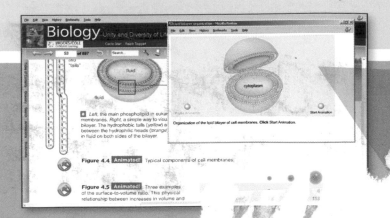

54

ANIMATIONS OF TEXT ART

Instead of simply studying content as you would in the printed textbook, the eBook gives you direct access to and animations, videos, and interactions. You can animation and then continue reading.

NOTES AND HIGHLIGHTS

Just like using a highlighter in a book, you can highlight important information in your eBook. You can also add annotations or notes to remind yourself to ask for help on a certain topic.

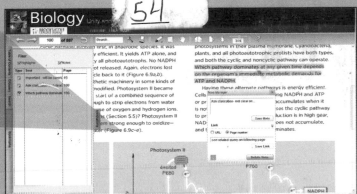

BOOKMARKS

You can bookmark specific pages that you want to return to later. From the Reference Panel you can view your bookmarks and link directly to any one of them.

ALSO AVAILABLE

Listen and learn about biology anywhere . . . with Audio Study Tools

These text-specific **Audio Study Tools** make it easy to learn biology's major concepts—such as mitosis and photosynthesis—anytime, anywhere. For each concept in the text, you can listen to:

- A one- to two-minute review
- Study tips (such as using word roots as a mnemonic device) and misconceptions (e.g., the misleading use of the term *dark reactions*)
- Questions to test your understanding
- 15–20 key terms, which are pronounced and defined

To purchase access to Audio Study Tools or the **eBook** for *Biology: The Unity and Diversity of Life*, 12th Edition, visit **www.iChapters.com**.

iChapters.com

Biology The Unity and Diversity of Life

Biology   The Unity and Diversity of Life Starr Taggart Evers Starr

Twelfth Edition

BROOKS/COLE
CENGAGE Learning

Australia • Brazil • Japan • Korea • Mexico • Singapore • Spain • United Kingdom • United States

BROOKS/COLE
CENGAGE Learning™

**Biology: The Unity and Diversity of Life,
Twelfth Edition**
Cecie Starr, Ralph Taggart, Christine Evers,
Lisa Starr

Publisher: Yolanda Cossio

Managing Development Editor: Peggy Williams

Assistant Editor: Elizabeth Momb

Editorial Assistant: Samantha Arvin

Technology Project Manager: Kristina Razmara

Marketing Manager: Amanda Jellerichs

Marketing Assistant: Katherine Malatesta

Marketing Communications Manager: Linda Yip

Project Manager, Editorial Production: Andy
    Marinkovich

Creative Director: Rob Hugel

Art Director: John Walker

Print Buyer: Karen Hunt

Permissions Editor: Bob Kauser

Production Service: Grace Davidson &
    Associates

Text and Cover Design: John Walker

Photo Researcher: Myrna Engler Photo
    Research Inc.

Copy Editor: Anita Wagner

Illustrators: Gary Head, ScEYEnce Studios,
    Lisa Starr

Compositor: Lachina Publishing Services

Cover Image: Biologist/photographer Tim
    Laman took these photos of mutualisms in
    Indonesia. *Top:* A wrinkled hornbill (*Aceros
    corrugatus*) eats fruits of a strangler fig (*Ficus
    stupenda*). The plant provides food for the
    bird, and the bird disperses its seeds. *Below:*
    Two species of sea anemone, each with its
    own species of anemone fish. Anemones
    provide a safe haven for anemonefish, who
    chase away other fish that would graze on
    the anemone's tentacles. www.timlaman.com

© 2009, 2006 Brooks/Cole, Cengage Learning

ALL RIGHTS RESERVED. No part of this work covered by the copyright herein
may be reproduced, transmitted, stored or used in any form or by any means
graphic, electronic, or mechanical, including but not limited to photocopying,
recording, scanning, digitizing, taping, Web distribution, information networks,
or information storage and retrieval systems, except as permitted under
Section 107 or 108 of the 1976 United States Copyright Act, without the prior
written permission of the publisher.

For product information and technology assistance, contact us at
Cengage Learning Customer & Sales Support, 1-800-354-9706.
For permission to use material from this text or product,
submit all requests online at **cengage.com/permissions**.
Further permissions questions can be emailed to
permissionrequest@cengage.com.

Library of Congress Control Number: 2008930411

ISBN-13: 978-0-495-55792-0

ISBN-10: 0-495-55792-7

Brooks/Cole
10 Davis Drive
Belmont, CA 94002
USA

Cengage Learning is a leading provider of customized learning solutions with
office locations around the globe, including Singapore, the United Kingdom,
Australia, Mexico, Brazil, and Japan. Locate your local office at:
international.cengage.com/region.

Cengage Learning products are represented in Canada by Nelson Education, Ltd.

For your course and learning solutions, visit **academic.cengage.com**.

Purchase any of our products at your local college store or at our preferred
online store **www.ichapters.com**.

Printed in the United States of America
1 2 3 4 5 6 7 12 11 10 09 08

CONTENTS IN BRIEF

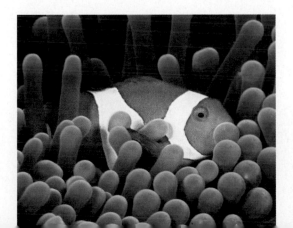

DETAILED CONTENTS

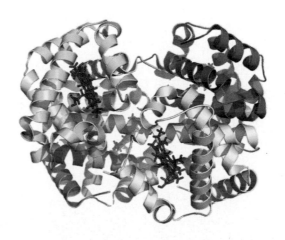

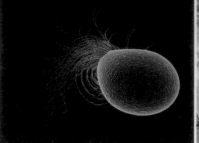
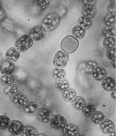
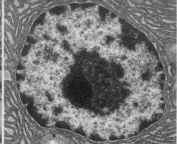

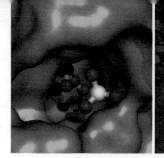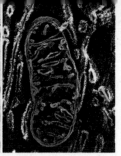

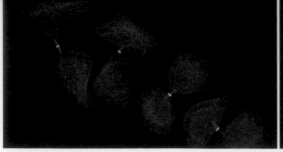
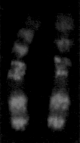

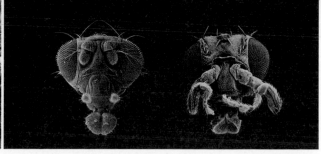

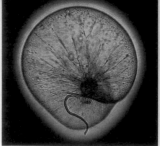

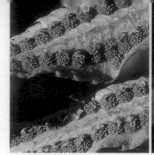

22 Protists—The Simplest Eukaryotes

23 The Land Plants

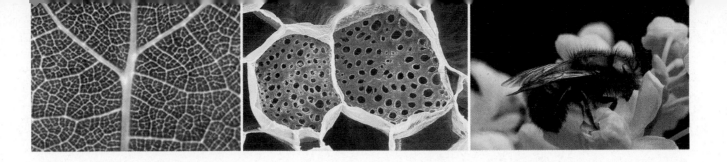

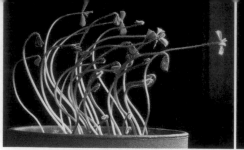

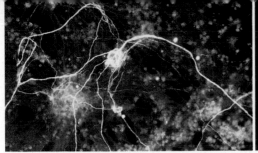
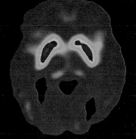

33 Neural Control

34 Sensory Perception

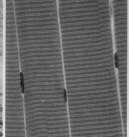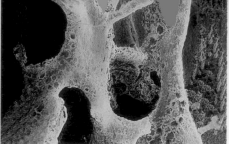

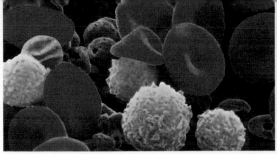
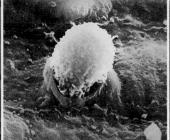

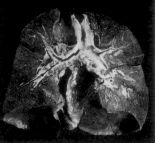
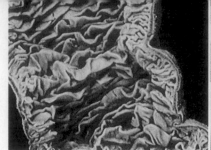

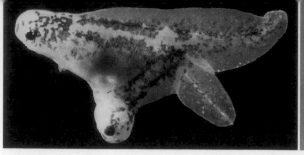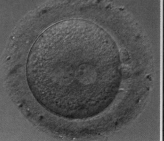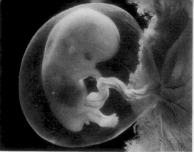

UNIT VII PRINCIPLES OF ECOLOGY

43 Animal Development

44 Animal Behavior

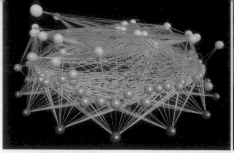
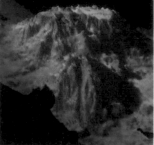

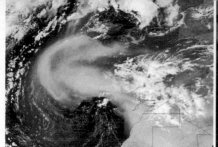

Preface

In preparation for this revision, we invited instructors who teach introductory biology for non-majors students to meet with with us and discuss the goals of their courses. The main goal of almost every instructor was something like this: "To provide students with the tools to make informed choices as consumers and as voters by familiarizing them with the way science works." Most students who use this book will not become biologists, and many will never take another science course. Yet for the rest of their lives they will have to make decisions that require a basic understanding of biology and the process of science.

Our book provides these future decision makers with an accessible introduction to biology. Current research, along with photos and videos of the scientists who do it, underscore the concept that science is an ongoing endeavor carried out by a diverse community of people. The research topics include not only what the researchers discovered, but also how the discoveries were made, how our understanding has changed over time, and what remains undiscovered. The role of evolution is a unifying theme, as it is in all aspects of biology.

As authors, we feel that understanding stems mainly from making connections, so we are constantly trying to achieve the perfect balance between accessibility and level of detail. A narrative with too much detail is inaccessible to the introductory student; one with too little detail comes across as a series of facts that beg to be memorized. Thus, we revised every page to make the text in this edition as clear and straightforward as possible, keeping in mind that English is a second language for many students. We also simplified many figures and added tables that summarize key points.

CHANGES IN THIS EDITION

Impacts, Issues   To make the *Impacts, Issues* essays more appealing, we shortened and updated them, and improved their integration throughout the chapters. Many new essays were added to this edition.

Key Concepts   Introductory summaries of the *Key Concepts* covered in the chapter are now enlivened with eye-catching graphics taken from relevant sections. The links to earlier concepts now include descriptions of the linked concepts in addition to the section numbers.

Take Home Message   Each section now concludes with a *Take Home Message* box. Here we pose a question that reflects the critical content of the section, and we also provide answers to the question in bulleted list format.

Figure It Out   *Figure It Out Questions* with answers allow students to check their understanding of a figure as they read through the chapter.

Data Analysis Exercise   To further strengthen a student's analytical skills and provide insight into contemporary research, each chapter includes a *Data Analysis Exercise*. The exercise includes a short text passage—

usually about a published scientific experiment—and a table, chart, or other graphic that presents experimental data. The student must use information in the text and graphic to answer a series of questions.

Chapter-Specific Changes   Every chapter was extensively revised for clarity; this edition has more than 250 new photos and over 300 new or updated figures. A page-by-page guide to content and figures is available upon request, but we summarize the highlights here.

• *Chapter 1, Invitation to Biology*   New essay about the discovery of new species. Greatly expanded coverage of critical thinking and the process of science; new section on sampling error.

• *Chapter 2, Life's Chemical Basis*   Sections on subatomic particles, bonding, and pH simplified; new pH art.

• *Chapter 3, Molecules of Life*   New essay about *trans* fats. Structural representations simplified and standardized.

• *Chapter 4, Cell Structure and Function*   New essay about foodborne *E. coli*; microscopy section updated; new section on cell theory and history of microscopy; two new focus essays on biofilms and lysosome malfunction.

• *Chapter 5, A Closer Look at Cell Membranes*   Membrane art reorganized; new figure illustrating cotransport.

• *Chapter 6, Ground Rules of Metabolism*   Energy and metabolism sections reorganized and rewritten; much new art, including molecular model of active site.

• *Chapter 7, Where It Starts—Photosynthesis*   New essay about biofuels. Sections on light-dependent reactions and carbon fixing adaptations simplified; new focus essay on atmospheric CO_2 and global warming.

• *Chapter 8, How Cells Release Chemical Energy*   All art showing metabolic pathways revised and simplified.

• *Chapter 9, How Cells Reproduce*   Updated micrographs of mitosis in plant and animal cells.

• *Chapter 10, Meiosis and Sexual Reproduction*   Crossing over, segregation, and life cycle art revised.

• *Chapter 11, Observing Patterns in Inherited Traits*   New essay about inheritance of skin color; mono- and dihybrid cross figures revised; new Punnett square for coat color in dogs; environmental effects on *Daphnia* phenotype added.

• *Chapter 12, Chromosomes and Human Inheritance*   Chapter reorganized; expanded discussion and new figure on the evolution of chromosome structure.

• *Chapter 13, DNA Structure and Function*   New opener essay on pet cloning; adult cloning section updated.

• *Chapter 14, From DNA to Protein*   New art comparing DNA and RNA, other art simplified throughout; new micrographs of transcription Christmas tree, polysomes.

• *Chapter 15, Controls Over Genes*   Chapter reorganized; eukaryotic gene control section rewritten; updated X chromosome inactivation photos; new lac operon art.

• *Chapter 16, Studying and Manipulating Genomes*   Text extensively rewritten and updated; new photos of *bt* corn, DNA fingerprinting; sequencing art revised.

• *Chapter 17, Evidence of Evolution*   Extensively revised, reorganized. Revised essay on evidence/inference; new focus essay on whale evolution; updated geologic time scale correlated with grand canyon strata.

• *Chapter 18, Processes of Evolution* Extensively revised, reorganized. New photos showing sexual selection in stalk-eyed flies, mechanical isolation in sage.

• *Chapter 19, Organizing Information About Species* Extensively revised, reorganized. New comparative embryology photo series; updated tree of life.

• *Chapter 20, Life's Origin and Early Evolution* Information about origin of agents of metabolism updated. New discussion of ribozymes as evidence for RNA world.

• *Chapter 21, Viruses and Prokaryotes* Opening essay about HIV moved here, along with discussion of HIV replication. New art of viral structure. New section describes the discovery of viroids and prions.

• *Chapter 22, Protists—The Simplest Eukaryotes* New opening essay about malaria. New figures show protist traits, how protists relate to other groups.

• *Chapter 23, The Land Plants* Evolutionary trends revised. More coverage of liverworts and hornworts.

• *Chapter 24, Fungi* New opening essay about airborne spores. More information on fungal uses and pathogens.

• *Chapter 25, Animal Evolution—The Invertebrates* New summary table for animal traits. Coverage of relationships among invertebrates updated.

• *Chapter 26, Animal Evolution—The Chordates* New section on lampreys. Human evolution updated.

• Material previously covered in the *Biodiversity in Prespective* chapter now integrated into other chapters.

• *Chapter 27, Plants and Animals—Common Challenges* New section about heat-related illness.

• *Chapter 28, Plant Tissues* Secondary structure section simplified; new essay on dendroclimatology.

• *Chapter 29, Plant Nutrition and Transport* Root function section rewritten and expanded; new translocation art.

• *Chapter 30, Plant Reproduction* Extensively revised. New essay on colony collapse disorder; new table showing flower specializations for specific pollinators; new section on flower sex; many new photos added.

• *Chapter 31, Plant Development* Sections on plant development and hormone mechanisms rewritten.

• *Chapter 32, Animal Tissues and Organ Systems* Essay on stem cells updated. New section on lab-grown skin.

• *Chapter 33, Neural Control* Reflexes integrated with coverage of spinal cord. Section on brain heavily revised.

• *Chapter 34, Sensory Perception* New art of vestibular apparatus, image formation in eyes, and accommodation. Improved coverage of eye disorders and disease.

• *Chapter 35, Endocrine Control* New section about pituitary disorders. Tables summarizing hormone sources now in appropriate sections, rather than at end.

• *Chapter 36, Structural Support and Movement* Improved coverage of joints and joint disorders.

• *Chapter 37, Circulation* Updated opening essay. New section about hemostasis. Blood cell diagram simplified. Blood typing section revised for clarity.

• *Chapter 38, Immunity* New essay on HPV vaccine; new focus essays on periodontal-cardiovascular disease and allergies; vaccines and AIDS sections updated.

• *Chapter 39, Respiration* Better coverage of invertebrate respiration and of Heimlich maneuver.

• *Chapter 40, Digestion and Human Nutrition* Nutritional information and obesity research sections updated.

• *Chapter 41, Maintaining the Internal Environment* New figure of fluid distribution in the human body. Improved coverage of kidney disorders and dialysis.

• *Chapter 42, Animal Reproductive Systems* New essay on intersex conditions. Coverage of reproductive anatomy, gamete production, intercourse, and fertilization.

• *Chapter 43, Animal Development* Information about principles of animal development streamlined.

• *Chapter 44, Animal Behavior* More on types of learning.

• *Chapter 45, Population Ecology* Exponential and logistic growth clarified. Human population material updated.

• *Chapter 46, Community Structure and Biodiversity* New table of species interactions. Competition section heavily revised.

• *Chapter 47, Ecosystems* New figures for food chain and food webs. Updated greenhouse gas coverage.

• *Chapter 48, The Biosphere* Improved coverage of lake turnover, ocean life, coral reefs, and threats to them.

• *Chapter 49, Human Impacts on the Biosphere* Covers extinction crisis, conservation biology, ecosystem degradation, and sustainable use of biological wealth.

Appendix V, Molecular Models New art and text explain why we use different types of molecular models.

Appendix VI, Closer Look at Some Major Metabolic Pathways New art shows details of electron transport chains in thylakoid membranes.

ACKNOWLEDGMENTS

No list can convey our thanks to the team of dedicated people who made this book happen. The professionals who are listed on the following page helped shape our thinking. Marty Zahn and Wenda Ribeiro deserve special recognition for their incisive comments on every chapter, as does Michael Plotkin for voluminous and excellent feedback. Grace Davidson calmly and tirelessly organized our efforts, filled in our gaps, and put all of the pieces of this book together. Paul Forkner's tenacious photo research helped us achieve our creative vision. At Cengage Learning, Yolanda Cossio and Peggy Williams unwaveringly supported us and our ideals. Andy Marinkovich made sure we had what we needed, Amanda Jellerichs arranged for us to meet with hundreds of professors, Kristina Razmara continues to refine our amazing technology package, Samantha Arvin helped us stay organized, and Elizabeth Momb managed all of the print ancillaries.

CECIE STARR, CHRISTINE EVERS, AND LISA STARR
June 2008

MARC C. ALBRECHT
University of Nebraska at Kearney

ELLEN BAKER
Santa Monica College

SARAH FOLLIS BARLOW
Middle Tennessee State University

MICHAEL C. BELL
Richland College

LOIS BREWER BOREK
Georgia State University

ROBERT S. BOYD
Auburn University

URIEL ANGEL BUITRAGO-SUAREZ
Harper College

MATTHEW REX BURNHAM
Jones County Junior College

P.V. CHERIAN
Saginaw Valley State University

WARREN COFFEEN
Linn Benton

LUIGIA COLLO
Universita' Degli Studi Di Brescia

DAVID T. COREY
Midlands Technical College

DAVID F. COX
Lincoln Land Community College

KATHRYN STEPHENSON CRAVEN
Armstrong Atlantic State University

SONDRA DUBOWSKY
Allen County Community College

PETER EKECHUKWU
Horry-Georgetown Technical College

DANIEL J. FAIRBANKS
Brigham Young University

MITCHELL A. FREYMILLER
University of Wisconsin - Eau Claire

RAUL GALVAN
South Texas College

NABARUN GHOSH
West Texas A&M University

JULIAN GRANIRER
URS Corporation

STEPHANIE G. HARVEY
Georgia Southwestern State University

JAMES A. HEWLETT
Finger lakes community College

JAMES HOLDEN
Tidewater Community College - Portsmouth

HELEN JAMES
Smithsonian Institution

DAVID LEONARD
Hawaii Department of Land and Natural Resources

STEVE MACKIE
Pima West Campus

CINDY MALONE
California State University - Northridge

KATHLEEN A. MARRS
Indiana University - Purdue University Indianapolis

EMILIO MERLO-PICH
GlaxoSmithKline

MICHAEL PLOTKIN
Mt. San Jacinto College

MICHAEL D. QUILLEN
Maysville Community and Technical College

WENDA RIBEIRO
Thomas Nelson Community College

MARGARET G. RICHEY
Centre College

JENNIFER CURRAN ROBERTS
Lewis University

FRANK A. ROMANO, III
Jacksonville State University

CAMERON RUSSELL
Tidewater Community College - Portsmouth

ROBIN V. SEARLES-ADENEGAN
Morgan State University

BRUCE SHMAEFSKY
Kingwood College

BRUCE STALLSMITH
University of Alabama - Huntsville

LINDA SMITH STATON
Pollissippi State Technical Community College

PETER SVENSSON
West Valley College

LISA WEASEL
Portland State University

DIANA C. WHEAT
Linn-Benton Community College

CLAUDIA M. WILLIAMS
Campbell University

MARTIN ZAHN
Thomas Nelson Community College

Introduction

Current configurations of the Earth's oceans and land masses—the geologic stage upon which life's drama continues to unfold. This composite satellite image reveals global energy use at night by the human population. Just as biological science does, it invites you to think more deeply about the world of life—and about our impact upon it.

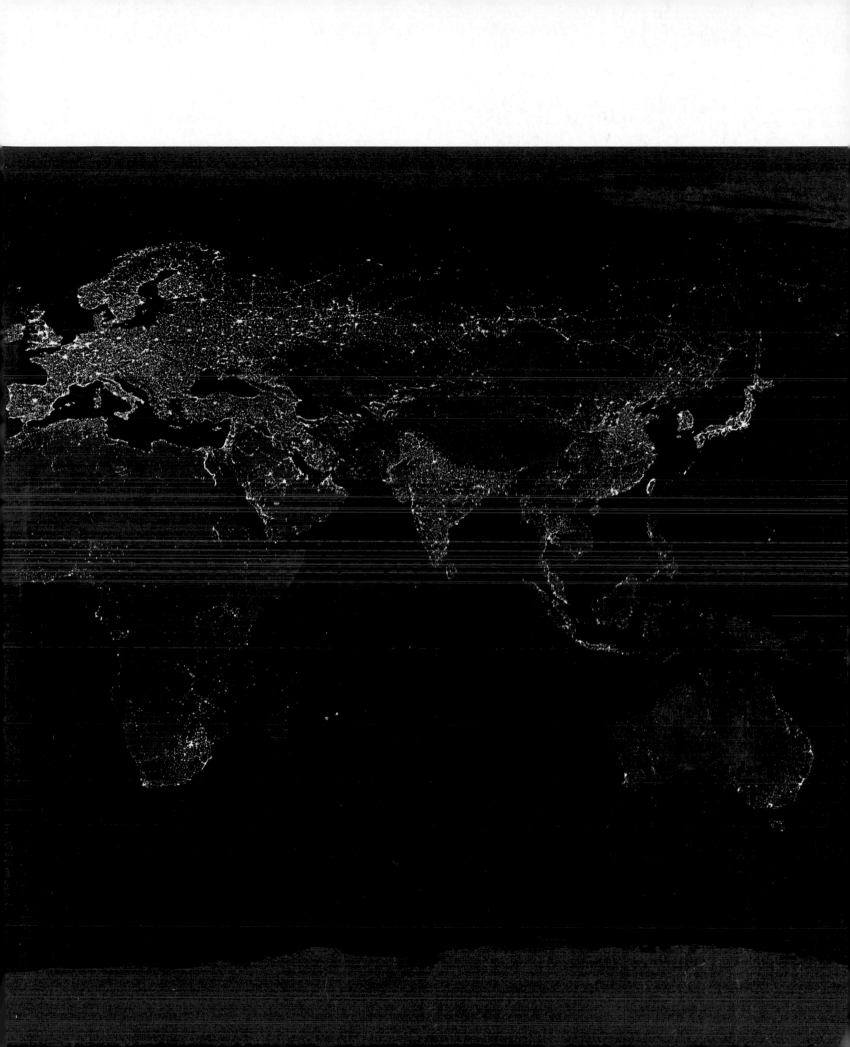

1 Invitation to Biology

Lost Worlds and Other Wonders

In this era of satellites, submarines, and global positioning systems, could there possibly be any more places on Earth that we have not explored? Well, yes. In 2005, for instance, helicopters dropped a team of biologists into a swamp in the middle of a vast and otherwise inaccessible tropical forest in New Guinea. Later, team member Bruce Beehler remarked, "Everywhere we looked, we saw amazing things we had never seen before. I was shouting. This trip was a once-in-a-lifetime series of shouting experiences."

The team discovered dozens of animals and plants that had been unknown to science, including a rhododendron with plate-sized flowers. They found animals that are on the brink of extinction in other parts of the world, and a bird that was supposedly extinct.

The expedition fired the imagination of people all over the world. It is not that finding new kinds of organisms is such a rare event. Almost every week, biologists discover many kinds of insects and other small organisms. However, the animals in this particular rain forest—mammals and birds especially—seem too big to have gone unnoticed. Had people just missed them? Perhaps not. No trails or other human disturbances cut through that part of the forest. The animals had never learned to be afraid of humans, so the team members could simply walk over and pick them up (Figure 1.1).

Many other animals have been discovered in the past few years, including lemurs in Madagascar, monkeys in India and Tanzania, cave-dwelling animals in two of California's national parks, carnivorous sponges near Antarctica, and whales and giant jellylike animals in the seas. Most came to light during survey trips similar to the New Guinea expedition—when biologists simply were attempting to find out what lives where.

Exploring and making sense of nature is nothing new. We humans and our immediate ancestors have been at it for thousands of years. We observe, come up with explanations about what the observations mean, and then test the explanations. Ironically, the more we learn about nature, the more we realize how much we have yet to learn.

You might choose to let others tell you what to think about the world around you. Or you might choose to develop your own understanding of it. Perhaps, like the New Guinea explorers, you are interested in animals and where they live. Maybe you are interested in aspects that affect your health, the food you eat, or your home and family. Whatever your focus may be, the scientific study of life—biology—can deepen your perspective on the world.

Throughout this book, you will find examples of how organisms are constructed, where they live, and what they do. These examples support concepts that, when taken together, convey what "life" is. This chapter gives you an overview of basic concepts. It sets the stage for upcoming descriptions of scientific observations and applications that can help you refine your understanding of life.

See the video! **Figure 1.1** Biologist Kris Helgen and a rare golden-mantled tree kangaroo in a tropical rain forest in the Foja Mountains of New Guinea. There, in 2005, explorers discovered forty previously unknown species.

Key Concepts

Levels of organization

We study the world of life at different levels of organization, which extend from atoms and molecules to the biosphere. The quality of "life" emerges at the level of cells. **Section 1.1**

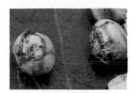

Life's underlying unity

All organisms consist of one or more cells, which stay alive through ongoing inputs of energy and raw materials. All sense and respond to change; all inherited DNA, a type of molecule that encodes information necessary for growth, development, and reproduction. **Section 1.2**

Life's diversity

Many millions of kinds of organisms, or species, have appeared and disappeared over time. Each kind is unique in some aspects of its body form or behavior. **Section 1.3**

Explaining unity in diversity

Theories of evolution, especially a theory of evolution by natural selection, help explain why life shows both unity and diversity. Evolutionary theories guide research in all fields of biology. **Section 1.4**

How we know

Biologists make systematic observations, predictions, and tests in the laboratory and in the field. They report their results so others may repeat their work and check their reasoning. **Sections 1.5–1.8**

Links to Earlier Concepts

■ This book parallels nature's levels of organization, from atoms to the biosphere. Learning about the structure and function of atoms and molecules primes you to understand the structure of living cells. Learning about processes that keep a single cell alive can help you understand how multicelled organisms survive, because their many living cells all use the same processes. Knowing what it takes for organisms to survive can help you see why and how they interact with one another and with their environments.

At the start of each chapter, we will use this space to remind you of such connections. Within chapters, cross-references will link you to relevant sections in earlier chapters

How would you vote? The discoverer of a new species usually is the one who gives it a scientific name. In 2005, a Canadian casino bought the right to name a monkey species. Should naming rights be sold? See CengageNOW for details, then vote online.

- We understand life by thinking about nature at different levels of organization.
- Nature's organization begins at the level of atoms, and extends through the biosphere.
- The quality of life emerges at the level of the cell.

Making Sense of the World

Most of us intuitively understand what nature means, but could you define it? **Nature** is everything in the universe except what humans have manufactured. It encompasses every substance, event, force, and energy —sunlight, flowers, animals, bacteria, rocks, thunder, humans, and so on. It excludes everything artificial.

Researchers, clerics, farmers, astronauts, children— anyone who is of the mind to do so attempts to make sense of nature. Interpretations differ, for no one can be expert in everything learned so far or have fore-knowledge of all that remains hidden. If you are reading this book, you are starting to explore how a subset of scientists, the biologists, think about things, what they found out, and what they are up to now.

A Pattern in Life's Organization

Biologists look at all aspects of life, past and present. Their focus takes them all the way down to atoms, and all the way up to global relationships among organisms and the environment. Through their work, we glimpse a great pattern of organization in nature.

The pattern starts at the level of atoms. **Atoms** are fundamental building blocks of all substances, living and nonliving (Figure 1.2*a*).

At the next level of organization, atoms join with other atoms, forming **molecules** (Figure 1.2*b*). Among the molecules are complex carbohydrates and lipids, proteins, and nucleic acids. Today, only living cells make these "molecules of life" in nature.

The pattern crosses the threshold to life when many molecules are organized as cells (Figure 1.2*c*). A **cell** is the smallest unit of life that can survive and reproduce on its own, given information in DNA, energy inputs, raw materials, and suitable environmental conditions.

An **organism** is an individual that consists of one or more cells. In larger multicelled organisms, trillions

B molecule

Two or more atoms joined in chemical bonds. In nature, only living cells make the molecules of life: complex carbohydrates and lipids, proteins, and nucleic acids.

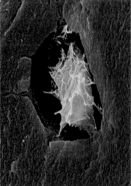

C cell

Smallest unit that can live and reproduce on its own or as part of a multicelled organism. A cell has DNA, an outermost membrane, and other components.

D tissue

Organized array of cells and substances that are interacting in some task. Bone tissue consists of secretions (*brown*) from cells such as this (*white*).

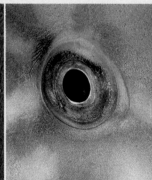

E organ

Structural unit of two or more tissues that interact in one or more tasks. This parrotfish eye is a sensory organ used in vision.

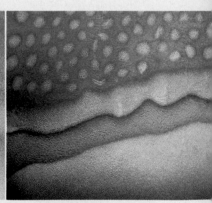

F organ system

Organs that interact in one or more tasks. The skin of this parrotfish is an organ system that consists of tissue layers, organs such as glands, and other parts.

single-celled organisms can form populations

A atom

Atoms are fundamental units of all substances. This image shows a model of a single hydrogen atom.

Figure 1.2 Animated Levels of organization in nature.

of cells organize into tissues, organs, and organ systems, all interacting in tasks that keep the whole body alive. Figure 1.2*d–g* defines these body parts.

Populations are at a greater level of organization. Each **population** is a group of individuals of the same kind of organism, or species, living in a specified area (Figure 1.2*h*). Examples are all heavybeak parrotfish living on Shark Reef in the Red Sea or all California poppies in California's Antelope Valley Poppy Reserve.

Communities are at the next level. A **community** consists of all populations of all species in a specified area. As an example, Figure 1.2*i* shows a sampling of the Shark Reef's species. This underwater community includes many kinds of seaweeds, fishes, corals, sea anemones, shrimps, and other living organisms that make their home in or on the reef. Communities may be large or small, depending on the area defined.

The next level of organization is the **ecosystem**: a community interacting with its physical and chemical environment. The most inclusive level, the **biosphere**, encompasses all regions of Earth's crust, waters, and atmosphere in which organisms live.

Bear in mind, life is more than the sum of its individual parts. In other words, some emergent property occurs at each successive level of life's organization. An **emergent property** is a characteristic of a system that does not appear in any of its component parts. For example, the molecules of life are themselves not alive. Considering them separately, no one would be able to predict that a particular quantity and arrangement of molecules will form a living cell. Life—an emergent property—appears first at the level of the cell but not at any lower level of organization in nature.

Take-Home Message

How does "life" differ from "nonlife"?

■ The building blocks—atoms—that make up all living things are the same ones that make up all nonliving things.

■ Atoms join as molecules. The unique properties of life emerge as certain kinds of molecules become organized into cells.

■ Higher levels of organization include multicelled organisms, populations, communities, ecosystems, and the biosphere.

G multicelled organism

Individual composed of different types of cells. Cells of most multicelled organisms, such as this parrotfish, form tissues, organs, and organ systems.

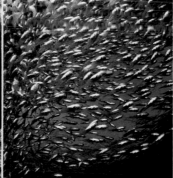

H population

Group of single-celled or multicelled individuals of a species in a given area. This is a population of one fish species in the Red Sea.

I community

All populations of all species in a specified area. These populations belong to a coral reef community in a gulf of the Red Sea.

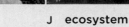
Red Sea

J ecosystem

A community that is interacting with its physical environment through inputs and outputs of energy and materials. Reef ecosystems flourish in warm, clear seawater throughout the Middle East.

K biosphere

All regions of Earth's waters, crust, and atmosphere that hold organisms. Earth is a rare planet. Life as we know it would be impossible without Earth's abundance of free-flowing water.

■ Continual inputs of energy and the cycling of materials maintain life's complex organization.

■ Organisms sense and respond to change.

■ DNA inherited from parents is the basis of growth and reproduction in all organisms.

Energy and Life's Organization

Eating supplies your body with energy and nutrients that keep it organized and functioning. **Energy** is the capacity to do work. A **nutrient** is a type of atom or molecule that has an essential role in growth and survival and that an organism cannot make for itself.

All organisms spend a lot of time acquiring energy and nutrients, although different kinds get such inputs from different sources. These differences allow us to classify organisms into one of two broad categories: producers or consumers.

Producers acquire energy and simple raw materials from environmental sources and make their own food. Plants are producers. By the process of **photosynthesis**, they use sunlight energy to make sugars from carbon dioxide and water. Those sugars function as packets of immediately available energy or as building blocks for larger molecules.

Consumers cannot make their own food; they get energy and nutrients indirectly—by eating producers and other organisms. Animals fall within the consumer category. So do decomposers, which feed on wastes or remains of organisms. We find leftovers of their meals in the environment. Producers take up the leftovers as sources of nutrients. Said another way, nutrients cycle between producers and consumers.

Energy, however, is not cycled. It flows through the world of life in one direction—from the environment, through producers, then through consumers. This flow maintains the organization of individual organisms, and it is the basis of life's organization within the biosphere (Figure 1.3). It is a one-way flow, because with each transfer, some energy escapes as heat. Cells do not use heat to do work. Thus, energy that enters the world of life ultimately leaves it—permanently.

Organisms Sense and Respond to Change

Organisms sense and respond to changes both inside and outside the body by way of receptors. A **receptor** is a molecule or cellular structure that responds to a specific form of stimulation, such as the energy of light or the mechanical energy of a bite (Figure 1.4).

Stimulated receptors trigger changes in activities of organisms. For example, after you eat, the sugars from

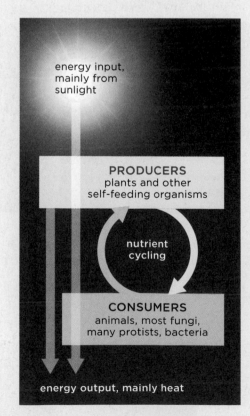

energy input, mainly from sunlight

PRODUCERS
plants and other self-feeding organisms

nutrient cycling

CONSUMERS
animals, most fungi, many protists, bacteria

energy output, mainly heat

A Energy inputs from the environment flow through producers, then consumers.

B Nutrients become incorporated into the cells of producers and consumers. Some nutrients released by decomposition cycle back to producers.

C All energy that enters an ecosystem eventually flows out of it, mainly as heat.

Figure 1.3 Animated The one-way flow of energy and cycling of materials through an ecosystem.

Figure 1.4 A roaring response to signals from pain receptors, activated by a lion cub flirting with disaster.

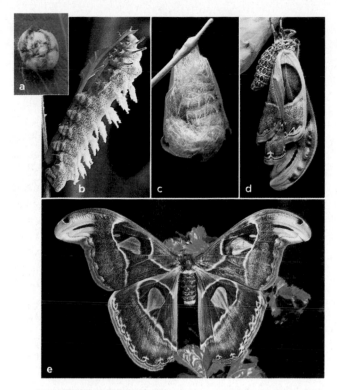

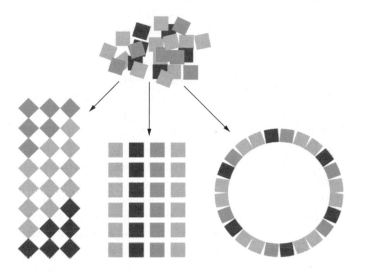

Figure 1.5 Development of the atlas moth. Instructions in DNA guide the development of this insect through a series of stages, from a fertilized egg (**a**), to a larval stage called a caterpillar (**b**), to a pupal stage (**c**), to the winged adult form (**d**,**e**).

Figure 1.6 Animated Three examples of objects assembled in different ways from the same materials.

your meal enter your bloodstream, and then your blood sugar level rises. The added sugars bind to receptors on cells of the pancreas (an organ). Binding sets in motion a series of events that causes cells throughout the body to take up sugar faster, so the sugar level in your blood returns to normal.

In multicelled organisms, the internal environment is all fluid inside of the body but outside of cells. Unless the composition of the internal environment is kept within certain ranges, body cells will die. By sensing and adjusting to change, organisms keep conditions in their internal environment within a range that favors cell survival. This process is called **homeostasis**, and it is a defining feature of life. All organisms, whether single-celled or multicelled, undergo homeostasis.

Organisms Grow and Reproduce

DNA, a nucleic acid, is the signature molecule of life. No chunk of rock has it. Why is DNA so important? It is the basis of growth, survival, and reproduction in all organisms. It is also the source of each individual's distinct features, or **traits**.

In nature, an organism inherits DNA—the basis of its traits—from parents. **Inheritance** is the transmission of DNA from parents to offspring. Moths look like moths and not like chickens because they inherited moth DNA, which differs from chicken DNA. **Reproduction** refers to actual mechanisms by which

parents transmit DNA to offspring. For all multicelled individuals, DNA has information that guides growth and **development**—the orderly transformation of the first cell of a new individual into an adult (Figure 1.5).

DNA contains instructions. Cells use some of those instructions to make proteins, which are long chains of amino acids. There are only 20 kinds of amino acids, but cells string them together in different sequences to make a tremendous variety of proteins. By analogy, just a few different kinds of tiles can be organized into many different patterns (Figure 1.6).

Different proteins have structural or functional roles. For instance, certain proteins are enzymes—functional molecules that make cell activities occur much faster than they would on their own. Without enzymes, such activities would not happen fast enough for a cell to survive. There would be no more cells—and no life.

Take-Home Message

How are all living things alike?

■ A one-way flow of energy and a cycling of nutrients through organisms and the environment sustain life, and life's organization.

■ Organisms maintain homeostasis by sensing and responding to change. They make adjustments that keep conditions in their internal environment within a range that favors cell survival.

■ Organisms grow, develop, and reproduce based on information encoded in their DNA, which they inherit from their parents.

1.3 Overview of Life's Diversity

■ Of an estimated 100 billion kinds of organisms that have ever lived on Earth, as many as 100 million are with us today.

Each time we discover a new **species**, or kind of organism, we assign it a two-part name. The first part of the name specifies the **genus** (plural, genera), which is a group of species that share a unique set of features. When combined with the second part, the name designates one species. Individuals of a species share one or more heritable traits, and they can interbreed successfully if the species is a sexually reproducing one.

Genus and species names are always italicized. For example, *Scarus* is a genus of parrotfish. The heavy-beak parrotfish in Figure 1.2*g* is called *Scarus gibbus*. A different species in the same genus, the midnight parrotfish, is *S. coelestinus*. Note that the genus name may be abbreviated after it has been spelled out one time.

We use various classification systems to organize and retrieve information about species. Most systems group species together on the basis of their observable characteristics, or traits. Table 1.1 and Figure 1.7 show a common system in which more inclusive groupings above the level of genus are phylum (plural, phyla), kingdom, and domain. Here, all species are grouped into domains Bacteria, Archaea, and Eukarya. Protists, plants, fungi, and animals make up domain Eukarya.

All **bacteria** (singular, bacterium) and **archaeans** are single-celled organisms. All of them are prokaryotic, which means they do not have a nucleus. In other organisms, this membrane-enclosed sac holds and protects a cell's DNA. As a group, prokaryotes have the most diverse ways of procuring energy and nutrients. They are producers and consumers in nearly all of the biosphere, including extreme environments such as frozen desert rocks, boiling sulfur-clogged lakes, and nuclear reactor waste. The first cells on Earth may have faced similarly hostile challenges to survival.

Cells of **eukaryotes** start out life with a nucleus. Structurally, **protists** are the simplest kind of eukaryote. Different protist species are producers or consumers. Many are single cells that are larger and more complex than prokaryotes. Some of them are tree-sized, multi-

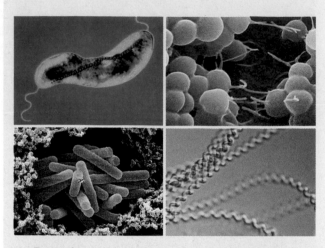

A Bacteria These prokaryotes tap more diverse sources of energy and nutrients than all other organisms. *Clockwise from upper left*, a magnetotactic bacterium has a row of iron crystals that acts like a tiny compass; bacteria that live on skin; spiral cyanobacteria; and *Lactobacillus* cells in yogurt.

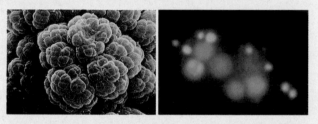

B Archaea Although they often appear similar to bacteria, these prokaryotes are evolutionarily closer to eukaryotes. *Left*, a colony of methane-producing cells. *Right*, two species from a hydrothermal vent on the seafloor.

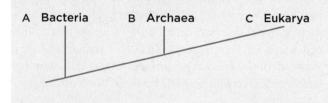

Figure 1.7 Animated Representatives of diversity from the three most inclusive branchings of the tree of life.

celled seaweeds. Protists are so diverse that they are now being reclassified into a number of separate major lineages based on emerging biochemical evidence.

Cells of fungi, plants, and animals are eukaryotic. Most **fungi**, such as the types that form mushrooms, are multicelled. Many are decomposers, and all secrete enzymes that digest food outside the body. Their cells then absorb the released nutrients.

Table 1.1	Comparison of Life's Three Domains
Bacteria	Single cells, prokaryotic (no nucleus). Most ancient lineage.
Archaea	Single cells, prokaryotic. Evolutionarily closer to eukaryotes.
Eukarya	Eukaryotic cells (with a nucleus). Single-celled and multicelled species categorized as protists, plants, fungi, and animals.

C Eukarya

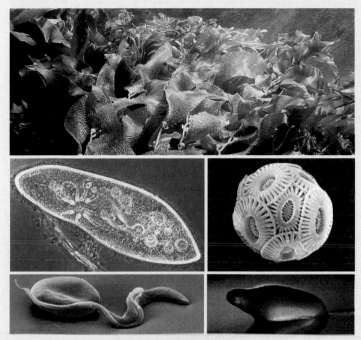

Protists are single-celled and multicelled eukaryotic species that range from the microscopic to giant seaweeds. Many biologists are now viewing the "protists" as many major lineages.

Plants are multicelled eukaryotes, most of which are photosynthetic. Nearly all have roots, stems, and leaves. Plants are the primary producers in land ecosystems. Redwood trees and flowering plants are examples.

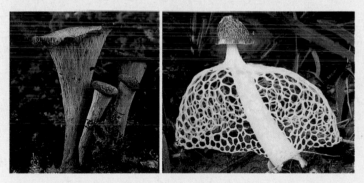

Fungi are eukaryotes. Most are multicelled. Different kinds are parasites, pathogens, or decomposers. Without decomposers such as fungi, communities would become buried in their own wastes.

Animals are multicelled eukaryotes that ingest tissues or juices of other organisms. Like this basilisk lizard, they actively move about during at least part of their life.

Plants are multicelled species. Most of them live on land or in freshwater environments. Nearly all plants are photosynthetic: They harness the energy of sunlight to drive the production of sugars from carbon dioxide and water. Besides feeding themselves, photosynthesizers also feed much of the biosphere.

The **animals** are multicelled consumers that ingest tissues or juices of other organisms. Herbivores graze, carnivores eat meat, scavengers eat remains of other organisms, and parasites withdraw nutrients from the tissues of a host. Animals grow and develop through a series of stages that lead to the adult form. Most kinds actively move about during at least part of their lives.

From this quick overview, can you get a sense of the tremendous range of life's variety—its diversity?

Take-Home Message

How do living things differ from one another?

■ Organisms differ in their details; they show tremendous variation in observable characteristics, or traits.

■ Various classification systems group species on the basis of shared traits.

■ A theory of evolution by natural selection is an explanation of life's diversity.

Individuals of a population are alike in certain aspects of their body form, function, and behavior, but the details of such traits differ from one individual to the next. For instance, humans (*Homo sapiens*) characteristically have two eyes, but those eyes come in a range of color among individuals.

Most traits are the outcome of information encoded in DNA, so they can be passed to offspring. Variations in traits arise through **mutations**, which are small-scale changes in DNA. Most mutations have neutral or negative effects, but some cause a trait to change in a way that makes an individual better suited to its environment. The bearer of such an **adaptive trait** has a better chance of surviving and passing its DNA to offspring than other individuals of the population. The naturalist Charles Darwin expressed the concept of "survival of the fittest" like this:

First, a natural population tends to increase in size. As it does, the individuals of the population compete more for food, shelter, and other limited resources.

Second, individuals of a population differ from one another in the details of shared traits. Such traits have a heritable basis.

Third, adaptive forms of traits make their bearers more competitive, so those forms tend to become more common over generations. The differential survival and reproduction of individuals in a population that differ in the details of their heritable traits is called **natural selection**.

Think of how pigeons differ in feather color and other traits (Figure 1.8*a*). Imagine that a pigeon breeder prefers black, curly-tipped feathers. She selects birds with the darkest, curliest-tipped feathers, and allows only those birds to mate. Over time, more and more pigeons in the breeder's captive population will have black, curly-tipped feathers.

Pigeon breeding is a case of artificial selection. One form of a trait is favored over others under contrived, manipulated conditions—in an artificial environment. Darwin saw that breeding practices could be an easily understood model for natural selection, a favoring of some forms of a given trait over others in nature.

Just as breeders are "selective agents" that promote reproduction of certain pigeons, agents of selection act on the range of variation in the wild. Among them are pigeon-eating peregrine falcons (Figure 1.8*b*). Swifter or better camouflaged pigeons are more likely to avoid falcons and live long enough to reproduce, compared with not-so-swift or too-flashy pigeons.

When different forms of a trait are becoming more or less common over successive generations, evolution is under way. In biology, **evolution** simply means change in a line of descent.

Take-Home Message

How did life become so diverse?

■ Individuals of a population show variation in their shared, heritable traits. Such variation arises through mutations in DNA.

■ Adaptive traits improve an individual's chances of surviving and reproducing, so they tend to become more common in a population over successive generations.

■ Natural selection is the differential survival and reproduction among individuals of a population that differ in the details of their shared, heritable traits. It and other evolutionary processes underlie the diversity of life.

wild rock pigeon

a

b

Figure 1.8 (**a**) Outcome of artificial selection: a few of the hundreds of varieties of domesticated pigeons descended from captive populations of wild rock pigeons (*Columba livia*). (**b**) A peregrine falcon (*left*) preying on a pigeon (*right*) is acting as an agent of natural selection in the wild.

■ Critical thinking means judging the quality of information.
■ Science is limited to that which is observable.

Thinking About Thinking

Most of us assume that we do our own thinking—but do we, really? You might be surprised to find out just how often we let others think for us. For instance, a school's job, which is to impart as much information as possible to students, meshes with a student's job, which is to acquire as much knowledge as possible. In this rapid-fire exchange of information, it is easy to forget about the quality of what is being exchanged. If you accept information without question, you allow someone else to think for you.

Critical thinking means judging information before accepting it. "Critical" comes from the Greek *kriticos* (discerning judgment). When you think this way, you move beyond the content of information. You look for underlying assumptions, evaluate the supporting statements, and think of possible alternatives (Table 1.2).

How does the busy student manage this? Be aware of what you intend to learn from new information. Be conscious of bias or underlying agendas in books, lectures, or online. Consider your own biases—what you want to believe—and realize those biases influence your learning. Respectfully question authority figures. Decide whether ideas are based on opinion or evidence. Such practices will help you decide whether to accept or reject information.

The Scope and Limits of Science

Because each of us is unique, there are as many ways to think about the natural world as there are people. **Science**, the systematic study of nature, is one way. It helps us be objective about our observations of nature, in part because of its limitations. We limit science to a subset of the world—only that which is observable.

Science does not address some questions, such as "Why do I exist?" Most answers to such questions are subjective; they come from within as an integration of the personal experiences and mental connections that shape our consciousness. This is not to say subjective answers have no value: No human society functions for very long unless its individuals share standards for making judgments, even if they are subjective. Moral, aesthetic, and philosophical standards vary from one society to the next, but all help people decide what is important and good. All give meaning to what we do.

Also, science does not address the supernatural, or anything that is "beyond nature." Science does not

Table 1.2   A Guide to Critical Thinking
What message am I being asked to accept?
What evidence supports the message? Is the evidence valid?
Is there another way to interpret the evidence?
What other evidence would help me evaluate the alternatives?
Is the message the most reasonable one based on the evidence?

assume or deny that supernatural phenomena occur, but scientists may still cause controversy when they discover a natural explanation for something that was thought to be unexplainable. Such controversy often arises when a society's moral standards have become interwoven with traditional interpretations of nature.

For example, Nicolaus Copernicus studied the planets centuries ago in Europe, and concluded that Earth orbits the sun. Today this conclusion seems obvious, but at the time it was heresy. The prevailing belief was that the Creator made Earth—and, by extension, humans—as the center of the universe. Galileo Galilei, another scholar, found evidence for the Copernican model of the solar system and published his findings. He was forced to publicly recant his publication, and to put Earth back at the center of the universe.

Exploring a traditional view of the natural world from a scientific perspective might be misinterpreted as questioning morality even though the two are not the same. As a group, scientists are no less moral, less lawful, or less compassionate than anyone else. As you will see in the next section, however, their work follows a particular standard: Explanations must be testable in the natural world in ways that others can repeat.

Science helps us communicate experiences without bias; it may be as close as we can get to a universal language. We are fairly sure, for example, that laws of gravity apply everywhere in the universe. Intelligent beings on a distant planet would likely understand the concept of gravity. We might well use such concepts to communicate with them—or anyone—anywhere. The point of science, however, is not to communicate with aliens. It is to find common ground here on Earth.

Take-Home Message

What is science?

■ Science is the study of the observable—those objects or events for which valid evidence can be gathered. It does not address the supernatural.

1.6 | How Science Works

■ Scientists make and test potentially falsifiable predictions about how the natural world works.

Observations, Hypotheses, and Tests

To get a sense of how science works, consider Table 1.3 and this list of common research practices:

1. Observe some aspect of nature.

2. Frame a question that relates to your observation.

3. Read about what others have discovered concerning the subject, then propose a **hypothesis**, a testable answer to your question.

4. Using the hypothesis as a guide, make a **prediction**: a statement of some condition that should exist if the hypothesis is not wrong. Making predictions is called the if–then process: "if" is the hypothesis, and "then" is the prediction.

5. Devise ways to test the accuracy of the prediction by conducting experiments or gathering information. Experiments may be performed on a **model**, or analogous system, if experimenting directly with an object or event is not possible.

6. Assess the results of the tests. Results that confirm the prediction are evidence—data—in support of the hypothesis. Results that disprove the prediction are evidence that the hypothesis may be flawed.

7. Report all the steps of your work, along with any conclusions you drew, to the scientific community.

Table 1.3   Example of a Scientific Approach

1. Observation	People get cancer.
2. Question	Why do people get cancer?
3. Hypothesis	Smoking cigarettes may cause cancer.
4. Prediction	If smoking causes cancer, then individuals who smoke will get cancer more often than those who do not.
5. Gather information	Conduct a survey of individuals who smoke and individuals who do not smoke. Determine which group has the highest incidence of cancers.
Laboratory experiment	Establish identical groups of laboratory rats (the model system). Expose one group to cigarette smoke. Compare the incidence of new cancers in each of the two groups.
6. Assess results	Compile test results and draw conclusions from them.
7. Report	Submit the results and the conclusions to the scientific community for review and publication.

You might hear someone refer to these practices as "the scientific method," as if all scientists march to the drumbeat of a fixed procedure. They do not. There are different ways to do research, particularly in biology (Figure 1.9). Some biologists do surveys; they observe without making hypotheses. Others make hypotheses and leave tests to others. Some stumble onto valuable information they are not even looking for. Of course, it is not only a matter of luck. Chance favors a mind that is already prepared, by education and experience, to recognize what the new information might mean.

Regardless of the variation, one thing is constant: Scientists do not accept information simply because someone says it is true. They evaluate the supporting evidence and find alternative explanations. Does this sound familiar? It should—it is critical thinking.

About the Word "Theory"

Most scientists avoid the word "truth" when discussing science. Instead, they tend to talk about evidence that supports or does not support a hypothesis.

Suppose a hypothesis has not been disproven even after years of tests. It is consistent with all of the evidence gathered to date, and it has helped us to make successful predictions about other phenomena. When any hypothesis meets these criteria, it is considered to be a **scientific theory**.

To give an example, observations for all of recorded history have supported the hypothesis that gravity pulls objects toward Earth. Scientists no longer spend time testing the hypothesis for the simple reason that, after many thousands of years of observation, no one has seen otherwise. This hypothesis is now a scientific theory, but it is not an "absolute truth." Why not? An infinite number of tests would be necessary to confirm that it holds under every possible circumstance.

A single observation or result that is *not* consistent with a theory opens that theory to revision. For example, if gravity pulls objects toward Earth, it would be logical to predict that an apple will fall down when dropped. However, a scientist might well see such a test as an opportunity for the prediction to fail. Think about it. If even one apple falls up instead of down, the theory of gravity would come under scrutiny. Like every other theory, this one remains open to revision.

A well-tested theory is as close to the "truth" as scientists will venture. Table 1.4 lists a few scientific theories. One of them, the theory of natural selection, holds after more than a century of testing. Like all other scientific theories, we cannot be sure that it will hold under all possible conditions, but we can say it

Figure 1.9 Scientists doing research in the laboratory and in the field. (**a**) Analyzing data with computers. (**b**) At the Centers for Disease Control and Prevention, Mary Ari testing a sample for the presence of dangerous bacteria. (**c**) Making field observations in an old-growth forest.

has a very high probability of not being wrong. If any evidence turns up that is inconsistent with the theory of natural selection, then biologists will revise it. Such a willingness to modify or discard even an entrenched theory is one of the strengths of science.

You may hear people apply the word "theory" to a speculative idea, as in the phrase "It's just a theory." Speculation is opinion or belief, a personal conviction that is not necessarily supported by evidence. A scientific theory is not an opinion: By definition, it must be supported by a large body of evidence.

Unlike theories, many beliefs and opinions cannot be tested. Without being able to test something, there is no way to disprove it. Even though personal conviction has tremendous value in our lives, it should not be confused with scientific theory.

Table 1.4 Examples of Scientific Theories

Atomic theory	All substances are composed of atoms.
Gravitation	Objects attract one another with a force that depends on their mass and how close together they are.
Cell theory	All organisms consist of one or more cells, the cell is the basic unit of life, and all cells arise from existing cells.
Germ theory	Microorganisms cause many diseases.
Plate tectonics	Earth's crust is cracked into pieces that move in relation to one another.
Evolution	Change occurs in lines of descent.
Natural selection	Variation in heritable traits influences differential survival and reproduction of individuals of a population.

Some Terms Used in Experiments

Careful observations are one way to test predictions that flow from a hypothesis. So are experiments. You will find examples of experiments in the next section. For now, just get acquainted with some of the important terms that researchers use:

1. **Experiments** are tests that can support or falsify a prediction.

2. Experiments are usually designed to test the effects of a single variable. A **variable** is a characteristic that differs among individuals or events.

3. Biological systems are an integration of so many interacting variables that it can be difficult to study one variable separately from the rest. Experimenters often test two groups of individuals, side by side. An **experimental group** is a set of individuals that have a certain characteristic or receive a certain treatment. This group is tested side by side with a **control group**, which is identical to the experimental group except for one variable—the characteristic or the treatment being tested. Ideally, the two groups have the same set of variables, except for the one being tested. Thus, any differences in experimental results between the two groups should be an effect of changing the variable.

Take-Home Message

How does science work?

■ Scientific inquiry involves asking questions about some aspect of nature, formulating hypotheses, making and testing predictions, and reporting the results.

■ Researchers design experiments to test the effects of one variable at a time.

■ A scientific theory is a long-standing, well-tested concept of cause and effect that is consistent with all evidence, and is used to make predictions about other phenomena.

■ Researchers unravel cause and effect in complex natural processes by changing one variable at a time.

Potato Chips and Stomach Aches

In 1996 the FDA approved Olestra®, a type of synthetic fat replacement made from sugar and vegetable oil, as a food additive. Potato chips were the first Olestra-laced food product on the market in the United States. Controversy soon raged. Some people complained of intestinal cramps after eating the chips and concluded that Olestra caused them.

Two years later, four researchers at Johns Hopkins University School of Medicine designed an experiment to test the hypothesis that this food additive causes cramps. They predicted that *if* Olestra causes cramps, *then* people who eat Olestra will be more likely to get cramps than people who do not.

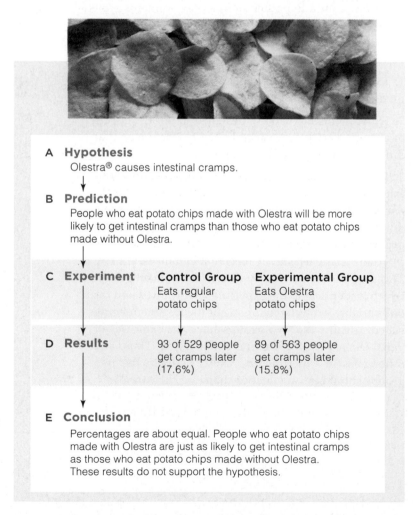

A Hypothesis
Olestra® causes intestinal cramps.

B Prediction
People who eat potato chips made with Olestra will be more likely to get intestinal cramps than those who eat potato chips made without Olestra.

C Experiment	Control Group	Experimental Group
	Eats regular potato chips	Eats Olestra potato chips
D Results	93 of 529 people get cramps later (17.6%)	89 of 563 people get cramps later (15.8%)

E Conclusion
Percentages are about equal. People who eat potato chips made with Olestra are just as likely to get intestinal cramps as those who eat potato chips made without Olestra. These results do not support the hypothesis.

Figure 1.10 Animated The steps in a scientific experiment to determine if Olestra causes cramps. A report of this study was published in the *Journal of the American Medical Association* in January of 1998.

To test the prediction, they used a Chicago theater as the "laboratory." They asked more than 1,100 people between ages thirteen and thirty-eight to watch a movie and eat their fill of potato chips. Each person got an unmarked bag that contained 13 ounces of chips. The individuals who got a bag of Olestra-laced potato chips were the experimental group. Individuals who got a bag of regular chips were the control group.

Afterward, researchers contacted all of the people and tabulated the reports of gastrointestinal cramps. Of 563 people making up the experimental group, 89 (15.8 percent) complained about problems. However, so did 93 of the 529 people (17.6 percent) making up the control group—who had munched on the regular chips! This simple experiment disproved the prediction that eating Olestra-laced potato chips at a single sitting can cause gastrointestinal cramps (Figure 1.10).

Butterflies and Birds

Consider the peacock butterfly, a winged insect that was named for the large, colorful spots on its wings. In 2005, researchers published a report on their tests to identify factors that help peacock butterflies defend themselves against insect-eating birds. The researchers made two observations. First, when a peacock butterfly rests, it folds its ragged-edged wings, so only the dark underside shows (Figure 1.11*a*). Second, when a butterfly sees a predator approaching, it repeatedly flicks its paired forewings and hindwings open and closed. At the same time, each forewing slides over the hindwing, which produces a hissing sound and a series of clicks.

The researchers asked this question, "Why does the peacock butterfly flick its wings?" After they reviewed earlier studies, they formulated three hypotheses that might explain the wing-flicking behavior:

1. When folded, the butterfly wings resemble a dead leaf. They may camouflage the butterfly, or help it hide from predators in its forest habitat.

2. Although the wing-flicking probably attracts predatory birds, it also exposes brilliant spots that resemble owl eyes (Figure 1.11*b*). Anything that looks like owl eyes is known to startle small, butterfly-eating birds, so exposing the wing spots might scare off predators.

3. The hissing and clicking sounds produced when the peacock butterfly rubs the sections of its wings together may deter predatory birds.

The researchers decided to test hypotheses 2 and 3. They made the following predictions:

a

b

c

Table 1.5   Results of Peacock Butterfly Experiment*

Wing Spots	Wing Sound	Total Number of Butterflies	Number Eaten	Number Survived
Spots	Sound	9	0	9 (100%)
No spots	Sound	10	5	5 (50%)
Spots	No sound	8	0	8 (100%)
No spots	No sound	10	8	2 (20%)

* Proceedings of the Royal Society of London, Series B (2005) 272: 1203–1207.

Figure 1.11 Peacock butterfly defenses against predatory birds. (**a**) With wings folded, a resting peacock butterfly looks like a dead leaf. (**b**) When a bird approaches, the butterfly repeatedly flicks its wings open and closed. This defensive behavior exposes brilliant spots. It also produces hissing and clicking sounds.

Researchers tested whether the behavior deters blue tits (**c**). They painted over the spots of some butterflies, cut the sound-making part of the wings on other butterflies, and did both to a third group; then the biologists exposed each butterfly to a hungry bird.

The results are listed in Table 1.5. **Figure It Out:** Which defense, wing spots or sounds, more effectively deterred the tits?

Answer: wing spots

1. *If* brilliant wing spots of peacock butterflies deter predatory birds, *then* individuals with no wing spots will be more likely to get eaten by predatory birds than individuals with wing spots.

2. *If* the sounds that peacock butterflies produce deter predatory birds, *then* individuals that do not make the sounds will be more likely to be eaten by predatory birds than individuals that make the sounds.

The next step was the experiment. The researchers painted the wing spots of some butterflies black, cut off the sound-making part of the hindwings of others, and did both to a third group. They put each butterfly into a large cage with a hungry blue tit (Figure 1.11c) and then watched the pair for thirty minutes.

Table 1.5 lists the results of the experiment. All of the butterflies with unmodified wing spots survived, regardless of whether they made sounds. By contrast, only half of the butterflies that had spots painted out but could make sounds survived. Most of the butterflies with neither spots nor sound structures were eaten quickly.

The test results confirmed both predictions, so they support the hypotheses. Birds are deterred by peacock butterfly sounds, and even more so by wing spots.

Asking Useful Questions

Researchers try to design single-variable experiments that will yield quantitative results, which are counts or some other data that can be measured or gathered objectively. Even so, they risk designing experiments and interpreting results in terms of what they want to find out. Particularly when studying humans, isolating a single variable is not often possible. For example, by thinking critically we may realize that the people who participated in the Olestra experiment were chosen randomly. That means the study was not controlled for gender, age, weight, medications taken, and so on. Such variables may well have influenced the results.

Scientists expect one another to put aside bias. If one individual does not, others will, because science works best when it is both cooperative and competitive.

Take-Home Message

Why do biologists do experiments?

■ Natural processes are often influenced by many interacting variables.

■ Experiments help researchers unravel causes of such natural processes by focusing on the effects of changing a single variable.

1.8 | Sampling Error in Experiments

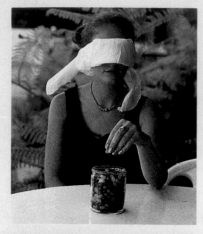

A Natalie, blindfolded, randomly plucks a jelly bean from a jar. There are 120 green and 280 black jelly beans in that jar, so 30 percent of the jelly beans in the jar are green, and 70 percent are black.

B The jar is hidden from Natalie's view before she removes her blindfold. She sees only one green jelly bean in her hand and assumes that the jar must hold only green jelly beans.

C Blindfolded again, Natalie picks out 50 jelly beans from the jar and ends up with 10 green and 40 black jelly beans.

D The larger sample leads Natalie to assume that one-fifth of the jar's jelly beans are green (20 percent) and four-fifths are black (80 percent). The sample more closely approximates the jar's actual green-to-black ratio of 30 percent to 70 percent. The more times Natalie repeats the sampling, the greater the chance she will come close to knowing the actual ratio.

■ Biology researchers experiment on subsets of a group.
■ Results from such an experiment may differ from results of the same experiment performed on the whole group.

Rarely can researchers observe all individuals of a group. For example, remember the explorers you read about in the chapter introduction? They did not survey the entire rain forest, which cloaks more than 2 million acres of New Guinea's Foja Mountains. Even if it were possible, doing so would take unrealistic amounts of time and effort. Besides, tromping about even in a small area can damage delicate forest ecosystems.

Given such constraints, researchers tend to experiment on subsets of a population, event, or some other aspect of nature that they select to represent the whole. They test the subsets, and then use the results to make generalizations about the whole population.

Suppose researchers design an experiment to identify variables that influence the population growth of golden-mantled tree kangaroos. They might focus only on the population living in one acre of the Foja Mountains. If they identify only 5 golden-mantled tree kangaroos in the specified acre, then they might extrapolate that there are 50 in every ten acres, 100 in every twenty acres, and so forth.

However, generalizing from a subset is risky because the subset may not be representative of the whole. If the only population of golden-mantled tree kangaroos in the forest just happens to be living in the surveyed acre, then the researchers' assumptions about the number of kangaroos in the rest of the forest will be wrong.

Sampling error is a difference between results from a subset and results from the whole. It happens most often when sample sizes are small. Starting with a large sample or repeating the experiment many times helps minimize sampling error (Figure 1.12). To understand why, imagine flipping a coin. There are two possible results: The coin lands heads up, or it lands tails up. You might predict that the coin will land heads up as often as it lands tails up. When you actually flip the coin, though, often it will land heads up, or tails up, several times in a row. If you flip the coin only a few times, the results may differ greatly from your prediction. Flip it many times, and you probably will come closer to having equal numbers of heads and tails.

Sampling error is an important consideration in the design of most experiments. The possibility that it occurred should be part of the critical thinking process as you read about experiments. Remember to ask: If the experimenters used a subset of the whole, did they select a large enough sample? Did they repeat the experiment many times? Thinking about these possibilities will help you evaluate the results and conclusions reached.

Figure 1.12 Animated Demonstration of sampling error.

Almost every week, another new species is discovered and we are again reminded that we do not yet know all of the organisms on our own planet. We don't even know how many to look for. The vast information about the 1.8 million species we do know about changes so quickly that collating it has been impossible—until now. A new web site, titled the Encyclopedia of Life, is intended to be an online reference source and database of species information maintained by collaborative effort. See it at www.eol.org.

How would you vote? Discovered in Madagascar in 2005, this tiny mouse lemur was named *Microcebus lehilahytsara* in honor of primatologist Steve Goodman (lehilahytsara is a combination of the Malagasy words for "good" and "man"). Should naming rights be sold? See CengageNOW for details, then vote online.

Summary

Section 1.1 There are **emergent properties** at each level of organization in **nature**. All matter consists of **atoms**, which combine as **molecules**. **Organisms** are one or more **cells**, the smallest units of life. A **population** is a group of individuals of a species in a given area; a **community** is all populations of all species in a given area. An **ecosystem** is a community interacting with its environment. The **biosphere** includes all regions of Earth that hold life.

- *Explore levels of biological organization with the interaction on CengageNOW.*

Section 1.2 All living things have similar characteristics (Table 1.6). All organisms require inputs of **energy** and **nutrients** to sustain themselves. **Producers** make their own food by processes such as **photosynthesis**; **consumers** eat producers or other consumers. By **homeostasis**, organisms use molecules and structures such as **receptors** to help keep the conditions in their internal environment within ranges that their cells tolerate. Organisms grow, **develop**, and **reproduce** using information in their **DNA**, a nucleic acid **inherited** from parents. Information encoded in DNA is the source of an individual's **traits**.

- *Use instructions with the animation on CengageNOW to see how different objects are assembled from the same materials. Also view energy flow and materials cycling.*

Section 1.3 Each type of organism is given a name that includes **genus** and **species** names. Classification systems group species by their shared, heritable traits. All organisms can be classified as **bacteria**, **archaea**, or **eukaryotes**. **Plants**, **protists**, **fungi**, and **animals** are eukaryotes.

- *Use the interaction on CengageNOW to explore characteristics of the three domains of life.*

Section 1.4 Information encoded in DNA is the basis of traits that an organism shares with others of its species. **Mutations** are the original source of variation in traits.

Some forms of traits are more adaptive than others, so their bearers are more likely to survive and reproduce. Over generations, such **adaptive traits** tend to become more common in a population; less adaptive forms of traits tend to become less common or are lost.

Thus, the traits that characterize a species can change over generations in evolving populations. **Evolution** is change in a line of descent. The differential survival and reproduction among individuals that vary in the details of their shared, heritable traits is an evolutionary process called **natural selection**.

Section 1.5 **Critical thinking** is judging the quality of information as one learns. **Science** is one way of looking at the natural world. It helps us minimize bias in our judgments by focusing on only testable ideas about observable aspects of nature.

Section 1.6 Researchers generally make observations, form **hypotheses** (testable assumptions) about it, then make **predictions** about what might occur if the hypothesis is correct. They test predictions with **experiments**, using **models**, **variables**, **experimental groups**, and **control groups**. A hypothesis that is not consistent with results of scientific tests (evidence) is modified or discarded. A **scientific theory** is a long-standing hypothesis that is used to make useful predictions.

Section 1.7 Scientific experiments simplify interpretations of complex biological systems by focusing on the effect of one variable at a time.

Section 1.8 Small sample size increases the likelihood of **sampling error** in experiments. In such cases, a subset may be tested that is not representative of the whole.

Table 1.6 Summary of Life's Characteristics

Shared characteristics that underlie life's unity

Organisms grow, develop, and reproduce based on information encoded in DNA, which is inherited from parents.

Ongoing inputs of energy and nutrients sustain all organisms, as well as nature's overall organization.

Organisms maintain homeostasis by sensing and responding to changes inside and outside of the body.

Basis of life's diversity

Mutations (heritable changes in DNA) give rise to variation in details of body form, the functioning of body parts, and behavior.

Diversity is the sum total of variations that have accumulated, since the time of life's origin, in different lines of descent. It is an outcome of natural selection and other processes of evolution.

Data Analysis Exercise

The photographs to the *right* represent the experimental and control groups used in the peacock butterfly experiment from Section 1.7.

See if you can identify each experimental group, and match it with the relevant control group(s). *Hint:* Identify which variable is being tested in each group (each variable has a control).

 **a** Wing spots painted out

 b Wing spots visible; wings silenced

 c Wing spots painted out; wings silenced

 d Wings painted but spots visible

 e Wings cut but not silenced

 f Wings painted but spots visible; wings cut but not silenced

Self-Quiz

Answers in Appendix III

1. _____ are fundamental building blocks of all matter.

2. The smallest unit of life is the _____ .

3. _____ move around for at least part of their life.

4. Organisms require _____ and _____ to maintain themselves, grow, and reproduce.

5. _____ is a process that maintains conditions in the internal environment within ranges that cells can tolerate.

6. Bacteria, Archaea, and Eukarya are three _____ .

7. DNA _____ .
   a. contains instructions for building proteins
   b. undergoes mutation
   c. is transmitted from parents to offspring
   d. all of the above

8. _____ is the transmission of DNA to offspring.
   a. Reproduction
   b. Development
   c. Homeostasis
   d. Inheritance

9. _____ is the process by which an organism produces offspring.

10. Science only addresses that which is _____ .

11. _____ are the original source of variation in traits.

12. A trait is _____ if it improves an organism's chances to survive and reproduce in its environment.

13. A control group is _____ .
   a. a set of individuals that have a certain characteristic or receive a certain treatment
   b. the standard against which experimental groups can be compared
   c. the experiment that gives conclusive results

14. Match the terms with the most suitable description.
   ___ emergent property
   ___ natural selection
   ___ scientific theory
   ___ hypothesis
   ___ prediction
   ___ species

   a. statement of what a hypothesis leads you to expect to see
   b. type of organism
   c. occurs at a higher organizational level in nature, not at levels below it
   d. time-tested hypothesis
   e. differential survival and reproduction among individuals of a population that vary in details of shared traits
   f. testable explanation

■ *Visit CengageNOW for additional questions.*

Critical Thinking

1. Why would you think twice about ordering from a cafe menu that lists only the second part of the species name (not the genus) of its offerings? *Hint:* Look up *Ursus americanus, Ceanothus americanus, Bufo americanus, Homarus americanus, Lepus americanus,* and *Nicrophorus americanus.*

2. How do prokaryotes and eukaryotes differ?

3. Explain the relationship between DNA and natural selection.

4. Procter & Gamble makes Olestra and financed the study described in Section 1.7. The main researcher was a consultant to Procter & Gamble during the study. What do you think about scientific information that comes from tests financed by companies with a vested interest in the outcome?

5. Once there was a highly intelligent turkey that had nothing to do but reflect on the world's regularities. Morning always started out with the sky turning light, followed by the master's footsteps, which was always followed by the appearance of food. Other things varied, but food always followed footsteps. The sequence of events was so predictable that it eventually became the basis of the turkey's theory about the goodness of the world. One morning, after more than 100 confirmations of the goodness theory, the turkey listened for the master's footsteps, heard them, and had its head chopped off.

   Any scientific theory is modified or discarded when contradictory evidence becomes available. The absence of absolute certainty has led some people to conclude that "facts are irrelevant—facts change." If that is so, should we stop doing scientific research? Why or why not?

6. In 2005 a South Korean scientist, Woo-suk Hwang, reported that he made immortal stem cells from eleven human patients. His research was hailed as a breakthrough for people affected by currently incurable degenerative diseases, because such stem cells might be used to repair a person's own damaged tissues. Hwang published his results in a respected scientific journal. In 2006, the journal retracted his paper after other scientists discovered that Hwang and his colleagues had faked their results. Does the incident show that the results of scientific studies cannot be trusted? Or does it confirm the usefulness of a scientific approach, because other scientists quickly discovered and exposed the fraud?

I PRINCIPLES OF CELLULAR LIFE

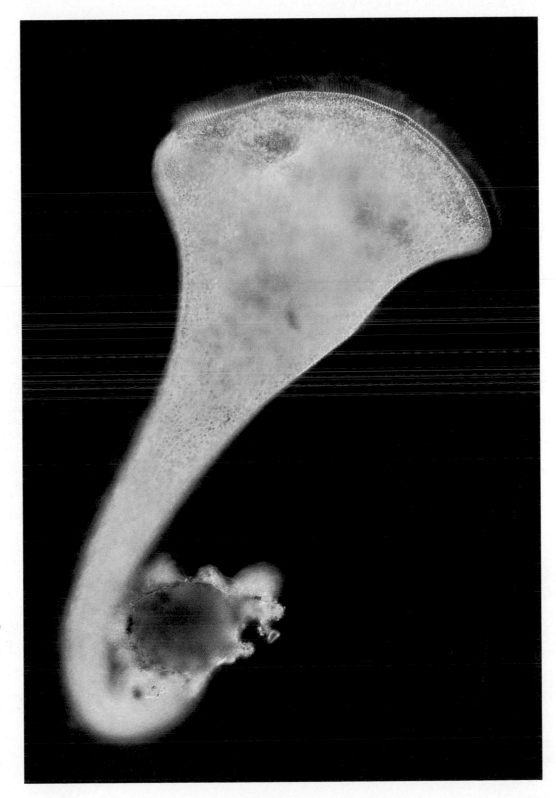

Staying alive means securing energy and raw materials from the environment. Shown here, a living cell of the genus *Stentor*. This protist has hairlike projections around an opening to a cavity in its body, which is about 2 millimeters long. Its "hairs" of fused-together cilia beat the surrounding water. They create a current that wafts food into the cavity.

2 Life's Chemical Basis

What Are You Worth?

Hollywood thinks actor Keanu Reaves is worth $30 million plus per movie, the Yankees think shortstop Alex Rodriguez is worth $252 million per decade, and the United States thinks the average public school teacher is worth $46,597 per year. How much is one human body really worth? You can buy the entire collection of ingredients that make up an average 70-kilogram (150-pound) body for about $118.63 (Figure 2.1). Of course, all you have to do is watch Keanu, Alex, or any teacher to know that a human body is far more than a collection of those ingredients. What makes us worth more than the sum of our parts?

The fifty-eight pure substances listed in Figure 2.1 are called elements. You will find the same elements that make up the human body in, say, dirt or seawater. However, the proportions of those elements differ between living and nonliving things. For example, a human body contains far more carbon. Seawater and most rocks have no more than a trace of it.

We are only starting to understand the processes by which a collection of elements becomes assembled as a living body. We do know that life's unique organization starts with the properties of atoms that make up certain elements. This is your chemistry. It makes you far more than the sum of your body's ingredients—a handful of lifeless chemicals.

Elements in a Human Body		
Element	Number of Atoms (x 10^{15})	Retail Cost
Hydrogen	41,808,044,129,611	$ 0.028315
Oxygen	16,179,356,725,877	0.021739
Carbon	8,019,515,931,628	6.400000
Nitrogen	773,627,553,592	9.706920
Phosphorus	151,599,284,310	68.198594
Calcium	150,207,096,162	15.500000
Sulfur	26,283,290,713	0.011623
Sodium	26,185,559,925	2.287748
Potassium	21,555,924,426	4.098737
Chlorine	16,301,156,188	1.409496
Magnesium	4,706,027,566	0.444909
Fluorine	823,858,713	7.917263
Iron	452,753,156	0.054600
Silicon	214,345,481	0.370000
Zinc	211,744,915	0.088090
Rubidium	47,896,401	1.087153
Strontium	21,985,848	0.177237
Bromine	19,588,506	0.012858
Boron	10,023,125	0.002172
Copper	6,820,886	0.012961
Lithium	6,071,171	0.024233
Lead	3,486,486	0.003960
Cadmium	2,677,674	0.010136
Titanium	2,515,303	0.010920
Cerium	1,718,576	0.043120
Chromium	1,620,894	0.003402
Nickel	1,538,503	0.031320
Manganese	1,314,936	0.001526
Selenium	1,143,617	0.037949
Tin	1,014,236	0.005387
Iodine	948,745	0.094184
Arsenic	562,455	0.023576
Germanium	414,543	0.130435
Molybdenum	313,738	0.001260
Cobalt	306,449	0.001509
Cesium	271,772	0.000016
Mercury	180,069	0.004718
Silver	111,618	0.013600
Antimony	98,883	0.000243
Niobium	97,195	0.000624
Barium	96,441	0.028776
Gallium	60,439	0.003367
Yttrium	40,627	0.005232
Lanthanum	34,671	0.000566
Tellurium	33,025	0.000722
Scandium	26,782	0.058160
Beryllium	24,047	0.000218
Indium	20,972	0.000600
Thallium	14,727	0.000894
Bismuth	14,403	0.000119
Vanadium	12,999	0.000322
Tantalum	6,654	0.001631
Zirconium	6,599	0.000830
Gold	6,113	0.001975
Samarium	2,002	0.000118
Tungsten	655	0.000007
Thorium	3	0.004948
Uranium	3	0.000103
Total	67,179,218,505,055 x 10^{15}	**$118.63**

See the video! Figure 2.1 Composition of an average-sized adult human body, by weight and retail cost. Manufacturers commonly add fluoride to toothpaste. Fluoride is a form of fluorine, one of several elements with vital functions—but only in trace amounts. Too much can be toxic.

Key Concepts

Atoms and elements

Atoms are particles that are the building blocks of all matter. They can differ in their numbers of component protons, electrons, and neutrons. Elements are pure substances, each consisting entirely of atoms that have the same number of protons. **Sections 2.1, 2.2**

Why electrons matter

Whether one atom will bond with others depends on the element, and the number and arrangement of its electrons. **Section 2.3**

Atoms bond

Atoms of many elements interact by acquiring, sharing, and giving up electrons. Ionic, covalent, and hydrogen bonds are the main interactions between atoms in biological molecules. **Section 2.4**

Water of life

Life originated in water and is adapted to its properties. Water has temperature-stabilizing effects, cohesion, and a capacity to act as a solvent for many other substances. These properties make life possible on Earth. **Section 2.5**

The power of hydrogen

Life is responsive to changes in the amounts of hydrogen ions and other substances dissolved in water. **Section 2.6**

Links to Earlier Concepts

■ With this chapter, we turn to the first of life's levels of organization—atoms and energy—so take a moment to review Section 1.1.

■ Life's organization requires continuous inputs of energy (1.2). Organisms store that energy in chemical bonds between atoms.

■ You will come across a simple example of how the body's built-in mechanisms maintain homeostasis (1.2).

How would you vote? Fluoride helps prevent tooth decay, but too much wrecks bones and teeth, and causes birth defects. A lot can kill you. Many communities in the United States add fluoride to drinking water. Do you want it in yours? See CengageNOW for details, then vote online.

Start With Atoms

■ The behavior of elements, which make up all living things, starts with the structure of individual atoms.

Characteristics of Atoms

Atoms are particles that are the building blocks of all substances. Even though they are about one billion times smaller than basketballs, atoms consist of even smaller subatomic particles called **protons** (p+), which carry a positive charge; **neutrons**, which carry no charge; and **electrons** (e−), which carry a negative charge. **Charge** is an electrical property that attracts or repels other subatomic particles. Protons and neutrons cluster in an atom's central core, or **nucleus**. Electrons move around the nucleus (Figure 2.2).

Atoms differ in the number of subatomic particles. The number of protons, which is the **atomic number**, determines the element. **Elements** are pure substances, each consisting only of atoms with the same number of protons. For example, a chunk of carbon contains only carbon atoms, all of which have six protons in their nucleus. The atomic number of carbon is 6. All atoms with six protons in their nucleus are carbon atoms, no matter how many electrons or neutrons they have. Each element has a symbol that is an abbreviation of its Latin name. Carbon's symbol, C, is from *carbo*, the Latin word for coal—which is mostly carbon.

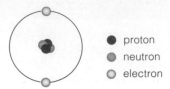

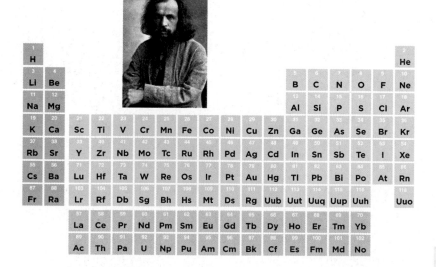

Figure 2.3 Periodic table of the elements and its creator, Dmitry Mendeleev. Until he came up with the table, Mendeleev was known mainly for his extravagant hair; he cut it only once per year.

Atomic numbers are shown above the element symbols. Some of the symbols are abbreviations for their Latin names. For instance, Pb (lead) is short for plumbum; the word "plumbing" is related—ancient Romans made their water pipes with lead. Appendix IV has a more detailed table.

Figure 2.2 Atoms. Electrons move about a nucleus of protons and neutrons. Models such as this do not show what an atom really looks like. A more accurate rendering would show electrons occupying fuzzy, three-dimensional shapes about 10,000 times larger than the nucleus.

All elements occur in different forms called **isotopes**. Atoms of isotopes have the same number of protons, but different numbers of neutrons. We refer to isotopes by **mass number**, which is the total number of protons and neutrons in their nucleus. The mass number of an isotope is shown as a superscript to the left of an element's symbol. For instance, the most common isotope of carbon is ^{12}C (six protons, six neutrons). Another is ^{13}C (six protons, seven neutrons).

The Periodic Table

Today, we know that the numbers of electrons, protons, and neutrons determine how an element behaves, but scientists were classifying elements by chemical behavior long before they knew about subatomic particles. In 1869, the chemist Dmitry Mendeleev arranged all of the elements known at the time into a table based on their chemical properties. He had constructed the first **periodic table of the elements**.

Elements are ordered in the periodic table by their atomic number (Figure 2.3). Those in each vertical column behave in similar ways. For instance, all of the elements in the far right column of the table are inert gases; their atoms do not interact with other atoms. In nature, such elements occur only as solitary atoms.

We find the first ninety-four elements in nature. The others are so unstable that they are extremely rare. We know they exist because they can be made, one atom at a time, for a fraction of a second. It takes a nuclear physicist to do this, because an atom's nucleus cannot be altered by heat or other ordinary means.

Take-Home Message

What are the basic building blocks of all matter?

■ Atoms are tiny particles, the building blocks of all substances.

■ Atoms consist of electrons moving around a nucleus of protons and (except for hydrogen) neutrons.

■ An element is a pure substance. Each kind consists only of atoms with the same number of protons.

2.2 | Putting Radioisotopes to Use

■ Some radioactive isotopes—radioisotopes—are used in research and in medical applications.

In 1896, Henri Becquerel made a chance discovery. He left some crystals of a uranium salt in a desk drawer, on top of a metal screen. Under the screen was an exposed film wrapped tightly in black paper. Becquerel developed the film a few days later and was surprised to see a negative image of the screen. He realized that "invisible radiations" coming from the uranium salts had passed through the paper and exposed the film around the screen.

Becquerel's images were evidence that uranium has **radioisotopes**, or radioactive isotopes. So do many other elements. The atoms of radioisotopes spontaneously emit subatomic particles or energy when their nucleus breaks down. This process, **radioactive decay**, can transform one element into another. For example, ^{14}C is a radioisotope of carbon. It decays when one of its neutrons spontaneously splits into a proton and an electron. Its nucleus emits the electron, and so an atom of ^{14}C (with eight neutrons and six protons) becomes an atom of ^{14}N (nitrogen 14, with seven neutrons and seven protons).

Radioactive decay occurs independently of external factors such as temperature, pressure, or whether the atoms are part of molecules. A radioisotope decays at a constant rate into predictable products. For example, after 5,730 years, we can predict that about half of the atoms in any sample of ^{14}C will be ^{14}N atoms. This predictability can be used to estimate the age of rocks and fossils by their radioisotope content. We return to this topic in Section 17.6.

Researchers and clinicians also introduce radioisotopes into living organisms. Remember, isotopes are atoms of the same element. All isotopes of an element generally have the same chemical properties regardless of the number of neutrons in their atoms. This consistent chemical behavior means that organisms use atoms of one isotope (such as ^{14}C) the same way that they use atoms of another (such as ^{12}C). Thus, radioisotopes can be used in tracers.

A **tracer** is any molecule with a detectable substance attached. Typically, a radioactive tracer is a molecule in which radioisotopes have been swapped for one or more atoms. Researchers deliver radioactive tracers into a biological system such as a cell or a multicelled body. Instruments that can detect radioactivity let researchers follow the tracer as it moves through the system.

For example, Melvin Calvin and his colleagues used a radioactive tracer to identify specific reaction steps of photosynthesis. The researchers made carbon dioxide with ^{14}C, then let green algae (simple aquatic organisms) take up the radioactive gas. Using instruments that detected the radioactive decay of ^{14}C, they tracked carbon through steps by which the algae—and all plants—make sugars.

Radioisotopes have medical applications as well. PET (short for *Positron-Emission Tomography*) helps us "see" cell activity. By this procedure, a radioactive sugar or other tracer is injected into a patient, who is then moved into a PET scanner (Figure 2.4*a*). Inside the patient's body, cells with differing rates of activity take up the tracer at different rates. The scanner detects radioactive decay wherever the tracer is, then translates that data into an image. Such images can reveal abnormal cell activity (Figure 2.4*b*).

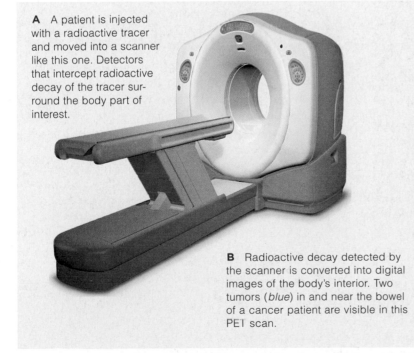

A A patient is injected with a radioactive tracer and moved into a scanner like this one. Detectors that intercept radioactive decay of the tracer surround the body part of interest.

B Radioactive decay detected by the scanner is converted into digital images of the body's interior. Two tumors (*blue*) in and near the bowel of a cancer patient are visible in this PET scan.

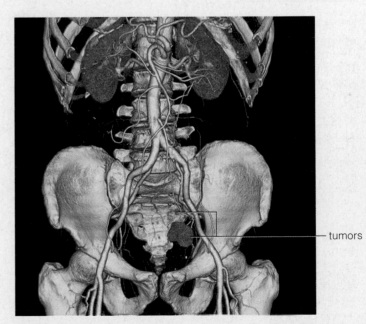

tumors

Figure 2.4 Animated PET scanning.

- Atoms acquire, share, and donate electrons.
- Whether an atom will interact with other atoms depends on how many electrons it has.

Electrons and Energy Levels

Electrons are really, really small: If they were as big as apples, you would be 3.5 times taller than our solar system is wide. Simple physics explains the motion of, say, an apple falling from a tree. Electrons are so tiny that everyday physics does not explain their behavior, but that behavior underlies interactions among atoms.

A typical atom has about as many electrons as protons, so a lot of electrons may be zipping around one nucleus. Those electrons never collide, despite moving at nearly the speed of light (300,000 kilometers per second, or 670 million miles per hour). Why not? They travel in different orbitals, which are defined volumes of space around the nucleus.

Imagine that an atom is a multilevel apartment building, with rooms available for rent by electrons. The nucleus is in the basement, and each "room" is an orbital. No more than two electrons can share a room at the same time. An orbital with only one electron has a vacancy, and another electron can move in.

Each floor in the apartment building corresponds to one energy level. There is only one room on the first floor: one orbital at the lowest energy level, closest to the nucleus. It fills up first. In hydrogen, the simplest atom, a single electron occupies that room. Helium has two electrons, so it has no vacancies at the lowest energy level. In larger atoms, more electrons rent the second-floor rooms. When the second floor fills, more electrons rent third-floor rooms, and so on. Electrons fill orbitals at successively higher energy levels.

The farther an electron is from the basement (the nucleus), the greater its energy. An electron in a first-floor room cannot move to the second or third floor, let alone the penthouse, unless an input of energy gives it a boost. Suppose an electron absorbs enough energy from sunlight to get excited about moving up. Move it does. If nothing fills that lower room, though, the electron immediately moves back down, emitting its extra energy as it does. In later chapters, you will see how some types of cells harvest that released energy.

Why Atoms Interact

Shells and Electrons We use a **shell model** to help us check an atom for vacancies (Figure 2.5). With this model, nested "shells" correspond to successive energy levels. Each shell includes all rooms on one floor of the

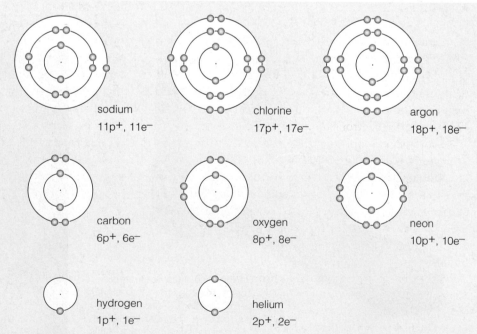

○ electron

C   Third shell This shell corresponds to the third energy level. It has four orbitals with room for eight electrons. Sodium has one electron in the third shell; chlorine has seven. Both have vacancies, so both form chemical bonds. Argon, with no vacancies, does not.

sodium 11p+, 11e−

chlorine 17p+, 17e−

argon 18p+, 18e−

B   Second shell This shell, which corresponds to the second energy level, has four orbitals—room for a total of eight electrons. Carbon has six electrons: two in the first shell and four in the second. It has four vacancies. Oxygen has two vacancies. Both carbon and oxygen form chemical bonds. Neon, with no vacancies, does not.

carbon 6p+, 6e−

oxygen 8p+, 8e−

neon 10p+, 10e−

A   First shell A single shell corresponds to the first energy level, which has a single orbital that can hold two electrons. Hydrogen has only one electron in this shell, so it has one vacancy. A helium atom has two electrons (no vacancies), so it does not form bonds.

hydrogen 1p+, 1e−

helium 2p+, 2e−

Figure 2.5 Animated Shell models, which help us check for vacancies in atoms. Each circle, or shell, represents all orbitals at one energy level. Atoms with vacancies in the outermost shell tend to form bonds. Remember, atoms do not look anything like these flat diagrams.

atomic apartment building. We draw an atom's shells by filling them with electrons (represented as dots or balls) from the innermost shell out, until there are as many electrons as the atom has protons.

If an atom's outermost shell is full of electrons, it has no vacancies. Atoms of such elements are chemically inactive; they are most stable as single atoms. Helium, neon, and the other inert gases in the right-hand column of the periodic table are like this.

If an atom's outermost shell has room for an extra electron, it has a vacancy. Atoms with vacancies tend to interact with other atoms; they give up, acquire, or share electrons until they have no vacancies in their outermost shell. Any atom is in its most stable state when it has no vacancies.

Atoms and Ions The negative charge of an electron cancels the positive charge of a proton, so an atom is uncharged only when it has as many electrons as protons. An atom with different numbers of electrons and protons is called an **ion**. An ion carries a charge; either it acquired a positive charge by losing an electron, or it acquired a negative charge by pulling an electron away from another atom.

Electronegativity is a measure of an atom's ability to pull electrons from other atoms. Whether the pull is strong or weak depends on the atom's size and how many vacancies it has; it is not a measure of charge

As an example, when a chlorine atom is uncharged, it has 17 protons and 17 electrons. Seven electrons are in its outer (third) shell, which can hold eight (Figure 2.6). It has one vacancy. An uncharged chlorine atom is highly electronegative—it can pull an electron away from another atom and fill its third shell. When that happens, the atom becomes a chloride ion (Cl⁻) with 17 protons, 18 electrons, and a net negative charge.

As another example, an uncharged sodium atom has 11 protons and 11 electrons. This atom has one electron in its outer (third) shell, which can hold eight. It has seven vacancies. An uncharged sodium atom is weakly electronegative, so it cannot pull seven electrons from other atoms to fill its third shell. Instead, it tends to lose the single electron in its third shell. When that happens, two full shells—and no vacancies—remain. The atom has now become a sodium ion (Na⁺), with 11 protons, 10 electrons, and a net positive charge.

From Atoms to Molecules Atoms do not like to have vacancies, and try to get rid of them by interacting with other atoms. A **chemical bond** is an attractive force that arises between two atoms when their electrons interact. A **molecule** forms when two or more atoms

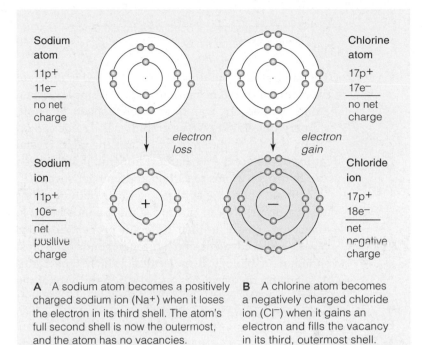

A A sodium atom becomes a positively charged sodium ion (Na⁺) when it loses the electron in its third shell. The atom's full second shell is now the outermost, and the atom has no vacancies.

B A chlorine atom becomes a negatively charged chloride ion (Cl⁻) when it gains an electron and fills the vacancy in its third, outermost shell.

Figure 2.6 Animated Ion formation.

of the same or different elements join in chemical bonds. The next section explains the main types of bonds in biological molecules.

Compounds are molecules that consist of two or more different elements in proportions that do not vary. Water is an example. All water molecules have one oxygen atom bonded to two hydrogen atoms. Whether water is in the seas, a waterfall, a Siberian lake, or anywhere else, its molecules have twice as many hydrogen as oxygen atoms. By contrast, in a **mixture**, two or more substances intermingle, and their proportions can vary because the substances do not bond with each other. For example, you can make a mixture by swirling sugar into water. The sugar dissolves, but no chemical bonds form.

Always two H for every O

Take-Home Message

Why do atoms interact?

■ An atom's electrons are the basis of its chemical behavior.

■ Shells represent all electron orbitals at one energy level in an atom. When the outermost shell is not full of electrons, the atom has a vacancy.

■ Atoms tend to get rid of vacancies by gaining or losing electrons (thereby becoming ions), or by sharing electrons with other atoms.

■ Atoms with vacancies can form chemical bonds. Chemical bonds connect atoms into molecules.

■ The characteristics of a bond arise from the properties of the atoms that take part in it.

The same atomic building blocks, arranged in different ways, make different molecules. For example, carbon atoms bonded one way form layered sheets of a soft, slippery mineral known as graphite. The same carbon atoms bonded a different way form the rigid crystal lattice of diamond—the hardest mineral. Bond oxygen and hydrogen atoms to carbon and you get sugar.

Although bonding applies to a range of interactions among atoms, we can categorize most bonds into distinct types based on their different properties. Three types—ionic, covalent, and hydrogen bonds—are most common in biological molecules. Which type forms depends on the vacancies and electronegativity of the atoms that take part in it. Table 2.1 compares different ways to represent molecules and their bonds.

Ionic Bonding

Remember from Figure 2.6, a strongly electronegative atom tends to gain electrons until its outermost shell is full. Then it is a negatively charged ion. A weakly electronegative atom tends to lose electrons until its outermost shell is full. Then it is a positively charged ion. Two atoms with a large difference in electronegativity may stay together in an **ionic bond**, which is a strong mutual attraction of two oppositely charged ions. Such bonds do not usually form by the direct transfer of an

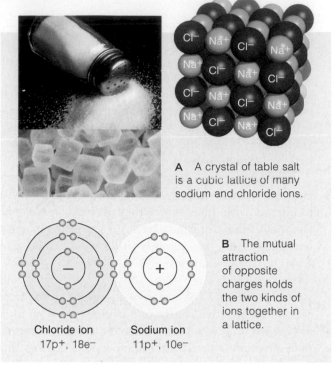

A A crystal of table salt is a cubic lattice of many sodium and chloride ions.

Chloride ion
17p+, 18e−

Sodium ion
11p+, 10e−

B The mutual attraction of opposite charges holds the two kinds of ions together in a lattice.

Figure 2.7 Animated Ionic bonds.

electron from one atom to another; rather, atoms that have already become ions stay close together because of their opposite charges.

Figure 2.7 shows crystals of table salt (sodium chloride, or NaCl). Ionic bonds in such solids hold sodium and chloride ions in an orderly, cubic arrangement.

Covalent Bonding

In a **covalent bond**, two atoms share a pair of electrons. Such bonds typically form between atoms with similar electronegativity and unpaired electrons. By sharing their electrons, each atom's vacancy becomes partially filled (Figure 2.8). Covalent bonds can be stronger than ionic bonds, but they are not always so.

Take a look at the structural formula in Table 2.1. Such formulas show how bonds connect the atoms. A line between two atoms represents a single covalent bond, in which two atoms share one pair of electrons. A simple example is molecular hydrogen (H_2), with one covalent bond between hydrogen atoms (H—H). Two lines between atoms represent a double covalent bond, in which two atoms share two pairs of electrons. Molecular oxygen (O=O) has a double covalent bond linking two oxygen atoms. Three lines indicate a triple covalent bond, in which two atoms share three pairs of electrons. A triple covalent bond links two nitrogen atoms in molecular nitrogen (N≡N).

Table 2.1 Different Ways To Represent the Same Molecule

Common name	Water	Familiar term.
Chemical name	Hydrogen oxide	Systematically describes elemental composition.
Chemical formula	H_2O	Indicates unvarying proportions of elements. Subscripts show number of atoms of an element per molecule. The absence of a subscript means one atom.
Structural formula	H—O—H	Represents each covalent bond as a single line between atoms. The bond angles may also be represented.
Structural model		Shows the positions and relative sizes of atoms.
Shell model		Shows how pairs of electrons are shared in covalent bonds.

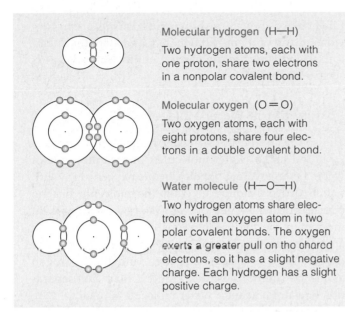

Molecular hydrogen (H—H)

Two hydrogen atoms, each with one proton, share two electrons in a nonpolar covalent bond.

Molecular oxygen (O=O)

Two oxygen atoms, each with eight protons, share four electrons in a double covalent bond.

Water molecule (H—O—H)

Two hydrogen atoms share electrons with an oxygen atom in two polar covalent bonds. The oxygen exerts a greater pull on the shared electrons, so it has a slight negative charge. Each hydrogen has a slight positive charge.

Figure 2.8 Animated Covalent bonds, in which atoms with unpaired electrons in their outermost shell become more stable by sharing electrons. Two electrons are shared in each covalent bond. When sharing is equal, the bond is nonpolar. When one atom exerts a greater pull on the electrons, the bond is polar.

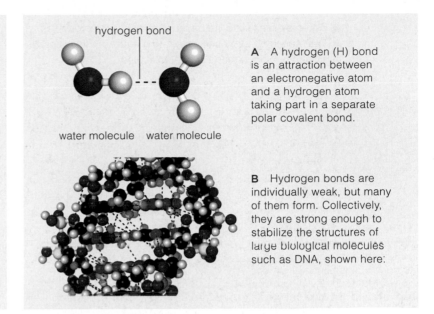

hydrogen bond

water molecule   water molecule

A A hydrogen (H) bond is an attraction between an electronegative atom and a hydrogen atom taking part in a separate polar covalent bond.

B Hydrogen bonds are individually weak, but many of them form. Collectively, they are strong enough to stabilize the structures of large biological molecules such as DNA, shown here:

Figure 2.9 Animated Hydrogen bonds. Hydrogen bonds form at a hydrogen atom taking part in a polar covalent bond. The hydrogen atom's slight positive charge weakly attracts an electronegative atom. As shown here, hydrogen (H) bonds can form between molecules or between different parts of the same molecule.

Some covalent bonds are **nonpolar**, meaning that the atoms participating in the bond are sharing electrons equally. There is no difference in charge between the two ends of such bonds. Nonpolar covalent bonds form between atoms with identical electronegativity. The molecular hydrogen (H_2), oxygen (O_2), and nitrogen (N_2) mentioned earlier are examples. These molecules are some of the gases that make up air.

Atoms participating in **polar** covalent bonds do not share electrons equally. Such bonds can form between atoms with a small difference in electronegativity. The atom that is more electronegative pulls the electrons a little more toward its "end" of the bond, so that atom bears a slightly negative charge. The atom at the other end of the bond bears a slightly positive charge.

For example, the water molecule shown in Table 2.1 has two polar covalent bonds (H—O—H). The oxygen atom carries a slight negative charge, but each of the hydrogen atoms carries a slight positive charge. Any such separation of charge into distinct positive and negative regions is called **polarity**. As you will see in the next section, the polarity of the water molecule is very important for the world of life.

Hydrogen Bonding

Hydrogen bonds form between polar regions of two molecules, or between two regions of the same molecule. A **hydrogen bond** is a weak attraction between a highly electronegative atom and a hydrogen atom taking part in a separate polar covalent bond.

Like ionic bonds, hydrogen bonds form by mutual attraction of opposite charges: The hydrogen atom has a slight positive charge and the other atom has a slight negative charge. However, unlike ionic bonds, hydrogen bonds do not make molecules out of atoms, so they are not chemical bonds.

Hydrogen bonds are weak. They form and break much more easily than covalent or ionic bonds. Even so, many of them form between molecules, or between different parts of a large one. Collectively, they are strong enough to stabilize the characteristic structures of large biological molecules (Figure 2.9).

- Water is essential to life because of its unique properties.
- The unique properties of water are a result of the extensive hydrogen bonding among water molecules.

Life evolved in water. All living organisms are mostly water, many of them still live in it, and all of the chemical reactions of life are carried out in water. What is so special about water?

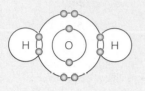

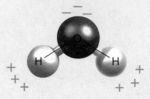

slight negative charge on the oxygen atom

slight positive charge on the hydrogen atoms

A The polarity of a water molecule arises because of the distribution of its electrons. The hydrogen atoms bear a slight positive charge, and the oxygen atom bears a slight negative charge.

Polarity of the Water Molecule

The special properties of water begin with the polarity of individual water molecules. In each molecule of water, polar covalent bonds join one oxygen atom with two hydrogen atoms. Overall, the molecule has no charge, but the oxygen pulls the shared electrons a bit more than the hydrogen atoms do. Thus, each of the atoms in a water molecule carries a slight charge: The oxygen atom is slightly negative, and the hydrogen atoms are slightly positive (Figure 2.10*a*). This separation of charge means a water molecule is polar.

The polarity of each water molecule attracts other water molecules, and hydrogen bonds form between them in tremendous numbers (Figure 2.10*b*). Extensive hydrogen bonding between water molecules imparts unique properties to water that make life possible.

B Many hydrogen bonds (dashed lines) that form and break rapidly keep water molecules clustered together in liquid water.

Water's Solvent Properties

A **solvent** is a substance, usually a liquid, that can dissolve other substances. Dissolved substances are **solutes**. Solvent molecules cluster around ions or molecules of a solute, thereby dispersing them and keeping them separated, or dissolved.

Water is a solvent. Clusters of water molecules form around the solutes in cellular fluids, tree sap, blood, the fluid in your gut, and most other fluids associated with life. When you pour table salt (NaCl) into a cup of water, the crystals of this ionically bonded solid separate into sodium ions (Na^+) and chloride ions (Cl^-). Salt dissolves in water because the negatively charged oxygen atoms of many water molecules pull on each Na^+, and the positively charged hydrogen atoms of many others pull on each Cl^- (Figure 2.11). The collective strength of many hydrogen bonds pulls the ions apart and keeps them dissolved.

Hydrogen bonds also form between water molecules and polar molecules such as sugars, so water easily dissolves polar molecules. Thus, polar molecules are **hydrophilic** (water-loving) substances. Hydrogen bonds do not form between water molecules and nonpolar molecules, such as oils, which are **hydrophobic** (water-dreading) substances. Shake a bottle filled with water and salad oil, then set it on a table. The water gathers together, and the oil clusters at the water's surface as new hydrogen bonds replace the ones broken

C Below 0°C (32°F), the hydrogen bonds hold water molecules rigidly in the three-dimensional lattice of ice. The molecules are less densely packed in ice than in liquid water, so ice floats on water.

The Arctic ice cap is melting because of global warming. It will probably be gone in fifty years, and so will polar bears. Polar bears must now swim farther between shrinking ice sheets, and they are drowning in alarming numbers.

Figure 2.10 Animated Water, a substance that is essential for life.

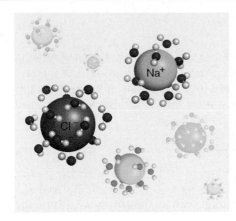

Figure 2.11 Animated Water molecules that surround an ionic solid pull its atoms apart, thereby dissolving them.

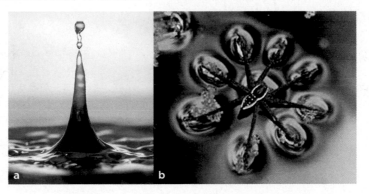

Figure 2.12 Cohesion of water. (**a**) After a pebble hits liquid water, individual molecules do not fly apart. Countless hydrogen bonds keep them together. (**b**) Cohesion keeps fishing spiders from sinking. (**c**) Water rises to the tops of plants because evaporation from leaves pulls cohesive columns of water molecules upward from the roots.

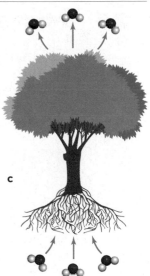

by shaking. The same interactions occur at the thin, oily membrane that separates water inside cells from water outside them. The organization of membranes—and life itself—starts with such interactions. You will read more about membranes in Chapter 5.

Water's Temperature-Stabilizing Effects

All molecules vibrate nonstop, and they move faster as they absorb heat. **Temperature** is a way to measure the energy of this molecular motion. The extensive hydrogen bonding in liquid water restricts the jiggling of water molecules. Thus, compared with other liquids, water absorbs more heat before it becomes measurably hotter. This property means that the temperature of water (and the air around it) stays relatively stable.

When the temperature of water is below its boiling point, hydrogen bonds form as fast as they break. As the water gets hotter, the molecules move faster, and individual molecules at the water's surface begin to escape into the air. By this process—**evaporation**—heat energy converts liquid water to a gas. The energy increase overcomes the attraction between water molecules, which break free.

It takes heat to convert liquid water to a gas, so the surface temperature of water decreases during evaporation. Evaporative water loss can help you and some other mammals cool off when you sweat in hot, dry weather. Sweat, which is about 99 percent water, cools the skin as it evaporates.

Below 0°C (32°F), water molecules do not jiggle enough to break hydrogen bonds, and become locked in the rigid, latticelike bonding pattern of ice (Figure 2.10c). Individual water molecules pack less densely in ice than they do in water, so ice floats on water. During cold winters, ice sheets may form near the surface of ponds, lakes, and streams. Such ice "blankets" insulate liquid water under them, so they help keep fish and other aquatic organisms from freezing.

Water's Cohesion

Another life-sustaining property of water is cohesion. **Cohesion** means that molecules resist separating from one another. You see its effect as surface tension when you toss a pebble into a pond (Figure 2.12a). Although the water ripples and sprays, individual molecules do not fly apart. Its hydrogen bonds collectively exert a continuous pull on the individual water molecules. This pull is so strong that the molecules stay together rather than spreading out in a thin film as other liquids do. Many organisms take special advantage of this unique property (Figure 2.12b).

Cohesion works inside organisms, too. For instance, plants continually absorb water as they grow. Water molecules evaporate from leaves, and replacements are pulled upward from roots (Figure 2.12c). Cohesion makes it possible for columns of liquid water to rise from roots to leaves inside narrow pipelines of vascular tissues. Section 29.3 returns to this topic.

Take-Home Message

Why is water essential to life?

■ Extensive hydrogen bonding among water molecules imparts unique properties to water that make life possible.

■ Water molecules hydrogen-bond with polar (hydrophilic) substances, dissolving them easily. They do not bond with nonpolar (hydrophobic) substances.

■ Ice is less dense than liquid water, so it floats. Ice insulates water beneath it.

■ The temperature of water is more stable than other liquids. Water also stabilizes the temperature of the air near it.

■ Cohesion keeps individual molecules of liquid water together.

■ Hydrogen ions have far-reaching effects because they are chemically active, and because there are so many of them.

■ Link to Homeostasis 1.2

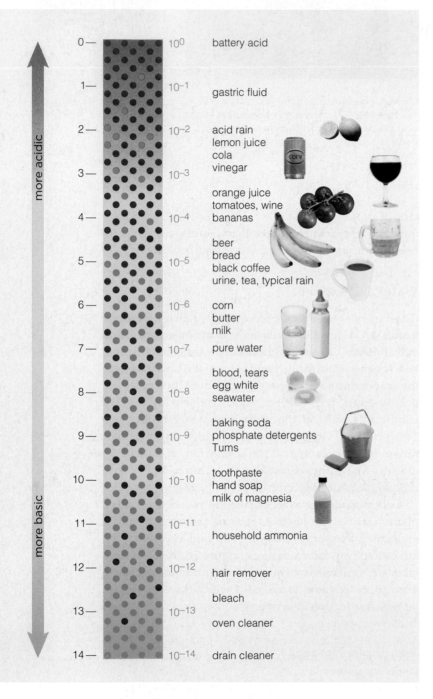

Figure 2.13 Animated A pH scale. Here, *red* dots signify hydrogen ions (H+) and *blue* dots signify hydroxyl ions (OH⁻). Also shown are approximate pH values for some common solutions.

This pH scale ranges from 0 (most acidic) to 14 (most basic). A change of one unit on the scale corresponds to a tenfold change in the amount of H+ ions (*blue* numbers).

Figure It Out: What is the approximate pH of cola?

Answer: 2.5

The pH Scale

At any given instant in liquid water, some of the water molecules are separated into hydrogen ions (H+) and hydroxide ions (OH⁻):

$$H_2O \rightleftharpoons H^+ + OH^-$$

water — hydrogen ions — hydroxyl ions

In chemical equations such as this, arrows indicate the direction of the reaction.

pH is a measure of the number of hydrogen ions in a solution. When the number of H+ ions is the same as the number of OH⁻ ions, the pH of the solution is 7, or neutral. The pH of pure water (not rainwater or tap water) is like this. The more hydrogen ions, the lower the pH. A one-unit decrease in pH corresponds to a tenfold increase in the amount of H+ ions, and a one-unit increase corresponds to a tenfold decrease in the amount of H+ ions. One way to get a sense of the difference is to taste dissolved baking soda (pH 9), distilled water (pH 7), and lemon juice (pH 2). A pH scale that ranges from 0 to 14 is shown in Figure 2.13.

Nearly all of life's chemistry occurs near pH 7. Most of your body's internal environment (tissue fluids and blood) is between pH 7.3 and 7.5.

How Do Acids and Bases Differ?

Substances called **acids** donate hydrogen ions as they dissolve in water. **Bases** accept hydrogen ions. Acidic solutions, such as lemon juice and coffee, contain more H+ than OH⁻, so their pH is below 7. Basic solutions, such as seawater and hand soap, contain more OH⁻ than H+. Basic, or alkaline, solutions have a pH greater than 7.

Acids and bases can be weak or strong. Weak acids, such as carbonic acid (H_2CO_3), are stingy H+ donors. Strong acids give up more H+ ions. One example is hydrochloric acid (HCl), which separates into H+ and Cl⁻ very easily in water:

$$HCl \rightleftharpoons H^+ + Cl^-$$

hydrochloric acid — hydrogen ions — chloride ions

Inside your stomach, the H+ from HCl makes gastric fluid acidic (pH 1–2). The acidity activates enzymes that digest proteins in your food.

Acids or bases that accumulate in ecosystems can kill organisms. For instance, fossil fuel emissions and nitrogen-containing fertilizers release strong acids into

Figure 2.14 Emissions of sulfur dioxide from a coal-burning power plant. Airborne pollutants such as sulfur dioxide dissolve in water vapor and form acidic solutions. They are a component of acid rain. The far-right photograph shows how acid rain can corrode stone sculptures.

the atmosphere. The acids lower the pH of rain (Figure 2.14). Some ecosystems are being damaged by such acid rain, which changes the composition of water and soil. Organisms in these regions are being harmed by the changes. We return to this topic in Section 48.2.

Salts and Water

A **salt** is a compound that dissolves easily in water and releases ions other than H^+ and OH^-. For example, when dissolved in water, the salt sodium chloride separates into sodium ions and chloride ions:

$$NaCl \longrightarrow Na^+ + Cl^-$$

sodium   sodium   chloride
chloride   ions   ions

Many ions are important components of cellular processes. For example, sodium, potassium, and calcium ions are critical for nerve and muscle cell function. As another example, potassium ions help plants minimize water loss on hot, dry days.

Buffers Against Shifts in pH

Cells must respond quickly to even slight shifts in pH because most enzymes and other biological molecules can function properly only within a narrow pH range. Even a slight deviation from that range can halt cellular processes completely.

Body fluids stay at a consistent pH because they are buffered. A **buffer system** is a set of chemicals, often a weak acid or base and its salt, that can keep the pH of a solution stable. It works because the two chemicals donate and accept ions that contribute to pH.

For example, when a base is added to an unbuffered fluid, the number of OH^- ions increases, so the pH rises. However, when a base is added to a buffered fluid, the acid component of the buffer releases H^+ ions. These combine with the extra OH^- ions, forming water, which does not affect pH. So, the buffered fluid's pH stays the same even when base is added.

Carbon dioxide, a gas that forms in many reactions, takes part in an important buffer system. It becomes carbonic acid when it dissolves in the water component of human blood:

$$H_2O + CO_2 \longrightarrow H_2CO_3$$

carbon dioxide   carbonic acid

The carbonic acid can separate into hydrogen ions and bicarbonate ions:

$$H_2CO_3 \longrightarrow H^+ + HCO_3-$$

carbonic acid   bicarbonate

This easily-reversed reaction constitutes the buffer system. Any excess OH^- combines with the H^+ to form water, which does not contribute to pH. Any excess H^+ combines with the bicarbonate; thus bonded, the hydrogen does not affect pH:

$$H^+ + HCO_3- \longrightarrow H_2CO_3$$

bicarbonate   carbonic acid

Together, these reactions keep the blood pH between 7.3 and 7.5—but only up to a point. A buffer system can neutralize only so many ions. Even slightly more than that limit causes the pH to swing widely.

A buffer system failure in a biological system can be catastrophic. In acute respiratory acidosis, carbon dioxide accumulates, and excess carbonic acid forms in blood. The resulting decline in blood pH may cause an individual to enter a coma, a level of unconsciousness that is dangerous. Alkalosis, a potentially lethal rise in blood pH, can also bring on coma. Even an increase to 7.8 can result in tetany, or prolonged muscle spasm.

Take-Home Message

Why are hydrogen ions important in biology?

■ Hydrogen ions contribute to pH. Acids release hydrogen ions in water; bases accept them. Salts release ions other than H^+ and OH^-.

■ Buffer systems keep the pH of body fluids stable. They are part of homeostasis.

What Are You Worth?

Contaminant or nutrient? An average human body contains highly toxic elements such as lead, arsenic, mercury, selenium, nickel, and even a few uranium atoms. The presence of these elements in the body is usually assumed to be the aftermath of environmental pollutants, but occasionally we discover that one of them has a vital function. For example, recently we found that having too little selenium can cause heart problems and thyroid disorders, so it may be part of some biological system we haven't yet unraveled.

The average body contains a substantial amount of fluorine, but as yet we know of no natural metabolic role for this element. Fluorine can substitute for other elements in biological molecules,

How would you vote?

When fluorine replaces calcium in teeth and bones, it changes the structural properties of these body parts. One effect is fewer cavities. Many communities add fluoride to drinking water. Do you want it in yours? See CengageNOW for details, then vote online.

but the substitution tends to make the molecules toxic. Several kinds of predator-deterring plant toxins are simple biological molecules with fluorine substituted for other elements.

Summary

Section 2.1 Most **atoms** have **electrons**, which have a negative **charge**. Electrons move around a **nucleus** of positively charged **protons** and, except in the case of hydrogen, uncharged **neutrons**. Atoms of an **element** have the same number of protons—the **atomic number** (Table 2.2). A **periodic table** lists all of the elements. We refer to **isotopes** of an element by their **mass number**.

Table 2.2 Summary of Players in the Chemistry of Life

Atom	Particles that are basic building blocks of all matter; the smallest unit that retains an element's properties
Element	Pure substance that consists entirely of atoms with the same, characteristic number of protons
Proton (p^+)	Positively charged particle of an atom's nucleus
Electron (e^-)	Negatively charged particle that can occupy a volume of space (orbital) around an atom's nucleus
Neutron	Uncharged particle of an atom's nucleus
Isotope	One of two or more forms of an element, the atoms of which differ in the number of neutrons
Radioisotope	Unstable isotope that emits particles and energy when its nucleus disintegrates
Tracer	Molecule that has a detectable substance (such as a radioisotope) attached
Ion	Atom that carries a charge after it has gained or lost one or more electrons
Molecule	Two or more atoms joined in a chemical bond
Compound	Molecule of two or more different elements in unvarying proportions (for example, water)
Mixture	Intermingling of two or more elements or compounds in proportions that can vary
Solute	Molecule or ion dissolved in a solvent
Acid	Substance that releases H^+ when dissolved in water
Base	Substance that accepts H^+ when dissolved in water
Salt	Substance that releases ions other than H^+ or OH^- when dissolved in water

Section 2.2 Researchers make **tracers** with detectable substances such as **radioisotopes**, which emit particles and energy as they **decay** spontaneously.

■ *Use the animation on CengageNOW to learn how radioisotopes are used in making PET scans.*

Section 2.3 We use **shell models** to view an atom's electron structure. Atoms with different numbers of electrons and protons are **ions**. Atoms with vacancies tend to interact with other atoms by donating, accepting, or sharing electrons. They form different **chemical bonds** depending on their **electronegativity**. A **compound** is a **molecule** of different elements. **Mixtures** are intermingled substances.

■ *Use the animation and interaction on CengageNOW to study electron distribution and the shell model.*

Section 2.4 An **ionic bond** is a very strong association between ions of opposite charge. Two atoms share a pair of electrons in a **covalent bond**, which may be **nonpolar** or **polar** (**polarity** is a separation of charge). **Hydrogen bonds** are weaker than either ionic or covalent bonds.

■ *Use the animation on CengageNOW to compare the types of chemical bonds in biological molecules.*

Section 2.5 **Evaporation** helps liquid water stabilize **temperature**. **Hydrophilic** substances dissolve easily in water; **hydrophobic** substances do not. **Solutes** are substances dissolved in water or another **solvent**. **Cohesion** keeps water molecules together.

■ *Use the animation on CengageNOW to view the structure of the water molecule and properties of liquid water.*

Section 2.6 **pH** reflects the number of hydrogen ions (H^+) in a solution. Typical pH scales range from 0 (most acidic) to 14 (most basic or alkaline). At neutral pH (7), the amounts of H^+ and OH^- ions are the same.

Salts are compounds that release ions other than H^+ and OH^- in water. **Acids** release H^+; **bases** accept H^+. A **buffer system** keeps a solution within a consistent range of pH. Most biological processes are buffered; they work only within a narrow pH range, usually near pH 7.

■ *Use the interaction on CengageNOW to investigate the pH of common solutions.*

Data Analysis Exercise

Living and nonliving things have the same kinds of atoms joined together as molecules, but those molecules differ in their proportions of elements and in how the atoms of those elements are arranged. The three charts in Figure 2.15 compare the proportions of some elements in the human body, Earth's crust, and seawater.

1. Which is the most abundant element in dirt? In the human body? In seawater?

2. What percentage of seawater is oxygen? Hydrogen? How many atoms of hydrogen are there for each atom of oxygen in seawater? In which molecule are hydrogen and oxygen found in that exact proportion?

3. How many atoms of chlorine are there for every atom of sodium in seawater? What common molecule has one atom of chlorine for every atom of sodium?

Human		Earth		Seawater	
Hydrogen	62.0%	Hydrogen	3.1%	Hydrogen	66.0%
Oxygen	24.0	Oxygen	60.0	Oxygen	33.0
Carbon	12.0	Carbon	0.3	Carbon	< 0.1
Nitrogen	1.2	Nitrogen	< 0.1	Nitrogen	< 0.1
Phosphorus	0.2	Phosphorus	< 0.1	Phosphorus	< 0.1
Calcium	0.2	Calcium	2.6	Calcium	< 0.1
Sodium	< 0.1	Sodium	< 0.1	Sodium	0.3
Potassium	< 0.1	Potassium	0.8	Potassium	< 0.1
Chlorine	< 0.1	Chlorine	< 0.1	Chlorine	0.3

Figure 2.15 Comparison of the abundance of some elements in a human, Earth's crust, and typical seawater. Each number is the percent of the total number of atoms in each source. For instance, 120 of every 1,000 atoms in a human body are carbon, compared with only 3 carbon atoms in every 1,000 atoms of dirt.

Self-Quiz
Answers in Appendix III

1. A(n) _____ is a molecule into which a radioisotope has been incorporated.

2. An ion is an atom that has _____ .
   a. the same number of electrons and protons
   b. a different number of electrons and protons
   c. a and b are correct

3. A(n) _____ forms when atoms of two or more elements bond covalently.

4. The measure of an atom's ability to pull electrons away from another atom is called _____ .

5. Atoms share electrons unequally in a(n) _____ bond.

6. Symbols for the elements are arranged according to _____ in the periodic table of the elements.

7. Liquid water has _____ .
   a. tracers
   b. a profusion of hydrogen bonds
   c. cohesion
   d. resistance to increases in temperature
   e. b through d
   f. all of the above

8. A(n) _____ substance repels water.

9. Hydrogen ions (H^+) are _____ .
   a. indicated by pH
   b. protons
   c. dissolved in blood
   d. all of the above

10. A(n) _____ is dissolved in a solvent.

11. When dissolved in water, a(n) _____ donates H^+.

12. A salt releases ions other than _____ in water.

13. A(n) _____ is a chemical partnership between a weak acid or base and its salt.

14. Match the terms with their most suitable description.
   ____ hydrophilic
   ____ atomic number
   ____ mass number
   ____ temperature
   a. measure of molecular motion
   b. number of protons in nucleus
   c. polar; readily dissolves in water
   d. number of protons and neutrons in nucleus

■ *Visit CengageNOW for additional questions.*

Critical Thinking

1. Alchemists were medieval scholars and philosophers who were the forerunners of modern-day chemists. Many spent their lives trying to transform lead (atomic number 82) into gold (atomic number 79). Explain why they never did succeed in that endeavor.

2. Meats are often "cured," or salted, dried, smoked, pickled, or treated with chemicals that can delay spoilage. Ever since the mid-1800s, sodium nitrite ($NaNO_2$) has been used in processed meat products such as hot dogs, bologna, sausages, jerky, bacon, and ham. Nitrites prevent growth of *Clostridium botulinum*. If ingested, this bacterium can cause a form of food poisoning called botulism.

   In water, sodium nitrite separates into sodium ions (Na^+) and nitrite ions (NO_2^-), which are called nitrites. Nitrites are rapidly converted to nitric oxide (NO), the compound that gives nitrites their preservative qualities. Eating preserved meats increases the risk of cancer, but nitrites may not be at fault. It turns out that nitric oxide has several important functions, including blood vessel dilation (for example, inside a penis during an erection), cell-to-cell signaling, and antimicrobial activities of the immune system. Draw a shell model for nitric oxide and then use it to explain why the molecule is so reactive.

3. Ozone is a chemically active form of oxygen gas. High in Earth's atmosphere, it forms a layer that absorbs about 98 percent of the sun's harmful rays. Oxygen gas consists of two covalently bonded oxygen atoms: $O=O$. Ozone has three covalently bonded oxygen atoms: $O=O-O$. Ozone reacts easily with many substances, and gives up an oxygen atom and releases gaseous oxygen ($O=O$). From what you know about chemistry, why do you suppose ozone is so reactive?

4. David, an inquisitive three-year-old, poked his fingers into warm water in a metal pan on the stove and did not sense anything hot. Then he touched the pan itself and got a nasty burn. Explain why water in a metal pan heats up far more slowly than the pan itself.

5. Some undiluted acids are more corrosive when diluted with water. That is why lab workers are told to wipe off splashes with a towel before washing. Explain.

3 | Molecules of Life

IMPACTS, ISSUES | Fear of Frying

The human body requires about one tablespoon of fat each day to remain healthy, but most of us eat far more than that. The average American consumes the equivalent of one stick of butter per day—100 pounds of fat per year—which may be part of the reason why the average American is overweight.

Being overweight increases one's risk for many diseases and health conditions. However, which type of fat we eat may be more important than how much fat we eat. Fats are more than just inert molecules that accumulate in strategic areas of our bodies if we eat too much of them. They are major constituents of cell membranes, and as such they have powerful effects on cell function.

The typical fat molecule has three tails—long carbon chains called fatty acids. Different fats have different fatty acid components. Those with a certain type of double bond in one or more of their fatty acids are called *trans* fats (Figure 3.1). Small amounts of *trans* fats occur naturally in red meat and dairy products, but most of the *trans* fats humans consume come from partially hydrogenated vegetable oil, an artificial food product.

Hydrogenation, a manufacturing process that adds hydrogen atoms to carbons, changes liquid vegetable oils into solid fats. Procter & Gamble Co. developed partially hydrogenated vegetable oil in 1908 as a substitute for the more expensive solid animal fats they were using to make candles. However, the demand for candles began to wane as more households in the United States became wired for electricity, and P & G began to look for another way to sell its proprietary fat. Partially hydrogenated vegetable oil looks a lot like lard, so in 1911 the company began marketing it as a revolutionary new food—a solid cooking fat with a long shelf life, mild flavor, and lower cost than lard or butter.

By the mid-1950s, hydrogenated vegetable oil had become a major part of the American diet. It was (and still is) found in a tremendous buffet of manufactured and fast foods: butter substitutes, cookies, crackers, cakes and pancakes, peanut butter, pies, doughnuts, muffins, chips, granola bars, breakfast bars, chocolate, microwave popcorn, pizzas, burritos, french fries, chicken nuggets, fish sticks, and so on.

For decades, hydrogenated vegetable oil was considered to be a more healthy alternative to animal fats. We now know that *trans* fats in hydrogenated vegetable oils raise the level of cholesterol in our blood more than any other fat, and they directly alter the function of our arteries and veins.

The effects of such changes are serious. Eating as little as 2 grams per day of hydrogenated vegetable oils increases a person's risk of atherosclerosis (hardening of the arteries), heart attack, and diabetes. A single serving of french fries made with hydrogenated vegetable oil contains about 5 grams of *trans* fats.

With this chapter, we introduce you to the chemistry of life. Although every living thing consists of the same basic kinds of molecules—carbohydrates, lipids, proteins, and nucleic acids—small differences in the way those molecules are put together often have big results.

trans fatty acid

See the video! Figure 3.1 *Trans* fats. The arrangement of hydrogen atoms around the carbon–carbon double bond in the middle of a *trans* fatty acid makes it a very unhealthy food. Consider skipping the french fries.

Key Concepts

Structure dictates function

We define cells partly by their capacity to build complex carbohydrates and lipids, proteins, and nucleic acids. All of these organic compounds have functional groups attached to a backbone of carbon atoms. **Sections 3.1, 3.2**

Carbohydrates

Carbohydrates are the most abundant biological molecules. They function as energy reservoirs and structural materials. Different types of complex carbohydrates are built from the same subunits of simple sugars, bonded in different patterns. **Section 3.3**

Lipids

Lipids function as energy reservoirs and as waterproofing or lubricating substances. Some are remodeled into other molecules. Lipids are the main structural component of all cell membranes. **Section 3.4**

Proteins

Structurally and functionally, proteins are the most diverse molecules of life. They include enzymes, structural materials, signaling molecules, and transporters. A protein's function arises directly from its structure. **Sections 3.5, 3.6**

Nucleotides and nucleic acids

Nucleotides have major metabolic roles and are building blocks of nucleic acids. Two kinds of nucleic acids, DNA and RNA, interact as the cell's system of storing, retrieving, and translating information about building proteins. **Section 3.7**

Links to Earlier Concepts

- Having learned about atoms, you are about to enter the next level of organization in nature: the molecules of life. Keep the big picture in mind by reviewing Section 1.1.

- You will be building on your understanding of how electrons are arranged in atoms (2.3) as well as the nature of covalent bonding and hydrogen bonding (2.4).

- Here again, you will consider one of the consequences of mutation in DNA (1.4), this time with sickle-cell anemia as the example.

How would you vote? All packaged foods in the United States now list *trans* fat content, but may be marked "zero grams of *trans* fats" even if a serving contains up to half a gram of it. Should hydrogenated vegetable oils be banned from all food? See CengageNOW for details, then vote online.

- All of the molecules of life are built with carbon atoms.
- We can use different models to highlight different aspects of the same molecule.

- Links to Elements 2.1, Covalent bonds 2.4

Carbon—The Stuff of Life

Living things are mainly oxygen, hydrogen, and carbon. Most of the oxygen and hydrogen are in the form of water. Put water aside, and carbon makes up more than half of what is left.

The carbon in living organisms is part of the molecules of life—complex carbohydrates, lipids, proteins, and nucleic acids. These molecules consist primarily of hydrogen and carbon atoms, so they are **organic**. The term is a holdover from a time when such molecules were thought to be made only by living things, as opposed to the "inorganic" molecules that formed by nonliving processes. The term persists, even though we now know that organic compounds were present on Earth long before organisms were, and we can also make them in laboratories.

Carbon's importance to life starts with its versatile bonding behavior. Each carbon atom can form covalent bonds with one, two, three, or four other atoms. Depending on the other elements in the resulting molecule, such bonds may be polar or nonpolar. Many organic compounds have a backbone—a chain of carbon atoms—to which other atoms attach. The ends of a backbone may join so that the carbon chain forms one or more ring structures (Figure 3.2). Such versatility means that carbon atoms can be assembled and remodeled into a variety of organic compounds.

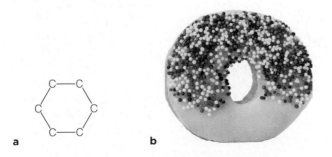

Figure 3.2 Carbon rings. (**a**) Carbon's versatile bonding behavior allows it to form a variety of structures, including rings. (**b**) Carbon rings form the framework of many sugars, starches, and fats, such as those found in doughnuts.

Representing Structures of Organic Molecules

Any molecule's structure can be depicted using different kinds of molecular models. Such models allow us to see different characteristics of the same molecule.

For example, structural models such as the one at *right* show how all the atoms in a molecule connect to one another. In such models, each line indicates one covalent bond. A double line (=) indicates a double bond; a triple line (≡) indicates a triple bond. Some of the atoms or bonds in a molecule may be implied but not shown. Hydrogen atoms bonded to a carbon backbone may also be omitted, and other atoms as well.

glucose

Carbon ring structures such as the ones that occur in glucose and other sugars are often represented as polygons. If no atom is shown at a corner or at the end of a bond, a carbon atom is implied there:

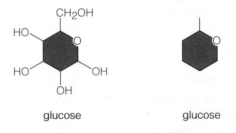

glucose   glucose

Ball-and-stick models such as the one at *right* show the positions of the atoms in three dimensions. Single, double, and triple covalent bonds are all shown as one stick connecting two balls, which represent atoms. Ball size reflects relative size of an atom. Elements are usually coded by color:

glucose

carbon   hydrogen   oxygen   nitrogen   phosphorus

Space-filling models such as the one at *right* show how atoms that are sharing electrons overlap. The elements in space-filling models are coded using the same color scheme as the ones in ball-and-stick models.

glucose

red blood cell

Figure 3.3 shows three different ways to represent the same molecule, hemoglobin, a protein that colors your blood red. Hemoglobin transports oxygen to tissues throughout the body of all vertebrates (animals that have a backbone). A ball-and-stick or space-filling model of such a large molecule can appear very complicated if all of the atoms are included. The space-filling model in Figure 3.3a is an example.

To reduce visual complexity, other types of models omit individual atoms. Surface models of large molecules can reveal large-scale features, such as folds or pockets, that can be difficult to see when individual atoms are shown. For example, in the surface model of hemoglobin in Figure 3.3b, you can see folds of the molecule that cradle two hemes. Hemes are complex carbon ring structures that often have an iron atom at their center. They are part of many important proteins that you will encounter in this book.

Very large molecules such as hemoglobin are often shown as ribbon models. Such models highlight different features of the structure, such as coils or sheets. In a ribbon model of hemoglobin (Figure 3.3c), you can see that the protein consists of four coiled components, each of which folds around a heme.

Such structural details are clues to how a molecule functions. For example, hemoglobin, which is the main oxygen-carrier in vertebrate blood, has four hemes. Oxygen binds at the hemes, so each hemoglobin molecule can carry up to four molecules of oxygen.

A A space-filling model of hemoglobin shows the complexity of the molecule.

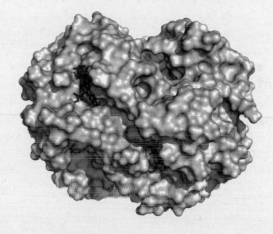

B A surface model of the same molecule reveals crevices and folds that are important for its function. Heme groups, in *red*, are cradled in pockets of the molecule.

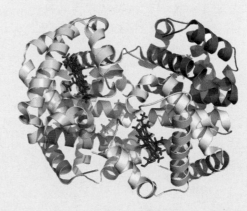

C A ribbon model of hemoglobin shows all four heme groups, also in *red*, held in place by the molecule's coils.

Figure 3.3 Visualizing the structure of hemoglobin, the oxygen-transporting molecule in red blood cells (*top left*). Models that show individual atoms usually depict them color-coded by element. Other models may be shown in various colors, depending on which features are highlighted.

Take-Home Message

How are all of the molecules of life alike?

■ Carbohydrates, lipids, proteins, and nucleic acids are organic molecules, which consist mainly of carbon and hydrogen atoms.

■ The structure of an organic molecule starts with its carbon backbone, a chain of carbon atoms that may form a ring.

■ We use different models to represent different characteristics of a molecule's structure. Considering a molecule's structural features gives us insight into how it functions.

From Structure to Function

- The function of organic molecules in biological systems begins with their structure.

- Links to Ions 2.3, Polarity 2.4, Acids and bases 2.6

All biological systems are based on the same organic molecules—a legacy of life's common origin—but the details of those molecules can differ among organisms. Remember, depending on the way carbon atoms bond together, they can form diamond, the hardest mineral, or graphite, one of the softest (Section 2.4). Similarly, the building blocks of carbohydrates, lipids, proteins, and nucleic acids bond together in different arrangements to form different molecules.

Functional Groups

An organic molecule that consists only of hydrogen and carbon atoms is called a hydrocarbon. Methane, the simplest hydrocarbon, is one carbon atom bonded to four hydrogen atoms. Most of the molecules of life have at least one **functional group**: a cluster of atoms covalently bonded to a carbon atom of an organic molecule. Functional groups impart specific chemical properties to a molecule, such as polarity or acidity. Figure 3.4 lists a few functional groups that are common in carbohydrates, lipids, proteins, and nucleic acids.

For example, alcohols are a class of organic compounds that have hydroxyl groups (—OH). These polar functional groups can form hydrogen bonds, so alcohols (at least the small ones) dissolve quickly in water. Larger alcohols do not dissolve as easily, because their long, nonpolar hydrocarbon chains repel water. Fatty acids also are like this, which is why lipids that have fatty acid tails do not dissolve easily in water.

Methyl groups impart nonpolar character. Reactive carbonyl groups (—C=O) are part of fats and carbohydrates. Carboxyl groups (—COOH) make amino acids and fatty acids acidic. Amine groups are basic. ATP releases chemical energy when it donates a phosphate group (PO_4) to another molecule. DNA and RNA also contain phosphate groups. Bonds between sulfhydryl

Group	Character	Location	Structure
hydroxyl	polar	amino acids; sugars and other alcohols	—OH
methyl	nonpolar	fatty acids, some amino acids	![methyl structure]
carbonyl	polar, reactive	sugars, amino acids, nucleotides	(aldehyde) (ketone)
carboxyl	acidic	amino acids, fatty acids, carbohydrates	(ionized)
amine	basic	amino acids, some nucleotide bases	—N—H —NH+ (ionized)
phosphate	high energy, polar	nucleotides (e.g., ATP); DNA and RNA; many proteins; phospholipids	—O—P—O− (P) icon
sulfhydryl	forms disulfide bridges	cysteine (an amino acid)	—SH —S—S— (disulfide bridge)

Figure 3.4 Animated Common functional groups in biological molecules, with examples of where they occur. Because such groups impart specific chemical characteristics to organic compounds, they are an important part of why the molecules of life function as they do.

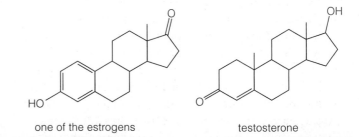

one of the estrogens   testosterone

female wood duck   male wood duck

Figure 3.5 Estrogen and testosterone, sex hormones that cause differences in traits between males and females of many species such as wood ducks (*Aix sponsa*). **Figure It Out:** Which functional groups differ between these hormones?   *Answer: The hydroxyl and carbonyl groups differ in position, and testosterone has an extra methyl group.*

Table 3.1 What Cells Do to Organic Compounds

Type of Reaction	What Happens
Condensation	Two molecules covalently bond into a larger one.
Cleavage	A molecule splits into two smaller ones. Hydrolysis is an example.
Functional group transfer	A functional group is transferred from one molecule to another.
Electron transfer	Electrons are transferred from one molecule to another.
Rearrangement	Juggling of covalent bonds converts one organic compound into another.

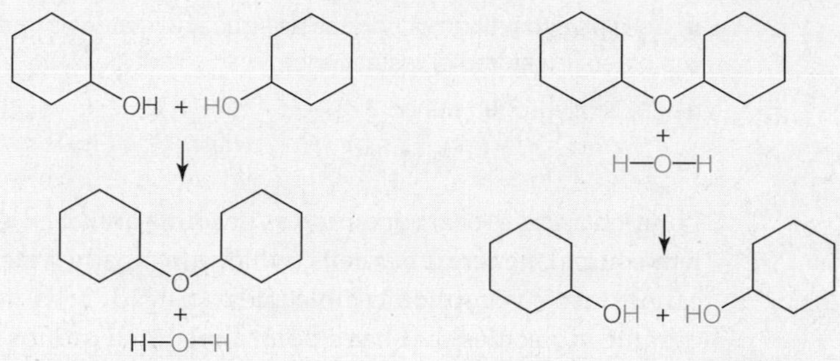

A Condensation. An —OH group from one molecule combines with an H atom from another. Water forms as the two molecules bond covalently.

B Hydrolysis. A molecule splits, then an —OH group and an H atom from a water molecule become attached to sites exposed by the reaction.

Figure 3.6 Animated Two examples of what happens to the organic molecules in cells. (**a**) In condensation, two molecules are covalently bonded into a larger one. (**b**) In hydrolysis, a water-requiring cleavage reaction splits a larger molecule into two smaller molecules.

groups (—SH) stabilize the structure of many proteins. Heat and some kinds of chemicals can temporarily break sulfhydryl bonds in human hair, which is why we can curl straight hair and straighten curly hair.

How much can one functional group do? Consider a seemingly minor difference in the functional groups of two structurally similar sex hormones (Figure 3.5). Early on, an embryo of a wood duck, human, or any other vertebrate is neither male nor female. If it starts making the hormone testosterone, a set of tubes and ducts will become male sex organs and male traits will develop. Without testosterone, those ducts and tubes become female sex organs, and hormones called estrogens will guide the development of female traits.

What Cells Do to Organic Compounds

Metabolism refers to activities by which cells acquire and use energy as they construct, rearrange, and split organic compounds. These activities help each cell stay alive, grow, and reproduce. They require enzymes—proteins that make reactions proceed faster than they would on their own. Some of the most common metabolic reactions are listed in Table 3.1. We will revisit these reactions in Chapter 6. For now, just start thinking about two of them.

With **condensation**, two molecules covalently bond into a larger one. Water usually forms as a product of condensation when enzymes remove an —OH group

from one of the molecules and a hydrogen atom from the other (Figure 3.6a). Some large molecules such as starch form by repeated condensation reactions.

Cleavage reactions split large molecules into smaller ones. One type of cleavage reaction, **hydrolysis**, is the reverse of condensation (Figure 3.6b). Enzymes break a bond by attaching a hydroxyl group to one atom and a hydrogen to the other. The —OH and —H are derived from a water molecule.

Cells maintain pools of small organic molecules—simple sugars, fatty acids, amino acids, and nucleotides. Some of these molecules are sources of energy. Others are used as subunits, or **monomers**, to build larger molecules that are the structural and functional parts of cells. These larger molecules, or **polymers**, are chains of monomers. When cells break down a polymer, the released monomers may be used for energy, or they may reenter cellular pools.

Take-Home Message

How do organic molecules work in living systems?

■ An organic molecule's structure dictates its function in biological systems.

■ Functional groups impart certain chemical characteristics to organic molecules. Such groups contribute to the function of biological molecules.

■ By reactions such as condensation, cells assemble large molecules from smaller subunits of simple sugars, fatty acids, amino acids, and nucleotides.

■ By reactions such as hydrolysis, cells split large organic molecules into smaller ones, and convert one type of molecule to another.

■ Carbohydrates are the most plentiful biological molecules in the biosphere.

■ Cells use some carbohydrates as structural materials; they use others for stored or instant energy.

■ Link to Hydrogen bonds 2.4

Long-chain hydrocarbons such as gasoline are an excellent source of energy, but cells (which are mostly water) cannot use hydrophobic molecules. Instead, cells use organic molecules that have polar functional groups— molecules that are easily assembled and broken apart inside a cell's watery interior.

Carbohydrates are organic compounds that consist of carbon, hydrogen, and oxygen in a 1:2:1 ratio. Cells use different kinds as structural materials and as sources of instant energy. The three main types of carbohydrates in living systems are monosaccharides, oligosaccharides, and polysaccharides.

Simple Sugars

"Saccharide" is from a Greek word that means sugar. Monosaccharides (one sugar unit) are the simplest of the carbohydrates. Common monosaccharides have a backbone of five or six carbon atoms, one ketone or aldehyde group, and two or more hydroxyl groups. Most monosaccharides are water soluble, so they are

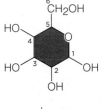

glucose

easily transported throughout the internal environments of all organisms.

Sugars that are part of DNA and RNA are monosaccharides with five carbon atoms. Glucose (at *left*) has six carbons. Cells use glucose as an energy source or as a structural material. They also use it as a precursor, or parent molecule, that

they remodel into other molecules. For example, vitamin C is derived from glucose.

Short-Chain Carbohydrates

An oligosaccharide is a short chain of covalently bonded monosaccharides (*oligo*– means a few). As examples, disaccharides consist of two sugar monomers. The lactose in milk is a disaccharide, with one glucose and one galactose unit. Sucrose, the most plentiful sugar in nature, has a glucose and a fructose unit (Figure 3.7). Sucrose extracted from sugarcane or sugar beets is our table sugar. Oligosaccharides with three or more sugar units are often attached to lipids or proteins that have important functions in immunity.

Complex Carbohydrates

The "complex" carbohydrates, or polysaccharides, are straight or branched chains of many sugar monomers —often hundreds or thousands. There may be one type or many types of monomers in a polysaccharide. The most common polysaccharides are cellulose, glycogen, and starch. All consist of glucose monomers, but they differ in their chemical properties. Why? The answer begins with differences in patterns of covalent bonding that link their glucose units (Figure 3.8).

For example, the covalent bonding pattern of starch makes the molecule coil like a spiral staircase (Figure 3.8*b*). Starch does not dissolve easily in water, so it resists hydrolysis. This stability is a reason why starch is used to store chemical energy in the watery, enzyme-filled interior of plant cells.

Most plants make much more glucose than they can use. The excess is stored as starch, in roots, stems, and leaves. However, because it is insoluble, starch

glucose   +   fructose   →   sucrose   +   water

Figure 3.7 Animated The synthesis of a sucrose molecule is an example of a condensation reaction. You are already familiar with sucrose—it is common table sugar.

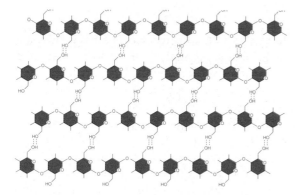

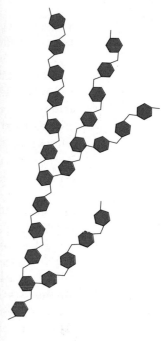

a Cellulose, a structural component of plants. Chains of glucose units stretch side by side and hydrogen bond at many —OH groups. The hydrogen bonds stabilize the chains in tight bundles that form long fibers. Very few types of organisms can digest this tough, insoluble material.

b In amylose, one type of starch, a series of glucose units form a chain that coils. Starch is the main energy reserve in plants, which store it in their roots, stems, leaves, fruits, and seeds (such as coconuts).

c Glycogen. In animals, this polysaccharide functions as an energy reservoir. It is especially abundant in the liver and muscles of active animals, including people.

Figure 3.8 Structure of (**a**) cellulose, (**b**) starch, and (**c**) glycogen, and their typical locations in a few organisms. All three carbohydrates consist only of glucose units, but the different bonding patterns that link the subunits result in substances with very different properties.

cannot be transported out of cells and distributed to other parts of the plant. When sugars are in short supply, hydrolysis enzymes nibble at the bonds between starch's sugar monomers. Cells make the disaccharide sucrose from the released glucose molecules. Sucrose is soluble and easily transported.

Cellulose, the major structural material of plants, may be the most abundant organic molecule in the biosphere. Glucose chains stretch side by side (Figure 3.8a). Hydrogen bonding between the chains stabilizes them in tight, sturdy bundles. Plant cell walls contain long cellulose fibers. Like steel rods inside reinforced concrete pillars, the tough fibers help tall stems resist wind and other forms of mechanical stress.

In animals, glycogen is the sugar-storage equivalent of starch in plants (Figure 3.8c). Muscle and liver cells store it. When the sugar level in blood falls, the liver cells break down glycogen, and the released glucose subunits enter the blood.

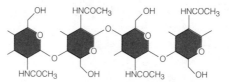

Figure 3.9 Chitin. This polysaccharide strengthens the hard parts of many small animals, such as crabs.

Chitin is a polysaccharide with nitrogen-containing groups on its many glucose monomers (Figure 3.9). Chitin strengthens hard parts of many animals, including the outer cuticle of crabs, beetles, and ticks. It also reinforces the cell wall of many fungi.

Take-Home Message

What are carbohydrates?

■ Subunits of simple carbohydrates (sugars), arranged in different ways, form various types of complex carbohydrates.

■ Cells use carbohydrates for energy, storage, or as structural materials.

3.4 | Greasy, Oily—Must Be Lipids

■ Lipids function as the body's major energy reservoir, and as the structural foundation of cell membranes.

Lipids are fatty, oily, or waxy organic compounds that are insoluble in water. Many lipids incorporate **fatty acids**: simple organic compounds that have a carboxyl group joined to a backbone of four to thirty-six carbon atoms (Figure 3.10).

Fats

Fats are lipids with one, two, or three fatty acids that dangle like tails from a small alcohol called glycerol. Most neutral fats, such as butter and vegetable oils, are triglycerides. **Triglycerides** are fats with three fatty acid tails linked to the glycerol (Figure 3.11). Triglycerides are the most abundant energy source in vertebrate bodies, and the richest. Gram for gram, triglycerides hold more than twice the energy of glycogen. Triglycerides are concentrated in adipose tissue that insulates and cushions parts of the body.

Figure 3.10 Examples of fatty acids. (**a**) The backbone of stearic acid is fully saturated with hydrogen atoms. (**b**) Oleic acid, with a double bond in its backbone, is unsaturated. (**c**) Linolenic acid, also unsaturated, has three double bonds. The first double bond occurs at the third carbon from the end of the tail, so oleic acid is called an omega-3 fatty acid. Omega-3 and omega-6 fatty acids are "essential fatty acids." Your body does not make them, so they must come from food.

The fatty acid tails of saturated fats have only single covalent bonds. Animal fats tend to remain solid at room temperature because their saturated fatty acid tails pack tightly. Fatty acid tails of unsaturated fats have one or more double covalent bonds. Such rigid bonds usually form kinks that prevent unsaturated fats from packing tightly (Figure 3.12*a*). Most vegetable oils are unsaturated, so they tend to remain liquid at room temperature. Partially hydrogenated vegetable oils are an exception. The double bond in these *trans* fatty acids keeps them straight. *Trans* fats pack tightly, so they are solid at room temperature (Figure 3.12*b*).

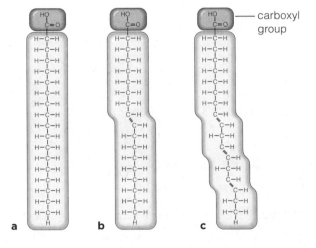

carboxyl group

a   b   c

Figure 3.11 Animated Triglyceride formation by the condensation of three fatty acids with one glycerol molecule. The photograph shows triglyceride-insulated emperor penguins during an Antarctic blizzard.

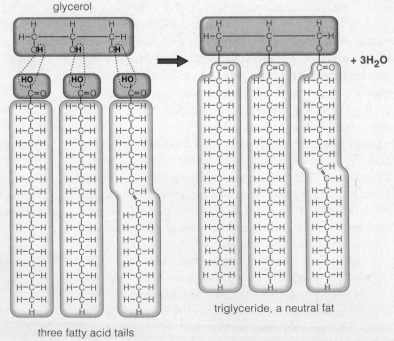

glycerol

$+ 3H_2O$

triglyceride, a neutral fat

three fatty acid tails

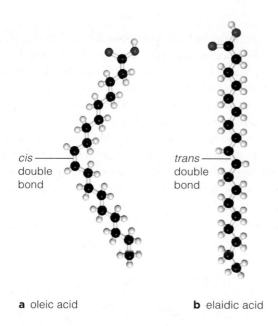

cis
double
bond

trans
double
bond

a oleic acid

b elaidic acid

Figure 3.12 The only difference between (**a**) oleic acid (a *cis* fatty acid) and (**b**) elaidic acid (a *trans* fatty acid) is the arrangement of hydrogens around one double bond. *Trans* fatty acids form during chemical hydrogenation processes.

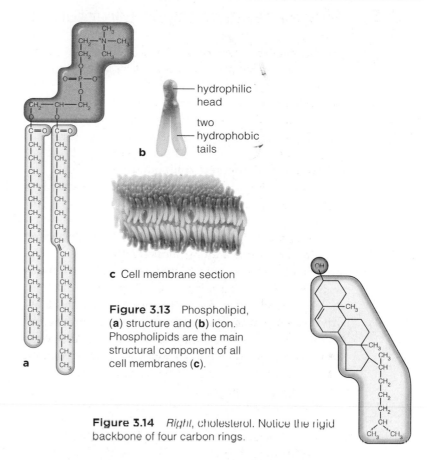

hydrophilic head

two hydrophobic tails

b

c Cell membrane section

Figure 3.13 Phospholipid, (**a**) structure and (**b**) icon. Phospholipids are the main structural component of all cell membranes (**c**).

a

Figure 3.14 *Right*, cholesterol. Notice the rigid backbone of four carbon rings.

Phospholipids

Phospholipids have a polar head with a phosphate in it, and two nonpolar fatty acid tails. They are the most abundant lipids in cell membranes, which have two phospholipid layers (Figure 3.13*a–c*). The heads of one layer are dissolved in the cell's watery interior, and the heads of the other layer are dissolved in the cell's fluid surroundings. All of the hydrophilic tails are sandwiched between the heads. You will read about membrane structure and function in Chapters 4 and 5.

Waxes

Waxes are complex, varying mixtures of lipids with long fatty acid tails bonded to long-chain alcohols or carbon rings. The molecules pack tightly, so the resulting substance is firm and water-repellent. Waxes in the cuticle that covers the exposed surfaces of plants help restrict water loss and keep out parasites and other pests. Other types of waxes protect, lubricate, and soften skin and hair. Waxes, together with fats and fatty acids, make feathers waterproof. Bees store honey and raise new generations of bees inside honeycomb, which they make from beeswax.

Cholesterol and Other Steroids

Steroids are lipids with a rigid backbone of four carbon rings and no fatty acid tails. They differ in the type, number, and position of functional groups. All eukaryotic cell membranes contain steroids. In animal tissues, cholesterol is the most common steroid (Figure 3.14). Cholesterol is remodeled into many molecules, such as bile salts (which help digest fats) and vitamin D (required to keep teeth and bones strong). Steroid hormones are derived from cholesterol. Estrogens and testosterone, hormones that govern reproduction and secondary sexual traits, are examples (Figure 3.5).

Take-Home Message

What are lipids?

■ Lipids are fatty, waxy, or oily organic compounds. They resist dissolving in water. The main classes of lipids are triglycerides, phospholipids, waxes, and steroids.

■ Triglycerides function as energy reservoirs in vertebrate animals.

■ Phospholipids are the main component of cell membranes.

■ Waxes are components of water-repelling and lubricating secretions.

■ Steroids are components of cell membranes, and precursors of many other molecules.

- Proteins are the most diverse biological molecule.
- Cells build thousands of different proteins by stringing together amino acids in different orders.

- Link to Covalent bonding 2.4

Proteins and Amino Acids

A **protein** is an organic compound composed of one or more chains of amino acids. An **amino acid** is a small organic compound with an amine group, a carboxyl group (the acid), and one or more atoms called an "R group." Typically, these groups are all attached to the same carbon atom (Figure 3.15). In water, the functional groups ionize: The amine group occurs as $—NH_3^+$, and the carboxyl group occurs as $—COO^-$.

Of all biological molecules, proteins are the most diverse. Structural proteins make up spiderwebs and feathers, hooves, hair, and many other body parts. Nutritious types abound in foods such as seeds and

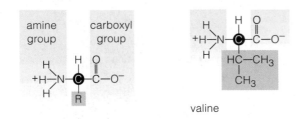

valine

Figure 3.15 Generalized structure of amino acids, and an example. *Green* boxes highlight R groups. Appendix V has models of all twenty of the common amino acids.

eggs. Most enzymes are proteins. Proteins move substances, help cells communicate, and defend the body. Amazingly, cells can synthesize thousands of different proteins from only twenty kinds of amino acids. The complete structures of those twenty amino acids are shown in Appendix V.

Protein synthesis involves bonding amino acids into chains called **polypeptides**. For each type of protein, instructions coded in DNA specify the order in which any of the twenty kinds of amino acids will occur at every place in the chain. A condensation reaction joins the amine group of an amino acid with the carboxyl group of the next in a peptide bond (Figure 3.16).

Levels of Protein Structure

Each type of protein has a unique sequence of amino acids. This sequence is known as the protein's primary structure (Figure 3.17*a*). Secondary structure emerges as the chain twists, bends, loops, and folds. Hydrogen bonding between amino acids makes stretches of the polypeptide chain form a sheet, or coil into a helix a bit like a spiral staircase (Figure 3.17*b*). The primary structure of each type of protein is unique, but similar patterns of coils and sheets occur in most proteins.

Much as an overly twisted rubber band coils back on itself, the coils, sheets, and loops of a protein fold up even more into compact domains. A "domain" is a part of a protein that is organized as a structurally stable unit. Such units are a protein's tertiary structure, its third level of organization. Tertiary structure

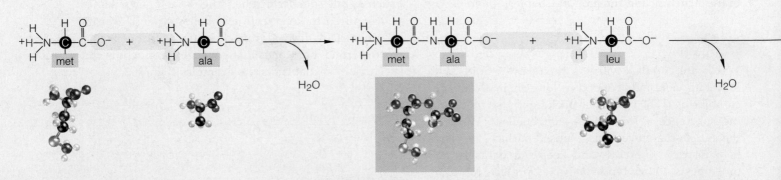

A DNA encodes the order of amino acids in a new polypeptide chain. Methionine (met) is typically the first amino acid.

B In a condensation reaction, a peptide bond forms between the methionine and the next amino acid, alanine (ala) in this example. Leucine (leu) will be next. Think about polarity, charge, and other properties of functional groups that become neighbors in the growing chain.

Figure 3.16 Animated Examples of peptide bond formation. Chapter 14 returns to protein synthesis.

makes a protein a working molecule. For instance, the barrel-shaped domains of some proteins function as tunnels through membranes (Figure 3.17c).

Many proteins also have a fourth level of organization, or quaternary structure: They consist of two or more polypeptide chains bonded together or in close association (Figure 3.17d). Most enzymes and many other proteins are globular, with several polypeptide chains folded into shapes that are roughly spherical. Hemoglobin, described shortly, is an example.

Enzymes often attach linear or branched oligosaccharides to polypeptide chains, forming glycoproteins such as those that impart unique molecular identity to a tissue or to a body.

Some proteins aggregate by many thousands into much larger structures, with their polypeptide chains organized into strands or sheets. Some of these fibrous proteins contribute to the structure and organization of cells and tissues. The keratin in your fingernails is an example. Other fibrous proteins, such as the actin and myosin filaments in muscle cells, are part of the mechanisms that help cells and cell parts move.

Take-Home Message

What are proteins?

■ Proteins consist of chains of amino acids. The order of amino acids in a polypeptide chain dictates the type of protein.

■ Polypeptide chains twist and fold into coils, sheets, and loops, which fold and pack further into functional domains.

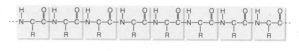

a Protein primary structure: Amino acids bonded as a polypeptide chain.

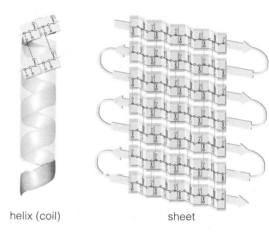

b Protein secondary structure: A coiled (helical) or sheetlike array held in place by hydrogen bonds (*dotted lines*) between different parts of the polypeptide chain.

helix (coil)   sheet

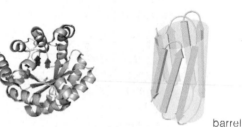

c Protein tertiary structure: A chain's coils, sheets, or both fold and twist into stable, functional domains such as barrels or pockets.

barrel

d Protein quaternary structure: two or more polypeptide chains associated as one molecule.

Figure 3.17 Four levels of a protein's structural organization.

C A peptide bond forms between the alanine and leucine. Tryptophan (trp) will be next. The chain is starting to twist and fold as atoms swivel around some bonds and attract or repel their neighbors.

D The sequence of amino acid subunits in this newly forming peptide chain is now met–ala–leu–trp. The process may continue until there are hundreds or thousands of amino acids in the chain.

3.6 Why Is Protein Structure So Important?

- When a protein's structure goes awry, so does its function.
- Links to Inheritance 1.2, Acids and bases 2.6

Just One Wrong Amino Acid . . .

Sometimes a protein's amino acid sequence changes, with drastic consequences. Let's use hemoglobin as an example. As blood moves through lungs, the hemoglobin inside red blood cells binds oxygen, then gives it up in regions of the body where oxygen levels are low. After giving up oxygen to tissues, the blood circulates back to the lungs, where the hemoglobin inside red blood cells binds more oxygen.

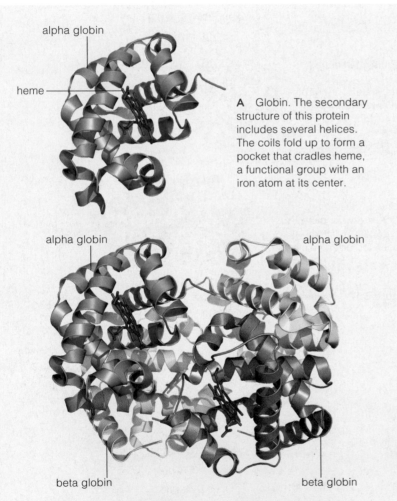

A Globin. The secondary structure of this protein includes several helices. The coils fold up to form a pocket that cradles heme, a functional group with an iron atom at its center.

B Hemoglobin is one of the proteins with quaternary structure. It consists of four globin molecules held together by hydrogen bonds. To help you distinguish among them, the two alpha globin chains are shown here in *green*, and the two beta globin chains are in *brown*.

Figure 3.18 Animated Globin and hemoglobin. (**a**) Globin, a coiled polypeptide chain that cradles heme, a functional group with an iron atom. (**b**) Hemoglobin, an oxygen-transport protein in red blood cells.

Hemoglobin's oxygen-binding properties depend on its structure. Each of the four globin chains in the protein forms a pocket that holds an iron-containing heme group (Figure 3.3 and 3.18). One oxygen molecule can bind to each heme in a hemoglobin protein.

Globin occurs in two slightly different forms, alpha globin and beta globin. In adult humans, two of each form fold up into a hemoglobin molecule. Negatively charged glutamic acid is normally the sixth amino acid in the beta globin chain, but a DNA mutation sometimes puts a different amino acid—valine—in the sixth position (Figure 3.19*a,b*). Valine is uncharged.

As a result of that one substitution, a tiny patch of the protein changes from polar to nonpolar—which in turn causes the protein's behavior to change slightly. Hemoglobin altered this way is called HbS. Under some conditions, molecules of HbS form large, stable, rod-shaped clumps. Red blood cells containing these clumps become distorted into a sickled shape (Figure 3.19*c*). Sickled cells tend to clog tiny blood vessels and disrupt blood circulation.

A human has two genes for beta globin, one inherited from each of two parents. (Genes are units of DNA that can encode proteins.) Cells use both genes to make beta globin. If one of a person's genes is normal and the other has the valine mutation, he or she makes enough normal hemoglobin to survive, but not enough to be completely healthy. Someone with two mutated globin genes can make only HbS hemoglobin. The outcome is sickle-cell anemia, a severe genetic disorder (Figure 3.19*d*).

Proteins Undone—Denaturation

The shape of a protein defines its biological activity: Globin cradles heme, an enzyme speeds a reaction, a receptor responds to some signal. These—and all other proteins—function as long as they stay coiled, folded, and packed in their correct three-dimensional shapes. Heat, shifts in pH, salts, and detergents can disrupt the hydrogen bonds that maintain a protein's shape. Without the bonds that hold them in their three-dimensional shape, proteins and other large biological molecules **denature**—their shape unravels and they no longer function.

Consider albumin, a protein in the white of an egg. When you cook eggs, the heat does not disrupt the covalent bonds of albumin's primary structure. But it destroys albumin's weaker hydrogen bonds, and so the protein unfolds. When the translucent egg white turns opaque, we know albumin has been altered. For a few proteins, denaturation might be reversed if and

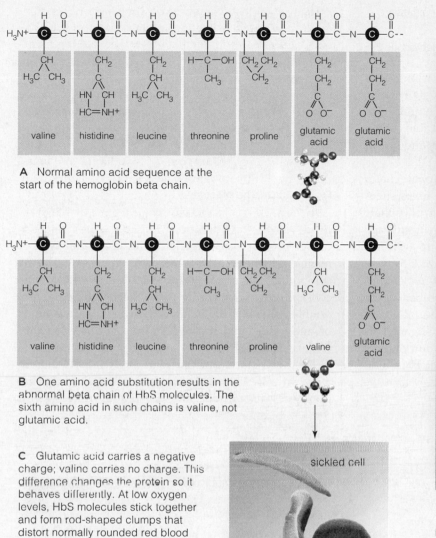

A Normal amino acid sequence at the start of the hemoglobin beta chain.

valine · histidine · leucine · threonine · proline · glutamic acid · glutamic acid

B One amino acid substitution results in the abnormal beta chain of HbS molecules. The sixth amino acid in such chains is valine, not glutamic acid.

valine · histidine · leucine · threonine · proline · valine · glutamic acid

C Glutamic acid carries a negative charge; valine carries no charge. This difference changes the protein so it behaves differently. At low oxygen levels, HbS molecules stick together and form rod-shaped clumps that distort normally rounded red blood cells into sickle shapes. (A sickle is a farm tool that has a crescent-shaped blade.)

sickled cell

normal cell

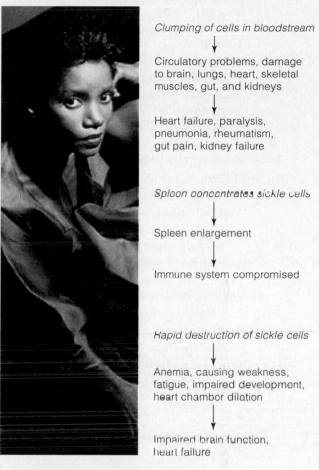

Clumping of cells in bloodstream

↓

Circulatory problems, damage to brain, lungs, heart, skeletal muscles, gut, and kidneys

↓

Heart failure, paralysis, pneumonia, rheumatism, gut pain, kidney failure

Spleen concentrates sickle cells

↓

Spleen enlargement

↓

Immune system compromised

Rapid destruction of sickle cells

↓

Anemia, causing weakness, fatigue, impaired development, heart chamber dilation

↓

Impaired brain function, heart failure

D Melba Moore is a celebrity spokesperson for sickle-cell anemia organizations. *Right*, range of symptoms for a person with two mutated genes for hemoglobin's beta chain.

Figure 3.19 Animated Sickle-cell anemia's molecular basis and symptoms. Section 18.6 explores evolutionary and ecological pressures that maintain this genetic disorder in human populations.

when normal conditions return, but albumin is not one of them. There is no way to uncook an egg.

A protein's structure dictates its function. Enzymes, hormones, transporters, hemoglobin—such proteins are critical for our survival. Their coiled, twisted and folded polypeptide chains form anchors, membrane-spanning barrels, or jaws that attack foreign proteins in the body. Mutations can alter the chains enough to block or enhance an anchoring, transport, or defense function. Sometimes the consequences are awful. Yet such changes also give rise to variation in traits, which is the raw material of evolution. Learn about protein structure and you are on your way to understanding life's richly normal and abnormal expressions.

Take-Home Message

Why is protein structure important?

■ A protein's function depends on its structure.

■ Mutations that alter a protein's structure may also alter its function.

■ Protein shape unravels if hydrogen bonds are disrupted.

Nucleic Acids

■ Nucleotides are subunits of DNA and RNA. Some have roles in metabolism.

■ Links to Inheritance 1.2, Diversity 1.4, Hydrogen bonds 2.4

Nucleotides are small organic molecules, various kinds of which function as energy carriers, enzyme helpers, chemical messengers, and subunits of DNA and RNA. Each nucleotide consists of a sugar with a five-carbon ring, bonded to a nitrogen-containing base and one or more phosphate groups.

The nucleotide **ATP** (adenosine triphosphate) has a row of three phosphate groups attached to its sugar (Figure 3.20). ATP transfers its outermost phosphate group to other molecules and so primes them to react. You will read about such phosphate-group transfers and their important metabolic role in Chapter 5.

Nucleic acids are polymers—chains of nucleotides in which the sugar of one nucleotide is joined to the phosphate group of the next. An example is **RNA**, or ribonucleic acid, named after the ribose sugar of its component nucleotides. RNA consists of four kinds of nucleotide monomers, one of which is ATP. RNA molecules are important in protein synthesis, which we will discuss in Chapter 14.

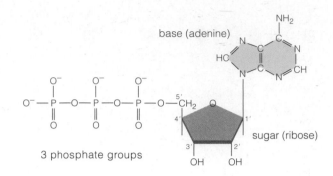

Figure 3.20 The structure of ATP.

DNA, or deoxyribonucleic acid, is another type of nucleic acid named after the deoxyribose sugar of its component nucleotides (Figure 3.21). A DNA molecule consists of two nucleotide chains twisted together as a double helix. Hydrogen bonds between the four kinds of nucleotide hold the two strands of DNA together (Figure 3.22).

Each cell starts out life with DNA inherited from a parent cell. That DNA contains all of the information necessary to build a new cell and, in the case of

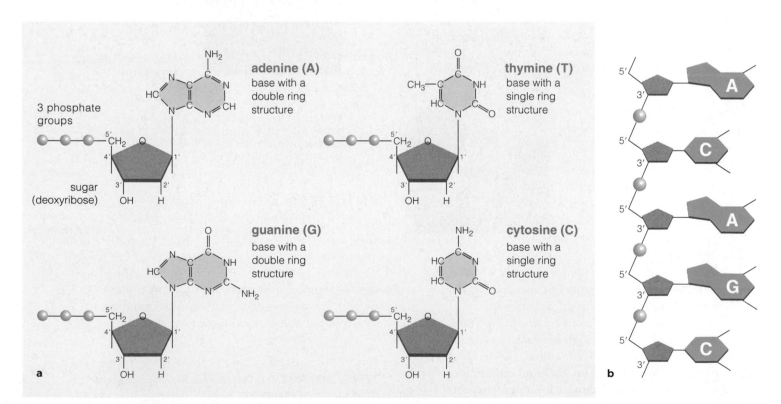

Figure 3.21 Animated (**a**) Nucleotides of DNA. The four kinds of nucleotides in DNA differ only in their component base, for which they are named. The carbon atoms of the sugar rings in nucleotides are numbered as shown. This numbering convention allows us to keep track of the orientation of a chain of nucleotides, as shown in (**b**).

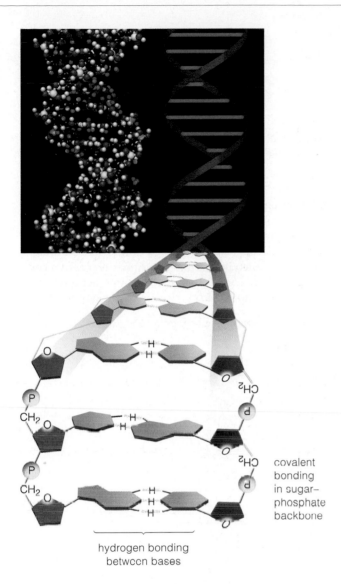

covalent
bonding
in sugar–
phosphate
backbone

hydrogen bonding
between bases

Figure 3.22 Models of the DNA molecule.

multicelled organisms, an entire individual. The cell uses the order of nucleotide bases in its DNA—the DNA sequence—to construct RNA and proteins. Parts of the sequence are identical or nearly so in all organisms. Other parts are unique to a species, or even to an individual. Chapter 13 returns to DNA structure and function.

Take-Home Message

What are nucleotides and nucleic acids?

■ Different nucleotides are monomers of the nucleic acids DNA and RNA, coenzymes, energy carriers, and messengers.

■ DNA's nucleotide sequence encodes heritable information.

■ Different types of RNA have roles in the processes by which a cell uses the heritable information in its DNA.

Summary

Section 3.1 Under present-day conditions in nature, only living cells make the molecules of life: complex carbohydrates and lipids, proteins, and nucleic acids.

The molecules of life differ, but all of them are **organic** compounds that consist mainly of carbon and hydrogen atoms. Carbon atoms can bond covalently with as many as four other atoms. Long carbon chains or rings form the backbone of the molecules of life.

Section 3.2 **Functional groups** attached to the carbon backbone influence the function of organic compounds. Table 3.2 (next page) summarizes the molecules of life and their functions. By the process of **metabolism**, cells acquire and use energy as they make, rearrange, and break down organic compounds.

Enzymatic reactions that are common in metabolism include **condensation**, which makes **polymers** from smaller **monomers**, and **hydrolysis**, which cleaves molecules into smaller ones.

■ *Use the animation on CengageNOW to explore functional groups, condensation, and hydrolysis.*

Section 3.3 Cells use **carbohydrates** as energy sources, transportable or storable forms of energy, and structural materials. The oligosaccharides and polysaccharides are polymers of monosaccharide monomers.

■ *Use the animation on CengageNOW to see how sucrose forms by condensation of glucose and fructose.*

Section 3.4 **Lipids** are greasy or oily nonpolar molecules, often with one or more **fatty acid** tails, and include **triglycerides** and other **fats**. **Phospholipids** are the main structural component of cell membranes. **Waxes** are part of water-repellent and lubricating secretions; **steroids** are precursors of other molecules.

■ *Use the animation on CengageNOW to see how a triglyceride forms by condensation.*

Section 3.5 **Proteins** are the most diverse molecules of life. Protein structure begins as a linear sequence of **amino acids** called a **polypeptide** chain (primary structure). The chains form sheets and coils (secondary structure), which may pack into functional domains (tertiary structure). Many proteins, including most enzymes, consist of two or more chains (quaternary structure). Fibrous proteins aggregate further into large chains or sheets.

■ *Use the animation on CengageNOW to explore amino acid structure and learn about peptide bond formation.*

■ *Read the InfoTrac article "Protein Folding and Misfolding," David Gossard, American Scientist, September 2002.*

Section 3.6 A protein's structure dictates its function. Sometimes a mutation in DNA results in an amino acid substitution that alters a protein's structure enough to compromise its function. Genetic diseases such as sickle-cell anemia may result.

Shifts in pH or temperature, and exposure to detergent or to salts may disrupt the many hydrogen bonds

Several countries are ahead of the United States in restricting the use of *trans* fats in food. In 2004, Denmark passed a law that prohibited importation of foods that contain partially hydrogenated vegetable oils. French fries and chicken nuggets the Danish import from the United States contain almost no *trans* fats; the same foods sold to consumers in the United States contain 5 to 10 grams of *trans* fats per serving.

How would you vote?

New York was the first U.S. city to ban *trans* fats from restaurant food. Should the use of *trans* fats in food be banned entirely? See CengageNOW for details, then vote online.

---

and other molecular interactions that hold a protein in its three-dimensional shape. If a protein unfolds so that it loses its three-dimensional shape (or **denatures**), it also loses its function.

■ *Use the animation on CengageNOW to learn more about hemoglobin structure and sickle-cell mutation.*

Section 3.7 **Nucleotides** are small organic molecules that consist of a sugar bonded to three phosphate groups and a nitrogen-containing base. **ATP** transfers phosphate groups to many kinds of molecules. Other nucleotides are coenzymes or chemical messengers. **DNA** and **RNA** are **nucleic acids**, each composed of four kinds of nucleotides. DNA encodes heritable information about a cell's proteins and RNAs. Different RNAs interact with DNA and with one another to carry out protein synthesis.

■ *Use the animation on CengageNOW to explore DNA.*

Table 3.2   Summary of the Main Organic Molecules in Living Things

Category	Main Subcategories	Some Examples and Their Functions	
CARBOHYDRATES ... contain an aldehyde or a ketone group, and one or more hydroxyl groups	**Monosaccharides** Simple sugars	Glucose	Energy source
	Oligosaccharides Short-chain carbohydrates	Sucrose	Most common form of sugar
	Polysaccharides Complex carbohydrates	Starch, glycogen	Energy storage
		Cellulose	Structural roles
LIPIDS ... are mainly hydrocarbon; generally do not dissolve in water but do dissolve in nonpolar substances, such as alcohols or other lipids	**Glycerides** Glycerol backbone with one, two, or three fatty acid tails (e.g., triglycerides)	Fats (e.g., butter), oils (e.g., corn oil)	Energy storage
	Phospholipids Glycerol backbone, phosphate group, another polar group; often two fatty acids	Lecithin	Key component of cell membranes
	Waxes Alcohol with long-chain fatty acid tails	Waxes in cutin	Conservation of water in plants
	Steroids Four carbon rings; the number, position, and type of functional groups differs	Cholesterol	Component of animal cell membranes; precursor of many steroids, vitamin D
PROTEINS ... are one or more polypeptide chains, each with as many as several thousand covalently linked amino acids	**Mostly fibrous proteins** Long strands or sheets of polypeptide chains; often strong, water-insoluble	Keratin	Structural component of hair, nails
		Collagen	Component of connective tissue
		Myosin, actin	Functional components of muscles
	Mostly globular proteins One or more polypeptide chains folded into globular shapes; many roles in cell activities	Enzymes	Great increase in rates of reactions
		Hemoglobin	Oxygen transport
		Insulin	Control of glucose metabolism
		Antibodies	Immune defense
NUCLEIC ACIDS AND NUCLEOTIDES ... are chains of units (or individual units) that each consist of a five-carbon sugar, phosphate, and a nitrogen-containing base	**Adenosine phosphates**	ATP	Energy carrier
		cAMP	Messenger in hormone regulation
	Nucleotide coenzymes	NAD^+, $NADP^+$, FAD	Transfer of electrons, protons (H^+) from one reaction site to another
	Nucleic acids Chains of nucleotides	DNA, RNAs	Storage, transmission, translation of genetic information

Data Analysis Exercise

Cholesterol does not dissolve in blood, so it is carried through the bloodstream by lipid–protein aggregates called lipoproteins. Lipoproteins vary in structure. Low-density lipoprotein (LDL) carries cholesterol to body tissues such as artery walls, where it can form health-endangering deposits. LDL is often called "bad" cholesterol. High-density lipoprotein (HDL) carries cholesterol away from tissues to the liver for disposal; it is often called "good" cholesterol.

In 1990, R.P. Mensink and M.B. Katan published a study that tested the effects of different dietary fats on blood lipoprotein levels. Their results are shown in Figure 3.23.

1. In which group was the level of LDL ("bad" cholesterol) highest?

2. In which group was the level of HDL ("good" cholesterol) lowest?

3. An elevated risk of heart disease has been correlated with increasing LDL-to-HDL ratios. In which group was the LDL:HDL ratio highest? Rank the three diets according to their potential effect on cardiovascular health.

	Main Dietary Fats			
	cis-fatty acids	*trans*-fatty acids	saturated fats	optimal level
LDL	103	117	121	<100
HDL	55	48	55	>40
ratio	1.87	2.43	2.2	<2

Figure 3.23 Effect of diet on lipoprotein levels. Researchers placed 59 men and women on a diet in which 10% of their daily energy intake consisted of *cis*-fatty acids, *trans*-fatty acids, or saturated fats. Blood LDL and HDL levels were measured after 3 weeks on the diet; averaged results are shown in mg/dL (milligrams per deciliter). All subjects were tested on each of the diets. The ratio of LDL to HDL is also shown.

Self-Quiz

Answers in Appendix III

1. Each carbon atom can share pairs of electrons with up to _____ other atom(s).

2. Sugars are a type of _____ .

3. _____ is a simple sugar (a monosaccharide).
a. Glucose   c. Ribose   e. both a and b
b. Sucrose   d. Chitin   f. both a and c

4. Unlike saturated fats, the fatty acid tails of unsaturated fats incorporate one or more _____ .

5. Is this statement true or false? Unlike saturated fats, all of the unsaturated fats are beneficial to health because their fatty acid tails bend and do not pack together.

6. Steroids are among the lipids with no _____ .

7. Which of the following is a class of molecules that encompasses all of the other molecules listed?
a. triglycerides   c. waxes   e. lipids
b. fatty acids   d. steroids   f. phospholipids

8. _____ are to proteins as _____ are to nucleic acids.
a. Sugars; lipids   c. Amino acids; hydrogen bonds
b. Sugars; proteins   d. Amino acids; nucleotides

9. A denatured protein has lost its _____ .
a. hydrogen bonds   c. function
b. shape   d. all of the above

10. _____ consist(s) of nucleotides.
a. sugars   b. DNA   c. RNA   d. b and c

11. _____ are the richest energy source in the body.
a. Sugars   b. Proteins   c. Fats   d. Nucleic acids

12. Match each molecule with its most suitable description.
_____chain of amino acids   a. carbohydrate
_____energy carrier in cells   b. phospholipid
_____glycerol, fatty acids, phosphate   c. polypeptide
_____two strands of nucleotides   d. DNA
_____one or more sugar monomers   e. ATP
_____richest source of energy   f. triglycerides

■ *Visit CengageNOW for additional questions.*

Critical Thinking

1. In the following list, identify the carbohydrate, the fatty acid, the amino acid, and the polypeptide:
a. $^{+}NH_3—CHR—COO^{-}$ c. $(glycine)_{20}$
b. $C_6H_{12}O_6$ d. $CH_3(CH_2)_{16}COOH$

2. Lipoproteins are relatively large, spherical clumps of protein and lipid molecules that circulate in the blood of mammals. They are like suitcases that move cholesterol, fatty acid remnants, triglycerides, and phospholipids from one place to another in the body. Given what you know about the insolubility of lipids in water, which of the four kinds of lipids would you predict to be on the outside of a lipoprotein clump, bathed in the fluid portion of blood?

3. In 1976, researchers were developing new insecticides by modifying sugars with chlorine (Cl_2) and other toxic gases. One young member of the team misunderstood instructions to "test" a new molecule. He thought he was supposed to "taste" it. Luckily, the molecule was not toxic, but it was sweet. It became the food additive sucralose.

Sucralose has three chlorine atoms substituted for three hydroxyl groups of sucrose. The highly electronegative chlorine atoms make sucralose strongly electronegative (Section 2.3). Sucralose binds so strongly to sweet-taste receptors on the tongue that our brain perceives it as 600 times sweeter than sucrose. The body does not recognize sucralose as a carbohydrate. Volunteers ate sucralose labeled with ^{14}C. Analysis of the radioactive molecules in their urine and feces showed that 92.8 percent of the sucralose passed unaltered through the body. Nonetheless, many are worried that the chlorine atoms impart toxicity to sucralose. How would you respond to that concern?

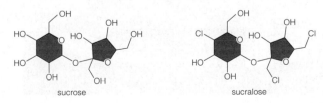

sucrose   sucralose

4 Cell Structure and Function

IMPACTS, ISSUES Food for Thought

We find bacteria at the bottom of the ocean, high up in the atmosphere, miles underground—essentially anywhere we look. Mammal intestines typically harbor fantastic numbers of them, but bacteria are not just stowaways there. Intestinal bacteria make vitamins that mammals cannot, and they crowd out more dangerous germs.

Escherichia coli is one of the most common intestinal bacteria of warm-blooded animals. Only a few of the hundreds of types, or strains, of *E. coli*, are harmful. One, O157:H7, makes a potent toxin that can severely damage the lining of the human intestine (Figure 4.1). After ingesting as few as ten O157:H7 cells, a person may become ill with severe cramps and bloody diarrhea that lasts up to ten days. In some people, complications of O157:H7 infection result in kidney failure, blindness, paralysis, and death. About 73,000 people in the United States become infected with *E. coli* O157:H7 each year, and more than 60 die.

E. coli O157:H7 lives in the intestines of other animals —mainly cattle, deer, goats, and sheep—apparently without sickening them. Humans are exposed to the bacteria when they come into contact with feces of animals that harbor it, for example, by eating contaminated ground beef. During slaughter, meat occasionally comes into contact with feces. Bacteria in the feces stick to the meat, then get thoroughly mixed into it during the grinding process. Unless contaminated meat is cooked to at least 71°C (160°F), live bacteria will enter the digestive tract of whoever eats it.

People also become infected by ingesting fresh fruits and vegetables that have come into contact with animal feces. For example, in 2006, at least 205 people became ill and 3 died after eating fresh spinach. The spinach was grown in a field close to a cattle pasture, and water contaminated with manure may have been used to irrigate the field. Washing contaminated produce with water does not remove *E. coli* O157:H7, because the bacteria are sticky.

The economic impact of such outbreaks, which occur with some regularity, extends beyond the casualties. Growers lost $50–100 million dollars recalling fresh spinach after the 2006 outbreak. In 2007, 5.7 million pounds of ground beef were recalled after 14 people were sickened. Food growers and processors are beginning to implement new procedures that they hope will reduce *E. coli* O157:H7 outbreaks. Some meats and produce are now tested for pathogens before sale, and improved documentation should allow a source of contamination to be pinpointed more quickly.

What makes bacteria sticky? Why do people but not cows get sick with *E. coli* O157:H7? You will begin to find answers to these and many more questions that affect your health in this chapter, as you learn about cells and how they work.

See the video! Figure 4.1 *E. coli* O157:H7 bacteria (*above, red*) on intestinal cells (*tan*) of a small child. This type of bacteria can cause a serious intestinal illness in people who eat foods contaminated with it, such as ground beef or fresh produce (*left*).

Key Concepts

What all cells have in common

Each cell has a plasma membrane, a boundary between its interior and the outside environment. The interior consists of cytoplasm and an innermost region of DNA. **Sections 4.1, 4.2**

Microscopes

Microscopic analysis supports three generalizations of the cell theory: Each organism consists of one or more cells and their products, a cell has a capacity for independent life, and each new cell is descended from another cell. **Section 4.3**

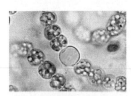

Prokaryotic cells

Archaeans and bacteria are prokaryotic cells, which have few, if any, internal membrane-enclosed compartments. In general, they are the smallest and structurally the simplest cells. **Sections 4.4, 4.5**

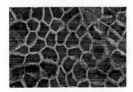

Eukaryotic cells

Cells of protists, plants, fungi, and animals are eukaryotic; they have a nucleus and other membrane-enclosed compartments. They differ in internal parts and surface specializations. **Sections 4.6–4.12**

A look at the cytoskeleton

Diverse protein filaments reinforce a cell's shape and keep its parts organized. As some filaments lengthen and shorten, they move cell structures or the whole cell. **Section 4.13**

Links to Earlier Concepts

■ Reflect on the overview of levels of organization in nature in Section 1.1. You will see how the properties of cell membranes emerge from the organization of lipids and proteins (3.4, 3.5).

■ What you know about scientific theory (1.6) will help you understand how scientific thought led to the development of the cell theory. This chapter also offers examples of the effects of mutation, and how researchers use tracers (2.2).

■ You will consider the cellular location of DNA (3.7) and the sites where carbohydrates (3.2, 3.3) are built and broken apart.

■ You will also expand your understanding of the vital roles of proteins in cell functions (3.6), and see how a nucleotide helps control cell activities (3.7).

How would you vote? Some think the safest way to protect consumers from food poisoning is by exposing food to high-energy radiation, which kills bacteria. Others think we should tighten food safety standards instead. Would you choose irradiated food? See CengageNOW for details, then vote online.

4.1 The Cell Theory

- The cell theory, a foundation of modern biology, states that cells are the fundamental units of all life.

- Link to Theory 1.6

Measuring Cells

Do you ever think of yourself as being about 3/2000 of a kilometer (1/1000 miles) tall? Probably not, yet that is how we measure cells. Use the scale bars in Figure 4.2 like a ruler and you can see that the cells shown are a few micrometers "tall." One micrometer (μm) is one-thousandth of a millimeter, which is one-thousandth of a meter, which is one-thousandth of a kilometer (0.62 miles). The cells are bacteria. Bacteria are among the smallest and structurally simplest cells on Earth. The cells that make up your body are generally larger and more complex than bacteria.

Animalcules and Beasties

Nearly all cells are so small that they are invisible to the naked eye. No one even knew cells existed until after the first microscopes were invented, around the end of the sixteenth century.

The first microscopes were not very sophisticated. Dutch spectacle makers Hans and Zacharias Janssen discovered that objects appear greatly enlarged (mag-nified) when viewed through a series of lenses. The father and son team created the first compound micro-scope (one that uses multiple lenses) in the year 1590, when they mounted two glass lenses inside a tube.

Given the simplicity of their instruments, it is amaz-ing that the pioneers in microscopy observed as much as they did. Antoni van Leeuwenhoek, a Dutch draper, had exceptional skill in constructing lenses and possi-bly the keenest vision. By the mid-1600s, he was spying on the microscopic world of rainwater, insects, fabric, sperm, feces—essentially any sample he could fit into his microscope (Figure 4.3a). He was fascinated by the tiny organisms he saw moving in many of his samples. For example, in scrapings of tartar from his teeth, Leeuwenhoek saw "many very small animalcules, the motions of which were very pleasing to behold." He (incorrectly) assumed that movement defined life, and (correctly) concluded that the moving "beasties" he saw were alive. Perhaps Leeuwenhoek was so pleased to behold his animalcules because he did not grasp the implications of what he was seeing: Our world, and our bodies, teem with microbial life.

Robert Hooke, a contemporary of Leeuwenhoek, added another lens that made the instrument easier to use. Many of the microscopes we use today are still based on his design. Hooke magnified a piece of thinly sliced cork from a mature tree and saw tiny compart-ments (Figure 4.3b). He named them cellulae—a Latin

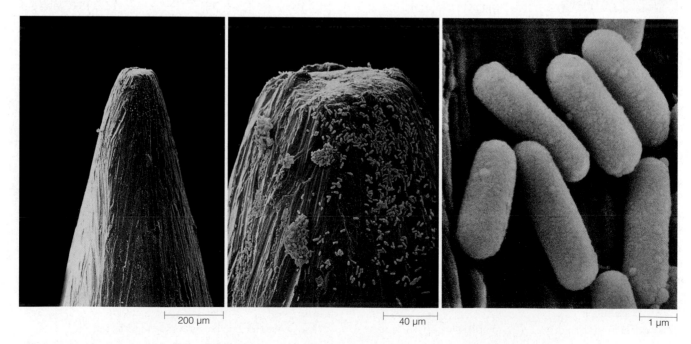

| 200 μm | 40 μm | 1 μm |

Figure 4.2 Rod-shaped bacterial cells on the tip of a household pin, shown at increasingly higher magnifications (enlargements). The "μm" is an abbreviation for micrometers, or 10^{-6} meters.
Figure It Out: About how big are these bacteria? *Answer: About 1 μm wide, and 5 μm long*

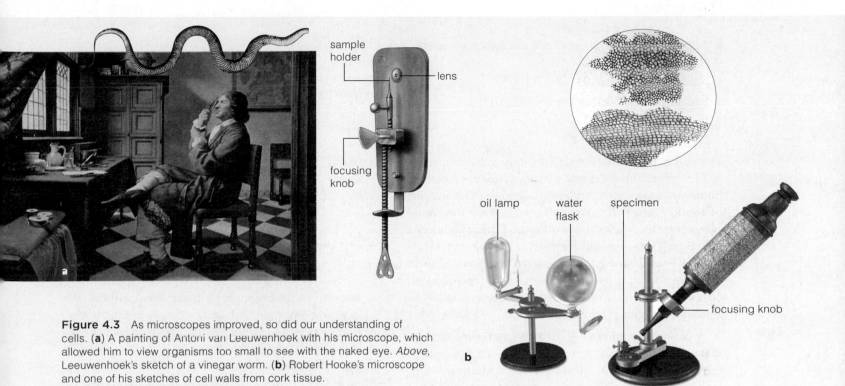

Figure 4.3 As microscopes improved, so did our understanding of cells. (**a**) A painting of Antoni van Leeuwenhoek with his microscope, which allowed him to view organisms too small to see with the naked eye. *Above*, Leeuwenhoek's sketch of a vinegar worm. (**b**) Robert Hooke's microscope and one of his sketches of cell walls from cork tissue.

word for the small chambers that monks lived in—and thus coined the term "cell." Actually they were dead plant cell walls, which is what cork consists of, but Hooke did not think of them as being dead because neither he nor anyone else knew cells could be alive. He observed cells "fill'd with juices" in green plant tissues but did not realize they were alive, either.

The Cell Theory Emerges

For nearly 200 years after their discovery, cells were thought to be part of a continuous membrane system in multicelled organisms, not separate entities. By the 1820s, vastly improved lenses brought cells into much sharper focus. Robert Brown, a botanist, was the first to identify a plant cell nucleus. Matthias Schleiden, another botanist, hypothesized that a plant cell is an independent living unit even when it is part of a plant. Schleiden compared notes with the zoologist Theodor Schwann, and both concluded that the tissues of animals as well as plants are composed of cells and their products. Together, the two scientists recognized that cells have a life of their own even when they are part of a multicelled body.

Another insight emerged from physiologist Rudolf Virchow, who studied how cells reproduce—that is, how they divide into descendant cells. Every cell, he realized, had descended from another living cell. These

and many other observations yielded four generalizations that today constitute the **cell theory**:

1. Every living organism consists of one or more cells.

2. The cell is the structural and functional unit of all organisms. A cell is the smallest unit of life, individually alive even as part of a multicelled organism.

3. All living cells come from division of other, preexisting cells.

4. Cells contain hereditary material, which they pass to their offspring during division.

The cell theory, first articulated in 1839 by Schwann and Schleiden and later revised, remains a foundation of modern biology. It was not always so. The theory was a radical new interpretation of nature that underscored life's unity. As with every scientific theory, it has remained (and always will be) open to revision if new data do not support it.

Take-Home Message

What is the cell theory?

■ All organisms consist of one or more cells.

■ A cell is the smallest unit with the properties of life.

■ Each new cell arises from division of another, preexisting cell.

■ Each cell passes hereditary material to its offspring.

What Is a Cell?

- All cells have a plasma membrane and cytoplasm, and all start out life with DNA.

- Links to Lipid structure 3.4, DNA 3.7

The Basics of Cell Structure

The **cell** is the smallest unit that shows the properties of life, which means it has a capacity for metabolism, homeostasis, growth, and reproduction. The interior of a **eukaryotic cell** is divided into various functional compartments, including a nucleus. **Prokaryotic cells** are usually smaller and simpler; none has a nucleus. Cells differ in size, shape, and activities. Yet, as Figure 4.4 suggests, all cells are similar in three respects. All cells start out life with a plasma membrane, a DNA-containing region, and cytoplasm:

1. A **plasma membrane** is the cell's outer membrane. It separates metabolic activities from events outside of the cell, but does not isolate the cell's interior. Water, carbon dioxide, and oxygen can cross it freely. Other substances cross only with the assistance of membrane proteins. Still others are kept out entirely.

2. All eukaryotic cells start life with a **nucleus**. This double-membraned sac holds a eukaryotic cell's DNA. The DNA inside prokaryotic cells is concentrated in a region of cytoplasm called the **nucleoid**.

3. **Cytoplasm** is a semifluid mixture of water, sugars, ions, and proteins between the plasma membrane and the region of DNA. Cell components are suspended in cytoplasm. For instance, **ribosomes**, structures on which proteins are built, are suspended in cytoplasm.

Diameter (cm)	2	3	6
Surface area (cm²)	12.6	28.2	113
Volume (cm³)	4.2	14.1	113
Surface-to-volume ratio	3:1	2:1	1:1

Figure 4.5 Animated Three examples of the surface-to-volume ratio. This physical relationship between increases in volume and surface area constrains cell size and shape.

Are any cells big enough to be seen without the help of a microscope? A few. They include the "yolks" of bird eggs, cells in watermelon tissues, and the eggs of amphibians and fishes. These cells can be relatively large because they are not very metabolically active. Most of their volume simply acts as a warehouse.

A physical relationship, the **surface-to-volume ratio**, strongly influences cell size and shape. By this ratio, an object's volume increases with the cube of its diameter, but its surface area increases only with the square. The ratio is important because the lipid bilayer can handle only so many exchanges between the cell's cytoplasm and the external environment.

Apply the surface-to-volume ratio to a round cell. As Figure 4.5 shows, when a cell expands in diameter during growth, its volume increases faster than its surface area does. Imagine that a round cell expands until it is four times its original diameter. The volume of the

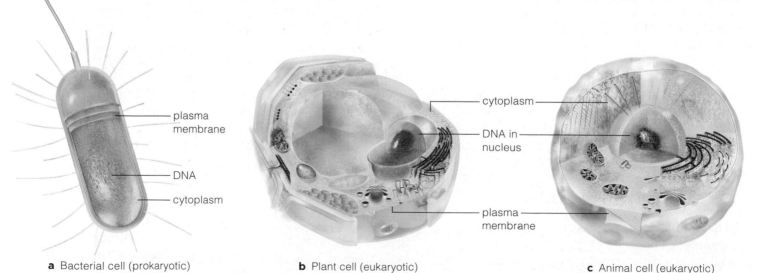

a Bacterial cell (prokaryotic)   plasma membrane   DNA   cytoplasm

b Plant cell (eukaryotic)   cytoplasm   DNA in nucleus   plasma membrane

c Animal cell (eukaryotic)

Figure 4.4 General organization of prokaryotic and eukaryotic cells. If the prokaryotic cell were drawn at the same scale as the other two cells, it would be about this big:

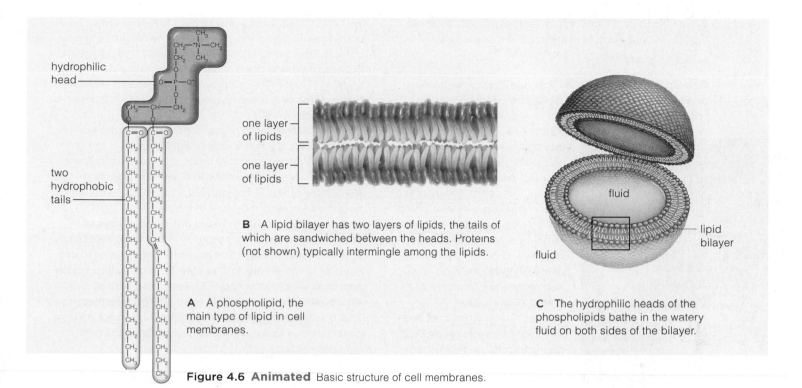

hydrophilic head

two hydrophobic tails

one layer of lipids

one layer of lipids

fluid

fluid

lipid bilayer

B A lipid bilayer has two layers of lipids, the tails of which are sandwiched between the heads. Proteins (not shown) typically intermingle among the lipids.

A A phospholipid, the main type of lipid in cell membranes.

C The hydrophilic heads of the phospholipids bathe in the watery fluid on both sides of the bilayer.

Figure 4.6 Animated Basic structure of cell membranes.

cell has increased 64 times (4^3), but its surface area has increased only 16 times (4^2). Each unit of plasma membrane must now handle exchanges with four times as much cytoplasm. If a cell's circumference gets too big, the inward flow of nutrients and outward flow of wastes will not be fast enough to keep the cell alive.

A big, round cell would also have trouble moving substances through its cytoplasm. Molecules disperse by their own random motions, but they move only so quickly. Nutrients or wastes would not get distributed fast enough to keep up with a large, round, active cell's metabolism. That is why many cells are long and thin, or frilly surfaced with folds that increase surface area. The surface-to-volume ratio of such cells is enough to sustain their metabolism. The amount of raw materials that cross the plasma membrane, and the speed with which they are distributed through cytoplasm, satisfy the cell's needs. Wastes are also removed fast enough to keep the cell from getting poisoned.

Surface-to-volume constraints also affect the body plans of multicelled species. For example, small cells attach end to end in strandlike algae, so each interacts directly with its surroundings. Muscle cells in your thighs are as long as the muscle in which they occur, but each is thin, so it exchanges substances efficiently with fluids in the tissue surrounding it.

Preview of Cell Membranes

The structural foundation of all cell membranes is the **lipid bilayer**, a double layer of lipids organized so that their hydrophobic tails are sandwiched between their hydrophilic heads (Figure 4.6).

Phospholipids are the most abundant type of lipid in cell membranes. Many different proteins embedded in a bilayer or attached to one of its surfaces carry out membrane functions. For example, some proteins form channels through a bilayer; others pump substances across it. In addition to a plasma membrane, many cells also have internal membranes that form channels or enclose sacs. These membranous structures compartmentalize tasks such as building, modifying, and storing substances. Chapter 5 offers a closer look at membrane structure and function.

Take-Home Message

How are all cells alike?

■ All cells start life with a plasma membrane, cytoplasm, and a region of DNA.

■ A lipid bilayer forms the structural framework of all cell membranes.

■ DNA of eukaryotic cells is enclosed by a nucleus. DNA of prokaryotic cells is concentrated in a region of cytoplasm called the nucleoid.

4.3 | How Do We See Cells?

■ We use different types of microscopes to study different aspects of organisms, from the smallest to the largest.

■ Link to Tracers 2.2

Modern Microscopes Like those early instruments mentioned in Section 4.1, many types of modern light microscopes still rely on visible light to illuminate objects. All

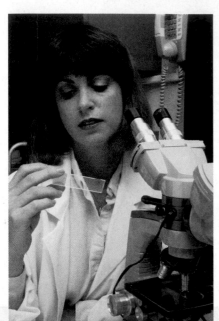

light travels in waves, a property that allows us to focus light with glass lenses. Light microscopes use visible light to illuminate a cell or some other specimen (Figure 4.7a). Curved glass lenses bend the light and focus it as a magnified image of the specimen. Photographs of images enlarged with any microscope are called micrographs (Figure 4.8).

Phase-contrast microscopes shine light through specimens, but most cells are nearly transparent. Their internal details may not be visible unless they

are first stained, or exposed to dyes that only some cell parts soak up. The parts that absorb the most dye appear darkest. The resulting increase in contrast (the difference between light and dark) allows us to see a greater range of detail (Figure 4.8a). Opaque samples are not stained; their surface details are revealed with reflected light microscopes (Figure 4.8b).

With a fluorescence microscope, a cell or a molecule is the light source; it fluoresces, or emits energy in the form of visible light, when a laser beam is focused on it. Some molecules, such as chlorophylls, fluoresce naturally (Figure 4.8c). More typically, researchers attach a light-emitting tracer to the cell or molecule of interest.

The wavelength of light—the distance from the peak of one wave to the peak behind it—limits the power of any light microscope. Why? Structures that are smaller than one-half of the wavelength of light are too small to scatter light waves, even after they have been stained. The smallest wavelength of visible light is about 400 nanometers. That is why structures that are smaller than about 200 nanometers across appear blurry under even the best light microscopes.

Other microscopes can reveal smaller details. For example, electron microscopes use electrons instead

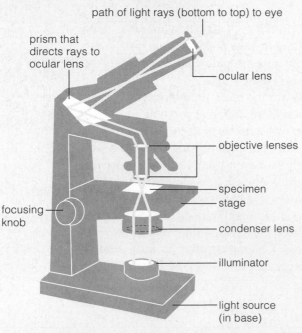

path of light rays (bottom to top) to eye

prism that directs rays to ocular lens

ocular lens

objective lenses

specimen

stage

condenser lens

focusing knob

illuminator

light source (in base)

A A compound light microscope has more than one glass lens.

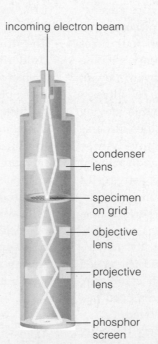

incoming electron beam

condenser lens

specimen on grid

objective lens

projective lens

phosphor screen

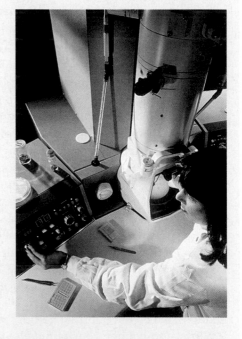

B Transmission electron microscope (TEM). Electrons passing through a thin slice of a specimen illuminate a fluorescent screen. Internal details of the specimen cast visible shadows, as in Figure 4.8d.

Figure 4.7 Animated Examples of microscopes.

   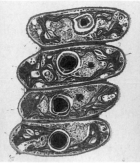 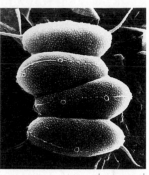

10 μm

a Light micrograph. A phase-contrast microscope yields high-contrast images of transparent specimens, such as cells.

b Light micrograph. A reflected light microscope captures light reflected from opaque specimens.

c Fluorescence micrograph. The chlorophyll molecules in these cells emitted red light (they fluoresced) naturally.

d A transmission electron micrograph reveals fantastically detailed images of internal structures.

e A scanning electron micrograph shows surface details of cells and structures. Often, SEMs are artificially colored to highlight certain details.

Figure 4.8 Different microscopes can reveal different characteristics of the same aquatic organism—a green alga (*Scenedesmus*). Try estimating the size of one of these algal cells by using the scale bar.

of visible light to illuminate samples (Figure 4.7b). Because electrons travel in wavelengths that are much shorter than those of visible light, electron microscopes can resolve details that are much smaller than you can see with light microscopes. Electron microscopes use magnetic fields to focus beams of electrons onto a sample.

With transmission electron microscopes, electrons form an image after they pass through a thin specimen. The specimen's internal details appear on the image as shadows (Figure 4.8d). Scanning electron microscopes direct a beam of electrons back and forth across a surface of a specimen, which has been coated with a thin layer of gold or another metal. The metal emits both electrons and x-rays, which are converted into an image of the surface (Figure 4.8e). Both types of electron microscopes can resolve structures as small as 0.2 nanometer.

Figure 4.9 compares the resolving power of light and electron microscopes with that of the unaided human eye.

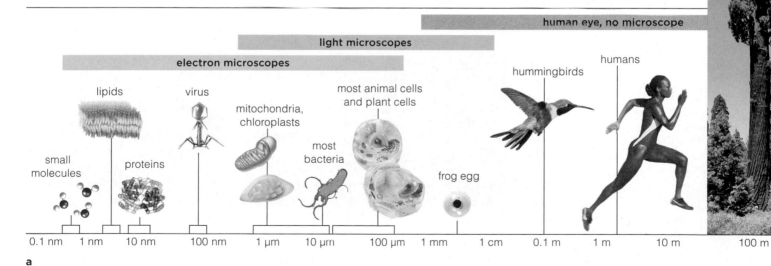

a

Figure 4.9 (a) Relative sizes of molecules, cells, and multicelled organisms. The diameter of most cells is in the range of 1 to 100 micrometers. Frog eggs, one of the exceptions, are 2.5 millimeters in diameter.

The scale shown here is exponential, not linear; each unit of measure is ten times larger than the unit preceding it. (b) Units of measure. See also Appendix IX. **Figure It Out:** Which is smallest, a protein, a lipid, or a water molecule?   *Answer: A water molecule*

b

1 centimeter (cm)	=	1/100 meter, or 0.4 inch
1 millimeter (mm)	=	1/1000 meter, or 0.04 inch
1 micrometer (μm)	=	1/1,000,000 meter, or 0.00004 inch
1 nanometer (nm)	=	1/1,000,000,000 meter, or 0.00000004 inch

1 meter $= 10^2$ cm $= 10^3$ mm $= 10^6$ μm $= 10^9$ nm

- Bacteria and archaea are the prokaryotes.
- Links to Polysaccharides 3.3, ATP 3.7

The word prokaryote means "before the nucleus," a reminder that the first prokaryotes evolved before the first eukaryotes. Prokaryotes are single-celled (Figure 4.10). As a group, they are the smallest and most metabolically diverse forms of life that we know about. Prokaryotes inhabit nearly all of Earth's environments, including some very hostile places.

Domains Bacteria and Archaea comprise all prokaryotes (Section 1.3). Cells of the two domains are alike in appearance and size, but differ in their structure and metabolic details (Figures 4.11 and 4.12). Some characteristics of archaeans indicate they are more closely related to eukaryotic cells than to bacteria. Chapter 21 revisits prokaryotes in more detail. Here we present an overview of their structure.

Most prokaryotic cells are not much wider than a micrometer. Rod-shaped species are a few micrometers long. None has a complex internal framework, but protein filaments under the plasma membrane impart shape to the cell. Such filaments also act as scaffolding for internal structures.

A rigid **cell wall** surrounds the plasma membrane of nearly all prokaryotes. Dissolved substances easily cross this permeable layer on the way to and from the plasma membrane. The cell wall of most bacteria consists of peptidoglycan, which is a polymer of cross-linked peptides and polysaccharides. The wall of most archaeans consists of proteins. Some types of eukaryotic cells (such as plant cells) also have a wall, but it is structurally different from a prokaryotic cell wall.

Sticky polysaccharides form a slime layer, or capsule, around the wall of many types of bacteria. The sticky capsule helps these cells adhere to many types of surfaces (such as spinach leaves and meat), and it also protects them from predators and toxins. A capsule can protect pathogenic (disease-causing) bacteria from host defenses.

Projecting past the wall of many prokaryotic cells are one or more **flagella** (singular, flagellum): slender cellular structures used for motion. A bacterial flagellum moves like a propeller that drives the cell through fluid habitats, such as a host's body fluids. It differs from a eukaryotic flagellum, which bends like a whip and has a distinctive internal structure.

Protein filaments called **pili** (singular, pilus) project from the surface of some bacterial species (Figure 4.12a). Pili help cells cling to or move across surfaces. One kind, a "sex" pilus, attaches to another bacterium and then shortens. The attached cell is reeled in, and genetic material is transferred from one cell to the other through the pilus.

The plasma membrane of all bacteria and archaeans selectively controls which substances move to and from the cytoplasm, as it does for eukaryotic cells. The plasma membrane bristles with transporters and receptors, and it also incorporates proteins that carry out important metabolic processes.

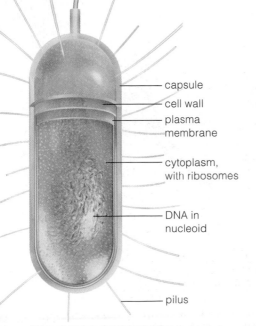

flagellum

capsule
cell wall
plasma membrane
cytoplasm, with ribosomes
DNA in nucleoid
pilus

Figure 4.10 Animated
Generalized body plan of a prokaryote.

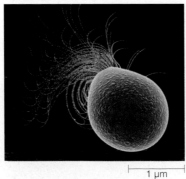

1 µm

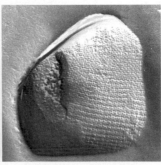

0.2 µm

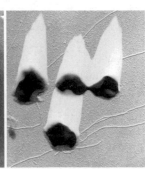

1 µm

a *Pyrococcus furiosus* was discovered in ocean sediments near an active volcano. It lives best at 100°C (212°F), and it makes a rare kind of enzyme that contains tungsten atoms.

b *Ferroglobus placidus* prefers superheated water spewing from the ocean floor. The unique composition of archaean lipid bilayers keeps these membranes intact at extreme heat and pH.

c *Metallosphaera prunae*, discovered in a smoking pile of ore at a uranium mine, prefers high temperatures and low pH. (*White* shadows are an artifact of electron microscopy.)

Figure 4.11 Some like it hot: many archaeans inhabit extreme environments. The cells in this example live without oxygen.

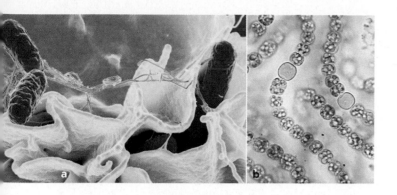

Figure 4.12 Bacteria. (**a**) Protein filaments, or pili, anchor bacterial cells to one another and to surfaces. Here, *Salmonella typhimurium* cells (*red*) use their pili to invade a culture of human cells. (**b**) Ball-shaped *Nostoc* cells stick together in a sheath of their own secretions. *Nostoc* are photosynthetic cyanobacteria. Other types of bacteria are shaped like rods or corkscrews.

For example, the plasma membrane of photosynthetic bacteria has arrays of proteins that capture light energy and convert it to the chemical energy of ATP (Section 3.7). The ATP is then used to build sugars. Similar metabolic processes occur in eukaryotes, but they take place at specialized internal membranes, not the plasma membrane.

The cytoplasm of prokaryotes contains thousands of ribosomes, structures upon which polypeptides are assembled. A prokaryotic cell's single chromosome, a circular DNA molecule, is located in an irregularly shaped region called the nucleoid. Most nucleoids are not enclosed by a membrane. Many prokaryotes also have plasmids in the cytoplasm. These small circles of DNA carry a few genes (units of inheritance) that can confer advantages, such as resistance to antibiotics.

One more intriguing point: There is evidence that all protists, plants, fungi, and animals evolved from a few ancient types of prokaryotes. For example, part of the plasma membrane of cyanobacteria folds into the cytoplasm. Pigments and other molecules that carry out photosynthesis are embedded in the membrane, just as they are in the inner membrane of chloroplasts—structures specialized for photosynthesis in eukaryotic cells. Section 20.4 returns to this topic.

Take-Home Message

What do all prokaryotic cells have in common?

■ All prokaryotes are single-celled organisms with no nucleus. These organisms inhabit nearly all regions of the biosphere.

■ Bacteria and archaeans are the only prokaryotes. Most kinds have a cell wall around their plasma membrane.

■ Prokaryotes have a relatively simple structure, but they are a diverse group of organisms.

4.5 | Microbial Mobs

■ Although prokaryotes are all single-celled, few live alone.

■ Link to Glycoproteins 3.5

Bacterial cells often live so close together that an entire community shares a layer of secreted polysaccharides and glycoproteins. Such communal living arrangements, in which single-celled organisms live in a shared mass of slime, are called **biofilms**. In nature, a biofilm typically consists of multiple species, all entangled in their own mingled secretions. It may include bacteria, algae, fungi, protists, and archaeans. Such associations allow cells living in a fluid to linger in a particular spot rather than be swept away by currents.

The microbial inhabitants of a biofilm benefit each other. Rigid or netlike secretions of some species serve as permanent scaffolding for others. Species that break down toxic chemicals allow more sensitive ones to thrive in polluted habitats that they could not withstand on their own. Waste products of some serve as raw materials for others.

Like a bustling metropolitan city, a biofilm organizes itself into "neighborhoods," each with a distinct microenvironment that stems from its location within the biofilm and the particular species that inhabit it (Figure 4.13). For example, cells that reside near the middle of a biofilm are very crowded and do not divide often. Those at the edges divide repeatedly, expanding the biofilm.

The formation and continuation of a biofilm is not random. Free-living cells sense the presence of other cells. Those that encounter a biofilm with favorable conditions switch their metabolism to support a more sedentary, communal lifestyle, and join in. Flagella disassemble, and sex pili form. If conditions become less favorable, the cells can revert to a free-living mode and swim away to find more hospitable accommodations.

0.2 cm

Figure 4.13 Biofilms. A single species of bacteria, *Bacillus subtilis*, formed this biofilm. Note the distinct regions.

4.6 Introducing Eukaryotic Cells

■ Eukaryotic cells carry out much of their metabolism inside organelles enclosed by membranes.

All eukaryotic cells start out life with a nucleus. *Eu–* means true; and *karyon*, meaning kernel, refers to the nucleus. A nucleus is a type of **organelle**: a structure that carries out a specialized function inside a cell. Many organelles, particularly those in eukaryotic cells, are bounded by membranes. Like all cell membranes, those around organelles control the types and amounts of substances that cross them. Such control maintains a special internal environment that allows an organelle to carry out its particular function. That function may be isolating a toxic or sensitive substance from the rest of the cell, transporting some substance through the cytoplasm, maintaining fluid balance, or providing a favorable environment for a reaction that could not occur in the cytoplasm. For example, a mitochondrion makes ATP after concentrating hydrogen ions inside its membrane system.

Much as interactions among organ systems keep an animal body running, interactions among organelles keep a cell running. Substances shuttle from one kind of organelle to another, and to and from the plasma membrane. Some metabolic pathways take place in a series of different organelles.

Table 4.1 lists common components of eukaryotic cells. These cells all start out life with certain kinds of organelles such as a nucleus and ribosomes. They also have a cytoskeleton, a dynamic "skeleton" of proteins (*cyto–* means cell). Specialized cells contain additional kinds of organelles and structures. Figure 4.14 shows two typical eukaryotic cells.

Table 4.1 Organelles of Eukaryotic Cells

Name	Function
Organelles with membranes	
Nucleus	Protecting, controlling access to DNA
Endoplasmic reticulum (ER)	Routing, modifying new polypeptide chains; synthesizing lipids; other tasks
Golgi body	Modifying new polypeptide chains; sorting, shipping proteins and lipids
Vesicles	Transporting, storing, or digesting substances in a cell; other functions
Mitochondrion	Making ATP by sugar breakdown
Chloroplast	Making sugars in plants, some protists
Lysosome	Intracellular digestion
Peroxisome	Inactivating toxins
Vacuole	Storage
Organelles without membranes	
Ribosomes	Assembling polypeptide chains
Centriole	Anchor for cytoskeleton

Take-Home Message

What do all eukaryotic cells have in common?

■ Eukaryotic cells start life with a nucleus and other membrane-enclosed organelles (structures that carry out specific tasks).

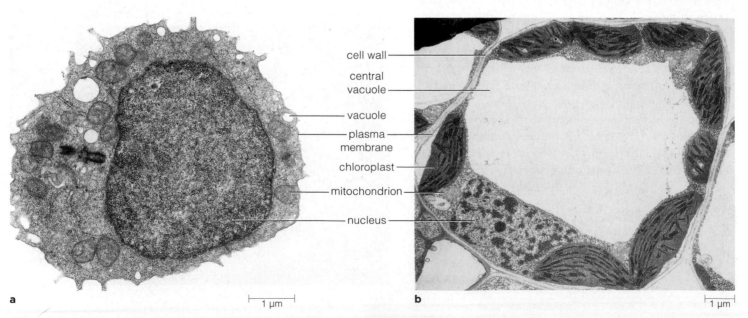

cell wall
central vacuole
vacuole
plasma membrane
chloroplast
mitochondrion
nucleus

a ⊢ 1 μm ⊣   b ⊢ 1 μm ⊣

Figure 4.14 Transmission electron micrographs of eukaryotic cells. (**a**) Human white blood cell. (**b**) Photosynthetic cell from a blade of timothy grass.

Visual Summary of Eukaryotic Cell Components

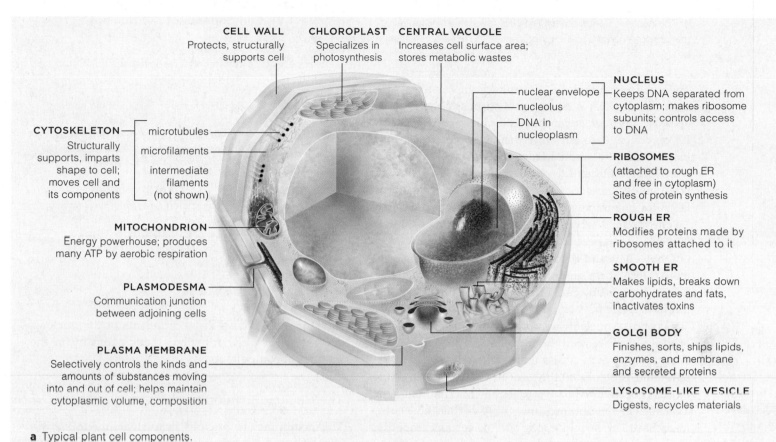

CELL WALL
Protects, structurally supports cell

CHLOROPLAST
Specializes in photosynthesis

CENTRAL VACUOLE
Increases cell surface area; stores metabolic wastes

CYTOSKELETON — microtubules
Structurally supports, imparts shape to cell; moves cell and its components
— microfilaments
— intermediate filaments (not shown)

MITOCHONDRION
Energy powerhouse; produces many ATP by aerobic respiration

PLASMODESMA
Communication junction between adjoining cells

PLASMA MEMBRANE
Selectively controls the kinds and amounts of substances moving into and out of cell; helps maintain cytoplasmic volume, composition

nuclear envelope
nucleolus
DNA in nucleoplasm

NUCLEUS
Keeps DNA separated from cytoplasm; makes ribosome subunits; controls access to DNA

RIBOSOMES
(attached to rough ER and free in cytoplasm) Sites of protein synthesis

ROUGH ER
Modifies proteins made by ribosomes attached to it

SMOOTH ER
Makes lipids, breaks down carbohydrates and fats, inactivates toxins

GOLGI BODY
Finishes, sorts, ships lipids, enzymes, and membrane and secreted proteins

LYSOSOME-LIKE VESICLE
Digests, recycles materials

a Typical plant cell components.

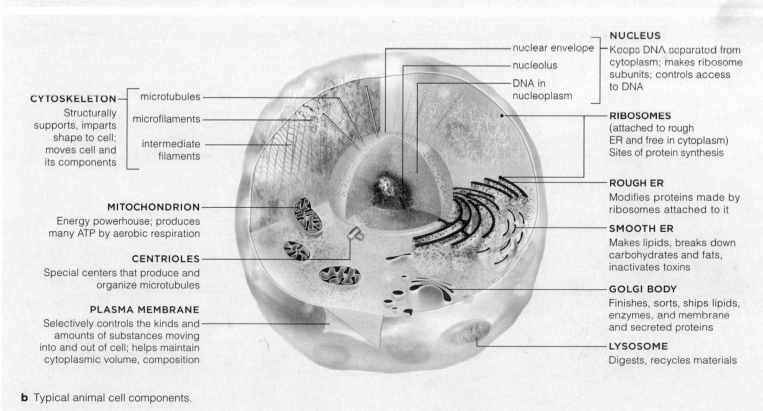

CYTOSKELETON — microtubules
Structurally supports, imparts shape to cell; moves cell and its components
— microfilaments
— intermediate filaments

MITOCHONDRION
Energy powerhouse; produces many ATP by aerobic respiration

CENTRIOLES
Special centers that produce and organize microtubules

PLASMA MEMBRANE
Selectively controls the kinds and amounts of substances moving into and out of cell; helps maintain cytoplasmic volume, composition

nuclear envelope
nucleolus
DNA in nucleoplasm

NUCLEUS
Keeps DNA separated from cytoplasm; makes ribosome subunits; controls access to DNA

RIBOSOMES
(attached to rough ER and free in cytoplasm) Sites of protein synthesis

ROUGH ER
Modifies proteins made by ribosomes attached to it

SMOOTH ER
Makes lipids, breaks down carbohydrates and fats, inactivates toxins

GOLGI BODY
Finishes, sorts, ships lipids, enzymes, and membrane and secreted proteins

LYSOSOME
Digests, recycles materials

b Typical animal cell components.

Figure 4.15 Animated Organelles and structures typical of (**a**) plant cells and (**b**) animal cells.

■ The nucleus keeps eukaryotic DNA away from potentially damaging reactions in the cytoplasm.
■ The nuclear envelope controls when DNA is accessed.

The nucleus contains all of a eukaryotic cell's DNA. A molecule of DNA is big to begin with, and the nucleus of most kinds of eukaryotic cells has many of them. If you could tease out all of the DNA molecules from the nucleus of a single human cell, unravel them, and stretch them out end to end, you would have a line of DNA about 2 meters (6–1/2 feet) long. That is a lot of DNA for one microscopic nucleus.

The nucleus serves two important functions. First, it keeps a cell's genetic material—its one and only copy of DNA—safe and sound. Isolated in its own compartment, DNA stays separated from the bustling activity of the cytoplasm, and from metabolic reactions that might damage it.

Second, a nuclear membrane controls the passage of molecules between the nucleus and the cytoplasm. For example, cells access their DNA when they make RNA and proteins, so the various molecules involved in this process must pass into the nucleus and out of it. The nuclear membrane allows only certain molecules to cross it, at certain times and in certain amounts. This control is another measure of safety for the DNA, and it is also a way for the cell to regulate the amount of RNA and proteins it makes.

Table 4.2 Components of the Nucleus

Nuclear envelope	Pore-riddled double membrane that controls which substances enter and leave the nucleus
Nucleoplasm	Semifluid interior portion of the nucleus
Nucleolus	Rounded mass of proteins and copies of genes for ribosomal RNA used to construct ribosomal subunits
Chromatin	Total collection of all DNA molecules and associated proteins in the nucleus; all of the cell's chromosomes
Chromosome	One DNA molecule and many proteins associated with it

Figure 4.16 shows the components of the nucleus. Table 4.2 lists their functions. Let's zoom in on the individual components.

The Nuclear Envelope

The membrane of a nucleus, or **nuclear envelope**, consists of two lipid bilayers folded together as a single membrane. As Figure 4.16 shows, the outer bilayer of the membrane is continuous with the membrane of

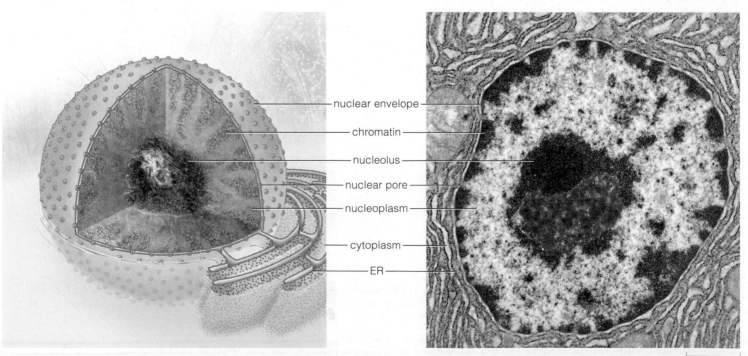

nuclear envelope
chromatin
nucleolus
nuclear pore
nucleoplasm
cytoplasm
ER

1 μm

Figure 4.16 The nucleus. TEM at *right*, nucleus of a mouse pancreas cell.

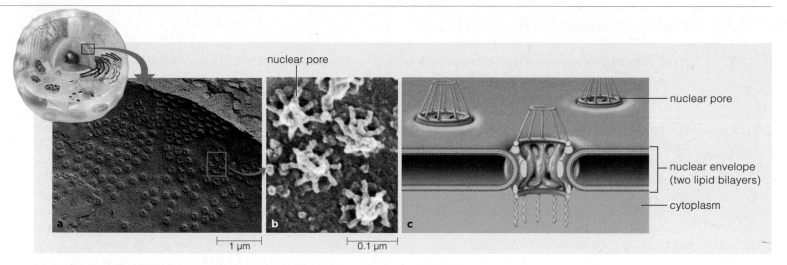

nuclear pore

nuclear pore

nuclear envelope
(two lipid bilayers)

cytoplasm

a

b

c

1 µm

0.1 µm

Figure 4.17 Animated Structure of the nuclear envelope. (**a**) The outer surface of a nuclear envelope was split apart, revealing the pores that span the two lipid bilayers. (**b**) Each nuclear pore is an organized cluster of membrane proteins that selectively allows certain substances to cross it on their way into and out of the nucleus. (**c**) Sketch of the nuclear envelope's structure.

another organelle, the ER. (We will discuss the ER in the next section.)

Different kinds of membrane proteins are embedded in the two lipid bilayers. Some are receptors and transporters; others aggregate into tiny pores that span the membrane (Figure 4.17). These molecules and structures work as a system to transport various molecules across the nuclear membrane. As with all membranes, water and gases cross nuclear membranes freely. All other substances can cross only through transporters and nuclear pores, both of which are selective about which molecules they allow through.

Fibrous proteins that attach to the inner surface of the nuclear envelope anchor DNA molecules and keep them organized. During cell division, these proteins help the cell parcel out the DNA into its offspring.

The Nucleolus

The nuclear envelope encloses **nucleoplasm**, a viscous fluid similar to cytoplasm. The nucleus also contains at least one **nucleolus** (plural, nucleoli), a dense, irregularly shaped region where subunits of ribosomes are assembled from proteins and RNA. The subunits pass through nuclear pores into the cytoplasm, where they join and become active in protein synthesis.

The Chromosomes

Chromatin is the name for all of the DNA, together with its associated proteins, in the nucleus. The genetic material of a eukaryotic cell is distributed among a specific number of DNA molecules. That number is characteristic of the type of organism and the type of cell, but it varies widely among species. For instance, the nucleus of a normal oak tree cell contains 12 DNA molecules, a human body cell, 46; and a king crab cell, 208. Each molecule of DNA, together with its many attached proteins, is called a **chromosome**.

Chromosomes change in appearance over the lifetime of a cell. When a cell is not dividing, its chromatin can appear grainy (as in Figure 4.16). Just before a cell divides, the DNA in each chromosome is copied, or duplicated. Then, during cell division, the chromosomes condense. As they do, they become visible in micrographs. The chromosomes first appear threadlike, then rodlike:

one chromosome
(one unduplicated
DNA molecule)

one chromosome
(one duplicated DNA
molecule, partially
condensed)

one chromosome
(one duplicated DNA
molecule, completely
condensed)

In later chapters, we will look in more detail at the dynamic structure and the function of chromosomes.

Take-Home Message

What is the function of the cell nucleus?

■ A nucleus protects and controls access to a eukaryotic cell's genetic material—its chromosomes.

■ The nuclear envelope is a double lipid bilayer. Proteins embedded in it control the passage of molecules between the nucleus and the cytoplasm.

The Endomembrane System

■ The endomembrane system is a set of organelles that makes, modifies, and transports proteins and lipids.

■ Links to Lipids 3.4, Proteins 3.5

The **endomembrane system** is a series of interacting organelles between the nucleus and the plasma membrane (Figure 4.18). Its main function is to make lipids, enzymes, and proteins for secretion or insertion into cell membranes. It also destroys toxins, recycles wastes, and has other specialized functions. The system's components vary among different types of cells, but here we present the most common ones.

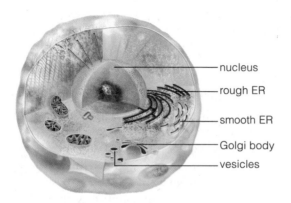

nucleus

rough ER

smooth ER

Golgi body

vesicles

The Endoplasmic Reticulum

Endoplasmic reticulum, or ER, is an extension of the nuclear envelope. It forms a continuous compartment that folds over and over into flattened sacs and tubes. Two kinds of ER are named for their appearance in electron micrographs. Many thousands of ribosomes attach to the outer surface of rough ER (Figure 4.18b). The ribosomes synthesize polypeptide chains, which extrude into the interior of the ER. Inside the ER, the proteins fold and take on their tertiary structure. Some of the proteins become part of the ER membrane itself; others are carried to different destinations in the cell.

Cells that make, store, and secrete a lot of proteins have a lot of rough ER. For example, ER-rich gland cells in the pancreas (an organ) make and secrete enzymes that help digest food in the small intestine.

Smooth ER has no ribosomes, so it does not make proteins (Figure 4.18d). Some of the polypeptides made in the rough ER end up in the smooth ER, as enzymes. These enzymes make most of the cell's membrane lipids. They also break down carbohydrates, fatty acids, and some drugs and poisons. In skeletal muscle cells, a special type of smooth ER called sarcoplasmic reticulum stores calcium ions and has a role in contraction.

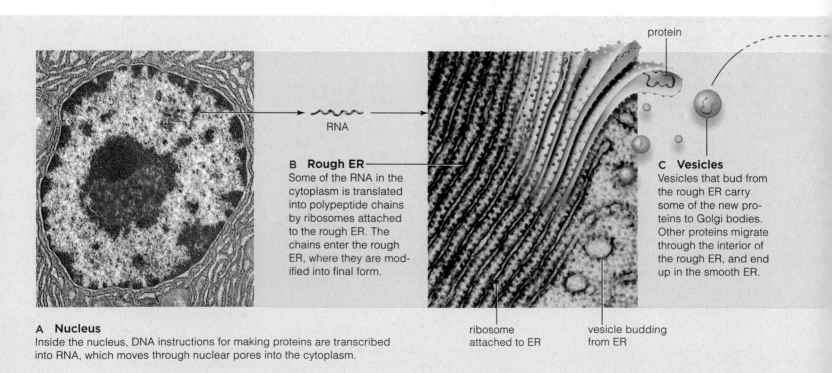

protein

RNA

A Nucleus
Inside the nucleus, DNA instructions for making proteins are transcribed into RNA, which moves through nuclear pores into the cytoplasm.

B Rough ER
Some of the RNA in the cytoplasm is translated into polypeptide chains by ribosomes attached to the rough ER. The chains enter the rough ER, where they are modified into final form.

C Vesicles
Vesicles that bud from the rough ER carry some of the new proteins to Golgi bodies. Other proteins migrate through the interior of the rough ER, and end up in the smooth ER.

ribosome attached to ER

vesicle budding from ER

Figure 4.18 Animated Endomembrane system, where lipids and many proteins are built, then transported to cellular destinations or to the plasma membrane. Chapter 14 describes transcription and translation.

Vesicles

Vesicles are small, membrane-enclosed, saclike organelles. They form in great numbers, and in a variety of types, either on their own or by budding from other organelles or the plasma membrane.

Many types of vesicles transport substances from one organelle to another, or to and from the plasma membrane (Figure 4.18c–f). Other kinds have different roles. For example, **peroxisomes** contain enzymes that digest fatty acids and amino acids. These vesicles form and divide on their own. Peroxisomes have a variety of functions, such as inactivating hydrogen peroxide, a toxic by-product of fatty acid breakdown. Enzymes in the peroxisomes convert hydrogen peroxide to water and oxygen, or they use it in reactions that break down alcohol and other toxins. Drink alcohol, and the peroxisomes in your liver and kidney cells degrade nearly half of it.

Plant and animal cells contain **vacuoles**. Although these vesicles appear empty under a microscope, they serve an important function. Vacuoles are like trash cans; they isolate and dispose of waste, debris, or toxic materials. A central vacuole, described in Section 4.11, helps a plant cell maintain its shape and size.

Golgi Bodies

Many vesicles fuse with and empty their contents into a **Golgi body**. This organelle has a folded membrane that typically looks like a stack of pancakes (Figure 4.18e). Enzymes in a Golgi body put finishing touches on polypeptide chains and lipids that have been delivered from the ER. They attach phosphate groups or sugars, and cleave certain polypeptide chains. The finished products—membrane proteins, proteins for secretion, and enzymes—are sorted and packaged into new vesicles that carry them to the plasma membrane or to lysosomes. **Lysosomes** are vesicles that contain powerful digestive enzymes. They fuse with vacuoles carrying particles or molecules for disposal, such as worn-out cell components. Lysosomal enzymes empty into the other vesicles and digest their contents into bits.

Take-Home Message

What is the endomembrane system?

■ The endomembrane system includes rough and smooth endoplasmic reticulum, vesicles, and Golgi bodies.

■ This series of organelles works together mainly to synthesize and modify cell membrane proteins and lipids.

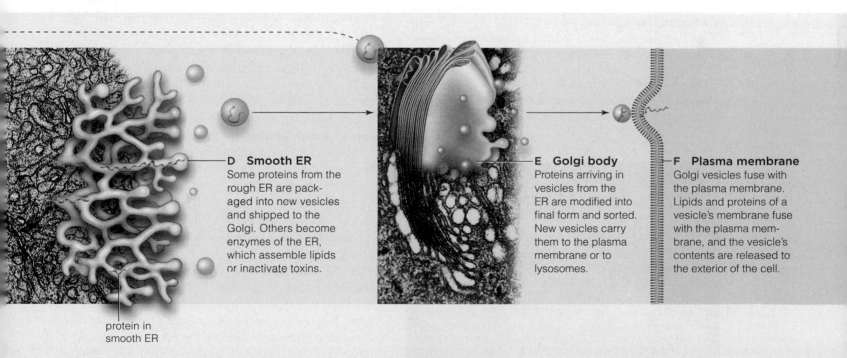

D Smooth ER
Some proteins from the rough ER are packaged into new vesicles and shipped to the Golgi. Others become enzymes of the ER, which assemble lipids or inactivate toxins.

protein in smooth ER

E Golgi body
Proteins arriving in vesicles from the ER are modified into final form and sorted. New vesicles carry them to the plasma membrane or to lysosomes.

F Plasma membrane
Golgi vesicles fuse with the plasma membrane. Lipids and proteins of a vesicle's membrane fuse with the plasma membrane, and the vesicle's contents are released to the exterior of the cell.

4.10 | Lysosome Malfunction

■ When lysosomes do not work properly, some cellular materials are not properly recycled, with devastating results.

■ Link to Mutations 1.4

Lysosomes serve as waste disposal and recycling centers. Enzymes inside them break large molecules into smaller subunits that the cell can use as building material or eliminate. Different kinds of molecules are broken down by different lysosomal enzymes.

In some people, a genetic mutation causes a deficiency or malfunction in one of the lysosomal enzymes. As a result, molecules that would normally get broken down accumulate instead. The result can be deadly.

For example, cells continually make, use, and break down gangliosides, a kind of lipid. This lipid turnover is especially brisk during early development. In Tay–Sachs disease, the enzyme responsible for ganglioside breakdown misfolds and is destroyed. Most commonly, affected infants seem normal for the first few months. Symptoms begin to appear as gangliosides accumulate to higher and higher levels inside their nerve cells. Within three to six months the child becomes irritable, listless, and may have seizures. Blindness, deafness, and paralysis follow. Affected children usually die by age five (Figure 4.19).

The mutation that causes Tay–Sachs is most prevalent in Jews of Eastern European descent. Cajuns and French Canadians also have a higher than average incidence, but Tay–Sachs occurs in all populations. The mutation can be detected in prospective parents by genetic screening, and in a fetus by prenatal diagnosis.

Researchers continue to explore options for treatment. Potential therapies involve blocking ganglioside synthesis, using gene therapy to deliver a normal version of the missing enzyme to the brain, or infusing normal blood cells from umbilical cords. All treatments are still considered experimental, and Tay–Sachs is still incurable.

Figure 4.19 Conner Hopf was diagnosed with Tay–Sachs disease at age 7–1/2 months. He died at 22 months.

4.11 | Other Organelles

■ Eukaryotic cells make most of their ATP in mitochondria.
■ Organelles called plastids function in storage and photosynthesis in plants and some types of algae.

■ Links to Metabolism 3.2, ATP 3.7

Mitochondria

The **mitochondrion** (plural, mitochondria) is a type of organelle that specializes in making ATP (Figure 4.20). Aerobic respiration, an oxygen-requiring series of reactions that proceeds inside mitochondria, can extract more energy from organic compounds than any other metabolic pathway. With each breath, you are taking in oxygen mainly for the mitochondria in your trillions of aerobically-respiring cells.

Typical mitochondria are between 1 and 4 micrometers in length; a few are as long as 10 micrometers. Some are branched. These organelles can change shape, split in two, and fuse together.

A mitochondrion has two membranes, one highly folded inside the other. This arrangement creates two compartments. Aerobic respiration causes hydrogen ions to accumulate between the two membranes. The buildup causes the ions to flow across the inner membrane, through the interior of membrane transport proteins. That flow drives the formation of ATP.

Nearly all eukaryotic cells have mitochondria, but prokaryotes do not (they make ATP in their cell walls and cytoplasm). The number of mitochondria varies by the type of cell and by the type of organism. For example, a single-celled yeast (a type of fungus) might have only one mitochondrion; a human skeletal muscle cell may have a thousand or more. Cells that have a very high demand for energy tend to have a profusion of mitochondria.

Mitochondria resemble bacteria, in size, form, and biochemistry. They have their own DNA, which is similar to bacterial DNA. They divide independently of the cell, and have their own ribosomes. Such clues led to a theory that mitochondria evolved from aerobic bacteria that took up permanent residence inside a host cell. By the theory of endosymbiosis, one cell was engulfed by another cell, or entered it as a parasite, but escaped digestion. That cell kept its plasma membrane intact and reproduced inside its host. In time, the cell's descendants became permanent residents that offered their hosts the benefit of extra ATP. Structures and functions once required for independent life were no longer needed and were lost over time. Later descendants evolved into mitochondria. We will explore evidence for the theory of endosymbiosis in Section 20.4.

Plastids

Plastids are membrane-enclosed organelles that function in photosynthesis or storage in plants and algal cells. Chloroplasts, chromoplasts, and amyloplasts are common types of plastids.

Photosynthetic cells of plants and many protists contain **chloroplasts**, organelles that are specialized for photosynthesis. Most chloroplasts have an oval or disk shape. Two outer membranes enclose a semifluid interior called the stroma (Figure 4.21). The stroma contains enzymes and the chloroplast's own DNA. Inside the stroma, a third, highly folded membrane forms a single compartment. The folds resemble stacks of flattened disks; the stacks are called grana (singular, granum). Photosynthesis takes place at this membrane, which is called the thylakoid membrane.

The thylakoid membrane incorporates many pigments and other proteins. The most abundant of the pigments are chlorophylls, which appear green. By the process of photosynthesis, the pigments and other molecules harness the energy in sunlight to drive the synthesis of ATP and the coenzyme NADPH. The ATP and NADPH are then used inside the stroma to build carbohydrates from carbon dioxide and water. We will describe the process of photosynthesis in more detail in Chapter 7.

In many ways, chloroplasts resemble photosynthetic bacteria, and like mitochondria they may have evolved by endosymbiosis.

Chromoplasts make and store pigments other than chlorophylls. They have an abundance of carotenoids, a pigment that colors many flowers, leaves, fruits, and roots red or orange. For example, as a tomato ripens, its green chloroplasts are converted to red chromoplasts, and the color of the fruit changes.

Amyloplasts are unpigmented plastids that typically store starch grains. They are notably abundant in cells of stems, tubers (underground stems), and seeds. Starch-packed amyloplasts are dense; in some plant cells, they function as gravity-sensing organelles.

The Central Vacuole

Amino acids, sugars, ions, wastes, and toxins accumulate in the water-filled interior of a plant cell's **central vacuole**. Fluid pressure in the central vacuole keeps plant cells—and structures such as stems and leaves—firm. Typically, the central vacuole takes up 50 to 90 percent of the cell's interior, with cytoplasm confined to a narrow zone in between this large organelle and the plasma membrane. Figure 4.14*b* has an example.

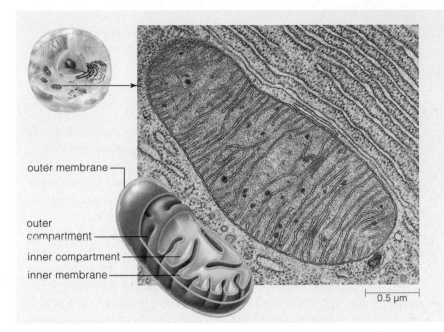

outer membrane
outer compartment
inner compartment
inner membrane

0.5 μm

Figure 4.20 Sketch and transmission electron micrograph of a mitochondrion. This organelle specializes in producing large quantities of ATP.

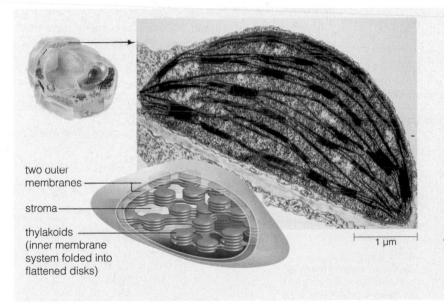

two outer membranes
stroma
thylakoids (inner membrane system folded into flattened disks)

1 μm

Figure 4.21 Animated The chloroplast, a defining character of photosynthetic eukaryotic cells. *Right*, transmission electron micrograph of a chloroplast from a tobacco leaf (*Nicotiana tabacum*). The lighter patches are nucleoids where DNA is stored.

Take-Home Message

What are some other specialized organelles of eukaryotes?

■ Mitochondria are eukaryotic organelles that produce ATP from organic compounds in reactions that require oxygen.

■ Chloroplasts are plastids that carry out photosynthesis.

■ Fluid pressure in a central vacuole keeps plant cells firm.

Cell Surface Specializations

- A wall or other protective covering often intervenes between a cell's plasma membrane and the surroundings.
- Link to Tissue 1.1

Eukaryotic Cell Walls

Like most prokaryotic cells, many types of eukaryotic cells have a cell wall around the plasma membrane. The wall is a porous structure that protects, supports, and imparts shape to the cell. Water and solutes easily cross it on the way to and from the plasma membrane. Cells could not live without such exchanges.

Animal cells do not have walls, but plant cells and many protist and fungal cells do. For example, a young plant cell secretes pectin and other polysaccharides onto the outer surface of its plasma membrane. The sticky coating is shared between adjacent cells, and it cements them together. Each cell then forms a **primary wall** by secreting strands of cellulose into the coating. Some of the coating remains as the middle lamella, a sticky layer in between the primary walls of abutting plant cells (Figure 4.22*a,b*).

Being thin and pliable, the primary wall allows the growing plant cell to enlarge. Plant cells with only a thin primary wall can change shape as they develop.

At maturity, cells in some plant tissues stop enlarging and begin to secrete material onto the primary wall's inner surface. These deposits form a firm **secondary wall**, of the sort shown in Figure 4.22*b*. One of the materials deposited is **lignin**, a complex polymer of alcohols that makes up as much as 25 percent of the secondary wall of cells in older stems and roots. Lignified plant parts are stronger, more waterproof, and less susceptible to plant-attacking organisms than younger tissues.

A **cuticle** is a protective body covering made of cell secretions. In plants, a semitransparent cuticle helps protect exposed surfaces of soft parts and limits water loss on hot, dry days (Figure 4.23).

Matrixes Between Cells

Most cells of multicelled organisms are surrounded and organized by **extracellular matrix** (ECM). This nonliving, complex mixture of fibrous proteins and polysaccharides is secreted by cells, and varies with the type of tissue. It supports and anchors cells, separates tissues, and functions in cell signaling.

Primary cell walls are a type of extracellular matrix, which in plants is mostly cellulose. The extracellular matrix of fungi is mainly chitin (Section 3.3). In most

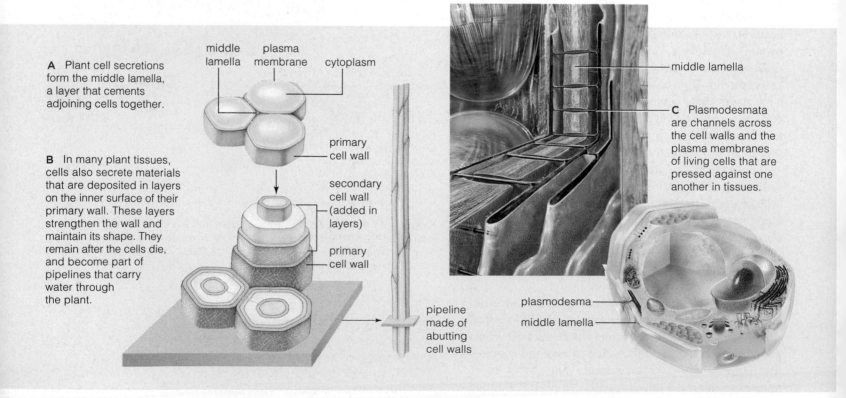

A Plant cell secretions form the middle lamella, a layer that cements adjoining cells together.

middle lamella   plasma membrane   cytoplasm

B In many plant tissues, cells also secrete materials that are deposited in layers on the inner surface of their primary wall. These layers strengthen the wall and maintain its shape. They remain after the cells die, and become part of pipelines that carry water through the plant.

primary cell wall

secondary cell wall (added in layers)

primary cell wall

pipeline made of abutting cell walls

middle lamella

C Plasmodesmata are channels across the cell walls and the plasma membranes of living cells that are pressed against one another in tissues.

plasmodesma

middle lamella

Figure 4.22 Animated Some characteristics of plant cell walls.

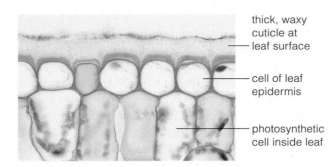

thick, waxy
cuticle at
leaf surface

cell of leaf
epidermis

photosynthetic
cell inside leaf

Figure 4.23 A plant cuticle is a waxy, waterproof covering secreted by living cells.

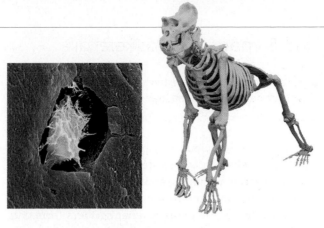

Figure 4.24 A living cell surrounded by hardened bone tissue, the main structural material in the skeleton of most vertebrates.

animals, extracellular matrix consists of various kinds of carbohydrates and proteins; it is the basis of tissue organization, and it provides structural support. For example, bone is mostly extracellular matrix (Figure 4.24). Bone ECM is mostly collagen, a fibrous protein, and it is hardened by mineral deposits.

Cell Junctions

A cell that is surrounded by a wall or other secretions is not isolated; it can still interact with other cells and with the surroundings. In multicelled species, such interaction occurs by way of **cell junctions**, which are structures that connect a cell to other cells and to the environment. Cells send and receive ions, molecules or signals through some junctions. Other kinds help cells recognize and stick to each other and to extracellular matrix.

In plants, channels called plasmodesmata (singular, plasmodesma) extend across the primary wall of two adjoining cells, connecting the cytoplasm of the cells (Figure 4.22c). Substances such as water, ions, nutrients, and signalling molecules can flow quickly from cell to cell through plasmodesmata.

Three types of cell-to-cell junctions are common in most animal tissues: tight junctions, adhering junctions, and gap junctions (Figure 4.25). Tight junctions link cells that line the surfaces and internal cavities of animals. These junctions seal the cells together tightly, so fluid cannot pass between them. Those in your gastrointestinal tract prevent gastric fluid from leaking out of your stomach and damaging your internal tissues. Adhering junctions anchor cells to each other and to extracellular matrix; they strengthen contractile tissues such as heart muscle. Gap junctions are open channels that connect the cytoplasm of adjoining cells; they are similar to plasmodesmata in plants. Gap junctions allow entire regions of cells to respond to a

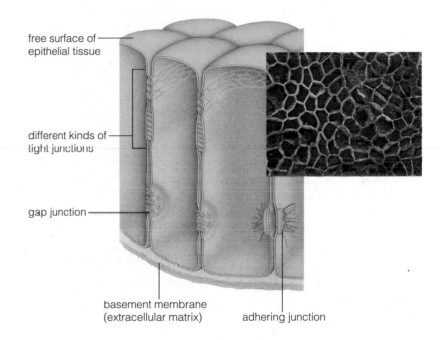

free surface of
epithelial tissue

different kinds of
tight junctions

gap junction

basement membrane
(extracellular matrix)

adhering junction

Figure 4.25 Animated Cell junctions in animal tissues. In the micrograph, a continuous array of tight junctions (*green*) seals the abutting surfaces of kidney cell membranes. DNA, which fills each cell's nucleus, appears *red*.

single stimulus. For example, in heart muscle, a signal to contract passes instantly from cell to cell through gap junctions, so all cells contract as a unit.

Take-Home Message

What structures form on the outside of eukaryotic cells?

■ Cells of many protists, nearly all fungi, and all plants, have a porous wall around the plasma membrane. Animal cells do not have walls.

■ Plant cell secretions form a waxy cuticle that helps protect the exposed surfaces of soft plant parts.

■ Cell secretions form extracellular matrixes between cells in many tissues.

■ Cells make structural and functional connections with one another and with extracellular matrix in tissues.

4.13 | The Dynamic Cytoskeleton

- Eukaryotic cells have an extensive and dynamic internal framework called a cytoskeleton.

- Links to Protein structure and function 3.5, 3.6

In between the nucleus and plasma membrane of all eukaryotic cells is a **cytoskeleton**—an interconnected system of many protein filaments. Parts of the system reinforce, organize, and move cell structures, and often the whole cell. Some are permanent; others form only at certain times. Figure 4.26 shows the main types.

Microtubules are long, hollow cylinders that consist of subunits of the protein tubulin. They form a dynamic scaffolding for many cellular processes, rapidly assembling when they are needed, disassembling when they are not. For example, some of the microtubules that assemble before a eukaryotic cell divides separate the cell's duplicated chromosomes, then disassemble. As another example, microtubules that form in the growing end of a young nerve cell support and guide its lengthening in a particular direction.

Microfilaments are fibers that consist primarily of subunits of the globular protein actin. They strengthen or change the shape of eukaryotic cells. Crosslinked, bundled, or gel-like arrays of them make up the **cell cortex**, a reinforcing mesh under the plasma membrane. Actin microfilaments that form at the edge of a cell drag or extend it in a certain direction (Figure 4.26). In muscle cells, microfilaments of myosin and actin interact to bring about contraction.

Intermediate filaments are the most stable parts of a cell's cytoskeleton. They strengthen and maintain cell and tissue structures. For example, some intermediate filaments called lamins form a layer that structurally supports the inner surface of the nuclear envelope.

All eukaryotic cells have similar microtubules and microfilaments. Despite the uniformity, both kinds of elements play diverse roles. How? They interact with accessory proteins, such as the **motor proteins** that can move cell parts in a sustained direction when they are repeatedly energized by ATP.

A cell is like a train station during a busy holiday, with molecules being transported through its interior. Microtubules and microfilaments are like dynamically assembled train tracks. Motor proteins are the freight engines that move along those tracks (Figure 4.27).

Some motor proteins move chromosomes. Others slide one microtubule over another. Some chug along tracks in nerve cells that extend from your spine to your toes. Many engines are organized in series, each moving some vesicle partway along the track before giving it up to the next in line. In plant cells, kinesins drag chloroplasts away from light that is too intense, or toward a light source under low-light conditions.

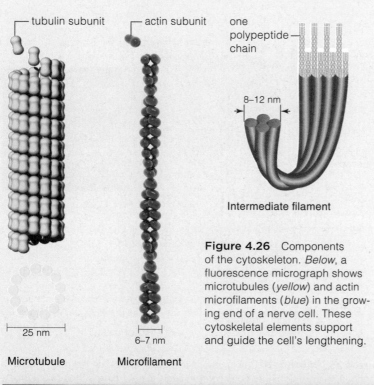

tubulin subunit

actin subunit

one polypeptide chain

8–12 nm

Intermediate filament

25 nm

6–7 nm

Microtubule

Microfilament

Figure 4.26 Components of the cytoskeleton. *Below*, a fluorescence micrograph shows microtubules (*yellow*) and actin microfilaments (*blue*) in the growing end of a nerve cell. These cytoskeletal elements support and guide the cell's lengthening.

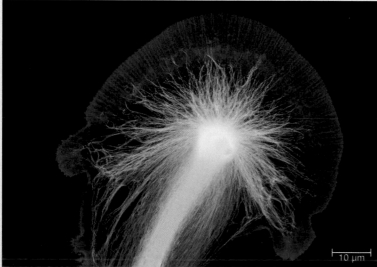

10 μm

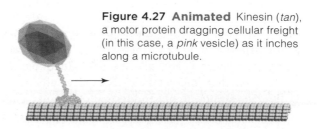

Figure 4.27 Animated Kinesin (*tan*), a motor protein dragging cellular freight (in this case, a *pink* vesicle) as it inches along a microtubule.

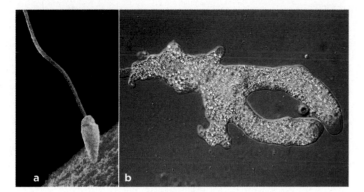

Figure 4.28 (**a**) Flagellum of a human sperm, which is about to penetrate an egg. (**b**) A predatory amoeba (*Chaos carolinense*) extending two pseudopods around its hapless meal: a single-celled green alga (*Pandorina*).

Cilia, Flagella, and False Feet

Organized arrays of microtubules occur in **eukaryotic flagella** (singular, flagellum) and **cilia** (cilium), which are whiplike structures that propel cells such as sperm through fluid (Figure 4.28a). Flagella tend to be longer and less profuse than cilia. The coordinated beating of cilia propels motile cells through fluid, and stirs fluid around stationary cells. For example, the coordinated motion of cilia on the thousands of cells lining your airways sweeps particles away from your lungs.

A special array of microtubules extends lengthwise through a flagellum or cilium. This 9+2 array consists of nine pairs of microtubules ringing another pair in the center (Figure 4.29). Protein spokes and links stabilize the array. The microtubules grow from a barrel-shaped organelle called the **centriole**, which remains below the finished array as a basal body.

Amoebas and other types of eukaryotic cells form **pseudopods**, or "false feet" (Figure 4.28b). As these temporary, irregular lobes bulge outward, they move the cell and engulf a target such as prey. Elongating microfilaments force the lobe to advance in a steady direction. Motor proteins that are attached to the microfilaments drag the plasma membrane along with them.

Take-Home Message

What is a cytoskeleton?

■ A cytoskeleton of protein filaments is the basis of eukaryotic cell shape, internal structure, and movement.

■ Microtubules organize the cell and help move its parts. Networks of microfilaments reinforce the cell surface. Intermediate filaments strengthen cells and tissues, and maintain their shape.

■ When energized by ATP, motor proteins move along tracks of microtubules and microfilaments. As part of cilia, flagella, and pseudopods, they can move the whole cell.

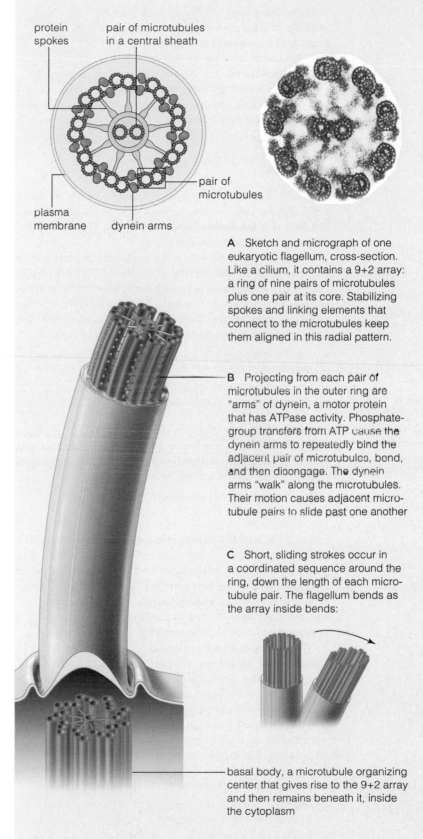

A Sketch and micrograph of one eukaryotic flagellum, cross-section. Like a cilium, it contains a 9+2 array: a ring of nine pairs of microtubules plus one pair at its core. Stabilizing spokes and linking elements that connect to the microtubules keep them aligned in this radial pattern.

B Projecting from each pair of microtubules in the outer ring are "arms" of dynein, a motor protein that has ATPase activity. Phosphate-group transfers from ATP cause the dynein arms to repeatedly bind the adjacent pair of microtubules, bond, and then disengage. The dynein arms "walk" along the microtubules. Their motion causes adjacent microtubule pairs to slide past one another

C Short, sliding strokes occur in a coordinated sequence around the ring, down the length of each microtubule pair. The flagellum bends as the array inside bends:

basal body, a microtubule organizing center that gives rise to the 9+2 array and then remains beneath it, inside the cytoplasm

Figure 4.29 Animated Eukaryotic flagella and cilia.

Irradiated meat, poultry, milk, and fruits are now available in supermarkets. By law, irradiated foods must be marked with the symbol on the right. Items that bear this symbol have been exposed to radiation, but are not themselves radioactive. Irradiating fresh foods kills bacteria and prolongs shelf life. However, some worry that the irradiation process may alter the food and produce harmful chemicals.

Whether health risks are associated with consuming irradiated foods is still unknown.

Summary

Sections 4.1–4.3 All organisms consist of one or more **cells**. By the **cell theory**, the cell is the smallest unit of life, and it is the basis of life's continuity. The **surface-to-volume** ratio limits cell size.

All cells start out life with a **plasma membrane**, a **nucleus** (in **eukaryotic cells**) or **nucleoid** (in **prokaryotic cells**), and **cytoplasm** in which structures such as **ribosomes** are suspended. The **lipid bilayer** is the foundation of all cell membranes. Different types of microscopes use light or electrons to reveal different details of cells.

- *Use the interactions on CengageNOW to investigate basic membrane structure and the physical limits on cell size.*

- *Use the animation on CengageNOW to learn how different types of microscopes function.*

Sections 4.4, 4.5 Bacteria and archaeans are prokaryotes (Table 4.3). None has a nucleus. Many have a **cell wall** and one or more **flagella** or **pili**. **Biofilms** are shared living arrangements among bacteria and other microbes.

- *Use the animation on CengageNOW to view prokaryotic cell structure.*

Sections 4.6–4.11 Eukaryotic cells start out life with a nucleus and other membrane-enclosed **organelles**. The nucleus contains **nucleoplasm** and **nucleoli**. **Chromatin** in the nucleus of a eukaryotic cell is divided into a characteristic number of **chromosomes**. Pores, receptors, and transport proteins in the **nuclear envelope** control the movement of molecules into and out of the nucleus.

The **endomembrane system** includes rough and smooth **endoplasmic reticulum**, **vesicles**, and **Golgi bodies**. This set of organelles functions mainly to make and modify lipids and proteins; it also recycles molecules and particles such as worn-out cell parts, and inactivates toxins.

Mitochondria produce ATP by breaking down organic compounds in the oxygen-requiring pathway of aerobic respiration. **Chloroplasts** are **plastids** that specialize in photosynthesis. Other organelles include **peroxisomes**, **lysosomes**, and **vacuoles** (including **central vacuoles**).

- *Use the interaction on CengageNOW to survey the major types of eukaryotic organelles.*

- *Use the animations on CengageNOW to view the nuclear membrane and the endomembrane system.*

- *Use the animation on CengageNOW to view a chloroplast.*

Section 4.12 Cells of most prokaryotes, protists, fungi, and all plant cells have a wall around the plasma membrane. Older plant cells secrete a rigid, **lignin**-containing **secondary wall** inside their pliable **primary wall**. Many eukaryotic cell types also secrete a **cuticle**. Plasmodesmata connect plant cells. **Cell junctions** connect animal cells to one another and to **extracellular matrix** (ECM).

- *Study the structure of cell walls and junctions with the animation on CengageNOW.*

Section 4.13 Eukaryotic cells have a **cytoskeleton**. The **cell cortex** consists of **intermediate filaments**. **Motor proteins** that are the basis of movement interact with **microfilaments** in **pseudopods**, or (in **cilia** and **eukaryotic flagella**) **microtubules** that grow from **centrioles**.

- *Learn more about cytoskeletal elements and their actions with the animation on CengageNOW.*

Self-Quiz   *Answers in Appendix III*

1. The _____ is the smallest unit of life.

2. True or false: Some protists are prokaryotes.

3. Cell membranes consist mostly of _____ .

4. Unlike eukaryotic cells, prokaryotic cells _____ .
   a. have no plasma membrane   c. have no nucleus
   b. have RNA but not DNA   d. a and c

5. Organelles enclosed by membranes are typical features of _____ cells.

6. The main function of the endomembrane system is building and modifying _____ and _____ .

7. Ribosome subunits are built inside the _____ .

8. No animal cell has a _____ .

9. Is this statement true or false? The plasma membrane is the outermost component of all cells. Explain.

10. Enzymes contained in _____ break down worn-out organelles, bacteria, and other particles.

11. Match each cell component with its function.
    ___mitochondrion   a. protein synthesis
    ___chloroplast   b. associates with ribosomes
    ___ribosome   c. ATP by sugar breakdown
    ___smooth ER   d. sorts and ships
    ___Golgi body   e. assembles lipids; other tasks
    ___rough ER   f. photosynthesis

- *Visit CengageNOW for additional questions.*

Data Analysis Exercise

An abnormal form of the motor protein dynein causes Kartagener syndrome, a genetic disorder characterized by chronic sinus and lung infections. Biofilms form in the thick mucus that collects in the airways, and the resulting bacterial activities and inflammation damage tissues.

Affected men can produce sperm but are infertile (Figure 4.30). Some have become fathers after a doctor injects their sperm cells directly into eggs. Review Figure 4.30, then explain how abnormal dynein could cause the observed effects.

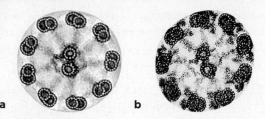

a   b

Figure 4.30 Cross-section of the flagellum of a sperm cell from (**a**) a human male affected by Kartagener syndrome and (**b**) an unaffected male.

Critical Thinking

1. In a classic episode of *Star Trek*, a gigantic amoeba engulfs an entire starship. Spock blows the cell to bits before it reproduces. Think of at least one problem a biologist would have with this particular scenario.

2. Many plant cells form a secondary wall on the inner surface of their primary wall. Speculate on the reason why the secondary wall does not form on the outer surface.

3. A student is examining different samples with a transmission electron microscope. She discovers a single-celled organism swimming in a freshwater pond (*below*).

Which of this organism's structures can you identify? Is it a prokaryotic or eukaryotic cell? Can you be more specific about the type of cell based on what you know about cell structure? Look ahead to Section 22.2 to check your answers.

Table 4.3 Summary of Typical Components of Prokaryotic and Eukaryotic Cells

Cell Component	Main Functions	Prokaryotic Bacteria, Archaea	Eukaryotic Protists	Fungi	Plants	Animals
Cell wall	Protection, structural support	✳	✳	✔	✔	—
Plasma membrane	Control of substances moving into and out of cell	✔	✔	✔	✔	✔
Nucleus	Physical separation of DNA from cytoplasm	—✳	✔	✔	✔	✔
DNA	Encodes hereditary information	✔	✔	✔	✔	✔
Nucleolus	Assembly of ribosome subunits	—	✔	✔	✔	✔
Ribosome	Protein synthesis	✔	✔	✔	✔	✔
Endoplasmic reticulum (ER)	Synthesis, modification of membrane proteins; lipid synthesis	—	✔	✔	✔	✔
Golgi body	Final modification of membrane proteins; sorting, packaging lipids and proteins into vesicles	—	✔	✔	✔	✔
Lysosome	Intracellular digestion	—	✔	✳	✳	✔
Centriole	Organization of cytoskeletal elements	★	✔	✔	✳	✔
Mitochondrion	ATP formation	—	✔	✔	✔	✔
Chloroplast	Photosynthesis	—	✳	—	✔	—
Central vacuole	Storage	—	—	✳	✔	—
Bacterial flagellum	Locomotion through fluid surroundings	✳	—	—	—	—
Flagellum or cilium with 9+2 microtubule array	Locomotion through or motion within fluid surroundings	—	✳	✳	✳	✔
Cytoskeleton	Cell shape; internal organization; basis of cell movement and, in many cells, locomotion	★	✳	✳	✳	✔

✔ Present in at least part of the life cycle of most or all groups.
✳ Known to be present in cells of at least some groups.
★ Occurs in a form unique to prokaryotes.
 * Some planctomycete bacteria have a double membrane around their DNA.

5 A Closer Look at Cell Membranes

IMPACTS, ISSUES | One Bad Transporter and Cystic Fibrosis

Every cell actively engages in the business of living. Think of how it has to move something as ordinary as water in one direction or the other across its plasma membrane. Water crosses a cell membrane freely. The cell has to be able to take in or send out water at different times in order to keep the cytoplasm from getting too concentrated or too dilute. If all goes well, the cell takes in or sends out water in just the right amounts—not too little, not too much.

Proteins called transporters move ions and molecules, including water, across cell membranes. Different transporters move different substances. One, called CFTR, is a transporter in the plasma membrane of epithelial cells. Sheets of these cells line the passageways and ducts of the lungs, liver, pancreas, intestines, reproductive system, and skin. CFTR pumps chloride ions out of these cells, and water follows the ions. A thin, watery film forms on the surface of the epithelial cell sheets. Mucus slides easily over the wet sheets of cells.

Sometimes a mutation changes the structure of CFTR. When epithelial cell membranes do not have enough working copies of the CFTR protein, chloride ion transport is disrupted. Not enough chloride ions leave the cells, and so not enough water leaves them either. The result is thick, dry mucus that sticks to the epithelial cell sheets.

In the respiratory tract, the mucus clogs airways to the lungs and makes breathing difficult. It is too thick for the ciliated cells lining the airways to sweep out, and bacteria thrive in it. Low-grade infections occur and may persist for years.

These symptoms—outcomes of mutation in the CFTR protein—characterize cystic fibrosis (CF), the most common fatal genetic disorder in the United States. Even with a lung transplant, most CF patients live no longer than thirty years, at which time their lungs usually fail. There is no cure.

More than 10 million people carry a mutated form of the CFTR gene. Some of them have sinus problems, but no other symptoms develop. Most do not know they carry the mutated gene. CF develops when a person inherits a mutated gene from both parents—an unlucky event that occurs in about 1 of 3,300 births (Figure 5.1). Think about it. A startling percentage of the human population can develop severe problems when even one kind of membrane protein does not work.

Your life depends on the functions of thousands of kinds of proteins and other molecules that keep cells working. Each cell functions properly only if it is responsive to conditions in the environments on both sides of its membranes. Cell membranes—these thin boundary layers make the difference between organization and chaos.

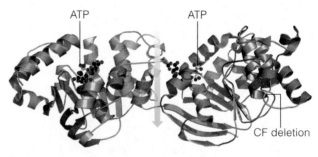

See the video! Figure 5.1 Cystic fibrosis. *Left*, a few of the many victims of cystic fibrosis, which occurs most often in people of northern European ancestry. At least one young person dies every day in the United States from complications of this disease.

Above, model of CFTR. The parts shown here are ATP-driven motors that widen or narrow a channel (*gray* arrow) across the plasma membrane. The tiny part of the protein that is lost in most cystic fibrosis mutations is shown on the ribbon in *green*.

Key Concepts

Membrane structure and function

Cell membranes have a lipid bilayer that is a boundary between the outside environment and the cell interior. Diverse proteins embedded in the bilayer or positioned at one of its surfaces carry out most membrane functions. **Sections 5.1, 5.2**

Diffusion and membrane transport

Gradients drive the directional movements of substances across membranes. Transport proteins work with or against gradients to maintain water and solute concentrations. **Sections 5.3, 5.4**

Membrane trafficking

Large packets of substances and engulfed cells move across the plasma membrane by the processes of endocytosis and exocytosis. Membrane lipids and proteins move to and from the plasma membrane during these processes. **Section 5.5**

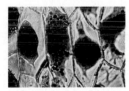

Osmosis

Water tends to diffuse across selectively permeable membranes, including cell membranes, to regions where its concentration is lower. **Section 5.6**

Links to Earlier Concepts

- Reflect again on the road map in Section 1.1. Here you will see how lipids (3.4) and proteins (3.5) become organized in cell membranes (4.2).

- In this chapter, you will consider examples of how a protein's function (3.6) arises from its structure. You will also learn more about the proteins that compose cell junctions (4.12).

- Lipids have both hydrophilic and hydrophobic properties (2.5), a duality that gives rise to the structural organization of all cell membranes.

- Your knowledge of charge (2.1) and the properties of water (2.5) will help you understand the movement of ions and molecules in response to gradients.

- You will revisit the endomembrane system (4.9) as you learn how the cytoskeleton (4.13) is involved in the cycling of membrane lipids and proteins.

- The movement of water into and out of cells is an important part of homeostasis (1.2). A review of what you know about plant cell walls (4.12) will help you understand how this movement affects growth in plants.

How would you vote? The ability to detect mutated genes that cause severe disorders such as cystic fibrosis raises ethical questions. Should we encourage the mass screening of prospective parents for mutations that cause CF? See CengageNow for details, then vote online.

■ The basic structure of all cell membranes is the lipid bilayer with many embedded proteins.

■ A membrane is a continuous, selectively permeable barrier.

■ Links to Emergent properties 1.1, Hydrophilic and hydrophobic 2.5, Lipids 3.4, Membranes 4.2, Tight junctions 4.12

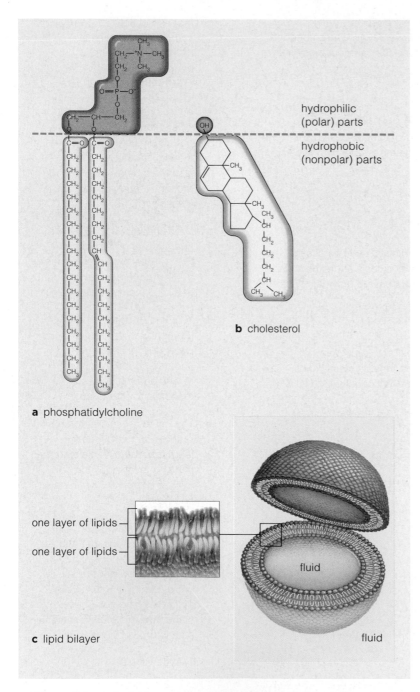

hydrophilic (polar) parts

hydrophobic (nonpolar) parts

b cholesterol

a phosphatidylcholine

one layer of lipids —

one layer of lipids —

fluid

c lipid bilayer   fluid

Figure 5.2 Cell membrane organization. (**a**) Phosphatidylcholine, the most common phospholipid component of animal cell membranes. (**b**) Cholesterol, the main steroid component of animal cell membranes. Phytosterols are its equivalent in plant cell membranes.

(**c**) Spontaneous organization of phospholipids into two layers (a lipid bilayer). When mixed with water, phospholipids aggregate into a bilayer, with their hydrophobic tails sandwiched between their hydrophilic heads.

Revisiting the Lipid Bilayer

Properties that are unique to cell membranes emerge when certain lipids—mainly phospholipids—interact. Each phospholipid molecule consists of a phosphate-containing head and two fatty acid tails (Figure 5.2a). The polar head is hydrophilic, which means it interacts with water molecules. The nonpolar tails are hydrophobic, so they do not interact with water molecules. The tails do, however, interact with the tails of other phospholipids. When swirled into water, phospholipids spontaneously assemble into two layers, with all of their nonpolar tails sandwiched between all of their polar heads. Such lipid bilayers are the framework of all cell membranes (Figure 5.2c).

The Fluid Mosaic Model

A **fluid mosaic model** describes the organization of cell membranes. By this model, a cell membrane is a mosaic, a mixed composition of mostly phospholipids, with steroids, proteins, and other molecules dispersed among them (Figure 5.3). The fluid part of the model refers to the behavior of phospholipids in membranes. The phospholipids remain organized as a bilayer, but they also drift sideways, they spin on their long axis, and their tails wiggle.

Variations

Differences in Membrane Composition Membranes differ in composition. The differences reflect their functions in cells. Even the two surfaces of a lipid bilayer are different. For example, carbohydrates attached to certain membrane proteins and lipids project from a plasma membrane but not into the cell. The kinds and numbers of attachments differ from one species to the next, and even among cells of the same body.

Different kinds of cells have different kinds of membrane phospholipids. For example, the fatty acid tails of membrane phospholipids vary in length and saturation. Usually, at least one of the two tails is unsaturated. An unsaturated fatty acid, remember, has one or more double covalent bonds in its carbon backbone (Section 3.4).

Differences in Fluidity We once thought that all proteins in a cell membrane were fixed in place, but key experiments proved otherwise. Two of those experiments are summarized in Figure 5.4. We now know that some proteins stay put, such as those that cluster as rigid pores. Protein filaments of the cytoskeleton

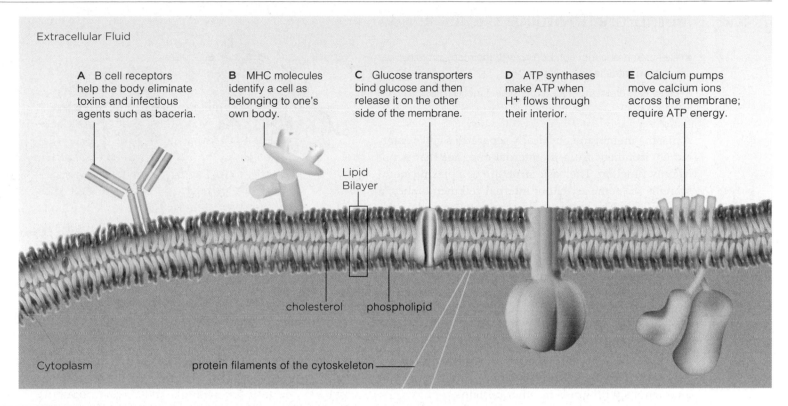

Extracellular Fluid

A B cell receptors help the body eliminate toxins and infectious agents such as baceria.

B MHC molecules identify a cell as belonging to one's own body.

C Glucose transporters bind glucose and then release it on the other side of the membrane.

D ATP synthases make ATP when H^+ flows through their interior.

E Calcium pumps move calcium ions across the membrane; require ATP energy.

Lipid Bilayer

cholesterol   phospholipid

Cytoplasm   protein filaments of the cytoskeleton

Figure 5.3 Animated Fluid mosaic model for the plasma membrane of an animal cell. Section 5.2 presents an overview of the main types of membrane proteins.

lock these and other proteins in place. Tight junctions that link the cytoskeletons of adjacent cells can keep the membrane proteins corraled to the upper or lower surfaces of cells in animal tissues. However, most of the proteins in bacterial and eukaryotic cell membranes drift around very quickly. Part of the reason that the membranes of these organisms are so fluid stems from the composition of the phospholipids in the lipid bilayer.

Archaeans do not build their phospholipids with fatty acids. Instead, they use molecules that have reactive side chains, so the tails of archaean phospholipids form covalent bonds among one another. As a result of this rigid crosslinking, archaean phospholipids do not drift, spin, or wiggle in a bilayer. Thus, the membranes of archaeans are far more rigid than those of bacteria or eukaryotes, a characteristic that may help these cells survive in extreme habitats.

Take-Home Message

What is the function of a cell membrane?

■ A cell membrane is a barrier that selectively controls exchanges between the cell and its surroundings. It is a mosaic of different kinds of lipids and proteins.

■ The foundation of cell membranes is the lipid bilayer—two layers of phospholipids, tails sandwiched between heads.

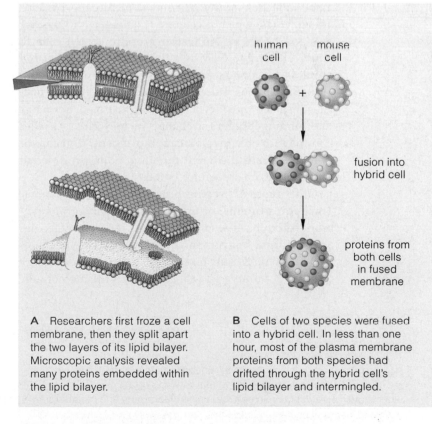

human cell   mouse cell

fusion into hybrid cell

proteins from both cells in fused membrane

A Researchers first froze a cell membrane, then they split apart the two layers of its lipid bilayer. Microscopic analysis revealed many proteins embedded within the lipid bilayer.

B Cells of two species were fused into a hybrid cell. In less than one hour, most of the plasma membrane proteins from both species had drifted through the hybrid cell's lipid bilayer and intermingled.

Figure 5.4 Animated Two studies of membrane structure, an observation and an experiment.

5.2 | Membrane Proteins

- Cell membrane function begins with the many proteins associated with the lipid bilayer.
- Links to Protein structure 3.5, Protein function 3.6, Cell junctions 4.12

A plasma membrane physically separates a cell's external environment from its internal one, but that is not its only function. The basic structure of a plasma membrane is the same as that of internal cell membranes: a lipid bilayer. The many kinds of proteins in and on the bilayer impart distinct functions to each membrane.

Membrane proteins can be assigned to one of two categories, depending on the way they associate with a membrane. Integral membrane proteins are permanently attached to a lipid bilayer. Some have transmembrane domains—hydrophobic regions that span the entire bilayer. Transmembrane domains anchor the protein in the bilayer, and some form channels all the way through it. Peripheral membrane proteins temporarily attach to one of the bilayer's surfaces by way of interactions with lipids or other proteins.

Each type of protein in a membrane imparts a specific function to it (Figure 5.5). Thus, different cell membranes can have different characteristics. For example, the plasma membrane has proteins that no other cell membrane has. Many peripheral membrane proteins are **enzymes**, which accelerate reactions without being changed by them. **Adhesion proteins** fasten cells to other cells and to ECM in animal tissues. **Recognition proteins** function as unique identity tags for each individual or species. **Receptor proteins** bind to a particular substance outside of the cell, such as a hormone. The binding triggers a change in the cell's activities that may involve metabolism, movement, division, or even cell death. Different receptors occur on different cells, but all are critical for homeostasis.

Other types of proteins occur on all cell membranes. **Transport proteins**, or transporters, are integral membrane proteins that move specific ions or molecules across a lipid bilayer. Some transporters are channels through which a substance diffuses; others use energy to actively pump a substance through the membrane.

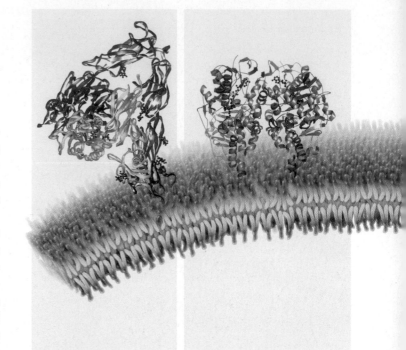

Adhesion Protein

Function Attachment of cells to one another and to extracellular matrix

Occurs only on plasma membranes

Membrane Attachment Integral

Example Integrins, including this one, are also receptors that mediate cell attachment, migration, differentiation, division, and survival.

Example Cadherins are part of adhering junctions between cells.

Example Selectins bind glycoproteins on the surface of cells that function in immunity.

Enzyme

Function Speeds a specific reaction

Membranes provide a relatively stable reaction site for enzymes, particularly those that work in series with other molecules. Arrays of membrane-bound enzymes and other proteins carry out important tasks such as photosynthesis and aerobic respiration.

Membrane Attachment Integral or peripheral

Example The enzyme shown here is a monoamine oxidase of mitochondrial membranes. It catalyzes a hydrolysis reaction that removes an ammonia group (NH_3) from amino acids.

Take-Home Message

What do membrane proteins do?

- Various membrane proteins impart functionality to a lipid bilayer.
- A plasma membrane, especially of multicelled species, has receptors and other proteins that function in self-recognition, adhesion, and metabolism.
- All cell membranes have transporters that passively and actively assist specific ions and molecules across the lipid bilayer.

Figure 5.5 Animated Major categories of membrane proteins, with descriptions and examples. You will see the icons above some of the descriptions again in this book.

Transporters span all cell membranes. The other proteins shown are components of plasma membranes. Organelle membranes also incorporate additional kinds of proteins.

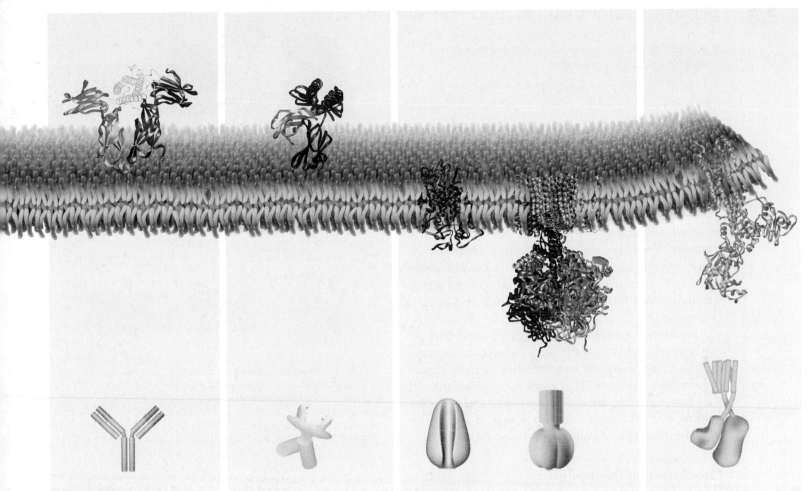

Receptor Protein

Function Binding signaling molecules

Binding causes a change in cell activity, such as gene expression, metabolism, movement, adhesion, division, or cell death.

Membrane Attachment Integral or peripheral

Example The B cell receptor shown here is a protein made only by white blood cells called B lymphocytes. B cell receptors are membrane-bound antibodies. These receptors are vital for immune responses. (We return to immunity in Chapter 38.)

Recognition Protein

Function Identifier of cell type, individual, or species

Membrane Attachment Integral

Example The MHC molecule shown here functions in vertebrate immunity. MHC molecules allow white blood cells called T lymphocytes to identify a cell as *nonself* (foreign) or *self* (belonging to one's own body). Fragments of invading organisms or other nonself particles bound to MHC molecules attract the attention of T lymphocytes. (We return to immunity in Chapter 38.)

Passive Transporter

Function Transport of molecules or ions

Does not require energy

Membrane Attachment Integral

Example On the *left*, a glucose transporter. When glucose binds to this transporter, the protein changes shape, and glucose is released to the other side of the membrane. Passive transporters that change shape are said to be "gated."

Example Other transporters, such as aquaporin, are open channels. Aquaporin transports water.

Example You will see the transporter shown on the *right* several more times in this book. When hydrogen ions flow through a channel in its interior, this molecule synthesizes ATP. Hence its name, ATP synthase.

Active Transporter

Function Transport of molecules or ions

Uses energy (usually in the form of ATP) to pump substances across the membrane

Membrane Attachment Integral

Example The calcium pump shown here uses ATP to pump calcium ions across a membrane.

Example In some contexts, ATP synthase works in reverse, using ATP to pump hydrogen ions across a membrane. In this role, the molecule is an active transporter.

■ Ions and molecules tend to move from one region to another, in response to gradients.

■ Links to Homeostasis 1.2, Charge 2.1, Water 2.5

Membrane Permeability

Any body fluid outside of cells is called extracellular fluid. Many different substances are dissolved in cytoplasm and in extracellular fluid, but the kinds and amounts of solutes in the two fluids differ. The ability of a cell to maintain these differences depends on a property of membranes called **selective permeability**: The membrane allows some substances but not others to cross it. This property helps the cell control which substances and how much of them enter and leave it (Figure 5.6).

Membrane barriers and crossings are vital, because metabolism depends on the cell's capacity to increase, decrease, and maintain concentrations of substances required for reactions. That capacity supplies the cell with raw materials, removes wastes, and maintains the volume and pH within tolerable ranges. It also serves these functions for membrane-enclosed sacs in cells.

Concentration Gradients

Concentration is the number of molecules (or ions) of a substance per unit volume of fluid. A difference in concentration between two adjacent regions is called a **concentration gradient**. Molecules or ions tend to move "down" their concentration gradient, from a region of higher concentration to one of lower concentration.

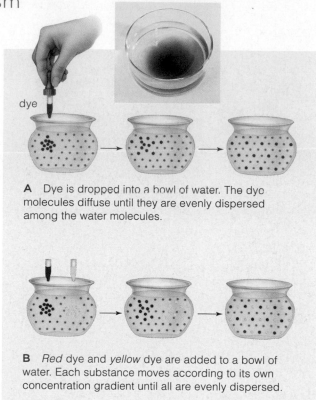

A Dye is dropped into a bowl of water. The dye molecules diffuse until they are evenly dispersed among the water molecules.

B *Red* dye and *yellow* dye are added to a bowl of water. Each substance moves according to its own concentration gradient until all are evenly dispersed.

Figure 5.7 Animated Examples of diffusion.

Why? Like individual atoms, molecules are always in motion. They collide at random and bounce off one another millions of times each second in both regions. However, the more crowded molecules are, the more often they collide. During any interval, more molecules are knocked out of a region of higher concentration than are knocked into it.

Diffusion is the net (or overall) movement of molecules or ions down a concentration gradient. It is an essential way in which substances move into, through, and out of cells. In multicelled species, diffusion also moves substances between cells in different regions of the body or between cells and the body's external environment. For instance, photosynthetic cells inside a leaf produce oxygen. The oxygen diffuses out of the cells and into air spaces inside the leaf, where its concentration is lower. Then it diffuses into the air outside the leaf, where its concentration is lower still.

Any substance tends to diffuse in a direction set by its own concentration gradient, not by the gradients of other solutes that may be sharing the same space. You can observe this tendency by squeezing a drop of dye into water. Dye molecules diffuse slowly into the region where they are less concentrated, regardless of the presence of other solutes (Figure 5.7).

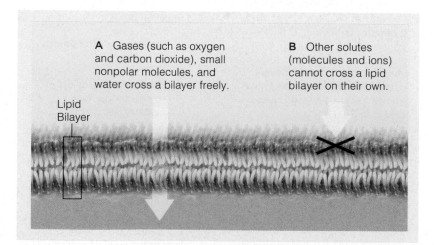

A Gases (such as oxygen and carbon dioxide), small nonpolar molecules, and water cross a bilayer freely.

B Other solutes (molecules and ions) cannot cross a lipid bilayer on their own.

Lipid Bilayer

Figure 5.6 Animated The selectively permeable nature of cell membranes. Small, nonpolar molecules, gases, and water molecules freely cross the lipid bilayer. Polar molecules and ions cross with the help of proteins that span the bilayer.

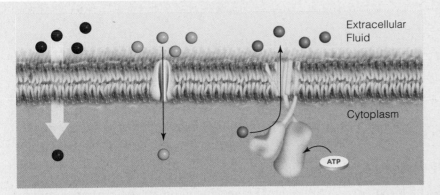

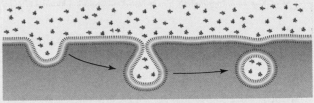

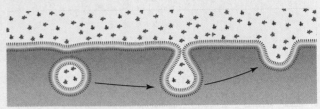

A Diffusion
A substance simply diffuses across lipid bilayer.

B Passive Transport
A solute moves across bilayer through interior of passive transporter; movement is driven by concentration gradient.

C Active Transport
Active transporter uses energy (often, ATP) to pump a solute through bilayer against its concentration gradient.

D Endocytosis Vesicle movement brings substances in bulk into cell.

E Exocytosis Vesicle movement ejects substances in bulk from cell.

Figure 5.8 Overview of membrane-crossing mechanisms.

The Rate of Diffusion

How quickly a solute diffuses depends on five factors:

1. Size. It takes less energy to move a smaller molecule, so smaller molecules diffuse faster.

2. Temperature. Molecules move faster at higher temperature, so they collide more often. Rebounds from the collisions propel them away from one another.

3. Steepness of the concentration gradient. The rate of diffusion is higher with steeper gradients. Again, molecules collide more often in a region of greater concentration. So, more molecules bounce out of a region of greater concentration than bounce into it.

4. Charge. Each ion dissolved in a fluid contributes to the fluid's overall electric charge. A difference in charge between two regions can affect the rate and direction of diffusion between them, because opposite charges attract and like charges repel. For example, positively charged substances, such as sodium ions, will move toward a region with a negative charge.

5. Pressure. Diffusion may be affected by a difference in pressure between two adjoining regions. Pressure squeezes molecules together; molecules that are more crowded collide and rebound more frequently.

How Substances Cross Membranes

Selective permeability is a property that arises from a membrane's structure. A lipid bilayer lets gases and nonpolar molecules cross freely, but it is impermeable to ions and large, polar molecules.

A passive transport protein allows a specific solute to follow its gradient across a membrane. The solute binds to the protein, and is released to the other side of the membrane. This process, which is called passive transport or facilitated diffusion, requires no energy input; the movement is driven by the solute's concentration gradient. Some molecules (such as water) that diffuse across a membrane on their own can also move through passive transport proteins.

An active transport protein pumps a specific solute across a membrane against its gradient. This mechanism, active transport, requires energy—typically in the form of ATP.

Other energy-requiring mechanisms move particles in bulk into or out of cells. In endocytosis, a patch of plasma membrane sinks inward, bringing with it molecules on the outside of the cell. In exocytosis, a vesicle in the cytoplasm fuses with the plasma membrane, so that its contents are released outside the cell.

Figure 5.8 shows an overview of these membrane-crossing mechanisms; the sections that follow describe them in detail.

Take-Home Message

What influences the movement of ions and molecules across cell membranes?

■ Diffusion is net movement of molecules or ions into an adjoining region where they are not as concentrated.

■ The steepness of a concentration gradient as well as temperature, molecular size, and electric and pressure gradients affect the rate of diffusion.

■ Substances move across cell membranes by diffusion, passive and active transport, endocytosis, and exocytosis.

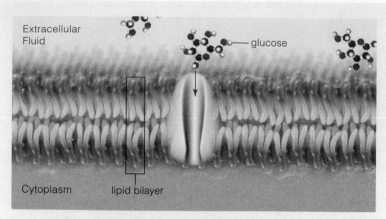

A A glucose molecule (here, in extracellular fluid) binds to a transport protein embedded in the lipid bilayer.

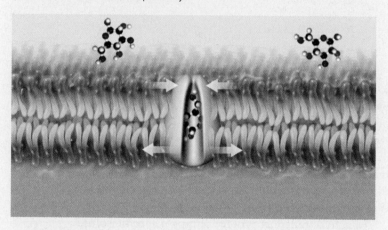

B Binding causes the protein to change shape.

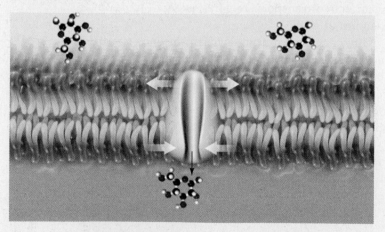

C The glucose molecule detaches from the transport protein on the other side of the membrane (here, in the cytoplasm), and the protein resumes its original shape.

Figure 5.9 Animated Passive transport. This model shows one of the glucose transporters that span the plasma membrane. Glucose crosses in both directions. The net movement of this solute is to the side of the membrane where it is less concentrated.

■ Many types of molecules and ions diffuse across a lipid bilayer only with the help of transport proteins.

Many solutes cross a membrane by associating with transport proteins. Each type of transport protein can move a specific ion or molecule across a membrane. Glucose transporters only transport glucose; calcium pumps only pump calcium; and so on. The specificity means that the amounts and types of substances that cross a membrane depend on which transport proteins are embedded in it.

Passive Transport

In **passive transport**, a concentration gradient drives the diffusion of a solute across a cell membrane, with the assistance of a transport protein. The protein does not require energy to assist the solute's movement; thus, passive transport is also called facilitated diffusion.

Some passive transporters are open channels; others are "gated." A gated transporter changes shape when a molecule binds to it, or in response to a change in electric charge. The protein's shape change moves the solute to the opposite side of the membrane, where it detaches. Then, the transporter reverts to its original shape. Figure 5.9 shows an example, a glucose transporter. Glucose molecules diffuse unassisted across a lipid bilayer, but the transporter increases the rate of diffusion by about 50,000 times.

The net movement of a particular solute through passive transporters tends to be toward the side of the membrane where the solute is less concentrated. This is because molecules or ions simply collide with the transporters more often on the side of the membrane where they are more concentrated.

Passive transport continues until the concentrations on both sides of the membrane are equal. However, such equilibrium rarely occurs in a living system. For example, cells use up glucose as fast as they get it. As soon as a glucose molecule enters a cell, it is broken down for energy or it is used to build other molecules. Thus, there is usually a concentration gradient across the membrane that favors uptake of more glucose.

Active Transport

Solute concentrations shift constantly in the cytoplasm and extracellular fluid. Maintaining a solute's concentration at a certain level often means transporting the solute against its gradient, to the side of a membrane where it is more concentrated. Such pumping does not occur without energy inputs, usually from ATP.

In **active transport**, a transport protein uses energy to pump a solute against its gradient across a cell membrane. Energy, often in the form of a phosphate-group transfer from ATP, changes the shape of the transporter. The change causes the transporter to release the solute to the other side of the membrane.

For example, **calcium pumps** are active transporters that move calcium ions across muscle cell membranes (Figure 5.10). Muscle cells contract when the nervous system causes calcium ions to flood out from a special organelle, the sarcoplasmic reticulum, which is wrapped around the muscle fiber. The flood clears out binding sites on motor proteins that make muscles contract (Section 4.13). Contraction ends after calcium pumps have moved most of the calcium ions back into the sarcoplasmic reticulum, against their concentration gradient. Calcium pumps keep the concentration of calcium in that compartment 1,000 to 10,000 times higher than it is in muscle cell cytoplasm.

The sodium–potassium pump is a **cotransporter**—it moves two substances at the same time (Figure 5.11). Nearly all of the cells in your body have these pumps, which transport sodium and potassium ions in opposite directions across a membrane. Sodium ions (Na^+) in the cytoplasm diffuse into the pump's open channel and bind to its interior. The pump changes shape after it receives a phosphate group from ATP. Its channel opens to the extracellular fluid, and it releases the Na^+. Then, potassium ions (K^+) from extracellular fluid diffuse into the channel and bind to its interior. The transporter releases the phosphate group, then reverts to its original shape. The channel opens to the cytoplasm, and the K^+ is released there.

Bear in mind, the membranes of all cells, not just those of animals, have active transporters. In Section 29.5, for example, you will learn how sugars made in a plant's leaves are pumped into tubes that distribute them through the plant body.

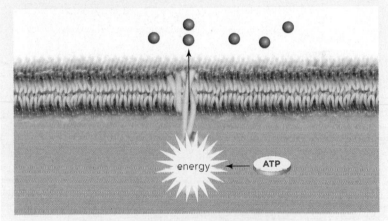

A Calcium ions bind to a calcium transporter (calcium pump).

B A phosphate group is transferred from ATP to the pump. The pump changes shape so that it ejects the calcium ions to the opposite side of the membrane, and then resumes its original shape.

Figure 5.10 Animated Active transport. This model shows a calcium transporter. After two calcium ions bind to the transporter, ATP transfers a phosphate group to it, thus providing energy that drives the movement of calcium against its concentration gradient across the cell membrane.

Take-Home Message

If a molecule or ion cannot diffuse through a lipid bilayer, how does it cross a cell membrane?

■ Transport proteins help specific molecules or ions to cross cell membranes. Which substances cross a membrane is mostly determined by the transport proteins embedded in it.

■ In passive transport, a solute binds a protein that releases it on the opposite side of the membrane. No energy is required; the net movement of solute is down its concentration gradient.

■ In active transport, a protein pumps a solute across a membrane, against its concentration gradient. The transporter must be activated, usually by an energy input from ATP.

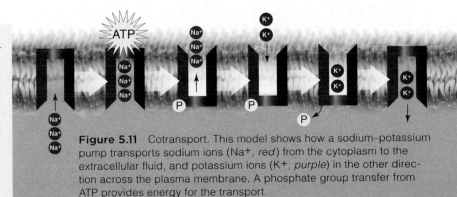

Figure 5.11 Cotransport. This model shows how a sodium–potassium pump transports sodium ions (Na^+, *red*) from the cytoplasm to the extracellular fluid, and potassium ions (K^+, *purple*) in the other direction across the plasma membrane. A phosphate group transfer from ATP provides energy for the transport.

■ By processes of exocytosis and endocytosis, cells take in and expel particles that are too big for transport proteins, as well as substances in bulk.

■ Links to Endomembrane system 4.9, Cytoskeleton 4.13

Endocytosis and Exocytosis

Think back on the structure of a lipid bilayer (Figure 5.2). When a bilayer is disrupted, as when part of the plasma membrane pinches off as a vesicle, it seals itself. Why? The disruption exposes the nonpolar fatty acid tails of the phospholipids to their watery surroundings. Remember, in water, phospholipids spontaneously rearrange themselves so that their tails stay together. When a patch of membrane buds, its phospholipid tails are repelled by water on both sides. The water "pushes" the phospholipid tails together, which helps round off the bud as a vesicle, and also seals the rupture in the membrane.

As part of vesicles, patches of membrane constantly move to and from the cell surface (Figure 5.12). The formation and movement of vesicles, which is called membrane trafficking, involves motor proteins and requires ATP (Section 4.13).

By **exocytosis**, a vesicle moves to the cell surface, and the protein-studded lipid bilayer of its membrane fuses with the plasma membrane. As the exocytic vesicle loses its identity, its contents are released to the surroundings (Figure 5.12).

There are three pathways of **endocytosis**, but they all take up substances near the cell's surface. A small patch of plasma membrane balloons inward, and then it pinches off after sinking farther into the cytoplasm. The membrane patch becomes the outer boundary of an endocytic vesicle, which delivers its contents to an organelle or stores them in a cytoplasmic region.

With receptor-mediated endocytosis, molecules of a hormone, vitamin, mineral, or another substance bind to receptors on the plasma membrane. A shallow pit forms in the membrane patch under the receptors. The pit sinks into the cytoplasm and closes back on itself, and in this way it becomes a vesicle (Figure 5.13).

Phagocytosis ("cell eating") is an endocytic pathway. Phagocytic cells such as amoebas engulf microorganisms, cellular debris, or other particles. In animals, macrophages and other phagocytic white blood cells engulf and digest pathogenic viruses and bacteria, cancerous body cells, and other threats.

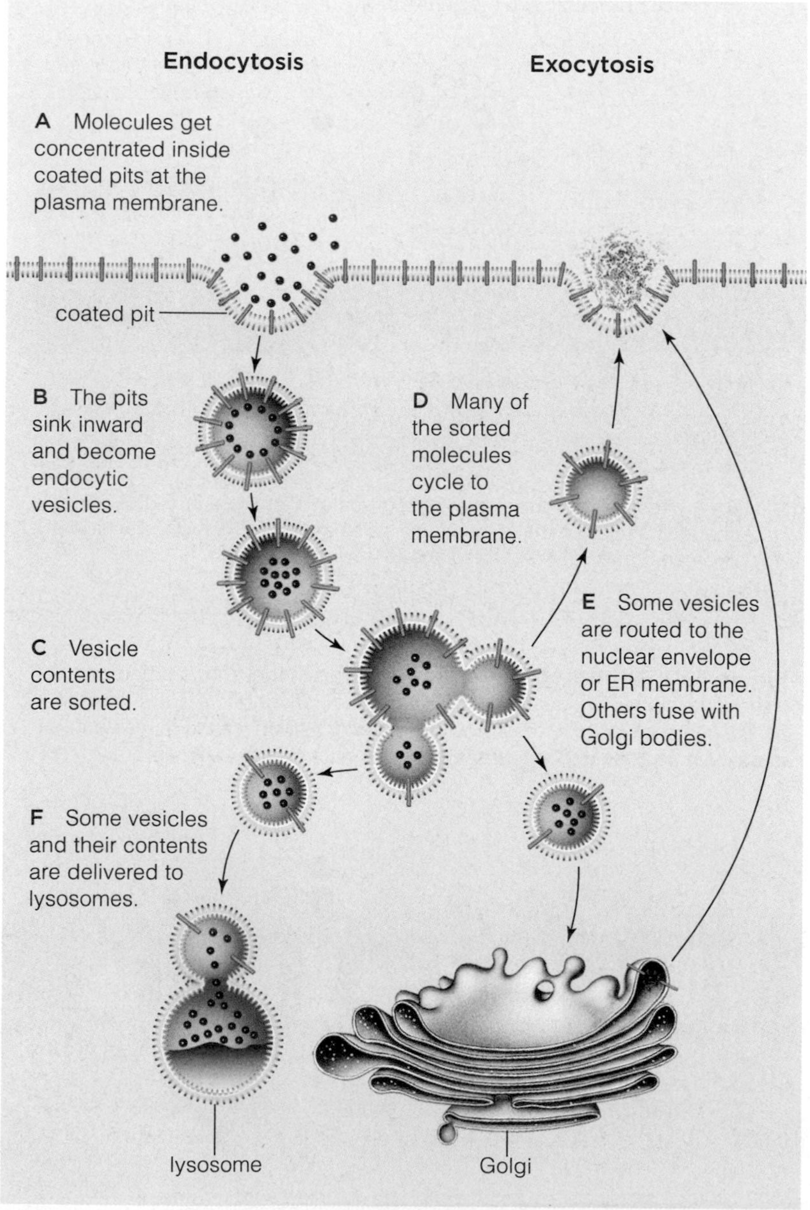

Endocytosis   **Exocytosis**

A Molecules get concentrated inside coated pits at the plasma membrane.

coated pit

B The pits sink inward and become endocytic vesicles.

C Vesicle contents are sorted.

D Many of the sorted molecules cycle to the plasma membrane.

E Some vesicles are routed to the nuclear envelope or ER membrane. Others fuse with Golgi bodies.

F Some vesicles and their contents are delivered to lysosomes.

lysosome   Golgi

Figure 5.12 Animated Endocytosis and exocytosis.

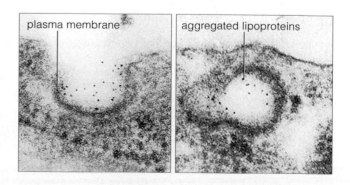

plasma membrane   aggregated lipoproteins

Figure 5.13 Endocytosis of lipoprotein aggregates.

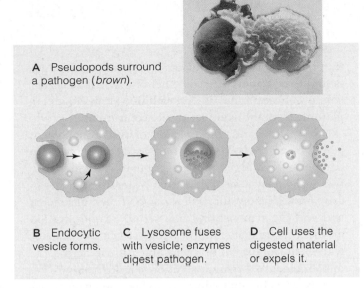

A Pseudopods surround a pathogen (*brown*).

B Endocytic vesicle forms.

C Lysosome fuses with vesicle; enzymes digest pathogen.

D Cell uses the digested material or expels it.

Figure 5.14 Animated Phagocytosis. (**a**) Micrograph of a phagocytic cell with its pseudopods (the extending lobes of cytoplasm) surrounding a pathogen.

(**b–d**) Diagram showing what happens inside a phagocytic cell after pseudopods (the extending lobes of cytoplasm) surround a pathogen. The plasma membrane above the bulging lobes fuses and forms an endocytic vesicle. Inside the cytoplasm, the vesicle fuses with a lysosome, which digests its contents.

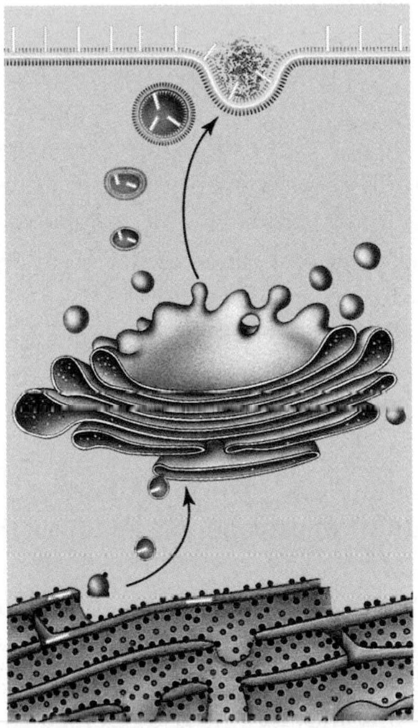

Figure 5.15 How membrane proteins become oriented to the inside or the outside of the cell.

Proteins of the plasma membrane are assembled in the ER, and finished inside Golgi bodies. The proteins become part of vesicle membranes that bud from the Golgi. The membrane proteins automatically become oriented in the proper direction when the vesicles fuse with the plasma membrane.

Figure It Out: What process does the upper arrow represent?

Answer: Exocytosis

We now know that receptor-mediated endocytosis is a misleading name, because receptors also function in phagocytosis. When these receptors bind to a target, they cause microfilaments to assemble in a mesh under the plasma membrane. The microfilaments contract, forcing some cytoplasm and plasma membrane above it to bulge outward as a lobe, or pseudopod (Figures 4.28*b* and 5.14). Pseudopods engulf a target and merge as a vesicle, which sinks into the cytoplasm and fuses with a lysosome (Section 4.9). Enzymes in the lysosome break down the vesicle's contents. Lysosomal enzymes digest the vesicle into fragments and smaller, reusable molecules.

Bulk-phase endocytosis is not as selective. A vesicle forms around a small volume of the extracellular fluid regardless of the kinds of substances dissolved in it.

Membrane Cycling

As long as a cell is alive, exocytosis and endocytosis are continually replacing and withdrawing patches of its plasma membrane, as in Figure 5.12.

The composition of a plasma membrane begins in the ER (Section 4.9). Membrane proteins and lipids are made and modified, and both become part of vesicles that transport them to Golgi bodies for final modification. The finished proteins and lipids are repackaged as new vesicles that travel to the plasma membrane and fuse with it. The lipids and proteins of the vesicle membrane become part of the plasma membrane. This is the process by which new plasma membrane forms.

Figure 5.15 shows what happens when an exocytic vesicle fuses with the plasma membrane. Golgi bodies package membrane proteins facing the inside of a vesicle, so after fusion the proteins face outside the cell.

In a cell that is no longer growing, the total area of the plasma membrane remains more or less constant. Membrane is lost as a result of endocytosis, but it is replaced by membrane arriving as exocytic vesicles.

Take-Home Message

How do cells take in large particles and bulk substances?

■ Exocytosis and endocytosis move materials in bulk across plasma membranes.

■ By exocytosis, a cytoplasmic vesicle fuses with the plasma membrane and releases its contents to the outside of the cell.

■ By endocytosis, a patch of plasma membrane sinks inward and forms a vesicle in the cytoplasm.

■ Receptor-mediated endocytosis and phagocytosis are two endocytic pathways that occur when specific substances bind to receptors. Bulk-phase endocytosis is not specific.

■ Plasma membrane lost during endocytosis is replaced by membrane that surrounds exocytic vesicles.

- Water diffuses across cell membranes by osmosis.
- Osmosis is driven by tonicity, and is countered by turgor.

- Links to Water 2.5, Plant cell walls 4.12

Osmosis

Like any other substance, water molecules tend to diffuse in response to their own concentration gradient. **Osmosis** is the name for this movement. As you read earlier, water crosses cell membranes on its own, and also through transport proteins.

You might be wondering: How can water be more or less concentrated? Think of water's concentration in terms of relative numbers of water molecules and solute molecules. The concentration of water depends on the total number of molecules or ions dissolved in it. The higher the solute concentration, the lower the water concentration.

For example, when you pour some sugar into a container that is partially filled with water, you increase the total volume of liquid. The number of water molecules is unchanged, but they are now dispersed in a larger total volume. As a result of the added solute, the number of water molecules per unit volume—the water concentration—has decreased.

Tonicity

Tonicity refers to the relative concentrations of solutes in two fluids that are separated by a selectively permeable membrane. When the solute concentrations

differ, the fluid with the lower concentration of solutes is said to be **hypotonic**. The other one, with the higher solute concentration, is **hypertonic**. Fluids that are **isotonic** have the same solute concentration.

Tonicity dictates the direction of water movement across membranes: Water diffuses from a hypotonic to a hypertonic fluid. Suppose a container is divided into two sections by a membrane that water, but not sugar, can cross. If you pour water into both compartments and add sugar to just one, you set up a concentration gradient. The sugar solution is hypertonic. By osmosis, water will follow its gradient and diffuse across the membrane into the sugar solution (Figure 5.16).

Now imagine that you have a sheet of a selectively permeable membrane that water, but not sucrose, can cross. You make a bag out of the membrane, then fill it with a 2 percent sucrose solution. If you drop the bag into a solution with 2 percent sucrose (an isotonic solution), the bag stays the same size (Figure 5.17a). If you drop it into a 10 percent sucrose solution (a hypertonic solution), the bag will shrink as water diffuses out of it. If you drop the bag into water with no sucrose in it (which is hypotonic with respect to the solution), it will swell up as water diffuses into it.

A cell is essentially a semipermeable membrane bag of fluid. What happens when the fluid outside of a cell is hypertonic? Water will follow its gradient and cross the membrane to the hypertonic side, and the volume of the cell will decrease as water diffuses out of it. If the outside fluid is very hypotonic, the volume of the cell will increase as water diffuses into it.

Most free-living cells can counter shifts in tonicity by selectively transporting solutes across the plasma membrane. Most cells of multicelled species cannot. In multicelled organisms, maintaining the tonicity of extracellular fluids is part of homeostasis. Thus, tissue fluid is normally isotonic with fluid inside cells (Figure 5.17b). If a tissue fluid were to become hypertonic, the cells would lose water, and they would shrivel (Figure 5.17c). If the fluid were to become hypotonic, too much water would diffuse into the cells, and they would burst (Figure 5.17d).

Effects of Fluid Pressure

Hydrostatic pressure, or as botanists say, **turgor**, often counters osmosis. Both terms refer to pressure that a volume of fluid exerts against a cell wall, membrane, tube, or any other structure that holds it. Cell walls in plants and many protists, fungi, and bacteria resist an increase in the volume of cytoplasm. The walls of

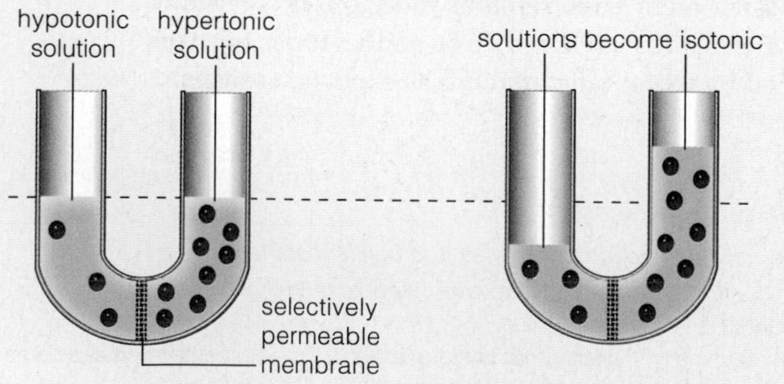

hypotonic solution   hypertonic solution

solutions become isotonic

selectively permeable membrane

A Initially, the volume of fluid is the same in the two compartments, but the solute concentration differs.

B The fluid volume in the two compartments changes as water follows its gradient and diffuses across the membrane.

Figure 5.16 Animated Experiment showing a change in fluid volume as an outcome of osmosis. A selectively permeable membrane separates two regions.

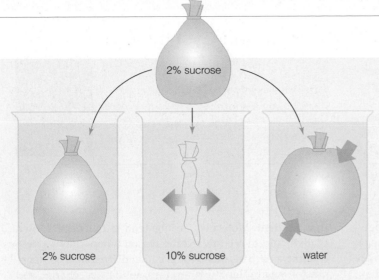

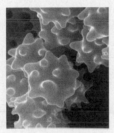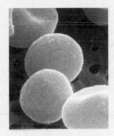

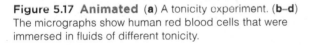

A What happens to a semipermeable membrane bag when it is immersed in an isotonic, a hypertonic, or a hypotonic solution?

B Red blood cells in an isotonic solution do not change in volume.

C Red blood cells in a hypertonic solution shrivel because water diffuses out of them.

D Red blood cells in a hypotonic solution swell because water diffuses into them.

Figure 5.17 Animated (**a**) A tonicity experiment. (**b–d**) The micrographs show human red blood cells that were immersed in fluids of different tonicity.

Figure 5.18 (**a**) A tomato plant undergoing osmotically induced wilting within thirty minutes after salty water was added to the soil in the pot. (**b**) Cells from an iris petal, plump with water. Their cytoplasm and central vacuole extend to the cell wall. (**c**) Cells from a wilted iris petal. Their cytoplasm and central vacuole shrank, and the plasma membrane moved away from the wall.

blood vessels resist an increase in blood volume. The amount of hydrostatic pressure that can stop water from diffusing into cytoplasmic fluid or other hypertonic solution is called **osmotic pressure**.

As one example, growing plant cells are hypertonic relative to water in soil (the cytoplasmic fluid usually has more solutes than soil water). Water diffusing into a young plant cell by osmosis exerts fluid pressure on the primary wall. The thin, pliable wall expands under pressure, which lets the cytoplasmic volume increase (Section 4.12). Expansion of the wall—and the cell—ends when the osmotic pressure inside the cell builds up enough to prevent the uptake of additional water.

Hydrostatic pressure also supports soft plant parts. When a plant with soft green leaves is growing well in soil that has enough water, hydrostatic pressure keeps the cells plump—and the plant erect. As the soil dries out, the concentration of salt in soil water increases. If the soil water becomes hypertonic with respect to

cytoplasmic fluid, water diffuses out of the plant's cells and hydrostatic pressure in them falls. The cytoplasm shrinks, and the plant wilts. Adding salt to the soil has the same effect. Figure 5.18 shows what happens when you pour salty water into soil around a tomato plant's roots. Within thirty minutes, the plant droops.

Take-Home Message

Why and how does water move into and out of cells?

■ Water moves in response to its own concentration gradient, which is influenced by solute concentration.

■ Osmosis is a net diffusion of water between two solutions that differ in water concentration and are separated by a selectively permeable membrane.

■ Water tends to move osmotically to regions of greater solute concentration (from a hypotonic to a hypertonic solution). No net diffusion occurs between isotonic solutions.

■ Fluid pressure that a solution exerts against a membrane or wall influences the osmotic movement of water.

CFTR is an active transporter of chloride ions. In about 90 percent of CF patients, loss of a single amino acid from the protein causes the disorder. The mutated CFTR proteins are functional, but enzymes destroy them before they reach the plasma membrane. Thus, cystic fibrosis is most often a result of impaired membrane trafficking of the CFTR protein.

How would you vote?

Lung tissue of a baby with cystic fibrosis; white patches are mucus. Should we screen prospective parents for CF mutations? See CengageNow for details, then vote online.

Summary

Sections 5.1, 5.2 A cell membrane is a **selectively permeable** barrier that separates an internal environment from an external one. Each is a mosaic of lipids (mainly phospholipids) and proteins. The lipids are organized as a double layer in which the nonpolar tails of both layers are sandwiched between the polar heads. Membranes of bacteria and eukaryotic cells can be described by a **fluid mosaic** model; those of archaeans are not fluid.

Proteins that are transiently or permanently associated with a membrane carry out most membrane functions. All membranes have **transport proteins**. Plasma membranes also incorporate **receptor proteins**, **adhesion proteins**, **enzymes**, and **recognition proteins** (Table 5.1).

■ *Use the animation on CengageNOW to learn about membrane structure and the experiments that elucidated it.*

■ *Use the animation on CengageNOW to familiarize yourself with the functions of receptor proteins.*

Section 5.3 A difference in the **concentration** of a substance between adjoining regions of fluid is a **concentration gradient**. Molecules or ions tend to follow their own gradient and move toward the region where they are less concentrated. This behavior is called **diffusion**. The steepness of the gradient, temperature, solute size, charge, and pressure influence the diffusion rate.

Gases, water, and small nonpolar molecules diffuse across a membrane. Most other molecules and ions cross only with the help of transport proteins.

■ *Use the interaction on CengageNOW to investigate diffusion across membranes.*

Section 5.4 Transport proteins move specific solutes across membranes. The types of transport proteins in a membrane determine which substances cross it. **Active transport** proteins such as **calcium pumps** use energy, usually from ATP, to move a solute against its concentration gradient. **Passive transport** proteins work without an energy input; solute movement is driven by the concentration gradient. **Cotransporters** move solutes in different directions across a membrane.

■ *Use the animation on CengageNOW to compare the processes of passive and active transport.*

Section 5.5 **Exocytosis**, **endocytosis**, and **phagocytosis** move bulk substances and large particles across plasma membranes. With exocytosis, a cytoplasmic vesicle fuses with the plasma membrane, and its contents are released to the outside of the cell. The vesicle's membrane lipids and proteins become part of the plasma membrane. With endocytosis, a patch of plasma membrane balloons into the cell, and forms a vesicle that sinks into the cytoplasm. Plasma membrane lost by endocytosis is replaced by exocytic vesicles.

■ *Use the animations on CengageNOW to discover how membrane components are cycled, and to explore phagocytosis.*

Section 5.6 **Osmosis** is the diffusion of water across a selectively permeable membrane, from the region with a lower solute concentration (**hypotonic**) toward the region with a higher solute concentration (**hypertonic**). There is no net movement of water between **isotonic** solutions. **Osmotic pressure** is the amount of **turgor** or **hydrostatic pressure** (fluid pressure against a cell membrane or wall) that stops osmosis.

■ *Use the animation on CengageNOW to explore osmosis.*

Table 5.1 Common Types of Membrane Proteins

Category	Function	Examples
Passive transporters	Allow ions or small molecules to cross a membrane to the side where they are less concentrated. Open or gated channels.	Porins; glucose transporter
Active transporters	Pump ions or molecules through membranes to the side where they are more concentrated. Require energy input, as from ATP.	Calcium pump; serotonin transporter
Receptors	Initiate change in a cell's activity by responding to an outside signal (e.g., by binding to a signaling molecule).	Insulin receptor; B cell receptor
Cell adhesion molecules	Help cells stick to one another and to extracellular matrix.	Integrins; cadherins
Recognition proteins	Identify cells as self (belonging to one's own body or tissue)	Histocompatibility molecules
Enzymes	Speed reactions without being altered by them.	Diverse hydrolases

Data Analysis Exercise

In most individuals with cystic fibrosis, the 508th amino acid of the CFTR protein (a phenylalanine) is missing. A CFTR protein with this change is synthesized correctly, and it can transport ions correctly, but it never reaches the plasma membrane to do its job.

Sergei Bannykh and his coworkers developed a procedure to measure the relative amounts of the CFTR protein localized in different regions of a cell. They compared the pattern of CFTR distribution in normal cells with the pattern in CFTR-mutated cells. A summary of their results is shown in Figure 5.19.

1. Which organelle contains the least amount of CFTR protein in normal cells? In CF cells? Which contains the most?

2. In which organelle is the amount of CFTR protein in CF cells closest to the amount in normal cells?

3. Where is the mutated CFTR protein getting held up?

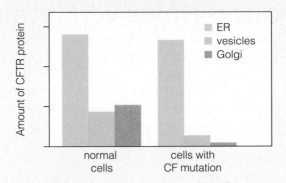

Figure 5.19 Comparison of the amounts of CFTR protein associated with endoplasmic reticulum (*blue*), vesicles traveling from ER to Golgi (*green*), and Golgi bodies (*orange*). The patterns of CFTR distribution in normal cells, and cells with the most common cystic fibrosis mutation, were compared.

Self-Quiz *Answers in Appendix III*

1. Cell membranes consist mainly of a _____ .
 a. carbohydrate bilayer and proteins
 b. protein bilayer and phospholipids
 c. lipid bilayer and proteins

2. In a lipid bilayer, _____ of all the lipid molecules are sandwiched between all the _____ .
 a. hydrophilic tails; hydrophobic heads
 b. hydrophilic heads; hydrophilic tails
 c. hydrophobic tails; hydrophilic heads
 d. hydrophobic heads; hydrophilic tails

3. By the _____ model, cell membranes are flexible structures composed of a mixture of many different types of molecules.

4. Most membrane functions are carried out by _____ .
 a. proteins   c. nucleic acids
 b. phospholipids   d. hormones

5. Organelle membranes incorporate _____ .
 a. transport proteins   c. recognition proteins
 b. adhesion proteins   d. all of the above

6. Some _____ proteins are also receptors.

7. Diffusion is the movement of ions or molecules from a region where they are _____ (more/less) concentrated to another where they are _____ (more/less) concentrated.

8. Name one molecule that can readily diffuse across a lipid bilayer.

9. Some sodium ions cross a cell membrane through transport proteins that first must be activated by an energy boost. This is an example of _____ .
 a. passive transport   c. facilitated diffusion
 b. active transport   d. a and c

10. Immerse a living cell in a hypotonic solution, and water will tend to _____ .
 a. move into the cell   c. show no net movement
 b. move out of the cell d. move in by endocytosis

11. Fluid pressure against a wall or cell membrane is called _____ .

12. Vesicles form by _____ .
 a. endocytosis   d. halitosis
 b. exocytosis   e. a and c
 c. phagocytosis   f. a through c

13. Put the following structures in order according to an exocytic trafficking pathway.
 a. plasma membrane   c. endoplasmic reticulum
 b. Golgi bodies   d. post-Golgi vesicles

14. Match the term with its most suitable description.
 ___ phagocytosis   a. identity protein
 ___ passive transport b. basis of diffusion
 ___ recognition protein c. important in membranes
 ___ active   d. one cell engulfs another
     transport   e. requires energy boost
 ___ phospholipid   f. docks for signals and
 ___ gradient   substances at cell surface
 ___ receptors   g. no energy boost required
                          to move solutes

■ *Visit CengageNOW for additional questions.*

Critical Thinking

1. Water moves osmotically into *Paramecium*, a single-celled aquatic protist. If unchecked, the influx would bloat the cell and rupture its plasma membrane, and the cell would die. An energy-requiring mechanism that involves contractile vacuoles (*right*) expels excess water. Water enters the vacuole's tubelike extensions and collects inside. A full vacuole contracts and squirts water out of the cell through a pore. Are *Paramecium*'s surroundings hypotonic, hypertonic, or isotonic?

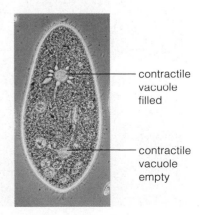

contractile vacuole filled

contractile vacuole empty

6 Ground Rules of Metabolism

A Toast to Alcohol Dehydrogenase

The next time someone asks you to have a drink, stop for a moment and think about the cells in your body that detoxify alcohol. It makes no difference whether you drink a bottle of beer, a glass of wine, or 1–1/2 ounces of vodka. Each holds the same amount of alcohol or, more precisely, ethanol (CH_3CH_2OH). Ethanol molecules move quickly from the stomach and small intestine into the bloodstream. Almost all of the ethanol someone drinks ends up in the liver, which has impressive numbers of alcohol-metabolizing enzymes. One of those enzymes, alcohol dehydrogenase, helps rid the body of ethanol and other toxic alcohols (Figure 6.1).

Detoxifying alcohol is hard on liver cells. It causes a slow-down in protein and glucose synthesis, and disrupts lipid and carbohydrate breakdown. Mitochondria use oxygen in ethanol metabolism—oxygen that normally would take part in the breakdown of fatty acids. Fatty acids accumulate as large fat globules in the tissues of heavy drinkers. As liver cells die of oxygen starvation, there are fewer and fewer cells for detoxification. One possible outcome is alcoholic hepatitis, a common disease characterized by inflammation and destruction of liver tissue. Alcoholic cirrhosis, another possibility, leaves the liver permanently scarred. (The word cirrhosis is from the Greek word *kirros*, or orange-colored, after the abnormal skin color of people with the disease.) Eventually, the liver just stops working, with dire health consequences.

The liver is the largest gland in the human body, weighing about 1.4 kg (3 pounds). It lies in the upper right side of the abdominal cavity. The liver has many important functions that affect the entire body. It helps digest fats and regulate the body's blood sugar level, and it breaks down many toxic compounds, not just ethanol. It also makes plasma proteins that circulate in blood. Plasma proteins are essential for blood clotting, immune function, and maintaining the solute balance of body fluids.

Now think about a self-destructive behavior known as binge drinking. The idea is to consume large amounts of alcohol in a brief period of time. Binge drinking is currently the most serious drug problem on college campuses throughout the United States. For example, a 2006 study showed that almost half of 4,580 undergraduate students surveyed are binge drinkers, meaning they consumed five or more alcoholic drinks in a two-hour period at least once during the year before the survey.

Binge drinking does far more than damage the liver. Aside from the related 500,000 injuries from accidents, the 600,000 assaults by intoxicated students, 100,000 cases of date rape, and 400,000 incidences of (whoops) unprotected sex among students, binge drinking kills upwards of 1,400 students every year. With this sobering example, we turn to metabolism, the cell's capacity to acquire and use energy.

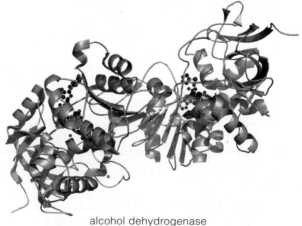

alcohol dehydrogenase

See the video! Figure 6.1 Alcohol dehydrogenase. This enzyme, which helps the body break down ethanol and other toxic alcohols, makes it possible for humans to drink beer, wine, and other alcoholic beverages.

Key Concepts

Energy flow in the world of life

Energy tends to disperse spontaneously. Each time energy is transferred, some of it disperses. Organisms maintain their organization only by continually harvesting energy. ATP couples reactions that release usable energy with reactions that require energy. **Sections 6.1, 6.2**

How enzymes work

Enzymes tremendously increase the rate of metabolic reactions. Environmental factors such as temperature, salt, and pH influence enzyme function. **Section 6.3**

The nature of metabolism

Metabolic pathways are energy-driven sequences of enzyme-mediated reactions. They concentrate, convert, or dispose of materials in cells. Controls over enzymes that govern key steps in metabolic pathways can shift cell activities fast. **Section 6.4**

Metabolism everywhere

Knowledge about metabolism, including how enzymes work, can help you interpret some natural phenomena. **Section 6.5**

Links to Earlier Concepts

- In this chapter, you will gain insight into how organisms tap into a one-way flow of energy to maintain their organization (1.2).

- Your knowledge of chemical bonding (2.4) and carbohydrates (3.3) will help you understand how cells store and retrieve energy in chemical bonds. You will also see how ATP (3.7) connects energy-requiring processes of metabolism (3.2) with energy-releasing ones.

- This chapter revisits the relationship between protein structure and function (3.5, 3.6), this time in the context of enzymes (1.2) and how they work. Factors such as temperature (2.5) and pH (2.6) affect enzyme function.

- You will start thinking about how cells harvest energy from organic molecules in sequences of electron (2.3) transfers, and the membrane proteins (5.2) that carry out such reactions.

- You will see an example of how scientists harness metabolic reactions to make tracers (2.2), and how such tracers help us better understand natural phenomena such as biofilms (4.5).

How would you vote? Some people have damaged their liver because they drank too much alcohol; others have had liver-damaging infections. Because there are not enough liver donors, should life-style be a factor in deciding who gets a transplant? See CengageNOW for details, then vote online.

- Assembly of the molecules of life starts with energy input into living cells.

- Links to Life's organization 1.2, Chemical bonding 2.4, Carbohydrates 3.3

Energy Disperses

Energy, remember, is the capacity to do work, but this definition is not perfect. Even the best physicists cannot say, exactly, what energy is. However, even without a perfect definition, we can understand energy just by thinking of familiar kinds such as light, electricity, pressure, heat, and motion (Figure 6.2).

We also understand that one form of energy can be converted to another form. For example, a light bulb can change electricity into light, and an automobile can change the chemical energy of gasoline into the energy of motion. What may not be obvious is that the total amount of energy in every such conversion stays the same. Energy does not appear from nowhere, and it does not vanish into nothing, a concept that is called the **first law of thermodynamics**.

Another concept describes the way energy behaves: It tends to disperse spontaneously. For example, heat flows from a hot pan to air in a cool kitchen until the temperature of both is the same. We never see cool air raising the temperature of a hot pan. Each form of energy—not just heat—tends to disperse until no part of a system holds more than another part.

Entropy is a measure of how much the energy of a particular system has become dispersed. Let's use the hot pan in a cool kitchen as an example of a system. As heat flows from the pan into the air, the entropy of the system increases. Entropy continues to increase until the heat is uniformly distributed throughout the kitchen, and there is no longer a net flow of heat from one area to another. Our system has now reached its maximum entropy with respect to heat (Figure 6.3).

When we say that energy disperses, we mean that a system tends to change toward a state of maximum entropy. The concept that entropy increases spontaneously is the **second law of thermodynamics**. If we see a decrease in entropy, we can expect that some energy change occurred to make it happen.

Biologists use the concept of entropy as it applies to chemical bonding, because energy flow in the world of life occurs primarily by the making and breaking of chemical bonds. How is entropy related to chemical bonding? Think about it just in terms of motion. Two unbound atoms can vibrate, spin, and rotate in every direction: They are at high entropy with respect to motion. A covalent bond between the atoms constrains them, so they move in fewer ways than they did before bonding. Thus, the entropy of two atoms decreases when a bond forms between them.

Entropy changes are part of the reason why some reactions occur spontaneously, and others require an energy input, as you will see in the next section.

Figure 6.2 Demonstration of a familiar type of energy—the energy of motion.

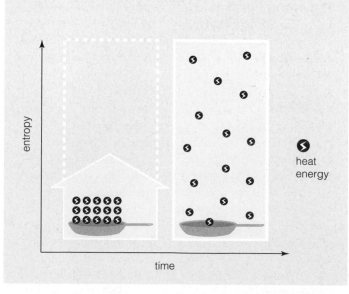

Figure 6.3 Entropy is the "mixedupness" of energy. Entropy tends to increase, but the total amount of energy always stays the same.

Figure 6.4 It takes more than 10,000 pounds of soybeans and corn to raise one 1,000-pound steer. Where do the other 9,000 pounds go? The animal's body breaks down molecules in the food to access energy stored in chemical bonds. Only about 10% of that energy goes toward building body mass. Some of the rest is used for activities (such as movement), but most is lost during energy conversions.

The One-Way Flow of Energy

Work occurs as energy is transferred from one place to another, and such energy transfers often involve the conversion of one form of energy to another. To a biologist, this statement means that all organisms use energy they harvest from the environment to drive cellular work. For example, photosynthetic cells of producers capture light energy from the sun by converting it to chemical energy stored in the bonds of carbohydrates. Most organisms access chemical energy stored in the bonds of carbohydrates by breaking those bonds. Both processes involve many energy transfers.

Some energy escapes with each transfer, usually in the form of heat. This is another way to interpret the second law: Energy transfers are never completely efficient. For example, the typical incandescent light bulb converts about 5% of the energy of electricity into light. The remaining energy, about 95% of it, ends up as heat that radiates from the bulb. Dispersed heat is not very useful for doing work, and it is not easily converted to a more useful form of energy (such as electricity). Because some of the energy in every transfer disperses as heat, and heat is not useful for doing work, we can say that the total amount of energy available for doing work in the universe is always decreasing.

Is life an exception to this depressing flow? An organized body is hardly dispersed. Energy becomes concentrated in each new organism as the molecules of life form and organize into cells. Even so, the second law applies. Living things constantly use energy to grow, to move, to acquire nutrients, to reproduce, and so on. Inevitable losses occur during the energy transfers that maintain life (Figure 6.4). Unless those losses are replenished with energy from another source, the complex organization of life will end. Most of the energy that fuels life on Earth is energy that has been lost from the sun, which has been losing energy since it formed 4.5 billion years ago.

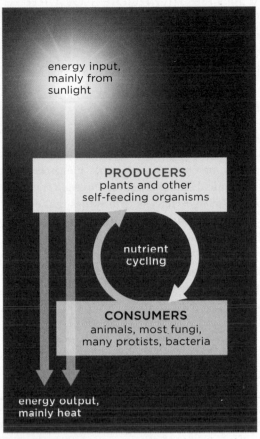

energy input, mainly from sunlight

PRODUCERS
plants and other self-feeding organisms

nutrient cycling

CONSUMERS
animals, most fungi, many protists, bacteria

energy output, mainly heat

ENERGY IN

Sunlight energy reaches environments on Earth. Producers of nearly all ecosystems secure some and convert it to stored forms of energy. They and all other organisms convert stored energy to forms that can drive cellular work.

ENERGY OUT

With each conversion, there is a one-way flow of a bit of energy back to the environment. Nutrients cycle between producers and consumers.

Figure 6.5 A one-way flow of energy into living organisms compensates for a one-way flow of energy out of them. Energy inputs drive a cycling of materials among producers and consumers.

In our world, energy flows from the sun, through producers, then consumers (Figure 6.5). During this journey, the energy changes form and changes hands many times. Each time, some energy escapes as heat until, eventually, all of it is irrevocably dispersed. However, the second law does not say how quickly the dispersal has to happen. Energy's spontaneous dispersal is resisted by chemical bonds. Think of all the bonds in the countless molecules that make up your skin, heart, liver, fluids, and other body parts. Those bonds hold the molecules, and you, together—at least for the time being.

Take-Home Message

What is energy?

■ Energy is the capacity to do work. It can be converted from one form to another, but it cannot be created or destroyed.

■ Energy tends to spread, or disperse, spontaneously.

■ Organisms can maintain their complex organization only as long as they replenish themselves with energy they harvest from someplace else.

6.2 Energy in the Molecules of Life

- All cells store and retrieve energy in chemical bonds of the molecules of life.

- Links to Bonding 2.4, Carbohydrates 3.3, Nucleotides 3.7

Energy In, Energy Out

You already know how chemical bonds join atoms into molecules. When molecules interact, chemical bonds can break, form, or both. A **reaction** is the process by which such chemical change occurs. During a chemical reaction, one or more **reactants** (molecules that enter a reaction) change into one or more **products** (molecules that remain at the reaction's end). A chemical reaction is typically shown as an equation (Figure 6.6).

Every chemical bond holds energy. The amount of energy that a particular bond holds depends on which elements are taking part in it. For example, the covalent bond between an oxygen and hydrogen atom in any water molecule always holds the same amount of energy. That is the amount of energy required to break the bond, and it is also the amount of energy released when the bond forms.

Bond energy and entropy both contribute to a molecule's **free energy**, which is the amount of energy that is available (free) to do work.

In most reactions, the free energy of the reactants differs from the free energy of the products. Reactions in which the reactants have less free energy than the products require a net energy input to proceed. Such reactions are **endergonic**, which means "energy in" (Figure 6.7a).

Cells store energy by running endergonic reactions. For example, energy (in the form of light) drives the overall reactions of photosynthesis, which convert car-

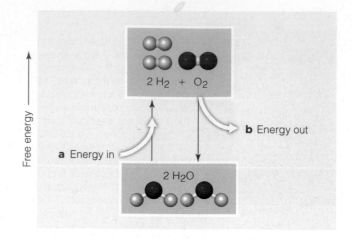

Figure 6.7 Energy inputs and outputs in chemical reactions.

(**a**) Endergonic reactions require an energy input because they convert molecules with lower free energy to molecules with higher free energy.

(**b**) Exergonic reactions end with an energy output because they convert molecules with higher free energy to molecules with lower free energy.

Figure It Out: Which law of thermodynamics explains energy inputs and outputs in chemical reactions?

Answer: The first law

bon dioxide and water to glucose and oxygen. Unlike light, glucose can be stored inside of a cell.

In other reactions, the reactants have greater free energy than the products. Such reactions are **exergonic**, which means "energy out," because they end with a net release of energy (Figure 6.7b). Cells access the free energy of molecules by running exergonic reactions. An example is the overall process of aerobic respiration, which converts glucose and oxygen to carbon dioxide and water for a net energy output.

Why the World Does Not Go Up in Flames

The molecules of life release energy when they combine with oxygen. For example, think of how a spark ignites tinder-dry wood in a campfire. Wood is mostly cellulose, which is a carbohydrate that consists of long chains of repeating glucose units (Section 3.3). A spark initiates a reaction that converts cellulose and oxygen to water and carbon dioxide. The reaction is exergonic, and it releases enough energy to initiate the same reaction with other cellulose and oxygen molecules. That is why a campfire keeps burning once it has been lit.

Earth is rich in oxygen—and in potential exergonic reactions. Why doesn't it burst into flames? Luckily, it takes energy to break the chemical bonds of reactants, even in an exergonic reaction. **Activation energy** is the minimum amount of energy that will get a chemical

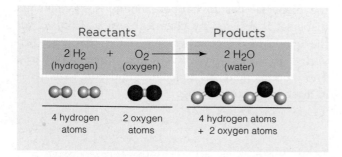

Figure 6.6 Chemical bookkeeping. In equations that represent chemical reactions, reactants are written to the left of an arrow that points to the products. A number before a formula indicates the number of molecules.

Atoms shuffle around in a reaction, but they never disappear: The same number of atoms that enter a reaction remain at the reaction's end.

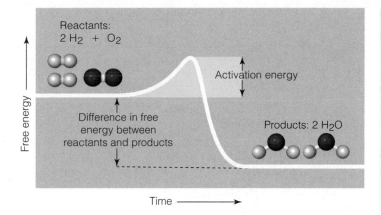

Figure 6.8 Activation energy. Most reactions will not proceed without an input of activation energy, which is shown here as a bump in an energy hill. In this example, the reactants have more free energy than the products. Activation energy keeps such exergonic reactions from running spontaneously.

reaction started (Figure 6.8). It is independent of any energy difference between reactants and products.

Both endergonic and exergonic reactions have activation energy, but the amount varies with the reaction. For example, guncotton, or nitrocellulose, is a highly explosive derivative of cellulose. Christian Schönbein accidentally discovered a way to manufacture it when he used a cotton apron to wipe up a nitric acid spill on his kitchen table, then hung it up to dry next to his oven. The apron exploded. Being a chemist in the 1800s, Schönbein had immediate hopes that he could market guncotton as a firearm explosive, but it proved to be too unstable. So little activation energy is needed to make guncotton react with oxygen that it explodes spontaneously. The substitute? Gunpowder, which has a higher activation energy for a reaction with oxygen.

ATP—The Cell's Energy Currency

 Cells pair reactions that require energy with reactions that release energy. ATP is part of that process for many reactions in cells. **ATP**, or adenosine triphosphate, is an energy carrier: It accepts energy released by exergonic reactions, and delivers energy to endergonic reactions. ATP is the main currency in a cell's energy economy, so we use a cartoon coin to symbolize it.

ATP is a nucleotide with three phosphate groups (Figure 6.9a). The bonds that link those phosphate groups hold a lot of energy. When a phosphate group is transferred from ATP to another molecule, energy is transferred along with it. That energy contributes to the "energy in" part of an endergonic reaction. A phosphate-group transfer is called **phosphorylation**.

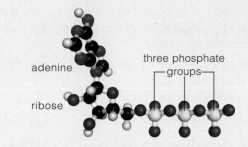

A Structure of ATP (adenine triphosphate).

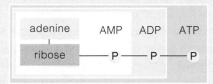

B The molecule is called ATP when it has three phosphate groups. After it loses one phosphate group, the molecule is called ADP (adenosine diphosphate); after losing two phosphate groups it is called AMP (adenosine monophosphate).

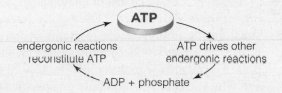

C ATP forms when an endergonic reaction drives the covalent bonding of ADP and phosphate. ATP energy is transferred to another molecule along with a phosphate group, and ADP forms again. Energy from such transfers drives endergonic reactions that are the stuff of cellular work, such as active transport and muscle contraction.

Figure 6.9 **Animated** ATP, the energy currency of all cells.

Cells constantly use up ATP to drive endergonic reactions, so they constantly replenish it. When ATP loses a phosphate, ADP (adenosine diphosphate) forms (Figure 6.9b). ATP forms again when ADP binds phosphate in an endergonic reaction. The cycle of using and replenishing ATP is called the **ATP/ADP cycle** (Figure 6.9c).

Take-Home Message

How do cells use energy?

■ Cells store and retrieve energy by making and breaking chemical bonds.

■ Activation energy is the minimum amount of energy required to start a chemical reaction.

■ Endergonic reactions cannot run without a net input of energy. Exergonic reactions end with a net release of energy.

■ ATP, the main energy carrier in all cells, couples reactions that release energy with reactions that require energy.

■ Enzymes make specific reactions occur much faster than they would on their own.

■ Links to Temperature 2.5, pH 2.6, Protein structure 3.5, Denaturation 3.6

How Enzymes Work

Centuries might pass before sugar would break down to carbon dioxide and water on its own, yet that same conversion takes just a few seconds inside your cells. Enzymes make the difference. **Enzymes** are catalysts, which are molecules that make chemical reactions proceed much faster than they would on their own. Most enzymes are proteins, but some are RNAs.

Most enzymes are not consumed or changed by participating in a reaction; they can work again and again. Each kind recognizes and alters specific reactants, or **substrates**. For instance, the enzyme thrombin cleaves a specific peptide bond in a protein called fibrinogen.

The polypeptide chains of enzymes are folded into one or more **active sites**. The sites are pockets where substrates bind and where reactions proceed (Figure 6.10). The active site is complementary in shape, size, polarity, and charge to the substrate. That fit is the reason why each enzyme acts only on specific substrates.

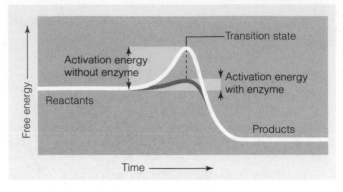

Figure 6.11 Animated An enzyme enhances the rate of a reaction by lowering its activation energy. **Figure It Out:** Is this reaction endergonic or exergonic?   *Answer: Exergonic*

Activation energy is a bit like a hill that reactants must climb before they can run down the other side to products. When we talk about activation energy, we are really talking about the energy required to break the bonds of the reactants. Depending on the reaction, that energy may force substrates close together, redistribute their charge, or cause some other change. The change brings on the **transition state**, when substrate bonds reach their breaking point and the reaction will run spontaneously to product. Enzymes can help bring on the transition state by lowering activation energy (Figure 6.11). They do this by the following four mechanisms, which work alone or in combination.

Helping Substrates Get Together The closer substrate molecules are to each other, the more likely they are to react. Binding at an active site is as effective as bringing substrates 10 millionfold closer together.

Orienting Substrates In Positions That Favor Reaction On their own, substrates collide from random directions. By contrast, binding at an active site positions substrates so they align appropriately for a reaction.

Inducing a Fit Between Enzyme and Substrate By the **induced-fit model**, a substrate is not quite complementary to an active site. The enzyme restrains the substrate, stretching or squeezing it into a shape that often puts it next to a reactive group or to another molecule. By forcing a substrate to fit into the active site, the enzyme ushers in the transition state.

Shutting Out Water Molecules Metabolism occurs in water-based fluids, but water molecules can interfere with certain reactions. The active sites of some enzymes repel water, and keep it away from the reactions.

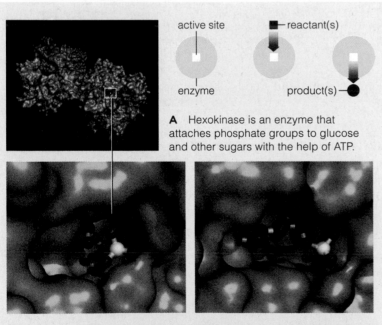

A Hexokinase is an enzyme that attaches phosphate groups to glucose and other sugars with the help of ATP.

B A glucose and a phosphate meet in hexokinase's active site, the microenvironment of which encourages these molecules to react.

C The glucose has bonded with the phosphate. The product of this reaction, glucose-6-phosphate, is shown leaving the active site.

Figure 6.10 The active site of an enzyme.

Effects of Temperature, pH, and Salinity

Adding energy in the form of heat boosts free energy, which is one reason why molecular motion increases with temperature (Section 2.5). The greater the free energy of reactants, the closer a reaction is to its activation energy. Thus, the rate of an enzymatic reaction typically increases with temperature, but only up to a point. An enzyme denatures above a characteristic temperature. Then, the reaction rate falls sharply as the shape of the enzyme changes and it stops functioning (Figure 6.12). For example, body temperatures above 42°C (107.6°F) adversely affect many of your enzymes, which is why such severe fevers are dangerous.

The pH tolerance of enzymes varies. In the human body, most enzymes work best at pH 6–8. For instance, the hexokinase molecule in Figure 6.10 is most active in areas of the small intestine where the pH is around 8. Some enzymes, like pepsin, work outside the typical range of pH. Pepsin functions only in stomach fluid, where it breaks down proteins in food. The fluid is very acidic, with a pH of about 2 (Figure 6.13).

An enzyme's activity is also influenced by the amount of salt in the surrounding fluid. Too much or too little salt can interfere with the hydrogen bonds that hold an enzyme in its three-dimensional shape.

Help From Cofactors

Cofactors are atoms or molecules (other than proteins) that associate with enzymes and are necessary for their function. Some are metal ions. Organic cofactors are called **coenzymes**. Almost all vitamins are coenzymes or precursors of them.

We can use an enzyme called catalase as an example of how cofactors work. Like hemoglobin (Section 3.6), catalase has four hemes. The iron atom at the center of each heme is a cofactor. Iron, like other metal atoms, affects electrons in nearby molecules. Catalase works by holding a substrate molecule close to one of its iron atoms. The iron pulls on the substrate's electrons, which brings on the transition state.

Catalase is an **antioxidant**, which means it neutralizes free radicals—atoms or molecules with one or more unpaired electrons. These dangerous leftovers of metabolic reactions attack the structure of biological molecules. Free radicals accumulate as we age, in part because the body makes fewer catalase molecules.

Some coenzymes are tightly bound to an enzyme. Others, such as NAD+ and NADP+, can diffuse freely through the cytoplasm. Unlike enzymes, many coenzymes become modified during a reaction.

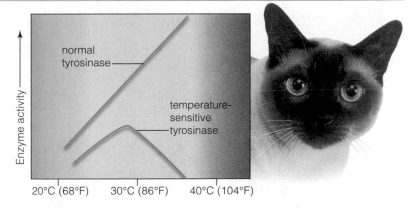

Figure 6.12 The enzyme tyrosinase is involved in the production of melanin, a black pigment in skin cells. Normally, tyrosinase activity increases with temperature between 20°C and 40°C. Mutations can cause tyrosinase activity to plummet at normal body temperatures. The Siamese mutation causes it to be inactive in warmer parts of a cat's body, which end up with less melanin, and lighter fur.

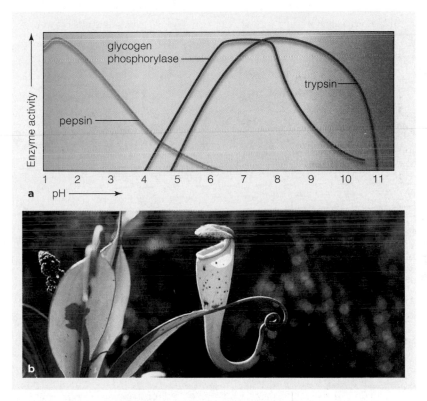

Figure 6.13 Enzymes and pH. (**a**) How pH values affect three enzymes. (**b**) Carnivorous plants of genus *Nepenthes* grow in nitrogen-poor habitats. They secrete acids and protein-digesting enzymes into fluid in a cup made of a modified leaf. The enzymes release nitrogen from small prey, such as insects, that are attracted to odors from the fluid and then drown in it. One of these pepsin-like enzymes functions best at pH 2.6.

Take-Home Message

How do enzymes work?

■ Enzymes greatly enhance the rate of specific reactions. Binding at an enzyme's active site causes a substrate to reach its transition state. In this state, the substrate's bonds are at the breaking point.

■ Each enzyme works best at certain temperatures, pH, and salt concentration.

■ Cofactors associate with enzymes and assist their function.

■ ATP, enzymes, and other molecules interact in organized pathways of metabolism.

■ Links to Electrons 2.3, Metabolism 3.2, Amino acids 3.5, Membrane proteins 5.2

Types of Metabolic Pathways

Metabolism, remember, refers to the activities by which cells acquire and use energy (Section 3.2). Any series of enzyme-mediated reactions by which a cell builds, rearranges, or breaks down an organic substance is called a **metabolic pathway**. Pathways that build mol-ecules from smaller ones are biosynthetic, or anabolic. Other pathways that break molecules apart are degradative, or catabolic.

Many metabolic pathways are linear, a straight line from reactants to products. Others are branched: Their intermediates can continue in more than one sequence of reactions. Still others are cyclic; the last step regenerates a reactant for the first step. For example, a cyclic pathway occurs during the second stage of photosynthesis. The entry point for the reactions is a molecule called RuBP; the last reaction of the pathway converts an intermediate to another molecule of RuBP.

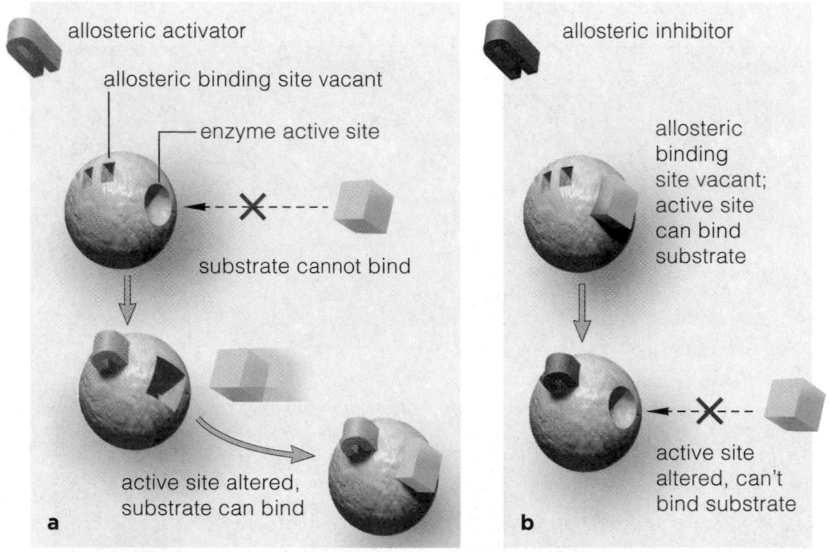

Controls Over Metabolism

Enzymatic reactions do not only run from reactants to products. Many also run in reverse at the same time, with some of the products being converted back to reactants. The rates of the forward and reverse reactions often depend on the concentrations of reactants and products: A high concentration of reactants pushes the reaction in the forward direction. A high concentration of products pushes it in the reverse direction.

Cells conserve energy and resources by making what they need—no more, no less—at any given moment. How does a cell adjust the types and amounts of molecules it produces? Feedback mechanisms help a cell maintain, raise, or lower its production of thousands of different substances. Some of these mechanisms adjust how fast enzyme molecules are made. Others activate or inhibit enzymes that have already been built.

In some cases, molecules that bind to an enzyme directly activate or inhibit it. Such regulatory molecules bind not at the active site, but rather at an allosteric site on the enzyme. An **allosteric** site is a region of an enzyme other than the active site that can bind regulatory molecules (*allo–* means other; *steric* means structure). Binding of an allosteric regulator alters the shape of the enzyme in a way that enhances or inhibits its function (Figure 6.14).

Allosteric effects can cause **feedback inhibition**, in which the end product of a series of enzymatic reactions inhibits the first enzyme in the series (Figure 6.15). For example, isoleucine inhibits its own synthesis, so cells make more of this amino acid when its concentration in the cytoplasm declines. Cells use up their stores of isoleucine and other amino acids—the building blocks of proteins—during protein synthesis (Section 3.5). When protein synthesis slows, less isoleucine gets incorporated into proteins, so the amino acid accumulates. Unused isoleucine binds to an allosteric site on an enzyme in its own synthesis pathway. The

allosteric activator

allosteric binding site vacant

enzyme active site

substrate cannot bind

active site altered, substrate can bind

a

allosteric inhibitor

allosteric binding site vacant; active site can bind substrate

active site altered, can't bind substrate

b

Figure 6.14 Animated Examples of allosteric control. (**a**) An active site becomes functional when an activator binds to an allosteric site. (**b**) An active site stops working when an inhibitor binds to an allosteric site.

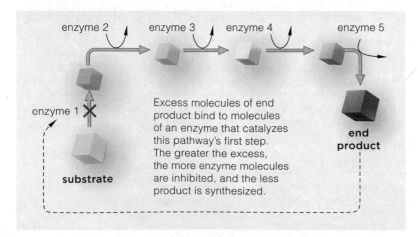

enzyme 2   enzyme 3   enzyme 4   enzyme 5

enzyme 1

Excess molecules of end product bind to molecules of an enzyme that catalyzes this pathway's first step. The greater the excess, the more enzyme molecules are inhibited, and the less product is synthesized.

substrate

end product

Figure 6.15 Animated Feedback inhibition. In this example, five kinds of enzymes act in sequence to convert a substrate to a product, which inhibits the activity of the first enzyme.

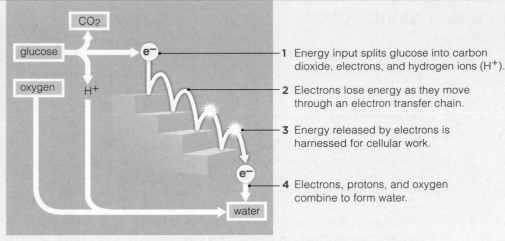

1 Energy input splits glucose into carbon dioxide, electrons, and hydrogen ions (H⁺).

2 Electrons lose energy as they move through an electron transfer chain.

3 Energy released by electrons is harnessed for cellular work.

4 Electrons, protons, and oxygen combine to form water.

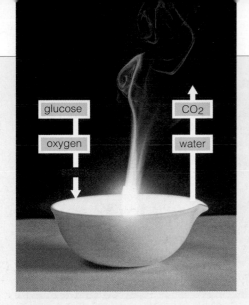

A Glucose and oxygen react when exposed to a spark. Energy is released all at once as CO_2 and water form.

B The same overall reaction occurs in small steps with an electron transfer chain. Energy is released in amounts that cells can harness for cellular work, such as muscle contraction or active transport.

Figure 6.16 Animated Uncontrolled versus controlled energy release.

binding changes the enzyme's shape, so less isoleucine forms. When the cell starts to make proteins again, it uses up the accumulated isoleucine until the allosteric sites on the enzyme molecules are freed up. Then, isoleucine synthesis begins again.

Redox Reactions

If a glucose molecule breaks apart into water and carbon dioxide all at once, it releases energy explosively (Figure 6.16a). Explosions are not good for cells. The only way cells can capture energy from glucose is to break down the molecule in small, manageable steps. Most of these steps are **oxidation–reduction reactions**. In each of these "redox" reactions, a molecule accepts electrons (it becomes *red*uced) from another molecule (which becomes *ox*idized). To remember what reduced means, think of how the negative charge of an electron "reduces" the charge of a recipient molecule. Think of the x in the word oxidation as a sideways + sign, which represents the increase in charge that occurs when a molecule loses an electron.

Coenzymes are among the many types of molecules that accept electrons in redox reactions, which are also called electron transfers. In the next two chapters, you will learn about the importance of redox reactions in electron transfer chains. An **electron transfer chain** is an organized series of reaction steps in which membrane-bound arrays of enzymes and other molecules give up and accept electrons in turn. Electrons are at a higher energy level when they enter a chain than when they leave. Think of the electrons as descending a staircase and losing a bit of energy at each step (Figure 6.16b).

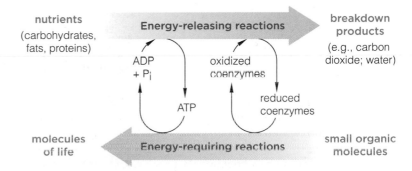

Figure 6.17 ATP forms in energy-releasing reactions, then delivers energy to energy-requiring reactions. Coenzymes (NAD⁺, NADP⁺, and FAD) accept electrons and hydrogen from energy-releasing reactions. The coenzymes (thus reduced to NADH, NADPH, and FADH₂) deliver their cargo of electrons and hydrogen to energy-requiring reactions.

Many coenzymes deliver electrons to electron transfer chains in photosynthesis and aerobic respiration. Energy released at certain steps in those chains helps drive the synthesis of ATP. Figure 6.17 is an overview of how ATP and coenzymes connect energy-releasing with energy-requiring pathways. These pathways will occupy our attention in chapters to come.

Take-Home Message

What are metabolic pathways?

■ Metabolic pathways are sequences of enzyme-mediated reactions. Some are biosynthetic; others are degradative.

■ Control mechanisms enhance or inhibit the activity of many enzymes. The adjustments help cells produce only what they require in any given interval.

■ Many metabolic pathways involve electron transfers, or oxidation–reduction reactions. Redox reactions occur in electron transfer chains. The chains are important sites of energy exchange in photosynthesis and aerobic respiration.

6.5 | Night Lights

- Bioluminescence is visible evidence of metabolism.
- Links to Tracers 2.2, Biofilms 4.5

Enzymes of Bioluminescence At night, in the warm waters of tropical seas or in the summer air above fields and gardens, you may see shimmers or flashes of light. The light, which is emitted from metabolic reactions in living organisms, is **bioluminescence** (from the Greek *bio–*, for life, and the Latin *lumen*, for shine). In different species, it helps attract mates or prey, or confuse predators.

Bioluminescent organisms emit light when enzymes called luciferases convert chemical bond energy to light energy (luciferase is a generic term that refers to many different enzymes). Figure 6.18 shows firefly luciferase, a temperature-sensitive enzyme that uses ATP to energize a light-emitting pigment molecule. Any substrate of a luciferase is called luciferin:

$$luciferin + ATP \rightarrow luciferin\text{-}ADP + P_i$$

Energized by the transfer, the modified luciferin spontanously releases its extra energy as light:

$$luciferin\text{-}ADP + O_2 \rightarrow oxyluciferin + AMP + CO_2 + light$$

Different luciferins emit colors across the spectrum of visible light—from red to orange, yellow, green, blue, and purple. Some even emit infrared or ultraviolet light.

A Research Connection Many species of protists, fungi, bacteria, insects, jellyfishes, and fishes are bioluminescent. Researchers can transfer genes for bioluminescence from one of these species into another, nonluminescent species,

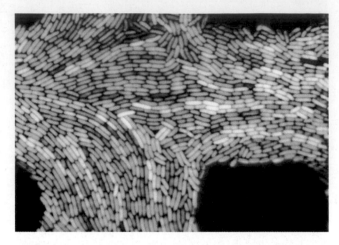

Figure 6.19 Bioluminescent biofilm. These bacteria have been altered to carry bioluminescence genes from a species of jellyfish.

so the recipient organisms light up under certain conditions. In itself, making organisms glow seems like a bizarre thing to do. However, the researchers are using the bioluminescence as a visible tracer in a variety of experiments.

For example, the *Escherichia coli* bacteria in Figure 6.19 are recipients of genes from a type of bioluminescent jellyfish. The bioluminescent light emitted by these cells indicates their metabolic activity. Differences in the intensity of light from individual cells reflect actual differences in metabolic activity among the cells in this biofilm. These bacteria are genetically identical; how could metabolic activity differ among them? The answer must be that each cell's metabolism depends on its location within the biofilm. Such research might help us discover why some bacterial cells, but not others, become resistant to antibiotics and are able to establish long-term infections in humans.

Figure 6.18 Bioluminescence. *Left*, a North American firefly (*Photinus pyralis*) emits a flash from its light organ, which contains peroxisomes packed with luciferase molecules. Firefly flashes may help potential mates find each other in the dark. *Right*, structure of firefly luciferase.

A Toast to Alcohol Dehydrogenase

In the human body, alcohol dehydrogenase (ADH) converts ethanol to acetaldehyde, an organic molecule even more toxic than ethanol and the most likely source of various hangover symptoms:

$$H-\overset{\overset{\displaystyle H}{|}}{\underset{\underset{\displaystyle H}{|}}{C}}-\overset{\overset{\displaystyle H}{|}}{\underset{\underset{\displaystyle H}{|}}{C}}-OH + NAD^+ \xrightarrow{\text{ADH}} H-\overset{\overset{\displaystyle H}{|}}{\underset{\underset{\displaystyle H}{|}}{C}}-\overset{\overset{\displaystyle O}{\|}}{C}-H + NADH$$

(ethanol)   (acetaldehyde)

A different enzyme, aldehyde dehydrogenase (ALDH), very quickly converts toxic acetaldehyde to nontoxic acetate:

$$H-\overset{\overset{\displaystyle H}{|}}{\underset{\underset{\displaystyle H}{|}}{C}}-\overset{\overset{\displaystyle O}{\|}}{C}-H + NAD^+ \xrightarrow{\text{ALDH}} H-\overset{\overset{\displaystyle H}{|}}{\underset{\underset{\displaystyle H}{|}}{C}}-\overset{\overset{\displaystyle O}{\|}}{C}-O^- + NADH + H^+$$

(acetaldehyde)   (acetate)

Thus, the overall pathway of ethanol metabolism in humans is:

$$\text{ethanol} \xrightarrow[\underset{NAD^+ \quad NADH}{}]{\text{ADH}} \text{acetaldehyde} \xrightarrow[\underset{NAD^+ \quad NADH}{}]{\text{ALDH}} \text{acetate}$$

In the average adult human body, this metabolic pathway can detoxify between 7 and 14 grams of ethanol per hour. The average alcoholic beverage contains between 10 and 20 grams of ethanol, which is why having more than one drink in any two-hour interval may result in a hangover.

Most organisms have alcohol dehydrogenase, which detoxifies the tiny quantities of alcohols that form in some metabolic pathways. In animals, the enzyme also detoxifies alcohols made by gut-inhabiting bacteria, and those in foods such as ripe fruit.

Despite the small amounts of alcohol that humans encounter naturally, our bodies make at least nine different kinds of alcohol dehydrogenase. It is interesting to speculate about why so many of them have evolved.

We do understand how some mutations in ADH affect our alcohol metabolism. For example, some mutations cause one of the ADH enzymes to be overactive, in which case acetaldehyde accumulates more quickly than ALDH can detoxify it:

$$\text{ethanol} \overset{\text{ADH}}{\underset{\xrightarrow{\hspace{1cm}}}{\overset{\xrightarrow{\hspace{1cm}}}{\xrightarrow{\hspace{1cm}}}}} \begin{array}{l}\text{acetaldehyde}\\ \text{acetaldehyde}\\ \text{acetaldehyde}\end{array} \xrightarrow{\text{ALDH}} \text{acetate}$$

People who carry such mutations become flushed and feel very ill after drinking even a small amount of alcohol. This unpleasant experience may be part of the reason that these people are less likely to become alcoholic than other people.

Different mutations that result in an underactive ALDH also cause acetaldehyde to accumulate:

$$\text{ethanol} \xrightarrow{\text{ADH}} \begin{array}{l}\text{acetaldehyde}\\ \text{acetaldehyde}\\ \text{acetaldehyde}\end{array} \xrightarrow{\quad\times\quad} \text{acetate}$$

These mutations are associated with the same effect—and the same protection from alcoholism—as mutations that cause an ADH to be overactive. Both types of mutations are common in people of Asian descent. For this reason, the alcohol flushing reaction is often called "Asian flush."

Mutations that disrupt the activity of an ADH enzyme have the opposite effect. Such mutations result in slowed alcohol metabolism, and people who carry them may not feel the ill effects of drinking alcoholic beverages as much as other people do. When these people drink alcohol, they have a tendency to become alcoholics. The study mentioned in the chapter opener showed that one-quarter of the undergraduate students who binged also had other signs of alcoholism.

Alcoholics will continue to drink despite the knowledge that doing so has tremendous negative consequences. In the United States, alcohol abuse is the leading cause of cirrhosis of the liver. The liver becomes so scarred, hardened, and filled with fat that it loses its function (Figure 6.20). It stops making the protein albumin, so the solute balance of body fluids is disrupted, and the legs and abdomen swell with watery fluid. It cannot remove drugs and other toxins from the blood, so they accumulate in the brain—which impairs mental functioning and alters personality. Restricted blood flow through the liver causes veins to enlarge and rupture, so internal bleeding is a risk. The damage to the body results in a heightened susceptibility to diabetes and liver cancer. Once cirrhosis has been diagnosed, a person has about a 50% chance of death within 10 years.

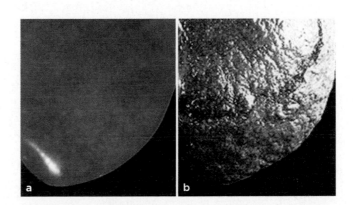

Figure 6.20 Alcoholic liver disease. (**a**) Normal human liver. (**b**) Enlarged, cirrhotic liver of an alcoholic. As few as 2 drinks per day can cause this disease.

Summary

Section 6.1 **Energy** is defined as a capacity to do work. Energy cannot be created or destroyed (**first law of thermodynamics**), but it can be converted from one form to another and thus transferred between objects or systems. Energy tends to disperse spontaneously (**second law of thermodynamics**). A bit disperses at each energy transfer, usually in the form of heat.

All living things maintain their organization only as long as they harvest energy from someplace else. Energy flows in one direction through the biosphere, starting mainly from the sun, then into and out of ecosystems. Producers and then consumers use energy to assemble, rearrange, and break down organic molecules that cycle among organisms throughout ecosystems.

Section 6.2 Cells store and retrieve **free energy** by making and breaking chemical bonds in metabolic **reactions**, in which **reactants** are converted to **products** (Table 6.1). **Activation energy** is the minimum energy required to start a reaction. **Endergonic** reactions require a net energy input. **Exergonic** reactions end with a net energy release.

ATP is an energy carrier between reaction sites in cells. It has three phosphate bonds; when a phosphate is transferred to another molecule, the energy of the bond is transferred along with it. Phosphate-group transfers (**phosphorylations**) to and from ATP couple reactions that release energy with reactions that require energy. Cells regenerate ATP by the **ATP/ADP cycle**.

■ *Use the animation on CengageNOW to learn about energy changes in chemical reactions and the role of ATP.*

Section 6.3 **Enzymes** are proteins or RNAs that greatly enhance the rate of a chemical reaction. Enzymes lower a reaction's activation energy by boosting local concentrations of **substrates**, orienting substrates in positions that favor reaction, inducing the fit between a substrate and the enzyme's **active site** (**induced-fit model**), and sometimes excluding water; all of which bring on a substrate's **transition state**. Each type of enzyme works best within a characteristic range of temperature, salt concentration, and pH. Most enzymes require the assistance of **cofactors**, which are metal ions or organic **coenzymes**. Cofactors in some **antioxidants** help them detoxify free radicals.

■ *Use the animation and interaction on CengageNOW to investigate how enzymes facilitate reactions.*

Section 6.4 Cells concentrate, convert, and dispose of most substances in enzyme-mediated reaction sequences called **metabolic pathways**. **Allosteric** sites are points of control by which a cell adjusts the types and amounts of substances it makes. **Feedback inhibition** is one example of enzyme control. **Oxidation–reduction** (redox) **reactions** in **electron transfer chains** allow cells to harvest energy in manageable increments.

■ *Use the animation on CengageNOW to compare the effects of controlled and uncontrolled energy release, and to observe mechanisms that exert control over enzymes.*

Section 6.5 **Bioluminescence** is light emitted by living organisms. Most bioluminescence is the product of enzyme-mediated reactions that often include ATP.

Self-Quiz

Answers in Appendix III

1. _____ is life's primary source of energy.
   a. Food   b. Water   c. Sunlight   d. ATP

2. Energy _____ .
   a. cannot be created or destroyed
   b. can change from one form to another
   c. tends to disperse spontaneously
   d. all of the above

3. Entropy _____ . (Choose all that are correct.)
   a. disperses   c. always increases, overall
   b. is a measure of disorder   d. is energy

4. If we liken a chemical reaction to an energy hill, then an _____ reaction is an uphill run.
   a. endergonic   c. ATP-assisted
   b. exergonic   d. both a and c

5. If we liken a chemical reaction to an energy hill, then activation energy is like _____ .
   a. a burst of speed   c. coasting downhill
   b. a bump at the hilltop   d. both a and b

6. _____ are always changed by participating in a reaction. (Choose all that are correct.)
   a. Enzymes   c. Reactants
   b. Cofactors   d. Intermediates

7. Enzymes _____ .
   a. are proteins, except for a few RNAs
   b. lower the activation energy of a reaction
   c. are changed by the reactions they catalyze
   d. a and b

Table 6.1	**Key Players in Metabolic Reactions**
Reactant	Substance that enters a metabolic reaction; also called a substrate of an enzyme
Intermediate	Substance that forms in a reaction or pathway between the reactants and products
Product	Substance that remains at the end of a reaction or pathway
Enzyme	Protein or RNA that greatly enhances the rate of a reaction, but is not changed by participating in it
Cofactor	Molecule or ion that assists enzymes; may carry electrons, hydrogen, or functional groups to other reaction sites
Energy carrier	Mainly ATP; couples reactions that release energy with reactions that require energy

Data Analysis Exercise

Ethanol is a toxin, so it makes sense that drinking it can cause various symptoms of poisoning—headache, stomach ache, nausea, fatigue, impaired memory, dizziness, tremors, and diarrhea, among other ailments. All are symptoms of hangover, the common word for what happens as the body is recovering from a bout of heavy drinking.

The most effective treatment for a hangover is to avoid drinking in the first place. Folk remedies (such as aspirin, coffee, bananas, more alcohol, honey, barley grass, pizza, milkshakes, glutamine, raw eggs, charcoal tablets, or cabbage) abound, but few have been studied scientifically. In 2003, Max Pittler and his colleagues tested one of them. The researchers gave 15 participants an unmarked pill containing either artichoke extract or a placebo (an inactive substance) just before or after drinking enough alcohol to cause a hangover. The results are shown in Figure 6.21.

1. How many participants experienced a hangover that was worse with the placebo than with the artichoke extract?

2. How many participants experienced a worse hangover with the artichoke extract?

3. Calculate the numbers you counted in questions 1 and 2 as a percentage of the total number of participants. How much difference is there between the percentages?

4. Does this data support the hypothesis that artichoke extract is an effective hangover treatment? Why or why not?

Participant (Age, Gender)	Severity of Hangover	
	Artichoke Extract	Placebo
1 (34, F)	1.9	3.8
2 (48, F)	5.0	0.6
3 (25, F)	7.7	3.2
4 (57, F)	2.4	4.4
5 (34, F)	5.4	1.6
6 (30, F)	1.5	3.9
7 (33, F)	1.4	0.1
8 (37, F)	0.7	3.6
9 (62, M)	4.5	0.9
10 (36, M)	3.7	5.9
11 (54, M)	1.6	0.2
12 (37, M)	2.6	5.6
13 (53, M)	4.1	6.3
14 (48, F)	0.5	0.4
15 (32, F)	1.3	2.5

Figure 6.21 Results of a study that tested artichoke extract as a hangover preventive. All participants were tested once with the placebo and once with the extract, with a week interval between. Each rated the severity of 20 hangover symptoms on a scale of 0 (not experienced) to 5 ("as bad as can be imagined"). The 20 ratings were averaged as a single, overall rating, which is listed here.

8. Which of the following statements is not correct? A metabolic pathway _____ .
 a. is a sequence of enzyme-mediated reactions
 b. may be biosynthetic or degradative
 c. generates heat
 d. can include an electron transfer chain
 e. none of the above

9. A molecule that donates electrons becomes _____ , and the one that accepts electrons becomes _____ .
 a. reduced; oxidized   c. oxidized; reduced
 b. reduced; reduced   d. oxidized; oxidized

10. A free radical is an atom or molecule that _____ .
 a. carries no charge   c. has an unpaired electron
 b. has too many electrons   d. has too few electrons

11. An antioxidant is a molecule that _____ .
 a. detoxifies free radicals   c. balances charge
 b. degrades toxins   d. oxidizes free radicals

12. Match each term with its most suitable description.
 ___ reactant   a. assists enzymes
 ___ enzyme   b. there at reaction's end
 ___ entropy   c. enters a reaction
 ___ product   d. increases spontaneously
 ___ redox reaction   e. energy cannot be created or destroyed
 ___ cofactor   f. a form of give and take
 ___ first law   g. usually unchanged by participating in a reaction

■ *Visit CengageNOW for additional questions.*

Critical Thinking

1. Beginning physics students are often taught the basic concepts of thermodynamics with two phrases: First, you can't win. Second, you can't break even. Explain.

2. Dixie Bee wanted to make JELL-O shots for her next party, but felt guilty about encouraging her guests to consume alcohol. She tried to compensate for the toxicity of the alcohol by adding pieces of healthy fresh pineapple to the shots, but when she did, the JELL-O never set up. What happened? Hint: JELL-O is mainly sugar and collagen, a protein.

3. Free radicals are atoms or molecules that are like ions with the wrong number of electrons. They form in many enzyme-catalyzed reactions, such as the digestion of fats and amino acids. They slip out of electron transfer chains. They form when x-rays and other kinds of ionizing radiation strike water and other molecules. Free radicals react easily with the molecules of life, and can destroy them.

Hydrogen peroxide (H_2O_2) forms in most organisms as a by-product of aerobic respiration. This toxic molecule can easily become an even more dangerous free radical, so cells must dispose of it fast or risk being damaged. One molecule of catalase can inactivate about 6 million hydrogen peroxide molecules per minute by combining them two at a time. Catalase also inactivates other toxins, including ethanol. Given that its active site specifically binds hydrogen peroxide, how can this enzyme act on other substances?

4. Catalase combines two hydrogen peroxide molecules ($H_2O_2 + H_2O_2$) to make two molecules of water. A gas also forms. What is the gas?

5. Hydrogen peroxide bubbles if dribbled on an open cut but does not bubble on unbroken skin. Explain why.

7 Where It Starts—Photosynthesis

IMPACTS, ISSUES Biofuels

Plants and other photosynthesizers harvest energy from sunlight, then store it in chemical bonds of organic molecules they make from carbon dioxide and water. That process is photosynthesis, and it feeds them, us, and most other life on Earth. Photosynthesis also satisfies almost all of our uniquely human needs for fuel—for energy that we can use to heat our homes, cook our food, and run our machines. Three hundred million years ago, photosynthesis supported the growth of vast swamp forests. Successive forests slowly decayed, compacted, and became fossil fuels that we now extract from the earth. Coal, petroleum, and natural gas are composed of molecules that were originally assembled by ancient photosynthesizers. As such, their supply is limited.

Where are we going to get our energy once Earth's supply of fossil fuels runs out? The sun gives off a lot of it, but unlike plants, we cannot capture usable energy from the sun in an economically feasible way. Luckily for us, photosynthesizers are still at it. The molecules they make end up in vegetation, agricultural products, and ultimately in animals and animal wastes—all biomass (organic matter that is not fossilized).

A lot of energy is locked up in biomass. We can burn it, but that is an inefficient and messy way to release its energy. We can, however, convert it to oils, gases, or alcohols that burn cleanly and yield more energy per unit volume than burning the biomass itself. Such biofuels are a renewable source of energy: they can be made from crops, weeds, or what we now consider waste. We now make biodiesel from oils that come from algae, soybeans, rapeseed, flaxseed, and even restaurant kitchens. Methane seeps from manure ponds, landfills, and cows; we just need to find an efficient way to collect it.

We can also make ethanol from biomass. First, the carbohydrates in the biomass must be broken down to their component sugars. Making ethanol from the sugars is the easy part: We simply feed them to microorganisms that can convert sugars to ethanol. It is much more difficult to break down the carbohydrates in biomass cost-effectively, and without fossil fuels. The more cellulose in the biomass, the more involved that process is. Cellulose is a tough, insoluble carbohydrate. Breaking the bonds between its component sugars uses a lot of chemicals and energy, which adds cost to the biofuel product.

Today, we make ethanol from sugar-rich food crops such as corn, sugar beets, and sugarcane. These crops can be expensive to grow, and using them to make biofuel competes with our food supply. Thus, researchers are working to find an inexpensive way to break down the abundant cellulose in fast-growing weeds like switchgrass (Figure 7.1), and agricultural wastes such as wood chips, wheat straw, cotton stalks, and rice hulls—all biomass that we now dump in landfills or burn.

See the video! Figure 7.1 Biofuels. (**a**) Switchgrass (*Panicum virgatum*) grows wild in North American prairies. (**b**) Researchers Ratna Sharma and Mari Chinn of North Carolina State University are working to make biofuel production from biomass like switchgrass and agricultural wastes economically feasible.

Key Concepts

The rainbow catchers

The flow of energy through the biosphere starts when chlorophylls and other photosynthetic pigments absorb the energy of visible light. **Sections 7.1, 7.2**

Making ATP and NADPH

Photosynthesis proceeds through two stages in the chloroplasts of plants and many types of protists. In the first stage, sunlight energy is converted to the chemical bond energy of ATP. The coenzyme NADPH forms in a pathway that also releases oxygen. **Sections 7.3–7.5**

Making sugars

The second stage is the "synthesis" part of photosynthesis. Sugars are assembled from CO_2. The reactions use ATP and NADPH that form in the first stage of photosynthesis. Details of the reactions vary among organisms. **Sections 7.6, 7.7**

Evolution and photosynthesis

The evolution of photosynthesis changed the composition of Earth's atmosphere. New pathways that detoxified the oxygen by-product of photosynthesis evolved. **Section 7.8**

Photosynthesis, CO_2, and global warming

Photosynthesis by autotrophs removes CO_2 from the atmosphere; metabolism by all organisms puts it back in. Human activities have disrupted this balance, and so have contributed to global warming. **Section 7.9**

Links to Earlier Concepts

- The concept of energy flow (Section 6.1) helps explain how photosynthesizers can harvest energy from the sun. A review of electron energy levels (2.3) and chemical bonding (2.4) may be useful.

- Photosynthesis is the primary function of chloroplasts (4.11). Surface specializations (4.12) indirectly support photosynthesis in plants.

- You will see how energy carriers such as ATP link energy-releasing metabolic reactions with energy-requiring ones (6.2), and how cells harvest energy with electron transfer chains (6.4).

- What you know about carbohydrates (3.3), membrane proteins (5.2), and concentration gradients (5.3) will help you understand the chemical processes of photosynthesis.

- You will see how free radicals (6.3) influenced the evolution of organisms that changed our atmosphere.

- An example of nutrient cycling (1.2) illustrates one of the ways that photosynthesis connects the biosphere with its inhabitants.

How would you vote? Ethanol and other fuels manufactured from crops cost more than gasoline, so they are renewable energy sources and have fewer emissions. Would you pay a premium to drive a vehicle that uses biofuels? If so, how much? See CengageNOW for details, then vote online.

■ Photosynthetic organisms use pigments to capture the energy of sunlight.

■ Links to Electrons 2.3, Bonding 2.4, Carbohydrates 3.3, Energy 6.1, Metabolism 6.4

Energy flow through nearly all ecosystems on Earth begins when photosynthesizers intercept energy from the sun. Let's turn now to the details of that process.

Properties of Light

Visible light is part of a spectrum of electromagnetic energy radiating from the sun. Such radiant energy travels in waves, undulating across space a bit like waves moving across an ocean. The distance between the crests of two successive waves of light is called **wavelength**, which we measure in nanometers (nm).

The electromagnetic energy of light is organized in packets called photons. A photon's energy and its wavelength are related, so all photons traveling at the same wavelength have the same amount of energy. Photons with the least amount of energy travel in longer wavelengths; those with the most energy travel in shorter wavelengths.

By the metabolic pathway of **photosynthesis**, organisms can harness the energy of light to build organic molecules from inorganic raw materials. Only light of wavelengths between 380 and 750 nanometers drives photosynthesis. Humans and many other organisms perceive light of all of these wavelengths combined as white, and particular wavelengths as different colors. White light separates into its component colors when it passes through a prism. The prism bends the longer wavelengths more than it bends the shorter ones, so a rainbow of colors forms (Figure 7.2).

Figure 7.2 also shows where visible light falls within the electromagnetic spectrum, which is the range of all wavelengths of radiant energy. Wavelengths of UV (ultraviolet) light, x-rays, and gamma rays are shorter than about 380 nanometers. They are energetic enough to alter or break the chemical bonds of DNA and other biological molecules, so they are a threat to life.

The Rainbow Catchers

Pigments are the molecular bridges between sunlight and photosynthesis. A **pigment** is an organic molecule that selectively absorbs light of specific wavelengths. Wavelengths of light that are not absorbed are reflected, and that reflected light gives each pigment its characteristic color. For example, a pigment that absorbs violet, blue, and green light reflects the remainder of the visible light spectrum: yellow, orange, and red light. This pigment would appear orange to us.

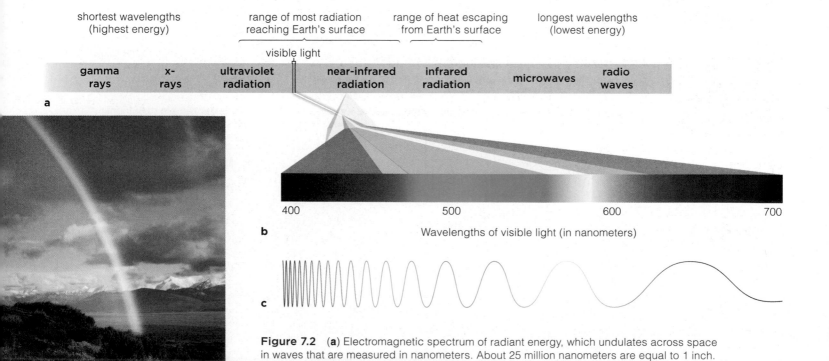

| shortest wavelengths (highest energy) | | range of most radiation reaching Earth's surface | range of heat escaping from Earth's surface | longest wavelengths (lowest energy) |

visible light

| gamma rays | x-rays | ultraviolet radiation | near-infrared radiation | infrared radiation | microwaves | radio waves |

a

400   500   600   700

b   Wavelengths of visible light (in nanometers)

c

Figure 7.2 (**a**) Electromagnetic spectrum of radiant energy, which undulates across space in waves that are measured in nanometers. About 25 million nanometers are equal to 1 inch. (**b**) Visible light is a very small part of the electromagnetic spectrum. A glass prism can break it into the bands we see in a rainbow. (**c**) The shorter the wavelength, the higher the energy.

Table 7.1 Some Pigments in Photosynthesizers

Pigment	Color	Plants	Protists	Bacteria	Archaeans
		Occurrence in Photosynthetic Organisms			
Chlorophyll *a*	green	✗	✗	✗	
Other chlorophylls	green	✗	✗	✗	
Phycobilins					
phycocyanobilin	blue		✗	✗	
phycoerythrobilin	red		✗	✗	
phycoviolobilin	purple		✗	✗	
phycourobilin	orange		✗	✗	
Carotenoids					
carotenes					
β-carotene	orange	✗	✗	✗	
α-carotene	orange	✗	✗	✗	
lycopene	red	✗	✗		
xanthophylls					
lutein	yellow				
zeaxanthin	yellow				
fucoxanthin	orange	✗	✗		
Anthocyanins	purple	✗	✗	✗	
Retinal	purple				✗

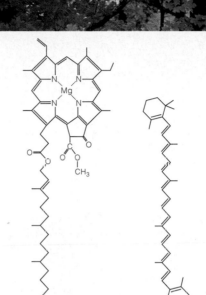

Figure 7.3 Structure of two photosynthetic pigments. Both structures are derived from evolutionary remodeling of the same synthesis pathway. The light-catching part of each is the array of single bonds alternating with double bonds.

In chlorophyll, the array is a ring structure almost identical to a heme group. Heme groups are part of hemoglobin, which is also a pigment (Figure 3.3).

chlorophyll *a*   β-carotene

Chlorophyll *a* is by far the most common photosynthetic pigment in plants, photosynthetic protists, and cyanobacteria. Chlorophyll *a* absorbs violet and red light, so it appears green. Accessory pigments absorb additional colors of light for photosynthesis. A few of the 600 or so known accessory pigments are listed in Table 7.1.

Most types of photosynthetic organisms use a mixture of pigments for photosynthesis. In leaves of typical plants, chlorophyll is usually so abundant that it masks the colors of all the other pigments. Thus, most leaves usually appear green. In autumn, however, pigment synthesis slows in many kinds of leafy plants, and chlorophyll breaks down faster than it is replaced. Other pigments tend to be more stable than chlorophyll, so the leaves of such plants turn red, orange, yellow, or purple as their chlorophyll content declines and accessory pigments become visible.

Collectively, photosynthetic pigments absorb nearly all of the wavelengths of visible light. Different kinds cluster in photosynthetic membranes. Together, they can absorb a broad range of wavelengths, like a radio antenna that can pick up different stations.

The light-trapping part of a pigment is an array of atoms in which single bonds alternate with double bonds (Section 2.4 and Figure 7.3). Electrons of these atoms occupy one large orbital that spans all of the atoms. Electrons in such arrays easily absorb photons, so pigment molecules are a bit like antennas that are specialized for receiving light energy.

Absorbing a photon excites electrons. Remember, an energy input can boost an electron to a higher energy level (Section 2.3). The excited electron returns quickly to a lower energy level by emitting the extra energy. As you will see in Section 7.4, photosynthetic cells can capture energy emitted from an electron by bouncing the energy like a super-speed volleyball among a team of photosynthetic pigments. When the energy reaches the team captain—a special pair of chlorophylls—the reactions of photosynthesis begin.

Take-Home Message

How do photosynthetic organisms absorb light?

■ Energy radiating from the sun travels through space in waves and is organized as packets called photons.

■ The spectrum of radiant energy from the sun includes visible light. Humans perceive light of certain wavelengths as different colors. The shorter the wavelength of light, the greater its energy.

■ Pigments absorb specific wavelengths of visible light. Photosynthetic organisms use chlorophyll *a* and other pigments to capture the energy of light. That energy is used to drive the reactions of photosynthesis.

7.2 Exploring the Rainbow

■ Photosynthetic pigments work together to harvest light of different wavelengths.

A Light micrograph of photosynthetic cells in a strand of *Chladophora*. Engelmann used this green alga to demonstrate that certain colors of light are best for photosynthesis.

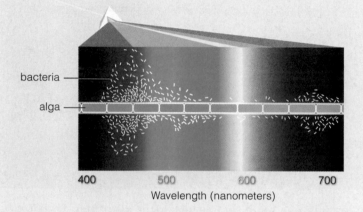

bacteria
alga

400   500   600   700
Wavelength (nanometers)

B Engelmann directed light through a prism so that bands of colors crossed a water droplet on a microscope slide. The water held a strand of *Chladophora* and oxygen-requiring bacteria. The bacteria clustered around the algal cells that were releasing the most oxygen—the ones that were most actively engaged in photosynthesis. Those cells were under red and violet light.

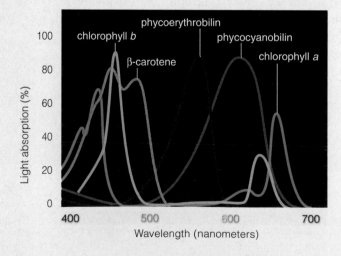

phycoerythrobilin
chlorophyll *b*
phycocyanobilin
β-carotene
chlorophyll *a*

Light absorption (%)

100 80 60 40 20 0

400   500   600   700
Wavelength (nanometers)

C Absorption spectra of a few photosynthetic pigments. Line color indicates the characteristic color of each pigment.

At one time, people thought that plants used only the substances in soil to grow. By 1882, a few chemists understood that there were more ingredients in that recipe: water, something in the air, and light. Botanist Theodor Engelmann designed an experiment to test his hypothesis that the color of light affects photosynthesis. It had long been known that photosynthesis releases oxygen, so Engelmann used the amount of oxygen released by photosynthetic cells as a measure of how much photosynthesis was occurring in them.

In his experiment, Engelmann used a prism to divide a ray of light into its component colors, then directed the resulting spectrum across a single strand of photosynthetic alga (Figure 7.4*a*) suspended in a drop of water. Oxygen-sensing equipment had not yet been invented, so Engelmann used oxygen-requiring bacteria to show him where the oxygen concentration in the water was highest. The bacteria moved through the water and gathered mainly where violet or red light fell across the algal strand (Figure 7.4*b*). Engelmann concluded that the algal cells illuminated by light of these colors were releasing the most oxygen—a sign that violet and red light are best at driving photosynthesis.

Engelmann's experiment allowed him to correctly identify the colors of light most efficient at driving photosynthesis in *Chladophora*. His results constituted an absorption spectrum—a graph that shows how efficiently the different wavelengths of light are absorbed by a substance. Peaks in the graph indicate wavelengths of light that the substance absorbs best (Figure 7.4*c*).

Engelmann's absorption spectrum represents the combined spectra of all the photosynthetic pigments in *Chladophora*. Most photosynthetic organisms use a combination of pigments to drive photosynthesis, and the combination differs by species. Why? Different proportions of wavelengths in sunlight reach different parts of Earth. The particular set of pigments in each species is an adaptation that allows an organism to absorb the particular wavelengths of light available in its habitat. For example, water absorbs light between wavelengths of 500 and 600 nm less efficiently than other wavelengths. Algae that live deep underwater have pigments that absorb light in the range of 500–600 nm, which is the range that water does not absorb very well. Phycobilins are the most common pigments in deep-water algae.

Figure 7.4 Animated Discovery that photosynthesis is driven by particular wavelengths of light. Theodor Engelmann used the green alga *Chladophora* (**a**) in an early photosynthesis experiment (**b**). His results constituted one of the first absorption spectra.

(**c**) Absorption spectra of chlorophylls *a* and *b*, β-carotene, and two phycobilins reveal the efficiency with which these pigments absorb different wavelengths of visible light.

Figure It Out: Which are the three main photosynthetic pigments in *Chladophora*?

Answer: chlorophyll a, chlorophyll b, and β-carotene

7.3 | Overview of Photosynthesis

■ Chloroplasts are organelles of photosynthesis in plants and other photosynthetic eukaryotes.

The **chloroplast** is an organelle that specializes in photosynthesis in plants and many protists (Figure 7.5a,b). Plant chloroplasts have two outer membranes, and are filled with a semifluid matrix called the **stroma**. Stroma contains the chloroplast's DNA, some ribosomes, and an inner, much-folded **thylakoid membrane**. The folds of a thylakoid membrane typically form stacks of disks (thylakoids) that are connected by channels. The space inside all of the disks and channels is a single, continuous compartment (Figure 7.5b).

Embedded in the thylakoid membrane are many clusters of light-harvesting pigments. These clusters absorb photons of different energies. The membrane also incorporates **photosystems**, which are groups of hundreds of pigments and other molecules that work as a unit to begin the reactions of photosynthesis. Chloroplasts contain two kinds of photosystems, type I and type II, which were named in the order of their discovery. Both types convert light energy into chemical energy.

Often, photosynthesis is summarized by this simple equation, from reactants to products:

$$6H_2O + 6CO_2 \xrightarrow[\text{enzymes}]{\text{light energy}} 6O_2 + C_6H_{12}O_6$$

water   carbon   oxygen   glucose
        dioxide

However, photosynthesis is actually a series of many reactions that occur in two stages. In the first stage, the **light-dependent reactions**, the energy of light gets converted to the chemical bond energy of ATP. Typically, the coenzyme $NADP^+$ accepts electrons and hydrogen ions, thus becoming NADPH. Oxygen atoms released from the breakdown of water molecules escape from the cell as O_2. The second stage, the **light-independent reactions**, runs on energy delivered by the ATP and NADPH formed in the first stage. That energy drives the synthesis of glucose and other carbohydrates from carbon dioxide and water (Figure 7.5c).

Take-Home Message

What are photosynthesis reactions and where do they occur?

■ In chloroplasts, the first stage of photosynthesis occurs at the thylakoid membrane. In these light-dependent reactions, light energy drives ATP and NADPH formation; oxygen is released.

■ The second stage of photosynthesis occurs in the stroma. In these light-independent reactions, ATP and NADPH drive the synthesis of sugars from water and carbon dioxide.

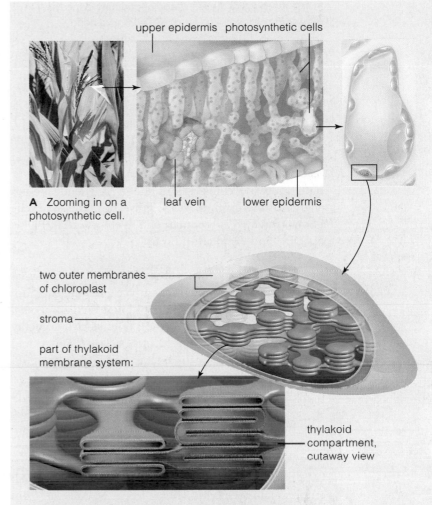

A Zooming in on a photosynthetic cell.

upper epidermis   photosynthetic cells
leaf vein   lower epidermis

two outer membranes of chloroplast
stroma
part of thylakoid membrane system:
thylakoid compartment, cutaway view

B Chloroplast structure. No matter how highly folded, its thylakoid membrane system forms a single, continuous compartment in the stroma.

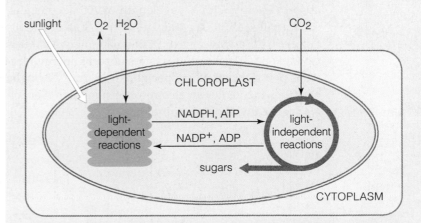

sunlight O_2 H_2O CO_2

CHLOROPLAST

light-dependent reactions → NADPH, ATP → light-independent reactions

$NADP^+$, ADP

sugars

CYTOPLASM

C In chloroplasts, ATP and NADPH form in the light-dependent stage of photosynthesis, which occurs at the thylakoid membrane. The second stage, which produces sugars and other carbohydrates, proceeds in the stroma.

Figure 7.5 Animated Sites of photosynthesis in a typical leafy plant.

■ The reactions of the first stage of photosynthesis convert the energy of light to the energy of chemical bonds.

■ Links to Electrons and energy levels 2.3, Chloroplasts 4.11, Membrane proteins 5.2, Membrane properties and gradients 5.3, Energy 6.1, Electron transfer chains 6.4

light-harvesting complex   photosystem

Figure 7.7 A view of some of the components of the thylakoid membrane from the stroma.

The first stage of photosynthesis is driven by light, so the collective reactions of this stage are said to be light-dependent. Two different sets of light-dependent reactions constitute a noncyclic and a cyclic pathway. Both pathways convert light energy to chemical bond energy in the form of ATP (Figure 7.6). The noncyclic pathway, which is the main one in chloroplasts, yields NADPH and O_2 in addition to ATP.

Capturing Energy for Photosynthesis Imagine what happens when a pigment absorbs a photon. The photon's energy boosts one of the pigment's electrons to a higher energy level (Section 2.3). The electron quickly emits the extra energy and drops back to its unexcited state. If nothing else were to happen, the energy would be lost to the environment.

In the thylakoid membrane, however, the energy of excited electrons is kept in play. Embedded in this membrane are millions of light-harvesting complexes (Figure 7.7). These circular clusters of photosynthetic pigments and proteins hold on to energy by passing it back and forth, a bit like volleyball players pass a ball among team members. The energy gets volleyed from cluster to cluster until a photosystem absorbs it.

At the center of each photosystem is a special pair of chlorophyll a molecules. The pair in photosystem I absorbs energy with a wavelength of 700 nanometers, so it is called P700. The pair in photosystem II absorbs energy with a wavelength of 680 nanometers, so it is called P680. When a photosystem absorbs energy, electrons pop right off of its special pair (Figure 7.8a). The electrons then enter an electron transfer chain (Section 6.4) in the thylakoid membrane.

Replacing Lost Electrons A photosystem can donate only a few electrons to electron transfer chains before it must be restocked with more. Where do the replacements come from? Photosystem II gets more electrons by pulling them off of water molecules. This reaction is so strong that it causes the water molecules to dissociate into hydrogen ions and oxygen (Figure 7.8b). The released oxygen diffuses out of the cell as O_2. The process by which any molecule becomes broken down by light energy is called **photolysis**.

Harvesting Electron Energy Light energy is converted to chemical energy when a photosystem donates electrons to an electron transfer chain (Figure 7.8c). Light does not take part in chemical reactions. Electrons do. In a series of redox reactions (Section 6.4), they pass from one molecule of the electron transfer chain to the next. With each reaction, the electrons release a bit of their extra energy.

The molecules of the electron transfer chain use the released energy to move hydrogen ions (H+) across the membrane, from the stroma to the thylakoid compartment (Figure 7.8d). Thus, the flow of electrons through electron transfer chains maintains a hydrogen ion gradient across the thylakoid membrane.

This gradient motivates hydrogen ions in the thylakoid compartment to diffuse back into the stroma. However, ions cannot simply diffuse through a lipid bilayer (Section 5.3). H+ can leave the thylakoid compartment only by flowing through membrane transport proteins called ATP synthases (Section 5.2). Hydrogen ion flow through an ATP synthase causes this protein to attach a phosphate group to ADP (Figure 7.8g,h). Thus, a hydrogen ion gradient across a thylakoid membrane drives the formation of ATP in the stroma.

Figure 7.6
Summary of the inputs and outputs of the noncyclic and cyclic light-dependent reactions of photosynthesis.

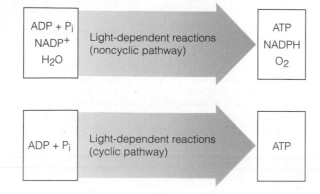

ADP + P$_i$
NADP$^+$
H$_2$O
→ Light-dependent reactions (noncyclic pathway) → ATP NADPH O$_2$

ADP + P$_i$ → Light-dependent reactions (cyclic pathway) → ATP

The Light-Dependent Reactions of Photosynthesis

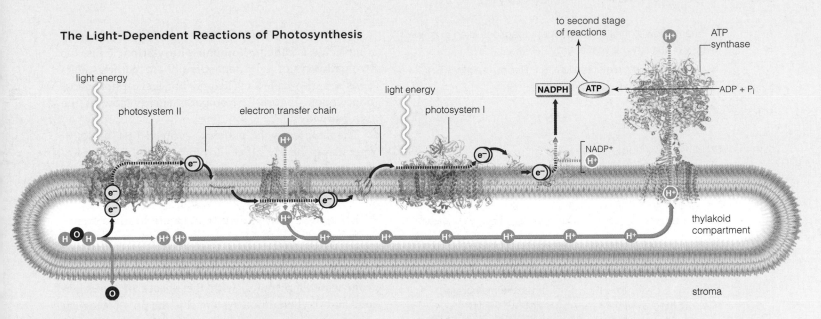

A Light energy drives electrons out of photosystem II.

B Photosystem II pulls replacement electrons from water molecules, which dissociate into oxygen and hydrogen ions (photolysis). The oxygen leaves the cell as O_2.

C Electrons from photosystem II enter an electron transfer chain.

D Energy lost by the electrons as they move through the chain causes H+ to be pumped from the stroma into the thylakoid compartment. An H+ gradient forms across the membrane.

E Light energy drives electrons out of photosystem I, which accepts replacement electrons from electron transfer chains.

F Electrons from photosystem I move through a second electron transfer chain, then combine with $NADP^+$ and H+. NADPH forms.

G Hydrogen ions in the thylakoid compartment are propelled through the interior of ATP synthases by their gradient across the thylakoid membrane.

H H+ flow causes the ATP synthases to attach phosphate to ADP, so ATP forms in the stroma.

Figure 7.8 Animated Noncyclic pathway of photosynthesis. Electrons that travel through two different electron transfer chains end up in NADPH, which delivers them to sugar-building reactions in the stroma. The cyclic pathway (not shown) uses a third type of electron transfer chain.

Accepting Electrons After electrons from a photosystem II have moved through an electron transfer chain, they are accepted by a photosystem I. The photosystem absorbs energy, and electrons pop off of its special pair of chlorophylls (Figure 7.8e). The electrons then enter a second, different electron transfer chain. At the end of this chain, $NADP^+$ accepts the electrons along with H+, so NADPH forms (Figure 7.8f):

$$NADP^+ + 2e^- + H^+ \rightarrow NADPH$$

ATP continues to form as long as electrons continue to flow through transfer chains in the thylakoid membrane. However, when NADPH is not being used, it accumulates in the stroma. The accumulation causes the noncyclic pathway to back up and stall. Then, the cyclic pathway runs independently in type I photosystems. This pathway allows the cell to continue making ATP even when the noncyclic pathway is not running.

The cyclic pathway involves photosystem I and an electron transfer chain that cycles electrons back to it.

The electron transfer chain that acts in the cyclic pathway uses electron energy to move hydrogen ions into the thylakoid compartment. The resulting hydrogen ion gradient drives ATP formation, just as it does in the noncyclic pathway. However, NADPH does not form, because electrons at the end of this chain are accepted by photosystem I, not $NADP^+$. Oxygen (O_2) does not form either, because photosystem I does not rely on photolysis to resupply itself with electrons.

Take-Home Message

What happens in the light-dependent reactions of photosynthesis?

■ In chloroplasts, ATP forms during the light-dependent reactions of photosynthesis, which may occur in a cyclic or a noncyclic pathway.

■ In the noncyclic pathway, electrons flow from water molecules, through two photosystems and two electron transfer chains, and end up in the coenzyme NADPH. This pathway releases oxygen and forms ATP.

■ In the cyclic pathway, electrons lost from photosystem I return to it after moving through an electron transfer chain. ATP forms, but NADPH does not. Oxygen is not released.

- Energy flow in the light-dependent reactions is an example of how organisms harvest energy.

- Links to Energy in metabolism 6.1, Redox reactions 6.4

Any light-driven reaction that attaches phosphate to a molecule is called **photophosphorylation**. Thus, the two pathways of light-dependent photosynthesis reactions are also called cyclic and noncyclic photophosphorylation. Figure 7.9 compares energy flow in the two pathways.

The simpler cyclic pathway evolved first, and still operates in nearly all photosynthesizers. Cyclic photophosphorylation yields ATP. No NADPH forms; no oxygen is released. Electrons lost from photosystem I are cycled back to it (Figure 7.9a).

Later, the photosynthetic machinery in some organisms became modified so that photosystem II became part of it. That modification was the beginning of a combined sequence of reactions that removes electrons from water molecules, with the release of hydrogen ions and oxygen. Photosystem II is the only biological system that is strong enough to oxidize—to pull electrons away from—water (Figure 7.9b).

Electrons that leave photosystem II do not return to it. They end up in NADPH, a powerful reducing agent (electron donor). NADPH delivers the electrons to sugar-producing reactions in the stroma.

In both cyclic and noncyclic photophosphorylation, molecules in the electron transfer chains use electron energy to shuttle H+ across the thylakoid membrane. Hydrogen ions accumulate in the thylakoid compartment, forming a gradient that powers ATP synthesis.

Today, the plasma membrane of different species of photosynthetic bacteria incorporates either type I or type II photosystems. Cyanobacteria, plants, and all photosynthetic protists use both types. Which of the two photophosphorylation pathways predominates at any given time depends on the organism's immediate metabolic demands for ATP and NADPH.

Having the alternate pathways is efficient, because cells can direct energy to producing NADPH and ATP or to producing ATP alone. NADPH accumulates when it is not being used. The excess backs up the noncyclic pathway, so the cyclic pathway predominates. The cell still makes ATP, but not NADPH. When sugar production is in high gear, NADPH is being used quickly. It does not accumulate, and the noncyclic pathway is the predominant one.

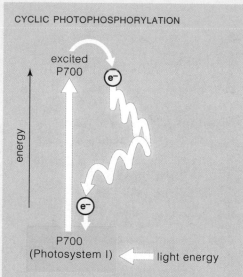

CYCLIC PHOTOPHOSPHORYLATION

A As long as electrons continue to pass through this electron transfer chain, H+ continues to be carried across the thylakoid membrane, and ATP continues to form. Light provides the energy boost that keeps the cycle going.

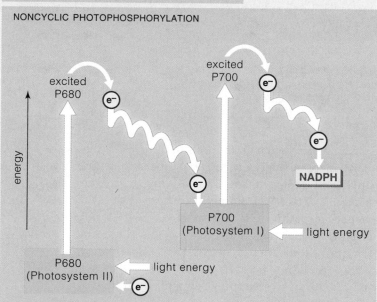

NONCYCLIC PHOTOPHOSPHORYLATION

B The noncyclic pathway is a one-way flow of electrons from water, to photosystem II, to photosystem I, to NADPH. As long as electrons continue to flow through the two electron transfer chains, H+ continues to be carried across the thylakoid membrane, and ATP and NADPH keep forming. Light provides the energy boosts that keep the pathway going.

Figure 7.9 Animated Energy flow in the light-dependent reactions of photosynthesis. The P700 in photosystem I absorbs photons of a 700-nanometer wavelength. The P680 of photosystem II absorbs photons of a 680-nanometer wavelength. Energy inputs boost P700 and P680 to an excited state in which they lose electrons.

Take-Home Message

How does energy flow in photosynthesis?

- Light provides energy inputs that keep electrons flowing through electron transfer chains. Energy lost by electrons as they flow through the chains sets up a hydrogen ion gradient that drives the synthesis of ATP alone, or ATP and NADPH.

Light-Independent Reactions: The Sugar Factory

- The cyclic, light-independent reactions of the Calvin–Benson cycle are the "synthesis" part of photosynthesis.

- Links to Carbohydrates 3.3, ATP as an energy carrier 6.2

The enzyme-mediated reactions of the **Calvin–Benson cycle** build sugars in the stroma of chloroplasts. These reactions are light-independent because light does not power them. Instead, they run on the bond energy of ATP and the reducing power of NADPH—molecules that formed in the light-dependent reactions.

The light-independent reactions build glucose from carbon dioxide (Figure 7.10). Extracting carbon atoms from an inorganic source and incorporating them into an organic molecule is called **carbon fixation**. In most plants, photosynthetic protists, and some bacteria, the enzyme **rubisco** fixes carbon by attaching CO_2 to five-carbon RuBP, or ribulose bisphosphate (Figure 7.11a).

The six-carbon intermediate that forms is unstable. It splits right away into two three-carbon molecules of PGA (phosphoglycerate). The PGAs receive a phosphate group from ATP, and hydrogen and electrons from NADPH (Figure 7.11b). Thus, two molecules of three-carbon PGAL (phosphoglyceraldehyde) form.

Glucose, remember, has six carbon atoms. To make one glucose molecule, six CO_2 must be attached to six RuBP molecules, so twelve PGAL intermediates form. Two of the PGAL combine to form a six-carbon sugar

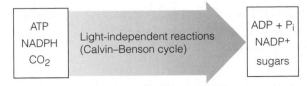

Figure 7.10 Summary of the inputs and outputs of the light-independent reactions of photosynthesis.

(Figure 7.11c). The ten remaining PGAL combine and regenerate the six RuBP (Figure 7.11d).

Plants can use the glucose they make in the light-independent reactions as building blocks for other organic molecules, or they can break it down to access the energy held in its bonds. However, most of the glucose is converted at once to sucrose or starch by other pathways that conclude the light-independent reactions. Excess glucose is stored in the form of starch grains inside the stroma of chloroplasts. When sugars are needed in other parts of the plant, the starch is broken down to sugar monomers and exported.

Take-Home Message

What happens during the light-independent reactions of photosynthesis?

- Driven by ATP energy, the light-independent reactions of photosynthesis use hydrogen and electrons (from NADPH), and carbon and oxygen (from CO_2) to build glucose and other sugars.

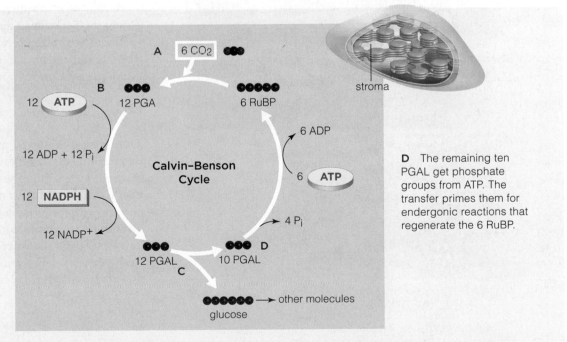

A Six CO_2 in air spaces inside of a leaf diffuse into a photosynthetic cell. Rubisco attaches each to a RuBP molecule. The resulting intermediates split, so twelve molecules of PGA form.

B Each PGA molecule gets a phosphate group from ATP, plus hydrogen and electrons from NADPH. Twelve intermediate molecules (PGAL) form.

C Two of the PGAL combine and form one molecule of glucose. The glucose may enter reactions that form other carbohydrates, such as sucrose and starch.

D The remaining ten PGAL get phosphate groups from ATP. The transfer primes them for endergonic reactions that regenerate the 6 RuBP.

Figure 7.11 Animated Light-independent reactions of photosynthesis which, in chloroplasts, occur in the stroma. The sketch is a summary of six cycles of the Calvin–Benson reactions and their product, one glucose molecule. *Black* balls signify carbon atoms. Appendix VI details the reaction steps.

- Environments differ, and so do details of photosynthesis.

- Links to Surface specializations 4.12, Controls over metabolic reactions 6.4

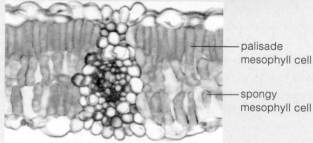

palisade mesophyll cell

spongy mesophyll cell

A C3 plant leaves. Chloroplasts are distributed evenly among two kinds of mesophyll cells in leaves of C3 plants such as basswood (*Tilia americana*). The light-dependent and light-independent reactions occur in both cell types.

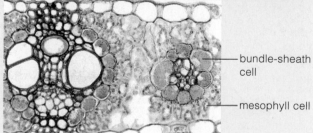

bundle-sheath cell

mesophyll cell

B C4 plant leaves. In C4 plants such as corn (*Zea mays*), carbon is fixed the first time in mesophyll cells, which are near the air spaces in the leaf, but have few chloroplasts. Specialized bundle-sheath cells ringing the leaf veins closely associate with mesophyll cells. Carbon fixation occurs for the second time in bundle-sheath cells, which are stuffed with rubisco-containing chloroplasts.

Rascally Rubisco

Plants that use only the Calvin–Benson cycle to fix carbon are called **C3 plants**, because *three*-carbon PGA is the first stable intermediate to form. Most plants use this pathway, but it can be inefficient in dry weather.

Plant surfaces that are exposed to air typically have a waxy, water-conserving cuticle (Section 4.12). They also have **stomata** (singular, stoma), which are small openings across the epidermal surfaces of leaves and green stems. Stomata close on dry days, which helps the plant minimize evaporative water loss from leaves and stems. However, like water, gases also enter and exit through stomata. When stomata are closed, CO_2 that is required for light-independent reactions cannot diffuse from air into leaves and stems, and the O_2 produced by the light-dependent reactions cannot diffuse out. Thus, when the light-dependent reactions run with stomata closed, oxygen builds up inside the plant. This buildup triggers an alternate pathway that reduces the cell's capacity to build sugars.

Remember that rubisco is the carbon-fixing enzyme of the Calvin–Benson cycle. At high O_2 levels, rubisco attaches oxygen (instead of carbon) to RuBP in a pathway called **photorespiration**. CO_2 is a product of photorespiration, so the cell loses carbon instead of fixing it. In addition, ATP and NADPH are used up to shunt the pathway's intermediates back to the Calvin–Benson cycle. So, sugar production in C3 plants becomes inefficient on dry days (Figures 7.12*a* and 7.13*a*).

Photorespiration can limit growth; C3 plants compensate for rubisco's inefficiency by making a lot of it. Rubisco is the most abundant protein on Earth.

C4 Plants

Over the past 50 to 60 million years, an additional set of reactions that compensates for rubisco's inefficiency evolved independently in many plant lineages. Plants that use the additional reactions also close stomata on dry days, but their sugar production does not decline. Examples are corn, switchgrass, and bamboo. We call these plants **C4 plants** because *four*-carbon oxaloacetate is the first stable intermediate to form in their carbon-fixation reactions (Figures 7.12*b* and 7.13*b*).

Figure 7.12 Different types of plants, different types of cells. Chloroplasts appear as green patches in the cross sections of the leaves. Purple areas are leaf veins.

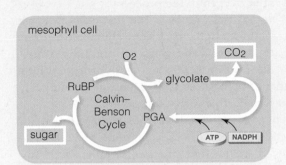

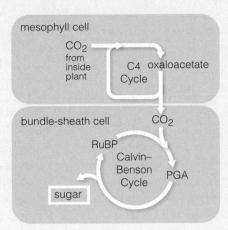

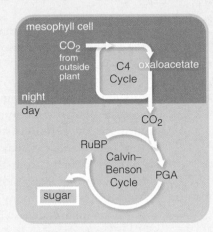

A C3 plants. On dry days, stomata close and oxygen accumulates to high concentration inside leaves. The excess causes rubisco to attach oxygen instead of carbon to RuBP. Cells lose carbon and energy as they make sugars.

B C4 plants. Oxygen also builds up inside leaves when stomata close during photosynthesis. An additional pathway in these plants keeps the CO_2 concentration high enough to prevent rubisco from using oxygen.

C CAM plants open stomata and fix carbon using a C4 pathway at night. When stomata are closed during the day, the organic compounds made during the night are converted to CO_2 that enters the Calvin–Benson cycle.

Figure 7.13 Light-independent reactions in three kinds of plants.

In C4 plants, the first set of light-independent reactions occurs in mesophyll cells. There, carbon is fixed by an enzyme that does not use oxygen even when the oxygen level is high. An intermediate is transported into bundle-sheath cells, where an ATP-requiring reaction converts it to CO_2. Rubisco fixes carbon for a second time as the CO_2 enters the Calvin–Benson cycle in the bundle-sheath cells. The C4 cycle keeps the CO_2 level near rubisco high, so it minimizes photorespiration. C4 plants use more ATP than C3 plants do, but on dry days they can make more sugar.

CAM Plants

Succulents, cactuses, and other **CAM plants** have an alternative carbon-fixing pathway that allows them to conserve water even in regions where the daytime temperatures can be extremely high. CAM stands for *Crassulacean Acid Metabolism*, after the Crassulaceae family of plants in which this pathway was first studied (Figure 7.14). Like C4 plants, CAM plants use a C4 cycle in addition to the Calvin–Benson cycle, but these two carbon-fixing cycles occur at different times rather than in different cells. The few stomata on a CAM plant open at night, when a C4 cycle fixes carbon from CO_2 in the air. The product of the cycle, a four-carbon acid, is stored in the cell's central vacuole. When the stomata close the next day, the acid moves out of the vacuole and becomes broken down to CO_2, which enters the Calvin–Benson cycle (Figure 7.13c).

Figure 7.14 A CAM plant: *Crassula argentea*, or jade plant.

Take-Home Message

How do carbon-fixing reactions vary?

■ When stomata are closed, oxygen builds up inside leaves of C3 plants. Rubisco then can attach oxygen (instead of carbon dioxide) to RuBP. This reaction, photorespiration, reduces the efficiency of sugar production, so it can limit growth.

■ Plants adapted to dry conditions limit photorespiration by fixing carbon twice. C4 plants separate the two sets of reactions in space; CAM plants separate them in time.

- The evolution of photosynthesis dramatically and permanently changed Earth's atmosphere.

- Link to Free radicals 6.3

Plants are the starting point for nearly all of the food (the carbon-based compounds) that you eat. They are **autotrophs**, or "self-nourishing" organisms. Like other autotrophs, plants can make their own food by securing energy directly from the environment, and they get their carbon from inorganic molecules (such as CO_2). Most bacteria, many protists, all fungi, and all animals are **heterotrophs**. These organisms get energy and carbon from organic molecules that have already been assembled by other organisms, for example by feeding on autotrophs, one another, or organic wastes or remains. *Hetero*– means other, as in "being nourished by others."

Plants are a kind of **photoautotroph**. By the process of photosynthesis, photoautotrophs make sugars from carbon dioxide and water using the energy of sunlight. Each year, plants collectively produce about 220

billion tons of sugar, enough to make about 300 quadrillion sugar cubes. That is a lot of sugar. They also release a lot of oxygen in the process.

It was not always this way. The first cells on Earth did not tap into sunlight. They were **chemoautotrophs** that extracted energy and carbon from simple molecules in the environment, such as hydrogen sulfide and methane. Both gases were plentiful in the nasty brew that was Earth's early atmosphere.

Ways of securing food did not change much for about a billion years. Then cyclic photophosphorylation evolved in the first photoautotrophs, and sunlight offered these organisms an essentially unlimited supply of energy. Not long afterward, the cyclic pathway became modified in some organisms. The new pathway, noncyclic photophosphorylation, split water molecules into hydrogen and oxygen. Molecular oxygen, which had previously been very rare in the atmosphere, began accumulating. From that time on, the world of life would never be the same (Figure 7.15).

The oxygen enrichment of Earth's early atmosphere exerted tremendous selection pressure on life. Oxygen reacts with metals, including enzyme cofactors, and free radicals form during those reactions. Free radicals, remember, are toxic (Section 6.3). Most early cells had no mechanism of detoxifying the oxygen radicals and became extinct. Only a few persisted in deep water, muddy sediments, and other oxygen-free habitats.

New pathways for detoxifying oxygen evolved, and one of them put oxygen's reactive properties to use. Oxygen accepts electrons at the end of electron transfer chains in these ATP-forming reactions, which are collectively called aerobic respiration.

Meanwhile, high in the ancient atmosphere, oxygen molecules were combining into ozone (O_3), a molecule that absorbs short-wavelength ultraviolet radiation in sunlight. The ozone layer that slowly formed in the upper atmosphere eventually shielded life from the sun's dangerous UV radiation. Only then did aerobic species emerge from the deep ocean, out from the mud and sediments, to diversify under the open sky.

Figure 7.15 Then and now—a view of how our atmosphere was irrevocably altered by photosynthesis. Photosynthesis is now the main pathway by which energy and carbon enter the web of life. Plants in this orchard are producing oxygen and carbon-rich parts—apples—at the Jerzy Boyz farm in Chelan, Washington.

Take-Home Message

How did photosynthesis affect Earth's atmosphere?

- The evolution of noncyclic photophosphorylation dramatically changed the oxygen content of Earth's atmosphere.

- In some organisms that survived the change, new pathways evolved for detoxifying oxygen radicals.

- Organisms did not live out in the open until after the ozone layer formed and began absorbing UV light from the sun.

7.9 | A Burning Concern

- Earth's natural atmospheric cycle of carbon dioxide is out of balance, mainly as a result of human activity.

- Link to Nutrient cycling 1.2

Have you ever wondered where all the atoms in your body came from? Think about just the carbon atoms. You eat other organisms to get the carbon atoms your body uses for energy and for raw materials. Those atoms may have passed through other heterotrophs before you ate them, but at some point they were part of photoautotrophic organisms. Photoautotrophs strip carbon from carbon dioxide, then use the atoms to build organic compounds. Your carbon atoms—and those of most other organisms—came from carbon dioxide.

Photosynthesis removes carbon dioxide from the atmosphere, and locks its carbon atoms inside organic compounds. When photosynthesizers and other aerobic organisms break down the organic compounds for energy, carbon atoms are released in the form of CO_2, which then reenters the atmosphere. Since photosynthesis evolved, these two processes have constituted a balanced cycle of the biosphere. You will learn more about the carbon cycle in Section 47.7. For now, know that the amount of carbon dioxide that photosynthesis removes from the atmosphere is roughly the same amount that organisms release back into it—at least it was, until humans came along.

As early as 8,000 years ago, humans began burning forests to clear land for agriculture. When trees and other plants burn, most of the carbon locked in their tissues is released into the atmosphere as carbon dioxide. Fires that occur naturally release carbon dioxide the same way.

Today, we are burning a lot more than our ancestors ever did. In addition to wood, we are burning fossil fuels—coal, petroleum, and natural gas—to satisfy our greater and greater demands for energy. As you will see in Section 23.5, fossil fuels are the organic remains of ancient organisms. When we burn these fuels, we release the carbon that has been locked inside of them for hundreds of millions of years back into the atmosphere—as carbon dioxide (Figure 7.16).

Researchers find pockets of our ancient atmosphere in Antarctica. Snow and ice have been accumulating in layers there, year after year, for the last 15 million years. Air and dust trapped in each layer reveal the composition of the atmosphere that prevailed when the layer formed. Thus, we now know that the atmospheric CO_2 level had been relatively stable for about 10,000 years before the industrial revolution. Since 1850, the CO_2 level has been steadily rising. In 2006, it was higher than it had been in *23 million years*.

Our activities have put Earth's atmospheric cycle of carbon dioxide out of balance. We are adding far more CO_2 to the atmosphere than photosynthetic organisms are removing from it. Today, we release around 26 billion tons of carbon dioxide into the atmosphere each year, more than ten times the amount we released in the year 1900. Most of it comes from burning fossil fuels. How do we know? Researchers can determine how long ago the carbon

Figure 7.16 Visible evidence of fossil fuel emissions in the atmosphere: the sky over New York City on a sunny day.

atoms in a sample of CO_2 were part of a living organism by measuring the ratio of different carbon isotopes in it. (You will read more about radioisotope dating techniques in Section 17.6.) These results are correlated with fossil fuel extraction, refining, and trade statistics.

The increase in atmospheric carbon dioxide is having dramatic effects on climate. CO_2 contributes to global warming, as you will read in Section 47.8. We are seeing a warming trend that mirrors the increase in CO_2 levels; Earth is now the warmest it has been for 12,000 years. The climate change is affecting biological systems everywhere. Life cycles are changing: Birds are laying eggs earlier; plants are flowering at the wrong times; mammals are hibernating for shorter periods. Migration patterns and habitats are also changing. The changes may be too fast, and many species may become extinct as a result.

Under normal circumstances, extra carbon dioxide stimulates photosynthesis, which means extra CO_2 uptake. However, the changes that we are already seeing in temperature and moisture patterns as a result of global warming are offsetting this benefit. Such changes are proving harmful to plants and other photosynthesizers.

Much research today targets development of energy sources that are not based on fossil fuels. For example, photosystem II catalyzes photolysis, the most efficient oxidation reaction in nature. Researchers are working to duplicate its catalytic function in artificial systems. If they are successful, then perhaps we too might be able to use light to split water into hydrogen, oxygen, and electrons—all of which can be used as clean sources of energy. Other research is focused on ways to remove carbon dioxide from the atmosphere—for example, by improving the efficiency of the enzyme rubisco in plants.

Corn and sugarcane are currently the top ethanol biofuel crops. These C4 plants flourish in hot, dry regions where photorespiration can limit the growth of C3 plants. They do not grow as well in areas where the growing season temperature averages less than 16°C (60°F), in part because the activity of rubisco decreases at this temperature (Section 6.3). C3 plants can compensate for the lowered activity by making more enzyme. C4 plants cannot. Their carbon-fixing cell specialization means there is less space in the

leaf for rubisco-containing chloroplasts. This space constraint limits the ability of C4 plants to make extra rubisco in cold climates.

Summary

Sections 7.1, 7.2   By metabolic pathways of **photosynthesis**, organisms capture the energy of light and use it to build sugars from water and carbon dioxide. **Pigments** such as **chlorophyll *a*** absorb visible light of particular **wavelengths** for photosynthesis.

■ *Use the animation on CengageNOW to see how Engelmann made an absorption spectrum for a photosynthetic alga.*

Section 7.3   In **chloroplasts**, the **light-dependent reactions** of photosynthesis occur at a much-folded **thylakoid membrane**, which incorporates two types of **photosystems**. The membrane forms a continuous compartment in the chloroplast's semifluid interior (**stroma**) where the **light-independent reactions** occur. The overall reactions of photosynthesis can be summarized as follows:

$$6H_2O + 6CO_2 \xrightarrow[enzymes]{light\ energy} 6O_2 + C_6H_{12}O_6$$

water · carbon dioxide · oxygen · glucose

■ *Use the animation on CengageNOW to view the sites where photosynthesis takes place.*

Sections 7.4, 7.5   Light-harvesting complexes in the thylakoid membrane absorb photons and pass the energy to photosystems, which then release electrons.

In noncyclic photophosphorylation, electrons released from photosystem II flow through an electron transfer chain. At the end of the chain, they enter photosystem I. Photon energy causes photosystem I to release electrons, which end up in NADPH. Photosystem II replaces lost electrons by pulling them from water, which then dissociates into H+ and O_2 (**photolysis**).

In cyclic **photophosphorylation**, the electrons released from photosystem I enter an electron transfer chain, then cycle back to photosystem I. NADPH does not form.

In both pathways, electrons flowing through electron transfer chains cause H+ to accumulate in the thylakoid compartment, and so a hydrogen ion gradient builds up across the thylakoid membrane. H+ flows back across the membrane through ATP synthases. This flow results in the formation of ATP in the stroma.

■ *Use the animation on CengageNOW to review pathways by which light energy is used to form ATP.*

Section 7.6   **Carbon fixation** occurs in light-independent reactions. Inside the stroma, the enzyme **rubisco** attaches

a carbon from CO_2 to RuBP to start the **Calvin–Benson cycle**. This cyclic pathway uses energy from ATP, carbon and oxygen from CO_2, and hydrogen and electrons from NADPH to make glucose.

■ *Use the animation on CengageNOW to see how glucose is produced in the light-independent reactions.*

Section 7.7   Environments differ, and so do details of sugar production in the light-independent reactions. On dry days, plants conserve water by closing their **stomata**, but O_2 from photosynthesis cannot escape. In **C3 plants**, the resulting high O_2 level in leaves causes rubisco to attach O_2 instead of CO_2 to RuBP. This pathway, called **photorespiration**, reduces the efficiency of sugar production. In **C4 plants**, carbon fixation occurs twice. The first reactions release CO_2 near rubisco, and thus limit photorespiration when stomata are closed. **CAM plants** open their stomata and fix carbon at night.

Section 7.8   **Autotrophs** make their own food using energy they get directly from the environment, and carbon from inorganic sources such as CO_2. **Heterotrophs** get energy and carbon from molecules that other organisms have already assembled.

Earth's early atmosphere held very little free oxygen, and **chemoautotrophs** were common. When the noncyclic pathway of photosynthesis evolved, oxygen released by **photoautotrophs** permanently changed the atmosphere, and was a selective force that favored evolution of new metabolic pathways, including aerobic respiration.

Section 7.9   Photoautotrophs remove CO_2 from the atmosphere; the metabolic activity of most organisms puts it back. Human activities disrupt this cycle by adding more CO_2 to the atmosphere than photoautotrophs are removing from it. The resulting imbalance is contributing to global warming.

Self-Quiz   *Answers in Appendix III*

1. Photosynthetic autotrophs use _____ from the air as a carbon source and _____ as their energy source.

2. Chlorophyll *a* absorbs mainly violet and red light, and it reflects mainly _____ light.
    a. violet and red   c. yellow
    b. green   d. blue

Data Analysis Exercise

Most corn is grown intensively in vast swaths, which means farmers who grow it use fertilizers and pesticides, both of which are often made from fossil fuels. Corn is an annual plant, and yearly harvests tend to cause runoff that depletes soil and pollutes rivers.

In 2006, David Tilman and his colleagues published the results of a 10-year study comparing the net energy output of various biofuels. The researchers grew a mixure of native perennial grasses without irrigation, fertilizer, pesticides, or herbicides, in sandy soil that was so depleted by intensive agriculture that it had been abandoned. They measured the usable energy in biofuels made from the grasses, from corn, and from soy. They also measured the energy it took to grow and produce each kind of biofuel (Figure 7.17).

1. About how much energy did ethanol produced from one hectare of corn yield? How much energy did it take to grow and produce that ethanol?

2. Which biofuel tested had the highest ratio of energy output to energy input?

3. Which of the three crops would require the least amount of land to produce a given amount of biofuel energy?

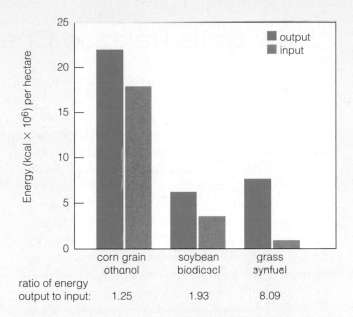

| ratio of energy output to input: | 1.25 | 1.93 | 8.09 |

Figure 7.17 Energy inputs and outputs of biofuels from different sources: corn and soy grown on fertile farmland, and grassland plants grown in infertile soil. One hectare is about 2.5 acres.

---

3. Light-dependent reactions in plants occur in the _____ .
a. thylakoid membrane   c. stroma
b. plasma membrane   d. cytoplasm

4. In the light-dependent reactions, _____ .
a. carbon dioxide is fixed   c. CO_2 accepts electrons
b. ATP forms   d. sugars form

5. What accumulates inside the thylakoid compartment during the light-dependent reactions?
a. glucose   b. RuBP   c. hydrogen ions   d. CO_2

6. When a photosystem absorbs light, _____ .
a. sugar phosphates are produced
b. electrons are transferred to ATP
c. RuBP accepts electrons
d. light-dependent reactions begin

7. Light-independent reactions proceed in the _____ .
a. cytoplasm   b. plasma membrane   c. stroma

8. The Calvin–Benson cycle starts when _____ .
a. light is available
b. carbon dioxide is attached to RuBP
c. electrons leave photosystem II

9. What substance is not part of the Calvin–Benson cycle?
a. ATP   d. PGAL
b. NADPH   e. O_2
c. RuBP   f. CO_2

10. A C3 plant absorbs a carbon radioisotope (as $^{14}CO_2$). In which stable, organic compound does the labeled carbon appear first? Which compound forms first if a C4 plant absorbs the same carbon radioisotope?

11. After noncyclic photophosphorylation evolved, its by-product, _____ , accumulated and changed the atmosphere.

12. A cat eats a bird, which ate a caterpillar that chewed on a weed. Which organisms are autotrophs? Heterotrophs?

13. Match each event with its most suitable description.
_____ PGAL formation   a. rubiscos function
_____ CO_2 fixation   b. water molecules split
_____ photolysis   c. ATP, NADPH required
_____ ATP forms; NADPH   d. electrons cycled back
     does not   to photosystem I

■ *Visit CengageNOW for additional questions.*

Critical Thinking

1. About 200 years ago, Jan Baptista van Helmont wanted to know where growing plants get the materials necessary for increases in size. He planted a tree seedling weighing 5 pounds in a barrel filled with 200 pounds of soil and then watered the tree regularly. After five years, the tree weighed 169 pounds, 3 ounces, and the soil weighed 199 pounds, 14 ounces. Because the tree had gained so much weight and the soil had lost so little, he concluded that the tree had gained all of its additional weight by absorbing the water he had added to the barrel. What really happened?

2. Only about eight classes of pigment molecules are known, but this limited group gets around. For example, photoautotrophs make carotenoids, which move through food webs, as when tiny aquatic snails graze on green algae and then flamingos eat the snails. Flamingos modify the carotenoids. Their cells split beta-carotene to form two molecules of vitamin A. This vitamin is the precursor of retinal, a visual pigment that converts light energy to electric signals in the eyes. Beta-carotene gets dissolved in fat under the skin. Cells that give rise to bright pink feathers take it up. Research another organism to identify sources for pigments that color its surfaces.

How Cells Release Chemical Energy

IMPACTS, ISSUES | When Mitochondria Spin Their Wheels

In the early 1960s, a Swedish physician, Rolf Luft, mulled over a patient's odd symptoms. The young woman felt weak and hot all the time. Even on the coldest winter days she could not stop sweating, and her skin was always flushed. She was thin, yet had a huge appetite. Luft inferred that his patient's symptoms pointed to a metabolic disorder: Her cells were very active, but much of their activity was being lost as metabolic heat. Luft checked the patient's rate of metabolism, the amount of energy her body was expending. Even while resting, her oxygen consumption was the highest that had ever been recorded! Examination of a tissue sample revealed that the patient's skeletal muscles had plenty of mitochondria, the cell's ATP-producing powerhouses. But there were too many of them, and they were abnormally shaped. The mitochondria were making very little ATP despite working at top speed.

The disorder, now called Luft's syndrome, was the first to be linked to defective mitochondria. The cells of someone with the disorder are like cities that are burning tons of coal in many power plants but not getting much usable energy output.

Skeletal and heart muscles, the brain, and other hardworking body parts with high energy demands are most affected.

More than forty disorders related to defective mitochondria are now known. One of them, Friedreich's ataxia, causes loss of coordination (ataxia), weak muscles, and serious heart problems. Many of those affected die when they are young adults (Figure 8.1).

Like the chloroplasts described in the previous chapter, mitochondria have an internal folded membrane system that allows them to make ATP. By the process of aerobic respiration, electron transfer chains in the mitochondrial membrane set up H^+ gradients that power ATP formation.

In Luft's syndrome, electron transfer chains in mitochondria work overtime, but too little ATP forms. In Friedreich's ataxia, a protein called frataxin does not work properly. This protein helps build some of the iron-containing enzymes of electron transfer chains. When it malfunctions, iron atoms that were supposed to be incorporated into the enzymes accumulate inside mitochondria instead.

Oxygen is present in mitochondria. As explained in Section 7.8, toxic free radicals form when oxygen reacts with metals. Too much iron in mitochondria means too many free radicals form, and these destroy the molecules of life faster than the cell can repair or replace them. Eventually, the mitochondria stop working, and the cell dies.

You already have a sense of how cells harvest energy in electron transfer chains: Details of the reactions vary from one type of organism to the next, but all life relies on this ATP-forming machinery. When you consider mitochondria in this chapter, remember that without them, you would not make enough ATP to even read about how they do it.

See the video! Figure 8.1 Sister, brother, and broken mitochondria. (**a**) Mitochondria are the body's ATP-producing powerhouses.

(**b**) Friedreich's ataxia, a genetic disorder, prevents mitochondria from making enough ATP. Leah (*left*) started to lose her sense of balance and coordination at age five. Six years later she was in a wheelchair; now she is diabetic and partially deaf. Her brother Joshua (*right*) could not walk by the time he was eleven, and is now blind. Both have heart problems; both had spinal fusion surgery. Special equipment allows them to attend school and work part-time. Leah is a professional model.

Key Concepts

Energy from carbohydrate breakdown

Various degradative pathways convert the chemical energy of glucose and other organic compounds to the chemical energy of ATP. Aerobic respiration yields the most ATP from each glucose molecule. In eukaryotes, it is completed inside mitochondria. **Section 8.1**

Glycolysis

Glycolysis is the first stage of aerobic respiration and of anaerobic fermentation pathways. Enzymes of glycolysis convert glucose to pyruvate. **Section 8.2**

How aerobic respiration ends

The final stages of aerobic respiration break down pyruvate to CO_2. Many coenzymes that become reduced deliver electrons and hydrogen ions to electron transfer chains. Energy released by electrons flowing through the chains is ultimately captured in ATP. Oxygen accepts electrons at the end of the chains. **Sections 8.3, 8.4**

How anaerobic pathways end

Fermentation pathways start with glycolysis. Substances other than oxygen accept electrons at the end of the pathways. Compared with aerobic respiration, the net yield of ATP from fermentation is small. **Sections 8.5, 8.6**

Other metabolic pathways

Molecules other than glucose are common energy sources. Different pathways convert lipids and proteins to substances that may enter glycolysis or the Krebs cycle. **Section 8.7**

Links to Earlier Concepts

- This chapter expands the picture of energy flow through the world of life (Section 6.1). It focuses on metabolic pathways (6.4) that make ATP (6.2) by degrading glucose and other molecules.

- The reactions of carbohydrate breakdown pathways occur either in the cytoplasm (4.2) or inside mitochondria (4.11). You may wish to review the structure of glucose and other carbohydrates (3.3).

- You will come across more examples of electron transfer chains (7.4) and reflect on the global connection between aerobic respiration and photosynthesis (7.8).

- Photoautotrophs make ATP during photosynthesis and use it to synthesize glucose and other carbohydrates.
- Most organisms, including photoautotrophs, make ATP by breaking down glucose and other organic compounds.

- Links to Energy 6.1 and 6.2, Metabolism 6.4

Organisms can stay alive only as long as they continue to resupply themselves with energy they get from their environment (Section 6.1). Plants and other photoautotrophs get their energy directly from the sun; heterotrophs eat autotrophs or one another. Regardless of its source, the energy must be converted to a form that can drive the diverse reactions necessary to sustain life. One such form is adenosine triphosphate (ATP), a common currency of energy expenditures in all cells.

Comparison of the Main Pathways

Most organisms make ATP by metabolic pathways that break down carbohydrates and other organic molecules. Some pathways are **aerobic** (they use oxygen); others are **anaerobic** (they occur in the absence of oxygen).

The main pathways by which cells harvest energy from organic molecules are called **aerobic respiration**, and anaerobic **fermentation**. Most types of eukaryotic cells either use aerobic respiration exclusively, or they use it most of the time. Many prokaryotes and protists in anaerobic habitats use alternative pathways. Some prokaryotes have their own unique version of aerobic respiration, but we focus on the pathway as it occurs in eukaryotic cells.

Fermentation and aerobic respiration begin with the same reactions in the cytoplasm. These reactions are **glycolysis**, and they convert one six-carbon molecule of glucose into two **pyruvate**, an organic molecule with a three-carbon backbone. After glycolysis, the pathways of fermentation and aerobic respiration diverge.

Aerobic respiration ends inside mitochondria, where oxygen accepts electrons at the end of electron transfer chains (Figure 8.2). Every breath you take provides your aerobically respiring cells with a fresh supply of oxygen. Fermentation ends in the cytoplasm, where a molecule other than oxygen accepts electrons at the end of the pathway.

Aerobic respiration is much more efficient than the fermentation pathways, which end with a net yield of two ATP per glucose. Aerobic respiration yields about thirty-six ATP per glucose. You and other multicelled organisms could not live without the higher yield of aerobic respiration.

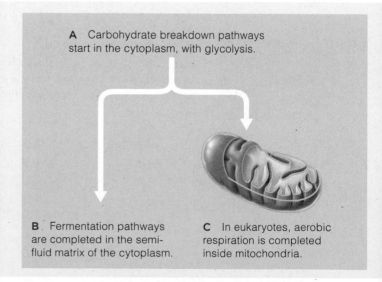

A Carbohydrate breakdown pathways start in the cytoplasm, with glycolysis.

B Fermentation pathways are completed in the semi-fluid matrix of the cytoplasm.

C In eukaryotes, aerobic respiration is completed inside mitochondria.

Figure 8.2 Animated Where the different pathways of carbohydrate breakdown start and end. Aerobic respiration alone can deliver enough ATP to sustain large multicelled organisms such as people, redwoods, and Canada geese.

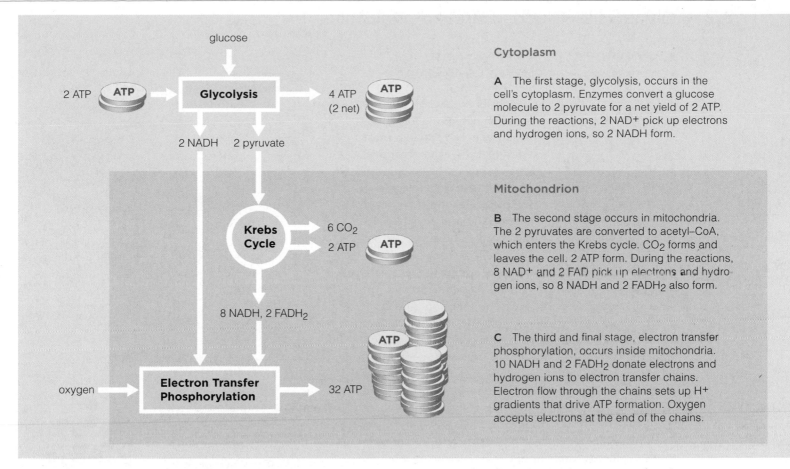

Cytoplasm

A The first stage, glycolysis, occurs in the cell's cytoplasm. Enzymes convert a glucose molecule to 2 pyruvate for a net yield of 2 ATP. During the reactions, 2 NAD+ pick up electrons and hydrogen ions, so 2 NADH form.

Mitochondrion

B The second stage occurs in mitochondria. The 2 pyruvates are converted to acetyl–CoA, which enters the Krebs cycle. CO_2 forms and leaves the cell. 2 ATP form. During the reactions, 8 NAD+ and 2 FAD pick up electrons and hydrogen ions, so 8 NADH and 2 $FADH_2$ also form.

C The third and final stage, electron transfer phosphorylation, occurs inside mitochondria. 10 NADH and 2 $FADH_2$ donate electrons and hydrogen ions to electron transfer chains. Electron flow through the chains sets up H+ gradients that drive ATP formation. Oxygen accepts electrons at the end of the chains.

Figure 8.3 Animated Overview of aerobic respiration. The reactions start in the cytoplasm and end inside mitochondria. The pathway's inputs and outputs are summarized on the *right*. **Figure It Out:** What is aerobic respiration's net yield of ATP?

Answer: 38 – 2 = 36 ATP per glucose

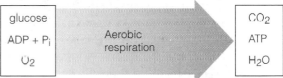

Overview of Aerobic Respiration

This equation summarizes aerobic respiration:

$$C_6H_{12}O_6 \; + \; O_2 \longrightarrow CO_2 \; + \; H_2O$$

glucose   oxygen   carbon dioxide   water

This equation only shows the substances at the start and end of the pathway, but not those at three intermediate stages (Figure 8.3). The first stage, glycolysis, converts glucose to pyruvate. During the second stage, the pyruvate is converted to acetyl–CoA, which enters the Krebs cycle. Carbon dioxide that forms in the second-stage reactions leaves the cell.

Electrons and hydrogen ions released by the reactions of the first two stages are picked up by two coenzymes, NAD+ (or nicotinamide adenine dinucleotide) and FAD (or flavin adenine dinucleotide). When these two coenzymes are carrying electrons and hydrogen, they are reduced (Section 6.4), and we refer to them as NADH and $FADH_2$.

Few ATP form during the first two stages. The big payoff occurs in the third stage after coenzymes give up electrons and hydrogen to electron transfer chains —the machinery of electron transfer phosphorylation. Operation of the transfer chains sets up hydrogen ion (H+) gradients that drive ATP formation. Oxygen in mitochondria accepts electrons and H+ at the end of the transfer chains, so water forms.

Take-Home Message

How do cells access the chemical energy in carbohydrates?

■ Most cells convert the chemical energy of carbohydrates to the chemical energy of ATP by aerobic respiration and fermentation pathways. These pathways start in the cytoplasm, with glycolysis.

■ Fermentation pathways end in the cytoplasm. They do not use oxygen. The net yield per glucose molecule is two ATP.

■ In eukaryotes, aerobic respiration ends in mitochondria. It uses oxygen, and the net yield per glucose molecule is thirty-six ATP.

8.2 Glycolysis—Glucose Breakdown Starts

- An energy investment of ATP starts glycolysis.
- Links to Hydrolysis 3.2, Glucose 3.3, Endergonic 6.2

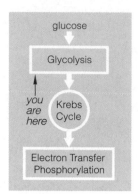

glucose

Glycolysis

you are here

Krebs Cycle

Electron Transfer Phosphorylation

Glycolysis is a series of reactions that begins carbohydrate breakdown pathways in most types of cells. The reactions, which occur in the cytoplasm, convert a glucose molecule to two pyruvates, for a net yield of two ATP and two NADH. The word "glycolysis" (from Greek *glyk–*, sweet; and *–lysis*, loosening) refers to the release of chemical energy from sugars. Different sugars can enter glycolysis, but for clarity we focus on glucose.

Glycolysis begins when a molecule of glucose enters a cell through a membrane transport protein (Section 5.2). The cell invests two ATP in the endergonic reactions that begin the pathway (Section 6.2). In the first reaction, an enzyme transfers a phosphate group from ATP to the glucose, thus forming glucose-6-phosphate (Figure 8.4a).

Unlike glucose, glucose-6-phosphate does not pass through glucose transporters in the plasma membrane, so it is trapped inside the cell. Almost all of the glucose that enters a cell is immediately converted to glucose-6-phosphate. This phosphorylation keeps the glucose concentration in the cytoplasm lower than it is in the fluid outside of the cell. By maintaining this concentration gradient across the plasma membrane, the cell favors uptake of even more glucose.

Glycolysis continues as glucose-6-phosphate accepts a phosphate group from another ATP, then splits in two (Figure 8.4b). PGAL is an abbreviation for the two three-carbon intermediates that result.

Enzymes attach a phosphate to each PGAL, forming two molecules of PGA (Figure 8.4c). In this reaction, two electrons and a hydrogen ion are transferred from each PGAL to NAD^+, so two NADH form. These reduced coenzymes will give up their cargo of electrons and hydrogen ions in reactions that follow glycolysis.

A phosphate group is transferred from each PGA to ADP, so two ATP form (Figure 8.4d). Two more ATP form when a phosphate is transferred from another pair of intermediates to two ADP (Figure 8.4e). Both reactions are **substrate-level phosphorylations**—direct transfers of a phosphate group from a substrate to ADP.

Remember, two ATP were invested to initiate the reactions of glycolysis. A total of four ATP form, so the *net* yield is two ATP per molecule of glucose that enters glycolysis (Figure 8.4f).

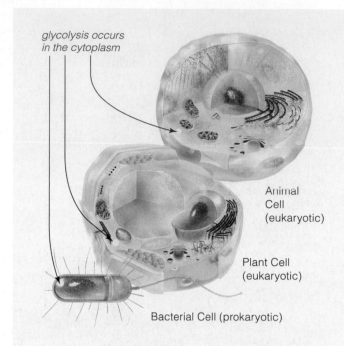

glycolysis occurs in the cytoplasm

Animal Cell (eukaryotic)

Plant Cell (eukaryotic)

Bacterial Cell (prokaryotic)

Figure 8.4 Animated Glycolysis. This first stage of carbohydrate breakdown starts and ends in the cytoplasm of prokaryotic and eukaryotic cells.

Glucose (*right*) is the reactant in the example shown on the *opposite page*; we track its six carbon atoms (*black* balls). Appendix VI shows the complete structures of the intermediates and the products.

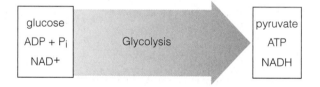

glucose
ADP + P_i
NAD^+

Glycolysis

pyruvate
ATP
NADH

Cells invest two ATP to start glycolysis, so the *net* energy yield from one glucose molecule is two ATP. Two NADH also form, and two pyruvate molecules are the end products (*above*).

Glycolysis ends with the formation of two three-carbon pyruvate molecules. These products may now enter the second stage reactions of aerobic respiration or fermentation.

Take-Home Message

What is glycolysis?

- Glycolysis is the first stage of carbohydrate breakdown in both aerobic respiration and fermentation.
- In glycolysis, one molecule of glucose is converted to two molecules of pyruvate, with a net energy yield of two ATP. Two NADH also form. The reactions occur in the cytoplasm.

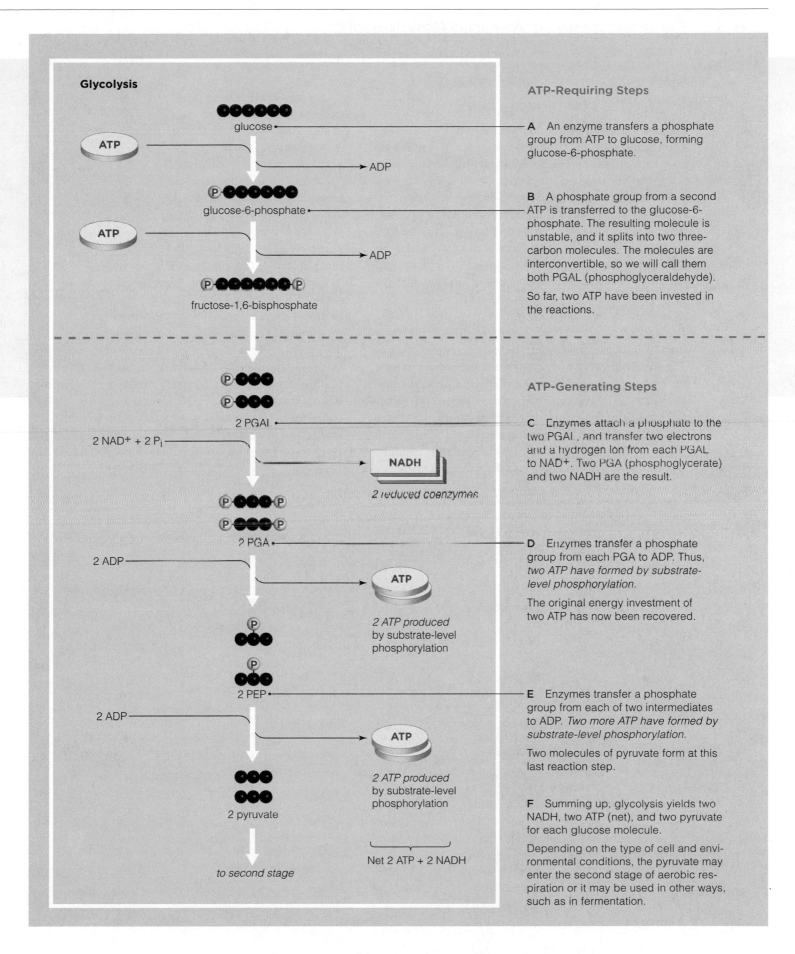

Glycolysis

ATP

glucose

ADP

ATP

glucose-6-phosphate

ADP

fructose-1,6-bisphosphate

A An enzyme transfers a phosphate group from ATP to glucose, forming glucose-6-phosphate.

B A phosphate group from a second ATP is transferred to the glucose-6-phosphate. The resulting molecule is unstable, and it splits into two three-carbon molecules. The molecules are interconvertible, so we will call them both PGAL (phosphoglyceraldehyde).

So far, two ATP have been invested in the reactions.

ATP-Generating Steps

2 PGAL

2 NAD+ + 2 Pᵢ

NADH

2 reduced coenzymes

2 PGA

2 ADP

ATP

2 ATP produced by substrate-level phosphorylation

2 PEP

2 ADP

ATP

2 ATP produced by substrate-level phosphorylation

2 pyruvate

Net 2 ATP + 2 NADH

to second stage

C Enzymes attach a phosphate to the two PGAL, and transfer two electrons and a hydrogen ion from each PGAL to NAD+. Two PGA (phosphoglycerate) and two NADH are the result.

D Enzymes transfer a phosphate group from each PGA to ADP. Thus, *two ATP have formed by substrate-level phosphorylation.*

The original energy investment of two ATP has now been recovered.

E Enzymes transfer a phosphate group from each of two intermediates to ADP. *Two more ATP have formed by substrate-level phosphorylation.*

Two molecules of pyruvate form at this last reaction step.

F Summing up, glycolysis yields two NADH, two ATP (net), and two pyruvate for each glucose molecule.

Depending on the type of cell and environmental conditions, the pyruvate may enter the second stage of aerobic respiration or it may be used in other ways, such as in fermentation.

8.3 Second Stage of Aerobic Respiration

■ The second stage of aerobic respiration finishes the breakdown of glucose that began in glycolysis.

■ Links to Mitochondria 4.11, Metabolism 6.4

The second stage of aerobic respiration occurs inside mitochondria. It includes two sets of reactions, acetyl–CoA formation and the **Krebs cycle**, that break down the pyruvate products of glycolysis (Figure 8.5). All of the carbon atoms that were once part of glucose end up in CO_2, which departs the cell. Only two ATP form. The big payoff is the formation of many reduced coenzymes that drive the third and final stage of aerobic respiration.

Acetyl–CoA Formation

The second-stage reactions start when two pyruvate formed by glycolysis enter the inner compartment of a mitochondrion. An enzyme splits each three-carbon pyruvate into one molecule of CO_2 and a two-carbon acetyl group (Figure 8.6a). The CO_2 diffuses out of the cell, and the acetyl group combines with coenzyme A (abbreviated CoA), forming acetyl–CoA. Electrons and hydrogen ions released by the reaction combine with the coenzyme NAD^+, so NADH forms.

The Krebs Cycle

The Krebs cycle breaks down acetyl–CoA to CO_2. The cycle is not a physical object, such as a wheel. It is a pathway, a sequence of enzyme-mediated reactions. It is called a cycle because the last reaction in the sequence regenerates the substrate of the first (Section 6.4). In the Krebs cycle, the substrate of the first reaction—and the product of the last—is four-carbon oxaloacetate.

Use Figure 8.6 to follow what happens during each cycle of Krebs reactions. First, the two carbon atoms of acetyl–CoA are transferred to four-carbon oxaloacetate, forming citrate, a form of citric acid (Figure 8.6b). The Krebs cycle is also called the citric acid cycle after this first intermediate. In later reactions, two CO_2 form and depart the cell; two NAD^+ accept hydrogen ions and electrons, so two NADH form (Figure 8.6c,d); ATP forms by substrate-level phosphorylation (Figure 8.6e); and FAD and another NAD^+ accept hydrogen ions and electrons (Figure 8.6f,g). The final steps of the pathway regenerate oxaloacetate (Figure 8.6h).

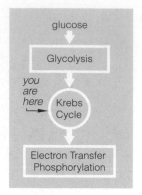

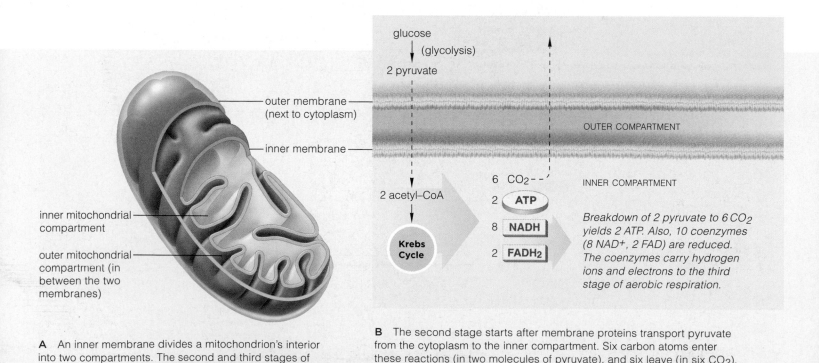

A An inner membrane divides a mitochondrion's interior into two compartments. The second and third stages of aerobic respiration take place at this membrane.

B The second stage starts after membrane proteins transport pyruvate from the cytoplasm to the inner compartment. Six carbon atoms enter these reactions (in two molecules of pyruvate), and six leave (in six CO_2). Two ATP form and ten coenzymes are reduced.

Figure 8.5 Animated Zooming in on aerobic respiration inside a mitochondrion.

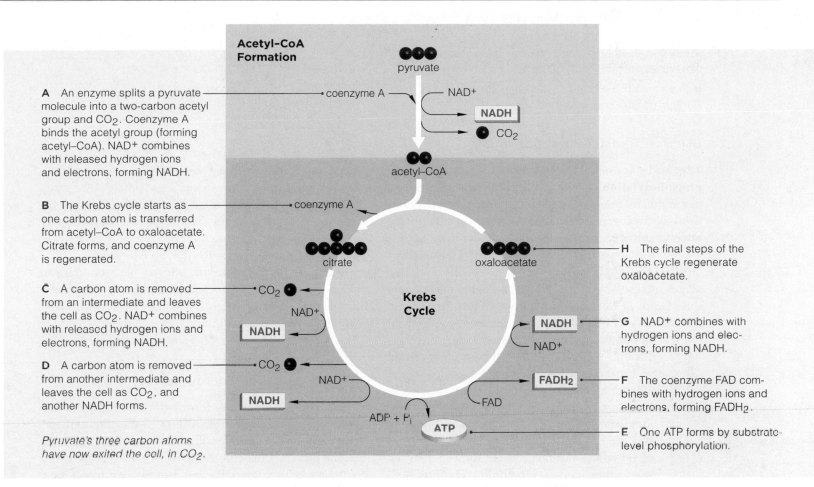

Acetyl–CoA Formation

A An enzyme splits a pyruvate molecule into a two-carbon acetyl group and CO_2. Coenzyme A binds the acetyl group (forming acetyl–CoA). NAD^+ combines with released hydrogen ions and electrons, forming NADH.

B The Krebs cycle starts as one carbon atom is transferred from acetyl–CoA to oxaloacetate. Citrate forms, and coenzyme A is regenerated.

C A carbon atom is removed from an intermediate and leaves the cell as CO_2. NAD^+ combines with released hydrogen ions and electrons, forming NADH.

D A carbon atom is removed from another intermediate and leaves the cell as CO_2, and another NADH forms.

Pyruvate's three carbon atoms have now exited the cell, in CO_2.

H The final steps of the Krebs cycle regenerate oxaloacetate.

G NAD^+ combines with hydrogen ions and electrons, forming NADH.

F The coenzyme FAD combines with hydrogen ions and electrons, forming $FADH_2$.

E One ATP forms by substrate-level phosphorylation.

pyruvate

coenzyme A · NAD^+ → NADH → CO_2

acetyl–CoA

coenzyme A

citrate · oxaloacetate

Krebs Cycle

NAD^+ → NADH

NAD^+ → NADH

NAD^+

NAD^+

FAD

$ADP + P_i$ → ATP

Figure 8.6 Animated Aerobic respiration's second stage: acetyl–CoA formation and the Krebs cycle. One set of second-stage reactions converts one pyruvate to three CO_2 for a yield of one ATP, four NADH, and one $FADH_2$ (*right*). The reactions occur in the mitochondrion's inner compartment. For simplicity, many of the pathway's intermediates are not shown. See Appendix VI for the detailed reactions.

pyruvate
$ADP + P_i$
NAD^+
FAD

Acetyl–CoA formation and the Krebs cycle

CO_2
ATP
NADH
$FADH_2$

Remember, glycolysis converted one glucose molecule to two pyruvate, and these were converted to two acetyl–CoA when they entered the inner compartment of a mitochondrion. There, the second stage reactions convert the two molecules of acetyl–CoA to six CO_2. At this point in aerobic respiration, one glucose molecule has been broken down completely: Six carbon atoms have left the cell, in six CO_2. Two ATP formed, which adds to the small net yield of glycolysis. However, six NAD^+ were reduced to six NADH, and two FAD were reduced to two $FADH_2$.

What is so important about reduced coenzymes? A molecule becomes reduced when it receives electrons (Section 6.4). Electrons carry energy that can be used to drive endergonic reactions. In this case, the electrons picked up by coenzymes during the first two stages of aerobic respiration carry energy that drives the reactions of the third stage.

In total, two ATP form and ten coenzymes (eight NAD^+ and two FAD) are reduced during acetyl–CoA formation and the Krebs cycle (the second stage of aerobic respiration). Add in the two NAD^+ reduced in glycolysis, and the full breakdown of each glucose molecule has a big potential payoff. Twelve reduced coenzymes will deliver electrons (and the energy they carry) to the third stage of aerobic respiration.

Take-Home Message

What happens during the second stage of aerobic respiration?

■ The second stage of aerobic respiration includes acetyl–CoA formation and the Krebs cycle. The reactions occur in the inner compartment of mitochondria.

■ The pyruvate that formed in glycolysis is converted to acetyl–CoA and carbon dioxide. The acetyl–CoA enters the Krebs cycle, which breaks it down to CO_2.

■ For each two pyruvates broken down in the second-stage reactions, two ATP form, and ten coenzymes (eight NAD^+ and two FAD) are reduced.

■ Many ATP are formed during the third and final stage of aerobic respiration.

■ Links to Cell membranes 5.1, Membrane proteins 5.2, Electron transfer chains 6.4

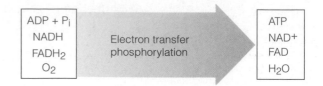

Electron Transfer Phosphorylation

The third stage of aerobic respiration, **electron transfer phosphorylation**, also occurs inside mitochondria. Its name means the flow of electrons through mitochondrial electron transfer chains ultimately results in the attachment of phosphate to ADP, which forms ATP.

The third stage begins with the coenzymes NADH and $FADH_2$, which became reduced in the first two stages of aerobic respiration. These coenzymes donate their cargo of electrons and hydrogen ions to electron transfer chains embedded in the inner mitochondrial membrane (Figure 8.7a). As the electrons pass through the chains, they give up energy little by little (Section 6.4). Some molecules of the transfer chains harness that energy to actively transport hydrogen ions from the inner mitochondrial compartment to the outer one. The ions accumulate in the outer compartment, so an H^+ concentration gradient forms across the inner mitochondrial membrane (Figure 8.7b).

This gradient attracts hydrogen ions back toward the inner mitochondrial compartment.

However, hydrogen ions cannot diffuse across a lipid bilayer without assistance. The ions can cross the inner mitochondrial membrane only by flowing through the interior of ATP synthases (Section 5.2 and Figure 8.7d). The flow causes these membrane transport proteins to attach phosphate groups to ADP, so ATP forms.

At the end of the mitochondrial electron transfer chains, oxygen accepts electrons and combines with H^+, forming water (Figure 8.7c). Aerobic respiration, which literally means "taking a breath of air," refers to oxygen as the final electron acceptor in this pathway.

Summing Up: The Energy Harvest

Thirty-two ATP typically form in the third stage of aerobic respiration. Add four ATP from the first and second stages, and the overall yield from the breakdown of one glucose molecule is thirty-six ATP (Figure 8.8).

Many factors affect the yield of aerobic respiration. For example, the two NADH from glycolysis cannot cross mitochondrial membranes; they transfer electrons

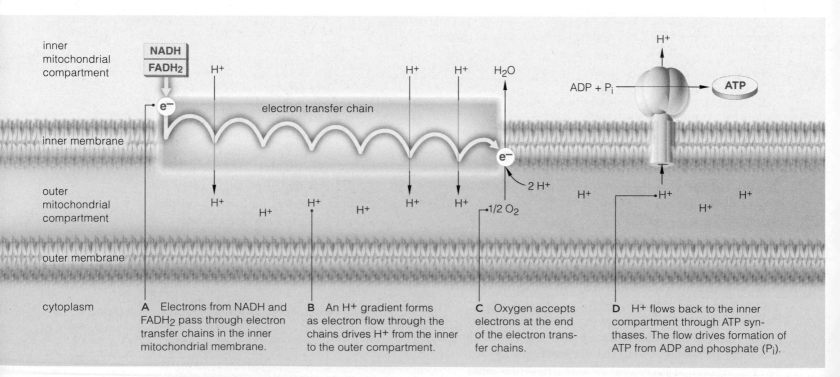

A Electrons from NADH and $FADH_2$ pass through electron transfer chains in the inner mitochondrial membrane.

B An H^+ gradient forms as electron flow through the chains drives H^+ from the inner to the outer compartment.

C Oxygen accepts electrons at the end of the electron transfer chains.

D H^+ flows back to the inner compartment through ATP synthases. The flow drives formation of ATP from ADP and phosphate (P_i).

Figure 8.7 Electron transfer phosphorylation, the third and final stage of aerobic respiration.

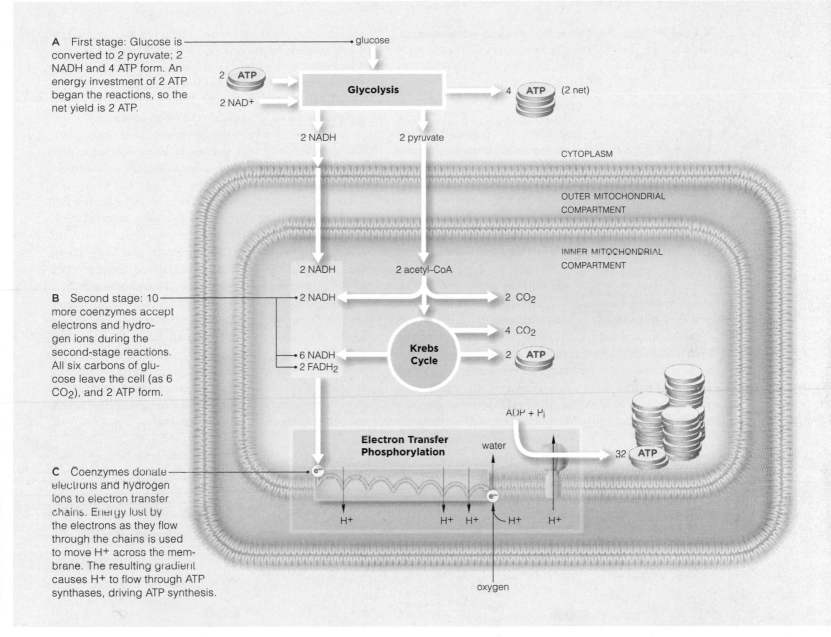

A First stage: Glucose is converted to 2 pyruvate; 2 NADH and 4 ATP form. An energy investment of 2 ATP began the reactions, so the net yield is 2 ATP.

B Second stage: 10 more coenzymes accept electrons and hydrogen ions during the second-stage reactions. All six carbons of glucose leave the cell (as 6 CO_2), and 2 ATP form.

C Coenzymes donate electrons and hydrogen ions to electron transfer chains. Energy lost by the electrons as they flow through the chains is used to move H+ across the membrane. The resulting gradient causes H+ to flow through ATP synthases, driving ATP synthesis.

glucose

2 ATP

2 NAD+

Glycolysis

4 ATP (2 net)

2 NADH   2 pyruvate

CYTOPLASM

OUTER MITOCHONDRIAL COMPARTMENT

INNER MITOCHONDRIAL COMPARTMENT

2 NADH   2 acetyl–CoA

2 NADH   2 CO_2

4 CO_2

6 NADH   **Krebs Cycle**   2 ATP
2 FADH$_2$

ADP + P$_i$

Electron Transfer Phosphorylation   water

32 ATP

H+   H+ H+   H+   H+

oxygen

Figure 8.8 Animated Summary of the steps in aerobic respiration. The typical overall yield of aerobic respiration is 36 ATP per glucose.

and hydrogen ions to molecules that can. After crossing the membranes, the intermediary molecules transfer the electrons to NAD+ or FAD in the inner compartment. This shuttling mechanism, which differs among cells, influences the ATP yield. In brain and skeletal muscle cells, the yield is thirty-eight ATP. In liver, heart, and kidney cells, it is thirty-six.

Remember that some energy dissipates with every transfer (Section 6.1). Even though aerobic respiration is a very efficient way of retrieving energy from carbohydrates, about 60 percent of the energy harvested in this pathway disperses as metabolic heat.

Take-Home Message

What is the third stage of aerobic respiration?

■ In aerobic respiration's third stage, electron transfer phosphorylation, energy released by electrons flowing through electron transfer chains is ultimately captured in the attachment of phosphate to ADP.

■ The reactions begin when coenzymes deliver electrons and hydrogen ions to electron transfer chains in the inner mitochondrial membrane.

■ Energy released by electrons as they pass through electron transfer chains pumps hydrogen ions from the inner mitochondrial compartment to the outer one. Thus, an H+ gradient forms.

■ The gradient drives H+ flow through ATP synthases, which results in ATP formation. The typical net yield of aerobic respiration is thirty-six ATP per glucose.

■ Fermentation pathways break down carbohydrates without using oxygen. The final steps in these pathways regenerate NAD+ but do not produce ATP.

Fermentation Pathways

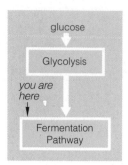

glucose

Glycolysis

you are here

Fermentation Pathway

Bacteria and single-celled protists that inhabit sea sediments, animal guts, improperly canned food, sewage treatment ponds, deep mud, and other anaerobic habitats are fermenters. Some of these organisms, including the bacteria that cause botulism, do not tolerate aerobic conditions. They will die when exposed to oxygen. Others, such as the single-celled fungi called yeasts, can switch between fermentation and aerobic respiration. Animal muscle cells can use both fermentation and aerobic respiration.

Glycolysis is the first stage of fermentation, just as it is for aerobic respiration (Figure 8.4). Again, two pyruvate, two NADH, and two ATP form in glycolysis. In the last stages of fermentation, the pyruvate is converted to other molecules, but it is not fully broken down to carbon dioxide and water. Electrons do not flow through transfer chains, so no more ATP forms. The final steps of fermentation only regenerate NAD+. Regenerating this coenzyme allows glycolysis—along with the small ATP yield it offers—to continue.

Fermentation yields enough energy to sustain many single-celled anaerobic species. It also helps some aerobic species produce ATP under anaerobic conditions.

Alcoholic Fermentation Pyruvate becomes converted to ethyl alcohol, or ethanol, in **alcoholic fermentation** (Figures 8.9a and 8.10). First, three-carbon pyruvate is split into two-carbon acetaldehyde and CO_2. Then, electrons and hydrogen are transferred from NADH to the acetaldehyde, forming NAD+ and ethanol.

Bakers mix one species of yeast, *Saccharomyces cerevisiae*, into dough. These cells break down carbohydrates in the dough, and release CO_2 in alcoholic fermentation. The dough expands (rises) as CO_2 forms bubbles in it. Some wild and cultivated strains of *Saccharomyces* are used to produce wine. Crushed grapes are left in vats along with large populations of yeast cells, which convert sugars in the juice to ethanol.

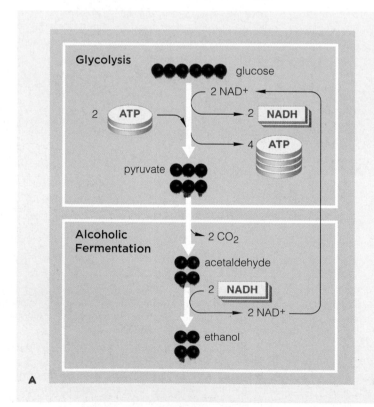

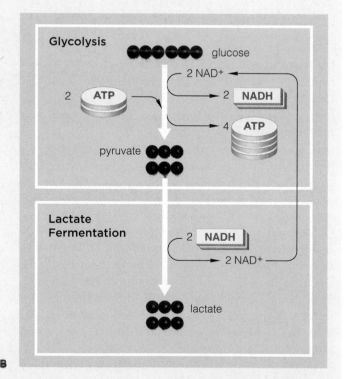

Figure 8.9 Animated (**a**) Alcoholic fermentation. (**b**) Lactate fermentation. In both pathways, the final steps do not produce ATP. They regenerate NAD+. The net yield is two ATP per molecule of glucose (from glycolysis).

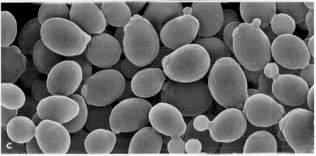

Figure 8.10 Animated Alcoholic fermentation in action.
(**a**) A vintner examines a fermentation product of *Saccharomyces*.
(**b**) A commercial vat of yeast dough rising with the help of yeast cells. (**c**) Scanning electron micrograph of yeast cells.

Lactate Fermentation In **lactate fermentation**, electrons and hydrogen ions are transferred from NADH directly to pyruvate. This reaction converts pyruvate to three-carbon lactate (lactic acid), and also converts NADH to NAD+ (Figure 8.9*b*).

Some lactate fermenters spoil food (Section 21.6), but others preserve it. For instance, we use *Lactobacillus acidophilus*, which digests lactose in milk, to ferment dairy products such as buttermilk, cheese, and yogurt. Other species of yeast ferment and preserve pickles, corned beef, sauerkraut, and kimchi.

Take-Home Message

What is fermentation?

■ ATP can form by carbohydrate breakdown in fermentation pathways, which are anaerobic.

■ The end product of lactate fermentation is lactate. The end product of alcoholic fermentation is ethanol.

■ Both pathways have a net yield of two ATP per glucose molecule. The ATP forms during glycolysis.

■ Fermentation reactions regenerate the coenzyme NAD+, without which glycolysis (and ATP production) would stop.

8.6 | The Twitchers

■ Both lactate fermentation and aerobic respiration yield ATP for muscles that partner with bones.

■ Links to Hemoglobin 3.6, Pigments 7.1

Skeletal muscles, which move bones, consist of cells fused as long fibers. The fibers differ in how they make ATP.

Slow-twitch muscle fibers have many mitochondria and produce ATP by aerobic respiration. They dominate during prolonged activity, such as long runs. Slow-twitch fibers are red because they have an abundance of myoglobin, a pigment related to hemoglobin. Myoglobin stores oxygen in muscle tissue.

Fast-twitch muscle fibers have few mitochondria and no myoglobin; they are pale. The ATP they make by lactate fermentation sustains only short bursts of activity, such as sprints or weight lifting (Figure 8.11). The pathway makes ATP quickly but not for long; it cannot support sustained activity. That is one reason chickens cannot fly very far. The flight muscles of a chicken are mostly fast-twitch fibers, which make up the "white" breast meat. Chickens fly only in short bursts, which they do to escape predators. More often, a chicken walks or runs. Its leg muscles are mostly slow-twitch muscle, the "dark meat."

Would you expect to find more light or dark breast muscles in a migratory duck? An ostrich? An albatross that can skim the ocean surface for months?

Most human muscles are a mix of fast-twitch and slow-twitch fibers, but the proportions vary among muscles and among individuals. Great sprinters tend to have more fast-twitch fibers. Great marathon runners tend to have more slow-twitch fibers. Section 33.5 offers a closer look at energy-releasing pathways in skeletal muscle.

Figure 8.11 Sprinters and lactate fermentation. The micrograph, a cross-section through a human thigh muscle, reveals two types of fibers. Light fibers sustain short, intense bursts of speed; they make ATP by lactate fermentation. Dark fibers contribute to endurance; they make ATP by aerobic respiration. They appear dark because the tissue was stained for the presence of ATP synthase.

Alternative Energy Sources in the Body

■ Pathways that break down molecules other than carbohydrates also keep organisms alive.

■ Links to Metabolism 3.2, Carbohydrates 3.3, Lipids 3.4, Proteins 3.5, Kilocalories 6.2

The Fate of Glucose at Mealtime and Between Meals

As you (and all other mammals) eat, glucose and other breakdown products of digestion are absorbed across the gut lining, and blood transports these small organic molecules throughout the body. The concentration of glucose in the bloodstream rises, and in response the pancreas (an organ) increases its rate of insulin secretion. The increase causes cells to take up glucose faster. Cells convert the glucose to glucose-6-phosphate, an intermediate of glycolysis (Figure 8.4).

When a cell takes in a lot of glucose, ATP-forming machinery goes into high gear. Unless the ATP is used quickly, its concentration rises in the cytoplasm. The high concentration of ATP causes glucose-6-phosphate to be diverted away from glycolysis and into a biosynthesis pathway that forms glycogen, a polysaccharide (Section 3.3). Liver and muscle cells especially favor the conversion of glucose to glycogen, and these cells maintain the body's largest stores of glycogen.

What happens if you eat too many carbohydrates? When the blood level of glucose gets too high, acetyl–CoA is diverted away from the Krebs cycle and into a pathway that makes fatty acids. That is why excess dietary carbohydrate ends up as fat (Table 8.1).

Between meals, the blood level of glucose declines. If the decline were not countered, that would be bad news for the brain, your body's glucose hog. At any time, your brain is taking up more than two-thirds of the freely circulating glucose. Why? Except in times of starvation, the brain's many nerve cells (neurons) use only this sugar, and they cannot store it.

The pancreas responds to low glucose levels in the blood by secreting glucagon. The hormone causes liver

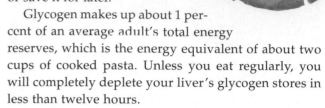

cells to convert stored glycogen to glucose. The cells release glucose into the bloodstream, so the glucose level rises, and brain cells keep working. Thus, hormones control whether cells use glucose as an energy source immediately or save it for later.

Glycogen makes up about 1 percent of an average adult's total energy reserves, which is the energy equivalent of about two cups of cooked pasta. Unless you eat regularly, you will completely deplete your liver's glycogen stores in less than twelve hours.

Energy From Fats

Of the total energy reserves in a typical adult who eats well, about 78 percent (about 10,000 kilocalories) is stored in body fat, and 21 percent in proteins.

How does a human body access its fat reservoir? A fat molecule, recall, has a glycerol head and one, two, or three fatty acid tails (Section 3.4). The body stores most fats as triglycerides, which have three fatty acid tails. Triglycerides accumulate in fat cells of adipose tissue. This tissue is an energy reservoir. It also insulates and pads the buttocks and other strategic areas of the body.

When the blood glucose level falls, triglycerides are tapped as an energy alternative. Enzymes in fat cells cleave the bonds between glycerol and the fatty acids, and both are released into the bloodstream. Enzymes in liver cells convert the glycerol to PGAL, which is an intermediate of glycolysis (Figure 8.4). Nearly all cells of your body can take up the fatty acids. Inside the cells, enzymes cleave the fatty acid backbones and convert the fragments to acetyl–CoA, which can enter the Krebs cycle (Figures 8.6 and 8.12).

Compared to carbohydrate breakdown, fatty acid breakdown yields more ATP per carbon atom. Between meals or during steady, prolonged exercise, fatty acid breakdown supplies about half of the ATP that muscle, liver, and kidney cells require.

Energy From Proteins

Some enzymes in your digestive system split dietary proteins into their amino acid subunits, which are then absorbed into the bloodstream. Cells use amino acids to build proteins or other molecules. Even so, when you eat more protein than your body needs, the amino

Table 8.1 Disposition of Organic Compounds

During meals	Excess glucose converted to glycogen or fat
Between meals	Glycogen degraded, glucose subunits enter glycolysis
	Fats degraded to fatty acids; some fragments enter glycolysis, others converted to acetyl–CoA
	Proteins degraded to amino acids, fragments become intermediates in Krebs cycle

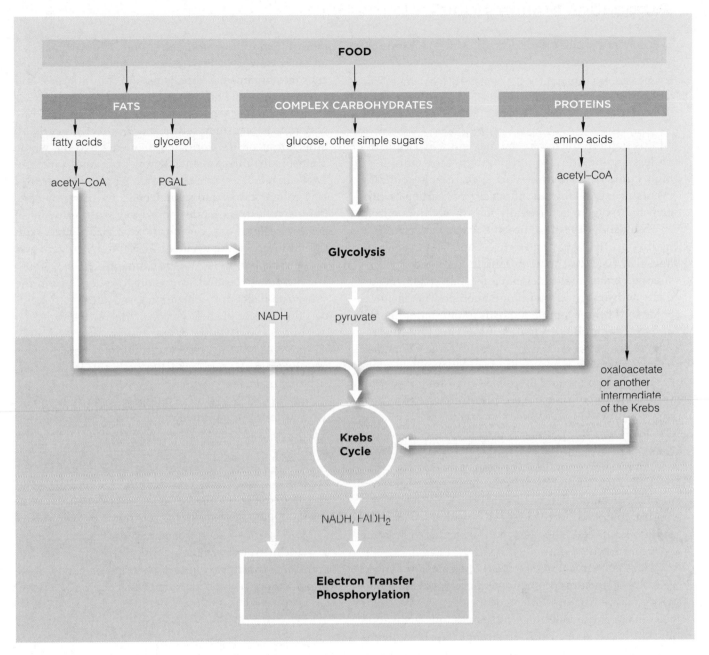

Figure 8.12 Animated Reaction sites where a variety of organic compounds enter aerobic respiration. Such compounds are alternative energy sources in the human body.

In humans and other mammals, complex carbohydrates, fats, and proteins from food do not enter the pathway of aerobic respiration directly. First, the digestive system, and then individual cells, break apart all of these molecules into simpler subunits. We return to this topic in Chapter 40.

acids become broken down further. Their NH_3^+ group is removed, and it becomes ammonia (NH_3). Their carbon backbone is split, and depending on the amino acid, acetyl–CoA, pyruvate, or an intermediate of the Krebs cycle forms. Your cells can divert any of these organic molecules into the Krebs cycle (Figure 8.12).

Maintaining and accessing energy reserves is complicated business. Controlling the use of glucose is important because it is the fuel of choice for the brain.

However, providing all of your cells with energy starts with the kinds and amounts of food you eat.

Take-Home Message

How are molecules other than glucose metabolized?

■ In humans and other mammals, the entrance of glucose or other organic compounds into an energy-releasing pathway depends on the kinds and proportions of carbohydrates, fats, and proteins in the diet.

■ Energy inputs drive the organization of molecules into units called cells.

■ Links to Life's organization 1.1, Water 2.5, Membrane organization 5.1, Selective permeability 5.3, Energy 6.1, Evolution of photosynthesis 7.8

At this point in the book, you may still have difficulty understanding the connections between yourself—a highly intelligent being—and such remote-sounding events as energy flow and the cycling of carbon, hydrogen, and oxygen. Is this really the stuff of humanity?

Think about the structure of a water molecule. Two hydrogen atoms sharing electrons with an oxygen may not seem very close to your daily life. Yet, through that sharing, water molecules have a polarity that makes them hydrogen-bond with one another. The chemical behavior of three simple atoms is a foundation for the organization of lifeless matter into living things.

Water also interacts with other molecules dispersed in it. Remember, phospholipids spontaneously organize into a two-layered film when they are mixed with water. Such lipid bilayers are the structural and functional foundation of all cell membranes.

Cells—and life—arose from such organization, but they continue by processes of metabolic control. With a membrane to contain them, metabolic reactions can proceed independently of conditions in the environment. With molecular functions built into their membranes, cells sense shifts in those conditions. Response mechanisms "tell" the cell what molecules to build or tear apart, and when to do it.

There is no mysterious force that creates proteins in cells. DNA, the double-stranded encyclopedia of inheritance, has a structure—a chemical message—that helps cells make and break down molecules, one generation after the next. Your own DNA strands tell your trillions of cells how to build proteins.

So yes, carbon, hydrogen, oxygen, and other atoms of organic molecules are the stuff of you, and us, and all of life. Yet life is more than molecules. It takes an ongoing flow of energy to turn molecules into cells, cells into organisms, organisms into communities, and so on through the biosphere (Section 1.1).

Photosynthesizers use energy from the sun and raw materials to feed themselves and, indirectly, nearly all other forms of life. Long ago they enriched the whole atmosphere with oxygen, a leftover of photosynthesis. That atmosphere favored aerobic respiration, a novel way to break down food molecules by using oxygen. Photosynthesizers made more food with leftovers of aerobic respiration—carbon dioxide and water. With this connection, the cycling of carbon, hydrogen, and oxygen through living things came full circle.

With few exceptions, infusions of energy from the sun sustain life's organization. And energy, remember, flows through the world of life in one direction (Section 6.1 and Figure 8.13). Only as long as energy lost from ecosystems is replaced with energy inputs—mainly from the sun—can life continue in all of its rich expressions.

In short, each new life is no more and no less than a marvelously complex system for prolonging order. Sustained with energy transfusions from the sun, life continues by its capacity for self-reproduction. With energy and the codes of inheritance in DNA, matter becomes organized, generation after generation. Even as individuals die, life elsewhere is prolonged. With each death, molecules are released and may be cycled as raw materials for new generations.

With this flow of energy and cycling of materials, each birth is affirmation of life's ongoing capacity for organization, each death a renewal.

Take-Home Message

What is the basis of life's unity and diversity?

■ Through biology, we have gained a profound insight into nature: Life's diversity, and its continuity, arise from unity at the level of molecules and energy.

energy in (mainly from sunlight)

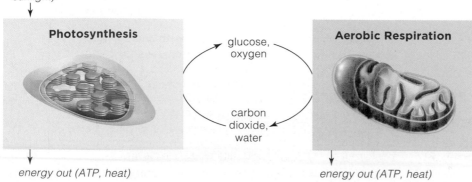

Photosynthesis

glucose, oxygen

carbon dioxide, water

Aerobic Respiration

energy out (ATP, heat)

energy out (ATP, heat)

Figure 8.13 Summary of links between photosynthesis (the main energy-requiring process) and aerobic respiration (the main energy-releasing process). Notice the one-way flow of energy.

Figure It Out: What does the middle circle represent?

Answer: The cycling of materials

When Mitochondria Spin Their Wheels

At least 83 proteins are directly involved in the electron transfer chains of electron transfer phosphorylation in mitochondria. A defect in any one of them—or in any of the thousands of other proteins used by mitochondria, such as frataxin (*right*)—can wreak havoc in the body. About one in 5,000 people suffer from a known mitochondrial disorder. New research is showing that mitochondrial defects may be involved in

many other illnesses such as diabetes, hypertension, Alzheimer's and Parkinson's disease, and even aging.

Summary

Section 8.1 Most organisms convert chemical energy of carbohydrates to the chemical energy of ATP. **Anaerobic and aerobic** pathways of carbohydrate breakdown start in the cytoplasm with the same set of reactions, **glycolysis**, which converts glucose and other sugars to **pyruvate**. **Fermentation** pathways end in the cytoplasm, do not use oxygen, and yield two ATP per molecule of glucose. Most eukaryotic cells use **aerobic respiration**, which uses oxygen and yields much more ATP than fermentation. In eukaryotes, it is completed in mitochondria.

■ *Use the animation on CengageNOW for an overview of aerobic respiration.*

Section 8.2 Enzymes of glycolysis use two ATP to convert one molecule of glucose or another six-carbon sugar to two molecules of pyruvate. In the reactions, electrons and hydrogen ions are transferred to two NAD^+, which are thereby reduced to NADH. Four ATP also form by **substrate-level phosphorylation**, the direct transfer of a phosphate group from a reaction intermediate to ADP.

The net yield of glycolysis is two pyruvate, two ATP, and two NADH per glucose molecule. The pyruvate may continue in fermentation in the cytoplasm, or it may enter mitochondria and the next steps of aerobic respiration.

■ *Use the animation on CengageNOW for a step-by-step journey through glycolysis.*

Section 8.3 The second stage of aerobic respiration, acetyl–CoA formation and the **Krebs cycle**, takes place in the inner compartment of mitochondria. The first steps convert two pyruvate from glycolysis to two acetyl–CoA and two CO_2. The acetyl–CoA enters the Krebs cycle. It takes two cycles to dismantle the two acetyl–CoA. At this stage, all of the carbon atoms in the glucose molecule that entered glycolysis have left the cell in CO_2.

During these reactions, electrons and hydrogen ions are transferred to NAD^+ and FAD, which are thereby reduced to NADH and $FADH_2$. ATP forms by substrate-level phosphorylation.

In total, the second stage of aerobic respiration results in the formation of six CO_2, two ATP, eight NADH, and two $FADH_2$ for every two pyruvates. Adding the yield from glycolysis, the total tally for the first two stages of aerobic respiration is twelve reduced coenzymes and four ATP for each glucose molecule.

■ *Use the animation on CengageNOW to explore a mitochondrion and observe the reactions inside it.*

Section 8.4 Aerobic respiration ends in mitochondria. In the third stage of reactions, **electron transfer phosphorylation**, reduced coenzymes deliver their electrons and H^+ to electron transfer chains in the inner mitochondrial membrane. Electrons moving through the chains release energy bit by bit; molecules of the chain use that energy to move H^+ from the inner to the outer compartment.

Hydrogen ions that accumulate in the outer compartment form a gradient across the inner membrane. The ions follow the gradient back to the inner compartment through ATP synthases. H^+ flow through these transport proteins drives ATP synthesis.

Oxygen combines with electrons and H^+ at the end of the transfer chains, thus forming water.

Overall, aerobic respiration typically yields thirty-six ATP for each glucose molecule.

■ *Use the animation on CengageNOW to see how each step in aerobic respiration contributes to a big energy harvest.*

Sections 8.5, 8.6 Fermentation pathways begin with glycolysis and finish in the cytoplasm. They do not use oxygen or electron transfer chains. The final steps oxidize NADH to NAD^+, which is required for glycolysis to continue, but produce no ATP. The end product of **lactate fermentation** is lactate. The end product of **alcoholic fermentation** is ethyl alcohol, or ethanol. Both pathways have a net yield of two ATP per glucose (from glycolysis).

Slow-twitch and fast-twitch skeletal muscle fibers can support different activity levels. Aerobic respiration and lactate fermentation proceed in different fibers that make up these muscles.

■ *Use the animation on CengageNOW to compare alcoholic and lactate fermentation.*

Section 8.7 In humans and other mammals, the simple sugars from carbohydrate breakdown, glycerol and fatty acids from fat breakdown, and carbon backbones of amino acids from protein breakdown may enter aerobic respiration at various reaction steps.

■ *Use the interaction on CengageNOW to follow the breakdown of different organic molecules.*

Section 8.8 The diversity and continuity of life arises from its unity at the level of molecules and energy.

Data Analysis Exercise

Tetralogy of Fallot (TF) is a genetic disorder characterized by four major malformations of the heart. The circulation of blood is abnormal, so TF patients have too little oxygen in their blood. Inadequate oxygen levels result in damaged mitochondrial membranes, which in turn cause cells to self-destruct. In 2004, Sarah Kuruvilla and her colleages looked at abnormalities in the mitochondria of heart muscle in TF patients. Some of their results are shown in Figure 8.14.

1. Which abnormality was most strongly associated with TF?

2. Can you make any correlations between blood oxygen content and mitochondrial abnormalities in these patients?

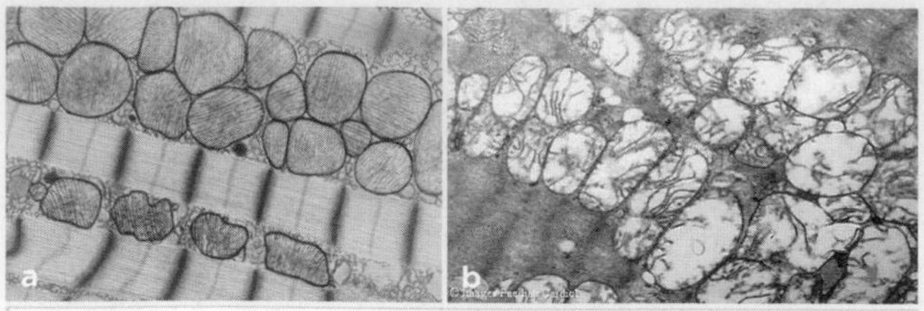

Figure 8.14 Mitochondrial changes in Tetralogy of Fallot (TF). (**a**) Normal heart muscle. Many mitochondria between the fibers provide muscle cells with ATP for contraction. (**b**) Heart muscle from a person with TF shows swollen, broken mitochondria.

(**c**) Mitochondrial abnormalities in TF patients. SPO$_2$ is oxygen saturation of the blood. A normal value of SPO$_2$ is 96%. Abnormalities seen are indicated by "+" signs.

Patient (age)	SPO$_2$ (%)	Mitochondrial Abnormalities in TF			
		Number	Shape	Size	Broken
1 (5)	55	+	+	−	−
2 (3)	69	+	+	−	−
3 (22)	72	+	+	−	−
4 (2)	74	I	+	−	−
5 (3)	76	+	+	−	+
6 (2.5)	78	+	+	−	+
7 (1)	79	+	+	−	−
8 (12)	80	+	−	+	−
9 (4)	80	+	+	−	−
10 (8)	83	+	−	+	−
11 (20)	85	+	+	−	−
12 (2.5)	89	+	−	+	−

c

Self-Quiz
Answers in Appendix III

1. Is the following statement true or false? Unlike animals, which make many ATP by aerobic respiration, plants make all of their ATP by photosynthesis.

2. Glycolysis starts and ends in the _____ .
a. nucleus
b. mitochondrion
c. plasma membrane
d. cytoplasm

3. Which of the following metabolic pathways require molecular oxygen (O$_2$)?
a. aerobic respiration
b. lactate fermentation
c. alcoholic fermentation
d. all of the above

4. Which molecule does not form during glycolysis?
a. NADH   b. pyruvate   c. FADH$_2$ d. ATP

5. In eukaryotes, aerobic respiration is completed in the _____ .
a. nucleus
b. mitochondrion
c. plasma membrane
d. cytoplasm

6. Which of the following reaction pathways is not part of the second stage of aerobic respiration?
a. electron transfer phosphorylation
b. acetyl–CoA formation
c. Krebs cycle
d. glycolysis
e. a and d

7. After the Krebs cycle runs _____ time(s), one glucose molecule has been completely oxidized.
a. one   b. two   c. three   d. six

8. In the third stage of aerobic respiration, _____ is the final acceptor of electrons from glucose.
a. water   b. hydrogen   c. oxygen   d. NADH

9. In alcoholic fermentation, _____ is the final acceptor of electrons stripped from glucose.
a. oxygen
b. pyruvate
c. acetaldehyde
d. sulfate

10. Fermentation makes no more ATP beyond the small yield from glycolysis. The remaining reactions _____ .
a. regenerate FAD
b. regenerate NAD$^+$
c. regenerate NADH
d. regenerate FADH$_2$

11. Your body cells can use _____ as an alternative energy source when glucose is in short supply.
a. fatty acids
b. glycerol
c. amino acids
d. all of the above

12. Match the event with its most suitable description.
___ glycolysis
___ fermentation
___ Krebs cycle
___ electron transfer phosphorylation

a. ATP, NADH, FADH$_2$, and CO$_2$ form
b. glucose to two pyruvates
c. NAD$^+$ regenerated, little ATP
d. H$^+$ flows through ATP synthases

■ *Visit CengageNOW for additional questions.*

Critical Thinking

1. At high altitudes, oxygen levels are low. Mountain climbers risk altitude sickness, which is characterized by shortness of breath, weakness, dizziness, and confusion. The early symptoms of cyanide poisoning resemble altitude sickness. Cyanide binds tightly to cytochrome *c* oxidase, a protein complex that is the last component of mitochondrial electron transfer chains. Cytochrome *c* oxidase with bound cyanide can no longer transfer electrons. Explain why cyanide poisoning starts with the same symptoms as altitude sickness.

2. As you learned, membranes impermeable to hydrogen ions are required for electron transfer phosphorylation. Membranes in mitochondria serve this function in eukaryotes. Prokaryotic cells do not have this organelle, but they can make ATP by electron transfer phosphorylation. How do you think they do it, given that they have no mitochondria?

II PRINCIPLES OF INHERITANCE

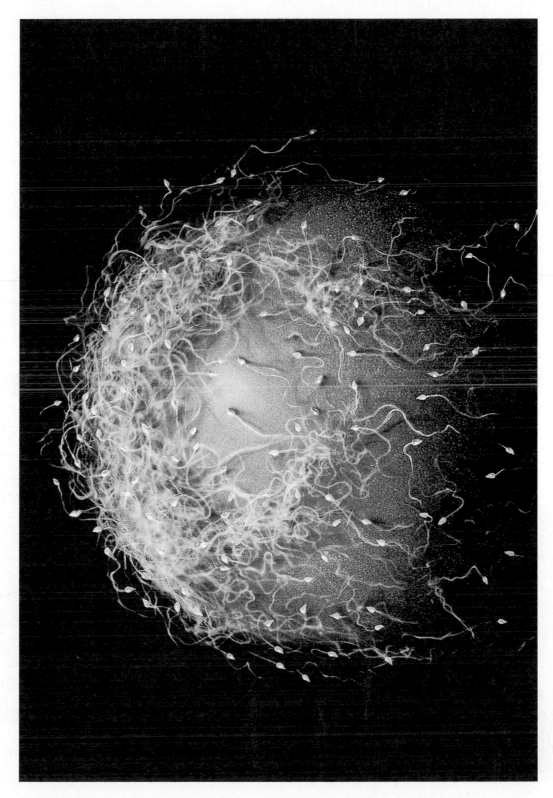

Human sperm, one of which will penetrate this mature egg and so set the stage for the development of a new individual in the image of its parents. This exquisite art is based on a scanning electron micrograph.

9 | How Cells Reproduce

IMPACTS, ISSUES | Henrietta's Immortal Cells

Each human starts out as a fertilized egg. By the time of birth, the human body consists of about a trillion cells, all descended from that single cell. Even in an adult, billions of cells divide every day and replace their damaged or aged predecessors. However, human cells cultured in the laboratory tend to divide a few times and die within weeks.

Since the mid-1800s, researchers had been trying to coax human cells to become immortal—to keep dividing outside of the body. Why? Many human diseases propagate only in human cells. Immortal cell lineages, or cell lines, would allow the researchers to study such diseases without experimenting on people. At Johns Hopkins University, George and Margaret Gey were among those researchers. They had been trying to culture human cells for almost thirty years when, in 1951, their assistant Mary Kubicek prepared one last sample of human cancer cells. Mary named the cells *HeLa*, after the first and last names of the patient from whom they were taken.

The HeLa cells began to divide. They divided again and again. Four days later, there were so many cells that the researchers had to transfer part of the population to more culture tubes. The cell populations increased at a phenomenal rate; cells were dividing every twenty-four hours and coating the inside of the tubes within days.

Sadly, cancer cells in the patient were dividing just as fast. Six months after she had been diagnosed with cancer, malignant cells had invaded tissues throughout her body.

Two months after that, Henrietta Lacks, a young woman from Baltimore, was dead.

Although Henrietta passed away, her cells lived on in the Geys' laboratory (Figure 9.1). The Geys used HeLa cells to distinguish among the viral strains that cause polio, which at the time was epidemic. They also shipped the cells to other laboratories all over the world. Researchers still use cell culture techniques developed by the Geys. They also continue to use HeLa cells to investigate cancer, viral growth, protein synthesis, the effects of radiation on cells, and more. Some HeLa cells even traveled into space for experiments on the *Discoverer XVII* satellite.

Henrietta Lacks was thirty-one, a wife and mother of four, when runaway cell divisions killed her. Decades later, her legacy continues to help humans all around the world, through her cells that are still dividing day after day.

Understanding cell division—and, ultimately, how new individuals are put together in the image of their parents—starts with answers to three questions. First, what kind of information guides inheritance? Second, how is that information copied inside a parent cell before being distributed to each of its descendant cells? Third, what kinds of mechanisms parcel out the information to descendant cells?

We will require more than one chapter to survey the nature of inheritance. In this chapter, we introduce the structures and mechanisms that cells use to reproduce.

See the video! Figure 9.1 *Left*, dividing HeLa cells—a cellular legacy of Henrietta Lacks, who was a young casualty of cancer. *Right*, Henrietta's son David holds a picture of his parents.

Key Concepts

Chromosomes and dividing cells

Individuals have a characteristic number of chromosomes in each of their cells. The chromosomes differ in length and shape, and they carry different portions of the cell's hereditary information. Division mechanisms parcel out the information into descendant cells. **Section 9.1**

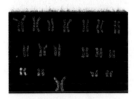

Where mitosis fits in the cell cycle

A cell cycle starts when a new cell forms by division of a parent cell, and ends when the cell completes its own division. A typical cell cycle proceeds through intervals of interphase, mitosis, and cytoplasmic division. **Section 9.2**

Stages of mitosis

Mitosis divides the nucleus, not the cytoplasm. It has four sequential stages: prophase, metaphase, anaphase, and telophase. A bipolar spindle forms and moves the cell's duplicated chromosomes into two parcels, which end up in two genetically identical nuclei. **Section 9.3**

How the cytoplasm divides

After nuclear division, the cytoplasm divides. Typically, one nucleus ends up in each of two new cells. The cytoplasm of an animal cell simply pinches in two. In plant cells, a cross-wall forms in the cytoplasm and divides it. **Section 9.4**

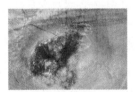

The cell cycle and cancer

Built-in mechanisms monitor and control the timing and rate of cell division. On rare occasions, the surveillance mechanisms fail, and cell division becomes uncontrollable. Tumor formation and cancer are outcomes. **Section 9.5**

Links to Earlier Concepts

- Before you start this chapter, think back on the changing appearance of chromosomes in the nucleus of eukaryotic cells (Section 4.8).

- You may also wish to review the introduction to microtubules and motor proteins (4.13). Doing so will help you understand how the mitotic spindle works.

- A review of plant cell walls (4.12) will help give you a sense of why plant cells do not divide by pinching their cytoplasm into two parcels, as animal cells do.

■ Individual cells or organisms produce offspring by the process of reproduction.

■ Link to Nucleus 4.8

When a cell reproduces, each of its cellular offspring inherits information encoded in parental DNA along with enough cytoplasm to start up its own operation. DNA contains protein-building instructions. Some of the proteins are structural materials; others are enzymes that speed construction of organic molecules. If a new cell does not inherit all of the information required to build proteins, it will not grow or function properly.

A parent cell's cytoplasm contains all the enzymes, organelles, and other metabolic machinery necessary for life. A descendant cell that inherits a blob of cytoplasm is getting start-up metabolic machinery that will keep it running until it can make its own.

Mitosis, Meiosis, and the Prokaryotes

In general, a eukaryotic cell cannot simply split in two, because only one of its descendant cells would get the nucleus—and thus, the DNA. A cell's cytoplasm splits only after its DNA has been packaged into more than one nucleus by way of mitosis or meiosis.

Mitosis is a nuclear division mechanism that occurs in the somatic cells (body cells) of multicelled eukaryotes. Mitosis and cytoplasmic division are the basis of increases in body size during development, and ongoing replacements of damaged or dead cells. Many species of plants, animals, fungi, and single-celled protists also make copies of themselves, or reproduce asexually, by mitosis (Table 9.1).

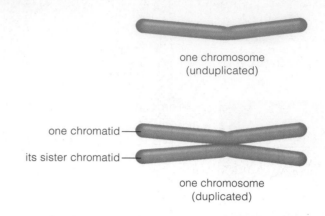

one chromosome
(unduplicated)

one chromatid ——
its sister chromatid ——

one chromosome
(duplicated)

Figure 9.2 A eukaryotic chromosome in the unduplicated state and duplicated state. Eukaryotic cells duplicate their chromosomes before mitosis or meiosis begins. After duplication, each chromosome consists of two sister chromatids.

Meiosis is a nuclear division mechanism that precedes the formation of gametes or spores, and it is the basis of sexual reproduction. In humans and all other mammals, the gametes called sperm and eggs develop from immature reproductive cells. Spores, which protect and disperse new generations, form during the life cycle of fungi, plants, and many kinds of protists.

As you will discover in this chapter and the next, meiosis and mitosis have much in common. Even so, their outcomes differ.

What about prokaryotes—bacteria and archaeans? Such cells reproduce asexually by prokaryotic fission, which is an entirely different mechanism. We consider prokaryotic fission later, in Section 21.5.

Key Points About Chromosome Structure

The genetic information of each eukaryotic species is distributed among some characteristic number of chromosomes that differ in length and shape (Section 4.8). Before a eukaryotic cell enters mitosis or meiosis, each of its chromosomes consists of one double-stranded DNA molecule (Figure 9.2). After the chromosomes are duplicated, each consists of *two* double-stranded DNA molecules. Those two molecules of DNA stay attached as a single chromosome until late in nuclear division. Until they separate, they are called **sister chromatids**.

During the early stages of mitosis and meiosis, each duplicated chromosome coils back on itself again and again, until it is highly condensed. Figure 9.3*a* shows an example of a duplicated human chromosome when it is most condensed. The structural organization of a chromosome arises from interactions between each DNA molecule and the proteins associated with it.

Table 9.1	Comparison of Cell Division Mechanisms
Mechanisms	**Functions**
Mitosis, cytoplasmic division	In all multicelled eukaryotes, the basis of: 1. Increases in body size during growth 2. Replacement of dead or worn-out cells 3. Repair of damaged tissues In single-celled and many multicelled species, also the basis of asexual reproduction
Meiosis, cytoplasmic division	In single-celled and multicelled eukaryotes, the basis of sexual reproduction; part of the processes by which gametes and sexual spores form (Chapter 10)
Prokaryotic fission	In bacteria and archaeans alone, the basis of asexual reproduction (Section 21.5)

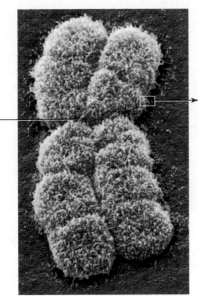

centromere

A Duplicated human chromosome in its most condensed form. If this chromosome were actually the size shown in the micrograph, its two DNA strands would stretch out about 800 meters (0.5 miles).

At regular intervals, a double-stranded DNA molecule winds twice around "spools" of proteins called **histones**. In a micrograph, these DNA–histone spools look like beads on a string (Figure 9.3d). Each "bead" is a **nucleosome**, the smallest unit of structural organization in eukaryotic chromosomes (Figure 9.3e).

When a duplicated chromosome condenses, its sister chromatids constrict where they attach to one another. This constricted region is called the **centromere** (Figure 9.3a). The location of a centromere differs for each type of chromosome. During nuclear division, a kinetochore forms at the centromere. Kinetochores are binding sites for microtubules that attach to chromatids.

What is the point of all this structural organization? It allows a huge amount of DNA to pack into a little nucleus. For example, the DNA from one of your body cells would stretch out to about 2 meters (6.5 feet)! That is a lot of DNA to pack into a nucleus that is typically less than 10 micrometers in diameter. The packing also serves a regulatory purpose. As you will see in Chapter 15, enzymes cannot access DNA that is tightly coiled.

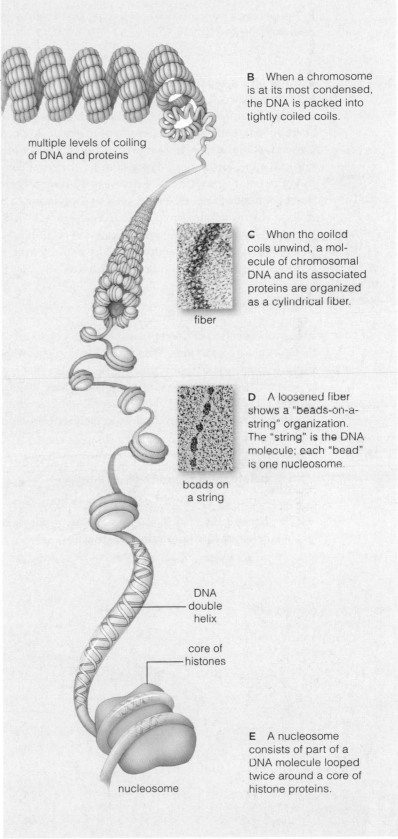

multiple levels of coiling of DNA and proteins

B When a chromosome is at its most condensed, the DNA is packed into tightly coiled coils.

fiber

C When the coiled coils unwind, a molecule of chromosomal DNA and its associated proteins are organized as a cylindrical fiber.

beads on a string

D A loosened fiber shows a "beads-on-a-string" organization. The "string" is the DNA molecule; each "bead" is one nucleosome.

DNA double helix

core of histones

nucleosome

E A nucleosome consists of part of a DNA molecule looped twice around a core of histone proteins.

Figure 9.3 Animated Chromosome structure. Several levels of structural organization allow a lot of DNA to pack very tightly into small nuclei.

Take-Home Message

What is cell division and why does it happen?

■ When a cell divides, each of its descendant cells receives a required number of chromosomes and some cytoplasm. In eukaryotic cells, the nucleus divides first, then the cytoplasm.

■ Mitosis is a nuclear division mechanism that is the basis of body size increases, cell replacements, and tissue repair in multicelled eukaryotes. Mitosis is also the basis of asexual reproduction in single-celled and some multicelled eukaryotes.

■ In eukaryotes, a nuclear division mechanism called meiosis precedes the formation of gametes and, in many species, spores. It is the basis of sexual reproduction.

Introducing the Cell Cycle

■ The cell cycle is a sequence of stages through which a cell passes during its lifetime.

■ Link to Microtubules 4.13

The life of a cell passes through a sequence of events between each cell division and the next (Figure 9.4). Interphase, mitosis, and cytoplasmic division are the stages of this **cell cycle**. The length of the cycle is about the same for all cells of the same type, but it differs from one cell type to the next. For example, the stem cells in your red bone marrow divide every 12 hours. Their descendants become red blood cells that replace 2 to 3 million worn-out ones in your blood each second. Cells in the tips of a bean plant root divide every 19 hours. In a sea urchin embryo, which develops rapidly from a fertilized egg, the cells divide every 2 hours.

The Life of a Cell

By a process called **DNA replication**, a cell copies all of its DNA before it divides. This work is completed during interphase, which is typically the longest interval of the cell cycle. **Interphase** consists of three stages, during which a cell increases its mass, roughly doubles the number of its cytoplasmic components, and replicates its DNA:

G1 Interval ("Gap") of cell growth and activity before the onset of DNA replication

S Time of "Synthesis" (DNA replication)

G2 Second interval (Gap), after DNA replication when the cell prepares for division

Gap intervals were named because outwardly they seem to be periods of inactivity. Actually, most cells going about their metabolic business are in G1. Cells preparing to divide enter S, when they copy their DNA. During G2, they make the proteins that will drive mitosis. Once S begins, DNA replication usually proceeds at a predictable rate and ends before the cell divides.

Control mechanisms work at certain points in the cell cycle. Some function as built-in brakes on the cell cycle. Apply the brakes that work in G1, and the cycle stalls in G1. Lift the brakes, and the cycle runs again.

Such controls are an important part of keeping the body functioning correctly. For example, the neurons (nerve cells) in most parts of your brain remain permanently in G1 of interphase; once they mature, they will never divide again. Experimentally driving them out of G1 causes them to die, not divide, because cells normally self-destruct if their cell cycle goes awry.

Cell suicide is is important because if controls over the cell cycle stop working, the body may be endangered. As you will see shortly, cancer begins this way. Crucial controls are lost, and the cell cycle spins out of control.

Mitosis and the Chromosome Number

After G2, a cell enters mitosis. Identical descendant cells result, each with the same number and kind of chromosomes as the parent. The **chromosome number** is the sum of all chromosomes in a cell of a given type. The body cells of gorillas have 48, those of human cells have 46. Pea plant cells have 14.

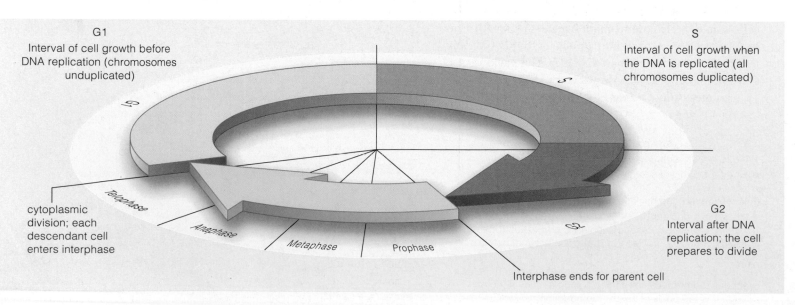

G1
Interval of cell growth before DNA replication (chromosomes unduplicated)

S
Interval of cell growth when the DNA is replicated (all chromosomes duplicated)

cytoplasmic division; each descendant cell enters interphase

Telophase

Anaphase

Metaphase

Prophase

G2
Interval after DNA replication; the cell prepares to divide

Interphase ends for parent cell

Figure 9.4 Animated Eukaryotic cell cycle. The length of each interval differs among cells.

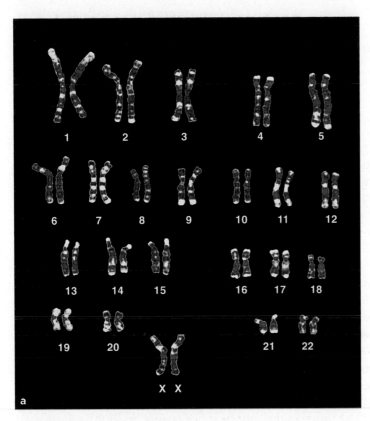

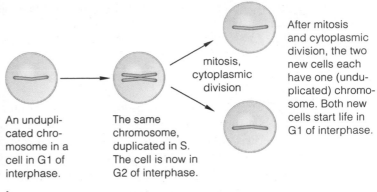

(a) An undupli-cated chro-mosome in a cell in G1 of interphase.

The same chromosome, duplicated in S. The cell is now in G2 of interphase.

mitosis, cytoplasmic division

After mitosis and cytoplasmic division, the two new cells each have one (undu-plicated) chromo-some. Both new cells start life in G1 of interphase.

b

Figure 9.5 Mitosis maintains the chromosome number.

(a) Human body cells are diploid—they have twenty-three pairs of chro-mosomes, for a total of forty-six. The last ones in this lineup of human chromosomes are a pair of sex chromosomes: Females have two X chro-mosomes; males have one X and one Y.

(b) What happens to each one of the forty-six chromosomes? Each time a human body cell undergoes mitosis and cytoplasmic division, its descend-dant cells end up with a complete set of forty-six chromosomes.

Figure It Out: Were the chromosomes in (a) taken from the cell of a male or a female? *Answer: Female*

Actually, human body cells have two of each type of chromosome: Their chromosome number is **diploid** (2n). The 46 are like two sets of books numbered 1 to 23 (Figure 9.5a). You have two volumes of each: a pair. Except for a pairing of sex chromosomes (XY) in males, members of each pair are the same length and shape, and they hold information about the same traits.

Think of them as two sets of books on how to build a house. Your father gave you one set. Your mother had her own ideas about wiring, plumbing, and so on. She gave you an alternate set that says slightly different things about many of those tasks.

With mitosis followed by cytoplasmic division, a diploid parent cell produces two diploid descendant cells. It is not just that each new cell gets forty-six or forty-eight or fourteen chromosomes. If only the total mattered, then one cell might get, say, two pairs of chromosome 22 and no pairs whatsoever of chro-mosome 9. Neither cell could function like its parent without two of each type of chromosome.

As the next section explains, a dynamic network of microtubules called the **bipolar spindle** forms dur-ing nuclear division. The spindle grows into the cyto-plasm from opposite ends, or poles, of the cell. During mitosis, some of the spindle's microtubules attach to the duplicated chromosomes. Microtubules from one pole connect to one chromatid of each chromosome; microtubules from the other pole connect to its sister:

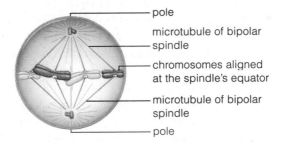

pole

microtubule of bipolar spindle

chromosomes aligned at the spindle's equator

microtubule of bipolar spindle

pole

The microtubules separate sister chromatids and move them to opposite ends of the cell. Two parcels of chro-mosomes form, and a nuclear membrane forms around each. The cytoplasm divides, and two new cells are the result. Figure 9.5b shows a preview of how mitosis maintains the parental chromosome number.

Take-Home Message

What is a cell cycle?

■ A cell cycle is a sequence of stages (interphase, mitosis, and cytoplasmic division) through which a cell passes during its lifetime.

■ During interphase, a new cell increases its mass, doubles the number of its cytoplasmic components, and duplicates its chromosomes. The cycle ends after the cell undergoes mitosis and then divides its cytoplasm.

- When a nucleus divides by mitosis, each new nucleus has the same chromosome number as the parent cell.
- There are four main stages of mitosis: prophase, metaphase, anaphase, and telophase.
- Link to Microtubules and motor proteins 4.13

A cell duplicates its chromosomes during interphase, so by the time mitosis begins, each chromosome consists of two sister chromatids joined at the centromere. During the first stage of mitosis, **prophase**, the chromosomes condense and become visible in micrographs (Figure 9.6a,b). "Mitosis" is from the Greek *mitos*, or thread, for the threadlike appearance of chromosomes during the nuclear division process.

Most animal cells have a centrosome, a region near the nucleus that organizes spindle microtubules while they are forming. The centrosome usually includes two barrel-shaped centrioles, and it is duplicated just before prophase begins. In prophase, one of the two centrosomes (along with its pair of centrioles) moves to the opposite pole of the nucleus. Microtubules that will form the bipolar spindle begin to grow from both centrosomes. (Plant cells do not have centrosomes, but they have other structures that guide spindle growth.) Motor proteins traveling along the microtubules help the spindle grow in the appropriate directions. Motor protein movement is driven by ATP (Section 4.13).

As prophase ends, the nuclear envelope breaks up and spindle microtubules penetrate the nuclear region (Figure 9.6c). Some microtubules from each spindle pole stop growing after they overlap in the middle of the cell. Others continue to grow until they reach a chromosome and attach to it.

One chromatid of each chromosome is tethered by microtubules extending from one spindle pole, and its sister chromatid is tethered by microtubules extending from the other spindle pole. The opposing sets of microtubules begin a tug-of-war by adding and losing tubulin subunits. As the microtubules grow and shrink, they push and pull the chromosomes. Soon, all the microtubules are the same length. At that point, they have aligned the chromosomes midway between the spindle poles (Figure 9.6d). The alignment marks **metaphase** (from ancient Greek *meta*, between).

Anaphase is the interval when sister chromatids of each chromosome separate and move toward opposite spindle poles (Figure 9.6e). Three cell activities bring this about. First, the spindle microtubules attached to each chromatid shorten. Second, motor proteins drag the chromatids along shrinking microtubules toward each spindle pole. Third, the microtubules that overlap midway between spindle poles begin to slide past one

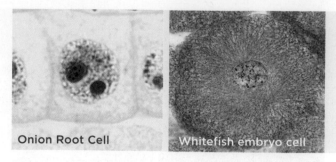

Onion Root Cell — **Whitefish embryo cell**

Figure 9.6 Animated Mitosis.

Opposite page, micrographs of onion root (plant) cells are shown on the *left*; whitefish embryo (animal) cells on the *right*.

The drawings show a diploid (2n) animal cell. For clarity, only two pairs of chromosomes are illustrated, but cells of nearly all eukaryotes have more than two. The two chromosomes of the pair inherited from one parent are coded *purple*; the two chromosomes inherited from the other parent are coded *blue*.

Above, interphase cells are shown for comparison, but interphase is not part of mitosis.

another. Motor proteins drive this movement, which pushes the spindle poles farther apart. Anaphase ends as each chromosome and its duplicate are heading to opposite spindle poles.

Telophase begins when the two clusters of chromosomes reach the spindle poles. Each cluster consists of the parental complement of chromosomes—two of each, if the parent cell was diploid. Vesicles derived from the old nuclear envelope fuse in patches around the clusters as the chromosomes decondense. Patch joins with patch until each set of chromosomes is enclosed by a new nuclear envelope. Thus, two nuclei form (Figure 9.6f). The parent cell in our example was diploid, so each new nucleus is diploid too. Once two nuclei have formed, telophase is over, and so is mitosis.

Take-Home Message

What happens during mitosis?

- Each chromosome in a cell's nucleus was duplicated before mitosis begins, so each consists of two sister chromatids.
- In prophase, chromosomes condense and microtubules form a bipolar spindle. The nuclear envelope breaks up. Some of the microtubules attach to the chromosomes.
- At metaphase, all duplicated chromosomes are aligned midway between the spindle poles.
- In anaphase, microtubules separate the sister chromatids of each chromosome, and pull them to opposite spindle poles.
- In telophase, two clusters of chromosomes reach the spindle poles. A new nuclear envelope forms around each cluster.
- Thus two new nuclei form. Each one has the same chromosome number as the parent cell's nucleus.

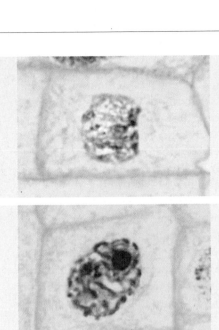

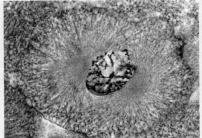

A Early Prophase

Mitosis begins. In the nucleus, the chromatin begins to appear grainy as it organizes and condenses. The centrosome is duplicated.

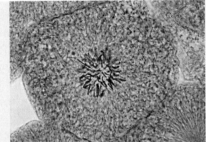

B Prophase

The chromosomes become visible as discrete structures as they condense further. Microtubules assemble and move one of the two centrosomes to the opposite side of the nucleus, and the nuclear envelope breaks up.

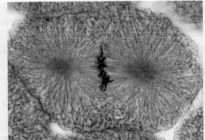

C Transition to Metaphase

The nuclear envelope is gone, and the chrompsomes are at their most condensed. Microtubules of the bipolar spindle assemble and attach sister chromatids to opposite spindle poles.

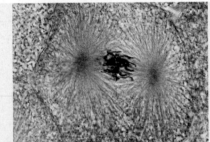

D Metaphase

All of the chromosomes are aligned midway between the spindle poles. Microtubules attach each chromatid to one of the spindle poles, and its sister to the opposite pole.

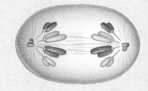

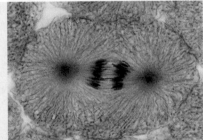

E Anaphase

Motor proteins moving along spindle microtubules drag the chromatids toward the spindle poles, and the sister chromatids separate. Each sister chromatid is now a separate chromosome.

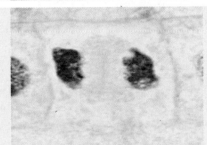

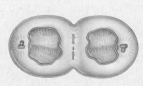

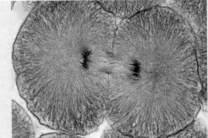

F Telophase

The chromosomes reach the spindle poles and decondense. A nuclear envelope begins to form around each cluster; new plasma membrane may assemble between them. Mitosis is over.

- In most kinds of eukaryotes, the cell cytoplasm divides between late anaphase and the end of telophase, but the mechanism of division differs.

- Links to Primary wall 4.12, Cytoskeleton 4.13

Division of Animal Cells

A cell's cytoplasm usually divides after mitosis. The process of cytoplasmic division, or **cytokinesis**, differs among eukaryotes. Typical animal cells partition their cytoplasm by pinching in two. The plasma membrane starts to sink inward as a thin indentation between the former spindle poles (Figure 9.7a). The indentation is called a cleavage furrow, and it is the first visible sign that the cytoplasm is dividing. The furrow advances until it extends around the cell. As it does, it deepens along a plane that corresponds to the former spindle equator (midway between the poles).

What is happening? The cell cortex, which is the mesh of cytoskeletal elements just under the plasma membrane, includes a band of actin and myosin filaments that wraps around the cell's midsection. ATP hydrolysis causes these filaments to interact, just as it does in muscle cells, and the interaction results in contraction. The band of filaments, which is called a **contractile ring**, is anchored to the plasma membrane. As it shrinks, the band drags the plasma membrane

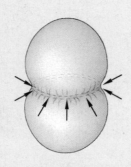
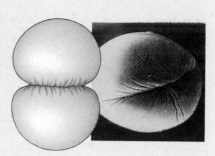
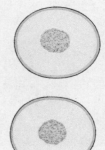

1 Mitosis is completed, and the bipolar spindle is starting to disassemble.

2 At the former spindle equator, a ring of actin filaments attached to the plasma membrane contracts.

3 This contractile ring pulls the cell surface inward as it continues to contract.

4 The contractile ring contracts until the cytoplasm is partitioned and the cell pinches in two.

A   Contractile Ring Formation

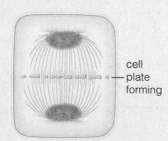
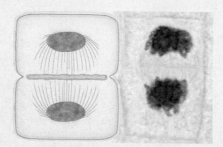

cell plate forming

1 The plane of division (and of the future cross-wall) was established by microtubules and actin filaments that formed and broke up before mitosis began. Vesicles cluster here when mitosis ends.

2 The vesicles fuse with each other and with endocytic vesicles bringing cell wall components and plasma membrane proteins from the cell surface. The fused materials form a cell plate along the plane of division.

3 The cell plate expands outward along the plane of division until it reaches the plasma membrane. When the cell plate attaches to the plasma membrane, it partitions the cytoplasm.

4 The cell plate matures as two new primary cell walls surrounding middle lamella material. The new walls join with the parent cell wall, so each daughter cell becomes enclosed by its own wall.

B   Cell Plate Formation

Figure 9.7 Animated Cytoplasmic division of an animal cell (**a**) and a plant cell (**b**).

inward until the cytoplasm (and the cell) is pinched in two (Figure 9.7a). Two new cells form this way. Each has a nucleus and some of the parent cell's cytoplasm, and each is enclosed in its own plasma membrane.

Division of Plant Cells

Dividing plant cells face a particular challenge. Unlike most animal cells, plant cells remain attached to one another and organized in tissues during development. Thus, plant growth occurs mainly in the direction of cell division, and the orientation of each cell's division is critical to the architecture of the plant.

Accordingly, plants have an extra step in cytokinesis. Microtubules under a plant cell's plasma membrane help orient the cellulose fibers in the cell wall. Before prophase, these microtubules disassemble, and then reassemble in a dense band around the nucleus along the future plane of division. The band, which also includes actin filaments, disappears as microtubules of the bipolar spindle form. An actin-depleted zone is left behind. The zone marks the plane in which cytoplasmic division will occur (Figure 9.7b).

The contractile ring mechanism that works for animal cells would not work for a plant cell. The contractile force of microfilaments is not strong enough to pinch through plant cell walls, which are stiff with cellulose and often lignin.

By the end of anaphase in a plant cell, a set of short microtubules has formed on either side of the division plane. These microtubules now guide vesicles derived from Golgi bodies and the cell's surface to the division plane. There, the vesicles and their wall-building contents start to fuse into a disk-shaped **cell plate**.

The plate grows outward until its edges reach the plasma membrane. It attaches to the membrane, and so partitions the cytoplasm. In time, the cell plate will develop into a primary cell wall that merges with the parent cell's wall. Thus, by the end of division, each of the descendant cells will be enclosed by its own plasma membrane and its own cell wall.

Appreciate the Process!

Take a moment to visualize the cells making up your palms, thumbs, and fingers. Now imagine the mitotic divisions that produced the many generations of cells before them while you were developing inside your mother (Figure 9.8). Be grateful for the precision of the mechanisms that led to the formation of your body parts at the right times, in the proper places. Why? An individual's survival depends on proper timing and

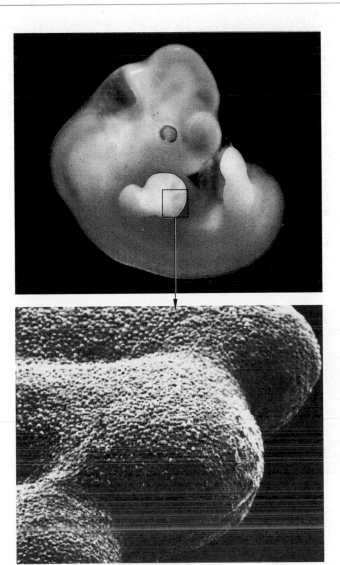

Figure 9.8 The paddlelike structure of a human embryo that develops into a hand by mitosis, cytoplasmic divisions, and other processes. The scanning electron micrograph reveals individual cells.

completion of cell cycle events. If a cell's cycle goes awry, the cell may begin to divide in an uncontrolled manner. Such unchecked divisions can destroy tissues and, ultimately, the individual.

Take-Home Message

How do cells divide?

■ After mitosis, the cytoplasm of the parent cell typically is partitioned into two descendant cells, each with its own nucleus. The process of cytoplasmic division, cytokinesis, differs among different kinds of eukaryotic cells.

■ In animal cells, a contractile ring partitions the cytoplasm. A band of actin filaments that rings the cell midsection contracts and pinches the cytoplasm in two.

■ In plant cells, a cell plate that forms midway between the spindle poles partitions the cytoplasm when it reaches and connects to the parent cell wall.

■ On rare occasions, controls over cell division are lost. Cancer may be the outcome.

■ Links to Receptors 5.2, Enzymes and free radicals 6.3, Ultraviolet radiation 7.1

The Cell Cycle Revisited Every second, millions of cells in your skin, bone marrow, gut lining, liver, and elsewhere are dividing and replacing their worn-out, dead, and dying predecessors. They do not divide at random. Many mechanisms control DNA replication and when cell division begins and ends.

What happens when something goes wrong? Suppose sister chromatids do not separate as they should during mitosis. As a result, one descendant cell ends up with too many chromosomes and the other with too few. Or sup-

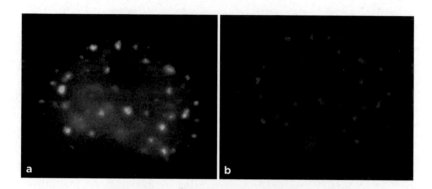

a   b

Figure 9.9 Protein products of checkpoint genes in action. A form of radiation damaged the DNA inside this nucleus. (**a**) *Green* dots pinpoint the location of a protein called *53BP1*, and (**b**) *red* dots pinpoint the location of another protein, *BRCA1*. Both proteins have clustered around the same chromosome breaks in the same nucleus. The integrated action of these proteins and others blocks mitosis until the DNA breaks are fixed.

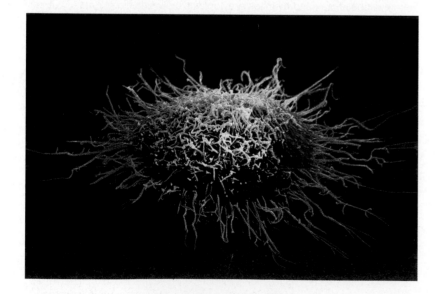

Figure 9.10 Scanning electron micrograph of the surface of a cervical cancer cell, the kind of malignant cell that killed Henrietta Lacks.

pose DNA gets damaged when a chromosome is being duplicated. A cell's DNA can also be damaged by free radicals (Section 6.3), chemicals, or environmental assaults such as ultraviolet radiation (Section 7.1). Such problems are frequent and inevitable, but a cell may not function properly unless they are countered quickly.

The cell cycle has built-in checkpoints that allow problems to be corrected before the cycle advances. Certain proteins, the products of checkpoint genes, can monitor whether a cell's DNA has been copied completely, whether it is damaged, and even whether nutrient concentrations are sufficient to support cell growth. Such proteins interact to advance, delay, or stop the cell cycle (Figure 9.9).

For example, some checkpoint gene products are kinases. This class of enzymes can activate other molecules by transferring a phosphate group to them. When DNA is broken or incomplete, the kinases activate certain proteins in a cascade of signaling events that ultimately stops the cell cycle or causes the cell to die.

As another example, checkpoint gene products called **growth factors** activate genes that stimulate cells to grow and divide. One kind, an epidermal growth factor, activates a kinase by binding to receptors on target cells in epithelial tissues. The binding is a signal to start mitosis.

Checkpoint Failure and Tumors Sometimes a checkpoint gene mutates so that its protein product no longer works properly. The result may be that the cell skips interphase, and division occurs over and over with no resting period. Or, DNA that has been damaged may be copied and packaged into descendant cells. In still other cases, the mutation alters signaling mechanisms that make an abnormal cell commit suicide (you will read more about apoptosis, the cellular self-destruct mechanism, in Section 27.6).

When all checkpoint mechanisms fail, a cell loses control over its cell cycle. The cell's descendants may form a **tumor**—an abnormal mass—in the surrounding tissue (Figures 9.10–9.12).

Mutated checkpoint genes are associated with an increased risk of tumor formation, and sometimes they run in families. Usually, one or more checkpoint gene products are missing in tumor cells. Checkpoint gene products that inhibit mitosis are called tumor suppressors because tumors form when they are missing. Checkpoint genes encoding proteins that stimulate mitosis are called proto-oncogenes (from the Greek *onkos*, or tumor); mutations that alter their products or the rate at which they are made can transform a normal cell into a tumor cell.

Moles and other tumors are **neoplasms**—abnormal masses of cells that lost control over how they grow and divide. Ordinary skin moles are among the noncancerous, or benign, neoplasms. They grow very slowly, and their cells retain the surface recognition proteins that keep them in their home tissue (Figure 9.11). Unless a benign neoplasm grows too large or becomes irritating, it poses no threat to the body.

Characteristics of Cancer A malignant neoplasm is one that is dangerous to health. **Cancer** occurs when the abnormally dividing cells of a malignant neoplasm disrupt body tissues, physically and metabolically. These typically disfigured cells can break loose from home tissues, slip into and out of blood vessels and lymph vessels, and invade other tissues where they do not belong (Figure 9.11). Cancer cells typically display the following three characteristics:

First, cancer cells grow and divide abnormally. Controls that usually keep cells from getting overcrowded in tissues are lost, so cancer cell populations may reach extremely high densities. The number of small blood vessels, or capillaries, that transport blood to the growing cell mass also increases abnormally.

Second, cancer cells often have an altered plasma membrane and cytoplasm. The membrane may be leaky and have altered or missing proteins. The cytoskeleton may be shrunken, disorganized, or both. The balance of metabolism is often shifted, as in an amplified reliance on ATP formation by fermentation rather than by aerobic respiration.

Third, cancer cells often have a weakened capacity for adhesion. Because their recognition proteins are altered or lost, they do not necessarily stay anchored in their proper tissues, and may break away and establish colonies in distant tissues. Metastasis is the name for this process of abnormal cell migration and tissue invasion.

Unless chemotherapy, surgery, or another procedure eradicates them, cancer cells can put an individual on a painful road to death. Each year, cancers cause 15 to 20 percent of all human deaths in developed countries alone. Cancers are not just a human problem. They are known to occur in most of the animal species studied to date.

Cancer is a multistep process. Researchers already know about many mutations that contribute to it. They are working to identify drugs that target and destroy cancer cells or stop them from dividing.

HeLa cells, for instance, were used in early tests of taxol, a drug that keeps microtubules from disassembling and so hampers mitosis. Frequent divisions of cancer cells make them more vulnerable to this poison than normal cells. Such research may yield drugs that put the brakes on cancer. We return to this topic in later chapters.

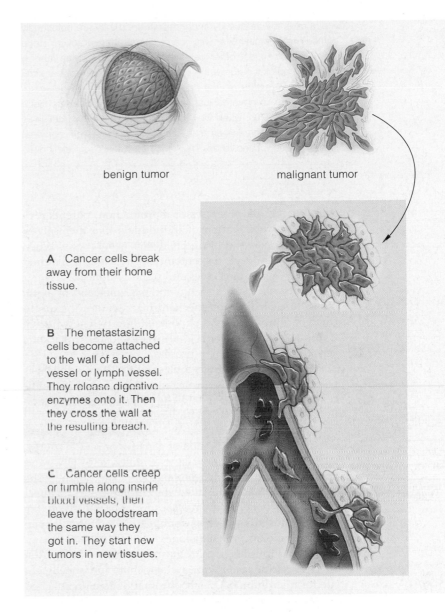

benign tumor   malignant tumor

A Cancer cells break away from their home tissue.

B The metastasizing cells become attached to the wall of a blood vessel or lymph vessel. They release digestive enzymes onto it. Then they cross the wall at the resulting breach.

C Cancer cells creep or tumble along inside blood vessels, then leave the bloodstream the same way they got in. They start new tumors in new tissues.

Figure 9.11 Animated Comparison of benign and malignant tumors. Benign tumors typically are slow-growing and stay in their home tissue. Cells of a malignant tumor migrate abnormally through the body and establish colonies even in distant tissues.

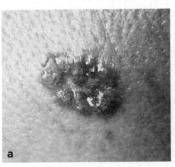
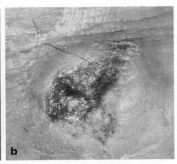
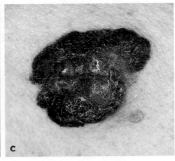

Figure 9.12 Skin cancers. (**a**) A basal cell carcinoma is the most common type. This slow-growing, raised lump is typically uncolored, reddish-brown, or black.

(**b**) The second most common form of skin cancer is a squamous cell carcinoma. This pink growth, firm to the touch, grows fast under the surface of skin exposed to sun.

(**c**) Malignant melanoma spreads fastest. Cells form dark, encrusted lumps. They may itch like an insect bite or bleed easily.

HeLa cells divide quickly and indefinitely, so they are notoriously difficult to contain. Even with careful laboratory practice, HeLa cells tend to infest other cell lines grown in the same laboratory, and quickly outgrow the other cells. Most cells appear similar in tissue culture, so the contamination may not be detected. Researchers discovered just how easy it is to propagate HeLa cells in the 1970s, when they found out that dozens of cell lines from various sources—as many as one in three—were not what they were supposed to be. The lines had been completely over-

How would you vote?

HeLa cells are sold worldwide by cell culture firms. Should the family of Henrietta Lacks (*right*) share in the profits of those sales? See CengageNOW for details, then vote online.

grown by HeLa cells. The finding undermined the significance of decades of research that had relied on the contaminated lines.

Summary

Section 9.1 By processes of reproduction, parents produce a new generation of individuals like themselves. Cell division is the bridge between generations. When a cell divides, each descendant cell receives a required number of DNA molecules and some cytoplasm.

Eukaryotic cells undergo mitosis, meiosis, or both. These nuclear division mechanisms partition the duplicated chromosomes of a parent cell into two new nuclei. Cytoplasm divides by a separate mechanism. Prokaryotic cells divide by a different process.

Mitosis followed by cytoplasmic division is the basis of growth, cell replacements, and tissue repair in multicelled species, and also the basis of asexual reproduction in many single-celled and multicelled species.

Meiosis, the basis of sexual reproduction in eukaryotes, precedes the formation of gametes or sexual spores.

A eukaryotic chromosome is a molecule of DNA and many **histones** and other proteins associated with it. The proteins structurally organize the chromosome and affect access to its genes. The smallest unit of organization, the **nucleosome**, is a stretch of double-stranded DNA looped twice around a spool of histones.

When duplicated, a chromosome consists of two **sister chromatids**, each with a kinetochore (an attachment site for microtubules). Sister chromatids remain attached at their **centromere** until late in mitosis (or meiosis).

- *Use the animation on CengageNOW to explore the structural organization of chromosomes.*

Section 9.2 A **cell cycle** starts when a new cell forms, runs through interphase, and ends when that cell reproduces by nuclear and cytoplasmic division. Most of a cell's activities occur in **interphase**, when it grows, roughly doubles the number of its cytoplasmic components, then duplicates its chromosomes.

Chromosome number is the sum of all chromosomes in cells of a specified type. For example, the chromosome number of human body cells is 46. These cells have two of each kind of chromosome, so they are **diploid**.

- *Use the interaction on CengageNOW to investigate the stages of the cell cycle.*

Section 9.3 Mitosis is a nuclear division mechanism that maintains the chromosome number. It proceeds in these four sequential stages:

Prophase. Duplicated chromosomes start to condense. Microtubules assemble and form a **bipolar spindle**, and the nuclear envelope breaks up. Some microtubules that extend from one spindle pole harness one chromatid of each chromosome; some that extend from the opposite spindle pole tether its sister chromatid. Other microtubules extend from both poles and grow until they overlap at the spindle's midpoint.

Metaphase. All chromosomes are aligned at the spindle's midpoint.

Anaphase. The sister chromatids of each chromosome detach from each other, and the spindle microtubules start moving them toward opposite spindle poles. Microtubules that overlap at the spindle's midpoint slide past each other, pushing the poles farther apart. Motor proteins drive all of these movements.

Telophase. A cluster of chromosomes that consists of a complete set of chromosomes reaches each spindle pole. A nuclear envelope forms around each cluster, forming two new nuclei. Both nuclei have the parental chromosome number.

- *Use the animation on CengageNOW to see how mitosis proceeds.*

Section 9.4 Most cells divide in two after their nucleus divides. Mechanisms of **cytokinesis**, or cytoplasmic division differ. In animal cells, a **contractile ring** of microfilaments that is part of the cell cortex pulls the plasma membrane inward until the cytoplasm is pinched in two. In plant cells, a band of microtubules and microfilaments that forms around the nucleus before mitosis marks the site at which a **cell plate** forms. The cell plate expands until it becomes a cross-wall, which partitions the cytoplasm when it fuses to the parent cell wall.

- *Compare the cytoplasmic division of plant and animal cells with the animation on CengageNOW.*

Section 9.5 Checkpoint gene products such as **growth factors** control the cell cycle. Mutated checkpoint genes can cause **tumors** (**neoplasms**) by disrupting the normal controls. **Cancer** is a multistep process involving altered cells that grow and divide abnormally. Cancer cells may metastasize, or break loose and colonize distant tissues.

- *Use the animation on CengageNOW to see how cancers spread through the body.*

Data Analysis Exercise

Despite their notorious ability to contaminate other cell lines, HeLa cells continue to be an extremely useful tool in cancer research. One early finding was that HeLa cells vary in chromosome number. The panel of chromosomes in Figure 9.13, originally published in 1989, shows all of the chromosomes in a single metaphase HeLa cell.

1. What is the chromosome number of this HeLa cell?

2. How many extra chromosomes do these cells have, compared to normal human cells?

3. Can you tell these cells came from a female? How?

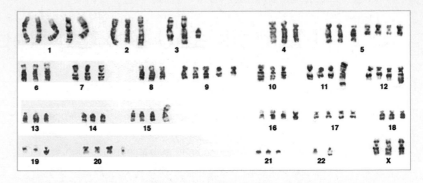

Figure 9.13 Chromosomes of a HeLa cell line.

Self-Quiz

Answers in Appendix III

1. Mitosis and cytoplasmic division function in _____ .
   a. asexual reproduction of single-celled eukaryotes
   b. growth and tissue repair in multicelled species
   c. gamete formation in prokaryotes
   d. both a and b

2. A duplicated chromosome has _____ chromatid(s).
   a. one   b. two   c. three   d. four

3. The basic unit that structurally organizes a eukaryotic chromosome is the _____ .
   a. higher order coiling   c. nucleosome
   b. bipolar mitotic spindle   d. microfilament

4. The chromosome number is _____ .
   a. the sum of all chromosomes in a cell of a given type
   b. an identifiable feature of each species
   c. maintained by mitosis
   d. all of the above

5. A somatic cell having two of each type of chromosome has a(n) _____ chromosome number.
   a. diploid   b. haploid   c. tetraploid   d. abnormal

6. Interphase is the part of the cell cycle when _____ .
   a. a cell ceases to function
   b. a cell forms its spindle apparatus
   c. a cell grows and duplicates its DNA
   d. mitosis proceeds

7. After mitosis, the chromosome number of the two new cells is _____ the parent cell's.
   a. the same as   c. rearranged compared to
   b. one-half of   d. doubled compared to

8. Name the intervals in the diagram of mitosis *below*.

9. Only _____ is not a stage of mitosis.
   a. prophase   b. interphase   c. metaphase   d. anaphase

10. Which of the following is a subset of the other two?
    a. cancer   b. neoplasm   c. tumor

11. Name one type of checkpoint gene product.

12. Match each stage with the events listed.
    ___metaphase   a. sister chromatids move apart
    ___prophase   b. chromosomes start to condense
    ___telophase   c. new nuclei form
    ___anaphase   d. all duplicated chromosomes are
                         aligned at the spindle equator

■ *Visit CengageNOW for additional questions.*

Critical Thinking

1. The anticancer drug taxol was first isolated from Pacific yews (*Taxus brevifolia*), which are slow-growing trees (*right*). Bark from about six yew trees provided enough taxol to treat one patient, but removing the bark killed the trees. Fortunately, taxol is now produced using plant cells that grow in big vats rather than in trees. What challenges do you think had to be overcome to get plant cells to grow and divide in laboratories?

2. Suppose you have a way to measure the amount of DNA in one cell during the cell cycle. You first measure the amount at the G1 phase. At what points in the rest of the cycle will you see a change in the amount of DNA per cell?

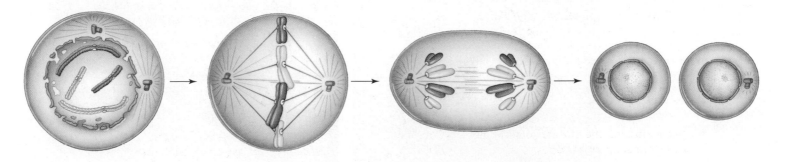

10 | Meiosis and Sexual Reproduction

IMPACTS, ISSUES Why Sex?

If the function of reproduction is the perpetuation of one's genetic material, then an asexual reproducer would seem to win the evolutionary race. In asexual reproduction, all of an individual's genetic information is passed to all of its off-spring. Sexual reproduction mixes up genetic information from two parents (Figure 10.1), so only about half of each parent's genetic information is passed to offspring.

So why sex? Variation in forms and combinations of heritable traits is typical of sexually reproducing populations. Remember from Section 1.4 that some forms of traits are more adaptive than others to conditions in the environment. If those conditions change, some of the diverse offspring of sexual reproducers may have forms of traits or combinations of them that help these individuals to survive the change. All offspring of asexual reproducers are adapted the same way to the environment—and equally vulnerable to changes in it.

Other organisms are part of the environment, and they, too, can change. Think of predator and prey—say, foxes and rabbits. If one rabbit is better than others at outrunning the foxes, it has a better chance of escaping, surviving, and passing on the genetic basis for its evasive ability to offspring. Thus, over many generations, the rabbits may get faster. If one fox is better than others at outrunning the faster rabbits, it has a

better chance of eating, surviving, and passing on the genetic basis for its predatory ability to offspring. Thus, over many generations, the foxes may tend to get faster. As one species changes, so does the other—an idea called the Red Queen hypothesis, after Lewis Carroll's book *Through the Looking Glass*. In the book, the Queen of Hearts tells Alice, "It takes all the running you can do, to keep in the same place."

An adaptive trait tends to spread more quickly through a sexually reproducing population than through an asexually reproducing one. Why? In asexual reproduction, new combinations of traits can arise only by mutation. An adaptive trait is perpetuated along with the same set of other traits—adaptive or not—until another mutation occurs. By contrast, sexual reproduction mixes up the genetic information of individuals that often have different forms of traits. It brings together adaptive traits, and separates adaptive traits from maladaptive ones, in far fewer generations than does mutation alone.

However, having a faster pace of achieving genetically diverse populations doesn't mean that sexual reproduction wins the evolutionary race. In terms of numbers of individuals and how long their lineages have endured, the most success-ful organisms on Earth are bacteria, which reproduce most often by just copying their DNA and dividing.

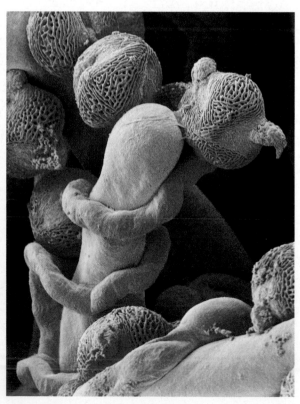

See the video!
Figure 10.1 Moments in the stages of sexual reproduction, a process that mixes up the genetic material of two organisms.

The photo on the *right* shows pollen grains (*orange*) germinating on flower carpels (*yellow*). Pollen tubes with male gametes inside are grow-ing from the grains down into tissues of the ovary, which house the flower's female gametes.

Key Concepts

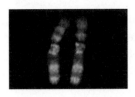

Sexual versus asexual reproduction

In asexual reproduction, one parent transmits its genetic information (DNA) to offspring. In sexual reproduction, offspring inherit DNA from two parents who usually differ in some number of alleles. Alleles are different forms of the same gene. **Section 10.1**

Stages of meiosis

Meiosis reduces the chromosome number. It occurs only in cells set aside for sexual reproduction. Meiosis sorts a reproductive cell's chromosomes into four haploid nuclei. **Sections 10.2, 10.3**

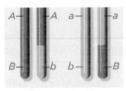

Chromosome recombinations and shufflings

During meiosis, each pair of maternal and paternal chromosomes swaps segments. Then, each chromosome is randomly segregated into one of the new nuclei. Both processes lead to novel combinations of alleles—and traits—among offspring. **Section 10.4**

Sexual reproduction in context of life cycles

Gametes form by different mechanisms in males and females. In most plants, spore formation and other events intervene between meiosis and gamete formation. **Section 10.5**

Mitosis and meiosis compared

Meiosis may have originated by evolutionary remodeling of mechanisms that already existed for mitosis and, before that, for repairing damaged DNA. **Section 10.6**

Links to Earlier Concepts

- This chapter returns to the concept of reproduction introduced in Section 1.2. Here, we detail the cellular basis of sexual reproduction, and begin to explore the far-reaching effects of gene shufflings—a process that introduces variations in traits among offspring (1.4).

- You will be revisiting the microtubules that move chromosomes (4.13, 9.3). Be sure you have a clear picture of the structural organization of chromosomes (9.1) and understand chromosome number (9.2).

- You will also draw on your understanding of cytoplasmic division (9.4) and checkpoint gene products (9.5) that monitor and repair chromosomal DNA during the cell cycle.

How would you vote? Japanese researchers successfully created a "fatherless" mouse with genetic material from the eggs of two females. The mouse is healthy and fertile. Should researchers be prevented from trying the same process with human eggs? See CengageNOW for details, then vote online.

10.1 | Introducing Alleles

■ Asexual reproduction produces genetically identical copies of a parent. By contrast, sexual reproduction introduces variation in the combinations of traits among offspring.

■ Link to Diversity 1.4

Each species has a unique set of **genes**: regions in DNA that encode information about traits. An individual's genes collectively contain the information necessary to make a new individual. With **asexual reproduction**, one parent produces offspring, so all of its offspring inherit the same number and kinds of genes. Mutations aside, then, all offspring of asexual reproduction are genetically identical copies of the parent, or **clones**.

Inheritance is far more complicated with **sexual reproduction**, the process involving meiosis, formation of mature reproductive cells, and fertilization. In most sexually reproducing multicelled eukaryotes, the first cell of a new individual has pairs of genes, on pairs of chromosomes. Typically, one chromosome of each pair is maternal and the other is paternal (Figure 10.2).

If the information in every gene of a pair were identical, then sexual reproduction would also produce clones. Just imagine: The entire human population might consist of clones, in which case everybody would look alike. But the two genes of a pair are often not identical. Why not? Inevitably, mutations accumulate in genes and permanently alter the information they carry. Thus, the two genes of any pair might "say" slightly different things about a particular trait. Each different form of a gene is called an **allele**.

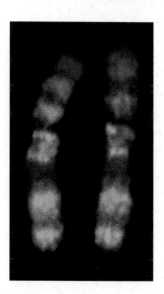

Figure 10.2 A maternal and a paternal chromosome pair. They appear identical in this micrograph, but any gene on one might differ slightly from its partner on the other.

Alleles influence differences in thousands of traits. For instance, whether your chin has a dimple in it or not depends on which alleles you inherited at one chromosome location. One allele says "dimple in the chin." A different allele says "no dimple in the chin." Alleles are one reason individuals of a sexually-reproducing species do not all look alike. The offspring of sexual reproducers inherit new combinations of alleles, which is the basis of new combinations of traits.

Take-Home Message

How does sexual reproduction introduce variation in traits?

■ Alleles are the basis of traits. Sexual reproduction bestows novel combinations of alleles—thus novel combinations of traits—on offspring.

10.2 | What Meiosis Does

■ Meiosis is a nuclear division mechanism that precedes cytoplasmic division of immature reproductive cells. It occurs only in sexually reproducing eukaryotic species.

■ Link to Chromosome number 9.2

Remember, the chromosome number is the total number of chromosomes in a cell of a given type (Section 9.2). A diploid cell has two copies of every chromosome; typically, one of each type was inherited from each of two parents. Except for a pairing of nonidentical sex chromosomes, the chromosomes of a pair are **homologous**, which means they have the same length, shape, and collection of genes (*hom*– means alike).

Mitosis maintains the chromosome number. **Meiosis**, a different nuclear division process, halves the chromosome number. Meiosis occurs in the immature reproductive cells—**germ cells**—of multicelled eukaryotes that reproduce sexually. In animals, meiosis of germ cells results in mature reproductive structures called **gametes**. (Plants have a slightly different process that we will discuss later.) A sperm cell is a type of male gamete; an egg is a type of female gamete. Gametes usually form inside special reproductive structures or organs (Figure 10.3).

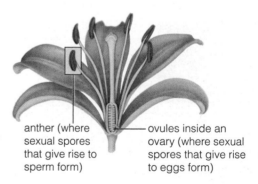

anther (where sexual spores that give rise to sperm form)

ovules inside an ovary (where sexual spores that give rise to eggs form)

a Flowering plant

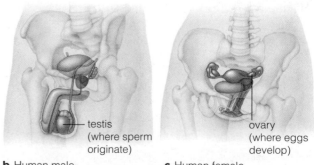

testis (where sperm originate)

ovary (where eggs develop)

b Human male   **c** Human female

Figure 10.3 Examples of reproductive organs, where cells that give rise to gametes originate.

Gametes have a single set of chromosomes, so they are **haploid** (*n*): their chromosome number is half of the diploid (2*n*) number. Human body cells are diploid, with 23 pairs of homologous chromosomes (Figure 10.4). Meiosis of a human germ cell normally produces gametes with 23 chromosomes: one of each pair. The diploid chromosome number is restored at fertilization, when two haploid gametes (one egg and one sperm) fuse to form a **zygote**, the first cell of a new individual.

Two Divisions, Not One

Meiosis is similar to mitosis in certain respects. A cell duplicates its DNA before the division process starts. The two DNA molecules and associated proteins stay attached at the centromere. For as long as they remain attached, they are sister chromatids (Section 9.1):

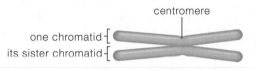

one chromosome in the duplicated state

As in mitosis, the microtubules of a spindle move the chromosomes to opposite poles of the cell. However, meiosis sorts the chromosomes into new nuclei twice. Two consecutive nuclear divisions form four haploid nuclei. There is typically no interphase between the two divisions, which are called meiosis I and II:

Interphase	Meiosis I	Meiosis II
DNA is replicated prior to meiosis I	Prophase I Metaphase I Anaphase I Telophase I	Prophase II Metaphase II Anaphase II Telophase II

In meoisis I, every duplicated chromosome aligns with its partner, homologue to homologue. After they are sorted and arranged this way, each homologous chromosome is pulled away from its partner:

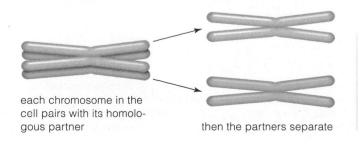

each chromosome in the cell pairs with its homologous partner

then the partners separate

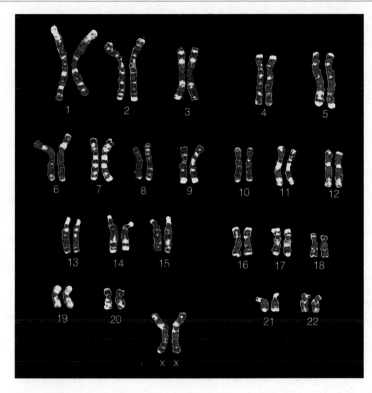

Figure 10.4 Twenty-three homologous pairs of human chromosomes. This example is from a human female, with two X chromosomes. Human males have a different pairing of sex chromosomes (XY).

After homologous chromosomes are pulled apart, each ends up in one of two new nuclei. The chromosomes are still duplicated—the sister chromatids are still attached. During meiosis II, the sister chromatids of each chromosome are pulled apart, so each becomes an individual, unduplicated chromosome:

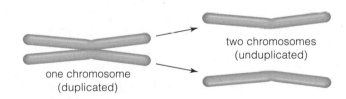

one chromosome (duplicated)

two chromosomes (unduplicated)

Meiosis distributes the duplicated chromosomes of a diploid nucleus (2*n*) into four new nuclei. Each new nucleus is haploid (*n*), with one unduplicated version of each chromosome. Typically, two cytoplasmic divisions accompany meiosis, so four haploid cells form. Figure 10.5 in the next section shows the chromosomal movements in the context of the stages of meiosis.

Take-Home Message

What is meiosis?

■ Meiosis is a nuclear division mechanism that occurs in immature reproductive cells of sexually reproducing eukaryotes. It halves a cell's diploid (2*n*) chromosome number, to the haploid number (*n*).

■ Links to Microtubules 4.13, Mitosis 9.2 and 9.3

Meiosis I

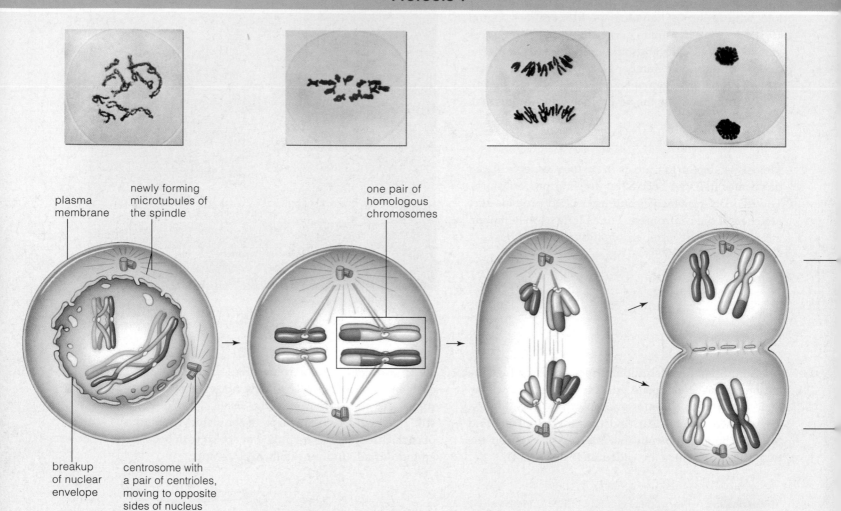

plasma membrane

newly forming microtubules of the spindle

one pair of homologous chromosomes

breakup of nuclear envelope

centrosome with a pair of centrioles, moving to opposite sides of nucleus

A Prophase I

The chromosomes were duplicated in interphase, so every chromosome now consists of two sister chromatids joined at the centromere. The nucleus is diploid (2n)—it contains two sets of chromosomes, one from each parent.

The chromosomes now condense. Homologous chromosomes pair up and swap segments (as indicated by color breaks). A bipolar spindle forms. The centrosome, with its two centrioles, gets duplicated; one centriole pair now moves to the opposite side of the cell as the nuclear envelope breaks up.

B Metaphase I

By the end of prophase I, spindle microtubules had connected the chromosomes to the spindle poles. Each chromosome is now attached to one spindle pole, and its homologue is attached to the other.

The microtubules grow and shrink, pushing and pulling the chromosomes as they do. When all of the microtubules are the same length, the chromosomes are aligned midway between spindle poles. This alignment marks metaphase I.

C Anaphase I

As spindle microtubules shorten, they pull each duplicated chromosome toward one of the spindle poles, so the homologous chromosomes separate.

Which chromosome (maternal or paternal) became attached to a particular spindle pole was random, so either may end up at a particular pole.

D Telophase I

One of each chromosome arrives at each spindle pole.

New nuclear envelopes form around the two clusters of chromosomes as they decondense. There are now two haploid (n) nuclei. The cytoplasm may divide at this point.

Figure 10.5 Animated Meiosis halves the chromosome number. Drawings show a diploid (2n) animal cell. For clarity, only two pairs of chromosomes are illustrated, but cells of nearly all eukaryotes have more than two. The two chromosomes of the pair inherited from one parent are in *purple;* the two inherited from the other parent are in *blue.* Micrographs show meiosis in a lily plant cell (*Lilium regale*).

Figure It Out: Are chromosomes in their duplicated or unduplicated state during metaphase II?

Answer: Duplicated

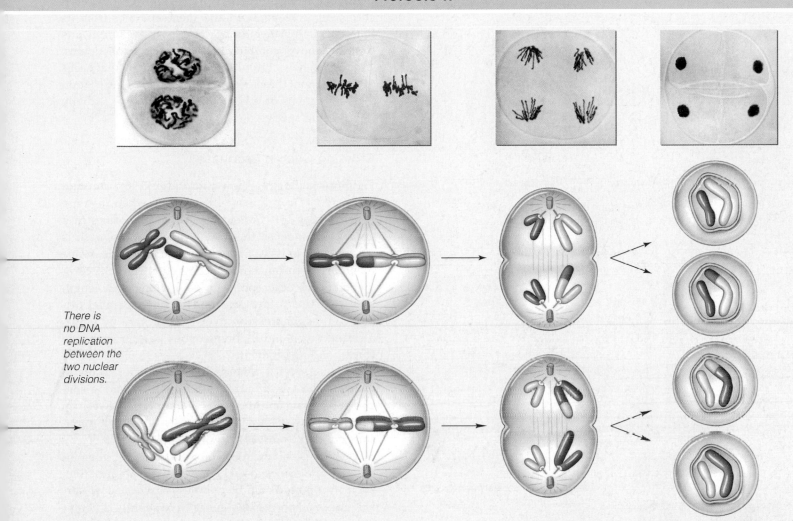

There is no DNA replication between the two nuclear divisions.

E Prophase II

Each nucleus contains one full set of chromosomes. Every chromosome is still duplicated—it consists of two sister chromatids joined at the centromere.

The chromosomes condense as a bipolar spindle forms. One centriole moves to the opposite side of each new nucleus, and the nuclear envelopes break up.

F Metaphase II

By the end of prophase II, spindle microtubules had connected the sister chromatids to the spindle poles. Each chromatid is now attached to one spindle pole, and its sister is attached to the other.

The microtubules grow and shrink, pushing and pulling the chromosomes as they do. When all of the microtubules are the same length, the chromosomes are aligned midway between spindle poles. This alignment marks metaphase II.

G Anaphase II

As spindle microtubules shorten, they pull each sister chromatid toward one of the spindle poles, so the sisters separate.

Which sister chromatid became attached to a particular spindle pole was random, so either may end up at a particular pole.

H Telophase II

Each chromosome now consists of a single, unduplicated molecule of DNA. One of each chromosome arrives at each spindle pole.

New nuclear envelopes form around each cluster of chromosomes as they decondense. There are now four haploid (*n*) nuclei. The cytoplasm may divide.

- Crossovers and the random sorting of chromosomes in meiosis result in new combinations of traits among offspring.

- Link to Chromosome structure 9.1

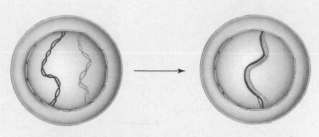

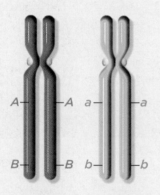

A Two homologous chromosomes, one maternal (*purple*) and one paternal (*blue*) are in their duplicated form: Each is two sister chromatids, joined at the centromere. Homologous chromosomes align and associate tightly during prophase I.

B Here, we focus on only two genes. One gene has alleles *A* and *a*; the other has alleles *B* and *b*.

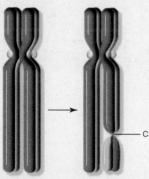

C Close contact between the homologous chromosomes promotes crossing over between nonsister chromatids, so paternal and maternal chromatids exchange segments.

crossover

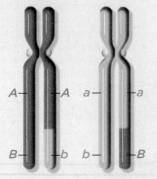

D Crossing over mixes up paternal and maternal alleles on homologous chromosomes.

The previous section mentioned briefly that duplicated chromosomes swap segments with their homologous partners during prophase I. It also showed how each chromosome aligns with and then separates from its homologous partner during anaphase I. Both events introduce novel combinations of alleles into gametes. Along with new chromosome combinations that occur in fertilization, these events contribute to the variation in combinations of traits among offspring of sexually reproducing species.

Crossing Over in Prophase I

Figure 10.6*a* illustrates one pair of duplicated chromosomes, early in prophase I of meiosis when they are in the process of condensing. All chromosomes in a germ cell condense this way. When they do, each is drawn close to its homologue. The chromatids of one homologous chromosome become stitched to the chromatids of the other, point by point along their length with little space in between. This tight, parallel orientation favors **crossing over**—the process by which a chromosome and its homologous partner exchange corresponding segments.

Crossing over is a normal and frequent process in meiosis. The rate of crossing over varies among species and among chromosomes; in humans, between 46 and 95 crossovers occur per meiosis, so each chromosome probably crosses over at least once.

Each crossover event is an opportunity for homologous chromosomes to exchange heritable information. Such swapping would be pointless if genes never varied, but remember, many genes have slightly different forms (alleles). Typically, a number of genes on one chromosome will not be identical to their partners on the homologous chromosome.

We will return to the impact of crossing over in later chapters. For now, remember that crossing over introduces novel combinations of alleles in both members of a pair of homologous chromosomes, which results in novel combinations of traits among offspring.

Figure 10.6 Animated Crossing over. *Blue* signifies a paternal chromosome, and *purple*, its maternal homologue.

For clarity, we show only one pair of homologous chromosomes and one crossover, but more than one crossover may occur in each chromosome pair.

Segregation of Chromosomes into Gametes

Normally, all new nuclei that form in meiosis I receive the same number of chromosomes, but which homologue ends up in which nucleus is random.

The process of chromosome segregation begins in prophase I. Suppose segregation is happening right now in one of your own germ cells. Crossovers have already made genetic mosaics of your chromosomes, but let's put crossing over aside to simplify tracking. Just call the twenty-three chromosomes you inherited from your mother the maternal ones, and the twenty-three you inherited from your father the paternal ones.

By metaphase I, microtubules emanating from both spindle poles have aligned all of the duplicated chromosomes at the spindle equator (Figure 10.5b). Have they attached all of the maternal chromosomes to one pole and all of the paternal chromosomes to the other? Probably not. Spindle microtubules attach to the kinetochores of the first chromosome they contact, regardless of whether it is maternal or paternal. Homologues get attached to opposite spindle poles. Thus, there is no pattern to the attachment of the maternal or paternal chromosomes to a particular spindle pole: Either homologous chromosome can end up at either pole.

Then, in anaphase I, each duplicated chromosome separates from its homologous partner and is pulled toward the pole to which it is attached.

Think of meiosis in a germ cell with just three pairs of chromosomes. By metaphase I, the three pairs would be attached to the spindle poles in one of four possible combinations (Figure 10.7). There would be eight (2^3) possible combinations of maternal and paternal chromosomes in new nuclei that form at telophase I.

By telophase II, each of the two nuclei would have divided and given rise to two new, identical haploid nuclei. (The nuclei would be identical because the sister chromatids of each duplicated chromosome were identical in our hypothetical example.) So, there would be eight possible combinations of maternal and paternal chromosomes in the four haploid nuclei that form by meiosis of that one germ cell.

Cells that give rise to human gametes have twenty-three pairs of homologous chromosomes, not three. Each time a human germ cell undergoes meiosis, the four gametes that form end up with one of 8,388,608 (or 2^{23}) possible combinations of homologous chromosomes! Remember, any number of genes may occur as different alleles on the maternal and paternal homologues. Are you getting an idea of why such fascinating combinations of traits show up among the generations of your own family tree?

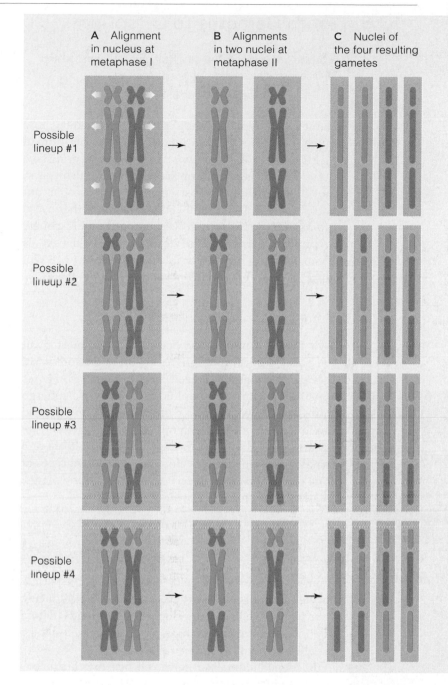

A Alignment in nucleus at metaphase I

B Alignments in two nuclei at metaphase II

C Nuclei of the four resulting gametes

Possible lineup #1

Possible lineup #2

Possible lineup #3

Possible lineup #4

Figure 10.7 Animated Hypothetical segregation of three pairs of chromosomes in meiosis I. Which chromosome of each pair gets packaged into which of the two new nuclei is random. *Left:* the four possible metaphase I lineups of three pairs of homologous chromosomes. *Right:* the resulting eight combinations of maternal (*purple*) and paternal (*blue*) chromosomes in the new nuclei.

Take-Home Message

How does meiosis introduce variation in combinations of traits?

■ Crossing over is recombination between nonsister chromatids of homologous chromosomes during prophase I. It makes new combinations of parental alleles.

■ Homologous chromosomes can be attached to either spindle pole in prophase I, so each homologue can be packaged into either one of the two new nuclei. Thus, the random assortment of homologous chromosomes increases the number of potential combinations of maternal and paternal alleles in gametes.

■ Aside from meiosis, the details of gamete formation and fertilization differ among plants and animals.

■ Link to Cytoplasmic division 9.4

Gametes are typically haploid, but do you know how much they differ in their details? For example, human sperm have one flagellum, roundworm sperm have none, and opossum sperm have two. Crayfish sperm look like pinwheels. Most eggs are microscopic, but an ostrich egg in its shell can weigh more than 2,200 grams (4.85 pounds). A flowering plant's male gamete is simply a sperm nucleus. We leave most of the details of sexual reproduction for later chapters, but you will need to know a few concepts before you get there.

Gamete Formation in Plants

Two kinds of multicelled bodies form in most plant life cycles. Typical **sporophytes** are diploid; spores form by meiosis in their specialized parts (Figure 10.8a). Spores consist of one or a few haploid cells. The cells undergo mitosis and give rise to a **gametophyte**, a multicelled haploid body inside which one or more gametes form. As an example, pine trees are sporophytes. Male and female gametophytes develop inside different types of pine cones that form on each tree. In flowering plants, gametophytes form in flowers.

Gamete Formation in Animals

Diploid germ cells give rise to animal gametes. In male animals, a germ cell develops into a primary spermatocyte. This large cell divides by meiosis, producing four haploid cells that develop into spermatids (Figure 10.9). Each spermatid matures as a male gamete, which is called a **sperm**.

In female animals, a germ cell becomes a primary oocyte, which is an immature egg. This cell undergoes

meiosis and division, as occurs with a primary spermatocyte. However, the cytoplasm of a primary oocyte divides unequally, so the four cells that result differ in size and function (Figure 10.10).

Two haploid cells form when the primary oocyte divides after meiosis I. One of the cells, the secondary oocyte, gets nearly all of the parent cell's cytoplasm. The other cell, a first polar body, is much smaller. Both cells undergo meiosis II and cytoplasmic division. One of the two cells that forms by division of the secondary oocyte develops into a second polar body. The other cell gets most of the cytoplasm and matures into a female gamete, which is called an ovum (plural, ova), or **egg**.

Polar bodies are not nutrient-rich or plump with cytoplasm, and generally do not function as gametes. In time they will degenerate. Their formation simply ensures that the egg will have a haploid chromosome number, and also will get enough metabolic machinery to support early divisions of the new individual.

More Shufflings at Fertilization

At **fertilization**, the fusion of two gametes produces a zygote. Fertilization restores the parental chromosome number. If meiosis did not precede fertilization, the chromosome number would double with every generation. If the chromosome number changes, so does the individual's set of genetic instructions. This set is like a fine-tuned blueprint that must be followed exactly, page by page, in order to build a body that functions normally. Changes in the blueprint can have serious—if not lethal—consequences, particularly in animals.

Fertilization also contributes to the variation that we see among offspring of sexual reproducers. Think about it in terms of human reproduction. The 23 pairs of homologous chromosomes are mosaics of genetic information after prophase I crossovers. Each gamete

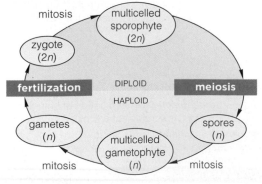

a Plant life cycle

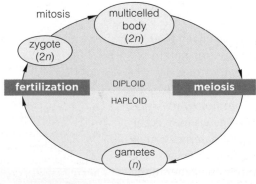

b Animal life cycle

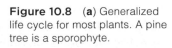

Figure 10.8 (**a**) Generalized life cycle for most plants. A pine tree is a sporophyte.

(**b**) Generalized life cycle for animals. The zygote is the first cell to form when the nuclei of two gametes, such as a sperm and an egg, fuse at fertilization.

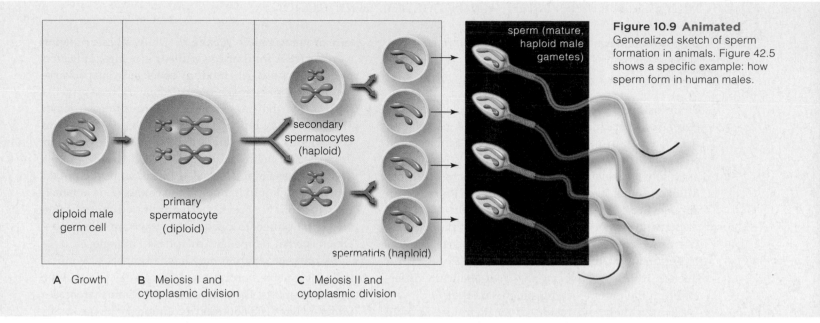

Figure 10.9 Animated Generalized sketch of sperm formation in animals. Figure 42.5 shows a specific example: how sperm form in human males.

sperm (mature, haploid male gametes)

secondary spermatocytes (haploid)

diploid male germ cell

primary spermatocyte (diploid)

spermatids (haploid)

A Growth

B Meiosis I and cytoplasmic division

C Meiosis II and cytoplasmic division

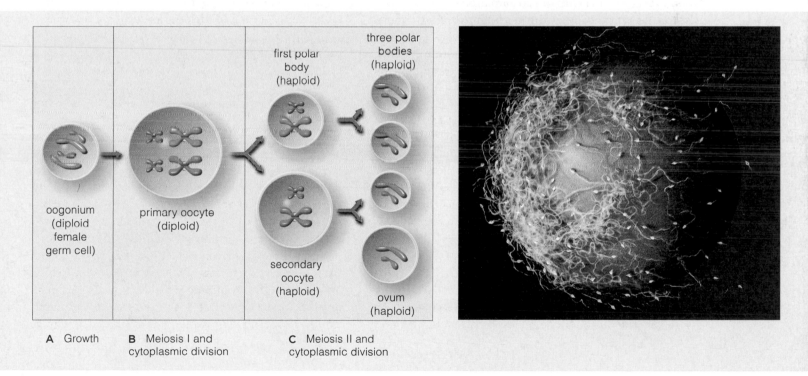

three polar bodies (haploid)

first polar body (haploid)

oogonium (diploid female germ cell)

primary oocyte (diploid)

secondary oocyte (haploid)

ovum (haploid)

A Growth

B Meiosis I and cytoplasmic division

C Meiosis II and cytoplasmic division

Figure 10.10 Animated Animal egg formation. Eggs are far larger than sperm and larger than the three polar bodies. The painting, based on a scanning electron micrograph, depicts human sperm surrounding an ovum. Figure 42.10 shows how human eggs form.

that forms receives one of millions of possible combinations of those chromosomes. Then, out of all the male and female gametes that form, which two actually get together at fertilization is a matter of chance. The sheer number of ways that parental genetic information can combine at fertilization is staggering!

Take-Home Message

Where does meiosis fit into the life cycle of plants and animals?

■ Meiosis and cytoplasmic division precede the development of haploid gametes in animals and spores in plants.

■ The union of two haploid gametes at fertilization results in a diploid zygote.

■ Though they have different results, mitosis and meiosis are fundamentally similar processes.

■ Links to Mitosis 9.3, Cell cycle controls 9.5

By mitosis and cytoplasmic division, one cell becomes two new cells. This process is the basis of growth and tissue repair in all multicelled species. Single-celled eukaryotes (and some multicelled ones) also reproduce asexually by way of mitosis and cytoplasmic division. Mitotic (asexual) reproduction results in clones, which are genetically identical copies of a parent.

By contrast, meiosis produces haploid parent cells, two of which fuse to form a diploid cell that is a new individual of mixed parentage. Meiotic (sexual) reproduction results in offspring that are genetically different from the parent—and from one another.

Though their end results differ, there are striking parallels between the four stages of mitosis and meiosis II (Figure 10.11). As one example, a bipolar spindle separates chromosomes during both processes. There are more similarities at the molecular level.

Long ago, the molecular machinery of mitosis may have been remodeled into meiosis. For example, cer-tain proteins repair breaks in DNA. These proteins monitor DNA for damage while it is being duplicated prior to mitosis. All modern species, from prokaryotes to mammals, make these proteins. Other proteins repair DNA that gets damaged during mitosis itself. This same set of repair proteins also seals up breaks in homologous chromosomes during crossover events in prophase I of meiosis.

In anaphase of mitosis, sister chromatids are pulled apart. What would happen if the connections between the sisters did not break? Each duplicated chromosome would be pulled to one or the other spindle pole—which is what happens in anaphase I of meiosis.

Sexual reproduction may have originated by mutations that affected processes of mitosis. As you will see in later chapters, the remodeling of existing processes into new ones is a common evolutionary theme.

Take-Home Message

Are the processes of mitosis and meiosis related?

■ Meiosis may have evolved by the remodeling of existing mechanisms of mitosis.

Meiosis I

one diploid nucleus ————————————→ two haploid nuclei

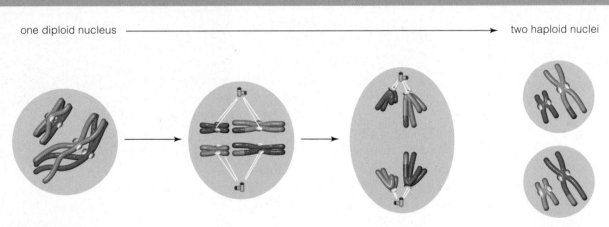

Prophase I
• Chromosomes condense.
• Homologous chromosomes pair.
• Crossovers occur.
• Bipolar spindle forms; it attaches chromosomes to spindle poles.
• Nuclear envelope breaks up.

Metaphase I
• Chromosomes align midway between spindle poles.

Anaphase I
• Homologous chromosomes separate as they are pulled toward spindle poles.

Telophase I
• Chromosome clusters arrive at spindle poles.
• New nuclear envelopes form.
• Chromosomes decondense.

Figure 10.11 Comparing mitosis and meiosis, beginning with a diploid cell containing two paternal and two maternal chromosomes.

Right, a bipolar spindle at metaphase, anaphase, and telophase of mitosis in a mouse cell. The *green* stain identifies the microtubules of the spindle. The *blue* stain identifies DNA in the cell's chromosomes.

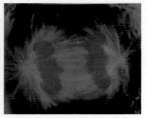

Mitosis

one diploid nucleus ⟶ two diploid nuclei

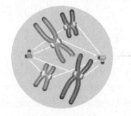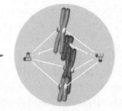

Prophase
- Chromosomes condense.
- Bipolar spindle forms; it attaches chromosomes to spindle poles.
- Nuclear envelope breaks up.

Metaphase
- Chromosomes align midway between spindle poles.

Anaphase
- Sister chromatids separate as they are pulled toward spindle poles.

Telophase
- Chromosome clusters arrive at spindle poles.
- New nuclear envelopes form.
- Chromosomes decondense.

Meiosis II

two haploid nuclei ⟶ four haploid nuclei

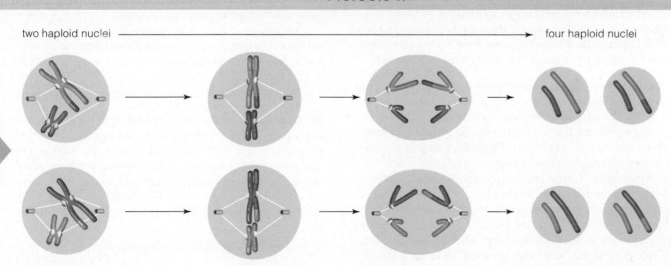

Prophase II
- Chromosomes condense.
- Bipolar spindle forms; it attaches chromosomes to spindle poles.
- Nuclear envelope breaks up.

Metaphase II
- Chromosomes align midway between spindle poles.

Anaphase II
- Sister chromatids separate as they are pulled toward spindle poles.

Telophase II
- Chromosome clusters arrive at spindle poles.
- New nuclear envelopes form.
- Chromosomes decondense.

There are a few all-female species of fishes, reptiles, and birds in nature, but not mammals. In 2004, researchers fused two mouse eggs in a test tube and made an embryo using no DNA from a male. The embryo developed into Kaguya, the world's first father-less mammal. The mouse grew up healthy, engaged in sex with a male mouse, and gave birth to offspring. The researchers wanted to find out if sperm was required for normal development.

How would you vote?

Researchers made a "fatherless" mouse (*right*) from two eggs. Should they be prevented from trying the process with human eggs? See CengageNOW for details, then vote online.

Summary

Section 10.1 Many eukaryotic life cycles have asexual and sexual phases. Offspring of **asexual reproduction** are genetically identical to their one parent—they are **clones**. The offspring of **sexual reproduction** differ from parents, and often from one another, in the details of shared traits. Meiosis in germ cells, haploid gamete formation, and fertilization occur in sexual reproduction. **Alleles** are different molecular forms of the same **gene**. Each specifies a different version of the gene's product. Meiosis shuffles parental alleles; thus, offspring inherit new combinations of alleles.

Section 10.2 **Meiosis**, a nuclear division mechanism that occurs in eukaryotic **germ cells**, precedes the formation of **gametes**. Meiosis halves the parental chromosome number. The fusion of two **haploid** gamete nuclei during fertilization restores the parental chromosome number in the **zygote**, the first cell of the new individual.

Offspring of most sexual reproducers inherit pairs of chromosomes, one of each pair from the mother and the other from the father. Except in individuals with non-identical sex chromosomes, the members of a pair are **homologous**: They have the same length, the same shape, and the same set of genes. The pairs interact at meiosis.

Section 10.3 All chromosomes are duplicated during interphase, before meiosis. Two divisions, meiosis I and II, halve the parental chromosome number.

In the first nuclear division, meiosis I, each duplicated chromosome lines up with its homologous partner; then the two move apart, toward opposite spindle poles.

Prophase I. Chromosomes condense and align tightly with their homologues. Each pair of homologues typically undergoes crossing over. Microtubules form the bipolar spindle. One of two pairs of centrioles is moved to the other side of the nucleus. The nuclear envelope breaks up, so microtubules growing from each spindle pole can penetrate the nuclear region. The microtubules then attach to one or the other chromosome of each homologous pair.

Metaphase I. A tug-of-war between the microtubules from both poles has positioned all pairs of homologous chromosomes at the spindle equator.

Anaphase I. Microtubules separate each chromosome from its homologue and move both to opposite spindle poles. As anaphase I ends, a cluster of duplicated chromosomes is nearing each spindle pole.

Telophase I. Two nuclei form; typically the cytoplasm divides. All of the chromosomes are still duplicated; each still consists of two sister chromatids.

The second nuclear division, meiosis II, occurs in both nuclei that formed in meiosis I. The chromosomes condense in **prophase II**, and align in **metaphase II**. Sister chromatids of each chromosome are pulled away from each other in **anaphase II**, so each becomes an individual chromosome. By the end of **telophase II**, there are four **haploid** nuclei, each with one set of chromosomes. The chromosomes are unduplicated at this stage.

■ *Use the animation on CengageNOW to explore what happens in the stages of meiosis.*

Section 10.4 Novel combinations of alleles arise by events in prophase I and metaphase I.

The *non*sister chromatids of homologous chromosomes undergo **crossing over** during prophase I: They exchange segments at the same place along their length, so each ends up with new combinations of alleles that were not present in either parental chromosome.

Crossing over during prophase I, and random segregation of maternal and paternal chromosomes into new nuclei, contribute to variation in traits among offspring.

■ *Use the animation on CengageNOW to see how crossing over and metaphase I alignments affect allele combinations.*

Section 10.5 Multicelled diploid and haploid bodies are typical in life cycles of plants and animals. A diploid **sporophyte** is a multicelled plant body that makes haploid spores. Spores give rise to **gametophytes**, or multicelled plant bodies in which haploid gametes form. Germ cells in the reproductive organs of most animals give rise to **sperm** or **eggs**. Fusion of a sperm and egg at **fertilization** results in a zygote.

■ *Use the animation on CengageNOW to see how gametes form.*

Section 10.6 Like mitosis, meiosis requires a bipolar spindle to move and sort duplicated chromosomes, but meiosis occurs only in cells that are set aside for sexual reproduction. Mitosis maintains the parental chromosome number. Meiosis halves the chromosome number, and it introduces new combinations of alleles into offspring. Some mechanisms of meiosis resemble those of mitosis, and may have evolved from them. For example, the same DNA repair enzymes act in both processes.

Data Analysis Exercise

In 1998, researchers at Case Western University were studying meiosis in mouse oocytes when they saw an unexpected and dramatic increase of abnormal meiosis events (Figure 10.12). Improper segregation of chromosomes during meiosis is one of the main causes of human genetic disorders, which we will discuss in Chapter 12.

The researchers discovered that the spike in meiotic abnormalities started immediately after the mouse facility's plastic cages and water bottles were washed in a new, alkaline detergent. The detergent had damaged the plastic, which began to leach bisphenol A (BPA). BPA is a synthetic chemical that mimics estrogen, a hormone. BPA is used to manufacture polycarbonate plastic items (including baby bottles and water bottles) and epoxies (including the coating on the inside of metal cans of food).

1. What percentage of mouse oocytes displayed abnormalities of meiosis with no exposure to damaged caging?

2. Which group of mice showed the most meiotic abnormalities in their oocytes?

3. What is abnormal about metaphase I as it is occurring in the oocytes shown in Figure 10.12b, c, and d?

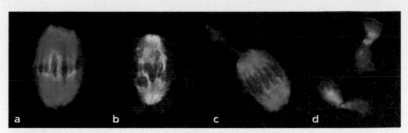

Caging materials	Total number of oocytes	Abnormalities
Control: New cages with glass bottles	271	5 (1.8%)
Damaged cages with glass bottles		
Mild damage	401	35 (8.7%)
Severe damage	149	30 (20.1%)
Damaged bottles	197	53 (26.9%)
Damaged cages with damaged bottles	58	24 (41.4%)

Figure 10.12 Meiotic abnormalities associated with exposure to damaged plastic caging. Fluorescent micrographs show nuclei of single mouse oocytes in metaphase I. (**a**) Normal metaphase; (**b–d**) examples of abnormal metaphase. Chromosomes are *red*; spindle fibers are *green*.

Self-Quiz

Answers in Appendix III

1. Meiosis and cytoplasmic division function in _____ .
a. asexual reproduction of single-celled eukaryotes
b. growth and tissue repair
c. sexual reproduction
d. both b and c

2. Sexual reproduction requires _____ .
a. meiosis   c. spore formation
b. fertilization   d. a and b

3. What is the name for alternative forms of the same gene?

4. Generally, a pair of homologous chromosomes _____ .
a. carry the same genes   c. are the same length, shape
b. interact at meiosis   d. all of the above

5. Sister chromatids are joined at the _____ .
a. kinetochore   c. centriole
b. spindle   d. centromere

6. Meiosis _____ the parental chromosome number.
a. doubles   c. maintains
b. halves   d. mixes up

7. Meiosis ends with the formation of _____ .
a. two cells   c. four cells
b. two nuclei   d. four nuclei

8. Sister chromatids of each duplicated chromosome separate during _____ .
a. prophase I   d. anaphase II
b. prophase II   e. both b and c
c. anaphase I

9. How does meiosis contribute to variation in traits among offspring of sexual reproducers?

10. The cell shown at *right* is in anaphase II. I know this because _____ .

11. Match each term with its description.

chromosome number	a. different molecular forms of the same gene
___alleles	b. maybe none between meiosis I, II
___metaphase I	c. all chromosomes aligned at spindle equator
___interphase	d. all chromosomes in a given type of cell

■ *Visit CengageNOW for additional questions.*

Critical Thinking

1. Explain why you can predict that meiosis gives rise to genetic differences between parent cells and descendant cells in fewer cell divisions than mitosis does.

2. Assume you can measure the amount of DNA in the nucleus of a primary oocyte, and then in the nucleus of a primary spermatocyte. Each gives you a mass *m*. What mass of DNA would you expect to find in the nucleus of each mature gamete (each egg and sperm) that forms after meiosis? What mass of DNA will be (1) in the nucleus of a zygote that forms at fertilization and (2) in that zygote's nucleus after the first DNA duplication?

3. The diploid chromosome numbers for the somatic cells of several eukaryotic species are listed at *right*. What is the number of chromosomes that normally ends up in gametes of each species? What would that number be after three generations if meiosis did not occur before gamete formation?

Fruit fly, *Drosophila melanogaster*	8
Garden pea, *Pisum sativum*	14
Frog, *Rana pipiens*	26
Earthworm, *Lumbricus terrestris*	36
Human, *Homo sapiens*	46
Amoeba, *Amoeba*	50
Dog, *Canis familiaris*	78
Vizcacha rat, *Tympanoctomys barrerae*	102
Horsetail, *Equisetum*	216

Observing Patterns in Inherited Traits

IMPACTS, ISSUES The Color of Skin

One of the most visible human traits is the color of skin, which can range from very pale to very dark brown. The color arises from melanosomes, organelles in skin cells that make red and brownish-black pigments called melanins. Most people have about the same number of melanosomes in their skin cells. Skin color variation occurs because the kinds and amounts of melanins made by the melanosomes varies among people.

Dark skin would have been adaptive under the intense sunlight of the African savannas where humans first evolved. Melanin protects skin cells exposed to sunlight because it absorbs ultraviolet (UV) radiation, which damages DNA and other biological molecules. Melanin-rich dark skin acts as a natural sunscreen, so it reduces the risk of certain cancers and other serious problems caused by overexposure to sunlight.

Early human groups that migrated to regions with colder climates were exposed to less sunlight. In these regions, lighter skin would have been adaptive. Why? UV radiation stimulates skin cells to make a molecule the body converts to essential vitamin D. Where sunlight exposure is minimal, UV radiation damage is less of a risk than vitamin D deficiency, which has serious health consequences for developing fetuses and children. People with dark, UV-shielding skin have a high risk of this deficiency in regions where sunlight exposure is minimal.

Like most other human traits, skin color has a genetic basis (Figure 11.1). More than 100 gene products affect the synthesis and deposition of melanin. Mutations in at least some of these genes may have contributed to adaptive variations of human skin color. For example, the *SLC24A5* gene on chromosome 15 encodes a membrane transport protein in melanosomes. Nearly all people of native African, American, or East Asian descent have the same version (allele) of this gene. By contrast, nearly all people of native European descent carry a particular mutation in the gene. The European allele results in less melanin, and lighter skin color, than the unmutated version.

Such genetic patterns offer clues about the past. For example, Chinese and Europeans do not share any skin pigmentation allele that does not also occur in other populations. However, most people of Chinese descent carry a particular allele of the *DCT* gene, the product of which helps convert tyrosine to melanin. Few people of European or African descent have this allele. Taken together, the distribution of the *SLC24A5* and *DCT* genes suggests that (1) an African population was ancestral to both the Chinese and Europeans, and (2) Chinese and European populations separated before their pigmentation genes mutated and their skin color changed.

Skin color is only one of many human traits that can vary because of mutations in single genes. The small scale of such differences is a reminder that all of us share a genetic legacy of common ancestry.

Figure 11.1 Skin color. Variations in skin color may have evolved as a balance between vitamin production and protection against harmful UV radiation.

Variation in skin color and in most other human traits begins with differences in alleles inherited from parents. Fraternal twin girls Kian and Remee were born in 2006 to parents Kylie (*left*) and Remi (*right*). Both Kylie's and Remi's mothers are of European descent, and have pale skin. Both of their fathers are of African descent, and have dark skin.

More than 100 genes affect skin color in humans. Kian and Remee inherited different alleles of some of those genes.

Key Concepts

Where modern genetics started

Gregor Mendel gathered the first experimental evidence of the genetic basis of inheritance. His meticulous work gave him clues that heritable traits are specified in units. The units, which are distributed into gametes in predictable patterns, were later identified as genes. **Section 11.1**

Insights from monohybrid experiments

Some experiments yielded evidence of gene segregation: When one chromosome separates from its homologous partner during meiosis, the alleles on those chromosomes also separate and end up in different gametes. **Section 11.2**

Insights from dihybrid experiments

Other experiments yielded evidence of independent assortment: Genes are typically distributed into gametes independently of other genes. **Section 11.3**

Variations on Mendel's theme

Not all traits appear in Mendelian inheritance patterns. An allele may be partly dominant over a nonidentical partner, or codominant with it. Multiple genes may influence a trait; some genes influence many traits. The environment also influences gene expression. **Sections 11.4–11.7**

Links to Earlier Concepts

■ Before starting this chapter, be sure you can generally define genes, alleles, and diploid versus haploid chromosome numbers (Sections 10.1 and 10.2).

■ You may want to scan the sections that introduce protein structure (3.5), enzymes (6.3), and pigments (7.1).

■ As you read, refer back to the visual road map of the stages of meiosis (10.3).

■ You will be considering experimental evidence of two major topics that were introduced earlier—the effects that crossing over and metaphase I alignments have on inheritance (10.4).

How would you vote? Traditionally, humans have been assigned to race categories based on physical attributes such as skin color, which have a genetic basis. Are twins such as Kian and Remee of different races? See CengageNOW for details, then vote online.

■ Recurring inheritance patterns are observable outcomes of sexual reproduction.

■ Links to Genes and alleles 10.1, Diploid and haploid 10.2

By the 1850s, most people had an idea that two parents contribute hereditary material to their offspring, but few suspected that the material is organized as units, or genes. Some thought that hereditary material must be fluid, with fluids from both parents blending at fertilization like milk into coffee.

Figure 11.2 Gregor Mendel, the founder of modern genetics.

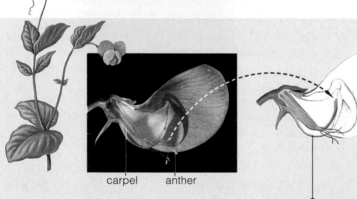

carpel   anther

A Garden pea flower, cut in half. Sperm form in pollen grains, which originate in male floral parts (anthers). Eggs develop, fertilization takes place, and seeds mature in female floral parts (carpels).

B Pollen from a plant that breeds true for purple flowers is brushed onto a floral bud of a plant that breeds true for white flowers. The white flower had its anthers snipped off. Artificial pollination is one way to ensure that a plant will not self-fertilize.

C Later, seeds develop inside pods of the cross-fertilized plant. An embryo in each seed develops into a mature pea plant.

D Each new plant's flower color is indirect but observable evidence that hereditary material has been transmitted from the parent plants.

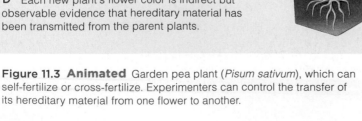

Figure 11.3 Animated Garden pea plant (*Pisum sativum*), which can self-fertilize or cross-fertilize. Experimenters can control the transfer of its hereditary material from one flower to another.

The idea of "blending inheritance" failed to explain the obvious. For example, many children who differ in their eye or hair color have the same two parents. If parental fluids blended, then the color would be some blended shade of the parental colors. If neither parent had freckles, freckled children would never pop up. A white horse bred with a black horse should always produce gray offspring, but offspring of such matings are not always gray. Blending inheritance did not explain the variation in traits that people could see with their own eyes.

Charles Darwin did not accept the idea of blending inheritance. However, though inheritance was central to his theory of natural selection, he could not quite see how it works. He saw that forms of traits often vary among individuals in a population. He realized that variations that help individuals survive and reproduce tend to appear more frequently in a population over generations. However, neither he nor anyone else at the time knew that hereditary material is divided into discrete units (genes). That insight is crucial to understanding how heredity works.

Even before Darwin presented his theory of natural selection, someone had been gathering evidence that would support it. Gregor Mendel, an Austrian monk (Figure 11.2), had been carefully breeding thousands of pea plants. By documenting how certain traits are passed from plant to plant, generation after generation, Mendel had been collecting indirect but observable evidence of how inheritance works.

Mendel's Experimental Approach

Mendel spent most of his adult life in Brno, a city near Vienna that is now part of the Czech Republic. He was not a man of narrow interests who just stumbled onto dazzling principles. He lived in a monastery close to European cities that were centers of scientific inquiry. Having been raised on a farm, Mendel was aware of agricultural principles and their applications. He kept abreast of current literature on breeding experiments. He was a dedicated member of an agricultural society, and he won awards for developing improved varieties of fruits and vegetables.

Just after Mendel entered the monastery at Brno, he took courses in mathematics, physics, and botany at the University of Vienna. Few scholars of his time were trained in both plant breeding and mathematics.

Shortly after his university training ended, Mendel started to study *Pisum sativum*, the garden pea plant. This plant is self-fertilizing: Its flowers produce both male and female gametes (sperm and eggs) that can

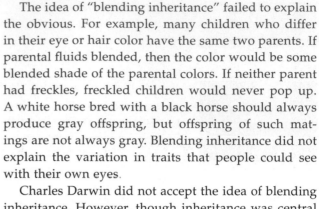

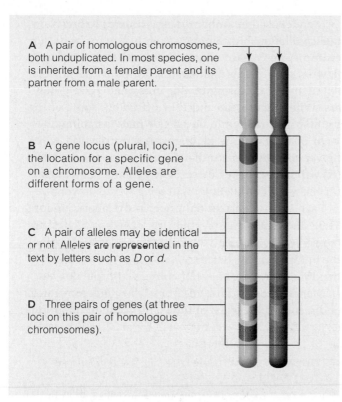

A A pair of homologous chromosomes, both unduplicated. In most species, one is inherited from a female parent and its partner from a male parent.

B A gene locus (plural, loci), the location for a specific gene on a chromosome. Alleles are different forms of a gene.

C A pair of alleles may be identical or not. Alleles are represented in the text by letters such as *D* or *d*.

D Three pairs of genes (at three loci on this pair of homologous chromosomes).

Figure 11.4 Animated A few genetic terms. Like other species with a diploid chromosome number, garden pea plants have pairs of genes, on pairs of homologous chromosomes.

Most genes come in slightly different forms called alleles. Different alleles may result in different versions of a trait. An allele at any given location on a chromosome may or may not be identical with its partner on the homologous chromosome.

come together and give rise to a new plant. Pea plants can "breed true" for certain traits such as white flowers. Breeding true for a trait means that, rare mutations aside, all offspring have the same form of the trait as the parent(s), generation after generation. For example, all offspring of pea plants that breed true for white flowers also have white flowers.

Breeders such as Mendel cross-fertilize plants when they transfer pollen from the flower of one plant to the flower of another. (Pollen grains are structures in which sperm develop. They form in anthers, which are the male parts of a flower.) For example, a breeder may open a flower bud of a true-breeding, white-flowered plant and snip out its anthers. Removing the anthers prevents the flower from fertilizing itself. The breeder then brushes the female parts of the flower with pollen from another plant, perhaps one that breeds true for purple flowers (Figure 11.3). Mendel discovered that the traits of the offspring of such cross-fertilized plants appear in predictable patterns.

Terms Used in Modern Genetics

In Mendel's time, no one knew about genes, meiosis, or chromosomes. As we follow his thinking, we will clarify the picture by substituting some modern terms, as stated here and in Figure 11.4.

1. **Genes** are heritable units of information about traits. Parents transmit genes to offspring. Each gene occurs at a specific location (**locus**) on a specific chromosome.

2. Cells with a diploid chromosome number ($2n$) have pairs of genes, on pairs of homologous chromosomes.

3. A **mutation** is a permanent change in a gene. It may cause a trait to change, as when a gene for flower color specifies purple and a mutated form specifies white. Such alternative forms of a gene are alleles.

4. All members of a lineage that breeds true for a specific trait have identical alleles for that trait. The offspring of a cross, or mating, between two individuals that breed true for different forms of a trait are **hybrids**. A hybrid has nonidentical alleles for the trait.

5. An individual with nonidentical alleles of a gene is **heterozygous** for the gene. An individual with identical alleles of a gene is **homozygous** for the gene.

6. An allele is **dominant** if its effect masks the effect of a **recessive** allele paired with it. Capital letters such as *A* signify dominant alleles; lowercase letters such as *a* signify recessive ones.

7. A **homozygous dominant** individual has a pair of dominant alleles (*AA*). A **homozygous recessive** individual has a pair of recessive alleles (*aa*). A heterozygous individual has a pair of nonidentical alleles (*Aa*). Heterozygotes are hybrids.

8. **Gene expression** is the process by which information in a gene is converted to a structural or functional part of a cell or body. Expressed genes determine traits.

9. Two terms help keep the distinction clear between genes and the traits they specify: **Genotype** refers to the particular alleles an individual carries; **phenotype** refers to an individual's traits.

10. F_1 stands for first-generation offspring of parents (P); F_2 for second-generation offspring. F is an abbreviation for filial (offspring).

Take-Home Message

What contribution did Gregor Mendel make to modern biology?

■ Mendel collected clues about how inheritance works among sexual reproducers by tracking observable traits through generations of pea plants.

11.2 | Mendel's Law of Segregation

- Garden pea plants inherit two "units" of information (genes) for a trait, one from each parent.

- Link to Sampling error 1.8

A **testcross** is a method of determining genotype. An individual of unknown genotype is crossed with one that is known to be homozygous recessive. The traits of the offspring may indicate that the individual is heterozygous or homozygous for a dominant trait.

Monohybrid experiments are testcrosses that check for a dominance relationship between two alleles at a single locus. Individuals with different alleles of a gene are crossed (or self-fertilized); traits of the offspring of such a cross may indicate whether one of the alleles is dominant over the other. A typical monohybrid experiment is a cross between individuals that are identically heterozygous at one gene locus ($Aa \times Aa$).

Mendel used monohybrid experiments to find dominance relationships among seven pea plant traits. For example, he crossed plants that bred true for purple flowers with plants that bred true for white flowers. All of the F_1 offspring of this cross had purple flowers. When he crossed those F_1 offspring, some of the F_2 offspring had white flowers! What was going on?

In pea plants, one gene governs purple and white flower color. Any plant that carries the dominant allele (A) will have purple flowers. Only plants homozygous for the recessive allele (a) will have white flowers.

Each gamete carries only one of the alleles (Figure 11.5). If plants homozygous for different alleles are crossed ($AA \times aa$), only one outcome is possible: All of the F_1 offspring are heterozygous (Aa). All of them carry the dominant allele A, so all will have purple flowers.

Mendel crossed hundreds of such F_1 heterozygotes, and recorded the traits of thousands of their offspring.

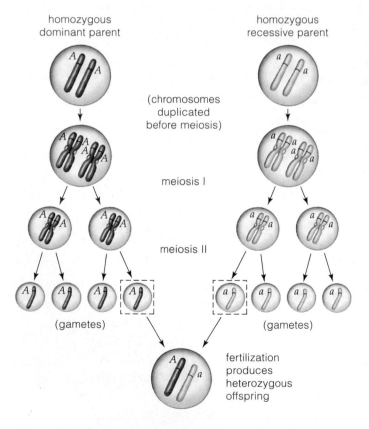

Figure 11.5 Segregation of a pair of alleles at a gene locus.

Figure 11.6 From some of Mendel's monohybrid experiments with pea plants, actual counts of F_2 offspring with certain phenotypes that reflect dominant or recessive hereditary "units" (alleles). All phenotype ratios in F_2 offspring were near 3 to 1.

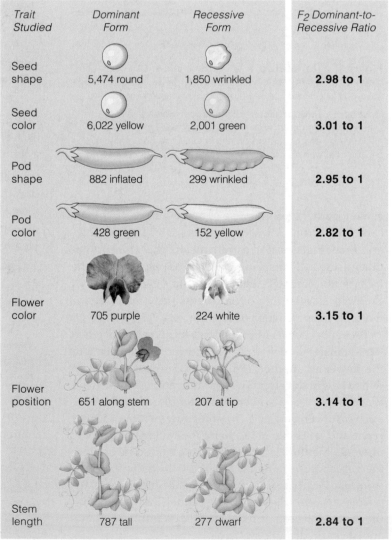

Trait Studied	Dominant Form	Recessive Form	F_2 Dominant-to-Recessive Ratio
Seed shape	5,474 round	1,850 wrinkled	**2.98 to 1**
Seed color	6,022 yellow	2,001 green	**3.01 to 1**
Pod shape	882 inflated	299 wrinkled	**2.95 to 1**
Pod color	428 green	152 yellow	**2.82 to 1**
Flower color	705 purple	224 white	**3.15 to 1**
Flower position	651 along stem	207 at tip	**3.14 to 1**
Stem length	787 tall	277 dwarf	**2.84 to 1**

About three of every four F₂ plants had the dominant trait, and about one of every four had the recessive trait (Figure 11.6).

Mendel's predictable results hinted that fertilization is a chance event with a finite number of possible outcomes. Mendel knew about **probability**, which is a measure of the chance that a particular outcome will occur. That chance depends on the total number of possible outcomes. For example, if you cross two Aa heterozygotes, the two types of gametes (A and a) can meet four different ways at fertilization:

Possible Event	Probable Outcome
Sperm A meets egg A	1 out of 4 offspring AA
Sperm A meets egg a	1 out of 4 offspring Aa
Sperm a meets egg A	1 out of 4 offspring Aa
Sperm a meets egg a	1 out of 4 offspring aa

Each of the offspring of this cross has 3 chances in 4 of inheriting at least one dominant A allele (and purple flowers). It has 1 chance in 4 of inheriting two recessive a alleles (and white flowers). Thus, the probability that an offspring of this cross will have purple or white flowers is 3 purple to 1 white, which we represent as a ratio of 3:1. We use grids called **Punnett squares** to calculate the probability of genotypes (and phenotypes) that will occur in offspring (Figure 11.7).

Mendel's observed ratios were not exactly 3:1, but he knew that deviations can arise from sampling error (Section 1.8). For example, if you flip a coin, it is just as likely to end up heads as tails (a probability of 1:1). But often it ends up heads, or tails, several times in a row. Thus, if you flip the coin only a few times, the observed ratio might differ greatly from the predicted ratio of 1:1. If you flip it many times, you are more likely to see that ratio. Mendel minimized his sampling error by maximizing his sample sizes.

The results from Mendel's monohybrid experiments became the basis of his law of **segregation**, which we state here in modern terms: Diploid cells have pairs of genes, on pairs of homologous chromosomes. The two genes of each pair are separated from each other during meiosis, so they end up in different gametes.

Take-Home Message

What is Mendel's law of segregation?

■ Diploid cells have pairs of genes, on pairs of homologous chromosomes. The two genes of each pair are separated from each other during meiosis, so they end up in different gametes.

■ Mendel discovered patterns of inheritance in pea plants by recording and analyzing the results of many testcrosses.

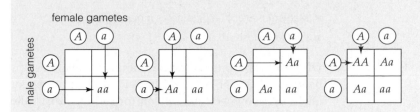

A From left to right, step-by-step construction of a Punnett square. Circles signify gametes, and letters signify alleles: A is dominant; a is recessive. The genotypes of the resulting offspring are inside the squares.

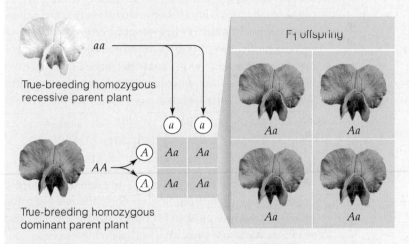

B A cross between two plants that breed true for different forms of a trait produces F₁ offspring that are identically heterozygous.

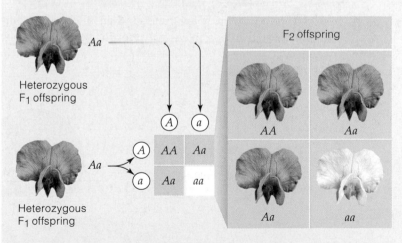

C A cross between the F₁ offspring is the monohybrid experiment. The phenotype ratio of F₂ offspring in this example is 3:1 (3 purple to 1 white).

Figure 11.7 Animated (a) Punnett-square method of predicting probable outcomes of genetic crosses. (b,c) One of Mendel's monohybrid experiments. On average, the ratio of dominant-to-recessive phenotypes among second-generation (F₂) plants of a monohybrid experiment is 3:1.

Figure It Out: How many possible genotypes are there in the F₂ generation?

Answer: Three: AA, Aa, and aa

11.3 Mendel's Law of Independent Assortment

- Many genes sort into gametes independently.
- Link to Meiosis 10.3

Dihybrid experiments test for dominance relationships between alleles at two loci. Individuals with different alleles of two genes are crossed (or self-fertilized); the ratio of traits in offspring offer clues about the alleles. Mendel analyzed the numerical results from dihybrid experiments, but given the prevailing understanding of heredity, he could only hypothesize that "units" specifying one trait (such as flower color) sort into gametes independently of "units" specifying other traits (such as plant height). He did not know that the units are genes, which occur in pairs on homologous chromosomes.

We can duplicate one of Mendel's dihybrid experiments by crossing two plants that breed true for two traits. Here, we track flower color (*A*, purple; *a*, white) and height (*B*, tall; *b*, short):

True-breeding parents:	$AABB$	$\times$	$aabb$
Gametes:	(AB, AB)		(ab, ab)
F_1 offspring, all dihybrids:		$AaBb$	

All F_1 offspring from this cross are tall with purple flowers. Remember, homologous chromosomes become attached to opposite spindle poles during meiosis, but which homologue gets attached to which pole is random (Section 10.3). Thus, four combinations of alleles are possible in the gametes of $AaBb$ dihybrids: AB, Ab, aB, and ab (Figure 11.8).

If two dihybrids are crossed, their alleles can combine in sixteen possible ways at fertilization (four types of gametes in one individual × four types of gametes in the other). In our example ($AaBb \times AaBb$), the sixteen combinations result in four different phenotypes (Figure 11.9). Nine of the sixteen are tall with purple flowers, three are short with purple flowers, three are tall with white flowers, and one is short with white flowers. The ratio of these phenotypes is 9:3:3:1. With more gene pairs, more combinations are possible. If the parents differ in, say, twenty gene pairs, 3.5 billion genotypes are possible!

Mendel published his results in 1866, but apparently his work was read by few and understood by no one. In 1871 he became abbot of his monastery, and his pioneering experiments ended. He died in 1884, never to know that his experiments would be the starting point for modern genetics.

In time, Mendel's hypothesis became known as the law of **independent assortment**. In modern terms, the law states that genes are sorted into gametes independently of other genes. The law holds for many genes in most organisms. However, it requires qualification because there are exceptions. For example, genes that are relatively close together on the same chromosome tend to stay together during meiosis (we will return to this topic in Section 11.5).

Take-Home Message

What is Mendel's law of independent assortment?

- A gene is distributed into gametes independently of how other genes are distributed.

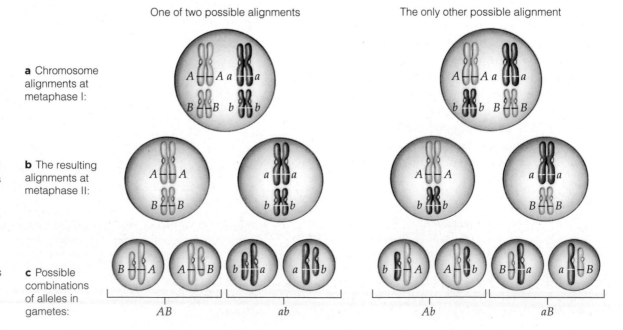

One of two possible alignments   The only other possible alignment

a Chromosome alignments at metaphase I:

b The resulting alignments at metaphase II:

Figure 11.8 Independent assortment at meiosis. This example shows just two pairs of homologous chromosomes in the nucleus of a diploid (2*n*) reproductive cell. Either chromosome of a pair may get attached to either pole. When two pairs are tracked, two different metaphase I alignments are possible.

c Possible combinations of alleles in gametes:

AB ab Ab aB

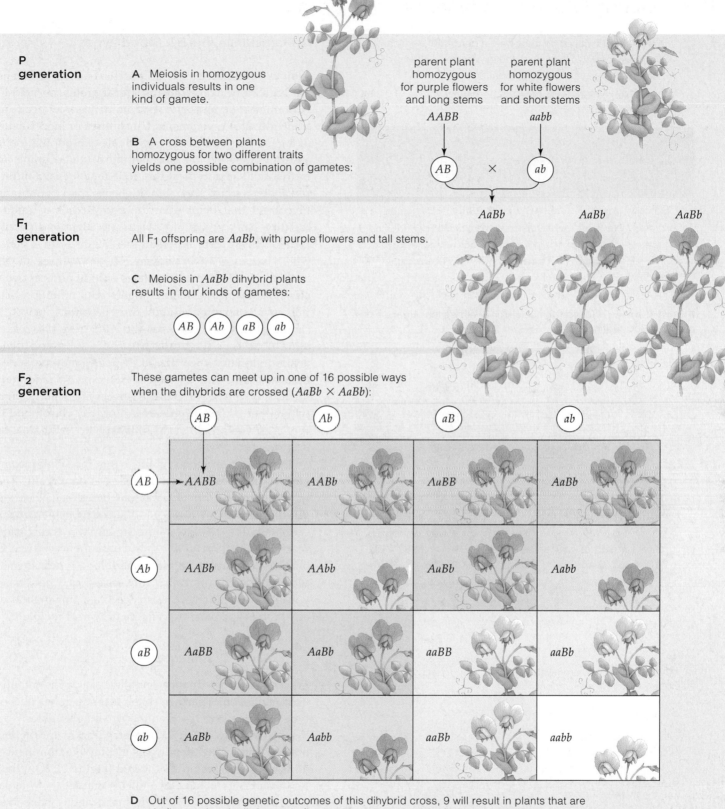

P generation

A Meiosis in homozygous individuals results in one kind of gamete.

parent plant homozygous for purple flowers and long stems

parent plant homozygous for white flowers and short stems

AABB

aabb

B A cross between plants homozygous for two different traits yields one possible combination of gametes:

AB × *ab*

F₁ generation

AaBb   *AaBb*   *AaBb*

All F₁ offspring are *AaBb*, with purple flowers and tall stems.

C Meiosis in *AaBb* dihybrid plants results in four kinds of gametes:

AB *Ab* *aB* *ab*

F₂ generation

These gametes can meet up in one of 16 possible ways when the dihybrids are crossed (*AaBb* × *AaBb*):

	AB	*Ab*	*aB*	*ab*
AB	*AABB*	*AABb*	*AaBB*	*AaBb*
Ab	*AABb*	*AAbb*	*AaBb*	*Aabb*
aB	*AaBB*	*AaBb*	*aaBB*	*aaBb*
ab	*AaBb*	*Aabb*	*aaBb*	*aabb*

D Out of 16 possible genetic outcomes of this dihybrid cross, 9 will result in plants that are purple-flowered and tall; 3, purple-flowered and short; 3, white-flowered and tall; and 1, white-flowered and short. The ratio of phenotypes of this dihybrid cross is 9:3:3:1.

Figure 11.9 Animated One of Mendel's dihybrid experiments. Here, *A* is an allele for purple flowers; *a*, white flowers; *B*, tall plants; *b*, short plants. **Figure It Out:** What do the flowers inside the boxes represent?

Answer: Phenotypes of the offspring

11.4 | Beyond Simple Dominance

■ Mendel focused on traits based on clearly dominant and recessive alleles. However, the expression patterns of genes for some traits are not as straightforward.

■ Links to Fibrous proteins 3.5, Pigments 7.1

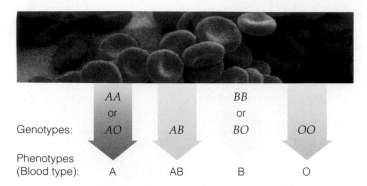

Genotypes:	AA or AO	AB	BB or BO	OO
Phenotypes (Blood type):	A	AB	B	O

Figure 11.10 Animated Combinations of alleles that are the basis of ABO blood typing.

homozygous parent (*RR*) × homozygous parent (*rr*) → heterozygous F$_1$ offspring (*Rr*)

A Cross a red-flowered with a white-flowered plant, and all of the F$_1$ offspring will be pink.

B Cross two F$_1$ plants, and the three phenotypes of the F$_2$ offspring will occur in a 1:2:1 ratio:

	R	r
R	RR	Rr
r	Rr	rr

Figure 11.11 Incomplete dominance in snapdragons.

Codominance in ABO Blood Types

With **codominance**, two nonidentical alleles of a gene are both fully expressed in heterozygotes, so neither is dominant or recessive. Codominance may occur in **multiple allele systems**, in which three or more alleles of a gene persist among individuals of a population.

ABO blood typing is a method of determining an individual's genotype at the *ABO* gene locus, a multiple allele system. The method checks for a membrane glycolipid that helps give the body's cells a unique identity. This glycolipid occurs in slightly different forms. Which form a person has begins with a gene, *ABO*, that encodes an enzyme. There are three alleles of this gene. The *A* and *B* alleles encode different versions of the enzyme. The *O* allele has a mutation that prevents its enzyme product from becoming active.

The alleles you carry for the *ABO* gene determine your blood type (Figure 11.10). The *A* and the *B* allele are codominant when paired. If your genotype is *AB*, then you have both versions of the enzyme, and your blood is type AB. The *O* allele is recessive when paired with either *A* or *B*. If you are *AA* or *AO*, your blood is type A. If you are *BB* or *BO*, it is type B. If you are *OO*, it is type O.

The reason for blood typing is that receiving incompatible blood cells in a transfusion is dangerous. The immune system attacks any red blood cells bearing glycolipids that are foreign to the body. Such an attack can cause the cells to clump or burst, a transfusion reaction with potentially fatal consequences. Type O blood is compatible with all other blood types, so people who have it are called universal blood donors. If you have AB blood, you can receive a transfusion of any blood type; you are called a universal recipient.

Incomplete Dominance

In **incomplete dominance**, one allele of a pair is not fully dominant over its partner, so the heterozygote's phenotype is somewhere between the two homozygotes.

A cross between two true-breeding snapdragons, one red and one white, reveals incomplete dominance: all F$_1$ offspring are pink-flowered (Figure 11.11). Why? Red snapdragons have two alleles that let them make a lot of red pigment. White snapdragons have two mutated alleles; they do not make any pigment at all, so their flowers are colorless. Pink snapdragons have a "red" allele and a "white" allele; such heterozygotes make only enough pigment to color the flowers pink. Cross two pink F$_1$ plants and you can expect to see red, pink, and white flowers in a 1:2:1 ratio in F$_2$ offspring.

Epistasis

Some traits are affected by interactions among different gene products, an effect called **epistasis**. Typically, one gene product suppresses the effect of another, so the resulting phenotype is somewhat unexpected. Epistatic interactions between two genes in chickens cause dramatic variations in their combs (Figure 11.12).

As another example, several genes affect Labrador retriever coat color, which can be black, yellow, or brown (Figure 11.13). A dog's coat color depends on how products of alleles at more than one locus make a dark pigment, melanin, and deposit it in tissues. Allele B (black) is dominant to b (brown). At a different locus, allele E promotes the deposition of melanin in fur, but two recessive alleles (ee) reduce it. A dog with two e alleles has yellow fur regardless of which alleles it has at the B locus.

Single Genes With a Wide Reach

One gene may influence multiple traits, an effect called **pleiotropy**. Genes encoding products used throughout the body are the ones most likely to be pleiotropic. For example, long fibers of fibrillin impart elasticity to the tissues of the heart, skin, blood vessels, tendons, and other body parts. Mutations in the fibrillin gene cause a genetic disorder called Marfan syndrome, in which tissues form with defective fibrillin or none at all. The largest blood vessel leading from the heart, the aorta, is particularly affected. In Marfan syndrome, muscle cells in the aorta's thick wall do not function very well, and the wall itself is not as elastic as it should be. The aorta expands under pressure, so the lack of elasticity eventually makes it thin and leaky. Calcium deposits accumulate inside. Inflamed, thinned, and weakened, the aorta can rupture abruptly during exercise.

Marfan syndrome can be very difficult to diagnose. Affected people are often tall, thin, and loose-jointed, but there are plenty of tall, thin, loose-jointed people without the disorder. Symptoms may not be apparent, so many people are not aware that they have Marfan. Until recently, it killed most of them before the age of fifty. Haris Charalambous was one (Figure 11.14).

Figure 11.12 Variation in combs, the fleshy, red crest on the head of chickens, is an outcome of interactions among products of alleles at two gene loci.

	EB	Eb	eB	eb
EB	EEBB black	EEBb black	EeBB black	EeBb black
Eb	EEBb black	EEbb chocolate	EeBb black	Eebb chocolate
eB	EeBB black	EeBb black	eeBB yellow	eeBb yellow
eb	EeBb black	Eebb chocolate	eeBb yellow	eebb yellow

Figure 11.13 *Left to right*, black, chocolate, and yellow Labrador retrievers. Epistatic interactions among products of two gene pairs affect coat color.

Figure 11.14 Rising basketball star Haris Charalambous, who died suddenly when his aorta burst during warmup exercises at the University of Toledo in 2006. He was 21.

Charalambous was very tall and lanky, with long arms and legs—traits that are valued in professional athletes such as basketball players. These traits are also associated with Marfan syndrome.

Like many other people, Charalambous did not know he had Marfan syndrome. An estimated 1 in 5,000 people are affected by Marfan worldwide.

Take-Home Message

Are all alleles clearly dominant or recessive?

■ An allele may be fully dominant, incompletely dominant, or codominant with its partner on a homologous chromosome.

■ In epistasis, two or more gene products influence a trait. In pleiotropy, one gene product influences two or more traits.

- The farther apart two genes are on a chromosome, the more often crossing over occurs between them.

- Links to Meiosis 10.3, Crossing over 10.4

As you learned in Section 11.3, alleles of genes on different chromosomes assort independently into gametes. What about genes on the same chromosome? Mendel studied seven genes in pea plants, which have seven pairs of chromosomes. Was he lucky enough to choose one gene on each of those seven chromosome pairs? Some surmised that if he had studied more genes, he would have discovered an exception to his law of independent assortment.

As it turns out, some of the genes Mendel studied *are* on the same chromosome. The genes are far enough apart that crossing over occurs between them very frequently—so frequently that they tend to assort into gametes independently, just as if they were on different chromosomes. By contrast, genes that are very close together on a chromosome do not tend to assort independently, because crossing over does not happen very often between them. Thus, gametes usually receive parental combinations of alleles of such genes. Genes that do not assort independently are said to be linked (Figure 11.15).

Alleles of some linked genes stay together during meiosis more than others do. The effect is simply due to the relative distance between genes: Genes that are closer together on a chromosome get separated less frequently by crossovers. For example, if genes *A* and *B* are twice as far apart on a chromosome as genes *C* and *D*, then we can expect crossovers to separate alleles of genes *A* and *B* more frequently than they separate alleles of genes *C* and *D:*

A B C D

Generalizing from this example, we can say the probability that a crossover event will separate alleles of two genes is proportional to the distance between those genes. In other words, the closer together any two genes are on a chromosome, the more likely gametes will be to receive parental combinations of alleles of those genes. Genes are said to be tightly linked if the distance between them is relatively small.

All genes on one chromosome are called a **linkage group**. Peas have 7 chromosomes, so they have 7 linkage groups. Humans have 23 chromosomes, so they have 23 linkage groups.

Human gene linkages were identified by tracking inheritance in families over several generations. One thing became clear: Crossovers are not at all rare, and may even be required in order for meiosis to run to completion. In many eukaryotes, at least two crossover events occur between every pair of homologous chromosomes during prophase I of meiosis.

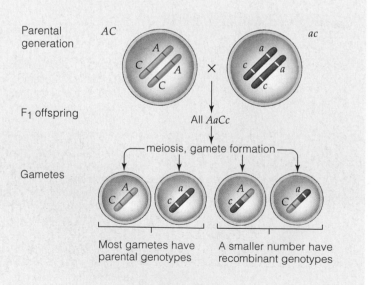

Mendel's blind luck in the genetics game?

Parental generation   *AC*   ×   *ac*

F₁ offspring   All *AaCc*

meiosis, gamete formation

Gametes

Most gametes have parental genotypes   A smaller number have recombinant genotypes

Figure 11.15 Animated Linkage and crossing over. Alleles of two genes on the same chromosome stay together when there is no crossover between them, and recombine when there is a crossover between them.

Take-Home Message

What is the effect of crossing over on inheritance?

- All genes on a chromosome are part of a linkage group.
- Crossing over disrupts linkage groups. The farther apart two genes are on a chromosome, the more often crossing over occurs between them.

11.6 | Genes and the Environment

- The environment can influence gene expression.
- Link to Enzyme function 6.3

Variations in traits are not always the result of differences in alleles. For example, in Section 6.3, you read about a heat-sensitive enzyme, tyrosinase, that affects the coat color of Siamese cats. This enzyme catalyzes one step in the synthesis of melanin, but it only works in cooler body regions, such as the legs, tail, and ears. Tyrosinase also affects coat color in Himalayan rabbits. The rabbits are homozygous for the c^h allele, which encodes a form of tyrosinase that stops working when the temperature in cells exceeds 33°C, or 91°F. Metabolic heat keeps the more massive body parts warm enough to deactivate the enzyme, so the fur is light on their surfaces. The ears and other slender appendages lose metabolic heat faster, so they are cooler, and melanin darkens them (Figure 11.16).

Yarrow plants offer another example of how environment influences phenotype. Yarrow is useful for experiments because it grows from cuttings. All cuttings of a plant have the same genotype, so experimenters know that genes are not the basis for any phenotypic differences among them. In one study, genetically identical yarrow plants had different phenotypes when grown at different altitudes (Figure 11.17).

Invertebrates, too, show phenotypic variation with environmental conditions. For instance, daphnias are microscopic freshwater relatives of shrimps. Aquatic insects prey on them. *Daphnia pulex* living in ponds with few predators have rounded heads, but those in ponds with many predators have more pointed heads (Figure 11.18). *Daphnia*'s predators emit chemicals that trigger the different phenotype.

The environment also affects human genes. One of our genes encodes a protein that transports serotonin across the membrane of brain cells. Serotonin lowers anxiety and depression during traumatic times. Some mutations in the serotonin transporter gene can reduce the ability to cope with stress. It is as if some of us are bicycling through life without an emotional helmet. Only when we crash does the mutation's phenotypic effect—depression—appear. Other human genes affect emotional state, but mutations in this one reduce our capacity to snap out of it when bad things happen.

Take-Home Message

Does the environment influence gene expression?

- Expression of some genes is affected by environmental factors such as temperature. The result may be variation in traits.

Figure 11.16 Animated Observable effect of the environment on gene expression. A Himalayan rabbit is homozygous for an allele that encodes a heat-sensitive form of an enzyme required for melanin synthesis. Cooler body parts, such as ears, are dark. The main body mass is warmer, and light. A patch of one rabbit's white fur was shaved off. An ice pack was tied above the fur-free patch. Fur grew back, but it was dark. The ice pack had cooled the patch enough for the enzyme to work, and melanin was produced.

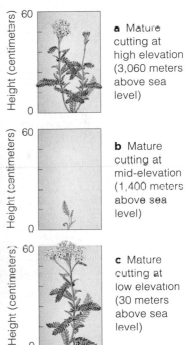

a Mature cutting at high elevation (3,060 meters above sea level)

b Mature cutting at mid-elevation (1,400 meters above sea level)

c Mature cutting at low elevation (30 meters above sea level)

Figure 11.17 Experiment showing environmental effects on phenotype in yarrow (*Achillea millefolium*). Cuttings from the same parent plant were grown in the same kind of soil at three different elevations.

Figure 11.18 (a) Light micrograph of a living daphnia. (b) Phenotypic effects of the presence of insects that prey on daphnias. The body form at the *left* develops when predators are absent or few. The form at the *right* develops when water contains chemicals emitted by the daphnia's insect predators. It has a longer tail spine and a pointed spine at the head.

■ Individuals of most species vary in some of their shared traits. Many traits show a continuous range of variation.

Continuous Variation

The individuals of a species typically vary in many of their shared traits. Some of those traits appear in two or three forms; others occur in a range of small differences that is called **continuous variation**. Continuous variation is an outcome of polygenic inheritance, in which multiple genes affect a single trait. The more genes and environmental factors that influence a trait, the more continuous is its variation.

Consider eye color. The colored part is the iris, a doughnut-shaped, pigmented structure. Several gene products contribute to iris color by making and distributing melanins. The more melanin deposited in the iris, the less light is reflected from it. Irises that are nearly black have dense melanin deposits that absorb almost all light, and reflect almost none. Melanin deposits are not as extensive in brown or hazel eyes, which reflect some incident light. Green, gray, and blue eyes have the least amount of melanin, so they reflect the most light.

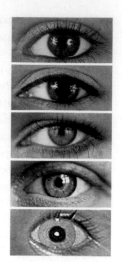

How do we determine whether a trait varies continuously? First,

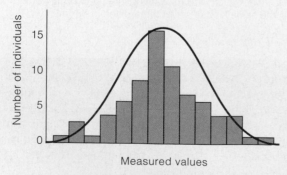

A Continuous variation in height among biology students at the University of Florida. Male (*above*) and female (*right*) students were divided into categories of one-inch increments in height.

B Bar graph showing the data from the male biology students in (**a**). Notice the rough bell-shaped distribution.

Figure 11.19 Animated Continuous variation. These examples show continuous variation in body height, one of the traits that help characterize human populations.

we divide the total range of phenotypes into measurable categories, such as inches of height. Next, we count how many individuals of a group fall into each category; this count gives the relative frequencies of phenotypes across our range of measurable values. Finally, we plot the data as a bar chart (Figure 11.19). In such charts, the shortest bars are categories with the fewest individuals, and the tallest bars are the categories with the most. A graph line around the top of the bars shows the distribution of values for the trait. If the line is a bell-shaped curve, or **bell curve**, the trait shows continuous variation.

Regarding the Unexpected Phenotype

Nearly all of the traits Mendel studied appeared in predictable ratios because the gene pairs happened to be on different chromosomes or far apart on the same chromosome. They tended to segregate independently. However, there is often far more variation in phenotypes, and not all of it is a result of crossing over.

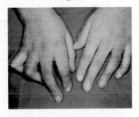

For example, certain mutations cause camptodactyly, in which finger shape and movement are abnormal. Any or all fingers on the left hand, right hand, or both hands may be bent and immobile (*right*).

What causes complex variation? Most organic molecules are synthesized in metabolic pathways that involve many enzymes. Genes encoding those enzymes can mutate in any number of ways, so their products may function within a spectrum of activity that ranges from excessive to not at all. Thus, the end product of a metabolic pathway can be produced within a range of concentration and activity. Environmental factors often add further variations on top of that.

Thus, phenotype results from complex interactions among gene products and the environment. We return to this topic in Chapter 18, as we consider some evolutionary consequences of variation in phenotype.

Take-Home Message

How does phenotype vary?

■ Some traits have a range of small differences, or continuous variation. The more genes and other factors that influence a trait, the more continuous the distribution of phenotype.

■ Enzymes and other gene products control steps of most metabolic pathways. Mutations, interactions among genes, and environmental conditions can affect one or more steps, and thus contribute to variation in phenotypes.

Summary

Section 11.1 By experimenting with pea plants, Mendel gathered evidence of patterns by which parents transmit genes to offspring. **Genes** are units of DNA that hold information about traits. Each has its own **locus**, or location, along the length of a chromosome. **Mutations** give rise to different forms of a gene (alleles).

Individuals that carry two identical alleles of a gene are **homozygous** for the gene: (AA) or (aa). Individuals that breed true for a trait are homozygous for alleles that affect the trait. Offspring of a cross between individuals homozygous for different alleles of a gene are **hybrids**, or **heterozygous**, with two nonidentical alleles (Aa).

A **dominant** allele masks the effect of a **recessive** allele partnered with it on the homologous chromosome. An individual with two dominant alleles (AA) is **homozygous dominant**. An individual with two recessive alleles is **homozygous recessive** (aa).

The alleles at any or all gene loci constitute an individual's **genotype**. **Gene expression** results in **phenotype**, which refers to an individual's observable traits.

■ *Learn how Mendel crossed garden pea plants, and the definitions of important genetic terms, on CengageNOW.*

Section 11.2 Crossing individuals that breed true for two forms of a trait ($AA \times aa$) yields identically heterozygous F_1 offspring (Aa). A cross between such F_1 offspring is a **monohybrid experiment**, which can reveal dominance relationships among the alleles. We use **Punnett squares** to calculate the **probability** of seeing certain phenotypes in F_2 offspring of such **testcrosses**:

	A	a
A	AA	Aa
a	Aa	aa

AA (dominant)
Aa (dominant)   } the expected
Aa (dominant)   phenotypic
aa (recessive)   } ratio of 3:1

Mendel's monohybrid experiment results led to his law of **segregation** (stated here in modern terms): Diploid organisms have pairs of genes, on pairs of homologous chromosomes. During meiosis, the genes of each pair separate, so each gamete gets one or the other gene.

■ *Use the interaction on CengageNOW to carry out monohybrid experiments and a testcross.*

Section 11.3 Crossing individuals that breed true for two forms of two traits ($AABB \times aabb$) yields identically heterozygous F_1 offspring ($AaBb$). A cross between these F_1 offspring is called a **dihybrid experiment**. The phenotype ratios in F_2 offspring of such testcrosses may reveal dominance relationships among the alleles. Mendel saw a 9:3:3:1 phenotype ratio in his dihybrid experiments:

9 dominant for both traits
3 dominant for A, recessive for b
3 dominant for B, recessive for a
1 recessive for both traits

These results led to the law of **independent assortment** (stated in modern terms): Meiosis assorts gene pairs on homologous chromosomes independently of other gene

A person of mixed ethnicity may make gametes that contain different mixtures of alleles for dark and light skin. It is fairly rare that one of those gametes contains all of the alleles for dark skin, or all of the alleles for light skin, but it happens, as evidenced by twins Kian and Remee.

In the case of the *SLC24A5* gene, a mutation that occurred between 6,000 and 10,000 years ago changed the 111th amino

acid of its protein product from alanine to threonine. This small change resulted in the European version of white skin.

How would you vote? Physical attributes such as skin color, which have a genetic basis, are often used to define race. Are twins such as Kian and Remee of different races? See CengageNOW for details, then vote online.

---

pairs on the other chromosomes. The random attachment of homologous chromosomes to opposite spindle poles during prophase I is the basis of this outcome.

■ *Use the interactions on CengageNOW to observe the results of a dihybrid cross.*

Section 11.4 Inheritance patterns frequently vary. With **incomplete dominance**, an allele is not fully dominant over its partner on a homologous chromosome, and both are expressed. The combination of alleles gives rise to an intermediate phenotype.

Codominant alleles are both expressed at the same time in heterozygotes, as in the **multiple allele system** underlying ABO blood typing. In **epistasis**, interacting products of one or more genes often affect the same trait. A **pleiotropic** gene affects two or more traits.

■ *Use the interactions on CengageNOW to explore patterns of non-Mendelian inheritance.*

Section 11.5 The farther apart two genes are on a chromosome, the greater the frequency of crossing over between them. Genes that are relatively close to each other on a chromosome tend to stay together during meiosis because few crossover events occur between them. Genes that are relatively far apart tend to assort independently into gametes. All of the genes on one chromosome constitute a **linkage group**.

Section 11.6 Various environmental factors may affect gene expression in individuals.

■ *Use the interactions on CengageNOW to see how the environment can affect phenotype.*

Section 11.7 A trait that is influenced by the products of multiple genes often occurs in a range of small increments of phenotype (**continuous variation**).

■ *Use the interaction on CengageNOW to plot the continuous distribution of height for a class.*

Self-Quiz
Answers in Appendix III

1. Alleles are _____ .
   a. different molecular forms of a gene
   b. different phenotypes
   c. self-fertilizing, true-breeding homozygotes

2. A bell curve indicates _____ in a trait.

3. A heterozygote has a _____ for a trait being studied.
   a. pair of identical alleles
   b. pair of nonidentical alleles
   c. haploid condition, in genetic terms

4. The observable traits of an organism are its _____ .
   a. phenotype   c. genotype
   b. sociobiology   d. pedigree

5. Second-generation offspring of a cross between parents who are homozygous for different alleles are the _____ .
   a. F_1 generation   c. hybrid generation
   b. F_2 generation   d. none of the above

6. F_1 offspring of the cross $AA \times aa$ are _____ .
   a. all AA c. all Aa
   b. all aa d. 1/2 AA and 1/2 aa

7. Refer to question 5. Assuming complete dominance, the F_2 generation will show a phenotypic ratio of _____ .
   a. 3:1   b. 9:1   c. 1:2:1   d. 9:3:3:1

8. A testcross is a way to determine _____ .
   a. phenotype   b. genotype   c. both a and b

9. Assuming complete dominance, crosses between two dihybrid F_1 pea plants, which are offspring from a cross $AABB \times aabb$, result in F_2 phenotype ratios of _____ .
   a. 1:2:1   b. 3:1   c. 1:1:1:1   d. 9:3:3:1

10. The probability of a crossover occurring between two genes on the same chromosome _____ .
    a. is unrelated to the distance between them
    b. decreases with the distance between them
    c. increases with the distance between them

11. Two genes that are close together on the same chromosome are _____ .
    a. linked   c. homologous   e. all of the
    b. identical alleles   d. autosomes   above

12. Match each example with the most suitable description.
    ___ dihybrid experiment   a. *bb*
    ___ monohybrid experiment   b. *AABB* × *aabb*
    ___ homozygous condition   c. *Aa*
    ___ heterozygous condition   d. *Aa* × *Aa*

■ *Visit CengageNOW for additional questions.*

Genetics Problems
Answers in Appendix III

1. Assuming that independent assortment occurs during meiosis, what type(s) of gametes will form in individuals with the following genotypes?
   a. *AABB*   b. *AaBB*   c. *Aabb*   d. *AaBb*

Data Analysis Exercise

A 2000 study measured average skin color of people native to more than fifty regions, and correlated them to the amount of UV radiation received in those regions. Some of their results are shown in Figure 11.20.

1. Which country receives the most UV radiation? The least?

2. The people native to which country have the darkest skin? The lightest?

3. According to this data, how does the skin color of indigenous peoples correlate with the amount of UV radiation incident in their native regions?

Country	Skin Reflectance	UVMED
Australia	19.30	335.55
Kenya	32.40	354.21
India	44.60	219.65
Cambodia	54.00	310.28
Japan	55.42	130.87
Afghanistan	55.70	249.98
China	59.17	204.57
Ireland	65.00	52.92
Germany	66.90	69.29
Netherlands	67.37	62.58

Figure 11.20
Skin color of indigenous peoples and regional incident UV radiation. Skin reflectance measures how much light of 685 nanometers wavelength is reflected from skin; UVMED is the annual average UV radiation received at Earth's surface.

---

2. Refer to problem 1. Determine the frequencies of each genotype among offspring from the following matings:
   a. $AABB \times aaBB$
   b. $AaBB \times AABb$
   c. $AaBb \times aabb$
   d. $AaBb \times AaBb$

3. Refer to problem 2. Assume a third gene has alleles C and c. For each genotype listed, what allele combinations will occur in gametes, assuming independent assortment?
   a. $AABBCC$
   b. $AaBBcc$
   c. $AaBBCc$
   d. $AaBbCc$

4. Sometimes the gene for tyrosinase mutates so its product is not functional. An individual who is homozygous recessive for such a mutation cannot make melanin. Albinism, the absence of melanin, results. Humans and many other organisms can have this phenotype (right). In the following situations, what are the probable genotypes of the father, the mother, and their children?

   a. Both parents have normal phenotypes; some of their children are albino and others are unaffected.
   b. Both parents are albino and have albino children.
   c. The woman is unaffected, the man is albino, and they have one albino child and three unaffected children.

5. Certain genes are vital for development. When mutated, they are lethal in homozygous recessives. Even so, heterozygotes can perpetuate recessive, lethal alleles. The allele *Manx* (M^L) in cats is an example. Homozygous cats ($M^L M^L$) die before birth. In heterozygotes ($M^L M$), the spine develops abnormally, and the cats end up with no tail (right).

   Two $M^L M$ cats mate. What is the probability that any one of their surviving kittens will be heterozygous?

6. Several alleles affect traits of roses, such as plant form and bud shape. Alleles of one gene govern whether a plant will be a climber (dominant) or shrubby (recessive). All F_1 offspring from a cross between a true-breeding climber and a shrubby plant are climbers. If an F_1 plant is crossed with a shrubby plant, about 50 percent of the offspring will be shrubby; 50 percent will be climbers. Using symbols A and a for the dominant and recessive alleles, make a Punnett-square diagram of the expected genotypes and phenotypes in F_1 offspring and in offspring of a cross between an F_1 plant and a shrubby plant.

7. Mendel crossed a true-breeding pea plant with green pods and a true-breeding pea plant with yellow pods. All the F_1 plants had green pods. Which color is recessive?

8. Suppose you identify a new gene in mice. One of its alleles specifies white fur, another specifies brown. You want to see if the two interact in simple or incomplete dominance. What sorts of genetic crosses would give you the answer?

9. In sweet pea plants, an allele for purple flowers (P) is dominant to an allele for red flowers (p). An allele for long pollen grains (L) is dominant to an allele for round pollen grains (l). Bateson and Punnett crossed a plant having purple flowers/long pollen grains with one having white flowers/round pollen grains. All F_1 offspring had purple flowers and long pollen grains. Among the F_2 generation, the researchers observed the following phenotypes:

   296 purple flowers/long pollen grains
   19 purple flowers/round pollen grains
   27 red flowers/long pollen grains
   85 red flowers/round pollen grains

What is the best explanation for these results?

10. Red-flowering snapdragons are homozygous for allele R^1. White-flowering snapdragons are homozygous for a different allele (R^2). Heterozygous plants ($R^1 R^2$) bear pink flowers. What phenotypes should appear among first-generation offspring of the crosses listed? What are the expected proportions for each phenotype?
   a. $R^1 R^1 \times R^1 R^2$
   b. $R^1 R^1 \times R^2 R^2$
   c. $R^1 R^2 \times R^1 R^2$
   d. $R^1 R^2 \times R^2 R^2$

(Incompletely dominant alleles are usually designated by superscript numerals, as shown, not by uppercase letters for dominance and lowercase letters for recessiveness.)

11. A single mutant allele gives rise to an abnormal form of hemoglobin (Hb^S, not Hb^A). Homozygotes ($Hb^S Hb^S$) develop sickle-cell anemia (Section 3.6). Heterozygotes ($Hb^A Hb^S$) have few symptoms. A couple who are both heterozygous for the Hb^S allele plan to have children. For each of the pregnancies, state the probability that they will have a child who is:
   a. homozygous for the Hb^S allele
   b. homozygous for the Hb^A allele
   c. heterozygous: $Hb^A Hb^S$

12 Chromosomes and Human Inheritance

IMPACTS, ISSUES Strange Genes, Tortured Minds

"This man is brilliant." That was the extent of a letter of recommendation from Richard Duffin, a professor of mathematics at Carnegie Mellon University. Duffin wrote the line in 1948 on behalf of John Forbes Nash, Jr. Nash was twenty years old at the time and applying for admission to Princeton University's graduate school. Over the next ten years, Nash made his reputation as one of the foremost mathematicians. He was socially awkward, but so are many highly gifted people. Nash showed no symptoms warning of the paranoid schizophrenia that eventually would debilitate him.

Full-blown symptoms emerged in his thirtieth year. Nash had to abandon his position at the Massachusetts Institute of Technology. Two decades passed before he was able to return to his pioneering work in mathematics.

Of every 100 people, 1 is affected by schizophrenia. This neurobiological disorder (NBD) is characterized by delusions, hallucinations, disorganized speech, and abnormal social behavior. Exceptional creativity often accompanies schizophrenia. It also accompanies other NBDs, including autism, chronic depression, and bipolar disorder, which manifests itself as jarring swings in mood and social behavior.

Certainly not every person with a high IQ has a neurobiological disorder, but a higher percentage of creative geniuses have NBDs than nongeniuses. In fact, emotionally healthy, highly creative people have more personality traits in common with people affected by NBDs than with individuals closer to the norm. For instance, both tend to be hypersensitive to environmental stimuli. Some may be on the edge of mental instability. Those who develop NBDs become part of a crowd that includes physicist Sir Isaac Newton, philosopher Socrates, composer Ludwig von Beethoven, painter Vincent van Gogh, psychiatrist Sigmund Freud, politician Winston Churchill, poets Edgar Allan Poe and Lord Byron, writers James Joyce and Ernest Hemingway, and many others (Figure 12.1).

We have not yet identified all the interactions among genes and environment that might tip such individuals one way or the other. But we do know about several mutations that predispose them to develop NBDs.

Indeed, NBDs tend to run in families. Many of the same families that produce creative geniuses also produce people affected by an NBD—sometimes more than one type. For decades, neuroscientists have been studying these families. They have identified several gene loci that, when mutated, are associated with a spectrum of neurobiological disorders.

With this intriguing connection, we invite you into the study of the chromosomal basis of human inheritance.

See the video! Figure 12.1 NBDs and creativity. Abraham Lincoln (*left*) suffered from chronic depression even as he changed the course of U.S. history. *Center*, Virginia Woolf's suicide after a mental breakdown is a tragic example of creative writers, who are, as a group, eighteen times more suicidal, ten times more likely to be depressed, and twenty times more likely to have bipolar disorder than the average person. Pablo Picasso (*right*) suffered from depression, and perhaps schizophrenia.

Key Concepts

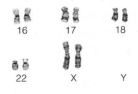

Autosomes and sex chromosomes

Animals have chromosomes called autosomes. The members of autosome pairs are identical in length, shape, and which genes they carry. In sexually reproducing species, the members of a pair of sex chromosomes differ between females and males. **Section 12.1**

Autosomal inheritance

Many genes on autosomes are expressed in Mendelian patterns of simple dominance. Some dominant or recessive alleles result in genetic disorders. **Sections 12.2, 12.3**

Sex-linked inheritance

Some traits are affected by genes on the X chromosome. Inheritance patterns of such traits differ between males and females. **Section 12.4**

Changes in chromosome structure or number

On rare occasions, a chromosome may undergo a large-scale, permanent change in its structure, or the number of autosomes or sex chromosomes may change. In humans, such changes usually result in a genetic disorder. **Sections 12.5, 12.6**

Human genetic analysis

Various analytical and diagnostic procedures often reveal genetic disorders. What an individual, and society at large, should do with the information raises ethical questions. **Sections 12.7, 12.8**

Links to Earlier Concepts

- In this chapter, you will be drawing upon your knowledge of chromosome structure (Section 9.1), mitosis (9.3), meiosis (10.3, 10.4), and gamete formation (11.3).

- Before you start, make sure you understand dominant and recessive alleles, and homozygous and heterozygous conditions (11.1).

- Sampling error (1.8), carbohydrates (3.3), protein structure (3.6), the cell cortex (4.13), pigments (7.1), glycolysis (8.2), and nutrition (8.7) will all turn up again, this time in the context of different genetic disorders.

How would you vote? Tests for predisposition to neurobiological disorders will be available soon. Insurance companies and employers may use the information to discriminate against otherwise healthy individuals. Do you support legislation governing the tests? See CengageNOW for details, then vote online.

- In humans, two sex chromosomes are the basis of sex. All other human chromosomes are autosomes.

- Links to Chromosome structure 9.1 and 9.3, Meiosis 10.3

Most animals, including humans, normally are either female or male. They also have a diploid chromosome number (2*n*), with pairs of homologous chromosomes in their body cells. Typically, all except one pair of those chromosomes are **autosomes**, which carry the same genes in both females and males. Autosomes of a pair have the same length, shape, and centromere location. Members of a pair of **sex chromosomes** differ between females and males. The differences determine an individual's sex.

The sex chromosomes of humans are called X and Y. Body cells of human females contain two X chromosomes (XX); those of human males contain one X and one Y chromosome (XY). The X and Y chromosomes differ in length, shape, and which genes they carry, but they interact as homologues during prophase I.

XX females and XY males are the rule among fruit flies, mammals, and many other animals, but there are other patterns. In butterflies, moths, birds, and certain fishes, males have two identical sex chromosomes, not females. Environmental factors (not sex chromosomes) determine sex in some species of invertebrates, turtles, and frogs. As an example, the temperature of the sand in which sea turtle eggs are buried determines the sex of the hatchlings.

Sex Determination

In humans, a new individual inherits a combination of sex chromosomes that dictates whether it will become a male or a female. All eggs made by a human female have one X chromosome. One-half of the sperm cells made by a male carry an X chromosome; the other half carry a Y chromosome. If an X-bearing sperm fertilizes an X-bearing egg, the resulting zygote will develop into a female. If the sperm carries a Y chromosome, the zygote will develop into a male (Figure 12.2*a*).

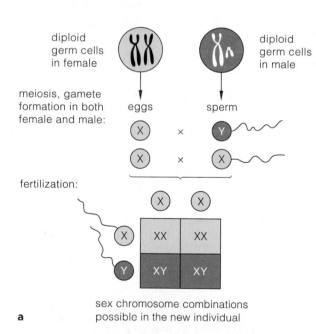

sex chromosome combinations possible in the new individual

a

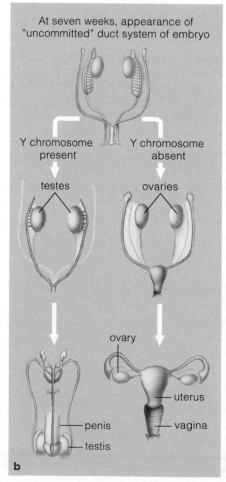

b

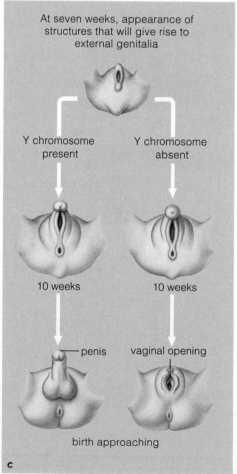

c

Figure 12.2 Animated (**a**) Punnett-square diagram showing the sex determination pattern in humans.

(**b**) An early human embryo appears neither male nor female. Then tiny ducts and other structures that can develop into male *or* female reproductive organs start forming. In an XX embryo, ovaries form in the absence of the Y chromosome and its *SRY* gene. In an XY embryo, the gene product triggers formation of testes, which secrete a hormone that initiates development of other male traits. (**c**) External reproductive organs in human embryos.

The human Y chromosome carries only 307 genes, but one of them is the *SRY* gene—the master gene for male sex determination. Its expression in XY embryos triggers the formation of testes, which are male gonads (Figure 12.2b). Some of the cells in these primary male reproductive organs make testosterone, a sex hormone that controls the emergence of male secondary sexual traits such as facial hair, increased musculature, and a deep voice. How do we know *SRY* is the male sex master gene? Mutations in this gene cause XY individuals to develop external genitalia that appear female.

An XX embryo has no Y chromosome, no *SRY* gene, and much less testosterone, so primary female reproductive organs (ovaries) form instead of testes. Ovaries make estrogens and other sex hormones that will govern the development of female secondary sexual traits, such as enlarged, functional breasts, and fat deposits around the hips and thighs.

The human X chromosome carries 1,336 genes. Some of those genes are associated with sexual traits, such as the distribution of body fat and hair. However, most of the genes on the X chromosome govern nonsexual traits such as blood clotting and color perception. Such genes are expressed in both males and females. Males, remember, also inherit one X chromosome.

Karyotyping

Sometimes the structure of a chromosome can change during mitosis or meiosis. Chromosome number can change also. A diagnostic tool called karyotyping helps us determine an individual's diploid complement of chromosomes (Figure 12.3). With this procedure, cells taken from an individual are put into a fluid growth medium that stimulates mitosis. The growth medium contains colchicine, a poison that binds tubulin and so interferes with assembly of mitotic spindles. The cells enter mitosis, but the colchicine prevents them from dividing, so their cell cycle stops at metaphase.

The cells and the medium are transferred to a tube. Then, the cells are separated from the liquid medium, and a hypotonic solution is added (Section 5.6). The cells swell up with water, so the chromosomes inside of them move apart. The cells are spread on a microscope slide and stained so the chromosomes become visible with a microscope.

The microscope reveals metaphase chromosomes in every cell. A micrograph of a single cell is digitally rearranged so the images of the chromosomes are lined up by centromere location, and arranged according to size, shape, and length. The finished array constitutes the individual's **karyotype**, which is compared with a

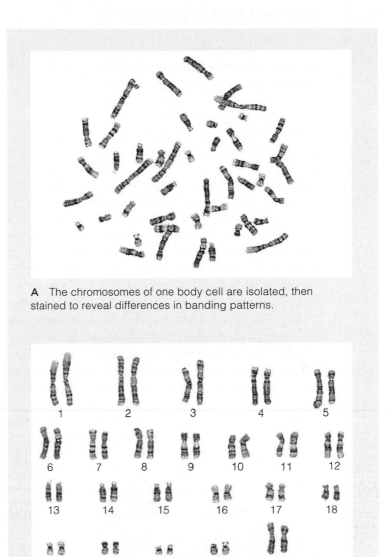

A The chromosomes of one body cell are isolated, then stained to reveal differences in banding patterns.

B The image is reassembled: The chromosomes are paired by size, centromere position, and other characteristics.

Figure 12.3 Animated Karyotyping, a diagnostic tool that reveals an image of a single cell's diploid complement of chromosomes. This human karyotype shows 22 pairs of autosomes and a pair of X chromosomes. **Figure It Out:** Was this cell taken from a male or female? *Answer: Female*

normal standard. The karyotype shows whether there are extra or missing chromosomes. Some other kinds of structural abnormalities are also visible.

Take-Home Message

What is the basis of sex determination in humans?

■ Members of a pair of sex chromosomes differ between males and females.

■ Other human chromosomes are autosomes—the same in males and females.

■ Many human traits can be traced to autosomal dominant or recessive alleles that are inherited in Mendelian patterns. Some of those alleles cause genetic disorders.

■ Links to Carbohydrates 3.3, Nutrition and aerobic respiration 8.7, Terms in genetics 11.1

Autosomal Dominant Inheritance

A dominant allele on an autosome (an autosomal dominant allele) is expressed in homozygotes and heterozygotes, so any trait it specifies tends to appear in every generation. When one parent is heterozygous, and the other is homozygous for the recessive allele, each of their children has a 50 percent chance of inheriting the dominant allele—and displaying the trait associated with it (Figure 12.4a).

An autosomal dominant allele causes achondroplasia, a genetic disorder that affects about 1 out of 10,000 people. Adult heterozygotes average about four feet, four inches tall, and have abnormally short arms and legs relative to other body parts (Figure 12.4a). While they were still embryos, the cartilage model on which a skeleton is constructed did not form properly. Most homozygotes die before or not long after birth.

A different autosomal dominant allele is responsible for Huntington's disease. With this genetic disorder, involuntary muscle movements increase as the nervous system slowly deteriorates. Typically, symptoms do not start until after age thirty; affected people die during their forties or fifties. The mutation that causes Huntington's alters a protein necessary for brain cell development. It is an expansion mutation, in which three nucleotides are duplicated many times. Hundreds of thousands of other expansion repeats occur harmlessly in and between genes on human chromosomes. This one alters the function of a critical gene product.

A dominant allele that causes severe problems can persist because expression of the allele does not interfere with reproduction. Achondroplasia is an example. With Huntington's and other late-onset disorders, people tend to reproduce before symptoms appear, so the allele may be passed unknowingly to children.

Autosomal Recessive Inheritance

Because autosomal recessive alleles are expressed only in homozygotes, traits associated with them may skip generations. Heterozygotes are carriers: They do not have the trait. Any child of two carriers has a 25 percent chance of inheriting the allele from both parents and being a homozygote with the trait (Figure 12.4b). All children of homozygous parents will be homozygous.

Galactosemia is a heritable metabolic disorder that affects about 1 in every 50,000 newborns. This case of autosomal recessive inheritance involves an allele for an enzyme that helps digest the lactose in milk or in milk products. The body normally converts lactose to glucose and galactose. Then, a series of three enzymes converts the galactose to glucose-6-phosphate (Figure 12.5). This intermediate can enter glycolysis or it can be converted to glycogen (Sections 3.3 and 8.7).

Figure 12.4 Animated (a) Example of autosomal dominant inheritance. A dominant allele (*red*) is fully expressed in heterozygotes. Achondroplasia, an autosomal dominant disorder, affects the three men shown above. At center, Verne Troyer (Mini Me in the Mike Myers spy movies), stands two feet, eight inches tall.

(b) An autosomal recessive pattern. In this example, both parents are heterozygous carriers of the recessive allele (*red*).

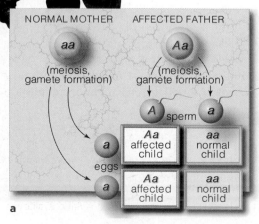

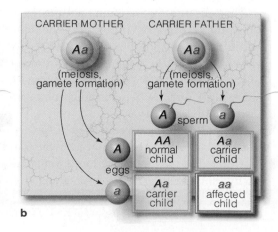

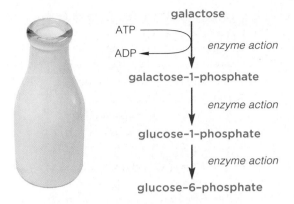

galactose

ATP ⟶
ADP ⟵ *enzyme action*

↓

galactose–1–phosphate

enzyme action

↓

glucose–1–phosphate

enzyme action

↓

glucose–6–phosphate

Figure 12.5 How galactose is normally converted to a form that can enter the breakdown reactions of glycolysis. A mutation that affects the second enzyme in the conversion pathway gives rise to galactosemia.

People with galactosemia do not make one of these three enzymes; they are homozygous recessive for a mutated allele. Galactose-1-phosphate accumulates to toxic levels in their body, and it can be detected in the urine. The condition leads to malnutrition, diarrhea, vomiting, and damage to the eyes, liver, and brain.

When they do not receive treatment, galactosemics typically die young. When they are quickly placed on a diet that excludes all dairy products, the symptoms may not be as severe.

What About Neurobiological Disorders?

Most of the neurobiological disorders mentioned in the chapter introduction do not follow simple patterns of Mendelian inheritance. In most cases, mutations in one gene do not give rise to depression, schizophrenia, or bipolar disorder. Multiple genes and environmental factors contribute to these outcomes. Nonetheless, it is useful to search for mutations that make some people more vulnerable to NBDs.

For example, researchers who conducted extensive family and twin studies have predicted that mutations in specific genes on chromosomes 1, 3, 5, 6, 8, 11–15, 18, and 22 increase an individual's chance of developing schizophrenia. Similarly, mutations in specific genes have been linked to bipolar disorder and depression.

Take-Home Message

How do we link traits to alleles on autosomal chromosomes?
■ Many traits, including some genetic disorders, can be traced to dominant or recessive alleles on autosomes because they are inherited in simple Mendelian patterns.

12.3 | Too Young to be Old

■ Progeria, genetic disorder that results in accelerated aging, is caused by mutations in an autosome.

Imagine being ten years old with a mind trapped in a body that is getting a bit more shriveled, more frail—old—every day. You are barely tall enough to peer over the top of a table. You weigh less than thirty-five pounds. Already you are bald and have a wrinkled nose. Possibly you have a few more years to live. Would you, like Mickey Hays and Fransie Geringer, still be able to laugh (Figure 12.6)?

On average, of every 8 million newborn humans, one will grow old far too soon. On one of its autosomes, that rare individual carries a mutated allele that gives rise to Hutchinson–Gilford progeria syndrome. While that new individual was still an embryo inside its mother, billions of DNA replications and mitotic cell divisions distributed the information encoded in that gene to each newly formed body cell. Its legacy will be an accelerated rate of aging and a sharply reduced life span.

The disorder arises by spontaneous mutation of a gene for lamin, a protein that normally makes up intermediate filaments in the nucleus (Section 4.13). The altered lamin is not processed properly. It builds up on the inner nuclear membrane and distorts the nucleus. How this buildup causes the symptoms of progeria is not yet known.

Those symptoms start before age two. Skin that should be plump and resilient starts to thin. Skeletal muscles weaken. Limb bones that should lengthen and grow stronger soften. Premature baldness is inevitable. Affected people do not usually live long enough to reproduce, so progeria does not run in families.

Most progeriacs can expect to die in their early teens as a result of strokes or heart attacks. These final insults are brought on by a hardening of the wall of arteries, a condition typical of advanced age. Fransie was seventeen when he died. Mickey died at age twenty.

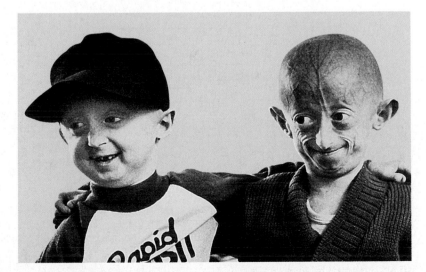

Figure 12.6 Mickey (*left*) and Fransie (*right*) met at a gathering of progeriacs at Disneyland, California. They were not yet ten years old.

■ X chromosome alleles give rise to phenotypes that reflect Mendelian patterns of inheritance. Some of those alleles cause genetic disorders.

■ Links to Cell cortex 4.13, Pigments 7.1

The X chromosome carries over 6 percent of all human genes. Mutations on this sex chromosome are known to cause or contribute to over 300 genetic disorders.

A recessive allele on an X chromosome (an X-linked recessive allele) leaves certain clues when it causes a genetic disorder. First, more males than females are affected by such X-linked recessive disorders. This is because heterozygous females have a second X chromosome that carries the dominant allele, which masks the effects of the recessive one. Heterozygous males have only one X chromosome, so they are not similarly protected (Figure 12.7). Second, an affected father cannot pass his X-linked recessive allele to a son because all children who inherit their father's X chromosome are female. Thus, a heterozygous female is the bridge between an affected male and his affected grandson.

X-linked dominant alleles that cause disorders are rarer than X-linked recessive ones, probably because they tend to be lethal in male embryos. Females, with two X chromosomes, most often have one functional allele that can dampen the effects of a mutated one.

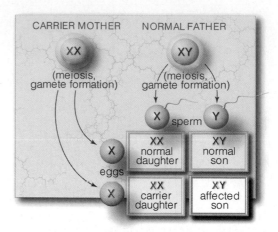

Figure 12.7 Animated X-linked recessive inheritance. In this case, the mother carries a recessive allele on one of her X chromosomes (*red*).

Hemophilia A

Hemophilia A is an X-linked recessive disorder that interferes with blood clotting. Most of us have a blood clotting mechanism that quickly stops bleeding from minor injuries. The mechanism involves protein products of genes on the X chromosome. Bleeding is prolonged in males who carry a mutated form of one of

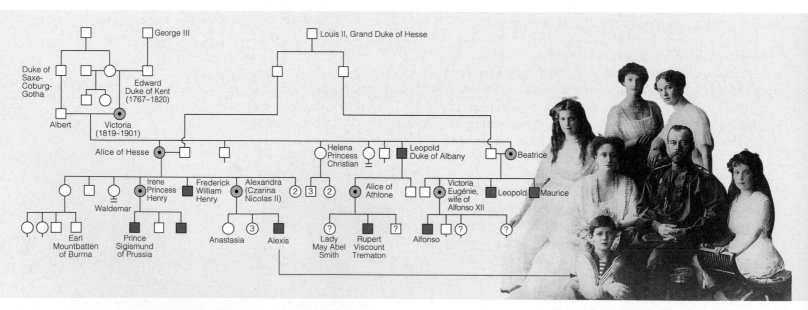

Figure 12.8 A classic case of X-linked recessive inheritance: a partial pedigree, or chart of genetic connections, of the descendants of Queen Victoria of England. At one time, the recessive X-linked allele that resulted in hemophilia was present in eighteen of Victoria's sixty-nine descendants, who sometimes intermarried.

Of the Russian royal family members shown, the mother (Alexandra Czarina Nicolas II) was a carrier. Through her obsession with the vulnerability of her son Alexis, a hemophiliac, she became involved in political intrigue that helped trigger the Russian Revolution of 1917.

○ female with normal alleles   □ male with normal allele
◉ female carrier   ■ affected male
③ three females   ⑦⑦ status unknown

Figure It Out: How many of Alexis' siblings were affected by hemophilia A?   *Answer: None*

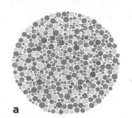
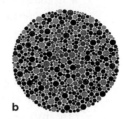

a   b

Figure 12.9 *Left,* what red–green color blindness means, using ripe red cherries on a green-leafed tree as an example. In this case, the perception of blues and yellows is normal, but the affected individual has difficulty distinguishing red from green.

Above, two of many Ishihara plates, which are standardized tests for different forms of color blindness. (**a**) You may have one form of red–green color blindness if you see the numeral 7 instead of 29 in this circle. (**b**) You may have another form if you see a 3 instead of an 8.

these X-linked genes, or in females who are homozygous for a mutation (clotting time is close to normal in heterozygous females). Affected people tend to bruise easily, and internal bleeding causes problems in their muscles and joints.

In the nineteenth century, the incidence of hemophilia A was relatively high in royal families of Europe and Russia, probably because the common practice of inbreeding kept the harmful allele circulating in their family trees (Figure 12.8). Today, about 1 in 7,500 people is affected, but that number may be rising because the disorder is now treatable. More affected people are living long enough to transmit the allele to children.

Red–Green Color Blindness

The pattern of X-linked recessive inheritance shows up among individuals who have some degree of color blindness. The term refers to a range of conditions in which an individual cannot distinguish among some or all colors in the spectrum of visible light. Mutated genes result in altered function of the photoreceptors (light-sensitive receptors) in the eyes.

Normally, humans can sense the differences among 150 colors. A person who is red–green color blind sees fewer than 25 colors: Some or all of the receptors that respond to red and green wavelengths are weakened or absent. Some people confuse red and green colors. Others see green as shades of gray, but perceive blues and yellows quite well (Figure 12.9). Two sections of a standard set of tests for color blindness are shown in Figure 12.9*a,b*.

Duchenne Muscular Dystrophy

Duchenne muscular dystrophy (DMD) is one of several X-linked recessive disorders that is characterized by muscle degeneration. DMD affects about 1 in 3,500 people, almost all of them boys.

A gene on the X chromosome encodes dystrophin, which is a protein that structurally supports the fused cells in muscle fibers by anchoring the cell cortex to the plasma membrane. When dystrophin is abnormal or absent, the cell cortex weakens and muscle cells die. The cell debris that remains in the tissues triggers chronic inflammation.

DMD is typically diagnosed in boys between the ages of three and seven. The rapid progression of this disorder cannot be stopped. When an affected boy is about twelve years old, he will begin to use a wheelchair. His heart muscles will start to break down. Even with the best of care, he will probably die before he is thirty years old, from a heart disorder or from respiratory failure (suffocation).

Take-Home Message

How do we link traits to alleles on sex chromosomes?

■ Mutated alleles on sex chromosomes cause or contribute to more than 300 genetic disorders. They are inherited in characteristic patterns.

■ A female heterozygous for one of those alleles may not show symptoms.

■ Males (XY) transmit an X-linked allele only to daughters, not to their sons.

■ On rare occasions, a chromosome's structure changes. Many of the alterations have severe or lethal outcomes.

■ Links to Protein structure 3.6, Meiosis 10.3

Large-scale changes in the structure of a chromosome may give rise to a genetic disorder. Such changes are rare, but they do occur spontaneously in nature. They can also be induced by exposure to certain chemicals or radiation. Either way, such alterations may be detected by karyotyping. Large-scale changes in chromosome structure include duplications, deletions, inversions, and translocations.

Duplication Even normal chromosomes have DNA sequences that are repeated two or more times. These repetitions are called **duplications**:

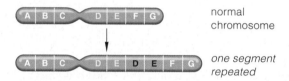

normal chromosome

one segment repeated

Duplications can occur through unequal crossovers at prophase I. Homologous chromosomes align side by side, but their DNA sequences misalign at some point along their length. The probability of misalignment is greater in regions where DNA has long repeats of the same sequence of nucleotides. A stretch of DNA gets deleted from one chromosome and is spliced into the partner chromosome. Some duplications, such as the expansion mutations that cause Huntington's, cause genetic abnormalities or disorders. Others have been evolutionarily important.

Deletion A **deletion** is the loss of some portion of a chromosome:

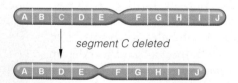

segment C deleted

In mammals, deletions usually cause serious disorders and are often lethal. The loss of genes results in the disruption of growth, development, and metabolism. For instance, a small deletion in chromosome 5 causes mental impairment and an abnormally shaped larynx. Affected infants tend to make a sound like the meow of a cat, hence the name of the disorder, cri-du-chat, which is French for "cat's cry" (Figure 12.10).

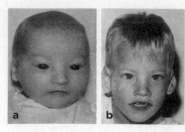

Figure 12.10 Cri-du-chat syndrome. (**a**) This infant's ears are low relative to his eyes. (**b**) Same boy, four years later. The high-pitched monotone of cri-du-chat children may persist into their adulthood.

Inversion With an **inversion**, part of the sequence of DNA within the chromosome becomes oriented in the reverse direction, with no molecular loss:

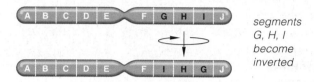

segments G, H, I become inverted

An inversion may not affect a carrier's health if it does not disrupt a gene region. However, it may affect an individual's fertility. Crossovers in an inverted region during meiosis may result in deletions or duplications that affect the viability of forthcoming embryos. Some carriers do not know that they have an inversion until they are diagnosed with infertility and their karyotype is tested.

Translocation If a chromosome breaks, the broken part may get attached to a different chromosome, or to a different part of the same one. This structural change is a **translocation**. Most translocations are reciprocal, or balanced; two chromosomes exchange broken parts:

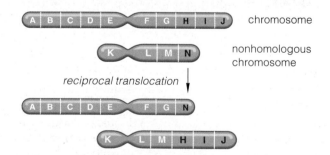

chromosome

nonhomologous chromosome

reciprocal translocation

A reciprocal translocation that does not disrupt a gene may have no adverse effect on its bearer. Many people do not even realize they carry a translocation until they have difficulty with fertility. The two translocated chromosomes pair abnormally with their non-translocated counterparts during meiosis. They segregate improperly about half of the time, so about half of the resulting gametes will carry major duplications or deletions. If one of these gametes unites with a normal gamete at fertilization, the resulting embryo almost always dies.

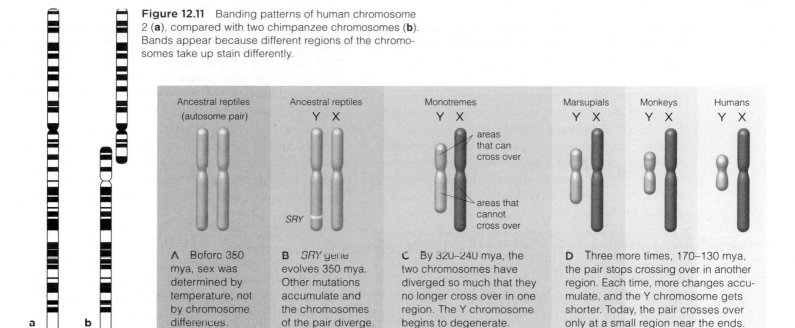

Figure 12.11 Banding patterns of human chromosome 2 (**a**), compared with two chimpanzee chromosomes (**b**). Bands appear because different regions of the chromosomes take up stain differently.

Ancestral reptiles (autosome pair)	Ancestral reptiles Y X	Monotremes Y X	Marsupials Y X	Monkeys Y X	Humans Y X

SRY

areas that can cross over

areas that cannot cross over

A Before 350 mya, sex was determined by temperature, not by chromosome differences.

B *SRY* gene evolves 350 mya. Other mutations accumulate and the chromosomes of the pair diverge.

C By 320–240 mya, the two chromosomes have diverged so much that they no longer cross over in one region. The Y chromosome begins to degenerate.

D Three more times, 170–130 mya, the pair stops crossing over in another region. Each time, more changes accumulate, and the Y chromosome gets shorter. Today, the pair crosses over only at a small region near the ends.

a   **b**

Figure 12.12 Evolution of the Y chromosome. Mya stands for million years ago.

Does Chromosome Structure Evolve?

As you see, alterations in chromosome structure may reduce fertility; individuals who are heterozygous for multiple changes may not be able to produce offspring at all. However, accumulation of multiple alterations in homozygous individuals can be the start of a new species. It may seem as if this outcome would be rare, but it can and does occur over generations in nature. Karyotyping studies show that structural alterations have been built into the DNA of nearly all species.

For example, certain duplications have allowed one copy of a gene to mutate while a different copy carries out its original function. The multiple and strikingly similar globin chain genes of humans and other primates apparently evolved by this process. Four globin chains associate in each hemoglobin molecule (Section 3.6). Which version of the globin chains participate in the association influences the oxygen-binding behavior of the resulting protein.

Some chromosome structure alterations contributed to differences among closely related organisms, such as apes and humans. Body cells of humans have twenty-three pairs of chromosomes, but those of chimpanzees, gorillas, and orangutans have twenty-four. Thirteen human chromosomes are almost identical with chimpanzee chromosomes. Nine more are similar, except for some inversions. One human chromosome matches up with two in chimpanzees and the other great apes (Figure 12.11). During human evolution, two chromosomes fused end to end and formed our chromosome 2. How do we know? The fused region contains remnants of a telomere, which is a special DNA sequence that caps the ends of chromosomes.

As another example, X and Y chromosomes were once homologous autosomes in reptilelike ancestors of mammals. In those organisms, ambient temperature probably determined sex, as it still does in turtles and some other modern reptiles. Then, about 350 million years ago, one of the two chromosomes underwent a structural change that interfered with crossing over in meiosis, and the two homologues diverged over evolutionary time. Eventually, the chromosomes became so different that they no longer crossed over at all in the changed region, which by that time held the *SRY* gene on the Y chromosome. One of the genes on the modern X chromosome is similar to *SRY*; it probably evolved from the same ancestral gene (Figure 12.12).

Take-Home Message

Does chromosome structure change?

■ A segment of a chromosome may be duplicated, deleted, inverted, or translocated. Such a change is usually harmful or lethal, but may be conserved in the rare circumstance that it has a neutral or beneficial effect.

■ Occasionally, new individuals end up with the wrong chromosome number. Consequences range from minor to lethal.

■ Links to Sampling error 1.8, Mitosis 9.3, Meiosis 10.3

Seventy percent of flowering plant species, and some insects, fishes, and other animals are **polyploid**—their cells have three or more of each type of chromosome. Changes in chromosome number may arise through **nondisjunction**, in which one or more pairs of chromosomes do not separate properly during mitosis or meiosis (Figure 12.13). Nondisjunction affects the chromosome number at fertilization. For example, suppose that a normal gamete fuses with an $n+1$ gamete (one that has an extra chromosome). The new individual will be trisomic ($2n+1$), having three of one type of chromosome and two of every other type. As another example, if an $n-1$ gamete fuses with a normal n gamete, the new individual will be $2n-1$, or monosomic.

Trisomy and monosomy are types of **aneuploidy**, a condition in which cells have too many or too few copies of a chromosome. Autosomal aneuploidy is usually fatal in humans, and it causes many miscarriages.

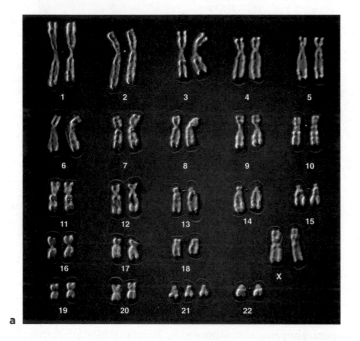

a

Autosomal Change and Down Syndrome

A few trisomic humans are born alive, but only those that have trisomy 21 will reach adulthood. A newborn with three chromosomes 21 will develop Down syndrome. This autosomal disorder is the most common type of aneuploidy in humans; it occurs once in 800 to 1,000 births and affects more than 350,000 people in the United States alone. Figure 12.13*a* shows a karyotype for a trisomic 21 female. The affected individuals have upward-slanting eyes, a fold of skin that starts at the inner corner of each eye, a deep crease across the sole of each palm and foot, one (instead of two) horizontal furrows on their fifth fingers, slightly flattened facial features, and other symptoms.

Not all of the outward symptoms develop in every individual. That said, trisomic 21 individuals tend to have moderate to severe mental impairment and heart problems. Their skeleton grows and develops abnormally, so older children have short body parts, loose joints, and misaligned bones of the fingers, toes, and hips. The muscles and reflexes are weak, and motor skills such as speech develop slowly. With medical care, trisomy 21 individuals live about fifty-five years. Early training can help affected individuals learn to care for themselves and to take part in normal activities. As a group, they tend to be cheerful.

The incidence of nondisjunction generally rises with the increasing age of the mother (Figure 12.14). Nondisjunction may occur in the father, although far less frequently. Trisomy 21 is one of the hundreds of conditions that can be detected easily through prenatal diagnosis (Section 12.8).

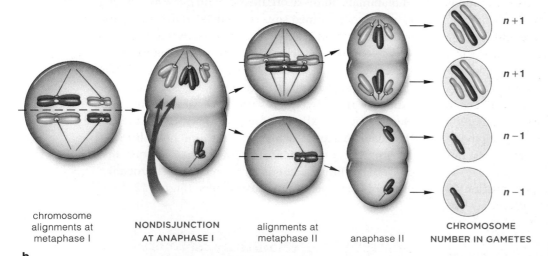

chromosome alignments at metaphase I

NONDISJUNCTION AT ANAPHASE I

alignments at metaphase II

anaphase II

$n+1$

$n+1$

$n-1$

$n-1$

CHROMOSOME NUMBER IN GAMETES

b

Figure 12.13 (**a**) A case of nondisjunction. This karyotype reveals the trisomic 21 condition of a human female. (**b**) One example of how nondisjunction arises. Of the two pairs of homologous chromosomes shown here, one fails to separate during anaphase I of meiosis. The chromosome number is altered in the gametes that form after meiosis.

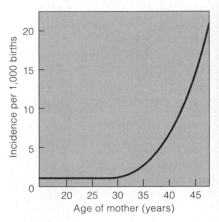

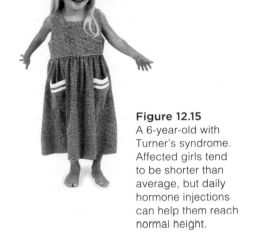

Figure 12.15
A 6-year-old with Turner's syndrome. Affected girls tend to be shorter than average, but daily hormone injections can help them reach normal height.

Figure 12.14 Relationship between the frequency of Down syndrome and mother's age at childbirth. The data are from a study of 1,119 affected children. The risk of having a trisomic 21 baby rises with the mother's age. About 80 percent of trisomic 21 individuals are born to women under thirty five, but these women have the highest fertility rates, and they have more babies.

Change in the Sex Chromosome Number

Nondisjunction also causes alterations in the number of X and Y chromosomes, with a frequency of about 1 in 400 live births. Most often, such alterations lead to difficulties in learning and impaired motor skills such as a speech delay, but problems may be so subtle that the underlying cause is never diagnosed.

Female Sex Chromosome Abnormalities Individuals with Turner syndrome have an X chromosome and no corresponding X or Y chromosome (XO). The condition occurs about 75 percent of the time because of nondisjunction originating with the father. About 1 in 2,500 to 10,000 newborn girls are XO (Figure 12.15). At least 98 percent of XO embryos will spontaneously abort early in pregnancy, so there are fewer cases compared with other sex chromosome abnormalities.

XO individuals are not as disadvantaged as other aneuploids. They grow up well proportioned but short (with an average height of four feet, eight inches). Most do not have functional ovaries, so they do not make enough sex hormones to become sexually mature. The development of secondary sexual traits such as breasts is also affected.

A female may inherit three, four, or five X chromosomes. The resulting XXX syndrome occurs in about 1 of 1,000 births. Only one X chromosome is typically active in female cells, so having extra X chromosomes usually does not result in physical or medical problems.

Male Sex Chromosome Abnormalities About 1 out of every 500 males has an extra X chromosome (XXY). Most cases are an outcome of nondisjunction during meiosis. The resulting disorder, Klinefelter syndrome, develops at puberty. XXY males tend to be overweight, tall, and within a normal range of intelligence. They make more estrogen and less testosterone than normal males, and this hormone imbalance has feminizing effects. Affected men tend to have small testes and prostate glands, low sperm counts, sparse facial and body hair, high-pitched voices, and enlarged breasts. Testosterone injections during puberty can reverse these feminized traits.

About 1 in 500 to 1,000 males has an extra Y chromosome (XYY). Adults tend to be taller than average and have mild mental impairment, but most are otherwise normal. XYY men were once thought to be predisposed to a life of crime. This misguided view was based on sampling error (too few cases in narrowly chosen groups such as prison inmates) and bias (the researchers who gathered the karyotypes also took the personal histories of the participants).

In 1976 a Danish geneticist reported results from his study of 4,139 tall males, all twenty-six years old, who had registered at their draft board. Besides their data from physical examinations and intelligence tests, the draft records offered clues to social and economic status, education, and any criminal convictions. Only twelve of the males studied were XYY, which meant that the "control group" had more than 4,000 males. The only findings? Mentally impaired, tall males who engage in criminal deeds are just more likely to get caught—irrespective of karyotype.

Take-Home Message

What are the effects of changes in the chromosome number?

■ Nondisjunction can change the number of autosomes or sex chromosomes in gametes. Such changes usually cause genetic disorders in offspring.

■ Sex chromosome abnormalities are usually associated with learning difficulties, speech delays, and motor skill impairment.

12.7 | Human Genetic Analysis

- Charting genetic connections with pedigrees reveals inheritance patterns for certain alleles.

- Links to Sampling error 1.8, Meiosis 10.4

Some organisms, including pea plants and fruit flies, are ideal for genetic analysis. They have few chromosomes, and reproduce fast in small spaces under controlled conditions. It does not take long to track a trait through many generations.

Humans, however, are a different story. Unlike flies grown in laboratories, we humans live under variable conditions, in different places, and we live as long as the geneticists who study us. Most of us select our own mates and reproduce if and when we want to. Most human families are not large, which means that there are typically not enough offspring to clarify any inheritance patterns.

Thus, to minimize sampling error (Section 1.8), geneticists gather information from multiple generations. If a trait follows a simple Mendelian inheritance pattern, geneticists can predict the probability of its recurrence

Figure 12.16 An intriguing pattern of inheritance. Eight percent of the men in central Asia carry nearly identical Y chromosomes, which implies descent from a shared ancestor. If so, then 16 million males living between northeastern China and Afghanistan—close to 1 of every 200 men alive today—may be part of a lineage that started with the warrior and notorious womanizer Genghis Khan.

in future generations. Some inheritance patterns are clues to past events (Figure 12.16). Those who analyze pedigrees use their knowledge of probability and patterns of Mendelian inheritance. Such researchers have traced many genetic abnormalities and disorders to a dominant or recessive allele and often to its location on an autosome or a sex chromosome (Table 12.1).

Table 12.1 Examples of Human Genetic Disorders and Genetic Abnormalities

Disorder or Abnormality	Main Symptoms	Disorder or Abnormality	Main Symptoms
Autosomal Recessive Inheritance		**X-Linked Recessive Inheritance**	
Albinism	Absence of pigmentation	Androgen insensitivity syndrome	XY individual but having some female traits; sterility
Hereditary methemoglobinemia	Blue skin coloration	Red–green color blindness	Inability to distinguish among some or all shades of red and green
Cystic fibrosis	Abnormal glandular secretions leading to tissue, organ damage		
Ellis–van Creveld syndrome	Dwarfism, heart defects, polydactyly	Fragile X syndrome	Mental impairment
Fanconi anemia	Physical abnormalities, bone marrow failure	Hemophilia	Impaired blood clotting ability
		Muscular dystrophies	Progressive loss of muscle function
Galactosemia	Brain, liver, eye damage	X-linked anhidrotic dysplasia	Mosaic skin (patches with or without sweat glands); other effects
Phenylketonuria (PKU)	Mental impairment		
Sickle-cell anemia	Adverse pleiotropic effects on organs throughout body	**Changes in Chromosome Structure**	
		Chronic myelogenous leukemia (CML)	Overproduction of white blood cells in bone marrow; organ malfunctions
Autosomal Dominant Inheritance		Cri-du-chat syndrome	Mental impairment; abnormally shaped larynx
Achondroplasia	One form of dwarfism		
Camptodactyly	Rigid, bent fingers	**Changes in Chromosome Number**	
Familial hypercholesterolemia	High cholesterol levels in blood; eventually clogged arteries		
Huntington's disease	Nervous system degenerates progressively, irreversibly	Down syndrome	Mental impairment; heart defects
		Turner syndrome (XO)	Sterility; abnormal ovaries, abnormal sexual traits
Marfan syndrome	Abnormal or no connective tissue		
Polydactyly	Extra fingers, toes, or both	Klinefelter syndrome	Sterility; mild mental impairment
Progeria	Drastic premature aging	XXX syndrome	Minimal abnormalities
Neurofibromatosis	Tumors of nervous system, skin	XYY condition	Mild mental impairment or no effect

Figure 12.17
A pedigree for Huntington's disease, a progressive degeneration of the nervous system.

Researcher Nancy Wexler and her team constructed this extended family tree for nearly 10,000 Venezuelans. Their analysis of unaffected and affected individuals revealed that a dominant allele on human chromosome 4 is the culprit.

Wexler has a special interest in the disorder; it runs in her family.

Inheritance patterns are often displayed as standardized charts of genetic connections called **pedigrees**. You already came across one pedigree in Figure 12.8. Figure 12.17 shows another.

What do we do with genetic information? The next section explores some of the options. When considering them, keep in mind some distinctions. First, a genetic abnormality is defined as a rare or uncommon version of a trait, such as when a person is born with six digits on each hand or foot instead of the usual five (Figure 12.18). Such abnormalities are not inherently life threatening, and how you view them is a matter of opinion. By contrast, a genetic disorder is an inherited condition that sooner or later causes mild to severe medical problems. A genetic disorder is characterized by a specific set of symptoms—a **syndrome**. By contrast, the term "disease" is usually reserved for an illness caused by infection or environmental factors.

One more point: Alleles that give rise to severe genetic disorders are generally rare in populations, because they put their bearers at risk. Why don't they disappear entirely? Rare mutations can reintroduce them. In heterozygotes, a normal allele may mask the effects of expression of a harmful recessive allele. Or, heterozygotes with a codominant allele may have an advantage in a particular environment. You will see an example of the latter case in Section 18.6.

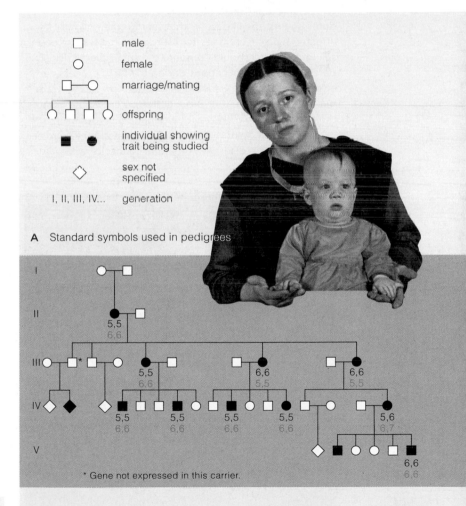

A Standard symbols used in pedigrees

* Gene not expressed in this carrier.

B A pedigree for *polydactyly*, which is characterized by extra fingers, toes, or both. The *black* numbers signify the number of fingers on each hand; the *blue* numbers signify the number of toes on each foot. Though it occurs on its own, polydactyly is also one of several symptoms of Ellis–van Creveld syndrome.

Figure 12.18 Animated Pedigrees.

Take-Home Message

How and why do we determine inheritance patterns in humans?

■ Geneticists track traits through a family tree using pedigrees.

■ Pedigrees reveal inheritance patterns for certain alleles. Such patterns may be used to determine the probability that future offspring will be affected by a genetic abnormality or disorder.

Prospects in Human Genetics

■ Genetic analyses can provide prospective parents with information about their future children.

With the first news of pregnancy, parents-to-be typically wonder if their baby will be normal. Quite naturally, they want their baby to be free of genetic disorders, and most babies are healthy. Many prospective parents have difficulty coming to terms with the possibility that a child of theirs might develop a severe genetic disorder, but sometimes that happens. What are their options?

Genetic Counseling Genetic counseling starts with diagnosis of parental genotypes, pedigrees, and genetic testing for known disorders. Using information gained from these tests, genetic counselors can predict a couple's probability of having a child with a genetic disorder.

Parents-to-be commonly ask genetic counselors to compare the risks associated with diagnostic procedures against the likelihood that their future child will be affected by a severe genetic disorder. At the time of counseling, they also should compare the small overall risk (3 percent) that complications during the birth process can affect any child. They should talk about how old they are, because the older either prospective parent is, the greater the risk of having a child with a genetic disorder.

As a case in point, suppose a first child or a close relative has a severe disorder. A genetic counselor will evaluate the pedigrees of the parents, and the results of any genetic tests. Using this information, counselors can predict risks for disorders in future children. The same risk will apply to each pregnancy.

Prenatal Diagnosis Doctors and clinicians commonly use methods of prenatal diagnosis to determine the sex of embryos or fetuses and to screen for more than 100 known genetic problems. Prenatal means before birth. Embryo is a term that applies until eight weeks after fertilization, after which fetus is appropriate.

Suppose a forty-five-year-old woman becomes pregnant and worries about Down syndrome. Between fifteen and twenty weeks after conception, she might opt for amniocentesis (Figure 12.19). In this diagnostic procedure, a physician uses a syringe to withdraw a small sample of fluid from the amniotic cavity. The "cavity" is a fluid-filled sac, enclosed by a membrane—the amnion—that in turn encloses the fetus. The fetus normally sheds some cells into the fluid. Cells suspended in the fluid sample can be analyzed for many genetic disorders, including Down syndrome, cystic fibrosis, and sickle-cell anemia.

Chorionic villi sampling (CVS) is a diagnostic procedure similar to amniocentesis. A physician withdraws a few cells from the chorion, which is a membrane that surrounds the amnion and helps form the placenta. The placenta is an organ that keeps the blood of mother and embryo separate, while allowing substances to be exchanged between them. Unlike amniocentesis, CVS can be performed as early as eight weeks into pregnancy.

It is now possible to see a live, developing fetus with fetoscopy. The procedure uses an endoscope, a fiber-optic device, to directly visualize and photograph the fetus, umbilical cord, and placenta with high resolution (Figure 12.20a). Characteristic physical effects of certain genetic abnormalities or disorders can be diagnosed, and sometimes corrected, by fetoscopy.

There are risks to a fetus associated with all three procedures, including punctures or infections. If the amnion does not reseal itself quickly, too much fluid may leak out of the amniotic cavity. Amniocentesis increases the risk of miscarriage by 1 to 2 percent. CVS occasionally disrupts the placenta's development and thus causes underdeveloped or missing fingers and toes in 0.3 percent of newborns. Fetoscopy raises the miscarriage risk by 2 to 10 percent.

Preimplantation Diagnosis Preimplantation diagnosis is a procedure associated with *in vitro* fertilization. Sperm and eggs from prospective parents are mixed in a sterile culture medium. One or more eggs may become fertilized. Then, mitotic cell divisions can turn the fertilized egg into a ball of eight cells within forty-eight hours (Figure 12.20b).

According to one view, the tiny, free-floating ball is a pre-pregnancy stage. Like all of the unfertilized eggs that a woman's body discards monthly during her reproductive years, it has not attached to the uterus. All of its cells have

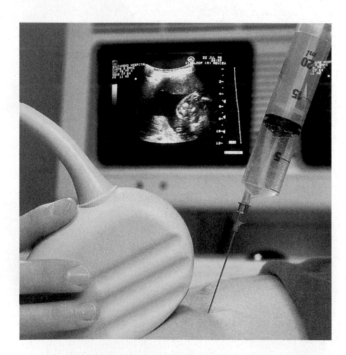

Figure 12.19 Animated Amniocentesis. A pregnant woman's doctor draws a sample of amniotic fluid into a syringe. The path of the needle is monitored by an ultrasound device. About 20 milliliters of fluid is withdrawn from the amniotic sac that holds the developing fetus. Amniotic fluid contains fetal cells and wastes that can be analyzed for genetic disorders.

Figure 12.20 Stages of human development. (**a**) Fetoscopy reveals a fetus in high resolution. (**b**) Eight-celled and (**c**) multicelled stages of human development.

the same genes, but they are not yet committed to being specialized one way or another. Doctors can remove one of these undifferentiated cells and analyze its genes. If it has no detectable genetic defects, the ball is inserted into the mother's uterus. The withdrawn cell will not be missed, and the ball may go on to develop into an embryo. Many of the resulting "test-tube babies" are born in good health. Some couples who are at risk of passing on the alleles for cystic fibrosis, muscular dystrophy, or some other genetic disorder have opted for this procedure.

Regarding Abortion What happens after prenatal diagnosis reveals a severe problem? Some prospective parents opt for an induced abortion. An *abortion* is an induced expulsion of a pre-term embryo or fetus from the uterus. We can only say here that individuals must weigh their awareness of the severity of the genetic disorder against their ethical and religious beliefs. We return to methods of inducing abortion in Section 42.9, in the context of human embryonic development.

Phenotypic Treatments Surgery, prescription drugs, hormone replacement therapy, and often dietary controls can minimize and in some cases eliminate the symptoms of many genetic disorders. For instance, strict dietary controls work in cases of phenylketonuria, or PKU. Individuals affected by this genetic disorder are homozygous for a recessive allele on an autosome. They cannot make a functional form of an enzyme that catalyzes the conversion

of one amino acid (phenylalanine) to another (tyrosine). Because the conversion is blocked, phenylalanine accumulates and is diverted into other metabolic pathways. The outcome is an impairment of brain function.

Affected people who restrict phenylalanine intake can lead essentially normal lives. They must avoid diet soft drinks and other products that are sweetened with aspartame, a compound that contains phenylalanine.

Genetic Screening Genetic screening is the widespread, routine testing for alleles associated with genetic disorders. It provides information on reproductive risks, and helps families that are already affected by a genetic disorder. If a genetic disorder is detected early enough, phenotypic treatments may minimize the damage it causes in some cases.

Hospitals routinely screen newborns for certain genetic disorders. For example, most newborns in the United States are routinely tested for PKU. Affected infants receive early treatment, so we now see fewer individuals with symptoms of the disorder. Besides helping individuals, the information from genetic screening can help us estimate the prevalence and distribution of harmful alleles in populations.

There are social risks that must be considered. How would you feel if you were labeled as someone that carries a "bad" allele? Would the knowledge invite anxiety? If you become a parent even though you know you have a "bad" allele, how would you feel if your child ends up affected by a genetic disorder? There are no easy answers here.

Strange Genes, Tortured Minds

Mutations that affect any of the steps in crucial metabolic pathways could impair brain chemistry, which in turn may result in NBDs. People with bipolar disorder or schizophrenia (such as John Nash, pictured at *right*) have altered gene expression, particularly in certain regions of the brain. Current research suggests that cells of people with these NBDs make too many or too few of the enzymes that carry out electron transfer phosphorylation. Remember, this stage of aero-

How would you vote?

Do you support legislation governing genetic testing that would identify individuals at risk for developing NBDs? See CengageNOW for details, then vote online.

bic respiration yields most of the body's ATP. In the brain, the disruption of electron transfer phosphorylation may alter cells in ways that boost creativity—but also invite illness.

Summary

Section 12.1 A human body cell has twenty-three pairs of homologous chromosomes. One is a pair of **sex chromosomes**. All of the others are pairs of **autosomes**. In both sexes, the two autosomes of each pair have the same length, shape, and centromere location, and they carry the same genes along their length.

Human females have identical sex chromosomes (XX) and males have nonidentical ones (XY). The *SRY* gene on the Y chromosome is the basis of male sex determination. Its expression causes an embryo to develop testes, which secrete testosterone. This hormone controls the development of male secondary sexual traits. An embryo with no Y chromosome (no *SRY* gene) develops into a female.

Karyotyping is a diagnostic tool that reveals missing or extra chromosomes, and some structural changes, in an individual's chromosomes. With this technique, a person's metaphase chromosomes are prepared for microscopy and then imaged. Images of the chromosomes are arranged by their defining features as a **karyotype**.

■ *Use the interaction on CengageNOW to see how sex is determined in humans.*

■ *Use the animation on CengageNOW to learn how to create a karyotype.*

Sections 12.2, 12.3 Certain dominant or recessive alleles on autosomes are associated with genetic abnormalities or genetic disorders.

■ *Use the interaction on CengageNOW to investigate autosomal inheritance.*

Section 12.4 Certain dominant and recessive alleles on the X chromosome are inherited in Mendelian patterns. Mutated alleles on the X chromosome contribute to more than 300 known genetic disorders. Males cannot transmit a recessive X-linked allele to their sons; a female passes such alleles to male offspring.

■ *Use the interaction on CengageNOW to investigate X-linked inheritance.*

Section 12.5 Rarely, a chromosome's structure becomes altered when part of it undergoes **duplication**, **deletion**, **inversion**, or **translocation**. Most alterations are harmful or lethal. Even so, many have accumulated in the chromosomes of all species over evolutionary time.

Section 12.6 The chromosome number of a cell can change permanently. Most often, such a change is an outcome of **nondisjunction**, which is the failure of one or more pairs of duplicated chromosomes to separate during meiosis. In **aneuploidy**, cells have too many or too few copies of a chromosome. In humans, the most common aneuploidy, trisomy 21, causes Down syndrome. Most other human autosomal aneuploids die before birth.

Polyploid individuals inherit three or more of each type of chromosome from their parents. About 70 percent of all flowering plants, and some insects, fishes, and other animals, are polyploid.

A change in the number of sex chromosomes usually results in impaired learning and motor skills. Problems can be so subtle that the underlying cause may not ever be diagnosed, as among XXY, XXX, and XYY children.

Sections 12.7, 12.8 A genetic abnormality is an uncommon version of a heritable trait that does not result in medical problems. A genetic disorder is a heritable condition that results a **syndrome** of mild or severe medical problems. Geneticists construct **pedigrees** to estimate the chance that a couple's offspring will inherit a genetic abnormality or disorder. Potential parents who may be at risk of transmitting a harmful allele to offspring have screening or treatment options.

■ *Use the animations on CengageNOW to examine a human pedigree, and to explore amniocentesis.*

Self-Quiz   *Answers in Appendix III*

1. The _____ of chromosomes in a cell are compared to construct karyotypes.
   a. length and shape   c. gene sequence
   b. centromere location d. both a and b

2. The _____ determines sex in humans.
   a. X chromosome   c. *SRY* gene
   b. *Dll* gene   d. both a and c

3. If one parent is heterozygous for a dominant autosomal allele and the other parent does not carry the allele, a child of theirs has a _____ chance of being heterozygous.
   a. 25 percent   c. 75 percent
   b. 50 percent   d. no chance; it will die

Data Analysis Exercise

A study in 1989 looked for a genetic relationship between mood disorders and intelligence. William Coryell and his colleagues found people with bipolar disorder, then tallied how many of their immediate family members had graduated from college. Some of the results are shown in Figure 12.21.

1. Which group of people, those with bipolar disorder or those without bipolar disorder, had the largest percentage of fathers with a college degree?

2. According to this data, if you had bipolar disorder, which one of your immediate relatives would be more likely to have a college degree? Which would be the least likely?

3. Are relatives of people with or without bipolar disorder more likely to graduate from college?

Proportion of College Graduates Among Relatives of People with Bipolar Disorder		
Relative	Nonbipolar (%)	Bipolar (%)
Father	16.1	27.3
Mother	10.5	18.2
Brother	23.9	38.7
Sister	16.1	31.8
Grandfather	7.6	13.0
Grandmother	5.1	7.1

Figure 12.21 Proportion of immediate relatives of nonbipolar and bipolar people that graduated from college.

4. Expansion mutations occur _____ within and between genes in human chromosomes.
   a. only rarely
   b. frequently
   c. not at all
   d. only in multiples of three

5. Name one X-linked recessive genetic disorder.

6. Men are about 16 times more likely to be affected by red–green color blindness than women. Why?

7. Is this statement true or false? A son can inherit an X linked recessive allele from his father.

8. Color blindness is a case of _____ inheritance.
   a. autosomal dominant
   b. autosomal recessive
   c. X-linked dominant
   d. X-linked recessive

9. A(n) _____ can alter chromosome structure.
   a. deletion
   b. duplication
   c. inversion
   d. translocation
   e. all of the above

10. Nondisjunction may occur during _____ .
   a. mitosis
   b. meiosis
   c. fertilization
   d. both a and b

11. Is this statement true or false? Body cells may inherit three or more of each type of chromosome characteristic of the species, a condition called polyploidy.

12. The karyotype for Klinefelter syndrome is _____ .
   a. XO
   b. XXX
   c. XXY
   d. XYY

13. A recognized set of symptoms that characterize a specific disorder is a _____ .
   a. syndrome
   b. disease
   c. pedigree

14. Match the terms appropriately.
   ___ polyploidy
   ___ deletion
   ___ aneuploidy
   ___ translocation
   ___ karyotype
   ___ nondisjunction during meiosis
   a. number and defining features of an individual's metaphase chromosomes
   b. segment of a chromosome moves to a nonhomologous chromosome
   c. extra sets of chromosomes
   d. gametes with the wrong chromosome number
   e. a chromosome segment lost
   f. one extra chromosome

■ *Visit CengageNOW for additional questions.*

Genetics Problems *Answers in Appendix III*

1. Human females are XX and males are XY.
   a. Does a male inherit the X from his mother or father?
   b. With respect to X-linked alleles, how many different types of gametes can a male produce?
   c. If a female is homozygous for an X-linked allele, how many types of gametes can she produce with respect to that allele?
   d. If a female is heterozygous for an X-linked allele, how many types of gametes might she produce with respect to that allele?

2. In Section 11.4, you read about a mutation that causes a serious genetic disorder, Marfan syndrome. A mutated allele responsible for the disorder follows a pattern of autosomal dominant inheritance. What is the chance that any child will inherit it if one parent does not carry the allele and the other is heterozygous for it?

3. Somatic cells of individuals with Down syndrome usually have an extra chromosome 21; they contain forty-seven chromosomes.
   a. At which stages of meiosis I and II could a mistake alter the chromosome number?
   b. A few individuals with Down syndrome have forty-six chromosomes: two normal-appearing chromosomes 21, and a longer-than-normal chromosome 14. Speculate on how this chromosome abnormality may have arisen.

4. As you read earlier, Duchenne muscular dystrophy is a genetic disorder that arises through the expression of a recessive X-linked allele. Usually, symptoms start to appear in childhood. Gradual, progressive loss of muscle function leads to death, usually by age twenty or so. Unlike color blindness, the disorder is nearly always restricted to males. Suggest why.

5. In the human population, mutation of two genes on the X chromosome causes two types of X-linked hemophilia (A and B). In a few cases, a woman is heterozygous for both mutated alleles (one on each of the X chromosomes). All of her sons should have either hemophilia A or B.
   However, on very rare occasions, one of these women gives birth to a son who does not have hemophilia, and his one X chromosome does not have either mutated allele. Explain how such an X chromosome could arise.

13 DNA Structure and Function

Here, Kitty, Kitty, Kitty, Kitty, Kitty

By now, we have mentioned repeatedly that DNA holds heritable information. Has anybody actually demonstrated that it does? Well, yes. One jarring demonstration occurred in 1997, when Scottish geneticist Ian Wilmut made a genetic copy—a clone—of a fully grown sheep. His team removed the nucleus (and the DNA it contained) from an unfertilized sheep egg. They replaced it with the nucleus of a cell taken from the udder of a different sheep. The hybrid egg became an embryo, and then a lamb. The lamb, whom the researchers named Dolly, was genetically identical to the sheep that had donated the udder cell.

At first, Dolly looked and acted like a normal sheep. But five years later, she was as fat and arthritic as a twelve-year-old sheep. The following year, Dolly contracted a lung disease that is typical of geriatric sheep, and was euthanized.

Dolly's telomeres hinted that she had developed health problems because she was a clone. Telomeres are short, repeated DNA sequences at the ends of chromosomes. They become shorter and shorter as an animal ages. When Dolly was only two years old, her telomeres were as short as those of a six-year-old sheep—the exact age of the adult animal that had been her genetic donor.

Since Dolly was born, mice, rabbits, pigs, cattle, goats, mules, deer, horses, cats, dogs, and a wolf have been cloned, but cloning mammals is far from routine. Not very many clonings end successfully. It usually takes hundreds of attempts to produce one embryo, and most embryos that do form die before birth or shortly after. About 25 percent of the clones that survive have health problems. For example, cloned pigs tend to limp and have heart problems. One never did develop a tail or, even worse, an anus.

What causes the problems? Even though all cells of an individual *inherit* the same DNA, an adult cell *uses* only a fraction of it compared to an embryonic cell. To make a clone from an adult cell, researchers must reprogram its DNA to function like the DNA of an egg. As Dolly's story reminds us, we still have a lot to learn about doing that.

Why do geneticists keep at it? The potential benefits are enormous. Cells cloned from people with incurable diseases may be grown as replacement tissues or organs in laboratories. Endangered animals might be saved from extinction; extinct animals may be brought back. Livestock and pet animals are already being cloned commercially (Figure 13.1).

Perfecting the methods to make healthy animal clones brings us closer to the possibility of cloning humans, technically and also ethically. For example, if cloning a lost cat for a grieving pet owner is acceptable, why would it not be acceptable to clone a lost child for a grieving parent?

With this chapter, we turn to our understanding of how DNA functions as the foundation of inheritance.

See the video! Figure 13.1 Demonstration that DNA holds heritable information—cloning of an adult. Compare the markings on Tahini, a Bengal cat (*above*) with those of Tabouli and Baba Ganoush, two of her clones (*right*). Eye color changes as a Bengal cat matures; both clones later developed the same eye color as Tahini's.

Key Concepts

Discovery of DNA's function

The work of many scientists over more than a century led to the discovery that DNA is the molecule that stores hereditary information about traits. **Section 13.1**

Discovery of DNA's structure

A DNA molecule consists of two long chains of nucleotides coiled into a double helix. Four kinds of nucleotides make up the chains, which are held together along their length by hydrogen bonds. **Section 13.2**

How cells duplicate their DNA

Before a cell begins mitosis or meiosis, enzymes and other proteins replicate its chromosome(s). Newly forming DNA strands are monitored for errors. Uncorrected errors may become mutations. **Section 13.3**

Cloning animals

Knowledge about the structure and function of DNA is the basis of several methods of making clones, which are identical copies of organisms. **Section 13.4**

The Franklin footnote

Science proceeds as a joint effort. Many scientists contributed to the discovery of DNA's structure. **Section 13.5**

Links to Earlier Concepts

- This chapter builds on our earlier introduction to radioisotopes (Section 2.2), hydrogen bonding (2.4), condensation reactions (3.2), carbohydrates (3.3), proteins (3.5), and nucleic acids (3.7).

- Your knowledge of enzyme specificity (6.3) and the cell cycle (9.2) will help you understand how DNA replication works. We also see an example of the importance of checkpoint genes (9.5).

- What you know about mitosis, meiosis, and asexual reproduction (10.1) will help you understand cloning procedures. The cells used for therapeutic cloning are no more developed than those at the eight-cell stage of human development (12.8).

How would you vote? Some view sickly or deformed clones as unfortunate but acceptable casualties of animal cloning research that also yields medical advances for human patients. Should animal cloning be banned? See CengageNOW for details, then vote online.

- Investigations that led to our understanding that DNA is the molecule of inheritance reveal how science advances.

- Links to Radioisotopes 2.2, Proteins 3.5

Early and Puzzling Clues

About the time Gregor Mendel was born, a Swiss medical student, Johann Miescher, was ill with typhus. The typhus left Miescher partially deaf, so becoming a doctor was no longer an option for him. He switched to organic chemistry instead. By 1869, he was collecting white blood cells from pus-filled bandages and sperm from fish so he could study the composition of the nucleus. Such cells do not contain much cytoplasm, which made isolating the substances in their nucleus easy. Miescher found that nuclei contain an acidic substance composed mostly of nitrogen and phosphorus. Later, that substance would be called **deoxyribonucleic acid**, or **DNA**.

Sixty years later, a British medical officer, Frederick Griffith, was trying to make a vaccine for pneumonia. He isolated two strains (types) of *Streptococcus pneumoniae*, a bacteria that causes pneumonia. He named one strain *R*, because it grows in *R*ough colonies. He named the other strain *S*, because it grows in *S*mooth colonies. Griffith used both strains in a series of experiments that did not lead to the development of a vaccine, but did reveal a clue about inheritance (Figure 13.2).

First, he injected mice with live *R* cells. The mice did not develop pneumonia. *The R strain was harmless.*

Second, he injected other mice with live *S* cells. The mice died. Blood samples from them teemed with live *S* cells. *The S strain was pathogenic; it caused pneumonia.*

Third, he killed *S* cells by exposing them to high temperature. *Mice injected with dead S cells did not die.*

Fourth, he mixed live *R* cells with heat-killed *S* cells and injected the mixture into mice. The mice died, *and blood samples drawn from them teemed with live S cells!*

What happened in the fourth experiment? If heat-killed *S* cells in the mix were not really dead, then mice injected with them in the third experiment would have died. If the harmless *R* cells had changed into killer cells, then mice injected with *R* cells in experiment 1 would have died.

The simplest explanation was that heat had killed the *S* cells, but had not destroyed their hereditary material, including whatever part specified "infect mice." Somehow, that material had been transferred from the dead *S* cells into the live *R* cells, which put it to use.

The transformation was permanent and heritable. Even after hundreds of generations, the descendants of transformed *R* cells were infectious. What had caused the transformation? *Which substance encodes the information about traits that parents pass to offspring?*

In 1940, Oswald Avery and Maclyn McCarty set out to identify that substance, which they termed the "transforming principle." They used a process of elimination that tested each type of molecular component of *S* cells.

Avery and McCarty repeatedly froze and thawed *S* cells. Ice crystals that form during this process disrupt membranes, thus releasing cell contents. The researchers then filtered any intact cells from the resulting slush. At the end of this process, the researchers had a fluid that contained lipid, protein, and nucleic acid components of *S* cells.

The *S* cell extract could still transform *R* cells after it had been treated with lipid- and protein-destroying enzymes. Thus, the transforming principle could not be lipid or protein. Carbohydrates had been removed during the purification process, so Avery and McCarty realized that the substance they were seeking must be nucleic acid—RNA or DNA. The *S* cell extract could still transform *R* cells after treatment with RNA-degrading enzymes, but not after treatment with DNA-degrading enzymes. DNA had to be the transforming principle.

The result surprised Avery and McCarty, who, along with most other scientists, had assumed that proteins were the substance of heredity. After all, traits are diverse, and

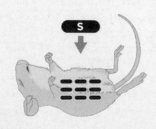

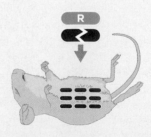

A Mice injected with live cells of harmless strain *R* do not die. Live *R* cells are in their blood.

B Mice injected with live cells of killer strain *S* die. Live *S* cells are in their blood.

C Mice injected with heat-killed *S* cells do not die. No live *S* cells are in their blood.

D Mice injected with live *R* cells plus heat-killed *S* cells die. Live *S* cells are in their blood.

Figure 13.2 Animated Fred Griffith's experiments, in which the hereditary material of harmful *Streptococcus pneumoniae* cells transformed cells of a harmless strain into killers.

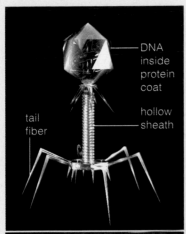

A *Top*, model of a bacteriophage. *Bottom*, micrograph of three viruses injecting DNA into an *E. coli* cell.

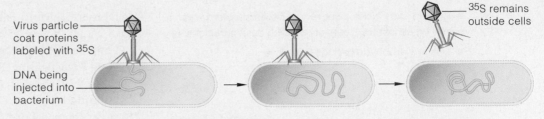

B In one experiment, bacteria were infected with virus particles labeled with a radioisotope of sulfur (^{35}S). The sulfur had labeled only viral proteins. The viruses were dislodged from the bacteria by whirling the mixture in a kitchen blender. Most of the radioactive sulfur was detected in the viruses, not in the bacterial cells. The viruses had not injected protein into the bacteria.

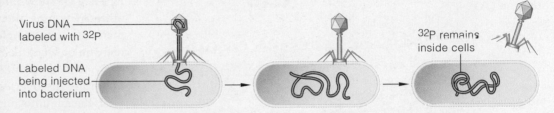

C In another experiment, bacteria were infected with virus particles labeled with a radioisotope of phosphorus (^{32}P). The phosphorus had labeled only viral DNA. When the viruses were dislodged from the bacteria, the radioactive phosphorus was detected mainly inside the bacterial cells. The viruses had injected DNA into the cells—evidence that DNA is the genetic material of this virus.

Figure 13.3 Animated The Hershey Chase experiments. Alfred Hershey and Martha Chase tested whether the genetic material injected by bacteriophage into bacteria is DNA, protein, or both. The experiments were based on the knowledge that proteins contain more sulfur (S) than phosphorus (P), and DNA contains more phosphorus than sulfur.

proteins were thought to be the most diverse biological molecules. Other molecules just seemed too uniform. The two scientists were so skeptical that they published their results only after they had convinced themselves, by years of painstaking experimentation, that DNA was indeed the hereditary material. They were also careful to point out that they had not proven DNA was the *only* hereditary material.

Confirmation of DNA's Function

By 1950, researchers had discovered **bacteriophage**, a type of virus that infects bacteria (Figure 13.3a). Like all viruses, these infectious particles carry hereditary information about how to make new viruses. After a virus infects a cell, the cell starts making new virus particles. Bacteriophages inject genetic material into bacteria, but was that material DNA, protein, or both?

Alfred Hershey and Martha Chase decided to find out by exploiting the long-known properties of protein (high sulfur content) and DNA (high phosphorus content). They cultured bacteria in growth medium containing an isotope of sulfur, ^{35}S (Section 2.2). In this medium, the protein (but not the DNA) of bacteriophage that infected the bacteria became labeled with ^{35}S.

Hershey and Chase allowed the labeled viruses to infect bacteria. They knew from electron micrographs that phages attach to bacteria by their slender tails. They reasoned it would be easy to break this precarious attachment, so they poured the virus–bacteria mixture into a Waring blender and turned it on. (A Waring blender was one of the kitchen appliances that was at the time a common piece of laboratory equipment.)

After blending, the researchers separated the bacteria from the virus-containing fluid, and measured the ^{35}S content of each separately. The fluid contained most of the ^{35}S. Thus, the viruses had not injected protein into the bacteria (Figure 13.3b).

Hershey and Chase repeated the experiment using an isotope of phosphorus, ^{32}P, which labeled the DNA (but not the proteins) of the bacteriophage. This time, they found that the bacteria contained most of the ^{32}P. The viruses had injected DNA into the bacteria (Figure 13.3c).

Both of these experiments—and many others—supported the hypothesis that DNA, not protein, is the material of heredity common to all life on Earth.

The Discovery of DNA's Structure

- Watson and Crick's discovery of DNA's structure was based on almost fifty years of research by other scientists.
- Link to Carbohydrate rings 3.3

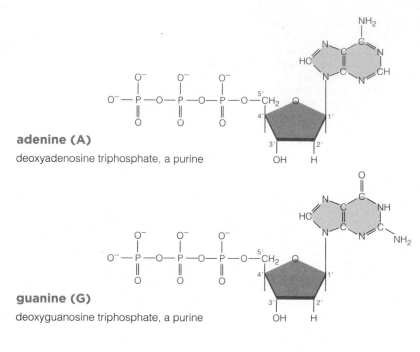

adenine (A)

deoxyadenosine triphosphate, a purine

guanine (G)

deoxyguanosine triphosphate, a purine

thymine (T)

deoxythymidine triphosphate, a pyrimidine

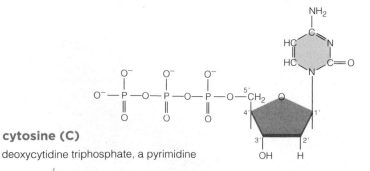

cytosine (C)

deoxycytidine triphosphate, a pyrimidine

Figure 13.4 Four kinds of nucleotides that are linked into strands of DNA. Each is nicknamed after its component base (in *blue*). Biochemist Phoebus Levene worked out the structure of these bases and how they are connected in DNA in the early 1900s. He worked with DNA for almost 40 years.

Numbering the carbons in the sugar rings (Section 3.7) allows us to keep track of the orientation of nucleotide chains, which is important in processes such as DNA replication. Compare Figure 13.6.

DNA's Building Blocks

Long before DNA's function was known, biochemists were investigating its composition. They had shown that DNA consists of only four kinds of nucleotide building blocks. A DNA **nucleotide** has a five-carbon sugar (deoxyribose), three phosphate groups, and one of four nitrogen-containing bases:

adenine	guanine	thymine	cytosine
A	G	T	C

Figure 13.4 shows the structures of these four nucleotides. Thymine and cytosine are called pyrimidines; their bases have single carbon rings. Adenine and guanine are purines; their bases have double carbon rings.

By 1952, biochemist Erwin Chargaff had made two important discoveries about the composition of DNA. First, the amounts of thymine and adenine in DNA are the same, as are the amounts of cytosine and guanine. Second, the proportion of adenine and guanine differs among species. We may show Chargaff's rules as:

$$A = T \quad \text{and} \quad G = C$$

The symmetrical proportions were an important clue to how nucleotides are arranged in DNA.

The first convincing evidence of that arrangement came from Rosalind Franklin, a researcher at King's College in London who specialized in x-ray crystallography. In this technique, x-rays are directed through a purified and crystallized substance. Atoms in the substance's molecules scatter the x-rays in a pattern that can be captured as an image. Researchers use the pattern to calculate the size, shape, and spacing between any repeating elements of the molecules—all of which are details of molecular structure.

Franklin made the first clear x-ray diffraction image of DNA in the form that occurs in cells. From the information in that image, she calculated that DNA is very long compared to its 2-nanometer diameter. She also identified a repeating pattern every 0.34 nanometers along its length, and another every 3.4 nanometers.

Franklin's image and data came to the attention of James Watson and Francis Crick, both at Cambridge University. Watson, an American biologist, and Crick, a British biophysicist, had been sharing their ideas about the structure of DNA. Biochemists Linus Pauling, Robert Corey, and Herman Branson had only recently described the alpha helix, a coiled pattern that occurs in many proteins (Section 3.5). Watson and Crick suspected that the DNA molecule was also a helix.

Watson and Crick spent many hours arguing about the size, shape, and bonding requirements of the four kinds of nucleotides that make up DNA. They pestered chemists to help them identify bonds they might have overlooked. They fiddled with cardboard cutouts, and made models from scraps of metal connected by suitably angled "bonds" of wire. Franklin's data provided them with the last piece of the puzzle. In 1953, Watson and Crick put together all of the clues that had been accumulating for the last fifty years and built the first accurate model of the DNA molecule (Figure 13.5).

Watson and Crick proposed that DNA's structure consists of two chains (or strands) of nucleotides, running in opposite directions and coiled into a double helix. Hydrogen bonds between the internally positioned bases hold the two strands together. Only two kinds of base pairings form: A to T, and G to C. Most scientists had assumed (incorrectly) that the bases had to be on the outside of the helix, because they would be more accessible to DNA-copying enzymes that way. You will see in Section 13.3 how those enzymes access the bases on the inside of the helix.

Patterns of Base Pairing

How do just two kinds of base pairings give rise to the stunning diversity of traits we see among living things? The answer is that the *order* in which one base pair follows the next—the DNA's **sequence**—is tremendously variable. For instance, a small piece of DNA from a petunia, a human, or any other organism might be:

one
base
pair
G A C T
C T G A
or
A T C G
T A G C
or
G G G C
C C C G

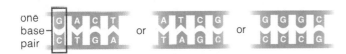

Notice how the two strands of DNA match up; each base on one is suitably paired with a partner base on the other. This bonding pattern (A to T, G to C) is the same in all molecules of DNA. However, which base pair follows the next in line differs among species, and among individuals of the same species. Thus, *DNA, the molecule of inheritance in every cell, is the basis of life's unity. Variations in its base sequence from one individual or one species to the next is the basis of life's diversity.*

Take-Home Message

What is the structure of DNA?

- A DNA molecule consists of two nucleotide chains (strands), running in opposite directions and coiled into a double helix.
- Internally positioned nucleotide bases hydrogen-bond between the two strands. A always pairs with T, and G with C. The sequence of bases is the genetic information.

Figure 13.5 Animated Structure of DNA, as illustrated by a composite of different models.

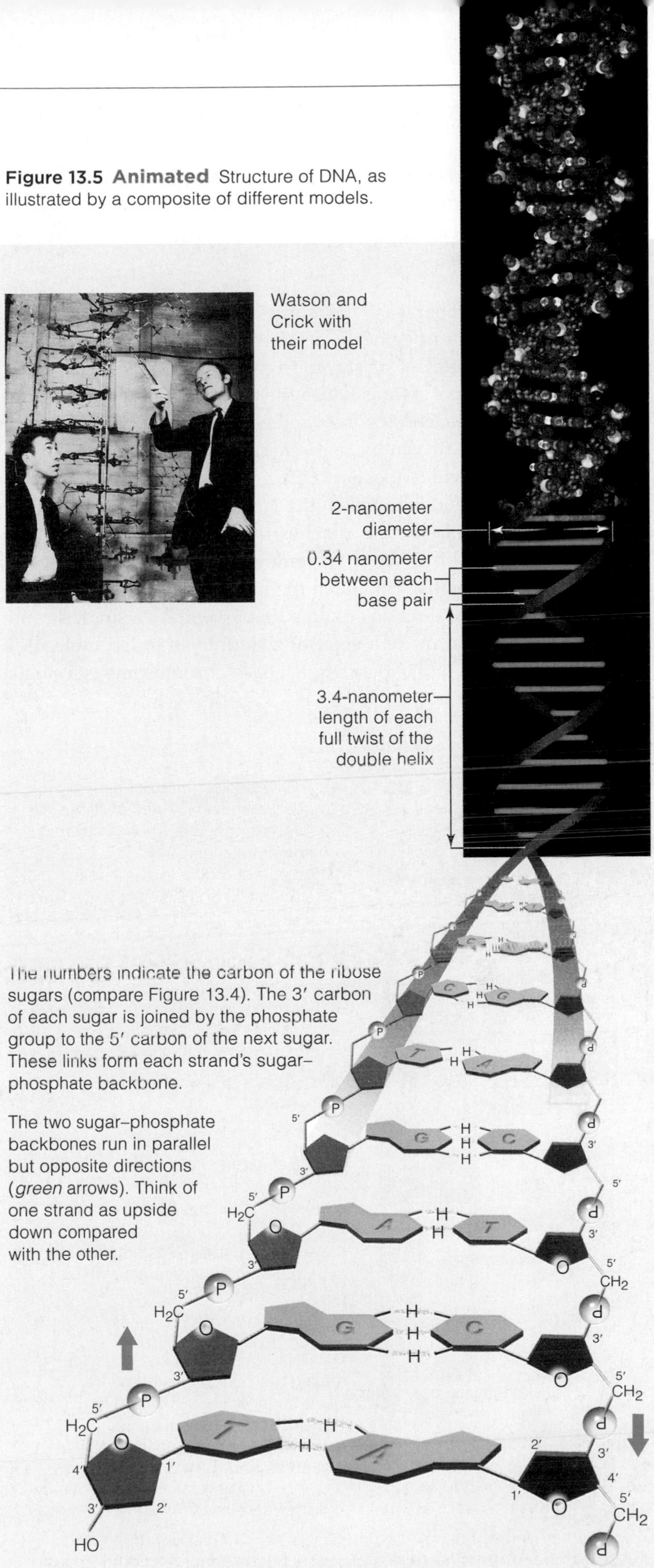
Watson and Crick with their model
2-nanometer diameter
0.34 nanometer between each base pair
3.4-nanometer length of each full twist of the double helix
The numbers indicate the carbon of the ribose sugars (compare Figure 13.4). The 3′ carbon of each sugar is joined by the phosphate group to the 5′ carbon of the next sugar. These links form each strand's sugar–phosphate backbone.
The two sugar–phosphate backbones run in parallel but opposite directions (*green* arrows). Think of one strand as upside down compared with the other.
P
5′
3′
H_2C
O
G
C
A
T
H
CH_2
4′
1′
2′
HO

13.3 DNA Replication and Repair

- A cell copies its DNA before mitosis or meiosis I.
- DNA repair mechanisms correct most replication errors.

- Links to Enzyme specificity 6.3, Cell cycle 9.2, Checkpoint genes 9.5

Remember, each cell copies its DNA before mitosis or meiosis I begins, so its descendant cells will inherit a complete set of chromosomes (Sections 9.2 and 10.1). **DNA polymerase** does the copying. This enzyme joins free nucleotides into a new strand of DNA. The process is driven by high-energy phosphate bonds in the nucleotides (Section 6.2). A free nucleotide has three phosphate groups, and DNA polymerase removes two of them when it attaches the nucleotide to a growing strand of DNA. That removal releases energy that the enzyme uses to attach the nucleotide to the strand.

How does an enzyme that assembles a single strand of DNA make a copy of a double-stranded molecule? Before replication begins, each chromosome is a single

molecule of DNA—one double helix. During DNA replication, an enzyme called DNA helicase breaks the hydrogen bonds that hold the helix together, so the two DNA strands unwind. Both strands are then replicated independently. Each new DNA strand winds up with its "parent" strand into a new double helix. Thus, after replication, there are two double-stranded molecules of DNA (Figure 13.6). One strand of each molecule is old and the other is new; hence the name of the process, **semiconservative replication** (Figure 13.7).

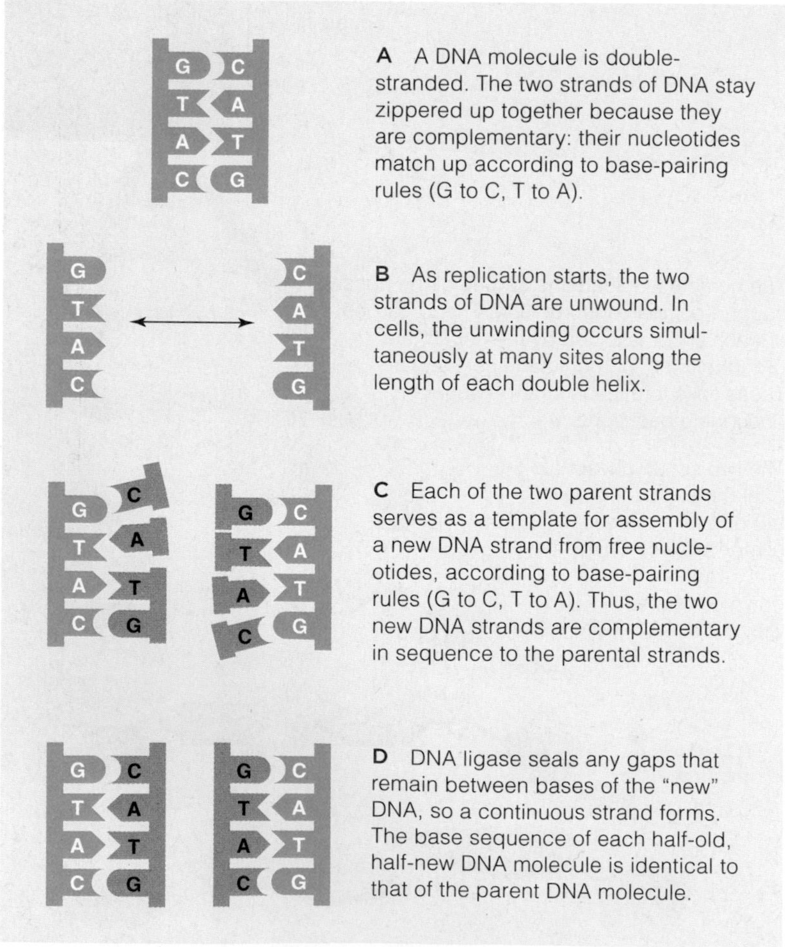

Numbering the carbons in nucleotides (Figure 13.5) allows us to keep track of the DNA strands in a double helix, because each strand has an unbonded 5' carbon at one end and an unbonded 3' carbon at the other:

DNA polymerase can attach free nucleotides only to a 3' carbon. Thus, it can replicate only one strand of a DNA molecule continuously (Figure 13.8). Synthesis of the other strand occurs in segments, in the direction opposite that of unwinding. Another enzyme that participates in DNA replication, **DNA ligase**, joins those segments into a continuous strand of DNA.

There are only four kinds of nucleotides in DNA, but the order in which those nucleotides occur is very important. The nucleotide sequence is a cell's genetic information; descendant cells must get an exact copy of it, or inheritance will go awry. As a DNA polymerase moves along a strand of DNA, it uses the sequence of bases as a template, or guide, to assemble a new strand of DNA. The base sequence of the new strand is complementary to that of the template, because DNA polymerase follows base-pairing rules.

For example, the polymerase adds a T to the end of the new DNA strand when it reaches an A in the parent DNA sequence; it adds a G when it reaches a C; and so on. Because each new strand of DNA is complementary in sequence to the parent strand, both double-stranded molecules that result from DNA replication are duplicates of the parent molecule.

Checking for Mistakes

A DNA molecule is not always replicated with perfect fidelity. Sometimes the wrong base is added to a growing DNA strand; at other times, bases get lost, or extra ones are added. Either way, the new DNA strand will no longer match up perfectly with its parent strand.

Some of these errors occur after the DNA becomes damaged by exposure to radiation or toxic chemicals. DNA polymerases do not copy damaged DNA very

A A DNA molecule is double-stranded. The two strands of DNA stay zipped up together because they are complementary: their nucleotides match up according to base-pairing rules (G to C, T to A).

B As replication starts, the two strands of DNA are unwound. In cells, the unwinding occurs simultaneously at many sites along the length of each double helix.

C Each of the two parent strands serves as a template for assembly of a new DNA strand from free nucleotides, according to base-pairing rules (G to C, T to A). Thus, the two new DNA strands are complementary in sequence to the parental strands.

D DNA ligase seals any gaps that remain between bases of the "new" DNA, so a continuous strand forms. The base sequence of each half-old, half-new DNA molecule is identical to that of the parent DNA molecule.

Figure 13.6 Animated DNA replication. Each strand of a DNA double helix is copied; two double-stranded DNA molecules result.

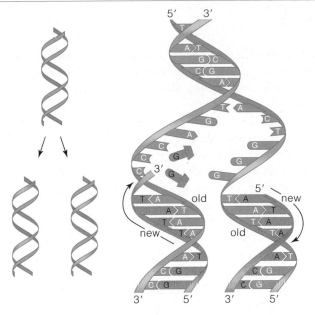

Figure 13.7 Semiconservative replication of DNA. Parent strands (*blue*) stay intact. A new strand (*purple*) is assembled on each parent in the direction shown by the arrows. The Y-shaped structure on the *right* is called a replication fork.

well. In most cases, **DNA repair mechanisms** fix DNA by enzymatically excising and replacing any damaged or mismatched bases before replication begins.

Most DNA replication errors occur simply because DNA polymerases catalyze a tremendous number of reactions very quickly—up to 1,000 bases per second. Mistakes are inevitable; some DNA polymerases make many of them. Luckily, most DNA polymerases proofread their own work. They correct any mismatches by immediately reversing the synthesis reaction to remove a mismatched nucleotide, and then resuming synthesis. If an error remains uncorrected, cellular controls may stop the cell cycle (Sections 9.2 and 9.5).

When proofreading and repair mechanisms fail, an error becomes a mutation—a permanent change in the DNA sequence. An individual or its offspring may not survive a mutation, because mutations can cause cancer in body cells. In cells that form eggs or sperm, they may lead to genetic disorders in offspring. However, not all mutations are dangerous. Some give rise to variations in traits that are the raw material of evolution.

Take-Home Message

How is DNA copied?

■ A cell replicates its DNA before mitosis or meiosis. Each strand of a DNA double helix serves as a template for synthesis of a new, complementary strand of DNA.

■ DNA repair mechanisms and proofreading maintain the integrity of a cell's genetic information. Unrepaired errors may be perpetuated as mutations.

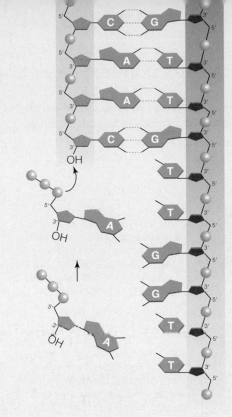

A Each DNA strand has two ends: one with a 5′ carbon, and one with a 3′ carbon. DNA polymerase can add nucleotides only at the 3′ carbon. In other words, DNA synthesis proceeds only in the 5′ to 3′ direction.

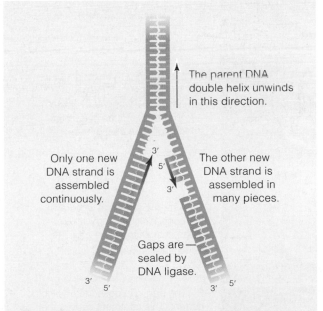

The parent DNA double helix unwinds in this direction.

Only one new DNA strand is assembled continuously.

The other new DNA strand is assembled in many pieces.

Gaps are sealed by DNA ligase.

B Because DNA synthesis proceeds only in the 5′ to 3′ direction, only one of the two new DNA strands can be assembled in a single piece.

The other new DNA strand forms in short segments, which are called Okazaki fragments after the two scientists who discovered them. DNA ligase joins the fragments into a continuous strand of DNA.

Figure 13.8 Discontinuous synthesis of DNA.
Figure It Out: What do the yellow balls represent?

Answer: Phosphate groups

■ Reproductive cloning is a reproductive intervention that results in an exact genetic copy of an adult individual.

■ Link to Asexual reproduction 10.1

A A cow egg is held in place by suction through a hollow glass tube called a micropipette. The polar body (Section 10.5) and chromosomes are identified by a *purple* stain.

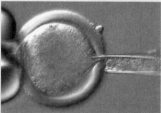

B A micropipette punctures the egg and sucks out the polar body and all of the chromosomes. All that remains inside the egg's plasma membrane is cytoplasm.

C A new micropipette prepares to enter the egg at the puncture site. The pipette contains a cell grown from the skin of a donor animal.

skin cell

D The micropipette enters the egg and delivers the skin cell to a region between the cytoplasm and the plasma membrane.

E After the pipette is withdrawn, the donor's skin cell is visible next to the cytoplasm of the egg. The transfer is complete.

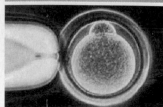

F The egg is exposed to an electric current. This treatment causes the foreign cell to fuse with and empty its nucleus into the cytoplasm of the egg. The egg begins to divide, and an embryo forms. After a few days, the embryo may be transplanted into a surrogate mother.

Figure 13.9 Animated Somatic cell nuclear transfer, using cattle cells. This series of micrographs was taken by scientists at Cyagra, a commercial company that specializes in cloning livestock.

The word "cloning" means making an identical copy of something. In biology, cloning can refer to a laboratory method by which researchers copy DNA fragments (we will discuss DNA cloning in Chapter 16). It can also refer to interventions in reproduction that result in an exact genetic copy of an organism.

Genetically identical organisms occur all the time in nature, and arise mainly by the process of asexual reproduction (Section 10.1). Embryo splitting, another natural process, results in identical twins. The first few divisions of a fertilized egg form a ball of cells that sometimes splits spontaneously. If both halves continue to develop independently, they become identical twins.

Embryo splitting has been a routine part of research and animal husbandry for decades. A ball of cells can be grown from a fertilized egg in a petri dish. If the ball is divided in two, each half will develop as a separate embryo. The embryos are implanted in surrogate mothers, which give birth to identical twins. Artificial twinning and any other technology that yields genetically identical individuals is called **reproductive cloning**.

Twins get their DNA from two parents, so they have a mixture of parental traits. Breeders that want no surprises may opt for a type of reproductive cloning that differs from embryo splitting. This process yields offspring with only one parent's traits; because it starts with nuclear DNA from an adult organism, it bypasses the genetic mixup of sexual reproduction (Section 10.5). All of the individuals that are produced by cloning an adult cell are genetically identical with the parent. However, the procedure presents more of a technical challenge than embryo splitting. A normal cell from an adult will not automatically start dividing as if it were a fertilized egg. It must first be tricked into rewinding its developmental clock.

All cells descended from a fertilized egg inherit the same DNA. As different cells in a developing embryo start using different subsets of their DNA, they differentiate, or become different in form and function. In animals, differentiation is usually a one-way path. Once a cell specializes, all of its descendant cells will be specialized the same way. By the time a liver cell, muscle cell, or other specialized cell forms, most of its DNA has been turned off, and is no longer used.

To clone an adult, scientists first transform one of its differentiated cells into an undifferentiated cell by turning its unused DNA back on. In **somatic cell nuclear transfer** (SCNT), a researcher removes the nucleus from an unfertilized egg, then inserts into the egg a nucleus from an adult animal cell (Figure 13.9). If all goes well, the egg's cytoplasm reprograms the transplanted DNA to direct the development of an embryo, which is then

Figure 13.10 Liz the cow and her clone. The clone was produced by somatic cell nuclear transfer, as in Figure 13.9.

implanted into a surrogate mother. The animal that is born to the surrogate is genetically identical with the donor of the nucleus (Figure 13.10). Dolly the sheep and the other animals described in the chapter introduction were produced using SCNT.

Adult cloning is now a common practice among people who breed prized livestock. Among other benefits, many more offspring can be produced in a given time frame by cloning than by traditional breeding methods, and offspring can be produced after a donor animal is castrated or even dead.

The controversial issue with adult cloning is not necessarily about livestock. The issue is that as the techniques become routine, cloning a human is no longer only within the realm of science fiction. Researchers are already using SCNT to produce human embryos for research, a practice called **therapeutic cloning**. The researchers harvest undifferentiated (stem) cells from the cloned human embryos. (We return to the topic of stem cells and their potential medical benefits in Chapter 32.) Reproductive cloning of humans is not the intent of such research, but somatic cell nuclear transfer would be the first step toward that end.

Take-Home Message

What is cloning?

■ Reproductive cloning technologies produce an exact copy of an individual—a clone.

■ Somatic cell nuclear transfer (SCNT) is a reproductive cloning method in which nuclear DNA of an adult donor is transferred to an enucleated egg. The hybrid cell develops into an embryo that is genetically identical to the donor individual.

■ Therapeutic cloning uses SCNT to produce human embryos for research purposes.

13.5 | Fame and Glory

■ In science, as in other professions, public recognition does not always include everyone who contributed to a discovery.

By the time she arrived at King's College, Rosalind Franklin was an expert x-ray crystallographer. She had solved the structure of coal, which is complex and unorganized (as are large biological molecules such as DNA), and she took a new mathematical approach to interpreting x-ray diffraction images. Like Pauling, she had built three-dimensional molecular models. Her assignment was to investigate DNA's structure. She did not know Maurice Wilkins was already doing the same thing just down the hall. Franklin had been told she would be the only one in the department working on the problem. When Wilkins proposed a collaboration with her, Franklin thought that Wilkins was oddly overinterested in her work and declined bluntly.

Wilkins and Franklin had been given identical samples of DNA, which had been carefully prepared by Rudolf Signer. Franklin's meticulous work with her sample yielded the first clear x-ray diffraction image of DNA as it occurs inside cells (Figure 13.11), and she gave a presentation on this work in 1952. DNA, she said, had two chains twisted into a double helix, with a backbone of phosphate groups on the outside, and bases arranged in an unknown way on the inside. She had calculated DNA's diameter, the distance between its chains and between its bases, the pitch (angle) of the helix, and the number of bases in each coil. Crick, with his crystallography background, would have recognized the significance of the work—if he had been there. Watson was in the audience but he was not a crystallographer, and he did not understand the implications of Franklin's x-ray diffraction image or her calculations.

Franklin started to write a research paper on her findings. Meanwhile, and perhaps without her knowledge, Watson reviewed Franklin's x-ray diffraction image with Wilkins, and Watson and Crick read a report containing Franklin's unpublished data. Crick, who had more experience with theoretical molecular modeling than Franklin, immediately understood what the image and the data meant. Watson and Crick used that information to build their model of DNA.

On April 25, 1953, Franklin's paper appeared third in a series of articles about the structure of DNA in the journal *Nature*. It supported with solid experimental evidence Watson and Crick's theoretical model, which appeared in the first article of the series.

Rosalind Franklin died at age 37, of ovarian cancer probably caused by extensive exposure to x-rays. Because the Nobel Prize is not given posthumously, she did not share in the 1962 honor that went to Watson, Crick, and Wilkins for the discovery of the structure of DNA.

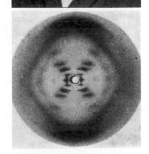

Figure 13.11 Rosalind Franklin and her famous x-ray diffraction image.

Human eggs are difficult to come by, so SCNT researchers are using adult human cells and enucleated cow eggs for therapeutic cloning. The nuclear DNA of the resulting hybrid eggs is human, and the cytoplasm is bovine. Remember, eukaryotic cytoplasm contains mitochondria, which have their own DNA and divide independently (Section 4.11). Thus, cells of embryos that develop from these hybrid eggs contain both human and cow DNA.

How would you vote?

Deformed or unhealthy clones, including Dolly (*right*) are unavoidable casualties of cloning research. Should cloning be banned? See CengageNOW for details, then vote online.

Summary

Section 13.1 Experiments with bacteria and **bacteriophage** offered solid evidence that **deoxyribonucleic acid (DNA)**, not protein, is hereditary material.

■ *Use the animation on CengageNOW to learn about experiments that revealed the function of DNA.*

Section 13.2 A DNA molecule consists of two strands of DNA coiled into a helix. Nucleotide monomers are joined to form each strand. A free **nucleotide** has a five-carbon sugar (deoxyribose), three phosphate groups, and one of four nitrogen-containing bases after which it is named: **adenine**, **thymine**, **guanine**, or **cytosine**.

Bases of the two DNA strands in a double helix pair in a consistent way: adenine with thymine (A–T), and guanine with cytosine (G–C). The order of the bases (the DNA **sequence**) varies among species and among individuals. The DNA of each species has unique sequences that set it apart from the DNA of all other species.

■ *Use the animation on CengageNOW to investigate the structure of DNA.*

Section 13.3 A cell replicates its DNA before mitosis or meiosis begins. By the process of **semiconservative replication**, one double-stranded molecule of DNA is copied, and two double-stranded DNA molecules identical to the parent are the result. One strand of each molecule is new, and the other is parental.

During the replication process, enzymes unwind the double helix at several sites along its length. **DNA polymerase** uses each strand as a template to assemble new, complementary strands of DNA from free nucleotides. DNA synthesis is discontinuous on one of the two strands of a DNA molecule. **DNA ligase** joins the segments into a continuous strand.

DNA repair mechanisms fix DNA damaged by chemicals or radiation. Proofreading by DNA polymerases corrects most base-pairing errors. Uncorrected errors can be perpetuated as mutations.

■ *Use the animation on CengageNOW to see how a DNA molecule is replicated.*

Section 13.4 Various **reproductive cloning** technologies produce genetically identical individuals (clones). In **somatic cell nuclear transfer** (SCNT), one cell from an adult is fused with an enucleated egg. The hybrid cell is treated with electric shocks or another stimulus that provokes the cell to divide and begin developing into a new individual. SCNT with human cells, which is called **therapeutic cloning**, produces embryos that are used for stem cell research.

■ *Use the animation on CengageNOW to observe the procedure used to create Dolly and other clones.*

Section 13.5 Sciences advances as a community effort. Ideally, individuals share their work and the recognition for achievement. As in all human endeavors, these ideals are not always achieved.

Self-Quiz
Answers in Appendix III

1. Bacteriophages are viruses that infect _____ .

2. Which is *not* a nucleotide base in DNA?
   a. adenine   c. uracil   e. cytosine
   b. guanine   d. thymine   f. All are in DNA.

3. What are the base-pairing rules for DNA?
   a. A–G, T–C   c. A–U, C–G
   b. A–C, T–G   d. A–T, G–C

4. One species' DNA differs from others in its _____ .
   a. sugars   c. base sequence
   b. phosphates   d. all of the above

5. When DNA replication begins, _____ .
   a. the two DNA strands unwind from each other
   b. the two DNA strands condense for base transfers
   c. two DNA molecules bond
   d. old strands move to find new strands

6. DNA replication requires _____ .
   a. template DNA   c. DNA polymerase
   b. free nucleotides   d. all of the above

7. DNA polymerase adds nucleotides to _____ (choose all that are correct).
   a. double-stranded DNA   c. double-stranded RNA
   b. single-stranded DNA   d. single-stranded RNA

8. Show the complementary strand of DNA that forms on this template DNA fragment during replication:

   5′–G G T T T C T T C A A G A G A–3′

9. _____ is an example of reproductive cloning.
   a. Somatic cell nuclear transfer (SCNT)
   b. Asexual reproduction
   c. Artificial embryo splitting
   d. a and c
   e. all of the above

Data Analysis Exercise

The graph in Figure 13.12 is reproduced from Alfred Hershey and Martha Chase's 1952 publication that showed DNA is the hereditary material of bacteriophage. The data are from the same two experiments described in Section 13.1, in which bacteriophage DNA and protein were labeled with radioactive tracers and allowed to infect bacteria. The virus–bacteria mixtures were whirled in a blender to dislodge the two, and the tracers were tracked inside and outside of the bacteria.

1. Before blending, what percentage of ^{35}S was outside the bacteria? What percentage was inside? What percentage of ^{32}P was outside the bacteria? What percentage was inside?

2. After 4 minutes in the blender, what percentage of ^{35}S was outside the bacteria? What percentage was inside? What percentage of ^{32}P was outside the bacteria? What percentage was inside?

3. How did the researchers know that the radioisotopes in the fluid came from outside the bacterial cells (extracellular) and not from bacteria that had broken apart?

4. The extracellular concentration of which isotope, ^{35}S or ^{32}P, increased the most with blending? DNA contains much more phosphorus than do proteins; proteins contain much more sulfur than do DNA. Do these results imply that the viruses inject DNA or protein into bacteria? Why?

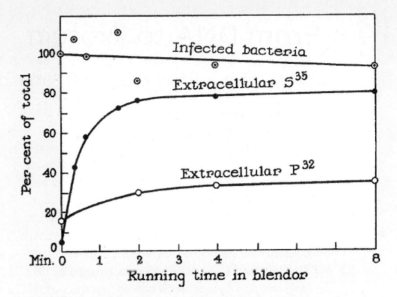

Figure 13.12 Detail of Alfred Hershey and Martha Chase's publication describing their experiments with bacteriophage. "Infected bacteria" refers to the percentage of bacteria that survived the blender.

From the Journal of General Physiology, *36(1), September 20, 1952: "Independent Functions of Viral Protein and Nucleic Acid in Growth of Bacteriophage."*

10. Match the terms appropriately.

___ bacteriophage	a. nitrogen-containing base, sugar, phosphate groups
___ clone	b. copy of an organism
___ nucleotide	c. nucleotide base with one carbon ring
___ purine	d. injects DNA into bacteria
___ DNA ligase	e. fills in gaps, seals breaks in a DNA strand
___ DNA polymerase	f. nucleotide base with two carbon rings
___ pyrimidine	g. adds nucleotides to a growing DNA strand

■ *Visit CengageNOW for additional questions.*

Critical Thinking

1. Matthew Meselson and Franklin Stahl's experiments supported the semiconservative model of replication. These researchers obtained "heavy" DNA by growing *Escherichia coli* with ^{15}N, a radioactive isotope of nitrogen. They also prepared "light" DNA by growing *E. coli* in the presence of ^{14}N, the more common isotope. An available technique helped them identify which of the replicated molecules were heavy, light, or hybrid (one heavy strand and one light). Use different colored pencils to draw the heavy and light strands of DNA. Starting with a DNA molecule having two heavy strands, show the formation of daughter molecules after replication in a ^{14}N-containing medium. Show the four DNA molecules that would form if the daughter molecules were replicated a second time in the ^{14}N medium. Would the resulting DNA molecules be heavy, light, or mixed?

2. Mutations are permanent changes in a cell's DNA base sequence, the original source of genetic variation and the raw material of evolution. How can mutations accumulate, given that cells have repair systems that fix changes or breaks in DNA strands?

3. There may be millions of woolly mammoths frozen in the ice of Siberian glaciers. These huge elephant-like mammals have been extinct for about 10,000 years, but a team of privately funded Japanese scientists is planning to resurrect one of them by cloning DNA isolated from frozen remains. What are some of the pros and cons, both technical and ethical, of cloning an extinct animal?

4. Xeroderma pigmentosum is an autosomal recessive disorder characterized by rapid formation of skin sores (*right*) that can develop into cancers. Affected individuals must avoid all forms of radiation—including sunlight and fluorescent lights. They have no mechanism for dealing with the damage that ultraviolet (UV) light can inflict on skin cells because they lack a DNA repair mechanism that corrects thymine dimers. When the nitrogen-containing bases in DNA absorb UV light, a covalent bond can form between two thymine bases in the same strand of DNA (*left*). The resulting thymine dimer makes a kink in the DNA strand. Propose what consequences might occur because of a thymine dimer during DNA replication.

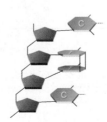

14 | From DNA to Protein

Ricin and Your Ribosomes

Ricin is a highly toxic protein. It is present in all tissues of the castor-oil plant (*Ricinus communis*), which is the source of the castor oil that is an ingredient in many plastics, cosmetics, paints, textiles, and adhesives. The oil—and the ricin—is most concentrated in castor-oil beans (Figure 14.1), but the ricin is discarded when the oil is extracted.

Injected, a dose of ricin as small as a few grains of salt can kill an adult. Inhaled or ingested, it takes more. Only plutonium and botulism toxin are more deadly. There is no antidote.

The lethal effects of ricin were known as long ago as 1888, but using ricin as a weapon is now banned by most countries under the Geneva Protocol. It takes no special skills or equipment to manufacture the toxin from easily obtained raw materials, so controlling its production is impossible. Thus, ricin appears periodically in the news.

For example, at the height of the Cold War, Georgi Markov, a Bulgarian writer, had defected to England and was working as a journalist for the BBC. As he made his way to a bus stop on a London street, an assassin used the tip of a modified umbrella to jam a small, ricin-laced ball into Markov's leg. Markov died in agony three days later.

In 2003, police acted on an intelligence tip and stormed a London apartment, where they found laboratory glassware and castor-oil beans. Traces of ricin were found in a United States Senate mailroom and State Department building, and also in an envelope addressed to the White House. In 2005, the FBI arrested a man who had castor-oil beans and an assault rifle stashed in his Florida home. Jars of banana baby food laced with ground castor-oil beans also made the news in 2005. In 2006, police found pipe bombs and a baby food jar full of ricin in a Tennessee man's shed. In 2008, castor beans, firearms, and several vials of ricin were found in a Las Vegas motel room after its occupant was hospitalized for ricin exposure.

Ricin is toxic because it inactivates ribosomes, the organelles upon which amino acids are assembled into proteins in all cells. Proteins are critical to all life processes. Cells that cannot make them die very quickly. Someone who inhales ricin typically dies from low blood pressure and respiratory failure within a few days of exposure.

This chapter details how the information encoded by a gene becomes converted to a gene product—an RNA or a protein. Though it is extremely unlikely that your ribosomes will ever encounter ricin, protein synthesis is nevertheless worth appreciating for how it keeps you—and all other organisms—alive.

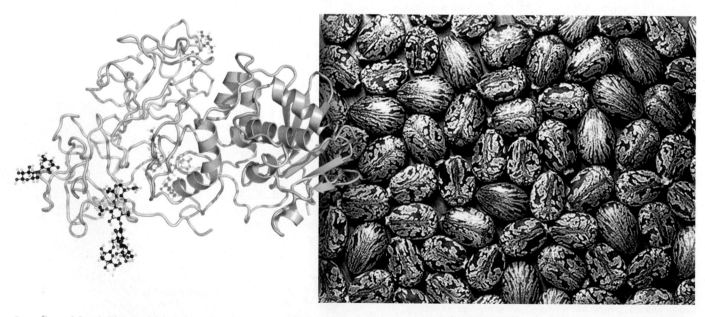

See the video! Figure 14.1 *Left*, model of ricin. One of its polypeptide chains (*green*) helps ricin penetrate a living cell. The other chain (*tan*) destroys the cell's capacity for protein synthesis. Ricin is a glycoprotein; sugars attached to the protein are shown. *Right*, seeds of the castor-oil plant, source of ribosome-busting ricin.

Key Concepts

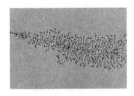

DNA to RNA to protein

Proteins consist of polypeptide chains. The chains are sequences of amino acids that correspond to sequences of nucleotide bases in DNA called genes. The path leading from genes to proteins has two steps: transcription and translation. **Section 14.1**

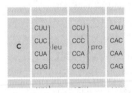

DNA to RNA: transcription

During transcription, one strand of a DNA double helix is a template for assembling a single, complementary strand of RNA (a transcript). Each transcript is an RNA copy of a gene. **Section 14.2**

RNA

Messenger RNA (mRNA) carries DNA's protein-building instructions. Its nucleotide sequence is read three bases at a time. Sixty-four mRNA base triplets—codons—represent the genetic code. Two other types of RNA interact with mRNA during translation of that code. **Section 14.3**

RNA to protein: translation

Translation is an energy-intensive process by which a sequence of codons in mRNA is converted to a sequence of amino acids in a polypeptide chain. **Section 14.4**

Mutations

Small-scale, permanent changes in the nucleotide sequence of DNA may result from replication errors, the activity of transposable elements, or exposure to environmental hazards. Such mutation can change a gene's product. **Section 14.5**

Links to Earlier Concepts

- This chapter builds on your understanding of enzymatic reactions (Section 3.2) and energy in metabolism (6.2). You will see how information coded in nucleic acids (3.7) becomes translated into proteins (3.5, 3.6).

- You will use what you know about genes (11.1) and base pairing (13.2) to understand transcription, which has many features in common with DNA replication (13.3).

- A review of peptide bond formation (3.5) will be helpful as you learn about translation.

- The last section of this chapter revisits mutations (1.4): their molecular basis, how environmental factors (2.3, 6.3, 7.1) cause them, and some of their consequences (9.5, 12.2).

How would you vote? Accidental exposure to ricin is unlikely, but terrorists may try to poison food or water supplies with it. Researchers have developed a vaccine against ricin. Do you want to be vaccinated? See CengageNOW for details, then vote online.

215

DNA, RNA, and Gene Expression

- Transcription converts information in a gene to RNA; translation converts information in an mRNA to protein.
- Links to Enzymatic reactions 3.2, Proteins 3.5, Nucleotides 3.7, Genes 11.1, DNA replication 13.3

The Nature of Genetic Information

A cell's DNA contains all of its genetic information, but how does the cell convert that information into structural and functional components? Let's start with the nature of the information itself.

DNA is like a book, an encyclopedia that contains all of the instructions for building a new individual. You already know the alphabet used to write the book: the four letters: A, T, G, and C, for the four nucleotide bases adenine, thymine, guanine, and cytosine. Each strand of DNA consists of a chain of those four kinds of nucleotides. The linear order, or sequence, of the four bases in the strand is the genetic information. That information occurs in subsets called genes, which are Mendel's "units of inheritance" that you read about in Chapter 11.

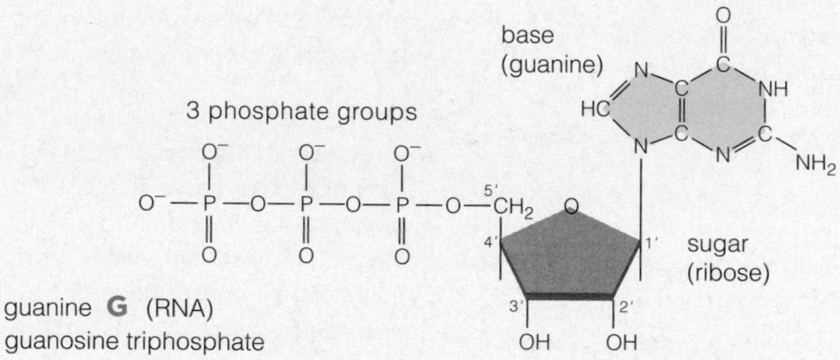

Converting a Gene to an RNA

Converting the information encoded by a gene into a product starts with RNA synthesis, or **transcription**. By this process, enzymes use the nucleotide sequence of a gene as a template to synthesize a strand of RNA (ribonucleic acid):

$$\text{DNA} \xrightarrow{\textit{transcription}} \text{RNA}$$

Except for the double-stranded RNA that is the genetic material of some types of viruses, RNA usually occurs in single-stranded form. A strand of RNA is structurally similar to a single strand of DNA. For example, both are chains of four kinds of nucleotides. Like a DNA nucleotide, an RNA nucleotide has three phosphate groups, a ribose sugar, and one of four bases. However, DNA and RNA nucleotides are slightly different. Three of the bases (adenine, cytosine, and guanine) are the same in DNA and RNA nucleotides, but the fourth base in RNA is uracil, not thymine, and the ribose sugar differs in RNA (Figure 14.2).

Despite these small differences in structure, DNA and RNA have very different functions (Figure 14.3). DNA's only role is to store a cell's heritable information. By contrast, a cell transcribes several kinds of RNAs, each of which has a different function. MicroRNAs are important in gene control, which is the subject of the next chapter. Three types of RNA have roles in protein synthesis. **Ribosomal RNA (rRNA)** is the main component of ribosomes, structures upon which polypeptide chains are built (Sections 4.4 and 4.6). **Transfer RNA (tRNA)** delivers amino acids to ribosomes, one by one, in the order specified by a **messenger RNA (mRNA)**.

Converting mRNA to Protein

mRNA is the only kind of RNA that carries a protein-building message. That message is encoded within the sequence of the mRNA itself by sets of three nucleotide bases, "genetic words" that follow one another along the length of the mRNA. Like the words of a sentence, a series of genetic words can form a meaningful parcel of information—in this case, the sequence of amino acids of a protein.

A Guanine, one of the four nucleotides in RNA. The others (adenine, uracil, and cytosine) differ only in their component bases. Three of the four bases in RNA nucleotides are identical to the bases in DNA nucleotides.

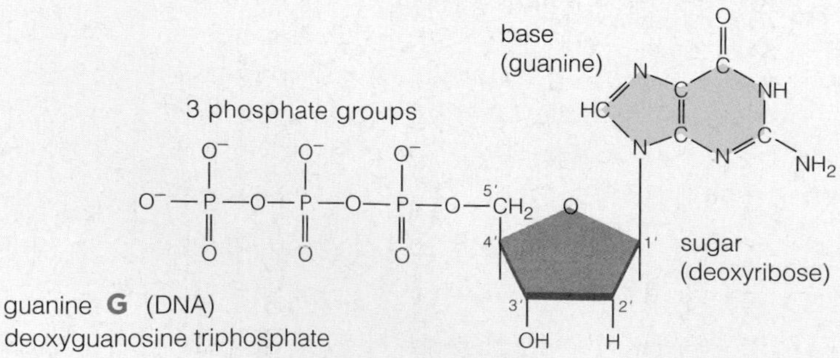

B Compare the DNA nucleotide guanine. The only structural difference between the RNA and DNA versions of guanine (or adenine, or cytosine) is the functional group on the 2' carbon of the sugar.

Figure 14.2 Ribonucleotides and nucleotides compared.

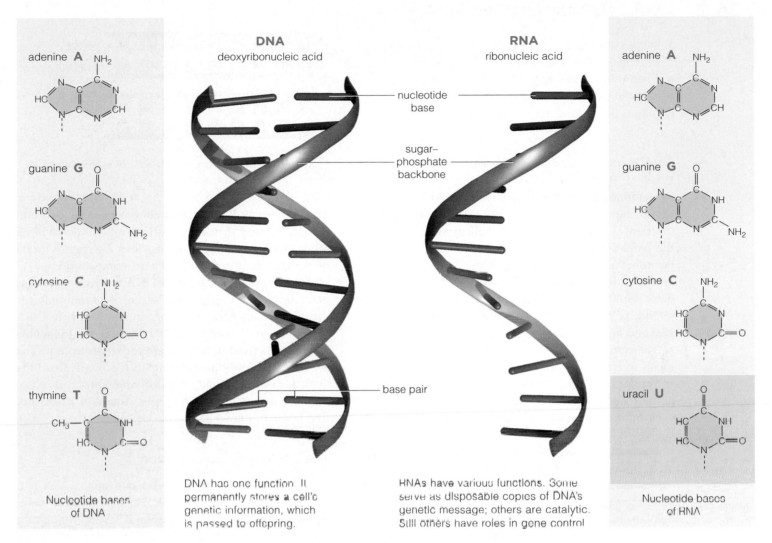

adenine **A**

guanine **G**

cytosine **C**

thymine **T**

Nucleotide bases
of DNA

DNA
deoxyribonucleic acid

nucleotide
base

sugar–
phosphate
backbone

base pair

DNA has one function. It
permanently stores a cell's
genetic information, which
is passed to offspring.

RNA
ribonucleic acid

RNAs have various functions. Some
serve as disposable copies of DNA's
genetic message; others are catalytic.
Still others have roles in gene control.

adenine **A**

guanine **G**

cytosine **C**

uracil **U**

Nucleotide bases
of RNA

Figure 14.3 DNA and RNA compared.

By the process of **translation**, the protein-building information in an mRNA is decoded (translated) into a sequence of amino acids. The result is a polypeptide chain that twists and folds into a protein:

$$\text{mRNA} \xrightarrow{\textit{translation}} \text{PROTEIN}$$

Sections 14.3 and 14.4 describe how rRNA and tRNA interact to translate the sequence of base triplets in an mRNA into the sequence of amino acids in a protein.

The processes of transcription and translation are part of **gene expression**, a multistep process by which genetic information encoded by a gene is converted into a structural or functional part of a cell or body:

$$\text{DNA} \xrightarrow{\textit{transcription}} \text{mRNA} \xrightarrow{\textit{translation}} \text{PROTEIN}$$

A cell's DNA sequence contains all of the information it needs to make the molecules of life. Each gene encodes an RNA, and different types of RNAs interact to assemble proteins from amino acids (Section 3.5). Proteins—enzymes—can assemble lipids and complex carbohydrates from simple building blocks (Section 3.2), replicate DNA (Section 13.3), and make RNA, as you will see in the next section.

Take-Home Message

What is the nature of genetic information carried by DNA?

■ The nucleotide sequence of a gene encodes instructions for building an RNA or protein product.

■ A cell transcribes the nucleotide sequence of a gene into RNA.

■ Although RNA is structurally similar to a single strand of DNA, the two types of molecules differ functionally.

■ A messenger RNA (mRNA) carries a protein-building code in its nucleotide sequence. rRNAs and tRNAs interact to translate that sequence into a protein.

- RNA polymerase links RNA nucleotides into a chain, in the order dictated by the base sequence of a gene.
- A new RNA strand is complementary in sequence to the DNA strand from which it was transcribed.

- Links to Base pairing 13.2, DNA replication 13.3

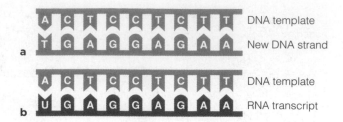

Figure 14.4 Base pairing during (**a**) DNA synthesis and (**b**) transcription.

DNA Replication and Transcription Compared

Remember that DNA replication begins with one DNA double helix and ends with two DNA double helices (Section 13.3). The two double helices are identical to the parent molecule because the process of DNA replication follows base-pairing rules. A nucleotide can be added to a growing strand of DNA only if it base-pairs with the corresponding nucleotide of the parent strand: G pairs with C, and A pairs with T (Section 13.2 and Figure 14.4*a*).

The same base-pairing rules also govern RNA synthesis in transcription. An RNA strand is structurally so similar to a DNA strand that the two can base-pair if their nucleotide sequences are complementary. In such hybrid molecules, G pairs with C; A pairs with U—uracil (Figure 14.4*b*).

During transcription, a strand of DNA acts as a template upon which a strand of RNA—a transcript—is assembled from RNA nucleotides. A nucleotide can be added to a growing strand of RNA only if it is complementary to the corresponding nucleotide of the parent strand of DNA: G pairs with C, and A pairs with U.

Thus, each new RNA is complementary in sequence to the DNA strand that served as its template. As in DNA replication, each nucleotide provides the energy for its own attachment to the end of a growing strand.

Transcription is similar to DNA replication in that one strand of a nucleic acid serves as a template for synthesis of another. However, in contrast with DNA replication, only part of one DNA strand, not the whole molecule, is used as a template for transcription. The enzyme **RNA polymerase**, not DNA polymerase, adds nucleotides to the end of a growing transcript. Also, transcription results in a single strand of RNA, not two DNA double helices.

The Process of Transcription

Transcription begins with a chromosome, which is a double helix molecule of DNA. The process gets under way when an RNA polymerase and several regulatory proteins attach to a specific binding site in the DNA

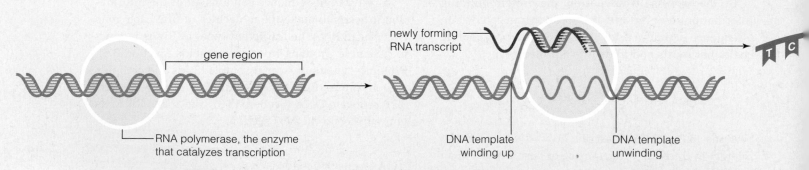

A RNA polymerase binds to a promoter in the DNA, along with regulatory proteins. The binding positions the polymerase near a gene in the DNA.

In most cases, the nucleotide sequence of the gene occurs on only one of the two strands of DNA. Only the complementary strand will be translated into RNA.

B The polymerase begins to move along the DNA and unwind it. As it does, it links RNA nucleotides into a strand of RNA in the order specified by the base sequence of the DNA.

The DNA double helix winds up again after the polymerase passes. The structure of the "opened" DNA molecule at the transcription site is called a transcription bubble, after its appearance.

Figure 14.5 Animated Transcription.

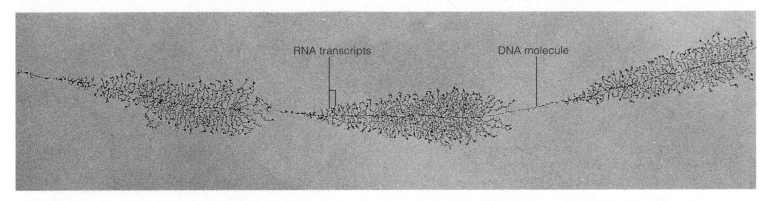

Figure 14.6 Typically, many RNA polymerases simultaneously transcribe the same gene, producing a conglomerate structure often called a "Christmas tree" after its shape. Here, three genes next to one another on the same chromosome are being transcribed.

Figure It Out: Are the polymerases transcribing this DNA molecule moving from left to right or from right to left? *Answer: Left to right*

called a **promoter** (Figure 14.5*a*). The binding positions the polymerase at a transcription start site close to a gene. The polymerase starts moving along the DNA, in the 5′ to 3′ direction over the gene (Figure 14.5*b*). As it moves, the polymerase unwinds the double helix just a bit so it can "read" the base sequence of the noncoding DNA strand. The polymerase joins free RNA nucleotides into a chain, in the order dictated by that DNA sequence. As in DNA replication, the synthesis is directional: An RNA polymerase adds nucleotides only to the 3′ end of a growing strand of RNA.

When the polymerase reaches the end of the gene, the DNA and the new RNA strand are released. RNA polymerase follows base-pairing rules, so the new RNA strand is complementary in base sequence to the DNA strand from which it was transcribed (Figure 14.5*c,d*). It is an RNA copy of a gene.

Typically, many polymerases transcribe a particular gene region at the same time, so many new RNA strands can be produced very quickly (Figure 14.6).

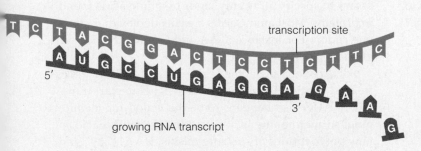

C What happened in the gene region? RNA polymerase catalyzed the covalent bonding of many nucleotides to one another to form an RNA strand. The base sequence of the new RNA strand is complementary to the base sequence of its DNA template—a copy of the gene.

D *Top*, at the end of the gene region, the last stretch of the new transcript unwinds and detaches from the DNA template. *Bottom*, a ball-and-stick model of a strand of RNA.

■ Base triplets in an mRNA are words in a protein-building message. Two other classes of RNA—rRNA and tRNA—translate those words into a polypeptide chain.

Post-Transcriptional Modifications

In eukaryotes, transcription takes place in the nucleus, where new RNA is modified before being shipped to the cytoplasm. Just as a dressmaker may snip off loose threads or add bows to a dress before it leaves the shop, so do eukaryotic cells tailor their RNA before it leaves the nucleus.

For example, most eukaryotic genes contain **introns**, nucleotide sequences that are removed from a new RNA. Introns intervene between **exons**, sequences that stay in the RNA (Figure 14.7). Introns are transcribed along with the exons, but are removed before the RNA leaves the nucleus. Either all exons remain in the mature RNA, or some are removed and the rest are spliced in various combinations. By such **alternative splicing**, one gene can encode different proteins.

New transcripts that will become mRNAs are further tailored after splicing. A modified guanine "cap" gets attached to the 5' end of each. Later, the cap will help the mRNA bind to a ribosome. A tail of 50 to 300 adenines is also added to the 3' end of a new mRNA; hence the name, poly-A tail.

mRNA—The Messenger

DNA stores heritable information about proteins, but making those proteins requires mRNA, tRNA, and rRNA. The three types of RNA interact to translate DNA's information into a protein.

An mRNA is a disposable copy of a gene; its job is to carry DNA's protein-building information to the other

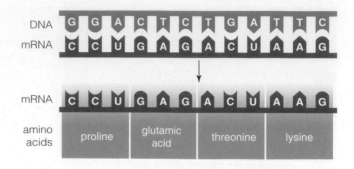

Figure 14.8 Example of the correspondence between DNA and proteins. A DNA strand is transcribed into mRNA, and the codons of the mRNA specify a chain of amino acids.

two types of RNA for translation. Like sentences, the genetic message carried by an mRNA can be understood by those who know the language. Each mRNA is a linear sequence of genetic "words," all spelled with an alphabet of just four nucleotides. Each "word" is three nucleotides long, and each is a code—a **codon**—for a particular amino acid. One codon follows the next along the length of an mRNA. Thus, the order of codons in an mRNA determines the order of amino acids in the polypeptide chain that will be translated from it (Figure 14.8).

With four different nucleotides possible in each of three positions, there are a total of sixty-four (or 4^3) mRNA codons. Collectively, the codons constitute the **genetic code** (Figure 14.9). Which of the four nucleotides is in the first, second, and third position of a triplet determines which amino acid the codon specifies. For instance, the codon AUG (adenine–uracil–guanine) encodes the amino acid methionine, and UGG encodes tryptophan. There are many more codons than are necessary to specify all twenty kinds of amino acids found in proteins. Most amino acids are encoded by more than one codon. For instance, GAA and GAG both code for glutamic acid.

Some codons signal the beginning and end of a gene. In most species, the first AUG is a signal to start translation. AUG also happens to be the codon for methionine, so methionine is always the first amino acid in new polypeptides of such organisms. UAA, UAG, and UGA do not specify an amino acid. They are signals that stop translation—stop codons. A stop codon marks the end of a coding sequence in an mRNA.

The genetic code is highly conserved, which means that many organisms use the same code and probably always have. Prokaryotes and some protists have a few codons that vary, as do mitochondria and chloroplasts. The variation was a clue that led to the theory of how organelles evolved, which we discuss in Section 20.4.

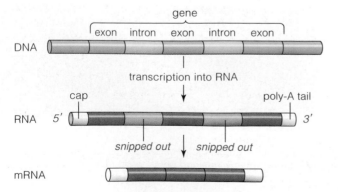

Figure 14.7 Animated Post-transcriptional modification of RNA in the nucleus. Introns are removed, exons may be spliced. An mRNA also gets a poly-A tail and modified guanine "cap."

first base	second base				third base
	U	**C**	**A**	**G**	
U	UUU ⎫ phe / UUC ⎭ / UUA ⎫ leu / UUG ⎭	UCU ⎫ / UCC ⎬ ser / UCA / UCG ⎭	UAU ⎫ tyr / UAC ⎭ / UAA **STOP** / UAG **STOP**	UGU ⎫ cys / UGC ⎭ / UGA **STOP** / UGG trp	U / C / A / G
C	CUU ⎫ / CUC ⎬ leu / CUA / CUG ⎭	CCU ⎫ / CCC ⎬ pro / CCA / CCG ⎭	CAU ⎫ his / CAC ⎭ / CAA ⎫ gln / CAG ⎭	CGU ⎫ / CGC ⎬ arg / CGA / CGG ⎭	U / C / A / G
A	AUU ⎫ / AUC ⎬ ile / AUA / AUG met	ACU ⎫ / ACC ⎬ thr / ACA / ACG ⎭	AAU ⎫ asn / AAC ⎭ / AAA ⎫ lys / AAG ⎭	AGU ⎫ ser / AGC ⎭ / AGA ⎫ arg / AGG ⎭	U / C / A / G
G	GUU ⎫ / GUC ⎬ val / GUA / GUG ⎭	GCU ⎫ / GCC ⎬ ala / GCA / GCG ⎭	GAU ⎫ asp / GAC ⎭ / GAA ⎫ glu / GAG ⎭	GGU ⎫ / GGC ⎬ gly / GGA / GGG ⎭	U / C / A / G

Figure 14.9 Animated The sixty four codons of the genetic code. The *left* column lists a codon's first base. The *top* row lists the second base. The *right* column lists the third. Appendix V shows the amino acids. **Figure It Out:** Which codons specify the amino acid lysine (lys)?

Answer: AAA and AAG

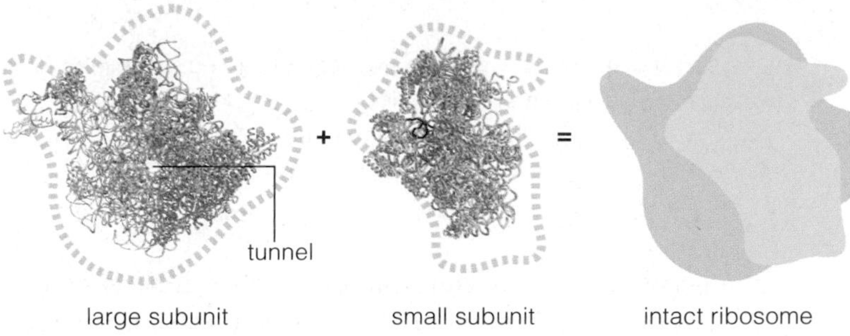

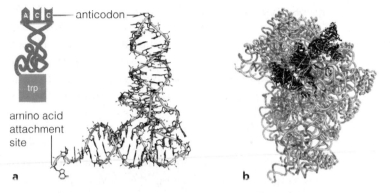

large subunit   small subunit   intact ribosome

Figure 14.10 The ribosome consists of a large and a small subunit. Notice the tunnel through the interior of the large subunit. rRNA components of the ribosome (*tan*) catalyze assembly of polypeptide chains, which thread through this tunnel as they form. We show an mRNA (*red*) attached to the small subunit.

anticodon

amino acid attachment site

a   **b**

Figure 14.11 tRNA. (**a**) Models of the tRNA that carries the amino acid tryptophan. Each tRNA's anticodon is complementary to an mRNA codon. Each also carries the amino acid specified by that codon. (**b**) During translation, tRNAs dock at an intact ribosome. Here, three tRNAs (*brown*) are docked at the small ribosomal subunit (the large subunit is not shown, for clarity). The anticodons of the tRNAs line up with complementary codons in an mRNA (*red*).

rRNA and tRNA—The Translators

A ribosome has a large and a small subunit. Each consists of proteins and rRNA (Figure 14.10). rRNA is one of the few examples of RNA with enzymatic activity: The rRNA of a ribosome, not the protein, catalyzes the formation of a peptide bond between amino acids.

As you will see in the next section, two ribosomal subunits converge as an intact ribosome on an mRNA during translation. tRNAs bring amino acids to this complex. A tRNA has two attachment sites: One is an **anticodon**, a triplet of nucleotides that base-pairs with an mRNA codon (Figure 14.11). The other binds to a free amino acid—the one specified by the codon.

Some tRNAs can base-pair with more than one type of codon. For example, the codons AUU, AUC, and AUA all specify isoleucine; a tRNA that carries isoleucine can base-pair with all of them.

As you will see in the next section, tRNAs deliver amino acids, one after the next, to a ribosome–mRNA complex during translation. The order of codons in the mRNA is the order in which tRNAs deliver their amino acid cargoes to the ribosome. As the amino acids are delivered, the ribosome joins them via peptide bonds into a new polypeptide chain (Section 3.5). Thus, the order of codons in an mRNA—DNA's protein-building message—is translated into a protein.

Take-Home Message

What are the functions of mRNA, tRNA, and rRNA?

■ Nucleotide bases in mRNA are "read" in sets of three during protein synthesis. Most of these base triplets (codons) code for amino acids. The genetic code comprises all sixty-four codons.

■ A tRNA has an anticodon complementary to an mRNA codon, and it has a binding site for the amino acid specified by that codon. tRNAs deliver amino acids to ribosomes.

■ Ribosomes, which consist of two subunits of rRNA and proteins, link amino acids into polypeptide chains.

Translation: RNA to Protein

■ Translation converts the information carried by an mRNA into a new polypeptide chain.
■ The order of the codons in the mRNA determines the order of the amino acids in the polypeptide chain.

■ Links to Peptide bonds 3.5, Energy in metabolism 6.2

Translation, the second part of protein synthesis, occurs in the cytoplasm of all cells. It has three stages: initiation, elongation, and termination.

The initiation stage begins when a small ribosomal subunit binds to an mRNA. Next, the anticodon of a special initiator tRNA base-pairs with the first AUG codon of the mRNA. Then, a large ribosomal subunit joins the small subunit. The cluster is now called an initiation complex (Figure 14.12a,b).

In the elongation stage, the ribosome assembles a polypeptide chain as it moves along the mRNA, threading the strand between its two subunits. The initiator tRNA carries the amino acid methionine, so the first amino acid of the new polypeptide chain is methionine. Other tRNAs bring successive amino acids to the complex as their anticodons base-pair with the codons in the mRNA, one after the next. The ribosome joins each amino acid to the end of the growing polypeptide chain by way of a peptide bond (Figure 14.12c–e and Section 3.5).

Termination occurs when the ribosome encounters a stop codon in the mRNA. Proteins called release factors recognize this codon and bind to the ribosome. The binding triggers enzyme activity that detaches the mRNA and the polypeptide chain from the ribosome (Figure 14.12f).

polysome

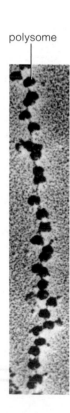

In cells that are making a lot of protein, new initiation complexes may form on an mRNA before other ribosomes finish translating it. Many ribosomes may simultaneously translate the same mRNA, in which case they are called polysomes (*left*). Transcription and translation both occur in the cytoplasm of prokaryotes, and these processes are closely linked in time and in space. Translation begins before transcription is done, so in these cells, a transcription Christmas tree (Figure 14.6) often appears decorated with polysome "balls."

Translation is a biosynthetic process that requires a lot of energy to run (Section 6.2). That energy is provided mainly in the form of phosphate-group transfers from the RNA nucleotide GTP (Figure 14.2a). GTP caps eukaryotic mRNAs, and its hydrolysis also fuels formation of the initiation complex, binding of tRNA to the ribosome, movement of the ribosome along the mRNA, formation of peptide bonds, and release of the ribosomal subunits from mRNA during termination. ATP is used to attach amino acids to free tRNAs.

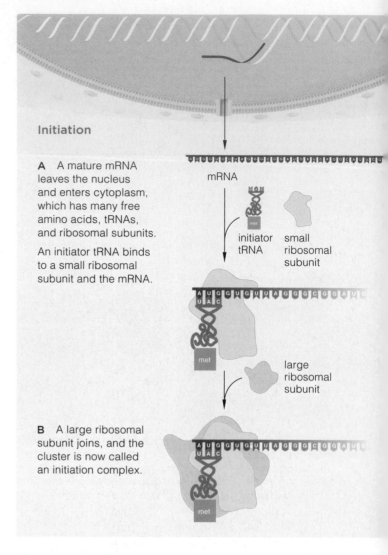

Initiation

A A mature mRNA leaves the nucleus and enters cytoplasm, which has many free amino acids, tRNAs, and ribosomal subunits.

An initiator tRNA binds to a small ribosomal subunit and the mRNA.

mRNA

initiator tRNA   small ribosomal subunit

met

large ribosomal subunit

B A large ribosomal subunit joins, and the cluster is now called an initiation complex.

met

Figure 14.12 Animated An example of translation as it occurs in eukaryotic cells.

(**a,b**) In initiation, an mRNA, an intact ribosome, and an initiator tRNA form an initiation complex.

(**c–e**) In elongation, the new polypeptide chain grows as the ribosome catalyzes the formation of peptide bonds between amino acids delivered by tRNAs.

(**f**) In termination, the mRNA and the new polypeptide chain are released, and the ribosome disassembles.

Take-Home Message

How is mRNA translated into protein?

■ Translation is an energy-requiring process that begins as an mRNA joins with an initiator tRNA and two ribosomal subunits.

■ Amino acids are delivered to the complex by tRNAs in the order dictated by successive mRNA codons. As they arrive, the ribosome joins each to the end of the polypeptide chain.

■ Translation ends when the ribosome encounters a stop codon in the mRNA.

Elongation

C An initiator tRNA carries the amino acid methionine, so the first amino acid of the new polypeptide chain will be methionine. A second tRNA binds the second codon of the mRNA (here, that codon is GUG, so the tRNA that binds carries the amino acid valine).

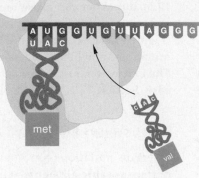
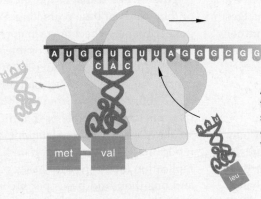

A peptide bond forms between the first two amino acids (here, methionine and valine).

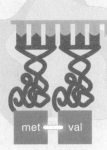
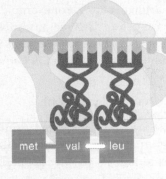

D The first tRNA is released and the ribosome moves to the next codon in the mRNA. A third tRNA binds to the third codon of the mRNA (here, that codon is UUA, so the tRNA carries the amino acid leucine).

A peptide bond forms between the second and third amino acids (here, valine and leucine).

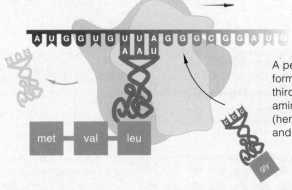
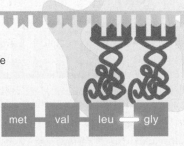

E The second tRNA is released and the ribosome moves to the next codon. A fourth tRNA binds the fourth mRNA codon (here, that codon is GGG, so the tRNA carries the amino acid glycine).

A peptide bond forms between the third and fourth amino acids (here, leucine and glycine).

Termination

F Steps **d** and **e** are repeated over and over until the ribosome encounters a stop codon in the mRNA. The mRNA transcript and the new polypeptide chain are released from the ribosome. The two ribosomal subunits separate from each other. Translation is now complete. Either the chain will join the pool of proteins in the cytoplasm or it will enter rough ER of the endomembrane system (Section 4.9).

14.5 Mutated Genes and Their Protein Products

■ If the nucleotide sequence of a gene changes, it may result in an altered gene product, with harmful effects.

■ Links to Mutation 1.4, Electrons 2.3, Protein structure 3.6, Free radicals 6.3, Radiant energy 7.1, Cancer 9.5, Huntington's disease 12.2, DNA replication 13.3

We have repeatedly mentioned mutations in reference to the harm they can cause, and also as the raw material of evolution. Mutations are small-scale changes in the nucleotide sequence of a cell's DNA. One or more nucleotides may be substituted for another or lost, or extra ones inserted. Such changes can alter the genetic instructions encoded in the DNA, and the result may be an altered gene product. Remember, more than one codon can specify the same amino acid, so cells have a margin of safety. For example, a mutation that changes a UCU to a UCC in an mRNA may not have further effects, because both codons specify serine. However, many mutations have negative consequences.

Common Mutations

A nucleotide mispaired during DNA replication may end up as a **base-pair substitution**, in which one nucleotide and its partner are replaced by a different base pair. A substitution may result in an amino acid change or a premature stop codon in a gene's protein product. Sickle-cell anemia is caused by a base-pair substitution in the hemoglobin beta chain gene (Figure 14.13b).

A **deletion** mutation, in which one or more bases is lost, is smaller than a chromosomal deletion (Section 12.5), but either can cause the reading frame of mRNA codons to shift. The shift garbles the genetic message (Figure 14.13c). Frameshifts are also caused by **insertion** mutations, in which extra bases are inserted into DNA. The expansion mutation that causes Huntington's disease (Section 12.2) is a type of insertion.

What Causes Mutations?

Insertion mutations are often caused by the activity of **transposable elements**, which are segments of DNA that can insert themselves anywhere in a chromosome (Figure 14.14). Transposable elements can be hundreds or thousands of base pairs long. When one interrupts a gene sequence, it becomes a major insertion that changes the gene's product. Transposable elements occur in the DNA of all species; about 45 percent of human DNA consists of them or their remnants. Certain kinds can move spontaneously from one place to another within the same chromosome, or to a different chromosome.

Many mutations occur spontaneously during DNA replication. That is not surprising, given the fast pace of replication (about twenty bases per second in humans, and one thousand bases per second in bacteria). DNA polymerases make mistakes at predictable rates, but most types fix errors as they occur (Section 13.3). Errors that remain uncorrected are mutations.

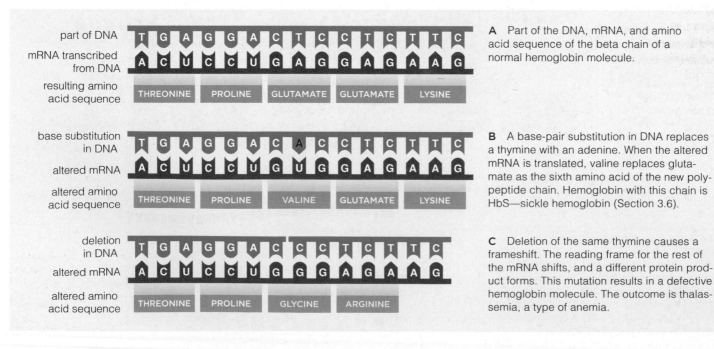

part of DNA	T G A G G A C T C C T C T T C
mRNA transcribed from DNA	A C U C C U G A G G A G A A G
resulting amino acid sequence	THREONINE PROLINE GLUTAMATE GLUTAMATE LYSINE

A Part of the DNA, mRNA, and amino acid sequence of the beta chain of a normal hemoglobin molecule.

base substitution in DNA	T G A G G A C A C C T C T T C
altered mRNA	A C U C C U G U G G A G A A G
altered amino acid sequence	THREONINE PROLINE VALINE GLUTAMATE LYSINE

B A base-pair substitution in DNA replaces a thymine with an adenine. When the altered mRNA is translated, valine replaces glutamate as the sixth amino acid of the new polypeptide chain. Hemoglobin with this chain is HbS—sickle hemoglobin (Section 3.6).

deletion in DNA	T G A G G A C C C T C T T C
altered mRNA	A C U C C U G G G A G A A G
altered amino acid sequence	THREONINE PROLINE GLYCINE ARGININE

C Deletion of the same thymine causes a frameshift. The reading frame for the rest of the mRNA shifts, and a different protein product forms. This mutation results in a defective hemoglobin molecule. The outcome is thalassemia, a type of anemia.

Figure 14.13 Animated Examples of mutation.

Figure 14.14 Barbara McClintock discovered transposable elements, which slip into and out of different locations in DNA. The curiously nonuniform coloration of individual kernels in Indian corn (*Zea mays*) sent her on the road to the discovery. She won a Nobel Prize for her research in 1983.

Several genes govern the formation and deposition of pigments in corn kernels, which are a type of seed. Interactions among these genes and their products result in yellow, white, red, orange, blue, or purple kernels. McClintock realized that unstable mutations in the genes cause streaks or spots of color in individual kernels.

The same pigment genes occur in all cells of a kernel, but those near a transposable element are inactive. Transposable elements move while a kernel's tissues are forming, so they can end up in different locations in the DNA of different cell lineages. Streaks and spots on the kernels are evidence of transposable element movement that inactivated and reactivated different pigment genes in different cell lineages.

Harmful environmental agents can cause mutations. For example, some forms of energy such as x-rays can ionize atoms by knocking electrons right out of them. Such ionizing radiation can break chromosomes into pieces, some of which may get lost during DNA replication (Figure 14.15*a*). Ionizing radiation also damages DNA indirectly when it penetrates living tissue, because it leaves a trail of destructive free radicals. Free radicals, remember, damage DNA (Section 6.3). That is why doctors and dentists use the lowest possible doses of x-rays on their patients.

Nonionizing radiation boosts electrons to a higher energy level, but not enough to knock them out of an atom. DNA absorbs one kind, ultraviolet (UV) light. Exposure to UV light can cause two adjacent thymine bases to bond covalently to one another. This bond, a thymine dimer, kinks the DNA (Figure 14.15*b*). During replication, the kinked part may be copied incorrectly, so a mutation is introduced into the DNA. Mutations that cause certain kinds of cancers begin with thymine dimers. They are the reason that exposing unprotected skin to sunlight increases the risk of skin cancer.

Some natural or synthetic chemicals can also cause mutations. For instance, certain chemicals in cigarette smoke transfer small hydrocarbon groups to the bases in DNA. The altered bases mispair during replication, or stop replication entirely.

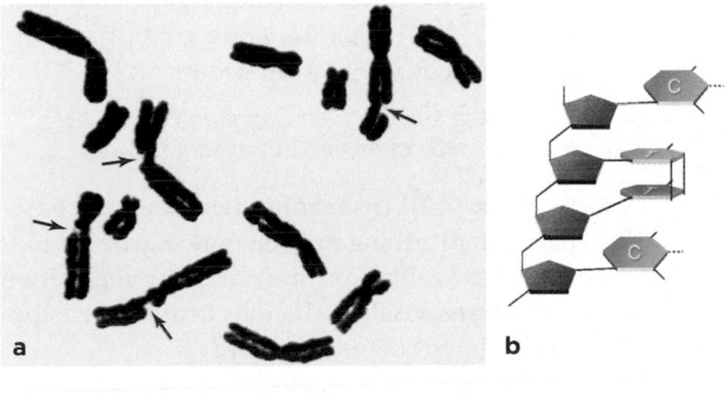

The Proof Is in the Protein

A mutation that occurs in a somatic cell of a sexually reproducing individual is not passed to the individual's offspring, so its effects do not endure. A mutation that arises in a germ cell or a gamete, however, may enter the evolutionary arena. It may also do so when passed on to offspring by asexual reproduction. Either way, an inherited mutation may affect an individual's capacity to function in its prevailing environment. The effects of uncountable mutations in millions of species have had spectacular evolutionary consequences—and that is a topic of later chapters.

Figure 14.15 Two types of DNA damage that can lead to mutations. (**a**) Chromosomes from a human cell after exposure to gamma rays (ionizing radiation). The broken pieces (*red arrows*) may get lost during DNA replication. The extent of damage in an exposed cell typically depends on how much radiation it absorbed. (**b**) A thymine dimer.

Take-Home Message

What is a mutation?

■ A mutation is a permanent small-scale change in the nucleotide sequence of DNA. A base-pair substitution, insertion, or deletion may alter a gene product.

■ Most mutations arise during DNA replication as a result of unrepaired DNA polymerase errors. Some mutations occur after exposure to harmful radiation or chemicals.

■ An inherited mutation may have positive or negative effects on an individual's capacity to function in its environment.

One of ricin's two polypeptide chains binds to a receptor on animal cell membranes that triggers endocytosis. The other chain is an enzyme; it removes a specific adenine base from one of the rRNA chains in the large ribosomal subunit. Once that happens, the ribosome stops working. A single molecule of ricin can inactivate about 1,500 ribosomes per minute. Protein synthesis grinds to a halt as ricin inactivates the rest of the cell's ribosomes.

How would you vote?

Terrorists may try to poison food or water supplies with ricin. Do you want to be vaccinated for ricin exposure? See CengageNOW for details, then vote online.

Summary

Section 14.1 The process of **gene expression** includes two steps, **transcription** and **translation** (Figure 14.16). It requires the participation of **messenger RNA (mRNA)**, **transfer RNA (tRNA)**, and **ribosomal RNA (rRNA)**.

Section 14.2 In eukaryotic cells, transcription occurs in the nucleus, and translation occurs in the cytoplasm. Both processes occur in the cytoplasm of prokaryotic cells.

In transcription, **RNA polymerase** binds to a **promoter** in the DNA near a gene, then assembles a strand of RNA by linking RNA nucleotides in the order dictated by the base sequence of the DNA.

■ *Use the animation on CengageNOW to explore transcription.*

Section 14.3 The RNA of eukaryotes is modified before it leaves the nucleus. **Introns** are removed. Some **exons** may be removed also, and the remaining ones spliced in different combinations (**alternative splicing**). A cap and a poly-A tail are also added to a new mRNA.

mRNA carries DNA's protein-building information. Its genetic message is written in **codons**, sets of three nucleotides. Sixty-four codons, most of which specify amino acids, constitute the **genetic code**. Variations occur among prokaryotes, organelles, and single-celled eukaryotes.

Each tRNA has an **anticodon** that can base-pair with a codon, and it binds to the kind of amino acid specified by the codon. Catalytic rRNA and proteins make up the two subunits of ribosomes.

■ *Use the interaction on CengageNOW to learn about transcript processing and the genetic code.*

Section 14.4 Genetic information carried by an mRNA directs the synthesis of a polypeptide chain during translation. First, an mRNA, an initiator tRNA, and two ribosomal subunits converge. The intact ribosome then catalyzes formation of a peptide bond between successive amino acids, which are are delivered by tRNAs in the order specified by the codons in the mRNA. Translation ends when the polymerase encounters a stop codon.

■ *Use the animation on CengageNOW to see the translation of an mRNA transcript.*

Section 14.5 Insertions, **deletions**, and **base-pair substitutions** may change a gene's product. These mutations may arise by replication error, **transposable element** activity, or exposure to environmental hazards.

■ *Use the animation on CengageNOW to investigate the effects of mutation.*

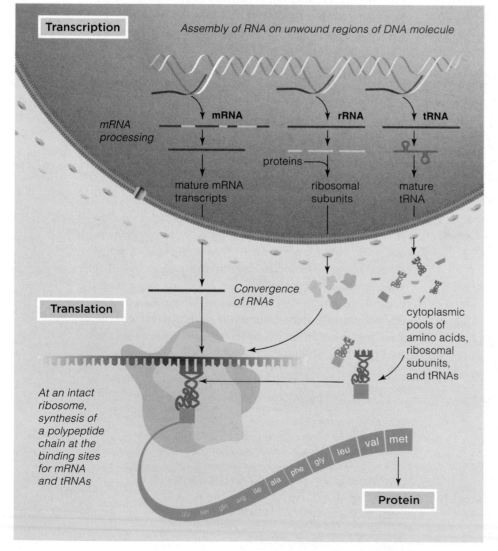

Transcription

Assembly of RNA on unwound regions of DNA molecule

mRNA processing

mRNA   **rRNA**   **tRNA**

proteins

mature mRNA transcripts   ribosomal subunits   mature tRNA

Convergence of RNAs

Translation

cytoplasmic pools of amino acids, ribosomal subunits, and tRNAs

At an intact ribosome, synthesis of a polypeptide chain at the binding sites for mRNA and tRNAs

gly ser gln arg ile ala phe gly leu val met

Protein

Figure 14.16 Animated Summary of protein synthesis as it occurs in eukaryotic cells.

Data Analysis Exercise

About one out of 3,500 people carry a mutation that affects the product of the NF1 gene, which is a tumor suppressor (Section 9.5). People who are heterozygous for one of these mutations have neurofibromatosis, an autosomal dominant genetic disorder (Section 12.2). Among other problems, soft, fibrous tumors (neurofibromas) form in the skin and nervous system. The homozygous condition may be lethal.

Most mutations associated with neurofibromatosis result in defective splicing of the gene's 60 exons. Each neurofibroma typically arises from a new mutation that disrupts the individual's one functional allele. In a 1997 study, Eduard Serra and his colleagues tested several tumors from an individual with the disorder for such mutations (Figure 14.17).

1. Which tumors are missing marker D17S250? Is this sequence inside or outside of the *NF1* gene?

2. In four of these six tumors, the entire large arm of chromosome 17 was deleted. Which four?

3. People affected by neurofibromatosis are 200 to 500 times more likely to develop malignant tumors than unaffected people. Why do you think that is the case?

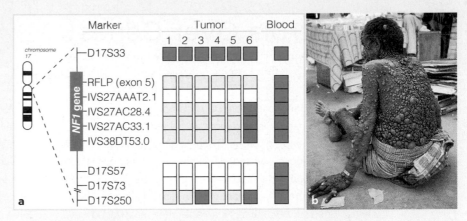

Figure 14.17 Neurofibromatosis. (**a**) Genetic analysis of six tumors from a single individual affected by neurofibromatosis. Each tumor was checked for the presence of nine nucleotide sequences (markers) in or near the *NF1* gene.

For each tumor (1–6), *green* boxes indicate that the marker is present; *yellow* boxes indicate the marker is missing; *white* boxes indicate inconclusive results. Blood was also tested as a control.

(**b**) An individual affected by neurofibromatosis.

Self-Quiz
Answers in Appendix III

1. A chromosome contains many genes that are transcribed into different _____ .
 a. proteins c. RNAs
 b. polypeptides d. a and b

2. A binding site for RNA polymerase is a _____ .

3. Energy that drives transcription is provided by _____ .

4. An RNA molecule is typically _____ -stranded.

5. RNAs form by _____ ; proteins form by _____ .
 a. replication; translation c. translation; transcription
 b. transcription; translation d. replication; transcription

6. _____ remain in mRNA.
 a. Introns b. Exons

7. How many codons constitute the genetic code?

8. Most codons specify a(n) _____ .
 a. protein c. amino acid
 b. polypeptide d. mRNA

9. Anticodons pair with _____ .
 a. mRNA codons c. RNA anticodons
 b. DNA codons d. amino acids

10. Energy that drives translation is provided by _____ .
 a. ATP c. UTP
 b. GTP d. a and b are correct

11. Using Figure 14.9, translate this nucleotide sequence into an amino acid sequence, starting at the first base:

5′—GGUUUCUUCAAGAGA—3′

12. Name one cause of mutations.

13. Match each term with the most suitable description.
 ___ genetic message a. protein-coding mRNA
 ___ sequence b. gets around
 ___ polysome c. read as base triplets
 ___ exon d. linear order of bases
 ___ genetic code e. occurs only in groups
 ___ intron f. set of 64 codons
 ___ transposable g. removed before translation
 element

■ *Visit CengageNOW for additional questions.*

Critical Thinking

1. Each position of a codon can be occupied by one of four (4) nucleotides. If codons were two (2) nucleotides long, they could encode a maximum of $4^2 = 16$ amino acids. What is the minimum number of nucleotides per codon necessary to specify all 20 biological amino acids?

2. Cigarette smoke contains at least fifty-five different chemicals identified as carcinogenic (cancer-causing) by the International Agency for Research on Cancer (IARC). When these carcinogens enter the bloodstream, enzymes convert them to a series of chemical intermediates that are easier to excrete. Some of the intermediates bind irreversibly to DNA. Propose a mechanism by which such binding causes cancer.

3. Termination of prokaryotic DNA transcription often depends on the structure of a newly forming RNA. Transcription stops where the mRNA folds back on itself, forming a hairpin-looped structure such as the one at *right*. How do you think this structure stops transcription?

15 | Controls Over Genes

Between You and Eternity

You are in college, your whole life ahead of you. Your risk of developing cancer is as remote as old age, an abstract statistic that is easy to forget. "There is a moment when everything changes—when the width of two fingers can suddenly be the total distance between you and eternity." Robin Shoulla wrote those words after being diagnosed with breast cancer. She was seventeen. At an age when most young women are thinking about school, parties, and potential careers, Robin was dealing with radical mastectomy—the removal of a breast, all lymph nodes under the arm, and skeletal muscles in the chest wall under the breast. She was pleading with her oncologist not to use her jugular vein for chemotherapy and wondering if she would survive to see the next year (Figure 15.1).

Robin's ordeal became part of a statistic—one of more than 200,000 new cases of breast cancer diagnosed in the United States each year. About 5,700 of those cases occur in women and men under thirty-four years of age.

Mutations in some genes predispose individuals to develop certain kinds of cancer. Tumor suppressor genes are named because tumors are more likely to occur when these genes mutate. Two examples are *BRCA1* and *BRCA2*. A mutated version of one or both of these genes is often found in breast and ovarian cancer cells. If a *BRCA* gene mutates in one of three especially dangerous ways, a woman has an 80 percent chance of developing breast cancer before the age of seventy.

Tumor suppressors are part of a system of stringent controls over gene expression that keeps the cells of multicelled organisms functioning normally. Such controls govern when and how fast specific genes are transcribed and translated. You will be considering the impact of gene controls in chapters throughout the book—and in some chapters of your life.

Robin Shoulla survived. Although radical mastectomy is rarely performed today (a modified procedure is less disfiguring), it is the only option when cancer cells invade muscles under the breast. It was Robin's only option. She may never know which mutation caused her cancer. Now, sixteen years later, she has what she calls a normal life—career, husband, children. Her goal as a cancer survivor: "To grow very old with gray hair and spreading hips, smiling."

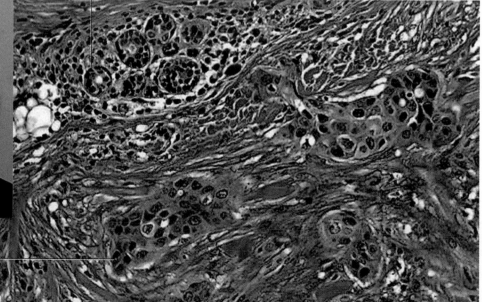

normal cells in organized clusters

irregular clusters of malignant cells

See the video! Figure 15.1 A case of breast cancer. *Right*, this light micrograph revealed irregular clusters of carcinoma cells that infiltrated milk ducts in breast tissue. *Above*, Robin Shoulla. Diagnostic tests revealed abnormal cells such as these in her body.

Key Concepts

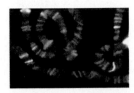

Overview of controls over gene expression

A variety of molecules and processes alter gene expression in response to changing conditions both inside and outside the cell. Selective gene expression also results in cell differentiation, by which different cell lineages become specialized. **Section 15.1**

Examples from eukaryotes

The orderly, localized expression of certain genes in embryos gives rise to the body plan of complex multicelled organisms. In female mammals, most of the genes on one of the two X chromosomes are inactivated in every cell. **Section 15.2**

Fruit fly development

Drosophila research revealed how a complex body plan emerges. All cells in a developing embryo inherit the same genes, but they use different subsets of those genes. **Section 15.3**

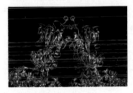

Examples from prokaryotes

Prokaryotic gene controls govern responses to short-term changes in nutrient availability and other aspects of the environment. The main gene controls bring about fast adjustments in the rate of transcription. **Section 15.4**

Links to Earlier Concepts

- A review of what you know about metabolic controls (Section 6.4) will be helpful as we revisit the concept of gene expression (14.1) in more detail. You may wish to review alleles (11.1), autosomal inheritance (12.2), and mutation (14.5).

- You will be applying what you know about the organization of chromosomal DNA (9.1, 9.2), and sex determination and X-linked inheritance in humans (12.1, 12.4), as we delve into controls over transcription (14.2), post-transcriptional processing (14.3), translation (14 4), and other processes that affect gene expression.

- You will revisit carbohydrates (3.3) and fermentation (8.5) as you learn about gene control in prokaryotes.

How would you vote? Some women at high risk of developing breast cancer opt for preventive surgical removal of their breasts before cancer develops. Many of those women never would have developed cancer. Should surgery be restricted to cancer treatment? See CengageNOW for details, then vote online.

■ Gene controls govern the kinds and amounts of substances that are present in a cell at any given interval.

■ Links to Histones 9.1, Gene expression 14.1, Transcription 14.2, Post-transcriptional modification 14.3, Translation 14.4

Which Genes Get Tapped?

All of the cells in your body are descended from the same fertilized egg, so they all contain the same DNA with the same genes. Some of the genes are transcribed by all cells; such genes affect structural features and metabolic pathways common to all cells.

enhancer

Figure 15.3 Hypothetical part of a chromosome that contains a gene. Molecules that affect the rate of transcription of the gene bind at promoter (*yellow*) and enhancer (*green*) sequences.

In other ways, however, nearly all of your body cells are specialized. **Differentiation**, the process by which cells become specialized, occurs as different cell lineages begin to express different subsets of their genes. Which genes a cell uses determines the molecules it will produce, which in turn determines what kind of cell it will be.

For example, most of your body cells express the genes that encode the enzymes of glycolysis, but only immature red blood cells use the genes that code for globin chains. Only your liver cells express genes for enzymes that neutralize certain toxins.

A cell rarely uses more than 10 percent of its genes at once. Which genes are expressed at any given time depends on many factors, such as conditions in the cytoplasm and extracellular fluid, and the type of cell. The factors affect controls governing all steps of gene expression, starting with transcription and ending with delivery of an RNA or protein product to its final destination. Such controls consist of processes that start, enhance, slow, or stop gene expression.

Control of Transcription Many controls affect whether and how fast certain genes are transcribed into RNA (Figure 15.2a). Those that prevent an RNA polymerase from attaching to a promoter near a gene also prevent transcription of the gene. Controls that help RNA polymerase bind to DNA also speed up transcription.

Some types of proteins affect the rate of transcription by binding to special nucleotide sequences in the DNA. For example, an **activator** speeds up transcription when it binds to a promoter. Activators also bind to DNA sequences called **enhancers**. An enhancer is not necessarily close to the gene it affects, and may even be on a different chromosome (Figure 15.3). As another example, a **repressor** slows or stops transcription when it binds to certain sites in DNA.

Regulatory proteins such as activators and repressors are called **transcription factors**. Whether and how fast a gene is transcribed depends on which transcription factors are bound to the DNA.

Interactions between DNA and the histone proteins it wraps around also affect transcription. RNA polymerase can only attach to DNA that is unwound from histones (Section 9.1). Attachment of methyl groups

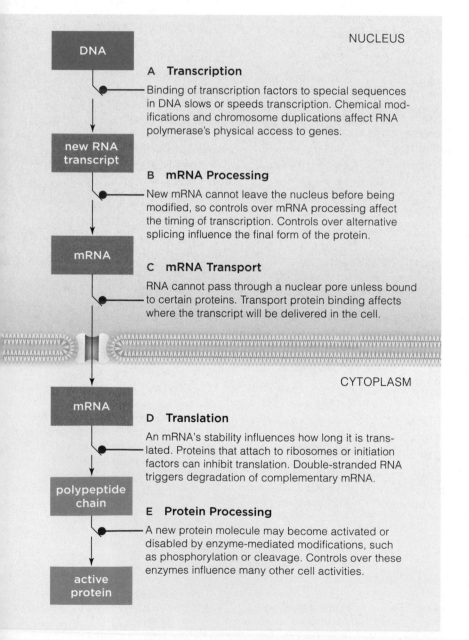

A Transcription

Binding of transcription factors to special sequences in DNA slows or speeds transcription. Chemical modifications and chromosome duplications affect RNA polymerase's physical access to genes.

B mRNA Processing

New mRNA cannot leave the nucleus before being modified, so controls over mRNA processing affect the timing of transcription. Controls over alternative splicing influence the final form of the protein.

C mRNA Transport

RNA cannot pass through a nuclear pore unless bound to certain proteins. Transport protein binding affects where the transcript will be delivered in the cell.

D Translation

An mRNA's stability influences how long it is translated. Proteins that attach to ribosomes or initiation factors can inhibit translation. Double-stranded RNA triggers degradation of complementary mRNA.

E Protein Processing

A new protein molecule may become activated or disabled by enzyme-mediated modifications, such as phosphorylation or cleavage. Controls over these enzymes influence many other cell activities.

Figure 15.2 Animated Points of control over eukaryotic gene expression.

promoter   exon1   intron   exon2   enhancer

transcription start site   transcription end

(—CH₃) causes DNA to wind tightly around histones; thus, methylation of DNA prevents its transcription.

The number of copies of a gene also affects how fast its product is made. For example, in some cells, DNA is copied repeatedly with no cytoplasmic division between replications. The result is a cell full of polytene chromosomes, each of which consists of hundreds or thousands of side-by-side copies of the same DNA molecule. All of the DNA strands carry the same genes. Translation of one gene, which occurs simultaneously on all of the identical DNA strands, produces a lot of mRNA, which is translated quickly into a lot of protein. Polytene chromosomes are common in the saliva gland cells of some insect larva and immature amphibian eggs (Figure 15.4).

mRNA Processing As you know, before eukaryotic mRNAs leave the nucleus, they are modified—spliced, capped, and finished with a poly-A tail (Section 14.3). Controls over these modifications can affect the form of a protein product and when it will appear in the cell (Figure 15.2b). For example, controls that determine which exons are spliced out of an mRNA affect which form of a protein will be translated from it.

mRNA Transport mRNA transport is another point of control (Figure 15.2c). For example, in eukaryotes, transcription occurs in the nucleus, and translation in the cytoplasm. A new RNA can pass through pores of the nuclear envelope only after it has been processed appropriately. Controls that delay the processing also delay an mRNA's appearance in the cytoplasm, and thereby delay its translation.

Controls also govern mRNA localization. A short base sequence near an mRNA's poly-A tail is like a zip code. Certain proteins that attach to the zip code drag the mRNA along cytoskeletal elements and deliver it to a particular organelle or area of the cytoplasm. Other proteins that attach to the zip code region prevent the mRNA from being translated before it reaches its destination. mRNA localization allows cells to grow or move in specific directions. It is also crucial for proper embryonic development.

Translational Control Most controls over eukaryotic gene expression affect translation (Figure 15.2d). Many govern the production or function of the various mole-

cules that carry out translation. Others affect mRNA stability: The longer an mRNA lasts, the more protein can be made from it. Enzymes begin to disassemble a new mRNA as soon as it arrives in the cytoplasm. The fast turnover allows cells to adjust their protein synthesis quickly in response to changing needs. How long an mRNA persists depends on its base sequence, the length of its poly-A tail, and which proteins are attached to it.

As a different example, microRNAs inhibit translation of other RNA. Part of a microRNA folds back on itself and forms a small double-stranded region. By a process called RNA interference, any double-stranded RNA (including a microRNA) is cut up into small bits that are taken up by special enzyme complexes. These complexes destroy every mRNA in a cell that can base-pair with the bits. So, expression of a microRNA complementary in sequence to a gene inhibits expression of that gene.

Post-Translational Modification

Many newly-synthesized polypeptide chains must get modified before they become functional (Figure 15.2e). For example, some enzymes become active only after they have been phosphorylated (another enzyme has attached a phosphate group to them). Such post-translational modifications can inhibit, activate, or stabilize many molecules, including the enzymes that participate in transcription and translation.

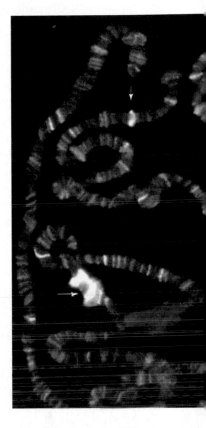

Figure 15.4 *Drosophila* polytene chromosomes. *Drosophila* larvae eat continuously, so they use a lot of saliva. In their salivary gland cells, giant polytene chromosomes form by repeated DNA replication.

Each of these chromosomes consists of hundreds or thousands of copies of the same DNA strand, aligned side by side. Transcription is visible as puffs, where the DNA has loosened (*arrows*).

Take-Home Message

What is gene expression control?

■ Most cells of multicelled organisms differentiate when they start expressing a unique subset of their genes. Which genes a cell expresses depends on the type of organism, its stage of development, and environmental conditions.

■ Various control processes regulate all steps between gene and gene product.

■ Many traits are evidence of selective gene expression.

■ Links to Chromosome number 9.2, Alleles 11.1, Sex chromosomes 12.1, X-linked inheritance 12.4, Mutation 14.5

X Chromosome Inactivation

Remember, in humans and other mammals, a female's cells each contain two X chromosomes, one inherited from her mother, the other one from her father (Section 12.1). One X chromosome is always tightly condensed, even during interphase (Figure 15.5a). We call the condensed X chromosomes "Barr bodies," after Murray Barr, who discovered them. RNA polymerase cannot access most of the genes on the condensed chromosome. **X chromosome inactivation** ensures that only one of the two X chromosomes in a female's cells is active.

X chromosome inactivation occurs when an embryo is a ball of about 200 cells. In humans and many other mammals, it occurs independently in every cell of a female embryo. The maternal X chromosome may get inactivated in one cell, and the paternal or maternal X chromosome may get inactivated in a cell next to it. Once the selection is made in a cell, all of that cell's descendants make the same selection as they continue dividing and forming tissues.

As a result of the X chromosome inactivation, an adult female mammal is a "mosaic" for the expression of X-linked genes. She has patches of tissue in which genes of the maternal X chromosome are expressed, and patches in which genes of the paternal X chromosome are expressed.

The homologous X chromosomes of most females have at least some alleles that are not identical. Thus, most females have variations in traits among patches of tissue. A female's mosaic tissues are visible if she is heterozygous for certain X chromosome mutations.

For example, incontinentia pigmenti is an X-linked disorder that affects the skin, teeth, nails, and hair. In heterozygous human females, mosaic tissues show up as lighter and darker patches of skin. The darker skin consists of cells in which the active X chromosome has the mutated allele; the lighter skin consists of cells in which the active X chromosome has the normal allele (Figure 15.5c).

Mosaic tissues are visible in other female mammals as well. For example, a gene on the X chromosomes of cats influences fur color. The expression of an allele (*O*) results in orange fur, and expression of another allele (*o*) results in black fur. Heterozygous cats (*Oo*) have patches of orange and black fur. Orange patches grow from skin cells in which the active X chromosome carries the *O* allele; black patches grow from skin cells in which the active X chromosome carries the *o* allele (Figure 15.6).

According to the theory of **dosage compensation**, X chromosome inactivation equalizes expression of X

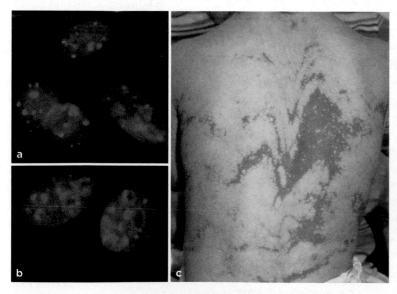

Figure 15.5 X chromosome inactivation. (**a**) Barr bodies (*red*) in the nucleus of four XX cells. (**b**) Compare the nucleus of two XY cells. (**c**) Mosaic tissues show up in human females who are heterozygous for mutations that cause incontinentia pigmenti. In darker patches of this girl's skin, the X chromosome with the mutation is active. In lighter skin, the X chromosome with the normal allele is active.

Figure 15.6 Animated Why is this cat "calico"? When she was an embryo, one or the other X chromosome was inactivated in each of her cells. The descendants of the cells formed mosaic patches of tissue. Orange or black fur results from expression of different alleles on the active X chromosome. (White patches are the outcome of a different gene, the product of which blocks synthesis of all pigment.)

A The pattern in which the floral identity genes *A*, *B*, and *C* are expressed affects differentiation of cells growing in whorls in the plant's tips. Their gene products guide expression of other genes in cells of each whorl; a flower results.

petals carpel

sepals stamens

B Mutations in *Arabidopsis* floral identity genes result in mutant flowers. *Top left*, *right*, some mutations lead to flowers with no petals. *Bottom left*, *B* gene mutations lead to flowers with sepals instead of petals. *Bottom right*, *C* gene mutations lead to flowers with petals instead of stamens and carpels. Compare the normal flower in (**a**).

Figure 15.7 Animated Control of flower formation, revealed by mutations in *Arabidopsis thaliana*.

chromosome genes between the sexes. The body cells of male mammals (XY) have one set of X chromosome genes. The body cells of female mammals (XX) have two sets, but only one is expressed. Normal development of female embryos depends on this control.

How does just one of two X chromosomes get inactivated? An X chromosome gene called *XIST* does the trick. This gene is transcribed on only one of the two X chromosomes. The gene's product, a large RNA, sticks to the chromosome that expresses the gene. The RNA coats the chromosome and causes it to condense into a Barr body. Thus, transcription of the *XIST* gene keeps the chromosome from transcribing other genes. The other chromosome does not express *XIST*, so it does not get coated with RNA; its genes remain available for transcription. It is still unknown how the cell chooses which chromosome will express *XIST*.

Flower Formation

When it is time for a plant to flower, populations of cells that would otherwise give rise to leaves instead differentiate into floral parts— sepals, petals, stamens, and carpels. How does the switch happen? Studies of mutations in the common wall cress plant, *Arabidopsis thaliana*, support the **ABC model**. This model explains how the specialized parts of a flower develop. Three sets of master genes—*A*, *B*, and *C*—guide the process. **Master genes** encode products that affect expression of many other genes. The expression of a master gene initiates cascades of expression of other genes, with the outcome being the completion of an intricate task such as the formation of a flower.

The master genes that control flower formation are switched on by environmental cues such as daylength, as you will see in Section 31.5. At the tip of a floral shoot (a modified stem), cells form whorls of tissue, one over the other like layers of an onion. Cells in each whorl give rise to different tissues depending on which of their *ABC* genes is activated. In the outer whorl, only the *A* genes are switched on, and their products trigger events that cause sepals to form. Cells in the next whorl express both *A* and *B* genes; they give rise to petals. Cells farther in express *B* and *C* genes; they give rise to male floral structures called stamens. The cells of the innermost whorl express only the *C* genes; they give rise to female floral structures called carpels (Figure 15.7*a*). Studies of the phenotypic effects of *ABC* gene mutations support this model (Figure 15.7*b*).

Take-Home Message

What are some examples of gene expression control?

■ Most genes on one X chromosome in female mammals (XX) are inactivated, which balances gene expression with males (XY).

■ Gene control also guides flower formation. ABC master genes are expressed differently in tissues of floral shoots.

There's a Fly in My Research

■ Research with fruit flies yielded the insight that body plans are a result of patterns of gene expression in embryos.

For about a hundred years, *Drosophila melanogaster* has been the subject of choice for many research experiments. Why? It costs almost nothing to feed this fruit fly, which is only about 3 millimeters long (*right*) and can live in bottles. *D. melanogaster* also reproduces fast and has a short life cycle. In addition, experimenting on insects that are considered nuisance pests presents few ethical dilemmas.

fruit fly, actual size

Many important discoveries about how gene expression guides development have come from *Drosophila* research. The discoveries help us understand similar processes in humans and other organisms, and provide clues to our shared evolutionary history.

Discovery of Homeotic Genes We now know of 13,767 genes on *Drosophila's* four chromosomes. As in most other eukaryotic species, some are **homeotic genes**: master genes that control formation of specific body parts (eyes, legs, segments, and so on) during the development of embryos. All homeotic genes encode transcription factors with a homeodomain, a region of about sixty amino acids that can bind to a promoter or some other sequence in DNA.

Localized expression of homeotic genes in tissues of a developing embryo gives rise to details of the adult body plan. The process begins long before body parts develop, as various master genes are expressed in local areas of the early embryo. The products of these master genes are transcription factors. They form in concentration gradients that span the entire embryo. Depending on where they are located within the gradients, embryonic cells begin to transcribe different homeotic genes. Products of these homeotic genes form in specific areas of the embryo.

— homeodomain

— DNA

The different products cause cells to differentiate into tissues that form specific structures such as wings or a head.

Researchers discovered homeotic genes by analyzing the DNA of mutant fruit flies that had body parts growing in the wrong places. As an example, the homeotic gene *antennapedia* is transcribed in embryonic tissues that give rise to a thorax, complete with legs. Normally, it is never transcribed in cells of any other tissue. Figure 15.8b shows what happens after a mutation causes *antennapedia* to be transcribed in embryonic tissue that gives rise to the head.

Over 100 homeotic genes have been identified. They control development by the same mechanisms in all eukaryotes, and many are interchangeable between species. Thus, we can expect that they evolved in the most ancient eukaryotic cells. Homeodomains often differ among species only in conservative substitutions—one amino acid has replaced another with similar chemical properties.

Knockout Experiments By controlling the expression of genes in *Drosophila* one at a time, researchers have made other important discoveries about how embryos of many organisms develop. In **knockout experiments**, researchers inactivate a gene by introducing a mutation into it. Then they observe how an organism that carries the mutation differs from normal individuals. The differences are clues to the function of the missing gene product.

Researchers name homeotic genes based on what happens in their absence. For instance, flies that have had their *eyeless* gene knocked out develop with no eyes. *Dunce* is required for learning and memory. *Wingless, wrinkled*, and *minibrain* genes are self-explanatory. *Tinman* is necessary for development of a heart. Flies with a mutated *groucho* gene have too many bristles above their eyes. One gene was named *toll*, after what a German researcher exclaimed upon seeing the disastrous effects of its mutation (*toll* is German for "cool!"). Figure 15.8 shows some mutant flies.

Humans, squids, mice, and many other animals have a homologue of the *eyeless* gene called *PAX6*. In humans, mutations in *PAX6* cause eye disorders such as aniridia—underdeveloped or missing irises. Altered expression of

Figure 15.8 Homeotic gene experiments. (**a**) Normal fly head. (**b**) Transcription of the *antennapedia* gene in the embryonic tissues of the thorax causes legs to form on the body. A mutation that causes *antennapedia* to be transcribed in the embryonic tissues of the head causes legs to form there too. The *antennapedia* homeodomain is modeled *above*, in *green*.

(**c**) Eyes form wherever the *eyeless* gene is expressed in fly embryos—here, on a wing.

(**d**) *Facing page*, more *Drosophila* mutations that yielded clues to homeotic gene function.

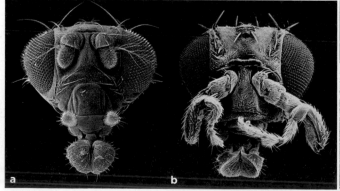

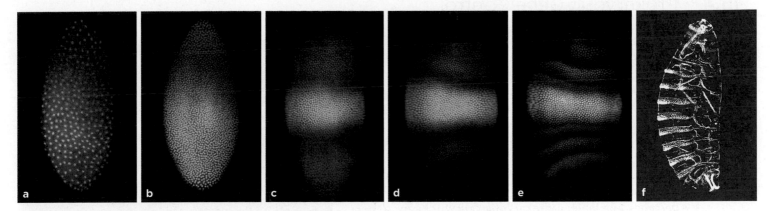

Figure 15.9 How gene expression control makes a fly, as illuminated by segmentation. The expression of different master genes is shown by different colors in fluorescence microscopy images of whole *Drosophila* embryos at successive stages of development. The bright dots are individual cell nuclei.

(**a**,**b**) The master gene *even-skipped* is expressed (in *red*) only where two maternal gene products (*blue* and *green*) overlap.

(**c**–**e**) The products of several master genes, including the two shown here in *green* and *blue*, confine the expression of *even-skipped* (*red*) to seven stripes. (**f**) One day later, seven segments develop that correspond to the position of the stripes.

the *eyeless* gene causes eyes to form not only on a fruit fly's head but also on its wings and legs (Figure 15.8*c*). *PAX6* works the same way in frogs—it causes eyes to form wherever it is expressed in tadpoles.

Researchers also discovered that *PAX6* is one of the homeotic genes that works across different species. If *PAX6* from a human, mouse, or squid is inserted into an *eyeless* mutant fly, it has the same effect as the *eyeless* gene: An eye forms wherever it is expressed. Such studies are evidence of a shared ancestor among these evolutionarily distant animals.

Filling In Details of Body Plans As an embryo develops, its differentiating cells form tissues, organs, and body parts. Some cells that alternately migrate and stick to other cells develop into nerves, blood vessels, and other structures that weave through the tissues. Events like these fill in the body's details, and all are driven by cascades of master gene expression. **Pattern formation** is the process by which a complex body forms from local processes

in an embryo. Patterning begins as maternal mRNAs are delivered to opposite ends of an unfertilized egg as it forms. The localized maternal mRNAs get translated right after the egg is fertilized, and their protein products diffuse away in gradients that span the entire embryo. Cells of the developing embryo begin to translate different master genes, depending on where they fall within those gradients. The products of those genes also form in overlapping gradients. Cells of the embryo translate still other master genes depending on where they fall within the gradients, and so on.

Such regional gene expression during development results in a three-dimensional map that consists of overlapping concentration gradients of master gene products. Which master genes are active at any given time changes, and so does the map. Some master gene products cause undifferentiated cells to differentiate, and specialized tissues are the outcome. The formation of body segments in a fruit fly embryo is an example of how pattern formation works (Figure 15.9). Section 43.4 returns to this topic.

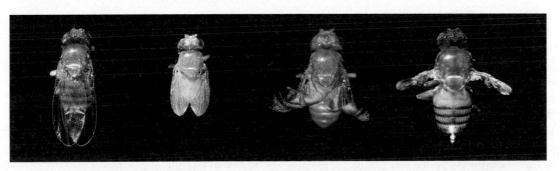

d *Left to right*, normal fly; yellow miniature; curly wings; vestigial wings; and ultrabithorax—a double thorax mutant.

15.4 | Prokaryotic Gene Control

■ Prokaryotes control gene expression mainly by adjusting the rate of transcription.

■ Links to Carbohydrates 3.3, Controls over metabolism 6.4, Lactate fermentation 8.5, Autosomal inheritance patterns 12.2

Prokaryotes do not undergo development and become multicelled organisms, so these cells do not use master genes. However, they do use gene controls. By adjusting gene expression, they can respond to environmental conditions. For example, when a certain nutrient becomes available, a prokaryotic cell will begin transcribing genes whose products allow the cell to use that nutrient. When the nutrient is not available, transcription of those genes stops. Thus, the cell does not waste energy and resources producing gene products that are not needed at a particular moment.

Prokaryotes control their gene expression mainly by adjusting the rate of transcription. Genes that are used together often occur together on the chromosome, one after the other. All of them are transcribed together into a single RNA strand, so their transcription is controllable in one step.

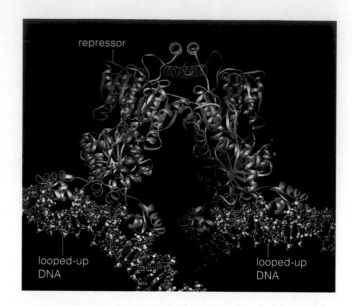

Figure 15.10 Model of the lactose operon repressor, shown here bound to operators. Binding twists the bacterial chromosome into a loop, which in turn prevents RNA polymerase from binding to the lac operon promoter.

The Lactose Operon

Escherichia coli lives in the gut of mammals, where it dines on nutrients traveling past. Its carbohydrate of choice is glucose, but it can make use of other sugars, such as the lactose in milk. *E. coli* cells can harvest the glucose subunit of lactose molecules with a set of three enzymes. However, unless there is lactose in the gut, *E. coli* cells keep the three genes for those enzymes turned off.

There is one promoter for all three genes. Flanking the promoter are two **operators**—DNA regions that are binding sites for a repressor. (Repressors, remember, stop transcription.) A promoter and one or more operators that together control the transcription of multiple genes are collectively called an **operon**.

When lactose is not present, lactose (lac) operon repressors bind *E. coli* DNA, and lactose-metabolizing genes stay switched off. One repressor molecule binds to both operators, and twists the DNA region with the promoter into a loop (Figure 15.10). RNA polymerase cannot bind to the twisted promoter, so it cannot transcribe the operon genes.

When lactose *is* in the gut, some of it is converted to another sugar, allolactose. Allolactose binds to the repressor and changes its shape. The altered repressor can no longer bind to the operators. The looped DNA unwinds, and the promoter is now free for RNA polymerase to begin transcription (Figure 15.11).

Note that *E. coli* cells use extra enzymes to metabolize lactose compared with glucose, so it is more efficient for them to use glucose. Accordingly, when both sugars are present, *E. coli* cells will use up all of the glucose before they switch to lactose metabolism.

How do *E. coli* shut off lactose metabolism in the presence of lactose? They have an additional level of control. Transcription of the lac operon genes occurs very slowly unless an activator binds to the promoter along with RNA polymerase. The activator consists of a protein with a bound nucleotide called cAMP (cyclic adenosine monophosphate). When glucose is plentiful, synthesis of cAMP is blocked, and the activator does not form. When glucose is scarce, cAMP is made. The activator forms and binds to the lac operon promoter. Lac operon genes are transcribed quickly, and lactose-metabolizing enzymes are produced at top speed.

Lactose Intolerance

Like infants of other mammals, human infants drink milk. Cells in the lining of the small intestine secrete lactase, an enzyme that cleaves the lactose in milk into its subunit monosaccharides. In most people, lactase production starts to decline at the age of five.

After that, it becomes more difficult to digest lactose in food—a condition called lactose intolerance.

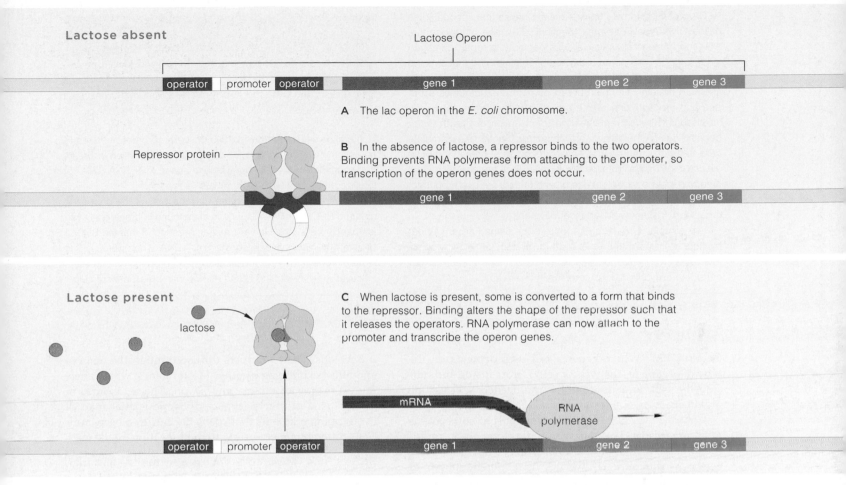

Lactose absent

Lactose Operon

| operator | promoter | operator | gene 1 | gene 2 | gene 3 |

A The lac operon in the *E. coli* chromosome.

Repressor protein

B In the absence of lactose, a repressor binds to the two operators. Binding prevents RNA polymerase from attaching to the promoter, so transcription of the operon genes does not occur.

| gene 1 | gene 2 | gene 3 |

Lactose present

lactose

C When lactose is present, some is converted to a form that binds to the repressor. Binding alters the shape of the repressor such that it releases the operators. RNA polymerase can now attach to the promoter and transcribe the operon genes.

mRNA

RNA polymerase

| operator | promoter | operator | gene 1 | gene 2 | gene 3 |

Figure 15.11 Animated Example of gene control in prokaryotes: the lactose operon on a bacterial chromosome. The operon consists of a promoter flanked by two operators, and three genes for lactose-metabolizing enzymes.

Figure It Out: What portion of the operon binds RNA polymerase when lactose is present?

Answer: The promoter

Lactose is not absorbed directly by the intestine. Thus, any that is not broken down in the small intestine ends up in the large intestine, which hosts *E. coli* and a variety of other prokaryotes. These resident organisms respond to the abundant sugar supply by switching on their lac operons. Carbon dioxide, methane, hydrogen, and other gaseous products of their various fermentation reactions accumulate quickly in the large intestine, distending its wall and causing pain. The other products of their metabolism (undigested carbohydrates) disrupt the solute–water balance inside the large intestine, and diarrhea results.

Not everybody is lactose intolerant. Many people carry a mutation in one of the genes responsible for the programmed lactase shutdown. The mutation is autosomal dominant (Section 12.2), so even heterozygotes make enough lactase to continue drinking milk without problems into adulthood.

Take-Home Message

Do prokaryotes have gene expression controls?

■ In prokaryotes, the main gene expression controls regulate transcription in response to shifts in nutrient availability and other outside conditions.

BRCA proteins promote transcription of genes that encode some of the DNA repair enzymes (Section 13.3). Any mutations that alter this function also alter a cell's capacity to repair damaged DNA. Other mutations are likely to accumulate, and that sets the stage for cancer (Section 14.5).

BRCA proteins also bind to receptors for the hormones estrogen and progesterone, which are abundant in breast and ovarian tissues. Binding regulates the transcription of growth factor genes. Among other things, growth factors (Section 9.5) stimulate cells to divide during normal, cyclic renewals of breast and ovarian tissues. When a mutation results in a BRCA protein that cannot bind to hormone receptors, the growth factors are overproduced. Cell division goes out of control, and tissue growth becomes disorganized. In other words, cancer develops.

Two groups of researchers, one at the Dana-Farber Cancer Institute at Harvard, the other at the University of Milan, recently found that XIST RNA localization is abnormal in breast cancer cells. In those cells, both X chromosomes are active.

How would you vote?

Some women at high risk of developing breast cancer opt for preventive breast removal. Many of them never would have developed cancer. Should the surgery be restricted to cancer treatment? See CengageNOW for details, then vote online.

It makes sense that having two active X chromosomes would have something to do with abnormal gene expression in breast and ovarian cancer cells, but why unmutated XIST RNA does not localize properly in such cells remains a mystery.

Mutations in the *BRCA1* gene may be part of the answer. A mutated *BRCA1* or *BRCA2* gene is often found in breast and ovarian cancer cells. The Harvard researchers found that the BRCA1 protein physically associates with XIST RNA. They were able to restore proper XIST RNA localization—and proper X chromosome inactivation—by rescuing BRCA1 function in breast cancer cells.

Summary

Section 15.1 Which genes a cell uses depends on the type of organism, the type of cell, factors inside and outside the cell, and, in complex multicelled species, the organism's stage of development.

Controls over gene expression are part of homeostasis in all organisms. They also drive development in multicelled eukaryotes. All cells of an embryo share the same genes. As different cell lineages use different subsets of genes during development, they become specialized, a process called **differentiation**. Specialized cells form tissues and organs in the adult.

Different molecules and processes govern every step between transcription of a gene and delivery of the gene's product to its final destination. Most controls operate at transcription; **transcription factors** such as **activators** and **repressors** influence transcription by binding to promoters, **enhancers**, or other sequences in DNA.

■ *Use the animation on CengageNOW to review the points of control for gene expression.*

Section 15.2 In female mammals, most genes on one of the two X chromosomes are permanently inaccessible. This **X chromosome inactivation** balances gene expression between the sexes (**dosage compensation**).

In plants, three sets of **master genes** guide cell differentiation in the whorls of a floral shoot (**ABC model**).

■ *Use the animation on CengageNOW to see how controls over gene expression affect eukaryotic development.*

Section 15.3 **Knockout experiments** involving **homeotic genes** in fruit flies (*Drosophila melanogaster*) revealed local controls over gene expression that govern the embryonic development of all complex, multicelled bodies, a process called **pattern formation**. Various master genes are expressed locally in different parts of an embryo as

it develops. Their products diffuse through the embryo and affect expression of other master genes, which affect the expression of others, and so on. These cascades of master gene products form a dynamic spatial map of overlapping gradients that spans the entire embryo body. Cells differentiate according to their location on the map.

Section 15.4 Most prokaryotic gene controls adjust transcription rates in response to environmental conditions, especially nutrient availability. The lactose **operon** governs expression of three genes active in lactose metabolism. Two **operators** that flank the promoter are binding sites for a repressor that blocks transcription.

■ *Use the animation on CengageNOW to explore the structure and function of the lactose operon.*

Self-Quiz   *Answers in Appendix III*

1. The expression of a given gene depends on _____ .
   a. the type of organism   c. the type of cell
   b. environmental conditions   d. all of the above

2. Gene expression in cells of multicelled eukaryotes changes in response to _____ .
   a. conditions outside the cell   c. operation of operons
   b. master gene products   d. a and b

3. Binding of _____ to _____ in DNA can increase the rate of transcription of specific genes.
   a. activators; promoters   c. repressors; operators
   b. activators; enhancers   d. both a and b

4. Proteins that influence gene expression by binding to DNA are called _____ .

5. Polytene chromosomes form in cells that _____ .
   a. have a lot of chromosomes   c. are polyploid
   b. are making a lot of protein d. b and c are correct

Data Analysis Exercise

Investigating a correlation between specific cancer-causing mutations and risk of mortality in humans is challenging, in part because each cancer patient is given the best treatment available at the time. There are no "untreated control" cancer patients, and the idea of what treatments are the best changes quickly as new drugs become available and new discoveries are made.

Figure 15.12 shows a study in which 442 women who had been diagnosed with breast cancer were checked for *BRCA* mutations, and their treatments and progress were followed over several years. All of the women in the study had at least two affected close relatives, so their risk of developing breast cancer due to an inherited factor was estimated to be greater than that of the general population.

1. According to this study, what is a woman's risk of dying of cancer if two of her close relatives have breast cancer?

2. What is her risk of dying of cancer if she carries a mutated *BRCA1* gene?

3. Is a *BRCA1* or *BRCA2* mutation more dangerous in breast cancer cases?

4. What other data would you have to see in order to make a conclusion about the effectiveness of preventive surgeries?

BRCA Mutations in Women Diagnosed With Breast Cancer				
	BRCA1	*BRCA2*	No *BRCA* Mutation	Total
Total number of patients	89	35	318	442
Avg. age at diagnosis	43.9	46.2	50.4	
Preventive mastectomy	6	3	14	23
Preventive oophorectomy	38	7	22	67
Number of deaths	16	1	21	38
Percent died	18.0	2.8	6.9	8.6

Figure 15.12 Results from a 2007 study investigating BRCA mutations in women diagnosed with breast cancer. All women in the study had a family history of breast cancer.

Some of the women underwent preventive mastectomy (removal of the noncancerous breast) during their course of treatment. Others had preventive oophorectomy (surgical removal of the ovaries) to prevent the possibility of getting ovarian cancer.

6. Controls over eukaryotic gene expression guide _____ .
  a. natural selection
  b. nutrient availability
  c. development
  d. all of the above

7. Incontinentia pigmenti is a rare example of _____ .
  a. autosomal dominant X-linked inheritance
  b. uneven pigmentation in humans

8. By the ABC model, _____ .
  a. *Antecedents* trigger *Behavior* that has *Consequences*
  b. three master gene sets (*A,B,C*) control flower formation
  c. gene *A* affects gene *B*, which affects gene *C*
  d. both b and c

9. During X chromosome inactivation, _____ .
  a. female cells shut down
  b. RNA coats chromosomes
  c. pigments form
  d. both a and b

10. A cell with a Barr body is _____ .
  a. prokaryotic
  b. a sex cell
  c. from a female mammal
  d. infected by Barr virus

11. Homeotic gene products _____ .
  a. flank a bacterial operon
  b. map out the overall body plan in embryos
  c. control the formation of specific body parts

12. Knockout experiments _____ genes.
  a. delete
  b. inactivate
  c. express
  d. either a or b

13. Gene expression in prokaryotic cells changes in response to _____ .
  a. activators; promoters
  b. activators; enhancers
  c. repressors; operators
  d. both a and c

14. A promoter and a set of operators that control access to two or more prokaryotic genes is a(n) _____ .

15. Match the terms with the most suitable description.
  ___ *ABC* genes
  ___ *XIST* gene
  ___ operator
  ___ Barr body
  ___ differentiation
  ___ methylation

  a. a big RNA is its product
  b. binding site for repressor
  c. cells become specialized
  d. —CH_3 additions to DNA
  e. inactivated X chromosome
  f. guide flower development

■ *Visit CengageNOW for additional questions.*

Critical Thinking

1. Unlike most rodents, guinea pigs are well developed at the time of birth. Within a few days, they can eat grass, vegetables, and other plant material.

Suppose a breeder decides to separate baby guinea pigs from their mothers three weeks after they were born. He wants to raise the males and the females in different cages. However, he has trouble identifying the sex of young guinea pigs. Suggest how a quick look through a microscope can help him identify the females.

2. Calico cats are almost always female. Male calico cats are rare, and usually they are sterile. Why?

3. Geraldo isolated an *E. coli* strain in which a mutation has hampered the capacity of the cAMP activator to bind the promoter of the lactose operon. How will this mutation affect transcription of the lactose operon when the *E. coli* cells are exposed to the following conditions?

  a. Lactose and glucose are both available.

  b. Lactose is available but glucose is not.

  c. Both lactose and glucose are absent.

Studying and Manipulating Genomes

IMPACTS, ISSUES Golden Rice or Frankenfood?

Vitamin A is necessary for good vision, growth, and immunity. A small child can get enough of it just by eating a carrot every few days, yet each year about 140 million children under the age of six suffer from serious health problems due to vitamin A deficiency. These children do not grow as they should, and they succumb easily to infection. As many as 500,000 of them go blind because of vitamin A deficiency, and half of them die within a year of losing their sight.

It is no coincidence that populations with the highest incidence of vitamin A deficiency also are the poorest. Most people in such populations tend to eat few animal products, vegetables, or fruits—all foods that are rich sources of vitamin A. Correcting and preventing vitamin A deficiency can be as simple as supplementing the diet with these foods, but changes in dietary habits are often limited by cultural traditions and poverty. Political and economic issues hamper long-term vitamin supplementation programs.

Geneticists Ingo Potrykus and Peter Beyer wanted to help such people by improving the nutritional value of rice. Why rice? Rice is the dietary staple for 3 billion people in impover-

ished countries around the world. Economies, traditions, and cuisines are based on growing and eating rice. So, growing and eating rice that happens to contain enough vitamin A to prevent disease would be compatible with prevailing methods of agriculture and traditional dietary preferences.

The body can easily convert beta-carotene, an orange photosynthetic pigment, into vitamin A. However, getting rice grains to make beta-carotene is beyond the scope of conventional methods of plant breeding. For example, corn seeds (kernels) make and store beta-carotene, but even the best gardener cannot induce rice plants to breed with corn plants.

Potrykus and Beyer genetically modified rice plants to make beta-carotene in their seeds—in the grains of Golden Rice (Figure 16.1). Like many other genetically modified organisms (GMOs), Golden Rice is transgenic, which means it carries genes from a different species. GMOs are made in laboratories, not on farms, but they are an extension of breeding practices used for many thousands of years to coax new plants and new breeds of animals from wild species.

No one wants children to suffer or die. However, many people are opposed to any GMO. Some worry that our ability to tinker with genetics has surpassed our ability to evaluate its impact. Should we be more cautious? Two people created a way to keep millions of children from dying. How much of a risk should we as a society take to help those children?

At this time, geneticists hold molecular keys to the kingdom of inheritance. As you will see, what they are unlocking is already having an impact on life in the biosphere.

See the video! Figure 16.1 Golden Rice, a miracle of modern science. (**a**) Vitamin A deficiency is common in Southeast Asia and other regions where people subsist mainly on rice.

(**b**) Rice plants with artificially inserted genes make and store the orange pigment beta-carotene in their seeds, or rice grains. The grains of this Golden Rice may help prevent vitamin A deficiency in developing countries. Compare unmodified rice grains in (**c**).

Key Concepts

DNA cloning

Researchers routinely make recombinant DNA by cutting and pasting together DNA from different species. Plasmids and other vectors can carry foreign DNA into host cells. **Section 16.1**

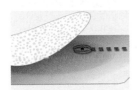

Needles in haystacks

Researchers manipulate targeted genes by isolating and making many copies of particular DNA fragments. **Section 16.2**

Deciphering DNA fragments

Sequencing reveals the linear order of nucleotides in a fragment of DNA. A DNA fingerprint is an individual's unique array of DNA sequences. **Sections 16.3, 16.4**

Mapping and analyzing whole genomes

Genomics is the study of genomes. Efforts to sequence and compare different genomes offer insights about our own genes. **Section 16.5**

Using the new technologies

Genetic engineering, the directed modification of an organism's genes, is now used in research, and it is being tested in medical applications. It continues to raise ethical questions. **Sections 16.6–16.10**

Links to Earlier Concepts

- This chapter builds on earlier explanations of DNA's structure (Section 13.2), and the molecules and processes that bring about DNA replication (13.3).

- We revisit mRNA (14.3) and bacteriophage (13.1) in the context of DNA cloning.

- Gene expression (14.1) and knockout experiments (15.3) are important concepts in genetic engineering, particularly as they apply to research on human genetic disorders (12.7).

- You may want to refer back to triglycerides (3.4) and lignin (4.1?).

- You will also see more examples of how researchers use light-emitting molecules (6.5) as tracers (2.2).

- Researchers cut up DNA from different sources, then paste the resulting fragments together.
- Cloning vectors can carry foreign DNA into host cells.
- Links to Bacteriophage 13.1, Base pairing 13.2, DNA ligase 13.3, mRNA 14.3, Introns 14.3

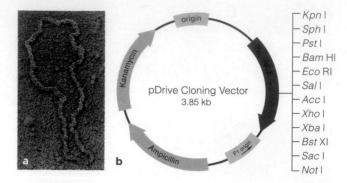

Figure 16.3 Cloning vectors. (**a**) Micrograph of a plasmid. (**b**) A commercial plasmid cloning vector. Restriction enzyme recognition sequences are indicated on the *right* by the name of the enzyme that cuts them. Researchers insert foreign DNA into the vector at these sites.

Bacterial genes help researchers identify host cells that take up a vector with inserted DNA. This vector carries two antibiotic resistance genes (*purple*) and the lactose operon (*red*).

Cut and Paste

In the 1950s, excitement over the discovery of DNA's structure gave way to frustration: No one could determine the order of nucleotides in a molecule of DNA. Identifying a single base among thousands or millions of others turned out to be a huge technical challenge.

A seemingly unrelated discovery offered a solution. Some types of bacteria resist infection by bacteriophage, which are viruses that inject their DNA into bacterial cells. Werner Arber, Hamilton Smith, and their coworkers discovered that special enzymes inside the bacteria chop up any injected bacteriophage DNA before it has a chance to integrate into the bacterial chromosome.

The enzymes restrict bacteriophage growth; hence their name, restriction enzymes. A **restriction enzyme** will cut DNA wherever a specific nucleotide sequence occurs. For example, the enzyme *Eco*RI (named after the organism from which it was isolated, *E. coli*) cuts DNA only at the sequence GAATTC (Figure 16.2*a*).

Other restriction enzymes cut different sequences. Many of them leave single-stranded tails, or "sticky ends," on DNA fragments (Figure 16.2*b*). Researchers realized that matching sticky ends base-pair together, regardless of the origin of the DNA (Figure 16.2*c*).

The enzyme DNA ligase speeds formation of covalent bonds between matching sticky ends in a mixture of DNA fragments (Figure 16.2*d*). Thus, using appropriate restriction enzymes and DNA ligase, research-

ers can cut and paste DNA from different organisms. The result, a hybrid molecule composed of DNA from two or more organisms, is **recombinant DNA**.

Making recombinant DNA is the first step in **DNA cloning**, a set of laboratory methods that uses living cells to make many copies of specific DNA fragments.

For example, researchers often insert specific DNA fragments into **plasmids**, small circles of bacterial DNA that are independent of the chromosome (Figure 16.3*a*). Before a bacterium divides, it copies its chromosome and any plasmids, so both descendant cells get one of each. If a plasmid carries a fragment of foreign DNA, that fragment gets copied and distributed to descendant cells along with the plasmid.

Thus, plasmids can be **cloning vectors**, molecules that carry foreign DNA into host cells (Figure 16.3*b*). A host cell that takes up a cloning vector can be grown in the laboratory to yield a huge population of geneti-

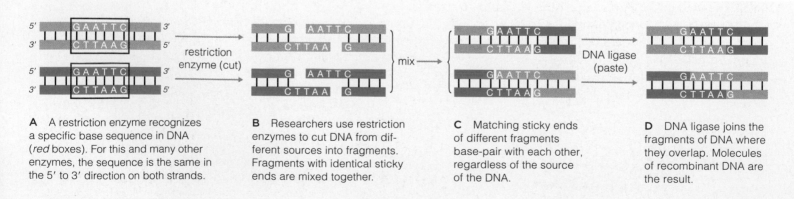

A A restriction enzyme recognizes a specific base sequence in DNA (*red* boxes). For this and many other enzymes, the sequence is the same in the 5′ to 3′ direction on both strands.

B Researchers use restriction enzymes to cut DNA from different sources into fragments. Fragments with identical sticky ends are mixed together.

C Matching sticky ends of different fragments base-pair with each other, regardless of the source of the DNA.

D DNA ligase joins the fragments of DNA where they overlap. Molecules of recombinant DNA are the result.

Figure 16.2 Making recombinant DNA.

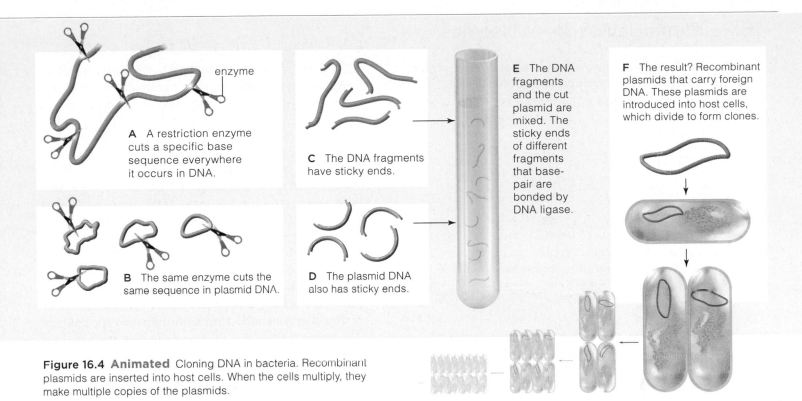

A A restriction enzyme cuts a specific base sequence everywhere it occurs in DNA.

B The same enzyme cuts the same sequence in plasmid DNA.

C The DNA fragments have sticky ends.

D The plasmid DNA also has sticky ends.

E The DNA fragments and the cut plasmid are mixed. The sticky ends of different fragments that base-pair are bonded by DNA ligase.

F The result? Recombinant plasmids that carry foreign DNA. These plasmids are introduced into host cells, which divide to form clones.

Figure 16.4 Animated Cloning DNA in bacteria. Recombinant plasmids are inserted into host cells. When the cells multiply, they make multiple copies of the plasmids.

cally identical cells called **clones**. Each clone contains a copy of the vector and the foreign DNA it carries (Figure 16.4). Researchers then harvest the DNA from the clones, and use it for experiments.

cDNA Cloning

Eukaryotic DNA, remember, contains introns. Unless you happen to be a eukaryotic cell, it is not easy to figure out which parts of the DNA encode gene products, and which do not. Researchers who study eukaryotic genes and their expression work with mRNA, because introns have already been snipped out (Section 14.3).

Messenger RNA cannot be cloned directly, because restriction enzymes and DNA ligase cut and paste only double-stranded DNA. However, mRNA can be used as a template to make double-stranded DNA in a test tube. **Reverse transcriptase**, a replication enzyme from certain types of viruses, transcribes mRNA into DNA. This enzyme uses the mRNA as a template to assemble a strand of complementary DNA, or **cDNA**:

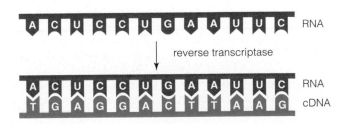

DNA polymerase added to the mixture strips the RNA from the hybrid molecule as it copies the cDNA into a second strand of DNA. The outcome is a double-stranded DNA copy of the original mRNA:

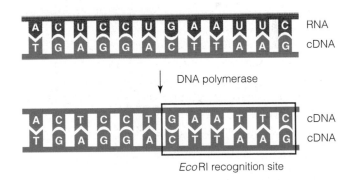

Like any other DNA, double-stranded cDNA may be cut with restriction enzymes, and the fragments can be pasted into a cloning vector using DNA ligase.

Take-Home Message

What is DNA cloning?

■ DNA cloning uses living cells to make identical copies of a particular fragment of DNA. Restriction enzymes cut DNA into fragments, then DNA ligase seals the fragments into cloning vectors. Recombinant DNA molecules result.

■ A cloning vector that holds foreign DNA can enter a living cell. The host cell can divide and give rise to huge populations of genetically identical cells (clones), each of which contains a copy of the foreign DNA.

- DNA libraries and the polymerase chain reaction (PCR) help researchers isolate particular DNA fragments.

- Links to Tracers 2.2, Base pairing 13.2

A Individual bacterial cells from a DNA library are spread over the surface of a solid growth medium. The cells divide repeatedly and form colonies—clusters of millions of genetically identical daughter cells.

B A piece of special paper pressed onto the surface of the growth medium will bind some cells from each colony.

C The paper is soaked in a solution that ruptures the cells and releases their DNA. The DNA clings to the paper in spots mirroring the distribution of colonies.

D A probe is added to the liquid bathing the paper. The probe hybridizes with (sticks to) only the spots of DNA that contain complementary base sequences.

E The bound probe makes a spot. Here, one radioactive spot darkens x-ray film. The position of the spot on the film is compared to the positions of all the original bacterial colonies. Cells from the colony that made the spot are cultured, and the DNA they contain is harvested.

The entire set of genetic material—the **genome**—of an organism typically comprises hundreds or thousands of genes. To study or manipulate one of those genes, researchers must first separate it from all of the others.

Researchers can isolate a gene by cutting an organism's DNA into pieces, and cloning all the pieces. The result is a genomic library, a set of clones that collectively host all of the DNA in a genome. Researchers can also harvest mRNA, make cDNA copies of it, and then clone the cDNA to make a cDNA library. A cDNA library represents only those genes being expressed at the time the mRNA was harvested.

Genomic and cDNA libraries are **DNA libraries**—sets of cells that host various cloned DNA fragments. In such libraries, a cell that hosts a particular gene of interest may be mixed up with thousands or millions of others that do not. All the cells look the same, so researchers get tricky to find that one clone among all of the others—the needle in the haystack.

Using a probe is one trick. A **probe** is a fragment of DNA labeled with a tracer (Section 2.2). Researchers design probes to match a targeted DNA sequence. For example, they might synthesize an oligomer (a short chain of nucleotides) based on a known DNA sequence, then attach a radioactive phosphate group to it.

The nucleotide sequence of a probe is complementary to that of the targeted gene, so the probe can base-pair with the gene. Base pairing between DNA (or DNA and RNA) from more than one source is called **nucleic acid hybridization**. A probe mixed with DNA from a library hybridizes with (sticks to) the targeted gene (Figure 16.5). Researchers pinpoint a clone that hosts the gene by detecting the label on the probe. The clone is cultured so a huge population of genetically identical cells forms. The DNA can then be extracted in bulk from the cells.

Big-Time Amplification: PCR

Researchers can isolate and mass-produce a particular DNA fragment without cloning. They do so with the *P*olymerase *C*hain *R*eaction, or **PCR**. This hot-and-cold cycled reaction uses a heat-tolerant DNA polymerase to copy a fragment of DNA by the billions.

PCR can transform one needle in a haystack—that one-in-a-million DNA fragment—into a huge stack of

Figure 16.5 Animated Nucleic acid hybridization. In this example, a radioactive probe helps identify a bacterial colony that contains a targeted sequence of DNA.

needles with a little hay in it (Figure 16.6). The starting material for PCR is a sample of DNA with at least one molecule of a target sequence. It might be DNA from a mixture of 10 million different clones, one sperm, a hair left at a crime scene, or a mummy. Essentially any sample that has DNA in it can be used for PCR.

First, the starting material is mixed with DNA polymerase, nucleotides, and primers. **Primers** are oligomers that base-pair with DNA at a certain sequence—here, on either end of the DNA to be amplified.

Researchers expose the reaction mixture to repeated cycles of high and low temperature. High temperature disrupts the hydrogen bonds that hold the two strands of a DNA double helix together (Section 13.2). During a high temperature cycle, every molecule of double-stranded DNA unwinds and becomes single-stranded. During a low temperature cycle, single DNA strands hybridize with complementary partners, and double-stranded DNA forms again.

Most DNA polymerases are destroyed by the high temperatures required to separate DNA strands. The kind that is used in PCR reactions, *Taq* polymerase, is from *Thermus aquaticus*. This bacterial species lives in superheated springs (Chapter 20 introduction), so its polymerase is necessarily heat-tolerant. Like other DNA polymerases, the *Taq* polymerase recognizes hybridized primers as places to start DNA synthesis. During a low temperature cycle, it starts synthesizing DNA where primers have hybridized with template. Synthesis proceeds along the template strand until the temperature rises and the DNA separates into single strands. The newly synthesized DNA is a copy of the target.

When the mixture cools, primers rehybridize, and DNA synthesis begins again. With each temperature cycle, the number of copies of target DNA can double. After about thirty PCR cycles, the number of template molecules may be amplified by about a billionfold.

A DNA template (*purple*) is mixed with primers (*red*), free nucleotides, and heat-tolerant *Taq* DNA polymerase.

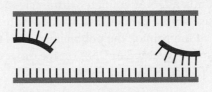

B When the mixture is heated, DNA strands separate. When it is cooled, some primers hydrogen-bond to the template DNA.

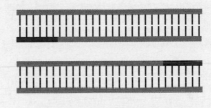

C *Taq* polymerase uses the primers to initiate synthesis, and complementary strands of DNA form. The first round of PCR is now complete.

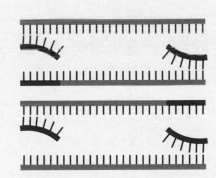

D The mixture is heated again, and all of the DNA separates into single strands. When the mixture is cooled, some of the primers hydrogen-bond to the DNA.

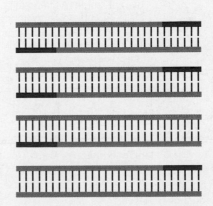

E *Taq* polymerase uses the primers to initiate DNA synthesis, and complementary strands of DNA form. The second round of PCR is complete.

Each round can double the number of DNA molecules. After 30 rounds, the mixture contains huge numbers of DNA fragments, all copies of the template DNA.

Take-Home Message

How do researchers study one gene in the context of many?

■ Researchers make DNA libraries or use PCR to isolate one gene from the many other genes in a genome.

■ Probes are used to identify one clone that hosts a DNA fragment of interest among many other clones in a DNA library.

■ PCR, the polymerase chain reaction, rapidly increases the number of molecules of a particular DNA fragment.

Figure 16.6 Animated Two rounds of PCR. Thirty cycles of this polymerase chain reaction may increase the number of starting DNA template molecules a billionfold.

DNA Sequencing

- DNA sequencing reveals the order of nucleotide bases in a fragment of DNA.

- Links to Tracers 2.2, Nucleotides 13.2, DNA replication 13.3

The order of the nucleotide bases in a DNA fragment is determined with **DNA sequencing**. The most commonly used method of sequencing DNA is similar to DNA replication, in that the DNA fragment is used as a template for DNA synthesis.

Researchers mix a DNA template with nucleotides, DNA polymerase, and a primer that hybridizes to the DNA. Starting at the primer, the polymerase joins free nucleotides into a new strand of DNA, in the order dictated by the sequence of the template.

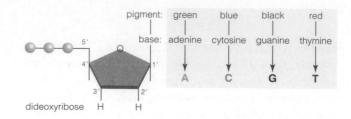

Figure 16.7 Structure of a dideoxynucleotide. Each of the four bases is labeled with a different colored pigment.

DNA polymerase joins a nucleotide to a DNA strand only at the hydroxyl group on the strand's 3' carbon (Section 13.3). The reaction mixture includes four kinds of *dideoxy*nucleotides, which have no hydroxyl group on their 3' carbon (Figure 16.7). During the sequencing reaction, a polymerase randomly adds either a regular nucleotide or a dideoxynucleotide to the end of a growing DNA strand. If it adds a dideoxynucleotide, the 3' carbon of the strand will not have a hydroxyl group, so synthesis of the strand ends there (Figure 16.8a,b).

After about 10 minutes, there are millions of DNA fragments of all different lengths; most are incomplete copies of the template DNA. All of the copies end with one of the four dideoxynucleotides (Figure 16.8c). For example, there will be many ten base-pair-long copies of the template in the mixture. If the tenth base in the template was adenine, every one of those fragments will end with a dideoxyadenine.

The fragments are then separated by **electrophoresis**. With this technique, an electric field pulls all the DNA fragments through a semisolid gel. DNA fragments of different sizes move through the gel at different rates. The shorter the fragment, the faster it moves, because shorter fragments slip through the tangled molecules of the gel faster than longer fragments do.

All fragments of the same length move through the gel at the same speed, so they gather into bands. All of the fragments in a given band have the same dideoxynucleotide at their ends. Each of the four types of dideoxynucleotides (A, C, G, or T) carries a different colored pigment label, and those tracers now impart distinct colors to the bands (Figure 16.8d). Each color designates one of the four dideoxynucleotides, so the order of colored bands in the gel represents the DNA sequence (Figure 16.8e).

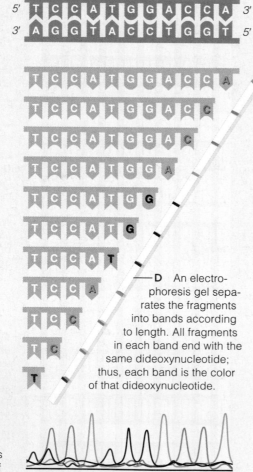

A The fragment of DNA to be sequenced is mixed with a primer, DNA polymerase, and nucleotides. The mixture also includes the four dideoxynucleotides labeled with four different colored pigments.

B The polymerase uses the DNA as a template to synthesize new strands again and again. Synthesis of each new strand stops when a dideoxynucleotide is added.

C At the end of the reaction, there are many truncated copies of the DNA template in the mixture.

D An electrophoresis gel separates the fragments into bands according to length. All fragments in each band end with the same dideoxynucleotide; thus, each band is the color of that dideoxynucleotide.

E A computer detects and records the color of each band on the gel. The order of colors of the bands represents the sequence of the template DNA.

Figure 16.8 Animated DNA sequencing. Researchers use a modified DNA replication reaction to determine the order of nucleotide bases in a fragment of DNA.

Take-Home Message

How is the order of nucleotides in DNA determined?

- With DNA sequencing, a strand of DNA is partially replicated. Electrophoresis is used to separate the resulting fragments of DNA, which are tagged with tracers, by length.

16.4 | DNA Fingerprinting

■ One individual can be distinguished from all others on the basis of DNA fingerprints.

Each human has a unique set of fingerprints. In addition, like members of other sexually reproducing species, each also has a **DNA fingerprint**—a unique array of DNA sequences. More than 99 percent of the DNA in all humans is the same, but the other fraction of 1 percent is unique to each individual. Some of these unique sequences are sprinkled throughout the human genome as **short tandem repeats**—many copies of the same 2- to 10-base-pair sequences, positioned one after the next along the length of a chromosome.

For example, one person's DNA might contain fifteen repeats of the bases TTTTC in a certain location. Another person's DNA might have TTTTC repeated two times in the same location. One person might have ten repeats of CGG; another might have fifty. Such repetitive sequences slip spontaneously into DNA during replication, and their numbers grow or shrink over generations. The mutation rate is relatively high around tandem repeat regions.

DNA fingerprinting reveals differences in the tandem repeats among individuals. With this technique, PCR is used to copy a region of a chromosome known to have tandem repeats of 4 or 5 nucleotides. The size of the copied DNA fragment differs among most individuals, because the number of tandem repeats in that region also differs.

Thus, the genetic differences between individuals can be detected by electrophoresis. As in DNA sequencing, the fragments form bands according to length as they migrate through a gel. Several regions of chromosomal DNA are typically tested. The resulting banding patterns on the electrophoresis gel constitute an individual's DNA fingerprint—which, for all practical purposes, is unique. Unless two people are identical twins, the chances that they have identical tandem repeats in even three regions of DNA is 1 in 1,000,000,000,000,000,000—or one in a quintillion—which is far more than the number of people that live on Earth.

A few drops of blood, semen, or cells from a hair follicle at a crime scene or on a suspect's clothing yield enough DNA to amplify with PCR for DNA fingerprinting (Figure 16.9). DNA fingerprints have been established as accurate and unambiguous, and are often used as evidence in court. For example, DNA fingerprints are now routinely submitted as evidence in paternity disputes. The technique is being widely used not only to convict the guilty, but also to exonerate the innocent: As of this writing, DNA fingerprinting evidence has helped release more than 160 innocent people from prison.

DNA fingerprint analysis has many applications. For instance, DNA fingerprinting was used to identify the remains of many individuals who died in the World Trade Center on September 11, 2001. It confirmed that human bones exhumed from a shallow pit in Siberia belonged to five individuals of the Russian imperial family, all shot to death in secrecy in 1918.

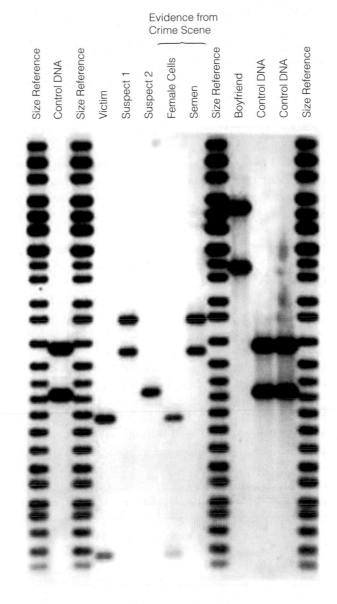

Figure 16.9 DNA fingerprinting in an investigation of sexual assault. A single short tandem repeat region was amplified from evidence found at the crime scene—the perpetrator's semen and the victim's cells. The two samples were compared with the same tandem repeat region amplified from DNA of the victim, her boyfriend, and two suspects (1 and 2).

The photo shows an x-ray film image of an electrophoresis gel from a forensics laboratory. The dark bands represent different sized DNA fragments that had been labeled with a radioactive tracer. Note the three samples of control DNA (to confirm that the assay was working correctly), and the four size reference samples. **Figure It Out:** Which suspect is guilty?

Answer: Suspect 1

Researchers also use DNA fingerprinting to study population dispersal in humans and other animals. Because only a tiny amount of DNA is necessary, such studies are not necessarily limited to living populations. Short tandem repeats on the Y chromosome are also used to determine genetic relationships among male relatives and descendants, and to trace an individual's ethnic heritage.

Studying Genomes

■ Comparing the sequence of our genome with that of other species is giving us insights into how the human body works.

■ Links to Triglycerides 3.4, Discovery of DNA structure 13.2, Knockout experiments 15.3

The Human Genome Project

Around 1986, people were arguing about sequencing the human genome. Many insisted that deciphering it would have enormous payoffs for medicine and pure research. Others said sequencing would divert funds from more urgent work that also had a better chance of success. At that time, sequencing 3 billion bases seemed like a daunting task: It would take at least 6 million sequencing reactions, all done by hand. Given the techniques available, the work would have taken more than fifty years to complete.

But techniques kept getting better, so more bases could be sequenced in less time. Automated (robotic) DNA sequencing and PCR had just been invented. Both of these techniques were still cumbersome and expensive, but many researchers sensed their potential. Waiting for faster technologies seemed the most efficient way to sequence 3 billion bases, but how fast did they need to be before the project could begin?

In 1987, private companies started to sequence the human genome. Walter Gilbert, one of the early inventors of DNA sequencing, started one that intended to sequence and patent the human genome. This development provoked widespread outrage, but it also spurred

commitments in the public sector. In 1988, the National Institutes of Health (NIH) effectively annexed the entire project by hiring James Watson (of DNA structure fame) to head the official Human Genome Project, and providing $200 million per year to fund it. A consortium formed between the NIH and international institutions that were sequencing different parts of the genome. Watson set aside 3 percent of the funding for studies of ethical and social issues arising from the research. He resigned later over a patent disagreement.

Amid ongoing squabbles over patent issues, Celera Genomics formed in 1998 (Figure 16.10). With Craig Venter at its helm, the company intended to commercialize genetic information. Celera started to sequence the genome using new, faster techniques, because the first to have the complete sequence had a legal basis for patenting it. The competition motivated the public consortium to move its efforts into high gear.

Then, in 2000, United States President Bill Clinton and British Prime Minister Tony Blair jointly declared that the sequence of the human genome could not be patented. Celera kept on sequencing anyway. Celera and the public consortium separately published about 90 percent of the sequence in 2001.

By 2003, fifty years after the discovery of the structure of DNA, the sequence of the human genome was officially completed. At this writing, about 99 percent of its coding regions—28,976 genes—have been identified. Researchers have not discovered what all of the genes encode, only where they are in the genome.

What do we do with this vast amount of data? The next step is to find out just what the sequence means.

Figure 16.10 Some of the bases of the human genome—and a few of the supercomputers used to sequence it—at Venter's Celera Genomics in Maryland.

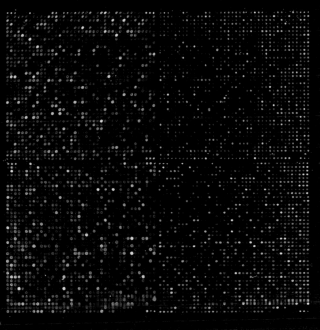

Genomics

Investigations into the genomes of humans and other species have converged into the new research field of **genomics**. Structural genomics focuses on determining the three-dimensional structure of proteins encoded by a genome. Comparative genomics compares genomes of different species; similarities and differences reflect evolutionary relationships.

The human genome sequence is a massive collection of seemingly cryptic data. Currently, the only way we are able to decipher it is by comparing it to genomes of other organisms, the premise being that all organisms are descended from shared ancestors, so all genomes are related to some extent. We see evidence of such genetic relationships just by comparing the sequence data. For example, the human and mouse sequences are about 78 percent identical; the human and banana sequences are about 50 percent identical.

Intriguing as these percentages might be, gene-by-gene comparisons offer more practical benefits. We have learned about the function of many human genes by studying their counterpart genes in other species. For example, researchers studying a human gene might disable the same gene in mice. The effects of the gene's absence on mice are clues to its function in humans.

These types of knockout experiments are revealing the function of many human genes. For example, researchers comparing the human and mouse genomes discovered a human version of the mouse gene *APOA5*. This gene encodes a protein that carries lipids in the blood. Mice with an *APOA5* knockout have four times the normal level of triglycerides in their blood. The researchers then looked for—and found—a correlation between *APOA5* mutations and high triglyceride levels in humans. High triglycerides are a risk factor for coronary artery disease.

Figure 16.11 (**a**) A DNA chip. (**b**) DNA chips are often used in gene expression research. Here, RNA from yeast cells carrying out fermentation was used to make cDNA, which was labeled with a green tracer; RNA from the same type of cells carrying out aerobic respiration was used to make cDNA labeled with a red tracer.

The probes were dropped onto a DNA chip with a 19-millimeter (3/4-inch) array of the complete yeast genome—around 6,000 genes. *Green* spots indicate the genes active during fermentation; *red* spots, genes active during aerobic respiration. *Yellow* spots are a combination of red and green; they indicate genes active in both pathways.

DNA Chips

Genomics researchers often use **DNA chips**, which are microscopic arrays (microarrays) of DNA samples that have been stamped in separate spots on small glass plates (Figure 16.11*a*). Typically, one microarray comprises hundreds or thousands of DNA fragments that collectively represent an entire genome.

Using DNA chips, researchers can compare the patterns of gene expression among cells—perhaps different types of cells from one individual, or the same cells at different times or under different conditions (Figure 16.11*b*). Which genes are expressed, at which times, in which cells, is information useful in research and other applications. For example, DNA chips have been used to determine which genes are deregulated in cancer cells. Soon, they may be used to quickly screen people for genetic predisposition to disease, to identify pathogens, and in forensic investigations.

Take-Home Message

What do we do with DNA sequence information?

■ Analysis of the human genome sequence is yielding new information about human genes and how they work. The information has practical applications in medicine, research, and other fields.

16.6 | Genetic Engineering

- The most common genetically modified organisms are bacteria and yeast.

- Links to Bioluminescence 6.5, Gene expression 14.1

Traditional cross-breeding methods can alter genomes, but only if individuals with the desired traits will interbreed. Genetic engineering takes gene-swapping to an entirely new level. **Genetic engineering** is a laboratory process by which deliberate changes are introduced into an individual's genome. A gene from one species may be transferred to another to produce a **transgenic** organism, or a gene may be altered and reinserted into an individual of the same species. Both methods result in **genetically modified organisms** (GMOs).

The most common GMOs are bacteria and yeast. These cells have the metabolic machinery to make complex organic molecules, and they are easily modified.

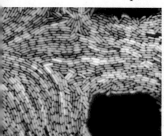

genetically modified bacteria expressing a jellyfish gene emit green light

Genetically modifying bacteria or yeast has practical applications. For example, the *E. coli* on the *left* have been modified to produce a fluorescent protein from jellyfish. The cells are genetically identical, so the visible variation in fluorescence among them reveals differences in gene expression. These GMOs may help us discover why some individual bacteria become dangerously resistant to antibiotics.

Some genetically modified bacteria and yeast are "factories" for medically important proteins. Diabetics were among the first beneficiaries of such organisms. Insulin for their injections was once extracted from animals, but it provoked an allergic reaction in some people. Human insulin, which does not provoke allergic reactions, has been produced by transgenic *E. coli* since 1982. Slight modifications of the gene have also yielded fast-acting and slow-release human insulin.

Engineered microorganisms also produce proteins used in food manufacturing. For example, cheese is traditionally made with an extract of calf stomachs, which contain the enzyme chymotrypsin. Most cheese manufacturers now use chymotrypsin made by genetically engineered bacteria. Other examples are GMO-made enzymes that improve the taste and clarity of beer and fruit juice, slow bread staling, or modify fats.

Take-Home Message

What is genetic engineering?

- Genetic engineering is the directed alteration of an individual's genome, and it results in a genetically modified organism (GMO).

- A transgenic organism carries a gene from a different species. Transgenic bacteria and yeast are used in research, medicine, and industry.

16.7 | Designer Plants

- Genetically engineered crop plants are widespread in the United States.

- Links to Lignin 4.12, Luciferase 6.5

Agrobacterium tumefaciens is a species of bacteria that infects many plants, including peas, beans, potatoes, and other important crops. Its plasmid contains genes that cause tumors to form on infected plants; hence the name Ti plasmid (for *T*umor-*i*nducing). Researchers use the Ti plasmid as a vector to transfer foreign or modified genes into plants. They remove the tumor-inducing genes from the plasmid, then insert desired genes into it. Whole plants can be grown from plant cells that take up the modified plasmid (Figure 16.12).

Modified *A. tumefaciens* bacteria are used to deliver genes into some food crop plants, including soybeans, squash, and potatoes. Researchers also transfer genes into plants by way of electric or chemical shocks, or by blasting them with DNA-coated pellets.

Genetically Engineered Plants

As crop production expands to keep pace with human population growth, it places unavoidable pressure on ecosystems everywhere. Irrigation leaves mineral and salt residues in soils. Tilled soil erodes, taking topsoil with it. Runoff clogs rivers, and fertilizer in it causes algae to grow so fast that fish suffocate. Pesticides can harm humans, other animals, and beneficial insects.

Pressured to produce more food at lower cost and with less damage to the environment, many farmers have begun to rely on genetically modified crop plants. Some of these modified plants carry genes that impart resistance to devastating plant diseases. Others offer improved yields, such as a strain of transgenic wheat that has double the yield of unmodified wheat.

GMO crops such as *Bt* corn and soy help farmers use smaller amounts of toxic pesticides. Organic farmers often spray their crops with spores of *Bt* (*Bacillus thuringiensis*), a bacterial species that makes a protein toxic only to insect larvae. Researchers transferred the gene encoding the *Bt* protein into plants. The engineered plants produce the *Bt* protein, but otherwise they are identical with unmodified plants. Insect larvae die shortly after eating their first (and only) GMO meal. Farmers can use much less pesticide on crops that make their own (Figure 16.13*a*).

Transgenic crop plants are also being developed for regions that are affected by severe droughts, such as Africa. Genes that confer drought tolerance and insect resistance are being transferred into crop plants such as corn, beans, sugarcane, cassava, cowpeas, banana,

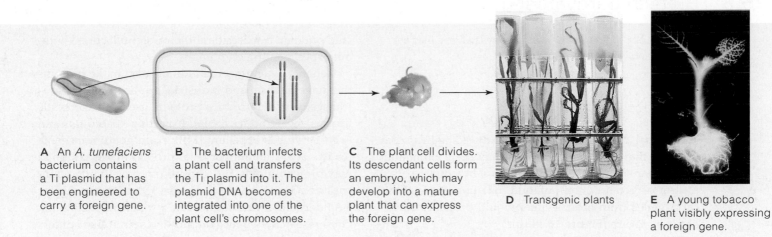

A An *A. tumefaciens* bacterium contains a Ti plasmid that has been engineered to carry a foreign gene.

B The bacterium infects a plant cell and transfers the Ti plasmid into it. The plasmid DNA becomes integrated into one of the plant cell's chromosomes.

C The plant cell divides. Its descendant cells form an embryo, which may develop into a mature plant that can express the foreign gene.

D Transgenic plants

E A young tobacco plant visibly expressing a foreign gene.

Figure 16.12 Animated (**a–d**) Ti plasmid transfer of an *Agrobacterium tumefaciens* gene to a plant cell. (**e**) Transgenic plant expressing a gene for the firefly enzyme luciferase (Section 6.5).

Figure 16.13 Examples of genetically modified plants. (**a**) Some GMO crops help farmers use less insecticide. *Top*, the *Bt* gene conferred insect resistance to the genetically modified plants that produced this corn. *Bottom*, unmodified corn is more vulnerable to pests.

(**b**) Lignin strengthens the secondary cell walls of many kinds of woody plants. Before paper can be made from wood, the lignin must be extracted from wood pulp. Paper products and clean-burning fuels such as ethanol may be easier to manufacture from the wood of trees engineered to produce less lignin.

Shown are a control plant (*left*) and three aspen seedlings (*right*) in which expression of a control gene in the lignin synthesis pathway was suppressed. The modified plants made normal lignin, but not as much of it.

and wheat. Such crops may help people that rely on agriculture for food and income in drought-stricken, impoverished regions of the world.

The USDA Animal and Plant Health Inspection Service (APHIS) regulates the introduction of GMOs into the environment. At this writing, the agency has deregulated seventy-three genetically modified plants, which means the plants are approved for unregulated use in the United States. Hundreds more are pending such deregulation.

The most widely planted GMO crops include corn, sorghum, cotton, soy, canola, and alfalfa engineered for resistance to glyphosate, an herbicide. Rather than tilling the soil to control weeds, farmers can spray their fields with glyphosate, which kills the weeds but not the engineered crops. However, weeds are becom-

ing resistant to glyphosate, so spraying it no longer kills the weeds in glyphosphate-resistant crop fields. The engineered gene is also appearing in wild plants and in nonengineered crops, which means that transgenes can—and do—escape into the environment.

Controversy raised by such GMO use invites you to read the research and form your own opinions. The alternative is to be swayed by media hype (the term "Frankenfood," for instance), or by reports from possibly biased sources (such as herbicide manufacturers).

Take-Home Message

Are there genetically engineered plants?

■ Plants with modified or foreign genes are now common farm crops.

■ Genetically engineered animals are invaluable in medical research and in other applications.

■ Link to Knockouts 15.3

Of Mice and Men

Traditional cross-breeding has produced animals so unusual that transgenic animals may seem mundane by comparison (Figure 16.14*a*). Cross-breeding is also a form of genetic manipulation, but transgenics that carry genes from other genera would probably never occur in nature (Figure 16.14*b,c*).

The first transgenic animals—mice—were produced in 1982. Researchers inserted a gene that codes for a rat growth hormone into a plasmid, then injected the recombinant plasmid into fertilized mouse eggs. The eggs were implanted in surrogate mother mice. A third of the mice that were born to the surrogates grew much larger than their littermates (Figure 16.15). The larger mice were transgenic: The rat gene had integrated into their chromosomes, and was being expressed.

Today, transgenic mice are commonplace, and they are invaluable in research into human genes. For example, researchers have discovered the function of many human genes by inactivating their counterparts in mice (Section 16.5).

Genetically modified animals are also used as models of many human diseases. For example, researchers inactivated the molecules involved in the control of glucose metabolism, one by one, in mice. Studying the effects of the knockouts has resulted in much of our current understanding of how diabetes works in humans. Genetically modified animals such as these mice are allowing researchers to

study human diseases (and their potential cures) without experimenting on humans.

Genetically engineered animals also make proteins that have medical and industrial applications. Various transgenic goats produce proteins used to treat cystic fibrosis, heart attacks, blood clotting disorders, and even nerve gas exposure. Milk from goats transgenic for lysozyme, an antibacterial protein in human milk, may protect infants and children in developing countries from acute diarrheal disease. Goats transgenic for a spider silk gene produce the protein in their milk. Once researchers figure out how to spin it like spiders do, the silk may be used to manufacture fashionable fabrics, bulletproof vests, sports equipment, and biodegradable medical supplies.

Rabbits make human interleukin-2, a protein that triggers divisions of immune cells. Genetic engineering has also given us dairy goats with heart-healthy milk, low-fat pigs, pigs with environmentally friendly low-phosphate manure, extra-large sheep, and cows that are resistant to mad cow disease.

Tinkering with the genes of animals raises ethical dilemmas. For example, many people view transgenic animal research as unconscionable. Many others see it as simply an extension of thousands of years of acceptable animal husbandry practices: The techniques have changed, but not the intent. We humans still have a vested interest in improving our livestock.

Knockout Cells and Organ Factories

Millions of people suffer with organs or tissues that are damaged beyond repair. In any given year, more than 80,000 of them are on waiting lists for an organ

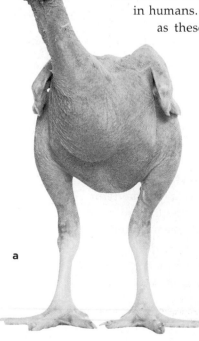

Figure 16.14 Genetically modified animals. (**a**) Featherless chicken developed by cross-breeding methods in Israel. Such chickens survive in deserts where cooling systems are not an option. (**b**) Mira, a goat transgenic for human antithrombin III (a protein that inhibits blood clotting). (**c**) The pig on the *left* is transgenic for a yellow fluorescent protein; its nontransgenic littermate is on the *right*.

Figure 16.15 Expression of a rat growth hormone gene in a transgenic mouse. These two mice are littermates. The mouse on the *left* weighed 29 grams (1 ounce); the one on the *right*, 44 grams (1.5 ounces).

transplant in the United States alone. Human donors are in such short supply that illegal organ trafficking is now a common problem.

Pigs are a potential source of organs for transplantation, because pig and human organs are about the same in both size and function. However, the human immune system battles anything it recognizes as nonself. It rejects a pig organ at once, because it recognizes a foreign glycoprotein on the plasma membrane of pig cells. Within a few hours, blood coagulates inside the organ's vessels and dooms the transplant. Drugs can suppress the immune response, but they also render organ recipients particularly vulnerable to infection.

Researchers have produced genetically engineered pigs that lack the offending glycoprotein on their cells. The human immune system may not reject tissues or organs transplanted from these pigs.

Transferring an organ from one species into another is called **xenotransplantation**. Critics of xenotransplantation are concerned that, among other things, pig-to-human transplants would invite pig viruses to cross the species barrier and infect humans, perhaps catastrophically. Their concerns are not unfounded. Evidence suggests that some of the worst pandemics arose because animal viruses adapted to new hosts—humans.

16.9 | Safety Issues

■ The first transfer of foreign DNA into bacteria ignited an ongoing debate about potential dangers of transgenic organisms that may enter the environment.

When James Watson and Francis Crick presented their model of the DNA double helix in 1953, they ignited a global blaze of optimism about genetic research. The very book of life seemed to be open for scrutiny. In reality, no one could read it. Scientific breakthroughs are not very often accompanied by the simultaneous discovery of the tools to study them. New techniques would have to be invented before that book would become readable.

Twenty years later, Paul Berg and his coworkers discovered how to make recombinant organisms by fusing DNA from two species of bacteria. By isolating DNA in manageable subsets, researchers now had the tools to be able to study its sequence in detail. They began to clone and analyze DNA from many different organisms. The technique of genetic engineering was born, and suddenly everyone was worried about it.

Researchers knew that DNA itself was not toxic, but they could not predict with certainty what would happen every time they fused genetic material from different organisms. Would they accidentally make a superpathogen? Could they make a new, dangerous form of life by fusing DNA of two normally harmless organisms? What if that new form escaped from the laboratory and transformed other organisms?

In a remarkably quick and responsible display of self-regulation, scientists reached a consensus on new safety guidelines for DNA research. Adopted at once by the NIH, these guidelines included precautions for laboratory procedures. They covered the design and use of host organisms that could survive only under the narrow range of conditions inside the laboratory. Researchers stopped using DNA from pathogenic or toxic organisms for recombination experiments until proper containment facilities were developed.

Now, all genetic engineering research is done under these laboratory guidelines. Releasing and importing genetically modified organisms is carefully regulated by the USDA. Such regulations are our best effort to minimize any risk involved in the research or as a result of it, but they are not a guarantee.

Take-Home Message

Why do we genetically engineer animals?

■ Animals that would be impossible to produce by traditional breeding methods are being created by genetic engineering. Such animals are used in research, medicine, and industry.

Take-Home Message

Is genetic engineering safe?

■ Rigorous safety guidelines for DNA research have been in place for decades in the United States and other countries. Researchers are expected to comply with these stringent standards.

16.10 | Modified Humans?

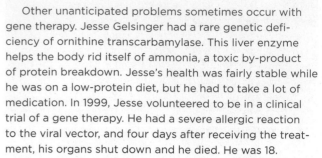

■ We as a society continue to work our way through the ethical implications of applying new DNA technologies.

■ The manipulation of individual genomes continues even as we are weighing the risks and benefits of this research.

■ Link to Human genetic disorders 12.7

Getting Better We know of more than 15,000 serious genetic disorders. Collectively, they cause 20 to 30 percent of infant deaths each year, and account for half of all mentally impaired patients and a fourth of all hospital admissions. They also contribute to many age-related disorders, including cancer, Parkinson's disease, and diabetes.

Drugs and other treatments can minimize the symptoms of some genetic diseases, but gene therapy is the only cure. **Gene therapy** is the transfer of recombinant DNA into an individual's body cells, with the intent to correct a genetic defect or treat a disease. The transfer, which occurs by way of viral vectors or lipid clusters, inserts an unmutated gene into an individual's chromosomes.

Human gene therapy is a compelling reason to embrace genetic engineering research. It is now being tested as a treatment for cystic fibrosis, hemophilia A, several types of cancer, inherited retinal disease, and inherited immune disorders, among other diseases. The results are encouraging.

For example, little Rhys Evans (*left*) was born with a severe immune disorder, SCID-X1. SCID-X1 stems from mutations in the *IL2RG* gene, which encodes a receptor for an immune signaling molecule. Children affected by this disorder can survive only in germ-free isolation tents, because they cannot fight infections.

In 1998, a viral vector was used to insert unmutated copies of *IL2RG* into cells taken from the bone marrow of eleven boys with SCID-X1. Each child's modified cells were infused back into his bone marrow. Months later, ten of the boys left their isolation tents for good. Their immune systems had been repaired by the gene therapy. Since then, gene therapy has freed many other SCID-X1 patients from life in an isolation tent. Rhys is one of them.

Getting Worse Manipulating a gene within the context of a living individual is unpredictable even when we know its sequence and where it is within the genome. No one, for example, can predict where a virus-injected gene will insert into chromosomes. Its insertion might disrupt other genes. If it interrupts a gene that is part of the controls over cell division, then cancer might be the outcome.

For example, three boys from the 1998 SCID-X1 clinical trial have since developed leukemia, and one of them died. The researchers had wrongly predicted that cancer related to the gene therapy would be rare. Research now implicates the very gene targeted for repair, especially when combined with the viral vector that delivered it.

Other unanticipated problems sometimes occur with gene therapy. Jesse Gelsinger had a rare genetic deficiency of ornithine transcarbamylase. This liver enzyme helps the body rid itself of ammonia, a toxic by-product of protein breakdown. Jesse's health was fairly stable while he was on a low-protein diet, but he had to take a lot of medication. In 1999, Jesse volunteered to be in a clinical trial of a gene therapy. He had a severe allergic reaction to the viral vector, and four days after receiving the treatment, his organs shut down and he died. He was 18.

Our understanding of how the human genome works clearly lags behind our ability to modify it.

Getting Perfect The idea of using human gene therapy to cure genetic disorders seems like a socially acceptable goal to most people. However, go one step further. Would it also be acceptable to modify genes of an individual who is within a normal range in order to minimize or enhance a particular trait? Researchers have already produced mice that have enhanced memory, bigger muscles, or improved learning abilities. Why not people?

The idea of selecting the most desired human traits is called eugenic engineering. Yet who decides which forms of traits are most desirable? Realistically, cures for many severe but rare genetic disorders will not be found, because the financial payback will not even cover the cost of the research. Eugenics, however, might just turn a profit. How much would potential parents pay to be sure that their child will be tall or blue-eyed? Would it be okay to engineer "superhumans" with breathtaking strength or intelligence? How about a treatment that can help you lose that extra weight, and keep it off permanently? The gray area between interesting and abhorrent can be very different depending on who you ask.

In a survey conducted in the United States, more than 40 percent of those interviewed said it would be fine to use gene therapy to make smarter and cuter babies. In one poll of British parents, 18 percent would be willing to use genetic enhancement to keep their child from being aggressive, and 10 percent would use it to keep a child from growing up to be homosexual.

Getting There Some people are adamant that we must never alter the DNA of anything. The concern is that gene therapy puts us on a slippery slope that may result in irreversible damage to ourselves and to nature. We as a society may not have the wisdom to know how to stop once we set foot on that slope. One is reminded of our peculiar human tendency to leap before we look.

And yet, something about the human experience allows us to dream of such things as wings of our own making, a capacity that carried us to the frontiers of space. In this brave new world, the questions before you are these: What do we stand to lose if serious risks are not taken? And, do we have the right to impose the consequences of taking such risks on those who would choose not to take them?

Beta-carotene is an orange photosynthetic pigment (Section 7.1) that is remodeled by cells of the small intestine into vitamin A. Potrykus and Beyer transferred two genes in the beta-carotene synthesis pathway into rice plants—one gene from corn and one from bacteria. All three genes were under the control of a promoter that works only in seeds. The transgenic rice plants began to make beta-carotene in their seeds—in the grains of Golden Rice. One cup of Golden Rice has enough beta-carotene to satisfy a child's daily recommended amount of vitamin A. The rice was ready in 2005, but is still not available for human consumption. The biosafety experiments required by regulatory agencies

How would you vote?

Should food distributors be required to identify food products made from genetically modified organisms? See CengageNOW for details, then vote online.

are far too expensive for a humanitarian agency in the public sector. Most of the transgenic organisms used for food today were carried through the deregulation process by private companies.

Summary

Section 16.1 **Recombinant DNA** consists of the fused DNA of different organisms. In **DNA cloning, restriction enzymes** cut DNA into pieces, then DNA ligase splices the pieces into **plasmids** or other **cloning vectors**. The resulting hybrid molecules are inserted into host cells such as bacteria. When a host cell divides, it forms huge populations of genetically identical descendant cells, or **clones**. Each clone has a copy of the foreign DNA.

RNA cannot be cloned directly. **Reverse transcriptase**, a viral enzyme, is used to convert single-stranded RNA into **cDNA** for cloning.

■ *Use the animation on CengageNOW to survey the tools of researchers who make recombinant DNA.*

Section 16.2 A **DNA library** is a collection of cells that host different fragments of DNA, often representing an organism's entire **genome**. Researchers can use **probes** to identify cells that host a specific fragment of DNA. Base-pairing between nucleic acids from different sources is called **nucleic acid hybridization**.

The polymerase chain reaction (**PCR**) uses **primers** and a heat-resistant DNA polymerase to rapidly increase the number of molecules of a DNA fragment.

■ *Use the interaction on CengageNOW to learn how researchers isolate and copy genes.*

Section 16.3 **DNA sequencing** can reveal the order of nucleotide bases in a fragment of DNA. DNA polymerase is used to partially replicate a DNA template. The reaction produces a mixture of DNA fragments of different lengths. **Electrophoresis** separates the fragments into bands.

■ *Use the animation on CengageNOW to investigate the technique of DNA sequencing.*

Section 16.4 **Short tandem repeats** are multiple copies of a short DNA sequence that follow one another along a chromosome. The number and distribution of short tandem repeats, unique in each individual, is revealed by electrophoresis as a **DNA fingerprint**.

■ *Use the animation on CengageNOW to observe the process of DNA fingerprinting.*

Section 16.5 The genomes of several organisms have been sequenced. **Genomics**, or the study of genomes, is providing insights into the function of the human genome. **DNA chips** are used to study gene expression.

Sections 16.6–16.8 Recombinant DNA technology and genome analysis are the basis of **genetic engineering**: directed modification of an organism's genetic makeup. Genes from one species are inserted into an individual of a different species to make a **transgenic** organism, or a gene is modified and reinserted into an individual of the same species. The result of either process is a **genetically modified organism** (GMO). GMO animals provide a source of organs for **xenotransplantation**.

■ *Use the animation on CengageNOW to see how the Ti plasmid is used to genetically engineer plants.*

Section 16.9 Rigorous safety procedures minimize potential risks to researchers in genetic engineering labs. Although these and other strict government regulations limit the release of genetically modified organisms into the environment, such laws are not guarantees against accidental releases or unforeseen environmental effects.

Section 16.10 With **gene therapy**, a gene is transferred into body cells to correct a genetic defect or treat a disease.

Self-Quiz   *Answers in Appendix III*

1. Researchers can cut DNA molecules at specific sites by using _____ .
   a. DNA polymerase   c. restriction enzymes
   b. DNA probes   d. reverse transcriptase

2. Fill in the blank: A _____ is a small circle of bacterial DNA that contains only a few genes and is separate from the bacterial chromosome.

3. By reverse transcription, _____ is assembled on a(n) _____ template.
   a. mRNA; DNA   c. DNA; ribosome
   b. cDNA; mRNA   d. protein; mRNA

4. For each species, all _____ in the complete set of chromosomes is the _____ .
   a. genomes; phenotype   c. mRNA; start of cDNA
   b. DNA; genome   d. cDNA; start of mRNA

Data Analysis Exercise

Autism is a neurobiological disorder with a range of symptoms that include impaired social interactions, stereotyped patterns of behavior such as hand-flapping or rocking, and, occasionally, greatly enhanced intellectual abilities.

Autism may have a genetic basis. Some autistic people have a mutation in neuroligin 3, a type of cell adhesion protein (Section 5.2) that connects brain cells. One mutation changes amino acid 451 from arginine to cysteine.

Mouse and human neuroligin 3 are very similar. In 2007, Katsuhiko Tabuchi and his colleagues genetically modified mice to carry the same arginine-to-cysteine substitution in their neuroligin 3. The mutation caused an increase in transmission of some types of signals between brain cells. Mice with the mutation had impaired social behavior, and, unexpectedly, enhanced spatial learning ability (Figure 16.16).

1. Did the modified or the unmodified mice learn the location of the platform faster in the first test?

2. Which mice learned faster the second time around?

3. Which mice showed the greatest improvement in memory between the first and the second test?

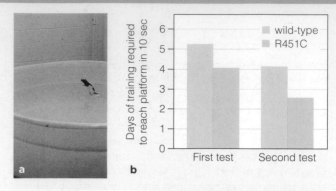

Figure 16.16 Enhanced spatial learning ability in mice with a mutation in neuroligin 3 (R451C), compared with unmodified (wild-type) mice. The mice were tested in a water maze, in which a platform is submerged a few millimeters below the surface of a deep pool of warm water (**a**). The platform, which is not visible to swimming mice, was moved for the second test.

Mice do not particularly enjoy swimming, so they locate a hidden platform as fast as they can. When tested again, they can remember its location by checking visual cues around the edge of the pool. How quickly they remember the platform's location is a measure of their spatial learning ability (**b**).

5. A set of cells that host various DNA fragments collectively representing an organism's entire set of genetic information is a _____ .

6. PCR can be used _____ .
  a. to increase the number of specific DNA fragments
  b. to make DNA fingerprints
  c. in a DNA sequencing reaction
  d. a and b are correct

7. Fragments of DNA can be separated by electrophoresis according to _____ .
  a. sequence   b. length   c. species

8. DNA sequencing relies on _____ .
  a. standard and labeled nucleotides
  b. primers and DNA polymerase
  c. electrophoresis
  d. all of the above

9. Which of the following can be used to carry foreign DNA into host cells? Choose all correct answers.
  a. RNA   e. lipid clusters
  b. viruses   f. blasts of pellets
  c. PCR   g. xenotransplantation
  d. plasmids   h. DNA microarrays

10. _____ can be used to correct a genetic defect.

11. Match the terms with the most suitable description.
  ___DNA fingerprint   a. carries a foreign gene
  ___Ti plasmid   b. slows bacteriophage growth
  ___nucleic acid   c. a person's unique collection
     hybridization   of short tandem repeats
  ___eugenic   d. base pairing of DNA or
     engineering   DNA and RNA from
  ___transgenic   different sources
  ___GMO   e. selecting "desirable" traits
  ___restriction enzyme   f. genetically modified
                          g. used in some gene transfers

■ *Visit CengageNOW for additional questions.*

Critical Thinking

1. The *FOXP2* gene encodes a transcription factor associated with vocal learning in mice, bats, birds, and humans. Mutations in *FOXP2* result in altered vocalizations in mice, and severe language disorders in humans. The chimpanzee, gorilla, and rhesus *FOXP2* proteins are identical; the human version differs in two of 715 amino acids. The change of two amino acids may have contributed to the development of language in humans. Would it be okay to transfer the human *FOXP2* gene into a nonhuman primate? What do you think might happen if the transgenic animal learned to speak?

2. Animal viruses can mutate so that they infect humans, occasionally with disastrous results. In 1918, an influenza pandemic that apparently originated with a strain of avian flu killed 50 million people. Researchers isolated samples of that virus, the influenza A(H1N1) strain, from bodies of infected people preserved in Alaskan permafrost since 1918. From the samples, the researchers reconstructed the DNA sequence of the viral genome, then reconstructed the virus. Being 39,000 times more infectious than modern influenza strains, the reconstructed A(H1N1) virus proved to be 100 percent lethal in mice.

Understanding how the A(H1N1) strain works can help us defend ourselves against other strains that may be like it. For example, researchers are using the reconstructed virus to discover which of its mutations made it so infectious and deadly in humans. Their work is urgent. A deadly new strain of avian influenza in Asia shares some mutations with the A(H1N1) strain. Even now, researchers are working to test the effectiveness of antiviral drugs and vaccines on the reconstructed virus, and to develop new ones.

Critics of the A(H1N1) reconstruction are concerned. If the virus escapes the containment facilities (even though it has not done so yet), it might cause another pandemic. Worse, terrorists could use the published DNA sequence and methods to make the virus for horrific purposes. Do you think this research makes us more or less safe?

III PRINCIPLES OF EVOLUTION

Two male frigate birds (*Fregata minor*) in the Galápagos Islands, far from the coast of Ecuador. Each male inflates a gular sac, a balloon of red skin at his throat, in a display that may catch the eye of a female. The males lurk together in the bushes, sacs inflated, until a female flies by. Then they wag their head back and forth and call out to her. Like other structures that males use only in courtship, the gular sac probably is an outcome of sexual selection—one of the topics you will read about in this unit.

Evidence of Evolution

Measuring Time

How do you measure time? Is your comfort level limited to your own generation? Probably you can relate to a few hundred years of human events, but how about a few million? Understanding the distant past requires an intellectual leap from the familiar to the unknown. Perhaps the idea of an asteroid slamming into Earth will help you make that leap. Asteroids are minor planets hurtling through space. They range in size from 1 to 1,500 kilometers (0.5 to 1,000 miles) wide. Millions of them orbit around the sun between Mars and Jupiter—cold, stony leftovers from the formation of our solar system.

Asteroids are hard to spot even with the best telescopes, because they do not emit light. Many cross Earth's orbit, but most of those pass us by before we know about them. Some have passed too close for comfort.

The mile-wide Barringer Crater in Arizona is hard to miss (Figure 17.1a). A 300,000-ton asteroid made this impressive pock in the sandstone when it slammed into Earth 50,000 years ago. The impact was 150 times more powerful than the bomb that leveled Hiroshima.

No human witnessed the impact, so how do we know what happened? Sometimes we have physical evidence of events that occurred before we were around to see them. In this case, geologists were able to infer the most probable cause of the Barringer Crater by analyzing tons of meteorites, melted sand, and other rocky clues at the site.

Similar evidence points to even larger asteroid impacts in the more distant past. For example, a mass extinction—a loss of major groups of organisms—occurred 65.5 million years ago, at what is known as the K–T boundary. An unusual layer of rock marks this boundary (Figure 17.1b). There are plenty of dinosaur fossils below this layer. Above it, in more recent rock layers, there are no dinosaur fossils anywhere. An impact crater near what is now the Yucatán dates to about 65.5 million years ago. Coincidence? Many scientists say no. They infer from the evidence that the impact of an asteroid that was 10 to 20 km (6 to 12 miles) wide caused a global catastrophe that wiped out the dinosaurs.

You are about to make an intellectual leap through time, to places that were not even known about a few centuries ago. We invite you to launch yourself from this premise: Natural phenomena that occurred in the past can be explained by the same physical, chemical, and biological processes that operate today. That premise is the foundation for scientific research into the history of life. The research represents a shift from experience to inference—from the known to what can only be surmised. It gives us astonishing glimpses into the past.

See the video! Figure 17.1 From evidence to inference. (**a**) What made the Barringer Crater? Rocky evidence points to a 300,000-ton asteroid that collided with Earth 50,000 years ago. (**b**) Bands that are part of a unique layer of rock that formed 65.5 million years ago, worldwide. The layer marks an abrupt transition in the fossil record that implies a mass extinction. The red pocketknife gives an idea of scale.

Key Concepts

Emergence of evolutionary thought

Long ago, naturalists started to catalog previously unknown species and think about the global distribution of all species. They discovered similarities and differences among major groups, including those represented as fossils in layers of sedimentary rock. **Sections 17.1, 17.2**

A theory takes form

Evidence of evolution, or changes in lines of descent, gradually accumulated. Charles Darwin and Alfred Wallace independently developed a theory of natural selection to explain how heritable traits that define each species evolve. **Sections 17.3, 17.4**

Evidence from fossils

The fossil record offers physical evidence of past changes in many lines of descent. We use the property of radioisotope decay to determine the age of rocks and fossils. **Sections 17.5–17.7**

Evidence from biogeography

Correlating evolutionary theories with geologic history helps explain the distribution of species, past and present. **Sections 17.8, 17.9**

Links to Earlier Concepts

- Section 1.4 sketched out the key premises of the theory of natural selection. Now you will consider evidence that led to its formulation. What you know about alleles and inheritance (11.1) will help you understand how natural selection works.

- This chapter explores one of the clashes that occurred between traditional belief systems and scientific thought. You may wish to review what you know about critical thinking (1.5) before you begin. Remember, science only concerns itself with the observable (1.6).

- Determining the age of ancient rocks and fossils depends on the properties of radioisotope decay (2.2) and compounds (2.3).

How would you vote? Many theories and hypotheses about events in the ancient past are necessarily based on traces left by those events, not on data collected by direct observations. Is indirect evidence ever enough to prove a theory about a past event? See CengageNOW for details, then vote online.

Early Beliefs, Confounding Discoveries

■ Belief systems are influenced by the extent of our knowledge. Those that are inconsistent with systematic observations of the natural world tend to change over time.

The seeds of biological inquiry were taking hold in the Western world more than 2,000 years ago. Aristotle, the Greek philosopher, had no books or instruments to guide him, and yet he was more than a collector of random observations. We can infer from his writings that he was making connections between his observations in an attempt to explain the order of the natural world. Like few others of his time, Aristotle viewed nature as a continuum of organization, from lifeless matter through complex plants and animals. Aristotle was one of the first **naturalists**—people who observe life from a scientific perspective.

By the fourteenth century, Aristotle's earlier ideas about nature had been transformed into a rigid view of life. By this view, a "great chain of being" extended from the lowest form (snakes), to humans, to spiritual beings. Each individual link in the chain was a kind of being, or species, and each was said to have been designed and forged at the same time in a perfect state. Once every link had been discovered and described, the meaning of life would be revealed.

European naturalists embarked on survey expeditions and brought back tens of thousands of plants and animals from Asia, Africa, North and South America, and the Pacific Islands. The newly discovered species were carefully catalogued as more links in the chain.

By the late 1800s, Alfred Wallace and a few other naturalists had moved beyond the idea of studying species just to catalogue them. They were seeing patterns in where species live and how they might be related, and had started to make hypotheses about the ecological and evolutionary forces that shape life on Earth. They were pioneers in **biogeography**, the study of patterns in the geographic distribution of species.

Some of the patterns they perceived raised questions that could not be answered within the framework of prevailing belief systems. For example, globe-trotting explorers had discovered plants or animals living in extremely isolated places. The isolated species looked suspiciously similar to plants or animals living across vast expanses of open ocean, or on the other side of impassable mountain ranges. Could different species be related? If so, how could the related species end up geographically isolated from one another?

The birds in Figure 17.2a–c, for example, share very similar features, although each lives on a different continent. All three flightless birds sprint about on long, muscular legs in flat, open grasslands about the same distance from the equator. All raise their long necks to watch for predators. Wallace thought that these birds

Figure 17.2 Species that resemble one another, but are native to distant geographic realms.

(**a**) South American rhea, (**b**) Australian emu, and (**c**) African ostrich. All three types of birds live in similar habitats. These related birds are unlike most other birds in several traits, including their long, muscular legs and their inability to fly.

Similar-looking, unrelated plants: a spiny cactus native to the hot deserts of the American Southwest (**d**), and a spiny spurge native to southwestern Africa (**e**).

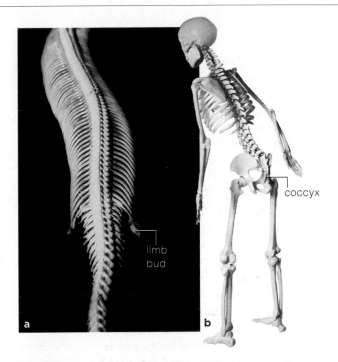

Figure 17.3 Vestigial body parts. (a) Pythons and boa constrictors have tiny leg bones, but snakes do not walk. (b) We humans use our legs, but not our coccyx bones.

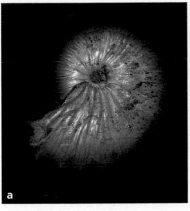

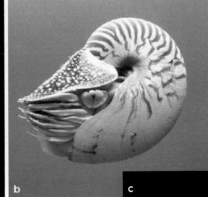

Figure 17.4 Fossil puzzles. (a) Fossilized ammonite that lived between 65 and 100 million years ago. This marine predator resembled the modern chambered nautilus (b).

(c) Fossilized foraminiferan shells, each from a different layer of sedimentary rock in a vertical sequence, and each a bit different from the ones in adjacent layers.

might be descended from an ancient common ancestor (and he was right), but he had no idea how they could have ended up on different continents.

Naturalists of the time also had trouble classifying organisms that are very similar in some features, but different in others. For example, the two plants shown in Figure 17.2d,e are native to different continents. Each lives about the same distance from the equator in the same kind of environment—hot deserts where water is seasonally scarce. Both have rows of sharp spines that deter herbivores, and both store water in their thick, fleshy stems. However, their reproductive parts are very different, so these plants cannot be as closely related as their outward appearance would suggest.

Comparisons like these are part of **comparative morphology**, the study of body plans and structures among groups of organisms. Organisms that are outwardly very similar may be quite different internally; think of fishes and porpoises. Others that differ greatly in outward appearance may be very similar in underlying structure. For example, a human arm, a porpoise flipper, an elephant leg, and a bat wing have comparable internal bones, as Section 19.2 will explain.

Comparative morphology revealed body parts that have no apparent function, which added to the confusion. According to prevailing beliefs, each organism's body plan had been created in a perfect state. If that were so, then why were there useless parts such as leg bones in snakes (which do not walk), or the vestiges of a tail in humans (Figure 17.3)?

Geologists that were mapping rock formations exposed by erosion or quarrying had discovered identical sequences of rock layers in different parts of the world. **Fossils** in the layers were recognized as stone-hard evidence of earlier forms of life, but some of the fossils were puzzling. For example, deep layers of rock had fossils of simple marine life. The layers above them held similar but more intricate fossils. In higher layers, fossils that were similar but even more intricate looked like modern species (Figure 17.4). What could those sequences in complexity mean? Fossils of gigantic animals that had no living representatives were also being unearthed. If the animals had been perfect at the time of creation, why were they now extinct?

Taken as a whole, findings from biogeography, comparative morphology, and geology did not fit with prevailing beliefs of the nineteenth century. If species had not been created in a perfect state (and fossil sequences and "useless" body parts implied they had not), then perhaps species had changed over time.

Take-Home Message

How have observations of the natural world changed our thinking?

■ Increasingly extensive observations of nature in the nineteenth century did not fit with prevailing belief systems.

■ The cumulative findings from biogeography, comparative morphology, and geology led to new ways of thinking about the natural world.

■ By the 1800s, many scholars realized that life on Earth had changed over time, and began to think about what could have caused the changes.

■ Links to Critical thinking 1.5, How science works 1.6

Squeezing New Evidence into Old Beliefs

In the nineteenth century, naturalists were faced with increasing evidence that life on Earth, and even Earth itself, had changed over time. Around 1800, Georges Cuvier, an expert in zoology and paleontology, was trying to make sense of all the new information. He had observed that there were abrupt changes in the fossil record and that many fossil species seemed to have no living counterparts. Given this evidence, he proposed a startling idea: many species that had once existed were now extinct.

Cuvier also knew about evidence that Earth's surface had changed. For example, he had seen fossilized seashells on mountainsides far from modern seas. Like most others of his time, he assumed Earth's age to be in the thousands, not millions, of years. He reasoned that geologic forces unlike those known today would have been necessary to raise sea floors to mountaintops in this short time span. Catastrophic geological events would have caused extinctions, after which surviving species repopulated the planet. Cuvier's idea came to be known as **catastrophism**. We now know it is incorrect; geologic processes have not changed over time.

Another scholar, Jean-Baptiste Lamarck, was thinking about processes that drive **evolution**—change in a line of descent. Lamarck thought that a species gradually improved over generations because of an inherent drive toward perfection, up the chain of being. This drive directed an unknown "fluida" into body parts needing change. By Lamarck's hypothesis, environmental pressures and internal needs cause changes in an individual's body, and offspring inherit the changes.

Try using Lamarck's hypothesis to explain why the giraffe's neck is very long. We might predict that some short-necked ancestor of the modern giraffe stretched its neck to browse on leaves beyond the reach of other animals. The stretches may have even made its neck a bit longer. By Lamarck's hypothesis, that animal's offspring would inherit a longer neck, and after many generations strained to reach ever loftier leaves, the modern giraffe would have been the result.

Lamark was correct in thinking that environmental factors affect a species' traits, but wrong about the inheritance of acquired characteristics. An individual's phenotype can change, as when a woman builds up large muscles by strength training. However, the child of an athletic mother is not born with bigger muscles than the child of a nonathletic mom. Offspring do not inherit traits a parent acquires during its lifetime; they inherit DNA. Under most circumstances, DNA passed to offspring is unaffected by a parent's activities.

Voyage of the *Beagle*

In 1831, the twenty-two year old Charles Darwin was wondering what to do with his life. Ever since he was

Figure 17.5 (**a**) Charles Darwin. (**b**) A replica of the *Beagle* sails off a rugged South American coastline. During one voyage, Darwin ventured into the Andes, where he found fossils of marine organisms in rock layers 3.6 kilometers (2.2 miles) above sea level.

(**c–e**) The Galápagos Islands are isolated in the ocean, far to the west of Ecuador. They arose by volcanic action on the sea floor about 5 million years ago. Winds and currents carried organisms to the once-lifeless islands. All of the native species are descended from those travelers. At *far right*, a blue-footed booby, one of many species Darwin observed during his voyage.

eight, he had wanted to hunt, fish, collect shells, or watch insects and birds—anything but sit in school. Later, at his father's insistence, he attempted to study medicine in college, but the crude, painful procedures used on patients in that era sickened him. His exasperated father urged him to become a clergyman, so Darwin packed for Cambridge, where he earned his degree in theology. Yet he spent most of his time with faculty members who embraced natural history.

John Henslow, a botanist, perceived Darwin's real interests. He arranged for Darwin to become a naturalist aboard the *Beagle*, a ship about to embark on a survey expedition to South America. The young man who had hated school and had no formal training in science quickly became an enthusiastic naturalist.

The *Beagle* set sail for South America in December, 1831 (Figure 17.5). During the ship's Atlantic crossing, Darwin read Henslow's parting gift, the first volume of Charles Lyell's *Principles of Geology*. What he learned gave him insights into the geological history of the regions he would encounter on his journey.

During the *Beagle*'s five-year voyage, Darwin found many unusual fossils. He saw diverse species living in environments that ranged from the sandy shores of remote islands to the plains high in the Andes.

He also started to mull over a radical theory. For many years, geologists had been chipping away at the sandstones, limestones, and other types of rocks that form after sediments slowly accumulate in lakebeds, river bottoms, and ocean floors. These rocks held evidence that gradual processes of geologic change operating in the present were the same ones that operated in the distant past. Lyell proposed that strange catastrophes were not necessary to explain Earth's surface. Over great spans of time, gradual, everyday geologic processes such as erosion could have sculpted Earth's current landscape.

The idea that gradual, repetitive change had shaped Earth became known as the **theory of uniformity**. It challenged the prevailing belief that Earth was 6,000 years old. According to traditional scholars, people had recorded everything that happened in those 6,000 years—and in all that time, no one had mentioned seeing a species evolve. Even so, by Lyell's calculations, it must have taken millions of years to sculpt Earth's surface. Was that not enough time for species to evolve? Darwin thought that it was. But *how* did they evolve? He would end up devoting the rest of his life to that burning question.

Take-Home Message

How did new evidence affect old beliefs?

■ In the 1800s, fossils and other evidence led some naturalists to propose that Earth and the species on it had changed over time. The naturalists also began to reconsider the age of Earth.

■ These ideas set the stage for Darwin's later insights about evolution.

route of *Beagle*

Darwin

EQUATOR

Wolf

Galápagos Islands

Pinta

Marchena   Genovesa

EQUATOR

Santiago
Bartolomé
Seymour
Rábida   Baltra
Fernandina   Pinzón
Santa Cruz
Santa Fe
Tortuga   San Cristóbal
Isabela
Española
Floreana

- Darwin's observations of species in different parts of the world helped him understand a driving force of evolution.

- Links to Natural selection 1.4, Alleles and traits 11.1

Old Bones and Armadillos

Darwin sent to England the thousands of specimens he had collected on his voyage. Among them were fossil glyptodons from Argentina. These armored mammals are extinct, but they have many traits in common with modern armadillos (Figure 17.6). For example, armadillos live only in places where glyptodons once lived. Like glyptodons, armadillos have helmets and protective shells that consist of unusual bony scales.

Could the odd shared traits mean that glyptodons were ancient relatives of armadillos? If so, perhaps traits of their common ancestor had changed in the line of descent that led to armadillos. But why would such changes occur?

A Key Insight—Variation in Traits

Back in England, Darwin pondered his notes and fossils. He also read an essay by one of his contemporaries, economist Thomas Malthus. Malthus had correlated increases in human population size with famine, disease, and war. He proposed that humans run out of food, living space, and other resources because they tend to reproduce beyond the capacity of their environment to sustain them. When that happens, the individuals of a population must either compete with one another for the limited resources, or develop technology to increase their productivity. Darwin realized that

Malthus's ideas had wider application: All populations, not just human ones, have the capacity to produce more individuals than their environment can support.

Darwin also reflected on species he had observed during his voyage. He knew that individuals of a species were not always identical. They had many traits in common, but they might vary in size, color, or other traits. It dawned on Darwin that having a particular version of a variable trait might give an individual an advantage over competing members of its species. The trait might enhance the individual's ability to secure limited resources—and to survive and reproduce—in its particular environment.

Darwin thought back to some of the bird species he saw on the Galápagos Islands (Figure 17.7). This island chain is separated from South America by 900 kilometers (550 miles) of open ocean, so he could assume that most of the bird species populating the islands had been isolated there for a long time. Different kinds of finches populate the coasts, dry lowlands, and mountain forests of the islands; each species has traits that suit its members to their particular habitat.

Darwin also knew about artificial selection, the process whereby humans choose traits that they favor in a domestic species. For example, he was familiar with dramatic variations in traits that pigeon breeders had produced through selective breeding (Section 1.4). He recognized that an environment could similarly select traits that make individuals of a population suited to it. For example, suppose a group of seed-eating birds lives in a dry environment where soft seeds are scarce. A bird is born with an extra-strong bill that allows it to crack open hard seeds that other members of the pop-

Figure 17.6 Ancient relatives? (**a**) A modern armadillo, about a foot long. (**b**) Fossil of a glyptodon, a car-sized mammal that lived between about 2 million and 15,000 years ago.

Glyptodons and armadillos are widely separated in time, but they share a restricted distribution and unusual traits, including a shell and helmet of keratin-covered bony plates—a material similar to crocodile and lizard skin. [The fossil in (**b**) is missing its helmet.] Their unique shared traits were a clue that helped Darwin develop a theory of evolution by natural selection.

Figure 17.7 Three of the thirteen species of finches native to the Galápagos Islands.

(**a**) A big-billed seed cracker, *Geospiza magnirostris.*

(**b**) *G. scandens* eats cactus fruit and insects in cactus flowers.

(**c**) *Camarhynchus pallidus* uses cactus spines and twigs to probe for wood-boring insects.

ulation cannot. Thus, the strong-billed bird can access an extra food source. All other things being equal, the strong-billed bird has a better chance of surviving and reproducing in this particular environment than other individuals in the population. Moreover, if bill hardness has a heritable basis, at least some of the bird's offspring may inherit the same advantage. After many generations, strong-billed birds would probably predominate in this population. Thus, over many generations, a population's environment may influence the traits shared by its individuals.

Natural Selection

These thoughts led Darwin to realize that variation in the details of shared traits make individuals of a population variably suited to their environment. In other words, the individuals of a natural population vary in **fitness**—adaptation to their environment as measured by relative genetic contribution to future generations. A trait that enhances an individual's fitness is called an evolutionary **adaptation**, or **adaptive trait**.

Individuals of a natural population tend to survive and reproduce with differing success depending on the details of their shared traits. Darwin understood that this process, which he called **natural selection**, could be a driving force of evolution. If an individual has a form of a trait that makes it better suited to an environment, then it is better able to compete for resources. If an individual is better able to compete for resources, then it has a better chance of surviving long enough to produce offspring. If individuals that bear an adaptive, heritable trait produce more offspring than those that do not, then the frequency of that trait will tend to increase in the population over successive generations. Table 17.1 summarizes this reasoning.

Darwin hypothesized that the process of evolution by natural selection could explain not only the variation within populations, but also the great diversity of species in the world and in the fossil record.

Table 17.1 Principles of Natural Selection

Observations about populations

■ Natural populations have an inherent reproductive capacity to increase in size over time.

■ As a population expands, resources that are used by its individuals (such as food and living space) eventually become limited.

■ When resources are limited, individuals of a population compete for them.

Observations about genetics

■ Individuals of a species share certain traits.

■ Individuals of a natural population vary in the details of their shared traits.

■ Traits have a heritable basis, in genes. Alleles (slightly different forms of a gene) arise by mutation.

Inferences

■ A certain form of a shared trait may make its bearer more competitive at securing a limited resource.

■ Individuals better able to secure a limited resource tend to leave more offspring than others of a population.

■ Thus, an allele associated with an adaptive trait tends to become more common in a population over generations.

Take-Home Message

What is natural selection?

■ Natural selection is the differential in survival and reproduction among individuals of a population that vary in the details of their shared, inherited traits.

■ Traits favored by natural selection are said to be adaptive. An adaptive trait increases the chances that an individual bearing it will survive and reproduce.

■ Darwin's insights into evolution were made possible by contributions of scientists who preceded him.
■ Alfred Wallace independently developed the idea of evolution by natural selection.

Darwin wrote out his ideas about natural selection, but let ten years pass without publishing them. In the meantime, Alfred Wallace, a naturalist who had been studying wildlife in the Amazon basin and the Malay Archipelago, wrote an essay and sent it to Darwin for advice. Wallace's essay had outlined Darwin's theory! Wallace had written earlier letters to Lyell and Darwin about patterns in the geographic distribution of species; he too had connected the dots. Wallace is now called the father of biogeography (Figure 17.8).

In 1858, just weeks after Darwin received Wallace's essay, their similar theories were presented jointly at a scientific meeting. Wallace was still in the field and knew nothing about the meeting, which Darwin did not attend. The next year, Darwin published *On the Origin of Species*, which laid out detailed evidence in support of his theory.

Many scholars readily accepted the idea of descent with modification, or evolution. However, there was a fierce debate over the idea that evolution occurs by natural selection. Decades would pass before experimental evidence from the field of genetics led to its widespread acceptance in the scientific community.

Figure 17.8 Alfred Wallace, codiscoverer of the process of evolution by natural selection.

Take-Home Message

What role did Alfred Wallace play in developing the theory of evolution by natural selection?

■ Wallace drew on his own observations of plant and animal species and proposed, like Darwin, that natural selection is a driving force of evolution.

■ Fossils are remnants or traces of organisms that lived in the past. They give us clues about evolutionary relationships.
■ The fossil record will always be incomplete.

How Do Fossils Form?

Most fossils are mineralized bones, teeth, shells, seeds, spores, or other hard body parts (Figure 17.9a,b). Trace fossils such as footprints and other impressions, nests, burrows, trails, bore holes, eggshells, or feces are evidence of an organism's activities (Figure 17.9c).

The process of fossilization begins when an organism or its traces become covered by sediments or volcanic ash. Water slowly infiltrates the remains, and metal ions and other inorganic compounds dissolved in the water gradually replace the minerals in the bones and other hard tissues. Sediments that accumulate on top of the remains exert increasing pressure on them. After a very long time, pressure and mineralization transform the remains into rock.

Most fossils are found in layers of sedimentary rock such as mudstone, sandstone, and shale (Figure 17.10). Sedimentary rock forms as rivers wash silt, sand, volcanic ash, and other particles from land to sea. The particles settle on sea floors in horizontal layers that vary in thickness and in composition. After hundreds of millions of years, layers of sediments became compacted into layers of rock.

We study the layers of sedimentary rock in order to understand the historical context of the fossils we find in them. Usually, the deeper layers in a stack were the first to form; those closest to the surface formed most recently. Thus, the deeper the layer of sedimentary rock, the older the fossils it contains. A layer's composition is also a clue about local or global events that were occurring as it formed; the K–T boundary layer discussed in the chapter introduction is one example. The relative thicknesses of the different layers provides other clues. For instance, the layers were thin during ice ages, when tremendous volumes of water froze and became locked in glaciers. Sedimentation slowed as rivers dried up. When the glaciers melted, sedimentation resumed and the layers became thicker.

The Fossil Record

We have fossils for more than 250,000 known species. Considering the current range of biodiversity, there must have been many millions more, but we will never know all of them. Why not?

The odds are against finding evidence of an extinct species because fossils are relatively rare. Most of the

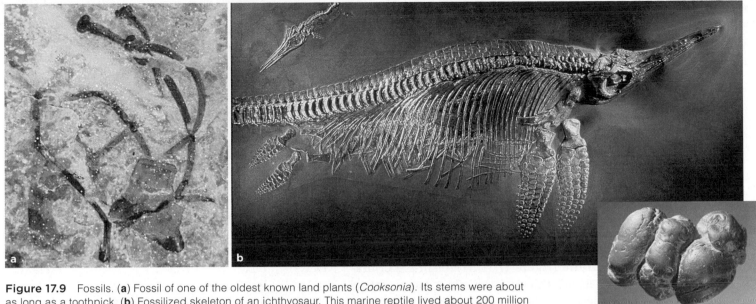

Figure 17.9 Fossils. (**a**) Fossil of one of the oldest known land plants (*Cooksonia*). Its stems were about as long as a toothpick. (**b**) Fossilized skeleton of an ichthyosaur. This marine reptile lived about 200 million years ago. (**c**) Coprolite. Fossilized food remains and parasitic worms inside such fossilized feces tell us about the diet and health of extinct species. A foxlike animal excreted this one.

time, an organism's remains are quickly obliterated by scavengers or decay. Organic materials decompose in the presence of oxygen, so remains endure only if they are encased in an air-excluding material such as sap, tar, ice, or mud. Remains that do become fossilized are often deformed, crushed, or scattered by erosion and other geologic assaults.

In order for us to know about an extinct species that existed long ago, we have to find a fossil of it. At least one specimen had to be buried before it decomposed or something ate it. The burial site had to escape geologic events, and it had to end up where we might be able to find it.

Most ancient species had no hard parts to fossilize, so we do not find much evidence of them. Unlike bony fishes or the hard-shelled mollusks, for example, jellyfishes and soft worms do not show up much in the fossil record, although they were probably much more common.

Also think about relative numbers of organisms. One plant might release millions of spores in a single season. The earliest humans lived in small bands and few offspring survived. What are the odds of finding even one fossilized human bone compared to the odds of finding a fossilized plant spore?

Finally, imagine one line of descent, a **lineage**, that vanished when its habitat on a remote volcanic island sank into the sea. Or imagine two lineages, one lasting only briefly and the other for billions of years. Which is more likely to be represented in the fossil record?

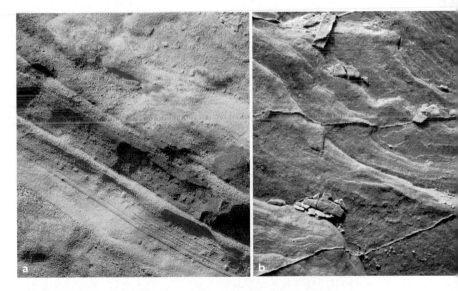

Figure 17.10 The two most common types of fossil-containing sedimentary rock: (**a**) sandstone, which consists mainly of compacted grains of sand or minerals, and (**b**) shale, which is compacted clay or mud. Both form in layers.

Take-Home Message

What are fossils?

■ Fossils are evidence of organisms that lived in the remote past, a stone-hard historical record of life. The oldest are usually in the deepest sedimentary rocks.

■ The fossil record will never be complete. Geologic events obliterated much of it. The rest of the record is slanted toward species that had hard parts, dense populations with wide distribution, and had persisted for a long time.

■ Even so, the fossil record is substantial enough to help us reconstruct patterns and trends in the history of life.

17.6 | Dating Pieces of the Puzzle

- Radiometric dating reveals the age of rocks and fossils.
- Links to Radioisotopes 2.2, Compounds 2.3

A radioisotope is a form of an element with an unstable nucleus (Section 2.2). Atoms of a radioisotope become atoms of other elements as their nucleus disintegrates. Such decay is not influenced by temperature, pressure, chemical bonding state, or moisture; it is only influenced by time. Like the ticking of a perfect clock, each type of radioisotope decays at a constant rate into predictable products called daughter elements.

For example, radioactive uranium 238 decays into thorium 234, which decays into something else, and so on until it becomes lead 206. The time it takes for half of a radioisotope's atoms to decay into a product is called **half-life** (Figure 17.11). The half-life of uranium 238 to lead 206 decay is 4.5 billion years.

The predictability of radioactive decay can be used to find the age of a volcanic rock—the date it cooled.

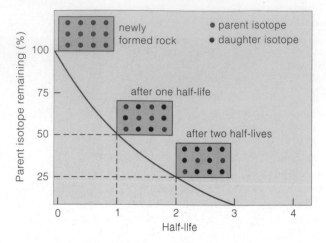

Figure 17.11 Animated Half-life—the time it takes for half of the atoms in a sample of radioisotope to decay. **Figure It Out:** How much of a parent radioisotope remains after two half-lives have elapsed?

Answer: 25 percent

Rock deep inside Earth is hot and molten; atoms swirl and mix in it. Rock that reaches the surface cools and hardens; as it does, minerals crystallize in it. Each kind of mineral has a characteristic structure and composition. For example, the mineral zircon (*right*) consists primarily of ordered arrays of zircon silicate molecules ($ZrSiO_4$). Some molecules in a zircon crystal have uranium atoms substituted for zirconium atoms, but never lead atoms, so new zircon that crystallizes from cooling, molten rock contains no lead. However, uranium decays into lead at a predictable rate. Thus, over time, uranium atoms disappear from a zircon crystal, and lead atoms accumulate in it. The ratio of uranium atoms to lead atoms in a zircon crystal can be measured precisely. That ratio can be used to calculate how long ago the crystal formed—its age.

zircon

We have just described **radiometric dating**, a method that can reveal the age of a material by determining its radioisotope and daughter element content. The oldest known terrestrial rock is a tiny zircon crystal from Jack Hills, Australia. It is 4.404 billion years old.

Recent fossils that still contain carbon can be dated by measuring their carbon 14 content (Figure 17.12). Most of the carbon 14 in a fossil will have decayed after about 60,000 years. The age of fossils older than that can only be estimated by dating volcanic rocks in lava flows above and below the fossil site.

A Long ago, trace amounts of ^{14}C and a lot more ^{12}C were incorporated into the tissues of a nautilus. The carbon atoms were part of organic molecules in the nautilus's food. As long as it was alive, the nautilus replenished its own tissues with carbon secured from food. Thus, the proportion of ^{14}C to ^{12}C in its tissues stayed the same.

B When the nautilus died, it stopped eating, so its body stopped gaining carbon. The ^{14}C in its body continued to decay, so the amount of ^{14}C decreased relative to the amount of ^{12}C. Half of the ^{14}C decayed in 5,370 years, half of what was left was gone after another 5,370 years, and so on.

C Fossil hunters discover the fossil. They measure its ^{14}C to ^{12}C ratio and use it to calculate the number of half-life reductions since death. In this example, the ratio is one-eighth of the ^{14}C to ^{12}C ratio in living organisms. Thus, this nautilus lived about 16,000 years ago.

Figure 17.12 Animated Using radiometric dating to find the age of a fossil. Carbon 14 (^{14}C) forms in the atmosphere and combines with oxygen to become carbon dioxide. Along with far greater quantities of the stable carbon isotope ^{12}C, trace amounts of ^{14}C enter food chains by way of photosynthesis. All living organisms incorporate carbon.

Take-Home Message

How do we determine the age of rocks and fossils?

- Researchers use the predictability of radioisotope decay to estimate the age of rocks and fossils.

17.7 | A Whale of a Story

■ New fossil discoveries are continually filling the gaps in our understanding of the ancient history of many lineages.

For some time, scientists have thought that the ancestors of whales probably walked on land, then took up life in the water again. However, evidence for this line of thinking was scanty. The skull and lower jaw of cetaceans—which include whales, dolphins, and porpoises—have distinctive features characteristic of some kinds of ancient carnivorous land animals. Molecular comparisons suggested that those animals were probably artiodactyls, hooved mammals with an even number of toes (two or four) on each foot; modern representatives of the lineage include hippopotamuses, camels, pigs, deer, sheep, and cows.

Until recently, gradual changes in skeletal features demonstrating the transition of whale lineages from terrestrial to aquatic life were missing from the fossil record. Researchers knew there were intermediate forms because they had found a representative fossil of a whale skull, but without a complete skeleton the rest of the story remained speculative. Then, in 2000, Philip Gingerich and his colleagues found two of the missing links in Pakistan, when they collected complete fossil skeletons of the ancient whales *Rodhocetus* and *Dorudon* (Figure 17.13).

The researchers knew these new fossil specimens represented intermediate forms in the whale lineage because intact, sheeplike ankle bones and ancient whalelike skull bones were present in the same skeletons.

The newly discovered fossils fill in many details of the story of the ancient history of whales. For example, the ankle bones of both *Rodhocetus* and *Dorudon* have distinctive features in common with the ankle bones of extinct as well as modern artiodactyls. Modern cetaceans do not have even a remnant of an ankle bone. With their in-between ankle bones, *Rodhocetus* and *Dorudon* were probably offshoots of the ancient artiodactyl-to-modern whale lineage as it transitioned back to life in water. The proportions of *Rodhocetus*'s limbs, skull, neck, and thorax indicate it swam with its feet, not its tail. Like modern whales, the 5-meter (16-foot) *Dorudon* was clearly a fully aquatic tail-swimmer: The entire hindlimb was only about 12 centimeters (5 inches) long, much too small to have supported the animal out of water.

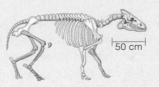

A A 30-million-year-old fossil of *Elomeryx*. This small terrestrial mammal was a member of the same artiodactyl group that gave rise to hippopotamuses, pigs, deer, sheep, cows, and whales.

B *Rodhocetus*, an ancient whale, lived about 47 million years ago. Its distinctive ankle bones point to a close evolutionary connection to artiodactyls. Inset: compare a *Rodhocetus* ankle bone (*left*) with that of modern artiodactyl, a pronghorn antelope (*right*).

C *Dorudon atrox*, an ancient whale that lived about 37 million years ago. Its artiodactyl-like ankle bones (*left*) were much too small to have supported the weight of its huge body on land, so this mammal had to be fully aquatic.

Figure 17.13 New links in the ancient lineage of whales.

You can think of each layer of sedimentary rock as a slice of geologic time; each holds clues to life on Earth during the period of time that it formed. Radiometric dating and fossils in the layers allow us to recognize similar sequences of sedimentary rock layers around the world. Transitions between layers mark the boundaries of great intervals of time in the **geologic time scale**, or chronology of Earth's history (Figure 17.14).

Eon	Era	Period	Epoch	mya	Major Geologic and Biological Events
PHANEROZOIC	CENOZOIC	QUATERNARY	Recent	0.01	Modern humans evolve. Major extinction event is now underway.
			Pleistocene	1.8	
		TERTIARY	Pliocene	5.3	Tropics, subtropics extend poleward. Climate cools; dry woodlands and grasslands emerge. Adaptive radiations of mammals, insects, birds.
			Miocene	23.0	
			Oligocene	33.9	
			Eocene	55.8	
			Paleocene	65.5 ◄	Major extinction event, perhaps precipitated by asteroid impact. Mass extinction of all dinosaurs and many marine organisms.
	MESOZOIC	CRETACEOUS	Late	99.6	
			Early		Climate very warm. Dinosaurs continue to dominate. Important modern insect groups appear (bees, butterflies, termites, ants, and herbivorous insects including aphids and grasshoppers). Flowering plants originate and become dominant land plants.
		JURASSIC		145.5	Age of dinosaurs. Lush vegetation; abundant gymnosperms and ferns. Birds appear. Pangea breaks up.
		TRIASSIC		199.6 ◄	Major extinction event
					Recovery from the major extinction at end of Permian. Many new groups appear, including turtles, dinosaurs, pterosaurs, and mammals.
	PALEOZOIC	PERMIAN		251 ◄	Major extinction event
					Supercontinent Pangea and world ocean form. Adaptive radiation of conifers. Cycads and ginkgos appear. Relatively dry climate leads to drought-adapted gymnosperms and insects such as beetles and flies.
		CARBONIFEROUS		299	High atmospheric oxygen level fosters giant arthropods. Spore-releasing plants dominate. Age of great lycophyte trees; vast coal forests form. Ears evolve in amphibians; penises evolve in early reptiles (vaginas evolve later, in mammals only).
		DEVONIAN		359 ◄	Major extinction event
					Land tetrapods appear. Explosion of plant diversity leads to tree forms, forests, and many new plant groups including lycophytes, ferns with complex leaves, seed plants.
		SILURIAN		416	Radiations of marine invertebrates. First appearances of land fungi, vascular plants, bony fish, and perhaps terrestrial animals (millipedes, spiders).
		ORDOVICIAN		443 ◄	Major extinction event
					Major period for first appearances. The first land plants, fish, and reef-forming corals appear. Gondwana moves toward the South Pole and becomes frigid.
		CAMBRIAN		488	Earth thaws. Explosion of animal diversity. Most major groups of animals appear (in the oceans). Trilobites and shelled organisms evolve.
PROTEROZOIC				542	Oxygen accumulates in atmosphere. Origin of aerobic metabolism. Origin of eukaryotic cells, then protists, fungi, plants, animals. Evidence that Earth mostly freezes over in a series of global ice ages between 750 and 600 mya.
ARCHAEAN AND EARLIER				2,500	3,800–2,500 mya. Origin of prokaryotes.
					4,600–3,800 mya. Origin of Earth's crust, first atmosphere, first seas. Chemical, molecular evolution leads to origin of life (from protocells to anaerobic prokaryotic cells).

A Transitions in sedimentary rock layers mark great time spans in Earth's history (not to the same scale). mya: millions of years ago. *Dates from International Commission on Stratigraphy data, 2007.*

B We can reconstruct some of the events in the history of life by studying rocky clues in the layers. Here, the *blue* triangles mark times of great mass extinctions. "First appearance" refers to appearance in the fossil record, not necessarily the first appearance on Earth; we often discover fossils that are significantly older than previously discovered specimens.

Figure 17.14 Animated The geologic time scale.

Kaibab Limestone

Toroweap Formation

Coconino Sandstone

Hermit Shale

Esplanade Sandstone

Wescogame Formation

Manakacha Formation

Watahomigi Formation

Redwall Limestone

Temple Butte Formation

Muav Limestone

Bright Angel Shale

Tapeats Sandstone*

Sixtymile Formation*

Chuar Group*

Nankoweap Formation*

Unkar Group*

Vishnu Basement Rocks*

C Sedimentary rock layers exposed by erosion in the Grand Canyon. Each layer has a characteristic composition and set of fossils (some are shown) that reflect events during its deposition. For example, Coconino Sandstone, which stretches from California to Montana, consists mainly of greatly weathered sand. Ripple marks and reptile tracks are the only fossils in it. Many think it is the remains of a vast sand desert, like the Sahara is today. *Layers not visible in this view of the Grand Canyon.

17.9 | Drifting Continents, Changing Seas

■ For billions of years, slow movements of Earth's outer layer and catastrophic events have changed the land, atmosphere, and oceans, with profound effects on the evolution of life.

When geologists first started to map vertical stacks of sedimentary rock, the theory of uniformity prevailed. The geologists knew that water, wind, fire, and other natural factors were continuously altering the surface of Earth. Eventually it became clear to them that such factors were part of a big picture of geologic change. Like life, Earth also changes dramatically. For instance, the Atlantic coasts of South America and Africa seemed to "fit" like jigsaw puzzle pieces. One model suggested that all continents were once part of a bigger supercontinent—**Pangea**—that had split into fragments and drifted apart. The model explained why the same types of fossils occur in sedimentary rock on both sides of the vast Atlantic Ocean.

At first, most scientists did not accept the model, which was called continental drift. To them, continents drifting about Earth seemed to be an outrageous idea, and no one knew what would drive such movement.

However, evidence that supported the model kept turning up. For instance, molten rock deep inside Earth wells up and solidifies on the surface. Some iron-rich minerals become magnetic as they solidify, and their magnetic poles align with Earth's poles when they do. If continents never moved, then all of these ancient rocky magnets would be aligned north-to-south, like compass needles. Indeed, the magnetic poles of rock formations on different continents are aligned—but not north-to-south. They point in many different directions. Either Earth's magnetic poles tilt off their north–south axis, or the continents wander.

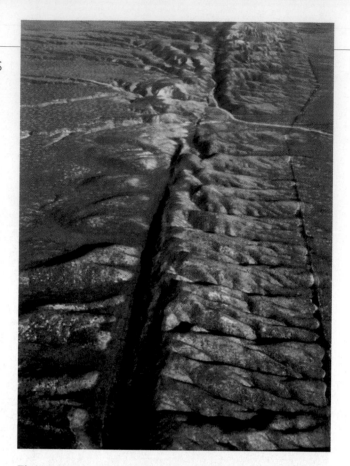

Figure 17.16 This aerial photo shows about 4.2 kilometers (2.6 miles) of the San Andreas Fault, which stretches a total of 1,300 km (800 miles) through California. The fault is the boundary between two tectonic plates sliding in opposite directions.

Then, deep-sea explorers discovered that ocean floors are not as static and featureless as had been assumed. Immense ridges stretch thousands of kilometers across

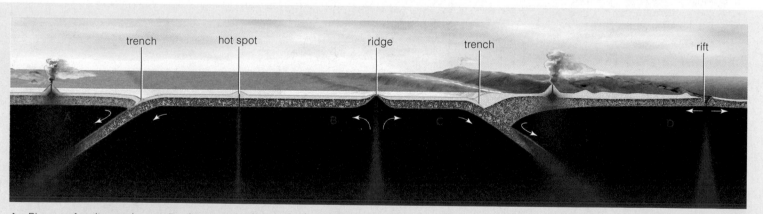

A Plumes of molten rock rupture a tectonic plate at what are called "hot spots." The Hawaiian Archipelago has been forming this way.

B At oceanic ridges, huge plumes of molten rock welling up from Earth's interior drive the movement of tectonic plates. New crust spreads laterally as it forms on the surface, forcing adjacent tectonic plates away from the ridge and into trenches elsewhere.

C At trenches, the advancing edge of one plate plows under an adjacent plate and buckles it. The Cascades, Andes, and other great coastal mountain ranges formed this way.

D At rifts, continents rupture in their interior as plates slide apart from each other.

Figure 17.15 Plate tectonics. Immense, rigid portions of Earth's outer layer of rock split, drift apart, and collide, all at a rate of less than 10 centimeters a year. As the plates move, they raft continents around the globe. The present configuration of Earth's tectonic plates is shown in Appendix VIII.

Figure 17.17 Animated A series of reconstructions of drifting continents. (**a**) The supercontinent Gondwana (*yellow*) had begun to break up by the Silurian. (**b**) The supercontinent Pangea formed during the Triassic, then (**c**) began to break up in the Jurassic. (**d**) K–T boundary. (**e**) The continents reached their modern configuration in the Miocene.

About 260 million years ago, seed ferns and other plants lived nowhere except on the area of Pangea that had once been Gondwana. So did mammal-like reptiles named therapsids. *Right*, fossilized leaf of one of the seed ferns (*Glossopteris*). *Far right*, a therapsid (*Lystrosaurus*) about 1 meter (3 feet) long. This tusked herbivore fed on fibrous plants in dry floodplains.

A 420 mya	B 237 mya	C 152 mya	D 65.5 mya	E 14 mya

the sea floor. Molten rock spewing from the ridges pushes old sea floor outward in both directions, then cools and hardens into new sea floor. Elsewhere, older sea floor plunges into deep trenches.

Such discoveries swayed the skeptics. Finally, there was a plausible mechanism for continental drift, which by then was named the theory of **plate tectonics**. By this theory, Earth's relatively thin outer layer of rock is cracked into immense plates, like a gigantic cracked eggshell. The plates grow from ridges and sink into trenches (Figure 17.15). As they do, they move like colossal conveyer belts, rafting continents on top of them to new locations. The movement is no more than about 10 centimeters (4 inches) a year, but that is enough to carry a continent all the way around the world after 40 million years or so. Evidence of the movement is all around us, in various geological features of our landscapes (Figure 17.16).

Researchers soon applied the plate tectonics theory to some long-standing puzzles. For example, fossils of a seed fern, *Glossopteris*, and of an early reptile, *Lystrosaurus*, occur in similar geologic formations in Africa, India, South America, and Australia (Figure 17.17). Why did these organisms occur on many continents, across vast expanses of oceans? The seeds of *Glossopteris* were too heavy to float or to be wind-blown over an ocean, and *Lystrosaurus* was too stocky to swim between continents. Researchers suspected that they both evolved on a supercontinent that

was even older than Pangea. This supercontinent, which they named **Gondwana**, must have existed about 300 million years ago. The researchers made this prediction: If Antarctica had once been part of Gondwana, then it must have the same geologic formations as well as *Glossopteris* and *Lystrosaurus* fossils.

At the time, Antarctica was mostly unexplored. Later expeditions found the formations and the fossils, in support of the prediction and the plate tectonics theory. Many modern species, including the ratite birds in Figure 17.2a–c, live only in places that were once part of Gondwana.

We now know that Gondwana drifted south, across the South Pole, then north until it merged with other land masses to form Pangea. We know that continents are always on the move. They collide, split into new continents, then collide all over again. Earth's outer layer of rock solidified 4.55 billion years ago. At least five times since then, a single supercontinent formed, with one ocean lapping at its coastline. All the while, the erosive forces of water and wind resculpted the land. So did the impacts and aftermaths of asteroids.

Such changes on land and in the ocean and atmosphere influenced life's evolution. Imagine early life in shallow, warm waters along continents. Shorelines vanished as continents collided and wiped out many lineages. Even as old habitats disappeared, new ones opened up for survivors—and evolution took off in new directions.

The K–T boundary layer consists of an unusual clay that formed 65 million years ago, worldwide (*right*). The clay is rich in iridium, an element rare on Earth's surface but common in asteroids. After finding the iridium, researchers looked for evidence of an asteroid big enough to cover the entire Earth with its debris. They found a crater that is about 65 million years old, buried under sediments off of the coast of Mexico's Yucatán Peninsula. It is so big—273.6 kilometers (170 miles) across and 1 kilometer (3,000 feet) deep—that no one had even noticed it before. This crater is evidence of an asteroid impact 40 million times more powerful than the one that

made the Barringer Crater—certainly big enough to have influenced life on Earth in a big way. Research since then estimates the size of the asteroid that made this impact was between 10 and 20 kilometers (6 to 12 miles) in diameter.

Summary

Section 17.1 By the 19th century, **naturalists** on globe-spanning survey expeditions were bringing back increasingly detailed observations of the natural world. **Fossils** were evidence of life in the distant past. Sudies of **biogeography** and **comparative morphology** led to new ways of thinking about the natural world.

■ *Use the animations on CengageNOW to learn more about fossil formation.*

Section 17.2 Prevailing belief systems may influence interpretation of the underlying cause of a natural event. Nineteenth-century naturalists tried to reconcile their traditional beliefs with physical evidence of **evolution**. **Catastrophism** and the **theory of uniformity** were two theories that surfaced during this time.

■ *Read the InfoTrac article "Typecasting a Bit Part," Stephen J. Gould, The Sciences, March 2000.*

Sections 17.3, 17.4 The observations of naturalist Alfred Wallace and Charles Darwin led to a theory of how species evolve. Here are the theory's main premises:

A population tends to grow until it begins to exhaust the resources of its environment. Individuals must then compete for food, shelter from predators, and so on.

Individuals with forms of traits that make them more competitive tend to produce more offspring.

Adaptive traits (adaptations) that impart greater **fitness** to an individual become more common in a population over generations, compared with less competitive forms.

Differential survival and reproduction of individuals of a population that vary in the details of shared traits is called **natural selection**. It is one of the processes that drives evolution.

■ *Read the InfoTrac article "What Darwin's Finches Can Teach Us About the Evolutionary Origin and Regulation of Biodiversity," B. Rosemary Grant and Peter Grant, Bioscience, March 2003.*

Section 17.5 Many fossils are found in stacked layers of sedimentary rock. Younger fossils usually occur in more recently deposited layers, on top of older fossils in older layers. Some **lineages** are represented as fossil series in sequential layers. Fossils are relatively scarce, so the fossil record will always be incomplete. Even so, it reveals much about life in the ancient past.

■ *Use the animations on CengageNOW to learn more about fossil formation.*

Section 17.6 The characteristic **half-life** of a radioisotope allows us to determine the age of rocks and fossils using a technique called **radiometric dating**.

■ *Use the animated interaction on CengageNOW to learn more about half-life.*

Section 17.7 Gaps in the fossil record are often filled by discoveries of new fossils. Such fossils add detail to our understanding of evolutionary history.

Section 17.8 Transitions in the fossil record became boundaries for great intervals of the **geologic time scale**. The scale is correlated with evolutionary events, and it includes dates obtained by radiometric dating.

■ *Use the animated interaction on CengageNOW to investigate the geologic time scale.*

Section 17.9 The discovery of the global distribution of land masses and fossils, magnetic rocks, and seafloor spreading from midoceanic ridges led to the theory of **plate tectonics**. By this theory, the movements of Earth's tectonic plates raft land masses to new positions. Several times in Earth's history, all land masses have converged as supercontinents. **Gondwana** and **Pangea** are examples. Such movements had profound impacts on evolution.

■ *Use the interaction on CengageNOW to learn more about drifting continents.*

Self-Quiz
Answers in Appendix III

1. Biogeographers study _____ .
   a. continental drift
   b. patterns in the world distribution of species
   c. mainland and island biodiversity
   d. both b and c are correct
   e. all are correct

Data Analysis Exercise

In the late 1970s, geologist Walter Alvarez was investigating the composition of the 1-centimeter-thick layer of clay that marks the K–T boundary all over the world. He asked his father, Nobel Prize–winning physicist Luis Alvarez, to help him analyze the elemental composition of the layer. The Alvarezes and their colleagues, chemists Frank Asaro and Helen Michel, tested the layer in Italy and in Denmark. The researchers discovered that the K–T boundary layer had a much higher iridium content than the surrounding rock layers (Figure 17.18).

Iridium belongs to a group of elements (Appendix IV) that are much more abundant in asteroids and other solar system materials than they are in Earth's crust. The Alvarez group concluded that the K–T boundary layer must have originated with extraterrestrial material. They calculated that an asteroid 14 kilometers (8.7 miles) in diameter would contain enough iridium to account for the iridium in the K–T boundary layer.

1. What was the iridium content of the K–T boundary layer?

2. How much higher was the iridium content of the boundary layer than the sample taken 0.7 meters above the layer?

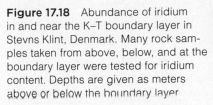

Sample Depth	Average Abundance of Iridium (ppb)
+ 2.7 m	< 0.3
+ 1.2 m	< 0.3
+ 0.7 m	0.36
boundary layer	41.6
– 0.5 m	0.25
– 5.4 m	0.30

Figure 17.18 Abundance of iridium in and near the K–T boundary layer in Stevns Klint, Denmark. Many rock samples taken from above, below, and at the boundary layer were tested for iridium content. Depths are given as meters above or below the boundary layer.

The iridium content of an average Earth rock contains 0.4 parts per billion (ppb) of iridium. An average meteorite contains about 550 parts per billion of iridium. Photo (above *right*): Luis and Walter Alvarez with a section of the iridium layer.

---

2. The bones of a bird's wing are similar to the bones in a bat's wing. This observation is an example of _____ .
   a. uniformity
   b. evolution
   c. comparative morphology
   d. a lineage

3. The number of species on an island depends on the size of the island and its distance from a mainland. This statement would most likely be made by _____ .
   a. an explorer
   b. a biogeographer
   c. a geologist
   d. a philosopher

4. Evolution is _____ .
   a. natural selection
   b. heritable change in a line of descent
   c. driven by natural selection
   d. b and c are correct

5. If the half-life of a radioisotope is 20,000 years, then a sample in which three-quarters of that radioisotope has decayed is _____ years old.
   a. 15,000
   b. 26,667
   c. 30,000
   d. 40,000

6. _____ has/have influenced the fossil record.
   a. Sedimentation and compaction
   b. Tectonic plate movements
   c. Radioisotope decay
   d. All of the above

7. Evidence suggests that life originated in the _____ .
   a. Archaean
   b. Proterozoic
   c. Phanerozoic
   d. Cambrian

8. The Cretaceous ended _____ million years ago.

9. Forces of geologic change include _____ (select all that are correct).
   a. erosion
   b. fossilization
   c. volcanoes
   d. evolution
   e. tectonic plate movements
   f. climate change
   g. asteroid impacts
   h. hot spots

10. Did Pangea or Gondwana form first?

11. Match the terms with the most suitable description.
   ___ fitness
   ___ fossils
   ___ natural selection
   ___ half-life
   ___ catastrophism
   ___ uniformity
   ___ sedimentary rock

   a. measured by reproductive success
   b. geologic change occurs continuously
   c. geologic change occurs in sudden major events
   d. good for finding fossils
   e. survival of the fittest
   f. characteristic of a radioisotope
   g. evidence of life in distant past

■ *Visit CengageNOW for additional questions.*

Critical Thinking

1. If you think of geologic time spans as minutes, life's history might be plotted on a clock such as the one shown on the *right*. According to this clock, the most recent epoch started after the last 0.1 second before noon. Where does that put you?

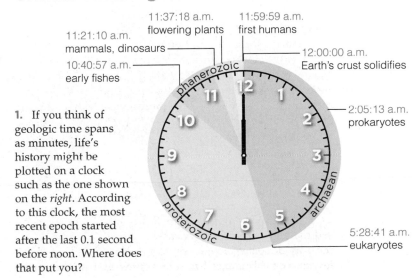

11:37:18 a.m. flowering plants
11:59:59 a.m. first humans
11:21:10 a.m. mammals, dinosaurs
10:40:57 a.m. early fishes
12:00:00 a.m. Earth's crust solidifies
2:05:13 a.m. prokaryotes
5:28:41 a.m. eukaryotes

IMPACTS, ISSUES Rise of the Super Rats

Slipping in and out of the pages of human history are rats—*Rattus*—the most notorious of mammalian pests. Rats thrive in urban centers, where garbage is plentiful and natural predators are not. The average U.S. city sustains about one rat for every ten people. Part of their success stems from an ability to reproduce very quickly; rat populations can expand within weeks to match the amount of garbage available for them to eat. Unfortunately for us, rats carry pathogens and parasites that cause bubonic plague, typhus, and other infectious diseases. They chew their way through walls and wires, and eat or foul 20 to 30 percent of our total food production (Figure 18.1). Rats cost us about $19 billion per year.

For years, people have been fighting back with dogs, traps, ratproof storage facilities, and poisons that include arsenic and cyanide. During the 1950s, they used baits laced with warfarin, a synthetic organic compound that interferes with blood clotting. Rats that ate the poisoned baits died within days after bleeding internally or losing blood through cuts or scrapes.

Warfarin was extremely effective, and compared to other rat poisons, it had a lot less impact on harmless species. It quickly became the rodenticide of choice. In 1958, however, a Scottish researcher reported that warfarin did not work against some rats. Similar reports from other European countries followed. About twenty years later, about 10 percent of rats caught in urban areas of the United States were resistant to warfarin. What happened?

To find out, researchers compared warfarin-resistant rats with still-vulnerable rats. They traced the difference to a gene on one of the rat chromosomes. Certain mutations in the gene were common among warfarin-resistant rat populations but rare among vulnerable ones. Warfarin binds to the gene's product—an enzyme that takes part in vitamin K–dependent synthesis of blood clotting factors. The mutations made the enzyme insensitive to warfarin.

"What happened" was evolution by natural selection. As warfarin exerted pressure on rat populations, the populations changed. The previously rare mutations became adaptive. Rats that had an unmutated gene died after eating warfarin. The lucky ones that had one of the mutations survived and passed it to their offspring. The rat populations recovered quickly, and a higher proportion of rats in the next generation carried the mutations. With each onslaught of warfarin, the frequency of the mutation in rat populations increased.

Selection pressures can and often do change. When warfarin resistance increased in rat populations, people stopped using warfarin. The frequency of the mutation in rat populations declined—probably because rats with the mutation are mildly vitamin K–deficient, so they are not as healthy as normal rats. Now, savvy exterminators in urban areas know that the best way to control a rat infestation is to exert another kind of selection pressure: Remove their source of food, which is usually garbage. Then the rats will eat each other.

See the video! Figure 18.1 *Left,* rats infesting 80,000 hectares (200,000 acres) of rice fields in the Philippine Islands ruin more than 20 percent of the crops. Rice is the main food for people in Southeast Asia. *Right,* rats thrive wherever people do. Dousing buildings and soil with rat poisons does not usually exterminate their populations, which recover quickly. It selects for rats that are resistant to the poisons.

Key Concepts

Populations evolve

Individuals of a population differ in which alleles they inherit, and so they differ in phenotype. Over generations, any allele may increase or decrease in frequency in a population. Such change is called micro-evolution. **Sections 18.1, 18.2**

Patterns of natural selection

Natural selection drives microevolution. Depending on the population and its environment, natural selection can shift or maintain the range of variation in heritable traits. **Sections 18.3–18.6**

Other microevolutionary processes

With genetic drift, change can occur in a line of descent by chance alone. Gene flow counters the evolutionary effects of mutation, natural selection, and genetic drift. **Sections 18.7, 18.8**

How species arise

Speciation varies in its details, but it typically starts after gene flow ends. Microevolutionary events that occur independently lead to genetic divergences, which are reinforced as reproductive isolation mechanisms evolve. **Sections 18.9–18.11**

Macroevolutionary patterns

Patterns of genetic change that involve more than one species are called macroevolution. Recurring patterns of macroevolution include exaptation, adaptive radiation, and extinction. **Section 18.12**

Links to Earlier Concepts

- This chapter builds on evidence for evolution that was introduced in the previous chapter (Sections 17.8, 17.9).

- Before starting, you may wish to review the premises of the theory of natural selection (17.3), the basic principles and terms of genetics (10.1, 11.1), and some of the cellular mechanisms of reproduction (4.13, 9.3, 10.3, 10.5).

- As you learn about the processes that drive evolution and their effects, you will be drawing upon your knowledge of the genetic basis of traits (10.4, 11.4, 11.6, 11.7) and the effects of genetic changes (3.6, 12.6, 12.7, 14.5, 15.3, 16.7).

- Your knowledge of sampling error (1.8) and probability (11.2) will help you understand the implications of experiments (1.7) that demonstrate evolution in action.

- Evolution starts with mutations in individuals.
- Mutation is the source of new alleles.
- Sexual reproduction can quickly spread a mutation through a population.

- Links to Alleles 10.1; Terms in modern genetics 11.1, Variations on simple dominance 11.4, Genes and the environment 11.6, Complex variation in traits 11.7, Mutation 14.5

Figure 18.2 Sampling phenotypic variation in (**a**) humans, and (**b**) a type of snail found on islands in the Caribbean. The variation in shared traits among individuals is an outcome of variations in alleles that influence those traits.

Variation in Populations

All individuals of a species share certain features. For example, giraffes have very long necks, brown spots on white coats, and so on. These are examples of morphological traits (*morpho–*, form). Individuals of a species also share physiological traits, such as metabolic activities. They also respond the same way to certain stimuli, as when hungry giraffes feed on tree leaves. These are behavioral traits.

However, individuals of a population vary in the details of their shared traits. A **population** is a group of individuals of the same species in a specified area. Think about the variations in the color and patterning of dog or cat fur. Figure 18.2*a* hints at the range of variations in human skin and eye color, and distribution, color, texture, and amount of hair. Almost every trait of any species may vary, and the variation can be quite dramatic (Figure 18.2*b*).

Many traits show qualitative differences; they have two or more different forms, or morphs, such as the purple or white pea plant flowers that Gregor Mendel studied. In addition, for many traits, individuals of a population show quantitative differences, or a range of incrementally small variations in a trait (Section 11.7).

The Gene Pool

Genes encode heritable information about traits. The individuals of a population inherit the same number and kind of genes (except for genes on nonidentical sex chromosomes). Together, the genes of a population comprise a **gene pool**—a pool of genetic resources.

In a sexually reproducing population, most genes in the pool have slightly different forms called **alleles** (Section 10.1). An individual carries two copies of each autosomal gene, and those copies may or may not be identical (Section 11.1). An individual's complement of alleles is its genotype. Alleles are the main source of variation in phenotype—the observable characteristics of an individual. For example, the color of your eyes is determined by the alleles you carry.

Some traits have only two distinct forms, such as male and female sexes. Having two forms is called a dimorphism (*di–*, two). Polymorphism (*poly–*, many) occurs when a gene has three or more alleles that persist in a population at high frequency (greater than 1 percent). The ABO alleles that determine human blood type are one example (Section 11.4).

You learned about patterns of inheritance in earlier chapters. Table 18.1 summarizes the key events involved. Mutation is the source of new alleles. Other events shuffle existing alleles into different combinations, but what a shuffle that is! There are $10^{116,446,000}$ possible combinations of human alleles. Not even 10^{10} people are living today. Unless you have an identical twin, it is unlikely that another person with your precise genetic makeup has ever lived or ever will.

One other point about the nature of the gene pool: Offspring inherit a genotype, not a phenotype. Section 11.6 described how environmental pressures can bring about variation in the range of phenotypes, but such effects last no longer than the individual.

Table 18.1 Genetic Events in Inheritance

Genetic Event	Effect
Mutation	Source of new alleles
Crossing over at meiosis I	Introduces new combinations of alleles into chromosomes
Independent assortment at meiosis I	Mixes maternal and paternal chromosomes
Fertilization	Combines alleles from two parents
Changes in chromosome number or structure	Transposition, duplication, or loss of chromosomes

Mutation Revisited

Being the original source of new alleles, mutations are worth another look—this time within the context of their impact on populations. We cannot predict when or in which individual a particular gene will mutate. We can, however, predict the average mutation rate of a species, which is the probability that a mutation will occur in a given interval. In humans, that rate is about 175 mutations per person per generation.

Many mutations give rise to structural, functional, or behavioral alterations that reduce an individual's chances of surviving and reproducing. Even one biochemical change may be devastating. For instance, the skin, bones, tendons, lungs, blood vessels, and other vertebrate organs incorporate the protein collagen. If the collagen gene mutates in a way that changes the protein's function, the entire body may be affected. A mutation that drastically changes phenotype is called a **lethal mutation** because it usually results in death.

A **neutral mutation** changes the base sequence in DNA, but the alteration has no effect on survival or reproduction. It neither helps nor hurts the individual. For instance, if you carry a mutation that keeps your earlobes attached to your head instead of swinging freely, attached earlobes should not in itself stop you from surviving and reproducing as well as anybody else. So, natural selection does not affect the frequency of this particular mutation in a population.

Occasionally, a change in the environment favors a previously neutral mutation, or a new beneficial mutation arises by chance. For example, a mutation that affects growth might make a corn plant grow larger or faster, and thereby give it better access to sunlight and nutrients. Even if a beneficial mutation bestows only a slight advantage, natural selection can increase its frequency in a population over time.

Mutations have been altering genomes for billions of years. Cumulatively, they have given rise to Earth's staggering biodiversity. Think about it. The reason you do not look like a bacterium or avocado or earthworm or even your neighbor who lives next door began with mutations that occurred in different lines of descent.

Stability and Change in Allele Frequencies

Researchers typically track **allele frequencies**: the relative abundances of alleles of a given gene among all individuals of a population. They start from a theoretical reference point, **genetic equilibrium**, which occurs when a population is not evolving with respect to the gene. Genetic equilibrium can only occur if every one of these five conditions are met: Mutations never happen; the population is infinitely large; the population stays isolated from all other populations of the same species; all individuals mate at random; and all individuals of the population survive and produce exactly the same number of offspring.

As you can imagine, all five conditions are never met in nature; thus, natural populations are never in equilibrium. **Microevolution**, or small-scale change in allele frequencies, is continually occurring in natural populations because processes that drive it are always in play. This chapter explores microevolutionary processes—mutation, natural selection, genetic drift, and gene flow—and their effects.

■ Researchers know whether or not a population is evolving by tracking deviations from a baseline of genetic equilibrium.

The Hardy–Weinberg Formula Early in the twentieth century, Godfrey Hardy (a mathematician) and Wilhelm Weinberg (a physician) independently applied the rules of probability to sexually reproducing populations. They perceived that gene pools can remain stable only when five conditions are being met:

1. Mutations do not occur.

2. The population is infinitely large.

3. The population is isolated from all other populations of the species (no gene flow).

4. Mating is random.

5. All individuals survive and produce the same number of offspring.

These conditions never occur all at once in nature. Thus, allele frequencies for any gene in the shared pool always change. However, we can think about a hypothetical situation in which the five conditions are being met and a population is not evolving.

Hardy and Weinberg developed a simple formula that can be used to track whether a population of any sexually reproducing species is in a state of genetic equilibrium. Consider tracking a hypothetical pair of alleles that affect butterfly wing color. A protein pigment is specified by dominant allele *A*. If a butterfly inherits two *AA* alleles, it will have dark-blue wings. If it inherits two recessive alleles (*aa*), it will have white wings. If it inherits one of each (*Aa*), the wings will be medium-blue (Figure 18.3).

At genetic equilibrium, the proportions of the wing-color genotypes are

$$p^2(AA) + 2pq(Aa) + q^2(aa) = 1.0$$

where *p* and *q* are the frequencies of alleles *A* and *a*. This equation became known as the Hardy–Weinberg equilibrium equation. It defines the frequency of a dominant allele (*A*) and a recessive allele (*a*) for a gene that controls a particular trait in a population.

The frequencies of *A* and *a* must add up to 1.0. To give a specific example, if *A* occupies half of all the loci for this gene in the population, then *a* must occupy the other half (0.5 + 0.5 = 1.0). If *A* occupies 90 percent of all the loci, then *a* must occupy 10 percent (0.9 + 0.1 = 1.0). No matter what the proportions,

$$p + q = 1.0$$

At meiosis, remember, paired alleles are assorted into different gametes. The proportion of gametes with the *A* allele is *p*, and the proportion with the *a* allele is *q*. The Punnett square on the next page shows the genotypes possible in the next generation (*AA*, *Aa*, and *aa*). Note that the frequencies of the three genotypes add up to 1.0:

$$p^2 + 2pq + q^2 = 1.0$$

490 *AA* butterflies
dark-blue wings

490 *AA* butterflies
dark-blue wings

490 *AA* butterflies
dark-blue wings

420 *Aa* butterflies
medium-blue wings

420 *Aa* butterflies
medium-blue wings

420 *Aa* butterflies
medium-blue wings

90 *aa* butterflies
white wings

90 *aa* butterflies
white wings

90 *aa* butterflies
white wings

Starting Population   **Next Generation**   **Next Generation**

Figure 18.3 Animated Finding out whether a population is evolving. The frequencies of wing-color alleles among all of the individuals in this hypothetical population of morpho butterflies are not changing; thus, the population is not evolving.

	p (A)	q (a)
p (A)	AA (p^2)	Aa (pq)
q (a)	Aa (pq)	aa (q^2)

Suppose that the population has 1,000 individuals and that each one produces two gametes:

490 *AA* individuals make 980 *A* gametes
420 *Aa* individuals make 420 *A* and 420 *a* gametes
90 *aa* individuals make 180 *a* gametes

The frequency of alleles *A* and *a* among 2,000 gametes is

$$A = \frac{980 + 420}{2,000 \text{ alleles}} = \frac{1,400}{2,000} = 0.7 = p$$

$$a = \frac{180 + 420}{2,000 \text{ alleles}} = \frac{600}{2,000} = 0.3 = q$$

At fertilization, gametes combine at random and start a new generation. If the population size stays constant at 1,000, there will be 490 *AA*, 420 *Aa*, and 90 *aa* individuals. The frequencies of the alleles for dark-blue, medium-blue, and white wings are the same as they were in the original gametes. Thus, dark-blue, medium-blue, and white wings occur at the same frequencies in the new generation.

As long as the assumptions that Hardy and Weinberg identified continue to hold, the pattern persists. If traits show up in different proportions from one generation to the next, though, one or more of the five assumptions is not being met. The hunt can begin for the evolutionary forces driving the change.

Applying the Rule How does the Hardy–Weinberg formula work in the real world? Researchers can use it to estimate the frequency of carriers of alleles that cause genetic traits and disorders.

As an example, hereditary hemochromatosis (HH) is the most common genetic disorder among people of Irish ancestry. Affected individuals absorb too much iron from food. The symptoms of this autosomal recessive disorder include liver problems, fatigue, and arthritis. A study in Ireland found the frequency for one allele that causes HH to be 0.14. If $q = 0.14$, then p is 0.86. Based on this study, the carrier frequency ($2pq$) can be calculated to be about 0.24. Such information is useful to doctors and to public health professionals.

Another example: A mutation in the *BRCA2* gene has been linked to breast cancer in adults. A deviation from the birth frequencies predicted by the Hardy–Weinberg formula suggests that this mutation can also have effects even before birth. In one study, researchers looked at the mutation's frequency among newborn girls. They found fewer homozygotes than expected, based on the number of heterozygotes and the Hardy–Weinberg formula. Thus, it seems that in homozygous form the mutation impairs the survival of female embryos.

18.3 Natural Selection Revisited

■ Natural selection is the most influential process of evolution.

■ Link to Natural selection theory 17.3

The remainder of this chapter explores the mechanisms and effects of processes that drive evolution, including natural selection. **Natural selection** is differential survival and reproduction among individuals of a population that vary in the details of their shared traits (Section 17.3). It influences the frequency of alleles in a population by operating on phenotypes that have a genetic basis.

We observe different patterns of natural selection, depending on the selection pressures and organisms involved. Sometimes, individuals with a trait at one extreme of a range of variation are selected against, and those at the other extreme are favored. We call this pattern directional selection. With stabilizing selection, midrange forms are favored, and the extremes are selected against. With disruptive selection, forms at the extremes of the range of variation are favored; intermediate forms are selected against. We will discuss these three modes of natural selection, which Figure 18.4 summarizes, in the following sections.

Section 18.6 explores sexual selection, a mode of natural selection that operates on a population by influencing mating success. This section also discusses balanced polymorphism, a particular case of natural selection in which heterozygous individuals have greater fitness in a certain environment than homozygous individuals.

Natural selection and other processes of evolution can alter a population or species so much that it becomes a new species. We discuss mechanisms of speciation in later sections of this chapter.

Remember, even though we can recognize patterns of evolution, none of them are purposeful. Evolution simply fills the nooks and crannies of opportunity.

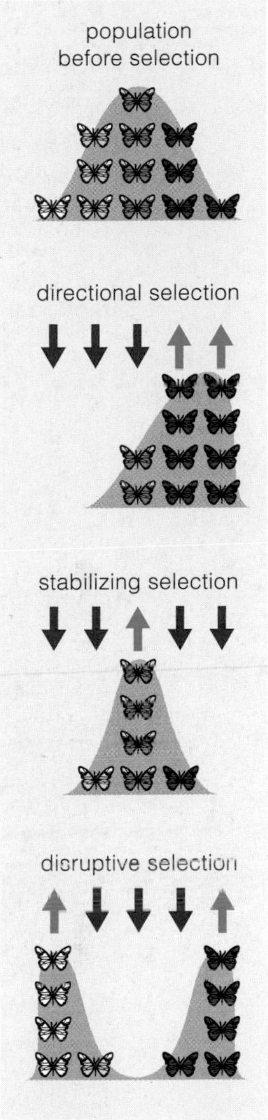

Figure 18.4 Three modes of natural selection. *Red* arrows indicate which forms of a trait are being selected against; *green*, forms that are being favored.

Take-Home Message

How does evolution occur?

■ Natural selection, one of the driving forces of evolution, occurs in recognizable patterns depending on the organisms and their environment.

■ Evolution is an opportunistic process.

■ Changing environmental conditions can result in a directional shift in allele frequencies.

■ Link to Experimental design 1.7

With **directional selection**, allele frequencies shift in a consistent direction, so forms at one end of a range of phenotypic variation become more common over time (Figure 18.5). The following examples show how field observations provide evidence of directional selection.

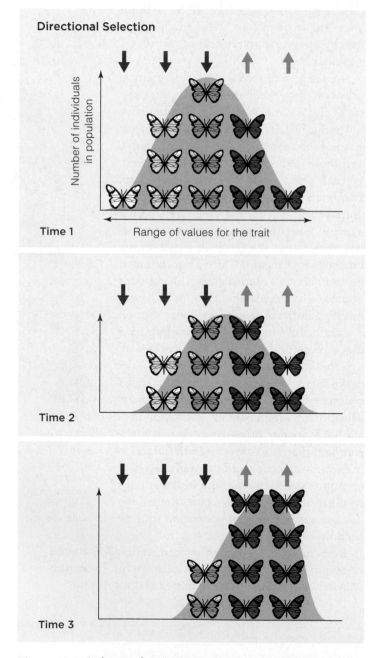

Figure 18.5 Animated Directional selection. These bell-shaped curves signify a range of continuous variation in a butterfly wing-color trait. *Red* arrows indicate which forms are being selected against; *green*, forms that are being favored.

Effects of Predation

The Peppered Moth Peppered moths (*Biston betularia*) feed and mate at night, and rest motionless on trees during the day. Their behavior and coloration camouflage them from day-flying, moth-eating birds.

Light-colored moths were the most common form in preindustrial England. A dominant allele that resulted in the dark color was rare. The air was clean, and light-gray lichens grew on the trunks and branches of most trees. Light moths were camouflaged when they rested on the lichens, but dark moths were not (Figure 18.6a).

By the 1850s, the dark moths were appearing more frequently. Why? The industrial revolution had begun, and smoke from coal-burning factories was beginning to change the environment. Air pollution was killing the lichens. Researchers hypothesized that dark moths were now better camouflaged from predators on the soot-darkened trees than light moths (Figure 18.6b).

In the 1950s, H. B. Kettlewell used a mark–release–recapture method to test the hypothesis. He bred both moth forms in captivity and marked hundreds so that they could be easily identified after being released in the wild. He released them near highly industrialized areas around Birmingham and near an unpolluted part of Dorset. His team recaptured more of the dark moths in the polluted area and more light ones near Dorset. They also observed predatory birds eating more light moths in Birmingham, and more dark moths in Dorset.

	Near Birmingham (pollution high)	Near Dorset (pollution low)
Light-Gray Moths		
Released	64	393
Recaptured	16 (25%)	54 (13.7%)
Dark-Gray Moths		
Released	154	406
Recaptured	82 (53%)	19 (4.7%)

Pollution controls went into effect in 1952. Tree trunks became free of soot, and lichens made a comeback. Moth phenotypes shifted in the reverse direction: Wherever pollution decreased, the frequency of dark moths decreased as well. Many other researchers since Kettlewell have confirmed the rise and fall of the dark-colored form of the peppered moth.

Pocket Mice Directional selection has been working among populations of rock pocket mice (*Chaetodipus intermedius*) in Arizona's Sonoran Desert. Rock pocket mice are small mammals that spend the day sleeping in underground burrows. They emerge to forage

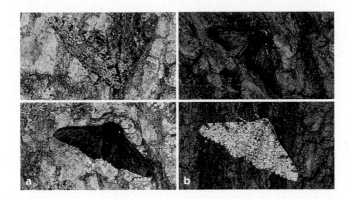

Figure 18.6 Natural selection of two forms of the same trait, body surface coloration, in two settings. (**a**) Light moths (*Biston betularia*) on a nonsooty tree trunk are hidden from predators. Dark ones stand out. (**b**) The dark color is more adaptive in places where soot darkens tree trunks.

Figure 18.7
Visible evidence of directional selection in populations of rock pocket mice.

(**a**) Rock pocket mice that have dark fur are more common in these areas of dark basalt rock.

(**b,c**) The two color types of rock pocket mice, each posed on the dark and light rocks of the area.

for seeds at night. The mice differ in coat color: Some are light brown; others are dark gray. Of about eighty genes that affect coat color in this species, one gene is responsible for the difference. One allele of the gene results in light fur; the other, dark.

The Sonoran Desert is dominated by outcroppings of light brown granite. There are also patches of dark basalt, the remains of ancient lava flows (Figure 18.7a). Most of the mice in populations that inhabit the light brown granite have light brown coats (Figure 18.7b). Most of the mice in populations that inhabit the dark rock have dark gray coats (Figure 18.7c).

Why? In each habitat, the individuals that match the rock color are camouflaged from their natural predators. Night-flying owls more easily see mice that do not match the rocks, so they preferentially eliminate nonmatching mice from each population. The owls are selective agents that directionally shift the frequency of coat color alleles in rock pocket mice populations.

Resistance to Antibiotics

Our attempts to control the environment can result in directional selection, as is the case with the warfarin-resistant rats. The use of antibiotics is another example. Antibiotics kill bacteria. Streptomycins, for example, block protein synthesis in certain bacteria. Penicillins disrupt formation of covalent bonds between the glycoproteins in bacterial cell walls. Cell walls that form in the presence of penicillin are weak, and they rupture.

When your grandparents were still young, scarlet fever, tuberculosis, and pneumonia caused one-fourth

of the annual deaths in the United States alone. Since the 1940s, we have been relying on antibiotics to fight these and other dangerous bacterial diseases. We also use them in other, less dire circumstances. Antibiotics are used preventively, both in humans and in livestock. They are part of the daily rations of millions of cattle, pigs, chickens, fish, and other animals that are raised on factory farms.

Bacteria evolve at an accelerated rate compared with humans, in part because they reproduce very quickly. For example, the common intestinal bacteria *E. coli* can divide every 17 minutes. Each new generation is an opportunity for mutation, so the gene pool of a bacterial population varies greatly. Thus, it is very likely that some cells will survive an antibiotic treatment. Then, natural selection takes over. A typical two-week course of antibiotics can potentially exert selection pressure on over a thousand generations of bacteria, and antibiotic resistant strains may be the outcome.

Antibiotic-resistant bacteria have plagued hospitals for many years, and now they are frequently found in schools. Even as researchers scramble to find new antibiotics, this trend is bad news for millions of people who contract cholera, tuberculosis, or another dangerous bacterial disease every year.

Take-Home Message

What is the effect of directional selection?

■ With directional selection, allele frequencies underlying a range of variation shift in a consistent direction in response to selection pressure.

Selection Against or in Favor of Extreme Phenotypes

■ Stabilizing selection is a form of natural selection that maintains an intermediate phenotype.

■ Disruptive selection is a form of natural selection that favors extreme forms of a trait.

Natural selection can bring about a directional shift in a population's range of phenotypes. Depending on the environment and the organisms involved, the process may also favor a midrange form of a trait, or it may eliminate the midrange form and favor extremes.

Stabilizing Selection

With **stabilizing selection**, an intermediate form of a trait is favored, and extreme forms are not. This mode of selection tends to preserve the midrange phenotypes in a population (Figure 18.8). For example, the body weight of sociable weavers (*Philetairus socius*) is subject to stabilizing selection (Figure 18.9). Weaver birds cooperate to build large communal nests in areas of the African savanna. Between 1993 and 2000, Rita Covas and her colleagues captured, tagged, weighed, and released birds living in communal nests before the breeding season began. The researchers then recaptured and weighed the surviving birds after the breeding season was over.

Covas's field studies indicated that body weight in sociable weavers is a trade-off between the risks of starvation and predation (Figure 18.9). Foraging is not easy in the sparse habitat of an African savanna, and

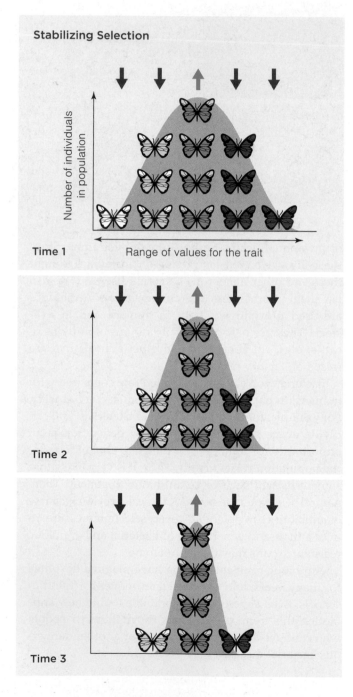

Figure 18.8 Animated Stabilizing selection eliminates extreme forms of a trait, and maintains the predominance of an intermediate phenotype in a population. *Red* arrows indicate which forms are being selected against; *green*, forms that are being favored. Compare the data set from a field experiment shown in Figure 18.9.

Figure 18.9 Stabilizing selection in sociable weavers. Graph shows the number of birds (out of 977) that survived a breeding season. **Figure It Out:** What is the optimal weight of a sociable weaver bird?

Answer: About 29 grams

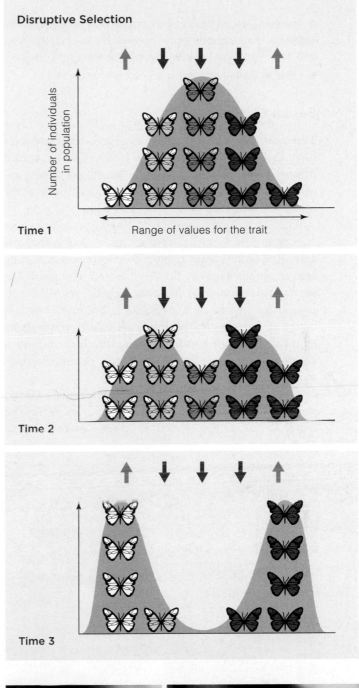

Figure 18.10 **Animated** Disruptive selection eliminates midrange forms of a trait, and maintains extreme forms.

Disruptive Selection

Time 1 — Range of values for the trait / Number of individuals in population

Time 2

Time 3

leaner birds do not store enough fat to avoid starvation. A meager food supply selects against birds with low body weight. Fatter birds may be more attractive to predators, and not as agile when escaping. Predators select against birds of high body weight. Thus, birds of intermediate weight have the selective advantage, and make up the bulk of sociable weaver populations.

Disruptive Selection

With **disruptive selection**, forms of a trait at both ends of a range of variation are favored, and intermediate forms are selected against (Figure 18.11). Consider the black-bellied seedcracker (*Pyrenestes ostrinus*) native to Cameroon, Africa. In these finches, there is a genetic basis for bills of a particular size. Females and males both have large or small bills—but no sizes between. Both forms occur in the same geographic range, and the birds breed randomly with respect to bill size. It is as if every human adult were four feet or six feet tall, with no one between.

Factors that affect *Pyrenestes* feeding performance maintain the dimorphism in bill size. The finches feed mainly on the seeds of two types of sedge, which is a grasslike plant. One sedge produces hard seeds; the other, soft seeds. Small-billed birds are better at opening the soft seeds, but large-billed birds are better at cracking the hard ones. All seeds are abundant during Cameroon's wet seasons, and all seedcrackers feed on both types of seeds. However, sedge seeds are scarce during the region's dry seasons. At those times, each bird focuses on eating the seeds that it opens most efficiently. Small-billed birds feed mainly on soft seeds, and large-billed birds feed mainly on hard seeds. Birds with intermediate-sized bills cannot open either type of seed as efficiently as the other birds, so they are less likely to survive the dry seasons.

Take-Home Message

What types of natural selection favor intermediate or extreme forms of traits?

■ With stabilizing selection, an intermediate phenotype is favored, and extreme forms are eliminated.

■ With disruptive selection, an intermediate form of a trait is selected against, and extreme phenotypes are favored.

lower bill 12 mm wide   lower bill 15 mm wide

Figure 18.11 In African seedcracker populations, birds with bills that are about 12 *or* 15 millimeters wide are favored. The difference is a result of competition for scarce food during dry seasons.

■ Natural selection theory helps explain diverse aspects of nature, including differences between males and females, and the relationship between sickle-cell anemia and malaria.

■ Links to Sickle-cell anemia 3.6, Codominance 11.4

Sexual Selection

The individuals of many sexually reproducing species have a distinct male or female phenotype, or **sexual dimorphism**. Individuals of one sex (often males) are often more colorful or larger than individuals of the other sex, and they tend to be more aggressive.

These adaptations seem puzzling because they take energy and time away from an individual's survival activities. Some are probably maladaptive because they attract predators. Why do they persist? The answer is **sexual selection**, in which the genetic winners out-reproduce others of a population because they are better at securing mates. In sexual selection, the most adaptive forms of a trait are those that help individuals defeat same-sex rivals for mates, or are the ones most attractive to the opposite sex.

By choosing among mates, a male or female acts as a selective agent on its own species. For example, the females of some species will shop for a mate among a congregation of males, which vary slightly in appearance and courtship behavior. The selected males pass the alleles for their attractive traits to the next generation of males. Females pass alleles that influence mate preference to the next generation of females.

Gerald Wilkinson and his colleagues demonstrated female preference for an exaggerated male trait in the stalk-eyed fly, *Cyrtodiopsis dalmanni*. The eyes of this Malaysian species form on long, horizontal eyestalks that provide no obvious adaptive advantage to their bearers, other than perhaps provoking sexual interest in other flies (Figure 18.12a). The researchers predicted that if eyestalk length is a sexually selected trait, then males with longer eyestalks would be more attractive to female flies than males with shorter eyestalks. They bred males with extra-long eyestalks, and found those males were indeed preferred by the female flies. Such experiments show how exaggerated traits can arise by sexual selection in nature.

Females of many species raise offspring with little help from males. In such species, males typically mate with any female, and females choose males that display species-specific cues, which often include flashy body parts or behaviors (Figure 18.12b). Flashy traits can be a physical hindrance and they may attract predators. However, a flashy male's survival despite his obvious handicap implies health and vigor, two traits that are likely to improve a female's chances of bearing healthy, vigorous offspring.

The males of species in which both sexes share parenting responsibilities usually are not overly flashy. Courtship behavior in such species may include demonstrations of the male's nurturing ability, such as offering nesting materials or food to the female.

Figure 18.12 Sexual selection. (**a**) Ooh, such sexy eyestalks! Stalk-eyed flies of the species *Cyrtodiopsis dalmanni* cluster on aerial roots to mate, and females cluster preferentially around males with the longest eyestalks. This photo, taken in Kuala Lumpur, Malaysia, shows a male with very long eyestalks (*top*) that has captured the interest of the three females below him.

(**b**) This male bird of paradise (*Paradisaea raggiana*) is engaged in a flashy courtship display. He caught the eye (and, perhaps, the sexual interest) of the smaller, less colorful female. The males of this species of bird compete fiercely for females, which are selective. (Why do you suppose the females are drab-colored?)

Balanced Polymorphism

With balancing selection, two or more alleles of a gene persist at high frequency in a population. Such **balanced polymorphism** may occur when conditions in the population's environment favor heterozygotes.

Consider the human *Hb* gene, which encodes the beta globin chain of hemoglobin (Section 3.6). Hemoglobin is the oxygen-transporting protein in blood. Hb^A is the normal allele of this gene; the Hb^S allele carries a particular mutation (Section 14.5) that causes homozygotes to develop sickle-cell anemia. Individuals homozygous for the Hb^S allele often die in their teens or early twenties. Despite being so harmful, the Hb^S allele persists at very high frequency among the human populations in tropical and subtropical regions of Asia and Africa.

Why? A clue to the answer is that populations with the highest frequency of the Hb^S allele also have the highest incidence of malaria (Figure 18.13). Mosquitoes transmit the parasitic agent of malaria, *Plasmodium*, to human hosts. The protozoan multiplies in the liver and then in red blood cells. The cells rupture and release new parasites during severe, recurring bouts of illness.

It turns out that Hb^A/Hb^S heterozygotes are more likely to survive malaria than people who make only normal hemoglobin. Several mechanisms are possible. For example, infected cells of heterozygotes take on a sickle shape. The abnormal shape brings infected cells to the attention of the immune system, which destroys them—along with the parasites they harbor. Infected cells of Hb^A/Hb^A homozygotes do not sickle, and the parasite may remain hidden from the immune system.

The persistence of the Hb^S allele may be a matter of relative evils. Malaria and sickle-cell anemia are both potentially deadly. In areas where malaria is common, Hb^A/Hb^S heterozygotes are more likely to survive and reproduce than Hb^A/Hb^A homozygotes. Heterozygotes are not completely healthy, but they do make enough normal hemoglobin to survive. Malaria or not, they are more likely to live long enough to reproduce than Hb^S/Hb^S homozygotes. The result is that nearly one-third of individuals that live in the most malaria-ridden regions of the world are Hb^A/Hb^S heterozygotes.

Take-Home Message

How does natural selection maintain variation?

■ With sexual selection, some version of a trait gives an individual an advantage over others in securing mates. Sexual dimorphism is one outcome of sexual selection.

■ Balanced polymorphism is a state in which natural selection maintains two or more alleles at relatively high frequencies.

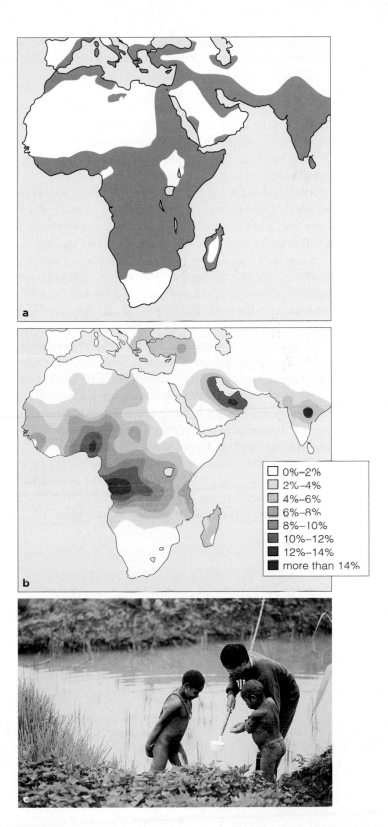

☐	0%–2%
☐	2%–4%
☐	4%–6%
☐	6%–8%
☐	8%–10%
☐	10%–12%
☐	12%–14%
■	more than 14%

Figure 18.13 Malaria and sickle-cell anemia. (**a**) Distribution of malaria cases (*green*) reported in Africa, Asia, and the Middle East in the 1920s, before the start of programs to control mosquitoes, which transmit *Plasmodium*. (**b**) Distribution (by percentage) of people that carry the sickle-cell allele. Notice the close correlation between the maps. (**c**) Physician searching for mosquito larvae in Southeast Asia.

18.7 Genetic Drift—The Chance Changes

- Especially in small populations, random changes in allele frequencies can lead to a loss of genetic diversity.

- Links to Sampling error 1.8, Probability 11.2, Ellis–van Creveld syndrome 12.7

Genetic drift is a random change in allele frequencies over time, brought about by chance alone. We explain genetic drift in terms of probability—the chance that some event will occur (Section 11.2).

We express the probability of an event occurring as a percentage. For instance, if 10 million people enter a drawing, each has the same probability of winning: 1 in 10 million, or a very improbable 0.00001 percent.

Remember, sample size is important in probability (Section 1.8). For example, every time you flip a coin, there is a 50 percent chance it will land heads up. With 10 flips, the proportion of times heads actually land up may be very far from 50 percent. With 1,000 flips, that proportion is more likely to be near 50 percent.

We can apply the same rule to populations. Because population sizes are not infinite, there will be random changes in allele frequencies. These random changes have a minor impact on large populations. However, such changes can lead to dramatic shifts in the allele frequencies of small populations.

Imagine two human populations. Population I has 10 individuals, and population II has 100. Say an allele *b* occurs in both populations at a 10 percent frequency. Only one person carries the allele in population I. If

that person does not reproduce, allele *b* will be lost from population I. However, ten people in population II carry the allele. All ten would have to die without reproducing for the allele to be lost from population II. Thus, the chance that population I will lose allele *b* is greater than that for population II. Steven Rich and his colleagues demonstrated this effect in populations of flour beetles (Figure 18.14).

Random change in allele frequencies can lead to the loss of genetic diversity and the homozygous condition. Both outcomes of genetic drift are possible in all populations, but they are more likely to occur in small ones. When all individuals of a population are homozygous for one allele, we say that **fixation** has occurred. Once an allele is fixed, its frequency will not change again unless mutation or gene flow introduces new alleles.

Bottlenecks and the Founder Effect

Genetic drift is pronounced when a few individuals rebuild a population or start a new one, such as occurs after a **bottleneck**—a drastic reduction in population size brought about by severe pressure. Suppose that a contagious disease, habitat loss, or overhunting nearly wipes out a population. Even if a moderate number of individuals do survive, allele frequencies will have been altered at random.

For example, northern elephant seals were once on the brink of extinction, with only twenty known sur-

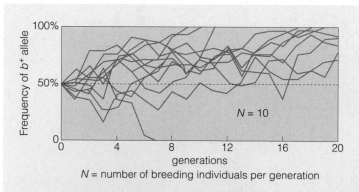

A The size of these populations of beetles was maintained at 10 breeding individuals. Allele *b+* was lost in one population (one graph line ends at 0).

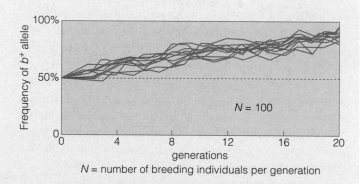

B The size of these populations was maintained at 100 individuals. Drift in these populations was less than in the small populations in (**a**).

Figure 18.14 Animated Effect of population size on genetic drift in flour beetles (*Tribolium castaneum*, shown *left* on a flake of cereal). Beetles homozygous for allele *b* were crossed with beetles homozygous for wild-type allele *b+*. F$_1$ individuals (*b+b*) were divided into populations of (**a**) 10 or (**b**) 100 randomly selected male and female beetles; population sizes were maintained for 20 generations.

Graph lines in (**b**) are smoother than in (**a**), indicating that drift was greatest in the sets of 10 beetles and least in the sets of 100. Notice that the average frequency of allele *b+* rose at the same rate in both groups, an indication that natural selection was at work too: Allele *b+* was weakly favored.
Figure It Out: In how many populations did allele *b+* become fixed?   *Answer: Six*

vivors in the world in the 1890s. Hunting restrictions implemented since then have allowed the population to recover to about 170,000 individuals. Every seal is homozygous for all of the genes analyzed to date.

The effects of genetic drift can be pronounced when a small group of individuals founds a new population. If the group is not representative of the original population in terms of allele frequencies, the new population will not be representative either. This outcome is called the **founder effect**. If the founding group was very small, the new population's genetic diversity may be quite reduced. For instance, imagine a patch of pink and yellow lily flowers on a mainland. By chance, a seabird lands on a yellow lily, and a few seeds stick to its feathers. The bird flies to a remote island and drops the seeds. Most of the seeds have the allele for yellow flowers. The seeds sprout, and a small, isolated population of lily plants establishes itself; most of the plants have yellow flowers. In the absence of gene flow or selection for flower color, genetic drift may fix the allele for yellow flowers in the island population.

The effects of genetic drift are also pronounced in inbred populations. **Inbreeding** is breeding or mating between close relatives, which share a large number of alleles. Inbreeding increases the frequency of homozygous individuals, so it lowers the genetic diversity of a population. Most societies discourage or forbid incest (mating between parents and children or between siblings), but more distant relatives such as cousins do often mate in some societies.

As an example, the Old Order Amish in Lancaster County, Pennsylvania, are moderately inbred. Amish people marry only other Amish people; intermarriage with other groups is not permitted. This population has an unusually high frequency of an allele associated with Ellis–van Creveld syndrome. Among other more serious problems, individuals affected by this syndrome have extra toes or fingers (Section 12.7). The particular allele that affects the Lancaster population has been traced to one man and his wife, who were among a group of 400 Amish who immigrated to the United States in the mid-1700s. As a result of the founder effect and inbreeding since then, about 1 of 8 people in the population is now heterozygous for the allele, and 1 in 200 is homozygous for it.

18.8 | Gene Flow

- Individuals, along with their alleles, move into and away from populations. The flow of alleles counters genetic change that tends to occur within a population.
- Link to Transgenic plants 16.7

Individuals of the same species do not always stay in the same geographic area, or in the same population. A population can lose alleles when individuals leave it permanently, an act called emigration. A population gains alleles when individuals permanently move in, an act called immigration. **Gene flow**, the movement of alleles among populations, occurs in both cases. Gene flow is a microevolutionary process that counters the evolutionary effects of mutations, natural selection, and genetic drift. It is most pronounced among populations of animals, which tend to be more mobile, but it also occurs in plant populations.

Consider the acorns that blue jays disperse when they gather nuts for the winter. Every fall, jays visit acorn-bearing oak trees repeatedly, then bury acorns in the soil of home territories that may be as much as a mile away (Figure 18.15). The jays transfer acorns—and the alleles inside them—among populations of oak trees that would otherwise be genetically isolated.

Figure 18.15 Blue jay, a mover of acorns that helps keep genes flowing between separate oak populations.

Many opponents of genetic engineering are concerned about gene flow from transgenic organisms into wild populations. The flow is already occurring: The *bt* gene and herbicide-resistance genes (Section 16.7) have been found in weeds and unmodified crop plants adjacent to test fields of transgenic plants. The long-term effects of this gene flow are currently unknown.

Take-Home Message

What is genetic drift?

- Genetic drift is a random change in allele frequencies over generations. The magnitude of its effect is greatest in small populations, such as one that endures a bottleneck.

Take-Home Message

What is gene flow?

- Gene flow is the physical movement of alleles into and out of a population, by way of immigration and emigration. It tends to counter the evolutionary effects of mutation, natural selection, and genetic drift on a population.

18.9 | Reproductive Isolation

■ Speciation differs in its details, but reproductive isolating mechanisms are always part of the process.

■ Links to Meiosis 10.3, Zygote 10.5

There are tremendous differences between species such as petunias and whales, beetles and emus, and so on. Such organisms look very different, so it is easy to tell them apart. Their separate lineages probably diverged so long ago that many changes accumulated in them. Organisms that share a more recent ancestor may be much more difficult to tell apart (Figure 18.16).

Evolutionary biologist Ernst Mayr defined a species as one or more groups of individuals that potentially can interbreed, produce fertile offspring, and do not interbreed with other groups. This "biological species concept" is useful for distinguishing species of sexual reproducers such as mammals, but it is not universally applicable. For example, not all populations of a species actually continue to interbreed. We may never know if two populations separated by a great distance could

Figure 18.16 Four butterflies, two species: Which are which? Two forms of the species *Heliconius melpomene* are on the *top* row; two of *H. erato* are on the *bottom* row.

interbreed even if they did get together. Also, populations often continue to interbreed even as they diverge into separate species. The point is, a "species" is a convenient but artificial construct of the human mind.

In nature, sexually reproducing species attain and maintain separate identities by **reproductive isolation**— the end of gene exchanges between populations. New species arise by the evolutionary process of **speciation**, which begins as gene flow ends between populations. Then the populations diverge genetically as mutation, natural selection, and genetic drift operate in each one independently. Speciation may occur after a very long period of divergence, or after one generation (as often occurs among flowering plants by polyploidy).

Reproductive isolating mechanisms arise as populations diverge (Figure 18.17). Such mechanisms are heritable aspects of body form, function, or behavior that prevent interbreeding among species. Prezygotic isolating mechanisms prevent successful pollination or mating, and postzygotic isolating mechanisms often result in weak or infertile hybrids. Both reinforce differences between diverging populations.

Prezygotic Isolating Mechanisms

Temporal Isolation Diverging populations cannot interbreed if the timing of their reproduction differs. Three species of periodical cicadas (*right*) feed on roots as they mature underground. Every 17 years, they emerge to reproduce. Each has a sibling species with nearly identical form and behavior. However, the siblings emerge on a 13-year cycle. A species and its sibling might interbreed—except they only get together once every 221 years!

Mechanical Isolation Body parts of a species may not physically match with those of other species that could otherwise be mates or pollinators. For example, *Salvia mellifera* (black sage) and *S. apiana* (white sage) grow

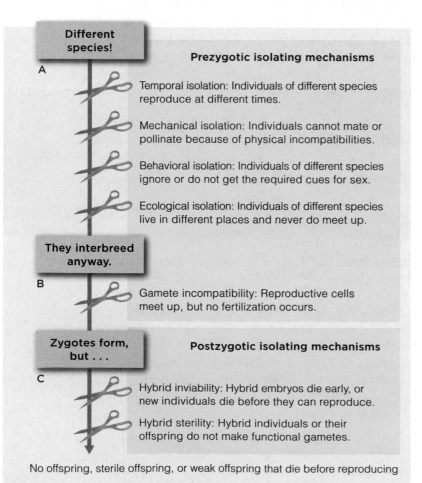

Different species!

A

Prezygotic isolating mechanisms

Temporal isolation: Individuals of different species reproduce at different times.

Mechanical isolation: Individuals cannot mate or pollinate because of physical incompatibilities.

Behavioral isolation: Individuals of different species ignore or do not get the required cues for sex.

Ecological isolation: Individuals of different species live in different places and never do meet up.

They interbreed anyway.

B

Gamete incompatibility: Reproductive cells meet up, but no fertilization occurs.

Zygotes form, but . . .

C

Postzygotic isolating mechanisms

Hybrid inviability: Hybrid embryos die early, or new individuals die before they can reproduce.

Hybrid sterility: Hybrid individuals or their offspring do not make functional gametes.

No offspring, sterile offspring, or weak offspring that die before reproducing

Figure 18.17 Animated Reproductive isolating mechanisms, which prevent interbreeding: barriers to (**a**) getting together, mating, or pollination; (**b**) successful fertilization; and (**c**) survival, fitness, or fertility of hybrid embryos or offspring.

Figure 18.18 Mechanical isolation. *Left*, the flowers of black sage, *Salvia mellifera*, are too small to support bumblebees; they are pollinated by smaller honeybees. *Right*, the pollen-bearing anthers of white sage flowers (*S. apiana*) are at the tips of filaments that project high above the petals. Honeybees that land on this flower are too small to reach the anthers, so only larger bees pollinate white sage.

in the same areas, but hybrids rarely form because the flowers of the two species have become specialized for distinct pollinators (Figure 18.18). Pollen-bearing parts (anthers) of white sage flowers are at the end of long filaments that extend far above the petals. The anthers are too high above the flower to brush small bees that land on the petals. Thus, small bees cannot pollinate white sage. This flower is pollinated mainly by bumblebees and other large bees. Large bees have difficulty finding footing on the tiny flowers of black sage; this species is pollinated mainly by small bees.

Behavioral Isolation Behavioral differences stop gene flow between related species. For instance, males and females of some bird species engage in courtship displays before sex (Figure 18.19). A female recognizes the singing, wing spreading, or bobbing of a male of her species as an overture to sex. Females of different species usually ignore this behavior.

Ecological Isolation Populations adapted to different microenvironments in the same area may be ecologically isolated. For example, in the Sierra Nevada, two species of manzanita (a plant) rarely hybridize. One that is better adapted for conserving water inhabits dry, rocky hillsides high in the foothills. The other lives on lower slopes where water stress is not as intense. The separation means cross-pollination is unlikely.

Figure 18.19 Animated Behavioral isolation. Species-specific courtship displays precede sex among many birds, including these albatrosses.

Gamete Incompatibility The reproductive cells of different species may have molecular incompatibilities, so fertilization cannot occur. This may be the primary speciation route of animals that fertilize their eggs by releasing free-swimming sperm in an aquatic habitat.

Postzygotic Isolating Mechanisms

Reduced Hybrid Viability When populations diverge, so do their genes. Even chromosomes of species that diverged recently may have major differences. Thus, a hybrid zygote may have extra or missing genes, or genes with incompatible products. In either case, its development probably will not proceed correctly and the resulting embryo will die prematurely. Hybrid offspring may have reduced fitness, such as occurs with ligers and tigons (offspring of lions and tigers), which have more health problems and a shorter life expectancy than individuals of either parent species.

Reduced Hybrid Fertility Some interspecies crosses produce robust but sterile offspring. For example, the offspring of a female horse (64 chromosomes) mated with a male donkey (62 chromosomes) is a mule. The mule's 63 chromosomes cannot pair up evenly during meiosis, so this animal makes few viable gametes.

Hybrid Breakdown Crossing fertile hybrids often produces offspring that have lower fitness with each successive generation. A mismatch between nuclear and mitochondrial DNA may be the cause (mitochondrial DNA is inherited from the mother only).

Take-Home Message

How do species attain and maintain separate identities?

■ Speciation is an evolutionary process by which new species form. It varies in its details and duration, but reproductive isolation is always part of the process.

■ Individuals of a species are most often reproductively isolated from individuals of other species.

- In the most common mode of speciation, a physical barrier arises and ends gene flow between populations.

- Link to Plate tectonics 17.9

Every species is a unique outcome of its own history and environment. Thus, speciation is not a predictable process such as a metabolic reaction; it happens in a unique way every time it happens. However, we can identify some underlying principles.

Genetic changes that lead to a new species usually begin with physical separation between populations, so **allopatric speciation** may be the most common way that new species form (*allo*– means different; *patria*, fatherland). By this speciation mode, a physical barrier separates two population and ends the gene flow between them. Then, reproductive isolating mechanisms arise, so even if the populations meet up again their individuals could not interbreed.

Whether a geographic barrier can block gene flow depends on an organism's means of travel or the way its gametes disperse. Populations of most species are separated by distance, and gene flow between them is usually intermittent. Barriers that arise abruptly may end the flow entirely. For example, the Great Wall of China cut off gene flow among wind-pollinated plants as it was being built. DNA analysis shows that populations of trees, shrubs, and herbs on either side of the wall are now diverging genetically.

The fossil record suggests that geographic isolation also happens slowly. For example, it happened after vast glaciers advanced into North America and Europe during the ice ages and cut off populations of plants and animals. After glaciers retreated, descendants of related populations met up again. Genetic divergences were not great between some separated populations; their descendants could still interbreed. Descendants of some other populations could no longer interbreed. Reproductive isolation had led to speciation.

Also, remember how Earth's crust is fractured into gigantic plates? Slow, colossal movements inevitably alter the configurations of land masses (Section 17.9). As Central America formed, part of an ancient ocean basin was uplifted, and it became a land bridge—now called the Isthmus of Panama. Some camelids crossed the bridge into South America. Geographic separation led to new species: llamas and vicunas (Figure 18.20).

The Inviting Archipelagos

An archipelago is an island chain some distance from a continent. The Florida Keys and other island chains are so close to a mainland that gene flow is more or less unimpeded, so there is little if any speciation. The Hawaiian Islands, the Galápagos Islands, and some other archipelagos are not close to a mainland. These remote, isolated islands are the tops of volcanoes that started building up on the sea floor, and, in time, broke the surface of the ocean. We can assume that their fiery surfaces were initially barren, with no life.

Winds or ocean currents carry a few individuals of some mainland species to such islands (Figure 18.21).

Figure 18.20 Allopatric speciations. The earliest camelids, no bigger than a rabbit, evolved in the Eocene grasslands and deserts of North America. By the end of the Miocene, they included the now-extinct *Procamelus*. The fossil record and comparative studies indicate that *Procamelus* may have been the common ancestral stock for llamas (**a**), vicunas (**b**), and camels (**c**). One descendant lineage dispersed into Africa and Asia and evolved into modern camels. A different lineage, ancestral to the llamas and vicunas, dispersed into South America after a land bridge formed between the two continents.

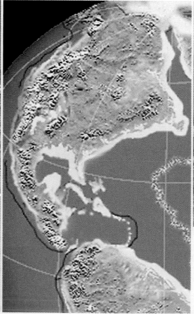

Left, Earth in the late Eocene, before a land bridge formed between North and South America. North America and Eurasia were still connected by a land bridge at the time.

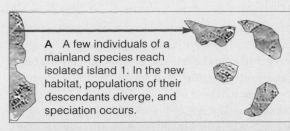  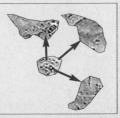

A A few individuals of a mainland species reach isolated island 1. In the new habitat, populations of their descendants diverge, and speciation occurs.

B Later, a few individuals of a new species colonize nearby island 2. Speciation follows genetic divergence in the new habitat.

C Genetically different descendants of the ancestral species may colonize islands 3 and 4 or even invade island 1. Genetic divergence and speciation may follow.

Akepa (*Loxops coccineus*)	Akekee (*L. caeruleirostris*)	Nihoa finch (*Telespiza ultima*)	Palila (*Loxioides bailleui*)	Maui parrotbill (*Pseudonestor xanthophrys*)	Apapane (*Himatione sanguinea*)
Insects, spiders from buds twisted apart by bill, some nectar; high mountain rain forest	Insects, spiders, some nectar; high mountain rain forest	Insects, buds, seeds, flowers, seabird eggs; rocky or shrubby slopes	Mamane seeds ripped from pods; buds, flowers, some berries, insects; high mountain dry forests	Rips dry branches for insect larvae, pupae, caterpillars; mountain forest with open canopy, dense underbrush	Nectar, especially of ohialehua flowers; caterpillars and other insects; spiders; high mountain forests

Poouli (*Melamprosops phaeosoma*)	Maui Alauahio (*Paroreomyza montana*)	Kauai Amakihi (*Hemignathus kaualensis*)	Akiapolaau (*Hemignathus munroi*)	Akohekohe (*Palmeria dolei*)	Iiwi (*Vestiaria coccinea*)
Tree snails, insects in understory; last known male died in 2004	Bark or leaf insects, some nectar; high mountain rain forest	Bark-picker, insects, spiders, nectar; high mountain rain forest	Probes, digs insects from big trees; high mountain rain forest	Mostly nectar from flowering trees, some insects, pollen; high mountain rain forest	Mostly nectar (ohia flowers, lobelias, mints), some insects; high mountain rain forest

Figure 18.21 Animated (a–c) Allopatric speciation on an isolated archipelago. *Above,* 12 of 57 known species and subspecies of Hawaiian honeycreepers, with some dietary and habitat preferences. Honeycreeper bills are adapted to feed on insects, seeds, fruits, nectar in floral cups, and other foods. Morphological studies, and comparisons of chromosomal and mitochondrial DNA sequences for proteins suggest that the ancestor of all Hawaiian honeycreepers resembled the housefinch (*Carpodacus*) shown at *left.*

Their descendants colonize other islands in the chain. Habitats and selection pressures that differ within and between the islands foster divergences that result in allopatric speciation. Later, new species may return to islands colonized by their ancestors.

The big island of Hawaii formed less than 1 million years ago. Its habitats range from old lava beds, rain forests, and grasslands to snow-capped volcanoes. The first birds to colonize it found a buffet of fruits, seeds, nectars, tasty insects—and few competitors for them. The near absence of competition in an abundance of vacant habitats spurred rapid speciation. Figure 18.21 hints at the variation that arose among the Hawaiian honeycreepers. Like thousands of other species, they are unique to Hawaii.

As another example of their speciation potential, the Hawaiian Islands make up only about 0.01 percent of the world's total land mass, yet 40 percent of the 1,450 known *Drosophila* (fruit fly) species arose there.

Take-Home Message

What is the most common mode of speciation?

■ With allopatric speciation, a physical barrier that intervenes between populations or subpopulations of a species prevents gene flow among them. As gene flow ends, genetic divergences give rise to new species.

■ Populations sometimes speciate even without a physical barrier that bars gene flow between them.

■ Links to Microtubules 4.13, Spindle 9.3, Polyploidy 12.6

Sympatric Speciation

In **sympatric speciation**, new species form within the home range of an existing species, in the absence of a physical barrier. *Sym–* means together.

Polyploidy Speciation can occur in an instant with a change in chromosome number. Sometimes, somatic cells duplicate their chromosomes but do not divide during mitosis. Or, nondisjunction in meiosis results in gametes with an unreduced chromosome number. Cells that result from such events are **polyploid**: They have three or more sets of chromosomes characteristic of their species (Section 12.6).

Polyploidy occurs spontaneously in plants, but it is somewhat rare in animals. Polyploid plants are often produced artificially by treating buds or seeds with colchicine. This microtubule poison prevents assembly of the spindle during mitosis or meiosis, so chromosomes cannot separate. Polyploid plants tend to be larger and more robust than diploids.

In some cases, a polyploid individual gives rise to an entire population. For example, polyploid cells that arise in a plant may proliferate to form shoots and flowers. If the flowers self-fertilize, a new species may result. The new species will be *auto*polyploid—it arose by chromosome multiplication in one parent species.

Allopolyploids have a combination of chromosome sets from different species. They originate after related species hybridize, and then the chromosome number multiplies in the offspring. For example, the ancestor of common bread wheat was a wild species, *Triticum monococcum*, that hybridized about 11,000 years ago with another wild species of *Triticum* (Figure 18.22). A spontaneous chromosome doubling in the resulting hybrid gave rise to *T. turgidum*, an allopolyploid species with two sets of chromosomes. Another hybridization led to the common bread wheat, *T. aestivum*.

Often, polyploids do not produce fertile offspring by breeding with the parent species; the mismatched chromosomes pair abnormally during meiosis. Some polyploid plants are backcrossed with diploid parents to make sterile offspring that are valued for agriculture. Seedless watermelons are produced this way.

About 95 percent of ferns originated by polyploidy; 70 percent of flowering plants are now polyploid, as well as a few conifers, insects and other arthropods, mollusks, fishes, amphibians, and reptiles.

Other Examples Speciation with no physical barrier to gene flow may occur with no chromosome number change. For example, sister species *Howea forsteriana* (thatch palm) and *H. belmoreana* (curly palm) diverged about 2 million years ago (Figure 18.23). The palms are still abundant in their native habitat, tiny Lord Howe Island. The island is so small that geographic isolation of the wind-pollinated palms is not possible, so their speciation may not have been allopatric.

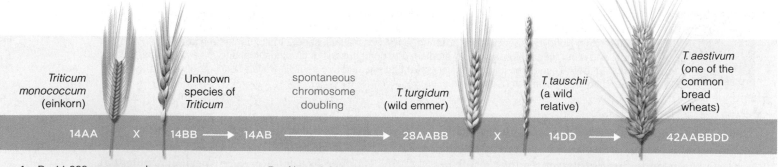

Triticum monococcum (einkorn)   Unknown species of *Triticum*   spontaneous chromosome doubling   *T. turgidum* (wild emmer)   *T. tauschii* (a wild relative)   *T. aestivum* (one of the common bread wheats)

14AA   X   14BB → 14AB → 28AABB   X   14DD → 42AABBDD

A By 11,000 years ago, humans were cultivating wild wheats. Einkorn has a diploid chromosome number of 14 (two sets of 7). It probably hybridized with another wild wheat species having the same number of chromosomes.

B About 8,000 years ago, the allopolyploid wild emmer originated from an AB hybrid wheat plant in which the chromosome number doubled. Wild emmer is tetraploid, or AABB; it has two sets of 14 chromosomes. There is recently renewed culinary interest in emmer, also called farro.

C AABB emmer probably hybridized with *T. tauschii*, a wild relative of wheat. Its diploid chromosome number is 14 (two sets of 7 DD). Common bread wheats have a chromosome number of 42 (six sets of 7 AABBDD).

Figure 18.22 Animated Presumed sympatric speciation in wheat. Wheat grains 11,000 years old and diploid wild wheats have been found in the Middle East, and chromosome analysis indicates that they hybridized. Later, in a self-fertilizing hybrid, homologous chromosomes failed to separate at meiosis, and it produced fertile polyploid offspring. A polyploid descendant hybridized with a wild species. We make bread from grains of their hybrid descendants.

Figure 18.23 Sympatric speciation in palms. **(a)** Lord Howe Island is so small that geographic isolation of native wind-pollinated palm species is impossible. **(b)** The thatch palm *Howea forsteriana* and **(c)** the curly palm *H. belmoreana* may have speciated in sympatry by reproductive isolation.

What ended gene flow between the two species? Differences in the pH of the island soils may be part of the answer. Most of the thatch palms grow in low-lying parts of the island, where the soil's pH is basic. Curly palms grow in volcanic soils, which are more acidic. Thatch palms growing in the basic soil flower six weeks earlier than palms of either species growing in acidic soil. We can expect that some individuals of an ancestral palm species that colonized low-lying regions of the island began to flower earlier. If the island was as small then as it is now, temporal reproductive isolation would have occurred without a physical barrier to gene flow. Disruptive selection would have reinforced the divergence of the two populations.

Sympatric speciation has also occurred in greenish warblers of Siberia (*Phylloscopus trochiloides*). A chain of populations of this bird forms a ring around Tibet. Adjacent populations do interbreed, but small genetic differences between them add up to major divergences between the populations at the ends. Such ring species present one of those paradoxes for those who like neat categories: Populations at the ends of the chain cannot interbreed; thus, they are technically distinct species. However, gene flow occurs continuously all around the ring; where should the line that divides the two species be drawn?

Isolation at Hybrid Zones

Parapatric speciation may occur when one population extends across a broad region encompassing diverse habitats. The different habitats exert distinct selection pressures on parts of the population, and the result may be divergences that lead to speciation. Hybrids that form in a contact zone between habitats are less fit than individuals on either side of it. An example is shown in Figure 18.24.

Take-Home Message

Can speciation occur with no physical barrier to end gene flow?

■ By a sympatric speciation model, new species arise from a population even in the absence of a physical barrier.

■ By a parapatric speciation model, populations maintaining contact along a common border evolve into distinct species.

Figure 18.24 Example of parapatric speciation. The habitats of two rare species, **(a)** the giant velvet walking worm, *Tasmanipatus barretti*, and **(b)** the blind velvet walking worm, *T. anophthalmus*, overlap in a hybrid zone on the island of Tasmania **(c)**. Hybrid offspring are sterile, which may be the main reason these two species are maintaining separate identities in the absence of a physical barrier between their habitats.

T. barretti
hybrid zone
T. anophthalmus

- Macroevolution includes patterns of change such as one species giving rise to multiple species, the origin of major groups, and major extinction events.

- Links to Homeotic genes 15.3, Geologic time scale 17.8, Plate tectonics 17.9

Patterns of Macroevolution

Microevolution describes genetic change within a single species or population. **Macroevolution** is our name for evolutionary patterns on a larger scale. Flowering plants evolved from seed plants, animals with four legs (tetrapods) evolved from fish, birds evolved from dinosaurs—all of these are examples of macroevolution that occurred over millions of years.

proboscis

nectar tube

10 cm

Figure 18.25 Coevolved species. The orchid *Angraecum sesquipedale*, discovered in Madagascar in 1852, stores its nectar at the base of a floral tube 30-centimeters (12 inches) long. Charles Darwin predicted that someone would eventually discover an insect in Madagascar with a proboscis long enough to reach the nectar and pollinate the flower. Decades later, the hawkmoth *Xanthopan morgani praedicta* was discovered in Madagascar. Its proboscis is 30–35 cm long.

Coevolution The process by which close ecological interactions between two species cause them to evolve jointly is **coevolution**. Each species acts as an agent of selection on the other, and each adapts to changes in the other. Over evolutionary time, the two species may become so interdependent that they can no longer survive without one another. We know of many coevolved species of predator and prey, host and parasite, pollinator and flower (Figure 18.25). Later chapters detail specific examples.

Stasis With the simplest macroevolutionary pattern, **stasis**, a lineage persists for millions of years with little or no change. For example, a type of ancient lobe-finned fish, the coelacanth, had been assumed extinct for at least 70 million years until a fisherman caught one in 1938. The modern coelacanth is almost identical to 100 million-year-old fossil specimens.

Exaptation A major evolutionary novelty typically stems from the adaptation of an existing structure for a completely different purpose. This macroevolutionary pattern is called preadaptation or **exaptation**. Some complex traits in modern species held different adaptive value in ancestral lineages. In other words, some traits were used for very different purposes than they are today. For example, feathers that allow modern birds to fly are derived from feathers that first evolved in some dinosaurs. Those dinosaurs could not have used their feathers for flight, but they probably did use them for insulation. Feathers in dinosaurs were a preadaptation to flight feathers in birds.

Adaptive Radiation A burst of divergences from a single lineage is called **adaptive radiation**, and it leads to many new species. This evolutionary pattern gave rise to the Hawaiian honeycreepers (Figure 18.21). Adaptive radiation only occurs when a lineage encounters a set of new niches. Think of a niche as a certain way of life, such as "burrowing in seafloor sediments" or "catching winged insects in the air at night." (You will learn more about niches in Chapter 46, within the context of community structure.)

A lineage that encounters a new set of niches tends to diversify over time. Genetic divergences give rise to many new species that fill the niches. A lineage may encounter new niches when some of its individuals gain physical access to a new habitat, for example by migrating to a different region. Or, geologic or climatic events sometimes change an existing habitat so that it becomes very different. For example, mammals were once distributed through tropical regions of the supercontinent Pangea, which later broke up into continents that drifted apart over millions of years (Section 17.9). Changes in habitats and resources on the continents as they drifted to different parts of the globe set the stage for adaptive radiations of the species they carried.

Some genetic changes may allow individuals of a lineage to enter new niches within their existing habitat. The niches had existed, but had been unavailable before the change. A **key innovation** is a structural or functional modification that bestows upon its bearer the opportunity to exploit a habitat more efficiently or in a novel way. For example, many new niches opened up for the ancestors of birds after they began to use their feathered forelimbs for flight.

Extinction By current estimates, more than 99 percent of all species that ever lived are now **extinct**, or irrevocably lost from Earth. In addition to continuing small-scale extinctions, the fossil record indicates that there have been more than twenty **mass extinctions**, which are losses of many lineages. These include five catastrophic events in which the majority of species on Earth disappeared (Section 17.8). After each event, adaptive radiations filled vacated niches with new species. Biodiversity recovered very slowly, over tens of millions of years (Figure 18.26).

Evolutionary Theory

Biologists do not doubt that macroevolution occurs, but many disagree about how it occurs. However we choose to categorize evolutionary processes, the very same genetic change may be at the root of all evolution—fast or slow, large-scale or small-scale. Dramatic jumps in morphology, if they are not artifacts of gaps in the fossil record, may be the result of mutations in homeotic or other regulatory genes. Macroevolution may include more processes than microevolution, or it may not. It may be an accumulation of many microevolutionary events, or it may be an entirely different process. Evolutionary biologists may disagree about these and other hypotheses, but all of them are trying to explain the same thing: how all species are related by descent from common ancestors.

Take-Home Message

What is macroevolution?

■ Macroevolution comprises large-scale patterns of evolutionary change such as adaptive radiations, the origin of major groups, and loss through extinction.

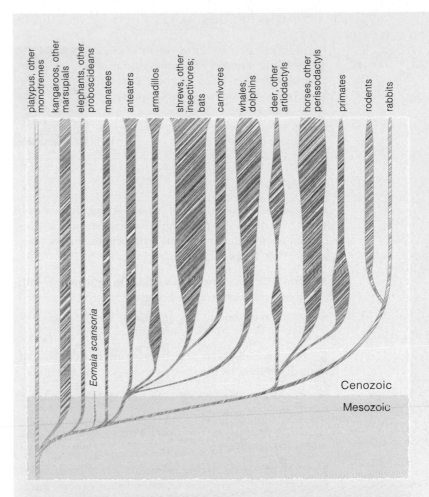

Figure 18.26 Evolutionary tree diagram showing the adaptive radiation of mammals following the K–T extinction event. Branch widths indicate the range of biodiversity in each group at different times. We show only a sample of modern mammals. The entire mammalian lineage includes more than 4,000 modern species.

The photograph shows a fossil of *Eomaia scansoria* (Greek for ancient mother climber), complete with the imprint of its fur. About 125 million years ago, this mouse-sized insect-eater crawled on low branches. It is thought to be an offshoot of the lineage that led to mammals.

Rise of the Super Rats

The rise of warfarin-resistant rat populations has led to the development of second-generation anticoagulants such as brodifacoum. Because these compounds do not kill rats immediately, poison-tainted rats can fall prey to owls, hawks, eagles, cats, and other predators, which can be sickened or killed as a result.

How would you vote?

One standard animal husbandry practice includes continually dosing healthy livestock with the same antibiotics prescribed for people. Should this practice stop? See CengageNOW for details, then vote online.

Summary

Section 18.1 Individuals of a **population** share a **gene pool**. Mutations are the original source of new **alleles**, but many are **lethal** or **neutral**. **Microevolution** is change in the **allele frequencies** of a population.

Section 18.2 Deviations from **genetic equilibrium** indicate that a population is evolving.

■ *Use the interaction on CengageNOW to investigate gene frequencies and genetic equilibrium.*

Section 18.3 **Natural selection** occurs in different patterns depending on the species and selection pressures.

Section 18.4 **Directional selection** shifts the range of variation in traits in one direction.

■ *Use the animation on CengageNOW to see how directional selection works.*

Section 18.5 **Stabilizing selection** favors intermediate forms of a trait. **Disruptive selection** favors extreme forms.

■ *Use the animation on CengageNOW to see how disruptive and stabilizing selection work.*

Section 18.6 **Sexual selection** leads to forms of traits that enhance reproductive success. **Sexual dimorphism** is one outcome. In **balanced polymorphism**, nonidentical alleles for a trait are maintained at relatively high frequencies.

Section 18.7 **Genetic drift** can lead to the loss of genetic diversity or **fixation**. It is pronounced in small or **inbreeding** populations, such as those that occur after an evolutionary **bottleneck**. A bottleneck is a type of **founder effect**.

■ *Use the interaction on CengageNOW to explore genetic drift.*

Section 18.8 **Gene flow** counters the effects of mutation, natural selection, and genetic drift.

Section 18.9 Individuals of sexually reproducing species can interbreed successfully under natural conditions, produce fertile offspring, and are reproductively isolated from other species. **Reproductive isolation** typically occurs after gene flow stops. Divergences then lead to **speciation** (Table 18.2).

■ *Use the animation on CengageNOW to explore how species become reproductively isolated.*

Section 18.10 In **allopatric speciation,** a geographic barrier interrupts gene flow between populations. Genetic divergences then give rise to new species.

■ *Use the animation on CengageNOW to learn more about speciation on an archipelago.*

Section 18.11 With **sympatric speciation**, populations in physical contact speciate. **Polyploid** species of many plants (and a few animals) originated by chromosome doublings and hybridizations. In **parapatric speciation**, populations in contact along a common border speciate.

■ *Use the animation on CengageNOW to explore the effects of sympatric speciation in wheat.*

Section 18.12 **Macroevolution** includes large-scale patterns of evolution. With **stasis**, a lineage changes very little over evolutionary time. In **exaptation**, a lineage uses a structure for a different purpose than its ancestor did. An **adaptive radiation** is rapid diversification into new species that occupy novel niches. A **key innovation** is a modification that allows exploitation of an environment in a new or more efficient way. **Coevolution** occurs when two species act as agents of selection upon one another. Most species that ever existed are **extinct**. **Mass extinctions** have occurred several times in the history of life.

Table 18.2 Different Speciation Models

	Allopatric	Sympatric	Parapatric
Original population	●	●	●
Initiating event	barrier arises	genetic change	new niche entered
Reproductive isolation	in isolation	within population	in new niche
New species	●●	●	●

Self-Quiz

Answers in Appendix III

1. Individuals don't evolve, _____ do.

2. Biologists define evolution as _____ .
   a. purposeful change in a lineage
   b. heritable change in a line of descent
   c. acquiring traits during the individual's lifetime

Data Analysis Exercise

Beginning in 1990, rat infestations in northwestern Germany started to intensify despite continuing use of rodenticides. In 2000, Michael H. Kohn and his colleagues analyzed the genetics of wild rat populations around Munich. For part of their research, they trapped wild rats in five towns, and tested those rats for resistance to warfarin and the second-generation anticoagulant bromadiolone. The results are shown in Figure 18.27.

1. In which of the five towns were most of the rats susceptible to warfarin?

2. Which town had the highest percentage of anticoagulant-resistant wild rats?

3. What percentage of rats in Olfen were resistant to warfarin?

4. In which town do you think the application of bromadiolone was most intensive?

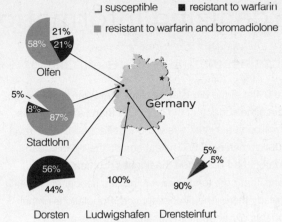

Figure 18.27 Resistance to anticoagulant rodenticides in wild populations of rats in Germany, 2000.

3. _____ is the original source of new alleles.
  a. Mutation   d. Gene flow
  b. Natural selection e. All are original sources of
  c. Genetic drift   new alleles

4. Natural selection can only occur in a population when there are _____ .
  a. differences in forms of heritable traits
  b. selection pressures
  c. both a and b

5. Stabilizing selection _____ (select all that apply).
  a. eliminates common forms of alleles
  b. eliminates uncommon forms of alleles
  c. favors intermediate forms of a trait
  d. favors extreme forms of a trait

6. Directional selection _____ (select all that apply).
  a. eliminates common forms of alleles
  b. eliminates uncommon forms of alleles
  c. favors intermediate forms of a trait
  d. favors extreme forms of a trait

7. Disruptive selection _____ (select all that apply).
  a. eliminates common forms of alleles
  b. eliminates uncommon forms of alleles
  c. favors intermediate forms of a trait
  d. favors extreme forms of a trait

8. Sexual selection, such as competition between males for access to fertile females, frequently influences aspects of body form and leads to _____ .
  a. aggression   c. sexual dimorphism
  b. sexual behavior   d. both b and c

9. The persistence of the sickle allele at high frequency in a population is a case of _____ .

10. _____ tends to keep different populations of a species similar to one another.

11. A fire devastates all trees in a wide swath of forest. Populations of a species of tree-dwelling frog on either side of the burned area diverge. This is an example of _____ .

12. Match the evolution concepts.
  ___ gene flow   a. leads to interdependent species
  ___ natural   b. changes in a population's allele
       selection   frequencies due to chance alone
  ___ mutation   c. allele frequencies change owing to
  ___ genetic   immigration, emigration, or both
       drift   d. survival of the fittest
  ___ adaptive   e. burst of divergences from one
       radiation   lineage into a set of niches
  ___ coevolution   f. source of new alleles

■ *Visit CengageNOW for additional questions.*

Critical Thinking

1. Rama the cama, a llama–camel hybrid, was born in 1997. The idea was to breed an animal that has the camel's strength and endurance and the llama's gentle disposition. However, instead of being large, strong, and sweet, Rama is smaller than expected and has a camel's short temper. He has his eye on Kamilah, a female cama born in early 2002. The question is, will any offspring from such a match be fertile? What might the offspring look like?

What does Rama's story tell you about the genetic changes required for irreversible reproductive isolation in nature? Explain why a biologist might not view Rama as evidence that llamas and camels are the same species.

2. Some theorists have hypothesized that many of our uniquely human traits arose by sexual selection. Over many thousands of years, women attracted to charming, witty men perhaps prompted the development of human intellect far beyond what was necessary for mere survival. Men attracted to women with juvenile features may have shifted the species as a whole to be less hairy and softer featured than any of our simian relatives. Can you think of a way to test this hypothesis?

IMPACTS, ISSUES Bye Bye Birdie

Kauai, the first of the big islands of the Hawaiian Archipelago, rose above the surface of the sea more than 5 million years ago. A million years later, a few finches reached it after traveling 4,000 kilometers (2,500 miles) across the open ocean. No predators had preceded the finches, but tasty insects and plants that bore tender leaves, nectar, seeds, and fruits were already there. The finches thrived. Populations of their descendants expanded into habitats along the coasts, through dry lowland forests, and into highland rain forests.

Between 1.8 million and 400,000 years ago, volcanic eruptions created the rest of the archipelago. Descendants of the first finches flew to the new islands, each of which had different foods and nesting sites. Over many generations, unique forms and behaviors evolved in many different lineages of birds—the Hawaiian honeycreepers. Such traits allowed the birds to exploit special opportunities presented by their island habitats.

The first Polynesians arrived on the Hawaiian Islands sometime before 1000 A.D., and Europeans followed in 1778. Hawaii's rich ecosystem was hospitable to all newcomers, including the settlers' dogs, cats, pigs, cows, goats, deer, and sheep. Escaped livestock began to eat and trample rain forest plants that had provided the honeycreepers with food and shelter. Entire forests were cleared to grow imported crops, and plants that escaped cultivation began to crowd out native plants. Mosquitoes introduced in 1826 spread diseases from imported birds such as chickens to native bird species. Stowaway rats and snakes ate their way through populations of native birds and their eggs. Mongooses deliberately imported to eat the rats and snakes preferred to eat birds and eggs.

Ironically, the very isolation that spurred adaptive radiations made the honeycreepers vulnerable to extinction. The birds had no built-in defenses against predators or diseases of the mainland. Specializations such as extravagantly elongated beaks became hindrances when the birds' habitats suddenly changed or disappeared.

Thus, at least 43 species of honeycreeper that had thrived on the islands before the arrival of humans were extinct by 1778. Today, 32 of the remaining 71 species are endangered, and 26 are extinct despite herculean efforts since the 1960s (Figure 19.1). Why? Invasive, non-native species of plants and animals are now established, and the rise in global temperatures is allowing disease-bearing mosquitoes to invade high-altitude habitats that had previously been too cold for them.

How do we know so much about the evolutionary history of Hawaiian honeycreepers? Evolutionary biologists are a bit like detectives. They use morphological and biochemical comparisons to discover relationships among species, then organize the resulting information using various classification schemes. Their methods are the focus of this chapter.

Figure 19.1 Three honeycreeper species: going, going, and gone.

(a) The palila (*Loxioides bailleui*) has an adaptation that allows it to feed mainly on the seeds of the mamane plant. The seeds are toxic to most other birds. The one remaining palila population is declining because mamane plants are being trampled by cows and gnawed to death by goats and sheep. The Hawaii Division of Forestry and Wildlife estimated that 3,862 palila existed as of 2007.

(b) The unusual lower bill of the akekee (*Loxops caeruleirostris*) points to one side, allowing this bird to open buds that harbor insects. Avian malaria carried by mosquitoes to higher altitudes is decimating the last population of this species. Between 2000 and 2007, the number of akekee plummeted from 7,839 birds to 3,536.

(c) This male poouli (*Melamprosops phaeosoma*)—rare, old, and missing an eye—died in 2004 from avian malaria. There were two other poouli alive at the time, but neither has been seen since then.

Key Concepts

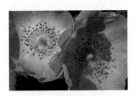

Taxonomy

Each species is given a two-part scientific name. Traditional classification schemes rank species into a hierarchy. Newer methods that group species by shared ancestry more appropriately reflect evolutionary history than do traditional ranking systems. **Section 19.1**

Comparing body form

Species may be grouped on the basis of similarities or differences in body form. Different lineages often have similar body parts, which may be evidence of descent from a shared ancestor. **Section 19.2**

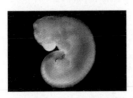

Comparing patterns of development

Species may be grouped on the basis of similarities or differences in patterns of development. Lineages with common ancestry often develop in similar ways. **Section 19.3**

```
RDVQFGWLIRNLHANG
RDVNYGWLIRYMHANG
RDVHYGWIIRYMHANG
RDVNYGWLIRNLHANG
RDVNYGWIIRYLHANG
RDVEGGWLLRYMHANG
TDVMNGWMVRSIHANG
```

Comparing biochemistry

Species may be grouped on the basis of similarities or differences in DNA and proteins. Molecular comparisons help us discover and confirm relationships among species and lineages. **Section 19.4**

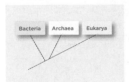

Making family trees

Evolutionary tree diagrams are based on the premise that all species interconnect through shared ancestors—some remote, others recent. A tree of life represents our best understanding of those connections. **Sections 19.5, 19.6**

Links to Earlier Concepts

- This chapter builds on the early introduction to species classification and diversity (Section 1.3). It explores how evolutionary connections among groups of organisms are deciphered, and how we organize the information about those connections. Organizing what we know about life's diversity allows us to communicate clearly with one another about it (1.5).

- You will learn about the types of data that are gathered in order to find evolutionary relationships. Before starting, you may wish to review what you learned about DNA (13.2), the genetic code (14.3), genomics (16.5), genetic equilibrium and neutral mutations (18.1), and genetic diversity (18.7).

- You will see how some master genes (15.2, 15.3) are evidence of shared ancestry.

- You will also come across another example of how we use DNA sequencing (16.3) and fingerprinting (16.4).

How would you vote? Often, when a species is on the brink of extinction, some individuals are captured and kept in zoos for captive breeding programs. Do you support captive breeding of highly endangered species? See CengageNOW for details, then vote online.

■ We group species based on what we know about their evolutionary relationships.

■ Links to Genus and species 1.3, Purpose of science 1.5

A Rose by Any Other Name . . .

Taxonomy, the science of naming and classifying species, began thousands of years ago. However, naming species in a consistent way became a priority in the eighteenth century. At the time, European explorers discovering the scope of life's diversity were having more and more trouble communicating about species, which often had multiple names. For example, one plant species native to Europe, Africa, and Asia was alternately known as the dog rose, briar rose, witch's briar, herb patience, sweet briar, dog berry, briar hip, eglantine gall, hep tree, hip fruit, hip rose, hip tree, hop fruit, and hogseed—and those are only the English names! A species often had several Latin names too. These names tended to be descriptive, but cumbersome. For example, the Latin name of the dog rose was *Rosa sylvestris inodora seu canina* (odorless woodland dog rose), and also *Rosa sylvestris alba cum rubore, folio glabro* (pinkish white woodland rose with smooth leaves).

The eighteenth century naturalist Carolus Linnaeus devised a much simpler naming system that we still use. By the Linnaean system, every species is given a unique two-part scientific name: The first part is the genus name, and together with the second part it designates the species. Thus, the dog rose now has one official scientific name: *Rosa canina*.

Linnaeus also ranked species into ever more inclusive categories. Each category, or **taxon** (plural, taxa) is an organism or group of them; the categories above species—genus, family, order, class, phylum, kingdom, and domain—are higher taxa (Figure 19.2). Each higher taxon consists of a group of the next lower taxon.

A species is usually assigned to higher taxa based on physical and molecular traits it shares with other species. That assignment may change as we discover more about the species and the traits involved. For example, Linnaeus grouped plants by the number and arrangement of reproductive parts, a scheme that resulted in odd pairings such as castor oil plants with pine trees. Today we place these plants in separate phyla.

Ranking Versus Grouping

Linnaeus devised his system of taxonomy before anyone knew about evolution. As we now know, evolution is a dynamic, extravagant, messy, and ongoing process that can be challenging for those who like their categories neat. For example, speciation does not usually occur at a precise moment in time: Individuals often

DOMAIN	Eukarya	Eukarya	Eukarya	Eukarya	Eukarya
KINGDOM	Plantae	Plantae	Plantae	Plantae	Plantae
PHYLUM	Magnoliophyta	Magnoliophyta	Magnoliophyta	Magnoliophyta	Magnoliophyta
CLASS	Magnoliopsida	Magnoliopsida	Magnoliopsida	Magnoliopsida	Magnoliopsida
ORDER	Apiales	Rosales	Rosales	Rosales	Rosales
FAMILY	Apiaceae	Cannabaceae	Rosaceae	Rosaceae	Rosaceae
GENUS	*Daucus*	*Cannabis*	*Malus*	*Rosa*	*Rosa*
SPECIES	*carota*	*sativa*	*domesticus*	*acicularis*	*canina*
COMMON NAME	carrot	marijuana	apple	arctic rose	dog rose

Figure 19.2 Linnaean classification of five species that are related at different levels. Each species has been assigned to ever more inclusive taxa—in this case, from genus to domain.
Figure It Out: Which of the plants shown here are in the same order?

Answer: *Marijuana, apple, arctic rose, and dog rose are all in the order Rosales.*

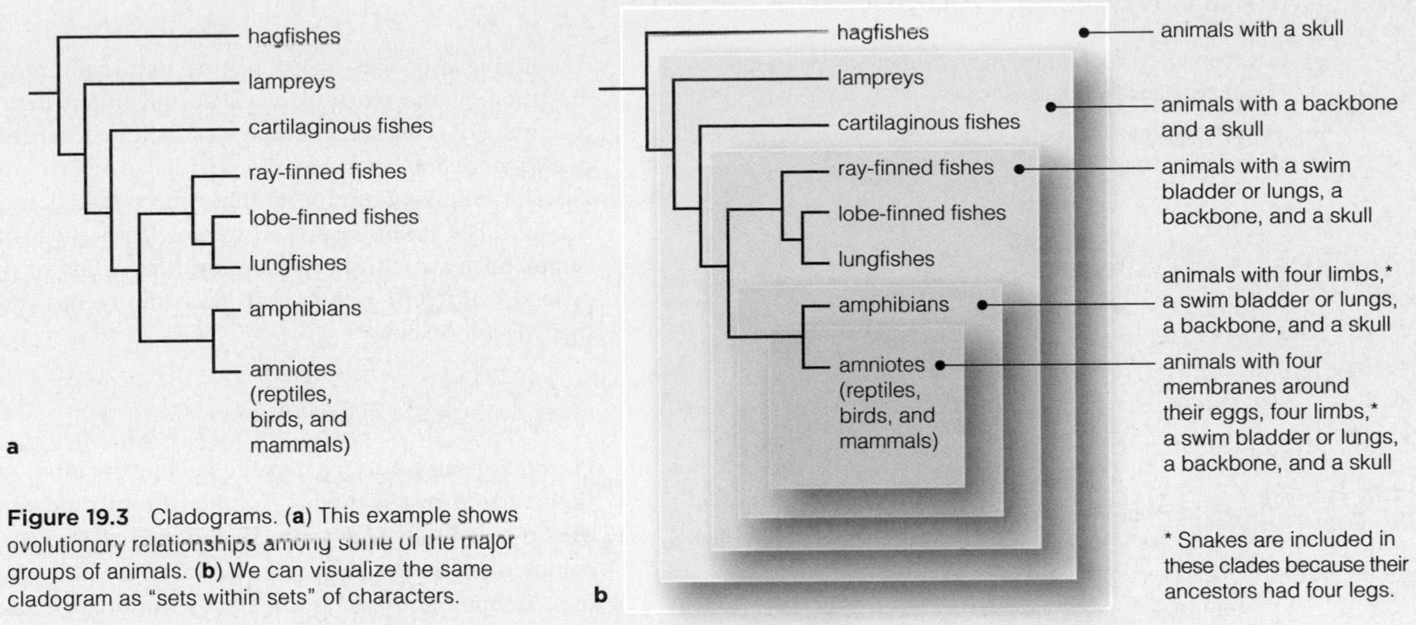
hagfishes
lampreys
cartilaginous fishes
ray-finned fishes
lobe-finned fishes
lungfishes
amphibians
amniotes (reptiles, birds, and mammals)
a
b
hagfishes
lampreys
cartilaginous fishes
ray-finned fishes
lobe-finned fishes
lungfishes
amphibians
amniotes (reptiles, birds, and mammals)
animals with a skull
animals with a backbone and a skull
animals with a swim bladder or lungs, a backbone, and a skull
animals with four limbs,* a swim bladder or lungs, a backbone, and a skull
animals with four membranes around their eggs, four limbs,* a swim bladder or lungs, a backbone, and a skull
* Snakes are included in these clades because their ancestors had four legs.

Figure 19.3 Cladograms. (**a**) This example shows evolutionary relationships among some of the major groups of animals. (**b**) We can visualize the same cladogram as "sets within sets" of characters.

continue to interbreed even as populations are diverging, and populations that have already diverged may come together and interbreed again.

Linnaean taxonomy can be problematic when species boundaries are fuzzy, but the bigger problem is that the rankings do not necessarily reflect evolutionary relationships, or **phylogeny**. Our increasing understanding of evolution is prompting a major, ongoing overhaul of the way biologists view life's diversity. Instead of trying to divide that tremendous diversity into a series of ranks, most biologists are now focusing on evolutionary connections. Each species is viewed not as a member or representative of a rank in a hierarchy, but rather as part of a bigger picture of evolution.

The central question of phylogeny is, "Who is related to whom?" Thus, methods of finding the answer to that question are an important part of phylogenetic classification systems. One method, **cladistics**, groups species on the basis of shared **characters**—quantifiable, heritable characteristics of the organisms of interest. A character can be a physical, behavioral, physiological, or molecular feature of an organism. Because each species has many characters, cladistic groupings may differ depending on which characters are used.

The result of a cladistic analysis is a **cladogram**, a diagram that shows a network of evolutionary relationships (Figure 19.3). Each line in a cladogram represents a lineage, which may branch into two lineages at a node. The node represents a common ancestor of the two lineages. Every branch ends with a **clade** (from *klados*, a Greek word for twig or branch), a group of species that share a set of characters. Ideally, a clade is a **monophyletic group** that comprises an ancestor and all of its descendants. Section 19.5 returns to cladograms and how they are constructed.

Cladograms and other **evolutionary tree diagrams** summarize our best data-supported hypotheses about how a group of species evolved. We use them to visualize evolutionary trends and patterns. For instance, the two lineages that emerge from a node on a cladogram are called **sister groups**. Sister groups are, by default, the same age. We may not know what that age is, but we can compare sister groups on a cladogram and say something about their relative rates of evolution.

Like other hypotheses, evolutionary tree diagrams are revised as new information is gathered. However, the diagrams are based on a solid premise: All species are interconnected by shared ancestry. Everything is related if you just go back far enough in time; evolutionary biologists' job is to figure out where the connections are. The following sections detail some of the types of comparative information they gather as evidence of evolutionary relationships.

Take-Home Message

How do we classify species?

- Taxonomy is a set of rules for naming organisms and classifying them into a series of ranks based on their traits. Though useful, taxonomic rankings do not necessarily reflect evolutionary relationships.
- Cladistics is a method of determining evolutionary relationships by grouping species that share a set of characters. A clade is a group of species that share a set of characters.
- Evolutionary tree diagrams show networks of evolutionary relationships. Such diagrams summarize our best understanding of the evolutionary history of a group of organisms.

19.2 | Comparing Body Form and Function

- Comparisons of body form and structures yields clues about evolutionary relationships.

- Links to Snake limbs 17.1, Evolution of whales 17.7

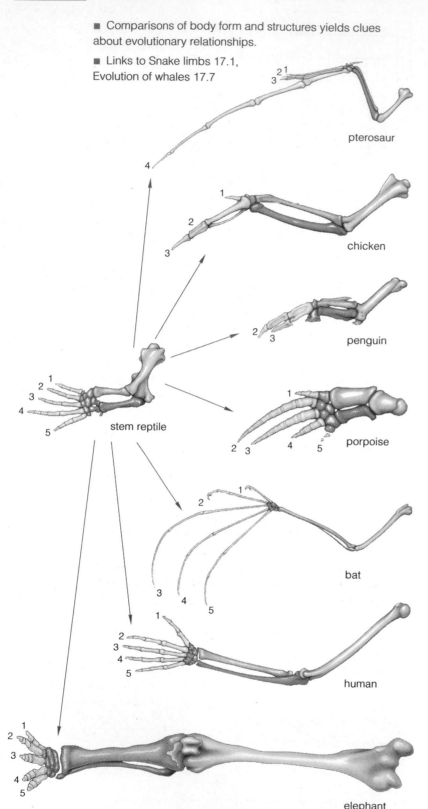

Classifying a species often begins with comparative morphology, the study of body forms and structures (Section 17.1), because similarities in the structure of one or more body parts is often evidence of a common ancestor. Similar body parts that reflect shared ancestry are called **homologous structures** (*hom–* means the same). Such structures may be used for different purposes in different groups, but the same genes direct their development.

Morphological Divergence

Populations of a species diverge genetically after gene flow ends between them (Chapter 18). In time, some of the morphological traits that define their species commonly diverge also. Change from the body form of a common ancestor is a macroevolutionary pattern called **morphological divergence**.

Even if the same body part of two related species evolved so that it has become dramatically different in the different lineages, some underlying aspect of form may remain similar. A careful look beyond the unique modifications can reveal shared heritage.

For example, we know from fossilized limb bones that all modern land vertebrates share an ancestor that crouched low to the ground on four legs. Descendants of this ancestral "stem reptile" diversified into many new habitats on land, and gave rise to the groups we call reptiles, birds, and mammals. A few lineages that had become adapted to walking on land even returned to life in the seas. The lineage that gave rise to whales is one example (Section 17.7).

The stem reptile's five-toed limbs were evolutionary clay. Over millions of years, they became molded into limbs with very different functions across many lineages (Figure 19.4). In penguins and porpoises, the limbs are now flippers useful for swimming. In modern horses, they are long and one-toed, suitable for fast running. Among elephants, they are strong and pillar-like, capable of supporting a great deal of weight. In extinct reptiles called pterosaurs, most birds, and bats, they have been modified for flight. They degenerated to nubs in pythons and boa constrictors (Section 17.1), and to nothing at all in other snakes.

The five-toed limb also became modified into the human arm and hand, in which the thumb evolved in opposition to the fingers. An opposable thumb was the basis of more precise motions and a firmer grip.

Even though vertebrate forelimbs are not the same in size, shape, or function from one group to the next, they clearly are alike in the structure and positioning of bony elements. They also are alike in the patterns of

Figure 19.4 Morphological divergence among vertebrate forelimbs, starting with the bones of a stem reptile. The number and position of many skeletal elements were preserved when these diverse forms evolved; notice the bones of the forearms. Certain bones were lost over time in some of the lineages (compare the digits numbered 1 through 5). The drawings are not to the same scale.

Figure 19.5 Morphological convergence. The flight surfaces of a bat wing (**a**), a bird wing (**b**), and an insect wing (**c**) are analogous structures. (**d**) This cladogram shows that wings evolved independently in the three separate lineages that led to bats, birds, and insects

nerves, blood vessels, and muscles that develop inside them. In addition, comparisons of the early embryos of different vertebrates reveal strong resemblances in patterns of bone development. Such similarities are evidence of shared ancestry.

Morphological Convergence

Similar body parts are not always homologous; they may have evolved independently in separate lineages as adaptations to the same environmental pressures. In this case, such parts are called analogous structures. **Analogous structures** look alike in different lineages but did not evolve in a shared ancestor; they evolved independently after the lineages diverged. Evolution of similar body parts in different lineages is known as **morphological convergence**.

We can sometimes identify analogous structures by studying their underlying form. For example, bird, bat, and insect wings all perform the same function: flight. However, several clues tell us that the flight surfaces of these wings are not homologous. The wing surfaces are adapted to the same physical constraints that govern flight, but the adaptations are different.

In the case of birds and bats, the limbs themselves are homologous, but the adaptations that make those limbs useful for flight are very different. The surface of a bat wing is a thin, membranous extension of the animal's skin. By contrast, the surface of a bird wing is a sweep of feathers, which are specialized structures derived from skin. Insect wings differ even more. An insect wing forms as a saclike extension of the body wall. Except at forked veins, the sac flattens and fuses into a thin membrane. The veins are reinforced with chitin, which structurally support the wing.

The unique adaptations for flight are evidence that wing surfaces of birds, bats, and insects are analogous structures—they evolved after the ancestors of these modern groups diverged (Figure 19.5).

Take-Home Message

Do similar body parts indicate an evolutionary relationship?

■ In morphological divergence, a body part inherited from a common ancestor becomes modified differently in different lines of descent. Such parts are called homologous structures.

■ In morphological convergence, body parts that appear alike evolved independently in different lineages, not in a common ancestor. Such parts are called analogous structures.

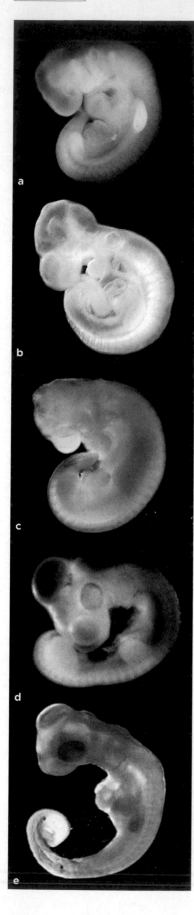

- Similar patterns of embryonic development may be evidence of evolutionary relationships.

- Links to Master genes in flower formation 15.2, Homeotic genes in development 15.3

The development of an embryo into the body of a plant or animal is orchestrated by layer after layer of master gene expression (Section 15.2). The failure of any single master gene to participate in this symphony of expression can result in a drastically altered body plan, typically with devastating consequences.

Because a mutation in a master gene typically unravels development, these genes tend to be highly conserved, which means they have changed very little or not at all over evolutionary time. Thus, a master gene with a similar sequence and function across different lineages is evidence that those lineages are related.

Similar Genes in Plants

The master genes called homeotic genes guide formation of specific body parts during development (Section 15.3). A mutation in one homeotic gene can disrupt details of the body's form. For example, any mutation that inactivates a floral identity gene, *Apetala1*, in *Brassica oleracea* plants (wild cabbage) results in mutated flowers.

Figure 19.6 Comparative embryology. Adult vertebrates are diverse, yet their embryos are similar in the early stages of development.

Compare the segmented backbone, four limb buds, and tail of these early embryos: (**a**) human, (**b**) mouse, (**c**) bat, (**d**) chicken, and (**e**) alligator.

Such flowers form with male reproductive structures (stamens) where petals are supposed to be. At least in the laboratory, these abundantly stamened flowers are exceptionally fertile, but such alterations usually are selected against in nature. *Apetala1* mutations in *Arabidopsis thaliana* plants (common wall cress) result in flowers that have no petals at all. The *Apetala1* gene affects the formation of petals across many different lineages, so it is very likely that this gene is the legacy of a shared ancestor.

Developmental Comparisons in Animals

How Many Legs? The embryos of many vertebrate species develop in similar ways. Their tissues form the same way, as embryonic cells divide, differentiate, and interact. For example, all vertebrates go through a stage in which they have four limb buds and a tail (Figure 19.6). How, then, did the adult forms of these lineages get to be so different? Part of the answer may lie with heritable changes in the onset, rate, or completion of crucial early steps in development. Very rarely, an altered body plan is advantageous.

For example, body appendages as diverse as crab legs, beetle legs, sea star arms, butterfly wings, fish fins, and mouse feet start out as clusters of cells that bud from the surface of the embryo. The buds form wherever the homeotic gene *Dlx* is expressed. The *Dlx* gene encodes a transcription factor that signals clusters of embryonic cells to "stick out from the body" and give rise to an appendage.

A master gene called *Hox* helps sculpt details of the body's form. It suppresses *Dlx* expression in all parts of an embryo that will not have appendages. Wherever *Hox* is expressed, *Dlx* is not, and appendages do not form. For example, *Hox* is expressed along the length of embryonic pythons, so *Dlx* is not expressed anywhere in this snake's body. As a result, an embryonic python's tiny limb buds never mature into back legs. In other organisms, appendages form wherever *Hox* is not expressed, and *Dlx* is expressed.

The *Dlx/Hox* gene control system operates across many phyla, which is strong evidence that it evolved a very long time ago. *Dlx* probably came first; in some Cambrian fossils, it looks like it was not suppressed at all (Figure 19.7a). *Hox* gene control over *Dlx* appears to have evolved later (Figure 19.7b–d).

Forever Young Observing development of the skull in chimpanzees and humans provides evidence that the two species are close relatives. At an early stage, a chimpanzee skull and a human skull appear quite

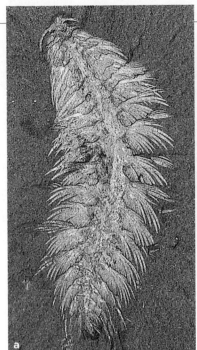

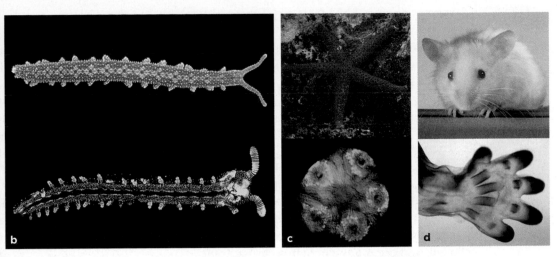

Figure 19.7 Master genes such as *Dlx* and *Hox* govern the formation of appendages across many lineages. Appendages form wherever *Dlx* is expressed, and *Dlx* is expressed wherever *Hox* is not. (**a**) Fossil animal that may be a case of unrestricted expression of *Dlx* in Cambrian times. Variations in *Dlx* expression are revealed by *green* fluorescence in the appendages of (**b**) a velvet walking worm embryo and (**c**) a sea star embryo; and by (**d**) *blue* dye in a mouse embryo's foot.

similar. As development continues, both skulls change shape as different parts grow at different rates (Figure 19.8). However, the human skull undergoes less pronounced differential growth than the chimpanzee skull does. As a result, a human adult has a rounder braincase, a flatter face, and a less protruding jaw compared with an adult chimpanzee.

In its proportions, a human adult skull is more like the skull of an infant chimpanzee than the skull of an adult chimpanzee. The similarity suggests that human evolution involved changes that slowed the rate of development, causing traits that were previously typical of juvenile stages to persist into adulthood.

Juvenile features also persist in other adult animals, notably salamanders called axolotls. The larvae of most species of salamander live in water and use external gills to breathe. Lungs that replace the gills as development continues allow the adult to breathe air and live on land. By contrast, axolotls never give up their aquatic life-style; their external gills and other larval traits persist into adulthood.

The closest relatives of axolotls are tiger salamanders. As you might expect, tiger salamander larvae resemble axolotls, although they are smaller.

Take-Home Message

Are similarities in development clues to shared ancestry?

■ Similarities in patterns of development are the result of master genes that have been conserved over evolutionary time.

■ Some differences between closely related species arose as a result of changes in the rate of development.

Figure 19.8 Animated Thinking about the morphological differences between two primates. These skulls are depicted as paintings on a rubber sheet divided into a grid. Stretching the sheets deforms the grid. Differences in how they are stretched are analogous to different growth patterns.

Shown here, proportional changes during skull development in (**a**) the chimpanzee and (**b**) the human. Chimpanzee skulls change more than human skulls, so the relative proportions in bones of adult and infant humans are more similar than those of adult and infant chimpanzees.

19.4 | Comparing DNA and Proteins

■ The kind and number of biochemical similarities among species are clues about evolutionary relationships.

■ Links to Nucleotide sequence 13.2, The genetic code 14.3, DNA sequencing 16.3, DNA fingerprinting 16.4, Genomics 16.5, Genetic equilibrium and neutral mutations 18.1

Each lineage has a unique set of characters that is a mixture of ancestral and novel traits, including biochemical features such as the nucleotide sequence of its DNA. Inevitable mutations change that sequence over time. The process of mutation is random, so the changes can occur anywhere in a chromosome.

Most of the mutations that accumulate in a lineage are neutral. Such mutations have little or no effect on an individual's survival or reproduction, so we can assume they accumulate in the DNA of a lineage at a constant rate. Neutral mutations alter the DNA of different lineages independently. Thus, the more recently two lineages diverged, the less time there has been for unique mutations to accumulate in the DNA of each one. That is why the DNA sequences of closely related species are more similar than those of distantly related ones—a general rule that can be used to estimate the relative times of divergence of different lineages.

The accumulation of neutral mutations in the DNA of a lineage can be likened to the predictable ticks of a **molecular clock**. Turn the hands of such a clock back, so the ticks wind back through the past. The last tick will be the time when the lineage embarked on its own unique evolutionary road.

How are molecular clocks calibrated? The number of differences in nucleotide or amino acid sequences between lineages can be correlated with the timing of morphological changes seen in the fossil record.

Identifying biochemical similarities and differences among species is now very fast and accurate, thanks to many advances in DNA sequencing and fingerprinting techniques (Sections 16.3 and 16.4). New sequences of genes and proteins from many genomes are compiled continually into online databases, which are accessible by anyone. Comparative genomics studies with such data have shown us (for example) that about 30 percent of the 6,609 genes of yeast cells have counterparts in the human genome. So do 50 percent of the 30,971 fruit fly genes and 40 percent of 19,023 roundworm genes.

Molecular Comparisons

Comparisons of protein primary structure (amino acid sequences) are often used to determine relationships among species. Two species with many identical proteins are likely to be close relatives. Two species with very few similar proteins probably have not shared an ancestor for a long time—long enough for many mutations to have accumulated in the DNA of their separate lineages.

Some essential genes have evolved very little; they are highly conserved across diverse species. One such gene encodes cytochrome *b*. This protein is an important component of electron transfer chains in mitochondria. In humans, its primary structure consists of 378 amino acids. Figure 19.9 compares part of the amino acid sequence of cytochrome *b* from several species.

In amino acid sequence comparisons, the number of differences among species can give us an idea of evolutionary relationships. The amino acids that differ are also clues. For example, a leucine to isoleucine change (a conservative amino acid substitution) may not affect the function of a protein very much, because

```
honeycreepers (10)...CRDVQFGWLIRNLHANGASFFFICIYLHIGRGIYYGSYLNK--ETWNIGVILLLTLMATAFVGYVLPWGQMSFWG...
      song sparrow...CRDVQFGWLIRNLHANGASFFFICIYLHIGRGIYYGSYLNK--ETWNVGIILLLALMATAFVGYVLPWGQMSFWG...
Gough Island finch...CRDVQFGWLIRNIHANGASFFFICIYLHIGRGLYYGSYLYK--ETWNVGILLLTLMATAFVGYVLPWGQMSFWG...
        deer mouse...CRDVNYGWLIRYMHANGASMFFICLFLHVGRGMYYGSYTFT--ETWNIGIVLLFAVMATAFMGYVLPWGQMSFWG...
 Asiatic black bear...CRDVHYGWIIRYMHANGASMFFICLFMHVGRGLYYGSYLLS--ETWNIGIILLFTVMATAFMGYVLPWGQMSFWG...
     bogue (a fish)...CRDVNYGWLIRNLHANGASFFFICIYLHIGRGLYYGSYLYK--ETWNIGVVLLLLVMGTAFVGYVLPWGQMSFWG...
             human...TRDVNYGWIIRYLHANGASMFFICLFLHIGRGLYYGSFLYS--ETWNIGIILLLATMATAFMGYVLPWGQMSFWG...
thale cress (a plant)...MRDVEGGWLLRYMHANGASMFLIVVYLHIFRGLYHASYSSPREFVWCLGVVIFLLMIVTAFIGYVLPWGQMSFWG...
      baboon louse...ETDVMNGWMVRSIHANGASWFFIMLYSHIFRGLWVSSFTQP--LVWLSGVIILFLSMATAFLGYVLPWGQMSFWG...
       baker's yeast...MRDVHNGYILRYLHANGASFFFMVMFMHMAKGLYYGSYRSPRVTLWNVGVIIFTLTIATAFLGYCCVYGQMSHWG...
```

Figure 19.9 Alignment of part of the amino acid sequence of mitochondrial cytochrome *b* from twenty species. This protein is a crucial component of mitochondrial electron transfer chains. The honeycreeper sequence is identical in ten species of honeycreeper; amino acids that differ in the other species are shown in *red*. Gaps in alignment are indicated by dashes. The amino acid abbreviations are spelled out in Appendix V. **Figure It Out:** Based on this comparison, which species is the most closely related to honeycreepers?

Answer: Song sparrow

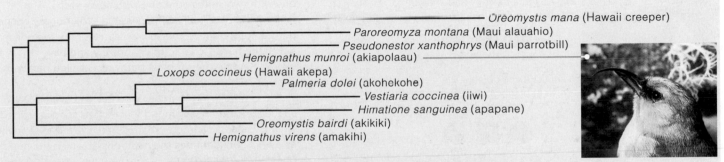

```
...GTAGCCCATATATGCCGCGACGTACAATTCGGCTGACTAATCCGCAACCT...Vestiaria coccinea (iiwi)
...GTAGCCCATGTATGCCGCGACGTACAATTCGGCTGACTAATCCGCAACCT...Himatione sanguinea (apapane)
...GTAGCCCACATATGCCGCGACGTACAATTCGGCTGACTAATCCGCAACCT...Palmeria dolei (akohekohe)
...GTTGCTCACATATGCCGTGACGTACAATTCGGCTGACTAATCCGCAACCT...Oreomystis mana (Hawaii creeper)
...GTAGCCCATATATGCCGCGACGTACAATTCGGCTGACTAATCCGCAACCT...Hemignathus virens (amakihi)
...GTAGCCCACATATGCCGCGACGTACAATTCGGCTGACTAATCCGCAACCT...Hemignathus munroi (akiapolaau)
...GTAGCCCACATATGCCGTGACGTACAGTTCGGCTGACTAATCCGCAACCT...Pseudonestor xanthophrys (Maui parrotbill)
...GTAGCCCACATATGCCGCGACGTACAATTCGGCTGACTAATCCGAAACCT...Paroreomyza montana (Maui alauahio)
...GTAGCCCACATATGCCGCGACGTACAATTCGGCTGACTAATCCGTAATCT...Oreomystis bairdi (akikiki)
...GTAGCCCACATATGCCGCGACGTACAATTCGGCTGACTAATCCGCAACCT...Loxops coccineus (Hawaii akepa)
```

A Part of the mitochondrial cytochrome *b* sequence compared across 10 species of honeycreeper. Differences are shown in *red*. Even though the amino acid sequence of cytochrome *b* is identical in all of these species, neutral mutations have accumulated in their separate lineages. The mutations did not change the amino acid sequence of the resulting protein.

Oreomystis mana (Hawaii creeper)
Paroreomyza montana (Maui alauahio)
Pseudonestor xanthophrys (Maui parrotbill)
Hemignathus munroi (akiapolaau)
Loxops coccineus (Hawaii akepa)
Palmeria dolei (akohekohe)
Vestiaria coccinea (iiwi)
Himatione sanguinea (apapane)
Oreomystis bairdi (akikiki)
Hemignathus virens (amakihi)

B A sequence comparison can be used to generate evolutionary trees. This one reflects a comparison of 790 bases of the mitochondrial cytochrome *b* DNA sequence from ten honeycreeper species. The length of the branches reflects the number of character changes (here, nucleotide differences), which in turn implies the relative length of time of divergence between the species. The arrangement of branches on such trees may differ depending on the data used to generate them.

Figure 19.10 Example of a DNA sequence comparison.

both amino acids are nonpolar, and both are about the same size. However, the substitution of a lysine (which is basic) for an aspartic acid (which is acidic) may dramatically change the character of a protein. Such nonconservative substitutions often affect phenotypic traits. Most mutations that affect phenotype are selected against, but occasionally one proves adaptive (Section 18.1). Thus, we are more likely to see nonconservative substitutions in lineages that diverged long ago. Nonconservative amino acid substitutions, deletions, and insertions are often at the root of differences in phenotype among divergent lineages.

The amino acid sequence of many proteins are identical in lineages that diverged relatively recently, such as the honeycreepers. We may be able to get an idea of evolutionary relationships among such lineages by looking for differences in the nucleotide sequences of their DNA (Figure 19.10). Even if the amino acid sequence of a protein is identical among lineages, the nucleotide sequence of the gene that encodes it may differ because of redundancy in the genetic code. For example, a nucleotide substitution that changes one codon from AAA to AAG in a protein-coding region would probably not affect the protein product, because both codons specify lysine. This base substitution is an example of a neutral mutation.

The DNA from nuclei, mitochondria, and chloroplasts of different species can be used in nucleotide comparisons. Mitochondrial DNA can also be used to compare different individuals of the same sexually reproducing animal species. Mitochondria are inherited intact from a single parent, usually the mother. They contain their own DNA; thus, any differences in mitochondrial DNA sequences between maternally related individuals are due to mutations, not genetic recombination during fertilization.

Take-Home Message

How does biochemistry reflect evolutionary history?

■ DNA and amino acid sequence differences are greatest among lineages that diverged long ago, and less among lineages that diverged more recently.

■ Evolutionary biologists reconstruct phylogeny by determining which pathways of change have the fewest number of steps between species.

To unravel evolutionary relationships, evolutionary biologists collect and analyze data such as the types described in the last three sections. They look for differences in specific characters—the nucleotide sequence of genomes, a set of morphological details, or some other measurable features that set apart the species of interest. They often use a combination of different kinds of data. Any character differences can be used for the analysis, but the bigger the dataset, the more solid the results.

The basic principle behind cladistics is a rule of simplicity: If there are different ways to change from one state to another, the way with the fewest number of steps is the one most likely to occur. Now apply this idea to evolution: When there are several possible evolutionary pathways, the shortest pathway is the one most likely to be correct.

For example, if an evolutionary change from state A to state D can pass through two additional states or five additional states, the correct pathway is more likely to be the one with two states—the one with fewer steps.

By determining all of the possible evolutionary connections among a set of species, we can identify the one in which the fewest character changes would have occurred overall. Finding the simplest pathway is called parsimony analysis (Figure 19.11).

For example, a researcher who has discovered DNA sequence differences among five related species will make what is called a character matrix. She arrays the sequences as a grid, and identifies the positions at which the nucleotides differ (in *red*):

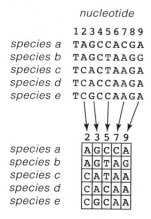

nucleotide

```
          123456789
species a  TAGCCACGA
species b  TAGCTAAGG
species c  TCACTAAGA
species d  TCACCAAGA
species e  TCGCCAAGA
```

```
           2 3 5 7 9
species a  A G C C A
species b  A G T A G
species c  C A T A A
species d  C A C A A
species e  C G C A A
```

A To get a sense of how parsimony analysis works, think about a few items that differ in measurable properties. For example, the following three objects differ in two characters, color and shape:

B If you shuffle these objects, there are only three different ways to put them next to one another:

or

or

C Now think about the total number of differences there are between each pair of adjacent objects. In this example, the middle arrangement has a total of two differences. The others have three:

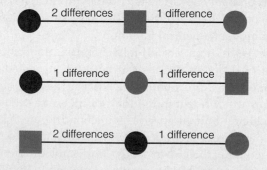

2 differences — 1 difference

1 difference — 1 difference

2 differences — 1 difference

D If we were to create these three arrangements by changing one object into the other two, one difference at a time, the middle arrangement would take the fewest number of steps.

She uses the differences in each column of the matrix for a parsimony analysis. Pairwise comparisons of the five sequences reveal three differences between species *a* and species *b;* three between species *b* and *c;* four between species *a* and *c,* and so on. The researcher (or her computer) makes an evolutionary tree for every possible way the five species might be connected, then adds up all of the nucleotide changes that would have to occur in each scenario. The tree with the lowest number of changes overall is the one that is most likely to be correct.

This method was used to prepare the evolutionary tree shown in Figure 19.12. Differences among skeletons of living and extinct honeycreepers were scored as character differences. Data from an outgroup (a species that is not monophyletic with the other species being studied) was added to the comparisons in order to "root" the tree.

Figure 19.11 A simple example of parsimony analysis. The evolutionary pathway with the least number of steps between states is the one with the highest probability of occurring.

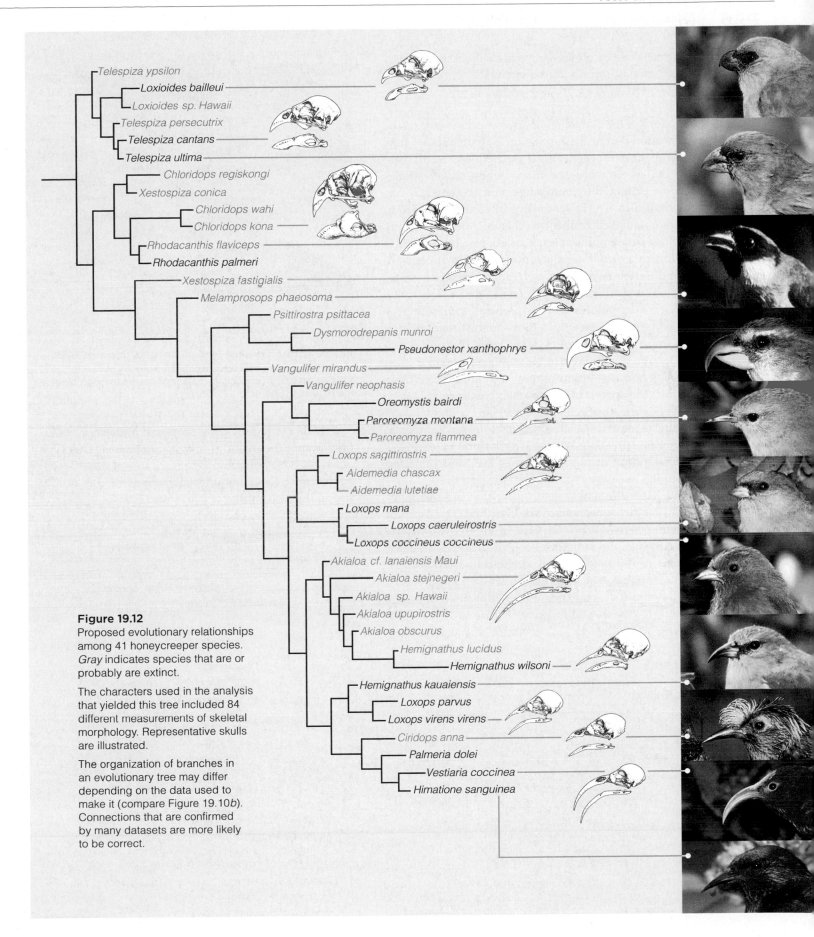

Figure 19.12
Proposed evolutionary relationships among 41 honeycreeper species. *Gray* indicates species that are or probably are extinct.

The characters used in the analysis that yielded this tree included 84 different measurements of skeletal morphology. Representative skulls are illustrated.

The organization of branches in an evolutionary tree may differ depending on the data used to make it (compare Figure 19.10*b*). Connections that are confirmed by many datasets are more likely to be correct.

Preview of Life's Evolutionary History

■ We can organize our knowledge about how species are related using diagrams such as a tree of life.

■ Links to Diversity 1.3, Genetic diversity 18.7

The story of the Hawaiian honeycreepers is a dramatic illustration of how evolution works. It also shows how finding ancestral connections can help species that are still living. As more and more honeycreeper species become extinct, the group's reservoir of genetic diversity dwindles. The lowered diversity means the group as a whole is less resilient to change, and more likely to suffer catastrophic species losses. Deciphering their phylogeny can tell us which honeycreeper species are most different from the others—and those are the ones most valuable in terms of preserving genetic diversity. Such research allows us to concentrate our resources and conservation efforts on those species that hold the best hope for the survival of the entire group (Figure 19.13). The poouli (Figure 19.1c) is one example of an outlying species. Unfortunately, that knowledge came too late; the species is now probably extinct.

Phylogeny research is ongoing for a tremendous number of species, including those in no immediate danger of extinction. With it, we continue to refine our understanding of how all species are interconnected by shared ancestry.

We have different ways to view that big picture of evolutionary connections. A **six-kingdom classification system** assigns all prokaryotes to kingdoms Bacteria and Archaea; the kingdom Protista includes the most ancient multicelled and all single-celled eukaryotes. Plants, fungi, and animals have their own kingdoms. A **three-domain system** sorts all life into three domains: Bacteria, Archaea, and Eukarya (Figure 19.14).

Figure 19.13 The range of Hawaiian honeycreeper diversity is dwindling along with their continued extinctions. Deciphering their evolutionary connections helps us in our efforts to preserve the remaining species.

Figure 19.15 shows one proposed pattern of evolution among the major groups of organisms. This type of evolutionary diagram is called a tree of life. We will be zooming in on individual sections of this family tree in the next unit.

Take-Home Message

What do we do with our knowledge of how species are related?

■ Phylogeny research is yielding an ever more specific and accurate picture of how all life is related by shared ancestry. A tree of life depicts these connections graphically.

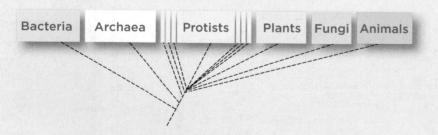

A This tree represents all life classified into six kingdoms. We have discovered that the kingdom of protists is not monophyletic, so some biologists now divide it up into a number of new kingdoms.

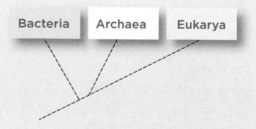

B This tree represents all life classified into three domains. The kingdoms protists, plants, fungi, and animals are subsumed into domain Eukarya. Compare Figure 19.15.

Figure 19.14 Animated Different ways of organizing the tree of life.

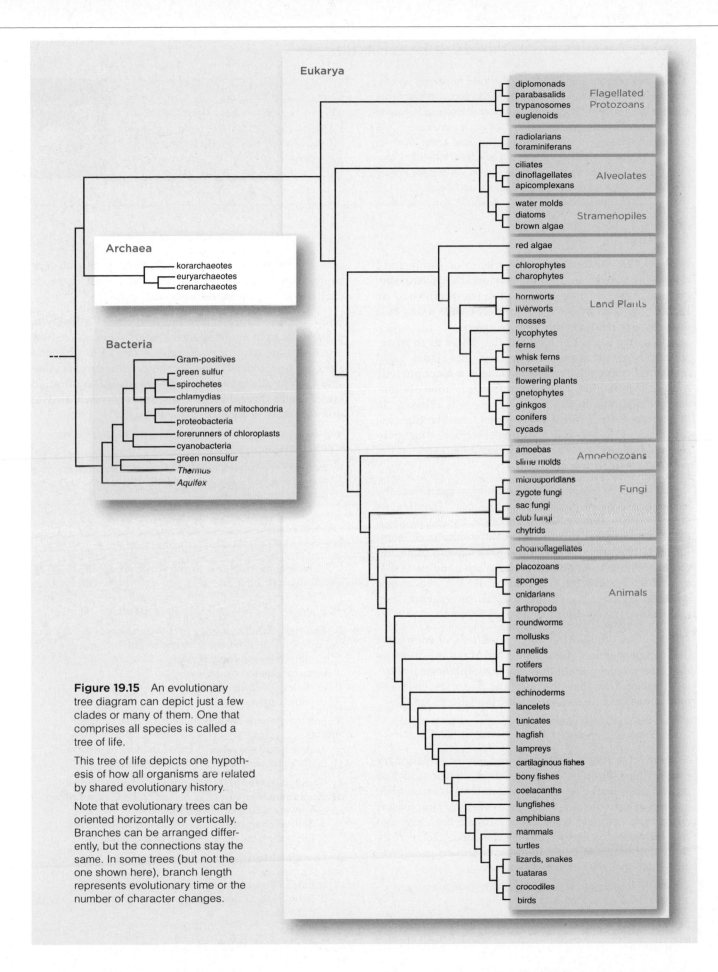

Figure 19.15 An evolutionary tree diagram can depict just a few clades or many of them. One that comprises all species is called a tree of life.

This tree of life depicts one hypothesis of how all organisms are related by shared evolutionary history.

Note that evolutionary trees can be oriented horizontally or vertically. Branches can be arranged differently, but the connections stay the same. In some trees (but not the one shown here), branch length represents evolutionary time or the number of character changes.

Bye Bye Birdie

In 2004, researchers captured one of the three remaining pooulis, with the intent of starting a captive breeding program before the species went extinct. They were unable to capture a female to mate with this male before it died in captivity a month later. Cells from this last bird were frozen, and may be used in the future for cloning. However, with no parents left to demonstrate the species' natural behavior to chicks, cloned birds will probably never be able to establish themselves as a natural population.

How would you vote?

Do you support removing individuals of highly endangered species from their natural habitat for captive breeding programs? See CengageNOW for details, then vote online.

Summary

Section 19.1 **Taxonomy** is the science of naming and classifying species. In traditional taxonomy systems, species are organized into a series of ranks (**taxa**) based on their traits. Such systems do not necessarily reflect true evolutionary relationships, or **phylogeny**.

Cladistics is a set of methods that allow us to reconstruct phylogeny. Species are grouped into **clades** based on shared **characters**. Ideally, a clade is a **monophyletic group**. The result of a cladistic analysis is an **evolutionary tree diagram**, in which a line represents a lineage. In evolutionary trees called **cladograms**, one line (lineage) can branch into two **sister groups** at a node, which represents a shared ancestor. One clade occurs at the end of every line in a cladogram.

Section 19.2 Comparative morphology can reveal the evolutionary connections among lineages. **Homologous structures** are similar body parts that, by **morphological divergence**, became modified differently in different lineages. Homologous structures are evidence of a common ancestor. **Analogous structures** are body parts that look alike in different lineages but did not evolve in a common ancestor. By **morphological convergence**, they evolved separately after the lineages diverged.

Section 19.3 Similarities among patterns of embryonic development reflect shared ancestry. Mutations in genes that affect development may cause morphological shifts in a lineage. Mutations that alter the rate of development may allow juvenile traits to persist into adulthood.

■ *Use the animated interaction on CengageNOW to explore proportional changes in embryonic development.*

Section 19.4 We can discover and clarify evolutionary relationships through comparisons of nucleic acid and protein sequences. Neutral mutations tend to accumulate in DNA at a predictable rate; like the ticks of a **molecular clock**, they can help researchers estimate how long ago two lineages diverged. Lineages that diverged recently share more nucleotide or amino acid sequences in common than ones that diverged long ago.

■ *Use the interaction on CengageNOW to learn more about amino acid comparisons.*

Section 19.5 Cladistic analysis is based on the premise that the most likely evolutionary pathway is the simplest one. The technique sorts out evolutionary relationships by finding the pathway in which the fewest character changes occurred from the ancestral species.

Section 19.6 Representing life's history as a tree with branchings from ancestral stems helps us visualize how organisms are related by descent. A tree of life summarizes our best understanding of the evolutionary relationships among all organisms. A **six-kingdom classification system** and a **three-domain classification system** are two different ways to organize life's diversity.

■ *Use the animation on CengageNOW to review different classification systems.*

■ *Read the InfoTrac article "How Taxonomy Helps Us Make Sense of the Natural World," Sue Hubbell, Smithsonian, May 1996.*

Self-Quiz   *Answers in Appendix III*

1. Homologous structures among major groups of organisms may differ in _____ .
   a. size   c. function
   b. shape   d. all of the above

2. Through _____ , a body part of an ancestor is modified differently in different lines of descent.
   a. morphological convergence
   b. morphological divergence
   c. analogous structures
   d. homologous structures

3. Some mutations are neutral because they do not affect _____ .
   a. amino acid sequence   c. the chances of survival
   b. nucleotide sequence   d. all of the above

4. By altering steps in the program by which embryos develop, a mutation in a _____ may lead to major differences between adults of related lineages.
   a. derived character   c. homologous structure
   b. homeotic gene   d. all of the above

5. Mitochondrial DNA may be used in cladistic comparisons of _____ .
   a. different species
   b. individuals of the same species
   c. different taxa
   d. a and b
   e. all of the above

Data Analysis Exercise

The poouli (*Melamprosops phaeosoma*) was discovered in 1973 by a group of students from the University of Hawaii. Its membership in the honeycreeper clade was—and continues to be—controversial, mainly because its appearance, odor, and behavior are so different from other honeycreepers. It particularly lacks the "old tent" scent of other honeycreepers.

A study published in 2001 by Robert Fleischer and his colleagues tried to clarify the relationships of *Melamprosops* to other honeycreepers by compared bone morphology and DNA sequences. Some of their data, and the resulting cladogram, are shown in Figure 19.16.

1. According to the cladogram, which species is/are most closely related to the poouli? The one(s) most closely related to the ancestor of the honeycreepers?

2. Is any honeycreeper species more closely related to the ancestral species than the poouli?

3. Not counting unresolved bases, how many differences are between the poouli and *Melospiza georgiana* sequences? The poouli and *Himatione sanguinea* sequences?

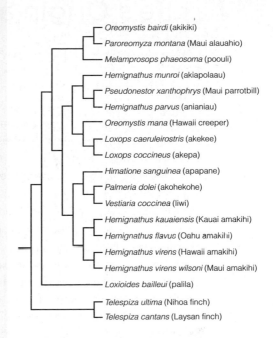

Figure 19.16 DNA sequence comparison of part of the cytochrome *b* control sequence from different honeycreeper species (*below*), and cladogram of honeycreeper phylogeny that resulted (*right*). The top sequence is from an outgroup species; N indicates an unresolved nucleotide. Differences from the *Melamprosops* sequence are indicated in *red*; gaps in the alignment are dashed.

Oreomystis bairdi (akikiki)
Paroreomyza montana (Maui alauahio)
Melamprosops phaeosoma (poouli)
Hemignathus munroi (akiapolaau)
Pseudonestor xanthophrys (Maui parrotbill)
Hemignathus parvus (anianiau)
Oreomystis mana (Hawaii creeper)
Loxops caeruleirostris (akekee)
Loxops coccineus (akepa)
Himatione sanguinea (apapane)
Palmeria dolei (akohekohe)
Vestiaria coccinea (iiwi)
Hemignathus kauaiensis (Kauai amakihi)
Hemignathus flavus (Oahu amakihi)
Hemignathus virens (Hawaii amakihi)
Hemignathus virens wilsoni (Maui amakihi)
Loxioides bailleui (palila)
Telespiza ultima (Nihoa finch)
Telespiza cantans (Laysan finch)

Melospiza georgiana (swamp sparrow)	TAGCCACGACACCTTATTATGAA-CCACTAGTGA-A-AACACTCCCGTAGGTATATTCAATAGATAG
Paroroeomyza montana	TCACCAAGACGATCTATTACGCTACACCAGGGAGGATGGCACTCCCACTGGTATATCCACTTGACAG
Loxioides bailleui	TCACTAAGACAGCTTACTACGCCAACTCACGCGAGAGAGCACTCCCGCTGGTATACTCACTTGATAG
Telespiza cantans	TCACCAAGACGGTTCACTATACCACACCAAGTGAGAGAGCACTCNNNNNGGTATACTCACTTTACAC
Hemignathus parvus	TCACCAAGACGACTTATTGTATCAA-CCAAGAGAGANNGCACTCCCAGTCGTAGATTCTCTTGACAG
Hemignathus kauaiensis	TCACTAAGACGACTTATTATACCCAACCAAGAGAAAAAGCACCCCCCACGGTAAGTCCTCTTGACCG
Himatione sanguinea	TCGCTTAGACGCCTTATTACGCTAAACCATCACAGANNGCACTACTGTTGGTCGATCCTCTTGACAG
Melamprosops phaeosoma	TCACTAAGACACCTTCTTATGTTCCATCAAGAGAGANNGCACNNNNNNGGTATAGCTTCTTGACAC

---

6. Molecular clocks are based on comparisons of the number of _____ mutations between species.

7. Cladistics is based on _____ .
 a. reconstructing phylogeny
 b. parsimony analysis of many clades
 c. character differences between species
 d. all of the above

8. In evolutionary trees, each branch point represents a(n) _____ .
 a. single lineage   c. divergence
 b. extinction   d. adaptive radiation

9. In cladograms, sister groups are _____ .
 a. inbred   c. represented by nodes
 b. the same age   d. members of the same family

10. Match the terms with the most suitable description.
 ___ phylogeny   a. sets within sets
 ___ cladogram   b. evolutionary history
 ___ homeotic genes   c. human arm and bird wing
 ___ homologous   d. similar across diverse taxa
 structures   e. neutral mutation measure
 ___ molecular clock   f. insect wing and bird wing
 ___ analogous
 structures

Visit CengageNOW for additional questions.

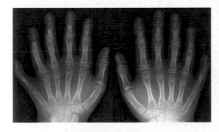

Figure 19.17 Polydactyly. Certain types of mutations result in extra fingers or toes. Most often, the extra digits are duplicates of another digit.

Critical Thinking

1. Some people still refer to species as "primitive" or "advanced." For example, they may say that mosses are primitive and flowering plants are advanced, or that crocodiles are primitive and mammals are advanced. Why is it incorrect to refer to a modern taxon as primitive?

2. Polydactyly is an inherited disorder characterized by extra digits on the hands or feet (Section 12.7 and Figure 19.17). Mutations in certain genes cause this disorder. In what family of genes do you think the mutations occur?

3. Construct a cladogram using the following objects:

CHAPTER 19   ORGANIZING INFORMATION ABOUT SPECIES   315

20 Life's Origin and Early Evolution

IMPACTS, ISSUES | Looking for Life in All the Odd Places

In the 1960s, microbiologist Thomas Brock began looking for signs of life in thermal springs and pools at Yellowstone National Park (Figure 20.1). He found microscopically small cells, including *Thermus aquaticus*. This prokaryote lives in extremely hot water, on the order of 80°C (176°F).

Brock's work had two unexpected results. First, it set researchers on paths that led to the discovery of a great domain of life—Archaea. Second, it led to a faster way to copy DNA and end up with useful amounts. *T. aquaticus* has a heat-resistant DNA polymerase that allows it to replicate its DNA at high temperatures that would denature most enzymes. Researchers now use a synthetic version of the *T. aquaticus* DNA polymerase in the polymerase chain reaction—PCR. As Section 16.2 explained, biotechnologists use PCR to make many copies of a specific piece of DNA.

Spurred by Brock's discovery, scientists began to explore harsh environments in search of new life forms. They found species that withstand extraordinary levels of temperature, pH, salinity, and pressure. For example, some prokaryotes are adapted to life in superheated water near hydrothermal vents on the seafloor. One of those species can grow at 121°C (249°F)! Other prokaryotes cling to life in glacial ice that never thaws. Still others live in acidic springs, where pH approaches zero, or in highly alkaline soda lakes.

Life also thrives deep below Earth's surface. *Bacillus infernus* lives at 75°C (167°F) on rocks 3 kilometers (a little less than 2 miles) beneath the soil surface in Virginia. The species name means "bacterium from hell." Other species of prokaryotes have been discovered at a similar depth on rocks in South African and Canadian mines.

Some eukaryotic cells also survive extreme conditions. Algal cells can color glacial ice red or grow in hot acidic springs. Photosynthetic cells called diatoms inhabit salty lakes that make most forms of life shrivel up and die. Flagellated protists called euglenoids swim in the waters of Berkeley Pit lake. This acidic metal-contaminated lake in Montana is one of the most toxic sites in the United States.

What is the point of these dramatic examples? Simply this: Life can adapt to nearly any environment that has sources of carbon and energy.

This chapter is your introduction to a slice through time, beginning with Earth's formation and moving on to life's chemical origins. The picture we paint here sets the stage for the next unit, which will take you along lines of descent to the present range of biodiversity.

Science cannot prove how life arose, but it can test hypotheses about what could have happened. As you will learn, life is a continuation of the physical and chemical history of the universe, and of the planet Earth.

See the video! Figure 20.1 Thomas Brock looking for life in a thermal pool at Yellowstone National Park. The inset micrograph shows one of the resident bacterial species—*Thermus aquaticus*. Recombinant DNA researchers make great use of its heat-resistant enzymes.

2 μm

Key Concepts

Origin of organic compounds

When Earth first formed more than 4 billion years ago, conditions were too harsh to support life. Over time, its crust cooled and seas formed. Organic compounds of the sort now found in living cells may have self-assembled in the seas or arrived in meteorites. **Section 20.1**

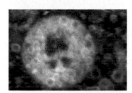

Origin of cells

In all living cells, proteins catalyze metabolic reactions, a plasma membrane encloses the cell, and DNA is the molecule of inheritance. Laboratory experiments provide insight into how cellular components and processes could have evolved. **Section 20.2**

Early evolution

The first cells were prokaryotic. Eukaryotes arose after evolution of the noncylic pathway of photosynthesis in some prokaryotes put oxygen into the air. Mitochondria and chloroplasts are descendants of bacteria that lived in other cells. **Sections 20.3–20.5**

Extraterrestrial life

Astrobiologists study the origin and evolution of life on Earth and elsewhere in the universe. **Section 20.6**

Links to Earlier Concepts

- In this chapter, you will use your knowledge of elements (2.1) and organic compounds (3.1), with a special emphasis on nucleic acids (3.7) and protein synthesis (14.1–14.4).

- You will consider the origins of prokaryotes and eukaryotes (4.4, 4.6) and features such as the cell membrane (5.1), the nucleus, and other organelles (4.8–4.11). You will be reminded of the mechanisms of photosynthesis (7.3).

- As we review the time line of Earth history (17.7), a knowledge of how fossils form (17.5, 17.6) and how molecular clocks (19.4) work will be helpful.

How would you vote? The abundance of life in extreme environments on Earth suggests there may be life beneath the surface of Mars. One way to find out is by sampling Martian soil and bringing it to Earth for study. Does such a plan pose a risk? See CengageNOW for details, then vote online.

20.1 | In the Beginning . . .

- Knowledge of modern chemistry and physics are the basis for scientific hypotheses about early events in Earth's history.

- Links to Elements 2.1, Organic compounds 3.1

Origin of the Universe and Our Solar System

According to the **big bang model**, the universe began in an instant, when all matter and energy suddenly was distributed outward from a single point. Evidence that all known galaxies (groups of stars) are moving away from one another supports this model. It is as if the universe is inflating like a balloon. By measuring movement of galaxies and then working backwards, scientists have estimated that the big bang occurred, and expansion began, 13 to 15 billion years ago.

The big bang model proposes that simple elements such as hydrogen and helium formed within minutes of the universe's birth. Then, over millions of years, gravity drew these gases together and they condensed as stars (Figure 20.2). Nuclear reactions inside these stars produced heavier elements. The great abundance of helium and hydrogen in the universe, together with the relative rarity of heavier elements, is consistent with predictions of the big bang model.

Explosions of early giant stars scattered materials from which galaxies formed. By one hypothesis, our galaxy began as a cloud of debris trillions of kilometers wide. Some of that debris served as the material for the galaxy's stars. Observations of stars and measurements of our sun's brightness and size, suggest that Earth's sun began burning about 5 billion years ago.

At first, a cloud of dust and debris surrounded the sun (Figure 20.3a). Planet-building began when rocks orbiting the sun (asteroids) collided, forming bigger rocky objects. The heavier these pre-planetary objects became, the more gravitational pull they exerted, and the more material they gathered. Eventually, by about 4.6 billion years ago, this process had assembled Earth and the other planets of our solar system.

Conditions on the Early Earth

Planet formation did not clear out all debris from orbit around the sun, so the early Earth received a constant hail of meteorites upon its still molten surface. More molten rock and gases spewed from volcanoes. Gases released by such volcanoes were the main source of the early atmosphere.

What was Earth's early atmosphere like? Studies of volcanic eruptions, meteorites, ancient rocks, and other planets provide clues. Such studies suggest that the air held water vapor, carbon dioxide, and gaseous hydrogen and nitrogen. However, Earth's early atmosphere was probably devoid of oxygen gas, since the geologic record shows that iron in rocks did not begin to combine with oxygen and form rust until much later in Earth's history.

If free oxygen had been abundant, then the organic compounds necessary for life could not have formed and persisted. Had oxygen gas been present, it would have reacted with and disabled organic compounds as quickly as they formed.

At first, any water falling on Earth's molten surface evaporated immediately. As the surface cooled, rocks formed. Later, rains washed mineral salts out of these rocks and the salty runoff pooled in early seas. It was in these seas that life first began (Figure 20.3b).

Figure 20.2 One site of ongoing star formation: columns of dust and gases in the Eagle Nebula, as photographed by the Hubble space telescope in 1995. The telescope was named for the astronomer Edwin Hubble. His discovery that the universe is expanding provided the earliest evidence in support of the big bang model.

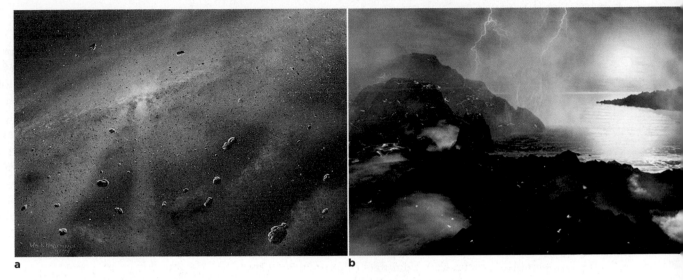

Figure 20.3
(a) What the cloud of dust, gases, rock, and ice around the early sun may have looked like. (b) An artist's depiction of early Earth.

a

b

Origin of the Building Blocks of Life

Until the early 1800s, chemists thought that organic molecules possessed a special "vital force" and could only be made inside living organisms. Then, in 1825, a German chemist synthesized urea, a molecule that is in urine. Later, another chemist synthesized alanine, an amino acid. These synthetic reactions proved that nonliving mechanisms could yield organic molecules.

Could amino acids and other building blocks of life have formed spontaneously on early Earth? In the 1950s, a graduate student named Stanley Miller did an experiment to test this hypothesis. He put gases then thought to be present in Earth's early atmosphere into a reaction chamber (Figure 20.4). He kept the mixture circulating and zapped it with sparks to simulate lightning. In less than one week, amino acids, sugars, and other organic compounds were present in the mix.

Since Miller's experiment, researchers have revised their ideas about which gases were present in Earth's early atmosphere. Miller and others have repeated his experiment using different gases and adding various ingredients to the mix. Amino acids form easily under some conditions. Adenine, a nucleotide base, has also been shown to form spontaneously.

Scientists continue to debate what conditions on early Earth were like. We will never be able to say for sure that building blocks of life formed spontaneously on this planet. We can only say that such a scenario is plausible, given what we know about chemistry.

By another hypothesis, simple organic compounds that served as building blocks for the first life formed in outer space. Support for this hypothesis comes from the presence of amino acids in interstellar clouds and in carbon-rich meteorites that land on Earth.

Regardless of whether the first small organic compounds formed on Earth or arrived with meteorites,

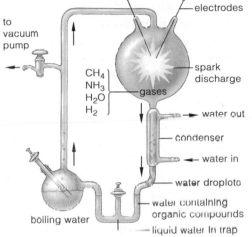

Figure 20.4 Animated
Diagram of the apparatus that Stanley Miller used to test whether organic compounds could have formed spontaneously on early Earth. Miller circulated water vapor, hydrogen gas (H_2), methane (CH_4), as well as ammonia (NH_3) in a glass chamber to simulate the first atmosphere. Sparks provided by an electrode simulated lightning.

questions remain. Where and how did small organic subunits assemble into proteins, phospholipids, and complex carbohydrates? Such polymers cannot form from low concentrations of subunits. What could have concentrated the subunits? We consider some possible answers to these questions in the next section.

Take-Home Message

What do scientists consider the best-supported hypotheses for the origins of the universe, our planet, and the building blocks of life?

■ The universe began in an instant, with a big bang about 13 to 15 billion years ago. It is still expanding.

■ Earth formed from material orbiting the sun. Earth's early atmosphere came from gas released by volcanoes. It was low in oxygen.

■ Small organic molecules that serve as the building blocks for living things can be formed by nonliving mechanisms. Amino acids form in reaction chambers designed to simulate early Earth and are also present in some meteorites.

■ We will never know for sure how the first cells came to be, but we can investigate the possible steps on the road to life.

■ Links to Nucleic acids 3.7, Membranes 5.1, Cofactors 6.3, Transcript modifications 14.3, Translation 14.4

Origin of Proteins and Metabolism

Metabolism and genetic replication boil down to this: One interacting group of molecules makes copies of itself over and over again using different molecules as "food" for the reactions.

Today, proteins called enzymes are the workhorse of metabolism. By one hypothesis, the first proteins assembled on clay-rich tidal flats (Figure 20.5a). Clay particles carry a slight negative charge, so positively charged amino acids dissolved in seawater will tend to stick to them. If clay became exposed during low tide, evaporation would have concentrated the amino acids even more. Energy from the sun could have caused amino acids to bond together. Some experiments do support this clay-template hypothesis. When exposed to conditions designed to simulate tidal flats, amino acids bond together, forming proteinlike chains.

By another hypothesis, simple metabolic pathways evolved at **hydrothermal vents**. Superheated, mineral-rich water gets forced out from these seafloor fissures under high pressure. Iron sulfides, hydrogen sulfides, and other minerals settle out of the water and build up as deposits near the vents (Figure 20.5b). Perhaps these minerals promoted formation of organic compounds from carbon dioxide and other molecules dissolved in water. Iron-sulfide cofactors act in this way, assisting enzymes found in modern cells (Section 6.3).

Researchers have carried out simulations to mimic conditions in rocks at hydrothermal vents. Such rocks are covered with tiny chambers about the size of cells (Figure 20.5c). Iron sulfide in the walls of these experimental chambers behaved like a cofactor. It donated hydrogen and electrons to dissolved carbon dioxide, forming organic molecules that became concentrated inside the chambers.

Origin of the Plasma Membrane

All modern cells have a plasma membrane composed of lipids and proteins (Section 5.1). We do not know whether the first agents of metabolism were enclosed in a membrane. By one hypothesis, a membrane did become the outer boundary of protocells. We define a **protocell** as any membrane-enclosed sac of molecules that captures energy, concentrates materials, engages in metabolism, and replicates itself.

Spontaneous formations of saclike structures under experimental conditions demonstrate how protocells may have formed. In simulations of sunbaked tidal flats, amino acids formed long chains. When moistened, the chains assembled as vesicle-like structures with fluid inside (Figure 20.6a). A mix of fatty acids and alcohols

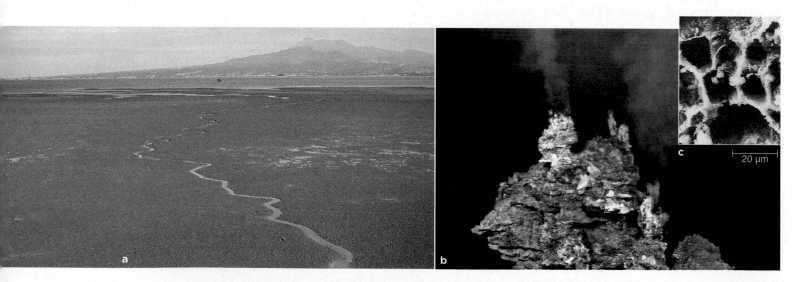

Figure 20.5 Where did the first complex organic compounds form? Two candidates are (**a**) clay templates in tidal flats, and (**b**) iron-sulfide rocks at hydrothermal vents on the deep ocean floor. (**c**) Laboratory simulations of conditions near the vents produce rocks riddled with cell-sized chambers. By one hypothesis, such chambers could have served as protected environments in which organic compounds accumulated and reactions took place. Iron-sulfide cofactors in living cells may be a legacy of such events.

Figure 20.6 Laboratory-formed models for protocells that may have preceded the emergence of cells. (**a**) Selectively permeable vesicles with an outer membrane of proteins were formed by heating amino acids, then wetting the resulting protein chains. (**b**) RNA-coated clay (stained *red*) enclosed in a simple membrane composed of fatty acids and alcohols (*green*). Mineral-rich clay promotes the assembly of such vesicle-like forms and catalyzes the formation of RNA strands from free nucleotides.

(**c**) One hypothesis for steps that led from lifeless chemicals to living cells.

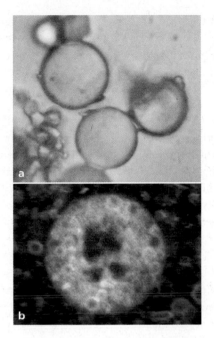

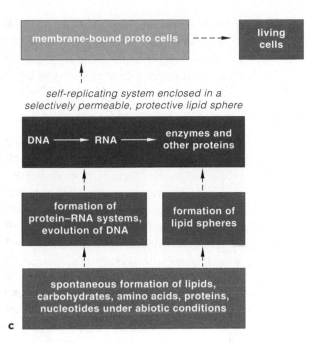

will self-assemble into sacs around bits of clay (Figure 20.6*b*). Finally, remember that when phospholipids and water mix, a lipid bilayer forms. Such a bilayer is the structural basis of all cell membranes.

Origin of Genetic Material

In all modern cells, DNA is the genetic material. Cells pass copies of their DNA to descendants, which use instructions encoded in that DNA to build proteins. Some of these proteins aid synthesis of new DNA, which is passed along to descendant cells, and so on. Protein synthesis depends on DNA, which in turn is built by proteins. How did this cycle begin?

In the 1960s, Francis Crick and Leslie Orgel took on this dilemma, suggesting that RNA may have been the first informational molecule. Since then, evidence for an early **RNA world**—a time when RNA both stored genetic information and functioned like an enzyme in protein synthesis—has accumulated.

Some RNAs still serve as enzymes in living cells. An rRNA in ribosomes catalyzes formation of peptide bonds during protein synthesis (Section 14.4). We know from DNA sequencing that ribosomal RNAs of eukaryotes resemble those of prokaryotes; ribosomes have not changed much over evolutionary time. This suggests that RNA's catalytic function evolved early in life's history.

The discovery of other catalytic RNAs, known as **ribozymes**, also supports the RNA world hypothesis. Natural ribozymes cut up and splice RNAs as part of transcript processing (Section 14.3). In the laboratory, researchers made synthetic, self-replicating ribozymes that copy themselves by assembling free nucleotides. Whether such ribozymes could have formed spontaneously on early Earth remains an open question.

Another question concerns the demise of the RNA world. If early self-replicating genetic systems were RNA based, why do all modern cells use DNA? The structure of DNA may hold the answer. Compared to single-stranded RNA, double-stranded DNA is less susceptible to breakage. A switch from RNA to DNA would have made larger, more stable genomes possible.

By another hypothesis, switching from RNA to DNA might have protected some early replicating systems from viruses that inserted themselves into RNA. Such viruses could not attack a DNA-based genome without evolving new enzymes. Until this viral evolution took place, DNA-based systems would be at an advantage.

Life's Early Evolution

■ Fossils and molecular comparisons among modern organisms inform us about the early history of life.

■ Links to Prokaryotic and eukaryotic cells 4.4, 4.6, Photosynthesis 7.3, Fossils 17.5, 17.6, Molecular clocks 19.4

The Golden Age of Prokaryotes

How old is life on Earth? Different methods provide different answers. Using accumulated mutations as a molecular clock (Section 19.4) indicates that the last universal common ancestor lived about 4.3 billion years ago. Microscopic filaments from Australia, which may be fossil cells, date back 3.5 billion years (Figure 20.7a). Microfossils from another Australian location indicate that cells were living around hydrothermal vents on the seafloor by 3.2 billion years ago.

The size and structure of early fossil cells suggests that they were prokaryotic (Section 4.4). In addition, gene sequence comparisons among living organisms put prokaryotes near the base of the tree of life. There was little oxygen in the air or seas 3 billion years ago, so early prokaryotes must have been anaerobic. They probably used dissolved carbon dioxide as a carbon source, and mineral ions as their source of energy.

The modern prokaryotes belong to two domains—archaeans and bacteria (Section 4.4). Based on gene differences between existing members of these groups, researchers estimate that they branched from a common ancestor by about 3.5 billion years ago.

After this divergence, one bacterial lineage evolved a new mode of nutrition: photosynthesis. How did the machinery of photosynthesis arise? By one hypothesis, some ancient bacteria had a pigment that detected heat energy. This pigment helped them locate the mineral-rich hydrothermal vents, upon which they depended. Later, mutations allowed the pigment to capture light energy, rather than detect heat. Hydrothermal vents emit a little light energy, which early photosynthesizers used to supplement their anaerobic metabolism. Later still, descendants of the vent-dwelling photosynthetic cells colonized sunlit waters. Here, they came to rely on photosynthesis and branched into many lineages.

What data support this hypothesis? Some modern bacteria do carry out photosynthesis using light from hydrothermal vents. Like most modern photosynthetic bacteria, these vent-dwellers form ATP by the cyclic pathway. This pathway does not produce oxygen.

The noncyclic pathway of photosynthesis, which does produce oxygen, evolved in a single bacterial lineage: the cyanobacteria (Figure 20.7b,c). Cyanobacteria are a relatively recent branch on the bacterial family tree, so noncyclic photosynthesis presumably arose through mutations that modified the cyclic pathway.

When the Proterozoic era began, 2.5 billion years ago, cyanobacteria and other photosynthetic bacteria were growing as dense mats in the seas. The mats trapped minerals and sediments. Over many years, continual cell growth and deposition of minerals formed large

Figure 20.7 Fossil prokaryotic cells. (**a**) A strand of what may be prokaryotic cells dates back 3.5 billion years. (**b,c**) Fossils of two types of cyanobacteria that lived 850 million years ago in Bitter Springs, Australia. (**d**) Artist's depiction of stromatolites in an ancient sea. (**e**) Cross-section through a fossilized stromatolite shows layers of material. Each layer formed when a mat of living prokaryotic cells trapped sediments. Descendant cells grew over the sediment layer, then trapped more sediment, forming the next layer.

Figure 20.8 Fossils of some early eukaryotes.

(**a**) One of the oldest known eukaryotic species, *Grypania spiralis*, which lived about 2.1 billion years ago. These fossil colonies are large enough to be visible to the naked eye.
(**b**) *Tawuia* probably was an early alga. (**c**) Fossils of a red alga, *Bangiomorpha pubescens*. This multicelled species lived 1.2 billion years ago. Cells were specialized; some formed a holdfast that anchored the body, others produced sexual spores.

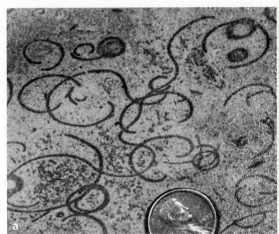

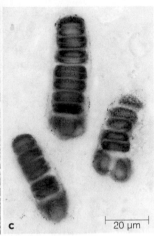

20 µm

dome-shaped, layered structures called **stromatolites** (Figure 20.7*d,e*). Such structures still form in some shallow seas today. The abundance of stromatolites soared during the Proterozoic era. Cyanobacterial populations increased, and so did their waste product: oxygen gas. Oxygen started to accumulate in Earth's waters and air. Sound familiar? Here we pick up the story that we began in Chapter 8.

An oxygen-rich atmosphere had three important consequences:

First, oxygen prevents the self-assembly of complex organic compounds, so life could no longer form spontaneously from nonliving materials.

Second, aerobic respiration evolved and became the main energy-releasing pathway. This pathway requires oxygen and produces ATP (Section 8.1). Compared to other energy-releasing pathways, aerobic respiration is far more efficient. It later met the high energy needs of multicelled eukaryotes.

Third, as oxygen enriched the atmosphere, an ozone layer formed. This layer keeps much of the sun's UV radiation from reaching Earth's surface. UV radiation can damage DNA. Without an ozone layer to protect it, life could not have moved onto land.

The Rise of Eukaryotes

In the Proterozoic era, forerunners of eukaryotic cells split from the archaean lineages. Rocks 2.8 billion years old have traces of lipids like those made by modern eukaryotes. The oldest eukaryotic fossils discovered so far date to 2.1 billion years ago (Figure 20.8*a,b*).

As you know, organelles are the defining features of eukaryotic cells (Section 4.6). What was their origin? The next section presents a few hypotheses. Another question: How do early eukaryotes fit in evolutionary trees? The earliest species that we can assign to any modern group is *Bangiomorpha pubescens*, a red alga that lived about 1.2 billion years ago (Figure 20.8*c*). This multicelled alga had specialized structures. Some of the cells helped anchor it. Others produced two types of sexual spores. Apparently, *B. pubescens* was one of the earliest sexually reproducing organisms.

After dominating the world's oceans for billions of years, stromatolites began to decline about 750 million years ago. The rise of algal competitors and changes in the mineral composition of seawater probably played a role. Newly evolved bacteria-eating organisms may have also contributed to the stromatolite decline.

So far, the earliest animal fossils known date back to 570 million years ago. These early animals were less than a millimeter across. They shared the oceans with bacteria, archaeans, fungi, and protists, including the lineage of green algae that would later give rise to land plants.

Animal diversity increased greatly during a great adaptive radiation in the Cambrian, 543 million years ago. When that period finally ended, all of the major animal lineages, including the vertebrates (animals with backbones), were represented in the seas.

Take-Home Message

What was early life like and how did it change Earth?

■ Life arose by 3–4 billion years ago; it was probably anaerobic and prokaryotic.

■ An early divergence separated ancestors of modern bacteria from the lineage that would lead to archaeans and eukaryotic cells.

■ The first photosynthetic cells were bacteria that used the cyclic pathway. Later, the oxygen-producing, noncyclic pathway evolved in cyanobacteria.

■ Oxygen accumulation in the air and seas halted spontaneous formation of the molecules of life, formed a protective ozone layer, and spurred the evolution of organisms that carried out the highly efficient pathway of aerobic respiration.

Where Did Organelles Come From?

- Eukaryotic cells have a composite ancestry, with different components derived from different prokaryotic ancestors.

- Links to Nucleus 4.8, Endomembrane system 4.9, Mitochondria and chloroplasts 4.11

Origin of the Nucleus, ER, and Golgi Body

Prokaryotic DNA lies in a region of cytoplasm, but the DNA of any eukaryotic cell resides inside a nucleus (Section 4.8). The outer boundary of the nucleus, the nuclear envelope, is continuous with the membrane of the endoplasmic reticulum (ER). How did eukaryotic internal cell membranes arise? Some prokaryotes may provide clues. Their plasma membrane folds into the cytoplasm, and it includes embedded enzymes, transport proteins, and other structures with roles in metabolic reactions (Figure 20.9a). Similar infoldings may have evolved in the ancestor of eukaryotic cells and become more elaborate over time.

What advantages could membrane infolding offer? Folds increase the membrane's surface area, so there is more space for any membrane-embedded metabolic machinery. Internal membranes allow specialization and a division of labor. A typical prokaryotic plasma membrane has to carry out all membrane functions. With infoldings, different parts of the membrane can become structurally and functionally specialized. As another advantage, infoldings create localized regions in which a cell can concentrate a specific substance.

Infoldings that extended around the DNA may have evolved into a nuclear envelope (Figure 20.9b). By one hypothesis, the nuclear envelope was favored because it kept genes safe from foreign DNA. Modern bacteria take up DNA from their surroundings and receive DNA injections from viruses. An alternative is that the nuclear envelope evolved after two prokaryotic cells fused. This membrane may have kept the incompatible genomes of the two cells separate.

Evolution of Mitochondria and Chloroplasts

Early in the history of life, cells became food for one another. Some cells engulfed and digested other cells. Intracellular parasites entered and dined inside their hosts. In some cases, the engulfed prey or parasites survived inside the predator or host. Inside this larger cell, they were protected, and had an ample supply of nutrients from their host's cytoplasm. Like their host, they continued to divide and reproduce.

Such interactions are the premise of the hypothesis of **endosymbiosis**, championed by Lynn Margulis and others. (Endo– means within; symbiosis means living together.) The symbiont lives out its life inside a host, and the interaction benefits one or both of them.

Most likely, mitochondria evolved after an aerobic bacterium entered and survived in a host cell (Figure 20.10a). The host cell may have been either an early eukaryote or an archaean. If it was an archaean, then a nuclear membrane might have evolved after the two cells fused. The membrane would have kept bacterial enzymes and genes from interfering with expression of the archaean host's genes.

In any case, the host began to use ATP produced by its aerobic symbiont while the symbiont began to rely on the host for raw materials. Over time, genes that specified the same or similar proteins in both the host and its symbiont were free to mutate. If a gene lost its function in one partner, a gene from the other could take up the slack. Eventually, the host and symbiont both became incapable of living independently.

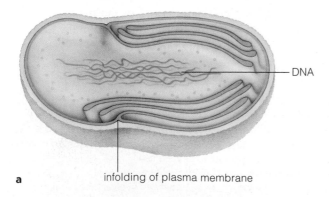

a

infolding of plasma membrane

DNA

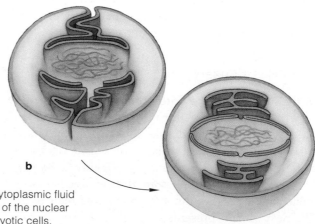

b

Figure 20.9 (**a**) Sketch of a soil bacterium, *Nitrobacter*. In this prokaryote, cytoplasmic fluid bathes permanent infoldings of the plasma membrane. (**b**) Model for the origin of the nuclear envelope and the endoplasmic reticulum. In the prokaryotic ancestors of eukaryotic cells, infoldings of the plasma membrane may have evolved into these organelles.

Evidence of Endosymbiosis

A chance discovery made by microbiologist Kwang Jeon supports the hypothesis that bacteria can evolve into organelles. In 1966, Jeon was studying *Amoeba proteus*, a species of single-celled protist. By accident, one of his cultures became infected by a rod-shaped bacterium. Some infected amoebas died immediately. Others survived, but growth slowed. Intrigued, Jeon maintained those infected cultures to see what would happen. Five years later, the descendant amoebas were host to many bacterial cells, yet they seemed healthy. When those amoebas received bacteria-killing drugs that usually do not harm amoebas, they died.

Experiments demonstrated that the amoebas had come to rely on the bacteria inside them. When amoebas in uninfected cultures were stripped of their nucleus and given the nucleus from an infected amoeba, they died. Something was missing. When bacteria were included with the nuclear transplant, recipient cells survived. Later studies showed that infected amoebas had lost the ability to make an essential enzyme. They depended on bacterial invaders to make it for them! The bacterial cells had become vital endosymbionts.

Mitochondria in living cells do resemble bacteria in size and structure (Section 4.11). A mitochondrion's inner membrane is like a bacterial plasma membrane. Like a bacterial chromosome, mitochondrial DNA is a circle with few noncoding regions between genes, and few or no introns. A mitochondrion does not replicate its DNA or divide at the same time as the cell.

Did chloroplasts, too, originate by endosymbiosis? By one scenario, a predatory early eukaryote engulfed photosynthetic cells. Those cells continued to function by absorbing nutrients from the host cytoplasm. The cells released oxygen and sugars into their aerobically respiring hosts, which benefited as a result.

In support of this hypothesis, genes and structures of modern chloroplasts resemble those of cyanobacteria. Also, a chloroplast does not replicate its DNA or undergo division at the same time as the cell.

Freshwater protists called glaucophytes offer more clues. Glauco– means pale green, and the name refers to the color of the protists' photosynthetic organelles, which resemble cyanobacteria (Figure 20.10b,c). Like cyanobacteria, these organelles have a cell wall.

However early eukaryotic cells arose, they had a nucleus, endomembrane system, mitochondria, and—in certain lineages—chloroplasts. These cells were the first protists. Over time, their many descendants came to include the modern protist lineages, as well as the plants, fungi, and animals. The next section provides a time frame for these pivotal evolutionary events.

Take-Home Message

How might the nucleus and other eukaryotic organelles have evolved?

- A nucleus and other organelles are defining features of eukaryotic cells.
- The nucleus and ER may have arisen through modification of infoldings of the plasma membrane.
- Mitochondria and chloroplasts descended from bacteria that were prey or parasites of early eukaryotic cells.

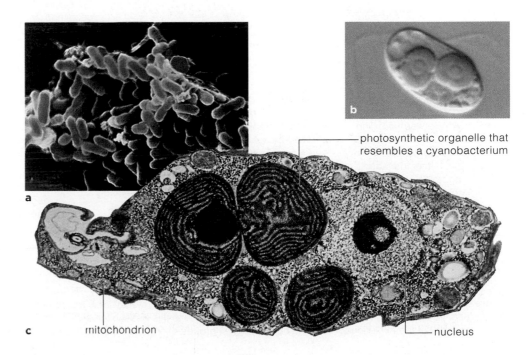

Figure 20.10 Examples of the many clues to ancient endosymbiotic interactions.

(**a**) What the ancestors of mitochondria may have looked like. Genes of the structurally simplest mitochondria we know about are similar to genes of *Rickettsia prowazekii*, a parasitic bacterium that causes typhus. Like mitochondria, *R. prowazekii* divides only inside the cytoplasm of eukaryotic cells. Enzymes in the host's cytoplasm catalyze the partial breakdown of organic compounds—a task that is completed inside aerobically respiring mitochondria.

(**b,c**) *Cyanophora paradoxa*, a glaucophyte. Its mitochondria resemble aerobic bacteria in size and structure. Its photosynthetic structures resemble cyanobacteria. They even have a wall similar in composition to the wall around a cyanobacterial cell.

photosynthetic organelle that resembles a cyanobacterium

mitochondrion

nucleus

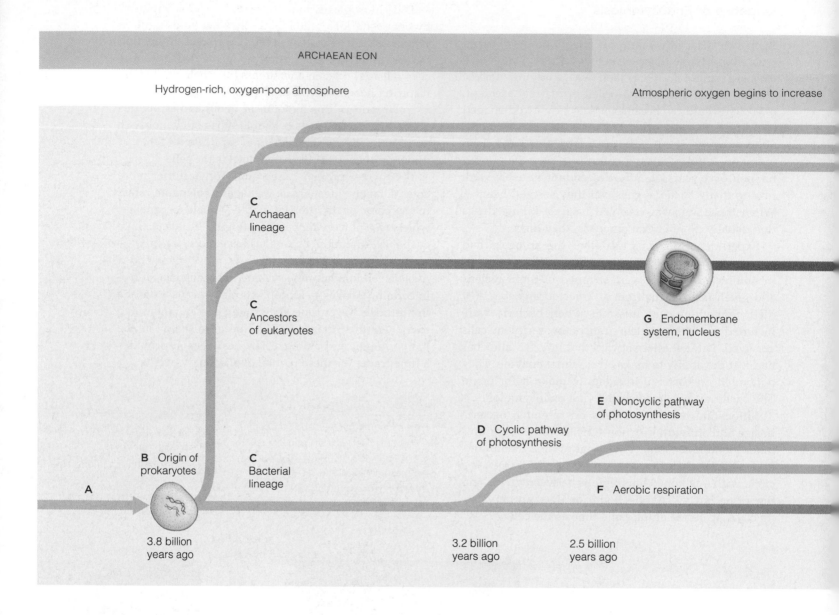

ARCHAEAN EON

Hydrogen-rich, oxygen-poor atmosphere

Atmospheric oxygen begins to increase

C
Archaean
lineage

C
Ancestors
of eukaryotes

G Endomembrane
system, nucleus

E Noncyclic pathway
of photosynthesis

D Cyclic pathway
of photosynthesis

B Origin of
prokaryotes

C
Bacterial
lineage

F Aerobic respiration

A

3.8 billion
years ago

3.2 billion
years ago

2.5 billion
years ago

Chemical and Molecular Evolution

A Between 5 billion and 3.8 billion years ago, lipids, proteins, nucleic acids, and complex carbohydrates formed from the simple organic compounds present on early Earth.

Origin of Prokaryotic Cells

B The first cells evolved by 3.8 billion years ago. They were prokaryotic; they did not have a nucleus or other organelles. Oxygen was scarce; the first cells made ATP by anaerobic pathways.

Three Domains of Life

C The first major divergence gave rise to bacteria and a common ancestor of archaeans and eukaryotic cells. Not long after that, the archaeans and eukaryotic cells parted ways.

Photosynthesis, Aerobic Respiration Evolve

D The cyclic pathway of photosynthesis evolved in one bacterial lineage.

E Noncyclic photosynthesis evolved in a branch from this lineage (cyanobacteria) and oxygen began to accumulate.

F Aerobic respiration evolved independently in some bacterial and archaean groups.

Origin of Endomembrane System, Nucleus

G Cell sizes and the amount of genetic information continued to expand in ancestors of what would become the eukaryotic cells. The endomembrane system, including the nuclear envelope, arose through the modification of cell membranes between 3 and 2 billion years ago.

Figure 20.11 Animated Milestones in the history of life, based on the most widely accepted hypotheses. As you read the next unit on life's past and present diversity, refer to this visual overview. It can serve as a simple reminder of the evolutionary connections among all groups of organisms. Time line not to scale.

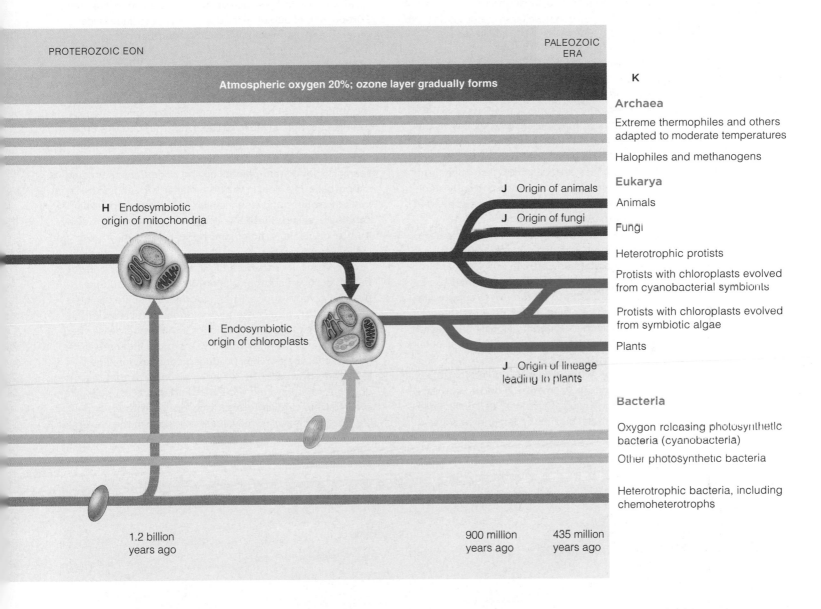

PROTEROZOIC EON

PALEOZOIC ERA

Atmospheric oxygen 20%; ozone layer gradually forms

K

Archaea

Extreme thermophiles and others adapted to moderate temperatures

Halophiles and methanogens

H Endosymbiotic origin of mitochondria

Eukarya

J Origin of animals

Animals

J Origin of fungi

Fungi

Heterotrophic protists

I Endosymbiotic origin of chloroplasts

Protists with chloroplasts evolved from cyanobacterial symbionts

Protists with chloroplasts evolved from symbiotic algae

Plants

J Origin of lineage leading to plants

Bacteria

Oxygen releasing photosynthetic bacteria (cyanobacteria)

Other photosynthetic bacteria

Heterotrophic bacteria, including chemoheterotrophs

1.2 billion years ago

900 million years ago

435 million years ago

Endosymbiotic Origin of Mitochondria

H Before about 1.2 billion years ago, an aerobic bacterium entered an anaerobic eukaryotic cell. Over generations, the two species established a symbiotic relationship. Descendants of the bacterial cell became mitochondria.

Endosymbiotic Origin of Chloroplasts

I By 1.5 billion years ago, a cyanobacterium entered a protist. Over generations, bacterial descendants evolved into chloroplasts. Later, some photosynthetic protists would evolve into chloroplasts inside other protist hosts.

Plants, Fungi, and Animals Evolve

J By 900 million years ago, representatives of all major lineages— including fungi, animals, and the algae that would give rise to plants—had evolved in the seas.

Lineages That Have Endured to the Present

K Today, organisms live in nearly all regions of Earth's waters, crust, and atmosphere. They are related by descent and share certain traits. However, each lineage encountered different selective pressures, and unique traits evolved in each one.

Figure It Out: From which prokaryotic lineage are mitochondria descended, bacteria or archaea?   *Answer: Bacteria*

20.6 | About Astrobiology

■ Studying conditions on other planets provides clues to how life arose on Earth. Similarly, what we learn about life on Earth informs our ideas about the possibility of life elsewhere.

What conditions allowed life to arise on Earth? What is necessary for life to persist? Could there be life elsewhere in the universe? The multidisciplinary field of **astrobiology** addresses these and other profound questions.

Among the most exciting topics in astrobiology is the possibility of life on Mars. Water, remember, is essential for life. Mars has ice at its poles and liquid water probably flowed across the planet's surface in the past. There is no ozone layer around Mars, so UV radiation currently sterilizes the surface. However, conditions might be somewhat more hospitable further down in the soil.

To practice soil sampling techniques that will be used during future unmanned Mars expeditions, scientists have turned to extreme habitats on Earth. In 2004, a team from the University of Arizona visited Chile's Atacama Desert. They sampled areas thought to be too dry to support life (Figure 20.12). By digging about 30 centimeters (1 foot) below the surface, scientists found previously undetected bacteria. Their advice to those planning the next sampling of Martian soil: Don't just scratch the surface.

An unmanned Mars mission now scheduled for 2013 will utilize this advice. A new instrument will drill up soil samples and test them for traces of amino acids, nucleotides, and other biological molecules that could indicate life.

In addition to Mars, astrobiologists are eager to explore Europa, one of Jupiter's moons. Europa's surface is ice-covered and astonishingly cold, but geothermal energy might be melting the ice in places deeper down. NASA is planning an unmanned mission to Europa in 20 to 30 years. The goal is to drill through the icy surface, then use a robot to explore any underlying oceans. A prototype robot has already been built and used to explore a deep well-like sinkhole in Mexico. The robot mapped this unseen realm and collected samples of bacteria from its depths.

Farther afield, earthbound and space telescopes are searching for distant planets having conditions that could support life. The goal is to find a planet with an Earth-like mass in orbit around a sun similar to our own. One likely candidate, called Gliese 581, lies about 20 light-years away. Its size, orbit, and type of sun suggest that it may have liquid water.

Suppose scientists do find evidence of past or present microbial life on another planet. Would it matter? Such a discovery would support the hypothesis that life can arise spontaneously as a consequence of chemical reactions. It would also make the possibility of intelligent life elsewhere in the universe seem more likely. The more life there is, the more likely evolution has produced complex, intelligent life forms elsewhere. Also, like the discoveries that the sun does not rotate around the Earth, and that all complex life evolved from simpler forms, discovery of extraterrestrial life would cause us to reevaluate our place in the universe.

Figure 20.12 Chile's Atacama Desert, which serves as a model and the testing ground for scientists interested in Martian soil. Jay Quade, a University of Arizona scientist, is visible in the distance at the right. He was part of a team that detected soil bacteria living underground in a part of the desert previously thought to be too dry to support life.

Looking for Life in All The Odd Places

Sometimes when researchers search for life, they are not quite sure what they have found. Consider nanobes: blobs and filaments found in deep rock layers and shown here. Nanobes have DNA and seem to grow, but they are only about one-tenth the size of a bacterial cell. Some researchers think nanobes are alive. Others think nanobes are too small to hold the machinery necessary for life and were

How would you vote?

Martian soil may contain microbes. Should we bring samples from Mars back to Earth for closer study? See CengageNOW for details, then vote online.

formed by nonbiological processes. Perhaps nanobes are something like the simple protocells that preceded life.

Summary

Section 20.1 According to the **big bang model**, the universe formed about 13 to 15 billion years ago and is still expanding (Table 20.1). Earth formed more than 4 billion years ago from rocky debris orbiting the sun. Its early atmosphere consisted mainly of volcanic gases, and held minimal oxygen.

Laboratory simulations provide indirect evidence that organic compounds such as amino acids and nucleotides self-assemble spontaneously under conditions like those thought to have prevailed on early Earth. Alternatively, the building blocks of life might have formed in deep space and reached Earth in meteorites.

■ *Use the animation on CengageNOW to see how Miller showed that organic compounds can form spontaneously.*

Section 20.2 Proteins that speed metabolic reactions might have first formed when amino acids stuck to clay, then bonded under the heat of the sun. Or, reactants could have begun interacting inside tiny holes in rocks near deep-sea **hydrothermal vents**, which emit scalding hot, mineral-rich water.

Membrane-like structures form when proteins or lipids are mixed with water. They serve as a model for **protocells**, which may have preceded cells.

An **RNA world**, a time in which RNA was the genetic material, may have preceded DNA-based systems. RNA still is a part of ribosomes that carry out protein synthesis in all organisms. Discovery of **ribozymes**, RNAs that act as enzymes, lends support to the RNA world hypothesis. Compared to DNA, RNA breaks apart easily. A switch from RNA to DNA would have made the genome more stable. It also might have offered a defense against viruses that attacked RNA-based cells.

Section 20.3 Fossils and gene sequencing suggest that the first life evolved 3 to 4 billion years ago. It was prokaryotic and, since oxygen levels were low, probably anaerobic. Early on, a divergence separated the ancestors of bacteria from the common ancestor of archaeans and eukaryotes.

Photosynthesis evolved in a bacterial lineage. The first pathway to evolve was cyclic. This photosynthesis pathway was modified in cyanobacteria; it became noncyclic and oxygen was released as a by-product.

Mats of photosynthetic bacteria dominated the seas for billions of years. Over generations these mats became

layered with sediments and formed **stromatolites**, dome-shaped structures which later became fossilized. Oxygen released by many cyanobacteria accumulated and changed Earth's atmosphere. The increased oxygen levels favored the evolution of aerobic respiration.

Protists were the first eukaryotic cells, and their fossils date back a little more than 2 billion years. By the start of the Cambrian, 570 million years ago, all major eukaryotic lineages were living in the seas.

Section 20.4 Internal membranes of eukaryotic cells such as a nuclear membrane and endoplasmic reticulum may have evolved by modification of infoldings of the plasma membrane. Mitochondria and chloroplasts most likely evolved by **endosymbiosis**. By this evolutionary process, one cell enters and lives in another. Then, over many generations, host and guest cells come to depend upon one another for essential metabolic processes.

Section 20.5 Evidence from many sources allows us to reconstruct the order of events and make a hypothetical time line for the history of life.

■ *Use the animated interaction on CengageNOW to investigate the history of life on Earth.*

Section 20.6 **Astrobiology** is a field of study concerned with the origins, evolution, and persistence of life on Earth as it relates to life in the universe. Astrobiologists study life in extreme habitats on Earth as their model for what might survive on other planets. In our own solar system, Mars and Europa (a moon of Jupiter) both have water and are considered possible cradles of life. Distant planets with Earth-like conditions may also have life.

Table 20.1 Major Events in the History of Life

Event	Estimated Time
Universe forms	13 to 15 billion years ago
Our sun forms	5 billion years ago
Earth forms	4.6 billion years ago
First prokaryotic cells	3.2 to 4.3 billion years ago
First eukaryotic cells	2.8 to 2 billion years ago

Data Analysis Exercise

Studies of ancient rocks and fossils can reveal changes that have taken place during Earth's existence. Figure 20.13 shows how asteroid impacts and composition of the atmosphere are thought to have changed over time. Use this figure and information in the chapter to answer the following questions.

1. Which occurred first, a decline in asteroid impacts, or a rise in the atmospheric level of oxygen?

2. How do modern levels of carbon dioxide and oxygen compare to those at the time when the first cells arose?

3. Which is now more abundant, oxygen or carbon dioxide?

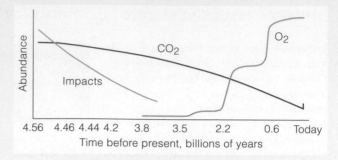

Figure 20.13 How asteroid impacts (*green*), atmospheric carbon dioxide concentration (*pink*), and oxygen concentration (*blue*) changed over geologic time.

Self-Quiz

Answers in Appendix III

1. According to the big bang model, _____ .
   a. Earth formed about 14 billion years ago
   b. the universe is continuing to expand
   c. carbon and oxygen were the first elements to form
   d. all of the above

2. An abundance of _____ in the atmosphere would have prevented the spontaneous assembly of organic compounds.

3. Stanley Miller's experiment demonstrated _____ .
   a. the great age of the Earth
   b. that amino acids can assemble spontaneously
   c. that oxygen is necessary for life
   d. all of the above

4. The prevalence of iron-sulfide cofactors in living organisms may be evidence that life arose _____ .
   a. in outer space   c. near deep-sea vents
   b. on tidal flats   d. in the upper atmosphere

5. According to one hypothesis, negatively-charged clay particles played a role in early _____ .
   a. protein formation   c. photosynthesis
   b. DNA replication   d. oxygen declines

6. An RNA that acts as an enzyme is a(n) _____ .

7. Certain pigments that evolved and were later used in photosynthesis may have first helped cells detect _____ .
   a. heat from hydrothermal vents
   b. iron-sulfide-rich rocks
   c. mineral-rich clays
   d. potential predators

8. The evolution of _____ resulted in an increase in the levels of atmospheric oxygen.
   a. sexual reproduction
   b. aerobic respiration
   c. the noncyclic pathway of photosynthesis
   d. the cyclic pathway of photosynthesis

9. Mitochondria are probably descendants of _____ .
   a. archaeans   c. cyanobacteria
   b. aerobic bacteria   d. anaerobic bacteria

10. Infoldings of the plasma membrane into the cytoplasm of some prokaryotes may have evolved into the _____ .
    a. nuclear envelope   c. primary cell wall
    b. ER membranes   d. both a and b

11. By the process of _____ , one cell lives inside another cell and the two become interdependent.

12. A _____ is a dome-shaped structure formed by mats of photosynthetic cells and sediments.

13. The first eukaryotes were _____ .
    a. fungi   c. protists
    b. plants   d. animals

14. Mars and Jupiter's moon Europa are considered possible candidates for life because they both _____ .
    a. have an ozone layer
    b. have ice and may have water
    c. are about the same temperature as Earth
    d. all of the above

15. Chronologically arrange the evolutionary events, with 1 being the earliest and 6 the most recent.
    ___1   a. emergence of the noncyclic
    ___2   pathway of photosynthesis
    ___3   b. origin of mitochondria
    ___4   c. origin of protocells
    ___5   d. emergence of the cyclic
    ___6   pathway of photosynthesis
                  e. origin of chloroplasts
                  f. the big bang

■ *Visit CengageNOW for additional questions.*

Critical Thinking

1. Researchers looking for fossils of the earliest life forms face many hurdles. For example, few sedimentary rocks date back more than 3 billion years. Review what you learned about plate tectonics (Section 17.9). Explain why so few remaining samples of these early rocks remain.

2. Craig Venter and Claire Fraser are working to create a "minimal organism." They are starting with *Mycoplasma genitalium*, a bacterium that has 517 genes. By disabling its genes one at a time, they discovered that 265–350 of them code for essential proteins. The scientists are synthesizing the essential genes and inserting them, one by one, into an engineered cell consisting only of a plasma membrane and cytoplasm. They want to see how few genes it takes to build a new life form. What properties would such a cell have to exhibit for you to conclude that it was alive?

IV EVOLUTION AND BIODIVERSITY

From the Green River Formation near Lincoln, Wyoming, the stunning fossilized remains of a bird trapped in time. During the Eocene, some 50 million years ago, sediments that had been gradually deposited in layers at the bottom of a large inland lake became its tomb. In this same formation, fossilized remains of sycamore, cattails, palms, and other plants suggest that the climate was warm and moist when the bird lived. Fossils from places all around the world yield clues to life's early history.

21 | Viruses and Prokaryotes

The Effects of AIDS

By the time Chedo Gowero (Figure 21.1) was thirteen years old, she had lost both her parents to AIDS, left school, and was working to support herself and her ten-year-old brother. Sadly, stories like Chedo's are common in Zimbabwe, a country where one-fifth of the population has AIDS.

AIDS is short for *Acquired Immune Deficiency Syndrome*, and the virus that causes it is called HIV, for *Human Immune Deficiency Virus*. Worldwide, more than 20 million people have died from AIDS and about 39 million are now infected with HIV. With rare exceptions, all of those infected either have AIDS or will develop it. In developed countries, antiviral drugs can help slow the progression of the disease. However, in less developed countries, only a miniscule percentage of the population has access to such drugs.

In the hardest hit regions, including sub-Saharan Africa, AIDS is shredding the cultural fabric. Orphans often drop out of school, endangering their nation's future, and some turn to prostitution or other crimes to survive. People weakened by AIDS cannot plant and care for crops, so food shortages—a chronic problem in sub-Saharan regions—keep getting worse.

The virus that causes all this misery was first isolated in the early 1980s. Since then, researchers have found that there are two strains, HIV-1 and HIV-2. HIV-1 is the most prevalent cause of AIDS. Sequencing of the HIV genome revealed that HIV closely resembles *Simian Immunodeficiency Virus* (SIV). SIV infects wild chimpanzees in Africa, so the first human infected probably was someone who butchered or ate meat from an SIV-infected chimp. We know that HIV has been in humans since at least 1959. Scientists recently detected HIV-1 in stored blood taken from an African man in 1959.

Most people now infected by HIV contracted the virus through sex with an infected partner. Anal, vaginal, and oral sex all can allow HIV to enter the body. A latex barrier such as a condom helps minimize the likelihood of such viral transfers. Infected mothers can transmit HIV to their child during the birth process, or in their milk. Exposure to HIV-infected blood by sharing of needles or transfusion also promotes infection.

Once inside a human body, HIV infects white blood cells that play a pivotal role in immune responses (a process we will discuss in Chapter 38). The virus takes over the cell's metabolic machinery and uses it to make more viral particles. Eventually, the infected white blood cell dies. Death of white blood cells as a result of HIV infection destroys the body's ability to defend itself. As a result, other viruses and disease-causing organisms run rampant, causing the symptoms of AIDS, and the problems that lead to death.

Despite international efforts, scientists have not yet come up with a vaccine that can prevent AIDS. Drugs that slow viral replication help HIV-infected people stay healthy, but the virus remains in their body. Therefore the drugs, which often have unpleasant side effects, must be taken for life.

HIV and other pathogens that endanger human health are a major focus of this chapter, but they are just part of the story. Most viruses and bacteria do us no harm. Some, such as the bacteria that live in our gut and synthesize essential vitamins, benefit us directly. Others benefit us indirectly. For example, bacteria put oxygen into the air and help crop plants grow by enriching the soil with nitrogen. Some viruses can kill harmful bacteria that could sicken us.

See the video! Figure 21.1 Chedo Gowero is one of the estimated 11 million African children orphaned by AIDS. HIV, the retrovirus that causes this disease (*far left*), first infected humans in Africa and continues to devastate that continent.

Key Concepts

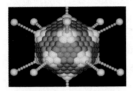

Viruses and other noncellular infectious particles

Viruses are noncellular particles made of protein and nucleic acid. They replicate by taking over the metabolic machinery of a host cell. Viroids are short sequences of infectious RNA. Prions are infectious misfolded versions of normal proteins. **Sections 21.1–21.3**

Features of prokaryotic cells

Prokaryotes are single-celled organisms that do not have a nucleus or the diverse cytoplasmic organelles found in most eukaryotic cells. Collectively, they show great metabolic diversity. They divide rapidly and exchange DNA by a variety of mechanisms. **Sections 21.4, 21.5**

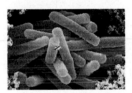

The bacteria

Bacteria are the most abundant prokaryotic cells on Earth. They perform important services such as degrading wastes, adding oxygen to the air, and providing essential nutrients to plants. Nearly all disease-causing prokaryotes are bacteria. **Section 21.6**

The archaeans

Archaeans are the more recently discovered, less studied prokaryotic group. Some show a remarkable ability to survive in extreme habitats, but others live in more ordinary places. They play important roles in ecosystems. **Section 21.7**

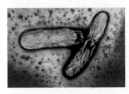

Evolution and disease

An immense variety of pathogens, or disease-causing agents, infect human hosts. Pathogens and their hosts coevolve; each acts as a selective agent on the other. **Section 21.8**

Links to Earlier Concepts

- In this chapter, we expand on our initial description of prokaryotes in Section 4.4. We reconsider classification of these cells (19.6), and methods used to determine their relationships to one another (16.5, 19.4). We discuss their role in Earth's early history (20.3) and think again about how they carry out photosynthesis (7.8).

  You will read again about the hypotheses that life emerged near hydrothermal vents (20.2) and that chloroplasts and mitochondria evolved from bacteria (20.4).

- Protein structure (3.5, 3.6) and protein synthesis (14.1–14.4), are discussed in the context of disease-causing agents.

- Chromosome structure (9.1) and some of the experiments that led to our understanding of DNA function (13.1) come up again, as do the tools of biotechnology (16.1, 16.2).

- Discussion of disease will provide new examples of directional selection (18.4) and coevolution (18.12).

How would you vote? Antiviral drugs extend lives of HIV-infected people but the drugs are patented and expensive. Cheap generic drugs infringe on these patents. Is it okay to ignore patents? Or would this discourage future drug research? See CengageNOW for details, then vote online.

- A virus consists of nucleic acid and protein. It is smaller than any cell and has no metabolic machinery of its own.

- Link to Discovery of DNA function 13.1

Viral Discovery and Traits

In the late 1800s, researchers studying tobacco plants discovered a new disease-causing agent, or **pathogen**. It was so small that it passed through screens that filtered out bacteria, and it could not be seen with a light microscope. The scientists called this unseen infectious entity a virus, a term that means "poison" in Latin.

Today, we define a **virus** as a noncellular infectious particle that cannot replicate on its own. A virus has a protein coat around genetic material that may be DNA or RNA. Some viruses also have a lipid envelope that covers their coat. A virus does not have ribosomes or other metabolic machinery of its own. To replicate, the virus must insert its genetic material into a cell of a specific organism, which we call its host. A viral infection is like a cellular hijacking; viral genes take over a host cell's machinery and direct it to synthesize viral proteins and nucleic acids. These components then self-assemble as viral particles. Table 21.1 summarizes viral characteristics.

Examples of Viruses

Virus structure varies, but a viral coat always consists of protein molecules arrayed in a repeating pattern. For example, tobacco mosaic virus, the first virus discovered, is rod-shaped, with coat proteins arranged in a helix around a strand of RNA (Figure 21.2a). In addition to tobacco, this virus infects tomatoes, peppers, petunias, and other plants. Infected plants are weak, with blotches of yellow or light green on their leaves. Similarly, brown blotches on orchid leaves are often symptoms of viral infection (Figure 21.3a). In tulips, a viral infection can produce streaks of color in

Table 21.1 Characteristics of a Virus

1. Noncellular; no cytoplasm, ribosomes, or other typical cell components

2. Genetic material may be DNA or RNA

3. Can only replicate inside a living host cell

4. Small (about 25 to 300 nanometers); nearly all are visible only with an electron microscope

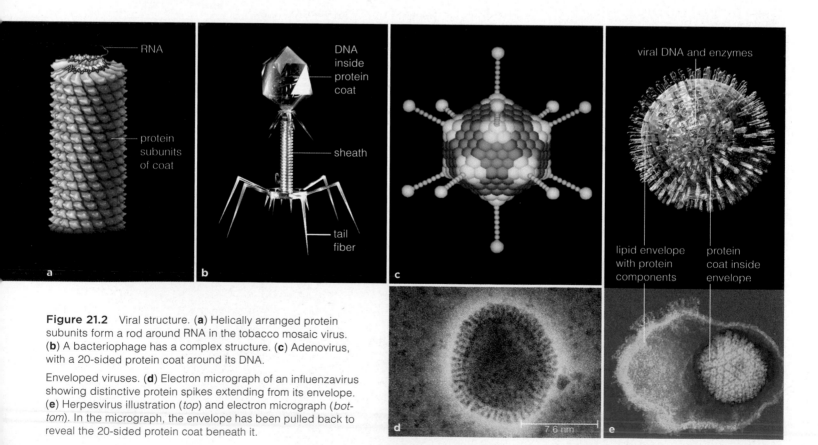

Figure 21.2 Viral structure. (**a**) Helically arranged protein subunits form a rod around RNA in the tobacco mosaic virus. (**b**) A bacteriophage has a complex structure. (**c**) Adenovirus, with a 20-sided protein coat around its DNA.

Enveloped viruses. (**d**) Electron micrograph of an influenzavirus showing distinctive protein spikes extending from its envelope. (**e**) Herpesvirus illustration (*top*) and electron micrograph (*bottom*). In the micrograph, the envelope has been pulled back to reveal the 20-sided protein coat beneath it.

flowers (Figure 21.3b). Plant cells have a thick wall, so plant viruses usually infect a plant only after insects or pruning create a wound that allows the virus in.

Bacteriophages are viruses that infect bacteria or archaeans. One well-studied group of bacteriophages has a complex structure (Figure 21.2b). Their DNA is in a "head" that has 20 triangular surfaces, a common coat shape for viruses. Attached to the head is a rod-like "tail" with fibers that help the virus attach to its host. Studies of these bacteriophages helped reveal DNA's function as genetic material (Section 13.1).

Adenoviruses are naked (non-enveloped) viruses that infect animals. Their 20-sided protein coat has a distinctive protein spike at each corner (Figure 21.2c). Human eye infections and upper respiratory ailments are often caused by adenoviruses. Other naked viruses cause hepatitis, polio, common colds, and warts.

More often, animal viruses have an envelope made of membrane derived from the host cell in which the virus self-assembled. For example, influenzaviruses, which can cause flu in humans and animals, are RNA viruses that have protein spikes radiating out through their envelope (Figure 21.2d). Other enveloped RNA viruses cause AIDS, rabies, rubella (German measles), bronchitis, mumps, measles, yellow fever, and West Nile encephalitis. Enveloped DNA viruses also cause disease. For example, herpesviruses are DNA viruses with a 20-sided coat underneath their envelope (Figure 21.2e). Various herpesviruses cause chicken pox, cold sores, genital herpes, or mononucleosis. An enveloped DNA virus also causes the deadly disease smallpox.

Impacts of Viruses

Viruses take a heavy toll on human health. They often are difficult to treat because they get inside cells, where medicines cannot reach them. We have many drugs that kill bacteria, but far fewer that eliminate viruses. The best medical defense against viruses is a vaccine that puts the immune system on the lookout for a specific virus and thus prevents infection. However, there are many viruses for which we do not yet have a vaccine.

Viruses can have devastating economic effects when they infect crop plants or domestic animals. For example, foot-and-mouth disease is a highly contagious viral disease of cattle, sheep, pigs, and goats. It is common in Asia, Africa, the Middle East, and South America. An outbreak in England in 2001 forced the slaughter of hundreds of thousands of animals. The United States has not had a case of foot-and-mouth disease since 1929, but meat imports and travelers who have visited foreign farms have potential to reintroduce the virus.

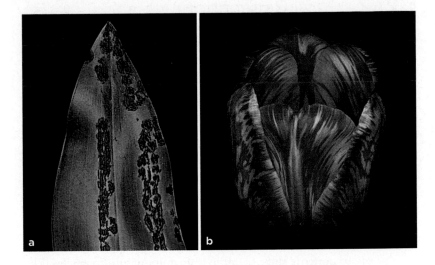

Figure 21.3 Effects of plant viruses. (**a**) Orchid fleck virus causes brown spots on leaves. (**b**) Another virus causes color streaks in this tulip's petals.

Some viruses benefit us indirectly, through effects on other species. Viruses infect organisms in all three domains of life. Generally, a viral infection decreases a host's ability to survive and reproduce. We benefit when viruses attack insects that eat our crops or bacteria that cause human disease. For example, meats can be treated with a bacteriophage-containing spray to kill bacteria that cause spoilage or food poisoning.

Viral Origins and Evolution

How did viruses originate and how are they related to cellular organisms? There are three main hypotheses. The first is that viruses are descendants of cells that were parasites inside other cells. Over time, most functions of these parasitic cells were delegated to the host, leaving the virus unable to survive on its own. A second hypothesis is that viruses are genetic elements that escaped from cells. The fact that some viral genes have counterparts in cellular organisms lends support to these two hypotheses.

The third hypothesis is that viruses represent a separate evolutionary branch, that they arose independently from the replicating molecules that preceded cells. This hypothesis would explain why most viral proteins are unlike any found in cellular organisms.

Take-Home Message

What are viruses and how do they affect us?

■ Viruses are noncellular infectious particles that multiply only inside living cells. Some viruses infect humans and harm us by causing disease. Others benefit us by controlling disease-causing organisms or pests.

■ Viruses may have evolved from cellular organisms or a part of their genome. Alternatively, viruses may represent a separate evolutionary branch.

21.2 Viindicates Viral Replication

- All viruses replicate only inside host cells, but the details of the process vary among viral groups.
- Links to Transcription 14.2, Translation 14.4

Steps in Replication

Viral multiplication cycles are varied, but nearly all go through the five steps listed in Table 21.2. The virus first attaches to an appropriate host cell by binding to a specific protein or proteins in the host's plasma membrane. The virus, or just its genetic material, enters the cell. Viral genes direct the cell to replicate the viral DNA or RNA and to build viral proteins. These components self-assemble to form new viral particles. The new viruses bud from the infected host cell or are released when the host bursts. A few examples will illustrate the process.

Bacteriophage Replication

There are two bacteriophage multiplication pathways (Figure 21.4). In the **lytic pathway**, the virus attaches to the host cell and injects DNA into it. Viral genes direct the host to make viral DNA and proteins, which assemble as viral particles. Soon the virus-filled cell lyses. Here, lysis refers to the disintegration of a host cell plasma membrane, wall, or both, which lets the cytoplasm—and the new viral particles—dribble out. Under direction of viral genes, the host makes a viral enzyme that initiates lysis, and causes its own death.

In the **lysogenic pathway**, a virus enters a latent state that extends the multiplication cycle. Viral genes get integrated into the host chromosome. The viral DNA is copied along with host DNA, and is passed along to all descendants of the host cell. Like miniature time bombs, the viral DNA inside these descendants awaits a signal to enter the lytic pathway.

Table 21.2 Steps in Most Viral Multiplication Cycles

1. Attachment Proteins on viral particle chemically recognize and lock onto specific receptors at the host cell surface.

2. Penetration Either the viral particle or its genetic material crosses the plasma membrane of a host cell and enters the cytoplasm.

3. Replication and synthesis Viral DNA or RNA directs host to make viral nucleic acids and viral proteins.

4. Assembly Viral components assemble as new viral particles.

5. Release The new viral particles are released from the cell.

Some bacteriophages can only replicate by the lytic pathway. They kill quickly and are not passed from one bacterial generation to the next. Others embark upon either the lytic or lysogenic pathway, depending on conditions in the host cell.

Replication of Herpes, an Enveloped DNA Virus

Like some bacteriophages, herpesviruses can enter a latent state. For example, most people are infected by Herpes simplex virus type 1 (HSV-1). After an initial infection, the virus remains latent in their nerve cells until sunburn or another stress reawakens it. Painful cold sores on or near the lips are a sign that the virus is replicating. A related strain of herpes, HSV-2 causes genital herpes, which we discuss in Chapter 42.

Herpesviruses are enveloped DNA viruses. Infection gets underway when the virus attaches to a host cell's plasma membrane. The membrane and viral envelope fuse, putting the viral DNA and protein into the host's cytoplasm. Viral DNA enters the nucleus and directs synthesis of new viral DNA and proteins. Each new viral particle self-assembles, then gets a bit of the host cell's inner nuclear membrane as its envelope. New viral particles exit the cell by exocytosis.

Replication of HIV, a Retrovirus

HIV is a retrovirus, an enveloped virus with RNA as its genetic material. It binds to receptors on certain white blood cells. The viral envelope fuses with the plasma membrane of the host cell, then viral protein and RNA enter the cell (Figure 21.5a).

For HIV to take over the host cell, viral RNA must be used to make viral DNA. The virus has an enzyme, **reverse transcriptase**, that catalyzes production of the DNA (Figure 21.5b). After the viral DNA forms, it gets integrated into a host's chromosome with the help of another viral enzyme (Figure 21.5c).

Once integrated into the host's chromosome, viral genes direct production of viral RNA and proteins (Figure 21.5d,e). RNA and viral proteins assemble into new viral particles. The particles bud from the host cell in an envelope derived from the host cell's plasma membrane (Figure 21.5f).

Drugs designed to fight HIV take aim at steps in viral replication. Some interfere with binding of HIV to a host cell. Others such as AZT slow reverse transcription of RNA. Integrase inhibitors prevent viral DNA from integrating into a human chromosome. Protease inhibitors prevents the processing of newly translated polypeptides into mature viral proteins.

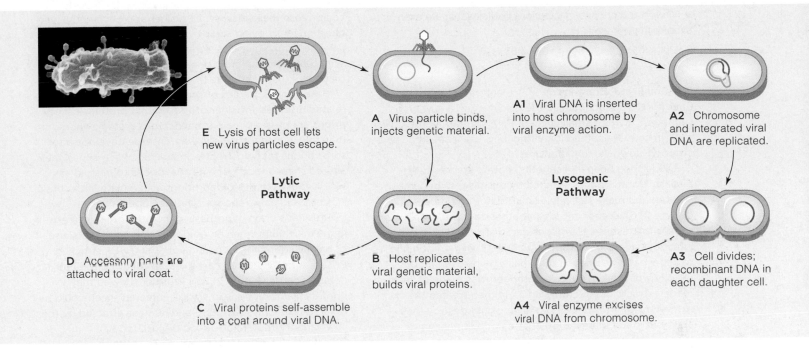

E Lysis of host cell lets new virus particles escape.

A Virus particle binds, injects genetic material.

A1 Viral DNA is inserted into host chromosome by viral enzyme action.

A2 Chromosome and integrated viral DNA are replicated.

Lytic Pathway

Lysogenic Pathway

D Accessory parts are attached to viral coat.

B Host replicates viral genetic material, builds viral proteins.

A3 Cell divides; recombinant DNA in each daughter cell.

C Viral proteins self-assemble into a coat around viral DNA.

A4 Viral enzyme excises viral DNA from chromosome.

Figure 21.4 Animated Pathways in the multiplication cycle of a bacteriophage.

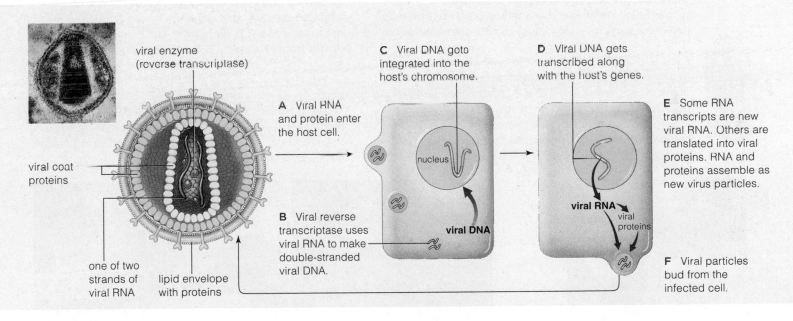

viral enzyme (reverse transcriptase)

C Viral DNA gets integrated into the host's chromosome.

D Viral DNA gets transcribed along with the host's genes.

A Viral RNA and protein enter the host cell.

E Some RNA transcripts are new viral RNA. Others are translated into viral proteins. RNA and proteins assemble as new virus particles.

viral coat proteins

nucleus

viral RNA viral proteins

B Viral reverse transcriptase uses viral RNA to make double-stranded viral DNA.

viral DNA

one of two strands of viral RNA

lipid envelope with proteins

F Viral particles bud from the infected cell.

Figure 21.5 Animated Multiplication cycle of HIV, the retrovirus that causes AIDS.

Drugs lower the number of HIV particles, so a person stays healthier. Less HIV in body fluids also means reduced risk of passing the virus to others. However, no drug eliminates the virus, all have unpleasant side effects, and all must be taken for life. Also, as you will learn shortly, HIV can become drug resistant.

Take-Home Message

How do viruses replicate?

■ A virus binds to a cell of its specific host, and the virus or its genetic material enters the host cell. Viral genes direct the production of viral components, which self-assemble as new viral particles.

21.3 | Viroids and Prions

■ Viroids and prions are infectious particles that are even simpler than viruses.

■ Links to Protein structure 3.5, 3.6

The Smallest Pathogens In 1971, plant pathologist Theodor Diener announced the discovery of a new type of pathogen. It was a small circle of RNA without a protective protein coat. He named it a **viroid**, because it seemed like a stripped-down version of a virus.

Diener had been investigating potato spindle tuber disease. Potato plants affected by this disease become stunted and make just a few small, deformed tubers (Figure 21.6). Diener initially hypothesized that a virus causes the disease. He reconsidered after realizing the infectious agent passed through filters that were too fine to allow passage of even the smallest virus. To find out exactly what this minuscule pathogen was made of, he made extracts of infected plants. He then treated the extracts with enzymes to see what would destroy their ability to infect plants. Extracts treated with enzymes that digested DNA or protein still infected plants. Only enzymes that digest RNA rendered the extracts harmless.

Plant pathologists have now described about thirty viroids, many of which cause disease in commercially valued plants. Only one viroid has been found to affect human health. It interacts with a virus in human liver cells, and these particles jointly cause the disease hepatitis D.

a

Fatal Misfoldings Neurologist Stanley Prusiner's research began after he watched helplessly as one of his patients died of Creutzfeldt–Jakob disease (CJD). This rare brain disease causes dementia and death. Prusiner knew that CJD resembled another rare disease, kuru, which occurred among members of a certain tribe in New Guinea who engaged in ritual cannibalism of their dead. What agent could cause these diseases? Prusiner approached the question by studying scrapie, a disease that affects sheep. Like the human diseases, scrapie causes neurological symptoms, and the brain becomes so riddled with holes that it looks like a sponge (Figure 21.7a).

Based on his studies of scrapie and his knowledge of related diseases, Prusiner proposed that proteins called **prions** are present in a normal nervous system, where they fold in a characteristic way. Disease develops after some prions fold incorrectly. In an unknown manner, their altered shape induces the normal prions to misfold as well. Deposits of misfolded prions accumulate in the brain, kill cells, and cause the sponge-like appearance.

Prusiner's prion hypothesis generated great interest in the mid-1980s when an epidemic of mad cow disease, or bovine spongiform encephalopathy (BSE), struck in Britain. A rise in human cases of CJD followed the epidemic in cattle (Figure 21.7b). Prusiner showed that a prion similar to the one in scrapie-infected sheep could be isolated from cows with BSE and humans affected by the new variant Creutzfeldt–Jakob disease (vCJD).

How did a prion get from sheep to cattle to people? The cattle ate feed that included remains of infected sheep, then the infected beef sickened humans. Use of animal parts in livestock feed is now banned, and the number of cases of BSE and vCJD has declined. Cattle with BSE still turn up, even in the United States, but are not a major threat to humans. In 2007, there were three deaths from vCJD, all in Britain. This brought the death toll for this disease since 1990 to 161.

Prusiner was awarded a Nobel Prize for his discovery of prions. He continues to study prion diseases, and he hopes to develop preventive treatments and cures. His research also addresses issues raised by skeptics. Some scientists suspect that an as yet unidentified virus or small nucleic acid has a role in these diseases.

b

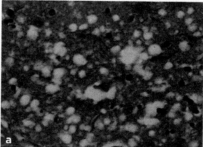

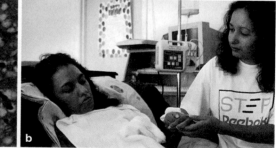

Figure 21.7 (**a**) Perforations in BSE-damaged brain tissue. *Right*, a model for a normal prion. The version that causes vCJD misfolds into a different three-dimensional shape and encourages normal versions to do the same.

(**b**) Charlene Singh being cared for by her mother. She is one of three people who developed symptoms of vCJD while living in the United States. Like the others, Singh most likely contracted the disease elsewhere; she spent her childhood in Britain. Loss of memory and sense of balance led to her diagnosis in 2001. She died in 2004.

Figure 21.6 (**a**) A potato produced by a plant infected by the potato spindle tuber viroid. (**b**) Yan Zhao and Rosemarie Hammond study plants infected by this viroid. They hope to learn how it gets into the nucleus of an infected plant cell.

21.4 | Prokaryotes—Enduring, Abundant, and Diverse

■ Prokaryotes evolved before eukaryotes. They still persist in enormous numbers and they show great metabolic diversity.

■ Links to Prokaryotes 4.4, Early life 20.3, Classification 19.6, Gene sequencing 16.3, Comparative biochemistry 19.4

Evolutionary History and Classification

The earliest cells, remember, had no nucleus; they were the first **prokaryotes**. (Pro– means before, karyon is taken to mean the nucleus.) Modern prokaryotes also lack a nucleus and they are structurally simple. Keep in mind that structural simplicity does not imply inferiority. Prokaryotes and eukaryotes have coexisted for more than a billion years, and prokaryotes still thrive. Some live in and feed on their larger, more complex eukaryotic neighbors. From an evolutionary viewpoint, prokaryotes and eukaryotes are both successful.

Traditionally, prokaryotic cells have been classified by numerical taxonomy. An unidentified prokaryotic cell is compared against a known group on the basis of shape, cell wall properties, metabolism, and other traits. The more traits that the cell shares with the known group, the closer is their inferred relatedness.

Such methods favored studies of prokaryotes that grew readily in the laboratory. These cells could be stained, examined under the microscope, and grown on different nutrients to characterize their metabolic traits. However, most prokaryotes cannot be cultured.

Automated gene sequencing and other methods of comparative biochemistry revolutionized the process of prokaryotic classification. They allow researchers to collect cell samples from nature, then compare genes to those of known species. Such comparisons revealed evidence of a divergence that occurred shortly after life began. One branching led to Bacteria. The other gave rise to Archaea and the ancestors of eukaryotes:

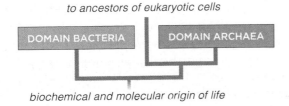

to ancestors of eukaryotic cells

DOMAIN BACTERIA   DOMAIN ARCHAEA

biochemical and molecular origin of life

Despite new techniques, relatively few prokaryotic species have been named. There are 1.4 million named eukaryotes, but less than 5,000 named prokaryotes. DNA analysis of soil and water samples suggests there may be millions of prokaryotic species.

In addition to discovering species, microbiologists identify **strains**, subgroups within a species that can be characterized by some identifiable trait or traits. For

Table 21.3 Prokaryotic Nutritional Modes

Mode of Nutrition	Carbon Source	Energy Source
Photoautotrophic	CO_2	Light
Chemoautotrophic	CO_2	Inorganic substances
Photoheterotrophic	Organic compounds	Light
Chemoheterotrophic	Organic compounds	Organic compounds

example, most *Escherichia coli* bacteria are harmless. However, one strain, *E. coli* O157:H7, makes a toxin that causes deadly food poisoning.

Abundance and Metabolic Diversity

In terms of reproductive success, the prokaryotes are unparalleled. Biologists at the University of Georgia once estimated that 5,000,000,000,000,000,000,000,000,000,000 bacterial cells were alive at that moment on Earth.

Metabolic diversity is important in the success of prokaryotes. There are four modes of nutrition and, as a group, prokaryotes utilize all of them (Table 21.3).

Photoautotrophs are photosynthetic; they use light energy to build organic compounds from carbon dioxide and water. This group includes nearly all plants, and some protists, as well as many prokaryotes.

Chemoautotrophs get energy by removing electrons from inorganic molecules such as sulfides or ammonia. They use this energy to build organic compounds from carbon dioxide and water. All are prokaryotes.

Photoheterotrophs are prokaryotes that use light energy and obtain carbon by breaking down organic compounds in the environment. This nutritional mode also occurs only in prokaryotes.

Chemoheterotrophs get both carbon and energy by breaking down organic compounds. Many prokaryotes are in this group, as are some protists, and all animals and fungi. Some chemoheterotrophs feed on living organisms and others are **saprobes**: organisms that break down wastes or remains. Saprobes play an important role as decomposers.

Take-Home Message

Why do biologists consider prokaryotic lineages successful?

■ Despite their structural simplicity, prokaryotic cells have persisted for billions of years. Prokaryotes gave rise to the eukaryotes early in the history of life, and the groups continue to coexist.

■ Prokaryotes are Earth's most abundant organisms. The group's metabolic diversity has contributed greatly to their success.

21.5 Prokaryotic Structure and Function

- Prokaryotic cells have many structural features that adapt them to their environment.

- Links to DNA function 13.1, Molecular toolkit 16.1

Cell Structure and Size

Modern prokaryotes are the bacteria and archaeans, single-celled organisms that do not enclose their DNA in a nucleus. Instead, their single prokaryotic chromosome lies in a region of the cytoplasm known as the **nucleoid** (Figure 21.8 and Table 21.4).

All prokaryotic cells have ribosomes, and some have infoldings of the plasma membrane. However, none has an endomembrane system like that of eukaryotes.

The typical prokaryotic cell cannot be seen without a light microscope. It is far smaller than a eukaryotic cell, about the size of a mitochondrion. In fact, there is evidence that certain bacteria were the ancestors of mitochondria (Section 20.4).

Prokaryotes can often be described by their shapes. A coccus is spherical, a bacillus is rod-shaped, and a spirillum is a spiral (Figure 21.8a).

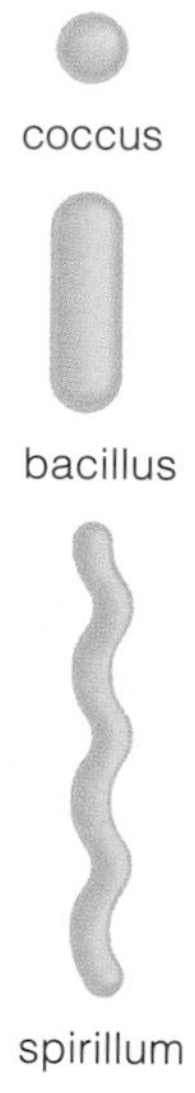

coccus

bacillus

spirillum

a

Table 21.4 Prokaryotic Cell Characteristics

1. No nucleus; chromosome in nucleoid

2. Generally a single chromosome (a circular DNA molecule); many species also contain plasmids

3. Cell wall present in most species

4. Ribosomes distributed in the cytoplasm

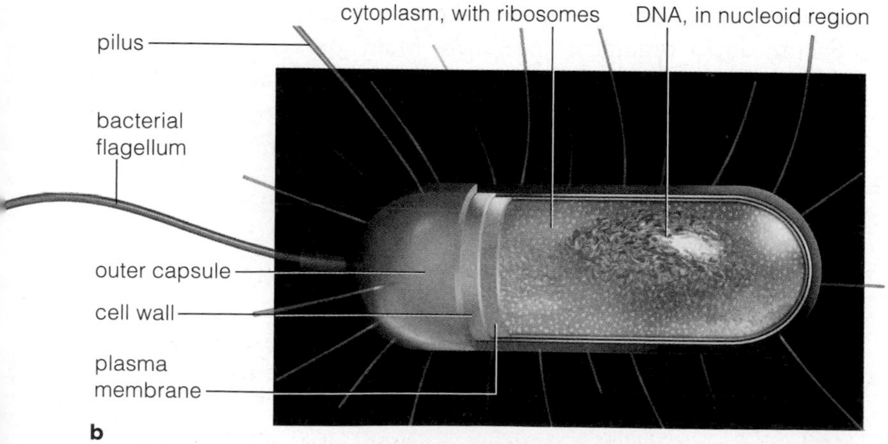

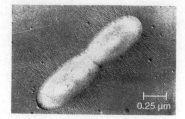

Figure 21.9 An *Escherichia coli* cell in the process of dividing by prokaryotic fission. **Figure It Out:** What are the many fine, threadlike structures extending from this cell? *Answer: pili*

0.25 μm

Nearly all prokaryotes have a semirigid, porous **cell wall** around the plasma membrane. The bacterial cell wall usually consists of peptidoglycan, a glycoprotein. Archean cell walls are made of other proteins.

Many prokaryotes have a secreted slime layer or a capsule outside the cell wall. Slime helps a cell adhere to surfaces. A capsule is tougher and helps some bacteria evade immune defenses of their hosts.

Many prokaryotic cells have one or more **flagella**. Unlike eukaryotic flagella, prokaryotic flagella do not contain microtubules and do not bend side to side. They instead rotate like a propeller.

Hairlike filaments called **pili** (singular, pilus) often extend from the cell surface. Some pili help a cell stick to a surface, such as river rocks or your teeth. Other cells glide from place to place using their pili as grappling hooks. The pilus is extended out to a surface, sticks to it, then shortens, drawing the cell forward. Other retractable pili help draw cells together prior to the exchange of genetic material, as described below.

Reproduction and Gene Transfers

Prokaryotes have staggering reproductive potential. Some types can divide every twenty minutes. One cell becomes two, then two become four, then four become eight, and so on. A cell nearly doubles in size before it divides. After division, each descendant cell has one **prokaryotic chromosome**—a circular, double-stranded molecule of DNA with a few proteins.

In some species, a descendant buds from a parent cell. More commonly, a cell reproduces by **prokaryotic fission** (Figures 21.9, 21.10). A parent cell replicates its single chromosome, and this DNA replica attaches to the plasma membrane adjacent to the parent molecule. The addition of more membrane moves the two DNA molecules apart. Eventually, the membrane and cell wall extend across the cell's midsection and divide the parent cell into two cells.

Besides inheriting DNA "vertically" from a parent cell, prokaryotes engage in **horizontal gene transfers**: They pick up genes from cells of the same or different species. One mechanism of horizontal gene transfer,

pilus

bacterial flagellum

outer capsule

cell wall

plasma membrane

cytoplasm, with ribosomes   DNA, in nucleoid region

b

Figure 21.8 (**a**) The three most common shapes among prokaryotic cells: spheres, rods, and spirals. (**b**) Body plan of a typical prokaryotic cell.

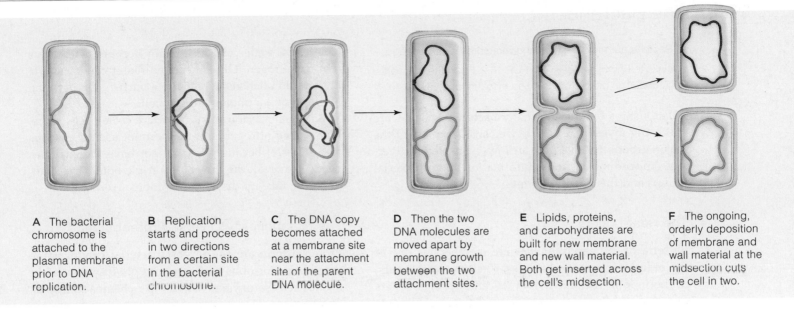

A The bacterial chromosome is attached to the plasma membrane prior to DNA replication.

B Replication starts and proceeds in two directions from a certain site in the bacterial chromosome.

C The DNA copy becomes attached at a membrane site near the attachment site of the parent DNA molecule.

D Then the two DNA molecules are moved apart by membrane growth between the two attachment sites.

E Lipids, proteins, and carbohydrates are built for new membrane and new wall material. Both get inserted across the cell's midsection.

F The ongoing, orderly deposition of membrane and wall material at the midsection cuts the cell in two.

Figure 21.10 Animated Prokaryotic fission, the reproductive mode of bacteria and archaeans.

conjugation, involves transfer of a plasmid between prokaryotic cells (Figure 21.11). A **plasmid** is a small circle of DNA that is separate from the bacterial chromosome (Section 16.1).

During conjugation, a special sex pilus draws two cells together. Then, one cell puts a copy of a plasmid and perhaps a few chromosomal genes into the other. Both bacteria and archaeans have plasmids and can engage in conjugation. Members of the two groups sometimes swap genes through this process.

Bacteriophages also move genes horizontally among prokaryotes, a process called **transduction**. The virus picks up DNA from a cell that it infects, then transfers that DNA to its next host.

Prokaryotes also acquire DNA by taking it up from the environment, a process called **transformation**. For example, Section 13.1 described how Frederick Griffith made harmless bacteria deadly by mixing them with dead cells of a harmful strain. The harmless bacteria picked up DNA that transformed them.

As you might expect, horizontal gene transfer can complicate attempts to reconstruct the evolutionary history of prokaryotes by comparing gene sequences. A history of gene transfers also makes defining boundaries between modern prokaryote species difficult.

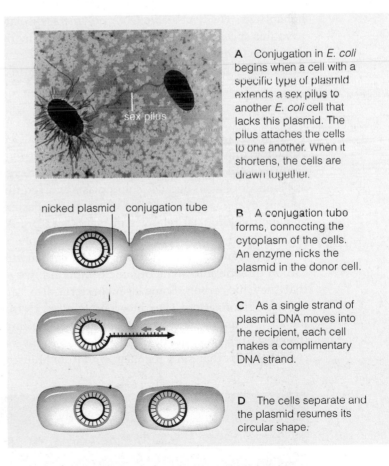

A Conjugation in *E. coli* begins when a cell with a specific type of plasmid extends a sex pilus to another *E. coli* cell that lacks this plasmid. The pilus attaches the cells to one another. When it shortens, the cells are drawn together.

nicked plasmid   conjugation tube

B A conjugation tube forms, connecting the cytoplasm of the cells. An enzyme nicks the plasmid in the donor cell.

C As a single strand of plasmid DNA moves into the recipient, each cell makes a complimentary DNA strand.

D The cells separate and the plasmid resumes its circular shape.

Figure 21.11 Animated Conjugation. (**a**) A sex pilus extending from one bacterial (*E. coli*) cell to another. The pilus will retract and draw the cells together, allowing formation of a conjugation tube between them.

(**b–d**) Steps in conjugation. For clarity, the plasmid's size has been greatly exaggerated and the bacterial chromosome is not shown. Conjugation requires two existing cells. It is a mechanism of gene transfer, not a mode of reproduction.

Take-Home Message

What is a typical prokaryote like?

■ The typical prokaryote is a walled cell with ribosomes, but no nucleus. It may stick to a surface or move about. It replicates by dividing in two, and can exchange genes with other prokaryotes.

- Bacteria are the oldest, most diverse prokaryotic lineage.
- Links to Photosynthesis 7.4, PCR 16.2, Hydrothermal vents 20.2, Evolution of chloroplasts and mitochondria 20.4

Relationships among the many bacterial lineages are still being investigated. Here we consider a few of the major groups to give you an idea of bacterial diversity. Appendix I has more information about bacterial groups and their defining traits.

The Heat Lovers

Bacteria of the genus *Aquifex* are members of one of the oldest prokaryotic lineages. They are heat-lovers, or thermophiles. Some live in volcanic springs, others near deep sea hydrothermal vents. Their ancient roots are taken as support for the hypothesis that life first arose near such vents (Section 20.2). *Thermus aquaticus*, shown earlier in Figure 20.1, is another inhabitant of hot springs. The heat-stable DNA polymerase isolated from this species was put to use in the first polymerase chain reactions. Biologists use these reactions to make many copies of a specific piece of DNA (Section 16.2).

The Cyanobacteria

Photosynthesis evolved in many bacterial lineages. However, only the **cyanobacteria** have the same light-capturing chlorophylls as plants, and release oxygen as plants do. This is because chloroplasts evolved from ancient cyanobacteria (Section 20.4). Cyanobacteria, and their chloroplast relatives in plants and protists, put nearly all of the oxygen into Earth's atmosphere.

When cyanobacteria incorporate the carbon from carbon dioxide into some organic compound, we say that they fix carbon. Some cyanobacteria also carry out **nitrogen fixation**: They incorporate nitrogen from the air into ammonia (NH_3). *Anabaena* is one example (Figure 21.12). This aquatic cyanobacterium forms long chains of cells that stick together. Under low-nitrogen conditions, some cells in a chain become heterocysts that fix nitrogen. Other nitrogen-fixing cyanobacteria are found in lichens, which are a partnership between a fungus and a photosynthetic cell.

Nitrogen fixation is an important ecological service. Plants need nitrogen, but they cannot use the gaseous form (N≡N) because they do not have the enzymes needed to break the molecule's triple bond. They can, however, take up dissolved ammonia from the soil.

The Metabolically Diverse Proteobacteria

Proteobacteria are the largest bacterial group. Some are photoautotrophs that carry out photosynthesis but do not release oxygen. Others are chemoautotrophs. One of these, *Thiomargarita namibiensis,* lives in marine sediments and is the largest prokaryote known (Figure 21.13a). It gets energy by stripping electrons from sulfur stored in a vacuole that makes up most of its volume.

Some other proteobacteria are chemoheterotrophs that live in the bodies of plants or animals. *Rhizobium* lives inside plant roots and fixes nitrogen. It helps the plant and gets shelter and sugars in return. *Escherichia coli* (Figures 21.9 and 21.11) lives in the mammalian gut. So does *Helicobacter pylori* (Figure 21.13b), the most common cause of stomach ulcers, and *Vibrio cholerae*, which causes cholera. The rickettsias are proteobacteria that cause typhus and Rocky Mountain spotted fever. Rickettsias are also the closest living relatives of cells that evolved into mitochondria (Section 20.4).

Some free-living, chemoautotrophic proteobacteria exhibit complex behavior. For example, magnetotactic bacteria contain particles of the iron-containing mineral magnetite and can detect Earth's magnetic field (Figure 21.13c). The bacteria are aquatic and use this information to navigate downward to deeper waters.

Myxobacteria are proteobacteria that glide about as a cohesive group and feed on other soil bacteria. When food dwindles, hundreds of thousands of cells form a multicelled fruiting body. In *Chondromyces crocatus*, the fruiting body is an elaborate, branched structure a few tenths of a millimeter high, with capsules at its tips (Figure 21.13d). Each capsule holds thousands of resting cells. Wind disperses capsules to new habitats.

The Gram-Positive Heterotrophs

Gram-positive bacteria are lineages with thick-walled cells that are tinted purple when prepared for microscopy by Gram-staining. Thin-walled bacteria such as cyanobacteria and proteobacteria are stained pink by this process, and are described as Gram-negative.

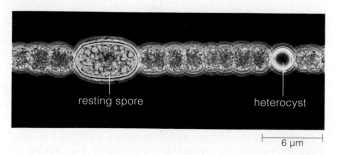

resting spore   heterocyst

6 µm

Figure 21.12 A chain of aquatic cyanobacteria (*Anabaena*), with a resting spore and a heterocyst that can fix nitrogen.

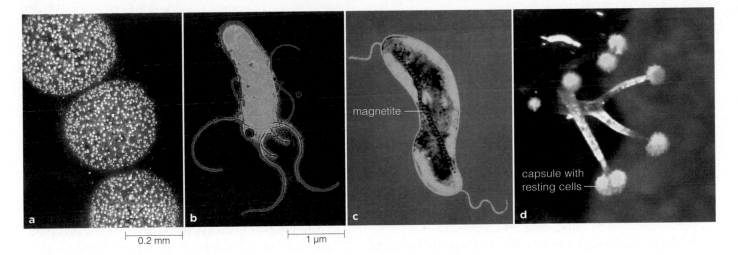

Figure 21.13 Proteobacteria. (**a**) *Thiomargarita namibiensis* lives in marine sediments and is so big it is visible without a microscope. White dots are sulfur in its vacuole. (**b**) *Helicobacter pylori*, the cause of most stomach ulcers. (**c**) A magnetotactic bacterium with a chain of magnetite particles. (**d**) Multicelled fruiting body of the myxobacteria *Chondromyces crocatus*. A capsule holds thousands of resting cells.

Most Gram-positive bacteria are chemoheterotrophs. *Lactobacillus* carries out the fermentation reactions that produce yogurt and other foods. *L. acidophilus* (Figure 21.14*a*) lives on skin and in the gut and vagina. The lactate it produces lowers the pH of it surroundings, which helps keep pathogenic bacteria in check.

The Gram-positive *Clostridium* and *Bacillus* species form **endospores** when conditions are unfavorable. An endospore holds a bacterial chromosome and a bit of cytoplasm (Figure 21.14*b*). It resists heat, boiling, irradiation, acid, and disinfectants. When conditions improve, the endospore germinates, releasing a cell.

Toxins made by some endospore-forming bacteria can be deadly. Inhale *Bacillus anthracis* endospores and you may get anthrax, a disorder in which the bacterial toxin interferes with breathing. *Clostridium tetani* endospores that germinate in wounds cause tetanus, in which toxins lock muscles in ongoing contraction. *C. botulinum* can grow in improperly canned foods and makes a toxin that, when ingested, causes a paralyzing food poisoning known as botulism.

Spirochetes and Chlamydias

Spirochetes resemble a stretched-out spring (Figure 21.15). Some are free-living; others live inside a host organism. One spirochete causes Lyme disease, which damages the heart, nervous system, and joints. Ticks are the vector for this disease. In microbiology, a **vector** is an organism that carries a pathogen between hosts.

Chlamydias are intracellular parasites of animals. Every year, *C. trachomatis* causes nearly a million cases of sexually transmitted disease in the United States.

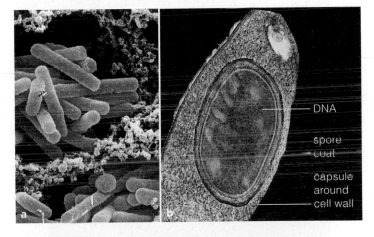

Figure 21.14 Gram-positive bacteria. (**a**) *Lactobacillus* cells in yogurt. (**b**) An endospore forming inside a *Clostridium tetani* cell.

Figure 21.15 *Borrelia burgdorferi*, a spirochete that causes Lyme disease. It moves from one host to another inside ticks.

Take-Home Message

What are bacteria?

■ Bacteria are the most abundant prokaryotes. Most are harmless or benefit us by releasing oxygen, fixing nitrogen, or otherwise cycling nutrients. A small minority of the bacterial chemoheterotrophs cause disease in humans.

■ Archaeans, the more recently discovered prokaryotic, lineage, are the closest prokaryotic relatives of eukaryotes.

■ Links to Pigments 7.1, Chromosome structure 9.1, Ribosomal RNA 14.3, Operons 15.4, Classification 19.6

The Third Domain

All prokaryotes were once put in the same kingdom, and it is easy to see why. Archaeans and bacteria are alike in size and shape. Neither has a nucleus. Both have a circular chromosome, with genes arranged as functional units called operons (Section 15.4).

The distinctive features of archaeans first became apparent in the 1970s. Molecular biologist Carl Woese began comparing the ribosomal RNAs of prokaryotes to find out how they are related to one another. Genes for rRNA are essential for protein synthesis (Section 14.3). However, certain sequences in those genes can mutate a bit without any loss of function. The longer two lineages have traveled on separate evolutionary roads, the more their genes for rRNA will differ.

Woese discovered that some prokaryotes fell into a distinct group. Their rRNA gene sequences positioned them between the bacteria and the eukaryotes. On the basis of this evidence, Woese proposed a three-domain classification system (Section 19.6).

The three-domain system is now widely accepted, and evidence that supports it keeps accumulating. As noted earlier, archaeans and bacteria have different cell wall components. Their membrane phospholipids are mirror images of each other and are synthesized in different ways. Like eukaryotic cells, archaeans spool their DNA around proteins called histones. Bacteria do not synthesize histones or structurally organize their DNA to the same extent.

Woese compares the discovery of archaeans to the discovery of a new continent, which he and others are now exploring. Already the explorers have identified several major subgroups.

Here, There, Everywhere

In their physiology, most archaeans are **methanogens** (methane producers), **extreme halophiles** (salt lovers), and **extreme thermophiles** (heat lovers). These three informal designations are not phylogenetic groupings. Some bacterial species are also methanogens, extreme halophiles, and extreme thermophiles. Archaeans and bacteria often coexist, and they often exchange genes.

Some methane-producing archaeans live in the gut of termites, cattle, and other animals (Figure 21.16a). Others live in marshes, Antarctic ice, seas, or rocks beneath Earth's surface (Figure 21.16b,c). Methanogens are strictly anaerobic; free oxygen kills them. They are chemoautotrophs that form ATP by pulling electrons from hydrogen gas or acetate. Methane (CH_4) gas forms as a product of these reactions.

Figure 21.16 Methanogenic archaeans.

(**a**) Researchers have isolated several kinds of methanogenic archaeans from the cattle gut.

(**b**) *Methanococcus jannaschii*, with two bundles of many flagella, grows in water heated to 85°C (185°F) near hydrothermal vents.

(**c**) Bubbles of methane gas almost 230 meters (750 feet) below sea level in the Black Sea. The methane is produced by archaeans far beneath the sea floor, and seeps up into deep, frigid ocean water. It combines with water to form deposits of methane hydrates.

0.5 µm

b

5 μm

By their metabolic activity, methanogens produce 2 billion tons of methane annually. The release of this carbon-containing gas into the air has a major impact on the global carbon cycle. As you will learn in Section 47.7, methane emissions contribute to global warming.

Extreme halophilic archaeans live in the Dead Sea, the Great Salt Lake, saltwater evaporation ponds, and other highly salty habitats (Figure 21.17a). Most make ATP by aerobic reactions but switch to photosynthesis when oxygen is low. They have a unique purple pigment called bacteriorhodopsin embedded in their plasma membrane. When excited by light, this protein pumps protons (H+) out of the cell. H+ flows back in through ATP synthases and so drives ATP formation.

Some extreme thermophilic archaeans live beside hydrothermal vents, where temperatures can exceed 110°C (230°F). Researchers came across *Nanoarchaeum equitans* as they were exploring hydrothermal vents near Iceland. Only 400 nanometers across, *N. equitans* is among the smallest known cells. It is a parasite, and its host is a slightly bigger archaean (Figure 21.17b).

Other heat-loving archaeans thrive in mineral-rich hot springs (Figure 21.17c,d). *Sulfolobus* species grow in well-oxygenated water at 80°C (176°F) at a pH of 3. The cells can act as chemoautotrophs that metabolize sulfur or switch to a heterotrophic mode and feed on carbon compounds.

Figure 21.17 Life in extreme environments. (**a**) In salty evaporation ponds in Utah's Great Salt Lake, extreme halophilic archaeans and red algae tint the water pink. (**b**) Parasitic *Nanoarchaeum equitans* (smaller *blue* spheres) grows as a parasite on another archaean, *Ignicoccus* (larger spheres). Both were isolated from 100°C (212°F) water near a hydrothermal vent.

(**c**) Membrane lipids typical of archaeans have been discovered in the Three Buddhas, a hot spring in Gerlach, Nevada. (**d**) Yellowstone's hot springs and pools are home to bacterial and archaeal extreme thermophiles.

Archaeans are common in the seas, and not just at hydrothermal vents. There may be as many archaeans in deep water as there are bacteria in water near the ocean's surface. Archaeans also occur in soil and fresh water, just about everywhere that bacteria live. Some archaeans thrive in the human gut, vagina, and mouth. In contrast to bacteria, few archaeans are thought to be human pathogens, although some that live in the mouth may contribute to gum disease.

Take-Home Message

What are archaeans?

■ Archaeans are the prokaryotic cells most closely related to eukaryotes. Many live in very hot or very salty habitats, but there are archaeans nearly everywhere. Unlike bacteria, hardly any cause human disease.

21.8 | Evolution and Infectious Disease

- Viruses, bacteria, and other pathogens evolve by natural selection, as do their hosts.

- Link to Directional selection 18.4

The Nature of Disease An infection occurs when some pathogen breaches the body's surface barriers, enters the internal environment, and multiplies. **Disease** follows when the body's defenses cannot be mobilized quickly enough to keep a pathogen's activities from interfering with normal body functions. Infectious diseases are spread by contact with tiny amounts of mucus, blood, or other body fluid that can contain a pathogen. In 2004, the World Health Organization estimated that about 19 percent of deaths were caused by infectious disease. Table 21.5 lists the most common causes of death in this category.

In an epidemic, a disease spreads fast through part of a population, then subsides. Sporadic diseases, such as whooping cough, occur irregularly and affect few people. Endemic diseases pop up more or less continually but do not spread far in large populations. Tuberculosis is like this. So is impetigo, a highly contagious bacterial infection that typically spreads no further than a single day-care center, or a similarly limited location.

In a pandemic, a disease breaks out and spreads worldwide. AIDS is a pandemic that has no end in sight. A 2002–2003 outbreak of SARS (severe acute respiratory

Table 21.5   Deaths From Infectious Diseases*

Disease	Type of Pathogen	Deaths per year
Acute respiratory infections	Bacteria, viruses	4 million
AIDS	Virus (HIV)	2.7 million
Diarrheas	Bacteria, viruses, protists	1.8 million
Tuberculosis	Bacteria	1.6 million
Malaria	Protists	1.3 million
Measles	Viruses	600,000
Whooping cough	Bacteria	294,000
Tetanus	Bacteria	204,000
Meningitis	Bacteria, viruses	173,000
Syphilis	Bacteria	157,000

* Deaths worldwide, based on The World Health Report for 2004.

syndrome) was a brief pandemic (Figure 21.18). It began in China, and travelers quickly carried it to countries around the world. Before government-ordered quarantines (isolation of those infected) halted its spread, about 8,000 people were sickened and about 10 percent of them died. Researchers quickly determined that a previously unknown coronavirus causes SARS. Other coronaviruses cause less dire respiratory infections such as colds.

There have been no reported cases of SARS since 2003. Is the pathogen gone for good? Only time will tell. Diseases sometimes disappear for years, then break out again without warning.

An Evolutionary Perspective Consider disease in terms of a pathogen's prospects for survival. A pathogen stays around only for as long as it has access to outside sources of energy and raw materials. To a microscopic organism or virus, a human is a treasure-house of both. With bountiful resources, a pathogen can reach amazing population sizes. Evolutionarily speaking, the pathogens that leave the most descendants win.

Two barriers prevent pathogens from evolving to a position of total dominance. First, any species that has a history of being attacked by a specific pathogen has coevolved with it and has built-in defenses against it. Second, if a pathogen kills its host too fast, the pathogen might disappear along with the host. Having a less-than-fatal effect can benefit a pathogen. Think of an infected host as a factory that makes and distributes pathogens. Killing the host would shut down this facility. The longer the infected individual survives, the more copies of the pathogen are made and dispersed.

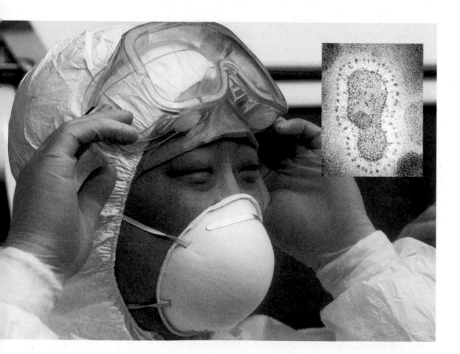

Figure 21.18 A health care worker in China dons protective gear during the SARS pandemic. The SARS virus, shown in the inset, is a coronavirus, like some viruses that causes common colds. Doctors and nurses who cared for SARS patients were among the casualties of the SARS pandemic.

Figure 21.19 Which is deadlier?
(**a**) The Ebola virus turns blood vessels to mush and kills up to 90 percent of those infected. So far, outbreaks have been infrequent and restricted to small areas in Africa. (**b**) *Mycobacterium tuberculosis* causes tuberculosis (TB). Without treatment, the virus kills about 50 percent of those infected, and it infects about 300 million worldwide.

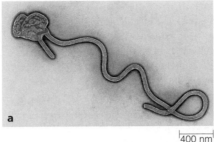
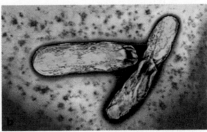

400 nm

0.5 µm

Pathogens most often kill hosts who are weakened by age or by the presence of other pathogens. Death may also occur after a pathogen infects a novel host that has not evolved any defenses against it.

Emerging Diseases As human populations have soared, people have moved farther and farther into jungles and other habitats previously considered marginal. In these remote regions, they eat the local animals and sometimes meet up with pathogens that have not coevolved with humans. As mentioned earlier, HIV is descended from SIV, a virus that infects African chimpanzees. These primates have evolved with SIV, and they are not killed by it.

Close contact with wild animals probably started the SARS epidemic. Chinese horseshoe bats (*Rhinolophus*) are a natural reservoir for a virus that has a sequence nearly identical to SARS. The bats and other wildlife are captured and sold live for food in Asian markets.

In Africa, bats serve as a reservoir for the highly deadly Ebola virus (Figure 21.19a). It often kills between 50 and 90 percent of those infected. The first symptoms are high fever and flu-like aches. Within a few days vomiting and diarrhea begin. Blood vessels are destroyed. Blood seeps into surrounding tissues and leaks out through all the body's orifices. Contacting body fluids of infected people can spread the disease. Understandably, at the start of an Ebola outbreak, government agencies throughout the world are notified. During 2007, independent outbreaks in Uganda and the Congo infected hundreds of people and killed more than one hundred.

World health officials are currently keeping a close eye on the H5N1 strain of avian influenza (bird flu). The first humans infected by this strain turned up in Hong Kong during 1997. Since then, human infections have been reported in other parts of Asia, Africa, the Pacific, Europe, and the Middle East. Half of those infected die.

So far, there seems to be little or no human-to-human transmission of H5N1. Nearly all people have become infected by direct contact with infected birds or bird feces. The virus has spread rapidly among birds and is expected to eventually affect them worldwide. With many infected birds, cases of bird-to-human transfer will no doubt rise. However, an even greater concern is the risk that a mutation will allow human-to-human

transfer. A 1918 flu pandemic that infected one third of the world population and killed 50 million people was caused by a distantly related strain of bird flu.

The Threat of Drug-Resistance As explained in Section 18.4, using antibiotics to treat any infectious disease results in directional selection. In a population of pathogens, those that are least affected by the drug are at a selective advantage. Drug-resistant individuals survive and have offspring while other drug-susceptible individuals die. As a result, the frequency of drug-resistant individuals increases over generations—which, for most pathogens, are short.

For example, *Streptococcus pneumoniae* is commonly transmitted among children at day care centers. It can cause pneumonia, meningitis, and chronic ear infections. Penicillin-resistant strains of *S. pneumoniae* first appeared in 1967. Today, about half of known strains are resistant.

Penicillin resistant strains of bacteria arise by mutations or horizontal gene transfers. In the case of *S. pneumoniae*, genetic comparisons showed that the genes that confer antibiotic resistance were transferred to an *S. pneumoniae* cell from a related species *S. mitis*.

Viruses are not cells, but they do have genes that can mutate, and so also can evolve by natural selection. For example, many strains of HIV are now resistant to one or more of the antiviral drugs designed to fight them.

It's a Small World As you know, HIV first infected humans in Africa, then spread worldwide. SARS broke out in Asia and became a global threat within months.

After the SARS pandemic, an international system of alerts was set up to help prevent people known to be infected by potentially deadly pathogens from spreading disease. However, in 2007, a young American man named Andrew Speaker made headlines by embarking on his honeymoon while infected with a drug-resistant strain of *Mycobacterium tuberculosis* (Figure 21.19b), the bacterium that causes tuberculosis. Speaker took commercial flights to several countries, then reentered the United States. Fortunately, he did not infect any of his fellow travelers. He did demonstrate how difficult it will be to contain the spread of dangerous pathogens in an age of global air travel and constant human migrations.

In developing countries, an estimated 80 percent of people who are HIV-positive do not know it. In the United States, it is estimated that a quarter of HIV-infected people have never been tested. Getting tested is the first step toward treatment that promotes health and lengthens the life span. The earlier treatment begins, the greater the chances of continued good health. If you think you could have been exposed to HIV, get tested as soon as possible.

Summary

Section 21.1 A **virus** is a noncellular infectious particle with a protein coat enclosing DNA or RNA. A virus cannot reproduce on its own. It infects a host cell, and takes over the host's mechanisms of replication and protein synthesis. A virus infects only a specific host type. For example, a **bacteriophage** infects only bacteria. Some viruses cause disease; they act as **pathogens** in humans. Others that infect certain nonhuman species benefit us.

Section 21.2 Nearly all viral multiplication cycles have five steps. The virus attaches to a host cell. The whole virus or just its genetic material enters the host cell. Viral genes and enzymes direct host mechanisms to replicate viral genetic material and synthesize viral proteins. Viral particles assemble and are released.

Bacteriophages may multiply by a **lytic pathway**, in which the new viral particles are made fast and released by lysis, or by a **lysogenic pathway**, in which viral DNA becomes part of the host chromosome.

Herpesviruses are enveloped viruses that can remain inactive in cells and periodically reawaken. HIV is an enveloped retrovirus with RNA. A viral enzyme, **reverse transcriptase**, uses RNA as a template to make DNA.

■ *Use the animation on CengageNOW to learn how a bacteriophage and enveloped viruses can multiply.*

Section 21.3 Viroids and prions are very small infectious agents. **Viroids** are circles of RNA without a protein coat. Many cause disease in plants. **Prions** are proteins that occur naturally in the vertebrate nervous system but can cause fatal disease when they misfold.

Sections 21.4, 21.5 Bacteria and archaeans are the only **prokaryotes**: single cells that do not have a nucleus or the other organelles that characterize eukaryotic cells. Many prokaryotic species include several **strains**, each with some distinctive trait. Prokaryotes are small, abundant, and—as a group—metabolically diverse. Autotrophs such as photosynthetic bacteria get carbon from carbon dioxide. Heterotrophs get carbon by breaking down organic compounds built by other organisms, as when **saprobes** feed on wastes and remains.

Most prokaryotes have a **cell wall** around their plasma membrane. **Pili** or **flagella** extend from many cells. The **prokaryotic chromosome** is a circular molecule of DNA that resides in a region of cytoplasm called the **nucleoid**. Many prokaryotes have one or more **plasmids**, circles of DNA that are separate from the chromosome and carry a few genes. Prokaryotes reproduce by **prokaryotic fission**: replication of the chromosome and division of a cell into two genetically identical descendant cells.

Three types of **horizontal gene transfer** move genes between prokaryotic cells. **Conjugation** transfers a plasmid and perhaps some chromosomal genes to another cell. Virus-assisted transfer of genes is **transduction**. With **transformation**, DNA is taken up from the environment.

■ *Use the animation on CengageNOW to observe prokaryotic fission and conjugation.*

Sections 21.6, 21.7 Bacteria are the oldest cell lineage and are the most abundant prokaryotes. Many bacteria are ecologically important. **Cyanobacteria** produce oxygen as a by-product of photosynthesis. Other bacteria carry out **nitrogen fixation**; they convert nitrogen gas to nitrogen-rich compounds plants can take up. A minority of bacteria are human pathogens. Some bacteria survive unfavorable conditions as **endospores**, which can survive boiling and other environmental assaults. Ticks are **vectors** for some bacterial pathogens, they carry the bacteria from one host to another.

Archaeans are prokaryotic, but they are like eukaryotic cells in certain features. Comparisons of structure, function, and genetic sequences position them in a separate domain, between eukaryotes and bacteria. Research is showing that archaeans are more diverse and widely distributed than was previously thought. They include **methanogens** (methane producers), **extreme halophiles** (salt lovers), and **extreme thermophiles** (heat lovers).

Section 21.8 **Disease** occurs when pathogens overrun a host. Hosts coevolve with pathogens. Selection favors host defenses that fight infection. Selection also favors pathogens that do not kill a host before it spreads the infection. Antibiotic use selects for antibiotic-resistant bacteria. Genes that convey drug resistance arise by mutation and may spread among bacteria by methods of horizontal gene transfer.

Diseases can be fatal if an individual is weakened by age or multiple pathogens, or lacks coevolved defenses. People do not have defenses against emerging pathogens.

Data Analysis Exercise

One of the sadder aspects of the AIDS pandemic is transmission of HIV from mother to child. Untreated mothers have a 15 to 30 percent chance of passing on the infection during pregnancy. The virus is also transmitted by breast-feeding.

Since the AIDS pandemic began, there have been more than 8,000 cases of mother-to-child HIV transmission in the United States. In 1993, American physicians began giving antiretroviral drugs to HIV-positive women during pregnancy and treating both mother and infant in the months after birth. Only about 10 percent of mothers were treated in 1993, but by 1999 more than 80 percent got the drugs.

Figure 21.20 shows the number of AIDS diagnoses among children in the United States. Use the information in this graph to answer the following questions.

1. How did the number of children diagnosed with AIDS change during the late 1980s?

2. What year did new AIDS diagnoses in children peak, and how many children were diagnosed that year?

3. How did the number of AIDS diagnoses change as the use of antiretrovirals in mothers and infants increased?

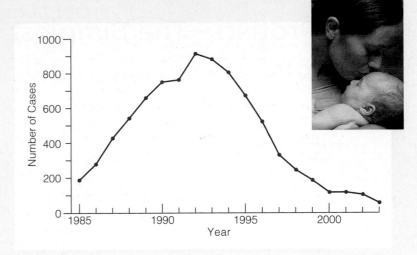

Figure 21.20 Number of new AIDS diagnoses in the United States per year among children who were exposed to HIV during pregnancy, birth, or by breast-feeding.

Self-Quiz
Answers in Appendix III

1. DNA or RNA may be the genetic material of _____ .
a. a bacterium   b. a prion   c. a virus   d. an archaean

2. A viroid consists entirely of _____ .
a. DNA   b. RNA   c. protein   d. lipids

3. Bacteriophages can multiply by _____ .
a. prokaryotic fission   c. a lysogenic pathway
b. a lytic pathway   d. both b and c

4. The genetic material of HIV is _____ .

5. Only _____ are prokaryotic.
a. archaeans   c. prions
b. bacteria   d. both a and b

6. Bacteria transfer plasmids by _____ .
a. prokaryotic fission   c. conjugation
b. transformation   d. the lytic pathway

7. All _____ are oxygen-releasing photoautotrophs.
a. spirochetes   c. cyanobacteria
b. chlamydias   d. proteobacteria

8. *E. coli* cells that live in your gut are _____ .
a. spirochetes   c. cyanobacteria
b. chlamydias   d. proteobacteria

9. All _____ are intracellular parasites of vertebrates.
a. spirochetes   c. cyanobacteria
b. chlamydias   d. proteobacteria

10. Some Gram-positive bacteria (e.g., *Bacillus anthracis*) survive harsh conditions by forming a(n) _____ .
a. pilus   c. endospore
b. heterocyst   d. plasmid

11. Only _____ reproduce by prokaryotic fission.
a. viruses   c. bacteria
b. archaeans   d. both b and c

12. A plasmid is a circle of _____ .

13. Which of the following infectious diseases kills the most people annually?
a. Ebola   b. AIDS   c. measles   d. syphilis

14. A worldwide outbreak of a disease is a(n) _____ .

15. Match the terms with their most suitable description.
____archaean
____bacteria
____virus
____plasmid
____extreme halophile
____prion
____sex pilus

a. infectious protein
b. nonliving infectious particle; nucleic acid core, protein coat
c. draws cells together
d. prokaryotes that most closely resemble eukaryotes
e. most common prokaryotic cells
f. small circle of bacterial DNA
g. salt lover

■ *Visit CengageNOW for additional questions.*

Critical Thinking

1. Like other organisms, plants suffer from viral diseases (Figure 21.3). Recall from Section 4.12 that plant parts are usually protected by a waxy cuticle and that each plant cell has a cellulose wall around its plasma membrane. Plant viruses often get into a host plant cell with assistance from plant-eating insects. Once inside a cell, the virus causes changes in the cell's plasmodesmata. Explain how altering these structures might benefit the virus.

2. When planting bean seeds, gardeners are advised to inoculate them first with a powder that contains nitrogen-fixing *Rhizobium* cells, which infect plant roots. How could the presence of these bacteria benefit the plants?

3. Viruses that do not have a lipid envelope tend to remain infectious outside the body for longer than enveloped viruses. "Naked" viruses are also less likely to be rendered harmless by soap and water. Can you explain why?

22 Protists—The Simplest Eukaryotes

IMPACTS, ISSUES | The Malaria Menace

Malaria is a leading cause of human death. Worldwide, it kills more than 1.3 million people annually. *Plasmodium*, a single-celled protist, is the cause of malaria. Mosquitoes carry the protist from one human host to another.

Plasmodium enters human blood and liver cells, where it feeds and multiplies. Infection by *Plasmodium* destroys oxygen-carrying red blood cells, causing weakness. Affected people have bouts of chills and fever. They often become jaundiced—waste materials accumulate in their body and tint them yellow. *Plasmodium*-infected blood cells that get into the brain can cause blindness, seizures, coma, and death. Children are especially susceptible (Figure 22.1).

Malaria was common in the United States, especially the South, until an aggressive campaign during the 1940s eradicated the disease. Swamps and ponds where mosquitoes breed were drained, and the insecticide DDT was sprayed inside millions of homes. Today, nearly all cases of malaria in the United States are in people who contracted the disease outside the country.

Malaria is mainly a tropical disease; *Plasmodium* cannot survive at low temperatures. Malaria remains common in parts of Mexico. South and Central America, as well as Asia and Pacific Islands, but it takes its greatest toll in Africa. One African child dies of malaria every 30 seconds.

As you learned in Section 18.6, malaria has been a potent selective force on humans in Africa. The allele responsible for sickle-cell anemia also reduces mortality from this disease.

Natural selection also acts on *Plasmodium*. The protist has recently become resistant to several antimalarial drugs. Over a longer time frame, it has evolved an amazing capacity to alter the behavior of its hosts. *Plasmodium* makes the mosquitoes that carry it more likely to feed several times a night, and thus more likely to bite several people. It also makes infected humans especially appetizing to a hungry mosquito. By manipulating its insect and human hosts, the protist maximizes the chances that its offspring will reach a new host.

With this chapter we turn to protists, a group sometimes referred to as "simple" eukaryotes. Most protists are structurally less complex than other eukaryotes, yet they still are exquisitely well adapted. The vast majority of protists do not cause disease, and many benefit us. Among other things, they decompose wastes, serve as food for larger organisms, take up some of the carbon dioxide that contributes to global warming, and help keep the pH of ocean waters stable.

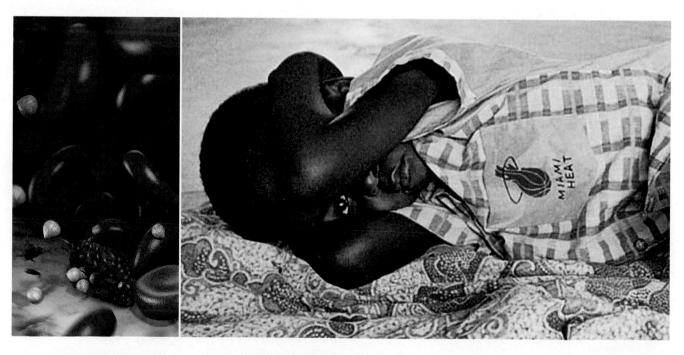

See the video! **Figure 22.1** A child in Mozambique awaits testing for malaria. At *left*, an artist's depiction of *Plasmodium* emerging from an infected red blood cell.

Key Concepts

Sorting out the protists

Protists include many lineages of single-celled eukaryotic organisms and their closest multicelled relatives. Gene sequencing and other methods are clarifying how protist lineages are related to one another and to plants, fungi, and animals. **Section 22.1**

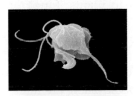

Flagellated protozoans and shelled cells

Flagellated protozoans include single-celled predators and some human parasites. Foraminiferans and radiolarians are shelled, single-celled heterotrophs. Most live in seas. **Sections 22.2, 22.3**

The alveolates

Ciliated protozoans, dinoflagellates, and apicomplexans are single-celled photoautotrophs, predators, and parasites. Their shared trait is a unique layer of sacs under the plasma membrane. **Sections 22.4–22.6**

The stramenopiles

Diatoms and brown algae are stramenopiles, most of which are photoautotrophs. The colorless water molds, which include major plant pathogens, are also stramenopiles. **Sections 22.7, 22.8**

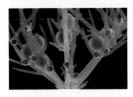

The closest relatives of land plants

Red algae and green algae are photosynthetic single cells and multi-celled seaweeds. One lineage of multicelled green algae includes the closest living relatives of land plants. **Sections 22.9, 22.10**

Relatives of fungi and animals

A great variety of amoeboid species formerly classified as members of separate lineages are now united as the amoebozoans. They are close relatives of fungi and animals. **Section 22.11**

Links to Earlier Concepts

- Section 20.3 introduced the earliest eukaryote lineage. Here we examine the diversity of their descendants and how they are classified (19.1). We consider again how organelles can evolve as a result of endosymbiosis (20.4).

- You will see how cells adapt to hypertonicity (5.6) and use anaerobic pathways (8.5) in low oxygen habitats. You will also see another example of bioluminescence (6.5).

- Organisms used in early studies of photosynthesis make an appearance (2.2, 7.2), as do photosynthetic pigments (7.1). You will also be reminded about plant plasmodesmata (4.12) and plant cell division (9.4).

- Variation in cell motility (4.13) is a big theme in this chapter. Another theme is the role protists play in their environment, as by buffering oceans (2.6) and affecting levels of carbon and oxygen in the atmosphere (7.9).

How would you vote? The insecticide DDT is highly effective against mosquitoes that transmit malaria, but it also harms wildlife and humans. Some people want to see DDT banned worldwide. Others think that with careful use it is an important weapon for the fight against malaria. Should DDT be banned worldwide? See CengageNow for details, then vote online.

- Protists include many lineages of mostly single-celled eukaryotes, some only distantly related to one another.

- Link to Classification 19.1

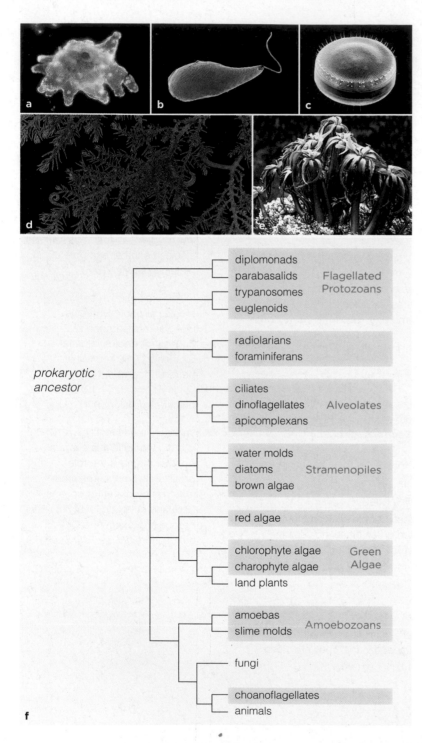

Figure 22.2 Single-celled protists include (**a**) amoebas, (**b**) euglenoids, and (**c**) diatoms. Most red algae (**d**) and all brown algae (**e**) are multicelled. (**f**) One proposed eukaryotic family tree with traditional protist groups indicated by *tan boxes*. Notice that the protists are not united as a single lineage.
Figure It Out: Are land plants more closely related to the red algae or the brown algae? *Answer: red algae*

Classification and Phylogeny

Protists are eukaryotic organisms that are not fungi, plants, or animals. No single trait is unique to protists; they are a collection of lineages, rather than a clade, or monophyletic group, as defined in Section 19.1.

As mentioned in earlier chapters, researchers have begun dividing the former kingdom Protista into many smaller groups. Gene sequencing plays a pivotal role in this work. It has shown that many protist groups are only distantly related to one another. In fact, some protists are more closely related to plants, fungi, or animals than to other protists.

Figure 22.2 shows examples of protists and where members of this group are currently thought to fit into the eukaryotic family tree. Tan boxes denote the protist groups discussed in this book. There are many additional protists, but this sampling of major groups will suffice to demonstrate protist diversity, ecological importance, and effects on human health.

Protist Organization and Nutrition

Most protist lineages include only single-celled species (Figure 22.2*a*–*c* and Table 22.1). However, some colonial protists exist, and multicellularity has evolved in several groups (Figures 22.2*d*,*e*). Some multicelled protists have large bodies that consist of many types of differentiated cells.

Many protists are heterotrophs in water or soil. They feed on decaying organic matter or prey on smaller organisms such as bacteria. Other heterotrophic protists live inside larger organisms, including humans. In some cases, protistan endosymbionts do no harm or even benefit their host. For example, flagellated protozoans that live in the termite gut give these insects the ability to digest wood. Other protists are parasites, and some infect humans.

Autotrophic protists have chloroplasts and carry out photosynthesis. Protist chloroplasts evolved by two slightly different mechanisms. Section 20.4 explained how cyanobacteria were engulfed by a heterotrophic cell and evolved into chloroplasts. We call this process primary endosymbiosis. The common ancestor of red and green algae got its chloroplasts this way. Later, various heterotrophic protists engulfed red or green algal cells. Some algal cells survived and evolved into chloroplasts. This process is secondary endosymbiosis.

A few protists are the only eukaryotes that function both as autotrophs and heterotrophs. These versatile "mixotrophs" switch between nutritional modes when environmental conditions favor one or the other.

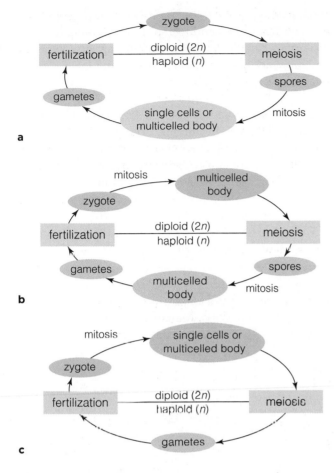

Figure 22.3 Examples of generalized protist life cycles.
(**a**) Haploid-dominated cycle, in which the only diploid cell is the zygote. (**b**) Alternation of generations, with haploid and diploid multicelled forms. (**c**) Diploid-dominated cycle.

Protist Life Cycles

Protists show great diversity in life cycles, and most can reproduce both sexually and asexually. Commonly, haploid cells dominate the life cycle; only the zygote is diploid (Figure 22.3a). The protistan parasites that cause malaria have this sort of cycle. In other groups, there is an **alternation of generations**, with multicelled haploid and diploid bodies (Figure 22.3b). Some algae have this kind of cycle, as do land plants. The spores produced in these cycles often are flagellated cells. In the final type of cycle, diploid cells dominate (Figure 22.3c). Diatoms have this kind of life cycle.

Often protists reproduce asexually as long as conditions favor growth. Sexual reproduction occurs when conditions become less favorable. Some single-celled protists also survive hard times by developing into a **cyst**, a walled, nonmotile structure that stays dormant until conditions favor growth.

Table 22.1 Characteristics of Some Protist Groups

Protist Group	Organization	Nutritional Mode
Flagellated Protozoans		
Diplomonads	Single cell	Heterotrophs; free-living or parasites of animals
Parabasalids	Single cell	Heterotrophs; free-living or parasites of animals
Kinetoplastids	Single cell	Heterotrophs; mostly parasites; some free-living
Euglenoids	Single cell	Most free-living heterotrophs; some autotrophs, mixotrophs
Radiolarians	Single cell	Free-living heterotrophs
Foraminiferans	Single cell	Free-living heterotrophs
Alveolates		
Ciliates	Single cell	Heterotrophs; most free-living, some parasites of animals
Dinoflagellates	Single cell	Autotrophs, mixotrophs, free-living or parasitic heterotrophs
Apicomplexans (Sporozoans)	Single cell	Heterotrophs; all parasites of animals
Stramenopiles		
Oomycotes	Single cell or multicelled	Heterotrophs; free-living or parasites of animals
Diatoms	Single cell	Mostly autotrophs, a few heterotrophs or mixotrophs
Brown Algae	Multicelled	Autotrophs
Red Algae	Most multicelled	Autotrophs
Green Algae	Single cell, colonial, or multicelled	Autotrophs
Amoebozoans		
Amoebas	Single cell	Heterotrophs; mostly free-living, a few parasites of animals
Slime Molds	Single-celled and aggregated stages in the life cycle	Heterotrophs

Take-Home Message

What are protists?

■ Protists are a collection of eukaryotic lineages. Most are single-celled, but there are some multicelled species.

■ Protists can be autotrophs or heterotrophs, and a few can switch between modes. Most reproduce both sexually and asexually.

■ Flagellated protozoans are single-celled species that swim in lakes, seas, and the body fluids of animals.

■ Links to Anaerobic pathways 8.5, Effects of tonicity 5.6

Flagellated protozoans are single, unwalled cells with one or more flagella. All groups are entirely or mostly heterotrophic. A **pellicle**, a layer of elastic proteins just beneath the plasma membrane, helps the cells retain their shape.

Haploid cells dominate the life cycle of these groups. The cells reproduce asexually by **binary fission**: A cell duplicates its DNA and organelles, then divides in half. The result is two identical cells.

The Anaerobic Flagellates

Diplomonads and parabasalids have multiple flagella and are among the few protists that can live in oxygen-poor waters. They lack typical mitochondria. Instead, they have hydrogenosomes, organelles that produce some ATP by an anaerobic pathway. Hydrogenosomes evolved from mitochondria and are an adaptation to anaerobic aquatic habitats. Free-living diplomonads and parabasalids thrive deep in seas and lakes. Others live in animal bodies. They may be harmful, helpful, or have no effect on their host.

The diplomonad *Giardia lamblia* causes giardiasis, a common disease in humans. The protist attaches to the intestinal lining, then sucks out nutrients (Figure 22.4*a,b*). Symptoms of giardiasis can include, cramps, nasusea, and severe diarrhea. They sometimes persist for weeks. Infected people and animals excrete cysts of G. *lamblia* in their feces. The cysts survive for months in water. Ingesting even a few can lead to infection.

The parabasalid *Trichomonas vaginalis* infects human reproductive tracts and causes trichomoniasis (Figure 22.4*c*). *T. vaginalis* does not make cysts, so it does not survive very long outside the human body. Fortunately for the parasite, sexual intercourse puts it directly into hosts. In the United States, about 6 million people are

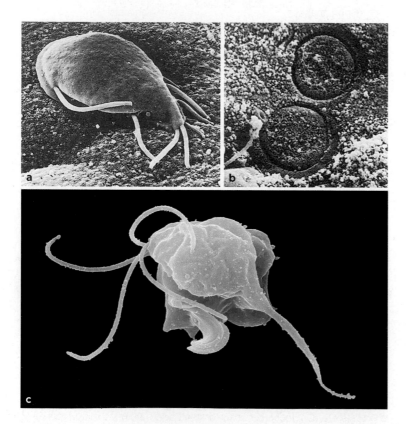

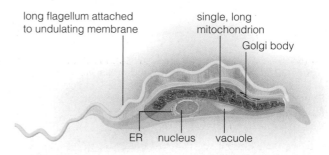

long flagellum attached to undulating membrane

single, long mitochondrion

Golgi body

ER   nucleus   vacuole

a

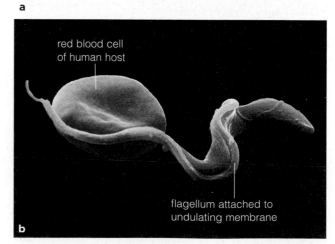

red blood cell of human host

flagellum attached to undulating membrane

b

Figure 22.4 (**a**) Electron micrographs of *Giardia lamblia* and (**b**) an example of the imprints that its sucking disk leaves on the intestinal lining. (**c**) Electron micrograph of *Trichomonas vaginalis*, which causes the sexually transmitted disease trichomoniasis.

Figure 22.5 (**a**) Sketch and (**b**) electron micrograph of the kinetoplastid *Trypanosoma brucei*. A bite of an infected tsetse fly delivers this trypanosome into the blood. It lives in the fluid portion of the blood (the plasma) and absorbs nutrients.

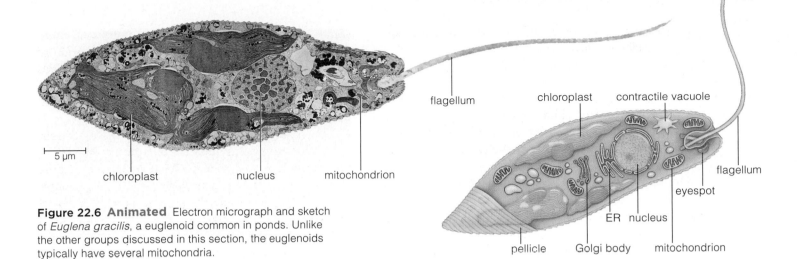

Figure 22.6 Animated Electron micrograph and sketch of *Euglena gracilis*, a euglenoid common in ponds. Unlike the other groups discussed in this section, the euglenoids typically have several mitochondria.

infected. In women, symptoms include vaginal soreness, itching, and a yellowish discharge. Infected males typically show no symptoms. Untreated infections damage the urinary tract, cause infertility, and increase risk of HIV infection. A dose of an antiprotozoal drug provides a quick cure. Both partners should be treated to prevent reinfection.

Trypanosomes and Other Kinetoplastids

Kinetoplastids are flagellated protozoans that have a single large mitochondrion. Inside the mitochondrion, near the base of the flagellum, is a clump of DNA. The clump is the kinetoplast for which the group is named.

Some kinetoplastids prey on bacteria in fresh water and seas, but **trypanosomes**, the largest subgroup, are parasitic; they live in plants and animals. Trypanosomes are long, tapered cells with an undulating membrane (Figure 22.5). A single flagellum that attaches to this membrane causes its characteristic wavelike motion.

Biting insects act as vectors for many trypanosomes that parasitize humans. Again, a vector is an insect or other animal that carries a pathogen between hosts.

Tsetse flies spread *Trypanosoma brucei*, which causes African trypanosomiasis, commonly known as African sleeping sickness. Infected people are drowsy during daytime and often cannot sleep at night. If untreated, the infection is fatal. Tsetse flies that spread *T. brucei* occur only in sub-Saharan Africa.

Bloodsucking bugs transmit *Trypanosoma cruzi*, the cause of Chagas disease. Untreated infection can harm the heart and digestive organs. Chagas disease is now prevalent in parts of Central and South America, and occurs at low frequency in the southern United States. Large influxes of immigrants from South and Central

America into the United States raised concerns about *T. cruzi* contamination of the blood supply. In 2006, blood banks began to test donated blood for *T. cruzi*.

The Euglenoids

Euglenoids are flagellated protists closely related to kinetoplastids. Most live in fresh water and none are human pathogens. The majority are tiny predators, a few are parasites of larger organisms, and about a third have chloroplasts (Figure 22.6). The structure of euglenoid chloroplasts and the pigments inside them indicate that these organelles evolved from a green alga by secondary endosymbiosis.

Photosynthetic euglenoids can detect light with an eyespot near the base of their long flagellum. They typically revert to heterotrophic nutrition if light levels decline, or conditions for photosynthesis become otherwise unfavorable.

A euglenoid is hypertonic relative to fresh water. As in other freshwater protists, one or more **contractile vacuoles** counter the tendency of water to diffuse into the cell. Excess water collects in contractile vacuoles, which contract and expel it to the outside.

Take-Home Message

What are flagellated protozoans?

■ Flagellated protozoans are single-celled protists with one or more flagella. They are typically heterotrophic and reproduce asexually by binary fission.

■ Diplomonads and parabasalids have adapted to life in oxygen-poor waters. Some members of these groups commonly infect humans and cause disease.

■ Trypanosomes also include human pathogens. They are transmitted by insects. Their relatives, the euglenoids, do not infect humans. Most prey on bacteria, but some have chloroplasts that evolved from green algae.

22.3 | Foraminiferans and Radiolarians

- Heterotrophic single cells with chalky or glassy shells live in great numbers in the world's oceans.

- Links to Buffers 2.6, Global warming 7.9, Fossils 17.1

The world's oceans contain countless foraminiferans and radiolarians. These protists live as single cells in shells of their own making. The shell is sieve-like, with numerous pores. Microtubule-reinforced threads of cytoplasm extend through the openings (Figure 22.7).

The Chalky-Shelled Foraminiferans

Foraminiferans (or forams) are single-celled protists with a calcium carbonate ($CaCO_3$) shell around their plasma membrane (Figure 22.7a). Most live on the sea floor, probing water and sediments for prey. Other forams are part of the marine **plankton**, the mostly microscopic organisms that drift or swim in the open sea. Planktonic forams often have photosynthetic protists such as diatoms or algae living inside them.

Long ago, the calcium-rich remains of forams and other protists with calicum carbonate shells began to accumulate on the sea floor. Over great spans of time, the deposits became limestone and chalk (Figure 22.8). Section 17.1 explained how fossil foram shells in such deposits help geologists match layers of rock in different geographic regions. Foraminiferan-rich limestones were used to construct the Egyptian pyramids and Sphinx, and are still important building materials.

Forams play a pivotal role in carbon cycling. They incorporate carbon dioxide (CO_2) in their shells. After

Figure 22.8 White chalk cliffs 90 meters (300 feet) high along the shore in Dover, England, were once sea floor. The chalk is the remains of foraminiferans and other organisms with calcium carbonate–rich shells.

a foram dies, its shell dissolves, releasing calcium and carbonate (CO_3^-) ions. The carbonate acts as buffer; it helps keep seas from getting too acidic (Section 2.6).

Rising levels of atmospheric carbon dioxide caused by human activities may disrupt this system. As more CO_2 enters the air, more dissolves in seawater. Some models suggest that if too much CO_2 gets into seas, the natural buffer system will be overwhelmed, allowing seawater to become acidic. An acidic ocean would prevent forams from building shells, thus leaving still more CO_2 in the atmosphere and water. Increasing atmospheric CO_2 is a problem because it contributes to global warming (Section 7.9).

The Glassy-Shelled Radiolarians

Radiolarians are heterotrophic protists with a glassy silica shell beneath their plasma membrane (Figure 22.7b). Most are part of the marine plankton. They live from surface waters to 5,000-meter depths. Numerous vacuoles in an outer zone of cytoplasm fill with air and keep the cells afloat.

Take-Home Message

What are foraminiferans and radiolarians?

- Foraminiferans and radiolarians are heterotrophic, single cells that live mainly in seawater.

- Forams make a shell of calcium carbonate and most live on the sea floor. Their activities help stabilize atmospheric carbon dioxide level and the pH of seawater.

- Radiolarians have a glassy silica shell; most are planktonic.

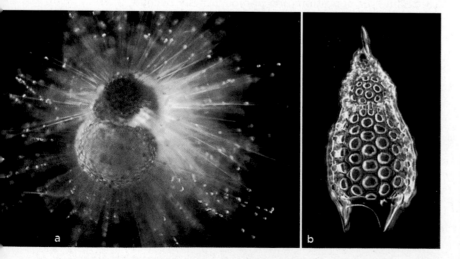

Figure 22.7 (a) Living foraminiferan extending thin pseudopods past its shell. The *yellow* "dots" are cells of algae. (b) A radiolarian's silica shell.

22.4 | The Ciliates

■ Ciliated cells hunt bacteria, other protists, and one another in freshwater habitats and the oceans.

■ Link to Meiosis 10.3

Three groups of protists—ciliates, dinoflagellates, and apicomplexans—are **alveolates**, members of a lineage characterized by a layer of sacs just under the plasma membrane. "Alveolus" means sac.

Ciliates, or ciliated protozoans, are highly diverse heterotrophs, with about 8,000 species. They occur just about anywhere there is water and most are predators (Figure 22.9). About a third live inside the bodies of animals. *Balantidium coli* is the only ciliate parasite of humans. It also infects pigs, and people usually become infected when *B. coli* cysts that are excreted in pig feces get into drinking water. Infection causes nausea and diarrhea.

Paramecium is a freshwater ciliate (Figure 22.10). Cilia cover its entire surface and beat in synchrony; they resemble a field of grass swaying in the wind. Starting at an oral groove at the cell surface, cilia sweep water laden with bacteria, algae, and other food particles into the gullet. Enzyme-filled vesicles digest the food. Contractile vacuoles squirt out excess water. Like other ciliates, *Paramecium* has organelles called trichocysts beneath its pellicle (Figure 22.10c). A

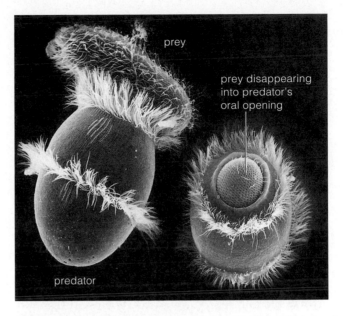

Figure 22.9 *Didinium* (*left*), a barrel-shaped ciliate with tufts of cilia, catching and engulfing *Paramecium* (*right*), another ciliate.

trichocyst contains a protein thread that can be expelled to help the cell capture prey or fend off a predator.

Ciliates reproduce asexually by binary fission, and most reproduce sexually by conjugation. A ciliate has a macronucleus that controls daily function, and one or more small micronuclei. During conjugation, cells pair up, four haploid micronuclei form by meiosis, and two are exchanged between cells. Each cell then forms a new macronucleus by combining one of its haploid micronuclei with a micronucleus from its partner.

Take-Home Message

What are ciliates?

■ Ciliates are heterotrophic single cells that move about with the help of cilia. Most are free-living predators, but some live inside animals.

■ Ciliates, along with dinoflagellates and apicomplexans, are alveolates. This group is characterized by having tiny sacs beneath their plasma membrane.

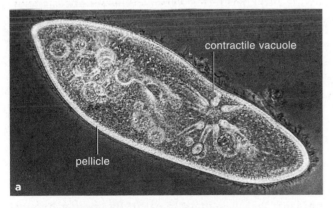

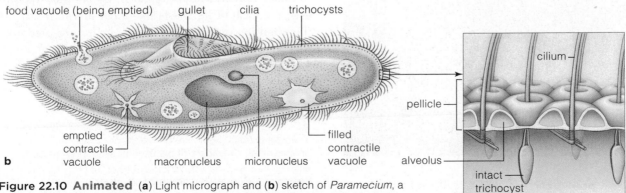

Figure 22.10 Animated (**a**) Light micrograph and (**b**) sketch of *Paramecium*, a typical ciliate. The close-up view (**c**) shows the alveoli (sacs) typical of alveolates.

22.5 | Dinoflagellates

- Dinoflagellates are single-celled heterotrophs and autotrophs. Most whirl about in the seas. Some live inside corals.

- Link to Bioluminescence 6.5

The name **dinoflagellate** means "whirling flagellate." These single-celled protists typically have two flagella; one extends out from the base of the cell, and the other wraps around the cell's middle like a belt (Figure 22.11a). Combined action of these two flagella causes the cell to rotate as it moves forward.

As noted in the previous section, dinoflagellates belong to the alveolate lineage. Most deposit cellulose in the alveoli (sacs) beneath their plasma membrane. The cellulose accumulates as thick but porous plates.

We find planktonic dinoflagellates in the sea and in places where salt water and fresh water mix. Cells are haploid and reproduce asexually most of the time.

Adverse conditions stimulate sexual reproduction; two dinoflagellates fuse and form a cyst that later undergoes meiosis.

About half of the dinoflagellates are heterotrophs. Most of these are predators, but some are parasites of fish and aquatic invertebrates.

Chloroplasts of most photosynthetic dinoflagellates evolved from red algae by secondary endosymbiosis. A few photosynthetic dinoflagellate species are themselves endosymbionts in cells of reef-building corals. A coral is an invertebrate animal, and its relationship with the protists is mutually beneficial. Dinoflagellates supply their host coral with sugar and oxygen for aerobic respiration. The coral provides the protists with nutrients, shelter, and the carbon dioxide necessary for photosynthesis.

Free-living photosynthetic dinoflagellates or other protists sometimes undergo great increases in population size, a phenomenon known as an **algal bloom**. In habitats enriched with nutrients, as from agricultural runoff, each liter of water may hold millions of cells. Blooms of certain species cause "red tides," during which an abundance of cells tints the water red.

Algal blooms can sicken humans and kill aquatic organisms (Figure 22.11b). Aerobic bacteria that feed on algal remains can use up the oxygen in the water, so that aquatic animals suffocate. Some dinoflagellate toxins also kill directly. *Karenia brevis* produces a toxin that binds to transport proteins in the plasma membrane of nerve cells. Eat shellfish contaminated with this toxin and you might end up dizzy and nauseated from neurotoxic shellfish poisoning. Symptoms usually develop hours after the meal and persist for a few days. Blooms of *K. brevis* occur almost every year in the Gulf of Mexico, but the severity and effects vary.

Some dinoflagellates are bioluminescent (Figure 22.12). Like fireflies, they have an enzyme (luciferase) that converts ATP energy to light energy (Section 6.5). Emitting light may protect a cell by startling a small predator that was about to eat it. By another hypothesis, the flash of light acts like a car alarm. It attracts the attention of other organisms, including predators that pursue the would-be dinoflagellate eaters.

Figure 22.11 (**a**) Micrograph of *Karenia brevis*, a dinoflagellate that makes a nerve toxin. (**b**) Results of a population explosion of *K. brevis* in waters near Naples, Florida. Poisoned fishes died and were washed ashore.

Figure 22.12 Tropical waters light up when *Noctiluca scintillans*, a bioluminescent dinoflagellate, is disturbed by something moving through the water.

Take-Home Message

What are dinoflagellates?

- Dinoflagellates are a group of mostly marine single-celled alveolate protists. Some are predators or parasites. Others are photosynthetic members of plankton or symbionts in corals.

- An algal bloom—a population explosion of protists—can harm aquatic organisms and endanger human health.

22.6 The Cell-Dwelling Apicomplexans

■ As mentioned in the chapter introduction, malaria is a major cause of human death. Here we consider the protist that causes malaria and another pathogenic relative.

Apicomplexans are parasitic alveolates that spend part of their life inside cells of their hosts. Their name refers to a complex of microtubules at their apical (top) end that allows them to enter a host cell. They are also sometimes called sporozoans.

Apicomplexans infect a variety of animals, from worms and insects to humans. Their life cycle may involve more than one host species. For example, Figure 22.13 shows the life cycle of *Plasmodium*, the agent of malaria. A female *Anopheles* mosquito transmits a motile infective stage (called the sporozoite) to a vertebrate host such as a human (Figure 22.13*a*). A sporozoite travels in blood vessels to liver cells, where it reproduces asexually (Figure 22.13*b*). Some offspring (merozoites) enter red blood cells and liver cells, where they divide asexually (Figure 22.13*c,d*). Other offspring develop into immature gametes, or gametocytes (Figure 22.13*e*).

When a mosquito bites an infected person, gametocytes are taken up with blood and mature in the mosquito gut (Figure 22.13*f,g*). Gametes fuse and form zygotes, which develop into new sporozoites. The sporozoites migrate to the insect's salivary glands, where they await transfer to a new vertebrate host.

Malaria symptoms usually start a week or two after a bite, when the infected liver cells rupture and release merozoites and cellular debris into blood. After the initial episode, symptoms may subside. However, continued infection damages the body and eventually kills the host. Again, malaria kills more than a million people each year.

Another apicomplexan disease is toxoplasmosis. It is caused by *Toxoplasma gondii*. Healthy people often harbor *T. gondii* inside their body without ill effect. However, in immune-suppressed people, such as those with AIDS, an infection can be fatal. Also, a maternal *T. gondii* infection during pregnancy can cause neurological birth defects.

Eating cysts in undercooked meat is a main cause of toxoplasmosis. *T. gondii* infects cattle, sheep, pigs, and poultry. Cats also can carry *T. gondii*. An infected cat excretes cysts in its feces (Figure 22.14). For this reason, pregnant women and people who have an impaired immune system are advised to avoid contact with cat feces. Cats that spend some time outdoors may be exposed to *T. gondii* cysts in soil or in wildlife and are most likely to be infected. Keeping a cat indoors and feeding it only commercially prepared food minimizes the risk of infection.

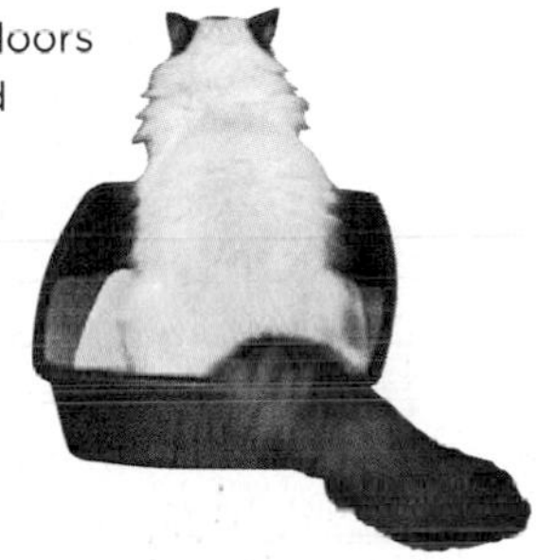

Figure 22.14 One potential source of the disease toxoplasmosis. Exposure to cysts in cat feces can lead to human infection by the apicomplexan parasite *Toxoplasma gondii*.

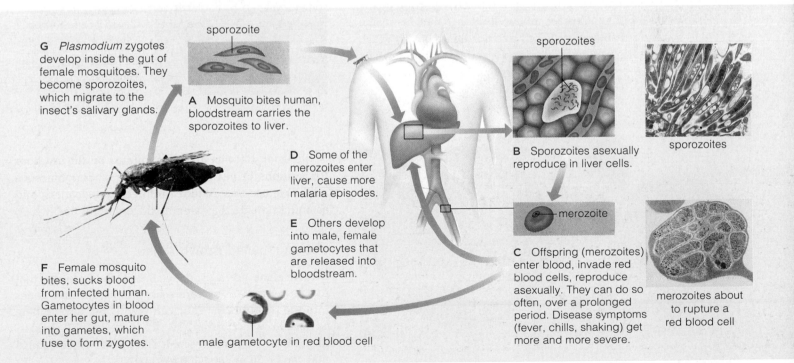

G *Plasmodium* zygotes develop inside the gut of female mosquitoes. They become sporozoites, which migrate to the insect's salivary glands.

sporozoite

A Mosquito bites human, bloodstream carries the sporozoites to liver.

D Some of the merozoites enter liver, cause more malaria episodes.

E Others develop into male, female gametocytes that are released into bloodstream.

F Female mosquito bites, sucks blood from infected human. Gametocytes in blood enter her gut, mature into gametes, which fuse to form zygotes.

male gametocyte in red blood cell

sporozoites

B Sporozoites asexually reproduce in liver cells.

sporozoites

merozoite

C Offspring (merozoites) enter blood, invade red blood cells, reproduce asexually. They can do so often, over a prolonged period. Disease symptoms (fever, chills, shaking) get more and more severe.

merozoites about to rupture a red blood cell

Figure 22.13 Animated Life cycle of one of the four *Plasmodium* species that cause malaria.

22.7 | The Stramenopiles

■ Colorless filamentous molds, photosynthetic single cells, and large seaweeds belong to the stramenopile lineage.

■ Link to Pigments 7.1

We turn now to another major protist lineage, known as the **stramenopiles**. The name means "straw-haired" and refers to a cell with a shaggy or hairy flagellum (*left*) that occurs during the life cycle of many members of this group. However, stramenopiles are defined mainly by the results of gene sequence comparisons, rather than by any visible characteristics.

The Diatoms

Diatoms are single-celled or colonial protists that have a two-part silica shell (Figures 22.2*c* and 22.15). Most are photosynthetic. Their chloroplasts and those of other photosynthetic stramenopiles are tinted brown by an accessory pigment called fucoxanthin. They evolved from a red alga by secondary endosymbiosis.

Diatoms live in seas, fresh water, and damp soils. In temperate seas, they are a major component of the phytoplankton, the photosynthetic portion of plankton. Diatoms are responsible for about 25 to 35 percent of all carbon taken up by photosynthetic organisms.

A diatom shell has two parts like a hat box or petri dish; a larger lid fits over a smaller bottom. When diatoms reproduce asexually, each new cell inherits half the parental shell. It uses the inherited portion as the "lid" and synthesizes a new "bottom" that fits inside it. As a result, average cell size declines a bit with each generation. When cells reach a certain minimum size, they reproduce sexually. Fertilization yields a zygote, which develops into a large cell.

Figure 22.16 Photo and diagram of the spore-bearing body (sporophyte) of a giant kelp. The stemlike stipe has leaflike blades. A holdfast anchors the kelp. Gas-filled bladders make stipes and blades buoyant. Spores form by meiosis on stipes. They divide and form small gamete-forming bodies. Fusion of gametes yields a zygote that develops into a new sporophyte.

Figure 22.17 *Fucus versiculosis*, a brown alga commonly known as bladderwrack or seawrack. It occurs on rocky shores along the North Atlantic coast of the United States and is harvested for use as an herbal supplement.

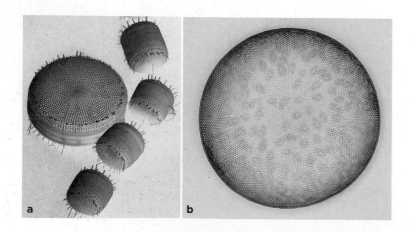

Figure 22.15 (**a**) Diatoms have a two-part silica shell. Average cell size decreases with each asexual division. (**b**) Fucoxanthin-tinted chloroplasts of this diatom are visible through its silica shell.

Shells of diatoms that accumulated on the sea floor for millions of years are the source of diatomaceous earth. We use this material in filters and cleaners, and as an insecticide that is nontoxic to vertebrates.

The Multicelled Brown Algae

Brown algae are a group of multicelled protists that live in temperate or cool seas, from intertidal zones into the open ocean. They can be olive-green, golden, or dark brown (Figures 22.2*e*, 22.16, and 22.17). In size, they range from microscopic filaments to giant kelps that can be 30 meters (100 feet) tall.

Giant kelps such as *Macrocystis* are the largest protists. The life cycle shows an alternation of generations, with multicellular haploid and diploid bodies (Figure 22.3*b*). The kelp sporophyte is the larger, longer-lived stage in the life cycle (Figure 22.16).

Kelps are of great ecological importance. Giant kelps form forestlike stands in costal waters of the Pacific Northwest. Like trees in a forest, kelps shelter a wide variety of other organisms. The Sargasso Sea in the North Altantic Ocean is named for its abundance of *Sargassum*. This kelp forms vast, floating mats up to 9 meters (30 feet) thick. The mats serve as habitat for fishes, sea turtles, and invertebrates.

Some brown algae are used commercially. Algins from kelp thicken many foods, beverages, cosmetics, and other products. Brown algae are also harvested for use as food, herbal supplements, and fertilizer.

The Heterotrophic Water Molds

The **water molds**, or oomycotes, are heterotrophs. They form a mesh of nutrient-absorbing filaments and were once classified as fungi, which have a similar growth pattern. Unlike fungi, water molds have cell walls of cellulose, not chitin, and their filaments are made up of diploid cells, rather than haploid ones.

Most water molds decompose organic matter in aquatic habitats, but some are aquatic parasites. For example, *Saprolegnia* often infects fish in aquariums, fish farms, and hatcheries (Figure 22.18). Other water molds live in damp places on land, or in plant tissues. Some that infect plants are important pathogens, as explained in the next section.

Figure 22.18 The water mold *Saprolegnia* grows as cottony filaments on an infected fish.

Take-Home Message

What are stramenopiles?

■ Stramenopiles include single-celled diatoms and multicelled brown algae, both important producers.

■ Heterotrophic water molds also belong to this group.

22.8 | The Plant Destroyers

■ The water molds include economically and ecologically important plant pathogens. They infect a wide variety of crop plants, as well as forest trees.

In the mid-1800s, an outbreak of a disease called late blight destroyed Ireland's potato crop. The resulting starvation, related illness, and emigration reduced Ireland's human population by about a third. The pathogen that caused the crop failure, *Phytophthora infestans*, is one of the water molds. Its genus name, *Phytophthora*, means "plant destroyer" and the group lives up to its name. Worldwide, *Phytophthora* species cause an estimated 5 billion dollars in crop losses every year. In addition to potatoes, they grow on and destroy cucumbers, squash, green beans, tomatoes, and other vegetables.

Phytophthora is now also assaulting some of North America's forests. In 1995, oak trees in northern California began to ooze sap, lose leaves, and die (Figure 22.19). The disease afflicting them became known as sudden oak death and *P. ramorum*, a pathogen previously unknown in North America, was identified as its cause. Scientists have since found *P. ramorum* infecting a wide variety of trees and shrubs, including rhododendrons, maples, firs, beeches, and redwoods. In addition to California, the pathogen has now been detected in Oregon, Washington, and the Canadian province of British Columbia. Once a tree becomes infected, there is no cure.

Containing *P. ramorum* is likely to prove difficult. In a forest, spores spread from tree to tree by rain and wind. Resting spores can survive in water and soil, so streams and hikers' boots can disperse them for longer distances. The shipping of nursery plants increases the threat. We know that at least one shipment of infected plants reached the East Coast, although it was quickly detected and destroyed. So far, surveys of eastern oak forests have turned up no sign of the disease.

Figure 22.19 Effects of *Phytophthora ramorum*. (**a**) Dead and dying trees near Big Sur, California, where *P. ramorum* is epidemic. Death of a tree's crown and open cankers that ooze sap (**b**) are early symptoms of infection.

- Green algae are the protists most like land plants. The group includes single-celled and multicelled species.

- Links to Engelmann's (7.2) and Calvin's (2.2) experiments, Plasmodesmata 4.12, Cell plate formation 9.4

"Green algae" is the informal name for about 7,000 photosynthetic species that range in size from microscopic cells to multicelled filamentous or branching forms more than a meter long.

Most green algae are **chlorophytes**, a large group that may be monophyletic. The others belong to one of several lineages collectively known as **charophyte algae**. One of these lineages is thought to be the sister group of the land plants.

All green algae resemble land plants in that they store sugars as starch and deposit cellulose fibers in their cell wall. Also like land plants, they have chloroplasts that evolved by primary endosymbiosis from cyanobacteria. These chloroplasts have a double membrane and contain chlorophylls *a* and *b*.

The Chlorophytes

Chlorophyte algae include single cells that live in the soil, on ice, or participate with fungi in formation of lichens. However, most single-celled chlorophytes are found in fresh water. Melvin Calvin used one of them, *Chlorella*, to study the light-independent reactions of photosynthesis (Section 2.2). Today *Chlorella* is grown commercially and sold as a health food.

Chlamydomonas is a single-celled species common in ponds (Figure 22.20). Haploid, flagellated cells (called spores) reproduce asexually, as long as nutrients and light are plentiful (Figure 22.20*a*). When conditions do not favor growth, gametes develop and fuse, forming a zygote with a thick, protective wall (Figure 22.20*b–f*). When conditions turn favorable, the zygote undergoes meiosis and then germinates, releasing the next generation of haploid, flagellated spores (Figure 22.20*g*).

Volvox is a colonial freshwater species. Hundreds to thousands of flagellated cells that resemble those of

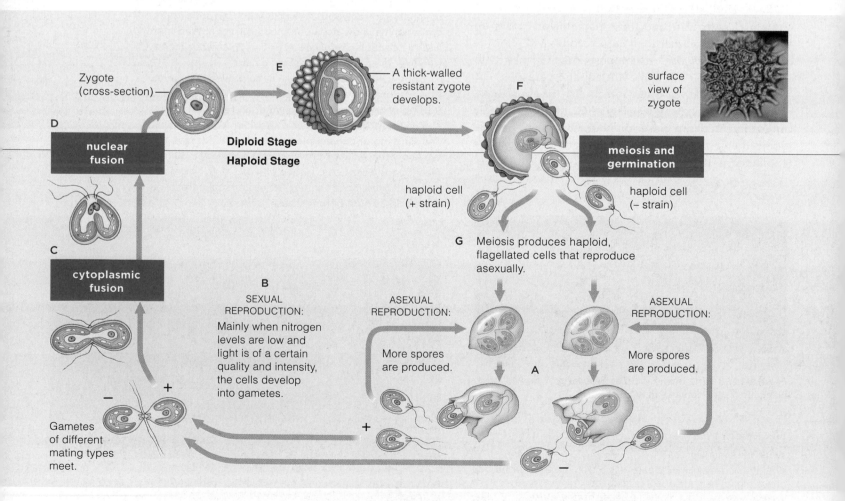

Figure 22.20 Animated Life cycle of *Chlamydomonas*, a single-celled, freshwater green alga.

Figure 22.21 Chlorophyte algae. (**a**) A *Volvox* colony, with flagellated cells joined by thin strands of cytoplasm. It ruptured and is releasing new colonies. (**b**) Thin sheets of sea lettuce (*Ulva*). (**c**) *Codium fragilis*, with spongy branches that can reach 1 meter in length.

Chlamydomonas are joined together by thin cytoplasmic strands to form a whirling, spherical colony (Figure 22.21a). Daughter colonies form inside the parental sphere, which eventually ruptures and releases them.

Other freshwater chlorophytes form long filaments. Theodor Engelmann used one of these, *Cladophora*, in his studies of photosynthesis (Section 7.2).

Some chlorophytes are common "seaweeds." Wispy sheets of *Ulva* cling to coastal rocks (Figure 22.21b). The sheets grow longer than your arm, but are usually less than 40 microns thick. *Ulva* is commonly known as sea lettuce and is a popular food in Scotland.

Codium fragilis, is a dark green, branching marine species, with the unappealing common name of Dead Man's Fingers (Figure 22.21c). Native to the Pacific, it was introduced to waters near Connecticut in the 1950s and is now prevalent along the Atlantic seaboard.

Charophyte Algae

In addition to the green alga–land plant similarities discussed above, the charophyte algae and land plants share other unique traits. As a result, botanists now consider these groups to be a clade.

Most charophytes live in freshwater. For example, desmids are a single-celled, freshwater group. Figure 22.22a shows two desmid cells formed by asexual cell division. Other charophytes, such as *Spirogyra*, form long, unbranched filaments. Still others form multi-celled disks.

Gene comparisons indicate that charales, or stoneworts, are the charophyte algae most closely related to land plants. Like plants they have plasmodemata

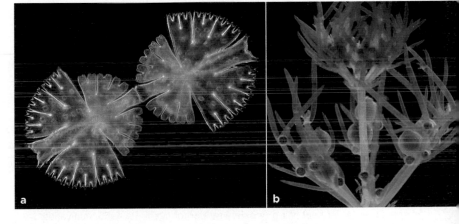

Figure 22.22 Charophyte algae. (**a**) Two newly formed desmid cells. The larger part of each cell is a remnant of the parent. The smaller portion is new. (**b**) *Chara*, a stonewort. Gametes, and then zygotes, form in protective jackets of cells on the "branches."

(Section 4.12), and divide their cytoplasm by cell plate formation (Section 9.4). *Chara*, native to Florida, is an example. Its haploid gametophyte grows as branching filaments in lakes and ponds (Figure 22.22b).

Take-Home Message

What are green algae?

■ Green algae are photosynthetic single-celled or multicelled protists. Like land plants, they have cellulose in their cell walls, store sugars as starch, and have chloroplasts descended from cyanobacteria.

■ Most green algae are chlorophytes. The smaller group of lineages known as charophyte algae form a clade with the land plants.

■ Red accessory pigments allow the red algae to survive at greater depths than other algae.

■ Link to Pigments 7.1

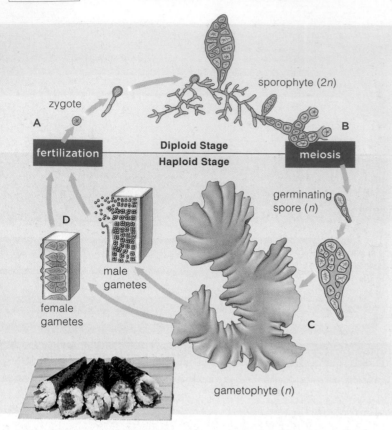

Figure 22.23 **Animated** Life cycle of a red alga (*Porphyra*). For centuries, Japanese fishermen cultivated and harvested a red alga in early fall. The rest of the year, it seemed to vanish. Kathleen Drew-Baker examined sheets of the alga grown in the lab. She saw gametes forming in packets near the sheet margins. She also studied gametes in a petri dish. After zygotes formed, individuals developed into tiny, branching filaments on bits of shell in the dish. That was how the alga spent most of the year!

Within a few years, researchers worked out the life cycle of *P. tenera*, a species used for seasoning or as a wrapper for sushi. By 1960, cultivation of *P. tenera* had become a billion-dollar industry.

Figure It Out: Are the cells in the sheets of *Porphyra* that are used to wrap sushi haploid or diploid? *Answer: haploid*

Of more than 4,000 species of **red algae**, nearly all live in warm marine currents and clear tropical seas. Of all photosynthetic protists, certain species of red algae live at the greatest depths. A few form a crust when they secrete calcium carbonate. Their remains become part of the structure of some coral reefs.

Chloroplasts of red algae contain chlorophyll *a* and pigments called phycobilins (Section 7.1). Phycobilins absorb blue-green and green light, and reflect red. Blue-green and green light penetrate deeper into water than other wavelengths, so having an ability to absorb this light allows red algae to thrive at greater depths. Shallow-water red algae tend to have little phycobilin and appear green. Deep dwellers are almost black.

Red algae and green algae share a common ancestor that had chloroplasts derived from cyanobacteria. Chloroplasts of green algae lost the ability to make phycobilins, but those of red algae retained it. Later, single-celled red algae evolved into plastids of apicomplexans, and chloroplasts of certain dinoflagellates.

Some single-celled species of red algae persist, but most are multicelled. They usually grow as sheets or in a branching pattern (Figures 22.23 and 22.24). Life cycles vary and are often complex, with both asexual and sexual phases. There is no flagellated stage.

Red algae have many commercial uses. Agar is a polysaccharide extracted from cell walls of some red algae. It keeps some baked goods and cosmetics moist, helps jellies set, and is used in the capsules that hold medicines. Carrageenan, another polysaccharide of red algae, is added to soy milk, dairy foods, and the fluid that is sprayed on airplanes to prevent ice formation. *Porphyra* is now cultivated worldwide as food (Figure 22.23). More than 130,000 tons are harvested annually.

Take-Home Message

What are red algae?

■ Red algae are mostly multicelled marine algae that live in clear, warm waters.

■ Pigments called phycobilins give red algae their red color and allow them to live at greater depths than other algae.

Figure 22.24 The red alga *Antithamnion plumula*. The filamentous, branching growth pattern is common among red algae.

22.11 | Amoeboid Cells at the Crossroads

- The amoebas and their social relatives, the slime molds, are shape-shifting heterotrophs.

- Link to Cell motility 4.13

Amoebozoa is one of the monophyletic groups now being carved out of the former hodgepodge kingdom Protista. Few amoebozoans have a cell wall, shell, or pellicle; nearly all undergo dynamic changes in shape. A compact blob of a cell can quickly send out pseudopods, move about, and capture food (Section 4.13).

The **amoebas** live as single cells. Figure 22.25a shows *Amoeba proteus*. Like most amoebas, it is a predator in freshwater habitats. Other amoebas live in the gut of humans and other animals. Each year, about 50 million people are affected by amoebic dysentery after drinking water contaminated with *Entamoeba histolytica* cysts.

Slime molds are "social amoebas." The plasmodial slime moles spend most of their life cycle as a plasmodium. This multinucleated mass arises from a diploid cell that undergoes mitosis many times without cytoplasmic division. A plasmodium streams out along the forest floor feeding on microbes and organic matter (Figure 22.25b). When food supplies dwindle, a plasmodium develops into spore-bearing fruiting bodies (Figure 22.25c).

Cellular slime molds, such as *Dictyostelium discoideum*, spend the bulk of their existence as individual amoeboid (amoeba-like) cells (Figure 22.26). Each cell eats bacteria and reproduces by mitosis. When this food runs out, thousands of cells come together. Often they form a "slug" that migrates in response to light and heat. When the slug reaches a suitable spot, it becomes a fruiting body. A stalk form and lengthens, and nonmotile spores form at its tip. Germination of a spore releases a diploid amoeboid cell that starts the cycle anew.

Dictyostelium and other amoebozoans provide clues to how signaling pathways of multicelled organisms evolved. Coordinated behavior—an ability to respond to stimuli as a unit—is a hallmark of multicellularity. It requires cell-to-cell communication, which may have originated in single-celled amoeboid ancestors. In *Dictyostelium*, a nucleotide called cyclic AMP is the signal that induces solitary amoeboid cells to stream together. It also triggers changes in gene expression. The changes cause some cells to differentiate into components of a stalk or into spores. Cyclic AMP also functions in signaling pathways of multicelled organisms. Intriguingly, molecular comparisons suggest that animals and fungi descended from an amoebozoan-like ancestor.

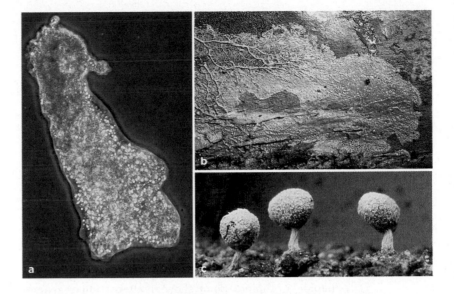

Figure 22.25 (**a**) *Amoeba proteus*, a free-living, freshwater amoeba. (**b**) *Physarum* plasmodium streaming across a rotting log. (**c**) *Physarum* fruiting bodies, which release haploid motile spores. When two spores fuse, they form a diploid cell. That cell undergoes repeated rounds of mitosis without cytoplasmic division, which forms a new plasmodium.

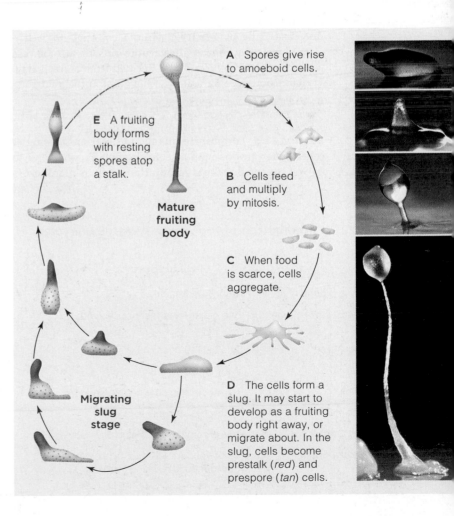

A Spores give rise to amoeboid cells.

B Cells feed and multiply by mitosis.

C When food is scarce, cells aggregate.

D The cells form a slug. It may start to develop as a fruiting body right away, or migrate about. In the slug, cells become prestalk (*red*) and prespore (*tan*) cells.

E A fruiting body forms with resting spores atop a stalk.

Mature fruiting body

Migrating slug stage

Figure 22.26 Animated Life cycle of *Dictyostelium discoideum*, a cellular slime mold. During aggregation, the cells secrete and respond to cyclic AMP.

The fight against malaria continues on many fronts. Scientists are working to synthesize a vaccine and looking into new drugs that interrupt the *Plasmodium* life cycle. People in affected areas are being screened for the disease and treated. Mosquito breeding areas are being eliminated. Bites are being prevented by use of insecticide-treated nets around sleeping areas and by spraying of insecticides, including DDT, inside homes.

How would you vote?

Should DDT be banned entirely, or should nations with malaria problems be allowed to spray it indoors? See CengageNOW for details, then vote online.

Summary

This chapter and the preceding one describe prokaryotes and protists, which are the simplest eukaryotes. Table 22.2 summarizes their similarities and differences.

Section 22.1 **Protists** are a collection of mostly single-celled eukaryotic lineages, some only distantly related to one another. The dominant stage of the life cycle may be haploid or diploid. Some show an **alternation of generations**, with haploid and diploid multicelled stages. Some protists survive adverse conditions by forming **cysts**.

Section 22.2 **Flagellated protozoans** are single-celled and mostly or entirely heterotrophic. Diplomonads and parabasalids lack mitochondria, but have organelles that make ATP by anaerobic pathways. Both include species that infect humans. Like many other protists, they reproduce asexually by **binary fission**. Most **euglenoids** live in freshwater; a **contractile vacuole** rids them of excess water. Some have chloroplasts derived from a green alga. **Trypanosomes** are parasites with a giant mitochondrion.

■ *Use the animation on CengageNOW to explore the structure of a euglenoid.*

Section 22.3 **Foraminiferans** and **radiolarians** are single-celled heterotrophs with a secreted shell. Forams tend to live on the sea floor, and **radiolarians** drift as **plankton**.

Sections 22.4–22.6 Tiny sacs (alveoli) beneath the plasma membrane characterize **alveolates**. **Ciliates** have many cilia and are aquatic heterotrophs. **Dinoflagellates** are aquatic heterotrophs and autotrophs with a cellulose covering. In nutrient-rich water, photosynthetic protists undergo population explosions known as **algal blooms**. **Apicomplexans** are intracellular parasites of animals.

■ *Use the animation and video on CengageNOW to explore ciliate structure and the* Plasmodium *life cycle.*

Sections 22.7, 22.8 **Stramenopiles** are named for a flagellum with hairlike filaments, but not all have this trait. **Water molds** are heterotrophs that grow as a mesh of absorptive filaments. Some are plant pathogens.

Diatoms are photosynthetic single cells with a two-part silica shell. Like brown algae, diatoms contain the pigment fucoxanthin. All **brown algae** are multicelled. They include tiny strands and giant kelps, which are the largest protists.

■ *Use the video on CengageNOW to visit a kelp forest.*

Section 22.9 Green algae are single-celled or multi-celled, mostly aquatic, autotrophs. Most are **chlorophytes**. **Charophyte algae** include the closest relatives of plants.

■ *Use the animation on CengageNOW to learn about the life cycle of a single-celled green alga.*

Table 22.2 Comparison of Prokaryotes With Eukaryotes

	Prokaryotes	Eukaryotes
Organisms represented:	Archaeans, bacteria	Protists, plants, fungi, and animals
Ancestry:	Two major lineages that evolved	Equally ancient prokaryotic species with a division of labor among specialized cells; complex types have tissues and organ systems
Typical cell size:	Small (1–10 micrometers)	Large (10–100 micrometers)
Cell wall:	Many with cell wall	Cellulose or chitin; none in animal cells
Organelles:	Rarely; no nucleus; no mitochondria	Typically profuse; nucleus present; mitochondria in most
Modes of metabolism:	Anaerobic, aerobic, or both	Aerobic modes predominate
Genetic material:	One chromosome; plasmids in some	Chromosomes of DNA and many associated proteins; in a nucleus
Mode of cell division:	Prokaryotic fission, mostly; some reproduce by budding	Nuclear division (mitosis, meiosis, or both) associated with one of various modes of cytoplasmic division, including binary fission

Data Analysis Exercise

Parasites sometimes alter their host's behavior in a way that increases their chances of transmission to another host. For example, *Toxoplasma gondii*, the cause of toxoplasmosis, infects rats and makes them less wary of cats.

Dr. Jacob Koella and his associates hypothesized that *Plasmodium* might benefit by making its human host more attractive to hungry mosquitoes when gametocytes were available in the host's blood. Recall that the gametocytes are the stage that can be taken up by the mosquito and mature into gametes inside its gut (Figure 22.13).

To test their hypothesis, the researchers recorded the response of mosquitoes to the odor of *Plasmodium*-infected children and uninfected children over the course of 12 trials on 12 separate days. Figure 22.27 shows their results.

1. On average, which group of children was most attractive to mosquitoes?

2. Which group of children averaged the fewest mosquitoes attracted?

3. What percentage of the total number of mosquitoes were attracted to the most attractive group?

4. Did the data support Dr. Koella's hypothesis?

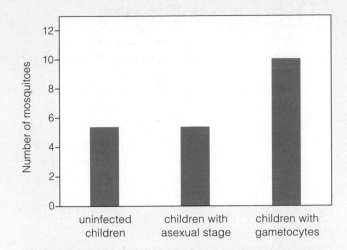

Figure 22. 27 Number of mosquitoes (out of 100) attracted to uninfected children, children harboring the asexual stage of *Plasmodium*, and children with gametocytes in their blood. The bars show the average number of mosquitoes attracted to that category of child over the course of 12 separate trials.

Section 22.10 Most **red algae** are multicelled. They can survive in deeper water than most photoautotrophs because their chloroplasts have phycobilins.

■ *Use the animation on CengageNOW to explore the life cycle of a red alga.*

Section 22.11 **Amoebozoans** include heterotrophic free-living **amoebas** and **slime molds**. The plasmodial slime molds feed as a multinucleated mass. Amoeba-like cells of cellular slime molds gather when food is scarce and form fruiting bodies that disperse resting spores.

■ *Use the animation on CengageNOW to learn about the life cycle of a cellular slime mold.*

Self-Quiz   *Answers in Appendix III*

1. True or false? Some protists are more related to plants than to other protists.

2. Diplomonads and parabasalids often live in anaerobic habitats and lack the _____ used in aerobic respiration.

3. Radiolarians and diatoms have a shell of _____ .

4. Which of the following might you find in seawater?
a. an apicomplexan   c. a dinoflagellate
b. a cellular slime mold d. a euglenoid

5. Diatoms are most closely related to the _____ .
a. dinoflagellates   c. green algae
b. water molds   d. red algae

6. Chloroplasts of green algae evolved from _____ .

7. Green algae are most closely related to _____ algae.

8. The kelp life cycle, with its haploid and diploid multicelled stages is an example of _____ .

9. Ciliates _____ through conjugation.
a. exchange genes   c. emit light
b. deter predators   d. infect cells

10. Which species does not cause human disease?
a. *Toxoplasma gondii*   c. *Dictyostelium discoideum*
b. *Entamoeba histolytica*   d. *Trichomonas vaginalis*

11. _____ is produced from red algae.
a. Diatomaceous earth   c. Carrageenan
b. Algin   d. both b and c

12. Match each term with its most suitable description.
___ diplomonad   a. protist population explosion
___ apicomplexan   b. silica-shelled producer
___ algal bloom   c. multinucleated motile mass
___ diatom   d. no mitochondria, anaerobic
___ brown alga   e. closest relative of land plants
___ red alga   f. multicelled, with fucoxanthin
___ green alga   g. agent of malaria
___ slime mold   h. deep dweller with phycobilins

■ *Visit CengageNOW for additional questions.*

Critical Thinking

1. Suppose you vacation in a developing country where sanitation is poor. Having read about parasitic flagellates in water and damp soil, what would you consider safe to drink? What foods might be best to avoid or which food preparation methods might make them safe to eat?

2. Runoff from highly fertilized cropland promotes algal blooms that can result in massive kills of aquatic species, birds, and other forms of wildlife. If you find the environmental cost unacceptable, what can you do personally to help stop this pollution?

The Land Plants

Beginnings and Endings

Change is the way of life. By the early Ordovician, some land plants were growing on the margins of continents. By 300 million years ago, some tree-sized ancestors of modern club mosses and horsetails dominated vast swamp forests. Then things changed. The global climate became cooler and drier, and moisture-loving plants declined. Hardier plants—cycads, ginkgos, and then conifers—rose to dominance. These were the gymnosperms, and they had a novel trait: They packaged their embryos in seeds.

Later, one branch of the gymnosperm lineage gave rise to flowering plants, causing yet another change. The flowering plants spread and soon became dominant in most regions (Figure 23.1a). However, conifers such as pines still retained their competitive edge in certain environments such as the high-latitude forests of the Northern Hemisphere.

Things changed yet again. About 11,000 years ago, humans began to cultivate crop plants. This advance helped human populations to soar. Over time, agricultural fields, homes, and eventually cities replaced many forests. Those that remained forests were logged to provide lumber and other products. Deforestation—removal of all trees from large tracts of land—was under way (Figure 23.1b).

The United States currently has more forest than it did a century ago, but there are regions of deforestation. Only about 4 percent of California's original coast redwood forest remains. In Maine, an area the size of Delaware was deforested in the past fifteen years. Deforestation may push some plant species toward extinction. Worldwide, about 350 of the 650 existing conifer species are threatened or endangered.

With this bit of perspective on change, we turn to the origins and adaptations of land plants. With few exceptions, they are photoautotrophs. These metabolic wizards produce organic compounds by absorbing energy from the sun, carbon dioxide from the air, and water and dissolved minerals from the soil. By the noncyclic pathway of photosynthesis, they split water molecules and release oxygen. Their oxygen output and carbon uptake sustain the atmosphere. Think of it—every atom of carbon in a redwood tree that stands a hundred meters high and weighs thousands of tons was taken up from the air.

In addition to their effects on the atmosphere, the 295,000 or so kinds of plants feed and shelter land animals. Without them, we humans and other land-dwelling animals never would have made it onto the evolutionary stage.

See the video! **Figure 23.1** Changing times. (**a**) Flowering plants arose during dinosaur times and displaced other plants in many habitats. (**b**) Deforestation of coniferous forests in British Columbia, Canada.

Key Concepts

Milestones in plant evolution

The earliest known plants date from 475 million years ago. Ever since then, environmental changes have triggered divergences, adaptive radiations, and extinctions. Structural and functional adaptations of lineages are responses to some of the changes. **Sections 23.1, 23.2**

Early-diverging plant lineages

Three plant lineages (mosses, hornworts, and liverworts) are commonly referred to as bryophytes, although they are not a natural group. The gamete-producing stage dominates their life cycle, and sperm reach the eggs by swimming through droplets or films of water. **Section 23.3**

Seedless vascular plants

Lycophytes, whisk ferns, horsetails, and ferns have vascular tissues but do not produce seeds. A large spore-producing body that has internal vascular tissues dominates the life cycle. As with bryophytes, sperm swim through water to reach eggs. **Sections 23.4, 23.5**

Seed-bearing vascular plants

Gymnosperms and, later, angiosperms radiated into higher and drier environments. Both produce pollen and seeds. Nearly all crop plants are seed plants. In angiosperms, flowers and fruits further enhanced reproductive success. **Sections 23.6–23.10**

Links to Earlier Concepts

- This chapter picks up the story of the evolution of plants from green algae (22.9). Reviewing gamete formation and fertilization (10.5) may help you put plant life cycles in perspective.

- We return again to the effects of continental movements and changes over the course of geologic time (17.8, 17.9).

- The chapter mentions the role of cell wall materials (4.12) and the importance of amino acids in the human diet (3.5).

- Understanding methods of classification (19.1) will help you get a handle on how plants are now grouped.

23.1 | Evolution on a Changing World Stage

■ Changes in atmospheric conditions and shifts in positions of continents affected the evolutionary history of land plants.

■ Links to Drifting continents 17.9, Green algae 22.9

Traditionally, red algae and green algae were grouped with the protists. However, as Section 22.9 explained, we now recognize one of the charophyte lineages of green algae as the sister group of land plants. Thus, many botanists now consider all members of the clade shown below to be in the plant kingdom.

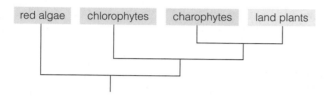

In this classification scheme, the clade of land plants is called the **embryophytes** or embryo-bearing plants.

The earliest fossil evidence of land plants dates back about 475 million years. By that time, enormous numbers of photosynthetic cells had come and gone, and oxygen-producing species had altered the composition of the atmosphere. High above Earth, the sun's energy had converted some oxygen into a dense ozone layer, which screened out ultraviolet radiation (Section 7.8). Before the protective ozone layer was in place, high doses of ultraviolet radiation would have destroyed DNA of any organisms that ventured onto land.

The three lineages commonly known as bryophytes were the first to branch from algal ancestors (Figures 23.2 and 23.3). Modern bryophytes include mosses

and the less familiar liverworts and hornworts (Table 23.1). Which evolved first? Debate continues, but gene comparisons among these groups suggest liverworts are the oldest lineage of land plants. Such comparisons also indicate that an offshoot from the hornwort lineage evolved into the first seedless vascular plants.

By about 430 million years ago, a seedless vascular plant called *Cooksonia* was growing in moist lowlands of the supercontinent Gondwana. It stood only a few centimeters high and had a simple branching pattern,

Table 23.1 Diversity of Modern Land Plants

Bryophytes	
Liverworts	9,000 species
Mosses	15,000 species
Hornworts	100 species
Seedless Vascular Plants	
Lycophytes	1,100 species
Whisk ferns	7 species
Horsetails	25 species
Ferns	12,000 species
Gymnosperms	
Cycads	130 species
Ginkgos	1 species
Conifers	600 species
Gnetopytes	70 species
Angiosperms (Flowering Plants)	
Basal groups (e.g., magnoliids)	9,200 species
Monocots	80,000 species
Eudicots	>180,000 species

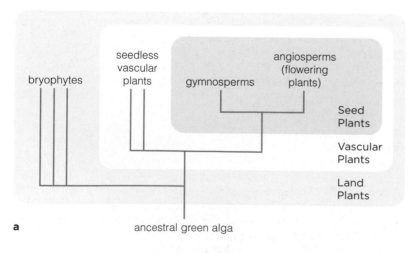

bryophyte (moss)   seedless vascular plant (fern)

b gymnosperm (conifer)   angiosperm (monocot)

Figure 23.2 (**a**) Evolutionary tree for land plants. Bryophytes and seedless vascular plants are not monophyletic groups. (**b**) Examples of land plants.

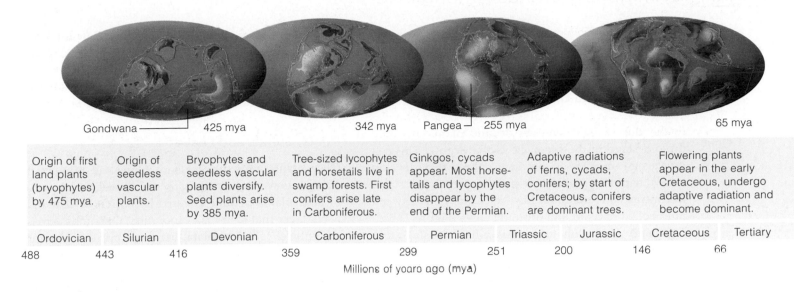

| Gondwana —— 425 mya | | 342 mya | Pangea ⌐ 255 mya | | 65 mya |

Origin of first land plants (bryophytes) by 475 mya.	Origin of seedless vascular plants.	Bryophytes and seedless vascular plants diversify. Seed plants arise by 385 mya.	Tree-sized lycophytes and horsetails live in swamp forests. First conifers arise late in Carboniferous.	Ginkgos, cycads appear. Most horsetails and lycophytes disappear by the end of the Permian.	Adaptive radiations of ferns, cycads, conifers; by start of Cretaceous, conifers are dominant trees.	Flowering plants appear in the early Cretaceous, undergo adaptive radiation and become dominant.

Ordovician	Silurian	Devonian	Carboniferous	Permian	Triassic	Jurassic	Cretaceous	Tertiary
488	443	416	359	299	251	200	146	66

Millions of years ago (mya)

Figure 23.3 A timeline for major events in the evolution of plants. These events took place on a changing global stage, as continents shifted, so did prevailing climates.

with no leaves or roots (Figure 23.4a). Spores formed at branch tips. *Psilophyton*, a taller seedless plant with a more complex branching structure appeared about 60 million years later (Figure 23.4b). Both *Cooksonia* and *Psilophyton* are known only from fossils; they are now extinct.

During the Carboniferous, other seedless vascular plants, including the relatives of modern club mosses and horsetails, grew to treelike size. Modern seedless vascular plants include club mosses and spike mosses, horsetails, and ferns.

The earliest fossil seeds date to 385 million years ago during the Devonian. Gymnosperms, a major seed-bearing lineage diversified during the Carboniferous. During the Permian, formation of the supercontinent Pangea caused a global shift toward a drier climate. This drying contributed to the demise of the tree-sized seedless vascular plants and favored a newly evolved gymnosperm lineage, the drought-tolerant conifers. Modern conifers include pines, firs, and spruces.

During the Triassic and Jurassic, as dinosaurs arose and diversified, the ferns, cycads, and conifers underwent their own adaptive radiation. By the end of the Jurassic, conifers were the dominant trees.

The flowering plants, or angiosperms, arose from a gymnosperm ancestor in the late Jurassic or early Cretaceous. In less than 40 million years, they would replace conifers and their relatives in most habitats.

The modifications in structure, function, and reproductive mode that occurred as different plant groups evolved are the focus of the next section.

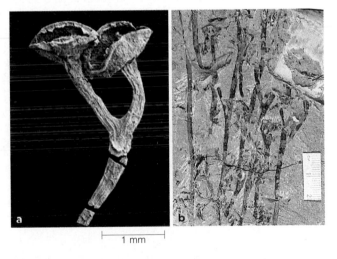

1 mm

Figure 23.4 Fossils of early seedless vascular plants. (a) *Cooksonia* stems always divided into two equal branches. It stood a few centimeters tall. (b) *Psilophyton* shows a more complex growth pattern. It branched unequally with a main stem and smaller branches to the side.

Take-Home Message

What events influenced plant evolution?

■ Land plants evolved from one lineage of charophyte algae after formation of an ozone layer had made life on land possible.

■ Bryophytes include three early-diverging land plant lineages. Nonvascular seedless plant lineages evolved next. The first seed plants were gymnosperms and angiosperms (flowering plants) branched from them.

■ Continental movements that caused the global climate to become drier favored groups that were better adapted to drought, such as seed plants.

■ Over time, the spore-producing bodies of plants became larger, more complex, and better adapted to dry habitats.

■ Links to Cell walls 4.12, Gametes 10.5, Green algae 22.9

From Haploid to Diploid Dominance

In the closest algal relatives of land plants, the only diploid stage in the life cycle is the zygote. This cell divides by meiosis to produce spores that develop into a haploid, multicelled gamete-producing stage. In contrast, all land plant life cycles alternate between multicelled haploid and diploid stages (Figure 23.5*a*).

A land plant **gametophyte** is a haploid stage that produces gametes by mitosis. Eggs become fertilized while still attached to the gametophyte. Mitosis of the zygote produces the multicelled embryo for which land plants are known. Further development yields the mature **sporophyte**: a diploid stage that forms spores by meiosis. A land plant spore is a nonmotile, haploid cell that divides by mitosis to form a gametophyte.

Biologists describe the land plant life cycle as an alternation of generations. This type of life cycle has advantages over formation of spores by meiotic division of the zygote. Forming a multicelled sporophyte increases the number cells that undergo meiosis and produce spores. Also a multicellular **sporangium**, or spore-forming structure, can protect developing spores and facilitate their dispersal.

The relative size, complexity, and longevity of the sporophyte and gametophyte stages varies among the land plants (Figure 23.5*b*). In bryophytes, sporophytes are small and short-lived relative to gametophytes.

Sporophytes became increasingly prominent in later-evolving lineages. Flowering plants have the largest and most complex sporophytes. For example, an oak tree is a sporophyte that stands many meters tall. Each attached oak gametophyte consists of only a few cells.

Roots, Stems, and Leaves

Once plants moved onto land, traits that prevented water loss were favored. Early on, aboveground parts of land plants became covered by **cuticle**, a secreted waxy layer that restricts evaporation. Openings across the cuticle, called **stomata**, became control points for balancing water conservation with the need to obtain carbon dioxide for photosynthesis (Figure 23.6).

Early land plants had structures that held them in place, but true roots evolved later. Roots anchored the plants and also took up water with dissolved mineral ions from the soil. Fungal symbionts in or on roots assisted in these tasks, as they still do today.

Moving substances taken up by roots to other body regions required **vascular tissues**, a system of internal pipelines (Figure 23.7). **Xylem** is the vascular tissue that distributes water and mineral ions. **Phloem** is the vascular tissue that distributes sugars made in photosynthetic cells. Of 295,000 or so modern plant species, more than 90 percent have xylem and phloem. These plants are members of the vascular plant lineage.

What made the vascular plants successful? For one thing, their vascular tissues are reinforced by **lignin**, an organic compound that lends structural support (Section 4.12). Lignified vascular tissues not only

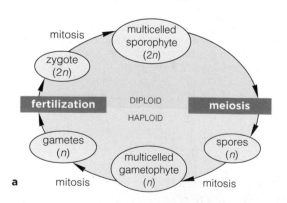

a

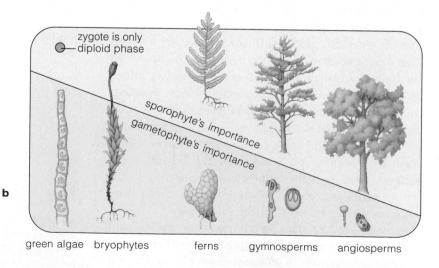

b

Figure 23.5 Animated (**a**) Generalized life cycle for land plants, as explained in Section 10.5. (**b**) One evolutionary trend in plant life cycles. Algae and bryophytes put the most energy into making gametophytes. Groups in seasonally dry habitats put the most energy into making sporophytes, which retain, nourish, and protect the new generation through harsh times.

Figure It Out: A pine tree is a gymnosperm. Which is the larger, more prominent phase in its life cycle, the sporophyte or gametophyte? *Answer: sporophyte*

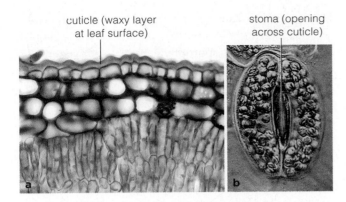

cuticle (waxy layer at leaf surface)

stoma (opening across cuticle)

Figure 23.6 Water-conserving adaptations. (**a**) Light micrograph showing the secreted waxy cuticle (stained *pink*) at the upper surface of an oleander leaf. (**b**) Light micrograph of a stoma, an opening across the cuticle. It can be opened to allow gas exchange, or closed to conserve water.

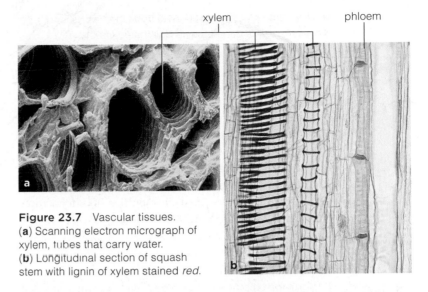

xylem

phloem

Figure 23.7 Vascular tissues. (**a**) Scanning electron micrograph of xylem, tubes that carry water. (**b**) Longitudinal section of squash stem with lignin of xylem stained *red*.

distributed materials, they also helped plants stand upright and allowed them to branch. Being tall and branched gave early vascular plants an advantage in spore dispersal. The most successful vascular lineages also evolved leaves, which increased a plant's surface area for intercepting sunlight and for gas exchange.

Pollen and Seeds

Reproductive traits also gave some vascular plants a competitive edge. All bryophytes, and some vascular plants such as the ferns, disperse by releasing spores (Figure 23.8*a*). Only the seed-bearing vascular plants release pollen grains and seeds.

A **pollen grain** is a walled, immature gametophyte that will give rise to sperm. After pollen grains are released, they travel to eggs on the same or another plant with the help of air currents or animals, most often insects. The ability to produce pollen gave seed plants an advantage in dry environments. Plants that do not make pollen require water to allow their sperm to swim to eggs. Pollen allowed seed plants to reproduce even when water was scarce.

Seeds also helped plants survive dry times. A **seed** is an embryo sporophyte and some nutritive tissue enclosed inside a waterproof seed coat. Many seeds have features that facilitate their dispersal away from the parent plant.

Gymnosperms and angiosperms are two lineages of seed-bearing plants. Cycads, conifers, and ginkgos are among the gymnosperms. Angiosperms, or flowering plants, branched off from a gymnosperm lineage, and they alone make flowers and fruits (Figure 23.8*b*). The great majority of modern plants are angiosperms.

Figure 23.8 Dispersal mechanisms. (**a**) Ferns disperse by releasing spores that form in clusters on the underside of leaves. (**b**) Seed plants release seeds, which contain embryo sporophytes. In flowering plants, such as a papaya, the seeds form inside floral tissue that develops into a fruit.

Take-Home Message

What adaptations contributed to plant diversification?

■ Plant life cycles shifted from a gametophyte-dominated cycle in bryophytes to a sporophyte-dominated cycle in other plants.

■ Life on land favored water-conserving features such as cuticle. In vascular plants, a system of vascular tissue—xylem and phloem—distributes material through the leaves, stems, and roots of sporophytes.

■ Bryophytes and seedless vascular plants release spores. Only seed plants release embryos inside protective seeds. Only in the flowering plants do seeds form inside floral tissue that later develops into a fruit.

■ Three land plant lineages—liverworts, hornworts, and mosses—have a gametophyte-dominated life-cycle.

■ Link to Nitrogen fixation 21.6

Bryophytes are not a monophyletic group, but rather a collection of three early-evolving land plant lineages: liverworts, hornworts, and mosses. Most of the 24,000 or so species live in places that are constantly moist. None make stem-stiffening lignin, so few stand more than 20 centimeters (8 inches) tall.

The gametophyte is the largest, most conspicuous phase of a bryophyte life cycle. Multicellular structures (gametangia) in or on the gametophyte surface enclose and protect developing gametes. Sperm are flagellated and swim to eggs. Insects and mites can assist sperm transfer where water does not form a continual path.

Sporophytes are unbranched and remain attached to the gametophyte even when mature. They produce wind-dispersed spores that withstand drought, making bryophytes important colonists of rocky places.

Liverworts

In most of the 9,000 or so liverwort species, the gametophyte has a ribbonlike part (a thallus) that attaches to soil or a surface by rootlike **rhizoids**. Rhizoids also wick up and store water but do not distribute it like roots of vascular plants.

The liverwort *Marchantia* is common on moist soil. It can reproduce asexually by producing gemmae—small clumps of cells—in cups on the gametophyte (Figure 23.9a,b). It also can reproduce sexually. In this genus, a gametophyte produces either eggs or sperm. Gametangia form atop a stalk that grows from the thallus (Figure 23.9c,d). Sperm swim to eggs and a zygote forms. The zygote develops into a sporophyte that hangs from the underside of the gametophyte.

Hornworts

sporophyte—

gametophyte

A pointy, hornlike sporophyte that can be several centimeters tall gives the hornworts their common name (*right*). The base of the sporophyte is embedded in gametophyte tissues and spores form in an upright sporangium, or capsule. When spores mature, the tip of the capsule splits, releasing them. The sporophyte grows continually from its base, so it can make and release spores over an extended period.

The sporophyte has chloroplasts and, in some cases, can survive even after the death of the gametophyte. These traits and certain genetic similarities suggest that hornworts may be close relatives of vascular plants.

The ribbonlike gametophyte holds nitrogen-fixing cyanobacteria (Section 21.6) in pores on its surface. The bacteria provide the plant with nitrogen compounds, and receive shelter in return.

Mosses

The moss sporophyte consists of a sporangium (the capsule) on a stalk embedded in gametophyte tissue (Figure 23.10a). Haploid spores form in the capsule by meiosis (Figure 23.10b). A spore germinates and develops into a gametophyte; sexes are usually separate (Figure 23.10c). Sperm form in a male gametangium, or antheridium, and eggs in a female gametangium, or archegonium (Figure 23.10d,e). After fertilization, the zygote develops into a sporophyte (Figure 23.10f,g).

Mosses frequently reproduce asexually by fragmentation. A bit of gametophyte breaks off and grows into a new plant.

Mosses are the most diverse group of bryophytes, with 15,000 or so species. Among these, 350 or so spe-

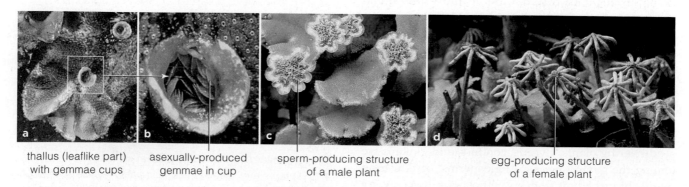

thallus (leaflike part) with gemmae cups

asexually-produced gemmae in cup

sperm-producing structure of a male plant

egg-producing structure of a female plant

Figure 23.9 (**a,b**) *Marchantia*, a liverwort, reproduces asexually by forming gemmae on the gametophyte surface. (**c,d**) *Marchantia* also reproduces sexually. Sexes are separate and gamete-producing structures form atop stalks.

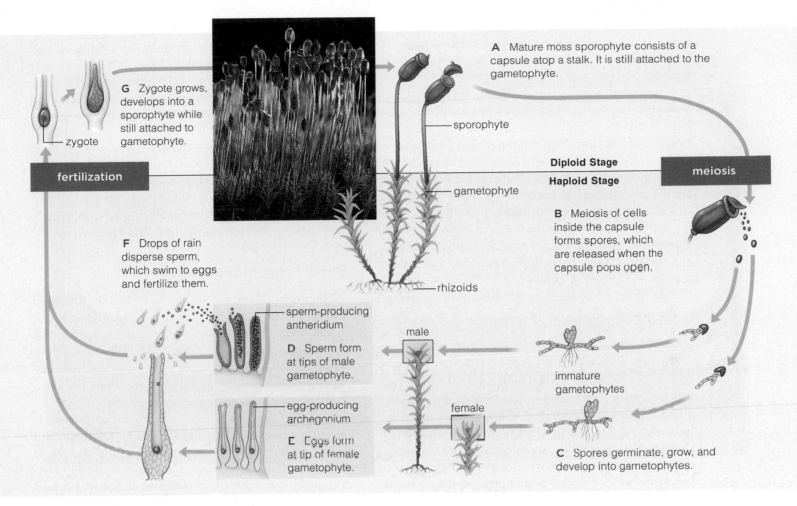

A Mature moss sporophyte consists of a capsule atop a stalk. It is still attached to the gametophyte.

G Zygote grows, develops into a sporophyte while still attached to gametophyte.

zygote

sporophyte

gametophyte

fertilization

Diploid Stage

Haploid Stage

meiosis

B Meiosis of cells inside the capsule forms spores, which are released when the capsule pops open.

rhizoids

F Drops of rain disperse sperm, which swim to eggs and fertilize them.

sperm-producing antheridium

D Sperm form at tips of male gametophyte.

male

egg-producing archegonium

female

E Eggs form at tip of female gametophyte.

immature gametophytes

C Spores germinate, grow, and develop into gametophytes.

Figure 23.10 Animated Life cycle of a moss (*Polytrichum*). The nonphotosynthetic sporophyte remains attached to and dependent on the gametophyte.

cies of peat mosses (*Sphagnum*) are of great ecological and commercial importance. *Sphagnum* is the dominant plant in bogs that cover more than 350 million acres in Europe, North Asia, and North America. Its remains accumulate as peat, which is harvested as a fuel (Figure 23.11) and used in planting mixes.

Soil in a peat bog can be as acidic as vinegar. Only acid-tolerant plants live alongside the mosses. Most bacteria and fungi do not grow well in this acidic habitat, so decomposition is leisurely. Well-preserved human remains more than a thousand years old have been found in European peat bogs. The high acidity kept the bodies from decomposing.

Take-Home Message

What are bryophytes?

■ Bryophyte is the common name for three plant lineages—liverworts, hornworts, and mosses.

■ Low-growing, haploid gametophytes dominate the bryophyte life cycle, and spores are the dispersal form.

sporophyte

gametophyte

Figure 23.11 (**a**) Peat bog in Ireland. This family is cutting blocks of peat and stacking them to dry as a home fuel source. Most peat is now harvested commercially and burned to generate electricity. (**b**) Peat moss (*Sphagnum*). You can clearly see four sporophytes (the *brown*, jacketed structures on the *white* stalks) attached to the pale gametophyte.

- A sporophyte with lignified vascular tissue is the dominant phase in the life cycle of the seedless vascular plants.
- Link to Gamete formation and fertilization 10.5

Some mosses have internal pipelines that transport fluid within their body. However, only vascular plants (tracheophytes) have lignin-strengthened vascular tissue with xylem and phloem. This innovation allowed the evolution of larger branching sporophytes, which are the predominant phase in vascular plant life cycles.

Vascular plants evolved from a bryophyte and like bryophytes early diverging lineages have flagellated sperm that swim to eggs. Also like bryophytes, they are seedless; they disperse by releasing their spores directly into the environment.

Two lineages of seedless vascular plants survived to the present. Lycophytes include club mosses, spike mosses, and quillworts. Monilophytes include whisk ferns, horsetails, and ferns.

These two lineages diverged before leaves and roots had evolved, and each developed these features in a different way. For example, lycophytes form spores along the sides of branches. Their leaves have one unbranched vein and probably evolved from a lateral sporangium. In contrast, monilophytes have spores at branch tips. Their leaves, which have branching veins, probably evolved from a branching network of stems.

Lycophytes

Most of the 1,200 modern lycophytes are club mosses. *Lycopodium* species are common in North America's hardwood forests, where they are known as ground pines (Figure 23.12a). *Lycopodium* spores form inside a **strobilus** (plural, strobili), a soft, cone shaped-structure made up of modified leaves. Many other kinds of vascular plants also have strolbili.

Lycopodium branches are sold in wreaths. The waxy spores ignite easily and were used in early flash photography. Spores also coated the inside of latex gloves and condoms until they were found to irritate skin.

Most spike mosses (*Selaginella*) live in moist tropical regions but some survive in deserts and are the most drought-tolerant vascular plants. Commonly known as resurrection plants, they curl up and turn brown when water is scarce. When rains return, they uncurl, turn green with new chlorophyll, and resume growth.

Whisk Ferns and Horsetails

Whisk ferns (*Psilotum*) are native to the southeastern United States. They have **rhizomes**, or underground stems, but no roots. The aboveground, photosynthetic stems appear leafless (Figure 23.12b). Spores form in fused sporangia at the tips of short branches. You may have noticed whisk ferns in bouquets. There is a commercial market for their unusual branches.

The 25 *Equisetum* species are known as horsetails or rushes (Figure 23.12c,d). They have rhizomes and hollow stems with tiny nonphotosynthetic leaves at the joints. Photosynthesis occurs in stems and in leaf-like branches. Deposits of silica in the stem support the plant and give stems a sandpapery texture. Before scouring powders and pads were widely available, people used some *Equisetum* stems as pot scrubbers, thus the common name "scouring rush."

Depending on the species, strobili form either at tips of photosynthetic stems or on specialized reproductive stems without chlorophyll. Each spore gives rise to a gametophyte not much bigger than a pinhead.

strobilus   sporangia   leaflike branch   strobilus

Figure 23.12 Seedless vascular plants. (**a**) A club moss (*Lycopodium*) about 20 centimeters (8 inches) tall. Spores form in strobili.

(**b**) Whisk fern (*Psilotum*) with sporangia at tips of short lateral branches.

Horsetails (*Equisetum*): (**c**) Photosynthetic stem and (**d**) spore-bearing strobilus at the tip of a nonphotosynthetic stem.

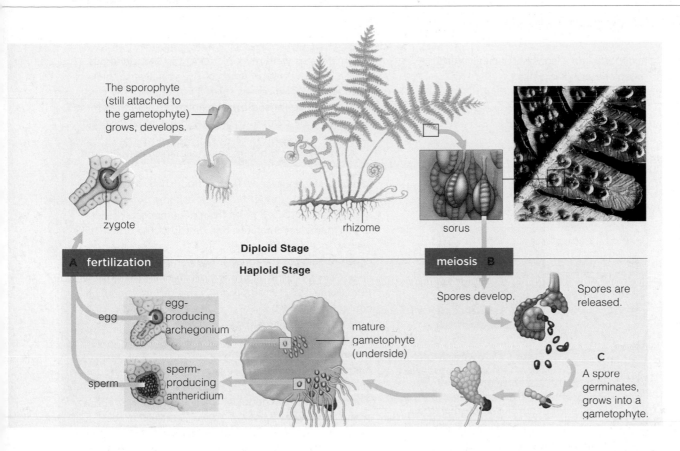

The sporophyte (still attached to the gametophyte) grows, develops.

zygote

rhizome

sorus

Diploid Stage

A   fertilization

Haploid Stage

meiosis   B

Spores develop.

Spores are released.

egg

egg-producing archegonium

mature gametophyte (underside)

sperm

sperm-producing antheridium

C

A spore germinates, grows into a gametophyte.

Figure 23.13 Animated Chain fern (*Woodwardia*) life cycle.

(**a**) After swimming sperm reach eggs, fertilization results in a diploid zygote. The zygote is the start of a sporophyte with a rhizome and many fronds.

(**b**) Many sori form on the underside of fronds. Each sorus is a cluster of sporangia in which spores form by way of meiosis.

(**c**) After the spores are released, they germinate and develop into small gametophytes that have a distinctive heart shape.

Ferns—No Seeds, But Much Diversity

With 12,000 or so species, ferns are the most diverse seedless vascular plants. All but 380 or so species live in the tropics. Most fern sporophytes have leaves and roots that grow out from rhizomes (Figure 23.13). Fern leaves, also known as fronds, often start out in a tight coil known as a "fiddlehead" before unfurling.

Sori (singular, sorus) are clusters of sporangia on the lower surface of fern fronds. Sori spring open and haploid spores pop out. After germination, a spore develops into a gametophyte that is typically bisexual and just a few millimeters across (Figure 23.13c).

Fern sporophytes vary greatly in structure and size (Figure 23.14). Some floating ferns have fronds only 1 millimeter long, but tree ferns can be 25 meters (80 feet) high. Fern fronds can be swordlike or divided into leaflets. Many tropical ferns are **epiphytes**. Such plants attach to and grow on a trunk or branch of another plant but do not withdraw nutrients from it.

Take-Home Message

What are seedless vascular plants?

■ Club mosses and relatives belong to one seedless vascular lineage. Ferns, horsetails and whisk ferns belong to the other.

■ A sporophyte with vascular tissues (xylem and phloem) dominates their life cycle, and spores are the dispersal form.

Figure 23.14   A sampling of fern diversity. (**a**) The floating fern *Azolla pinnata*. The whole plant is not as wide as a finger. Chambers in the leaves shelter nitrogen-fixing cyanobacteria. Southeast Asian farmers grow this species in rice fields as a natural alternative to chemical fertilizers. (**b**) Bird's nest fern (*Asplenium nidus*), one of the epiphytes. (**c**) Lush forest of tree ferns (*Cyathea*) in Australia's Tarra-Bulga National Park.

23.5 | Ancient Carbon Treasures

■ Existing coal deposits are a legacy of ancient forests dominated by seedless vascular plants.

■ Link to Geologic time scale 17.8

When climates were mild, plants of the Carboniferous put on growth through much of the year. Dense thickets of underground rhizomes spread fast and far. The club mosses, horsetails, and other plants with lignin-reinforced tissues had the competitive edge, and some evolved into tall, massively stemmed giants (Figure 23.15).

After the forests had formed, climates changed, and the sea level rose and fell many times. When the waters receded, steamy swamp forests flourished. After the sea moved back in, submerged trees became buried in sediments that protected them from decomposers. Layers of sediments accumulated one on top of the other. Their weight squeezed the water out of the saturated, undecayed remains of the forests. The compaction generated heat. Over time, the increasing pressure and heat transformed the compacted organic remains into seams of **coal** (Figure 23.16).

With its high percentage of carbon, coal is rich in stored energy and one of our premier "fossil fuels." It took a staggering amount of photosynthesis, burial, and compaction to form each major seam of coal in the ground. It has taken us only a few centuries to deplete much of the world's known coal deposits. Often you will hear about annual production rates for coal or some other fossil fuel. How much do we really produce each year? We produce nothing. We simply extract it from the ground. Coal is a nonrenewable source of energy.

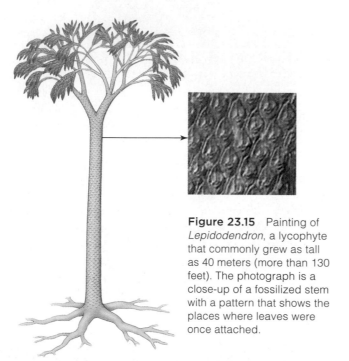

Figure 23.15 Painting of *Lepidodendron*, a lycophyte that commonly grew as tall as 40 meters (more than 130 feet). The photograph is a close-up of a fossilized stem with a pattern that shows the places where leaves were once attached.

stem of a giant lycophyte (*Lepidodendron*), which grew 40 meters (131 feet) tall

Medullosa, one of the early seed plants

stem of a giant horsetail (*Calamites*), which was almost 20 meters (66 feet) tall

Figure 23.16 Reconstruction of a Carboniferous forest. *Right*, photograph of part of a seam of coal.

23.6 | Seed-Bearing Plants

■ Seeds and pollen allowed gymnoperms and flowering plants to survive and thrive in drier habitats.

■ Link to Geologic time scale 17.8

Rise of the Seed Plants

The first seed-bearing plants evolved during the late Devonian. One lineage gave rise to cycads and other gymnosperms. The period that many people call the Age of Dinosaurs, botanists call the Age of Cycads. In the early Cretaceous, angiosperms (flowering plants), branched from a gymnosperm ancestor.

Modifications in spore production contributed to seed plant success. All land plants produce spores by meiosis. In some bryophytes and nonvascular plants, one type of spore forms. It develops into a bisexual gametophyte that produces both eggs and sperm. In other bryophytes and nonvascular plants—and in all seed plants—two types of spores form. **Microspores** develop into male gametophyes. **Megaspores** develop into female gametophytes.

In seed plants alone, the male gametophyte that develops from a microspore is a pollen grain. A pollen grain consists of a few cells, one of which produces sperm. Evolution of pollen put the seed plants at an advantage in dry habitats. Pollen can travel long distances on the wind or on insect bodies. Thus, a film of water was no longer necessary for reproduction.

Another unique trait of seed plants is the **ovule**, a specialized sporangium enclosed within a protective layer of cells called the integument. Inside an ovule, megaspores form by meiosis and develop into egg-producing gametophytes.

Fertilization occurs and an embryo forms inside the ovule. The seed is a mature ovule. A seed coat derived from ovule tissues encloses the embryo sporophyte. Nutrient-rich tissue stored inside the seed supports the embryo's growth.

Human Uses of Seed Plants

Many seed plants receive human help in dispersing seeds. By 10,000 years ago, we had domesticated some seed plants as sources of food. We now recognize 3,000 or so plant species as edible and grow about 150 as food crops (Figure 23.17). We use others, most notably conifers, as sources of lumber. Still other seed plants provide medicines, as when extracts from yew trees slow the growth of cancers. People grow tobacco, marijuana, opium poppies, and coca (the source of cocaine) for their mind-altering properties. They grow flax, cotton, and hemp for use in fabrics, carpets, and ropes. They often dye these products with pigments extracted from other seed plants. Humans may be contributing to the demise of many seed plants, but those that assist us continue to prosper.

pine
pollen grains

Take-Home Message

What factors contributed to the success of seed plants?

■ Seed plants release pollen grains, which allow fertilization to occur even in the absence of environmental water. They form eggs inside ovules on the body of the parental sporophyte. A seed is a mature ovule with an embryo sporophyte and some nutritive tissue inside it.

■ Humans depend heavily on cultivated seed plants and have contributed to the widespread dispersal of seed plants that they favor.

Figure 23.17 Edible treasures from flowering plants. (**a**) Some of the nearly 100 varieties of apples (*Malus domestica*) grown in the United States. (**b**) Mechanized harvesting of wheat, *Triticum*. (**c**) Hand picking shoots of tea plants (*Camellia sinensis*) in Indonesia. Leaves of plants on hillsides in moist, cool regions have the best flavor. (**d**) In Hawaii, a field of sugarcane, *Saccharum officinarum*. We make sugar and syrup by boiling down sap extracted from its stems.

■ Gymnosperms are one of the two modern lineages of seed plants. Conifers, mentioned in the chapter introduction, are the best known gymnosperms.

Gymnosperms are vascular seed plants that produce seeds on the surface of ovules. Seeds are said to be "naked," because unlike those of angiosperms they are not inside a fruit. (*Gymnos* means naked and *sperma* is taken to mean seed.) However, many gymnosperms enclose their seeds in a fleshy or papery covering.

Conifers

The 600 or so species of **conifers** are woody trees and shrubs. Seeds form in female cones. Male cones release pollen, which wind carries to female cones. Conifers typically have needlelike or scalelike leaves. Leaves often have a thick cuticle, and conifers tend to be more resistant to drought and cold than flowering plants. Most conifers shed some leaves steadily but remain evergreen. A few deciduous species shed all leaves at once seasonally. The tallest trees (redwoods) and oldest trees (bristlecone pine) are conifers (Figure 23.18*a*).

Lesser Known Gymnosperms

Cycads and ginkgos were most diverse in dinosaur times. They are the only modern seed plants that have motile sperm. Sperm emerge from pollen grains, then swim in fluid produced by the plant's ovule.

About 130 species of **cycads** exist, mainly in the dry tropics and subtropics. Cycads look like palms or ferns but are not close relatives of them (Figure 23.18*b*). The "sago palms" commonly used in landscaping and as houseplants are actually cycads.

The only living **ginkgo** species is *Ginkgo biloba*, the maidenhair tree (Figure 23.18*c–f*). It is one of the few deciduous gymnosperms. *G. biloba* is native to China, but its attractive fan-shaped leaves and resistance to insects, disease, and air pollution make it a popular street tree in urban areas. Usually only male trees are planted because the seeds produced by female trees give off a strong, unpleasant odor when they decay. Some studies indicate that dietary supplements made from ginkgo leaves may slow memory loss in people who have Alzheimer's disease.

Gnetophytes include tropical trees, leathery vines, and desert shrubs. Extracts from the stems of *Ephedra* (Figure 23.18*g*) are sold as an herbal stimulant and a weight loss aid. Such supplements can be dangerous; a few people have died while using them.

The strange-looking gnetophyte, *Welwitschia*, lives only in Africa's Namib desert. It has a taproot and a woody stem with strobili. Two straplike leaves grow 5 meters (16 feet) long. These leaves split lengthwise repeatedly as the plant matures (Figure 23.18*h*).

Figure 23.18 A sampling of gymnosperms. (**a**) Bristlecone pine (*Pinus longaeva*) on a mountaintop in the Sierra Nevada. (**b**) *Cycas armstrongii*, an Australian cycad with its seeds. *Ginkgo biloba*: (**c**) fleshy seeds, (**d**) leaves, (**e**) fossil leaf, and (**f**) fall foliage. Two gnetophytes: (**g**) *Ephedra viridis*; (**h**) *Welwitschia mirabilis*, with strappy leaves and seed-bearing strobili.

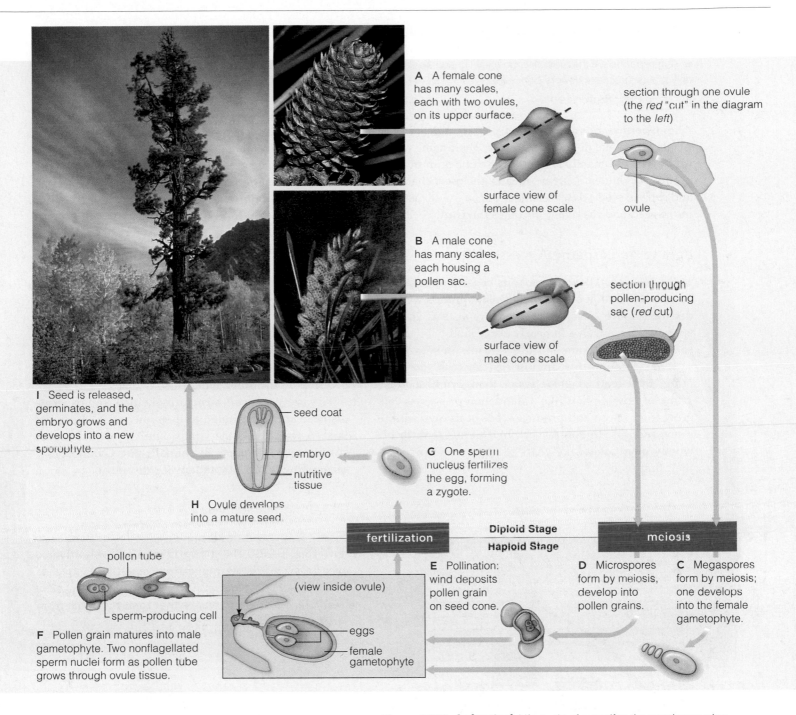

A A female cone has many scales, each with two ovules, on its upper surface.

section through one ovule (the *red* "cut" in the diagram to the *left*)

surface view of female cone scale

ovule

B A male cone has many scales, each housing a pollen sac.

section through pollen-producing sac (*red* cut)

surface view of male cone scale

I Seed is released, germinates, and the embryo grows and develops into a new sporophyte.

seed coat

embryo

nutritive tissue

G One sperm nucleus fertilizes the egg, forming a zygote.

H Ovule develops into a mature seed.

Diploid Stage	
fertilization	meiosis
Haploid Stage	

pollen tube

sperm-producing cell

F Pollen grain matures into male gametophyte. Two nonflagellated sperm nuclei form as pollen tube grows through ovule tissue.

(view inside ovule)

eggs

female gametophyte

E Pollination: wind deposits pollen grain on seed cone.

D Microspores form by meiosis, develop into pollen grains.

C Megaspores form by meiosis; one develops into the female gametophyte.

Figure 23.19 Animated Life cycle of a conifer, the ponderosa pine.

A Representative Life Cycle

A pine tree is a sporophyte, and its life cycle is typical of conifers (Figure 23.19). Ovules form on the upper surfaces of scales in female cones. An egg-producing female gametophyte develops in each ovule. In male cones, microspores become winged pollen grains.

Millions of tiny pollen grains are released and drift with the winds. Pollination occurs when pollen lands on an ovule. The pollen grain germinates and some cells of the male gametophyte begin to grow, forming the pollen tube (Figure 23.19*f*). After about a year, the sperm tube reaches the egg and the nucleus of a sperm cell in the tube fuses with the egg nucleus, forming a zygote. The zygote develops into an embryo sporophyte, which, with the ovule tissues, becomes a seed.

Take-Home Message

What are gymnosperms?

■ Gymnosperms include conifers, ginkgos, and some nonwoody plants.

■ These vascular plants release pollen and seeds, which form in strobili or, in the case of conifers, in woody cones.

- Angiosperms are the most diverse plant lineage and the only plants that make flowers and fruits.

- Link to Classification 19.1

Angiosperms are vascular seed plants, and the only plants that make flowers and fruits. Their name refers to **ovaries**, the chambers that enclose one or more egg-producing ovules. (*Angio–* means enclosed chamber, and *sperma*, seed.) After fertilization, an ovule matures into a seed and the ovary becomes the **fruit**.

Keys to Angiosperm Success

In the Mesozoic, flowering plants began a spectacular adaptive radiation even as other plant groups were in decline (Figure 23.20). Now there are at least 260,000 species. They survive in nearly every land habitat and some make their home in lakes, streams, or seas.

What accounts for angiosperm success? For one thing, they tend to grow faster than gymnosperms. Think of how a plant like a dandelion or a grass can grow from a seed and produce seeds of its own within a few months. In contrast, gymnosperms tend to be woody plants that take years to mature and produce their first seeds.

Another factor was the **flower**, a specialized reproductive shoot (Figure 23.21). After pollen-producing plants evolved, some insects began feeding on pollen. Plants gave up some pollen but gained a reproductive edge. How? Insects moved pollen from male parts of one flower to female parts of another.

Some flowering plants evolved traits that attracted specific **pollinators**, animals that move pollen of one plant species onto female reproductive structures of the same species. Insects are the most common pollinators, but birds, bats, and other vertebrates also act in this role (Figure 23.22 *a–c*). Big colorful flowers, sugary nectar, or a strong fragrance help attract pollinators to specific plants. Wind pollinated plants tend to have small, unscented flowers that lack nectar.

Over time, plants coevolved with their animal pollinators. **Coevolution** refers to two or more species jointly evolving as a result of their close ecological interactions. Heritable changes in one exert selection pressure on the other, which also evolves.

A variety of fruit structures helped angiosperms disperse and contributed to their success. Some fruits float in water, ride the winds, stick to animal fur, or survive a trip through an animal's gut. Gymnosperm seeds show fewer adaptations for dispersal.

Flowering Plant Diversity

Nearly 90 percent of all modern species of plants are flowering plants. The group is tremendously diverse even in size. Species range from aquatic duckweeds 1 millimeter long to *Eucalyptus* trees 100 meters (330 feet) tall. A few are parasites that steal nutrients from other plants (Figure 23.22*f*). Pitcher plants and some

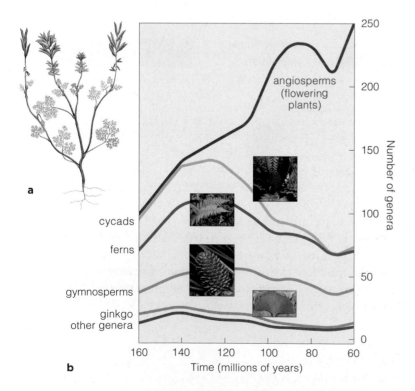

Figure 23.20 (**a**) Sketch of *Archaefructus sinensis*, one of the earliest known flowering plants. It probably grew in shallow lakes. (**b**) Diversity of vascular plants in Mesozoic times. Conifers and other gymnosperms started to decline even before flowering plants started their major adaptive radiation.

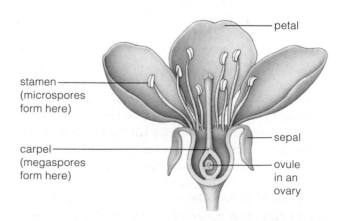

Figure 23.21 Structures of a typical modern flower. It has male and female parts (stamens and carpels) and accessory parts (petals and sepals).

Figure 23.22 (**a,b**) Flowers are modified shoots. Their colors, fragrances, and shapes are adaptations that attract pollinators, primarily insects. (**c**) This hummingbird pollinator has a long bill that fits the long, delicate nectar tube of the flower of a columbine (*Aquilegia*).

(**d**) A water lily (*Nymphaea*), a member of one of the early lineages, with spirally arranged petals.
(**e**) Pansies (*Viola*) are among the familiar eudicots.

(**f**) Dwarf mistletoe (*Arceuthobium*) is a highly specialized eudicot that has a reduced amount of chlorophyll and parasitizes conifers. Colorless flowers produce droplets of nectar that lure insects.

(**g**) Evolutionary tree diagram for flowering plants.

others in nitrogen-poor habitats attract, trap, and dissolve insects, then absorb the nutrients (Figure 6.12).

Flowering plants were once divided into only two groups based on their number of cotyledons, or seed leaves, that form on embryo sporophytes. Plants with one cotyledon were named monocots; those with two were named dicots. It now appears that the monocots branched from a more ancient dicot lineage.

Researchers have now identified the oldest lineages of flowering plants. There are three, as represented by their modern descendants: water lilies, star anise, and *Amborella* (Figure 23.22g).

Genetic divergences gave rise to other groups that became dominant: magnoliids, eudicots (true dicots), and monocots. Among the 9,200 or so magnoliids are magnolias and avocado trees. The most diverse group, eudicots, has about 170,000 species. It includes

most herbaceous (nonwoody) plants such as lettuces, cabbages, dandelions, daisies, and cactuses. Most flowering shrubs or trees, such as roses, maples, oaks, elms, and fruit trees, are eudicots. Among the 80,000 named monocot species are palms, lilies, grasses, and orchids. Sugarcane and cereal grasses—especially rice, wheat, corn, oats, and barley—are the most important monocot crop plants. For a more detailed classification of plants, refer to Appendix I.

Take-Home Message

What are the features of angiosperms?

■ Angiosperms are seed plants in which seeds develop inside the ovaries of flowers. After pollination, an ovary becomes a fruit.

■ Angiosperms are the most successful plants. Short life-cycles, coevolution with insect pollinators, and diverse fruit structures enhanced their success.

■ Flowering plants make fruits and supply their embryo sporophytes with endosperm, a nutritive tissue.

Figure 23.23 shows a flowering plant life cycle. The female gametophye forms in a flower's ovary. Pollen forms inside stamens. After pollination, a pollen tube delivers two sperm to the ovary and double fertilization occurs. One sperm fertilizes the egg. The other fertilizes a cell with two nuclei, forming a triploid cell that divides and becomes **endosperm**, a nutrient-rich tissue that is unique to angiosperm seeds. The endosperm nourishes the developing embryo.

Ovary tissue matures into a fruit that encloses the seed. Fruits help disperse seeds by attracting animals with sugary flesh, by sticking to fur or feathers, or by catching breezes with wings and other extensions.

Take-Home Message

What is unique about the flowering plant life cycle?
■ Flowering plants form eggs in ovaries and pollen in stamens.
■ The seed contains endosperm and is enclosed within a fruit.

Figure 23.23
Animated Life cycle of a lily (*Lilium*), one of the monocots.

(**a**) The sporophyte dominates this life cycle. (**b**) Pollen forms in pollen sacs. (**c**) Eggs develop in an ovule within an ovary. (**d**) Pollination occurs and a tube grows from the pollen grain into the ovule, delivering two sperm.

(**e**) Double fertilization occurs in all flowering plant life cycles. One sperm fertilizes the haploid egg. The other fertilizes a diploid cell. The resulting triploid cell divides repeatedly and forms endosperm, a tissue that will nourish the embryo sporophyte.

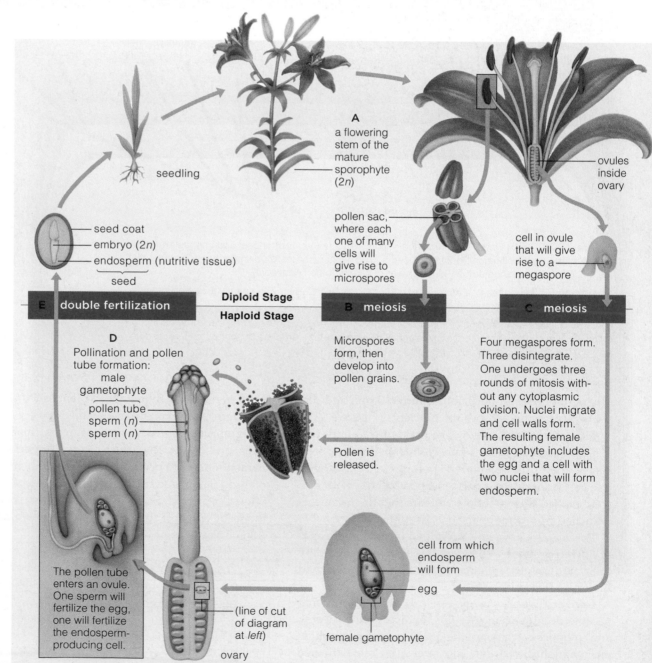

seedling

A
a flowering stem of the mature sporophyte (2n)

ovules inside ovary

seed coat
embryo (2n)
endosperm (nutritive tissue)
seed

pollen sac, where each one of many cells will give rise to microspores

cell in ovule that will give rise to a megaspore

E double fertilization	**Diploid Stage**	**B** meiosis	**C** meiosis
	Haploid Stage		

D
Pollination and pollen tube formation: male gametophyte
pollen tube
sperm (n)
sperm (n)

Microspores form, then develop into pollen grains.

Four megaspores form. Three disintegrate. One undergoes three rounds of mitosis without any cytoplasmic division. Nuclei migrate and cell walls form. The resulting female gametophyte includes the egg and a cell with two nuclei that will form endosperm.

Pollen is released.

The pollen tube enters an ovule. One sperm will fertilize the egg, one will fertilize the endosperm-producing cell.

(line of cut of diagram at *left*)

ovary

cell from which endosperm will form

egg

female gametophyte

23.10 | The World's Most Nutritious Plant

■ Botanists use knowledge of plant biology and genetics to find new ways to feed a hungry world.

■ Link to Amino acids and proteins 3.5

Alejandro Bonifacio was raised in poverty in rural Bolivia. As a child he spoke a language that predates the Incas. He learned Spanish before attending college. There, he earned a bachelor's degree in agriculture and became a plant breeder for Bolivia's department of agriculture.

His research interest is *Chenopodium quinoa*, a plant that originated in the Andes. Quinoa (pronounced keen-wah) is a leafy eudicot, a distant relative of spinach and beets. Its nutritious seeds are not a cereal grain, but for many thousands of years, they have been a staple of Latin American diets. Together with corn and potatoes, quinoa helped feed the great Inca civilization.

Quinoa seeds are 16 percent protein, on average. Some varieties contain more. Wheat seeds are about 12 percent protein, and rice seeds, 8 percent. More importantly, quinoa has all amino acids that humans require, while wheat and rice proteins are deficient in the amino acid lysine. Quinoa also has more iron than most cereal grains, and a good deal of calcium, phosphorus, and many B vitamins.

In addition, quinoa plants are easy to grow. They are highly resistant to drought, frost, and salty soils. Quinoa is the only food crop that can grow in the salt deserts that prevail in much of Bolivia.

Far to the north, even before Bonifacio received his college scholarship, Daniel Fairbanks became a botanist at Brigham Young University. Fairbanks, too, saw quinoa's potential to feed millions in Peru and Bolivia, the most impoverished places in Latin America. Many families in

these countries are subsistence farmers. Protein-deficiency is common. It causes skin problems, muscle loss and fatigue, and impairs growth and development.

In 1991, Bonifacio and Fairbanks met at a conference about Andean crops and they became friends. Later, Bonifacio received a fellowship to study in the United States, and Fairbanks became his advisor. Bonifacio earned his PhD and learned his third language—English.

The two are now codirectors of an international research program with a holistic approach to improving quinoa production for farmers locked in poverty. They collect quinoa strains and look for ways to conserve, improve, and use genetic diversity. They are identifying the traits of quinoa strains and researching the best way of preserving seeds for future study. They are developing a genetic map of quinoa.

Today, more than twenty scientists in four countries take part in this program. They research the economic impact of new strains and agricultural technologies. They investigate substitutes for chemical pesticides to control quinoa moths. They keep in mind cultural preferences for seeds of particular sizes and colors. Farmers and home cooks help evaluate new varieties.

Thousands of Bolivian families now grow more food, thanks to the new quinoa strains. Children who would have been dead or sickened by protein deficiency are now attending school.

In a recent letter, Fairbanks told us that he learned more from Bonifacio than Bonifacio learned from him. He appended a photo of his colleague in a research field, standing next to one of his new quinoa varieties, so that we can put a face with the name (Figure 23.24).

Figure 23.24 Alejandro Bonifacio checks genetically improved quinoa plants.

The 2004 Nobel Peace Prize was awarded to Wangari Maathai of Kenya (*right*), the founder of the Green Belt Movement. Maathai warns that environmental destruction leads to shortages that can threaten peace and notes that small positive actions by many individuals can have a great collective effect. At her urging, members—mostly poor rural women—have planted more than 25 million trees.

How would you vote?

It's estimated that about 40 percent of the trees harvested each year end up in paper. Processing costs raise the price of recycled paper. Are you willing to pay more for products made from recycled paper? See CengageNOW for details, then vote online.

Summary

Sections 23.1, 23.2 The land plants, or **embryophytes**, evolved from charophytes, a kind of green algae. Nearly all are photoautotrophs. Groups listed in Figure 23.25 reflect these trends: A **gametophyte** dominates bryophyte life cycles. A **sporophyte** dominates in all other groups. Features that contributed to success on land include a **sporangium** that protects spores, **cuticle** and **stomata** that minimize water loss, **xylem** and **phloem** (two types of **vascular tissues**), and **lignin** in cell walls. In seed plants, **pollen grains** allowed reproduction without standing water, and embryo sporophytes were protected in **seeds**.

■ *Use the animation on CengageNOW to investigate plant life cycles.*

Section 23.3 Mosses, liverworts, and hornworts are **bryophytes**. They are nonvascular (no xylem or phloem). Sperm swim through water to eggs. The sporophyte forms on and is nourished by the gametophyte. **Rhizoids** attach a gametophyte to soil or another surface.

■ *Use the animation on CengageNOW to observe the life cycle of a moss.*

Sections 23.4, 23.5 Club mosses and spike mosses are one lineage of seedless vascular plants. Horsetails, whisk ferns, and true ferns are another. In both, the life cycle is dominated by the sporophyte. Roots and above-ground stems grow from **rhizomes**. Spore-bearing structures include the **strobili** of horsetails and the **sori** of ferns. Many ferns live as **epiphytes**, attached to another plant. Sperm swim through water to reach eggs. Energy-rich, compacted remains of Carboniferous swamp forests that were dominated by giant lycophytes became **coal**.

■ *Use the animation on CengageNOW to observe the life cycle of a fern.*

Section 23.6 Gymnosperms and flowering plants, or angiosperms, are seed-bearing vascular plants.

Seed plants produce **microspores** that become pollen grains, which are sperm-producing male gametophytes. They make **megaspores** that develop into egg-producing female gametophytes inside **ovules**. A seed is a mature ovule. It includes nutritive tissue and a tough seed coat that protects the embryo sporophyte inside the seed from harsh conditions.

Section 23.7 **Gymnosperms** include **conifers**, **cycads**, **ginkgos**, and **gnetophytes**. Many are well adapted to dry climates. Their ovules form on strobili or, in the case of conifers, on woody cones.

■ *Use the animation on CengageNOW to observe the life cycle of a pine.*

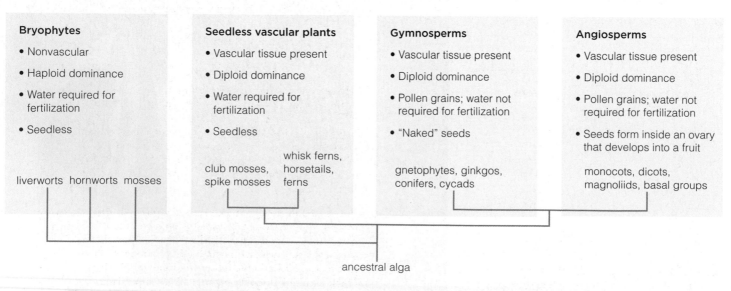

Bryophytes

• Nonvascular

• Haploid dominance

• Water required for fertilization

• Seedless

liverworts hornworts mosses

Seedless vascular plants

• Vascular tissue present

• Diploid dominance

• Water required for fertilization

• Seedless

club mosses, whisk ferns,
spike mosses horsetails, ferns

Gymnosperms

• Vascular tissue present

• Diploid dominance

• Pollen grains; water not required for fertilization

• "Naked" seeds

gnetophytes, ginkgos, conifers, cycads

Angiosperms

• Vascular tissue present

• Diploid dominance

• Pollen grains; water not required for fertilization

• Seeds form inside an ovary that develops into a fruit

monocots, dicots, magnoliids, basal groups

ancestral alga

Figure 23.25 Summary of plant evolutionary trends. All groups shown have living representatives.

Data Analysis Exercise

The U.N. Food and Agriculture Organization (UNFAO) recognizes the importance of forests to human populations and keeps track of forest abundance. Figure 23.26 shows UNFAO data on the amount of forest throughout the world in 1990, 2000, and 2005.

1. How many hectares of forested land were there in North America in 2005?

2. In which region(s) did the amount of forested land increase between 1990 and 2005?

3. How many hectares of forest did the world lose between 1990 and 2005?

4. In 2002, China embarked on an ambitious campaign to add 76 million hectares of trees over a ten-year period. Do you see any indication that this campaign is successful?

	Forested Area (in millions of hectares)		
Region	1990	2000	2005
Africa	699	656	635
Asia	574	567	572
Central America	28	24	22
Europe	989	988	1001
North America	678	678	677
Oceania	233	208	206
South America	891	853	832
World total	4,077	3,988	3,952

Figure 23.26 Changes in forested area by region from 1990 to 2005. One hectare is 2.47 acres. The full report on the world's forests is online at www.fao.org/forestry/en/.

Sections 23.8–23.10 **Angiosperms** are the most diverse plants. They have **flowers** and **coevolved** with animal **pollinators**. Ovules enclosed in the floral ovary mature into **fruits**. Double fertilization produces **endosperm** in seeds.

■ *Use the animation on CengageNOW to observe the life cycle of a flowering plant.*

Self-Quiz
Answers in Appendix III

1. The first land plants were _____ .
  a. gnetophytes   c. bryophytes
  b. gymnosperms   d. lycophytes

2. Lignin is not found in stems of _____ .
  a. mosses   b. ferns   c. monocots   d. a and b

3. A waxy cuticle helps land plants _____ .
  a. conserve water   c. reproduce
  b. take up carbon dioxide   d. stand upright

4. True or false? Ferns produce seeds inside strobili.

5. _____ attach mosses to soil and absorb water.
  a. rhizoids   c. roots
  b. rhizomes   d. microphylls

6. Bryophytes alone have a relatively large _____ and an attached, dependent _____ .
  a. sporophyte; gametophyte
  b. gametophyte; sporophyte

7. Club mosses, horsetails, and ferns are _____ plants.
  a. multicelled aquatic   c. seedless vascular
  b. nonvascular seed   d. seed-bearing vascular

8. Coal consists primarily of compressed remains of the _____ that dominated Carboniferous swamp forests.
  a. seedless vascular plants   c. flowering plants
  b. conifers   d. hornworts

9. The sperm of _____ swim to eggs.
  a. mosses   b. ferns   c. conifers   d. a and b

10. A seed is a(n) _____ .
  a. female gametophyte   c. mature pollen tube
  b. mature ovule   d. immature microspore

11. True or false? Only seed plants produce pollen.

12. Which angiosperm lineage includes the most species?
  a. magnoliids   c. monocots
  b. eudicots   d. water lilies

13. Match the terms appropriately.
  ___ bryophyte   a. seeds, but no fruits
  ___ seedless   b. has flowers and fruits
       vascular plant   c. xylem and phloem, but
  ___ gymnosperm   no ovules
  ___ angiosperm   d. gametophyte dominates

14. Match the terms appropriately.
  ___ ovule   a. gamete-producing body
  ___ cuticle   b. spore-producing body
  ___ gametophyte   c. where eggs form
  ___ sporophyte   d. underground stem
  ___ fruit   e. mature ovary
  ___ endosperm   f. nutritive tissue in seed
  ___ rhizome   g. where fern spores form
  ___ sorus   h. waxy layer

■ *Visit CengageNOW for additional questions.*

Critical Thinking

1. Early botanists admired ferns but found their life cycle perplexing. In the 1700s, they learned to propagate ferns by sowing what appeared to be tiny dustlike "seeds" from the undersides of fronds. Despite many attempts, the scientists could not find the pollen source, which they assumed must stimulate these "seeds" to develop. Imagine you could write to one of these botanists. Compose a note that would clear up their confusion.

2. The dominant stage in most plants is diploid. By one hypothesis, diploid dominance was favored because it allowed a greater level of genetic diversity. Suppose that a recessive mutation arises. It is mildly disadvantageous now, but it will be useful in some future environment. Explain why such a mutation would be more likely to persist in a diploid dominant plant than in a haploid dominant one.

24 Fungi

High-Flying Fungi

Fungi are not known for their mobility. You probably don't think of mushrooms and their relatives as world travelers, but they do get around. Fungi produce microscopic spores that can stick in crevices on tiny particles. When these particles get carried aloft by the wind, the spores go along for the ride. Some fungal spores travel surprising distances in this way, riding the winds that swirl high above Earth.

Dust storms in deserts of Africa and Asia, for example, launch fungus-laden particles into the atmosphere. Every year, hundreds of millions of tons of dust blow from Africa across the Atlantic, bringing spores of *Aspergillus sydowii* with them (Figure 24.1). When these spores land in Caribbean waters, they can germinate and infect sea fans (a type of coral). Trans-Atlantic transport of African dust has more than doubled since the 1970s as a result of drought in the Sahel region. Researchers suspect that fungal passengers on that dust may have contributed to declines in Caribbean reefs during the same period.

Airborne transport of fungal spores can also affect human health. On days when winds convey a lot of African dust to Caribbean nations, the number of fungal spores in air samples increases, as does the number of hospital admissions for asthma. Fungi that cause allergies, respiratory problems, and skin diseases have been found on the African dust.

In the southwestern United States, dust storms put spores of *Coccidioides immitis* into the air. The spores can cause outbreaks of valley fever (coccidioidomycosis). Most people fight off the infection with minor or no symptoms. However, people with a weakened immune system and pregnant women can be severely affected.

The constant rain of fungal spores is just one aspect of the biology of fungi. As you will learn in this chapter, most fungi live in the soils and are decomposers, not pathogens. They serve an important ecological role—breaking down organic wastes and remains, and making nutrients available to plants. Other fungi partner with photosynthetic cells, forming lichens. Fungi serve as food for many animals. Humans value them for their medicinal properties and as food. Single-celled fungi help us make bread and beer, and countless mushrooms end up on our pizzas and in our salads and sauces.

See the video! Figure 24.1 U.S. Geological Survey (USGS) scientist Ginger Garrison and a colleague analyzing dust samples in Cape Verde, an island nation off the west coast of Africa. *Inset photo*, a fungus (*Aspergillus sydowii*) that crosses the Atlantic on windborne African dust and causes disease in some Caribbean corals.

Key Concepts

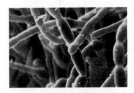

Traits and classification

Fungi are single-celled and multicelled heterotrophs. They secrete digestive enzymes onto organic matter, then absorb the released nutrients. They reproduce sexually and asexually by producing spores. Zygote fungi, club fungi, and sac fungi are major groups. **Section 24.1**

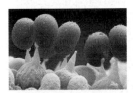

The major groups

In zygote fungi, which include many molds, the single-celled zygote produces spores by meiosis. Many sac fungi and club fungi make complex spore-bearing structures such as mushrooms. Meiosis in cells on these structures produces spores. **Sections 24.2–24.5**

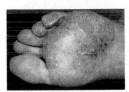

Living together

Many fungi live on, in, or with other species. Some live inside plant leaves, stems, or roots. Others form lichens by living with algae or cyanobacteria. **Section 24.6**

Fungal pathogens

A minority of fungi are parasites and some of these species cause disease in humans. Fungi also make toxins that can be deadly when eaten. **Section 24.7**

Links to Earlier Concepts

- Many fungi play important roles in nutrient cycling, a topic introduced in Section 1.2. Others are pathogens that cause disease (21.8). We use the fermentation reactions (8.5) of others to produce foods and drinks.

- In this chapter, you will learn more about fungal cell walls (4.12), and the structural material chitin (3.3). You will consider again the eukaryotic flagellum (4.13). The lignin in plants is also relevant here (4.12, 23.2).

- You will learn how fungi interact with many other organisms, including the cyanobacteria (21.6), green algae (22.9), and land plants (Chapter 23).

How would you vote? Spraying spores of fungi that infect plants might help reduce illicit crops such as the opium poppies used to make heroin. Do ecological and health risks outweigh the benefits? See CengageNOW for details, then vote online.

■ Fungi are heterotrophs that obtain nutrition by extracellular digestion, and disperse by producing spores.

■ Links to Nutrient cycle 1.2, Carbohydrates 3.3

Characteristics and Ecology

Fungi are spore-producing heterotrophs that include chitin, a nitrogen-containing polysaccharide, in their cell wall (Section 3.3). Some fungi live as single cells; they are commonly called yeasts. However, most are multicelled. Molds and mushrooms are the most familiar examples of multicelled fungi (Figure 24.2*a*,*b*).

A multicelled fungus grows as a mesh of branching filaments collectively called the **mycelium** (plural, mycelia). Each filament is one **hypha** (plural, hyphae). It consists of cells arranged end-to-end (Figure 24.2*c*). Depending on the fungal group, there may or may not be cross-walls between cells of a hypha.

All fungi feed by absorbing nutrients from their environment. As fungal cells grow in or over organic matter, the cells secrete digestive enzymes and absorb breakdown products. This nutritional mode is known as extracellular digestion and absorption. Most fungi are free-living saprobes: organisms that feed on and decompose organic wastes and remains. In this role, they help keep nutrients cycling in ecosystems. Other fungi live in or on other organisms. Some of these are parasites. Others benefit their host or have no effect.

Fungi form mutually beneficial partnerships with many organisms, expecially with plants. In fact, most plants have beneficial fungi growing inside or on their roots. Fungi also associate with photosynthetic cells, forming what we call a lichen. Other fungi live in the gut of some herbivores. The fungi help their host digest plant material.

Fungi parasitize a diverse array of organisms, ranging from algae, to plants, to insects, to mammals. They are important as pathogens of crop plants and a small number threaten human health.

Overview of Fungal Life Cycles

In fungi, as in some protists, the diploid stage is the least conspicuous part of the life cycle. Depending on the fungal group, either a haploid stage or a dikaryotic stage dominates the cycle. "Dikaryotic" means that a cell contains two genetically different nuclei ($n+n$).

Fungi disperse by producing spores. A fungal spore is a cell or cluster of cells, often with a thick wall that allows it to survive harsh conditions. With the exception of one group, fungal spores are nonmotile; they cannot move themselves from place to place. Spores may form by mitosis (asexual spores) or by meiosis (sexual spores). Scientists have traditionally classified fungi largely on the basis of the distinctive structures, in which they produce their sexual spores.

Phylogeny and Classification

Comparisons of gene sequences show that fungi are more closely related to animals than to plants:

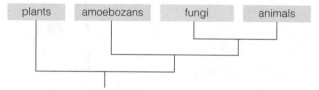

Chytrids, zygote fungi (zygomycetes), and glomeromycetes are small groups that are not monophyletic (Table 24.1). They do not have a dikaryotic stage, and each hypha is a tube with few or no cross-walls, or

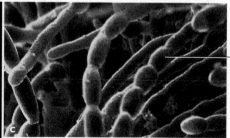

one cell (part of one hypha of the mycelium)

Figure 24.2 Multicelled fungi (**a**) Green mold (*Penicillium digitatum*) growing on grapefruit. (**b**) Scarlet hood mushroom (*Hygrophorus*) on the floor of a Virginia forest.

Molds and mushrooms are two examples of a mycelium, a multicelled body made up of individual threadlike hyphae (**c**). Material flows easily among the cells of a hypha.

Table 24.1 Major Groups of Fungi

Groups with no dikaryotic stage, few or no cross-walls (septae) between cells of hyphae:

Chytrids

1,000 species. Make spores asexually and sexually. Spores and gametes flagellated. Live in seawater, fresh water, moist soil, and in or on other organisms.

Zygote Fungi (Zygomycetes)

1,100 species. Make spores asexually and sexually. Live in soil, and inside or on other organisms. Some species are human pathogens.

Glomeromycetes

150 species. Not known to reproduce sexually. All live inside plant roots without harming the plant.

Groups with a dikaryotic mycelium that has regular cross-walls between cells of hyphae:

Sac Fungi (Ascomycetes)

More than 32,000 species. Make spores asexually and sexually. Live in soil, and inside or on other organisms. Some are human pathogens. Many partner with photosynthetic cells and form lichens.

Club Fungi (Basidiomycetes)

More than 26,000 species. Make spores asexually and sexually. Include species with the largest and most complex spore-bearing structures. Live in soil, and inside or on other organisms.

septae (singular, septa). Relationships among these groups, and their connection to the two main groups of fungi, are still being investigated.

The two largest monophyletic fungal groups are sac fungi (ascomycetes) and the club fungi (basidiomycetes). Members of both make a dikaryotic mycelium and the cells of their hyphae are separated by septae.

What made sac fungi and club fungi successful? For one thing, having a dikaryotic mycelium increased the genetic diversity of their sexually produced spores. Also, septate hyphae are advantageous in dry habitats. Without septae, an injury to one cell of a hypha can cause the entire hypha to dry out and die.

Take-Home Message

What are characteristics of fungi?

■ Fungi are heterotrophs that absorb nutrients from their environment. They live as single cells or as a multicelled mycelium. They disperse by producing spores.

24.2 | The Flagellated Fungi

■ Chytrids are the only modern fungi with a life cycle that includes flagellated cells.

■ Link to Flagella 4.13

Chytrids are an ancient fungal lineage. Their flagellated spores and gametes swim about in ponds, seas, damp soil, and the bodies of some animals. The chytrid flagellum has the same sort of structure seen in other eukaryotes (Section 4.13). This consistency suggests that the common ancestor of all modern eukaryotes was flagellated. Genetic comparisons suggest it was some sort of protist.

Most chytrids feed on organic wastes and remains, thus helping to recycle materials. A few kinds swim about in the gut of sheep, cattle, and other herbivores and help them digest cellulose. Others are parasites.

The chytrid *Batrachochytrium dendrobatidis* is a parasite of amphibians (Figure 24.3). It was discovered in the late 1990s when scientists investigated sudden population declines of frogs in Australia and South America. Since then, *B. dendrobatidis* has been detected in wild frogs in North America, South America, Europe, Africa, and Asia.

Transport of amphibians for sale as pets has probably aided the spread of this parasite. The first Asian infection was reported in late 2006 in Tokyo, Japan, by a collector who had purchased imported frogs. Since then, infections have also been detected in wild frogs in Japan.

The worldwide spread of *B. dendrobatidis* is a cause for great concern among ecologists. Frogs help control populations of harmful insects and also serve as food for many other animals. Chytrid infection may push species already threatened for other reasons to extinction.

Figure 24.3 Frogs and fungus. (**a**) Harlequin frog, one of the many species infected by the chytrid *B. dendrobatidis*. (**b**) Cross-section of skin from a chytrid-infected frog. The arrows indicate flask-shaped structures that contain fungal spores. Spores of *B. dendrobatidis* can survive in water for as long as seven weeks before infecting a new host.

24.3 | Zygote Fungi and Relatives

- Zygote fungi form a branching haploid mycelium on organic material, and inside living plants and animals.

- Link to Cell division mechanisms 9.1

Typical Zygote Fungi

Only **zygote fungi** (zygomycetes) produce a zygospore during sexual reproduction. Most of their life cycle is spent as a haploid mycelium that has few or no cross-walls between cells. There are no dikaryotic hyphae. Most zygote fungi are saprobes, but some parasitize animals, protists, or other fungi. Others associate with plant roots in a mutually beneficial way.

Rhizopus stolonifer, black bread mold, is a zygote fungus with a typical life cycle (Figure 24.4). It reproduces both asexually and sexually. There are two mating strains, plus (+) and minus (–). Sexual reproduction occurs when hyphae of these two strains meet. After contact, a structure called a gametangium forms at the

tip of each hypha. Cytoplasmic fusion of gametangia is followed by fusion of their nuclei. The result is a diploid zygospore with a thick, protective wall (Figure 24.4e). Meiosis occurs as the zygospore germinates. A hypha emerges, bearing a sac with haploid spores at its tip. After these spores are released, they germinate and each gives rise to a haploid mycelium. This mycelium grows rapidly and forms spores by mitosis at the tips of raised hyphae.

In addition to spoiling bread, *Rhizopus* species turn post-harvest fruits and vegetables to mush. *Rhizopus oryzae* can infect people who have a weakened immune system. Hyphae of this fungus proliferate inside blood vessels and cause zygomycosis, an often fatal disease. "Mycosis" is a general term for any infectious disease caused by a fungus.

Pilobolus, another zygomycete, is common on horse dung (Figure 24.5). Spores pass through the horse gut and end up in the feces. The spores germinate and

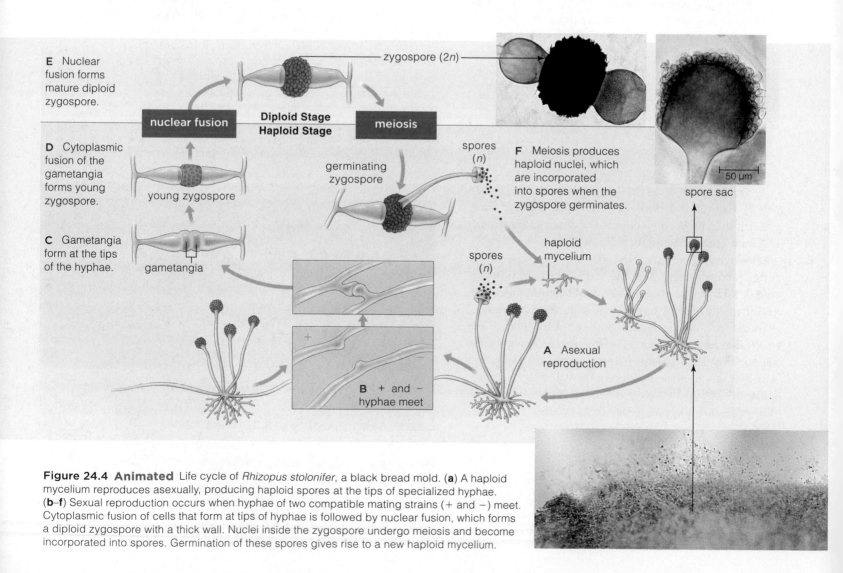

Figure 24.4 Animated Life cycle of *Rhizopus stolonifer*, a black bread mold. (**a**) A haploid mycelium reproduces asexually, producing haploid spores at the tips of specialized hyphae. (**b**–**f**) Sexual reproduction occurs when hyphae of two compatible mating strains (+ and −) meet. Cytoplasmic fusion of cells that form at tips of hyphae is followed by nuclear fusion, which forms a diploid zygospore with a thick wall. Nuclei inside the zygospore undergo meiosis and become incorporated into spores. Germination of these spores gives rise to a new haploid mycelium.

Figure 24.5 Spore-bearing structures of *Pilobolus*. The name means "hat-thrower." The dark "hats" are spore sacs.

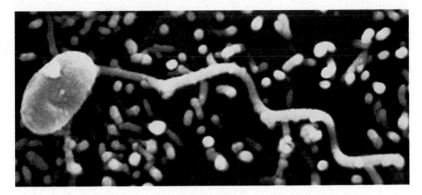

Figure 24.6 Scanning electron micrograph of a microsporidian spore, with its polar tube extruded.

produce a mycelium that produces specialized spore-bearing hyphae. At the tip of each of these hypha is a dark-walled, spore-containing sac. Below this sac, the stalk balloons outward; it is swollen with a fluid-filled central vacuole. During the day, the stalk bends so the spore-sac is directed toward the sun. Fluid pressure builds up inside the central vacuole until the vesicle ruptures. The forceful blast can propel the spore sacs up to 2 meters (about 6.5 feet)—which is impressive, since the stalk is less than 10 millimeters tall.

Microsporidians—Intracellular Parasites

Microsporidians are intracellular parasites of nearly all animals. They were long considered protists, but gene comparisons indicate that they are closely related to zygote fungi. Some biologists place them within this group; others consider them a separate phylum. Like some parasitic protists, microsporidians do not have mitochondria. They rely on the host cell for ATP.

A microsporidian spore has a long polar tube that is stored coiled up in the cytoplasm. When the spore contacts an appropriate host cell, the tube uncoils and enters that cell (Figure 24.6). Infectious contents of the spore then flow through the tube into the host.

At least fourteen species of microsporidians infect humans. *Enterocytozoon bieneusi* infection is the most common. Spores can enter the human body in food or drink, or by inhalation. People who have AIDS or other immune suppressing conditions are at the greatest risk for developing microsporidian disease. The parasites most often take up residence inside the intestine, where they cause diarrhea, abdominal cramping, and nausea. Microsporidians can also live inside cells of skin, eyes, kidneys, and the brain. If untreated, a microsporidian infection can be fatal.

Figure 24.7 Glomeromycete hypha branching inside a plant cell.

Glomeromycetes—Plant Symbionts

Glomeromycetes were previously placed among the zygote fungi, but are now considered a separate group. They are not known to reproduce sexually. All associate with plant roots. A hypha grows into a root and branches inside the wall of a root cell, where it shares space with the plant cell (Figure 24.7). Having a fungal roommate does not harm a root cell; the fungus shares nutrients from the soil with its host. We discuss fungus–plant associations again in Section 24.6.

Take-Home Message

What are zygote fungi and their relatives?

■ Zygote fungi form a thick-walled diploid spore when they reproduce sexually. Some spoil food or cause disease. Microsporidians are a subgroup that lives inside animal cells.

■ Glomeromycetes, a related group, associate with and benefit plants.

■ Sac fungi are the most diverse fungal group. There are single-celled and multicelled forms.

■ Link to Fermentation 8.5

Sac fungi, or ascomycetes, include more than 32,000 named species. They include single-celled yeasts and multicelled species. The hyphae have cross-walls at regular intervals and often intertwine as elaborate

spore-producing bodies. Septate hyphae evolved in the common ancestor of sac fungi and club fungi. The cross-walls contributed to the success of both groups. Hyphae strengthened with cross-walls can form larger spore-producing bodies. Cross-walls also divided the cytoplasm, so damage to one part of a hypha did not cause the whole hypha to dry out and die. That is one reason why sac fungi and club fungi are usually more prevalent than zygote fungi in dry environments.

Most fungi that partner with photosynthetic cells in lichens are sac fungi, as are many fungal pathogens of plants. The coral-killing *Aspergillus* species shown in Figure 24.1 and the mold shown in Figure 24.2 are sac fungi. Sac fungi also are the group that most often causes disease in humans, a topic we will return to in the final section of this chapter.

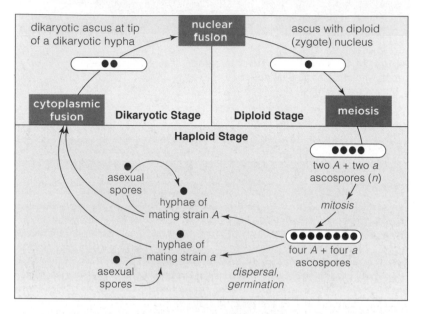

Figure 24.8 A haploid mycelium that makes spores by mitosis dominates the life cycle of *Neurospora*. Sexual reproduction occurs when the hyphae of different mating strains meet. Cytoplasmic fusion produces dikaryotic hyphae that, together with haploid hyphae, form the ascocarp. Nuclear fusion occurs in asci, saclike cells inside the ascocarp. The resulting zygote undergoes meiosis, forming four haploid spores. The haploid spores divide by mitosis, yielding eight ascospores.

Sexual Reproduction

Not all sac fungi reproduce sexually. In those that do, spores typically form inside a saclike cell called an ascus (plural, asci). Figure 24.8 shows the life cycle of *Neurospora crassa* (red bread mold). This species is often used in genetic research because it can be grown in the laboratory and results of genetic crosses are easily observable. Sexual reproduction begins when the hyphae of two compatible types meet and form dikaryotic (*n+n*) hyphae. Nuclear fusion, followed by meiosis, occurs in asci that form at the tips of hyphae.

Multicelled sac fungi often produce asci on a fruiting body, or ascocarp (Figure 24.9). It is typically made of interwoven haploid and dikaryotic hyphae.

Figure 24.9 Ascocarps. (**a**) *Sarcoscypha coccinea*, scarlet cup fungus. The cup is an ascocarp. Asci, each containing eight ascospores, form on its inner surface. (**b**) Morels, the edible ascocarps of *Morchella esculenta*. (**c**) A basket of truffles. These ascocarps form underground and the spores are inside them. Truffles are a highly valued gourmet food.

Asexual Reproduction

Most yeasts are single-celled sac fungi. For example, *Candida* is a sac fungus that causes "yeast infections" of the mouth and vagina. Yeasts often can reproduce asexually by budding (Figure 24.10a). Multicelled sac fungi also reproduce asexually. They produce haploid spores called conidia or conidiospores at the tips of specialized hyphae. Figure 24.10b shows an example.

Human Uses of Sac Fungi

We put sac fungi to a great variety of uses. As already mentioned, *Neurospora* is used in genetic studies.

Morels (Figure 24.9b) and truffles (Figure 24.9c) are among the edible ascocarps. A truffle forms underground. When spores mature, the fungus gives off a scent like that of an amorous male pig. Female pigs that catch a whiff disperse truffle spores as they root through the soil in search of a seemingly subterranean suitor. Dogs can also be trained to snuffle out truffles. Searching for truffles can be worthwhile. In 2006, a single 1.5 kilogram (about 3 pound) Italian truffle sold for $160,000.

Fermentation reactions in sac fungi help us make foods and beverages. A packet of baking yeast holds spores of *Saccharomyces cerevisiae*. When bread dough is set out to rise in a warm place, spores germinate and release cells that reproduce by budding. Carbon dioxide, a by-product of the fermentation reactions in these cells, causes dough to expand. Fermentation by *S. cerevisiae* also helps produce beer and wine (Section 8.5). One *Aspergillus* species ferments soy beans and wheat for soy sauce. Another makes citric acid that is used as a preservative and to flavor soft drinks. *Penicillium roquefortii* adds those tangy blue veins to blue cheeses such as Roquefort and Gorgonzola.

Some sac fungi are sources of drugs. Most famously, the initial source of the antibiotic penicillin was the soil fungus, *Penicillium chrysogenum*. Another antibiotic, cephalosporin, was first isolated from *Cephalosporium*. Statins from *Aspergillus* help lower cholesterol levels, and cyclosporin from *Trichoderma* helps prevent the rejection of transplanted organs.

Sac fungi that infect plant or animal pests can be used as natural herbicides or pesticides. For example, *Arthrobotrys* is a predatory sac fungus. It makes special hyphae with loops that tighten and trap roundworms (Figure 24.11). After feeding on a worm, the fungus makes asexual spores. Researchers hope to control roundworms that damage crops by spreading *Arthrobotrys* spores in agricultural fields.

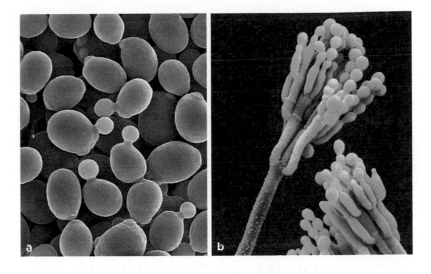

Figure 24.10 Asexual reproduction in sac fungi. (**a**) Cells of the yeast *Candida albicans*. Notice the small cells budding from the larger ones. (**b**) Conidia (asexual spores) of *Eupenicillium*. "Conidia" means dust.

part of one hypha that forms a nooselike ring   roundworm

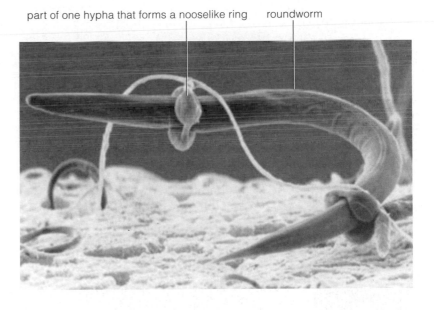

Figure 24.11 Animated A predatory fungus (*Arthrobotrys*) that captures and feeds on roundworms. Rings that form on the hyphae constrict and entrap the worms, then hyphae grow into the captive and digest it.

Take-Home Message

What are sac fungi?

■ Sac fungi are the largest fungal group. Some are single cells, but in most a haploid mycelium dominates the life cycle. Sac fungi that reproduce sexually typically form spores inside an ascus. Yeasts reproduce asexually by budding, multicelled species by formation of conidia.

■ We use sac fungi as sources of food and beverages, as drugs, and as control agents for pest organisms.

Club Fungi—Basidiomycetes

- Club fungi make the largest and most elaborate fruiting bodies; some familiar mushrooms are examples.
- Link to Plant adaptations to life on land 23.2

Club fungi are most often multicelled. The dikaryotic (n+n) phase is predominant in their life cycle, and they form sexual spores inside club-shaped cells. Typically, these cells develop on a fruiting body, or basidiocarp, composed of interwoven dikaryotic hyphae.

As one example, button mushrooms in markets and on pizzas are usually spore-bearing parts of *Agaricus bisporus*. Haploid hyphae of *A. bisporus* grow underground. When the hyphae of two mating strains meet and fuse, the result is a dikaryotic mycelium (Figure 24.12a,b). This mycelium grows through the soil, and forms mushrooms whenever conditions favor sexual reproduction. Hanging beneath each mushroom's cap are thin tissue sheets (gills) fringed with club-shaped cells. The two nuclei in these dikaryotic cells fuse and form a diploid zygote (Figure 24.12c,d). The zygote undergoes meisis, forming four haploid spores, These spores are dispersed by the wind, germinate, and start a new cycle (Figure 24.12e,f).

Club fungi play an important role as decomposers of forest plants; they are the only fungi capable of breaking down the lignin that stiffens many plant stems (Section 23.2). Some forest fungi are long-lived giants. For example, in one Oregon forest, the mycelium of a honey mushroom (*Armillaria ostoyae*) extends through more than 2,000 acres of soil. By one estimate, this fungus is 2,400 years old. This species helps break down stumps and logs, but it also attacks living trees and can kill them.

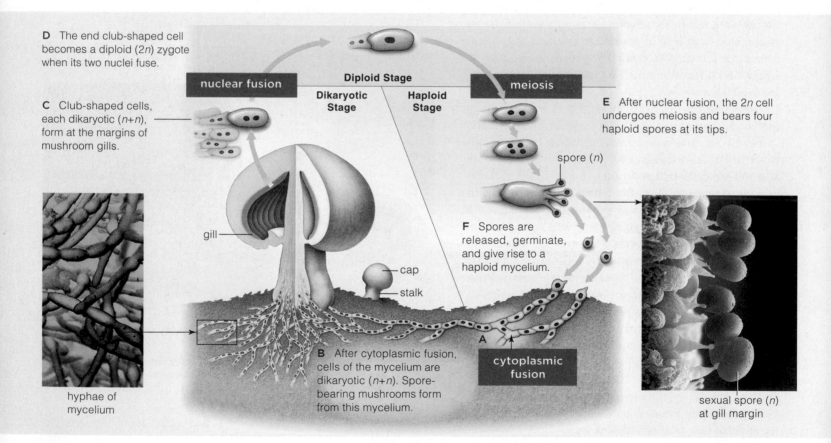

D The end club-shaped cell becomes a diploid (2n) zygote when its two nuclei fuse.

C Club-shaped cells, each dikaryotic (n+n), form at the margins of mushroom gills.

E After nuclear fusion, the 2n cell undergoes meiosis and bears four haploid spores at its tips.

F Spores are released, germinate, and give rise to a haploid mycelium.

B After cytoplasmic fusion, cells of the mycelium are dikaryotic (n+n). Spore-bearing mushrooms form from this mycelium.

nuclear fusion

Diploid Stage

Dikaryotic Stage

Haploid Stage

meiosis

spore (n)

cytoplasmic fusion

gill

cap

stalk

A

hyphae of mycelium

sexual spore (n) at gill margin

Figure 24.12 Animated Life cycle typical of a club fungus having two mating strains of hyphae.
(**a**) Haploid hyphal cells of two compatible strains meet. Their cytoplasm fuses, but the nuclei do not.
(**b**) Mitotic cell divisions form a mycelium in which each cell has two nuclei. Under favorable conditions, many hyphae of the mycelium intertwine and form a mushroom. (**c,d**) Club-shaped structures develop on the mushroom gills. The end cell on the "club" becomes diploid when two nuclei fuse. (**e**) Meiosis results in four haploid spores, which migrate into four short cytoplasmic extensions at the tip of the club. (**f**) The spores drift away from the gills. (**g**) Each may germinate and give rise to a new mycelium.

Figure It Out: What are the *blue* and *red* dots in this figure?   Answer: Genetically different nuclei

Smuts and rusts are also plant pathogens. Unlike most of the club fungi, they do not produce a large fruiting body. Wheat stem rust is an example (Figure 24.13*a*). Asexually produced, rust-colored spores can spread the infection quickly among plants, reducing crop yield by up to 70 percent.

Other club fungi include puffballs, shelf fungi, coral fungi, and chanterelles (Figure 24.13). Chanterelles are edible, but some species that have a similar appearance are poisonous. Other edible wild mushrooms also have poisonous look-alikes. For instance, most puffballs can be eaten while they are young and white. However, when *Amanita phalloides*, the death cap, first emerges from the ground it can look like a lot like a puffball to the untrained eye. Only later does it develop the distinctive cap (Figure 24.13*f*). Eating an *Amanita* species can cause nausea and abdominal cramping, followed by liver and kidney failure, and death.

Figure 24.13 Club fungus diversity. (**a**) *Puccinia graminis*, stem rust of wheat. Windborne spores spread this disease. The life cycle is complicated and requires two different plant species as hosts.

Examples of basidiocarps. (**b**) An immature puffball (*Calvatia*). Spores form inside. When mature, it turns brown and spores escape through an opening at the top or a crack in the covering. The largest puffballs can be more than a meter in diameter. (**c**) Chanterelles, one of the tasty wild mushrooms. (**d**) One of the coral fungi (*Ramaria*), with a highly branched structure. (**e**) Sulfur shelf fungus (*Laetiporus*) is a pathogen. Its hyphae grow into a host tree and digest internal tissues.

(**f**) Death cap (*Amanita phalloides*). The stem, cap, and spores are toxic. Even with treatment, about a third of poisonings are fatal. Worldwide *Amanita* species cause about 90 percent of mushroom poisonings.

Take-Home Message

What are club fungi?

■ Club fungi are fungi in which a dikaryotic mycelium dominates the life cycle. They are important decomposers of wood and have the largest, most complex fruiting bodies of all fungi.

- Fungi form associations with plants and with single-celled photosynthetic species.

- Links to Cyanobacteria 21.6, Green algae 22.9

Lichens

Most **lichens** are a symbiotic interaction between a sac fungus and a green alga or a cyanobacterium. Some club fungi also form lichens.

A lichen forms after the tip of a fungal hypha binds to a suitable photosynthetic cell. Both cells lose their wall and divide. The result is a multicelled body that may be flattened, erect, leaflike, or pendulous. Some lichens have a layered organization (Figure 24.14).

The fungus makes up most of the lichen's mass. Fungal tissues shelter a photosynthetic species, which shares nutrients with the fungus. Is a lichen a case of mutualism? **Mutualism** is a symbiotic interaction that benefits both partners. However, in another view the fungus might be exploiting a photosynthetic species that it holds captive within its tissues. The degree to which each partner benefits may vary among species.

Lichens reproduce asexually by fragmentation. The fungal partner may also release spores. To survive, a newly germinated fungus must make contact with the appropriate photosynthetic partner.

Lichens can colonize places that are too hostile for most organisms. For example, when a glacier retreats, lichens colonize newly exposed bedrock. By releasing acids and holding water that freezes and thaws, they break down the rock. When soil conditions improve, plants move in and take root. Millions of years ago, lichens may have preceded plants onto the land.

Today, some lichens are threatened by air pollution. They absorb pollutants and cannot break them down.

Fungal Endophytes

Endophytic fungi are mostly sac fungi that reside in the leaves and stems of the vast majority of plants. Usually, the interaction neither helps nor harms the plant. Some hosts benefit when the fungus produces chemicals that deter herbivores. For example, a fungus that lives inside tall fescue (a kind of grass) makes alkaloids that can sicken grazers. Once sickened, an animal will avoid the grass. Other endophytes protect the host from pathogens, including other fungi or oomycotes, such as *Phytophthora* (Section 22.8).

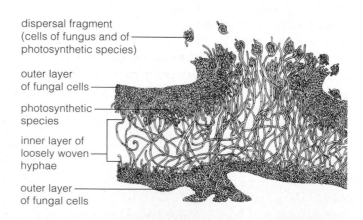

dispersal fragment
(cells of fungus and of photosynthetic species)

outer layer of fungal cells

photosynthetic species

inner layer of loosely woven hyphae

outer layer of fungal cells

Figure 24.14 (**a**) Leaflike lichen on a birch tree. (**b**) *Usnea*, or old man's beard, is one of the pendant lichens. (**c**) Encrusting lichens on granite. (**d**) Organization of one stratified lichen, as it would look in cross-section.

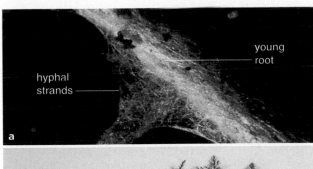

Figure 24.15 (**a**) Mycorrhiza formed by a fungus and its partner, a young hemlock tree. (**b**) Effects of the presence or absence of mycorrhizae on plant growth in phosphorus-poor, sterilized soil. The juniper seedlings at left were controls; they grew without a fungus. The six-month-old seedlings at right, the experimental group, were grown with a partner fungus.

Mycorrhizae—The Fungus-Roots

Many soil fungi, including truffles, live in or on tree roots in a partnership that is known as a **mycorrhiza** (plural, mycorrhizae). In some cases, the hyphae form a dense net around roots but do not penetrate them. Club fungi often take part in mycorrhizae with tree roots in temperate forests. Most forest mushrooms are reproductive structures of these fungi. In other cases, hyphae of the fungus penetrate root cells, as shown in Figure 24.7. About 80 percent of vascular plants form such a partnership with a glomeromycete fungus.

Hyphae of both kinds of mycorrhizae grow through soil and functionally increase the absorptive surface area of their partner. Fungal hyphae are thin. They are better at growing between soil particles than even the smallest plant roots. The fungus concentrates nutrients and shares them with the plant. The plant gives up sugars to the fungus. It is a beneficial trade; many plants do poorly without mycorrhizae (Figure 24.15).

Take-Home Message

What types of symbiotic relationships do fungi form?
■ A lichen consists of a fungus and photosynthetic cells.
■ Fungi also form mutually beneficial partnerships with plants; the fungus can live in stems, leaves, or roots.

24.7 | An Unloved Few

■ Although most fungi are harmless and provide beneficial ecological services, a small minority can harm human health.

Fungi often infect the human skin. Most often the fungus causes flaking, redness, and itching, but does not pose a serious health threat to an otherwise healthy person. For example, a variety of fungi can take up residence in the thin skin between your toes, causing what is commonly called "athlete's foot" (Figure 24.16a). Such infections can usually be cured with over-the-counter medications. To prevent athlete's foot, do not go barefoot in public showers or other places where infected people may have walked and shed fungal spores. Also, keep your feet dry; skin fungi grow best in continually damp places.

Sac fungi of the genus *Candida* often occur in low numbers in the vagina, but their overgrowth can cause fungal vaginitis, or a vaginal yeast infection. Symptoms usually include itching or burning sensations and a thick, odorless, whitish vaginal discharge. Intercourse is often painful. Disrupting the normal populations of bacteria in the vagina by douching or using antibiotics increases the risk of fungal vaginitis, as does use of oral contraceptives. Nonprescription medications placed into the vagina will usually control the infection. If such a treatment is not effective, a woman should see a doctor.

Histoplasmosis is a common fungal disease in the midwestern and south central United States, where the soil holds spores of *Histoplasma capsulatum*. Most people that inhale such spores have no symptoms or a brief episode of coughing, but no long-term effect. However, in some individuals—usually the elderly or immune-suppressed the fungus may spread from the lungs, through the blood, and into other organs, with fatal results.

Similarly, soils in the American Southwest hold spores of *Coccidioides*, which can cause coccidioidomycosis, or valley fever. Like histoplasmosis, this malady can be fatal to the elderly or to those with an impaired immune system.

As a final example of fungal effects on human health, consider *Claviceps purpurea*. It is not a human pathogen, but rather a parasite of rye and other cereal grains (Figure 24.16b). Alkaloids made by the fungus can taint flour and cause a type of poisoning called ergotism. Symptoms include vomiting, visual and auditory hallucinations, and convulsions. Severe ergotism may be fatal.

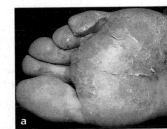

Ergotism may have played a role in the witch hunts in early American colonies such as Salem, Massachusetts. Symptoms reported by the "bewitched" such as tremors and hearing voices are among those caused by ergotism.

Figure 24.16 (**a**) A case of athlete's foot, caused by *Epidermophyton floccosum*. (**b**) Spores of *Claviceps purpurea*, on an infected rye plant. Alkaloids from this fungus cause ergotism.

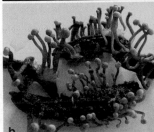

Fusarium, a sac fungus, can be a high flier. David Schmale of Virginia Tech (*left*) has collected the spores of more than a dozen *Fusarium* species from the air. Many infect plants and some cause human disease. In 2006, *Fusarium* spores got into contact lens solution and caused eye infections worldwide. Of 122 infections in the United States, a third were so bad that patients required surgery to replace the eye's clear layer, the cornea.

How would you vote?

One *Fusarium* strain kills opium poppies. Should fungal spores be sprayed in Afghanistan to reduce supplies of opium and heroin? See CengageNOW for details, then vote online.

Summary

Section 24.1 All **fungi** are heterotrophs that secrete digestive enzymes on organic matter and absorb released nutrients. Most are saprobes that feed on organic remains. Other fungi are harmless or beneficial symbionts, or parasites. Fungi are more closely related to animals than to plants. They include single-celled yeasts and multicelled species. In the multicelled species, spores germinate and give rise to filaments called **hyphae**. The filaments typically grow as an extensive mesh called a **mycelium**.

Sections 24.2, 24.3 **Chytrids** are an ancient fungal group and the only fungi with flagellated spores and gametes. Chytrids that infect amphibians are a matter of concern worldwide.

Zygote fungi include common molds. The hyphae are continuous tubes with few or no cross-walls. A thick-walled, diploid zygospore forms during sexual reproduction. Meiosis of cells inside the zygospore produces haploid spores that germinate and produce a haploid mycelium. Mycelia also produce asexual spores.

Microsporidians are zygote fungi that live inside animal cells. Like some other zygote fungi, they can cause human disease. **Glomeromycetes**, close relatives of zygote fungi, live inside plant roots.

■ *Use the animation on CengageNOW to observe the life cycle of black bread mold (Rhizopus), a zygote fungus.*

Section 24.4 **Sac fungi** are the most diverse fungal group. They include single-celled yeasts and multicelled species that have hyphae with cross-walls. Many sac fungi produce asexual spores, or conidia. Sexual spores are produced in asci. In multicelled species, these saclike structures form on an ascocarp consisting of dikaryotic hyphae. Many sac fungi are economically important.

■ *View the video on CengageNOW and watch a nematode-trapping fungus in action.*

Section 24.5 The mostly multicelled **club fungi** have hyphae with cross-walls. This group produces the largest and most complex fruiting bodies (basidiocarps). Many are important decomposers in forest habitats.

Typically, a dikaryotic mycelium dominates the life cycle. It grows by mitosis and, in some species, extends through a vast volume of soil. When conditions favor reproduction, a basidiocarp, also made up of dikaryotic hyphae, develops. A mushroom is an example. Haploid spores form by meiosis at the tips of club-shaped cells.

■ *Use the the animation on CengageNOW to learn about the life cycle of a club fungus.*

Section 24.6 Many fungi are symbionts that spend all or part of their life cycle in or on another species. **Endophytic fungi** live in many stems and leaves without harming the host plant. Some protect hosts from herbivores or from plant pathogens. This is an example of a **mutualism**, a mutually beneficial interaction.

A **lichen** is a composite organism that consists of a fungal symbiont and one or more photoautotrophs, such as green algae or cyanobacteria. The fungus makes up the bulk of the lichen and obtains a supply of nutrients from its photosynthetic partner.

A **mycorrhiza** (fungus-root) is a symbiotic interaction between a fungus and a plant. Fungal hyphae surround or penetrate the roots and supplement their absorptive surface area. The fungus shares some absorbed mineral ions with the plant and gets sugars in return.

Section 24.7 A number of pathogenic fungi can cause diseases in humans.

Self-Quiz   *Answers in Appendix III*

1. All fungi _____ .
   a. are multicelled
   b. form flagellated spores
   c. are heterotrophs
   d. all of the above

2. Saprobic fungi derive nutrients from _____ .
   a. nonliving organic matter
   b. living plants
   c. living animals
   d. photosynthesis

3. In _____ , a hypha has few or no cross-walls.
   a. all fungi
   b. zygote fungi
   c. sac fungi
   d. club fungi

4. A slice of white bread contains the remains of many yeast cells, one type of _____ .
   a. chytrid
   b. zygote fungus
   c. sac fungus
   d. club fungus

5. In many _____ , an extensive dikaryotic mycelium is the longest-lived phase of the life cycle.
   a. chytrids
   b. zygote fungi
   c. sac fungi
   d. club fungi

6. A mushroom is _____ .
   a. the food-absorbing part of a chytrid
   b. the only part of the fungal body not made of hyphae
   c. a reproductive structure that releases sexual spores
   d. produced by meiosis in a zygospore

Data Analysis Exercise

The club fungus *Armillaria ostoyae* infects living trees and acts as a parasite, withdrawing nutrients from them. If the tree dies, the fungus continues to dine on its remains. Fungal hyphae grow out from the roots of infected trees and roots of dead stumps. If these hyphae contact roots of a healthy tree, they can invade and cause a new infection.

Canadian forest pathologists hypothesized that removing stumps after logging could help prevent tree deaths. To test this hypothesis, they carried out an experiment. In half of a forest they removed stumps after logging. In a control area, they left stumps behind. For more than 20 years, they recorded tree deaths and whether *A. ostoyae* caused them. Figure 24.17 shows the results.

1. Which tree species was most often killed by *A. ostoyae* in control forests? Which was least affected by the fungus?

2. For the most affected species, what percentage of deaths did *A. ostoyae* cause in control and in experimental forests?

3. Looking at the overall results, do the data support the hypothesis? Does stump removal reduce effects of *A. ostoyae*?

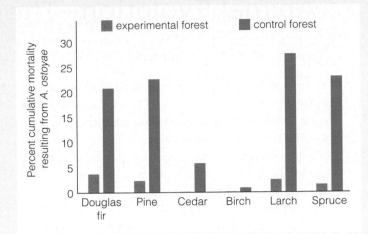

Figure 24.17 Results of a long-term study of how logging practices affect tree deaths caused by the fungus *A. ostoyae*. In the experimental forest, whole trees—including stumps—were removed (*brown* bars). The control half of the forest was logged conventionally, with stumps left behind (*blue* bars).

7. Spores released from a mushroom's gills are _____ .
   a. club-shaped
   b. dikaryotic
   c. haploid
   d. both a and c

8. The antibiotic penicillin was isolated from a _____ .
   a. chytrid
   b. zygote fungus
   c. sac fungus
   d. club fungus

9. Some green algae partner with a fungus and form a _____ .
   a. lichen
   b. mycorrhiza
   c. hypha
   d. zygospore

10. A long-term interspecific interaction that benefits both participants is a _____ .

11. All glomeromycetes _____ .
   a. cause human disease
   b. associate with roots
   c. are club fungi
   d. are part of a lichen

12. True or false? Only sac fungi form mycorrhizae.

13. Histoplasmosis is an example of a(n) _____ .
   a. endophyte
   b. lichen
   c. mycorrhiza
   d. mycosis

14. Single-celled sac fungi known as yeasts can reproduce asexually by _____ .
   a. zygospore formation
   b. conjugation
   c. budding
   d. fragmentation

15. Match the terms appropriately.
   ____chytrid
   ____sac fungus
   ____lichen
   ____club fungus
   ____zygospore fungus
   ____mycorrhiza
   ____microsporidian

   a. forms spores in an ascus
   b. produces flagellated spores
   c. lives in animal cells
   d. can digest lignin
   e. forms thick-walled diploid spore
   f. fungus and photosynthetic cells
   g. fungus and plant root

■ *Visit CengageNOW for additional questions.*

Critical Thinking

1. Certain poisonous mushrooms have bright, distinctive colors that mushroom-eating animals learn to recognize. Once sickened, an animal avoids these species. Other toxic mushrooms are as dull-looking as edible ones, but they have an unusually strong odor. Some scientists think the strong odors aid in defense against mushroom-eating animals that are active at night. Explain their reasoning.

2. Chances are, a dermatophytic (skin-living) fungus has taken up residence in you or in someone you know. *Trichophyton*, *Microsporum*, and *Epidermophyton* are the main culprits. They cause diseases known as tineas, and health professionals refer to each kind according to which body tissues are infected. As Table 24.2 shows, dermatophytes thrive on almost all body surfaces. They feed on the outer, dead layers of skin by secreting enzymes that dissolve keratin, the main skin protein, and other skin components. Infected areas commonly become raised, red, and itchy.

Dermatophyte diseases are persistent. Ointments and creams may not reach the deepest infected skin layers. There are fewer antifungal drugs than antibacterial ones and antifungals often have bad side effects. Reflect on the evolutionary relationships among bacteria, fungi, and humans. Why it is harder to fight fungi than bacteria?

Table 24.2 Common Dermatophyte Diseases

Disease	Infected Body Parts
Tinea corporis (ringworm)	Trunk, limbs
Tinea pedis (athlete's foot)	Feet, toes
Tinea capitis	Scalp, eyebrows, eyelashes
Tinea cruris (jock itch)	Groin, perianal area
Tinea barbae	Bearded areas
Tinea unguium	Toenails, fingernails

Animal Evolution—The Invertebrates

IMPACTS, ISSUES Old Genes, New Drugs

East of Australia, small, reef-fringed islands dot the vast expanse of the South Pacific Ocean. Shelled animals abound in the warm, nearshore waters of the islands, which include Samoa, Fiji, Tonga, and Tahiti. Among them are more than 500 kinds of predatory mollusks called cone snails (*Conus*), which have endured for millions of years. Humans find them tasty as well as beautiful (Figure 25.1).

Cone snails fascinate biologists for different reasons. The snails are stealthy hunters. They lie in wait, often buried in sediment, sniffing the water for the odor of prey such as fish or other invertebrates. When prey come along, the snail shoots out a harpoon loaded with conotoxins. This venom can paralyze a small fish within seconds by disrupting the signals flowing through its nervous system. It occasionally kills even larger animals. People who have been stung by the snails have died; paralysis of chest muscles halted their breathing.

Each *Conus* species makes a unique mix of 100 to 300 conotoxins that affect different membrane proteins. The broad range of specific effects makes the snail toxins potential sources of new drugs. For example, one conotoxin stops cells from releasing signaling molecules that contribute to the sense of pain. Ziconotide, a synthetic version of this toxin, relieves severe chronic pain. The nonaddictive drug is 1,000 times more potent than morphine.

While studying *C. geographicus* (Figure 25.1), University of Utah researchers found that a gene involved in conotoxin synthesis has ancient roots. In cone snails, the gene encodes the enzyme gamma-glutamyl carboxylase (GGC). The gene also occurs in fruit flies and humans, which means it has been around for at least 500 million years. It must have arisen in a common ancestor of snails, insects, and vertebrates. When these groups went their separate ways, the gene mutated independently in each lineage, and its product diverged in function. GGC helps repair blood vessels in humans. We have yet to figure out what it does in fruit flies.

This example supports an organizing principle in the study of life. Look back through time, and you discover that all organisms are related. At each branch point in the animal family tree, mutations gave rise to changes in biochemistry, body plans, or behavior. The mutations were the source of unique traits that help define each lineage.

This chapter describes the unique traits of the major invertebrate lineages. Of about 2 million named animals, only about 50,000 are vertebrates—animals with a backbone. The vast majority, including cone snails, are invertebrates. Do not assume that the invertebrates are "primitive." Invertebrates arose long before the vertebrates and their longevity attests to how well they are adapted to their environment.

See the video! Figure 25.1 (**a**) The mollusk *Conus geographicus*, engulfing a small fish. The tubelike structure extended straight up in this photograph is a siphon. It can detect small amounts of chemicals in water, as when small fishes and other prey swim within range. This cone snail impaled its prey, a fish, with a harpoon-like device. The snail then pumped paralyzing conotoxins into its prey. (**b**) A small sampling of the diverse patterns of *Conus* shells.

Key Concepts

Introducing the animals

Animals are multicelled heterotrophs that actively move about during all or part of the life cycle. Early animals were small and structurally simple. Their descendants evolved a more complex structure and greater integration among specialized parts. **Sections 25.1, 25.2**

The structurally simple invertebrates

Placozoans and sponges have no body symmetry or tissues. The radially symmetrical cnidarians such as jellyfish have two tissue layers and unique stinging cells used in feeding and in defense. **Sections 25.3–25.5**

The major invertebrate lineages

One major lineage of animals with tissues includes the flatworms, annelids, mollusks, nematodes, and arthropods. All are bilaterally symmetrical. The arthropods, which include the insects, are the most diverse of all animal groups. **Sections 25.6–25.17**

On the road to vertebrates

Echinoderms are on the same branch of the animal family tree as the vertebrates. They are invertebrates with bilateral ancestors, but adults now have a decidedly radial body plan. **Section 25.18**

Links to Earlier Concepts

- This chapter draws on your understanding of levels of organization (1.1), and of adaptive traits (17.3) and exaptation (18.12). We return again to the topic of how to classify organisms (19.1, 19.6).

- We discuss comparative genomic studies (16.5), homeotic genes (15.3) and patterns of development (19.3). You will learn about how membrane proteins (5.2) played a role in evolution of multicelled animals.

- You may wish to refer back to the geological time scale (17.8), to put events into perspective.

- Earlier chapters discussed diseases caused by bacteria (21.6, 21.8) and by protists (22.2, 22.6). Here you will learn a bit about their animal vectors. You will also be reminded of the interaction between dinoflagellates (22.5) and corals.

How would you vote? Marine invertebrates are ecologically important and a source of human food. Some make chemicals that are used as drugs. Bottom trawling, a type of fishing, destroys invertebrate habitat. Should it be banned? See CengageNOW for details, then vote online.

Animal Traits and Body Plans

- All animals are multicelled heterotrophs, and the overwhelming majority are invertebrates.

- Link to Ratio of surface area-to-volume 4.2

What Is an Animal?

Animals are multicelled heterotrophs that move about during part or all of the life cycle. Their body cells do not have a wall and are typically diploid. Table 25.1 introduces animal phyla. Appendix I provides more details about taxonomy.

Most animals are invertebrates; they do not have a backbone. This chapter covers invertebrate diversity. Chapter 26 covers vertebrates (animals with a backbone) and their closest invertebrate relatives.

Variation in Animal Body Plans

Organization All animals are multicelled. As Table 25.1 shows, the most ancient animal lineages such as

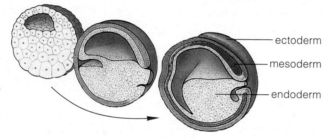

Figure 25.2
How a three-layer animal embryo forms. Most animals have this type of embryo.

ectoderm
mesoderm
endoderm

sponges are organized as aggregations of cells. There are different cell types that carry out different tasks. All animals show such an internal division of labor. Later, in the common ancestor of most animals, cells became organized as tissues. A tissue consists of cells of a particular type organized in a specific pattern.

Tissue formation begins in an embryo. At first, animal embryos had two tissue layers: outer ectoderm and an inner endoderm. Jellyfish and flatworms still have this organization. Later still, a third embryonic layer evolved. Called mesoderm, it lay between the inner and outer layers (Figure 25.2). The evolution of mesoderm allowed an increase in complexity. Most animal groups have many organs derived from mesoderm.

Body Symmetry The structurally simplest animals such as sponges are asymmetrical; you cannot divide their body into halves that are mirror images. Jellyfish and their relatives the hydras have **radial symmetry**; body parts are repeated around a central axis, like the spokes of a wheel (Figure 25.3*a*). Most animals have **bilateral symmetry**; many parts are paired, with one on each side of the body (Figure 25.3*b*). Most bilateral animals have undergone **cephalization**; nerve cells have become concentrated at the head end. In some lineages, this concentration of cells evolved into a brain.

Gut and Body Cavity Most animals have a **gut**: a digestive sac or tube that opens at the body surface.

Table 25.1 Animal Groups Surveyed in Chapters 25 and 26

Animal Phylum	Representative Groups	Living Species	Organization	Body Symmetry	Digestion	Circulation
Placozoa	Placozoans (*Trichoplax*)	1	Connected cells	None	Extracellular	Diffusion
Porifera	Barrel sponges, encrusting sponges	8,000	Connected cells	None	Intracellular	Diffusion
Cnidaria	Sea anemones, jellyfishes, corals	11,000	2 tissue layers	Radial	Saclike gut	Diffusion
Platyhelminthes	Planarians, tapeworms, flukes	15,000	2 tissue layers, organs	Bilateral	Saclike gut	Diffusion
Annelida	Polychaetes, earthworms, leeches	15,000	3 tissue layers, organs	Bilateral	Complete gut	Closed system
Mollusca	Snails, slugs, clams, octopuses	110,000	3 tissue layers, organs	Bilateral	Complete gut	Open in most, closed in some
Rotifera	Rotifers	2,150	3 tissue layers, organs	Bilateral	Complete gut	Diffusion
Tardigrada	Water bears	950	3 tissue layers, organs	Bilateral	Complete gut	Diffusion
Nematoda	Pinworms, hookworms	20,000	3 tissue layers, organs	Bilateral	Complete gut	Closed system
Arthropoda	Spiders, crabs, millipedes, insects	1,113,000	3 tissue layers, organs	Bilateral	Complete gut	Open system
Echinodermata	Sea stars, sea urchins	6,000	3 tissue layers, organs	Larvae bilateral; adults radial	Complete gut	Open system
Chordata	Invertebrate chordates	2,100	3 tissue layers, organs	Bilateral	Complete gut	Closed system
	Vertebrates (fishes, amphibians, reptiles, birds, mammals)	4,500	3 tissue layers, organs	Bilateral	Complete gut	Closed system

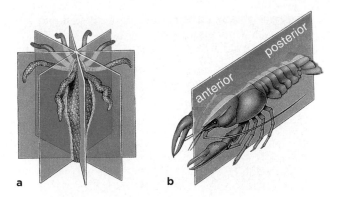

a   **b**

Figure 25.3 (**a**) Radial body symmetry of *Hydra*, a cnidarian. (**b**) Bilateral symmetry of a lobster, an arthropod. The anterior end is the head end; the posterior end is the tail end.

A saclike gut is an incomplete digestive system; food enters and wastes leave through the same body opening. A tubular gut is a complete digestive system, with a mouth at one end and an anus at the other.

A complete digestive system has advantages. Parts of the tube can became specialized for taking in food, digesting food, absorbing nutrients, or compacting the wastes. Unlike a saclike gut, a complete gut can carry out all of these tasks simultaneously.

A flatworm's gut is enclosed by a more or less solid mass of tissues and organs (Figure 25.4*a*). However, in most animals, a fluid-filled body cavity surrounds the gut (Figure 25.4*b*,*c*). If this cavity has a lining of tissue derived from mesoderm, it is called a **coelom** (Figure 25.4*c*). A cavity incompletely lined by mesodermal tissue is a **pseudocoel**, which means false coelom.

A fluid-filled coelom or pseudocoel provided three advantages. First, materials could diffuse through the fluid to body cells. Second, muscles could redistribute the fluid to alter the body shape and aid locomotion. Finally, organs were not hemmed in by a mass of tissue, so they could grow larger and move more freely.

The two major lineages of bilateral animals differ in how their digestive system and coelom form. In **protostomes**, the first opening that appears on the embryo becomes the mouth, and the second becomes the anus. In **deuterostomes**, the first opening develops into the anus and the second becomes the mouth.

Circulation In small animals, gases and nutrients can diffuse through a body. However, diffusion alone cannot move substances fast enough to keep a big animal alive (Section 4.2). In most animals, a circulatory system speeds distribution of materials. In a closed circulatory system, a heart or hearts propel blood through a continuous system of vessels. Materials diffuse out of vessels and into cells. In an open circulatory system,

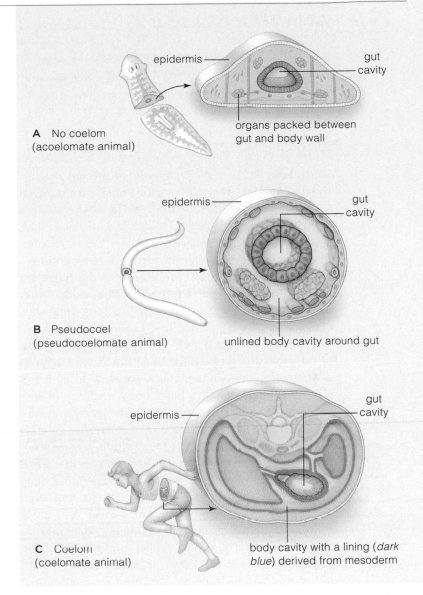

A No coelom (acoelomate animal)

organs packed between gut and body wall

epidermis — gut cavity

B Pseudocoel (pseudocoelomate animal)

unlined body cavity around gut

epidermis — gut cavity

C Coelom (coelomate animal)

body cavity with a lining (*dark blue*) derived from mesoderm

Figure 25.4 Animated (**a**) A flatworm is acoelomate; it has no body cavity. (**b**) A roundworm has a partially lined cavity (a pseudocoel). (**c**) All vertebrates are coelomate. Peritoneum, a tissue derived from mesoderm (and shown here in *dark blue*) lines the vertebrate coelom.

blood leaves vessels and exchanges materials directly with tissues before returning to the heart. A closed system allows for faster blood flow than an open one.

Segmentation Many bilateral animals are segmented; similar units are repeated along the length of the body. As you will see, repetition opened the way to specialization. When many segments perform the same task, some can change and take on new functions.

Take-Home Message

What traits characterize animals?

■ Animals are multicelled heterotrophs that usually ingest food. Their unwalled, diploid body cells are usually organized as tissues, but body plans vary.

- Fossils and gene comparisons among modern species provide insights into how animals arose and diversified.

- Links to Membrane proteins 5.2, Homeotic genes 15.3, Speciation 18.10, Exaptation 18.12, Classification 19.1,19.6

Becoming Multicellular

According to the colonial theory of animal origins, the first animals evolved from a colonial protist. What was that protist like? **Choanoflagellates**, the modern protists most closely related to animals, provide clues. Their name means "collared flagellate." Each choano-flagellate cell has a collar of microvilli surrounding a flagellum (Figure 25.5a). Movement of the flagellum directs food-laden water past microvilli, which filter out food. As you will see, sponges feed the same way.

Some choanoflagellates live as single cells, while others form colonies (Figure 25.5b). A colony is a group of cells that all carry out the same functions. Each cell can survive independently if separated. In contrast, a multicelled organism has a body made up of several types of cells that carry out different tasks and are arranged in a specific pattern. The cells must interact to survive and only some of them produce gametes.

Studies of choanoflagellates have shown that some proteins associated with multicellularity have deep evolutionary roots. These protists have proteins that are similar to proteins involved in adhesion or inter-cellular signaling proteins in animals. What role do these proteins serve in single-celled protists? Adhesion proteins may help cells stick together during sexual reproduction. Proteins like those used in animal signaling pathways may help the protists detect molecules associated with food or pathogens. As Section 18.12 explained, traits that evolve as adaptations in one context often change over time and later become adaptive in a somewhat different or entirely different context in a descendant group.

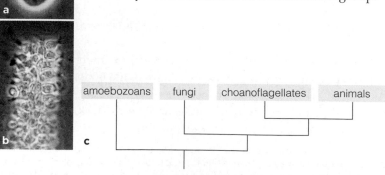

Figure 25.5 (a) Free-living choanoflagellate. A collar of microvilli rings its flagellum. (b) Colony of choanoflagellates. Some researchers view it as a model for the origin of animals. (c) Relationships among animals, choanoflagellates, and related groups. The amoebozoans and choanoflagellates are both protists.

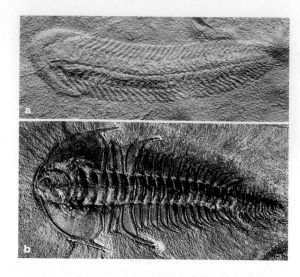

Figure 25.6 Fossil animals. (**a**) *Spriggina*, an Ediacaran, lived about 570 million years ago. It was about 3 centimeters (1 inch) long. By one hypothesis, it was a soft-bodied ancestor of arthropods, such as trilobites (**b**) that arose during the Cambrian.

A Great Adaptive Radiation

We do not know exactly when the very first animals evolved. They were certainly small and soft-bodied, and thus unlikely to leave conspicuous fossils. We do know that by about 570 million years ago, a diverse collection of multicelled organisms, including some early animals, lived in the seas. The animals are called Ediacarans because their fossils were first discovered in Australia's Ediacaran hills. They include multicelled species that range from tiny blobs to frondlike forms that stood more than a meter high. Most Ediacaran lineages do not have any living descendants. However, early representatives of some modern animal groups may have been among them (Figure 25.6).

Animals underwent a dramatic adaptive radiation during the Cambrian (542–488 million years ago). By the end of this period, all major animal lineages were present in the seas. What caused this Cambrian explosion in diversity? Rising oxygen levels and changes in global climate may have played a role. Also, supercontinents were breaking up. Movement of land masses isolated populations, thus increasing opportunities for allopatric speciation (Section 18.10). Biological factors also encouraged speciation. Once the first predators arose, mutations that produced protective hard parts would have been favored. Evolution of new genes that regulate body plans may also have sped things along (Section 15.3). Mutations in these genes would have allowed adaptive changes to body form in response to predation or altered habitat conditions.

Relationships and Classification

Animals have traditionally been classified on the basis of their morphology (body form) and developmental patterns. More recently, gene sequence comparisons have been used to investigate relationships. Results from the two methods sometimes differ. Figure 25.7 compares the traditional classification scheme with a new one based on gene comparisons. In both, animals are placed into a series of nested groups, represented here by boxes of different colors. However, animals with a three-layer embryo (the pink boxes) are subdivided differently in the two classification schemes.

The traditional classification scheme (Figure 25.7a) puts great emphasis on the possession of a body cavity and the features of this cavity. Animals that have a three-layer embryo are sorted into three groups. Acoelomate animals have no body cavity. Coelomate animals have a cavity fully lined with tissue derived from mesoderm. Pseudocoelomate animals have a body cavity partially lined with mesodermal tissue. In this scheme, roundworms and rotifers are grouped together because both have a pseudocoelom. Coelomate animals are further divided into protostomes and deuterostomes based on aspects of their development.

The newer scheme (Figure 25.7b) puts all animals with a three-layer embryo into the protostome group or deuterostome group. Within protostomes, there are two lineages. Ecdysozoa includes animals that **molt**, or periodically shed, a body covering as they grow. Lophotrochozoans do not molt and have their own unique traits. This new scheme puts roundworms and rotifers, both pseudocoelomate, in separate lineages.

Which is correct? Critics of the newer scheme argue that it is unlikely that a coelom could have evolved independently in two lineages. However, a common ancestor of all animals with a three-layer embryo might have had a coelom. By this scenario, flatworms lost the coelom and roundworms and rotifers modified theirs. Traits are often modified or lost over time. We organize this chapter around the relationships shown in Figure 25.7b, realizing that new information may yet modify it.

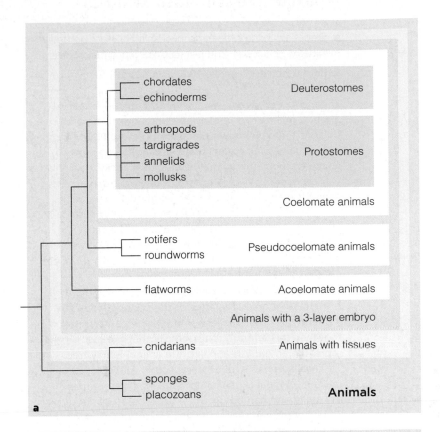

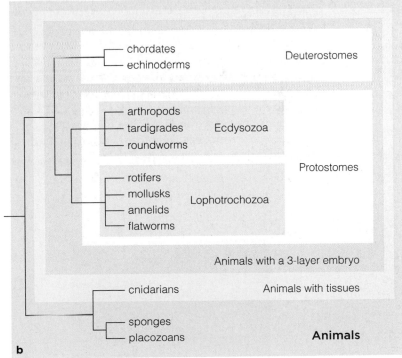

Take-Home Message

What do we know about animal origins and diversification?

■ Animals probably evolved from a choanoflagellate-like protist. Most modern groups arose in an adaptive radiation during the Cambrian. We continue to investigate how groups are related.

Figure 25.7 Proposed evolutionary trees for animals. Colored boxes show subgroupings. (**a**) Traditional classification based mainly on morphology. (**b**) A more recently proposed tree based on gene and protein comparisons. The two trees differ in how they divide up the animals with a three-layer embryo. As more genomes are sequenced, scientists will be better able to distinguish the true relationships. **Figure It Out:** What group of acoelomate animals is considered protostomes in the new (b) classification scheme? *Answer: The flatworms*

25.3 | The Simplest Living Animal

■ Placozoans, the simplest known animals, have no body symmetry, no tissues, and just four different types of cells.

■ Link to Homeotic genes 15.3

Trichoplax adhaerens, an asymmetrical marine animal about 2 millimeters in diameter and 2 micrometers thick, is the only known **placozoan** (Figure 25.8). *Tricho-* means hairy, *plax* means plate, and *adhaerens* means sticky. In short, the animal resembles a sticky, hairy plate.

T. adhaerens lives in coastal waters of tropical seas, where it feeds on bacteria and single-celled algae. Its four types of cells form two layers. A ciliated surface allows the animal to glide from place to place. When *T. adhaerens* happens upon food, gland cells in its lower layer secrete enzymes and absorb breakdown products. Cells also take in bits by phagocytosis.

Gene sequence comparisons show that *T. adhaerens* is the closest animal relative of the choanoflagellates. Its genome is the smallest of any known animal. Taken together, this information suggests that the placozoans represent an early branch on the animal family tree.

T. adhaerens's evolutionary history and small genome make it an ideal organism for studies. It reproduces asexually and can be grown in the laboratory. Studying *T. adhaerens* can reveal the history of human genes. As explained in the chapter introduction, a gene that evolved in one context often mutates and takes on different or additional functions in descendant lineages.

Scientists have already discovered that although *T. adhaerens* has no nerve cells, it has genes like those that encode signaling molecules in human nerves. It also has a gene similar to the homeotic genes that regulate development in more complex animals (Section 15.3). Researchers often discover that genes that now take part in complex structures had other functions in the simpler animals that evolved earlier.

Figure 25.8 *Trichoplax adhaerens*, the only known species of placozoan. This specimen was grown in the laboratory at Yale University. Its red color comes from the red algae cells on which it had been feeding.

25.4 | The Sponges

■ Sponges are simple but successful. They have survived in the seas since Precambrian times.

■ Links to Flagella 4.13, Phagocytosis 5.5

Characteristics and Ecology

Sponges (phylum Porifera) are aquatic animals with no symmetry, tissues, or organs. They resemble a colony of choanoflagellates but with more kinds of cells and a greater division of labor. Most sponges live in tropical seas, but some species occur in arctic seas or in fresh water. Sponges attach to the seafloor or other surfaces. A few are big enough to sit in; others would fit on a fingertip. Shapes range from sprawling and flat, to lobed, compact, or tubelike (Figure 25.9).

The phylum name, Porifera, means pore-bearing, and a typical sponge has many pores and one or more larger openings. Flattened, nonflagellated cells cover the sponge's outer surface; flagellated collar cells line its inner one; and a jellylike matrix lies between the cell layers (Figure 25.10).

Most sponges feed by filtering bacteria from water. As in choanoflagellates, motion of the flagella drives movement of food-laden water. Water is drawn into the sponge through the many pores in the body wall and flows out through one or more larger openings. As water flows past collar cells in the sponge's inner surface, villi of these cells trap food and engulf it by phagocytosis (Section 5.5). Digestion is intracellular. Amoeba-like cells move throughout the matrix. They receive food-filled vesicles from collar cells and then distribute food to other cells throughout the body.

Sponges cannot run away from predators, but they have other defenses. In many species, cells inside the matrix secrete fibrous proteins or glassy spikes called spicules (Figure 25.9b). The coarse or glassy materials make sponges difficult for most predators to eat. Also, some sponges secrete slime or chemicals that repel predators. In addition, such chemicals can help fend off competitors for living space.

Sponges themselves can serve as habitat for marine worms, arthropods, echinoderms, and other invertebrates. Some sponges receive sugars from single-celled algae or photosynthetic bacteria living in their tissues. Bacterial cells may make up as much as 40 percent of such a sponge's body mass.

Sponges have been gathered for use by humans since ancient times. At present, about $40 million worth of sponges are harvested each year. Disease and overharvesting have caused population declines of the most widely sought species, so many bath sponges are now grown on undersea farms.

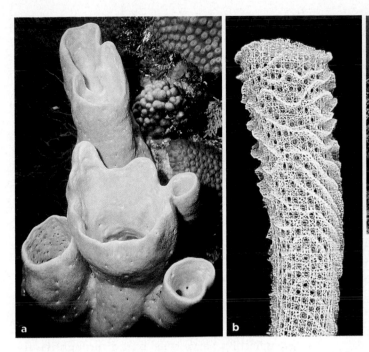

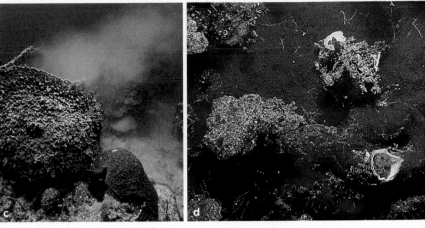

Figure 25.9 (**a**) A vase-shaped sponge. (**b**) Structural framework of Venus's flower basket (*Euplectella*). In this marine sponge, fused-together silica spicules form a rigid network. A thin layer of flattened cells stretches over its outer surface. A tuft of spicules anchors the sponge to a surface. (**c**) A basket sponge releasing a cloud of sperm. (**d**) Encrusting sponge growing on a ledge in a temperate sea.

Sponge Reproduction and Dispersal

A typical sponge is a **hermaphrodite**: an individual that produces both eggs and sperm. A sponge releases its sperm into the water, but holds onto its eggs (Figure 25.9c). After fertilization occurs, a zygote forms and develops into a ciliated larva. A **larva** (plural, larvae) is a free-living, sexually immature stage in an animal life cycle. Sponge larvae exit the parental body, swim about briefly, then settle and develop into adults.

Many sponges can also reproduce asexually. New individuals bud from existing ones or fragments break away and grow into new sponges.

Some freshwater sponges can survive unfavorable conditions by producing gemmules: tiny clusters of cells encased in a hardened coat. Gemmules survive freezing, extreme heat, and drying out. When conditions get better, gemmules grow into a new sponge.

Sponge Self-Recognition

Sponges show cell adhesion and self-recognition. In some species, individual cells rejoin to form a sponge after being broken apart. The separated cells do not hook up at random. If cells from different individual sponges are mixed together, the cells sort themselves out. In more complex animals, such a capacity for self-recognition serves as the foundation for immune responses to pathogens.

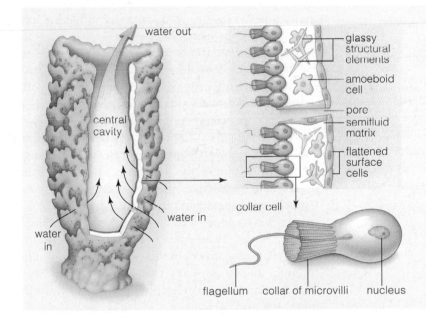

Figure 25.10 Animated Body plan of a simple sponge. Flattened cells cover the outer surface and line pores. Flagellated collar cells line inner canals and chambers. Microvilli of these cells act like a sieve that strains food from the water. Amoeboid cells in the matrix distribute nutrients and secrete structural elements.

Take-Home Message

What are sponges?

■ Sponges are typically marine animals with no body symmetry or tissues. Larvae swim briefly, but adults stay put. They pull water through their body and filter out food. Toxins and fibrous or sharp body materials deter predators.

■ Cnidarians are radial animals with two tissue layers. They have a long history; fossils date back to Precambrian times.

■ Link to Dinoflagellates 22.5

General Features

Cnidarians (phylum Cnidaria) include 10,000 species of radially symmetrical animals such as corals, sea anemones, and jellyfishes. Nearly all are marine. There are two cnidarian body shapes—medusa and polyp (Figure 25.11a,b). In both, a tentacle ringed mouth opens onto a saclike gastrovascular cavity that serves in digestion and gas exchange.

Medusae such as jellyfishes are shaped like a bell or umbrella, with a mouth on the lower surface. Most swim or drift about. Polyps such as sea anemones are tubular and one end usually attaches to a surface.

Both medusae and polyps consist of two tissues. The outer epidermis develops from ectoderm, and the inner gastrodermis from endoderm. Mesoglea, a jellylike, acellular secreted matrix, fills the space between the two tissue layers. Medusae tend to have a lot of mesoglea; polyps usually have less.

The name Cnidaria is from *cnidos*, the Greek word for nettle, a kind of stinging plant. Cnidarian tentacles have stinging cells with unique organelles called **nematocysts**. Nematocysts help capture prey and also function as a defense (Figure 25.12). Like a jack-in-the-box, a nematocyst holds a coiled thread beneath a hinged lid. When something brushes against the nematocyst's trigger, the lid opens. The thread inside pops out and entangles prey or sticks a barb into it. Unlucky swimmers that brush up against jellyfish and trigger this response receive painful, and occasionally deadly, stings. More often, nematocyst-covered tentacles snag tiny invertebrates or fish. Food is pulled through the mouth into the gastrovascular cavity. Gland cells of the gastrodermis secrete enzymes that digest the prey.

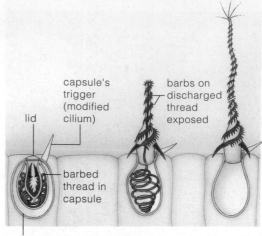

nematocyst (capsule at free surface of epidermal cell)

Figure 25.12 Animated Example of nematocyst action. Mechanical stimulation causes the thread coiled up inside the capsule to spring out and penetrate prey.

Interconnecting nerve cells extend through the tissues, forming a **nerve net**, a simple nervous system. Body parts move when nerve cells signal contractile cells. Such contractions redistribute mesoglea, just as a water-filled balloon changes shape when you squeeze it. A fluid-filled cavity or cellular mass that contractile cells exert force on is called a **hydrostatic skeleton**.

Diversity and Life Cycles

We divide the cnidarians into four classes: hydrozoans, anthozoans, cubozoans, and scyphozoans. *Obelia* is a small marine hydrozoan with a life cycle that includes polyp, medusa, and larval stages (Figure 25.13). The cnidarian larva, called a planula, is ciliated and bilateral. It develops into a polyp that reproduces asexually by budding. Gamete-producing medusae develop on tips of specialized polyps. Each medusa is less than a centimeter in diameter.

Hydra, another hydrozoan, lives in freshwater. The predatory polyp stands up to 20 millimeters (3/4 inch) high (Figure 25.14a,b). There is no medusa stage and reproduction usually occurs asexually by budding.

Anthozoans such as corals and sea anemones also do not have a medusa stage (Figure 25.14c,d). Gametes form on polyps. Coral reefs are colonies of polyps that enclose themselves in a skeleton of secreted calcium carbonate. In a mutually beneficial relationship, photosynthetic dinoflagellates (Section 22.5) live inside the polyp's tissues. The protists get shelter and carbon dioxide from the coral, which gets sugars and oxygen in return. If a reef-building coral loses its protist symbionts, an event called "coral bleaching," it may die.

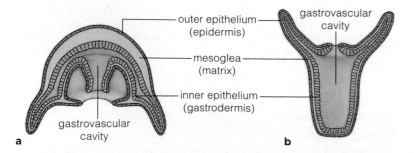

Figure 25.11 Animated Two cnidarian body plans: (**a**) medusa and (**b**) polyp, cutaway views. Both are saclike, with two thin tissue layers—an outer epidermis, and an inner gastrodermis. Jellylike, secreted mesoglea lies between the two.

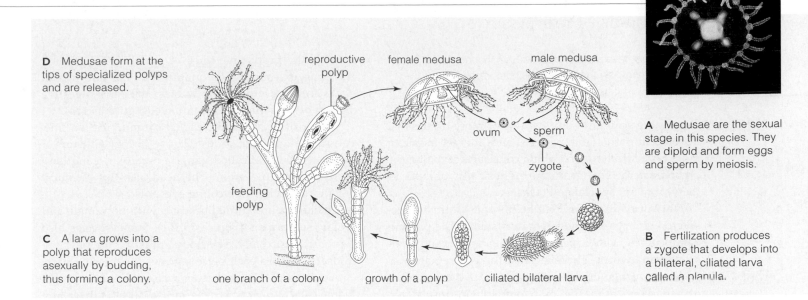

D Medusae form at the tips of specialized polyps and are released.

reproductive polyp

female medusa

male medusa

ovum   sperm

zygote

feeding polyp

A Medusae are the sexual stage in this species. They are diploid and form eggs and sperm by meiosis.

C A larva grows into a polyp that reproduces asexually by budding, thus forming a colony.

one branch of a colony

growth of a polyp

ciliated bilateral larva

B Fertilization produces a zygote that develops into a bilateral, ciliated larva called a planula.

Figure 25.13 Animated Life cycle of *Obelia*, a hydrozoan.

Box jellies, best known for their potentially deadly sting, are among the cubozoans (Figure 25.14e). They have surprisingly complex eyes around the rim of their bell. The polyp is tiny and develops into a medusa, rather than producing and releasing medusae.

Scyphozoans are sometimes called "true jellyfish." They include most of the species that commonly wash up on beaches. Some scyphozoans are harvested and dried as food, especially in Asia. A Portuguese man-of-war (*Physalia*) is a colonial type. Beneath a float are many specialized polyps and medusae (Figure 25.14f). Tentacles of the polyps can extend for several meters.

Figure 25.14 Cnidarian diversity. (**a,b**) One of the few freshwater cnidarians, a hydroid (*Hydra*), capturing a water flea and digesting it. (**c**) Sea anemone and (**d**) polyps of a reef-building coral, both anthozoans. (**e**) The box jelly, *Chironex*, is a cubozoan that makes a toxin that can kill a person. (**f**) A Portuguese man-of-war (*Physalia*), is a colony of scyphozoans. The purplish-blue, air-filled float is a modified polyp that keeps the colony at the surface of the water.

Take-Home Message

What are cnidarians?

■ Cnidarians are radial animals such as jellyfishes, corals, and sea anemones, with unique organelles called nematocysts. Medusae and polyps are the two common body shapes. A nerve net and hydrostatic skeleton allow movement.

25.6 Flatworms—Simple Organ Systems

- Flatworms have a three-layer embryo that develops into an adult with many organ systems but no coelom.
- Link to Levels of organization 1.1

Organs are structural units of two or more tissues that develop in predictable patterns and interact in one or more tasks. Each organ system consists of two or more organs interacting chemically, physically, or both as they carry out specialized tasks.

Flatworms (phylum Platyhelminthes) form a three-layered embryo and have organ systems. The phylum name comes from Greek; *platy–* means flat, and *helminth* means worm. Turbellarians, flukes (trematodes), and tapeworms (cestodes) are the main classes. Most turbellarians are marine, but some live in fresh water, and a few live in damp places on land. Flukes and tapeworms are parasites of animals.

Flatworms are bilateral and cephalized. Although they lack a coelom, they have genes that resemble those that regulate coelom development in other animals. Tapeworms have distinct segments and turbellarians, though not externally segmented, do have internally repeated organs. By one hypothesis, the ancestor of all flatworms was segmented and coelomate, and these traits were lost as the lineages evolved.

Structure of a Free-Living Flatworm

The planarians are free-living turbellarians that glide about in ponds and streams. Cilia at the body surface provide the propulsive force. A muscular tube called the **pharynx** connects the mouth with the gut. It both sucks in food and expels wastes; thus the digestive system of planarians is incomplete (Figure 25.15*a*).

A pair of **nerve cords**, each a communication line, runs the length of the body (Figure 25.15*b*). Clusters of nerve cell bodies, called **ganglia** (singular, ganglion), serve as a simple brain. The head also has chemical receptors and light-detecting eyespots.

Planarians are hermaphrodites with both male and female sex organs (Figure 25.15*c*). Some species also reproduce asexually. The body splits in two near the midsection, then each piece regrows the missing part.

A system of tubes regulates water and solute levels. Flame cells appear to be "flickering" as cilia drive any excess water into these tubes, which open to the body surface at a pore (Figure 25.15*d*).

Flukes and Tapeworms—The Parasites

Flukes and tapeworms are parasites of many animals. Often, immature stages spend time in one or more intermediate hosts, then reproduction takes place in a definitive host. For example, Figure 25.16 shows the life cycle of a blood fluke (*Schistosoma*). Aquatic snails serve as the intermediate hosts, but reproduction can only take place inside a mammal, such as a human.

Ancestors of tapeworms probably had a gut and mouth. But inside the vertebrate gut, a habitat rich in predigested food, these features were unnecessary and

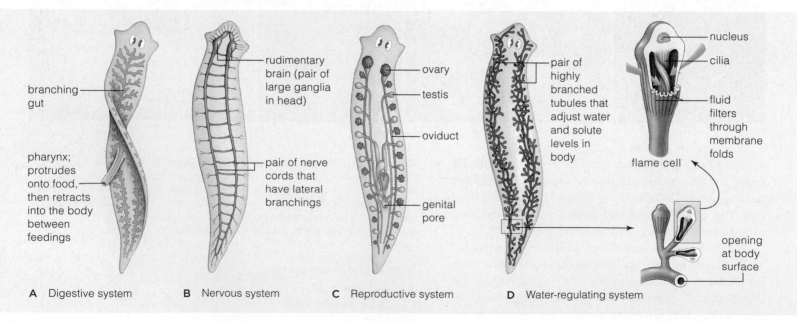

A Digestive system **B** Nervous system **C** Reproductive system **D** Water-regulating system

branching gut

pharynx; protrudes onto food, then retracts into the body between feedings

rudimentary brain (pair of large ganglia in head)

pair of nerve cords that have lateral branchings

ovary

testis

oviduct

genital pore

pair of highly branched tubules that adjust water and solute levels in body

flame cell

nucleus

cilia

fluid filters through membrane folds

opening at body surface

Figure 25.15 Animated Organ systems of a planarian, one of the flatworms. Repetition of organs along the length of the body suggests that an ancestor was segmented.

—over many generations—were lost. Modern species latch onto the intestinal wall with a scolex, a structure with hooks or suckers at the head end. Nutrients reach cells by diffusing across the tapeworm body wall.

A tapeworm body consists of **proglottids**. It grows as these repeating body units form and bud from the region behind the scolex. The tapeworm can fertilize itself, because each proglottid is hermaphroditic. The sperm from one can fertilize eggs in another. Older proglottids (farthest from the scolex) contain fertilized eggs. The oldest proglottids break off, then exit the body in feces. Fertilized eggs can survive for months on their own before reaching an intermediate host.

Some tapeworms parasitize humans. Larvae enter the body when a person eats undercooked meat or fish that contains larvae. For example, Figure 25.17 shows the life cycle of the beef tapeworm.

Take-Home Message

What are flatworms?

■ Flatworms develop from a three-layer embryo, are bilateral, and have organs. Some are free-living, others parasites.

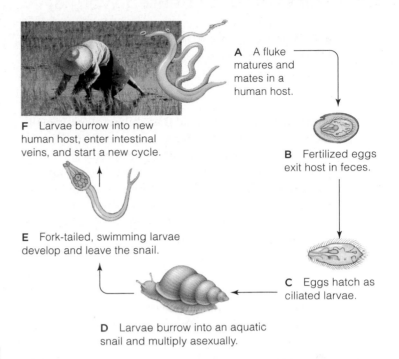

A A fluke matures and mates in a human host.

B Fertilized eggs exit host in feces.

C Eggs hatch as ciliated larvae.

D Larvae burrow into an aquatic snail and multiply asexually.

E Fork-tailed, swimming larvae develop and leave the snail.

F Larvae burrow into new human host, enter intestinal veins, and start a new cycle.

Figure 25.16 Life cycle of *Schistosoma japonicum*, found mainly in China, Indonesia, and the Philippines. This blood fluke parasitizes humans. Early symptoms of the resulting disease, schistosomiasis, are not obvious. Later, side effects of immune responses to fluke eggs damage internal organs. Worldwide, an estimated 200 million people are currently infected by some kind of schistosome.

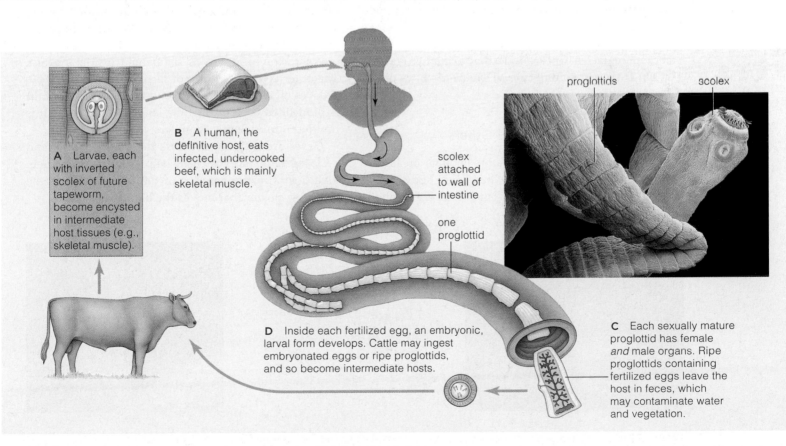

A Larvae, each with inverted scolex of future tapeworm, become encysted in intermediate host tissues (e.g., skeletal muscle).

B A human, the definitive host, eats infected, undercooked beef, which is mainly skeletal muscle.

scolex attached to wall of intestine

one proglottid

C Each sexually mature proglottid has female *and* male organs. Ripe proglottids containing fertilized eggs leave the host in feces, which may contaminate water and vegetation.

D Inside each fertilized egg, an embryonic, larval form develops. Cattle may ingest embryonated eggs or ripe proglottids, and so become intermediate hosts.

proglottids   scolex

Figure 25.17 Animated Life cycle of a beef tapeworm (*Taenia saginata*). Adult worms can grow 7 meters (22 feet) long. The photograph shows the pork tapeworm, *T. solium*.

■ An annelid body is coelomate and segmented; it consists of many repeated units.

■ Link to Fluid pressure 5.6

Annelids (phylum Annelida) are bilateral worms with a coelom and a body that is segmented, inside and out. The majority of the 12,000 or so species are marine worms called polychaetes. The other two groups are oligochaetes (which include earthworms) and leeches. Except in leeches, nearly all segments bear chaetae, or chitin-reinforced bristles. Hence the names polychaete and oligochaete (*poly*–, many; *oligo*–, few).

The Marine Polychaetes

The best known of the polychaetes are the sandworms (*Nereis*) (Figure 25.18*a*). They are often sold as bait for saltwater fishing. These active predators have chitin-strengthened jaws that they use to capture other soft-bodied invertebrates. Each body segment has a pair of paddlelike appendages called parapodia that help the worm burrow in sediments and pursue prey.

Other polychaetes have modifications of this basic body plan. The fan worms and feather duster worms live inside a tube made of secreted mucus and sand grains. The head end protrudes from the tube and its elaborate tentacles capture food that drifts by (Figure 25.18*b*). The worms do not crawl much and have only tiny parapodia.

Leeches—Bloodsuckers and Others

Leeches occur in the ocean, damp habitats on land, and—most commonly—fresh water. Their body lacks conspicuous bristles and has a sucker at either end.

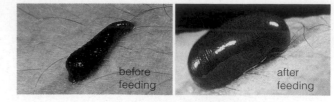

Figure 25.19 The leech *Hirudo medicinalis* feeding on human blood. A leech sticks to the skin with suckers at either end of the body, then draws blood with chitin-hardened jaws.

Many leeches are scavengers and predators of small invertebrates. Others attach to a vertebrate, pierce its skin, and suck blood (Figure 25.19). Their saliva has a protein that keeps blood from clotting while the leech feeds. For this reason, doctors who reattach a severed finger or ear often apply leeches to the reattached part. As they feed, the leeches prevent unwanted clots from forming inside blood vessels of the reattached part.

The Earthworm—An Oligochaete

Oligochaetes include marine and freshwater worms, but the land-dwelling earthworms are most familiar. We consider their body in detail as our example of annelid structure (Figure 25.20).

An earthworm body is segmented inside and out. The outer layer is a cuticle of secreted proteins. Visible grooves on its surface correspond to internal partitions. A fluid-filled coelom runs the length of the body. It is divided into coelomic chambers, one per segment.

Gases are exchanged across the body surface, and a closed circulatory system helps distribute oxygen. Five hearts in the anterior of the worm provide the pumping power that moves the blood.

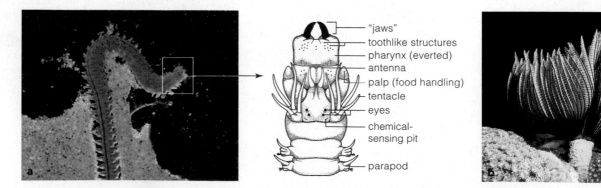

"jaws"
toothlike structures
pharynx (everted)
antenna
palp (food handling)
tentacle
eyes
chemical-sensing pit
parapod

Figure 25.18 Polychaetes. (**a**) The sandworm (*Nereis vexillosa*) burrows into sediment on marine mudflats using its many parapodia. It is an active predator with hard jaws. (**b**) A feather duster worm (*Eudistylia*) lives in a tube and filters food from the water with its tentacles.

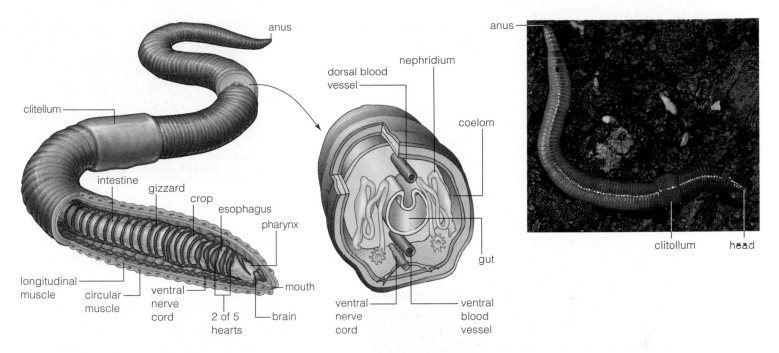

Figure 25.20 **Animated** Earthworm body plan. Each segment contains a coelomic chamber full of organs. A gut, ventral nerve cord, and dorsal and ventral blood vessels run through all coelomic chambers.

A complete digestive system also extends through all coelomic chambers. An earthworm eats its way through the soil, digesting the organic debris in it. The undigested bits are eliminated through the anus. You can buy these earthworm "castings" as a fertilizer.

The solute composition and volume of coelomic fluid is regulated by **nephridia** (singular, nephridium) that occur in nearly all segments. Each nephridium collects coelomic fluid, adjusts its composition, then expels waste through a pore in the next segment.

An earthworm has a rudimentary "brain," a fused pair of ganglia, that coordinates activities. The brain sends signals via a pair of nerve cords. In response to the nervous commands, muscles contract in ways that put pressure on fluid inside coelomic chambers. This fluid is a hydrostatic skeleton.

Two sets of muscles—longitudinal muscles parallel to the body's long axis and circular ones that ring the body—work in opposition. When a segment's longitudinal muscles contract, the segment gets shorter and fatter. When circular muscles contract, a segment gets longer and thinner. Together these two sets of muscles redistribute fluid, causing body segments to change shape in ways that propel the worm (Figure 25.21).

Earthworms are hermaphrodites. A secretory region, the clitellum, produces mucus that glues two worms together while they swap sperm. Later, the clitellum secretes a silky case that encloses the fertilized eggs.

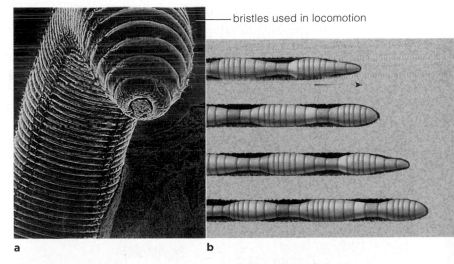

Figure 25.21 How earthworms move through soil. (**a**) Bristles on sides of the body extend and withdraw as muscle contractions act on coelomic fluid inside each segment. (**b**) Bristles are extended when a segment's diameter is at its widest (when circular muscle is relaxed and longitudinal muscle is contracted). They retract as the segment gets long and thin. A worm's front end is pushed forward, then bristles anchor it and the back of the body is pulled up behind it.

Take-Home Message

What are annelids?

■ Annelids are bilateral, coelomate, segmented marine worms, earthworms, and leeches. They have digestive, nervous, excretory, and circulatory systems.

- The ability to secrete a protective shell gave mollusks an advantage over other soft-bodied invertebrates.

- Link to Patterns of development 19.3

General Characteristics

Mollusks (phylum Mollusca) are bilaterally symmetrical invertebrates with a reduced coelom. Most live in seas, but some live in freshwater or on land. All have a **mantle**, a skirtlike extension of the upper body wall that covers a mantle cavity. Aquatic mollusks typically have one or more respiratory organs called **gills** inside their fluid-filled mantle cavity. Cilia on the surface of the gills cause water to flow through the cavity. In mollusks with a shell, the shell is made of a calcium-rich, bonelike material secreted by the mantle.

Many mollusks feed using a **radula**—a tonguelike organ hardened with chitin (*right*). The mollusk digestive system is always complete.

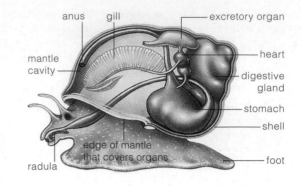

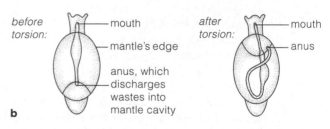

Figure 25.23 Animated Gastropod body plan. (**a**) Body plan of an aquatic snail. (**b**) Torsion, a developmental process unique to gastropods, twists the body relative to the root.

Mollusk Diversity

With more than 100,000 living species, mollusks are second only to arthropods in level of diversity. There are four main classes: chitons, gastropods, bivalves, and cephalopods (Figure 25.22).

Chitons probably are the most like ancestral mollusks. All are marine and have a dorsal shell made of

Figure 25.22 Mollusk groups. (**a**) A chiton with a shell of overlapping plates. (**b**) An aquatic snail, a gastropod, uses its big "foot" to crawl on the glass of an aquarium. (**c**) A bivalve (scallop) with a two-part hinged shell and many eyes (*blue* dots) around the edge of its mantle. (**d**) A squid, a cephalopod.

eight plates (Figure 25.22*a*). Chitons cling to rocks and scrape up algae with their radula. They do not have a distinct head and do not move fast or far. When disturbed, a chiton simply holds on tight and relies on its shell for protection.

With about 60,000 species of snails and slugs, the gastropods are the most diverse mollusks. Their name means "belly foot." Most species glide about on the broad muscular foot that makes up most of the lower body mass (Figures 25.22*b* and 25.23*a*). A gastropod shell, when present, is one-piece and often coiled.

Gastropods have a distinct head that usually has eyes and sensory tentacles. In many aquatic species, a part of the mantle forms an inhalant siphon, a tube through which water is drawn into the mantle cavity. The cone snails discussed in the chapter introduction use the siphon to sniff out their prey. Cone snails are predatory, and their radula is modified as a harpoon, but most gastropods are herbivores.

During development, gastropods undergo a unique rearrangement of body parts called **torsion**. The body mass twists, putting previously posterior parts, including the anus, up above the head (Figure 25.23*b*).

Gastropods include the only terrestrial mollusks. In land-dwelling snails and slugs (Figure 25.24*a*,*b*), a lung replaces the gill. Glands on the foot continually secrete mucus that protects the animal as it moves across dry, abrasive surfaces. Most mollusks have separate sexes, but land-dwellers tend to be hermaphrodites. Unlike

Figure 25.24 Variations on the gastropod body plan. Land snails (**a**) and slugs (**b**) are adapted to life in a dry habitat. They have a lung in place of gills. The slime trail left behind after they move over a surface is mucus secreted by their big foot.

(**c**) Two Spanish shawl nudibranchs (*Flabellina iodinea*). These sea slugs feed on cnidarians and store undischarged nematocysts inside their bright red respiratory organs.

mantle   eye

opening that leads to lung   sensory tentacle

foot   b   c

other mollusks, which produce a swimming larva, the embryos of these groups develop directly into adults.

Slugs and sea slugs lack a shell. They also undergo detorsion: Like other mollusks, they rotate the body early in development. Later, they rotate body parts again, so their anus ends up farther back. Wouldn't it be simpler to just skip torsion? Perhaps, but evolution occurs by small changes that result from random mutations; it does not proceed purposefully.

Lacking a shell, slugs and sea slugs must defend themselves in other ways. Some make and secrete distasteful substances. Certain sea slugs eat cnidarians such as jellyfish and store undischarged nematocysts that serve as a defense. For example, frilly extensions on the back of a Spanish shawl nudibranch function in gas exchange and hold nematocysts (Figure 25.24c).

Bivalves include many of the mollusks that end up on our dinner plates, including mussels, oysters, clams, and scallops (Figure 25.22c). All bivalves have a hinged, two-part shell. Powerful adductor muscles hold the valves together (Figure 25.25). Contraction of these muscles pulls the two valves shut, enclosing the body and protecting it from predation or drying out. Some scallops can "swim" by repeatedly opening and closing their shell. As the shell shuts, the force of the expelled water causes the scallop to scoot backwards. A bivalve has a reduced head, but eyes arrayed around the edge of its mantle alert it to danger.

Bivalves have a large triangular foot commonly used in burrowing. For example, a clam burrows beneath the sand and extends its siphons into the water above. Like other bivalves, it does not have a radula. It feeds by drawing water into its mantle cavity and trapping bits of food in mucus on its gills. Movement of cilia

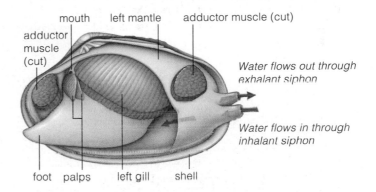

mouth   left mantle   adductor muscle (cut)

adductor muscle (cut)

Water flows out through exhalant siphon

Water flows in through inhalant siphon

foot   palps   left gill   shell

Figure 25.25 Animated Body plan of a clam, a bivalve.

directs particle-laden mucus to the mouth. A pair of labial palps sorts out particles, and sweeps food into the mouth.

The **cephalopods** are the fourth major group. They are all aquatic and include the squids (Figure 25.22d), octopuses, and relatives. Compared to other mollusks, cephalopods are faster, smarter, and generally larger. They are the only mollusks that have a closed circulatory system. We take a closer look at this group in the next section.

Take-Home Message

What are mollusks?

■ Mollusks are invertebrates with a bilateral body plan, a reduced coelom, and a mantle that drapes over their internal organs. In most species, the mantle secretes a protective hardened shell.

■ Most mollusks are aquatic, but some gastropods have adapted to life on land. In addition to gastropods, mollusks include chitons, bivalves, and cephalopods.

25.9 | Cephalopods—Fast and Brainy

■ *Cephalopod* means "head foot," and tentacles attached to the head are evolutionary modifications of the foot. They surround the mouth, which has a hard, horny beak.

Five hundred million years ago, during the Ordovician, cephalopods were the top predators of the open seas (Figure 25.26*a*). All lived inside a shell that had multiple chambers. Except for a few species of nautiluses, their modern descendants have a highly reduced shell or none at all (Figure 25.26*b–d*).

Why were shells reduced? Jawed fishes started an adaptive radiation about 400 million years ago (Section 18.12). Fishes that hunted cephalopods or competed with them for prey became swifter and larger. In what appears to have been a long-term race for speed, most cephalopods lost their external shell. They became streamlined, fast, and surprisingly smart.

For cephalopods, jet propulsion became the name of the game. They moved faster by shooting a jet of water out from the mantle cavity, through a funnel-shaped siphon. All modern cephalopods do the same. The brain controls siphon activity and governs the direction in which the body moves.

Increased speed was accompanied by increasingly complex eyes. Cephalopods, like vertebrates, have an eye with a lens that focuses incoming light. Speed also required change in respiratory and circulatory systems. Of all groups of mollusks, only cephalopods have a closed circulatory system. Blood pumped by the main heart gives up carbon dioxide and picks up oxygen in two gills. Two accessory hearts keep blood moving quickly to and from all body tissues.

Cephalopods include the fastest (squids), biggest (giant squid), and smartest (octopuses) invertebrates. Of all invertebrates, octopuses have the largest brain relative to body size, and show the most complex behavior. Captive octopuses easily learn to navigate mazes or to unscrew the lid of a jar that holds tasty prey.

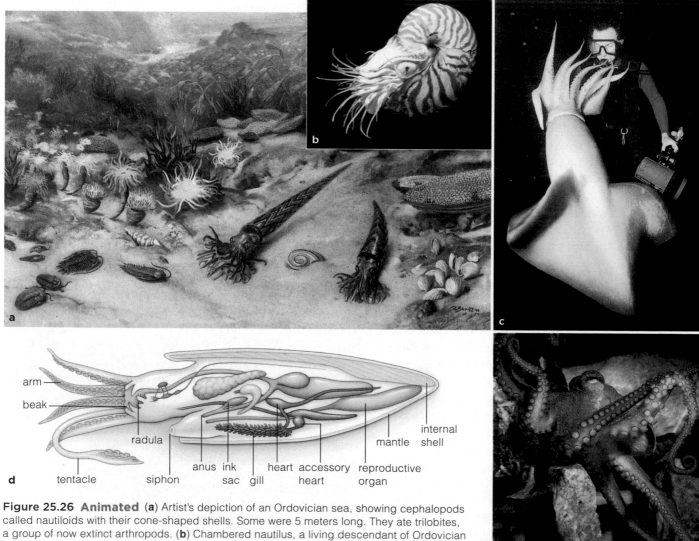

arm
beak
radula
tentacle
anus ink sac
siphon
gill
heart
accessory heart
reproductive organ
mantle
internal shell

Figure 25.26 Animated (**a**) Artist's depiction of an Ordovician sea, showing cephalopods called nautiloids with their cone-shaped shells. Some were 5 meters long. They ate trilobites, a group of now extinct arthropods. (**b**) Chambered nautilus, a living descendant of Ordovician nautiloids. (**c**) A diver with a squid (*Dosidicus*). (**d**) Body plan of a cuttlefish. (**e**) An octopus.

25.10 | Rotifers and Tardigrades—Tiny and Tough

■ Rotifers and tardigrades are among the smallest animals. Their positions on the animal family tree are still in question.

The 2,150 species of **rotifers** (phylum Rotifera) live in freshwater and in damp land habitats. Most are less than one millimeter long. The group name is Latin for "wheel bearer." It refers to the constantly moving cilia on the head, which direct food to the mouth and look like turning wheels (Figure 25.27). There are excretory organs (protonephridia) and a complete digestive system, but no circulatory or respiratory organs.

Digestive and excretory organs are located inside a pseudocoelom. Traditionally, the rotifers and roundworms were grouped together as pseudocoelomates. However, gene comparisons suggest that rotifers are more closely related to annelids and mollusks.

Some rotifers glue themselves to a surface by their toes, but most swim or crawl about. Some species are all female. New individuals develop from unfertilized eggs—a process called parthenogenesis. Other species produce males seasonally or have two sexes.

Tardigrades (phylum Tardigrada) are similarly tiny animals that often live beside rotifers in damp moss and temporary ponds. Commonly called water bears, they waddle about on four pairs of stubby legs (Figure 25.28). "Tardigrada" means slow walker.

About 950 tardigrades have been named. Most suck juices from plants or algae. Some, including the one in Figure 25.28a, are predators. They eat roundworms, rotifers, and one another. The digestive system is complete and there are excretory organs, but no circulatory or respiratory organs. The coelom is reduced.

Like roundworms and insects, tardigrades have an external body covering that they molt (shed periodically) as they grow. The molting and gene sequence data suggest tardigrades belong in Ecdysozoa, but relationships within this group are poorly understood.

Tardigrades and rotifers living in habitats that often dry up completely have evolved a remarkable ability. They can survive dry periods by entering a sort of suspended animation. As the habitat dries up, sugar replaces water in their tissues and metabolism slows to a nearly nonexistent pace. In tardigrades, the water content of the body can drop to 1 percent of normal.

Dormant tardigrades can withstand extraordinary heat and cold. They have survived for a few days at –200°C (–328°F) and a few minutes at 151°C (304°F). Also, a tardigrade can remain dormant for years, then revive within a few hours of being placed in water. For all of these reasons, tardigrades are often said to be the toughest animals.

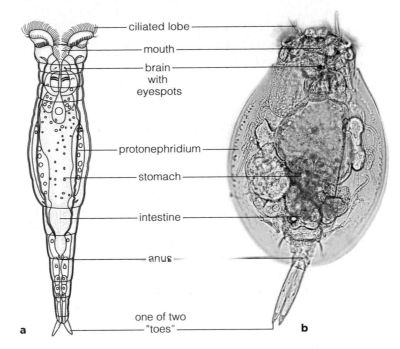

Figure 25.27 (**a**) Body plan of a bdelloid rotifer. (**b**) Micrograph of *Euchlanis*, which secretes a transparent covering around its body.

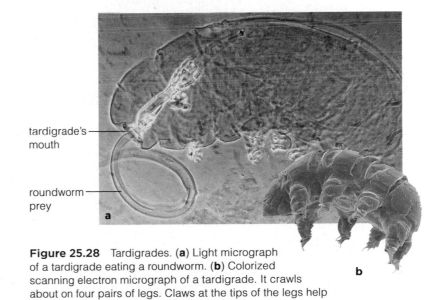

Figure 25.28 Tardigrades. (**a**) Light micrograph of a tardigrade eating a roundworm. (**b**) Colorized scanning electron micrograph of a tardigrade. It crawls about on four pairs of legs. Claws at the tips of the legs help the animal grip surfaces, such as strands of dampened moss.

Take-Home Message

What are rotifers and tardigrades?

■ Rotifers and tardigrades (water bears) are tiny bilateral animals. Most live in damp habitats or fresh water. Some have adapted to an environment that often dries out by evolving the ability to enter a dormant state.

■ Rotifers have a pseudocoelom, but genetic comparisons suggest they are closest to annelids and mollusks. Tardigrades have a coelom and molt; they are probably relatives of the roundworms and insects.

■ Roundworms are among the most abundant animals. A shovelful of rich soil can contain millions.

■ Link to Genomics 16.5

Roundworms, or nematodes (phylum Nematoda), are bilateral, unsegmented worms with a cuticle-covered, cylindrical body (Figure 25.29). A muscular pharynx sucks up food and the digestive system is complete. Nearly all of the 22,000 or so named species are less than 5 millimeters long; but one that lives as a parasite inside sperm whales can be 13 meters long.

Roundworms have a collagen-rich, pliable cuticle that is repeatedly molted as the animal grows. The roundworms were traditionally grouped with rotifers as pseudocoelomates. However, many small roundworms do not have a body cavity. In addition, gene similarities and the shared trait of a cuticle that is molted suggest that roundworms are closer to insects.

The roundworm *Caenorhabditis elegans* is a favorite for genetic experiments. It has the same tissue types as complex animals, but it is transparent, has only 959 body cells, and reproduces fast. Its genome is about 1/30 the size of ours. With such traits, each cell's fate is easy to monitor during development.

Most roundworms feed on organic matter in soil or water, but some are parasites inside plants or animals.

A few parasitic roundworms impair human health. For example, eating undercooked pork or wild game can lead to an infection by *Trichinella spiralis*. The resulting disease, trichinosis, can be fatal. The roundworm moves from the intestines, through the blood, and into muscles, where it forms cysts (Figure 25.30a).

Ascaris lumbricoides, a large roundworm, currently infects more than 1 billion people, primarily in Asia and in Latin America (Figure 25.30b). People become infected when they eat its eggs, which survive in soil and get on hands and into food. When enough adults occupy a host, they can clog the digestive tract.

Hookworms, too, infect more than 1 billion people. Juveniles in the soil cut into human skin and migrate through blood vessels to the lungs. They climb up the windpipe, then enter the digestive tract when the host swallows. Once inside the small intestine, hookworms attach to the intestinal wall and suck blood.

Wuchereria bancrofti and certain other roundworms cause lymphatic filariasis. Repeated infections injure lymph vessels, so lymph pools inside the legs and feet (Figure 25.30c). Elephantiasis, the common name for this disease, refers to fluid-swollen, elephant-like legs. Mosquitoes carry larval roundworms to new hosts.

Pinworms (*Enterobius vermicularis*) commonly infect children. Female worms less than a millimeter long leave the rectum at night and lay eggs near the anus. The migration causes itching, and scratching puts eggs under fingernails. From there, they get into food and onto toys. Swallowing eggs causes a new infection.

pharynx   intestine   eggs in uterus   gonad

false coelom (unlined body cavity)   muscular body wall   anus

Figure 25.29 Animated Body plan and micrograph of *Caenorhabditis elegans*, a free-living roundworm. Sexes are separate and this is a female.

Take-Home Message

What are roundworms?

■ Roundworms are unsegmented, pseudocoelomate worms with a secreted cuticle that is molted. Most roundworms are decomposers, but some are parasites of humans.

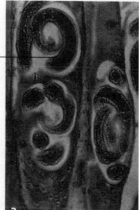
larva in a muscle fiber, longitudinal section

Figure 25.30 (**a**) *Trichinella spiralis* larvae in muscle tissue of a host animal. (**b**) Live roundworms (*Ascaris lumbricoides*). These intestinal parasites cause stomach pain, vomiting, and appendicitis. (**c**) A case of elephantiasis that resulted from an infection by the roundworm *Wuchereria bancrofti*.

25.12 | Arthropods—Animals With Jointed Legs

- There are more than a million species of the jointed legged animals we call arthropods.

- Links to Chiton 3.3, Adaptive traits 17.3

Arthropods (phylum Arthropoda) are bilateral, with a reduced coelom. They have a hard, jointed external skeleton, a complete digestive system, an open circulatory system, and respiratory and excretory organs. If we use number of species as our measure, arthropods are the most successful animals. One main lineage, trilobites, is extinct (Figure 25.6b). Modern subgroups are chelicerates, crustaceans, and myriapods and insects. Table 25.2 provides examples of each group. We survey these groups in sections to follow. Here we begin by thinking about the five key adaptations that continue to contribute to their great evolutionary success.

Key Arthropod Adaptations

Hardened Exoskeleton Arthropods secrete a cuticle of chitin (Section 3.3), proteins, and waxes. It is their **exoskeleton**, a hard, external skeleton. It helps fend off predators, and muscles that attach to it move body parts. Among land arthropods, the exoskeleton helps conserve water and support an animal's weight.

A hardened exoskeleton does not restrict growth, because—like the roundworms—arthropods molt their cuticle after each growth spurt. Hormones regulate molting. They cause the formation of a new cuticle under the old one, which is then shed (Figure 25.31a).

Jointed Appendages If an arthropod's cuticle were uniformly hard and thick like a plaster cast, it would prevent movement. Arthropod cuticles thin at joints: the regions where two hard body parts meet. Body parts move at joints; "arthropod" means jointed leg (Figure 25.31b). Such legs often became modified for specialized tasks.

Highly Modified Segments In early arthropods, body segments were distinct and all body appendages were similar. In many of their descendants, segments became fused into structural units such as a head, a thorax (midsection), and an abdomen (hind section). Appendages became modified for special tasks. For example, among insects, thin extensions of the wall of some segments evolved into wings (Figure 25.31c).

Sensory Specializations Most arthropods have one or more pairs of eyes: organs that sample the visual world. In insects and crustaceans, eyes are compound, with many lenses. With the exception of chelicerates,

Figure 25.31 (a) Centipede molting its old exoskeleton (*gray*). (b) Jointed legs of a crab. (c) A wing attached to the thorax of a fly. (d) Larva of a monarch butterfly, a specialized stage that feeds on plant leaves.

Table 25.2 Living Arthropod Subgroups

Group	Representatives	Named Species
Chelicerates	Horseshoe crabs	4
	Arachnids (scorpions, spiders, ticks, mites)	70,000
Crustaceans	Crabs, shrimp, lobsters, barnacles, pill bugs	42,000
Myriapods	Millipedes and centipedes	2,800
Insects	Beetles, ants, butterflies, flies	>1 million

most arthropods also have paired **antennae** that can detect touch and waterborne or airborne chemicals.

Specialized Developmental Stages The body plan of many arthropods changes during the life cycle. Individuals often undergo **metamorphosis**: Tissues get remodeled as juveniles become adults. Each stage is specialized for a different task. For instance, wingless, plant-eating caterpillars metamorphose into winged butterflies that disperse and find mates (Figure 25.31d). Having such different bodies also prevents adults and juveniles from competing for the same resources.

Take-Home Message

What are arthropods?

- Arthropods are the most diverse animal phylum. A jointed exoskeleton, a segmented body plan with specialized segments, sensory specializations, and a life cycle that often includes metamorphosis contributed to their success.

- Trilobites are an extinct arthropod group. Modern arthropod groups include horseshoe crabs, spiders, ticks, crabs, lobsters, centipedes, and insects.

■ Chelicerates include the oldest living arthropod lineage (horseshoe crabs) and other arthropods without antennae.

■ Link to Disease-causing bacteria 21.6

Chelicerates include horseshoe crabs, scorpions, spiders, ticks, and mites (Figure 25.32). The body has a cephalothorax (fused head and thorax) and abdomen. There are four pairs of walking legs. The head has eyes, but no antennae. Near the mouth are paired feeding appendages called chelicerae and pedipalps.

Horseshoe crabs live in seas, eat clams and worms, and have a hard shield over the cephalothorax (Figure 25.32*a*). A spinelike segment (the telson) acts as a rudder when they swim. Horseshoe crab eggs, laid onshore in spring, are essential food for some migratory birds.

All land chelicerates, including spiders, scorpions, ticks, and mites, are arachnids. Scorpions and spiders are predators that subdue prey with venom. Scorpions dispense venom through a stinger on the telson (Figure 25.32*b*). Spiders deliver venom with a bite. Their fang-like chelicerae have poison glands (Figures 25.32*c,d* and 25.33). Of 38,000 spider species, about 30 produce venom that can harm humans. Most spiders indirectly help us by eating insect pests.

The spider abdomen has paired spinners that eject silk for webs and nests. An open circulatory system allows blood to mingle with tissue fluids. **Malpighian tubules** move excess water and nitrogen-rich wastes from the tissues to the gut for disposal. In many species, gas exchange occurs at leaflike "book lungs."

All ticks suck blood from vertebrates (Figure 25.32*e*). Some can transmit bacteria that cause Lyme disease or other diseases (Section 21.6). The 45,000 or so species of mites include parasites, predators, and scavengers (Figure 25.32*f*). Most are less than a millimeter long.

Take-Home Message

What are chelicerates?

■ Chelicerates are arthropods that do not have antennae. Horseshoe crabs are a small, marine lineage. The far more diverse arachnids live mostly on land.

Figure 25.32 Chelicerates. (**a**) A horseshoe crab (*Limulus*). All horseshoe crabs are marine. They are the closest living relatives of the extinct trilobites.

Members of the arachnid subgroup: (**b**) A scorpion. Some scorpion stings can be fatal to humans. (**c**) A jumping spider. It does not make a web. It pounces on its prey. (**d**) A web-weaving black widow spider (*Latrodectus*) has venom that can be fatal to humans. Only the females bite. They have a red hourglass marking on their abdomen. (**e**) Tick swollen after a blood meal. (**f**) A dust mite.

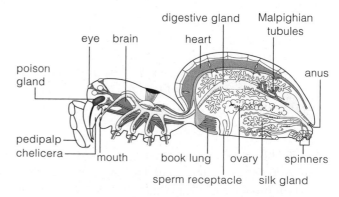

Figure 25.33 Body plan of a spider.

25.14 The Mostly Marine Crustaceans

■ Crustaceans are the only arthropod group that is mainly aquatic. Most crustaceans live in the seas.

Crustaceans are a group of mostly marine arthropods with two pairs of antennae. Some live in freshwater. A few such as wood lice (Figure 25.34a) live on land.

Small crustaceans reach great numbers in the seas and are an important food source for larger animals. Krill (euphausids) have a shrimplike body a few centimers long and swim in upper ocean waters (Figure 25.34b). Most copepods are also marine zooplankton, but others live in freshwater (Figure 25.34c). Some copepod parasites of fish or whales can be large; a few of these are as long as your forearm.

Larval barnacles swim, but adults are enclosed in a calcified shell and live attached to piers, rocks, and even whales (Figure 25.34d). They filter food from the water with feathery legs. As adults, they cannot move about, so you would think that mating might be tricky. But barnacles tend to settle in groups, and most are hermaphrodites. An individual extends a penis, often several times its body length, out to neighbors.

Lobsters, crayfish, crabs, and shrimps all belong to the same crustacean subgroup (the decapods). All are bottom feeders with five pairs of walking legs (Figure 25.35). In some lobsters, crayfish, and crabs, the first pair of legs has become modified into a pair of claws.

Like all arthropods, crabs molt as they grow (Figure 25.36). Some spider crabs do quite a bit of growing. With legs that can reach more than a meter in length, these crabs are the largest living arthropods.

Take-Home Message

What are crustaceans?

■ Crustaceans are mostly marine arthropods that have two pairs of antennae. They are ecologically important as a food source and include the largest living arthropods.

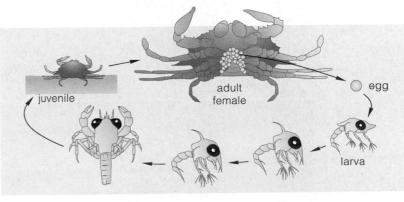

Figure 25.34 Representative crustaceans. (**a**) A wood louse (also known as a pill bug, sow bug, or roly-poly) is a scavenger on land. (**b**) Antarctic krill (*Euphausia superba*) can be up to 6 centimeters long. Populations can reach densities of 10,000 individuals per cubic meter of seawater. (**c**) A free-living female copepod (*Macrocyclops albidus*) from the Great Lakes is about one millimeter long. (**d**) Goose barnacle. Adults cement themselves to one spot head down, and filter food from the water with feathery jointed legs.

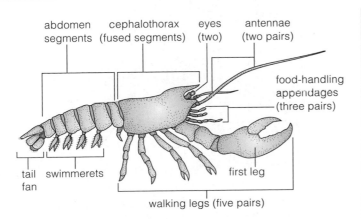

Figure 25.35 Body plan of a lobster (*Homarus americanus*).

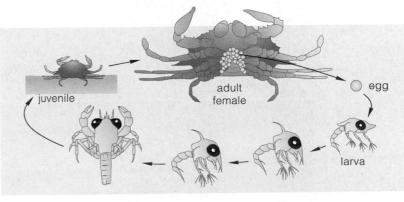

Figure 25.36 Animated Crab life cycle. Larval and juvenile stages molt repeatedly and grow in size before they are mature adults. Adults continue to molt. A female carries her fertilized eggs under her abdomen until they hatch.

■ Centipedes and millipedes use their many legs to walk about on land, hunting prey or scavenging.

Myriapod means "many feet," and aptly describes the centipedes and millipedes. Both have a long body with many similar segments (Figure 25.37). The head has a pair of antennae and two simple eyes. Myriapods are ground-dwellers that move about at night and hide under rocks and leaves during the day.

Centipedes have a low-slung, flattened body with a single pair of legs per segment, for a total of 30 to 50. They are fast-moving predators. Their first pair of legs has become modified as fangs that inject paralyzing venom. Most centipedes prey on insects, but some big tropical species eat small vertebrates (Figure 25.37a).

Millipedes are slower moving animals that feed on decaying vegetation. Their body is rounded and has two pairs of legs on most segments, with 250 or so pairs in total (Figure 25.37b).

Take-Home Message

What are myriapods?

■ Myriapods are land-dwelling arthropods with two antennae and an abundance of body segments. Centipedes are predators and millipedes are scavengers.

■ Arthropods are the most successful animal phylum, and insects are the most successful arthropods.

■ Link to Genomics 16.5

Insect Characteristics

With more than a million species, insects are the most diverse arthropod group. They are also breathtakingly abundant. By some estimates, the ants alone make up about 10 percent of the world's animal biomass (the total weight of all living animals).

Insects have a three-part body plan, with a head, thorax, and abdomen (Figure 25.38). The head has one pair of antennae and two compound eyes. Such eyes consist of many individual units, each with a lens. Near the mouth are jawlike mandibles and other feeding appendages. Insects feed in a variety of ways and their mouthparts reflect their speciality (Figure 25.39).

Three pairs of legs extend out from the insect thorax. In some groups, the thorax also has one or two pairs of wings. Insects are the only winged invertebrates.

A few insects spend some time in the water, but the group is overwhelmingly terrestrial. A respiratory system consisting of tracheal tubes carries air from openings at the body surface to tissues deep inside the body. Like all other arthropods, insects have an open circulatory system.

Figure 25.37 (**a**) A Southeast Asian centipede feeds on its frog prey. The false antennae on the last segment may prevent predators from attacking the centipede's head. (**b**) A millipede.

— abdomen

thorax with six legs

head with two eyes, and two antennae

Figure 25.38 A bed bug (*Cimex lectularius*) illustrates the basic insect body plan: a head, thorax, and abdomen. The bug is 7 millimeters long and feeds on human blood.

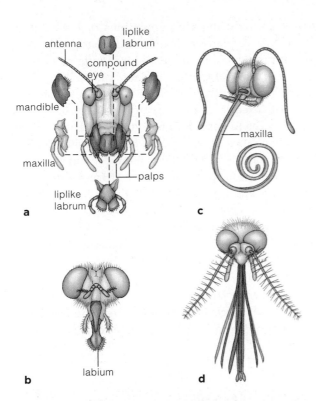

antenna

liplike
labrum

compound
eye

mandible

maxilla

liplike
labrum

palps

a

b labium

maxilla

c

d

Figure 25.39 Animated Examples of insect appendages. Head parts of (**a**) grasshoppers, which chew fibrous plant parts; (**b**) flies, which sop up nutrients; (**c**) butterflies, which siphon nectar from flowering plants; and (**d**) mosquitoes, which pierce hosts and suck up blood.

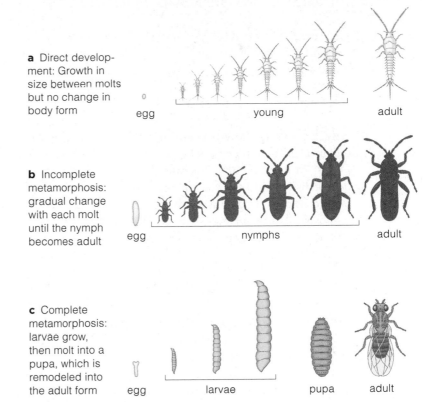

a Direct development: Growth in size between molts but no change in body form

egg   young   adult

b Incomplete metamorphosis: gradual change with each molt until the nymph becomes adult

egg   nymphs   adult

c Complete metamorphosis: larvae grow, then molt into a pupa, which is remodeled into the adult form

egg   larvae   pupa   adult

Figure 25.40 Insect development. (**a**) Silverfish show direct development. Young simply change in size with each molt. (**b**) Bugs, including bedbugs, undergo *incomplete* metamorphosis. Small changes occur with each molt. (**c**) Fruit flies show *complete* metamorphosis. A larva develops into a pupa, which is remodeled into an adult.

Insects have a complete digestive system divided into a foregut, a midgut where food is digested, and a hindgut, where water is reabsorbed. As in spiders and other land-dwelling arthropods, Malpighian tubules inside the abdomen function in excretion. Nitrogen-rich wastes produced by digestion of proteins diffuse from blood into these tubes. There, enzymes convert the waste to crystals of uric acid, which an insect excretes. Malpighian tubules help insects eliminate toxic metabolic wastes without losing precious water.

An insect abdomen also contains sex organs. Sexes are separate. Depending on the group, a fertilized egg either hatches into a small version of the adult, or a juvenile that will later undergo **metamorphosis**. During metamorphosis, tissues of a juvenile are reorganized (Figure 25.40). Incomplete metamorphosis means that the changes in body form take place a bit at a time. Juveniles called nymphs, change a little with each molt. Complete metamorphosis is more dramatic. In this case, the juvenile, called a larva, grows and molts with no change in body plan. Then it is transformed into a pupa, which undergoes the tissue remodeling that produces the adult.

Insect Origins

Until recently, insects were thought to be close relatives of the myriapods. Both groups have a single pair of antennae and unbranched legs. Then—as we have seen so many times—new information made scientists rethink the connections. The current hypothesis holds that the insects are most closely related to crustaceans. Specifically, insects are thought to be descended from freshwater crustaceans, with silverfish (Figure 25.40*a*) as an early insect lineage. If this hypothesis is correct, then insects are the crustaceans of the land.

Take-Home Message

What are insects?

■ Insects are the most diverse and abundant animals. They have a three-part body plan. The head has compound eyes, a pair of antennae, and specialized mouthparts. A thorax has three pairs of legs and, in some lineages, wings.

■ Insects are adapted to life on land. A system of tracheal tubes delivers air to their tissues. Malpighian tubules in their abdomen allow them to expel waste while minimizing water loss.

■ By the most recent hypothesis, insects evolved from a crustacean lineage.

■ It would be hard to overestimate the importance of insects, for either good or ill.

■ Links to Flagellate protozoans 22.2, Malaria 22.6

A Sampling of Insect Diversity

Again, insects show tremendous diversity, with more than a million species. The representatives in Figure 25.41 provide a glimpse of the variety. Of these, only silverfish (Figure 25.41a) undergo direct development. In addition to bugs, insects with incomplete metamorphosis include earwigs, lice, cicadas, damselflies, termites, and grasshoppers. Earwigs are scavengers that have a flattened body (Figure 25.41b). Like silverfish, they sometimes end up in our basements and garages. Lice are wingless and suck blood from warm-blooded animals (Figure 25.41c). Cicadas (Figure 25.41d) and the related leafhoppers and aphids, are winged and suck juices from plants. Damselflies (Figure 25.41e) and the related dragonflies are agile aerial predators of other insects. Termites live in big family groups. They have prokaryotic and protistan symbionts in their gut that allow them to digest wood (Figure 25.41f). They are unwelcome when they devour buildings or decks, but are important decomposers. Grasshoppers cannot eat wood, but they do chew their way through tough, nonwoody plant parts (Figure 25.41g).

The four most successful insect lineages all have wings and undergo complete metamorphosis. There are approximately 150,000 species of flies, or dipterans (Figure 25.41h), and at least as many beetles, or coleopterans (Figure 25.41i,j). The wasp in Figure 25.41k is one of about 130,000 hymenopterans. This group also includes the bees and ants. Lepidopterans—moths and butterflies (Figure 25.41l)—weigh in with about 120,000 species. As a comparison, consider that there are about 4,500 species of mammals.

Ecological Services

As you learned in Section 23.8, the flowering plants coevolved with insect pollinators. The vast majority of these plants are pollinated by members of one of the four most successful insect groups. The other groups contain few or no pollinators. By one hypothesis, the close interactions between pollinating insect groups and flowering plants contributed to an increased rate of speciation in both.

Today, declines in populations of insect pollinators are a matter of concern. Development of natural areas, use of pesticides, and the spread of newly introduced diseases are reducing the populations of insects that pollinate native plants and agricultural crops. We discuss this problem in more detail in Chapter 30.

Insects are also important as food for wildlife. Most songbirds nourish their nestlings on a diet consisting largely of insects. Many migratory songbirds travel long distances to nest and raise young in areas where insect abundance is seasonally high. Aquatic larvae of insects such as dragonflies, mayflies, and mosquitoes serve as food for trout and other freshwater fish. Amphibians and reptiles feed mainly on insects. Even humans eat insects. In many cultures, they are considered a tasty source of protein.

Insects dispose of wastes and remains. Flies and beetles are quick to discover an animal corpse or a pile of feces. They lay their eggs in or on this organic material, and the larvae that hatch devour it. By their actions, these insects keep organic wastes and remains from piling up, and help distribute nutrients through the ecosystem.

Competitors for Crops

Insects are our main competitors for food and other plant products. It is estimated that about a quarter to a third of all crops grown in the United States are lost to insects. Also, in an age of global trade and travel, we have more than just home-grown pests to worry about. Consider the Mediterranean fruit fly (Figure 25.41h). The Med fly, as it is known, lays eggs in citrus and other fruits, as well as many vegetables. Damage done to plants and fruits by larvae of the Med fly can cut crop yield in half. Med flies are not native to the United States and there is an ongoing inspection program for imported produce, but some Med flies still slip in. So far, eradication efforts have been successful, but they have cost hundreds of millions of dollars. Still, this amount is probably a bargain. If the Med fly were to become permanently established, losses would likely run into the billions.

Vectors for Disease

What is the deadliest animal? It may be the mosquito. As you learned earlier, certain mosquito species transmit malaria, which kills more than a million people each year (Section 22.6). Mosquitoes are also vectors for viruses and roundworms that cause disease. Other biting insects can spread other pathogens. Biting flies transmit African sleeping sickness; biting bugs spread

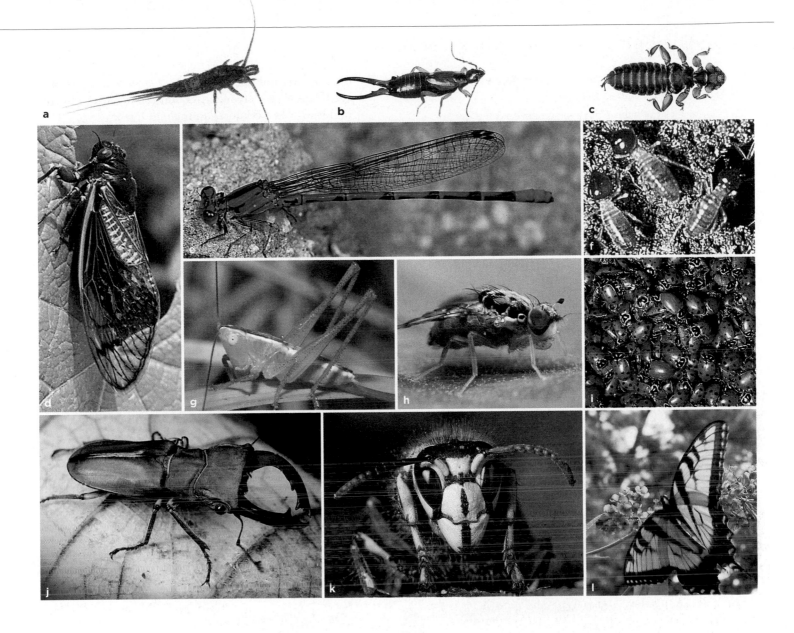

Chagas disease (Section 22.2). Fleas that bite rats and then bite humans can transmit bubonic plague. Body lice can transmit typhus.

So far as we know, bedbugs like the one in Figure 25.38 do not cause disease. However, a heavy bedbug infestation can cause weakness as a result of blood loss, especially in children.

Take-Home Message

What effect do insects have?

■ There are many groups of insects. The four most diverse groups all include members that are pollinators of flowering plants. As such, insects help provide us with food crops. Insects also have important ecological roles as food for animals and as agents of waste disposal.

■ A small number of insect species compete with us for crops or carry pathogens.

Figure 25.41 A sampling of insect diversity. (**a**) One of the silverfish, the only insect group with direct development.

Insects with incomplete metamorphosis: (**b**) European earwig, a common household pest. Curved pincers at the tail end indicate this is a male. In females, pincers are straight. (**c**) Duck louse. It eats bits of feathers and skin. (**d**) A cicada. Male cicadas are among the loudest of all insects. They have specialized sound-producing organs that they use to attract females. (**e**) A damselfly, one of the insects that has aquatic larvae. (**f**) Sterile soldier termites with glue-squirting heads ready to defend their colony. (**g**) A grasshopper.

Members of the four most diverse groups. All are winged and undergo complete metamorphosis. (**h**) Mediterranean fruit fly. Larvae of this insect destroy citrus fruit and other crops. (**i**) Ladybird beetles with a distinctive red and black spotted wing cover. (**j**) Staghorn beetle from New Guinea. Males, such as this one, have huge mandibles. Females have smaller mandibles. (**k**) A bald-faced hornet, a wasp, is a hymenopteran. This is a fertile female, or queen. She lives in a papery nest with her many offspring. (**l**) A swallowtail butterfly, a lovely lepidopteran, shown acting as pollinator.

- Echinoderms begin life as bilateral larvae and develop into spiny-skinned, radial adults.

- Link to Patterns of development 19.3

The Protostome–Deuterostome Split

In Section 25.1 we introduced the two major lineages of animals, protostomes and deuterostomes. Thus far, all of the animals with a three-layer embryo that we have discussed—from flatworms to arthropods—have been protostomes. This section begins our survey of deuterostome lineages. Echinoderms are the largest group of invertebrate deuterostomes. We will discuss other invertebrate deuterostomes, and the vertebrates (also deuterostomes), in the next chapter.

Echinoderm Characteristics and Body Plan

Echinoderms (phylum Echinodermata) include about 6,000 marine invertebrates. Their name means "spiny-skinned" and refers to interlocking spines and plates of calcium carbonate embedded in their skin. Adults have a radial body plan, with five parts (or multiples

of five), around a central axis. The larvae, however, are bilateral, which suggests that the ancestor of echinoderms was a bilateral animal.

Sea stars (also called starfish) are the most familiar echinoderms, and we will use them as our example of the echinoderm body plan (Figure 25.42). Sea stars do not have a brain, but they do have a decentralized nervous system. Eye spots at the tips of arms detect light and movement.

A typical sea star is an active predator that moves about on tiny, fluid-filled tube feet. Tube feet are part of a **water–vascular system** unique to echinoderms. Figure 25.42*a* shows the system of fluid-filled canals in each arm of a sea star. Side canals deliver coelomic fluid into muscular ampullae that function like the bulb on a medicine dropper (Figure 25.42*b*). Contraction of an ampulla forces fluid into the attached tube foot, extending the foot. A sea star glides along smoothly as coordinated contraction and relaxation of the ampullae redistributes fluid among hundreds of tube feet.

Sea stars often feed on bivalve mollusks. They can slide their stomach out through their mouth and into the bivalve's shell. The stomach secretes acid and

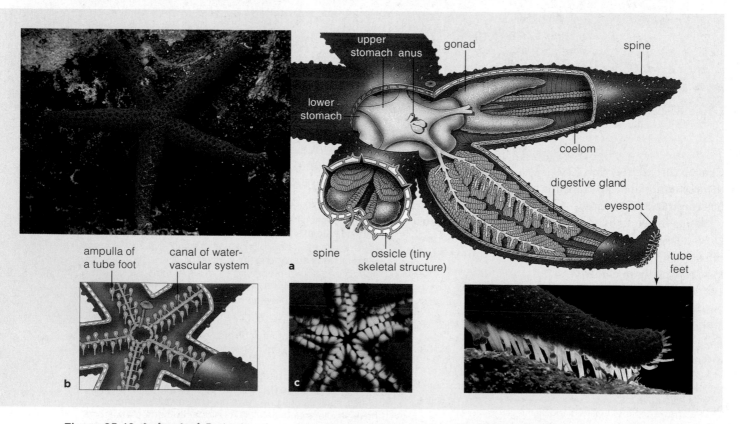

Figure 25.42 Animated Body plan of a sea star. (**a**) Major components of the central body and the radial arms, with a close-up of its little tube feet. (**b**) Organization of the water–vascular system. In combination with many tube feet, it is the basis of locomotion. (**c**) A sea star's toothy feeding apparatus.

enzymes that kill the mollusk and begin to digest it. Partially digested food is taken into the stomach and digestion is completed with the aid of digestive glands in the arms.

Gas exchange occurs by diffusion across the tube feet and tiny skin projections at the body surface. There are no specialized excretory organs.

Sexes are separate. Either male or female gonads are in the arms. Eggs and sperm are released into the water. Fertilization produces an embryo that develops into a ciliated, bilateral larva. The larva swims about and develops into the adult, nonswimming form.

Sea stars and other echinoderms have a remarkable ability to regenerate lost body parts. If a sea star is cut into pieces, any portion with some of the central disk can regrow the missing body parts.

Echinoderm Diversity

Brittle stars are the most diverse and abundant echinoderms (Figure 25.43a). They are less familiar than sea stars because they generally live in deeper water. They have a central disk and highly flexible arms that move about in a snakelike way. Most brittle stars are scavengers on the sea floor.

In sea urchins, calcium carbonate plates form a stiff, rounded cover from which spines protrude (Figure 25.43b). The spines provide protection and are used in movement. Some urchins graze on algae. Others act as scavengers or prey on invertebrates. Sea urchin roe (eggs) are used in some sushi. Overharvesting for markets in Asia threatens species that produce the most highly prized roe.

In sea cucumbers, hardened parts have been reduced to microscopic plates embedded in a soft body. Some species such as the one in Figure 25.43c filter food from the seawater. Others have a wormlike body and, like earthworms, they feed by eating their way through sediments and digesting any organic material.

Lacking spines or sharp plates, sea cucumbers have an alternative defense. When threatened, they expel a sticky mass of specialized threads and internal organs out through their anus. If this maneuver successfully distracts the predator, the sea cucumber escapes, and its missing parts grow back.

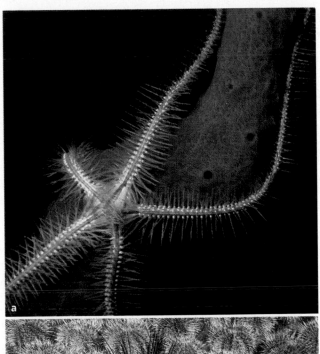

Take-Home Message

What are echinoderms?

■ Echinoderms are deuterostome invertebrates that have a radial body as adults. They are brainless and have a unique water–vascular system that functions in locomotion.

Figure 25.43 (a) Brittle star. Its slender arms (rays) make fast, snakelike movements. (b) Underwater "forest" of sea urchins, which can move about on spines and a few tube feet. (c) Sea cucumber, with rows of tube feet along its soft body.

Marine invertebrates are important components of ecosystems, a source of food, and a treasure trove of molecules with potential for use in industrial applications or as medicines. Various species of cone snails, sponges, corals, crabs, and sea cucumbers make compounds that show promise as drugs. However, even as we begin to explore this potential, marine biodiversity is on the decline as a result of habitat destruction and overharvesting.

How would you vote?

Bottom trawling helps keep seafood prices low, but can destroy invertebrate habitats. Should it be banned? See CengageNOW for details, then vote online.

Summary

Section 25.1 **Animals** are multicelled heterotrophs with unwalled cells. Some animals have no body symmetry or have **radial symmetry**, like a wheel. Most have **bilateral symmetry** and show **cephalization**, a concentration of nerves and sensory structures at the head end. Most digest food in a **gut**. The gut may be surrounded by tissues or inside a fluid-filled cavity. The cavity may be a fully lined **coelom**, or a partially lined **pseudocoel**.

Two major branches of bilateral animals, protostomes and deuterostomes, have a coelom and a complete gut. In **protostomes**, the first opening on the embryo becomes a mouth. In **deuterostomes**, an anus forms first.

- *Use the animation on CengageNOW to familiarize yourself with terms necessary to understand animal body plans.*

Sections 25.2, 25.3 Animals most likely evolved from a colony similar to **choanoflagellates**, a type of protist. **Placozoans** are the structurally simplest modern animals.

The oldest animal fossils, called the Ediacarans, date back about 600 million years. A great adaptive radiation during the Cambrian gave rise to most modern lineages.

Relationships among animal groups are still being investigated. For example, recent genetic studies suggest that all invertebrates that **molt** are closely related.

Section 25.4 **Sponges** are asymmetrical and do not have tissues or organs. They filter food from water and are **hermaphrodites**: each makes eggs and sperm. Adults stay put, but immature forms, called **larvae**, swim.

- *Use the animation on CengageNOW to explore the body plan of a sponge.*

Section 25.5 **Cnidarians**, such as jellyfishes, corals, and sea anemones, are radially symmetrical. They alone make **nematocysts**, which they use to catch prey and to defend themselves. They have two tissues with a jellylike layer that functions as a **hydrostatic skeleton** between them. A **nerve net** controls movements. A gastrovascular cavity functions in both respiration and digestion.

- *Use the animation on CengageNOW to compare cnidarian body plans and life cycles.*

Section 25.6 **Flatworms**, such as planarians, are bilateral protostomes and the simplest animals to have organ systems. **Nerve cords** connect to **ganglia** in the head that serve as a control center. The gut is saclike and a **pharynx** takes in food and expels waste. Tapeworms are parasitic flatworms with a body made of units called **proglottids**. Flukes are also parasites.

- *Use the animation on CengageNOW to learn about flatworm organ systems and life cycles.*

Section 25.7 **Annelids** are segmented worms (such as earthworms and polychaetes) and leeches. Circulatory, digestive, solute-regulating, and nervous systems extend through all coelomic chambers. **Nephridia** regulate the composition of body fluid.

- *Use the animation on CengageNOW to investigate the body plan of an earthworm.*

Sections 25.8, 25.9 **Mollusks** have a sheetlike **mantle**. Most have respiratory **gills** in the mantle cavity and feed using a food-scraping **radula**. Examples are **chitons**; **gastropods** (such as snails) which undergo **torsion**; **bivalves** (such as clams); and **cephalopods**.

- *Use the animation on CengageNOW to compare molluscan body plans.*

Section 25.10 **Rotifers** and **tardigrades** are tiny animals of damp or aquatic habitats. Rotifers have a ciliated head and a pseudocoelom. Tardigrades, or water bears, have a reduced coelom and molt. Both groups can dry out and survive long periods of adverse conditions.

Section 25.11 The **roundworms** (nematodes) have an unsegmented body, a cuticle that is molted, a complete gut, and a false coelom. Some are parasites of humans.

- *Use the animation on CengageNOW to learn about the roundworm body plan.*

Sections 25.12–25.17 **Arthropods**, the largest phylum of animals, have a jointed **exoskeleton**, or external skeleton. Most have one or more pairs of sensory **antennae**. **Malpighian tubules** expel waste in land-dwelling groups.

Chelicerates include the marine horseshoe crabs and the arachnids (spiders, scorpions, ticks, and mites). The mostly marine **crustaceans** include wood lice, crabs, lobsters, barnacles, krill, and copepods. **Myriapods** are predatory centipedes and scavenging millipedes. Insects, the most successful arthropods, include the only winged invertebrates. Most insects undergo **metamorphosis**, a change in body form between larval and adult stages. Insects pollinate plants, dispose of wastes, and serve as food, but some eat crops or transmit disease.

- *Use the animation on CengageNOW to learn about arthropod life cycles and body plans.*

Data Analysis Exercise

Atlantic horseshoe crabs, *Limulus polyphemus,* have long been ecologically important. For more than a million years, their eggs have fed migratory shorebirds. More recently, people began to harvest horseshoe crabs for use as bait. More recently still, people started using horseshoe crab blood to test injectable drugs for potentially deadly bacterial toxins. To keep horseshoe crab populations stable, blood is extracted from captured animals, which are then returned to the wild. Concerns about the survival of animals after bleeding led researchers to do an experiment. They compared survival of animals captured and maintained in a tank with that of animals captured, bled, and kept in a similar tank. Figure 25.44 shows the results.

1. In which trial did the most control crabs die? In which did the most bled crabs die?

2. Looking at the overall results, how did the mortality of the two groups differ?

3. Based on these results, would you conclude that bleeding harms horseshoe crabs more than capture alone does?

| | Control Animals | | Bled Animals | |
Trial	Number of crabs	Number that died	Number of crabs	Number that died
1	10	0	10	0
2	10	0	10	3
3	30	0	30	0
4	30	0	30	0
5	30	1	30	6
6	30	0	30	0
7	30	0	30	2
8	30	0	30	5
Total	200	1	200	16

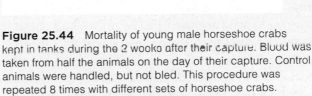

Figure 25.44 Mortality of young male horseshoe crabs kept in tanks during the 2 weeks after their capture. Blood was taken from half the animals on the day of their capture. Control animals were handled, but not bled. This procedure was repeated 8 times with different sets of horseshoe crabs.

Section 25.18 Echinoderms, such as sea stars, are invertebrate members of the deuterostome lineage. They have skin with spines, spicules, or plates of calcium carbonate. A **water–vascular system** with tube feet helps most glide about. Adults are radial, but bilateral ancestry is evident in their larval stages and other features.

■ *Use the animation on CengageNOW to examine a sea star body plan and observe tube feet in action.*

Self-Quiz   *Answers in Appendix III*

1. True or false? Animal cells do not have walls.

2. A body cavity fully lined with tissue derived from mesoderm is a _____ .

3. The modern protist group most closely related to animals is the _____ .

4. A _____ filters food from the water and has no tissues or organs.
   a. sponge   c. cnidarian
   b. roundworm   d. flatworm

5. Cnidarians alone have _____ .
   a. nematocysts   c. a hydrostatic skeleton
   b. a mantle   d. Malpighian tubules

6. Flukes are most closely related to _____ .
   a. tapeworms   c. arthropods
   b. roundworms   d. echinoderms

7. Nephridia have the same functional role as _____ .
   a. gemmules of sponges   c. flame cells of planarians
   b. mandibles of insects d. tube feet of echinoderms

8. Which invertebrate phylum includes the most species?
   a. mollusks   c. arthropods
   b. roundworms   d. flatworms

9. A radula is used to _____ .
   a. detect light   c. produce silk
   b. scrape up food   d. eliminate excess water

10. Barnacles are shelled _____ .
    a. gastropods   c. crustaceans
    b. cephalopods   d. copepods

11. The _____ include the only winged invertebrates.
    a. cnidarians   c. arthropods
    b. echinoderms   d. placozoans

12. The _____ have a coelom and are radial as adults.

13. Match the organisms with their descriptions.
    ___choanoflagellates   a. complete gut, pseudocoelom
    ___placozoan   b. sister group to animals
    ___sponges   c. simplest organ systems
    ___cnidarians   d. no tissues, filters out food
    ___flatworms   e. jointed exoskeleton
    ___roundworms   f. mantle over body mass
    ___annelids   g. segmented worms
    ___arthropods   h. tube feet, spiny skin
    ___mollusks   i. nematocyst producers
    ___echinoderms   j. simplest known animal

■ *Visit CengageNOW for additional questions.*

Critical Thinking

1. Many different species of flatworms, roundworms, and annelids are parasites of mammals. There are no such parasites among sponges, cnidarians, mollusks, and echinoderms. Propose a plausible explanation for this difference.

2. A massive die-off of lobsters in the Long Island Sound was blamed on pesticides sprayed to control the mosquitoes that carry West Nile virus. Why might a chemical designed to kill insects also harm lobsters?

Animal Evolution—The Chordates

Transitions Written in Stone

By Charles Darwin's time, all major groups of organisms had been identified. One objection to acceptance of Darwin's theory of evolution by natural selection was the apparent lack of transitional forms between groups. If new species evolve from older ones, then where were the "missing links," species with traits intermediate between two groups?

In fact, workmen at a limestone quarry in Germany had already unearthed one such link. The pigeon-sized fossil resembled a small dinosaur. It had teeth, three long clawed fingers on a pair of forelimbs, and a long bony tail. Later, diggers found another specimen. Later still, someone noticed feathers. If they were fossilized birds, then why did they have teeth and a bony tail? If dinosaurs, what were they doing with feathers? The specimen was named *Archaeopteryx*, meaning ancient winged one (Figure 26.1*a*).

So far, a total of eight *Archaeopteryx* fossils have been excavated, all from German limestone. Radiometric dating (Section 17.6) revealed that *Archaeopteryx* lived about 150 million years ago, in the late Jurassic. What is now limestone was once sediments in a shallow lagoon near the shore of

the supercontinent Pangea. When bodies of organisms fell into this lagoon, fine sediments quickly covered them. Over time, the sediments compacted and hardened. They became a stony tomb for more than 600 species, including marine invertebrates, dinosaurs, and *Archaeopteryx*.

No human witnessed the transitions that led to modern animal diversity. However, fossils are physical evidence of changes, and radiometric dating assigns the fossils to places in time. The structure, biochemistry, and gene sequences of living organisms provide information about branchings.

The theory of evolution by natural selection provides the best explanation for the observed genetic similarities and differences between species and for the transitional forms we observe in the fossil record. Evolutionists often argue over how to interpret data and which of the known mechanisms can best explain life's history. At the same time, they eagerly look to new evidence to support or disprove hypotheses. As you will see, fossils and other evidence form the foundation for this chapter's account of vertebrate evolution, including the story of our own origins.

See the video! Figure 26.1 Placing *Archaeopteryx* in time. (**a**) One of the *Archaeopteryx* fossils from Germany. It clearly shows feathers, a long bony tail, and teeth. No modern bird has a bony tail or teeth. (**b**) Painting based on fossils of plants and animals that lived in a Jurassic forest. Foreground, two gliding *Archaeopteryx*. Behind them a huge herbivorous *Apatosaurus* (a herbivore) is pursued by *Saurophaganax* ("king of reptile eaters"). In the distant background are *Camptosaurus* (*left*) and *Stegosaurus* (*right*).

Key Concepts

Characteristics of chordates

A unique set of four traits characterizes the chordates: a supporting rod (notochord); a hollow, dorsal nerve cord; a pharynx with gill slits in the wall; and a tail extending past an anus. Certain invertebrates and all vertebrates belong to this group. **Section 26.1**

Trends among vertebrates

In vertebrate lineages, a backbone replaced the notochord. Jaws and fins evolved in water. Fleshy fins with skeletal supports evolved into limbs that allowed some vertebrates to walk onto land. On land, lungs replaced gills and circulation changed in concert. **Section 26.2**

Transition from water to land

Vertebrates evolved in the seas, where cartilaginous and bony fishes still live. Of all vertebrates, modern bony fishes are most diverse. One group gave rise to aquatic tetrapods (four-legged walkers), the descendants of which moved onto dry land. **Sections 26.3–26.6**

The amniotes

Amniotes—reptiles, birds, and mammals—have waterproof skin and eggs, highly efficient kidneys, and other traits that adapt them to a life that is typically lived entirely on land. Reptiles and birds belong to one amniote lineage, and mammals to another. **Sections 26.7–26.11**

Early humans and their ancestors

Changes in climate and available resources were selective forces that shaped the anatomy and behavior of early humans and their primate ancestors. Behavioral and cultural flexibility helped humans disperse from Africa throughout the world. **Sections 26.12–26.14**

Links to Earlier Concepts

■ This chapter continues the story of the deuterostome lineage first described in Section 25.1.

■ Be sure you understand the processes of gene duplications (12.5), convergent evolution (19.2), adaptation (17.3), allopatric speciation (18.10), and adaptive radiation (18.12). They come up repeatedly. Knowledge of cladistics (19.1, 19.5) will also be important.

■ You will see how physical factors such as asteroids striking Earth (Chapter 17 introduction) and plate tectonics (17.9) influenced animal evolution and distribution. You may find it useful to refer back to the geologic time scale (17.8).

■ We will return to the story of amphibian declines (24.2). In considering vertebrate body plans, we contrast the vertebrate endoskeleton with the exoskeleton (25.12) of arthropods.

How would you vote? Some private collectors have purchased rare and valuable vertebrate fossils. Private trade raises purchase costs for museums and encourages theft from protected fossil beds. Is the private sale of significant vertebrate fossils unethical? See CengageNOW for details, then vote online.

26.1 | The Chordate Heritage

- Chordates are the most diverse lineage of deuterostomes. Some are invertebrates, but most are vertebrates.
- Link to Animal classification 25.1

Chordate Characteristics

The preceding chapter ended with the echinoderms, a phylum of invertebrate deuterostomes. The majority of deuterostomes are **chordates** (phylum Chordata). Chordate embryos have four defining traits: (1) A rod of stiff but flexible connective tissue, a **notochord**, extends the length of the body and supports it. (2) A dorsal, hollow nerve cord parallels the notochord. (3) Gill slits open across the wall of the pharynx (throat region). (4) A muscular tail extends beyond the anus. Depending on the chordate group, some, all, or none of these traits persist in the adult.

Chordates are bilateral and coelomate (Section 25.1). They show cephalization (sensory structures are concentrated at the head end) and segmentation (paired structures such as muscles are repeated along either side of the long body axis). They have a complete digestive system and a closed circulatory system.

Most of the 50,000 or so chordates are **vertebrates** (subphylum Vertebrata), animals that have a backbone (Table 26.1). The bulk of this chapter describes their traits and evolution. Here we begin our survey with tunicates and lancelets, two groups of marine invertebrate chordates. We also take a brief look at hagfishes, another in-between group.

The Invertebrate Chordates

Lancelets (subphylum Cephalochordata) are invertebrate, fish-shaped chordates 3 to 7 centimeters long

Table 26.1 Modern Chordate Groups

Group	Named Species
Invertebrate chordates:	
Lancelets	30
Tunicates	2,150
Craniates:	
Hagfishes (jawless fishes)	60
Vertebrates:	
Lampreys (jawless fishes)	41
Jawed fishes:	
Cartilaginous fishes	1,160
Bony fishes	26,000
Amphibians	4,900
Reptiles	8,200
Birds	8,600
Mammals	4,500

For details of chordate classification, see Appendix I.

(Figure 26.2). They retain all characteristic chordate traits as adults. A dorsal nerve cord extends into the head. A single eyespot at the end of the nerve cord detects light, but the head has no brain, braincase, or paired sensory organs like those of fishes.

A lancelet wiggles backward into sediments until it is buried up to its mouth, then filters food from the water. Movement of cilia causes water to flow in through the mouth, into the pharynx, then out of the body through gill slits. Cilia also move food particles that get trapped in mucus on the pharynx to the gut.

Like vertebrates, lancelets have segmented muscles. Contractile units in muscle cells run parallel with the body's long axis. The force that muscles direct against the notochord produces a side-to-side motion that allows lancelets to burrow and swim short distances.

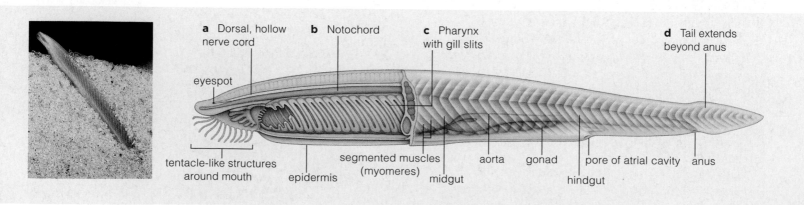

a Dorsal, hollow nerve cord   **b** Notochord   **c** Pharynx with gill slits   **d** Tail extends beyond anus

eyespot

tentacle-like structures around mouth — epidermis — segmented muscles (myomeres) — midgut — aorta — gonad — hindgut — pore of atrial cavity — anus

Figure 26.2 Animated Photo and body plan of a lancelet, a small filter-feeder. Like other chordates, it has a dorsal, hollow nerve cord (**a**), a supporting notochord (**b**), a pharynx with gill slits (**c**), and a tail that extends past the anus (**d**).

As you will see, that is how fish swim and how the first land vertebrates walked.

Tunicates (subphylum Urochordata) are invertebrates in which larvae have typical chordate traits, but adults retain only the pharynx with gill slits (Figure 26.3). Larvae swim about briefly, then undergo metamorphosis. The tail breaks down and other parts become rearranged into the adult body form.

A secreted carbohydrate-rich covering or "tunic" encloses the adult body and gives the group its common name. Most tunicates are sea squirts that live attached to an undersea surface. When disturbed, they squirt water. Other tunicates, known as salps, drift or swim in the open seas. Both groups filter food from the water. Water flows in an oral opening and past gill slits, where the food sticks to mucus and gets sent to a gut. Water leaves through another body opening.

Until recently, lancelets were considered the closest invertebrate relatives of vertebrates. An adult lancelet certainly looks more like fish than an adult tunicate does, but such superficial similarities are sometimes deceiving. New studies of developmental processes and gene sequences indicate that tunicates are the closest living relatives of vertebrates.

Keep in mind that neither tunicates nor lancelets are ancestors of vertebrates. These groups share a recent common relative, but each has unique traits that put it onto a separate branch of the animal family tree.

A Braincase but No Backbone

Fishes, amphibians, reptiles, birds, and mammals are **craniates**. A cranium—a braincase of cartilage or bone—encloses and protects their brain, and they have paired eyes and other sensory structures on the head.

Hagfishes are the only modern chordates that have a cranium, but no backbone (Figure 26.4). Like lancelets, these soft-bodied, jawless fishes have a notochord that supports the body. Like other craniates, a hagfish has paired ears that detect vibrations and a pair of eyes. However, their eye has no lens so their vision is poor. Sensory tentacles near the mouth respond to touch and dissolved chemicals. They help a hagfish find its food—soft invertebrates and dead or dying fish. There are no fins. A hagfish moves with a wriggling motion, similar to that of a lancelet.

Hagfishes are sometimes called slime eels because, when threatened, they can secrete a gallon of slimy mucus. Exuding slime is a useful defense for a soft-bodied animal and it deters most predators. However, it has not kept humans from harvesting hagfish. Most of what is sold as "eelskin" is actually hagfish skin.

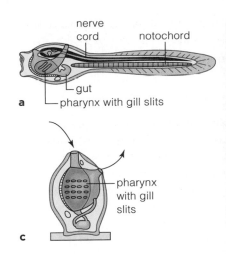
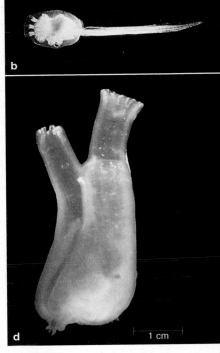

Figure 26.3 (**a,b**) Tunicate larva. It swims briefly, then glues its head to a surface and metamorphoses. Tissues of its tail, notochord, and much of the nervous system are remodeled. (**c,d**) Adult tunicate. Arrows in (**c**) indicate direction of water flow: in one opening, into the pharynx, through gill slits, then out through another opening.

Figure 26.4 Hagfish body plan. The two photographs show a hagfish before and after it coated its body with slimy mucous secretions.

Take-Home Message

What traits characterize the chordates?

■ We define chordates based on traits seen in their embryos. Only in one group of invertebrate chordates, lancelets, do these traits persist in adults. Tunicates are the other group of invertebrate chordates.

■ Hagfishes are the only craniates that are not vertebrates.

26.2 Vertebrate Traits and Trends

- A supportive backbone, a larger brain, and hardened jaws contributed to vertebrate success.
- Link to External skeleton 25.12

An Internal Skeleton and a Big Brain

Vertebrates have an **endoskeleton**, an internal skeleton, consisting of cartilage and (in most groups) bone. The endoskeleton encloses and protects internal organs. It also interacts with skeletal muscles to move the body and its parts. Compared with an external skeleton, an internal one provides less protection, but it has other advantages. It consists of living cells, so it grows and does not have to be molted. It allows greater flexibility and speed of movement. It also provides relatively lightweight support that allows animals to grow big. All large land animals are vertebrates.

The notochord of a vertebrate embryo develops into a **vertebral column**, or backbone. This flexible but sturdy structure is made of many individual skeletal elements called **vertebrae**. It encloses and protects the spinal cord that develops from the embryonic nerve cord. The anterior end of that nerve cord develops into a brain, which is protected by a cranium.

Vertebrate brains are larger and more complex than those of invertebrate chordates. Paired eyes relay information to the brain, as do paired ears. In fishes, paired ears help maintain balance and detect pressure waves in water. When vertebrates moved onto land, ears became modified to detect pressure waves in air.

With the exception of fishes called lampreys, all modern vertebrates have jaws (Figure 26.5a). **Jaws** are hinged skeletal elements used in feeding. The earliest vertebrates were jawless fishes (Figure 26.5b). Jawed fishes called placoderms appeared during the Silurian. They had bony plates on their head and body. Their jaws were expansions of hard parts that structurally supported the gill slits (Figure 26.6).

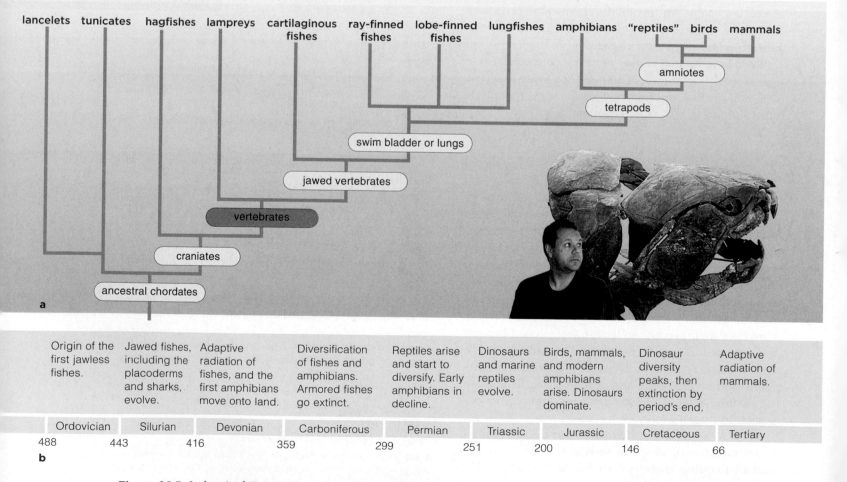

Figure 26.5 Animated The chordate family tree. (**a**) Compare the size of the human with *Dunkleosteus*, an extinct placoderm. (**b**) Time line for events in vertebrate evolution. Numbers indicate millions of years ago. Periods are not to scale.

Figure It Out: Which tetrapods are not also amniotes?

Answer: Amphibians

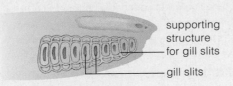 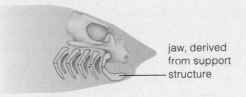 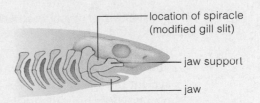

A In early jawless fishes, supporting elements reinforced a series of gill slits on both sides of the body.

B In early jawed fishes (e.g., placoderms), the first elements were modified and served as jaws. Cartilage reinforced the mouth's rim.

C Sharks and other modern jawed fishes have strong jaw supports.

Figure 26.6 Animated Comparison of gill-supporting structures.

The evolution of jaws started an arms race between predators and prey. Fishes with a bigger brain that could better plan pursuit or escape had an advantage, as did those that were fast and maneuverable. Fishes evolved **fins**, body appendages that help them swim. The fins go by these names:

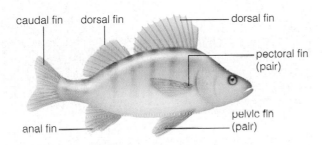

In the Devonian, fishes underwent a great adaptive radiation. Groups with heavy armor died out and a lineage of fishes with bones in the pelvic and pectoral fins arose. This lineage gave rise to amphibians, the first animals with paired limbs, and the vertebrates began to move onto land.

Circulatory and Respiratory Systems

In lancelets and tunicates, some gas exchange occurs at gill slits, but most gases just diffuse across the body wall. Paired gills evolved in early vertebrates. **Gills** are respiratory organs with moist, thin folds that are richly supplied with blood vessels. Gills enhance the exchange of gases and thereby support higher levels of activity than diffusion alone. The force of a beating heart drives blood flow through vessels in fish gills.

Gills became more efficient in larger, more active fishes. But gills cannot function out of water. In fishes ancestral to land vertebrates, two small outpouchings on the side of the gut wall evolved into **lungs**: moist, internal sacs that serve in gas exchange.

Vertebrates have a closed circulatory system. Such systems allow faster blood flow than open systems (Section 25.1). Vertebrate circulatory systems evolved in concert with the respiratory system. In fishes, a two-chambered heart pumps blood through one circuit: from the heart, to gills, through the body, and back to the heart. In most land vertebrates, the heart is divided into four chambers and pumps blood through two separate circuits. One circuit carries oxygen-poor blood from the heart to the lungs and returns oxygen-enriched blood to the heart. The other circuit then pumps this blood to body tissues. Together, lungs and a two-circuit circulatory system enhance the rate of gas exchange and thus sustain a high level of activity.

Other Organ Systems

Vertebrates have a pair of **kidneys**, organs that filter blood and adjust the volume and composition of the extracellular fluid. On land, highly efficient kidneys that help conserve water proved advantageous.

Vertebrates reproduce sexually and sexes are usually separate. Fishes and amphibians typically release eggs and sperm into water. The reptiles, birds, and mammals have organs that allow fertilization to take place inside a female, and eggs that resist water loss.

Vertebrates have a well-developed immune system. Specialized white blood cells allow this system to recognize, remember, and respond fast to pathogens.

Take-Home Message

What are vertebrates?

■ Vertebrates are chordates with an internal skeleton that includes a backbone. Most also have jaws. Compared to invertebrate chordates, vertebrates have a larger, more complex brain.

■ Paired fins in one lineage of fish were the evolutionary predecessors of limbs of land vertebrates. Moving to land also involved modification of circulatory and respiratory systems, a more efficient kidney, and internal fertilization.

26.3 | The Jawless Lampreys

■ Lampreys are vertebrates, but they do not have jaws or paired fins as the jawed fishes do.

The 50 or so species of lampreys are an evolutionarily ancient lineage of fishes. Fossils show that their body plan has been basically unchanged since the Devonian. Like hagfishes, lampreys do not have jaws or fins, but lampreys do have a backbone made of cartilage.

Lampreys are among the few fishes that undergo metamorphosis. Larval lampreys live in fresh water and, like lancelets, they burrow into sediments and filter food from the water. After several years, body tissues get remodeled into the adult form. About half of the lamprey species remain in fresh water and do not feed as adults. The other half are parasites. Some of these stay in fresh water; others migrate to the sea.

Figure 26.7 shows the distinctive mouth of an adult parasitic lamprey. It has an oral disk with horny teeth made of the protein keratin. A parasitic lamprey uses its oral disk to attach to another fish. Once attached, it secretes enzymes and uses a tooth-covered tongue to scrape up bits of the host's tissues. The host fish often dies from blood loss or a resulting infection.

In the early 1800s, sea lampreys invaded the Great Lakes of North America. They probably entered the Hudson River, then made their way through newly built canals. By 1946, lampreys were established in all the Great Lakes. Their arrival caused local extinction of many native fish species. Today, attempts to reduce the lamprey population cost millions of dollars each year and, so far, have had little success.

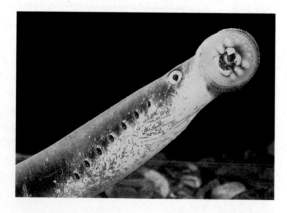

Figure 26.7 Adult parasitic lamprey with eight gill slits on each side of its body and an impressive oral disk. The lamprey latches on to another fish and feeds on its tissues.

Take-Home Message

What are lampreys?

■ Lampreys are a lineage of jawless fishes that undergo metamorphosis. As adults, about half are ecologically important parasites of other fishes.

26.4 | The Jawed Fishes

■ Jawed fishes come in a great variety of shapes and sizes. Nearly all have paired fins and a body covered with scales.

■ Link to Gene duplications 12.5

Most jawed fishes have paired fins and **scales**: hard, flattened structures that grow from and often cover the skin. Scales and an internal skeleton make a fish body denser than water and thus prone to sinking. Fish that are highly active swimmers have fins with a shape that helps lift them, something like the way that wings help lift up an airplane. Water resists movements through it, so speedy swimmers typically have a streamlined body that reduces friction.

There are two groups of jawed fishes: cartilaginous fishes and bony fishes.

Figure 26.8 Cartilaginous fishes. (**a**) Manta ray. Two fleshy projections on its head unfurl and funnel plankton to its mouth. (**b**) Galápagos sharks. Notice the gill slits on both the ray and the shark. (**c**) The cavernous mouth of a whale shark. The shark is as long as a city bus. Like the manta ray, a whale shark is primarily a plankton feeder.

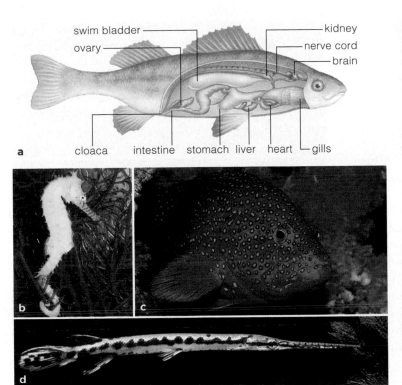

Figure 26.10 An Australian lungfish, a bony fish. In oxygen-poor waters it fills its lungs by rising to the surface and inhaling air.

Figure 26.11 A coelacanth (*Latimeria*), a lobe-finned fish. It is a bony fish that has skeletal elements in its pelvic and pectoral fins.

Figure 26.9 Ray-finned bony fishes. (**a**) Body plan of a perch. (**b**) Sea horse. (**c**) Coral grouper. (**d**) Long-nose gar, a fast predator.

Cartilaginous Fishes

Cartilaginous fishes (Chondrichthyes) include about 850 species of mostly marine sharks and rays. All have a skeleton of cartilage, and five to seven gill slits (Figure 26.8). Their teeth are modified scales hardened with bone and dentin. The teeth grow in rows and are continually shed and replaced.

Rays have a flattened body with large pectoral fins. Manta rays filter plankton from water and some are 6 meters (20 feet) wide (Figure 26.8*a*). Stingrays are bottom feeders. Their barbed tail has a venom gland.

Sharks include predators that swim in upper ocean waters (Figure 26.8*b*), plankton feeders (Figure 26.8*c*), and bottom feeders that suck up invertebrates and act as scavengers.

Bony Fishes

In **bony fishes** (Osteichthyes) bone replaces cartilage in much of the skeleton. Unlike most cartilaginous fishes, bony fishes have a cover, or operculum, that protects their gills. Bony fishes also typically have a **swim bladder**: a gas-filled flotation device. By adjusting the volume of gas inside its swim bladder, a bony fish can stay suspended in water at different depths.

The three bony fish subgroups are ray-finned fishes, lungfishes, and lobe-finned fishes.

Ray-finned fishes (Figure 26.9) have thin, flexible fin supports derived from skin. With 21,000 species, they are the most diverse fishes. Teleosts, the largest ray-finned group, includes the fishes in Figure 26.9*a–c*, as well as most fishes we eat. Long ago, the entire teleost genome was duplicated. Mutations in copied genes may have facilitated diversification of this group.

Lungfishes (Figure 26.10) are bony fishes that have gills and lunglike sacs—modified outpouchings of the gut wall. They fill the sacs by surfacing and gulping air, then oxygen diffuses from the sacs into the blood.

Coelacanths (*Latimeria*) are the only modern group of lobe-finned fishes. The two populations we know about may be separate species. Their ventral fins are fleshy extensions of the body wall and have internal skeletal elements (Figure 26.11). Lobe-finned fishes are the fish most closely related to amphibians.

Take-Home Message

What are the characteristics of jawed fishes?

■ Jawed fishes are cartilaginous fishes and bony fishes. Both groups typically have scales. The ray-finned lineage of bony fishes is the most diverse group of vertebrates. Lobe-finned fishes are the fish closest to the amphibians.

■ Amphibians spend part of their life on land, but most still return to water to breed.

■ Link to Homologous structures 19.2

Adapting to Life on Land

Amphibians are land-dwelling vertebrates that need water to breed and have a three-chambered heart. Their lineage branched from that of lobe-finned fishes during the Devonian. Fossils show how the skeleton was modified as fishes adapted to swimming evolved into four-legged walkers, or **tetrapods** (Figure 26.12). Bones of a fish's pelvic and pectoral fins are homologous with amphibian limb bones (Section 19.2).

The transition to land was not simply a matter of skeletal changes. Division of the heart into three chambers allowed flow in two circuits, one to the body and one to the increasingly important lungs. Changes to the inner ear improved detection of airborne sounds. Eyes became protected from drying out by eyelids.

What was the selective advantage to living on land? An ability to survive out of water would have been favored in seasonally dry places. Also, on land, individuals escaped aquatic predators and had new food—insects, which also evolved during the Devonian.

Modern Amphibians

The three subgroups of modern amphibians are the salamanders, the caecilians, and the frogs and toads. All are carnivores as adults. Amphibians release eggs

Figure 26.13 (**a**) Red-spotted salamander, with equal-sized forelimbs and hindlimbs. (**b**) A legless caecilian.

and sperm into water. Their aquatic larvae have gills. Larvae feed and grow until hormonal changes cause them to metamorphose into adults. Most species lose their gills and develop lungs during this transition. However, a few salamanders retain gills as adults. Others lose gills and exchange gases across their skin.

The 535 species of salamanders and related newts live mainly in North America, Europe, and Asia. In body form, they are the modern group most like early

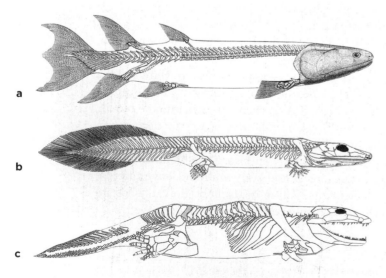

Figure 26.12 Skeleton of a Devonian lobe-finned fish (**a**), and two early amphibians, *Acanthostega* (**b**), and *Ichthyostega* (**c**). The painting (**d**) shows what *Acanthostega* (foreground), and *Ichthyostega* (background) may have looked like.

Figure 26.14 (**a**) Adult frog displaying the power of its well developed hindlimbs. (**b**) A larval frog, or tadpole.

tetrapods. Forelimbs and back limbs are of similar size and there is a long tail (Figure 26.13*a*). As salamanders walk, their body bends from side to side, like the body of a swimming fish. Their ancestors that first ventured onto land probably moved in a similar way.

Caecilians are close relatives of salamanders that have adapted to a burrowing way of life. They include about 165 limbless, blind species (Figure 26.13*b*). Most caecilians burrow through soil and use their senses of touch and smell to pursue invertebrate prey.

Frogs and toads belong to the most diverse amphibian lineage; there are more than 5,000 modern species. Muscular, elongated hindlegs allow the tailless adults to swim, hop, and make leaps that can be spectacular, given their body size (Figure 26.14*a*). The forelegs are much smaller and help absorb the impact of landings.

Larvae of salamanders and caecilians have a body shape more or less like that of an adult, except for the presence of gills. In contrast, the larvae of frogs and toads are markedly different from adults. The larvae have gills and a tail, but no limbs. They are commonly known as tadpoles (Figure 26.14*b*).

Take-Home Message

What are amphibians?

■ Amphibians are vertebrates with a three-chambered heart. They start life in water as gilled larvae, then undergo metamorphosis. Adults typically have lungs and are carnivores on land.

26.6 | Vanishing Acts

■ Amphibians depend on access to standing water to breed and have a thin skin unprotected by scales. These features make them vulnerable to habitat loss, disease, and pollution.

■ Links to Chytrid fungi 24.2, Flukes 25.6

There is no question that amphibians are in trouble. Of about 5,500 known species, population sizes of at least 200 are plummeting. The alarming declines have been best documented in North America and Europe, but the changes are happening worldwide.

At this writing, six frog species, four toad species, and eleven salamander species are considered threatened or endangered in the United States and Puerto Rico. One, the California red-legged frog (*Rana aurora*), inspired Mark Twain's well-known short story, "The Celebrated Jumping Frog of Calaveras County." This species is the largest frog native to the western United States.

Researchers correlate many declines with shrinking or deteriorating habitats. Developers and farmers commonly fill in low-lying ground that once collected seasonal rains and formed pools of standing water. Nearly all amphibians need to deposit their eggs and sperm in water, and their larvae must develop in water.

Also contributing to declines are introductions of new species in amphibian habitats, long-term shifts in climate, increases in ultraviolet radiation, and the spread of certain pathogens and parasites. Section 24.2 discussed chytrid infections of amphibians and Figure 26.15 provides an example of the effects of a parasitic fluke (a subgroup of flatworms). Chemical pollution of aquatic habitats also harms amphibians. We will consider the negative effects of one agricultural chemical on frogs in Chapter 35.

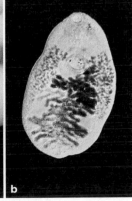

Figure 26.15 (**a**) Example of frog deformities. (**b**) A parasitic fluke (*Ribeiroia*). It burrows into limb buds of frog tadpoles and physically or chemically alters individual cells. Infected tadpoles grow extra legs or none at all. Where *Ribeiroia* populations are most dense, the number of tadpoles that successfully complete metamorphosis is low. Nutrient enrichment of water by fertilizers and pesticide contamination make frogs more easily infected.

26.7 The Rise of Amniotes

- Amniotes took waterproofing to a new level with their skin and eggs, making them well adapted to dry habitats.

- Links to Geologic time scale 17.8, Cladistics 19.5

In the late Carboniferous, one amphibian lineage gave rise to the "stem" reptiles, the first amniotes. **Amniotes** make eggs having four unique membranes that allow embryos to develop away from water (Figures 26.16a,b and 26.21). Amniotes have waterproof skin and a pair of efficient kidneys. Nearly all fertilize eggs in the female's body. These traits adapt them to life on land.

One early branching of the amniote lineage led to synapsids: mammals and extinct mammal-like species (Figure 26.16c). A now extinct synapsid subgroup, the therapsids, included ancestors of mammals, as well as *Lystrosaurus*, a tusked herbivore shown in Figure 17.17.

Three other branches of the amniote lineage have survived. One branch led to turtles, one to lizards and snakes, and the third to crocodilians and birds. As you can see, the traditional division of birds and "reptiles" into separate classes does not reflect phylogeny; the reptiles are not a clade (Section 19.5). Nevertheless, the term **reptile** persists as a way to refer to amniotes that lack the defining traits of birds or mammals. That is how we use it in this book.

The earliest reptiles had a lizardlike body. With well-muscled jaws and sharp teeth, they could seize and kill their prey with more force than amphibians. Waterproof scales rich in the protein keratin covered the body and suited reptiles to drier habitats, but such scales also prevented gas exchange across the skin. Compared to amphibians, the early reptiles had larger, more efficient lungs. They also had larger brains that allowed more complex behavior.

Figure 26.16 Amniote eggs and phylogeny.

(**a**) A painting of a nest of a duck-billed dinosaur (*Maiasaura*) that lived about 80 million years ago in what is now Montana. Like the modern crocodilians and birds, this dinosaur protected its eggs in a nest and may have cared for the hatchlings. (**b**) Two eastern hognose snakes emerging from amniote eggs.

(**c**) Family tree for amniotes. Snakes, lizards, tuataras, birds, crocodilians, turtles, and mammals are modern amniote groups.

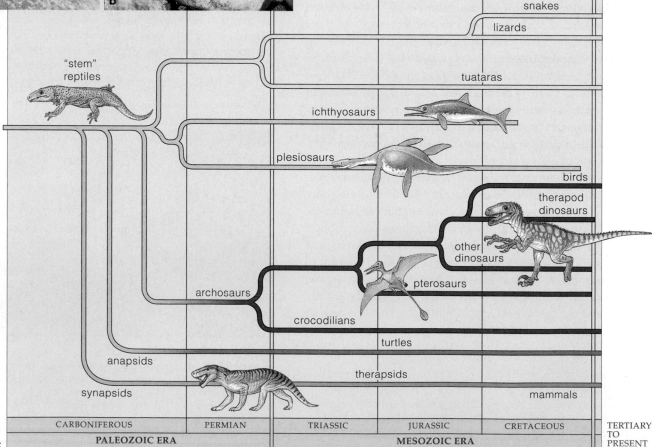

Figure 26.17 *Temnodontosaurus.* This ichthyosaur hunted large squids, ammonites, and other prey in the warm, shallow seaways of the early Jurassic. Fossils that measure 30 feet (9 meters) long have been found in England and Germany.

Biologists define the reptiles known as **dinosaurs** by certain skeletal features, such as the configuration of their pelvis and hips. Dinosaurs evolved by the late Triassic. Early species were the size of a turkey and ran on two legs. Adaptive zones opened up for this lineage as the Jurassic began, after fragments from an asteroid or comet hit what are now France, Quebec, Manitoba, and North Dakota. Nearly all the animals that survived these asteroid impacts were small, had high metabolic rates, and could tolerate big changes in temperature.

Surviving dinosaur groups, such as those shown in Figure 26.1, became the "ruling reptiles." For 125 million years they dominated the land even as other groups, including the ichthyosaurs, flourished in the seas (Figure 26.17). Many kinds of dinosaurs were lost in a mass extinction that ended the Jurassic. Others died off during the Cretaceous.

As the Cretaceous ended, another asteroid impact wiped out many groups. Feathered dinosaurs ancestral to birds survived, as did ancestors of modern reptiles: crocodilians, turtles, tuataras, snakes, and lizards.

Take-Home Message

What are amniotes?

■ Amniotes are animals whose embryos develop inside a waterproof egg. They also have waterproof skin and highly efficient kidneys that reduce water loss.

■ Dinosaurs are extinct amniotes and birds are their descendants. Reptiles and mammals are the other modern amniotes.

26.8 | So Long, Dinosaurs

■ The effects of an asteroid impact on life on Earth are vividly illustrated by the story of the dinosaurs' demise.

■ Link to Asteroid impacts (Chapter 17 introduction)

Chapter 17 made reference to a mass extinction that defines the Cretaceous–Tertiary (K–T) boundary. After methodically analyzing the elemental composition of soils, maps of gravitational fields, and other evidence from around the world, Walter Alvarez and Luis Alvarez developed a hypothesis: A direct hit by an enormous asteroid caused the K–T extinction event. This came to be known as the **K–T asteroid impact hypothesis.**

Later, researchers discovered an enormous impact crater on the seafloor in the Gulf of Mexico. Known as the Chicxulub Crater, it is 9.6 kilometers deep and about 300 kilometers across. By one estimate, to make a crater that big, the impact would have blasted least 200,000 cubic kilometers of dense gases and debris into the sky.

Did this impact cause the K–T extinction event? Many researchers think so. However, Gerta Keller and others hold that the Chicxulub Crater was formed 300,000 years before the K–T extinction. They hypothesize that a series of asteroid impacts occurred and that the crater formed by the impact at the K–T boundary is yet to be discovered.

Researchers also debate the mechanism by which an asteroid impact could have caused the known extinctions. Some argue that atmospheric debris must have blocked sunlight for months, causing a deep, dark freeze that killed plants and starved animals. They are trying to explain the fossil record, which shows that land plant and animal species died off. Others think that the volume of debris blasted aloft would not have been great enough to cause the widespread extinctions recorded in the fossil record.

An alternative scenario was proposed after the comet Shoemaker–Levy 9 slammed into Jupiter in 1994. Debris blasted into the Jovian atmosphere and triggered intense heating. That event led Jay Melosh and his colleagues to propose that an enormous asteroid impact raised Earth's atmospheric temperature by thousands of degrees. In one terrible hour, the world erupted in flames. Any animals out in the open—including nearly all dinosaurs—were broiled alive.

Not every living thing disappeared. Snakes, lizards, crocodiles, and turtles survived, as did birds and mammals. The proponents of Melosh's hypothesis argue that smaller species may have escaped the firestorm by burrowing underground. Critics point out that most birds are ill equipped to burrow. Also, many of the invertebrate species that lived on the ocean floor disappeared, too. How could they have "broiled" deep underwater?

In short, one or more asteroids are implicated in the K–T extinctions. Where they hit and exactly what happened next remains an open question.

An asteroid impact ended the golden age of dinosaurs.

■ Reptile have a scale-covered body. Most have four limbs of approximately equal size, but the snakes are limbless.

General Characteristics

"Reptile" is derived from the Latin *repto*, which means to creep. Some reptiles do creep. Others swim or race or lumber about. Modern reptiles include about 8,160 species. Figure 26.18 shows a typical body plan.

Like fishes, reptiles have scales. However, reptile scales develop from the outer layer of skin (epidermis), whereas fish scales arise from a deeper layer (dermis).

Like amphibians and fishes, reptiles have a **cloaca**, an opening that expels digestive and urinary waste and functions in reproduction. All male reptiles, except tuataras, have a penis and fertilize eggs in a female's body. In most groups, females lay eggs which develop on land. In some lizards and snakes, eggs are held in a female's body and young are born fully developed.

Also like amphibians and fishes, all modern reptiles are **ectotherms**; their body temperature is determined by the temperature of their surroundings. Reptiles in temperate regions spend the cold season inactive in a burrow on land, or in the case of some freshwater turtles, beneath the mud at a lake bottom.

Major Groups

Turtles The unique feature of the 300 or so species of turtles and tortoises is a bony, scale-covered shell that connects to the backbone (Figure 26.19*a,b*). Turtles do not have teeth; a "beak" made of keratin covers their jaws. Some feed on plants and others are predators.

Many sea turtles are endangered. Adults travel back to the same tropical beach where they were hatched to mate and lay their eggs. An increasing human presence on these beaches threatens these species.

Lizards With 4,710 species, lizards are the most diverse reptiles. The smallest fits on a dime (*left*). The largest, the Komodo dragon, can reach 3 meters (10 feet) in length. It is an ambush predator that snags prey with its peglike teeth. Its saliva contains deadly pathogenic bacteria. Chameleons are lizards that catch prey with a sticky tongue that can be longer than their body. Iguanas are herbivorous lizards.

Lizards have interesting defenses to avoid becoming prey themselves. Some try to outrun a predator or startle it (Figure 26.19*c,d*). Many can detach their tail. The detached tail wriggles briefly, which may distract a predator from its fleeing, tailless owner.

Tuataras The two species of tuatara that live on small islands near the coast of New Zealand are all that remains of a lineage that thrived during the Triassic. Tuatara means "peaks on the back" in the language of New Zealand's native Maori people. This name refers to a spiny crest (Figure 26.19*e*). Tuataras are reptiles but walk like salamanders and have amphibian-like brain structures. Also, a third eye develops under the skin of the forehead. It becomes covered by scales in adults and its function, if any, is unclear.

Snakes During the Cretaceous, snakes evolved from short-legged, long-bodied lizards. Some of the 2,995 modern snakes still have bony remnants of hindlimbs, but most are limbless. All are carnivores. Many have flexible jaws that help them swallow prey whole. All snakes have teeth; not all have fangs. Rattlesnakes and other fanged types bite and subdue prey with venom made in modified salivary glands (Figure 26.19*f*). On average, only about 2 of the 7,000 snake bites reported annually in the United States are fatal.

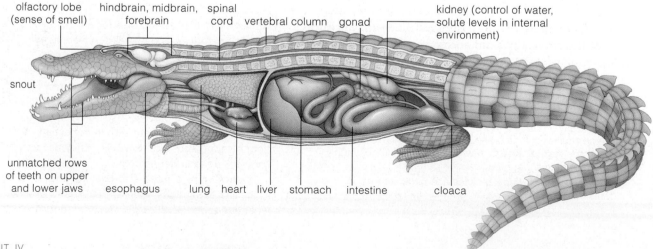

Figure 26.18 Animated Body plan of a crocodile. The heart has four chambers, so blood flows through two entirely separate circuits. This keeps oxygen-poor blood returning from the body from mixing with oxygen-rich blood from the lungs.

olfactory lobe (sense of smell)

hindbrain, midbrain, forebrain

spinal cord

vertebral column

gonad

kidney (control of water, solute levels in internal environment)

snout

unmatched rows of teeth on upper and lower jaws

esophagus

lung

heart

liver

stomach

intestine

cloaca

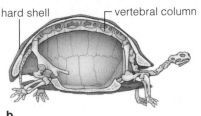

b

Figure 26.19 (**a**) Galápagos tortoise. (**b**) Turtle shell and skeleton. Most turtles can pull their head into their shell when threatened, however the shell is reduced in some sea turtles.

(**c**) A lizard fleeing and (**d**) lizard confronting a threat. (**e**) Tuatara (*Sphenodon*). (**f**) Rattlesnake. (**g**) Spectacled caiman, a crocodilian, showing its peglike teeth. Upper and lower teeth do not align as they do in mammals.

Crocodilians Almost a dozen species of crocodiles, alligators, and caimans are the closest living relatives of birds. All are predators in or near water. They have powerful jaws, a long snout, and sharp teeth (Figures 26.18 and 26.19*g*). They clench prey, drag it under water, tear it apart, and then gulp down the chunks.

Crocodilians are the only reptiles that have a four-chambered heart, as mammals and birds do. Such a heart prevents the mixing of oxygen-poor blood from tissues with oxygen-rich blood from the lungs.

Crocodilians are the closest living relatives of birds, and like birds they display complex parental behavior. For example, they build and guard a nest, then feed and care for the hatchlings.

Take-Home Message

What are the modern reptiles like?

■ Reptiles range in size from tiny lizards to giant crocodiles. Some are aquatic, but most live on land. All lay eggs on land. There are herbivores, but the majority are carnivores.

- In one group of dinosaurs, the scales became modified as feathers. Birds are modern descendants of this group.

- Links to Beak morphology 17.3, 18.10

From Dinosaurs to Birds

Birds are the only living animals that have feathers. Feathers are modified reptilian scales. *Sinosauropteryx prima*, a small carnivorous dinosaur that lived in the late Jurassic, was covered with fine, downy feathers (Figure 26.20*a*). Similar feathers give juvenile birds a fluffy appearance (Figure 26.20*b*). Downy feathers do not allow flight, but they provide insulation.

Archaeopteryx, described and shown in the chapter introduction, was like modern birds in having both short, downy feathers and flight feathers. However, this early bird had teeth and a long bony tail.

Confuciusornis sanctus is the earliest known bird with a beak like that of modern birds (Figure 26.20*c,d*). Its tail was short, with long feathers. Still, its dinosaur ancestry remains apparent—it had grasping, clawed digits at the front tips of its wings.

Figure 26.20 (**a**) A feathered dinosaur, *Sinosauropteryx prima*. It was covered with downy feathers like those of a modern chick (**b**). The early bird *Confuciusornis sanctus* (**c**), lived about the same time as *Archaeopteryx*. It had a short tail with long feathers, wings with grasping digits and claws, and a toothless beak, similar to that of a modern bird such as this cardinal (**d**).

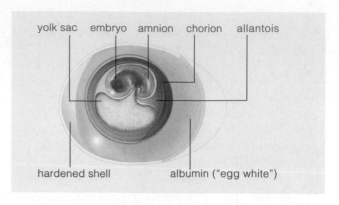

Figure 26.21 Animated A bird egg, a type of amniote egg with four membranes outside the embryo. The chorion assists in gas exchange; the amnion secretes fluid that keeps the embryo moist; the allantois stores waste; and the yolk stack holds yolk that nourishes the developing embryo.

General Characteristics

Like other amniotes, birds produce eggs with internal membranes (Figure 26.21). In birds, a shell hardened with calcium encloses the egg. Fertilization is internal. Males do not have a penis; sperm is transferred from the male's cloaca to the female's.

Birds do not have teeth. Instead, jaw bones covered with layers of the protein keratin form a horny beak. Beak shape varies, with different types of beaks suited to different diets (Sections 17.3, 18.10).

Birds are **endotherms**, which means "heated from within." Physiological mechanisms allow endotherms to maintain their body temperature within a limited range. A bird's downy feathers slow the loss of metabolic heat. Feathers also act as a water-shedding body covering, play a role in courtship, and allow flight.

Flight feathers are just one of the adaptations that helps birds fly. Birds also have a lightweight skeleton, powerful flight muscles, and highly efficient respiratory and circulatory systems.

A wing is a modified forelimb, with feathers that extend outward, increasing its surface area (Figure 26.22*a,b*). The feathers give the wing a shape that helps lift the bird as air passes over it.

Air cavities inside bones keep body weight low, and make it easier for a bird to get and remain airborne. The flight muscles connect an enlarged breastbone, or sternum, to bones of the upper limb (Figure 26.22*c*).

Flight requires a lot of energy, which is provided by aerobic respiration (Section 8.1). To ensure adequate oxygen supply, birds have a unique system of air sacs that keeps air flowing continually through their lungs. A four-chamber heart pumps blood through two fully separated circuits.

Flight also requires a great deal of coordination. Much of the bird brain controls movement. Birds also have excellent vision, including color vision.

Bird Diversity and Behavior

The approximately 9,000 named species of birds vary in size, proportions, coloration, and capacity for flight. A bee hummingbird weighs 1.6 grams (0.6 ounces). The ostrich, a flightless sprinter, weighs 150 kilograms (330 pounds).

More than half of all bird species belong to the subgroup of perching birds. Among them are familiar sparrows, jays, starlings, swallows, finches, robins, warblers, orioles, and cardinals (Figure 26.20d).

We see one of the most impressive forms of behavior among birds that migrate with the changing seasons. Migration is a recurring movement from one region to another in response to some environmental rhythm. Seasonal change in daylength is one cue for internal timing mechanisms called "biological clocks." Such a clock triggers physiological and behavioral changes that induce migratory birds to fly between breeding grounds and wintering grounds. Many types of birds migrate long distances. They use the sun, stars, and Earth's magnetic field as directional cues. Arctic terns make the longest migrations. They spend summers in the arctic and winters in the antarctic.

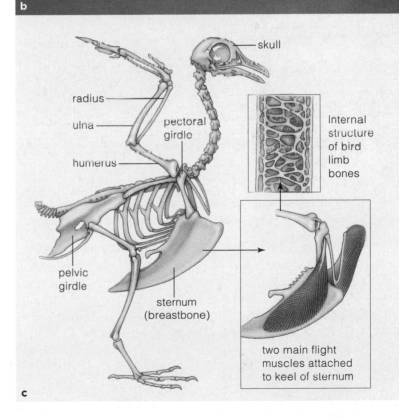

Figure 26.22 Animated Adaptations for flight. (**a**) Birds fly by flapping their wings. The downstroke provides lift. (**b**) Some birds, like this Laysan albatross, have wings that let them glide long distances. With wings more than 2 meters (6.5 feet) across, this bird weighs less than 10 kilograms (22 pounds). It is so at home in the air that it sleeps while aloft.

(**c**) A bird's skeleton is made up of lightweight bones with internal air pockets. The wing is a modified forelimb (see Figure 19.5). Powerful flight muscles attach to a large breastbone, or sternum.

Take-Home Message

What are birds?

■ Birds are the only living animals with feathers. They evolved from dinosaurs and have a body adapted for flight. Bones are lightweight; air sacs increase the efficiency of respiration; and a four-chambered heart keeps blood moving rapidly.

■ Mammals scurried about while dinosaurs dominated the land, then radiated once they were gone.

■ Links to Morphological convergence 19.2, Plate tectonics 17.9, Adaptive radiation 18.12

Mammalian Traits

Mammals are animals in which females nourish their offspring with milk they secrete from mammary glands (Figure 26.23a). The group name is derived from the Latin *mamma*, meaning breast. Milk is a nutrient-rich food source that also contains immune system proteins that help protect offspring from disease.

Mammals are the only animals that have hair or fur. Both are modifications of scales. Like birds, mammals are endotherms. A coat of fur or head of hair helps them maintain their core temperature. Most mammals have whiskers, stiffened hairs on the face that serve a sensory purpose. Mammals are the only animals that sweat, although not all mammals can do so.

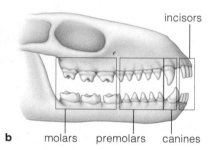

b   molars   premolars   canines

Figure 26.23 Distinctly mammalian traits. (**a**) A human baby, already with a mop of hair, being nourished by milk secreted from the mammary gland in a breast. (**b**) Four types of teeth and a single lower jaw bone.

Only mammals have four different kinds of teeth (Figure 26.23b). In other vertebrates, an individual's teeth may vary a bit in size, but they are all the same shape. Mammals have incisors that can be used to gnaw, canines that tear and rip flesh, and premolars and molars that grind and crush hard foods. Not all mammals have all four tooth types, but most have some combination. Having a variety of different kinds of teeth allows mammals to eat a wider variety of foods than most other vertebrates.

Like birds and crocodilians, mammals have a four-chambered heart that pumps their blood through two fully separate circuits. Gas exchange occurs in a pair of well-developed lungs.

Mammalian Evolution

As noted earlier (Figure 26.16), mammals belong to the synapsid branch of the amniote lineage. The earliest mammals appeared when the dinosaurs were becoming dominant. **Monotremes** (egg-laying mammals) and **marsupials** (pouched mammals) both evolved during the Jurassic. **Placental mammals** evolved a bit later, in the Cretaceous. Placental mammals are named for their **placenta**, an organ that allows materials to pass between a mother and an embryo developing inside her body. Placental embryos grow faster than those of other mammals. Also the offspring are born more fully formed and thus are less vulnerable to predation.

Continental movements affected the evolution and dispersion of mammal groups. Because monotremes and marsupials evolved while Pangea was intact, they dispersed across this supercontinent (Figure 26.24a). Placental mammals evolved after Pangea had begun breaking up (Figure 26.24b). As a result, monotremes

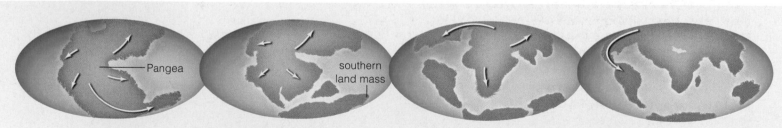

A About 150 million years ago, during the Jurassic, the first monotremes and marsupials evolved and migrated through the supercontinent Pangea.

B Between 130 and 85 million years ago, during the Cretaceous, placental mammals arose and began to spread. Monotremes and marsupials that lived on the southern land mass evolved in isolation from placental mammals.

C Starting about 65 million ago, mammals expanded in range and diversity. Marsupials and early placental mammals displaced monotremes in South America.

D About 5 million years ago, in the Pliocene, advanced placental mammals invaded South America. They drove most marsupials and the early placental species to extinction.

Figure 26.24 Animated Effects of continental drift on the evolution and distribution of mammalian lineages.

Figure 26.25 Paleocene mammals in a sequoia forest in what is now Wyoming. With the exception of the marsupial on the tree branch, all are members of now extinct mammalian lineages.

Figure 26.26 *Indricotherium*, the "giraffe rhinoceros." At 15 tons and 5.5 meters (18 feet) high at the shoulder, it is the largest land mammal we know about. It lived in Asia during the Oligocene and is a relative of the rhinoceros.

Figure 26.27 Example of convergent evolution. (**a**) Australia's spiny anteater, one of only three modern species of monotremes. (**b**) Africa's aardvark and (**c**) South America's giant anteater. Compare the ant-snuffling snouts.

a   egg-laying mammal   **b**   pouched mammal   **c**   placental mammal

and marsupial mammals on land masses that broke early from Pangea, lived for millions of years in the absence of placental mammals. For example, Australia split off from Pangea early on and so does not have native placental mammals.

Australia remains a separate continent, but continental movement reunited other land masses. When placental mammals entered regions where they were previously unknown, the native monotreme and marsupial populations declined. The new arrivals often outcompeted them and, in many cases, drove them to local extinction (Figure 26.24*c,d*).

After the dinosaurs disappeared at the end of the Cretaceous, mammals underwent the great adaptive radiation illustrated earlier in Figure 18.26. Figures 26.25 and 26.26 provide examples of some of the resulting diversity.

Members of different mammal lineages adapted to similar habitats on different continents. For example, Australia's spiny anteater, South America's giant anteater, and Africa's aardvark all hunt ants using their long snout (Figure 26.27). The similar snouts are an example of morphological convergence (Section 19.2).

Take-Home Message

What are mammals?

■ Mammals are animals that nourish young w[...] or fur. Their four kinds of teeth allow them to eat ma[...]. Mammals originated in the Jurassic, then unde[...] after dinosaurs died out. Continental movement[...]ion.

■ Mammals successfully established themselves on every continent and in the seas. What are the existing species like?

Figure 26.28 Female platypus, a monotreme, with two young that hatched from eggs with a rubbery shell. She has a beaverlike tail, a ducklike bill, and webbed feet. Sensory receptors in the bill help a platypus find prey under water. Platypuses burrow into riverbanks using claws exposed when they retract the webbing on their feet. Both males and females have spurs on their hind feet. The male's spurs deliver venom, making them the only venomous mammals.

Fig
Aust
of nat
teeth in marsupials. (a) A koala, *Phascolarctos cinereus*, from
in the wil eucalyptus trees and is threatened by destruction
offspring lopment. (b) A young Tasmanian devil shows its
It is the only carnivorous marsupial surviving
um with her four genetically identical
e embryo splits in early development.

Egg-Laying Monotremes

Three species of monotremes still exist. Two are spiny anteaters, and one of these is shown in Figure 26.27a. The third species is the platypus (Figure 26.28).

All female monotremes lay and incubate eggs that have a leathery shell like that of reptiles. Offspring hatch in a relatively undeveloped state—tiny, hairless, and blind. Young cling to the mother or are held in a skin fold on her belly. Milk oozes from openings on the mother's skin; monotremes do not have nipples.

Pouched Marsupials

Most of the 240 modern species of marsupials live in Australia and on nearby islands. Groups include kangaroos, the koala (Figure 26.29a), and the Tasmanian devil (Figure 26.29b). The opossum (Figure 26.29c) is the only marsupial native to North America.

Young marsupials develop briefly in their mother's body, nourished by egg yolk and by nutrients that diffuse from maternal tissues. They are born at an early developmental stage, and crawl along their mother's body to a permanent pouch on her ventral surface. They attach to a nipple in the pouch, suckle, and grow.

Placental Mammals

Compared to other mammals, the placental mammals develop to a far more advanced stage inside their mother's body. An organ called the placenta allows materials to pass between maternal and embryonic bloodstreams (Figure 26.30a). The placenta transfers nutrients more efficiently than diffusion does, allowing the embryo to grow faster. After birth, young suckle milk from nipples on the mother's ventral surface.

Placental mammals are now the dominant mammals in most land habitats, and the only ones that live in the seas. Figure 26.30b–i shows a few of the more than 4,000 species. Appendix I lists major groups.

Nearly half of mammal species are rodents and, of those, about half are rats. The next most diverse group is the bats, with about 375 species. Bats are the only flying mammals. Although some may resemble flying mice, bats are more closely related to carnivores such as wolves and foxes than to rodents.

Take-Home Message

What are living mammals like?

■ Most mammals living today are placental mammals. Of these, the rodents and bats are the most diverse groups.

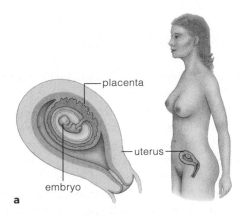

Figure 26.30 Placental mammals. (**a**) Location of the placenta in a pregnant human female. (**b**) Blue whale and (**c**) the size of its skeleton relative to a human. At 200 tons, an adult is the largest living animal.

(**d**) A Florida manatee eats plants in warm coastal waters and rivers. (**e**) A camel traverses hot deserts. (**f**) Harp seals pursue prey in frigid waters and rest on ice.

(**g**) Flying squirrel, really just a glider. The only flying mammals are bats (**h**); this one is a Kittl's hog-nosed bat. (**i**) Red fox hiding in blue spruce. Thick, insulating fur protects it from winter cold.

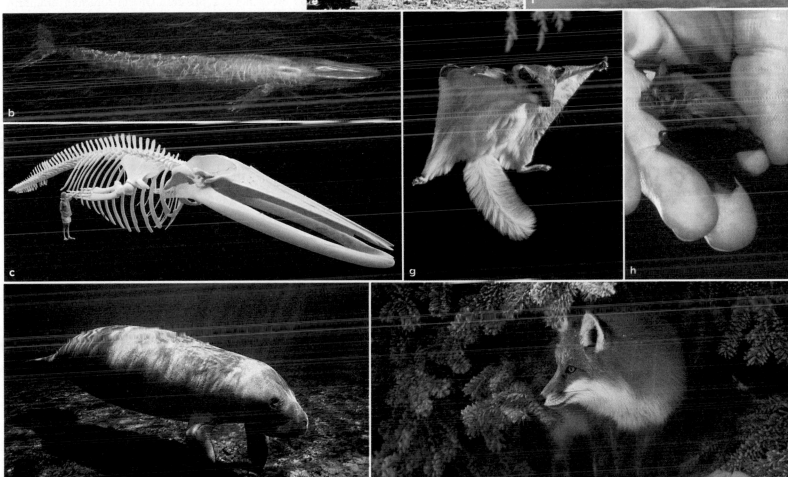

■ The primates are the mammalian subgroup to which humans and our closest relatives belong.

■ Link to Adaptive traits 17.3

Primates include 260 species of prosimians, monkeys, apes, and humans (Figure 26.31). Prosimians ("before monkeys") evolved first. Modern prosimians include tarsiers and lemurs in Africa, Asia, and Madagascar (Page 17 shows a newly discovered lemur species). Anthropoids include monkeys, apes, and humans; all are widely distributed. Hominoids include apes and humans. Our closest living relatives are chimpanzees and bonobos (previously called pygmy chimpanzees). Humans and extinct humanlike species are **hominids**. Table 26.2 summarizes primate subgroups.

Table 26.2   Primate Classification

Prosimians	Lemurs, tarsiers
Anthropoids	New World monkeys (e.g., spider monkeys)
	Old World monkeys (e.g., baboons, macaques)
	Hominoids:
	Hylobatids (gibbons, siamangs)
	Pongids (orangutans, gorillas, chimpanzees, bonobos)
	Hominids (humans, extinct humanlike species)

Overview of Key Trends

Five trends that led to uniquely human traits began in early tree-dwelling species. They came about through modifications to the eyes, bones, teeth, and brain.

Enhanced daytime vision. Early primates had an eye on each side of a mouse-shaped head. Later on, some had a more upright, flattened face, with eyes up front. The ability to focus both eyes on an object improved depth perception. Also, eyes became more sensitive to variations in light intensity and in color. During this time, the sense of smell declined in importance.

Upright walking. Humans are **bipedal**: their skeleton and muscles are adapted for upright standing and walking. For example, their S-shaped backbone keeps the head and torso centered over the feet, and arms are shorter than legs. By contrast, prosimians and monkeys move about on four legs, all about the same length. Gorillas walk on two legs while leaning on the knuckles of longer arms (Figure 26.31*c,e*).

How can we determine whether a fossil primate was bipedal? The position of the foramen magnum, an opening in the skull, is one clue. This opening allows the brain to connect with the spinal cord. In animals that walk on all fours, the foramen magnum is located at the back of the skull. In upright walkers, it is near the center of the skull's base (Figure 26.32).

Better grips. Early mammals spread their toes apart to support their weight as they walked or ran on four legs. In ancient tree-dwelling primates, hands became

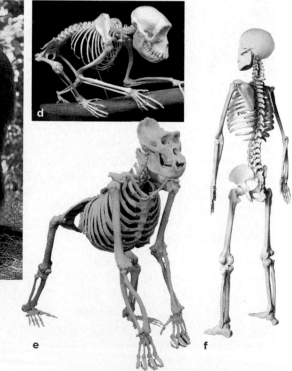

Figure 26.31 Primates. (**a**) Tarsier, a prosimian climber and leaper. (**b**) Spider monkey, an agile climber. (**c**) Male gorilla, using its forearms to support its weight as it walks on its knuckles and two legs. Comparisons of the skeletal structure of (**d**) a monkey, (**e**) a gorilla, and (**f**) a human. Skeletons are not to the same scale.

a Hole at back of skull; the backbone is habitually parallel with ground or a plant stem

b Hole close to center of base of skull; the backbone is habitually perpendicular to ground

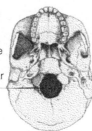

Figure 26.32 A hole in the head, the foramen magnum, in (**a**) a four-legged walker and (**b**) an upright walker. Position of this hole helps us determine if a fossil species was bipedal.

modified. Fingers could curl around things (prehensile movements), and the thumb could touch all fingertips (opposable movements). In time, hands were freed from load-bearing functions and were modified in ways that allowed powerful or precision gripping:

power grip   precision grip

Having a capacity for versatile hand positions gave the ancestors of humans the ability to make and use tools. Refined prehensile and opposable movements led to the development of technologies and culture.

Modified jaws and teeth. Modifications to the jaws correlate with a shift from eating insects, to fruits and leaves, to a mixed, or omnivorous, diet. Rectangular jaws and long canine teeth evolved in monkeys and apes. A bow-shaped jaw and smaller, more uniformly sized teeth evolved in the early hominids.

Brain, behavior, and culture. The braincase and the brain increased in size and complexity. As brain size increased, so did length of pregnancy and extent of maternal care. Compared to early primates, later groups had fewer offspring and invested more in them.

Early primates were solitary. Later, some started to live in small groups. Social interactions and cultural traits started to affect reproductive success. **Culture** is the sum of all learned behavioral patterns transmitted among members of a group and between generations.

Origins and Early Divergences

The first primates arose in the tropical forests of East Africa by about 65.5 million years ago (mya). Early species resembled modern tree shrews (Figure 26.33*a,b*). They foraged at night among low branches for insects and seeds. They had a long snout and eyes located toward the sides of the head.

a

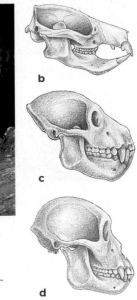

b

c

d

Figure 26.33 (**a**) Southeast Asian tree shrew (*Tupaia*), a close relative of modern primates. Skull comparisons: (**b**) *Plesiadapis*, an early, shrewlike primate. (**c**) Monkey-sized *Aegyptopithecus*, one of the Oligocene anthropoids. (**d**) *Proconsul africanus*. This early hominoid was the size of a four-year-old child.

We know from fossils that the prosimians had evolved by the Eocene. Skeletal changes adapted them to life among the treetops. They had a shorter snout and front-facing eyes. Their brain was bigger than that of early primates. Climbers and leapers had to estimate body weight, distance, wind speed, and suitable destinations. Adjustments had to be quick for a body in motion far above the ground.

By 36 million years ago, tree-dwelling anthropoids arose (Figure 26.33*c*). Between 23 and 18 million years ago, in tropical rain forests, they gave rise to the first hominoids: early apes (Figure 26.33*d*).

Hominoids dispersed through Africa, Asia, and Europe as climates were changing due to shifts in land masses. During this time, Africa became cooler and drier. Tropical forests, with their abundance of edible soft fruits and leaves, were replaced by open woodlands and, later, grasslands. Food became drier, harder, and more difficult to find. Hominoids that had evolved in moist forests either moved into new adaptive zones or died out. Most species died out, but not the shared ancestor of apes and humans. By 6 million years ago, hominids had emerged.

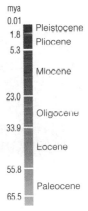

mya
0.01 — Pleistocene
1.8
5.3 — Pliocene

Miocene

23.0
— Oligocene
33.9

Eocene

55.8

Paleocene
65.5

Take-Home Message

What trends shaped the primate lineage ancestral to humans?

■ Early primates were long-snouted animals that clambered about near the ground. Later species were climbers with a skeleton and brain that better adapted them to this new way of life.

- We have fossil evidence of many hominids, but do not know exactly how they are related to one another.

- Link to Gene duplication 12.5

Early Hominids

Genetic comparisons indicate that hominids diverged from apelike ancestors about 6 to 8 million years ago. Fossils that may be hominids are about 6 million years old. *Sahelanthropus tchadensis* had a hominid-like flat face, prominent brow, and small canines, but its brain was the size of a chimpanzee's (Figure 26.34a). *Orrorin tugenensis* and *Ardipithecus ramidus* also had hominid-like teeth. Some researchers suspect that those species stood upright; others disagree. More fossils will have to be discovered to clarify the picture.

An indisputably bipedal hominid, *Australopithecus afarensis*, was established in Africa by about 3.9 million years ago. Remarkably complete skeletons reveal that it habitually walked upright (Figure 26.35). About 3.7 million years ago, two *A. afarensis* individuals walked across a layer of newly deposited volcanic ash. Soon thereafter, a light rain fell, and transformed the powdery ash they had just crossed into stone, preserving their footprints (Figure 26.35c,d).

A. afarensis was one of the **australopiths,** or "southern apes." This informal group includes *Australopithecus* and *Paranthropus* species. *Australopithecus* species were petite; they had a narrow jaw and small teeth (Figure 26.34b). One or more species probably are ancestral to modern humans. In contrast, *Paranthropus* species had a stockier build, a wider face, and larger molars. Jaw muscles attached to a pronounced bony crest at the top of their skull (Figure 26.34c). Large molars and strong jaw muscles indicate that fibrous, difficult-to-chew plant parts accounted for a large part of the diet. *Paranthropus* died out about 1.2 million years ago.

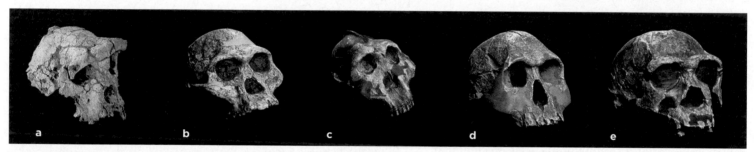

| *Sahelanthropus tchadensis* | *Australopithecus africanus* | *Paranthropus boisei* | *Homo habilis* | *Homo erectus* |
| 6 million years ago | 3.2–2.3 million years ago | 2.3–1.2 million years ago | 1.9–1.6 million years ago | 1.9 million to 53,000 years ago |

Figure 26.34 A sampling of fossilized hominid skulls from Africa, all to the same scale.

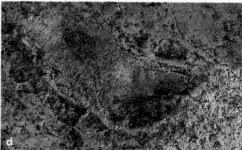

Figure 26.35 (**a**) Fossilized bones of Lucy, a female australopith (*Australopithecus afarensis*). This bipedal hominid lived in Africa 3.2 million years ago.

Chimpanzees and other apes have a splayed-out big toe (**b**). Early hominids did not. How do we know? (**c,d**) At Laetoli in Tanzania, Mary Leakey discovered footprints made in soft, damp volcanic ash 3.7 million years ago. The arch, big toe, and heel marks of these footprints are signs of bipedal hominids.

Early Humans

What do the fossilized fragments of early hominids tell us about human origins? The record is still too incomplete for us to be sure how all the diverse forms were related, let alone which might have been our ancestors. Besides, exactly which traits should we use to define **humans**—members of the genus *Homo*?

Well, what about brains? Our brain is the basis of unsurpassed analytical skills, verbal skills, complex social behavior, and technological innovations. How did an early hominid make the evolutionary leap to becoming human?

Comparing the brains of modern primates can give us clues. We know that genes for some brain proteins underwent repeated duplication (Section 12.5) as the primate lineage evolved. Further studies of how these proteins function may provide additional insight into how our uniquely human mental traits arose.

Until then, we are left to speculate on the evidence of physical traits among diverse fossils. They include a skeleton that permitted bipedalism, a smaller face, larger cranium, and smaller teeth with more enamel. These traits emerged during the late Miocene and can be observed in *Homo habilis*. The name of this early human means "handy man" (Figure 26.36).

Most of the early known forms of *Homo* are from the East African Rift Valley. Fossil teeth indicate that these early humans ate hard-shelled nuts, dry seeds, soft fruits, leaves, and insects. *H. habilis* may have enriched its diet by scavenging carcasses left behind by carnivores such as saber-tooth cats, but it did not have teeth adapted to a diet rich in meat.

Our close relatives, the chimpanzees and bonobos, use sticks and other natural objects as tools (Section 44.6). They smash nuts open with rocks and use sticks to dig into termite nests and capture insects. Early hominids most likely did the same.

By 2.5 million years ago, some hominids had begun modifying rocks in ways that made them better tools. Pieces of volcanic rock chipped to a sharp edge were found with animal bones that show evidence that they were scraped by such tools.

The layers of Tanzania's Olduvai Gorge document refinements in toolmaking abilities (Figure 26.37). The layers that date to about 1.8 million years ago hold crudely chipped pebbles. More recent layers contain more complex tools, such as knifelike cleavers.

Olduvai Gorge also holds hominid fossils. At the time of their discovery, these fossils were classified as *Homo erectus*. This name means "upright man." Today, some researchers reserve that name for fossils in Asia.

Figure 26.36 Painting of a band of *Homo habilis* in an East African woodland. Two australopiths are shown in the distance at the left.

Figure 26.37 A sample of stone tools from Olduvai Gorge in Africa. From left to right, crude chopper, more refined chopper, hand ax, and cleaver.

They prefer to call the African fossils *H. ergaster*. In our discussions, we will adopt a traditional approach using "*H. erectus*" in reference to African populations and to descendant populations that, over generations, made their way into Europe and Asia.

H. erectus adults averaged about 1.5 meters (5 feet) tall, and had a larger brain than *H. habilis*. Improved hunting skills may have helped *H. erectus* get the food needed to maintain a large body and brain. Also, *H. erectus* built fires, so cooking probably broadened their diet by softening previously inedible hard foods.

Take-Home Message

What were the now extinct hominids like?

■ Australopiths and certain hominids that preceded them walked upright. *Homo habilis*, the earliest known human species, also walked upright. *Homo erectus* had a larger brain and dispersed out of Africa.

- Modern humans first evolved in Africa and relatively recently spread from there throughout the world.

- Link to Allopatric speciation 18.10

Branchings of the Human Lineage

By 1.7 million years ago, *Homo erectus* populations had become established in places as far away from Africa as the island of Java and eastern Europe. At the same time, African populations continued to thrive. Over thousands of generations, geographically separated groups adapted to local conditions. Some populations became so different from parental *H. erectus* that we call them new species: *H. neanderthalensis* (Neandertals), *H. floresiensis*, and *H. sapiens*, or fully modern humans (Figure 26.38).

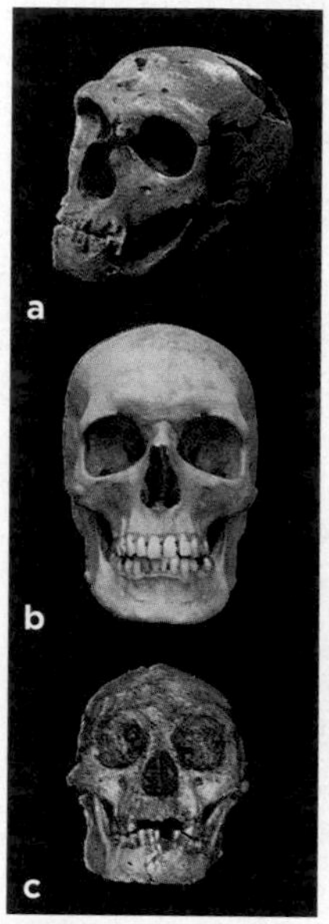

We know from one fossil found in Ethiopia that *Homo sapiens* had evolved by 195,000 years ago. Compared to *H. erectus*, *H. sapiens* had smaller teeth, facial bones, and jawbones. *H. sapiens* also had a higher, rounder skull, a larger brain, and a capacity for spoken language.

From 200,000 to 30,000 years ago, Neandertals lived in Africa, the Middle East, Europe, and Asia. They were stocky enough to endure colder climates. A stocky body has a lower ratio of surface area to volume than a thin one, so it loses heat less quickly.

Neandertals had a big brain. Did they have a spoken language? We do not know. They vanished when *H. sapiens* entered the same regions. The new arrivals may have driven Neadertals to extinction through warfare or by outcompeting them for resources. Members of the two species may have occasionally mated, but comparisons

Figure 26.38
Recent *Homo* species. (**a**) *H. neanderthalensis*, (**b**) *H. sapiens* (modern human), (**c**) *H. floresiensis*.

between DNA from modern humans and DNA from Neandertal remains indicate that Neandertals did not contribute to the gene pool of modern *Homo sapiens*.

In 2003, human fossils about 18,000 years old were discovered on the Indonesian island of Flores. Like *H. erectus*, they had a heavy brow and a relatively small brain for their body size. Adults would have stood a meter tall. Scientists who found the fossils assigned them to a new species, *H. floresiensis*. Not everyone is convinced. Some think the fossils belong to *H. sapiens* individuals who had a disease or disorder.

Where Did Modern Humans Originate?

Neandertals evolved from *H. erectus* populations in Europe and western Asia. *H. floresiensis* evolved from *H. erectus* in Indonesia. Where did *H. sapiens* originate? Two major models agree that *H. sapiens* evolved from *H. erectus* but differ over where and how fast. Both attempt to explain the distribution of *H. erectus* and *H. sapiens* fossils, as well as genetic differences among modern humans who live in different regions.

Multiregional Model By the **multiregional model**, populations of *H. erectus* in Africa and other regions evolved into populations of *H. sapiens* gradually, over more than a million years. Gene flow among populations maintained the species through the transition to fully modern humans (Figure 26.39*a*).

By this model, some of the genetic variation now seen among modern Africans, Asians, and Europeans began to accumulate soon after their ancestors branched from an ancestral *H. erectus* population. The model is based on interpretation of fossils. For example, faces of *H. erectus* fossils from China are said to look more like modern Asians than those of *H. erectus* that lived in Africa. The idea is that much variation seen among modern *H. sapiens* evolved long ago, in *H. erectus*.

Replacement Model By the more widely accepted **replacement model**, *H. sapiens* arose from a single *H. erectus* population in sub-Saharan Africa within the past 200,000 years. Later, bands of *H. sapiens* entered regions already occupied by *H. erectus* populations, and drove them all to extinction (Figure 26.39*b*). If this model is correct, then the regional variations observed among modern *H. sapiens* populations arose relatively recently. This model emphasizes the enormous degree of genetic similarity among living humans.

Fossils support the replacement model. *H. sapiens* fossils date back to 195,000 years ago in East Africa and 100,000 years ago in the Middle East. In Australia,

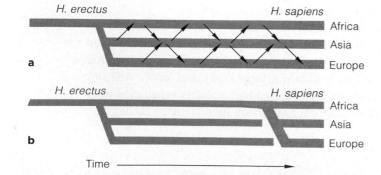

Figure 26.39 Two models for the origin of *H. sapiens*. (**a**) Multiregional model. *H. sapiens* slowly evolves from *H. erectus* in many regions. Arrows represent ongoing gene flow among populations. (**b**) Replacement model. *H. sapiens* rapidly evolves from one *H. erectus* population in Africa, then disperses and replaces *H. erectus* populations in all regions.

the oldest such fossils date to 60,000 years ago and, in Europe, they date to 40,000 years ago. Global comparisons of markers in mitochondrial DNA, and in the X and Y chromosomes, place the modern Africans closest to the root of the family tree. They also reveal that the most recent common ancestor of all humans now alive lived in Africa approximately 60,000 years ago.

Leaving Home

Long-term shifts in the global climate drove human bands away from Africa (Figure 26.40). About 120,000 years ago, Africa's interior was getting cooler and drier. As patterns and amounts of rainfall changed, so did the distribution of herds of grazing animals and the humans who hunted them. A few hunters may have journeyed north from East Africa into Israel, where fossils 100,000 years old were found inside a cave. These populations apparently died out. Eruption of Mount Toba in Indonesia 73,000 years ago may have killed them, along with other ancient travelers. The enormous eruption released 10,000 times more ash than the 1981 eruption of Mount St. Helens in Washington State. The resulting cloud of debris had a devastating impact on the global climate.

Later waves of travelers had better luck, as some individuals left established groups and ventured into new territory. Successive generations continued along the coasts of Africa, then Australia and Eurasia. In the Northern Hemisphere, much of Earth's water became locked in vast ice sheets, which lowered the sea level by hundreds of meters. Previously submerged land was drained off between some regions. About 15,000 years ago, one small band of humans crossed such a land bridge from Siberia into North America.

Deserts and mountains influenced the dispersal routes (Figure 26.40b). Until about 100,000 years ago, enough rain fell in northern Africa to sustain plants and herds of grazing animals. By 45,000 years ago, blazing hot sand stretched for more than 3,200 kilometers (2,000 miles). Humans whose ancestors had passed through this region no longer had the option of moving back to the grasslands of central Africa. The newly enlarged desert blocked their way.

With return to Africa no longer an option, groups moved east into central Asia, where the towering Himalayas and other peaks of the Hindu Kush forced some to detour north, into western China, and others south, into India. Descendants of humans that moved

Figure 26.40 (a) Some dispersal routes for small bands of *Homo sapiens*. This map shows ice sheets and deserts that prevailed about 60,000 years ago (ya). It is based on clues from sedimentary rocks and ice core drillings. Times when modern humans appeared in these regions are based on fossils studies of genetic markers in mitochondrial DNA and Y chromosomes from 10,000 individuals around the world:

Africa	by 195,000 years ago
Israel	100,000
Australia	60,000
China	50,000
Europe	40,000
North America	11,000

(b) Global climate changes caused expansion and contraction of deserts in Africa and the Middle East. Resulting changes in food sources may have encouraged migrations of small groups out of Africa. Locations of ice sheets, deserts, and tall mountain ranges influenced migration routes.

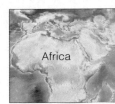

Africa

120,000 ya

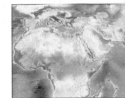

60,000 ya

30,000 ya

into Asia eventually reached Siberia, then traveled into North America. Colonists from central Asia moved west, across cold grasslands. Some crossed mountains in the Balkans, and continued on into Europe.

With each step of their journey, humans faced and overcame extraordinary hardships. During this time, they devised cultural means to survive in inhospitable environments. Unrivaled capacities for modifying the habitat and for language served them well. Cultural evolution is ongoing. Hunters and gatherers persist in a few parts of the world, while others of us live in the age of "high tech." This coexistence of such diverse groups is a tribute to the great behavioral flexibility of the human species.

b   present

Take-Home Message

What do fossils and DNA studies tell us about the evolution of modern humans?

■ Fossils and genetic evidence indicate that modern humans, *H. sapiens*, evolved from a *H. erectus* population in Africa.

■ Modern humans dispersed out of Africa during a time when long-term shifts in climate influenced their options.

Sale of vertebrate fossils is big business for rock shops, auction houses, and websites. Most such fossils are not particularly important to scientists, but some are. For example, one of the few *Archaeopteryx* fossils in existence is privately held. It shows details of the bird's feet that are not visible in other fossils. Some scientists argue that private ownership of such fossils thwarts research and endangers an irreplaceable legacy.

Summary

Section 26.1 Four features help define **chordates**: a notochord, a dorsal hollow nerve cord, a pharynx with gill slits, and a tail extending past the anus. All features form in embryos and may or may not persist in adults. Invertebrate chordates include **tunicates** and **lancelets**, both marine filter-feeders. **Craniates** are chordates with a braincase of cartilage or bone. Structurally, a jawless fish called the hagfish is the simplest modern craniate. Most craniates are **vertebrates**.

- *Use the animation on CengageNOW to examine the body plan and chordate features of a lancelet.*

Section 26.2 Vertebrates have an **endoskeleton** with a **vertebral column** (backbone) of cartilaginous or bony **vertebrae**. **Jaws** and paired **fins** evolved in early fishes. In lineages that moved onto land, **gills** were replaced by **lungs**, **kidneys** became better at conserving water, and the circulatory system became more efficient.

- *Use the animation on CengageNOW to explore the chordate family tree and see how jaws evolved.*

Sections 26.3, 26.4 Lampreys are jawless fishes with a backbone. Jawed fishes include the **cartilaginous fishes** and **bony fishes**. Both have **scales** on their skin. A **swim bladder** helps bony fishes regulate their buoyancy.

Sections 26.5, 26.6 **Tetrapods**, or four-legged walkers, evolved from lobe-finned bony fishes. **Amphibians** are tetrapods that live on land, but typically return to water to reproduce. Many amphibians now face extinction.

Sections 26.7, 26.8 **Amniotes**, the first vertebrates able to complete their life cycle on dry land, have water-conserving skin and kidneys, and amniote eggs. **Reptiles** (including extinct **dinosaurs**) and birds are one amniote lineage; modern mammals are another. The **K–T asteroid hypothesis** proposes that an asteroid impact led to the extinction of dinosaurs.

Section 26.9 Reptiles are **ectotherms** (cold-blooded animals) with scales. Eggs are laid on land and fertilization is usually internal. A **cloaca** functions in excretion and reproduction. Lizards and snakes are the most diverse groups. Crocodilians are the closest relatives of birds.

- *Use the animation on CengageNOW to explore the body plan of a crocodile.*

Section 26.10 **Birds** are **endotherms** (warm-blooded animals) and the only living animals with feathers. The body plan of most has been highly modified for flight.

- *Use the animation on CengageNOW to see what is inside a bird egg and how birds are adapted for flight.*

Sections 26.11, 26.12 **Mammals** nourish young with milk secreted by mammary glands, have fur or hair, and have more than one kind of tooth. Three lineages are egg-laying mammals (**monotremes**), pouched mammals (**marsupials**), and **placental mammals**, the most diverse group. A **placenta** is an organ that facilitates exchange of substances between the embryonic and maternal blood.

- *Use the animation on CengageNOW to see how the current distribution of mammalian groups arose.*

Figure 26.41 Estimated dates for the origin and extinction of three hominid genera. Purple lines show one view of how the human species relate to one another. Number of species, which fossils belong to each species, and how species relate remains a matter of debate.

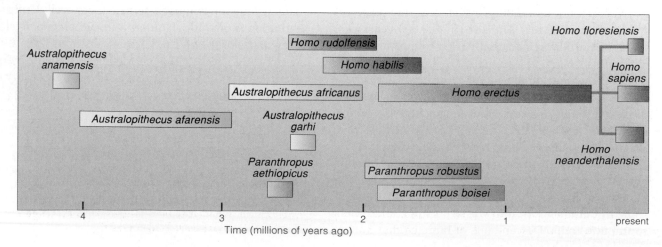

Data Analysis Exercise

As Section 26.13 mentioned, one trend in primate evolution involved changes in life history traits, such as length of infancy and the time it takes to reach adulthood.

Figure 26.42 compares five primate lineages, from most ancient to most recent. It graphs life spans and years spent as "infants" when ongoing maternal care is required. It shows time spent as subadults, when individuals are no longer dependent on their mother for care, but have not yet begun to breed. It shows the length of the reproductive years, and the length of time lived after reproductive years have passed.

1. What is the average life span for a lemur? A gibbon?

2. Which group reaches adulthood most quickly?

3. Which group has the longest expanse of reproductive years?

4. Which groups survive past their reproductive years?

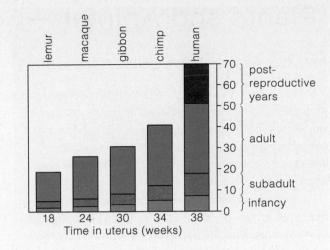

Figure 26.42 Trend toward longer life spans and greater dependency of offspring on adults for five primate lineages.

Sections 26.13–26.15 **Primates** include prosimians such as tarsiers and anthropoids such as monkeys, apes, and **hominids**—humans and extinct humanlike forms. Early primates were shrewlike. **Bipedalism**, improved daytime vision, refined hand movements, smaller teeth, bigger brains, social complexity, extended parental care, and, later, **culture** evolved in some lineages.

All hominids, including **australopiths** and **humans**, originated in Africa. The human lineage (*Homo*) arose by 2 million years ago (Figure 26.41). By the **multiregional model**, *H. sapiens* evolved in many separate regions. The **replacement model** postulates that *H. sapiens* evolved in Africa, then dispersed. It is now the favored model.

Self-Quiz

Answers in Appendix III

1. List the four distinguishing chordate traits.

2. Which of these traits are retained by an adult lancelet?

3. Vertebrate jaws evolved from _____ .
   a. gill supports   b. ribs   c. scales   d. teeth

4. Lampreys and sharks both have _____ ,
   a. jaws   d. a swim bladder
   b. a bony skeleton   e. a four-chambered heart
   c. a cranium   f. lungs

5. Which group of bony fish gave rise to tetrapods?

6. Reptiles and birds belong to one major lineage of amniotes, and _____ belong to another.
   a. sharks   c. mammals
   b. frogs and toads   d. salamanders

7. Reptiles are adapted to life on land by _____ .
   a. tough skin   d. amniote eggs
   b. internal fertilization e. both a and c
   c. efficient kidneys   f. all of the above

8. The closest modern relatives of birds are _____ .
   a. crocodilians   b. prosimians   c. tuataras   d. lizards

9. Only birds have _____ .
   a. a cloaca   c. feathers
   b. a four-chambered heart d. amniote eggs

10. An australopith is a _____ .
    a. craniate   d. amniote
    b. vertebrate   e. placental mammal
    c. hominoid   f. all of the above

11. Match the organisms with the appropriate description.
    ___ lancelets   a. pouched mammals
    ___ fishes   b. invertebrate chordates
    ___ amphibians   c. feathered amniotes
    ___ reptiles   d. egg-laying mammals
    ___ birds   e. humans and close relatives
    ___ monotremes   f. cold-blooded amniotes
    ___ marsupials   g. first land tetrapods
    ___ hominids   h. most diverse vertebrates

■ *Visit CengageNOW for additional questions.*

Critical Thinking

1. In 1798, a stuffed platypus specimen was delivered to the British Museum. Reports that it laid eggs added to the confusion. To modern biologists, a platypus is clearly a mammal. It has fur and the females produce milk. Young animals have typical mammalian teeth that are replaced by hardened pads as the animal matures. Why do you think modern biologists can more easily accept that a mammal can have some seemingly reptilian traits?

2. The cranial volume of early *H. sapiens* averaged 1,200 cubic centimeters. It now averages 1,400 cubic centimeters. By one hypothesis, females chose the cleverest mates, the advantage being offspring with genes that favorably affect intelligence. What types of data might a researcher gather to test this sexual selection hypothesis?

A Cautionary Tale

A cell can only survive within a certain range of conditions. As explained in Section 6.3, changes in acidity, salinity, or temperature can inactivate the enzymes that catalyze the many reactions necessary for life. To remain alive, any multicelled organism must keep conditions inside its body within the range its cells can tolerate.

Heat stroke is an example of what can happen when internal conditions get out of balance. It can be deadly. For example, Korey Stringer, a football player for the Minnesota Vikings, collapsed of heat stroke during a practice (Figure 27.1). He and his team were working out in full uniform on a day when temperature and humidity were high.

Stringer was rushed to the hospital with an internal body temperature of 108.8°F (42.7°C), and a blood pressure too low to measure. Doctors immersed him in a bath of ice water to bring his temperature down, but irreparable damage had

already been done. Stringer's blood clotting mechanism shut down and he started to bleed internally. Then his kidneys faltered. He stopped breathing and was attached to a respirator, but his heart gave out. Less than twenty-four hours after the football practice had started, Stringer was pronounced dead. He was twenty-seven years old.

The human body functions best when internal temperature remains between about 97°F (36°C) and 100°F (38°C). Above 104°F (40°C), blood flow is increasingly diverted from internal organs to the skin. Heat is transferred from skin to air, as long as a body is warmer than its surroundings. Sweating helps get rid of heat, but it is less effective on humid days.

When internal temperature climbs above 105°F (40.6°C), normal cooling processes fail and heat stroke occurs. The body stops sweating, and its core temperature begins to shoot up. The heart beats faster; fainting or confusion follow. Without prompt treatment, brain damage or death can occur.

We use this sobering example as our introduction to anatomy and physiology. *Anatomy* is the study of body form. *Physiology* is the study of how the body's parts are put to use. This information can help you understand what is going on inside your own body. More broadly, it can also help you appreciate how all organisms survive.

We discuss the anatomy and physiology of plants and animals separately in later chapters. In this chapter, we provide an overview of the processes and structural traits that the two groups share in common.

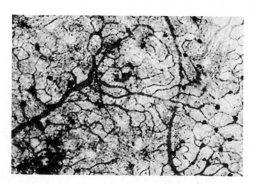

See the video! Figure 27.1 *Left*, Korey Stringer, during his last practice with his team. When the body's temperature rises, profuse sweating increases evaporative cooling. Also, blood is directed to capillaries of the skin (*above*), which radiate heat into the air. In Stringer's case, homeostatic control mechanisms were no match for strenuous activity on a hot, humid day.

Key Concepts

Many levels of structure and function

Cells of plants and animals are organized in tissues. Tissues make up organs, which work together in organ systems. This organization arises as the plant or animal grows and develops. Interactions among cells and among body parts keep the body alive. **Section 27.1**

Similarities between animals and plants

Animals and plants exchange gases with their environment, transport materials through their body, maintain volume and composition of their internal environment, and coordinate cell activities. They also respond to threats and to variations in available resources. **Section 27.2**

Homeostasis

Homeostasis is the process of keeping conditions in the body's internal environment stable. The feedback mechanisms that often play a role in homeostasis involve receptors that detect stimuli, an integrating center, and effectors that carry out responses. **Sections 27.3–27.5**

Cell communication in multicelled bodies

Cells of tissues and organs communicate by secreting chemical molecules into extracellular fluid, and by responding to signals secreted by other cells. **Section 27.6**

Links to Earlier Concepts

- With this chapter, we return to the concept of levels of organization introduced in Section 1.1. We also explore some examples of sensing and responding to stimuli (1.2), one of the signature traits of life.

- You will learn how constraints imposed by the ratio of surface area to volume (4.2) affect body structures.

- Cellular structures such as cell junctions (4.12) and membrane proteins (5.2) also come into play, as do cellular processes such as transport (5.3) and energy-releasing pathways (8.1).

- We discuss the ability of plants and animals to fight infectious disease (21.8) and how their bodies are adapted to life on land (23.1, 26.5, 26.7).

How would you vote? The interior of a vehicle heats up fast on even a mild day. Each year children left in vehicles die as a result of heat stroke. Some states have made it a crime to leave a child alone in a parked car. Do you support such laws? See CengageNOW for details, then vote online.

- Earlier chapters covered plant and animal diversity. Here we begin to consider how their bodies are organized.

- Links to Levels of organization 1.1, Natural selection 17.3, Land plants 23.1, Land animals 26.5 and 26.7

From Cells to Multicelled Organisms

The body of any plant or animal consists of hundreds to hundreds of trillions of cells. In all but the simplest bodies, cells become organized as tissues, organs, and organ systems, each capable of specialized functions. Said another way, there is a division of labor among parts of a plant or animal body (Section 23.1).

A **tissue** consists of one or more cell types—and often an extracellular matrix—that collectively perform a specific task or tasks. Each tissue is characterized by the types of cells it includes and their proportions. For example, nervous tissue has different types and proportions of cells than muscle tissue or bone tissue.

An **organ** consists of two or more tissues that occur in specific proportions and interact in carrying out a specific task or tasks. For example, a leaf is an organ that serves in gas exchange and photosynthesis (Figure 27.2); lungs are organs of gas exchange (Figure 27.3).

Organs that interact in one or more tasks form an **organ system**. Leaves and stems are components of a plant's gas exchange system. Lungs and airways are organs of the respiratory system of land vertebrates.

Growth Versus Development

A plant or animal becomes structurally organized as it grows and develops. For any multicelled species, **growth** refers to an increase in the number, size, and volume of cells. We describe it in quantitative terms. **Development** is a series of stages in which specialized tissues, organs, and organ systems form in heritable patterns. We describe it in qualitative terms; usually by describing the stages. For example, both plants and animals have an early stage called the embryo.

Evolution of Form and Function

All anatomical and physiological traits have a genetic basis and thus have been affected by natural selection. The traits we see in modern species are the outcome of differences in survival and reproduction among many generations of individuals who varied in their traits. Only traits that proved adaptive in the past have been passed along to modern generations.

For example, Section 23.1 discussed how plants adapted to life on dry land. As plants radiated out of the aquatic environment onto land, they faced a new challenge—they had to keep from drying out in air. We see solutions to this challenge in the anatomy of roots, stems, and leaves (Figure 27.2). Internal pipes called xylem convey water that roots absorb from soil upward to leaves. The epidermal tissue that covers leaves and stems of vascular plants secretes a waxy cuticle that reduces evaporative water loss. Stomata, small gaps across a leaf's epidermis, can open to allow gas exchange or close to prevent water loss.

Flower, a reproductive organ

Cross-section of a leaf, an organ of photosynthesis and gas exchange

shoot system (aboveground parts)

root system (belowground parts, mostly)

Cross-section of a stem, an organ of structural support, storage, and distribution of water and food

Figure 27.2 Animated Anatomy of a tomato plant. Its vascular tissues (*purple*) conduct water, dissolved mineral ions, and organic compounds. Another tissue makes up the bulk of the plant body. A third covers all external surfaces. Organs such as flowers, leaves, stems, and roots are each made up of all three tissues.

Figure 27.3 Parts of the human respiratory system. Cells making up the tissues of this system carry out specialized tasks.

Airways to paired lungs are lined with epithelial tissue. Ciliated cells in this tissue whisk any bacteria and particles that might cause infections away from the lungs.

Lungs are organs of gas exchange. Inside them are air sacs lined with continually moist epithelial tissue. Tiny vessels (capillaries) filled with blood surround the air sacs and interact with them in the task of gas exchange.

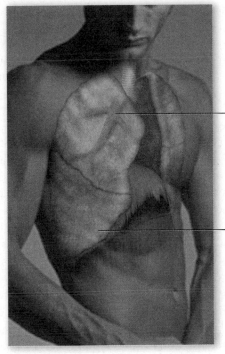

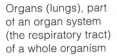

Organs (lungs), part of an organ system (the respiratory tract) of a whole organism

Ciliated cells and mucus-secreting cells of a tissue that lines respiratory airways

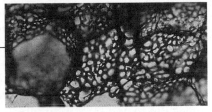

Lung tissue (tiny air sacs) laced with blood capillaries—one-cell-thick tubular structures that hold blood, which is a fluid connective tissue

Similarly, animals evolved in water and faced new challenges when they moved onto land (Sections 26.5 and 26.7). Gases can only move into and out of an animal's body by moving across a moist surface. That is not a problem for aquatic organisms, but on land, evaporation can cause moist surfaces to dry out. The evolution of respiratory systems allowed land animals to exchange gases with air across a moist surface deep inside their body.

In land vertebrates, the respiratory system typically includes airways and paired lungs (Figure 27.3). The tissue that lines the airways leading to lungs includes ciliated cells that can capture airborne particles and pathogens. Deep inside the lungs gases are exchanged between air and blood across the thin, continually moistened tissue of tiny air sacs.

The Internal Environment

A single-celled organism gets necessary nutrients and gases from the fluid around it. Plant and animal cells are also surrounded by fluid. This **extracellular fluid (ECF)** is like an internal environment in which body cells live. To keep cells alive, a body's parts work in concert in ways that maintain the volume and composition of the extracellular fluid.

A Body's Tasks

The next two units describe how a plant or an animal carries out the following essential functions:

- Maintains favorable conditions for its cells
- Acquires and distributes water, nutrients, and other raw materials; disposes of wastes
- Defends itself against pathogens
- Reproduces
- Nourishes and protects gametes and (in many species) embryos

Each living cell engages in metabolic activities that keep it alive. At the same time, integrated activities of cells in tissues, organs, and organ systems sustain the body as a whole. Their interactions keep conditions in the internal environment within tolerable limits—a process we call **homeostasis**.

Take-Home Message

How are plant and animal bodies organized?

■ Plant and animal bodies typically consist of cells organized as tissues, organs, and organ systems. The ways in which body parts are arranged and function have a genetic basis and have been shaped by natural selection.

■ Collectively, cells, tissues, and organs maintain conditions inside the body.

- Although plants and animals differ in many ways, they share some common challenges.

- Links to Surface area-to-volume ratio 4.2, Diffusion and transport mechanisms 5.3, Energy-releasing pathways 8.1

Gas Exchange

To begin thinking about the processes that occur in both plants and animals, consider how the golfer Tiger Woods is like a tulip (Figure 27.4). Cells inside both bodies release energy by carrying out aerobic respiration (Section 8.1). This pathway requires oxygen and produces carbon dioxide. Some tulip cells also carry out photosynthesis, an energy-storing process that requires carbon dioxide and produces oxygen.

All multicelled species respond, structurally and functionally, to this common challenge: Quickly move molecules to and from individual cells.

By the process of **diffusion**, ions or molecules of a substance move from a place where they are concentrated to one where they are more scarce (Section 5.3). Plants and animals keep gases diffusing in directions most suitable for metabolism and cell survival. How? That question will lead you to stomata at leaf surfaces (Section 28.4) and to the circulatory and respiratory systems of animals (Chapters 37 and 39).

Internal Transport

Diffusion is most effective over small distances. As an object's diameter increases, its ratio of surface area to volume decreases (Section 4.2). This means that as the diameter of a body part becomes larger, interior cells get farther and farther from the body surface, and there is less body surface per cell.

As a result of this constraint, plants and animals that rely on diffusion alone to move materials through their body tend to be small and flat. Flatworms and liverworts are two examples (Figure 27.5a,b). Both are just a few cell layers thick.

Most plants and animals that are not small and flat have vascular tissues—systems of tubes through which substances move to and from cells. A leaf vein in a vascular plant consists of long strands of xylem and phloem, the two types of vascular tissue (Figure 27.5c). Human blood vessels such as veins and capillaries are our vascular tissues (Figure 27.5d).

In both plants and animals, vascular tissue carries water, nutrients, and signaling molecules. In animals, this tissue also distributes gases. Gases move into and through a plant by diffusion. Components of animal blood fight infection. Similarly, phloem of vascular plants carries chemicals made in response to injury.

Maintaining the Water–Solute Balance

Plants and animals continually gain and lose water and solutes. Even so, to stay alive they must maintain the volume and composition of their extracellular fluid within limited ranges. How do they do this? Plants and animals differ hugely in this respect, yet you can still find common responses by zooming down to the level of molecules.

At the surface of a body or an organ, cells in sheets of tissue carry out active and passive transport. Recall that in **passive transport**, a solute moves down its concentration gradient with the assistance of a transport protein. In **active transport**, a protein pumps one specific solute from a region of low concentration to one of higher concentration (Section 5.3).

Active transport by cells in plant roots helps control which solutes move into the plant. In leaves, active transport puts sugars made by photosynthesis into phloem, which distributes them through the plant.

In animals, active transport moves nutrients from food inside the gut into body cells. In vertebrates, active transport allows kidneys to eliminate wastes and excess solutes and water in the urine.

Cell-to-Cell Communication

Plants and animals have another crucial similarity: Both depend on communication among cells. Many types of specialized cells release signaling molecules

Figure 27.4 What do Tiger and the tulips have in common?

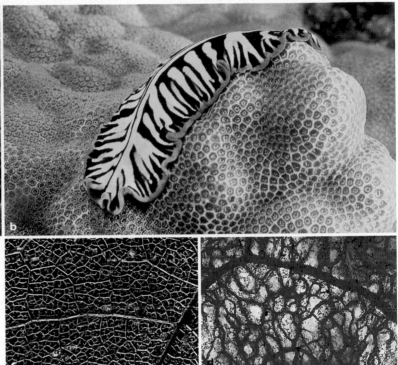

Figure 27.5 Having a flattened body allows a liverwort (**a**) and a flatworm (**b**) to do just fine without vascular tissues. All their cells lie close to the body surface.

Evolution of vascular tissues such as (**c**) leaf veins in a dicot and (**d**) blood vessels in a human allow these organisms to grow much larger and have thicker body parts.

that help coordinate and control events in the body as a whole. Signaling mechanisms guide how the plant or animal body grows, develops, and maintains itself, and also reproduces.

On Variations in Resources and Threats

A habitat is a place where members of a species typically live. Each habitat has a specific set of resources and poses a unique set of challenges. Each has unique physical characteristics. Water and nutrients may be plentiful or scarce. The habitat may be brightly lit, a bit shady, or dark. It may be whipped by winds or still. Temperature may vary a lot or a little over the course of a day. Similarly, conditions may change with the season or stay more or less constant.

Biotic (living) components of the habitat vary as well. Different producers, predators, prey, pathogens, or parasites may be present. Competition for resources and reproductive partners may be minimal or fierce. Variation in these factors promotes diversity in form and function.

Even with all the diversity, we may still see similar responses to similar challenges. Sharp cactus spines or porcupine quills deter most animals that might eat a cactus or porcupine (Figure 27.6). Modified epidermal cells give rise to both spines and quills that defend the body against potential predators.

Figure 27.6 Protecting body tissues from predation: (**a**) Cactus spines. (**b**) Quills of a porcupine (*Erethizon dorsatum*).

Take-Home Message

How are plant and animal bodies similar?

■ Plants and animals carry out aerobic respiration and exchange gases with the environment.

■ Most plants and most animals have vascular tissues that function in transport.

■ Plants and animals keep their internal environment stable by regulating which substances enter their body and which are eliminated.

- Detecting and responding to changes is a characteristic trait of all living things and the key to homeostasis.

- Link to Sensing and responding to change 1.2

Detecting and Responding to Changes

In animals, homeostasis involves interactions among receptors, integrators, and effectors (Figure 27.7). A **receptor** is a cell or cell component that changes in response to specific stimuli. Some receptors such as those in eyes, ears, and skin respond to external stimuli such as light, sound, or touch. Receptors involved in homeostasis function like internal watchmen. They detect changes inside the body. For example, some receptors detect blood pressure changes, others detect changes in the level of carbon dioxide in the blood, and still others detect changes in internal temperature.

Information from sensory receptors throughout the body flows to an **integrator**: a collection of cells that receives and processes information about stimuli. In vertebrates, this integrator is the brain.

In response to the signals it receives, the integrator sends a signal to **effectors**—muscles, glands, or both—that carry out responses to the stimulation.

Sensory receptors, integrators, and effectors often interact in feedback systems. In such systems, some stimulus causes a change from a set point, which then "feeds back" and affects the original stimulus.

Negative Feedback

In a **negative feedback mechanism**, a change leads to a response that reverses that change. Think of how a furnace with a thermostat operates. A user sets the thermostat to a desired temperature. When the temperature decreases below this preset point, the furnace turns on and emits heat. When the temperature rises to the desired level, the thermostat turns off the heat.

Similar feedback mechanisms help keep a human's internal body temperature near 98.6°F (37°C) despite changes in the temperature of the surroundings.

STIMULUS Sensory input into the system

| **Receptor**
such as a free nerve ending in the skin | → | **Integrator**
such as the brain or the spinal cord | → | **Effector**
a muscle or a gland |

Figure 27.7 The three types of components that interact in homeostasis in animal bodies.

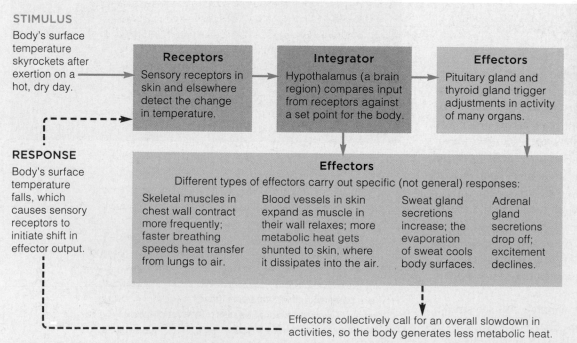

STIMULUS

Body's surface temperature skyrockets after exertion on a hot, dry day.

RESPONSE

Body's surface temperature falls, which causes sensory receptors to initiate shift in effector output.

Receptors
Sensory receptors in skin and elsewhere detect the change in temperature.

Integrator
Hypothalamus (a brain region) compares input from receptors against a set point for the body.

Effectors
Pituitary gland and thyroid gland trigger adjustments in activity of many organs.

Effectors
Different types of effectors carry out specific (not general) responses:

Skeletal muscles in chest wall contract more frequently; faster breathing speeds heat transfer from lungs to air.

Blood vessels in skin expand as muscle in their wall relaxes; more metabolic heat gets shunted to skin, where it dissipates into the air.

Sweat gland secretions increase; the evaporation of sweat cools body surfaces.

Adrenal gland secretions drop off; excitement declines.

Effectors collectively call for an overall slowdown in activities, so the body generates less metabolic heat.

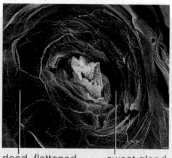

dead, flattened skin cell   sweat gland pore

Scanning electron micrograph of a sweat gland pore at the skin surface. Such glands are among the effectors for this control pathway.

Figure 27.8 Animated Major homeostatic controls over a human body's internal temperature. *Solid* arrows signify the main control pathways. *Dashed* arrows signify the feedback loop.

Consider what happens when you exercise on a hot day. During exercise, muscles increase their metabolic rate. Because metabolic reactions generate heat, body temperature rises. Receptors sense the increase and trigger changes that affect the whole body (Figure 27.8). Blood flow shifts, so more blood from the body's hot interior flows to the skin. This maximizes the amount of heat that dissipates to the surrounding air. At the same time, glands in the skin increase their secretion of sweat. Sweat is mostly water and as it evaporates, it helps cool the body surface. Breathing rate and the volume of each breath increase, speeding the transfer of heat from the blood flowing through your lungs to the air. Levels of excitatory hormones decline, so you feel more sluggish. As your activity level slows, and your rate of heat loss to the environment rises, your temperature falls. Thus, the stimulus (high body temperature) that triggered these responses is reversed by the responses.

For most people, most of the time, this feedback mechanism will prevent overheating. The heat illness that occurs when negative feedback mechanisms fail is the topic of the next section.

Positive Feedback

Positive feedback mechanisms also operate in a body, although they are less common than negative feedback ones. These mechanisms spark a chain of events that intensify change from an original condition. In living organisms, intensification eventually leads to a change that ends feedback.

For example, when a woman is giving birth, muscles of her uterus contract and force the fetus against the wall of this organ. The resulting pressure on the uterine wall induces secretion of a signaling molecule (oxytocin) that causes stronger contractions. In a positive feedback loop, as contractions get more forceful, pressure on the uterine wall increases, thus causing still stronger contractions. The positive feedback cycle continues until the child is born.

Take-Home Message

What types of mechanisms operate in animal homeostasis?
■ Change-detecting receptors, an information-processing brain, and muscles and glands controlled by the brain interact in homeostasis.
■ Negative feedback mechanisms can reverse changes to conditions within the body.
■ Positive feedback is less common than negative feedback. It causes a temporary intensification of a change in the body.

27.4 | Heat-Related Illness

■ Heat stroke is a failure of homeostasis that can cause irreversible brain damage or death.

In a typical year, about 175 Americans die as a direct result of heat exposure. To avoid heat-related problems, listen to your body. Most heat-related deaths in young, healthy adults occur when people continue to exert themselves despite clear warnings that something is amiss.

Social pressure to continue an activity often plays a role in exertion-induced heat stress. An attempt to impress a coach or peers, or to satisfy a boss, can push a healthy person beyond safe limits. Symptoms of heat exhaustion include dizziness, blurred vision, muscle cramping, weakness, nausea, and vomiting. Korey Stringer vomited repeatedly during his final practice, but did not stop working out. Similarly, a young firefighter recruit in Florida complained of weakness and blurred vision. Yet he ran until he collapsed with a body temperature of 108°F. Immediate treatment by fellow firefighters and quick hospitalization could not save him; he died nine days later.

Part of the problem is that heat exhaustion can impair judgment. Profuse sweating causes loss of water and salts, changing the concentration of the extracellular fluid. Blood flow to the gut and liver decreases. Starved of nutrients and oxygen they need, these organs release toxins into the blood. The toxins interfere with function of the nervous system, as well as other organ systems. As a result, a person may be incapable of recognizing and responding to seemingly obvious signs of danger.

To stay safe outside on a hot day, drink plenty of water and avoid excessive exercise. If you must exert yourself, take frequent breaks and monitor how you feel. Wear light-colored, lightweight, breathable clothing. Stay in the shade, or if you must be in direct sunlight wear a hat and use a strong sunscreen. Sunburn impairs the skin's ability to transfer heat to the air.

Keep in mind that high humidity adds to the danger. Evaporation slows when there is more water in the air, so sweating is less effective on humid days. A 95°F (35°C) day with 90 percent humidity puts more heat stress on the body than a 100°F (37.8°C) day accompanied by 55 percent humidity.

Responses to heat can vary with age and certain medical conditions. Pregnant women, the elderly, and people with heart problems or diabetes are at an elevated risk for heat stroke and should be especially careful. Use of alcohol, blood pressure medications, antidepressants, and other drugs also make heat-related problems more likely. Also, people can become acclimated to a high external temperature; those who are not used to living with heat are at an increased risk for heat-related problems.

If you suspect someone is suffering from heat stroke, call for medical help immediately. Give the heat-stroke victim water to drink, then have them lie down with their feet slightly elevated. Spray or sponge them with cool water and, if possible, place ice packs under their armpits.

- Plants too must maintain internal conditions within a range that their cells can tolerate.

- Link to Infectious disease 21.8

Directly comparing plants and animals is not always possible. For example, as a plant grows, new tissues arise only at particular sites in roots and shoots. In animal embryos, tissues form all through the body. Plants do not have the equivalent of an animal brain. But they do have some decentralized mechanisms that influence the internal environment and keep the body functioning properly. Two simple examples illustrate the point; chapters to follow include more.

Walling Off Threats

Unlike people, trees consist mostly of dead and dying cells. Also unlike people, trees cannot run away from attacks. When a pathogen infiltrates their tissues, trees cannot unleash infection-fighting white blood cells in response, because they have none. However, plants do have **systemic acquired resistance**: a defense response to infections and injured tissues. Cells in an affected tissue release signaling molecules. The molecules cause synthesis and release of organic compounds that will protect the plant against attacks for days or months to come. Some protective compounds are so effective that synthetic versions are being used to boost disease resistance in crop plants and ornamental plants.

Most trees also have another defense that minimizes effects of pathogens. When wounded, such trees wall off the damaged tissue, release phenols and other toxic compounds, and often secrete resins. A heavy flow of gooey compounds saturates and protects the bark and wood at the wound. It also seeps into the soil around roots. Some of these toxins are so potent that they can kill cells of the tree itself. Compartments form around injured, infected, or poisoned tissues, and new tissues grow right over them. This plant response to wounds is called **compartmentalization**.

Drill holes into a tree species that makes a strong compartmentalization response and the wound gets walled off fast (Figure 27.9). In a species that makes a moderate response, decomposers cause the decay of more wood surrounding the holes. Drill into a weak compartmentalizer, and decomposers cause massive decay deep into the trunk.

Even strong compartmentalizers live only so long. If too much tissue gets walled off, flow of water and solutes to living cells slows and the tree begins to die. What about the bristlecone pine, which grows high in mountain regions (Section 23.7)? One tree we know of is almost 5,000 years old. These trees live under harsh conditions in remote places where pathogens are few. The trees spend most of each year dormant beneath a blanket of snow, and grow slowly during a short, dry summer. This slow growth makes a bristlecone pine's wood so dense that few insects can bore into it.

Sand, Wind, and the Yellow Bush Lupine

If you have ever walked barefoot across beach sand on a sunny summer day you know how hot it can get. Sandy soil also tends to drain quickly, and to be low in nutrients. Few plants are adapted to survive in this habitat, but the yellow bush lupine, *Lupinus arboreus*, thrives here (Figure 27.10). This shrubby plant is native to coastal dunes of central and southern California.

Several factors contribute to the lupine's success in its challenging coastal environment. It is a legume and, like other members of this plant family, it shelters

A Strong

B Moderate

C Weak

Figure 27.9 Animated Results of an experiment in which holes were drilled into living trees to test compartmentalization responses. From top to bottom, decay patterns (*green*) in trunks of three species of trees that made strong, moderate, and weak compartmentalization responses, respectively.

Figure 27.10 Yellow bush lupine, *Lupinus arboreus*, in a sandy shore habitat. On hot, windy days, its leaflets fold up longitudinally along the crease that runs down their center. This helps minimize evaporative water loss.

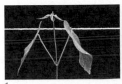  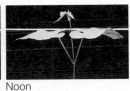 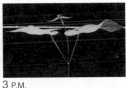 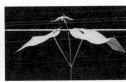 

1 A.M. | 6 A.M. | Noon | 3 P.M. | 10 P.M. | Midnight

Figure 27.11 Animated Observational test of rhythmic leaf movements by a young bean plant (*Phaseolus*). Physiologist Frank Salisbury kept the plant in darkness for twenty-four hours. Despite the lack of light cues, the leaves kept on folding and unfolding at sunrise (6 A.M.) and sunset (6 P.M.).

nitrogen-fixing bacteria inside its young roots (Section 24.6). The bacteria share some nitrogen with their host plant, thus giving it a competitive edge in nitrogen poor soil.

Another environmental challenge near the beach is the lack of fresh water. Leaves of a yellow bush lupine are structurally adapted for water conservation. Each leaf has a dense array of fine epidermal hairs that project above it, particularly on the leaf's lower surface. Collectively, these hairs trap moisture that evaporates from the stomata. The dampened hairs keep humidity around the stomata high, which helps minimize water losses to the air.

The yellow bush lupine also makes a homeostatic response. It folds its leaves lengthwise when conditions are hot and windy (Figure 27.10). This folding shelters stomata from the wind and further raises the humidity around them. When winds are strong and the potential for water loss is greatest, the leaves fold tightly. The least-folded leaves are close to the plant's center or on the side most sheltered from the wind. Folding is a response to heat as well as to wind. When air temperature is highest during the day, leaves fold at an angle that helps minimizes the amount of light they intercept, and the amount of heat they absorb.

Rhythmic Leaf Folding

Another example of a plant response is rhythmic leaf folding (Figure 27.11). A bean plant holds its leaves horizontally during the day but folds them close to its stem at night. A plant exposed to constant light or darkness for a few days will continue to move its leaves in and out of the "sleep" position at the time of sunrise and sunset. The response might help reduce heat loss at night, when air cools, and so maintain the plant's internal temperature within tolerable limits.

Rhythmic leaf movements are just one example of a **circadian rhythm**: a biological activity pattern that recurs with an approximately 24-hour cycle. Circadian means "about a day." Both plants and animals, as well as other organisms, have circadian rhythms.

Take-Home Message

How does homeostasis in plants differ from that animals?

■ Control mechanisms that function in homeostasis in plants are not centrally controlled as they are in most animals.

■ Systemic acquired resistance, compartmentalization, and leaf movements in response to environmental changes are examples of these mechanisms.

27.6 | How Cells Receive and Respond to Signals

- Coordinated action requires communication among body cells. Signaling mechanisms are essential to that integration.

- Links to Cell junctions 4.12, Membrane proteins 5.2

Cells in any multicelled body communicate with their neighbors and often with cells farther away. Section 4.12 described how plasmodesmata in plants and gap junctions in animals allow substances to pass quickly between adjoining cells. Communication among more distant cells involves special molecules. Some molecular signals diffuse from one cell to another through the fluid between them. Others travel in blood vessels or in a plant's vascular tissues.

Molecular mechanisms by which cells "talk" to one another evolved early in the history of life. They often have three steps: signal reception, signal transduction, and a cellular response (Figure 27.12*a*).

During signal reception, a specific receptor is activated, as by reversibly binding a signaling molecule. The receptors are often membrane proteins of the sort shown in Section 5.2.

Next, the signal is transduced, or converted to a form that acts inside the signal-receiving cell. Some signal receptor proteins are enzymes that undergo a shape change when a signaling molecule binds. Once activated in this way, the enzyme catalyzes formation of a molecule that then acts as an intracellular signal.

Finally, the cell responds to the signal. For example, it may alter its growth or which genes it expresses.

Consider one example, a signaling pathway that occurs as an animal develops. As part of development, many cells heed calls to self-destruct at a particular time. **Apoptosis** is a process of programmed cell death. It often starts when certain molecular signals bind to receptors at the cell surface (Figure 27.12*b*). A chain of reactions leads to the activation of self-destructive enzymes. Some of these enzymes chop up structural proteins, such as cytoskeleton proteins and histones that organize DNA. Others snip apart nucleic acids.

An animal cell undergoing apoptosis shrinks away from its neighbors. Its membrane bubbles inward and outward. The nucleus and then the whole cell break apart. Phagocytic white blood cells that patrol tissues engulf the dying cells and their remnants. Enzymes in the phagocytes digest the engulfed bits.

Many cells committed suicide as your hands were developing. Each hand starts as a paddlelike structure. Normally, apoptosis in vertical rows of cells divides the paddle into individual fingers within a few days (Figure 27.13). When the cells do not die on cue, the paddle does not split properly (Figure 27.14).

Besides helping to sculpt certain developing body parts, apoptosis also removes aged or damaged cells from a body. For example, keratinocytes are the main cells in your skin. Normally they live for three weeks or so, then undergo apoptosis. Formation of new cells balances out the death of old ones, so your skin stays

Signal Reception	→	Signal Transduction	→	Cellular Response
Signal binds to a receptor, usually at the cell surface.		Binding brings about changes in cell properties, activities, or both.		Changes alter cell metabolism, gene expression, or rate of division.

a

Signal to die docks at receptor.

Signal leads to activation of protein-destroying enzymes.

b

Figure 27.12 (**a**) Signal transduction pathway. A signaling molecule docks at a receptor. The signal activates enzymes or other cytoplasmic components that cause changes inside the cell. (**b**) An artist's fanciful depiction of what happens during apoptosis, the process by which a body cell self-destructs.
Figure It Out: What are the blue objects with sharp blades? *Answer: Protein-destroying enzymes*

A Cautionary Tale

A parked car can heat up quickly even on a mild day. Children's bodies do not regulate temperature as well as adults' bodies do. Together, these facts can add up to tragedy. Between 1997 and 2007, 339 children who were left alone in cars died of heat stroke. In some cases, an adult unknowingly left the child behind, but about 20 percent of deaths occurred after an adult deliberately left an infant or child in the car.

How would you vote?

Children left alone in cars have died of heat stroke. Should it be illegal to leave a child in a car for even a minute? See CengageNOW for details, then vote online.

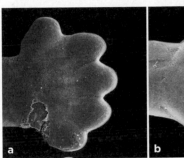

Figure 27.13 Animated Formation of human fingers. (**a**) Forty-eight days after fertilization, tissue webs connect embryonic digits. (**b**) Three days later, after apoptosis by cells making up the tissue webs, the digits are separated.

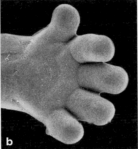

Figure 27.14 Digits remained attached when embryonic cells did not commit suicide on cue.

uniformly thick. If you spend too much time in the sun, cells enter apoptosis ahead of schedule, so your skin peels. Peeling is bad news for individual cells but it helps protect your body. Cells exposed to excess UV radiation often end up with damaged DNA and are more likely to become cancerous.

Some walled plant cells also die on cue. They get emptied of cytoplasm, and the walls of abutting ones act as pipelines for water. Cells that attach leaves to a stem die in response to seasonal change or stress, and leaves are shed. When a plant tissue is wounded or attacked by a pathogen, signals may trigger the death of nearby cells, which form a wall around the threat, as described in the previous section.

Take-Home Message

How do cells in a multicelled body communicate?

■ Cell communication involves binding of signaling molecules to membrane receptors, transduction of that signal, and the cellular response to it.

Summary

Section 27.1 Anatomy is the scientific study of body form, and physiology is the study of body functions. Structural and functional organization emerges during the **growth** and **development** of an individual.

Bodies have levels of organization. Each cell carries out metabolic tasks that keep it alive. At the same time, individual cells interact in **tissues**, and often, in **organs** and **organ systems**. Together cells, tissues, and organs maintain conditions in the **extracellular fluid (ECF)**, the fluid outside of cells. Maintaining the ECF is an aspect of **homeostasis**: the process of keeping the conditions inside a body within a range that body cells can tolerate.

■ *Use the animation on CengageNOW to investigate the structural organization of a tomato plant.*

Section 27.2 Plants and animals have adapted to some of the same environmental challenges. Small plants and animals rely on **diffusion** of material through their body. Larger ones have vascular tissues. **Active transport** and **passive transport** maintain water and solute concentrations inside both plants and animals. Both groups have mechanisms that allow them to respond to signals from other cells, as well as to environmental changes.

Sections 27.3, 27.4 In animal bodies, **receptors** detect stimuli and send signals to an **integrator** such as a brain. Signals from the integrator cause **effectors** (muscles and glands) to respond. With **negative feedback mechanisms**, receptors detect a change, then effectors respond and reverse the change. Such mechanisms act in homeostasis. With **positive feedback mechanisms**, detection of a change leads to a response that intensifies the change.

Heat stroke is an example of the consequences of a failure of homeostasis.

■ *Use the animation on CengageNOW to observe the effects of negative feedback on temperature control in humans.*

Section 27.5 Plants do not have a brain, but they do have decentralized mechanisms of homeostasis, such as **systemic acquired resistance** to pathogens and an ability to wall off a wound (**compartmentalization**). Plants respond to changes in their environment when they fold leaves in ways that minimize water loss or help maintain temperature. Rhythmic leaf folding is a type of **circadian rhythm**, an event repeated on a 24-hour cycle.

■ *Use the animation on CengageNOW to learn about plant defense mechanisms.*

Data Analysis Exercise

As part of ongoing efforts to prevent heat-related illness, the National Weather Service has devised a heat index (HI) to alert people to the risks of high temperature coupled with high humidity. The heat index is sometimes referred to as the "apparent temperature." It tells you what the temperature feels like, given the level of relative humidity. The higher the HI value, the higher the heat disorder risk with prolonged exposure or with exertion.

Figure 27.15 shows the heat index chart. The maximum possible value is 137. *Gold* indicates temperatures near the danger level, *orange* indicates danger, and *pink* means extreme danger.

1. What is the heat index on a day when the temperature is 96°F and the relative humidity is 45 percent?

2. What is the heat index on a day when the temperature is 96°F and the relative humidity is 75 percent?

3. How does the danger level indicated by these two heat index values compare?

4. What is the lowest temperature that, when coupled with 100% relative humidity, can cause extreme danger?

Temp (°F)	Relative humidity (%)												
	40	45	50	55	60	65	70	75	80	85	90	95	100
110	136												
108	130	137											
106	124	130	137										
104	119	124	131	137									
102	114	119	124	130	137								
100	109	114	118	124	129	136							
98	105	109	113	117	123	128	134						
96	101	104	108	112	116	121	126	132					
94	97	100	103	106	110	114	119	124	129	135			
92	94	96	99	101	105	108	112	116	121	126	131		
90	91	93	95	97	100	103	106	109	113	117	122	127	132
88	88	89	91	93	95	98	100	103	106	110	113	117	121
86	85	87	88	89	91	93	95	97	100	102	105	108	112
84	83	84	85	86	88	89	90	92	94	96	98	100	103
82	81	82	83	84	84	85	86	88	89	90	91	93	95
80	80	80	81	81	82	82	83	84	84	85	86	86	87

Figure 27.15 Heat index (HI) chart.

Section 27.6 Communication between cells involves signal reception, signal transduction, and a response by a target cell. Many signals are transduced by membrane proteins that trigger reactions in the cell. Reactions may alter gene expression or metabolic activities. An example is a signal that unleashes the protein-cleaving enzymes of **apoptosis**, the programmed self-destruction of a cell.

■ *Use the animation on CengageNOW to see how a human hand forms.*

Self-Quiz

Answers in Appendix III

1. Fill in the blank. An increase in the number, size, and volume of plant cells or animal cells is called _____ .

2. A leaf is an example of _____ .
a. a tissue
b. an organ
c. an organ system
d. none of the above

3. A substance moves spontaneously to a region of lower concentration by the process of _____ .
a. diffusion
b. active transport
c. passive transport
d. a and c

4. Aerobic respiration occurs in _____ .
a. plants
b. animals
c. both plants and animals
d. neither

5. A plant's xylem and phloem are _____ tissues.
a. vascular
b. sensory
c. respiratory
d. digestive

6. An animal's muscles and glands are _____ .
a. integrators
b. receptors
c. effectors
d. all are correct

7. Fill in the blank: With _____ feedback, a change in conditions triggers a response that intensifies that change.

8. Systemic acquired resistance _____ .
a. helps protect plants from infections
b. is an example of a circadian response
c. requires white blood cells
d. all are correct

9. When a signal is transduced, it is _____ .
a. heightened
b. dampened
c. converted to a new form
d. ignored

10. The process of _____ sculpts a developing hand from a paddlelike form.
a. apoptosis
b. transduction
c. positive feedback
d. diffusion

11. Match the terms with their most suitable description.
____ circadian rhythm
____ homeostasis
____ apoptosis
____ integrator
____ effectors
____ negative feedback

a. programmed cell death
b. 24-hour or so cyclic activity
c. central command center
d. stable internal environment
e. muscles and glands
f. an activity changes some condition, then the change triggers its own reversal

■ *Visit CengageNOW for additional questions.*

Critical Thinking

1. The Arabian oryx (*Oryx leucoryx*), an endangered antelope, lives in the harsh deserts of the Middle East. Most of the year there is no free water, and temperatures routinely reach 47°C (117°F). The most common tree in the region is the umbrella thorn tree (*Acacia tortilis*). List the common challenges faced by the oryx and acacia that are unlike those faced by plants and animals in other environments.

2. Eating a heavy, protein-rich meal on a hot day can increase the risk of heat illness. Why?

V HOW PLANTS WORK

The sacred lotus, *Nelumbo nucifera*, busily doing what its ancestors did for well over 100 million years—flowering spectacularly during the reproductive phase of its life cycle.

Plant Tissues

Droughts Versus Civilization

The more we dig up records of past climates, the more we wonder about what is happening now. In any given year, places around the world have severe droughts—far less rainfall than we expect to see. In themselves, droughts are not that unusual, but some have been severe enough to cause mass starvation, cripple economies, and invite conflicts over dwindling resources. What is the long-term forecast? As global warming changes Earth's weather patterns, heat waves are expected to be more intense, and droughts more frequent and more severe.

Humans built the whole of modern civilization on a vast agricultural base. Today we reel from droughts that last two, five, seven years or so. Imagine one lasting 200 years! It happened. About 3,400 years ago, rainfall dried up and brought an end to the Akkadian civilization in northern Mesopotamia. We know about the drought from ice cores. Researchers take such samples by drilling a long pipe down through deep ice, then pulling it out. The ice core inside the pipe holds dust and air bubbles trapped in layers of snow that fell year in, year out. The ice in some regions is more than 3,000 meters (9,800 feet) thick, and has layers that have accumulated over the last 200,000 years. These layers hold clues to past atmospheric conditions, and they point to recurring climate changes that may have brought an end to many societies around the world.

A catastrophic drought contributed to the collapse of the Mayan civilization centuries ago (Figure 28.1). More recently, Afghanistan was scorched by seven years of drought—the worst in the past century. The vast majority of Afghans are subsistence farmers; the drought wiped out their harvests, dried up their wells, and killed their livestock. Despite relief efforts, starvation was rampant. Desperate rural families sold their land, their possessions, and their daughters. As of this writing, extreme drought is affecting southern China and about one-third of the continental United States; Australia is in the middle of the worst drought in 1,000 years.

This unit focuses on seed-bearing vascular plants, especially the flowering types that are integral to our lives. You will be looking at how these plants function and at their patterns of growth, development, and reproduction. You will consider how they are adapted to withstand a variety of stressful conditions and why prolonged water deprivation kills them.

The vulnerability of the agricultural base for societies around the world will impact your future. Which nations will stumble during long-term climate change? Which ones will make it through a severe drought that does not end any time soon?

See the video! Figure 28.1
We depend on adaptations by which plants get and use resources, which include water. Directly or indirectly, plants make the food that sustains nearly all forms of life on Earth.

Left, mute reminder of the failed Mayan civilization. *Above*, from a Guatemalan field, a stunted corncob—a reminder of prolonged drought and widespread crop failures.

Key Concepts

Overview of plant tissues

Seed-bearing vascular plants have a shoot system, which includes stems, leaves, and reproductive parts. Most also have a root system. Such plants have ground, vascular, and dermal tissues. Plants lengthen or thicken only at active meristems. **Sections 28.1, 28.2**

Organization of primary shoots

Ground, vascular, and dermal tissues are organized in characteristic patterns in stems and leaves. The patterns differ between monocots and eudicots. Stem and leaf specializations maximize sunlight interception, water conservation, and gas exchange. **Sections 28.3, 28.4**

Organization of primary roots

Ground, vascular, and dermal tissues are organized in a characteristic pattern in roots. The pattern differs between monocots and eudicots. Roots absorb water and minerals, and anchor the plant. **Section 28.5**

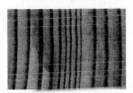

Secondary growth

In many plants, older branches and roots put on secondary growth that thickens them during successive growing seasons. Wood is extensive secondary growth. **Sections 28.6, 28.7**

Modified stems

Certain types of stem modifications are adaptations for storing water or nutrients, or for reproduction. **Section 28.8**

Links to Earlier Concepts

- This chapter builds on what you learned in Sections 23.1, 23.8, and 27.1, which introduced plant structure and growth, and correlated them with present and past functions.

- You will revisit some structural specializations of plant cells (4.12, 7.7, 23.2), and see how water-conserving adaptations (27.5) function in plant homeostasis (27.1, 27.2). You will also see how secondary growth is part of compartmentalization (27.5).

■ The unique organization of tissues in flowering plants is part of the reason why they are the dominant group of the plant kingdom.

■ Links to Plant evolution 23.1, Angiosperms 23.8, Evolution of plant structure 27.1

The Basic Body Plan

Figure 28.2 shows the body plan of a typical flowering plant. It has **shoots**: aboveground parts such as stems, leaves, and flowers. Stems support upright growth, a bonus for cells that intercept energy from the sun. They also connect the leaves and flowers with **roots**, which are structures that absorb water and dissolved minerals as they grow down and outward in the soil. Roots often anchor the plant. All root cells store food for their own use, and some types also store it for the rest of the plant body.

Shoots and roots consist of three tissue systems. The **ground tissue system** functions in several tasks, such as photosynthesis, storage, and structural support of other tissues. Pipelines of the **vascular tissue system** distribute water and mineral ions that the plant takes up from its surroundings. They also carry sugars produced by photosynthetic cells to the rest of the plant. The **dermal tissue system** covers and protects exposed surfaces of the plant.

The ground, vascular, and dermal tissue systems consist of cells that are organized as simple and complex tissues. Simple tissues are constructed primarily of one type of cell; examples include parenchyma, collenchyma, and sclerenchyma. Complex tissues have two or more types of cells. Xylem, phloem, and epidermis are examples. You will learn more about all of these tissues in the next section.

Eudicots and Monocots—Same Tissues, Different Features

The same tissues form in all flowering plants, but they do so in different patterns. Consider **cotyledons**, which are leaflike structures that contain food for a plant embryo. These "seed leaves" wither after the seed germinates and the developing plant begins to make its own food by photosynthesis. Cotyledons consist of the same types of tissues in all plants that have them, but the seeds of **eudicots** have two cotyledons and those of **monocots** have only one. Figure 28.3 shows other differences between these two types of flowering plants. Most shrubs and trees, such as rose bushes and maple trees, are eudicots. Lilies, orchids, and corn are typical monocots.

Introducing Meristems

All plant tissues arise at **meristems**, each a region of undifferentiated cells that can divide rapidly. Portions of the descendant cells differentiate and mature into

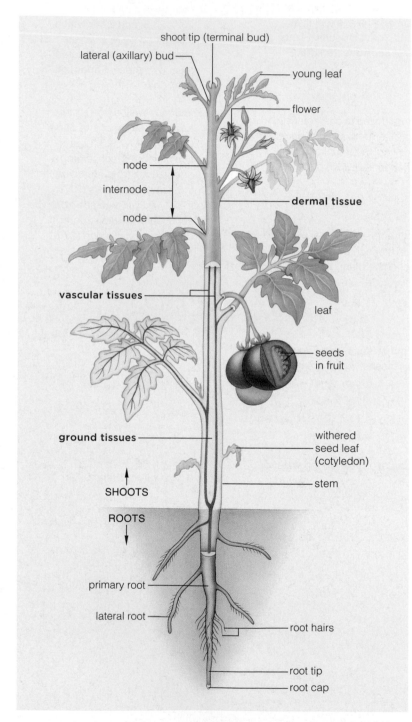

shoot tip (terminal bud)
lateral (axillary) bud
young leaf
flower
node
internode
node
dermal tissue
vascular tissues
leaf
seeds in fruit
ground tissues
withered seed leaf (cotyledon)
stem
SHOOTS
ROOTS
primary root
lateral root
root hairs
root tip
root cap

Figure 28.2 Animated Body plan of a tomato plant (*Lycopersicon esculentum*). Its vascular tissues (*purple*) conduct water, dissolved minerals, and organic substances. They thread through ground tissues that make up most of the plant. Epidermis, a type of dermal tissue, covers root and shoot surfaces.

A Characteristics of Eudicots

In seeds, two cotyledons
(seed leaves of embryo)

Flower parts in fours or fives
(or multiples of four or five)

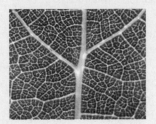

Leaf veins usually forming
a netlike array

Pollen grains with three
pores or furrows

Vascular bundles
organized in a ring
in ground tissue

B Characteristics of Monocots

In seeds, one cotyledon
(seed leaf of embryo)

Flower parts in threes
(or multiples of three)

Leaf veins usually running
parallel with one another

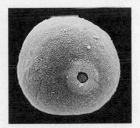

Pollen grains with one
pore or furrow

Vascular bundles
throughout
ground tissue

Figure 28.3 Animated Comparison of eudicots and monocots.

Figure 28.4 *Right*, locations of apical and lateral meristems.

specialized tissues. New, soft plant parts lengthen by activity at **apical meristems** in the tips of shoots and roots. The seasonal lengthening of young shoots and roots is called **primary growth** (Figure 28.4a).

Some plants also undergo **secondary growth**—their stems and roots thicken over time. In woody eudicots and in gymnosperms such as pine trees, secondary growth occurs when cells of a thin cylindrical layer called the **lateral meristem** divide (Figure 28.4b).

shoot apical meristem
(new cells forming)

cells dividing,
differentiating

three tissue
systems
developing

three tissue
systems
developing

cells dividing,
differentiating

root apical meristem
(new cells forming)

a Many cellular descendants of *apical meristems* are the start of lineages of differentiated cells that grow, divide, and lengthen shoots and roots.

Take-Home Message

What is the basic structure of flowering plants?

■ Plants typically have aboveground shoots, such as stems, leaves, and flowers. All have ground, vascular, and dermal tissue systems.

■ The patterns in which plant tissues are organized differ between eudicots and monocots.

■ Plants lengthen, or put on primary growth, at soft shoot and root tips. Many plants put on secondary growth; older stems and roots thicken over successive growing seasons.

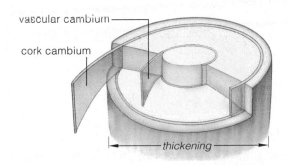

vascular cambium

cork cambium

thickening

b In woody plants, the activity of two *lateral meristems—vascular cambium* and *cork cambium*—result in secondary growth that thickens older stems and roots.

■ Different plant tissues form just behind shoot and root tips, and on older stem and root parts.

■ Links to Plant cell surface specializations 4.12, Stomata 7.7, Lignin in plant evolution 23.2, Growth 27.1

Table 28.1 summarizes the common plant tissues and their functions. Some of these tissues are visible in the micrograph shown in Figure 28.5. Plant parts are typically cut along standard planes like this cross-section in order to simplify our interpretation of micrographs (Figure 28.6).

Simple Tissues

Parenchyma tissue makes up most of the soft primary growth of roots, stems, leaves, and flowers, and it also has storage and secretion functions. **Parenchyma** is a

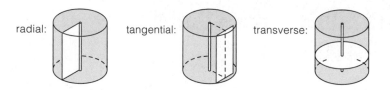

Figure 28.6 Terms that identify how tissue specimens are cut from a plant. Longitudinal cuts along a stem or root radius give *radial* sections. Cuts at right angles to the radius give *tangential* sections. Cuts perpendicular to the long axis of a stem or root give *transverse* sections—that is, cross-sections.

Table 28.1	Overview of Flowering Plant Tissues	
Tissue Type	**Main Components**	**Main Functions**
Simple Tissues		
Parenchyma	Parenchyma cells	Photosynthesis, storage, secretion, tissue repair, other tasks
Collenchyma	Collenchyma cells	Pliable structural support
Sclerenchyma	Fibers or sclereids	Structural support
Complex Tissues		
Vascular		
Xylem	Tracheids, vessel members; parenchyma cells; sclerenchyma cells	Water-conducting tubes; reinforcing components
Phloem	Sieve-tube members, parenchyma cells; sclerenchyma cells	Tubes of living cells that distribute organic compounds; supporting cells
Dermal		
Epidermis	Undifferentiated as well as specialized cells (e.g., guard cells)	Secretion of cuticle; protection; control of gas exchange and water loss
Periderm	Cork cambium; cork cells; parenchyma	Forms protective cover on older stems, roots

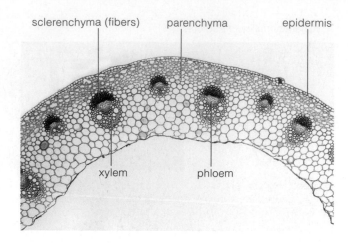

Figure 28.5 Some tissues in a buttercup stem (*Ranunculus*).

simple tissue that consists mainly of parenchyma cells, which are typically thin-walled, flexible, and many-sided. These cells are alive in mature tissue, and they can continue to divide. Plant wounds are repaired by dividing parenchyma cells. **Mesophyll**, the only photosynthetic tissue, is a type of parenchyma.

Collenchyma is a simple tissue that consists mainly of collenchyma cells, which are elongated and alive in mature tissue. This stretchable tissue supports rapidly growing plant parts, including young stems and leaf stalks (Figure 28.7a). Pectin, a polysaccharide, imparts flexibility to a collenchyma cell's primary wall, which is thickened where three or more of the cells abut.

Cells of **sclerenchyma** are variably shaped and dead at maturity, but the lignin-rich walls that remain help this tissue resist compression. Remember, lignin is the organic compound that structurally supports upright plants, and helped them evolve on land (Section 23.2). Lignin also deters some fungal attacks.

Fibers and sclereids are typical sclerenchyma cells. Fibers are long, tapered cells that structurally support the vascular tissues in some stems and leaves (Figure 28.7b). They flex and twist, but resist stretching. We use certain fibers as materials for cloth, rope, paper, and other commercial products. The far stubbier and often branched sclereids strengthen hard seed coats, such as peach pits, and make pear flesh gritty (Figure 28.7c).

Complex Tissues

Vascular Tissues Xylem and phloem are vascular tissues that thread through ground tissue. Both consist of elongated conducting tubes that are often sheathed in sclerenchyma fibers and parenchyma. **Xylem**, which conducts water and mineral ions, consists of two types of cells, **tracheids** and **vessel members**, that are dead at maturity (Figure 28.8a,b). The secondary walls of these cells are stiffened and waterproofed with lignin. They

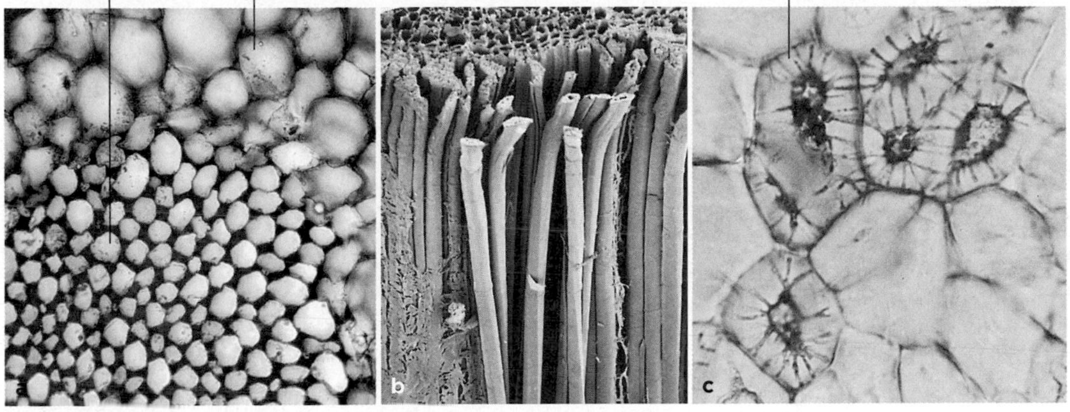

collenchyma   parenchyma

lignified secondary wall

Figure 28.7 Simple tissues.
(**a**) Collenchyma and parenchyma
from a supporting strand inside of
a celery stem, transverse section.

Sclerenchyma: (**b**) Fibers from a
strong flax stem, tangential view.
(**c**) Stone cells, a type of sclereid
in pears, transverse section.

interconnect to form conducting tubes, and they also
lend structural support to the plant. The perforations
in adjoining cell walls align, so fluid moves laterally
between the tubes as well as upward through them.

Phloem conducts sugars and other organic solutes.
Its main cells, sieve-tube members, are alive in mature
tissue. They connect end to end at sieve plates, forming **sieve tubes** that distribute sugars to all parts of
the plant (Figure 28.8c). Phloem's **companion cells** are
parenchyma cells that load sugars into the sieve tubes.

Dermal Tissues The first dermal tissue to form on a
plant is **epidermis**, which usually is a single layer of
cells. Secretions deposited on the outward-facing cell
walls form a cuticle. Plant cuticle is rich in deposits of
cutin, a waxy substance. It helps the plant conserve
water and repel pathogens (Figures 28.5 and 28.9).

The epidermis of leaves and young stems includes
specialized cells. For example, a stoma is a small gap
across epidermis; it opens when the pair of guard cells
around it swells (Section 7.7). Diffusion of water vapor,
oxygen, and carbon dioxide gases across the epidermis
is controlled at stomata. Periderm, a different tissue,
replaces epidermis in woody stems and roots

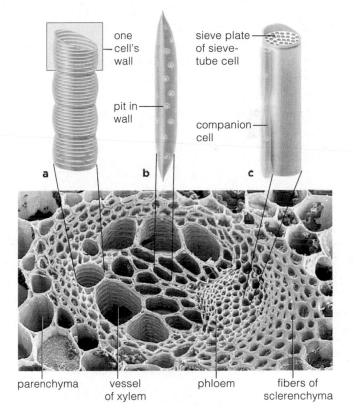

one
cell's
wall

pit in
wall

sieve plate
of sieve-
tube cell

companion
cell

a   b   c

parenchyma   vessel   phloem   fibers of
                of xylem   sclerenchyma

Figure 28.8 Simple and complex tissues in a stem. In xylem, (**a**) part
of a column of vessel members, and (**b**) a tracheid. (**c**) One of the living
cells that interconnect as sieve tubes in phloem.

Take-Home Message

What are the main types of plant tissues?

■ Cells of parenchyma have diverse roles, such as secretion,
storage, photosynthesis, and tissue repair. Collenchyma and
sclerenchyma support and strengthen plant parts.

■ Xylem and phloem are vascular tissues that thread through
the ground tissue. In xylem, water and ions flow through tubes
of dead tracheid and vessel member cells. In phloem, sieve
tubes that consist of living cells distribute sugars.

■ Epidermis covers all young plant parts exposed to the surroundings. Periderm that forms on older stems and roots
replaces epidermis of younger stems.

leaf surface   cuticle   epidermal cell   photosynthetic cell

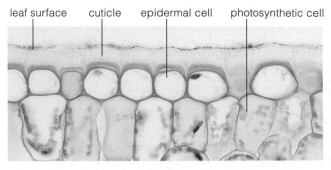

Figure 28.9
A typical plant
cuticle, with
many epider-
mal cells and
photosyn-
thetic cells
under it.

■ Inside the soft, young stems and leaves of both eudicots and monocots, the ground, vascular, and dermal tissue systems are organized in predictable patterns.

Behind the Apical Meristem

The structural organization of a new flowering plant has become mapped out by the time it is an embryo sporophyte inside a seed coat. As you will read later, a tiny primary root and shoot have already formed as part of the embryo. Both are poised to resume growth and development as soon as the seed germinates.

Terminal buds are a shoot's main zone of primary growth. Just beneath a terminal bud's surface, cells of shoot apical meristem divide continually during the growing season. Some of the descendants divide and differentiate into specialized tissues. Each descendant cell lineage divides in particular directions, at different rates, and the cells go on to differentiate in size, shape, and function. Figure 28.10 shows an example.

Buds may be naked or encased in modified leaves called bud scales. Small regions of tissue bulge out near the sides of a bud's apical meristem; each is the start of a new leaf. As the stem lengthens, the leaves form and mature in orderly tiers, one after the next. A region of stem where one or more leaves form is called a node; the region between two successive nodes is called an internode (Figure 28.2).

Lateral buds, or axillary buds, are dormant shoots of mostly meristematic tissue. Each one forms inside a leaf axil, the point at which the leaf is attached to the stem. Different kinds of axillary buds are the start of side branches, leaves, or flowers. A hormone secreted by a terminal bud can keep lateral buds dormant, as Section 31.2 will explain.

Inside the Stem

In most flowering plants, cells of primary xylem and phloem are bundled together as long, multistranded

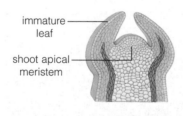

immature leaf
shoot apical meristem

Figure 28.10 Stem of *Coleus*, a eudicot. (**a–c**) Successive stages of the stem's primary growth, starting with the shoot apical meristem.

(**d**) The light micrograph shows a longitudinal cut through the stem's center. The tiers of leaves in the photograph below it formed in this linear pattern of development.

Figure It Out:
What is the transparent layer of cells on the outer surface of b and c?

Answer: Epidermis

a Sketch of the shoot tip in the micrograph at right, tangential cut. The descendant meristematic cells are color-coded *orange*.

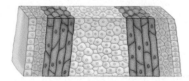

b Same tissue region later on, after the shoot tip lengthened above it

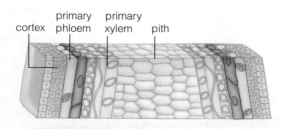

cortex primary phloem primary xylem pith

c Same tissue region later still, with lineages of cells lengthening and differentiating

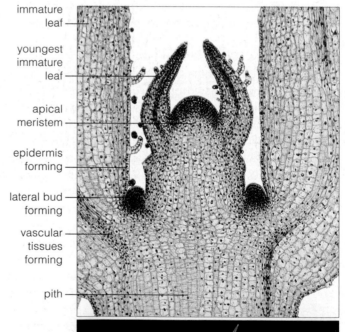

immature leaf
youngest immature leaf
apical meristem
epidermis forming
lateral bud forming
vascular tissues forming
pith

d

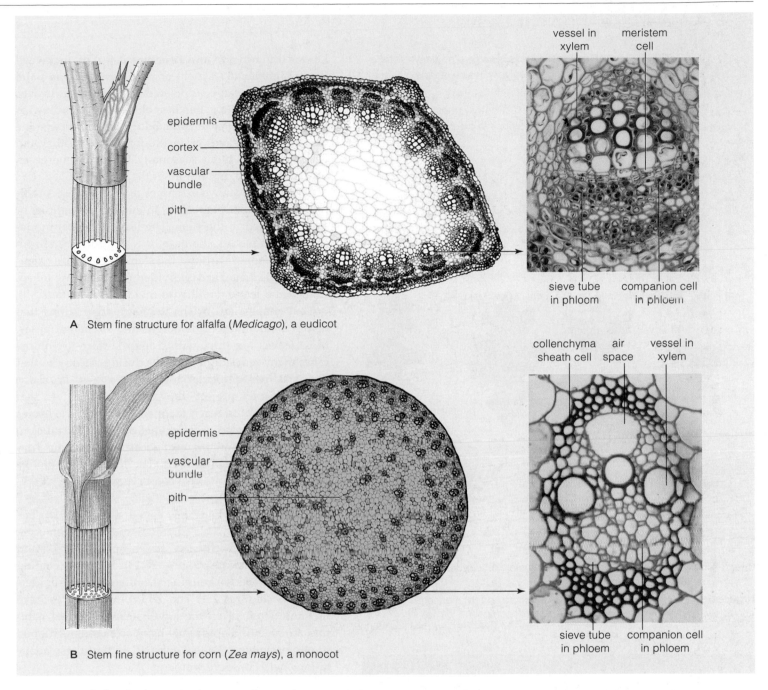

A Stem fine structure for alfalfa (*Medicago*), a eudicot

B Stem fine structure for corn (*Zea mays*), a monocot

Figure 28.11 Animated Zooming in on a eudicot and a monocot stem.

cords in the same cylindrical sheath of cells. The cords are called **vascular bundles**, and they thread lengthwise through the ground tissue system of all shoots.

Vascular bundles form in two distinct patterns. The vascular bundles of most eudicots form in a cylinder that runs parallel with the long axis of the shoot. Figure 28.11*a* shows how the cylinder divides the parenchyma of ground tissue into cortex (parenchyma between the vascular bundles and the epidermis) and pith (parenchyma inside the cylinder of vascular bundle).

Most monocot and some magnoliids have a different arrangement. Vascular bundles in stems of these plants are distributed all throughout the ground tissue (Figure 28.11*b*). In the next chapter, you will see how these vascular tissues take up, conduct, and give up water and solutes throughout the plant.

Take-Home Message

How are plant tissues organized inside stems?

■ Buds are the main zones of primary growth in shoots. Ground, vascular, and dermal tissues form in organized patterns.

■ The arrangement of vascular bundles, which are multistranded cords of vascular tissue, differs between eudicot and monocot stems.

■ All leaves are metabolic factories where photosynthetic cells churn out sugars, but they vary in size, shape, surface specializations, and internal structure.

■ Links to Plasmodesmata 4.12, Photosynthesis in leaf cells 7.7, Water conservation adaptations in plants 27.5

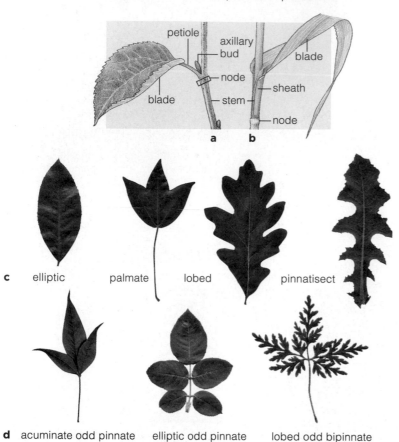

c elliptic   palmate   lobed   pinnatisect

d acuminate odd pinnate   elliptic odd pinnate   lobed odd bipinnate

Figure 28.12 Common leaf forms of (**a**) eudicots and (**b**) monocots, and a few examples of (**c**) simple leaves and (**d**) compound leaves.

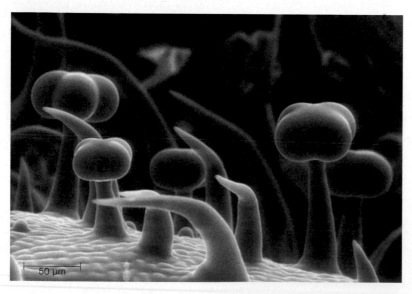

Leaves differ in size and structure. A leaf of duckweed is 1 millimeter (0.04 inch) across; leaves of one palm (*Raphia regalis*) can be 25 meters (82 feet) long. Leaves are shaped like cups, needles, blades, spikes, tubes, or feathers. They differ in color, odor, and edibility (some make toxins). Leaves of deciduous species wither and drop from their stems seasonally. Leaves of evergreen plants also drop, but not all at the same time.

Figure 28.12 shows examples of leaf shapes. A typical leaf has a flat blade and, in eudicots, a petiole, or stalk, attached to the stem. The leaves of most monocots are flat blades, the base of which forms a sheath called a coleoptile around the stem. Grasses are examples. Simple leaves are undivided, but many are lobed. Compound leaves are blades divided into leaflets.

Leaf shapes and orientations are adaptations that help a plant intercept sunlight and exchange gases. Most leaves are thin, with a high surface-to-volume ratio; many reorient themselves during the day so that they stay perpendicular to the sun's rays. Typically, adjacent leaves project from a stem in a pattern that allows sunlight to reach them all. However, the leaves of plants native to arid regions may stay parallel to the sun's rays, reducing heat absorption and thus conserving water (Section 27.5). The thick or needlelike leaves of some plants also conserve water.

Leaf Epidermis Epidermis covers every leaf surface exposed to the air. This surface tissue may be smooth, sticky, or slimy, with hairs, scales, spikes, hooks, and other specializations (Figure 28.13). A cuticle coating restricts water loss from the sheetlike array of epidermal cells (Figures 28.9 and 28.14). Most leaves have far more stomata on the lower surface. In arid habitats, stomata and epidermal hairs often are positioned in depressions in the leaf surface. Both of these adaptations help conserve water.

Mesophyll—Photosynthetic Ground Tissue Each leaf has mesophyll, a photosynthetic parenchyma with air spaces between cells (Section 7.7 and Figure 28.14). Carbon dioxide reaches the cells by diffusing into the leaf through stomata, and oxygen released by photo-

Figure 28.13 Example of leaf cell surface specialization: hairs on a tomato leaf. The lobed heads are glandular structures that occur on the leaves of many plants; they secrete aromatic chemicals that deter plant-eating insects. Those on marijuana plants secrete the psychoactive chemical tetrahydrocannabinol (THC).

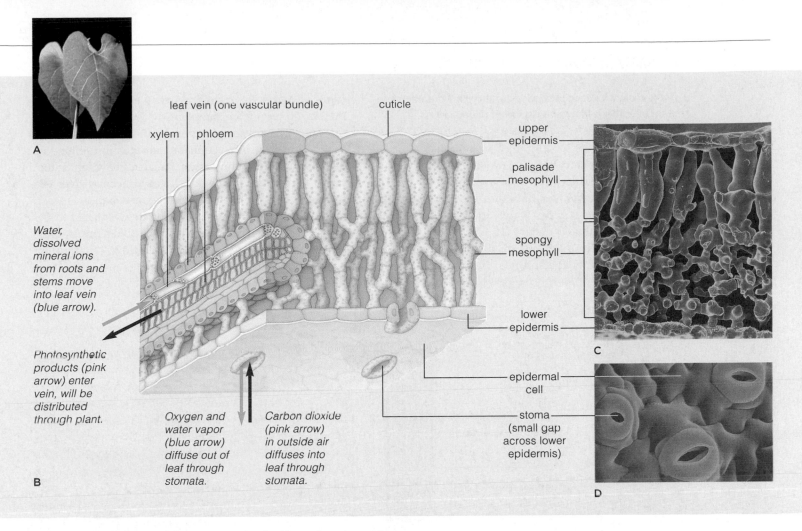

leaf vein (one vascular bundle)

cuticle

xylem   phloem

upper epidermis

palisade mesophyll

spongy mesophyll

lower epidermis

epidermal cell

stoma (small gap across lower epidermis)

A

Water, dissolved mineral ions from roots and stems move into leaf vein (blue arrow).

Photosynthetic products (pink arrow) enter vein, will be distributed through plant.

B

Oxygen and water vapor (blue arrow) diffuse out of leaf through stomata.

Carbon dioxide (pink arrow) in outside air diffuses into leaf through stomata.

C

D

Figure 28.14 Animated Leaf organization for *Phaseolus*, a bean plant. (**a**) Foliage leaves. (**b**–**d**) Leaf fine structure.

synthesis diffuses out the same way. Plasmodesmata connect the cytoplasm of adjacent cells. Substances can flow rapidly across the walls of adjoining cells through these cell junctions (Section 4.12).

Leaves oriented perpendicular to the sun have two layers of mesophyll. Palisade mesophyll is attached to the upper epidermis. The elongated parenchyma cells of this tissue have more chloroplasts than cells of the spongy mesophyll layer below (Figure 28.14). Blades of grass and other monocot leaves that grow vertically can intercept light from all directions. The mesophyll in such leaves is not divided into two layers.

Veins—The Leaf's Vascular Bundles Leaf **veins** are vascular bundles typically strengthened with fibers. Inside the bundles, continuous strands of xylem rapidly transport water and dissolved ions to mesophyll. Continuous strands of phloem rapidly transport the products of photosynthesis (sugars) away from mesophyll. In most eudicots, large veins branch into a network of minor veins embedded in mesophyll. In most monocots, all veins are similar in length and run parallel with the leaf's long axis (Figure 28.15).

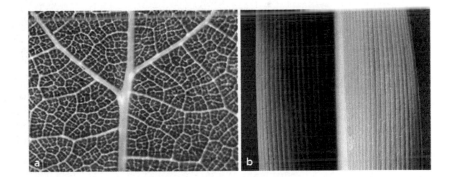

Figure 28.15 Typical vein patterns in flowering plants. (**a**) The netlike array in this grape leaf is common among eudicots. A stiffened midrib runs from the petiole to the leaf tip. Ever smaller veins branch from it. (**b**) The strong parallel orientation of veins in an *Agapanthus* leaf is typical of monocots. Like umbrella ribs, stiffened veins help maintain leaf shape.

Take-Home Message

How does a leaf's structure contribute to its function?

■ A leaf's shape, orientation, and structure typically function in sunlight interception, gas exchange, and distribution of water and solutes to and from living cells. Its epidermis encloses mesophyll and veins.

28.5 | Primary Structure of Roots

- Roots mainly function to provide plants with a large surface area for absorbing water and dissolved mineral ions.
- Link to Homeostasis in plants 27.1 and 27.2

Unless tree roots start to buckle a sidewalk or clog a sewer line, flowering plant root systems tend not to occupy our thoughts. Yet these are dynamic systems that actively mine soil for water and minerals. Most grow no deeper than 5 meters (16 feet). However, the roots of one hardy mesquite shrub grew 53.4 meters (175 feet) down into the soil near a streambed. Some types of cactus have shallow roots that can radiate 15 meters (50 feet) from the plant.

Someone measured the roots of a young rye plant that had been growing for four months in 6 liters (1.6 gallons) of soil. If the surface area of that root system were laid out as one sheet, it would occupy over 600 square meters, or close to 6,500 square feet!

A root's structural organization begins in a seed. As the seed germinates, a primary root pokes through the seed coat. In nearly all eudicot seedlings, that young root thickens.

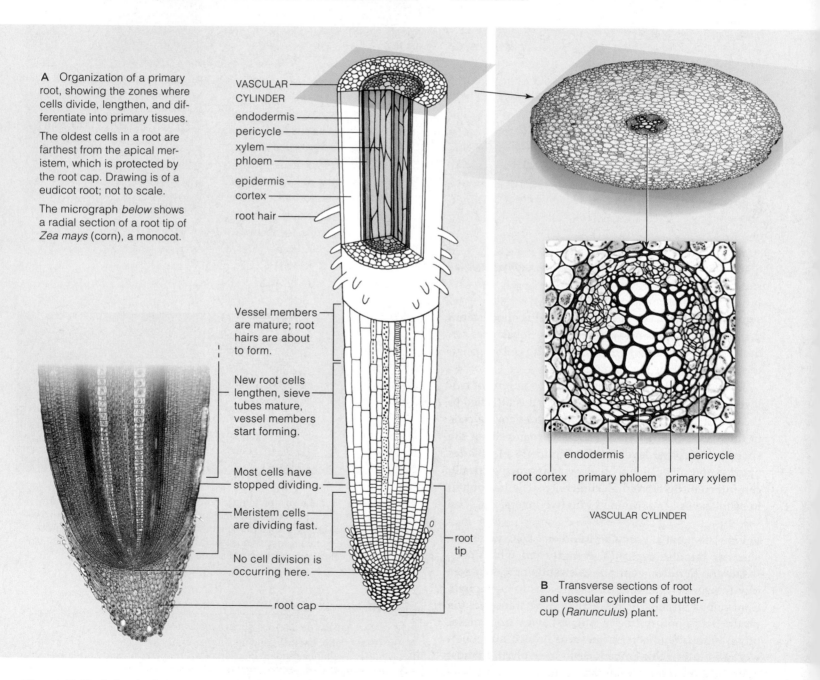

A Organization of a primary root, showing the zones where cells divide, lengthen, and differentiate into primary tissues.

The oldest cells in a root are farthest from the apical meristem, which is protected by the root cap. Drawing is of a eudicot root; not to scale.

The micrograph *below* shows a radial section of a root tip of *Zea mays* (corn), a monocot.

VASCULAR CYLINDER
endodermis
pericycle
xylem
phloem
epidermis
cortex
root hair

Vessel members are mature; root hairs are about to form.

New root cells lengthen, sieve tubes mature, vessel members start forming.

Most cells have stopped dividing.

Meristem cells are dividing fast.

No cell division is occurring here.

root cap

root tip

endodermis   pericycle

root cortex   primary phloem   primary xylem

VASCULAR CYLINDER

B Transverse sections of root and vascular cylinder of a buttercup (*Ranunculus*) plant.

Figure 28.16 Animated Tissue organization of a typical root.

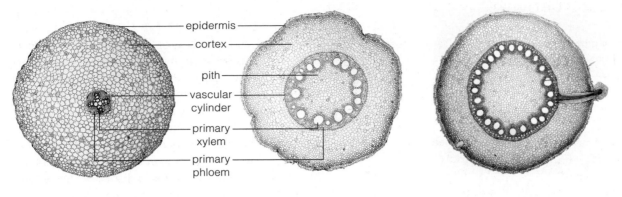

epidermis
cortex
pith
vascular cylinder
primary xylem
primary phloem

a eudicot root structure   **b** monocot root structure   **c** lateral root growing from pericycle

Figure 28.17 Comparison of root structure of (**a**) a eudicot (buttercup, *Ranunculus*) and (**b**) a monocot (corn, *Zea mays*). In corn and some other monocots, the vascular cylinder divides the ground tissue into cortex and pith. (**c**) A lateral root forms and branches from the pericycle of *Zea mays*.

Look at the root tip in Figure 28.16*a*. Some descendants of root apical meristem give rise to a root cap, a dome-shaped mass of cells that protects the soft, young root as it grows through soil. Other descendants give rise to lineages of cells that lengthen, widen, or flatten when they differentiate as part of the dermal, ground, and vascular tissue systems.

Ongoing divisions push cells away from the active root apical meristem. Some of their descendants form epidermis. The root epidermis is the plant's absorptive interface with soil. Many of its specialized cells send out fine extensions called **root hairs**, which collectively increase the surface area available for taking up soil water, dissolved oxygen, and mineral ions. Chapter 29 looks at the role of root hairs in plant nutrition.

Descendants of meristem cells also form the root's **vascular cylinder**, a central column of conductive tissue. The root vascular cylinder of typical eudicots is mainly primary xylem and phloem (Figure 28.17*a*); that of typical monocots divides the ground tissue into two zones, cortex and pith (Figure 28.17*b*). The vascular cylinder is sheathed by a pericycle, an array of parenchyma cells one or more layers thick (Figure 28.16*b*). These cells are differentiated, but they still divide repeatedly in a direction perpendicular to the axis of the root. Masses of cells erupt through the cortex and epidermis as the start of new, lateral roots (Figure 28.17*c*).

As you will see in Chapter 29, water entering a root moves from cell to cell until it reaches the endodermis, a layer of cells that encloses the pericycle. Wherever endodermal cells abut, their walls are waterproofed. Water must pass through the cytoplasm of endodermal cells to reach the vascular cylinder. Transport proteins in the plasma membrane control the uptake of water and dissolved substances.

Root primary growth results in one of two kinds of root systems. The **taproot system** of eudicots consists of a primary root and its lateral branchings. Carrots, oak trees, and poppies are among the plants that have a taproot system (Figure 28.18*a*). By comparison, the primary root of most monocots is quickly replaced by adventitious roots that grow outward from the stem. Lateral roots that are similar in diameter and length branch from adventitious roots. Together, the adventitious and lateral roots of such plants form a **fibrous root system** (Figure 28.18*b*).

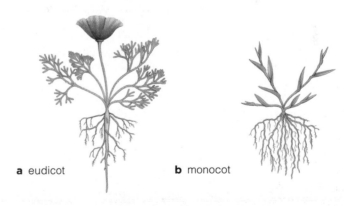

a eudicot   **b** monocot

Figure 28.18 Different types of root systems. (**a**) Taproot of the California poppy, a eudicot. (**b**) Fibrous roots of a grass plant, a monocot.

Take-Home Message

What is the function of plant roots?

■ Roots provide a plant with a tremendous surface area for absorbing water and solutes. Inside each is a vascular cylinder, with long strands of primary xylem and phloem.

■ Taproot systems consist of a primary root and lateral branchings. Fibrous root systems consist of adventitious and lateral roots that replace the primary root.

■ Secondary growth occurs at two types of lateral meristem, vascular cambium and cork cambium.

■ Link to Compartmentalization 27.5

Each spring, as primary growth resumes at buds, secondary growth thickens the girth of stems and roots of some plants. Figure 28.19 shows a typical pattern of secondary growth at the **vascular cambium**. This lateral meristem forms a cylinder, a few cells thick, inside older stems and roots. Divisions of vascular cambium cells produce secondary xylem on the cylinder's inner surface, and secondary phloem on its outer surface. As the core of xylem thickens, it also displaces the vascular cambium toward the surface of the stem. The displaced cells of the vascular cambium divide in a widening circle, so the tissue's cylindrical form is maintained.

Vascular cambium consists of two types of cells. Long, narrow cells give rise to the secondary tissues that extend lengthwise through a stem or root: tracheids, fibers, and parenchyma in secondary xylem; and sieve tubes, companion cells, and fibers in secondary phloem. Small, rounded cells that divide perpendicularly to the axis of the stem give rise to "rays" of parenchyma, radially oriented like spokes of a bicycle wheel. Secondary xylem and phloem of the rays conduct water and solutes radially through the stems and roots of older plants.

A core of secondary xylem, or **wood**, contributes up to 90 percent of the weight of some plants. Thin-walled, living parenchyma cells and sieve tubes of secondary phloem lie in a narrow zone outside of the vascular cambium. Bands of thick-walled reinforcing fibers are often interspersed through this secondary phloem. The only living sieve tubes are within a centimeter or so of the vascular cambium; the rest are dead, but they help protect the living cells behind them.

As seasons pass, the expanding inner core of xylem continues to direct pressure toward the stem or root surface. In time, it ruptures the cortex and the outer

A Secondary growth (thickening of older stems and roots) occurs at two lateral meristems. Vascular cambium gives rise to secondary vascular tissues; cork cambium gives rise to periderm.

cork cambium   vascular cambium

pith   cortex

stem surface

primary xylem   primary phloem

vascular cambium

B In spring, primary growth resumes at terminal and lateral buds. Secondary growth resumes at vascular cambium. Divisions of meristem cells in the vascular cambium expand the inner core of xylem, which displaces the vascular cambium (*orange*) toward the surface of the stem or root.

secondary xylem   secondary phloem

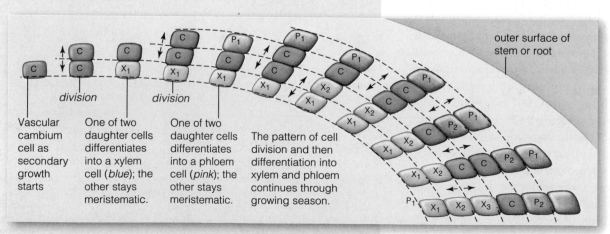

outer surface of stem or root

| Vascular cambium cell as secondary growth starts | One of two daughter cells differentiates into a xylem cell (*blue*); the other stays meristematic. | One of two daughter cells differentiates into a phloem cell (*pink*); the other stays meristematic. | The pattern of cell division and then differentiation into xylem and phloem continues through growing season. |

C Overall pattern of growth at vascular cambium.

Figure 28.19 Animated Secondary growth.

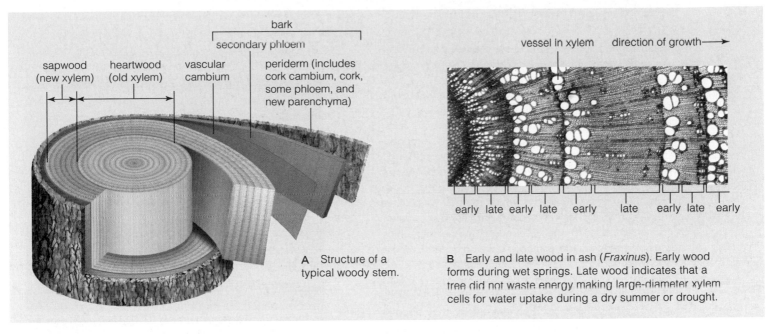

bark

secondary phloem

sapwood (new xylem) | heartwood (old xylem) | vascular cambium | periderm (includes cork cambium, cork, some phloem, and new parenchyma)

vessel in xylem   direction of growth →

early late early late   early   late   early late early

A Structure of a typical woody stem.

B Early and late wood in ash (*Fraxinus*). Early wood forms during wet springs. Late wood indicates that a tree did not waste energy making large-diameter xylem cells for water uptake during a dry summer or drought.

Figure 28.20 Animated Structure of wood.

secondary phloem. Then, another lateral meristem, the **cork cambium**, forms and gives rise to **periderm**. This dermal tissue consists of parenchyma and cork, as well as the cork cambium that produces them. What we call **bark** is secondary phloem and periderm. Bark consists of all of the living and dead tissues outside of the vascular cambium (Figure 28.20*a*).

The **cork** component of bark has densely packed rows of dead cells, the walls of which are thickened with a fatty substance called suberin. Cork protects, insulates, and waterproofs the stem or root surface. Cork also forms over wounded tissues. When leaves drop from the plant, cork forms at the places where petioles had attached to stems.

Wood's appearance and function change as a stem or root ages. Metabolic wastes, such as resins, tannins, gums, and oils, clog and fill the oldest xylem so much that it no longer is able to transport water and solutes. These substances often darken and strengthen the wood, which is called **heartwood**.

Sapwood is moist, still-functional xylem between heartwood and vascular cambium. In trees of temperate zones, dissolved sugars travel from roots to buds through sapwood's secondary xylem in spring. The sugar-rich fluid is sap. Each spring, New Englanders collect maple tree sap to make maple syrup.

Vascular cambium is inactive during cool winters or long dry spells. When the weather warms or moisture returns, the vascular cambium gives rise to early wood, with large-diameter, thin-walled cells. Late wood, with small-diameter, thick-walled xylem cells, forms in dry summers. A transverse cut from older trunks reveals alternating bands of early and late wood (Figure 28.20*b*). Each band is a growth ring, or "tree ring."

Trees native to regions in which seasonal change is pronounced tend to add one growth ring each year. Those in desert regions may add more than one ring of early wood in response to a single season of plentiful rain. In the tropics, seasonal change is almost nonexistent, so growth rings are not a feature of tropical trees.

Oak, hickory, and other eudicot trees that evolved in temperate and tropical zones are hardwoods, with vessels, tracheids, and fibers in xylem. Pines and other conifers are softwoods because they are weaker and less dense than the hardwoods. Their xylem has tracheids and parenchyma rays but no vessels or fibers.

Like other organisms, plants compete for resources. Plants with taller stems or broader canopies that defy the pull of gravity also intercept more light energy streaming from the sun. By tapping a greater supply of energy for photosynthesis, they have the metabolic means to produce large root and shoot systems. The larger its root and shoot systems, the more competitive the plant can be in acquiring resources.

Take-Home Message

What is secondary growth in plants?

■ Secondary growth thickens the stems and roots of older plants.

■ Wood is mainly accumulated secondary xylem.

■ Secondary growth occurs at two types of lateral meristem: vascular cambium and cork cambium. Secondary vascular tissues form at a cylinder of vascular cambium. A cylinder of cork cambium gives rise to periderm, which is part of a protective covering of bark.

28.7 | Tree Rings and Old Secrets

■ The number and relative thickness of a tree's rings hold clues to environmental conditions during its lifetime.

Tree rings can be used to estimate average annual rainfall; to date archaeological ruins; to gather evidence of wildfires, floods, landslides, and glacier movements; and to study the ecology and effects of parasitic insect populations. How? Some tree species, such as redwoods and bristlecone pines, lay down wood over centuries, one ring per year. Count an old tree's rings, and you have an idea of its age. If you know the year in which the tree was cut, you can find out which ring formed in what year by counting them backwards from the outer edge. Compare the thicknesses of the rings, and you have clues to events in those years (Figure 28.21).

For instance, In 1587, about 150 English settlers arrived at Roanoke Island off of the coast of North Carolina. When ships arrived in 1589 to resupply the colony, they discovered that the island had been abandoned. Searches up and down the coast failed to turn up the missing colonists.

About twenty years later, the English established a colony at Jamestown, Virginia. Although this colony survived, the initial years were difficult. In the summer of 1610 alone, more than 40 percent of the colonists died, many of them from starvation.

Researchers examined wood cores from bald cypress trees (*Taxodium distichum*) that had been growing at the time the Roanoke and Jamestown colonies were founded. Differences in the thicknesses of the trees' growth rings revealed that the colonists were in the wrong place at the wrong time (Figure 28.22). The settlers arrived at Roanoke just in time for the worst drought in 800 years. Nearly a decade of severe drought struck Jamestown. We know that the corn crop of the Jamestown colony failed. Drought-related crop failures probably occurred at Roanoke as well. The settlers also had difficulty finding fresh water. Jamestown was established at the head of an estuary; when the river levels dropped, their drinking water supply mixed with ocean water and became salty. Piecing together these bits of evidence gives us an idea of what life must have been like for the early settlers.

direction of growth ⟶

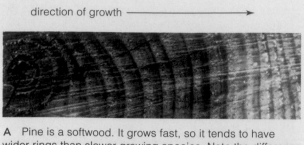

A Pine is a softwood. It grows fast, so it tends to have wider rings than slower growing species. Note the difference between the appearance of heartwood and sapwood.

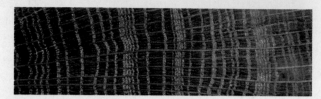

B The rings of this oak tree show dramatic differences in yearly growth patterns over its lifetime.

C An elm made this series between 1911 and 1950.

Figure 28.21 Animated Tree rings. In most species, each ring corresponds to one year, so the number of rings indicate the age of the tree. Relative thickness of the rings can be used to estimate data such as average annual rainfall long before such records were kept.

year: 1 2 3

Figure 28.22 (**a**) Location of two of the early American colonies. (**b**) Rings of a bald cypress tree, transverse section. This tree was living when English colonists first settled in North America. Narrower annual rings mark years of severe drought.

Jamestown Colony

Virginia

North Carolina

Lost Colony (Roanoke Island)

a

b   1587–1589   1606–1612

28.8 | Modified Stems

■ Many plants have modified stem structures that function in storage or reproduction.

The structure of a typical stem is shown in Figure 28.2, but there are many variations on that structure in different types of plants. Most serve special reproductive or storage functions.

Stolons Stolons, often called runners, are stems that branch from the main stem of the plant, typically on or near the surface of the soil. Stolons may look like roots, but they have nodes; roots do not have nodes. Adventitious roots and leafy shoots that sprout from the nodes develop into new plants (Figure 28.23*a*).

Rhizomes Rhizomes are fleshy, scaly stems that typically grow under the soil and parallel to its surface. A rhizome is the main stem of the plant, and it also serves as the plant's primary storage tissue. Branches that sprout from nodes grow aboveground for photosynthesis and flowering. Examples include ginger, irises, many ferns, and some grasses (Figure 28.23*b*).

Bulbs A bulb is a short section of underground stem encased by overlapping layers of thickened, modified leaves called scales. The scales contain starch and other substances that a plant holds in reserve when conditions in the environment are unfavorable for growth. When favorable conditions return, the plant uses these stored substances to sustain rapid growth. The scales develop from a basal plate, as do roots. A dry, paperlike outermost scale of many bulbs serves as a protective covering. An onion is an example (Figure 28.23*c*).

Corms A corm is a thickened underground stem that stores nutrients. Like a bulb, a corm has a basal plate from which roots grow. Unlike a bulb, a corm is solid rather than layered, and it has nodes from which new plants develop (Figure 28.23*d*).

Tubers Tubers are thickened portions of underground stolons; they are the plant's primary storage tissue. Tubers are like corms in that they have nodes from which new shoots and roots sprout, but they do not have a basal plate. Potatoes are tubers; their "eyes" are the nodes (Figure 28.23*e*).

Cladodes Cactuses and other succulents have photosynthetic stems called cladodes: flattened stems that store water. New plants form at the nodes. The cladodes of some plants appear quite leaflike, but most are unmistakably fleshy (Figure 28.23*f*).

Figure 28.23 Variations on a stem. Counterclockwise from *top:* (**a**) plants such as this aquatic eelgrass (*Vallisneria*) propagate themselves by sending out stolons. New plants develop at nodes in the stolons. (**b**) The main stems of turmeric plants (*Curcuma longa*) are undergound rhizomes. (**c**) Clearly visible scales of an onion (*Allium cepa*) surround the stem at the center of the bulb.

(**d**) Taro, also known as arrowroot, is a corm of *Colocasia esculenta* plants. Corms, unlike bulbs, do not have layers of scales. (**e**) Potatoes are tubers that grow on stolons of *Solanum tuberosum* plants. (**f**) The stems of prickly pear (*Opuntia*) are spiky cladodes. These paddlelike structures store water, allowing the plant to survive in very dry regions.

Take-Home Message

Are all stems alike?

■ Many plants have modified stems that function in storage or reproduction. Stolons, rhizomes, bulbs, corms, tubers, and cladodes are examples.

Even a short drought reduces photosynthesis and crop yields. Like other plants, crop plants conserve water by closing stomata, which of course also stops carbon dioxide from moving in. Without a continuous supply of carbon dioxide, the plant's photosynthetic cells cannot continue to make sugars. Drought-stressed flowering plants make fewer flowers or stunted ones. Even if flowers get pollinated, fruits may fall off the plant before ripening.

How would you vote?

Should cities restrict urban growth? Should farming be restricted to areas with sufficient rainfall to sustain agriculture? See CengageNOW for details, then vote online.

Summary

Section 28.1 Most flowering plants have aboveground **shoots**, including stems, photosynthetic leaves, and flowers. Most kinds also have **roots**. Shoots and roots consist of **ground**, **vascular**, and **dermal tissue systems**. Ground tissues store materials, function in photosynthesis, and structurally support the plant. Tubes in vascular tissues conduct substances to all living cells. Dermal tissues protect plant surfaces.

Monocots and **eudicots** consist of the same tissues organized in different ways. For example, monocots and eudicots differ in how xylem and phloem are distributed through ground tissue, in the number of petals in flowers, and in the number of **cotyledons**. All plant tissues originate at **meristems**, which are regions of undifferentiated cells that retain their ability to divide. **Primary growth** (or lengthening) arises from **apical meristems**. **Secondary growth** (or thickening) arises from **lateral meristems**.

■ *Use the animation on CengageNOW to explore a plant body plan and to compare monocot and eudicot tissues.*

Section 28.2 The simple plant tissues are **parenchyma**, **collenchyma**, and **sclerenchyma**. The living, thin-walled cells in parenchyma have diverse roles in ground tissue. Photosynthetic parenchyma is called **mesophyll**. Living cells in collenchyma have sturdy, flexible walls that support fast-growing plant parts. Cells in sclerenchyma die at maturity, but their lignin-reinforced walls remain and support the plant.

Vascular tissues (**xylem and phloem**) and dermal tissues (epidermis and periderm) are examples of complex plant tissues. **Vessel members** and **tracheids** of xylem are dead at maturity; their perforated, interconnected walls conduct water and dissolved minerals. Phloem's **sieve-tube members** remain alive at maturity. These cells interconnect to form tubes that conduct sugars. **Companion cells** load sugars into the sieve tubes. **Epidermis** covers and protects the outer surfaces of primary plant parts. **Periderm** replaces epidermis on woody plants, which have extensive secondary growth.

Section 28.3 Stems of most species support upright growth, which favors interception of sunlight. **Vascular bundles** of xylem and phloem thread through them. New shoots form at **terminal buds** and **lateral buds** on stems.

In most herbaceous and young woody eudicot stems, a ring of bundles divides the ground tissue into cortex and

pith. In woody eudicot stems, the ring becomes bands of different tissues. Monocot stems often have vascular bundles distributed throughout ground tissue.

■ *Use the animation on CengageNOW to look inside stems.*

Section 28.4 Leaves are photosynthesis factories that contain mesophyll and vascular bundles (**veins**) between their upper and lower epidermis. Air spaces around mesophyll cells allow gas exchange. Water vapor and gases cross the cuticle-covered epidermis at stomata.

■ *Use the animation on CengageNOW to explore the structure of a leaf.*

Section 28.5 Roots absorb water and mineral ions for the rest of the plant. Inside each is a **vascular cylinder** with primary xylem and phloem. **Root hairs** increase the surface area of roots. Most eudicots have a **taproot system**; many monocots have a **fibrous root system**.

■ *Use the animation on CengageNOW to learn about root structure and function.*

Sections 28.6, 28.7 Activity at **vascular cambium** and **cork cambium**, both lateral meristems, thickens the older stems and roots of many plants. **Wood** is classified by its location and function, as in **heartwood** or **sapwood**. **Bark** is secondary phloem and periderm. The **cork** in **periderm** protects and waterproofs woody stems and roots.

■ *Use the animation on CengageNOW to learn about the structure of wood.*

Section 28.8 Stem modifications in many types of plants function in storage or reproduction.

Self-Quiz   *Answers in Appendix III*

1. Which of the following two distribution patterns for vascular tissues is common among eudicots? Which is common among monocots?

Data Analysis Exercise

Douglas fir trees (*Pseudotsuga menziesii*) are exceptionally long-lived, and particularly responsive to rainfall levels. Researcher Henri Grissino-Mayer sampled Douglas firs in El Malpais National Monument, in west central New Mexico. Pockets of vegetation in this site have been surrounded by lava fields for about 3,000 years, so they have escaped wildfires, grazing animals, agricultural activity, and logging. Grissino-Mayer compiled tree ring data from old, living trees, and dead trees and logs to generate a 2,129-year annual precipitation record (Figure 28.24).

1. The Mayan civilization began to suffer a massive loss of population around 770 A.D. Do these tree ring data reflect a drought condition at this time? If so, was that condition relatively more or less severe than the "dust bowl" drought?

2. One of the worst population catastrophes ever recorded occurred in Mesoamerica between 1519 and 1600 A.D., when approximately 22 million people native to the region died. According to these data, which period between 137 B.C. and 1992 had the most severe drought? How long did that particular drought last?

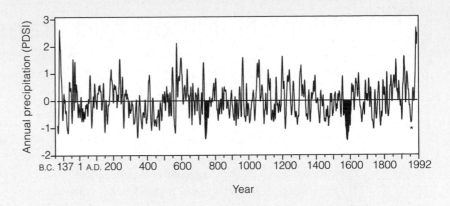

Figure 28.24 A 2,129-year annual precipitation record complied from tree rings in El Malpais National Monument, New Mexico. Data was averaged over 10-year intervals; graph correlates with other indicators of rainfall collected in all parts of North America. PDSI: Palmer Drought Severity Index: 0, normal rainfall; increasing numbers mean increasing excess of rainfall; decreasing numbers mean increasing severity of drought.

* A severe drought contributed to a series of catastrophic dust storms that turned the midwestern United States into a "dust bowl" between 1933 and 1939.

2. Roots and shoots lengthen through activity at _____ .
  a. apical meristems   c. vascular cambium
  b. lateral meristems   d. cork cambium

3. In many plant species, older roots and stems thicken by activity at _____ .
  a. apical meristems   c. vascular cambium
  b. cork cambium   d. both b and c

4. Bark is mainly _____ .
  a. periderm and cork   c. periderm and phloem
  b. cork and wood   d. cork cambium and phloem

5. _____ conducts water and minerals throughout a plant, and _____ conducts sugars.
  a. Phloem; xylem   c. Xylem; phloem
  b. Cambium; phloem   d. Xylem; cambium

6. Mesophyll consists of _____ .
  a. waxes and cutin   c. photosynthetic cells
  b. lignified cell walls d. cork but not bark

7. In phloem, organic compounds flow through _____ .
  a. collenchyma cells   c. vessels
  b. sieve tubes   d. tracheids

8. Xylem and phloem are _____ tissues.
  a. ground   b. vascular   c. dermal   d. both b and c

9. In early wood, cells have _____ diameters, _____ walls.
  a. small; thick   c. large; thick
  b. small; thin   d. large; thin

10. Match each plant part with a suitable description.
  ___apical meristem   a. massive secondary growth
  ___lateral meristem   b. source of primary growth
  ___xylem   c. distribution of sugars
  ___phloem   d. source of secondary growth
  ___vascular cylinder   e. distribution of water
  ___wood   f. central column in roots

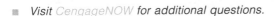
Visit CengageNOW for additional questions.

Critical Thinking

1. Is the plant with the yellow flower *above* a eudicot or a monocot? What about the plant with the purple flower?

2. Oscar and Lucinda meet in a tropical rain forest and fall in love, and he carves their initials into the bark of a tiny tree. They never do get together, though. Ten years later, still heartbroken, Oscar searches for the tree. Given what you know about primary and secondary growth, will he find the carved initials higher relative to ground level? If he goes berserk and chops down the tree, what kinds of growth rings will he see?

3. Are the structures shown *below left* stolons, rhizomes, bulbs, corms, or tubers? (Hint: Notice where the shoots are growing from.) What about the structures shown *below right*?

IMPACTS, ISSUES | Leafy Cleanup Crews

From World War I until the 1970s, the United States Army tested and disposed of weapons at J-Field, Aberdeen Proving Ground in Maryland (Figure 29.1a). Obsolete chemical weapons and explosives were burned in open pits, together with plastics, solvents, and other wastes. Lead, arsenic, mercury, and other metals heavily contaminated the soil and groundwater. So did highly toxic organic compounds, including trichloroethylene (TCE). TCE damages the nervous system, lungs, and liver, and can cause coma and death. Today, the toxic groundwater is seeping toward nearby marshes and the Chesapeake Bay.

There was too much contaminated soil at J-Field to remove, so the Army and the Environmental Protection Agency turned to phytoremediation: the use of plants to take up and concentrate or degrade environmental contaminants. They planted hybrid poplar trees (*Populus trichocarpa* × *deltoides*) that cleanse groundwater by taking up TCE and other organic compounds from it (Figure 29.1b).

How? The roots of the hybrid poplars take up water from the soil. Along with the water come dissolved nutrients and chemical contaminants, including TCE. The trees break down some of the TCE, and release some of it into the atmosphere. Airborne TCE is the lesser of two evils: TCE persists for a long time in groundwater, but it breaks down quickly in air that is polluted with other chemicals.

In other types of phytoremediation, groundwater contaminants accumulate in tissues of the plants, which are then harvested for safer disposal elsewhere.

The best plants for phytoremediation take up many contaminants, grow fast, and grow big. Not very many species can tolerate toxic substances, but genetically engineered ones may increase our number of choices for this purpose. For example, alpine pennycress (*Thlaspi caerulescens*) absorbs zinc, cadmium, and other potentially toxic minerals dissolved in soil water. Unlike typical cells, the cells of pennycress plants store zinc and cadmium inside a central vacuole. Isolated inside these organelles, the toxic elements are kept safely away from the rest of the cells' activities. Pennycress is a small, creeping plant, so its usefulness for phytoremediation is limited. Researchers are working to transfer a gene that confers its toxin-storing capacity to larger plants.

Many adaptations that help the toxin-busters cleanse contaminated areas are the same ones that absorb and distribute water and solutes through the plant body. When considering these adaptations, remember that many details of plant physiology are adaptations to limited environmental resources. In nature, plants rarely have unlimited supplies of the resources they require to nourish themselves, and nowhere except in overfertilized gardens does soil water contain lavish amounts of dissolved minerals.

See the video! Figure 29.1 Phytoremediation in action. (**a**) J-Field, once a weapons testing and disposal site. (**b**) Today, hybrid poplars are helping to remove substances that contaminate the field's soil and groundwater.

Key Concepts

Plant nutrients and soil

Many plant structures are adaptations to limited amounts of water and essential nutrients. The amount of water and nutrients available for plants to take up depends on the composition of soil. Soil is vulnerable to leaching and erosion. **Section 29.1**

Water uptake and movement through plants

Certain specializations help roots of vascular plants take up water and nutrients. Xylem distributes absorbed water and solutes from roots to leaves. **Sections 29.2, 29.3**

Water loss versus gas exchange

A cuticle and stomata help plants conserve water, a limited resource in most land habitats. Closed stomata stop water loss but also stop gas exchange. Some plant adaptations are trade offs between water conservation and gas exchange. **Section 29.4**

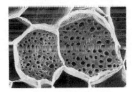

Sugar distribution through plants

Phloem distributes sucrose and other organic compounds from photosynthetic cells in leaves to living cells throughout the plant. Organic compounds are actively loaded into conducting cells, then unloaded in growing tissues or storage tissues. **Section 29.5**

Links to Earlier Concepts

- In this chapter, you will be taking a closer look at how fluids move through plants. This movement depends on hydrogen bonding in water (Section 2.4), membrane transporters (5.2–5.4), and osmosis and turgor (5.6).

- It will help to review what you learned about nutrients (1.2), ions (2.3), water (2.5), and carbohydrates (3.3), as well as photosynthesis (7.3, 7.6) and aerobic respiration (8.4).

- You will use your knowledge of vascular tissues (28.2), leaves (28.4), and roots (28.5). You will also see more examples of plant symbionts (24.6).

- We will revisit some adaptations of land plants (23.2), including the cuticle (4.12) and stomata (7.7). You will see an example of how cell signaling (27.6) is part of homeostasis in plants.

How would you vote? Transgenic plants may be more efficient at cleaning up contaminated sites than unmodified plants. Do you support using genetically engineered plants for phytoremediation? See CengageNOW for details, then vote online.

29.1 | Plant Nutrients and Availability in Soil

■ Plants require elemental nutrients from soil, water, and air.
■ Different types of soil affect the growth of different plants.

■ Links to Nutrients 1.2, Ions 2.3

The Required Nutrients

A **nutrient** is an element or molecule with an essential role in an organism's growth and survival. Plants require sixteen nutrients, all elements available in water and air, or as minerals that have dissolved as ions in the water. Examples include calcium and potassium. Nine of the elements are macronutrients, which means that they are required in amounts greater than 0.5 percent of the plant's dry weight (its weight after all of the water has been removed). Seven other elements are micronutrients, which make up traces (typically a few parts per million) of the plant's dry weight. A deficiency in any one of these nutrients may affect plant growth (Table 29.1).

Table 29.1 Plant Nutrients and Deficiency Symptoms

Type of Nutrient	Deficiency Symptoms
MACRONUTRIENT	
Carbon, oxygen, hydrogen	None; all are available in abundance from water and carbon dioxide
Nitrogen	Stunted growth; chlorosis (leaves turn yellow and die because of insufficient chlorophyll)
Potassium	Reduced growth; curled, mottled, or spotted older leaves, leaf edges brown; weakened plant
Calcium	Terminal buds wither; deformed leaves; stunted roots
Magnesium	Chlorosis; drooped leaves
Phosphorus	Purplish veins; stunted growth; fewer seeds, fruits
Sulfur	Light-green or yellowed leaves; reduced growth
MICRONUTRIENT	
Chlorine	Wilting; chlorosis; some leaves die
Iron	Chlorosis; yellow, green striping in leaves of grasses
Boron	Buds die; leaves thicken, curl, become brittle
Manganese	Dark veins, but leaves whiten and fall off
Zinc	Chlorosis; mottled or bronzed leaves; abnormal roots
Copper	Chlorosis; dead spots in leaves; stunted growth
Molybdenum	Pale green, rolled or cupped leaves

Properties of Soil

Soil consists of mineral particles mixed with variable amounts of decomposing organic material, or **humus**. The particles form by the weathering of hard rocks. Humus forms from dead organisms and organic litter: fallen leaves, feces, and so on. Water and air occupy spaces between the particles and organic bits.

Soils differ in their proportions of mineral particles and how compacted they are. The particles, which differ in size, are primarily sand, silt, and clay. The biggest sand grains are 0.05 to 2 millimeters in diameter. You can see individual grains by sifting beach sand through your fingers. Individual particles of silt are too small to see; they are only 0.002 to 0.05 millimeters in diameter. Particles of clay are even smaller.

Each clay particle consists of thin, stacked layers of negatively charged crystals. Sheets of water molecules alternate between the layers. Because of its negative charge, clay can temporarily bind positively charged mineral ions dissolved in the soil water. Clay latches onto dissolved nutrients that would otherwise trickle past roots too quickly to be absorbed.

Even though they do not bind mineral ions as well as clay, sand and silt are necessary for growing plants. Without enough sand and silt to intervene between the tiny particles of clay, the soil packs so tightly that air is excluded. Without air spaces in the soil, root cells cannot secure enough oxygen for aerobic respiration.

Soils and Plant Growth Soils with the best oxygen and water penetration are **loams**, which have roughly equal proportions of sand, silt, and clay. Most plants grow best in loams.

Humus also affects plant growth because it releases nutrients, and its negatively charged organic acids can trap the positively charged mineral ions in soil water. Humus swells and shrinks as it absorbs and releases water, and these changes in size aerate soil by opening spaces for air to penetrate.

Most plants grow well in soils that contain between 10 and 20 percent humus. Soil with less than 10 percent humus may be nutrient-poor. Soil with more than 90 percent humus stays so saturated with water that air (and the oxygen in it) is excluded. The soil in swamps and bogs contains so much organic matter that very few kinds of plants can grow in them.

How Soils Develop Soils develop over thousands of years. They are in different stages of development in different regions. Most form in layers, or horizons, that are distinct in color and other properties (Figure 29.2).

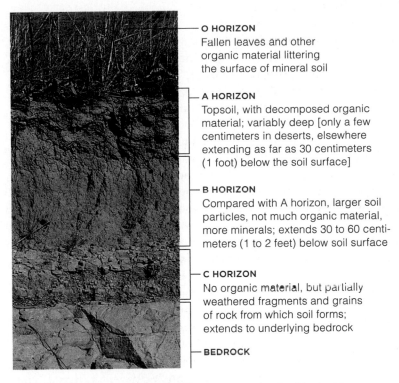

O HORIZON
Fallen leaves and other organic material littering the surface of mineral soil

A HORIZON
Topsoil, with decomposed organic material; variably deep [only a few centimeters in deserts, elsewhere extending as far as 30 centimeters (1 foot) below the soil surface]

B HORIZON
Compared with A horizon, larger soil particles, not much organic material, more minerals; extends 30 to 60 centimeters (1 to 2 feet) below soil surface

C HORIZON
No organic material, but partially weathered fragments and grains of rock from which soil forms; extends to underlying bedrock

BEDROCK

Figure 29.2 From a habitat in Africa, an example of soil horizons.

Figure 29.3 *Right:* Runaway erosion in Providence Canyon, Georgia, is the result of poor farming practices combined with soft soil. Settlers that arrived in the area around 1800 plowed the land straight up and down the hills. The furrows made excellent conduits for rainwater, which proceeded to carve out deep crevices that made even better rainwater conduits. The area became useless for farming by 1850. It now consists of about 445 hectares (1,100 acres) of deep canyons that continue to expand at the rate of about 2 meters (6 feet) per year.

The layers help us characterize soil in a given place, and compare it with soils in other places. For instance, the A horizon is **topsoil**. This layer typically contains the greatest amount of organic matter, so the roots of most plants grow most densely in it. Topsoil is deeper in some places than in others. Section 48.5 shows soil profiles for some major classes of ecosystems on land.

Leaching and Erosion

Minerals, salts, and other molecules dissolve in water as it filters through soil. **Leaching** is the process by which water removes soil nutrients and carries them away. Leaching is fastest in sandy soils, which do not bind nutrients as well as clay soils. During heavy rains, more leaching occurs in forests than in grasslands. Why? Grass plants absorb water more quickly than trees.

Soil erosion is a loss of soil under the force of wind and water. Strong winds, fast-moving water, sparse vegetation, and poor farming practices cause the great-est losses (Figure 29.3). For example, each year, about 25 billion metric tons of topsoil erode from croplands in the midwestern United States. The topsoil enters the Mississippi River, which then dumps it into the Gulf of Mexico. Nutrient losses because of this erosion affect not only plants that grow in the region, but also the other organisms that depend on them for survival.

Take-Home Message

From where do plants get the nutrients they require?

■ Plants require nine macronutrients and seven micronutrients, all elements. All are available from water, air, and soil.

■ Soil consists mainly of mineral particles: sand, silt, and clay. Clay attracts and reversibly binds dissolved mineral ions.

■ Soil contains humus, a reservoir of organic material rich in organic acids.

■ Most plants grow best in loams (soils with equal proportions of sand, silt, and clay) and between 10 and 20 percent humus.

■ Leaching and erosion remove nutrients from soil.

- Root specializations such as hairs, mycorrhizae, and nodules help the plant absorb water and nutrients.

- Links to Plasmodesmata 4.12, Aquaporins 5.2, Transport proteins 5.3, Osmosis 5.6, Nitrogen fixation 21.6, Fungal symbionts 24.6, Root structure 28.5

Figure 29.4 Examples of root specializations.

(**a**) The hairs on this root of a white clover plant (*Trifolium repens*) are about 0.2 mm long. (**b**) Mycorrhizae (*white* hairs) extending from the tip of these roots (*tan*) greatly enhance their surface area for absorbing scarce minerals from the soil.

(**c**) Root nodules on this soybean plant fix nitrogen from the air, and share it with the plant. (**d**) A nodule forms where bacteria infect the root. (**e**) Soybean plants growing in nitrogen-poor soil show the effect of root nodules on growth. Only the plants in the rows at *right* were inoculated with *Rhizobium* bacteria and formed nodules. **Figure It Out:** Are *Rhizobium* bacteria parasites or mutualists?

Answer: Mutualists

In actively growing plants, new roots infiltrate different patches of soil as they replace old roots. The new roots are not "exploring" the soil. Rather, their growth is simply greater in areas where the water and nutrient concentrations best match the requirements of the particular plant.

Certain specializations help plants take up water and nutrients from both soil and air. In roots, mycorrhizae and root hairs help plants absorb water and ions from soil, and root nodules help certain plants absorb additional nitrogen from the air.

Root Hairs As most plants put on primary growth, their root tips sprout many **root hairs** (Figure 29.4*a*). Collectively, these thin extensions of root epidermal cells enormously increase the surface area available for absorbing water and dissolved mineral ions. Root hairs are fragile structures no more than a few millimeters long. They do not develop into new roots, and live only a few days. New ones constantly form just behind the root tip (Section 28.5).

Mycorrhizae As Section 24.6 explains, a **mycorrhiza** (plural, mycorrhizae) is a form of mutualism between a young root and a fungus. Both species benefit from the association. The fungal hyphae grow as a velvety covering around the root or penetrate its cells (Figure 29.4*b*). Collectively, hyphae have a far greater surface area than the root itself, so they can absorb scarce minerals from a larger volume of soil. The root's cells give some sugars and nitrogen-rich compounds to the fungus, and the fungus gives some of the minerals it mines to the plant.

Root Nodules Certain types of bacteria in soil are mutualists with clover, peas, and other legumes. Like all other plants, legumes require nitrogen for growth. Nitrogen gas ($N{\equiv}N$, or N_2) is abundant in the air, but plants do not have enzymes that can break it apart. The bacteria do. Their enzymes convert nitrogen gas to ammonia (NH_3). The metabolic conversion of nitrogen gas to ammonia is an energy-intensive process called **nitrogen fixation** (Section 21.6). Other types of soil bacteria convert ammonia to nitrate (NO_3^-), the form of nitrogen that plants can use most easily. You will read more about nitrogen fixation in Section 47.9.

Root nodules are swollen masses of bacteria-infected root cells (Figure 29.4*c*). The bacteria (*Rhizobium* and *Bradyrhizobium*, both anaerobic) fix nitrogen and share it with the plant. In return, the plant provides the bacteria with an oxygen-free environment, and shares its photosynthetically produced sugars with them.

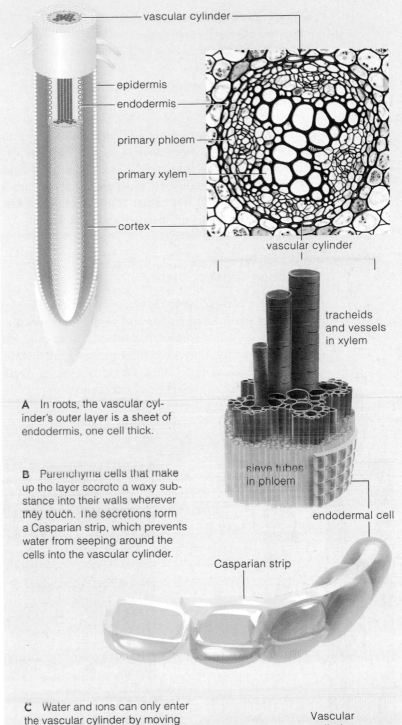

Figure 29.5 Animated In most flowering plants, transport proteins in the plasma membranes of root cells control the plant's uptake of water and dissolved mineral ions from the soil.

vascular cylinder

epidermis
endodermis
primary phloem
primary xylem
cortex

vascular cylinder

tracheids and vessels in xylem

sieve tubes in phloem

endodermal cell

Casparian strip

A In roots, the vascular cylinder's outer layer is a sheet of endodermis, one cell thick.

B Parenchyma cells that make up the layer secrete a waxy substance into their walls wherever they touch. The secretions form a Casparian strip, which prevents water from seeping around the cells into the vascular cylinder.

C Water and ions can only enter the vascular cylinder by moving through cells of the endodermis. They enter the cells via plasmodesmata or via transport proteins in the cells' plasma membranes.

Water and ions must cross at least one lipid bilayer before entering a vascular cylinder. Thus, plasma membrane transport proteins control the movement of these substances into the rest of the plant.

Vascular cylinder

Casparian strip

water and nutrients

Cortex

How Roots Control Water Uptake

Osmosis drives the movement of soil water into a root, then into the walls of parenchyma cells that make up the root cortex. Some of the nutrient-laden water stays in the cell walls; it permeates the cortex by diffusing around the cells' plasma membranes. Water molecules enter the cells' cytoplasm by diffusing across plasma membranes directly or through aquaporins (Section 5.2). Active transporters in the membranes pump dissolved mineral ions into the cells. After moving into cytoplasm, the water and ions diffuse from cell to cell through plasmodesmata (Section 4.12).

A vascular cylinder is separated from the root cortex by endodermis, a tissue composed of a single layer of parenchyma cells (Figure 29.5a). These cells secrete a waxy substance into their walls wherever they abut. The substance forms a **Casparian strip**, a waterproof band between the plasma membranes of endodermal cells (Figure 29.5b). The Casparian strip prevents the water that is seeping around the cells in the root's cortex from passing through endodermal cell walls into the vascular cylinder.

Water and ions enter a root's vascular cylinder by moving through plasmodesmata, or by crossing endodermal cell plasma membranes. Either way, they have to cross at least one plasma membrane. Thus, plasma membrane transport proteins can control the amount of water, and the amount and types of ions, that move from the root cortex into the vascular cylinder (Figure 29.5c). The selectivity of these proteins also offers protection against toxins that may be in soil water.

The roots of many plants also have an **exodermis**, a layer of cells just beneath their surface. Exodermal cells often deposit their own Casparian strip that functions like the one next to the vascular cylinder.

Take-Home Message

How do roots take up water and nutrients?

■ Root hairs, mycorrhizae, and root nodules greatly enhance a root's ability to take up water and nutrients.

■ Transport proteins in root cell plasma membranes control the uptake of water and ions into the vascular cylinder.

- Evaporation from leaves and stems drives the upward movement of water through pipelines of xylem inside a plant.
- Water's cohesion allows it to be pulled from roots into all other parts of the plant.
- Links to Hydrogen bonding 2.4, Properties of water 2.5, Xylem 28.2, Root structure 28.5

Soil water moves into roots and then into the plant's aboveground parts. How does water move all the way from roots to leaves that may be more than 100 meters (330 feet) above the soil? The movement does not occur by active pumping, but rather is driven by two features of water that you learned about in Section 2.5: evaporation and cohesion.

Cohesion–Tension Theory

In vascular plants, water moves inside xylem. Section 28.2 introduced the **tracheids** and **vessel members** that make up its water-conducting tubes. These cells are dead at maturity; only their lignin-impregnated walls are left behind (Figure 29.6). Obviously, being dead, the cells are not expending any energy to pump water against gravity.

The botanist Henry Dixon explained how water is transported in plants. By his **cohesion–tension theory**, water inside xylem is pulled upward by air's drying power, which creates a continuous negative pressure called tension. The tension extends continuously from leaves to roots. Figure 29.7 illustrates the theory.

First, air's drying power causes **transpiration**: the evaporation of water from aboveground plant parts. Most of the water a plant takes up is lost by evaporation, typically from stomata on the plant's leaves and stems. Transpiration creates negative pressure inside the conducting tubes of xylem. In other words, the evaporation of water from leaves and stems pulls on the water that remains in the xylem.

Second, the continuous columns of fluid inside the narrow conductive tubes of xylem resist breaking into droplets. Remember from Section 2.5 that the collective strength of many hydrogen bonds among water molecules imparts cohesion to liquid water. Because water molecules are all connected to one another by hydrogen bonds, a pull on one also pulls on the others. Thus, the negative pressure created by transpiration exerts tension on the entire column of water that fills a xylem tube. That tension extends from leaves

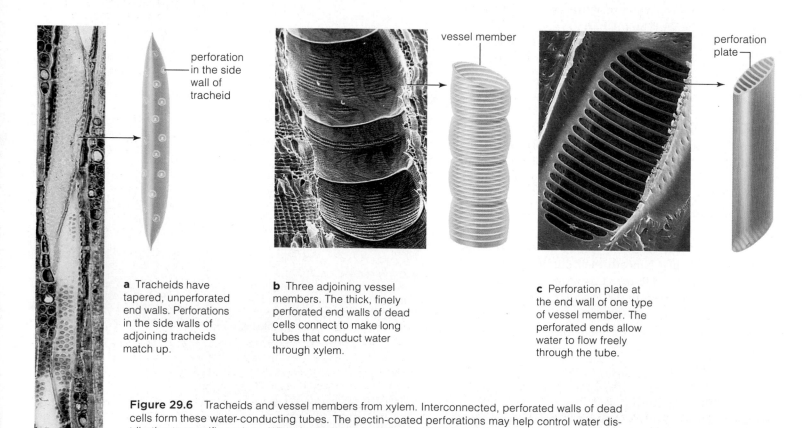

a Tracheids have tapered, unperforated end walls. Perforations in the side walls of adjoining tracheids match up.

b Three adjoining vessel members. The thick, finely perforated end walls of dead cells connect to make long tubes that conduct water through xylem.

c Perforation plate at the end wall of one type of vessel member. The perforated ends allow water to flow freely through the tube.

Figure 29.6 Tracheids and vessel members from xylem. Interconnected, perforated walls of dead cells form these water-conducting tubes. The pectin-coated perforations may help control water distribution to specific regions. When hydrated, the pectins swell and stop the flow. During dry periods, they shrink, and water moves freely through open perforations toward leaves.

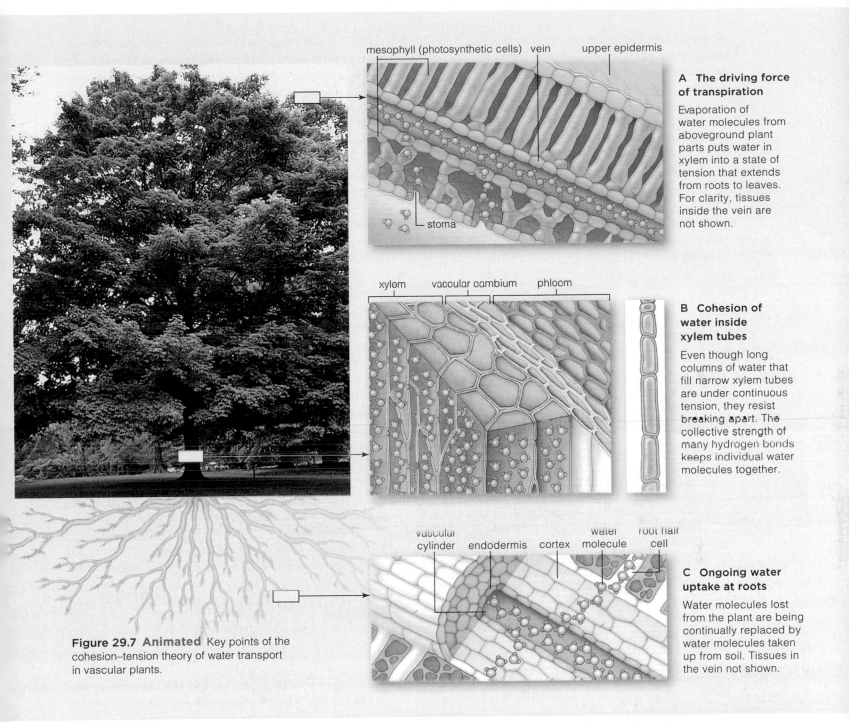

mesophyll (photosynthetic cells) vein upper epidermis

stoma

A The driving force of transpiration

Evaporation of water molecules from aboveground plant parts puts water in xylem into a state of tension that extends from roots to leaves. For clarity, tissues inside the vein are not shown.

xylem vascular cambium phloem

B Cohesion of water inside xylem tubes

Even though long columns of water that fill narrow xylem tubes are under continuous tension, they resist breaking apart. The collective strength of many hydrogen bonds keeps individual water molecules together.

vascular cylinder endodermis cortex water molecule root hair cell

C Ongoing water uptake at roots

Water molecules lost from the plant are being continually replaced by water molecules taken up from soil. Tissues in the vein not shown.

Figure 29.7 Animated Key points of the cohesion–tension theory of water transport in vascular plants.

that may be hundreds of feet in the air, down through stems, and on into young roots where water is being absorbed from the soil.

The movement of water through plants is driven mainly by transpiration. However, evaporation is only one of many other processes in plants that involve the loss of water molecules. Such processes all contribute to the negative pressure that results in water movement. Photosynthesis is an example.

Take-Home Message

What makes water move inside plants?

■ Transpiration is the evaporation of water from leaves, stems, and other plant parts.

■ By a cohesion–tension theory, transpiration puts water in xylem into a continuous state of tension from leaves to roots.

■ Tension pulls columns of water in xylem upward through the plant. The collective strength of many hydrogen bonds (cohesion) keeps the water from breaking into droplets as it rises.

■ Water is an essential resource for all land plants. Thus, water-conserving structures and processes are key to the survival of these plants.

■ Links to Plant cuticle 4.12, Osmosis 5.6, Gases in photosynthesis 7.3, Stomata 7.7, Gases in aerobic respiration 8.4, Land plant adaptations 23.2, Cell signaling 27.6, Leaf structure 28.4

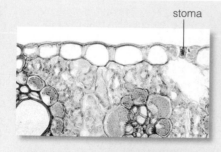

stoma

A Cuticle (*gold*) and stoma on a leaf. Each stoma is formed by two guard cells, which are specialized epidermal cells.

20 µm

guard cells

B This stoma is open. When the guard cells swell with water, they bend so that a gap opens between them.

The gap allows the plant to exchange gases with air. The exchange is necessary to keep metabolic reactions running.

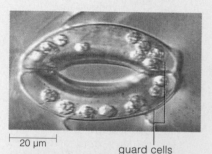

C This stoma is closed. The guard cells, which are not plump with water, are collapsed against each other so there is no gap between them.

A closed stoma limits water loss, but it also limits gas exchange, so photosynthesis and respiration reactions slow.

solutes water

D How do stoma open and close? When a stoma is open, the guard cells are maintaining a relatively high concentration of solutes by pumping solutes into their cytoplasm. Water diffuses into the hypertonic cytoplasm and keeps the cells plump.

ABA signal   solutes   water

E When water is scarce, a hormone (ABA) activates a pathway that lowers the concentrations of solutes in guard cell cytoplasm. Water follows its gradient and diffuses out of the cells, and the stoma closes.

In land plants, at least 90 percent of the water transported from roots to a leaf evaporates right out. Only about 2 percent is used in metabolism, but that amount must be maintained or photosynthesis, growth, membrane functions, and other processes will shut down.

If a plant is running low on water, it cannot move around to seek out more, as most animals can. A cuticle and stomata (Sections 4.12 and 23.2) help the plant conserve the water it already holds in its tissues. Both of these structures restrict the amount of water vapor that diffuses out of the plant's surfaces.

However, the cuticle and stomata also restrict gas exchanges between the plant and the air. Why is that important? The concentrations of carbon dioxide and oxygen gases in air spaces inside the plant affect the rate of critical metabolic pathways (such as photosynthesis and aerobic respiration) in the plant's cells. If a plant were entirely impermeable to water vapor and gases, it could not take in enough carbon dioxide to run photosynthesis. Neither could it sustain aerobic respiration for very long, because too much oxygen would build up in its tissues. Thus, water-conserving structures and mechanisms must balance the plant's needs for water with its needs for gas exchanges.

The Water-Conserving Cuticle

Even mildly water-stressed plants would wilt and die without a cuticle. This water-impermeable layer coats the walls of all plant cells exposed to air (Figure 29.8*a*). It consists of epidermal cell secretions: a mixture of waxes, pectin, and cellulose fibers embedded in cutin, an insoluble lipid polymer. The cuticle is translucent, so it does not prevent light from reaching photosynthetic tissues.

Controlling Water Loss at Stomata

A pair of specialized epidermal cells defines each stoma. When these two **guard cells** swell with water, they bend

Figure 29.8 Water-conserving structures in plants.
(**a**) Cuticle and stoma in a cross-section of basswood (*Tilia*) leaf.

(**b**–**e**) Stomata in action. Whether a stoma is open or closed depends on how much water is plumping up these guard cells. The amount of water in guard cell cytoplasm is influenced by hormonal signals.

The round structures inside the cells are chloroplasts. Guard cells are the only type of epidermal cell with these organelles.

Figure 29.9 Stomata at the leaf surface of a holly plant growing in a smoggy, industrialized region. Airborne pollutants not only block sunlight from photosynthetic cells, they also clog stomata, and can damage them so much that they close permanently.

slightly so a gap forms between them. The gap is the stoma. When the cells lose water, they collapse against each other, so the gap closes (Figure 29.8b,c).

Environmental cues such as water availability, the level of carbon dioxide inside the leaf, and light intensity affect whether stomata open or close. These cues trigger osmotic pressure changes in the cytoplasm of guard cells. For example, when the sun comes up, the light causes guard cells to begin pumping solutes (in this case, potassium ions) into their cytoplasm. The resulting buildup of potassium ions causes water to enter the cells by osmosis. The guard cells plump up, so the gap between them opens. Carbon dioxide from the air diffuses into the plant's tissues, and photosynthesis begins.

As another example, root cells release the hormone abscisic acid (ABA) when soil water becomes scarce. ABA travels through the plant's vascular system to leaves and stems, where it binds to receptors on guard cells. The binding causes solutes to exit these cells. Water follows by osmosis, the guard cells lose plumpness and collapse against each other, and the stomata close (Figure 29.8e).

Most stomata close at night, in most plants. Water is conserved, and carbon dioxide builds up in leaves as cells make ATP by aerobic respiration. The stomata of CAM plants, including most cactuses, open at night, when the plant takes in and fixes carbon from carbon dioxide. During the day, they close, and the plant uses the carbon that it fixed during the night for photosynthesis (Section 7.7).

Stomata also close in response to some of the chemicals in polluted air. The closure protects the plant from chemical damage, but it also prevents the uptake of carbon dioxide for photosynthesis, and so inhibits growth. Think about it on a smoggy day (Figure 29.9).

Take-Home Message

How do land plants conserve water?

■ A waxy cuticle covers all epidermal surfaces of the plant exposed to air. It restricts water loss from plant surfaces.

■ Plants conserve water by closing their many stomata. Closed stomata also prevent gas exchanges necessary for photosynthesis and aerobic respiration.

■ A stoma stays opens when the guard cells that define it are plump with water. It closes when the cells lose water and collapse against each other.

■ Xylem distributes water and minerals through plants, and phloem distributes the organic products of photosynthesis.

■ Links to Carbohydrates 3.3, Active transport 5.4, Osmosis and turgor 5.6, Photosynthetic products 7.6, Plant vascular tissues 28.2

Phloem is a vascular tissue with organized arrays of conducting tubes, fibers, and strands of parenchyma cells. Unlike conducting tubes of xylem, sieve tubes in phloem consist of living cells. Sieve-tube cells are positioned side by side and end to end, and their abutting end walls (sieve plates) are porous. Dissolved organic compounds flow through the tubes (Figure 29.10a,b).

Companion cells that are pressed against the sieve tubes actively transport the organic products of photosynthesis into them. Some of the molecules are used in the cells that make them, but the rest travel through the sieve tubes to the other parts of the plant: roots, stems, buds, flowers, and fruits.

Plants store their carbohydrates mainly as starch, but starch molecules are too big and too insoluble to transport across plasma membranes. Cells break down starch molecules to sucrose and other small molecules that are easily transported through the plant.

Some experiments with plant-sucking insects demonstrated that sucrose is the main carbohydrate transported in phloem. Aphids feeding on the juices in the conducting tubes of phloem were anesthetized with high levels of carbon dioxide (Figure 29.11). Then their bodies were detached from their mouthparts, which remained attached to the plant. Researchers collected and analyzed fluid exuded from the aphids' mouthparts. For most of the plants studied, sucrose was the most abundant carbohydrate in the fluid.

Pressure Flow Theory

Translocation is the formal name for the process that moves sucrose and other organic compounds through phloem of vascular plants. Phloem translocates photosynthetic products along declining pressure and solute concentration gradients. The **source** of the flow is any region of the plant where organic compounds are being loaded into sieve tubes. A common source is photosynthetic mesophyll in leaves. The flow ends at a **sink**, which is any plant region where the products are being used or stored. For instance, while flowers and fruits are forming on the plant, they are sinks.

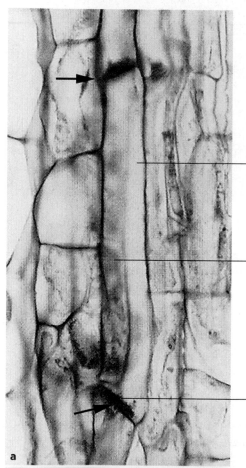

one of a series of living cells that abut, end to end, and form a sieve tube

companion cell (in the background, pressed tightly against sieve tube)

perforated end plate of sieve-tube cell, of the sort shown in (**b**)

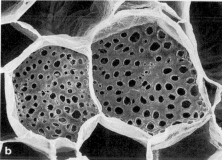

Figure 29.10 (**a**) Part of a sieve tube inside phloem. Arrows point to perforated ends of individual tube members. (**b**) Scanning electron micrograph of the sieve plates on the ends of two side-by-side sieve-tube members.

Figure 29.11 Honeydew exuding from an aphid after this insect's mouthparts penetrated a sieve tube. High pressure in phloem forced this droplet of sugary fluid out through the terminal opening of the aphid gut.

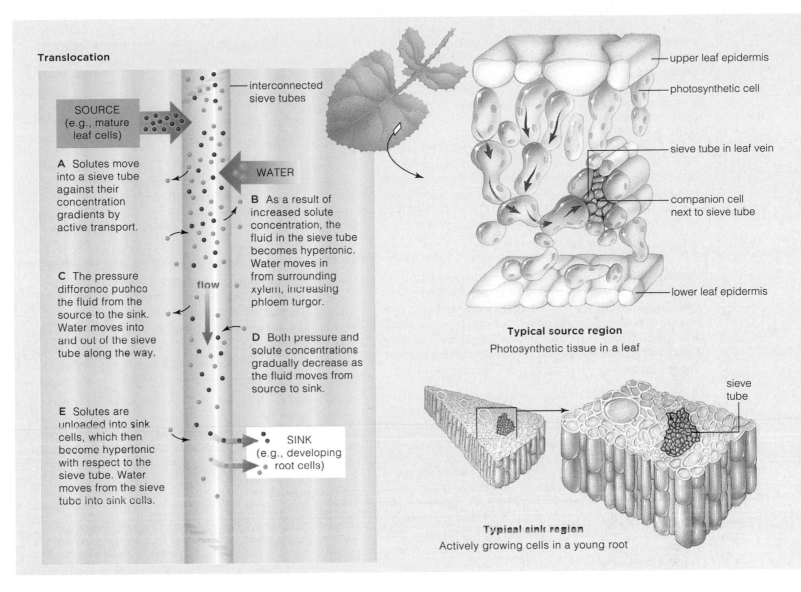

Figure 29.12 Animated Translocation of organic compounds. Review Section 7.6 to get an idea of how translocation relates to photosynthesis in vascular plants.

Why do organic compounds in phloem flow from source to sink? High fluid pressure drives the movement of fluid in phloem (Section 5.6). According to the **pressure flow theory**, internal pressure builds up in sieve tubes at a source. The pressure can be five times higher than the air pressure inside an automobile tire. A pressure gradient pushes solute-rich fluid to a sink, where the solutes are removed from the phloem.

Use Figure 29.12 to track what happens to sugars and other organic solutes as they move from the photosynthetic cells into small leaf veins. Companion cells in veins actively transport the solutes into sieve-tube members. When the solute concentration increases in the tubes, water also moves into them by osmosis. The increase in fluid volume exerts extra pressure (turgor) on the walls of the sieve tubes.

Phloem in a sink region has a lower internal pressure than that of a source region. Sucrose is unloaded at a sink, and water is diffusing out of phloem there by osmosis. The difference in fluid pressure between sources and sinks moves the sugar-laden fluid inside phloem through the plant.

Take-Home Message

How do organic molecules move through plants?

■ Plants store carbohydrates as starch, and distribute them as sucrose and other small, water-soluble molecules.

■ Concentration and pressure gradients in the sieve-tube system of phloem force organic compounds to flow to different parts of the plant.

■ The gradients are set up by companion cells moving organic molecules into sieve tubes at sources, and the unloading of the molecules at sinks.

With elemental pollutants such as lead or mercury, the best phytoremediation strategies use plants that absorb and then store these toxins in aboveground tissues, which can be harvested for safe disposal. Researchers have genetically modified such plants to enhance their absorptive and storage capacity. Dr. Kuang-Yu Chen, pictured at *right*, is analyzing zinc and cadmium levels in plants that can tolerate these elements.

In the case of organic toxins such as TCE, the best phytoremediation strategies use plants with biochemical pathways that break down the compounds to less-toxic molecules. Phytoremediation researchers are beefing up these pathways in many plants. Some

How would you vote?

Do you support the use of transgenic plants with an enhanced capacity to take up or detoxify pollutants for phytoremediation? See CengageNOW for details, then vote online.

are transferring genes from bacteria or animals into plants; others are enhancing expression of genes that encode molecular participants in the plants' own detoxification pathways.

Summary

Section 29.1 Plant growth requires steady sources of water and **nutrients** obtainable from carbon dioxide and soil (Figure 29.13). The availability of water and nutrients in **soil** is largely determined by its proportions of sand, silt, and clay; and its **humus** content. **Loams** have roughly equal proportions of sand, silt, and clay. **Leaching** and **soil erosion** deplete nutrients in soil, particularly **topsoils**.

Section 29.2 **Root hairs** greatly increase roots' surface area for absorption. Fungi are symbionts with young roots in **mycorrhizae**, which enhance a plant's ability to absorb mineral ions from soil. **Nitrogen fixation** by bacteria in **root nodules** gives a plant extra nitrogen. In both cases, the symbionts receive some of the plant's sugars.

Roots control the movement of water and dissolved mineral ions into the vascular cylinder. Endodermal cells that form a layer around the cylinder deposit a waterproof band, a **Casparian strip**, in their abutting walls. The strip keeps water from diffusing around the cells. Water and nutrients enter a root vascular cylinder only by moving through the plasma membrane of parenchyma cells. The uptake is controlled by active transport proteins embedded in the membranes. Some plants also have an **exodermis**, an additional layer of cells that deposit a second Casparian strip just inside the root surface.

■ *Use the animation on CengageNOW to see how vascular plant roots control nutrient uptake.*

Section 29.3 Water and dissolved mineral ions flow through conducting tubes of xylem. The interconnected, perforated walls of **tracheids** and **vessel members** (cells that are dead at maturity) form the tubes.

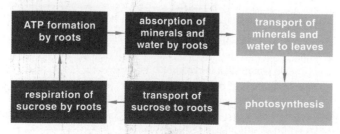

Figure 29.13 Summary of processes that sustain plant growth.

Transpiration is the evaporation of water from plant parts, mainly at stomata, into air. By a **cohesion–tension theory**, transpiration pulls water upward by creating a continuous negative pressure (or tension) inside xylem from leaves to roots. Hydrogen bonds among water molecules keep the columns of fluid continuous inside the narrow vessels.

■ *Use the animation on CengageNOW to learn about water transport in vascular plants.*

Section 29.4 A cuticle and stomata balance a plant's loss of water with its needs for gas exchange. Stomata are gaps across the cuticle-covered epidermis of leaves and other plant parts. Each is defined by a pair of **guard cells**. Closed stomata limit the loss of water, but also prevent the gas exchange required for photosynthesis and aerobic respiration.

Environmental signals, including pollution, can cause stomata to open or close. Hormonal signals trigger guard cells to pump ions into or out of their cytoplasm; water follows the ions (by osmosis). Water moving into guard cells plumps them, which opens the gap between them. Water diffusing out of the cells causes them to collapse against each other, so the gap closes.

Section 29.5 Organic compounds become distributed through a plant by **translocation**. Companion cells actively transport sugars and other organic products of photosynthesis into sieve tubes of phloem at **source** regions. The molecules are unloaded from the tubes at **sink** regions. By the **pressure flow theory**, the movement of fluid through phloem is driven by pressure and solute gradients.

■ *Use the animation on CengageNOW to observe how vascular plants distribute organic compounds.*

Self-Quiz   *Answers in Appendix III*

1. Carbon, hydrogen, and oxygen are plant _____ .
   a. macronutrients   d. essential elements
   b. micronutrients   e. both a and d
   c. trace elements

2. A(n) _____ strip between abutting endodermal cell walls forces water and solutes to move through these cells rather than around them.

Data Analysis Exercise

Plants used for phytoremediation take up organic pollutants from the soil or air, then transport the chemicals to plant tissues, where they are stored or broken down. Researchers are now designing transgenic plants with enhanced ability to take up or break down toxins.

In 2007, Sharon Doty and her colleagues published the results of their efforts to design plants useful for phytoremediation of soil and air containing organic solvents. The researchers used *Agrobacterium tumefaciens* (Section 16.7) to deliver a mammalian gene into poplar plants. The gene encodes cytochrome P450, a type of heme-containing enzyme involved in the metabolism of a range of organic molecules, including solvents such as TCE. The results of one of the researchers' tests on these transgenic plants are shown in Figure 29.14.

1. How many transgenic plants did the researchers test?

2. In which group did the researchers see the slowest rate of TCE uptake? The fastest?

3. On day 6, what was the difference between the TCE content of air around transgenic plants and that around vector control plants?

4. Assuming no other experiments were done, what two explanations are there for the results of this experiment? What other control might the researchers have used?

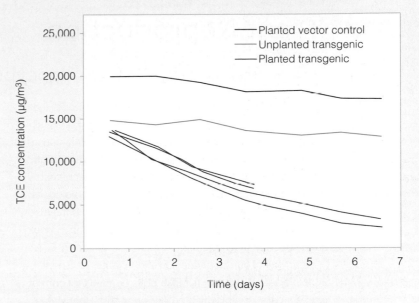

Figure 29.14 Results of tests on transgenic poplar trees. Planted trees were incubated in sealed containers with an initial 15,000 micrograms of TCE (trichloroethylene) per cubic meter of air. Samples of the air in the containers were taken daily and measured for TCE content. Controls included a tree transgenic for a *Ti* plasmid with no cytochrome P450 in it (vector control), and a bare-root transgenic tree (one that was not planted in soil).

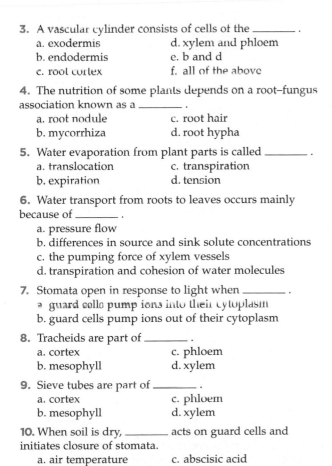

3. A vascular cylinder consists of cells of the _____ .
- a. exodermis
- b. endodermis
- c. root cortex
- d. xylem and phloem
- e. b and d
- f. all of the above

4. The nutrition of some plants depends on a root–fungus association known as a _____ .
- a. root nodule
- b. mycorrhiza
- c. root hair
- d. root hypha

5. Water evaporation from plant parts is called _____ .
- a. translocation
- b. expiration
- c. transpiration
- d. tension

6. Water transport from roots to leaves occurs mainly because of _____ .
- a. pressure flow
- b. differences in source and sink solute concentrations
- c. the pumping force of xylem vessels
- d. transpiration and cohesion of water molecules

7. Stomata open in response to light when _____ .
- a. guard cells pump ions into their cytoplasm
- b. guard cells pump ions out of their cytoplasm

8. Tracheids are part of _____ .
- a. cortex
- b. mesophyll
- c. phloem
- d. xylem

9. Sieve tubes are part of _____ .
- a. cortex
- b. mesophyll
- c. phloem
- d. xylem

10. When soil is dry, _____ acts on guard cells and initiates closure of stomata.
- a. air temperature
- b. humidity
- c. abscisic acid
- d. oxygen

11. Match the concepts of plant nutrition and transport.
- ___ stomata
- ___ plant nutrient
- ___ sink
- ___ root system
- ___ hydrogen bonds
- ___ transpiration
- ___ translocation

- a. evaporation from plant parts
- b. harvests soil water and nutrients
- c. balance water loss with gas exchange
- d. cohesion in water transport
- e. sugars unloaded from sieve tubes
- f. organic compounds distributed through the plant body
- g. essential element

■ *Visit CengageNOW for additional questions.*

Critical Thinking

1. Successful home gardeners, like farmers, make sure that their plants get enough nitrogen from either nitrogen-fixing bacteria or fertilizer. Which biological molecules incorporate nitrogen? Nitrogen deficiency stunts plant growth; leaves yellow and then die. How would nitrogen deficiency cause these symptoms?

2. When moving a plant from one location to another, the plant is more likely to survive if some native soil around the roots is transferred along with the plant. Formulate a hypothesis that explains that observation.

3. If a plant's stomata are made to stay open at all times, or closed at all times, it will die. Why?

4. Allen is studying the rate at which tomato plants take up water from soil. He notices that several environmental factors, including wind and relative humidity, affect the rate. Explain how they might do so.

IMPACTS, ISSUES | Plight of the Honeybee

In the fall of 2006, commercial beekeepers in Europe, India, and North America began to notice something was amiss in their honeybee hives. The bees were dying off in unusually high numbers. Many colonies did not survive through the winter that followed. By spring, the phenomenon had a name: colony collapse disorder. Farmers and biologists began to worry about what would happen if the honeybee populations continued to decline. Honey production would suffer, but many commercial crops would fail too.

Nearly all of our crops are flowering plants. As Chapter 23 explained, these plants make pollen grains that consist of a few cells, one of which produces sperm. Honeybees are pollinators; they carry pollen from one plant to another, pollinating flowers as they do. Typically, a flower will not

develop into a fruit unless it receives pollen from another flower. Even plants with flowers that can self-pollinate tend to make bigger fruits and more of them when they are cross-pollinated (Figure 30.1).

Many types of insects pollinate plants, but honeybees are especially efficient pollinators of a variety of plant species. They are also the only ones that tolerate living in man-made hives that can be loaded onto trucks and carted wherever crops require pollination. Loss of their portable pollination service is a huge threat to our agricultural economy.

We do not know what causes colony collapse disorder. Honeybees can be infected by a variety of pests and diseases that may be part of the problem. For example, Israeli acute paralysis virus has been detected in many affected hives. Pesticides may also be taking a toll. In the past few years, neonicotinoids have become the most widely used insecticides in the United States. These chemicals are systemic insecticides, which means they are taken up by all plant tissues, including the nectar and pollen that honeybees collect. Neonicotinoids are highly toxic to honeybees.

Colony collapse disorder is currently in the spotlight because it affects our food supply. However, other pollinator populations are also dwindling. Habitat loss is probably the main factor, but pesticides that harm honeybees also harm other pollinators.

Flowering plants rose to dominance in part because they coevolved with animal pollinators. Most flowers are specialized to attract and be pollinated by a specific species or type of pollinator. Those adaptations put the plants at risk of extinction if coevolved pollinator populations decline. Wild animal species that depend on the plants for fruits and seeds will also be affected. Recognizing the prevalence and importance of these interactions is our first step toward finding workable ways to protect them.

See the video! Figure 30.1 Importance of insect pollinators. (**a**) Honeybees are efficient pollinators of a variety of flowers, including berries. (**b**) Raspberry flowers can pollinate themselves, but the fruit that forms from a self-fertilized flower is of lower quality than that of a cross-pollinated flower. The two berries on the *left* formed from self-pollinated flowers. The one on the *right* formed from an insect-pollinated flower.

Key Concepts

Links to Earlier Concepts

- A review of what you know about plant tissue organization (Sections 28.2, 28.3, 28.8) and plant life cycles (10.5, 23.2) will be helpful as we examine in detail some of the reproductive adaptations that contributed to the evolutionary success of flowering plants (23.8, 23.9).

- This chapter revisits some of the evolutionary processes (18.11, 18.12) that resulted in the current spectrum of structural diversity in flowering plants.

- You will draw upon your understanding of membrane proteins (5.2) as you learn more about cell signaling (27.6) and development (15.2) in plant reproduction.

- We also revisit meiosis (10.3), Mendelian inheritance (11.1), cloning (13.4), radiometric dating (17.6), aneuploidy (12.6), and polyploidy in plants (18.11) within the context of plant asexual reproduction (10.1).

Structure and function of flowers

Flowers are shoots that are specialized for reproduction. Modified leaves form their parts. Gamete-producing cells develop in their reproductive structures; other parts such as petals are adapted to attract and reward pollinators. **Sections 30.1, 30.2**

Gamete formation and fertilization

Male and female gametophytes develop inside the reproductive parts of flowers. In flowering plants, pollination is followed by double fertilization. As in animals, signals are key to sex. **Sections 30.3, 30.4**

Seeds and fruits

After fertilization, ovules mature into seeds, each an embryo sporophyte together with tissues that nourish and protect it. As seeds develop, tissues of the ovary and often other parts of the flower mature into fruits, which function in seed dispersal. **Sections 30.5, 30.6**

Asexual reproduction in plants

Many species of plants reproduce asexually by vegetative reproduction. Humans take advantage of this natural tendency by propagating plants asexually for agriculture and research. **Section 30.7**

How would you vote? Systemic insecticides get into the nectar and pollen of flowering plants and thus can poison honeybees and other insect pollinators. To protect pollinators, should the use of these chemicals on flowering plants be restricted? See CengageNOW for details, then vote online.

Reproductive Structures of Flowering Plants

■ Specialized reproductive shoots called flowers consist of whorls of modified leaves.

■ Links to Plant life cycles 10.5 and 23.2, ABC model of flowering 15.2, Lateral buds 28.3

The sporophyte dominates the life cycle of flowering plants. A **sporophyte** is a diploid spore-producing plant body that grows by mitotic cell divisions of a fertilized egg (Sections 10.5 and 23.2).

Flowers are the specialized reproductive shoots of angiosperm sporophytes. Spores that form by meiosis inside flowers develop into haploid **gametophytes**, or structures in which haploid gametes form by mitosis.

Anatomy of a Flower

A flower forms when a lateral bud along the stem of a sporophyte develops into a short, modified branch called a receptacle. Master genes that become active in the apical meristem of the branch direct the formation of a flower (Section 15.2).

The petals and other parts of a typical flower are modified leaves that form in four spirals or four rings (whorls) at the end of the floral shoot. The outermost whorl develops into a calyx, which is a ring of leaflike sepals (Figure 30.2*a*). The sepals of most flowers are photosynthetic and inconspicuous; they serve to protect the flower's reproductive parts.

Just inside the calyx, petals form in a whorl called the corolla (from the Latin *corona*, or crown). Petals are usually the largest and most brightly colored parts of a flower. They function mainly to attract pollinators.

A whorl of stamens forms inside the ring of petals. **Stamens** are the male parts of a flower. In most flowers, they consist of a thin filament with an anther at the tip. Inside a typical anther are two pairs of elongated pouches called pollen sacs. Meiosis of diploid cells in each sac produces haploid, walled spores. The spores differentiate into **pollen grains**, which are immature male gametophytes. The durable coat of a pollen grain is a bit like a suitcase that carries and protects the cells inside on their journey to meet an egg.

The innermost whorl of modified leaves are folded and fused into **carpels**, the female parts of a flower. Carpels are sometimes called pistils. Many flowers have one carpel; others have several carpels, or several

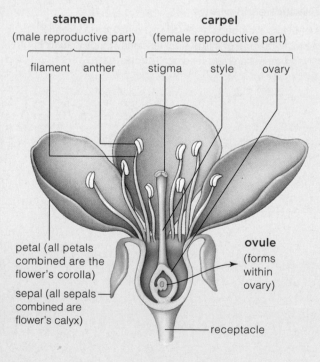

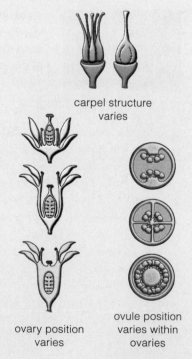

A Like many flowers, a cherry blossom (*Prunus*) has several stamens and one carpel. The male reproductive parts are stamens, which consist of pollen-bearing anthers atop slender filaments. The female reproductive part is the carpel, which consists of stigma, style, and ovary.

B Flower structure varies among different plant species.

Figure 30.2 Animated Structure of flowers.

groups of carpels, that may be fused (Figure 30.2*b*). The upper region of a carpel, a sticky or hairy stigma, is specialized to trap pollen grains. Often, the stigma sits on top of a slender stalk called a style. The lower, swollen region of a carpel is the **ovary**, which contains one or more ovules. An **ovule** is a tiny bulge of tissue inside the ovary. A cell in the ovule undergoes meiosis and develops into the haploid female gametophyte.

At fertilization, a diploid zygote forms when male and female gametes meet inside an ovary. The ovule then matures into a seed. The life cycle of the plant is completed when the seed germinates, and a new sporophyte forms and matures (Figure 30.3). We return to fertilization and seed development in later sections.

Diversity of Flower Structure

Remember that mutations in some master genes give rise to dramatic variations in flower structure (Section 15.2). We see many such variations in the range of diversity of flowering plants.

Regular flowers are symmetric around their center axis: If the flower were cut like a pie, the pieces would be roughly identical (Figure 30.4*a*). Irregular flowers are not radially symmetric (Figure 30.4*b*). Flowers may form as single blossoms, or in clusters called inflorescences. Some species, like sunflowers (*Helianthus*), have inflorescences that are actually composites of many flowers grouped into a single head. Other types of inflorescence include umbrella-like forms (Figure 30.4*c*) or elongated spikes (Figure 30.4*d*).

A cherry blossom (Figure 30.2) has all four sets of modified leaves (sepals, petals, stamens, and carpels), so it is called a complete flower. Incomplete flowers lack one or more of these structures (Figure 30.4*e*). Cherry blossoms are also called perfect flowers, because they have both stamens and carpels.

Perfect flowers may be fertilized by pollen from other plants, or they can self-pollinate. Self-pollination can be adaptive in situations where plants are widely spaced, such as in newly colonized areas. However, in general, offspring of self-pollinated flowers or plants tend to be less vigorous than those of cross-pollinated plants. Accordingly, adaptations of many plant species encourage or even require cross-pollination.

For example, pollen may be released from a flower's anthers only after its stigma is no longer receptive to being fertilized by pollen. As another example, the imperfect flowers of some species have either stamens or carpels, but not both. Depending on the species, the separate male and female flowers form on different plants, or on the same plant.

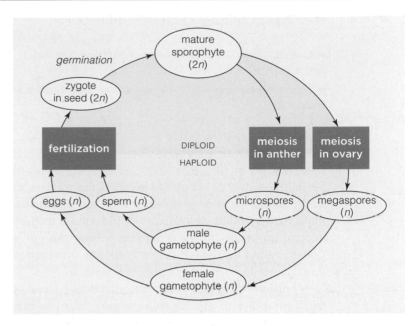

Figure 30.3 Animated Typical flowering plant life cycle.

Figure 30.4 Examples of structural variation in flowers. (**a**) Arctic rose (*Rosa acicularis*), a regular flower; (**b**) white sage (*Salvia apiana*), an irregular flower; (**c**) carrot (*Daucus carota*), an umbrella-like inflorescence; (**d**) yucca (*Yucca sp.*), an elongated inflorescence, and (**e**) meadow-rue (*Thalictrum pubescens*), an incomplete flower that has stamens but no petals.

Take-Home Message

What are flowers?

■ Flowers are short reproductive branches of sporophytes. The different parts of a flower (sepals, petals, stamens, and carpels) are modified leaves.

■ The male parts of flowers are stamens, which typically consist of a filament with an anther at the tip. Pollen forms inside anthers.

■ The female parts of flowers are carpels, which typically consist of stigma, style, and ovary. Haploid, egg-producing female gametophytes form in an ovule inside the ovary.

■ Flowers vary in structure. Many of the variations are adaptations that maximize the plant's chance of cross-pollination.

■ Flowering plants coevolved with pollination vectors that help them reproduce sexually.

■ Links to Coevolution 18.12, Coevolution of flowers and pollinators 23.8

Getting By With a Little Help From Their Friends

Sexual reproduction in plants involves the transfer of pollen, typically from one plant to another. Unlike animals, plants cannot move about to find a mate, so they depend on factors in the environment that can move pollen around for them (Section 23.8). The diversity of flower form in part reflects that dependence.

A **pollination vector** is an agent that delivers pollen from an anther to a compatible stigma. Many plants are pollinated by wind, which is entirely nonspecific in

Figure 30.6 *Opposite*, flowers of a giant saguaro cactus (*Carnegia gigantea*). Birds and insects sip nectar from these large, white flowers by day, and bats sip by night. The flowers offer a sweet nectar.

where it dumps pollen. Such plants often release pollen grains by the billions, insurance in numbers that some of their pollen will reach a receptive stigma.

Other plants enlist the help of **pollinators**—living pollination vectors—to transfer pollen among individuals of the same species. An insect, bird, or other animal that is attracted to a particular flower often picks up pollen on a visit, then inadvertently transfers it to the flower of a different plant on a later visit. The more specific the attraction, the more efficient the transfer of pollen among plants of the same species. Given the selective advantage for flower traits that attract specific pollinators, it is not surprising that about 90 percent of flowering plants have coevolved animal pollinators.

A flower's shape, pattern, color, and fragrance are adaptations that attract specific animals (Table 30.1). For example, the petals of flowers pollinated by bees usually are bright white, yellow, or blue, typically with pigments that reflect ultraviolet light. Such UV-reflecting pigments are often distributed in patterns that bees can recognize as visual guides to nectar (Figure 30.5). We see these patterns only with special camera filters; our eyes do not have receptors that respond to UV light.

Pollinators such as bats and moths have an excellent sense of smell, and can follow concentration gradients of airborne chemicals to a flower that is emitting them (Figure 30.6). Not all flowers smell sweet; odors like dung or rotting flesh beckon beetles and flies.

An animal's reward for a visit to the flower may be **nectar** (a sweet fluid exuded by flowers), oils, nutritious pollen, or even the illusion of having sex (Figure 30.7). Nectar is the only food for most adult butterflies, and it is the food of choice for hummingbirds. Honeybees collect nectar and convert it to honey, which helps feed the bees through the winter. Pollen is an even richer food, with more vitamins and minerals than nectar.

Many flowers have specializations that exclude non-pollinators. For example, nectar at the bottom of a long floral tube or spur is often accessible only to a certain

Figure 30.5 Bees as pollinators. (**a**) The blueberry bee (*Osmia ribifloris*) is an efficient pollinator of a variety of plants, including this barberry (*Berberis*). (**b**) How we see a gold-petaled marsh marigold. (**c**) Bee-attracting pattern of the same flower. We can see this UV-reflecting pattern only with special camera filters.

Table 30.1 Common Traits of Flowers Pollinated by Specific Animal Vectors

Floral Trait	Bats	Bees	Beetles	Birds	Butterflies	Flies	Moths
				Vector			
Color:	Dull white, green, purple	Bright white, yellow, blue, UV	Dull white or green	Scarlet, orange, red, white	Bright, such as red, purple	Pale, dull, dark brown or purple	Pale/dull red, pink, purple, white
Odor:	Strong, musty, emitted at night	Fresh, mild, pleasant	None to strong	None	Faint, fresh	Putrid	Strong, sweet, emitted at night
Nectar:	Abundant, hidden	Usually	Sometimes, not hidden	Ample, deeply hidden	Ample, deeply hidden	Usually absent	Ample, deeply hidden
Pollen:	Ample	Limited, often sticky, scented	Ample	Modest	Limited	Modest	Limited
Shape:	Regular, bowl-shaped, closed during the day	Shallow with landing pad; tubular	Large, bowl-shaped	Large funnel-shaped cups, strong perch	Narrow tube with spur; wide landing pad	Shallow, funnel-shaped or trap-like and complex	Regular; tube-shaped with no lip
Examples:	Banana, agave	Larkspur, violet	Magnolia, dogwood	Fuschia, hibiscus	Phlox	Skunk cabbage, philodendron	Tobacco, lily, some cactuses

pollinator that has a matching feeding device (Figure 18.25). Often, stamens adapted to brush against a pollinator's body or lob pollen onto it will function only when triggered by that pollinator. Such relationships are to both species' mutual advantage: A flower that captivates the attention of an animal has a pollinator that spends its time seeking out (and pollinating) only those flowers; the animal receives an exclusive supply of the reward offered by the plant.

Take-Home Message

What is the purpose of the nonreproductive traits of flowers?

■ The shape, pattern, color, and fragrance of flowers attract coevolved pollinators.

■ Pollinators are often rewarded for visiting a flower by obtaining nutritious pollen or sweet nectar.

Figure 30.7 Intimate connections. (**a**) Female burnet moths (*Zygaena filipendulae*) perch on purple flowers—preferably those of field scabious (*Knautia arvensis*)—when they are ready to mate. The visual combination attracts males. (**b**) A zebra orchid (*Caladenia cairnsiana*) mimics the scent of a female wasp. Male wasps follow the scent to the flower, then try to copulate with and lift the dark red mass of tissue on the lip. The wasp's movements trigger the lip to tilt upward, which brushes the wasp's back against the flower's stigma and pollen.

- In flowering plants, fertilization has two outcomes: It results in a zygote, and it is the start of endosperm, which is a nutritious tissue that nourishes the embryo sporophyte.

- Links to Evolution of seed-bearing plants 23.8, Life cycle of flowering plants 23.9, Cell signaling 27.6

Microspore and Megaspore Formation

Figure 30.8 zooms in on a flowering plant life cycle. On the male side, masses of diploid, spore-producing cells form by mitosis in the anthers. Typically, walls develop around the cell masses to form four pollen sacs (Figure 30.8a). Each cell inside the sacs undergoes meiosis, forming four haploid **microspores** (Figure 30.8b).

Mitosis and differentiation of microspores produce pollen grains. Each pollen grain consists of a durable coat that surrounds two cells, one inside the cytoplasm of the other (Figure 30.8c). After a period of dormancy, the pollen sacs split open, and pollen is released from the anther (Figure 30.8d).

On the female side, a mass of tissue—the ovule—starts growing on the inner wall of an ovary (Figure 30.8e). One cell in the middle of the mass undergoes meiosis and cytoplasmic division, forming four haploid **megaspores** (Figure 30.8f).

Three of the four megaspores typically disintegrate. The remaining megaspore undergoes three rounds of mitosis without cytoplasmic division. The outcome is a single cell with eight haploid nuclei (Figure 30.8g). The cytoplasm of this cell divides unevenly, and the result is a seven-celled embryo sac that constitutes the female gametophyte (Figure 30.8h). The gametophyte is enclosed and protected by cell layers, called integuments, that developed from ovule tissue. One of the cells in the gametophyte, the **endosperm mother cell**, has two nuclei ($n + n$). Another cell is the egg.

Pollination and Fertilization

Pollination refers to the arrival of a pollen grain on a receptive stigma. Interactions between the two structures stimulate the pollen grain to resume metabolic activity (germinate). One of the two cells in the pollen grain then develops into a tubular outgrowth called a pollen tube. The other cell undergoes mitosis and cytoplasmic division, producing two sperm cells (the male gametes) within the pollen tube. A pollen tube together with its contents of male gametes constitutes the mature male gametophyte (Figure 30.8d).

The pollen tube grows from its tip down through the carpel and ovary toward the ovule, carrying with it the two sperm cells. Chemical signals secreted by

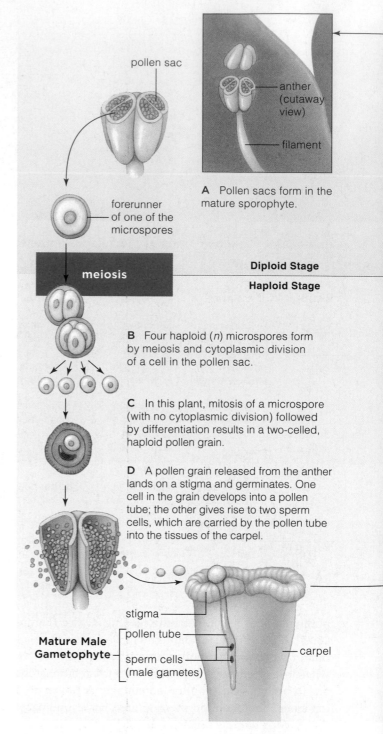

A Pollen sacs form in the mature sporophyte.

pollen sac

anther (cutaway view)

filament

forerunner of one of the microspores

meiosis

Diploid Stage

Haploid Stage

B Four haploid (*n*) microspores form by meiosis and cytoplasmic division of a cell in the pollen sac.

C In this plant, mitosis of a microspore (with no cytoplasmic division) followed by differentiation results in a two-celled, haploid pollen grain.

D A pollen grain released from the anther lands on a stigma and germinates. One cell in the grain develops into a pollen tube; the other gives rise to two sperm cells, which are carried by the pollen tube into the tissues of the carpel.

Mature Male Gametophyte —
stigma
pollen tube
sperm cells (male gametes)
carpel

Figure 30.8 Animated Life cycle of cherry (*Prunus*), a eudicot. **Figure It Out:** What structure gives rise to a pollen grain by mitosis?
Answer: A microspore

the female gametophyte guide the tube's growth to the embryo sac within the ovule. Many pollen tubes may grow down into a carpel, but only one typically penetrates an embryo sac. The sperm cells are then released into the sac (Figure 30.8i). Flowering plants undergo **double fertilization**: One of the sperm cells from the

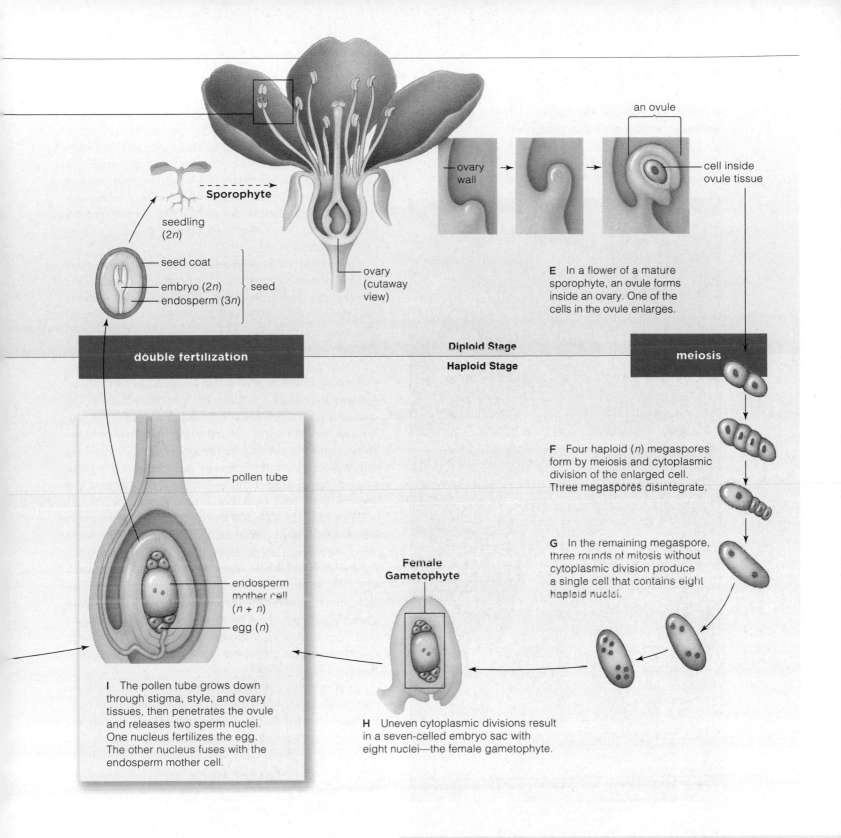

Sporophyte

seedling
(2n)

seed coat
embryo (2n) seed
endosperm (3n)

ovary
(cutaway
view)

an ovule

ovary
wall

cell inside
ovule tissue

E In a flower of a mature sporophyte, an ovule forms inside an ovary. One of the cells in the ovule enlarges.

Diploid Stage

Haploid Stage

double fertilization

meiosis

F Four haploid (n) megaspores form by meiosis and cytoplasmic division of the enlarged cell. Three megaspores disintegrate.

pollen tube

G In the remaining megaspore, three rounds of mitosis without cytoplasmic division produce a single cell that contains eight haploid nuclei.

endosperm
mother cell
(n + n)

egg (n)

**Female
Gametophyte**

I The pollen tube grows down through stigma, style, and ovary tissues, then penetrates the ovule and releases two sperm nuclei. One nucleus fertilizes the egg. The other nucleus fuses with the endosperm mother cell.

H Uneven cytoplasmic divisions result in a seven-celled embryo sac with eight nuclei—the female gametophyte.

pollen tube fuses with (fertilizes) the egg and forms a diploid zygote. The other fuses with the endosperm mother cell, forming a triploid (3n) cell. This cell will give rise to triploid **endosperm**, a nutritious tissue that forms only in seeds of flowering plants. Right after a seed germinates, endosperm will sustain the rapid growth of the sporophyte seedling until true leaves form and photosynthesis begins.

Take-Home Message

How does fertilization occur in flowering plants?

■ In flowering plants, male gametophytes form in pollen grains; female gametes form in ovules. Pollination occurs when pollen arrives on a receptive stigma.

■ A pollen grain germinates on a receptive stigma as a pollen tube containing male gametes. The pollen tube grows into the carpel and enters an ovule. Double fertilization occurs when one of the male gametes fuses with the egg, the other with the endosperm mother cell.

30.4 Flower Sex

- Interactions between pollen grain and stigma govern pollen germination and pollen tube growth.

- Links to Recognition and adhesion proteins 5.2, Cell signaling 27.6, Plant epidermis 28.2

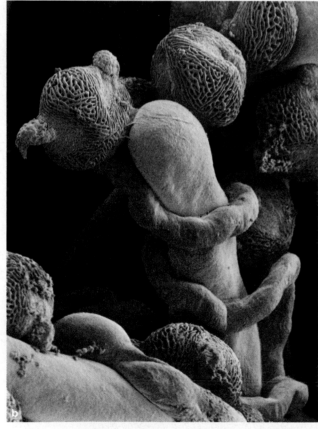

Figure 30.9 Pollen. (**a**) Pollen grains from several species. Elaborately sculpted pollen coats are adapted to cling to insect bodies; smooth coats are adapted for wind dispersal.

(**b**) Pollen tubes grow from pollen grains (*orange*) that germinated on stigmas (*yellow*) of prairie gentian (*Gentiana*). Molecular cues guide a pollen tube's growth through carpel tissues to the egg.

The main function of a pollen grain's coat is to protect the two cells inside of it on what may be a long, turbulent ride to a stigma. Pollen grains make terrific fossils because the outer layer of the coat consists primarily of sporopollenin, an extremely hard, durable mixture of long-chain fatty acids and other organic molecules. In fact, sporopollenin is so resistant to degradation by enzymes and harsh chemicals that we still don't know exactly what it is.

Given the coat's toughness, how does a pollen grain "know" when to germinate? How does a microscopic pollen tube that grows through centimeters of tissue find its way to a single cell deep inside of the carpel? The answers to such questions involve cell signaling (Section 27.6).

Sex in plants, like sex in animals, involves an interplay of signals. It begins when recognition proteins on epidermal cells of a stigma bind to molecules in the coat of a pollen grain. Within minutes, lipids and proteins in the pollen grain's coat begin to diffuse onto the stigma, and the pollen grain becomes tightly bound via adhesion proteins in stigma cell membranes. The specificity of recognition proteins means that a stigma can preferentially bind pollen of its own species.

Pollen is very dry, and the cells inside are dormant. These adaptations make the grains light and portable. After a pollen grain attaches to a stigma, nutrient-rich fluid begins to diffuse from the stigma into the grain. The fluid stimulates the cells inside to resume metabolism, and a pollen tube that contains the male gametes grows out of one of the furrows or pores in the pollen's coat (Figure 30.9). Gradients of nutrients (and perhaps other molecules) direct the growth of the pollen tube down through the style.

Cells of the female gametophyte secrete chemical signals that guide the growth of the pollen tube from the bottom of the style to the egg. These signals are species-specific; pollen tubes of different species do not recognize them, and will not reach the ovule. In some species, the signals are also part of mechanisms that can keep a flower's pollen from fertilizing its own stigma. Only pollen from another flower (or another plant) can give rise to a pollen tube that recognizes the female gametophyte's chemical guidance.

Take-Home Message

What constitutes sex in plants?

- Species-specific molecular signals stimulate pollen germination and guide pollen tube growth to the egg.

- In some species, the specificity of the signaling also limits self-pollination.

30.5 | Seed Formation

■ After fertilization, mitotic cell divisions transform a zygote into an embryo sporophyte encased in a seed.

The Embryo Sporophyte Forms

In flowering plants, double fertilization produces a zygote and a triploid (3n) cell. Both begin mitotic cell divisions; the zygote develops into an embryo sporophyte, and the triploid cell develops into endosperm (Figure 30.10a–c). When the embryo approaches maturity, the integuments of the ovule separate from the ovary wall and become layers of the protective seed coat. The embryo sporophyte, its reserves of food, and the seed coat have now become a mature ovule, a self-contained package called a **seed** (Figure 30.10d). The seed may enter a period of dormancy until it receives signals that conditions in the environment are appropriate for germination.

Seeds as Food

As an embryo is developing, the parent plant transfers nutrients to the ovule. These nutrients accumulate in endosperm mainly as starch with some lipids, proteins, or other molecules. Eudicot embryos transfer nutrients in endosperm to their two cotyledons before germination occurs. The embryos of monocots tap endosperm only after seeds germinate.

The nutrients in endosperm and cotyledons nourish seedling sporophytes. They also nourish humans and other animals. Rice (*Oryza sativa*), wheat (*Triticum*), rye (*Secale cereale*), oats (*Avena sativa*), and barley (*Hordeum vulgare*) are among the grasses commonly cultivated for their nutritious seeds, or grains. The embryo (the germ) of a grain contains most of the seed's protein and vitamins, and the seed coat (the bran) contains most of the minerals and fiber. Milling removes bran and germ, leaving only the starch-packed endosperm.

Maize, or corn (*Zea mays*), is the most widely grown grain crop. Popcorn pops because the moist endosperm steams when heated; pressure builds inside the seed until it bursts. Cotyledons of bean and pea seeds are valued for their starch and protein; those of coffee (*Coffea*) and cacao (*Theobroma cacao*), for their stimulants.

Take-Home Message

What is a seed?

■ After fertilization, the zygote develops into an embryo, the endosperm becomes enriched with nutrients, and the ovule's integuments develop into a seed coat.

■ A seed is a mature ovule. It contains an embryo sporophyte.

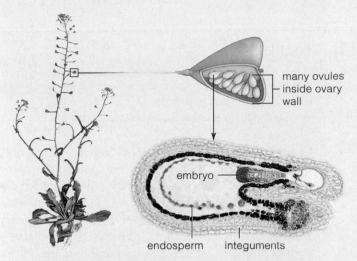

many ovules inside ovary wall

embryo

endosperm   integuments

A After fertilization, a *Capsella* flower's ovary develops into a fruit. Surrounded by integuments, an embryo forms inside each of the ovary's many ovules.

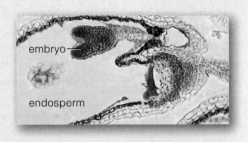

embryo

endosperm

B The embryo is heart-shaped when cotyledons start forming. Endosperm tissue expands as the parent plant transfers nutrients into it.

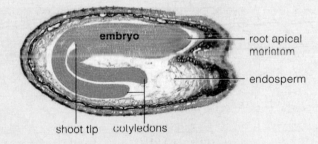

embryo

root apical meristem

endosperm

shoot tip   cotyledons

C The developing embryo is torpedo-shaped when the enlarging cotyledons bend inside the ovule.

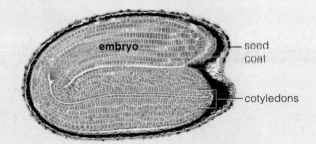

embryo

seed coat

cotyledons

D A layered seed coat that formed from the layers of integuments surrounds the mature embryo sporophyte. In eudicots like *Capsella*, nutrients have been transferred from endosperm into two cotyledons.

Figure 30.10 Animated Embryonic development of shepherd's purse (*Capsella*), a eudicot.

■ As embryos develop inside the ovules of flowering plants, tissues around them form fruits.

■ Water, wind, and animals disperse seeds in fruits.

Only flowering plants form seeds in ovaries, and only they make fruits. A **fruit** is a seed-containing mature ovary, often with fleshy tissues that develop from the ovary wall (Figure 30.11). In some plants, fruit tissues develop from parts of the flower other than the ovary wall (such as petals, sepals, stamens, or receptacles). Apples, oranges, and grapes are familiar fruits, but so are many "vegetables" such as beans, peas, tomatoes, grains, eggplant, and squash.

An embryo or seedling can use the nutrients stored in endosperm or cotyledons, but not in fruit. The function of fruit is to protect and disperse seeds. Dispersal increases reproductive success by minimizing competition for resources among parent and offspring, and by expanding the area colonized by the species.

Just as flower structure is adapted to certain pollination vectors, so are fruits adapted to certain dispersal vectors: environmental factors such as water or wind, or mobile organisms such as birds or insects.

Water-dispersed fruits have water-repellent outer layers. The fruits of sedges (*Carex*) native to American

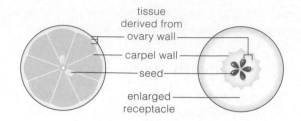

Figure 30.11 Parts of a fruit develop from parts of a flower. *Left*, the tissues of an orange (*Citrus*) develop from the ovary wall. *Right*, the flesh of an apple is an enlarged receptacle.

Figure It Out: How many carpels were there in the flower that gave rise to this orange?

Answer: Eight

marshlands have seeds encased in a bladderlike envelope that floats (Figure 30.12*a*). Buoyant fruits of the coconut palm (*Cocos nucifera*) have thick, tough husks that can float for thousands of miles in seawater.

Many plant species use wind as a dispersal agent. Part of a maple fruit (*Acer*) is a dry outgrowth of the ovary wall that extends like a pair of thin, lightweight wings (Figure 30.12*b*). The fruit breaks in half when it drops from the tree; as the halves drop to the ground, wind currents that catch the wings spin the attached seeds away. Tufted fruits of thistle, cattail, dandelion,

Figure 30.12 Examples of adaptations that aid fruit dispersal. (**a**) Air-filled bladders that encase the seeds of certain sedges (*Carex*) allow the fruits to float in their marshy habitats.

(**b**) Wind lifts the "wings" of maple (*Acer*) fruits, which spin the seeds away from the parent tree.

(**c**) Wind that catches the hairy modified sepals of a dandelion fruit (*Taraxacum*) lifts the seed away from the parent plant.

(**d**) Curved spines make cocklebur (*Xanthium*) fruits stick to the fur of animals (and clothing of humans) that brush past it.

(**e**) The fruits of the California poppy (*Eschscholzia californica*) are long, dry pods that split open suddenly. The movement jettisons the seeds.

(**f**) The red, fleshy fruit of crabapples attracts cedar waxwings.

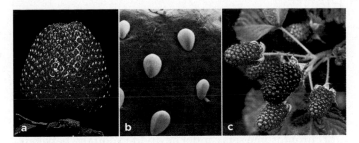

Figure 30.13 Aggregate fruits. (**a**) A strawberry (*Fragaria*) is not a berry. The flower's carpels turn inside out as the fruits form. The red, juicy flesh is an expanded receptacle; the hard "seeds" on the surface are individual dry fruits (**b**).

(**c**) Boysenberries and other *Rubus* species are not berries, either. Each is an aggregate fruit of many small drupes.

and milkweed may be blown as far as 10 kilometers (6 miles) from the parent plant (Figure 30.12*c*). The fruits of cocklebur, bur clover, and many other plants have hooks or spines that stick to the feathers, feet, fur, or clothing of more mobile species (Figure 30.12*d*). The dry, podlike fruit of plants such as California poppy (*Eschscholzia californica*) propel their seeds through the air when they pop open explosively (Figure 30.12*e*).

Colorful, fleshy, fragrant fruits attract insects, birds, and mammals that disperse seeds (Figure 30.12*f*). The animal may eat the fruit and discard the seeds, or eat the seeds along with the fruit. Abrasion of the seed coat by digestive enzymes in an animal's gut can facilitate germination after the seed departs in feces.

Botanists categorize fruits by how they originate, their tissues, and appearance (Table 30.2). Simple fruits, such as pea pods, acorns, and *Capsella*, are derived from one ovary. Strawberries and other aggregate fruits form from separate ovaries of one flower; they mature as a cluster of fruits. Multiple fruits form from fused ovaries of separate flowers. The pineapple is a multiple fruit that forms from fused ovary tissues of many flowers.

Fruits also may be categorized in terms of which tissues they incorporate. True fruits such as cherries consist only of the ovary wall and its contents. Other floral parts, such as the receptacle, expand along with the ovary in accessory fruits. Most of the flesh of an apple, an accessory fruit, is an enlarged receptacle.

To categorize a fruit based on appearance, the first step is to describe it as dry or juicy (fleshy). Dry fruits are dehiscent or indehiscent. If dehiscent, the fruit wall splits along definite seams to release the seeds inside. California poppy fruits and pea pods are examples.

A dry fruit is indehiscent if the wall does not split open; seeds are dispersed inside intact fruits. Acorns and grains (such as corn) are dry indehiscent fruits, as are the fruits of sunflowers, maples, and strawberries. Strawberries are not berries and their fruits are not

juicy. A strawberry's red flesh is an accessory to the dry indehiscent fruits on its surface (Figure 30.13*a,b*).

Drupes, berries, and pomes are three types of fleshy fruits. Drupes have a pit, a hard jacket around the seed. Cherries, apricots, almonds, and olives are drupes, as are the individual fruits of boysenberries and other *Rubus* species (Figure 30.13*c*).

A berry forms from a compound ovary. It has one to many seeds, no pit, and fleshy fruit. Grapes and tomatoes are berries. Lemons, oranges, and other citrus fruits (*Citrus*) are a type of berry called a hesperidium: An oily, leathery peel encloses juicy pulp. Each "section" of the pulp started out as an ovary of a partially fused carpel. Pumpkins, watermelons, and cucumbers are pepos, berries in which a hard rind of accessory tissues forms over the somewhat slippery true fruit.

A pome has seeds in a core derived from the ovary; fleshy tissues derived from the receptacle enclose the core. Two familiar pomes are apples and pears.

Table 30.2 Three Ways To Classify Fruits

How did the fruit originate?

Simple fruit	One flower, single or fused carpels
Aggregate fruit	One flower, several unfused carpels; becomes cluster of several fruits
Multiple fruit	Individually pollinated flowers grow and fuse

What is the fruit's tissue composition?

True fruit	Only ovarian wall and its contents
Accessory fruit	Ovary and other floral parts, such as receptacle

Is the fruit dry or fleshy?

Dry		
	Dehiscent	Dry fruit wall splits on seam to release seeds
	Indehiscent	Seeds dispersed inside intact, dry fruit wall
Fleshy		
	Drupe	Fleshy fruit around hard pit surrounding seed
	Berry	Fleshy fruit, often many seeds, no pit
		Pepo: Hard rind on ovary wall
		Hesperidium: Leathery rind on ovary wall
	Pome	Fleshy accessory tissues, seeds in core tissue

Take-Home Message

What is a fruit?

■ A mature ovary, with or without accessory tissues that develop from other parts of a flower, is a fruit.

■ We can categorize a fruit in terms of how it originated, its composition, and whether it is dry or fleshy.

30.7 | Asexual Reproduction of Flowering Plants

■ Many plants also reproduce asexually, which permits rapid production of genetically identical offspring.

■ Links to Asexual versus sexual reproduction 10.1, Meiosis 10.3, Mendelian inheritance 11.1, Aneuploidy 12.6, Cloning 13.4, Radiometric dating 17.6, Speciation by polyploidy in plants 18.11, Modified stems 28.8

Plant Clones

Unlike most animals, most flowering plants can reproduce asexually. By **vegetative reproduction**, new roots and shoots grow from extensions or fragments of a parent plant. Each new plant is a clone, a genetic replica of its parent.

You already know that new roots and shoots sprout from nodes on modified stems (Section 28.8). This is one example of vegetative reproduction. As another example, "forests" of quaking aspen (*Populus tremuloides*) are actually stands of clones that grew from root suckers, which are shoots that sprout from the aspens' shallow, cordlike lateral roots. Suckers sprout after aboveground parts of the aspens are damaged or removed. One stand in Utah consists of about 47,000 shoots and stretches for 107 acres (Figure 30.14).

No one knows how old those aspen clones are. As long as conditions in the environment favor growth, such clones are as close as any organism gets to being immortal. The oldest known plant is a clone: the one and only population of King's holly (*Lomatia tasmanica*), which consists of several hundred stems growing along 1.2 kilometers (0.7 miles) of a river gully in Tasmania. Radiometric dating of the plant's fossilized leaf litter show that the clone is at least 43,600 years old—predating the last ice age!

The ancient species of *Lomatia* is triploid. With three sets of chromosomes, it is sterile—it can only reproduce asexually. Why? During meiosis, an odd number of chromosome sets cannot be divided equally between the two spindle poles. If meiosis does not fail entirely, unequal segregation of chromosomes during meiosis results in aneuploid offspring, which are rarely viable.

Agricultural Applications

Cuttings and Grafting For thousands of years, we humans have been taking advantage of the natural capacity of plants to reproduce asexually. Almost all houseplants, woody ornamentals, and orchard trees are clones that have been grown from stem fragments (cuttings) of a parent plant.

Propagating some plants from cuttings may be as simple as jamming a broken stem into the soil. This method uses the plant's natural ability to form roots and new shoots from stem nodes. Other plants must be grafted. Grafting means inducing a cutting to fuse with the tissues of another plant. Often, the stem of a desired plant is spliced onto the roots of hardier one.

Propagating a plant from cuttings ensures that offspring will have the same desirable traits as the parent plant. For example, domestic apple trees (*Malus*) are typically grafted because they do not breed true for fruit color, flavor, size, or texture. Even trees grown

Figure 30.14 Quaking aspen (*Populus tremuloides*). A single plant gave rise to this stand of shoots by asexual reproduction. Such clones are connected by underground lateral roots, so water can travel from roots near a lake or river to those in drier soil some distance away.

Figure 30.15 Apples (*Malus*). (**a**) Commercial growers must plant grafted apple trees in order to reap consistent crops. (**b**) Fruit of 21 wild apple trees.

(**c**) Gennaro Fazio (*left*) and Phil Forsline (*right*) are part of an effort to maintain the genetic diversity of apple trees in the United States. Cross-breeding is yielding new apples with the palatability of commercial varieties, and the disease resistance of wild trees.

from seeds of the same fruit produce fruits that vary, sometimes dramatically so. The genus is native to central Asia, where apple trees grow wild in forests. Each tree in the forests is different from the next, and very few of the fruits are palatable (Figure 30.15).

In the early 1800s, the eccentric humanitarian John Chapman (known as Johnny Appleseed) planted millions of apple seeds in the midwestern United States. He sold the trees to homesteading settlers, who would plant orchards and make hard cider from the apples. About one of every hundred trees produced fruits that could be eaten out of hand. Its lucky owner would graft the tree and patent it. Most of the apple varieties sold in American grocery stores are clones of these trees, and they are still propagated by grafting.

Grafting is also used to increase the hardiness of a desirable plant. In 1862, the plant louse *Phylloxera* was accidentally introduced into France via imported American grapevines. European grapevines had little resistance to this tiny insect, which attacks and kills the root systems of the vines. By 1900, *Phylloxera* had destroyed two-thirds of the vineyards in Europe, and devastated the wine-making industry. Today, French vintners routinely graft their prized grapevines onto the roots of *Phylloxera*-resistant American vines.

Tissue Culture An entire plant may be cloned from a single cell with **tissue culture propagation**, by which a somatic cell is induced to divide and form an embryo (Section 13.4). The method can yield millions of genetically identical plants from a single specimen. The technique is being used in research intended to improve food crops. It is also used to propagate rare or hybrid ornamental plants such as orchids.

Seedless Fruits In some plants such as figs, blackberries, and dandelions, fruits may form even in the absence of fertilization. In other species, fruit may continue to form after ovules or embryos abort. Seedless grapes and navel oranges are the result of mutations that result in arrested seed development. These plants are sterile, so they are propagated by grafting.

Seedless bananas are triploid (*3n*). In general, plants tolerate polyploidy better than animals do. Triploid banana plants are robust, but sterile: They are propagated by adventitious shoots that sprout from corms.

Despite their ubiquity in nature (Section 12.6), polyploid plants rarely arise spontaneously. Plant breeders often use the microtubule poison colchicine to artificially increase the frequency of polyploidy in plants (Section 18.11). Tetraploid (*4n*) offspring of colchicine-treated plants are then backcrossed with diploid parent plants. The resulting triploid offspring are sterile: They make seedless fruit after pollination (but not fertilization) by a diploid plant, or on their own. Seedless watermelons are produced this way.

Take-Home Message

How do plants reproduce asexually?

■ Many plants propagate asexually when new shoots grow from a parent plant or pieces of it. Offspring of such vegetative reproduction are clones.

■ Humans propagate plants asexually for agricultural or research purposes by grafting, tissue culture, or other methods.

Theobroma cacao (*right*) is a species of flowering plant that is native to the deep tropical rainforests of middle and south America. The bumpy, football-sized fruits of *T. cacao* contain 40 or so black, bitter seeds. We make chocolate by processing those seeds, but the tree has proven difficult to cultivate outside of rainforests.

Why? *T. cacao* trees do not produce very many seeds when they are grown in typically sun-drenched cultivated plantations. As plantation owners found out, *T. cacao* has a preferred pollinator: midges. These tiny, flying insects live and breed only in

damp, rotting leaf litter of tropical rain forest floors. The flowers of *T. cacao* trees form low to the ground, directly on the woody trunk. This is an adaptation that encourages pollination by—not surprisingly—insects that live in the damp, rotting leaf litter of rain forest floors. Thus, no forests, no midges. No midges, no chocolate.

Summary

Section 30.1 Flowers consist of modified leaves (sepals, petals, **stamens**, and **carpels**) at the ends of specialized branches of angiosperm **sporophytes**. An **ovule** develops from a mass of **ovary** wall tissue inside carpels. Spores produced by meiosis in ovules develop into female **gametophytes**; those produced in anthers develop into immature male gametophytes (**pollen grains**). Adaptations of many flowers restrict self-pollination.

■ *Use the animation on CengageNOW to investigate a flowering plant life cycle and floral structure.*

Section 30.2 A flower's shape, pattern, color, and fragrance typically reflect an evolutionary relationship with a particular **pollination vector**, often a coevolved animal. Coevolved **pollinators** receive **nectar**, pollen, or another reward for visiting a flower.

Sections 30.3, 30.4 Meiosis of diploid cells inside pollen sacs of anthers produces haploid microspores. Each **microspore** develops into a pollen grain.

Mitosis and cytoplasmic division of a cell in an ovule produces four **megaspores**, one of which gives rise to the female gametophyte. One of the seven cells of the gametophyte is the egg; another is the **endosperm mother cell**.

Pollination is the arrival of pollen grains on a receptive stigma. A pollen grain germinates and forms a pollen tube that contains two sperm cells. Species-specific molecular signals guide the tube's growth down through carpel tissues to the egg. In **double fertilization**, one of the sperm cells in the pollen tube fertilizes the egg, forming a zygote; the other fuses with the endosperm mother cell and gives rise to **endosperm**.

■ *Use the animation on CengageNOW to take a closer look at the life cycle of a eudicot.*

Section 30.5 As a zygote develops into an embryo, the endosperm collects nutrients from the parent plant, and the ovule's protective layers develop into a seed coat. A **seed** is a mature ovule: an embryo sporophyte and endosperm enclosed within a seed coat.

Eudicot embryos transfer nutrients from endosperm to their two cotyledons. Carbohydrates, lipids, and proteins stored in endosperm or cotyledons make seeds a nutritious food source for humans and other animals.

Section 30.6 As an embryo sporophyte develops, the ovary wall and sometimes other tissues mature into a **fruit** that encloses the seeds. Fruit functions in the protection and dispersal of seeds.

■ *Use the animation on CengageNOW to see how an embryo sporophyte develops in a eudicot seed.*

Section 30.7 Many species of flowering plants reproduce asexually by **vegetative reproduction**. The offspring produced by asexual reproduction are clones of the parent. Many agriculturally valuable plants are produced by grafting or **tissue culture propagation**.

Self-Quiz   *Answers in Appendix III*

1. The _____ of a flower contains one or more ovaries in which eggs develop, fertilization occurs, and seeds mature.
   a. pollen sac   c. receptacle
   b. carpel   d. sepal

2. Seeds are mature _____; fruits are mature _____.
   a. ovaries; ovules   c. ovules; ovaries
   b. ovules; stamens   d. stamens; ovaries

3. Meiosis of cells in pollen sacs forms haploid _____.
   a. megaspores   c. stamens
   b. microspores   d. sporophytes

4. After meiosis in an ovule, _____ megaspores form.
   a. two   b. four   c. six   d. eight

5. The seed coat forms from the _____.
   a. ovule wall   c. endosperm
   b. ovary   d. residues of sepals

6. Cotyledons develop as part of _____.
   a. carpels   c. embryo sporophytes
   b. accessory fruits   d. petioles

Data Analysis Exercise

Massonia depressa is a low-growing succulent plant native to the desert of South Africa. The dull-colored flowers of this monocot develop at ground level, have tiny petals, emit a yeasty aroma, and produce a thick, jelly-like nectar. These features led researchers to suspect that desert rodents such as gerbils pollinate this plant (Figure 30.17).

To test their hypothesis, the researchers trapped rodents in areas where *M. depressa* grows and checked them for pollen. They also put some plants in wire cages that excluded mammals, but not insects, to see whether fruits and seeds would form in the absence of rodents. The results are shown in Figure 30.18.

1. How many of the 13 captured rodents showed some evidence of pollen from *M. depressa*?

2. Would this evidence alone be sufficient to conclude that rodents are the main pollinators for this plant?

3. How did the average number of seeds produced by caged plants compare with that of control plants?

4. Do these data support the hypothesis that rodents are required for pollination of *M. depressa*? Why or why not?

Figure 30.18 *Right,* results of experiments testing rodent pollination of *M. depressa*. (**a**) Evidence of visits to *M. depressa* by rodents. (**b**) Fruit and seed production of *M. depressa* with and without visits by mammals. Mammals were excluded from plants by wire cages with openings large enough for insects to pass through. 23 plants were tested in each group.

Figure 30.17 The dull, petal-less, ground-level flowers of *Massonia depressa* are accessible to rodents, who push their heads through the stamens to reach the nectar. Note the pollen on the gerbil's snout.

a

Type of rodent	Number caught	# with pollen on snout	# with pollen in feces
Namaqua rock rat	4	3	2
Cape spiny mouse	3	2	2
Hairy-footed gerbil	4	2	4
Cape short-eared gerbil	1	0	1
African pygmy mouse	1	0	0

b

	Mammals allowed access to plants	Mammals excluded from plants
Percentage of plants that set fruit	30.4	4.3
Average number of fruits per plant	1.39	0.47
Average number of seeds per plant	20.0	1.95

7. Name one reward that a pollinator may receive in return for a visit to a flower of its coevolved plant partner.

8. By _____ , a new plant forms from a tissue or structure that drops or is separated from the parent plant.
   a. parthenogenesis   c. vegetative reproduction
   b. exocytosis   d. nodal growth

9. Wanting to impress friends with her sophisticated knowledge of botany, Dixie Bee prepares a plate of tropical fruits for a party and cuts open a papaya (*Carica papaya*). Soft skin and soft fleshy tissue enclose many seeds in a slimy tissue (Figure 30.16a). Knowing her friends will ask her how to categorize this fruit, she panics, runs to her biology book, and opens it to Section 30.6. What does she find out?

10. Having succeeded in spectacularly impressing her friends, Dixie Bee prepares a platter of peaches (Figure 30.16b) for her next party. How will she categorize this fruit?

11. Match the terms with the most suitable description.
   ___ovule
   ___receptacle
   ___double fertilization
   ___anther
   ___carpel
   ___mature female gametophyte
   ___mature male gametophyte

   a. pollen tube together with its contents
   b. embryo sac of seven cells, one with two nuclei
   c. starts out as cell mass in ovary; may become a seed
   d. female reproductive part
   e. pollen sacs inside
   f. base of floral shoot
   g. formation of zygote and first cell of endosperm

■ *Visit CengageNOW for additional questions.*

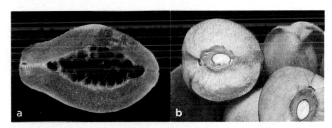

Figure 30.16 Tangential sections reveal seeds of two mature fruits: (**a**) papaya (*Carica papaya*) and (**b**) peach (*Prunus*).

Critical Thinking

1. Would you expect winds, bees, birds, bats, butterflies, or moths to pollinate the flower pictured to the *left*? Explain your choice.

2. All but one species of large-billed birds native to New Zealand's tropical forests are now extinct. Numbers of the surviving species, the kereru, are declining rapidly due to the habitat loss, poaching, predation, and interspecies competition that wiped out the other native birds. The kereru remains the sole dispersing agent for several native trees that produce big seeds and fruits. One tree, the puriri (*Vitex lucens*), is New Zealand's most valued hardwood. Explain, in terms of natural selection, why we might expect to see no new puriri trees in New Zealand.

Plant Development

IMPACTS, ISSUES Foolish Seedlings, Gorgeous Grapes

In 1926, researcher Ewiti Kurosawa was studying what Japanese call *bakane*, the "foolish seedling" effect. The stems of rice seedlings infected with a fungus, *Gibberella fujikuroi*, grew twice the length of uninfected seedlings. The abnormally elongated stems were weak and spindly, and eventually toppled. Kurosawa discovered that he could cause the lengthening experimentally by applying extracts of the fungus to seedlings. Many years later, other researchers purified the substance from fungal extracts that brought about the lengthening. They named it gibberellin, in reference to the name of the fungus.

Gibberellins, as we now know, are a major class of plant hormones. Hormones are secreted signaling molecules that stimulate some response in target cells. Cells that bear molecular receptors for a hormone may be in the same tissue as the hormone-secreting cell, or in a distant tissue.

Researchers have isolated more than eighty different forms of gibberellin from seeds of flowering plants and fungi. These signaling molecules cause young cells in stems to elongate, and the collective elongation lengthens plant parts. In nature,

gibberellins also help dormant seeds and buds resume growth in spring.

Applications of synthetic gibberellins make celery stalks longer and crispier. They prevent the rind of navel oranges in orchard groves from ripening before pickers can get to them. Walk past plump seedless grapes in produce bins of grocery stores and marvel at how fleshy fruits of the grape plant (*Vitis*) grow in dense clusters along stems. Seedless grapes tend to be smaller than seeded varieties, because their undeveloped seeds do not produce normal amounts of gibberellin. Farmers spray their seedless grape plants with synthetic gibberellin, which increases the size of the resulting fruit (Figure 31.1). Gibberellin also makes the stems elongate between nodes, which opens up space between individual grapes. Improved air circulation between the fruit reduces infections by fruit-damaging fungi.

Gibberellin and other plant hormones control the growth and development of plants. Plant cells secrete hormones in response to environmental cues, as when warm spring rains arrive after a cold winter, and the hours of daylight increase.

With this chapter, we complete our survey of plant structure and function. So far, you read about the tissue organization of primary and secondary growth in flowering plants. You considered the tissue systems by which plants acquire and distribute water and solutes that sustain their growth. You learned how flowering plants reproduce, from gamete formation and pollination on through the formation of a mature embryo sporophyte inside a protective seed coat.

At some point after its dispersal from a parent plant, remember, a seed germinates and growth resumes. In time, the mature sporophyte typically forms flowers, then seeds of its own. Depending on the species, it may drop old leaves throughout the year or all at once, in autumn.

Continue now with the internal mechanisms that govern plant development, and the environmental cues that turn the mechanisms on or off at different times.

Figure 31.1 Seedless grapes radiate market appeal. The hormone gibberellin causes grape stems to lengthen, which improves air circulation around individual grapes and gives them more room to grow. The fruit also enlarges, which makes growers happy (grapes are sold by weight).

Key Concepts

Patterns of plant development

Plant development includes seed germination and all events of the life cycle, such as root and shoot development, flowering, fruit formation, and dormancy. These activities have a genetic basis, but are also influenced by environmental factors. **Section 31.1**

Mechanisms of hormone action

Cell-to-cell communication is essential to development and survival of all multicelled organisms. In plants, such communication occurs by hormones. **Sections 31.2, 31.3**

Responses to environmental cues

Plants respond to environmental cues, including gravity, sunlight, and seasonal shifts in night length and temperatures, by altering patterns of growth. Cyclic patterns of growth are responses to changing seasons and other recurring environmental patterns. **Sections 31.4–31.6**

Links to Earlier Concepts

- This chapter revisits hormones (Section 27.2), homeostasis (27.5), and signaling pathways (27.6) in the context of plant physiology. In plants, development depends on cell-to-cell communication, just as animal development does (15.3).

- Plant hormones are involved in gene expression and control (15.1), and the function of structures such as meristems (28.3) and stomata (29.4).

- As you learn about plant responses to environmental stimuli, you will be drawing upon your understanding of carbohydrates (3.2, 3.3); how turgor (5.6) pushes on plant cell walls (4.12); light (7.1); and photosynthesis (7.4, 7.6). You will also revisit cell components, including plastids (4.11), the cytoskeleton (4.13), and membrane transport proteins (5.2).

How would you vote? 1-Methylcyclopropene, or MCP, is a gas that keeps ethylene from binding to cells in plant tissues. It is used to prolong the shelf life of cut flowers and the storage time for fruits. Should produce treated with MCP be labeled to alert consumers? See CengageNOW for details, then vote online.

■ Patterns of development in plants have a genetic basis, and they are also influenced by the environment.

■ Links to Carbohydrates 3.3, Plant cell walls 4.12, Gene control 15.1, Hormones 27.2, Meristems 28.3

In Chapter 30, we left the embryo sporophyte after its dispersal from the parent plant. What happens next? An embryonic plant complete with shoot and root apical meristems formed as part of the embryo (Figure 31.2). However, the seed dried out as it matured, and the desiccation caused the embryo's cells to stop dividing. The embryo entered a period of temporarily suspended development called dormancy.

An embryo may idle in its protective seed coat for years before it resumes metabolic activity. **Germination** is the process by which a mature embryo sporophyte resumes growth. The process begins with water seeping into a seed. The water activates enzymes that start to hydrolyze stored starches into sugar monomers. It also swells tissues inside the seed, so the coat splits open and oxygen enters. Meristem cells in the embryo begin to use the sugars and the oxygen for aerobic respiration as they start dividing rapidly. The embryonic plant begins to grow from the meristems. Germination ends when the first part of the embryo—the embryonic root, or radicle—breaks out of the seed coat.

Seed dormancy is a climate-specific adaptation that allows germination to occur when conditions in the environment are most likely to support the growth of a seedling. For example, the weather in regions near the equator does not vary by season, so seeds of most plants native to such regions do not enter dormancy; they can germinate as soon as they are mature. By contrast, the seeds of many annual plants native to colder regions are dispersed in autumn. If they germinated immediately, the seedlings would not survive the cold winter. Instead, the seeds stay dormant until spring, when milder temperatures and longer daylength are more suitable for tender seedlings.

How does a dormant embryo sporophyte "know" when to germinate? The triggers, other than the presence of water, differ by species, and all have a genetic basis. For example, some seed coats are so dense that they must be abraded or broken (by being chewed, for example) before water can even enter the seed. Seeds of some species of lettuce (*Lactuca*) must be exposed to bright light. The germination of wild California poppy seeds (*Eschscholzia californica*) is inhibited by light and enhanced by smoke. The seeds of some species of pine (*Pinus*) will not germinate unless they have been previously burned. The seeds of many cool-climate plants require exposure to freezing temperatures.

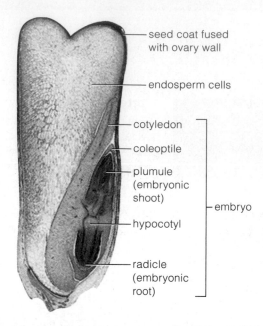

seed coat fused with ovary wall

endosperm cells

cotyledon

coleoptile

plumule (embryonic shoot)

embryo

hypocotyl

radicle (embryonic root)

Figure 31.2 Anatomy of a corn seed (*Zea mays*). During germination, cell divisions resume mainly at apical meristems of the plumule (the embryonic shoot) and radicle (the embryonic root).

A plumule consists of an apical meristem and two tiny leaves. In grasses such as corn, the growth of this delicate structure through soil is protected by a sheathlike coleoptile.

Germination is just one of many patterns of development in plants. As a sporophyte grows and matures, its tissues and parts develop in other patterns characteristic of its species (Figures 31.3 and 31.4). Leaves form in predictable shapes and sizes, stems lengthen and thicken in particular directions, flowering occurs at a certain time of year, and so on. As in germination, these patterns have a genetic basis, but they also have an environmental component.

Development includes **growth**, which is an increase in cell number and size. Plant cells are interconnected by shared walls, so they cannot move about within the organism. Thus, plant growth occurs primarily in the direction of cell division—and cell division occurs primarily at meristems. Behind meristems, cells differentiate and form specialized tissues. However, unlike animal cell differentiation, plant cell differentiation is often reversible, as when new shoots form on mature roots, or when new roots sprout from a mature stem.

Take-Home Message

What is plant development?

■ In plants, growth and differentiation results in the formation of tissues and parts in predictable patterns.

■ Germination and other patterns of plant development are an outcome of gene expression and environmental influences.

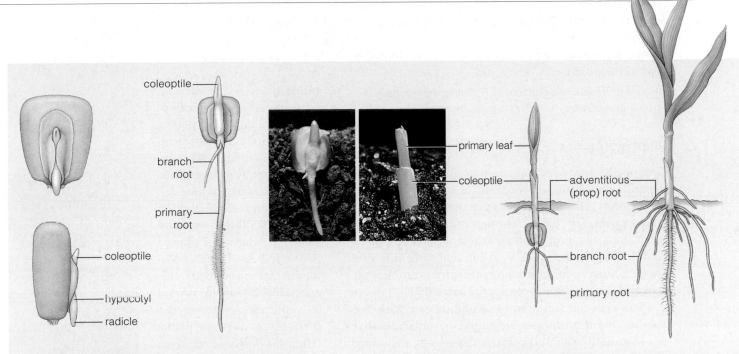

A After a corn grain (seed) germinates, its radicle and coleoptile emerge. The radicle develops into the primary root. The coleoptile grows upward and opens a channel through the soil to the surface, where it stops growing.

B The plumule develops into the seedling's primary shoot, which pushes through the coleoptile and begins photosynthesis. In corn plants, adventitious roots that develop from the stem afford additional support for the rapidly growing plant.

Figure 31.3 Animated Early growth of corn (*Zea mays*), a monocot.

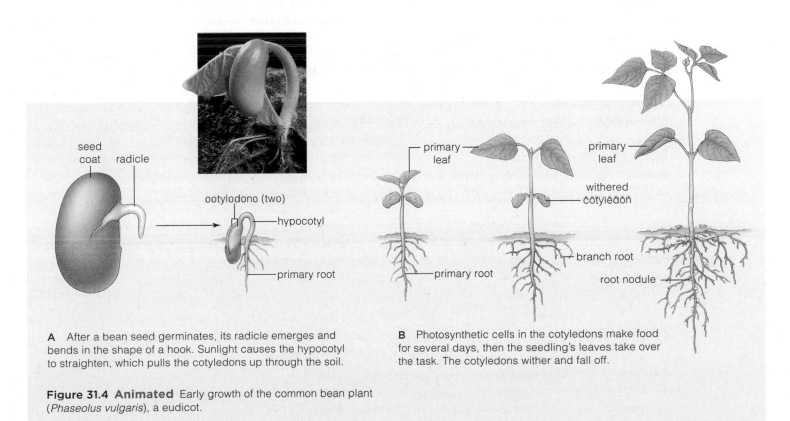

A After a bean seed germinates, its radicle emerges and bends in the shape of a hook. Sunlight causes the hypocotyl to straighten, which pulls the cotyledons up through the soil.

B Photosynthetic cells in the cotyledons make food for several days, then the seedling's leaves take over the task. The cotyledons wither and fall off.

Figure 31.4 Animated Early growth of the common bean plant (*Phaseolus vulgaris*), a eudicot.

Plant Hormones and Other Signaling Molecules

■ Plant development depends on cell-to-cell communication, which is mediated by plant hormones.

■ Links to Transcription factors 15.1, Cell communication in animal development 15.3, Function of stomata 29.4

Plant Hormones

You may be surprised to learn that plant development depends on extensive coordination among individual cells, just as it does in animals (Section 15.3). A plant is an organism, not just a collection of cells, and as such it develops as a unit. Cells in different parts of a plant coordinate their activities by communicating with one another. Such communication means, for example, that root and shoot growth occur at the same time.

Plant cells use hormones to communicate with one another. **Plant hormones** are signaling molecules that can stimulate or inhibit plant development, including growth. Environmental cues such as the availability of water, length of night, temperature, and gravity influence plants by triggering the production and dispersal of hormones. When a plant hormone binds to a target cell, it may modify gene expression, solute concentrations, enzyme activity, or activate another molecule in the cytoplasm. Later sections give examples.

Five types of plant hormones—gibberellins, auxins, abscisic acid, cytokinins, and ethylene—all interact to orchestrate plant development (Table 31.1).

Gibberellins Growth and other processes of development in all flowering plants, gymnosperms, mosses, ferns, and some fungi are regulated in part by **gibberellins**. These hormones induce cell division and elongation in stem tissue; thus, they cause stems to lengthen between the nodes. As mentioned in the chapter introduction, this effect can be demonstrated by application of gibberellin to the leaves of young plants (Figure 31.5). The short stems of Mendel's dwarf pea plants (Section 11.3) are the result of a mutation that reduces the rate of gibberellin synthesis in these plants. Gibberellins are also involved in breaking dormancy of seeds, seed germination, and the induction of flowering in biennials and some other plants.

Figure 31.5 Foolish cabbages! The three tall cabbage plants were treated with gibberellins. The two short plants in front of the ladder were not treated.

Table 31.1 Major Plant Hormones and Some of Their Effects

Hormone	Primary Source	Effect	Site of Effect
Gibberellins	Stem tip, young leaves	Stimulates cell division, elongation	Stem internode
	Embryo	Stimulates germination	Seed
	Embryo (grass)	Stimulates starch hydrolysis	Endosperm
Auxins	Stem tip, young leaves	Stimulates cell elongation	Growing tissues
		Initiates formation of lateral roots	Roots
		Inhibits growth (apical dominance)	Axillary buds
		Stimulates differentiation of xylem	Cambium
		Inhibits abscission	Leaves, fruits
	Developing embryos	Stimulates fruit development	Ovary
Abscisic acid	Leaves	Closes stomata	Guard cells
		Stimulates formation of dormant buds	Stem tip
	Ovule	Inhibits germination	Seed coat
Cytokinins	Root tip	Stimulates cell division	Stem tip, axillary buds
		Inhibits senescence (aging)	Leaves
Ethylene	Damaged or aged tissue	Inhibits cell elongation	Stem
		Stimulates senescence (aging)	Leaves
		Stimulates ripening	Fruits

Figure 31.6 Effect of rooting powders that contain auxin. Cuttings of winter honeysuckle (*Lonicera fragrantissima*) that were treated with a lot of auxin (*right*), some auxin (*middle*), and no auxin (*left*).

Table 31.2 Some Commercial Uses of Plant Hormones
Gibberellins Increase fruit size; delay citrus fruit ripening; synthetic forms can make some dwarf mutants grow tall
Synthetic auxins Promote root formation in cuttings; induce seedless fruit production before pollination; keep mature fruit on trees until harvest time; widely used as herbicides against broadleaf weeds in agriculture
ABA Induces nursery stock to enter dormancy before shipment to minimize damage during handling
Cytokinins Tissue culture propagation; prolong shelf life of cut flowers
Ethylene Allows shipping of green, still-hard fruit (minimizes bruises and rotting). Carbon dioxide application stops ripening of fruit in transit to market, then ethylene is applied to ripen distributed fruit quickly

Auxins **Auxins** are plant hormones that promote or inhibit cell division and elongation, depending on the target tissue. Auxins that are produced in apical meristems result in elongation of shoots. They also induce cell division and differentiation in vascular cambium, fruit development in ovaries, and lateral root formation in roots (Figure 31.6). Auxins also have inhibitory effects. For example, auxin produced in a shoot tip prevents the growth of lateral buds along a lengthening stem, an effect called **apical dominance**. Gardeners routinely pinch off shoot tips to make a plant bushier. Pinching the tips ends the supply of auxin in a main stem, so lateral buds give rise to branches. Auxins also inhibit abscission, which is the dropping of leaves, flowers, and fruits from the plant.

Abscisic Acid **Abscisic acid** (ABA) is a hormone that was misnamed; it inhibits growth, and has little to do with abscission. ABA is part of a stress response that causes stomata to close (Section 29.4). It also diverts photosynthetic products from leaves to seeds, an effect that overrides growth-stimulating effects of other hormones as the growing season ends. ABA inhibits seed germination in some species, such as apple (*Malus*). Such seeds do not germinate before most of the ABA they contain has been broken down, for example by a long period of cold, wet conditions.

Cytokinins Plant **cytokinins** form in roots and travel via xylem to shoots, where they induce cell divisions in the apical meristems. These hormones also release lateral buds from apical dominance, and inhibit the normal aging process in leaves. Cytokinins signal to shoots that roots are healthy and active. When roots stop growing, they stop producing cytokinins, so shoot growth slows and leaves begin to deteriorate.

Ethylene The only gaseous hormone, **ethylene**, is produced by damaged cells. It is also produced in autumn in deciduous plants, or near the end of the life cycle as part of a plant's normal process of aging. Ethylene inhibits cell division in stems and roots. It also induces fruit and leaves to mature and drop. Ethylene is widely used to artificially ripen fruit that has been harvested while still green (Table 31.2).

Other Signaling Molecules

As we now know, other signaling molecules have roles in various aspects of plant development. For example, brassinosteroids stimulate cell division and elongation; stems remain short in their absence. FT protein is part of a signaling pathway in flower formation. Salicylic acid, a molecule similar to aspirin, interacts with nitric oxide in regulating transcription of gene products that help plants resist attacks by pathogens. Systemin is a polypeptide that forms as insects feed on plant tissues; it enhances transcription of genes that encode insect toxins. Jasmonates, derived from fatty acids, interact with other hormones in control of germination, root growth, and tissue defense. You will see an example of how jasmonates help defend plant tissues in the next section.

Take-Home Message

What regulates growth and development in plants?

■ Plant hormones are signaling molecules that influence plant development.

■ The five main classes of plant hormones are gibberellins, auxins, cytokinins, abscisic acid, and ethylene.

■ Interactions among hormones and other kinds of signaling molecules stimulate or inhibit cell division, elongation, differentiation, and other events.

■ Plant hormones are involved in signal perception, transduction, and response.

■ Links to Carbohydrates 3.2 and 3.3, Membrane proteins 5.2, Turgor 5.6, Plant cell walls 4.12, Rubisco 7.6, Gene expression 15.1, Signal transduction 27.6

Gibberellin and Germination

During germination, water absorbed by a barley seed causes cells of the embryo to release gibberellin (Figure 31.7). The hormone diffuses into the aleurone, a protein-rich layer of cells surrounding the endosperm. In the aleurone, gibberellin induces transcription of the gene for amylase, an enzyme that hydrolyzes starch into sugar monomers (Sections 3.2 and 3.3). The amylase is released into the endosperm's starchy interior, where it proceeds to break down stored starch molecules into sugars. The embryo takes up the sugars and uses them for aerobic respiration, which fuels rapid cell divisions at the embryo's meristems.

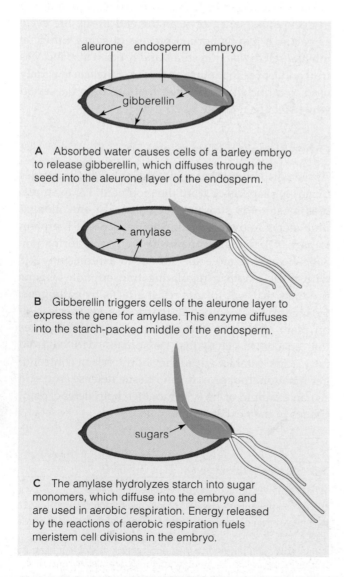

A Absorbed water causes cells of a barley embryo to release gibberellin, which diffuses through the seed into the aleurone layer of the endosperm.

B Gibberellin triggers cells of the aleurone layer to express the gene for amylase. This enzyme diffuses into the starch-packed middle of the endosperm.

C The amylase hydrolyzes starch into sugar monomers, which diffuse into the embryo and are used in aerobic respiration. Energy released by the reactions of aerobic respiration fuels meristem cell divisions in the embryo.

Figure 31.7 Action of gibberellin in barley seed germination.

Auxin Augmentation

There are a few naturally occurring auxins, but the one with the majority of effects is indole-3-acetic acid (IAA). This molecule plays a critical role in all aspects of plant development, starting with the first division of the zygote. It is involved in polarity and tissue patterning in the embryo, formation of plant parts (primary leaves, shoot tips, stems, and roots), differentiation of vascular tissues, formation of lateral roots (and adventitious roots in some species), and, as you will see in the next sections, responses to environmental stimuli.

How can one molecule have so many roles? Part of the answer is that IAA has multiple effects on plant cells. For example, it causes cells to expand by increasing the activity of proton pumps, which are membrane transporter proteins that pump hydrogen ions from the cytoplasm into the cell wall. The resulting increase in acidity causes the wall to become less rigid. Turgor pushing on the softened wall from the inside stretches the cell irreversibly. IAA also affects gene expression by binding to certain regulatory molecules. The binding results in the degradation of repressor proteins that block transcription of specific genes (Section 15.1).

IAA can exert different effects at different concentrations. Although present in almost all plant tissues, IAA is unevenly distributed through them. In a sporophyte, IAA is made mainly in shoot tips and young leaves, and its concentration is highest there. It forms gradients in plant tissues by moving away from these developing parts, but the movement is more complicated than diffusion alone can explain.

IAA is transported in phloem over long distances, such as from shoots to roots. Over shorter distances, it moves by a cell-to-cell transport system that involves active transport. IAA diffuses into a cell, but it also is actively transported through membrane proteins located on the top of the cell. It moves out of the cell only through efflux carriers, which are active transport proteins present on the bottom of the cell. In other words, IAA moves into a cell on the top, and out of it on the bottom. Thus, it tends to be transported in a polar fashion through local tissues, from the tip toward the base of a stem (Figure 31.8). A different mechanism moves auxin molecules upward from the root tip to the shoot–root junction.

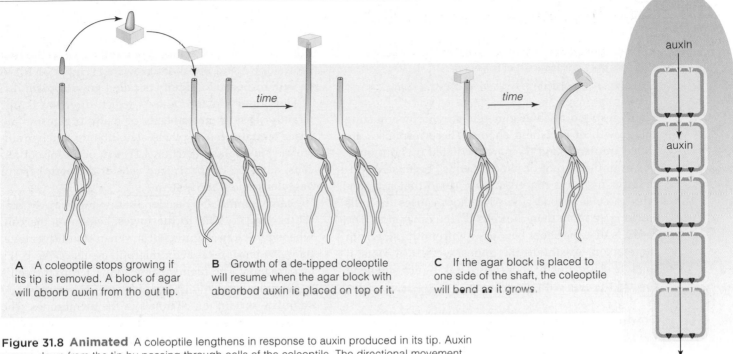

A A coleoptile stops growing if its tip is removed. A block of agar will absorb auxin from the cut tip.

B Growth of a de-tipped coleoptile will resume when the agar block with absorbed auxin is placed on top of it.

C If the agar block is placed to one side of the shaft, the coleoptile will bend as it grows.

Figure 31.8 Animated A coleoptile lengthens in response to auxin produced in its tip. Auxin moves down from the tip by passing through cells of the coleoptile. The directional movement is driven by different types of active transporters positioned at the top and bottom of the cells' plasma membranes (*right*).

Jeopardy and Jasmonates

Many plants protect themselves with thorns or nasty tasting chemicals that deter herbivores (plant-eating animals). Some get help from wasps.

Damage to a leaf, such as occurs when an herbivore chews on it, triggers a stress response in the plant. The wounding results in the cleavage of certain peptides (such as systemin) in mesophyll cells. Thus activated, the peptides stimulate synthesis of jasmonates, which turn on transcription of a variety of genes.

Some of the resulting gene products break down molecules used in normal activities, such as rubisco (Section 7.6), so growth temporarily slows. Other gene products produce chemicals that the plant releases into the air. The chemicals are detected by wasps that parasitize herbivores (Figure 31.9). The signaling is quite specific: A leaf releases a different set of chemicals depending on which herbivore is chewing on it. Certain wasp species recognize these chemicals as a signal leading to preferred prey. They follow airborne concentration gradients of the chemicals back to the plant, where they attack the herbivores.

Take-Home Message

What are some examples of plant hormone effects?

■ Gibberellin affects expression of genes for nutrient utilization in germination; auxin causes cell lengthening; and jasmonates are involved in plant defensive signaling.

Figure 31.9 Jasmonates in plant defenses. (**a**) Consuelo De Moraes studies chemical signaling in plants. (**b**) A caterpillar chewing on a tobacco leaf (*Nicotiana*) triggers a chemical response from the leaf's cells. The cells release certain chemicals into the air. (**c,d**) A parasitoid wasp follows the chemicals back to the stressed leaves, then attacks a caterpillar and deposits an egg inside it. When the egg hatches, it will release a caterpillar-munching larva.

De Moraes discovered that such interactions are highly specific: Leaf cells release different chemicals in response to different caterpillar species. Each chemical attracts only the wasps that parasitize the particular caterpillar that triggered the chemical's release.

■ Plants alter growth in response to environmental stimuli. Hormones are typically part of this effect.

■ Links to Plastids 4.11, Cytoskeleton 4.13, Pigments 7.1

Plants respond to environmental stimuli by adjusting the growth of roots and shoots. These responses are called **tropisms**, and they are mediated by hormones. For example, a root or shoot "bends" because of differences in auxin concentration. Auxin that accumulates in cells on one side of a shoot causes the cells to elongate more than the cells on the other side. The result is that the shoot bends away from the side with more auxin. Auxin has the opposite effect in roots: It inhibits elongation of root cells. Thus, a root will bend toward the side with more auxin.

Gravitropism No matter how a seed is positioned in the soil when it germinates, the radicle always grows down, and the primary shoot always grows up. Even if a seedling is turned upside down just after germina-

tion, the primary root and shoot will curve so the root grows down and the shoot grows up (Figure 31.10). A growth response to gravity is called **gravitropism**.

How does a plant "know" which direction is up? Gravity-sensing mechanisms of many organisms are based on **statoliths**. In plants, statoliths are starch-grain-stuffed amyloplasts (Section 4.11) that occur in root cap cells, and also in specialized cells at the periphery of vascular tissues in the stem.

Starch grains are heavier than cytoplasm, so statoliths tend to sink to the lowest region of the cell, wherever that is (Figure 31.11). When statoliths move, they put tension on actin microfilaments of the cell's cytoskeleton. The filaments are connected to the cell's membranes, and the change in tension is thought to stimulate certain ion channels in the membranes. The result is that the cell's auxin efflux carriers move to the new "bottom" of the cell within minutes of a change in orientation. Thus, auxin is always transported to the down-facing side of roots and shoots.

A Gravitropism of a corn seedling. No matter what the orientation of a seed in the soil, a seedling's primary root grows down, and its primary shoot grows up.

B These seedlings were rotated 90° counterclockwise after they germinated. The plant adjusts to the change by redistributing auxin, and the direction of growth shifts as a result.

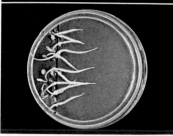

C In the presence of auxin transport inhibitors, seedlings do not adjust their direction of growth after a 90° counterclockwise rotation. Mutations in genes that encode auxin transport proteins have the same effect.

Figure 31.10 Gravitropism.

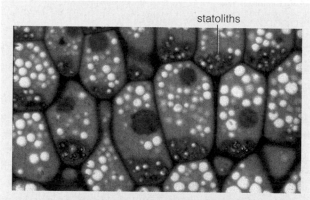

statoliths

A Heavy, starch-packed statoliths are settled on the bottom of gravity-sensing cells in a corn root cap.

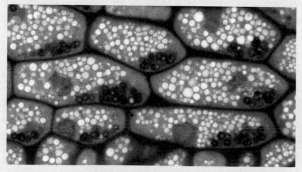

B Ten minutes after the root was rotated, the statoliths settled to the new "bottom" of the cells. The redistribution causes auxin redistribution, and the root tip curves down.

Figure 31.11 Animated Gravity, statoliths, and auxin.
Figure It Out: In which direction was this root rotated?

Answer: 90° counterclockwise

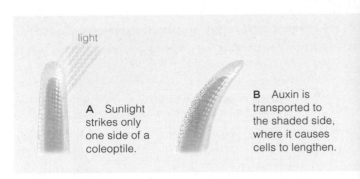

A Sunlight strikes only one side of a coleoptile.

B Auxin is transported to the shaded side, where it causes cells to lengthen.

Figure 31.12 Animated Phototropism. (**a,b**) Auxin-mediated differences in cell elongation between two sides of a coleoptile induce bending toward light. The photo shows shamrock (*Oxalis*) responding to a directional light source.

Phototropism Light streaming in from one direction causes a stem to curve toward its source. This response, **phototropism**, orients certain parts of the plant in the direction that will maximize the amount of light intercepted by its photosynthetic cells.

Phototropism in plants occurs in response to blue light. Nonphotosynthetic pigments called phototropins absorb blue light, and translate its energy into a cascade of intracellular signals. The ultimate effect of this cascade is that auxin is redistributed to the shaded side of a shoot or coleoptile. As a result, cells on the shaded side elongate faster than cells on the illuminated side. Differences in growth rates between cells on opposite sides of a shoot or coleoptile causes the entire structure to bend toward the light (Figure 31.12).

Thigmotropism A plant's contact with a solid object may result in a change in the direction of its growth, a response called **thigmotropism**. The mechanism that gives rise to this response is not well understood, but it involves the products of calcium ions and at least five genes called *TOUCH*.

We see thigmotropism when a vine's tendril touches an object. The cells near the area of contact stop elongating, and the cells on the opposite side of the shoot keep elongating. The unequal growth rates of cells on opposite sides of the shoot cause it to curl around the object (Figure 31.13). A similar mechanism causes roots to grow away from contact, which allows them to "feel" their way around rocks and other impassable objects in the soil.

Mechanical stress, as inflicted by wind or grazing animals, inhibits stem lengthening in a touch response related to thigmotropism (Figure 31.14).

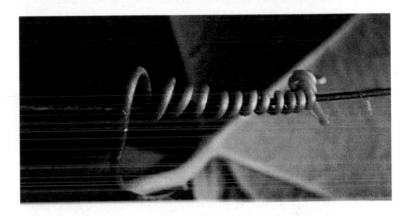

Figure 31.13 Passion flower (*Passiflora*) tendril twisting thigmotropically around a wire support.

Figure 31.14 Effect of mechanical stress on tomato plants. (**a**) This plant, the control, was not shaken. (**b**) This plant was mechanically shaken for thirty seconds each day, for twenty-eight days. (**c**) This one had two shakings each day. All plants were the same age.

Take-Home Message

How do plants respond to environmental stimuli?

■ Plants adjust the direction and rate of growth in response to environmental stimuli that include gravity, light, contact, and mechanical stress.

- Seasonal shifts in night length, temperature, and light trigger seasonal shifts in plant development.

- Links to Photosynthesis 7.4 and 7.6, Master genes in flowering 15.2, Homeostasis in plants 27.5

Biological Clocks

Most organisms have a **biological clock**—an internal mechanism that governs the timing of rhythmic cycles of activity. Section 27.5 showed a bean plant changing the light-intercepting position of its leaves over twenty-four hours even when it was kept in the dark.

A cycle of activity that starts anew every twenty-four hours or so is called a **circadian rhythm** (Latin *circa*, about; *dies*, day). In the circadian response called **solar tracking**, a leaf or flower changes position in response to the changing angle of the sun throughout the day. For example, a buttercup stem swivels so the flower on top of it always faces the sun. Unlike a phototropic response, solar tracking does not involve redistribution of auxin and differential growth. Instead, the absorption of blue light by photoreceptor proteins increases fluid pressure in cells on the sunlit side of a stem or petiole. The cells change shape, which bends the stem.

Similar mechanisms cause flowers of some plants to open only at certain times of day. For example, the flowers of many bat-pollinated plants unfurl, secrete nectar, and release fragrance only at night. Closing flowers periodically protects the delicate reproductive parts when the likelihood of pollination is low.

Setting the Clock

Like a mechanical clock, a biological one can be reset. Sunlight resets biological clocks in plants by activating

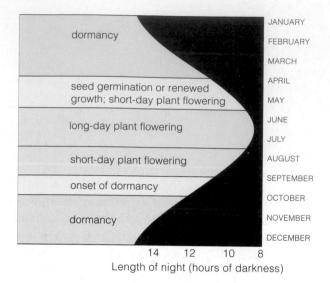

Figure 31.16 Plant growth and development correlated with seasonal climate changes in northern temperate zones.

and inactivating photoreceptors called **phytochromes**. These blue-green pigments are sensitive to red light (660 nanometers) and far-red light (730 nanometers). The relative amounts of these wavelengths in sunlight that reaches a given environment vary during the day and with the season. Red light causes phytochromes to change from an inactive form to an active form. Far-red light causes them to change back to their inactive form (Figure 31.15).

Active phytochromes bring about transcription of many genes, including some that encode components of rubisco, photosystem II, ATP synthase, and other proteins used in photosynthesis; phototropin for phototropic responses; and molecules involved in flowering, gravitropism, and germination.

When to Flower?

Photoperiodism is an organism's response to changes in the length of night relative to the length of day. Except at the equator, night length varies with the season. Nights are longer in winter than in summer, and the difference increases with latitude (Figure 31.16).

You have probably noticed that different species of plants flower at different times of the year. In these plants, flowering is photoperiodic. *Long-day* plants such as irises flower only when the hours of darkness fall below a critical value (Figure 31.17a). Chrysanthemums and other *short-day* plants flower only when the hours of darkness are greater than some critical value (Figure 31.17b). Sunflowers and other *day-neutral* plants flower when they mature, regardless of night length.

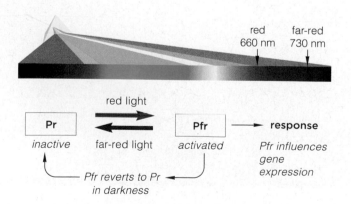

Figure 31.15 Animated Phytochromes. Red light changes the structure of a phytochrome from inactive to active form; far-red light changes it back to the inactive form. Activated phytochromes control important processes such as germination and flowering.

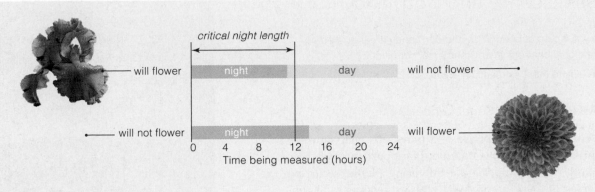

A Long-day plants flower only when hours of darkness are *less* than the critical value for the species. Irises will flower only when night length is less than 12 hours.

B Short-day plants flower only when hours of darkness are *greater* than the critical value for the species. Chrysanthemums will flower only when night length exceeds 12 hours.

Figure 31.17 Animated Different plant species flower in response to different night lengths. Each horizontal bar represents 24 hours.

Figure 31.18 shows two experiments that demonstrated how phytochromes play a role in photoperiodism. In the first experiment, a long-day and a short-day plant were exposed to long "nights," interrupted by a brief pulse of red light (which activates phytochrome). Both plants responded in their typical way to a season of short nights. In the second experiment, the pulse of red light (which activates phytochrome) was followed by a pulse of far-red light (which deactivates phytochrome). Both plants responded in their typical way to a season of long nights.

Leaves detect night length and produce signals that travel through the plant. In one experiment, a single leaf was left on a cocklebur, a short-day plant. The leaf was shielded from light for 8–1/2 hours every day, which is the threshold amount of darkness required for flowering. The plant flowered. Later, the leaf was grafted onto another cocklebur plant that had *not* been exposed to long hours of darkness. After grafting, the recipient plant flowered, too.

How does a compound produced by leaves cause flowering? In response to night length and other cues, leaf cells transcribe more or less of a flowering gene. The transcribed mRNA migrates from leaves to shoot tips, where it is translated. Its protein product helps activate the master genes that control the formation of flowers (Section 15.2).

The length of night is not the only cue for flowering. Some biennials and perennials flower only after exposure to cold winter temperatures (Figure 31.19). This process is called **vernalization** (from Latin *vernalis*, which means "to make springlike").

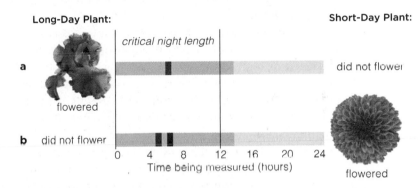

Figure 31.18 Phytochrome plays a role in flowering. (**a**) An flash of red light interrupting a long night causes plants to respond as if the night were short. Long-day plants flower. (**b**) A pulse of far-red light, which inactivates phytochrome, cancels the effect of the red flash: Short-day plants flower.

Figure 31.19 Local effect of cold on dormant buds of lilac (*Syringa*). For this experiment, a single branch was positioned to protrude from a greenhouse through a cold winter. The rest of the plant was kept inside and exposed only to warm temperatures. Only buds exposed to the low outside temperatures resumed growth and flowered in springtime.

Take-Home Message

Do plants have biological clocks?

■ Flowering plants respond to recurring cues from the environment with recurring cycles of development.

■ The main environmental cue for flowering is the length of night relative to the length of day, which varies by the season in most places. Low winter temperatures stimulate the flowering of many plant species in spring.

31.6 Senescence and Dormancy

■ Dropping of plant parts and dormancy are triggered by seasonal changes in environmental conditions.

■ Link to Plant extracellular matrix 4.12

Abscission and Senescence

Senescence is the phase of a plant life cycle between full maturity and the death of plant parts or the whole plant. In many species of flowering plants, recurring cycles of growth and inactivity are responses to conditions that vary seasonally. Such plants are typically native to regions that are too dry or too cold for optimal growth during part of the year. Plants may drop leaves during such unfavorable intervals. The process by which plant parts are shed is **abscission**. It occurs in deciduous plants in response to shortening daylight hours, and year-round in evergreen plants. Abscission may also be induced by injury, water or nutrient deficiencies, or high temperatures.

Let's use deciduous plants as an example. As leaves and fruits grow in early summer, their cells produce auxin. The auxin moves into the stems, where it helps maintain growth. By midsummer, the nights are getting longer. Plants begin to divert nutrients away from their leaves, stems, and roots, and into flowers, fruits, and seeds. As the growing season comes to a close, nutrients are routed to twigs, stems, and roots, and auxin production declines in leaves and fruits.

The auxin-deprived structures release ethylene that diffuses into nearby abscission zones—twigs, petioles, and fruit stalks. The ethylene is a signal for cells in the

control (pods not removed)   experimental plant (pods removed)

Figure 31.21 Experiment in which seed pods removed from a soybean plant as soon as they formed delayed senescence.

zone to produce enzymes that digest their own walls and the middle lamella (Section 4.12). The cells bulge as their walls soften, and separate from one another as their middle lamella—the layer that cements them together—dissolves. Tissue in the zone weakens, and the structure above it drops (Figure 31.20).

If the seasonal diversion of nutrients into flowers, seeds, and fruits is interrupted, leaves and stems stay on a deciduous plant longer (Figure 31.21).

Dormancy

For many species, growth stops in autumn as a plant enters **dormancy**, a period of arrested growth that is triggered by (and later ended by) environmental cues. Long nights, cold temperatures, and dry, nitrogen-poor soil are strong cues for dormancy in many plants.

Dormancy-breaking cues usually operate between fall and spring. Dormant plants do not resume growth until certain conditions in the environment occur. A few species require exposure of the dormant plant to many hours of cold temperature. More typical cues include the return of milder temperatures and plentiful water and nutrients. With the return of favorable conditions, life cycles begin to turn once more as seeds germinate and buds resume growth.

Figure 31.20 Horse chestnut (*Aesculus hippocastanum*) leaves changing color in autumn. The horseshoe-shaped leaf scar at *right* is all that remains of an abscission zone that formed before a leaf detached from the stem.

Take-Home Message

What triggers dropping of plant parts and dormancy?

■ Abscission and dormancy are triggered by environmental cues such as seasonal changes in temperature or daylength.

Foolish Seedlings, Gorgeous Grapes

Fruit ripening is a type of senescence. Like wounded tissues, senescing tissues (including ripening fruit) release ethylene gas. This plant hormone stimulates the production of enzymes such as amylase. These enzymes convert stored starches and acids to sugars, and soften the cell

ethylene

walls of fleshy fruits—sweetening and softening effects that we associate with ripening. Ethylene emitted by one fruit can stimulate the ripening—and over-ripening—of nearby fruits.

Fruit that is harvested at the peak of ripeness can be stored for months or even years after treatment with MCP. MCP binds per-

manently to ethylene receptors on fruit, but unlike ethylene, does not stimulate them. Thus, ripe fruit treated with MCP becomes insensitive to ethylene, so it will not over-ripen. MCP treatment is marketed as SmartFresh technology.

Summary

Section 31.1 Gene expression and cues from the environment coordinate plant development, which is the formation and **growth** of tissues and parts in predictable patterns (Figure 31.22). **Germination** is one pattern of development in plants.

■ *Use the animation on CengageNOW to compare monocot and eudicot growth and development.*

Sections 31.2, 31.3 Like animal hormones, **plant hormones** secreted by one cell alter the activity of a different cell. Plant hormones can promote or arrest growth of a plant by stimulating or inhibiting cell division, differentiation, elongation, and reproduction.

Gibberellins lengthen stems, break dormancy in seeds and buds, and stimulate flowering.

Auxins lengthen coleoptiles, shoots, and roots by promoting cell enlargement.

Cytokinins stimulate cell division, release lateral buds from **apical dominance**, and inhibit senescence.

Ethylene promotes senescence and abscission. It also inhibits growth of roots and stems.

Abscisic acid promotes bud and seed dormancy, and it limits water loss by causing stomata to close.

■ *Use the animation on CengageNOW to observe the effect of auxin on plant growth.*

Section 31.4 In **tropisms**, plants adjust the direction and rate of growth in response to environmental cues.

In **gravitropism**, roots grow down and stems grow up in response to gravity. **Statoliths** are part of this response. In **phototropism**, stems and leaves bend toward or away from light. Blue light is the trigger for such phototropic responses. In some plants, the direction of growth changes in response to contact (**thigmotropism**). Growth may also be affected by mechanical stress.

■ *Use the animation on CengageNOW to investigate plant tropisms.*

Sections 31.5, 31.6 Internal timing mechanisms such as **biological clocks** (including **circadian rhythms**) are set by daily and seasonal variations in environmental con-

ditions. **Solar tracking** is one type of circadian rhythm. Another, **photoperiodism**, is a response to changes in length of night relative to length of day. Light-detection in plants involves nonphotosynthetic pigments called **phytochromes** (in photoperiodism) and phototropins (in phototropism).

Short-day plants flower in spring or fall, when nights are long. Long-day plants flower in summer, when nights are short. Day-neutral plants flower whenever they are mature enough to do so. Some plants require exposure to cold before they can flower, a process called **vernalization**.

Dormancy is a period of arrested growth that does not end until specific environmental cues occur. Dormancy is typically preceded by **abscission**. **Senescence** is the part of the plant life cycle between maturity and death of the plant or plant parts.

■ *Use the animation on CengageNOW to learn how plants respond to night length.*

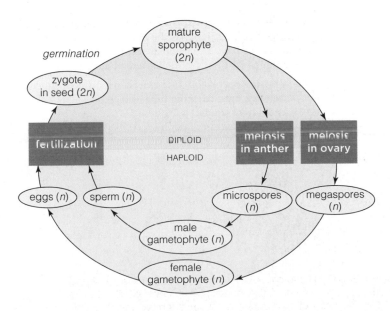

Figure 31.22 Summary of development in the life cycle of a typical eudicot.

Data Analysis Exercise

In 2007, researchers Casey Delphia, Mark Mescher, and Consuelo De Moraes (pictured in Figure 31.9a) published a study on the production of different volatile chemicals by tobacco plants (*Nicotiana tabacum*) in response to predation by two types of insects: western flower thrips (*Frankliniella occidentalis*) and tobacco budworms (*Heliothis virescens*). Their results are shown in Figure 31.23.

1. Which treatment elicited the greatest production of volatiles?

2. Which volatile chemical was produced in the greatest amount? What was the stimulus?

3. Which one of the chemicals tested is most likely produced by tobacco plants in a non-specific response to predation?

4. Are there any chemicals produced in response to predation by budworms, but not in response to predation by thrips?

Volatile Compound Produced	Treatment					
	C	T	W	WT	HV	HVT
Myrcene	0	0	0	0	17	22
β-Ocimene	0	433	15	121	4,299	5,315
Linalool	0	0	0	0	125	178
Indole	0	0	0	0	74	142
Nicotine	0	0	233	160	390	538
β-Elemene	0	0	0	0	90	102
β-Caryophyllene	0	100	40	124	3,704	6,166
α-Humulene	0	0	0	0	123	209
Sesquiterpene	0	7	0	0	219	268
α-Farnesene	0	15	0	0	293	457
Caryophyllene oxide	0	0	0	0	89	166
Total	0	555	288	405	9,423	13,563

Figure 31.23 Volatile compounds produced by tobacco plants (*Nicotiana tabacum*) in response to predation by different insects. Groups of plants were untreated (C), attacked by thrips (T), mechanically wounded (W), mechanically wounded and attacked by thrips (WT), attacked by budworms (HV), or attacked by budworms and thrips (HVT). Values are indicated in nanograms/day.

Self-Quiz
Answers in Appendix III

1. Which of the following statements is false?
   a. Auxins and gibberellins promote stem elongation.
   b. Cytokinins promote cell division, retard leaf aging.
   c. Abscisic acid promotes water loss and dormancy.
   d. Ethylene promotes fruit ripening and abscission.

2. Plant hormones _____.
   a. may have multiple effects
   b. are influenced by environmental cues
   c. are active in plant embryos within seeds
   d. are active in adult plants
   e. all of the above

3. _____ is the strongest stimulus for phototropism.
   a. Red light   c. Green light
   b. Far-red light   d. Blue light

4. _____ light makes phytochrome switch from inactive to active form; _____ light has the opposite effect.
   a. Red; far-red   c. Far-red; red
   b. Red; blue   d. Far-red; blue

5. The following oat coleoptiles have been modified: either cut or placed in a light-blocking tube. Which ones will still bend toward a light source?

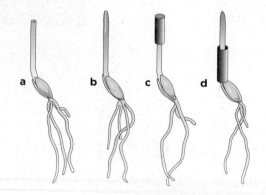

6. In some plants, flowering is a _____ response.
   a. phototropic   c. photoperiodic
   b. gravitropic   d. thigmotropic

7. Match the observation with the hormone most likely to be its cause.
   ____ethylene
   ____cytokinin
   ____auxin
   ____gibberellin
   ____abscisic acid

   a. Your cabbage plants bolt (they form elongated flowering stalks).
   b. The *philodendron* in your room is leaning toward the window.
   c. The last of your apples is getting really mushy.
   d. The seeds of your roommate's marijuana plant do not germinate no matter what he does to them.
   e. Lateral buds on your *Ficus* plant are sprouting branch shoots.

■ *Visit CengageNOW for additional questions.*

Critical Thinking

1. Reflect on Chapter 28. Would you expect hormones to influence primary growth only? What about secondary growth in, say, a hundred-year-old oak tree?

2. Photosynthesis sustains plant growth, and inputs of sunlight sustain photosynthesis. Why, then, do seedlings that germinated in a fully darkened room grow taller than different seedlings of the same species that germinated in full sun?

3. Belgian scientists discovered that certain mutations in common wall cress (*Arabidopsis thaliana*) cause excess auxin production. Predict the impact on the plant's phenotype.

4. Beef cattle typically are given somatotropin, an animal hormone that makes them grow bigger (the added weight means greater profits). There is concern that such hormones may have unforeseen effects on beef-eating humans. Do you think plant hormones can affect humans? Why or why not?

VI HOW ANIMALS WORK

How many and what kinds of body parts does it take to function as a lizard in a tropical forest? Make a list of what comes to mind as you start reading Unit VI, then see how resplendent the list can become at the unit's end.

32 Animal Tissues and Organ Systems

IMPACTS, ISSUES Open or Close the Stem Cell Factories?

Imagine being able to grow new body parts to replace lost or diseased ones. This dream motivates researchers who study stem cells. Stems cells are self-renewing; they divide and produce more stem cells. In addition, some descendants of stem cells differentiate into the specialized cells that make up specific body parts. In short, all cells in your body "stem" from stem cells.

Cell types that your body continually replaces, such as blood and skin, arise from adult stem cells. Such stem cells are specialists that normally differentiate into a limited variety of cells. For example, stem cells in adult bone marrow can become blood cells, but not muscle cells or brain cells.

Embryos have stem cells that are more versatile. After all, these cells are the source of all tissue types in the new body. Embryonic stem cells are formed soon after fertilization when cell division produces a pinhead-sized ball of cells. By birth, embryonic stem cells have disappeared.

Stem cells that can become nerve cells or muscle cells are rare in adults. Thus, unlike skin and blood cells, nerves and muscles are not replaced if they get damaged or die. This is why an injury to the nerves of the spinal cord can cause permanent paralysis.

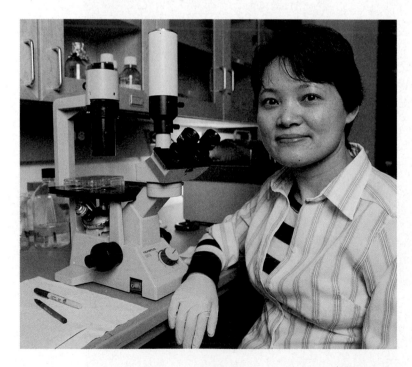

See the video! Figure 32.1 Junying Yu at the University of Wisconsin-Madison is part of a research team that developed a method of turning a newborn's skin cells into cells that behave like embryonic stem cells.

In theory, embryonic stem cell treatments could provide new nerve cells for paralyzed people. Treatments might also help treat other nerve and muscle disorders such as heart disease, muscular dystrophy, and Parkinson's disease.

Despite the promise of embryonic stem cell research, some people oppose it. They are troubled by the original source of the cells—early human embryos. The embryos typically come from fertility clinics that would otherwise have destroyed them and are donated by their parents.

So far, scientists have not found any adult stem cells that have the same potential of embryonic stem cells. However, they may be able to genetically engineer such cells. For example, James Thompson and Junying Yu (Figure 32.1) used viruses to insert genes from embryonic cells into skin cells of a newborn boy. The result was easy-to-grow cells that showed the same features as embryonic stem cells in culture. A research team in Japan achieved similar results by using viruses to insert genes into adult skin cells.

Does that mean using embryonic stem cells will become unnecessary? Possibly, but there are still obstacles. First, the retroviruses used to insert the genes can cause cancer. Thus cells created by this method cannot safely be placed in a human body. Second, while the engineered cells seem to behave like embryonic stem cells in the lab, they might behave differently once implanted in a person. Further research will be necessary to see whether stem cells can be engineered in a safer way, and if they actually have the same potential as embryonic stem cells in a clinical context.

Stem cells, the source of all tissues and organs, are a fitting introduction to this unit. The unit deals with animal anatomy (how a body is put together) and physiology (how a body works). In this unit, you will return repeatedly to a concept outlined in Chapter 27. Cells, tissues, and organs interact smoothly when the body's internal environment is maintained within a range that individual cells can tolerate. In most kinds of animals, blood and interstitial fluid are the internal environment. The processes involved in maintaining this environment are collectively called homeostasis.

Regardless of the species, the body parts must interact and perform the following tasks :

1. Coordinate and control activities of its individual parts.

2. Acquire and distribute raw materials to individual cells and dispose of wastes.

3. Protect tissues against injury or attack.

4. Reproduce and, in many species, nourish and protect offspring through early growth and development.

Key Concepts

Animal organization

All animals are multicelled, with cells joined by cell junctions. Typically, cells are organized in four tissues: epithelial tissue, connective tissue, muscle tissue, and nervous tissue. Organs, which consist of a combination of tissues, interact in organ systems. **Section 32.1**

Types of animal tissues

Epithelial tissue covers the body's surface and lines its internal tubes. Connective tissue provides support and connects body parts. Muscle tissue moves the body and its parts. Nervous tissue detects internal and external stimuli and coordinates responses. **Sections 32.2–32.5**

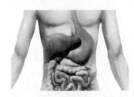

Organ systems

Vertebrate organ systems compartmentalize the tasks of survival and reproduction for the body as a whole. Different systems arise from ectoderm, mesoderm, and endoderm, the primary tissue layers that form in the early embryo. **Section 32.6**

A closer look at skin

Skin is an example of an organ system. It includes epithelial layers, connective tissue, adipose tissue, glands, blood vessels, and sensory receptors. It helps protect the body, conserve water, control temperature, excrete wastes, and detect external stimuli. **Sections 32.7, 32.8**

Links to Earlier Concepts

- With this chapter, we begin to consider the tissue and organ system levels of organization in animals (Section 1.1). You will also learn more about the cells involved in sensing and responding to stimuli (1.2).

- This chapter expands on the nature of animal body plans (25.1) and trends in vertebrate evolution (26.2).

- You will think again about the importance of diffusion across cell membranes (5.3), aerobic respiration (8.1), and the structure and metabolism of lipids (3.4, 8.7). The protein hemoglobin (3.6) comes up as we discuss blood.

- Cancer (9.5) and the effects of UV radiation (14.5) are revisited in the context of skin and sunlight exposure.

■ Cells of animal bodies are united by cell junctions, and typically organized as tissues, organs, and organ systems.

■ Link to Levels of organization 1.1

From Tissue to Organs to Organ Systems

All animals are multicelled, and nearly all have cells organized as tissues. A **tissue** consists of interacting cells and extracellular substances that carry out one or more specialized tasks.

Four types of tissue occur in all vertebrate bodies. Epithelial tissues cover body surfaces and line internal cavities. Connective tissues hold body parts together and provide structural support. Muscle tissues move the body and its parts. Nervous tissues detect stimuli and relay information. We will consider each type of tissue in detail in the sections that follow.

Typically, animal tissues are organized into organs. An **organ** is a structural unit of two or more tissues organized in a specific way and capable of carrying out specific tasks. Your heart is an organ that consists of all four types of tissues in certain proportions and arrangements. In **organ systems**, two or more organs and other components interact physically, chemically, or both in a common task, as when the force generated by a beating heart moves blood through the body.

A body's cells, tissues, and organs interact smoothly when the internal environment stays within a range that the cells can tolerate. In most animals, blood and interstitial fluid (fluid between cells) are the internal environment. **Homeostasis** is the process of maintaining the internal environment (Section 27.1).

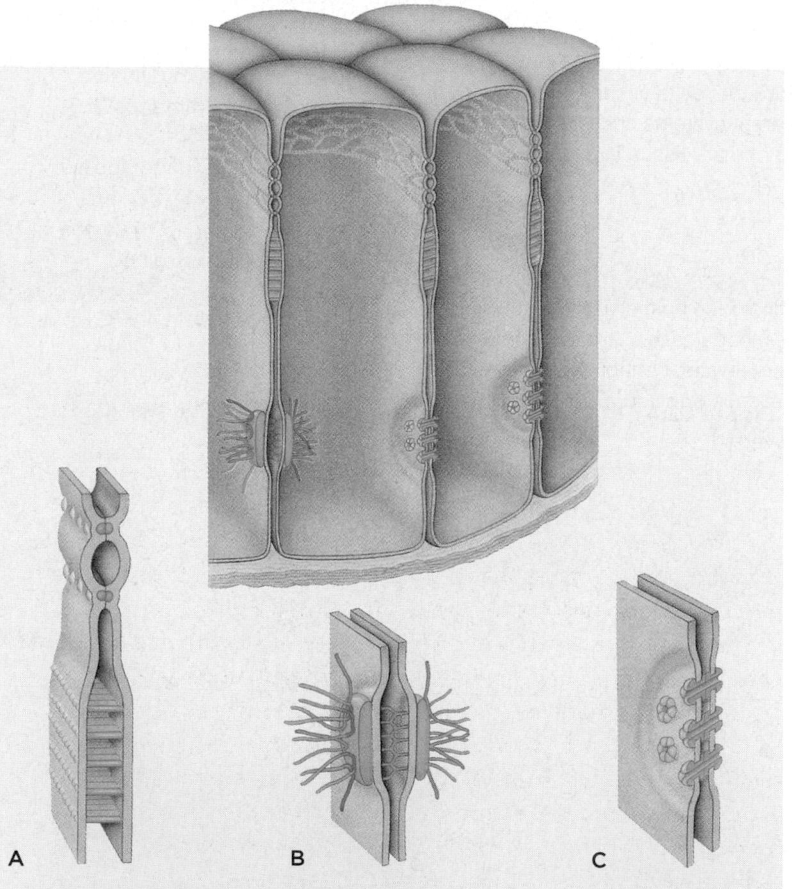

Cell Junctions

Cells in most animal tissues connect to their neighbors by way of one or more types of cell junctions.

In epithelial tissues, rows of proteins that form **tight junctions** between plasma membranes of adjacent cells prevent fluid from seeping between these cells. To cross an epithelium, a fluid must pass through the epithelial cells. Transport proteins in cell membranes control which ions and molecules cross the epithelium (Section 5.2).

An abundance of tight junctions in the lining of the stomach normally keeps acidic fluid from leaking out. If a bacterial infection damages this lining, acid and enzymes can erode the underlying connective tissue and muscle layers. The result is a painful peptic ulcer.

Adhering junctions hold cells together at distinct spots, like buttons hold a shirt closed (Figure 32.2b). Skin and other tissues that are subject to abrasion or stretching are rich in adhering junctions.

Gap junctions permit ions and small molecules to pass from the cytoplasm of one cell to another (Figure 32.2c). Heart muscle and other tissues in which the cells perform some coordinated action have many of these communication channels.

A	B	C
Tight junctions	**Adhering junction**	**Gap junction**
Rows of proteins that run parallel with the free surface of a tissue; stop leaks between adjoining cells	A mass of interconnected proteins that welds two cells together; anchored under the plasma membrane by intermediate filaments of cytoskeleton	Cylindrical arrays of proteins spanning the plasma membrane of adjoining cells, paired as open channels

Figure 32.2 Animated Examples of cell junctions in animal tissues.

Take-Home Message

How is an animal body organized?

■ Nearly all animals have cells united by cell junctions and organized into tissues, organs, and organ systems.

■ All body parts work together in homeostasis, the process of keeping internal conditions within the range cells can tolerate.

32.2 | Epithelial Tissue

■ Sheets of epithelial tissue cover the body's outer surface and line its internal ducts and cavities.

■ Link to Diffusion 5.3

General Characteristics

An **epithelium** (plural, epithelia), or **epithelial tissue**, is a sheet of cells that covers an outer body surface or lines an internal cavity. One surface of the epithelium faces the outside environment or a body fluid. A secreted extracellular matrix, known as the **basement membrane**, attaches the epithelium's opposite surface to an underlying tissue (Figure 32.3).

Epithelial tissues are described in terms of the shape of the constituent cells and the number of cell layers. A simple epithelium is one cell thick; a stratified epithelium has multiple cell layers. Squamous epithelium cells are flattened or platelike. Cells of cuboidal epithelium are short cylinders that look like cubes when viewed in cross-section. Cells in columnar epithelium are taller than they are wide. Figure 32.4 shows these shapes in the three types of simple epithelium.

Different kinds of epithelia are suited to different tasks. Simple squamous epithelium is the thinnest type. It lines blood vessels and the tiny air sacs inside lungs. Because it is thin, gases and nutrients diffuse across it easily. In contrast, thicker stratified squamous epithelium has a protective function. The outer layer of your skin consists of this tissue.

Cells of cuboidal and columnar epithelium act in absorption and secretion. In some tissues, such as the lining of the kidneys and small intestine, fingerlike projections called **microvilli** extend from the free surface of epithelial cells. These projections increase the surface area across which substances are absorbed. In other tissues, such as the upper airways and oviducts, the free surface is ciliated. Action of the cilia helps move the mucus secreted by the epithelium

Glandular Epithelium

Only epithelial tissue contains gland cells. These cells produce and secrete substances that function outside

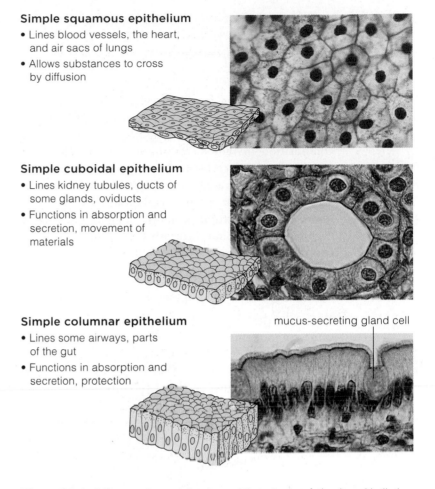

Simple squamous epithelium
• Lines blood vessels, the heart, and air sacs of lungs
• Allows substances to cross by diffusion

Simple cuboidal epithelium
• Lines kidney tubules, ducts of some glands, oviducts
• Functions in absorption and secretion, movement of materials

Simple columnar epithelium
• Lines some airways, parts of the gut
• Functions in absorption and secretion, protection

mucus-secreting gland cell

Figure 32.4 Micrographs and drawings of three types of simple epithelia in vertebrates, with examples of their functions and locations.

the cell. In most animals, secretory cells are clustered inside **glands**, organs that release substances onto the skin, or into a body cavity or the interstitial fluid.

Exocrine glands have ducts or tubes that deliver their secretions onto an internal or external surface. Exocrine secretions include mucus, saliva, tears, milk, digestive enzymes, and earwax.

Endocrine glands have no ducts. They secrete their products, hormones, directly into the interstitial fluid between cells. Hormone molecules diffuse into blood, which carries them to target cells.

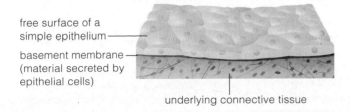

free surface of a simple epithelium

basement membrane (material secreted by epithelial cells)

underlying connective tissue

Figure 32.3 Generalized structure of a simple epithelium.

Take-Home Message

What are epithelial tissues?

■ Epithelial tissues are sheetlike layers of cells attached by a basement layer to an underlying tissue. They cover body surfaces and line cavities and ducts.

■ Some epithelial cells are ciliated or have microvilli that aid absorption.

■ Secretory epithelium forms endocrine and exocrine glands.

■ Connective tissues connect body parts and provide structural and functional support to other body tissues.

■ Links to Lipids 3.4, Hemoglobin 3.6, Storage of excess sugars as fats 8.7

Connective tissues consist of cells in an extracellular matrix of their own secretions. Connective tissues are classified by the cell types that they include and the composition of their extracellular matrix. There are two kinds of soft connective tissues: loose and dense. In both, fibroblasts are the main type of cell. Fibroblasts secrete a matrix of complex carbohydrates with long fibers of the structural proteins collagen and elastin. Cartilage, bone tissue, adipose tissue, and blood are specialized connective tissues.

Soft Connective Tissues

Loose and dense connective tissues are made up of the same components but in different proportions. Loose connective tissue has fibroblasts and fibers dispersed widely through its matrix. Figure 32.5a is an example. This tissue, the most common type in the vertebrate body, helps hold organs and epithelia in place.

In dense, irregular connective tissue, the matrix is packed full of fibroblasts and collagen fibers that are oriented every which way, as in Figure 32.5b. Dense, irregular connective tissue makes up deep skin layers. It supports intestinal muscles and also forms capsules around organs that do not stretch, such as kidneys.

Dense, regular connective tissue has fibroblasts in orderly rows between parallel, tightly packed bundles of fibers (Figure 32.5c). This organization helps keep the tissue from being torn apart when placed under mechanical stress. Tendons and ligaments are mainly dense, regular connective tissue. The tendons connect skeletal muscle to bones. Ligaments attach one bone to another and are stretchier than tendons. Elastic fibers in their matrix facilitate movements around joints.

Specialized Connective Tissues

All vertebrate skeletons include **cartilage**, which has a matrix of collagen fibers and rubbery glycoproteins. Cartilage cells (chondrocytes) secrete the matrix, which eventually imprisons them (Figure 32.5d). When you were an embryo, cartilage formed a model for your developing skeleton; then bone replaced most of it. Cartilage still supports the outer ears, nose, and throat. It cushions joints and acts as a shock absorber between vertebrae. Blood vessels do not extend through cartilage, so nutrients and oxygen must diffuse from vessels in nearby tissues. Also, unlike cells of other connective tissues, cartilage cells do not divide often in adults.

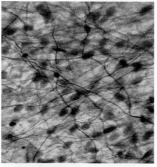
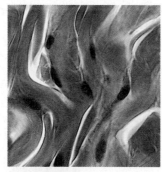
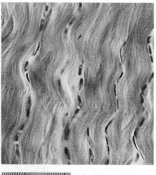
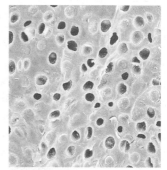

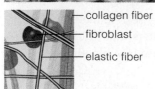
— collagen fiber
— fibroblast
— elastic fiber

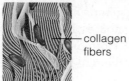
— collagen fibers

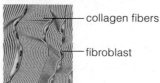
— collagen fibers
— fibroblast

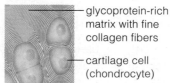
— glycoprotein-rich matrix with fine collagen fibers
— cartilage cell (chondrocyte)

a Loose connective tissue
• Underlies most epithelia
• Provides elastic support and serves as a fluid reservoir

b Dense, irregular connective tissue
• In deep skin layers, around intestine, and in kidney capsule
• Binds parts together, provides support and protection

c Dense, regular connective tissue
• In tendons connecting muscle to bone and ligaments that attach bone to bone
• Provides stretchable attachment between body parts

d Cartilage
• Internal framework of nose, ears, airways; covers the ends of bones
• Supports soft tissues, cushions bone ends at joints, provides a low-friction surface for joint movements

Figure 32.5 Micrographs and drawings of connective tissues.

Adipose tissue is the body's main energy reservoir. Most cells can convert excess sugars and lipids into fats (Section 8.7). However, only the cells of adipose tissue bulge with so much stored fat that the nucleus gets pushed to one side and flattened (Figure 32.5e). Adipose cells have little matrix between them. Small blood vessels run through the tissue and carry fats to and from cells. In addition to its energy-storage role, adipose tissue cushions and protects body parts, and a layer under the skin functions as insulation.

Bone tissue is a connective tissue in which living cells (osteocytes) are imprisoned in a calcium-hardened matrix that they secreted (Figure 32.5f). Bone tissue is the main component of bones, organs that interact with muscles to move a body. Bones also support and protect internal organs. Figure 32.6 shows a femur, a leg bone that is structurally adapted to bear weight. Blood cells form in the spongy interior of some bones.

Blood is considered a connective tissue because its cells and platelets are descended from stem cells in bone (Figure 32.7). Red blood cells filled with hemoglobin transport oxygen (Section 3.6). White blood cells help defend the body against dangerous pathogens. Platelets are cell fragments that function in clot formation. Cells and platelets drift in plasma, a fluid extracellular matrix consisting mostly of water, with dissolved nutrients and other substances.

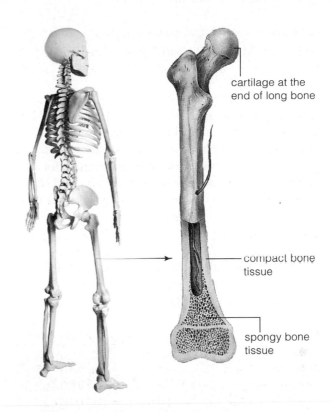

cartilage at the end of long bone

compact bone tissue

spongy bone tissue

Figure 32.6 Locations of cartilage and bone tissue. Spongy bone tissue has hard parts with spaces between. Compact bone tissue is more dense. The bone shown here is the femur, the largest and strongest bone in the human body.

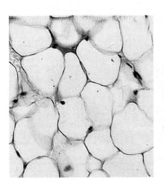

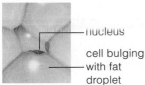

nucleus

cell bulging with fat droplet

e Adipose Tissue

• Underlies skin and occurs around heart and kidneys

• Serves in energy storage, provides insulation, cushions and protects some body parts

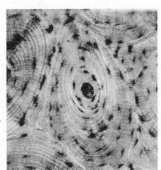

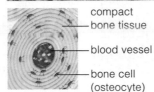

compact bone tissue

blood vessel

bone cell (osteocyte)

f Bone Tissue

• Makes up the bulk of most vertebrate skeletons

• Provides rigid support, attachment site for muscles, protects internal organs, stores minerals, produces blood cells

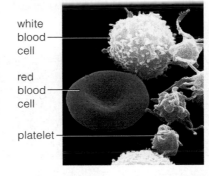

white blood cell

red blood cell

platelet

Figure 32.7 Cellular components of human blood. Cells and cell fragments (platelets) drift along in plasma, the fluid portion of the blood. Plasma consists of water with dissolved proteins, salts, and nutrients.

Take-Home Message

What are connective tissues?

■ Connective tissues consist of cells in a secreted extracellular matrix.

■ Various soft connective tissues underlie epithelia, form capsules around organs, and connect muscle to bones or bones to one another.

■ Cartilage is a specialized connective tissue with a rubbery extracellular matrix.

■ Adipose tissue is a specialized connective tissue with fat-filled cells.

■ Bone is a specialized connective tissue with a calcium-hardened matrix.

■ Blood is considered a connective tissue because blood cells form in bone. The cells are carried along by plasma, the fluid portion of the blood.

- Muscle tissue is made up of cells that can contract.
- Links to Cytoskeletal proteins 4.13, Aerobic respiration 8.1

Cells of muscle tissues contract—or forcefully shorten—in response to signals from nervous tissue. Muscle tissues consist of many cells arranged in parallel with one another, in tight or loose arrays. Coordinated contractions of layers or rings of muscles move the whole body or its parts. Muscle tissue occurs in most animals, but we focus here on the kinds found in vertebrates.

Skeletal Muscle Tissue

Skeletal muscle tissue, the functional partner of bone (or cartilage), helps move and maintain the positions of the body and its parts. Skeletal muscle tissue

has parallel arrays of long, cylindrical muscle fibers (Figure 32.8a). The fibers form during development, when embryonic cells fuse together, so each fiber has multiple nuclei. A fiber is filled with myofibrils—long strands with row after row of contractile units called sarcomeres. The rows of sarcomeres are so regular that skeletal muscle has a striated, or striped, appearance.

Each sarcomere consists of parallel arrays of the proteins actin and myosin (Section 4.13). ATP-powered interactions between the actin and myosin filaments shorten sarcomeres and brings about muscle contraction. We describe this process in detail in Section 36.7.

Skeletal muscle tissue makes up 40 percent or so of the weight of an average human. Reflexes activate it, but we also can cause it to contract when we want to move a body part. That is why skeletal muscles are commonly called "voluntary" muscles.

Cardiac Muscle Tissue

Cardiac muscle tissue occurs only in the heart wall (Figure 32.8b). Like skeletal muscle tissue, it contains sarcomeres and looks striated. Unlike skeletal muscle tissue, it consists of branching cells. Cardiac muscle cells are attached at their ends by adhering junctions that prevent them from being ripped apart during forceful contractions. Signals to contract pass swiftly from cell to cell at gap junctions that connect cells along their length. Rapid flow of signals ensures that all cells in cardiac muscle tissue contract as a unit.

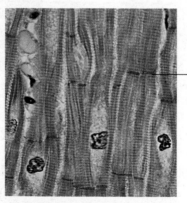
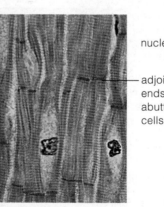
nucleus

adjoining ends of abutting cells

nucleus

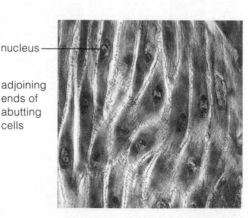

a Skeletal muscle

- Long, multinucleated, cylindrical cells with conspicuous striping (striations)
- Interacts with bone to bring about movement, maintain posture
- Reflex activated, but also under voluntary control

b Cardiac muscle

- Striated cells attached end to end, each with a single nucleus
- Occurs only in the heart wall
- Contraction is not under voluntary control

c Smooth muscle

- Cells with a single nucleus, tapered ends, and no striations
- Found in the walls of arteries, the digestive tract, the reproductive tract, the bladder, and other organs
- Contraction is not under voluntary control

Figure 32.8 Micrographs of muscle tissues, and a photo of skeletal muscles in action.

Compared to other muscle tissues, cardiac muscle has far more mitochondria, which provide the beating heart with a dependable supply of ATP from aerobic respiration. Unlike skeletal muscle, cardiac muscle has little stored glycogen. If blood flow to cardiac cells is interrupted, cells run out of glucose and oxygen fast, so aerobic respiration slows. A heart attack interrupts blood flow, and cardiac muscle dies as a result.

Cardiac muscle and smooth muscle tissue occur in "involuntary" muscle, so named because most people cannot make it contract just by thinking about it.

Smooth Muscle Tissue

We find layers of **smooth muscle tissue** in the wall of many soft internal organs, such as the stomach, uterus, and bladder. This tissue's unbranched cells contain a nucleus at their center and are tapered at both ends (Figure 32.8c). Contractile units are not arranged in an orderly repeating fashion, as they are in skeletal and cardiac muscle tissue, so smooth muscle tissue is not striated. Even so, cells of this tissue contain actin and myosin filaments, which are anchored to the plasma membrane by intermediate filaments.

Smooth muscle tissue contracts more slowly than skeletal muscle, but its contractions can be sustained longer. Smooth muscle contractions propel material through the gut, shrink the diameter of blood vessels, and close sphincters. A sphincter is a ring of muscle in a tubular organ.

32.5 | Nervous Tissue

■ Nervous tissue detects internal and external stimuli, and coordinates responses to these stimuli.

■ Link to Sensing and responding 1.2

Nervous tissue consists of specialized signaling cells called **neurons**, and the cells that support them. A neuron has a cell body that holds its nucleus and other organelles (Figure 32.9). Projecting from the cell body are long cytoplasmic extensions that allow the cell to receive and send electrochemical signals.

When a neuron receives sufficient stimulation, an electrical signal travels along its plasma membrane to the ends of certain of its cytoplasmic extensions. Here, the electrical signal causes release of chemical signaling molecules. These molecules diffuse across a small gap to an adjacent neuron, muscle fiber, or gland cell, and alter that cell's behavior.

Your nervous system has more than 100 billion neurons. There are three types. Sensory neurons are excited by specific stimuli, such as light or pressure. Interneurons receive and integrate sensory information. They store information and coordinate responses to stimuli. In vertebrates, interneurons occur mainly in the brain and spinal cord. Motor neurons relay commands from the brain and spinal cord to glands and to muscle cells, as in Figure 32.10.

Neuroglial cells keep neurons positioned where they should be and provide metabolic support. They constitute a significant portion of the nervous tissue. More than half of your brain volume is neuroglia.

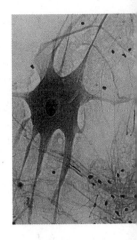

Figure 32.9 Micrograph of a motor neuron. It has a cell body with a nucleus (visible as a dark spot), and long cytoplasmic extensions.

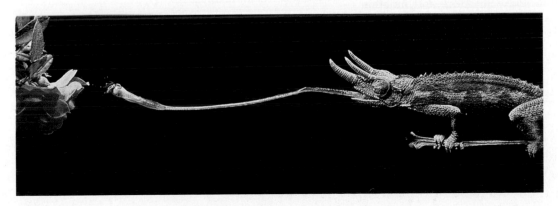

Figure 32.10 One example of the coordinated interaction between skeletal muscle tissue and nervous tissue.

Interneurons in the brain of this lizard, a chameleon, calculate the distance and the direction of a tasty fly.

In response to this stimulus, signals from the interneurons flow along certain motor neurons and reach muscle fibers inside the lizard's long, coiled-up tongue. The tongue uncoils swiftly and precisely to reach the very spot where the fly is perched.

Overview of Major Organ Systems

- Interacting tissues form organs and organ systems.
- Links to Animal body plans 25.1, Trends in vertebrate evolution 26.2

Development of Tissues and Organs

How do tissues of a vertebrate body develop? After fertilization, mitotic cell divisions form a ball of cells that arrange themselves as three **germ layers**, or primary tissue layers (Figure 25.2). Growth and differentiation of these germ layers yields all adult tissues. **Ectoderm**, the outermost germ layer, becomes the nervous tissue and the epithelium of skin. **Mesoderm**, the middle germ layer, gives rise to muscle, connective tissue, and the lining of body cavities derived from the coelom. The innermost germ layer, **endoderm**, forms epithelium of the gut and also organs—such as lungs—that evolved from outpocketings of the gut.

As noted in the chapter introduction, **stem cells** are self-renewing cells; some of their descendants are stem cells, while others differentiate to form specific tissues. An embryonic stem cell that develops before the germ layers form can give rise to any adult tissue. Stem cells of later embryos or after birth are more specialized; each gives rise to only specific tissue types.

Vertebrate Organ Systems

Like other vertebrates, humans are bilateral and have a lined body cavity known as a coelom (Section 25.1). A sheet of smooth muscle, the diaphragm, divides the coelom into an upper thoracic cavity and a cavity that has abdominal and pelvic regions (Figure 32.11a). The heart and lungs are in the thoracic cavity. The stomach, intestines, and liver lie inside the abdominal cavity. The bladder and reproductive organs are in the pelvic cavity. A cranial cavity in the head and spinal cavity in the back are not derived from the coelom.

Figure 32.12 introduces organ systems that divide up the necessary tasks that ensure survival and reproduction of a vertebrate body. Structure and function of these systems is the topic of the remaining chapters in this unit. Figure 32.11b,c introduces some anatomical terms we will use in these discussions.

Take-Home Message

How do vertebrate organ systems arise and function?

- In vertebrates, organs arise from three embryonic germ layers: ectoderm, mesoderm, and endoderm.
- All vertebrates have a set of organ systems that compartmentalize the many specialized tasks required for survival and reproduction of a body.

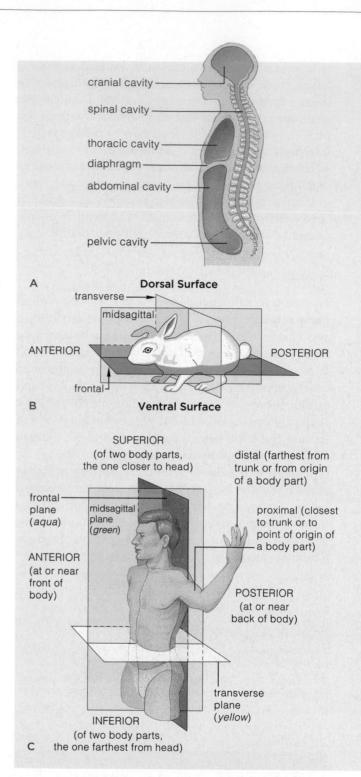

Figure 32.11 Animated (**a**) Main body cavities in humans. (**b,c**) Directional terms and planes of symmetry for the body. For vertebrates that keep their main body axis parallel with Earth's surface, *dorsal* refers to the upper surface (back) and *ventral* to the lower surface. For upright walkers, *anterior* (the front) corresponds to ventral and *posterior* (the back) to dorsal.

Figure 32.12 Animated *Facing page,* human organ systems and their functions.

Integumentary System

Protects body from injury, dehydration, and some pathogens; controls its temperature; excretes certain wastes; receives some external stimuli.

Nervous System

Detects external and internal stimuli; controls and coordinates responses to stimuli; integrates all organ system activities.

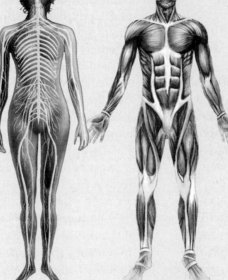

Muscular System

Moves body and its internal parts; maintains posture; generates heat by increases in metabolic activity.

Skeletal System

Supports and protects body parts; provides muscle attachment sites; produces red blood cells; stores calcium, phosphorus.

Circulatory System

Rapidly transports many materials to and from interstitial fluid and cells; helps stabilize internal pH and temperature.

Endocrine System

Hormonally controls body functioning; with nervous system integrates short- and long-term activities. (Male testes added.)

Lymphatic System

Collects and returns some tissue fluid to the bloodstream; defends the body against infection and tissue damage.

Respiratory System

Rapidly delivers oxygen to the tissue fluid that bathes all living cells; removes carbon dioxide wastes of cells; helps regulate pH.

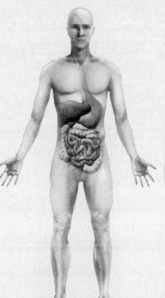

Digestive System

Ingests food and water; mechanically, chemically breaks down food and absorbs small molecules into internal environment; eliminates food residues.

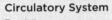

Urinary System

Maintains the volume and composition of internal environment; excretes excess fluid and bloodborne wastes.

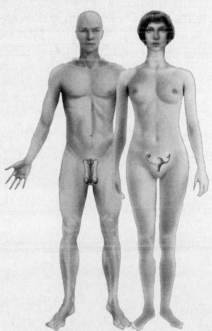

Reproductive System

Female: Produces eggs; after fertilization, affords a protected, nutritive environment for the development of new individuals. *Male:* Produces and transfers sperm to the female. Hormones of both systems also influence other organ systems.

Vertebrate Skin—Example of an Organ System

- Skin is the body's interface with the environment.
- Links to Cancer 9.5, UV radiation and mutations 14.5

Structure and Function of Skin

The integumentary system, or skin, is the vertebrate organ system with the largest surface area. It includes sensory receptors that detect changes in external conditions. Skin forms a barrier that helps defend a body against pathogens. It helps control internal temperature and, in land vertebrates, it helps conserve water. In humans, it helps make vitamin D.

Skin consists of two layers, an outer **epidermis** and a deeper **dermis** (Figure 32.13). Underlying the dermis is a layer of connective tissue called the hypodermis.

The dermis consists of dense connective tissue with stretch-resistant collagen fibers. Blood vessels, lymph vessels, and sensory neurons run through the dermis. Nutrients delivered to the dermis by blood vessels diffuse up to cells in the epidermis. There are no blood vessels in this upper layer.

The epidermis is stratified squamous epithelium. Its structure varies among vertebrate groups. Evolution of a thick layer of keratinocytes—cells that make the waterproof protein keratin—accompanied the move onto land. Ongoing mitotic divisions in the deepest epidermal layers push newly formed keratinocytes toward the skin's surface. As cells move toward the surface, they become flattened, lose their nucleus, and die. Dead cells at the skin surface form an abrasion-resistant layer that helps prevent water loss. Surface cells are continually abraded or flake off.

The epidermis is the body's first line of defense against pathogens. Phagocytic dendritic cells prowl through it. These white blood cells engulf pathogens and alert the immune system to these threats.

As vertebrate lineages evolved, some keratinocytes became specialized and keratin-rich structures such as claws, nails, and beaks evolved. Hair and fur of mammals consist of dead keratinocytes. Hair follicles lie in the dermis, but are of epidermal origin. An average human scalp has about 100,000 hairs. Genes, nutrition, and hormones all affect hair growth.

Epidermally derived gland cells also lie in the dermis. In humans, these include about 2.5 million sweat glands. Sweat glands help humans and many other mammals dissipate heat. Sweat is mostly water, with dissolved salts. Most regions of the mammalian dermis also have oil glands (sebaceous glands). The oily secretions lubricate and soften hair and skin, and deter bacterial growth.

Amphibians do not have sweat glands, but most have mucous glands that help keep their surface moist. Many also have glands that secrete distasteful substances or poisons. Pigmented cells in the dermis give some highly poisonous frogs a distinctive coloration that predators learn to avoid (Figure 32.14).

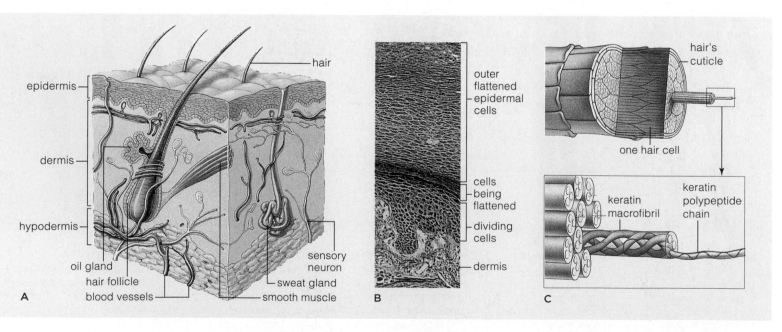

Figure 32.13 Animated (a) Skin structure. (b) Section through human skin. (c) Structure of a hair. It arises from a hair follicle derived from epidermal cells that have sunk into the dermis.

Figure It Out: How many polypeptide chains are in a keratin macrofibril? *Answer: Three*

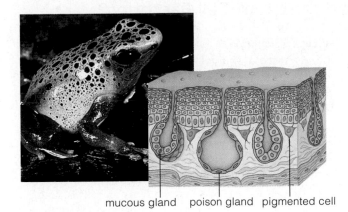

Figure 32.14 Skin of a frog (*Dendrobates azureus*). The dermis contains epidermally derived glands that secrete mucus and poison. Pigment cells in the dermis give the frog its distinctive color and warn predators that it is poisonous.

Sunlight and Human Skin

As the Chapter 11 introduction explained, skin color variation has a genetic basis. Color variations arise from differences in the distribution and activity of melanocytes. These cells make the brownish pigment called **melanin** and donate it to keratinocytes. In pale skin, little melanin is formed. Such skin appears pink because the red color of the iron in hemoglobin shows through thin-walled blood vessels and the epidermis.

Melanin has a protective function. It absorbs ultraviolet (UV) radiation that might otherwise damage underlying skin layers. Exposure to sunlight causes increased production of melanin, producing a "tan."

A bit of UV radiation is a good thing; it stimulates melanocytes to make a molecule that the body later converts to vitamin D. We need this vitamin to absorb calcium ions from food. However, excessive UV exposure damages collagen and causes elastin fibers to clump. Chronically tanned skin gets less resilient and becomes leathery. UV also damages DNA, increasing the risk of skin cancer (Section 9.5).

As we age, epidermal cells divide less often. Skin thins and becomes less elastic as collagen and elastin fibers become sparse. Glandular secretions that kept it soft and moist dwindle. Wrinkles deepen. Many people needlessly accelerate the aging process by tanning or smoking, which shrinks the skin's blood supply.

Take-Home Message

What are the properties of vertebrate skin?

■ Vertebrate skin consists of all four tissue types arranged in two layers, an outer epidermis and a deeper dermis.

■ Skin's keratinized, melanin-containing cells provide a waterproof barrier that protects internal body cells.

| 32.8 | Farming Skin |

■ Commercially grown skin substitutes are already in use for treatment of chronic wounds.

■ Skin may be a source of stem cells that could be used to grow other organs.

Adults make few new muscle cells or nerve cells, but they do constantly renew their skin cells. Each day you lose skin cells, and new ones move up to replace them. The whole epidermis is renewed every month, and an adult sheds about 0.7 kilogram (1.5 pounds) of skin each year.

Skin cells are already being cultured for medical uses (Figure 32.15). Commercially available cultured skin substitutes are made using infant foreskins that were removed during routine circumcisions. The foreskins (a tissue that covers the tip of the penis) provide a rich source of keratinocytes and fibroblasts. These cells are grown in culture with other biological materials, and the resulting products are used to close chronic wounds, help burns heal, and cover sores on patients with epidermolysis bullosa.

Epidermolysis bullosa (EB) is a rare inherited disorder caused by mutations in the structural proteins of skin, such as keratin, collagen, or laminin. The protein defect causes skin layers to separate easily, so upper layers blister and slough off. Affected people are covered with open sores and must avoid touch. Even the friction of clothing on their skin can open a wound. Use of cultured skin substitutes cannot cure EB, but it does help wounds heal faster, thus reducing pain and the risk of life-threatening infections.

Unlike real skin, cultured skin substitutes do not include melanocytes, sweat glands, oil glands, and other differentiated structures. Use of adult epidermal stem cells may one day allow production of cultured skin as complex as real skin. Stem cells, recall, divide and produce more stem cells, as well as specialized cells that make up specific tissues.

As noted in the chapter introduction, researchers also have more ambitious hopes for epidermal cells. If these cells could be genetically engineered, and their differentiation controlled, they might provide starting material to replace other types of tissues, without the controversy raised by use of embryonic stem cells.

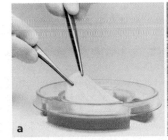

Figure 32.15 (a) A commercially available cultured skin substitute called Apligraf. It has a two-layered structure, with living keratinocytes on top, and fibroblasts below. (b) When placed over a wound, as shown here, the cultured skin cells can help prevent infection while encouraging faster healing.

In vitro fertilization—uniting egg and sperm outside the body—is a common practice in fertility clinics. It produces a cell cluster smaller than a grain of sand. The cluster is implanted in a woman's uterus or frozen for later use. An estimated 500,000 such "embryos" are now frozen and many will never be implanted in their mother. They are a potential source of stem cells, or a potential child—if a woman is willing to carry them to term.

How would you vote?

Should embryos unwanted by parents and stored in fertility clinics be used as a source of stem cells for research? See CengageNOW for details, then vote online.

Summary

Section 32.1 Animal cells are organized as **tissues**, aggregations of cells and intercellular substances that interact in specific tasks. Animal tissues have a variety of cell junctions. **Tight junctions** stop fluid from leaking across an epithelium. **Adhering junctions** hold neighboring cells together. **Gap junctions** are open channels that connect the cytoplasm of abutting cells and permit rapid transfer of ions and small molecules between them.

Tissues are organized into **organs**, which interact as components of **organ systems**. Together, all body parts maintain **homeostasis**—they keep conditions in the internal environment stable and suitable for life.

■ *Use the animation on CengageNOW to compare the structure and function of the main animal cell junctions.*

Section 32.2 **Epithelial tissues** cover the body surface and line its internal spaces. They have one free surface exposed to a body fluid or the environment. A secreted **basement membrane** connects the epithelium to underlying tissue. **Microvilli** increase the free surface area of epithelia that absorb substances. Epithelia may also be ciliated or secretory. Gland cells and secretory **glands** are derived from epithelium. **Endocrine glands** secrete hormones into blood. **Exocrine glands** secrete products such as sweat or digestive enzymes through ducts.

Section 32.3 **Connective tissues** "connect" tissues to one another, both functionally and structurally. Different types bind, organize, support, strengthen, protect, and insulate other tissues. All contain cells scattered in a secreted matrix. Soft connective tissue underlies skin, holds internal organs in place, and connects muscle to bone, or bones to one another. The different types of soft connective tissues all have the same components (fibroblasts and a matrix with elastin and collagen fibers) but in different proportions. Rubbery **cartilage**, calcium-hardened **bone tissue**, lipid-storing **adipose tissue**, and **blood** are specialized connective tissues.

Section 32.4 Muscle tissues contract and move a body or its parts. Muscle contraction is a response to signals from the nervous system and it requires ATP energy. The three types of muscle are **skeletal muscle, cardiac muscle**, and **smooth muscle** tissue. Only skeletal muscle and cardiac muscle tissues appear striated. Only skeletal muscle is under voluntary control.

Skeletal muscle is the functional partner of bones and consists of long cells with many nuclei. Cardiac muscle occurs only in the heart wall. Its cells are joined together end to end. Smooth muscle occurs in walls of hollow and tubular organs such as blood vessels and the bladder.

Section 32.5 **Nervous tissue** makes up the communication lines that extend through the body. **Neurons** are cells that can become excited and relay messages along their plasma membrane. Sensory neurons detect stimuli. Interneurons integrate information and call for responses. Motor neurons deliver commands to muscles and glands that carry out responses. Nervous tissue also contains a diverse collection of **neuroglial cells**. Neuroglia protect and support the neurons.

Section 32.6 An organ system consists of two or more organs that interact chemically, physically, or both in tasks that help keep individual cells as well as the whole body functioning. Most vertebrate organ systems contribute to homeostasis; they help maintain conditions in the internal environment within tolerable limits and so benefit individual cells and the body as a whole.

All tissues and organs of an adult animal arise from three primary tissue layers, or **germ layers**, that form in early embryos: **ectoderm, mesoderm**, and **endoderm**. Cells in all tissues are derived from **stem cells**. Stem cells in early embryos—before germ layers form—can become any tissue. Stem cells in later stages are more specialized and produce only a limited number of tissues.

■ *Use the animation on CengageNOW to investigate the function of vertebrate organ systems and learn terms that describe their locations.*

Sections 32.7, 32.8 The skin is an organ system that functions in protection, temperature control, detection of shifts in external conditions, vitamin production, and defense. It has two-layers, the outer **epidermis** and the deeper **dermis**. Hair, fur, and nails are rich in keratin and derived from epidermal cells. A brownish pigment called **melanin** protects the skin from ultraviolet radiation that can damage DNA.

Skin is continually renewed. Some kinds of skin cells are already being cultured for medical uses.

■ *Use the animation on CengageNOW to explore the structure of human skin.*

Data Analysis Exercise

Diabetes is a disorder in which the blood sugar level is not properly controlled. Among other effects, this disorder reduces blood flow to the lower legs and feet. As a result, about 3 million diabetes patients have ulcers, or open wounds that do not heal, on their feet. Each year, about 80,000 require amputations.

Several companies provide cultured cell products designed to promote the healing of diabetic foot ulcers. Figure 32.16 shows the results of a clinical experiment that tested the effect of the cultured skin product shown in Figure 32.15 versus standard treatment for diabetic foot wounds. Patients were randomly assigned to either the experimental treatment group or the control group and their progress was monitored for 12 weeks.

1. What percentage of wounds had healed at 8 weeks when treated the standard way? When treated with cultured skin?

2. What percentage of wounds had healed at 12 weeks when treated the standard way? When treated with cultured skin?

3. How early was the healing difference between the control and treatment groups obvious?

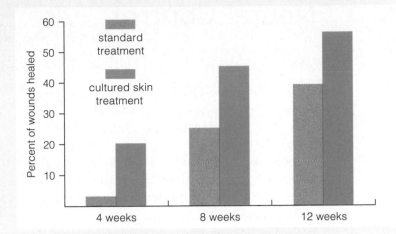

Figure 32.16 Results of a multicenter study of the effects of standard treatment versus use of a cultured cell product for diabetic foot ulcers. Bars show the percentage of foot ulcers that had completely healed.

Self-Quiz   *Answers in Appendix III*

1. _____ tissues are sheetlike with one free surface.

2. _____ function in cell-to-cell communication.
   a. Tight junctions   c. Gap junctions
   b. Adhering junctions   d. all of the above

3. In most animals, glands are formed of _____ tissue.
   a. epithelial   c. muscle
   b. connective   d. nervous

4. A sweat gland is an _____ gland.
   a. endocrine   b. exocrine

5. Most _____ have many collagen and elastin fibers.
   a. epithelial tissues   c. muscle tissues
   b. connective tissues   d. nervous tissues

6. What is the fluid portion of the blood called?

7. Your body converts excess carbohydrates and proteins to fats. _____ specializes in storing the fats.
   a. Epithelial tissue   c. Adipose tissue
   b. Dense connective tissue   d. both b and c

8. Only cells of _____ can shorten (contract).
   a. epithelial tissue   c. muscle tissue
   b. connective tissue   d. nervous tissue

9. _____ detects and integrates information about changes and controls responses to those changes.
   a. Epithelial tissue   c. Muscle tissue
   b. Connective tissue   d. Nervous tissue

10. Which type of muscle can be voluntarily controlled?

11. Which type of neuron delivers signals to muscles?

12. Exposure to sunlight causes increased production of _____ , which shields against harmful UV radiation.
   a. melanin   c. keratin
   b. hemoglobin   d. collagen

13. The main cell type in the epidermis is _____ .
   a. neuroglia   c. keratinocytes
   b. motor neurons   d. osteocytes

14. Match the terms with the most suitable description.
   ___ exocrine gland   a. strong, pliable; like rubber
   ___ endocrine gland   b. secretion through duct
   ___ endoderm   c. outermost primary tissue
   ___ ectoderm   d. contracts, not striated
   ___ cartilage   e. innermost primary tissue
   ___ smooth muscle   f. muscle of the heart wall
   ___ cardiac muscle   g. cements cells together
   ___ blood   h. fluid connective tissue
   ___ adhering   i. ductless secretion
        junction

■ *Visit CengageNOW for additional questions.*

Critical Thinking

1. Many people oppose the use of animals for testing the safety of cosmetics. They say alternative test methods are available, such as the use of lab-grown tissues in some cases. Given what you learned in this chapter, speculate on the advantages and disadvantages of tests that use specific lab-grown tissues as opposed to living animals.

2. Porphyria is a name for a set of rare genetic disorders. Affected people lack one of the enzymes in the metabolic pathway that forms heme, the iron-containing group of hemoglobin. As a result, intermediates of heme synthesis (porphyrins) accumulate. When porphyrins are exposed to sunlight, they absorb energy and release energized electrons. Electrons careening around the cell can break bonds and cause damaging free radicals to form. In the most extreme cases, gums and lips can recede, which makes some front teeth—the canines—look more fanglike.

Affected individuals must avoid sunlight, and garlic can exacerbate their symptoms. By one hypothesis, people who were affected by the most extreme forms of porphyria may have been the source for vampire stories. Would you consider this hypothesis plausible? What other kinds of historical data might support or disprove it?

Ecstasy, an illegal drug, can make you feel socially accepted, less anxious, and more aware of your surroundings and of sensory stimuli. It also can leave you dying in a hospital, foaming at the mouth and bleeding from all orifices as your temperature skyrockets. It can send your family and friends spiraling into horror and disbelief as they watch you stop breathing. Lorna Spinks ended life that way when she was nineteen years old (Figure 33.1).

Her anguished parents released these photographs because they wanted others to know what their daughter did not: Ecstasy can kill.

Ecstasy is a psychoactive drug; it alters brain function. The active ingredient, MDMA (3,4-methylenedioxymethamphet-amine), is a type of amphetamine, or "speed." As one effect, it makes neurons release an excess of the signaling molecule serotonin. The serotonin saturates receptors on target cells and cannot be cleared away, so cells cannot be released from overstimulation.

The abundance of serotonin promotes feelings of energy, empathy, and euphoria. But the unrelenting stimulation calls for rapid breathing, dilated eyes, restricted urine formation,

and a racing heart. Blood pressure soars, and the body's internal temperature can rise out of control. Spinks became dizzy, flushed, and incoherent after taking just two Ecstasy tablets. She died because her increased temperature caused her organ systems to shut down.

Few Ecstasy overdoses end in death. Panic attacks and fleeting psychosis are more common short-term effects. We do not know much about the drug's long-term effects; users are unwitting guinea pigs for unscripted experiments.

We know that Ecstasy use depletes the brain's store of serotonin and that this shortage can last for some time. In animals, multiple doses of MDMA alter the structure and number of serotonin-secreting neurons. This is a matter of concern because, low serotonin levels in humans are associated with inability to concentrate, memory loss, and depression.

Human MDMA users do have memory loss, and the more often a person uses the drug, the worse their memory gets. Fortunately, at least over the short term, capacity for memory seems to be restored when Ecstasy use stops. However, undoing the neural imbalances often takes many months.

Think about it. The nervous system evolved as a way to sense and respond fast to changing conditions inside and outside the body. Vision and taste, hunger and passion, fear and rage—awareness of stimulation starts with a flow of information along communication lines of the nervous system. Even before you were born, excitable cells called neurons started organizing in newly forming tissues and chattering among themselves. All through your life, in moments of danger or reflection, excitement or sleep, their chattering has continued and will continue for as long as you do.

Each of us possesses a complex nervous system, a legacy of millions of years of evolution. Its architecture and its functions give us an unparalleled capacity for learning and sharing experiences with others. Perhaps the saddest consequence of drug abuse is the implicit denial of this legacy—the denial of self when we choose not to assess how drugs can harm our brain, or cease to care.

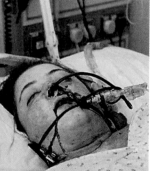

See the video! Figure 33.1 Photos of Lorna Spinks alive (*left*), and minutes after her death (*right*). She died after taking two Ecstasy tablets. If you suspect someone is having a bad reaction to Ecstasy or any other drug, get medical help fast and be honest about the cause of the problem. Immediate, informed medical action may save a life.

Key Concepts

How animal nervous tissue is organized

In radially symmetrical animals, excitable neurons interconnect as a nerve net. Most animals are bilaterally symmetrical with a nervous system that has a concentration of neurons at the anterior end and one or more nerve cords running the length of the body. **Section 33.1**

How neurons work

Messages flow along a neuron's plasma membrane, from input to output zones. Chemicals released at a neuron's output zone may stimulate or inhibit activity in an adjacent cell. Psychoactive drugs interfere with the information flow between cells. **Sections 33.2–33.7**

Vertebrate nervous system

The central nervous system consists of the brain and spinal cord. The peripheral nervous system includes many pairs of nerves that connect the brain and spinal cord to the rest of the body. The spinal cord and peripheral nerves interact in spinal reflexes. **Sections 33.8, 33.9**

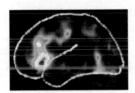

About the brain

The brain develops from the anterior part of the embryonic nerve cord. A human brain includes evolutionarily ancient tissues and newer regions that provide the capacity for analytical thought and language. Neuroglia make up the bulk of the brain. **Sections 33.10–33.13**

Links to Earlier Concepts

- In this chapter, you will find many examples of the cell processes covered in Unit One. Nervous signals involve receptor proteins (5.2) and transport mechanisms (5.3, 5.4, 5.5). They depend on ion gradients, a type of potential energy (6.1).

- You will reconsider trends in animal evolution (25.1, 26.2) and chordate traits (26.1) with emphasis on nervous systems.

- You will also revisit some health applications such as cancer (9.5), alcohol abuse (Chapter 6 introduction), and stem cell research (Chapter 32 introduction).

- You will see examples of PET scans, a technique that uses radioisotopes, as explained in Section 2.2.

■ Interacting neurons give animals a capacity to respond to stimuli in the environment and inside their body.

■ Link to Trends in animal evolution 25.1

Of all multicelled organisms, animals respond fastest to external stimuli. Activities of neurons are the key to these quick responses. A **neuron** is a cell that can relay electrical signals along its plasma membrane and can communicate with other cells by way of specific chemical messages. Cells called **neuroglia** functionally and structurally support neurons in most animals.

A typical animal has three types of neurons. **Sensory neurons** detect internal or external stimuli and signal interneurons or motor neurons. **Interneurons** process information received from sensory neurons or other interneurons, then send signals along to interneurons or motor neurons. **Motor neurons** signal and control muscles and glands.

The Cnidarian Nerve Net

Cnidarians such as the hydras and jellyfishes, are the simplest animals with neurons. These radial, aquatic animals have a nerve net that allows them to respond to food or threats that arrive from all directions (Figure 33.2*a*). A **nerve net** is a mesh of interconnected neurons.

Information can flow in any direction among cells of the nerve net, and there is no centralized, controlling organ that functions like a brain. By causing cells in the body wall to contract, the nerve net can alter the size of the animal's mouth, change the body shape, or shift the position of tentacles.

Bilateral, Cephalized Nervous Systems

Most animals have a bilaterally symmetrical body (Section 25.1). Evolution of bilateral body plans was accompanied by **cephalization**, the concentration of neurons that detect and process information at the body's anterior, or head, end.

Planarians and the other flatworms are the simplest animals with a bilateral, cephalized nervous system. A planarian's head end has a pair of ganglia (Figure 33.2*b*). A **ganglion** (plural, ganglia), is a cluster of neuron cell bodies that functions as an integrating center. A planarian's ganglia receive signals from eye spots and chemical-detecting cells on its head. The ganglia also connect to a pair of nerve cords that run the length of the body. The cords have no ganglia. Nerves cross the body between the cords, giving the nervous system a ladderlike appearance. The cross connections help coordinate activities of the two sides of the body.

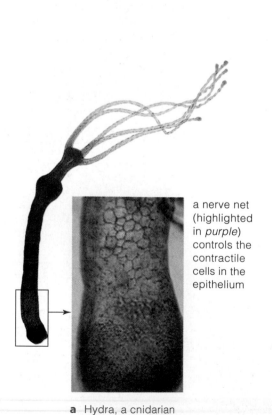

a nerve net (highlighted in *purple*) controls the contractile cells in the epithelium

a Hydra, a cnidarian

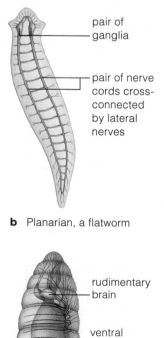

pair of ganglia

pair of nerve cords cross-connected by lateral nerves

b Planarian, a flatworm

rudimentary brain

ventral nerve cord

ganglion

c Earthworm, an annelid

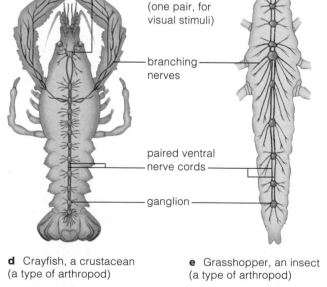

brain

brain

optic lobe (one pair, for visual stimuli)

branching nerves

paired ventral nerve cords

ganglion

d Crayfish, a crustacean (a type of arthropod)

e Grasshopper, an insect (a type of arthropod)

Figure 33.2 (**a**) Hydras and other cnidarians have a nerve net. (**b**) A planarian has a ladderlike nervous system with two nerve cords and a pair of ganglia in the head. (**c,d,e**) Annelids and arthropods have paired ventral nerve cords with ganglia in each segment. The nerve cords connect to a simple brain.

Annelids and arthropods have paired ventral nerve cords that connect to a simple brain (Figure 33.2c–e). In addition, a pair of ganglia in each body segment provides local control over that segment's muscles.

Chordates have a single, dorsal nerve cord (Section 26.1). In vertebrates, the anterior region of this cord evolved into a brain. Bigger brains gave some animals a competitive edge in finding resources and reacting to danger. Also, among vertebrates that moved onto land, certain brain centers became modified and expanded in ways that helped animals better move about and respond to stimuli in their new environment.

The Vertebrate Nervous System

The nervous system of vertebrates has two functional divisions (Figure 33.3). Most interneurons are located in the **central nervous system**—the brain and spinal cord. Nerves that extend through the rest of the body make up the **peripheral nervous system**. These nerves are further classified as autonomic or somatic, based on which organs they are associated with.

Figure 33.4 shows the location of the human brain, spinal cord, and some peripheral nerves. As you will learn, each nerve contains long extensions, or axons, of sensory neurons, motor neurons, or both. Afferent axons carry sensory signals into the central nervous system; efferent axons relay commands for response out of it. For instance, you have a sciatic nerve in each of your legs. These nerves swiftly relay signals from sensory receptors in leg muscles, joints, and skin in toward the spinal cord. At the same time, they relay signals from the spinal cord to leg muscles.

In sections to follow, you will consider the kinds of messages that flow along these communication lines.

Take-Home Message

What are the features of animal nervous systems?

■ Most animals have three types of interacting neurons—sensory neurons, interneurons, and motor neurons.

■ The simplest animals that have neurons are cnidarians. Their neurons are arranged as a nerve net.

■ Most animals are bilaterally symmetrical and have a nervous system with a concentration of nerve cells at their head end.

■ Bilateral invertebrates usually have a pair of ventral nerve cords. In contrast, the chordates have a dorsal nerve cord.

■ Cnidarians do not have a central information-processing organ. Flatworms have a pair of ganglia that serve this function. Other invertebrates have larger and more complex brains.

■ The vertebrate nervous system includes a well-developed brain, a spinal cord, and peripheral nerves.

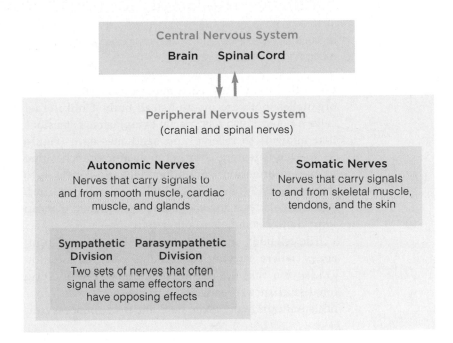

Figure 33.3 Functional divisions of vertebrate nervous systems. The spinal cord and brain are its central portion. The peripheral nervous system includes spinal nerves, cranial nerves, and their branches, which extend through the rest of the body. Peripheral nerves carry signals to and from the central nervous system. Section 33.6 explains the functional divisions of the peripheral system.

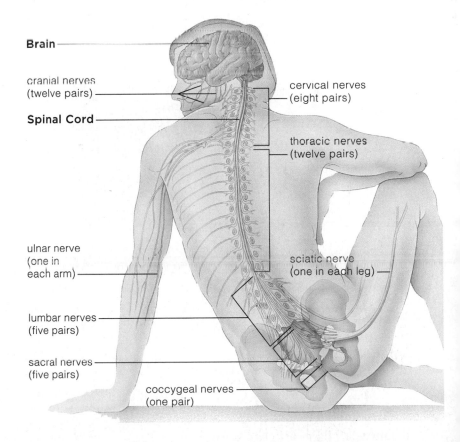

Figure 33.4 Some of the major nerves of the human nervous system.

■ Neurons have cytoplasmic extensions specialized for receiving and sending signals.

Like other body cells, each neuron has a nucleus and organelles; both are inside its cell body. Unlike other cells, a neuron also has special cytoplasmic extensions that allow it to receive and send messages (Figure 33.5). **Dendrites** are short, cytoplasmic branches that receive information from other cells and convey it to the cell body. A neuron usually has several dendrites. A neuron also has an **axon**, a longer extension that can send chemical signals to other cells.

The cell body and dendrites function as signal input zones, where arriving signals alter ion concentration gradients across the plasma membrane. The resulting ion disturbance spreads into a trigger zone, which connects with the axon. From here, the disturbance is con-ducted along the axon to axon terminals. When it reaches these output zones, the disturbance causes release of signaling molecules.

Information usually flows from sensory neurons, to interneurons, to motor neurons (Figure 33.6). The three types of neurons differ somewhat in the type and arrangement of their cytoplasmic extensions. A sensory neuron typically has no dendrites. One end of its axon has receptor endings that can detect a specific stimulus (Figure 33.6a). Axon terminals at the other end send chemical signals, and the cell body lies in between. An interneuron has many signal-receiving dendrites and one axon (Figure 33.6b). In vertebrates, nearly all interneurons reside in the central nervous system and some have many thousands of dendrites. A motor neuron also has multiple dendrites and one axon (Figure 33.6c).

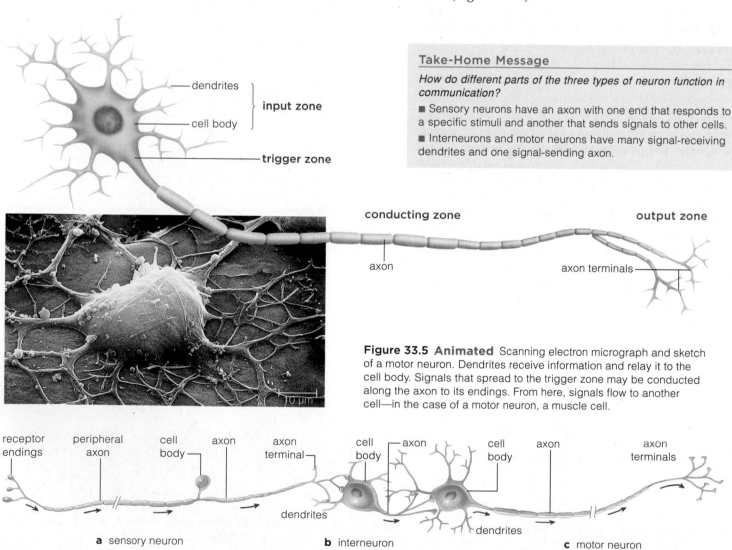

dendrites

input zone

cell body

trigger zone

conducting zone

output zone

axon

axon terminals

Take-Home Message

How do different parts of the three types of neuron function in communication?

■ Sensory neurons have an axon with one end that responds to a specific stimuli and another that sends signals to other cells.

■ Interneurons and motor neurons have many signal-receiving dendrites and one signal-sending axon.

Figure 33.5 Animated Scanning electron micrograph and sketch of a motor neuron. Dendrites receive information and relay it to the cell body. Signals that spread to the trigger zone may be conducted along the axon to its endings. From here, signals flow to another cell—in the case of a motor neuron, a muscle cell.

receptor endings   peripheral axon   cell body   axon   axon terminal   cell body   axon   cell body   axon   axon terminals

dendrites

a sensory neuron

dendrites

b interneuron

dendrites

c motor neuron

Figure 33.6 The three types of neurons. Arrows indicate direction of information flow. (**a**) Sensory neurons detect stimuli and signal other cells. (**b**) Interneurons relay signals between neurons. (**c**) Motor neurons signal effectors—muscle or gland cells.

33.3 | Membrane Potentials

- Properties of the neuron membrane affect ion movement.
- Links to Transport mechanisms 5.3, 5.4, Potential energy 6.1

Resting Potential

All cells have an electric gradient across their plasma membrane. The cytoplasmic fluid near this membrane has more negatively charged ions and proteins than the interstitial fluid outside the cell does. As in a battery, these separated charges have potential energy. We call the voltage difference across a cell membrane a membrane potential and measure it in thousandths of a volt, or millivolts (mV). An unstimulated neuron has a **resting membrane potential** of about –70 mV.

Distributions of three kinds of ions are important in generating the resting potential. First, the cytoplasm of a neuron includes many negatively charged proteins that are not present in the interstitial fluid. Being large and charged, these proteins cannot diffuse across the lipid bilayer of the cell membrane.

The other two important ions are positively charged potassium ions (K^+) and positively charged sodium ions (Na^+). These ions move in and out of the neuron with the assistance of transport proteins (Section 5.3).

Sodium–potassium pumps (Figure 33.7a and Section 5.4) use energy from a molecule of ATP to transport two potassium ions into the cell and three sodium ions out. Since the pump moves more positive charges out of the cell than in, its action increases the charge gradient across the neuron membrane. Action of the pump also contributes to concentration gradients for sodium and potassium across this membrane.

Nearly all sodium pumped out of the neuron stays out—as long as the cell is at rest. In contrast, some potassium ions flow down their concentration gradient (out of the cell) through channel proteins (Figure 33.7b). Leaking of potassium (K^+) outward increases the number of unbalanced negative ions in the cell.

In summary, the cytoplasm of a resting neuron has negatively charged proteins that the interstitial fluid lacks. It also has fewer sodium ions (Na^+) and more potassium ions (K^+). We can show the relative concentrations of the relevant ions this way, with the green ball representing negatively charged proteins:

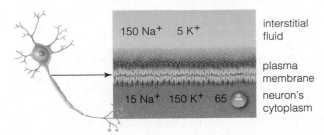

interstitial fluid

plasma membrane

neuron's cytoplasm

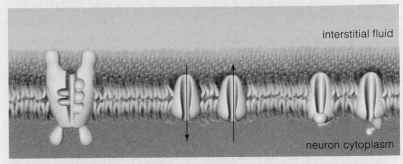

A Sodium–potassium pumps actively transport 3 Na^+ out of a neuron for every 2 K^+ they pump in.

B Passive transporters allow K^+ ions to leak across the plasma membrane, down their concentration gradient.

C In a resting neuron, gates of voltage-sensitive channels are shut (*left*). During action potentials, the gates open (*right*), allowing Na^+ or K^+ to flow through them.

Figure 33.7 Animated Icons for protein channels and pumps that span a neuron's plasma membrane. (**a**) Sodium–potassium pumps (Na^+/K^+ pumps) and (**b**) open potassium (K^+) channels contribute to the resting potential. (**c**) Voltage-gated channels are required for action potentials.

Action Potentials

Neurons and muscle cells are said to be "excitable" because, when properly stimulated, they undergo an **action potential**—an abrupt reversal in the electric gradient across the plasma membrane. Channels with gates that open at a particular voltage, or membrane potential, are essential to action potentials.

Neurons have such voltage-gated channels in the membrane of their trigger zone and conducting zone (Figure 33.7c). Some of these voltage-gated channels let potassium ions diffuse across the membrane through their interior. Others let sodium ions move across. The voltage-gated channels are shut in a neuron at rest, but they swing open during an action potential.

With this bit of background on membrane proteins and ion gradients, you are ready to look at how an action potential arises at a neuron's trigger zone and propagates itself, undiminished, to an output zone.

Take-Home Message

How do gradients across a neuron membrane contribute to neuron function?

- The interior of a resting neuron is more negative than the fluid outside the cell. The presence of negatively charged proteins and activity of transport proteins contribute to this charge difference, or resting membrane potential.
- A resting neuron also has concentration gradients for sodium and potassium across its membrane, with more sodium outside and more potassium inside.
- When properly stimulated, a neuron undergoes an action potential. Voltage-gated channels open and the membrane potential briefly reverses.

■ Movement of sodium and potassium ions through gated channels causes a brief reversal of the membrane potential.

■ Link to Transport mechanisms 5.3, 5.4

Approaching Threshold

A small alteration in the ion concentration gradients across the plasma membrane of a neuron can shift the membrane potential. We call the resulting change a local, graded potential. "Local" means it only spreads out for a millimeter or so. "Graded" means that the change in potential can vary in size. A local potential occurs when ions enter a region of neuron cytoplasm and change the membrane potential in that region. For example, a little sodium entering may shift membrane potential in a region from –70 millivolts to –66 mV.

Stimulation of a neuron's input zone can cause a local, graded potential. If the stimulus is sufficiently intense or long-lasting, ions diffuse from the input zone into the adjacent trigger zone. The membrane here includes sodium channels with voltage-sensitive gates (Figure 33.8a). When the difference in charge across the membrane increases to a specific level, the **threshold potential**, the gated sodium channels in the trigger zone open and start an action potential.

Opening of these voltage-gated channels allows sodium to flow down its electrical and concentration gradients into the neuron (Figure 33.8b). In an example of positive feedback (Section 27.3), gated sodium channels open in an accelerating way after threshold is reached. As sodium starts to flow in, it makes the neuron cytoplasm more positive, so more sodium channels open. Now the stimulus that brought the neuron to threshold becomes unimportant. Sodium rushing into the neuron—not diffusion of ions from the input zone—drives the feedback cycle:

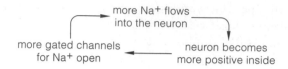

An All-or-Nothing Spike

Researchers can study changes in membrane potentials by inserting one electrode into an axon and another into the fluid just outside of it (Figure 33.9). They connect these electrodes to a device that shows membrane potential. Figure 33.10 shows what a recording looks like before, during, and after an action potential. Once threshold level is reached, membrane potential always rises to the same level as an action potential peak. Thus, an action potential is an all-or-nothing event.

The reversal of charge during an action potential lasts only milliseconds. Above a certain voltage, gates on sodium channels swing shut. About the same time, gates on potassium (K+) channels open (Figure 33.8c). The resulting outflow of positively charged potassium makes cytoplasm once again more negative than the interstitial fluid. Diffusion of ions quickly restores the Na+ and K+ ion gradients to match those set up by action of sodium–potassium pumps (Figure 33.8d).

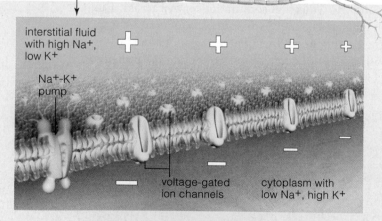

A Close-up of the trigger zone of a neuron. One sodium–potassium pump and some of the voltage-gated ion channels are shown. At this point, the membrane is at rest and the voltage-gated channels are closed. The cytoplasm's charge is negative relative to interstitial fluid.

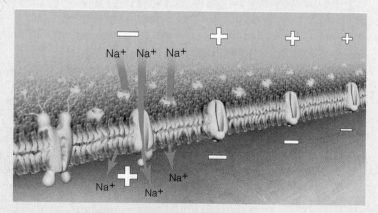

B Arrival of a sufficiently large signal in the trigger zone raises the membrane potential to threshold level. Gated sodium channels open and sodium (Na+) flows down its concentration gradient into the cytoplasm. Sodium inflow reverses the voltage across the membrane.

Figure 33.8 Animated Propagation of an action potential along part of a motor neuron's axon.

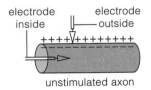

electrode inside | electrode outside

+++++V+++++++

unstimulated axon

Figure 33.9 How membrane potentials can be investigated. Electrodes placed inside and outside an axon allow researchers to measure membrane potential. Figure 33.10 shows the record this method produces when a neuron is stimulated enough to produce an action potential.

A Resting membrane potential is −70 mV.

B Stimulation causes an influx of positive ions and a rise in the membrane potential.

C Once potential exceeds threshold (−60 mV), the sodium (Na+) gates begin to open, and Na+ rushes in. This causes more gates to open, and so on. Voltage shoots up rapidly as a result.

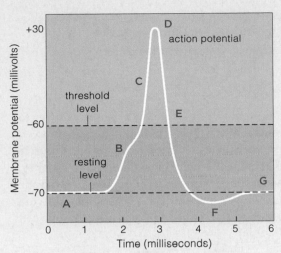

D Every action potential peaks at +33 mV; no more, no less. At this point, Na+ gates have closed and potassium (K+) gates have opened.

E Flow of K+ out of the neuron causes the potential to fall.

F So much K+ exits that potential declines below resting potential.

G Na+–K+ pump action restores resting potential.

Figure 33.10 Animated How membrane potential changes during an action potential.
Figure It Out: How long does the increase in potential last?   *Answer: About 2 milliseconds*

Direction of Propagation

Each action potential is self-propagating. Some of the sodium that enters one region of an axon diffuses into an adjoining region, driving that region to threshold and opening sodium gates. As these gates swing open in one region after the next, the action potential moves toward the axon terminals without weakening.

Once sodium gates close, another action potential cannot occur right away. The brief refractory period limits the maximum speed of signals and causes them to move one way, toward axon terminals. Diffusion of ions from a region undergoing an action potential can only open gated channels that did not already open.

Take-Home Message

What happens during an action potential?

■ An action potential begins in the neuron's trigger zone. A strong stimulus decreases the voltage difference across the membrane. This causes gated sodium channels to open, and the voltage difference reverses.

■ An action potential travels along an axon as consecutive patches of membrane undergo reversals in membrane potential.

■ At each patch of membrane, an action potential ends when potassium ions flow out of the neuron, and voltage difference across the membrane is restored.

■ Action potentials move in one direction, toward axon terminals, because gated sodium channels are briefly inactivated after an action potential.

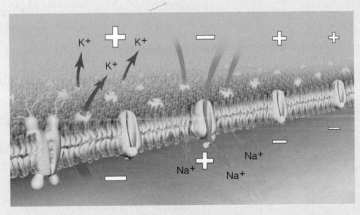

C The charge reversal makes gated Na+ channels shut and gated K+ channels open. The K+ outflow restores the voltage difference across the membrane. The action potential is propagated along the axon as positive charges spreading from one region push the next region to threshold.

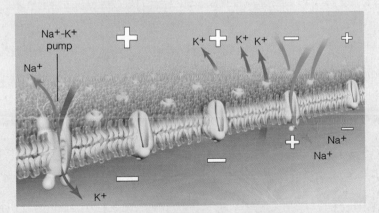

D After an action potential, gated Na+ channels are briefly inactivated, so the action potential moves one way only, toward axon terminals. Na+ and K+ gradients disrupted by action potentials are restored by diffusion of ions that were put into place by activity of sodium–potassium pumps.

33.5 | How Neurons Send Messages to Other Cells

- Action potentials do not pass directly from a neuron to another cell; chemicals carry the signals between cells.
- Links to Receptor proteins 5.2, Exocytosis 5.5

Chemical Synapses

An action potential travels along a neuron's axon to axon terminals at its tips. The region where an axon terminal sends chemical signals to a neuron, a muscle fiber, or a gland cell is called a **synapse**. At a synapse, the signal-sending neuron is called the presynaptic cell. A fluid-filled space about 20 nanometers wide separates it from the input zone of a postsynaptic cell that receives the signal. Figure 33.11 shows a synapse between a motor neuron and a skeletal muscle fiber. Such a synapse is called a **neuromuscular junction**.

Action potentials arrive at a neuromuscular junction by traveling along the axon of a motor neuron to axon terminals (Figure 33.11a,b). Inside the axon terminals are vesicles with molecules of **neurotransmitter**, a type of signaling molecule that relays messages between presynaptic and postsynaptic cells.

Release of the neurotransmitter requires an influx of calcium ions (Ca++). The plasma membrane of an axon terminal has gated channels for these ions. In a resting neuron, these gates are closed and calcium

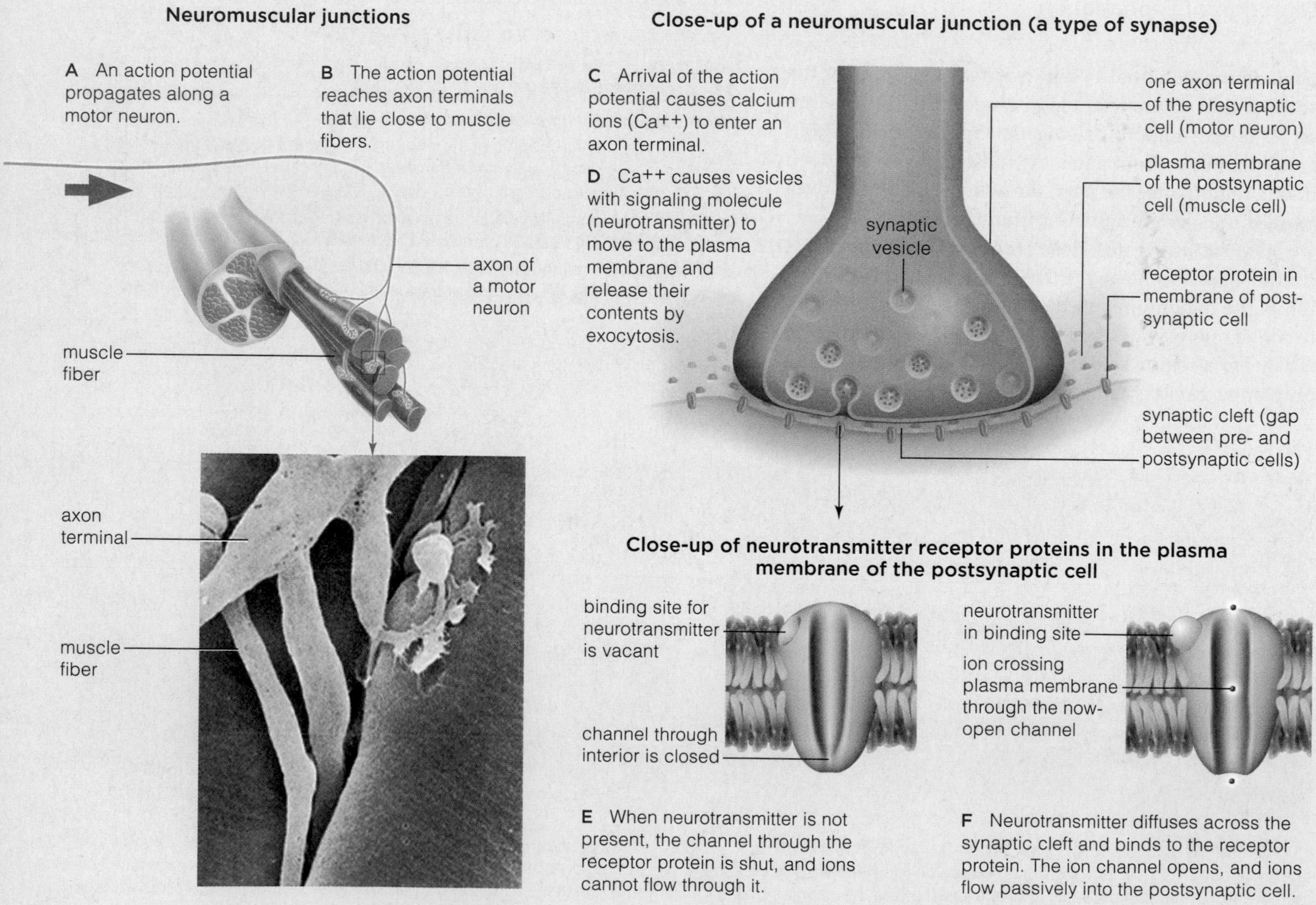

Neuromuscular junctions

A An action potential propagates along a motor neuron.

B The action potential reaches axon terminals that lie close to muscle fibers.

axon of a motor neuron

muscle fiber

axon terminal

muscle fiber

Close-up of a neuromuscular junction (a type of synapse)

C Arrival of the action potential causes calcium ions (Ca++) to enter an axon terminal.

D Ca++ causes vesicles with signaling molecule (neurotransmitter) to move to the plasma membrane and release their contents by exocytosis.

synaptic vesicle

one axon terminal of the presynaptic cell (motor neuron)

plasma membrane of the postsynaptic cell (muscle cell)

receptor protein in membrane of post-synaptic cell

synaptic cleft (gap between pre- and postsynaptic cells)

Close-up of neurotransmitter receptor proteins in the plasma membrane of the postsynaptic cell

binding site for neurotransmitter is vacant

channel through interior is closed

neurotransmitter in binding site

ion crossing plasma membrane through the now-open channel

E When neurotransmitter is not present, the channel through the receptor protein is shut, and ions cannot flow through it.

F Neurotransmitter diffuses across the synaptic cleft and binds to the receptor protein. The ion channel opens, and ions flow passively into the postsynaptic cell.

Figure 33.11 Animated How information is transmitted at a neuromuscular junction, a synapse between a motor neuron and a skeletal muscle fiber. The micrograph shows several such junctions.

pumps actively transport calcium out of the cell. As a result, there are fewer calcium ions in the neuron cytoplasm than in the interstitial fluid. Arrival of an action potential opens gated calcium channels, and calcium flows into the axon terminal. The resulting increase in calcium concentration causes exocytosis; vesicles filled with neurotransmitter move to the plasma membrane and fuse with it. This releases neurotransmitter into the synaptic cleft (Figure 31.11c,d). At a neuromuscular junction, the neurotransmitter released by the motor neuron is acetylcholine (ACh).

The plasma membrane of a postsynaptic cell has receptors that bind neurotransmitter (Figure 31.11e). When ACh binds to receptors in the membrane of a skeletal muscle fiber, channels for sodium ions open (Figure 33.11f). Sodium ions stream passively through these channels into the muscle cell.

Like a neuron, a muscle fiber is excitable; it can undergo an action potential. The rise in sodium caused by the binding of ACh drives the fiber's membrane toward threshold. Once threshold is reached, action potentials stimulate muscle contraction by a process described in detail in Section 36.8.

Some neurotransmitters bind to more than one type of postsynaptic cell, causing a different result in each. For example, ACh stimulates contraction in skeletal muscle but it slows contraction in cardiac muscle.

Synaptic Integration

Typically, a neuron or effector cell gets messages from many neurons at the same time. Certain interneurons in the brain are on the receiving end of synapses with 10,000 neurons! An incoming signal may be excitatory and push the membrane potential closer to threshold. Or it may be inhibitory and nudge the potential away from threshold.

How does a postsynaptic cell respond to all of this information? Through **synaptic integration**, a neuron sums all inhibitory and excitatory signals arriving at its input zone. Incoming synaptic signals can amplify, dampen, or cancel one another's effects. Figure 33.12 illustrates how an excitatory signal and an inhibitory signal of differing sizes that arrive at a synapse at the same time are integrated.

Competing signals cause the membrane potential at the postsynaptic cell's input zone to rise and fall. When the excitatory signals outweigh inhibitory ones, ions diffuse from the input zone into the trigger zone and drive the postsynaptic cell to threshold. Gated sodium channels swing open, and an action potential occurs as described in the preceding section.

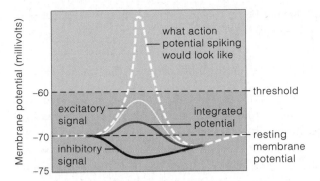

Figure 33.12 Synaptic integration. Excitatory and inhibitory signals arrive at a postsynaptic neuron's input zone at the same time. The graph lines show a postsynaptic cell's response to an excitatory signal (*yellow*), to an inhibitory signal (*purple*) and to both at once (*red*). In this example, summation of the two signals did not lead to an action potential (*white* waveform).

Neurons also integrate signals that arrive in quick succession from a single presynaptic cell. An ongoing stimulus can trigger a series of action potentials in a presynaptic cell, which will bombard a postsynaptic cell with waves of neurotransmitter.

Cleaning the Cleft

After signaling molecules do their work, they must be removed from synaptic clefts to make way for new signals. Some diffuse away. Membrane pumps move others back into presynaptic cells or neuroglial cells. Secreted enzymes break down specific kinds, as when the enzyme acetylcholinesterase breaks down ACh.

When neurotransmitter accumulates in a synaptic cleft, it disrupts the signaling pathways. That is how nerve gases such as sarin exert their deadly effects. After being inhaled, they bind to acetylcholinesterase and thus inhibit ACh breakdown. ACh accumulates, causing skeletal muscle paralysis, confusion, headaches, and, when the dosage is high enough, death.

Take-Home Message

How does information pass between cells at a synapse?

■ Action potentials travel to a neuron's output zone. There they stimulate release of neurotransmitters—chemical signals that affect another cell.

■ Neurotransmitters are signaling molecules secreted into a synaptic cleft from a neuron's output zone. They may have excitatory or inhibitory effects on a postsynaptic cell.

■ Synaptic integration is the summation of all excitatory and inhibitory signals arriving at a postsynaptic cell's input zone at the same time.

■ For a synapse to function properly, neurotransmitter must be cleared from the synaptic cleft after the chemical signal has served its purpose.

A Smorgasbord of Signals

- Different types of neurons release different neurotransmitters.
- Link to PET scans 2.2

Neurotransmitter Discovery and Diversity

In the early 1920s, Austrian scientist Otto Loewi was working to find out what controls the heart's beating. He surgically removed a frog heart—with the nerve that adjusts its rate still attached—and put it in saline solution. The heart continued to beat and, when Loewi stimulated the nerve, the heartbeat slowed a bit.

Loewi suspected stimulation of the nerve caused release of a chemical signal. To test this hypothesis, he put two frog hearts into a saline-filled chamber and stimulated the nerve connected to one of them. Both

hearts started to beat more slowly. As expected, the nerve had released a chemical that not only affected the attached heart, but also diffused through the liquid and slowed the beating of the second heart.

Loewi had discovered one of the responses to ACh, the neurotransmitter you read about in the preceding section. ACh acts on skeletal muscle, smooth muscle, the heart, many glands, and the brain. In myasthenia gravis, an autoimmune disease, the body mistakenly attacks its skeletal muscle receptors for ACh. Eyelids droop first, then other muscles weaken.

Interneurons in the brain also use ACh as a signaling molecule. A low ACh level in the brain contributes to memory loss in Alzheimer's disease. Affected people often can recall long-known facts, such as a childhood address, but have trouble remembering recent events.

There are many other neurotransmitters (Table 33.1). Norepinephrine and epinephrine (commonly known as adrenaline) prepare the body to respond to stress or to excitement. They are made from the amino acid tyrosine. So is dopamine, a neurotransmitter that influences reward-based learning and fine motor control.

Parkinson's disease involves impairment or death of dopamine-secreting neurons in a brain region that governs motor control (Figure 33.13). Hand tremors are often the earliest symptom. Later, sense of balance may be affected, and any movement can be difficult.

The neurotransmitter serotonin affects memory and mood. The drug fluoxetine (Prozac) lifts depression by raising serotonin levels. GABA (gamma-aminobutyric acid) inhibits release of neurotransmitters by other neurons. Diazapam (Valium) and alprazolam (Xanax) are drugs that lower anxiety by boosting GABA's effects.

Table 33.1 Major Neurotransmitters and Their Effects

Neurotransmitter	Examples of Effects
Acetylcholine (ACh)	Induces skeletal muscle contraction, slows cardiac muscle contraction rate, affects mood and memory
Epinephrine and norepinephrine	Speed heart rate; dilate the pupils and airways to lungs; slow gut contractions; increase anxiety
Dopamine	Dampens excitatory effects of other neurotransmitters; has roles in memory, learning, fine motor control
Serotonin	Elevates mood; role in memory
GABA	Inhibits release of other neurotransmitters

The Neuropeptides

Some neurons also make neuropeptides that serve as **neuromodulators**, molecules that influence the effects of neurotransmitters. One neuromodulator, substance P, enhances pain perception. Neuromodulators called enkephalins and endorphins are natural painkillers. They are secreted in response to strenuous activity or injuries and inhibit release of substance P. Endorphins also are released when people laugh, reach orgasm, or get a comforting hug or a relaxing massage.

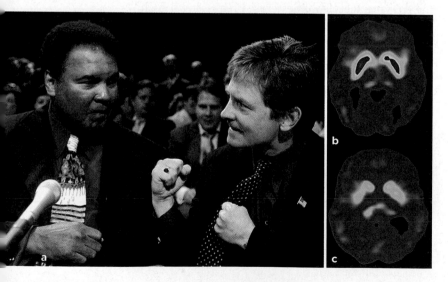

Figure 33.13 Battling Parkinson's disease. (**a**) This neurological disorder affects former heavyweight champion Muhammad Ali, actor Michael J. Fox, and about half a million other people in the United States. (**b**) A normal PET scan and (**c**) one from an affected person. *Red* and *yellow* indicate high metabolic activity in dopamine-secreting neurons. Section 2.2 explains PET scans.

Take-Home Message

What kinds of signaling molecules do neurons make?

- Neurons make neurotransmitters that signal other neurons or effector cells. Some neurons also make neuromodulators that can influence a neurotransmitter's effects on other cells.

33.7 | Drugs Disrupt Signaling

■ Psychoactive drugs exert their effects by interfering with the action of neurotransmitters.

■ Link to Alcohol's effects Chapter 6 introduction

People take psychoactive drugs, both legal and illegal, to alleviate pain, relieve stress, or feel pleasure. Many drugs are habit-forming, and users often develop tolerance; it takes larger or more frequent doses of the drug to obtain the desired effect.

Habituation and tolerance can lead to **drug addiction**, by which a drug takes on a vital biochemical role. Table 33.2 lists the main warning signs of addiction. Three or more signs may be cause for concern.

All major addictive drugs stimulate release of dopamine, a neurotransmitter with a role in reward-based learning. In just about all animals with a nervous system, dopamine release provides pleasurable feedback when an animal engages in behavior that enhances survival or reproduction. This response is adaptive; it helps animals learn to repeat the behaviors that benefit them. When drugs cause dopamine release, they tap into this ancient learning pathway. Drug users inadvertently teach themselves that the drug is essential to their well-being.

Stimulants Stimulants make users feel alert but also anxious, and they can interfere with fine motor control. Nicotine is a stimulant that blocks brain receptors for ACh. The caffeine in coffee, tea, and many soft drinks is also a stimulant. It blocks receptors for adenosine, which acts as a signaling molecule to suppress brain cell activity.

Cocaine, a powerful stimulant, is inhaled or smoked. Users feel elated and aroused, then become depressed and exhausted. Cocaine stops the uptake of dopamine, serotonin, and norepinephrine, from synaptic clefts. When norepinephrine is not cleared away, blood pressure soars. Overdoses may cause strokes or heart attacks that can end in death. Cocaine is highly addictive. Heavy cocaine use remodels the brain so that only cocaine can bring about a sense of pleasure (Figure 33.14).

Amphetamines reduce appetite and energize users by increasing secretion of serotonin, norepinephrine, and dopamine in the brain. Various types of amphetamine are ingested, smoked, or injected. The chapter introduction focused on the synthetic amphetamine found in Ecstasy. Crystal meth is another widely abused amphetamine. As with cocaine, users require more and more to get high or just to feel okay. Long-term use shrinks the brain areas involved in memory and emotions.

Depressants Depressants such as alcohol (ethyl alcohol) and barbiturates slow motor responses by inhibiting ACh output. Alcohol stimulates the release of endorphins and GABA, so users typically experience a brief euphoria followed by depression. Combining alcohol with barbiturates can be deadly. As the introduction to Chapter 6 explains, alcohol abuse damages the brain, liver, and other organs. Alcoholics deprived of the drug undergo tremors, seizures, nausea, and hallucinations.

Analgesics Analgesics mimic a body's natural painkillers—endorphins and enkephalins. The narcotic analgesics, such as morphine, codeine, heroin, fentanyl, and oxycodone, suppress pain. They cause a rush of euphoria and are highly addictive. Ketamine and PCP (phencyclidine) belong to a different class of analgesics. They give users an out-of-body experience and numb the extremities, by slowing the clearing of synapses. Use of either drug can lead to seizures, kidney failure, and fatal heat stroke. PCP can induce a violent, agitated psychosis that sometimes lasts more than a week.

Hallucinogens Hallucinogens distort sensory perception and bring on a dreamlike state. LSD (lysergic acid diethylamide) resembles serotonin and binds to receptors for it. Tolerance develops, but LSD is not addictive. However, users can get hurt, and even die, because they do not perceive and respond to hazards, such as oncoming cars. Flashbacks, or brief distortions of perceptions, may occur years after the last intake of LSD. Two related drugs, mescaline and psilocybin, have weaker effects.

Marijuana consists of parts of *Cannabis* plants. Smoking a lot of marijuana can cause hallucinations. More often, users become relaxed and sleepy as well as uncoordinated and inattentive. The active ingredient, THC (delta-9-tetrahydrocannabinol), alters levels of dopamine, serotonin, norepinephrine, and GABA. Chronic use can impair short-term memory and decision-making ability.

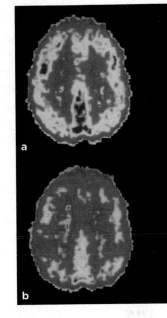

Figure 33.14 PET scans revealing (**a**) normal brain activity and (**b**) cocaine's long-term effect. *Red* shows areas of most activity, and *yellow*, *green*, and *blue* show successively reduced activity.

Table 33.2 Warning Signs of Drug Addiction

1. Tolerance; takes increasing amounts of the drug to get the same effect.

2. Habituation; takes continued drug use over time to maintain the self-perception of functioning normally.

3. Inability to stop or curtail drug use, even if desire to do so persists.

4. Concealment; not wanting others to know of the drug use.

5. Extreme or dangerous actions to get and use a drug, as by stealing, by asking more than one doctor for prescriptions, or by jeopardizing employment by using drugs at work.

6. Deterioration of professional and personal relationships.

7. Anger and defensiveness if someone suggests there may be a problem.

8. Drug use preferred over previous favored activities.

■ Peripheral nerves run through your body and carry information to and from the central nervous system.

Axons Bundled as Nerves

In humans, the peripheral nervous system includes 31 pairs of spinal nerves that connect to the spinal cord and 12 pairs of cranial nerves that connect directly to the brain. Each peripheral nerve consists of axons of many neurons bundled together inside a connective tissue sheath (Figure 33.15a). All spinal nerves include axons from both sensory and motor neurons. Cranial nerves may include axons of motor neurons, axons of sensory neurons, or axons of both sensory and motor neurons. Interneurons, remember, are not part of the peripheral nervous system.

The neuroglial cells called Schwann cells wrap like jelly rolls around the axons of most peripheral nerves (Figure 33.15b). The Schwann cells collectively form an insulating **myelin sheath** that makes action potentials flow faster. Ions cannot cross a sheathed neural membrane. As a result, ion disturbances associated with an action potential spread through an axon's cytoplasm until they reach a node, a small gap between Schwann cells. At each node, the membrane contains numerous gated sodium channels. When these gates open, the voltage difference reverses abruptly. By jumping from node to node in long axons, a signal can move as fast as 120 meters per second. In unmyelinated axons, the maximum speed is about 10 meters per second.

Functional Subdivisions

We subdivide the peripheral system into the somatic nervous system and the autonomic nervous system.

Somatic and Autonomic Systems The sensory part of the **somatic nervous system** conducts information about external conditions from sensory neurons to the central nervous system. The motor part of the somatic system relays commands from the brain and spinal cord to the skeletal muscles. It is the only part of the nervous system normally under voluntary control. The **autonomic nervous system** is concerned with signals to and from internal organs and glands.

Sympathetic and Parasympathetic Divisions The nerves of the autonomic system are in two categories: sympathetic and parasympathetic. Both service most organs and work antagonistically, meaning the signals from one type oppose signals from the other (Figure

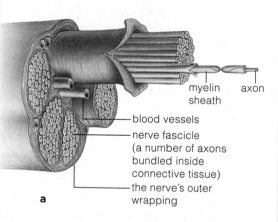

a

myelin sheath

axon

blood vessels

nerve fascicle (a number of axons bundled inside connective tissue)

the nerve's outer wrapping

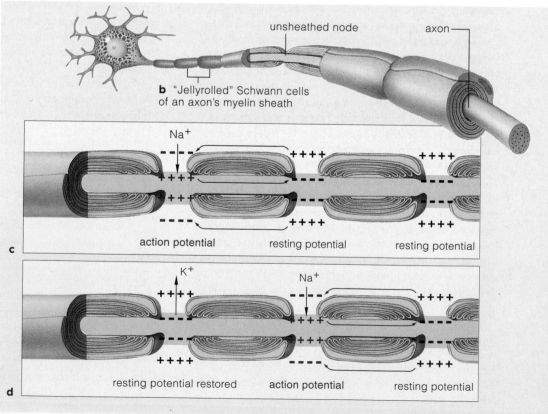

unsheathed node

axon

b "Jellyrolled" Schwann cells of an axon's myelin sheath

c

Na⁺

action potential   resting potential   resting potential

d

K⁺   Na⁺

resting potential restored   action potential   resting potential

Figure 33.15 Animated (**a**) Structure of one type of nerve. (**b–d**) In axons with a myelin sheath, ions flow across the neural membrane at nodes, or small gaps between the cells that make up the sheath. Many gated channels for sodium ions are exposed to extracellular fluid at the nodes. When excitation caused by an action potential reaches a node, the gates open and sodium rushes in, starting a new action potential. Excitation spreads rapidly to the next node, where it triggers a new action potential, and so on down the axon to the output zone.

Figure 33.16
Animated

(**a**) Sympathetic and (**b**) parasympathetic nerves of the autonomic system. Each half of the body has nerves of the same type.

Ganglia containing the cell bodies of sympathetic neurons lie near the spinal cord. Ganglia of the autonomic neurons lie in or near the organ they control.

Figure It Out: Which parasympathetic nerve has branches that send signals to the heart, stomach, and kidneys?

Answer: Vagus nerve

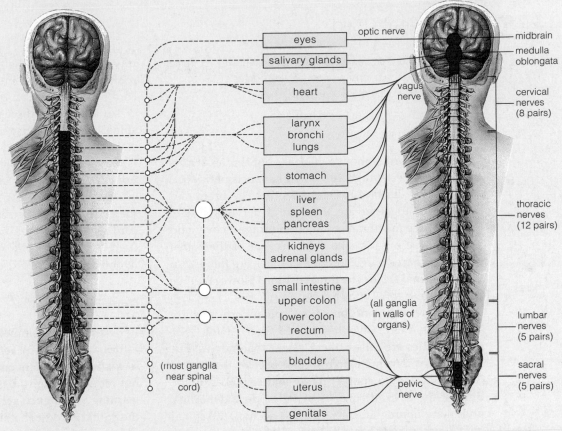

A   Sympathetic outflow from spinal cord

Some responses to sympathetic outflow:
• Heart rate increases.
• Pupils of eyes dilate (widen, let in more light).
• Glandular secretions decrease in airways to lungs.
• Salivary gland secretions thicken.
• Stomach and intestinal movements slow down.
• Sphincters contract.

B   Parasympathetic outflow from spinal cord and brain

Some responses to parasympathetic outflow:
• Heart rate decreases.
• Pupils of eyes constrict (keep more light out).
• Glandular secretions increase in airways to lungs.
• Salivary gland secretions become more watery.
• Stomach and intestinal movements increase.
• Sphincters relax.

33.16). **Sympathetic neurons** are most active in times of stress, excitement, and danger. Their axon terminals release norepinephrine. **Parasympathetic neurons** are most active in times of relaxation. Release of ACh by their axon terminals promotes housekeeping tasks, such as digestion and urine formation.

What happens when something startles or scares you? Parasympathetic input falls. Sympathetic signals increase. When unopposed, sympathetic signals raise your heart rate and blood pressure, make you sweat more and breathe faster, and induce adrenal glands to secrete epinephrine. The signals put you in a state of intense arousal, so you are primed to fight or make a fast getaway. Hence the term **fight–flight response**.

Opposing sympathetic and parasympathetic signals govern most organs. For instance, both act on smooth muscle cells in the gut wall. As sympathetic neurons are releasing norepinephrine at synapses with these cells, parasympathetic neurons are releasing ACh at other synapses with the same muscle cells. One signal tells the gut to slow down contractions; the other calls for increased activity. The outcome is finely adjusted through synaptic integration.

Take-Home Message

What is the peripheral nervous system?

■ The peripheral nervous system includes nerves that connect the body with the central nervous system. A nerve consists of the bundled axons of many neurons. Typically each axon is wrapped in a myelin sheath that increases the speed of action potential transmission.

■ Neurons of the somatic part of the peripheral system control skeletal muscle and convey information about the external environment to the central nervous system.

■ The autonomic system carries information to and from smooth muscle, cardiac muscle, and glands. Signals from its two divisions—sympathetic and parasympathetic—have opposing effects on effectors.

- The spinal cord serves as an information highway for traffic to and from the brain, and also as a reflex center.
- Spinal reflexes do not involve the brain.

An Information Highway

Your **spinal cord** is about a thick as your thumb. It runs through the vertebral column and connects peripheral nerves with the brain (Figure 33.17). The brain and spinal cord together are the central nervous system (CNS). Three membranes, called **meninges**, cover and protect these organs. The central canal of the spinal cord and spaces between the meninges are filled with **cerebrospinal fluid**. The fluid cushions blows and thus protects central nervous tissue.

The outermost portion of the spinal cord is **white matter**: bundles of myelin-sheathed axons. In the CNS, such bundles are called tracts, rather than nerves. The tracts carry information from one part of the central nervous system to another. **Gray matter** makes up the bulk of the CNS. It consists of cell bodies, dendrites, and many neuroglial cells. In cross-section, the spinal cord's gray matter has a butterfly-like shape.

Spinal nerves of the peripheral nervous system connect to the spinal cord at dorsal and ventral "roots." Remember, all spinal nerves have sensory and motor components. Sensory information travels to the spinal cord through a dorsal root. Cell bodies of sensory neurons are found in dorsal root ganglia. Motor signals travel away from the spinal cord through a ventral root. Cell bodies of motor neurons are in the spinal cord's gray matter.

An injury that disrupts the signal flow through the spinal cord can cause a loss of sensation and paralysis. Symptoms depend on what portion of the cord is damaged. Nerves carrying signals to and from the upper body lie higher in the cord than nerves that govern the lower body. An injury to the lumbar region of the cord often paralyzes the legs. An injury to higher cord regions can paralyze all limbs, as well as muscles used in breathing. More that 250,000 Americans now live with a spinal cord injury.

Reflex Pathways

Reflexes are the simplest and most ancient paths of information flow. A **reflex** is an automatic response to a stimulus, a movement or other action that does not require thought. Basic reflexes do not require any learning. With such reflexes, sensory signals flow to the spinal cord or the brain stem, which then calls for a response by way of motor neurons.

For example, the stretch reflex is one of the spinal reflexes. It causes a muscle to contract after gravity or some other force stretches it. Suppose you hold a bowl as someone drops fruit into it. The increased load makes your hand drop a bit, which stretches the biceps muscle in your arm. Stretching of the muscle

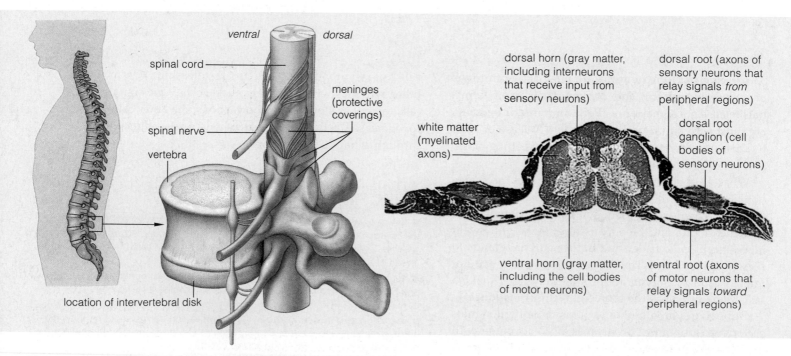

Figure 33.17 Animated Location and organization of the spinal cord.

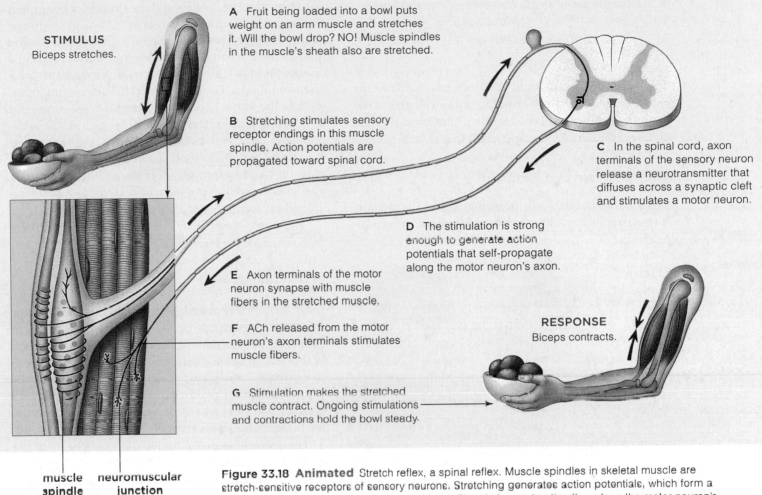

STIMULUS
Biceps stretches.

A Fruit being loaded into a bowl puts weight on an arm muscle and stretches it. Will the bowl drop? NO! Muscle spindles in the muscle's sheath also are stretched.

B Stretching stimulates sensory receptor endings in this muscle spindle. Action potentials are propagated toward spinal cord.

C In the spinal cord, axon terminals of the sensory neuron release a neurotransmitter that diffuses across a synaptic cleft and stimulates a motor neuron.

D The stimulation is strong enough to generate action potentials that self-propagate along the motor neuron's axon.

E Axon terminals of the motor neuron synapse with muscle fibers in the stretched muscle.

F ACh released from the motor neuron's axon terminals stimulates muscle fibers.

G Stimulation makes the stretched muscle contract. Ongoing stimulations and contractions hold the bowl steady.

RESPONSE
Biceps contracts.

muscle spindle   neuromuscular junction

Figure 33.18 Animated Stretch reflex, a spinal reflex. Muscle spindles in skeletal muscle are stretch-sensitive receptors of sensory neurons. Stretching generates action potentials, which form a synapse with a motor neuron in the spinal cord. Signals for contraction flow along the motor neuron's axon, from spinal cord back to the stretched muscle. The muscle contracts, steadying the arm.

causes muscle spindles between the muscle fibers to stretch. Muscle spindles are sensory organs that house receptor endings of sensory neurons (Figure 33.18).

The more the biceps muscle stretches, the greater the frequency of action potentials along axons of the muscle spindle neurons. Inside the spinal cord, these axons synapse with motor neurons that control the stretched muscle. Signals from the sensory neurons cause action potentials in the motor neurons, which release ACh at the neuromuscular junction. In response to this signal, the biceps contracts and steadies the arm against the added load.

The knee-jerk reflex is another stretch reflex. A tap just below the knee stretches the thigh muscle. The stretch is detected by muscle spindles in this muscle. The muscle spindles send signals to the spinal cord, where motor neurons become excited. As a result, signals flow from the spinal cord back to the leg, and the leg jerks in response.

Another spinal reflex, the withdrawal reflex, allows quick action when you touch something hot. Touch a hot surface and signals flow to the spinal cord. Unlike the stretch reflex, the withdrawal response involves an interneuron of the spinal cord. A heat-detecting sensory neuron sends signals to the spinal interneuron, which then relays the signal to motor neurons. Before you know it, your biceps has contracted, pulling your hand away from the potentially damaging heat.

Take-Home Message

What are the functions of the spinal cord?

■ Tracts of the spinal cord relay information between peripheral nerves and the brain. The axons involved in these pathways make up the bulk of the cord's white matter. Cell bodies, dendrites, and neuroglia make up gray matter.

■ The spinal cord also has a role in some simple reflexes, automatic responses that occur without conscious thought or learning. Signals from sensory neurons enter the cord through the dorsal root of spinal nerves. Commands for responses go out along the ventral root of these nerves.

33.10 | The Vertebrate Brain

- The brain is part of the central nervous system and is the body's main information integrating organ.

- Link to Trends in vertebrate evolution 26.2

In all vertebrates, the embryonic neural tube develops into a spinal cord and brain. During development the brain becomes organized as three functional regions: the forebrain, midbrain, and hindbrain (Figure 33.19).

The Hindbrain and Midbrain

The hindbrain sits atop the spinal cord. The portion just above the cord, the **medulla oblongata**, influences the strength of heartbeats and the rhythm of breathing. It also controls reflexes such as swallowing, vomiting, and sneezing. Above the medulla oblongata lies the **pons**, which assists in regulation of breathing. Pons means "bridge," and tracts extend through the pons to the midbrain. The **cerebellum**, the largest hindbrain region, lies at the back of the brain and serves mainly to coordinate voluntary movements.

Fishes and amphibians have the most pronounced midbrain (Figure 33.20). It sorts out sensory input and initiates motor responses. In primates, the midbrain is the smallest of the three brain regions and plays an important role in reward-based learning.

The pons, medulla, and midbrain are collectively referred to as the **brain stem**.

The Forebrain

Early vertebrates relied heavily on their forebrain's olfactory lobes; odors provided essential information about the environment. Paired outgrowths from the brain stem integrated olfactory input and responses to it. Especially among land vertebrates, these outgrowths expanded into the two halves of the **cerebrum**, the two cerebral hemispheres. Most sensory signals destined for the cerebrum pass through the adjacent **thalamus**.

The **hypothalamus** ("under the thalamus") is the center for homeostatic control of the internal environment. It regulates behaviors related to internal organ activities, such as thirst, sex, and hunger and governs temperature. The hypothalamus is also an endocrine gland. It interacts with the adjacent pituitary gland to control hormone secretions. Another endocrine gland, the pineal gland, lies deep in the forebrain. We discuss endocrine function in detail in Chapter 35.

Also in the forebrain is a group of structures that we refer to collectively as the limbic system. We discuss the role of the human system in the next section.

Protection at the Blood–Brain Barrier

The neural tube's lumen—the space inside it—persists in adult vertebrates as a system of cavities and canals filled with cerebrospinal fluid. This clear fluid forms

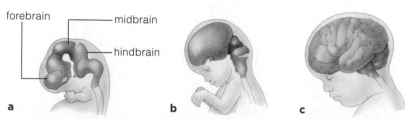

forebrain — midbrain — hindbrain

a b c

FOREBRAIN	Cerebrum	Localizes, processes sensory inputs; initiates, controls skeletal muscle activity; governs memory, emotions, abstract thought in the most complex vertebrates
	Olfactory lobe	Relays sensory input from nose to olfactory areas of cerebrum
	Thalamus	Relays sensory signals to and from cerebral cortex; has a role in memory
	Hypothalamus	With pituitary gland, functions in homeostatic control. Adjusts volume, composition, temperature of internal environment; governs organ-related behaviors (e.g., sex, thirst, hunger), and expression of emotions
	Limbic system	Governs emotions; has roles in memory
	Pituitary gland (Chapter 35)	With hypothalamus, provides endocrine control of metabolism, growth, development
	Pineal gland (Chapter 35)	Helps control some circadian rhythms; also has role in mammalian reproduction
MIDBRAIN	Roof of midbrain (tectum)	In fishes and amphibians, it coordinates sensory input (as from optic lobes) with motor responses. In mammals, it is reduced and mainly relays sensory input to the forebrain
HINDBRAIN	Pons	Tracts bridge cerebrum and cerebellum, also connect spinal cord with forebrain. With the medulla oblongata, controls rate and depth of respiration
	Cerebellum	Coordinates motor activity for moving limbs and maintaining posture, and for spatial orientation
	Medulla oblongata	Tracts relay signals between spinal cord and pons; functions in reflexes that affect heart rate, blood vessel diameter, and respiratory rate. Also involved in vomiting, coughing, other vital functions

Figure 33.19 Neural tube to brain. The human neural tube at (**a**) 7 weeks of embryonic development. The brain at (**b**) 9 weeks, and (**c**) at birth. The chart lists and describes major components in the three regions of the adult vertebrate brain.

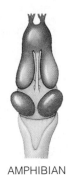
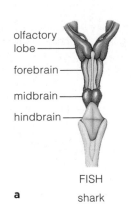

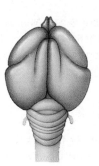
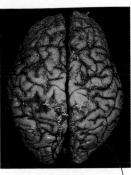

olfactory lobe
forebrain
midbrain
hindbrain

a

FISH
shark

AMPHIBIAN
frog

REPTILE
alligator

BIRD
goose

MAMMAL
human

when water and small molecules are filtered out of the blood into brain cavities called ventricles. The fluid then seeps out and bathes the brain and spinal cord. It returns to the bloodstream by entering veins.

A **blood–brain barrier** protects the spinal cord and brain from harmful substances. The barrier is formed by the walls of blood capillaries that service the brain. In most parts of the brain, tight junctions form a seal between adjoining cells of the capillary wall, so water-soluble substances must pass through the cells to reach the brain. Transport proteins in the plasma membrane of these cells allow essential nutrients to cross. Oxygen and carbon dioxide diffuse across the barrier, but most waste urea cannot breach it.

No other portion of extracellular fluid has solute concentrations maintained within such narrow limits. Even changes brought on by eating and exertion are limited. Why? Hormones and other chemicals in blood affect neural function. Also, changes in ion concentrations can alter the threshold for action potentials.

The blood–brain barrier is not perfect; some toxins such as nicotine, alcohol, caffeine, and mercury slip across. Also, inflammation or a traumatic blow to the head can damage it and compromise neural function.

The Human Brain

The average human brain weighs 1,330 grams, or 3 pounds. It contains about 100 billion interneurons, and neuroglia makes up more than half of its volume. The human midbrain is relatively smaller than that of other vertebrates. A human cerebellum is the size of a fist and has more interneurons than all other brain regions combined. As in other vertebrates, the cerebellum plays a role in the sense of balance, but it took on added functions as humans evolved. It affects learning of motor and some mental skills, such as language.

A deep fissure divides the forebrain's cerebrum into two halves, the cerebral hemispheres (Figure 33.20).

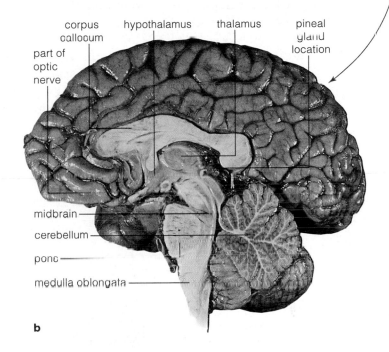

corpus callosum
hypothalamus
thalamus
pineal gland location
part of optic nerve
midbrain
cerebellum
pons
medulla oblongata

b

Each half deals mainly with input from the opposite side of the body. For instance, signals about pressure on the right arm reach the left hemisphere. Activity of the hemispheres is coordinated by signals that flow both ways across the **corpus callosum**, a thick band of nerve tracts. The next section focuses on the cerebral cortex, the thin outer layers of the cerebrum.

Take-Home Message

What are the structural and functional divisions of the vertebrate brain?

■ We recognize three regions, the forebrain, midbrain, and hindbrain, based on the embryonic tissue from which they develop. The brain stem, which includes parts of the hindbrain and the midbrain, is the most evolutionarily ancient region of brain tissue. It is involved in reflex behaviors.

■ The forebrain includes the cerebrum, which evolved as an expansion of the olfactory lobe and is now the main processing center in humans. It also includes the hypothalamus, which has important roles in thirst, temperature regulation, and other responses related to homeostasis.

■ Our capacity for language and conscious thought arises from the activity of the cerebral cortex.

■ The cortex interacts with other brain regions in shaping our emotional responses and memories.

Functions of the Cerebral Cortex

Each half of the cerebrum, or cerebral hemisphere, is divided into frontal, temporal, occipital, and parietal lobes (Figure 33.21). The **cerebral cortex,** the outermost gray matter on each lobe, contains distinct areas that receive and process diverse signals.

The cerebral hemispheres overlap in function, but there are differences. Most often, mathematical skills and language arise mainly from activity in the left hemisphere. The right hemisphere interprets music, judges spatial relations, and assesses visual inputs.

The body is spatially mapped out in the primary motor cortex of each frontal lobe, which controls and

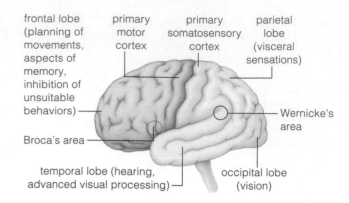

Figure 33.21 Animated Lobes of the brain, with primary receiving and integrating centers of the human cerebral cortex.

coordinates the movements of skeletal muscles on the opposite side of the body. Much of the motor cortex is devoted to finger, thumb, and tongue muscles, which can make fine movements. Figure 33.22 depicts the proportions of the motor cortex that are devoted to controlling different body parts.

The premotor cortex of each frontal lobe regulates complex movements and learned motor skills. Swing a golf club, play the piano, or type on a keyboard, and the premotor cortex coordinates the activity of many different muscle groups.

Broca's area in the frontal lobe helps us translate thoughts into speech. It controls the tongue, throat, and lip muscles and gives humans our capacity to speak complex sentences. In most people, Broca's area is in the left hemisphere. Damage to Broca's area often prevents normal speech, although an affected individual can still understand language.

The primary somatosensory cortex is at the front of the parietal lobe. Like the motor cortex, it is organized as a map that corresponds to body parts. It receives sensory input from the skin and joints, and one part has a role in taste perception (Section 34.3).

The perceptions of sound and odor arise in sensory areas of each temporal lobe. Wernicke's area, in this lobe, functions in the comprehension of spoken and

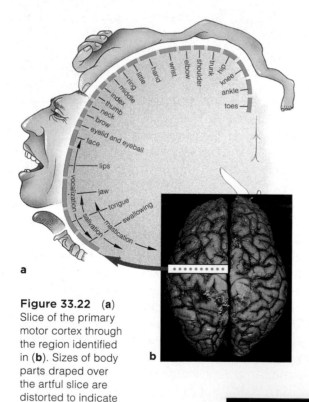

Figure 33.22 **(a)** Slice of the primary motor cortex through the region identified in (**b**). Sizes of body parts draped over the artful slice are distorted to indicate which ones get the most precise control.

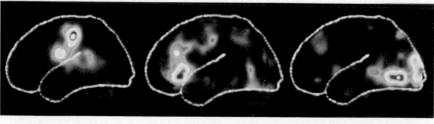

Figure 33.23 Three PET scans that identify which brain areas were active when a person performed three kinds of tasks. *Yellow* and *orange* indicate high activity.

Motor cortex activity when speaking　　Prefrontal cortex activity when generating words　　Visual cortex activity when seeing written words

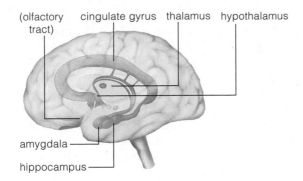

(olfactory tract) cingulate gyrus thalamus hypothalamus

amygdala

hippocampus

Figure 33.24 Limbic system components.

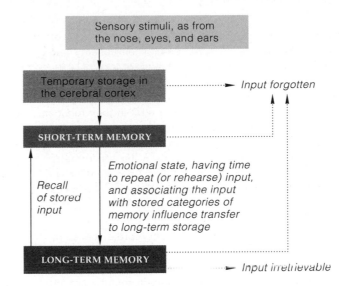

Sensory stimuli, as from the nose, eyes, and ears

Temporary storage in the cerebral cortex → *Input forgotten*

SHORT-TERM MEMORY

Emotional state, having time to repeat (or rehearse) input, and associating the input with stored categories of memory influence transfer to long-term storage

Recall of stored input

LONG-TERM MEMORY → *Input irretrievable*

Figure 33.25 Stages in memory processing.

written language, including Braille, a written language for the blind.

A primary visual cortex at the back of each occipital lobe receives sensory input from both eyes.

Association areas are scattered through the cortex, but not in the primary motor and sensory areas. Each integrates diverse inputs (Figure 33.23). For instance, one visual association area around the primary visual cortex compares what we see with visual memories.

Connections With the Limbic System

The **limbic system** encircles the upper brain stem. It governs emotions, assists in memory, and correlates organ activities with self-gratifying behavior such as eating and sex. That is why the limbic system is known as our emotional visceral brain. "Gut reactions" called up by the limbic system can often be overridden by the cerebral cortex.

The hypothalamus, hippocampus, amygdala, and cingulate gyrus are part of the limbic system (Figure 33.24). The hypothalamus is the major control center for homeostatic responses and it correlates emotions with visceral activities. The hippocampus helps store memories and access memories of earlier threats. The almond-shaped amygdala helps interpret social cues, and contributes to the sense of self. It is highly active during episodes of fear and anxiety, and often it is overactive in people afflicted with panic disorders. The cingulate gyrus has a role in attention and in emotion. It is often smaller and less active than normal in people with schizophrenia.

Evolutionarily, the limbic system is related to the olfactory lobes. Olfactory input causes signals to flow to the hippocampus, amygdala, and hypothalamus as well as to the olfactory cortex. That is one reason why specific odors can call up emotionally significant memories. Information about taste also travels to the limbic system and can trigger emotional responses.

Making Memories

The cerebral cortex receives information continually, but only a fraction of it becomes memories. Memory forms in stages. Short-term memory lasts seconds to hours. This stage holds a few bits of information, a set of numbers, words of a sentence, and so forth. In long-term memory, larger chunks of information get stored more or less permanently (Figure 33.25).

Different types of memories are stored and brought to mind by different mechanisms. Repetition of motor tasks can create skill memories, which are highly persistent. Once you learn to ride a bicycle, drive a car, dribble a basketball, or play an accordion, you seldom forget how. Skill memories involve the cerebellum, which controls motor activity.

Declarative memory stores facts and impressions of events, as when it helps you remember how a lemon smells or that a quarter is worth more than a dime. It starts when the sensory cortex signals the amygdala, a gatekeeper to the hippocampus. A memory will be retained only if signals loop repeatedly in the sensory cortex, hippocampus, and thalamus.

Emotions influence memory retention. For instance, epinephrine released during times of stress helps place short-term memories into long-term storage.

Take-Home Message

What are the functions of the cerebral cortex?

■ The cerebral cortex controls voluntary activity, sensory perception, abstract thought, and language and speech. It receives information and processes some of it into memories. It also oversees the limbic system, the brain's center of emotional responses.

33.12 | The Split Brain

■ Investigations by Roger Sperry into the importance of information flow between the cerebral hemisphere showed that the two halves of the brains have a division of labor.

As mentioned in the preceding section, the two cerebral hemispheres look alike but differ a bit in their functions. The differences first became apparent in the mid-1800s, through studies of people who had brain injuries. For instance, damage to Broca's area in the left frontal cortex interfered with the ability to vocalize words. Injury to Wernicke's area in the left temporal lobe did not interfere with the capacity to say words, but the affected person could not put words into sentences.

Fast-forward to the 1960s. Ever more evidence of the importance of the left hemisphere continued to flow in. Researchers began wondering what role, if any, the right hemisphere plays in the advanced functions of typical right-handed people. Roger Sperry and his coworkers decided to find out.

Sperry became interested in "split-brain" patients. These people had undergone surgery to sever their corpus callosum, a thick band of nerves that connects the two cerebral hemispheres. At the time, this was an experimental way to treat severe epilepsy. Epileptic seizures are like electrical storms in the brain. Surgeons severed a patient's corpus callosum to prevent flow of disturbed electrical signals from one hemisphere to the other. After a brief recovery, patients were able to lead what seemed to be normal lives, with fewer seizures.

But were those patients really normal? The surgery had stopped the flow of information across 200 million or so axons in the corpus callosum. Surely something had to be different. Something was.

Sperry designed elegant experiments to examine the split-brain experience. He devised a mechanism of presenting the two halves of affected patients with two different parts of a visual stimulus. At the time, researchers already knew that the visual connections to and from one hemisphere are mainly concerned with the opposite half of the visual field, as in Figure 33.26. Sperry projected a word—say, COWBOY—onto a screen so that COW fell in the left half of the visual field, and BOY fell in the right (Figure 33.27).

The subjects of this experiment reported seeing the word BOY. The left hemisphere, which controls language, recognized the word. However, when asked to write the word with the left hand—which was hidden from view—the subject wrote COW. The right hemisphere "knew" the other half of the word (COW) and had directed the left hand's motor response. But it could not tell the left hemisphere what was going on because of the severed corpus callosum. The subject knew a word was being written but could not say what it was!

"The surgery," Sperry reported, "left these people with two separate minds, two spheres of consciousness." Sperry concluded that both hemispheres contribute to normal perception by sharing information that shapes the experience we call consciousness.

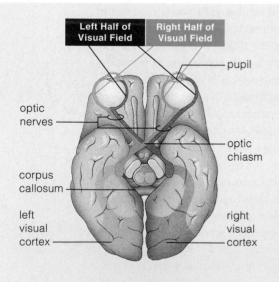

A Pathway by which sensory input about visual stimuli reaches the visual cortex of the human brain.

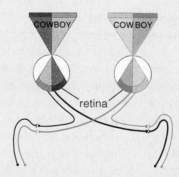

B Each eye gathers visual information at the retina, a thin layer of densely packed photoreceptors at the back of the eyeball (Section 34.7).

Light from the *left* half of the visual field strikes receptors on the right side of both retinas. Parts of two optic nerves carry signals from the receptors to the right cerebral hemisphere.

Light from the *right* half of the visual field strikes receptors on the left side of both retinas. Parts of the optic nerves carry signals from them to the left hemisphere.

Figure 33.26 Animated Visual information and the brain.

Figure 33.27 One example of the response of a split-brain patient to visual stimuli. As described in the text, this type of experiment demonstrated the importance of the corpus callosum in coordinating activities between the two cerebral hemispheres.

33.13 | Neuroglia—The Neurons' Support Staff

■ Although we focus on the neurons, neuroglial cells make up the bulk of the brain and have important roles too.

■ Links to Cell Cycle 9.2, Cancer 9.5

Types of Neuroglia

Neuroglial cells, or neuroglia, outnumber neurons in a human brain by about 10 to 1. Neuroglia act as a framework that holds neurons in place; *glia* means glue in Latin. While a nervous system is developing, new neurons migrate along highways of neuroglia to reach their final destination.

An adult brain has four main types of neuroglial cells: oligodendrocytes, microglia, astrocytes, and ependymal cells. The oligodendrocytes make myelin sheaths that insulate axons in the central nervous system. As mentioned earlier, Schwann cells are neuroglia that perform this same function for peripheral nerves.

Multiple sclerosis (MS) is an autoimmune disorder in which white blood cells wrongly attack and destroy the myelin sheaths of oligodendrocytes. The myelin is replaced by scar tissue and the conduction ability of the affected axons declines. Certain genes increase the likelihood of MS, but a viral infection might set it in motion. Once it begins, information flow is disrupted. Dizziness, numbness, muscle weakness, fatigue, visual problems, and other symptoms commonly follow. MS affects at least 300,000 people in the United States.

Microglia are, as the name implies, the smallest of the neuroglial cells. They continually survey the brain. If brain tissue is injured or infected, microglia become active, motile cells that engulf dead or dying cells and debris. They also produce chemical signals that alert the immune system to the threat.

Star-shaped astrocytes are the most abundant cells in the brain (Figure 33.28). They have diverse roles. They wrap around blood vessels that supply the brain and stimulate formation of the blood-brain barrier, take up neurotransmitters released by neurons, assist in immune defense, make lactate that fuels activities of neurons, and synthesize nerve growth factor. A growth factor is a molecule that is secreted by one cell and causes division or differentiation of another cell. Neurons do not divide; they are stopped in G1 of the cell cycle (Section 9.2). But nerve growth factor causes a neuron to form new synapses with its neighbors.

Ependymal cells are neuroglia that line the brain's fluid-filled cavities (ventricles) and the spinal cord's central canal. Some ependymal cells are ciliated and the action of their cilia keeps the cerebrospinal fluid flowing in a consistent direction through the system of cavities and canals.

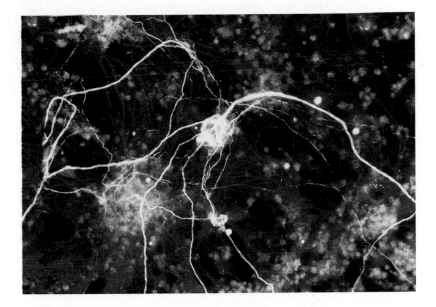

Figure 33.28 Astrocytes (*orange*) and a neuron (*yellow*) in brain tissue. The cells in this light micrograph were made visible by immunofluorescence. This procedure attaches fluorescent dye molecules to antibodies that then bind to specific molecules on a cell.

About Brain Tumors

Neurons do not divide, so they do not give rise to tumors. However, sometimes neuroglial cells divide uncontrollably, and the result is a glioma. This is the most common kind of primary brain tumor—a tumor that arises from cells in the brain. Brain tumors also arise from uncontrolled division of cells in meninges, or as a result of metastasis—arrival of cancerous cells from elsewhere in the body (Section 9.5).

Men are more prone to brain tumors than women. Exposure to ionizing radiation, such as x-rays, or to chemical carcinogens increases risk. What about the radio waves from cell phones? No study has shown that use of a cell phone causes brain cancer. However, cell phones are a relatively recent invention and brain tumors can take years to develop. To be cautious, some doctors recommend use of a headset, which keeps the wave-emitting part of the phone away from the brain.

Take-Home Message

What are the functions of neuroglia?

■ Neuroglial cells make up the bulk of the brain. They provide a framework for neurons, insulate neuron axons, assist neurons metabolically, and protect the brain from injury and disease.

■ Unlike neurons, neuroglia continue to divide in adults. Thus, neuroglia can be a source of brain tumors.

Now that you know a bit more about how a brain functions, take a moment to reconsider effects of MDMA, the active ingredient in Ecstasy. MDMA harms and possibly kills brain interneurons that produce the neurotransmitter serotonin. Remember, neurons do not divide, so damaged ones are not replaced. MDMA also impairs the blood–brain barrier, so it allows larger than normal molecules to pass into the brain for as long as 10 weeks after use.

How would you vote?

Should people who are caught using illegal drugs be offered addiction treatment as an alternative to jail time? See CengageNOW for details, then vote online.

Summary

Section 33.1 **Neurons** are electrically excitable cells that signal other cells by means of chemical messages. **Sensory neurons** detect stimuli. **Interneurons** relay signals between neurons. **Motor neurons** signals effectors (muscles and glands). **Neuroglia** support the neurons.

Radially symmetrical animals have a **nerve net**. Most animals have a bilateral nervous system with **cephalization**; they have paired **ganglia** (clusters of neuron cell bodies) or a brain at the head end.

The vertebrate **central nervous system** is a brain and spinal cord. The **peripheral nervous system** includes all nerves that run through the body.

Sections 33.2–33.4 A neuron's **dendrites** receive signals and its **axon** transmits signals. Neurons maintain a **resting membrane potential**, a slight voltage difference across their plasma membrane. An **action potential** is a brief reversal of the membrane potential. It occurs only if membrane potential increases to the **threshold potential**.

An action potential occurs when opening of voltage-gated sodium channels allows sodium to flow down its concentration gradient into the neuron. Then, opening of voltage-gated potassium channels allows potassium ions to flow out of the neuron.

All action potentials are the same size and travel in one direction only, away from the cell body and toward the axon terminals.

■ *Use the animation on CengageNOW to learn about a neuron's structure and its membrane properties and to view an action potential step by step.*

Sections 33.5–33.7 Neurons send chemical signals to cells at **synapses**. A synapse between a motor neuron and a muscle fiber is a **neuromuscular junction**. Arrival of an action potential at a presynaptic cell's axon terminals triggers the release of **neurotransmitter**, a type of chemical signal. Neurotransmitter diffuses to receptors on a postsynaptic cell and binds to them. A postsynaptic cell's response is determined by **synaptic integration** of all messages arriving at the same time.

Neuromodulators are chemicals secreted by neurons that can alter neurotransmitter effects.

Psychoactive drugs disrupt neurotransmitter-based signaling. Some cause **drug addiction**, a dependence on the drug that interferes with normal functioning.

■ *Use the animation on CengageNOW and learn about a synapse between a motor neuron and a muscle cell.*

Section 33.8 Nerves are bundles of axons that carry signals through the body. **Myelin sheaths** enclose most axons and increase signal conduction rates.

The peripheral nervous system is functionally divided into the **somatic nervous system**, which controls skeletal muscles, and the **autonomic nervous system**, which controls internal organs and glands.

Signals from **sympathetic neurons** of the autonomic system increase in times of stress or danger. The signals cause a **fight–flight response**. During less stressful times, signals from **parasympathetic neurons** dominate. Organs receive signals from both types of neurons.

■ *Use the animation on CengageNOW to explore the structure of a nerve and to compare the effects of sympathetic and parasympathetic stimulation.*

Section 33.9 Like the brain, the **spinal cord** consists of **white matter** (with myelinated axons) and **gray matter** (with cell bodies, dendrites, and neuroglia). The spinal cord and brain are enclosed by membranous **meninges** and cushioned by **cerebrospinal fluid**. Spinal reflexes involve peripheral nerves and the spinal cord. A **reflex** is an automatic response to stimulation; it does not require conscious thought.

■ *Use the animation on CengageNOW to explore the spinal cord and see what happens during a stretch reflex.*

Sections 33.10–33.12 The neural tube of a vertebrate embryo develops into the spinal cord and brain. The **brain stem** is the evolutionarily oldest brain tissue. It includes the **pons** and **medulla oblongata**, which control reflexes involved in breathing and other essential tasks. The **cerebellum** acts in motor control. The **thalamus** and **hypothalamus** function in homeostasis. A **blood–brain barrier** protects the brain from many harmful chemicals.

The **cerebral cortex**, the most recently evolved brain region, governs complex functions. It has specific areas that receive different types of sensory input or control voluntary movements. The cerebral cortex interacts with the **limbic system** in emotions and memory. Activity of the two halves of the **cerebrum** is coordinated by means of the **corpus callosum** that connects them.

■ *Use the animation on CengageNOW to learn about the structure and function of the human brain.*

Section 33.13 Neuroglial cells make up the bulk of the brain. Unlike neurons, they continue to divide in adults.

Data Analysis Exercise

Animal studies are often used to assess effects of prenatal exposure to illicit drugs. For example, Jack Lipton used rats to study the behavioral effect of prenatal exposure to MDMA, the active ingredient in Ecstasy. He injected female rats with either MDMA or saline solution when they were 14 to 20 days pregnant. This is the period when their offsprings' brains were forming. When those offspring were 21 days old, Lipton tested their ability to adjust to a new environment. He placed each young rat in a new cage and used a photobeam system to record how much each rat moved around before settling down. Figure 33.29 shows his results.

1. Which rats moved around most (caused the most photobeam breaks) during the first 5 minutes in a new cage, those prenatally exposed to MDMA or the controls?

2. How many photobeam breaks did the MDMA-exposed rats make during their second 5 minutes in the new cage?

3. Which rats moved around the most during the last 5 minutes of the study?

4. Does this study support the hypothesis that MDMA affects a developing rat's brain?

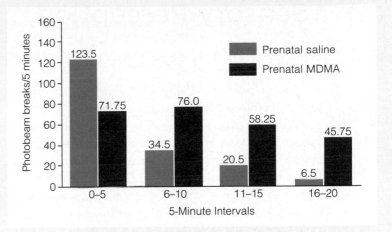

Figure 33.29 Effect of prenatal exposure to MDMA on activity levels of 21-day-old rats placed in a new cage. Movements were detected when the rat interrupted a photobeam. Rats were monitored at 5-minute intervals for a total of 20 minutes. *Blue* bars are results for rats whose mothers received saline, *red* bars are rats whose mothers received MDMA.

Self-Quiz

Answers in Appendix III

1. _____ relay messages from the brain and spinal cord to muscles and glands.
   a. Motor neurons b. Interneurons c. Sensory neurons

2. When a neuron is at rest, _____ .
   a. it is at threshold potential
   b. gated sodium channels are open
   c. the sodium–potassium pump is operating
   d. both a and c

3. Action potentials occur when _____ .
   a. a neuron receives adequate stimulation
   b. more and more sodium gates open
   c. sodium–potassium pumps kick into action
   d. both a and b

4. True or false? Action potentials vary in their size.

5. Neurotransmitters are released by _____ .
   a. axon terminals   c. dendrites
   b. the cell body   d. the myelin sheath

6. What chemical is released by axon terminals of a motor neuron at a neuromuscular junction?
   a. ACh   b. serotonin   c. dopamine   d. epinephrine

7. Which neurotransmitter is important in reward-based learning and drug addiction?
   a. ACh   b. serotonin   c. dopamine   d. epinephrine

8. Skeletal muscles are controlled by _____ .
   a. sympathetic signals   c. somatic nerves
   b. parasympathetic signals   d. both a and b

9. When you sit quietly on the couch and read, output from _____ neurons prevails.
   a. sympathetic   b. parasympathetic

10. Cell bodies of the sensory neurons that deliver signals to the spinal cord are in the _____ .
   a. white matter b. gray matter c. dorsal root ganglia

11. Which of the following are not in the brain?
   a. Schwann cells   b. astrocytes   c. microglia

12. True or false? Neurons do not divide in adults.

13. Match each item with its description.
   ___muscle spindle
   ___neurotransmitter
   ___limbic system
   ___corpus callosum
   ___cerebral cortex
   ___neural tube
   ___neuroglia
   ___white matter
   ___blood–brain barrier

   a. start of brain, spinal cord
   b. connects the hemispheres
   c. protects brain and spinal cord from some toxins
   d. type of signaling molecule
   e. support team for neurons
   f. stretch-sensitive receptor
   g. roles in emotion, memory
   h. most complex integration
   i. myelinated axons of neurons

■ *Visit CengageNOW for additional questions.*

Critical Thinking

1. In humans, the axons of some motor neurons extend more than a meter, from the base of the spinal cord to the big toe. What are some of the functional challenges involved in the development and maintenance of such impressive cellular extensions?

2. Some survivors of disastrous events develop post-traumatic stress disorder (PTSD). Symptoms include nightmares about the experience and suddenly feeling as if the event is recurring. Brain-imaging studies of people with PTSD showed that their hippocampus was shrunken and their amygdala unusually active. Given these changes, what other brain functions might be disrupted in PTSD?

3. In human newborns, especially premature ones, the blood–brain barrier is not yet fully developed. Why is this one reason to pay careful attention to the diet of infants?

34 Sensory Perception

A Whale of a Dilemma

Imagine yourself in the sensory world of a whale, 200 meters (650 feet) beneath the ocean surface. Almost no sunlight penetrates this deep, so the whale sees little as its moves through water. Many fishes detect motion with a lateral line system, which responds to differences in water pressure. Fishes also use dissolved chemicals as navigational cues. However, a whale has no lateral line, and it has a very poor sense of smell. How does it know where it is going?

All whales use sounds—acoustical cues. Water is an ideal medium for transmitting sound waves, which move five times faster in water than in air. Unlike humans, whales do not have a pair of ear flaps that collect sound waves. Some whales do not even have a canal leading to ear components inside their head. Others have ear canals packed with wax. How, then, do whales hear? Their jaws pick up vibrations traveling through water. The vibrations are transmitted from the jaws, through a layer of fat, to a pair of pressure-sensitive middle ears.

Whales use sound to communicate, locate food, and find their way around underwater. Killer whales and some other species of toothed whales use echolocation. The whale emits high-pitched sounds and then listens as the echoes bounce off objects, including prey. Its ears are especially sensitive to sounds of high frequencies. Baleen whales, including the humpback whale, make very low-pitched sounds that can travel across an entire ocean basin. Their ears are adapted to detect those sounds.

The ocean is becoming a lot noisier, and the superb acoustical adaptations of whales now put them at risk. For example, in 2001 some whales beached themselves near an area where the United States Navy was testing a sonar system (Figure 34.1). This system emits loud low-frequency sounds and uses their echoes to locate submarines. Humans cannot hear the sonar sounds. Whales can.

As autopsies later revealed, the beached whales had blood in their ears and in acoustic fat. Apparently the intense sounds emitted by the sonar made them race to the surface in fear. Rapid change in pressure damaged internal tissues.

Sonar testing continues because the threat of stealth submarine attacks against the United States is real. Also, noise from commercial shipping may be a worse problem for whales. Massive tankers generate low-frequency sounds that frighten whales or drown out acoustical cues. Realistically, global shipping of oil and other resources that industrial nations require is not going to stop. If research shows that whales are at risk, will those same nations be willing to design and deploy newer, more expensive tankers that are quieter?

In this chapter, we turn to sensory systems. Using these organ systems, animals detect stimuli inside and outside their body and become aware of touches, sounds, sights, odors, and other sensations. As you will learn, animals differ in their type and number of sensory receptors that sample the environment, and thus also differ in their perception of that environment.

See the video! Figure 34.1 A few children drawn to one of the whales that stranded itself during military testing of a new sonar system. Of sixteen stranded whales, six died on the beach. Volunteers pushed the others out to sea. Their fate is unknown.

Key Concepts

How sensory pathways work

Sensory receptors detect specific stimuli. Different animals have receptors for different stimuli. Information from sensory receptors becomes encoded in the number and frequency of action potentials sent to the brain along particular nerve pathways. **Section 34.1**

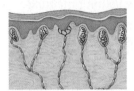

Somatic and visceral senses

Somatic sensations such as touch are easily localized and arise from receptors in the skin, muscles, or near joints. Visceral sensations, such as a feeling of fullness in your stomach, are less easily pinpointed. They arise from receptors in the walls of internal organs. **Section 34.2**

Chemical senses

The senses of smell and taste require chemoreceptors, which bind molecules of specific substances dissolved in the fluid bathing them. **Section 34.3**

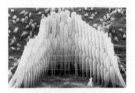

Balance and hearing

Organs in the ear function in balance and in hearing. The inner ear's vestibular apparatus detects body position and motion. The outer and middle ear collect and amplify sound waves. Mechanoreceptors in the inner ear send signals about sound to the brain. **Sections 34.4–34.6**

Vision

Most organisms have light-sensitive pigments, but vision requires eyes. Vertebrates have an eye that operates like a film camera. Their retina, which has photoreceptors, is analogous to the film. A sensory pathway starts at the retina and ends in the visual cortex. **Sections 34.7–34.10**

Links to Earlier Concepts

- This chapter builds heavily on the previous one. You will see examples of action potentials (Section 33.3), and learn more about neuromodulators (33.6), the stretch reflex (33.9), and the limbic system and cerebral cortex (33.11).

- Our discussions of evolution of sensory organs will refer to earlier sections about morphological convergence (19.2), vertebrate evolution (26.2), and primate evolution (26.13) in particular.

- In discussing vision, we return to the topic of pigments (7.1), and to the effects of Vitamin A deficiency (Chapter 16 introduction).

- You will also learn about how pathogenic amoebas (?? 11) and roundworms (25.11) can harm vision.

How would you vote? Maritime activities such as shipping cause an underwater ruckus. Would you support a ban on activities that generate excessive noise levels from territorial waters of the United States and other nations? See CengageNOW for details, then vote online.

■ An animal's sensory receptors determine what features of the environment it can detect and respond to.

■ Links to Action potentials 33.3, Stretch reflex 33.9

As the previous chapter explained, an animal's sensory neurons detect specific stimuli, or forms of energy, in the internal or external environment. Stimulation of the receptor endings of a sensory neuron causes action potentials that travel along the plasma membrane.

Sensory Receptor Diversity

All animals that have neurons have sensory neurons. However, the types of stimuli these neurons detect vary among animal groups. We can classify sensory neurons based on the kinds of stimuli to which they respond.

Mechanoreceptors are sensory endings that respond to mechanical energy. Some detect a body's position or acceleration. For example, a jellyfish can tell which way is up because it has cells with statoliths. A statolith is a dense object that shifts position when a cell's orientation changes. Shifts trigger action potentials.

Other mechanoreceptors fire off action potentials in response to touch or to stretching of a body part. The muscle spindles involved in the human stretch reflex (Section 33.9) are a type of mechanoreceptor.

Still other mechanoreceptors respond to vibrations caused by pressure waves. Hearing involves this type of receptor. As the chapter introduction noted, different animals detect sound waves of different frequencies. Whales detect ultra-low frequencies that humans cannot hear. Bats emit and respond to sounds too high for humans to perceive (Figure 34.2a).

Pain receptors, also called nociceptors, detect tissue damage. They have a protective function and are often involved in reflexes that minimize further harm.

Some **thermoreceptors** respond to a specific temperature; others fire in response to a temperature change. Pythons and some other snakes have thermoreceptors concentrated in pits on their head (Figure 34.2b). These receptors help a snake detect warm-blooded prey.

Chemoreceptors detect specific solutes dissolved in a fluid. Nearly all animals have chemoreceptors that help them locate chemical nutrients and avoid taking in poisons. Chemoreceptors also function in smell.

Osmoreceptors detect a change in the concentration of solutes in a body fluid, such as blood.

Figure 34.2 Examples of sensory receptors. (**a**) Mechanoreceptors inside a bat's inner ear allow the animal to detect high-pitched, or ultrasonic, pressure waves. (**b**) Thermoreceptors in pits above and below a python's mouth allow it to detect body heat, or infrared energy, of nearby prey.

Figure 34.3 A marsh marigold looks yellow to humans (**a**), but photographing it with UV-sensitive film reveals a dark area around the reproductive parts (**b**). This pattern is caused by UV-absorbing pigment and is visible to insect pollinators.

Photoreceptors detect light energy. Humans detect only visible light, but insects and some other animals, including rodents, also respond to ultraviolet light. Flowers often have UV-absorbing pigments arranged in patterns that are invisible to us, but obvious to their insect pollinators (Figure 34.3).

From Sensing to Sensation

In animals that have a brain, the processing of sensory signals gives rise to sensation: awareness of a stimulus. Sensation is different than perception, which refers to a conscious understanding of what a sensation means.

Sensory receptors in skin, skeletal muscles, or near the joints give rise to somatic sensations. Sensations of touch and warmth are examples. Visceral sensations, such as the feeling that your bladder or stomach is full, arise from receptors in internal organs. Sensory receptors restricted to specific sensory organs, such as eyes or ears, function in special senses—vision, smell, balance, hearing, and taste.

For example, stretch receptors in a gymnast's arm and leg muscles keep the brain informed of changes in muscle length (Figure 34.4a). The gymnast's brain integrates this sensory input with signals from eyes and the organs of balance in the inner ear, then issues commands that cause muscles to adjust their length and help maintain balance and posture.

Stimulation of a sensory receptor produces action potentials which, remember, are always the same size (Section 33.3). The brain gets additional information about stimuli by noting which nerve pathways carry the action potentials, the frequency of action potentials traveling on each axon in the pathway, and the number of axons recruited by the stimulus.

First, an animal's brain is prewired, or genetically programmed, to interpret action potentials in certain ways. That is why you may "see stars" after an eye gets poked, even in a dark room. Photoreceptors in the eye that are mechanically disturbed send signals along one of two optic nerves to the brain. The brain interprets all signals from an optic nerve as "light."

Second, a strong signal makes receptors fire action potentials more often and longer than a weak signal. The same receptors are stimulated by a whisper and a whoop. Your brain interprets the difference by variations in frequency of signals (Figure 34.4b).

Third, a stronger stimulus recruits more sensory receptors, compared to a weak stimulus. A gentle tap on the arm activates fewer receptors than a slap.

Stimulus duration also affects response. In **sensory adaptation**, sensory neurons cease firing in spite of

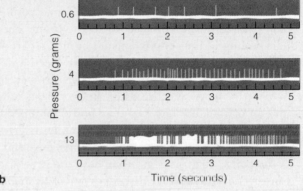

Figure 34.4 Animated (a) A young gymnast benefiting from information flowing from his muscle spindles and other sensory receptors to his brain.

(b) Recordings of action potentials from a pressure receptor with endings in a human hand. The graphs chart the variations in stimulus strength. A thin rod was pressed against skin with varying amounts of pressure. Vertical bars above each thick horizontal line record individual action potentials. Frequency of action potentials rises with each increase in stimulus strength.

continued stimulation. Put on a sock and you briefly feel it against your skin, but you quickly lose your awareness of it. Mechanoreceptors in the skin adapt to this stimulus, allowing you to focus on other things.

Take-Home Message

How do animals detect and process sensory stimuli?

■ Sensory neurons undergo action potentials in response to specific stimuli. Different kinds of sensory receptors respond to different types of stimuli.

■ In animals with a brain, input from sensory neurons can give rise to sensation.

■ Action potentials are all the same size, but which axons are responding, how many are responding, and the frequency of action potentials provides the brain with information about stimulus location and strength.

- Signals from receptors in the skin, joints, muscles, and internal organs flow through the spinal cord to the brain.

- Links to Neuromodulators 33.6, Cerebral cortex 33.11

Sensory neurons responsible for somatic sensations are located in skin, muscle, tendons, and joints. **Somatic sensations** are easily localized to a specific part of the body. In contrast, **visceral sensations**, which arise from neurons in the walls of soft internal organs, are often difficult to pinpoint. It is easy to determine exactly where someone is touching you, but less easy to say exactly where you feel a stomachache.

The Somatosensory Cortex

Signals from the sensory neurons involved in somatic sensation travel along axons to the spinal cord, then along tracts in the spinal cord to the brain. The signals end up in the **somatosensory cortex**, a part of the cerebral cortex. Like the motor cortex (Section 33.11), the somatosensory cortex has neurons arrayed like a map of the body (Figure 34.5). Body parts shown as disproportionately large in the "body" mapped onto this brain correspond to body regions with the most sensory receptors, such as the fingertips, face, and lips. Body parts, such as legs, that have relatively fewer sensory neurons appear disproportionately small.

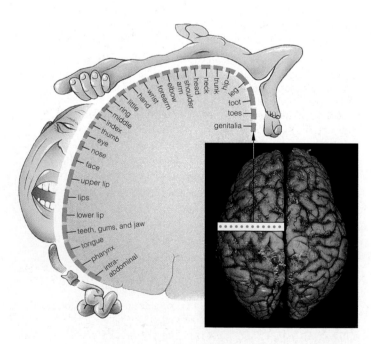

Figure 34.5 A map showing where the different body regions are represented in the human primary somatosensory cortex. This brain region is a narrow strip of the cerebral cortex that runs from the top of the head to just above each ear. Compare Figure 33.21.

Receptors Near the Body Surface

As an example of the types of receptors that report to the somatosensory cortex, consider those in the human skin (Figure 34.6). Free nerve endings that coil around the roots of hairs in the dermis detect even the slightest pressure. Other free nerve endings detect temperature changes or tissue damage. Free nerve endings also occur in skeletal muscles, tendons, joints, and walls of internal organs. Here, they give rise to sensations that range from itching, to a dull ache, to sharp pain.

Other skin receptors are surrounded by a capsule and are named for the scientists who first described them. Meissner's corpuscles and Pacinian corpuscles are the main receptors that detect touch and pressure in hairless skin regions such as fingertips, palms, and the soles of feet. Small Meissner's corpuscles lie in the upper dermis and detect light touches. Pacinian are larger and respond to stronger pressure. They lie deeper in the dermis and also occur near joints and in the wall of some organs. Concentric layers of connective tissue wrap around their sensory endings.

Either pressure or a specific temperature can cause other encapsulated receptors to respond. Ruffini endings adapt more slowly than Meissner's and Pacinian corpuscles. If you hold a stone in your hand, Ruffini endings inform your brain that the stone is still there even after other receptors have adapted and stopped responding. Ruffini endings also fire when temperature exceeds 45°C (113°F). The bulb of Krause, also an encapsulated receptor, responds to touch and cold. It is found in skin and certain mucous membranes.

Muscle Sense

Remember those stretch receptors in muscle spindle fibers (Section 33.9)? The more a muscle stretches, the more frequently stretch receptors fire. In concert with receptors in tendons and near movable joints, they inform the brain about positions of the body's limbs.

The Sense of Pain

Pain is the perception of a tissue injury. Somatic pain is a response to signals from pain receptors in skin, skeletal muscles, joints, and tendons. Visceral pain is associated with organs inside body cavities. It occurs as a response to a smooth muscle spasm, inadequate blood flow to an internal organ, over-stretching of a hollow organ, and other abnormal conditions.

Injured or distressed body cells release chemicals that stimulate nearby pain receptors. Signals from the

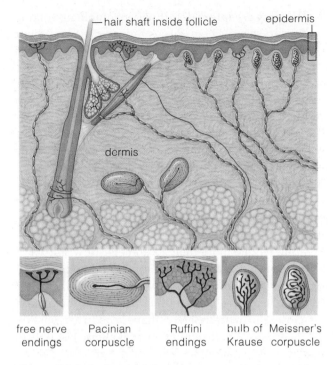

Figure 34.6 **Animated** Sensory receptors in human skin.

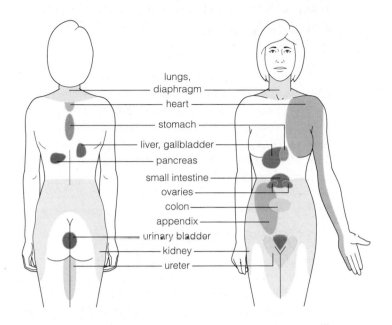

Figure 34.7 **Animated** Sites of referred pain. Colored regions indicate the area that the brain interprets as affected when specific internal organs are actually distressed.

pain receptors then travel along the axons of sensory neurons to the spinal cord. Here, the sensory axons synapse with the spinal interneurons that relay signals about pain to the brain. The signals proceed through the brain to the cerebral cortex, where they are assessed and the appropriate responses are set in motion.

Numerous substances affect signal transmission at the synapse between a pain-detecting sensory neuron and a spinal interneuron. For example, substance P (a neuromodulator) makes the interneurons more likely to send signals to the sensory cortex. In contrast, the natural opiates—endorphins and enkephalins (Section 33.6)—impair flow of signals along the pain pathway.

Pain relievers, or analgesics, interfere with steps in the pain pathway. For example, aspirin reduces pain by slowing production of prostaglandins. These local signaling molecules, which are released by damaged tissues, increase the sensitivity of pain receptors to stimulation. As another example, synthetic opioids such as morphine mimic the activity of endorphins.

The drug ziconotide is a chemical first discovered in the venom of a cone snail (Chapter 24 introduction). When injected into the spinal cord, ziconotide blocks calcium channels in axon terminals of pain receptor neurons. Because calcium ion inflow is necessary for neurotransmitter release (Section 33.5), preventing it keeps signals from reaching spinal interneurons that normally convey pain signals to the brain.

Sometimes, the brain mistakenly interprets signals about a visceral problem as if the signals were coming from the skin or joints. The result is called **referred pain**. The classic example is a pain that radiates from chest across the shoulder and down the left arm during a heart attack (Figure 34.7). Tissue in the heart, not the arm, is affected so why does the arm hurt?

The answer lies in the construction of the nervous system. Each level of the spinal cord receives sensory input from the skin as well as from some of the organs. The skin encounters more painful stimuli than the organs do, so its signals more often flow along the pathway to the brain. The brain sometimes attributes signals that arrive along a pathway to their most common source—skin—even if they originate elsewhere.

Take-Home Message

How do somatic and visceral sensations arise?

■ Somatic sensations are signals from sensory receptors in skin, skeletal muscle, and joints. They travel along sensory neuron axons to the spinal cord, then to the somatosensory cortex.

■ Visceral sensations begin with the stimulation of sensory neurons in the walls of organs inside the body. These signals are relayed to the spinal cord, and then the brain.

■ Pain is the sensation associated with tissue damage. Because pain signals originate most often with somatic sources, the brain sometimes misinterprets visceral pain as if it were caused by a problem in the skin or a joint.

Sampling the Chemical World

■ Both smell and taste begin with chemoreceptors.

■ Link to Limbic system 33.11

Sense of Smell

Olfaction, a sense of smell, starts with chemoreceptors that bind specific substances. A stimulus can trigger action potentials that olfactory nerves transmit to the cerebral cortex. Messages also travel to the limbic system, which integrates them with emotional state and stored memories (Section 33.11).

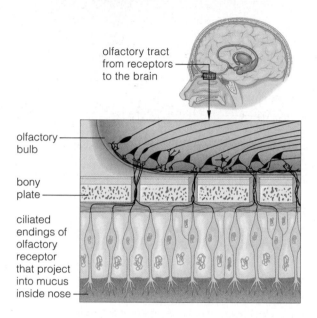

olfactory tract from receptors to the brain

olfactory bulb

bony plate

ciliated endings of olfactory receptor that project into mucus inside nose

Figure 34.8 Pathway from the sensory endings of olfactory receptors in the human nose to the cerebral cortex and limbic system. Axons of these sensory receptors pass through holes in a bony plate between the lining of the nasal cavities and the brain.

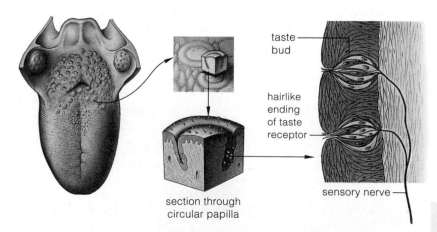

taste bud

hairlike ending of taste receptor

sensory nerve

section through circular papilla

Figure 34.9 Taste receptors in the human tongue. Taste buds are clusters of receptor cells and supporting cells inside special epithelial papillae. One type, a circular papilla, is shown in section here. The tongue has about 5,000 taste buds, each enclosing as many as 150 taste receptor cells.

Olfactory receptors detect water-soluble or volatile (easily vaporized) chemicals. A human nose has about 5 million olfactory receptors; a bloodhound nose has 200 million. Receptor axons send action potentials to two olfactory bulbs. These small brain structures sort out components of a scent, then signal the cerebrum for further processing (Figure 34.8).

Many animals use olfactory cues to find their way, locate food, and communicate socially. A **pheromone** is a type of signaling molecule that is secreted by one individual and affects the behavior of other members of its species. For example, female silk moths secrete a sex pheromone. Male silk moths have antennae with olfactory receptors that help them detect a pheromone-secreting female more than a kilometer upwind.

Reptiles and most mammals, have a **vomeronasal organ**, a collection of sensory neurons in the nasal cavity that is sensitive to pheromones. Humans and our closest primate relatives have a reduced version of this organ. Whether humans make and respond to pheromones remains a matter of debate. We discuss the role of pheromones in more detail in Chapter 44.

Sense of Taste

Taste receptors are also chemoreceptors that detect chemicals dissolved in fluid, but they have a different structure and location than olfactory receptors. Taste receptors help animals locate food and avoid poisons. An octopus "tastes" with receptors in suckers on its tentacles; a fly "tastes" using receptors in its antennae and feet. In humans, many taste buds are embedded in the upper surface of the tongue (Figure 34.9). These sensory organs are located in specialized epithelial structures, or papillae, that look like raised bumps or red dots on the tongue surface.

You perceive many tastes, but all are a combination of five main sensations: *sweet* (elicited by glucose and the other simple sugars), *sour* (acids), *salty* (sodium chloride or other salts), *bitter* (plant toxins, including alkaloids), and *umami* (elicited by amino acids such as glutamate which, as in aged cheese and aged meat, has a savory taste). The food additive MSG (monosodium glutamate) can enhance flavor by stimulating the taste receptors that contribute to the sensation of umami.

Take-Home Message

How are the senses of smell and taste similar?

■ Smell and taste begin with the stimulation of chemoreceptors by the binding of specific dissolved molecules.

34.4 | Sense of Balance

■ Organs inside your inner ear are essential to maintaining posture and a sense of balance.
■ Somatic sensory receptors also contribute to balance.

Organs of equilibrium are parts of sensory systems that monitor the body's positions and motions. Each vertebrate ear includes such organs inside a fluid-filled sensory structure called the **vestibular apparatus**. The organs are located in three semicircular canals and in two sacs, the saccule and utricle (Figure 34.10a).

Organs of the vestibular apparatus have **hair cells**, a type of mechanoreceptor with modified cilia at one end. Fluid pressure inside the canals and sacs makes the cilia bend. The mechanical energy of this bending deforms the hair cell plasma membrane just enough to let ions slip across and stimulate an action potential. A vestibular nerve carries the sensory input to the brain. As you will see, other hair cells function in hearing.

The three semicircular canals are oriented at right angles to one another, so rotation of the head in any combination of directions—front/back, up/down, or left/right—moves the fluid inside them. An organ of equilibrium rests on the bulging base of each canal. The cilia of its hair cells are embedded in a jellylike mass (Figure 34.10b). When fluid moves in the canal, it pushes against the mass and generates the pressure required for initiating action potentials.

The brain receives signals from semicircular canals on both sides of the head. By comparing the number and frequency of action potentials coming from each side of the head, the brain senses dynamic equilibrium: the angular movement and rotation of the head. Among other things, this sense allows you to keep your eyes locked on an object even when you swivel your head or nod.

Organs in the saccule and utricle act in the sense of static equilibrium. These organs help the brain keep track of the head's position and how fast it is moving in a straight line. They also help keep the head upright and maintain posture.

Inside the saccule and utricle, is a jellylike mass weighted with calcite statoliths. This mass lies on top of mechanoreceptors (hair cells). When you tilt your head, or start or stop moving, the weighted mass shifts, bending hair cells and altering their rate of action potentials.

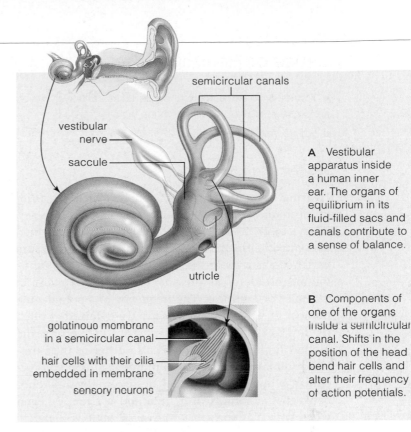

A Vestibular apparatus inside a human inner ear. The organs of equilibrium in its fluid-filled sacs and canals contribute to a sense of balance.

B Components of one of the organs inside a semicircular canal. Shifts in the position of the head bend hair cells and alter their frequency of action potentials.

Figure 34.10 Animated Organs of equilibrium in the inner ear.

The brain also takes into account information from the eyes, and from receptors in the skin, muscles, and joints. Integration of the signals provides awareness of the body's position and motion in space, as shown by figure skater Sarah Hughes at left.

A stroke, an inner ear infection, or loose particles in the semicircular canals can cause vertigo, a sensation that the world is moving or spinning around. Vertigo also arises from conflicting sensory inputs, as when you stand at a height and look down. The vestibular apparatus reports that you are motionless, but your eyes report that your body is floating in space.

Mismatched signals can cause motion sickness. On a curvy road, passengers in a car experience changes in acceleration and direction that scream "motion" to their vestibular apparatus. At the same time, signals from their eyes about objects inside the car tell their brain that the body is at rest. Driving can minimize motion sickness because the driver focuses on sights outside the car such as scenery rushing past, so the visual signals are consistent with vestibular signals.

Take-Home Message

What gives us our sense of balance?

■ Mechanoreceptors in the fluid-filled vestibular apparatus of the inner ear detect the body's position in space, and when we start or stop moving.

■ Your ears collect, amplify, and sort out sound waves, which are pressure waves traveling through the air.

■ Link to Vertebrate evolution 26.2

Properties of Sound

Hearing is the perception of sound, which is a form of mechanical energy. A sound arises when vibration of an object causes pressure variations in air, water, or some other medium. We can represent the pressure variations as waveforms. The amplitude of a sound—the magnitude of its pressure waves—determines its intensity or loudness. The frequency of a sound—the number of wave cycles per second—determines pitch (Figure 34.11). The more wave cycles per second, the higher the frequency. Sounds also differ in their timbre or quality. Differences in timbre can help you recognize people by their voices, or discern the difference between the sounds of a flute and a trumpet, even when both play the same note at the same volume.

The Vertebrate Ear

Water readily transfers vibrations to body tissues, so fishes do not require elaborate ears to detect sounds. When vertebrates left water for land, their capacity to collect and amplify vibrations evolved in response to a new environmental challenge: transfer of sound waves from the air to body tissues is inefficient. The structure of human ears helps maximize the efficiency of transfer.

As Figure 34.12*a* indicates, the **outer ear** of humans and most other mammals is adapted to gathering sounds from the air. The pinna, a skin-covered flap of cartilage projecting from the side of the head, collects sound waves and directs them into the auditory canal. The canal conveys sounds to the middle ear.

The **middle ear** amplifies and transmits air waves to the inner ear. An **eardrum**, or tympanic membrane, first evolved in early reptiles as a shallow depression on each side of the head. Pressure waves cause this thin membrane to vibrate. Behind the eardrum is an air-filled cavity and three small bones known as the hammer, anvil, and stirrup (Figure 34.12*b*). These bones transmit the force of sound waves from the eardrum

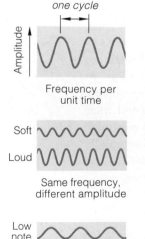

one cycle

Amplitude

Frequency per
unit time

Soft

Loud

Same frequency,
different amplitude

Low
note

High
note

Same amplitude,
different frequency

Figure 34.11 Animated
Wavelike properties of sound.

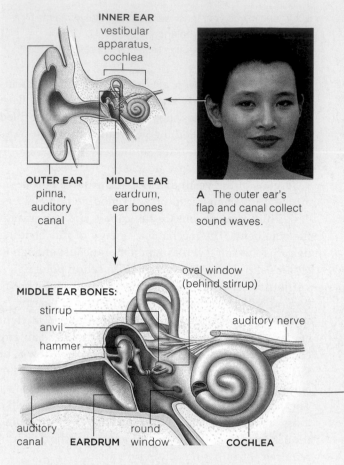

INNER EAR
vestibular
apparatus,
cochlea

OUTER EAR
pinna,
auditory
canal

MIDDLE EAR
eardrum,
ear bones

A The outer ear's
flap and canal collect
sound waves.

oval window
(behind stirrup)

MIDDLE EAR BONES:

stirrup

anvil

hammer

auditory nerve

auditory
canal

EARDRUM

round
window

COCHLEA

B The eardrum and middle ear bones amplify sound.

Figure 34.12 Animated How humans hear.

to the smaller surface of the oval widow. This flexible membrane is the boundary between the middle ear and the inner ear.

The **inner ear**, remember, has a vestibular apparatus that functions in the sense of balance (Section 34.4). It also has a **cochlea**, which in humans is a pea-sized, fluid-filled structure that resembles a coiled snail shell (the Greek word *koklias* means snail).

If you could straighten out a cochlea and look inside it, you would notice two fluid-filled compartments (Figure 34.12*c*). One compartment bends in a U-shape. Its two arms are known as the vestibular duct and tympanic duct. The other compartment, the cochlear duct, lies between the arms of the "U."

When sound waves make the three tiny bones of the middle ear vibrate, the stirrup pushes against the oval window. The oval window bows inward, creating a fluid pressure wave. The wave travels through fluid of the vestibular and tympanic ducts, until it reaches the round window, which bows outward in response.

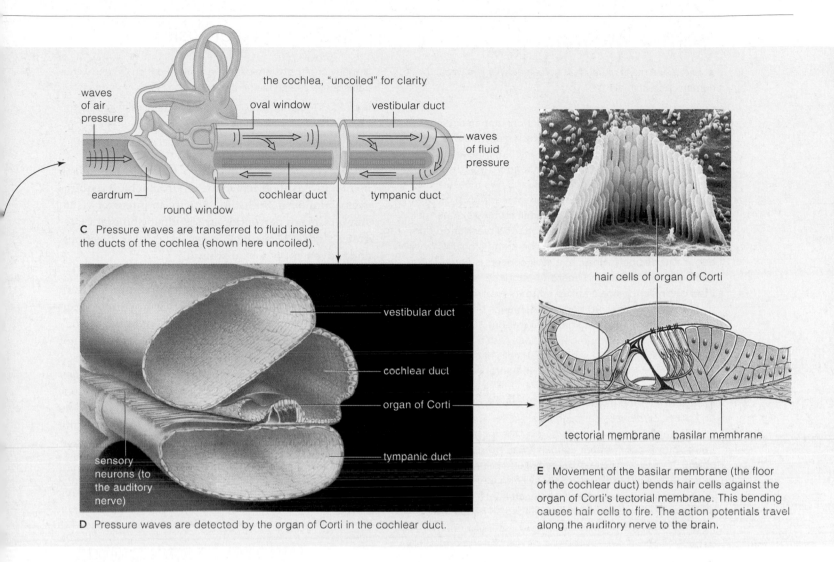

waves of air pressure

the cochlea, "uncoiled" for clarity

oval window

vestibular duct

waves of fluid pressure

eardrum

round window

cochlear duct

tympanic duct

C Pressure waves are transferred to fluid inside the ducts of the cochlea (shown here uncoiled).

vestibular duct

cochlear duct

organ of Corti

sensory neurons (to the auditory nerve)

tympanic duct

D Pressure waves are detected by the organ of Corti in the cochlear duct.

hair cells of organ of Corti

tectorial membrane   basilar membrane

E Movement of the basilar membrane (the floor of the cochlear duct) bends hair cells against the organ of Corti's tectorial membrane. This bending causes hair cells to fire. The action potentials travel along the auditory nerve to the brain.

As fluid shifts back and forth between the round window and oval window, pressure waves cause the lower wall of the cochlear duct to begin vibrating up and down. This lower wall is the basilar membrane (Figure 34.12*d,e*). Sitting on top of the membrane is the **organ of Corti**, an acoustical organ with arrays of hair cells. A hair cell is a mechanoreceptor with a tuft of modified cilia at one end. The cilia project into a tectorial membrane that drapes over them. Movement of the basilar membrane pushes cilia against the tectorial membrane. When the cilia bend, the hair cells undergo action potentials, which then travel along an auditory nerve to the brain.

The number of hairs cells that fire and the frequency of their signals inform the brain how loud a sound is. The louder a sound, the more action potentials flow along the auditory nerve to the brain.

The brain can determine the pitch of a sound by assessing which part of the basilar membrane is vibrating most. The basilar membrane is not uniform along its length. It is stiff and narrow near the oval window, and broader and more flexible deeper into the coil. High-pitched sounds make the stiff, narrow, closer-in part of the basilar membrane vibrate most. Low-pitched sounds cause vibrations mainly in the wide flexible part close to the membrane's tip. More vibrations make more hair cells in that region fire.

Hearing loss or deafness can occur because sound waves do not reach the inner ear, as when an eardrum is ruptured or ear bones do not move properly. It can also occur because of auditory nerve damage or hair cell loss. Some antibiotic drugs can kill hair cells. So can loud noise, a topic we consider in the next section.

Take-Home Message

How do vertebrates hear?

■ Human ears collect pressure waves from the surroundings and convert them to pressure waves in fluid inside the inner ear. Pressure waves in this fluid stimulate hair cells, which are auditory receptors that send action potentials along auditory nerves to the brain.

34.6 | Noise Pollution

■ Excessive noise caused by human activity is a threat to humans and animals.

As detailed in the chapter introduction, human activities have made the world's oceans a noisy place. This noise alters the sensory world through which marine animals move, alters their behavior, and endangers their health.

Things are not much quieter on land. We measure the intensity of a sound in decibels. An increase of 10 on this scale means an increase of ten fold in loudness. A normal conversation is about 60 decibels, a food blender operating at high speed is about 90 decibels, and a chain saw is about 100 decibels. Music at a rock concert is about 120 decibels. So is the sound heard through the earbuds of an iPod or similar device cranked up to its maximum volume.

Noise louder than 90 decibels damages hair cells in the cochlea (Figure 34.13). Humans have about 30,000 such cells at birth, and the number declines with age. Exposure to loud noise accelerates loss of hair cells and of hearing.

In humans, a high level of environmental noise also impairs concentration and interferes with sleep patterns. It raises anxiety and increases the risk of high blood pressure and other cardiovascular problems.

Land animals are also affected by the increasing din. Loud sounds can frighten animals away from food or young. It can also distract them, making them vulnerable to predators. In birds that rely heavily on auditory signals during courtship, man-made noise can interfere with the ability to find and secure a mate. Canadian researchers recently reported the effects of noisy compressors used to extract oil and gas on ovenbirds, a type of song bird. Birds that share their habitat with the noisy machinery have 15 percent fewer offspring than those in quiet forest habitat.

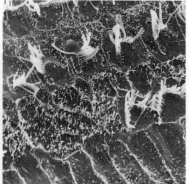

Figure 34.13 Results of an experiment on the effect of intense sound on the inner ear. *Left,* from a guinea pig ear, two rows of hair cells that normally project into the tectorial membrane in the organ of Corti. *Right,* hair cells inside the same organ after twenty-four hours of exposure to noise levels comparable to extremely loud music.

34.7 | Sense of Vision

■ Many organisms are sensitive to light, but only those with a camera eye see an image as you do.

■ Links to Morphological convergence 19.2, Primates 26.13

Requirements for Vision

Vision is detection of light in a way that provides a mental image of objects in the environment. It requires eyes and a brain with the capacity to interpret visual stimuli. Image perception arises when the brain integrates signals regarding shapes, brightness, positions, and movement of visual stimuli.

Eyes are sensory organs that hold photoreceptors. Pigment molecules inside the photoreceptors absorb light energy. That energy is converted to the excitation energy in action potentials that are sent to the brain.

Certain invertebrates, such as earthworms, do not have eyes, but they do have photoreceptors dispersed under the epidermis or clustered in parts of it. They use light as a cue to orient the body, detect shadows, or adjust biological clocks, but they do not have a true sense of vision. Detecting visual detail requires many photoreceptors, and many invertebrate eyes do not have many such receptors.

The quality of the image formed by an eye improves with a **lens**, a transparent body that bends light rays

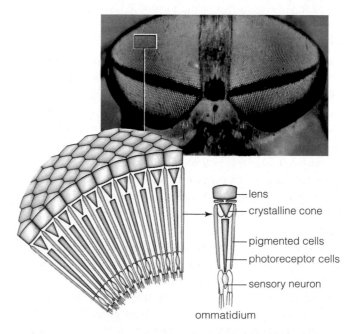

Figure 34.14 The compound eye of a deerfly, with many densely packed, identical units called ommatidium. Each unit has a lens that focuses light on photoreceptor cells. Although the mosaic image produced by such an eye is fuzzy, the eye is very good at detecting movement.

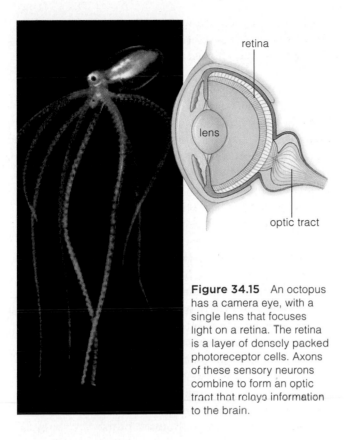

retina

lens

optic tract

Figure 34.15 An octopus has a camera eye, with a single lens that focuses light on a retina. The retina is a layer of densely packed photoreceptor cells. Axons of these sensory neurons combine to form an optic tract that relays information to the brain.

Figure 34.16 In owls, eyes face forward and photoreceptors are concentrated near the top of the inner eyeball. Such birds mainly look down for prey. When on the ground, they must turn their heads almost upside down to see something above their head.

from any point in the visual field so that rays converge on photoreceptors. Light rays bend at boundaries between substances of different densities.

Insects have **compound eyes** with many lenses, each in a separate unit known as an ommatidium (Figure 34.14). The brain constructs images based on the light intensities detected by the different units. Compound eyes do not provide the clearest vision, but they are highly sensitive to movement.

Cephalopod mollusks such as squids and octopuses have the most complex eyes of any invertebrate (Figure 34.15). Their **camera eyes** have an adjustable opening that allows light to enter a dark chamber. Each eye's single lens focuses incoming light onto a **retina**, a tissue densely packed with photoreceptors. The retina of a camera eye is analogous to the light-sensitive film used in a traditional film camera. Signals from the photoreceptors in each eye travel along one of the two optic tracts to the brain. Compared to compound eyes, camera eyes yield a more sharply defined and detailed image.

Vertebrates also have camera eyes, and because they are distant relatives of cephalopod mollusks, camera eyes are presumed to have evolved independently in the two lineages. This is an example of morphological convergence (Section 19.2).

Many animals have eyes placed on either side of the head, which maximizes the visible area. Predators, including owls, tend to have two eyes that face forward (Figure 34.16). Having two eyes that both survey the same area supplies the brain with overlapping information that enhances depth perception. The brain can compare information from the eyes to determine how far apart objects are.

Primates have good depth perception. As Section 26.13 explained, primates evolved from a shrewlike ancestor that had eyes on either side of its head. The enhanced depth perception from forward-facing eyes may have provided an advantage when early primates began living in and moving through the treetops.

Take-Home Message

How do animal visual systems differ?

■ Some animals such as earthworms have photoreceptors that detect light, but do not form any sort of image.

■ Other animals, including insects, have compound eyes. A compound eye has many individual units, each with its own lens. It produces a mosaic image that is fuzzy, but highly sensitive to movement.

■ A camera eye with an adjustable opening and a lens that focuses light on a photoreceptor-rich retina provides a richly detailed image. Camera eyes evolved independently in cephalopod mollusks and vertebrates.

■ The human eye is a multilayered structure with a light-bending cornea, a focusing lens, and a photoreceptor-rich retina. The eye is surrounded by protective structures.

Anatomy of the Eye

Each human eyeball sits inside a protective, cuplike, bony cavity called the orbit. Skeletal muscles that run from the rear of the eye to the bones of the orbit move the eyeball up and down or side to side.

Eyelids, eyelashes, and tears all help protect delicate eye tissues. Periodic blinking is a reflex that spreads a film of tears over the eyeball's exposed surface. Tears are secreted by glands in the eyelids and consist of water, lipids, salts, and proteins. Among the proteins are enzymes that break down bacterial cell walls and thus help prevent eye infections.

A protective mucous membrane, the **conjunctiva**, lines the inner surface of the eyelids and folds back to cover most of the eye's outer surface. Conjunctivitis, commonly called pinkeye, is an inflammation of this membrane. A viral or bacterial infection can cause it.

The eyeball is spherical, and has a three-layered structure (Figure 34.17). The front portion of each eye is covered by a **cornea** made of transparent crystalline proteins. A dense, white, fibrous **sclera** covers the rest of the eye's outer surface.

The eye's middle layer includes the choroid, iris, and the ciliary body. The blood vessel–rich **choroid** is darkened by the brownish pigment melanin. This dark layer prevents light reflection within the eyeball. Attached to the choroid, and suspended behind the cornea, is a muscular, doughnut-shaped **iris**. It too has melanin. Whether your eyes are blue, brown, or green depends on the amount of melanin in your iris.

Light enters the eye's interior through the **pupil**, an opening at the center of the iris. Muscles of the iris can adjust pupil diameter in response to light conditions. Bright light causes the iris muscle encircling the pupil to contract, so the pupil contracts (shrinks). In low light, the spoke-like radial muscle contracts and the pupil dilates (widens).

A ciliary body of muscle, fibers, and secretory cells, attaches to the choroid. The ciliary body holds the lens in its proper place, just behind the pupil. The stretchable, transparent lens is about 1 centimeter (1/2 inch) in diameter and bulges outward on both sides.

The eye has two internal chambers. The ciliary body produces the fluid that fills the anterior chamber. Called aqueous humor, this fluid bathes the iris and lens. A jellylike vitreous body fills the larger chamber behind the lens. The innermost layer of the eye, the retina, is at the back of this chamber. The retina contains the light-detecting photoreceptors.

The cornea and lens both bend incoming light so that rays converge at the back of the eye, on the retina. The image formed on the retina is upside down and the mirror image of the real world (Figure 34.18). The

---

Wall of eyeball (three layers)

Outer layer	*Sclera.* Protects eyeball	
	Cornea. Focuses light	
Middle layer	*Pupil.* Serves as entrance for light	
	Iris. Adjusts diameter of pupil	
	Ciliary body. Its muscles control the lens shape; its fine fibers hold lens in place	
	Choroid. Its blood vessels nutritionally support wall cells; its pigments stop light scattering	
	Start of optic nerve. Carries signals to brain	
Inner layer	*Retina.* Absorbs, transduces light energy	

Interior of eyeball

Lens	Focuses light on photoreceptors
Aqueous humor	Transmits light, maintains fluid pressure
Vitreous body	Transmits light, supports lens and eyeball

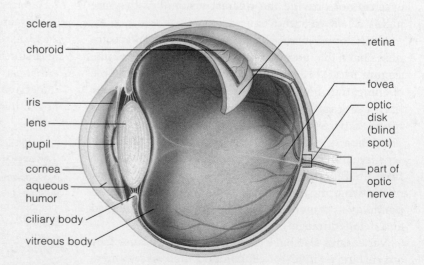

Figure 34.17 Animated Components and structure of the human eye.

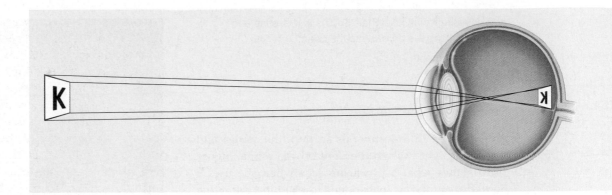

Figure 34.18 Animated Pattern of retinal stimulation in the human eye. The curved, transparent cornea changes the trajectory of light rays that enter the eye. As a result, light rays that fall on the retina produce a pattern that is upside down and inverted left to right.

brain makes the necessary adjustments so you perceive the correct orientation when you view an object.

Focusing Mechanisms

With **visual accommodation**, the shape or position of a lens adjusts so that incoming light rays fall on the retina, not in front of it or behind it. Without these adjustments, only objects at a fixed distance would stimulate retinal photoreceptors in a focused pattern. Objects closer or farther away would appear fuzzy.

Fishes and reptiles have eyes with a lens that can be shifted forward or back, but lens shape is constant. Extending or decreasing the distance between the lens and retina keeps light focused on the retina.

In birds and mammals, the lens is elastic; pulling on the lens changes its shape. A ring-shaped **ciliary muscle** (part of the ciliary body) encircles the lens and attaches to it by short fibers. Contraction of this muscle adjusts the shape of the lens. When the ciliary muscle is relaxed, fibers are taut, the lens is under tension, and it flattens (Figure 34.19a). When the ciliary muscle contracts, fibers attached to the lens slacken allowing the lens to become more round (Figure 34.19b).

The curvature of the lens determines the extent to which light rays will bend, and thus where they will fall in the eye. A flat lens will focus light from a distant object onto the retina. However, the lens must be rounder to focus light from nearby objects. When you read a book, ciliary muscle contracts and fibers that connect this muscle to the lens slacken. The decreased tension on the lens allows it to round up enough to focus light from the page onto your retina. Gaze into the distance and ciliary muscle around the lens relaxes, allowing the lens to flatten. Continual viewing of a close object, such as a computer screen or book, keeps ciliary muscle contracted. To reduce eyestrain, take breaks and focus on more distant objects.

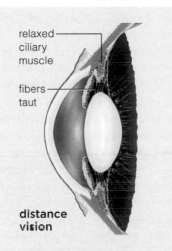
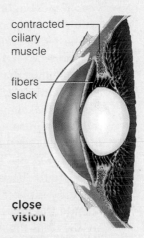

relaxed ciliary muscle

fibers taut

distance vision

contracted ciliary muscle

fibers slack

close vision

A Relaxed ciliary muscle pulls fibers taut, the lens is stretched into a flatter shape that focuses light from a distant object on the retina.

B Contracted ciliary muscle allows fibers to slacken; the lens rounds up and focuses light from a close object on the retina.

Figure 34.19 Animated How the eye varies its focus. The lens is encircled by ciliary muscle. Elastic fibers attach the muscle to the lens. The shape of the lens is adjusted by contracting or relaxing the ciliary muscle, increasing or decreasing the tension on the fibers, and thus changing the shape of the lens. **Figure It Out:** The thicker a lens, the more it bends light. Does the lens bend light more with distance vision or close vision? *Answer: Close vision*

Take-Home Message

How is the structure of the human eye related to its function?

■ The eye consists of delicate tissues that are surrounded by a bony orbit and constantly bathed in infection-fighting tears.

■ The cornea at the front of the eye bends light rays, which then enter the eye's interior through the pupil. The diameter of the pupil can be regulated depending on the amount of available light.

■ Behind the pupil, the lens focuses light on the retina, the eye's innermost photoreceptor-containing layer. Muscle contractions can alter the shape of the lens to focus light from near or distant objects.

■ Processing of visual signals begins in the retina and continues along the pathway to the brain.

■ Link to Pigments 7.1

Structure of the Retina

As explained in the previous section, the cornea and lens bend light rays so they fall on the retina. Figure 34.20 shows what a physician sees when she uses a lighted magnifying instrument to examine the retina inside the eyeball. The **fovea**, the area of the retina that is richest in photoreceptors, appears as a reddish spot in an area relatively free of blood vessels. With normal vision, most light rays are focused on the fovea. Also visible in this photo is the start of the optic nerve.

The retina consists of multiple cell layers. Nearest the source of light are several layers of interneurons such as amacrine cells, horizontal cells, and bipolar cells (Figure 34.21). These cells are involved in processing of visual signals. The two types of photoreceptors, rod cells and cone cells, lie in the deepest retinal layer, the one closest to the choroid.

Rod cells are photoreceptors that detect dim light. They are the basis for coarse perception of movement and for peripheral vision. They are the most abundant outside the fovea. **Cone cells** detect bright light and are the basis for sharp vision and for color perception. The fovea has the greatest density of cone cells.

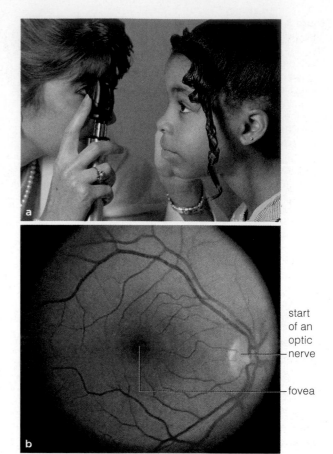

Figure 34.20 (**a**) Examining the retina. (**b**) View of the retina, showing the fovea and start of the optic nerve.

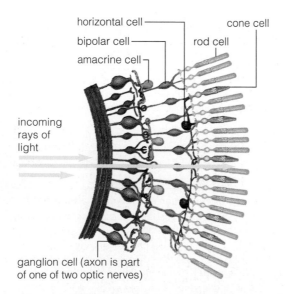

Figure 34.21 Animated Organization of the retina. The light-sensitive rods and cones lie beneath and send signals to interneurons involved in visual processing.

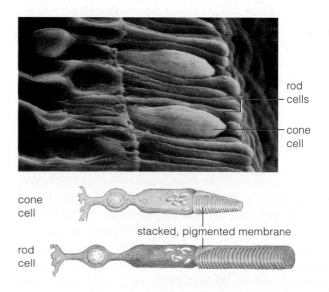

Figure 34.22 Scanning electron micrograph and diagrams of rod cells and cone cells. There are three types of cones. Each responds to a different wavelength of light.

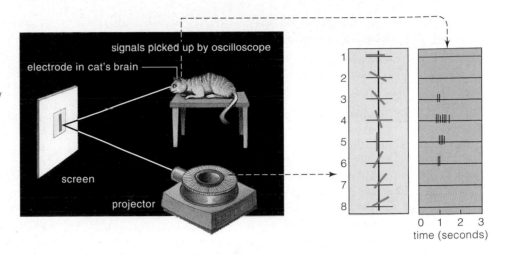

Figure 34.23 Animated An experiment into the response of cells of the visual cortex. David Hubel and Torsten Wiesel implanted an electrode in a cat's brain. They placed the cat in front of a screen upon which different patterns of light were projected, here, a hard-edged bar. Light or shadow falling on part of the screen excited or inhibited signals sent to a single neuron in the visual cortex. Tilting the bar at different angles, as shown in the *tan* box, produced changes in the neuron's activity, shown in the *purple* box. A vertical bar image produced the strongest signal (*numbered 5 in the sketch*). When the bar image tilted slightly, signals were less frequent. When the bar was tilted past a certain angle, signals stopped.

How Photoreceptors Work

Stacks of membranous disks fill much of the interior of a rod cell (Figure 34.22). Each membranous disk holds molecules of rhodopsin. Rhodopsin consists of a protein (opsin) and retinal, a light-absorbing pigment synthesized from vitamin A.

As long as rod cells are in the dark, they undergo action potentials and release an inhibitory neurotransmitter at their synapses with bipolar cells. Exposure to blue-green light causes rhodopsin to change shape, and halts release of the inhibitory neurotransmitter. With this inhibition lifted, the bipolar cells are free to signal other interneurons in the retina. Eventually, this signaling causes action potentials that travel along the optic nerve to the brain.

Humans have three types of cone cells—red, green, and blue—each with a slightly different kind of opsin. Differences in opsins affect which wavelength of light a cone absorbs. As in rods, photon absorption by cones leads indirectly to action potentials in other cells.

Visual Processing

Interneurons that connect to photoreceptors receive, process, and begin to integrate visual signals. Input from hundreds of rods and cones converges on each bipolar cell. Information also flows laterally among the amacrine cells and horizontal cells of the retina. Eventually, all of the signals converge on about one million ganglion cells. These are the output neurons; their axons are the start of an optic nerve.

The region where the optic nerve exits the eye is known as the **blind spot** because it does not have photoreceptors. You do not normally notice your blind spots because the visual fields of your eyes overlap. The portion of the visual field that is missed because of the blind spot in one eye is seen by the other eye.

Figure 34.24 Flow of information from the retina to processing centers in the brain. Signals from both eyes reach both of the brain's two hemispheres. The signals from the left half of the visual field end up in the brain's right visual cortex. Signals from the right half of the visual field end up in the left cortex.

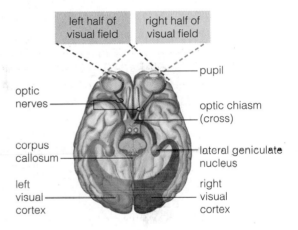

Different neurons inside the brain's visual cortex respond to different visual patterns. Figure 34.23 shows an experiment that demonstrated this mechanism.

Signals from the right visual field of each eye travel to the left hemisphere. Signals from the left visual field go to the right hemisphere (Figure 34.24). Each optic nerve ends in a brain region (lateral geniculate nucleus) that processes signals. From here, the signals travel to the visual cortex where the final integration process produces visual sensations.

Take-Home Message

How does the retina function?

■ The retina's deepest layer, closest to the choroid, contains photoreceptors: rod cells that work in dim light and cone cells that allow sharp color vision.

■ Interneurons that overlie the photoreceptors receive signals from them.

■ Signal processing begins in the brain and is completed in the visual cortex.

■ Genetic conditions, age-related changes, nutritional deficits, and infectious agents can impair vision.

■ Links to X-linked inheritance 12.4, Vitamin A deficiency Chapter 16 introduction, Amoebas 22.11, Roundworms 25.11

Color Blindness Color blindness arises when one or more types of cones fail to develop or do not work properly. With the most common type, an affected person has trouble distinguishing reds from greens. This X-linked recessive trait affects about 7 percent of men in the United States. As is the case for other X-linked traits, it shows up predominantly in males (Section 12.4). Only 0.4 percent of women are affected.

Lack of Focus About 150 million Americans have disorders in which light rays do not converge as they should. Astigmatism results from an unevenly curved cornea, which cannot properly focus incoming light on the lens.

Nearsightedness occurs when the distance from the front to the back of the eye is longer than normal or when ciliary muscles react too strongly. With either disorder, images of distant objects get focused in front of the retina instead of on it (Figure 34.25*a*).

In farsightedness, the distance from front to back of the eye is unusually short or ciliary muscles are too weak. Either way, light rays from nearby objects get focused behind the retina (Figure 34.25*b*). Also, the lens loses its flexibility as a person ages. That is why most people who are over age forty have relatively impaired close vision.

Glasses, contact lenses, or surgery can correct some focusing problems. About 1.5 million Americans undergo laser surgery (LASIK) annually. Typically, LASIK eliminates the need for glasses during most activities, although older adults usually continue to require reading glasses.

Macular Degeneration In the United States, an estimated 13 million people have age-related macular degeneration (AMD). The macula is the cone-rich region that surrounds and includes the fovea. Destruction of photoreceptors in the macula clouds the center of the visual field more than the periphery (Figure 34.26*b*).

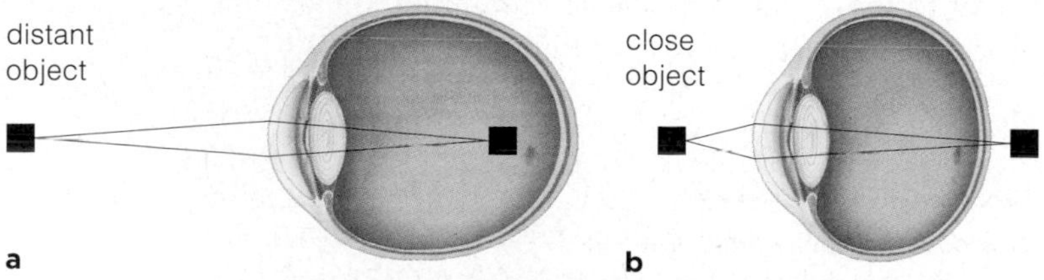

Mutations in certain genes can increase the risk of AMD. So do smoking, obesity, and high blood pressure. A diet rich in vegetables seems to protect against it. Damage caused by AMD usually cannot be reversed, but drug treatments and laser therapy can slow its progression.

Glaucoma With glaucoma, too much aqueous humor builds up inside the eyeball. The increased fluid pressure damages blood vessels and ganglion cells. It also can interfere with peripheral vision and visual processing. Although we often associate chronic glaucoma with old age, the conditions that give rise to the disorder start to develop long before symptoms show up. When doctors detect the increased fluid pressure before the damage becomes severe, they can manage the disorder with medication, surgery, or both.

Cataracts A cataract is a clouding of the lens. It typically develops slowly. The cloudy lens reduces the amount and focusing of light that reaches the retina. Early symptoms are poor night vision and blurred vision (Figure 34.26*c*). Vision ends after the lens becomes fully opaque. Excessive exposure to ultraviolet radiation, use of steroids, and some diseases such as diabetes can promote the onset and development of cataracts. An artificial implant can replace a badly clouded lens. Millions of people in developed countries undergo cataract surgery each year. Worldwide, about 16 million are currently blind as a result of cataracts.

Nutritional Blindness Each year, as many as half a million children worldwide go blind because they do not have enough vitamin A in their diet. Among other things, the body needs vitamin A to make retinal, the pigment in both rods and cones. The Chapter 16 introduction described efforts to genetically engineer rice to contain vitamin A, as a partial solution to vitamin A deficiency. This vitamin can be obtained as part of a balanced diet that includes meat, eggs, and yellow and orange vegetables.

Infectious Agents The bacterium *Chlamydia trachomatis* causes the disease trachoma. The bacteria infect the conjunctiva, the membrane that lines the eyelids and covers the sclera (the white part of the eye). Repeated infections cause corneal scarring and lead to blindness. About 6 million people have been blinded by trachoma in Africa, Asia, the Middle East, Latin America, and the Pacific Islands. It is the leading cause of infectious blindness.

Roundworms (Section 25.11) cause onchocerciasis, the second most common type of infectious blindness. It is also called "river blindness" because the biting flies that transmit it are most common around African rivers.

Other bacterial diseases and viral diseases, including syphilis, can also cause blindness. So can infection by certain kind of amoebas (Section 22.11). These amoebas have turned up in batches of certain contact lens solutions, as have eye-damaging fungi.

distant object

close object

a

b

Figure 34.25 Focusing problems. (**a**) In nearsightedness, light rays from distant objects converge in front of the retina. (**b**) In farsightedness, light rays from close objects have not yet converged when they arrive at the retina.

a Normal vision

b Vision with macular degeneration

c Vision with cataracts

Figure 34.26 Photos that simulate how normal vision (**a**) compares with vision of a person with age-related macular degeneration (**b**) or cataracts (**c**). Macular degeneration obscures the center of the visual field. Cataracts lessen the amount of the light that reaches the retina and scatter that light so the resulting image appears fuzzy.

Summary

Section 34.1 The types of sensory receptors that an animal has determine the types of stimuli it detects and can respond to. Stimulation of a sensory receptor causes action potentials. **Mechanoreceptors** respond to mechanical energy such as touch. **Pain receptors** respond to tissue damage. **Thermoreceptors** are sensitive to temperature. **Chemoreceptors** fire in response to dissolved chemicals. **Osmoreceptors** sense and respond to water concentration. **Photoreceptors** respond to light.

The brain evaluates action potentials from sensory receptors based on which of the body's nerves delivers them, their frequency, and the number of axons firing in any given interval. Continued stimulation of a receptor may lead to a diminished response (**sensory adaptation**).

The **somatic sensations** arise from sensory receptors located in skin, or near muscles or joints. **Visceral sensations** arise from receptors near organs in body cavities. The receptors for special senses—taste, smell, hearing, balance, and vision—are in specific sensory organs.

■ *Use the animation on CengageNOW to see how stimulus intensity affects action potential frequency.*

Section 34.2 Signals from free nerve endings, encapsulated receptors, and stretch receptors in the skin, skeletal muscles, and joints reach the **somatosensory cortex**. Interneurons in this part of the cerebral cortex are laid out like a map of the body surface.

Pain is the perception of tissue damage. In vertebrates, a variety of neuromodulators enhance or lessen signals about pain. With **referred pain**, the brain mistakenly attributes signals that come from an internal organ to the skin or muscles.

■ *Use the animation on CengageNOW to learn about sensory receptors in human skin and referred pain.*

Section 34.3 The senses of taste and smell (**olfaction**) involve chemoreceptors and pathways to the cerebral cortex and limbic system. In humans, **taste receptors** are concentrated in taste buds on the tongue and walls of the mouth. **Olfactory receptors** line human nasal passages. **Pheromones** are chemical signals that act as social cues among many animals. A **vomeronasal** organ functions in detection of pheromones in many vertebrates.

Section 34.4 **Organs of equilibrium** detect gravity, acceleration, and other forces that affect body positions and motions. The **vestibular apparatus** is a system of fluid-filled sacs and canals in the inner ear. The sense of dynamic equilibrium arises when body movements cause shifts in the fluid, which causes cilia of **hair cells** to bend. Static equilibrium depends on signals from hair cells that lie beneath a weighted, jellylike mass. A shift in head position or a sudden stop or start shifts the mass, bends the hair cells, and makes these cells fire.

■ *Use the animation on CengageNOW to explore static and dynamic equilibrium.*

A Whale of a Dilemma

Animal sensory systems evolved over countless generations in a world without human activity. Now, we have dramatically altered the sensory landscape for many animals. The world has become noisier and more brightly lit. Our communication systems fill the air with radio waves. How do these changes affect the species with which we share the planet? How much harm do the changes do? We do not know the answers to these questions.

How would you vote?

Excessive noise can harm marine organisms. Should we regulate the maximum allowable noise level underwater? See CengageNOW for details, then vote online.

Sections 34.5, 34.6 **Hearing** is the perception of sound, which is a form of mechanical energy. Sound waves are pressure waves. We perceive variations in the amplitude of the waves as differences in loudness. We perceive variations in wave frequency as differences in pitch.

Human ears have three functional regions. The skin-covered flap of the **outer ear** collects sound waves. The **middle ear** contains the **eardrum** and a set of tiny bones that amplify sound waves and transmit them to the inner ear. The **inner ear** is where pressure waves elicit action potentials inside a **cochlea**. This coiled structure with fluid-filled ducts holds the mechanoreceptors responsible for hearing in its **organ of Corti**.

Pressure waves traveling through the fluid inside the cochlea bend hair cells of the organ of Corti. The brain gauges the loudness of a sound by the number of signals the sound elicits. It determines a sound's pitch by which part of the cochlea's coil the signals arrive from.

Hearing loss may be caused by nerve problems, damaged hair cells, or failure of signals to reach the inner ear. Exposure to loud noise can damage hair cells. Noise also disrupts human health and animal behavior.

■ *Use the animation on CengageNOW to learn about the properties of sound and the human sense of hearing.*

Section 34.7 Most organisms can respond to light, but **vision** requires eyes and brain centers capable of processing the visual information. An **eye** is a sensory organ that contains a dense array of photoreceptors.

Insects have a **compound eye**, with many individual units. Each unit has a **lens**, a structure that bends light rays so they fall on the photoreceptors. Like squids and octopuses, humans have **camera eyes**, with an adjustable opening that lets in light, and a single lens that focuses the light on a photoreceptor-rich **retina**.

In animals with eyes that face forward, the brain gets overlapping information about the viewed area. This allows more accurate depth perception.

Sections 34.8–34.10 A human eye is protected by eyelids lined by the **conjunctiva**. This membrane also covers the **sclera**, or white of the eye. The clear, curved **cornea** at the front of the eye bends incoming light. Light enters the eye's interior through the **pupil**, an adjustable opening in the center of the muscular, doughnut-shaped **iris**. Light that enters the eye falls on the retina. The retina sits on a pigmented **choroid** that absorbs light so it is not reflected inside the eye.

With **visual accommodation**, the **ciliary muscle** adjusts the shape of the lens so that light from a near or distant object falls on the retina's photoreceptors. Humans have two types of photoreceptors. **Rod cells** detect dim light and are important in coarse vision and peripheral vision. **Cone cells** detect bright light and colors; they provide a sharp image. The greatest concentration of cones is in the portion of the retina called the **fovea**. The rods and cones interact with other cells in the retina that start processing visual information before sending it to the brain. Visual signals travel to the cerebral cortex along two optic nerves. There are no photoreceptors in the eye's **blind spot**, the area where the optic nerve begins.

Abnormalities in eye shape, in the lens, and in cells of the retina can impair vision.

■ *Use the animation on CengageNOW to investigate the structure, function, and organization of the eye and retina.*

Self-Quiz
Answers in Appendix III

1. A stimulus is a specific form of energy in the outside environment that is detected by _____ .
    a. a sensory neuron   c. a motor neuron
    b. an interneuron   d. all of the above

2. _____ is defined as a decrease in the response to an ongoing stimulus.
    a. Perception   c. Sensory adaptation
    b. Visual accommodation   d. Somatic sensation

3. Which is a somatic sensation?
    a. taste   c. touch   e. a through c
    b. smell   d. hearing   f. all of the above

4. Chemoreceptors play a role in the sense of _____ .
    a. taste   c. touch   e. both a and b
    b. smell   d. hearing   f. all of the above

5. In the _____ , interneurons are arranged like maps that correspond to different parts of the body surface.
    a. somatosensory cortex   c. basilar membrane
    b. retina   d. all of the above

6. Mechanoreceptors in the _____ send signals to the brain about the body's position relative to gravity.
    a. eyes   b. ears   c. tongue   d. nose

7. The middle ear functions in _____ .
    a. detecting shifts in body position
    b. amplifying and transmitting sound waves
    c. sorting sound waves out by frequency

8. The organ of Corti responds to _____ .
    a. sound   b. light   c. heat   d. pheromones

Data Analysis Exercise

Frequent exposure to noise of a particular pitch can cause loss of hair cells in the part of the cochlea's coil that responds to that pitch. Many workers are at risk for such frequency-specific hearing loss because they work with or around noisy machinery. Taking precautions such as using ear plugs to reduce sound exposure is important. Noise-induced hearing loss can be prevented, but once it occurs it is irreversible. Dead or damaged hair cells are not replaced.

Figure 34.27 shows the threshold decibel levels at which sounds of different frequencies can be detected by an average 25-year-old carpenter, a 50-year-old carpenter, and a 50-year-old who has not been exposed to on-the-job noise. Sound frequencies are given in hertz (cycles per second). The more cycles per second, the higher the pitch.

1. Which sound frequency was most easily detected by all three people?

2. How loud did a 1,000-hertz sound have to be for the 50-year-old carpenter to detect it?

3. Which of the three people had the best hearing in the range of 4,000 to 6,000 hertz? Which had the worst?

4. Based on this data, would you conclude that the hearing decline in the 50-year-old carpenter was caused by age or by job-related noise exposure?

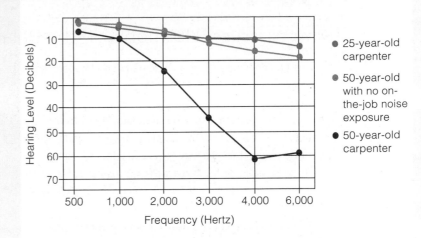

Figure 34.27 Effects of age and occupational noise exposure. The graph shows the threshold hearing capacities (in decibels) for sounds of different frequencies (given in hertz) in a 25-year-old carpenter (*blue*), a 50-year-old carpenter (*red*), and a 50-year-old who did not have any on-the-job noise exposure (*brown*).

9. Color vision begins with signals from _____ .
   a. rods   b. cones   c. hair cells   d. the blind spot

10. When you view a close object, your lens gets _____ .
   a. more rounded   c. more flattened
   b. cloudier   d. more transparent

11. Bright light causes the _____ to shrink.
   a. lens   b. pupil   c. fovea   d. blind spot

12. Label the parts of the human eye in this diagram:

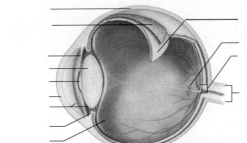

13. Match each structure with its description.
   ___ fovea   a. sensitive to vibrations
   ___ cochlea   b. functions in balance
   ___ lens   c. type of photoreceptor cell
   ___ hair cell   d. has most cone cells
   ___ rod cell   e. contains chemoreceptors
   ___ taste bud   f. focuses rays of light
   ___ vestibular   g. sorts out sound waves
       apparatus   h. helps brain assess heat,
   ___ free nerve ending   pressure, pain

■ *Visit CengageNOW for additional questions.*

Critical Thinking

1. Laura loves to eat broccoli and brussels sprouts. Lionel cannot stand them. Everyone has the same five kinds of taste receptors, so what is going on? Is Lionel just being difficult? Perhaps not. The number and distribution of receptors that respond to bitter substances vary among individuals of a population—and studies now indicate that some of this variation is heritable.

People who have the greatest number of receptors for bitter substances find many fruits and vegetables highly unpalatable. These supertasters make up about 25 percent of the general population. They tend to be slimmer than average but are more likely to develop colon polyps and colon cancer. How might Lionel's highly sensitive taste buds put him at increased risk for colon cancer?

2. Are organs of dynamic equilibrium, static equilibrium, or both activated during a roller-coaster ride?

3. The strength of Earth's magnetic field and its angle relative to the surface vary with latitude. Diverse species sense these differences and use them as cues for assessing their location and direction of movement. Behavioral experiments have shown that sea turtles, salamanders, and spiny lobsters use information from Earth's magnetic field during their migrations. Whales and some burrowing rodents also seem to have a magnetic sense. Evidence about humans is contradictory. Suggest an experiment to test whether humans can detect a magnetic field.

4. After a leg injury, pain makes a person avoid putting too much weight on the affected leg. An injured insect shows no such shielding response and does not make natural pain-relieving chemicals. Is this sufficient evidence to conclude that insects do not have a sense of pain?

35 Endocrine Control

IMPACTS, ISSUES Hormones in the Balance

Atrazine has been used widely as an herbicide for more than forty years. Each year in the United States, about 76 million pounds are sprayed, mostly to kill weeds in cornfields. From there, atrazine gets into soil and water. Atrazine molecules break down within a year but they still turn up in ponds, wells, groundwater, and rain. Do they have bad effects? Tyrone Hayes, a University of California biologist, thinks so. His data suggest that atrazine is an endocrine disruptor: a synthetic compound that alters the action of natural hormones and adversely affects health and development (Figure 35.1).

Hayes studied atrazine's effects on African clawed frogs (*Xenopus laevis*) and leopard frogs (*Rana pipiens*). He found that exposing male tadpoles to atrazine in the laboratory caused some to develop both female and male reproductive organs. This effect occurred even at atrazine levels far below those allowed in drinking water.

Does atrazine have similar effects in the wild? To find out, Hayes collected leopard frogs from ponds and ditches across the Midwest. Male frogs from every contaminated pond had abnormal sex organs. In the pond with the most atrazine, 92 percent of males had ovary tissue.

Other scientists have also reported that atrazine causes or contributes to frog deformities. The Environmental Protection Agency found the data intriguing. Among other tasks, this agency regulates chemical applications in agriculture. It called for further study of atrazine's effects on amphibians and is encouraging farmers to minimize atrazine-laden runoff from their fields.

Numerous hormone disruptors infiltrate aquatic habitats. For instance, the estrogens in birth control pills are excreted in urine and cannot be removed by standard wastewater treatments. In streams or rivers, estrogen-tainted water causes male fish to develop female traits.

An excess of estrogen-like chemicals may lower sperm counts. Estrogen is a sex hormone. Both men and women produce it and have receptors for it, although females make much more. In males, estrogen docks at receptors on target cells in reproductive organs and helps sperm to mature. Other synthetic chemicals, including kepone and DDT, bind to estrogen receptors, thus blocking estrogen's actions, including its role in sperm maturation. Both chemicals are now banned in the United States.

This chapter focuses on the hormones—their sources, targets, effects, and interactions. All vertebrates have similar hormone-secreting glands and systems. Keep this point in mind when you think about the endocrine disruptors. What you learn in this chapter will help you evaluate the costs and benefits of synthetic chemicals that affect hormone action.

See the video! Figure 35.1 Benefits and costs of herbicide applications. *Left*, atrazine can keep cornfields nearly weed-free; no need for constant tilling that causes soil erosion. Tyrone Hayes (*right*) suspects that the chemical scrambles amphibian hormonal signals.

Key Concepts

Signaling mechanisms

Hormones and other signaling molecules function in communication among body cells. A hormone travels through the blood and acts on any cell that has receptors for it. The receptor may be at a target cell's surface or inside the cell. **Sections 35.1, 35.2**

A master integrating center

In vertebrates, the hypothalamus and pituitary gland are connected structurally and functionally. Together, they coordinate activities of many other glands. Pituitary hormones affect growth, reproductive functions, and composition of extracellular fluid. **Sections 35.3, 35.4**

Other hormone sources

Negative feedback loops to the hypothalamus and pituitary control secretions from many glands. Signals from the nervous system and internal solute concentrations also influence hormone secretion. **Sections 35.5–35.12**

Invertebrate hormones

Hormones control molting and other events in invertebrate life cycles. Vertebrate hormones and receptors for them first evolved in ancestral lineages of invertebrates. **Section 35.13**

Links to Earlier Concepts

- This chapter continues the story of cell signaling that began in Section 27.6. You will see many examples of feedback mechanisms (27.3). We will also revisit gap junctions (32.1) and glandular epithelium (32.2).

- Knowing the properties of steroids (3.4), proteins (3.5), and the function of the plasma membrane (5.4) will help you understand how different types of hormones interact with cells.

- The nervous and endocrine systems work together. You will hear again about action potentials (33.3), synapses (33.5), sympathetic neurons (33.8), the anatomy of the brain (33.10), and visual processing (34.9).

- You will see how hormones affect metabolism of glucose (6.7), gamete formation (10.5), and molting (25.11).

- Genetics concepts relevant to this chapter include gene duplications (12.5), gene expression (14.1), the role of promoters (14.2), introns (14.3), and techniques of genetic engineering (16.6).

How would you vote? Some widely used agricultural chemicals may disrupt hormone action in untargeted species. Should potentially harmful chemicals be kept on the market while researchers investigate them? See CengageNOW for details, then vote online.

■ Animal cells communicate with one another by way of a variety of short-range and long-range chemical signals.

■ Links to Gap junctions 32.1, Glandular epithelium 32.2, Synapses 33.5

Intercellular Signaling in Animals

In all animals, cells constantly signal one another in response to changes in the internal and external environments. Receiving such signals can influence a cell's metabolic activity, division, or gene expression.

Gap junctions allow signals to move directly from the cytoplasm of one cell to that of an adjacent cell (Section 32.1). Other cell–cell communication involves signaling molecules that are secreted into interstitial fluid (the fluid between cells). These molecules exert effects only when they bind to a receptor on or inside another cell. We refer to a cell that has receptors that bind and respond to a specific signaling molecule as a "target" of that molecule.

Some secreted signaling molecules diffuse a short distance through interstitial fluid and bind to nearby cells. For example, neurons secrete signaling molecules called **neurotransmitters** into the synaptic cleft that separates them from a target cell. Neurotransmitter diffuses the short distance across the cleft and binds to the target (Section 33.5).

Only neurons release neurotransmitters, but many cells secrete **local signaling molecules** that affect their neighbors. Prostaglandins are one type of local signal. When released by injured cells, they activate pain receptors and increase local blood flow. The enhanced blood flow delivers more infection-fighting proteins and white blood cells to the injured region.

Animal hormones are longer-range communication molecules. After being secreted into interstitial fluid, they enter capillaries and are distributed throughout the body. Compared to neurotransmitters or local signaling molecules, hormones last longer, travel farther, and exert their effects on a greater number of cells.

Some animals produce intercellular communication signals called **pheromones** that diffuse through water or air and bind to target cells in other individuals. Pheromones help integrate social behavior. We discuss them in Chapter 44, in the context of social behavior. For the rest of this chapter, our focus is hormones.

Overview of the Endocrine System

The word "hormone" dates back to the early 1900s. Physiologists W. Bayliss and E. Starling were trying to determine what triggers the secretion of pancreatic juices when food travels through a dog's gut. As they knew, acids mix with food in the stomach. Arrival of the acidic mixture inside the small intestine triggers pancreatic secretions that reduce the acidity. Was the nervous system stimulating this pancreatic response, or was some other signaling mechanism at work?

To find an answer, Bayliss and Starling blocked the nerves—but not blood vessels—to the small intestine of a laboratory animal. The pancreas still responded when acidic food from the stomach entered the small intestine. The pancreas even responded to extracts of cells from the intestinal lining, which is a glandular epithelium (Section 32.2). Apparently, some substance produced by glandular cells signaled the pancreas to start its secretions.

That substance is now called secretin. Identifying its mode of action supported a hypothesis that dated back centuries: The blood carries internal secretions that influence the activities of the body's organs.

Starling coined the term "hormone" for glandular secretions (the Greek word *hormon* means to set in motion). Later on, researchers identified many other hormones and their sources. Glands and other hormone sources are collectively referred to as an animal's **endocrine system**. Figure 35.2 surveys major sources of hormones in the human endocrine system.

Nervous–Endocrine Interactions

The endocrine system and nervous system are closely linked. Both neurons and endocrine cells are derived from an embryo's ectodermal layer. Both respond to the hypothalamus, a command center in the forebrain (Section 33.10). Most organs receive and respond to both nervous signals and hormones.

Hormones influence the development of the brain, both before and after birth. Hormones can also affect nervous processes such as sleep/wake cycles, emotion, mood, and memory. Conversely, the nervous system affects hormone secretion. For example, in a stressful situation, nervous signals call for increased secretion of some hormones and decreased secretion of others.

Take-Home Message

How do cells of an animal body communicate with one another?

■ Animals cells communicate through gap junctions and by release of molecules that bind to receptors in or on other cells.

■ Neurotransmitters and local signaling molecules disperse by diffusion and affect only nearby cells. Hormones enter the blood and are distributed throughout the body, so they have wider reaching effects.

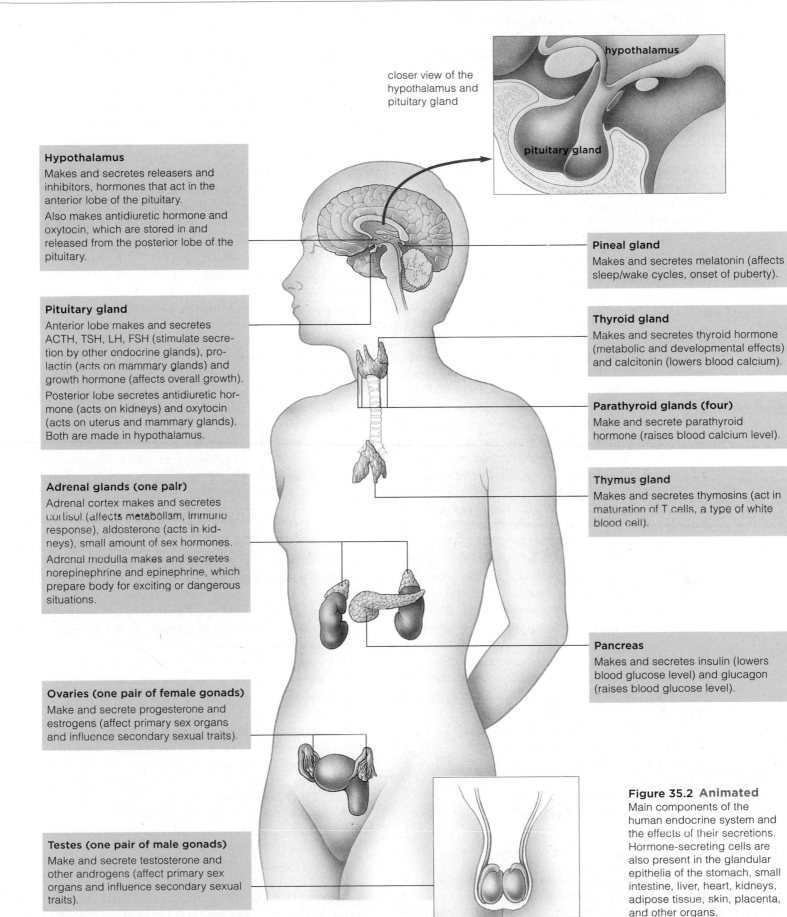

closer view of the hypothalamus and pituitary gland

hypothalamus

pituitary gland

Hypothalamus

Makes and secretes releasers and inhibitors, hormones that act in the anterior lobe of the pituitary.

Also makes antidiuretic hormone and oxytocin, which are stored in and released from the posterior lobe of the pituitary.

Pituitary gland

Anterior lobe makes and secretes ACTH, TSH, LH, FSH (stimulate secretion by other endocrine glands), prolactin (acts on mammary glands) and growth hormone (affects overall growth).

Posterior lobe secretes antidiuretic hormone (acts on kidneys) and oxytocin (acts on uterus and mammary glands). Both are made in hypothalamus.

Adrenal glands (one pair)

Adrenal cortex makes and secretes cortisol (affects metabolism, immune response), aldosterone (acts in kidneys), small amount of sex hormones.

Adrenal medulla makes and secretes norepinephrine and epinephrine, which prepare body for exciting or dangerous situations.

Ovaries (one pair of female gonads)

Make and secrete progesterone and estrogens (affect primary sex organs and influence secondary sexual traits).

Testes (one pair of male gonads)

Make and secrete testosterone and other androgens (affect primary sex organs and influence secondary sexual traits).

Pineal gland

Makes and secretes melatonin (affects sleep/wake cycles, onset of puberty).

Thyroid gland

Makes and secretes thyroid hormone (metabolic and developmental effects) and calcitonin (lowers blood calcium).

Parathyroid glands (four)

Make and secrete parathyroid hormone (raises blood calcium level).

Thymus gland

Makes and secretes thymosins (act in maturation of T cells, a type of white blood cell).

Pancreas

Makes and secretes insulin (lowers blood glucose level) and glucagon (raises blood glucose level).

Figure 35.2 Animated

Main components of the human endocrine system and the effects of their secretions. Hormone-secreting cells are also present in the glandular epithelia of the stomach, small intestine, liver, heart, kidneys, adipose tissue, skin, placenta, and other organs.

35.2 The Nature of Hormone Action

- For a hormone to have an effect, it must bind to receptors on or inside a target cell.

- Links to Steroids 3.4, Proteins 3.5, Cell membranes 5.4, Sex determination 12.1, Promoters 14.2, Cell signaling 27.6

From Signal Reception to Response

Cell communication involves three steps (Section 27.6). A signal activates a target cell receptor, the signal is transduced (changed into form that affects target cell behavior), and the cell makes a response:

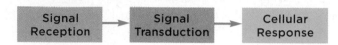

Enzymes make hormones from a variety of sources. Steroid hormones are derived from cholesterol. Amine hormones are modified amino acids. Peptide hormones are short chains of amino acids; protein hormones are longer chains. Table 35.1 lists a few examples of each.

Hormones initiate responses in different ways. In all cases, binding to a receptor is reversible and the effect of the hormone declines over time. The decline occurs as the body breaks the hormones down so they no longer bind to receptors and elicit a response.

Intracellular Receptors Steroid hormones are made from cholesterol and, like other lipids, they easily diffuse across a plasma membrane. Once inside a cell, steroid hormones form a hormone–receptor complex by binding to a receptor in the cytoplasm or nucleus. Most often, this hormone–receptor complex binds to and activates a promoter (Section 14.2). Activation of the promoter allows binding of RNA polymerase, which then transcribes an adjacent gene or genes. Transcription and translation produce a protein product, such as an enzyme, that carries out the target cell's response to the signal. Figure 35.3*a* is a simple illustration of this type of steroid hormone action.

Receptors at the Plasma Membrane Most amine hormones, and all peptide or protein hormones, are too big and polar to diffuse across a membrane. They bind to receptors that span a target cell's plasma membrane. Often, this binding activates an enzyme that converts ATP to cAMP (cyclic adenosine monophosphate). The cyclic AMP then functions as a **second messenger**: a molecule that forms inside a cell in response to an external signal and affects that cell's activity.

For example, when there is too little glucose in the blood, certain cells in the pancreas secrete the peptide hormone glucagon. When glucagon binds to receptors in the plasma membrane of target cells, it causes formation of cAMP inside them (Figure 35.3*b*). The cAMP activates an enzyme that activates a different enzyme, setting into motion a cascade of reactions. The last enzyme activated catalyzes breakdown of glycogen into glucose and thus raises the blood glucose level.

Some cells have receptors for steroid hormones at their plasma membrane. Binding of a steroid hormone to such a receptor does not influence gene expression. Instead, it triggers a faster response by way of a second messenger or by affecting the membrane. For example, when the steroid hormone aldosterone binds to receptors at the surface of kidney cells, the membrane of these cells becomes more permeable to sodium ions.

Receptor Function and Diversity

A cell can only respond to a hormone for which it has appropriate and functional receptors. All hormone receptors are proteins and gene mutations can make them less efficient or even nonfunctional. In this case, even though the hormone that targets the mutated receptor is present in normal amounts, the hormone will have a lesser or no effect.

For example, typical male genitals will not form in an XY embryo without testosterone, one of the steroid hormones (Section 12.1). XY individuals who have androgen insensitivity syndrome secrete testosterone, but a mutation alters their receptors for it. Without functional receptors, it is as if testosterone is not present. As a result, the embryo forms testes, but they do not descend into the scrotum, and the genitals appear female. Such individuals are often raised as females, as discussed in more detail in Chapter 42.

Variations in receptor structure also affect responses to hormones. Different tissues often have receptor proteins that respond in different ways to binding of the same hormone. For example, in Chapter 41, you will learn how ADH (antidiuretic hormone) from the posterior lobe of the pituitary acts on kidney cells and

Table 35.1 Categories and Examples of Hormones

Steroids	Testosterone and other androgens, estrogens, progesterone, aldosterone, cortisol
Amines	Melatonin, epinephrine, thyroid hormone
Peptides	Glucagon, oxytocin, antidiuretic hormone, calcitonin, parathyroid hormone
Proteins	Growth hormone, insulin, prolactin, follicle-stimulating hormone, luteinizing hormone

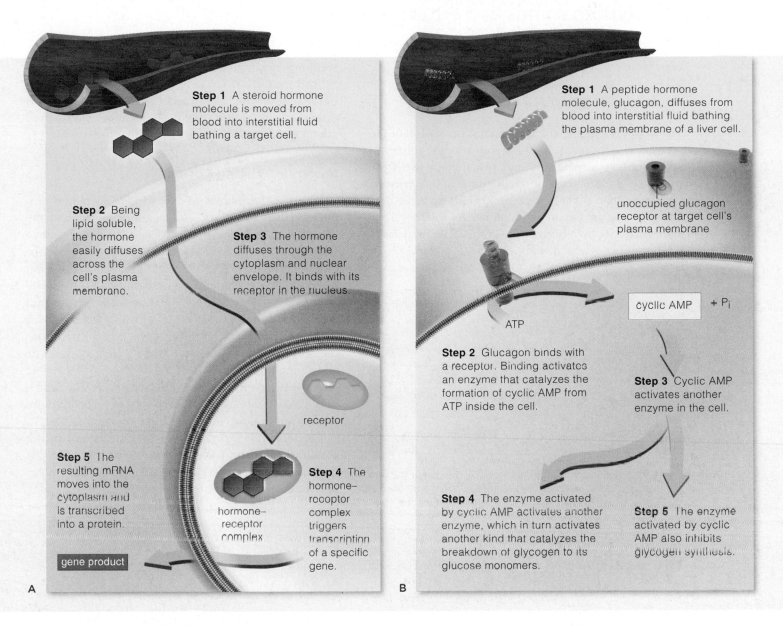

Step 1 A steroid hormone molecule is moved from blood into interstitial fluid bathing a target cell.

Step 2 Being lipid soluble, the hormone easily diffuses across the cell's plasma membrane.

Step 3 The hormone diffuses through the cytoplasm and nuclear envelope. It binds with its receptor in the nucleus.

receptor

Step 5 The resulting mRNA moves into the cytoplasm and is transcribed into a protein.

hormone–receptor complex

Step 4 The hormone–receptor complex triggers transcription of a specific gene.

gene product

A

Step 1 A peptide hormone molecule, glucagon, diffuses from blood into interstitial fluid bathing the plasma membrane of a liver cell.

unoccupied glucagon receptor at target cell's plasma membrane

cyclic AMP $+ P_i$

ATP

Step 2 Glucagon binds with a receptor. Binding activates an enzyme that catalyzes the formation of cyclic AMP from ATP inside the cell.

Step 3 Cyclic AMP activates another enzyme in the cell.

Step 4 The enzyme activated by cyclic AMP activates another enzyme, which in turn activates another kind that catalyzes the breakdown of glycogen to its glucose monomers.

Step 5 The enzyme activated by cyclic AMP also inhibits glycogen synthesis.

B

Figure 35.3 Animated (a) Typical steroid hormone action inside a target cell. (b) Typical peptide hormone action at the plasma membrane. Cyclic AMP, which serves as the second messenger, relays a signal from a plasma membrane receptor into the cell.

Figure It Out: Where does the second messenger form after glucagon binds to a cell? *Answer: In the cytoplasm*

helps maintain solute concentrations in the internal environment. ADH is sometimes referred to as vasopressin, because it also binds to receptors in the wall of blood vessels and causes these vessels to narrow. In many mammals, ADH helps maintain blood pressure. ADH also binds to brain cells and influences sexual and social behavior, as we will discuss in Section 44.1. This diversity of responses to a single hormone is an outcome of variations in ADH receptors. In each kind of cell, a different kind of receptor summons up a different cellular response.

Take-Home Message

How do hormones exert their effects on target cells?

■ Hormones exert their effects by binding to protein receptors, either inside a cell or at the plasma membrane.

■ Most steroid hormones bind to a promoter inside the nucleus and alter the expression of specific genes.

■ Peptide and protein hormones usually bind to a receptor at the plasma membrane. They trigger formation of a second messenger, a molecule that relays a signal into the cell.

■ Variations in receptor structure affect how a cell responds to a hormone.

■ The hypothalamus and pituitary gland deep inside the brain interact as a central command center.

■ Links to Feedback controls 27.3, Action potentials 33.3, Human brain 33.10, Exocrine glands 32.2

The **hypothalamus** is the main center for control of the internal environment. It lies deep inside the forebrain and connects, structurally and functionally, with the **pituitary gland** (Figure 35.4). In humans, this gland is no bigger than a pea. Its posterior lobe secretes hormones made in the hypothalamus. Its anterior lobe synthesizes its own hormones. Table 35.2 summarizes the hormones released from the pituitary gland.

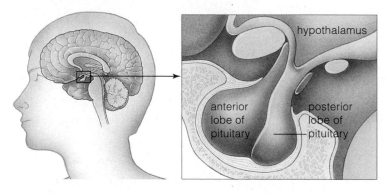

Figure 35.4 Location of the hypothalamus and pituitary gland. The two lobes of the pituitary (anterior and posterior) release different hormones.

The hypothalamus signals the pituitary by way of secretory neurons that make hormones, rather than neurotransmitters. These neurons have their cell body in the hypothalamus. Axons of some of these neurons extend into the pituitary's posterior lobe. Axons from others end in the stalk just above the pituitary.

Posterior Pituitary Function

Antidiuretic hormone and oxytocin are hormones produced in the cell bodies of secretory neurons of the hypothalamus (Figure 35.5a). These hormones move through axons to axon terminals inside the posterior pituitary (Figure 35.5b). Arrival of an action potential (Section 33.3) at the axon terminals causes these terminals to release hormone. The hormone diffuses into capillaries (small blood vessels) inside the posterior pituitary (Figure 35.5c). From here, blood distributes the hormone throughout the body, where it exerts its effect on target cells (Figure 35.5d).

Antidiuretic hormone (**ADH**) affects certain kidney cells. The hormone causes these cells to reabsorb more water, thus making the urine more concentrated.

Oxytocin (**OT**) triggers muscle contractions during childbirth. It also makes milk move into the ducts of mammary glands when a female is nursing her young, and it affects social behavior in some species.

Table 35.2 Primary Actions of Hormones Released From the Human Pituitary Gland

Pituitary Lobe	Secretions	Designation	Main Targets	Primary Actions
Posterior Nervous tissue (extension of hypothalamus)	Antidiuretic hormone (vasopressin)	ADH	Kidneys	Induces water conservation as required to maintain extracellular fluid volume and solute concentrations
	Oxytocin	OT	Mammary glands Uterus	Induces milk movement into secretory ducts Induces uterine contractions during childbirth
Anterior Glandular tissue, mostly	Adrenocorticotropic hormone	ACTH	Adrenal glands	Stimulates release of cortisol, an adrenal steroid hormone
	Thyroid-stimulating hormone	TSH	Thyroid gland	Stimulates release of thyroid hormones
	Follicle-stimulating hormone	FSH	Ovaries, testes	In females, stimulates estrogen secretion, egg maturation; in males, helps stimulate sperm formation
	Luteinizing hormone	LH	Ovaries, testes	In females, stimulates progesterone secretion, ovulation, corpus luteum formation; in males, stimulates testosterone secretion, sperm release
	Prolactin	PRL	Mammary glands	Stimulates and sustains milk production
	Growth hormone (somatotropin)	GH	Most cells	Promotes growth in young; induces protein synthesis, cell division; roles in glucose, protein metabolism in adults

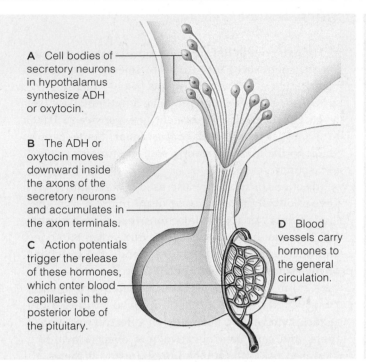

A **Cell bodies of secretory neurons in hypothalamus synthesize ADH or oxytocin.**

B **The ADH or oxytocin moves downward inside the axons of the secretory neurons and accumulates in the axon terminals.**

C **Action potentials trigger the release of these hormones, which enter blood capillaries in the posterior lobe of the pituitary.**

D **Blood vessels carry hormones to the general circulation.**

Figure 35.5 Animated Interactions between the pituitary gland's posterior lobe and the hypothalamus.

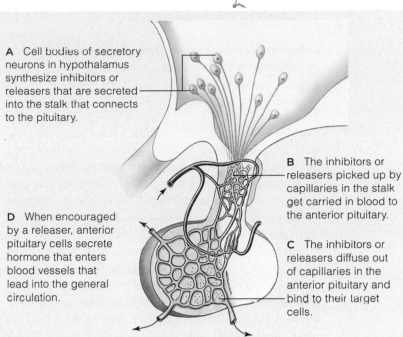

A **Cell bodies of secretory neurons in hypothalamus synthesize inhibitors or releasers that are secreted into the stalk that connects to the pituitary.**

B **The inhibitors or releasers picked up by capillaries in the stalk get carried in blood to the anterior pituitary.**

C **The inhibitors or releasers diffuse out of capillaries in the anterior pituitary and bind to their target cells.**

D **When encouraged by a releaser, anterior pituitary cells secrete hormone that enters blood vessels that lead into the general circulation.**

Figure 35.6 Animated Interactions between the pituitary gland's anterior lobe and the hypothalamus.

Anterior Pituitary Function

The anterior pituitary produces hormones of its own, but hormones from the hypothalamus control their secretion. Most hypothalamic hormones that act on the anterior pituitary are **releasers**; they encourage secretion of hormones by target cells. Hypothalamic **inhibitors** call for a reduction in target cell secretions.

Hypothalamic releasers and inhibitors are secreted into the stalk that connects the hypothalamus to the pituitary (Figure 35.6a). They diffuse into blood and are carried to the anterior lobe of the pituitary (Figure 35.6b). Here, they diffuse out of capillaries and bind to target cells (Figure 35.6c). When stimulated by a releaser, the target cell releases an anterior pituitary hormone into the blood (Figure 35.6d).

The target cells of some anterior pituitary hormones are inside other glands:

Adrenocorticotropic hormone (ACTH) stimulates the release of hormones by adrenal glands.

Thyroid-stimulating hormone (TSH) regulates the secretion of thyroid hormone by the thyroid gland.

Follicle-stimulating hormone (FSH) and **luteinizing hormone (LH)** affect sex hormone secretion and production of gametes by gonads—a male's testes or a female's ovaries.

Prolactin (PRL) targets the mammary glands, which are exocrine glands (Section 32.2). It stimulates and sustains milk production after childbirth.

Growth hormone (GH) has targets in most tissues. It triggers secretions of signals that promote growth of bone and soft tissues in the young. It also influences metabolism in adults.

Feedback Controls of Hormone Secretion

The hypothalamus and pituitary are involved in many feedback controls. With positive feedback mechanisms, a stimulus causes a response, such as hormone secretion, that increases the intensity of the stimulus. For example, Section 27.3 described how the stretching of muscles during childbirth causes oxytocin secretion, which causes more stretching, and so on.

Negative feedback mechanisms are more common. In this case, a stimulus elicits a response that decreases the stimulus. Several examples of negative feedback mechanisms that involve the hypothalamus and pituitary gland are described later in this chapter.

Take-Home Message

How do the hypothalamus and pituitary gland interact?

■ Some secretory neurons of the hypothalamus make hormones (ADH, OT) that move through axons into the posterior pituitary, which releases them.

■ Other hypothalamic neurons produce releasers and inhibitors that are carried by the blood into the anterior pituitary. These hormones regulate the secretion of anterior pituitary hormones (ACTH, TSH, LH, FSH, PRL, and GH).

- Disturbances of growth hormone production or function can cause excessive or reduced growth.
- Link to Genetic engineering 16.6

Growth hormone (GH) secreted by the anterior pituitary affects target cells throughout the body. Among other effects, GH calls for production of cartilage and bone and increases muscle mass. Normally, GH production surges during teenage years, causing a growth spurt. Level of the hormone then declines with age.

Excessive secretion of GH during childhood causes gigantism. Affected people have a normally proportioned body, but are unusually large (Figure 35.7a).

Excess production of GH during adulthood causes acromegaly. Bones can no longer lengthen, and instead become thicker. The hands, feet, and facial bones are most often visibly affected (Figure 35.7b). The Greek word *acro* means extremities, and *megas* means large.

Both gigantism and acromegaly usually arise as the result of a benign (noncancerous) pituitary tumor.

Pituitary dwarfism occurs when the body produces too little GH or receptors do not respond to it properly during childhood. Affected individuals are short but normally proportioned (Figure 35.7c). Pituitary dwarfism can be inherited, or it can result from a pituitary tumor or injury.

Human growth hormone can now be made through genetic engineering (Section 16.6). Injections of recombinant human growth hormone (rhGH) increase the growth rate of children who have a naturally low GH level. However, such treatment is expensive ($10,000 to $20,000 a year) and controversial. Some people object to the idea of treating short stature as a defect to be cured.

Injections of rhGH are also used to treat adults who have a low GH level because of pituitary or hypothalamic tumors or injury. Injections that restore normal GH level can help affected individuals maintain bone and muscle mass, while reducing body fat. Injections of rhGH have also been touted by some as a way to slow normal aging or to boost athletic performance. However, such uses are not approved by regulatory agencies, have not been shown effective in clinical trials, and can have negative side effects, including increased risk of high blood pressure and diabetes.

Take-Home Message

What are the effects of too much or too little growth hormone?

- Excessive growth hormone causes faster than normal bone growth. When the excess occurs during childhood, the result is gigantism. In adults, the result is acromegaly.
- A deficiency of GH during childhood can cause dwarfism.

Figure 35.7 Examples of effects of disrupted growth hormone function.

(**a**) Standing 6 feet 5 inches tall, this 12-year-old boy with pituitary gigantism towers over his mother.

(**b**) A woman before and after she became affected by acromegaly. Notice how her chin elongated.

(**c**) Dr. Hiralal Maheshwari, *right*, with two men from a village in Pakistan where a heritable form of dwarfism is common. The men of the village average 130 centimeters (a little over 4 feet) tall. Dr. Maheshwari found that these men make less than the typical amount of GH because their pituitary gland does not respond to the hypothalamic releaser that normally stimulates GH secretion.

age 16

age 52

a

b

c

■ A cell in a vertebrate body is a target for a diverse array of hormones from endocrine glands and secretory cells.

The next few sections of this chapter describe effects of the main vertebrate hormones that are released by endocrine glands other than the pituitary. Table 35.3 provides an overview of this information.

In addition to major endocrine glands, vertebrates have hormone-secreting cells in some organs. As noted earlier, cells of the small intestine make secretin, which acts on the pancreas. Parts of the gut also secrete other hormones that affect appetite and digestion. In addition, adipose (fat) tissue makes leptin, a hormone that acts in the brain and suppresses appetite.

When oxygen level in blood falls, kidneys secrete erythropoietin, a hormone that stimulates maturation and production of oxygen-transporting red blood cells. Even the heart makes a hormone: atrial natriuretic peptide. It stimulates water and salt excretion by kidneys.

As you learn about the effects of specific hormones, keep in mind that cells in most tissues have receptors for more than one hormone. The response called up by one hormone may oppose or reinforce that of another. For example, every skeletal muscle fiber has receptors for glucagon, insulin, cortisol, epinephrine, estrogen, testosterone, growth hormone, somatostatin, and thyroid hormone, as well as others. Thus, blood levels of all of these hormones affect the muscles.

Take-Home Message

What are the sources and effects of vertebrate hormones?

■ In addition to the pituitary gland and hypothalamus, endocrine glands and endocrine cells secrete hormones. The gut, kidneys, and heart are among the organs that are not considered glands, but do include hormone-secreting cells.

■ Most cells have receptors for multiple hormones, and the effect of one hormone can be enhanced or opposed by that of another.

Table 35.3 Sources and Actions of Vertebrate Hormones Discussed in Sections 35.6 to 35.12

Source	Examples of Secretion(s)	Main Target(s)	Primary Actions
Thyroid	Thyroid hormone	Most cells	Regulates metabolism; has roles in growth, development
	Calcitonin	Bone	Lowers calcium level in blood
Parathyroids	Parathyroid hormone	Bone, kidney	Elevates calcium level in blood
Pancreatic islets	Insulin	Liver, muscle, adipose tissue	Promotes cell uptake of glucose; thus lowers glucose level in blood
	Glucagon	Liver	Promotes glycogen breakdown; raises glucose level in blood
	Somatostatin	Insulin-secreting cells	Inhibits digestion of nutrients, hence their absorption from gut
Adrenal cortex	Glucocorticoids (including cortisol)	Most cells	Promotes breakdown of glycogen, fats, and proteins as energy sources; thus help raise blood level of glucose
	Mineralocorticoids (including aldosterone)	Kidney	Promotes sodium reabsorption (sodium conservation); help control the body's salt–water balance
Adrenal medulla	Epinephrine (adrenaline)	Liver, muscle, adipose tissue	Raises blood level of sugar, fatty acids; increases heart rate and force of contraction
	Norepinephrine	Smooth muscle of blood vessels	Promotes constriction or dilation of certain blood vessels; thus affects distribution of blood volume to different body regions
Gonads			
Testes (in males)	Androgens (including testosterone)	General	Required in sperm formation; development of genitals; maintenance of sexual traits; growth, development
Ovaries (in females)	Estrogens	General	Required for egg maturation and release; preparation of uterine lining for pregnancy and its maintenance in pregnancy; genital development; maintenance of sexual traits; growth, development
	Progesterone	Uterus, breasts	Prepares, maintains uterine lining for pregnancy; stimulates development of breast tissues
Pineal gland	Melatonin	Brain	Influences daily biorhythms, seasonal sexual activity
Thymus	Thymosins	T lymphocytes	Poorly understood regulatory effect on T lymphocytes

- The thyroid regulates metabolic rate, and the adjacent parathyroids regulate calcium levels.

- Link to Feedback mechanisms 27.3

The Thyroid Gland

The human **thyroid gland** is at the base of the neck, and attaches to the trachea (Figure 35.8). The gland secretes two iodine-containing molecules (triiodothyronine and thyroxine) that we refer to collectively as thyroid hormone. Thyroid hormone increases the metabolic activity of tissues throughout the body. The thyroid gland also secretes calcitonin, a hormone that causes deposition of calcium in the bones of growing children. Normal adults produce little calcitonin.

The anterior pituitary gland and hypothalamus regulate thyroid hormone secretion by way of a negative feedback loop. Figure 35.9 shows what happens when the level of thyroid hormone in the blood declines. In response to this decline, the hypothalamus secretes a releasing hormone (TRH) that acts in the anterior lobe of the pituitary. The releaser causes the pituitary to secrete thyroid-stimulating hormone (TSH). TSH in turn induces the thyroid gland to release thyroid hormone. As a result, the blood level of thyroid hormone rises back to its set point. Once that point is reached, the secretion of TRH and TSH slows.

Thyroid hormone includes iodine, a nutrient that humans obtain from their food. Thus, too little iodine in the diet is one cause of hypothyroidism—a low level of thyroid hormone. A goiter, or enlarged thyroid, is often a symptom (Figure 35.10a). The thyroid enlarges because the feedback loop illustrated in Figure 35.9 is disrupted and the gland receives constant stimulation to increase its output. Use of iodized salt is an easy, inexpensive way to ensure adequate iodine intake, but such salt is not available everywhere.

Hypothyroidism can cause developmental problems. If a mother lacks iodine during her pregnancy, or a child has a genetic defect that interferes with thyroid hormone production, the child's nervous system may not form properly. A low level of thyroid hormone during infancy or early childhood also stunts growth and impairs mental ability.

Hypothyroidism sometimes arises in adults as the result of an injury or an immune disorder that affects the thyroid or pituitary. Regardless of the cause, symptoms of insufficient thyroid hormone often include

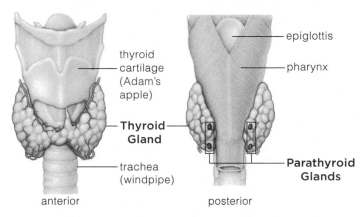

Figure 35.8 Location of human thyroid and parathyroid glands.

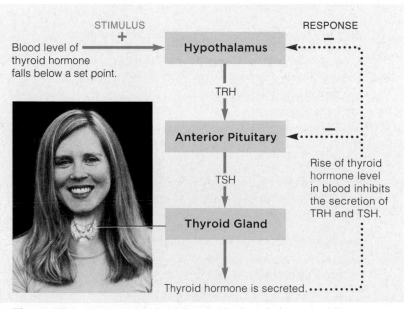

Figure 35.9 Negative feedback loop to the hypothalamus and the pituitary's anterior lobe that governs thyroid hormone secretion.

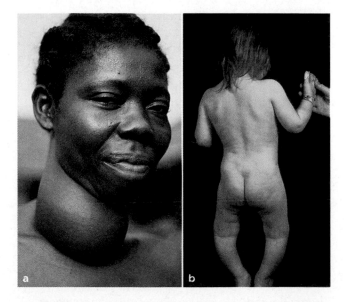

Figure 35.10 (a) A goiter caused by a diet that includes too little iodine. (b) A child with rickets caused by a lack of vitamin D has characteristic bowed legs.

weight gain, sluggishness, forgetfulness, depression, joint pain, weakness, and increased sensitivity to cold. The use of synthetic thyroid hormone can eliminate symptoms, but treatment must be continued for life.

A goiter can also be a symptom of Graves' disease. In this case, an immune malfunction causes the thyroid to produce an excess of thyroid hormone. The resulting hyperthyroidism, causes anxiety, insomnia, heat intolerance, protruding eyes, weight loss, and tremors. Drugs, surgery, or radiation can be used to reduce thyroid hormone level in the blood.

The Parathyroid Glands

Four **parathyroid glands**, each about the size of a grain of rice, are located on the thyroid's posterior surface (Figure 35.8). The glands release parathyroid hormone (PTH) in response to a decline in the level of calcium in blood. Calcium ions have roles in neuron signaling, blood clotting, muscle contraction, and other essential physiological processes.

PTH targets bone cells and kidney cells. In bones, it induces specialized cells called osteoclasts to secrete bone-digesting enzymes. Calcium and other minerals released from the bone enter the blood. In the kidneys, PTH stimulates tubule cells to reabsorb more calcium. It also stimulates secretion of enzymes that activate vitamin D, transforming it to calcitriol. Calcitriol is a steroid hormone that encourages cells in the intestinal lining to absorb more calcium from food.

A nutritional disorder known as rickets occurs in children who do not get enough vitamin D. Without adequate vitamin D, the child does not absorb much calcium, so formation of new bone slows. At the same time, low calcium in the blood triggers PTH secretion. As PTH rises, the child's body breaks down existing bones. Bowed legs and deformities in pelvic bones are common symptoms of rickets (Figure 35.10b).

Tumors and other conditions that cause excessive PTH secretion also weaken bone, and they increase risk of kidney stones, because calcium released from bone ends up in the kidney. Disorders that reduce PTH output lower blood calcium. The resulting seizures and unrelenting muscle contractions can be deadly.

Take-Home Message

What are the functions of the thyroid and parathyroid glands?

■ The thyroid gland has roles in regulation of metabolism and in development. Iodine is required to make thyroid hormone.

■ The parathyroid glands are the main regulators of blood calcium level.

35.7 | Twisted Tadpoles

■ Impaired thyroid function in frogs is another indication of hormone disruptors in the environment.

A tadpole is an aquatic larva of a frog. It undergoes a major remodeling in body form—a metamorphosis—when it makes the transition to an adult. For instance, it sprouts legs, lungs replace its gills, and its tail disappears. A surge in thyroid hormone triggers these changes. A tadpole keeps growing if its thyroid tissue is removed, but it will never undergo metamorphosis or take on adult form.

Some water pollutants may be the chemical equivalent of thyroid removal. For one study, investigators exposed embryos of African clawed frogs (*X. laevis*) to water drawn from lakes in Minnesota and Vermont. Half of the water samples came from lakes where deformity rates were low. The other half came from "hot spots," places where the water has as many as twenty kinds of dissolved pesticides and where deformity rates are high.

The embryos that were raised in hot-spot water often developed into tadpoles that had a bent spine and other abnormalities, as in Figure 35.11. Some tadpoles never did undergo metamorphosis and change into adult form. Control embryos raised in water from other lakes developed normally.

To find out if something in the water was interfering with thyroid hormone, the researchers added thyroid hormone to hot-spot water. Embryos raised in this mix developed into tadpoles that had fewer deformities or none at all. This result suggested that something in the water impaired normal thyroid hormone action.

Frogs are highly sensitive to disturbances in thyroid function, and thyroid disruptions are easy to detect. That is why toxicologists use laboratory frogs to test whether chemicals are thyroid disruptors. These scientists also use frogs to determine exactly how disruptive chemicals exert their effects.

Among the chemicals under study are perchlorates, which are widely used in explosives, propellants, and batteries. Perchlorates can interfere with the metabolism of iodine. As little as 5 parts per billion in water may stop a frog's forelimbs from developing.

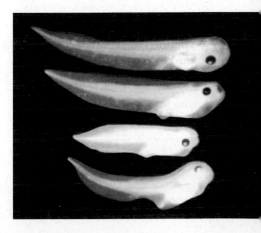

Figure 35.11 Evidence that pollutants affect frog development. The uppermost *Xenopus laevis* tadpole in this photographic series was raised in water from a lake with few deformed frogs. Tadpoles below it developed in water taken from three "hot-spot" lakes with increasingly higher concentrations of dissolved chemical compounds. As later tests showed, supplemental thyroid hormone can lessen or eliminate hot-spot deformities.

Pancreatic Hormones

■ Two pancreatic hormones with opposing effects work together to regulate the level of sugar in the blood.

■ Link to Endocrine and exocrine glands 32.2

The **pancreas** is an organ that lies in the abdominal cavity, behind the stomach (Figure 35.12) and has both endocrine and exocrine functions. Its exocrine cells secrete digestive enzymes into the small intestine. Its endocrine cells are in clusters called pancreatic islets.

Alpha cells of the pancreatic islets secrete the hormone **glucagon**. Glucagon targets cells in the liver and causes the activation of enzymes that break glycogen into glucose subunits. By its action, glucagon raises the level of glucose in blood.

Beta cells of the islets secrete the hormone **insulin**. This hormone's main targets are liver, fat, and skeletal muscle cells. Insulin stimulates muscle and fat cells to take up glucose. In all target cells, insulin activates enzymes that function in protein and fat synthesis, and it inhibits the enzymes that catalyze protein and fat breakdown. As a result of its actions, insulin lowers the level of glucose in the blood.

As you can see, glucagon and insulin have opposing effects on blood glucose level. Together, their actions keep blood glucose within the range that body cells can tolerate. When blood glucose level rises above a set point, alpha cells secrete less glucagon and beta cells secrete more insulin (Figure 35.12*a–c*). As glucose is taken up and stored inside cells, blood glucose declines (Figure 35.12*d,e*). In contrast, any decline in blood glucose below the set point turns up glucagon secretion and slows insulin secretion (Figure 35.12*f–h*). The resulting release of glucose from the liver causes blood glucose to rise (Figure 35.12*i,j*).

Take-Home Message

How do the actions of pancreatic hormones help maintain the level of blood glucose within a range body cells can tolerate?

■ Insulin is secreted in response to high blood glucose and it increases glucose uptake and storage by cells.

■ Glucagon is secreted in response to low blood glucose and it increases breakdown of glycogen to glucose.

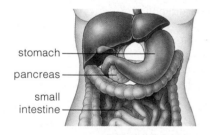

stomach
pancreas
small intestine

Figure 35.12 Animated *Above*, location of the pancreas. *Right*, how cells that secrete insulin and glucagon react to shifts in the blood level of glucose. Insulin and glucagon work antagonistically to regulate glucose level, an example of homeostasis.

(**a**) *After* a meal, glucose enters blood faster than cells can take it up. Its level in blood increases. (**b,c**) In the pancreas, the increase stops alpha cells from secreting glucagon and stimulates beta cells to secrete insulin. (**d**) In response to insulin, muscle and adipose cells take up and store glucose, and liver cells synthesize more glycogen. (**e**) The outcome? Insulin *lowers* the glucose blood level.

(**f**) *Between* meals, the glucose level in blood declines. (**g,h**) This stimulates alpha cells to secrete glucagon and stops beta cells from secreting insulin. (**i**) In the liver, glucagon causes cells to break glycogen down into glucose, which enters the blood. (**j**) The outcome? Glucagon *raises* the amount of glucose in blood.

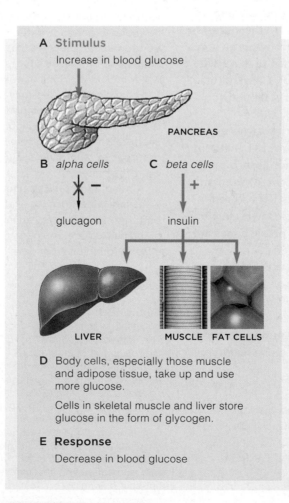

A Stimulus
Increase in blood glucose

PANCREAS

B *alpha cells* —

glucagon

C *beta cells* +

insulin

LIVER   MUSCLE   FAT CELLS

D Body cells, especially those muscle and adipose tissue, take up and use more glucose.

Cells in skeletal muscle and liver store glucose in the form of glycogen.

E Response
Decrease in blood glucose

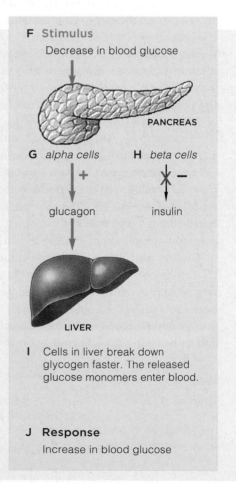

F Stimulus
Decrease in blood glucose

PANCREAS

G *alpha cells* +

glucagon

H *beta cells* —

insulin

LIVER

I Cells in liver break down glycogen faster. The released glucose monomers enter blood.

J Response
Increase in blood glucose

35.9 | Blood Sugar Disorders

■ Glucose is the main energy source for brain cells and the only one for red blood cells. Having too much or too little glucose in blood causes problems throughout the body.

Diabetes mellitus is a metabolic disorder in which cells do not take up glucose as they should. As a result, sugar accumulates in blood and in urine. Complications develop throughout the body (Table 35.4). Excess sugar in the urine encourages growth of pathogenic bacteria, and it damages small blood vessels in the kidneys. Diabetes is the most common cause of permanent kidney failure.

Uncontrolled diabetes also damages blood vessels and nerves elsewhere, especially in the arms, hands, legs, and feet. Diabetics account for more than 60 percent of lower limb amputations.

Type 1 Diabetes There are two main types of diabetes mellitus. Type 1 develops after the body has mounted an autoimmune response against its insulin-secreting beta cells. Certain white blood cells wrongly identify the cells as foreign (nonself) and destroy them. Environmental factors add to a genetic predisposition to the disorder. Symptoms usually start to appear during childhood and adolescence, which is why this metabolic disorder is also known as juvenile-onset diabetes. Individuals with type 1 diabetes require injections of insulin, and must monitor their blood sugar level carefully (Figure 35.13).

Type 1 diabetes accounts for only 5 to 10 percent of all reported cases, but it is the most dangerous in the short term. Insulin discourages metabolism of fats and proteins, so too little insulin causes excessive fat and protein breakdown. Two outcomes are weight loss and accumulation of ketones in blood and urine. Ketones are normal acidic products of fat breakdown, but when too many build up, the result is ketoacidosis. The altered acidity and solute levels can interfere with brain function. Extreme cases may lead to coma or death.

Type 2 Diabetes Type 2 diabetes is by far the most common form of the disorder. Insulin levels are normal or even high. However, target cells do not respond to the hormone as they should, and blood sugar levels remain elevated. Symptoms typically start to develop in middle age, when insulin production declines. Genetics also is a factor, but obesity increases the risk.

Diet, exercise, and oral medications can control most cases of type 2 diabetes. However, if glucose levels are not lowered by these means, pancreatic beta cells receive continual stimulation. Eventually they falter, and insulin production declines. When that happens, a type 2 diabetic may require insulin injections.

Worldwide, rates of type 2 diabetes are soaring. By one estimate, more than 150 million people are now affected. Western diets and sedentary life-styles are contributing factors. The prevention of diabetes and its complications is acknowledged to be among the most pressing public heath priorities around the world.

Hypoglycemia In hypoglycemia, the level of blood glucose falls low enough to disrupt normal body functions. Rare insulin-secreting tumors can cause it, but most cases occur after an insulin-dependent diabetic miscalculates and injects a bit too much insulin to balance food intake. The result is insulin shock. The brain stalls as its fuel source dwindles. Common symptoms are dizziness, confusion, and difficulty speaking. Insulin shock can be lethal, but an injection of glucagon quickly reverses the condition.

Table 35.4	Some Complications of Diabetes
Eyes	Changes in lens shape and vision; damage to blood vessels in retina; blindness
Skin	Increased susceptibility to bacterial and fungal infections; patches of discoloration; thickening of skin on the back of hands
Digestive system	Gum disease; delayed stomach emptying that causes heartburn, nausea, vomiting
Kidneys	Increased risk of kidney disease and failure
Heart and blood vessels	Increased risk of heart attack, stroke, high blood pressure, and atherosclerosis
Hands and feet	Impaired sensations of pain; formation of calluses, foot ulcers; poor circulation in feet especially sometimes leads to tissue death that can only be treated by amputation

Figure 35.13 A diabetic checks his blood glucose by placing a blood sample into a glucometer. Compared with Caucasians, Hispanics and African Americans are about 1.5 times more likely to be diabetic. Native Americans and Asians are at even greater risk. Proper diet helps control blood sugar, even in type 1 diabetics.

- Atop each kidney is an adrenal gland with two parts. Each part produces and releases different hormones.

- Links to Alternative energy sources 8.7, Sympathetic neurons 33.8

There are two **adrenal glands**; one perches above each kidney. (In Latin *ad–* means near, and *renal* refers to the kidney.) Each adrenal gland is about the size of a big grape. Its outer layer is the **adrenal cortex** and its inner portion is the **adrenal medulla**. The two parts of the gland are controlled by different mechanisms, and they secrete different hormones.

Hormonal Control of the Adrenal Cortex

The adrenal cortex secretes three steroid hormones. One of these, **aldosterone**, controls sodium and water reabsorption in the kidneys. Chapter 41 explains its function in great detail. The adrenal cortex also produces and secretes small amounts of both male and female sex hormones, which we discuss in Section 35.12 and Chapter 42. For now, we focus on **cortisol**, an adrenal hormone that has wide-reaching effects on metabolism and immunity.

A negative feedback loop governs the cortisol level in blood (Figure 35.14). A decrease in cortisol triggers secretion of CRH (corticotropin-releasing hormone) by the hypothalamus. CRH then stimulates secretion of ACTH (adrenocorticotropic hormone). This anterior pituitary hormone causes the adrenal cortex to release cortisol. The blood level of cortisol keeps increasing until it reaches a set point. Then the hypothalamus and anterior pituitary slow their release of CRH and ACTH, and cortisol secretion also winds down.

Cortisol has many effects. It induces liver cells to break down their store of glycogen, and it suppresses uptake of glucose by other cells. Cortisol also prompts adipose cells to degrade fats, and skeletal muscles to degrade proteins. Breakdown products of fats and proteins function as alternative energy sources (Section 8.7). Cortisol also suppresses immune responses.

With injury, illness, or anxiety, the nervous system overrides the feedback loop, and cortisol in blood can soar. In the short term, this response helps get enough glucose to the brain when food supplies are likely to be low. Cortisol also suppresses inflammatory responses. As the next section explains, long-term stress and elevation of cortisol level can cause health problems.

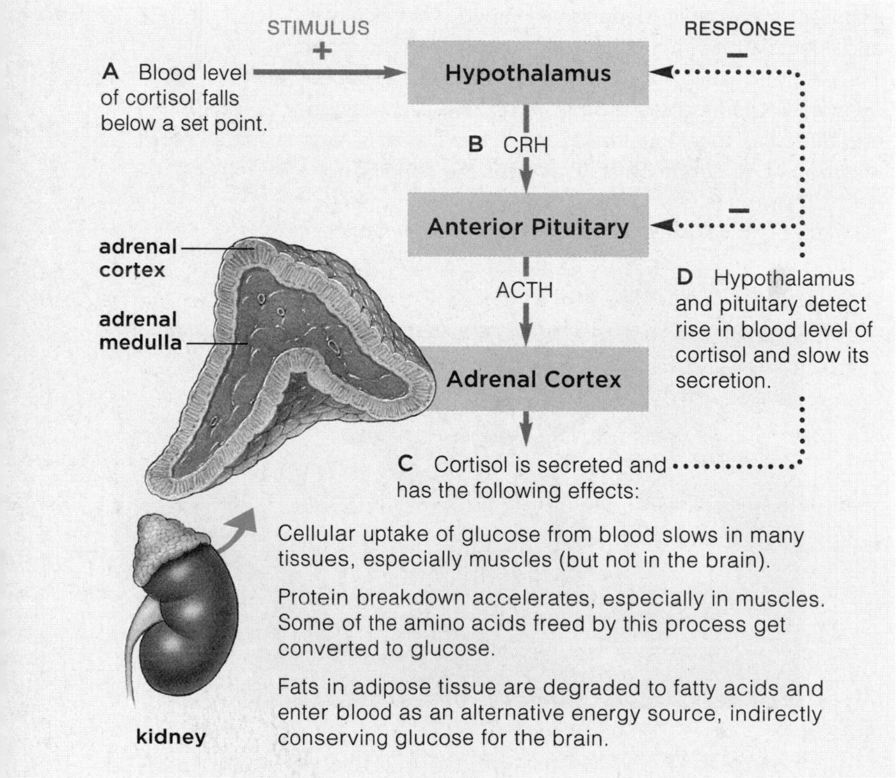

Nervous Control of the Adrenal Medulla

The adrenal medulla contains specialized neurons of the sympathetic division (Section 33.8). Like other sympathetic neurons, those in the adrenal medulla release norepinephrine and epinephrine. However, in this case, the norepinephrine and epinephrine enter the blood and function as hormones, rather than acting as neurotransmitters at a synapse. Epinephrine and norepinephrine released into the blood have the same effect on a target organ as direct stimulation by a sympathetic nerve.

Remember that sympathetic stimulation plays a role in the fight–flight response. Epinephrine and norepinephrine dilate the pupils, increase breathing, and make the heart beat faster. They prepare the body to deal with an exciting or dangerous situation.

STIMULUS + / RESPONSE −

A Blood level of cortisol falls below a set point.

Hypothalamus

B CRH

Anterior Pituitary

ACTH

Adrenal Cortex

D Hypothalamus and pituitary detect rise in blood level of cortisol and slow its secretion.

C Cortisol is secreted and has the following effects:

adrenal cortex

adrenal medulla

Cellular uptake of glucose from blood slows in many tissues, especially muscles (but not in the brain).

Protein breakdown accelerates, especially in muscles. Some of the amino acids freed by this process get converted to glucose.

Fats in adipose tissue are degraded to fatty acids and enter blood as an alternative energy source, indirectly conserving glucose for the brain.

kidney

Figure 35.14 Animated Structure of the human adrenal gland. An adrenal gland rests on top of each kidney. The diagram shows a negative feedback loop that governs cortisol secretion.

Take-Home Message

What is the function of the adrenal glands?

- The adrenal cortex secretes aldosterone, cortisol, and small amounts of sex hormones. Aldosterone affects urine concentration and cortisol affects metabolism and the stress response.

- The adrenal medulla releases epinephrine and norepinephrine, which prepare the body for excitement or danger.

35.11 | Too Much or Too LIttle Cortisol

■ Short-term responses to stress help us function in hard times, but chronic stress is unhealthy.

■ Link to Memory 33.11

Chronic Stress and Elevated Cortisol

Each summer, a troop of olive baboons (*Papio anubis*) on East Africa's Serengeti plains has visitors. For more than twenty years, neurobiologist Robert Sapolsky and his Kenyan colleagues have been studying how these baboons interact and how a baboon's social position influences its hormone levels and health.

Remember, when the body is stressed, commands from the nervous system trigger secretion of cortisol, epinephrine, and norepinephrine. As these secretions find their targets, they help the body deal with the immediate threat by diverting resources from longer-term tasks. This stress response is highly adaptive for short bursts of activity, as when it diverts blood flow to muscles of an animal fleeing from a predator.

Sometimes stress does not end. The baboons live in big troops with a clearly defined dominance hierarchy. Those on top of the hierarchy get first access to food, grooming, and sexual partners. Those at the bottom must relinquish resources to a higher ranking baboon or face attack (Figure 35.15). Not surprisingly, the low-ranking baboons tend to have elevated cortisol levels.

Physiological responses to chronic stress interfere with growth, the immune system, sexual function, and cardiovascular function. Chronically high cortisol levels also harm cells in the hippocampus, a brain region central to memory and learning (Section 33.11).

We also see the impact of long-term elevated cortisol levels in humans affected by Cushing's syndrome, or hypercortisolism. This rare metabolic disorder might be triggered by an adrenal gland tumor, oversecretion of ACTH by the anterior pituitary, or ongoing use of the drug cortisone. Doctors often prescribe cortisone to relieve chronic pain, inflammation, or other health problems. The body converts it to cortisol.

The symptoms of hypercortisolism include a puffy, rounded "moon face" and increased fat deposition around the torso. Blood pressure and blood glucose become unusually high. White blood cell counts are low, so affected people are more prone to infections. Thin skin, decreased bone density, and muscle loss are common. Wounds may be slow to heal. Women's menstrual cycles are erratic or nonexistent. Men may be impotent. Often, the hippocampus shrinks. Patients with the highest cortisol level also have the greatest reduction in the volume of the hippocampus, and the most impaired memory.

Figure 35.15 A dominant baboon (*right*) raising the stress level—and cortisol level—of a less dominant member of its troop.

Can status-related social stress affect human health? People who are low in a socioeconomic hierarchy do tend to have more health problems—obesity, hypertension, and diabetes—than those who are better off. These differences persist even after researchers factor out the obvious causes, such as variations in diet and access to health care. By one hypothesis, a heightened cortisol level caused by low social status may be one of the links between poverty and poor health.

Low Cortisol Level

Tuberculosis and other infectious diseases can damage the adrenal glands, and slow or halt cortisol secretion. The result is Addison's disease, or hypocortisolism. In developed countries, this hormonal disorder more often arises after autoimmune attacks on the adrenal glands. President John F. Kennedy had this form of the disorder. Symptoms often include fatigue, weakness, depression, weight loss, and darkening of the skin. If cortisol levels get too low, blood sugar and blood pressure can fall to life-threatening levels. Addison's disease is treated with a synthetic form of cortisone.

Take-Home Message

What are the effects of abnormal cortisol levels?

■ High cortisol levels, produced by chronic stress or an endocrine disorder, impair growth, healing, sexual function, and memory. Blood pressure and blood sugar are higher than normal.

■ With low cortisol levels, blood pressure and blood sugar fall. If they decline too far, the result can be life threatening.

35.12 | Other Endocrine Glands

- Outputs from the gonads, pineal gland, and thymus all change as an individual enters puberty.

- Links to Gamete formation 10.5, Visual signals 34.9

The Gonads

The **gonads**, or primary reproductive organs, produce gametes (eggs or sperm) as well as sex hormones. The gonads of male vertebrates are testes (singular, testis) and the main hormone they secrete is **testosterone**, the male sex hormone. The female gonads are the ovaries. They secrete mainly **estrogens** and **progesterone**, the female sex hormones. Figure 35.16 shows the location of the human gonads.

Puberty is a post-embryonic stage of development when the reproductive organs and structures mature. At puberty, a female mammal's ovaries increase their estrogen production, which causes breasts and other female secondary sexual traits to develop. Estrogens and progesterone control egg formation and ready the uterus for pregnancy. In males, a rise in testosterone output triggers the onset of sperm formation and the development of secondary sexual traits.

The hypothalamus and anterior pituitary control the secretion of sex hormones (Figure 35.17). In both males and females, the hypothalamus produces GnRH (gonadotropin-releasing hormone). This releaser causes the anterior pituitary to secrete follicle-stimulating hormone (FSH) and luteinizing hormone (LH). FSH and LH cause the gonads to secrete sex hormones.

Testes secrete mostly testosterone, but they also make a little bit of estrogen and progesterone. The estrogen is necessary for sperm formation. Similarly, a female's ovaries make mostly estrogen and progesterone, but also a little testosterone. The presence of testosterone contributes to libido—the desire for sex.

We discuss the role of sex hormones in gamete formation, the menstrual cycle, and development in detail in Chapter 42.

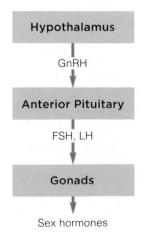

Figure 35.17
Generalized diagram showing control of sex hormone secretion.

The Pineal Gland

Deep in the vertebrate brain is the **pineal gland**. This small, pine cone–shaped gland secretes **melatonin**, a hormone that serves as part of an internal timing mechanism, or biological clock. Melatonin secretion declines when the retina detects light and sends signals along the optic nerve to the brain (Section 34.9).

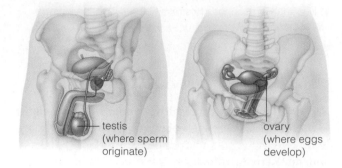

Figure 35.16 Location of human gonads, which produce gametes and secrete sex hormones.

Melatonin may affect human gonads. A decline in the production of this hormone starts at puberty and may help trigger it. Some pineal gland disorders are known to accelerate or delay puberty.

Melatonin also targets neurons that can lower body temperature and make us drowsy in dim light. The blood level of melatonin peaks in the middle of the night. Exposure to bright light sets a biological clock that controls sleeping versus arousal. Travelers who cross many time zones are advised to spend time in the sun after reaching a destination. Doing so helps them reset their biological clock and minimize jet lag.

In winter, seasonal affective disorder, also called "winter blues," causes some people to be depressed, to binge on carbohydrates, and to crave sleep. Bright artificial light in the morning decreases pineal gland activity and can improve mood.

The Thymus

The **thymus** lies beneath the breastbone. It secretes thymosins, hormones that help the infection-fighting white blood cells called T cells mature. The thymus grows until puberty, when it is about the size of an orange. Then, the surge in sex hormones causes it to shrink, and its secretions decline. However, the thymus enhances immune function even in adults.

Take-Home Message

What are the roles of the gonads, pineal gland, and thymus?

- A female's ovaries or a male's testes are gonads that make sex hormones as well as gametes.

- The pineal gland is inside the brain and produces melatonin, which influences sleep-wake cycles and onset of puberty.

- The thymus is in the chest and it secretes thymosins that are necessary for the maturation of white blood cells called T cells.

■ Genes that encode hormone receptors and enzymes involved in hormone synthesis have evolved over time.

■ Links to Gene duplication 12.5, Introns 14.3, Molting 25.11

Evolution of Hormone Diversity

We can trace the evolutionary roots of some vertebrate hormones and receptors back to signaling molecules in invertebrates. For example, receptors for the hormones FSH, LH, and TSH all have a similar structure. The genes that encode these receptors have a similar sequence and have introns (noncoding DNA) in the same places. The slightly different forms of receptor most likely evolved when a gene was duplicated, then copies mutated over time (Section 12.5).

When did the ancestral gene arise? Sea anemones do not have an endocrine system, but they do have a receptor protein gene like that for FSH. This suggests that the ancestral receptor gene existed long ago in a common ancestor of sea anemones and vertebrates.

Estrogen receptors may also have a long history. Sea slugs (Figure 35.18), a kind of mollusk, have receptors that are similar to vertebrate estrogen receptors.

Hormones and Molting

Other hormones are unique to invertebrates. For example, arthropods, which include crabs and insects, have a hardened external cuticle that they periodically shed as they grow (Section 25.11). Shedding of the old cuticle is called molting. A soft new cuticle forms beneath an old one before the animal molts. Although details vary among arthropod groups, molting is generally under the control of **ecdysone**, a steroid hormone.

The arthropod molting gland produces and stores ecdysone, then releases it for distribution throughout the body when conditions favor molting. Hormone-secreting neurons inside the brain control ecdysone's release. The neurons respond to internal signals and environmental cues, including light and temperature.

Figure 35.19 is an example of the control steps in crabs and other crustaceans. In response to seasonal cues, secretion of a molt-inhibiting hormone declines and ecdysone secretion rises. Ecdysone causes changes in the animal's structure and physiology. The existing cuticle separates from the epidermis and the muscles. Inner layers of the old cuticle break down. At the same time, cells of the epidermis secrete the new cuticle.

The steps in molting differ a bit in insects, which do not have a molt-inhibiting hormone. Rather, stimulation of the insect brain sets in motion a cascade of signals that trigger the production of molt-inducing

Figure 35.18 The sea hare (*Aplysia*), a type of mollusk. Some receptors in its plasma membrane are similar to vertebrate receptors that bind the steroid hormone estrogen.

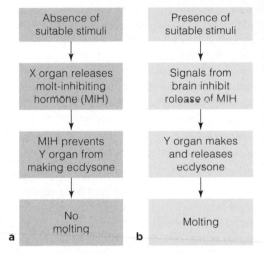

Figure 35.19 Hormonal control of molting in crustaceans such as crabs. Two hormone-secreting organs play a role. The X organ is in the eye stalk. The Y organ is at the base of the crab's antennae.

(a) In the absence of environmental cues for molting, secretions from the X organ prevent molting. **(b)** When stimulated by proper environmental cues, the brain sends nervous signals that inhibit X organ activity. With the X organ suppressed, the Y organ releases the ecdysone that stimulates molting.

(c) A newly molted blue crab with its old shell. The new shell remains soft for a about 12 hours, making it a "soft-shelled crab." In this state, the crab is highly vulnerable to predators, including human seafood lovers.

ecdysone. Chemicals that mimic ecdysone or interfere with its function are used as insecticides. When such insecticides run off from fields and get into water, they can affect ecdysone-related responses in other arthropods, such as crayfish, crabs, or shrimps.

Take-Home Message

What types of hormone systems do we see in invertebrates?

■ We can trace the evolutionary roots of the vertebrate endocrine system in invertebrates. Cnidarians such as sea anemones, and mollusks such as sea slugs, have receptors that resemble those that bind vertebrate hormones.

■ Invertebrates also have hormones with no vertebrate counterparts. Hormones that control molting in arthropods are an example.

Testosterone and estrogens have a very similar structure and enzymes can convert one to the other. The enzyme aromatase converts testosterone to estrogens. When human cells growing in culture are exposed to the herbicide atrazine, their aromatase activity rises, so more testosterone gets converted to estrogen. Atrazine may have the same effect in frogs, which would explain the altered sex organs first reported by Tyrone Hayes.

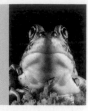

How would you vote?

Should atrazine use be continued while its health and environmental effects are studied in more detail? See CengageNOW for details, then vote online.

Summary

Section 35.1 **Hormones, neurotransmitters, local signaling molecules,** and **pheromones** are chemicals that are secreted by one cell type and that adjust the behavior of other, target cells. Any cell is a target if it has receptors for a signaling molecule.

All vertebrates have an **endocrine system** of secretory glands and cells. In most cases, the hormonal secretions travel through the bloodstream to nonadjacent targets.

■ *Use the animation on CengageNOW to learn about the main sources of hormones in the human body.*

Section 35.2 Some steroid hormones enter a target cell and bind to receptors inside it. Others bind to the cell's plasma membrane and alter the membrane properties.

The peptide and protein hormones bind to plasma membrane receptors. Binding may lead to formation of a **second messenger,** which relays a signal into the cell.

■ *Use the animation on CengageNOW to compare the mechanisms of steroid and protein hormone action.*

Sections 35.3, 35.4 The **hypothalamus,** a forebrain region, is structurally and functionally linked with the **pituitary gland** as a major center for homeostatic control.

The posterior pituitary releases two hormones made by neurons of the hypothalamus. **Antidiuretic hormone** acts in kidneys to concentrate urine. **Oxytocin** acts on the uterus and milk ducts. Other hypothalamic neurons secrete **releasers** and **inhibitors** that encourage or slow the secretion of anterior pituitary hormones.

The anterior pituitary produces several hormones that regulate other glands. **Adrenocorticotropic hormone** acts on the adrenal glands. **Follicle-stimulating hormone** and **luteinizing hormone** regulate the gonads. The thyroid is stimulated by **thyroid-stimulating hormone.** Mammary glands are stimulated by **prolactin.** The anterior pituitary also makes **growth hormone,** which affects cells throughout the body and stimulates bone growth. Gigantism, dwarfism, and acromegaly result from mutations that affect growth hormone function.

■ *Use the animation on CengageNOW to study how the hypothalamus and pituitary interact.*

Section 35.5 In addition to major endocrine glands, there are hormone-secreting cells in tissues and organs throughout the body. Most cells have receptors for, and are influenced by, many different hormones.

Sections 35.6, 35.7 A feedback loop to the anterior pituitary and hypothalamus governs the **thyroid gland** in the base of the neck. The thyroid affects metabolic rate and development. Iodine is required for thyroid function.

Four **parathyroid glands** make a hormone that acts on bone and kidney cells and raises blood calcium level.

Sections 35.8, 35.9 The **pancreas** in the abdominal cavity has exocrine and endocrine functions. Beta cells secrete **insulin** when blood glucose level is high. Insulin stimulates uptake of glucose by muscle and liver cells. When blood glucose is low, alpha cells secrete **glucagon,** which calls for glycogen breakdown and glucose release by the liver. The two hormones work in opposition to keep blood glucose levels within the optimal range.

Diabetes occurs when the body does not make insulin or its cells do not respond to it.

■ *Use the animation on CengageNOW to see how the actions of insulin and glucagon regulate blood sugar.*

Sections 35.10, 35.11 There is an **adrenal gland** on each kidney. The **adrenal cortex** secretes **aldosterone** which targets the kidney, and **cortisol,** the stress hormone. Cortisol secretion is governed by a negative feedback loop to the anterior pituitary gland and hypothalamus. In times of stress, the nervous system overrides feedback controls.

Norepinephrine and epinephrine released by neurons of the **adrenal medulla** influence organs as sympathetic stimulation does; they cause a fight–flight response.

■ *Watch the animation on CengageNOW to see how cortisol levels are maintained by negative feedback.*

Section 35.12 The **gonads** (ovaries or testes) secrete sex hormones. Ovaries secrete mostly **estrogens** and **progesterone.** Testes secrete mostly **testosterone.** Sex hormones control gamete formation and, in **puberty,** regulate the development of secondary sexual traits.

Light suppresses secretion of **melatonin** by the **pineal gland** in the brain. Melatonin affects biological clocks—internal timing mechanisms.

The **thymus** in the chest produces hormones that help some white blood cells (T cells) mature.

Section 35.13 Some vertebrate hormone receptor proteins resemble similar receptor proteins in invertebrates. This suggests the receptors evolved in a common ancestor of both groups. The steroid hormone **ecdysone** affects molting in arthropods and has no vertebrate counterpart.

Data Analysis Exercise

Contamination of water by agricultural chemicals affects the reproductive function of some animals. Are there effects on humans? Epidemiologist Shanna Swann and her colleagues studied sperm collected from men in four cities in the United States (Figure 35. 20). The men were partners of women who had become pregnant and were visiting a prenatal clinic, so all were fertile. Of the four cities, Columbia, Missouri, is located in the county with the most farmlands. New York City in New York is in an area with no agriculture.

1. In which cities did researchers record the highest and lowest sperm counts?

2. In which cities did samples show the highest and lowest sperm motility (ability to move)?

3. Aging, smoking, and sexually transmitted diseases adversely affect sperm. Could differences in any of these variables explain the regional differences in sperm count?

4. Do these data support the hypothesis that living near farmlands can adversely affect male reproductive function?

	Location of clinic			
	Columbia, Missouri	Los Angeles, California	Minneapolis, Minnesota	New York, New York
Average age	30.7	29.8	32.2	36.1
Percent nonsmokers	79.5	70.5	85.8	81.6
Percent with history of STD	11.4	12.9	13.6	15.8
Sperm count (million/ml)	58.7	80.8	98.6	102.9
Percent motile sperm	48.2	54.5	52.1	56.4

Figure 35.20 Data from a study of sperm collected from men who were partners of pregnant women that visited prenatal health clinics in one of four cities. STD stands for sexually transmitted disease.

Self-Quiz
Answers in Appendix III

1. _____ are signaling molecules that travel through the blood and affect distant cells in the same individual.
  a. Hormones
  b. Neurotransmitters
  c. Pheromones
  d. Local signaling molecules
  e. both a and b
  f. a through d

2. A _____ is synthesized from cholesterol and can diffuse across the plasma membrane.
  a. steroid hormone
  b. pheromone
  c. peptide hormone
  d. all of the above

3. Match each pituitary hormone with its target.
  ___ antidiuretic hormone
  ___ oxytocin
  ___ luteinizing hormone
  ___ growth hormone
  a. gonads (ovaries, testes)
  b. mammary glands, uterus
  c. kidneys
  d. most body cells

4. Releasers secreted by the hypothalamus cause the secretion of hormones by the _____ pituitary lobe.
  a. anterior
  b. posterior

5. In adults, too much _____ can cause acromegaly.
  a. growth hormone
  b. cortisol
  c. insulin
  d. melatonin

6. A diet lacking in iodine can cause _____ .
  a. rickets
  b. a goiter
  c. diabetes
  d. gigantism

7. Low blood calcium triggers secretion by _____ .
  a. adrenal glands
  b. parathyroid glands
  c. ovaries
  d. the thyroid gland

8. _____ lowers blood sugar levels; _____ raises it.
  a. Glucagon; insulin
  b. Insulin; glucagon

9. The _____ has endocrine and exocrine functions.
  a. hypothalamus
  b. pancreas
  c. pineal gland
  d. parathyroid gland

10. Secretion of _____ suppresses immune responses.
  a. melatonin
  b. antidiuretic hormone
  c. thyroid hormone
  d. cortisol

11. Exposure to bright light lowers blood _____ levels.
  a. glucagon
  b. melatonin
  c. thyroid hormone
  d. parathyroid hormone

12. True or false? Some heart cells and kidney cells secrete hormones.

13. True or false? Only women make follicle-stimulating hormone (FSH); only men make luteinizing hormone (LH).

14. True or false? All hormones secreted by arthropods such as crabs and insects are also secreted by vertebrates.

15. Match the term listed at left with the most suitable description at right.
  ___ adrenal medulla
  ___ thyroid gland
  ___ posterior pituitary gland
  ___ pancreatic islets
  ___ pineal gland
  ___ prostaglandin
  a. affected by day length
  b. a local signaling molecule
  c. secretes hormones made in the hypothalamus
  d. source of epinephrine
  e. secrete insulin, glucagon
  f. hormones require iodine

■ *Visit CengageNOW for additional questions.*

Critical Thinking

1. A large study of nurses suggests that night shift work may raise the risk of breast cancer. Changes in melatonin level may contribute to the increased risk. There is evidence that this hormone can slow the rate of cancer cell division. Nurses who work night shifts tend to have lower melatonin levels than those working days. Why is secretion of this hormone especially likely to be reduced by night work?

2. Sex hormone secretion is governed by a negative feedback loop to the hypothalamus and pituitary, similar to that for thyroid hormone or cortisol. Because of this, a veterinarian can tell whether or not a female dog has been neutered with a blood test. Dogs that still have their ovaries have a lower blood level of luteinizing hormone (LH) than dogs that have been neutered. Explain why removing a dog's ovaries would result in an elevated level of LH.

IMPACTS, ISSUES Pumping Up Muscles

The male sex hormone testosterone has anabolic effects; it encourages protein synthesis and thus increases muscle mass. That's one reason why men, who naturally make a lot of testosterone, tend to be more muscular than women, who make far less (Figure 36.1). It is also why some body builders and athletes turn to anabolic steroids (synthetic derivatives of testosterone), or to supplements that claim to raise natural testosterone levels.

For example, in the late 1990s, androstenedione, or "andro," soared in popularity after a baseball player, Mark McGwire, said he had used it during his successful attempt to break Major League Baseball's single-season home-run record. Andro forms naturally in the body as an intermediate in the synthesis of the sex hormone testosterone.

Does taking andro as a dietary supplement improve athletic performance? Results from the few controlled studies are mixed. Moreover, andro, like all anabolic steroids, has side effects. It increases a man's level of the female hormone estrogen, which can also be formed from andro. Estrogen has feminizing effects on males, including shrunken testicles, formation of female-like breasts, and hair loss. Also, like all anabolic steroids, andro increases risk of liver damage and cardiovascular attack. In 2004, the U.S. Food and Drug Administration announced that, in light of these side effects, it was banning the sale of andro. Even with all the negative publicity, some athletes continued to use anabolic steroids, risking both their health and reputation.

Athletes also use approved nutritional supplements such as creatine, which is a short chain of amino acids. The body makes some creatine and obtains more from food. When muscles must contract hard and fast, they normally turn first to phosphorylated creatine as an instant energy source.

Does creatine work? In some controlled studies, creatine improved performance during brief, high-intensity exercise. Nevertheless, excessive creatine intake puts a strain on the kidneys, and it is too soon to know whether creatine supplements have any long-term side effects. Also, no regulatory agency checks to see how much creatine is actually present in any commercial product.

With this chapter, we turn to the skeletal and muscular systems. What you learn here can help you evaluate how far both systems can and should be pushed in the pursuit of enhanced performance.

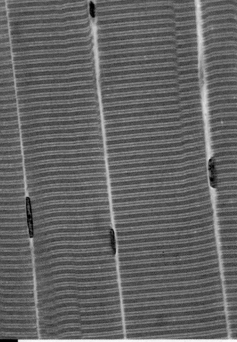

See the video! Figure 36.1 *Left*, a male with an abundance of skeletal muscle tissue, which has parallel rows of muscle fibers (*above*).

Key Concepts

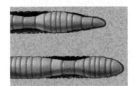

Invertebrate skeletons

Contractile force exerted against a skeleton moves animal bodies. In many invertebrates a fluid-filled body cavity is a hydrostatic skeleton. Others have an exoskeleton of hard structures at the body surface. Still others have a hard internal skeleton, or endoskeleton. **Section 36.1**

Vertebrate skeletons

Vertebrates have an endoskeleton of cartilage, bone, or both. Bones interact with muscles to move the body. They also protect and support organs, and store minerals. Blood cells form in some bones. A joint is a place where bones meet; there are several kinds. **Sections 36.2–36.5**

The muscle-bone partnership

Skeletal muscles are bundles of muscle fibers that interact with bones and with one another. Some cause movements by working as pairs or groups. Others oppose or reverse the action of a partner muscle. Tendons attach skeletal muscles to bones. **Section 36.6**

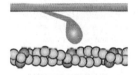

Skeletal muscle function

Muscle fibers contract in response to signals from a motor neuron. A muscle fiber contains many myofibrils, each divided crosswise into sarcomeres. ATP-driven interactions between protein filaments shorten sarcomeres, causing muscle contraction. **Sections 36.7–36.11**

Links to Earlier Concepts

- This chapter elaborates on some of the animal traits and evolutionary trends you learned about in Chapters 25 and 26.

- You will also build on your knowledge of connective (32.3) and muscle (32.4) tissues.

- You will learn more about the X-linked disorder muscular dystrophy (12.4), and how bacterial endospores (21.6) can affect muscles.

- You will see examples of active transport (5.4) and revisit the filaments involved in cell movement (4.13).

- Nervous control of muscle (33.5) and the effects of some hormones (35.6) are also discussed again.

- A skeleton can be internal or external.
- Links to Cnidarians 25.5, Annelids 25.7, Arthropods 25.12, Echinoderms 25.18

When you think of a skeleton, you probably picture an internal framework of bones, but this is just one type of skeleton. In other animals, a skeleton consists of a fluid or of external hard parts. Animal body parts move when muscles interact with the skeleton.

Hydrostatic Skeletons

Cnidarians and annelids are among the animals with a **hydrostatic skeleton**: a fluid-filled closed chamber or chambers that muscles act against. For example, a sea anemone's body is inflated by water that flows in through its mouth and fills its gastrovascular cavity (Figure 36.2). Beating of cilia causes the inward flow of water. Contraction of a ring of muscle around the mouth traps the water inside the body. Contractions of other muscles can redistribute the water and alter body shape. By analogy, think about how squeezing or pulling on a water-filled balloon changes its shape.

An anemone has circular muscles that ring its body and longitudinal ones that run from its top to bottom. Contracting circular muscles and relaxing longitudinal ones makes an anemone taller and thinner. When circular muscles relax and longitudinal ones contract, the anemone gets shorter and fatter. The animal can also open its mouth, contract both sets of muscles, and draw in its tentacles. This action forces most fluid from the gastrovascular cavity out of the body, and the body shrinks into a protective resting position (Figure 36.2b).

In earthworms, a coelom divided into many fluid-filled segments is the hydrostatic skeleton (Section 25.7). Longitudinal and circular muscles put pressure on the coelomic fluid in each segment, causing it to become long and narrow or short and wide. Waves of contraction that run the length of the body move the worm through the soil (Figure 36.3).

Exoskeletons

An **exoskeleton** is a stiff body covering to which the muscles attach. For example, bivalve mollusks such as clams and scallops have a hinged two-part shell.

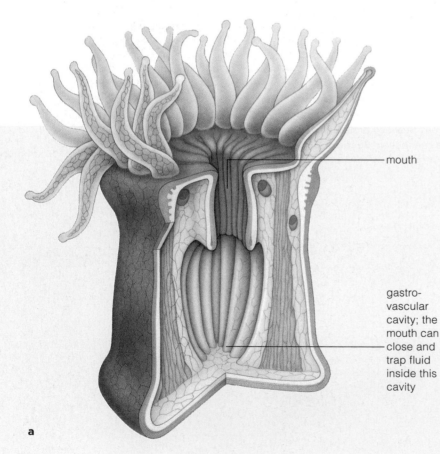

mouth

gastro-vascular cavity; the mouth can close and trap fluid inside this cavity

a

b

Figure 36.2 Animated Hydrostatic skeleton of a sea anemone. (**a**) Water is drawn into the gastrovascular cavity through the mouth. When the cavity is filled and mouth is closed, muscles can act on the trapped fluid and alter body shape. There are two sets of muscles: circular muscles ring the body; longitudinal ones run the length of the body. (**b**) One anemone inflated with water (*left*) and another that has expelled water from the gastrovascular cavity and pulled in its tentacles (*right*).

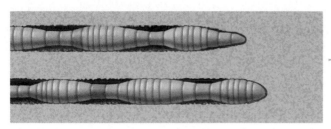

Figure 36.3 How an earthworm moves through the soil. Muscles act on coelomic fluid in individual body segments, causing the segments to change shape. A segment narrows when circular muscle ringing it contracts and longitudinal muscle running its length relaxes. The segment widens when circular muscle relaxes and longitudinal muscle contracts.

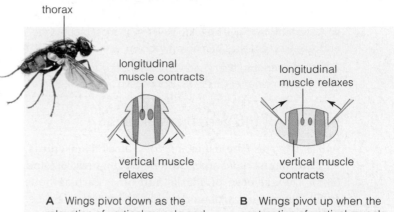

A Wings pivot down as the relaxation of vertical muscle and the contraction of longitudinal muscle pulls in sides of thorax.

B Wings pivot up when the contraction of vertical muscle and relaxation of longitudinal muscle flattens the thorax.

Figure 36.4 Animated Fly wing movement. Wings attach to the thorax at pivot points. When muscles inside the thorax contract and relax, the thorax changes shape and the wings pivot up and down at their attachment point.

A powerful muscle attached to the two halves of the shell can pull them together, shutting the shell. Some scallops can swim through the water by opening and closing their shell. Each time the shell is pulled shut, forcing water out, the scallop scoots backwards a bit.

Crabs, spiders, insects, and other arthropods have a hinged exoskeleton with attachment sites for sets of muscles that pull on the hardened parts. For example, a fly's wings flap when muscles attached to its thorax alternately contract and relax (Figure 36.4).

Redistribution of body fluid also has a role in some arthropod movements. In spiders, muscles attached to the exoskeleton contract and pull the legs inward, but there are no opposing muscles to pull legs out again. Instead, a large muscle of the thorax contracts, which causes blood to surge into the hind legs (Figure 36.5). Similarly, redistribution of fluid extends the proboscis of a moth or butterfly, allowing the insect to sip nectar.

Figure 36.5 Side view of a jumping spider making a leap. When a large muscle in the thorax contracts, volume of the thoracic cavity decreases, forcing blood into the hind legs. The resulting surge of high fluid pressure extends the legs. Some jumping spiders can leap a distance 25 times the length of their body.

Endoskeletons

An **endoskeleton** is an internal framework of hardened elements to which the muscles attach. Echinoderms and vertebrates have an endoskeleton. The skeleton of echinoderms such as sea stars (Figure 36.6) and sea urchins consists of calcium-carbonate plates embedded in the body wall.

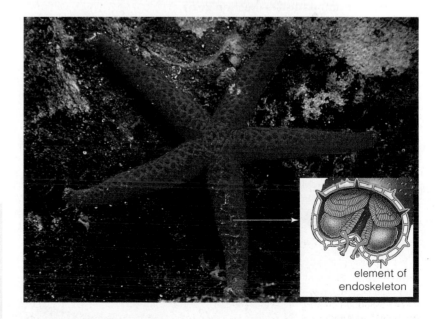

element of endoskeleton

Figure 36.6 A sea star. The sketch shows a cross-section through one arm. Hard plates embedded in the body wall form an endoskeleton.

Take-Home Message

What kinds of skeletons do invertebrates have?

■ Soft-bodied animals such as sea anemones and earthworms have a hydrostatic skeleton, which is an enclosed fluid that contracting muscles act upon.

■ Some mollusks and all arthropods have a hardened external skeleton, or exoskeleton.

■ Echinoderms have an endoskeleton, or internal skeleton.

36.2 | The Vertebrate Endoskeleton

■ All vertebrates have an endoskeleton. In most groups, the endoskeleton consists primarily of bones.

■ Links to Vertebrate evolution 26.2, Transition to life on land 26.5, Human bipedalism 26.13, Connective tissues 32.3

Features of the Vertebrate Skeleton

All vertebrates (the fishes, reptiles, amphibians, birds, and mammals) have an endoskeleton (Figures 36.7 and 36.8). The skeleton of sharks and other cartilaginous fishes consists of cartilage, a rubbery connective tissue. Other vertebrate skeletons include some cartilage, but consist mostly of bone tissue (Section 32.3).

The term "vertebrate" refers to the **vertebral column**, or backbone, a feature common to all members of this group. The backbone supports the body, serves as an attachment point for muscles, and protects the spinal cord, which runs through a canal inside it. Bony segments called **vertebrae** (singular, vertebra) make up the backbone. **Intervertebral disks** of cartilage between vertebrae act as shock absorbers and flex points.

The vertebral column, along with the bones of the head and rib cage, constitute the **axial skeleton**. The **appendicular skeleton** consists of the pectoral (shoulder) girdle, the pelvic (hip) girdle, and limbs (or bony fins) attached to them.

You learned earlier how vertebrate skeletons have evolved over time. For example, jaws are derived from the gill supports of ancient jawless fishes (Section 26.2). As another example, bones in the limbs of land vertebrates are homologous to those in fins of lobe-finned fishes (Section 26.5).

The Human Skeleton

For a closer look at vertebrate skeletal features, think about a human skeleton. The human skull's flattened cranial bones fit together to form the braincase that surrounds and protects the brain (Figure 36.8a). The brain and spinal cord connect through an opening called the **foramen magnum**. In upright walkers such as humans, this opening lies at the base of the skull (Section 26.13). Facial bones include cheekbones and other bones around the eyes, the bone that forms the bridge of the nose, and bones of the jaw.

Both males and females have twelve pairs of ribs (Figure 36.8b). Ribs and the breastbone, or sternum, form a protective cage around the heart and lungs.

The vertebral column extends from the base of the skull to the pelvic girdle (Figure 36.8c). In humans, natural selection favored an ability to walk upright and led to modification of the backbone. Viewed from the side, our backbone has an S shape that keeps our head and torso centered over our feet (Section 26.13).

Maintaining an upright posture requires that vertebrae and intervertebral disks stack one on top of the other, rather than being parallel to the ground, as in four-legged walkers. The stacking puts additional pressure on disks and, as people age, their disks often slip out of place or rupture, causing back pain.

The scapula (shoulder blade), and clavicle (collarbone) are bones of the human pectoral girdle (Figure 36.8d). The thin clavicle transfers force from the arms to the axial skeleton. When a person falls on an outstretched arm, the excessive force transferred to the clavicle frequently causes it to fracture or break.

The upper arm has one bone, the humerus. The forearm has two bones, the radius and ulna. Carpals are bones of the wrist, metacarpals are bones of the palm, and phalanges (singular, phalanx) are finger bones.

The pelvic girdle consists of two sets of fused bones, one set on each side of the body. It protects organs inside the pelvic cavity and supports the weight of the upper body when you stand upright (Figure 36.8e).

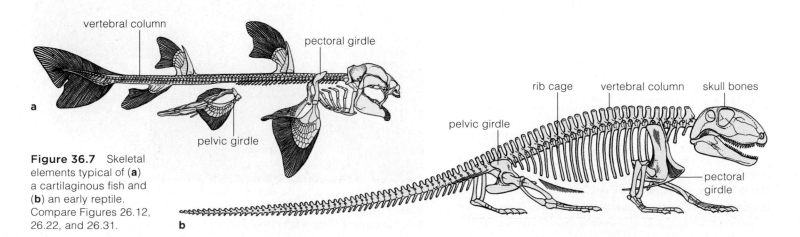

Figure 36.7 Skeletal elements typical of (**a**) a cartilaginous fish and (**b**) an early reptile. Compare Figures 26.12, 26.22, and 26.31.

vertebral column

pectoral girdle

pelvic girdle

a

pelvic girdle

rib cage   vertebral column   skull bones

pectoral girdle

b

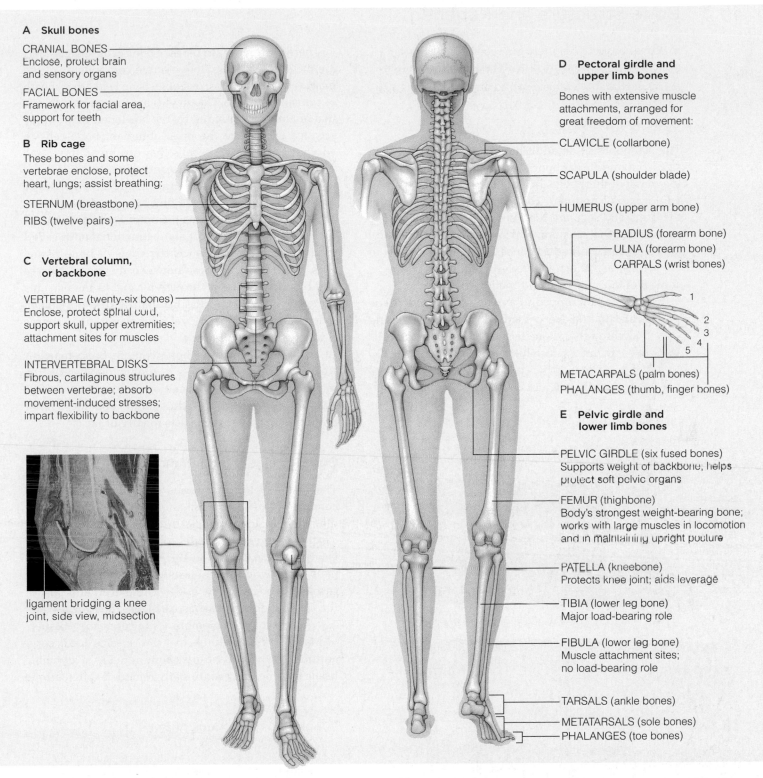

Figure 36.8 Animated Bone (*tan*) and cartilage (*light blue*) elements of the human skeleton. *Left*, labels for the axial portion, and (*right*) for the appendicular portion.

Bones of the leg include the femur (thighbone), the patella (kneecap), and the tibia and fibula (bones of the lower leg). Tarsals are ankle bones, and metatarsals are bones of the sole of the foot. Like the bones of the fingers, those of the toes are called phalanges.

Take-Home Message

What type of skeleton is present in humans and other vertebrates?

■ The endoskeleton of vertebrates usually consists mainly of bone. Its axial portion includes the skull, vertebral column, and ribs. Its appendicular part includes a pectoral girdle, a pelvic girdle, and the limbs.

■ Some features of the human skeleton such as an S-shaped backbone are adaptations to upright posture and walking.

36.3 | Bone Structure and Function

- Bones consist of living cells in a secreted extracellular matrix. Proper diet and exercise will help keep them healthy.
- Links to Extracellular matrix 4.12, Parathyroid glands 35.6

Bone Anatomy

The 206 bones of an adult human's skeleton range in size from middle ear bones as small as a grain of rice to the massive thighbone, or femur, which weighs about a kilogram (2 pounds). The femur and other bones of the arms and legs are long bones. Other bones, such as the ribs, the sternum, and most bones of the skull are flat bones. Still other bones, such as the carpals in the wrists, are short and roughly squarish in shape. Table 36.1 summarizes the functions of bones.

Each bone is wrapped in a dense connective tissue sheath that has nerves and blood vessels running through it. Bone tissue consists of bone cells in an extracellular matrix (Section 4.12). The matrix is mainly collagen (a protein) with calcium and phosphorus salts.

There are three main types of bone cells. **Osteoblasts** are the bone builders; they secrete components of the matrix. In adult bones, osteoblasts lie beneath a sheath of connective tissue. **Osteocytes** are former osteoblasts that are now surrounded by the hardened matrix they secreted. These are the most abundant bone cells in adults. **Osteoclasts** are cells that can break down the matrix by secreting enzymes and acids.

A long bone such as a femur includes two types of bone tissue, compact bone and spongy bone (Figure 36.9). Compact bone forms the outer layer and shaft of the femur. It is made up of many functional units called osteons, each having concentric rings of bone tissue, with bone cells in spaces between the rings. Nerves and blood vessels run through a canal in the osteon's center. Spongy bone fills the shaft and knobby ends of long bones. It is strong yet lightweight; open spaces riddle its hardened matrix.

The cavities inside a bone contain bone marrow. **Red marrow** fills the spaces in spongy bone and is the major site of blood cell formation. **Yellow marrow** fills the central cavity of an adult femur and most other mature long bones. It consists mainly of fat.

Bone Formation and Remodeling

The first skeleton that forms in a vertebrate embryo consists of cartilage. It remains cartilage in sharks and other cartilaginous fishes. In other vertebrates, early cartilage serves as a model for an adult skeleton that is largely bone (Figure 36.10). Most bones in these animals form when osteoblasts move into and replace cartilage models. A few bones in the head and part of the clavicle do not start as cartilage; they form when osteoblasts colonize membranes of connective tissue.

Many bones continue to grow in size until early adulthood. Even in adults, bone remains a dynamic tissue that the body continually remodels. Microscopic

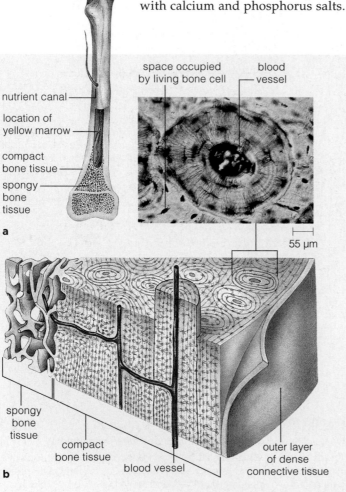

a

space occupied by living bone cell

blood vessel

nutrient canal

location of yellow marrow

compact bone tissue

spongy bone tissue

55 µm

spongy bone tissue

compact bone tissue

blood vessel

outer layer of dense connective tissue

b

Figure 36.9 Animated (**a**) Structure of a human femur, or thighbone, and (**b**) a section through its spongy and compact bone tissues.

Table 36.1 Functions of Bone
1. Movement. Bones interact with skeletal muscle and change or maintain positions of the body and its parts.
2. Support. Bones support and anchor muscles.
3. Protection. Many bones form hardened chambers or canals that enclose and protect soft internal organs.
4. Mineral storage. Bones are a reservoir for calcium and phosphorus ions. Deposits and withdrawals of these ions help maintain their concentrations in body fluids.
5. Blood cell formation. Only certain bones contain the tissue where blood cells form.

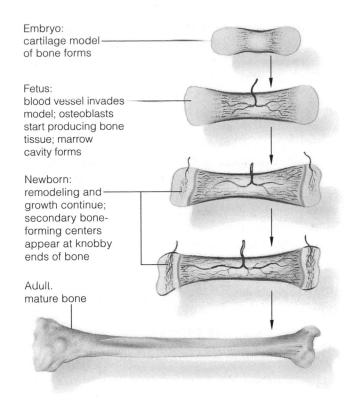

Embryo:
cartilage model
of bone forms

Fetus:
blood vessel invades
model; osteoblasts
start producing bone
tissue; marrow
cavity forms

Newborn:
remodeling and
growth continue;
secondary bone-
forming centers
appear at knobby
ends of bone

Adult:
mature bone

Figure 36.10 Long bone formation, starting with osteoblast activity in a cartilage model formed earlier in the embryo. The bone-forming cells are active first in the shaft region, then at the knobby ends. In time, cartilage is left only at the ends.

fractures that result from normal body movements are repaired. In response to hormonal signals, osteoclasts dissolve portions of the matrix, releasing stored mineral ions into the blood. Osteoblasts secrete new matrix, which replaces that broken down by osteoclasts.

Bones and teeth contain most of the body's calcium. Hormones regulate calcium concentration in blood by affecting calcium uptake from the gut and calcium release from bone. When the blood calcium level is too high, the thyroid gland secretes calcitonin. This hormone slows the release of calcium into blood by inhibiting osteoclast action. When blood has too little calcium, the parathyroid glands release parathyroid hormone, or PTH (Section 35.6). This hormone stimulates osteoclast activity. It also decreases calcium loss in urine and helps activate vitamin D. The vitamin stimulates cells in the gut lining to absorb calcium.

Other hormones also affect bone turnover. The sex hormones estrogen and testosterone encourage bone deposition. Cortisol, the stress hormone, slows it.

Until an individual is about twenty-four years old, osteoblasts secrete more matrix than osteoclasts break down, so bone mass increases. Bones become denser

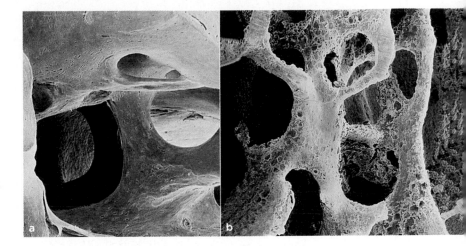

Figure 36.11 (a) Normal bone tissue. (b) Bone weakened by osteoporosis. The term osteoporosis means "porous bones."

and stronger. Later in life, as osteoblasts become less active, bone mass gradually declines.

About Osteoporosis

Osteoporosis is a disorder in which bone loss outpaces bone formation. As a result, the bones become weaker and more likely to break (Figure 36.11). Osteoporosis is most common in postmenopausal woman because they no longer produce the sex hormones that encourage bone deposition. However, about 20 percent of osteoporosis cases occur in men.

To reduce your risk of osteoporosis, ensure that your diet provides adequate levels of vitamin D and calcium. A premenopausal woman requires 1,000 milligrams of calcium daily; a post-menopausal woman requires 1,500 milligrams a day. Avoid smoking and excessive alcohol intake, which slow bone deposition. Get regular exercise to encourages bone renewal and avoid an excessive intake of cola soft drinks. Several studies have shown that women who drink more than two such soft drinks a day have a slightly lower than normal bone density.

Take-Home Message

What are the structural and functional features of bones?

■ Bones have a variety of shapes and sizes.

■ A sheath of connective tissues encloses the bone, and the bone's inner cavity contains marrow. Red marrow produces blood cells.

■ All bones consist of bone cells in a secreted extracellular matrix. A bone is continually remodeled; osteoclasts break down the matrix of old bone and ostoeoblasts lay down new bone. Hormones regulate this process.

36.4 Skeletal Joints—Where Bones Meet

■ Bones interact with one another at joints. Depending on the type, they allow no, little, or much range of motion.

A **joint** is an area of contact or near contact between bones. There are three types of joints: fibrous joints, cartilaginous joints, and synovial joints (Figure 36.12a).

At fibrous joints, bones are held securely in place by dense, fibrous connective tissue. Fibrous joints hold teeth in their sockets in the jaw.

Pads or disks of cartilage connect bones at cartilaginous joints. The flexible connection allows just a bit of movement. Cartilaginous joints connect vertebrae to one another and connect some ribs to the sternum.

Synovial joints are the most common kind of joints. They include joints of knees, hips, shoulders, wrists,

and ankles. At these joints, bones are separated by a small cavity and smooth cartilage covers their ends, reducing friction. Cords of dense connective tissue called **ligaments** hold bones in place at a synovial joint. Some ligaments form a capsule that encloses the joint. The capsule's lining secretes a lubricating synovial fluid. *Synovial* means "egglike" in Latin, and describes the thick consistency of the fluid.

Different synovial joints allow different kinds of movements. For example, joints at the shoulders and hips are ball-and-socket joints that allow a wide range of rotational motion. At other joints, including some in the wrists and ankles, bones glide past one another. Joints at the elbows and knees function like a hinged door; they allow the bones to move back and forth in one plane only.

Figure 36.12b shows some of the ligaments that hold the fibula and tibia together at the knee joint. The knee also is stabilized by wedges of cartilage called menisci (singular, meniscus).

Take-Home Message

What are joints?

■ Joints are areas where bones meet and interact.

■ In the most common type, synovial joints, the bones are separated by a small fluid-filled space and are held together by ligaments of fibrous connective tissue.

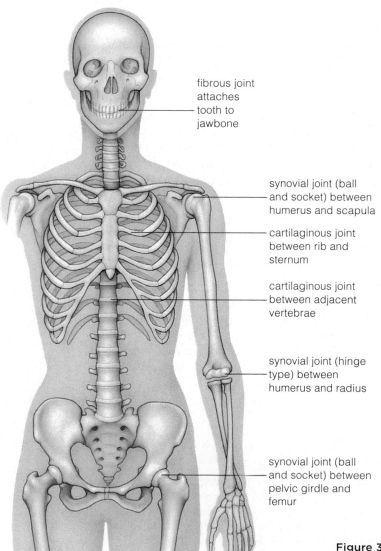

fibrous joint attaches tooth to jawbone

synovial joint (ball and socket) between humerus and scapula

cartilaginous joint between rib and sternum

cartilaginous joint between adjacent vertebrae

synovial joint (hinge type) between humerus and radius

synovial joint (ball and socket) between pelvic girdle and femur

a

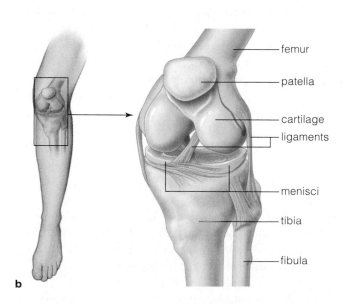

femur

patella

cartilage

ligaments

menisci

tibia

fibula

b

Figure 36.12 (**a**) Examples of the three types of joints. (**b**) Simplified diagram of the structure of the left knee with muscles stripped away. Several ligaments attach the femur to the tibia and chunks of cartilage called menisci help keep the bones properly aligned. Compare the photo in Figure 36.8.

36.5 | Those Aching Joints

■ We ask a lot of our joints when we engage in sports, carry out repetitive tasks, or slip on a pair of high heels.

Common Injuries A sprained ankle is the most common joint injury. It occurs when one or more of the ligaments that hold bones together at the ankle joint overstretches or tears. A sprained ankle is usually treated immediately with rest, application of ice, compression with an elastic bandage, and elevation of the affected area. After the ankle heals, exercises may help strengthen muscles that stabilize the joint and prevent future sprains.

A tear of the cruciate ligaments in the knee joint may require surgery. Cruciate means cross, and these short ligaments cross one another in the center of the joint. They are visible in Figure 36.12*b*. The cruciate ligaments stabilize the knee and when they are torn completely, bones may shift so the knee gives out when a person tries to stand. A blow to the lower leg, as often occurs in football, can injure a cruciate ligament, but so can a fall or misstep. Female athletes are at a higher risk for cruciate ligament tears than men who play the equivalent sport. For example, female soccer players tear these ligaments four times as often as male soccer players do.

Another common knee injury is a torn meniscus. A meniscus is a C-shaped wedge of cartilage that reduces friction between the bones, cushions them, and helps keep them in place. Each knee has two menisci. A minor tear at the edge of the meniscus may heal on its own, but cartilage repairs itself only very slowly. If a chunk of meniscus cartilage gets torn off, it can drift about in the synovial fluid of the joint and end up jammed into a spot where it interferes with normal function.

A dislocation means that bones of a joint are out of place. It is usually highly painful and requires immediate treatment. The bones must be placed back into proper position and immobilized for a time to allow healing.

Arthritis and Bursitis Arthritis means inflammation of a joint. As you will learn in Chapter 38, inflammation is the body's normal response to injury. However, with arthritis, inflammation—and the associated pain and swelling—become chronic.

The most common type of arthritis is osteoarthritis. It usually appears in old age, after cartilage wears down at a frequently used joint. It affects different joints in different people. For example, women who habitually wear high-heeled shoes increase their risk of osteoarthritis of the knees (Figure 36.13). Such shoes put added pressure on the cartilage that cushions the knee joint, increasing the chances that it will wear down and fail.

Rheumatoid arthritis is an autoimmune disorder; the immune system mistakenly attacks the fluid-secreting lining of synovial joints. It can occur at any age and women are two to three times more likely than men to be affected.

Gout is another form of arthritis. It occurs when crystals of uric acid accumulate in certain joints, most notably those of the big toes. The resulting pain can be chronic and excruciating. Uric acid is a natural product of protein breakdown, but certain genes, excess alcohol intake, or obesity can cause blood levels to rise.

Arthritis can be treated with drugs that relieve pain and minimize inflammation. Joints affected by osteoarthritis can also be replaced with artificial, or prosthetic, joints. Knee and hip replacements are now common and allow a person to resume normal activities.

With bursitis, a bursa becomes inflamed. A **bursa** (as shown in Figure 36.16*b*) is a fluid filled sac that functions as a cushion between parts in many joints. Repeating a movement that puts pressure on a particular bursa usually causes the inflammation. For example, swinging a tennis racket or golf club can lead to inflammation of a bursa in the shoulder or elbow. Continually leaning on an elbow, kneeling to work on something on the floor, or even sitting or standing a certain way can also cause bursitis.

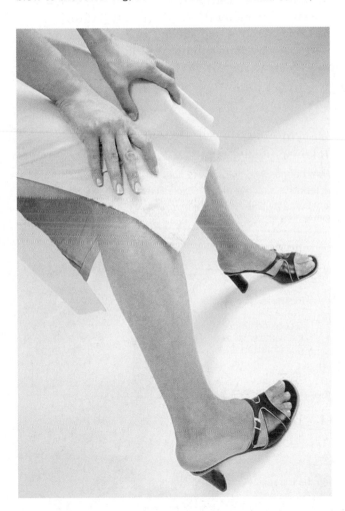

Figure 36.13 High heels now, may lead to aching knees later. A study by researchers at Tufts University showed that shoes with heels 2.7 inches high increased pressure on the knee joint by 20 to 25 percent over barefoot walking. Wide heels increased pressure on knees more than narrow ones, perhaps because women walked more confidently in them.

- Only skeletal muscles attach to and pull on bones.
- Link to Types of muscle 32.4

Skeletal muscles consist of bundles of muscle fibers sheathed in dense connective tissue. A **muscle fiber** is a long, cylindrical cell with multiple nuclei that holds contractile filaments. It has several nuclei because it is descended from a group of cells that fused together in the developing embryo.

Most muscles and bones interact as a lever system, in which a rod is attached to a fixed point and moves about it. The bone is a rigid rod near a joint (the fixed point). Muscle contraction transmits force to the bone and makes it move, as in Figure 36.14.

Fully extend your right arm, place your left hand over the upper arm, and slowly bend your elbow, as in Figure 36.15a. Can you feel the muscle contracting? By causing this muscle to shorten a bit, you caused the bone attached to the muscle to move a large distance. Besides acting on bone, skeletal muscles also can interact with one another. Some work in pairs

or groups to bring about a movement. Muscles can only pull on bones; they cannot push. Often two muscles work in opposition; action of one resists or reverses action of another. For example, the biceps in the upper arm opposes the triceps. Such pairings are the case for most muscles in the limbs (Figures 36.14 and 36.15).

Bear in mind, only *skeletal* muscle moves bones. As you read in Section 32.4, smooth muscle is mainly a component of soft internal organs, such as the stomach. Cardiac muscle is found only in the heart wall. Later chapters consider the structure and function of smooth muscle and cardiac muscle.

The human body has close to 700 skeletal muscles, some near the surface, others in the body wall (Figure 36.16). A straplike **tendon** of connective tissue attaches skeletal muscles to bone. As an example, the Achilles tendon attaches calf muscles to the heel bone and is the largest tendon in the body (Figure 36.16a).

Later chapters explain the roles skeletal muscles play in respiration and in blood circulation. We turn now to mechanisms that bring about muscle contraction.

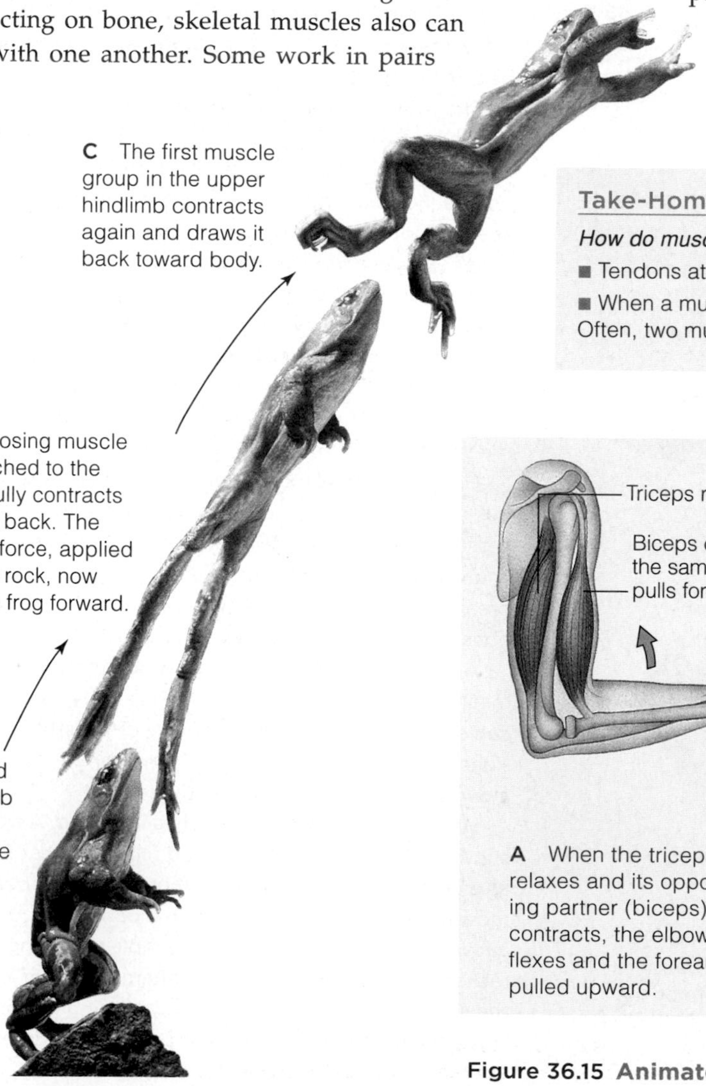

C The first muscle group in the upper hindlimb contracts again and draws it back toward body.

B An opposing muscle group attached to the limb forcefully contracts and pulls it back. The contractile force, applied against the rock, now propels the frog forward.

A A muscle attached to each upper hindlimb contracts and pulls it slightly forward relative to main body axis.

Figure 36.14 A frog on a rock demonstrating how small contractions and the action of opposing muscles can cause big movements.

Take-Home Message

How do muscles interact with bones?

- Tendons attach skeletal muscles to bone.
- When a muscle contracts, it pulls on the attached bone. Often, two muscles attached to a bone have opposing actions.

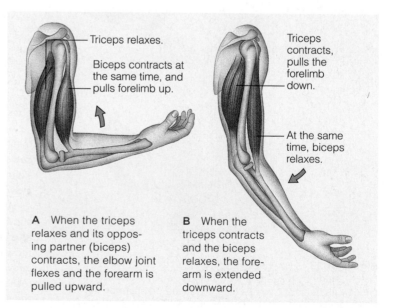

Triceps relaxes.

Biceps contracts at the same time, and pulls forelimb up.

Triceps contracts, pulls the forelimb down.

At the same time, biceps relaxes.

A When the triceps relaxes and its opposing partner (biceps) contracts, the elbow joint flexes and the forearm is pulled upward.

B When the triceps contracts and the biceps relaxes, the forearm is extended downward.

Figure 36.15 Animated Two opposing muscle groups in human arms.

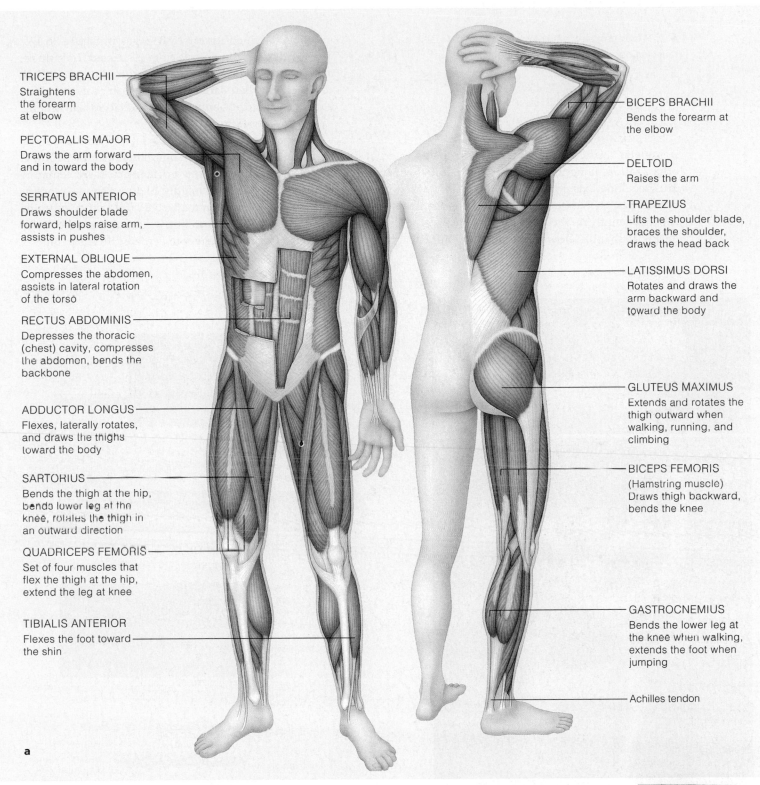

TRICEPS BRACHII
Straightens the forearm at elbow

PECTORALIS MAJOR
Draws the arm forward and in toward the body

SERRATUS ANTERIOR
Draws shoulder blade forward, helps raise arm, assists in pushes

EXTERNAL OBLIQUE
Compresses the abdomen, assists in lateral rotation of the torso

RECTUS ABDOMINIS
Depresses the thoracic (chest) cavity, compresses the abdomen, bends the backbone

ADDUCTOR LONGUS
Flexes, laterally rotates, and draws the thighs toward the body

SARTORIUS
Bends the thigh at the hip, bends lower leg at the knee, rotates the thigh in an outward direction

QUADRICEPS FEMORIS
Set of four muscles that flex the thigh at the hip, extend the leg at knee

TIBIALIS ANTERIOR
Flexes the foot toward the shin

BICEPS BRACHII
Bends the forearm at the elbow

DELTOID
Raises the arm

TRAPEZIUS
Lifts the shoulder blade, braces the shoulder, draws the head back

LATISSIMUS DORSI
Rotates and draws the arm backward and toward the body

GLUTEUS MAXIMUS
Extends and rotates the thigh outward when walking, running, and climbing

BICEPS FEMORIS
(Hamstring muscle) Draws thigh backward, bends the knee

GASTROCNEMIUS
Bends the lower leg at the knee when walking, extends the foot when jumping

Achilles tendon

a

Figure 36.16 Animated (**a**) Muscles of the human musculoskeletal system. These are the skeletal muscles that gym enthusiasts are familiar with; many more are not shown. Also labeled is the Achilles tendon, the largest tendon in the body and the most frequently injured. It attaches muscles in the calf to the heel bone.

(**b**) Tendons at a synovial joint. Bursae form between tendons and bones or some other structure. These fluid-filled sacs help reduce friction between adjacent tissues.

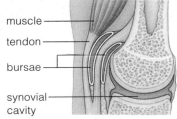

muscle

tendon

bursae

synovial cavity

b

36.7 How Does Skeletal Muscle Contract?

- ATP-fueled movements of protein filaments inside a muscle fiber result in muscle contraction.
- Link to Cytoskeleton 4.13

Fine Structure of Skeletal Muscle

A skeletal muscle's function arises from its internal organization. Long muscle fibers run parallel with the muscle's long axis. The muscle fibers are packed with **myofibrils**, each a bundle of contractile filaments that run the length of the fiber (Figure 36.17a). Light-to-dark crossbands show up along the entire length of myofibrils stained for microscopy, as in Figure 36.17b. The bands give the muscle fiber a striated, or striped, appearance. These bands define the units of muscle contraction called **sarcomeres**. A mesh of cytoskeletal elements called Z bands anchors adjacent sarcomeres to one another (Figure 36.17c).

The sarcomere has parallel arrays of thin and thick filaments (Figure 36.18a). Thin filaments attached to Z bands extend inward, toward the sarcomere center. A thin filament consists mainly of two chains of **actin**, a globular protein (Figure 36.17d). Two other proteins associate with the actin, but we can ignore their role for now. Thick filaments are centered in a sarcomere.

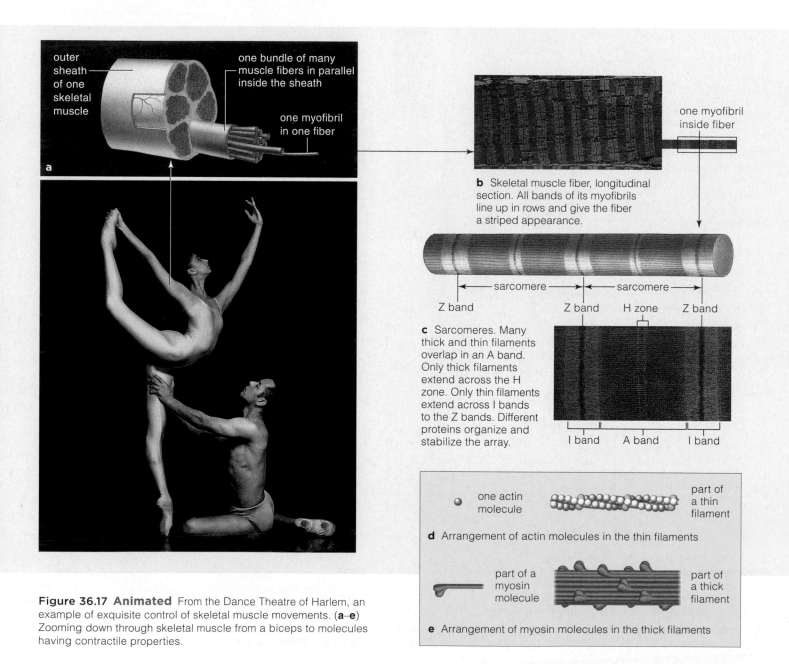

a

outer sheath of one skeletal muscle

one bundle of many muscle fibers in parallel inside the sheath

one myofibril in one fiber

one myofibril inside fiber

b Skeletal muscle fiber, longitudinal section. All bands of its myofibrils line up in rows and give the fiber a striped appearance.

sarcomere — sarcomere

Z band   Z band   H zone   Z band

c Sarcomeres. Many thick and thin filaments overlap in an A band. Only thick filaments extend across the H zone. Only thin filaments extend across I bands to the Z bands. Different proteins organize and stabilize the array.

I band   A band   I band

one actin molecule — part of a thin filament

d Arrangement of actin molecules in the thin filaments

part of a myosin molecule — part of a thick filament

e Arrangement of myosin molecules in the thick filaments

Figure 36.17 Animated From the Dance Theatre of Harlem, an example of exquisite control of skeletal muscle movements. (**a–e**) Zooming down through skeletal muscle from a biceps to molecules having contractile properties.

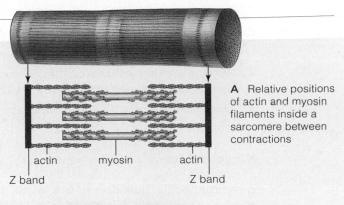

Each consists of **myosin**, a motor protein with a club-like head (Figure 36.17e). The head is positioned just a few nanometers away from a thin filament.

Muscle fibers, myofibrils, thin filaments, and thick filaments all run parallel with a muscle's long axis. As a result, all sarcomeres in all fibers of a muscle work together and pull in the same direction.

The Sliding-Filament Model

The **sliding-filament model** explains how interactions between thick and thin filaments bring about muscle contraction. According to this model, filaments do not change length and myosin filaments do not change position. Instead, myosin heads bind to actin filaments and slide them toward the center of a sarcomere. As actin filaments are pulled inward, Z bands attached to them are drawn closer together, and the sarcomere shortens (Figure 36.18a,b).

Part of the myosin head can bind ATP and break it into ADP and phosphate. This reaction readies myosin for action (Figure 36.18c). Muscle contraction occurs when signals from the nervous system cause calcium levels around filaments to rise, a process we consider in the next section. For now, it is enough to know that a rise in calcium allows myosin heads to bind actin, forming a cross-bridge between the actin and myosin filaments (Figure 36.18d).

After binding actin, each myosin head tilts toward the sarcomere center, and the ADP and phosphate are released (Figure 36.18e). Movement of the myosin head slides the attached actin filament toward the center of the sarcomere. The collective sliding of many myosin heads pulls the Z bands toward one another.

Binding of a new ATP frees the myosin head from actin, and the head goes back to its original position (Figure 36.18f). The head attaches to another binding site on the actin, tilts in another stroke, and so on as long as calcium and ATP are available. Hundreds of myosin heads perform a series of repeated strokes all along the length of the actin filaments.

Take-Home Message

What is the sliding-filament model for muscle contraction?

■ The sliding-filament model explains how interactions among protein filaments within a muscle fiber's individual contractile units (its sarcomeres) bring about muscle contractions.

■ By this model, a sarcomere shortens when actin filaments are pulled toward the center of the sarcomere by ATP-fueled interactions with myosin filaments.

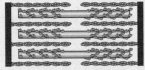

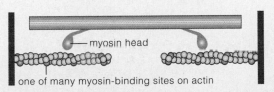

A Relative positions of actin and myosin filaments inside a sarcomere between contractions

actin   myosin   actin

Z band   Z band

B Relative positions of actin and myosin filaments in the same sarcomere, contracted

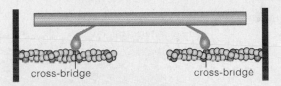

— myosin head

one of many myosin-binding sites on actin

C Myosin in a muscle at rest. Earlier, all myosin heads were energized by binding ATP, which they hydrolyzed to ADP and inorganic phosphate.

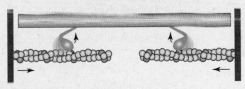

cross-bridge   cross-bridge

D A rise in the local concentration of calcium exposes binding sites for myosin on actin filaments, so cross-bridges form.

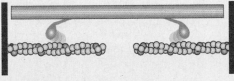

E Binding makes each myosin head tilt toward the sarcomere's center and slide the bound actin along with it. ADP and phosphate are released as the myosin heads drag the actin filaments inward, which pulls the Z bands closer.

F New ATP binds to myosin heads, which detach from actin. ATP is hydrolyzed, which returns myosin heads to their original positions.

Figure 36.18 Animated A sliding-filament model for the contraction of a sarcomere in skeletal muscle. (**a,b**) Organized, overlapping arrays of actin and myosin filaments interact and reduce each sarcomere's width. (**c–f**) For clarity, we show the action of two myosin heads only. Each head binds repeatedly to an actin filament and slides it toward the center of the sarcomere. Collective action of many myosin heads makes the sarcomere shorten (contract).

■ Like neurons, muscle cells are excitable. Action potentials in muscle trigger calcium release that allows contraction.

■ Links to Active transport 5.4, Neuromuscular junctions 33.5

Nervous Control of Contraction

A neuromuscular junction is a synapse between a motor neuron and a muscle fiber (Section 33.5 and Figure 36.19a,b). For a skeletal muscle to contract, an action potential must first travel to a neuromuscular junction and cause the release of acetylcholine (ACh) from a motor neuron's axon terminals. Like a neuron, a muscle fiber is excitable, and the binding of ACh to receptors at its plasma membrane causes an action potential. The action potential travels along the muscle plasma membrane, then down T tubules that extend from this membrane. The T tubules deliver the action potential to the **sarcoplasmic reticulum**, a special type of smooth endoplasmic reticulum that wraps around myofibrils and stores calcium ions (Figure 36.19c).

The arrival of action potentials opens voltage-gated channels in the sarcoplasmic reticulum, allowing calcium ions to flow out, down their concentration gradient. This raises the calcium concentration around the actin and myosin filaments, allowing them to interact, and muscle contraction occurs.

When contraction ends, calcium pumps of the type described and illustrated in Section 5.4 transport the calcium ions back into the sarcoplasmic reticulum. The muscle fiber is ready for another signal.

The Roles of Troponin and Tropomyosin

How does the release of calcium from the sarcoplasmic reticulum allow actin and myosin to interact? Calcium affects troponin and tropomyosin, two proteins that regulate binding of myosin to actin filaments.

Figure 36.20a,b shows a single thin filament in a muscle fiber at rest. Under these circumstances, there is little calcium in the fluid around the thin filament. Tropomyosin, a fibrous protein, wraps around actin and covers myosin-binding sites, preventing myosin from binding. Troponin, a globular protein attached to the tropomyosin, has a site that can reversibly bind calcium ions.

When an action potential causes release of calcium from the sarcoplasmic reticulum, some of the calcium binds to troponin (Figure 36.20c). As a result, troponin changes shape and pulls tropomyosin—to which it is attached—away from the myosin-binding site on actin (Figure 36.20d). With this binding site cleared, myosin can bind to actin, and the sliding action described in the previous section takes place (Figure 36.20e,f).

So, to summarize events of muscle contraction, a signal (ACh) from a motor neuron causes an action potential in a muscle fiber, which opens calcium gates in the sarcoplasmic reticulum. Some released calcium ions bind to troponin, which pulls tropomyosin away from the myosin-binding site on actin. Cross-bridges form, sarcomeres shorten, and the muscle contracts.

Afterwards, calcium pumps transport calcium ions back into the sarcoplasmic reticulum. As the calcium level in the muscle fiber declines, troponin resumes its resting shape, tropomyosin settles back into place over the myosin-binding site, and the muscle relaxes.

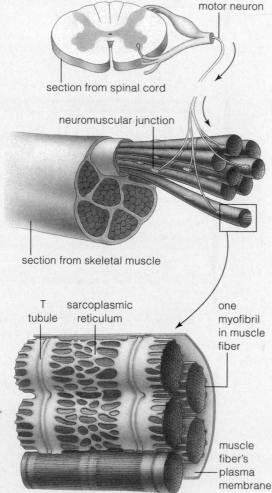

A A signal travels along the axon of a motor neuron, from the spinal cord to a skeletal muscle.

motor neuron

section from spinal cord

B The signal is transferred from the motor neuron to the muscle at neuromuscular junctions. Here, ACh released by the neuron's axon terminals diffuses into the muscle fiber and causes action potentials.

neuromuscular junction

section from skeletal muscle

C Action potentials propagate along a muscle fiber's plasma membrane down to T tubules, then to the sarcoplasmic reticulum, which releases calcium ions. The ions promote interactions of myosin and actin that result in contraction.

T tubule   sarcoplasmic reticulum

one myofibril in muscle fiber

muscle fiber's plasma membrane

Figure 36.19 Animated Pathway by which the nervous system controls skeletal muscle contraction. A muscle fiber's plasma membrane encloses many individual myofibrils. Tubelike extensions of the membrane connect with part of the sarcoplasmic reticulum, which wraps lacily around the myofibrils.

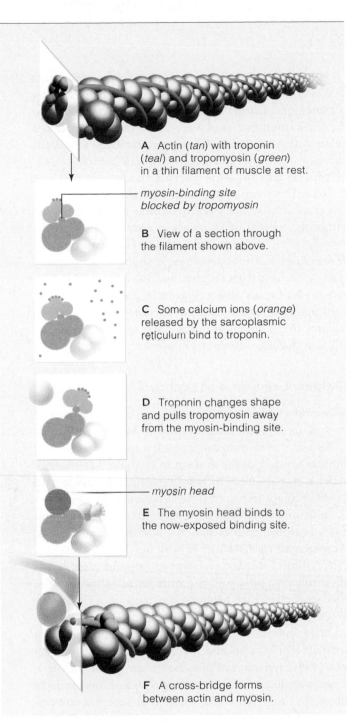

A Actin (*tan*) with troponin
(*teal*) and tropomyosin (*green*)
in a thin filament of muscle at rest.

*myosin-binding site
blocked by tropomyosin*

B View of a section through
the filament shown above.

C Some calcium ions (*orange*)
released by the sarcoplasmic
reticulum bind to troponin.

D Troponin changes shape
and pulls tropomyosin away
from the myosin-binding site.

myosin head

E The myosin head binds to
the now-exposed binding site.

F A cross-bridge forms
between actin and myosin.

Figure 36.20 Animated The interactions among actin,
tropomyosin, and troponin in a skeletal muscle cell.

Take-Home Message

*What initiates muscle contraction? What role does calcium play in
muscle contraction?*

■ A skeletal muscle contracts in response to a signal from
a motor neuron. Release of ACh at a neuromuscular junction
causes an action potential in the muscle cell.

■ An action potential results in release of calcium ions, which
affect proteins attached to actin. Resulting changes in the
shape and location of these proteins open the myosin-binding
site on actin, allowing cross-bridge formation.

| 36.9 | Energy for Contraction |

■ Multiple metabolic pathways can supply the ATP required
for muscle contraction.

■ Links to Energy-releasing pathways 8.1, Fermentation 8.5

The availability of ATP affects whether and how long
a muscle can contract. ATP is the first energy source a
muscle uses, but cells store little ATP. Once that ATP
gets used up, the muscle turns to creatine phosphate.
Phosphate transfers from creatine phosphate to ADP
can produce more ATP (Figure 36.21), and thus keep
a muscle going until ATP output from other pathways
increases. This is why taking creatine supplements, as
described in the chapter introduction, may enhance
athletic feats that require short bursts of activity.

Most of the ATP used during prolonged, moderate
activity is produced by aerobic respiration. Glucose
derived from stored glycogen fuels five to ten minutes
of activity. Next, glucose and fatty acids that the blood
delivers to muscle fibers are broken down. Fatty acids
fuel activities that last more than half an hour.

Not all fuel is broken down aerobically. Even in
resting muscle, some pyruvate is converted to lactate
by fermentation. Lactate production rises with exercise. This pathway does not yield much ATP, but it
can operate even when oxygen is low.

Take-Home Message

What is the source of ATP that powers muscle contraction?

■ Muscles first use any stored ATP, then transfer phosphate from creatine phosphate to ADP to form ATP.

■ With ongoing exercise, aerobic respiration and lactate fermentation yield the
ATP that supplies the energy for muscle contraction.

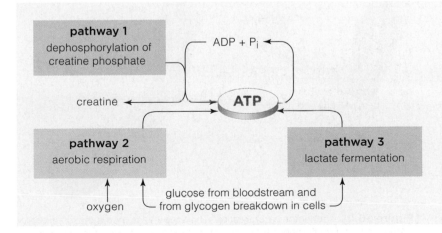

Figure 36.21 Animated Three metabolic pathways by which muscles
obtain the ATP molecules that fuel their contraction.

■ So far, we have been concentrating on individual muscle fibers, but in bodies many fibers respond as a unit.

Motor Units and Muscle Tension

A motor neuron has many axon terminals that synapse on different fibers in a muscle. One motor neuron and all of the muscle fibers it synapses with constitute one **motor unit**. Briefly stimulate a motor neuron, and the fibers of its motor unit contract for a few milliseconds. That contraction is a **muscle twitch** (Figure 36.22a).

A new stimulus that occurs before a response ends makes the fibers twitch again. Repeatedly stimulating a motor unit during a short interval makes all of the twitches run together in a sustained contraction called **tetanus** (Figure 36.22c). Force generated by tetanus is three or four times the force of a single twitch.

Muscle tension is the mechanical force exerted by a muscle. The more motor units stimulated, the greater the muscle tension. Opposing muscle tension is a load, either the weight of an object or gravity's pull on the muscle. Only when muscle tension exceeds opposing forces does a stimulated muscle shorten. Isotonically contracting muscles shorten and move some load, as when you lift an object (Figure 36.23a). Isometrically contracting muscles tense but do not shorten, as when you try to lift an object but fail because it is too heavy (Figure 36.23b).

Fatigue, Exercise, and Aging

When unrelenting stimulation keeps a skeletal muscle in tetanus, muscle fatigue follows. **Muscle fatigue** is a decrease in a muscle's capacity to generate force; muscle tension declines despite ongoing stimulation. After a few minutes of rest, the fatigued muscle will contract again in response to stimulation.

In humans, all muscle fibers form before birth and exercise does not stimulate the addition of new ones. Aerobic exercise—low intensity, but long duration—makes muscles more resistant to fatigue. It increases their blood supply and the number of mitochondria, the organelles that produce the bulk of ATP during aerobic respiration.

Brief, intense exercise such as weight lifting results in synthesis of actin and myosin. This helps a muscle exert more tension but does not improve endurance.

As people age, the number and size of their muscle fibers decline. The tendons that attach muscle to bone stiffen and are more likely to tear. Older people may exercise intensely for long periods, but their muscle mass can no longer increase as much. Even so, aerobic exercise does improve blood circulation, and modest strength training can slow the loss of muscle tissue.

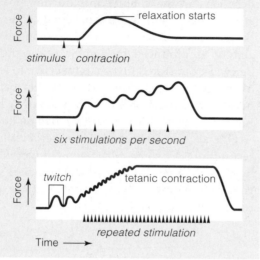

A A single, brief stimulus causes a twitch, a rapid contraction followed by immediate relaxation.

relaxation starts

stimulus   contraction

B Repeated stimuli over a short time have an additive effect; they increase the force of contraction.

six stimulations per second

C Sustained stimulation causes tetanus, a sustained contraction with several times the force of a twitch.

twitch   tetanic contraction

repeated stimulation

Time ⟶

Figure 36.22 Animated Recordings of twitches in a muscle fiber when the motor neuron controlling is artificially stimulated. **Figure It Out:** Which graph allows you to compare the force generated by a twitch and tetanus? *Answer: C*

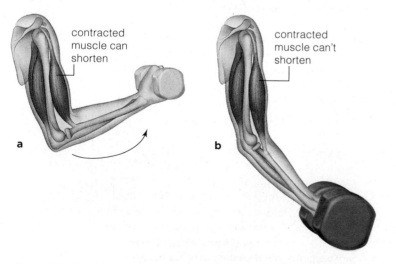

contracted muscle can shorten

contracted muscle can't shorten

a

b

Figure 36.23 (**a**) Isotonic contraction. The load is less than a muscle's peak capacity to contract. The muscle can contract, shorten, and lift the load. (**b**) Isometric contraction. The load exceeds a muscle's peak capacity, so the muscle contracts but cannot shorten.

Take-Home Message

How do whole muscles respond to stimulation and exercise?

■ Brief stimulation of a muscle causes a twitch; ongoing stimulation results in a more forceful contraction called tetanus.

■ Exercise cannot add muscle fibers, but it can increase the number of protein filaments and mitochondria in existing ones.

36.11 | Disruption of Muscle Contraction

■ Some genetic disorders, diseases, or toxins can cause muscles to contract too little or too much.

■ Links to X-linked inheritance 12.4, Endospores 21.6

Muscular Dystrophies Muscular dystrophies are a class of genetic disorders in which skeletal muscles progressively weaken. With Duchenne muscular dystrophy, symptoms begin to appear in childhood. Myotonic muscular dystrophy is the most common kind in adults.

A mutation of a gene on the X chromosome causes Duchenne muscular dystrophy. This affected gene encodes dystrophin, a protein found in the plasma membrane of muscle fibers. A mutant form of dystrophin allows foreign material to enter a muscle fiber, which causes the fiber to break down (Figure 36.24). Muscular dystrophy arises in about 1 in 3,500 males. Like other X-linked disorders, it rarely causes symptoms in females, who nearly always have a normal version of the gene on their other X chromosome. Affected boys usually begin to show signs of weakness by the time they are three years old, and require a wheelchair in their teens. Most die in their twenties of the respiratory failure that occurs when the skeletal muscles involved in breathing stop functioning.

Motor Neuron Disorders When motor neurons cannot signal muscles to contract, or signaling is impaired, skeletal muscles weaken or become paralyzed. For example, poliovirus can infect and kill motor neurons. Children are most frequently infected; those who survive an infection may be paralyzed or have a weakened voluntary muscle response as a result. Polio vaccines have been available since the 1950s, so the disease is on the decline. No new cases have been reported in the United States since 1979. However, infections continue to occur in less-developed countries. Also, some people who had polio as children develop post-polio syndrome as adults. Fatigue and progressive muscle weakness are the main symptoms. There are at least 250,000 polio survivors in the United States, and may be as many as a million.

Amyotrophic lateral sclerosis (ALS) also kills motor neurons. It is sometimes called Lou Gehrig's disease, after a famous baseball player whose career was cut short by the disease in the late 1930s. ALS usually causes death by respiratory failure within three to five years of diagnosis, but some people survive much longer. For example, the astrophysicist Stephen Hawking was diagnosed with ALS in 1963. Though now confined to a wheelchair and unable to speak, he continues to write and to lecture with the assistance of a voice synthesizer.

Botulism and Tetanus Bacteria of the genus *Clostridium* produce toxins that disrupt the flow of signals from nerves to muscles. Resting spores (endospores) of *C. botulinum* sometimes are in canned food. When spores germinate, the bacteria that grow make botulinum, an odorless toxin. When a person eats tainted food, botulinum enters motor

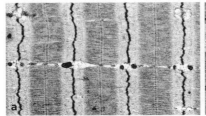

Figure 36.24 Electron micrographs of (**a**) normal skeletal muscle tissue and (**b**) muscle tissue of a person affected by muscular dystrophy.

Figure 36.25 An 1809 painting showing a casualty of a battle wound as he lay dying of tetanus in a military hospital.

neurons and keeps them from releasing acetylcholine (ACh). Muscles cannot contract without this neurotransmitter. Affected people can die if skeletal muscles with roles in breathing become paralyzed.

A related bacterium, *C. tetani*, lives in the gut of cattle, horses, and other grazing animals, and even some people. Its endospores can last for years in soil. *C. tetani* spores sometimes get into a deep wound and germinate there. The bacteria grow and produce a toxin that blood or nerves deliver to the spinal cord and brain.

In the spinal cord, the bacterial toxin blocks release of neurotransmitters such as GABA (Section 33.6) that exert inhibitory control over motor neurons. Without these controls, nothing dampens signals to contract, so symptoms of the disease known as tetanus begin. Overstimulated muscles stiffen and cannot be released from contraction. Fists and the jaw may stay clenched; lockjaw is a common name for the disease. The backbone may become locked in an abnormally arching curve (Figure 36.25). Death occurs when respiratory and cardiac muscles become locked in contraction.

Vaccines have eradicated tetanus in the United States. Worldwide, the annual death toll is over 200,000. Most are newborns infected during an unsanitary delivery.

In a 2004 article, scientists in Germany reported on their study of an unusual child. Even at birth the boy had bulging biceps and thigh muscles. Investigation showed that he has a mutation in the gene for myostatin, a regulatory protein that slows muscle growth. He apparently makes little or no myostatin. Gene mutations that decrease myostatin levels may give some athletes a natural edge in putting on muscle mass. The boy's mother is a sprinter.

How would you vote?

Should dietary supplement makers have to prove that their products are both safe and effective? See CengageNOW for details, then vote online.

Summary

Sections 36.1, 36.2 Nearly all animals move by applying the force of muscle contraction to their skeletal elements. A **hydrostatic skeleton** is a confined fluid upon which muscle contractions act. An **exoskeleton** consists of hardened parts at the body surface. An **endoskeleton** consists of hardened parts inside the body.

Humans, like other vertebrates, have an endoskeleton. The **axial skeleton** consists of skull bones, a **vertebral column** (backbone), and a rib cage. The **appendicular skeleton** is the pelvic girdle, pectoral girdle, and paired limbs. The vertebral column consists of individual segments called **vertebrae**, with **intervertebral disks** between them. The spinal cord runs through the vertebral column and connects with the brain through the **foramen magnum**, a hole in the base of the skull. Placement of this hole, and other features of the human skeleton, are adaptations for upright walking in humans.

■ *Use the animation on CengageNOW to learn about the skeletal systems of invertebrates and humans.*

Sections 36.3–36.5 Bones are organs rich in collagen, calcium, and phosphorus. In addition to having a role in movement, they store minerals and protect organs. Some have **red marrow** that makes blood cells; most have **yellow marrow**. In a human embryo, bones develop from a cartilage model. Even in adults, bones are continually remodeled. **Osteoblasts** are cells that synthesize bone, whereas **osteoclasts** break bone down. **Osteocytes** are former osteoblasts enclosed in a matrix of their secretions.

A **joint** is an area of close contact between bones. One or more **ligaments** hold bones together at most joints. Bits of cartilage and fluid-filled **bursae** cushion joints.

■ *Use the animation on CengageNOW to study the structure of a human femur.*

Section 36.6 A **muscle fiber** is a long, cylindrical cell with multiple nuclei. In a skeletal muscle, muscle fibers are bundled inside a dense connective tissue sheath that extends beyond the fibers. **Tendons** are extensions of this sheath. They attach most skeletal muscles to bones.

When skeletal muscles contract, they transmit force to bones and move them. Some muscles work together, and others work as opposing pairs.

■ *Use the animation on CengageNOW to review the location and function of human skeletal muscles.*

Section 36.7 The internal organization of a skeletal muscle promotes a strong, directional contraction. Many **myofibrils** make up a skeletal muscle fiber. A myofibril consists of **sarcomeres**, units of muscle contraction, lined up along its length. Each sarcomere has parallel arrays of **actin** and **myosin** filaments. The **sliding-filament model** describes how ATP-driven sliding of actin filaments past myosin filaments shortens the sarcomere. Shortening of all sarcomeres in all myofibrils of all muscle fibers of a muscle bring about the muscle's contraction.

■ *Use the animation on CengageNOW to explore muscle structure and observe muscle contraction.*

Sections 36.8, 36.9 Signals from motor neurons result in action potentials in muscle fibers, which in turn cause the **sarcoplasmic reticulum** to release stored calcium. The flow of this calcium into the cytoplasm makes accessory proteins associated with the thin filaments shift in such a way that actin and myosin heads can interact and bring about a muscle contraction.

Muscle fibers produce the ATP needed for contraction by way of three pathways: dephosphorylation of creatine phosphate, aerobic respiration, and lactate fermentation.

■ *Use the animation on CengageNOW to observe how the nervous system controls muscle contraction, and how a muscle gets the energy for contraction.*

Sections 36.10, 36.11 A motor neuron and all the muscle fibers it controls are a **motor unit**. Brief stimulation of a motor unit causes a **twitch**. Repeated stimulation causes a **tetanus**, or sustained contraction. **Muscle tension** is the force exerted by a contracting muscle. **Muscle fatigue** is a decline in muscle tension despite ongoing stimulation.

Genetic disorders that affect muscle structure impair muscle function. So do some diseases and toxins that affect motor neurons.

■ *Use the animation on CengageNOW to observe how a muscle fiber responds to stimulation of a motor neuron.*

Self-Quiz   *Answers in Appendix III*

1. A hydrostatic skeleton consists of _____ .
   a. a fluid in an enclosed space
   b. hardened plates at the surface of a body
   c. internal hard parts
   d. none of the above

Data Analysis Exercise

Tiffany, shown in Figure 36.26, was born with multiple fractures in her arms and legs. By age six, she had undergone surgery to correct more than 200 bone fractures. Her fragile, easily broken bones are symptoms of osteogenesis imperfecta (OI), a genetic disorder caused by a mutation in a gene for collagen. As bones develop, collagen forms a scaffold for deposition of mineralized bone tissue. The scaffold forms improperly in children with OI. Figure 36.26 also shows the results of an experimental test of a new drug. Treated children, all less than two years old, were compared to similarly affected children of the same age who were not treated with the drug.

1. An increase in vertebral area during the 12-month period of the study indicates bone growth. How many of the treated children showed such an increase?

2. How many of the untreated children showed an increase in vertebral area?

3. How did the rate of fractures in the two groups compare?

4. Do the results shown support the hypothesis that giving young children who have OI this drug, which slows bone breakdown, can increase bone growth and reduce fractures?

Treated child	Vertebral area in cm² (Initial)	(Final)	Fractures per year
1	14.7	16.7	1
2	15.5	16.9	1
3	6.7	16.5	6
4	7.3	11.8	0
5	13.6	14.6	6
6	9.3	15.6	1
7	15.3	15.9	0
8	9.9	13.0	4
9	10.5	13.4	4
Mean	11.4	14.9	2.6

Control child	Vertebral area in cm² (Initial)	(Final)	Fractures per year
1	18.2	13.7	4
2	16.5	12.9	7
3	16.4	11.3	8
4	13.5	7.7	5
5	16.2	16.1	8
6	18.9	17.0	6
Mean	16.6	13.1	6.3

Figure 36.26 Results of a a clinical trial of a drug treatment for osteogenesis imperfecta (OI), which affects the child shown at *right*. Nine children with OI received the drug. Six others were untreated controls. Surface area of certain vertebrae was measured before and after treatment. Fractures occurring during the 12 months of the trial were also recorded.

2. Bones are _____ .
a. mineral reservoirs
b. skeletal muscle's partners
c. sites where blood cells form (some bones only)
d. all of the above

3. Bones move when _____ muscles contract.
a. cardiac
b. skeletal
c. smooth
d. all of the above

4. A ligament connects _____ .
a. bones at a joint
b. a muscle to a bone
c. a muscle to a tendon
d. a tendon to bone

5. Parathyroid hormone stimulates _____ .
a. osteoclast activity
b. bone deposition
c. red blood cell formation
d. all of the above

6. The _____ attaches to the pelvic girdle.
a. radius
b. sternum
c. femur
d. tibia

7. The _____ is the basic unit of contraction.
a. osteoblast
b. sarcomere
c. twitch
d. myosin filament

8. In sarcomeres, phosphate-group transfers from ATP activate _____ .
a. actin   b. myosin   c. both   d. neither

9. A sarcomere shortens when _____ .
a. thick filaments shorten
b. thin filaments shorten
c. both thick and thin filaments shorten
d. none of the above

10. ATP for muscle contraction can be formed by _____ .
a. aerobic respiration
b. lactate fermentation
c. creatine phosphate breakdown
d. all of the above

11. A virus causes _____ .
a. polio   b. botulism   c. muscular dystrophy

12. A motor unit is _____ .
a. a muscle and the bone it moves
b. two muscles that work in opposition
c. the amount a muscle shortens during contraction
d. a motor neuron and the muscle fibers it controls

13. Match the words with their defining feature.
____osteoblast
____muscle twitch
____muscle tension
____joint
____myosin
____red marrow
____metacarpals
____myofibrils
____muscle fatigue
____foramen magnum
____sarcoplasmic reticulum

a. stores and releases calcium
b. all in the hands
c. blood cell production
d. decline in tension
e. bone-forming cell
f. motor unit response
g. force exerted by cross bridges
h. area of contact between bones
i. muscle fiber's threadlike parts
j. actin's partner
k. hole in the head

■ *Visit CengageNOW for additional questions.*

Critical Thinking

1. Compared to most people, long-distance runners have far more mitochondria in skeletal muscles. In sprinters, skeletal muscle fibers have more of the enzymes required for glycolysis but not as many mitochondria. Suggest why.

2. Zachary's younger brother Noah had Duchenne muscular dystrophy and died at the age of 16. Zachary is now 26 years old, healthy, and planning to start a family of his own. However, he worries that his sons might be at high risk for muscular dystrophy. His wife's family has no history of this genetic disorder. Review Sections 12.4 and 36.11 and decide whether Zachary's concerns are well founded.

37 Circulation

And Then My Heart Stood Still

The heart is the body's most durable muscle. It starts to beat during the first month of human development, and keeps on going for a lifetime. Each heartbeat is set in motion by an electrical signal generated by a natural pacemaker in the heart wall. In some people, this pacemaker malfunctions. Electrical signaling gets disrupted, the heart stops beating, and blood flow halts. This is called sudden cardiac arrest. In the United States, it strikes more than 300,000 people per year. An inborn heart defect causes most cardiac arrests in people under age 35. In older people, heart disease usually causes the heart to stop functioning.

The chance of surviving sudden cardiac arrest rises by 50 percent when cardiopulmonary resuscitation (CPR) is started within four to six minutes of the arrest. With this technique, a person alternates mouth-to-mouth respiration with chest compressions that keep the victim's blood moving.

CPR cannot restart the heart. That requires a defibrillator, a device that delivers an electric shock to the chest and resets the natural pacemaker. You have probably seen this procedure depicted in hospital dramas.

Matt Nader (Figure 37.1a) learned about the importance of CPR and defibillation when he went into sudden cardiac arrest while playing in a high school football game. Nader's parents, who were watching the game, rushed from their seats and begin CPR on their son. At the same time, someone ran to get the school's automated external defibrillator (AED). This device is about the size of a laptop computer (Figure 37.1b). It provides simple voice commands about how to attach electrodes to a person in distress, then checks for a heartbeat and, if required, shocks the heart.

The AED restarted Nader's heart, and he went on to testify before the Texas Legislature about his experience. Thanks in part to his efforts, Texas passed a law requiring all high schools to have AEDs at athletic events and practices.

Because most cardiac arrests do not occur in a hospital, the presence of a bystander willing to carry out CPR and use an AED often means the difference between life and death. Yet studies show only about 15 percent of sudden cardiac arrest victims get CPR before trained personnel arrive. The problem is most people do not know how to administer CPR or use an AED. A half-day course given by the American Red Cross or another community health organization can teach you both skills. Taking the time to learn these skills is something we can all do for one another.

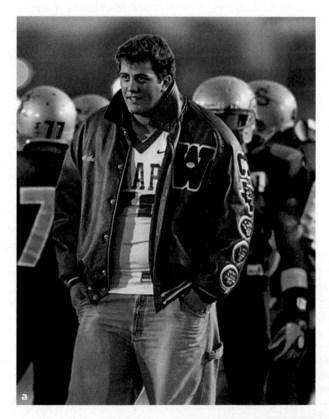

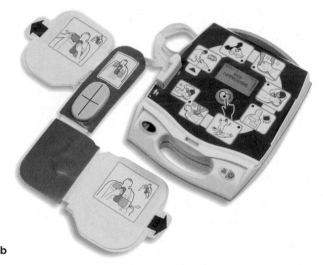

b

See the video! Figure 37.1 Surviving sudden cardiac arrest. (**a**) Matt Nader, a talented high school football player, discovered he had a heart defect when his heart stopped during a game. CPR and quick defibrillation saved his life.

(**b**) One type of automated external defibrillator. Such devices are designed to be simple enough to be used by a trained layperson. AEDs are increasingly available in public places, but they only make a difference if someone uses them.

a

Key Concepts

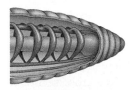

Overview of circulatory systems

Many animals have either an open or a closed circulatory system that transports substances to and from all body tissues. All vertebrates have a closed circulatory system, in which blood is always contained within the heart or blood vessels. **Section 37.1**

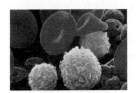

Blood composition and function

Vertebrate blood is a fluid connective tissue. It consists of red blood cells, white blood cells, platelets, and plasma (the transport medium). Red blood cells function in gas exchange; white blood cells defend tissues, and platelets function in clotting. **Sections 37.2–37.4**

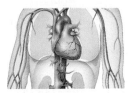

The human heart and two flow circuits

The four-chambered human heart pumps blood through two separate circuits of blood vessels. One circuit extends through all body regions, the other through lung tissue only. Both circuits loop back to the heart. **Sections 37.5, 37.6**

Blood vessel structure and function

The heart pumps blood rhythmically, on its own. Adjustments at arterioles regulate how blood volume is distributed among tissues. Exchange of gases, wastes, and nutrients between the blood and tissues takes place at capillaries. **Sections 37.7, 37.8**

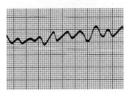

When the system breaks down

Cardiovascular problems include clogged blood vessels or abnormal heart rhythms. Some problems have a genetic basis; most are related to age or life-style. **Section 37.9**

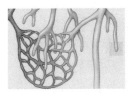

Links with the lymphatic system

A lymph vascular system delivers excess fluid that collects in tissues to the blood. Lymphoid organs cleanse blood of infectious agents and cellular debris. **Section 37.10**

Links to Earlier Concepts

- In this chapter, you will see examples of the role that diffusion (Section 5.3) plays in exchange of substances. You will revisit metabolism of alcohol (Chapter 6 introduction), and how glucose gets stored as glycogen (8.7).

- You will learn more about blood as a connective tissue (32.3) and how the muscle of the heart contracts (36.7, 36.8). You will also see how cell junctions (32.1) play a role in this contraction.

- ABO blood typing (11.4) and membrane proteins (5.2) are discussed again, as are hemoglobin and sickle-cell anemia (3.6, 18.6), hemophilia (12.4), and thalassemia (14.5)

- You will be reminded again of how diabetes affects the circulatory system (35.9), the role of the thymus gland (35.12), the effects of autonomic stimulation (33.8), and the overriding importance of homeostasis (27.1, 27.3).

- Evolutionary changes to the circulatory system (25.1, 26.2) also receive additional attention here.

How would you vote? Should public high schools in your state require all students to take a course in CPR? Is such a course worth diverting time and resources from the basic curriculum? See CengageNOW for details, then vote online.

- A circulatory system distributes materials throughout the body of some invertebrates and all vertebrates.

- Links to Animal evolution 25.1, Vertebrate evolution 26.2, Diffusion 5.3

From Structure to Function

A **circulatory system** moves substances into and out of cellular neighborhoods. **Blood**, its transport medium, typically flows inside tubular vessels under pressure generated by a **heart**, a muscular pump. Blood makes exchanges with **interstitial fluid**—fluid that fills spaces between cells. Interstitial fluid in turn exchanges substances with cells.

Blood and interstitial fluid serve as the body's internal environment. Interactions among organ systems keep the composition and volume of this environment within ranges that cells can tolerate (Section 25.1).

Structurally, there are two main kinds of circulatory systems. Arthropods and most mollusks have an **open circulatory system**. Their blood moves through hearts and large vessels but also mixes with interstitial fluid (Figure 37.2*a*). Annelids and vertebrates have a **closed circulatory system**. Their blood remains inside a heart or blood vessel at all times (Figures 37.2*b* and 37.3).

In a closed circulatory system, the blood volume moves continually through large and small vessels. Blood moves fastest where it is confined in a few big vessels and it slows in **capillaries**, the vessels with the smallest diameter. The slowdown in capillaries gives the blood and interstitial fluid time to exchange substances by diffusion (Section 5.3).

Blood slows in capillaries not because these vessels are small, but because of their huge numbers. Your body has billions, and their collective cross-sectional area is much greater than that of the far fewer, larger vessels that deliver blood to them. When blood enters capillaries, its speed declines, as if a narrow river (the few larger vessels) were delivering water to a wide lake (the many capillaries). Figure 37.3*d* illustrates the concept. Velocity picks up again in the larger, but far fewer, vessels that return blood to the heart. Similarly, water picks up speed when it flows from a wide lake into a narrow river.

Evolution of Circulation in Vertebrates

All vertebrates have a closed circulatory system, but fishes, amphibians, birds, and mammals differ in their pumps and plumbing. These differences evolved over

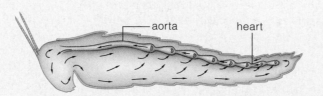

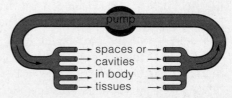

A In a grasshopper's open system, a heart (not like yours) pumps blood through a vessel, a type of aorta. From there, blood moves into tissue spaces, mingles with interstitial fluid, then reenters the heart at openings in the heart wall.

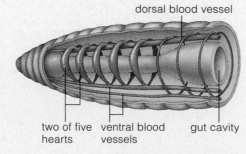

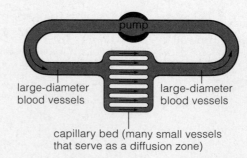

B The closed system of an earthworm confines blood inside pairs of muscular hearts near the head end and inside many blood vessels.

Figure 37.2 Animated Comparison of open and closed circulatory systems.

hundreds of millions of years after some vertebrates left the water for land. The earliest vertebrates, recall, had gills. Like other respiratory structures, gills have a thin, moist surface, across which oxygen and carbon dioxide diffuse. In time, internally moistened sacs called lungs evolved and supported the move to dry land. Other modifications helped blood flow faster in a loop between the heart and lungs (Section 26.2).

In most fishes, blood flows in one circuit (Figure 37.3a). The contractile force of a two-chambered heart drives it through a capillary bed inside each gill. From there, blood flows into a large vessel, then through capillary beds in body tissues and organs, and back to the heart. The blood is not under much fluid pressure when it leaves the gill capillaries, so it moves slowly through the single circuit back to the heart.

In amphibians, the heart is partitioned into three chambers, with two atria emptying into one ventricle. Oxygenated blood flows from the lungs to the heart in one circuit, then a forceful contraction pumps it through the rest of the body in a second circuit. Still, the oxygenated blood and oxygen-poor blood mix a bit in the ventricle (Figure 37.3b).

In birds and mammals, the heart has fully separate right and left halves, each with two chambers, and it pumps blood in two **separate** circuits (Figure 37.3c). In the **pulmonary circuit**, oxygen-poor, carbon dioxide–rich blood flows from the right half of the heart to the lungs. There, blood picks up oxygen, gives up carbon dioxide, and flows into the left half of the heart.

In the longer **systemic circuit**, the heart's left half pumps oxygenated blood to tissues where oxygen is used and carbon dioxide forms in aerobic respiration. Blood gives up oxygen and picks up carbon dioxide at tissues, then flows to the heart's right half.

With two fully separate circuits, blood pressure can be regulated independently in each circuit. Strong contraction of the heart's left ventricle provides sufficient force to keep blood moving fast through the long systemic circuit. Less forceful contraction of the right ventricle protects delicate lung capillaries that would be blown apart by high pressure.

Take-Home Message

What are the two types of animal circulatory systems?

■ Some animals, including insects, have an open circulatory system in which blood leaves vessels and mingles with the interstitial fluid.

■ Other animals, including annelids and all vertebrates, have a closed circulatory system, in which materials are exchanged across the walls of small blood vessels.

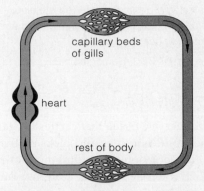

A In fishes, the heart has two chambers: one atrium and one ventricle. Blood flows through one circuit. It picks up oxygen in the capillary beds of the gills, and delivers it to capillary beds in all body tissues. Oxygen-poor blood then returns to the heart.

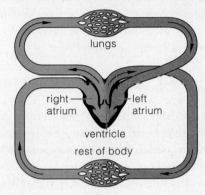

B In amphibians, the heart has three chambers: two atria and one ventricle. Blood flows along two partially separated circuits. The force of one contraction pumps blood from the heart to the lungs and back. The force of a second contraction pumps blood from the heart to all body tissues and back to the heart.

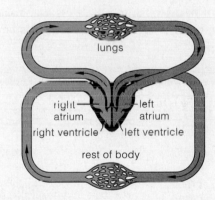

C In birds and mammals, the heart has four chambers: two atria and two ventricles. The blood flows through two fully separated circuits. In one circuit, blood flows from the heart to the lungs and back. In the second circuit, blood flows from the heart to all body tissues and back.

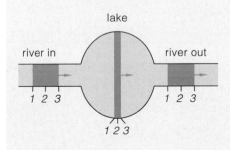

D Why flow slows in capillaries. Picture a volume of water in two fast rivers flowing into and out of a lake. The flow rate is constant, with an identical volume moving from points 1 to 3 in the same interval. However, flow velocity decreases in the lake. Why? The volume spreads out through a larger cross-sectional area and flows forward a shorter distance during the specified interval.

Figure 37.3 Animated (**a–c**) Comparison of flow circuits in the closed circulatory systems of fishes, amphibians, birds, and mammals. *Red* indicates oxygenated blood; *blue*, oxygen-poor blood. (**d**) Analogy illustrating why blood flow slows in capillaries.

■ Tumbling along in the fluid plasma of a vertebrate blood-stream are cells that distribute oxygen through the body and defend the body from pathogens.

■ Link to Connective tissues 32.3, Hemoglobin 3.6

Functions of Blood

Blood is the fluid connective tissue that carries oxygen, nutrients, and other solutes to cells and picks up their metabolic wastes and secretions, including hormones. Blood helps stabilize internal pH. It is a highway for cells and proteins that protect and repair tissues. In birds and mammals, it helps keep body temperature within tolerable limits by moving excess heat to skin, which can give up heat to the surroundings.

Blood Volume and Composition

Body size and the concentrations of water and solutes dictate the blood volume. Average-sized humans hold about 5 liters (a bit more than 10 pints), which is 6 to 8 percent of the total body weight. In vertebrates, blood is a viscous fluid that is thicker than water, and slower flowing. Blood's fluid portion is **plasma**. Its cellular portion consists of blood cells and platelets that arise from stem cells in bone marrow (Section 32.3). A stem

cell is an unspecialized cell that retains a capacity for mitotic cell division. Some portion of its daughter cells divide and differentiate into specialized cell types.

Plasma About 50 to 60 percent of the blood's total volume is plasma (Figure 37.4). Plasma is 90 percent water. Besides being the transport medium for blood cells and platelets, it acts as a solvent for hundreds of different plasma proteins. Some proteins transport lipids and fat-soluble vitamins; others have a role in blood clotting or immunity. Plasma also holds sugars, lipids, amino acids, vitamins, and hormones, as well as the gases oxygen, carbon dioxide, and nitrogen.

Red Blood Cells Erythrocytes, or **red blood cells**, transport oxygen from lungs to aerobically respiring cells and help carry carbon dioxide wastes from them. In all mammals, red blood cells lose their nucleus, mitochondria, and other organelles as they mature. Mature red blood cells are flexible disks with a depression at their center. They slip easily through narrow blood vessels and the flattened shape facilitates gas exchange.

Most oxygen that diffuses into your blood binds to hemoglobin in red blood cells. You learned about this protein in Section 3.6. Stored hemoglobin fills about 98 percent of the interior of human red blood cells. It

Components	Amounts	Main Functions
Plasma Portion (50–60% of total blood volume)		
1. Water	91–92% of total plasma volume	Solvent
2. Plasma proteins (albumins, globulins, fibrinogen, etc.)	7–8%	Defense, clotting, lipid transport, extracellular fluid volume controls
3. Ions, sugars, lipids, amino acids, hormones, vitamins, dissolved gases, etc.	1–2%	Nutrition, defense, respiration, extracellular fluid volume controls, cell communication, etc.
Cellular Portion (40–50% of total blood volume; numbers per microliter)		
1. Red blood cells	4,600,000–5,400,000	Oxygen, carbon dioxide transport to and from lungs
2. White blood cells:		
Neutrophils	3,000–6,750	Fast-acting phagocytosis
Lymphocytes	1,000–2,700	Immune responses
Monocytes (macrophages)	150–720	Phagocytosis
Eosinophils	100–380	Killing parasitic worms
Basophils	25–90	Anti-inflammatory secretions
3. Platelets	250,000–300,000	Roles in blood clotting

red blood cell   white blood cell   platelet

Figure 37.4 Typical components of human blood. Numbers for cellular components are all per microliter. The sketch of a test tube shows what happens when you prevent a blood sample from clotting. The sample separates into straw-colored plasma, which floats on a reddish cellular portion. The scanning electron micrograph shows these components.

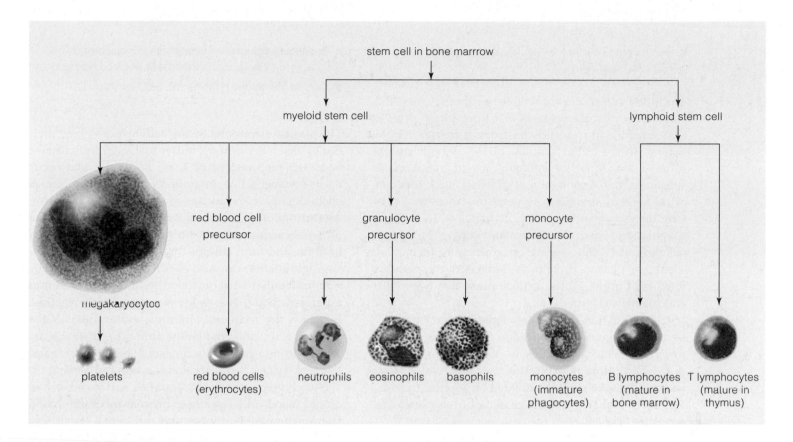

Figure 37.5 Main cellular components of mammalian blood and how they originate.

makes the cells and oxygenated blood appear bright red. Oxygen-poor blood is dark red, but it looks blue through blood vessel walls near the body surface.

In addition to hemoglobin, a mature red blood cell has enough stored glucose and enzymes to last about 120 days. In a healthy person, ongoing replacements keep red blood cell numbers at a fairly stable level. A **cell count** measures the quantity of cells of one type per microliter of blood. Men typically have a higher red blood cell count than women of reproductive age, who lose blood during menstruation.

White Blood Cells Leukocytes, or **white blood cells**, carry out ongoing housekeeping tasks and function in defense. The cells differ in their size, nuclear shape, and staining traits (Figure 37.5), as well as function.

Neutrophils, basophils, and eosinophils all develop from one type of precursor cell. They are sometimes collectively referred to as the granulocytes, because their cytoplasm contains granules that can be stained by specific dyes. Neutrophils are the most abundant white cells; they are phagocytes that engulf bacteria and debris. Eosinophils attack larger parasites, such as worms, and have a role in allergies. Basophils secrete chemicals that have a role in inflammation.

Monocytes circulate in the blood for a few days, then move into the tissues, where they develop into phagocytic cells known as macrophages. As you will see in the next chapter, macrophages interact with lymphocytes to bring about immune responses. There are two types of lymphocytes, B cells and T cells. B cells mature in bone, whereas T cells mature in the thymus. Both protect the body against specific threats.

Platelets Megakaryocytes are ten to fifteen times bigger than other blood cells that form in bone marrow. They break up into membrane-wrapped fragments of cytoplasm called **platelets**. After a platelet forms, it will last five to nine days. When activated, it releases substances needed for blood clotting.

Take-Home Message

What are the components of human blood and what are their functions?

■ Blood consists mainly of plasma, a protein-rich fluid that carries wastes, gases, and nutrients.

■ Blood cells and platelets form in bone marrow and are transported in plasma. Red blood cells contain hemoglobin that carries oxygen from lungs to tissues. White cells help defend the body from pathogens. Platelets are cell fragments that have a role in clotting.

- Plasma proteins and platelets interact in clotting.
- Link to Hemophilia 12.4

The blood vessels are vulnerable to ruptures, cuts, and similar injuries. **Hemostasis** is a three-phase process that stops blood loss and constructs a framework for repairs. In the initial vascular phase, smooth muscle in the damaged vessel wall contracts in an automatic spasm. In the second phase, platelets stick together at the injured site. They release substances that prolong the spasm and attract more platelets. In the final coagulation phase, plasma proteins convert blood to a gel and form a clot. During clot formation, fibrinogen, a soluble plasma protein, is converted to insoluble threads of fibrin. Fibrin forms a mesh that traps cells and platelets (Figure 37.6).

Clot formation involves a cascade of enzyme reactions. Fibrinogen is converted to fibrin by the enzyme thrombin, which circulates in blood as the inactive precursor prothrombin. Prothrombin is activated by an enzyme (factor X) that is activated by another enzyme, and so on. What starts the cascade of reactions? The exposure of collagen in the damaged vessel wall.

If a mutation affects any one of the enzymes that acts in the cascade of clotting reactions, the blood may not clot properly. Such mutations cause the genetic disorder hemophilia (Section 12.4).

Stimulus
A blood vessel is damaged.

Phase 1 response
A vascular spasm constricts the vessel.

Phase 2 response
Platelets stick together plugging the site.

Phase 3 response
Clot formation starts:

1. Enzyme cascade results in activation of Factor X.
2. Factor X converts prothrombin in plasma to thrombin.
3. Thrombin converts fibrinogen, a plasma protein, to fibrin threads.
4. Fibrin forms a net that entangles cells and platelets, forming a clot.

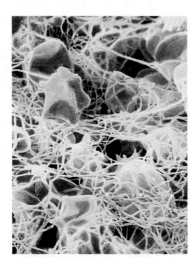

Figure 37.6 The three-phase process of hemostasis. The micrograph shows the result of the final clotting phase—blood cells and platelets in a fibrin net.

Take-Home Message

How does the body respond to blood vessel damage and halt bleeding?

- The vessel constricts, platelets accumulate, and cascading enzyme reactions involving protein components of plasma cause clot formation.

- Genetically determined differences in molecules on the surface of red blood cells are the basis of blood typing.
- Links to Membrane proteins 5.2, ABO genetics 11.4

The plasma membrane of any cell includes many molecules that vary among individuals. An individual's body ignores versions of these molecules that occur on its own cells, but unfamiliar cell surface molecules elicit defensive responses by the immune system. **Agglutination** is a normal response in which plasma proteins called antibodies bind foreign cells, such as bacteria, and form clumps that attract phagocytes.

Agglutination can also occur when red blood cells with unfamiliar surface molecules are transfused into a person's body. The result is a transfusion reaction, in which the recipient's immune system attacks the donated cells, causing them to clump together (Figure 37.7). The clumps of cells clog small blood vessels and damage tissues. A transfusion reaction can be fatal.

Blood typing—analysis of specific surface molecules on red blood cells—can help prevent mixing of blood from incompatible donors and recipients. It can also put physicians on the alert for blood-related problems that can arise during some pregnancies.

ABO Blood Typing

ABO blood typing analyzes variations in one type of glycolipid on the surface of red blood cells. Section 11.4 describes the genetics of these variations. People who have one form of the molecule have type A blood. Those with a different form have type B blood. People with both forms of the molecule have type AB blood. Those who have neither form are type O. See below.

ABO Type	Glycolipid(s) on Red Cells	Antibodies Present
A	A	Anti-B
B	B	Anti-A
AB	Both A and B	None
O	Neither A nor B	Anti-A, Anti-B

If you are blood type O, your immune system treats both type A and type B cells as foreign. You can accept blood only from people who are type O (Figure 37.8). However, you can donate blood to anyone. If you are blood type A, your body will recognize type B cells as foreign. If you are type B, your blood will react against type A cells. If you are blood type AB, your immune system treats both type A and type B as "self," so you can receive blood from anyone.

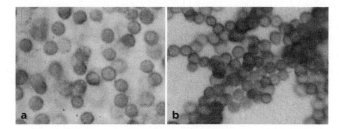

Figure 37.7 Light micrographs showing (**a**) an absence of agglutination in a mixture of two different yet compatible blood types and (**b**) agglutination in a mixture of incompatible types.

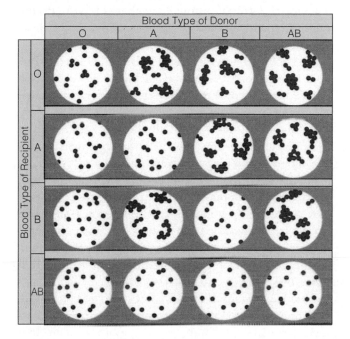

Figure 37.8 Animated Results of mixing blood of the same or differing ABO blood types. **Figure It Out:** How many incompatible combinations are shown? *Answer: Seven*

Rh Blood Typing

Rh blood typing is based on the presence or absence of the Rh protein (first identified in blood of *Rh*esus monkeys). If you are type Rh+, your blood cells bear this protein. If you are type Rh−, they do not.

Normally, Rh− individuals do not have antibodies against the Rh protein. However, they will produce such antibodies if they are exposed to Rh+ blood. This can happen during some pregnancies. If an Rh+ man impregnates an Rh− woman, the resulting fetus may be Rh+. The first time that an Rh− woman carries an Rh+ fetus, she will not have antibodies against the Rh protein (Figure 37.9a). However, fetal red blood cells may get into her blood during childbirth, causing her to form anti-Rh+ antibodies. If the woman gets pregnant again, these antibodies cross the placenta and get into fetal blood. If a fetus is Rh+, the antibodies attack

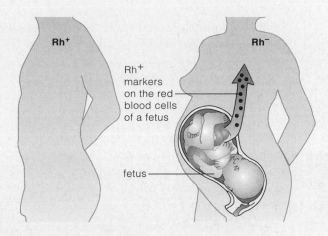

A An Rh+ man and an Rh− woman carrying his Rh+ child. This is the mother's first Rh+ pregnancy, so she has no anti-Rh+ antibodies. But during birth, some of the child's Rh+ cells get into her blood.

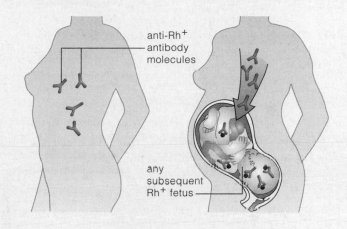

B The foreign marker stimulates antibody formation. If this woman gets pregnant again and if her second fetus (or any other) carries the Rh+ protein, her anti-Rh+ antibodies may attack the fetal red blood cells.

Figure 37.9 Animated How Rh differences can complicate pregnancy.

its red blood cells and can kill the fetus (Figure 37.9b). To prevent any problems, an Rh− mother who has just given birth to an Rh+ child should be injected with a drug that blocks production of antibodies that could cause problems during future pregnancies.

ABO blood types do not cause a similar condition because maternal antibodies for A and B molecules do not cross the placenta and attack the fetal cells.

Take-Home Message

What is a blood type?

■ Blood type refers to the kind of surface molecules on red blood cells. Genes determine which form of these molecules is present in a particular individual.

■ When blood of incompatible types mixes, the immune system attacks the unfamiliar molecules, with results that can be fatal.

37.5 | Human Cardiovascular System

- The term "cardiovascular" comes from the Greek *kardia* (for heart) and Latin *vasculum* (vessel).

- Links to Alcohol metabolism Chapter 6 introduction, Glycogen storage 8.7, Homeostasis 27.1 and 27.3

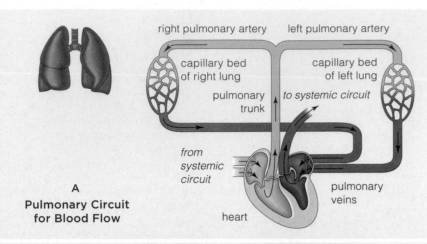

right pulmonary artery | left pulmonary artery

capillary bed of right lung

capillary bed of left lung

pulmonary trunk

to systemic circuit

from systemic circuit

pulmonary veins

heart

A
Pulmonary Circuit for Blood Flow

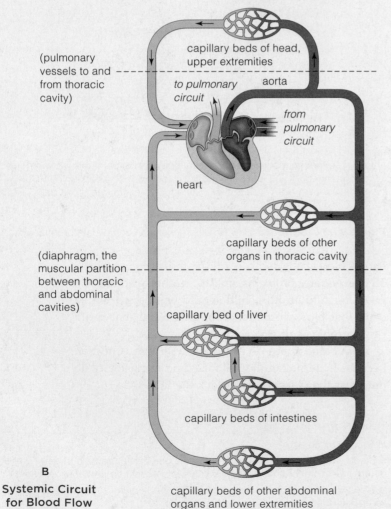

(pulmonary vessels to and from thoracic cavity)

capillary beds of head, upper extremities

to pulmonary circuit

aorta

from pulmonary circuit

heart

capillary beds of other organs in thoracic cavity

(diaphragm, the muscular partition between thoracic and abdominal cavities)

capillary bed of liver

capillary beds of intestines

B
Systemic Circuit for Blood Flow

capillary beds of other abdominal organs and lower extremities

In humans, as in all mammals, the heart is a double pump that propels blood through two cardiovascular circuits. Each circuit extends from the heart, through arteries, arterioles, capillaries, venules, and veins, and reconnects with the heart (Figures 37.10 and 37.11).

A short loop, the pulmonary circuit, oxygenates blood (Figure 37.10*a*). It leads from the heart's right half to capillary beds in the lungs. Blood is oxygenated in the lungs, then flows to the heart's left half.

The systemic circuit is a longer loop (Figure 37.10*b*). The heart's left half pumps oxygenated blood into the main artery in the body: the **aorta**. That blood gives up oxygen in all tissues, then the oxygen-poor blood flows back to the heart's right half.

In the systemic circuit, most blood flows through one capillary bed, then returns to the heart. However, blood that passes through the capillaries in the small intestine then flows through the hepatic portal vein to a capillary bed in the liver. This arrangement allows the blood to pick up glucose and other substances absorbed from the gut, and deliver them to the liver. The liver stores some of the absorbed glucose as glycogen (Section 8.7). It also breaks down some absorbed toxins, including alcohol (Chapter 6 introduction).

As Figure 37.12 shows, the cardiovascular system distributes nutrients, gases, and other substances that enter the body by way of the digestive system and respiratory system. It moves carbon dioxide and other metabolic wastes to the respiratory and urinary systems for disposal. These are the main systems that keep operating conditions of the internal environment within tolerable ranges, a process we call homeostasis (Sections 27.1 and 27.3).

Take-Home Message

What are the two circuits of the human circulatory system?

- In the pulmonary circuit, oxygen-poor blood flows from the heart, through a pair of lungs, then back to the heart. It takes up oxygen and gives up carbon dioxide in the lungs.

- In the systemic circuit, oxygenated blood flows from the heart to capillary beds of all tissues. There it gives up oxygen and takes up carbon dioxide, then flows back to the heart.

Figure 37.10 Animated (**a**,**b**) Pulmonary and systemic circuits of the human cardiovascular system. Blood vessels carrying oxygenated blood are shown in *red*. Those that hold oxygen-poor blood are color-coded *blue*.

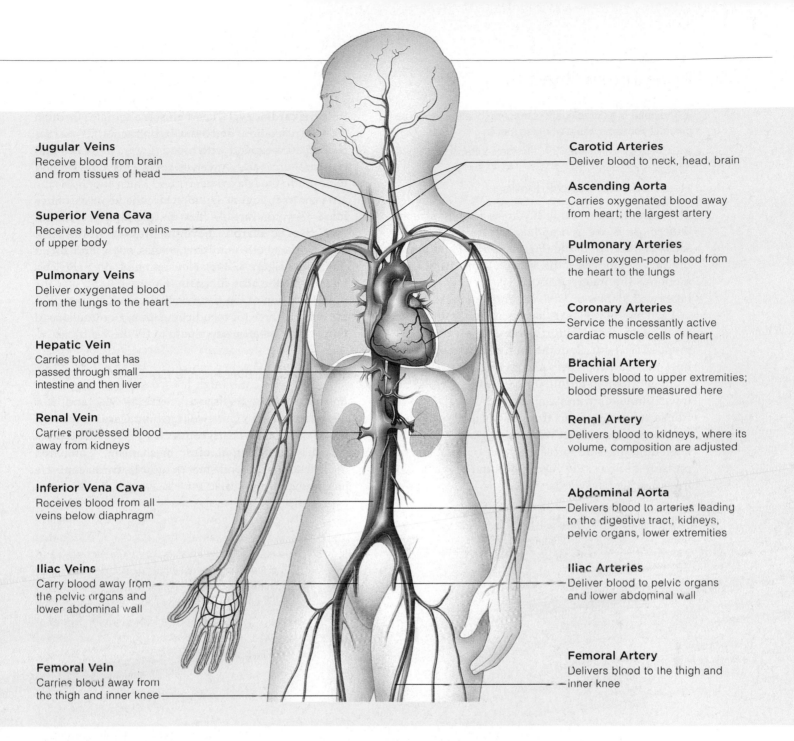

Jugular Veins
Receive blood from brain and from tissues of head

Superior Vena Cava
Receives blood from veins of upper body

Pulmonary Veins
Deliver oxygenated blood from the lungs to the heart

Hepatic Vein
Carries blood that has passed through small intestine and then liver

Renal Vein
Carries processed blood away from kidneys

Inferior Vena Cava
Receives blood from all veins below diaphragm

Iliac Veins
Carry blood away from the pelvic organs and lower abdominal wall

Femoral Vein
Carries blood away from the thigh and inner knee

Carotid Arteries
Deliver blood to neck, head, brain

Ascending Aorta
Carries oxygenated blood away from heart; the largest artery

Pulmonary Arteries
Deliver oxygen-poor blood from the heart to the lungs

Coronary Arteries
Service the incessantly active cardiac muscle cells of heart

Brachial Artery
Delivers blood to upper extremities; blood pressure measured here

Renal Artery
Delivers blood to kidneys, where its volume, composition are adjusted

Abdominal Aorta
Delivers blood to arteries leading to the digestive tract, kidneys, pelvic organs, lower extremities

Iliac Arteries
Deliver blood to pelvic organs and lower abdominal wall

Femoral Artery
Delivers blood to the thigh and inner knee

Figure 37.11 Animated Major blood vessels of the human cardiovascular system. This art is simplified for clarity. For example, the arteries or veins labeled for one arm occur in both arms.

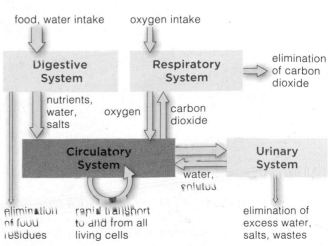

Figure 37.12 Functional links between the circulatory system and other organ systems with major roles in maintaining the internal environment.

The Human Heart

- The heart is a durable, spontaneously beating, muscular pump. It contracts with a wringing motion.

- Links to Cell junctions 32.1, Muscle contraction 36.7, 36.8

Heart Structure and Function

The human heart's durability arises from its structure. Outermost is the *pericardium*, a tough, two-layered sac of connective tissue (Figure 37.13). Fluid between the layers lubricates the heart during its wringing motions. The inner pericardial layer attaches to the heart wall, or *myocardium*, of cardiac muscle.

Each half of the heart has an **atrium** (plural, atria), an entrance chamber for blood, and a **ventricle** that pumps blood out. Endothelium, a kind of epithelium, lines the heart chambers and all blood vessels.

To get from an atrium into a ventricle, blood must travel through an atrioventricular (AV) valve. To flow from a ventricle into an artery, it has to pass through a semilunar valve. Heart valves are like one-way doors. High fluid pressure forces the valve open. When fluid pressure declines, the valve shuts and prevents blood from flowing backwards.

In the **cardiac cycle**, heart muscle alternates through *diastole* (relaxation) and *systole* (contraction). First, the relaxed atria expand with blood (Figure 37.14*a*). Fluid pressure forces the AV valves open. This allows blood to flow into the relaxed ventricles, which expand as the atria contract (Figure 37.14*b*). Once the ventricles have filled, they contract. As they do, the fluid pressure in them rises so sharply above the pressure in the great arteries that both semilunar valves open, and blood flows out (Figure 37.14*c*). Now emptied, the ventricles relax while the atria fill again (Figure 37.14*d*).

Contraction of the thick-walled ventricles provides the driving force for blood circulation. Contractions of thinner-walled atria serve only to fill the ventricles.

How Does Cardiac Muscle Contract?

Cardiac Muscle Revisited Sections 36.7 and 36.8 describe skeletal muscle contraction. **Cardiac muscle**, found only in the heart, contracts by the same type of ATP-driven sliding-filament mechanism. Compared to skeletal muscle and smooth muscle, cardiac muscle has more mitochondria.

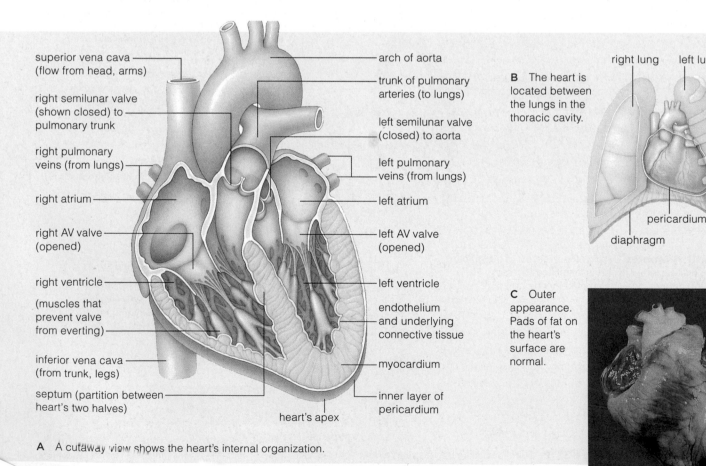

superior vena cava
(flow from head, arms)

right semilunar valve
(shown closed) to
pulmonary trunk

right pulmonary
veins (from lungs)

right atrium

right AV valve
(opened)

right ventricle

(muscles that
prevent valve
from everting)

inferior vena cava
(from trunk, legs)

septum (partition between
heart's two halves)

heart's apex

arch of aorta

trunk of pulmonary
arteries (to lungs)

left semilunar valve
(closed) to aorta

left pulmonary
veins (from lungs)

left atrium

left AV valve
(opened)

left ventricle

endothelium
and underlying
connective tissue

myocardium

inner layer of
pericardium

A A cutaway view shows the heart's internal organization.

B The heart is located between the lungs in the thoracic cavity.

right lung   left lung

ribs
1–8

pericardium

diaphragm

C Outer appearance. Pads of fat on the heart's surface are normal.

Figure 37.13 Animated The human heart.

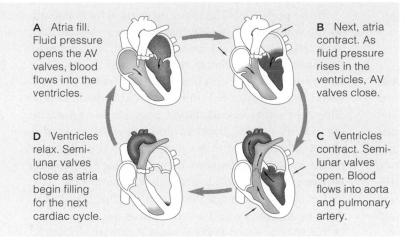

A Atria fill. Fluid pressure opens the AV valves, blood flows into the ventricles.

B Next, atria contract. As fluid pressure rises in the ventricles, AV valves close.

C Ventricles contract. Semilunar valves open. Blood flows into aorta and pulmonary artery.

D Ventricles relax. Semilunar valves close as atria begin filling for the next cardiac cycle.

Figure 37.14 Animated Cardiac cycle. You can hear the cycle through a stethoscope as a "lub-dup" near the chest wall. At each "lub," the heart's AV valves are closing as its ventricles are contracting. At each "dup," the heart's semilunar valves are closing as its ventricles are relaxing.

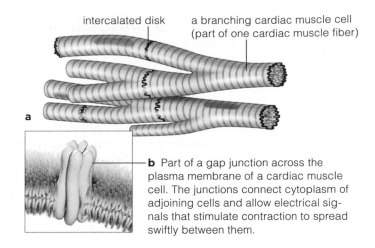

intercalated disk   a branching cardiac muscle cell (part of one cardiac muscle fiber)

a

b Part of a gap junction across the plasma membrane of a cardiac muscle cell. The junctions connect cytoplasm of adjoining cells and allow electrical signals that stimulate contraction to spread swiftly between them.

Figure 37.15 (**a**) Cardiac muscle cells. Compare Figure 32.8*b*. Many adhering junctions in intercalated disks at the ends of cells hold adjacent cells together, despite the mechanical stress caused by the heart's wringing motions. (**b**) The sides of cardiac muscle cells are subject to less mechanical stress than the ends. The sides have a profusion of gap junctions across the plasma membrane.

Sarcomeres arranged along the length of each cell give cardiac muscle a striated appearance. The cells attach end to end at intercalated disks, regions with many adhering junctions (Figure 37.15*a*). Neighboring cells communicate through gap junctions. These gap junctions allow waves of excitation to wash swiftly over the entire heart (Section 32.1 and Figure 37.15*b*).

How the Heart Beats In cardiac muscle, some specialized cells do not contract. Instead, they are part of the **cardiac conduction system**, which initiates and distributes signals that tell other cardiac muscle cells to contract. As Figure 37.16 shows, the system consists of a sinoatrial (SA) node and an atrioventricular (AV) node, functionally linked by junctional fibers. These fibers are bundles of long, thin cardiac muscle cells.

The SA node, a clump of noncontracting cells in the right atrium's wall, is the **cardiac pacemaker**. Its cells have specialized membrane channels that allow them to fire action potentials about seventy times per minute. The brain does not have to direct the SA node to fire; this natural pacemaker has spontaneous action potentials. Nervous signals from the brain only adjust the rate and strength of contractions. Even if a heart is removed from the body, it will keep beating for a short time.

A signal from the SA node starts the cardiac cycle. The signal spreads through the atria, causing them to contract. Simultaneously, the signal excites junctional fibers, which conduct it to the AV node. This clump of cells is the only electric bridge to the ventricles. The

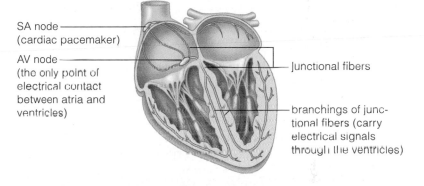

SA node (cardiac pacemaker)

AV node (the only point of electrical contact between atria and ventricles)

Junctional fibers

branchings of junctional fibers (carry electrical signals through the ventricles)

Figure 37.16 Animated The cardiac conduction system.

time it takes for a signal to cross this bridge is enough to keep ventricles from contracting before they fill.

From the AV node, a signal travels along a bundle of fibers. These junctional fibers branch in the septum, between the heart's left and right ventricles. Branching fibers extend down to the heart's lowest point and up the ventricle walls. Ventricles contract from the bottom up, with a twisting motion.

Take-Home Message

How is the human heart structured and how does it function?

■ The four-chambered heart is partitioned into two halves, each with an atrium and a ventricle. Contraction of ventricles drives blood circulation.

■ The SA node is the cardiac pacemaker. Its spontaneous, rhythmically repeated signals make cardiac muscle fibers of the heart wall contract in a coordinated fashion.

Pressure, Transport, and Flow Distribution

■ Contracting ventricles put pressure on the blood, forcing it through a series of vessels.

■ Link to Autonomic nervous system 33.8

Figure 37.17 compares the structure of blood vessels. **Arteries** are rapid-transport vessels for blood pumped out of the heart's ventricles. They deliver blood to the **arterioles**: smaller vessels where controls over distribution of blood flow operate. Arterioles branch into **capillaries**, small, thin-walled vessels that substances diffuse into and out of easily. **Venules** are small vessels located between capillaries and veins. **Veins** are large vessels that deliver blood back to the heart and serve as blood volume reservoirs.

Blood pressure is the pressure exerted by the blood on the walls of the vessels that enclose it. Ventricular contractions put blood under pressure, and because the right ventricle contracts less forcefully than the left ventricle, blood entering the pulmonary circuit is under less pressure than blood entering the systemic circuit. In either circuit, blood pressure is highest in the arteries and declines as blood flows through the circuit (Figure 37.18). The rate of flow between two points in a circuit depends on the pressure difference between those points, and the resistance to flow. The wider and smoother a vessel is, the less resistance there is, and the faster fluid can move through it.

Rapid Transport in Arteries

With their large diameter and low resistance to flow, arteries are efficient rapid transporters of oxygenated blood. They also are pressure reservoirs that smooth out pressure differences during every cardiac cycle. Their thick, muscular, elastic wall bulges whenever a heartbeat forces a large volume of blood into them. Between contractions, the wall recoils.

Flow Distribution at Arterioles

No matter how active you are, all blood from the right half of your heart flows to your lungs, and all blood from the left half is distributed to other tissues along the systemic circuit. The brain gets a constant supply of blood, but flow to other organs varies with activity. When you are resting, the blood flow is distributed as shown in Figure 37.19.

When you exercise, less blood flows to the kidneys and gut, and more flows to skeletal muscles in your legs. Like traffic cops, your arterioles guide the flow

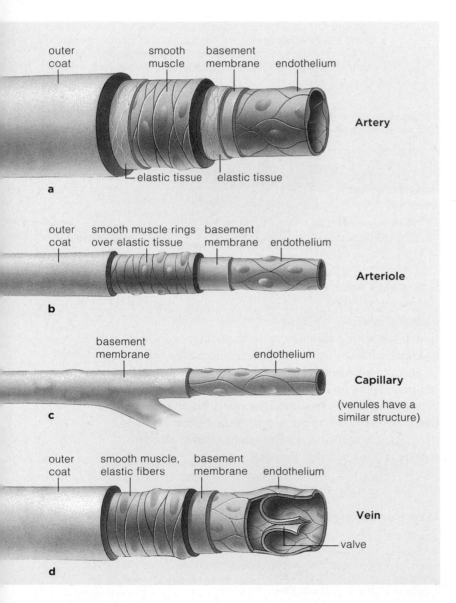

Figure 37.17 Structural comparison of human blood vessels. The drawings are not to the same scale.

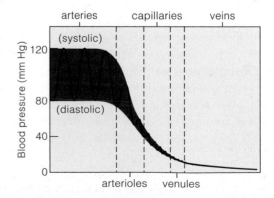

Figure 37.18 Plot of fluid pressure for a volume of blood as it flows through the systemic circuit. Systolic pressure occurs when ventricles contract, diastolic when ventricles are relaxed.

based on orders from the autonomic nervous system (Section 33.8) and endocrine system. Signals from both act on rings of smooth muscle cells in arteriole walls (Figure 37.17b). Some signals cause dilation, or widening, of a blood vessel by causing the smooth muscle cells in its wall to relax. Other signals decrease blood vessel diameter by causing the smooth muscle in its wall to contract. When arterioles that supply a particular organ widen, more blood flows to that organ.

Arterioles also respond to shifts in concentrations of substances in a tissue. As an example, when you exercise, your skeletal muscle cells use up oxygen, and carbon dioxide concentration around them rises. Arterioles in the muscle widen in response to these localized changes. As a result, more oxygenated blood flows through the tissue, and more metabolic waste products are carried away. When skeletal muscles relax, they require less oxygen. The concentration of oxygen rises locally, and the arterioles narrow.

Controlling Blood Pressure

We generally measure blood pressure at the brachial artery in an upper arm (Figure 37.20). In each cardiac cycle, *systolic* (peak) pressure occurs when contracting ventricles force blood into arteries. *Diastolic* pressure, the lowest pressure, occurs when ventricles are most relaxed. Blood pressure is measured in millimeters of mercury (mm Hg) and recorded as "systolic pressure over diastolic pressure," as in 120/80 mm Hg.

Blood pressure depends on the total blood volume, how much blood the ventricles pump out (the cardiac output), and whether the arterioles are constricted or dilated. Receptors in the aorta and in the carotid arteries of the neck send signals to a control center in the medulla (a portion of the brain stem) when blood pressure increases or decreases. In response, this brain region calls for changes in cardiac output and arteriole diameter. This reflex response is a short-term control over blood pressure. Over the longer term, kidneys influence blood pressure by adjusting fluid loss and thus altering the total blood volume. The greater the blood volume, the higher the blood pressure.

Take-Home Message

What determines blood pressure and distribution?

■ The rate and strength of heartbeats and resistance to flow through blood vessels dictates blood pressure. Pressure is greatest in contracting ventricles and at the start of arteries.

■ How much blood flows to specific tissues varies over time and is altered by adjustments to the diameter of arterioles.

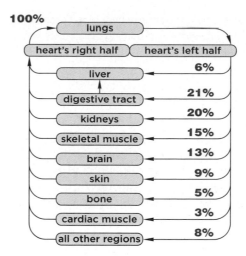

100%

lungs	
heart's right half / heart's left half	
liver	6%
digestive tract	21%
kidneys	20%
skeletal muscle	15%
brain	13%
skin	9%
bone	5%
cardiac muscle	3%
all other regions	8%

Figure 37.19 Distribution of the heart's output in a resting person. How much blood flows through a given tissue can be adjusted by selectively narrowing and widening arterioles all along the systemic circuit.

Figure It Out: What percentage of the brain's blood supply arrives from the heart's right half?

Answer: None

Figure 37.20 Animated Measuring blood pressure. *Left*, a hollow inflatable cuff attached to a pressure gauge is wrapped around the upper arm. A stethoscope is placed over the brachial artery, just below the cuff.

The cuff is inflated with air to a pressure above the highest pressure of the cardiac cycle, when ventricles contract. Above this pressure, you will not hear sounds through the stethoscope, because no blood is flowing through the vessel.

Air in the cuff is slowly released until the stethoscope picks up soft tapping sounds. Blood flowing into the artery under the pressure of the contracting ventricles—the systolic pressure—causes the sounds. When these sounds start, a gauge typically reads about 120 mm Hg. That amount of pressure will force mercury (Hg) to move up 120 millimeters in a glass column of a standard diameter.

More air is released from the cuff. Eventually the sounds stop. Blood is now flowing continuously, even when the ventricles are the most relaxed. The pressure when the sounds stop is the lowest during a cardiac cycle. This diastolic pressure is usually about 80 mm Hg.

Right, compact monitors are now available that automatically record the systolic/diastolic blood pressure.

■ A capillary bed is a diffusion zone, where blood exchanges substances with the interstitial fluid bathing cells before veins carry it back to the heart.

■ Links to Epithelium 32.2, Diffusion 5.3, Endocytosis 5.5

Capillary Function

A capillary is a cylinder of endothelial cells, one cell thick, wrapped in basement membrane (Section 32.2). Figure 37.21 shows a few of the 10 billion to 40 billion capillaries that service a human body. Collectively, they offer a huge surface area for exchange of substances with interstitial fluid. In nearly all tissues, cells are very close to one or more capillaries. Proximity is essential. Diffusion distributes molecules and ions so slowly that it is effective only over small distances.

Red blood cells, which are about 8 micrometers in diameter, have to squeeze in single file through the capillaries. The squeeze puts oxygen-transporting red blood cells and solutes in the plasma in direct or near contact with the exchange surface—the capillary wall.

To move between the blood and interstitial fluid, a substance must cross a capillary wall. Oxygen, carbon dioxide, and small lipid-soluble molecules can diffuse across endothelial cells of a capillary. Proteins are too big to diffuse across plasma membranes, but some do enter endothelial cells by endocytosis, diffuse through the cell, then escape by exocytosis on the opposite side. Also, fluid with small solutes and ions leaks out of capillaries through spaces between adjacent cells.

Compared to other capillaries in the body, those in the brain are much less leaky. Brain endothelial cells adhere so tightly to one another that plasma does not leak between them. This property of brain capillaries creates the blood–brain barrier (Section 33.10).

As blood flows through a typical capillary bed, it is subject to two opposing forces. Hydrostatic pressure, an outward-directed force, results from contraction of

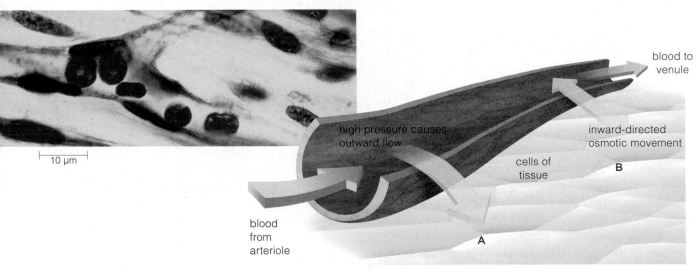

Figure 37.21 Fluid movement at a capillary bed. Fluid crosses a capillary wall by way of ultrafiltration and reabsorption. (**a**) At the capillary's arteriole end, a difference between blood pressure and interstitial fluid pressure forces out plasma, but few plasma proteins, through clefts between endothelial cells of the capillary wall. Ultrafiltration is the outward flow of fluid across the capillary wall as a result of hydrostatic pressure.

(**b**) Reabsorption is the osmotic movement of some interstitial fluid into the capillary. It happens when the water concentration between interstitial fluid and the plasma differs. Plasma, with its dissolved proteins, has a greater solute concentration and therefore a lower water concentration. Reabsorption near the end of a capillary bed tends to balance ultrafiltration at the start of it. Normally, there is only a small net filtration of fluid, which vessels of the lymphatic system return to blood (Section 37.10).

Arteriole end of capillary bed	
Outward-Directed Pressure:	
Hydrostatic pressure of blood in capillary:	35 mm Hg
Osmosis due to interstitial proteins:	28 mm Hg
Inward-Directed Pressure:	
Hydrostatic pressure of interstitial fluid:	0
Osmosis due to plasma proteins:	3 mm Hg
Net Ultrafiltration Pressure:	
(35 – 0) – (28 – 3) = 10 mm Hg	
Ultrafiltration favored	

Venule end of capillary bed	
Outward-Directed Pressure:	
Hydrostatic pressure of blood in capillary:	15 mm Hg
Osmosis due to interstitial proteins:	28 mm Hg
Inward-Directed Pressure:	
Hydrostatic pressure of interstitial fluid:	0
Osmosis due to plasma proteins:	3 mm Hg
Net Reabsorption Pressure:	
(15 – 0) – (28 – 3) = –10 mm Hg	
Reabsorption favored	

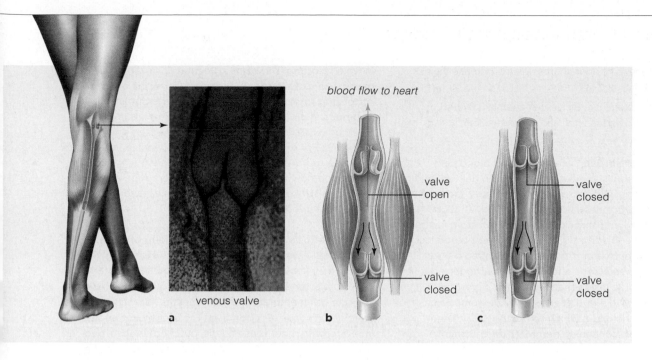

blood flow to heart

venous valve

a

valve open

valve closed

b

valve closed

valve closed

c

Figure 37.22 Venous valve action.

(**a**) Valves in medium-sized veins prevent the backflow of blood.

Adjacent skeletal muscles helps raise fluid pressure inside a vein. (**b**) These muscles bulge into a vein as they contract. Pressure inside the vein rises and helps keeps blood flowing forward. (**c**) When muscles relax, the pressure that they exerted on the vein is lifted. Venous valves shut and cut off backflow.

the ventricles. Osmotic pressure, an inward-directed force, results from differences in solute concentration between blood and interstitial fluid.

At the arterial end of a capillary bed, hydrostatic pressure is high. It forces fluid out between cells of the capillary wall, into interstitial fluid (Figure 37.21a). This process is **ultrafiltration**. The fluid forced out has high levels of oxygen, ions, and nutrients such as glucose. Ultrafiltration moves large quantities of essential substances from blood into interstitial fluid.

As the blood continues on to the venous end of the capillary bed, hydrostatic pressure drops and osmotic pressure predominates (Figure 37.21b). Water is drawn by osmosis from interstitial fluid into the protein-rich plasma. This process is **capillary reabsorption**.

Normally, there is a small net outward flow of fluid from capillaries, which lymph vessels return to blood. If high blood pressure causes too much fluid to flow out or something interferes with fluid return, interstitial fluid collects in tissues. The resulting swelling is called edema. Roundworm infections that damage lymph vessels also cause severe edema (Section 25.11).

Venous Pressure

Blood from several capillaries flows into each venule. These thin-walled vessels join together to form veins, the large-diameter, low-resistance transport tubes that carry blood to the heart. Many veins, especially in the legs, have flaplike valves that help prevent backflow (Figure 37.22). These valves automatically shut when blood in the vein starts to reverse direction.

Sometimes venous valves lose their elasticity. Then veins become enlarged and bulge near the surface of skin. This elasticity loss commonly occurs in veins of the legs; they become varicose veins. Failure of valves in veins around the anus causes hemorrhoids.

The vein wall can bulge quite a bit under pressure, much more so than an arterial wall. Thus, veins act as reservoirs for great volumes of blood. When you rest, they hold about 60 percent of the total blood volume.

During exercise, fluid pressure in veins increases, and less blood collects inside them. Veins have a bit of smooth muscle inside their wall, and exercise-induced signals from the nervous system make it contract. The contraction causes veins to stiffen so they cannot hold as much blood, and the pressure inside them rises. At the same time, skeletal muscles that move limbs bulge and press against veins, squeezing blood toward the heart (Figure 37.22b,c).

Exercise-induced deep breathing also raises venous pressure. As the chest expands, organs get squeezed and press against adjacent veins. The pressure assists in moving blood toward the heart.

37.9 Blood and Cardiovascular Disorders

■ High blood pressure and atherosclerosis increase the risk of both heart attack and stroke.

■ Links to Cholesterol 3.4, Sickle-cell anemia 3.6, 18.6, Hemophilia 12.4, Thalassemia 14.5, Diabetes 35.9

Red Blood Cell Disorders In anemias, red blood cells are few or compromised. As a result, oxygen delivery and metabolism falter. Shortness of breath, fatigue, and chills follow. Hemorrhagic anemias result from sudden blood loss, as from a wound; chronic anemias result from low red blood cell production or a slight but persistent blood loss.

Bacteria and protozoans that replicate in red blood cells cause some hemolytic anemias. The pathogens get into red cells, divide inside them, and then cause the cell to break apart and die. A diet with too little iron causes iron-deficiency anemia, in which red blood cells cannot make enough iron-containing heme. Sickle-cell anemia arises from a mutation that alters hemoglobin and allows cells to change shape (Section 3.6).

Beta-thalassemias occur when mutations disrupt or stop synthesis of a globin chains of hemoglobin (Section 14.5). Few red blood cells form and those that do are thin and fragile. Polycythemia is an excess of red blood cells. This increases oxygen delivery, but also makes blood more viscous and elevates blood pressure.

White Blood Cell Disorders Epstein–Barr virus can cause infectious mononucleosis. The virus infects B lymphocytes and the body produces large numbers of monocytes in response. Symptoms typically last weeks and include a sore throat, fatigue, muscle aches, and low-grade fever.

Leukemias are cancers that originate in cells of the bone marrow. They cause overproduction of abnormally formed white blood cells that do not function properly.

Lymphomas are cancers that originate from B or T lymphocytes. Division of the cancerous lymphocytes produces tumors in lymph nodes and other parts of the lymphatic system.

Clotting Disorders Too much or too little clotting can cause health problems. Hemophilia is a genetic disorder in which clotting is impaired (Section 12.4). Other disorders cause clots to form spontaneously inside a vessel. A clot that forms inside a vessel and stays put is called a thrombus. A clot that breaks loose and travels in blood is an embolus. Both clot types can block vessels and cause problems. For example, a stroke occurs when a vessel in the brain ruptures or gets blocked by an embolus. Either way, blood flow to brain cells is disrupted. A person who survives a stroke often has impairments caused by death of blood-starved brain cells.

Atherosclerosis In atherosclerosis, buildup of lipids in the arterial wall narrows the lumen, or space inside the vessel. As you may know, cholesterol plays a role in this "hardening of the arteries." The human body requires cholesterol to make cell membranes, myelin sheaths, bile salts, and steroid hormones (Section 3.4). The liver makes enough cholesterol to meet these needs, but more is absorbed from food in the gut. Genetics affects how different people's bodies deal with an excess of dietary cholesterol.

Most of the cholesterol dissolved in blood is bound to protein carriers. The complexes are known as low density lipoproteins, or LDLs, and most cells can take them up. A lesser amount is bound up in high density lipoproteins, or HDLs. Cells in the liver metabolize HDLs, using them in the formation of bile, which the liver secretes into the gut. Eventually, the bile leaves the body in the feces.

When the LDL level in blood rises, so does the risk of atherosclerosis. The first sign of trouble is a buildup of lipids in an artery's endothelium (Figure 37.23). Fibrous connective tissue forms over the entire mass. The mass, an atherosclerotic plaque, bulges into the vessel's interior, narrowing its diameter and slowing blood flow.

A hardened plaque can rupture an artery wall, thereby triggering clot formation. A heart attack occurs when a cardiac artery is completely blocked, most commonly by a clot. If the blockage is not removed fast, cardiac muscle cells die. Clot-dissolving drugs can restore blood flow if they are given within an hour of the onset of an attack, so a suspected heart attack should receive prompt attention.

In coronary bypass surgery, doctors open a person's chest and use a blood vessel from elsewhere in the body

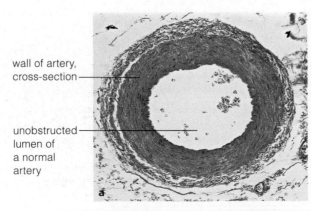
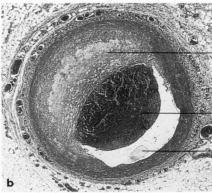

wall of artery, cross-section

unobstructed lumen of a normal artery

atherosclerotic plaque

blood clot sticking to plaque

narrowed lumen

Figure 37.23 Sections from (**a**) a normal artery and (**b**) an artery with a lumen narrowed by an atherosclerotic plaque. A clot clogged this one.

(usually a leg vein) to divert blood around the clogged coronary artery. (Figure 37.24). In laser angioplasty, laser beams vaporize plaques. In balloon angioplasty, doctors inflate a small balloon in a blocked artery to flatten the plaques. A wire mesh tube called a stent is then inserted to keep the vessel open.

Hypertension—A Silent Killer Hypertension refers to chronically high blood pressure (above 140/90). Often the cause is unknown. Heredity is a factor, and African Americans have an elevated risk. Diet also plays a role; in some people high salt intake causes water retention that raises blood pressure. Hypertension is sometimes described as a silent killer, because people often are unaware they have it. Hypertension makes the heart work harder than normal. This can cause the heart to enlarge and to function less efficiently. High blood pressure also increases risk of atherosclerosis. An estimated 180,000 Americans die each year as a result of hypertension.

Rhythms and Arrhythmias As you read in Section 37.6, the SA node controls the rhythmic beating of the heart. Electrocardiograms, or ECGs, record the electrical activity during the cardiac cycle (Figure 37.25a).

ECGs can reveal arrhythmias, which are abnormal heart rhythms (Figure 37.25b–d). Arrhythmias are not always dangerous. For example, endurance athletes commonly experience bradycardia, a below-average resting heart rate. Ongoing exercise has made their heart more efficient, and the nervous system has adjusted the firing rate of the cardiac pacemaker downward. Tachycardia, a faster than normal heart rate, can be caused by exercise, stress, or some underlying heart problem.

In atrial fibrillation, the atria do not contract normally. They quiver, which increases the risk of blood clots and stroke. Ventricular fibrillation is the most dangerous type of arrhythmia. It causes the ventricles to flutter, and their pumping action falters or stops. Blood flow halts, leading to loss of consciousness and death. A shock administered by a defibrillator such as the new AEDs mentioned in the chapter introduction can restore a heart's normal rhythm. It does so by resetting the heart's natural pacemaker, the SA node.

Risk Factors Cardiovascular disorders are the leading cause of death in the United States. Each year, they affect about 40 million people, and about 1 million die. Tobacco smoking tops the list of risk factors. Other factors include a family history of such disorders, hypertension, a high cholesterol level, diabetes mellitus, and obesity (Section 35.9). Age also is a factor. The older you get, the greater the risk of cardiovascular disorders. Physical inactivity, too, increases the risk. Regular exercise helps lower the risk of cardiovascular disorders even when the exercise is not particularly strenuous. Gender is another factor; until about age fifty, males are at greater risk.

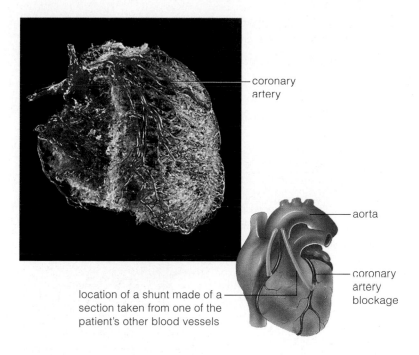

coronary artery

aorta

coronary artery blockage

location of a shunt made of a section taken from one of the patient's other blood vessels

Figure 37.24 The photo shows coronary arteries and other blood vessels that service the heart. Resins were injected into them. Then the cardiac tissues were dissolved to make an accurate, three-dimensional corrosion cast. The sketch shows two coronary bypasses (color-coded *green*), which extend from the aorta past two clogged parts of the coronary arteries

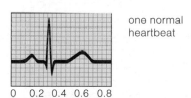

one normal heartbeat

0   0.2 0.4 0.6 0.8
a time (seconds)

Figure 37.25 (**a**) ECG of one normal beat of the human heart. (**b–d**) Recordings that identified three types of arrhythmias.

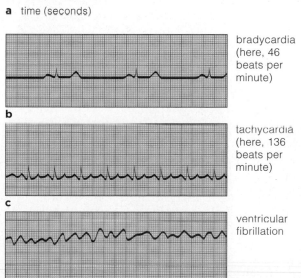

bradycardia (here, 46 beats per minute)

b

tachycardia (here, 136 beats per minute)

c

ventricular fibrillation

d

■ Vessels and organs of the lymphatic system interact closely with the circulatory system.

■ Link to Thymus gland 35.12

Lymph Vascular System

A portion of the lymphatic system, called the **lymph vascular system**, consists of vessels that collect water and solutes from interstitial fluid, then deliver them to the circulatory system. The lymph vascular system includes lymph capillaries and vessels (Figure 37.26). Fluid that moves through these vessels is the **lymph**.

The lymph vascular system serves three functions. First, its vessels are drainage channels for water and plasma proteins that leaked out of capillaries and that must be returned to the circulatory system. Second, it delivers fats absorbed from food in the small intestine to the blood. Third, it transports cellular debris, pathogens, and foreign cells to lymph nodes, which serve as disposal sites.

The lymph vascular system extends to capillary beds. There, excess fluid enters lymph capillaries. These capillaries have no obvious entrance;

Tonsils
Defense against bacteria and other foreign agents

Right Lymphatic Duct
Drains right upper portion of the body

Thymus Gland
Site where certain white blood cells acquire means to chemically recognize specific foreign invaders

Thoracic Duct
Drains most of the body

Spleen
Major site of antibody production; disposal site for old red blood cells and foreign debris; site of red blood cell formation in the embryo

Some Lymph Vessels
Return excess interstitial fluid and reclaimable solutes to the blood

Some Lymph Nodes
Filter bacteria and many other agents of disease from lymph

Bone Marrow
Marrow in some bones is production site for infection-fighting blood cells (as well as red blood cells and platelets)

a

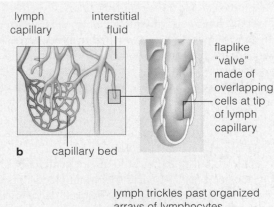

lymph capillary   interstitial fluid

flaplike "valve" made of overlapping cells at tip of lymph capillary

b   capillary bed

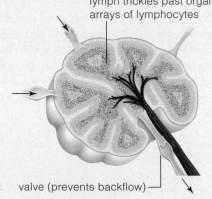

lymph trickles past organized arrays of lymphocytes

c   valve (prevents backflow)

Figure 37.26 Animated (**a**) Components of the human lymphatic system and their functions. Not shown are patches of lymphoid tissue in the small intestine and in the appendix. (**b**) Diagram of lymph capillaries at the start of a drainage network, the lymph vascular system. (**c**) Cutaway view of a lymph node. Its inner compartments are packed with organized arrays of infection-fighting white blood cells.

water and solutes move into clefts between cells. As you can see from Figure 37.26b, endothelial cells overlap, forming flaplike valves. Lymph capillaries merge into larger diameter lymph vessels, which have smooth muscle in their wall and valves that prevent backflow. Finally, lymph vessels converge onto collecting ducts, which drain into veins in the lower neck.

Lymphoid Organs and Tissues

The other portion of the lymphatic system has roles in the body's defense responses to injury and attack. It includes the lymph nodes, tonsils, adenoids, spleen, and thymus, as well as some patches of tissue in the wall of the small intestine and appendix.

Lymph nodes are located at intervals along lymph vessels (Figure 37.26c). Lymph filters through at least one node before it enters the blood. Large numbers of lymphocytes (B and T cells) that formed in the bone marrow take up stations inside the nodes. When they identify pathogens in the lymph, they sound the alarm that summons up an immune response, as described in detail in the next chapter.

Tonsils are two patches of lymphoid tissue at the back of the throat. Adenoids are similar tissue clumps at the rear of the nasal cavity. Tonsils and adenoids help the body respond fast to inhaled pathogens.

The **spleen** is the largest lymphoid organ, about the size of a fist in an average adult. In embryos only, it functions as a site of red blood cell formation. After birth, the spleen filters pathogens, worn-out red blood cells and platelets from the many blood vessels that branch through it. The spleen has white blood cells that engulf and digest pathogens and altered body cells. It also holds antibody-producing B cells. People can survive removal of their spleen, but they become more vulnerable to infections.

In the **thymus gland**, T lymphocytes differentiate and become capable of recognizing and responding to particular pathogens. The thymus gland also makes the hormones that influence these actions. It is central to immunity, the focus of the next chapter.

Take-Home Message

What are the functions of the lymphatic system?

■ The lymph vascular system consists of tubes that collect and deliver excess water and solutes from interstitial fluid to blood. It also carries absorbed fats to the blood, and delivers disease agents to lymph nodes.

■ The system's lymphoid organs, including lymph nodes, have specific roles in body defenses.

Summary

Section 37.1 A **circulatory system** moves substances to and from interstitial fluid faster than diffusion alone could move them. **Interstitial fluid** fills spaces between cells. It exchanges substances with cells and with **blood**, a fluid transport medium.

Some invertebrates have an **open circulatory system**, in which blood spends part of the time mingling with tissue fluids. In vertebrates, a **closed circulatory system** confines blood inside a **heart**, a type of muscular pump, and blood vessels, the smallest of which are **capillaries**.

As lungs took on added importance in vertebrates on land, the circulatory system also evolved, making gas exchange more efficient. In birds and mammals, the heart has four chambers, so blood travels in two fully separated circuits. The **systemic circuit** carries blood from the heart to body tissues, then returns it to the heart. Blood in the **pulmonary circuit** moves from the heart to the lungs, then back to the heart.

■ *Use the animation on CengageNOW to compare animal circulatory systems.*

Sections 37.2, 37.3 Blood is a fluid connective tissue that consists of plasma, blood cells, and platelets. **Plasma** is mostly water in which diverse ions and molecules are dissolved. **Red blood cells**, or erythrocytes, contain the hemoglobin that functions in rapid transport of oxygen and, to a lesser extent, carbon dioxide. They do not have a nucleus when mature. A variety of **white blood cells**, or leukocytes, have roles in day-to-day tissue maintenance and repair and in defenses against pathogens. Cell fragments called **platelets** interact with blood cells and plasma proteins in **hemostasis** after a vessel is damaged. Platelets and all blood cells arise from stem cells in bone marrow. A **cell count** is the number of blood cells of a specific type in a given volume.

Section 37.4 Among the molecules on the surface of red blood cells are glycolipids and proteins that can be used to type an individual's blood. The body mounts an attack against any cells that bear unfamiliar molecules, causing **agglutination**, or a clumping of cells. **ABO blood typing** helps match the blood of donors and recipients to avoid blood transfusion problems. **Rh blood typing** and the appropriate treatment prevent problems that can arise when maternal and fetal Rh blood types differ.

■ *Use the animation on CengageNOW to learn about blood types and blood transfusions.*

Section 37.5 The human heart is a four-chambered muscular pump, the contraction of which forces blood through two separate circuits. In the pulmonary circuit, oxygen-poor blood from the heart's right half flows to the lungs, picks up oxygen, then flows to the heart's left half. In the systemic circuit, the oxygen-rich blood flows from the heart's left half, out the **aorta**, and to body tissues. Oxygen-poor blood returns to the heart's right half.

Traditional CPR alternates blowing into a person's mouth to inflate their lungs with chest compressions. The requirement for mouth-to-mouth contact makes many people reluctant to use this method on strangers. A new method called CCR (cardiocerebral resuscitation) relies on chest compressions alone. This method may be as good as or even better than traditional CPR as treatment for most people who have sudden cardiac arrest or a heart attack.

How would you vote?

Knowledge of CPR can save lives. Should high schools require students to learn CPR? See CengageNOW for details, then vote online.

Most blood flows through only one capillary system, but blood in intestinal capillaries will later flow through liver capillaries. The liver metabolizes or stores nutrients and neutralizes some bloodborne toxins.

■ *Use the animation on CengageNOW to explore the human cardiovascular system.*

Section 37.6 A human heart is a double pump that consists mainly of **cardiac muscle**. It is partitioned into two halves, each with two chambers: an **atrium** that receives blood and a **ventricle** that expels it.

During one **cardiac cycle**, all heart chambers undergo rhythmic relaxation (diastole) and contraction (systole). When a cycle starts, each atrium expands as blood fills it. Both ventricles already are filling as the atria contract. When ventricles contract, they force blood into the aorta and pulmonary arteries. Ventricular contraction provides the force that powers movement of blood through blood vessels. Atrial contraction simply fills the ventricles.

A **cardiac conduction system** produces and distributes electrical signal that cause the heart's beating. It consists of an SA node in the right atrium that is functionally linked by conducting fibers to an AV node.

The SA node, the **cardiac pacemaker**, spontaneously generates the action potentials that set the pace for cardiac contractions. The nervous system does not initiate heartbeats; it only adjusts their rate and strength. Waves of excitation wash over the heart's atria, down fibers in its septum, then up the walls of the ventricles.

■ *Use the animation on CengageNOW to learn about the structure and function of the human heart.*

Section 37.7 **Blood pressure** varies in the circulatory system. It is highest in contracting ventricles. It declines as blood travels through **arteries, arterioles, capillaries, venules,** and **veins** of the systemic or pulmonary circuit. It is lowest in relaxed atria. The speed of flow depends on heartbeat strength and rate, and on resistance to flow in the blood vessels. Adjusting the diameter of arterioles that supply different parts of the body redistributes the blood volume as necessary. In any interval, when a tissue needs more blood, the arterioles that supply it widen, allowing increased blood flow.

■ *Use the animation on CengageNOW to see how blood pressure is measured.*

Section 37.8 Substances move between the blood and interstitial fluid at capillary beds. **Ultrafiltration** pushes

a small amount of fluid out of capillaries. Fluid moves back in by **capillary reabsorption**. Normally, inward and outward directed forces are nearly balanced, but there is a small net outward flow from a capillary bed.

Several capillaries drain into each venule. Veins are transport vessels that serve as a blood volume reservoir where the flow volume back to the heart is adjusted.

Section 37.9 In a blood disorder, an individual has too many, too few, or abnormal red or white blood cells. Formation of blood clots inside vessels can cause health problems. Common circulatory disorders include atherosclerosis, hypertension (chronic high blood pressure), heart attacks, strokes, and certain arrhythmias. Regular exercise, maintaining normal body weight, and not smoking lower risk for these disorders.

Section 37.10 Some fluid that leaves capillaries enters the **lymph vascular system**. The fluid, now called **lymph**, is filtered by **lymph nodes**. White blood cells in the nodes attack any pathogens. The spleen and thymus are organs of the lymphatic system. The **spleen** filters the blood and removes any old red blood cells. The **thymus gland** produces hormones and is the site where T lymphocytes (a kind of white blood cell) mature.

■ *Learn about the human lymphatic system with the animation on CengageNOW.*

Self-Quiz   *Answers in Appendix III*

1. The velocity of blood flow _____ when blood enters capillaries.
     a. increases   b. decreases   c. stays the same

2. All vertebrates have _____ .
     a. an open circulatory system
     b. a closed circulatory system
     c. a four-chambered heart
     d. both b and c

3. Which are not found in the blood?
     a. plasma
     b. blood cells and platelets
     c. gases and dissolved substances
     d. All of the above are found in blood.

4. A person who has type O blood _____ .
     a. can receive a transfusion of blood of any type
     b. can donate blood to a person of any blood type
     c. can donate blood only to a person of type O
     d. cannot be a blood donor

Data Analysis Exercise

Risk of death by stroke is not distributed evenly across the United States. Epidemiologists refer to a swath of states in the Southeast as the "stroke belt" because of the increased incidence of stroke deaths there. By one hypothesis, the high rate of deaths from stroke in this region results largely from a relative lack of access to immediate medical care. Compared to other parts of the country, more stroke-belt residents live in rural settings with few medical services.

Figure 37.27 compares the rate of stroke deaths in stroke-belt states (Alabama, Arkansas, Georgia, Mississippi, North Carolina, South Carolina, and Tennessee) with that of New York State. It also breaks down the death risk in each region by ethnic group and by sex.

1. How does the rate of stroke deaths among blacks living in the stroke-belt compare with whites in the same region?

2. How does the rate of stroke deaths among blacks living in New York compare with whites in the same region?

3. Which group has the higher rate of stroke deaths, blacks living in New York, or whites living in the stroke belt?

4. Do these data support the hypothesis that poor access to care causes the high rate of death by stroke in the stroke belt?

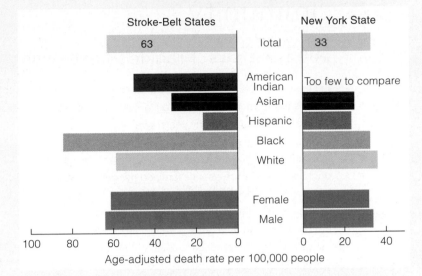

Figure 37.27 Comparison of the age-adjusted rate of deaths by stroke in Southeastern "stroke-belt" states and in New York State. Source: National Vital Statistics System—Mortality (NVSS-M), NCHS, CDC.

5. In the blood, most oxygen is transported _____ .
a. in red blood cells
b. in white blood cells
c. bound to hemoglobin
d. both a and c

6. Which has a more muscular wall? _____ .
a. right atrium
b. left ventricle

7. Blood flows directly from the left atrium to _____ .
a. the aorta
b. the left ventricle
c. the right atrium
d. the pulmonary arteries

8. All blood cells descend from stem cells in _____ .
a. the spleen
b. the left ventricle
c. the right atrium
d. bone marrow

9. Contraction of _____ drives the flow of blood through the aorta and pulmonary arteries.
a. atria
b. arterioles
c. ventricles
d. skeletal muscle

10. Blood pressure is highest in the _____ and lowest in the _____ .
a. arteries; veins
b. arterioles; venules
c. veins; arteries
d. capillaries; arterioles

11. At rest, the largest volume of blood is in _____ .
a. arteries
b. capillaries
c. veins
d. arterioles

12. At the start of a capillary bed (closest to arterioles), ultrafiltration moves _____ .
a. proteins into the capillary
b. interstitial fluid into the capillary
c. proteins into the interstitial fluid
d. water, ions, and small solutes into interstitial fluid

13. Which is not a function of the lymphatic system?
a. filters out pathogens
b. returns fluid to the circulatory system
c. helps certain white blood cells mature
d. distributes oxygen to the tissues

14. Match the components with their functions.
____ capillary bed
____ lymph node
____ blood
____ ventricle
____ SA node
____ veins
____ aorta
a. filters out pathogens
b. cardiac pacemaker
c. main blood volume reservoir
d. largest artery
e. fluid connective tissue
f. zone of diffusion
g. contractions drive blood circulation

■ *Visit CengageNOW for additional questions.*

Critical Thinking

1. The highly publicized deaths of a few airline travelers led to warnings about economy-class syndrome. The idea is that sitting motionless for long periods on flights allows blood to pool and clots to form in legs. More recent studies suggest that long-distance flights cause problems in about 1 percent of air travelers, and that the risk is the same regardless of whether a person is in a first-class seat or an economy seat. Physicians suggest that air travelers drink plenty of fluids and periodically get up and walk around the cabin. Given what you know about blood flow in the veins, explain why these precautions can lower the risk of clot formation.

2. Mitochondria occupy about 40 percent of the volume of human cardiac muscle but only about 12 percent of the volume of skeletal muscle. Explain this difference.

3. In some people the valve between an atrium and ventricle does not close properly. This condition can be diagnosed by listening carefully to the heart. The listener will hear a whooshing sound called a murmur when the ventricle of the affected chamber contracts. What causes this sound?

IMPACTS, ISSUES Frankie's Last Wish

In October of 2000, Frankie McCullough had known for a few months that something was not quite right. She hadn't had an annual checkup in many years; after all, she was only 31 and had been healthy her whole life. It never occurred to her to doubt her own invincibility until the moment she saw the doctor's face change as he examined her cervix. Frankie had cervical cancer.

The cervix is the lowest part of the uterus, or womb. Epithelial or endocrine cells of the cervix can become cancerous, but the process is usually slow. The cells pass through several precancerous stages that are detectable by routine Pap tests (Figure 38.1). Precancerous and even early-stage cancerous cells can be removed from the cervix before they spread to other parts of the body. However, plenty of women like Frankie do not take advantage of regular exams. Those who end up at the gynecologist's office with pain or bleeding may be experiencing symptoms of advanced cervical cancer, the treatment of which offers only about a 9% chance of survival. About 3,600 women die of cervical cancer each year in the United States; many more than that die in places where routine gynecological testing is not common.

What causes cancer? At least in the case of cervical cancer, we know the answer to that question: Healthy cervical cells are transformed into cancerous ones by infection with human papillomavirus (HPV). HPV is a DNA virus that infects skin and mucous membranes. There are about 100 different types of HPV; a few cause warts on the hands or feet, or in the mouth. About 30 others that infect the genital area some-times cause genital warts, but usually there are no symptoms of infection. Genital HPV is spread very easily by sexual contact. At least 80% of women have been infected by age 50.

A genital HPV infection usually goes away on its own, but not always. A persistent infection with one of about 10 strains is the main risk factor for cervical cancer. Types 16 and 18 are particularly dangerous: One of the two is found in more than 70% of all cervical cancers. In 2006, the FDA approved Gardasil, a vaccine against four types of genital HPV, including 16 and 18. The vaccine prevents cervical cancer caused by these HPV strains. It is most effective in girls who have not yet become sexually active, because they are least likely to have become infected with any of those four strains of HPV.

The HPV vaccine came too late for Frankie McCullough. Despite radiation treatments and chemotherapy, her cervical cancer spread quickly. She died on September 16, 2001, leaving a wish for other young women: awareness. "If there is one thing I could tell a young woman to convince them to have a yearly exam, it would be not to assume that your youth will protect you. Cancer does not discriminate; it will attack at random, and early detection is the answer." She was right; almost all of the women with newly diagnosed invasive cervical cancer have not had a Pap test in at least five years, and many of them have never had one.

Pap tests, HPV vaccines, and all other medical tests and treatments are direct benefits of our increasing understanding of the interplay of the human body with its pathogens, an interaction that we call immunity.

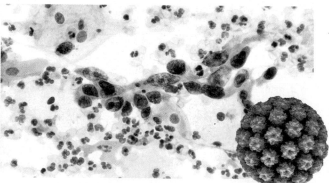

See the video! Figure 38.1 HPV and cervical cancer. *Left*, Frankie McCullough (waving) died of cervical cancer in 2001. *Above*, a Pap test reveals cancer cells (with enlarged, irregularly shaped nuclei) among normal squamous epithelial cells of the cervix. Cells with multiple nuclei are indicative of HPV infection. The *orange ball* is a model of an HPV16 virus.

Key Concepts

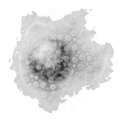

Overview of body defenses

The vertebrate body has three lines of immune defenses. Surface barriers prevent invasion by ever-present pathogens; general innate responses rid the body of most pathogens; adaptive responses specifically target pathogens and cancer cells. **Section 38.1**

Surface barriers

Skin, mucous membranes, and secretions at the body's surfaces function as barriers that exclude most microbes. **Sections 38.2, 38.3**

Innate immunity

Innate immune responses involve a set of general, immediate defenses against invading pathogens. Innate immunity includes phagocytic white blood cells, plasma proteins, inflammation, and fever. **Section 38.4**

Adaptive immunity

In an adaptive immune response, white blood cells destroy specific pathogens or altered cells. Some make antibodies in an antibody-mediated immune response; others destroy ailing body cells in a cell-mediated response. **Sections 38.5–38.8**

Immunity in our lives

Vaccines are an important part of any health program. Failed or faulty immune mechanisms can result in allergies, immune deficiencies, or autoimmune disorders. The immune system itself is a target of human immunodeficiency virus (HIV). **Sections 38.9–38.12**

Links to Earlier Concepts

- In this chapter you will be integrating what you have learned about disease-causing agents and their hosts (Section 21.8). You will apply your knowledge of prokaryotic cells and viruses (4.4, 4.5, 16.1, 21.1, 21.2) as you learn about their interactions with eukaryotic cells.

- You will revisit what you know about protein structure (3.5), the endomembrane system (4.9), membrane proteins (5.2), endocytosis and phagocytosis (5.5), osmosis (5.6), fever (6.3), alternative splicing (14.3), cell junctions (32.1), and apoptosis (27.6) to understand the immune defenses of vertebrates.

- This chapter has several examples of what happens when pathogens invade internal environments (27.1), including the human nervous system (33.13), joints (36.5), and the cardiovascular system (37.9).

- Earlier sections on cell signaling (27.3, 33.6, 35.1) gave you the background to understand immune signaling mechanisms. You will see how body systems, including exocrine glands (32.2), skin (32.7), the circulatory system (37.2, 37.8), and the lymphatic system (37.10) work together to fight infection.

How would you vote? Clinical trials of some vaccines take place in underdeveloped countries that have fewer regulations governing human testing than the United States. Should clinical trials be held to the same ethical standards no matter where they take place? See CengageNOW for details, then vote online.

Integrated Responses to Threats

- In vertebrates, the innate and adaptive immune systems work together to combat infection and injury.

- Links to Phagocytosis 5.5, Coevolution of pathogens and hosts 21.8, Neuropeptides 33.6, White blood cells 37.2

Evolution of the Body's Defenses

Humans continually cross paths with a tremendous array of viruses, bacteria, fungi, parasitic worms, and other pathogens, but you need not lose sleep over this. Humans coevolved with these pathogens, so you have defenses that protect your body from them. **Immunity**, an organism's capacity to resist and combat infection, began well before multicelled eukaryotes evolved from free-living cells. Mutations in membrane protein genes introduced new patterns in the proteins, patterns that were unique in cells of a given type. As multicellularity evolved, so did mechanisms of identifying the patterns as self, or belonging to one's own body.

By 1 billion years ago, nonself recognition had also evolved. Cells of all modern multicelled eukaryotes

Figure 38.2 A physical barrier to infection: mucus and the mechanical action of cilia keep pathogens from getting a foothold in the airways to the lungs. Bacteria and other particles get stuck in mucus secreted by goblet cells (*gold*). Cilia (*pink*) on other cells sweep the mucus toward the throat for disposal.

Table 38.1 Innate and Adaptive Immunity Compared

	Innate Immunity	Adaptive Immunity
Response time	Immediate	About a week
How antigen is detected	Fixed set of receptors for molecular patterns found on pathogens	Random recombinations of gene sequences generates billions of receptors
Specificity of response	None	Specific antigens targeted
Persistence	None	Long-term

Table 38.2 Some Chemical Weapons in Immunity

Substance	Functions
Complement	Direct cell lysis; enhancement of lymphocyte responses
Cytokines	Cell-to-cell and cell–tissue communication:
Interleukins	Inflammation, T cell and B cell proliferation and differentiation, bone marrow stem cell stimulation, neutrophil chemotaxis, NK cell activation, fever
Interferons	Resistance to virus infection, NK cell activation
TNFs	Inflammation; tumor cell destruction
Other chemicals	
(enzymes, peptides, clotting factors, toxins, hormones, protease inhibitors)	Antimicrobial activities, cell lysis, complement activation and binding, coagulation, signaling, other diverse functions

bear a set of receptors that collectively can recognize around 1,000 different nonself cues, which are called pathogen-associated molecular patterns (PAMPs). As their name suggests, PAMPs occur mainly on or in pathogens. They include some components of prokaryotic cell walls, bacterial flagellum and pilus proteins, double-stranded RNA unique to some viruses, and so on. When a cell's receptors bind to a PAMP, they trigger a set of immediate, general defense responses. In mammals, for example, binding triggers activation of complement. **Complement** is a set of proteins that circulate in inactive form throughout the body. Activated complement can destroy microorganisms or flag them for phagocytosis (Section 5.5).

Pattern receptors and the responses they initiate are part of **innate immunity**, a set of fast, general defenses against infection. All multicelled organisms start out life with these defenses, which do not change within the individual's lifetime.

Vertebrates have another set of defenses carried out by interacting cells, tissues, and proteins. This **adaptive immunity** tailors immune defenses to a vast array of specific pathogens that an individual may encounter during its lifetime. It is triggered by **antigen**: a PAMP or any other molecule or particle recognized by the body as nonself. Most antigens are polysaccharides, lipids, and proteins typically present on viruses, bacteria or other foreign cells, tumor cells, toxins, and allergens.

Three Lines of Defense

The mechanisms of adaptive immunity evolved within the context of innate immunity. The two systems were once thought to operate independently of each other, but we now know they function together.

We describe both systems together in terms of three lines of defense. The first line comprises the physical, chemical, and mechanical barriers that keep pathogens on the outside of the body (Figure 38.2). Innate immunity, the second line of defense, begins after tissue is damaged, or after a PAMP is detected inside the body. Its general response mechanisms rid the body of many different kinds of invaders before populations of them become established in the internal environment.

Activation of innate immunity triggers the third line of defense, adaptive immunity. White blood cells form huge populations that target a specific antigen and destroy anything bearing it. Some of the cells persist after infection ends. If the same antigen returns, these memory cells mount a secondary response. Adaptive immunity can specifically target billions of antigens. Table 38.1 compares innate and adaptive immunity.

The Defenders

White blood cells (Figure 38.3) carry out all immune responses. Many kinds circulate through the body in blood and lymph; others populate the lymph nodes, spleen, and other tissues. Some white blood cells are phagocytic; all are secretory. Their secretions include cell-to-cell signaling molecules called **cytokines**. These peptides and proteins coordinate all aspects of immunity. Vertebrate cytokines include interleukins, interferons, and tumor necrosis factors (Table 38.2).

Different types of white blood cells are specialized for specific tasks, such as phagocytosis. **Neutrophils** are the most abundant of the circulating phagocytes. **Macrophages** that patrol tissue fluids are mature monocytes, which patrol the blood. **Dendritic cells** alert the adaptive immune system to the presence of antigen.

Some white blood cells contain secretory vesicles: granules that hold cytokines, enzymes, or pathogen-busting toxins. **Eosinophils** target parasites too big for phagocytosis. **Basophils** circulating in blood and **mast cells** anchored in tissues secrete substances contained by their granules in response to injury or antigen. Often associated with nerves, mast cells also respond to neuropeptides (Section 33.6), so they link the nervous and immune systems.

Lymphocytes are a special category of white blood cells that are central to adaptive immunity. **B** and **T lymphocytes** (B and T cells) have the capacity to collectively recognize billions of specific antigens. There are several kinds of T cells, including some that target infected or cancerous body cells. **Natural killer cells** (NK cells) can destroy infected or cancerous body cells that are undetectable by cytotoxic T cells.

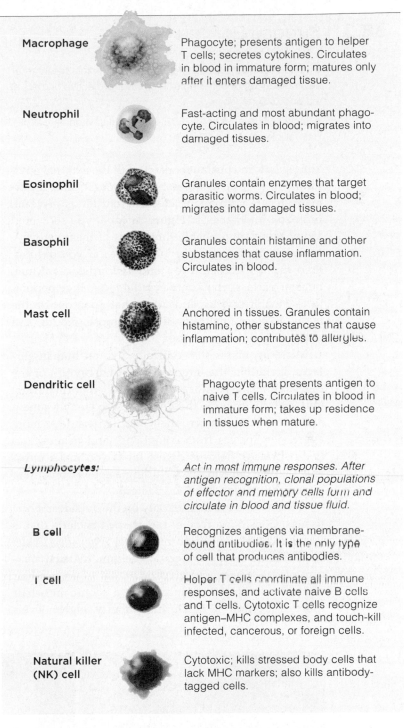

Macrophage	Phagocyte; presents antigen to helper T cells; secretes cytokines. Circulates in blood in immature form; matures only after it enters damaged tissue.
Neutrophil	Fast-acting and most abundant phagocyte. Circulates in blood; migrates into damaged tissues.
Eosinophil	Granules contain enzymes that target parasitic worms. Circulates in blood; migrates into damaged tissues.
Basophil	Granules contain histamine and other substances that cause inflammation. Circulates in blood.
Mast cell	Anchored in tissues. Granules contain histamine, other substances that cause inflammation; contributes to allergies.
Dendritic cell	Phagocyte that presents antigen to naive T cells. Circulates in blood in immature form; takes up residence in tissues when mature.
Lymphocytes:	*Act in most immune responses. After antigen recognition, clonal populations of effector and memory cells form and circulate in blood and tissue fluid.*
B cell	Recognizes antigens via membrane-bound antibodies. It is the only type of cell that produces antibodies.
T cell	Helper T cells coordinate all immune responses, and activate naive B cells and T cells. Cytotoxic T cells recognize antigen–MHC complexes, and touch-kill infected, cancerous, or foreign cells.
Natural killer (NK) cell	Cytotoxic; kills stressed body cells that lack MHC markers; also kills antibody-tagged cells.

Figure 38.3 White blood cells (leukocytes). Staining shows details such as cytoplasmic granules that contain enzymes, toxins, and signaling molecules.

Take-Home Message

What is immunity?

■ The innate immune system is a set of general defenses against a fixed number of antigens. It acts immediately to prevent infection.

■ Vertebrate adaptive immunity is a system of defenses that can specifically target billions of different antigens.

■ White blood cells are central to both systems; signaling molecules such as cytokines integrate their activities.

■ A pathogen can cause infection only if it enters the internal environment by penetrating skin or other protective barriers at the body's surfaces.

■ Links to Bacterial cell walls 4.4, Internal environment 27.1, Hair follicles and skin 32.7

Your skin is in constant contact with the external environment, so it picks up many microorganisms. It normally teems with about 200 different kinds of yeast, protozoa, and bacteria (Figure 38.4a). If you showered today, there are probably thousands of them on every square inch of your external surfaces. If you did not, there may be billions. They tend to flourish in warmer, moister parts, such as between the toes. Huge populations inhabit cavities and tubes that open out on the body's surface, including the eyes, nose, mouth, and anal and genital openings.

Microorganisms that typically live on human surfaces, including the interior tubes and cavities of the digestive and respiratory tracts, are called **normal flora**. Our surfaces provide them with a stable environment and nutrients. In return, their populations deter more aggressive species from colonizing (and penetrating) body surfaces; help us digest food; and make nutrients that we depend on, including a cobalt-containing vitamin (B_{12}) made only by bacteria.

Normal flora are helpful only on the outside of body tissues. Consider a type of rod-shaped bacteria that is a major constituent of normal flora, *Propionibacterium acnes* (Figure 38.4b). It feeds on sebum, a greasy mixture of fats, waxes and glycerides that lubricates hair and skin. Sebaceous glands secrete sebum into hair follicles (Section 32.7). During puberty, higher levels

Table 38.3	Vertebrate Surface Barriers
Physical	Intact skin and epithelia that line tubes and cavities such as the gut and eye sockets; established populations of normal flora
Mechanical	Mucus; broomlike action of cilia; flushing action of tears, saliva, urination, diarrhea
Chemical	Secretions (sebum, other waxy coatings); low pH of urine, gastric juices, urinary and vaginal tracts; lysozyme

of steroid hormones trigger sebaceous glands to make more sebum than before. Excess sebum combines with dead, shed skin cells and so blocks the openings of hair follicles. *P. acnes* can survive on the surface of the skin, but far prefer anaerobic habitats such as the interior of blocked hair follicles. There, they multiply to tremendous numbers. Secretions of the flourishing *P. acnes* populations leak into internal tissues, attracting neutrophils that initiate inflammation in the tissue around the follicles. The resulting pustules are called acne.

Normal flora can cause serious illness if they invade tissues. The bacterial agent of tetanus, *Clostridium tetani*, passes through our intestines so often that we consider it a normal inhabitant. The bacteria responsible for diphtheria, *Corynebacterium diphtheriae*, was normal skin flora before widespread use of the vaccine eradicated the disease. *Staphylococcus aureus*, a resident of human skin, nasal membranes, and intestines, is also a leading cause of human bacterial disease (Figure 38.4c). Normal flora cause or worsen pneumonia; ulcers; colitis; whooping cough; meningitis; abscesses of the lung and brain; and colon, stomach, and intestinal cancers.

 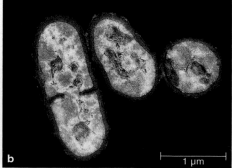 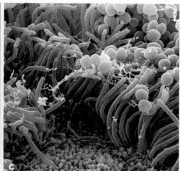

Figure 38.4 Some microbial inhabitants of human surfaces. (**a**) *Staphylococcus epidermidis*, the most common colonizer of human skin. (**b**) *Propionibacterium acnes*, the bacterial cause of acne.

(**c**) *Staphylococcus aureus* cells (*yellow*) adhering to mucus-coated cilia of human nasal epithelial cells. *S. aureus* is a common inhabitant of human skin and linings of the mouth, nose, throat, and intestines. It is also the leading cause of bacterial disease in humans. Antibiotic-resistant strains of *S. aureus* are now widespread. A particularly dangerous kind (MRSA) that is resistant to all penicillins is now endemic in most hospitals around the world. MRSA is called a "superbug."

skin surface

epithelial
cells die and
become filled
with keratin
as they are
pushed toward
skin surface

epidermis

dividing
epithelial
cells

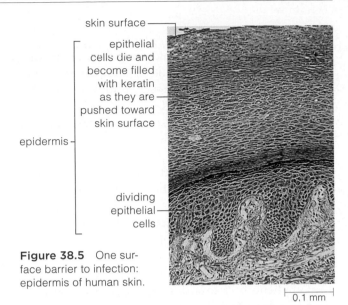

Figure 38.5 One surface barrier to infection: epidermis of human skin.

├─ 0.1 mm ─┤

In contrast to body surfaces, the blood and tissue fluids of healthy people are typically microorganism-free. Physical, chemical, and mechanical barriers keep microorganisms on the outside of body tissues (Table 38.3). For example, healthy, intact skin is an effective physical barrier. Vertebrate skin has a tough outer layer (Figure 38.5). Microorganisms flourish on this waterproof, oily surface, but rarely penetrate it.

Sticky mucus that coats the surfaces of many epithelial linings can trap microorganisms. Broomlike cilia on cells of the linings sweep the trapped microorganisms toward the outside of the body (Figure 38.4c). Mucus also contains **lysozyme**, an enzyme that chops up the polysaccharides in bacterial cell walls and so unravels their structure. Lysozyme ensures that bacteria stuck in the mucus do not survive long enough to breach the walls of the sinuses and lower respiratory tract.

Normal flora in the mouth resist lysozyme in saliva. Most microorganisms that enter the stomach are killed by gastric fluid, a potent brew of protein-digesting enzymes and acid. Most of those that survive to reach the small intestine are killed by bile salts. The hardy ones that make it to the large intestine must compete with about 500 resident species. Any that displace normal flora there are typically flushed out by diarrhea.

Lactic acid produced by *Lactobacillus* helps keep the vaginal pH outside the range of tolerance of most fungi and other bacteria. Urination's flushing action usually stops pathogens from colonizing the urinary tract.

Take-Home Message

What prevents ever-present microorganisms from entering the body's internal environment?

■ Surface barriers keep microorganisms that contact or inhabit vertebrate surfaces from invading the internal environment.

38.3 | Remember to Floss

■ Nine of every ten cardiovascular disease patients have serious periodontal disease. There is a connection.

■ Links to Biofilms 4.5, Cell junctions 32.1, Cardiovascular disease 37.9

Your mouth is a particularly inviting habitat for microorganisms, offering plenty of nutrients, warmth, moisture, and surfaces for colonization. Accordingly, it harbors huge populations of various species of *Streptococcus*, *Lactobacillus*, *Staphylococcus*, and other bacteria.

A few of the 400 or so species of microorganisms that normally live in the mouth cause dental **plaque**, a thick biofilm of various bacteria and occasional archaea, their extracellular products, and saliva glycoproteins. Plaque sticks tenaciously to teeth (Figure 38.6). Some bacteria that live in it are fermenters. They break down bits of carbohydrate that stick to teeth and then secrete organic acids, which etch away tooth enamel and make cavities.

In young, healthy people, tight junctions (Section 32.1) between the gum epithelium and teeth form a barrier that keeps oral microorganisms out of the internal environment. As we age, the connective tissue beneath gum epithelium thins, and the barrier becomes vulnerable. Deep pockets form between the teeth and gums, and a very nasty gang of anaerobic bacteria and archaea accumulates in these pockets. Their noxious secretions, including destructive enzymes and acids, cause inflammation of surrounding gum tissues—a condition called periodontitis.

Porphyromonas gingivalis is one of those anaerobic species. Along with every other species of oral bacteria associated with periodontitis, *P. gingivalis* also occurs in atherosclerotic plaque (Section 37.9). Periodontal wounds are an open door to the circulatory system and its arteries.

Atherosclerosis is now known to be a disease of inflammation. Macrophages and T cells are attracted to lipid deposits in the vessel walls. Their secretions initiate inflammation that further attracts lipids, and the lesion grows as the immune cells die and become part of the deposits. What role the oral microorganisms play in this scenario is not yet clear, but one thing is certain—they contribute to the inflammation that fuels coronary artery disease.

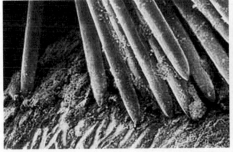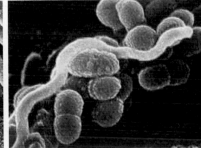

Figure 38.6 Plaque. *Left*, micrograph of toothbrush bristles scrubbing plaque on a tooth surface. *Right*, the main cause of plaque, *Streptococcus mutans*.

■ Innate immune mechanisms nonspecifically protect animals from pathogens that invade internal tissues.

■ Links to Osmosis 5.6, Fever 6.3, Lysis 21.2, Effectors 27.3, Prostaglandins 35.1, Blood 37.2, Capillary function 37.8

What happens if a pathogen slips by surface defenses and enters the body's internal environment? All animals are normally born with a set of fast-acting, off-the-shelf immune defenses that can keep an invading pathogen from establishing a population in the body's internal environment. These innate immune defenses include phagocyte and complement action, inflammation, and fever—all general mechanisms that normally do not change much over an individual's lifetime.

Phagocytes and Complement Macrophages are large phagocytes that engulf and digest essentially everything except undamaged body cells. They patrol the interstitial fluid, so they are often the first white blood cells to encounter an invading pathogen. When receptors on a macrophage bind to antigen, the cell begins to secrete cytokines. These signaling molecules attract more macrophages, neutrophils, and dendritic cells to the site of invasion.

Antigen also triggers complement activation (Figure 38.7a,b). In vertebrates, about 30 different types of complement protein circulate in inactive form throughout the blood and interstitial fluid. Some become activated when they encounter antigen, or an antibody bound to antigen (we will return to antibodies in Section 38.6). The activated complement proteins are enzymes that cut other inactive complement proteins, which thereby become activated and cut other inactive complement proteins, and so on. These cascading reactions quickly produce tremendous concentrations of activated complement localized at the site of invasion.

Activated complement attracts phagocytic cells. Like snuffling bloodhounds, these cells can follow complement gradients back to an affected tissue. Some complement proteins attach directly to pathogens. Phagocytes have complement receptors, so a pathogen coated with complement is recognized and engulfed faster than an uncoated pathogen. Other activated complement proteins self-assemble into complexes that puncture bacterial cell walls or plasma membranes (Figure 38.7c–e).

Activated complement proteins also work in adaptive immunity, by guiding the maturation of immune cells and mediating some interactions among them.

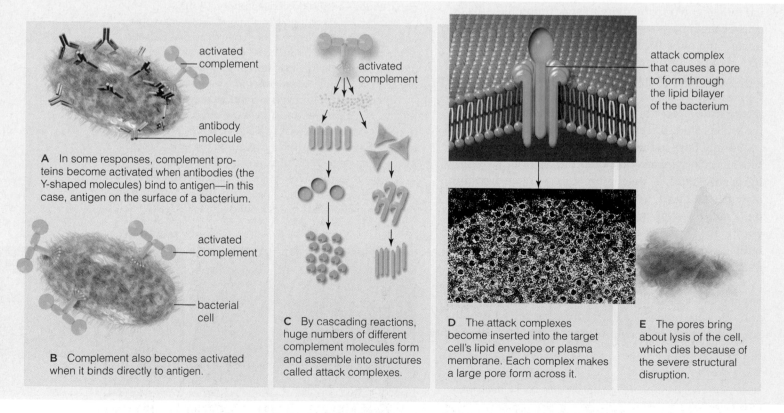

A In some responses, complement proteins become activated when antibodies (the Y-shaped molecules) bind to antigen—in this case, antigen on the surface of a bacterium.

activated complement

antibody molecule

activated complement

bacterial cell

B Complement also becomes activated when it binds directly to antigen.

activated complement

C By cascading reactions, huge numbers of different complement molecules form and assemble into structures called attack complexes.

attack complex that causes a pore to form through the lipid bilayer of the bacterium

D The attack complexes become inserted into the target cell's lipid envelope or plasma membrane. Each complex makes a large pore form across it.

E The pores bring about lysis of the cell, which dies because of the severe structural disruption.

Figure 38.7 Animated One effect of complement protein activation. Activation causes lysis-inducing pore complexes to form. The micrograph shows holes in a pathogen's surface that were made by membrane attack complexes.

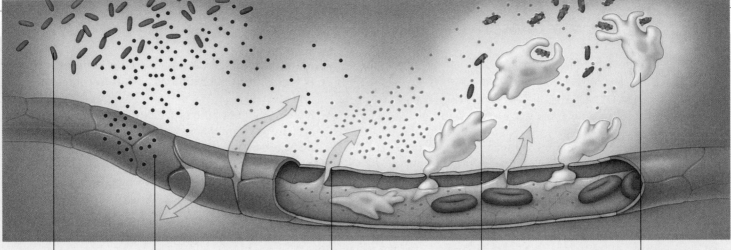

A Bacteria invade a tissue and release toxins or metabolic products that damage tissue.

B Mast cells in tissue release histamine, which widens arterioles (causing redness and warmth) and increases capillary permeability.

C Fluid and plasma proteins leak out of capillaries; localized edema (tissue swelling) and pain result.

D Complement proteins attack bacteria. Clotting factors also wall off inflamed area.

E Neutrophils and macrophages engulf invaders and debris. Macrophage secretions kill bacteria, attract more lymphocytes, and initiate fever.

Figure 38.8 Animated Inflammation in response to bacterial infection. *Above*, in this example, white blood cells and plasma proteins enter a damaged tissue. *Right*, the micrograph shows a phagocyte squeezing through a blood vessel wall.

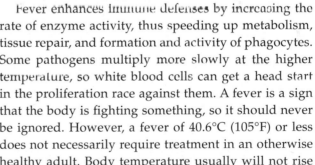

Inflammation Activated complement and cytokines trigger **inflammation**, a local response to tissue damage. The outward symptoms include redness, warmth, swelling, and pain. Inflammation begins when pattern receptors on basophils, mast cells, or neutrophils bind to antigen, or when mast cells directly bind to activated complement. In response to the binding, the cells release prostaglandins, histamines, and other substances into the affected tissue (Section 35.1).

These substances have two effects. First, they cause nearby arterioles to widen. As a result, blood flow to the area increases, reddening and warming the tissue. The increased flow speeds the arrival of more phagocytes, which are attracted to the cytokines. Second, the signaling molecules cause spaces between cells in capillary walls to widen, so they make capillaries in an affected tissue "leakier." Phagocytes and plasma proteins squeeze between the cells, out of the blood vessel and into interstitial fluid (Figure 38.8). The transfer changes the osmotic balance across the capillary wall, so more water diffuses from the blood into tissue. The tissue swells with fluid, putting pressure on free nerve endings and thus giving rise to sensations of pain.

Fever **Fever** is a temporary rise in body temperature above the normal 37°C (98.6°F) that often occurs in response to infection. Some cytokines stimulate brain cells to make and release prostaglandins, which act on the hypothalamus to raise the body's internal temperature set point. As long as the temperature of the body is below the new set point, the hypothalamus signals effectors (Section 27.3) to give rise to a sensation of cold, to constrict blood vessels in the skin, and to trigger shivering, or "chills." All of these responses help raise the internal temperature of the body.

Fever enhances immune defenses by increasing the rate of enzyme activity, thus speeding up metabolism, tissue repair, and formation and activity of phagocytes. Some pathogens multiply more slowly at the higher temperature, so white blood cells can get a head start in the proliferation race against them. A fever is a sign that the body is fighting something, so it should never be ignored. However, a fever of 40.6°C (105°F) or less does not necessarily require treatment in an otherwise healthy adult. Body temperature usually will not rise above that value, but if it does, immediate hospitalization is recommended because a fever of 42°C (107.6°F) can result in brain damage or death.

Take-Home Message

What is innate immunity?

■ Innate immunity is the body's built-in set of general immune defenses.

■ Complement, phagocytes, inflammation, and fever quickly eliminate most invaders from the body before their populations become established.

■ Vertebrate adaptive immunity is defined by self/nonself recognition, specificity, diversity, and memory.

■ Links to Lysosomes 4.9, Recognition proteins 5.2, Phagocytosis 5.5, Lymphatic system 37.10

If innate immune mechanisms do not quickly rid the body of an invading pathogen, populations of pathogenic cells may become established in body tissues. By that time, long-lasting adaptive immune mechanisms have begun to target the invaders specifically.

Tailoring Responses to Specific Threats

Life is so diverse that the number of different antigens is essentially unlimited. No system can recognize all of them, but vertebrate adaptive immunity comes close. Unlike innate immunity, the adaptive immune system changes: It "adapts" to different antigens an individual encounters during its lifetime. Lymphocytes and phagocytes interact to effect the four defining characteristics of adaptive immunity: self/nonself recognition, specificity, diversity, and memory.

MHC marker

Self versus nonself recognition starts with the molecular patterns that give each kind of cell or virus a unique identity. The plasma membrane of your cells bears **MHC markers** (*left*), which are self-recognition proteins named after the genes that encode them. Your T cells also bear antigen receptors called **T cell receptors**, or **TCRs**. Part of a TCR recognizes MHC markers as self; part also recognizes an antigen as nonself.

Specificity means that defenses are tailored to target specific antigens.

Diversity refers to the antigen receptors on a body's collection of B and T cells. There are potentially billions of different antigen receptors, so an individual has the potential to counter billions of different threats.

Memory refers to the capacity of the adaptive immune system to "remember" an antigen. It take a few days for B and T cells to respond in force the first time they encounter an antigen. If the same antigen shows up again, they make a faster, stronger response. That is why we do not get as sick the second time around.

First Step—The Antigen Alert

Recognition of a specific antigen is the first step of the adaptive immune response. A new B or T cell is naive, which means that no antigen has bound to its receptors yet. Once it binds to an antigen, it begins to divide by mitosis, and tremendous populations form.

T cell receptors do not recognize antigen unless it is presented by an antigen-presenting cell. Macrophages, B cells, and dendritic cells do the presenting. First, they engulf something antigenic (Figure 38.9*a*). Vesicles that contain the antigenic particle form in the cells' cytoplasm and fuse with lysosomes. Lysosomal enzymes digest the particle into bits (Sections 4.9 and 5.5).

The lysosomes also contain MHC markers that bind to some of the antigen bits. The resulting antigen–MHC complexes become displayed at the cell's surface when the vesicles fuse with (and become part of) the plasma membrane (Figure 38.9*b*). The display of MHC markers paired with antigen fragments is a call to arms.

Any T cell that bears a receptor for this antigen will bind the antigen–MHC complex. The T cell then starts secreting cytokines, which signal all other B or T cells with the same antigen receptor to divide again and again. Huge populations of B and T cells form after a few days; all of the cells recognize the same antigen. Most are **effector cells**, differentiated lymphocytes that

Figure 38.9 Antigen processing. (**a**) A macrophage ingests a foreign cell.

(**b**) From encounter to display, what happens when a B cell, macrophage, or dendritic cell engulfs an antigenic particle—in this case, a bacterium. These cells engulf, process, and then display antigen bound to MHC markers. The displayed antigen is presented to T cells.

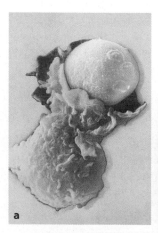

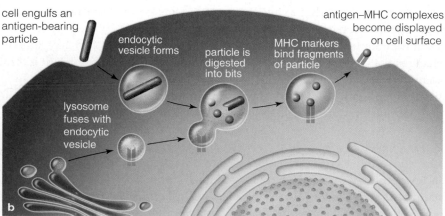

cell engulfs an antigen-bearing particle

endocytic vesicle forms

particle is digested into bits

MHC markers bind fragments of particle

antigen–MHC complexes become displayed on cell surface

lysosome fuses with endocytic vesicle

a

b

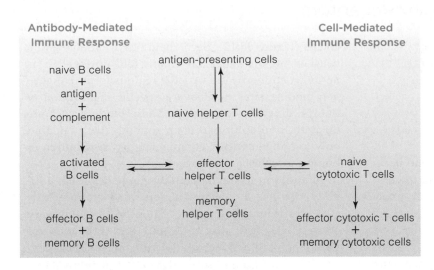

Antibody-Mediated Immune Response

naive B cells
+
antigen
+
complement

↓

activated B cells

↓

effector B cells
+
memory B cells

Cell-Mediated Immune Response

antigen-presenting cells

↕

naive helper T cells

↓

effector helper T cells
+
memory helper T cells

naive cytotoxic T cells

↓

effector cytotoxic T cells
+
memory cytotoxic cells

Figure 38.10 Overview of key interactions between antibody-mediated and cell-mediated responses—the two arms of adaptive immunity. A "naive" cell is one that has not made contact with its specific antigen.

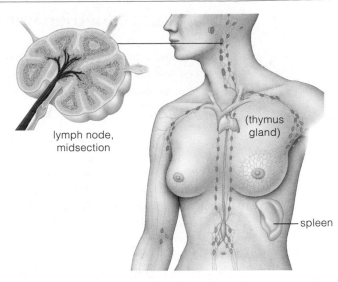

lymph node, midsection

(thymus gland)

spleen

Figure 38.11 Battlegrounds of adaptive immunity. Lymph nodes along lymph vascular highways hold macrophages, dendritic cells, B cells, and T cells. The spleen filters antigenic particles from blood.

act at once. Some are **memory cells**, long-lived B and T cells reserved for future encounters with the antigen.

Two Arms of Adaptive Immunity

Like a boxer's one-two punch, adaptive immunity has two separate arms: the antibody-mediated and the cell-mediated immune responses (Figure 38.10). These two responses work together to eliminate diverse threats.

Not all threats present themselves in the same way. For example, bacteria, fungi, or toxins can circulate in blood or interstitial fluid. These cells are intercepted quickly by B cells and other phagocytes that interact in the **antibody-mediated immune response**. In this response, B cells produce antibodies, which are proteins that can bind to specific antigen-bearing particles. We return to antibodies in the next section.

Some kinds of threats are not targeted by B cells. For example, B cells cannot detect body cells altered by cancer. As another example, some viruses, bacteria, fungi, and protists can hide and reproduce inside body cells; B cells can detect them only briefly, when they slip out of one cell to infect others. Such intracellular pathogens are targeted primarily by the **cell-mediated immune response**, which does not involve antibodies. In this response, cytotoxic T cells and NK cells detect and destroy altered or infected body cells.

Intercepting and Clearing Out Antigen

After engulfing an antigen-bearing particle, a dendritic cell or macrophage migrates to a lymph node (Section 37.10), where it will present antigen to many T cells

that filter through the node (Figure 38.11). Every day, about 25 billion T cells pass through each node. T cells that recognize and bind to antigen presented by a phagocyte initiate an adaptive response.

Antigen-bearing particles in interstitial fluid flow through lymph vessels to a lymph node, where they meet up with arrays of resident B cells, dendritic cells, and macrophages. These phagocytes engulf, process, and present antigen to T cells that are passing through the node. Any antigenic particle that escapes a lymph node to enter blood is taken up by the spleen.

During an infection, the lymph nodes swell because T cells accumulate inside of them. When you are ill, you may notice your swollen lymph nodes as tender lumps under the jaw or elsewhere.

The tide of battle turns when the effector cells and their secretions destroy most antigen-bearing agents. With less antigen present, fewer immune fighters are recruited. Complement proteins assist in the cleanup by binding antibody–antigen complexes, forming large clumps that can be quickly cleared from the blood by the liver and spleen. Immune responses subside after the antigenic particles are cleared from the body.

Take-Home Message

What is the adaptive immune system?

■ Phagocytes and lymphocytes interact to bring about vertebrate adaptive immunity, which has four defining characteristics: self/nonself recognition, specificity, diversity, and memory.

■ The two arms of adaptive immunity work together. Antibody-mediated responses target antigen in blood or interstitial fluid; cell-mediated responses target altered body cells.

- Antigen receptors give lymphocytes the potential to recognize billions of different antigens.

- Links to Protein structure 3.5, Membrane proteins 5.2, Alternative splicing 14.3, Exocrine glands 32.2

Antibody Structure and Function

If we liken B cells to assassins, then each one has a genetic assignment to liquidate one particular target—an antigen-bearing extracellular pathogen or toxin. Antibodies are their molecular bullets. **Antibodies** are proteins, Y-shaped antigen receptors made only by B cells. Each can bind to the antigen that prompted its synthesis. Many antibodies circulate in blood and enter interstitial fluid during inflammation, but they do not kill pathogens directly. Instead, they activate complement, facilitate phagocytosis, prevent pathogens from attaching to body cells, and neutralize toxins.

An antibody molecule consists of four polypeptides: two identical "light" chains and two identical "heavy"

chains (Figure 38.12). Each chain has a variable and a constant region. When the chains fold up together as an intact antibody, the variable regions form two antigen binding sites that have a specific distribution of bumps, grooves, and charge. These binding sites are the antigen receptor part of an antibody: they can bind only to antigen with a complementary distribution of bumps, grooves, and charge.

In addition to the antigen-binding sites, each antibody also has a constant region that determines its structural identity, or class. There are five antibody classes: IgG, IgA, IgE, IgM, and IgD (Ig stands for immunoglobulin, which is another name for antibody). The different classes serve different functions (Table 38.4). Most of the antibodies circulating in the bloodstream and tissue fluids are IgG, which binds pathogens, neutralizes toxins, and activates complement. IgG is the only antibody that can cross the placenta to protect a fetus before its own immune system is active. IgA is the main antibody in mucus and other exocrine

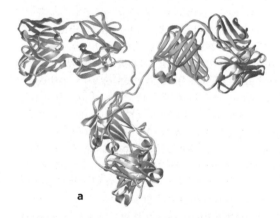

a

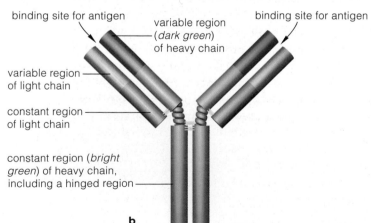

binding site for antigen

variable region
(*dark green*)
of heavy chain

binding site for antigen

variable region
of light chain

constant region
of light chain

constant region (*bright
green*) of heavy chain,
including a hinged region

b

Figure 38.12 Antibody structure. (**a**) An antibody molecule has four polypeptide chains joined in a Y-shaped configuration. In this ribbon model, the two heavy chains are shown in *green*, and the two light chains are *teal*. (**b**) Each chain has a variable and a constant region.

Table 38.4	Structural Classes of Antibodies

Secreted antibodies

IgG		Main antibody in blood; activates complement, neutralizes toxins; protects fetus and is secreted in early milk.
IgA		Abundant in exocrine gland secretions (e.g., tears, saliva, milk, mucus), where it occurs in dimeric form (*shown*). Interferes with binding of bacteria and viruses to body cells.

Membrane-bound antibodies

IgE		Anchored to surface of basophils, mast cells, eosinophils, and some dendritic cells. IgE binding to antigen induces anchoring cell to release histamines and cytokines. Factor in allergies and asthma.
IgD		B cell receptor.
IgM		B cell receptor, as a monomer. Also is secreted as pentamer (group of five, *shown*).

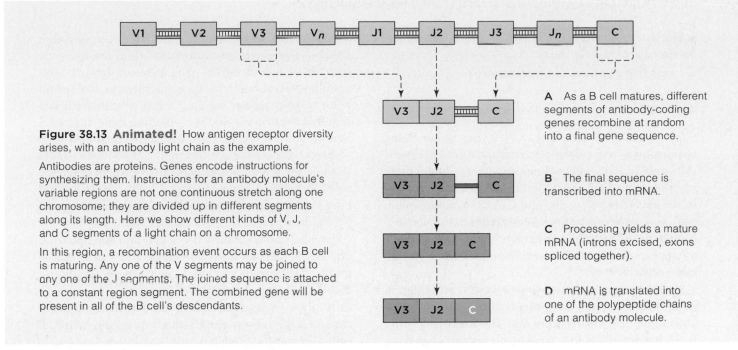

Figure 38.13 Animated! How antigen receptor diversity arises, with an antibody light chain as the example.

Antibodies are proteins. Genes encode instructions for synthesizing them. Instructions for an antibody molecule's variable regions are not one continuous stretch along one chromosome; they are divided up in different segments along its length. Here we show different kinds of V, J, and C segments of a light chain on a chromosome.

In this region, a recombination event occurs as each B cell is maturing. Any one of the V segments may be joined to any one of the J segments. The joined sequence is attached to a constant region segment. The combined gene will be present in all of the B cell's descendants.

A As a B cell matures, different segments of antibody-coding genes recombine at random into a final gene sequence.

B The final sequence is transcribed into mRNA.

C Processing yields a mature mRNA (introns excised, exons spliced together).

D mRNA is translated into one of the polypeptide chains of an antibody molecule.

gland secretions (Section 32.2). Bound to antigen, it interacts with mast cells, basophils, macrophages, and NK cells to initiate inflammation. IgA is secreted as a dimer (two antibodies bound together), which makes it stable enough to patrol harsh environments such as the interior of the digestive tract. There, IgA encounters pathogens before they contact body cells. IgE is incorporated into the plasma membrane of mast cells, basophils, and some types of dendritic cells. Binding of antigen to IgE triggers the anchoring cell to release histamines and cytokines. A new B cell bristles with **B cell receptors**, which are membrane-bound IgM or IgD antibodies. Secreted IgM pentamers (polymers of five) efficiently bind antigen and activate complement.

The Making of Antigen Receptors

Most humans can make about 2.5 billion unique antigen receptors. This diversity arises because the genes that encode the receptors do not occur in a continuous stretch on one chromosome; instead, they occur in several segments on different chromosomes, and there are several different versions of each segment. The segments are spliced together during B and T cell differentiation, but which version of each segment gets spliced into the antigen receptor gene of a particular cell is random (Section 14.3 and Figure 38.13). As a B or T cell differentiates, it ends up with one out of about 2.5 billion different combinations of gene segments.

Before a new B cell leaves bone marrow, it already is synthesizing its unique antigen receptors. The constant region of each receptor is embedded in the lipid bilayer of the cell's plasma membrane, and the two arms project above the membrane. In time the B cell bristles with more than 100,000 antigen receptors. It is now a "naive" B cell, meaning it has not yet met its antigen.

T cells also form in bone marrow, but they mature only after they take a tour in the thymus gland (Section 37.10). There, they encounter hormones that stimulate them to make MHC receptors and T cell receptors.

Because of the random splicing of antigen receptor gene segments, the TCRs of some new T cells bind body proteins instead of antigen, and most do not recognize MHC markers. So how does an individual end up with a working set of T cells that does not attack its own body? Thymus cells have a built-in quality control that weeds out "bad" TCRs. They snip small peptides from a variety of body proteins and attach them to MHC markers. T cells that bind to a peptide–MHC complex have TCRs that recognize a self protein; those that do not bind any complex do not recognize MHC markers. Both types of cells die. Thus, any T cell that leaves the thymus to begin its journey through the circulatory system bristles with functional TCRs.

Take-Home Message

What are antigen receptors?

■ The adaptive immune system has the potential to recognize about 2.5 billion different antigens via receptors on B cells and T cells.

■ Antibodies are secreted or membrane-bound antigen receptors. They are made only by B cells.

- In an antibody-mediated immune response, B cells are stimulated to produce antibodies targeting a specific antigen.

    - Link to Receptor-mediated endocytosis 5.5

An Antibody-Mediated Response

Suppose that you accidentally nick your finger. Being opportunists, some *Staphylococcus aureus* cells on your skin invade your internal environment. Complement in interstitial fluid quickly attaches to carbohydrates in the bacterial cell walls, and cascading complement activation reactions begin. Within an hour, complement-coated bacteria tumbling along in lymph vessels reach a lymph node in your elbow. There they filter past an army of naive B cells.

As it happens, one of the naive B cells in that lymph node makes antigen receptors that recognize a poly-saccharide in *S. aureus* cell walls. This and every other B cell has receptors that recognize a complement coating on bacteria. Binding to antigen and complement together stimulates the B cell to engulf one of the bacteria by receptor-mediated endocytosis (Section 5.5). The B cell is now activated (Figure 38.14*a*).

Meanwhile, more *S. aureus* cells have been secreting metabolic products into interstitial fluid around your cut. The secretions attract phagocytes. A dendritic cell engulfs several bacteria, then migrates to the lymph node in your elbow. By the time it gets there, it has digested the bacteria and is displaying their fragments bound to MHC markers on its surface (Figure 38.14*b*).

Each hour, about 500 different naive T cells travel through the lymph node, inspecting resident dendritic cells. In this case, one of those T cells has TCRs that bind the *S. aureus* antigen–MHC complexes displayed by the dendritic cell.

For the next 24 hours, the T cell and the dendritic cell interact. When they disengage, the T cell returns to the circulatory system and begins to divide (Figure 38.14*c*). A huge population of genetically identical T cells forms; each cell has receptors that can bind the *S. aureus* antigen. These clones differentiate into helper T cells and memory T cells.

By the theory of clonal selection, the T cell was "selected" because its receptors bind to the *S. aureus* antigen. T cells with receptors that do not bind the antigen do not divide to form huge clonal populations.

A   The B cell receptors on a naive B cell bind to a specific antigen on the surface of a bacterium. The bacterium's complement coating triggers the B cell to engulf it. Fragments of the bacterium bind MHC markers, and the complexes become displayed at the surface of the now-activated B cell.

B   A dendritic cell engulfs the same kind of bacterium that the B cell encountered. Digested fragments of the bacterium bind to MHC markers, and the complexes become displayed at the dendritic cell's surface. The dendritic cell is now an antigen-presenting cell.

C   The antigen–MHC complexes on the antigen-presenting cell are recognized by antigen receptors on a naive T cell. Binding causes the T cell to divide and differentiate into effector and memory helper T cells.

D   Antigen receptors of one of the effector helper T cells bind antigen–MHC complexes on the B cell. Binding makes the T cell secrete cytokines.

E   The cytokines induce the B cell to divide, giving rise to many identical B cells. The cells differentiate into effector B cells and memory B cells.

F   The effector B cells begin making and secreting huge numbers of IgA, IgG, or IgE, all of which recognize the same antigen as the original B cell receptor. The new antibodies circulate throughout the body and bind to any remaining bacteria.

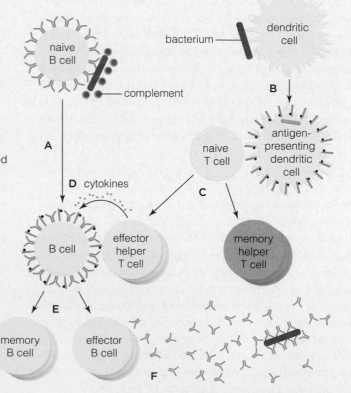

Figure 38.14 Animated Example of an antibody-mediated immune response.

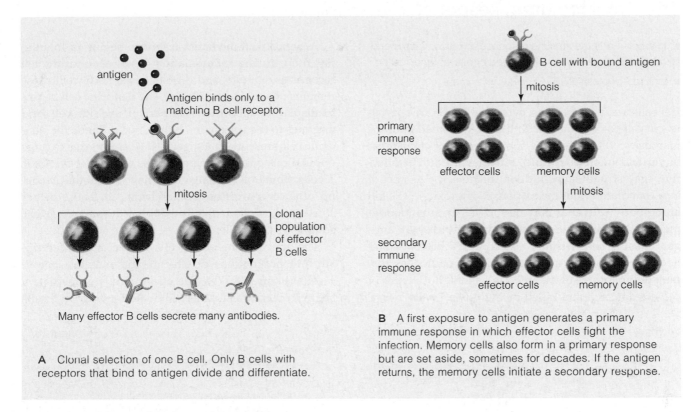

antigen

Antigen binds only to a matching B cell receptor.

mitosis

clonal population of effector B cells

Many effector B cells secrete many antibodies.

A Clonal selection of one B cell. Only B cells with receptors that bind to antigen divide and differentiate.

B cell with bound antigen

mitosis

primary immune response

effector cells   memory cells

mitosis

secondary immune response

effector cells   memory cells

B A first exposure to antigen generates a primary immune response in which effector cells fight the infection. Memory cells also form in a primary response but are set aside, sometimes for decades. If the antigen returns, the memory cells initiate a secondary response.

Figure 38.15 **Animated** B cell maturation.

Let's go back to that B cell in the lymph node. By now, it has digested the bacterium, and it is displaying bits of *S. aureus* bound to MHC molecules on its plasma membrane. One of the new helper T cells recognizes the antigen–MHC complexes displayed by the B cell. Like long-lost friends, the B cell and the helper T cell stay together for a while and communicate.

One of the messages that is communicated consists of cytokines secreted by the helper T cell. The cytokines stimulate the B cell to begin mitosis after the two cells disengage (Figure 38.14d). The B cell divides again and again to form a huge population of genetically identical cells, all with receptors that can bind to the *S. aureus* antigen (Figure 38.15a). These clones differentiate into effector and memory B cells (Figure 38.14e).

The effector cells start working immediately. They switch antibody classes, which means they begin to produce and secrete IgG, IgA, or IgE instead of making membrane-bound B cell receptors. The new antibody molecules recognize the same *S. aureus* antigen as the original B cell receptor. Antibodies now circulate throughout the body and attach themselves to any remaining bacterial cells. An antibody coating prevents bacteria from attaching to body cells, and flags them for phagocytosis and disposal (Figure 38.14f).

Memory B and T cells also form, but these do not act right away. They persist long after the initial infection

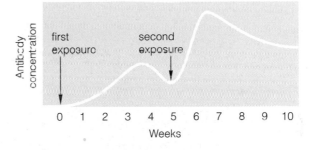

Antibody concentration

first exposure

second exposure

0 1 2 3 4 5 6 7 8 9 10
Weeks

Figure 38.16 Antibody levels in a primary and secondary immune response. A secondary immune response is faster and stronger than the primary response that preceded it.

ends. If the same antigen enters the body at a later time, these memory cells will initiate a secondary response (Figures 38.15b and 38.16). In the secondary response, larger populations of effector cell clones form much more quickly than they did in the primary response, so more antibodies can be produced in a shorter time.

Take-Home Message

What happens during an antibody-mediated immune response?

■ Antigen presenting cells, T cells, and B cells interact in an antibody-mediated immune response targeting a specific antigen.

■ Populations of B cells form; these make and secrete antibodies that recognize and bind the antigen.

- In a cell-mediated immune response, cytotoxic T cells and NK cells are stimulated to kill infected or altered body cells.

- Link to Apoptosis 27.6

If B cells are like assassins, cytotoxic T cells are specialists in cell-to-cell combat. Antibody-mediated immune responses efficiently target pathogens that circulate in blood and interstitial fluid, but they are not as effective against pathogens hidden inside cells. As part of a cell-mediated immune response, cytotoxic T cells kill ailing body cells that may be missed by an antibody-mediated response. Such cells typically display antigen: Cancer cells display altered body proteins, and body cells infected with intracellular pathogens display polypeptides of the infecting agent. Both types of cell are detected and killed by cytotoxic T cells.

A typical cell-mediated response begins in interstitial fluid during inflammation when a dendritic cell recognizes, engulfs, and digests a sick body cell or the remains of one (Figure 38.17a). The dendritic cell begins to display antigen that was part of the sick cell, and migrates to the spleen or a lymph node. There, the dendritic cell presents its antigen–MHC complexes to huge populations of naive helper T cells and naive cytotoxic T cells. Some of the naive cells have TCRs that recognize the complexes on the dendritic cell. Those helper T cells and cytotoxic T cells that bind the antigen–MHC complexes on the dendritic cell become activated.

The activated helper T cells divide and differentiate into populations of effector and memory helper T cells (Figure 38.17b). The effector cells immediately begin to secrete cytokines. Activated cytotoxic T cells

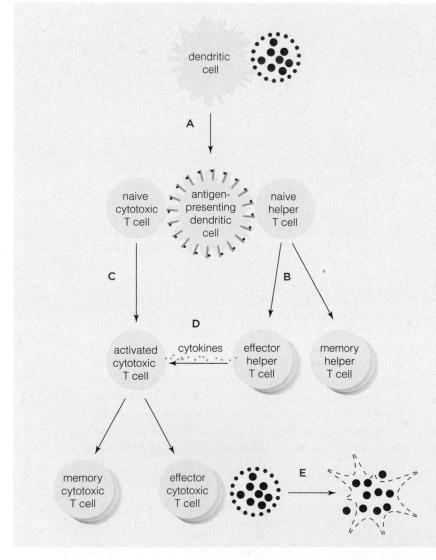

A A dendritic cell engulfs a virus-infected cell. Digested fragments of the virus bind to MHC markers, and the complexes are displayed at the dendritic cell's surface. The dendritic cell, now an antigen-presenting cell, migrates to a lymph node.

B Receptors on a naive helper T cell bind to antigen–MHC complexes on the dendritic cell. The interaction activates the helper T cell, which then begins to divide. A large population of descendant cells forms. Each cell bears T cell receptors that recognize the same antigen. The cells differentiate into effector and memory cells.

C Receptors on a naive cytotoxic T cell bind to the antigen–MHC complexes on the surface of the dendritic cell. The interaction activates the cytotoxic T cell.

D The activated cytotoxic T cell recognizes cytokines secreted by the effector helper T cells as signals to divide. A large population of descendant cells forms. Each cell bears T cell receptors that recognize the same antigen. The cells differentiate into effector and memory cells.

E The new cytotoxic T cells circulate through the body. They recognize and touch-kill any body cell that displays the viral antigen–MHC complexes on its surface.

Figure 38.17 Animated Example of a primary cell-mediated immune response.
Figure It Out: What do the large red spots represent?
Answer: Viruses

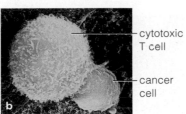

cytotoxic T cell

cancer cell

Figure 38.18 T cell receptor function. (**a**) A TCR (*green*) on a T cell binds to an MHC marker (*tan*) on an antigen-presenting cell. An antigen (*red*) is bound to the MHC marker.

(**b**) A cytotoxic T cell caught in the act of touch-killing a cancer cell.

recognize the cytokines as signals to divide and differentiate, and tremendous populations of effector and memory cytotoxic T cells form (Figure 38.17c,d). All of them recognize and bind the same antigen—the one displayed by that first ailing cell. As in an antibody-mediated response, the memory cells that form in a primary cell-mediated response will mount a secondary response if the antigen returns at a later time.

The effector cytotoxic T cells start working immediately. They circulate throughout blood and interstitial fluid, and bind to any other body cell displaying the original antigen together with MHC markers (Figure 38.18a). After it is bound to an ailing cell, a cytotoxic T cell releases perforin and proteases. These toxins poke holes in the sick cell and induce it to die by apoptosis (Figures 38.17e and 38.18b).

Cytotoxic T cells also recognize the MHC markers of foreign body cells (cytotoxic T cells are responsible for rejection of transplanted organs). They must recognize MHC molecules on the surface of a body cell in order to kill it. However, some infections or cancer can alter a cell so that it is missing part or all of its MHC markers. NK ("natural killer") cells are crucial for fighting such cells. Unlike cytotoxic T cells, NK cells can kill body cells that lack MHC markers. Cytokines secreted by helper T cells (Figure 38.17d) also stimulate NK cell division. The resulting populations of effector NK cells attack body cells tagged for destruction by antibodies. They also recognize certain proteins displayed by body cells under stress. Stressed body cells with normal MHC markers are not killed; only those with altered or missing MHC markers are destroyed.

Take-Home Message

What happens during a cell-mediated immune response?

■ Antigen-presenting cells, T cells, and NK cells interact in a cell-mediated immune response targeting body cells that have been altered by cancer or infected.

38.9 | Allergies

■ An immune response to a typically harmless substance is an allergy. Allergies can be annoying or life-threatening.

In millions of people, exposure to harmless substances stimulates an immune response. Any substance that is ordinarily harmless yet provokes such responses is an **allergen**. Sensitivity to an allergen is called an **allergy**. Drugs, foods, pollen, dust mites, fungal spores, poison ivy, and venom from bees, wasps, and other insects are among the most common allergens.

Some people are genetically predisposed to having allergies. Infections, emotional stress, and changes in air temperature can trigger reactions. A first exposure to an allergen stimulates the immune system to make IgE, which becomes anchored to mast cells and basophils. With later exposures, antigen binds to the IgE. Binding triggers the anchoring cell to secrete histamine and cytokines that initiate inflammation. If this reaction occurs at the lining of the respiratory tract, a copious amount of mucus is secreted and the airways constrict; sneezing, stuffed-up sinuses, and a drippy nose result (Figure 38.19a). Contact with an allergen that penetrates the skin's outer layers causes the skin to redden, swell, and become itchy.

Antihistamines relieve allergy symptoms by dampening the effects of histamines. These drugs act on histamine receptors, and also inhibit the release of cytokines and histamines from basophils and mast cells.

Some people are hypersensitive to drugs, insect stings, foods, or vaccines. A second exposure to the allergen can result in anaphylactic shock, a severe, whole-body allergic reaction. Huge amounts of cytokines and histamines released in all parts of the body provoke an immediate, systemic reaction. Fluid leaking from blood into tissues causes the blood pressure to drop too much (shock), and tissues to swell. Swelling tissue constricts airways and may block them. Anaphylactic shock is rare but life-threatening and requires immediate treatment (Figure 38.19c). It may occur at any time, upon exposure to even a tiny amount of allergen. Risks include a prior allergic reaction of any kind.

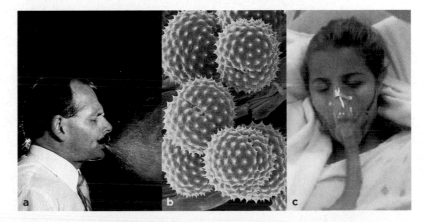

Figure 38.19 Allergies. (**a**) A mild allergy may cause upper respiratory symptoms. (**b**) Ragweed pollen, a common allergen. (**c**) Anaphylactic shock is a severe allergic reaction that requires immediate treatment.

38.10 | Vaccines

■ Vaccines are designed to elicit immunity to a disease.

Immunization refers to processes designed to induce immunity. In active immunization, a preparation that contains antigen—a **vaccine**—is administered orally or injected. The first immunization elicits a primary immune response, just as an infection would. A second immunization, or booster, elicits a secondary immune response for enhanced immunity.

In passive immunization, a person receives antibodies purified from the blood of another individual. The treatment offers immediate benefit for someone who has been exposed to a potentially lethal agent, such as tetanus or rabies, Ebola virus, or a venom or toxin. Because the antibodies were not made by the recipient's lymphocytes, memory cells do not form, so benefits last only as long as the injected antibodies do.

The first vaccine was the result of desperate attempts to survive smallpox epidemics that swept repeatedly through cities all over the world. Smallpox is a severe disease that kills up to one-third of the people it infects (Figure 38.20). Before 1880, no one knew what caused infectious diseases or how to protect anyone from getting them, but there were clues. In the case of smallpox, survivors seldom contracted the disease a second time. They were immune, or protected from infection.

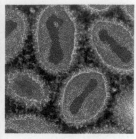

Figure 38.20 Young survivor and the cause of her disease, smallpox viruses. Worldwide use of the vaccine eradicated naturally occurring cases of smallpox; vaccinations for it ended in 1972.

The idea of acquiring immunity to smallpox was so appealing that people had been risking their lives on it for two thousand years. For example, many people poked into their skin bits of scabs or threads soaked in pus. Some survived the crude practices and became immune to smallpox, but many others did not.

By the late 1700s, it was widely known that dairymaids did not get smallpox if they had already recovered from cowpox (a mild disease that affects cattle as well as humans). In 1796, Edward Jenner, an English physician, injected liquid from a cowpox sore into the arm of a healthy boy. Six weeks later, Jenner injected the boy with liquid from a smallpox sore. Luckily, the boy did not get smallpox. Jenner's experiment showed directly that the agent of cowpox elicits immunity to smallpox. Jenner named his procedure "vaccination," after the Latin word for cowpox (*vaccinia*). The use of Jenner's vaccine spread quickly through Europe, then to the rest of the world. The last known case of naturally occurring smallpox was in 1977, in Somalia. The vaccine had eradicated the disease.

We now know that the cowpox virus is an effective vaccine for smallpox because the antibodies it elicits also recognize smallpox virus antigens. Our knowledge of how the immune system works has allowed us to develop many other vaccines that save millions of lives every year. These vaccines are an important part of worldwide public health programs (Table 38.5).

Table 38.5 Recommended Immunization Schedule

Vaccine	Age of Vaccination
Hepatitis B	Birth to 2 months
Hepatitis B boosters	1–4 months and 6–18 months
Rotavirus	2, 4, and 6 months
DTP: diphtheria, tetanus, and pertussis (whooping cough)	2, 4, and 6 months
DTP boosters	15–18 months, 4–6 years, and 11–12 years
HiB (*Haemophilus influenzae*)	2, 4, and 6 months
HiB booster	12–15 months
Pneumococcal	2, 4, and 6 months
Pneumococcal booster	12–15 months
Inactivated poliovirus	2 and 4 months
Inactivated poliovirus boosters	6–18 months and 4–6 years
Influenza	Yearly, 6 months–18 years
MMR (measles, mumps, rubella)	12–15 months
MMR booster	4–6 years
Varicella (chicken pox)	12–15 months
Varicella booster	4–6 years
Hepatitis A series	1–2 years
HPV series	11–12 years
Meningococcal	11–12 years

Source: Centers for Disease Control and Prevention (CDC), 2008

Take-Home Message

How does immunization work?

■ Immunization is the administration of an antigen-bearing vaccine designed to elicit immunity to a specific disease.

38.11 | Immunity Gone Wrong

■ The immune system of some people does not function properly. The outcome is often severe or lethal.

■ Links to Multiple sclerosis 33.13, Arthritis 36.5

Despite the redundancies of immune system functions and built-in quality controls, immunity does not always work as well as it should. Its sheer complexity is part of the problem, because there are so many points at which it could go wrong. Autoimmune disorders occur when an immune response is misdirected against the person's own body cells. In immunodeficiency, the immune response is insufficient to protect a person from disease.

Autoimmune Disorders

Sometimes lymphocytes and antibody molecules fail to discriminate between self and nonself. When that happens, they mount an **autoimmune response**, or an immune response that targets one's own tissues.

For example, autoimmunity occurs in rheumatoid arthritis, a disease in which self antibodies form and bind to the soft tissue in joints. The resulting inflammation leads to eventual disintegration of bone and cartilage in the joints (Section 36.5).

Antibodies to self proteins may bind to hormone receptors, as in Graves' disease. Self antibodies that bind stimulatory receptors on the thyroid gland cause it to produce excess thyroid hormone, which quickens the body's overall metabolic rate. Antibodies are not part of the feedback loops that normally regulate thyroid hormone production. So, antibody binding continues unchecked, the thyroid continues to release too much hormone, and the metabolic rate spins out of control. Symptoms of Graves' disease include uncontrollable weight loss; rapid, irregular heartbeat; sleeplessness; pronounced mood swings; and bulging eyes.

A neurological disorder, multiple sclerosis, occurs when self-reactive T cells attack the myelin sheaths of axons in the central nervous system (Section 33.13). Symptoms range from weakness and loss of balance to paralysis and blindness. Specific MHC gene alleles increase susceptibility, but a bacterial or viral infection may trigger the disorder.

Immune responses tend to be stronger in women than in men, and autoimmunity is far more frequent in women. We know that estrogen receptors are part of gene expression controls throughout the body. T cells have receptors for estrogens, so these hormones may enhance T cell activation in autoimmune diseases. Women's bodies have more estrogen, so interactions between their B cells and T cells may be amplified.

Figure 38.21 A case of severe combined immunodeficiency (SCID). Cindy Cutshwall was born with a deficient immune system. She carries a mutated gene for adenosine deaminase (ADA). Without this enzyme, her cells cannot break down adenosine completely, so a reaction product that is toxic to white blood cells accumulated in her body. High fevers, severe ear and lung infections, diarrhea, and an inability to gain weight were outcomes.

In 1991, when Cindy was nine years old, she and her parents consented to one of the first human gene therapies. Genetic engineers spliced the normal ADA gene into the genetic material of a harmless virus. The modified virus delivered copies of the normal gene into her bone marrow cells. Some cells incorporated the gene in their DNA and started making the missing enzyme.

Now in her twenties, Cindy is doing well. She still requires weekly injections to supplement her ADA production. Other than that, she is able to live a normal life. She is a strong advocate of gene therapy.

Immunodeficiency

Impaired immune function is dangerous and sometimes lethal. Immune deficiencies render individuals vulnerable to infections by opportunistic agents that are typically harmless to those in good health. Primary immune deficiencies, which are present at birth, are the outcome of mutations. Severe combined immunodeficiencies (SCIDs) are examples. A genetic disorder called adenosine deaminase (ADA) deficiency is a type of SCID (Figure 38.21). Secondary immune deficiency is the loss of immune function after exposure to an outside agent, such as a virus. AIDS (acquired immunodeficiency syndrome, described in the next section) is the most common secondary immune deficiency.

Take-Home Message

What happens when the immune system does not function as it should?

■ Misdirected or compromised immunity, which sometimes occurs as a result of mutation or environmental factors, can have severe or lethal outcomes.

■ AIDS is an outcome of interactions between the HIV virus and the human immune system.

■ Links to cDNA 16.1, Viruses 21.1, HIV replication 21.2

Acquired immune deficiency syndrome, or **AIDS**, is a constellation of disorders that occur as a consequence of infection with HIV, the human immunodeficiency virus (Figure 38.22*a*). This virus cripples the immune system, so it makes the body very susceptible to infections and rare forms of cancer. Worldwide, approximately 39.5 million individuals currently have AIDS (Table 38.6 and Figure 38.22*b*).

There is no way to rid the body of the HIV virus, no cure for those already infected. At first, an infected person appears to be in good health, perhaps fighting "a bout of flu." But symptoms eventually emerge that foreshadow AIDS: fever, many enlarged lymph nodes, chronic fatigue and weight loss, and drenching night sweats. Then, infections caused by normally harmless microorganisms strike. Yeast infections of the mouth, esophagus, and vagina often occur, as well as a form of pneumonia caused by the fungus *Pneumocystis carinii*. Colored lesions erupt. These lesions are evidence of Kaposi's sarcoma, a type of cancer that is common among AIDS patients (Figure 38.22*c*).

HIV Revisited HIV is a retrovirus that has a lipid envelope. Remember, this type of envelope is a small piece of plasma membrane a virus particle acquires as it buds from a cell (Section 21.2). Proteins jut out from the enve-

lope, span it, and line its inner surface. Just beneath the envelope, more viral proteins enclose two RNA strands and copies of reverse transcriptase. When a virus particle infects a cell, the reverse transcriptase copies the viral RNA into DNA, which becomes integrated into the host cell's DNA.

A Titanic Struggle HIV mainly infects macrophages, dendritic cells, and helper T cells. When virus particles enter the body, dendritic cells engulf them. The dendritic cells then migrate to lymph nodes, where they present processed HIV antigen to naive T cells. An army of HIV-neutralizing IgG antibodies and HIV-specific cytotoxic T cells forms.

We have just described a typical adaptive immune response. It rids the body of most—but not all—of the virus. In this first response, HIV infects a few helper T cells in a few lymph nodes. For years or even decades, the IgG antibodies keep the level of HIV in the blood low, and the cytotoxic T cells kill HIV-infected cells.

Patients are contagious during this stage, although they might show no symptoms of AIDS. HIV viruses persist in a few of their helper T cells, in a few lymph nodes. Eventually, the level of virus-neutralizing IgG in the blood plummets, and T cell production slows. Why IgG decreases is still a major topic of research, but its effect is certain: The adaptive immune system

Table 38.6 Global HIV and AIDS Cases

Region	AIDS Cases	New HIV Cases
Sub-Saharan Africa	22,500,000	1,700,000
South/Southeast Asia	4,000,000	340,000
Central Asia/East Europe	1,600,000	150,000
Latin America	1,600,000	100,000
North America	1,300,000	46,000
East Asia	800,000	92,000
Western/Central Europe	760,000	31,000
Middle East/North Africa	380,000	35,000
Caribbean Islands	230,000	17,000
Australia/New Zealand	75,000	14,000
Worldwide total	33,200,000	2,500,000

Source: Joint United Nations Programme HIV/AIDS, 2007 data

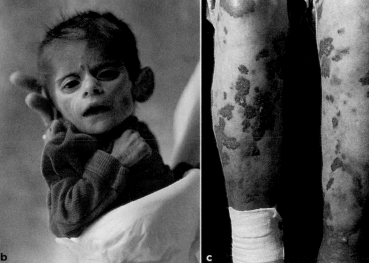

Figure 38.22 AIDS. (**a**) A human T cell (*blue*), infected with HIV (*red*). (**b**) This Romanian baby contracted AIDS from his mother's breast milk. He did not live long enough to develop lesions of Kaposi's sarcoma (**c**), a cancer that is a common symptom of HIV infection in older AIDS patients.

becomes less and less effective at fighting the virus. The number of virus particles rises; up to 1 billion HIV viruses are built each day. Up to 2 billion helper T cells become infected. Half of the viruses are destroyed and half of the helper T cells are replaced every two days. Lymph nodes begin to swell with infected T cells.

Eventually, the battle tilts as the body makes fewer replacement helper T cells and the body's capacity for adaptive immunity is destroyed. Other types of viruses make more particles in a day, but the immune system eventually wins. HIV demolishes the immune system. Secondary infections and tumors kill the patient.

Transmission HIV is transmitted most frequently by having unprotected sex with an infected partner. The virus occurs in semen and vaginal secretions, and can enter a partner through epithelial linings of the penis, vagina, rectum, and the mouth. The risk of transmission increases by the type of sexual act; for example, anal sex carries 50 times the risk of oral sex.

Infected mothers can transmit HIV to a child during pregnancy, labor, delivery, or breast-feeding. HIV also travels in tiny amounts of infected blood in the syringes shared by intravenous drug abusers, or by patients in hospitals of poor countries. HIV is not transmitted by casual contact.

Testing Most AIDS tests check blood, saliva, or urine for antibodies that bind to HIV antigens. These antibodies are detectable in 99 percent of infected people within three months of exposure to the virus. One test can detect viral RNA at about eleven days after exposure. Currently, the only reliable tests are performed in clinical laboratories; home test kits may result in false negatives, which may cause an infected person to unknowingly transmit the virus.

Drugs and Vaccines Drugs cannot cure AIDS, but they can slow its progress. Of the twenty or so FDA-approved AIDS drugs, most target processes unique to retroviral replication. For example, RNA nucleotide analogs such as AZT are called reverse transcriptase inhibitors. They interrupt HIV replication when they substitute for normal nucleotides in the viral RNA-to-DNA synthesis process (Sections 16.1 and 21.2). Other drugs such as protease inhibitors affect different parts of the viral replication cycle.

A three-drug "cocktail" of one protease inhibitor plus two reverse transcriptase inhibitors is currently the most successful AIDS therapy, and has changed the course of the disease from a short-term death sentence to a long-term, often manageable illness.

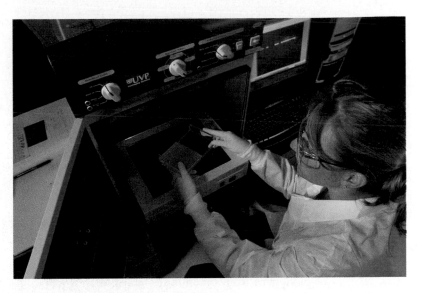

Figure 38.23 At the Global AIDS Program's International Laboratory Branch of the National Center for HIV/AIDS, Viral Hepatitis, STD, and TB Prevention, researcher Amanda McNulty examines a DNA electrophoresis gel. She is investigating HIV drug resistance in people from Africa, Vietnam, and Haiti.

Researchers are using several strategies to develop an HIV vaccine. At this writing, organizations around the world are testing 42 different HIV vaccines. Most of them consist of isolated HIV proteins or peptides, and many deliver the antigens in viral vectors. Live, weakened HIV virus is an effective vaccine in chimpanzees, but the risk of HIV infection from the vaccines themselves far outweighs their potential benefits in humans. Other types of HIV vaccines are notoriously ineffective. IgG antibody exerts selective pressure on the virus, which has a very high mutation rate because it replicates so fast. The human immune system just cannot produce antibodies fast enough to keep up with the mutations (Figure 38.23).

At present, our best option for halting the spread of HIV is prevention, by teaching people how to avoid being infected. The best protection against AIDS is to avoid unsafe behaviors. In most circumstances, HIV infection is the consequence of a choice: either to have unprotected sex, or to use a shared needle for intravenous drugs. Educational programs around the world are having an effect on the spread of the virus: In many (but not all) countries, the incidence of new cases of HIV each year is beginning to slow. Overall, however, our global battle against AIDS is not being won.

Take-Home Message

What is AIDS?

■ AIDS occurs as a result of infection by HIV, a virus that infects lymphocytes and so cripples the human immune system.

The Gardasil HPV vaccine consists of viral capsid proteins that self-assemble into virus-like particles (VLPs). These proteins are produced by a recombinant yeast, *Saccharomyces cerevisiae*. The yeast carries genes for one capsid protein from each of four strains of HPV, so the VLPs carry no viral DNA. Thus, the VLPs are not infectious, but the antigenic proteins they consist of elicit an immune response at least as strong as infection with HPV virus.

How would you vote?

Should clinical trials of potential vaccines be held to the same ethical standards no matter where they take place? See CengageNOW for details, then vote online.

Summary

Section 38.1 Three lines of immune defense protect vertebrates from infection. An **antigen**-bearing pathogen that breaches surface barriers triggers **innate immunity**, a set of general defenses that usually prevents populations of pathogens from becoming established in the internal environment. **Adaptive immunity**, which can specifically target billions of different antigens, follows. **Complement** and signaling molecules such as **cytokines** coordinate the activities of white blood cells (**dendritic cells**, **macrophages**, **neutrophils**, **basophils**, **mast cells**, **eosinophils**, **B** and **T lymphocytes**, and **NK cells**) in **immunity**.

Sections 38.2, 38.3 Vertebrates can fend off pathogens such as those that cause dental **plaque** at body surfaces with physical, mechanical, and chemical barriers (including **lysozyme**). Most **normal flora** do not cause disease unless they penetrate inner tissues.

Section 38.4 An innate immune response includes fast, general responses that can eliminate invaders before an infection can become established. Complement attracts phagocytes, and punctures some invaders. **Inflammation** begins when mast cells in tissue release histamine, which increases blood flow and also makes capillaries leaky to phagocytes and plasma proteins. **Fever** fights infection by increasing the metabolic rate.

- *Use the animation on CengageNOW to investigate inflammation and the action of complement.*

Section 38.5 Adaptive immunity is characterized by self/nonself recognition, target specificity, diversity (the capacity to intercept billions of different pathogens), and memory. B and T cells carry out adaptive responses.

The **antibody-mediated immune response** and the **cell-mediated immune response** work together to rid the body of a specific pathogen. Macrophages, dendritic cells, and B cells engulf and digest viruses or bacteria into bits. The phagocytes then present the antigenic bits on their surfaces bound to **MHC markers** (self markers). T cells that recognize the complexes via **T cell receptors** (TCRs) initiate the formation of many **effector cells** that target other antigen-bearing particles. **Memory cells** that are reserved for later encounters with the same antigen also form.

Sections 38.6, 38.7 B cells, assisted by T cells and signaling molecules, carry out antibody-mediated immune responses. B cells make **antibodies** that bind to specific antigens. Antigen receptors—T cell receptors and **B cell receptors** (a type of antibody)—recognize specific antigens. These receptors are the basis of the immune system's capacity to recognize billions of different antigens.

- *Use the animations on CengageNOW to see an antibody-mediated immune response, how antigen receptor diversity is generated, and clonal selection of B cells.*

Section 38.8 Antigen-presenting cells, T cells, and NK cells interact in cell-mediated responses. They target and kill body cells altered by infection or cancer.

- *Use the animation on CengageNOW to observe a cell-mediated immune response.*

Sections 38.9–38.11 **Allergens** are normally harmless substances that induce an immune response; sensitivity to an allergen is called an **allergy**. **Immunization** with **vaccines** designed to elicit immunity to specific diseases saves millions of lives each year. In an **autoimmune response**, a body's own cells are inappropriately recognized as foreign and attacked. Immune deficiency is a reduced capacity to mount an immune response.

Section 38.12 **AIDS** is caused by HIV, a virus that destroys the immune system mainly by infecting helper T cells. At present, AIDS cannot be cured.

Self-Quiz   *Answers in Appendix III*

1. _____ is/are the first line of defense against threats.
   a. Skin, mucous membranes   d. Resident bacteria
   b. Tears, saliva, gastric fluid   e. a through c
   c. Urine flow   f. all of the above

2. Complement proteins _____ .
   a. form pore complexes   c. neutralize toxins
   b. promote inflammation   d. a and b

3. _____ trigger immune responses.
   a. Cytokines   d. Antigens
   b. Lysozymes   e. Histamines
   c. Immunoglobulins   f. all of the above

4. Name one defining characteristic of innate immunity.

5. Name one defining characteristic of adaptive immunity.

6. Antibodies are _____ .
   a. antigen receptors   c. proteins
   b. made only by B cells   d. all of the above

7. _____ binding antigen triggers allergic responses.
   a. IgA   b. IgE   c. IgG   d. IgM   e. IgD

Data Analysis Exercise

In 2003, Michelle Khan and her coworkers published their findings on a 10-year study in which they followed cervical cancer incidence and HPV status in 20,514 women. All women who participated in the study were free of cervical cancer when the test began. Pap tests were taken at regular intervals, and the researchers used a DNA probe hybridization test to detect the presence of specific types of HPV in the women's cervical cells.

The results are shown in Figure 38.24 as a graph of the incidence rate of cervical cancer by HPV type. Women who are HPV positive are often infected by more than one type, so the data were sorted into groups based on the women's HPV status ranked by type: either positive for HPV16; or negative for HPV16 and positive for HPV18; or negative for HPV16 and 18 and positive for any other cancer-causing HPV; or negative for all cancer-causing HPV.

1. At 110 months into the study, what percentage of women who were not infected with any type of cancer-causing HPV had cervical cancer? What percentage of women who were infected with HPV16 also had cervical cancer?

2. In which group would women infected with both HPV16 and HPV18 fall?

3. Is it possible to estimate from this graph the overall risk of cervical cancer that is associated with infection of cancer-causing HPV of any type?

4. Do these data support the conclusion that being infected with HPV16 or HPV18 raises the risk of cervical cancer?

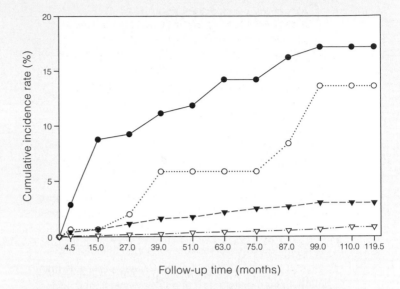

Figure 38.24 Cumulative incidence rate of cervical cancer correlated with HPV status in 20,514 women aged 16 years and older.

The data were grouped as follows: HPV16 positive (closed circles), or else HPV18 positive (open circles), or else all other cancer-causing HPV types combined (closed triangles). Open triangles: no cancer-causing HPV type was detected.

8. Antibody-mediated responses work against _____ .
a. intracellular pathogens
b. extracellular pathogens
c. cancerous cells
d. both a and b
e. both b and c
f. a, b, and c

9. Cell-mediated responses work against _____ .
a. intracellular pathogens
b. extracellular pathogens
c. cancerous cells
d. both a and b
e. both a and c
f. a, b, and c

10. _____ are targets of cytotoxic T cells.
a. Extracellular virus particles in blood
b. Virus-infected body cells or tumor cells
c. Parasitic flukes in the liver
d. Bacterial cells in pus
e. Pollen grains in nasal mucus

11. Allergies occur when the body responds to _____ .
a. pathogens
b. normally harmless substances
c. toxins
d. all of the above

12. Match the immunity concepts.
___ anaphylactic shock
___ antibody secretion
___ phagocyte
___ immune memory
___ autoimmunity
___ antigen receptor
___ inflammation

a. neutrophil
b. effector B cell
c. general defense
d. immune response against own body
e. secondary response
f. B cell receptor
g. hypersensitivity to an allergen

■ *Visit CengageNOW for additional questions.*

Critical Thinking

1. As described in Section 38.10, Edward Jenner was lucky. He performed a potentially harmful experiment on a boy who managed to survive the procedure. What would happen if a would-be Jenner tried to do the same thing today in the United States?

2. Elena developed chicken pox when she was in first grade. Later in life, when her children developed chicken pox, she remained healthy even though she was exposed to countless virus particles daily. Explain why.

3. Before each flu season, you get a flu shot, an influenza vaccination. This year, you get "the flu" anyway. What happened? There are at least three explanations.

4. Monoclonal antibodies are made by immunizing a mouse with a particular antigen, then removing its spleen. Individual B cells producing mouse antibodies specific for the antigen are isolated from the mouse's spleen and fused with cancerous B cells from a myeloma cell line.

The resulting hybrid myeloma cells—hybridoma cells—are cloned, or grown in tissue culture as separate cell lines. Each line produces and secretes antibodies that recognize the antigen to which the mouse was immunized. These monoclonal antibodies can be purified and used for research or other purposes.

Monoclonal antibodies are sometimes used in passive immunization. They tend to be effective, but only in the immediate term. IgG produced by one's own immune system can last up to about six months in the bloodstream, but monoclonals delivered in passive immune therapy typically last for less than a week. Why the difference?

39 Respiration

IMPACTS, ISSUES Up in Smoke

Each day, 3,000 or so teenagers join the ranks of habitual smokers in the United States. Most are not even fifteen years old. When they first light up, they cough and choke on the irritants in the smoke. Most become dizzy and nauseated, and develop headaches. Sound like fun? Hardly. Why, then, do they ignore signals about threats to the body and work so hard to be a smoker? Mainly to fit in. To many adolescents, a misguided perception of social benefits overwhelms the seemingly remote threats to health (Figure 39.1).

Despite teenage perceptions, changes that can make the threat a reality start right away. Ciliated cells keep many pathogens and pollutants that enter airways from reaching the lungs. These cells can be immobilized for hours by the smoke from a single cigarette. Smoke also kills white blood cells that patrol and defend respiratory tissues. Pathogens multiply in the undefended airways. The result is more colds, more asthma attacks, and more bronchitis.

The highly addictive stimulant nicotine constricts blood vessels, which increases blood pressure. The heart has to work harder to pump blood through the narrowed tubes. Nicotine also triggers a rise in "bad" cholesterol (LDL) and a decline in the "good" kind (HDL) in blood. It makes blood stickier, encouraging clots that can block blood vessels.

Tobacco smoke has more than forty known carcinogens and 80 percent of lung cancers occur in smokers. Women who smoke are more susceptible to cancers than men. On average, women develop cancers earlier, and with lower exposure to tobacco. Fewer than 15 percent of women diagnosed with lung cancer survive five years. Smoking also increases breast cancer risk; females who start to smoke as teenagers are about 70 percent more likely to get breast cancer than those women who never smoked. Therefore, the trend of increased smoking among women in less-developed countries especially troubling.

Families, coworkers, and friends get unfiltered doses of the carcinogens in tobacco smoke. Each year in the United States, lung cancers arising from secondhand smoke kill about 3,000. Children exposed to secondhand smoke also are more likely to develop chronic middle ear infections, asthma, and other respiratory problems later in life.

This chapter samples a few respiratory systems. All exchange gases with the outside environment. They also contribute to homeostasis—maintaining the body's internal operating conditions within ranges that cells can tolerate. If you or someone you know smokes, you might use the chapter as a guide to smoking's impact on health. For a more graphic preview, find out what goes on every day with smokers in hospital emergency rooms or intensive care units. There's no glamor there. It is not cool, and it is not pretty.

See the video! Figure 39.1 Learning to smoke is easy, compared with trying to quit. In one survey, two-thirds of female smokers who were sixteen to twenty-four wanted to give up smoking entirely. Of those who tried to quit, only about 3 percent remained nonsmokers for an entire year.

Key Concepts

Principles of gas exchange

Respiration is the sum of processes that move oxygen from air or water in the environment to all metabolically active tissues and move carbon dioxide from those tissues to the outside. Oxygen levels are more stable in air than in water. **Sections 39.1, 39.2**

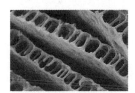

Gas exchange in invertebrates

Gas exchange occurs across the body surface or gills of aquatic invertebrates. In large invertebrates on land, it occurs across a moist, internal respiratory surface or at fluid-filled tips of branching tubes that extend from the surface to internal tissues. **Section 39.3**

Gas exchange in vertebrates

Gills or paired lungs are gas exchange organs in most vertebrates. The efficiency of gas exchange is improved by mechanisms that cause blood and water to flow in opposite directions at gills, and by muscle contractions that move air into and out of lungs. **Sections 39.4–39.7**

Respiratory problems

Respiration can be disrupted by damage to respiratory centers in the brain, physical obstructions, infectious disease, and inhalation of pollutants, including cigarette smoke. **Section 39.8**

Gas exchange in extreme environments

At high altitudes, the human body makes short-term and long-term adjustments to the thinner air. Built-in respiratory mechanisms and specialized behaviors allow sea turtles and diving marine mammals to stay under water, at great depths, for long periods. **Section 39.9**

Links to Earlier Concepts

- Understanding diffusion (Section 5.3) and aerobic respiration (8.1) will help you understand the need for gas exchange and the process by which it occurs. You will also revisit the role of red blood cells (37.2) and the hemoglobin they hold (3.6).

- You will learn about the role of the brain stem (33.10), autonomic nervous system (33.8), and chemoreceptors (34.1) in the regulation of breathing. You will also be reminded of the role of the respiratory system in temperature regulation (27.3).

- You will see how adaptations of animal body plans (25.1, 17.1) and the evolutionary changes that accompanied the move of vertebrates onto land (26.5) allow respiration in specific environments.

- Respiratory effects of algal blooms (22.5), tuberculosis (21.8), and marijuana use (33.7) are also discussed.

How would you vote? Tobacco is a threat to health and a profitable product for American companies. As tobacco use declines in the United States, should the United States government encourage international efforts to reduce tobacco use around the world? See CengageNOW for details, then vote online.

The Nature of Respiration

■ All animals must supply their cells with oxygen and rid their body of carbon dioxide.

■ Links to Diffusion 5.3, Aerobic respiration 8.1

All animals move their body or body parts during at least some part of their life cycle. This movement requires energy, which is usually supplied by ATP. The most efficient way to make ATP is aerobic respiration, a pathway that requires oxygen and releases carbon dioxide as a by-product (Section 8.1). How does an animal supply its cells with the oxygen necessary for aerobic respiration and rid itself of carbon dioxide waste? In animals that have organ systems, a respiratory system carries out these tasks. In humans and other vertebrates, the respiratory system interacts with other organ systems as shown in Figure 39.2.

The Basis of Gas Exchange

Respiration is the physiological process by which an animal exchanges oxygen and carbon dioxide with its environment. Respiration depends upon the tendency of gaseous oxygen (O_2) and carbon dioxide (CO_2) to diffuse down their concentration gradients—or, as we say for gases, their pressure gradients—between the external and internal environments.

Aquatic animals live in an environment where the availability of O_2 can vary widely from place to place and change over time. Air is a more reliable source of oxygen. Earth's atmosphere is 78 percent nitrogen,

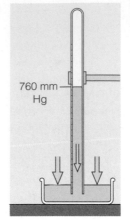

Figure 39.3 How a mercury barometer measures atmospheric pressure. That pressure makes mercury (Hg), a viscous liquid, rise or fall in a narrow tube. At sea level, it rises 760 millimeters (29.91 inches) from the tube's base. Atmospheric pressure varies with altitude. On the top of Mount Everest, atmospheric pressure is only about one-third the pressure at sea level.

760 mm Hg

21 percent oxygen, 0.04 percent carbon dioxide, and 0.06 percent other gases. Total atmospheric pressure as measured by a mercury barometer is 760 mm at sea level (Figure 39.3). Oxygen's contribution to the total, its **partial pressure**, is 21 percent of 760, or 160 mm Hg. "Hg" is the symbol for mercury.

Gases enter and leave the internal environment by crossing a **respiratory surface**: a moist layer thin enough for gases to diffuse across. The surface has to be moist because gases can only diffuse quickly across a membrane if they first dissolve in fluid.

Factors Affecting Diffusion Rates

Several factors affect how much gas diffuses across a respiratory surface. For example, the steeper the partial pressure gradient, the faster the rate of diffusion.

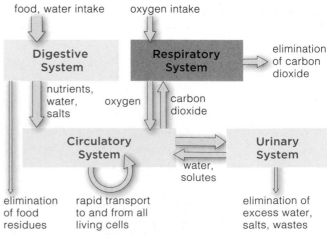

food, water intake   oxygen intake

Digestive System   **Respiratory System**   elimination of carbon dioxide

nutrients, water, salts   oxygen   carbon dioxide

Circulatory System   **Urinary System**

water, solutes

elimination of food residues   rapid transport to and from all living cells   elimination of excess water, salts, wastes

Figure 39.2 A dog's breathing helps meet its cells' need for oxygen. In dogs and other vertebrates, the respiratory system interacts with other organ systems that contribute to homeostasis.

Surface-to-Volume Ratio The greater the area of a respiratory surface, the more molecules can cross it in any given interval. Remember, as an animal grows, its volume increases faster than its surface area does (Section 4.2). If an animal does not have specialized respiratory organs, it usually has a small, flattened body. In such animals, diffusion alone delivers enough oxygen to cells, because no cell is more than a few millimeters away from gases outside the body.

Ventilation Moving air or water past a respiratory surface keeps the pressure gradient across the surface high and thus increases the rate of gas exchange. For example, frogs and humans breathe in and out, which ventilates their lungs. Breathing forces stale air with waste CO_2 away from the respiratory surface in the lungs, and draws in fresh air with more O_2. Fish and other animals that live in water have mechanisms that keep water moving across their respiratory surface.

Respiratory Proteins **Respiratory proteins** contain one or more metal ions that reversibly bind oxygen atoms. Oxygen atoms bind to these proteins when the partial pressure of oxygen is high, and are released when the partial pressure of oxygen declines. By reversibly binding oxygen, respiratory proteins help maintain a steep partial pressure gradient for oxygen between cells and the blood. The gradient is steepened because any oxygen that is bound to a molecule in solution does not contribute to the partial pressure of O_2 in that solution.

Hemoglobin, an iron-containing respiratory protein, fills vertebrate red blood cells (Sections 3.6 and 37.2). It also circulates in the blood of annelids, mollusks, and crustaceans, which do not have red blood cells. The respiratory proteins hemerythrin (with iron) and hemocyanin (with copper) also aid oxygen transport in some invertebrates. **Myoglobin** a heme-containing respiratory protein, is found in muscle of vertebrates and some invertebrates. It helps stabilize the oxygen level inside muscle cells.

Take-Home Message

What is respiration and what factors influence it?

■ Respiration supplies cells with oxygen for aerobic respiration and removes carbon dioxide wastes.

■ Gases are exchanged by diffusion across a respiratory surface: a thin, moist membrane.

■ The area of a respiratory surface and the partial pressure gradients across it influence the rate of exchange. Ventilation and respiratory proteins help keep partial pressure gradients steep and thus enhance gas exchange.

39.2 | Gasping for Oxygen

■ Rising water temperatures, slowing streams, and organic pollutants reduce the oxygen available for aquatic species.

■ Link to Algal bloom 22.5

Any animal can tolerate only a limited range of environmental conditions. For aquatic animals, dissolved oxygen content of water (DO) is one of the most important factors affecting their survival. More oxygen dissolves in cooler, fast-flowing water than in warmer, still water. When water temperature increases or water becomes stagnant, aquatic species that have high oxygen needs suffocate (Figure 39.4).

As oxygen levels in water fall, so does biodiversity. Pollution can cause DO to decline. A lake enriched with runoff that contains manure or sewage offers a nutrition boost to aerobic bacteria living on the lake bottom. The bacteria are decomposers. As their populations soar, they use up lots of oxygen, so the amount available to other species plummets. The same thing can happen after phosphate-rich or nitrogen-rich fertilizers cause an algal bloom—a population explosion of protists such as dinoflagellates (Section 22.5). The protists multiply rapidly, then die. Their decomposition depletes the water of oxygen.

In freshwater lakes and streams, aquatic larvae of mayflies and stoneflies are the first invertebrates to disappear when oxygen levels fall. These insect larvae are active predators that demand considerable oxygen. Gilled snails disappear, too. Such invertebrate declines have cascading effects on fishes that feed on them. Some fish are more directly affected. Trout and salmon are especially intolerant of low oxygen. Carp (including koi and goldfish) are among the most tolerant of oxygen declines; they survive even in warm algae-rich ponds or tiny goldfish bowls.

When oxygen levels fall below 4 parts per million, no fishes can survive. Leeches thrive as most competing invertebrates disappear. In waters with the lowest oxygen concentration, annelids called sludge worms (*Tubifex*) often are the only animals. They are colored red by large amounts of hemoglobin. Compared to the hemoglobin in most organisms, the *Tubifex* hemoglobin is better at binding oxygen when oxygen levels are low. A high affinity for oxygen allows these worms to exploit low-oxygen habitats such as sediments in deep lakes, where food is plentiful and competitors and predators are scarce.

Figure 39.4 A fish kill. When the oxygen level in water declines, fishes and other aquatic organisms can suffocate.

■ Invertebrates arose in water but some groups evolved respiratory organs that allow them to breathe air.

■ Link to Animal body plans 25.1

Integumentary Exchange

Some invertebrates do not have any respiratory organs (Figure 39.5a,b). Sponges, cnidarians, flatworms, and earthworms are examples. Such animals live in aquatic or continually damp land environments and rely on **integumentary exchange**: the diffusion of gases across their outer body surface, or integument. Animals that depend on this method of gas exchange usually are small and flat, or when larger, have cells arranged in thin layers. Integumentary exchange also supplements the effects of respiratory organs in many invertebrates that have gills, and even some vertebrates.

Invertebrate Gills

Gills are filamentous respiratory organs that increase the surface area available for gas exchange in many aquatic animals. Blood vessels in gill filaments pick up oxygen and distribute it throughout the body.

Most aquatic mollusks draw water into their mantle cavity, where it flows over a gill (as shown earlier in Figures 25.23, 25.25, and 25.26). In some sea slugs, gills are visible on the body surface (Figure 39.5c).

Many aquatic arthropods such as lobsters and crabs have feathery gills inside their exoskeleton, where the delicate tissues are protected from damage. The gills evolved from walking legs.

Snails With Lungs

Snails and slugs that spend some time on land have a lung instead of, or in addition to, their gill. A **lung** is a saclike respiratory organ. Inside it, branching tubes deliver air to a respiratory surface serviced by many blood vessels. In snails and slugs, a pore at the side of the body can be opened to allow air into the lung, and it can be shut to conserve water (Figure 39.6).

Tracheal Tubes and Book Lungs

The most successful air-breathing land invertebrates are insects and arachnids, such as spiders. They have a hard integument that helps conserve water but also

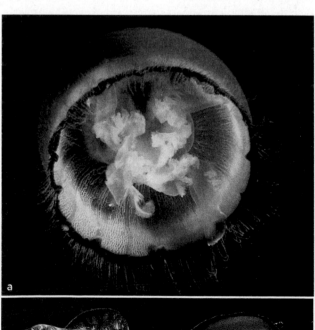

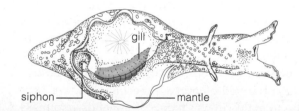

Figure 39.5 Respiration in water. (**a**) A jellyfish and (**b**) a marine flatworm do not have respiratory organs. All of the cells in these animals lie close to the body surface, and gas exchange takes place by diffusion across that surface.

(**c**) Gill of the marine sea slug *Aplysia*, a mollusk. Having a gill increases the surface area for gas exchange. Blood vessels that run through the gill carry gases to and from body tissues.

Figure 39.6 A land snail (*Helix aspersa*) with the opening that leads to its lung visible at the left. Compare Figure 25.24.

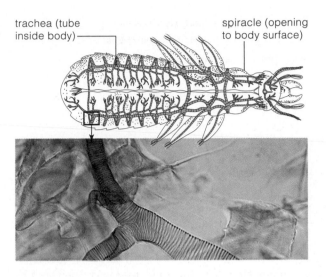

trachea (tube inside body)

spiracle (opening to body surface)

Figure 39.7 Insect tracheal system. Chitin rings reinforce branching, air-filled tubes in such respiratory systems.

blocks gas exchange. Insects and some spiders have a **tracheal system** that consists of repeatedly branching, air-filled tubes reinforced with chitin.

Tracheal tubes start at spiracles—small openings across the integument (Figure 39.7). There is usually a pair of spiracles per segment; one on each side of the body. They can be opened or closed to regulate the amount of oxygen that enters the body. Substances that clog spiracles are used as insecticides. For example, horticultural oils sprayed on fruit trees kill scale bugs, aphids, and mites by clogging their spiracles.

At the tips of the finest tracheal branches is a bit of fluid in which gases dissolve. The tips of insect tracheal tubes are adjacent to body cells, and oxygen and carbon dioxide diffuse between these tubes and the tissues. Because tracheal tubes end next to cells, insects have no need for a respiratory protein such as hemoglobin to carry gases.

Some insects can force air into and out of tracheal tubes. For example, when a grasshopper's abdominal muscles contract, organs press on the pliable tracheal tubes and force air out of them. When these muscles relax, pressure on tracheal tubes decreases, the tubes widen, and air rushes in.

Some spiders have one or two **book lungs** in addition to or instead of tracheal tubes. In a book lung, air and blood exchange gases across thin sheets of tissue (Figure 39.8). Hemocyanin in a spider's blood picks up oxygen and turns blue-green as it passes through a book lung. It gives up oxygen and becomes colorless in body tissues.

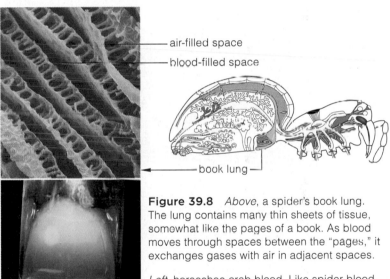

air-filled space

blood-filled space

book lung

Figure 39.8 *Above*, a spider's book lung. The lung contains many thin sheets of tissue, somewhat like the pages of a book. As blood moves through spaces between the "pages," it exchanges gases with air in adjacent spaces.

Left, horseshoe crab blood. Like spider blood, it contains the respiratory pigment hemocyanin, which turns blue-green when carrying oxygen.

Take-Home Message

How do invertebrates exchanges gases with their environment?

■ Some invertebrates do not have respiratory organs and exchange gases across the body wall. This process also supplements the action of gills in many invertebrates.

■ Gills are filamentous organs that increase the surface area for gas exchange in aquatic habitats. Blood vessels run through gill filaments.

■ Some land snails have a lung in their mantle cavity. Land arthropods have tracheal tubes or book lungs, respiratory organs that bring air deep inside their body.

■ Fishes use gills to extract oxygen from water; land vertebrates obtain it from air that enters their lungs.

■ Link to Move to land 26.5

The Gills of Fishes

gill cover

All fishes have gill slits that open across the pharynx (their throat region). In jawless fishes and cartilaginous fishes, the gill slits are visible from the outside, but bony fishes have a gill cover that hides them (Figure 39.9a).

In all fishes, respiration occurs when water flows into the mouth, enters the pharynx, then moves out of the body through the gill slits. Some sharks swim constantly with their mouth open, so water flows passively over their gills. However, most fish actively draw water over their gills. A bony fish sucks water inward by opening its mouth, closing its gill covers, and contracting muscles that enlarge the oral cavity

(Figure 39.9b). Water is forced out when the fish closes its mouth, opens the gill cover, and contracts muscles that make the oral cavity smaller (Figure 39.9c).

If you could remove the gill cover of a bony fish, you would see that the gills themselves consist of bony gill arches, each with many gill filaments attached (Figure 39.10a,b). Each gill filament holds many capillary beds where gases are exchanged with blood.

Blood in a gill capillary and water flowing past gill filaments move in opposite directions (Figure 39.10c). The result is a **countercurrent exchange**, in which two fluids exchange substances while flowing in opposite directions. Oxygen-poor blood enters a capillary and travels past water with an increasing oxygen content. Because these fluids flow in opposite directions, their oxygen content can never equalize, as it would if they flowed in the same direction. As a result, oxygen diffuses from water into the blood all along the capillary.

Evolution of Paired Lungs

The first vertebrate lungs evolved from outpouchings of the gut wall in some bony fishes. Such lungs may have helped these fishes survive short trips on land. Gills would have been useless in air: Without water to buoy them up and keep them moist, gills would collapse under their own weight and dry out. Lungs became increasingly important as aquatic tetrapods spent more time on land (Section 26.5).

Amphibian larvae have external gills. Most often, as the animal develops, these gills disappear and are replaced by paired lungs. Amphibians also exchange some gases across their thin-skinned body surface. In all amphibians, most carbon dioxide that forms during aerobic respiration leaves the body across the skin.

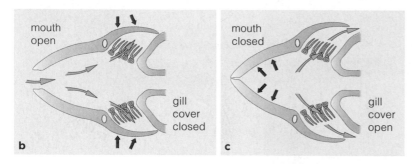

Figure 39.9 (**a**) Location of the gill cover of a bony fish. (**b**) Water is sucked into the mouth and over the gills when a fish closes its gill covers, opens its mouth, and expands its oral cavity. (**c**) The water moves out when the fish closes its mouth, opens its gill covers, and squeezes the water past its gills.

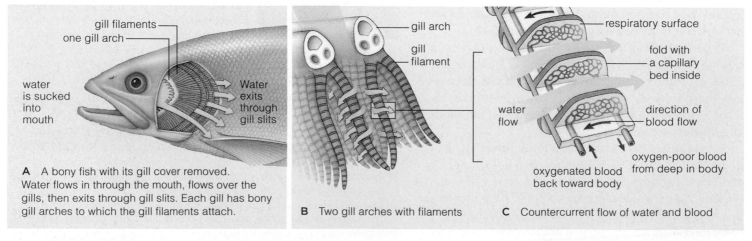

A A bony fish with its gill cover removed. Water flows in through the mouth, flows over the gills, then exits through gill slits. Each gill has bony gill arches to which the gill filaments attach.

B Two gill arches with filaments

C Countercurrent flow of water and blood

Figure 39.10 Animated Structure and function of the gills of a bony fish.

A Lowering the floor of the mouth draws air inward through nostrils.

B Closing nostrils and raising the floor of the mouth pushes air into lungs.

C Rhythmically raising and lowering the floor of the mouth assists gas exchange.

D Contracting chest muscles and raising the floor of the mouth forces air out of lungs, and the frog exhales.

Figure 39.11 Animated How a frog breathes.

Frogs have paired lungs. They inhale by lowering the floor of the mouth, which draws air in through their nostrils. Then they close their nostrils and lift the floor of the mouth and throat, pushing air into the lungs (Figure 39.11).

Reptiles, birds, and mammals—the amniotes—have waterproof skin and no gills as adults. Gas exchange occurs in their two well-developed lungs. Contraction of chest muscles draws air through airways and into the lungs.

In reptiles and mammals, gas exchange occurs in sacs at the ends of the smallest airways. In birds, there are no such "dead ends" inside the lung. Birds have small, inelastic lungs that do not expand and contract when the bird breathes. Instead, air sacs attached to the lungs inflate and deflate. It takes two breaths to move air through this system (Figure 39.12). Oxygen-rich air flows through tiny tubes in the lung during both inhalations and exhalations. The lining of these tubes is the respiratory surface. Continual movement of air past this surface greatly increases the efficiency of gas exchange.

We turn next to the human respiratory system. Its operating principles apply to most vertebrates, even though lungs evolved differently among them.

A Inhalation 1
Muscles expand chest cavity, drawing air in through nostrils. Some of the air flowing in through the trachea goes to lungs and some goes to posterior air sacs.

B Exhalation 1
Anterior air sacs empty. Air from posterior air sacs moves into lungs.

C Inhalation 2
Air in lungs moves to anterior air sacs and is replaced by newly inhaled air.

D Exhalation 2
Air in anterior air sacs moves out of the body and air from posterior sacs flows into the lungs.

trachea

anterior air sacs

lung

posterior air sacs

Figure 39.12 Animated Respiratory system of a bird. Large, stretchy air sacs attach to two small, inelastic lungs. Contraction and expansion of chest muscles cause air to flow into and then out of this system.

Air flows in through many air tubes inside the lung, and into posterior air sacs. The lining of the tiniest air tubes, sometimes called air capillaries, is the site of gas exchange—the respiratory surface.

It takes more than one breath for air to flow through the system, but air flows continuously through the lungs and over the respiratory surface. This unique ventilating system supports the high metabolic rates that birds require for flight and other energy-demanding activities.

Right, this scanning electron micrograph of lung tissue shows the tubes through which air flows to and from air sacs. Gas exchange takes place across the lining of these tubes.

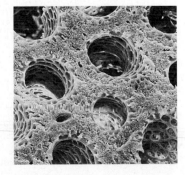

Take-Home Message

What kind of respiratory systems do vertebrates have?

■ Most fish exchange gases with water that flows over their gills. The direction of blood flow in gill capillaries is opposite that of water flow. This countercurrent flow aids gas exchange.

■ Amphibians exchange gases across their skin and (usually) at the respiratory surface of paired lungs.

■ Reptiles, birds, and mammals do not exchange any gases across the skin. They rely on paired lungs. Birds have the most efficient vertebrate lungs. A system of air sacs ensures that air moves constantly through a bird's lung.

Human Respiratory System

- The human respiratory system functions in gas exchange, but also in speech, in the sense of smell, and in homeostasis.

- Link to Temperature regulation 27.3

The System's Many Functions

Figure 39.13 shows the human respiratory system and lists the functions of its parts. It also shows skeletal muscles that assist in respiration. Rhythmic contraction and relaxation of these muscles cause air to move into and out of the lungs.

The respiratory system functions in gas exchange, but it has a wealth of additional roles. We can speak, sing, or shout by controlling vibrations as air moves past our vocal cords. We have a sense of smell because airborne molecules stimulate olfactory receptors in the nose. Cells lining nasal passages and other airways of the system help defend the body; they intercept and neutralize airborne pathogens. The respiratory system contributes to the body's acid–base balance by getting rid of carbon dioxide wastes. Controls over breathing even help maintain body temperature, because water evaporating from airways has a cooling effect.

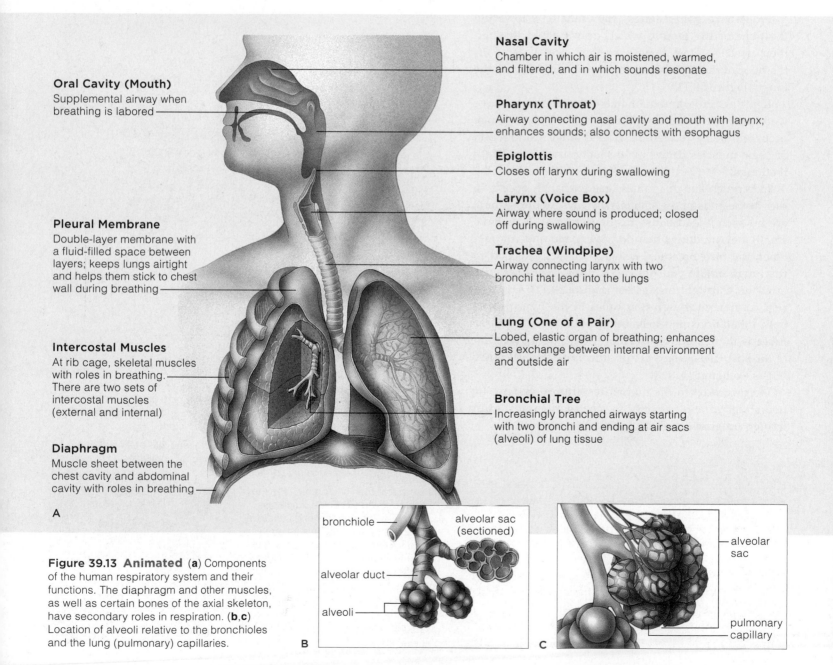

Oral Cavity (Mouth)
Supplemental airway when breathing is labored

Pleural Membrane
Double-layer membrane with a fluid-filled space between layers; keeps lungs airtight and helps them stick to chest wall during breathing

Intercostal Muscles
At rib cage, skeletal muscles with roles in breathing. There are two sets of intercostal muscles (external and internal)

Diaphragm
Muscle sheet between the chest cavity and abdominal cavity with roles in breathing

Nasal Cavity
Chamber in which air is moistened, warmed, and filtered, and in which sounds resonate

Pharynx (Throat)
Airway connecting nasal cavity and mouth with larynx; enhances sounds; also connects with esophagus

Epiglottis
Closes off larynx during swallowing

Larynx (Voice Box)
Airway where sound is produced; closed off during swallowing

Trachea (Windpipe)
Airway connecting larynx with two bronchi that lead into the lungs

Lung (One of a Pair)
Lobed, elastic organ of breathing; enhances gas exchange between internal environment and outside air

Bronchial Tree
Increasingly branched airways starting with two bronchi and ending at air sacs (alveoli) of lung tissue

A

bronchiole
alveolar sac (sectioned)
alveolar duct
alveoli

alveolar sac
pulmonary capillary

Figure 39.13 Animated (**a**) Components of the human respiratory system and their functions. The diaphragm and other muscles, as well as certain bones of the axial skeleton, have secondary roles in respiration. (**b**,**c**) Location of alveoli relative to the bronchioles and the lung (pulmonary) capillaries.

B

C

From Airways to Alveoli

The Respiratory Passageways Take a deep breath. Now look at Figure 39.13 to get an idea of where the air traveled in your respiratory system.

If you are healthy and sitting quietly, air probably entered through your nose, rather than your mouth. As air moves through your nostrils, tiny hairs filter out any large particles. Mucus secreted by cells of the nasal lining captures most fine particles and airborne chemicals. Ciliated cells in the nasal lining also help remove any inhaled contaminants.

Air from the nostrils enters the nasal cavity, where it gets warmed and moistened. It flows next into the **pharynx**, or throat. It continues to the **larynx**, a short airway commonly known as the voice box because a pair of vocal cords projects into it (Figure 39.14). Each vocal cord is skeletal muscle with a cover of mucus-secreting epithelium. Contraction of the vocal cords changes the size of the **glottis**, the gap between them.

When the glottis is wide open, air flows through it silently. When muscle contraction narrows the glottis, flow of air outward through the tighter gap makes vocal cords vibrate and produces sounds. The tension on the cords and the position of the larynx determine the sound's pitch. To get a feel for how this works, place one finger on your "Adam's apple," the laryngeal cartilage that sticks out most at the front of your neck. Hum a low note, then a high one. You will feel the vibration of your vocal cords and how laryngeal muscles shift the position of your larynx.

In laryngitis, overuse or infection has inflamed the vocal cords. The swollen cords cannot vibrate as they should, which makes speaking difficult.

At the entrance to the larynx is an **epiglottis**. When this tissue flap points up, air moves into the **trachea**, or windpipe. When you swallow, the epiglottis flops over, points down, and covers the larynx entrance, so food and fluids enter the esophagus. The esophagus connects the pharynx to the stomach.

The trachea branches into two airways, one to each lung. Each airway is a **bronchus** (plural, bronchi). Its epithelial lining has many ciliated and mucus-secreting cells that fend off respiratory tract infections. Bacteria and airborne particles stick to the mucus. Cilia sweep the mucus toward the throat for expulsion.

The Paired Lungs Human lungs are cone-shaped organs in the thoracic cavity, one on each side of the heart. The rib cage encloses and protects the lungs. A two-layer-thick pleural membrane covers each lung's outer surface and lines the inner thoracic cavity wall.

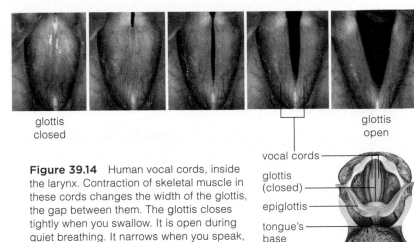

glottis closed

glottis open

vocal cords
glottis (closed)
epiglottis
tongue's base

Figure 39.14 Human vocal cords, inside the larynx. Contraction of skeletal muscle in these cords changes the width of the glottis, the gap between them. The glottis closes tightly when you swallow. It is open during quiet breathing. It narrows when you speak, so that air flow causes the cords to vibrate.

Once inside a lung, air moves through finer and finer branchings of a "bronchial tree." The branches are called **bronchioles**. At the tips of the finest bronchioles are respiratory **alveoli** (singular, alveolus), little air sacs where gases are exchanged (Figure 39.13b,c). Each alveolus has a wall that is only one cell thick. Collectively, the many alveoli provide an extensive surface for gas exchange. If all 6 million alveoli in your lungs could be stretched out in a single layer, they would cover half of a tennis court!

Air in alveoli exchanges gases with blood flowing through pulmonary capillaries (Latin *pulmo*, lung). At this point, a different organ system gets involved. The circulatory system transports oxygen to body tissues and carries carbon dioxide away from them.

Muscles and Respiration A broad sheet of smooth muscle beneath the lungs, the **diaphragm**, partitions the coelom into a thoracic cavity and an abdominal cavity. Of all smooth muscle, it alone can be controlled voluntarily. You can make it contract by deliberately inhaling. The diaphragm and **intercostal muscles**, the skeletal muscles between the ribs, interact to change the volume of the thoracic cavity during breathing.

Take-Home Message

What roles do the components of the human respiratory system play?

■ In addition to gas exchange, the human respiratory system acts in the sense of smell, voice production, body defenses, acid–base balance, and temperature regulation.

■ Air enters through the nose or mouth. It flows through the pharynx (throat) and larynx (voice box) to a trachea that branches into two bronchi, one to each lung. Inside each lung, additional branching airways deliver air to alveoli, where gases are exchanged with pulmonary capillaries.

■ Rhythmic signals from the brain cause muscle contractions that cause air to flow into the lungs.

■ Links to Autonomic signals 33.8, Brain stem 33.10, Chemoreceptors 34.1

The Respiratory Cycle

A **respiratory cycle** is one breath in (inhalation) and one breath out (exhalation). Inhalation is always active; muscle contractions drive it. Changes in the volume of the lungs and thoracic cavity during a respiratory cycle alter pressure gradients between air inside and outside the respiratory tract (Figures 39.15 and 39.16).

When you inhale, the diaphragm flattens, moving downward. External intercostal muscles contract and lift the rib cage up and outward (Figure 39.15*a*). As the thoracic cavity expands, so do the lungs. Pressure in the alveoli falls below atmospheric pressure, and air flows down the pressure gradient, into the airways.

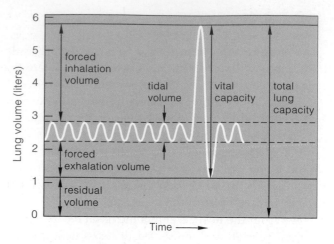

Figure 39.16 Animated Respiratory volumes. In normal breathing, lungs hold 2.7 liters at the end of inhalation and 2.2 liters at the end of exhalation; the tidal volume of air entering and leaving is 0.5 liter. Lungs never deflate completely. When air flows out and lung volume is low, the wall of the smallest airways collapses and prevents further air loss.

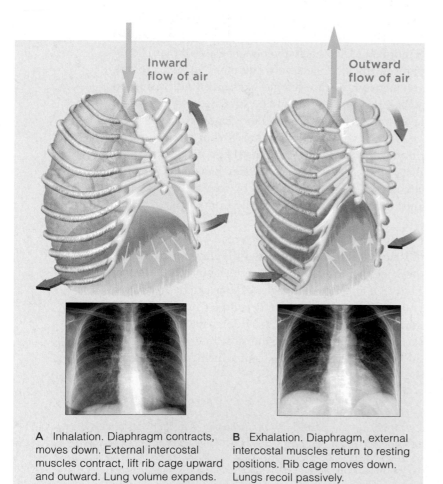

A Inhalation. Diaphragm contracts, moves down. External intercostal muscles contract, lift rib cage upward and outward. Lung volume expands.

B Exhalation. Diaphragm, external intercostal muscles return to resting positions. Rib cage moves down. Lungs recoil passively.

Figure 39.15 Animated Changes in the size of the thoracic cavity during a single respiratory cycle. The x-ray images reveal how inhalation and expiration change the lung volume.

Exhalation is usually passive. When muscles that caused inhalation relax, the lungs passively recoil and lung volume decreases. This compresses alveolar sacs, raising air pressure inside them. Air moves down the pressure gradient, out of the lungs (Figure 39.15*b*).

Exhalation is active only when you exercise vigorously or consciously attempt to expel more air. During active exhalation, internal intercostal muscles contract, pulling the thoracic wall inward and downward. At the same time, muscles of the abdominal wall contract. Abdominal pressure increases and exerts an upward-directed force on the diaphragm. The volume of the thoracic cavity decreases more than normal, and a bit more air is forced out.

Upward-directed force on the diaphragm is also the reason that the **Heimlich maneuver** works (Figure 39.17). Performing this procedure can save the life of a person who is choking. A choking person has food lodged in their trachea. By making upward thrusts into the choker's upper abdomen, a rescuer raises the intra-abdominal pressure, which forces the choker's diaphragm upward. The force of air rushing out of the lungs into the trachea can dislodge the food, allowing the victim to resume breathing.

Respiratory Volumes

The maximum volume of air that the lungs can hold, total lung volume, averages 5.7 liters in men and 4.2 liters in women. Usually lungs are less than half full. **Vital capacity**, the maximum volume that can move in and out in one cycle, is one measure of lung health. **Tidal volume**—the volume that moves in and out in a normal respiratory cycle—is about 0.5 liter (Figure 39.16). Your lungs never fully deflate; thus air inside them always is a mix of freshly inhaled air and "stale air" that was left behind during the previous exhalation. Even so, there is plenty of oxygen for exchange.

Control of Breathing

Neurons in the medulla oblongata of the brain stem serve as a control center for respiration. When you rest, these neurons fire spontaneous action potentials 10 to 14 times per minute. Nerves carry these signals to the diaphragm and intercostal muscles, causing the contractions that result in inhalation. Between action potentials, the muscles relax and you exhale.

Breathing patterns change with activity level. When you are more active, muscle cells increase their rate of aerobic respiration and produce more CO_2. This CO_2 enters blood, where it combines with water and forms carbonic acid (Section 39.7). The acid dissociates and H^+ levels rise in the blood and in cerebrospinal fluid. Chemoreceptors inside the medulla oblongata and in carotid artery and aorta walls detect the change. These receptors signal the respiratory center, which calls for changes in the breathing pattern (Figure 39.18).

Chemoreceptors in the carotid arteries also signal the medulla oblongata when the O_2 partial pressure in arterial blood falls below a life-threatening 60 mm Hg. Ordinarily, the O_2 partial pressure does not fall that low. This control mechanism has survival value only at high altitudes and during severe lung diseases.

Reflexes such as swallowing or coughing can briefly halt breathing. Breathing patterns can also be deliberately altered, as when you hold your breath to dive, or break normal breathing rhythm to talk. In addition, commands from sympathetic nerves make you breathe faster when you are frightened (Section 33.8).

Take-Home Message

What happens when we breathe?

■ Inhalation is always an active process. Contraction of the diaphragm and external intercostal muscles increase the volume of the thoracic cavity. This reduces air pressure in alveoli below atmospheric pressure, so air moves inward.

■ Exhalation is usually passive. As muscles relax, the thoracic cavity shrinks back down, air pressure in alveoli rises above atmospheric pressure, and air moves out.

■ Only some of the air in the lungs is replaced with each breath. The lungs are never fully emptied of air.

■ The brain controls the rate and depth of breathing.

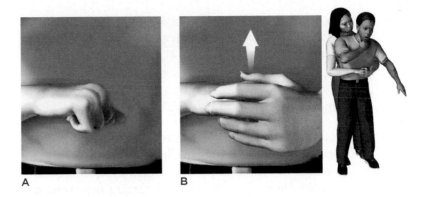

Figure 39.17 Animated How to perform the Heimlich maneuver on an adult who is choking.

1. Determine that the person is actually choking; a person who has an object lodged in their trachea cannot cough or speak.

2. Stand behind the person and place one fist below his or her rib cage, just above the navel, with your thumb facing inward as in (**a**).

3. Cover the fist with your other hand and thrust inward and upward with both fists as in (**b**). Repeat until the object is expelled.

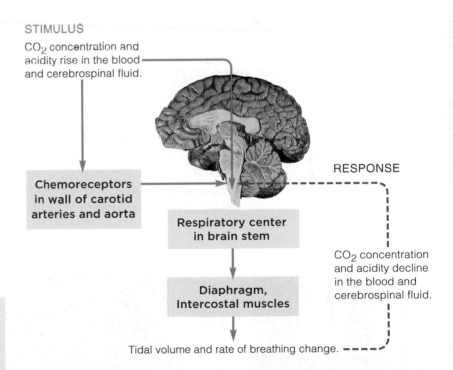

STIMULUS

CO_2 concentration and acidity rise in the blood and cerebrospinal fluid.

Chemoreceptors in wall of carotid arteries and aorta

Respiratory center in brain stem

RESPONSE

CO_2 concentration and acidity decline in the blood and cerebrospinal fluid.

Diaphragm, Intercostal muscles

Tidal volume and rate of breathing change.

Figure 39.18 Respiratory response to increased activity levels. An increase in activity raises the CO_2 output. It also makes the blood and cerebrospinal fluid more acidic. Chemoreceptors in blood vessels and the medulla sense the changes and signal the brain's respiratory center, also in the brain stem.

In response, the respiratory center signals the diaphragm and intercostal muscles. The signals call for alterations in the rate and depth of breathing. Excess CO_2 is expelled, which causes the level of this gas and acidity to decline. Chemoreceptors sense the decline and signal the respiratory center, so breathing is adjusted accordingly.

- Gases are exchanged by diffusion in alveoli.
- Red blood cells play a role in transport of both oxygen and carbon dioxide.

- Links to Hemoglobin 3.6, Red blood cells 37.2

The Respiratory Membrane

Gases diffuse between an alveolus and a pulmonary capillary at the lung's **respiratory membrane**. This thin membrane consists of alveolar epithelium, capillary endothelium, and the fused basement membranes of the alveolus and capillary (Figure 39.19). Secretions keep the alveolar side of the respiratory membrane moist so that gases can diffuse quickly across it.

O_2 and CO_2 diffuse passively across the respiratory membrane. Therefore, the net direction of movement of these gases depends upon their partial pressure gradients across the membrane. Air flow into and out of the lungs and blood flow through pulmonary capillaries keep O_2 and CO_2 partial pressure gradients steep.

Oxygen Transport

The inhaled air that reaches alveoli contains a great deal of O_2 compared to the blood in pulmonary capillaries. As a result, O_2 in the lungs tends to diffuse into blood plasma inside the pulmonary capillaries, and then into red blood cells.

As many as 30 trillion red blood cells circulate in your blood. Each holds many millions of hemoglobin molecules. Again, the hemoglobin molecule consists of four polypeptide chains, each associated with one heme group (Figure 39.20a). Each heme group includes one iron atom that reversibly binds O_2. Hemoglobin with oxygen bonded to it is **oxyhemoglobin**, or HbO_2.

About 98.5 percent of the oxygen you inhale gets bound to heme groups of hemoglobin. The amount of HbO_2 that forms in a given interval depends on the partial pressure of O_2. The higher the partial pressure of O_2, the more HbO_2 will form.

Heme binds O_2 only weakly. It releases O_2 in places where the partial pressure of O_2 is much lower than that in the alveoli. This is true in metabolically active tissues, as the boxes color-coded pink in Figure 39.21 show. Other factors that encourage release of O_2 from heme, including high temperature, low pH, and high CO_2 partial pressure, also are typical of these tissues.

Myoglobin, also an iron-containing respiratory protein, helps cardiac muscle and some skeletal muscles store oxygen. Structurally, myoglobin resembles the globin in hemoglobin, but it holds more tightly onto oxygen (Figure 39.20b). The O_2 that hemoglobin gives up near a cardiac muscle cell diffuses into the cell and binds to myoglobin inside it. When blood flow cannot keep up with a cell's increased O_2 needs, as during periods of intense exercise, the myoglobin releases O_2, which allows mitochondria to keep on making ATP.

Carbon Dioxide Transport

Carbon dioxide diffuses into blood capillaries in any tissue where its partial pressure is higher than it is in blood. This is the case in metabolically active tissues, as the boxes color-coded blue in Figure 39.21 show.

Carbon dioxide is transported to the lungs in three forms. About 10 percent remains dissolved in plasma. Another 30 percent reversibly binds with hemoglobin and forms carbaminohemoglobin ($HbCO_2$). However, most CO_2 that diffuses into the plasma—60 percent— is transported as bicarbonate (HCO_3^-).

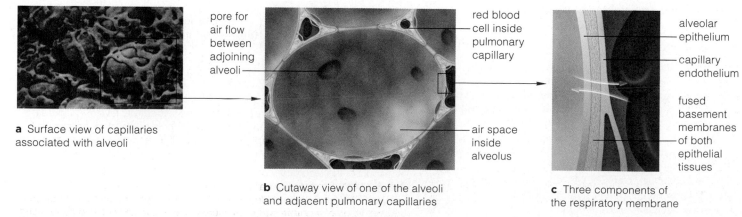

a Surface view of capillaries associated with alveoli

pore for air flow between adjoining alveoli

b Cutaway view of one of the alveoli and adjacent pulmonary capillaries

red blood cell inside pulmonary capillary

air space inside alveolus

alveolar epithelium

capillary endothelium

fused basement membranes of both epithelial tissues

c Three components of the respiratory membrane

Figure 39.19 Zooming in on the respiratory membrane in human lungs.

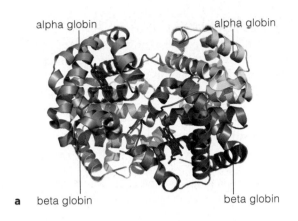 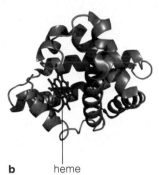

alpha globin   alpha globin

a   beta globin   beta globin   **b**   heme

Figure 39.20 (**a**) Structure of hemoglobin, the oxygen-transporting protein of red blood cells. It consists of four globin chains, each associated with an iron-containing heme group, color-coded *red*.

(**b**) Myoglobin, an oxygen-storing protein in muscle cells. Its single chain associates with a heme group. Compared to hemoglobin, myoglobin has a higher affinity for oxygen, so it helps speed the transfer of oxygen from blood to muscle cells.

How does bicarbonate form? Carbon dioxide first combines with water, forming carbonic acid (H_2CO_3). This compound separates into bicarbonate and H^+:

$$CO_2 + H_2O \rightleftharpoons H_2CO_3 \rightleftharpoons HCO_3^- + H^+$$

carbonic acid   bicarbonate

Red blood cells have **carbonic anhydrase**, an enzyme that catalyzes the above reaction. The bicarbonate that forms in red blood cells diffuses into plasma, whereas most of the H^+ binds to hemoglobin.

When red blood cells reach the alveolar capillaries—where the CO_2 partial pressure is relatively low—the reactions reverse, forming water and CO_2. The CO_2 diffuses into the air in an alveolus and is exhaled.

The Carbon Monoxide Threat

Carbon monoxide (CO) is a colorless, odorless gas. It is present in the smoke from cigarettes and fossil fuel combustion. Hemoglobin has a higher affinity for CO than for O_2. When CO builds up in the air, it fills O_2 binding sites on hemoglobin, preventing transport of O_2 and causing carbon monoxide poisoning. Nausea, headache, confusion, dizziness, and weakness set in as tissues are starved of oxygen. In the United States, accidental CO poisoning kills about 500 people each year. To minimize your risk, be sure that fuel-burning appliances have been properly vented to the outside, and install a carbon monoxide detector.

Take-Home Message

How are gases transported in blood?

■ Most oxygen in blood is bound to hemoglobin, which binds oxygen in alveoli where oxygen partial pressure is high, and releases it in tissues where oxygen partial pressure is lower.

■ Most carbon dioxide is transported in blood in the form of bicarbonate, nearly all of which forms by enzyme action inside red blood cells.

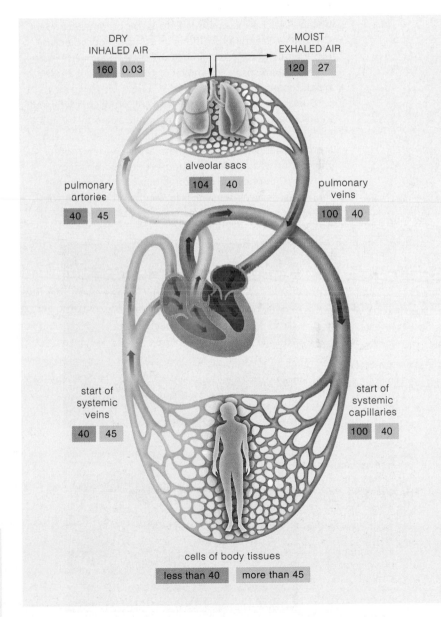

DRY INHALED AIR   160   0.03

MOIST EXHALED AIR   120   27

alveolar sacs   104   40

pulmonary arteries   40   45

pulmonary veins   100   40

start of systemic veins   40   45

start of systemic capillaries   100   40

cells of body tissues   less than 40   more than 45

Figure 39.21 Animated Partial pressures (in mm Hg) for oxygen (*pink* boxes) and carbon dioxide (*blue* boxes) in the atmosphere, blood, and tissues.
Figure It Out: What is the partial pressure of oxygen in arteries that carry blood to systemic capillary beds? *Answer: 100 mm Hg*

■ Genetic disorders, infectious disease, and lifestyle choices can increase the risk of respiratory problems.

■ Links to Tuberculosis 21.8, Marijuana's effects 33.7

Interrupted Breathing A tumor or other damage to the brain stem's medulla oblongata can affect respiratory controls. It can cause apnea, a disorder in which breathing repeatedly stops and restarts spontaneously, especially during sleep. More often, sleep apnea occurs when the tongue, tonsils, or another soft tissue obstructs the upper airways. Breathing stops for up to several seconds many times each night. Interrupted sleep patterns and daytime fatigue follow. The risk for heart attacks and strokes rises, because each time breathing stops, blood pressure soars. Changes in sleeping positions or using a mouthpiece or other kinds of devices can help mild sleep apnea. Severe cases require surgical removal of the soft tissues that block the airways.

Sudden infant death syndrome (SIDS) occurs when an infant does not awaken from an apneic episode. Infants who sleep on their back are less vulnerable to SIDS than stomach sleepers. They are more at risk if their mother smoked or was exposed to smoke during pregnancy.

Hannah Kinney of Harvard Medical School reported that an underlying weakness in the respiratory control center may be fatal when combined with environmental stresses. She compared brains of infants who died of SIDS with those of infants who died of other causes. The SIDS babies had fewer receptors for serotonin in their medulla oblongata. This neurotransmitter carries signals between neurons (Section 33.6). Weak signaling may impair the responses to potentially deadly respiratory stress.

Potentially Deadly Infections About one-third of the human population is infected by *Mycobacterium tuberculosis*, the cause of tuberculosis (Section 21.8). These bacteria colonize the lungs, but infection does not always result in disease. Carriers can be identified by a TB skin test. If untreated, about 10 percent of them eventually will develop the disease. They start to cough and may have chest pain. They may have trouble breathing and cough up bloody mucus. Antibiotics cure TB, but only if they are taken diligently for at least six months. An active, untreated infection can be fatal.

Lungs also get infected by bacteria, viruses, and—less commonly—fungi that cause pneumonia. Pneumonia is not one disease; it is a general term for lung inflammation caused by an infectious organism. Coughing, an aching chest, shortness of breath, and fever are usual symptoms. An x-ray can reveal infected tissues filled with fluid and immune cells instead of air. The treatment and outcome depend on the type of pathogen.

Chronic Bronchitis and Emphysema Facing the lumen of your bronchioles is a ciliated, mucus-producing epithelium (Figure 39.22). It is one of many defenses that protect you from respiratory infections. Chronic irritation of the lining may lead to bronchitis. With this respiratory disease, epithelial cells become irritated and secrete too much mucus. Excessive mucus triggers coughing, and provides a moist, nutrient-rich place for pathogens to grow.

Early attacks of bronchitis are treatable. When the aggravation continues, bronchioles become chronically inflamed as bacteria, chemical agents, or both attack the lining of these airways. The lining's ciliated cells die, and mucus-secreting cells multiply. Fibrous scar tissue forms. Over time, scarring narrows or obstructs the airways. Breathing becomes labored and difficult.

Chronic bronchitis can lead to emphysema. With this condition, tissue-destroying bacterial enzymes digest the thin, stretchable alveolar wall. As walls deteriorate, inelastic fibrous tissue builds up around them. Alveoli enlarge, and gas exchange becomes less efficient. In time, the lungs become distended and inelastic, so the balance between air flow and blood flow is compromised. It becomes hard even to catch a breath.

About 2 million people in the United States currently have emphysema, and it causes or contributes to about 100,000 deaths every year.

A number of individuals are genetically predisposed to develop emphysema. They do not have a workable gene for antitrypsin, an enzyme that inhibits bacterial attacks

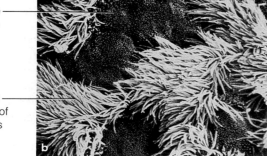

free surface of a mucus-secreting cell

free surface of a cluster of ciliated cells

Figure 39.22 (**a**) Cigarette smoke about to enter bronchi that lead to the lungs. Smoke irritates ciliated and mucus-secreting cells that line the airways (**b**) and can exacerbate bronchitis.

Risks Associated With Smoking	Reduction in Risks by Quitting
Shortened life expectancy Nonsmokers live about 8.3 years longer than those who smoke two packs a day from their midtwenties on.	Cumulative risk reduction; after 10–15 years, the life expectancy of ex-smokers approaches that of nonsmokers.
Chronic bronchitis, emphysema Smokers have 4–25 times higher risk of dying from these diseases than do nonsmokers.	Greater chance of improving lung function and slowing down rate of deterioration.
Cancer of lungs Cigarette smoking is the major cause.	After 10–15 years, risk approaches that of nonsmokers.
Cancer of mouth 3–10 times greater risk among smokers.	After 10–15 years, risk is reduced to that of nonsmokers.
Cancer of larynx 2.9–17.7 times more frequent among smokers.	After 10 years, risk is reduced to that of nonsmokers.
Cancer of esophagus 2–9 times greater risk of dying from this.	Risk proportional to amount smoked; quitting should reduce it.
Cancer of pancreas 2–5 times greater risk of dying from this.	Risk proportional to amount smoked; quitting should reduce it.
Cancer of bladder 7–10 times greater risk for smokers.	Risk decreases gradually over 7 years to that of nonsmokers.
Cardiovascular disease Cigarette smoking a major contributing factor in heart attacks, strokes, and atherosclerosis.	Risk for heart attack declines rapidly, for stroke declines more gradually, and for atherosclerosis it levels off.
Impact on offspring Women who smoke during pregnancy have more stillbirths, and the weight of liveborns is lower than the average (which makes babies more vulnerable to disease and death).	When smoking stops before fourth month of pregnancy, risk of stillbirth and lower birth weight eliminated.
Impaired immunity More allergic responses, destruction of white blood cells (macrophages) in respiratory tract.	Avoidable by not smoking.
Bone healing Surgically cut or broken bones may take 30 percent longer to heal in smokers, perhaps because smoking depletes the body of vitamin C and reduces the amount of oxygen delivered to tissues. Reduced vitamin C and reduced oxygen interfere with formation of collagen fibers in bone (and many other tissues).	Avoidable by not smoking.

a

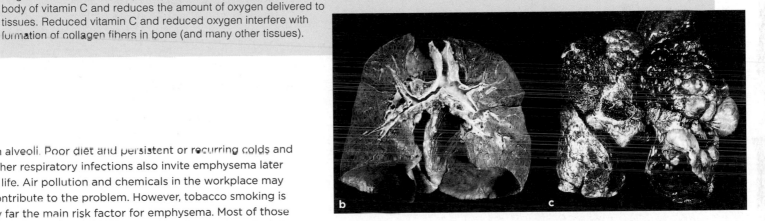

on alveoli. Poor diet and persistent or recurring colds and other respiratory infections also invite emphysema later in life. Air pollution and chemicals in the workplace may contribute to the problem. However, tobacco smoking is by far the main risk factor for emphysema. Most of those affected are over age 50. Twenty or thirty years of smoke exposure leave lungs looking like those in Figure 39.23c.

Smoking's Impact Globally, cigarette smoking kills 4 million people each year. By 2030, the number may rise to 10 million, with about 70 percent of the deaths occurring in developing countries. In the United States, the direct medical costs of treating smoke-induced disorders drains $22 billion a year from the economy. As G. H. Brundtland—a medical doctor and the former director of the World Health Organization—points out, tobacco is the only legal consumer product that kills half of its regular users. If you are a smoker, you may wish to reflect on the information in Figure 39.23a.

Cigarettes also do more than sicken and kill smokers. Nonsmokers die of cancers and disease brought on by breathing secondhand smoke. Children who breathe cigarette smoke at home have a heightened risk for developing lung problems. Smoking while pregnant increases risk of miscarriage and low birth weight.

Figure 39.23 (**a**) From the American Cancer Society, a list of major risks incurred by smoking and the benefits of quitting. (**b**) Appearance of normal lung tissue in humans. (**c**) Appearance of lung tissues from someone who was affected by emphysema.

Smoking marijuana (*Cannabis*) also poses significant respiratory risks. Although marijuana contains fewer toxic particles, or "tar," than tobacco, marijuana is usually smoked without a filter. Also, people smoking marijuana tend to inhale more deeply than tobacco smokers, to hold hot smoke in their lungs for longer periods, and to smoke their cigarettes down to stubs, where tar accumulates. As a result, long-term marijuana smokers have an increased risk of respiratory problems, and they tend to show lung damage earlier than cigarette smokers. On the other hand, unlike tobacco, marijuana has not been shown to increase the risk of lung cancer.

- Specialized features of some respiratory systems adapt organisms to high altitude or deep dives.

- Links to Evolutionary adaptation 17.1, Hypertension 37.9

Respiration at High Altitudes

Atmospheric pressure decreases with altitude. Above 5,500 meters, or about 18,000 feet, it is 380 mm Hg—half of what it is at sea level. Oxygen still is 21 percent of the total pressure, so there is about half as much oxygen as there is at sea level.

Llamas are animals that live at high altitudes in the Andes (Figure 39.24). Their hemoglobin helps them survive in the "thin air," with its lower oxygen level. Compared to the hemoglobin of humans and most other mammals, llama hemoglobin binds oxygen more efficiently. Also, the lungs and the heart of a llama are unusually large relative to the animal's body size.

Most people live at lower altitudes where there is plenty of oxygen. When they ascend too fast to high altitudes, the transport of oxygen to cells plummets. Hypoxia, or cellular oxygen deficiency, is the result.

In an acute compensatory response to hypoxia, the brain commands the heart and respiratory muscles to work harder. People breathe faster and more deeply than usual; they hyperventilate. As a result CO_2 is exhaled faster than it forms and ion balances in the cerebrospinal fluid get skewed. Shortness of breath, a pounding heart, dizziness, nausea, and vomiting are symptoms of the resulting altitude sickness.

Compared to people at low elevations, people who grew up at high altitudes have more alveoli and blood vessels in their lungs. Their heart has larger ventricles and pumps greater volumes of blood.

A healthy person who is unaccustomed to life at a high altitude can become physiologically adjusted to such an environment. Through **acclimatization**, the body makes long-term adjustments in cardiac output, and the rate and magnitude of breathing. Hypoxia also stimulates kidney cells to secrete more **erythropoietin**. This hormone induces stem cells in the bone marrow to divide repeatedly, and induces descendant cells to develop as red blood cells. Under typical conditions, the body produces 2 million to 3 million red blood cells per second to replace those that die off. Under extreme oxygen deprivation, increased erythropoietin secretion can result in a six-fold rise in red blood cell formation. Increased numbers of circulating red blood cells improve the oxygen-delivery capacity of blood.

However, an altitude-induced increase in red blood cell count can put a strain on the heart. Having more blood cells thickens blood, so the heart has to work harder to pump blood through the circulatory system. Stronger contractions increase blood pressure, putting a person at risk for the health problems associated with chronic hypertension (Section 37.9).

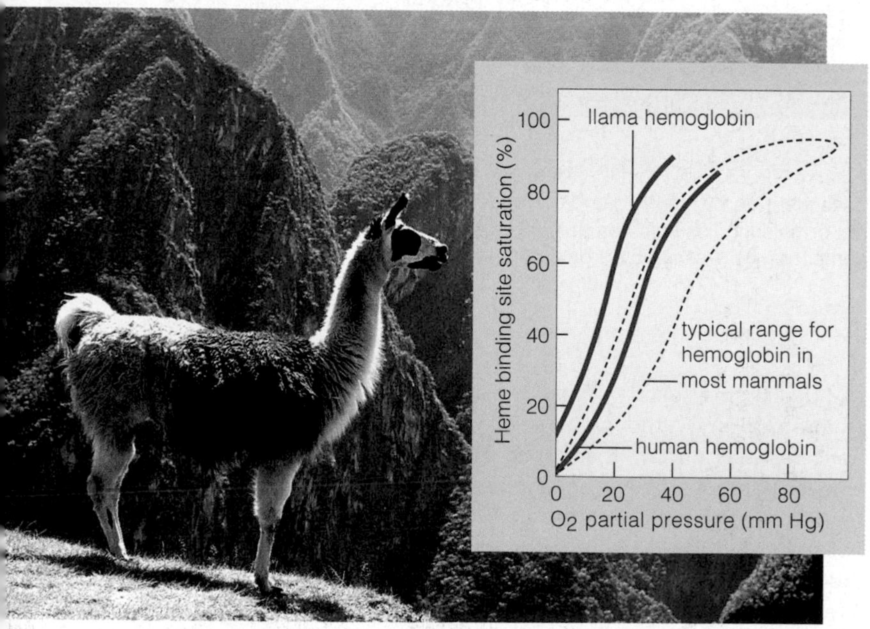

Deep-Sea Divers

Water pressure increases with depth. Human divers using tanks of compressed air risk nitrogen narcosis, sometimes called "raptures of the deep." The deeper a diver goes, the more gaseous nitrogen (N_2) dissolves in interstitial fluid. N_2 affects the lipid bilayer of cell membranes. In neurons, this dissolved nitrogen can disrupt signaling, causing a diver to feel euphoric and drowsy. The deeper divers descend, the more weakened and clumsy they become.

Returning to the surface from a deep dive also has risks. As a diver ascends, pressure falls and N_2 moves from interstitial fluid into blood and is exhaled. If a diver rises too fast, N_2 bubbles form inside the body. The resulting decompression sickness, also known as "the bends," usually begins with joint pain. Bubbles of N_2 can slow the flow of blood to organs. If bubbles form in the brain or lungs, the result can be fatal.

Humans who train to dive without oxygen tanks can remain submerged for about three minutes. So far, the human free diving record is 210 meters. Compare that with the impressive depth records for species

Figure 39.24 Saturation curve for hemoglobin of humans, llamas, and other mammals. **Figure It Out:** At what partial pressure of oxygen do half the heme groups in human blood have oxygen bound?   *Answer: 30 mm Hg*

Species	Maximum Depth
Sperm whale (*Physeter macrocephalus*)	2,200 meters
Leatherback turtle (*Dermochelys coriacea*)	1,200 meters
Southern elephant seal (*Mirounga leonina*)	1,620 meters
Weddell seal (*Leptonychotes weddelli*)	741 meters
Bottlenose dolphin (*Tursiops truncatus*)	>600 meters
Emperor penguin (*Aptenodytes forsteri*)	565 meters

Figure 39.25 (**a**) Two Atlantic leatherback sea turtles returning to the sea after laying eggs. The leathery shell is adapted for deep diving; it bends rather than breaks under extreme pressure. (**b**) Bottlenose dolphins. The chart at *left* lists a few diving records.

listed in Figure 39.25. What types of adaptations make deep dives possible?

Leatherback sea turtles leave water only to lay eggs (Figure 39.25*a*). They spend the rest of their time in open oceans diving for jellyfishes, their main prey. As a turtle or other air-breathing animal dives deeper and deeper, the weight of more and more water presses down onto the body. Lungs filled with air would collapse inward, but most diving animals move air out of the lungs and into cartilage-reinforced airways before they dive too deep. Also, the pressure at great depths could crack a typical turtle's hard shell, but the leatherback's soft shell bends and flexes under such pressure.

Making a deep dive means spending long intervals without access to air. The longest dive recorded for a leatherback turtle lasted for a little more than an hour. Sperm whales can stay submerged for two hours.

If a diving animal's lungs are emptied of air and if it has no access to the surface, then how does it meet its oxygen requirements? It does so in four ways.

First, before it dives, it breathes deeply. A sperm whale blows out about 80–90 percent of the air in its lungs with each exhalation; you exhale only about 15 percent. The deep breaths keep oxygen pressure inside alveoli high, so more oxygen diffuses into the blood.

Second, diving animals can store great amounts of oxygen inside their blood and muscles. They tend to have a large blood volume relative to their body size, a high red blood cell count, and considerable amounts of myoglobin in their muscles. A skeletal muscle of a bottlenose dolphin (Figure 39.25*b*) has about 3.5 times the amount of myoglobin that a comparable skeletal

muscle in a dog has. A muscle in a sperm whale has 7 times as much as the dog muscle.

Third, more oxygen gets distributed to the heart, brain, and other organs that require an uninterrupted supply of ATP for a deep dive. The blood volume and dissolved gases are stored and distributed efficiently with the assistance of valves and plexuses—meshes of blood vessels in local tissues. Metabolic rate and heart rate also decrease. So do oxygen uptake and carbon dioxide formation.

Fourth, whenever possible, a diving animal makes the most of its oxygen stores by sinking and gliding instead of actively swimming. It conserves energy by avoiding unnecessary movements.

Take-Home Message

What are some adaptations that aid respiration in extreme environments?

■ Hemoglobin with a high affinity for oxygen adapts some animals to life at high altitudes where oxygen partial pressure is low.

■ A high red blood cell count, large amount of myoglobin, and other traits allow some animals to hold their breath for long, deep dives.

In the United States, tobacco use is declining. Smoking is banned from airline cabins and airports. Many states and cities ban it in theaters, restaurants, and other enclosed spaces. Cigarette sales to minors are prohibited, as is cigarette advertising on television or near schools. However, in most developing countries smoking is largely unrestricted and the proportion of smokers continues to rise, especially among women.

How would you vote?

Should the United States support efforts to reduce the sale of tobacco products worldwide? See CengageNOW for details, then vote online.

Summary

Section 39.1 **Respiration** is the physiological process by which O_2 enters the internal environment and CO_2 leaves by diffusing across a **respiratory surface**. Each gas moves down its own **partial pressure** gradient into or out of animal bodies. Constraints imposed by the surface-to-volume ratio shape respiratory structures and ventilation mechanisms. **Respiratory proteins** such as **hemoglobin** in red blood cells and **myoglobin** in muscle bind oxygen and help maintain gradients that favor gas exchange.

Section 39.2 Oxygen content of water can vary and affects the survival of aquatic species.

Section 39.3 Some invertebrates do not have special respiratory organs and rely on **integumentary exchange**, diffusion of gases across the body surface. **Gills** enhance respiration in other aquatic invertebrates. On land, **lungs**, **book lungs**, and **tracheal systems** aid gas exchange.

Section 39.4 Water flowing over fish gills exchanges gases with blood flowing in the opposite direction inside gill capillaries. This **countercurrent exchange** is highly efficient. Most amphibians have lungs, and also exchange gases across the skin. Reptiles, birds, and mammals rely on lungs for gas exchange. In birds, air sacs connected to lungs keep air flowing continually through them.

■ *Use the animation on CengageNOW to compare various vertebrate respiratory systems.*

Section 39.5 In humans, air flows through two nasal cavities and a mouth into the **pharynx** (throat), then the **larynx** (voice box). A flap of tissue called the **epiglottis** directs air through the **glottis**, the opening to the **trachea** (windpipe). The trachea branches into two **bronchi** that enter the lungs. In the lungs, bronchi lead to finely branching **bronchioles** that have **alveoli** at their tips. Gases are exchanged at these thin-walled air sacs.

Contractions of the dome-shaped **diaphragm** and the **intercostal muscles** between the ribs alter the volume of the thoracic cavity during breathing.

■ *Use the animation on CengageNOW to explore the human respiratory system.*

Section 39.6 Each **respiratory cycle** consists of one inhalation and one exhalation. Inhalation is always an active process. As muscle contractions expand the chest cavity, pressure in lungs decreases below atmospheric pressure, and air flows into the lungs. These events are reversed during exhalation, which normally is passive.

If a person is choking, the **Heimlich maneuver** can be used to expel food from their trachea.

Tidal volume is normally far less than **vital capacity**. The medulla oblongata in the brain stem adjusts the rate and magnitude of breathing.

■ *Use the animation on CengageNOW to learn about the respiratory cycle and the Heimlich maneuver.*

Section 39.7 In human lungs, the alveolar wall, the wall of a pulmonary capillary, and their fused basement membranes form a thin **respiratory membrane** between air inside an alveolus and the internal environment. O_2 following its partial pressure gradient diffuses across the respiratory membrane, into the plasma of the blood, and finally into red blood cells.

Red blood cells are filled with hemoglobin that binds O_2 where its partial pressure is high, forming **oxyhemoglobin**. In metabolically active tissue, O_2 released from hemoglobin diffuses out of capillaries, through interstitial fluid, and into cells.

CO_2 diffuses from cells to blood. Most CO_2 reacts with water inside red blood cells, to form bicarbonate. The enzyme **carbonic anhydrase** catalyzes this reaction, which is reversed in the lungs. There, CO_2 and water vapor form and are expelled in exhalations.

Carbon monoxide (CO) is a dangerous gaseous pollutant that binds to hemoglobin more strongly than oxygen.

■ *Use the animation on CengageNOW to compare partial pressures of gases in different body regions.*

Section 39.8 Respiratory disorders include apnea and sudden infant death syndrome (SIDS). Respiratory diseases include tuberculosis, pneumonia, bronchitis, and emphysema. Smoking worsens or increases risk of many respiratory problems. Worldwide, smoking remains a leading cause of debilitating diseases and deaths.

Section 39.9 Air's oxygen concentration declines with altitude. Short-term physiological changes that occur in response to high altitude are called **acclimatization**. They include altered breathing patterns and an increase in **erythropoietin**, a hormone that stimulates red blood cell formation. Specialized mechanisms and behaviors allow some turtles and marine mammals to dive deeply for long intervals.

Data Analysis Exercise

Radon is a colorless, odorless gas emitted by many rocks and soils. It is formed by the radioactive decay of uranium and is itself radioactive (Section 2.2). There is some radon in the air almost everywhere, but routinely inhaling a lot of it raises the risk of lung cancer. Radon also seems to increase cancer risk far more in smokers than in nonsmokers. Figure 39.26 is an estimate of how radon in homes affects risk of lung cancer mortality. Note that this data shows only the death risk for radon-induced cancers. Smokers are also at risk from lung cancers that are caused by tobacco.

1. If 1,000 smokers were exposed to a radon level of 1.3 pCi/L over a lifetime (the average indoor radon level) how many would die of a radon-induced lung cancer?

2. How high would the radon level have to be to cause approximately the same number of cancers among 1,000 nonsmokers?

3. The risk of dying in a car crash is about 7 out of 1,000. Is a smoker in a home with an average radon level (1.3 pCi/L), more likely to die in a car crash or of radon-induced cancer?

Radon Level (pCi/L)	Risk of Cancer Death From Lifetime Radon Exposure	
	Never Smoked	Current Smokers
20	36 out of 1,000	260 out of 1,000
10	18 out of 1,000	150 out of 1,000
8	15 out of 1,000	120 out of 1,000
4	7 out of 1,000	62 out of 1,000
2	4 out of 1,000	32 out of 1,000
1.3	2 out of 1,000	20 out of 1,000
0.4	>1 out of 1,000	6 out of 1,000

Figure 39.26 Estimated risk of lung cancer death as a result of lifetime radon exposure. Radon levels are measured in picocuries per liter (pCi/L). The Enviromental Protection Agency considers a radon level above 4 pCi/Liter to be unsafe. For information about testing your home for radon and what to do if the radon level is high, visit the EPA's Radon Information Site at www.epa.gov/radon.

Self-Quiz
Answers in Appendix III

1. The most abundant gas in the atmosphere is _____ .
a. nitrogen   c. oxygen
b. carbon dioxide d. hydrogen

2. Respiratory proteins such as hemoglobin _____
a. contain metal ions
b. occur only in vertebrates
c. increase the efficiency of oxygen transport
d. both a and c

3. In insects, most gas exchange occurs at _____ .
a. the tips of tracheal tubes   c. gills
b. the body surface   d. paired lungs

4. Countercurrent flow of water and blood increases the efficiency of gas exchange in _____ .
a. fishes   c. birds
b. amphibians   d. all of the above

5. In human lungs, gas exchange occurs at the _____ .
a. two bronchi   c. alveolar sacs
b. pleural sacs   d. both b and c

6. When you breathe quietly, inhalation is _____ and exhalation is _____ .
a. passive; passive   c. passive; active
b. active; active   d. active; passive

7. During inhalation, _____ .
a. the thoracic cavity expands
b. the diaphragm relaxes
c. atmospheric pressure declines
d. both a and c

8. True or false? Human lungs hold some air, even after a forced exhalation.

9. Most oxygen being transported in blood _____ .
a. is bound to hemoglobin
b. combines with carbon to form carbon dioxide
c. is in the form of bicarbonate
d. is dissolved in the plasma

10. At high altitudes, _____ .
a. nitrogen bubbles out of the blood
b. hemoglobin has fewer oxygen-binding sites
c. atmospheric pressure is lower than at sea level
d. both b and c

11. Myoglobin helps muscles to _____ .
a. synthesize hemoglobin
b. store oxygen
c. form bicarbonate
d. both b and c

12. True or false? Hemoglobin has a higher affinity for carbon dioxide than for oxygen.

13. Match the words with their descriptions.
____ trachea   a. muscle of respiration
____ pharynx   b. gap between vocal cords
____ alveolus   c. between bronchi and alveoli
____ hemoglobin   d. windpipe
____ bronchus   e. respiratory protein
____ bronchiole   f. site of gas exchange
____ glottis   g. airway leading to lung
____ diaphragm   h. throat

■ *Visit CengageNOW for additional questions.*

Critical Thinking

1. The red blood cell enzyme carbonic anhydrase contains the metal zinc. Humans obtain zinc from their diet, especially from red meat and some seafoods. A zinc deficiency does not reduce the number of red blood cells, but it does impair respiratory function by reducing carbon dioxide output. Explain why a zinc deficiency has this effect.

2. Look again at Figure 39.21. Notice that the oxygen and carbon dioxide content of blood in pulmonary veins is the same as at the start of the systemic capillaries. Notice also that systemic veins and pulmonary arteries have equal partial pressures. Explain the reason for these similarities.

IMPACTS, ISSUES | Hormones and Hunger

Like other mammals, humans have adipose tissue with cells that store fat. This energy warehouse served our early hominid ancestors well. As foragers, they could seldom be certain of where their next meal was coming from. Filling their adipose cells with fat when food was abundant helped them survive when food became scarce.

Lean pickings are not a problem for most Americans. With 60 percent of adults overweight or obese, they are among the fattest people in the world. "Obesity" means there is too much fat in adipose tissue. It increases the risk of heart disease, diabetes, and some cancers. Many people try to lose weight, but extra pounds are tough to shed. Why? For one thing, hormones are involved.

When you take in more calories than you burn, your fat-storing cells plump up and increase their secretion of leptin. This hormone acts on a brain region that affects appetite. Mutant mice that cannot make leptin eat and eat, until they look like inflated balloons (Figure 40.1a). Inject an obese mutant mouse with leptin, and it eats less and slims down.

However, lack of leptin or leptin receptors is extremely rare in humans. Having more fat, obese people make more leptin than slim ones, but for unknown reasons an obese person's body does not heed leptin's call to stop eating.

Ghrelin, another hormone, increases appetite. Some cells in the stomach lining and the brain secrete ghrelin when the stomach is empty. Secretions slow after a big meal. In one study of ghrelin's effects, a group of obese volunteers stayed on a low-fat, low-calorie diet for six months. They lost weight, but the concentration of ghrelin in their bloodstream climbed dramatically—they were hungrier than ever!

Some extremely obese people undergo gastric bypass surgery, which effectively reduces the size of the stomach and small intestine. The surgery makes people feel full faster. It also reduces the amount of nutrients that they absorb from food. The results can be dramatic (Figure 40.1b). However, the surgery raises risk for vitamin and mineral deficiencies.

Gastric bypass is more effective than standard weight loss methods. Post-bypass patients are far less likely to regain pounds. One reason may be that these patients secrete less ghrelin after the bypass surgery, so they feel less hungry.

Discussion of food intake and body weight lead us into the world of nutrition. The word encompasses all the processes by which an animal ingests and digests food, then absorbs the released nutrients as energy sources and building blocks for cells. When all works well, inputs balance the outputs, and weight remains within a healthy range.

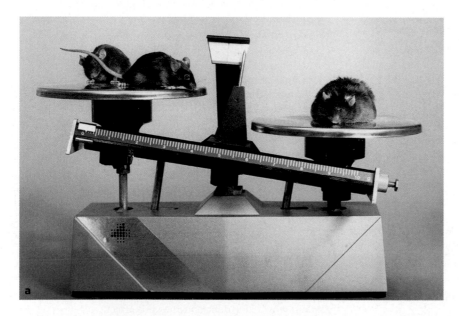  

See the video! Figure 40.1 Examples of hormonal effects on appetite. (**a**) Two normal mice (*left*) weigh less than one mutant mouse (*right*) that cannot synthesize leptin. This hormone that acts in the brain to suppress appetite. Compared with normal mice, a leptin-deficient mutant eats and weighs a lot more.

(**b**) A young woman before (*left*) and after gastric bypass surgery (*right*). This surgery reduces the amount of food a person can take in and the amount of ghrelin she secretes.

Key Concepts

Overview of digestive systems

Some animal digestive systems are saclike, but most are a tube with two openings. In complex animals, a digestive system interacts with other organ systems in the distribution of nutrients and water, disposal of residues and wastes, and homeostasis. **Section 40.1**

Human digestive system

Human digestion starts in the mouth, continues in the stomach, and is completed in the small intestine. Secretions of the salivary glands, liver, and pancreas aid digestion. Most nutrients are absorbed in the small intestine. The large intestine concentrates wastes. **Sections 40.2–40.6**

Organic metabolism and nutrition

Nutrients absorbed from the gut are raw materials used in synthesis of the body's complex carbohydrates, lipids, proteins, and nucleic acids. A healthy diet normally provides all nutrients, vitamins, and minerals necessary to support metabolism. **Sections 40.7–40.9**

Balancing caloric inputs and outputs

Maintaining body weight requires balancing calories taken in with calories burned in metabolism and physical activity. **Section 40.10**

Links to Earlier Concepts

- You already know about the structure of carbohydrates (Section 3.3), lipids (3.4), and proteins (3.5). In this chapter you will learn about how your body digests these molecules.

  You will also learn how the body obtains vitamins and minerals required to make coenzymes (6.3), electron transfer chain components (6.4), hemoglobin (37.2), and certain hormones.

- You will discover how low pH (2.6) and the action of enzymes (6.3) help break down food, and how the products of digestion (3.3–3.5) cross cell membranes.

- Characteristics of epithelium (32.2) and smooth muscle (32.4), as well as the sense of taste (34.3) and the action of the autonomic nervous system (33.8), are discussed again, as is the anatomy of the throat (39.5).

- You will be reminded of the variety of animal body plans (25.1), and the way in which natural selection affects traits related to feeding (17.3, 18.5, and 18.10).

40.1 The Nature of Digestive Systems

■ All animals are heterotrophs that ingest food, break it down, and absorb its nutrient subunits into their body.

■ Links to Animal body plans 25.1, Bill shape and natural selection 17.3, 18.5, and 18.10

An animal's **digestive system** is a body cavity or a tube that mechanically and chemically breaks food down to small particles, then to molecules that can be absorbed into the internal environment. The digestive system also expels any unabsorbed residues. Together with other organ systems, it plays an essential role in homeostasis (Figure 40.2).

Incomplete and Complete Systems

Remember from Section 25.1 that some invertebrates have an **incomplete digestive system**. Food enters their saclike gut through an opening at the body surface. Wastes leave through the same opening. In flatworms, a saclike, branching gut cavity opens at the start of the pharynx, a muscular tube (Figure 40.3*a*). Food that enters the sac is digested, its nutrients are absorbed, then wastes are expelled. This two-way movement of material does not favor specialization of gut regions for specific tasks.

Most groups of invertebrates and all vertebrates have a **complete digestive system**: a tubular gut with an opening at both ends. Along the tube's length are regions that specialize in processing food, absorbing nutrients, or concentrating wastes. Figure 40.3*b* shows the complete digestive system of a frog. The tubular portion consists of the mouth, pharynx, esophagus, stomach, small intestine, large intestine, and anus. A liver, gallbladder, and pancreas are accessory organs that assist digestion by secreting enzymes and other products into the small intestine.

A complete digestive system carries out five tasks:

1. *Mechanical processing and motility:* movements that break up, mix, and directionally propel food material

2. *Secretion:* release of substances, especially digestive enzymes, into the lumen (the space inside the tube)

3. *Digestion:* breakdown of food into particles, then into nutrient molecules small enough to be absorbed

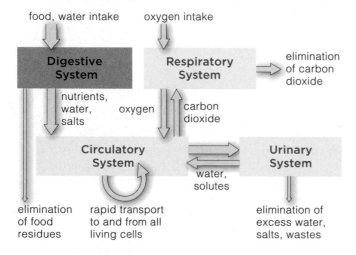

Figure 40.2 Organ systems with key roles in the uptake, processing, and distribution of nutrients and water in complex animals.

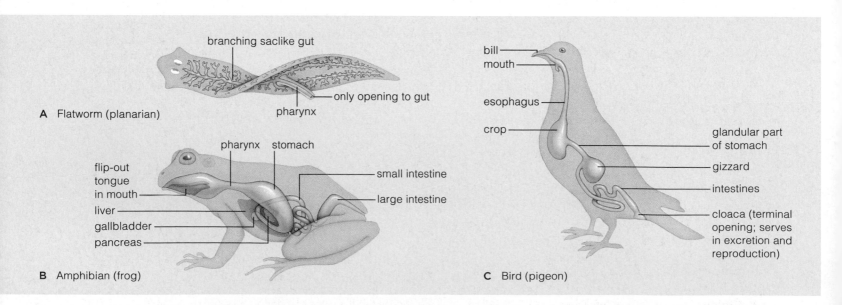

A Flatworm (planarian)

B Amphibian (frog)

C Bird (pigeon)

Figure 40.3 Animated (**a**) Incomplete digestive system. (**b,c**) Two complete digestive systems.

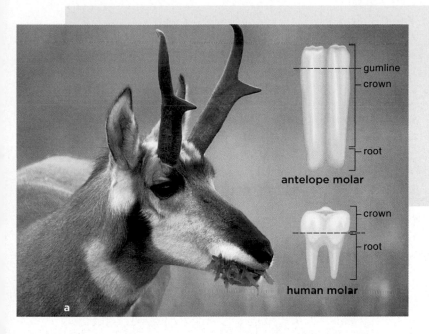

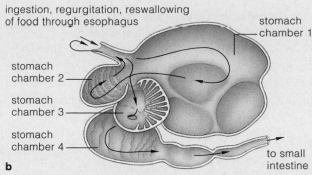

Figure 40.4 Animated (**a**) Human and pronghorn antelope molars. (**b**) An antelope's multiple stomach chambers. In the first two, food is mixed with fluid and exposed to microbes (prokaryotes, protists, and fungi) that engage in fermentation. Some of the microbes degrade cellulose; others synthesize organic compounds, fatty acids, and vitamins. Partly digested food is regurgitated into the mouth, chewed, then swallowed. It enters the third chamber and is digested again before entering the last chamber.

4. *Absorption:* uptake of digested nutrients and water across the gut wall, into extracellular fluid

5. *Elimination:* expulsion of undigested or unabsorbed solid residues

Dietary Adaptations

In birds, the size and shape of the bill are diet-related traits shaped by natural selection (Sections 17.3, 18.5, and 18.10). So are other traits. For example, a pigeon (Figure 40.3c) uses its bill to pick up small seeds from the ground. Like other seed-eating birds, a pigeon has a large crop, a saclike food-storing region above the stomach. The bird quickly fills its crop with seeds, then flies off and digests them later. This eat-and-run strategy reduces the amount of time that the bird is on the ground, where it is most vulnerable to predators.

Birds do not have teeth. They grind up food inside a gizzard: a stomach chamber lined with hard protein particles. Compared to hawks and other meat-eating birds, seed eaters have larger gizzards relative to their body size. Also, seed eaters have a relatively longer intestinal tract, because seeds require more processing time than easier-to-digest meat. In all birds, undigested residues collect in a cloaca before being expelled.

Mammalian teeth are adaptations to specific diets. For example, pronghorn antelope browse on grass and nibble on shrubby plants. Antelope molars (cheek teeth) have a flattened crown that serves as a grinding platform (Section 26.11). The crown on your molars is proportionately much smaller (Figure 40.4a). Why? You do not brush your mouth against dirt as you eat,

but an antelope does. Abrasive soil particles mix with the animal's food, so the crown of an antelope molar gets a lot of wear. An enlarged crown is an adaptation that keeps the molars from wearing down to nubs.

The antelope's gut also shows specializations for a diet of plant material. Like cattle, goats, and sheep, antelopes are **ruminants**, hoofed mammals that have multiple stomach chambers (Figure 40.4b). Microbes that live inside the first two stomach chambers carry out fermentation reactions that break down cellulose in plant cell walls. Solids accumulate in the second chamber, forming "cud" that is regurgitated—moved back into the mouth for a second round of chewing. Nutrient-rich fluid moves from the second chamber to the third and fourth chambers, and finally to the intestine. This system allows ruminants to maximize the amount of nutrients they extract from plant foods rich in cellulose. Cellulose is so tough and insoluble that most animals cannot digest it.

Take-Home Message

What are digestive systems and how do they vary among animal groups?

■ Digestive systems mechanically and chemically degrade food into small molecules that can be absorbed, along with water, into the internal environment. These systems also expel the undigested residues from the body.

■ Digestive systems may be incomplete or complete. Incomplete digestive systems are a saclike cavity with one opening. A complete digestive system is a tube with two openings and regional specializations in between.

■ Some digestive traits, such as the shape of teeth or the length of different portions of the digestive tract, are adaptations that allow an animal to exploit a particular type or types of foods.

Overview of the Human Digestive System

■ If the tubular gut of an adult human were fully stretched out, it would extend up to 9 meters (30 feet).

■ Accessory organs along the length of the gut secrete enzymes and other substances that aid the breakdown of food into its component molecules.

■ Links to Epithelium 32.2, Taste 34.3, Trachea 39.5

Humans have a complete digestive system, a tubular gut with two openings (Figure 40.5). Mucus-covered epithelium (Section 32.2) lines the tube, and different parts of the tube specialize in digesting food, absorbing released nutrients, or concentrating and storing the unabsorbed waste. The salivary glands, the pancreas,

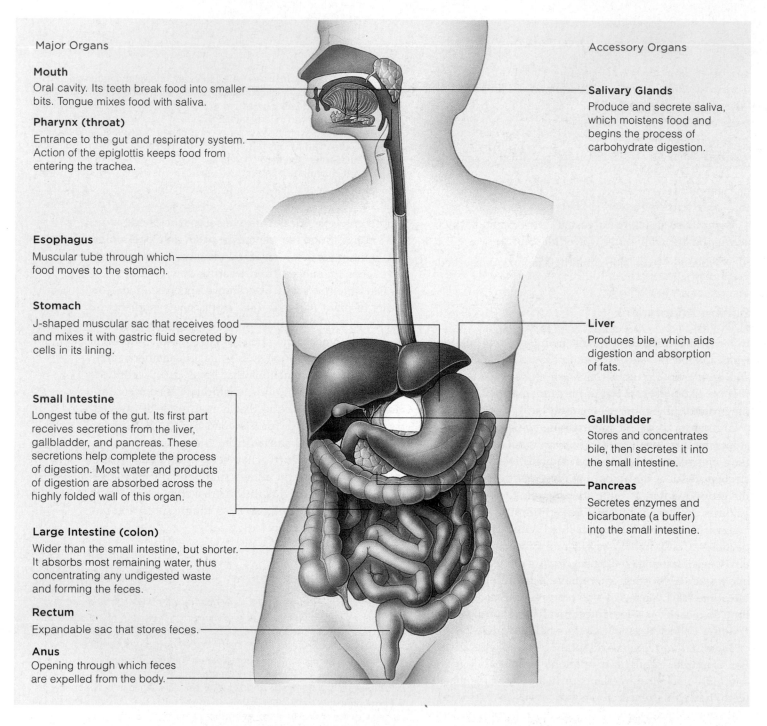

Major Organs

Mouth
Oral cavity. Its teeth break food into smaller bits. Tongue mixes food with saliva.

Pharynx (throat)
Entrance to the gut and respiratory system. Action of the epiglottis keeps food from entering the trachea.

Esophagus
Muscular tube through which food moves to the stomach.

Stomach
J-shaped muscular sac that receives food and mixes it with gastric fluid secreted by cells in its lining.

Small Intestine
Longest tube of the gut. Its first part receives secretions from the liver, gallbladder, and pancreas. These secretions help complete the process of digestion. Most water and products of digestion are absorbed across the highly folded wall of this organ.

Large Intestine (colon)
Wider than the small intestine, but shorter. It absorbs most remaining water, thus concentrating any undigested waste and forming the feces.

Rectum
Expandable sac that stores feces.

Anus
Opening through which feces are expelled from the body.

Accessory Organs

Salivary Glands
Produce and secrete saliva, which moistens food and begins the process of carbohydrate digestion.

Liver
Produces bile, which aids digestion and absorption of fats.

Gallbladder
Stores and concentrates bile, then secretes it into the small intestine.

Pancreas
Secretes enzymes and bicarbonate (a buffer) into the small intestine.

Figure 40.5 Animated Overview of the components of the human digestive system, together with a brief description of their primary functions in digestion.

the liver, and the gallbladder are accessory organs involved in secreting substances into the tube. Food enters the mouth, and travels through the pharynx and esophagus to the gut. A human gut, or **gastrointestinal tract**, starts at the stomach and extends through the intestines to the tube's terminal opening.

Food is partially processed inside the mouth, or oral cavity. The tongue is a bundle of membrane-covered skeletal muscle attached to the floor of the mouth. The tongue positions food so it can be swallowed, and the many chemoreceptors in taste buds at the tongue's surface contribute to our sense of taste (Section 34.3).

Swallowing forces food into the pharynx. A human **pharynx**, or throat, is the entrance to the digestive and respiratory tracts (Section 39.5). The presence of food at the back of the throat triggers a swallowing reflex. When you swallow, the flaplike epiglottis flops down and the vocal cords constrict, so the route between the pharynx and larynx is blocked. This reflex keeps food from getting stuck in an airway and choking you.

A muscular tube called the **esophagus** connects the pharynx with the stomach. The esophagus undergoes **peristalsis**, rhythmic muscle contractions that propel food or liquid through a tubular digestive organ. The **stomach** is a stretchable sac that stores food, secretes acid and enzymes, and mixes them all together.

Between the esophagus and stomach is a **sphincter**. Like all sphincters, this ring of smooth muscle blocks the flow of substances past it when it has contracted. In people who have gastroesophageal reflux disease (GERD), this sphincter does not shut properly. As a result, acidic stomach fluids splash back and irritate esophageal tissues causing burning pain (heartburn).

The stomach leads to the **small intestine**, the part of the gut where most carbohydrates, lipids, and proteins are digested and where most of the released nutrients and water are absorbed. Secretions from the liver and pancreas assist the small intestine in these tasks.

The **large intestine** absorbs most remaining water and ions, thus compacting wastes. Wastes are briefly stored in a stretchable tube, the **rectum**, before being expelled from the gut's terminal opening, or **anus**.

Take-Home Message

What type of digestive system do humans have?

■ Humans have a complete digestive system with a muscular, mucosa-lined gut.

■ Accessory organs positioned adjacent to the gut secrete substances into its interior. These substances aid digestion or absorption of food.

■ Chewing your food begins the process of digestion.

Mechanical digestion begins when teeth rip and crush food. Each tooth is embedded in the jaw at a fibrous joint and consists mostly of bonelike **dentin** (Figure 40.6a). Dentin-secreting cells reside in a central pulp cavity. These cells are serviced by nerves and blood vessels that extend through the tooth's root. **Enamel**—the hardest material in the body—covers the tooth's exposed crown and reduces wear.

Human adults have thirty-two teeth of four types (Figure 40.6b). Chisel-shaped incisors shear off bits of food. Cone-shaped canines tear up meats. Premolars and molars have broad bumpy crowns that serve as platforms for grinding and crushing food.

Chemical digestion begins when food mixes with saliva from **salivary glands**. Saliva is mostly water, with bicarbonate, enzymes, and mucins. Bicarbonate, a buffer, keeps the pH in the mouth from becoming too acidic. The enzyme **salivary amylase** hydrolyzes starch, breaking it into disaccharides. Mucin proteins combine with water and form mucus that makes food pieces stick together in easy-to-swallow clumps.

Take-Home Message

How does the mouth function in digestion?

■ Digestion begins when teeth mechanically break food into smaller bits and salivary amylase chemically breaks starch into disaccharides.

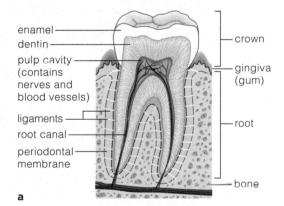

enamel
dentin
pulp cavity (contains nerves and blood vessels)
ligaments
root canal
periodontal membrane

crown
gingiva (gum)
root
bone

a

Figure 40.6 Human teeth. (**a**) Cross-section of a human tooth. The crown is the portion extending above the gum; the root is embedded in the jaw. Tiny ligaments attach the tooth to the jawbone.

(**b**) The four types of teeth in adult. Molars and premolars grind up food. Incisors and canines rip and tear off bits.

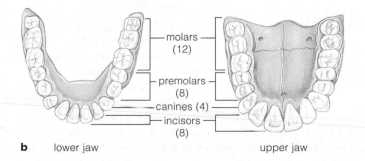

molars (12)
premolars (8)
canines (4)
incisors (8)

b   lower jaw   upper jaw

40.4 Food Breakdown in the Stomach and Small Intestine

■ In the stomach and small intestine, smooth muscle contractions mix food with digestive enzymes.

■ Links to pH 2.6, Enzymes 6.3, Smooth muscle 32.4, Autonomic nervous system 33.8

Carbohydrate breakdown, again, starts in the mouth. Protein breakdown begins in the stomach. Digestion of both is completed in the small intestine. Lipids are also digested in the small intestine. Digestion occurs as contractions of smooth muscle in the gut wall mix food with enzymes (Figure 40.7 and Table 40.1).

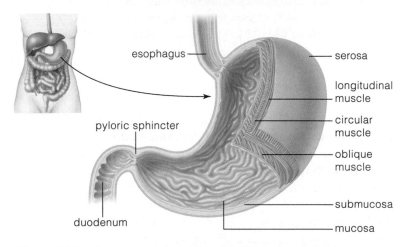

Figure 40.7 Structure of the stomach wall. The outermost layer, the serosa, is connective tissue covered by epithelium. Beneath the serosa, three layers of smooth muscle differ in their orientation and direction of contraction. Their coordinated action mixes stomach contents with gastric fluid secreted by the mucosa that lines the stomach's interior.

Digestion in the Stomach

The stomach is a muscular, stretchable sac with three functions. First, the stomach stores food and controls the rate of passage to the small intestine. Second, it pummels and mechanically breaks down food. Third, it secretes substances that aid in chemical digestion.

A mucus-secreting epithelium—the **mucosa**—lines the inner gut wall. In the stomach, cells of the mucosa secrete about two liters of **gastric fluid** each day. This fluid includes mucus, hydrochloric acid, and enzymes such as pepsinogens. Acid lowers the pH to about 2. When food enters the stomach, endocrine cells in the stomach lining secrete the hormone gastrin into blood. Gastrin binds to secretory cells of the mucosa, causing them to step up secretion of acid and pepsinogens.

Rhythmic contraction of smooth muscle in the stomach wall mixes gastric fluid and food into a semiliquid mass called **chyme**. Eventually, the contractions propel chyme through the pyloric sphincter that connects the stomach with the small intestine (Figure 40.7).

The acidity of chyme makes proteins unfold, exposing their peptide bonds. Acid also causes pepsinogens to become pepsins, enzymes that break peptide bonds. The strong acidity kills most bacteria, but acid-tolerant *Helicobacter pylori* sometimes infects the lining of the stomach and the upper intestine. A chronic *H. pylori* infection can harm the lining and expose the underlying tissues to acid, causing a painful ulcer. Antibiotics are now routinely used to treat such ulcers.

Table 40.1 Summary of Chemical Digestion

Location	Enzymes Present	Enzyme Source	Enzyme Substrate	Main Breakdown Products
Carbohydrate Digestion				
Mouth, stomach	Salivary amylase	Salivary glands	Polysaccharides	Disaccharides
Small intestine	Pancreatic amylase	Pancreas	Polysaccharides	Disaccharides
	Disaccharidases	Intestinal lining	Disaccharides	**Monosaccharides*** (such as glucose)
Protein Digestion				
Stomach	Pepsins	Stomach lining	Proteins	Protein fragments
Small intestine	Trypsin, chymotrypsin	Pancreas	Proteins	Protein fragments
	Carboxypeptidase	Pancreas	Protein fragments	**Amino acids***
	Aminopeptidase	Intestinal lining	**Amino acids***	
Lipid Digestion				
Small intestine	Lipase	Pancreas	Triglycerides	**Free fatty acids, monoglycerides***
Nucleic Acid Digestion				
Small intestine	Pancreatic nucleases	Pancreas	DNA, RNA	Nucleotides
	Intestinal nucleases	Intestinal lining	Nucleotides	**Nucleotide bases, monosaccharides***

* Breakdown products small enough to be absorbed into the internal environment.

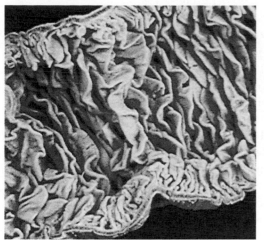

a A section of highly folded mucosa

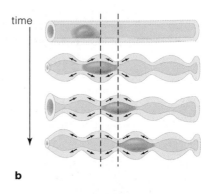

Figure 40.8 (**a**) Structure of the small intestine. Its wall has a highly folded inner lining, the mucosa. (**b**) Rings of circular muscle inside the wall contract and relax in a pattern. Back-and-forth movement propels, mixes, and forces chyme up against the wall, enhancing digestion and absorption.

Digestion in the Small Intestine

Chyme from the stomach and various secretions from the pancreas enter the duodenum, the first part of the small intestine. Pancreatic enzymes break down large organic compounds in chyme into monosaccharides, monoglycerides, fatty acids, amino acids, nucleotides, and nucleotide bases (Table 40.1). Bicarbonate from the pancreas buffers acids, thus protecting the intestinal lining and ensuring intestinal enzymes work properly.

In addition to enzymes, fat digestion requires bile. **Bile** is a mixture of salts, pigments, cholesterol, and lipids. It is made in the liver, and is concentrated and stored in the **gallbladder**. A fatty meal stimulates the gallbladder to contract, forcing bile out through a duct that leads to the small intestine.

Bile salts enhance fat digestion by **emulsification**, a process which disperses any droplets of fat in a fluid. Water-insoluble triglycerides from food tend to clump together and form fat globules. Movement of the small intestine counteracts this tendency. Rings of smooth muscle in the intestinal wall contract in an oscillating pattern (Figure 40.8*b*). These contractions mix the chyme and break up fat globules into small droplets that become coated with bile salts. This coating of bile salts keeps the droplets separated. The smaller drops present a greater surface area to enzymes that break up fats into fatty acids and monoglycerides.

Gallstones, hard pellets of cholesterol and bile salts, can form in the gallbladder. Most are harmless. If they block the bile duct or otherwise interfere with function of the gallbladder, they can be removed surgically.

The breakdown products of digestion are absorbed across the epithelial lining of the small intestine, into the internal environment. How each kind gets across is the focus of the next section.

Controls Over Digestion

The nervous system, endocrine system, and nerves in the gut wall control digestion. Arrival of food in the stomach causes signals to flow along reflex pathways to gut muscles and glands. Other pathways alert the brain. In response, gut muscles contract and glands secrete hormones into the blood (Table 40.2). A large meal stimulates more forceful contractions than a small one. Composition of a meal also has an effect. Stomach emptying takes longer after a high-fat meal than after a meal lower in fat.

With stress or exercise, sympathetic neurons signal gut muscles to contract more slowly (Section 33.8). This is why chronic stress or exercising immediately after a meal can cause digestive problems.

Table 40.2 Main Hormonal Controls of Digestion

Hormone	Source	Effects on Digestive System
Gastrin	Stomach	Stimulates stomach acid secretion
Cholecystokinin (CCK)	Small intestine	Stimulates pancreatic enzyme secretion and gallbladder contraction
Secretin	Small intestine	Stimulates pancreas to secrete bicarbonate and slows contractions of small intestine

Take-Home Message

Where and how does digestion occur?

■ Digestion begins in the mouth and continues in the stomach, but the bulk of it occurs in the small intestine.

■ Enzyme activity, acidity, and mechanical processes break food into small organic molecules that can be absorbed.

- The small intestine is the main site of absorption for the products of digestion.

- Links to Organic monomers 3.3–3.5, Lysozyme 38.2

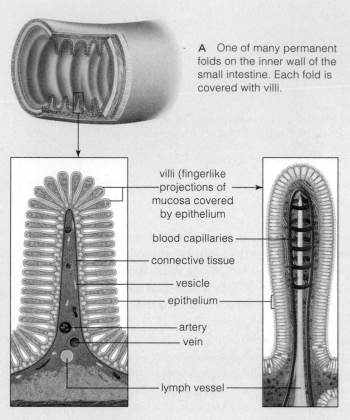

A One of many permanent folds on the inner wall of the small intestine. Each fold is covered with villi.

villi (fingerlike projections of mucosa covered by epithelium)

blood capillaries

connective tissue

vesicle

epithelium

artery

vein

lymph vessel

B At the free surface of each mucosal fold are many fingerlike absorptive structures called villi.

C A villus is covered with specialized epithelial cells. It also contains blood capillaries and lymph vessels.

D Epithelial cells in the intestinal mucosa. The four types shown *below* are color-coded enlargements of cells on the surface of the villus shown in (**c**).

Absorptive brush border cells are the most abundant cells on a villus. Their crown of microvilli extends into the intestinal lumen. The small-intestinal enzymes discussed in the previous section are built into brush border cell plasma membranes. Other cells of the mucosa secrete mucus, hormones, or lysozyme (an enzyme that digests bacterial cell walls).

secretes lysozyme

secretes hormones

secretes mucus

absorbs nutrients

brush border cell

From Structure to Function

The small intestine is "small" only in its diameter—about 2.5 cm (1 inch). It is the longest segment of the gut. Uncoiled, it would extend for about 5 to 7 meters (16 to 23 feet). Water and nutrients cross the lining of this long tube to reach the internal environment.

Three features of the small intestine lining enhance absorption. First, this lining is folded (Figure 40.9*a*). Second, millions of multicelled, fingerlike absorptive structures called **villi** (singular, villus) extend out from each of the folds (Figure 40.9*b*). Each villus houses a lymph vessel and blood vessels (Figure 40.9*c*). Third, most cells on the villus surface are **brush border cells** (Figure 40.9*d*). These specialized cells have membrane extensions called **microvilli** (singular, microvillus) that project into the lumen. Collectively, all of the folds and projections make the surface area of intestinal mucosa about the size of half a tennis court!

The brush border cells function in both digestion and absorption. Digestive enzymes at the surface of the microvilli break down sugars, protein fragments, and nucleotides as listed in Table 40.1. Also at the microvillus surface are many transport proteins that act in absorption, as explained below.

In addition to brush border cells, the lining of the small intestine includes secretory cells (Figure 40.9*d*). These cells secrete hormones, mucus, and bacteria-killing chemicals, such as lysozyme (Section 38.2).

How Are Materials Absorbed?

Water and Solute Absorption Each day, eating and drinking puts 1 to 2 liters of fluid into the small intestine. Secretions from the stomach, accessory glands, and intestinal lining contribute another 6 to 7 liters. About 80 percent of the water in that fluid is absorbed across the small intestinal lining and into the internal

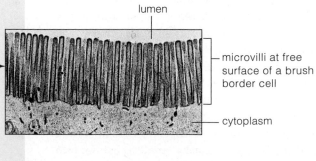

lumen

microvilli at free surface of a brush border cell

cytoplasm

Figure 40.9 Animated The lining of the small intestine.

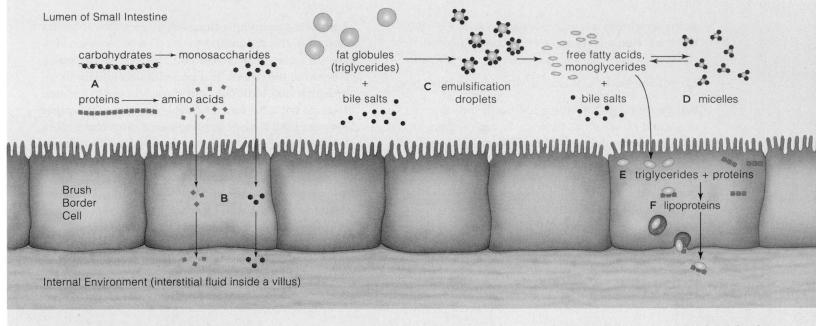

Lumen of Small Intestine

carbohydrates → monosaccharides

A

proteins → amino acids

fat globules
(triglycerides)

+

bile salts

C emulsification
droplets

free fatty acids,
monoglycerides

+

bile salts

D micelles

Brush
Border
Cell

B

E triglycerides + proteins

F lipoproteins

Internal Environment (interstitial fluid inside a villus)

A Enzymes secreted by the pancreas and cells of the intestinal mucosa complete the digestion of carbohydrates to monosaccharides, and proteins to amino acids.

B Monosaccharides and amino acids are actively transported across the plasma membrane of brush border cells in the intestinal lining, then out of the same cells and into the internal environment.

C Movements of the intestinal wall break up fat globules into small droplets. Bile salts coat the droplets, so that globules cannot form again. Pancreatic enzymes digest the droplets to fatty acids and monoglycerides.

D Micelles form when bile salts combine with products of fat digestion: monoglycerides and fatty acids These products slip into and out of micelles.

E Concentrating monoglycerides and fatty acids in micelles enhances diffusion of these substances into brush border cells. These lipids diffuse across the plasma membrane's lipid bilayer, into the cells.

F In a brush border cell, the products of fat digestion form triglycerides, which associate with proteins. The resulting lipoproteins are then expolled by exocytosis into the interstitial fluid inside the villus.

Figure 40.10 Animated Summary of digestion and absorption in the small intestine.
Figure It Out: What do the purple dots in the micelles represent? *Answer: Bile salts*

environment by osmosis (Section 5.6). Transport proteins in the plasma membrane of brush border cells move salts, sugars, and amino acids from the intestinal lumen into these cells. Other transport proteins then move these solutes from the brush border cells into interstitial fluid inside a villus (Figure 40.10*b*). This movement of solutes creates an osmotic gradient, so water moves in the same direction.

From the interstitial fluid, water, salts, sugars, and amino acids enter the blood capillary inside the villus. The blood then distributes them throughout the body.

Fat Absorption Being lipid soluble, the fatty acids and monoglycerides released by fat digestion enter a villus by diffusing across the lipid bilayer of brush border cells. Remember, bile salts aid fat digestion by coating fatty droplets (Section 40.4 and Figure 40.10*c*).

Bile salts also combine with the products of fat digestion (fatty acids and monoglycerides) to form tiny droplets called micelles (Figure 40.10*d*). When a micelle contacts a brush border cell, the micelle's fatty

acids and monglycerides diffuse into that cell (Figure 40.10*e*). The bile salts that were in the micelle remain in the intestinal lumen, where they will become part of new micelles.

Inside the brush border cells, monoglycerides and fatty acids form triglycerides that join with proteins. The resulting lipoproteins move by exocytosis into the interstitial fluid inside a villus (Figure 40.10*f*). From the interstitial fluid, triglycerides enter lymph vessels. Lymph—and triglycerides—eventually drain into the bloodstream (Section 37.10).

Take-Home Message

How are substances absorbed from the small intestine?

■ With a folded mucosa, villi, and microvilli, the small intestine has a vast surface area for absorbing water and nutrients.

■ Substances are absorbed through the brush border cells that line the free surface of each villus. Passive and active transport mechanisms help water and solutes cross; micelle formation helps lipid-soluble products cross.

■ The large intestine is wider than the small intestine, but also much shorter—only about 1.5 meters (5 feet) long.

Structure and Function of the Large Intestine

Not everything that enters the small intestine can be or should be absorbed. Muscular contractions propel indigestible material, dead bacteria and mucosal cells, inorganic substances, and some water from the small intestine into the large intestine. As the wastes travel through the large intestine, they become compacted as **feces**. Compaction occurs as the large intestine actively pumps sodium ions out of the lumen, into the internal environment. Water follows by osmosis.

A cup-shaped cecum, is the first part of the large intestine (Figure 40.11*a*). A wormlike pouch, called the **appendix** extends from it. From the cecum, material enters the ascending colon, which extends up along the wall of the abdominal cavity. The transverse colon extends across this cavity, and the descending colon connects to the rectum (Figures 40.5 and 40.11).

Contraction of the smooth muscle in the colon wall mixes its contents and propels them along its length. Compared with other gut regions, wastes move more slowly through the colon, which also has a moderate pH. These conditions favor growth of bacteria such as *Escherichia coli*. The bacteria make vitamins K and B12, which are absorbed across the colon lining.

After a meal, gastrin and signals from autonomic nerves cause much of the colon to contract forcefully and propel feces to the rectum. The rectum stretches, which activates a defecation reflex to expel feces. The nervous system can override the reflex by calling for contraction of a sphincter at the anus.

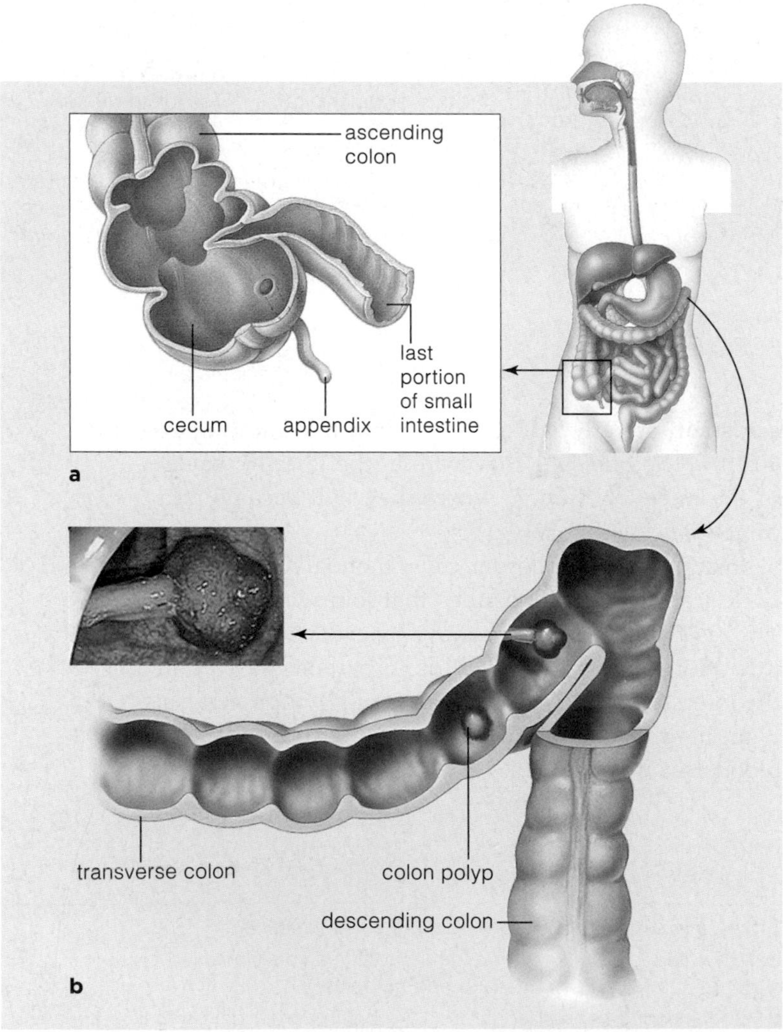

Disorders of the Large Intestine

Healthy adults typically defecate about once a day, on average. Emotional stress, a diet low in fiber, minimal exercise, dehydration, and some medications can lead to constipation, in which defecation occurs fewer than three times a week, is difficult, and yields small, hardened, dry feces. Occasional constipation usually goes away on its own. A chronic problem should be discussed with a doctor. Diarrhea—frequent passing of watery feces—can result from bacterial infection or problems with nervous controls. If prolonged, it can cause dehydration and disrupt blood solute levels.

Appendicitis—an inflammation of the appendix—requires prompt treatment. Removing the inflamed appendix prevents it from bursting and releasing large numbers of bacteria into the abdominal cavity. Such a rupture could cause a life-threatening infection.

Some people are genetically predisposed to develop colon polyps, small growths on the colon wall (Figure 40.11*b*). Most polyps are benign, but some can become cancerous. If detected in time, colon cancer is highly curable. Blood in feces and dramatic changes in bowel habits may be symptoms of colon cancer and should be reported to a doctor. Also, anyone over the age of 50 should have a colonoscopy, in which clinicians use a camera to examine the colon for polyps or cancer.

Figure 40.11 (**a**) Location of cecum and appendix of the large intestine. (**b**) Sketch and photo of polyps in the transverse colon.

Take-Home Message

What is the function of the large intestine?

■ The large intestine completes the process of absorption, then concentrates, stores, and eliminates wastes.

■ Most absorbed organic compounds are broken down for energy, stored, or used to build larger organic compounds.

■ Links to Glycogen 3.3, Alcohol metabolism Chapter 6 introduction, Systemic circulation 37.5

Figure 40.12a shows the main routes by which organic molecules from food are shuffled and reshuffled in the body. Living cells constantly recycle some carbohydrates, lipids, and proteins by breaking them apart. They use breakdown products as energy sources and building blocks. The nervous and endocrine systems regulate this turnover.

The **liver** is a large organ that functions in digestion, metabolism, and homeostasis (Figure 40.12b). All blood from the capillaries in the small intestine enters the hepatic portal vein, which delivers it to the liver. The blood flows through capillaries in the liver before returning to the heart (Section 37.5).

The liver helps protect the body against dangerous substances that were ingested or formed as a result of digestion. For example, Chapter 6 explained the role of the liver in detoxifying alcohol, and how alcohol abuse can damage this essential organ. As another example, ammonia (NH_3) is a toxic product of amino acid breakdown. The liver converts ammonia to urea, a much less toxic compound. Urea is carried by the blood to kidneys and is excreted in the urine.

Most of the body's fat-soluble vitamins such as vitamins A and D are stored in the liver. The liver also stores glucose. After a meal, liver and muscle cells take up glucose and convert it to glycogen (Section 3.3). Excess carbohydrates and proteins are also converted to fats, which are stored mainly in adipose tissue.

In between meals, the brain takes up much of the glucose circulating in the blood. The brain cannot use fats or proteins as an energy source. Other body cells dip into their stores of glycogen and fat. Adipose cells degrade fats to glycerol and fatty acids, which enter blood. Liver cells break down glycogen and release glucose, which also enters blood. Body cells take up the released fatty acids and glucose and use them to fuel ATP production.

Take-Home Message

What happens to compounds absorbed from the gut?

■ Absorbed compounds are carried by the blood to the liver. The liver detoxifies dangerous substances and stores vitamins and glucose. The glucose is stored as glycogen.

■ Adipose tissue takes up absorbed carbohydrates and proteins and converts them to fats.

■ In between meals, the liver breaks down stored glycogen, and releases its glucose subunits into the blood. This ensures that the brain, which can only use carbohydrates as fuel, always has an adequate supply of energy.

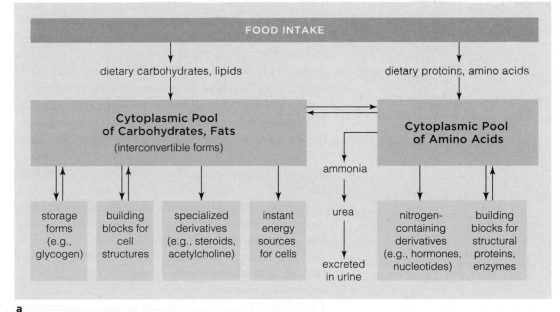

Liver Functions

Forms bile (assists fat digestion), rids body of excess cholesterol and blood's respiratory pigments

Controls amino acid levels in the blood; converts potentially toxic ammonia to urea

Controls glucose level in blood; major reservoir for glycogen

Removes hormones that served their functions from blood

Removes ingested toxins, such as alcohol, from blood

Breaks down worn-out and dead red blood cells, and stores iron

Stores some vitamins

Figure 40.12 (a) Summary of major pathways of organic metabolism. Cells continually synthesize and tear down carbohydrates, fats, and proteins. Most urea forms in the liver, an organ that is at the crossroads of organic metabolism (b).

■ Eating provides your cells with a source of energy and a supply of essential building materials.

■ Links to Trans fats Chapter 3 introduction, Carbohydrates 3.3, Lipids 3.4, Proteins 3.5, Quinoa 23.10

USDA Dietary Recommendations

Scientists at the Department of Agriculture and other United States government agencies research diets that may help prevent diabetes, cancers, and other health problems. They periodically update their nutritional guidelines. In 2005, they replaced their traditional one-size-fits-all food pyramid with a new Internet-based program that generates recommendations specific for

USDA Nutrition Guidelines	
Food Group	Amount Recommended
Vegetables	2.5 cups/day
Dark green vegetables	3 cups/week
Orange vegetables	2 cups/week
Legumes	3 cups/week
Starchy vegetables	3 cups/week
Other vegetables	6.5 cups/week
Fruits	2 cups/day
Milk Products	3 cups/day
Grains	6 ounces/day
Whole grains	3 ounces/day
Other grains	3 ounces/day
Fish, poultry, lean meat	5.5 ounces/day
Oils	24 grams/day

Figure 40.13 Example of nutritional guidelines from the United States Department of Agriculture (USDA). These recommendations are for females between ages ten and thirty who get less than 30 minutes of vigorous exercise daily. Portions add up to a 2,000-kilocalorie daily intake.

a person's age, sex, height, weight, and activity level (Figure 40.13). To generate your own healthy eating plan, visit the USDA web site: www.mypyramid.gov.

In stark contrast to the diet of a typical American, the new guidelines recommend lowering the intake of refined grains, saturated fats, *trans*-fatty acids, added sugar or caloric sweeteners, and salt (no more than a teaspoon per day). They also recommend eating more vegetables and fruits with a high potassium and fiber content, fat-free or low-fat milk products, and whole grains. About 55 percent of daily caloric intake should come from carbohydrates.

Energy-Rich Carbohydrates

Fresh fruits, whole grains, and vegetables—especially legumes such as peas and beans—provide abundant complex carbohydrates (Section 3.3). The body breaks the starch in these foods into glucose, your primary source of energy. These foods also provide essential vitamins and fiber. Eating foods high in soluble fiber helps lower one's cholesterol level and may reduce the risk of heart disease. A diet high in insoluble fiber helps prevent constipation.

Foods that contain a lot of processed carbohydrates such as white flour, refined sugar, and corn syrup are sometimes said to be full of "empty calories." This is a way of saying that these foods provide little in the way of vitamins or fiber.

Good Fat, Bad Fat

You cannot stay alive without lipids. Cell membranes incorporate phospholipids and cholesterol, one of the sterols. Fats serve as energy reserves, insulation, and cushioning. They also help store fat-soluble vitamins.

Linoleic acid and alpha-linolenic acid are **essential fatty acids**. The human body cannot synthesize them, so you must get them from your diet. Both are polyunsaturated fats; their long carbon tails include two or more double bonds (Table 40.3). Unsaturated fats are liquid at room temperature (Section 3.4).

We divide the polyunsaturated fatty acids into two categories: omega-3 fatty acids and omega-6 fatty acids. Omega-3 fatty acids, the main fat in oily fish such as sardines, seem to have special health benefits. Studies suggest that a diet high in omega-3 fatty acids can reduce the risk of cardiovascular disease, lessen the inflammation associated with rheumatoid arthritis, and help diabetics control their blood glucose.

Oleic acid, the main fat in olive oil, may also have health benefits. It is monounsaturated, which means

its carbon tails have only one double bond. A diet in which olive oil is substituted for saturated fats helps prevent heart disease.

Dairy products and meats are rich in saturated fats and cholesterol. Overindulging in these foods increases risk for heart disease, stroke, and some cancers.

Trans fatty acids, or *trans* fats, are manufactured from vegetable oils. However, they have a molecular structure that makes them even worse for the heart than saturated fats (Chapter 3 introduction).

Body-Building Proteins

Amino acids are building blocks of proteins (Section 3.5). Your cells can make some amino acids but you must get eight **essential amino acids** from food. The eight essential amino acids are methionine (or cysteine, its metabolic equivalent), isoleucine, leucine, lysine, phenylalanine, threonine, tryptophan, and valine.

Most proteins in meat are "complete"; their amino acid ratios match a human's nutritional needs. Nearly all plant proteins are incomplete, in that they lack one or more amino acids essential for humans. Proteins of quinoa (*Chenopodium quinoa*) are a notable exception (Section 23.10).

To get required amino acids from a vegetarian diet, one must combine plant foods so that the amino acids missing from one component are present in some others. As an example, rice and beans together provide all necessary amino acids, but rice alone or beans alone do not. You do not have to eat the two complementary foods at the same meal, but both should be consumed within a 24-hour period.

About Low-Carb/High-Protein Diets

Many people turn to diets lower in carbohydrates and higher in proteins and fats to promote rapid weight loss. The long-term effectiveness and health effect of these diets is controversial. We know that increased

Table 40.3 Main Types of Dietary Lipids

Polyunsaturated Fatty Acids: Liquid at room temperature; essential for health.
   Omega-3 fatty acids
      Alpha-linoleinc acid and its derivatives
      Sources: Nut oils, vegetable oils, oily fish
   Omega-6 fatty acids
      Linoleic acid and its derivatives
      Sources: Nut oils, vegetable oils, meat

Monounsaturated Fatty Acids: Liquid at room temperature. Main dietary source is olive oil. Beneficial in moderation.

Saturated Fatty Acids: Solid at room temperature. Main sources are meat and dairy products, palm and coconut oils. Excessive intake may raise risk of heart disease.

Trans **Fatty Acids (Hydrogenated Fats):** Solid at room temperature. Manufactured from vegetable oils and used in many processed foods. Excessive intake may raise risk of heart disease.

protein intake increases ammonia production (Section 40.7). Enzymes in the liver convert ammonia to urea, which kidneys filter from blood and excrete in urine. Also, when a body uses fats rather than carbohydrates as its main source of energy, large amounts of acidic metabolic wastes called ketones form. Ketones must be filtered from blood and excreted. Thus, high-fat, high-protein diets make kidneys work harder, raising the risk of kidney problems. Anyone with impaired kidney function should avoid such a diet.

Take-Home Message

What are the main types of nutrients that humans require?

■ Carbohydrates are broken down to glucose, the body's main energy source. Foods rich in complex carbohydrates also supply fiber and vitamins.

■ Fats are burned for energy and used as building materials. Polyunsaturated and monounsaturated fats should provide most of your fat calories. Excessive consumption of saturated fats and *trans* fats raises risk of heart disease.

■ Proteins are the source of amino acids used to build your body's own proteins. Meat provides all essential amino acids. Most plant foods lack one or more amino acids, but when combined correctly these foods can meet all human amino acid needs.

■ In addition to major nutrients, the body requires certain organic and inorganic substances to function properly.

■ Links to Electron transfer chains 6.4, Coenzymes 6.3, Thyroid hormones 35.6, Blindness 34.10, Hemoglobin 37.2

Vitamins are organic substances that are essential in very small amounts; no other substance can carry out their metabolic functions. At a minimum, human cells require the thirteen vitamins listed in Table 40.4. Each has specific roles. For instance, the B vitamin niacin is modified to make NAD, a coenzyme (Section 6.3).

Minerals are inorganic substances that are essential for growth and survival because no other substance can serve their metabolic functions (Table 40.5). As an example, all of your cells use iron as a component of electron transfer chains (Section 6.4). Red blood cells require iron to make oxygen-transporting hemoglobin (Section 37.2). Iodine is essential for development of a

Table 40.4 Major Vitamins: Sources, Functions, and Effects of Deficiencies or Excesses*

Vitamin	Common Sources	Main Functions	Effects of Chronic Deficiency	Effects of Extreme Excess
Fat-Soluble Vitamins				
A	Its precursor comes from beta-carotene in yellow fruits, yellow or green leafy vegetables; also in fortified milk, egg yolk, fish, liver	Used in synthesis of visual pigments, bone, teeth; maintains epithelia	Dry, scaly skin; lowered resistance to infections; night blindness; permanent blindness	Malformed fetuses; hair loss; changes in skin; liver and bone damage; bone pain
D	Inactive form made in skin, activated in liver, kidneys; in fatty fish, egg yolk, fortified milk products	Promotes bone growth and mineralization; enhances calcium absorption	Bone deformities (rickets) in children; bone softening in adults	Retarded growth; kidney damage; calcium deposits in soft tissues
E	Whole grains, dark green vegetables, vegetable oils	Counters effects of free radicals; helps maintain cell membranes; blocks breakdown of vitamins A and C in gut	Lysis of red blood cells; nerve damage	Muscle weakness; fatigue; headaches; nausea
K	Enterobacteria form most of it; also in green leafy vegetables, cabbage	Blood clotting; ATP formation via electron transport	Abnormal blood clotting; severe bleeding (hemorrhaging)	Anemia; liver damage and jaundice
Water-Soluble Vitamins				
B_1 (thiamin)	Whole grains, green leafy vegetables, legumes, lean meats, eggs	Connective tissue formation; folate utilization; coenzyme action	Water retention in tissues; tingling sensations; heart changes; poor coordination	None reported from food; possible shock reaction from repeated injections
B_2 (riboflavin)	Whole grains, poultry, fish, egg white, milk	Coenzyme action (FAD)	Skin lesions	None reported
B_3 (niacin)	Green leafy vegetables, potatoes, peanuts, poultry, fish, pork, beef	Coenzyme action (NAD^+)	Contributes to pellagra (damage to skin, gut, nervous system, etc.)	Skin flushing; possible liver damage
B_6	Spinach, tomatoes, potatoes, meats	Coenzyme in amino acid metabolism	Skin, muscle, and nerve damage; anemia	Impaired coordination; numbness in feet
Pantothenic acid	In many foods (meats, yeast, egg yolk especially)	Coenzyme in glucose metabolism, fatty acid and steroid synthesis	Fatigue; tingling in hands; headaches; nausea	None reported; may cause diarrhea occasionally
Folate (folic acid)	Dark green vegetables, whole grains, yeast, lean meats; enterobacteria produce some folate	Coenzyme in nucleic acid and amino acid metabolism	A type of anemia; inflamed tongue; diarrhea; impaired growth; mental disorders	Masks vitamin B_{12} deficiency
B_{12}	Poultry, fish, red meat, dairy foods (not butter)	Coenzyme in nucleic acid metabolism	A type of anemia; impaired nerve function	None reported
Biotin	Legumes, egg yolk; colon bacteria produce some	Coenzyme in fat, glycogen formation and in amino acid metabolism	Scaly skin (dermatitis); sore tongue; depression; anemia	None reported
C (ascorbic acid)	Fruits and vegetables, especially citrus, berries, cantaloupe, cabbage, broccoli, green pepper	Collagen synthesis; possibly inhibits effects of free radicals; structural role in bone, cartilage, and teeth; used in carbohydrate metabolism	Scurvy; poor wound healing; impaired immunity	Diarrhea, other digestive upsets; may alter results of some diagnostic tests

* Guidelines for appropriate daily intakes are being worked out by the Food and Drug Administration.

healthy nervous system and to make thyroid hormone (Section 35.6).

Healthy people can get all the vitamins and minerals they need from a well-balanced diet. In most cases, vitamin and mineral supplements are necessary only for strict vegetarians, the elderly, and people who are chronically ill or taking a medicine that interferes with nutrient absorption.

In addition to vitamins and minerals, a healthy diet should include a variety of **phytochemicals**, also known as phytonutrients. These organic molecules are found in plant foods and while not essential, they may reduce the risk of certain disorders. For example, eating leafy green vegetables ensures adequate intake of the plant pigments lutein and zeaxanthin. A diet low in these phytochemicals raises the risk of macular degeneration, a major cause of blindness (Section 34.10). As another example, isoflavones in soy products can help lower cholesterol level in the blood and protect against heart diseases.

Keep this in mind: The more colors you see among the vegetables on your plate, the greater the variety of beneficial phytochemicals in your food.

Take-Home Message

What roles do vitamins, minerals, and phytonutrients play?

■ Vitamins are organic molecules with an essential role in metabolism.

■ Minerals are inorganic substances with an essential role.

■ Phytochemicals are plant molecules that are not essential but may reduce the risk of certain disorders.

Table 40.5 Major Minerals: Sources, Functions, and Effects of Deficiencies or Excesses*

Mineral	Common Sources	Main Functions	Effects of Chronic Deficiency	Effects of Extreme Excess
Calcium	Dairy products, dark green vegetables, dried legumes	Bone, tooth formation; blood clotting; neural and muscle action	Stunted growth; fragile bones; nerve impairment; muscle spasms	Impaired absorption of other minerals; kidney stones in susceptible people
Chloride	Table salt (usually too much in diet)	HCl formation in stomach; contributes to body's acid–base balance; neural action	Muscle cramps; impaired growth, poor appetite	Contributes to high blood pressure in certain people
Copper	Nuts, legumes, seafood, drinking water	Used in synthesis of melanin, hemoglobin, and some transport chain components	Anemia; changes in bone and blood vessels	Nausea; liver damage
Fluorine	Fluoridated water, tea, seafood	Bone, tooth maintenance	Tooth decay	Digestive upsets; mottled teeth and deformed skeleton in chronic cases
Iodine	Marine fish, shellfish, iodized salt, dairy products	Thyroid hormone formation	Enlarged thyroid (goiter) with metabolic disorders	Toxic goiter
Iron	Whole grains, green leafy vegetables, legumes, nuts, eggs, lean meat, molasses, dried fruit, shellfish	Formation of hemoglobin and cytochrome (transport chain component)	Iron-deficiency anemia; impaired immune function	Liver damage; shock; heart failure
Magnesium	Whole grains, legumes, nuts, dairy products	Coenzyme role in ATP–ADP cycle; roles in muscle, nerve function	Weak, sore muscles; impaired neural function	Impaired neural function
Phosphorus	Whole grains, poultry, red meat	Component of bone, teeth, nucleic acids, ATP, phospholipids	Muscular weakness; loss of minerals from bone	Impaired absorption of minerals into bone
Potassium	Diet alone provides ample amounts	Muscle and neural function; roles in protein synthesis and body's acid–base balance	Muscular weakness	Muscular weakness; paralysis; heart failure
Sodium	Table salt; diet provides ample to excessive amounts	Key role in body's salt–water balance; roles in muscle and neural function	Muscle cramps	High blood pressure in susceptible people
Sulfur	Proteins in diet	Component of body proteins	None reported	None likely
Zinc	Whole grains, legumes, nuts, meats, seafood	Component of digestive enzymes; roles in normal growth, wound healing, sperm formation, and taste and smell	Impaired growth; scaly skin; impaired immune function	Nausea, vomiting, diarrhea; impaired immune function and anemia

* Guidelines for appropriate daily intakes are being worked out by the Food and Drug Administration.

40.10 | Weighty Questions, Tantalizing Answers

■ Fat cells do not increase in number after birth. Putting on weight simply fills existing fat cells with more fat.

■ Links to Fat tissue 32.3, Limbic system 33.11, Insulin 35.8, Diabetes 35.9, Inflammation 38.4

Weight and Health Being overweight has a negative effect on health. Among other things, it increases the risk of type 2 diabetes, high blood pressure, heart disease, breast and colon cancer, arthritis, and gallstones.

Why does excess weight have ill effects? As Section 8.7 explained, triglycerides in fat cells are the body's main form of energy storage. Fat cells of people who are at a healthy weight hold a moderate amount of triglycerides and function normally. In obese people, an excess of these molecules distends fat cells and impairs their function. Like cells damaged in other ways, the overstuffed fat cells respond by sending out signals that summon up an inflammatory response (Section 38.4). The resulting chronic inflammation harms organs throughout the body and increases risk of cancer.

Overstuffed fat cells also increase secretion of signals that interfere with the action of insulin. Remember that this hormone encourages cells to take up sugar from the blood (Section 35.8). When insulin becomes ineffective, the result is type 2 diabetes (Section 35.9).

Armed with an understanding of how weight impairs health, researchers are looking for ways to dampen or offset harmful signals secreted by fat cells. One day, it may be possible to keep fat cells from causing inflammation or interfering with insulin function. But for now, the only way to prevent these effects is by losing the excess weight.

What Is the "Right" Body Weight? Figure 40.14 shows one of the widely accepted weight guidelines for women and men. The body mass index (BMI) is another guideline. It is a measurement designed to help assess increased health risk associated with weight gains. You can calculate your body mass index with this formula:

$$BMI = \frac{weight\ (pounds) \times 703}{height\ (inches)^2}$$

Generally, individuals with a BMI of 25 to 29.9 are said to be overweight. A score of 30 or more indicates **obesity**: an overabundance of fat in adipose tissue that may lead to severe health problems. How body fat gets distributed also helps predict the risks. Fat deposits just above the belt, as in a "beer belly," are associated with the greatest risk of heart problems. Fat deposits just below the skin of arms and legs, commonly referred to as "cellulite," have less of an effect on the heart.

If your BMI is too high, dieting alone will probably not lower it to a healthy level. When you simply eat less than normal, your body slows its metabolic rate to conserve energy. So how do you lose weight? You must decrease your caloric intake and increase your energy output. For most people, this means eating reasonable portions of low-calorie, nutritious foods and exercising regularly.

Energy stored in food is expressed as kilocalories or Calories (with a capital C). One kilocalorie equals 1,000 calories, which are units of heat energy.

Here is a way to calculate how many kilocalories you should take in daily to maintain a preferred weight. First, multiply the weight (in pounds) by 10 if you are not active physically, by 15 if you are moderately active, and by 20 if

Weight Guidelines for Women		Weight Guidelines for Men	
Starting with an ideal weight of 100 pounds for a woman who is 5 feet tall, add five additional pounds for each additional inch of height. Examples:		Starting with an ideal weight of 106 pounds for a man who is 5 feet tall, add six additional pounds for each additional inch of height. Examples:	
Height (feet)	Weight (pounds)	Height (feet)	Weight (pounds)
5' 2"	110	5' 2"	118
5' 3"	115	5' 3"	124
5' 4"	120	5' 4"	130
5' 5"	125	5' 5"	136
5' 6"	130	5' 6"	142
5' 7"	135	5' 7"	148
5' 8"	140	5' 8"	154
5' 9"	145	5' 9"	160
5' 10"	150	5' 10"	166
5' 11"	155	5' 11"	172
6'	160	6'	178

Figure 40.14 How to estimate "ideal" weights for adults. Values shown are consistent with a long-term Harvard study into the link between excess weight and risk of cardiovascular disorders. The "ideal" varies. It is influenced by specific factors such as having a small, medium, or large skeletal frame; bones are heavy.

you are highly active. Next, subtract one of the following amounts from the multiplication result:

Age:	Subtract:
25–34	0
35–44	100
45–54	200
55–64	300
Over 65	400

For example, if you are 25 years old, are highly active, and weigh 120 pounds, you will require $120 \times 20 = 2{,}400$ kilocalories daily to maintain weight. If you want to gain weight you will require more; to lose, you will require less. The amount is only a rough estimate. Other factors, such as height, must be considered. A person who is 5 feet, 2 inches tall and is active does not require as much energy as an active 6-footer whose body weight is the same.

Genes, Hormones, and Obesity Numerous studies have explored the role that genetics plays in obesity. As one example, Claude Bouchard studied experimental overeating by twelve pairs of male twins. All were lean young men in their early twenties. For 100 days they did not exercise, and they adhered to a diet that provided 6,000 more kilocalories a week than usual.

They all gained weight, but some gained three times as much as others. Members of each set of twins tended to gain a similar amount, which suggests that genes affect the response to overfeeding. For another test, Bouchard put sets of obese twins on a low-calorie diet. Once again, each set of twins lost a similar amount.

As the chapter introduction indicated, we are learning more about how genes that encode hormones contribute to obesity. Figure 40.15 details how researchers uncovered the role of the appetite-suppressing hormone leptin in mice. Researchers have now also identified the leptin gene in humans; it is on chromosome 7 (Appendix VII).

Leptin deficiency of the sort seen in mice is extremely rare in humans. However, three cousins in a Turkish family have been found to be entirely leptin deficient. All three were greatly obese. When UCLA researchers gave them daily leptin injections, the leptin-deficient men lost an average of 50 percent of their body weight without even trying to diet. The injections apparently caused changes in their brains. Scans showed increases in the gray matter of the cingulate gyrus, a portion of the limbic system known from other research to affect cravings (Section 33.11).

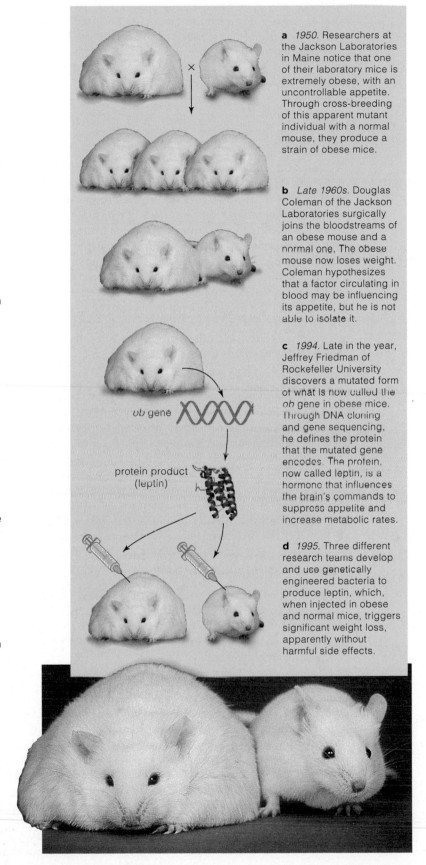

ob gene

protein product (leptin)

a *1950.* Researchers at the Jackson Laboratories in Maine notice that one of their laboratory mice is extremely obese, with an uncontrollable appetite. Through cross-breeding of this apparent mutant individual with a normal mouse, they produce a strain of obese mice.

b *Late 1960s.* Douglas Coleman of the Jackson Laboratories surgically joins the bloodstreams of an obese mouse and a normal one. The obese mouse now loses weight. Coleman hypothesizes that a factor circulating in blood may be influencing its appetite, but he is not able to isolate it.

c *1994.* Late in the year, Jeffrey Friedman of Rockefeller University discovers a mutated form of what is now called the *ob* gene in obese mice. Through DNA cloning and gene sequencing, he defines the protein that the mutated gene encodes. The protein, now called leptin, is a hormone that influences the brain's commands to suppress appetite and increase metabolic rates.

d *1995.* Three different research teams develop and use genetically engineered bacteria to produce leptin, which, when injected in obese and normal mice, triggers significant weight loss, apparently without harmful side effects.

Figure 40.15 Chronology of research developments that identified leptin as a heritable factor that affects body weight.

Americans are eating fewer and fewer meals at home. An ever-increasing array of fast-food outlets benefits from this trend. However, frequent fast-food meals increase the risk of obesity and diabetes. One part of the problem is over-sized portions. Another is that people simply do not make healthy choices. Many fast-food restaurants now offer salads or veggie burgers, but most diners prefer higher fat, higher calorie options.

How would you vote?

Most fast-food meals are high in saturated fats and in calories. Should these foods carry warning labels? See CengageNOW for details, then vote online.

Summary

Section 40.1 A **digestive system** breaks food down into molecules that are small enough to be absorbed into the internal environment. It also stores and eliminates any unabsorbed materials, and promotes homeostasis by its interactions with other organ systems. Some invertebrates have an **incomplete digestive system**: a saclike gut with a single opening. Most animals, and all vertebrates, have a **complete digestive system**: a tube with two openings (mouth and anus) and specialized areas between them.

Features of the digestive system may adapt an animal to a particular diet. For example, the multiple stomach chambers of cattle and other **ruminants** allow them to maximize the nutrients they get from plant food.

■ *Use the animation on CengageNOW to compare vertebrate digestive systems.*

Section 40.2 The human **pharynx** is the entrance to the digestive and respiratory systems. **Peristalsis** moves food down the **esophagus** and through a **sphincter** (a ring of muscle that can close off an opening) into the **stomach**, the start of the **gastrointestinal tract**. From the stomach, material moves to the **small intestine**. Most digestion occurs and most nutrients and water are absorbed here. The **large intestine** concentrates undigested wastes, which are stored in the **rectum** until expelled through the **anus**.

■ *Use the animation on CengageNOW to explore the components of the human digestive system.*

Section 40.3 Teeth are mostly bonelike **dentin**, with a covering of hard **enamel**. They break food into bits that become coated with saliva from **salivary glands**. Saliva contains the enzyme **salivary amylase**, which begins the process of starch digestion.

Section 40.4 Protein digestion starts in the stomach, where cells in its lining (the **mucosa**) release **gastric fluid**. This fluid contains protein-digesting enzymes and acid. It mixes with food and forms the semiliquid **chyme**.

Most digestion is completed in the small intestine, which receives a variety of digestive enzymes from the pancreas. **Bile**, which assists in fat digestion, is made in the liver and stored in the **gallbladder**. Delivery of bile into the small intestine, causes the **emulsification** of fats, breaking them into smaller, more easily digested droplets.

The nervous and endocrine systems respond to the volume and composition of food in the gut. They cause changes in muscle activity and in the rate of secretion of hormones and enzymes.

Section 40.5 The lining of the small intestine is highly folded. Multicelled, absorptive structures called **villi** are on each fold. Most cells at the surface of each villus are **brush border cells** that have **microvilli** on their surface. Brush border cells function in digestion and absorption. Their many membrane proteins transport salts, simple sugars, and amino acids from the intestinal lumen into the villus interior. A blood vessel inside each villus takes up absorbed sugars and amino acids.

Monoglycerides and fatty acids diffuse into a brush border cell, where they combine with proteins. The result is lipoproteins, which move by exocytosis into interstitial fluid, then enter lymph vessels that deliver them to blood.

■ *Use the animation on CengageNOW to learn about the small intestine's structure and how it absorbs nutrients.*

Section 40.6 The large intestine absorbs water and ions, thus compacting undigested solid wastes as **feces**. The **appendix** is a thin extension of the first part of the large intestine.

Section 40.7 The small organic compounds absorbed from the gut are stored, used in biosynthesis or as energy sources, or excreted by other organ systems. Blood that flows through the small intestine travels next to the **liver**, whic eliminates ingested toxins and stores some excess glucose as glycogen.

Sections 40.8, 40.9 Food must provide both energy and raw materials, including **essential amino acids** and **essential fatty acids**. It must also include two additional types of compounds needed for metabolism: **vitamins**, which are organic, and **minerals**, which are inorganic. **Phytochemicals** are plant molecules that are not essential, but may improve health or prevent certain disorders.

Section 40.10 An unhealthy overabundance of fat, or **obesity**, stresses fat cells and increases the risk of many disorders. To maintain your body weight, energy (caloric) intake must balance with energy output. Genetic factors influence how difficult it is for a person to reach and maintain a healthy weight. Hormones can influence both appetite and metabolic rate.

■ *Use the interaction on CengageNOW to calculate your body mass index.*

Data Analysis Exercise

The human *AMY-1* gene encodes salivary amylase, an enzyme that breaks down starch. The number of copies of this gene varies, and people who have more copies generally make more enzyme. In addition, the average number of *AMY-1* copies differs among cultural groups.

George Perry and his colleagues hypothesized that duplications of the *AMY-1* gene would confer a selective advantage in cultures in which starch is a large part of the diet. To test this hypothesis, the scientists compared the number of copies of the *AMY-1* gene among members of seven cultural groups that differed in their traditional diets. Figure 40.16 shows their results.

1. Starchy tubers are a mainstay of Hadza hunter–gatherers in Africa, whereas fishing sustains Siberia's Yakut. Almost 60 percent of Yakut had fewer than 5 copies of the *AMY1* gene. What percent of the Hadza had fewer than 5 copies?

2. None of the Mbuti (rain-forest hunter–gathers) had more than 10 copies of *AMY-1*. Did any European Americans?

3. Do these data support the hypothesis that a starchy diet favors duplications of the *AMY-1* gene?

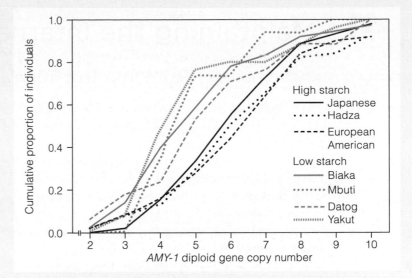

Figure 40.16 Number of copies of the *AMY-1* gene among members of cultures with traditional high-starch or low-starch diets. The Hadza, Biaka, Mbuti, and Datog are tribes in Africa. The Yakut live in Siberia.

Self-Quiz
Answers in Appendix III

1. A digestive system functions in _____ .
  a. secreting enzymes   c. eliminating wastes
  b. absorbing compounds   d. all of the above

2. Protein digestion begins in the _____ .
  a. mouth   c. small intestine
  b. stomach   d. colon

3. Most nutrients are absorbed in the _____ .
  a. mouth   c. small intestine
  b. stomach   d. colon

4. Bile has roles in _____ digestion and absorption.
  a. carbohydrate   c. protein
  b. fat   d. amino acid

5. Monosaccharides and amino acids absorbed from the small intestine enter _____ .
  a. blood vessels   c. fat droplets
  b. lymph vessels   d. the large intestine

6. The largest number of bacteria thrive in the _____ .
  a. stomach   c. large intestine
  b. small intestine   d. esophagus

7. The pH is lowest in the _____ .
  a. stomach   c. large intestine
  b. small intestine   d. esophagus

8. Most water that enters the gut is absorbed across the lining of the _____ .
  a. stomach   c. large intestine
  b. small intestine   d. esophagus

9. _____ are inorganic substances with essential metabolic roles that no other substance can fulfill.
  a. Phytonutrients   c. Vitamins
  b. Minerals   d. both a and c

10. True or false? Glucose-rich blood flows from the small intestine to the liver, which stores glucose as glycogen.

11. Ammonia is a toxic product of the digestion of _____ .
  a. fats   b. proteins   c. carbohydrates   d. vitamins

12. Ammonia is converted to less toxic urea by the _____ .
  a. liver   b. stomach   c. gallbladder   d. rectum

13. The essential fatty acids are _____ .
  a. *trans* fats   c. polyunsaturated fats
  b. saturated fats   d. lysine and methionine

14. Match each organ with a digestive function.
  ___ gallbladder   a. makes bile
  ___ large   b. compacts undigested residues
      intestine   c. secretes most digestive enzymes
  ___ liver   d. absorbs most nutrients
  ___ small   e. secretes gastric fluid
      intestine   f. stores, secretes bile
  ___ stomach
  ___ pancreas

■ *Visit CengageNOW for additional questions.*

Critical Thinking

1. Anorexia nervosa is an eating disorder in which people, most often women, starve themselves. Although the name means "nervous loss of appetite," most affected people are obsessed with food and continually hungry. Anorexia nervosa has complex causes, including some recently discovered genetic factors. Reported incidence of anorexia has soared during the past 20 years. Is it likely that a rise in the frequency of alleles that put people at risk for anorexia has caused this rise in reported cases?

2. Starch and sugar have the same number of calories per gram. However, not all vegetables are equally calorie dense. For example, a serving of boiled sweet potato provides about 1.2 calories per gram, while a serving of kale yields only 0.3 calories per gram. What could account for the difference in the calories your body obtains from these two foods?

41 Maintaining the Internal Environment

Truth in a Test Tube

Light or dark? Clear or cloudy? A lot or a little? Asking about and examining urine is an ancient art (Figure 41.1). About 3,000 years ago in India, the pioneering healer Susruta reported that some patients formed an excess of sweet-tasting urine that attracted insects. In time, the disorder was named diabetes mellitus, which loosely translates as "passing honey-sweet water." Doctors still diagnose it by testing the sugar level in urine, although they have replaced the taste test with chemical analysis.

Today, physicians routinely check the pH and solute concentrations of urine to monitor their patients' health. Acidic urine suggests metabolic problems. Alkaline urine can indicate an infection. Damaged kidneys will produce urine high in proteins. An abundance of some salts can result from dehydration or trouble with the hormones that control kidney function. Special urine tests detect chemicals produced by cancers of the kidney, bladder, and prostate gland.

Do-it-yourself urine tests are popular. If a woman is hoping to become pregnant, she can use one test to keep track of the amount of luteinizing hormone, or LH, in her urine. About midway through a menstrual cycle, LH triggers ovulation, the release of an egg from an ovary. Another over-the-counter urine test can reveal whether she has become pregnant. Still other tests help older women check for declining hormone levels in urine, a sign that they are entering menopause.

Not everyone is in a hurry to have their urine tested. Olympic athletes can be stripped of their medals when mandatory urine tests reveal they use prohibited drugs. Major League Baseball players agreed to urine tests only after repeated allegations that certain star players took prohibited steroids. The National Collegiate Athletic Association (NCAA) tests urine samples from about 3,300 student athletes per year for any performance-enhancing substances as well as for "street drugs."

If you use marijuana, cocaine, Ecstasy, or other kinds of psychoactive drugs, urine tells the tale. After the active ingredient of marijuana enters blood, the liver converts it to another compound. As kidneys filter blood, they add the compound to newly forming urine. It can take as long as ten days for all molecules of the compound to become fully metabolized and removed from the body. Until that happens, urine tests can detect it.

It is a tribute to the urinary system that urine is such a remarkable indicator of health, hormonal status, and drug use. Each day, a pair of fist-sized kidneys filter all of the blood in an adult human body, and they do so more than forty times. When all goes well, kidneys rid the body of excess water and excess or harmful solutes, including a variety of metabolites, toxins, hormones, and drugs.

So far in this unit, you have considered several organ systems that work to keep cells supplied with oxygen, nutrients, water, and other substances. Turn now to the kinds that maintain the composition, volume, and even the temperature of the internal environment.

See the video! Figure 41.1 *This page,* a seventeenth-century physician and a nurse examining a urine specimen. Urine's consistency, color, odor, and—at least in the past—taste provide clues to health conditions. Urine forms inside kidneys, and it provides clues to abnormal changes in the volume and composition of blood and interstitial fluid. *Facing page,* testing for the presence of drugs in urine samples.

Key Concepts

Maintaining the extracellular fluid

Animals continually produce metabolic wastes. They continually gain and lose water and solutes. Yet the overall composition and volume of extracellular fluid must be kept within a narrow range. Most animals have organs that accomplish this task. **Sections 41.1–41.3**

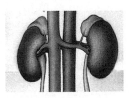

The human urinary system

The human urinary system consists of two kidneys, two ureters, a bladder, and a urethra. Inside a kidney, millions of nephrons filter water and solutes from the blood. Most of this filtrate is returned to the blood. Water and solutes not returned leave the body as urine. **Section 41.4**

What kidneys do

Urine forms by filtration, reabsorption, and secretion. Its content is adjusted continually by hormonal and behavioral responses to shifts in the internal environment. Hormones, as well as a thirst mechanism, influence whether urine is concentrated or dilute. **Sections 41.5–41.8**

Adjusting the core temperature

Heat losses to the environment and heat gains from the environment and from metabolic activity determine an animal's body temperature. Adaptations in body form and behavior help maintain core temperature within a tolerable range. **Sections 41.9, 41.10**

Links to Earlier Concepts

- In this chapter, you will see how osmosis (Section 5.6) affects water gain and loss in animal bodies, and learn about an animal group that has contractile vacuoles (22.2). You will also learn more about the efficient kidneys of amniotes (26.7).

  You will be reminded that aerobic respiration (8.1) produces water, and protein metabolism (40.7) yields ammonia, which is why a high-protein diet (40.8) can stress kidneys.

- Your knowledge of pH and buffer systems (2.6) will help you understand acid–base balance in the body.

- You will learn the roles of osmoreceptors (34.1), the hypothalamus (33.10), the pituitary gland (35.3), the adrenal glands (35.10) and the autonomic nervous system (33.8), in regulating body fluids. You will also learn about another spinal reflex (33.9).

- The discussion of body temperature will refer back to properties of water (2.5), forms of energy (6.1), feedback controls (27.3), heat illness (27.4), sweat glands (32.7), and fever (38.4).

How would you vote? Prospective employees are sometimes screened for drug and alcohol use by testing their urine. Should an employer be allowed to require a urine test before hiring an individual, or are such tests an invasion of privacy? See CengageNOW for details, then vote online.

- All animals constantly acquire and lose water and solutes, yet they must keep the volume and the composition of their internal environment—the extracellular fluid—stable.

- Links to Osmosis 5.6, Aerobic respiration 8.1, Ammonia 40.7

By weight, all organisms consist mostly of water, with dissolved salts and other solutes. Fluid outside cells—the extracellular fluid (ECF)—functions as the body's internal environment. In humans and other vertebrates, extracellular fluid consists mostly of interstitial fluid, which fills the spaces between cells, and plasma, the fluid portion of the blood (Figure 41.2).

Keeping the solute composition and volume of the extracellular fluid within the range that living cells can tolerate is a major aspect of homeostasis. Water and solute gains need to be balanced by water and solute losses. An animal can lose water and solutes in feces and urine, in exhalations, and in secretions. The animal gains water by eating and drinking. In aquatic animals, water also moves into or out of the body by osmosis across the body surface (Section 5.6).

In all animals, metabolic reactions put water and solutes into the ECF. The most abundant molecules of metabolic waste are carbon dioxide and ammonia. Aerobic respiration produces carbon dioxide and water (Section 8.1). Ammonia forms when amino acids or nucleic acids are broken down (Section 40.7). Carbon dioxide diffuses out across the body surface or leaves with the help of respiratory organs. In most animals, excretory organs rid the body of ammonia and other unwanted solutes, as well as excess water.

- Most invertebrates regulate the volume and composition of their body fluid through the action of excretory organs.

- Link to Contractile vacuole 22.2

Sponges are among the simplest invertebrates; they do not have tissues or organs (Section 25.4). A sponge excretes metabolic wastes at the cellular level. All of a sponge's cells are located close to the body surface, so metabolic wastes can simply diffuse from that surface into the surrounding water.

Freshwater sponges face a challenge common to all freshwater animals. Their body fluid contains a higher concentration of solutes than the surrounding water. As a result, water constantly moves into the body by osmosis. In freshwater sponges, this inward flow is countered by action of **contractile vacuoles** similar to those of freshwater protists (Section 22.2). Fluid accumulates inside this organelle, which then contracts and expels the fluid to the outside through a pore.

In planarians, a group of freshwater flatworms, a pair of branching, tubular, excretory organs run the length of the body (Figure 41.3). Along the tubes are **flame cells**, so called because the cells contain a tuft of cilia that looks like a flickering flame when viewed with a microscope. Movement of the cilia draws interstitial fluid into the tubes, propels it along, and forces it out of the body through pores at the body surface.

An earthworm is a segmented annelid worm with a fluid-filled body cavity (a coelom) and a closed circulatory system (Section 25.7). Most body segments have a

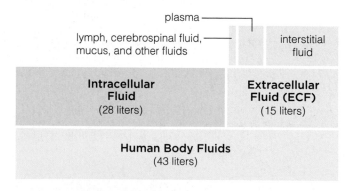

Figure 41.2 Fluid distribution in the human body.

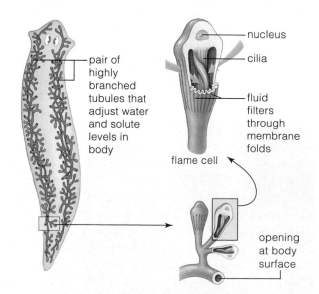

Figure 41.3 Planarian excretory organs. Action of cilia in flame cells drives flow of interstitial fluid into branching tubes, then out of the body through pores at the body surface.

Take-Home Message

What is the function of excretory organs?

- Excretory organs help maintain the volume and composition of the extracellular fluid by getting rid of water and certain solutes.

funnel where coelomic fluid enters the nephridium (coded *green*)

pore where ammonia-rich fluid leaves the body

body wall

storage bladder

loops where blood vessels take up solutes

one body segment of an earthworm

Figure 41.4 Excretory system of an earthworm. Most body segments have a pair of nephridia. One nephridium is shown in the diagram in *green*. Coelomic fluid enters a nephridium through a ciliated funnel in the segment just anterior to it. As fluid travels through the nephridium, essential solutes leave this tube and enter into adjacent blood vessels (shown in *red*). This process yields an ammonia-rich fluid that exits the body through a pore.

pair of tubular excretory organs called **nephridia**. The anterior end of each nephridium is a ciliated funnel that collects coelomic fluid from the adjacent segment (Figure 41.4). As the fluid flows through the tubular portion of the nephridium, essential solutes and some water leave the tubes and are reabsorbed by adjacent blood vessels, but wastes remain in the tubule. The ammonia-rich fluid that forms through this process is stored in a bladderlike organ before leaving the body through a pore.

Land-dwelling arthropods such as insects, spiders, and centipedes do not excrete ammonia. Instead, some enzymes in their blood convert ammonia to **uric acid**. Uric acid and other solutes are actively transported into **Malpighian tubules**. These tubules are long, thin excretory organs that connect to and empty into a region of the gut (Figure 41.5). Solutes are pumped from the blood into Malpighian tubules, then water follows by osmosis. Both the water and solutes move through the tubules and enter the gut.

Unlike ammonia, uric acid need not be dissolved in a large amount of water in order to be excreted from the body. Thus, nearly all of the water taken up by Malpighian tubules can be reabsorbed into the blood across the wall of the rectum. The uric acid is then excreted from the rectum in the form of crystals mixed with just a tiny bit of water to produce a thick paste.

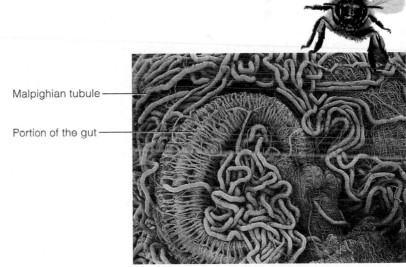

Malpighian tubule

Portion of the gut

Figure 41.5 Scanning, colorized electron micrograph of Malpighian tubules (*gold*) in a honeybee. The tubules are outpouchings of the gut (*pink*). They are bathed by the bee's blood, and take up substances from it.

Take-Home Message

How do invertebrates regulate the volume and composition of their body fluid?

■ Sponges are simple animals with no excretory organs. Wastes diffuse out across the body wall and excess water is expelled by contractile vacuoles.

■ Flatworms and earthworms have tubular excretory organs that deliver fluid with dissolved ammonia to a pore at the body surface.

■ Insects convert ammonia to uric acid, which Malpighian tubules deliver to the gut. Excreting uric acid rather than ammonia reduces water loss.

■ All vertebrates have paired kidneys—excretory organs that filter metabolic wastes and toxins out of the blood and adjust the level of solutes.

■ Links to Osmosis 5.6, Amniote traits 26.7

Vertebrates have a **urinary system** that filters water and solutes from their blood, then reclaims or excretes water and certain solutes as needed to maintain the volume and composition of the extracellular fluid. A pair of organs called **kidneys** filter blood. The urinary system interacts with other vertebrate organ systems as illustrated in Figure 41.6.

Fluid Balance in Fishes and Amphibians

Most marine invertebrates have body fluids that have the same concentration of solutes as seawater. As a result, there is no net movement of water into or out of their body as a result of osmosis. Body fluids of sharks and other cartilaginous fishes are also isotonic with seawater, although the fluids have different types of solutes. The fishes maintain a high internal solute concentration by retaining large amounts of urea, a solute that is scarce in seawater.

Bony fishes have body fluids that are less salty than seawater, but saltier than fresh water. Thus, wherever they live, they face an osmotic challenge. A marine bony fish loses water by osmosis across its body surfaces, especially its gills. To replace this lost water, the fish gulps seawater, then pumps the unwanted salts out through its gills (Figure 41.7a). It produces a small amount of urine that contains some salts.

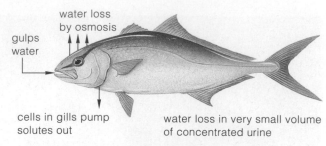

a Marine bony fish: Body fluids are less salty than the surrounding water; they are hypotonic.

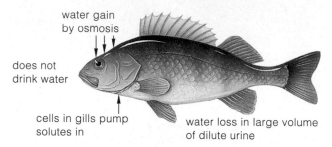

b Freshwater bony fish: Body fluids are less salty than the surrounding water; they are hypertonic.

Figure 41.7 Fluid and solute balance in bony fishes.

In contrast, a freshwater bony fish continually gains water. It does not drink and still produces a large volume of dilute urine. Solutes lost in the urine are offset by solutes absorbed from the gut, and by sodium ions pumped in across the gills.

When in water, amphibians face the same challenge as freshwater bony fishes. Water moves inward across their skin. Most keep their body fluid from becoming too dilute by pumping ions in across the skin.

On land, amphibians tend to lose water when it evaporates across their skin. Most amphibians excrete either ammonia or urea as adults, but some that spend much of their time in dry habitats excrete uric acid. Converting urea to uric acid takes energy, but this cost is offset by the benefit of reducing the amount of water required for excretion.

Fluid Balance in Reptiles, Birds, and Mammals

Waterproof skin and a pair of highly efficient kidneys are among the features that adapt amniotes—reptiles, birds, and mammals—to life on land (Section 26.7).

Reptiles and birds convert ammonia to uric acid, while mammals convert it to **urea**. It takes twenty to thirty times more water to excrete 1 gram of urea than to excrete 1 gram of uric acid. Thus, a typical mammal requires more water than a bird or reptile of similar body size. Even so, some mammals have adaptations

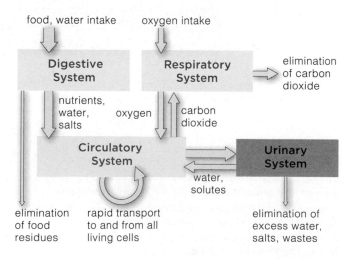

Figure 41.6 Functional links between the urinary, digestive, respiratory, and circulatory systems. Guided by the nervous and endocrine systems, these systems help maintain homeostasis.

	Kangaroo Rat	Human
Daily Water Gain (milliliters):	60 ml	2600 ml
By ingesting solids	10%	33%
By ingesting liquids	0%	54%
By metabolism	90%	13%
Daily Water Loss (milliliters):	60 ml	2600 ml
In urine	23%	58%
In feces	4%	8%
By evaporation	73%	34%

Figure 41.8 Water gains and losses in two mammals, a human and a kangaroo rat. In both, water intake must balance water losses.

Figure It Out: Which species loses a greater percentage of its water to evaporation? *Answer: Kangaroo rat*

that allow them to get along with very little water. For example, the kangaroo rat (*Dipodomys deserti*) is a small mammal that lives in the New Mexico desert where standing water is scarce, except during a brief rainy season. The rat conserves water by sheltering in its burrow during the heat, then foraging at night for dry seeds and bits of plants.

The kangaroo rat hops rapidly and far as it searches for seeds and flees from predators. All that activity requires ATP energy. Aerobic respiration (Section 8.1) of compounds in food provides energy and produces carbon dioxide and water. Each day, the "metabolic water" derived from this and other reactions makes up 90 percent of a kangaroo rat's water intake. In contrast, metabolic water accounts for about 13 percent of a human's daily water gain (Figure 41.8).

A kangaroo rat conserves and recycles water when it rests in its cool burrow. It moistens and warms the air that it inhales. When it exhales, water condenses in its cooler nose, and some diffuses back into the body. Seeds emptied from a kangaroo rat's cheek pouches soak up water dripping from the nose. The kangaroo rat reclaims water when it eats dripped-on seeds.

A kangaroo rat has no sweat glands and its feces contain only half the water that human feces do. Like a human, the kangaroo rat must eliminate metabolic wastes in urine, but the rat's highly efficient kidneys minimize urinary water loss. A kangaroo rat produces urine that can be as much as three to five times more concentrated than human urine.

As another example of how mammalian kidneys can help an animal adapt to an unusual habitat, think about whales and dolphins. These marine mammals had land-dwelling ancestors, so solute concentrations in their blood are like those of other land mammals. Yet whales and dolphins eat highly salty food and do not drink fresh water. How do they rid their body of ingested salts and obtain the water needed to maintain proper solute concentration in their body fluid?

The kidneys of marine mammals tend to be larger than those of land mammals of a similar size, and marine mammal kidneys are divided into many small lobes that increase their surface area. Having large, highly efficient kidneys allows whales and dolphins to make and excrete urine that is saltier than seawater. As for meeting their water needs, like kangaroo rats, whales and dolphins conserve nearly all the water released by digestion and metabolism of their food.

Take-Home Message

How do vertebrates regulate volume and composition of their body fluid?

■ All vertebrates have a urinary system with two kidneys that filter the blood and adjust its solute concentration.

■ Fish and amphibians also adjust their internal solute concentration by pumping solutes across their gills or skin.

■ Reptiles and birds excrete uric acid but mammals excrete urea, which requires more water to excrete.

■ Some mammals have highly efficient kidneys and other adaptations that allow them to live in habitats where fresh water is scarce.

■ The human urinary system forms urine, stores it, and then expels it from the body.

■ Link to Reflexes 33.9

Components of the Urinary System

As in all other vertebrates, the human urinary system includes two kidneys, two ureters, a urinary bladder, and a urethra (Figure 41.9a). Kidneys filter blood and form urine. The other organs collect and store urine, and channel it to the body surface for excretion.

Each human kidney is a bean-shaped organ about the size of an adult fist. The kidneys are located at the rear of the abdominal cavity, with one on each side of the backbone (Figure 41.9a,b). Kidneys lie beneath the peritoneum, the tissue that lines the abdominal cavity.

The outermost layer of a kidney is a renal capsule that consists of fibrous connective tissue (Figure 41.9c). The Latin *renal* means "relating to the kidneys." The bulk of tissue inside the renal capsule is divided into two zones: the outer **renal cortex** and the inner **renal medulla**. A renal artery carries blood to each kidney and a renal vein transports blood away from it.

Urine collects in the renal pelvis, a central cavity inside each kidney. A tubular **ureter** conveys the fluid from a kidney into the **urinary bladder**. This muscular organ stores urine until a sphincter at its lower end opens and urine flows into the **urethra**.

As the bladder fills with urine, it stretches and a reflex action occurs. Stretch receptors in the bladder wall signal neurons in the spinal cord. These neurons then command the smooth muscle in the bladder wall to contract. As the bladder contracts, sphincters that encircle the urethra relax, so urine can flow out of the body. After age two or three, the brain overrides this spinal reflex and prevents urine from flowing through the urethra at inconvenient moments.

In males, the urethra runs the length of the penis. Urine and semen flow through it, but a sphincter cuts off urine flow during erections. In females, the urethra opens onto the body surface between the vagina and the clitoris. A female's urethra is a relatively short tube (about 4 centimeters, or 1.5 inches long), so pathogens move more easily through it to the urinary bladder. That is one reason why women get bladder infections more often than men do.

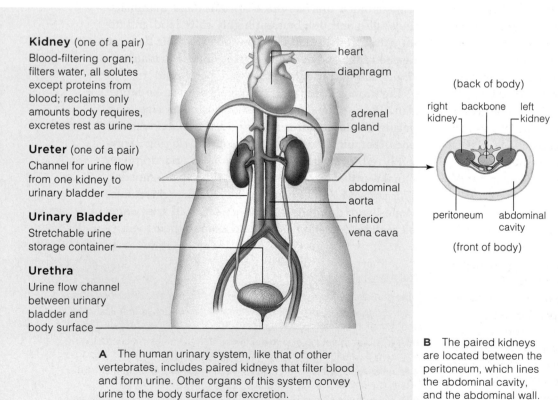

Kidney (one of a pair)
Blood-filtering organ; filters water, all solutes except proteins from blood; reclaims only amounts body requires, excretes rest as urine

Ureter (one of a pair)
Channel for urine flow from one kidney to urinary bladder

Urinary Bladder
Stretchable urine storage container

Urethra
Urine flow channel between urinary bladder and body surface

heart
diaphragm
adrenal gland
abdominal aorta
inferior vena cava

A The human urinary system, like that of other vertebrates, includes paired kidneys that filter blood and form urine. Other organs of this system convey urine to the body surface for excretion.

(back of body)
right kidney   backbone   left kidney
peritoneum   abdominal cavity
(front of body)

B The paired kidneys are located between the peritoneum, which lines the abdominal cavity, and the abdominal wall.

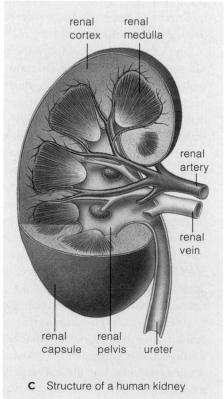

renal cortex   renal medulla
renal artery
renal vein
renal capsule   renal pelvis   ureter

C Structure of a human kidney

Figure 41.9 Animated Human urinary system.

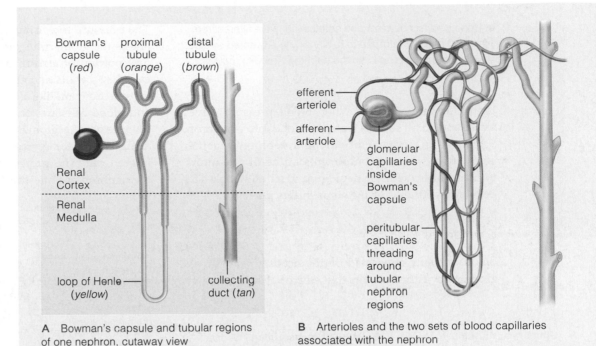

Figure 41.10 Animated
(**a**) The structure of a nephron. Nephrons are functional units of a kidney. They interact with neighboring blood vessels to form the urine.

(**b**) The arterioles and blood capillaries associated with each nephron. Large gaps between cells in the walls of glomerular capillaries make the capillaries about a hundred times more permeable than any others in the body.

Only a thin basement membrane separates each capillary wall from cells of the innermost layer of Bowman's capsule. Cells of this inner layer have long extensions that interdigitate with one another, like interlaced fingers. Fluid flows through the narrow slits between these extensions.

Bowman's capsule (*red*) proximal tubule (*orange*) distal tubule (*brown*)

Renal Cortex

Renal Medulla

loop of Henle (*yellow*)

collecting duct (*tan*)

A Bowman's capsule and tubular regions of one nephron, cutaway view

efferent arteriole

afferent arteriole

glomerular capillaries inside Bowman's capsule

peritubular capillaries threading around tubular nephron regions

B Arterioles and the two sets of blood capillaries associated with the nephron

Nephrons—The Functional Units of the Kidney

In the section to follow, you will be taking a look at three processes that rid the body of excess water and solutes in the form of urine. Tracking the steps of the processes will be simpler if you first acquaint yourself with the structures that carry out these functions.

Overview of Nephron Structure A kidney has more than 1 million **nephrons**—microscopically small tubes with a wall only one cell thick. Each nephron begins in the renal cortex, where its wall balloons outward and folds back on itself, to form a cup-shaped **Bowman's capsule** (Figure 41.10*a*). Past the capsule, the nephron twists a bit and straightens out as a **proximal tubule** (the part nearest the beginning of the nephron). After extending down into the renal medulla, the nephron makes a hairpin turn, the **loop of Henle**. It reenters the cortex and it twists again, as the **distal tubule** (the farthest from the start of the nephron), which drains into a **collecting duct**. Up to eight nephrons drain into each duct. Many collecting ducts extend through the kidney medulla and open onto the renal pelvis.

Blood Vessels Around Nephrons Inside each kidney, the renal artery branches into many afferent arterioles. Each arteriole branches into a **glomerulus**, a capillary bed inside Bowman's capsule (Figure 41.10*b*). As the next section explains, these capillaries interact with Bowman's capsule as a blood-filtering unit.

As blood passes through the glomerulus, a portion of it is filtered into Bowman's capsule. The rest enters an efferent arteriole. This arteriole quickly branches to become **peritubular capillaries**, which thread lacily around the nephron (*peri*–, around). Blood inside these capillaries continues into venules, and then through a vein leading out of the kidney.

Urine forms by three physiological processes that involve all the nephrons, glomerular capillaries, and peritubular capillaries. The processes are glomerular filtration, tubular reabsorption, and tubular secretion. They are the topic of the next section.

Each minute, nephrons of both kidneys collectively filter about 125 milliliters (1/2 cup) of fluid from the blood flowing past, which amounts to 180 liters (about 47.5 gallons) per day. At this rate of flow, the kidneys filter the entire volume of blood about 40 times a day!

Take-Home Message

How do the components of the human urinary system function?

■ Kidneys filter water and solutes from blood. The body reclaims most of the filtered fluid, or filtrate. The rest flows as urine through ureters into a bladder that stores it. Urine flows out of the body through the urethra.

■ The functional unit of the human kidneys is the nephron, a microscopic tube that interacts with two systems of capillaries to filter blood and form urine.

- Urine consists of water and solutes that were filtered from the blood and not returned to it, along with unwanted solutes secreted from the blood into the nephron's tubular regions.

- Link to Autonomic nervous system 33.8

Urine formation begins when blood pressure drives water and small solutes from the blood into a nephron. Variations in permeability along a nephron's tubular parts affect whether components of the filtrate return to blood or leave in urine. Figure 41.11 and Table 41.1 provide overviews of the steps in this process.

Glomerular Filtration

Blood pressure generated by the beating heart drives **glomerular filtration**, the first step of urine formation.

The pressure forces about 20 percent of the fluid that enters the glomerulus out across its wall and into the first portion of a nephron. Collectively, a glomerular capillary's walls and the inner wall of the Bowman's capsule function like a filter. Plasma proteins, platelets, and blood cells are too large to go through this filter. They leave the glomerulus via the efferent arteriole, along with the 80 percent of the fluid that did not get filtered out. The protein-free plasma that does enter the nephron becomes the filtrate:

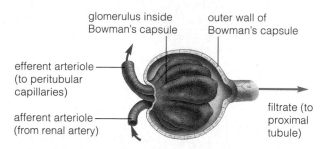

A Glomerular filtration Driven by pressure from a beating heart, water and solutes are forced across the wall of glomerular capillaries and into Bowman's capsule.

A Glomerular filtration
Occurs at glomerular capillaries in Bowman's capsule. Glomerular filtration nonselectively moves water, ions, and solutes from blood into Bowman's capsule.

B Tubular reabsorption Occurs all along a nephron's tubular parts. Most of the filtrate leaks or is transported out of the nephron's tubular parts into interstitial fluid, then is selectively reabsorbed into blood.

Tubular Reabsorption

Only a small fraction of the filtrate will be excreted. Most water and solutes are reclaimed during **tubular reabsorption**. By this process transport proteins move sodium ions (Na^+), chloride ions (Cl^-), bicarbonate, glucose, and other substances across the tubule wall and into peritubular capillaries. Movement of these solutes causes water to follow by osmosis:

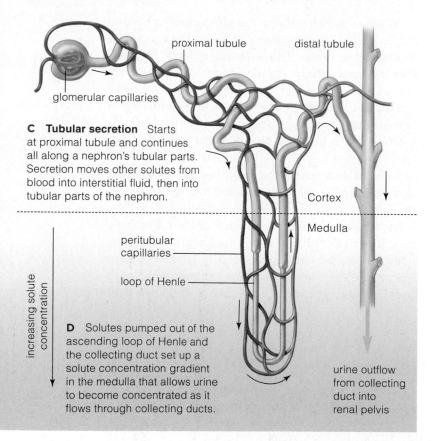

C Tubular secretion Starts at proximal tubule and continues all along a nephron's tubular parts. Secretion moves other solutes from blood into interstitial fluid, then into tubular parts of the nephron.

D Solutes pumped out of the ascending loop of Henle and the collecting duct set up a solute concentration gradient in the medulla that allows urine to become concentrated as it flows through collecting ducts.

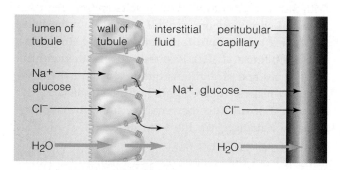

B Tubular reabsorption As the filtrate flows through the proximal tubule, ions and nutrients are actively and passively transported from the filrate into interstitial fluid. Water follows by osmosis. Cells of peritubular capillaries transport ions and nutrients into blood. Water again follows by osmosis.

Tubular reabsorption returns close to 99 percent of the water that enters a nephron to the blood. It also returns all the glucose and amino acids, most Na^+ and bicarbonate, and about half of the urea.

Figure 41.11 Animated How urine forms and becomes concentrated. The lettered in-text art in this section provides a closer look at each of the processes designated by a letter in this diagram.

Tubular Secretion

A build-up of excess hydrogen (H^+) ions, potassium (K^+) ions, or wastes such as urea can harm the body. By **tubular secretion**, transport proteins in the walls of peritubular capillaries actively transport urea and excess ions into the interstitial fluid. Then active transport proteins in a nephron's wall pump urea and ions into the filtrate, so they may be excreted in the urine:

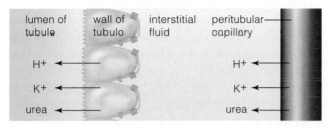

C **Tubular secretion** Transport proteins actively transport H^+, K^+, and urea out of peritubular capillaries and into filtrate.

As Section 41.7 explains, secretion of H^+ is essential to maintenance of the body's acid–base balance.

Concentrating the Urine

Sip soda all day and your urine will be dilute; sleep eight hours and it will be concentrated. Urine often has far more solutes than either plasma or most interstitial fluid. What concentrates the urine? Urine gets concentrated when water moves out of a nephron by osmosis. For urine to become concentrated, interstitial fluid surrounding the nephron must be saltier than the filtrate inside it. Only in the renal medulla does an outward-directed solute concentration gradient form, with the interstitial fluid saltiest deep in the medulla.

This concentration gradient is established as filtrate flows though the loop of Henle which extends into the medulla. The loop's two arms are close together and differ in permeability:

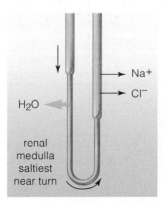

D The ascending limb of the loop of Henle actively pumps out salt, but is not permeable to water. Pumping salt outward creates a concentration gradient, with the saltiest interstitial fluid in the deepest part of the medulla. The descending part of the loop is permeable to water, but not to salt. As filtrate flows through the loop, it first loses water by osmosis, then loses salt by active transport.

Filtrate becomes concentrated as it flows through the descending part of the loop of Henle and loses water by osmosis. It then becomes less concentrated when salt is actively transported out in the ascending part of the loop. As a result, filtrate entering the distal tubule is less concentrated than normal body fluid.

The distal tubule delivers filtrate to the collecting duct, which—like the descending loop of Henle—extends down into the medulla. In the deepest part of the medulla, urea is pumped out of the collecting duct, making the interstitial fluid nearby even saltier. As urine passes through the collecting duct, the increasing saltiness of the interstitial fluid around it favors flow of water out of the duct by osmosis.

The body can adjust how much water is reabsorbed at distal tubules and collecting ducts. When it needs to conserve water, distal tubules and collecting ducts become more permeable to water, so less leaves in urine. When the body needs to rid itself of excess water, the distal tubule and collecting ducts become less permeable to water and the urine remains dilute. As the next section explains, hormones adjust the permeability of the distal tubule and collecting duct.

Table 41.1 Processes of Urine Formation

Process	Characteristics
Glomerular filtration	Pressure generated by heartbeats drives water and small solutes (not proteins) out of leaky glomerular capillaries and into Bowman's capsule, the entrance to the nephron.
Tubular reabsorption	Most water and solutes in the filtrate move from a nephron's tubular portions, into interstitial fluid around the nephron, then into blood inside the peritubular capillaries.
Tubular secretion	Urea, H^+, and some other solutes move out of peritubular capillaries, into interstitial fluid, then into the filtrate inside the nephron for excretion in urine.

Take-Home Message

How does urine form and become concentrated?

■ The force of the beating heart drives protein-free plasma out of glomerular capillaries and into the nephron's tubular portion as filtrate.

■ Nearly all of the water and solutes that leave the blood as filtrate later leave the tubule and return to the blood in peritubular capillaries.

■ Water and solutes that remain in the tubule, and solutes secreted into the tubule along its length, become the urine.

■ Concentration of urine as it flows down through the loop of Henle sets up a solute concentration gradient in the surrounding interstitial fluid of the renal medulla. The existence of this gradient allows urine to become concentrated as it flows through the collecting duct to the renal pelvis.

■ Urine consists of water and solutes that were filtered from the blood and not returned to it, along with solutes secreted from the blood into the nephron.

■ Links to Hypothalamus 33.10, Osmoreceptors 34.1, Pituitary hormones 35.3, Adrenal glands 35.10

Regulating Thirst

When you do not drink enough fluid to make up for normal fluid losses, the concentration of sodium and other solutes in your blood rises. You make less saliva and your dry mouth stimulates nerve endings that signal the **thirst center**, a region of the hypothalamus. At the same time, the thirst center receives input from osmoreceptors that detect the level of solutes inside the brain (Section 34.1). The thirst center responds by notifying the cerebral cortex, which in turn compels you to search for and drink fluid.

While thirst mechanisms call for uptake of water, hormonal controls act to conserve water already inside the body. The hormones exert their effects mainly at distal tubules and collecting ducts.

Effects of Antidiuretic Hormone

When internal sodium levels rise, the hypothalamus stimulates the pituitary gland to secrete **antidiuretic hormone** (ADH). ADH binds to cells of distal tubules and collecting ducts, and causes them to become more permeable to water. As a result, water moves out of the filtrate more freely, peritubular capillaries reabsorb more of it, and less departs in the urine (Figure 41.12). In time, solute levels decline because the volume of extracellular fluid rises, and ADH secretion slows.

Other factors also stimulate ADH secretion. With heavy blood loss, receptors in the atria sense a decline in blood pressure and call for increased ADH. Stress, heavy exercise, or vomiting also cause internal changes that trigger a rise in ADH output.

ADH increases water reabsorption by stimulating the insertion of proteins called aquaporins into the plasma membrane of the distal tubules and collecting ducts. An aquaporin is a porelike passive transport protein that selectively allows water to cross the membrane. When ADH binds to cells of distal tubules and collecting ducts, vesicles that hold aquaporin subunits move toward the cells' plasma membrane. As these vesicles fuse with the plasma membrane, the subunits assemble themselves into functional aquaporins. Once in place, aquaporins facilitate the rapid flow of water out of the filtrate, back into the interstitial fluid.

Effects of Aldosterone

Any decrease in the volume of extracellular fluid also activates some cells in the arterioles that deliver blood to the nephrons. These cells release renin, an enzyme that sets in motion a complex chain of reactions.

Stimulus

a Water loss lowers the blood volume Sensory receptors in the hypothalamus detect a big deviation from the set point.

b The hypothalamus stimulates the pituitary gland to step up its secretion of ADH.

c ADH circulates in blood, reaches nephrons in the kidneys. By acting on cells of distal tubules and collecting ducts, it makes the tube walls more permeable to water.

hypothalamus

Response

f Sensory receptors in the hypothalamus detect the increase in blood volume. Signals calling for ADH secretion slow down.

pituitary gland

e The blood volume rises.

d More water is reabsorbed by peritubular capillaries around the nephrons, so less water is lost in urine.

Figure 41.12 Feedback control of ADH secretion, one of the negative feedback loops from kidneys to the brain that helps adjust the volume of extracellular fluid. Nephrons in the kidneys reabsorb more water when we do not take in enough water or lose too much, as by profuse sweating.

Renin coverts angiotensinogen, a protein secreted by the liver into the blood, into angiotensin I. Another enzyme converts angiotensin I to angiotensin II, which acts on the adrenal cortex. The cortex is the outer part of the adrenal gland that sits atop the kidney. The adrenal cortex responds to angiotensin II by secreting the hormone **aldosterone** into the blood. Aldosterone acts on the kidney's collecting ducts, increasing the activity of sodium–potassium pumps so that more sodium gets reabsorbed. Water follows the sodium by osmosis, and the urine becomes more concentrated.

Thus, both ADH and aldosterone cause urine to become more concentrated, although they do so by different mechanisms.

Atrial natriuretic peptide (ANP) is a hormone that makes urine more dilute. Muscle cells in the heart's atria release ANP when high blood volume causes the atrial walls to stretch. ANP directly inhibits secretion of aldosterone by acting on the adrenal cortex. It also acts indirectly by inhibiting renin release. In addition, ANP increases the glomerular filtration rate, so more fluid enters kidney tubules.

Hormonal Disorders and Fluid Balance

The metabolic disorder diabetes insipidus arises if the pituitary gland secretes too little ADH, ADH receptors do not respond to ADH, or aquaporins are impaired or missing. A large volume of highly dilute urine, and unquenchable thirst are symptoms of the disorder.

Some cancers, infections, and medications such as antidepressants, stimulate ADH oversecretion. With an excess of ADH, the kidneys retain too much water. Solute concentrations in the interstitial fluid decrease, which is bad news, especially for brain cells; they are highly sensitive to solute concentrations. If untreated, ADH oversecretion can be fatal.

An adrenal gland tumor may cause oversecretion of aldosterone, or hyperaldosteronism. The excess of aldosterone causes fluid retention, which can increase the blood pressure to dangerous levels.

41.7 | Acid–Base Balance

■ The kidneys help maintain the pH of body fluids. They are the only organs that can selectively rid the body of H^+ ions.

■ Link to pH and buffer systems 2.6

Metabolic reactions such as protein breakdown and lactate fermentation add H^+ to the extracellular fluid. Despite these continual additions, a healthy body can maintain its H^+ concentration within a tight range; a state known as **acid–base balance**. Buffer systems, and adjustments to the activity of respiratory and urinary systems are essential to this balance.

A **buffer system** involves substances that reversibly bind and release H^+ or OH^-. Such a system minimizes pH changes as acidic or basic molecules enter or leave a solution (Section 2.6).

The pH of human extracellular fluid usually stays between 7.35 and 7.45. In the absence of any buffer, adding acids to ECF could make its pH decrease. But excess hydrogen ions react with buffers, including the bicarbonate–carbonic acid buffer system:

$$H^+ + HCO_3^- \rightleftharpoons H_2CO_3 \rightleftharpoons CO_2 + H_2O$$

bicarbonate   carbonic acid

Adjustments in the rate and depth of breaths help offset changes in pH. When the blood pH decreases, breathing quickens and deepens, so CO_2 is expelled faster than it forms. As you can tell from the equation above, less CO_2 means less carbonic acid can form, so the pH rises. Slower, shallower breathing allows CO_2 to accumulate, so more carbonic acid can form.

Control of bicarbonate reabsorption and secretion of H^+ can adjust the pH inside kidneys. Reabsorbed bicarbonate moves into peritubular capillaries, where it buffers excess acid. H^+ secreted into tubule cells combines with phosphate or ammonia ions and forms compounds that are excreted in the urine.

When the kidney's secretion of H^+ falters, or excess H^+ is formed by metabolic reactions, or not enough bicarbonate is reabsorbed, the pH of body fluids can fall below 7.1, a condition called acidosis.

41.8 | When Kidneys Fail

■ Kidney failure can be treated with dialysis, but only a kidney transplant can fully restore function.

■ Link to High-protein diets 40.8

Causes of Kidney Failure The vast majority of kidney problems occur as complications of diabetes mellitus or high blood pressure. These disorders damage small blood vessels, including capillaries that interact with nephrons. Some people are genetically predisposed to infections or conditions that damage kidneys. Kidneys also fail after filtering lead, arsenic, pesticides, or other toxins from the blood. Occasionally, repeated high doses of aspirin or other drugs irreversibly damage the kidneys.

High-protein diets force the kidneys to work overtime to dispose of nitrogen-rich breakdown products (Section 40.8). Such diets also increase the risk for kidney stones. These hardened deposits form when uric acid, calcium, and other wastes settle out of urine and collect in the renal pelvis. Most kidney stones are washed away in urine, but sometimes one becomes lodged in a ureter or the urethra and causes severe pain. Any stone that blocks urine flow raises risk of infections and permanent kidney damage.

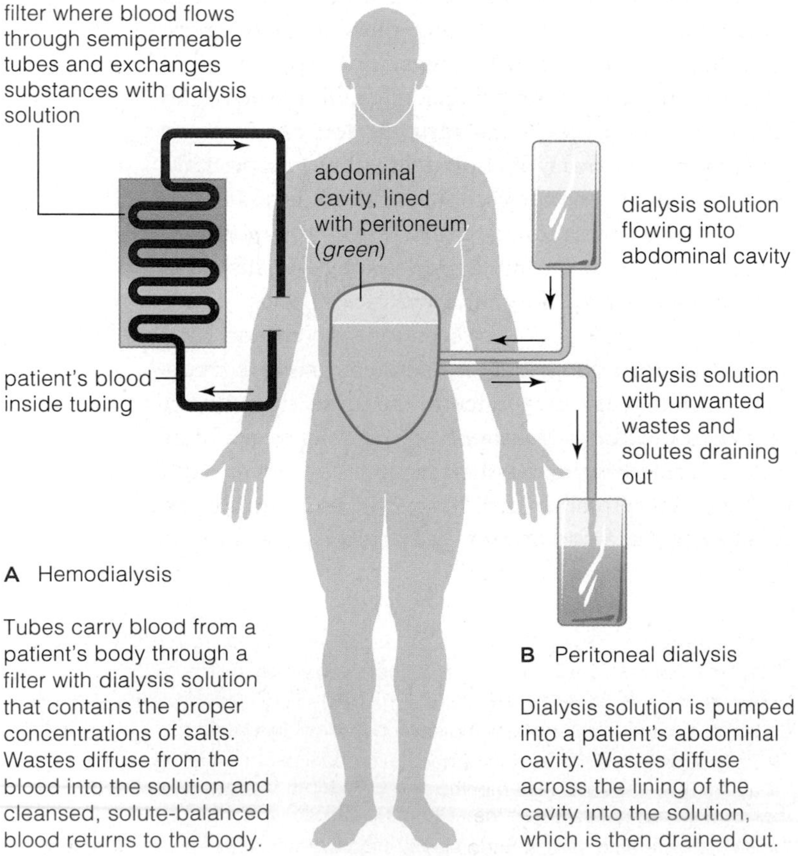

We usually measure kidney function in terms of the rate of filtration through glomerular capillaries. Kidney failure occurs when the filtration rate falls by half, regardless of whether it is caused by low blood flow to the kidneys or by damaged tubules or blood vessels. Kidney failure can be fatal. Wastes build up in the blood and interstitial fluid. The pH rises and changes in the concentrations of other ions, most notably Na^+ and K^+, interfere with metabolism.

Kidney Dialysis Kidney dialysis is used to restore proper solute balances in a person with kidney failure. "Dialysis" refers to exchanges of solutes across a semipermeable membrane between two solutions.

With hemodialysis, a dialysis machine is connected to a patient's blood vessel (Figure 41.13a). The machine pumps a patient's blood through semipermeable tubes submerged in a warm solution of salts, glucose, and other substances. As the blood flows through the tubes, wastes dissolved in the blood diffuse out and solute concentrations return to normal levels. Cleansed, solute-balanced blood is returned to the patient's body. Typically a person has hemodialysis three times a week at an outpatient dialysis center. Each treatment takes several hours.

Peritoneal dialysis can be done at home. Each night, dialysis solution is pumped into a patient's abdominal cavity (Figure 41.13b). Wastes diffuse across the peritoneal lining into the fluid, which is drained out the next morning. Thus this body lining serves as the dialysis membrane.

Kidney dialysis can keep a person alive through an episode of temporary kidney failure. When kidney damage is permanent, dialysis must be continued for the rest of a person's life, or until a donor kidney becomes available for transplant surgery.

Kidney Transplants Each year in the United States, about 12,000 people are recipients of kidney transplants. More than 40,000 others remain on a waiting list because there is a shortage of donated kidneys. The National Kidney Foundation estimates that every day, 17 people die of kidney failure while waiting for a transplant.

Most kidneys used as transplants come from people who had arranged to be organ donors after their death. However, an increasing number of kidneys are removed from a living donor, most often a relative. A kidney transplant from a living donor has a better chance of success than one from a deceased person. One kidney is adequate to maintain good health, so the risks to a living donor are mainly related to the surgery—unless a donor's remaining kidney fails.

The benefits of organs from living donors, a lack of donated organs, and high dialysis costs have led some to suggest that people should be allowed to sell a kidney. Critics argue that it is unethical to tempt people to risk their health for money. Section 16.8 describes another potential alternative. Some day, genetically modified pigs may become organ factories.

filter where blood flows through semipermeable tubes and exchanges substances with dialysis solution

abdominal cavity, lined with peritoneum (*green*)

patient's blood inside tubing

dialysis solution flowing into abdominal cavity

dialysis solution with unwanted wastes and solutes draining out

A Hemodialysis

Tubes carry blood from a patient's body through a filter with dialysis solution that contains the proper concentrations of salts. Wastes diffuse from the blood into the solution and cleansed, solute-balanced blood returns to the body.

B Peritoneal dialysis

Dialysis solution is pumped into a patient's abdominal cavity. Wastes diffuse across the lining of the cavity into the solution, which is then drained out.

Figure 41.13 Two types of kidney dialysis.

41.9 | Heat Gains and Losses

- Maintaining the body's core temperature is another aspect of homeostasis. Some animals expend more energy than others to keep their body warm.
- Links to Properties of water 2.5, Forms of energy 6.1

How the Core Temperature Can Change

Metabolic reactions release heat, so the heat generated by metabolism affects the temperature of an animal's internal core. Animals also gain heat from, and lose heat to, their surroundings. An animal's internal core temperature is stable only when the metabolic heat produced and the heat gained from the environment balance any heat losses to the environment:

$$\text{change in body heat} = \text{heat produced} + \text{heat gained} - \text{heat lost}$$

Figure 41.14 (a) Sidewinder, an ectotherm. (b) Pine grosbeak, an endotherm, using fluffed feathers as insulation against winter cold.

Heat is gained or lost at body surfaces by radiation, conduction, convection, and evaporation.

Thermal radiation is emission of heat from a warm object into the space around it. Just as the sun radiates heat energy into space, an animal radiates metabolically produced heat. A typical human at rest gives off about as much heat as a 100-watt light bulb.

In **conduction**, heat is transferred within an object or among objects that contact one another. An animal loses heat when it contacts a cooler object, and gains heat when it contacts a warmer one.

In **convection**, heat is transferred by the movement of heated air or water away from the source of heat. As air or water heats up, it rises and moves away from the object, such as a body, that warmed it.

In **evaporation**, heat energy converts a liquid into a gas, a process that cools any remaining liquid (Section 2.5). When that liquid is water at a body surface, this cooling helps decrease body temperature. Evaporative cooling is most effective with dry air and a breeze; high humidity and still air slow it.

Endotherms, Ectotherms, and Heterotherms

Fishes, amphibians, and reptiles are **ectotherms**, which means that they are "heated from the outside"; their body temperature fluctuates with the temperature of their environment. Ectotherms typically have a low metabolic rate and little insulation; they lack fur, hair, or feathers. They regulate their internal temperature by altering their position, rather than their metabolism. A rattlesnake (Figure 41.14a) is an example. When its body is cold, the snake basks in the sun. When the snake becomes too hot, it moves into shade.

Most birds and mammals are **endotherms**, which means "heated from within." Compared to ectotherms, endotherms have relatively high metabolic rates. For example, a mouse uses thirty times more energy than a lizard of the same body weight. An ability to produce a large amount of metabolic heat helps endotherms remain active in a wider range of temperatures than ectotherms. Fur, hair, or feathers insulate endotherms and minimize heat transfers (Figure 41.14b).

Some birds and mammals are **heterotherms**. They can maintain a fairly constant core temperature some of the time, but let it shift at other times. For example, hummingbirds have a very high metabolic rate when foraging for nectar during the day. At night, metabolic activity decreases so much that the bird's body may become almost as cool as the surroundings.

Warm climates favor ectotherms, which do not have to spend as much energy as endotherms do to maintain core temperature. Thus, in tropical regions, reptiles exceed mammals in numbers and diversity. In all cool or cold regions, however, most vertebrates tend to be endotherms. About 130 species of mammals and 280 species of birds occur in the arctic, but fewer than 5 species of reptiles are native to this region.

Take-Home Message

How do animals regulate their body temperature?

- Animals can gain heat from the environment, or lose heat to it. They can also generate heat by metabolic reactions.
- Fishes, amphibians, and reptiles are ectotherms that warm themselves mostly by heat gained from the environment.
- Birds and mammals are endotherms that maintain body temperature with their own metabolic heat.

■ A variety of mechanisms help mammals keep their core temperature from fluctuating with that of the environment.

■ Links to Feedback control of temperature 27.3, Heat illness 27.4, Sweat glands 32.7, Fever 38.4

The hypothalamus is the main regulatory center for the control of mammalian body temperature. This brain region receives signals from thermoreceptors (Section 34.1) in the skin, as well as from others located deep inside the body. When the temperature deviates from a set point, the hypothalamus integrates the responses of skeletal muscles, smooth muscle of arterioles in the skin, and sweat glands. Negative feedback loops back to the hypothalamus inhibit the responses when core temperature returns to the set point (Section 27.3).

Most mammals maintain body temperature within a few degrees. Dromedary camels are an exception; they can adjust their hypothalamic set point (Figure 41.15). Over the course of a day, their body temperature can vary from 34°C to 41.7°C (93°F to 107°F).

Responses to Heat Stress

When any mammal becomes too hot, the temperature control centers in the hypothalamus issue commands that widen the diameter of blood vessels in the skin. Increased blood flow to the skin delivers more metabolic heat to the body surface, where it can be given up by radiation to the surroundings (Table 41.2).

Another response to heat stress, evaporative heat loss, occurs at moist respiratory surfaces and across skin. Animals that sweat lose some water this way. For instance, humans and some other mammals have sweat glands that release water and solutes through pores at the skin's surface (Section 32.7). An average human adult has more than 2 million sweat glands.

For each liter of sweat produced, the body loses about 600 kilocalories of heat energy through evaporative cooling. During strenuous exercise, sweating helps the body rid itself of the extra heat produced by increased metabolic activity of skeletal muscles.

Sweat dripping from skin dissipates little heat. The body cools greatly when sweat evaporates. On humid days, the air's high water content slows evaporation, so sweating is less effective at cooling the body.

Not all mammals sweat. Many drool, lick their fur, or pant to speed cooling. "Panting" refers to shallow, rapid breathing. It assists evaporative water loss from the respiratory tract, nasal cavity, mouth, and tongue.

Sometimes peripheral blood flow and evaporative heat loss cannot counter heat stress, so the body's core temperature increases above normal, a condition called hyperthermia. In humans, an increase in core temperature above 40.6°C (105°F) is dangerous (Section 27.4).

A **fever** is an increase in body temperature that most often occurs as a response to infection (Section 38.4). Chemicals released by an infectious agent or by white blood cells that detect the infection influence the hypothalamus. In response to these chemicals, the hypothalamus allows the core temperature to rise a bit above normal set point. Increased temperature makes the body less hospitable for pathogens and encourages immune responses. Generally, the hypothalamus does not let the core temperature rise above 41.5°C (105°F). When a fever exceeds that point or lasts more than a few days, the condition causing it is life threatening and medical evaluation is essential.

Responses to Cold Stress

A mammal responds to the cold by redistributing its blood flow, fluffing its hair or fur, and shivering.

Figure 41.15 Short-term adaptation to desert heat stress. Dromedary camels let their core temperature rise during the hottest hours of the day. A hypothalamic mechanism adjusts their internal thermostat, so to speak. By allowing their temperature set point to rise, camels minimize their sweat production and thus can conserve water.

Table 41.2 Heat and Cold Stress Compared

Stimulus	Main Responses	Outcome
Heat stress	Widening of blood vessels in skin; behavioral adjustments; in some species, sweating, panting	Dissipation of heat from body
	Decreased muscle action	Heat production decreases
Cold stress	Narrowing of blood vessels in skin; behavioral adjustments (e.g., minimizing surface parts exposed)	Conservation of body heat
	Increased muscle action; shivering; nonshivering heat production	Heat production increases

Figure 41.16 Two responses to cold stress.

(**a**) Polar bears (*Ursus maritimus*, "bear of the sea"). A polar bear is active even during severe arctic winters. It does not get too chilled after swimming because the coarse, hollow guard hairs of its coat shed water quickly. Thick, soft underhair traps heat. An insulating layer of brown adipose tissue about 11.5 centimeters (4–1/2 inches) thick helps generate metabolic heat.

(**b**) In 1912, the *Titanic* collided with an immense iceberg on her maiden voyage. It took about 2–1/2 hours for the *Titanic* to sink and rescue ships arrived in less than two hours. Even so, 1,517 people died. Many died in lifeboats or while afloat in life vests. Hypothermia killed them.

Thermoreceptors in the skin signal the hypothalamus when conditions get chilly. The hypothalamus then causes smooth muscle in arterioles that deliver blood to the skin to constrict. For example, when your fingers or toes are chilled, all but 1 percent of the blood that would usually flow to skin is diverted to other body regions. Constriction of arterioles that supply the skin lessens movement of metabolic heat to the body surface, where it would be lost to the surroundings.

As another response to cold, reflex contractions of smooth muscle in the skin cause fur (or hair) to "stand up." This response creates a layer of still air next to skin, thus reducing heat lost by convection and thermal radiation. Minimizing exposed body surfaces can also prevent heat loss, as when polar bear cubs curl up and cuddle against their mother (Figure 41.16*a*).

With prolonged cold exposure, the hypothalamus commands skeletal muscles to contract ten to twenty times each second. Although this **shivering response** increases heat production, it has a high energy cost.

Long-term or severe cold exposure also leads to an increase in thyroid activity that raises the rate of metabolism. Thyroid hormone binds to cells of brown adipose tissue, causing **nonshivering heat production**. By this process, mitchondria in cells of brown adipose tissue carry out reactions that release energy as heat, rather than storing it in ATP.

Brown adipose tissue occurs in mammals that live in cold regions and in the young of many species. In human infants, this tissue makes up about 5 percent of body weight. Unless cold exposure is ongoing, the tissue disappears after childhood ends.

Failure to protect against cold causes hypothermia, a condition in which the core temperature falls. In humans, a decline to 35°C (95°F) alters brain functions. "Stumbles, mumbles, and fumbles" are said to be the symptoms of early hypothermia. Severe hypothermia causes loss of consciousness, disrupts heart rhythm, and can be fatal (Figure 41.16*b* and Table 41.3).

Table 41.3 Impact of Increases in Cold Stress

Core Temperature	Physiological Responses
36°–34°C (about 95°F)	Shivering response; faster breathing, metabolic heat output. Peripheral vasoconstriction, more blood deeper in body. Dizziness, nausea.
33°–32°C (about 91°F)	Shivering response ends. Metabolic heat output declines.
31°–30°C (about 86°F)	Capacity for voluntary motion is lost. Eye and tendon reflexes inhibited. Consciousness lost. Cardiac muscle action becomes irregular.
26°–24°C (about 77°F)	Ventricular fibrillation sets in (Section 37.9). Death follows.

Take-Home Message

How do mammals maintain their body temperature?

■ Temperature shifts are detected by thermoreceptors that send signals to an integrating center in the hypothalamus. This center serves as the body's thermostat and calls for adjustments that maintain core temperature.

■ Mammals respond to cold with reduced blood flow to skin, fluffing up of fur or hair, increased muscle activity, shivering, and nonshivering heat production.

■ Mammals counter heat stress by increasing blood flow to the skin, sweating and panting, and by reducing their activity level.

Solutes and nutrients the body needs are reabsorbed from the filtrate that enters kidney tubules. Water-soluble drugs and toxins are generally not reabsorbed, so they end up in the urine. How quickly the kidneys clear a substance from the blood depends in part on the efficiency of the kidneys, which can vary with age and health. A healthy 35-year-old eliminates drugs from the body about twice as fast as a healthy 85-year-old.

How would you vote?

Should employers be allowed to require potential employees to pass a urine test as a condition of employment? See CengageNOW for details, then vote online.

Summary

Section 41.1 Plasma and interstitial fluid are the main components of extracellular fluid. Maintaining the volume and composition of extracellular fluid is an essential aspect of homeostasis. All organisms balance solute and fluid gains with solute and fluid losses, and all eliminate metabolic wastes. Most have excretory organs that rid the body of ammonia and other unwanted solutes.

Section 41.2 Sponges are simple animals in which excretion occurs at a cellular level. In freshwater sponges and other freshwater animals, water flows into the body by osmosis. Like some protists, sponge cells eliminate excess water using organelles called **contractile vacuoles**.

In flatworms, the action of ciliated **flame cells** draws interstitial fluid into a system of tubes that delivers it to the body surface. Earthworms have excretory organs called **nephridia** that take up coelomic fluid, and deliver wastes to a pore at the body surface.

In insects and spiders, **Malpighian tubules** take up fluid, uric acid, and solutes from the blood and deliver them to the gut. Uric acid is formed from ammonia, but requires less water to be excreted.

Section 41.3 Vertebrates have a **urinary system** that interacts with other organ systems in homeostasis. A pair of **kidneys** filter water and solutes from their blood.

Cartilaginous fishes retain urea in their body, so they do not lose or gain water by osmosis. Marine bony fish constantly gain water by osmosis, while those that live in fresh water lose water by osmosis.

On land, the main challenge is avoiding dehydration. Birds and reptiles save water by eliminating nitrogen-rich wastes as uric acid crystals. Mammals excrete **urea**, which must be dissolved in a lot of water.

Section 41.4 The human urinary system consists of two kidneys, a pair of **ureters**, a **urinary bladder**, and the **urethra**. Kidney **nephrons** are small, tubular structures that interact with nearby capillaries to form urine. Each nephron starts as **Bowman's capsule** in a kidney's outer region, or **renal cortex**. The nephron continues as a **proximal tubule**, a **loop of Henle** that descends into and ascends from the **renal medulla**, and a **distal tubule** that drains into a **collecting duct**.

Bowman's capsule and capillaries of the **glomerulus** that it cups around serve as a blood-filtering unit. Most filtrate that enters Bowman's capsule is reabsorbed into **peritubular capillaries** around the nephron. The portion of the filtrate not returned to blood is excreted as urine.

- *Use the animation on CengageNOW to explore the anatomy of the human urinary system.*

Sections 41.5, 41.6 Blood pressure drives **glomerular filtration**, which puts protein-free plasma into the kidney tubules. Most water and solutes return from these tubules to the blood by **tubular reabsorption**. Substances move from the blood into the tubules by **tubular secretion**.

A part of the hypothalamus serves as a **thirst center**. The hypothalamus signals the pituitary gland to release **antidiuretic hormone**, which increases the reabsorption of water. **Aldosterone**, a hormone secreted by the adrenal cortex, increases sodium reabsorption. Both antidiuretic hormone and aldosterone concentrate the urine. **Atrial natriuretic peptide**, a hormone made by the heart, slows secretion of aldosterone and makes urine more dilute.

- *Use the animation on CengageNOW to learn about the three processes of urine formation.*

Section 41.7 The urinary system helps regulate **acid–base balance** by eliminating H+ in urine and reabsorbing bicarbonate, which has a role in the main **buffer system**.

Section 41.8 When kidneys fail, frequent dialysis or a kidney transplant is required to sustain life.

Section 41.9 Animals produce metabolic heat. They also gain or lose heat by **thermal radiation**, **conduction**, and **convection**; and lose it by **evaporation**.

Ectotherms such as reptiles control core temperature mainly by behavior. **Endotherms** (most mammals and birds) regulate temperature mainly by controlling production and loss of metabolic heat. **Heterotherms** control core temperature only part of the time.

Section 41.10 In mammals, the hypothalamus is the main center for temperature control. A **fever** is elevation of body temperature as a defensive response to infection.

Widening of blood vessels in the skin, sweating, and panting are responses to heat. Mammals alone can sweat, but not all mammals have this ability.

Exposure to cold causes constriction of blood vessels in skin, makes hair (or fur) stand upright, and elicits a **shivering response**. Long-term exposure to cold can alter metabolism and encourage **nonshivering heat production**, in which brown adipose tissue produces heat.

Data Analysis Exercise

Products labeled as "organic" fill an increasing amount of space on supermarket shelves. What does this label mean? A food that carries the USDA's organic label must be produced without pesticides such as malathion and chlorpyrifos, which conventional farmers typically use on fruits, vegetables, and many grains.

Does eating organic food significantly affect the level of pesticide residues in a child's body? Chensheng Lu of Emory University used urine testing to find out (Figure 41.17). For fifteen days, the urine of twenty-three children (aged 3 to 11) was monitored for breakdown products of pesticides. During the first five days, children ate their standard, nonorganic diet. For next five days, they ate organic versions of the same types of foods and drinks. Then, for the final five days, the children returned to their nonorganic diet.

1. During which phase of the experiment did the children's urine contain the lowest level of the malathion metabolite?

2. During which phase of the experiment was the maximum level of the chlorpyrifos metabolite detected?

3. Did switching to an organic diet lower the amount of pesticide residues excreted by the children?

Study Phase	No. of Samples	Malathion Metabolite		Chlorpyrifos Metabolite	
		Mean (µg/liter)	Maximum (µg/liter)	Mean (µg/liter)	Maximum (µg/liter)
1. Nonorganic	87	2.9	96.5	7.2	31.1
2. Organic	116	0.3	7.4	1.7	17.1
3. Nonorganic	156	4.4	263.1	5.8	25.3

Figure 41.17 *Above*, levels of metabolites (breakdown products) of malathion and chlorpyrifos detected in the urine of children taking part in a study of the effects of an organic diet. The difference in the mean level of metabolites in the organic and inorganic phases of the study was statistically significant. *Right*, the USDA organic food label.

4. Even in the nonorganic phases of this experiment, the highest pesticide metabolite levels detected were far below those known to be harmful. Given this data, would you spend more to buy organic foods?

Self-Quiz

Answers in Appendix III

1. An insect's _____ deliver nitrogen wastes to its gut.
  a. nephridia   c. Malpighian tubules
  b. nephrons   d. contractile vacuoles

2. Body fluids of a marine bony fish have a solute concentration that is _____ its surroundings.
  a. higher than   b. lower than   c. equal to

3. Bowman's capsule, the start of the tubular part of a nephron, is located in the _____ .
  a. renal cortex   c. renal pelvis
  b. renal medulla   d. renal artery

4. Fluid that enters Bowman's capsule flows directly into the _____ .
  a. renal artery   c. distal tubule
  b. proximal tubule   d. loop of Henle

5. Blood pressure forces water and small solutes into Bowman's capsule during _____ .
  a. glomerular filtration   c. tubular secretion
  b. tubular reabsorption   d. both a and c

6. Kidneys return most of the water and small solutes back to blood by way of _____ .
  a. glomerular filtration   c. tubular secretion
  b. tubular reabsorption   d. both a and b

7. ADH binds to receptors on distal tubules and collecting ducts, making them _____ permeable to _____ .
  a. more; water   c. more; sodium
  b. less; water   d. less; sodium

8. Increased sodium reabsorption _____ .
  a. will make urine more concentrated
  b. will make urine more dilute
  c. is stimulated by aldosterone
  d. both a and c

9. True or false? Increased secretion of H+ into kidney tubules helps lower the pH of the blood.

10. Match each structure with a function.
  ___ ureter   a. start of nephron
  ___ Bowman's capsule   b. delivers urine to
  ___ urethra   body surface
  ___ collecting duct   c. carries urine from
  ___ pituitary   kidney to bladder
      gland   d. secretes ADH
               e. target of aldosterone

11. The main control center for maintaining the temperature of the mammalian body is in the _____ .
  a. anterior pituitary   c. adrenal gland
  b. renal cortex   d. hypothalamus

12. An animal with a low metabolism that maintains its temperature mainly by adjusting its behavior is _____ .
  a. an endotherm   b. an ectotherm

13. True or false? Exposure to cold increases blood flow to your skin, thus warming the skin.

■ *Visit CengageNOW for additional questions.*

Critical Thinking

1. The kangaroo rat kidney efficiently excretes a very small volume of urine (Section 41.3). Compared to a human, its nephrons have a loop of Henle that is proportionally much longer. Explain how a long loop helps the rat conserve water.

2. In cold habitats, ectotherms are few and endotherms often show morphological adaptations to cold. Compared to closely related species that live in warmer areas, cold dwellers tend to have smaller appendages. Also, animals adapted to cool climates tend to be larger than relatives in warmer places. The largest bear is the polar bear and the largest penguin is Antarctica's emperor penguin.

Think about heat transfers between animals and their habitat, then explain why smaller appendages and larger overall body size are advantageous in very cold climates.

42 Animal Reproductive Systems

Male or Female? Body or Genes?

Athlete Santhi Soundarajan was born in a rural area in India in 1981. She overcame poverty and malnutrition to become a competitive runner, and in 2006 she represented her country in the Pan Asian Games (Figure 42.1). She won a silver medal, but her triumph was short-lived. A few days after the close of the games, the Olympic Council of Asia announced that Soundarajan had been stripped of her medal. Although she had been raised as a female, she has a Y chromosome, rather than a typical woman's two X chromosomes.

The International Olympics Committee (IOC) began a program of gender testing in 1968. At first, they required that athletes "prove" their femaleness by undergoing a physical exam. In the early 1970s, the committee switched to a less intrusive method. Experts examined a few of an athlete's cells under a microscope for evidence of two X chromosomes. In 1992 the committee upgraded its methods again, this time to a test that detects the *SRY* gene. *SRY* is the gene on the Y chromosome that normally causes development of testes in a human XY embryo (Section 12.1).

The Olympic testing program did not turn up any men deliberately pretending to be women. It did detect athletes who had been brought up as females and thought of them-

selves as women, but had a Y chromosome. In the 1996 Summer Olympics, 8 of 3,387 women athletes tested positive for an *SRY* gene. Further tests revealed that each of them had some sort of genetic abnormality. Because these genetic conditions prevented testosterone from exerting muscle-building effects, the women were not considered to have any unfair advantage and they were allowed to compete.

The IOC and most other groups that govern competitive athletic events have now banned gender testing. They did so in response to geneticists and physicians who spoke out against the practice. These professionals argued that disqualifying athletes on the basis of such tests is a form of discrimination that can cause great hardship to athletes with genetic abnormalities.

Generally, when a child is born, a quick look at their genitals (external sex organs) reveals their sex. Males have a penis; females, a vagina. Chromosomal sex (XX or XY) determines which gonads (ovaries or testes) form. Hormones secreted by the gonads then shape the genitals and other phenotypic aspects of sex. However, mutations can result in ambiguous genitals. A boy may be born with a tiny penis and with testes deep within his abdomen. Or, a girl may have a large clitoris and no opening to her vagina. In other cases, a child who has typical female genitals is actually a genetic male whose body either does not make or does not respond to the male sex hormone testosterone. Such a female lacks ovaries and a uterus, so she will not menstruate, but in terms of her body shape and strength, she is typically female.

Such intersex conditions challenge our thinking about what it means to be male or female. In the United States, children who have unusual genitals have traditionally been operated on within their first year to make them appear as normal as possible. Sometimes the best cosmetic outcome is obtained by assigning a child to the opposite genetic sex. Some doctors and some intersex individuals who underwent genital surgery as infants now argue against early surgery. They advocate accepting a child's unusual appearance and putting off any surgery until after puberty. Postponing surgery until this time allows affected individuals to make their own decision about what type of surgery, if any, they want to have.

In this chapter and the next, we consider the structure of reproductive systems and their normal function. Unlike the other organ systems, a reproductive system is not necessary for an individual's survival. It is, however, the key to passing on genes and thus to ensuring the survival of one's lineage. In humans, it is also an important component of our self-identity.

Figure 42.1 Indian athlete Santhi Soundarajan rests on the track after the 800-meter race for which she won a silver medal at the Pan Asian Games in 2006. She lost the medal after testing indicated that she has a Y chromosome.

Key Concepts

Modes of animal reproduction

Some animals reproduce asexually, but sexual reproduction predominates in most animals. Some sexual reproducers make both eggs and sperm, but most are either male or female. Living on land favored fertilization of eggs inside the female body. **Section 42.1**

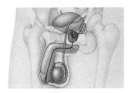

Male reproductive function

A human male has a pair of testes that make sperm and secrete the sex hormone testosterone. Sperm mixes with secretions from other glands and leaves the body through ducts. **Sections 42.2, 42.3**

Female reproductive function

A human female has a pair of ovaries that produce eggs and sex hormones. An approximately monthly hormonal cycle causes release of eggs. Ducts carry eggs toward the uterus, where offspring develop. The vagina receives sperm and is the birth canal. **Sections 42.4–42.7**

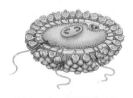

Intercourse and fertilization

Sexual intercourse requires coordinating nervous and hormonal signals. It can load to pregnancy, which humans use a variety of methods to prevent, promote, or terminate. **Sections 42.8, 42.9**

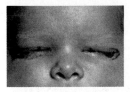

Sexually-transmitted diseases

A variety of pathogens make their home in the human reproductive tract. They are passed between partners by sexual interactions and may be transmitted to offspring during childbirth. Effects of sexually transmitted diseases range from discomfort to death. **Section 42.10**

Links to Earlier Concepts

■ Section 10.1 introduced the concepts of sexual and asexual reproduction, which we expand upon here. Gamete formation (10.5) is also explained in more detail.

■ This chapter draws upon your knowledge of human sex determination (12.1), and revisits the subject of prenatal diagnosis (12.8).

■ You will learn more about how the hypothalamus and pituitary (35.3) affect sexual organs, and about the sex hormones (35.12). You will also see how the autonomic nervous system (33.8) affects intercourse.

■ In considering reproductive health, we revisit tumors (9.5) and effects of prostaglandins (35.1). We end the chapter with a look at the infectious diseases (21.8) that are transmitted sexually, including AIDS (Chapter 21 introduction, 21.2).

How would you vote? Children born with intersex disorders have traditionally had surgery early in life. Some people think such surgery should be delayed until after puberty so a child can choose or reject it. Would you delay surgery if your child was thus affected? See CengageNOW for details, then vote online.

Modes of Animal Reproduction

- Sexual reproduction dominates the life cycle of most animals, including many that can also reproduce asexually.
- Link to Sexual and asexual reproduction 10.1

Asexual Reproduction in Animals

With **asexual reproduction**, a single individual makes offspring that are genetically identical to it, so a parent has all its genes represented in each offspring. Asexual reproduction can be advantageous in a stable environment. Gene combinations that make the parent successful can be expected to do the same for offspring.

Many invertebrates reproduce asexually. Some can reproduce by fragmentation—a piece breaks off and grows into a new individual. New hydras bud from existing ones (Figure 42.2a). Some insects and rotifers produce offspring from unfertilized eggs, a process called parthenogenesis. Most animals that reproduce asexually can also switch to sexual reproduction.

Among the vertebrates, some fishes, amphibians, and lizards can form offspring from unfertilized eggs. However, no mammals reproduce asexually.

Costs and Benefits of Sexual Reproduction

With **sexual reproduction**, two parents make gametes that combine at fertilization to produce offspring with gene combinations unlike either parent (Section 10.4).

Sexual reproducers incur higher genetic and energetic costs than asexual reproducers. On average, only half of a sexually reproducing parent's genes end up in each offspring. Producing gametes, and finding and courting an appropriate mate also has costs. What benefits offset these costs? Most animals live where resources and threats change over time. In such environments, production of offspring that differ from both parents and from one another can be advantageous. By reproducing sexually, a parent increases the likelihood that some of its offspring will have a gene combination that suits them to their new environment.

Variations on Sexual Reproduction

Some animals that reproduce sexually produce both eggs and sperm; they are **hermaphrodites**. Tapeworms and some roundworms are simultaneous hermaphrodites. They produce eggs and sperm at the same time, and can fertilize themselves. Earthworms and slugs are simultaneous hermaphrodites too, but they require a partner. So do hamlets, a type of marine fish (Figure 42.2b). During a bout of mating, hamlet partners take turns in the "male" and "female" roles. Other fishes are sequential hermaphrodites. They switch from one sex to another during the course of a lifetime. More typically, vertebrates have separate sexes that remain fixed for life; an individual is either male or female.

Figure 42.2 Examples of animal reproduction. (**a**) A hydra reproducing asexually by budding. (**b**) Barred hamlets mating. The fish are hermaphrodites that fertilize eggs externally. During mating, each fish alternates between laying eggs and fertilizing its partner's eggs. (**c**) A male elephant inserting his penis into his female partner. The eggs will be fertilized and the offspring will develop inside the mother's body, nourished by nutrients delivered by her bloodstream.

Figure 42.3 A look at where some invertebrate and vertebrate embryos develop, how they are nourished, and how (if at all) parents protect them.

(**a**) Most snails lay eggs and abandon them, (**b**) Spider eggs develop in a silk egg sac. Females often die soon after they make the sac, but some species guard the sac, then cart spiderlings about for a few days while they feed them.

(**c**) Ruby-throated hummingbirds and all other birds lay fertilized eggs with big yolk reserves. The eggs develop and hatch outside the mother. One or both parent birds typically expend energy feeding and caring for the young.

(**d**) Embryos of some sharks, lizards, and snakes develop in their mother, receive nourishment continuously from yolk reserves, and are born in a well-developed state. Shown here, live birth of a lemon shark.

Embryos of most mammals draw nutrients from maternal tissues and are born live. (**e**) In kangaroos and other marsupials, embryos are born "unfinished." They complete embryonic development inside a pouch on the mother's ventral surface. (**f**) Juveniles (joeys) continue to be nourished with milk secreted by mammary glands inside the pouch.

A human female (**g**) retains a fertilized egg inside her uterus. Her own tissues nourish the developing individual until birth.

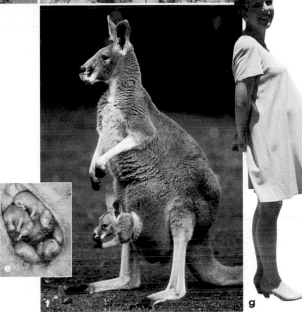

Most aquatic invertebrates, fishes, and amphibians release gametes into the water, where they combine during **external fertilization**. Most land animals have **internal fertilization**; sperm and egg meet inside the female's body. A specialized organ is typically used to deliver sperm. In mammals, a male's penis serves this purpose (Figure 42.2c).

Internally fertilized eggs may be laid outside the body and abandoned (Figure 42.3a), or a parent may lay and protect the eggs and later the young (Figure 42.3b,c). In other animals, offspring develop from eggs held inside the mother's body (Figure 42.3d–g).

Most female animals make some investment in **yolk**, a thick fluid rich in proteins and lipids that nourish the developing individual. The amount of yolk varies among species. Sea urchins release tiny eggs that hold little yolk. Not much yolk is needed because a fertilized sea urchin egg becomes a self-feeding, swimming larva in less than a day. In contrast, birds make eggs with a large quantity of yolk. Yolk is the embryo's only nourishment during its time in an eggshell. Kiwi birds

have the longest incubation time, about 11 weeks, and their eggs have an unusually large amount of yolk. A typical bird's egg is about one-third yolk; whereas the kiwi's egg is two-thirds yolk.

A human mother nourishes her offspring through nine months of development from a nearly yolkless, fertilized egg. Nutrients in the mother's blood diffuse into an offspring's blood and support its development (Figure 42.3g). You will learn more about how humans develop and nourish their young in Chapter 43.

Take-Home Message

How do animal reproductive systems vary?

■ Many invertebrates and some vertebrates can reproduce asexually. Most of these species can also reproduce sexually.

■ Animals that reproduce sexually have genetically variable offspring. Sexual reproducers may produce eggs and sperm at the same time, produce both at different times in their life, or always produce only one or the other.

■ Fertilization may occur in the mother's body, or outside it. Internally fertilized eggs may be laid in the environment or develop in a mother's body.

■ A human male's reproductive system produces hormones and sperm, which it delivers to a female's reproductive tract.

■ Link to Human sex determination 12.1

The Male Gonads

Human gametes form in primary reproductive organs, or **gonads**. Males have a pair of gonads called **testes** (singular, testis) that produce sperm. Testes also make and secrete the sex hormone **testosterone**. In addition to gonads, the male reproductive system includes a system of ducts and accessory glands (Table 42.1 and Figure 42.4).

Earlier, Figure 12.2 showed how two testes form on the wall of an XY embryo's abdominal cavity. Before birth, testes descend into the scrotum, a pouch of loose skin suspended below the pelvic girdle. Inside this pouch, smooth muscle encloses the testes. Contraction and relaxation of this muscle in response to threats or temperature adjusts the position of the testes. When a man feels cold or afraid, reflexive muscle contractions draw his testes closer to his body. When he feels warm, relaxation of muscle in the scrotum allows his testes to hang lower, so the sperm-making cells do not overheat. These cells function best when they are just a bit below normal body temperature.

Table 42.1 Human Male Reproductive System

Reproductive organs

Testes (2)	Sperm, sex hormone production
Epididymides (2)	Site of sperm maturation and subsequent storage
Vasa deferentia (2)	Rapid transport of sperm
Ejaculatory ducts (2)	Conduction of sperm to penis
Penis	Organ of sexual intercourse

Accessory glands

Seminal vesicles (2)	Secrete most fluid in semen
Prostate gland	Secretes some fluid in semen
Bulbourethral glands (2)	Secrete a lubricating mucus

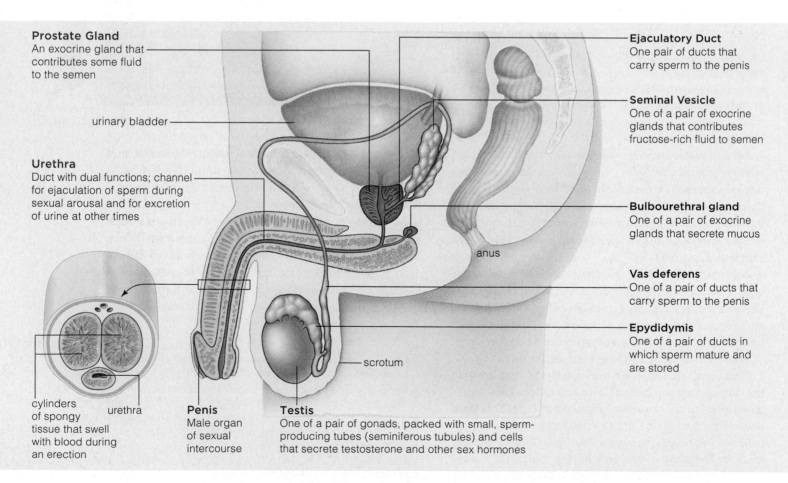

Prostate Gland
An exocrine gland that contributes some fluid to the semen

urinary bladder

Urethra
Duct with dual functions; channel for ejaculation of sperm during sexual arousal and for excretion of urine at other times

cylinders of spongy tissue that swell with blood during an erection

urethra

Penis
Male organ of sexual intercourse

Testis
One of a pair of gonads, packed with small, sperm-producing tubes (seminiferous tubules) and cells that secrete testosterone and other sex hormones

scrotum

anus

Ejaculatory Duct
One pair of ducts that carry sperm to the penis

Seminal Vesicle
One of a pair of exocrine glands that contributes fructose-rich fluid to semen

Bulbourethral gland
One of a pair of exocrine glands that secrete mucus

Vas deferens
One of a pair of ducts that carry sperm to the penis

Epydidymis
One of a pair of ducts in which sperm mature and are stored

Figure 42.4 Animated Components of the human male reproductive system and their functions.

A male enters **puberty**—the stage of development when reproductive organs mature—sometime between the ages of 11 and 16 years. Testes enlarge and sperm production begins. Secretion of testosterone increases and leads to development of secondary sexual traits: thickened vocal cords that deepen the voice; increased growth of hair on the face, chest, armpits, and pubic region; and an altered distribution of fat and muscle.

Reproductive Ducts and Accessory Glands

Sperm form by meiosis in the testes, a process we discuss in the next section. Here we consider the path sperm travel to the body surface. The journey begins when the immature, nonmotile sperm are pushed by cilia action from the testis into the epididymis (plural, epididymides), a coiled duct perched on a testis. The Greek *epi–* means upon and *didymos* means twins. In this context, the "twins" refers to the two testes. Secretions from the epididymis wall nourish the sperm and help them mature.

The last region of each epididymis stores mature sperm and is continuous with the first portion of a vas deferens (plural, vasa deferentia). In Latin, *vas* means vessel, and *deferens*, to carry away. A vas deferens is a duct that carries sperm away from an epididymis, and to a short ejaculatory duct. Ejaculatory ducts deliver sperm to the urethra, the duct that extends through a male's penis to open at the body surface.

The **penis** is the male organ of intercourse and also has a role in urination. Beneath its outer layer of skin, connective tissue encloses three elongated cylinders of spongy tissue. When a male becomes sexually excited, nervous signals cause blood to flow into the spongy tissue faster than it flows out. As fluid pressure rises, the normally limp penis becomes erect.

Sperm stored in the epididymides and first part of the vasa deferentia typically continue their journey toward the body surface only when a male reaches the peak of sexual excitement and ejaculates. During ejaculation, smooth muscle in the walls of the epididymides and vasa deferentia undergoes rhythmic contractions that propel sperm and accessory gland secretions out of the body as a thick, white fluid called **semen**.

Semen is a complex mix of sperm, proteins, nutrients, ions, and signaling molecules. Sperm constitute less than 5 percent of semen's volume; the bulk of it is secretions from accessory glands. Seminal vesicles, exocrine glands near the base of the bladder, secrete fructose-rich fluid into the vasa deferentia. Sperm use fructose (a sugar) as their energy source. The prostate gland, which encircles the urethra, is the other major contributor to semen volume. Its secretions help raise the pH of the female reproductive tract, so sperm can swim more efficiently. Both seminal vesicles and the prostate gland also secrete prostaglandins, which are local signaling molecules.

The two pea-sized bulbourethral glands secrete a lubricating mucus into the urethra. This mucus helps clear the urethra of urine prior to ejaculation.

Prostate and Testicular Problems

In a young, healthy man, the prostate gland is about the size of a walnut. However, inflammation or age can cause this gland to enlarge. Because the urethra runs through the prostate gland, even benign prostate enlargement can narrow this duct and cause difficulty urinating. Medication, laser treatments, and surgery are used to relieve symptoms.

Prostate enlargement can be a symptom of prostate cancer. This cancer is a leading cause of death for men, surpassed only by lung cancers. In the United States, more than 200,000 men are diagnosed with prostate cancer each year, and about 35,000 die. Many prostate cancers grow relatively slowly, but some grow fast and spread easily to other parts of the body. Risk factors for prostate cancer include advancing age, a diet rich in animal fats, smoking, and a couch-potato life-style. Genes also play a role. If a man has an affected father or brother, his own risk of prostate cancer doubles.

Doctors can diagnose prostate cancer by blood tests that detect increases in prostate-specific antigen (PSA) and by physical examination. Surgery and radiation therapy can cure cancers that are detected early.

Testicular cancer is relatively rare, with 7,000 cases a year in the United States. Even so, it is the most common cancer among men aged 15 to 34. Once a month, after a warm shower or bath, a male should examine his testes for lumps, enlargement, or hardening. The treatment of testicular cancer is usually successful if the cancer is detected early, before it has spread.

Take-Home Message

What are the functions of the male reproductive organs?

■ A pair of testes, the primary reproductive organs in human males, produce sperm. They also make and secrete the sex hormone testosterone.

■ Sperm and secretions from accessory glands form semen. During sexual arousal, semen is propelled through a series of ducts and leaves the body through an opening in the penis.

■ One accessory gland, the prostate, frequently becomes enlarged with age. It is also a common site for male cancers.

42.3 | Sperm Formation

- In his reproductive years, a male continually produces new germ cells, which undergo meiosis to produce sperm.
- Sperm formation is controlled by hormones.
- Links to Gamete formation 10.5, Sex hormones 35.12

From Germ Cells to Mature Sperm

Although smaller than a golf ball, a testis holds coiled seminiferous tubules that would extend for 125 meters (longer than a football field) if stretched out (Figure 42.5a). Leydig cells that cluster between these tubules secrete the hormone testosterone (Figure 42.5b).

Male germ cells, or spermatogonia (singular, spermatogonium) line the inner wall of each semiferous tubule. In a sexually mature male, these diploid cells undergo mitosis again and again. With each division, the youngest descendants force older ones farther into the interior of the tubule. The displaced older cells are primary spermatocytes. Sertoli cells, another type of cell inside the tubules, provide metabolic support to the spermatocytes.

Primary spermatocytes enter meiosis while they are being displaced—but their cytoplasm does not quite divide. Thin cytoplasmic bridges keep them connected to one another during the nuclear divisions. Signaling molecules cross the bridges freely and induce them to mature at the same rate.

The completion of meiosis I yields two secondary spermatocytes (Figure 42.5c). These are haploid cells with duplicated chromosomes (Section 10.5). The sister chromatids of each chromosome will move apart during meiosis II, which produces immature sperm, or spermatids. As spermatids mature into sperm, the bridges of cytoplasm between them break down.

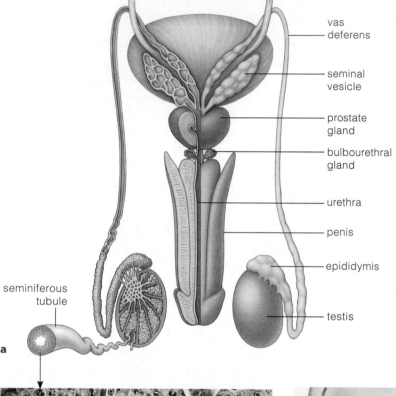

a

seminiferous tubule

vas deferens
seminal vesicle
prostate gland
bulbourethral gland
urethra
penis
epididymis
testis

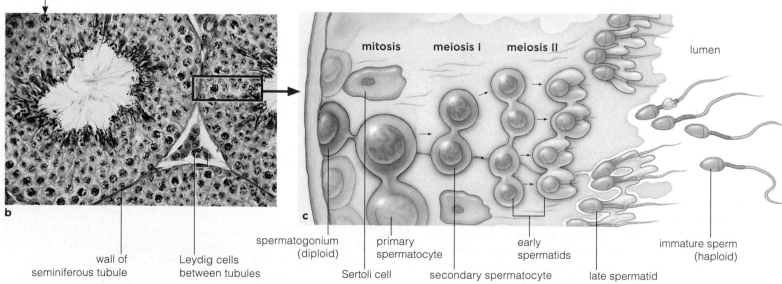

b

wall of seminiferous tubule

Leydig cells between tubules

c

mitosis   meiosis I   meiosis II   lumen

spermatogonium (diploid)

primary spermatocyte

Sertoli cell

secondary spermatocyte

early spermatids

late spermatid

immature sperm (haploid)

Figure 42.5 Animated Where and how sperm form. (**a**) Male reproductive tract, posterior view. (**b**) Light micrograph of cells in three adjacent seminiferous tubules, cross-section. Testosterone-secreting Leydig cells, occupy spaces between the tubules. (**c**) Diploid germ cells (spermatogonia) line a seminiferous tubule. These cells undergo mitosis to form primary spermatocytes, which undergo meiosis to form spermatids. Spermatids mature into sperm.

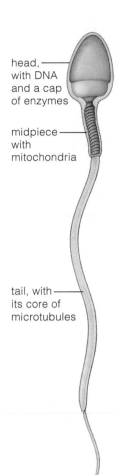

head, with DNA and a cap of enzymes

midpiece with mitochondria

tail, with its core of microtubules

Figure 42.6 Structure of a mature sperm, a male gamete.

A spermatozoan, or mature sperm, is a haploid, flagellated cell (Figure 42.6). A sperm uses its flagellum, or "tail," to swim toward an egg. Mitochondria in the adjacent midpiece supply the energy required for flagellar movement. A sperm's "head" is packed full of DNA and tipped by an enzyme-containing cap. The enzymes can help a sperm penetrate an oocyte by partly digesting away its outer layer.

Sperm formation takes about 100 days, from start to finish. An adult male makes sperm on an ongoing basis, so that many millions of cells are in different stages of development on any given day.

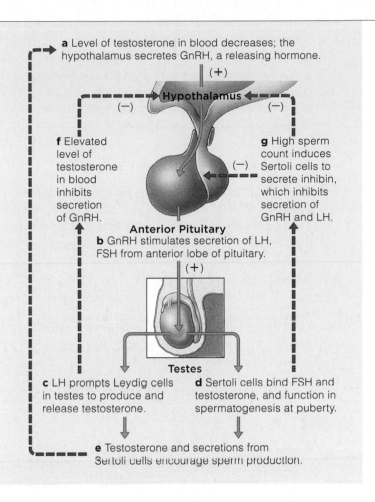

a Level of testosterone in blood decreases; the hypothalamus secretes GnRH, a releasing hormone.

(+)

Hypothalamus

(−)   (−)

f Elevated level of testosterone in blood inhibits secretion of GnRH.

g High sperm count induces Sertoli cells to secrete inhibin, which inhibits secretion of GnRH and LH.

(−)

Anterior Pituitary
b GnRH stimulates secretion of LH, FSH from anterior lobe of pituitary.

(+)

Testes

c LH prompts Leydig cells in testes to produce and release testosterone.

d Sertoli cells bind FSH and testosterone, and function in spermatogenesis at puberty.

e Testosterone and secretions from Sertoli cells encourage sperm production.

Figure 42.7 Signaling pathways in sperm formation. Negative feedback loops control hormonal secretions of the hypothalamus, the anterior lobe of the pituitary gland, and the testes.

Hormonal Control of Sperm Formation

Four hormones—GnRH, LH, FSH, and testosterone—are part of the signaling pathways that control sperm formation (Figure 42.7).

Gonadotropin-releasing hormone (GnRH) is one of the hypothalamic hormones that targets the pituitary gland (Figure 42.7a and Section 35.3). GnRH stimulates anterior pituitary cells to secrete **luteinizing hormone (LH)** and **follicle-stimulating hormone (FSH)** (Figure 42.7b). As you will learn, these two hormones have a role in both male and female reproductive function.

In males, both LH and FSH target cells inside the testes. LH binds to the Leydig cells that lie in between the seminiferous tubules, stimulating them to secrete testosterone (Figure 42.7c). FSH targets Sertoli cells, inside seminiferous tubules. FSH, in combination with testosterone, prompts Sertoli cells to produce growth factors and other molecular signals (Figure 42.7d). These substances bathe neighboring male germ cells and encourage the development and maturation of sperm (Figure 42.7e).

A negative feedback loop regulates testosterone secretion and sperm formation. A high concentration of testosterone in the blood slows secretion of GnRH by the hypothalamus (Figure 42.7f). The decrease in GnRH then lowers the output of LH and FSH by the testes. In addition, a high sperm count encourages the Sertoli cells to release the hormone inhibin (Figure 42.7g). Like testosterone, inhibin calls for a slowdown in GnRH and FSH secretion.

Take-Home Message

How do sperm form and what role do hormones play?

■ Meiosis in germ cells in seminiferous tubules of the testes produces sperm—the haploid male gametes.

■ Hormonal control of the process begins with GnRH from the hypothalamus. GnRH causes secretion of the hormones FSH and LH by the pituitary gland.

■ FSH and LH act on the testes, where they stimulate release of testosterone and other factors needed for formation and development of sperm.

Reproductive System of Human Females

■ The reproductive system of human females functions in the production of gametes and sex hormones.

■ The system receives sperm, and has a chamber in which developing offspring are protected and nourished until birth.

Components of the System

Figures 42.8 and 42.9 show the reproductive organs of a human female, and Table 42.2 lists their functions. The gonads are a pair of **ovaries** that produce oocytes (immature eggs) and secrete sex hormones on a cyclic

basis. Upon its release, an oocyte enters into one of the pair of **oviducts**, or Fallopian tubes.

Fertilization most often occurs in the oviduct. The fertilized egg tumbles into the **uterus**, a hollow, pear-shaped organ above the urinary bladder. An embryo forms and development is completed in the uterus. A thick layer of smooth muscle, the myometrium, makes up most of the uterine wall. Endometrium lines the uterus and consists of glandular epithelium, connective tissues, and blood vessels. The narrowed-down, lowest portion of the uterus is the cervix, which opens into the vagina.

The **vagina**, a muscular, mucosa-lined tube, extends from the cervix to the body's surface. It is lubricated by its own mucus secretions, and it functions as the female organ of intercourse. The vagina also functions as the birth canal in childbirth. Two pairs of skin folds enclose the surface openings of the vagina and urethra. Adipose tissue fills the pair of outer folds (the labia majora). Thin inner folds (the labia minora) have a rich blood supply and swell during sexual arousal.

The tip of the clitoris, a highly sensitive sex organ, is positioned between the two inner folds, just in front of the urethra. The clitoris and penis develop from the

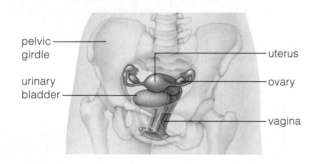

Figure 42.8 Location of the human female reproductive system relative to the pelvic girdle and the urinary bladder.

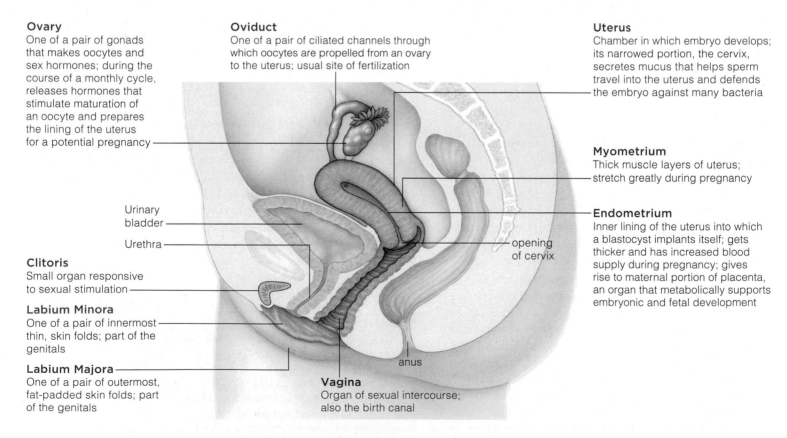

Ovary
One of a pair of gonads that makes oocytes and sex hormones; during the course of a monthly cycle, releases hormones that stimulate maturation of an oocyte and prepares the lining of the uterus for a potential pregnancy

Oviduct
One of a pair of ciliated channels through which oocytes are propelled from an ovary to the uterus; usual site of fertilization

Uterus
Chamber in which embryo develops; its narrowed portion, the cervix, secretes mucus that helps sperm travel into the uterus and defends the embryo against many bacteria

Myometrium
Thick muscle layers of uterus; stretch greatly during pregnancy

Endometrium
Inner lining of the uterus into which a blastocyst implants itself; gets thicker and has increased blood supply during pregnancy; gives rise to maternal portion of placenta, an organ that metabolically supports embryonic and fetal development

Clitoris
Small organ responsive to sexual stimulation

Labium Minora
One of a pair of innermost thin, skin folds; part of the genitals

Labium Majora
One of a pair of outermost, fat-padded skin folds; part of the genitals

Vagina
Organ of sexual intercourse; also the birth canal

Figure 42.9 Animated Components of the human female reproductive system and their functions.

Table 42.2 Female Reproductive Organs

Ovaries (2)	Oocyte production and maturation, sex hormone production
Oviducts (2)	Tubes between the ovaries and the uterus; fertilization normally takes place here
Uterus	Chamber in which new individual develops
Cervix	Entrance to the uterus; secretes mucus that enhances sperm travel into uterus and reduces embryo's risk of infection
Vagina	Organ of sexual intercourse; birth canal

same embryonic tissue. Both have an abundance of highly sensitive touch receptors, and both swell with blood and become erect during sexual arousal.

Overview of the Menstrual Cycle

Females of most mammalian species follow an estrous cycle, meaning they are fertile and "in heat" (sexually receptive to males) only at certain times. Females of humans and some other primates follow a **menstrual cycle**. Their fertile periods are cyclic, intermittent, and not tied to sexual receptivity. In other words, they can get pregnant only during certain times in their cycle but may be receptive to sex at any time.

Human females typically begin to menstruate at about age twelve. Section 42.6 describes the menstrual cycle in detail, but here is an overview: Every twenty-eight days or so, an oocyte matures in an ovary, and is released. During a two-week interval, the uterus gets primed for pregnancy. If the oocyte does not get fertilized, blood and bits of endometrium flow out through the vagina. This menstrual flow indicates the start of a new cycle.

A woman goes through such monthly cycles until she reaches her late forties or early fifties, when her sex hormone output dwindles. The decline in hormone secretions correlates with the onset of **menopause**, the twilight of a female's fertility.

Take-Home Message

What are the main female reproductive organs?

■ Ovaries are female gonads; they produce eggs and secrete sex hormones.

■ Eggs travel through oviducts to the uterus, the chamber where offspring develop.

■ The vagina receives sperm and serves as the birth canal.

42.5 | Female Troubles

■ Hormonal changes cause premenstrual symptoms, menstrual pain, and hot flashes.

■ Links to Prostaglandins 35.1, Benign tumors 9.5

PMS Many women regularly experience discomfort a week or so before they menstruate. Body tissues swell because premenstrual changes influence aldosterone secretion. This adrenal gland hormone stimulates reabsorption of sodium and, indirectly, water (Section 41.6). Breasts may become tender because hormones cause their milk ducts to enlarge. Cycle-induced changes also cause depression, irritability, or anxiety. Headaches and sleeping problems are common.

The regular recurrence of these symptoms is known as premenstrual syndrome (PMS). A balanced diet and regular exercise make PMS less likely and less severe. Taking oral contraceptives minimizes hormone swings and therefore PMS. In some cases, drugs that completely suppress the secretion of sex hormones can help.

Menstrual Pain Prostaglandins secreted during menstruation stimulate contractions of smooth muscle in the uterine wall. Many women do not notice the contractions, but others experience a dull ache or sharp pain. Women who secrete high levels of prostaglandins are more likely to feel uncomfortable while menstruating.

Endometriosis, the growth of endometrial tissue in the wrong regions of the pelvis, affects about 15 percent of women and can cause pain during menstruation. Hormones cause misplaced tissue to bleed, then heal, forming scars that can be painful. Hormone suppression methods help, but only surgery can provide a cure.

More than one-third of women over age thirty have benign uterine tumors called fibroids. Most fibroids cause no symptoms, but some result in pain, long menstrual periods, and excessive bleeding. A woman who needs to change pads or tampons on an hourly basis should discuss this condition with her doctor. Surgical removal of fibroids halts the excessive bleeding and the pain.

Hot Flashes, Night Sweats Three-fourths of the women entering menopause have hot flashes. They get abruptly and uncomfortably hot, flushed, and sweaty as blood surges to their skin. When episodes occur at night, they disrupt sleep. Hormone replacement therapy can relieve these symptoms, but the therapy raises risk of breast cancer and stroke, especially if continued for more than a few years. Exercising, avoiding alcohol, and eating soy-based products can also help reduce symptoms.

42.6 Preparations for Pregnancy

- A fertile woman undergoes hormonal changes and releases eggs in an approximately monthly cycle.

- Links to Gamete formation 10.5, Sex hormones 35.12

The Ovarian Cycle

At birth, a girl has about 2 million **primary oocytes**, immature eggs that have entered meiosis but stopped short in prophase I. Beginning with her first menstrual cycle, these oocytes mature, typically one at a time, in approximately a 28-day cycle. Figure 42.10 depicts the events of this cycle in an ovary.

A primary oocyte and the cells that surround it constitute an ovarian follicle (Figure 42.10*a*). In the first part of the ovarian cycle—the follicular phase—cells around the oocyte divide repeatedly, while the oocyte enlarges and secretes glycoproteins. These secreted glycoproteins form a noncellular layer known as the **zona pellucida** (Figure 42.10*b*). As the follicle matures, a fluid-filled cavity opens in the cell layer around the

oocyte (Figure 42.10*c*). More than one follicle often starts to mature during the follicular phase, but typically only one goes on to become fully mature.

Follicle maturation requires about 14 days and is under hormonal control. As the follicular phase starts, the hypothalamus secretes GnRH. This hormone stimulates cells in the anterior pituitary to increase their secretion of FSH and LH (Figure 42.11*a*). Rising levels of FSH and LH in the blood allow follicle maturation and stimulate follicle cells to secrete **estrogens**, a type of sex hormone (Figure 42.11*b,c*).

The pituitary detects the rising level of estrogens in blood and responds with an outpouring of LH. The LH surge encourages the primary oocyte to complete meiosis I and undergo cytoplasmic division. One of the resulting haploid cells, the **secondary oocyte**, gets most of the cytoplasm. The other haploid cell is the first polar body, a cell that will eventually degenerate (Figure 42.10*d*). The LH surge also causes the follicle to swell and eventually to burst. The secondary oocyte,

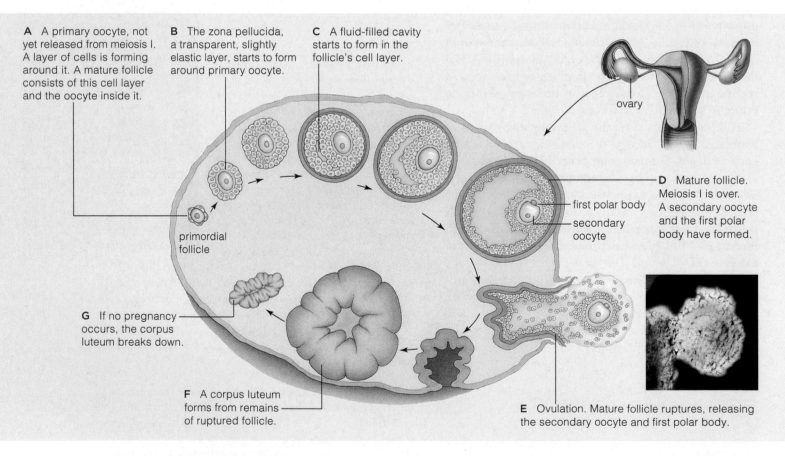

A A primary oocyte, not yet released from meiosis I. A layer of cells is forming around it. A mature follicle consists of this cell layer and the oocyte inside it.

B The zona pellucida, a transparent, slightly elastic layer, starts to form around primary oocyte.

C A fluid-filled cavity starts to form in the follicle's cell layer.

ovary

primordial follicle

first polar body

secondary oocyte

D Mature follicle. Meiosis I is over. A secondary oocyte and the first polar body have formed.

G If no pregnancy occurs, the corpus luteum breaks down.

F A corpus luteum forms from remains of ruptured follicle.

E Ovulation. Mature follicle ruptures, releasing the secondary oocyte and first polar body.

Figure 42.10 Animated Cyclic events in a human ovary, cross-section. The follicle does not "move around" as in this diagram, which simply shows the *sequence* of events. All of these structures form in the same place during one menstrual cycle. In the cycle's first phase, a follicle grows and matures. At ovulation, the second phase, the mature follicle ruptures and releases a secondary oocyte. In the third phase, a corpus luteum forms from the follicle's remnants.

still surrounded by the zona pellucida and some follicle cells, is released into an oviduct. Thus, the midcycle surge of LH is the trigger for **ovulation**, the release of a secondary oocyte from an ovary (Figure 42.10*e*).

Ovulation is followed by the luteal phase of the ovarian cycle. During this phase, the ruptured follicle becomes a yellowish glandular structure known as the **corpus luteum** (Figure 42.10*f*). In Latin, *corpus* means body, and *luteum* means yellow.

The corpus luteum secretes a large amount of the sex hormone progesterone, and a lesser amount of estrogens. The high progesterone level feeds back to the brain and reduces secretion of LH and FSH, so a new follicle does not develop.

If pregnancy does not occur, the corpus luteum lasts no more than 12 days. In the final days of the luteal phase, a decline in LH causes it to break down (Figure 42.10*g*). Then a new follicular phase begins.

Correlating Events in the Ovary and Uterus

Menstruation, the flow of blood and endometrial tissue out of the uterus and through the vagina, coincides with the beginning of the follicular phase in the ovary (Figure 42.11*c*,*d*). Menstruation usually lasts for 1 to 5 days. Then, as the follicular phase goes on, estrogens secreted by a maturing follicle encourage the uterine lining to repair itself and thicken.

After ovulation, in the luteal phase, estrogens and progesterone secreted by the corpus luteum act on the endometrium. These hormones encourage the growth of blood vessels and of glands that secrete glycogen. The uterus is now ready to sustain a pregnancy.

If no pregnancy occurs, the corpus luteum breaks down and progesterone and estrogen levels plummet. Blood vessels supplying the endometrium wither and the endometrium starts to break down. As the bloody tissue is shed, a new follicular phase begins.

Take-Home Message

What cyclic changes occur in the ovary and uterus?

■ Every 28 days or so FSH and LH stimulate maturation of an ovarian follicle.

■ A midcycle surge of LH triggers ovulation—the release of a secondary oocyte into an oviduct.

■ Estrogen secreted by a maturing follicle causes the endometrium to thicken. After ovulation, progesterone secreted by the corpus luteum encourages secretion by endometrial glands.

■ If pregnancy does not occur, the corpus luteum breaks down, hormone levels drop, the endometrial lining is shed, and the cycle begins again.

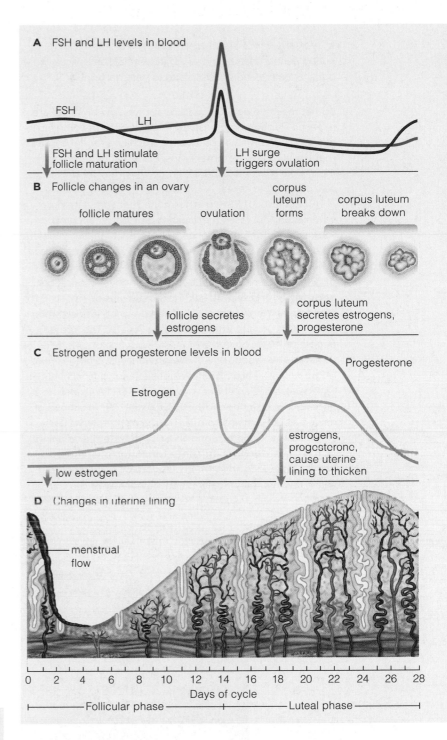

Figure 42.11 Animated Changes in a human ovary and uterus correlated with changing hormone levels. We start with the onset of menstrual flow on day one of a twenty-eight-day menstrual cycle.

(**a**,**b**) Prompted by GnRH from the hypothalamus, the anterior pituitary secretes FSH and LH, which stimulate a follicle to grow and an oocyte to mature in an ovary. A midcycle surge of LH triggers ovulation and the formation of a corpus luteum. A decline in FSH after ovulation stops more follicles from maturing.

(**c**,**d**) Early on, estrogen from a maturing follicle calls for repair and rebuilding of the endometrium. After ovulation, the corpus luteum secretes some estrogen and more progesterone that primes the uterus for pregnancy. If pregnancy occurs, the corpus luteum will persist, and its secretions will stimulate the maintenance of the uterine lining.

42.7 | FSH and Twins

■ Typically, only a single egg matures and gets released during each menstrual cycle. Abundant FSH can cause two eggs to mature and possibly lead to fraternal twins.

Sometimes two oocytes mature at the same time and are released during one menstrual cycle. If both become fertilized, the outcome will be two genetically different zygotes that develop into fraternal twins. Fraternal twins are no more alike than any other siblings. They may be the same sex, or different sexes.

A high level of FSH, the hormone that stimulates egg maturation, increases the likelihood of fraternal twins. FSH level and the prevalence of fraternal twinning varies among families and among ethnic groups. A woman who is herself a fraternal twin has double the average chances of giving birth to fraternal twins. If she does so once, her odds triple for a second set. Fraternal twins are most common among women of African descent, less common among Caucasians, and rare among Asians. The Yoruba people of Africa have the highest incidence of twin or triplet births—about one in every twenty-two pregnancies (Figure 42.12). They also have unusually high levels of FSH.

Age also has an effect. A woman's FSH levels rise from puberty through her midthirties, causing her likelihood of having fraternal twins to rise. Thus, a trend toward later childbearing is contributing to a rise in fraternal twinning.

FSH level does not influence formation of identical twins. Such twins arise when a zygote or early embryo splits, and two genetically identical individuals develop. A split is a chance event; a tendency to produce identical twins does not run in families and such twins are equally likely among women of all ethnic groups and ages.

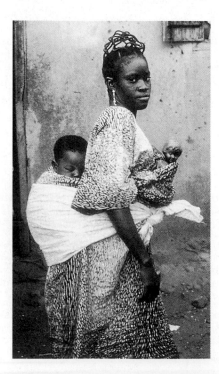

Figure 42.12 Yoruba mother. The rate of twin births among the Yoruba is the world's highest, but the mortality rate also is high; half of the twins die shortly after birth. Grieving mothers use a carving (Ere Ibeji) as a ritual point of contact with lost infants. Commercially produced plastic dolls are now being substituted for traditional wood carvings.

42.8 | When Gametes Meet

■ When a female and a male engage in sexual intercourse, the excitement of the moment may obscure what can happen if a secondary oocyte is in an oviduct.

■ Links to Autonomic signals 33.8, Pituitary hormones 35.3

Internal fertilization involves coordinated changes in the physiology of two individuals and then additional interactions between their gametes. It all begins with sexual intercourse, or coitus.

Sexual Intercourse

Physiology of Sex For males, intercourse requires an erection. Long cylinders of spongy tissue make up the bulk of the penis (Figure 42.4). When a male is not sexually aroused, his penis remains limp, because the large blood vessels that transport blood to the spongy tissue are constricted. When a male becomes aroused, parasympathetic signals induce the vessels that supply the spongy tissue to widen. Inward flow of blood now exceeds outward flow, and the increase in fluid pressure expands the internal chambers. As a result, the penis enlarges and stiffens, so it can be inserted into a female's vagina.

During intercourse, pelvic thrusting stimulates the mechanoreceptors in the male's penis and the female's clitoris. The female's vaginal wall, labia, and clitoris swell with blood.

In both partners, the heart rate and breathing rate rise. The posterior pituitary steps up its secretion of oxytocin, which inhibits signals from the brain center that controls fear and anxiety (the amygdala). When stimulation continues, oxytocin surges at orgasm.

At orgasm, oxytocin causes rhythmic contractions of smooth muscle of the reproductive tract. At the same time, endorphin release in the brain evokes feelings of pleasure. In a male, orgasm is usually accompanied by ejaculation, in which contracting muscles force the semen out of the penis. You may have been told that a female will not become pregnant as long as she does not reach orgasm. Do not believe it.

Regarding Viagra The ability to get and sustain an erection peaks during the late teens. As a male grows older, he may have episodes of erectile dysfunction. With this disorder, the penis cannot stiffen enough for intercourse. Men who have circulatory problems are most often affected. Smoking also increases risk. The National Institutes of Health estimates that 30 million men are affected in the United States alone. Viagra and similar drugs prescribed for erectile dysfunction cause blood vessels that carry blood into the penis to

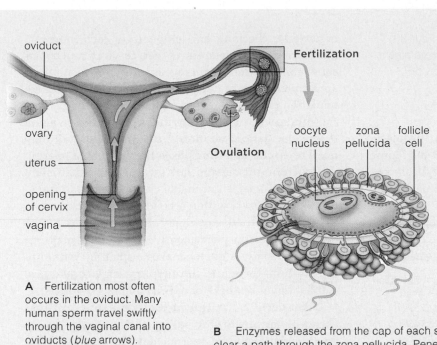

A Fertilization most often occurs in the oviduct. Many human sperm travel swiftly through the vaginal canal into oviducts (*blue* arrows).

Inside an oviduct, the sperm surround a secondary oocyte that was released by ovulation.

B Enzymes released from the cap of each sperm clear a path through the zona pellucida. Penetration of the secondary oocyte by a sperm causes the oocyte to releases substances that harden the zona pellucida and prevent other sperm from binding.

C The oocyte nucleus completes meiosis II, forming a nucleus with a haploid maternal genome. The sperm's tail and other organelles degenerate. Its DNA is enclosed by a membrane, forming a haploid nucleus with paternal genes.

Later, the two nuclear membranes will break down and paternal and maternal chromosomes will become arranged on a bipolar spindle in preparation for the first mitotic division.

Figure 42.13 Animated Events in human fertilization. The light micrograph shows a fertilized human oocyte.

Figure It Out: In the micrograph, what are the small cells on the right, just beneath the zona pellucida? *Answer: The polar bodies*

widen and deliver more blood. Such drugs can cause headaches and (rarely) sudden hearing loss. They can also interact with other drugs, and should never be taken without a prescription.

Fertilization

On average, an ejaculation can put 150 million to 350 million sperm in the vagina. Less than thirty minutes later, hundreds of them reach the oviducts. The sperm swim toward the ovaries. As the sperm travel, they undergo changes that prepare them to bind to and penetrate an oocyte.

Fertilization most often occurs in an oviduct (Figure 42.13a). Sperm bind to an oocyte's zona pellucida, and this binding triggers the release of acrosomal enzymes from the cap on a sperm's head. The enzymes digest the zona pellucida, clearing a passage to the oocyte's plasma membrane (Figure 42.13b). Usually only one sperm enters the secondary oocyte. The sperm's tail and other organelles break down.

Penetration of an oocyte by a sperm has two major effects. First, penetration causes changes in the oocyte that prevent other sperm from entering it. Second, the penetration causes the oocyte to complete meiosis II

and divide (Figure 42.13c). This division produces a mature egg, or ovum, and a polar body. This polar body, along with the one formed earlier by meiosis I, will degenerate.

In most animals, egg and sperm nuclei fuse to form a diploid nucleus in the zygote, the first cell of a new individual. In humans and other mammals, nuclei do not fuse. Instead, the nuclear membranes of the egg and the sperm break down. The maternal and paternal chromosomes then become oriented on a mitotic spindle for the first cell division. This division is the first step in development, a process explained in detail in the next chapter.

Take-Home Message

What happens during intercourse and fertilization?

■ Sexual arousal involves nervous and hormonal signals.

■ Ejaculation releases millions of sperm into the vagina. Sperm travel through the uterus toward the oviducts, the site where fertilization most often occurs.

■ Penetration of a secondary oocyte by a single sperm causes the oocyte to complete meiosis II, and prevents additional sperm from penetrating it.

■ Sperm organelles disintegrate. The DNA of the sperm, along with that of the oocyte, become the genetic material of the zygote.

■ There are many options for people who wish to put off reproducing or improve their chances of becoming a parent.

■ Link to Prenatal diagnosis 12.8

Birth Control Options

Emotional and economic factors often lead people to seek ways to control their fertility. Table 42.3 and Figure 42.14 list common fertility control options and compare their effectiveness. Most effective is abstinence—no sex—which may take great self-discipline.

Rhythm methods are forms of abstinence; a woman simply avoids sex in her fertile period. She calculates when she is fertile by recording how long menstrual cycles last, checking her temperature each morning, monitoring the thickness of her cervical mucus, or some combination of these methods. Miscalculations are frequent. Sperm deposited in the vagina just before ovulation may live long enough to meet an egg.

Withdrawal, or removing the penis from the vagina before ejaculation, requires great willpower and may fail. Pre-ejaculation fluids from the penis hold sperm.

Douching, or rinsing out the vagina immediately after intercourse, is unreliable. Some sperm can travel through the cervix within seconds of ejaculation.

Surgical methods are highly effective, but are meant to make a person permanently sterile. Men may opt for a vasectomy. A doctor makes a small incision into the scrotum, then cuts and ties off each vas deferens. A tubal ligation blocks or cuts a woman's oviducts.

Other fertility control methods use physical and chemical barriers to stop sperm from reaching an egg. Spermicidal foam and spermicidal jelly poison sperm. They are not always reliable, but their use with a condom or diaphragm reduces the risk of pregnancy.

A diaphragm is a flexible, dome-shaped device that is positioned inside the vagina so it covers the cervix. A diaphragm is relatively effective if it is first fitted by a doctor and used correctly with a spermicide. A cervical cap is a similar but smaller device.

Condoms are thin, tight-fitting sheaths worn over the penis during intercourse. Good brands may be as much as 95 percent effective when used correctly with a spermicide. Only condoms made of latex offer protection against sexually transmitted diseases (STDs). However, even the best ones can tear or leak.

An intrauterine device, or IUD, is inserted into the uterus by a physician. Some IUDs make cervical mucus thicken so sperm cannot swim through it. Others shed copper, which interferes with implantation.

The birth control pill is the most common fertility control method in developed countries. "The Pill" is a mixture of synthetic estrogens and progesterone-like hormones that prevents both maturation of oocytes and ovulation. When used correctly, the pill is at least 94 percent effective. It can reduce menstrual cramps, but sometimes causes nausea, headaches, and weight gain. Its use lowers risk of ovarian and uterine cancer but raises risk of breast, cervical, and liver cancer.

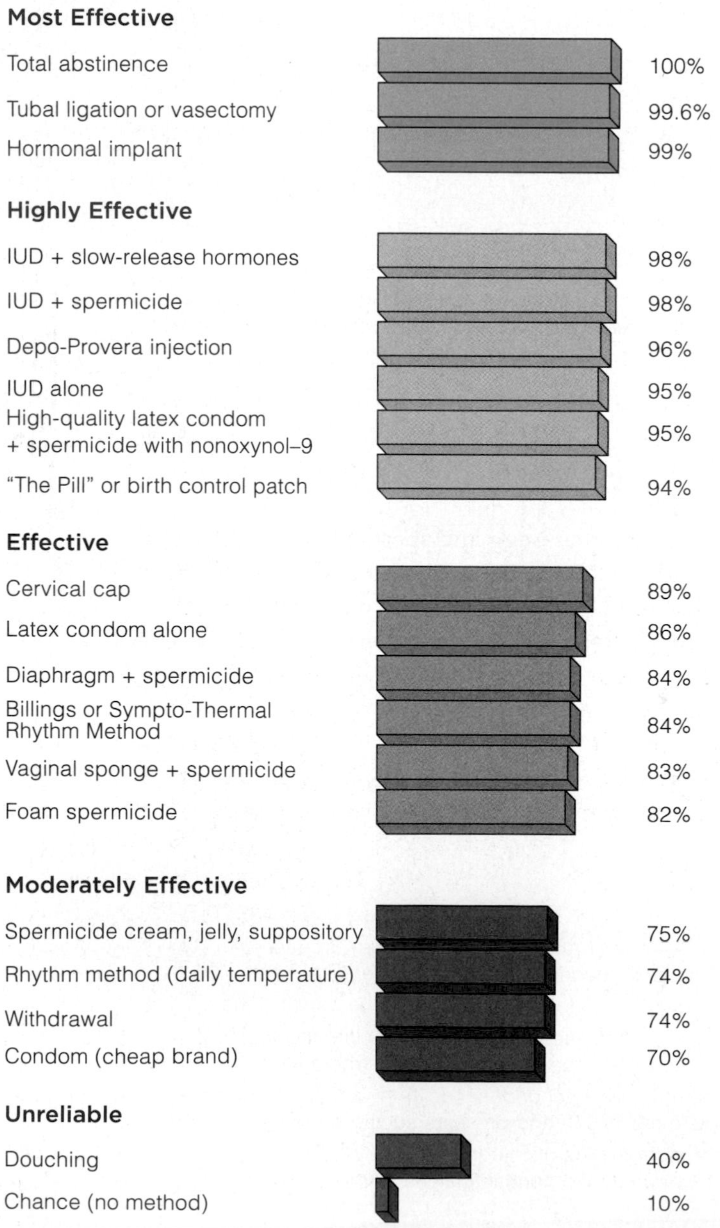

Most Effective

Total abstinence	100%
Tubal ligation or vasectomy	99.6%
Hormonal implant	99%

Highly Effective

IUD + slow-release hormones	98%
IUD + spermicide	98%
Depo-Provera injection	96%
IUD alone	95%
High-quality latex condom + spermicide with nonoxynol–9	95%
"The Pill" or birth control patch	94%

Effective

Cervical cap	89%
Latex condom alone	86%
Diaphragm + spermicide	84%
Billings or Sympto-Thermal Rhythm Method	84%
Vaginal sponge + spermicide	83%
Foam spermicide	82%

Moderately Effective

Spermicide cream, jelly, suppository	75%
Rhythm method (daily temperature)	74%
Withdrawal	74%
Condom (cheap brand)	70%

Unreliable

Douching	40%
Chance (no method)	10%

Figure 42.14 Comparison of the effectiveness of some methods of contraception. These percentages also indicate the number of unplanned pregnancies per 100 couples who use only that method of birth control for a year. For example, "94% effectiveness" for oral contraceptives means that 6 of every 100 females will still become pregnant, on average.

A birth control patch is a small, flat adhesive patch applied to skin. The patch delivers the same mixture of hormones as an oral contraceptive, and it blocks ovulation the same way. Like birth control pills, it is not for everyone. Some women, especially those who smoke, can develop dangerous blood clots and other serious cardiovascular disorders.

Hormone injections or implants prevent ovulation. Injections act for several months, whereas the implant Implanon lasts for three years. Both methods are quite effective, but may cause sporadic, heavy bleeding.

Some women turn to emergency contraception after a condom tears, or after unprotected consensual sex or rape. Such "morning-after pills" are now available without a prescription to women over age 18. They prevent ovulation and work best if taken immediately after intercourse. However, they can be effective up to five days later. The pills are not meant to be used on a regular basis. Nausea, vomiting, abdominal pain, headache, and dizziness are side effects.

About Abortion

About 10 percent of detected pregnancies end in a spontaneous abortion, or miscarriage. Many other pregnancies end without ever having been detected. By some estimates, as many as 50 percent of all pregnancies are cut short by some genetic problem. Risk of miscarriage increases with maternal age.

Induced abortion is the deliberate dislodging and removal of an embryo or fetus from the uterus. In the United States, about half of all unplanned pregnancies end in induced abortion. Parents who find out through genetic tests that an embryo has a genetic abnormality may decide to terminate the pregnancy. About 80 percent of embryos diagnosed with Down syndrome are aborted (Section 12.8).

From a clinical standpoint, abortion usually is a brief, low-risk procedure, especially during the first trimester of pregnancy. Mifepristone (RU-486) and similar drugs can induce abortion during the first nine weeks. They interfere with how the body sustains the uterine lining for the pregnancy. Use of a suction device terminates pregnancies as late as fourteen weeks. Later abortions require more difficult surgical procedures.

Assisted Reproductive Technology

About 15 percent of couples in the United States are infertile; either the woman does not become pregnant or repeatedly miscarries. When a couple make normal sperm and oocytes but cannot conceive naturally, they

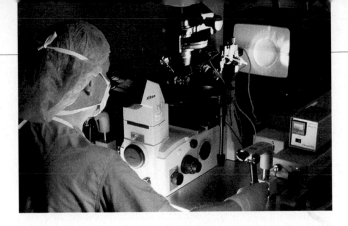

Figure 42.15 *In vitro* fertilization. A fertility doctor uses a micromanipulator to insert a human sperm into an oocyte. The video screen shows the view through the microscope.

may turn to technology for help. With *in vitro* fertilization, a doctor combines eggs and sperm outside the body (Figure 42.15). The resulting zygotes are allowed to divide, then one or more small clusters of cells are transferred to a woman's uterus to undergo development. Assisted reproduction attempts are costly and most fail. In 30-year-old women, about one-third of *in vitro* attempts result in a birth. In 40-year-olds, only one in six attempts succeeds.

Take-Home Message

What methods do humans use to control their fertility?

- Physical barriers and hormonal treatments can prevent pregnancy.
- Spontaneous or induced abortion ends an existing pregnancy.
- Assisted reproductive technology helps some couples who are having trouble conceiving

Table 42.3 Common Methods of Contraception

Method	Description
Abstinence	Avoid intercourse entirely
Rhythm method	Avoid intercourse in female's fertile period
Withdrawal	End intercourse before male ejaculates
Douche	Wash semen from vagina after intercourse
Vasectomy	Cut or close off male's vasa deferentia
Tubal ligation	Cut or close off female's oviducts
Condom	Enclose penis, block sperm entry to vagina
Diaphragm, cervical cap	Cover cervix, block sperm entry to uterus
Spermicides	Kill sperm
Intrauterine device	Prevent sperm entry to uterus or prevent implantation of embryo
Oral contraceptives	Prevent ovulation
Hormone patches, implants, or injections	Prevent ovulation
Emergency contraception pill	Prevent ovulation

- Sex acts transfer body fluids in which some human pathogens travel from one host to another.

- Links to Infectious disease 21.8, HIV 21.2

Consequences of Infection

Each year, the pathogens that cause **sexually transmitted diseases**, or STDs, infect about 15 million people in the United States (Table 42.4). Two-thirds of those infected are under age twenty-five, one-fourth are teenagers. Over 65 million Americans now live with an incurable STD. Treating these diseases and their complications costs about $8.4 billion in an average year.

The social consequences of STDs are sobering. Women become infected more easily than men, and develop more complications. Each year, about 1 million American women develop pelvic inflammatory disease (PID), a complication of some bacterial STDs. PID scars the reproductive tract, can cause chronic pain and infertility, and increases the risk of a tubal pregnancy (Figure 42.16a).

A mother can transmits an STD to her child. Herpes simplex type 2 virus kills about 50 percent of the embryos it infects and causes neural defects in many survivors. Exposure to *Chlamydia* during childbirth can lead to an infection of the newborn's throat or eyes (Figure 42.16b).

Table 42.4 New STD Cases Annually

STD	U.S. Cases	Global Cases
HPV infection	6,200,000	400,000,000
Trichomoniasis	5,000,000	174,000,000
Chlamydia	3,000,000	92,000,000
Genital herpes	1,000,000	20,000,000*
Gonorrhea	650,000	62,000,000
Syphilis	70,000	12,000,000
AIDS	40,000	4,900,000

* Global data on genital herpes last compiled in 1997.

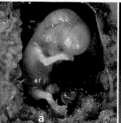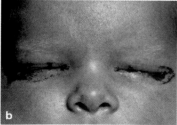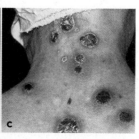

Figure 42.16 A few downsides of unsafe sex. (**a**) Increased risk of tubal pregnancy. Scarring caused by STDs can cause an embryo to implant itself in an oviduct, rather than the uterus. Untreated tubal pregnancies can rupture an oviduct and cause bleeding, infection, and death. (**b**) An infant with chlamydia-inflamed eyes. The child's mother passed on the bacterial pathogen during the birth process. (**c**) Chancres (open sores) caused by syphilis.

Major Agents of Sexually Transmitted Disease

HPV Human papillomavirus (HPV) infection is the most widespread and fastest growing STD in the United States. At least 20 million are already infected. Some of the 100 or so HPV strains can cause genital warts: bumpy growths on external genitals and the area around the anus. A few HPV strains are the main cause of cervical cancer. Sexually active females should have an annual Pap smear to check for cervical changes. A vaccine can prevent HPV infection if given before viral exposure (Chapter 38 introduction).

Trichomoniasis The flagellated protist *Trichomonas vaginalis* causes the disease trichomoniasis (Section 22.2). In women, symptoms typically include vaginal soreness, itching, and a yellowish discharge. Infected males often show no symptoms. Untreated infections damage the urinary tract, cause infertility, and invite HIV infection. A single dose of an antiprotozoal drug quickly cures an infection. Both sexual partners should be treated.

Chlamydia Chlamydial infection is primarily a young person's disease. Forty percent of those infected are between ages fifteen and nineteen; 1 in 10 sexually active teenage girls is infected. *Chlamydia trachomatis* causes the disease (Figure 42.17a). Antibiotics can quickly kill this bacterium. Most infected females remain undiagnosed; they have no symptoms. Between 10 and 40 percent of those who are untreated will develop pelvic inflammatory disease. Half of infected males have symptoms, such as abnormal discharges from the penis and painful urination. Untreated males risk an inflamed reproductive tract and infertility.

Genital Herpes About 45 million Americans have genital herpes, caused by Herpes simplex type 2 virus. Transmission to new hosts requires direct contact with active herpesviruses or with sores that contain them. Mucous membranes of the mouth and genitals are most vulnerable. Early symptoms are often mild or absent. Small, painful blisters may form on the genitals. Within three weeks, the virus enters latency. Blisters crust over and heal, but viral particles remain hidden in the body.

Sporadic reactivation of herpesvirus typically causes painful blisters at or near the original infection site. Sexual intercourse, menstruation, emotional stress, or other types of infection trigger flare-ups. An antiviral drug decreases healing time and pain, but genital herpes is incurable.

Gonorrhea The STD gonorrhea is caused by *Neisseria gonorrhoeae* (Figure 42.17b). This bacterium can cross the mucous membranes of the urethra, cervix, or anal canal during sexual intercourse. An infected female may notice a vaginal discharge or burning sensation while urinating. If the bacterium enters her oviducts, it may cause cramps, fever, vomiting, and scarring that can lead to sterility. Less than one week after a male is infected, yellow pus oozes

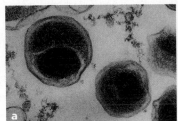 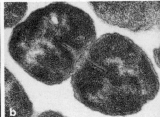 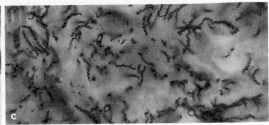

Figure 42.17 Light micrographs of bacteria that cause (**a**) chlamydia, (**b**) gonorrhea, and (**c**) syphilis. All can be killed with antibiotic drugs.

from his penis. Urination becomes more frequent, and it also may be painful.

Prompt treatment with antibiotics quickly cures this disease, which is rampant. Many ignore early symptoms or wrongly believe infection confers immunity. A person can contract gonorrhea repeatedly, probably because there are now at least sixteen strains of *N. gonorrhoeae*.

Syphilis The spirochete bacterium *Treponema pallidum* causes syphilis, a dangerous STD (Figure 42.17c). During sex with an infected partner, these bacteria get onto the genitals or into the cervix, vagina, or oral cavity. They then slip into the body through tiny cuts. One to eight weeks later, many *T. pallidum* cells are twisting about inside a flattened, painless chancre, a localized ulcer.

The chancre is a sign of the primary stage of syphilis. It usually heals, but treponemes multiply inside the spinal cord, brain, eyes, bones, joints, and mucous membranes. In the infectious secondary stage, a skin rash develops and more chancres form (Figure 42.16c). In about half of the cases, immune responses succeed and symptoms subside or disappear. In the remainder of cases, lesions and scars appear in the skin and liver, bones, and other organs. Few treponemes form during this tertiary stage, but the host's immune system is hypersensitive to them. Chronic immune reactions may damage the brain and spinal cord, and cause paralysis.

Possibly because the symptoms are so alarming, more people seek early treatment for syphilis than they do for gonorrhea. Later stages require prolonged treatment.

AIDS Infection by HIV, human immunodeficiency virus, can lead to AIDS—acquired immune deficiency syndrome (Chapter 21 introduction). At first, a person may not know that he or she is infected. Over time, the virus begins to destroy the immune system, and the set of chronic disorders that characterize AIDS develop. Some normally harmless bacteria already living in and on the body are the first to take advantage of the lowered resistance. This infection can open the door to other, more dangerous pathogens. Over time, these agents can overwhelm the compromised immune system and cause death.

Most often, HIV spreads by way of anal, vaginal, and oral intercourse, and through intravenous drug use. Virus particles in blood, semen, urine, or vaginal secretions enter a new host through cuts and abrasions of the penis, vagina, rectum, or mouth. Oral sex is least likely to cause

infection. Anal sex is 5 times more dangerous than vaginal sex and 50 times more dangerous than oral sex.

Most health care workers advocate safe sex, although there is confusion over what "safe" means. The use of high-quality latex condoms together with a nonoxynol-9 spermicide helps prevent viral transmission. However, as mentioned earlier, this practice still carries a slight risk. Open-mouth kissing with an HIV-positive individual is risky. Caressing is not, as long there are no lesions or cuts where HIV-laden body fluids can enter the body. Skin lesions caused by any other sexually transmitted disease can serve as points of entry for the virus.

Confidential testing for HIV exposure is now widely available, and early diagnosis saves lives. It keeps a person from unknowingly infecting others and allows treatment to be started when it is most effective. An HIV infection cannot be cured, but drug therapies can extend the life of those infected (Figure 42.18). When early diagnosis and treatment are followed by ongoing medical care, an HIV-positive person can have a nearly normal life span.

However, once infected, a person can always infect others. Also, the drugs that keep people alive often have unpleasant side effects, including nausea, fatigue, diarrhea, and bone loss. These side effects cause many HIV-positive people to risk their lives by stopping treatment or taking less than the recommended amount of their medication.

Figure 42.18 Basketball legend Magic Johnson, one of the torch bearers of the 2002 Winter Olympics. He was diagnosed as HIV positive in 1991. He contracted the virus through heterosexual sex, and credits his survival to AIDS drugs and informed medical care. Johnson continues to campaign to educate others about the risk of AIDS.

Male or Female? Body or Genes?

Parents who have a child with ambiguous genitals face a difficult choice. Surgery can make their child appear more normal, but it can harm nerves and impair sexual function. The best cosmetic result may even require sex reassignment, as when a boy with a micropenis is surgically altered and reared as a female. On the other hand, parents who opt to avoid surgery worry about the psychological trauma that having an unusual body may cause.

How would you vote?

Should parents of a child who has unusual genitals wait and allow the child to choose or decline normalizing surgery? See CengageNOW for details, then vote online.

Summary

Section 42.1 **Asexual reproduction** produces genetic copies of the parent. **Sexual reproduction** is energetically more costly, and a parent does not have as many of its genes represented among the offspring. However, sexual reproduction produces variable offspring, which may be advantageous in environments where conditions fluctuate from one generation to the next.

Most animals reproduce sexually and have separate sexes, but some are **hermaphrodites** that produce both eggs and sperm. With **external fertilization**, gametes are released into water. Most animals on land have **internal fertilization**; gametes meet in a female's body. Offspring may develop inside or outside the maternal body. **Yolk** helps nourish developing young.

Sections 42.2, 42.3 The human reproductive system consists of primary reproductive organs, or **gonads**, and accessory organs and ducts. Male gonads are **testes**, which produce sperm and the sex hormone **testosterone**. Testosterone influences reproduction, as well as development of gender-specific secondary sexual traits that emerge when sexual organs mature at **puberty**.

Gonadotropin-releasing hormone (GnRH) released by the hypothalamus causes the pituitary gland to secrete **luteinizing hormone (LH)** and **follicle-stimulating hormone (FSH)**. These hormones affect gamete formation in both males and females. Sperm form in a series of ducts. Glands that empty into these ducts supply components of the **semen**. The **penis** is the male organ of intercourse.

■ *Use the animation on CengageNOW to learn about the reproductive system of human males and how sperm form.*

Sections 42.4–42.7 **Ovaries**, the female gonads, produce eggs and secrete **progesterone** and **estrogens**. Eggs are released into **oviducts** that connect to the **uterus**, the chamber where offspring develop. The **vagina** serves as the female organ of intercourse and as the birth canal.

A **menstrual cycle** is an approximately monthly cycle of fertility. Feedback loops from ovaries to the hypothalamus and the anterior pituitary gland control it. In the cycle's follicular phase, FSH stimulates maturation of a **primary oocyte** and cells that surround it. Women who have high FSH levels are more likely to release more than one egg at a time and have fraternal twins. FSH and LH also prompt ovaries to secrete estrogens that cause thickening of the lining of the uterus. A midcycle surge

in LH triggers **ovulation**, release of a **secondary oocyte** from an ovary. During the luteal phase, a **corpus luteum** forms from cells that surrounded the egg. Its hormonal secretions, mainly progesterone, cause the uterine wall to thicken. If fertilization does not occur, the corpus luteum degenerates and menstrual fluid flows out of the vagina as the cycle starts again. Menstrual cycles continue until a woman's fertility ends at **menopause**.

■ *Use the animation on CengageNOW to learn about the female reproductive system, cyclic changes in an ovary, and hormonal changes during the menstrual cycle.*

Sections 42.8–42.10 Hormones and nerves govern the physiological changes that occur during arousal and intercourse. Millions of sperm are ejaculated, but usually only one penetrates the secondary oocyte. Fertilization forms a zygote, which will develop into a new individual.

Humans prevent pregnancy by abstinence, surgery, physical or chemical barriers, and by influencing female sex hormones. Unsafe sex and other behaviors promote the spread of pathogens that cause **sexually transmitted diseases**, or STDs.

■ *Use the animation on CengageNOW to see what happens during fertilization.*

Self-Quiz   *Answers in Appendix III*

1. Sexual reproduction _____ .
   a. requires internal fertilization
   b. produces offspring that vary in their traits
   c. is more efficient than asexual reproduction
   d. puts all of a parent's genes in each offspring

2. Testosterone is secreted by the _____ .
   a. testes   c. prostate gland
   b. hypothalamus   d. all of the above

3. Semen contains secretions from the _____ .
   a. adrenal gland   c. prostate gland
   b. pituitary gland   d. all of the above

4. Male germ cells undergo meiosis in the _____ .
   a. urethra   c. prostate gland
   b. seminiferous tubules   d. vasa deferentia

5. The female _____ is derived from the same embryonic tissue as the male penis.
   a. cervix   b. clitoris   c. vagina   d. oviduct

Data Analysis Exercise

Adrenal glands normally make a little testosterone, but a mutation in the gene for the enzyme 21-hydroxylase causes excess production of this hormone. A female child who has a 21-hydroxylase deficiency is exposed to abnormally high levels of testosterone during development. This hormone can enlarge her clitoris and cause her labia to fuse, giving her genitals a more male appearance.

The drug dexamethasone slows the adrenal glands' testosterone production. Figure 42.19 shows data from a study in which doctors gave this drug to pregnant women carrying daughters with 21-hydroxylase deficiency. Sixteen of these women had previously given birth to a daughter with 21-hydroxylase deficiency. These daughters (sisters of the treated newborns) serve as a point of comparison.

1. How many daughters produced by dexamethasone-treated pregnancies had normal female genitals?

2. How many phenotypically normal girls had the women's earlier untreated pregnancies produced?

3. How many women who previously had girls with level 4 or 5 masculinization saw an improvement with treatment?

4. Do the data support the hypothesis that giving the drug dexamethasone to a pregnant woman can reduce the effects of her developing daughter's 21-hydrolase deficiency?

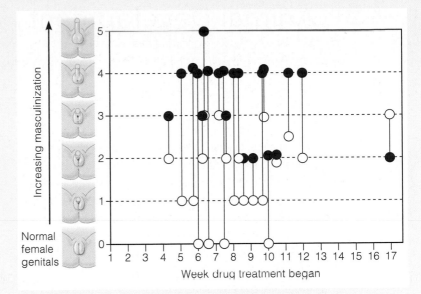

Figure 42.19 Degree of masculinization of 21-hydroxylase deficient females exposed to dexamethasone in the womb (*open* circles), compared to that of older affected sisters who were untreated during development (*dark* circles). Graphics along the side depict appearance of the newborn's genital area.

6. The cervix is the entrance to the _____ .
a. oviducts b. vagina c. uterus d. clitoris

7. During a menstrual cycle, a midcycle surge of _____ triggers ovulation.
a. estrogens b. progesterone c. LH d. FSH

8. The corpus luteum develops from _____ and secretes hormones that cause the lining of the uterus to thicken.
a. follicle cells   c. a primary oocyte
b. polar bodies   d. a secondary oocyte

9. A male has an erection when _____ .
a. muscles running the length of the penis contract
b. Leydig cells release a surge of testosterone
c. the posterior pituitary releases oxytocin
d. spongy tissue inside the penis fills with blood

10. Birth control pills deliver synthetic _____ .
a. estrogens and progesterone
b. LH and FSH
c. testosterone
d. oxytocin and prostaglandins

11. Match each hormone with its source.
___FSH and LH   a. pituitary gland
___GnRH   b. ovaries
___estrogens   c. hypothalamus
___testosterone   d. testes

12. Match each disease with the type of agent that causes it. The choices can be used more than once.
___chlamydial infection   a. bacteria
___AIDS   b. protist
___syphilis   c. virus
___genital warts
___gonorrhea
___genital herpes
___trichomoniasis

13. Match each structure with its description.
___testis   a. conveys sperm out of body
___epididymis   b. secretes semen components
___labia majora   c. stores sperm
___urethra   d. produces testosterone
___vagina   e. produces estrogens and
___ovary   progesterone
___oviduct   f. usual site of fertilization
___prostate gland   g. lining of uterus
___endometrium   h. fat-padded skin folds
                              i. birch canal

■ *Visit CengageNOW for additional questions.*

Critical Thinking

1. Drugs that inhibit signals of sympathetic neurons may be prescribed for males who have high blood pressure. How might such drugs interfere with sexual performance?

2. In most groups of birds, males do not have a penis. Both males and females have a single opening, called a cloaca, through which wastes leave the body. The male's sperm also exit through this opening. During mating, a male perches on a female's back and bends his abdomen under, so his cloaca covers hers. This action is referred to as a "cloacal kiss." Some birds even carry out this feat in midair. Flightless birds such as ostriches and kiwis do have a penis. Did the common reptile ancestor of all birds have a penis or not? What types of information would help you answer this question?

3. Some sperm mitochondria do get into an egg during fertilization, but they do not persist. As sperm mature, their mitochondria become tagged with a protein (ubiquitin) that signals the egg to destroy them. What organelle would you expect to be involved in this destruction process?

43 Animal Development

IMPACTS, ISSUES Mind-Boggling Births

In December of 1998, Nkem Chukwu of Houston, Texas, gave birth to six girls and two boys. They were the first set of human octuplets to be born alive (Figure 43.1). The births were premature. In total, all eight newborns weighed a bit more than 4.5 kilograms (10 pounds). Odera, the smallest, weighed about 300 grams (less than 1 pound), and six days later she died when her heart and lungs gave out. Two others required surgery. All seven survivors had to spend months in the hospital before going home, but now are in good health.

Why did octuplets form in the first place? Chukwu had trouble getting pregnant. Her doctors gave her hormone injections, which caused many of her eggs to mature and be released at the same time. When the doctors realized that she was carrying a large number of embryos, they suggested reducing the number. Chukwu chose instead to try to carry all of them to term.

Her first child was thirteen weeks premature. The others were surgically delivered two weeks later.

Over the past two decades, the incidence of multiple births has increased by almost 60 percent. There have been four times as many higher order multiple births—triplets or more. What is going on? A woman's fertility peaks in her midtwenties. By thirty-nine, her chance of conceiving naturally has declined by about half. Yet the number of first-time mothers who are more than forty years old doubled in the past decade. Many had turned to reproductive intervention, including fertility drugs and *in vitro* fertilization.

Weigh the rewards against risks. Carrying more than one embryo increases the risk of miscarriage, premature delivery, or stillbirth. Multiple-birth newborns weigh less than normal and are more likely to have birth defects, including cleft lip, heart malformations, and disorders in which the bladder or spinal cord is exposed at the body surface.

With this example, we turn to one of life's most amazing dramas—the development of complex animals. How does a single fertilized egg of a human—or frog or bird or any other animal—give rise to so many specialized kinds of cells? How does development yield an adult with all the complex tissues and organs discussed throughout this unit?

Answers to these questions will emerge as we consider the developmental processes common to all animals. You will see how experiments helped scientists approach these questions and how these experimental studies led to our current understanding of developmental processes.

We will also continue the story of human reproduction and the human life cycle, which we began in the previous chapter. We will see how humans develop from a single cell to an adult body with trillions of specialized cells.

See the video! Figure 43.1 Testimony to the potency of fertility drugs—seven survivors of a set of octuplets. Besides manipulating so many other aspects of nature, humans are now manipulating their own reproduction.

Key Concepts

Principles of animal embryology

Animals develop through cleavage, gastrulation, organ formation, and then growth and tissue specialization. Cleavage parcels out material stored in different parts of the egg cytoplasm into different cells, thus starting the process of cell specialization. **Sections 43.1–43.5**

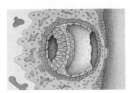

Human development begins

A pregnancy starts with fertilization and implantation of a blastocyst in the uterus. After implantation, a three-layered embryo forms and organ formation begins. All organs have formed by the end of the eighth week. **Sections 43.6–43.8**

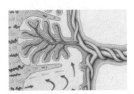

Function of the placenta

The placenta allows substances to diffuse between bloodstreams of a mother and her developing child. It also produces hormones that help sustain the pregnancy. **Section 43.9**

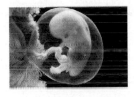

Later human development

By the time the fetal period begins, the developing individual appears distinctly human. Harmful substances that get into a mother's blood can cross the placenta and cause birth defects in the developing embryo or fetus. **Sections 43.10, 43.11**

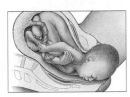

Birth and lactation

Positive feedback control plays a role in the process of labor, or childbirth. After birth, the newborn is nourished by milk secreted by mammary glands. **Section 43.12**

Links to Earlier Concepts

- This chapter builds on our discussions of cleavage (Section 9.4), and gamete formation and fertilization (10.5, 42.3, 42.6, 42.8). We revisit RNA localization (15.3), as well as the cell differentiation and master genes that influence it (15.1–15.3, 19.3).

- You will learn more about primary tissue layers of embryos (25.1, 32.6), and see more examples of feedback controls (27.3) and cell signaling (27.6).

- You will build upon your understanding of the evolution of vertebrate body plans (25.1, 26.1, 26.12) and of the two main animal lineages (25.7).

- The effects of thyroid hormone (35.6), and carbon monoxide (38.7) on an embryo are discussed, as is the protective effect of maternal antibodies (38.6).

■ Animals as different as sea stars and sea otters pass through the same stages in their developmental journey from a single, fertilized egg to a multicelled adult.

■ Links to Gamete formation 10.5, 42.3, 42.6, Animal body plans 25.1, Germ layers 32.6, Fertilization 42.8

Figure 43.2 shows six sequential processes that occur in reproduction and development of all animals with tissues and organs. This group includes most invertebrates and all vertebrates (Sections 25.1 and 26.1).

In the first process, gamete formation, eggs or sperm arise from germ cells in the parental body (Sections 10.5, 42.3, and 42.6). During fertilization (Section 42.8), the first cell of a new individual—the zygote—forms after a sperm penetrates a mature egg.

Cleavage carves up the zygote by repeated mitotic cell divisions. The number of cells increases, but the

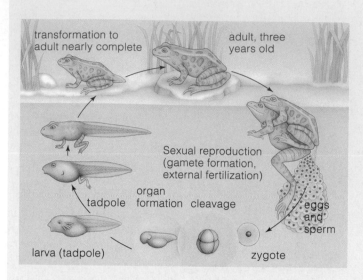

A We zoom in on the life cycle as a female releases her eggs into the water and a male releases sperm over the eggs.

A frog zygote forms at fertilization. About one hour after fertilization, a surface feature called the gray crescent appears on this type of embryo. It establishes the frog's head-to-tail axis. Gastrulation will start at the gray crescent.

Figure 43.3 Animated Reproduction and development in the life cycle of the leopard frog, *Rana pipiens*.

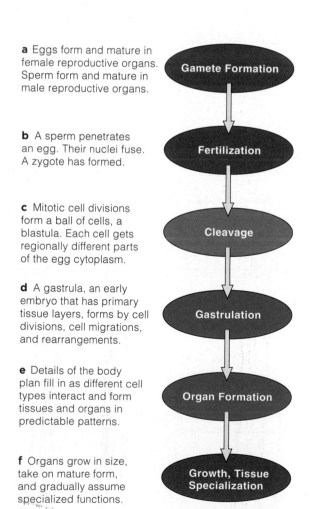

a Eggs form and mature in female reproductive organs. Sperm form and mature in male reproductive organs.

Gamete Formation

b A sperm penetrates an egg. Their nuclei fuse. A zygote has formed.

Fertilization

c Mitotic cell divisions form a ball of cells, a blastula. Each cell gets regionally different parts of the egg cytoplasm.

Cleavage

d A gastrula, an early embryo that has primary tissue layers, forms by cell divisions, cell migrations, and rearrangements.

Gastrulation

e Details of the body plan fill in as different cell types interact and form tissues and organs in predictable patterns.

Organ Formation

f Organs grow in size, take on mature form, and gradually assume specialized functions.

Growth, Tissue Specialization

Figure 43.2 Overview of reproductive and developmental processes that occur in animals with tissues and organs. We discussed gamete formation and fertilization in Chapter 42.

zygote's original volume does not. Cells become more numerous but smaller (Figure 43.3*b,c*). Cells formed during cleavage are called blastomeres. They typically are arranged as a **blastula**: a ball of cells that enclose a cavity (blastocoel) filled with their own secretions.

In the fourth stage, **gastrulation**, cells self-organize as an early embryo—a **gastrula**—that has two or three primary tissue layers. The tissues are the **germ layers** of the new individual. Germ layers, remember, are the forerunners of the adult animal's tissues and organs (Section 32.6).

During organ formation, tissues become arranged into organs. Many organs incorporate tissues derived from more than one germ layer.

Growth and tissue specialization is the final process of animal development. The tissues and organs continue to grow, and they slowly take on their final sizes, shapes, proportions, and functions. Growth and tissue specialization will continue into adulthood.

Figure 43.3 shows examples of the stages for one vertebrate, the leopard frog (*Rana pipiens*). A female releases eggs into the water and a male releases sperm onto them. Fertilization is external. The zygote formed by fertilization undergoes cleavage (Figure 43.3*b*). The

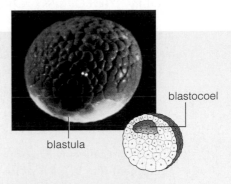

B Here we show the first three divisions of cleavage, a process that carves up the zygote's cytoplasm. In this species, cleavage results in a blastula, a ball of cells with a fluid-filled cavity.

C Cleavage is over when the blastula forms.

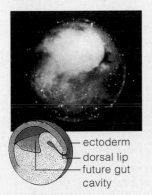

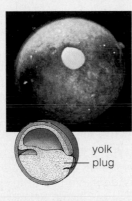

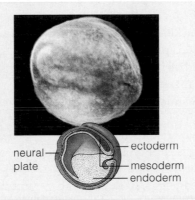

D The blastula becomes a three-layered gastrula—a process called gastrulation. At the dorsal lip, a fold of ectoderm above the first opening that appears in the blastula, cells migrate inward and start rearranging themselves.

E Organs begin to form as a primitive gut cavity opens up. A neural tube, then a notochord and other organs form from the primary tissue layers.

Tadpole, a swimming larva with segmented muscles and a notochord extending into a tail.

Limbs grow and the tail is absorbed during metamorphosis to the adult form.

Sexually mature, four-legged adult leopard frog.

F The frog's body form changes as it grows and its tissues specialize. The embryo becomes a tadpole, which metamorphoses into an adult.

repeated mitotic divisions form a blastula consisting of several thousand cells (Figure 43.3c).

The blastula undergoes gastrulation, which forms the three germ layers (Figure 43.3d). After the three primary tissues have formed, tissue specialization and organ formation begins. A typical vertebrate's neural tube and notochord form (Figure 43.3e). In frogs, as in some other animals, a larva (in this case as tadpole) undergoes metamorphosis, the remodeling of tissues into the adult form (Figure 43.3f).

Each stage in the development process builds on the one that precedes it.

Take-Home Message

What are the stages in reproduction and development in a typical animal?

■ Most animal life cycles start with gamete formation and fertilization. Development involves cleavage, gastrulation, organ formation, and then growth and tissue specialization.

■ The location of materials in an egg and distribution of those materials to descendant cells affects early development.

■ Links to Egg formation 10.5, RNA localization 15.3, Protostome and deuterostome lineages 25.7

Information in the Cytoplasm

A sperm, recall, consists of paternal DNA and a bit of equipment that helps it swim to and penetrate an egg. An oocyte, or immature egg, has far more cytoplasm (Section 10.5). Its cytoplasm has yolk proteins that will nourish a new embryo, mRNA transcripts for proteins that will be translated in early development, tRNAs and ribosomes to translate the mRNA transcripts, and proteins required to build mitotic spindles.

Certain components are not distributed all through the egg cytoplasm; they are localized in one particular region or another. This **cytoplasmic localization** is a feature of all oocytes (Section 15.3).

Cytoplasmic localization gives rise to the polarity that characterizes all animal eggs. In a yolk-rich egg, the vegetal pole has most of the yolk and the animal pole has little. In some amphibian eggs, dark pigment molecules accumulate in the cell cortex, a cytoplasmic region just beneath the plasma membrane. Pigment is the most concentrated close to the animal pole. After a sperm penetrates the egg at fertilization, the cortex rotates. Rotation reveals a gray crescent, a region of the cell cortex that is lightly pigmented (Figure 43.4a).

Early in the 1900s, experiments by Hans Spemann showed that some substances essential to development are localized in the gray crescent. In one experiment, he separated the first two blastomeres that formed at cleavage. Each blastomere had half of the gray crescent and developed into an embryo (Figure 43.4b). In the next experiment, Spemann altered the cleavage plane (Figure 43.4c). One blastomere received all of the gray crescent, and developed normally. The other, with no gray crescent, formed only a ball of cells.

Cleavage Divides Up the Maternal Cytoplasm

Once an oocyte is fertilized, the resulting zygote enters cleavage. By this process, a ring of microfilaments just beneath the plasma membrane contracts and pinches the cell in two (Section 9.4). The zygote's cytoplasm does not grow in size during cleavage; the repeated cuts divide its volume into ever smaller blastomeres.

Simply by virtue of where the cuts are made, different blastomeres receive different portions of the

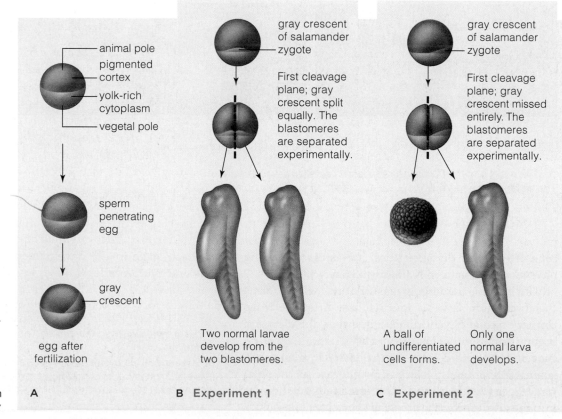

Figure 43.4 Animated
Experimental evidence of cytoplasmic localization in an amphibian oocyte.

(**a**) Many amphibian eggs have dark pigment concentrated in cytoplasm near the animal pole. At fertilization, the cytoplasm shifts, and exposes a gray crescent-shaped region just opposite the sperm's entry point. With normal first cleavage, each resulting cell gets half of the gray crescent.

(**b**) In one experiment, the first two cells formed by normal cleavage were physically separated from each other. Each developed into a normal larva.

(**c**) In another experiment, a zygote was manipulated so the first cleavage plane missed the gray crescent. Only one of the descendant cells received gray crescent material, and only it developed normally.

Figure It Out: Is the gray crescent region required for normal amphibian development?   *Answer: Yes*

animal pole
pigmented cortex
yolk-rich cytoplasm
vegetal pole

sperm penetrating egg

gray crescent

egg after fertilization

A

gray crescent of salamander zygote

First cleavage plane; gray crescent split equally. The blastomeres are separated experimentally.

Two normal larvae develop from the two blastomeres.

B Experiment 1

gray crescent of salamander zygote

First cleavage plane; gray crescent missed entirely. The blastomeres are separated experimentally.

A ball of undifferentiated cells forms.

Only one normal larva develops.

C Experiment 2

a Early protostome embryo. Its four cells are undergoing spiral cleavage, oblique to the anterior–posterior axis:

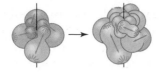

b Early deuterostome embryo. Its four cells are undergoing radial cleavage, *parallel with* and *perpendicular to* the anterior–posterior axis:

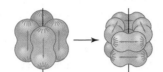

Figure 43.5 Examples of the two cleavage patterns most common in the two main lineages of bilateral animals.

a Sea urchin egg, with little yolk. Cleavage is complete. First cells formed are equally sized.

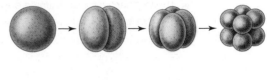

b Frog egg, with moderate amount of yolk. Yolk slows cleavage so lower cells are larger.

c Fish egg, with a large amount of yolk. Cleavage is restricted to the layer of cytoplasm on top of the yolk.

Two cells formed by first cleavage

mass of yolk

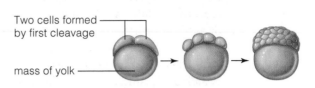

Figure 43.6 Comparison of cleavage patterns amoung deuterostomes that have different amounts of yolk in their eggs. Yolk slows division.

maternal cytoplasm. Orientation of the cell divisions is not random and has major implications for future development. The pattern of cleavage determines how much and which portion of the maternal cytoplasm a blastomere will receive. As a result of cytoplasmic localization of material inside the egg, cleavage distributes different kinds and quantities of materials into different blastomeres. For example, cleavage may put a specific maternal mRNA into one blastomere but not others. Thus, cleavage creates cell lineages that differ in the contents of their cytoplasm. Later, possessing different maternal materials will cause different cell lineages to express different genes, forming specialized tissues.

Variations in Cleavage Patterns

The details of cleavage vary among species. Differences start with the first division, which determines whether the first two cells will be equal or unequal in size and what part of the egg cytoplasm they receive.

There are two major animal lineages, protostomes and deuterostomes (Section 25.1), and they differ in their cleavage pattern. Most bilateral invertebrates are protostomes, which undergo spiral cleavage (Figure 43.5a). Echinoderms and all vertebrates are deuterostomes, and typically undergo radial cleavage (Figure 43.5b). Mammals, however, have a somewhat different pattern called rotational cleavage. The first cleavage divides the zygote along a plane that runs from top to bottom. Next, one cell divides the same way and the other divides in half at the cell equator.

The amount of yolk stored inside an egg also affects cleavage patterns. When there is little yolk, cleavage is complete; the first cut divides all the cytoplasm. An abundance of yolk will impede divisions, so cleavage is incomplete. Sea urchin eggs have little yolk, so their cleavage is complete and all blastomeres are similar in size (Figure 43.6a). The same is true for the nearly

yolkless eggs of mammals. Frogs and other amphibians also undergo complete cleavage, but it proceeds more slowly at the yolk-rich vegetal pole than at the yolk-free animal pole. As a result, cells vary a bit in their size (Figure 43.6b). Eggs of reptiles, birds, and most fishes are so yolky that cuts are exceedingly slow or blocked entirely, except in the small, disk-shaped region that has the least yolk (Figure 43.6c).

Structure of the Blastula

Collectively, the cells produced by cleavage constitute the blastula. Tight junctions hold the loose collection of cells together. Structure of the blastula varies with a species' cleavage pattern. In sea urchins, complete cleavage produces a blastula that is a hollow ball of cells. In animals with highly yolky eggs, such as birds and many fish, a disk-shaped collection of cells, called a blastodisc, forms atop the yolk. There is no large fluid-filled space. A mammal's blastula is a **blastocyst**, with outer cells that secrete fluid into the ball's cavity and other cells clustered in a mass against the cavity wall. The inner cells will develop into the embryo.

Take-Home Message

What are the effects of cytoplasmic localization and cleavage?

■ In an unfertilized egg, many enzymes, mRNAs, yolk, and other materials are localized in specific parts of the cytoplasm. This cytoplasmic localization helps guide development.

■ Cleavage divides a fertilized egg into a number of small cells but does not increase its original volume. The cells (blastomeres) inherit different parcels of cytoplasm that will make them behave differently, later in development.

From Blastula to Gastrula

■ The first tissues of an animal body form during gastrulation, when cells of the blastula rearrange themselves.

■ Link to Primary tissues 32.6

Hundreds to thousands of cells form during cleavage, depending on the species. Starting with gastrulation, cells begin to migrate and rearrange themselves. Figure 43.7 provides an example. Mechanisms of gastrulation vary among species. For example, an entire sheet of cells may bend inward, individual cells may migrate, or rows of cells may bend back on themselves.

In most animals, gastrulation produces a gastrula with three primary tissue layers: an outermost layer of **ectoderm**, a middle layer of **mesoderm**, and an inner layer of **endoderm** (Section 32.6).

What initiates gastrulation? Hilde Mangold, one of Spemann's students, discovered the answer. She knew that during gastrulation, some cells of a

salamander blastula move inward through an opening on its surface. Cells in the dorsal (upper) lip of the opening are descended from cells in a zygote's gray crescent. Mangold hypothesized that signals from dorsal lip cells caused gastrulation. She predicted that a transplant of dorsal lip material from one embryo to another would cause gastrulation at the recipient site. Mangold carried out many transplants (Figure 43.8*a*), and the results supported her prediction. Cells migrated inward at the transplant site, as well as at the usual location (Figure 43.8*b*). A salamander larva with two joined sets of body parts developed (Figure 43.8*c*). Apparently, signals from transplanted cells had caused their new neighbors to develop in a novel way.

This experiment also explained the results shown in Figure 43.4*c*. Without any gray crescent cytoplasm, an embryo does not have cells that would normally become the dorsal lip. In the absence of the signals produced by these cells, development stops short.

The effect of cells of a salamander gastrula's dorsal lip region on nearby cells is an example of **embryonic induction**. By this process, the fate of one group of embryonic cells is affected by its proximity to another group of cells. In this case, the cells of the dorsal lip alter the behavior of their neighbors.

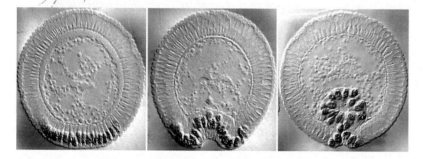

Figure 43.7 Gastrulation in a fruit fly (*Drosophila*). In these insects, cleavage is restricted to the outermost region of cytoplasm; the interior is filled with yolk. The series of photographs, all cross-sections, shows sixteen cells (stained *gold*) migrating inward. The opening that the cells move in through will become the fly's mouth. Descendants of the stained cells will form mesoderm. Movements shown in the photos occur during a period of less than 20 minutes.

Take-Home Message

What is gastrulation and how is it controlled?

■ Gastrulation is the developmental process during which cells rearrange themselves into primary tissue layers.

■ Gastrulation occurs when certain cells of the blastula make and release short-range signals that cause nearby cells to move about, either singly or as a cohesive group. This process is an example of embryonic induction.

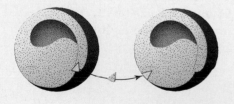

A Dorsal lip excised from donor embryo, grafted to novel site in another embryo.

B Graft induces a second site of inward migration.

C The embryo develops into a "double" larva, with two heads, two tails, and two bodies joined at the belly.

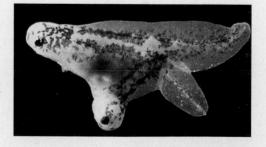

Figure 43.8 Animated Experimental evidence that signals from dorsal lip cells start amphibian gastrulation. A dorsal lip region of a salamander embryo was transplanted to a different site in another embryo. A second set of body parts started to form.

43.4 | Specialized Tissues and Organs Form

■ Cell differentiation lays the groundwork for formation of specialized tissues and organs.

■ Links to Cell differentiation 15.1, Master genes 15.2, Apoptosis 27.6

Cell Differentiation

From gastrulation onward, selective gene expression occurs: Different cell lineages express different genes. That is the start of **cell differentiation**, the process by which cell lineages become specialized (Section 15.1).

Intercellular signals can encourage differentiation, as during induction. In addition, **morphogens**, signalling molecules encoded by master genes, diffuse out from their source and form a concentration gradient in the embryo. A morphogen's effects on target cells depends on its concentration. Cells close to the source of a morphogen are exposed to a high concentration and turn on different genes than distant cells exposed to a lower morphogen concentration.

Morphogenesis and Pattern Formation

Cellular signals help bring about **morphogenesis**, the process by which tissues and organs form. During morphogenesis, some cells migrate to new locations. For example, neurons in the center of the developing brain creep along extensions of glial cells or the axons of other neurons until they reach their final position. Sheets of cells change shape, forming organs such as the neural tube, the forerunner of the vertebrate brain and spinal cord (Figure 43.9). Some cells even die on cue. By the process of **apoptosis**, signals from cells cause others to self-destruct. Apoptosis sculpts human fingers from a paddlelike body part (Section 27.6).

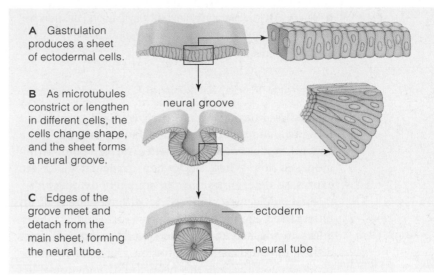

A Gastrulation produces a sheet of ectodermal cells.

B As microtubules constrict or lengthen in different cells, the cells change shape, and the sheet forms a neural groove.

neural groove

C Edges of the groove meet and detach from the main sheet, forming the neural tube.

ectoderm

neural tube

Figure 43.9 Animated Neural tube formation. Microtubule changes alter cell shape, causing the sheet of ectoderm to fold into a tubular form.

Why does a hand form at the end of an arm? Why not a foot? **Pattern formation** is the process by which body parts form in a specific place. For example, a tissue called AER (apical ectodermal ridge) forms at tips of a chick's limb buds and induces mesoderm beneath it to develop as a limb (Figure 43.10a). Whether a wing or a leg forms depends on positional information set down earlier in development (Figure 43.10b).

Take Home Message

What processes produce specialized cells, tissues, and organs?

■ Selective gene expression is the basis of cell differentiation. Signaling molecules contribute to differentiation. Morphogens diffuse through an embryo and have different effects depending on their concentration in each region.

■ Organs take shape as cells migrate, fold as sheets, and die on cue.

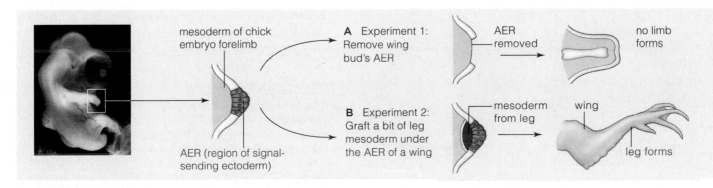

mesoderm of chick embryo forelimb

A Experiment 1: Remove wing bud's AER

AER removed

no limb forms

AER (region of signal-sending ectoderm)

B Experiment 2: Graft a bit of leg mesoderm under the AER of a wing

mesoderm from leg

wing

leg forms

Figure 43.10 Animated Control of limb formation in a chick. (**a**) Cells at the tip of a limb bud tell mesoderm under it to form a limb. Remove these AER cells and no limb forms. (**b**) Whether a limb becomes a wing or a leg depends on positional signals that the mesoderm received earlier.

■ Similarities in developmental pathways among animals are evidence of common ancestry.

■ Link to Homeotic genes 15.3

A General Model for Animal Development

Through studies of animals such as roundworms, fruit flies, fish, and mice, researchers have come up with a general model for development. The key point of the model is this: Where and when particular genes are expressed determines how an animal body develops.

First, molecules confined to different areas of an unfertilized egg induce localized expression of master genes in the zygote. Products of these master genes diffuse outward, so concentration gradients for these products form along the head-to-tail and dorsal-to-ventral axes of the developing embryo.

Second, depending on where they fall within these concentration gradients, cells in the embryo activate or suppress other master genes. The products of these genes become distributed in gradients, which affect other genes, and so on.

Third, this positional information affects expression of **homeotic genes**, genes that regulate development of specific body parts, as introduced in Section 15.3. All animals have similar homeotic genes. For example, a mouse's *eyeless* gene guides development of its eyes. Introduce the mouse version of this gene into a fruit fly, and eyes will form in tissues where the introduced gene is expressed.

Developmental Constraints and Modifications

The developmental model just described helps explain why we only see certain types of animal body plans. We know that body plans are influenced by physical constraints such as the surface-to-volume ratio. An animal cannot evolve large size unless it has circulatory and respiratory mechanisms to service body cells that reside far from the body surface.

There are also architectural constraints. These are constraints imposed by the existing body framework. For example, the first vertebrates on land had a body plan with four limbs. Evolution of wings in birds and bats occurred through modification of existing forelimbs, not by sprouting new limbs. Although it might be advantageous to have both wings and arms, no vertebrate with both has ever been discovered.

Finally, there are phyletic constraints on body plans. These constraints are imposed by interactions among genes that regulate development in a lineage. Once master genes evolved, their interactions determined the basic body form. Mutations that dramatically alter effects of these master genes are often lethal.

For example, vertebrates have paired bones and skeletal muscles along the body's head-to-tail axis. This pattern arises early in development, when the mesoderm on either side of the embryo's neural tube becomes divided into blocks of cells called **somites** (Figure 43.11). The somites will later develop into bones and skeletal muscles. A complex pathway involving many genes governs somite formation. Any mutation that disrupts this pathway and halts somite formation is lethal during development. Thus, we do not find vertebrates with an unsegmented body plan, although the number of somites does vary among species.

In short, mutations that affect development led to a variety of forms among animal lineages. These mutations brought about morphological changes through the modification of existing developmental pathways, rather than by blazing entirely new genetic trails.

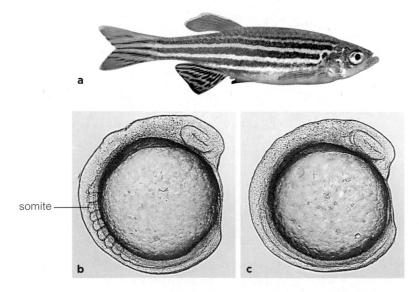

Figure 43.11 (**a**) An adult zebrafish. (**b**) Normal zebrafish embryo with somites that give rise to bone and muscle. (**c**) Mutant embryo that cannot form somites. It will die in early development.

Take-Home Message

Why are developmental processes and body plans similar among animal groups?

■ In all animals, cytoplasmic localization sets the stage for cell signaling. The signals activate sets of master genes shared by most animal groups. The products of these genes cause embryonic cells to form tissues and organs at certain locations.

■ Once a developmental pathway evolves, drastic changes to genes that govern this pathway are generally not favored.

43.6 | Overview of Human Development

- Like all animals, humans begin life as a single cell and go through a series of developmental stages.

- Links to Placental mammals 26.12, Human fertilization 42.8

Chapter 42 introduced the structure and function of human reproductive organs, and explained how an egg and sperm meet at fertilization to form a zygote (Section 42.8). The remaining sections of this chapter will continue this story, with an in-depth look at human development. In this section, we provide an overview of the process and define the stages that we will discuss. Prenatal (before birth) and postnatal (after birth) stages are listed in Table 43.1.

It takes about five trillion mitotic divisions to go from the single cell of a zygote to the ten trillion or so cells of an adult human. The process gets underway during a pregnancy that typically lasts an average of thirty-eight weeks from the time of fertilization.

The first cleavage occurs about 12 to 24 hours after fertilization. It takes about one week for a blastocyst to form. Again, a blastocyst is a mammalian blastula. In humans and other placental mammals, a blastocyst embeds itself in its mother's uterus. As the offspring develops, nutrients diffusing from the maternal bloodstream across the placenta sustain it (Section 26.12).

All major organs, including the sex organs, form during the embryonic period, which ends after eight weeks. The bones of the developing skeleton are laid down as cartilage models, which are then invaded by bone cells that convert the cartilage to bone.

At the end of the embryonic period, the developing individual is referred to as a **fetus**. In the fetal period, from the start of the ninth week until birth, organs grow and become specialized.

We divide the prenatal period into three trimesters. The first trimester includes months one through three; the second trimester is months four through six; the third trimester is months seven through nine.

Births before 37 weeks are considered premature. A fetus born earlier than 22 weeks rarely survives because its lungs are not yet fully mature. About half of births that occur before 26 weeks result in some sort of long-term disability.

After birth, the human body continues to grow and its body parts continue to change in proportion. Figure 43.12 shows the proportional changes during development. Postnatal growth is most rapid between 13 and 19 years. Sexual maturation occurs at puberty, and bones stop growing shortly thereafter. The brain is the last organ to become fully mature: Portions of it continue to develop until the individual is about 19 to 22 years old.

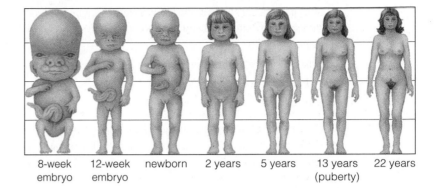

| 8-week embryo | 12-week embryo | newborn | 2 years | 5 years | 13 years (puberty) | 22 years |

Figure 43.12 Observable, proportional changes in prenatal and postnatal periods of human development. Changes in overall physical appearance are slow but noticeable until the teens.

Table 43.1 Stages of Human Development

Prenatal period

Zygote	Single cell resulting from fusion of sperm nucleus and egg nucleus at fertilization.
Blastocyst (blastula)	Ball of cells with surface layer, fluid-filled cavity, and inner cell mass.
Embryo	All developmental stages from two weeks after fertilization until end of eighth week.
Fetus	All developmental stages from ninth week to birth (about 38 weeks after fertilization).

Postnatal period

Newborn	Individual during the first two weeks after birth.
Infant	Individual from two weeks to fifteen months.
Child	Individual from infancy to about ten or twelve years.
Pubescent	Individual at puberty; secondary sexual traits develop. Girls, between 10 and 15 years; boys, between 11 and 16 years.
Adolescent	Individual from puberty until about 3 or 4 years later; physical, mental, emotional maturation.
Adult	Early adulthood (between 18 and 25 years); bone formation and growth finished. Changes proceed slowly after this.
Old age	Aging processes result in expected tissue deterioration.

Take-Home Message

How does human development proceed?

- Humans are placental mammals, so offspring develop in the mother's uterus.
- By the end of the second week, the blastocyst is embedded in the uterus.
- By the end of the eighth week, the embryo has all the typical human organs.
- Most of a pregnancy is taken up with the fetal period, during which organs grow and take on their specialized functions.

■ After a human blastocyst forms, it burrows into the wall of its mother's uterus and a system of membranes forms outside the embryo.

■ Link to Human fertilization 42.8

Cleavage and Implantation

Fertilization of a human egg typically occurs in one of the oviducts. Cleavage gets underway within a day or two of fertilization, as the zygote travels through the oviduct toward the uterus (Figure 43.13a,b). By the time it reaches the uterus, the zygote has become a cluster of sixteen cells called a **morula** (Figure 43.13c).

A blastocyst of a few hundred cells forms by the fifth day. It consists of an outer layer of cells, a cavity filled with their secretions (a blastocoel), and an inner cell mass (Figure 43.13d). The embryo develops from the inner cell mass. The outer cells will help form membranes that surround the developing embryo.

About six days after fertilization, the blastocyst is usually in the uterus. It now expands by cell divisions and uptake of fluid. This increase in size ruptures the noncellular zona pellucida, allowing the blastocyst to slip out of this enclosing layer. **Implantation** begins when the blastocyst attaches to the endometrium and

burrows into it. During implantation, the inner cell mass develops into two flattened layers of cells called the embryonic disk (Figure 43.13e,f).

In an ectopic pregnancy, the blastocyst implants in tissue other than the uterus—most commonly an oviduct. Such a pregnancy cannot be carried to term and must be removed surgically to protect the life of the mother. Use of birth control pills, a history of sexually transmitted disease, and certain inflammatory disorders increase the risk of ectopic pregnancy.

Extraembryonic Membranes

Membranes start forming outside the embryo during implantation (Table 43.2). A fluid-filled amniotic cavity opens up between the embryonic disk and part of the blastocyst surface (Figure 43.13f). Many cells migrate around the wall of the cavity and form the **amnion**, a membrane that will enclose the embryo. Fluid in the cavity will function as a buoyant cradle in which an embryo can grow, move freely, and be protected from abrupt temperature changes and any potentially jarring impacts.

As the amnion forms, other cells migrate around the inner wall of the blastocyst and form a lining that

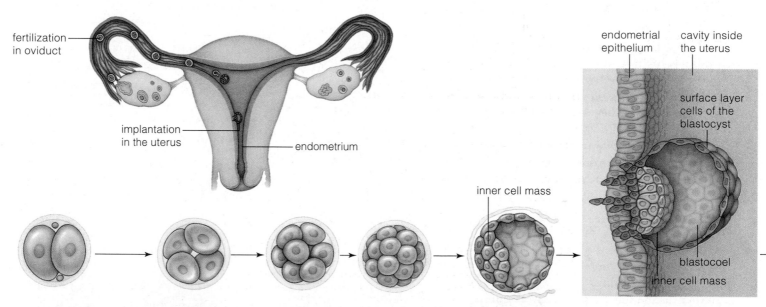

fertilization in oviduct

implantation in the uterus

endometrium

inner cell mass

endometrial epithelium

cavity inside the uterus

surface layer cells of the blastocyst

blastocoel

inner cell mass

A DAYS 1–2. The first cleavage furrow extends between the two polar bodies. Later cuts are angled, so cells become asymmetrically arranged. Until the eight-cell stage forms, they are loosely organized, with space between them.

B DAY 3. After the third cleavage, cells abruptly huddle into a compacted ball, which tight junctions among the outer cells stabilize. Gap junctions formed along the interior cells enhance intercellular communication.

C DAY 4. By 96 hours there is a ball of sixteen to thirty-two cells shaped like a mulberry. It is a morula (after *morum*, Latin for mulberry). Cells of the surface layer will function in implantation and will give rise to a membrane, the chorion.

D DAY 5. A blastocoel (fluid-filled cavity) forms in the morula as a result of surface cell secretions. By the thirty-two-cell stage, differentiation is occurring in an inner cell mass that will give rise to the embryo proper. This embryonic stage is the blastocyst.

E DAYS 6–7. Some of the blastocyst's surface cells attach themselves to the endometrium and start to burrow into it. Implantation has started.

actual size

becomes the yolk sac. In reptiles and birds, this sac holds yolk. In humans, cells of the yolk sac give rise to the embryo's blood cells and to germ cells.

Before a blastocyst is fully implanted, spaces that open in maternal tissues become filled with blood seeping in from ruptured capillaries. In the blastocyst, a new cavity opens up around the amnion and yolk sac. The lining of this cavity becomes the **chorion**, a membrane that is folded into many fingerlike projections that extend into blood-filled maternal tissues. It will become part of the placenta. The **placenta** is an organ that functions in exchanges of materials between the bloodstreams of a mother and her developing child.

After the blastocyst is implanted, an outpouching of the yolk sac will become the fourth extraembryonic membrane—the **allantois**. It gives rise to the urinary bladder and the placenta's blood vessels.

Early Hormone Production

Once implanted, a blastula releases **human chorionic gonadotropin (HCG)**. This hormone causes the corpus luteum to keep secreting progesterone and estrogens. These hormones prevent menstruration and maintain the uterine lining. After about three months, the placenta takes over the secretion of HCG.

Table 43.2 Human Extraembryonic Membranes

Amnion	Encloses, protects embryo in a fluid-filled, buoyant cavity
Yolk sac	Becomes site of red blood cell formation; germ cell source
Chorion	Lines amnion and yolk sac, becomes part of placenta
Allantois	Source of urinary bladder and blood vessels for placenta

HCG can be detected in a mother's urine as early as the third week of pregnancy. At-home pregnancy tests include a treated "dipstick" that changes color when exposed to urine that contains HCG.

Take-Home Message

What occurs during the first two weeks of human development?

■ Cleavage produces a morula and then a blastocyst, which slips out of the zona pellucida and implants itself in the endometrium, the lining of the uterus.

■ During implantation, projections from the blastocyst grow into maternal tissues. Connections that will support the developing embryo begin to form.

■ The inner cell mass of the blastocyst will become the embryo. Other portions of the blastocyst give rise to four external membranes. The outermost of these is the amnion, which encloses and protects the embryo in a fluid-filled cavity.

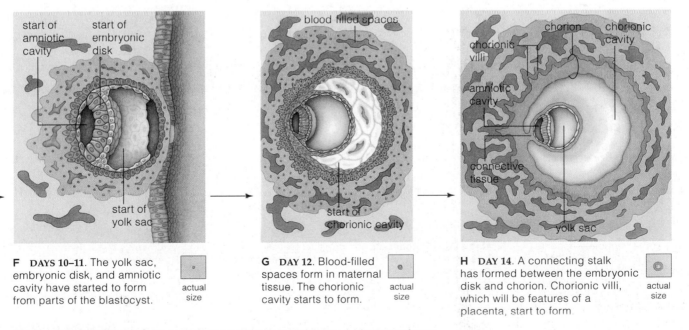

F **DAYS 10–11.** The yolk sac, embryonic disk, and amniotic cavity have started to form from parts of the blastocyst.

G **DAY 12.** Blood-filled spaces form in maternal tissue. The chorionic cavity starts to form.

H **DAY 14.** A connecting stalk has formed between the embryonic disk and chorion. Chorionic villi, which will be features of a placenta, start to form.

Figure 43.13 Animated From fertilization through implantation. A blastocyst forms, and its inner cell mass becomes an embryonic disk two cells thick. It will later become the embryo. Three extraembryonic membranes (the amnion, chorion, and yolk sac) start forming. A fourth membrane (allantois) forms after implantation is over.

43.8 | Emergence of the Vertebrate Body Plan

■ Gastrulation occurs in the third week as the embryo sets off down the pathway of typical vertebrate development.

■ Link to Coelomic cavity 25.1

By two weeks after fertilization, the inner cell mass of a blastocyst is a two-layered embryonic disk. During gastrulation in the third week, cells migrate inward along a depression, the primitive streak, that forms on the disk's surface (Figure 43.14a). The resulting three germ layers of the gastrula are the forerunners of all tissues (Table 43.3). The primitive streak's location establishes the body's head-to-tail axis.

Many master genes are now being expressed and the tissues and organs are beginning to take shape. For example, by the eighteenth day after fertilization,

the embryonic disk has two folds that will merge into a neural tube, which will develop into the spinal cord and brain (Figure 43.14b). Mesodermal folding also forms a notochord, which acts as a structural model for the bony segments of the vertebral column.

Spina bifida is a birth defect in which the neural tube and one or more vertebrae do not form as they should. As a result, the spinal cord protrudes out of the vertebral column at birth.

By the end of the third week, somites form. These paired segments of mesoderm will develop into bones, skeletal muscles of the head and trunk, and overlying dermis of the skin. Pharyngeal arches (Figure 43.14c) that start to form at this time will later contribute to the pharynx, larynx, and the face, neck, mouth, and nose. Small spaces begin to open up in certain parts of the mesoderm; these spaces will eventually interconnect as a coelomic cavity (Section 25.1).

Table 43.3 Derivatives of Human Germ Layers

Ectoderm (outer layer)	Outer layer (epidermis) of skin; nervous tissue
Mesoderm (middle layer)	Connective tissue of skin; skeletal, cardiac, smooth muscle; bone; cartilage; blood vessels; urinary system; gut organs; peritoneum (coelom lining); reproductive tract
Endoderm (inner layer)	Lining of gut and respiratory tract, and organs derived from these linings

Take-Home Message

What happens during weeks three and four of a pregnancy?

■ Gastrulation takes place, producing a three-layered embryo.

■ The neural tube and notochord form.

■ Somites appear on either side of the neural tube.

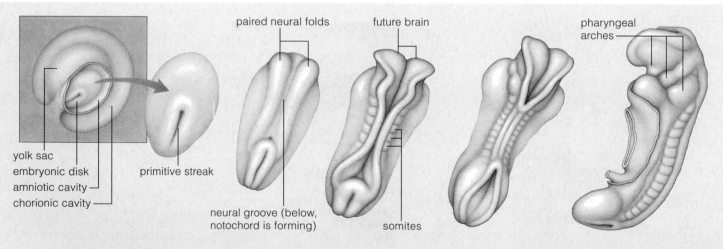

A DAY 15. A faint band appears around a depression along the axis of the embryonic disk. This band is the primitive streak, and it marks the onset of gastrulation in vertebrate embryos.

B DAYS 18–23. Organs start to form through cell divisions, cell migrations, tissue folding, and other events of morphogenesis. Neural folds will merge to form the neural tube. Somites (bumps of mesoderm) appear near the embryo's dorsal surface. They will give rise to most of the skeleton's axial portion, skeletal muscles, and much of the dermis.

C DAYS 24–25. By now, some embryonic cells have given rise to pharyngeal arches. These will contribute to the formation of the face, neck, mouth, nasal cavities, larynx, and pharynx.

Figure 43.14 Hallmarks of the embryonic period of humans and other vertebrates. A primitive streak and then a notochord form. Neural folds, somites, and pharyngeal arches form later. (**a,b**) Dorsal views of the embryo's back. (**c**) Side view.

43.9 | The Function of the Placenta

■ The placenta allows transfer of substances between a mother and her developing child without mixing their blood.

All exchange of materials between an embryo and its mother takes place by way of the placenta, a pancake-shaped, blood-engorged organ that consists of uterine lining and extraembryonic membranes. At full term, the placenta covers about one-fourth of the uterus's inner surface (Figure 43.15).

The placenta begins forming early in pregnancy. By the third week, maternal blood has begun to pool in spaces in the endometrial tissue. Chorionic villi—tiny fingerlike projections from the chorion—extend into the pools of maternal blood.

Embryonic blood vessels extend outward through the umbilical cord to the placenta, and then into the chorionic villi. Embryonic blood exchanges substances with maternal blood, but the two bloodstreams do not mix. If they mixed, some maternal antibodies could attack the embryo. Oxygen and nutrients diffuse from maternal blood and into embryonic blood vessels in the villi. Wastes diffuse the other way, and the mother's body disposes of them.

After the third month, the placenta produces large amounts of HCG, progesterone, and estrogens. These hormones encourage the ongoing maintenance of the uterine lining.

Take-Home Message

What is the function of the placenta?

■ Vessels of the embryo's circulatory system extend through the umbilical cord to the placenta, where they run through pools of maternal blood.

■ Maternal and embryonic blood do not mix; substances diffuse between the maternal and embryonic bloodstreams.

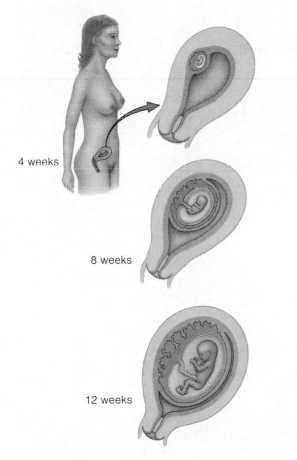

4 weeks

8 weeks

12 weeks

Figure 43.15 Relationship between fetal and maternal blood circulation in a full-term placenta. Blood vessels extend from the fetus, through the umbilical cord, and into chorionic villi. Maternal blood flows into spaces between villi. However, the two bloodstreams do not intermingle. Oxygen, carbon dioxide, and other small solutes diffuse across the placental membrane surface.

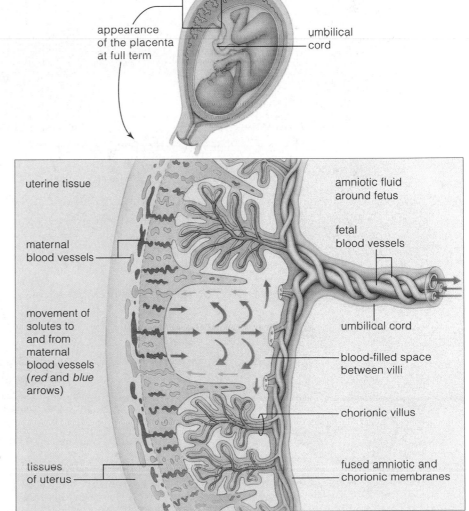

appearance of the placenta at full term

umbilical cord

uterine tissue

maternal blood vessels

movement of solutes to and from maternal blood vessels (*red* and *blue* arrows)

tissues of uterus

amniotic fluid around fetus

fetal blood vessels

umbilical cord

blood-filled space between villi

chorionic villus

fused amniotic and chorionic membranes

43.10 Emergence of Distinctly Human Features

- A human embryo's tail and pharyngeal arches label it as a chordate. The features disappear during fetal development.
- Link to Sex organ formation 12.1

When the fourth week ends, the embryo is 500 times the size of a zygote, but still less than 1 centimeter long. Growth slows as details of organs begin to fill in. Limbs form; paddles are sculpted into fingers and toes. The umbilical cord and the circulatory system develop. Growth of the head now surpasses that of all other regions (Figure 43.16). Reproductive organs begin forming, as described in Section 12.1. At the end of the eighth week, all organ systems have formed and we define the individual as a human fetus.

In the second trimester, reflexive movements begin as developing nerves and muscles connect. Legs kick, arms wave about, and fingers grasp. The fetus frowns,

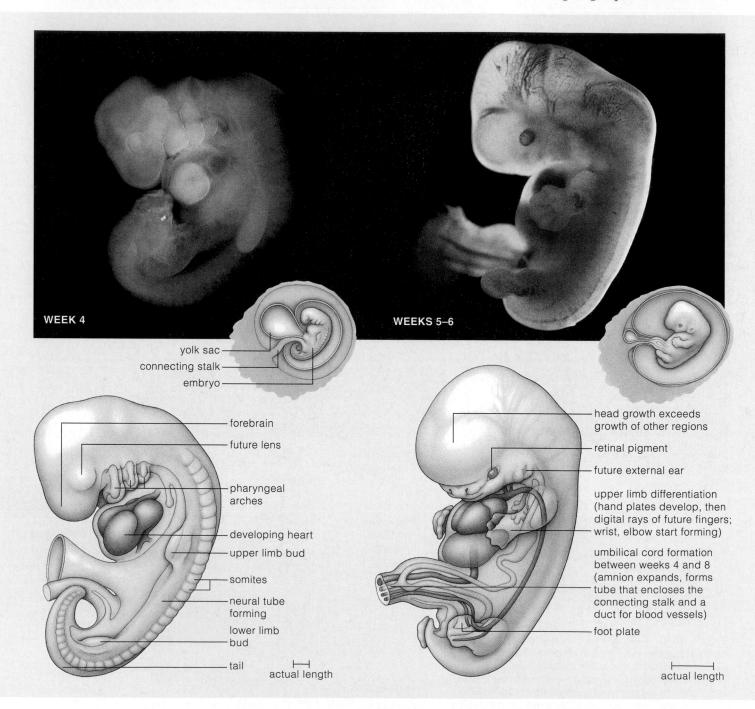

WEEK 4

yolk sac
connecting stalk
embryo

forebrain
future lens
pharyngeal arches
developing heart
upper limb bud
somites
neural tube forming
lower limb bud
tail

actual length

WEEKS 5–6

head growth exceeds growth of other regions
retinal pigment
future external ear
upper limb differentiation (hand plates develop, then digital rays of future fingers; wrist, elbow start forming)
umbilical cord formation between weeks 4 and 8 (amnion expands, forms tube that encloses the connecting stalk and a duct for blood vessels)
foot plate

actual length

Figure 43.16 Human embryo at successive stages of development.

squints, puckers its lips, sucks, and hiccups. When a fetus is five months old, its heartbeat can be heard clearly through a stethoscope positioned on the mother's abdomen. The mother can sense movements of fetal arms and legs.

By now, soft, fetal hair (lanugo) covers the skin; most will be shed before birth. A thick, cheesy coating (vermix) protects the skin from abrasion. In the sixth month, eyelids and eyelashes form. Eyes open during the seventh month, the start of the final trimester. By this time, all portions of the brain have formed and have begun to function.

Take-Home Message

What occurs during the late embryonic and the fetal periods?

■ The embryo takes on its human appearance by week eight but remains tiny. During the fetal period, organs begin functioning and growth is rapid.

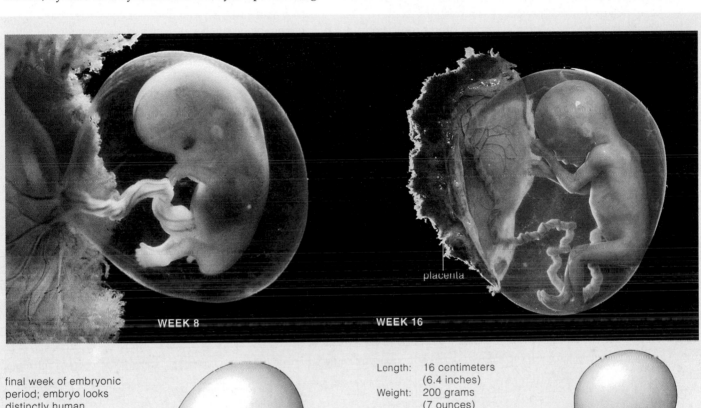

WEEK 8   WEEK 16   placenta

final week of embryonic period; embryo looks distinctly human compared to other vertebrate embryos

upper and lower limbs well formed; fingers and then toes have separated

primordial tissues of all internal, external structures now developed

tail has become stubby

actual length

Length: 16 centimeters (6.4 inches)
Weight: 200 grams (7 ounces)

WEEK 29
Length: 27.5 centimeters (11 inches)
Weight: 1,300 grams (46 ounces)

WEEK 38 (full term)
Length: 50 centimeters (20 inches)
Weight: 3,400 grams (7.5 pounds)

During fetal period, length measurement extends from crown to heel (for embryos, it is the longest measurable dimension, as from crown to rump).

Mother as Provider and Protector

■ An embryo depends on its mother to supply nutrients and is subjected to toxins or pathogens to which she is exposed.

■ Links to Thyroid hormone 35.6, Antibodies 38.6, Carbon monoxide 38.7

According to the Centers for Disease Control, about 3 percent of children born in the United States have some sort of birth defect. The defects include visible problems such as a cleft lip or a club foot, as well as internal problems such as heart or intestinal malformations. Some birth defects have a genetic basis, but others result from an environmental factor such as poor nutrition or exposure to a **teratogen**. A teratogen is a toxin or infectious agent that interferes with development. Figure 43.17 shows the periods when specific organs are the most vulnerable to damage by exposure to teratogens.

Nutritional Considerations A pregnant woman who eats a well-balanced diet supplies her future child with all of the proteins, carbohydrates, and lipids it needs for growth and

development. Her own demands for vitamins and minerals increase, but both are absorbed preferentially across the placenta and taken up by the embryo. Taking B-complex vitamins in early pregnancy reduces the embryo's risk of neural tube defects. Folate (folic acid) is especially critical in this respect.

Dietary deficiencies affect many developing organs. For example, if a mother does not get enough iodine, her new-born may be affected by cretinism, a disorder that affects brain function and motor skills (Section 35.6).

A diabetic woman who does not control her blood sugar during pregnancy provides excess sugar to her fetus. This excess can cause birth defects. Also, the fetus converts the extra sugar to fat and becomes unusually large. An oversized fetus can cause problems during delivery.

About Morning Sickness About two-thirds of pregnant women begin to have episodes of nausea with or without vomiting around the sixth week of pregnancy. Although commonly known as morning sickness, the symptoms can occur at any time of day. They typically end by the twelfth

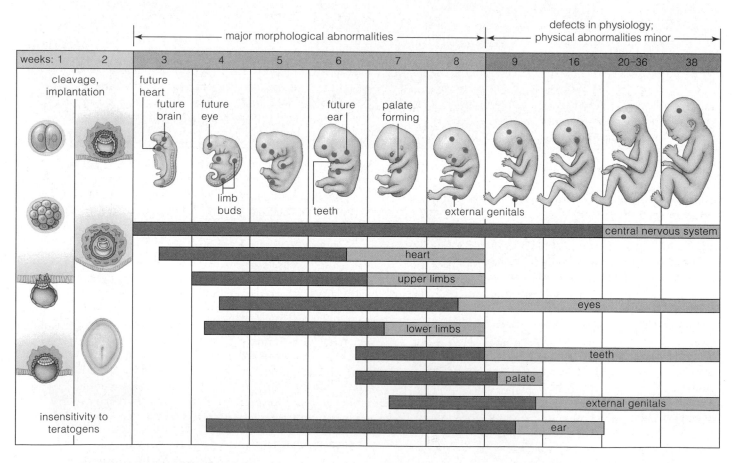

Figure 43.17 Teratogen sensitivity. Teratogens are drugs, infectious agents, and environmental factors that cause birth defects. *Dark blue* signifies the highly sensitive period for an organ or body part; *light blue* signifies periods of less severe sensitivity. For example, the upper limbs are most sensitive to damage during weeks 4 through 6, and somewhat sensitive during weeks 7 and 8.

Figure It Out: Is teratogen exposure in the 16th week more likely to affect the heart or the genitals? *Answer: Genitals*

week. Morning sickness generally does not cause problems and may have an adaptive function. Morning sickness most often occurs during the period when organs of the child she is carrying are developing and are most vulnerable to teratogens. Women who have morning sickness are less likely to miscarry than women who are not affected, and women who vomited were more likely to carry a child to term than those who only felt nauseous. The foods women who have morning sickness report they are most likely to avoid—fish, poultry, meat, and eggs—are the ones most likely to be tainted by dangerous microorganisms.

Infectious Agents Some antibodies in a pregnant woman's blood cross the placenta and protect an embryo or fetus from bacterial infections (Section 38.6). But some viral diseases can be dangerous in the early weeks after fertilization. Rubella, or German measles, is an example. A woman may sidestep the risk of passing on the rubella virus by getting vaccinated before she becomes pregnant.

A relative of the protist that causes malaria sometimes lurks in garden soil, cat feces, and undercooked meat. It causes toxoplasmosis. The disease often does not cause symptoms, so a pregnant woman may become infected and not realize it. If the parasite crosses the placenta, it can infect her child and lead to developmental problems, a miscarriage, or stillbirth. To minimize the risk, pregnant women should eat well-cooked meat and avoid cat feces.

Alcohol and Caffeine Alcohol passes across the placenta, so when a pregnant woman drinks alcohol, her embryo or fetus feels the effects. Alcohol exposure can cause fetal alcohol syndrome, or FAS. A small head and brain, facial abnormalities, slow growth, mental impairment, heart problems, and poor coordination characterize affected infants (Figure 43.18). The damage is permanent. Children affected by FAS never catch up, physically or mentally.

Most doctors now advise women who are pregnant or attempting to become pregnant to avoid alcohol entirely. Even before a woman knows she is pregnant, tissues of the embryonic nervous system have begun forming, and alcohol can damage them. Even moderate drinking during pregnancy increases risk of miscarriage and stillbirth.

Laboratory studies have shown that caffeine interferes with nervous system development in animals, and physicians have suspected that it may also harm human embryos. A recent study supports this hypothesis. The study showed that women who took in 200 milligrams a day of caffeine (the equivalent of one and a half cups of coffee), had twice as many miscarriages as those who avoided caffeine. The study's authors advise pregnant women to chose decaffeinated or caffeine-free beverages.

Smoking Smoking or exposure to secondhand smoke increases the risk of miscarriage and adversely affects fetal growth and development. Carbon monoxide in the smoke can outcompete oxygen for the binding sites on

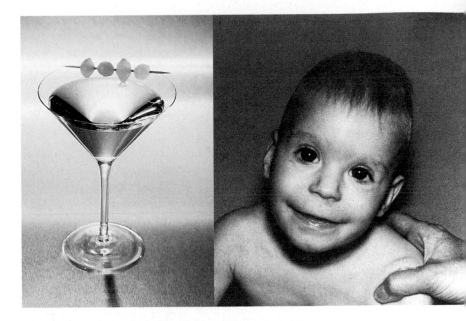

Figure 43.18 An child with fetal alcohol syndrome—FAS. Obvious symptoms are low and prominently positioned ears, improperly formed cheekbones, and an abnormally wide, smooth upper lip. Growth-related complications, heart problems, and nervous system abnormalities are also common.

hemoglobin (Section 38.7), so the embryo or fetus of a smoker gets less oxygen than that of a nonsmoker. In addition, levels of the addictive stimulant nicotine in amniotic fluid can be higher than those in the mother's blood.

Effects of maternal smoking persist long after birth. One British study tracked a group of children born in the same week over the course of seven years. More children of smokers died of postdelivery complications, and the survivors were smaller, with twice as many heart defects. When the study ended, the children of smokers were nearly half a year behind the normal reading age.

Prescription Drugs Some medications cause birth defects. For example, thalidomide was routinely prescribed to treat morning sickness during the 1960s in Europe. Infants of some of the women who used it during the first trimester had severely deformed arms and legs or none at all. The FDA never approved use of thalidomide for pregnant women in the United States.

Isotretinoin (Accutane) is widely used in the United States and elsewhere. This highly effective treatment for severe acne is often prescribed for young women. If taken early in a pregnancy, it can cause heart problems or facial and cranial deformities in the embryo.

Certain antidepressants increase the risk of birth defects. Paroxetine (Paxil) and related drugs inhibit the reuptake of serotonin. Use of these drugs during early pregnancy increases the likelihood of heart malformations. Taking them later in pregnancy increases risk that an infant will have fatal heart and lung disorders.

- As in other placental mammals, human fetuses are born live and are nourished with nutritious milk secreted from the mother's mammary glands. Shifts in the levels of hormones help control these processes.

- Links to Pituitary hormones 35.3, Positive feedback 27.3

Giving Birth

A mother's body changes as her fetus nears full term, at about 38 weeks after fertilization. Until the last few weeks, her firm cervix helped prevent the fetus from slipping out of her uterus prematurely. Now cervical connective tissue becomes thinner, softer, and more flexible. These changes will allow the cervix to stretch enough to permit the fetus to pass out of the body.

The birth process is known as **labor**. Typically, the amnion ruptures right before birth, so amniotic fluid drains out from the vagina. The cervical canal dilates. Strong contractions propel the fetus through it, then out through the vagina (Figure 43.19).

A positive feedback mechanism operates during labor. When the fetus nears full term, it typically shifts position so that its head puts pressure on the mother's cervix. Receptors inside the cervix sense pressure and signal the hypothalamus, which signals the posterior lobe of the pituitary to secrete **oxytocin**. In a positive feedback loop, oxytocin binds to smooth muscle of the uterus, causing uterine contractions that push the fetus against the cervix. The added pressure triggers more oxytocin secretion, which causes more contractions and more cervical stretching. Forceful uterine contractions continue until the fetus is forced through the cervix and out of the mother's body.

Strong muscle contractions also detach and expel the placenta from the uterus as the "afterbirth." The umbilical cord that connects the newborn to this mass of expelled tissue is clamped, cut short, and tied. The short stump of cord left in place withers and falls off. The navel marks the former attachment site.

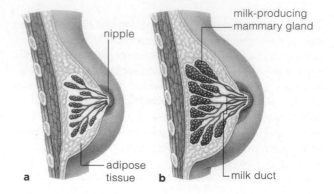

Figure 43.20 Cutaway views of (**a**) a breast of a woman who is not pregnant and (**b**) a breast of a lactating woman.

Nourishing the Newborn

Before a pregnancy, a woman's breast tissue is mostly adipose tissue. Milk ducts and mammary glands are small and inactive (Figure 43.20). During pregnancy, these structures enlarge in preparation for **lactation**, or milk production. **Prolactin**, a hormone secreted by the mother's anterior pituitary, triggers milk synthesis.

After birth, a decline in progesterone and estrogens causes milk production to go into high gear. The stimulus of a newborn's suckling causes the release of oxytocin. The hormone stimulates muscles around the milk glands to contract and force milk into the ducts.

Besides being nutrient-rich, human breast milk has antibodies that protect a newborn from some viruses and bacteria. Nursing mothers should remember that drugs, alcohol, and other toxins end up in milk.

Take-Home Message

What roles do hormones play in birth and lactation?

- During birth, the hormone oxytocin stimulates muscle contractions that force a fetus out of its mother's body.

- Prolactin stimulates milk production and oxytocin causes milk secretion from milk ducts.

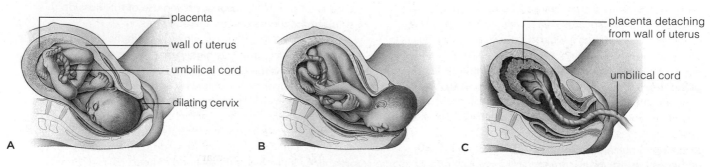

Figure 43.19 Animated Expulsion of (**a,b**) a human fetus and (**c**) afterbirth during labor. The afterbirth consists of the placenta, tissue fluid, and blood.

Multiple births that result from use of fertility drugs not only put off spring at risk, they also threaten the health of the mother. Among other things, carrying two or more fetuses requires a greater blood volume, which puts a strain on the mother's heart and raises her risk of high blood pressure. Also, such pregnancies require more placental area, raising the risk of dangerous blood loss when the placenta detaches after birth.

How would you vote?

Should use of fertility drugs be discouraged to prevent higher risk multiple pregnancies? See CengageNOW for details, then vote online.

Summary

Section 43.1 Most animal life cycles have six stages of development. Gametes form, then fertilization takes place. **Cleavage** produces a **blastula**. **Gastrulation** results in an early embryo (a **gastrula**) that has two or three primary tissue layers, or **germ layers**. Finally, organs form, and tissues and organs become specialized.

■ *Use the animation on CengageNOW to track development of a frog.*

Section 43.2, 43.3 **Cytoplasmic localization**, storage of different substances in different parts of the cytoplasm, is a feature of all oocytes. Cleavage distributes different portions of the egg cytoplasm to different cells. Patterns of cleavage vary among animal lineages. Cleavage ends with formation of a blastula. The mammalian blastula is a **blastocyst**, which has a fluid-filled cavity and an inner cell mass. The inner cell mass will become the embryo.

During gastrulation, cell rearrangement produces layers of tissue. Most often, three tissue layers form: outer **ectoderm**, inner **endoderm**, and **mesoderm** in between ectoderm and endoderm. Gastrulation is controlled by signal-sending cells that cause movement of neighboring cells. This type of signalling interaction is an example of **embryonic induction**.

■ *Use the animation on CengageNOW to learn about cytoplasmic localization and control of gastrulation.*

Sections 43.4, 43.5 Selective gene expression leads to **cell differentiation**: cells become specialized by activating different subsets of their genome. **Morphogens**, products of master genes, act as long-range signals that diffuse out from a source and form a concentration gradient. This gradient affects which genes a cell turns on or off.

Morphogenesis, the formation of tissues and organs, occurs as cells migrate, change shape, and undergo programmed cell death (**apoptosis**). Development of organs and limbs in particular places is **pattern formation**. Cues about position play a role in pattern formation.

A general model for animal development is based on comparative studies. By this model, cytoplasmic localization in an oocyte causes localized expression of master genes in the zygote. Diffusion of morphogens products of these master genes—creates gradients that cause the differential expression of other genes such as **homeotic genes**, which govern the formation of specific body parts.

Master genes are similar among all major animal groups. Developmental changes are constrained by interactions among master genes, as well as by physical and architectural factors. For example, in all vertebrates, paired blocks of mesoderm called **somites** form and give rise to paired skeletal muscles and bone.

■ *Use the animation on CengageNOW to learn how the neural tube forms and how a chick wing develops.*

Section 43.6 Human prenatal development takes nine months. Organs take shape during the embryonic period, which is over at the end of the eighth week. For the remainder of the pregnancy, the **fetus** grows larger and organs take on their specialized roles. Growth and development continue after birth (in the postnatal period).

Sections 43.7–43.11 Human fertilization usually occurs inside an oviduct. Cleavage produces a **morula**, then a blastocyst. During **implantation**, a blastocyst buries itself in the uterine wall. Membranes form outside the blastocyst and support its development. The **amnion** encloses and protects the embryo in a fluid-filled sac. The **chorion** and **allantois** become part of the **placenta**, the organ that allows exchange of substances between the maternal and fetal bloodstreams. An implanted blastula makes **human chorionic gonadotropin**, a hormone that prevents menstruation and thus maintains the pregnancy.

Gastrulation occurs after implantation. The first organ to form, the neural tube, later becomes the brain and spinal cord. Somites form on either side of the neural tube.

By the end of the eighth week, the embryo has lost its tail and pharyngeal arches and has a distinctly human appearance. It continues to grow in size and its organs continue to mature during the fetal period.

Nutrients and antibodies move across the placenta from mother to embryo or fetus, as do **teratogens**, which can cause birth defects.

■ *Use the animation on CengageNOW to observe the events of human development.*

Section 43.12 Hormones typically induce **labor** at about 38 weeks. Positive feedback controls secretion of **oxytocin**, a hormone that causes contractions that expel the fetus and then the afterbirth. **Prolactin** regulates the maturation of the mammary glands and then oxytocin causes **lactation**.

■ *Use the animation on CengageNOW to observe labor.*

Data Analysis Exercise

People considering fertility treatments should be aware that such treatments raise the risk of multiple births, and that multiple pregnancies are associated with an increased risk of some birth defects.

Figure 43.21 shows the results of Yiwei Tang's study of birth defects reported in Florida from 1996 to 2000. Tang compared the incidence of various defects among single and multiple births. She calculated the relative risk for each type of defect based on type of birth, and corrected for other differences that might increase risk such as maternal age, income, race, and medical care during pregnancy. A relative risk of less than 1 means that multiple births pose less risk of that defect occurring. A relative risk greater than 1 means multiples are more likely to have a defect.

1. What was the most common type of birth defect in the single-birth group?

2. Was that type of defect more or less common among the multiple-birth newborns than among single births?

3. Tang found that multiples have more than twice the risk of single newborns for one type of defect. Which type?

4. Does a multiple pregnancy increase the relative risk of chromosomal defects in offspring?

	Prevalence of Defect		Relative Risk
	Multiples	Singles	
Total birth defects	358.50	250.54	1.46
Central nervous system defects	40.75	18.89	2.23
Chromosomal defects	15.51	14.20	0.93
Gastrointestinal defects	28.13	23.44	1.27
Genital/urinary defects	72.85	58.16	1.31
Heart defects	189.71	113.89	1.65
Musculoskeletal defects	20.92	25.87	0.92
Fetal alcohol syndrome	4.33	3.63	1.03
Oral defects	19.84	15.48	1.29

Figure 43.21 Prevalence, per 10,000 live births, of various types of birth defects among multiple and single births. Relative risk for each defect is given after researchers adjusted for the mother's age, race, previous adverse pregnancy experience, education, Medicaid participation during pregnancy, as well as the infant's sex and number of siblings.

Self-Quiz   *Answers in Appendix III*

1. The typical end product of cleavage is a _____ .
   a. zygote   c. gastrula
   b. blastula   d. gamete

2. Is this statement true or false? Materials are randomly distributed in egg cytoplasm, so cleavage parcels out same kinds of cytoplasmic components to all cells.

3. Cells differentiate as a direct result of _____ .
   a. selective gene expression   c. gastrulation
   b. morphogenesis   d. all of the above

4. _____ help bring about morphogenesis.
   a. Cell migrations   c. Cell suicide
   b. Changes in cell shape   d. all of the above

5. Match each term with the most suitable description.
   ___ apoptosis   a. blastomeres form
   ___ embryonic induction   b. cellular rearrangements
   ___ cleavage   form primary tissues
   ___ gastrulation   c. cells die on cue
   ___ pattern formation   d. cells influence neighbors
                            e. tissues, organs emerge in
                               the correct places

6. A _____ implants in the lining of the human uterus.
   a. zygote   b. gastrula   c. blastocyst   d. fetus

7. The _____ , a fluid-filled sac, surrounds and protects an embryo and keeps it from drying out.
   a. amnion   b. allantois   c. yolk sac   d. chorion

8. At full term, a placenta _____ .
   a. is composed of extraembryonic membranes alone
   b. directly connects maternal and fetal blood vessels
   c. keeps maternal and fetal blood vessels separated

9. During the second trimester of pregnancy, _____ .
   a. gastrulation ends   b. eyes open   c. heartbeats start

10. _____ stimulates milk synthesis in mammary glands.
   a. HCG   c. Testosterone
   b. Prolactin   d. Oxytocin

11. Number these events in human development in the correct order.
   ___ gastrulation occurs
   ___ blastocyst forms
   ___ morula forms
   ___ zygote forms
   ___ neural tube forms
   ___ pharyngeal arches form

12. _____ gives rise to skeletal muscle and bone.
   a. Mesoderm   c. Ectoderm
   b. Endoderm   d. all of the above

■ *Visit CengageNOW for additional questions.*

Critical Thinking

1. By UNICEF estimates, each year 110,000 people are born with birth defects as a result of prenatal rubella infections. Deafness and blindness only occur if the mother becomes infected during the first trimester of pregnancy. Why?

2. The most common ovarian tumors in young women are ovarian teratomas. The name comes from the Greek word *teraton*, which means monster. What makes these tumors "monstrous" is the presence of well-differentiated tissues, most commonly bones, teeth, fat, and hair. Early physicians suggested that teratomas arose as a result of nightmares, witchcraft, or intercourse with the devil. Unlike all other tumors, which arise from somatic cells, teratomas arise from germ cells. Explain why a tumor derived from a germ cell is able to produce more differentiated cell types than one derived from a somatic cell.

VII PRINCIPLES OF ECOLOGY

Lioness and her cub at sunset on the African savanna. What are the consequences of their interactions with each other, with other kinds of organisms, and with their environment? By the end of this last unit, you might find worlds within worlds in such photographs.

44 Animal Behavior

My Pheromones Made Me Do It

One spring day as Toha Bergerub was walking down a street near her Las Vegas home, she felt a sharp pain above her right eye—then another, and another. Within a few seconds, hundreds of stinging bees covered the upper half of her body. Firefighters in protective gear rescued her, but not before she was stung more than 500 times. Bergerub, who was seventy-seven years old at the time, spent a week in the hospital, but recovered fully.

Bergerub's attackers were Africanized honeybees, a hybrid between gentle European honeybees and a more aggressive subspecies native to Africa (Figure 44.1). Bee breeders had imported African bees to Brazil in the 1950s. The breeders thought cross-breeding might yield a mild-tempered but more active pollinator for commercial orchards. However, some African bees escaped and mated with European honeybees that had become established in Brazil before them.

Then, in a grand example of geographic dispersal, some descendants of the hybrids buzzed all the way from Brazil to Mexico and on into the United States. So far, they have settled in Texas, New Mexico, Nevada, Utah, California, Oklahoma, Louisiana, Alabama, and Florida.

Africanized honeybees became known as "killer bees," although they rarely kill humans. They have been in the United States since 1990, yet no more than fifteen people have died after being attacked.

All honeybees defend their hives by stinging. Each can sting only once, and all make the same kind of venom. Even so, compared with European honeybees, Africanized ones get riled up more easily, attack in greater numbers, and stay agitated longer. Some are known to have chased people for more than a quarter of a mile.

What makes Africanized bees so testy? Part of the answer is that they have a heightened response to alarm pheromone. A pheromone is a social cue, a type of chemical signal that is emitted by one individual and influences another individual of the same species. For instance, when a honeybee worker guarding the entrance to a hive senses an intruder, it releases alarm pheromone. Pheromone molecules diffuse through the air and excite other bees, which fly out and sting the intruder.

In one study, researchers tested hundreds of colonies of Africanized honeybees and European honeybees to quantify their responses to alarm pheromone. The researchers positioned a seemingly threatening object, such as a scrap of black cloth, near the entrance of each hive. Then they released a small quantity of an artificial alarm pheromone. The Africanized bees flew out of their hive and zeroed in on the perceived threat much faster. Those bees plunged six to eight times as many stingers into the target.

The two strains of honeybees also show other behavioral differences. Africanized bees are less picky about where they establish a colony. They are more likely to abandon their hive after a disturbance. Of greater concern to beekeepers, the Africanized bees are less interested in storing large amounts of honey.

Such differences among honeybees lead us into the world of animal behavior—to coordinated responses that animal species make to stimuli. We invite you to reflect first on behavior's genetic basis, which is the foundation for its instinctive and learned mechanisms. Along the way, you will also come across examples of the adaptive value of behavior.

See the video! Figure 44.1 Two Africanized honeybees stand guard at their hive entrance. If a threat appears, they will release an alarm pheromone that stimulates hivemates to join an attack.

Key Concepts

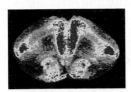

Foundations for behavior

Behavioral variations within or among species often have a genetic basis. Behavior can also be modified by learning. When behavioral traits have a heritable basis, they can evolve by way of natural selection. **Sections 44.1–44.3**

Animal communication

Interactions between members of a species depend on evolved modes of communication. Communication signals hold clear meaning for both the sender and the receiver of signals. **Section 44.4**

Mating and parental care

Behavioral traits that affect the ability to attract and hold a mate are shaped by sexual selection. Males and females are subject to different selective pressure. Parental care can increase reproductive success, but it has energetic costs. **Section 44.5**

Costs and benefits of social behavior

Life in social groups has reproductive benefits and costs. Self-sacrificing behavior has evolved among a few kinds of animals that live in large family groups. Human behavior is influenced by evolutionary factors, but humans alone make moral choices. **Sections 44.6–44.8**

Links to Earlier Concepts

- This chapter builds on your knowledge of sensory and endocrine systems (Sections 34.1, 35.3). We will discuss the role of hormones in lactation (43.12) and other behaviors. We will also look in more detail at pheromones 35.1.

- You may wish to review the concepts of adaptation (17.3) and sexual selection (18.6). You will see another example of the use of knockout experiments (15.3).

- You will be reminded again of the limits of science (1.5), and the rise of cultural traits (26.13).

How would you vote? Africanized bees are expanding their range in North America. Learning more about them may help us devise ways to protect ourselves. Should research into the genetic basis of their behavior be a high priority? See CengageNOW for details, then vote online.

- Variations in behavior within or among species often have their basis in genetic differences.

- Links to Knockout experiments 15.3, Sensory systems 34.1, Pituitary hormones 35.3, Lactation 43.12

How Genes Affect Behavior

Animal behavior requires a capacity to detect stimuli. A **stimulus**, recall, is some type of information about the environment that a sensory receptor has detected (Section 34.1). Which types of stimuli an animal is able to detect and the types of responses it can make start with the structure of its nervous system. Differences in genes that affect the structure and activity of the nervous system cause many differences in behavior.

Keep in mind, however, that mutations that affect metabolism or structural traits also influence behav-

Figure 44.2
(a) Banana slug, the food of choice for adult garter snakes of coastal California. (b) A newborn garter snake from a coastal population, tongue-flicking at a cotton swab that had been drenched with fluids from a banana slug.

Characteristics	Rover	Sitter
Foraging behavior	Switches feeding area frequently	Tends to feed in one area
Genotype	*FF* or *Ff*	*ff*
PKG (enzyme) level	Higher	Lower
Speed of learning olfactory cues	Faster	Slower
Long-term memory for olfactory cues	Shorter	Longer

Figure 44.3 Characteristics of rovers and sitters, two behavioral phenotypes that occur in wild fruit fly populations. The two types differ in foraging behavior, learning, and memory, but not in general activity level. When food is not present, rovers and sitters are equally likely to move about.

ior. For example, suppose you notice that some birds routinely eat large seeds and others focus on small seeds. Those that eat large seeds might do so because they cannot detect the smaller seeds. Or, they might see but ignore small seeds because the structure of their beaks allows them to easily open larger ones.

Studying Variation Within a Species

One way to investigate the genetic basis of behavior is to examine behavioral differences among members of a single species. For example, Stevan Arnold studied feeding behavior in two populations of garter snakes. Some garter snakes live in coastal forests of the Pacific Northwest and their preferred food is banana slugs, which are common on the forest floor (Figure 44.2a). Farther inland, there are no banana slugs and the garter snakes prefer to eat fishes and tadpoles. Were these prey preferences inborn? To find out, Arnold offered newborn garter snakes of both populations a banana slug as their first meal. Most offspring of coastal snakes ate it. Offspring of inland snakes usually ignored it.

Newborn coastal snakes also flicked their tongue more often at a cotton swab soaked in slug juices, as in Figure 44.2b. (Tongue-flicking pulls molecules into the mouth.) Arnold hypothesized that inland snakes lack the genetically determined ability to associate the scent of slugs with "FOOD!" He predicted that if coastal garter snakes were crossed with inland snakes, the resulting offspring would make an intermediate response to slug odors. Results from his experimental crosses confirmed this prediction. Hybrid baby snakes tongue-flick at cotton swabs with slug juices more than newborn inland snakes do, but not as often as newborn coastal snakes do. Exactly which gene or genes underlie this difference has not been determined.

We do know about one gene that influences feeding behavior in fruit flies (*Drosophila melanogaster*). Marla Sokolowski showed that in wild fruit fly populations about 70 percent of the flies are "rovers"; they tend to move from place to place when food is present. About 30 percent of flies are "sitters"; they tend to feed in one place. Genotype at the *foraging* (*for*) locus determines whether a fly is rover or a sitter. Flies that have the dominant allele (*F*) are rovers. Those homozygous for the recessive allele (*f*) are sitters.

Sokolowski went on to uncover the molecular basis for the observed differences in behavior. She showed that the *for* gene encodes a cGMP-dependent protein kinase (PKG). This enzyme activates other molecules by donating a phosphate group to them, and it plays a role in many intercellular signaling pathways. Rovers make a bit more PKG than sitters. Having more PKG in the brain allows rovers to learn about new odors faster than sitters, but it also makes rovers forget what they learned faster. Figure 44.3 summarizes genotypes and behaviors of the rover and sitter phenotypes.

Examples such as this one, in which researchers can point to a single gene as the predominant cause of natural variations in behavior, are extremely rare. More typically, differences in many genes and exposure to different environmental factors cause members of a species to differ in their behavior.

Comparisons Among Species

Comparing behavior of related species can sometimes help clarify the genetic basis of a behavior. For instance, all mammals secrete the pituitary hormone oxytocin (OT), which acts in labor and lactation (Section 35.3). In many mammals, OT also influences pair bonding, aggression, territoriality, and other forms of behavior.

Among small rodents called prairie voles (*Microtus ochrogaster*), OT is the hormonal key that unlocks the female's heart. The female bonds with a male after a night of repeated matings, and she mates for life. In one experimental test of OT's influence, researchers injected pair-bonded female prairie voles with a drug that blocks OT action. Females that got the injection immediately dumped their partners.

Genetic differences in the number and distribution of OT receptors may help explain differences in mating systems among vole species. For example, prairie voles, which are monogamous and mate for life, have more OT receptors than mountain voles (*M. montanus*), which are highly promiscuous (Figure 44.4).

Compared to males of promiscuous vole species, males of monogamous species also have more antidiuretic hormone (ADH) receptors in their forebrain. To test the effect of this difference, scientists isolated the gene for the ADH receptor in prairie voles. They then used a virus to add copies of this gene into the forebrain of some naturally promiscuous male meadow voles (*M. pennsylvanicus*). Results confirmed the role of ADH receptors in monogamy. Experimentally treated males preferred a female with whom they had mated over a new one. Control males that received the gene in a different brain region or virus with a different gene showed no preference for a familiar partner.

Knockouts and Other Mutations

Study of mutations can also help researchers understand behavior. As an example, fruit fly males with

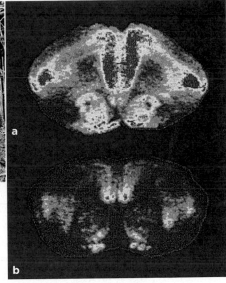
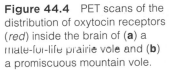

Figure 44.4 PET scans of the distribution of oxytocin receptors (*red*) inside the brain of (**a**) a male-for-life prairie vole and (**b**) a promiscuous mountain vole.

a mutation in the *fruitless* (*fru*) gene do not perform normal courtship movements and they court males in addition to females. When researchers compared the brains of male *fru* mutants to brains of normal males, they found the mutants—like normal females—lacked a certain set of neurons. Apparently development of that set of neurons has an integral role in governing typical male mate preference and courtship behavior.

As another example, knockout experiments (Section 15.3) confirmed the importance of oxytocin in mouse maternal behavior. Researchers produced female mice in which the gene for the OT receptor was knocked out. Lacking a functional receptor for OT, these mice could not respond to this hormone. As expected, these females did not lactate; oxytocin is required for contraction of milk ducts (Section 43.12). Knockout females also were less likely than normal mice to retrieve pups that researchers moved out of the nest. Based on these results, researchers concluded that oxytocin is required for normal maternal behavior in mice.

Take-Home Message

How do researchers study the effect of genes on animal behavior?

■ Studying variations in behavior within a species or among related species allows researchers to determine whether the variation has a genetic basis. Such differences are rarely caused by variation in a single gene; many genes affect behavior.

■ Researchers sometimes can determine the effect of a gene on a specific behavior by studying individuals in which the gene is nonfunctional.

- Some behaviors are inborn and can be performed without any practice.
- Most behaviors are modified as a result of experience.

Instinctive Behavior

All animals are born with the capacity for **instinctive behavior**—an innate response to a specific and usually simple stimulus. A newborn coastal garter snake behaves instinctively when it attacks a banana slug. A male fruit fly instinctively waves its wings during courtship of a female.

The life cycle of the cuckoo bird provides several examples of instinct at work. This European bird is a social parasite. Females lay eggs in nests of other birds. A newly hatched cuckoo is blind, but contact with an egg laid by its foster parent stimulates an instinctive response. That hatchling maneuvers the egg onto its back, then shoves it out of the nest (Figure 44.5a). This behavior removes any potential competition for the foster parent's attention.

A cuckoo's egg-dumping response is a **fixed action pattern**: a series of instinctive movements, triggered by a specific stimulus, that—once started—continues to completion without the need for further cues. Such fixed behavior has survival advantages when it permits a fast response to an important stimulus. However, a fixed response to simple stimuli has limitations. For example, the cuckoo's foster parents are not equipped to note color and size of offspring. A simple stimulus —a chick's gaping mouth—induces the fixed action pattern of parental feeding behavior (Figure 44.5b).

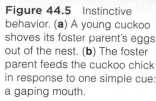

Figure 44.5 Instinctive behavior. (**a**) A young cuckoo shoves its foster parent's eggs out of the nest. (**b**) The foster parent feeds the cuckoo chick in response to one simple cue: a gaping mouth.

Time-Sensitive Learning

Learned behavior is behavior that is altered by experience. Some instinctive behavior can be modified with learning. A garter snake's initial strikes at prey are instinctive, but the snake learns to avoid dangerous or unpalatable prey. Learning may occur throughout an animal's life, or be restricted to a critical period.

Imprinting is a form of learning that occurs during a genetically determined time period. For example, baby geese learn to follow the large object that bends over them in response to their first peep (Figure 44.6). With rare exceptions, this object is their mother. When mature, the geese will seek out a sexual partner that is similar to the imprinted object.

A genetic capacity to learn, combined with actual experiences in the environment, shapes most forms of behavior. For example, a male songbird has an inborn capacity to recognize his species' song when he hears older males singing it. The young male uses these overheard songs as a guide to fill in details of his own song. Males reared alone sing a simplified version of their species' song. So do males exposed only to the songs of other species.

Many birds must learn their species-specific song during a limited period early in life. For example, a male white-crowned sparrow will not sing normally if he does not hear a male "tutor" of his own species during his first 50 or so days. Hearing a same-species tutor later in life will not influence his singing.

Most birds must also practice their song to perfect it. In one experiment, researchers temporarily paralyzed throat muscles of zebra finches who were beginning to sing. After being temporarily unable to practice, these

Figure 44.6 Nobel laureate Konrad Lorenz with geese that imprinted on him. The smaller photograph shows results of a more typical imprinting episode.

birds never mastered their song. In contrast, temporary paralysis of throat muscles in very young birds or adults did not impair later song production. Thus, in this species, there is a critical period for song practice, as well as for song learning.

Conditioned Responses

Nearly all animals are lifelong learners. Most learn to associate certain stimuli with rewards and others with negative consequences.

With **classical conditioning**, an animal's involuntary response to a stimulus becomes associated with another stimulus that is presented at the same time. In the most famous example, Ivan Pavlov rang a bell whenever he fed a dog. Eventually, the dog's reflexive response to food—increased salivation—was elicited by the sound of the bell alone.

With **operant conditioning**, an animal modifies its voluntary behavior in response to consequences of that behavior. This type of learning was first described for conditions in the lab. For example, a rat that presses a lever in a laboratory cage and is rewarded with a food pellet becomes more likely to press the lever again. A rat that receives a shock when it enters a particular area of a cage will quickly learn to avoid that area.

Other Types of Learned Behavior

With **habituation**, an animal learns by experience not to respond to a stimulus that has neither positive nor negative effects. For example, pigeons in cities learn not to flee from the large numbers of people who walk past them.

Many animals learn about the landmarks in their environment and form a sort of mental map. This map may be put to use when the animal needs to return home. For example, a fiddler crab foraging up to 10 meters (30 feet) away from its burrow is able to scurry straight home when it perceives a threat.

Many animals also learn the details of their social landscape; they learn to recognize mates, offspring, or competitors by appearance, calls, odor, or some combination of cues. For example, when two male lobsters meet up for the first time they will fight (Figure 44.7). Later, they will recognize one another by scent and behave accordingly, with the loser actively avoiding the winner. A lobster also recognizes its mate's scent.

With **observational learning**, an animal imitates the behavior of another individual. For example, Ludwig Huber and Bernhard Voelkel allowed marmoset monkeys to watch another marmoset demonstrate how to

Figure 44.7 Getting to know one another. Two male lobsters battle at their first meeting. Later, the loser will remember the odor of the winner and avoid him. Without another meeting, memory of the defeat lasts up to two weeks.

Figure 44.8 Observational learning. A marmoset opens a container using its teeth. After watching one individual successfully perform this maneuver, other marmosets used the same technique. Analysis of videos of their movements showed that the observers closely imitated the behavior they had seen earlier.

open a plastic container and retrieve the treat inside it. Marmosets who had seen the demonstrator open the container with its hands imitated this behavior, using their hands in the same way. In contrast, those who had watched a demonstrator open the box with its teeth attempted to do the same (Figure 44.8).

Take-Home Message

How do instinct and learning shape behavior?

■ Instinctive behavior can initially be performed without any prior experience, as when a simple cue triggers a fixed action pattern. Even instinctive behavior may be modified by experience.

■ Certain types of learning can only occur at particular times in the life cycle.

■ Learning affects both voluntary and involuntary behaviors.

- If a behavior varies and some of that variation has a genetic basis, then it will be subject to natural selection.
- Link to Adaptive traits 17.3

Behavior that increases an individual's reproductive success is adaptive. For example, Larry Clark and Russell Mason studied the nest decorating behavior of starlings. These birds tuck sprigs of aromatic plants such as wild carrot into their nests. Clark and Mason suspected that the plant bits control parasitic mites that feed on nestlings. To test their hypothesis, the researchers replaced natural starling nests with man-made ones that either had wild carrot sprigs or were sprig-free. They predicted that the decorated nests would have fewer mites than undecorated ones.

After the starling chicks left the nests, Clark and Mason recorded the number of mites left behind. The number was greater in sprig-free nests (Figure 44.9). Why? As it turns out, one organic compound in the leaves of wild carrot prevents mites from maturing.

Mason and Clark concluded that decorating a nest with sprigs deters bloodsucking mites. They inferred that this nest-decorating behavior is adaptive because it promotes nestling survival, increasing reproductive success for the nest-decorating birds.

As you will learn in Section 44.7, some behavior that increases the reproductive success of relatives at the expense of the individual can also be adaptive.

Take-Home Message

What makes a behavior adaptive?

- Most behavior is adaptive because it increases the reproductive success of the individual performing it. Some is adaptive because it benefits relatives.

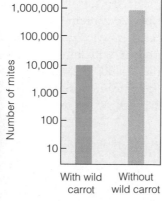

Figure 44.9 Results of an experiment to test the effect of wild carrot sprigs on the number of mites in starling nests. Nests with wild carrot pieces had significantly fewer mites than those with no greenery. There may be a selective advantage to using wild carrot and other aromatic plants as nest materials.

- Cooperating to mate or in other ways requires individuals to share information about themselves and their environment.
- Link to Pheromones 35.1

Communication signals are cues for social behavior between members of a species. Chemical, acoustical, visual, and tactile signals transmit information from signalers to signal receivers.

Pheromones are chemical cues. Signal pheromones make a receiver alter its behavior fast. The honeybee alarm pheromone is an example. So are sex attractants that help males and females find each other. Priming pheromones cause longer-term responses, as when a chemical dissolved in the urine of certain male mice triggers ovulation in females of the same species.

Many acoustical signals, such as bird song, attract mates or define a territory. Others are alarm signals, such as a prairie dog's bark that warns of a predator.

One visual signal is a male baboon threat display, which communicates readiness to fight a rival (Figure

Figure 44.10 Visual signals. (**a**) A male baboon shows his teeth in a threat display. (**b**) Penguins engaged in a courtship display. (**c**) A wolf's play bow tells another wolf that behavior that follows is play, not aggression.

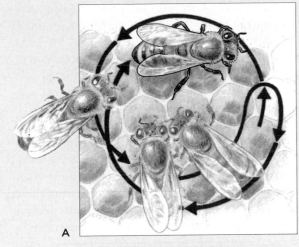

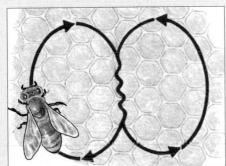

A

B

Figure 44.11 Animated Honeybee dances, an example of a tactile display. **(a)** Bees that have visited a source of food close to their hive return and perform a *round* dance on the hive's vertically oriented honeycomb. The bees that maintain contact with the dancer later fly out and search for food near the hive.

(b) A bee that visits a feeding source more than 100 meters (110 yards) from her hive performs a *waggle* dance. Orientation of an abdomen-waggling dancer in the straight run of her dance informs other bees about the direction of the food.

(c) If the food is in line with the sun, the dancer's waggling run proceeds straight up the honeycomb. **(d)** If food is in the opposite direction from the sun, the dancer's waggle run is straight down. **(e)** If food is 90 degrees to the right of the direction of the sun, the waggle run is offset by 90 degrees to the right of vertical.

The speed of the dance and the number of waggles in the straight run provide information about distance to the food. A dance inspired by food that is 200 meters away is much faster and has more waggles per straight run, than a dance inspired by a food source that is 500 meters away.

Figure It Out: Do the dances shown in parts c–e indicate different distances from the hive?   *Answer: No. The number of waggles in the straight run does not vary.*

When bee moves straight up comb, recruits fly straight toward the sun

When bee moves straight down comb, recruits fly to source directly away from the sun

When bee moves to right of vertical, recruits fly at 90° angle to right of the sun

C

D

E

44.10*a*). Visual signals are part of courtship displays that often precede mating in birds (Figure 44.10*b*). Unambiguous signals work best, so movements often get exaggerated and body form evolves in ways that draw attention to the movements.

With tactile displays, information is transmitted by touch. For example, after discovering food, a foraging honeybee worker returns to the hive and performs a complex dance. The bee moves in a defined pattern, jostling a crowd of other bees that surround her. The signals give other bees information about the distance and the direction of the food source (Figure 44.11).

The same signal sometimes functions in more than one context. For example, dogs and wolves solicit play behavior with a play bow (Figure 44.10*c*). A play bow informs an animal's prospective playmate that signals that follow, which would ordinarily be construed as aggressive or sexual, are friendly play behavior.

A communication signal evolves and persists only if it benefits both sender and receiver. If the signal has disadvantages, then natural selection will tend to favor individuals that do not send or respond to it. Other factors can also select against signalers. For example,

male tungara frogs attract females with complex calls, which also make it easier for frog-eating bats to zero in on the caller. When bats are near, male frogs call less, and usually with less flair. The subdued signal is a trade-off between locating a partner for mating and the need for immediate survival.

There are illegitimate signalers, too. For example, fireflies attract mates by producing flashes of light in a characteristic pattern. Some female fireflies prey on males of other species. When a predatory female sees the flash from a male of the prey species, she flashes back as if she were a female of his own species. If she lures him close enough, she captures and eats him.

Take-Home Message

What are the benefits and costs of communication signals?

■ A communication signal transfers information from one individual to another individual of the same species. Such signals benefit both the signaler and the receiver.

■ Signals have a potential cost. Some individuals of a different species benefit by intercepting signals or by mimicking them.

■ In studying behavior, we expect that each sex will evolve in ways that maximize its benefits, and minimize its costs, which can lead to conflicts.

■ Link to Sexual selection 18.6

Sexual Selection and Mating Behavior

Males or females of a species often compete for access to mates, and many are choosy about their partners. Both situations lead to **sexual selection**. As explained in Section 18.6, this microevolutionary process favors characteristics that provide a competitive advantage in attracting and often holding on to mates.

But whose reproductive success is it—the male's or the female's? Male animals, remember, produce many small sperm, and females produce far larger but fewer eggs. For the male, success generally depends on how many eggs he can fertilize. For the female, it depends more on how many eggs she produces or how many offspring she can raise. Usually, the most important factor in a female's sexual preference is the quality of the mate, not the quantity of partners.

Female hangingflies (*Harpobittacus*) will mate only with males that supply food. A male hunts and kills a moth or some other insect. Then he releases a sex pheromone, which attracts females to him and his "nuptial gift" (Figure 44.12*a*). The female begins to eat the male's offering and copulation begins. Only after the female has been eating for five minutes or so does she start to accept sperm from her partner. Even after mating begins, a female can break off from her suitor, if she finishes eating his gift. If she does end the mating, she will seek out a new male and his sperm will replace the first male's. Thus, the larger the male's gift, the greater the chance that mostly his sperm will actually end up fertilizing the eggs of his mate.

Females of certain species shop around for males who have appealing traits. Consider the fiddler crabs that live along many sandy shores. One of the male's two claws is enlarged; it often accounts for more than half his total body weight (Figure 44.12*b*). During their breeding season, hundreds of males excavate mating burrows near one another. Each male stands next to his burrow, waving his oversized claw. Female crabs stroll along, checking out males. If a female likes what she sees, she inspects her suitor's burrow. Only when a burrow has the right location and dimensions does she mate with its owner and lay eggs in his burrow.

Some female birds are similarly choosy. Male sage grouse (*Centrocercus urophasianus*) converge at a lek, a type of communal display ground, where each stakes out a few square meters. With tail feathers erect, the males emit booming calls by puffing and deflating big neck pouches (Figure 44.12*d*). As they do, they stamp about on their patch of prairie. Females tend to select and mate with one male sage grouse. Afterward, they

Figure 44.12 (**a**) Male hangingfly dangling a moth as a nuptial gift for a potential mate. Females of some hangingfly species choose sexual partners that offer the largest gift to them. By waving his enlarged claw, a male fiddler crab (**b**) may attract the eye of a female fiddler crab (**c**). A male sage grouse (**d**) showing off as he competes for female attention at a communal display ground.

go off to nest and raise any offspring by themselves. Often, many females favor the same few males, and most males never have an opportunity to mate.

In another behavioral pattern, the sexually receptive females of some species cluster in defendable groups. Where you come across such a group, you are likely to observe males competing for access to the clusters. Competition for ready-made harems has resulted in combative male lions, sheep, elk, elephant seals, and bison, to name a few examples (Figure 44.13).

Parental Care

When females fight for males, we can predict that the males provide more than sperm delivery. Some, such as the male midwife toad, help with parenting. The male holds strings of fertilized eggs around his legs until the eggs hatch (Figure 44.14a). Once her eggs are being cared for, a female can mate with other males, if she can find some that are not already caring for eggs. Late in the breeding season, males without strings of eggs are rare, and females fight for access to them. The females even attempt to pry mating pairs apart.

Parental behavior uses up time and energy, which parents otherwise might spend on living long enough to reproduce again. However, for some animals, the benefit of increased survival of the young outweighs the cost of parenting.

Few reptiles provide care for young. Crocodilians, the reptiles most closely related to birds, are a notable exception. Crocodile parents bury their eggs in a nest. When young are ready to hatch, they call and parents dig them out and care for them for some time.

Most birds are monogamous, and both parents often care for the young (Figure 44.14b). In mammals, males typically leave after mating. Females raise the young alone, and males attempt to mate again or conserve energy for the next breeding season (Figure 44.14c). Mammalian species in which males help care for the young tend to be monogamous, at least over the course of a breeding season. Only about 5 percent of mammals are monogamous.

Figure 44.13 Male bison locked in combat during the breeding season.

Figure 44.14 (**a**) Male midwife toad with developing eggs wrapped around his legs. (**b**) A pair of Caspian terns cooperate in the care of their chick. (**c**) A female grizzly will care for her cub for as long as two years. The male takes no part in the cub's upbringing.

Take-Home Message

How does natural selection affect mating systems?

■ Males and females each behave in ways that will maximize their own reproductive success.

■ Most males compete for females and mate with more than one. Monogamy and male parental care are not common.

■ Survey the animal kingdom and you find evolutionary costs and benefits across a range of social groups.

■ Link to Culture 26.13

Defense Against Predators

In some groups, cooperative responses to predators reduce the net risk to all. Vulnerable individuals can be on the alert for predators, join a counterattack, or engage in more effective defenses (Figure 44.15).

Birds, monkeys, meerkats, prairie dogs, and many other animals make alarm calls, as in Figure 44.15*a*. A prairie dog makes a particular bark when it sights an eagle and a different signal when it sights a coyote. Others dive into burrows to escape an eagle's attack or stand erect and observe the coyote's movements.

Sawfly caterpillars feed in clumps on branches and benefit by coordinated repulsion of predatory birds. When a potential predator approaches, the caterpillars rear up and vomit partly digested eucalyptus leaves (Figure 44.15*b*). Birgitta Sillén-Tullberg demonstrated that predatory birds prefer individual caterpillars to a wiggling group. When offered caterpillars one at a time, the birds ate an average of 5.6. Birds offered a cluster of twenty caterpillars ate an average of 4.1.

Whenever animals cluster, some individuals shield others from predators. Preference for the center of a group can create a **selfish herd**, in which individuals hide behind one another. Selfish-herd behavior occurs in bluegill sunfishes. A male sunfish builds a nest by scooping out a depression in mud on the bottom of a lake. Females lay eggs in these nests, and snails and fishes prey on eggs. Competition for the safest sites is greatest near the center of a group, with large males taking the innermost locations. Smaller males cluster around them and bear the brunt of the egg predation. Even so, the nests of small males are safer at the edge of the group than they would be alone in the open.

Improved Feeding Opportunities

Many mammals, including wolves, lions, wild dogs, and chimpanzees, live in social groups and cooperate in hunts (Figure 44.16). Are cooperative hunters more efficient than solitary ones? Often, no. In one study, researchers observed a solitary lion that caught prey about 15 percent of the time. Two lions cooperatively hunting caught prey twice as often but had to share it, so the amount of food per lion balanced out. When more lions joined a hunt, the success rate per lion fell. Wolves show a similar pattern. Among carnivores that hunt cooperatively, hunting success does not seem to be the major advantage of group living. Individuals hunt together, but they also may fend off scavengers, care for one another's young, and protect territory.

Group living also allows transmission of cultural traits, or behaviors learned by imitation. For example, chimpanzees make and use simple tools by stripping leaves from branches. They use thick sticks to make holes in a termite mound, then insert long, flexible "fishing sticks" into the holes (Figure 44.17). The long stick agitates the termites, which attack and cling to it.

Figure 44.15 Group defenses. (**a**) Black-tailed prairie dogs bark an alarm call that warns others of predators. Does this call put the caller at risk? Not much. Prairie dogs usually act as sentries only after they finish feeding and happen to be standing next to their burrows. (**b**) Australian sawfly caterpillars form clumps and regurgitate a fluid (the *yellow* blobs) that predators find unappealing. (**c**) Musk oxen adults (*Ovibos moschatus*) form a ring of horns, often around their young.

Chimps withdraw the stick and lick off termites, as a high-protein snack. Different groups of chimpanzees use slightly different tool-shaping and termite-fishing methods. Youngsters of each group learn by imitating the adults.

Dominance Hierarchies

In many social groups, subordinate individuals do not get an equal share of resources. Most wolf packs, for instance, have one dominant male that breeds with just one dominant female. The others are nonbreeding brothers and sisters, aunts and uncles. All hunt and carry food back to individuals that guard the young in their den.

Why would a subordinate give up resources and often breeding privileges? It might get injured or die if it challenges a strong individual. It might not be able to survive on its own. A subordinate might even get a chance to reproduce if it lives long enough or if its dominant peers are taken out by a predator or old age. As one example, some subordinate wolves move up the social ladder when the opportunity arises.

Regarding the Costs of Group Living

If social behavior is advantageous, then why are there so few social species? In most habitats, costs outweigh benefits. For instance, when individuals are crowded together they compete more for resources. Cormorants and other seabirds form dense breeding colonies, as in Figure 44.18. All compete for space and food.

Large social groups also attract more predators. If individuals are crowded together, they are vulnerable to parasites and contagious diseases that jump from host to host. Individuals may also be at risk of being killed or exploited by others. Given the opportunity, a pair of breeding herring gulls will cannibalize the eggs and even the chicks of their neighbors.

Figure 44.16 Members of a wolf pack (*Canis lupus*). Wolves cooperate in hunting, caring for young, and defending territory. Benefits are not distributed equally. Only the highest ranking individuals, the alpha male and alpha female, breed.

Figure 44.17 Chimpanzees (*Pan troglodytes*) using sticks as tools for extracting tasty termites from a nest. This behavior is learned by imitation.

Figure 44.18 Nearly uniform spacing in a crowded cormorant colony.

Take-Home Message

What are the benefits and costs of social groups?

■ Living in a social group can provide benefits, as through cooperative defenses or shielding against predators.

■ Group living has costs: increased competition, increased vulnerability to infections, and exploitation by others.

■ Extreme cases of sterility and self-sacrifice have evolved in only a few groups of insects and one group of mammals. How are genes of the nonreproducers passed on?

Social Insects

Animals that are eusocial live together for generations in a group that has a reproductive division of labor. Eusocial insects include the honeybees, termites, and ants. In all of these groups, sterile workers care cooperatively for the offspring produced by just a few breeding individuals. Such workers often are highly specialized in their form and function (Figure 44.19).

A queen honeybee is the only fertile female in her hive. She is larger than other females, partly because of her enlarged ovaries (Figure 44.20a). She secretes a pheromone that makes all other female bees sterile.

All of the 30,000 to 50,000 worker bees are females that develop from fertilized eggs laid by the queen. They feed the larvae, maintain the hive, and construct honeycomb from wax they secrete. Workers also gather nectar and pollen that feeds the colony. They guard the hive and will sacrifice themselves to repel intruders.

In spring and summer, the queen lays unfertilized eggs that develop into drones. These male bees are stingless and subsist on food gathered by their worker sisters. Each day, drones fly in search of a mate. If one is lucky, he will meet a virgin queen on her one flight away from a colony. He dies after mating. A young queen mates with many males, and stores their sperm for use over her lifetime of several years.

Like honeybees, termites live in enormous family groups with a queen specialized for producing eggs (Figure 44.20b). Unlike the honeybee hive, a termite mound holds sterile individuals of both sexes. A king

supplies the female with sperm. Winged reproductive termites of both sexes develop seasonally.

Social Mole-Rats

Sterility and extreme self-sacrifice are uncommon in vertebrates. The only eusocial mammals are African mole-rats. The best studied is *Heterocephalus glaber*, the naked mole-rat. Clans of this nearly hairless rodent build and occupy burrows in dry parts of East Africa.

A mole rat clan consists of a reproductive "queen" (Figure 44.20c), the one to three "kings," with whom she mates, and their nonbreeding worker offspring. Workers care for the queen, the king(s), and the young. Some workers serve as diggers that excavate tunnels and chambers. When a digger finds an edible root, it hauls a bit back to the main chamber and chirps. Its chirps recruit other workers to help carry food back to the chamber. Still other workers function as guards. When a predator appears, they chase and attack it at great risk to themselves.

Evolution of Altruism

A sterile worker in a social insect colony or a naked mole-rat clan shows **altruistic behavior**: behavior that enhances another individual's reproductive success at the altruist's expense. How did this behavior evolve? According to William Hamilton's **theory of inclusive fitness**, genes associated with altruism are selected if they lead to behavior that promotes the reproductive success of an altruist's closest relatives.

A sexually reproducing, diploid parent caring for offspring is not helping exact genetic copies of itself.

Figure 44.19 Specialized ways of serving and defending the colony. (**a**) An Australian honeypot ant worker. This sterile female is a living container for her colony's food reserves. (**b**) Army ant soldier (*Eciton burchelli*) with formidable mandibles. (**c**) Eyeless soldier termite (*Nasutitermes*). It bombards intruders with a stream of sticky goo from its nozzle-shaped head.

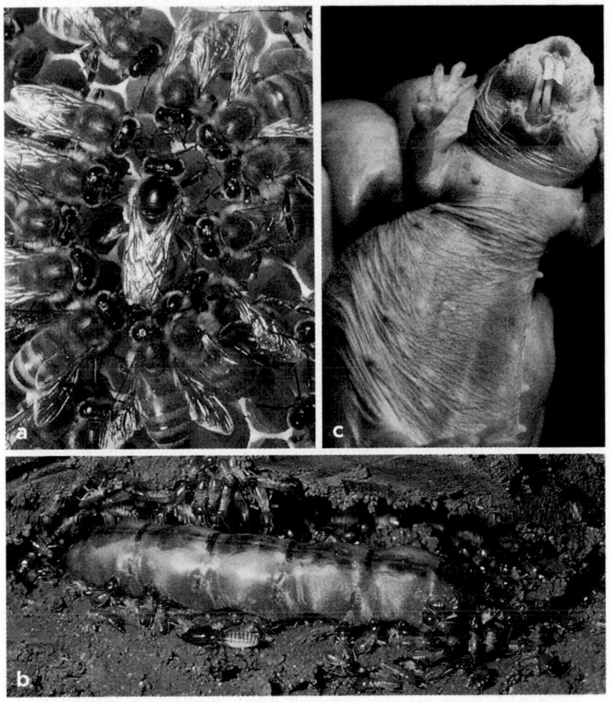

44.8 | Human Behavior

- Evolutionary forces shaped human behavior—but humans alone can make moral choices about their actions.
- Link to Limits of science 1.5

Hormones and Pheromones Are humans, too, influenced by hormones that contribute to bonding behavior in other mammals? Perhaps. Consider that autism, a developmental disorder in which people have trouble making social contacts, is often associated with low oxytocin levels. Oxytocin is known to affect bonding behavior in other mammals.

Pheromones in sweat may also affect human behavior. Women who live together often have synchronized menstrual cycles and experiments have shown that a woman's menstrual cycle will lengthen or shorten after she has been exposed to sweat from a woman who was in a different phase of the cycle. Other experiments have shown that exposure to male sweat can alter a woman's cortisol level.

Morality and Behavior If we are comfortable with studying the evolutionary basis of behavior of termites, naked mole-rats, and other animals, why do some people resist the idea of analyzing the evolutionary basis of human behavior? A common fear is that an objectionable behavior will be defined as "natural." To evolutionary biologists, however, "adaptive" does not mean "morally right." It simply means a behavior increases reproductive success. Scientific studies do not address moral issues (Section 1.5).

For example, infanticide is morally repugnant. Is it unnatural? No. It happens in many animal groups and all human cultures. Male lions often kill the offspring of other males when they take over a pride. Thus deprived of parenting tasks, the lionesses can now breed with the infanticidal male and increase that male's reproductive success.

Biologists would predict that unrelated human males are a threat to infants. Evidence supports the prediction. The absence of a biological father and the presence of an unrelated male increases risk of death for an American child under age two by more than sixty times.

What about parents who kill their own offspring? In her book on maternal behavior, primatologist Sarah Blaffer Hrdy cites a study of one village in Papua New Guinea in which parents killed about 40 percent of the newborns. As Hrdy argues, when resources or social support are hard to come by, a mother's fitness might increase if a newborn who is unlikely to survive is killed. The mother can allocate child-rearing energy to her other offspring or save it for children she may have in the future.

Do most of us find such behavior appalling? Yes. Can considering the possible evolutionary advantages of the behavior help us prevent it? Perhaps. An analysis of the conditions under which infanticide occurs tells us this: When mothers lack the resources they need to care for their children, they are more likely to harm them. We as a society can act upon such information.

Figure 44.20 Three queens. (**a**) Queen honeybee with her sterile daughters. (**b**) A termite queen (*Macrotermes*) dwarfs her offspring and mate. Ovaries fill her enormous abdomen. (**c**) A naked mole-rat queen.

Each of its gametes, and each of its offspring, inherits one-half of its genes. Other individuals of the social group that have the same ancestors also share genes. Siblings (brothers or sisters) are as genetically similar as a parent and offspring. Nephews and nieces share about one-fourth of their uncle's genes.

Sterile workers promote genes for self-sacrifice by helping close relatives survive and reproduce. In honeybee, termite, and ant colonies, sterile workers assist fertile relatives with whom they share genes. A guard bee will die after she stings, but her sacrifice preserves many copies of her genes in her hivemates.

Inbreeding increases the genetic similarity among relatives and may play a role in mole-rat sociality. A clan is highly inbred as a result of many generations of sibling, mother–son, and father–daughter matings. Dry habitats and patchy food sources also may favor cooperation in digging, locating food, and fending off competitors and predators.

Take-Home Message

How can altruistic behavior be selectively advantageous?

- Altruistic behavior may be favored when individuals pass on genes indirectly, by helping relatives survive and reproduce.

When a European queen bee mates with an Africanized drone, her worker offspring are just as aggressive as workers in a pure Africanized colony. In contrast, a cross between an Africanized queen and a European drone yields workers with an intermediate level of aggression. Unfortunately, European queen–Africanized male pairings occur far more frequently than the reciprocal cross. Africanized males outcompete European males for matings.

How would you vote?

Africanized honeybees continue to increase their range. Should study of their genetics be a high priority? See CengageNOW for details, then vote online.

Summary

Section 44.1 Behavior refers to coordinated responses that an animal makes to a **stimulus**. Genes that affect the nervous system often affect behavior, but other genes may also influence it. Studies of natural behavioral variations within and among species provide information about the genetic basis for behaviors, as does the study of induced or natural mutations.

Section 44.2 **Instinctive behavior** can occur without having been learned by experience. A **fixed action pattern** is an instinctive series of responses to a simple cue.

Learned behavior is altered by experience. **Imprinting** is one form of learning that happens only during a sensitive period early in life. With **classical conditioning**, an animal learns to associate an involuntary response to one stimulus with another stimulus. With **operant conditioning**, an animal modifies a voluntary behavior in response to the behavior's consequences. With **habituation**, an animal stops responding to an ongoing stimulus. With **observational learning**, it imitates another's actions.

Section 44.3 A behavior that has a genetic basis is subject to evolution by natural selection. Adaptive forms of behavior evolved as a result of individual differences in reproductive success in past generations.

Section 44.4 **Communication signals** allow animals of the same species to share information. Such signals evolve and persist only if they benefit both senders and receivers of the signal.

Chemical signals such as **pheromones** have roles in social communication, as do acoustical signals, visual signals that are part of courtship and threat displays, and tactile signals.

■ *Use the animation on CengageNOW to explore the honeybee dance language.*

Section 44.5 **Sexual selection** favors traits that give an individual a competitive edge in attracting and often holding on to mates. The females of many species select males that have traits or engage in behaviors they find attractive. When large numbers of females cluster in a defensible area, males may compete with one another to control the areas.

Parental care has reproductive costs in terms of future reproduction and survival. It is adaptive when benefits to a present set of offspring offset the costs.

Section 44.6 Animals that live in social groups may benefit by cooperating in predator detection, defense, and rearing the young. A **selfish herd** forms when animals hide behind one another. Benefits of group living are often distributed unequally. Species that live in large groups incur costs, including increased disease and parasitism, and increased competition for resources.

Section 44.7 Ants, termites, and some other insects as well as two species of mole-rats are eusocial. They live in colonies with overlapping generations and have a reproductive division of labor. Most colony members do not reproduce; they assist their relatives instead. According to the **theory of inclusive fitness**, such **altruistic behavior** is perpetuated because altruistic individuals share genes with their reproducing relatives. Altruistic individuals help perpetuate the genes that led to their altruism by promoting the reproductive success of close relatives that also carry copies of these genes.

Section 44.8 Hormones and possibly pheromones influence human behavior. A behavior that is adaptive in the evolutionary sense may still be judged by society to be morally wrong. Science does not address morality.

Self-Quiz   *Answers in Appendix III*

1. Genes affect the behavior of individuals by _____ .
   a. influencing the development of nervous systems
   b. affecting the kinds of hormones in individuals
   c. determining which stimuli can be detected
   d. all of the above

2. Stevan Arnold offered slug meat to newborn garter snakes from different populations to test his hypothesis that the snakes' response to slugs _____ .
   a. was shaped by indirect selection
   b. is an instinctive behavior
   c. is based on pheromones
   d. is adaptive

3. A behavior is defined as adaptive if it _____ .
   a. varies among individuals of a population
   b. occurs without prior learning
   c. increases an individual's reproductive success
   d. is widespread across a species

4. The honeybee dance language transmits information about distance to food by way of _____ signals.
   a. tactile   c. acoustical
   b. chemical   d. visual

Data Analysis Exercise

Honeybees disperse by forming new colonies. An old queen leaves the hive along with a group of workers. These bees fly off, find a new nest site, and set up a new hive. Meanwhile, at the old hive, a new queen emerges, mates, and takes over. A new hive can be several kilometers from the old one.

Africanized honeybees form new colonies more often than European ones, a trait that contributes to their spread. Africanized bees also spread by taking over existing hives of European bees. In addition, in areas where European and Africanized hives coexist, European queens are more likely to mate with Africanized males, thus introducing Africanized traits into the colony. Figure 44.21 shows the counties in the United States in which Africanized honeybees became established from 1990 through 2006.

1. Where in the United States did Africanized bees first become established?

2. In what states did Africanized bees first appear in 2005?

3. Why is it likely that human transport of bees contributed to the spread of Africanized honeybees to Florida?

4. Based on this map, would you expect Africanized honeybees to colonize additional states in the next five years?

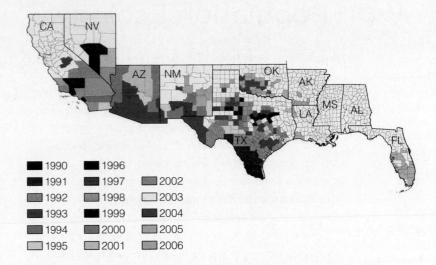

Figure 44.21 The spread of Africanized honeybees in the United States, from 1990 through 2006. The USDA adds a county to this map only when the state officially declares bees in that county Africanized. Bees can be identified as Africanized on the basis of morphological traits or analysis of their DNA.

5. A _____ is a chemical that conveys information between individuals of the same species.
   a. pheromone   c. hormone
   b. neurotransmitter   d. all of the above

6. In _____ , males and females typically cooperate in care of the young.
   a. mammals   c. amphibians
   b. birds   d. all of the above

7. Generally, living in a social group costs the individual in terms of _____ .
   a. competition for food, other resources
   b. vulnerability to contagious diseases
   c. competition for mates
   d. all of the above

8. Social behavior evolves because _____ .
   a. social animals are more advanced than solitary ones
   b. under some conditions, the costs of social life to individuals are offset by benefits to the species
   c. under some conditions, the benefits of social life to an individual offset the costs to that individual
   d. under most conditions, social life has no costs to an individual

9. Eusocial insects _____ .
   a. live in extended family groups
   b. include termites, honeybees, and ants
   c. show a reproductive division of labor
   d. a and c
   e. all of the above

10. Helping other individuals at a reproductive cost to oneself might be adaptive if those helped are _____ .
    a. members of another species
    b. competitors for mates
    c. close relatives
    d. illegitimate signalers

11. True or false? Some mammals live in colonies and act as sterile workers that serve close relatives.

12. Match the terms with their most suitable description.
    ___ fixed action pattern
    ___ altruism
    ___ basis of instinctive and learned behavior
    ___ imprinting
    ___ pheromone

    a. time-dependent form of learning requiring exposure to key stimulus
    b. genes plus actual experience
    c. series of responses that runs to completion independently of feedback from environment
    d. assisting another individual at one's own expense
    e. one communication signal

■ *Visit CengageNOW for additional questions.*

Critical Thinking

1. For billions of years, the only bright objects in the night sky were stars or the moon. Night-flying moths used them to navigate in a straight line. Today, the instinct to fly toward bright objects causes moths to exhaust themselves fluttering around streetlights and banging against brightly lit windowpanes. This behavior is not adaptive, so why does it persist?

2. Damaraland mole-rats are relatives of naked mole-rats (Figure 44.19). In their clans, too, nonbreeding individuals of both sexes cooperatively assist one breeding pair. Even so, breeding individuals in wild Damaraland mole-rat colonies usually are unrelated, and few subordinates move up in the hierarchy to breeding status. Researchers suspect that ecological factors, not genetic ones, were the more important selective force in Damaraland mole-rat altruism. Explain why.

45 Population Ecology

The Numbers Game

In 1722, on Easter morning, a European explorer landed on a small volcanic island in the South Pacific and discovered a few hundred hungry, skittish people living in caves. He noticed withered grasses and scorched, shrubby plants— and the absence of trees. He wondered about the hundreds of massive stone statues near the coast and 500 unfinished, abandoned ones in inland quarries (Figure 45.1). Some weighed 100 tons and stood 10 meters (33 feet) high.

Easter Island, as it came to be called, is no larger than 165 square kilometers (64 square miles). Archaeologists have determined that voyagers from the Marquesas discovered this eastern outpost of Polynesia more than 1,650 years ago. The place was a paradise. Its volcanic soil supported dense forests and lush grassland. The colonists used long, straight palms to build canoes that were strengthened with rope made of fibers from hauhau trees. They used wood as fuel to cook fishes and dolphins. They cleared forests to plant crops. They had many children.

By 1440, as many as 15,000 people were living on the island. Crop yields declined; ongoing harvests and erosion had depleted the soil of nutrients. Fish vanished from the waters close to the island, so fishermen had to sail farther and farther out on the open ocean.

Those in power built statues to appeal to the gods. They directed others to carve images of unprecedented size and move the new statues to the coast. Wars broke out and by 1550, no one ventured offshore to fish. They could not build any more canoes because there were no more trees.

As central authority crumbled, the dwindling numbers of islanders retreated to caves and launched raids against one another. Winners ate the losers and tipped over statues. Even if the survivors had wanted to, they had no way to get off the island. The once-flourishing population collapsed.

Any natural population has the capacity to increase in number, given the right conditions. In North America, white-tailed deer are behaving like early settlers on Easter Island. With plenty of food and few predators, deer numbers are soaring. Deer overpopulation harms forests, damages crops, and increases the incidence of highway accidents.

With this chapter, we begin a survey of principles that govern the growth and sustainability of all populations. The principles are the bedrock of ecology—the systematic study of how organisms interact with one another and with their environment. Those interactions start within and between populations and extend to communities, ecosystems, and the biosphere.

See the video! Figure 45.1 Row of massive statues on Easter Island. Islanders set them up long ago, apparently as a plea for help after their once-large population wreaked havoc on their tropical paradise. Their plea had no effect whatsoever on reversing the loss in biodiversity on the island and in the surrounding sea. The human population did not recover, either.

Key Concepts

The vital statistics

Ecologists explain population growth in terms of population size, density, distribution, and number of individuals in different age categories. Field studies allow ecologists to estimate population size and density. **Sections 45.1, 45.2**

Exponential rates of growth

A population's size and reproductive base influence its rate of growth. When the population is increasing at a rate proportional to its size, it is undergoing exponential growth. **Section 45.3**

Limits on increases in number

Over time, an exponentially growing population typically overshoots the carrying capacity—the maximum number of individuals of a species that environmental resources can sustain. Some populations stabilize after a big decline. Others never recover. **Section 45.4**

Patterns of survival and reproduction

Resource availability, disease, and predation are major factors that can restrict population growth. These limiting factors differ among species and shape their life history patterns. **Sections 45.5, 45.6**

The human population

Human populations sidestepped limits to growth by way of global expansion into new habitats, cultural interventions, and innovative technology. Even so, no population can continue to expand indefinitely. **Sections 45.7–45.10**

Links to Earlier Concepts

- Earlier chapters defined and explored the evolutionary history and genetic nature of populations, including those of humans (Sections 18.1 and 26.15). Now you will consider factors that limit population growth, including contraception (42.9).

- You will be reminded of the effects of infectious disease (Chapter 21 introduction, 21.8), and the stunning reproductive capacity of prokaryotes (21.5).

- Gene flow (18.8) and directional selection (18.4) are discussed in the context of evolving populations. We also consider how sampling error (1.8) affects population studies.

How would you vote? Soaring numbers of white-tailed deer threaten forest plants and the animals that depend on them. Is encouraging deer hunting in regions where their overabundance is a threat to other species the best solution? See CengageNOW for details, then vote online.

45.1 Population Demographics

- A population's size, density, distribution, and age structure are shaped by ecological factors, and may shift over time.

- Link to Population genetics 18.1

Ecologists typically use the term "population" to refer to all members of a species within an area defined by the researcher. Studies of population ecology start with **demographics**: statistics that describe population size, age structure, density, distribution, and other factors.

Population size is the number of individuals in the population. **Age structure** is the number of individuals in each of several age categories. Individuals are often grouped as pre-reproductive, reproductive, or post-reproductive. Those in the pre-reproductive category

have the capacity to produce offspring when mature. Together with individuals in the reproductive group, they make up the population's **reproductive base**.

Population density is the number of individuals in a specified portion of a habitat. A habitat, remember, is the type of place where a species lives. We characterize a habitat by its physical and chemical features, and its array of species.

Density refers to how many individuals are in an area but not how they are dispersed through it. Even a habitat that looks uniform, such as a sandy shore, has variations in light, moisture, and many other variables. A population may live in only a small part of the habitat, and it may do so all of the time or only some of the time.

The pattern in which individuals are dispersed in their habitat is the **population distribution**. It may be clumped, nearly uniform, or random (Figure 45.2).

A clumped distribution is most common, for several reasons. First, the conditions and resources that most species require tend to be patchy. Animals cluster at a water hole, seeds sprout only in moist soil, and so on. Second, most seeds and some animal offspring cannot disperse far from their parents. Third, some animals spend their lives in social groups that offer protection and other advantages.

With a nearly uniform distribution, individuals are more evenly spaced than we would expect on the basis of chance alone. Such distribution is relatively rare. It happens when competition for resources or territory is fierce, as in a nesting colony of seabirds.

We observe random distribution only when habitat conditions are nearly uniform, resource availability is fairly steady, and individuals of a population or pairs of them neither attract nor avoid one another. Each wolf spider does not hunt far from its burrow, which can be almost anywhere in forest soil (Figure 45.2b).

The scale of the study area and timing of a study can influence the observed pattern of distribution. For example, seabirds often are spaced almost uniformly at a nesting site, but nesting sites are clustered along a shoreline. Also, these birds crowd together during the breeding season, but disperse when breeding is over.

Figure 45.2 Three patterns of population distribution: (**a**) clumped, as in squirrelfish schools; (**b**) random, as when wolf spiders dig their burrows almost anywhere in forest soil; and (**c**) more or less uniform, as in a royal penguin nesting colony.

Take-Home Message

How do we describe a natural population?

- Each population has characteristic demographics, such as size, density, distribution pattern, and age structure.

- Environmental conditions and species interactions shape these characteristics, which may change over time.

45.2 | Elusive Heads to Count

■ Ecologists carry out field studies to test hypotheses about populations and to monitor the status of populations that are threatened or endangered.

■ Link to Sampling error 1.8

Many white-tailed deer (*Odocoileus virginianus*) live in the forests, fields, and suburbs of North America. How could you find out how many deer live in a particular region?

A full count would be a careful measure of absolute population density. In the United States, census takers attempt such a count of human populations every ten years, although not everyone answers the door. Ecologists sometimes make counts of large species in small areas, such as fur seals at their breeding grounds, and sea stars in a tidepool.

More often, a full count would be impractical, so they sample part of a population and estimate its total density. For instance, you could divide a map of your county into small plots, or quadrats. **Quadrats** are sampling areas of the same size and shape, such as rectangles, squares, and hexagons. You could count individual deer in several plots and, from that, extrapolate the average number for the county as a whole. Ecologists often make such estimates for plants and other species that stay put (Figure 45.3). Such estimates run the risk of sampling error (Section 1.8), if the number of sampled plots is not large.

Ecologists use **capture–recapture methods** to estimate the population sizes of deer and other animals that do not stay put. First, they trap and mark some individuals. Deer get collars, squirrels get tattoos, salmon get tags, birds get leg rings, butterflies get wing markers, and so forth (Figure 45.4). Marked animals are released at time 1. At time 2, traps are reset. The proportion of marked animals in the second sample is then taken to be representative of the proportion marked in the whole population:

$$\frac{\text{marked individuals in sampling at time 2}}{\text{total captured in sampling 2}} = \frac{\text{marked individuals in sampling at time 1}}{\text{total population size}}$$

Ideally, both marked and unmarked individuals of the population are captured at random, no marked animal is overlooked, and marking does not affect whether animals die or otherwise depart during the study interval.

In the real world, recaptured individuals might not be a random sample; they might over- or underrepresent their population. Squirrels marked after being attracted to bait in boxes might now be trap-happy or trap-shy. Instead of mailing tags of marked fish to ecologists, a fisherman may keep them as souvenirs. Birds lose leg rings.

Estimates of population size may also vary depending on the time of year they are made. The distribution of a population may change seasonally. Many types of animals move between different parts of their range in response to seasonal changes in resource abundance.

As with other population data, the accuracy of size estimates can be increased by repeated samplings. The more data that can be accumulated, the lower the risk of sampling error.

Figure 45.3 Easy-to-count creosote bushes near the eastern base of the Sierra Nevada. They are an example of a relatively uniform distribution pattern. Individual plants compete for scarce water in this desert, which has extremely hot, dry summers and mild winters.

Figure 45.4 Two individuals marked for population studies. (**a**) Florida Key deer and (**b**) Costa Rican owl butterfly (*Caligo*).

■ Populations are dynamic units. They are continually adding and losing individuals. All populations have a capacity to increase in number.

■ Link to Bacterial reproduction 21.5

Gains and Losses in Population Size

Populations continually change size. They increase in size because of births and **immigration**, the arrival of new residents from other populations. They decrease in size because of deaths and **emigration**, departure of individuals that then take up permanent residence elsewhere. For example, a freshwater turtle population changes size in the spring when young turtles emigrate from their home pond. The young emigrants typically become immigrants at another pond some distance away.

What about the individuals of species that migrate daily or seasonally? A **migration** is a recurring round-trip between regions, usually in response to expected shifts or gradients in environmental resources. Some or all members of a population leave an area, spend time in another area, then return. For our purposes, we may ignore these recurring gains and losses, because we can assume that they balance out over time.

From Zero to Exponential Growth

Zero population growth is an interval during which the number of births is balanced by an equal number of deaths. Population size remains stable, with no net increase or decrease in the number of individuals.

We can measure births and deaths in terms of rates per individual, or per capita. *Capita* means head, as in a head count. Subtract a population's per capita death rate (d) from its per capita birth rate (b) and you have the **per capita growth rate**, or r:

$$
\underset{\substack{\text{(per capita} \\ \text{growth rate)}}}{r} = \underset{\substack{\text{(per capita} \\ \text{birth rate)}}}{b} - \underset{\substack{\text{(per capita} \\ \text{death rate)}}}{d}
$$

As long as r remains constant and greater than zero, **exponential growth** will continue: Population size will increase by the same proportion in every successive time interval.

Imagine a population of 2,000 mice living in a field. If 1,000 mice are born each month, the birth rate is 0.5 per mouse per month (1,000 births/2,000 mice). If 200 mice die each month, the death rate is 200/2,000 = 0.1 per mouse per month. Given these birth and death rates, r is 0.5 − 0.1 = 0.4 per mouse per month. In other words, the mouse population grows by 4 percent each

Figure 45.5 Animated

(**a**) Net monthly increases in a hypothetical population of mice when the per capita rate of growth (r) is 0.4 per mouse per month and the starting population size is 2,000.

(**b**) Graph these numerical data and you end up with a J-shaped growth curve.

		Starting Population Size	Net Monthly Increase	New Population Size
$G = r \times$		2,000 =	800	2,800
$r \times$		2,800 =	1,120	3,920
$r \times$		3,920 =	1,568	5,488
$r \times$		5,488 =	2,195	7,683
$r \times$		7,683 =	3,073	10,756
$r \times$		10,756 =	4,302	15,058
$r \times$		15,058 =	6,023	21,081
$r \times$		21,081 =	8,432	29,513
$r \times$		29,513 =	11,805	41,318
$r \times$		41,318 =	16,527	57,845
$r \times$		57,845 =	23,138	80,983
$r \times$		80,983 =	32,393	113,376
$r \times$		113,376 =	45,350	158,726
$r \times$		158,726 =	63,490	222,216
$r \times$		222,216 =	88,887	311,103
$r \times$		311,103 =	124,441	435,544
$r \times$		435,544 =	174,218	609,762
$r \times$		609,762 =	243,905	853,667
$r \times$		853,667 =	341,467	1,195,134

A

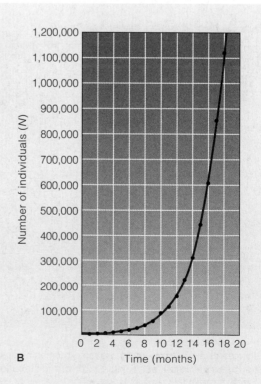

B

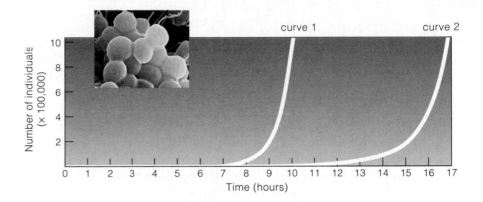

Figure 45.6 Effect of deaths on the rate of increase for two hypothetical populations of bacteria. Plot the population growth for bacterial cells that reproduce every half hour and you get growth curve 1. Next, plot the population growth of bacterial cells that divide every half hour, with 25 percent dying between divisions, and you get growth curve 2. Deaths slow the rate of increase, but as long as the birth rate exceeds the death rate and is constant, exponential growth will continue.

month. We can calculate the population growth (G) for each interval based on the per capita growth rate (r) and the number of individuals (N):

$$
\underset{\substack{\text{(population growth} \\ \text{per unit time)}}}{G} = \underset{\substack{\text{(per capita} \\ \text{growth rate)}}}{r} \times \underset{\substack{\text{(number of} \\ \text{individuals)}}}{N}
$$

After one month, 2,800 mice are scurrying about in the field (Figure 45.5a). A net increase of 800 fertile mice has made the reproductive base larger. They all reproduce, so the population size expands, for a net increase of 0.4 × 2,800 = 1,120. Population size is now 3,920. At this growth rate, the number of mice would rise from 2,000 to more than 1 million in two years! Plot the increases against time and you end up with a J-shaped curve that is characteristic of exponential growth (Figure 45.5b).

With exponential growth, a population grows faster and faster, although the per capita growth rate stays the same. It is like the compounding of interest on a bank account. The annual interest rate remains fixed, yet every year the amount of interest paid increases. Why? The annual interest paid into the account adds to the size of the balance, and the next interest payment will be calculated based on that larger balance.

In exponentially growing populations, r is like the interest rate. Although r remains constant, population growth accelerates as the population size increases. When 6,000 individuals reproduce, population growth is three times higher than it was when there were only 2,000 reproducers.

As another example, think of a single bacterium in a culture flask. After thirty minutes, the cell divides in two. Those two cells divide, and so on every thirty minutes. If no cells die between divisions, then the population size will double in every interval—from 1 to 2, then 4, 8, 16, 32, and so on. The time it takes for a population to double in size is its **doubling time**.

Consider how doubling time works in our flask of bacteria. After 9–1/2 hours, or nineteen doublings, there are more than 500,000 bacterial cells. After ten hours, or twenty doublings, there are more than one

million. Curve 1 in Figure 45.6 is a plot of this change over time.

The size of r affects the speed of exponential growth. Suppose 25 percent of the bacteria in our hypothetical flask die every 30 minutes. Under these conditions, it would take 17 hours, rather than 10, for the population to reach 1 million (curve 2 in Figure 45.6). The higher death rate decreases r, so exponential growth occurs more slowly. However, as long as r is greater than zero and constant, growth plots out as a J-shaped curve.

What Is the Biotic Potential?

Now imagine a population living in an ideal habitat, free of all threats such as predators and pathogens. Every individual has plenty of shelter, food, and other vital resources. Under such conditions, a population would reach its **biotic potential**: the maximum possible per capita rate of increase for its species.

All species have a characteristic biotic potential. For many bacteria, it is 100 percent every half hour or so. For humans, it is about 2 to 5 percent per year.

The actual growth rate depends on many factors. A population's age distribution, how often its individuals reproduce, and how many offspring an individual can produce are examples. The human population has not reached its biotic potential, but it is growing exponentially. We will return to the topic of the human population later in the chapter.

Take-Home Message

What determines the size of a population and its growth rate?

■ The size of a population is influenced by its rates of births, deaths, immigration, and emigration.

■ Subtract the per capita death rate from the per capita birth rate to get r, the per capita growth rate of a population. As long as r is constant and greater than zero, a population will grow exponentially. With exponential growth, the number of individuals increases faster and faster over time.

■ The biotic potential of a species is its maximum possible population growth rate under optimal conditions.

- Natural populations seldom continue to grow unchecked.
- Competition and crowding can slow growth.

Environmental Limits on Growth

Most of the time, a population cannot fulfill its biotic potential because of environmental limits. That is why sea stars—the females of which could make 2,500,000 eggs each year—do not fill the oceans with sea stars.

Any essential resource that is in short supply is a **limiting factor** on population growth. Food, mineral ions, refuge from predators, and safe nesting sites are examples (Figure 45.7). Many factors can potentially limit population growth. Which specific factor is the first to be in short supply and thus limit growth varies from one environment to another.

To get a sense of the limits on growth, start again with a bacterial cell in a culture flask, where you can control the variables. First, enrich the culture medium with glucose and other nutrients bacteria require for growth. Next, let the cells reproduce.

Initially, growth may be exponential. Then it slows, and population size remains relatively stable. After a brief stable period, population size plummets until all the bacterial cells are dead. What happened? The larger population required more nutrients. Over time, nutrient levels declined, and the cells could no longer divide. Even after cell division stopped, existing cells kept taking up and using nutrients. When the nutrient supply was exhausted, the last cells died out.

Suppose you continued adding nutrients to the flask. Population growth would still slow and then halt. As before, the bacteria would eventually die. Why? Like other organisms, bacteria generate metabolic wastes. Over time, this waste would accumulate and poison the habitat preventing further growth. No population can grow exponentially forever. Remove one limiting factor and another one becomes limiting.

Carrying Capacity and Logistic Growth

Carrying capacity refers to the maximum number of individuals of a population that a given environment can sustain indefinitely. Ultimately, it means that the sustainable supply of resources determines population size. We can use the pattern of **logistic growth**, shown in Figure 45.8, to reinforce this point. By this pattern, a small population starts growing slowly in size, then it grows rapidly, then its size levels off as the carrying capacity is reached.

Figure 45.7 One example of a limiting factor. (**a**) Wood ducks build nests only inside cavities of specific dimensions. With the clearing of old growth forests, the access to natural cavities of the correct size and position is now a limiting factor on wood duck population size. (**b**) Artificial nesting boxes are being placed in preserves to help ensure the health of wood duck populations.

Figure 45.8 Animated Idealized S-shaped curve characteristic of logistic growth. After a rapid growth phase (time B to C), growth slows and the curve flattens as carrying capacity is reached (time C to D).

In the real world, population size often declines when a change in the environment lowers carrying capacity (time D to E). That happened to the human population of Ireland in the mid-1800s. Late blight, a disease caused by a water mold, destroyed the potato crop that was the mainstay of Irish diets (Section 22.8).

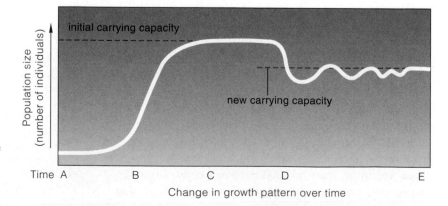

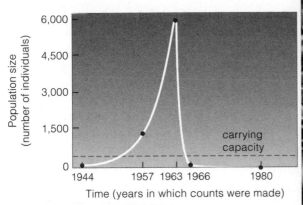

Graphing logistic growth yields an S-shaped curve, as shown in Figure 45.8 (A to C). In equation form,

$$\begin{array}{c} \text{population} \\ \text{growth per} \\ \text{unit time} \end{array} = \begin{array}{c} \text{maximum} \\ \text{per capita} \\ \text{population} \\ \text{growth rate} \end{array} \times \begin{array}{c} \text{number} \\ \text{of} \\ \text{individuals} \end{array} \times \begin{array}{c} \text{proportion} \\ \text{of resources} \\ \text{not yet used} \end{array}$$

An S-shaped curve is simply an approximation of what takes place in nature. Often a population that is growing fast overshoots its carrying capacity. Figure 45.9 shows what happened to a small population of reindeer. As the population size increased, more and more individuals competed for resources such as food and shelter, so each reindeer received a smaller share. More individuals died of starvation and fewer young were born. Deaths began to outnumber births. Finally, the death rate soared and the birth rate plummeted.

Two Categories of Limiting Factors

Density-dependent factors lower reproductive success and appear or worsen with crowding. Competition for limited resources leads to density-dependent effects, as does disease. Pathogens and parasites can spread more easily when hosts are crowded. As one example, human populations in cities support huge numbers of rats that can carry bubonic plague, typhus, and other deadly infectious diseases.

Density-dependent factors control population size through negative feedback. High density causes these factors to come into play, then their effects act to lower population density. A logistic growth pattern results from this feedback effect.

Density-independent factors decrease reproductive success too, but their likelihood of occurring and their magnitude of effect are unaffected by crowding. Fires, snow storms, earthquakes, and other natural disasters affect crowded and uncrowded populations alike. For

Figure 45.9 Graph of changes in a reindeer population that exceeded its habitat's carrying capacity (*blue* dashed line) and did not recover.

In 1944, during World War II, a United States Coast Guard crew established a station on St. Matthew, an island 320 kilometers (200 miles) west of Alaska in the Bering Sea. They brought in 29 reindeer as a backup food source. Reindeer eat lichens. Thick mats of lichens cloaked the island, which is no more than 51 kilometers long and 6.4 kilometers (32 miles by 4 miles) across. World War II drew to a close before any reindeer were shot. The Coast Guard pulled out, leaving behind seabirds, arctic foxes, voles—and a herd of healthy reindeer with no predators big enough to hunt them.

In 1957, biologist David Klein visited St. Matthew. On a hike from one end of the island to the other, he counted 1,350 well-fed reindeer and saw trampled and overgrazed lichens. In 1963, Klein and three other biologists returned to the island. They counted 6,000 reindeer. They could not help but notice the profusion of reindeer tracks and feces, and a lot of trampled, dead lichens.

Klein returned to St. Matthew in 1966. Bleached-out reindeer bones littered the island. Forty-two reindeer were still alive. Only one was a male, it had abnormal antlers, which made it unlikely to reproduce. There were no fawns. Klein figured out that thousands of reindeer had starved to death during the unusually harsh winter of 1963–1964. By the 1980s, there were no reindeer on the island at all.

example, in December of 2004, a powerful tsunami (a giant wave caused by an earthquake) hit Indonesia. It killed about 250,000 people. The degree of crowding did not make the tsunami any more or less likely to happen, or to strike any particular island. The logistic growth equation cannot be used to predict effects of density-independent factors.

Take-Home Message
How do limiting factors affect population growth?

■ Carrying capacity is the maximum number of individuals of a population that can be sustained indefinitely by the resources in a given environment.

■ With logistic growth, population growth is fastest when density is low, slows as the population approaches carrying capacity, and then levels off.

■ Density-dependent factors such as disease result in a pattern of logistic growth. Density-independent factors such as natural disasters also affect population size.

- Life span, age at maturity, and the number of offspring produced vary widely among organisms. Natural selection influences these life history traits.

So far, you have looked at populations as if all of their members are identical with regard to age. For most species, however, individuals that make up a group are at many different stages of development. Often, those stages require different resources, as when cat-erpillars that eat leaves later develop into butterflies that sip nectar. In addition, individuals might be more or less vulnerable to danger at different stages.

In short, each species has a **life history pattern**. It has a set of adaptations that affect when an individual starts reproducing, how many offspring it has at one time, how often it reproduces, and other traits. In this section and the next, we will consider variables that underlie these age-specific patterns.

Life Tables

Each species has a characteristic life span, but only a few individuals survive to the maximum age possible. Death is more likely at some ages. Individuals tend to reproduce during an expected age interval and to be most likely to die during another interval.

Age-specific patterns in populations are useful to life insurance and health insurance companies as well as ecologists. Such investigators focus on a **cohort**—a group of individuals born during the same interval—from their time of birth until the last one dies.

Ecologists often divide a natural population into age classes and record the age-specific birth rates and mortality. The resulting data is summarized in a life table (Table 45.1). Such tables inform decisions about how changes, such as harvesting a species or altering its environment, might affect the species' numbers. Birth and death schedules for the northern spotted owl are one case in point. They were cited in federal court rulings that halted mechanized logging in the owl's habitat—old-growth forests of the Pacific Northwest.

Human life tables are usually not based on a real cohort. Instead, information about current conditions is used to predict the births and deaths for a hypothetical group. Table 45.2 is such a life table for humans based on conditions in the United States during 2003.

Survivorship Curves

A **survivorship curve** is a graph line that emerges when you plot a cohort's age-specific survival in its habitat. Each species has a characteristic survivorship curve. Three types are common in nature.

A type I curve indicates survivorship is high until late in life. Populations of large animals that bear one or, at most, a few offspring at a time and give these young extended parental care show this pattern (Figure 45.10a). For example, a female elephant has one calf at a time and cares for it for several years. Type I curves are typical of human populations when individuals have access to good health care.

Table 45.1 Life Table for an Annual Plant Cohort*

Age Interval (days)	Survivorship (number surviving at start of interval)	Number Dying During Interval	Death Rate (number dying/ number surviving)	"Birth" Rate During Interval (number of seeds from each plant)
0–63	996	328	0.329	0
63–124	668	373	0.558	0
124–184	295	105	0.356	0
184–215	190	14	0.074	0
215–264	176	4	0.023	0
264–278	172	5	0.029	0
278–292	167	8	0.048	0
292–306	159	5	0.031	0.33
306–320	154	7	0.045	3.13
320–334	147	42	0.286	5.42
334–348	105	83	0.790	9.26
348–362	22	22	1.000	4.31
362–	0	0	0	0
		996		

* *Phlox drummondii*; data from W. J. Leverich and D. A. Levin, 1979.

Table 45.2 Life Table for Humans in the United States (based on 2003 conditions)

Age Interval	Number at Start of Interval	Number Dying During Age Interval	Life Expectancy (Years Remaining) at Start of Interval	Reported Live Births
0–1	100,000	687	77.5	
1–5	99,313	124	77.0	
5–10	99,189	73	73.1	
10–15	99,116	95	68.2	6,781
15–20	99,022	328	63.2	415,262
20–25	98,693	474	58.4	1,034,454
25–30	98,219	467	53.7	1,104,485
30–35	97,752	542	48.9	965,633
35–44	97,210	767	45.2	475,606
44–45	96,444	1,157	39.5	103,679
45–50	95,287	1,702	35.0	5,748
50–55	93,585	2,441	30.6	374
55–60	91,185	3,425	26.3	
60–65	87,760	5,092	22.2	
65–70	82,668	7,133	18.4	
70–75	75,535	9,825	14.9	
75–80	65,710	12,969	11.8	
80–85	52,741	15,753	9.0	
85–90	36,988	15,648	6.8	
90–95	21,344	12,363	5.0	
95–100	8,977	6,614	3.6	
100+	2,363	2,363	2.6	

A type II curve indicates that death rates do not vary much with age (Figure 45.10b). In lizards, small mammals, and big birds, old individuals are about as likely to die of disease or predation as young ones.

A type III curve indicates that the death rate for a population peaks early in life. It is typical of species that produce many small offspring and provide little or no parental care. Figure 45.10c shows how the curve plummets for sea urchins, which release great numbers of eggs. Sea urchin larvae are soft and tiny, so fish, snails, and sea slugs devour most of them before protective hard parts can develop. A type III curve is common for marine invertebrates, insects, fishes, fungi, and for annual plants such as phlox (Table 45.1).

Reproductive Strategies

Some organisms such as bamboo and Pacific salmon reproduce just once, then die. Others such as oak trees, mice, and humans reproduce repeatedly. A one-shot strategy is favored when an individual is unlikely to have a second chance to reproduce. For Pacific salmon, reproduction requires a life-threatening journey from the sea to a stream. For bamboo, environmental conditions that favor reproduction occur only sporadically.

Population density may also influence the optimal reproductive strategy. At low density, there will be little competition for resources, so individuals who turn resources into offspring fast are at an advantage. Such individuals reproduce while still young, produce many small offspring, and invest very little in parental care. Selection that favors traits that maximize number of offspring is called *r*-selection. When population density nears the carrying capacity, outcompeting others for resources becomes more important. Big individuals that reproduce later in life and produce fewer, higher quality offspring have the advantage in this scenario. Selection for traits that improve offspring quality is *K*-selection. Some organisms have traits associated mainly with *r*-selection or with *K*-selection, but most have a mixture of these traits.

a Elephants have type I survivorship, with low mortality until advanced age.

b Snowy egrets are type II populations, with a fairly constant death rate.

c Sea urchins are type III populations. Spines protect this adult, but the larvae are tiny, soft-bodied, and vulnerable to predation.

Figure 45.10 Three generalized survivorship curves and examples.

Take-Home Message

How do researchers study and describe life history patterns?

■ Tracking a cohort (a group of individuals) from their birth until the last one dies reveals patterns of reproduction, death, and migrations.

■ Survivorship curves reveal differences in age-specific survival among species or among populations of the same species.

■ Different environments and population densities can favor different reproductive strategies.

- Predation can serve as a selection pressure that shapes life history patterns.

- Links to Directional selection 18.4, Gene flow 18.8

Predation on Guppies in Trinidad Several years ago, two evolutionary biologists drenched with sweat and clutching fishnets were wading through a stream. John Endler and David Reznick were in the mountains of Trinidad, an island in the southern Caribbean Sea. They wanted to capture guppies (*Poecilia reticulata*), small fishes that live in the shallow freshwater streams (Figure 45.11). The biologists were beginning what would become a long-term study of guppy traits, including life history patterns.

Male guppies are usually smaller and more colorful than female guppies of the same age. A male's colors serve as visual signals during courtship rituals. The drabber females are less conspicuous to predators and, unlike males they continue to grow after reaching sexual maturity.

Reznick and Endler were interested in how predators influence the life history of guppies. For their study sites, they decided on streams with many small waterfalls. These waterfalls are barriers that prevent guppies in one part of a stream from moving easily to another. As a result, each stream holds several populations of guppies, and very little gene flow occurs among those populations (Section 18.8).

The waterfalls also keep guppy predators from moving into different parts of the stream. In this habitat, the main

a *Right*, guppy that shared a stream with killifishes (*below*).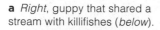

b *Right*, guppy that shared a stream with cichlids (*below*).

Figure 45.11 (**a,b**) Guppies and two guppy eaters, a killifish and a cichlid. (**c**) Biologist David Reznick contemplating interactions among guppies and their predators in a freshwater stream in Trinidad.

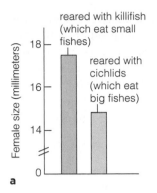
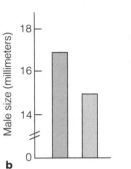
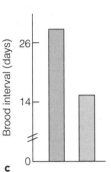
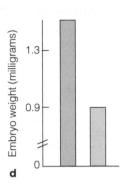

a   b   c   d

Figure 45.12 Experimental evidence of natural selection among guppy populations subject to different predation pressures. Compared to the guppies raised with killifish (*green* bars), guppies raised with cichlids (*tan* bars) differed in body size and in the length of time between broods.

guppy predators are killifish and cichlids. These two types of predatory fish differ in size and prey preferences. The killifish is relatively small and preys mostly on immature guppies. It ignores the larger adults. The cichlids are large fish. They tend to pursue mature guppies and ignore the small ones. Some parts of the streams hold one type of predator but not the other, so different guppy populations face different predation pressures.

As Reznick and Endler discovered, guppies in streams with cichlids grow faster and are smaller at maturity than those in streams with killifish (Figure 45.12). Also, guppies hunted by cichlids reproduce earlier, have more offspring at a time, and breed more frequently.

Were the differences in life history traits genetic, or did environmental differences cause them? To find out, the scientists collected guppies from cichlid-dominated and killifish-dominated streams. They reared these two groups in separate aquariums under identical conditions, with no predators present. Two generations later, the life history traits of these groups still differed, as they had in natural populations. Apparently, the differences in life history traits observed in the wild do have a genetic basis.

Reznick and Endler hypothesized that predators serve as selective agents that influence guppy life history traits. The scientists made a prediction: If life history traits are adaptive responses to predation, then these traits will change when a population is exposed to a new predator.

To test their prediction, Reznick and Endler found a stream region above a waterfall that had killifish but no guppies or cichlids. They brought in some guppies from a region below the waterfall where there were cichlids but no killifish. At the experimental site, the guppies that had previously lived only with cichlids were now exposed to killifish. The control site was the downstream region below the waterfall, where relatives of the transplanted guppies still coexisted with cichlids.

Reznik and Endler revisited the stream over the course of eleven years and thirty-six generations of guppies. They monitored traits of guppies above and below the waterfall. Their data showed that guppies at the upstream experimental site were evolving. Exposure to a novel predator had caused big changes in their rate of growth, age at

first reproduction, and other life history traits. By contrast, guppies at the control site showed no such changes. As Reznick and Endler concluded, life history traits in guppies can evolve rapidly in response to the selective pressure exerted by predation.

Overfishing and the Atlantic Cod The evolution of life history traits in response to predation pressure is not merely interesting. It has commercial importance. Just as guppies evolved in response to predators, the North Atlantic codfish (*Gadus morhua*) evolved in response to fishing pressure. North Atlantic codfish can be big (*below*). From the mid-1980s to early 1990s, the number of fisherman pursuing codfish rose. Fishermen kept the largest fish, and threw smaller ones back. This human behavior put codfish that became sexually mature when they were still small at an advantage, and such fish became increasingly common. As codfish numbers declined, smaller and smaller fish were kept.

Looking back, a rapid decline in age at first reproduction was a sign that the cod population was under great pressure. In 1992, Canada banned cod fishing in some areas. That ban, and later restrictions, came too late to stop the Atlantic cod population from plummeting. The population still has not recovered from this decline.

Had biologists recognized the life history changes as a warning sign, they might have been able to save this fishery and protect the livelihood of thousands of workers. Monitoring the life history data for other economically important fishes may help prevent over-fishing of other species in the future.

Human Population Growth

- The size of the human population is at its highest level ever and is expected to continue to increase.

- Links to Infectious disease 21.8, Human dispersal 26.15

The Human Population Today

In 2008, the estimated average rate of increase for the human population was 1.16 percent per year. As long as birth rates continue to exceed death rates, annual additions will drive a larger absolute increase each year into the foreseeable future.

Although many people enjoy abundant resources, about a fifth of the human population lives in severe poverty, and more than 800 million are malnourished (Figure 45.13). More than 1 billion people lack access to clean drinking water. More than 2 billion people face a shortage in fuelwood, which they depend on to heat their homes and cook their food. Rising populations will only increase pressure on limited resources.

Extraordinary Foundations for Growth

How did we get into this predicament? For most of its history, the human population grew very slowly. The growth rate began to increase about 10,000 years ago, and during the past two centuries, growth rates soared (Figure 45.14). Three trends promoted the large increases. First, humans were able to migrate into new habitats and expand into new climate zones. Second, humans developed new technologies that increased the carrying capacity of existing habitats. Third, humans sidestepped some limiting factors that tend to restrain the growth of other species.

Geographic Expansion Early humans evolved in the dry woodlands of Africa, then moved into the savannas. We assume they subsisted mainly on plant foods, but they probably also scavenged bits of meat. Bands of hunter–gatherers moved out of Africa about 2 million years ago. By 44,000 years ago, their descendants were established in much of the world (Section 26.15).

Few species can expand into such a broad range of habitats, but the early humans had large brains that allowed them to develop the necessary skills. They learned how to start fires, build shelters, make clothing, manufacture tools, and cooperate in hunts. With the advent of language, knowledge of such skills did not die with the individual. Compared to most species, humans displayed a greater capacity to disperse fast over long distances and to become established in physically challenging new environments.

Increased Carrying Capacity Beginning about 11,000 years ago, bands of hunter–gatherers were shifting to agriculture. Instead of counting on the migratory game herds, they were settling in fertile valleys and other regions that favored seasonal harvesting of fruits and grains. They developed a more dependable basis for life. A pivotal factor was the domestication of wild grasses, including species ancestral to modern wheat and rice. Now people harvested, stored, and planted seeds all in one place. They domesticated animals as sources of food and to pull plows. They dug irrigation ditches and diverted water to croplands.

Agricultural productivity became a basis for increases in population growth rates. Towns and cities formed. Later, food supplies increased yet again. Farmers started to use chemical fertilizers, herbicides, and pesticides to protect their crops. Transportation and food distribution improved. Even at its simplest, the management of food supplies through agricultural practices increased the carrying capacity for the human population.

Sidestepped Limiting Factors Until about 300 years ago, malnutrition and infectious diseases

Banks of corn silos in Wisconsin

Figure 45.13 Far from well-fed humans in highly developed countries, an Ethiopian child shows the effects of starvation. Ethiopia is one of the poorest developing countries, with an annual per capita income of $120. Average caloric intake is more than 25 percent below the minimum necessary to maintain good health. Malnutrition stunts the growth, weakens the body, and impairs the brain development of about half of Ethiopia's children. Despite ongoing food shortages, Ethiopia's population has one of the highest annual rates of increase in the world. If growth continues at its current rate, the population of 75 million will double in less than 25 years.

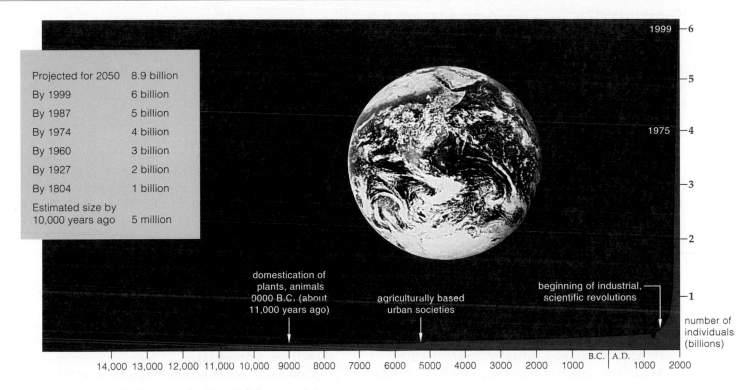

Projected for 2050	8.9 billion
By 1999	6 billion
By 1987	5 billion
By 1974	4 billion
By 1960	3 billion
By 1927	2 billion
By 1804	1 billion
Estimated size by 10,000 years ago	5 million

domestication of plants, animals 0000 B.C. (about 11,000 years ago)

agriculturally based urban societies

beginning of industrial, scientific revolutions

number of individuals (billions)

Figure 45.14 Growth curve (*red*) for the world human population. The *blue* box indicates how long it took for the human population to increase from 5 million to 6 billion. The dip between years 1347 and 1351 marks the time when 60 million people died during a pandemic that may have been a bubonic plague.

kept death rates high enough to more or less balance birth rates. Infectious diseases are density-dependent controls. Plagues swept through crowded cities. In the mid-1300s, one third of Europe's population was lost to a pandemic known as the Black Death. Waterborne diseases such as cholera that are associated with poor sanitation ran rampant. Then plumbing improved and vaccines and medications began to cut the death toll from disease. Births increasingly outpaced deaths—*r* became larger and exponential growth accelerated.

The industrial revolution took off in the middle of the eighteenth century. People had discovered how to harness the energy of fossil fuels, starting with coal. Within decades, cities of western Europe and North America became industrialized. World War I sparked the development of more technologies. After the war, factories turned to mass production of cars, tractors, and other affordable goods. Advances in agricultural practices meant that fewer farmers were required to support a larger population.

In sum, by controlling disease agents and tapping into fossil fuels—a concentrated source of energy—the human population sidestepped many factors that had previously limited its rate of increase.

Where have the far-flung dispersals and ongoing advances in technology and infrastructure gotten us? It took more than 100,000 years for the human population size to reach 1 billion. As Figure 45.14 shows, it took just 123 years to reach 2 billion, 33 more to reach 3 billion, 14 more to reach 4 billion, and then 13 more to get to 5 billion. It took only 12 more years to arrive at 6 billion! No doubt new technology will continue to increase Earth's human carrying capacity, but growth cannot be sustained indefinitely.

Why not? Ongoing increases in population size will cause density-dependent controls to exert their effects. For instance, globe-hopping travelers can carry pathogens to dense urban areas all around the world in a matter of weeks (Section 21.8). Also, limited resources cause economic hardship and civil strife.

Take-Home Message

Why have human populations grown so much, and what can we expect?

■ Through expansion into new habitats, cultural interventions, and technological innovations, the human population has temporarily skirted environmental resistance to growth.

■ Without technological breakthroughs, density-dependent controls will kick in and slow human population growth.

■ Acknowledgment of the risks posed by rising populations has led to increased family planning in almost every region.

■ Links to AIDS Chapter 21 introduction, Contraception 42.9

Some Projections

Most governments recognize that population growth, resource depletion, pollution, and quality of life are interconnected. Many offer family planning programs, and the United Nations Population Division estimates that about 60 percent of the world's married women now use some sort of contraception.

An increase in contraceptive use is contributing to a global decline in birth rate. Death rates are also falling in most regions. Improved diet and health care are lowering the infant mortality rate (the number of infants per 1,000 who die in their first year). On the other hand, AIDS has caused the death rate to soar in some African countries (Chapter 21 introduction).

World population is expected to peak at 8.9 billion by 2050, and possibly to decline as the century ends. Think of all the resources that will be required. We will have to boost food production, and find more energy and fresh water to meet even the most basic needs of billions more people. Utilizing natural resources on a larger scale will intensify pollution.

We expect to see the most growth in India, China, Pakistan, Nigeria, Bangladesh, and Indonesia, in that order. China (with 1.3 billion people) and India (with 1.09 billion) dwarf other countries; together, they hold 38 percent of the world population. Next in line is the United States, with 294 million.

Shifting Fertility Rates

The **total fertility rate** (TFR) is the average number of children born to the women of a population during their reproductive years. In 1950, the worldwide TFR averaged 6.5. Currently it is 2.7, which is still above the replacement level of 2.1—or the average number of children a couple must bear to keep the population at a constant level, given current death rates.

TFRs vary among countries. TFRs are at or below replacement levels in many developed countries; the developing countries in western Asia and Africa have the highest. Figure 45.15 has some examples of the disparities in demographic indicators.

Comparing age structure diagrams is revealing. In Figure 45.16, focus on the reproductive age category for the next fifteen years. Women generally bear children when they are 15 to 35 years old. We can expect populations that have a broad base to grow faster. The United States population has a relatively narrow base below a wide area that represents the 78 million baby-boomers (Figure 45.16c). This cohort began forming in 1946 when American soldiers came home after World War II and started to raise families.

Global increases in population seem certain. Even if every couple from this time forward has no more than two children, population growth cannot slow for sixty years. About 1.9 billion are about to enter their reproductive years. More than one-third of the world population is in the broad pre-reproductive base.

China has the most wide-reaching family planning program. Its government discourages premarital sex. It urges people to delay marriage and limit families to one or two children. It offers abortions, contraceptives, and sterilization at no cost to married couples, which mobile units and paramedics provide even in remote areas. Couples who follow guidelines get more food, free medical care, better housing, and salary bonuses.

Population in 2006	298 million	
	188 million	
	132 million	
Population in 2025 (projected)	349 million	
	211 million	
	206 million	
Population under age 15	20%	
	26%	
	42%	
Population above age 65	13%	
	6%	
	3%	
Total fertility rate (TFR)	2.1	
	1.9	
	5.5	
Infant mortality rate	6 per 1,000 live births	
	29 per 1,000 live births	
	97 per 1,000 live births	
Life expectancy	78 years	
	72 years	
	47 years	
Per capita income	$43,740	
	$3,460	
	$560	

Figure 45.15 Key demographic indicators for three countries, mainly in 2006. The United States (*brown* bar) is highly developed, Brazil (*red* bar) is moderately developed, and Nigeria (*beige* bar) is less developed. **Figure It Out:** What is the difference in life expectancy between the United States and Nigeria?   *Answer: 31 years*

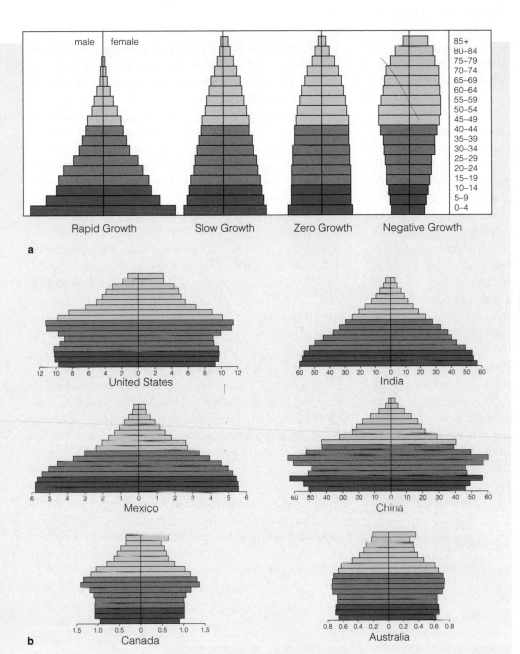

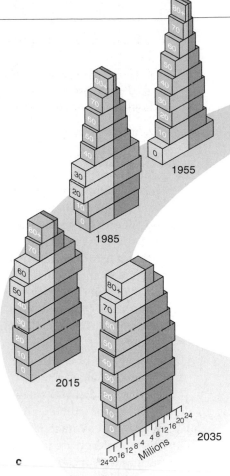

Figure 45.16 Animated (a) General age structure diagrams for countries with rapid, slow, zero, and negative rates of population growth. The pre-reproductive years are the *green* bars; reproductive years, *purple*; post-reproductive years, *light blue*. A vertical axis divides each graph into males (*left*) and females (*right*). Bar widths correspond to the proportions of individuals in each age group.

(b) 1997 age structure diagrams for six nations. Population sizes are measured in millions.

(c) Sequential age structure diagrams for the United States population. *Gold* bars track the baby-boomer generation.

Their offspring get free tuition and special treatment when they enter the job market. Parents having more than two children lose benefits and pay more taxes.

Since 1972, China's TFR has fallen sharply, from 5.7 to 1.75. An unintended consequence has been a shift in the country's sex ratio. Traditional cultural preference for sons, especially in rural areas, led some parents to abort female fetuses or commit infanticide. Worldwide, 1.06 boys are born for every girl. However, among those under age 15 in China, there are 1.134 boys for every girl. More than 100,000 girls are abandoned each year. The government is offering additional cash and tax incentives to the parents of girls. In the meantime,

the population time bomb keeps on ticking in China. About 150 million of its young females now make up the pre-reproductive age category.

■ The most developed countries have the slowest growth rates and use the most resources. As more countries become industrialized, pressure on Earth's resources will increase.

Demographic Transitions

The **demographic transition model** describes how the population growth rate changes as a country becomes more developed (Figure 45.17). Living conditions are harsh in the preindustrial stage, before technological and medical advances spread. Birth and death rates are both high, so the rate of population growth is low. In the transitional stage, industrialization begins. Food production and health care improve, and the death rate slows. Not surprisingly, in agricultural societies where families are expected to help in the fields, the birth rate is high. The annual growth rates in such societies are between 2.5 to 3 percent. When living conditions improve, the birth rate starts to fall and the population size levels off.

In the industrial stage, population growth slows. Cities filled with employment opportunities attract people, and average family size declines. Large numbers of children are no longer required to work a farm, and higher survival means it is not necessary to have many offspring to ensure that a few live.

In the postindustrial stage, the population growth rate becomes negative. The birth rate falls below the death rate, and the population size slowly decreases.

The United States, Canada, Australia, the bulk of western Europe, Japan, and much of the former Soviet Union have reached the industrial stage. Developing countries such as Mexico are now in the transitional stage, with people continuing to migrate to cities from agricultural regions.

Many currently developing countries are expected to enter the industrial stage in the next few decades. However, there are concerns that the continued rapid population growth in these countries will overwhelm their economic growth, food production, and health care systems.

The demographic transition model was developed to describe what happened when western Europe and North America became industrialized. It may not be relevant to today's less developed countries, which receive aid from existing highly developed countries, and must also compete against these countries in a global market.

There are also regional differences in how well the transition to an industrial stage is proceeding. In Asia, rising affluence is bringing higher life expectancy and lowered birth rates, as predicted. However, in sub-Saharan Africa, the AIDS epidemic is keeping some countries from moving out of the lowest stage of economic development.

Resource Consumption

Industrialized nations use the most resources. As an example, the United States accounts for about 4.6 percent of the world's population, yet it uses about 25

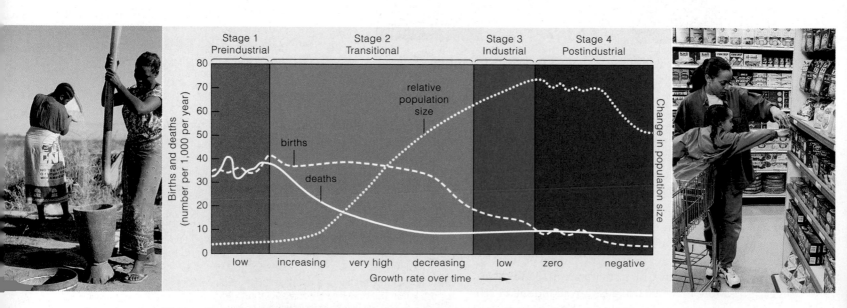

Figure 45.17 Animated Demographic transition model for changes in population growth rates and sizes, correlated with long-term changes in the economy.

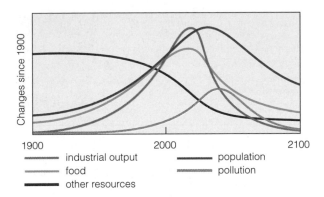

Figure 45.18 Computer-based projection of what might happen if human population size continues to skyrocket without dramatic policy changes and technological innovation. The assumptions were that the population has already overshot the carrying capacity and current trends will continue unchanged.

— industrial output   — population
— food   — pollution
— other resources

percent of the world's minerals and energy supplies. Billions of people living in India, China, and other less developed nations dream of owning the same kinds of consumer goods as people in developed countries. Earth does not have enough resources to make that possible. For everyone now alive to have a lifestyle like an average American would require four times the resources present on Earth.

What will happen if the human population keeps on increasing as predicted? How will we find the food, energy, water, and other basic resources needed to sustain so many people? Can we provide the necessary education, housing, medical care, and other social services? Some models suggest not (Figure 45.18). Other analysts claim we can adapt to a more crowded world if innovative technologies improve crop yields, if people rely less on meat for protein, and if resources are shared more equitably among regions. We have made great strides in increasing our agricultural output, but have been less successful in getting food to the people who need it.

Take-Home Message

How does industrialization affect population growth and resource consumption?

■ Differences in population growth and resource consumption among countries can be correlated with levels of economic development. Growth rates are typically greatest during the transition to industrialization.

■ Global conditions have changed so that the demographic transition model may no longer apply to modern nations.

■ An average person living in a highly developed nation uses far more resources than a person in a less-developed nation.

45.10 | Rise of the Seniors

■ While some countries face overpopulation, others have declining birth rates and an increasing average age.

In some developed countries, the decreasing total fertility rate and increasing life expectancy have resulted in a high proportion of older adults. In Japan, people over 65 currently make up about 20 percent of the population. In the United States, the proportion of people over 65 is projected to reach this level by 2030 (Figure 45.19). In 2050, there could be as many as 31 million Americans over age 85.

The aging of a population has social implications. Older individuals have traditionally been supported by a younger workforce. In the United States, most older people receive social security payments and government-subsidized medical care. As a result of inflation and increases in life expectancy, the benefits being distributed to current seniors exceed the contributions these people paid into the program. When baby boomers begin to receive benefits, the deficit will skyrocket. Keeping the system going will require ever greater contributions from the younger, still-working population. Increasing numbers of debilitated seniors will also challenge the health care system. Thus, finding ways to keep people healthy later in life is both a social and an economic priority.

Figure 45.19 Two of the 37 million Americans over age 65.

Take-Home Message

How does slowing population growth affect age distribution?

■ When population growth slows, the proportion of older individuals rises.

The Numbers Game

Many states are struggling to control rising numbers of white-tailed deer. In Ohio, the number has risen from 17,000 deer in 1970 to more than 700,000. In West Virginia, deer are overbrowsing plants that grow on the forest floor, including wild ginseng, which is an important export crop. Biologist James McGraw argues that controlling deer and saving West Virginia's forests will require either reintroducing big predators or increasing deer hunting.

How would you vote?

Without natural predators, deer numbers are soaring. Is encouraging deer hunting the best solution? See CengageNOW for details, then vote online.

Summary

Sections 45.1, 45.2 Each population is a group of individuals of the same species. Its growth is affected by its **demographics**. These include **population size** and **age structure**, such as the size of the **reproductive base**. They also include **population density** and **population distribution**. Most populations in nature have a clumped distribution pattern.

Counting the number of individuals in **quadrats** is a way to estimate the density of a population in a specified area. **Capture–recapture methods** can be used to estimate the population density for mobile animals.

- *Use the interaction on CengageNOW to learn how to estimate population size.*

Section 45.3 **Immigration** and **emigration** permanently affect population size, but **migration** does not. The per capita birth rate minus the per capita death rate gives us *r*, the population's **per capita growth rate**. When births equal deaths we have **zero population growth**.

In cases of **exponential growth**, a population's growth is proportional to its size. The population size increases at a fixed rate in any given interval. The time required for a population to double is the **doubling time**. The maximum possible rate of increase is a species' **biotic potential**.

- *View the animation on CengageNOW to observe a pattern of exponential growth.*

Section 45.4 **Limiting factors** constrain population increases. With **logistic growth**, a small population starts growing slowly, then grows rapidly, then levels off once **carrying capacity** is reached. **Density-dependent factors** are conditions or events that lower reproductive success and have an increasing effect with crowding. **Density-independent factors** are conditions or events that can lower reproductive success, but their effect does not vary with crowding.

- *Watch the animation on CengageNOW to learn about logistic growth.*

Sections 45.5, 45.6 The time to maturity, number of reproductive events, number of offspring per event, and life span are aspects of a **life history pattern**. A **cohort** is a group of individuals that were born at the same time. Three types of **survivorship curves** are common: a high death rate late in life, a constant rate at all ages, or a high rate early in life. Life histories have a genetic basis

and are subject to natural selection. At low population density, *r*-**selection** favors quickly producing as many offspring as possible. At a higher population density, *K*-**selection** favors investing more time and energy in fewer, higher quality offspring. Most populations have a mixture of both *r*-selected and *K*-selected traits.

Section 45.7 The human population has surpassed 6.6 billion. Expansion into new habitats and agriculture allowed early increases. Later, medical and technological innovations raised the carrying capacity and sidestepped many limiting factors.

Section 45.8 A population's **total fertility rate** (TFR) is the average number of children born to women during their reproductive years. The global TFR is declining and most countries have family planning programs of some sort. Even so, the pre-reproductive base of the world population is so large that population size will continue to increase for at least sixty years.

- *Use the interaction on CengageNOW to compare age structure diagrams.*

Section 45.9 The **demographic transition model** predicts how human population growth rates will change with industrialization. Generally, the death rate and birth rate both fall with rising industrialization, but conditions in countries can vary in ways that affect this trend.

Developed nations have a much higher per capita consumption of resources than developing nations. Earth does not have enough resources to support the current population in the style of the developed nations.

- *Use the interaction on CengageNOW to learn about the demographic transition model.*

Section 45.10 Slowing population growth leads to an increase in the proportion of elderly in the population.

Self-Quiz   *Answers in Appendix III*

1. Most commonly, individuals of a population show a _____ distribution through their habitat.
   a. clumped   c. nearly uniform
   b. random   d. none of the above

2. The rate at which population size grows or declines depends on the rate of _____ .
   a. births   c. immigration   e. a and b
   b. deaths   d. emigration   f. all of the above

Data Analysis Exercise

In 1989, Martin Wikelski started a long-term study of marine iguana populations in the Galápagos Islands (Section 17.2). He marked the iguanas on two islands— Genovesa and Santa Fe—and collected data on how their body size, survival, and reproductive rates varied over time. The iguanas eat algae and have no predators, so deaths are usually the result of food shortages, disease, or old age. His studies showed that numbers decline during El Niño events, when the surrounding waters heat up.

In January 2001, an oil tanker ran aground and leaked a small amount of oil into the waters near Santa Fe—Figure 45.20 shows the number of marked iguanas that Wikelski and his team counted in their census of study populations just before the spill and about a year later.

1. Which island had more marked iguanas at the time of the first census?

2. How much did the population size on each island change between the first and second census?

3. Wikelski concluded that changes on Santa Fe were the result of the oil spill, rather than sea temperature or other climate factors common to both islands. How would the census numbers be different from those he observed if an adverse event had affected both islands?

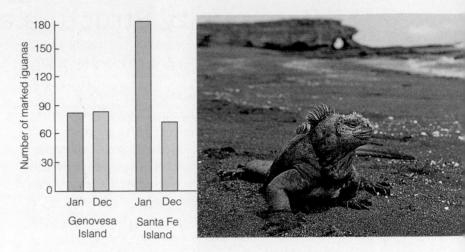

Figure 45.20 Shifting numbers of marked marine iguanas on two Galápagos islands. An oil spill occurred near Santa Fe just before the January 2001 census (*green* bars). A second census was carried out in December 2001 (*tan* bars).

3. Suppose 200 fish are marked and released in a pond. The following week, 200 fish are caught and 100 of them have marks. There are about _____ fish in this pond.
   a. 200   b. 300   c. 400   d. 2,000

4. A population of worms is growing exponentially in a compost heap. Thirty days ago there were 400 worms and now there are 800. How many worms will there be thirty days from now, assuming conditions remain constant?
   a. 1,200   b. 1,600   c. 3,200   d. 6,400

5. For a given species, the maximum rate of increase per individual under ideal conditions is its _____ .
   a. biotic potential   c. environmental resistance
   b. carrying capacity   d. density control

6. _____ is a density-independent factor that influences population growth.
   a. Resource competition   c. Predation
   b. Infectious disease   d. Harsh weather

7. A life history pattern for a population is a set of adaptations that influence the individual's _____ .
   a. longevity   c. age at reproductive maturity
   b. fertility   d. all of the above

8. The human population is now over 6.6 billion. It was about half that in _____ .
   a. 2004   b. 1960   c. 1802   d. 1350

9. Compared to the less developed countries, the highly developed ones have a higher _____ .
   a. death rate   c. total fertility rate
   b. birth rate   d. resource consumption rate

10. _____ population growth increases the proportion of older individuals in a population.
   a. Slowing   b. Accelerating

11. Match each term with its most suitable description.
   ___ carrying capacity
   ___ exponential growth
   ___ biotic potential
   ___ limiting factor
   ___ logistic growth

   a. maximum rate of increase per individual under ideal conditions
   b. population growth plots out as an S-shaped curve
   c. maximum number of individuals sustainable by the resources in a given environment
   d. population growth plots out as a J-shaped curve
   e. essential resource that restricts population growth when scarce

■ *Visit CengageNOW for additional questions.*

Critical Thinking

1. Think back to Section 45.6. When researchers moved guppies from populations preyed on by cichlids to a habitat with killifish, the life histories of the transplanted guppies evolved. They came to resemble those of guppy populations preyed on by killifish. Males became gaudier; some scales formed larger, more colorful spots. How might a decrease in predation pressure on sexually mature fish favor this change?

2. The age structure diagrams for two hypothetical populations are shown at right. Describe the growth rate of each population and discuss the current and future social and economic problems that each is likely to face.

Community Structure and Biodiversity

IMPACTS, ISSUES Fire Ants in the Pants

Step on a nest of red imported fire ants, *Solenopsis invicta* (Figure 46.1*a*), and you will be sorry. The ants are quick to defend their nest. Ants stream out from the ground and inflict a series of stings. Venom injected by the stinger causes burning pain and results in the formation of a pus-filled bump that is slow to heal. Multiple stings can cause nausea, dizziness, and—rarely—death.

S. invicta arrived in the United States from South America in the 1930s, probably as stowaways on a ship. The ants spread out from the Southeast and have been found as far west as California and as far north as Kansas and Delaware.

Like many introduced species, the ants disrupt natural communities. They attack livestock, pets, and wildlife. They also outcompete native ants and may be contributing to the decline of other native wildlife. For example, the Texas horned lizard vanished from most of its home range when *S. invicta* moved in and displaced the native ants—the lizard's food of choice. The horned lizard cannot tolerate eating the imported fire ants.

Invicta means "invincible" in Latin and *S. invicta* is living up to its species name. Pesticides have not managed to halt the foreign ant's spread. The chemicals might even be facilitating dispersal by preferentially wiping out native ant populations.

Ecologists are enlisting biological controls. Phorid flies control *S. invicta* in its native habitat (Figure 46.1*b*). The flies are parasitoids, a type of parasite that kills its host in a rather gruesome way. A female fly pierces the cuticle of an adult ant, then lays an egg in the ant's soft tissues. The egg hatches into a larva, which grows and eats its way through tissues to the ant's head. After the larva gets big enough, it causes the ant's head to fall off (Figure 46.1*c*). The larva develops into an adult within the detached head.

Several phorid fly species have now been introduced in various southern states. The flies are surviving, reproducing, and increasing their range. They probably will never kill off all *S. invicta* in affected areas, but they are expected to reduce the density of colonies.

This example introduces community structure: patterns in the number of species and their relative abundances. As you will see, species interactions and disturbances to the habitat can shift community structure in small and large ways—some predictable, others unexpected.

See the video! Figure 46.1 (**a**) Red imported fire ant (*S. invicta*) mounds. (**b**) A phorid fly that lays its eggs on the ants. (**c**) An ant that lost its head after the larva of a phorid fly moved into it.

Key Concepts

Community characteristics

A community consists of all species in a habitat. Each species has a niche—the sum of its activities and relationships. A habitat's history, its biological and physical characteristics, and the interactions among species in the habitat affect community structure. **Section 46.1**

Types of species interactions

Commensalism, mutualism, competition, predation, and parasitism are types of interspecific interactions. They influence the population size of participating species, which in turn influences the community's structure. **Sections 46.2–46.7**

Community stability and change

Communities have certain elements of stability, as when some species persist in a habitat. Communities also change, as when new species move into the habitat and others disappear. Physical characteristics of the habitat, species interactions, disturbances, and chance events affect how a community changes over time. **Sections 46.8–46.10**

Global patterns in community structure

Biogeographers identify regional patterns in species distribution. They have shown that tropical regions hold the greatest number of species, and also that characteristics of islands can be used to predict how many species an island will hold. **Section 46.11**

Links to Earlier Concepts

- In this chapter, you will see how natural selection (Section 17.3) and coevolution (18.12) shape traits of species in communities.

- You will revisit examples of interspecific interactions such as bacteria that live inside protists (20.4), plant–pollinator interactions (23.8, 30.2), lichens (24.6), and root nodules and mycorrhizae (29.2).

- You will consider again the evolution of prey defenses such as ricin (Chapter 14 introduction), nematocysts (25.5), and the way that evolution affects pathogens (21.8).

- Knowledge of biogeography (17.1) will help you understand how communities in different regions differ.

Which Factors Shape Community Structure?

- Community structure refers to the number and relative abundances of species in a habitat. It changes over time.

- Link to Coevolution 18.12

The type of place where a species normally lives is its **habitat**, and all species living in a habitat represent a **community**. A community has a dynamic structure. It shows shifts in its species diversity—the number and relative abundances of species.

Many factors influence community structure. First, climate and topography influence a habitat's features, including temperature, soil, and moisture. Second, a habitat has only certain kinds and amounts of food and other resources. Third, species themselves have traits that adapt them to certain habitat conditions, as in Figure 46.2. Fourth, the species interact in ways that cause shifts in their numbers and abundances. Finally, the timing and history of disturbances, both natural and human-induced, affect community structure.

The Niche

All species of a community share the same habitat—the same "address"—but each also has a "profession," or unique ecological role, that sets it apart. This role is the species' **niche**, which we describe in terms of

Table 46.1	Direct Two-Species Interactions	
Type of Interaction	Effect on Species 1	Effect on Species 2
Commensalism	Helpful	None
Mutualism	Helpful	Helpful
Interspecific competition	Harmful	Harmful
Predation	Helpful	Harmful
Parasitism	Helpful	Harmful

the conditions, resources, and interactions necessary for survival and reproduction. Aspects of an animal's niche include temperatures it can tolerate, the kinds of foods it can eat, and the types of places it can breed or hide. A description of a plant's niche would include its soil, water, light, and pollinator requirements.

Categories of Species Interactions

Species in a community interact in a variety of ways (Table 46.1) **Commensalism** benefits one species and does not affect the other. Most bacteria in your gut are commensal. They benefit by living inside you, but do not help or harm you. **Mutualism** provides benefits to both species. **Interspecific competition** hurts both species. **Predation** and **parasitism** help one species at another's expense. Predators are free-living organisms that kill their prey. Parasites live on or in a host and usually do not kill it.

Parasitism, commensalism, and mutualism can all be types of **symbiosis**, which means "living together." Symbiotic species, or symbionts, spend most or all of their life cycle in close association with each other. An endosymbiont is a species that lives inside its partner.

Regardless of whether one species helps or hurts another, two species that interact closely for extended periods may coevolve. With **coevolution**, each species is a selective agent that shifts the range of variation in the other (Section 18.12).

Figure 46.2 Three of twelve fruit-eating pigeon species in Papua New Guinea's tropical rain forests: (**a**) pied imperial pigeon, (**b**) superb crowned fruit pigeon, and (**c**) the turkey-sized Victoria crowned pigeon. The forest's trees differ in the size of fruit and fruit-bearing branches. The big pigeons eat big fruit. Smaller ones, with smaller bills, cannot peck open big, thick-skinned fruit. They eat the small, soft fruit on branches too spindly to hold big pigeons.

Trees feed the birds, which help the trees. Seeds in fruit resist digestion in the bird gut. Flying pigeons disperse seed-rich droppings, often some distance from mature trees that would outcompete new seedlings for water, minerals, and sunlight. With dispersal, seedlings have a better chance of surviving.

Take-Home Message

What is a biological community?

- A community consists of all species in a habitat, each with a unique niche, or ecological role.

- Species in a community interact and may benefit, harm, or have no net effect on one another. Some are symbionts; they associate closely for most or all of their life cycle.

46.2 | Mutualism

- A mutualistic interaction benefits both partners.

- Links to Endosymbiosis and organelles 20.4, Pollination 23.8 and 30.2, Lichens 24.6, Plant mutualisms 29.2

Mutualists are common in nature. For example, birds, insects, bats, and other animals serve as pollinators of flowering plants (Sections 23.8 and 30.2). Pollinators feed on energy-rich nectar and pollen. In return, they transfer pollen between plants, facilitating pollination. Similarly, pigeons take food from rain forest trees but disperse their seeds to new sites (Figure 46.2).

In some mutualisms, neither species can complete its life cycle without the other. Yucca plants and the moths that pollinate them show such interdependence (Figure 46.3). In other cases, the mutualism is helpful but not a life-or-death requirement. Most plants, for example, use more than one pollinator.

Mutualists help most plants take up mineral ions (Section 29.2). Nitrogen-fixing bacteria living on roots of legumes such as peas provide the plant with extra nitrogen. Mycorrhizal fungi living in or on plant roots enhance the plant's mineral uptake.

Other fungi partner with photosynthetic bacteria or algae, thus forming lichens (Section 24.6). In all mutualisms, there is some conflict between partners. In a lichen, the fungus would do best by obtaining as much sugar as possible from its photosynthetic partner. That partner would do best by keeping as much sugar as possible for its own use.

Some mutualists defend one another. For example, most fishes avoid sea anemones, which have stinging cells called nematocysts in their tentacles. However, an anemone fish can nestle among those tentacles (Figure 46.4). A mucus layer shields the anemone fish from stings, and the tentacles keep it safe from predatory fish. The anemone fish repays its partner by chasing off the few fishes that feed on sea anemone tentacles.

Finally, reflect on a theory outlined in Section 20.4, whereby certain aerobic bacteria became mutualistic endosymbionts of early eukaryotic cells. The bacteria received nutrients and shelter. In time, they evolved into mitochondria and provided the "host" with ATP. Cyanobacteria living inside eukaryotic cells evolved into chloroplasts by a similar process.

Figure 46.3 Mutualism in the high desert of Colorado.

Each species of *Yucca* plant is pollinated by one species of yucca moth, which cannot complete its life cycle with any other plant. The moth matures when yucca plants flower. A female moth collects yucca pollen and rolls it into a ball. She flies to another flower and pierces the floral ovary, and lays eggs inside. As she crawls out, she pushes a ball of pollen onto the flower's pollen-receiving platform.

After pollen grains germinate, they give rise to pollen tubes, which grow through the ovary tissues and deliver sperm to the plant's eggs. Seeds develop after fertilization.

Meanwhile, moth eggs develop into larvae that eat a few seeds, then gnaw their way out of the ovary. Seeds that larvae do not eat give rise to new yucca plants.

Figure 46.4 The sea anemone *Heteractis magnifica*, which shelters about a dozen fish species. It has a mutualistic association with the pink anemone fish (*Amphiprion perideraion*). This tiny but aggressive fish chases away predatory butterfly fishes that would bite off tips of anemone tentacles. The fish cannot survive and reproduce without the protection of an anemone. The anemone does not need a fish to protect it, but it does better with one.

Take-Home Message

What is mutualism?

- Mutualism is a species interaction in which each species benefits by associating with the other.

- In some cases the mutualism is necessary for both species; more often it is not essential for one or both partners.

■ Resources are limited and individuals of different species often compete for access to them.

■ Links to Natural selection 17.3, Limiting factor 45.4

As Charles Darwin understood, intense competition for resources among individuals of the same species leads to evolution by natural selection (Section 17.3). Competitive interactions between different species—interspecific competition—is not usually as intense. Why not? The requirements of two species might be similar, but they can never be as close as they are for individuals of the same species.

With interference competition, one species actively prevents another from accessing some resource. As an example, one species of scavenger will often chase

Figure 46.5 Interspecific competition among scavengers. (**a**) A golden eagle and a red fox face off over a moose carcass. (**b**) In a dramatic demonstration of interference competition, the eagle attacks the fox with its talons. After this attack, the fox retreated, leaving the eagle to exploit the carcass.

another away from a carcass (Figure 46.5). As another example, some plants use chemical weapons against potential competition. Aromatic chemicals that ooze from tissues of sagebrush plants, black walnut trees, and eucalyptus trees seep into the soil around these plants. The chemicals prevent other kinds of plants from germinating or growing.

In exploitative competition, species do not interact directly; each reduces the amount of resources available to the other by using that resource. For example, deer and blue jays both eat acorns in oak forests. The more acorns the birds eat, the fewer there are for the deer.

Effects of Competition

Deer and blue jays share a fondness for acorns, but each also has other sources of food. Any two species differ in their resource requirements. Species compete most intently when the supply of a shared resource is the main limiting factor for both (Section 45.4).

In the 1930s, G. Gause conducted experiments with two species of ciliated protists (*Paramecium*) that compete for bacterial prey. When cultured separately, the growth curves for these species were about the same. When grown together, growth of one species outpaced the other, and drove it to extinction (Figure 46.6).

Experiments by Gause and others are the basis for the concept of **competitive exclusion**: Whenever two species require the same limited resource to survive or reproduce, the better competitor will drive the less competitive species to extinction in that habitat.

Competitors can coexist when their resource needs are not exactly the same, however, competition generally supresses population growth of both species. For instance, Gause also studied two *Paramecium* species with differing food preferences. When grown together, one fed on bacteria suspended in culture tube liquid. The other ate yeast cells near the bottom of the tube. When grown together, population growth rates fell for both species, but they continued to coexist.

Experiments by Nelson Hairston showed the effects of competition between slimy salamanders (*Plethodon glutinosus*) and Jordan's salamanders (*P. jordani*). The salamanders coexist in wooded habitats (Figure 46.7). Hairston removed all slimy salamanders from certain test plots and Jordan's salamanders from others. He left a final group of plots unaltered as controls.

After five years, the numbers and abundances of the two species had not changed in the control plots. In the plots with slimy salamanders alone, population density had soared. Numbers also increased in plots with Jordan's salamanders alone. Hairston concluded

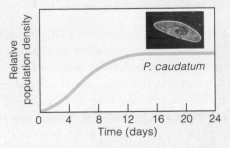

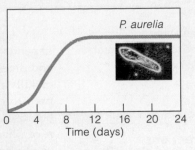

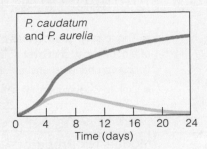

A *Paramecium caudatum* and *P. aurelia* grown in separate culture flasks established stable populations. The S-shaped graph curves indicate logistic growth and stability.

B For this experiment, the two species were grown together. *P. aurelia* (*brown* curve) drove *P. caudatum* toward extinction (*green* curve).

Figure 46.6 Animated Results of competitive exclusion between two related species that compete for the same food. Two species cannot coexist indefinitely in the same habitat *when they require identical resources.*

that whenever these salamanders coexist, competitive interactions suppress the population growth of both.

Resource Partitioning

Think back on those fruit-eating pigeon species. They all require fruit, but each eats fruits of a certain size. Their preferences are a case of **resource partitioning**: a subdividing of an essential resource, which reduces the competition among species that require it.

Similarly, three annual plant species live in the same field. They all require minerals and water, but their roots take them up at different depths (Figure 46.8).

When species with very similar requirements share a habitat, competition puts selective pressure on them. In each species, individuals who differ most from the competing species are favored. The outcome may be **character displacement**: Over the generations, a trait of one species diverges in a way that lowers the intensity of competition with the other species. Modification of the trait promotes partitioning of a resource.

Figure 46.7 Two species of salamanders, *Plethodon glutinosus* (*top*) and *P. jordani* (*bottom*), that compete in areas where their habitats overlap.

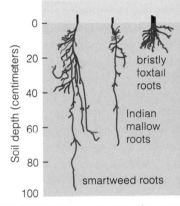

Figure 46.8 A case of resource partitioning among three annual plant species in a plowed but abandoned field. Roots of each species take up water and mineral ions from a different soil depth. This reduces competition among them and allows them to coexist.

For example, researchers Peter and Rosemary Grant demonstrated a change in beak size in the Galápagos finch *Geospiza fortis*. It occurred after a larger finch, *G. magnirostris*, moved onto the island where *G. fortis* had previously been alone. Arrival of *G. magnirostris* put big-beaked *G. fortis* individuals at a disadvantage. They now had to compete with *G. magnirostris* for big seeds. Small-beaked *G. fortis* had no such competition, and enjoyed higher reproductive success. As a result, the average beak size of *G. fortis* declined over time.

Take-Home Message

What happens when species compete for resources?

■ In some interactions, one species actively blocks another's access to a resource. In other interactions, one species is simply better than another at exploiting a shared resource.

■ When two species compete, selection favors individuals whose needs are least like those of the competing species.

■ The relative abundances of predator and prey populations of a community shift over time in response to species interactions and changing environmental conditions.

■ Link to Coevolution 18.12

Models for Predator–Prey Interactions

Predators are consumers that get energy and nutrients from **prey**, which are living organisms that predators capture, kill, and eat. The quantity and types of prey species affect predator diversity and abundance, and predator types and numbers do the same for prey.

The extent to which a predator species affects prey numbers depends in part on how individual predators respond to changes in prey density. Figure 46.9a compares models for the three main predator responses to increases in density.

In a type I response, the proportion of prey killed is constant, so the number killed in any given interval depends solely on prey density. Web-spinning spiders and other passive predators tend to show this type of response. As the number of flies in an area increases, more and more become caught in each spider's web. Filter-feeding predators also show a type I response.

In a type II response, the number of prey killed depends on the capacity of predators to capture, eat, and digest prey. When prey density increases, the rate of kills rises steeply at first because there are more prey to catch. Eventually, the rate of increase slows, because each predator is exposed to more prey than it can handle at one time. Figure 46.9b is an example of this type of response, which is common in nature. A wolf that just killed a caribou will not hunt another until it has eaten and digested the first one.

In a type III response, the number of kills increases slowly until prey density exceeds a certain level, then rises rapidly, and finally levels off. This response is common in nature in three situations. In some cases, the predator switches among prey, concentrating its efforts on the species that is most abundant. In other cases, the predators need to learn how to best capture each prey species; they get more lessons when more prey are around. In still other cases, the number of hiding places for prey is limited. Only after prey density rises and some individual prey have no place to hide, does the number of kills increase.

Knowing which type of response a predator makes to prey helps ecologists predict long-term effects of predation on a prey population.

The Canadian Lynx and Snowshoe Hare

In some cases, a time lag in the predator's response to prey density leads to cyclic changes in abundance of predators and prey. When prey density becomes low, the number of predators declines. As a result, prey are safer and their number increases. This increase allows predators to increase. Then predation causes another prey decline, and the cycle begins again.

Consider a ten-year oscillation in populations of a predator, the Canadian lynx, and the snowshoe hare

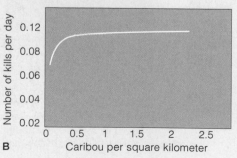

Figure 46.9 Animated (**a**) Three models for responses of predators to prey density. Type I: Prey consumption rises linearly as prey density rises. Type II: Prey consumption is high at first, then levels off as predator bellies stay full. Type III: When prey density is low, it takes longer to hunt prey, so the predator response is low. (**b**) A type II response in nature. For one winter month in Alaska, B. W. Dale and his coworkers observed four wolf packs (*Canis lupus*) feeding on caribou (*Rangifer tarandus*). The interaction fit the type II model for the functional response of predators to the prey density.

A — Prey population density
Number of prey killed per predator per unit time
I
II
III

B — Caribou per square kilometer
Number of kills per day
0.12
0.08
0.06
0.04
0.02
0   0.5   1   1.5   2   2.5

Figure 46.10 Graph of the abundances of Canadian lynx (*dashed* line) and snowshoe hares (*solid* line), based on counts of pelts sold by trappers to Hudson's Bay Company during a ninety-year period. Charles Krebs observed that predation causes heightened alertness among snowshoe hares, which continually look over their shoulders during the declining phase of each cycle. The photograph at *right* supports the Krebs hypothesis that there is a three-level interaction going on, one that involves plants.

The graph may be a good test of whether you tend to accept someone else's conclusions without questioning their basis in science. Remember those sections in Chapter 1 that introduced the nature of scientific methods?

What other factors may have had an impact on the cycle? Did the weather vary, with more severe winters imposing greater demand for hares (to keep lynxes warmer) and higher death rates? Did the lynx compete with other predators, such as owls? Did the predators turn to alternative prey during low points of the hare cycle?

that is its main prey (Figure 46.10). To determine the causes of this pattern, Charles Krebs and coworkers tracked hare population densities for ten years in the Yukon River Valley of Alaska. They set up one-square-kilometer control plots and experimental plots. They used fences to keep predatory mammals out of some plots. Extra food or fertilizers that helped plants grow were used in other plots. The researchers captured and put radio collars on more than 1,000 snowshoe hares, lynx, and other animals, and then released them.

In predator-free plots, the hare density doubled. In plots with extra food, it tripled. In plots having extra food and fewer predators, it increased elevenfold.

The experimental manipulations delayed the cyclic declines in population density but did not stop them. Why not? Owls and other raptors flew over the fences. Only 9 percent of the collared hares starved to death; predators killed some of the others. Krebs concluded that a simple predator–prey or plant–herbivore model did not fully explain his results. Other variables were at work, in a multilevel interaction.

Coevolution of Predators and Prey

Interactions among predators and prey can influence characteristic species traits. If a certain genetic trait in a prey species helps it escape predation, that trait will increase in frequency. If some predator characteristic helps overcome a prey defense, it too will be favored. Each defensive improvement selects for a countering improvement in predators, which selects for another defensive improvement, and so on, in a never-ending arms race. The next section describes some outcomes.

Take-Home Message

How do predator and prey populations change over time?

■ Predator populations show three general patterns of response to changes in prey density. Population levels of prey may show recurring oscillations.

■ The numbers in predator and prey populations often vary in complex ways that reflect the multiple levels of interaction in a community.

■ Predator and prey populations exert selective pressures on one another.

| An Evolutionary Arms Race

■ Predators select for better prey defenses, and prey select for more efficient predators.

■ Links to Ricin Chapter 14 introduction, Coevolution 18.12, Nematocysts 25.5

Prey Defenses

Earlier chapters, including Chapter 25, introduced some examples of prey defenses. Many species have hard parts that make them difficult to eat. Spikes in a sponge body, clam and snail shells, lobster and crab exoskeletons, sea urchin spines—all of these traits help deter predators and thereby contribute to evolutionary success.

Also, many heritable traits function in **camouflage**: body shape, color pattern, behavior, or a combination of factors make an individual blend with its surroundings. Predators cannot eat prey they cannot find. Section 18.4 explains how alleles that improved the camouflage of a prey species, the desert pocket mouse, were adaptive in particular habitats.

Camouflage is widespread. Marsh birds called bitterns live among tall reeds. When threatened, a bittern points its beak skyward and blends with the reeds (Figure 46.11a). On a breezy day, the bird enhances the effect by swaying slightly. A caterpillar with mottled color patterns appears to be a bird dropping (Figure 46.11b). Desert plants of the genus *Lithops* usually look like rocks (Figure 46.11c). They flower only during a brief rainy season, when plenty of other plants tempt herbivores.

Many prey species contain chemicals that taste bad or sicken predators. Some produce toxins through metabolic processes. Others use chemical or physical weapons that they get from their prey. For instance, after sea slugs dine on a sea anemone or a jellyfish, they can store its stinging nematocysts in their own tissues (Figure 25.24c).

Leaves, stems, and seeds of many plants contain bitter, hard-to-digest, or toxic chemicals. Remember the Chapter 14 introduction? It explains how ricin acts to kill or sicken animals. Ricin evolved in castor bean seeds as a defense against herbivores. Caffeine in coffee beans and nicotine in tobacco leaves evolved as defenses against insects.

Many prey species advertise their bad-tasting or toxin-laden properties by **warning coloration**. They have flashy patterns and colors that predators learn to recognize and avoid. For instance, a toad might catch a yellow jacket once. But a painful sting from this wasp teaches the toad that black and yellow stripes mean *AVOID ME!*

Mimicry is an evolutionary convergence in body form; species come to resemble one another. In some cases, two or more well-defended organisms end up looking alike.

Figure 46.11 Prey camouflage. (**a**) What bird? When a predator approaches its nest, the least bittern stretches its neck (which is colored like the surrounding withered reeds), points its bill upward, and sways like reeds in the wind. (**b**) An inedible bird dropping? No. This caterpillar's body coloration and its capacity to hold its body in a rigid position help camouflage it from predatory birds. (**c**) Find the plants (*Lithops*) hiding in the open from herbivores with the help of their stonelike form, pattern, and coloration.

a A dangerous model   **b** One of its edible mimics   **c** Another edible mimic   **d** And another edible mimic

Figure 46.12 Examples of mimicry. Edible insect species often resemble toxic or unpalatable species that are not at all closely related. (**a**) A yellow jacket can deliver a painful sting. It might be the model for nonstinging wasps (**b**), beetles (**c**), and flies (**d**) of strikingly similar appearance.

In others, a tasty, harmless prey species evolves the same warning coloration as an unpalatable or well-defended one (Figure 46.12). Predators may avoid the mimic after experiencing the disgusting taste, irritating secretion, or painful sting of the species it resembles.

When an animal is cornered or under attack, survival may depend on a last-chance trick. Opossums "play dead." Other animals startle predators. Section 1.7 describes an experiment that tested the peacock butterfly defenses—a show of eye-like spots and hissing. Other species puff up, bare sharp teeth, or flare neck ruffs (Figure 26.19*d*). When cornered, many animals, including skunks, some snakes, many toads, and certain insects, secrete or squirt stinky or irritating repellents (Figure 46.13*a*).

Adaptive Responses of Predators

A predator's evolutionary success hinges on eating prey. Stealth, camouflage, and ways of avoiding repellents are countermeasures to prey defenses. For example, some edible beetles spray noxious chemicals at their attackers.

A grasshopper mouse grabs the beetle and plunges the sprayer end into the ground, and then chews on the tasty, unprotected head (Figure 46.13*b*). Some evolved traits in herbivores are responses to plant defenses. The digestive tract of koalas can handle tough, aromatic eucalyptus leaves that would sicken other herbivorous mammals.

Also, a speedier predator catches more prey. Consider the cheetah, the world's fastest animal on land. One was clocked at 114 kilometers (70 miles) per hour. Compared with other big cats, a cheetah has longer legs relative to body size and nonretractable claws that act like cleats to increase traction. Thomson's gazelle, its main prey, can run longer but not as fast (80 kilometers per hour). Without a head start, the gazelle is likely to be outrun.

Camouflaging helps predators as well as prey. Think of white polar bears stalking seals on ice, striped tigers crouched in tall-stalked, golden grasses, and scorpionfish on the sea floor (Figure 46.13*c*). Camouflage can be quite stunning among predatory insects (Figure 46.13*d*). Even so, with each new, improved camouflaging trait, predators select for enhanced predator-detecting ability in prey.

Figure 46.13 Predator responses to prey defenses. (**a**) Some beetles spray noxious chemicals at attackers, which deters them some of the time. (**b**) Grasshopper mice plunge the chemical-spraying tail end of their beetle prey into the ground and feast on the head end. (**c**) This leaf scorpionfish, is a venomous predator with camouflaging fleshy flaps, multiple colors, and many spines. (**d**) Where do the pink flowers end and the pink praying mantis begin?

- Predators have only a brief interaction with prey, but parasites live on or in their hosts.

- Link to Evolution and disease 21.8

Parasites and Parasitoids

Parasites spend all or part of their life living in or on other organisms, from which they steal nutrients. Although most parasites are small, they can have a major impact on populations of their hosts. Many parasites are pathogens; they cause disease in their hosts. For example, *Myxobolus cerebralis* is a parasite of trout, salmon and related fishes. Following infection, a host fish develops deadly whirling disease (Figure 46.14).

Even when a parasite does not cause such dramatic symptoms, infection can weaken the host so it is more vulnerable to predation or less attractive to potential mates. Some parasitic infections cause sterility. Others shift the sex ratio of their host species. Parasites affect host numbers by altering birth and death rates. They also indirectly affect species that compete with their host. The decline in trout caused by whirling disease allows competing fish populations to increase.

Sometimes the gradual drain of nutrients during a parasitic infection indirectly leads to death. The host is so weak that it cannot fight off secondary infections. A rapid death is rare. Usually death happens only after a parasite attacks a novel host—one with no coevolved defenses—or after the body is overwhelmed by a huge population of parasites.

In evolutionary terms, killing the host too quickly is bad for the parasite. Ideally, a host will live long enough to give the parasite time to produce plenty of offspring. The longer the host survives, the more offspring the parasite can produce. That is why we can predict that natural selection will favor parasites with less-than-fatal effects on hosts (Section 21.8).

Unit Four describes many parasites. Some spend their entire life in or on a single host species. Others have different hosts during different stages of the life cycle. Insects and other arthropods can act as **vectors**: organisms that convey a parasite from host to host.

Even a few plants are parasitic. Nonphotosynthetic species such as dodders obtain energy and nutrients from a host plant (Figure 46.15). Other species carry out photosynthesis but steal nutrients and water from their host. Most mistletoe are like this; their modified roots tap into the vascular tissues of host trees.

Many tapeworms, flukes, and certain roundworms are parasitic invertebrates (Figure 46.16). So are ticks, many insects, and some crustaceans.

Parasitoids are insects that lay eggs in other insects. Larvae hatch, develop in the host's body, eat its tissue, and eventually kill it. The fire ant–killing phorid flies described in this chapter's introduction do this. As many as 15 percent of all insects may be parasitoids.

Social parasites are animals that take advantage of the behavior of a host to complete their life cycle. Cuckoos and North American cowbirds, as explained shortly, are social parasites.

Figure 46.14 (**a**) A young trout with a twisted spine and darkened tail caused by whirling disease, which damages cartilage and nerves. Jaw deformities and whirling movements are other symptoms. (**b**) Spores of *Myxobolus cerebralis*, the parasite that causes the disease. The disease now occurs in many lakes and streams in western and northeastern states.

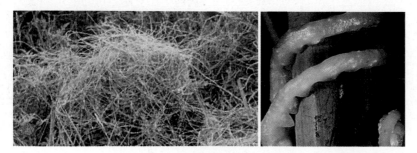

Figure 46.15 Dodder (*Cuscuta*), also known as strangleweed or devil's hair. This parasitic flowering plant has almost no chlorophyll. Leafless stems twine around a host plant during growth. Modified roots penetrate the host's vascular tissues and absorb water and nutrients from them.

Figure 46.16 Adult roundworms (*Ascaris*), an endoparasite, inside the small intestine of a host pig. Sections 25.6 and 25.11 show more examples of parasitic worms.

Figure 46.17 Biological control agent: a commercially raised parasitoid wasp about to deposit an egg in an aphid. After the egg it laid hatches, a wasp larva will devour the aphid from the inside.

Biological Control Agents

Some parasites and parasitoids are now raised commercially for use as biological control agents. Use of such agents is promoted as an alternative to pesticides. For example, some parasitoid wasps attack aphids, which are widespread plant pests (Figure 46.17).

Effective biological control agents are adapted to a specific host species and to its habitat. They are good at finding the hosts. Their population growth rate is high compared to the host's. Their offspring are good at dispersing. Also, they make a type III response to changes in prey density (Section 46.4), without much lag time after the prey or host population size shifts.

Biological control is not without risks of its own. Releasing multiple species of biological control agent in an area may allow competition among them, and lower their effectiveness against an intended target. Also, introduced parasites sometimes go after nontargeted species in addition to, or instead of, those species they were introduced to control.

For example, parasitoids deliberately introduced to the Hawaiian Islands attacked the wrong target. They were brought in to control stinkbugs that are pests of Hawaii's crops. Instead, the parasitoids decimated the population of koa bugs, Hawaii's largest native bug. Introduced parasitoids also have been implicated in ongoing declines of many native Hawaiian butterfly and moth populations.

Take-Home Message

What are parasites, parasitoids, and social parasites?

■ Parasitic species feed on another species but generally do not kill their host.

■ Parasitoids are insects that eat other insects from inside out.

■ Social parasites manipulate the social behavior of another species to their own benefit.

46.7 | Strangers in the Nest

■ The brown-headed cowbird's genus name (*Molothrus*) means intruder in Latin. They intrude into other birds' nests and lay their eggs there.

Brown-headed cowbirds (*Molothrus ater*) evolved in the Great Plains of North America and they were commensal with bison. Great herds of these hefty ungulates stirred up plenty of tasty insects as they migrated through the grasslands, and, being insect-eaters, cowbirds wandered around with them (Figure 46.18a).

Cowbirds are social parasites that lay their eggs in the nests constructed by other birds, so young cowbirds are reared by foster parents. Many species became "hosts" to cowbirds; they did not have the capacity to recognize the differences between cowbird eggs and their own eggs. Concurrently, cowbird hatchlings became innately wired for hostile takeovers. They demand to be fed by unwitting, and often smaller, foster parents (Figure 46.18b). For thousands of years, cowbirds have perpetuated their genes at the expense of hosts.

When American pioneers moved west, many cleared swaths of woodlands for pastures. Cowbirds now moved in the other direction. They adapted easily to a life with new ungulates—cattle—in the man-made grasslands; hence their name. They started to penetrate adjacent woodlands and exploit novel species. Today, brown-headed cowbirds parasitize at least fifteen kinds of native North American birds. Some of those birds are threatened or endangered.

Besides being successful opportunists, cowbirds are big-time reproducers. A female can lay an egg a day for ten days, give her ovaries a rest, do the same again, and then again in one season. As many as thirty eggs in thirty nests—that is a lot of cowbirds.

Figure 46.18 (a) Brown-headed cowbirds (*Molothrus ater*) originally evolved as commensalists with bison of the North American Great Plains. (b) Cowbirds are social parasites. The large nestling at the *left* is a cowbird. The smaller foster parent is rearing the cowbird in place of its own offspring.

46.8 Ecological Succession

- Which species are present in a community depends on physical factors such as climate, biotic factors such as which species arrived earlier, and the frequency of disturbances.

- Links to Mosses 23.3, Lichens 24.6, Nitrogen-fixing bacteria 29.2

Figure 46.19 One observed pathway of primary succession in Alaska's Glacier Bay region. (**a**) As a glacier retreats, meltwater leaches minerals from the rocks and gravel left behind. (**b**) Pioneer species include lichens, mosses, and some flowering plants such as mountain avens (*Dryas*), which associate with nitrogen-fixing bacteria. Within 20 years, alder, cottonwood, and willow seedlings take hold. Alders also have nitrogen-fixing symbionts. (**c**) Within 50 years, alders form dense, mature thickets in which cottonwood, hemlock, and a few evergreen spruce grow. (**d**) After 80 years, western hemlock and spruce crowd out alders. (**e**) In areas deglaciated for more than a century, tall Sitka spruce are the predominant species.

Successional Change

Species composition of a community can change over time. Species often alter the habitat in ways that allow other species to come in and replace them. We call this type of change ecological succession.

The process of succession starts with the arrival of **pioneer species**, which are opportunistic colonizers of new or newly vacated habitats. Pioneers species have high dispersal rates, grow and mature fast, and produce many offspring. Later, other species replace the pioneers. Then replacements are replaced, and so on.

Primary succession is a process that begins when pioneer species colonize a barren habitat with no soil, such as a new volcanic island or land exposed by the retreat of a glacier (Figure 46.19). The earliest pioneers to colonize a new habitat are often mosses and lichens (Sections 23.3 and 24.6). They are small, have a brief life cycle, and can tolerate intense sunlight, extreme temperature changes, and little or no soil. Some hardy, annual flowering plants with wind-dispersed seeds are also among the pioneers.

Pioneers help build and improve the soil. In doing so, they may set the stage for their own replacement. Many pioneer species partner with nitrogen-fixing bacteria, so they can grow in nitrogen-poor habitats. Seeds of later species find shelter inside mats of the pioneers. Organic wastes and remains accumulate and, by adding volume and nutrients to soil, this material helps other species take hold. Later successional species often shade and eventually displace earlier ones.

In **secondary succession**, a disturbed area within a community recovers. If improved soil is still present, secondary succession can be fast. It commonly occurs in abandoned fields, burned forests, and tracts of land where plants were killed by volcanic eruptions.

Factors Affecting Succession

When the concept of ecological succession was first developed in the late 1800s, it was thought to be a predictable and directional process. Physical factors such as climate, altitude, and soil type were considered to be the main determinants of which species appeared in what order during succession. Also by this view, succession culminates in a "climax community," an array of species that will persist over time and will be reconstituted in the event of a disturbance.

Ecologists now realize that the species composition of a community changes frequently, in unpredictable ways. Communities do not journey along a well-worn path to some predetermined climax state.

Figure 46.20 A natural laboratory for succession after the 1980 Mount Saint Helens eruption (**a**). The community at the base of this Cascade volcano was destroyed. (**b**) In less than a decade, pioneer species came in. (**c**) Twelve years later, seedlings of a dominant species, Douglas firs, took hold.

Random events can determine the order in which species arrive in a habitat and thus affect the course of succession. Arrival of a certain species may make it easier or more difficult for others to take hold. As an example, surf grass can only grow along a shoreline if algae have already colonized that area. The algae act as an anchoring site for the grass. In contrast, when sagebrush gets established in a dry habitat, chemicals it secretes into the soil keep most other plants out.

Ecologists had an opportunity to investigate these factors after the 1980 eruption of Mount Saint Helens leveled about 600 square kilometers (235 square miles) of forest in Washington State (Figure 46.20). Ecologists recorded the natural pattern of colonization. They also carried out experiments in plots inside the blast zone. They added seeds of certain pioneer species to some plots and left other plots seedless. The results showed that some pioneers helped other later arriving plants become established. Different pioneers kept the same late arrivals out.

Disturbances also can influence the species composition in communities. According to the **intermediate disturbance hypothesis**, species richness is greatest in communities where disturbances are moderate in their intensity or frequency. In such habitats, there is enough time for new colonists to arrive and become established but not enough for competitive exclusion to cause extinctions:

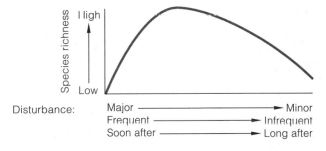

In short, the modern view of succession holds that the species composition of a community is affected by (1) physical factors such as soil and climate, (2) chance events such as the order in which species arrive, and (3) the extent of disturbances in a habitat. Because the second and third factors may vary even between two geographically close regions, it is generally difficult to predict exactly what any given community will look like at any point in the future.

Take-Home Message

What is succession?

■ Succession, a process in which one array of species replaces another over time. It can occur in a barren habitat (primary succession), or a region in which a community previously existed (secondary succession).

■ Chance events make successional changes difficult to predict.

- The loss or addition of even one species may destabilize the number and abundances of species in a community.

- Link to Sudden oak death 22.8

The Role of Keystone Species

As you read earlier, short-term physical disturbances can influence the species composition of a community. Long-term changes in climate or some other environmental variable also have an effect. In addition, a shift in species interactions can dramatically alter the community by favoring some species and harming others.

The uneasy balance of forces in a community comes into focus when we observe the effects of a keystone species. A **keystone species** has a disproportionately large effect on a community relative to its abundance. Robert Paine was the first to describe the effect of a keystone species after his experiments on the rocky shores of California's coast. Species living in the rocky

intertidal zone withstand pounding surf by clinging to rocks. A rock to cling to is a limiting factor. Paine set up control plots with the sea star *Pisaster ochraceus* and its main prey—chitons, limpets, barnacles, and mussels. In experimental plots he removed all sea stars.

Mussels (*Mytilus*) happen to be the prey of choice for sea stars. In the absence of sea stars, they took over Paine's experimental plots; they became the strongest competitors and crowded out seven other species of invertebrates. In this intertidal zone, predation by sea stars normally keeps the number of prey species high because it restricts competitive exclusion by mussels. Remove all the sea stars, and the community shrinks from fifteen species to eight.

The impact of a keystone species can vary between habitats that differ in their species arrays. Periwinkles (*Littorina littorea*) are alga-eating snails that live in the intertidal zone. Jane Lubchenco found removing them can increase *or* decrease the diversity of algal species, depending on the habitat (Figure 46.21).

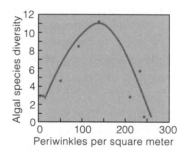

d Algal diversity in tidepools

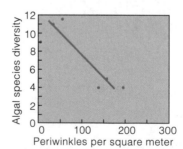

e Algal diversity on rocks that become exposed at high tide

Figure 46.21 Effect of competition and predation in an intertidal zone. (**a**) Grazing periwinkles (*Littorina littorea*) affect the number of algal species in different ways in different marine habitats. (**b**) *Chondrus* and (**c**) *Enteromorpha*, two kinds of algae in their natural habitats. (**d**) By grazing on the dominant alga in tidepools (*Enteromorpha*), the periwinkles promote the survival of less competitive algal species that would otherwise be overgrown. (**e**) *Enteromorpha* does not grow on rocks. Here, *Chondrus* is dominant. Periwinkles find *Chondrus* tough and dine instead on less competitive algal species. By doing so, periwinkles decrease the algal diversity on the rocks.

Table 46.2   Outcomes of Some Species Introductions Into the United States

Species Introduced	Origin	Mode of Introduction	Outcome
Water hyacinth	South America	Intentionally introduced (1884)	Clogged waterways; other plants shaded out

Dutch elm disease:
  Ophiostoma ulmi (fungus)
  Bark beetle (vector)
	Asia (by way
	of Europe)
		Accidental; on infected elm timber (1930)
		Accidental; on unbarked elm timber (1909)
			Millions of mature elms destroyed

Chestnut blight fungus	Asia	Accidental; on nursery plants (1900)	Nearly all eastern American chestnuts killed
Zebra mussel	Russia	Accidental; in ballast water of ship (1985)	Clogged pipes and water intake valves of power plants; displaced native bivalves in Great Lakes
Japanese beetle	Japan	Accidental; on irises or azaleas (1911)	Close to 300 plant species (e.g., citrus) defoliated
Sea lamprey	North Atlantic	Accidental; on ship hulls (1860s)	Trout, other fish species destroyed in Great Lakes
European starling	Europe	Intentional release, New York City (1890)	Outcompetes native cavity-nesting birds; crop damage; swine disease vector
Nutria	South America	Accidental release of captive animals being raised for fur (1930)	Crop damage, destruction of levees, overgrazing of marsh habitat

In tidepools, periwinkles prefer to eat a certain alga (*Enteromorpha*) which can outgrow other algal species. By keeping that alga in check, periwinkles help other, less competitive algal species survive. On rocks of the lower intertidal zone, *Chondrus* and other tough, red algae dominate. Here, periwinkles preferentially graze on competitively weaker algae. Periwinkles promote species richness in tidepools but reduce it on rocks.

Not all keystone species are predators. For example, beavers can be a keystone species. These large rodents cut down trees by gnawing through their trunks. Some of the felled trees are used to build dams that create a pool where only a shallow stream would otherwise exist. Thus the presence of beavers affects which types of fish and aquatic invertebrates are present.

Species Introductions Can Tip the Balance

Instabilities are also set in motion when residents of an established community move out from their home range, then successfully take up residence elsewhere. This type of directional movement, called geographic dispersal, happens in three ways.

First, over a number of generations, a population might expand its home range by slowly moving into any outlying regions that prove hospitable. Second, a population might be moved away from a home range by continental drift, at an almost imperceptibly slow pace over long spans of time. Third, some individuals might be rapidly transported across great distances, an event called jump dispersal. Birds that travel long distances facilitate such jumps by carrying seeds of plants. For some time now, humans have been a major cause of jump dispersal. They have introduced species that benefit them, as by bringing crop plants from the Americas to Europe. They have also unknowingly transported stowaways, as when Asian long-horned beetles were imported along with wood products.

When you hear someone speaking enthusiastically about exotic species, you can safely bet the speaker is not an ecologist. An **exotic species** is a resident of an established community that dispersed from its home range and became established elsewhere. Unlike most imports, which never do take hold outside the home range, an exotic species permanently insinuates itself into a new community.

In its new locale, the exotic species is often untroubled by competitors, predators, parasites, and diseases that kept it in check back home. Freed from its usual constraints, the exotic species can often outcompete similar species native to its new habitat.

You have already learned how some imports are affecting community structure. The chapter introduction described how red imported fire ants that arrived from South America outcompete North American ant species. Sudden oak death, described in Section 22.8, is caused by a protist from Asia. A parasite from Europe is the cause of whirling disease in trout. The list of detrimental exotic species is depressingly long. Table 46.2 lists some well-known imports, and the next section describes four others in some detail.

Take-Home Message

How can a single species affect community structure?

■ A keystone species is one that has a major effect on species richness and relative abundances in a habitat.

■ Removal of a keystone species or introductions of an exotic species can affect the types and abundances of species in a community.

■ Nonnative species introduced by human activities are affecting native communities on every continent.

■ Link to Green algae 22.9

Battling Algae

The long, green, feathery branches of *Caulerpa taxifolia* look great in saltwater aquariums, so researchers at the Stuttgart Aquarium in Germany developed a sterile strain of this green alga and shared it with other marine institutions. Was it from Monaco's Oceanographic Museum that the hybrid strain escaped into the wild? Some say yes, Monaco says no.

In any case, a small patch of the aquarium strain was found growing in the Mediterranean near Monaco in 1984. Boat propellers and fishing nets dispersed the alga, and it now blankets tens of thousands of acres of sea floor in the Mediterranean and Adriatic (Figure 46.22*a*).

Just how bad is *C. taxifolia?* The aquarium strain can thrive on sandy or rocky shores and in mud. It can live ten days after being discarded in meadows. Unlike its tropical parents, it can also survive in cool water and polluted water. It has the potential to displace endemic algae, overgrow reefs, and destroy marine food webs. Its success is due in part to production of a toxin (Caulerpenyne) that poisons invertebrates and fishes, including algae eaters that keep other algae in check.

In 2000, scuba divers discovered *C. taxifolia* growing near the southern California coast. Someone might have drained water from a home aquarium into a storm drain or into the lagoon itself. The government and private groups quickly sprang into action. So far, eradication and surveillance programs have worked, but at a cost of more than $3.4 million.

Importing *C. taxifolia* or any closely related species of *Caulerpa* into the United States is now illegal. To protect native aquatic communities, aquarium water should never be dumped into storm drains or waterways. It should be discarded into a sink or toilet so wastewater treatment can kill any algal spores (Section 22.9).

The Plants That Overran Georgia

In 1876, kudzu (*Pueraria montana*) was introduced to the United States from Japan. In its native habitat, this perennial vine is a well-behaved legume with an extensive root system. It *seemed* like a good idea to use it for forage and to control erosion on slopes. But kudzu grew faster in the American Southeast. No native herbivores or pathogens were adapted to attack it. Competing plant species posed no serious threat to it.

With nothing to stop it, kudzu can grow 60 meters (200 feet) per year. Its vines now blanket streambanks, trees, telephone poles, houses, and almost anything else in their path (Figure 46.22*b*). Kudzu withstands burning, and grows back from its deep roots. Grazing goats and herbicides help. But goats eat most other plants along with it, and herbicides taint freshwater supplies. Kudzu invasions now stretch from Connecticut down to Florida and are reported in Arkansas. It crossed the Mississippi River into Texas. Thanks to jump dispersal, it is now an invasive species in Oregon.

Figure 46.22 (**a**) Aquarium strain of *Caulerpa taxifolia* suffocating yet another richly diverse marine ecosystem.

(**b**) Kudzu (*Pueraria montana*) taking over part of Lyman, South Carolina. This vine has become invasive in many states from coast to coast. Ruth Duncan of Alabama (*above*), who makes 200 kudzu vine baskets a year, can't keep up.

Figure 46.23 Rabbit-proof fence? Not quite. This photo shows part of a fence built in 1907 to hold back rabbits that were wreaking havoc with the vegetation in Australia. The fence did not solve the rabbit problem, but it did restrict movements of native wildlife such as kangaroos and emus.

On the bright side, Asians use a starch extracted from kudzu in drinks, herbal medicines, and candy. A kudzu processing plant in Alabama may export this starch to Asia, where the demand currently exceeds the supply. Also, kudzu may help save forests; it can be an alternative source for paper and other wood products. Today, about 90 percent of Asian wallpaper is kudzu-based.

The Rabbits That Ate Australia

During the 1800s, British settlers in Australia just could not bond with koalas and kangaroos, and so they imported familiar animals from home. In 1859, in what would be the start of a major ecological disaster, a landowner in northern Australia imported and released two dozen European rabbits (*Oryctolagus cuniculus*). Good food and great sport hunting—that was the idea. An ideal rabbit habitat with no natural predators—that was the reality.

Six years later, the landowner had killed 20,000 rabbits and was besieged by 20,000 more. The rabbits displaced livestock and caused the decline of native wildlife. Now 200 million to 300 million are hippity-hopping through the southern half of the country. They graze on grasses in good times and strip bark from shrubs and trees during droughts. Thumping hordes turn shrublands as well as grasslands into eroded deserts. Their burrows undermine the soil and set the stage for widespread erosion.

Rabbits have been shot and their warrens fumigated, plowed under, and dynamited. The first assaults killed 70 percent of them, but the rabbits rebounded in less than a year. When a fence 2,000 miles long was built to protect western Australia, rabbits made it from one side to the other before workers could finish the job (Figure 46.23).

In 1951, the government introduced a myxoma virus that normally infects South American rabbits. The virus causes myxomatosis. This disease has mild effects on its coevolved host but nearly always kills *O. cuniculus*. Fleas and mosquitoes transmit the virus to new hosts. With no coevolved defenses against the import, European rabbits died in droves. But natural selection has since favored a rise in rabbit populations resistant to the imported virus.

In 1991, on an uninhabited island in Australia's Spencer Gulf, researchers released rabbits that were injected with a calicivirus. The rabbits died from blood clots in their lungs, heart, and kidneys. Then, in 1995, the test virus escaped from the island to the mainland, perhaps on insect vectors.

The combination of the two imported viruses, along with traditional control methods has brought the rabbit population under control. There still are some rabbits, but vegetation is growing back and native herbivores are increasing in numbers.

Gray Squirrels Versus Red Squirrels

The eastern gray squirrel (*Sciurus carolinensis*) is native to eastern North America, where it is a welcome sight in forests, yards, and parks. It has become similarly common throughout Britain and parts of Italy where it has been introduced. Here, the squirrel is considered an exotic pest that has thrived at the expense of Europe's native red squirrel (*Sciurus vulgaris*). In Britain, the imported grays now outnumber the native reds 66 to 1.

The gray squirrels are at an advantage over their European cousins because they excel at detecting and stealing nuts that red squirrels stored for the winter. In addition, gray squirrels carry and spread a virus that kills Britain's red squirrels, but are not themselves affected by the virus.

To protect the remaining red squirrels, the British have begun trapping and killing gray squirrels. Efforts are also under way to develop a contraceptive drug that would be effective against grays, but not the native reds.

■ The richness and relative abundances of species differ from one habitat or region of the world to another.

■ Link to Biogeography 17.1

Biogeography is the scientific study of how species are distributed in the natural world (Section 17.1). We see patterns that correspond with differences in sunlight, temperature, rainfall, and other factors that vary with latitude, elevation, or water depth. Still other patterns relate to the history of a habitat and the species in it. Each species has its own unique physiology, capacity for dispersal, resource requirements, and interactions with other species.

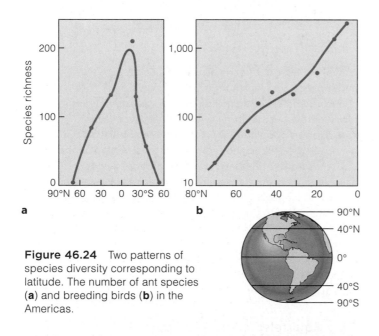

Figure 46.24 Two patterns of species diversity corresponding to latitude. The number of ant species (**a**) and breeding birds (**b**) in the Americas.

Mainland and Marine Patterns

Perhaps the most striking pattern of species richness corresponds with distance from the equator. For most major plants and animal groups, the number of species is greatest in the tropics and declines from the equator to the poles. Figure 46.24 illustrates two examples of this pattern. Consider just a few factors that help bring about such a pattern and maintain it.

First, for reasons explained in Section 48.1, tropical latitudes intercept more intense sunlight and receive more rainfall, and their growing season is longer. As one outcome, resource availability tends to be greater and more reliable in the tropics than elsewhere. One result is a degree of specialized interrelationships not possible where species are active for shorter periods.

Second, tropical communities have been evolving for a long time. Some temperate communities did not start forming until the end of the last ice age.

Third, species richness may be self-reinforcing. The number of species of trees in tropical forests is much greater than in comparable forests at higher latitudes. Where more plant species compete and coexist, more species of herbivores also coexist, partly because no single herbivore species can overcome all the chemical defenses of all plants. In addition, more predators and parasites can evolve in response to more kinds of prey and hosts. The same principles apply to tropical reefs.

Island Patterns

As you saw in Section 45.4, islands are laboratories for population studies. They have also been laboratories for community studies. For instance, in the mid-1960s volcanic eruptions formed a new island 33 kilome-

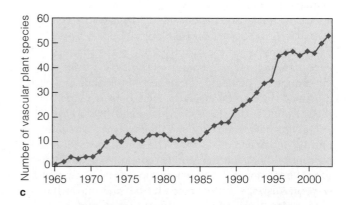

Figure 46.25 Surtsey, a volcanic island, at the time of its formation (**a**) and in 1983 (**b**). The graph (**c**) shows the number of vascular plant species found in yearly surveys. Sea gulls began nesting on the island in 1986.

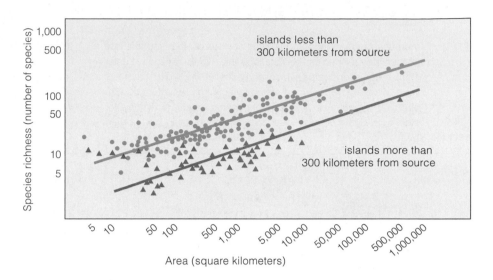

Figure 46.26 Island biodiversity patterns.

Distance effect: Species richness on islands of a given size declines as distance from a source of colonists rises. *Green* circles are values for islands less than 300 kilometers from the colonizing source. *Orange* triangles are values for islands more than 300 kilometers (190 miles) from a source of colonists.

Area effect: Among islands the same distance from a source of colonists, larger islands tend to support more species than smaller ones.

Figure It Out: Which is likely to have more species, a 100-km² island more than 300 km from a colonizing source or a 500-km² island less than 300 km from a colonist source? *Answer: The 500-km² island*

ters (21 miles) from the coast of Iceland. The island was named Surtsey (Figure 46.25). Bacteria and fungi were early colonists. The first vascular plant became established on the island in 1965. Mosses appeared two years later and thrived (Figure 46.25*b*). The first lichens were found five years after that. The rate of arrivals of new vascular plants picked up considerably after a seagull colony became established in 1986 (Figure 46.25*c*). This example illustrates the important role birds play in introducing species to islands.

The number of species on Surtsey will not continue increasing forever. Can we estimate how many species there will be when the number levels off? The **equilibrium model of island biogeography** addresses this question. According to this model, the number of species living on any island reflects a balance between immigration rates for new species and extinction rates for established ones. The distance between an island and a mainland source of colonists affects immigration rates. An island's size affects both immigration rates and extinction rates.

Consider first the **distance effect:** Islands far from a source of colonists receive fewer immigrants than those closer to a source. Most species cannot disperse very far, so they will not turn up far from a mainland.

Species richness also is shaped by the **area effect:** Big islands tend to support more species than small ones. More colonists will happen upon a larger island simply by virtue of its size. Also, big islands are more likely to offer a variety of habitats, such as high and low elevations. These options make it more likely that a new arrival will find a suitable habitat. Finally, big islands can support larger populations of species than small islands. The larger a population, the less likely it is to become locally extinct as the result of some random event.

Figure 46.26 illustrates how interactions between the distance effect and the area effect can influence the number of species on islands.

Robert H. MacArthur and Edward O. Wilson first developed the equilibrium model of island biogeography in the late 1960s. Since then it has been modified and its use has been expanded to help scientists think about habitat islands—natural settings surrounded by a "sea" of degraded habitat. Many parks and wildlife preserves fit this description. Island-based models can help estimate the size of an area that must be set aside as a protected reserve to ensure survival of a species.

One more note about island communities: An island often differs from its source of colonists in physical aspects, such as rainfall and soil type. It also differs with regard to species array; not all species reach the island. As a result of these differences, a population on an island often faces different selection pressures than its same-species relatives on the mainland and evolves in a different way as a result.

In a pattern that is the opposite of character displacement, a species may find itself on an island that lacks a major competitor found on the mainland. In the absence of this competition, traits of the island population may become more like those of the competitor that it left behind.

Take-Home Message

What are some biogeographic patterns in species richness?

■ Generally, species richness is highest in the tropics and lowest at the poles. Tropical habitats have conditions that more species can tolerate, and tropical communities have often been evolving for longer than temperate ones.

■ When a new island forms, species richness rises over time and then levels off. The size of an island and its distance from a colonizing source influence its species richness.

Increased global trade and faster ships are contributing to a rise in the rate of species introductions into North America. Faster ships mean shorter trips, which increases the likelihood that pests will survive a voyage. Wood-eating insects from Asia turn up with alarming frequency in the wood of packing crates and spools for steel wire. Some of these insects, such as the Asian long-horned beetle, now pose a serious threat to North America's forests.

How would you vote?

Is inspecting more imported goods to detect potentially harmful exotic species worth the added cost? See CengageNOW for details, then vote online.

Summary

Section 46.1 Each species occupies a certain **habitat** characterized by physical and chemical features and by the array of other species living in it. All populations of all species in a habitat are a **community**. Each species in a community has its own **niche**, or way of living. Species interactions between members of a community include **commensalism**, which does not help or harm either species, **mutualism**, which benefits both species, **interspecific competition**, which harms both species, and **parasitism** and **predation**, in which one species benefits at the expense of another. Commensalism, mutualism, and parasitism may be a **symbiosis**, in which species live together. Interacting species undergo **coevolution**.

Section 46.2 In a mutualism, two species interact and both benefit. Some mutualists cannot complete their life cycle without the interaction.

Section 46.3 By the process of **competitive exclusion**, one species outcompetes a rival with the same resource needs, driving it to extinction. **Character displacement** makes competing species less similar, which facilitates **resource partitioning**.

■ *Use the animation on CengageNOW to learn about competitive interactions.*

Sections 46.4, 46.5 **Predators** are free-living and usually kill their **prey**. Predator and prey numbers often fluctuate in cycles. Carrying capacity, predator behavior, and availability of other prey affect these cycles. Predators and their prey exert selection pressure on one another. Evolutionary results of such selection include **warning coloration**, **camouflage**, and **mimicry**.

■ *Use the interaction on CengageNOW to learn about three alternative models for predator responses to prey density.*

Sections 46.6, 46.7 **Parasites** live in or on a host and withdraw nutrients from its tissues. Hosts may or may not die as a result. An animal **vector** often carries the parasite between hosts. **Parasitoids** lay eggs on a host, then their larvae devour the host. **Social parasites** manipulate some aspect of a host's behavior.

Section 46.8 Ecological succession is the sequential replacement of one array of species by another over time. **Primary succession** happens in new habitats. **Secondary**

succession occurs in disturbed ones. The first species of a community are **pioneer species**. The pioneers may help, hinder, or have no effect on later colonists.

The older idea that all communities eventually reach a predictable climax state has been replaced by models that emphasize the role of chance and disturbances. The **intermediate disturbance hypothesis** holds that disturbances of moderate intensity and frequency maximize species diversity.

Sections 46.9, 46.10 Community structure reflects an uneasy balance of forces that operate over time. Major forces are competition and predation. **Keystone species** are especially important in maintaining the composition of a community. The removal of a keystone species or introduction of an **exotic species**—one that evolved in a different community—can alter community structure in ways that may be permanent.

Section 46.11 Species richness, the number of species in a given area, varies with latitude, elevation, and other factors. Tropical regions tend to have more species than higher latitude regions. The **equilibrium model of island biogeography** helps ecologists estimate the number of species that will become established on an island. The **area effect** is the tendency of large islands to have more species than small islands. The **distance effect** is the tendency of islands near a source of colonists to have more species than distant islands.

■ *Learn about the area effect and distance effect with the interaction on CengageNOW.*

Self-Quiz   *Answers in Appendix III*

1. A habitat _____ .
   a. has distinguishing physical and chemical features
   b. is where individuals of a species normally live
   c. is occupied by various species
   d. all of the above

2. A species' niche includes its _____ .
   a. habitat requirements
   b. food requirements
   c. reproductive requirements
   d. all of the above

3. Which cannot be a symbiosis?
   a. mutualism   c. commensalism
   b. parasitism   d. interspecific competition

Data Analysis Exercise

Ant-decapitating phorid flies are just one of the biological control agents used to battle imported fire ants. Researchers have also enlisted the help of *Thelohania solenopsae*, another natural enemy of the ants. This microsporidian is a parasite that infects ants and shrinks the ovaries of the colony's egg-producing female (the queen). As a result, a colony dwindles in numbers and eventually dies out.

Are these biological controls useful against imported fire ants? To find out, USDA scientists treated infested areas with either traditional pesticides or pesticides plus biological controls (both flies and the parasite). The scientists left some plots untreated as controls. Figure 46.27 shows the results.

1. How did population size in the control plots change during the first four months of the study?

2. How did population size in the two types of treated plots change during this same interval?

3. If this study had ended after the first year, would you conclude that biological controls had a major effect?

4. How did the two types of treatment (pesticide alone versus pesticide plus biological controls) differ in their longer-term effects?

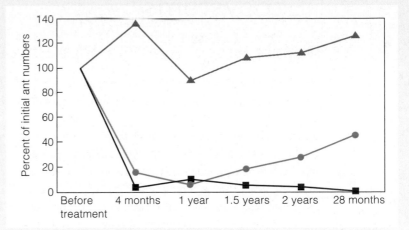

Figure 46.27 Effects of two methods of controlling red imported fire ants. The graph shows the numbers of red imported fire ants over a 28-month period. *Orange* triangles represent untreated control plots. *Green* circles are plots treated with pesticides alone. *Black* squares are plots treated with pesticide and biological control agents (phorid flies and a microsporidian parasite).

---

4. Lizards and songbirds that share a habitat and both eat flies are an example of _____ competition.
- a. exploitative
- b. interference
- c. intraspecific
- d. interspecific
- e. both a and d

5. With character displacement, two competing species become _____ .
- a. more alike
- b. less alike
- c. symbionts
- d. extinct

6. Predator and prey populations _____ .
- a. always coexist at relatively stable levels
- b. may undergo cyclic or irregular changes in density
- c. cannot coexist indefinitely in the same habitat
- d. both b and c

7. Match the terms with the most suitable descriptions.
- ___ predation
- ___ mutualism
- ___ commensalism
- ___ parasitism
- ___ interspecific competition

- a. one free-living species feeds on another and usually kills it
- b. two species interact and both benefit by the interaction
- c. two species interact and one benefits while the other is neither helped nor harmed
- d. one species feeds on another but usually does not kill it
- e. two species attempt to utilize the same resource

8. By a currently favored hypothesis, species richness of a community is greatest between physical disturbances of _____ intensity or frequency.
- a. low
- b. intermediate
- c. high
- d. variable

9. True or false? Parasitoids usually live inside their host without killing it.

10. Match the terms with the most suitable descriptions.
- ___ geographic dispersal
- ___ area effect
- ___ pioneer species
- ___ climax community
- ___ keystone species
- ___ exotic species
- ___ resource partitioning

- a. opportunistic colonizer of barren or disturbed habitat
- b. greatly affects other species
- c. individuals leave home range, become established elsewhere
- d. more species on large islands than small ones at same distance from the source of colonists
- e. array of species at the end of successional stages in a habitat
- f. allows competitors to coexist
- g. often outcompete, displace native species of established community

■ *Visit CengageNOW for additional questions.*

Critical Thinking

1. With antibiotic resistance rising, researchers are looking for ways to reduce use of these drugs. Some cattle once fed antibiotic-laced food now get probiotic feed that can bolster populations of helpful bacteria in the animal's gut. The idea is that if a large population of beneficial bacteria is in place, then harmful bacteria cannot become established or thrive. Which ecological principle is guiding this research?

2. Flightless birds that live on islands often have relatives on the mainland that can fly. The island species presumably evolved from fliers that, in the absences of predators, lost their ability to fly. Many flightless birds on islands are now declining because rats and other predators have been introduced to their previously isolated island. Despite the change in selective pressure, no flightless island bird has yet regained the ability to fly. Why is this unlikely to happen?

47 Ecosystems

Bye-Bye, Blue Bayou

Each Labor Day, the coastal Louisiana town of Morgan City celebrates the Louisiana Shrimp and Petroleum Festival. The state is the nation's top shrimp harvester and the third-largest producer of petroleum, which is refined into gasoline and other fossil fuels. But the petroleum industry's success may be contributing indirectly to the decline of the state's fisheries. Why? The lower atmosphere is warming up, and fossil fuel burning is one of the causes (Section 7.9). As the climate heats up, the ocean's surface waters get warmer and expand, glaciers melt, and sea level rises.

If current trends continue, some coastal lowlands will be submerged. With more than 40 percent of the nation's salt-water wetlands, Louisiana has the most to lose. This state's coastal marshes, or bayous, are already in danger. Dams and levees keep back sediments that would normally be deposited in the marshes. Since the 1940s, Louisiana has lost an area of marshland the size of Rhode Island (Figure 47.1).

Louisiana's marshes are an ecological treasure. Millions of migratory birds overwinter there. The marshes are also the source of more than $3.5 billion worth of fish, shrimp, and shellfish. If the marshes disappear, so will the revenue.

Equally troubling is what will happen to low-lying towns and cities along the coasts after the marshes are gone. Then, there will be nothing to buffer devastating storm surges that threaten the coasts during hurricanes.

In 2005, the category 5 hurricane Katrina slammed into the Gulf Coast. High winds and flooding ruined countless buildings, and more than 1,700 people died. Climate change models suggest that if temperatures continue to rise, more hurricanes are likely to reach category 5 status.

The models also indicate that warming seas will promote overgrowth of algae, which can kill fish. Warmer water can encourage growth of many types of pathogenic bacteria, so more people are expected to become sick after swimming in contaminated water, or eating shellfish harvested from it.

Inland, heat waves are becoming more intense as global temperatures rise, and more people are dying of heat stroke. Fueled by rising temperatures and extended dry seasons, wildfires are becoming more frequent and more devastating. Disease-spreading mosquitoes are now spreading into regions that were too cold for them even a few years ago.

This chapter is about the flow of energy and nutrients through ecosystems. It will give you the tools to do some of your own critical thinking about human impacts on Earth's environments. We have become major players in the global flows of energy and nutrients even before we fully understand how ecosystems work. Decisions we make today about global climate change and other environmental issues are likely to shape Earth's environments—and the quality of human life—far into the future.

See the video! Figure 47.1 *Left*, Fishing camp in Louisiana. It was built in a once-thriving marsh that has since given way to the open waters of Barataria Bay.

Above, a marsh restoration project in Louisiana's Sabine National Wildlife Refuge. In marshland that has become open water, sediments are barged in and marsh grasses are planted on them.

Key Concepts

Organization of ecosystems

An ecosystem consists of a community and its physical environment. A one-way flow of energy and a cycling of raw materials among its interacting participants maintain it. It is an open system, with inputs and outputs of energy and nutrients. **Section 47.1**

Food webs

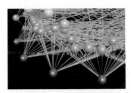

Food chains are linear sequences of feeding relationships. Food chains cross-connect as food webs. Most of the energy that enters a food web returns to the environment, mainly as metabolic heat. Nutrients are recycled within the food web. **Section 47.2**

Energy and materials flow

Ecosystems differ in how much energy their producers capture and how much is stored in each trophic level. Some toxins that enter an ecosystem can become increasingly concentrated as they pass from one trophic level to another. **Sections 47.3, 47.4**

Cycling of water and nutrients

The availability of water, carbon, nitrogen, phosphorus, and other substances influences primary productivity. These substances move slowly in global cycles, from environmental reservoirs, into food webs, then back to reservoirs. **Sections 47.5–47.10**

Links to Earlier Concepts

- This chapter builds on your understanding of the laws of thermodynamics (Section 6.1). We discuss ecological roles of producers such as phytoplankton (22.7), and of decomposers (21.6 and 24.5).

- You will be reminded of the importance of water to the world of life (2.5) and how transpiration works (29.3). We also revisit the effects of acid rain (2.6) and the role of water in leaching nutrients (29.1).

- You will see how nitrogen fixation (21.6 and 29.2) plays an essential role in nutrient cycles and how excess nitrogen contributes to algal blooms (22.5).
  You will also learn more about carbon imbalances (7.9), and be reminded that carbon is stored in peat bogs (23.3) and the shells of protists such as foraminiferans (22.3). You will also hear again about attempts to control the protist-caused disease malaria (22.6).

- Discussions of nutrient cycles will also draw on your knowledge of tectonic plates (17.9).

47.1 | The Nature of Ecosystems

■ In an ecosystem, energy and nutrients from the environment flow among a community of species.

■ Links to Laws of thermodynamics 6.1, Leaching 29.1

Overview of the Participants

Diverse natural systems abound on Earth's surface. In climate, soil type, array of species, and other features, prairies differ from forests, which differ from tundra and deserts. Reefs differ from the open ocean, which differs from streams and lakes. Yet, despite all these differences, all systems are alike in many aspects of their structure and function.

We define an **ecosystem** as an array of organisms and a physical environment, all interacting through a one-way flow of energy and a cycling of nutrients. It is an open system, because it requires ongoing inputs of energy and nutrients to endure (Figure 47.2).

All ecosystems run on energy captured by **primary producers**. These autotrophs, or "self-feeders," obtain energy from a nonliving source—generally sunlight—and use it to build organic compounds from carbon dioxide and water. Plants and phytoplankton are the main producers. Chapter 7 explains how they capture energy from the sun to assemble sugars from carbon dioxide and water, by the process of photosynthesis.

Consumers are heterotrophs that get energy and carbon by feeding on tissues, wastes, and remains of producers and one another. We can describe consumers by their diets. Herbivores eat plants. Carnivores eat the flesh of animals.

Parasites live inside or on a living host and feed on its tissues. Omnivores devour both animal and plant materials. **Detritivores**, such as earthworms and crabs, dine on small particles of organic matter, or detritus. **Decomposers** feed on organic wastes and remains and break them down into inorganic building blocks. The main decomposers are bacteria and fungi.

Energy flows one way—into an ecosystem, through its many living components, then back to the physical environment (Section 6.1). Light energy captured by producers is converted to bond energy in organic molecules, which is then released by metabolic reactions that give off heat. This is a one-way process because heat energy cannot be recycled; producers cannot convert heat into chemical bond energy.

In contrast, many nutrients are cycled within an ecosystem. The cycle begins when producers take up hydrogen, oxygen, and carbon from inorganic sources, such as the air and water. They also take up dissolved nitrogen, phosphorus, and other minerals necessary for biosynthesis. Nutrients move from producers into the consumers who eat them. After an organism dies, decomposition returns nutrients to the environment, from which producers take them up again.

Not all nutrients remain in an ecosystem; typically there are gains and losses. Mineral ions are added to an ecosystem when weathering processes break down rocks, and when winds blow in mineral-rich dust from elsewhere. Leaching and soil erosion remove minerals (Section 29.1). Gains and losses of each mineral tend to balance out over time in a healthy ecosystem.

Trophic Structure of Ecosystems

All organisms of an ecosystem take part in a hierarchy of feeding relationships called **trophic levels** ("troph" means nourishment). When one organism eats another, energy is transferred from the eaten to the eater. All organisms at the same trophic level in an ecosystem are the same number of transfers away from the energy input into that system.

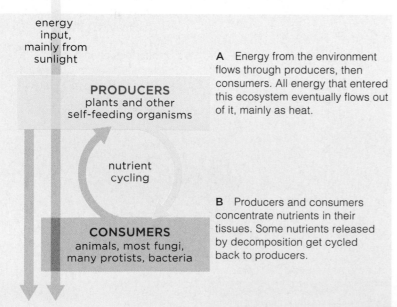

energy input, mainly from sunlight

PRODUCERS
plants and other self-feeding organisms

A Energy from the environment flows through producers, then consumers. All energy that entered this ecosystem eventually flows out of it, mainly as heat.

nutrient cycling

B Producers and consumers concentrate nutrients in their tissues. Some nutrients released by decomposition get cycled back to producers.

CONSUMERS
animals, most fungi, many protists, bacteria

Figure 47.2 Animated Model for ecosystems on land, in which energy flow starts with autotrophs that capture energy from the sun. Energy flows one way, into and out of the ecosystem. Nutrients get cycled among producers and heterotrophs.

Figure 47.3 Example of a food chain and corresponding trophic levels in tallgrass prairie, Kansas.

hawk

Fourth Trophic Level

carnivore
(third-level consumer)

sparrow

Third Trophic Level

carnivore
(second-level consumer)

grasshopper

Second Trophic Level

herbivore
(primary consumer)

big
bluestem
grass

First Trophic Level

autotroph
(primary producer)

A **food chain** is a sequence of steps by which some energy captured by primary producers is transferred to organisms at successively higher trophic levels. For example, big bluestem grass and other plants are the major primary producers in a tallgrass prairie (Figure 47.3). They are at this ecosystem's first trophic level. In one food chain, energy flows from bluestem grass to grasshoppers, to sparrows, and finally to hawks. Grasshoppers are primary consumers; they are at the second trophic level. Sparrows that eat grasshoppers are second-level consumers and at the third trophic level. Hawks are third-level consumers, and they are at the fourth trophic level.

At each trophic level, organisms interact with the same sets of predators, prey, or both. Omnivores feed at several levels, so we would partition them among different levels or assign them to a level of their own.

Identifying one food chain is a simple way to start thinking about who eats whom in ecosystems. Bear in mind, many different species usually are competing for food in complex ways. Tallgrass prairie producers (mainly flowering plants) feed grazing mammals and herbivorous insects. But many more species interact in the tallgrass prairie and in most other ecosystems, particularly at lower trophic levels. A number of food chains cross-connect with one another—as food webs —and that is the topic of the next section.

Take-Home Message

What is the trophic structure of an ecosystem?

■ An ecosystem includes a community of organisms that interact with their physical environment by a one-way energy flow and a cycling of materials.

■ Autotrophs tap into an environmental energy source and make their own organic compounds from inorganic raw materials. They are the ecosystem's primary producers.

■ Autotrophs are at the first trophic level of a food chain, a linear sequence of feeding relationships that proceeds through one or more levels of heterotrophs, or consumers.

47.2 | The Nature of Food Webs

■ All food webs consist of multiple interconnecting food chains. Ecologists who untangled the chains of many food webs discovered patterns of organization. The patterns reflect environmental constraints and the inefficiency of energy transfers from one trophic level to the next.

Interconnecting Food Chains

A **food web** diagram illustrates trophic interactions among species in one particular ecosystem. Figure 47.4 shows a small sampling of the participants in an arctic food web. Nearly all food webs include two types of food chains. In a **grazing food chain**, the energy stored

human (Inuk)   **arctic wolf**   **arctic fox**

HIGHER TROPHIC LEVELS

A sampling of carnivores that feed on herbivores and one another

gyrfalcon   **snowy owl**   **ermine**

SECOND TROPHIC LEVEL

Major parts of the buffet of primary consumers (herbivores)

vole   **arctic hare**   **lemming**

mosquito   **flea**

Parasitic consumers feed at more than one trophic level.

FIRST TROPHIC LEVEL

This is just part of the buffet of primary producers.

grasses, sedges   **purple saxifrage**   **arctic willow**

Detritivores and Decomposers (nematodes, annelids, saprobic insects, protists, fungi, bacteria)

Figure 47.4 Animated A very small sampling of organisms in an arctic food web on land.

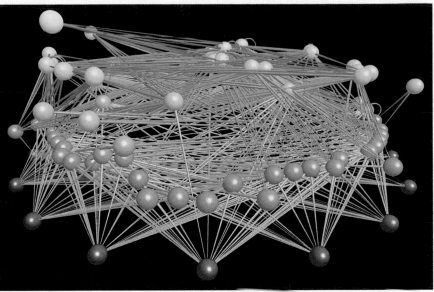

Figure 47.5 Computer model for a food web in East River Valley, Colorado. Balls signify species. Their colors identify trophic levels, with producers (coded *red*) at the bottom and predators (*yellow*) at top. The connecting lines thicken, starting from an eaten species to the eater.

in producer tissues flows to herbivores, which tend to be relatively large animals. In a **detrital food chain**, the energy in producers flows to detritivores, which tend to be smaller animals, and to decomposers.

In most land ecosystems, the bulk of the energy that becomes stored in producer tissues moves through detrital food chains. For example, in an arctic ecosystem, grazers such as voles, lemmings, and hares graze on some plant parts. However, far more plant matter becomes detritus. Bits of dead plant material sustain detritivores such as nematodes and soil-dwelling insects, and decomposers such as soil bacteria and fungi.

Grazing food chains tend to predominate in aquatic ecosystems. Zooplankton (heterotrophic protists and tiny animals that drift or swim) consume most of the phytoplankton. A smaller amount of phytoplankton ends up on the ocean floor as detritus.

Detrital food chains and grazing food chains interconnect to form the overall food web. For example, animals at higher trophic levels often eat both grazers and detritivores. Also, after grazers die, the energy in their tissues flows to detritivores and decomposers.

How Many Transfers?

When ecologists looked at food webs for a variety of ecosystems, they discovered some common patterns. For example, the energy captured by producers usually passes through no more than four or five trophic levels. Even in ecosystems with many species, the number of transfers is limited. Remember that energy transfers are not that efficient (Section 6.1). Energy losses limit the length of a food chain.

Field studies and computer simulations of aquatic and land food ecosystems reveal more patterns. Food chains tend to be shortest in habitats where conditions vary widely over time. Chains tend to be longer in stable habitats, such as the ocean depths. The most complex webs tend to have a large variety of herbivores, as in grasslands. By comparison, the food webs with fewer connections tend to have more carnivores.

Diagrams of food webs help ecologists predict how ecosystems will respond to change. Neo Martinez and his colleagues constructed the one shown in Figure 47.5. By comparing different food webs, they realized that trophic interactions connect species more closely than people thought. On average, each species in any food web was two links away from all other species. Ninety-five percent of species were within three links of one another, even in large communities with many species. As Martinez concluded in a paper discussing his findings, "Everything is linked to everything else." He cautioned that extinction of any species in a food web may have an impact on many other species.

Take-Home Message

How does energy flow affect food chains and food webs?

■ Tissues of living plants and other producers are the basis for grazing food chains. Remains of producers are the basis for detrital food webs.

■ Nearly all ecosystems include both grazing food chains and detrital food chains that interconnect as the system's food web.

■ The cumulative energy losses from energy transfers between trophic levels limits the length of food chains.

■ Even when an ecosystem has many species, trophic interactions link each species with many others.

- Primary producers capture energy and take up nutrients, which then move to other trophic levels.

- Link to Phytoplankton 22.7

Capturing and Storing Energy

The flow of energy through an ecosystem begins with **primary production**: the rate at which producers (most often plants or photosynthetic protists) capture and store energy. The amount of energy captured by all producers in the ecosystem is defined as the system's gross primary production. The portion of energy that producers invest in growth and reproduction (rather than in maintenance) is net primary production.

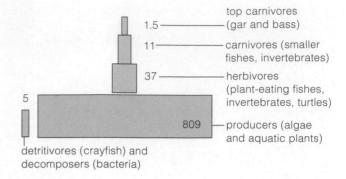

top carnivores
1.5 ——— (gar and bass)

11 ——— carnivores (smaller fishes, invertebrates)

37 ——— herbivores (plant-eating fishes, invertebrates, turtles)

5

809 ——— producers (algae and aquatic plants)

detritivores (crayfish) and decomposers (bacteria)

Figure 47.7 Biomass (in grams per square meter) for Silver Springs, a freshwater aquatic ecosystem in Florida. In this system, primary producers make up the bulk of the biomass.

Factors such as temperature and the availability of water and nutrients affect producer growth, and thus influence primary production. As a result, the primary production varies among habitats and may also vary seasonally (Figure 47.6). Per unit area, the net primary production on land tends to be higher than that in the oceans. However, because oceans cover about 70 percent of Earth's surface, they contribute nearly half of the global net primary productivity.

Ecological Pyramids

Ecologists often represent the trophic structure of an ecosystem in the form of ecological pyramids. In such diagrams, primary producers collectively form a base for successive tiers of consumers above them.

A **biomass pyramid** illustrates the dry weight of all organisms at each trophic level in an ecosystem. Figure 47.7 shows the biomass pyramid for Silver Springs, an aquatic ecosystem in Florida.

Most commonly, primary producers make up most of the biomass in a pyramid, and top carnivores make up very little. If you visited Silver Springs, you would see a lot of aquatic plants but very few gars (the main top predator in this ecosystem). Similarly, when you walk through a prairie, you would see more grams of grass than of hawks.

However, if producers are small and reproduce rapidly, a biomass pyramid can have its smallest tier at the bottom. For example, producers in the open ocean are

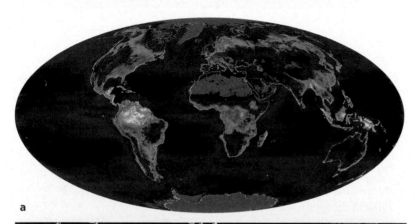

a

North America

Atlantic Ocean in Winter   Africa

b

North America

Atlantic Ocean in Spring   Africa

c

Figure 47.6 Primary productivity. (**a**) Summary of satellite data on net primary production during 2002. Productivity is coded as *red* (highest) down through *orange, yellow, green, blue,* and *purple* (lowest). (**b,c**) Satellite data showing seasonal shifts in net primary productivity for the North Atlantic Ocean.

single-celled protists that devote most energy that they harness to rapid reproduction, rather than to building a big body. They get eaten as fast as they reproduce, so a smaller biomass of phytoplankton can support a greater biomass of zooplankton and bottom feeders.

An **energy pyramid** illustrates how the amount of usable energy diminishes as it is transferred through an ecosystem. Sunlight energy is captured at the base (the primary producers) and declines with successive levels to its tip (the top carnivores). Energy pyramids are always "right-side-up," with their largest tier at the bottom. Such pyramids depict energy flow per unit of water (or land) per unit of time. Figure 47.8 shows the energy pyramid for the Silver Springs ecosystem and the energy flow that this pyramid represents.

Ecological Efficiency

Anywhere between 5 and 30 percent of the energy in the tissues of organisms at one trophic level ends up in the tissues of those at the next trophic level. Several factors influence the efficiency of transfers. First, not all energy harvested by consumers is used to build biomass. Some is lost as metabolic heat. Second, not all biomass can be digested by most consumers. Few herbivores have the ability to break down the lignin and cellulose that reinforce bodies of most land plants. Similarly, many animals have some biomass tied up in an internal or external skeleton. Hair, feathers, and fur are also part of the biomass that is difficult to digest.

The ecological efficiency of energy transfers is usually higher in aquatic ecosystems than on land. Algae lack lignin, and so are more easily digested than land plants. Also, aquatic ecosystems usually have a higher proportion of ectotherms (cold-blooded animals), such as fish, than land ecosystems do. Ectotherms lose less energy as heat than endotherms (warm-blooded animals) so more is transferred to the next level. Higher efficiencies of transfers allow for longer food chains.

Take-Home Message

How does energy flow through ecosystems?

■ Primary producers capture energy and convert it into biomass. We measure this process as primary production.

■ A biomass pyramid depicts dry weight of organisms at each trophic level in an ecosystem. Its largest tier is usually producers, but the pyramid for some aquatic systems is inverted.

■ An energy pyramid depicts the amount of energy that enters each level. Its largest tier is always at the bottom (producers).

■ Efficiency of transfers tends to be greatest in aquatic systems, where primary producers usually lack lignin and consumers tend to be ectotherms.

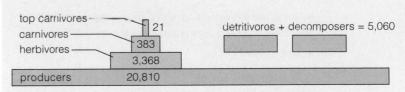

A Energy pyramid for the Silver Springs ecosystem. The size of each step in the pyramid represents the amount of energy that enters that trophic level annually, as shown in detail below.

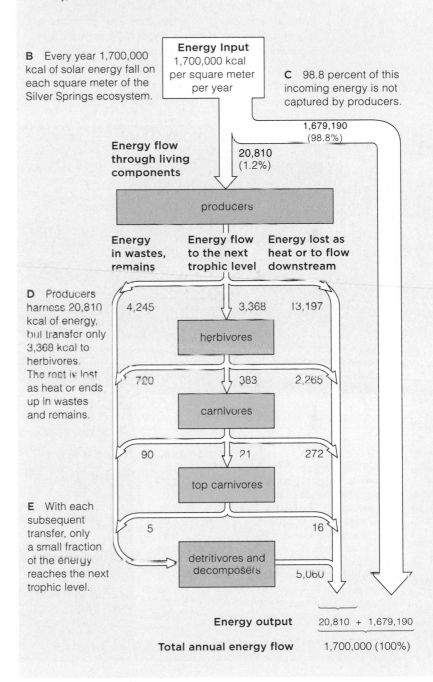

B Every year 1,700,000 kcal of solar energy fall on each square meter of the Silver Springs ecosystem.

Energy Input
1,700,000 kcal per square meter per year

C 98.8 percent of this incoming energy is not captured by producers.

D Producers harness 20,810 kcal of energy, but transfer only 3,368 kcal to herbivores. The rest is lost as heat or ends up in wastes and remains.

E With each subsequent transfer, only a small fraction of the energy reaches the next trophic level.

Energy output 20,810 + 1,679,190

Total annual energy flow 1,700,000 (100%)

Figure 47.8 Animated Annual energy flow in Silver Springs measured in kilocalories (kcal) per square meter per year. **Figure It Out:** What percent of the energy carnivores received from herbivores was later passed on to top carnivores?

Answer: 21/383 × 100 = 5.5 percent

47.4 | Biological Magnification

■ Some harmful substances become more and more concentrated as they pass from one trophic level to the next.

■ Link to Malaria 22.6

DDT and Silent Spring The synthetic pesticide dichloro-
diphenyl-trichloroethane, or DDT, was invented in the
late 1800s and came into widespread use in the 1940s.
Spraying DDT saved many human lives by killing lice
that spread typhus, and mosquitoes that carried malaria.
Farmers also embraced this new chemical that increased
crop yields by killing common agricultural pests. In the
1950s, swelling numbers of suburbanites turned to DDT
to keep their shrubbery free of leaf-munching insects.

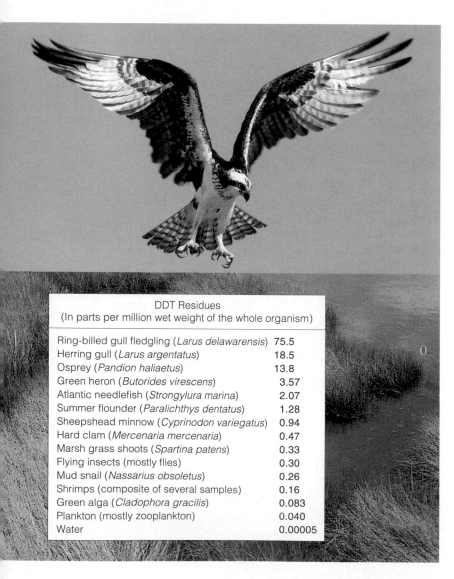

DDT Residues
(In parts per million wet weight of the whole organism)

Ring-billed gull fledgling (*Larus delawarensis*)	75.5
Herring gull (*Larus argentatus*)	18.5
Osprey (*Pandion haliaetus*)	13.8
Green heron (*Butorides virescens*)	3.57
Atlantic needlefish (*Strongylura marina*)	2.07
Summer flounder (*Paralichthys dentatus*)	1.28
Sheepshead minnow (*Cyprinodon variegatus*)	0.94
Hard clam (*Mercenaria mercenaria*)	0.47
Marsh grass shoots (*Spartina patens*)	0.33
Flying insects (mostly flies)	0.30
Mud snail (*Nassarius obsoletus*)	0.26
Shrimps (composite of several samples)	0.16
Green alga (*Cladophora gracilis*)	0.083
Plankton (mostly zooplankton)	0.040
Water	0.00005

Figure 47.9 Biological magnification in an estuary on Long Island, New
York, as reported in 1967 by George Woodwell, Charles Wurster, and Peter
Isaacson. Effects of DDT vary among species. Ospreys such as the one in the
upper photo are highly sensitive. At 4 ppm of DDT, osprey eggs are fragile and
unlikely to hatch. Gulls tolerate far higher doses of DDT without eggshell effects.

Unfortunately, DDT also affected nonpest species.
Where DDT was sprayed to control Dutch elm disease,
songbirds died. In forests sprayed to kill budworm larvae,
DDT got into streams and killed fishes.

Rachel Carson, who had worked for the U.S. Fish and
Wildlife Service, began compiling information about the
harmful effects of pesticide use. She published her findings
in 1962 as the book *Silent Spring*. The public embraced
Carson's ideas but the pesticide industry mounted a campaign to discredit her. At the time, Carson was battling
terminal breast cancer. Yet she vigorously defended her
position until her death in 1964.

After Carson's death, study of DDT's impact increased.
Researchers showed that DDT, like some other synthetic
chemicals, undergoes **biological magnification**. By this
process, a chemical that degrades slowly or not at all
becomes increasingly concentrated in tissues of organisms
as it moves up a food chain (Figure 47.9). In birds that
are top carnivores such as ospreys, brown pelicans, bald
eagles, and peregrine falcons, high DDT levels made eggs
fragile, causing population sizes to plummet.

In recognition of the ecological effects of DDT, the
United States has banned its use and export. Predatory
bird populations in this country have largely recovered.
Some countries still use DDT to fight malaria-carrying
mosquitoes, but application is limited to indoor spraying.
Even this use is controversial; some people would like to
see a worldwide ban on the chemical. In additional to the
environmental concerns, they cite studies indicating that
maternal exposure to DDT during pregnancy may cause
premature births and affects a child's mental development.

The Mercury Menace Birds bore the brunt of DDT's
effects but fish get the spotlight when it comes to mercury
pollution. Coal-burning power plants and some industrial
processes put mercury into the air, then rain washes it into
aquatic habitats. In some regions, runoff from abandoned
or operating mines also contributes to aquatic mercury.

Like DDT, mercury accumulates as it moves up through
food chains. Mercury adversely affects development of
the human nervous system, so children and women who
are pregnant or nursing should not eat fish that are top
carnivores. Shark, swordfish, king mackerel, and tilefish are
riskiest. You should also avoid these high-mercury fish if
you are planning on becoming pregnant in the near future.
Once mercury settles into your tissues, it can take a year
for your body to get rid of it.

Everyone should avoid making fish that can have a
high mercury content a major part of their diet. You can
receive the health benefits of eating fish by choosing other
species that are lower in mercury. For example, catfish,
salmon, sardines, pollack, and canned light tuna are good
choices. If you fish and plan to eat what you catch, check
for local advisories about contaminants. The EPA website
www.epa.gov/waterscience/fish/states.htm can link you to
the appropriate agency.

47.5 | Biogeochemical Cycles

- Nutrients move from nonliving environmental reservoirs into living organisms, then back into those reservoirs.
- Links to Tectonic plates 17.9, Nitrogen fixation 21.6

In a **biogeochemical cycle**, an essential element moves from one or more nonliving environmental reservoirs, through living organisms, then back to the reservoirs (Figure 47.10). As explained in the Chapter 2 introduction, oxygen, hydrogen, carbon, nitrogen, and phosphorus are some of the elements essential to all forms of life. We refer to these and other required elements as nutrients.

Depending on the element, environmental reservoirs may include Earth's rocks and sediments, waters, and atmosphere. Chemical and geologic processes move elements to and from these reservoirs. For example, elements that had been locked in rocks become part of the atmosphere as a result of volcanic activity. Uplifting elevates rocks where they are exposed to erosive forces of wind and rain. The rocks slowly dissolve; elements in them enter rivers, and eventually seas.

Elements enter the living part of an ecosystem by way of primary producers. Photosynthetic organisms take up essential ions dissolved in water. Land plants also take up carbon dioxide from the air.

Some bacteria fix nitrogen gas (Section 21.6). Their action makes this nutrient available to producers.

Nutrients move through food webs when organisms eat one another. Fungi and prokaryotes speed nutrient cycling within an ecosystem by decomposing remains and wastes of other organisms, so elements that were tied up in those materials are once again available to primary producers.

The next sections describe the four biogeochemical cycles that affect the most abundant elements in living organisms. In the water cycle, oxygen and hydrogen move on a global scale as part of molecules of water. In atmospheric cycles, a gaseous form of a nutrient such as carbon or nitrogen moves through ecosystems. A nutrient that does not often occur as a gas, such as phosphorus, moves in sedimentary cycles. Such nutrients accumulate on the ocean floor, then return to land by slow movements of Earth's crust (Section 17.9).

Take-Home Message

What are biogeochemical cycles?

- Biogeochemical cycles describe the continual flow of nutrients between nonliving environmental reservoirs and living organisms.
- Prokaryotes play a pivotal role in transfers between the living and nonliving portions of the cycle.
- Elements that occur in gases move through atmospheric cycles. Elements that do not normally occur as a gas move in sedimentary cycles.

Figure 47.10 Generalized biogeochemical cycle. In such cycles, a nutrient moves among nonliving environmental reservoirs and into and out of the living portion of an ecosystem. For all nutrients, the portion tied up in environmental reservoirs far exceeds the amount in living organisms.

- All organisms are mostly water and the cycling of this essential resource has implications for all life.
- Links to Properties of water 2.5, Leaching and erosion 29.1, Transpiration 29.3

How and Where Water Moves

The world ocean holds most of Earth's water (Table 47.1). As Figure 47.11 shows, in the **water cycle**, water moves among the atmosphere, the oceans, and environmental reservoirs on land. Sunlight energy drives evaporation, the conversion of water from liquid form to a vapor. Transpiration, explained in Section 29.3, is evaporation of water from plant parts. In cool upper layers of the atmosphere, condensation of water vapor into droplets gives rise to clouds. Later, clouds release the water as precipitation—as rain, snow, or hail.

A **watershed** is an area from which all precipitation drains into a specific waterway. It may be as small as a valley that feeds a stream, or as large as the Mississippi River Basin, which covers about 41 percent of the continental United States.

Most precipitation falling in a watershed seeps into the ground. Some collects in **aquifers**, permeable rock layers that hold water. **Groundwater** is water in soil and aquifers. When soil gets saturated, water becomes **runoff**; it flows over the ground into streams.

Table 47.1 Environmental Water Reservoirs

Main Reservoirs	Volume (10^3 cubic kilometers)
Ocean	1,370,000
Polar ice, glaciers	29,000
Groundwater	4,000
Lakes, rivers	230
Soil moisture	67
Atmosphere (water vapor)	14

Flowing water moves dissolved nutrients into and out of a watershed. Experiments in New Hampshire's Hubbard Brook watershed illustrated that vegetation helps slow nutrient losses. Experimental deforestation caused a spike in loss of mineral ions (Figure 47.12).

A Global Water Crisis

Our planet has plenty of water, but most of it is too salty to drink or use for irrigation. If all Earth's water filled a bathtub, the amount of fresh water that could be used sustainably in a year would fill a teaspoon.

Of the fresh water we use, about two-thirds goes to agriculture, but irrigation can harm soil. Piped-in water

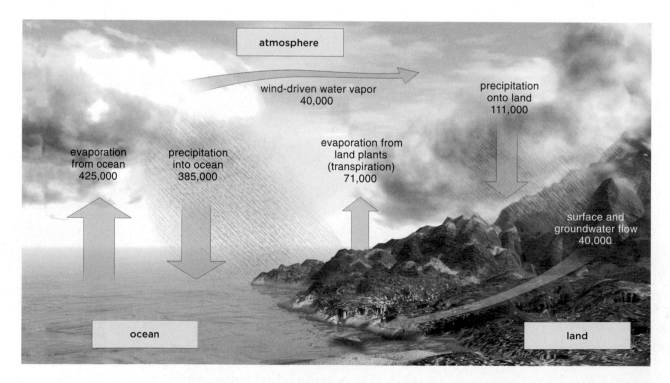

Figure 47.11 Animated The water cycle. Arrows identify processes that move water. The numbers shown indicate the amounts moved, as measured in cubic kilometers per year.

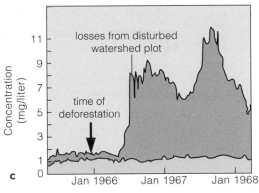

Figure 47.12 Hubbard Brook experimental watershed. (**a**) Runoff in this watershed is collected by concrete basins for easy monitoring. (**b**) This plot of land was stripped of all vegetation as an experiment. (**c**) After experimental deforestation, calcium levels in runoff increased sixfold (*medium blue*). A control plot in the same watershed showed no similar increase during this time (*light blue*).

often has high concentrations of salts. **Salinization**, the buildup of mineral salts in soil, stunts crop plants and decreases yields.

Groundwater supplies drinking water to about half of the United States population. Pollution of this water now poses a threat. Chemicals leaching from landfills, hazardous waste facilities, and underground storage tanks often contaminate it. Unlike flowing rivers and streams, which can recover fast, polluted groundwater is difficult and expensive to clean up.

Water overdrafts are also common; water is drawn from aquifers faster than natural processes replenish it. When too much fresh water is withdrawn from an aquifer near the coast, salt water moves in and replaces it. Figure 47.13 highlights regions of aquifer depletion and saltwater intrusion in the United States.

Overdrafts have now depleted half of the Ogallala aquifer, which extends from South Dakota into Texas. This aquifer supplies the irrigation water for about 20 percent of the nation's crops. For the past thirty years, withdrawals have exceeded replenishment by a factor of ten. What will happen when water runs out?

Contaminants such as sewage, animal wastes, and agricultural chemicals make water in rivers and lakes unfit to drink. In addition, pollutants disrupt aquatic ecosystems, and in some cases they drive vulnerable species to local extinction.

Desalinization, the removal of salt from seawater, may help increase freshwater supplies. However, the process requires a lot of fossil fuel. Desalinization is feasible mainly in Saudi Arabia and other places that have small populations and very large fuel reserves. In addition, the process produces mountains of waste salts that must be disposed of.

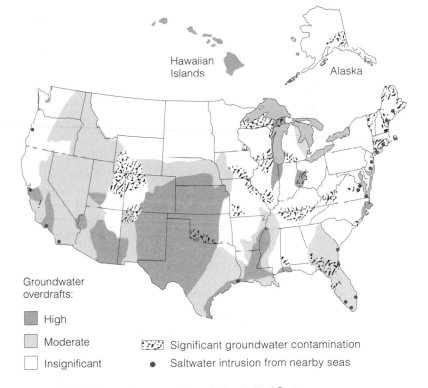

Groundwater overdrafts:

■ High

■ Moderate   ▨ Significant groundwater contamination

□ Insignificant   ● Saltwater intrusion from nearby seas

Figure 47.13 Groundwater problems in the United States.

Take-Home Message

What is the water cycle and how do humans affect it?

■ In the water cycle, water moves on a global scale. It moves slowly from the world ocean—the main reservoir—through the atmosphere, onto land, then back to the ocean.

■ Of the fresh water that human populations use, about two-thirds sustains agriculture.

■ Aquifers that supply much of the world's drinking water are becoming polluted and depleted.

■ Carbon dioxide in air makes the carbon cycle an atmospheric cycle, but most carbon is in sediments and rocks.

■ Links to Carbon fixation 7.6, Foraminiferans 22.3, Peat bogs 23.3

In the **carbon cycle**, carbon moves through the lower atmosphere and all food webs on its way to and from its largest reservoirs (Figure 47.14). Earth's crust holds the most carbon—66 million to 100 million gigatons. A gigaton is a billion tons. There are 4,000 gigatons of carbon in the known fossil fuel reserves.

Organisms contribute to Earth's carbon deposits. Single-celled protists such as foraminiferans (Section 22.3) produce shells rich in calcium carbonate. Over hundreds of millions of years, uncountable numbers of these cells died, sank, and were buried in seafloor sediments. The carbon in their remains cycles slowly, as movements of Earth's crust uplift portions of the sea floor, making it part of a land ecosystem.

Most of the annual carbon movement takes place between the ocean and atmosphere. The ocean holds 38,000–40,000 gigatons of dissolved carbon, primarily in the form of bicarbonate and carbonate ions. The air holds about 766 gigatons of carbon, mainly combined with oxygen in the form of carbon dioxide (CO_2).

On land, detritus in soil holds 1,500–1,600 gigatons of carbon. Peat bogs and the permafrost, a perpetually frozen layer of soil that underlies arctic regions, are major reservoirs. Another 540–610 gigatons is present in biomass, or tissues of organisms.

Ocean currents move carbon from upper ocean waters into deep sea reservoirs. Carbon dioxide enters warm surface waters and is converted to bicarbonate. Then, prevailing winds and regional differences in density drive the flow of bicarbonate-rich seawater in a gigantic loop from the surface of the Pacific and Atlantic oceans down to the Atlantic and Antarctic sea floors. Here, bicarbonate moves into cold, deep storage

Figure 47.14 Animated *Right,* carbon cycling in (**a**) marine ecosystems and (**b**) land ecosystems. *Gold* boxes highlight the most important carbon reservoirs. The vast majority of carbon atoms are in sediments and rocks, followed by lesser amounts in seawater, soil, the atmosphere, and biomass (in that order). Typical annual fluxes in global distribution of carbon, in gigatons, are:

From atmosphere to plants by carbon fixation	120
From atmosphere to ocean	107
To atmosphere from ocean	105
To atmosphere from plants	60
To atmosphere from soil	60
To atmosphere from fossil fuel burning	5
To atmosphere from net destruction of plants	2
To ocean from runoff	0.4
Burial in ocean sediments	0.1

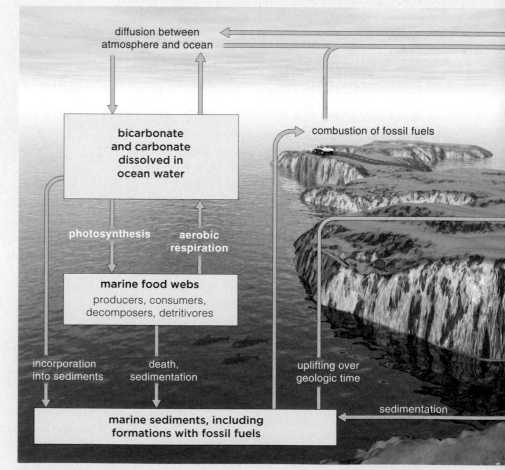

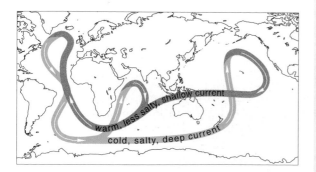

Figure 47.15 Loop that moves carbon dioxide to carbon's deep ocean reservoir. The loop sinks in the cold, salty North Atlantic. It rises in the warmer Pacific.

reservoirs before water loops back up (Figure 47.15). Storage of carbon in the deep sea helps dampen any short-term effects of increases in atmospheric carbon.

Biologists sometimes refer to the global cycling of carbon in the form of carbon dioxide and bicarbonate as a carbon–oxygen cycle. Plants, phytoplankton, and some bacteria fix carbon when they engage in photosynthesis (Section 7.6). Each year, they tie up billions of metric tons of carbon in sugars and other organic compounds. Breakdown of those compounds by aerobic respiration releases carbon dioxide into the air. More carbon dioxide escapes into the air when fossil fuels or forests burn and when volcanoes erupt.

The time that an ecosystem holds a given carbon atom varies. Organic material decomposes rapidly in tropical forests, so carbon does not build up at the soil surface. By contrast, bogs and other anaerobic habitats do not favor decomposition, so material accumulates, as in peat bogs (Section 23.3).

Humans are altering the carbon cycle. Each year, we withdraw 4 to 5 gigatons of fossil fuel from environmental reservoirs. Our activities put about 6 gigatons more carbon in the air than can be moved into ocean reservoirs by natural processes. Only about 2 percent of the excess carbon entering the atmosphere becomes dissolved in ocean water. Carbon dioxide in the air traps heat, so increased outputs of it may be a factor in global climate change. The next section looks at this possibility and some environmental implications.

Take-Home Message

What is the carbon cycle?

■ In the carbon–oxygen cycle, carbon moves into and out of ecosystems mainly combined with oxygen, as in carbon dioxide, bicarbonate, and carbonate.

■ Earth's crust is the largest carbon reservoir, followed by the world ocean. Most of the annual cycling of carbon occurs between the ocean and atmosphere.

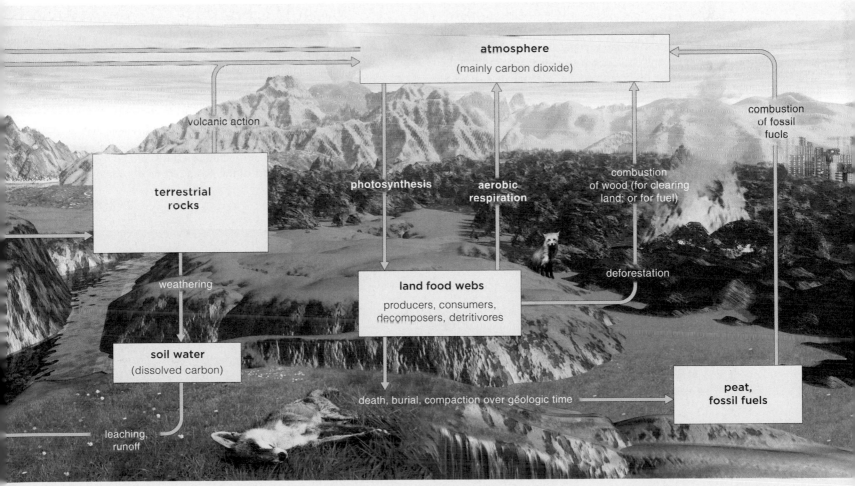

B

47.8 | Greenhouse Gases and Climate Change

- Concentrations of gases in Earth's atmosphere help determine the temperature near Earth's surface. Human activities are altering gas concentrations and causing climate change.

- Link to Carbon imbalances 7.9

Concentrations of various gaseous molecules profoundly influence the average temperature of the atmosphere near Earth's surface. That temperature, in turn, has far-reaching effects on global and regional climates.

Atmospheric molecules of carbon dioxide, water, nitrous oxide, methane, and chlorofluorocarbons (CFCs) are among the main players in interactions that can shift global temperatures. Collectively, the gases trap heat a bit like a greenhouse does, hence the familiar name "greenhouse gases."

Radiant energy from the sun passes through the atmosphere and is absorbed by Earth's surface. The energy warms the surface, which means that the surface emits infrared radiation (heat). The infrared energy radiates back toward space, but greenhouse gases in the atmosphere interfere with its progress. How? The gases absorb some of the infrared energy, and then emit a portion of it back toward Earth's surface (Figure 47.16). Without this process, which is called the **greenhouse effect**, Earth's surface would be so cold that very little life would survive.

In the 1950s, researchers at a laboratory on Hawaii's highest volcano began to measure the atmospheric concentrations of greenhouse gases. That remote site is almost free of local airborne contamination. It also is representative of atmospheric conditions for the Northern Hemisphere. What did they find? Briefly, concentrations of CO_2 follow annual cycles of primary production. They decline in summer, when the rates of photosynthesis are highest. They rise in winter, when photosynthesis declines but aerobic respiration and fermentation continue.

Figure 47.17 *Facing page*, graphs of recent increases in four categories of atmospheric greenhouse gases. A key factor is the sheer number of gasoline-burning vehicles in large cities. *Above*, Mexico City on a smoggy morning. With 10 million residents, it is the world's largest city.

The alternating troughs and peaks along the graph line in Figure 47.17a are annual lows and highs of global CO_2 concentrations. For the first time, researchers saw the effects of carbon dioxide fluctuations for the entire hemisphere. Notice the midline of the troughs and peaks in the cycle. It shows that carbon dioxide concentration is steadily increasing—as are concentrations of other major greenhouse gases.

Atmospheric levels of greenhouse gases are far higher than they were for most of the past. Carbon dioxide may

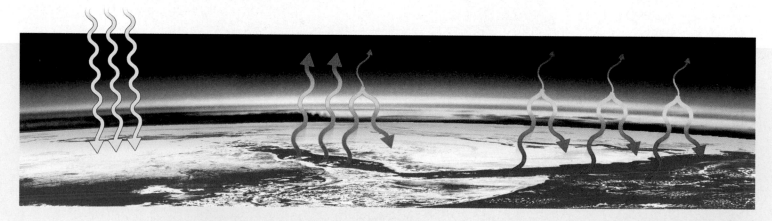

A Radiant energy from the sun penetrates the lower atmosphere, and it warms Earth's surface.

B The warmed surface radiates heat (infrared radiation) back toward space. Greenhouse gases absorb some of the infrared energy, and then emit a portion of it back toward Earth.

C Increased concentrations of greenhouse gases trap more heat near Earth's surface. Sea surface temperatures rise, so more water evaporates into the atmosphere. Earth's surface temperature rises.

Figure 47.16 Animated The greenhouse effect.

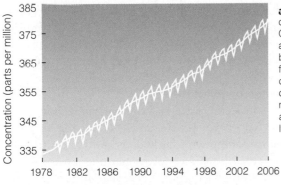

a Carbon dioxide (CO_2). Of all human activities, the burning of fossil fuels and deforestation contribute the most to rising atmospheric levels.

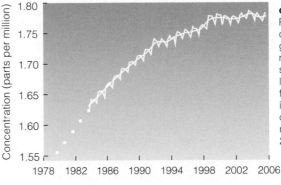

c Methane (CH_4). Production and distribution of natural gas as fuel adds to methane released by some bacteria that live in swamps, rice fields, landfills, and in the digestive tract of cattle and other ruminants (Section 21.7).

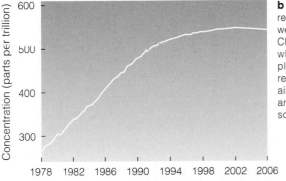

b CFCs. Until restrictions were in place, CFCs were widely used in plastic foams, refrigerators, air conditioners, and industrial solvents.

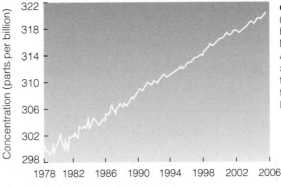

d Nitrous oxide (N_2O). Denitrifying bacteria produce N_2O in metabolism. Also fertilizers and animal waste from large-scale feedlots release large amounts.

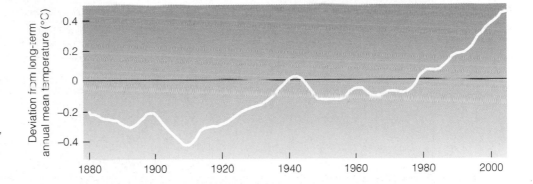

Figure 47.18 Recorded changes in the global mean temperature over land and sea between 1880 and 2005, given as degrees above or below average temperature during 1960–1990.

be at its highest level since 470,000 years ago, possibly since 20 million years ago. There is scientific consensus that human activities—mainly the burning of fossil fuels—are contributing significantly to the current increases in greenhouse gases. The big worry is that the increase may have far-reaching environmental consequences.

The increase in greenhouse gases may be a factor in **global warming**, a long-term increase in temperature near Earth's surface (Figure 47.18). In the past thirty years, the global surface temperature increased at a faster rate, to 1.8°C (3.2°F) per century. Warming is most dramatic at the upper latitudes of the Northern Hemisphere.

Data from satellites, weather stations and balloons, research ships, and computer programs suggest that some irreversible climate changes are already under way. Water

expands as it is heated, and heating also melts glaciers and other ice. Together, thermal expansion and addition of meltwater will cause sea level to rise. In the past century, the sea level may have risen as much as 20 centimeters (8 inches) and the rate of rise appears to be accelerating.

Scientists expect continued temperature increases to have far-reaching effects on climate. An increased rate of evaporation will alter global rainfall patterns. Intense rains and flooding probably will become more frequent in some regions, while droughts increase in others. Hurricanes probably will become more intense.

It bears repeating: As investigations continue, a key research goal is to investigate all of the variables in play. With respect to consequences of climate change, the most crucial variable may be the one we do not know.

■ Gaseous nitrogen makes up about 80 percent of the lower atmosphere, but most organisms can't use this gaseous form.

■ Links to Acid rain 2.6, Nitrogen fixation 21.6 and 29.2, Algal blooms 22.5, Decomposers 21.6 and 24.5, Leaching 29.1

Inputs Into Ecosystems

Nitrogen moves in an atmospheric cycle known as the **nitrogen cycle** (Figure 47.19). Gaseous nitrogen makes up about 80 percent of the atmosphere. Triple covalent bonds hold its two atoms of nitrogen together as N_2, or $N \equiv N$. Plants cannot use gaseous nitrogen, because they do not make the enzyme that can break its triple bond. Volcanic eruptions and lightning can convert some N_2 into forms that enter food webs. Far more is converted through **nitrogen fixation**. By this process, bacteria break all three bonds in N_2, then incorporate the N atoms into ammonia (NH_3). Ammonia gets converted into ammonium (NH_4^+) and nitrate (NO_3^-). These two nitrogen salts dissolve readily in water and are taken up by plant roots.

Many species of bacteria fix nitrogen (Section 21.6). Nitrogen-fixing cyanobacteria live in aquatic habitats, soil, and as components of lichens. Another nitrogen-fixing group, *Rhizobium*, forms nodules on the roots of peas and other legumes. Each year, nitrogen-fixing bacteria collectively take up about 270 million metric tons of nitrogen from the atmosphere.

The nitrogen incorporated into plant tissues moves up through trophic levels of ecosystems. It ends up in

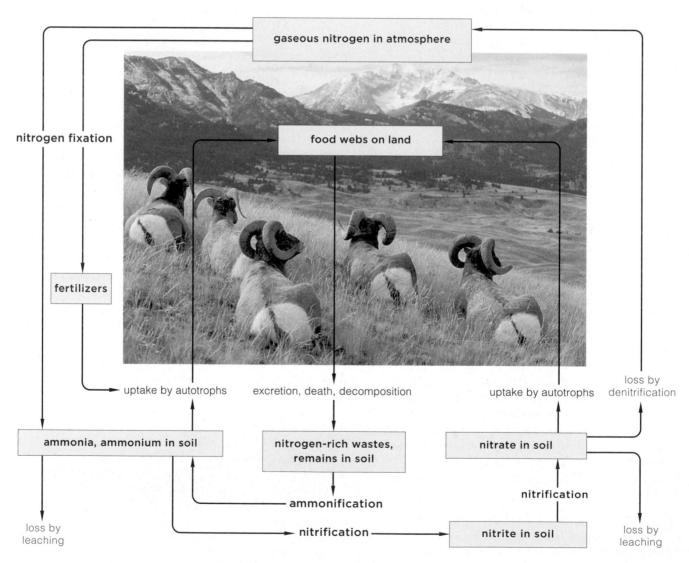

Figure 47.19 Animated Nitrogen cycle in an ecosystem on land. Nitrogen becomes available to plants through the activities of nitrogen-fixing bacteria. Other bacterial species cycle nitrogen to plants. They break down organic wastes to ammonium and nitrates.

nitrogen-rich wastes and remains, which bacteria and fungi decompose (Sections 21.6 and 24.5). By the process of **ammonification**, these organisms break apart proteins and other nitrogen-containing molecules and produce ammonium. Some of the ammonium product gets released into the soil, where plants and nitrifying bacteria take it up. **Nitrification** begins when bacteria convert ammonium to nitrite (NO_2^-). Other nitrifying bacteria then use the nitrite in reactions that end with the formation of nitrate. Nitrate, like ammonium, can be taken up by plant roots.

Natural Losses From Ecosystems

Ecosystems lose nitrogen through **denitrification**. By this process, denitrifying bacteria convert nitrate or nitrite to gaseous nitrogen or to nitrogen oxide (NO_2). Denitrifying bacteria are typically anaerobes that live in waterlogged soils and aquatic sediments.

Ammonium, nitrite, and nitrate also are lost from a land ecosystem in runoff and by leaching, the removal of some nutrients as water trickles down through the soil (Section 29.1). Nitrogen-rich runoff enters streams and other aquatic ecosystems.

Disruptions by Human Activities

Deforestation and conversion of grassland to farmland also causes nitrogen losses from an ecosystem. With each clearing and harvest of plants, nitrogen stored in plant tissues is removed. Plant removal also makes soil more vulnerable to erosion and leaching.

Farmers can counter nitrogen depletion by rotating their crops. For example, they plant corn and soybeans in the same field in alternating years. Nitrogen-fixing bacteria that associate with legumes such as soybeans add nitrogen to the soil (Section 29.2).

In developed countries, most farmers also spread synthetic nitrogen-rich fertilizers. High temperature and pressure converts nitrogen and hydrogen gases to ammonia fertilizers. Although the manufactured fertilizers improve crop yields, they also modify soil chemistry. Adding ammonium to the soil increases the concentration of hydrogen ions, as well as nitrogen. High acidity encourages ion exchange: Nutrient ions bound to particles of soil get replaced by hydrogen ions. As a result, calcium and magnesium ions needed for plant growth seep away in soil water.

Burning of fossil fuel in power plants and by vehicles releases nitrogen oxides. These gases contribute to global warming and acid rain (Section 2.6). Winds frequently carry gaseous pollutants far from their sources.

Figure 47.20 Dead and dying trees in Great Smoky Mountains National Park. Forests are among the casualties of nitrogen oxides and other forms of air pollution.

By some estimates, pollutants blowing into the Great Smoky Mountains National Park have increased the amount of nitrogen in the soil sixfold (Figure 47.20).

Nitrogen in acid rain can have the same effects as use of manufactured fertilizers. Different plant species respond in different ways to increased nitrogen level. Changes in soil nitrogen disrupt the balance among competing species in a community, causing diversity to decline. The impact can be especially pronounced in forests at high elevations or at high latitudes, where soils tend to be naturally nitrogen-poor.

Some human activities disrupt aquatic ecosystems through nitrogen enrichment. For instance, about half of the nitrogen in fertilizers applied to fields runs off into rivers, lakes, and estuaries. More nitrogen enters waters in sewage from cities and in animal wastes. As one result, nitrogen inputs promote algal blooms (Section 22.5). Phosphorus in fertilizers has the same negative effects, as explained in the next section.

Take-Home Message

What is the nitrogen cycle?

■ The ecosystem phase of the nitrogen cycle starts with nitrogen fixation. Bacteria convert gaseous nitrogen in the air to ammonia and then to ammonium, which is a form that plants easily take up.

■ By ammonification, bacteria and fungi make additional ammonium available to plants when they break down nitrogen-rich organic wastes and remains.

■ By nitrification, bacteria convert nitrites in soil to nitrate, which also is a form that plants easily take up.

■ The ecosystem loses nitrogen when denitrifying bacteria convert nitrite and nitrate back to gaseous nitrogen, and when nitrogen is leached from soil.

■ Unlike carbon and nitrogen, phosphorus seldom occurs as a gas. Like nitrogen, it can be taken up by plants only in ionized form, and it, too, is often a limiting factor on plant growth.

In the **phosphorus cycle**, phosphorus passes quickly through food webs as it moves from land to ocean sediments, then slowly back to dry land. Earth's crust is the largest reservoir of phosphorus.

Phosphorus in rocks is mainly in the form of phosphate (PO_4^{3-}). Weathering and erosion put phosphate ions from rocks into streams and rivers, which deliver them to oceans (Figure 47.21). There, the phosphates accumulate as underwater deposits along the edges of continents. After millions of years, movements of Earth's crust result in uplifting of parts of the sea floor. Once uplifted, the rocky phosphate deposits on land are subject to weathering and erosion, which release phosphates from the rocks and start the phosphorus cycle over again.

Phosphates are required building blocks for ATP, phospholipids, nucleic acids, and other compounds. Plants take up dissolved phosphates from soil water. Herbivores get them by eating plants; carnivores get them by eating herbivores. Animals lose phosphate in urine and in feces. Bacterial and fungal decomposers release phosphate from organic wastes and remains, then plants take them up again.

The water cycle helps move phosphorus and other minerals through ecosystems. Water evaporates from the ocean and falls on land. As it flows back to the ocean, it transports silt and dissolved phosphates that the primary producers require for growth.

Of all minerals, phosphorus most frequently acts as the limiting factor for plant growth. Only newly weathered, young soil has an abundance of phosphorus. Many tropical and subtropical ecosystems that are already low in phosphorus are likely to be further depleted by human actions. In an undisturbed forest, decomposition releases phosphorus stored in biomass. When forest is converted to farmland, the ecosystem loses phosphorus that had been stored in trees. Crop yields soon decline. Later, after the fields are abandoned, regrowth remains sparse. Spreading finely ground, phosphate-rich rock can help restore fertility, but many developing countries lack this resource.

Many developed countries have a different problem. Phosphorous in runoff from heavily fertilized fields pollutes water. Sewage from cities and factory farms also contain phosphorus. Dissolved phosphorus that

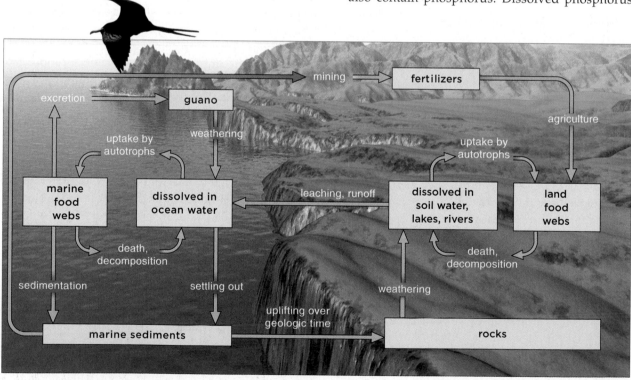

Figure 47.21 Animated Phosphorus cycle. In this sedimentary cycle, phosphorus moves mainly in the form of phosphate ions (PO_4^{3-}) to the ocean. It moves through phytoplankton of marine food webs, then to fishes that eat plankton. Seabirds eat the fishes, and their droppings (guano) accumulate on islands. Humans collect and use guano as a phosphate-rich fertilizer.

gets into aquatic ecosystems can promote destructive algal blooms. Like the plants, algae require nitrogen, phosphorus, and other ions to keep growing. In many freshwater ecosystems, nitrogen-fixing bacteria keep the nitrogen levels high, so phosphorus becomes the limiting factor. When phosphate-rich pollutants pour in, algal populations soar and then crash. As aerobic decomposers break down remains of dead algae, the water becomes depleted of the oxygen that fishes and other organisms require.

Eutrophication refers to nutrient enrichment of any ecosystem that is otherwise low in nutrients. It can occur naturally, but human activities often accelerate it, as the experiment shown in Figure 47.22 demonstrated. Eutrophication of a lake is difficult to reverse. It can take years for excess nutrients that encourage algal growth to be depleted.

Take-Home Message

What is the phosphorus cycle?

■ The phosphorus cycle is a sedimentary cycle that moves this element from its main reservoir (Earth's crust), through soils and sediments, aquatic habitats, and bodies of living organisms.

Figure 47.22 A eutrophication experiment. Researchers put a plastic curtain across a channel between two basins of a natural lake. They added nitrogen, carbon, and phosphorus to the water on one side of the curtain (here, the *lower* part of the lake) and added nitrogen and carbon to the water on the other side. Within months, the basin with phosphorous was eutrophic, with a dense algal bloom (*green*) covering its surface.

Summary

Section 47.1 An **ecosystem** consists of an array of organisms along with nonliving components of their environment. There is a one-way flow of energy into and out of an ecosystem, and a cycling of materials among resident species. All ecosystems have inputs and outputs of energy and nutrients.

Sunlight supplies energy to most ecosystems. **Primary producers** convert sunlight energy into chemical bond energy. They also take up the nutrients that they, and all consumers, require. Herbivores, carnivores, omnivores, **decomposers**, and **detritivores** are **consumers**.

Energy moves from organisms at one **trophic level** to organisms at another. Organisms are at the same trophic level if they are an equal number of steps away from the energy input into the ecosystem. A **food chain** shows one path of energy and nutrient flow among organisms. It depicts who eats whom.

■ *Use the animation on CengageNOW to learn about energy flow and nutrient cycling.*

Section 47.2 Food chains interconnect as **food webs**. The efficiency of energy transfers is always low, so most ecosystems have no more than four or five trophic levels. In a **grazing food chain**, most energy captured by producers flows to herbivores. In **detrital food chains**, most energy flows from producers directly to detritivores and decomposers. Both types of food chains interconnect in nearly all ecosystems.

■ *Use the animation on CengageNOW to explore a food web.*

Section 47.3 A system's **primary production** is the rate at which producers capture and store energy in their tissues. It varies with climate, seasonal changes, nutrient availability, and other factors.

Energy pyramids and **biomass pyramids** depict how energy and organic compounds are distributed among the organisms of an ecosystem. All energy pyramids are largest at their base. If producers get eaten as fast as they reproduce, the biomass of consumers can exceed that of producers, so the biomass pyramid is upside down.

■ *Use the animation on CengageNOW to see how energy flows through one ecosystem.*

Section 47.4 With **biological magnification**, a chemical substance is passed from organisms at each trophic level to those above and becomes increasingly concentrated in body tissues.

Section 47.5 In a **biogeochemical cycle**, water or some nutrient moves from an environmental reservoir, through organisms, then back to the environment.

Section 47.6 In the **water cycle**, evaporation, condensation, and precipitation move water from its main reservoir—oceans—into the atmosphere, onto land, then back to oceans. **Runoff** is water that flows over ground into streams. A **watershed** is an area where all precipitation drains into a specific waterway. Water in **aquifers** and in the soil is **groundwater**. Use of irrigation can cause

Bye-Bye, Blue Bayou

In 2006, China overtook the United States as the country that emits the most carbon dioxide. Still, an average American life-style causes about 20 tons of carbon emissions per year. That's more than four times the emissions of an average person in China. It's also more than twice that of people in western Europe. Automotive emissions are one factor; fuel efficiency standards in both China and Europe are more stringent than they are in the United States.

How would you vote?

Should the United States increase fuel efficiency standards for cars and trucks to lower carbon dioxide output? See CengageNOW for details, then vote online.

---

salinization—salt buildup—in soil. **Desalinization** is an energy-intensive method of obtaining fresh water from salt water.

- *Use the animation on CengageNOW to learn about the water cycle.*

Section 47.7 The **carbon cycle** moves carbon from reservoirs in rocks and seawater, through its gaseous forms (methane and CO_2) in the air, and through ecosystems. Deforestation and the burning of wood and fossil fuels are adding more carbon dioxide to the atmosphere than the oceans can absorb.

- *Use the animation on CengageNOW to observe the flow of carbon through its global cycle.*

Section 47.8 The **greenhouse effect** refers to the ability of certain gases to trap heat in the lower atmosphere. It warms Earth's surface. Human activities are putting larger than normal amounts of greenhouse gases, including carbon dioxide, into the atmosphere. The rise in these gases correlates with a rise in global temperatures (**global warming**) and other climate changes.

- *Use the animation on CengageNOW to explore the greenhouse effect and global warming.*

Section 47.9 The **nitrogen cycle** is an atmospheric cycle. Air is the main reservoir for N_2, a gaseous form of nitrogen that plants cannot use. In **nitrogen fixation**, certain bacteria take up N_2 and form ammonia. **Ammonification** releases ammonia from organic remains. **Nitrification** involves conversion of ammonium to nitrite and then nitrate, which plants are able to take up. Some nitrogen is lost to the atmosphere by **denitrification** carried out by bacteria. Human activities add nitrogen to ecosystems; for example, through fossil fuel burning (which releases nitrogen oxides) and application of fertilizers. The added nitrogen can disrupt ecosystem processes.

- *Use the animation on CengageNOW to learn how nitrogen is cycled in an ecosystem.*

Section 47.10 The **phosphorus cycle** is a sedimentary cycle; Earth's crust is the largest reservoir and there is no major gaseous form. Phosphorus is often the factor that limits population growth of plant and algal producers. Excessive inputs of phosphorus to an aquatic ecosystem can accelerate **eutrophication**.

- *Use the animation on CengageNOW to learn how phosphorus is cycled in an ecosystem.*

Self-Quiz *Answers in Appendix III*

1. In most ecosystems, the primary producers use energy from _____ to build organic compounds.
   a. sunlight
   b. heat
   c. breakdown of wastes and remains
   d. breakdown of inorganic substances in the habitat

2. Organisms at the lowest trophic level in a tallgrass prairie are all _____ .
   a. at the first step away from the original energy input
   b. autotrophs   d. both a and b
   c. heterotrophs   e. both a and c

3. Decomposers are commonly _____ .
   a. fungi   b. plants   c. bacteria   d. a and c

4. All organisms at the first trophic level _____ .
   a. capture energy from a nonliving source
   b. obtain carbon from a nonliving source
   c. would be at the bottom of an energy pyramid
   d. all of the above

5. Primary productivity on land is affected by _____ .
   a. nutrient availability   c. temperature
   b. amount of sunlight   d. all of the above

6. If biological magnification occurs, the _____ will have the highest levels of toxins in their systems.
   a. producers   c. primary carnivores
   b. herbivores   d. top carnivores

7. Most of Earth's fresh water is _____ .
   a. in lakes and streams   c. frozen as ice
   b. in aquifers and soil   d. in bodies of organisms

8. Earth's largest carbon reservoir is _____ .
   a. the atmosphere   c. seawater
   b. sediments and rocks   d. living organisms

9. Carbon is released into the atmosphere by _____ .
   a. photosynthesis   c. burning fossil fuels
   b. aerobic respiration   d. b and c

10. Greenhouse gases _____ .
    a. slow the escape of heat energy from Earth into space
    b. are produced by natural and human activities
    c. are at higher levels than they were 100 years ago
    d. all of the above

11. The _____ cycle is a sedimentary cycle.
    a. water   c. nitrogen
    b. carbon   d. phosphorus

12. Earth's largest phosphorus reservoir is _____ .
    a. the atmosphere   c. sediments and rocks
    b. guano   d. living organisms

Data Analysis Exercise

To assess the impact of human activity on the carbon dioxide level in Earth's atmosphere, it helps to take a long view. One useful data set comes from deep core samples of Antarctic ice. The oldest ice core that has been fully analyzed dates back a bit more than 400,000 years. Air bubbles trapped in the ice provide information about the gas content in Earth's atmosphere at the time the ice formed. Combining ice core data with more recent direct measurements of atmospheric carbon dioxide—as in Figure 47.23—can help scientists put current changes in the atmospheric carbon dioxide into historical perspective.

1. What was the highest carbon dioxide level between 400,000 B.C. and 0 A.D.?

2. During this period, how many times did carbon dioxide reach a level comparable to that measured in 1980?

3. The industrial revolution occurred around 1800. What was the trend in carbon dioxide level in the 800 years prior to this event? What about in the 175 years after it?

4. Was the rise in the carbon dioxide level between 1800 and 1975 larger or smaller than the rise between 1980 and 2007?

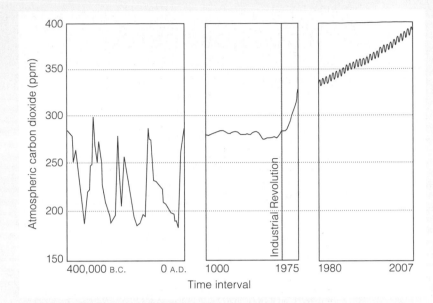

Figure 47.23 Changes in atmospheric carbon dioxide levels (in parts per million). Direct measurements began in 1980. Earlier data are based on ice cores.

13. Plant growth requires _____ uptake from the soil.
   a. nitrogen   d. both a and c
   b. carbon   e. all of the above
   c. phosphorus

14. Nitrogen fixation converts _____ to _____ .
   a. nitrogen gas; ammonia   d. ammonia; nitrates
   b. nitrates; nitrites   e. nitrogen gas; nitrogen
   c. ammonia; nitrogen gas   oxides

15. Match each term with its most suitable description.
   ____producers   a. steps from energy source
   ____herbivores   b. feed on small bits of
   ____decomposers   organic matter
   ____detritivores   c. degrade organic
   ____trophic level   wastes and remains to
   ____biological   inorganic forms
      magnification   d. capture sunlight energy
                   e. feed on plants
                   f. toxins accumulate

■ *Visit CengageNOW for additional questions.*

Critical Thinking

1. Marguerite has a vegetable garden in Maine. Eduardo has one in Florida. What are some of the variables that influence primary production in each place?

2. Where does your water come from? A well, a reservoir? Beyond that, what area is included within your watershed and what are the current flows like? Visit the Science in Your Watershed site at water.usgs.gov/wsc and research these questions.

3. Look around you and name all of the objects, natural or manufactured, that might be contributing to amplification of the greenhouse effect.

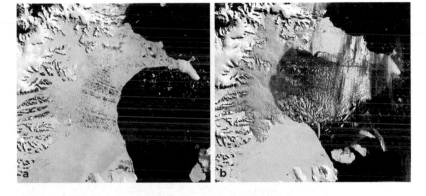

Figure 47.24 Antarctica's Larsen B ice shelf in (**a**) January and (**b**) March 2002. About 720 billion tons of ice broke from the shelf, forming thousands of icebergs. Some of the icebergs project 25 meters (82 feet) above the surface of the ocean. About 90 percent of an iceberg's volume is hidden underwater.

4. Polar ice shelves are vast, thickened sheets of ice that float on seawater. In March 2002, 3,200 square kilometers (1,250 square miles) of Antarctica's largest ice shelf broke free from the continent and shattered into thousands of icebergs (Figure 47.24). Scientists knew the ice shelf was shrinking and breaking up, but this event was the single largest loss ever observed at one time. Why should this concern people who live in more temperate climates?

5. Nitrogen-fixing bacteria live throughout the ocean, from its sunlit upper waters to 200 meters (650 feet) beneath its surface. Recall that nitrogen is a limiting factor in many habitats. What effect would an increase in populations of marine nitrogen-fixers have on primary productivity in the waters? What effect would that change have on carbon uptake in those waters?

48 The Biosphere

IMPACTS, ISSUES Surfers, Seals, and the Sea

Professional surfer Ken Bradshaw has ridden a lot of waves, but one in particular stands out. In January of 1998, he found himself off the coast of Hawaii riding the biggest wave he had ever seen (Figure 48.1). It towered more than 12 meters (39 feet) high and gave him the ride of a lifetime.

That wave was one manifestation of a climate event that happens about every three to seven years. During such an event, Pacific waters along the west coast of South America and westward become warmer than normal. This change in water temperature leads to shifts in marine currents and wind patterns, and causes wave-generating winter storms.

The rise in water temperature also disrupts currents that normally carry nutrients from the deep ocean toward western coasts of the Americas. The resulting nutrient shortage slows the growth of marine primary producers, causing cascading effects throughout marine food webs. One effect, which most often begins around Christmas, is a shortage of fish in waters near the coast of Peru. Peruvian fisherman noted this pattern and named the periodic climate effect El Niño, meaning "the baby boy," in reference to the birth of Jesus.

The decline in fish populations during an El Niño can have devastating effects on marine mammals that normally feed on those fish. During the 1997–1998 El Niño, about half of the sea lions on the Galápagos Islands starved to death. California's population of northern fur seals also suffered a sharp decline.

The temperature change in Pacific waters during the 1997–1998 El Niño was the largest on record, and it affected climates around the world. Giant waves, including the one that Bradshaw rode, battered eastern Pacific coasts. Heavy rains caused massive flooding and landslides in California and Peru. At the same time, less rain than normal fell in Australia and Indonesia, leading to crop failures and wildfires.

As you will learn in this chapter, the circulation pattern of water in Earth's oceans is just one of the physical factors that affect the distribution of species through the biosphere. We define the biosphere as all the places where we find life on Earth. It includes the hydrosphere (the ocean, ice caps, and other bodies of water, liquid and frozen), the lithosphere (Earth's rocks, soils, and sediments), and the lower portions of the atmosphere (gases and particles that envelop Earth).

See the video! Figure 48.1 A powerful El Niño caused this enormous wave in the Pacific. It also affected fish populations, causing sea lion pups (photo at *left*) and seals to starve.

Key Concepts

Air circulation patterns

Air circulation patterns start with regional differences in energy inputs from the sun, Earth's rotation and orbit, and the distribution of land and seas. These factors give rise to the great weather systems and regional climates. **Sections 48.1, 48.2**

Ocean circulation patterns

Interactions among ocean currents, air circulation patterns, and landforms produce regional climates, which affect where different organisms can live. **Section 48.3**

Land provinces

Biogeographic realms are vast regions characterized by species that evolved nowhere else. They are divided into biomes characterized mainly by the dominant vegetation. Sunlight intensity, moisture, soil, and evolutionary history vary among biomes. **Sections 48.4–48.11**

Water provinces

Water provinces cover more than 71 percent of Earth's surface. All freshwater and marine ecosystems have gradients in light availability, temperature, and dissolved gases that vary daily and seasonally. The variations influence primary productivity. **Sections 48.12–48.16**

Applying the concepts

Understanding interactions among the atmosphere, ocean, and land can lead to discoveries about specific events—in one case, recurring cholera epidemics—that impact human life. **Section 48.17**

Links to Earlier Concepts

- With this chapter, you reach the highest level of organization in nature (Section 1.1).

- You will learn more about soils (29.1), distribution of primary productivity (47.3), carbon-fixing pathways (7.7), and the effects of deforestation (Chapter 23 introduction).

- Our discussions of aquatic provinces will draw on your knowledge of properties of water (2.5), acid rain (2.6, 47.9), the water cycle (47.6), and eutrophication (47.10).
  You will learn more about coral reefs (25.5) and life at hydrothermal vents (20.2).

- You will be reminded of the effects of fossil fuel use (23.5), including global warming (47.8). You will learn about threats to the ozone layer (20.3).

- The chapter ends with an example of a scientific approach to problem solving (1.6, 1.7).

How would you vote? We cannot stop an El Niño from happening, but we might be able to minimize its severity. Would you support the use of taxpayer dollars to fund research into the causes and effects of El Niño? See CengageNOW for details, then vote online.

■ How much solar energy reaches Earth's surface varies from place to place and with the season.

■ Link to Fossil fuels 23.5

Air Circulation and Regional Climates

Climate refers to average weather conditions, such as cloud cover, temperature, humidity, and wind speed, over time. Regional climates differ because the factors that influence winds and ocean currents—intensity of sunlight, the distribution of land masses and seas, and elevation—vary from place to place.

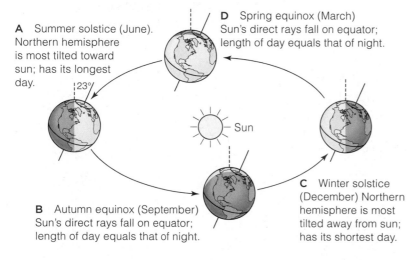

A Summer solstice (June). Northern hemisphere is most tilted toward sun; has its longest day.

23°

Sun

D Spring equinox (March) Sun's direct rays fall on equator; length of day equals that of night.

B Autumn equinox (September) Sun's direct rays fall on equator; length of day equals that of night.

C Winter solstice (December) Northern hemisphere is most tilted away from sun; has its shortest day.

Figure 48.2 Animated Earth's tilt and yearly rotation around the sun cause seasonal effects. The 23° tilt of Earth's axis causes the Northern Hemisphere to receive more intense sunlight and have longer days in summer than in winter.

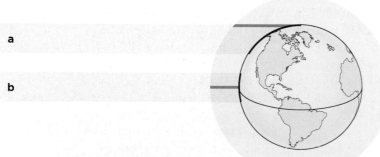

a

b

Figure 48.3 Variation in intensity of solar radiation with latitude. For simplicity, we depict two equal parcels of incoming radiation on an equinox, a day when incoming rays are perpendicular to Earth's axis.

Rays that fall on high latitudes (**a**) pass through more atmosphere (*blue*) than those that fall near the equator (**b**). Compare the length of the *green* lines. Atmosphere is not to scale.

Also, energy in the rays that fall at the high latitude is spread over a greater area than energy that falls on the equator. Compare the length of the *red* lines.

Each year, Earth rotates around the sun in an elliptical path (Figure 48.2). Seasonal changes arise because Earth's axis is not perpendicular to the plane of this ellipse, but rather is tilted about 23 degrees. In June, when the Northern Hemisphere is angled toward the sun, it receives more intense sunlight and has longer days than the Southern Hemisphere (Figure 48.2a). In December, the opposite occurs (Figure 48.2c). Twice a year—on spring and autumn equinoxes—Earth's axis is perpendicular to incoming sunlight. On these days, every place on Earth receives 12 hours of daylight and 12 hours of darkness (Figure 48.2b,d).

On any particular day, equatorial regions get more sunlight energy than higher latitudes for two reasons (Figure 48.3). First, fine particles of dust, water vapor, and greenhouse gases absorb some solar radiation or reflect it back into space. Because sunlight traveling to high latitudes passes through more atmosphere to reach Earth's surface than light traveling to the equator, less energy reaches the ground. Second, energy in any incoming parcel of sunlight is spread out over a smaller surface area at the equator than at the higher latitudes. As a result of these factors, Earth's surface warms more at the equator than at the poles.

This regional difference in surface warming is the start of global air circulation patterns (Figure 48.4). Warm air can hold more moisture than cooler air and is less dense, so it rises. Near the equator, air warms, picks up moisture from the oceans, and rises (Figure 48.4a). Air cools when it rises to higher altitudes and flows north and south, releasing moisture as rain that supports lush tropical rain forests. Deserts often form at latitudes of about 30°, where the drier and cooler air descends (Figure 48.4b). Farther north and south, the air picks up moisture again. It rises, and then releases moisture at latitudes of about 60° (Figure 48.4c). In the polar regions cold air that holds little moisture descends (Figure 48.4d). Precipitation is sparse, and polar deserts form.

Prevailing winds do not blow directly north and south because Earth's rotation and curvature influence the air circulation pattern. Air masses are not attached to Earth's surface, so as an air mass moves north or south this surface rotates beneath it, rotating faster at the equator than the poles. As a result, when viewed from Earth's surface, air masses that move north or south will seem to be deflected east or west, with the deflection greatest at high latitude (Figure 48.4e,f).

Regional winds occur where the presence of land masses cause differences in air pressure near Earth's surface. Because land absorbs and releases heat faster than water does, air rises and falls faster over land

D At the poles, cold air sinks and moves toward lower latitudes.

C Air rises again at 60° north and south, where air flowing poleward meets air coming from the poles.

B As the air flows toward higher latitudes, it cools and loses moisture as rain. At around 30° north and south latitude, the air sinks and flows north and south along Earth's surface.

A Warmed by energy from the sun, air at the equator picks up moisture and rises. It reaches a high altitude, and spreads north and south.

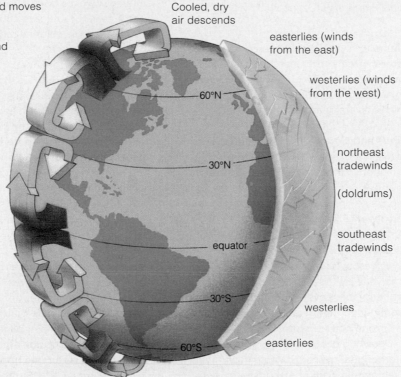

Cooled, dry air descends

easterlies (winds from the east)

westerlies (winds from the west)

60°N

30°N

northeast tradewinds

(doldrums)

southeast tradewinds

equator

30°S

westerlies

60°S

easterlies

E Major winds near Earth's surface do not blow directly north and south because of Earth's rotation. Winds deflect to the right of their original direction in the Northern Hemisphere and to the left in the Southern Hemisphere.

F For example, air moving from 30° south toward the equator is deflected to the left (west), as the southeast trade winds. The winds are named by the direction from which they blow.

Figure 48.4 Animated Global air circulation patterns and their effects on climate.

Figure It Out: What is the direction of prevailing winds in the central United States? *Answer: Winds blow from west to east.*

than it does over the ocean. Air pressure is lowest where air rises and greatest where air sinks.

Harnessing the Sun and Wind

The need for energy to support human activities continues to increase. Fossil fuels, including gasoline and coal, are nonrenewable energy sources (Section 23.5). Solar and wind energy are renewable. The amount of solar energy that Earth receives per year is about 10 times the energy of all fossil fuel reserves combined.

Solar energy can be harnessed directly to heat air or water that can then be pumped through buildings to heat them. Solar energy can also be captured by photovoltaic cells and used to generate electricity. The electricity can be used directly, stored in a battery, or used to form oxygen and hydrogen gases from water. Proponents of solar–hydrogen energy argue that it could end smog, oil spills, and acid rain without any of the risks of nuclear power. Hydrogen gas can fuel cars and heat buildings. However, hydrogen is a small molecule that leaks easily from pipelines or containers.

How increased leakage of hydrogen into the air would affect the environment is unknown.

We use solar energy indirectly by harnessing winds. Wind energy is only practical where winds blow faster than 8 meters per second (18 miles per hour). Winds seldom blow constantly, but wind energy can charge batteries to supply power even on still days. Energy from winds of North and South Dakota alone could meet 80 percent of the United States' energy needs.

Wind farms have drawbacks. Turbine blades can be noisy and can kill birds and bats. Large facilities may alter local weather patterns. Also, some people see wind farms as a form of "visual pollution" that ruins otherwise scenic views and lowers property values.

Take-Home Message

What causes global air circulation patterns and differences in climate?

■ Longitudinal differences in the amount of solar radiation reaching Earth produce global air circulation patterns.

■ Earth's shape and rotation also affect air circulation patterns.

■ Particles and gases act as air pollutants that endanger human health and disrupt ecosystems.

■ Links to Acid rain 2.6 and 47.9, Ozone 20.3, CFCs 47.8

A **pollutant** is a natural or synthetic substance released into soil, air, or water in greater than natural amounts; it disrupts normal processes because organisms evolved in its absence, or are adapted to lower levels of it. Today, air pollution threatens biodiversity and human health.

Swirling Polar Winds and Ozone Thinning High in Earth's atmosphere, molecules of ozone (O_3) absorb most of the ultraviolet (UV) radiation in incoming sunlight. Between 17 and 27 kilometers above sea level (10.5 and 17 miles), the ozone concentration is so great that scientists refer to this region as the **ozone layer** (Figure 48.5a).

In the mid-1970s, scientists started to notice that the ozone layer was getting thinner. Its thickness had always varied a bit with the season, but now there was steady decline from year to year. By the mid-1980s, the spring ozone thinning over Antarctica was so pronounced that people were calling it an "ozone hole" (Figure 48.5b).

Declining ozone quickly became an international concern. With a thinner ozone layer, people would be exposed to more UV radiation and get more skin cancers (Section 14.5). Higher UV levels also harm wildlife, which do not have the option of rubbing on more sunscreen. Higher UV levels might even harm plants and other producers, slowing rates of photosynthesis and release of oxygen into the atmosphere.

Chlorofluorocarbons, or CFCs, are the main ozone destroyers. These odorless gases were once widely used as propellants in aerosol cans, as coolants, and in solvents and plastic foam. CFCs interact with ice crystals and UV light in the stratosphere. These reactions release chlorine radicals that degrade ozone. A single chlorine radical can break apart thousands of ozone molecules.

Ozone thins the most at the poles because swirling winds concentrate CFCs in this region during dark, cold polar winters. In the spring, increasing daylight and the presence of ice clouds allow a surge in the formation of chlorine radicals from the highly concentrated CFCs.

In response to the potential threat posed by ozone thinning, developed countries agreed in 1992 to phase out the production of CFCs and other ozone destroyers. As a result of that agreement, the concentrations of CFCs in the atmosphere are now starting to decline (Section 47.8). However, they are expected to stay high enough to significantly affect the ozone layer for the next twenty years.

No Wind, Lots of Pollutants, and Smog Often, weather conditions cause a thermal inversion: A layer of cool, dense air becomes trapped under a warm, less dense layer. Trapped air sets the stage for **smog**, an atmospheric condition in which air pollutants accumulate to high concentration. The accumulation occurs because winds cannot disperse pollutants trapped under a thermal inversion layer (Figure 48.6). Thermal inversions have contributed to some of the highest recorded air pollution levels.

Industrial smog forms as a gray haze over cities that burn a lot of coal and other fossil fuels during cold, wet winters. Photochemical smog forms above big cities in warm climate zones. Photochemical smog is most dense over cities in natural topographic basins, such as Los Angeles and Mexico City. Exhaust fumes from vehicles contain nitric oxide, a pollutant that combines with oxygen and forms nitrogen dioxide. Exhaust fumes also contain

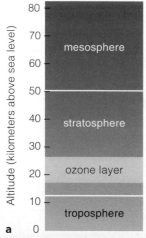

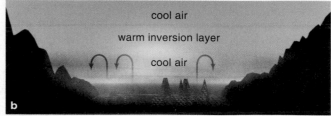

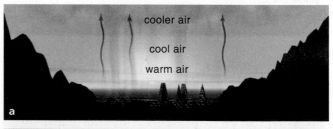

Figure 48.5 Animated
(**a**) The atmospheric layers. Ozone concentrated in the stratosphere helps shield life from UV radiation. (**b**) Seasonal ozone thinning above Antarctica in 2001. *Dark blue* represents the low ozone concentration, at the ozone hole's center.

Figure 48.6 (**a**) Normal air circulation in smog-forming regions. (**b**) Air pollutants trapped under a thermal inversion layer.

hydrocarbons that react with nitrogen dioxide to form ozone and other photochemical oxidants. A high ozone level in the lower atmosphere harms plants and animals.

Winds and Acid Rain Coal-burning power plants, smelters, and factories emit sulfur dioxides. Vehicles, power plants that burn gas and oil, and nitrogen-rich fertilizers emit nitrogen oxides. In dry weather, airborne oxides coat dust particles and fall as dry acid deposition. In moist air, they form nitric acid vapor, sulfuric acid droplets, and sulfate and nitrate salts. Winds typically disperse these pollutants far from their source. They fall to Earth in rain and snow. We call this a wet acid deposition, or **acid rain**.

The pH of typical rainwater is about 5 (Section 2.6). Acid rain can be 10 to 100 times more acidic—as potent as lemon juice! It corrodes metals, marble, rubber, plastics, nylon stockings, and other materials. It alters soil pH and can kill trees (Section 47.9) and other organisms.

Rain in much of eastern North America is thirty to forty times more acidic than it was even a few decades ago (Figure 48.7a). The heightened acidity has caused fish populations to vanish from more than 200 lakes in the Adirondack Mountains of New York (Figure 48.7b). It also is contributing to the decline of forests.

Windborne Particles and Health Pollen, fungal spores, and other natural particles are carried aloft by winds, along with pollutant particles of many sizes. Inhaling small particles can irritate nasal passages, the throat, and lungs. It triggers asthma attacks and can increase their severity. The smallest particles are most likely to reach the lungs, where they can interfere with respiratory function.

Exhaust from vehicles is a major source of particulate pollution. Diesel-fueled engines are the worst offenders because they emit more of the smallest, most dangerous particles than their gasoline-fueled counterparts.

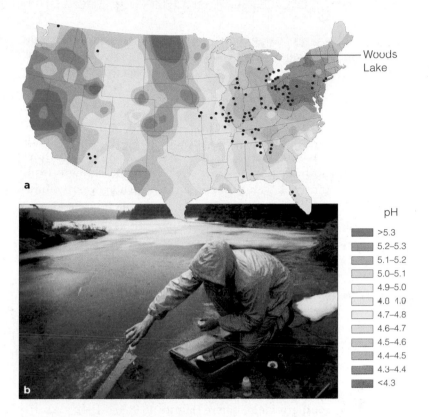

Figure 48.7 Animated (a) Average precipitation acidities in the United States in 1998. (b) Biologist measuring the pH of New York's Woods Lake. In 1979, the lake water's pH was 4.8. Since then, experimental addition of calcite to soil around the lake has successfully raised the pH of the water to more than 6.

Regardless of their source, air pollutants travel on the winds across continents and the open ocean. As Figure 48.8 shows, airborne pollutants do not stop at national borders. We all share the same air.

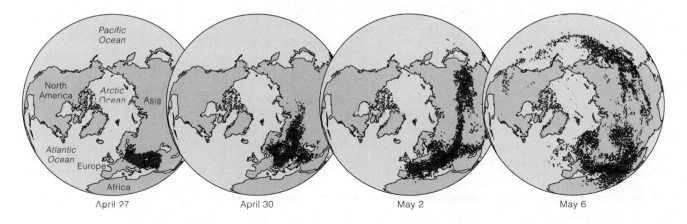

Figure 48.8 Global distribution of radioactive fallout released during the 1986 meltdown of the Chernobyl nuclear power plant in Ukraine. The meltdown allowed radioactive particles to enter the air, then winds dispersed them around the world. The incidence of thyroid cancers in Ukraine and neighboring Belarus continues to rise, a legacy of childhood exposure to high radiation levels.

■ The ocean, a continuous body of water, covers more than 71 percent of Earth's surface. Driven by solar heat and wind friction, its upper 10 percent moves in currents that distribute nutrients through marine ecosystems.

Ocean Currents and Their Effects

Latitudinal and seasonal variations in sunlight warm and cool water. At the equator, where vast volumes of water warm and expand, the sea level is about 8 centimeters (3 inches) higher than at either pole. The volume of water in this "slope" is enough to get sea surface water moving in response to gravity, most often toward the poles. The moving water warms air above it. At midlatitudes, oceans transfer 10 million billion calories of heat energy per second to the air!

Enormous volumes of water flow as ocean currents. The force of major winds, Earth's rotation, and topography determine the directional movement of these currents. Surface currents circulate clockwise in the Northern Hemisphere and counterclockwise in the Southern Hemisphere (Figure 48.9).

Swift, deep, and narrow currents of nutrient-poor water flow away from the equator along the east coast of continents. Along the east coast of North America, warm water flows north, as the Gulf Stream. Slower, shallower, broader currents of cold water parallel the west coast of continents and flow toward the equator.

Ocean currents affect climates. Coasts in the Pacific Northwest are cool and foggy in summer because the cold California current chills the air, so water condenses out as droplets. Boston and Baltimore are muggy in summer because air masses pick up heat and moisture from the warm Gulf Stream, then deliver it to these cities.

Ocean circulation patterns shift over geologic time as land masses move (Section 17.9). Some worry that global warming could also alter these patterns.

Rain Shadows and Monsoons

Mountains, valleys, and other surface features of the land affect climate. Suppose you track a warm air mass after it picks up moisture off California's coast. It moves inland, as wind from the west, and piles up

Figure 48.9 Animated Major climate zones correlated with surface currents of the world ocean. Warm surface currents start moving from the equator toward the poles, but prevailing winds, Earth's rotation, gravity, the shape of ocean basins, and landforms influence the direction of flow. Water temperatures, which differ with latitude and depth, contribute to the regional differences in air temperature and rainfall.

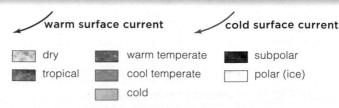

warm surface current   cold surface current

- dry
- tropical
- cold
- warm temperate
- cool temperate
- subpolar
- polar (ice)

A Prevailing winds move moisture inland from the Pacific Ocean.

B Clouds pile up and rain forms on side of mountain range facing prevailing winds.

C Rain shadow on side facing away from the prevailing winds makes arid conditions.

4,000/ 75
3,000/ 85   2,000/ 25
1,800/ 125   1,000/ 25
1,000/ 85
moist habitats
15/ 25

Figure 48.10 Animated Rain shadow effect. On the side of mountains facing away from prevailing winds, rainfall is light. *Black* numbers signify annual precipitation, in centimeters, averaged on both sides of the Sierra Nevada, a mountain range. *White* numbers signify elevations, in meters.

against the Sierra Nevada. This high mountain range parallels the distant coast. The air cools as it rises in altitude and loses moisture as rain (Figure 48.10). The result is a **rain shadow**—a semiarid or arid region of sparse rainfall on the leeward side of high mountains. "Leeward" is the side facing away from the wind. The Himalayas, Andes, Rockies, and other great mountain ranges cause vast rain shadows.

Differences in the heat capacity of water and land give rise to coastal breezes. In the daytime, water does not warm as fast as the land. Air heated by the warm land rises, and cool offshore air moves in to replace it (Figure 48.11*a*). After sundown, land becomes cooler than the water, so the breezes reverse (Figure 48.11*b*).

Differential heating of water and land also causes **monsoons**, winds that change their direction seasonally. For example, the continental interior of Asia heats up in the summer, so air rises above it. The resulting low pressure draws in moisture from over the warm Indian Ocean to the south, and these north-blowing winds deliver heavy rains. In the winter, the continental interior is cooler than the ocean. As a result, cool, dry winds blowing from the north toward southern coasts, cause a seasonal drought.

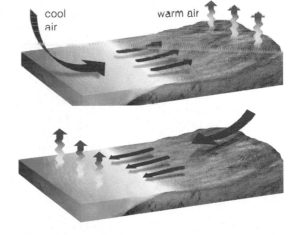

A Afternoon; land is warmer than sea; breeze blows onto the shore

cool air

warm air

B Evening; sea is warmer than land; the breeze blows out to sea

Figure 48.11 Animated Coastal breezes.

Take-Home Message

How do ocean currents arise and how do they affect regional climates?

■ Surface ocean currents, which are set in motion by latitudinal differences in solar radiation, are affected by winds and by Earth's rotation.

■ Collective effects of air masses, oceans, and landforms determine regional temperature and moisture levels.

- Regions with different physical conditions support different types of organisms.

- Links to Biogeography 17.1, Plate tectonics 17.9

Suppose you live in the coastal hills of California and decide to tour the Mediterranean coast, the southern tip of Africa, and central Chile. In each region, you see highly branched, tough-leafed woody plants that look a lot like the highly branched, tough-leafed chaparral plants back home. Vast geographic and evolutionary distances separate the plants. Why are they alike?

You decide to compare their locations on a global map and discover that American and African desert plants live about the same distance from the equator. Chaparral plants and their distant look-alikes all grow along the western and southern coasts of continents between latitudes 30° and 40°. You have noticed one of many patterns in the global distribution of species.

Early naturalists divided Earth's land masses into six **biogeographic realms**—vast expanses where they could expect to find communities of certain types of plants and animals (Figure 48.12). For example, palm trees and camels live in the Ethiopian realm. In time, the six classic realms became subdivided.

Biomes are finer subdivisions of the land realms, but they are still identifiable on a global scale. Most biomes occur on more than one continent. For instance, dry forest (coded orange in Figure 48.12) covers vast regions of South America, India, and Asia. Similarly, the North American prairie, South American pampa, southern Africa veld, and Eurasian steppe are all types of temperate grasslands (Figure 48.13).

The distribution of biomes is influenced by climate (especially temperature and patterns of rainfall), soil type, and interactions among the array of species that make up their communities. Consumers are adapted to the dominant vegetation. Each species, remember,

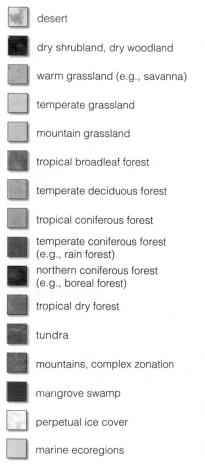

desert

dry shrubland, dry woodland

warm grassland (e.g., savanna)

temperate grassland

mountain grassland

tropical broadleaf forest

temperate deciduous forest

tropical coniferous forest

temperate coniferous forest (e.g., rain forest)

northern coniferous forest (e.g., boreal forest)

tropical dry forest

tundra

mountains, complex zonation

mangrove swamp

perpetual ice cover

marine ecoregions

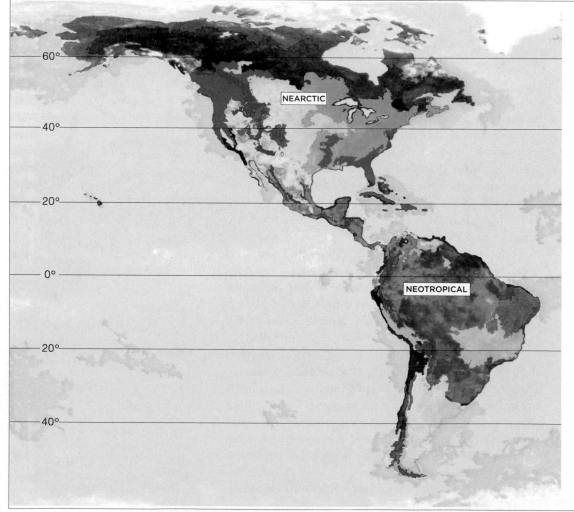

Figure 48.12 Animated Global distribution of major categories of biomes and marine ecoregions.

shows adaptations in its form, function, behavior, and life history pattern.

Distribution of biomes has also been influenced by evolutionary history. For example, species that evolved together on Pangea ended up on different land masses after this supercontinent broke up (Section 17.9).

Similarly, environmental features and evolutionary history helped shape the distribution of species in the seas. Figure 48.12 shows the key marine ecoregions as well as Earth's biomes.

Take-Home Message

What are biomes?

■ Biomes are vast expanses of land dominated by distinct kinds of plants that support characteristic communities.

■ The global distribution of biomes is a result of topography, climate, and evolutionary history.

Figure 48.13 Two examples of temperate grassland biome. *Top*, Argentine pampa. *Bottom*, Mongolian steppe. See also Figure 48.16.

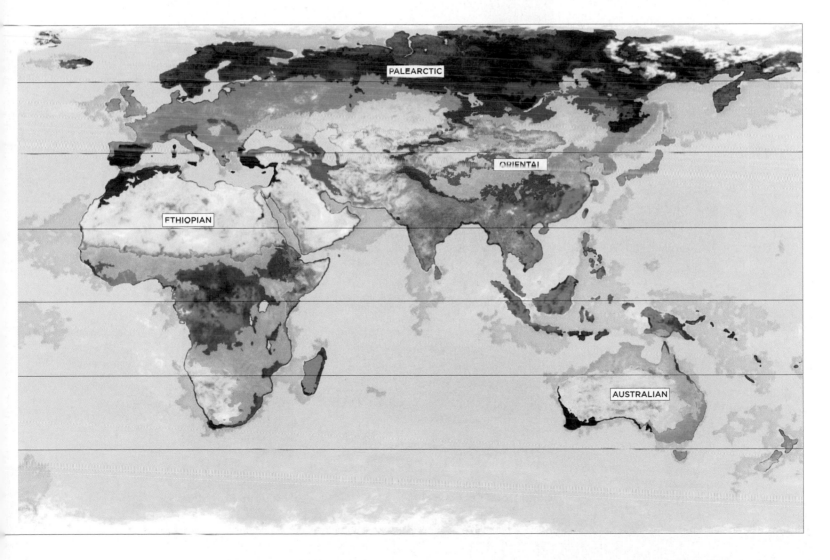

Soils of Major Biomes

■ Plants obtain necessary nutrients from soil. As a result, properties of soil have a big impact on primary production.

■ Link to Soil profiles 29.1

Plants obtain water and dissolved mineral ions from soil. Soil consists of mineral particles and decomposing organic matter called humus (Section 29.1). Water and air fill spaces between soil particles.

The properties of soils vary. Clay is richest in minerals, but its fine, close-packed particles drain poorly; there is little air space for roots to take up oxygen. In gravelly or sandy soils, leaching draws off water and mineral ions. Most plants grow best in a soil that is a mixture of different-sized particles and has a moderate amount of humus.

Each biome has a **soil profile**, a layered structure that develops over time (Figures 29.2 and 48.14). The top layers are surface litter (the O horizon) and topsoil (the A horizon). Topsoil is the most important layer for plant growth. Deserts have little topsoil, and their soil is nutrient-poor and high in salts. Grasslands have the deepest topsoil; it can be more than one meter thick. This is why grasslands are favored for conversion to agriculture. In tropical rain forests, decomposition is rapid, so little topsoil can accumulate above the poorly draining lower layers of soil. In coniferous and deciduous temperate forests, decomposition proceeds more slowly, so leaf litter accumulates and upper soil layers tend to be richer.

Take-Home Message

How do soils affect biome characteristics?

■ Each biome has a characteristic soil profile, with different amounts of inorganic and organic components.

■ Properties of topsoil are the most important for plant growth.

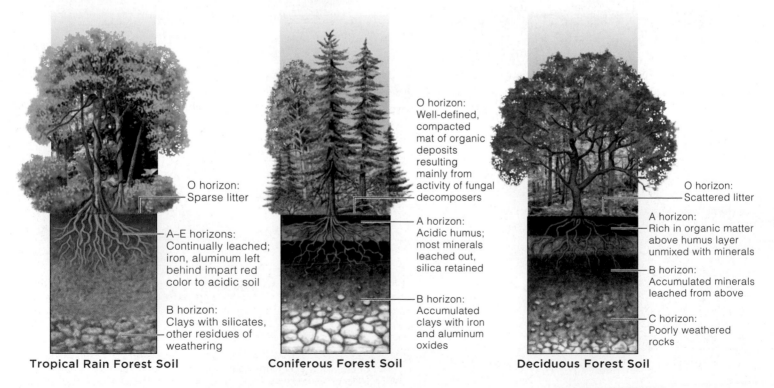

O horizon: Pebbles, little organic matter
A horizon: Shallow, poor soil
B horizon: Evaporation causes salt build-up; leaching removes nutrients
C horizon: Rock fragments from uplands
Desert Soil

A horizon: Alkaline, deep, rich in humus
B horizon: Percolating water enriches layer with calcium carbonates
Grassland Soil

O horizon: Sparse litter
A–E horizons: Continually leached; iron, aluminum left behind impart red color to acidic soil
B horizon: Clays with silicates, other residues of weathering
Tropical Rain Forest Soil

O horizon: Well-defined, compacted mat of organic deposits resulting mainly from activity of fungal decomposers
A horizon: Acidic humus; most minerals leached out, silica retained
B horizon: Accumulated clays with iron and aluminum oxides
Coniferous Forest Soil

O horizon: Scattered litter
A horizon: Rich in organic matter above humus layer unmixed with minerals
B horizon: Accumulated minerals leached from above
C horizon: Poorly weathered rocks
Deciduous Forest Soil

Figure 48.14 Soil profiles for some major biomes. The A horizon, or topsoil, is the most important source of nutrients for plant growth.

48.6 | Deserts

■ With this section we begin a survey of the major biomes. Our first stop is deserts, which are defined by low rainfall.

■ Links to Carbon-fixing pathways 7.7, Atacama Desert 20.6, Desert kangaroo rat 41.3

Deserts

Deserts are regions that receive an average of less than 10 centimeters (4 inches) of rain per year. Most are located at about 30° north and south latitude, where air depleted of moisture sinks. In these regions, the low humidity allows much sunlight to reach the soil surface, so the ground heats fast during the day. Low humidity also causes the ground to cool fast at night. Soils are typically nutrient poor and somewhat salty. Despite these forbidding conditions, some plants and animals survive, especially in areas where moisture is available in more than one season (Figure 48.15).

Many desert plants have adaptations that reduce water loss. Light-colored spines or hairs can help keep humidity around the stomata high and also reflect sunlight. Alternative carbon-fixing pathways also help desert plants conserve water (Section 7.7). Cactuses and agaves are CAM plants and open their stomata only at night. Many annuals that live in deserts are C4 plants. Woody desert shrubs such as mesquite and creosote have extensive, efficient root systems that take up the little water that is available. Mesquite roots have been found as deep as 60 meters beneath the soil surface.

Animals also have adaptations that allow them to conserve water. The desert kangaroo rat discussed in Section 41.3 is a resident of the Sonoran Desert. So are the animals shown in Figure 48.15.

The driest of all deserts may be Chile's cool Atacama Desert, which lies in a rain shadow behind the Andes. Parts of this area are so dry that they were thought to be entirely lifeless. However, scientists recently found bacteria deep in the soil (Section 20.6).

Take-Home Message

What features characterize desert biomes?

■ A desert gets very little rain and has low humidity. There is plenty of sunlight, but poor soil and lack of water prevent most plants from surviving here.

■ Many plants and animals in deserts have adaptations that minimize their need for water.

Figure 48.15 Two parts of the same biome—the Sonoran Desert in Arizona. The sun's rays are just as intense in desert lowlands (**a**) as in the uplands (**b**), but differences in water availability, temperature, and soil types influence plant growth. Creosote bush (*Larrea*) dominates lowlands. A greater variety of plants survive in the somewhat wetter and cooler uplands.

Examples of desert animals. (**c**) The Sonoran desert tortoise escapes the heat by burrowing. (**d**) Lesser long-nosed bats spend spring and summer in the Sonoran Desert, where they avoid the daytime heat in caves and abandoned mine shafts. The bats are important pollinators of cactuses and agaves.

Grasslands, Shrublands, and Woodlands

- Where there is more rain than in deserts, grasses take hold. In areas with a bit more rain still, shrubs take hold.

- Links to Soil erosion 29.1, Tallgrass prairie 47.1

Grassland, shrublands, and woodlands

Grasslands

Grasslands form in the interior of continents between deserts and temperate forests (Figure 48.16). Summers are warm, and winters are cold. Annual rainfall of 25 to 100 centimeters (10–40 inches) keeps desert from forming, but is too little water to support forest. Low-growing primary producers tolerate strong winds, sparse and infrequent rain, and intervals of drought. Growth tends to be seasonal. Constant trimming by grazing animals, along with periodic fires, keeps trees and most shrubs from taking hold.

Shortgrass and tallgrass prairie (Figure 48.16a,b) are North America's main grasslands. Perennial grasses that fix carbon by the water-conserving C4 pathway dominate these biomes. Grass roots extend profusely through the topsoil and help hold it in place, preventing erosion by the constant winds.

During the 1930s, much of the shortgrass prairie of the American Great Plains was plowed under to grow wheat. The strong winds, a prolonged drought, and unsuitable farming practices turned much of the region into what the newspapers of that time called the Dust Bowl. John Steinbeck's historical novel *The Grapes of Wrath*, eloquently describes the human costs of this environmental disaster.

Tallgrass prairie (Section 47.1) once covered 140 million acres, mostly in Kansas. Tall grasses, legumes, and herbaceous plants such as daisies thrived in the continent's interior, which had somewhat richer topsoil and slightly more frequent rainfall than the shortgrass prairie. Nearly all tallgrass prairie has been converted to cropland. The Tallgrass Prairie National Preserve was created in 1996 to protect the little that remains.

Figure 48.16 Three examples of grasslands. (**a**) Tallgrass prairie in eastern Kansas. See also Figure 47.3. (**b**) Bison grazing in shortgrass prairie in South Dakota. (**c**) A herd of wildebeest graze in the African savanna. Figure 48.13 shows additional grasslands.

Figure 48.17 California chaparral. (**a,b**) Dominant plants are mostly branching, woody evergreens less than 2 meters (6 feet) tall, with leathery leaves. The leaves often contain oils that deter herbivores and also make the plants highly flammable.

(**c**) A firestorm in the chaparral-covered hills above Malibu. Today, most fires that occur in this biome are caused by humans. (**d**) Toyon (*Heteromeles arbutifolia*), a fire-adapted chaparral shrub, resprouting from its roots after a fire.

Savannas are broad belts of grasslands with a few scattered shrubs and trees. Savannas lie between the tropical forests and hot deserts of Africa, India, and Australia. Temperatures are warm year-round. During the rainy season 90–150 centimeters (35–60 inches) of rain falls. Fires that occur during the dry season help keep forest from replacing the grassland.

Africa's savanna's are famous for their abundant wildlife (Figure 48.16c). Herbivores include giraffes, zebras, elephants, a variety of antelopes, and immense herds of wildebeests. Lions and hyenas are carnivores that eat the grazers.

Dry Shrublands and Woodlands

Dry shrublands receive less than 25 to 60 centimeters (10–24 inches) of rain annually. We see them in South Africa, in Mediterranean regions, and in California, where they are known as chaparral. California has about 6 million acres of chaparral (Figure 48.17a,b).

Rains occur seasonally, and lightning-sparked fires sometimes sweep through shrublands during the dry season. In California, where homes are often built near chaparral, the fires frequently cause property damage (Figure 48.17c). Foliage of many chaparral shrubs is highly flammable. However, the plants have adapted to occasional fires. Some grow back from root crowns

after a fire (Figure 48.17d). Seeds of other chaparral species germinate only after they are exposed to heat or smoke, ensuring that the seeds sprout only when young seedlings face little competition.

Dry woodlands prevail where the annual rainfall is 40 to 100 centimeters (16–40 inches). Drought-tolerant trees are often tall, but do not form a continuous canopy. Examples include the eucalyptus forests of Australia, and the oak forests of California and Oregon.

Take-Home Message

What are grasslands, dry shrublands, and woodlands?

■ Grasslands form in the interior of continents. Grasses and other short, nonwoody plants predominate. Activity of grazing animals and occasional fires help prevent trees and shrubs from taking hold.

■ Dry shrublands such as California's chaparral also include fire-adapted species—predominantly short, woody shrubs.

■ Dry woodlands are dominated by trees that are adapted to withstand seasonal drought.

More Rain, Broadleaf Forests

- Broadleaf (angiosperm) trees dominate moist forests in both temperate and equatorial regions.

- Links to Pigments 7.1, Plant nutrients 29.1

Broadleaf forests

Semi-Evergreen and Deciduous Broadleaf Forests

Semi-evergreen forests occur in the humid tropics of Southeast Asia and India. These forests include a mix of broadleaf trees that retain leaves year round, and deciduous broadleaf trees. Deciduous trees or shrubs shed leaves once a year, prior to the season when cold or dry conditions would not favor growth. Deciduous trees in a semi-evergreen forest shed their leaves in preparation for the dry season.

Where less than 2.5 centimeters (1 inch) of rain falls in the dry season, **tropical deciduous forests** form. In tropical deciduous forests, most trees shed leaves at the start of the dry season.

Temperate deciduous forests form in parts of eastern North America, western and central Europe, and parts of Asia, including Japan. About 50 to 150 centimeters (about 20–60 inches) of precipitation fall throughout the year. Leaves turn brilliant red, orange, and yellow before dropping in autumn (Figure 48.18 and Section 7.1). Having discarded their leaves, the trees become dormant during the cold winter, when water is locked in snow and ice. In the spring, when conditions again favor growth, deciduous trees flower and new leaves appear. Also during the spring, leaves shed the prior autumn decay and form a rich humus. Rich soil and a somewhat open canopy that lets sunlight through allows many understory plants to flourish.

Tropical Rain Forests

Evergreen broadleaf forests form between latitudes 10° north and south in equatorial Africa, the East Indies, Malaysia, Southeast Asia, South America, and Central America. Yearly rainfall averages 130 to 200 centimeters (50 to 80 inches). Regular rains, combined with an average temperature of 25°C (77°F) and high humidity, support tropical rain forests of the sort shown in the next section. In structure and diversity, these biomes are the most complex. Some trees are 30 meters (100 feet) tall. Many form a closed canopy that stops most sunlight from getting to the forest floor. Vines and epiphytes (plants that grow on another plant, but do not withdraw nutrients from it) grow in the shady canopy.

Decomposition and mineral cycling happen fast in these forests, so litter does not accumulate. Soils are highly weathered, heavily leached, and are very poor nutrient reservoirs.

Figure 48.18 North American temperate deciduous forest. The series above shows changes in a deciduous tree's foliage from winter (*far left*) through spring, summer, and fall.

Take-Home Message

What is a broadleaf forest?

- Conditions in broadleaf forests favor dense stands of trees that form a continuous canopy.

- Deciduous broadleaf trees shed leaves seasonally. Evergreen broadleafs drop them in small numbers throughout the year.

48.9 | You and the Tropical Forests

■ The Chapter 23 introduction discussed the deforestation of northern conifer forests. We turn here to factors that currently threaten the once-vast tropical forests.

Southeast Asia, Africa, and Latin America stretch across the tropical latitudes. Developing nations on these continents have the fastest-growing populations and high demands for food, fuel, and lumber. Of necessity, people turn to forests (Figure 48.19). Most of these tropical forests may vanish within our lifetime. That possibility concerns people in highly developed nations, who use most of the world's resources, including forest products.

On purely ethical grounds, the destruction of so much biodiversity is a concern. Tropical rain forests have the greatest variety and numbers of insects, and the world's largest ones. They are homes to the most species of birds and to plants with the largest flowers (*Rafflesia*). Forest canopies and understories support monkeys, tapirs, and jaguars in South America; and apes, leopards, and okapis in Africa. Massive vines twist around tree trunks. Orchids, mosses, lichens, and other organisms grow on branches, absorbing minerals from rains. Communities of microbes, insects, spiders, and amphibians live, breed, and die in small pools of water that collect in furled leaves.

Also, products provided by rain forest species save and enhance human lives. Analysis of compounds in rain forest species can point the way toward new drugs. Quinine, an antimalarial drug, was first derived from an extract of *Cinchona* bark from a tree in the Amazonian rain forest. Two chemotherapy drugs, vincristine and vinblastine, were extracted from the rosy periwinkle (*Catharanthus roseus*), a low-growing plant native to Madagascar's rain forests. Today, these drugs help fight leukemia, lymphoma, breast cancer, and testicular cancer. Many ornamental plants, spices, and foods, including cinnamon, chocolate, and coffee, originated in tropical forests. So did the latex, gums, resins, dyes, waxes, and oils used in tires, shoes, toothpaste, ice cream, shampoo, and condoms.

Conservation biologists decry the loss of forest species and their essential natural services. Yet tropical rain forest loss keeps accelerating. The amount of temperate forest is increasing in North America, Europe, and China, but this rise is overshadowed by the staggering losses of tropical forests elsewhere.

The disappearance of rain forests could influence the atmosphere. The forests take up and store carbon, and they release oxygen. Burning enormous tracts of tropical forest to make way for agriculture releases carbon dioxide, which contributes to global warming (Section 47.8).

Ironically, concern about greenhouse gas release from fossil fuels may encourage rain forest destruction. Areas of rainforest in the Amazon and Indonesia are being cleared to make way for plantations that grow soybeans or palms. Oils from these plants are exported, mainly to Europe, where they are used to produce biodiesel fuel.

Rafflesia

Jaguars

Figure 48.19 Tropical rain forests in Southeast Asia and Latin America.

48.10 | Coniferous Forests

- Compared to broadleaf trees, conifers are more tolerant of cold and drought, and can withstand poorer soils. Where these conditions occur, coniferous forests prevail.

- Link to Conifers 23.7

Coniferous forests

Conifers—evergreen gymnosperms with seed-bearing cones—dominate **coniferous forests**. Their leaves are typically needle-shaped, with a thick cuticle. Stomata are sunk below the leaf surface. These adaptations help conifers conserve water during drought or times when soil water is frozen. As a group, conifers tolerate poorer soils and drier habitats than broadleaf trees.

In the Northern Hemisphere, montane coniferous forests extend southward through the great mountain ranges (Figure 48.20a). Spruce and fir dominate at the highest elevations, with firs and pines taking over as you move farther down the slopes.

Boreal, or northern, coniferous forests occur in Asia, Europe, and North America in formerly glaciated areas where lakes and streams abound (Figure 48.20b). These forests are dominated by pine, fir, and spruce. They are also known as taigas, which means "swamp forests." Most rain falls in summer. Winters are long, cold, and dry. Moose are dominant grazers in this biome.

Conifers also dominate temperate lowlands along the Pacific coast from Alaska into northern California. These coniferous forests hold the world's tallest trees, Sitka spruce to the north and coast redwoods to the south. Large tracts have been logged (Chapter 23).

We find other conifer-dominated ecosystems in the eastern United States. About a quarter of New Jersey is pine barrens, a mixed forest of pitch pines and scrub oaks that grow in sandy, acidic soil. Pine forest covers about one-third of the Southeast. Fast-growing loblolly pines dominate these forests and are a major source of lumber and wood pulp. The pines survive periodic fires that kill most hardwood species. If fires are suppressed, hardwoods will replace the pines.

Take-Home Message

What are coniferous forests?

- Coniferous forests consist of hardy evergreen trees able to withstand conditions that most broadleaf trees cannot.

a

b

Moose

Figure 48.20 (a) Montane coniferous forest near Mount Rainier, Washington. (b) Taiga in Alberta, Canada.

48.11 | Tundra

- Low-growing plants tolerate cold and wind in tundra, which forms at high altitudes and latitudes.
- Link to Global warming 47.8

Arctic tundra

Figure 48.22 Alpine tundra in Washington's Cascade range.

Arctic tundra forms between the polar ice cap and the belts of boreal forests in the Northern Hemisphere. Most is in northern Russia and Canada. It is Earth's youngest biome; it appeared about 10,000 years ago, when glaciers retreated at the end of the last ice age.

Arctic tundra is blanketed with snow for as long as nine months of the year. Annual snow and rain is usually less than 25 centimeters (10 inches). During a brief summer, plants grow rapidly under the nearly continuous sunlight. Lichens and shallow-rooted, low-growing plants are a base for food webs that include voles, arctic hares, caribou, arctic foxes, wolves, and brown bears (Figure 48.21). Enormous numbers of migratory birds nest here in the summer, when the air is thick with mosquitoes.

Only the surface layer of tundra soil thaws during summer. Below that lies the **permafrost**, a frozen layer 500 meters (1,600 feet) thick in places. Permafrost acts as a barrier that prevents drainage, so the soil above it remains perpetually waterlogged. Cool, anaerobic conditions slow decay, so organic remains can build up. Organic matter in the permafrost makes the arctic tundra one of Earth's greatest stores of carbon.

As global temperatures rise, the amount of frozen soil that melts each summer is increasing. With warmer temperatures, much of the snow and ice that would otherwise reflect sunlight is disappearing. As a result, newly exposed dark soil absorbs heat from the sun's rays, which encourages more melting.

Alpine tundra occurs at high altitudes throughout the world (Figure 48.22). Even in the summer, some patches of snow persist in shaded areas, but there is no permafrost. The alpine soil is well drained, but thin and nutrient-poor. As a result, primary productivity is low. Grasses, heaths, and small-leafed shrubs grow in patches where soil has accumulated to a greater depth. These low-growing plants can withstand strong winds that discourage the growth of trees.

Take-Home Message

What is tundra?

- Arctic tundra prevails at high latitudes, where short, cold summers alternate with long, cold winters. Lichens and short plants grow above a frozen soil layer, the permafrost, which is a reservoir for carbon.
- Alpine tundra, also dominated by short plants, prevails at high altitudes.

Brown bear

Figure 48.21 Arctic tundra in the summer.

■ Freshwater and saltwater provinces cover more of Earth's surface than all land biomes combined. Here we begin our survey of these watery realms.

■ Links to Properties of water 2.5, Aquatic respiration 39.2, Food chains 47.2, Water cycle 47.6, Eutrophication 47.10

Lakes

A **lake** is a body of standing fresh water. If it is sufficiently deep, it can be divided into zones that differ in their physical characteristics and species composition (Figure 48.23). Near shore is the littoral zone, from the Latin *litus* for shore. Here, sunlight penetrates all the way to the lake bottom; aquatic plants and algae that attach to the bottom are the primary producers. The lake's open waters include an upper, well-lit limnetic zone, and—if the lake is deep—a dark profundal zone where light does not penetrate. Primary producers in the limnetic zone can include aquatic plants, green algae, diatoms, and cyanobacteria. These organisms serve as food for rotifers, copepods, and other types of zooplankton. In the profundal zone, where there is not enough light for photosynthesis, consumers feed on organic debris that drifts down from above.

Nutrient Content and Succession Like a habitat on land, a lake undergoes succession; it changes over time (Section 46.8). A newly formed lake is oligotrophic: deep, clear, and nutrient-poor, with low primary productivity (Figure 48.24). Later, sediments accumulate and plants take root. The lake becomes eutrophic.

Eutrophication refers to processes, either natural or artificial, that enrich a body of water with nutrients (Section 47.10).

Seasonal Changes Temperate-zone lakes undergo a seasonal variation in temperature gradients, from the surface to the bottom. During winter, a layer of ice forms at the lake surface. Unlike most substances, water is denser as a liquid than as a solid (ice). As water cools, its density increases, until it reaches 4°C (39°F). Below this temperature, additional cooling decreases

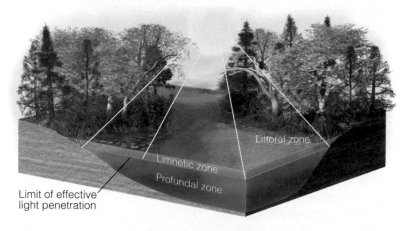

Limnetic zone
Profundal zone
Littoral zone

Limit of effective light penetration

Figure 48.23 Lake zonation. A lake's littoral zone extends around the shore to a depth where rooted aquatic plants stop growing. Its limnetic zone is the open waters where light penetrates and photosynthesis occurs. Below that lies the cool, dark profundal zone, where detrital food chains predominate.

Oligotrophic Lake	Eutrophic Lake
Deep, steeply banked	Shallow with broad littoral
Large deep-water volume relative to surface-water volume	Small deep-water volume relative to surface-water volume
Highly transparent	Limited transparency
Water blue or green	Water green to yellow- or brownish-green
Low nutrient content	High nutrient content
Oxygen abundant through all levels throughout year	Oxygen depleted in deep water during summer
Not much phytoplankton; green algae and diatoms dominant	Abundant, thick masses of phytoplankton; and cyanobacteria dominant
Aerobic decomposers favored in profundal zone	Anaerobic decomposers in profundal zone
Low biomass in profundal zone	High biomass in profundal zone

Figure 48.24 Crater Lake, an oligotrophic lake in a collapsed volcano that filled with snow melt. The chart compares oligotrophic and eutrophic lakes.

water's density—which is why ice floats on water (Section 2.5). In an ice-covered lake, water just under the ice is near its freezing point and at its lowest density. The densest (4°C) water is at the lake bottom (Figure 48.25a).

In spring, the air warms and the ice melts. When the temperature of the meltwater rises to 4°C, it sinks. This causes a **spring overturn**, during which oxygen-rich water in surface water moves downward while nutrient-rich water from the lake's depths moves up (Figure 48.25b). Winds aid in the overturn.

In the summer, a lake has three layers that differ in their temperature and oxygen content (Figure 48.25c). The upper layer is warm and oxygen-rich. It overlies the **thermocline**, a thin layer where temperature falls rapidly. Beneath the thermocline is the coolest water. The thermocline acts as a barrier that keeps the upper and lower layers from combining. During the summer, decomposers use up the oxygen in the lakes deepest waters, and nutrients from the depths cannot escape into surface waters. In autumn, the upper layer cools and sinks, and the thermocline vanishes. During the **fall overturn**, oxygen-rich water moves down while nutrient-rich water moves up (Figure 48.25d).

Overturns influence primary productivity. After a spring overturn, longer daylength and an abundance of nutrients support the greatest primary productivity. During the summer, vertical mixing ceases. Nutrients do not move up, and photosynthesis slows. By late summer, nutrient shortages limit growth. Fall overturn brings nutrients to the surface and favors a brief burst of photosynthesis. The burst ends as winter brings shorter days and the amount of sunlight declines.

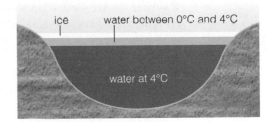

A Winter. Ice covers the thin layer of slightly warmer water just below it. Densest (4°) water is at bottom. Winds do not affect water under the ice, so there is little circulation.

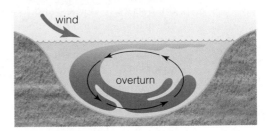

B Spring. Ice thaws. Upper water warms to 4°C and sinks. Winds blowing across water create vertical currents that help overturn water, bringing nutrients up from bottom.

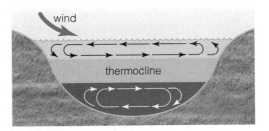

C Summer. Sun warms the upper water, which floats on a thermocline, a layer across which temperature changes abruptly. Upper and lower water do not mix because of this thermal boundary.

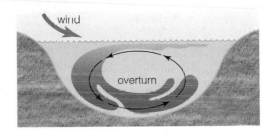

D Fall. Upper water cools and sinks downward, eliminating the thermocline. Vertical currents mix water that was separated during the summer.

Figure 48.25 Seasonal changes in a temperate zone lake.

Streams and Rivers

Streams are flowing-water ecosystems that begin as freshwater springs or seeps. As they flow downslope, they grow and merge to form rivers. Rainfall, snowmelt, geography, altitude, and shade cast by plants affect flow volume and temperature.

Properties of a stream or river vary along its length. Streambed composition affects solute concentrations, as when limestone rocks dissolve and add calcium. Water that flows rapidly over rocks mixes with air and holds more oxygen than slower-moving, deeper water. Also, cold water holds more oxygen than warm water. As a result, different parts of a stream or river support species with different oxygen needs (Section 39.2).

A stream imports nutrients into many food webs. In forests, trees cast shade and hinder photosynthesis, but dead leaves sustain detrital food chains (Section 47.2). Aquatic species take up and release nutrients as water flows downstream. Nutrients move upstream in the tissues of migratory fish and other animals. The nutrients cycle between aquatic organisms and water as it flows on a one-way course to the sea.

Take-Home Message

What factors affect life in freshwater provinces?

■ Lakes have gradients in light, dissolved oxygen, and nutrients.

■ Primary productivity varies with a lake's age and—in temperate zones—with the season.

■ Different conditions along the length of a stream or river favor different organisms.

48.13 | "Fresh" Water?

■ All water is recycled. What happens to the water that you send down your drain, or flush down your toilet?

■ Link to Groundwater 47.6

Pollutants flow into rivers, lakes, and groundwater from countless sources. Pollutants include sewage, animal wastes, industrial chemicals, fertilizers, and pesticides. Runoff from roads adds engine oil and antifreeze that dripped from vehicles, and rubber residues from tire wear. Leaky underground fuel tanks allow gasoline and other fuels to seep into groundwater.

How do we keep these poisons out of our drinking water? One safeguard is wastewater treatment. There are three stages of treatment. In primary treatment, screens and settling tanks remove large bits of organic material (sludge), which is dried, and burned or dumped in landfills. In secondary treatment, microbes break down any organic matter that remained after the primary treatment. Water is then treated with chlorine or exposed to ultraviolet light to kill disease-causing microorganisms. By now, most organic wastes are gone—but not all nitrogen, phosphorus, toxins, and heavy metals. Tertiary treatment uses chemical filters to remove these contaminants from water but it adds to the cost of treatment. In the United States most water is discharged after secondary treatment.

One variation on standard wastewater treatment is a solar-aquatic system such as the one constructed by biologist John Todd (Figure 48.26). Sewage enters tanks in which aquatic plants grow. Decomposers degrade the wastes and release nutrients that promote plant growth. Heat from sunlight speeds the decomposition. Water next flows through an artificial marsh that filters out algae and organic wastes. Then it flows through other tanks filled with living organisms, including plants that take up metals. After ten days, water flows into a second artificial marsh for final filtering and cleansing. Versions of this system are now used to treat both sewage and industrial wastes.

48.14 | Coastal Zones

■ Where sea meets shore we find regions of high primary productivity.

■ Link to Grazing and detrital food chains 47.2

Wetlands and the Intertidal Zone

Like freshwater ecosystems, estuaries and mangrove wetlands have distinct physical and chemical features, including their depth, water temperature, salinity, and light penetration. An **estuary** is an enclosed coastal region where seawater mixes with nutrient-rich fresh water from rivers and streams (Figure 48.27u). Water inflow continually replenishes nutrients, which is one reason estuaries are highly productive.

Primary producers include algae and other types of phytoplankton, and plants that tolerate being submerged at high tide. Detrital food chains are common (Section 47.2). Estuaries are marine nurseries; many larval and juvenile invertebrates and fishes develop in them. Migratory birds use estuaries as rest stops.

Estuaries can be broad and shallow like Chesapeake Bay, Mobile Bay, and San Francisco Bay, or narrow and deep like the fjords of Norway. Many face threats. Fresh water that should refresh them is diverted for human uses. Rivers deliver harmful substances, such as pesticides and fertilizers that entered streams in the runoff from agricultural fields.

In tidal flats at tropical latitudes, we find nutrient-rich mangrove wetlands. "Mangrove" is the common term for certain salt-tolerant woody plants that live in sheltered areas along tropical coasts. The plants have prop roots that extend out from their trunk (Figure 48.27b). Specialized cells at the surface of some roots allow gas exchange with air.

Increasing human populations along tropical coasts threaten mangrove wetlands. People have traditionally cut these trees for firewood. A more recent threat is conversion of mangrove wetlands to shrimp farms. The shrimp mainly end up on dinner plates in the United States, Japan, and western Europe. Disappearance of mangrove wetlands threatens the fishes and migratory birds that depend upon them for shelter and food.

Figure 48.26 John Todd in the experimental solar-aquatic wastewater treatment facility he designed. Unlike traditional treatments, Todd's system does not require toxic chemicals or emit unpleasant odors. Bacteria, fungi, plants, invertebrates, and fish break down wastes. Solar-aquatic treatment systems are now in use in eight countries around the world.

Figure 48.27 Wetlands. (**a**) South Carolina salt marsh. Marsh grass (*Spartina*) is the major producer. (**b**) In the Florida Everglades, a mangrove wetland lined with red mangroves (*Rhizophora*).

Intertidal zone's upper littoral, submerged only at highest tide of lunar cycle

midlittoral; submerged at each highest regular tide and exposed at lowest tide

lower littoral; exposed only at low tide of lunar cycle

Figure 48.28 Contrasting coasts. (**a,b**) Algae-rich rocky shores where invertebrates abound. (**c**) A sandy shore in Australia shows fewer signs of life. Invertebrates burrow in its sediments.

Rocky and Sandy Coastlines

Rocky and sandy coastlines support ecosystems of the intertidal zone. Biologists divide a shoreline into three vertical zones that differ in physical characteristics and diversity. The upper littoral zone is submerged only at the highest tide of a lunar cycle. It holds the fewest species. The midlittoral zone is submerged during the highest average tide and exposed at the lowest tide. The lower littoral zone, exposed only during the lowest tide of the lunar cycle, has the most diversity.

You can easily see the zonation along a rocky shore (Figure 48.28*a,b*). Algae clinging to rocks are primary producers for the prevailing grazing food chains. The primary consumers include a variety of snails.

Zonation is less obvious on sandy shores where detrital food chains start with material washed ashore (Figure 48.28*c*). Some crustaceans eat detritus in the upper littoral zone. Nearer to the water, other invertebrates feed as they burrow through the sand.

Take-Home Message

What kinds of ecosystems occur along coastlines?

■ We find estuaries where rivers empty into seas. The rivers deliver nutrients that foster high productivity.

■ Mangrove wetlands are common along shorelines in tropical latitudes.

■ Rocky and sandy shores show zonation, with different zones exposed during different phases of the tidal cycle. Diversity is highest in the zone that is submerged most of the time.

The Once and Future Reefs

- Coral reefs are highly productive, and greatly threatened.
- Links to Dinoflagellates 22.5, Corals 25.5

Coral reefs are wave-resistant formations that consist primarily of calcium carbonate secreted by generations of coral polyps (Section 25.5). Reef-forming corals live mainly in clear, warm waters between latitudes 25° north and 25° south (Figure 48.29d). The mineral-hardened cell walls of red algae such as the one shown at *left* contribute to the structural framework of many reefs. The resulting reef is home to a remarkably diverse array of vertebrate and invertebrate species.

coralline alga

Australia's Great Barrier Reef parallels Queensland for 2,500 kilometers (1,550 miles), and is the largest example of biological architecture. Actually it is a string of reefs, some 150 kilometers (95 miles) across. It supports 500 coral species, 3,000 fish species, 1,000 kinds of mollusks, and 40 kinds of sea snakes. Figure 48.29e shows a wealth of warning colors, tentacles, and stealthy behavior—all signs of fierce competition for resources among species jostling for the limited space.

Reefbuilding corals have photosynthetic dinoflagellates (Section 22.5) in their tissues. These protists give a coral its color and provide it with oxygen and sugars. When a coral is stressed, it expels the protists and loses its color, an event called coral bleaching (Figure 48.30). If the coral remains stressed for more than a few months, protists do not return and the coral dies.

Figure 48.30 Coral bleaching on Australia's Great Barrier Reef.

Abnormal, widespread bleaching in the Caribbean and the tropical Pacific began in the 1980s. So did increases in sea surface temperature, which might be a key stress factor. Is the damage one outcome of global warming? If so, as marine biologists Lucy Bunkley-Williams and Ernest Williams suggest, the future looks grim for reefs, which may be destroyed within three decades.

Also, people can directly destroy reefs, as by sewage discharges into nearshore waters of populated islands. Massive oil spills, commercial dredging operations, and mining for coral rock have catastrophic impact. Logging of areas adjacent to reefs increases runoff of nutrients and silt, which can harm reef species.

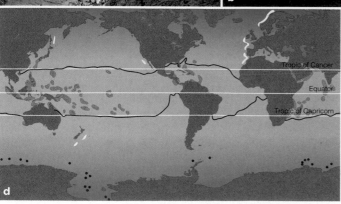

Figure 48.29 Coral reef formations. (**a**) *Fringing* reefs form near land when rainfall and runoff are light, as on the downwind side of young volcanic islands. Many reefs in the Hawaiian Islands and Tahiti are like this.

(**b**) Ring-shaped *atolls* consist of coral reefs and coral debris. They fully or partly enclose a shallow lagoon, often with a channel to open ocean. Biodiversity is not great in the shallow water, which can get too hot for corals. (**c**) *Barrier* reefs parallel the shore of continents and volcanic islands, as in Bora Bora. Behind them are calm lagoons.

(**d**) Distribution map for coral reefs (*orange*) and coral banks (*yellow*). Nearly all reef-building corals live in warm seas, here enclosed in dark lines. Past latitudes 25° north and south, solitary and colonial corals (*red*) form coral banks in temperate seas and in cold seas above continental shelves. (**e**) *Facing page*, a sampling of coral reef biodiversity.

Fishing nets can break pieces off corals, but some fishermen prefer even more destructive practices such as dropping dynamite in the water. Fish hiding in the coral are blasted out and float to the surface, some dead, but others only stunned. The practice is outlawed nearly everywhere, but enforcement is often lacking.

Capture of fish for the pet trade also has harmful effects. In some places, sodium cyanide is squirted into the water to stun fish, which float to the surface. Most fish that survive being stunned with cyanide are shipped off for sale in pet stores in the United States or Europe.

Invasive species also threaten reefs. In Hawaii, reefs are being overgrown by exotic algae, including several species imported for cultivation during the 1970s.

Reef biodiversity is in danger around the world, from Australia and Southeast Asia to the Hawaiian Islands, the Galápagos Islands, the Gulf of Panama, Florida, and Kenya. For example, the biodiversity on the coral reef off Florida's Key Largo has been reduced by 33 percent since 1970.

LIONFISH

PART OF A FIJIAN CORAL REEF

MORAY EEL

NUDIBRANCH

LONGNOSE HAWKFISH AND RED SEA FAN

e   BANDED CORAL SHRIMP

PURPLE TUBE SPONGE

GREEN CORAL POLYP

- Earth's vast oceans are still largely unexplored. We are only beginning to catalogue the diversity they contain.

- Links to Hydrothermal vents 20.2, Chemoautotrophs 21.4

Oceanic Zones and Habitats

Like a lake, an ocean shows gradients in light, nutrient availability, temperature, and oxygen concentration. We refer to the open waters of the ocean as the **pelagic province** (Figure 48.31*a*). These open waters include the neritic zone—the water over continental shelves—together with the more extensive oceanic zone farther offshore. The neritic zone receives nutrients in runoff from land, and is the zone of greatest productivity.

In the ocean's upper, brightly lit waters, photosynthetic microorganisms are the primary producers, and grazing food chains predominate. Depending on the region, some light may penetrate as far as 1,000 meters beneath the sea surface. Below that, organisms live in darkness, and organic material that drifts down from above is the basis of detrital food chains. At the top of food webs, carnivores range from the familiar sharks and squids to giant colonial cnidarians and the bizarre deep-sea angler fishes (Figure 48.32*a,b*). In what may be the greatest circadian migrations, many species rise thousands of feet at night to feed in upper waters, then move down in the morning.

The **benthic province** is the ocean bottom—its rocks and sediments. Benthic biodiversity is greatest on the margins of continents, or the continental shelves. The benthic province also includes some largely unexplored concentrations of biodiversity on seamounts and at hydrothermal vents.

Seamounts are undersea mountains that stand 1,000 meters or more tall, but are still below the sea surface (Figure 48.31*b*). They attract large numbers of fishes and are home to many marine invertebrates (Figure 48.32*c*). Like islands, seamounts often are home to species that evolved there and are found nowhere else.

The abundance of life at seamounts makes them attractive to commercial fishing vessels. Fish and other organisms are often harvested by trawling, a fishing technique in which a large net is dragged along the bottom, capturing everything in its path. The process is ecologically devastating; trawled areas are stripped bare of life, and silt stirred up by the giant, weighted nets suffocates filter-feeders in adjacent areas.

Superheated water that contains dissolved minerals spews out from the ocean floor at **hydrothermal vents**. When this heated, mineral-rich water mixes with cold seawater, the minerals settle out as extensive deposits. Chemoautotrophic prokaryotes (Section 21.4) can get energy from these deposits. The prokaryotes serve as primary producers for food webs that include diverse invertebrates, such as tube worms and brittle stars (Figure 48.32*d–f*). As explained in Section 20.2, one hypothesis holds that life originated on the sea floor in such heated, nutrient-rich places.

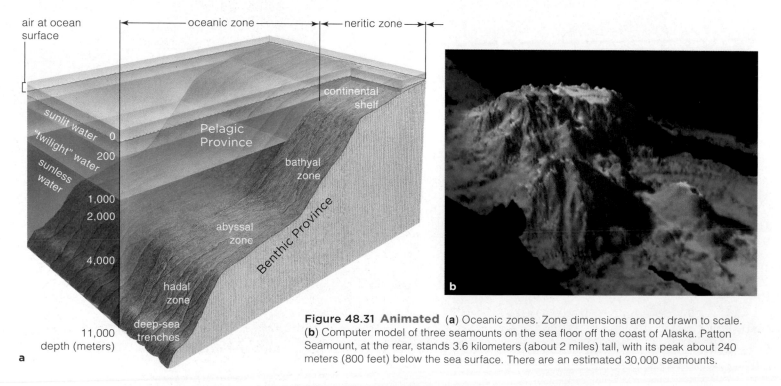

Figure 48.31 Animated (**a**) Oceanic zones. Zone dimensions are not drawn to scale. (**b**) Computer model of three seamounts on the sea floor off the coast of Alaska. Patton Seamount, at the rear, stands 3.6 kilometers (about 2 miles) tall, with its peak about 240 meters (800 feet) below the sea surface. There are an estimated 30,000 seamounts.

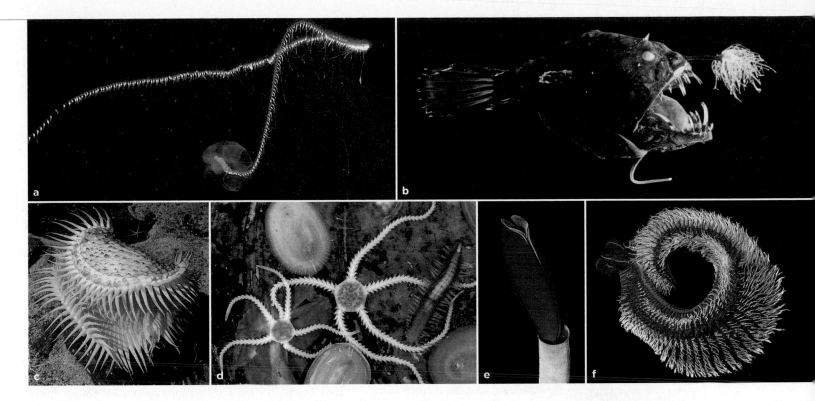

Figure 48.32 What lies beneath: a vast, largely unexplored world of marine life. (a) A siphonophore, *Praya dubia*, one of the colonial relatives of corals and jellyfish can be 40 meters (130 feet) long. (b) Deep-sea angler fish with bioluminescent lures.

(c) A flytrap anemone, from Davidson Seamount, just off the California coast.

Residents of hydrothermal vent communities: (d) brittle stars, limpets, and a polychaete worm; (e) tube worm, a polychaete; (f) Pompei-worm, another polychaete.

Upwelling—A Nutrient Delivery System

Deep, cold ocean waters are rich in nutrients. By the process of **upwelling**, this nutrient-laden water moves upward along the coasts of continents. Winds set the coastal waters in motion. For example, in the Northern Hemisphere, prevailing winds that blow from north to south parallel to the west coasts of continents start the surface waters moving (Figure 48.33). Upwelling occurs as Earth's rotation deflects the masses of slow-moving water away from a coast and cold, deep water moves up vertically in its place.

In the Southern Hemisphere, winds from the south tug surface water away from the coast. Cold, deeper water of the Humboldt Current moves in to replace it. Nutrients in this water sustain phytoplankton that are the basis for a rich fishery.

Every three to seven years, surface waters of the western equatorial Pacific Ocean warm up, causing a change in the direction of the wind. This warming most often happens around Christmas, so fishermen

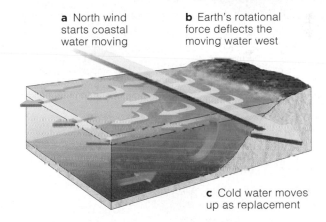

a North wind starts coastal water moving

b Earth's rotational force deflects the moving water west

c Cold water moves up as replacement

Figure 48.33 Coastal upwelling in the Northern Hemisphere.

in Peru named this event **El Niño**, as discussed in the chapter introduction. The name became part of a more inclusive term: the El Niño Southern Oscillation, or ENSO. The next section takes a closer look at some of the consequences of this recurring event.

Take-Home Message

What factors affect life in ocean provinces?

■ Oceans have gradients in light, dissolved oxygen, and nutrients. Nearshore and well lit zones are the most productive and species-rich.

■ On the sea floor, pockets of diversity occur on seamounts and around hydrothermal vents.

■ Upwelling brings nutrient-rich water from deep regions of the sea to surface waters along coasts.

■ Events in the atmosphere and oceans, and on land, interconnect in ways that profoundly affect the world of life.

■ Link to Copepods 25.14

An El Niño Southern Oscillation, or ENSO, is defined by changes in sea surface temperatures and in the air circulation patterns. "Southern oscillation" refers to a seesawing of the atmospheric pressure in the western equatorial Pacific—Earth's greatest reservoir of warm water and warm air. It is the source of heavy rainfall, which releases enough heat energy to drive global air circulation patterns.

Between ENSOs, the warm waters and heavy rains move westward (Figure 48.34*a*). During an ENSO, the prevailing surface winds over the western equatorial Pacific pick up speed and "drag" surface waters east (Figure 48.34*b*). As they do, the westward transport of water slows down. Sea surface temperatures rise, evaporation accelerates, and air pressure falls. These changes affect weather worldwide.

El Niño episodes persist for 6 to 18 months. Often they are followed by a **La Niña** episode in which the Pacific waters become cooler than usual. Other years, waters are neither warmer nor colder than average.

As noted in the chapter introduction, 1997 ushered in the most powerful El Niño event of the century. The average sea surface temperatures in the eastern Pacific rose by 5°C (9°F). This warmer water extended 9,660 kilometers (6,000 miles) west from the coast of Peru.

The 1997–1998 El Niño/La Niña roller-coaster had extraordinary effects on the primary productivity in the equatorial Pacific. With the massive eastward flow of nutrient-poor warm water, photoautotrophs were almost undetectable in satellite photos that measure primary productivity (Figure 48.35*a*).

During the La Niña rebound, cooler, nutrient-rich water welled up to the sea surface and was displaced westward all along the equator. As satellite images revealed, upwelling had sustained an algal bloom that stretched across the equatorial Pacific (Figure 48.35*b*).

During the 1997–1998 El Niño event, 30,000 cases of cholera were reported in Peru alone, compared with only 60 cases from January to August in 1997. People knew that water contaminated by *Vibrio cholerae* causes epidemics of cholera (Figure 48.36*b*). The disease agent triggers severe diarrhea. Bacteria-contaminated feces enter the water supply and individuals who use the tainted water become infected.

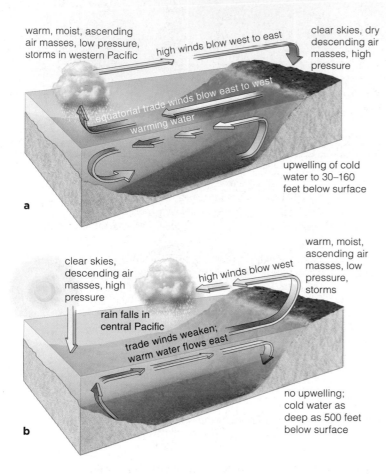

a — warm, moist, ascending air masses, low pressure, storms in western Pacific · high winds blow west to east · clear skies, dry descending air masses, high pressure · equatorial trade winds blow east to west · warming water · upwelling of cold water to 30–160 feet below surface

b — clear skies, descending air masses, high pressure · high winds blow west · warm, moist, ascending air masses, low pressure, storms · rain falls in central Pacific · trade winds weaken; warm water flows east · no upwelling; cold water as deep as 500 feet below surface

Figure 48.34 Animated (**a**) Westward flow of cold, surface water between ENSOs. (**b**) Eastward dislocation of warm water during El Niño.

Figure 48.35 Satellite data on primary productivity in the equatorial Pacific Ocean. The concentration of chlorophyll in the water was used as the measure.

(**a**) During the 1997–1998 El Niño episode, a massive amount of nutrient-poor water moved to the east, and so photosynthetic activity was negligible.

(**b**) During a subsequent La Niña episode, massive upwelling and westward displacement of nutrient-rich water led to a vast algal bloom that stretched all the way to the coast of Peru.

a Near-absence of phytoplankton in the equatorial Pacific during an El Niño.

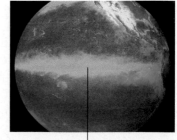

b Huge algal bloom in the equatorial Pacific in the La Niña rebound event.

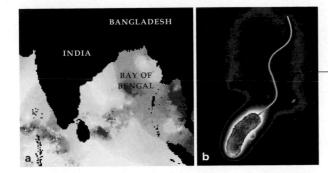

What people did not know was where *V. cholerae* remained between cholera outbreaks. It could not be found in humans or in water supplies. Even so, the cholera would often break out simultaneously in far apart places—usually coastal cities where the urban poor draw water from rivers near the sea.

Marine biologist Rita Colwell had been thinking about the fact that humans are not the host between outbreaks. Was there an environmental reservoir for the pathogen? Maybe, but nobody had detected it in water samples subjected to standard culturing.

Then Colwell had a flash of insight: What if no one could find the pathogen because it changes its form and enters a dormant stage between outbreaks?

During one cholera outbreak in Louisiana, Colwell realized that she could use an antibody-based test to detect a protein unique to *V. cholerae*'s surface. Later, tests in Bangladesh revealed bacteria in fifty-one of fifty-two samples of water. Standard culture methods had missed it in all but seven samples.

V. cholerae survives in rivers, estuaries, and seas. As Colwell knew, plankton also thrive in these aquatic environments. She decided to restrict her search for the unknown host to warm waters near Bangladesh, where outbreaks of cholera occur seasonally (Figure 48.36). It was here that Colwell discovered a dormant *V. cholerae* stage inside copepods, a type of tiny marine crustacean (Section 25.14). Copepods eat phytoplankton, so the abundance of copepods—and of *V. cholerae* cells inside them—increases and decreases with the abundance of phytoplankton.

Colwell suspected that water temperature changes in the Bay of Bengal were tied to cholera outbreaks, so she looked at medical reports for the 1990–1991 and 1997–1998 El Niño episodes. She found that the number of reported cholera cases rose four to six weeks after an El Niño event began. El Niño brings warmer water with more nutrients to the Bay of Bengal, encouraging the growth of phytoplankton. This added food then increases the number of cholera-carrying copepods.

Figure 48.36 (**a**) Satellite data on rising sea surface temperatures in the Bay of Bengal. *Red* shows warmest summer temperatures. (**b**) *Vibrio cholerae*, the agent of cholera. Copepods host a dormant stage of this bacterium that waits out adverse environmental conditions that do not favor its growth and reproduction. (**c**) A typical Bangladeshi waterway from which water samples were drawn for analysis. (**d**) In Bangladesh, Rita Colwell comparing samples of unfiltered and filtered drinking water.

Today, Colwell and Anwarul Huq, a Bangladeshi scientist, are investigating salinity and other factors that may relate to outbreaks. Their goal is to design a model for predicting where cholera will occur next. They advised women in Bangladesh to use sari cloth as a filter to remove *V. cholerae* cells from the water (Figure 48.36*d*). The copepod hosts are too big to pass through the thin cloths, which can be rinsed in clean water, sun-dried, and used again. This inexpensive, simple method has cut cholera outbreaks by half.

Take-Home Message

What occurs during an El Niño event?

■ During an El Niño event, changes in ocean temperatures and winds alter currents, affecting weather, marine food webs, and human health.

It is becoming increasingly clear that "normal" weather depends on what time frame you consider. Cycles of heating and cooling in the Pacific Ocean alter conditions over the course of 3 to 7 years. Evidence of longer-term cycles is also emerging. Some cycles seem to have a period as long as 50 to 70 years. These findings suggest that long-term plans that are based on current weather and climate conditions may be short-sighted.

How would you vote?

Is supporting studies of El Niño and the other long term climate cycles a good use of government funds? See CengageNOW for details, then vote online.

Summary

Sections 48.1, 48.2 Global air circulation patterns affect **climate** and the distribution of communities. The patterns are set into motion by latitudinal variations in incoming solar radiation. Air circulation patterns are influenced by Earth's daily rotation and annual path around the sun, the distribution of landforms and seas, and elevations of landforms. Solar energy, and the winds that it causes, are renewable, clean sources of energy.

Human put **pollutants** into the atmosphere. Use of CFCs depletes the **ozone layer** in the upper atmosphere and allows more UV radiation to reach Earth's surface.

Smog, a form of air pollution, occurs when fossil fuels are burned in warm, still air above cities. Coal-burning power plants also are big contributors to **acid rain**, which alters habitats and kills many organisms.

- *Use the animation on CengageNOW to learn how Earth's tilt affects seasons, how sunlight drives air circulation, how CFCs destroy ozone, and how acid rain forms.*

Section 48.3 Latitudinal and seasonal variations in sunlight warm sea surface water and start currents. The currents distribute heat energy worldwide and influence the weather patterns. Ocean currents, air currents, and landforms interact in shaping global temperature zones, as when the presence of coastal mountains causes a **rain shadow** or **monsoon** rains fall seasonally.

- *Use the animation on CengageNOW to learn about ocean currents, rain shadows, and coastal breezes.*

Sections 48.4, 48.5 **Biogeographic realms** are vast areas with communities of plants and animals found nowhere else. **Biomes** are somewhat smaller regions with a particular type of dominant vegetation. Regional variations in climate, elevation, **soil profiles**, and evolutionary history affect the distribution of biomes.

- *Use the animation on CengageNOW to see the distribution of biomes and compare some of their soil profiles.*

Sections 48.6–48.11 **Deserts** form around latitudes 30° north and south. Vast **grasslands** form in the interior of midlatitude continents. Slightly moister southern or western coastal regions support **dry woodlands** and **dry shrublands**.

From the equator to latitudes 10° north and south, high rainfall, high humidity, and mild temperatures can support **evergreen broadleaf forests**.

Semi-evergreen forests and **tropical deciduous forests** form between latitudes 10° and 25°, depending on how much of the annual rainfall occurs in a prolonged dry season. **Temperate deciduous forests** form at higher latitudes. Where a cold, dry season alternates with a cold, rainy season, **coniferous forests** dominate. Conifers also are favored in temperate areas with poor soil.

Low-growing, hardy plants of the **arctic tundra** occur at high latitudes, where there is a layer of **permafrost**. At high altitudes, similar plants grow as **alpine tundra**.

Sections 48.12–48.15 Most **lakes**, streams, and other aquatic ecosystems have gradients in the penetration of sunlight, water temperature, and in dissolved gases and nutrients. These characteristics vary over time and affect primary productivity.

In temperate-zone lakes, a **spring overturn** and a **fall overturn** cause vertical mixing of waters and trigger a burst of productivity. In summer, a **thermocline** prevents upper and lower waters from mixing.

Coastal zones support diverse ecosystems. Among these, the coastal wetlands, **estuaries**, and **coral reefs** are especially productive.

Sections 48.16, 48.17 Life persists throughout the ocean. Diversity is highest in sunlit waters at the top of the **pelagic province**, or ocean waters. In the **benthic province**—the seafloor—diversity is high near deep-sea **hydrothermal vents** and on **seamounts**.

Upwelling is an upward movement of deep, cool, often nutrient-rich ocean water, typically along the coasts of continents. An **El Niño** event is a warming of eastern Pacific waters that triggers changes in rainfall and other weather patterns around the world. A **La Niña** is a cooling of these same waters; it too influences global weather patterns.

- *Use the interaction on CengageNOW to learn about oceanic zones and to observe how an El Niño event affects ocean currents and upwelling.*

Self-Quiz   *Answers in Appendix III*

1. Solar radiation drives the distribution of weather systems and so influences _____ .
   a. temperature zones   c. seasonal variations
   b. rainfall distribution   d. all of the above

Data Analysis Exercise

To try to predict the effect of El Niño or La Niña events in the near future, the National Oceanographic and Atmospheric Administration collects information about sea surface temperature (SST) and atmospheric conditions. They compare monthly temperature averages in the eastern equatorial Pacific Ocean to historical data and calculate the difference (the degree of anomaly) to determine if El Niño conditions, La Niña conditions, or neutral conditions are developing. El Niño is a rise in the average SST above 0.5°C. A decline of the same amount is La Niña. Figure 48.37 shows data for nearly 39 years.

1. When did the greatest positive temperature deviation occur during this time period?

2. What type of event, if any, occurred during the winter of 1982–1983? What about the winter of 2001–2002?

3. During a La Niña event, less rain than normal falls in the American West and Southwest. In the time interval shown, what was the longest interval without a La Niña event?

4. What type of conditions were in effect in the fall of 2007 when California suffered severe wildfires?

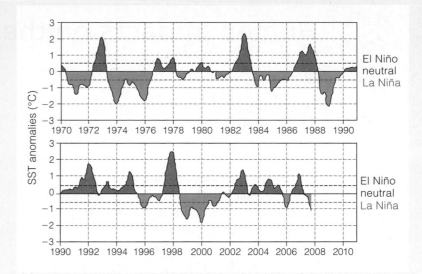

Figure 48.37 Sea surface temperature anomalies (differences from the historical mean) in the eastern equatorial Pacific Ocean. A rise above the dotted *red* line is an El Niño event, a decline below the *blue* line is La Niña.

---

2. _____ shields living organisms against the sun's UV wavelengths.
   a. A thermal inversion   c. The ozone layer
   b. Acid precipitation   d. The greenhouse effect

3. Regional variations in the global patterns of rainfall and temperature depend on _____ .
   a. global air circulation   c. topography
   b. ocean currents   d. all of the above

4. A rain shadow is a reduction in rainfall _____ .
   a. on the inland side of a coastal mountain range
   b. during an El Niño event
   c. that occurs seasonally in the tropics

5. Air masses rise _____ .
   a. at the equator   c. as air cools
   b. at the poles   d. all of the above

6. Biomes are _____ .
   a. water provinces   d. partly characterized
   b. water and land zones   by dominant plants
   c. land regions   e. both c and d

7. Biome distribution depends on _____ .
   a. climate   c. soils
   b. elevation   d. all of the above

8. Grasslands most often predominate _____ .
   a. near the equator   c. in interior of continents
   b. at high altitudes   d. b and c

9. Permafrost underlies _____ , and is a vast store of carbon.
   a. arctic tundra   c. coniferous forests
   b. alpine tundra   d. all of the above

10. During _____ , deeper, often nutrient-rich water moves to the surface of a body of water.
   a. spring overturns   c. upwellings
   b. fall overturns   d. all of the above

11. Chemoautotrophic prokaryotes are the primary producers for food webs _____ .
   a. in grasslands   c. on coral reefs
   b. in deserts   d. at hydrothermal vents

12. Match the terms with the most suitable description.
   ___tundra   a. equatorial broadleaf forest
   ___chaparral   b. partly enclosed by land where
   ___desert   fresh water and seawater mix
   ___savanna   c. type of grassland with trees
   ___estuary   d. has low-growing plants at
   ___boreal forest   high latitudes or elevations
   ___tropical rain   e. at latitudes 30° north and south
      forest   f. mineral-rich, superheated
   ___hydrothermal   water supports communities
      vents   g. conifers dominate
                      h. dry shrubland

■ *Visit CengageNOW for additional questions.*

Critical Thinking

1. London, England, is at the same latitude as Calgary in Canada's province of Alberta. However, the mean January temperature in London is 5.5°C (42°F), whereas in Calgary it is minus 10°C (14°F). Compare the locations of these two cities and suggest a reason for this temperature difference.

2. Increased industrialization in China has environmentalists worried about air quality elsewhere. Are air pollutants from Beijing more likely to end up in eastern Europe or the western United States? Why?

3. The use of off-road recreational vehicles may double in the next twenty years. Enthusiasts would like increased access to government-owned deserts. Some argue that it's the perfect place for off-roaders because "There's nothing there." Explain whether you agree, and why.

49 Human Impacts on the Biosphere

A Long Reach

We began this book with the story of biologists who ventured into a remote forest in New Guinea, and their excitement at the many previously unknown species that they encountered. At the far end of the globe, a U.S. submarine surfaced in Arctic waters and discovered polar bears hunting on the ice-covered sea (Figure 49.1). The bears were about 270 miles from the North Pole and 500 miles from the nearest land.

Even such seemingly remote regions are no longer beyond the reach of human explorers—and human influence. You already know that increasing levels of greenhouse gases are raising the temperature of Earth's atmosphere and seas. In the Arctic, the warming is causing sea ice to thin and to break up earlier in the spring. This raises the risk that polar bears hunting far from land will become stranded and unable to return to solid ground before the ice thaws.

Polar bears are top predators and their tissues contain a surprisingly high amount of mercury and organic pesticides. The pollutants entered the water and air far away, in more temperate regions. Winds and ocean currents deliver them to polar realms. Contaminants also travel north in the tissues of migratory animals such as seabirds that spend their winters in temperate regions and nest in the Arctic.

In places less remote than the Arctic, effects of human populations have a more direct effect. As we cover more and more of the world with our dwellings, factories, and farms, less appropriate habitat remains for other species. We also put species at risk by competing with them for resources, overharvesting them, and introducing nonnative competitors.

It would be presumptuous to think that we alone have had a profound impact on the world of life. As long ago as the Proterozoic, photosynthetic cells were irrevocably changing the course of evolution by enriching the atmosphere with oxygen. Over life's existence, the evolutionary success of some groups assured the decline of others. What is new is the increasing pace of change and the capacity of our own species to recognize and affect its role in this increase.

A century ago, Earth's physical and biological resources seemed inexhaustible. Now we know that many practices put into place when humans were largely ignorant of how natural systems operate take a heavy toll on the biosphere. The rate of species extinctions is on the rise and many types of biomes are threatened. These changes, the methods scientists use to document them, and the ways that we can address them, are the focus of this chapter.

Figure 49.1 Three polar bears investigate an American submarine that surfaced in ice-covered Arctic waters.

Key Concepts

The newly endangered species

Human activities have accelerated the rate of extinctions. Habitat loss, degradation, and fragmentation lead to extinctions, as do species introductions and overharvesting. **Sections 49.1, 49.2**

Assessing biodiversity

Our knowledge of species is biased toward large land animals. Conservation biologists assess the state of ecosystems and their biodiversity, with the goal of preserving as much of it as possible. **Sections 49.3, 49.4**

Harmful practices

Building homes, using energy, purchasing products, raising crops, and discarding trash all have harmful environmental effects that endanger species and ecosystems. **Sections 49.5–49.7**

Sustainable solutions

All nations have biological wealth that can benefit human populations. Recognizing the value of biodiversity and putting it to use in sustainable ways is good for Earth and all of its species. **Section 49.8**

Links to Earlier Concepts

- You already know about extinction (Section 18.12) and how mass extinctions were used to create the geologic time scale (17.8). Here we take a look at how human population growth and resource use (45.7, 45.9), including use of fossil fuel (23.5) are accelerating extinctions.

- You will learn how human activities can cause inbreeding by disrupting gene flow (18.8, 18.9). You will also be reminded of the effects of aquifer depletion (47.6), acid rain (47.9), soil erosion (29.1), and greenhouse gas emissions (47.8). You will see how transpiration (29.3) affects local rainfall patterns.

- We will look at the story of lichens and pollution (18.4) from another perspective, and see another example of the effects of pathogenic oomycotes (22.8).

How would you vote? The Arctic contains reserves of gas, oil, and minerals. The United States has a claim to some Arctic territory. Should it push for protection of the Arctic rather than exploitation of these resources? See CengageNOW for details, then vote online.

The Extinction Crisis

- Extinction is a natural process, but we are accelerating it.
- Links to Geologic time scale 17.8, Extinction 18.12

Era	Period	Major extinction under way
CENOZOIC	QUATERNARY —1.8 mya TERTIARY	With high population growth rates and cultural practices (e.g., agriculture, deforestation), humans become major agents of extinction.
	—65.5	*Major extinction event*
MESOZOIC	CRETACEOUS —145.5 JURASSIC —199.6 TRIASSIC	Slow recovery after Permian extinction, then adaptive radiations of some marine groups and plants and animals on land. Asteroid impact at K–T boundary, 85% of all species disappear from land and seas.
	—251	*Major extinction event*
PALEOZOIC	PERMIAN —299 CARBONIFEROUS	Pangea forms; land area exceeds ocean surface area for first time. Asteroid impact? Major glaciation, colossal lava outpourings, 90%–95% of all species lost.
	—359	*Major extinction event*
	DEVONIAN —416 SILURIAN	More than 70% of marine groups lost. Reef builders, trilobites, jawless fishes, and placoderms severely affected. Meteorite impact, sea level decline, global cooling?
	—443	*Major extinction event*
	ORDOVICIAN —488 CAMBRIAN	Second most devastating extinction in seas; nearly 100 families of marine invertebrates lost.
	—542	*Major extinction event*
	(Precambrian)	Massive glaciation; 79% of all species lost, including most marine microorganisms.

a

Mass Extinctions and Slow Recoveries

Extinction, like speciation, is a natural process (Section 18.12). Species arise and become extinct on an ongoing basis. Based on several lines of evidence, scientists estimate that 99 percent of all species that have ever lived are now extinct.

The rate of extinction picks up dramatically during a mass extinction, when many kinds of organisms in many different habitats all become extinct in a relatively short period. Five great mass extinctions mark the boundaries for the geologic time periods (Section 17.8). With each mass extinction event, biodiversity plummeted both on land and in the oceans. Afterwards, the surviving species underwent adaptive radiations. Each time, biodiversity recovered extremely slowly. It took at least 10 million years for diversity to return to the level that preceded the extinction event. Figure 49.2a reviews the major extinctions and recoveries.

This pattern of extinctions is a composite of what happened to the major taxa. However, lineages differ in their time of origin, their tendency to branch and give rise to new species, and how long they endure. If we consider the number of species as the measure of success for any lineage, not all lineages are equally successful. Figure 49.2b, illustrates how the number of species changed over time in some major lineages. Expansion of one lineage sometimes occurred at the same time as contraction of another, as when a decline in gymnosperms accompanied the adaptive radiation of angiosperms.

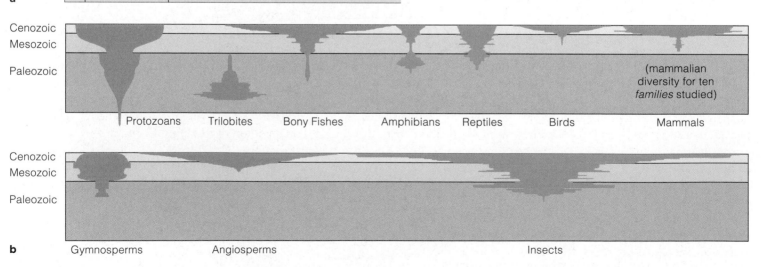

b

Figure 49.2 Animated (**a**) Dates of the five greatest mass extinctions and recoveries in the past. Compare Figure 17.14. (**b**) Species diversity over time for a sampling of taxa. The width of each *blue* shape represents the number of species in that lineage. Notice the variation among lineages.

Figure 49.3 Drawing of a dodo (*Raphus cucullatus*). Extinct since the late 1600s, it was larger than a turkey and flightless.

Figure 49.4 Living or extinct? Colorized photo of an ivory-billed woodpecker (*Campephilus principalis*). It is, or was, North America's largest woodpecker and a native of the Southeastern states.

The Sixth Great Mass Extinction

We are currently in the midst of a mass extinction. The current extinction rate is estimated to be 100 to 1,000 times above the typical background rate, putting it on a par with the five major extinction events. Unlike those early events, this one cannot be blamed on some natural catastrophe such as an asteroid impact. Rather, this mass extinction is the outcome of the success of a single species—humans—and their effect on Earth.

The ongoing extinction event may have begun as early as 60,000 years ago. The estimated arrival time of humans in Australia and North America correlates with a rise in the extinction rate for large mammals. Climate change certainly played a role in the declines, but hunting may have been a contributing factor.

It is easier to pin the blame for recent extinctions on humans. The World Conservation Union has compiled a list of more than 800 documented extinctions that occurred since 1500. As one example, the dodo (Figure 49.3) was a big, flightless bird that lived on the island of Mauritius in the Indian Ocean. Dodos were plentiful in 1600, when Dutch sailors first arrived on the island, but 80 or so years later the birds were extinct. Some were eaten by the sailors. However, destruction of nests and habitat by rats, cats, and pigs that accompanied the humans probably had a greater effect.

Extinctions of animals tend to garner more press than those of plants. Disappearances of large land animals, especially birds and mammals, are usually well documented. We know less about the losses of small animals, especially invertebrates. Historically, the losses of microorganisms, protists, and fungi have been almost entirely undocumented.

It can be difficult to determine whether a species is entirely extinct. As its numbers dwindle, sightings will become rare, but a few individuals may survive in isolated pockets of habitat. Consider, for example, the ivory-billed woodpecker, a spectacular bird that is native to swamp forests of the American Southeast (Figure 49.4). Lumbering of these forests caused the species' decline, and it was believed to have become extinct in the 1940s. A possible sighting in Arkansas in 2004 led to an extensive hunt for evidence of the bird's survival. By the end of 2007, this search had produced some blurry photos, snippets of video, and a few recordings of what may or may not be ivory-bill calls and knocks. Definitive proof that the bird still lives remains elusive.

If the ivory-billed woodpecker is still around, it is an **endangered species**, a species that has population levels so low that it faces extinction in all or part of its range. A **threatened species** is one that is likely to become endangered in the near future. Nearly all species that are currently endangered or threatened owe their precarious position to human influences, as detailed in the next section.

Take-Home Message

How are humans affecting the pattern of extinctions?

■ Humans are causing a rise in the rate of extinction.

■ Previous mass extinctions occurred as a result of global catastrophes. Species diversity takes millions of years to recover after a mass extinction.

■ Many species are currently endangered or threatened as a result of human activity, in what is being called the sixth great mass extinction.

■ Expansion of human populations and the accompanying industrialization threatens countless species.

■ Links to Inbreeding and gene flow 18.8 and 18.9, Aquifer depletion 47.6, Greenhouse gases 47.8

Habitat Loss, Fragmentation, and Degradation

Each species requires a particular type of habitat, and any loss, degradation, or fragmentation of that habitat reduces population numbers. An **endemic species**, one that is confined to the limited area in which it evolved, is more likely to go extinct than a species with a more widespread distribution.

Species with highly specific resource requirements are particularly vulnerable to habitat alterations. For example, giant pandas (Figure 49.5) are endemic to China's bamboo forests and feed mainly on this plant. As China's human population soared, bamboo was cut for building materials and to make room for farms. As the bamboo forests disappeared, so did pandas. Their numbers, which may have once been as high as 100,000, fell to 1,000 or so animals in the wild.

In addition to habitat loss, pandas are affected by habitat fragmentation; suitable panda habitat is now limited to widely separated patches atop mountains. Because of this fragmentation, pandas facing adverse conditions in one area cannot move to new site. The fragmentation also hampers the dispersal of young females. This decreases gene flow, effectively dividing the population into smaller subunits. The small group size encourages inbreeding and reduces the genetic diversity of the species as a whole. Current efforts to save giant pandas involve protecting existing habitat, creating corridors of suitable habitat to connect now isolated preserves, and captive breeding programs.

In the United States, habitat loss affects nearly all of the more than 700 species of threatened or endangered flowering plants. For example, the conversion of prairies and meadows to farms and housing developments has put both the eastern and western species of prairie fringed orchids (*Platanthera*) on the federal threatened species list (Figure 49.6a).

Humans also degrade habitat in less direct ways. For example, Edwards Aquifer in Texas consists of water-filled, underground, limestone formations that supply drinking water to the city of San Antonio. Excessive withdrawals of water from this aquifer, along with pollution of the water that recharges it, endanger the species that live in the aquifer, such as the Texas blind salamander (Figure 49.6b). Biologists find this species of interest because it demonstrates the evolutionary effects of many generations of life in total darkness.

Figure 49.5 Giant panda (*Ailuropoda melanoleuca*), one of the best-known endangered species. The panda's diet consists almost entirely of bamboo. Destruction and fragmentation of China's bamboo forests threaten its survival.

Acid rain, pesticide residues, fertilizer runoff, and emissions of greenhouse gases also degrade habitats and contribute to the decline of species. The chapter introduction explained how melting of polar ice may harm polar bears. Recognition of this threat may lead to the listing of this species as endangered.

Overharvesting and Poaching

When European settlers first arrived in North America, they found 3 billion to 5 billion passenger pigeons. In the 1800s, commercial hunting caused a steep decline in the bird's numbers. The last time anyone saw a wild passenger pigeon was 1900—and he shot it. The last captive bird died in 1914.

We are still overharvesting species. The crash of the Atlantic codfish population, described in Section 45.6, is one recent example. As another, the white abalone was the first marine invertebrate listed as threatened in the United States. Commercial harvesting of white abalones accelerated during the 1970s. By 1990, only 1 percent of the population remained.

Biologist Boris Worm estimates that populations of about 29 percent of commercially harvested marine fish and invertebrates have already collapsed—annual catch for these species is now less than 10 percent of the recorded maximum. If current trends continue, all populations of marine species that we now harvest for commercial sale could collapse by 2050.

Poaching, the illegal harvest of species, is another threat, especially in less-developed countries. People who have few other sources of protein will kill and eat

Figure 49.6 Two North American species under threat. (**a**) Habitat destruction threatens the eastern fringed prairie orchid (*Platanthera leucophaea*). (**b**) Aquifer depletion and pollution endanger Texas blind salamanders, *Typhlomolge rathbuni*. Generations of life in a dark aquifer, where there is no selection against mutations that impair eye development, have reduced this species' eyes to tiny black spots.

local animals, regardless of the animals' endangered status. Endangered species are also collected or killed for profit. It is a sad commentary on human nature that the rarer a species becomes, the higher the price it fetches on the black market. Globalization means that species can be sold to high bidders anywhere in the world. For example, rhino horn from endangered animals in Africa ends up as a traditional medicine in Asia and as knife handles in Yemen.

Species Introductions

Exotic predators (Section 46.9) are another threat. For example, rats that expanded their range by stowing away on ships now endanger many island species. Rats eat bird eggs and nestlings. They also devour other small animals such as snails. Humans also unintentionally dispersed the brown tree snake, which is native to Samoa. The arrival of this snake in Guam resulted in the extinction of most birds endemic to the island, and endangers the three that remain.

Exotic species often outcompete native ones. In the American Southeast, introduced vines such as kudzu (Figure 46.22*b*) and Japanese honeysuckle overgrow and threaten native low-growing plants. In California's mountain streams, competition from European brown trout and eastern brook trout—both introduced for sport fishing—endangers native golden trout.

Exotic pathogens also cause species declines. For example, avian malaria was unknown in Hawaii until it was carried to the islands by introduced birds and dispersed by introduced mosquitoes. Avian malaria is contributing to the extinction of native honeycreepers (birds described in the Chapter 19 introduction).

Interacting Effects

A species most often becomes endangered because of a number of concurrent threats. Often, the decline or loss of one species endangers another. For example, running buffalo clover (*Trifolium stoloniferum*) and the buffalo, or bison, that grazed on it were once common in the Midwest. The plants thrived in the open woodlands that the buffalo favored. Here, the soil was enriched by the hefty herbivore's droppings and periodically disturbed by its hooves. Buffalo helped to disperse the clover's seeds, which survive passage through the animal's gut. When buffalo were hunted to near extinction, clover populations declined. Now listed as an endangered species, the clover is further threatened by conversion of habitat for human use, competition from introduced plants, and attacks by introduced insects and pathogens.

Take-Home Message

How do human activities endanger existing species?

■ Species decline when humans destroy or fragment natural habitat by converting it to human use, or degrade it through pollution or withdrawal of an essential resource.

■ Humans also directly cause declines by overharvesting species and by poaching.

■ Global travel and trade can introduce exotic species that harm native ones.

■ Most endangered species are affected by multiple threats.

49.3 | The Unknown Losses

■ We have only begun to evaluate the threats to many groups of species, especially the microbial ones.

Endangered species listings have historically focused on vertebrates. Biologists have just begun to evaluate the threats to invertebrates and plants. Our impact on protists and fungi is essentially unknown, and the World Conservation Union's IUCN Red List of Threatened Species (Table 49.1) does not even address prokaryotes.

In a 2006 article, microbiologist Tom Curtis made a plea for increased research on microbial ecology and microbial diversity. He argued that we have barely begun to comprehend the vast number of microbial species and to understand their importance.

Curtis concluded by writing, "I make no apologies for putting microorganisms on a pedestal above all other living things. For if the last blue whale choked to death on the last panda, it would be disastrous but not the end of the world. But if we accidentally poisoned the last two species of ammonia-oxidizers, that would be another matter. It could be happening now and we wouldn't even know . . . " The ammonia-oxidizing bacteria are essential because they make nitrogen available to plants.

Table 49.1 Global List of Threatened Species (2007)*

	Described Species	Evaluated for Threats	Found to Be Threatened
Vertebrates			
Mammals	5,416	4,863	1,094
Birds	9,956	9,956	1,217
Reptiles	8,240	1,385	422
Amphibians	6,199	5,915	1,808
Fishes	30,000	3,119	1,201
Invertebrates			
Insects	959,000	1,255	623
Mollusks	81,000	2,212	978
Crustaceans	40,000	553	460
Corals	2,175	13	5
Others	130,200	83	42
Land Plants			
Mosses	15,000	92	79
Ferns and allies	13,025	211	139
Gymnosperms	980	909	321
Angiosperms	258,650	10,771	7,899
Protists			
Green algae	3,715	2	0
Red algae	5,956	58	9
Brown algae	2,849	15	6
Fungi			
Lichens	10,000	2	2
Mushrooms	16,000	1	1

* IUCN–WCU Red List, available online at www.iucnredlist.org

49.4 | Assessing Biodiversity

■ Conservation biologists are busy surveying and seeking ways to protect the world's existing biodiversity.

■ Links to Lichens and pollution 18.4, Oomycotes 22.8

Conservation Biology

Biologists recognize three levels of **biodiversity**: genetic diversity, species diversity, and ecosystem diversity. The rate of decline in biodiversity is accelerating at all three levels. Conservation biology addresses these declines. The goals of this relatively new field of biology are (1) to survey the range of biodiversity, (2) to investigate the evolutionary and ecological origins of biodiversity, and (3) to find ways to maintain and use biodiversity in ways that benefit human populations. The objective is to conserve as much biodiversity as possible by using it in sustainable ways.

Monitoring Indicator Species

Habitat damage and loss can affect different species in different ways. An **indicator species** is a species that alerts biologists to habitat degradation and impending loss of diversity when its populations decline. As one example, biologists can assess the health of a stream by monitoring certain fish and invertebrates. A decline in a trout population can be an early sign of problems in a freshwater habitat because trout do not tolerate pollutants or low oxygen levels.

Lichens function as indicators of habitat quality on land. Because lichens absorb mineral ions from dust in the air, they are harmed by air pollution. The lichens absorb toxic metals such as mercury and lead, and cannot get rid of them. Section 18.4 described how, with the onset of the industrial revolution, decline of lichens in England's forests selected for a particular coloration pattern among forest moths.

Identifying Regions at Risk

With so many species at risk, conservation biologists are working to identify the **hot spots**, habitats that are rich in endemic species and under great threat. The idea is that once identified, hot spots can take priority in worldwide conservation efforts.

The identification of a hot spot involves making an inventory of the organisms in some limited area, such as an isolated valley. Sampling quadrats and capture-mark–recapture studies identify species present in the area, and allow an estimation of their population size (Section 45.2). The Chapter 1 introduction highlights one exploratory survey in New Guinea.

Critical or endangered ecoregion
Vulnerable ecoregion
Stable or intact ecoregion
No information available

Figure 49.7 The location and current conservation status of the land ecoregions deemed most important by the World Wildlife Fund.

On a broader scale, conservation biologists define **ecoregions**, which are land or aquatic regions characterized by climate, geography, and the species found within them. The most widely used ecoregion system was developed by conservation scientists of the World Wildlife Fund. These scientists defined 867 distinctive land ecoregions. Figure 49.7 shows the locations and conservation status of ecoregions that are considered the top priority for conservation efforts.

The goal of prioritizing ecoregions is to save representative examples of all of Earth's existing biomes. By focusing on hotspots and critical ecoregions, rather than on individual endangered species, scientists hope to maintain ecosystem processes that naturally sustain biological diversity.

Table 49.2 lists the critical/endangered ecoregions located partially or entirely in the United States. Each has a large number of endemic species and is under threat. As one example, the Klamath-Siskiyou forest in southwestern Oregon and northwestern California is home to many rare conifers. Logging is the main threat to this region. However, a newly introduced conifer pathogen, *Phytophthora lateralis*, is also a concern. It is a relative of the oomycote protist that causes sudden oak death (Section 22.8). Two endangered birds, the northern spotted owl and the marbled murlet, nest in old-growth portions of the forest. Endangered coho salmon breed in streams that run through the forest.

Table 49.2 Critical or Endangered Ecoregions in the U.S.

Ecoregion	Area (sq. km)	Major Threats
Northern prairie	700,000	Conversion to pasture or farms; oil and gas development
Klamath-Siskiyou coniferous forest	50,300	Logging, exotic root disease spread by road building
Pacific temperate rain forest	295,000	Logging
Sierra Madre pine-oak forests	289,000	Overgrazing, logging, overuse for recreation
California chaparral and woodlands	121,000	Establishment of exotic species, overgrazing, fire suppression
Nevada coniferous forest	53,000	Logging, urban expansion
Southeastern coniferous and broadleaf forests	585,000	Logging, suppression of fire, urban expansion

Take-Home Message

How do conservation biologists help protect biodiversity?

■ Conservation biologists assess Earth's species richness and create systems for prioritizing conservation efforts.

■ Hot spots are areas that include many endemic species and face a high degree of threat. Ecoregions are larger areas characterized by physical factors and species composition.

49.5 | Effects of Development and Consumption

- As human populations soar, their need for energy and other resources puts pressure on native species.
- Links to Acid rain 47.9, Resource use 45.9, Fossil fuel 23.5

Effects of Urban and Suburban Development

When homes, factories, and shopping centers replace undisturbed habitat, biodiversity declines. Worldwide, people continue to migrate from rural areas into cities at an ever accelerating pace (Figure 49.8).

In the United States, expansion of urban and suburban areas is a factor in the declines of many species. The threatened Florida sandhill crane is being pressured by the expanding city of Orlando. In Nevada, a small population of relict leopard frogs, once thought to be extinct, is hanging on in rapidly growing Clark County. Another endangered amphibian, the Houston toad, now survives only between the growing cities of Austin and Houston. In northern California, new housing developments near San Francisco may harm the endangered Mission Blue butterfly.

Proximity to human development affects different species in different ways. Exotic plants introduced to beautify suburban yards can release seeds that become established in the wild and outcompete natives such as California's chapparal species. Dogs and cats that are allowed to roam can kill wild animals or change their behavior in ways that interfere with breeding. Roads interrupt and restrict land animal movement and so hamper gene flow. Lighting at night also has negative impacts. For example, light from cities along tropical beaches can disorient endangered sea turtle hatchlings as they attempt to make their way to the sea. Night-flying migratory birds that use light to navigate tend to collide with tall, well-lit buildings.

Effects of Resource Consumption

The life-style of people in industrial nations requires large quantities of resources, and the extraction and delivery of these resources affects biodiversity. In the United States, the size of the average family has declined since the 1950s, but the size of the average home has doubled. Larger homes require more lumber to build and to furnish, which encourages logging. Big homes also require more energy to heat and to cool.

Most of the energy used in developed countries is supplied by fossil fuels—petroleum, natural gas, and coal (Figure 49.9). You already know that use of these nonrenewable fuels contributes to global warming and acid rain. In addition, extraction and transportation of these fuels have negative impacts. Oil harms many species when it leaks from pipelines or from ships. Strip mining for coal degrades the immediate area and it often lowers the water quality of nearby streams.

Figure 49.8 Cities displace wild species and require huge amounts of resources. In 2008, for the first time, a majority of the human population was city dwellers.

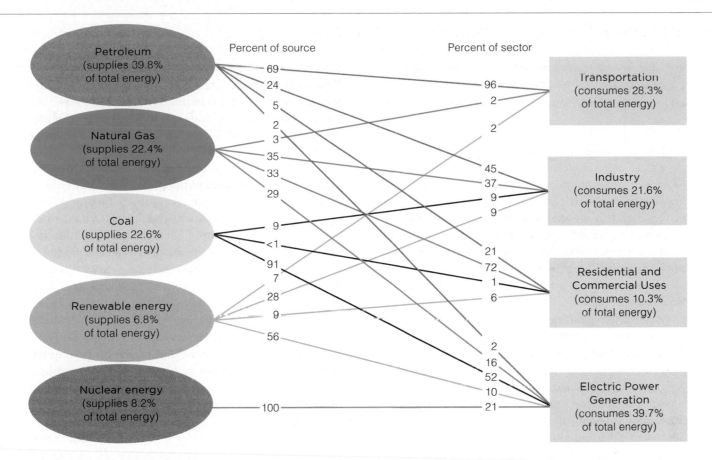

Percent of source   Percent of sector

Petroleum (supplies 39.8% of total energy)
69
24
5

Natural Gas (supplies 22.4% of total energy)
2
3
35
33
29

Coal (supplies 22.6% of total energy)
9
<1
91

Renewable energy (supplies 6.8% of total energy)
7
28
9
56

Nuclear energy (supplies 8.2% of total energy)
100

Transportation (consumes 28.3% of total energy)
96
2
2

Industry (consumes 21.6% of total energy)
45
37
9
9

Residential and Commercial Uses (consumes 10.3% of total energy)
21
72
1
6

Electric Power Generation (consumes 39.7% of total energy)
2
16
52
10
21

Figure 49.9 United States energy consumption in 2006 by source and sector of use. For example, 69 percent of the petroleum used in the United States goes to transportation, and petroleum supplies 96 percent of the transportation sector's energy needs.

Figure It Out: What percent of coal used in the United States goes to generate electricity? What percentage the country's electric power is derived from coal?

Answer: 91%, 52%

For example, the runoff from coal mines in Tennessee is poisoning endangered freshwater mussels that live in water downstream from the mines.

Even renewable energy sources can cause problems. For example, dams in rivers of the Pacific Northwest generate renewable hydroelectric power. However, the dams also prevent endangered salmon from returning to streams above the dam to breed. As these salmon populations have declined, so have endangered killer whales (orcas) that feed on adult salmon in the ocean. Such effects are not exactly what people have in mind when they think about "green energy."

The reality is that all commercially produced energy has some kind of negative environmental impact. The best way to minimize that impact is to use less energy.

Extracting and delivering materials that sustain the economies of developed countries also has environmental costs. Petroleum is used not only as fuel, but also as raw material for plastics, a topic we will return to later in this chapter. Surface mines extract essential minerals, such as the copper we use in computers and other electronic products, and the manganese used to

manufacture steel for building, cars, and appliances. Surface mining strips an area of vegetation and soil, creating an ecological dead zone. It puts dust into the air, creates mountains of rocky waste, and sometimes contaminates nearby waterways.

Where do raw materials in manufactured products you buy come from? With globalization, it's hard to know. Mines in developing countries operate under regulations that are usually less strict than those in the United States or less stringently enforced, so their environmental impact is even greater.

Take-Home Message

How do development and resource use affect biodiversity?

■ Expansion of cities and suburbs has a negative impact on biodiversity. Developed areas displace wild species and also harm them indirectly, as by introducing competing plants or causing light pollution.

■ Processes that extract or capture energy, both renewable and nonrenewable, can destroy or degrade habitat.

■ Obtaining the raw materials used in consumer products frequently involves degradation of the environment, which can reduce biodiversity.

■ Human activities have the potential not only to harm individual species, but also to transform entire biomes.

■ Links to Soil erosion 29.1, Transpiration 29.3

As human populations increase, greater numbers of people are forced to farm in areas that are ill-suited to agriculture. Others allow livestock to overgraze in grasslands. **Desertification**, a conversion of grassland or woodlands to desertlike conditions, is one result.

Deserts naturally expand and contract over time as climate conditions vary (Section 26.15). However, poor agricultural practices that encourage soil erosion can sometimes lead to rapid shifts from grassland or woodland to desert.

For example, during the mid-1930s, large portions of prairie on the southern Great Plains were plowed under to plant crops. This plowing exposed the deep prairie topsoil to the force of the region's constant winds. Coupled with a drought, the result was an economic and ecological disaster. Winds carried more than a billion tons of topsoil aloft as sky-darkening dust clouds turned the region into what came to be known as the Dust Bowl (Figure 49.10). Tons of displaced soil fell to earth as far away as New York City and Washington, D.C.

Desertification now threatens vast areas. In Africa, the Sahara Desert is expanding south into the Sahel region. Overgrazing in this region strips grasslands of their vegetation and allows winds to erode the soil. Winds carry the soil aloft and westward. Soil particles land as far away as the southern United States and the Caribbean (Figure 49.11). As described in the Chapter 24 introduction, pathogens traveling on dust particles threaten Caribbean corals.

In China's northwestern regions, overplowing and overgrazing has expanded the Gobi Desert so that dust clouds periodically darken skies above Beijing. Winds carry some of this dust across the Pacific to the United States. In an effort to hold back the desert, China has now planted billions of trees as a "Green Wall."

Drought encourages desertification, which results in more drought in a positive feedback cycle. Plants cannot thrive in a region where the topsoil has blown away. With less transpiration (Section 29.3), less water enters the atmosphere, so local rainfall decreases.

The best way to prevent desertification is to avoid farming in areas subject to high winds and periodic drought. If these areas must be utilized, methods that do not repeatedly disturb the soil can minimize the risk of desertification.

Take-Home Message

What is desertification?

■ Desertification turns productive grassland or woodland into a desertlike region in which little grows.

Figure 49.10 A giant dust cloud about to descend on a farm in Kansas during the 1930s. A large portion of the southern Great Plains was then known as the Dust Bowl. Drought and poor agricultural practices allowed winds to strip tons of topsoil from the ground and carry it aloft.

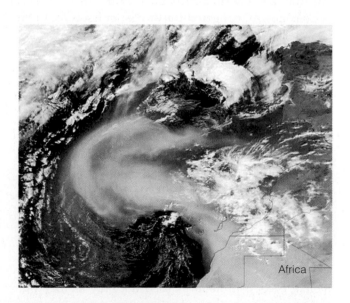

Africa

Figure 49.11 Modern-day dust cloud blowing from Africa's Sahara Desert out into the Atlantic ocean. The Sahara's area is increasing as a result of a long-term drought, overgrazing, and stripping of woodlands for firewood.

49.7 | The Trouble With Trash

■ Recycling saves limited resources, and it also keeps dangerous trash out of habitats where it can do harm.

■ Link to Groundwater and aquifers 47.6

Six billion people use and discard a lot of stuff. Where does all the waste go? Historically, unwanted material was simply buried in the ground or dumped out at sea. Trash was out of sight, and also out of mind. We now know that chemicals in buried trash can contaminate groundwater and aquifers. Waste dumped into the sea harms marine life. For example, seabirds eat floating bits of plastic and feed them to their chicks, with deadly results (Figure 49.12).

In 2006, the United States generated 251 million tons of garbage, which averages out to 2.1 kilograms (4.6 pounds) per person per day. By weight, about a third of that material was recycled, but there is plenty of room for improvement. Two-thirds of plastic soft drink bottles and three-quarters of glass bottles were not recycled. Nonrecycled trash now gets burned in high-temperature incinerators or placed in engineered landfills lined with material that minimizes the risk of groundwater contamination. No solid municipal waste can legally be dumped at sea.

Nevertheless, plastic and other garbage constantly enters our coastal waters. Foam cups and containers from fast-food outlets, plastic shopping bags, plastic water bottles, and other material discarded as litter ends up in storm drains. From there it is carried to streams and rivers, which can convey it to the sea. A seawater sample taken near the mouth of the San Gabriel River in southern California had 128 times as much plastic as plankton by weight.

Once in the ocean, trash can persist for a surprisingly long time. Components of a disposable diaper will last for more than 100 years, as will fishing line. A plastic bag will be around for more than 50 years, and a cigarette filter for more than 10.

To reduce the impact of plastic trash, choose more durable objects over disposable ones, and avoid buying plastic when other, less environmentally harmful alternatives exist. If you use plastic, be sure to recycle or dispose of it properly.

Take-Home Message

What are the ecological effects of trash?

■ Trash, especially plastics, often ends up in the oceans where it harms marine life.

■ You can minimize your environmental impact by avoiding disposable goods and by recycling.

Figure 49.12 (**a**) A recently deceased Laysan albatross chick, dissected to reveal the contents of its gut. (**b**) Scientists found more than 300 pieces of plastic inside the bird. One of the pieces had punctured its gut wall, resulting in its death. The chick was fed the plastic by its parents, who gathered the material from the ocean surface, mistaking it for food.

■ Managing biodiversity can sustain biological wealth, while also providing economic opportunity.

■ Link to Human population growth 45.7

Bioeconomic Considerations

Every nation enjoys three forms of wealth—material, cultural, and biological wealth. Its biological wealth— biodiversity—can be a source of food, medicine, and other products. However, protecting biological wealth is often a tricky proposition. Even in developed countries, people often oppose environmental protections because they fear such measures will have adverse economic consequences. However, taking care of the environment can make good economic sense. People can both preserve and profit from their biological wealth. We now turn to some success stories.

Sustainable Use of Biological Wealth

Using Genetic Diversity One particularly observant Mexican college student discovered *Zea diploperennis*, a wild maize long believed to be extinct. It had disappeared from most of its range, but a relic population clung to life in a 900-acre region of mountain terrain near Jalisco. Unlike domesticated corn, *Z. diploperennis* is perennial and resistant to most viruses. Gene transfers from this wild species into crop plants may boost production of corn for hungry people in Mexico and elsewhere. In recognition of the potential value of *Z. diploperennis* species, the Mexican government has set aside its mountainous habitat as a biological reserve, the first ever created to protect a wild relative of an important crop plant.

Discovering Useful Chemicals Many species make chemical compounds that could serve as medicines or other commercial products. Most developing countries do not have laboratories to test species for potential products, but large pharmaceutical companies in other countries do. The National Institute of Biodiversity of Costa Rica collects and identifies species that look promising, and sends away extracts of these species for chemical analyses. If a product of one of the species is marketed, Costa Rica will share in the profits, which are earmarked for conservation programs.

Ecotourism Setting aside species-rich preserves and encouraging tourists to visit them can have biological and economic benefits. For example, during the 1970s, George Powell was studying birds in the Monteverde Cloud Forest in Costa Rica. This forest was rapidly being cleared and Powell got the idea to buy part of it as a nature sanctuary. His efforts inspired individuals and conservation groups to donate funds, and much of the forest is now protected as a private nature reserve.

Figure 49.13 Strip logging. The practice may protect biodiversity as it permits logging on tropical slopes. A narrow corridor paralleling the land's contours is cleared. A roadbed is made at the top to haul away logs. After a few years, saplings grow in the cleared corridor. Another corridor is cleared above the roadbed. Nutrients leached from exposed soil trickle into the first corridor. There they are taken up by saplings, which benefit from all the nutrient input by growing faster. Later, a third corridor is cut above the second one—and so on in a profitable cycle of logging, which the habitat sustains over time.

uncut forest

cut 1 year ago

dirt road

cut 3–5 years ago

cut 6–10 years ago

uncut forest

stream in watershed

The reserve's plants and animals include more than 100 mammalian species, 400 bird species, and 120 species of amphibians and reptiles. The reserve is one of the few habitats left for the jaguar, ocelot, puma, and their relatives.

More than 50,000 tourists now visit the Monteverde Cloud Forest Reserve each year. Ecotourism centered on this reserve provides employment to local people and has other beneficial effects. For example, a nonprofit school set up inside the reserve helps educate the area's children.

Sustainable Logging A tropical forest yields wood for local needs and for export to developed countries. However, severe erosion often follows the logging of forested slopes. Gary Hartshorn devised a method to minimize erosion. As explained in Figure 49.13, strip logging allows cycles of logging. It yields sustainable economic benefits for local loggers, while minimizing effects of erosion and maximizing forest biodiversity.

Responsible Ranching Developed countries are also implementing conservation practices that sustain biological wealth. For example, **riparian zones** are narrow corridors of vegetation along a stream or river. They are of great ecological importance. Plants in a riparian zone act as a line of defense against flood damage by sponging up water during spring runoffs and summer storms. Shade cast by a canopy of taller shrubs and trees in a riparian zone helps conserve water during droughts. A riparian zone provides wildlife with food, shelter, and shade, particularly in arid and semiarid regions. In the western United States, 67 to 75 percent of the endemic species spend all or part of their life cycle in riparian zones. Among them are 136 kinds of songbirds, some of which will nest only in the plants of a riparian zone.

In ranch country, cattle tend to congregate near rivers and streams, which are often their only source of water. There, the animals trample and feed until grasses and herbaceous shrubs are gone. It takes only a few head of cattle to destroy a riparian zone. All but 10 percent of the riparian vegetation of Arizona and New Mexico is already gone, mainly into the stomachs of grazing cattle.

To preserve the biodiversity in riparian zones, cattle on some ranches are now kept away from riverbanks and provided with an alternate water source. Figure 49.14 shows how excluding cattle from the riparian zone can make a difference. Once cattle are excluded, native vegetation regrows fast and the biodiversity of the habitat is restored.

Figure 49.14 Riparian zone restoration. The photos show Arizona's San Pedro River, before restoration (*above*) and after (*below*).

These examples show how we can put our knowledge of biological principles to use. The health of our planet depends on our ability to recognize that the principles of energy flow and of resource limitation, which govern the survival of all systems of life, do not change. It is our biological and cultural imperative to come to terms with these principles, and to ask ourselves this: What will be our long-term effect on the world of life?

Take-Home Message

How do we meet human needs and sustain biodiversity?

■ Any nation has biological wealth, which people will tend to protect if they recognize its value.

■ Sustainable practices allow people to benefit economically from biological resources without destroying them.

Global warming is thawing ice in the Arctic and opening access to this continent, which was previously protected from development by the lack of shipping lanes. Eight countries, including the United States, Canada, and Russia, control parts of the Arctic and have rights to its oil, gas, and mineral deposits. Conservationists worry that exploitation of these resources will put more pressure on arctic species already vulnerable to extinction.

How would you vote?

Should conservation of the Arctic receive priority over exploitation of its resources? See CengageNOW for details, then vote online.

Summary

Section 49.1 The current rate of species loss is high enough to suggest that an extinction crisis is under way. After other mass extinctions, it has taken millions of years for biodiversity to recover to its previous level.

Human-caused extinctions may have begun when humans first entered the Americas and Australia. Many more-recent extinctions definitely resulted from human activity. **Endangered species** currently face a high risk of extinction. **Threatened species** are likely to become endangered in the future.

■ *Use the animation on CengageNOW to view extinction patterns for different taxa.*

Section 49.2 **Endemic species**, which evolved in one place and are present only in that habitat, are highly vulnerable to extinction. Species with highly specialized resource needs are also especially vulnerable.

Humans cause habitat loss, degradation of habitat, and habitat fragmentation, all of which can endanger a species. Humans also directly reduce populations and endanger species by overharvesting. Species introductions also cause declines of native species. In most cases, a species becomes endangered because of multiple factors. Sometimes, a decline in one species as a result of human activity leads to decline of another species.

Sections 49.3, 49.4 Our knowledge of existing species is limited and biased toward vertebrates. We know little about the abundance and diversity of microbial species that carry out essential ecosystem processes.

We recognize three levels of **biodiversity**: genetic diversity, species diversity, and ecosystem diversity. All are threatened. The field of conservation biology surveys the range of biodiversity, investigates its origins, and identifies ways to maintain and use it in ways that benefit human populations.

An **indicator species** is one that is especially sensitive to environmental change and can be monitored to determine the health of an ecosystem.

Given that resources are limited, biologists attempt to identify **hot spots**, regions rich in endemic species and under a high level of threat. Biologists also identify **ecoregions**, larger regions characterized by their physical characteristics as well as the species in them.

The biologists prioritize ecoregions, with the goal of identifying those whose conservation will ensure that a representative sample of all of Earth's current biomes remains intact. The United States holds part or all of several ecoregions that are considered critical or endangered by international conservation organizations.

Section 49.5 Growth of cities and suburbs displaces wild species. Proximity to humans can also put stress on some species, as by the effects of introduced competitors or predators, or by ill effects of nighttime lighting.

People in developed countries contribute to species extinctions by their pattern of resource consumption. Their use of large amounts of fossil fuels contributes to global warming and also has other adverse effects on the environment. Renewable energy sources can also degrade habitat. Extracting fuel and mineral resources requires mining and other processes that pollute and otherwise make habitat unsuitable for native organisms.

Section 49.6 **Desertification** is the conversion of grassland or woodland to desertlike conditions. In the 1930s, a drought and poor agricultural practices caused desertification of a portion of the Great Plains, which became known as the Dust Bowl.

Desertification is currently a problem in China and in Africa. Some effects of desertification are felt far from the problem site because winds can pick up soil and carry it for long distances.

Section 49.7 Production of large amounts of trash is another threat to biodiversity. Plastic that enters oceans is particularly harmful and persistent.

Section 49.8 All nations have biological wealth; they have unique species that are of value to humans. People tend to preserve biological wealth when they recognize and benefit economically from its existence.

Preserving areas and using them for ecotourism can benefit local people while protecting endangered species.

Resources can also be harvested sustainably, as by strip logging of tropical forests on mountain slopes. This method minimizes erosion and ensures that there is always tree cover.

Developed nations also benefit by using their biological wealth in a sustainable fashion. For example, cattle ranching can have adverse effects on riparian zones, which are areas of high species diversity on river banks. Responsible ranchers exclude cattle from riparian zones to sustain biodiversity.

Data Analysis Exercise

Winds carry chemical contaminants produced and released at temperate latitudes to the Arctic, where the chemicals enter food webs. By the process of biological magnification (Section 47.4), top carnivores in arctic food webs—such as polar bears and people—end up with high doses of these chemicals. For example, indigenous arctic people who eat a lot of local wildlife tend to have unusually high levels of polychlorinated biphenyls, or PCBs, in their bodies. The Arctic Monitoring and Assessment Programme studies the effects of these chemicals on health and reproduction. Figure 49.15 shows the effect of PCBs on the sex ratio at birth in indigenous populations in the Russian Arctic.

1. Which sex is was most common in offspring of women with less than one microgram per millileter of PCB in serum?

2. At what PCB concentrations were women more likely to have daughters?

3. In some Greenland villages, nearly all recent newborns are female. Would you expect PCB levels in those villages to be above or under 4 micrograms per milliliter?

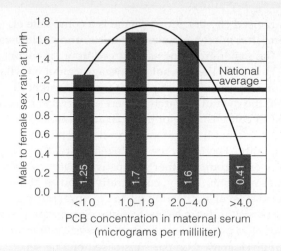

Figure 49.15 Effect of maternal PCB concentration on sex ratio of newborns in indigenous populations in the Russian Arctic. The red line indicates the average sex ratio for births in Russia—1.06 males per female.

Self-Quiz

Answers in Appendix III

1. True or false? Most species that evolved have already become extinct.

2. Dodos were driven to extinction _____ .
   a. when humans arrived in North America
   b. by overharvesting and introduced species
   c. as a result of global warming
   d. both a and b

3. An _____ species has population levels so low it is at great risk of extinction in the near future.
   a. endemic   c. indicator
   b. endangered   d. exotic

4. Gene flow among populations is hampered by _____ .
   a. habitat fragmentation   c. poaching
   b. species introductions   d. all of the above

5. _____ , native to the United States, have now been driven to extinction.
   a. Dodos   c. Pandas
   b. Passenger pigeons   d. Bison (buffalo)

6. Which of the following has the most representatives among the known endangered species?
   a. bacteria   c. vertebrates
   b. fungi   d. invertebrates

7. An _____ species can be monitored to gauge the health of its environment.
   a. endemic   c. indicator
   b. endangered   d. exotic

8. A(n) _____ is an area that conservation biologists consider a high priority for preservation.
   a. hot spot   c. biome
   b. ecoregion   d. biogeographic province

9. True or false? Artificial lighting harms some species.

10. Dams built to provide renewable hydroelectric power have caused declines in populations of _____ .
   a. salmon   c. sea turtles
   b. killer whales   d. both a and b

11. In the United States most plastic enters oceans by way of _____ .
   a. littering   c. offshore drilling
   b. careless boaters d. municipal dumping

12. Match the organisms with their descriptions.
   ___ riparian zone   a. evolved and found in one area
   ___ hot spot   b. cause of some dust storms
   ___ endemic species   c. locals benefit from visitors
   ___ indicator species   d. many species, under threat
   ___ desertification   e. less erosion, sustains forest
   ___ ecotourism   f. species-rich area near river
   ___ strip logging   g. highly sensitive to changes
   ___ conservation   h. assesses and seeks ways to
       biology   preserve biodiversity

■ *Visit CengageNOW for additional questions.*

Critical Thinking

1. Many biologists think that global climate change resulting from greenhouse gas emissions is the single greatest threat to existing biodiversity. List some negative effects that climate change could have on native species in your area.

2. In one seaside community in New Jersey, the U.S. Fish and Wildlife Service suggested removing feral cats (domestic cats that live in the wild) in order to protect some endangered wild birds (plovers) that nested on the town's beaches. Many residents were angered by the proposal, arguing that the cats have just as much right to exist as the birds. Do you agree?

Appendix I. Classification System

This revised classification scheme is a composite of several that microbiologists, botanists, and zoologists use. The major groupings are agreed upon, more or less. However, there is not always agreement on what to name a particular grouping or where it might fit within the overall hierarchy. There are several reasons why full consensus is not possible at this time.

First, the fossil record varies in its completeness and quality. Therefore, the phylogenetic relationship of one group to other groups is sometimes open to interpretation. Today, comparative studies at the molecular level are firming up the picture, but the work is still under way. Also, molecular comparisons do not always provide definitive answers to questions about phylogeny. Comparisons based on one set of genes may conflict with those comparing a different part of the genome. Or comparisons with one member of a group may conflict with comparisons based on other group members.

Second, ever since the time of Linnaeus, systems of classification have been based on the perceived morphological similarities and differences among organisms. Although some original interpretations are now open to question, we are so used to thinking about organisms in certain ways that reclassification often proceeds slowly.

A few examples: Traditionally, birds and reptiles were grouped in separate classes (Reptilia and Aves); yet there are compelling arguments for grouping the lizards and snakes in one group and the crocodilians, dinosaurs, and birds in another. Many biologists still favor a six-kingdom system of classification (archaea, bacteria, protists, plants, fungi, and animals). Others advocate a switch to the more recently proposed three-domain system (archaea, bacteria, and eukarya).

Third, researchers in microbiology, mycology, botany, zoology, and other fields of inquiry inherited a wealth of literature, based on classification systems that have been developed over time in each field of inquiry. Many are reluctant to give up established terminology that offers access to the past.

For example, botanists and microbiologists often use *division*, and zoologists *phylum*, for taxa that are equivalent in hierarchies of classification.

Why bother with classification frameworks if we know they only imperfectly reflect the evolutionary history of life? We do so for the same reasons that a writer might break up a history of civilization into several volumes, each with a number of chapters. Both are efforts to impart structure to an enormous body of knowledge and to facilitate retrieval of information from it. More importantly, to the extent that modern classification schemes accurately reflect evolutionary relationships, they provide the basis for comparative biological studies, which link all fields of biology.

Bear in mind that we include this appendix for your reference purposes only. Besides being open to revision, it is not meant to be complete. Names shown in "quotes" are polyphyletic or paraphyletic groups that are undergoing revision. For example, "reptiles" comprise at least three and possibly more lineages.

The most recently discovered species, as from the mid-ocean province, are not listed. Many existing and extinct species of the more obscure phyla are also not represented. Our strategy is to focus primarily on the organisms mentioned in the text or familiar to most students. We delve more deeply into flowering plants than into bryophytes, and into chordates than annelids.

PROKARYOTES AND EUKARYOTES COMPARED

As a general frame of reference, note that almost all bacteria and archaea are microscopic in size. Their DNA is concentrated in a nucleoid (a region of cytoplasm), not in a membrane-bound nucleus. All are single cells or simple associations of cells. They reproduce by prokaryotic fission or budding; they transfer genes by bacterial conjugation.

Table A lists representative types of autotrophic and heterotrophic prokaryotes. The authoritative reference, *Bergey's Manual of Systematic Bacteriology*, has called this a time of taxonomic transition. It references groups mainly by numerical taxonomy (Section 19.1) rather than by phylogeny. Our classification system does reflect evidence of evolutionary relationships for at least some bacterial groups.

The first life forms were prokaryotic. Similarities between Bacteria and Archaea have more ancient origins relative to the traits of eukaryotes.

Unlike the prokaryotes, all eukaryotic cells start out life with a DNA-enclosing nucleus and other membrane-bound organelles. Their chromosomes have many histones and other proteins attached. They include spectacularly diverse single-celled and multicelled species, which can reproduce by way of meiosis, mitosis, or both.

DOMAIN BACTERIA	DOMAIN ARCHAEA	DOMAIN EUKARYA				
Kingdom Bacteria	Kingdom Archaea	Kingdom Protista	Kingdom Fungi	Kingdom Plantae	Kingdom Animalia

DOMAIN OF BACTERIA

KINGDOM BACTERIA   The largest, and most diverse group of prokaryotic cells. Includes photosynthetic autotrophs, chemosynthetic autotrophs, and heterotrophs. All prokaryotic pathogens of vertebrates are bacteria.

PHYLUM AQIFACAE   Most ancient branch of the bacterial tree. Gram-negative, mostly aerobic chemoautotrophs, mainly of volcanic hot springs. *Aquifex*.

PHYLUM DEINOCOCCUS-THERMUS   Gram-positive, heat-loving chemoautotrophs. *Deinococcus* is the most radiation resistant organism known. *Thermus* occurs in hot springs and near hydrothermal vents.

PHYLUM CHLOROFLEXI   Green nonsulfur bacteria. Gram-negative bacteria of hot springs, freshwater lakes, and marine habitats. Act as nonoxygen-producing photoautotrophs or aerobic chemoheterotrophs. *Chloroflexus*.

PHYLUM ACTINOBACTERIA   Gram-positive, mostly aerobic heterotrophs in soil, freshwater and marine habitats, and on mammalian skin. *Propionibacterium, Actinomyces, Streptomyces*.

PHYLUM CYANOBACTERIA   Gram-negative, oxygen-releasing photoautotrophs mainly in aquatic habitats. They have chlorophyll *a* and photosystem I. Includes many nitrogen-fixing genera. *Anabaena, Nostoc, Oscillatoria*.

PHYLUM CHLOROBIUM   Green sulfur bacteria. Gram-negative nonoxygen-producing photosynthesizers, mainly in freshwater sediments. *Chlorobium*.

PHYLUM FIRMICUTES   Gram-positive walled cells and the cell wall-less mycoplasmas. All are heterotrophs. Some survive in soil, hot springs, lakes, or oceans. Others live on or in animals. *Bacillus, Clostridium, Heliobacterium, Lactobacillus, Listeria, Mycobacterium, Mycoplasma, Streptococcus*.

PHYLUM CHLAMYDIAE   Gram-negative intracellular parasites of birds and mammals. *Chlamydia*.

PHYLUM SPIROCHETES   Free-living, parasitic, and mutualistic gram-negative spring-shaped bacteria. *Borelia, Pillotina, Spirillum, Treponema*.

PHYLUM PROTEOBACTERIA   The largest bacterial group. Includes photoautotrophs, chemoautotrophs, and heterotrophs, free-living, parasitic, and colonial groups. All are gram-negative.

Class Alphaproteobacteria. *Agrobacterium, Azospirillum, Nitrobacter, Rickettsia, Rhizobium*.

Class Betaproteobacteria. *Neisseria*.

Class Gammaproteobacteria. *Chromatium, Escherichia, Haemopilius, Pseudomonas, Salmonella, Shigella, Thiomargarita, Vibrio, Yersinia*.

Class Deltaproteobacteria. *Azotobacter, Myxococcus*.

Class Epsilonproteobacteria. *Campylobacter, Helicobacter*.

DOMAIN OF ARCHAEA

KINGDOM ARCHAEA   Prokaryotes that are evolutionarily between eukaryotic cells and the bacteria. Most are anaerobes. None are photosynthetic. Originally discovered in extreme habitats, they are now known to be widely dispersed. Compared with bacteria, the archaea have a distinctive cell wall structure and unique membrane lipids, ribosomes, and RNA sequences. Some are symbiotic with animals, but none are known to be animal pathogens.

PHYLUM EURYARCHAEOTA   Largest archean group. Includes extreme thermophiles, halophiles, and methanogens. Others are abundant in the upper waters of the ocean and other more moderate habitats. *Methanocaldococcus, Nanoarchaeum*.

PHYLUM CRENARCHAEOTA   Includes extreme theromophiles, as well as species that survive in Antarctic waters, and in more moderate habitats. *Sulfolobus, Ignicoccus*.

PHYLUM KORARCHAEOTA   Known only from DNA isolated from hydrothermal pools. As of this writing, none have been cultured and no species have been named.

DOMAIN OF EUKARYOTES

KINGDOM "PROTISTA"   A collection of single-celled and multicelled lineages, which does not constitute a monophyletic group. Some biologists consider the groups listed below to be kingdoms in their own right.

PARABASALIA   Parabasalids. Flagellated, single-celled anaerobic heterotrophs with a cytoskeletal "backbone" that runs the length of the cell. There are no mitochondria, but a hydrogenosome serves a similar function. *Trichomonas, Trichonympha*.

DIPLOMONADIDA   Diplomonads. Flagellated, anaerobic single-celled heterotrophs that do not have mitochondria or Golgi bodies and do not form a bipolar spindle at mitosis. May be one of the most ancient lineages. *Giardia*.

EUGLENOZOA   Euglenoids and kinetoplastids. Free-living and parasitic flagellates. All with one or more mitochondria. Some photosynthetic euglenoids with chloroplasts, others heterotrophic. *Euglena, Trypanosoma, Leishmania*.

RHIZARIA   Formaminiferans and radiolarians. Free-living, heterotrophic amoeboid cells that are enclosed in shells. Most live in ocean waters or sediments. *Pterocorys, Stylosphaera*.

ALVEOLATA   Single cells having a unique array of membrane-bound sacs (alveoli) just beneath the plasma membrane.

Ciliata. Ciliated protozoans. Heterotrophic protists with many cilia. *Paramecium, Didinium*.

Dinoflagellates. Diverse heterotrophic and photosynthetic flagellated cells that deposit cellulose in their alveoli. *Gonyaulax, Gymnodinium, Karenia, Noctiluca*.

Apicomplexans. Single-celled parasites of animals. A unique microtubular device is used to attach to and penetrate a host cell. *Plasmodium*.

STRAMENOPHILA   Stramenophiles. Single-celled and multicelled forms; flagella with tinsel-like filaments.

Oomycotes. Water molds. Heterotrophs. Decomposers, some parasites. *Saprolegnia, Phytophthora, Plasmopara*.

Chrysophytes. Golden algae, yellow-green algae, diatoms, coccolithophores. Photosynthetic. *Emiliania, Mischococcus*.

Phaeophytes. Brown algae. Photosynthetic; nearly all live in temperate marine waters. All are multicellular. *Macrocystis, Laminaria, Sargassum, Postelsia*.

RHODOPHYTA   Red algae. Mostly photosynthetic, some parasitic. Nearly all marine, some in freshwater habitats. Most multicellular. *Porphyra, Antithamion*.

CHLOROPHYTA   Green algae. Mostly photosynthetic, some parasitic. Most freshwater, some marine or terrestrial. Single-celled, colonial, and multicellular forms. Some biologists place the chlorophytes and charophytes with the land plants in a kingdom called the Viridiplantae. *Acetabularia, Chlamydomonas, Chlorella, Codium, Udotea, Ulva, Volvox*.

CHAROPHYTA   Photosynthetic. Closest living relatives of plants. Include both single-celled and multicelled forms. Desmids, stoneworts. *Micrasterias, Chara, Spirogyra*.

AMOEBOZOA   True amoebas and slime molds. Heterotrophs that spend all or part of the life cycle as a single cell that uses pseudopods to capture food. *Amoeba, Entoamoeba* (amoebas), *Dictyostelium* (cellular slime mold), *Physarum* (plasmodial slime mold).

Nearly all multicelled eukaryotic species with chitin-containing cell walls. Heterotrophs, mostly saprobic decomposers, some parasites. Nutrition based upon extracellular digestion of organic matter and absorption of nutrients by individual cells. Multicelled species form absorptive mycelia and reproductive structures that produce asexual spores (and sometimes sexual spores).

PHYLUM CHYTRIDIOMYCOTA Chytrids. Primarily aquatic; saprobic decomposers or parasites that produce flagellated spores. *Chytridium.*

PHYLUM ZYGOMYCOTA Zygomycetes. Producers of zygospores (zygotes inside thick wall) by way of sexual reproduction. Bread molds, related forms. *Rhizopus, Philobolus.*

PHYLUM ASCOMYCOTA Ascomycetes. Sac fungi. Sac-shaped cells form sexual spores (ascospores). Most yeasts and molds, morels, truffles. *Saccharomycetes, Morchella, Neurospora, Claviceps, Candida, Aspergillus, Penicillium.*

PHYLUM BASIDIOMYCOTA Basidiomycetes. Club fungi. Most diverse group. Produce basidiospores inside club-shaped structures. Mushrooms, shelf fungi, stinkhorns. *Agaricus, Amanita, Craterellus, Gymnophilus, Puccinia, Ustilago.*

"IMPERFECT FUNGI" Sexual spores absent or undetected. The group has no formal taxonomic status. If better understood, a given species might be grouped with sac fungi or club fungi. *Arthobotrys, Histoplasma, Microsporum, Verticillium.*

"LICHENS" Mutualistic interactions between fungal species and a cyanobacterium, green alga, or both. *Lobaria, Usnea.*

KINGDOM PLANTAE Most photosynthetic with chlorophylls *a* and *b*. Some parasitic. Nearly all live on land. Sexual reproduction predominates.

BRYOPHYTES (NONVASCULAR PLANTS)

Small flattened haploid gametophyte dominates the life cycle; sporophyte remains attached to it. Sperm are flagellated; require water to swim to eggs for fertilization.

PHYLUM HEPATOPHYTA Liverworts. *Marchantia.*

PHYLUM ANTHOCEROPHYTA Hornworts.

PHYLUM BRYOPHYTA Mosses. *Polytrichum, Sphagnum.*

SEEDLESS VASCULAR PLANTS

Diploid sporophyte dominates, free-living gametophytes, flagellated sperm require water for fertilization.

PHYLUM LYCOPHYTA Lycophytes, club mosses. Small single-veined leaves, branching rhizomes. *Lycopodium, Selaginella.*

PHYLUM MONILOPHYTA

Subphylum Psilophyta. Whisk ferns. No obvious roots or leaves on sporophyte, very reduced. *Psilotum.*

Subphylum Sphenophyta. Horsetails. Reduced scalelike leaves. Some stems photosynthetic, others spore-producing. *Calamites* (extinct), *Equisetum.*

Subphylum Pterophyta. Ferns. Large leaves, usually with sori. Largest group of seedless vascular plants (12,000 species), mainly tropical, temperate habitats. *Pteris, Trichomanes, Cyathea* (tree ferns), *Polystichum.*

SEED-BEARING VASCULAR PLANTS

PHYLUM CYCADOPHYTA Cycads. Group of gymnosperms (vascular, bear "naked" seeds). Tropical, subtropical. Compound leaves, simple cones on male and female plants. Plants usually palm-like. Motile sperm. *Zamia, Cycas.*

PHYLUM GINKGOPHYTA Ginkgo (maidenhair tree). Type of gymnosperm. Motile sperm. Seeds with fleshy layer. *Ginkgo.*

PHYLUM GNETOPHYTA Gnetophytes. Only gymnosperms with vessels in xylem and double fertilization (but endosperm does not form). *Ephedra, Welwitchia, Gnetum.*

PHYLUM CONIFEROPHYTA Conifers. Most common and familiar gymnosperms. Generally cone-bearing species with needle-like or scale-like leaves. Includes pines (*Pinus*), redwoods (*Sequoia*), yews (*Taxus*).

PHYLUM ANTHOPHYTA Angiosperms (the flowering plants). Largest, most diverse group of vascular seed-bearing plants. Only organisms that produce flowers, fruits. Some families from several representative orders are listed:

BASAL FAMILIES

Family Amborellaceae. *Amborella.*
Family Nymphaeaceae. Water lilies.
Family Illiciaceae. Star anise.

MAGNOLIIDS

Family Magnoliaceae. Magnolias.
Family Lauraceae. Cinnamon, sassafras, avocados.
Family Piperaceae. Black pepper, white pepper.

EUDICOTS

Family Papaveraceae. Poppies.
Family Cactaceae. Cacti.
Family Euphorbiaceae. Spurges, poinsettia.
Family Salicaceae. Willows, poplars.
Family Fabaceae. Peas, beans, lupines, mesquite.
Family Rosaceae. Roses, apples, almonds, strawberries.
Family Moraceae. Figs, mulberries.
Family Cucurbitaceae. Squashes, melons, cucumbers.
Family Fagaceae. Oaks, chestnuts, beeches.
Family Brassicaceae. Mustards, cabbages, radishes.
Family Malvaceae. Mallows, okra, cotton, hibiscus, cocoa.
Family Sapindaceae. Soapberry, litchi, maples.
Family Ericaceae. Heaths, blueberries, azaleas.
Family Rubiaceae. Coffee.
Family Lamiaceae. Mints.
Family Solanaceae. Potatoes, eggplant, petunias.
Family Apiaceae. Parsleys, carrots, poison hemlock.
Family Asteraceae. Composites. Chrysanthemums, sunflowers, lettuces, dandelions.

MONOCOTS

Family Araceae. Anthuriums, calla lily, philodendrons.
Family Liliaceae. Lilies, tulips.
Family Alliaceae. Onions, garlic.
Family Iridaceae. Irises, gladioli, crocuses.
Family Orchidaceae. Orchids.
Family Arecaceae. Date palms, coconut palms.
Family Bromeliaceae. Bromeliads, pineapples.
Family Cyperaceae. Sedges.
Family Poaceae. Grasses, bamboos, corn, wheat, sugarcane.
Family Zingiberaceae. Gingers.

KINGDOM ANIMALIA Multicelled heterotrophs, nearly all with tissues and organs, and organ systems, that are motile during part of the life cycle. Sexual reproduction occurs in most, but some also reproduce asexually. Embryos develop through a series of stages.

PHYLUM PORIFERA Sponges. No symmetry, tissues.

PHYLUM PLACOZOA Marine. Simplest known animal. Two cell layers, no mouth, no organs. *Trichoplax.*

PHYLUM CNIDARIA Radial symmetry, tissues, nematocysts.
Class Hydrozoa. Hydrozoans. *Hydra, Obelia, Physalia, Prya.*
Class Scyphozoa. Jellyfishes. *Aurelia.*
Class Anthozoa. Sea anemones, corals. *Telesto.*

PHYLUM PLATYHELMINTHES Flatworms. Bilateral, cephalized; simplest animals with organ systems. Saclike gut.

  Class Turbellaria. Triclads (planarians), polyclads. *Dugesia*.
  Class Trematoda. Flukes. *Clonorchis, Schistosoma*.
  Class Cestoda. Tapeworms. *Diphyllobothrium, Taenia*.

PHYLUM ROTIFERA Rotifers. *Asplancha, Philodina*.

PHYLUM MOLLUSCA Mollusks.

  Class Polyplacophora. Chitons. *Cryptochiton, Tonicella*.

  Class Gastropoda. Snails, sea slugs, land slugs. *Aplysia, Ariolimax, Cypraea, Haliotis, Helix, Liguus, Limax, Littorina*.

  Class Bivalvia. Clams, mussels, scallops, cockles, oysters, shipworms. *Ensis, Chlamys, Mytelus, Patinopectin*.

  Class Cephalopoda. Squids, octopuses, cuttlefish, nautiluses. *Dosidiscus, Loligo, Nautilus, Octopus, Sepia*.

PHYLUM ANNELIDA Segmented worms.

  Class Polychaeta. Mostly marine worms. *Eunice, Neanthes*.

  Class Oligochaeta. Mostly freshwater and terrestrial worms, many marine. *Lumbricus* (earthworms), *Tubifex*.

  Class Hirudinea. Leeches. *Hirudo, Placobdella*.

PHYLUM NEMATODA Roundworms. *Ascaris, Caenorhabditis elegans, Necator* (hookworms), *Trichinella*.

PHYLUM ARTHROPODA

  Subphylum Chelicerata. Chelicerates. Horseshoe crabs, spiders, scorpions, ticks, mites.

  Subphylum Crustacea. Shrimps, crayfishes, lobsters, crabs, barnacles, copepods, isopods (sowbugs).

  Subphylum Myriapoda. Centipedes, millipedes.

  Subphylum Hexapoda. Insects and sprintails.

PHYLUM ECHINODERMATA Echinoderms.

  Class Asteroidea. Sea stars. *Asterias*.

  Class Ophiuroidea. Brittle stars.

  Class Echinoidea. Sea urchins, heart urchins, sand dollars.

  Class Holothuroidea. Sea cucumbers.

  Class Crinoidea. Feather stars, sea lilies.

  Class Concentricycloidea. Sea daisies.

PHYLUM CHORDATA Chordates.

  Subphylum Urochordata. Tunicates, related forms.

  Subphylum Cephalochordata. Lancelets.

CRANIATES

  Class Myxini. Hagfishes.

VERTEBRATES (SUBGROUP OF CRANIATES)

  Class Cephalaspidomorphi. Lampreys.

  Class Chondrichthyes. Cartilaginous fishes (sharks, rays, skates, chimaeras).

  Class "Osteichthyes." Bony fishes. Not monophyletic (sturgeons, paddlefish, herrings, carps, cods, trout, seahorses, tunas, lungfishes, and coelocanths).

TETRAPODS (SUBGROUP OF VERTEBRATES)

  Class Amphibia. Amphibians. Require water to reproduce.
    Order Caudata. Salamanders and newts.
    Order Anura. Frogs, toads.
    Order Apoda. Apodans (caecilians).

AMNIOTES (SUBGROUP OF TETRAPODS)

  Class "Reptilia." Skin with scales, embryo protected and nutritionally supported by extraembryonic membranes.

  Subclass Anapsida. Turtles, tortoises.

  Subclass Lepidosaura. *Sphenodon*, lizards, snakes.

  Subclass Archosaura. Crocodiles, alligators.

Class Aves. Birds. In some classifications birds are grouped in the archosaurs.

  Order Struthioniformes. Ostriches.

  Order Sphenisciformes. Penguins.

  Order Procellariiformes. Albatrosses, petrels.

  Order Ciconiiformes. Herons, bitterns, storks, flamingoes.

  Order Anseriformes. Swans, geese, ducks.

  Order Falconiformes. Eagles, hawks, vultures, falcons.

  Order Galliformes. Ptarmigan, turkeys, domestic fowl.

  Order Columbiformes. Pigeons, doves.

  Order Strigiformes. Owls.

  Order Apodiformes. Swifts, hummingbirds.

  Order Passeriformes. Sparrows, jays, finches, crows, robins, starlings, wrens.

  Order Piciformes. Woodpeckers, toucans.

  Order Psittaciformes. Parrots, cockatoos, macaws.

Class Mammalia. Skin with hair; young nourished by milk-secreting mammary glands of adult.

  Subclass Prototheria. Egg-laying mammals (monotremes; duckbilled platypus, spiny anteaters).

  Subclass Metatheria. Pouched mammals or marsupials (opossums, kangaroos, wombats, Tasmanian devils).

  Subclass Eutheria. Placental mammals.

    Order Edentata. Anteaters, tree sloths, armadillos.

    Order Insectivora. Tree shrews, moles, hedgehogs.

    Order Chiroptera. Bats.

    Order Scandentia. Insectivorous tree shrews.

    Order Primates.

    Suborder Strepsirhini (prosimians). Lemurs, lorises.

    Suborder Haplorhini (tarsioids and anthropoids).

      Infraorder Tarsiiformes. Tarsiers.

      Infraorder Platyrrhini (New World monkeys).

        Family Cebidae. Spider monkeys, howler monkeys, capuchin.

      Infraorder Catarrhini (Old World monkeys and hominoids).

        Superfamily Cercopithecoidea. Baboons, macaques, langurs.

        Superfamily Hominoidea. Apes and humans.
          Family Hylobatidae. Gibbon.
          Family "Pongidae." Chimpanzees, gorillas, orangutans.
          Family Hominidae. Existing and extinct human species (*Homo*) and humanlike species, including the australopiths.

    Order Lagomorpha. Rabbits, hares, pikas.

    Order Rodentia. Most gnawing animals (squirrels, rats, mice, guinea pigs, porcupines, beavers, etc.).

    Order Carnivora. Carnivores (wolves, cats, bears, etc.).

    Order Pinnipedia. Seals, walruses, sea lions.

    Order Proboscidea. Elephants, mammoths (extinct).

    Order Sirenia. Sea cows (manatees, dugongs).

    Order Perissodactyla. Odd-toed ungulates (horses, tapirs, rhinos).

    Order Tubulidentata. African aardvarks.

    Order Artiodactyla. Even-toed ungulates (camels, deer, bison, sheep, goats, antelopes, giraffes, etc.).

    Order Cetacea. Whales, porpoises.

Appendix II. Annotations to A Journal Article

This journal article reports on the movements of a female wolf during the summer of 2002 in northwestern Canada. It also reports on a scientific process of inquiry, observation and interpretation to learn where, how and why the wolf traveled as she did. In some ways, this article reflects the story of "how to do science" told in section 1.5 of this textbook. These notes are intended to help you read and understand how scientists work and how they report on their work.

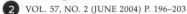

(1) ARCTIC

(2) VOL. 57, NO. 2 (JUNE 2004) P. 196–203

(3) Long Foraging Movement of a Denning Tundra Wolf

(4) Paul F. Frame,[1,2] David S. Hik,[1] H. Dean Cluff,[3] and Paul C. Paquet[4]

(5) *(Received 3 September 2003; accepted in revised form 16 January 2004)*

(6) **ABSTRACT** Wolves (*Canis lupus*) on the Canadian barrens are intimately linked to migrating herds of barren-ground caribou (*Rangifer tarandus*). We deployed a Global Positioning System (GPS) radio collar on an adult female wolf to record her movements in response to changing caribou densities near her den during summer. This wolf and two other females were observed nursing a group of 11 pups. She traveled a minimum of 341 km during a 14-day excursion. The straight-line distance from the den to the farthest location was 103 km, and the overall minimum rate of travel was 3.1 km/h. The distance between the wolf and the radio-collared caribou decreased from 242 km one week before the excursion to 8 km four days into the excursion. We discuss several possible explanations for the long foraging bout.

(7) *Key words:* wolf, GPS tracking, movements, *Canis lupus*, foraging, caribou, Northwest Territories

(8) **RÉSUMÉ** Les loups (*Canis lupus*) dans la toundra canadienne sont étroitement liés aux hardes de caribous des toundras (*Rangifer tarandus*). On a équipé une louve adulte d'un collier émetteur muni d'un système de positionnement mondial (GPS) afin d'enregistrer ses déplacements en réponse au changement de densité du caribou près de sa tanière durant l'été. On a observé cette louve ainsi que deux autres en train d'allaiter un groupe de 11 louveteaux. Elle a parcouru un minimum de 341 km durant une sortie de 14 jours. La distance en ligne droite de la tanière à l'endroit le plus éloigné était de 103 km, et la vitesse minimum durant tout le voyage était de 3,1 km/h. La distance entre la louve et le caribou muni du collier émetteur a diminué de 242 km une semaine avant la sortie à 8 km quatre jours après la sortie. On commente diverses explications possibles pour ce long épisode de recherche de nourriture.

Mots clés: loup, repérage GPS, déplacements, *Canis lupus*, recherche de nourriture, caribou, Territoires du Nord-Ouest

Traduit pour la revue *Arctic* par Nésida Loyer.

(9) Introduction

Wolves (*Canis lupus*) that den on the central barrens of mainland Canada follow the seasonal movements of their main prey, migratory barren-ground caribou (*Rangifer tarandus*) (Kuyt, 1962; Kelsall, 1968; Walton et al., 2001). However, most wolves do not den near caribou calving grounds, but select sites farther south, closer to the tree line (Heard and Williams, 1992). Most caribou migrate beyond primary wolf denning areas by mid-June and do not return until mid-to-late July (Heard et al., 1996; Gunn et al., 2001). Conse-

quently, caribou density near dens is low for part of the summer.

During this period of spatial separation from the main caribou herds, wolves must either search near the homesite for scarce caribou or alternative prey (or both), travel to where prey are abundant, or use a combination of these strategies. **(10)**

Walton et al. (2001) postulated that the travel of tundra wolves outside their normal summer ranges is a response to low caribou availability rather than a pre-dispersal exploration like that observed in terri torial wolves (Fritts and Mech, 1981; Messier, 1985). The authors postulated this because most such travel was directed toward caribou calving grounds. We report details of such a long-distance excursion by a breeding female tundra wolf wearing a GPS radio collar. We discuss the relationship of the excursion to movements of satellite-collared caribou (Gunn et al., 2001), supporting the hypothesis that tundra wolves make directional, rapid, long-distance movements in response to seasonal prey availability. **(11)**

[1] Department of Biological Sciences, University of Alberta, Edmonton, Alberta T6G 2E9, Canada
[2] Corresponding author: pframe@ualberta.ca
[3] Department of Resources, Wildlife, and Economic Development, North Slave Region, Government of the Northwest Territories, P.O. Box 2668, 3803 Bretzlaff Dr., Yellowknife, Northwest Territories X1A 2P9, Canada; Dean_Cluff@gov.nt.ca
[4] Faculty of Environmental Design, University of Calgary, Calgary, Alberta T2N 1N4, Canada; current address: P.O. Box 150, Meacham, Saskatchewan S0K 2V0, Canada

© The Arctic Institute of North America

196

Annotation notes (right margin):

1 Title of the journal, which reports on science taking place in Arctic regions.

2 Volume number, issue number and date of the journal, and page numbers of the article.

3 Title of the article: a concise but specific description of the subject of study—one episode of long-range travel by a wolf hunting for food on the Arctic tundra.

4 Authors of the article: scientists working at the institutions listed in the footnotes below. Note #2 indicates that P. F. Frame is the *corresponding author*—the person to contact with questions or comments. His email address is provided.

5 Date on which a draft of the article was received by the journal editor, followed by date one which a revised draft was accepted for publication. Between these dates, the article was reviewed and critiqued by other scientists, a process called peer review. The authors revised the article to make it clearer, according to those reviews.

6 ABSTRACT: A brief description of the study containing all basic elements of this report. First sentence summarizes the *background* material. Second sentence encapsulates the *methods* used. The rest of the paragraph sums up the *results*. Authors introduce the main *subject* of the study—a female wolf (#388) with pups in a den—and refer to later *discussion* of possible explanations for her behavior.

7 Key words are listed to help researchers using computer databases. Searching the databases using these key words will yield a list of studies related to this one.

8 RÉSUMÉ: The French translation of the abstract and key words. Many researchers in this field are French Canadian. Some journals provide such translations in French or in other languages.

9 INTRODUCTION: Gives the background for this wolf study. This paragraph tells of known or suspected wolf behavior that is important for this study. Note that (a) major species mentioned are always accompanied by scientific names, and (b) statements of fact or *postulations* (claims or assumptions about what is likely to be true) are followed by references to studies that established those facts or supported the postulations.

10 This paragraph focuses directly on the wolf behaviors that were studied here.

11 This paragraph starts with a statement of the *hypothesis* being tested, one that originated in other studies and is supported by this one. The hypothesis is restated more succinctly in the last sentence of this paragraph. This is the *inquiry* part of the scientific process—asking questions and suggesting possible answers.

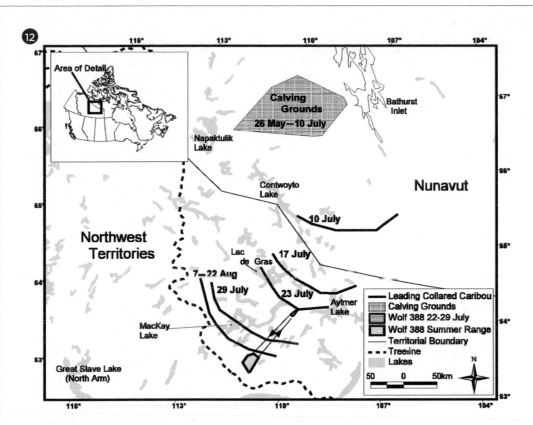

12 This map shows the study area and depicts wolf and caribou locations and movements during one summer. Some of this information is explained below.

13 STUDY AREA: This section sets the stage for the study, locating it precisely with latitude and longitude coordinates and describing the area (illustrated by the map in Figure 1).

14 Here begins the story of how prey (caribou) and predators (wolves) interact on the tundra. Authors describe movements of these nomadic animals throughout the year.

15 We focus on the denning season (summer) and learn how wolves locate their dens and travel according to the movements of caribou herds.

Figure 1. Map showing the movements of satellite radio-collared caribou with respect to female wolf 388's summer range and long foraging movement, in summer 2002.

13 Study Area

Our study took place in the northern boreal forest–low Arctic tundra transition zone (63° 30′ N, 110° 00′ W; Figure 1, Timoney et al., 1992). Permafrost in the area changes from discontinuous to continuous (Harris, 1986). Patches of spruce (*Picea mariana, P. glauca*) occur in the southern portion and give way to open tundra to the northeast. Eskers, kames, and other glacial deposits are scattered throughout the study area. Standing water and exposed bedrock are characteristic of the area.

14 Details of the Caribou-Wolf System

The Bathurst caribou herd uses this study area. Most caribou cows have begun migrating by late April, reaching calving grounds by June (Gunn et al., 2001;

Figure 1). Calving peaks by 15 June (Gunn et al., 2001), and calves begin to travel with the herd by one week of age (Kelsall, 1968). The movement patterns of bulls are less known, but bulls frequent areas near calving grounds by mid-June (Heard et al., 1996; Gunn et al., 2001). In summer, Bathurst caribou cows generally travel south from their calving grounds and then, parallel to the tree line, to the northwest. The rut usually takes place at the tree line in October (Gunn et al., 2001). The winter range of the Bathurst herd varies among years, ranging through the taiga and along the tree line from south of Great Bear Lake to southeast of Great Slave Lake. Some caribou spend the winter on the tundra (Gunn et al., 2001; Thorpe et al., 2001).

In winter, wolves that prey on Bathurst caribou do not behave territorially. Instead, they follow the herd throughout its winter range (Walton et al., 2001; Musiani, 2003). However, during denning (May– **15**

16 Other variables are considered—prey other than caribou and their relative abundance in 2002.

17 METHODS: There is no one scientific method. Procedures for each and every study must be explained carefully.

18 Authors explain when and how they tracked caribou and wolves, including tools used and the exact procedures followed.

19 This important subsection explains what data were calculated (average distance ...) and how, including the software used and where it came from. (The calculations are listed in Table 1.) Note that the behavior measured (traveling) is carefully defined.

20 RESULTS: The heart of the report and the *observation* part of the scientific process. This section is organized parallel to the Methods section.

21 This subsection is broken down by periods of observation. Pre-excursion period covers the time between 388's capture and the start of her long-distance travel. The investigators used visual observations as well as telemetry (measurements taken using the global positioning system (GPS)) to gather data. They looked at how 388 cared for her pups, interacted with other adults, and moved about the den area.

Table 1. Daily distances from wolf 388 and the den to the nearest radio-collared caribou during a long excursion in summer 2002.

Date (2002)	Mean distance from caribou to wolf (km)	Daily distance from closest caribou to den
12 July	242	241
13 July	210	209
14 July	200	199
15 July	186	180
16 July	163	162
17 July	151	148
18 July	144	137
19 July[1]	126	124
20 July	103	130
21 July	73	130
22 July[2]	40	110
23 July[2]	9	104
29 July[3]	16	43
30 July	32	43
31 July	28	44
1 August	29	46
2 August[4]	54	52
3 August	53	53
4 August	74	74
5 August	75	75
6 August	74	75
7 August	72	75
8 August	76	75
9 August	79	79

[1] Excursion starts.
[2] Wolf closest to collared caribou.
[3] Previous five days' caribou locations not available.
[4] Excursion ends.

August, parturition late May to mid-June), wolf movements are limited by the need to return food to the den. To maximize access to migrating caribou, many wolves select den sites closer to the tree line than to caribou calving grounds (Heard and Williams, 1992). Because of caribou movement patterns, tundra denning wolves are separated from the main caribou herds by several hundred kilometers at some time during summer (Williams, 1990:19; Figure 1; Table 1).

Muskoxen do not occur in the study area (Fournier and Gunn, 1998), and there are few moose there (H.D. Cluff, pers. obs.). Therefore, alternative prey for wolves includes waterfowl, other ground-nesting birds, their eggs, rodents, and hares (Kuyt, 1972; Williams, 1990:16; H.D. Cluff and P.F. Frame, unpubl. data). During 56 hours of den observations, we saw no ground squirrels or hares, only birds. It appears that the abundance of alternative prey was relatively low in 2002.

Methods

Wolf Monitoring

We captured female wolf 388 near her den on 22 June 2002, using a helicopter net-gun (Walton et al., 2001). She was fitted with a releasable GPS radio collar (Merrill et al., 1998) programmed to acquire locations at 30-

minute intervals. The collar was electronically released (e.g., Mech and Gese, 1992) on 20 August 2002. From 27 June to 3 July 2002, we observed 388's den with a 78 mm spotting scope at a distance of 390 m.

Caribou Monitoring

In spring of 2002, ten female caribou were captured by helicopter net-gun and fitted with satellite radio collars, bringing the total number of collared Bathurst cows to 19. Eight of these spent the summer of 2002 south of Queen Maud Gulf, well east of normal Bathurst caribou range. Therefore, we used 11 caribou for this analysis. The collars provided one location per day during our study, except for five days from 24 to 28 July. Locations of satellite collars were obtained from Service Argos, Inc. (Landover, Maryland).

Data Analysis

Location data were analyzed by ArcView GIS software (Environmental Systems Research Institute Inc., Redlands, California). We calculated the average distance from the nearest collared caribou to the wolf and the den for each day of the study.

Wolf foraging bouts were calculated from the time 388 exited a buffer zone (500 m radius around the den) until she re-entered it. We considered her to be traveling when two consecutive locations were spatially separated by more than 100 m. Minimum distance traveled was the sum of distances between each location and the next during the excursion.

We compared pre- and post-excursion data using Analysis of Variance (ANOVA; Zar, 1999). We first tested for homogeneity of variances with Levene's test (Brown and Forsythe, 1974). No transformations of these data were required.

Results

Wolf Monitoring

Pre-Excursion Period: Wolf 388 was lactating when captured on 22 June. We observed her and two other females nursing a group of 11 pups between 27 June and 3 July. During our observations, the pack consisted of at least four adults (3 females and 1 male) and 11 pups. On 30 June, three pups were moved to a location 310 m from the other eight and cared for by an uncollared female. The male was not seen at the den after the evening of 30 June.

Before the excursion, telemetry indicated 18 foraging bouts. The mean distance traveled during these bouts was 25.29 km (± 4.5 SE, range 3.1–82.5 km). Mean greatest distance from the den on foraging

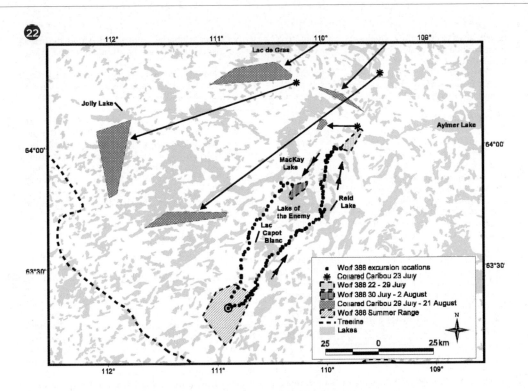

Figure 2. Details of a long foraging movement by female wolf 388 between 19 July and 2 August 2002. Also shown are locations and movements of three satellite radio-collared caribou from 23 July to 21 August 2002. On 23 July, the wolf was 8 km from a collared caribou. The farthest point from the den (103 km distant) was recorded on 27 July. Arrows indicate direction of travel.

22 The key in the lower right-hand corner of the map shows areas (shaded) within which the wolves and caribou moved, and the dotted trail of 388 during her excursion. From the results depicted on this map, the investigators tried to determine when and where 388 might have encountered caribou and how their locations affected her traveling behavior.

23 The wolf's excursion (her long trip away from the den area) is the focus of this study. These paragraphs present detailed measurements of daily movements during her two-week trip—how far she traveled, how far she was from collared caribou, her time spent traveling and resting, and her rate of speed. Authors use the phrase "minimum distance traveled" to acknowledge they couldn't track every step but were measuring samples of her movements. They knew that she went at least as far as they measured. This shows how scientists try to be exact when reporting results. Results of this study are depicted graphically in the map in Figure 2.

bouts was 7.1 km (± 0.9 SE, range 1.7–17.0 km). The average duration of foraging bouts for the period was 20.9 h (± 4.5 SE, range 1–71 h).

The average daily distance between the wolf and the nearest collared caribou decreased from 242 km on 12 July, one week before the excursion period, to 126 km on 19 July, the day the excursion began (Table 1).

Excursion Period: On 19 July at 2203, after spending 14 h at the den, 388 began moving to the northeast and did not return for 336 h (14 d; Figure 2). Whether she traveled alone or with other wolves is unknown. During the excursion, 476 (71%) of 672 possible locations were recorded. The wolf crossed the southeast end of Lac Capot Blanc on a small land bridge, where she paused for 4.5 h after traveling for 19.5 h (37.5

km). Following this rest, she traveled for 9 h (26.3 km) onto a peninsula in Reid Lake, where she spent 2 h before backtracking and stopping for 8 h just off the peninsula. Her next period of travel lasted 16.5 h (32.7 km), terminating in a pause of 9.5 h just 3.8 km from a concentration of locations at the far end of her excursion, where we presume she encountered caribou. The mean duration of these three movement periods was 15.7 h (± 2.5 SE), and that of the pauses, 7.3 h (± 1.5). The wolf required 72.5 h (3.0 d) to travel a minimum of 95 km from her den to this area near caribou (Figure 2). She remained there (35.5 km2) for 151.5 h (6.3 d) and then moved south to Lake of the Enemy, where she stayed (31.9 km^2) for 74 h (3.1 d) before returning to her den. Her greatest distance from the den, 103 km, was recorded 174.5 h (7.3 d) after the excursion began, at 0433 on 27 July. She was 8 km from a collared caribou on 23 July, four days after the excursion began (Table 1).

The return trip began at 0403 on 2 August, 318 h (13.2 d) after leaving the den. She followed a relatively direct path for 18 h back to the den, a distance of 75 km.

The minimum distance traveled during the excursion was 339 km. The estimated overall minimum travel rate was 3.1 km/h, 2.6 km/h away from the den and 4.2 km/h on the return trip.

24 **Post-Excursion Period:** We saw three pups when recovering the collar on 20 August, but others may have been hiding in vegetation.

Telemetry recorded 13 foraging bouts in the post-excursion period. The mean distance traveled during these bouts was 18.3 km (+ 2.7 SE, range 1.2–47.7 km), and mean greatest distance from the den was 7.1 km (+ 0.7 SE, range 1.1–11.0 km). The mean duration of these post-excursion foraging bouts was 10.9 h (+ 2.4 SE, range 1–33 h).

When 388 reached her den on 2 August, the distance to the nearest collared caribou was 54 km. On 9 August, one week after she returned, the distance was 79 km (Table 1).

Pre- and Post-Excursion Comparison

25 We found no differences in the mean distance of foraging bouts before and after the excursion period (F = 1.5, df = 1, 29, p = 0.24). Likewise, the mean greatest distance from the den was similar pre- and post-excursion (F = 0.004, df = 1, 29, p = 0.95). However, the mean duration of 388's foraging bouts decreased by 10.0 h after her long excursion (F = 3.1, df = 1, 29, p = 0.09).

26 *Caribou Monitoring*

Summer Movements: On 10 July, 5 of 11 collared caribou were dispersed over a distance of 10 km, 140 km south of their calving grounds (Figure 1). On the same day, three caribou were still on the calving grounds, two were between the calving grounds and the leaders, and one was missing. One week later (17 July), the leading radio-collared cows were 100 km farther south (Figure 1). Two were within 5 km of each other in front of the rest, who were more dispersed. All radio-collared cows had left the calving grounds by this time. On 23 July, the leading radio-collared caribou had moved 35 km farther south, and all of them were more widely dispersed. The two cows closest to the leader were 26 km and 33 km away, with 37 km between them. On the next location (29 July), the most southerly caribou were 60 km

farther south. All of the caribou were now in the areas where they remained for the duration of the study (Figure 2).

A Minimum Convex Polygon (Mohr and Stumpf, 1966) around all caribou locations acquired during the study encompassed 85 119 km^2.

Relative to the Wolf Den: The distance from the **27** nearest collared caribou to the den decreased from 241 km one week before the excursion to 124 km the day it began. The nearest a collared caribou came to the den was 43 km away, on 29 and 30 July. During the study, four collared caribou were located within 100 km of the den. Each of these four was closest to the wolf on at least one day during the period reported.

28 **Discussion**

Prey Abundance

Caribou are the single most important prey of tundra **29** wolves (Clark, 1971; Kuyt, 1972; Stephenson and James, 1982; Williams, 1990). Caribou range over vast areas, and for part of the summer, they are scarce or absent in wolf home ranges (Heard et al., 1996). Both the long distance between radio-collared caribou and the den the week before the excursion and the increased time spent foraging by wolf 388 indicate that caribou availability near the den was low. Observations of the pups' being left alone for up to 18 h, presumably while adults were searching for food, provide additional support for low caribou availability locally. Mean foraging bout duration decreased by 10.0 h after the excursion, when collared caribou were closer to the den, suggesting an increase in caribou availability nearby.

Foraging Excursion

One aspect of central place foraging theory (CPFT) **30** deals with the optimality of returning different-sized food loads from varying distances to dependents at a central place (i.e., the den) (Orians and Pearson, 1979). Carlson (1985) tested CPFT and found that the predator usually consumed prey captured far from the central place, while feeding prey captured nearby to dependants. Wolf 388 spent 7.2 days in one area near caribou before moving to a location 23 km back towards the den, where she spent an additional 3.1 days, likely hunting caribou. She began her return trip from this closer location, traveling directly to the den. While away, she may have made one or more successful kills and spent time meeting her own energetic needs before returning to the den. Alternatively, it may have taken several attempts to make a kill, which she then fed on before beginning her return trip. We do not know if she returned food to the pups, but such behavior would be supported by CPFT.

[31] Other workers have reported wolves' making long round trips and referred to them as "extraterritorial" or "pre-dispersal" forays (Fritts and Mech, 1981; Messier, 1985; Ballard et al., 1997; Merrill and Mech, 2000). These movements are most often made by young wolves (1–3 years old), in areas where annual territories are maintained and prey are relatively sedentary (Fritts and Mech, 1981; Messier, 1985). The long excursion of 388 differs in that tundra wolves do not maintain annual territories (Walton et al., 2001), and the main prey migrate over vast areas (Gunn et al., 2001).

Another difference between 388's excursion and those reported earlier is that she is a mature, breeding female. No study of territorial wolves has reported reproductive adults making extraterritorial movements in summer (Fritts and Mech, 1981; Messier, 1985; Ballard et al., 1997; Merrill and Mech, 2001). However, Walton et al. (2001) also report that breeding female tundra wolves made excursions.

Direction of Movement

[32] Possible explanations for the relatively direct route 388 took to the caribou include landscape influence and experience. Considering the timing of 388's trip and the locations of caribou, had the wolf moved northwest, she might have missed the caribou entirely, or the encounter might have been delayed.

A reasonable possibility is that the land directed 388's route. The barrens are crisscrossed with trails worn into the tundra over centuries by hundreds of thousands of caribou and other animals (Kelsall, 1968; Thorpe et al., 2001). At river crossings, lakes, or narrow peninsulas, trails converge and funnel towards and away from caribou calving grounds and summer range. Wolves use trails for travel (Paquet et al., 1996; Mech and Boitani, 2003; P. Frame, pers. observation). Thus, the landscape may direct an animal's movements and lead it to where cues, such as the odor of caribou on the wind or scent marks of other wolves, may lead it to caribou.

[33] Another possibility is that 388 knew where to find caribou in summer. Sexually immature tundra wolves sometimes follow caribou to calving grounds (D. Heard, unpubl. data). Possibly, 388 had made such journeys in previous years and killed caribou. If this were the case, then in times of local prey scarcity she might travel to areas where she had hunted successfully before. Continued monitoring of tundra wolves may answer questions about how their food needs are met in times of low caribou abundance near dens.

[34] Caribou often form large groups while moving south to the tree line (Kelsall, 1968). After a large aggregation of caribou moves through an area, its scent can linger for weeks (Thorpe et al., 2001:104). It is conceivable that 388 detected caribou scent on the wind, which was blowing from the northeast on 19–21 July (Environment Canada, 2003), at the same time her excursion began. Many factors, such as odor strength and wind direction and strength, make systematic study of scent detection in wolves difficult under field conditions (Harrington and Asa, 2003). However, humans are able to smell odors such as forest fires or oil refineries more than 100 km away. The olfactory capabilities of dogs, which are similar to wolves, are thought to be 100 to 1 million times that of humans (Harrington and Asa, 2003). Therefore, it is reasonable to think that under the right wind conditions, the scent of many caribou traveling together could be detected by wolves from great distances, thus triggering a long foraging bout.

Rate of Travel

[35] Mech (1994) reported the rate of travel of Arctic wolves on barren ground was 8.7 km/h during regular travel and 10.0 km/h when returning to the den, a difference of 1.3 km/h. These rates are based on direct observation and exclude periods when wolves moved slowly or not at all. Our calculated travel rates are assumed to include periods of slow movement or no movement. However, the pattern we report is similar to that reported by Mech (1994), in that homeward travel was faster than regular travel by 1.6 km/h. The faster rate on return may be explained by the need to return food to the den. Pup survival can increase with the number of adults in a pack available to deliver food to pups (Harrington et al., 1983). Therefore, an increased rate of travel on homeward trips could improve a wolf's reproductive fitness by getting food to pups more quickly.

Fate of 388's Pups

[36] Wolf 388 was caring for pups during den observations. The pups were estimated to be six weeks old, and were seen ranging as far as 800 m from the den. They received some regurgitated food from two of the females, but were unattended for long periods. The excursion started 16 days after our observations, and it is improbable that the pups could have traveled the distance that 388 moved. If the pups died, this would have removed parental responsibility, allowing the long movement.

Our observations and the locations of radio-collared caribou indicate that prey became scarce in

---

31 Here our authors note other possible explanations for wolves' excursions presented by other investigators, but this study does not seem to support those ideas.

32 Authors discuss possible reasons for why 388 traveled directly to where caribou were located. They take what they learned from earlier studies and apply it to this case, suggesting that the lay of the land played a role. Note that their description paints a clear picture of the landscape.

33 Authors suggest that 388 may have learned in traveling during previous summers where the caribou were. The last two sentences suggest ideas for future studies.

34 Or maybe 388 followed the scent of the caribou. Authors acknowledge difficulties of proving this, but they suggest another area where future studies might be done.

35 Authors suggest that results of this study support previous studies about how fast wolves travel to and from the den. In the last sentence, they speculate on how these observed patterns would fit into the theory of evolution.

36 Authors also speculate on the fate of 388's pups while she was traveling. This leads to . . .

37 Discussion of cooperative rearing of pups and, in turn, to speculation on how this study and what is known about cooperative rearing might fit into the animal's strategies for survival of the species. Again, the authors approach the broader theory of evolution and how it might explain some of their results.

38 And again, they suggest that this study points to several areas where further study will shed some light.

39 In conclusion, the authors suggest that their study supports the hypothesis being tested here. And they touch on the implications of increased human activity on the tundra predicted by their results.

40 ACKNOWLEDGEMENTS: Authors note the support of institutions, companies and individuals. They thank their reviewers ad list permits under which their research was carried on.

41 REFERENCES: List of all studies cited in the report. This may seem tedious, but is a vitally important part of scientific reporting. It is a record of the sources of information on which this study is based. It provides readers with a wealth of resources for further reading on this topic. Much of it will form the foundation of future scientific studies like this one.

the area of the den as summer progressed. Wolf 388 may have abandoned her pups to seek food for herself. However, she returned to the den after the excursion, where she was seen near pups. In fact, she foraged in a similar pattern before and after the excursion, suggesting that she again was providing for pups after her return to the den.

37 A more likely possibility is that one or both of the other lactating females cared for the pups during 388's absence. The three females at this den were not seen with the pups at the same time. However, two weeks earlier, at a different den, we observed three females cooperatively caring for a group of six pups. At that den, the three lactating females were observed providing food for each other and trading places while nursing pups. Such a situation at the den of 388 could have created conditions that allowed one or more of the lactating females to range far from the den for a period, returning to her parental duties afterwards. However, the pups would have been weaned by eight weeks of age (Packard et al., 1992), so nonlactating adults could also have cared for them, as often happens in wolf packs (Packard et al., 1992; Mech et al., 1999).

Cooperative rearing of multiple litters by a pack could create opportunities for long-distance foraging movements by some reproductive wolves during summer periods of local food scarcity. We have recorded multiple lactating females at one or more tundra wolf dens per year since 1997. This reproductive strategy may be an adaptation to temporally and **38** spatially unpredictable food resources. All of these possibilities require further study, but emphasize both the adaptability of wolves living on the barrens and their dependence on caribou.

Long-range wolf movement in response to caribou **39** availability has been suggested by other researchers (Kuyt, 1972; Walton et al., 2001) and traditional ecological knowledge (Thorpe et al., 2001). Our report demonstrates the rapid and extreme response of wolves to caribou distribution and movements in summer. Increased human activity on the tundra (mining, road building, pipelines, ecotourism) may influence caribou movement patterns and change the interactions between wolves and caribou in the region. Continued monitoring of both species will help us to assess whether the association is being affected adversely by anthropogenic change.

40 Acknowledgements

This research was supported by the Department of Resources, Wildlife, and Economic Development, Government of the Northwest Territories; the Department of Biological Sciences at the University of Alberta; the Natural Sciences and Engineering Research Council of Canada; the Department of Indian and Northern Affairs Canada; the Canadian Circumpolar Institute; and DeBeers Canada, Ltd. Lorna Ruechel assisted with den observations. A. Gunn provided caribou location data. We thank Dave Mech for the use of GPS collars. M. Nelson, A. Gunn, and three anonymous reviewers made helpful comments on earlier drafts of the manuscript. This work was done under Wildlife Research Permit – WL002948 issued by the Government of the Northwest Territories, Department of Resources, Wildlife, and Economic Development.

41 References

BALLARD, W.B., AYRES, L.A., KRAUSMAN, P.R., REED, D.J., and FANCY, S.G. 1997. Ecology of wolves in relation to a migratory caribou herd in northwest Alaska. Wildlife Monographs 135. 47 p.

BROWN, M.B., and FORSYTHE, A.B. 1974. Robust tests for the equality of variances. Journal of the American Statistical Association 69:364–367.

CARLSON, A. 1985. Central place foraging in the red-backed shrike (*Lanius collurio* L.): Allocation of prey between forager and sedentary consumer. Animal Behaviour 33:664–666.

CLARK, K.R.F. 1971. Food habits and behavior of the tundra wolf on central Baffin Island. Ph.D. Thesis, University of Toronto, Ontario, Canada.

ENVIRONMENT CANADA. 2003. National climate data information archive. Available online: http://www.climate.weatheroffice.ec.gc.ca/Welcome_e.html

FOURNIER, B., and GUNN, A. 1998. Musk ox numbers and distribution in the NWT, 1997. File Report No. 121. Yellowknife: Department of Resources, Wildlife, and Economic Development, Government of the Northwest Territories. 55 p.

FRITTS, S.H., and MECH, L.D. 1981. Dynamics, movements, and feeding ecology of a newly protected wolf population in northwestern Minnesota. Wildlife Monographs 80. 79 p.

GUNN, A., DRAGON, J., and BOULANGER, J. 2001. Seasonal movements of satellite-collared caribou from the Bathurst herd. Final Report to the West Kitikmeot Slave Study Society, Yellowknife, NWT. 80 p. Available online: http://www.wkss.nt.ca/HTML/08_ProjectsReports/PDF/Seasonal MovementsFinal.pdf

HARRINGTON, F.H., and ASA, C.S. 2003. Wolf communication. In: Mech, L.D., and Boitani, L., eds. Wolves: Behavior, ecology, and conservation. Chicago: University of Chicago Press. 66–103.

HARRINGTON, F.H., MECH, L.D., and FRITTS, S.H. 1983. Pack size and wolf pup survival: Their relationship under varying ecological conditions. Behavioral Ecology and Sociobiology 13:19–26.

HARRIS, S.A. 1986. Permafrost distribution, zonation and stability along the eastern ranges of the cordillera of North America. Arctic 39(1):29–38.

HEARD, D.C., and WILLIAMS, T.M. 1992. Distribution of wolf dens on migratory caribou ranges in the Northwest

Territories, Canada. Canadian Journal of Zoology 70:1504–1510.

HEARD, D.C., WILLIAMS, T.M., and MELTON, D.A. 1996. The relationship between food intake and predation risk in migratory caribou and implication to caribou and wolf population dynamics. Rangifer Special Issue No. 2:37–44.

KELSALL, J.P. 1968. The migratory barren-ground caribou of Canada. Canadian Wildlife Service Monograph Series 3. Ottawa: Queen's Printer. 340 p.

KUYT, E. 1962. Movements of young wolves in the Northwest Territories of Canada. Journal of Mammalogy 43:270–271.

———. 1972. Food habits and ecology of wolves on barren-ground caribou range in the Northwest Territories. Canadian Wildlife Service Report Series 21. Ottawa: Information Canada. 36 p.

MECH, L.D. 1994. Regular and homeward travel speeds of Arctic wolves. Journal of Mammalogy 75:741–742.

MECH, L.D., and BOITANI, L. 2003. Wolf social ecology. In: Mech, L.D., and Boitani, L., eds. Wolves: Behavior, ecology, and conservation. Chicago: University of Chicago Press. 1–34.

MECH, L.D., and GESE, E.M. 1992. Field testing the Wildlink capture collar on wolves. Wildlife Society Bulletin 20:249–256.

MECH, L.D., WOLFE, P., and PACKARD, J.M. 1999. Regurgitative food transfer among wild wolves. Canadian Journal of Zoology 77:1192–1195.

MERRILL, S.B., and MECH, L.D. 2000. Details of extensive movements by Minnesota wolves (Canis lupus). American Midland Naturalist 144:428–433.

MERRILL, S.B., ADAMS, L.G., NELSON, M.E., and MECH, L.D. 1998. Testing releasable GPS radiocollars on wolves and white-tailed deer. Wildlife Society Bulletin 26:830–835.

MESSIER, F. 1985. Solitary living and extraterritorial movements of wolves in relation to social status and prey abundance. Canadian Journal of Zoology 63:239–245.

MOHR, C.O., and STUMPF, W.A. 1966. Comparison of methods for calculating areas of animal activity. Journal of Wildlife Management 30.290–304.

MUSIANI, M. 2003. Conservation biology and management of wolves and wolf-human conflicts in western North America. Ph.D. Thesis, University of Calgary, Calgary, Alberta, Canada.

ORIANS, G.H., and PEARSON, N.E. 1979. On the theory of central place foraging. In: Mitchell, R.D., and Stairs, G.F., eds. Analysis of ecological systems. Columbus: Ohio State University Press. 154–177.

PACKARD, J.M., MECH, L.D., and REAM, R.R. 1992. Weaning in an arctic wolf pack: Behavioral mechanisms. Canadian Journal of Zoology 70:1269–1275.

PAQUET, P.C., WIERZCHOWSKI, J., and CALLAGHAN, C. 1996. Summary report on the effects of human activity on gray wolves in the Bow River Valley, Banff National Park, Alberta. In: Green, J., Pacas, C., Bayley, S., and Cornwell, L., eds. A cumulative effects assessment and futures outlook for the Banff Bow Valley. Prepared for the Banff Bow Valley Study. Ottawa: Department of Canadian Heritage.

STEPHENSON, R.O., and JAMES, D. 1982. Wolf movements and food habits in northwest Alaska. In: Harrington, F.H., and Paquet, P.C., eds. Wolves of the world. New Jersey: Noyes Publications. 223–237.

THORPE, N., EYEGETOK, S., HAKONGAK, N., and QITIRMIUT ELDERS. 2001. The Tuktu and Nogak Project: A caribou chronicle. Final Report to the West Kitikmeot/Slave Study Society, Ikaluktuuttiak, NWT. 160 p.

TIMONEY, K.P., LA ROI, G.H., ZOLTAI, S.C., and ROBINSON, A.L. 1992. The high subarctic forest-tundra of northwestern Canada: Position, width, and vegetation gradients in relation to climate. Arctic 45(1):1–9.

WALTON, L.R., CLUFF, H.D., PAQUET, P.C., and RAMSAY, M.A. 2001. Movement patterns of barren-ground wolves in the central Canadian Arctic. Journal of Mammalogy 82:867–876.

WILLIAMS, T.M. 1990. Summer diet and behavior of wolves denning on barren-ground caribou range in the Northwest Territories, Canada. M.Sc. Thesis, University of Alberta, Edmonton, Alberta, Canada.

ZAR, J.H. 1999. Biostatistical analysis. 4th ed. New Jersey: Prentice Hall. 663 p.

Appendix III. Answers to Self-Quizzes and Genetics Problems

Italicized numbers refer to relevant section numbers

CHAPTER 1

1. Atoms — *1.1*
2. cell — *1.1*
3. Animals — *1.3*
4. energy, nutrients — *1.2*
5. Homeostasis — *1.2*
6. Domains — *1.3*
7. d — *1.2*
8. d — *1.2*
9. Reproduction — *1.2*
10. observable — *1.5*
11. Mutations — *1.4*
12. adaptive — *1.4*
13. b — *1.6*
14. c — *1.1*
    e — *1.4*
    d — *1.6*
    f — *1.6*
    a — *1.6*
    b — *1.3*

CHAPTER 2

1. Tracer — *2.2*
2. b — *2.3*
3. compound — *2.3*
4. electronegativity — *2.3*
5. polar covalent — *2.4*
6. atomic number — *2.1*
7. e — *2.5*
8. hydrophobic — *2.5*
9. d — *2.6*
10. solute — *2.5*
11. acid — *2.6*
12. hydrogen ions (H+) or hydroxyl ions (OH−) — *2.6*
13. buffer system — *2.6*
12. c — *2.5*
    b — *2.1*
    d — *2.1*
    a — *2.5*

CHAPTER 3

1. four — *3.1*
2. carbohydrate — *3.3*
3. f — *3.3, 3.7*
4. double covalent bonds — *3.4*
5. False — *3.4*
6. fatty acid tails — *3.4*
7. e — *3.4*
8. d — *3.3, 3.5*
9. d — *3.6*
10. d — *3.7*
11. c — *3.4*
12. c — *3.5*
    e — *3.7*
    b — *3.4*
    d — *3.7*
    a — *3.3*
    f — *3.4*

CHAPTER 4

1. cell — *4.2*
2. False (all protists are eukaryotes) — *4.6*
3. phospholipids — *4.2*
4. c — *4.2*
5. eukaryotic — *4.6*
6. lipds, proteins — *4.9*
7. nucleus — *4.8*
8. cell wall — *4.12*
9. False (cell walls enclose the plasma membrane of many cells) — *4.12*
10. lysosomes — *4.9*
11. c — *4.11*
    f — *4.11*
    a — *4.2*

e — *4.9*
d — *4.9*
b — *4.9*

CHAPTER 5

1. c — *5.1*
2. c — *5.1*
3. fluid mosaic — *5.1*
4. a — *5.2*
5. a — *5.2*
6. adhesion — *5.2*
7. more, less — *5.3*
8. oxygen (CO2, water, etc.) — *5.3*
9. b — *5.4*
10. a — *5.6*
11. hydrostatic pressure (or turgor) — *5.6*
12. e — *5.5*
13. d, b, e, a — *5.5*
14. d — *5.5*
    g — *5.4*
    a — *5.2*
    e — *5.4*
    c — *5.1*
    b — *5.3*
    f — *5.2*

CHAPTER 6

1. c — *6.1*
2. d — *6.1*
3. b, c — *6.1*
4. d — *6.2*
5. b — *6.2*
6. c, d — *6.2, 6.4*
7. d — *6.3*
8. e — *6.4*
9. c — *6.4*
10. c — *6.3*
11. a — *6.3*
12. c — *6.2*
    g — *6.3*
    d — *6.1*
    b — *6.2*
    f — *6.4*
    a — *6.3*
    e — *6.1*

CHAPTER 7

1. carbon dioxide, light (or sunlight) — *7.1, 7.8*
2. b — *7.1*
3. a — *7.3*
4. b — *7.3*
5. c — *7.3*
6. d — *7.4*
7. c — *7.3*
8. b — *7.6*
9. e — *7.6*
10. PGA; oxaloacelate — *7.7*
11. oxygen gas (O2) — *7.8*
12. The cat, bird, and caterpillar are heterotrophs. The weed is an autotroph. — *7.8*
13. c — *7.6*
    a — *7.6*
    b — *7.4*
    d — *7.4*

CHAPTER 8

1. False (plants make ATP by aerobic respiration too) — *8.1*
2. d — *8.2*
3. a — *8.1*
4. c — *8.2*
5. b — *8.1*

6. e — *8.3*
7. b — *8.3*
8. c — *8.4*
9. c — *8.5*
10. b — *8.5*
11. d — *8.7*
12. b — *8.1*
    c — *8.5*
    a — *8.3*
    d — *8.4*

CHAPTER 9

1. d — *9.1*
2. b — *9.1*
3. c — *9.1*
4. d — *9.2*
5. a — *9.2*
6. c — *9.2*
7. a — *9.2*
8. See Figure 9.6 — *9.3*
9. b — *9.2*
10. a — *9.5*
11. kinase, growth factor, epidermal growth factor, tumor suppressor are all mentioned in this chapter — *9.5*
12. d — *9.3*
    b — *9.3*
    c — *9.3*
    a — *9.3*

CHAPTER 10

1. c — *10.1*
2. d — *10.1*
3. alleles — *10.1*
4. d — *10.1*
5. d — *10.2*
6. b — *10.2*
7. d — *10.2*
8. d — *10.3*
9. meiosis gives rise to nonparental combinations of alleles — *10.1, 10.4*
10. sister chromatids have separated — *10.3*
11. d — *10.2*
    a — *10.1*
    c — *10.3*
    b — *10.2*

CHAPTER 11

1. a — *11.1*
2. continuous variation — *11.7*
3. b — *11.1*
4. a — *11.1*
5. b — *11.1*
6. c — *11.2*
7. a — *11.4*
8. b — *11.2*
9. d — *11.3*
10. c — *11.5*
11. a — *11.5*
12. b — *11.3*
    d — *11.2*
    a — *11.1*
    c — *11.1*

CHAPTER 12

1. d — *12.1*
2. c — *12.1*
3. b — *12.2*
4. b — *12.2*
5. the three mentioned in text are hemophilia, red–green color blindness, DMD (Duchenne muscular dystrophy) — *12.4*

6. Genes for red and green light photoreceptors are located on the X chromosome — *12.4*
7. False — *12.4*
8. d — *12.4*
9. e — *12.5*
10. d — *12.6*
11. True — *12.6*
12. c — *12.6*
13. a — *12.7*
14. c — *12.6*
    e — *12.5*
    f — *12.6*
    b — *12.5*
    a — *12.1*
    d — *12.6*

CHAPTER 13

1. bacteria — *13.1*
2. c — *13.2*
3. d — *13.2*
4. c — *13.2*
5. a — *13.3*
6. d — *13.3*
7. b — *13.3*
8. 3'-CCAAAGAAGTTCTCT-5' — *13.3*
9. d — *13.4*
10. d — *13.1*
    b — *13.4*
    a — *13.2*
    f — *13.2*
    e — *13.3*
    g — *13.3*
    c — *13.2*

CHAPTER 14

1. c — *14.1*
2. promoter — *14.2*
3. high-energy phosphate bonds of free nucleotides — *14.1*
4. single — *14.1*
5. b — *14.1*
6. b — *14.3*
7. 64 — *14.3*
8. c — *14.3*
9. a — *14.3*
10. d — *14.4*
11. gly-phe-phe-lys-arg — *14.3*
12. replication error, ionizing or nonionizing radiation, transposable elements, and toxic chemicals are mentioned in the text — *14.5*
13. c — *14.3*
    d — *14.1*
    e — *14.4*
    a — *14.3*
    f — *14.3*
    g — *14.3*
    b — *14.5*

CHAPTER 15

1. d — *15.1*
2. b — *15.1*
3. d — *15.1*
4. transcription factors — *15.1*
5. b — *15.1*
6. c — *15.1*
7. a — *15.2*
8. b — *15.2*
9. b — *15.2*
10. c — *15.2*
11. c — *15.3*
12. d — *15.3*
13. c — *15.4*

14. operon — *15.4*
15. f — *15.2*
    a — *15.2*
    b — *15.4*
    e — *15.2*
    c — *15.1*
    d — *15.1*

CHAPTER 16

1. c — *16.1*
2. plasmid — *16.1*
3. b — *16.1*
4. b — *16.2*
5. DNA library — *16.2*
6. d — *16.2*
7. b — *16.3*
8. d — *16.3*
9. b — *16.10*
    d — *16.1*
    e — *16.10*
    f — *16.7*
10. Gene therapy — *16.10*
11. c — *16.4*
    g — *16.7*
    d — *16.2*
    e — *16.10*
    a — *16.6*
    f — *16.6*
    b — *16.1*

CHAPTER 17

1. d — *17.1*
2. c — *17.1*
3. b — *17.1*
4. d — *17.2–17.4*
5. d — *17.6*
6. d — *17.5, 17.6, 179*
7. a — *17.7*
8. 65.5 — *17.7*
9. a, c, e, f, g, h — *Chapter introduction, 17.2, 17.8*
10. Gondwana — *17.8*
11. a — *17.3*
    g — *17.1*
    e — *17.3*
    f — *17.6*
    c — *17.2*
    b — *17.2*
    d — *17.5*

CHAPTER 18

1. populations — *18.1*
2. b — *18.1*
3. a — *18.1*
4. c — *18.3*
5. b, c — *18.5*
6. d — *18.4*
7. a, d — *18.5*
8. c — *18.6*
9. balanced polymorphism — *18.6*
10. Gene flow — *18.8*
11. allopatric speciation — *18.10*
12. c — *18.8*
    d — *18.3*
    f — *18.1*
    b — *18.7*
    e — *18.12*
    a — *18.12*

CHAPTER 19

1. d — *19.2*
2. b — *19.2*
3. c — *19.4*
4. b — *19.3*
5. d — *19.4*
6. neutral — *19.4*
7. c — *19.1*

8. c — 19.1
9. b — 19.1
10. b — 19.1
b — 19.1
d — 19.3
c — 19.2
e — 19.4
f — 19.2

CHAPTER 20
1. b — 20.1
2. oxygen — 20.1
3. b — 20.1
4. c — 20.2
5. a — 20.2
6. ribozyme — 20.2
7. a — 20.3
8. c — 20.3
9. b — 20.4
10. d — 20.4
11. endosymbiosis — 20.4
12. stromatolite — 20.3
13. c — 20.3
14. b — 20.6
15. f — 20.1
c — 20.2
d — 20.5
a — 20.5
b — 20.5
e — 20.5

CHAPTER 21
1. c — 21.1
2. b — 21.3
3. d — 21.2
4. RNA — 21.2
5. d — 21.5
6. c — 21.5
7. c — 21.6
8. d — 21.6
9. b — 21.6
10. c — 21.6
11. c — 21.5
12. DNA — 21.5
13. b — 21.8
14. pandemic — 21.8
15. d — 21.5
e — 21.5
b — 21.1
f — 21.5
g — 21.7
a — 21.3
c — 21.5

CHAPTER 22
1. True — 22.1
2. mitocondria — 22.2
3. silica — 22.3, 22.7
4. c — 22.5
5. b — 22.7
6. cyanobacteria — 22.1, 22.9
7. red — 22.1
8. alternation of generations — 22.7
9. a — 22.4
10. c — 22.11
11. c — 22.10
12. d — 22.2
g — 22.6
a — 22.5
b — 22.7
f — 22.7
h — 22.10
e — 22.9
c — 22.11

CHAPTER 23
1. c — 23.1
2. a — 23.2
3. a — 23.2
4. False — 23.4
5. a — 23.3
6. b — 23.3
7. c — 23.4
8. a — 23.5
9. d — 23.3, 23.4
10. b — 23.6
11. True — 23.6
12. b — 23.8
13. d — 23.3
c — 23.4
c — 23.7
b — 23.8
14. c — 23.6
h — 23.2
a — 23.2
b — 23.2
e — 23.8
f — 23.9
d — 23.4
g — 23.4

CHAPTER 24
1. c — 24.1
2. a — 24.1
3. b — 24.3
4. c — 24.4
5. d — 24.5
6. c — 24.5
7. c — 24.5
8. c — 24.4
9. a — 24.6
10. mutualism — 24.6
11. b — 24.3
12. False — 24.6
13. d — 24.7
14. c — 24.4
15. b — 24.2
a — 24.4
f — 24.6
d — 24.5
e — 24.3
g — 24.6
c — 24.3

CHAPTER 25
1. True — 25.1
2. coelom — 25.1
3. choanoflagellates — 25.2
4. a — 25.4
5. a — 25.5
6. a — 25.6
7. c — 25.6, 25.7
8. c — 25.12
9. b — 25.8
10. c — 25.4
11. c — 25.12, 25.17
12. echinoderms — 25.18
13. b — 25.2
j — 25.3
d — 25.4
i — 25.5
c — 25.6
a — 25.11
g — 25.7
e — 25.12
f — 25.8
h — 25.18

CHAPTER 26
1. notochord, dorsal nerve cord, pharynx with gill slits, tail extending beyond anus — 26.1
2. All of them — 26.1
3. a — 26.2
4. c — 26.3, 26.4
5. lobe-finned fish — 26.4
6. c — 26.7
7. f — 26.7
8. a — 26.9
9. c — 26.10
10. f — 26.14
11. b — 26.1
h — 26.4
g — 26.5
f — 26.9
c — 26.10
d — 26.11
a — 26.11
e — 26.13

CHAPTER 27
1. growth — 27.1
2. b — 27.1
3. d — 27.2
4. c — 27.2
5. a — 27.2
6. c — 27.3
7. positive — 27.3
8. a — 25.5
9. c — 27.6
10. a — 27.6
11. b — 27.5
d — 27.1
a — 27.6
c — 27.3
e — 27.3
f — 27.3

CHAPTER 28
1. left, eudicots; right, monocots — 28.1
2. a — 28.1
3. d — 28.6
4. c — 28.6
5. c — 28.2
6. c — 28.2
7. b — 28.2
8. b — 28.2
9. d — 28.6
10. b — 28.1
d — 28.6
e — 28.2
c — 28.2
f — 28.5
a — 28.6

CHAPTER 29
1. e — 29.1
2. Casparian — 29.2
3. e — 29.2
4. b — 29.2
5. c — 29.3
6. d — 29.3
7. a — 29.4
8. d — 29.3
9. c — 29.5
10. c — 29.4
11. c — 29.4
g — 29.1
e — 29.5
b — 29.2
d — 29.3
a — 29.3
f — 29.5

CHAPTER 30
1. b — 30.1
2. c — 30.5, 30.6
3. b — 30.3
4. b — 30.3
5. a — 30.5
6. c — 30.5
7. e.g., pollen or nectar — 30.2
8. c — 30.7
9. A papaya is a berry. — 30.6
10. A peach is a drupe. — 30.6
11. c — 30.3, 30.5
f — 30.1
g — 30.3
e — 30.3
d — 30.1
b — 30.3
a — 30.3

CHAPTER 31
1. c — 31.2
2. e — 31.2
3. d — 31.4
4. a — 31.6
5. b and d — 31.3, 31.4
6. c — 31.5
7. c — 31.2
e — 31.2
b — 31.2, 31.4
a — 31.2
d — 31.2

CHAPTER 32
1. Epithelial — 32.1
2. c — 32.1
3. a — 32.2
4. b — 32.2
5. b — 32.3
6. plasma — 32.3
7. c — 32.3
8. c — 32.4
9. d — 32.5
10. skeletal muscle — 32.4
11. motor neuron — 32.5
12. a — 32.7
13. c — 32.7
14. b — 32.2
i — 32.2
e — 32.6
c — 32.6
a — 32.3
d — 32.4
f — 32.4
h — 32.3
g — 32.1

CHAPTER 33
1. a — 33.1
2. c — 33.3
3. d — 33.3
4. False — 33.4
5. a — 33.5
6. a — 33.5
7. c — 33.6, 33.7
8. c — 33.8
9. b — 33.8
10. c — 33.9
11. a — 33.8, 33.13
12. True — 33.13
13. f — 33.9
d — 33.6
g — 33.11
b — 33.10, 33.12
h — 33.11
a — 33.10
e — 33.13
i — 33.9
c — 33.10

CHAPTER 34
1. a — 34.1
2. c — 34.1
3. c — 34.2
4. e — 34.3
5. a — 34.2
6. b — 34.4
7. b — 34.5
8. a — 34.5
9. b — 34.9
10. a — 34.8
11. b — 34.8
12. See Figure 34.17
13. d — 34.9
g — 34.5
f — 34.7
a — 34.5
c — 34.9
e — 34.3
b — 34.4
h — 34.2

CHAPTER 35
1. a — 35.1
2. a — 35.2
3. c, b, a, d — 35.3
4. a — 35.3
5. a — 35.4
6. b — 35.6
7. b — 35.6
8. b — 35.8
9. b — 35.8
10. d — 35.10
11. b — 35.12
12. True — 35.5
13. False — 35.12
14. False — 35.13
15. d — 35.10
f — 35.6
c — 35.3
e — 35.8
a — 35.12
b — 35.1

CHAPTER 36
1. a — 36.1
2. d — 36.3
3. b — 36.6
4. a — 36.4
5. a — 36.3
6. c — 36.2
7. b — 36.7
8. b — 36.7
9. b — 36.7
10. d — 36.9
11. a — 36.11
12. d — 36.10
13. e — 36.3
f — 36.10
g — 36.10
h — 36.4
j — 36.7
c — 36.3
b — 36.2
i — 36.7
d — 36.10
k — 36.2
a — 36.8

CHAPTER 37
1. b — 37.1
2. b — 37.1
3. d — 37.2
4. b — 37.4
5. d — 37.2
6. b — 37.6
7. b — 37.6
8. d — 37.2
9. c — 37.6
10. a — 37.7
11. c — 37.7
12. d — 37.8
13. d — 37.10
14. f — 37.1, 37.8
a — 37.10
g — 37.2
c — 37.6
c — 37.7
d — 37.5

CHAPTER 38
1. f — 38.2
2. d — 38.4
3. d — 38.1
4. fixed, general, immediate, limited to about 1000 antigens — 38.4
5. self/nonself recognition, specificity, diversity, memory — 38.5
6. d — 38.6
7. b — 38.9
8. b — 38.7
9. e — 38.8
10. b — 38.8
11. b — 38.9
12. g — 38.8
b — 38.7
a — 38.1
e — 38.7
d — 38.10

f	38.6							
c	38.4							

CHAPTER 39

1.	a	39.1
2.	d	39.1
3.	a	39.3
4.	a	39.4
5.	c	39.5
6.	d	39.6
7.	a	39.6
8.	True	39.6
9.	a	39.7
10.	c	39.9
11.	b	39.1, 39.9
12.	False	39.7
13.	d	39.5
	h	39.5
	f	39.5
	e	39.1
	g	39.5
	c	39.5
	b	39.5
	a	39.5

CHAPTER 40

1.	d	40.1
2.	b	40.4
3.	c	40.5
4.	b	40.5
5.	a	40.5
6.	c	40.6
7.	a	40.4
8.	b	40.5
9.	b	40.9
10.	True	40.7
11.	b	40.7
12.	a	40.7
13.	c	40.8
14.	f	40.4
	b	40.6
	a	40.4
	d	40.4
	e	40.4
	c	40.4

CHAPTER 41

1.	c	41.2
2.	b	41.3
3.	a	41.4
4.	b	41.4
5.	a	41.5
6.	b	41.5
7.	a	41.6
8.	d	41.6
9.	False	41.7
10.	c	41.4
	a	41.4
	b	41.4
	e	41.4
	d	41.6
11.	d	41.10
12.	b	41.9
13.	False	41.10

CHAPTER 42

1.	b	42.1
2.	a	42.2
3.	c	42.2
4.	b	42.3
5.	b	42.4
6.	c	42.4
7.	c	42.5
8.	a	42.5
9.	d	42.8
10.	a	42.9
11.	a	42.3
	c	42.3
	b	42.4
	d	42.2
12.	a, c, a, c, a, c, b	42.10
13.	d	42.2
	c	42.2
	h	42.2
	a	42.2
	i	42.4
	e	42.4
	f	42.8
	g	42.4

CHAPTER 43

1.	b	43.1
2.	False	43.2
3.	a	43.4
4.	d	43.4
5.	c	43.4
	d	43.3
	a	43.1
	b	43.1
	e	43.4
6.	c	43.7
7.	a	43.7
8.	c	43.9
9.	c	43.10
10.	b	43.12
11.	4	43.8
	3	43.7
	2	43.7
	1	43.7
	5	43.8
	6	43.8
12.	a	43.8

CHAPTER 44

1.	d	44.1
2.	b	44.1
3.	c	44.3
4.	a	44.4
5.	a	44.4
6.	b	44.5
7.	d	44.6
8.	c	44.6
9.	e	44.7
10.	c	44.7
11.	True	44.7
12.	c	44.2
	d	44.7
	b	44.1, 44.2
	a	44.2
	e	44.4

CHAPTER 45

1.	a	45.1
2.	f	45.1
3.	c	45.2
4.	b	45.3
5.	a	45.3
6.	d	45.4
7.	d	45.5
8.	b	45.7
9.	d	45.9
10.	a	45.10
11.	c	45.4
	d	45.3
	a	45.3
	e	45.4
	b	45.4

CHAPTER 46

1.	d	46.1
2.	d	46.1
3.	d	46.1
4.	e	46.3
5.	b	46.3
6.	b	46.4
7.	a	46.4
	b	46.2
	c	46.1
	d	46.6
	e	46.3
8.	b	46.8
9.	False	46.6
10.	c	46.9
	d	46.11
	a	46.8
	e	46.8
	b	46.9
	g	46.9
	f	46.3

CHAPTER 47

1.	a	47.1
2.	d	47.1
3.	d	47.1
4.	d	47.1
5.	d	47.3
6.	d	47.4
7.	c	47.6
8.	b	47.7
9.	d	47.7
10.	d	47.8
11.	d	47.10
12.	c	47.10
13.	d	47.9, 47.10
14.	a	47.9
15.	d	47.1
	e	47.1
	c	47.1
	b	47.1
	a	47.1
	f	47.4

CHAPTER 48

1.	d	48.1
2.	c	48.2
3.	d	48.3
4.	a	48.3
5.	a	48.1
6.	e	48.4
7.	e	48.4
8.	c	48.7
9.	a	48.11
10.	d	48.12, 48.16
11.	d	48.16
12.	d	48.11
	h	48.7
	e	48.6
	c	48.7
	b	48.14
	g	48.10
	a	48.8
	f	48.16

CHAPTER 49

1.	True	49.1
2.	b	49.1
3.	b	49.1
4.	a	49.2
5.	b	49.2
6.	c	49.3
7.	c	49.4
8.	a	49.4
9.	True	49.5
10.	d	49.5
11.	a	49.7
12.	f	49.8
	d	49.4
	a	49.1
	g	49.4
	b	49.6
	c	49.8
	e	49.8
	h	49.4

CHAPTER 11: GENETIC PROBLEMS

1. **a.** *AB*
   b. *AB, aB*
   c. *Ab, ab*
   d. *AB, Ab, aB, ab*

2. **a.** All offspring will be *AaBB*.
   b. 1/4 *AABB*   (25% each genotype)
   1/4 *AABb*
   1/4 *AaBB*
   1/4 *AaBb*
   c. 1/4 *AaBb*   (25% each genotype)
   1/4 *Aabb*
   1/4 *aaBb*
   1/4 *aabb*
   d. 1/16 *AABB* (6.25% of genotype)
   1/8 *AaBB*   (12.5%)
   1/16 *aaBB*   (6.25%)
   1/8 *AABb*   (12.5%)
   1/4 *AaBb*   (25%)
   1/8 *aaBb*   (12.5%)
   1/16 *AAbb*   (6.25%)
   1/8 *Aabb*   (12.5%)
   1/16 *aabb*   (6.25%)

3. **a.** *ABC*
   b. *ABC, aBC*
   c. *ABC, aBC, ABc, aBc*
   d. *ABC*
   aBC
   AbC
   abC
   ABc
   aBc
   Abc
   abc

4. **a.** Both parents are heterozygotes (*Aa*). Their children may be albino (*aa*) or unaffected (*AA* or *Aa*).
   b. All are homozygous recessive (*aa*).
   c. Homozygous recessive (*aa*) father, and heterozygous (*Aa*) mother. The albino child is *aa*, the unaffected children *Aa*.

5. A mating of two M^L cats yields 1/4 MM, 1/2 M^LM, and 1/4 M^LM^L. Because M^LM^L is lethal, the probability that any one kitten among the survivors will be heterozygous is 2/3.

6. Possible outcomes of an experimental cross between F_1 rose plants heterozygous for height (Aa):

	A	a
A	AA climber	Aa climber
a	Aa climber	aa shrubby

3:1 possible ratio of genotypes and phenotypes in F_2 generation

Possible outcomes of a testcross between an F_1 rose plant heterozygous for height and a shrubby rose plant:

Gametes F_1 hybrid:

Gametes shrubby plant:	A	a
a	Aa climber	aa shrubby
a	Aa climber	aa shrubby

1:1 possible ratio of genotypes and phenotypes in F_2 generation

7. Yellow is recessive. Because F_1 plants have a green phenotype and must be heterozygous, green must be dominant over the recessive yellow.

8. A mating between a mouse from a true-breeding, white-furred strain and a mouse from a true-breeding, brown-furred strain would provide you with the most direct evidence. Because true-breeding strains of organisms typically are homozygous for a trait being studied, all F_1 offspring from this mating should be heterozygous. Record the phenotype of each F_1 mouse, then let them mate with one another. Assuming only one gene locus is involved, these are possible outcomes for the F_1 offspring:

a. All F_1 mice are brown, and their F_2 offspring segregate: 3 brown : 1 white. *Conclusion:* Brown is dominant to white.

b. All F_1 mice are white, and their F_2 offspring segregate: 3 white : 1 brown. *Conclusion:* White is dominant to brown.

c. All F_1 mice are tan, and the F_2 offspring segregate: 1 brown : 2 tan : 1 white. *Conclusion:* The alleles at this locus show incomplete dominance.

9. The data reveal that these genes do not assort independently because the observed ratio is very far from the 9:3:3:1 ratio expected with independent assortment. Instead, the results can be explained if the genes are located close to each other on the same chromosome, which is called linkage.

10. a. <u>1/2</u> red   <u>1/2</u> pink   _____ white
    b. _____ red   <u>All</u> pink   _____ white
    c. <u>1/4</u> red   <u>1/2</u> pink   <u>1/4</u> white
    d. _____ red   <u>1/2</u> pink   <u>1/2</u> white

11. Because both parents are heterozygotes (Hb^AHb^S), the following are the probabilities for each child:

    a. 1/4 Hb^SHb^S

    b. 1/4 Hb^AHb^A

    c. 1/2 Hb^AHb^S

CHAPTER 12: GENETIC PROBLEMS

1. a. Human males (XY) inherit their X chromosome from their mother.

b. A male can produce two kinds of gametes. Half carry an X chromosome and half carry a Y chromosome. All the gametes that carry the X chromosome carry the same X-linked allele.

c. A female homozygous for an X-linked allele produces only one kind of gamete.

d. Fifty percent of the gametes of a female who is heterozygous for an X-linked allele carry one of the two alleles at that locus; the other fifty percent carry its partner allele for that locus.

2. Because Marfan syndrome is a case of autosomal dominant inheritance and because one parent bears the allele, the probability that any child of theirs will inherit the mutant allele is 50 percent.

3. a. Nondisjunction might occur during anaphase I or anaphase II of meiosis.

b. As a result of translocation, chromosome 21 may get attached to the end of chromosome 14. The new individual's chromosome number would still be 46, but its somatic cells would have the translocated chromosome 21 in addition to two normal chromosomes 21.

4. A daughter could develop this muscular dystrophy only if she inherited two X-linked recessive alleles— one from each parent. Males who carry the allele are unlikely to father children because they develop the disorder and die early in life.

5. In the mother, a crossover between the two genes at meiosis generates an X chromosome that carries neither mutant allele.

6. The phenotype appeared in every generation shown in the diagram, so this must be a pattern of autosomal dominant inheritance.

7. There is no scientific answer to this question, which simply invites you to reflect on the difference between a scientific and a subjective interpretation of this individual's condition.

Appendix IV. Periodic Table of the Elements

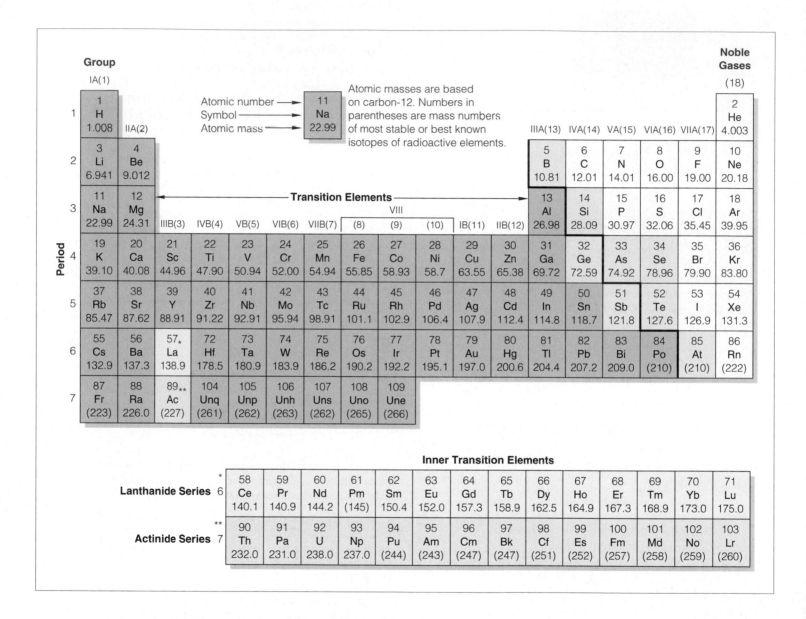

Group

IA(1)

Atomic number → 11
Symbol → Na
Atomic mass → 22.99

Atomic masses are based on carbon-12. Numbers in parentheses are mass numbers of most stable or best known isotopes of radioactive elements.

Noble Gases

(18)

Period	IA(1)	IIA(2)				Transition Elements							IIIA(13)	IVA(14)	VA(15)	VIA(16)	VIIA(17)	(18)
1	1 H 1.008																	2 He 4.003
2	3 Li 6.941	4 Be 9.012											5 B 10.81	6 C 12.01	7 N 14.01	8 O 16.00	9 F 19.00	10 Ne 20.18
3	11 Na 22.99	12 Mg 24.31	IIIB(3)	IVB(4)	VB(5)	VIB(6)	VIIB(7)	VIII (8)	(9)	(10)	IB(11)	IIB(12)	13 Al 26.98	14 Si 28.09	15 P 30.97	16 S 32.06	17 Cl 35.45	18 Ar 39.95
4	19 K 39.10	20 Ca 40.08	21 Sc 44.96	22 Ti 47.90	23 V 50.94	24 Cr 52.00	25 Mn 54.94	26 Fe 55.85	27 Co 58.93	28 Ni 58.7	29 Cu 63.55	30 Zn 65.38	31 Ga 69.72	32 Ge 72.59	33 As 74.92	34 Se 78.96	35 Br 79.90	36 Kr 83.80
5	37 Rb 85.47	38 Sr 87.62	39 Y 88.91	40 Zr 91.22	41 Nb 92.91	42 Mo 95.94	43 Tc 98.91	44 Ru 101.1	45 Rh 102.9	46 Pd 106.4	47 Ag 107.9	48 Cd 112.4	49 In 114.8	50 Sn 118.7	51 Sb 121.8	52 Te 127.6	53 I 126.9	54 Xe 131.3
6	55 Cs 132.9	56 Ba 137.3	57* La 138.9	72 Hf 178.5	73 Ta 180.9	74 W 183.9	75 Re 186.2	76 Os 190.2	77 Ir 192.2	78 Pt 195.1	79 Au 197.0	80 Hg 200.6	81 Tl 204.4	82 Pb 207.2	83 Bi 209.0	84 Po (210)	85 At (210)	86 Rn (222)
7	87 Fr (223)	88 Ra 226.0	89** Ac (227)	104 Unq (261)	105 Unp (262)	106 Unh (263)	107 Uns (262)	108 Uno (265)	109 Une (266)									

Inner Transition Elements

Lanthanide Series *6	58 Ce 140.1	59 Pr 140.9	60 Nd 144.2	61 Pm (145)	62 Sm 150.4	63 Eu 152.0	64 Gd 157.3	65 Tb 158.9	66 Dy 162.5	67 Ho 164.9	68 Er 167.3	69 Tm 168.9	70 Yb 173.0	71 Lu 175.0
Actinide Series **7	90 Th 232.0	91 Pa 231.0	92 U 238.0	93 Np 237.0	94 Pu (244)	95 Am (243)	96 Cm (247)	97 Bk (247)	98 Cf (251)	99 Es (252)	100 Fm (257)	101 Md (258)	102 No (259)	103 Lr (260)

Appendix V. Molecular Models

A molecule's structure can be depicted by different kinds of molecular models. Such models allow us to visualize different characteristics of the same structure.

Structural models show how atoms in a molecule connect to one another:

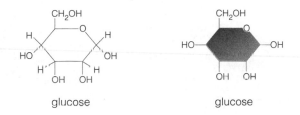

methane   glucose

In such models, each line indicates one covalent bond: Double bonds are shown as two lines; triple bonds as three lines. Some atoms or bonds may be implied but not shown. For example, carbon ring structures such as those of glucose and other sugars are often represented as polygons. If no atom is shown at the corner of a polygon, a carbon atom is implied. Hydrogen atoms bonded to one of the atoms in the carbon backbone of a molecule may also be omitted:

glucose   glucose

Ball-and-stick models show the relative sizes of the atoms and their positions in three dimensions:

methane   glucose

All types of covalent bonds (single, double, or triple) are shown as one stick. Typically, the elements in such models are coded in standardized colors:

carbon   hydrogen   oxygen   nitrogen

Space-filling models show the outer boundaries of the atoms in three dimensions:

methane   glucose

A model of a large molecule can be quite complex if all the atoms are shown. This space-filling model of hemoglobin is an example:

To reduce visual complexity, other types of models omit individual atoms. Surface models of large molecules can show features such as an active site crevice (Figure 5.7). In this surface model of hemoglobin, you can see two heme groups (red) nestled in pockets of the protein:

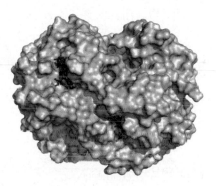

Large molecules such as proteins are often shown as ribbon models. Such models highlight secondary structure such as coils or sheets. In this ribbon model of hemoglobin, you can see the four coiled polypeptide chains, each of which folds around a heme group:

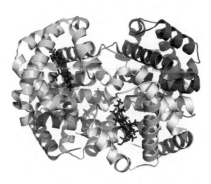

Such structural details are clues to how a molecule functions. Hemoglobin is the main oxygen carrier in vertebrate blood. Oxygen binds at the hemes, so one hemoglobin molecule can hold four molecules of oxygen.

The Amino Acids

Neutral, nonpolar side group

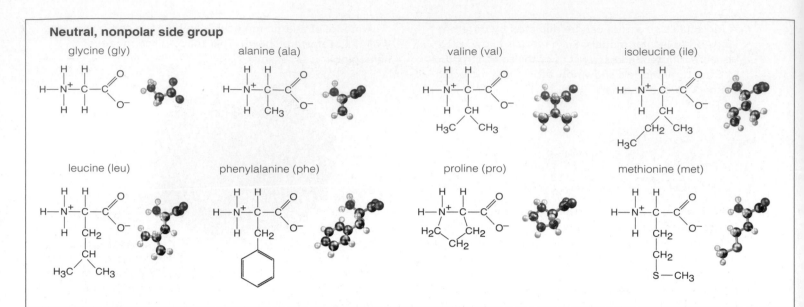

glycine (gly)   alanine (ala)   valine (val)   isoleucine (ile)

leucine (leu)   phenylalanine (phe)   proline (pro)   methionine (met)

Neutral, polar side group

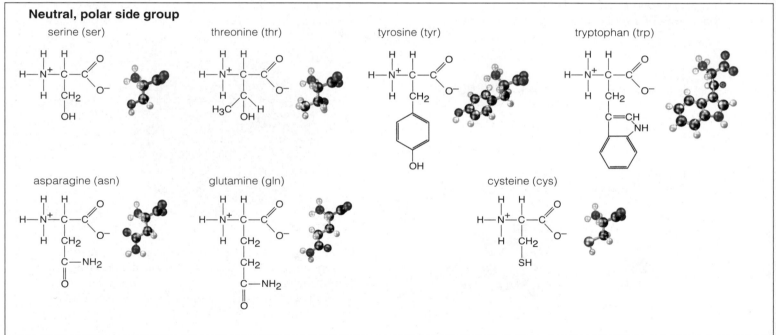

serine (ser)   threonine (thr)   tyrosine (tyr)   tryptophan (trp)

asparagine (asn)   glutamine (gln)   cysteine (cys)

Acidic side group

aspartic acid (asp)   glutamic acid (glu)

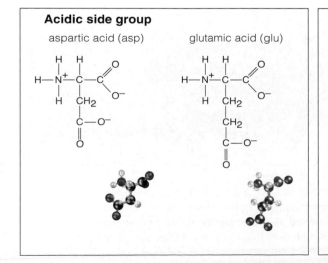

Basic side group

lysine (lys)   arginine (arg)   histidine (his)

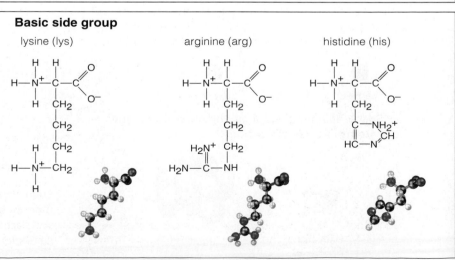

Appendix VI. Closer Look at Some Major Metabolic Pathways

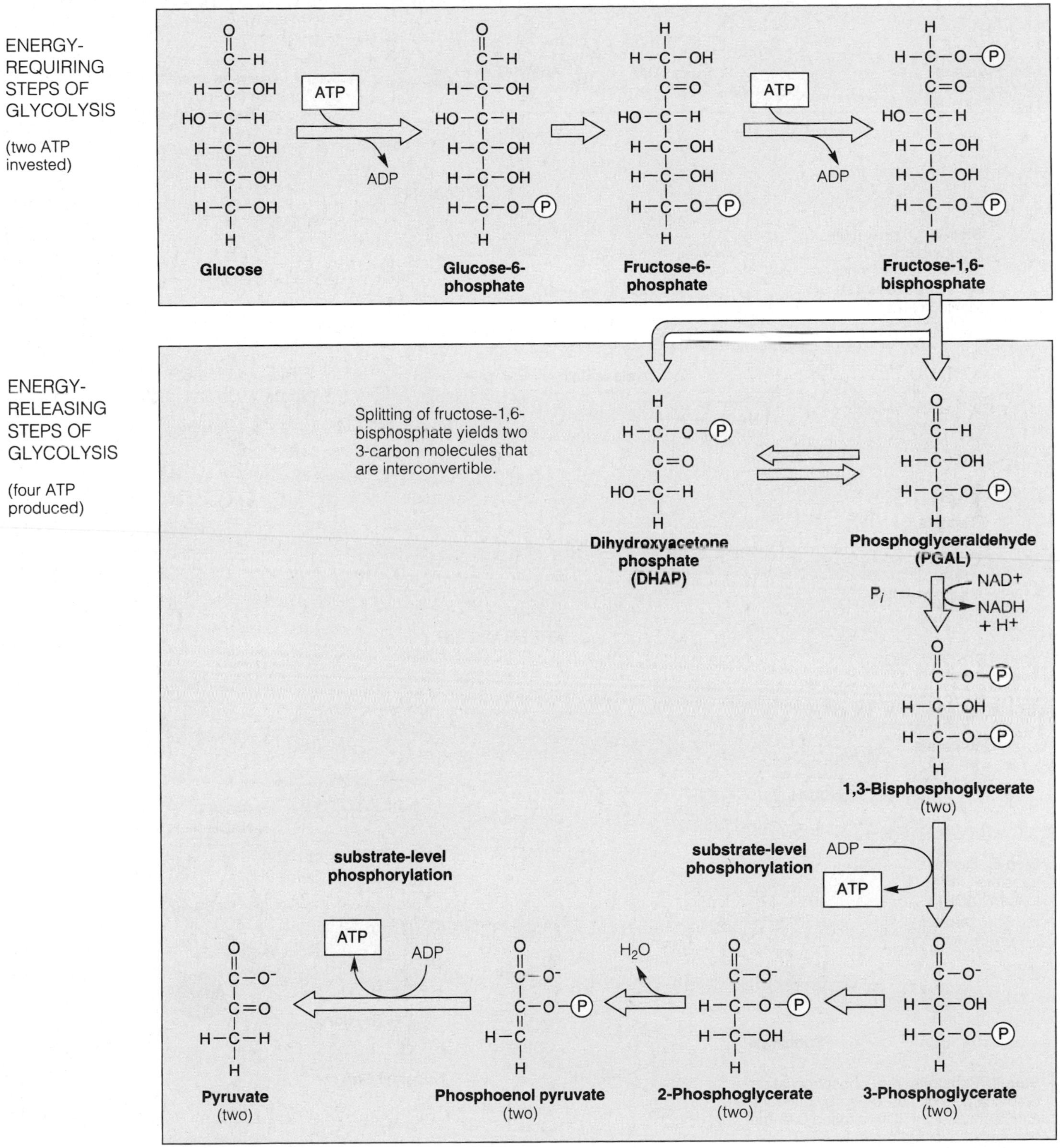
ENERGY-REQUIRING STEPS OF GLYCOLYSIS
(two ATP invested)
ATP
ADP
ATP
ADP
Glucose
Glucose-6-phosphate
Fructose-6-phosphate
Fructose-1,6-bisphosphate
ENERGY-RELEASING STEPS OF GLYCOLYSIS
(four ATP produced)
Splitting of fructose-1,6-bisphosphate yields two 3-carbon molecules that are interconvertible.
Dihydroxyacetone phosphate (DHAP)
Phosphoglyceraldehyde (PGAL)
Pi
NAD+
NADH + H+
1,3-Bisphosphoglycerate (two)
substrate-level phosphorylation
ADP
ATP
3-Phosphoglycerate (two)
2-Phosphoglycerate (two)
H2O
Phosphoenol pyruvate (two)
substrate-level phosphorylation
ATP
ADP
Pyruvate (two)

Figure A Glycolysis, ending with two 3-carbon pyruvate molecules for each 6-carbon glucose molecule entering the reactions. The *net* energy yield is two ATP molecules (two invested, four produced).

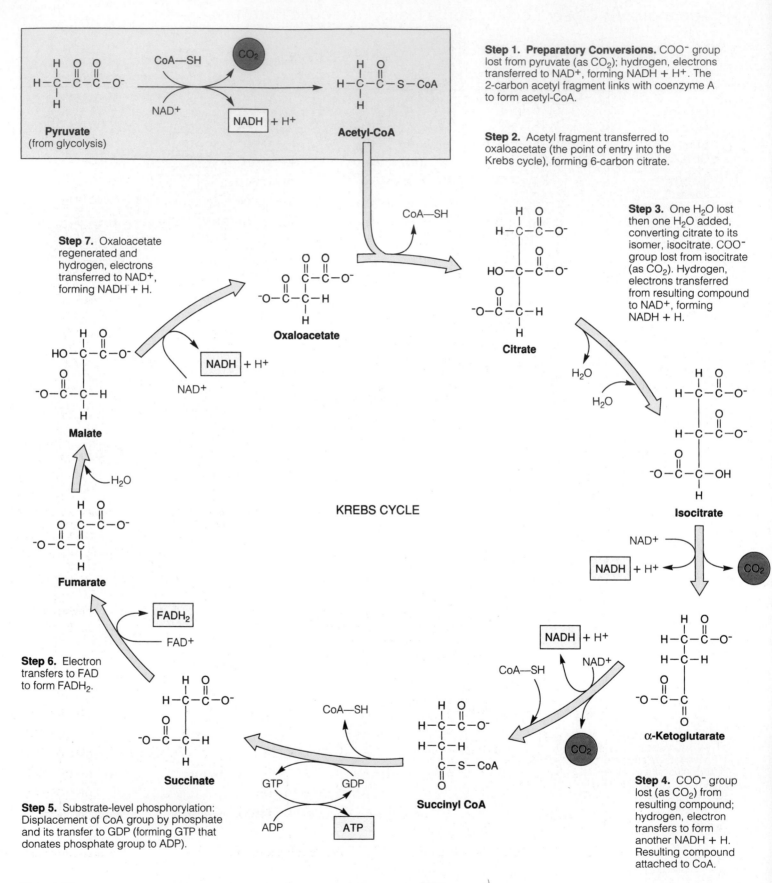

Step 1. Preparatory Conversions. COO^- group lost from pyruvate (as CO_2); hydrogen, electrons transferred to NAD^+, forming $NADH + H^+$. The 2-carbon acetyl fragment links with coenzyme A to form acetyl-CoA.

Step 2. Acetyl fragment transferred to oxaloacetate (the point of entry into the Krebs cycle), forming 6-carbon citrate.

Step 3. One H_2O lost then one H_2O added, converting citrate to its isomer, isocitrate. COO^- group lost from isocitrate (as CO_2). Hydrogen, electrons transferred from resulting compound to NAD^+, forming $NADH + H$.

Step 7. Oxaloacetate regenerated and hydrogen, electrons transferred to NAD^+, forming $NADH + H$.

Step 6. Electron transfers to FAD to form $FADH_2$.

Step 5. Substrate-level phosphorylation: Displacement of CoA group by phosphate and its transfer to GDP (forming GTP that donates phosphate group to ADP).

Step 4. COO^- group lost (as CO_2) from resulting compound; hydrogen, electron transfers to form another $NADH + H$. Resulting compound attached to CoA.

KREBS CYCLE

Figure B Krebs cycle, also known as the citric acid cycle. *Red* identifies carbon atoms entering the cyclic pathway (by way of acetyl-CoA) and leaving (by way of carbon dioxide). These cyclic reactions run twice for each glucose molecule that has been degraded to two pyruvate molecules.

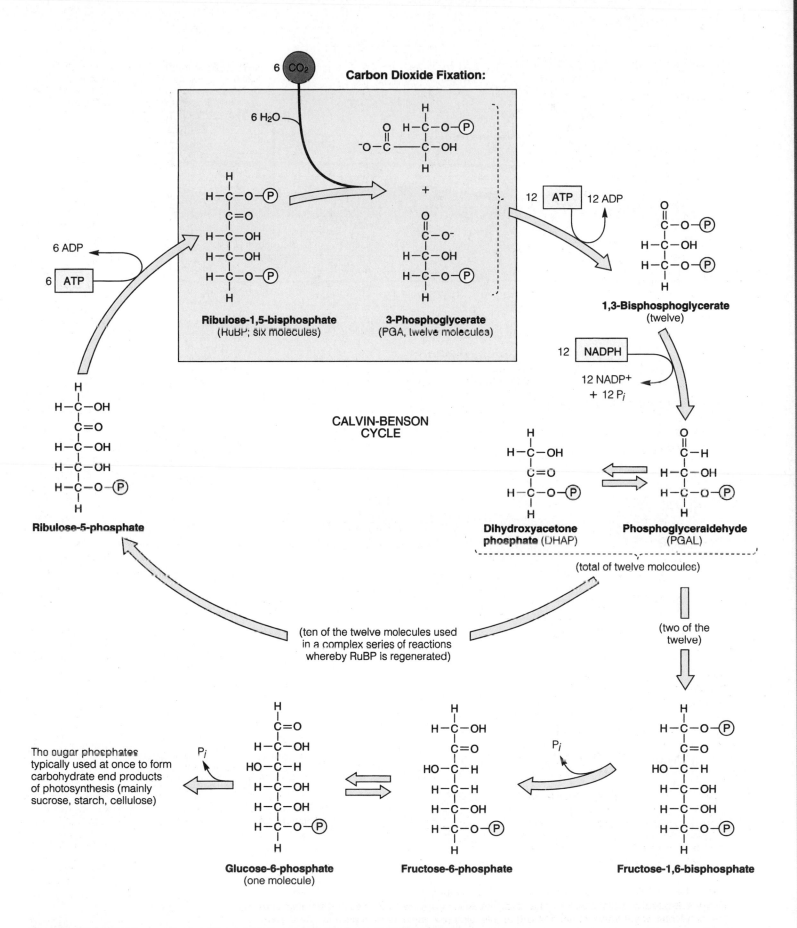

Figure C Calvin–Benson cycle of the light-independent reactions of photosynthesis.

Noncyclic photophosphorylation

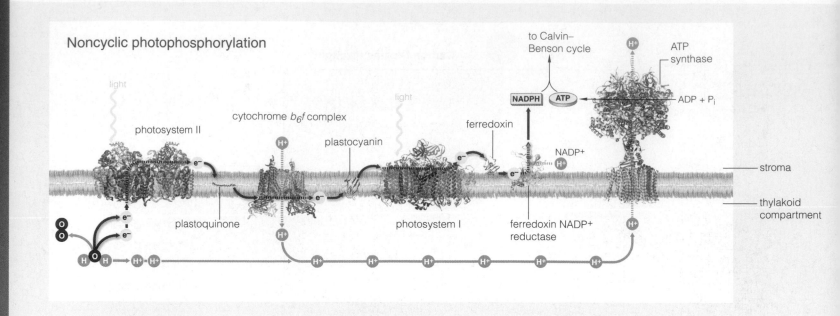

Cyclic photophosphorylation

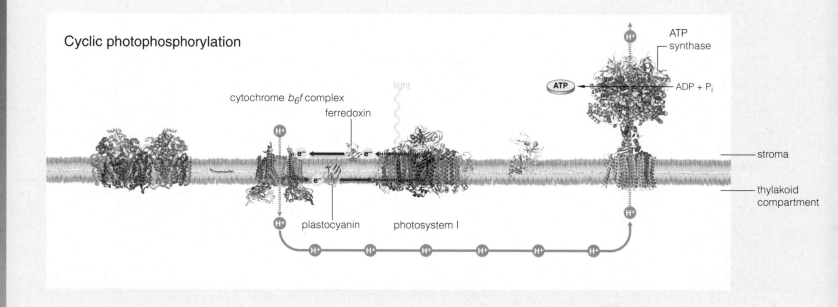

The arrangement of electron transfer chain components in highly folded thylakoid membranes maximizes the efficiency of ATP production. ATP synthases are positioned only on the outer surfaces of the thylakoid stacks, in contact with the stroma and its supply of NADP+ and ADP.

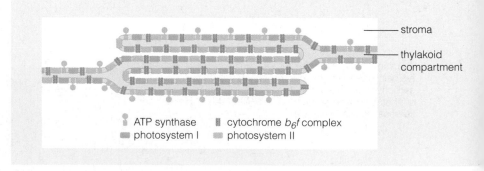

ATP synthase   cytochrome b_6f complex
photosystem I   photosystem II

Figure D Electron transfer in the light-dependent reactions of photosynthesis. Members of the electron transfer chains are densely packed in thylakoid membranes; electrons are transferred directly from one molecule to the next. For clarity, we show the components of the chains widely spaced.

Appendix VII. A Plain English Map of the Human Chromosomes

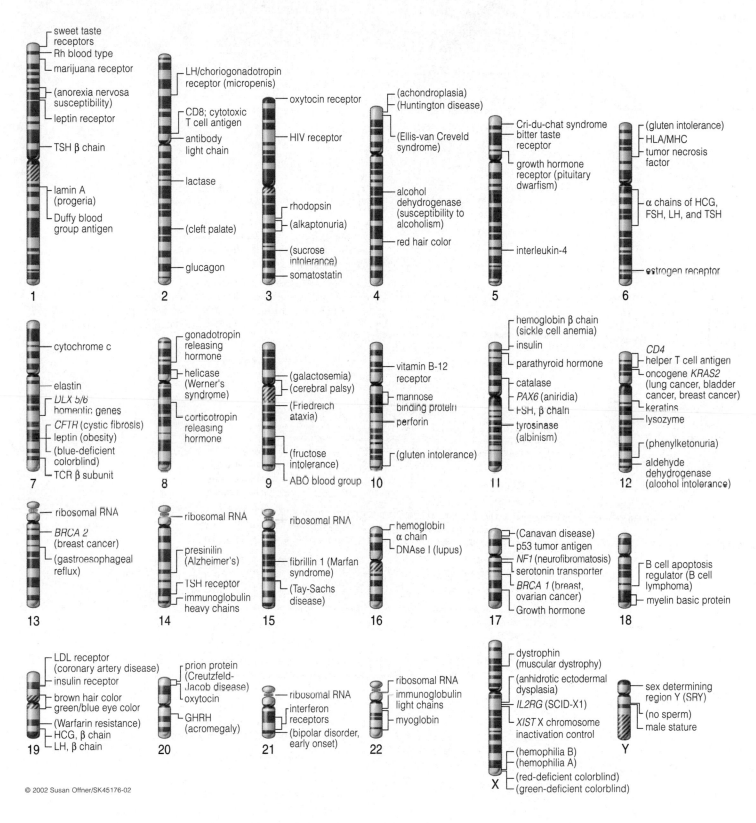

1
- sweet taste receptors
- Rh blood type
- marijuana receptor
- (anorexia nervosa susceptibility)
- leptin receptor
- TSH β chain
- lamin A (progeria)
- Duffy blood group antigen

2
- LH/choriogonadotropin receptor (micropenis)
- CD8; cytotoxic T cell antigen
- antibody light chain
- lactase
- (cleft palate)
- glucagon

3
- oxytocin receptor
- HIV receptor
- rhodopsin
- (alkaptonuria)
- (sucrose intolerance)
- somatostatin

4
- (achondroplasia)
- (Huntington disease)
- (Ellis-van Creveld syndrome)
- alcohol dehydrogenase (susceptibility to alcoholism)
- red hair color

5
- Cri-du-chat syndrome
- bitter taste receptor
- growth hormone receptor (pituitary dwarfism)
- interleukin-4

6
- (gluten intolerance)
- HLA/MHC
- tumor necrosis factor
- α chains of HCG, FSH, LH, and TSH
- estrogen receptor

7
- cytochrome c
- elastin
- DLX 5/6 homeotic genes
- CFTR (cystic fibrosis)
- leptin (obesity)
- (blue-deficient colorblind)
- TCR β subunit

8
- gonadotropin releasing hormone
- helicase (Werner's syndrome)
- corticotropin releasing hormone

9
- (galactosemia)
- (cerebral palsy)
- (Friedreich ataxia)
- (fructose intolerance)
- ABO blood group

10
- vitamin B-12 receptor
- mannose binding protein
- perforin
- (gluten intolerance)

11
- hemoglobin β chain (sickle cell anemia)
- insulin
- parathyroid hormone
- catalase
- PAX6 (aniridia)
- FSH, β chain
- tyrosinase (albinism)

12
- CD4
- helper T cell antigen
- oncogene KRAS2 (lung cancer, bladder cancer, breast cancer)
- keratins
- lysozyme
- (phenylketonuria)
- aldehyde dehydrogenase (alcohol intolerance)

13
- ribosomal RNA
- BRCA 2 (breast cancer)
- (gastroesophageal reflux)

14
- ribosomal RNA
- presinilin (Alzheimer's)
- TSH receptor
- immunoglobulin heavy chains

15
- ribosomal RNA
- fibrillin 1 (Marfan syndrome)
- (Tay-Sachs disease)

16
- hemoglobin α chain
- DNAse I (lupus)

17
- (Canavan disease)
- p53 tumor antigen
- NF1 (neurofibromatosis)
- serotonin transporter
- BRCA 1 (breast, ovarian cancer)
- Growth hormone

18
- B cell apoptosis regulator (B cell lymphoma)
- myelin basic protein

19
- LDL receptor (coronary artery disease)
- insulin receptor
- brown hair color
- green/blue eye color
- (Warfarin resistance)
- HCG, β chain
- LH, β chain

20
- prion protein (Creutzfeld-Jacob disease)
- oxytocin
- GHRH (acromegaly)

21
- ribosomal RNA
- interferon receptors
- (bipolar disorder, early onset)

22
- ribosomal RNA
- immunoglobulin light chains
- myoglobin

X
- dystrophin (muscular dystrophy)
- (anhidrotic ectodermal dysplasia)
- IL2RG (SCID-X1)
- XIST X chromosome inactivation control
- (hemophilia B)
- (hemophilia A)
- (red-deficient colorblind)
- (green-deficient colorblind)

Y
- sex determining region Y (SRY)
- (no sperm)
- male stature

Haploid set of human chromosomes. The banding patterns characteristic of each type of chromosome appear after staining with a reagent called Giemsa. The locations of some of the 20,065 known genes (as of November, 2005) are indicated. Also shown are locations that, when mutated, cause some of the genetic diseases discussed in the text.

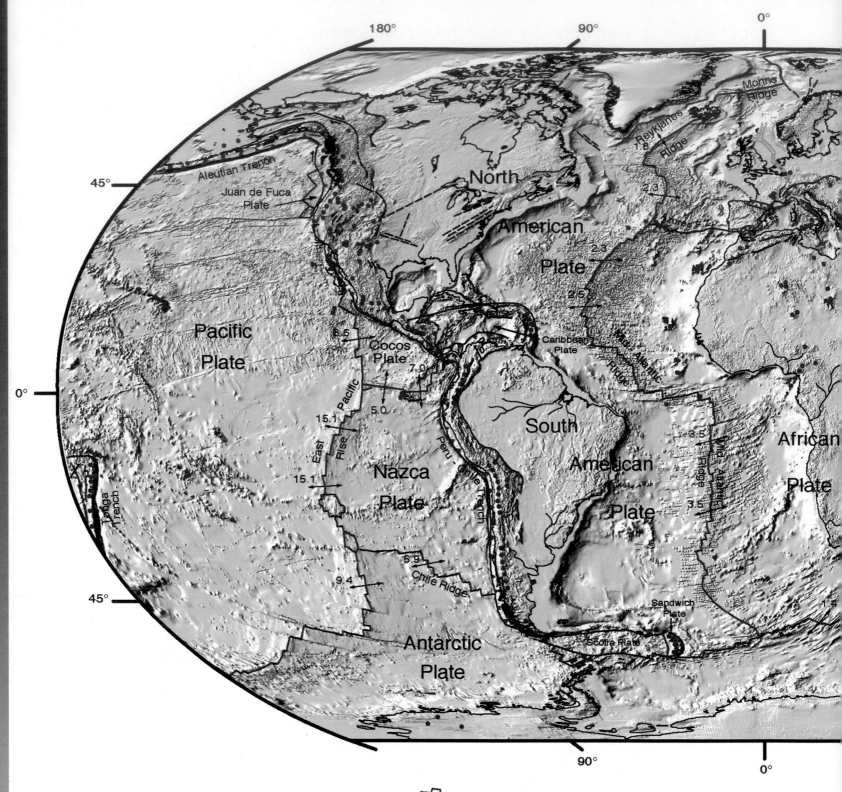

180° 90° 0°

Mohns Ridge

Reykjanes Ridge
7.8

45°

Aleutian Trench

Juan de Fuca Plate

North

2.3

American

2.3

Plate

2.5

Pacific

8.5

Cocos Plate

Caribbean Plate

Mid-Atlantic Ridge

Plate

7.0

0°

5.0

15.1

East Pacific Rise

Peru-Chile Trench

South

African

3.5

Nazca

American

Mid-Atlantic Ridge

Plate

Plate

Plate

15.1

3.5

Tonga Trench

5.9

Chile Ridge

Sandwich Plate

1.4

9.4

45°

Antarctic

Scotia Plate

Plate

90° 0°

Appendix VIII.
Restless Earth—Life's Changing Geologic Stage

This NASA map summarizes the tectonic and volcanic activity of Earth during the past 1 million years. The reconstructions at far right indicate positions of Earth's major land masses through time.

Actively-spreading ridges and transform faults

Total spreading rate, cm/year
1.4

Major active fault or fault zone; dashed where nature, location, or activity uncertain

Normal fault or rift; hachures on downthrown side

Reverse fault (overthrust, subduction zones); generalized; barbs on upthrown side

Volcanic centers active within the last one million years; generalized. Minor basaltic centers and seamounts omitted.

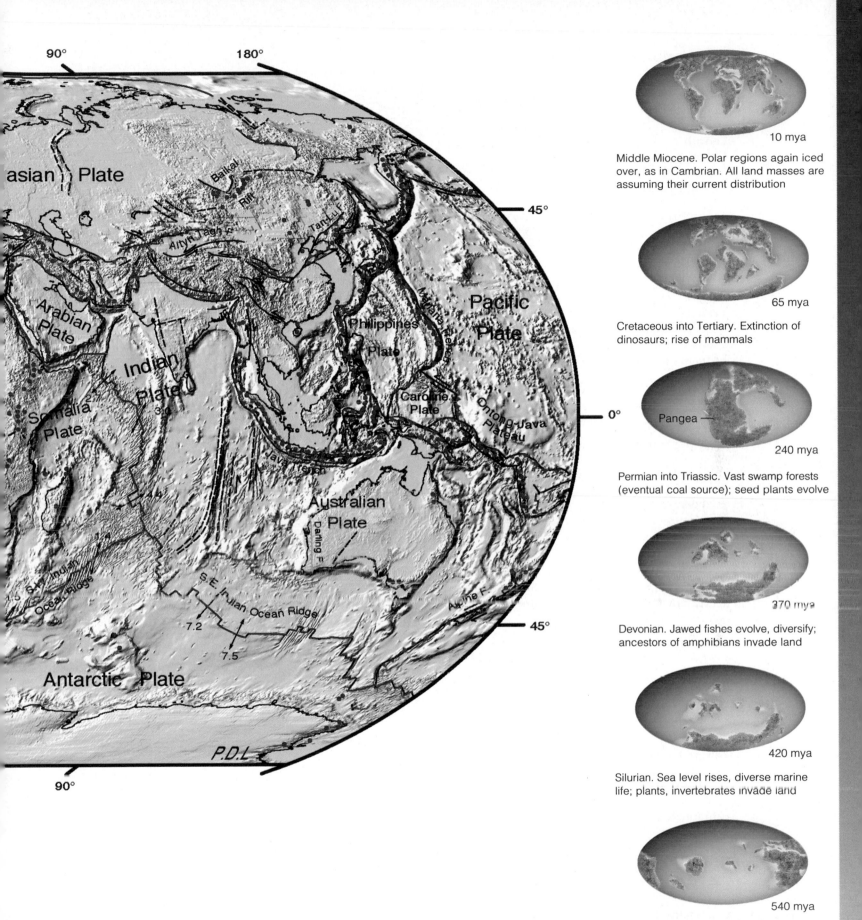

90°   180°

45°

asian   Plate

Baikal

Rift

Arabian
Plate

Altyn Tagh F.

Tarim

45°

Indian
Plate

Somalia
Plate

Philippines
Plate

Pacific
Plate

Mariana Trench

3.7

0°

Caroline
Plate

Ontong-Java
Plateau

Java Trench

4.4

Australian
Plate

Darling F.

1.4

S.W. Indian
Ocean Ridge

S.E. Indian Ocean Ridge

Alpine F.

45°

7.2

7.5

Antarctic Plate

P.D.L

90°

10 mya

Middle Miocene. Polar regions again iced
over, as in Cambrian. All land masses are
assuming their current distribution

65 mya

Cretaceous into Tertiary. Extinction of
dinosaurs; rise of mammals

Pangea

240 mya

Permian into Triassic. Vast swamp forests
(eventual coal source); seed plants evolve

370 mya

Devonian. Jawed fishes evolve, diversify;
ancestors of amphibians invade land

420 mya

Silurian. Sea level rises, diverse marine
life; plants, invertebrates invade land

540 mya

Cambrian. Fragments of Rodinia, the first
supercontinent. Major adaptive radiations
in equatorial seas; icy polar regions

Appendix VIII

Appendix IX. Units of Measure

Length

1 kilometer (km) = 0.62 miles (mi)
1 meter (m) = 39.37 inches (in)
1 centimeter (cm) = 0.39 inches

To convert	multiply by	to obtain
inches	2.25	centimeters
feet	30.48	centimeters
centimeters	0.39	inches
millimeters	0.039	inches

Area

1 square kilometer = 0.386 square miles
1 square meter = 1.196 square yards
1 square centimeter = 0.155 square inches

Volume

1 cubic meter = 35.31 cubic feet
1 liter = 1.06 quarts
1 milliliter = 0.034 fluid ounces = 1/5 teaspoon

To convert	multiply by	to obtain
quarts	0.95	liters
fluid ounces	28.41	milliliters
liters	1.06	quarts
milliliters	0.03	fluid ounces

Weight

1 metric ton (mt) = 2,205 pounds (lb) = 1.1 tons (t)
1 kilogram (kg) = 2.205 pounds (lb)
1 gram (g) = 0.035 ounces (oz)

To convert	multiply by	to obtain
pounds	0.454	kilograms
pounds	454	grams
ounces	28.35	grams
kilograms	2.205	pounds
grams	0.035	ounces

Temperature

Celcius (°C) to Fahrenheit (°F) :
$$°F = 1.8 \, (°C) + 32$$

Fahrenheit (°F) to Celsius:
$$°C = \frac{(°F - 32)}{1.8}$$

	°C	°F
Water boils	100	212
Human body temperature	37	98.6
Water freezes	0	32

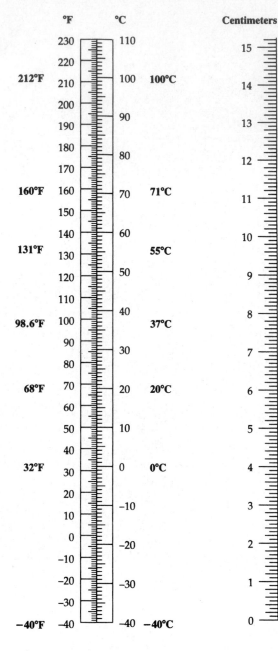

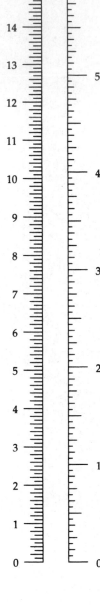

Appendix X. A Comparative View of Mitosis in Plant and Animal Cells

For step-by-step description of the stages of mitosis, refer to Figure 9.6.

Mitosis in a generalized animal cell. For simplicity, only two chromosomes are shown.

 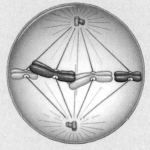 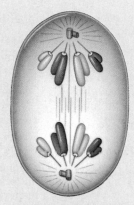 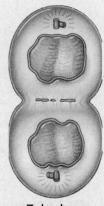

| Prophase | Metaphase | Anaphase | Telophase |

Mitosis in a white-fish cell.

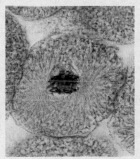  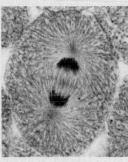 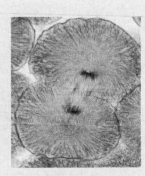

| Prophase | Metaphase | Anaphase | Telophase |

Mitosis in a lily cell.

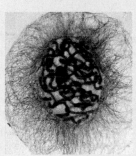  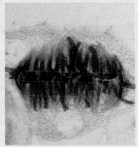  

| Prophase | Metaphase | Anaphase | Telophase |

Glossary of Biological Terms

ABC model Model of the genetic basis of flower formation; products of three master genes (*A, B, C*) control the development of sepals, petals, and stamens and carpels. 233

ABO blood typing Method of identifying certain glycoproteins (A or B) on an individual's red blood cells; the absence of either type is designated O. 642

abscisic acid Plant hormone; stimulates stomata to close in response to water stress; induces dormancy in buds and seeds. 527

abscission Plant parts are shed in response to seasonal change, drought, injury, or some nutrient deficiency. 534

acclimatization A body adjusts to a new environment; e.g., after moving from sea level to a high-altitude habitat. 696

acid Any substance that releases hydrogen ions in water. 30

acid–base balance Outcome of control over solute concentrations; extracellular fluid is neither too acidic nor too basic. 731

acid rain Rain or snow made acidic by airborne oxides of sulfur or nitrogen. 865

actin Globular protein; roles in cell shape, cell motility, and muscle contraction. 628

action potential A brief, self-propagating reversal in the voltage difference across the membrane of a neuron or muscle cell. 557

activation energy Minimum amount of energy required to start a reaction; enzymes lower it in metabolic reactions. 96

activator A regulatory protein that increases the rate of transcription when it binds to a promoter or enhancer. 230

active site Chemically stable crevice in an enzyme where substrates bind and a reaction can be catalyzed repeatedly. 98

active transport Mechanism by which a solute is moved across a cell membrane against its concentration gradient, through a transport protein. Requires energy input, as from ATP. 85, 464

adaptation A heritable trait that enhances an individual's fitness; an adaptive trait. 265

adaptive immunity Set of vertebrate immune responses characterized by self/nonself recognition, antigen specificity, antigen receptor diversity, and immune memory. Includes antibody-mediated and cell-mediated responses. 660

adaptive radiation A burst of genetic divergences from a lineage gives rise to many new species. 296

adaptive trait A heritable trait that enhances an individuals fitness; an evolutionary adaptation. 10, 265

adenine (A) A type of nitrogen-containing base in nucleotides; also, a nucleotide with an adenine base. Base-pairs with thymine in DNA and uracil in RNA. 206

adhering junction Cell junction composed of adhesion proteins; anchors cells to each other or to extracellular matrix. 540

adhesion protein In multicelled species, a membrane protein that helps cells stick to each other or to extracellular matrix. 80

adipose tissue Specialized connective tissue made up of fat-storing cells. 543

adrenal cortex Outer zone of an adrenal gland; secretes steroid hormones, including aldosterol and cortisol. 610

adrenal gland Endocrine gland located on top of the kidney; major role in stress response; affects glucose metabolism. 610

adrenal medulla Innermost zone of an adrenal gland; secretes epinephrine and norepinephrine. 610

adrenocortocotropic hormone (ACTH) Anterior pituitary hormone; stimulates release of cortisol by adrenal glands. 603

aerobic Oxygen-requiring. 124

aerobic respiration Metabolic pathway that breaks down carbohydrates to produce ATP by using oxygen. Typical yield: 36 ATP per molecule of glucose. 124

age structure Of a population, the number of individuals in each age category. 798

agglutination Clumping of foreign cells, such as red blood cells, after antibodies bind to antigens on their surface. 642

AIDS Acquired immune deficiency syndrome. A collection of diseases that develops after a virus (HIV) weakens the immune system. 676

alcoholic fermentation Anaerobic pathway that breaks down glucose, forms ethanol and ATP. Begins with glycolysis; end reactions regenerate NAD^+ so glycolysis continues. Net yield: 2 ATP per glucose. 132

aldosterone Hormone secreted by adrenal cortex; acts in kidneys to promote sodium reabsorption; concentrates urine. 610, 731

algal bloom Large increase in population size of single-celled photosynthetic protist as a result of nutrient enrichment of body of water. 358

allantois An extraembryonic membrane of amniotes. In reptiles, birds, some mammals, it exchanges gases and stores wastes; in humans, it helps form a placenta. 769

allele One of two or more forms of a gene; alleles arise by mutation and encode slightly different versions of the same gene product. 156, 278

allele frequency At a specific locus, the abundance of one allele relative to others among individuals of a population. 279

allergen A normally harmless substance that provokes an immune response in some people. 673

allergy Sensitivity to an allergen. 673

allopatric speciation Speciation route in which a physical barrier that separates members of a population ends gene flow between them. 292

allosteric A region of an enzyme other than the active site that can bind regulatory molecules. 100

alpine tundra Biome prevailing at high altitudes throughout the world; even in the summer snow persists in shaded areas, but there is no permafrost. 877

alternation of generations Alternation of haploid (gamete-producing) and diploid (spore-producing) multicelled phases in the life cycle of an organism. 353

alternative splicing mRNA processing event in which some exons are removed or joined in various combinations. By this process, one gene can specify two or more slightly different proteins. 220

altruistic behavior Social behavior that can lower an individual's reproductive success but improve that of others. 792

alveolate A type of single-celled eukaryote with many tiny, membrane-bound sacs just beneath the plasma membrane; e.g., ciliate, apicomplexan, or dinoflagellate. 357

alveolus, plural **alveoli** In a vertebrate lung, one of many tiny, thin-walled sacs where air exchanges gases with blood. 689

amino acid A small organic compound with a carboxylic acid group, an amino group, and a characteristic side group (R); monomer of polypeptide chains. 44

ammonification Process by which bacteria and fungi break down nitrogen-containing organic material and release ammonia and ammonium ions. 855

amnion Extraembryonic membrane of amniotes; outer layer of a fluid-filled sac inside which the embryo develops. 768

amniote Member of a vertebrate lineage that produces eggs having four extra-embryonic membranes (chorion, allantois, yolk sac, and amnion). Modern groups are reptiles, birds, and mammals. 442

amoeba A solitary amoebozoan protist that moves about on pseudopods. All are predatory or parasitic. **365**

amphibian A thin-skinned vertebrate that spends time on land but lays eggs in water; e.g., a frog, a toad, a salamander. **440**

anaerobic Occurring in the absence of oxygen. **124**

analogous structures Similar structures that evolved separately in different lineages; e.g., the flight surfaces of bat wings and fly wings. **305**

anaphase Stage of mitosis in which sister chromatids separate and move to opposite spindle poles. **146**

aneuploidy A chromosome abnormality in which there are too many or too few copies of a particular chromosome; e.g., having three copies of chromosome 21, which causes Down syndrome. **194**

angiosperm A flowering plant; it forms seeds inside a floral ovary, which develops into a fruit. **382**

animal A multicelled heterotroph with unwalled cells. It develops through a series of embryonic stages and is motile during part or all of the life cycle. **9, 404**

animal hormone Intercellular communication molecule secreted by an endocrine gland or cell. It enters the blood and is distributed throughout the body. **598**

annelid Bilateral, coelomate invertebrate with a highly segmented body; major groups are polychaetes, oligochaetes, and leeches. **414**

antennae In some arthropods, paired sensory appendages on the head that act in touch, smell, taste, and in detection of vibrations and temperature. **421**

antibody Y-shaped antigen receptor protein made only by B cells. **668**

antibody-mediated immune response One of two arms of adaptive immunity in which antibodies are produced in response to a specific antigen. **667**

anticodon Set of three nucleotides in a tRNA; base-pairs with mRNA codon. **221**

antidiuretic hormone (ADH) Hormone released by posterior pituitary; induces water reabsorption by kidneys. **602, 730**

antigen A molecule or particle that the immune system recognizes as nonself; triggers an immune response. **660**

antioxidant Substance that neutralizes free radicals or other strong oxidizers. **99**

anus Waste-expelling, terminal opening of a complete digestive system. **705**

aorta The main artery of human systemic circulation; receives blood from the left ventricle. **644**

apical dominance Growth-inhibiting effect on lateral (axillary) buds, mediated by auxin produced in shoot tips. **527**

apical meristem In shoot and root tips, mass of undifferentiated cells, the division of which lengthens plant parts. **477**

apicomplexan A parasitic alveolate protist that penetrates the host cell using a unique microtubular structure; e.g., malaria-causing *Plasmodium* species. **359**

apoptosis Programmed cell death. A cell commits suicide in response to molecular signals; part of a program of development and maintenance of an animal body. **470, 765**

appendicular skeleton Bone structure consisting of the pectoral (shoulder) girdle, the pelvic (hip) girdle, and limbs (or bony fins) attached to them. **620**

aquifer Permeable rock layers that hold water. **848**

archaean A member of the prokaryotic domain Archaea. Members have some unique features but also share some traits with bacteria and other traits with eukaryotic species. **8**

arctic tundra Biome prevailing at high latitudes, where short, cool summers alternate with long, cold winters; it forms between the polar ice cap and the belts of boreal forests in the Northern Hemisphere. **877**

area effect Biogeographical pattern; larger islands support more species than smaller ones at equivalent distances from sources of colonizing species. **835**

arteriole A blood vessel that carries blood from an artery to a capillary bed. **648**

artery A thick-walled, muscular vessel that carries blood away from the heart. **648**

arthropod Type of invertebrate having a hardened exoskeleton and specialized segments with jointed appendages; e.g., millipedes, spiders, lobsters, insects. **421**

asexual reproduction Any reproductive mode by which offspring arise from one parent and inherit that parent's genes only; e.g., prokaryotic fission, transverse fission, budding, vegetative propagation. **156, 740**

astrobiology A field of study concerned with the origins, evolution, and persistence of life on Earth as it relates to life in the universe. **328**

atom Particle that is a fundamental building block of matter; consists of varying numbers of electrons, protons, and neutrons. **4, 22**

atomic number The number of protons in the nucleus of atoms of a given element. **22**

ATP Adenosine triphosphate. Nucleotide that consists of an adenine base, the five-carbon sugar ribose, and three phosphate groups. The main energy carrier between reaction sites in cells. **48, 97**

ATP/ADP cycle How a cell regenerates its ATP supply. ADP forms when ATP loses a phosphate group, then ATP forms as ADP gains a phosphate group. **97**

atrial natriuretic peptide Hormone secreted by the heart in response to high blood volume; its effects makes the urine more dilute. **731**

atrium One of the two upper chambers in the heart that receive blood from veins. **646**

australopith Member of one of many now-extinct species classified as hominids, but not as members of the genus *Homo*. **454**

autoimmune response Immune response that targets one's own tissues. **675**

autonomic nervous system Portion of the peripheral system that carries signals related to smooth muscle, cardiac muscle, and glands of the viscera. **564**

autosome Any chromosome other than a sex chromosome. **186**

autotroph Organism that makes its own food using carbon from inorganic molecules such as CO_2, and energy from light or chemical reactions. **118**

auxin A plant hormone; stimulates cell division and elongation; role in gravitropism and phototropism. **527**

axial skeleton The skull, backbone, ribs, and breastbone (sternum) **670**

axon A neuron's signal-conducting zone; action potentials typically self-propagate away from the cell body along it. **556**

B cell receptor Membrane-bound IgM or IgD antibody on a naive B cell. **669**

B lymphocyte B cell. Type of white blood cell that makes antibodies. **661**

bacteria Members of the prokaryotic domain Bacteria; the most diverse and most ancient prokaryotic lineage. **8**

bacteriophage Type of virus that infects bacteria. **205, 335**

balanced polymorphism The maintenance of two or more alleles for a trait in some populations, as a result of natural selection against homozygotes. **287**

bark In woody plants, secondary phloem and periderm. **487**

base A substance that accepts hydrogen ions as it dissolves in water. **30**

basement membrane Noncellular, secreted material that attaches epithelium to an underlying tissue. **541**

base-pair substitution Type of mutation; a single base-pair change. **224**

basophil White blood cell that circulates in blood; role in inflammation. **661**

bell curve Curve that typically results when range of variation for a continuous trait is plotted against frequency in the population. **181**

benthic province Oceanic zone comprised of the ocean bottom—its rocks, and sediments. **884**

big bang model Model describing the origin of universe, as a nearly instantaneous distribution of all matter and energy through space. **318**

bilateral symmetry Body plan in which many appendages and organs are paired, one to each side of the main body axis. **404**

bile Mix of salts, cholesterol, and pigments made in the liver, stored by the gallbladder, and used in fat digestion. **707**

binary fission Asexual reproductive mode of some protists. **354**

biodiversity Variety of forms of life, in terms of genetic diversity, species diversity, and ecosystem diversity. **896**

biofilm Community of different types of microorganisms living within a shared mass of slime. **61**

biogeochemical cycle Slow movement of an element from environmental reservoirs, through food webs, then back. **847**

biogeographic realm One of many vast expanses of land defined by the presence of certain types of plants and animals. **868**

biogeography Study of patterns in the geographic distribution of species and communities. **260, 834**

biological clock Internal time-measuring mechanism by which individuals adjust their activities seasonally, daily, or both, in response to environmental cues. **532**

biological magnification A pesticide or other chemical that becomes increasingly concentrated in the tissues of organisms at higher trophic levels. **846**

bioluminescence Light emitted as a result of reactions in a living organism. **102**

biomass pyramid Chart in which tiers of a pyramid depict biomass (dry weight) in each of an ecosystem's trophic levels. **844**

biome A subdivision of a biogeographic realm; usually described in terms of the dominant plants; e.g., tropical broadleaf forest, grassland, tundra. **868**

biosphere All regions of Earth's waters, crust, and air where organisms live. **5**

biotic potential The maximum rate of increase per individual for a population growing under ideal conditions. **801**

bipedal Habitually standing upright on two legs. **452**

bipolar spindle In a eukaryotic cell, a dynamically assembled and dissasembled array of microtubules that moves chromosomes during mitosis or meiosis. **145**

bird A warm-blooded, feathered amniote descended from certain dinosaurs. **446**

bivalve Member of a mollusk subgroup that has a headless body enclosed in a hinged two part shell. **417**

blastocyst A type of blastula with a surface layer of blastomeres, a cavity filled with their secretions, and an inner cell mass that develops into the embryo. **763**

blastula A ball of cells and a cavity filled with their own secretions; outcome of the cleavage stage of animal development. **760**

blind spot Small area at the back of the retina where the optic nerve exits the eye and there are no photoreceptors. **591**

blood Fluid connective tissue that is the transport medium of circulatory systems. In vertebrates, consists of plasma, blood cells, and platelets. **543, 638**

blood–brain barrier Blood capillaries that protect the brain and spinal cord by exerting close control over which solutes enter cerebrospinal fluid. **569**

blood capillary *See* capillary, blood.

blood pressure Fluid pressure generated by heartbeats; causes blood circulation. **648**

bone tissue In vertebrates, a specialized connective tissue with a matrix hardened by calcium and other mineral ions. **543**

bony fish Fish with an endoskeleton that consists mostly of bone tissue. A lungfish, lobe-finned fish, or ray-finned fish. **439**

book lung Respiratory organ of some spiders; air and blood exchange gases across thin pagelike sheets of tissue. **685**

bottleneck Severe reduction in population size; can reduce genetic diversity. **288**

Bowman's capsule Cup-shaped first part of a nephron; the water and solutes filtered out of glomerular capillaries enter it. **727**

brain stem The most evolutionarily ancient nerve tissue in a vertebrate brain. **568**

bronchiole One of many tiny airways that deliver air to the alveoli in a lung. **689**

bronchus, plural **bronchi** An airway that delivers air from the trachea to a lung. **689**

brown alga A stramenopile; a multicelled marine autotroph with an abundance of the pigment fucoxanthin; e.g., kelps. **360**

brush border cell Cell type specialized for absorption; found on the sides and tip of a villus in the small intestine. **708**

bryophyte Nonvascular land plant. The haploid stage dominates its life cycle, and its sperm require standing water to reach eggs. A moss, liverwort, or hornwort. **374**

buffer system Set of chemicals that can keep the pH of a solution stable by alternately donating and accepting ions that contribute to pH. **31, 731**

bursa Fluid filled sac that functions as a cushion between parts in many joints. **625**

C3 plant Type of plant that uses only the Calvin–Benson cycle to fix carbon. **116**

C4 plant Type of plant that minimizes photorespiration by fixing carbon twice, using a C4 pathway in addition to the Calvin–Benson cycle. **116**

calcium pump Active transport protein; pumps calcium ions across a cell membrane against their concentration gradient. **85**

Calvin–Benson cycle Light-independent reactions of photosynthesis; cyclic pathway that forms glucose from CO_2. **115**

CAM plant Type of C4 plant that conserves water by opening stomata only at night, when it fixes carbon by a C4 pathway. **117**

camera eye Eye in which light enters through a small opening and is focused by a lens on a photoreceptor-rich retina. Evolved independently in cephalopods and vertebrates. **587**

camouflage Body coloration, patterning, form, or behavior that helps predators or prey blend with the surroundings and possibly escape detection. **824**

cancer Disease that occurs when a malignant neoplasm physically and metabolically disrupts body tissues. **151**

capillary, blood Smallest diameter blood vessel; blood exchanges substances with interstitial fluid across its wall, which is only one cell thick. **638, 648**

capillary reabsorption The process by which water moves by osmosis from the interstitial fluid into protein-rich plasma at the venous end of a capillary bed. **651**

capture–recapture method Individuals of a mobile species are captured (or selected) at random, marked, then released so they can mix with unmarked individuals. One or more samples are taken. The ratio of marked to unmarked individuals is used to estimate total population size. **799**

carbohydrate Organic molecule that consists primarily of carbon, hydrogen, and oxygen atoms in a 1:2:1 ratio. **40**

carbon cycle Atmospheric cycle. Carbon moves from its environmental reservoirs (sediments, rocks, the ocean), through the atmosphere (mostly as CO_2), food webs, and back to the reservoirs. **850**

carbon fixation Process by which carbon from an inorganic source such as CO_2 is incorporated into an organic compound. Occurs in the light-independent reactions of photosynthesis. **115**

carbon monoxide (CO) A colorless, odorless gas released by combustion of fossil fuels. **693**

carbonic anhydrase Enzyme in red blood cells that speeds the interconversion of CO_2 and water into bicarbonate. **693**

cardiac conduction system Specialized cardiac muscle cells that initiate and send signals that make other cardiac muscle cells contract. SA node, AV node, and junctional fibers that link them. **647**

cardiac cycle A recurring sequence of muscle contraction and relaxation that corresponds to one heartbeat. **616**

cardiac muscle Muscle tissue of the heart. **646**

cardiac muscle tissue A contractile tissue present only in the heart wall. **544**

cardiac pacemaker Sinoatrial (SA) node; a cluster of self-excitatory cardiac muscle cells that set the normal heart rate. **647**

carpel Female reproductive structure of a flower; a sticky or hairlike stigma, often stalked, above a chamber (ovary) that contains one or more ovules. **508**

carrying capacity Maximum number of individuals of a species that a particular environment can sustain. **802**

cartilage Specialized connective tissue with fine collagen fibers in a rubbery matrix that resists compression. **542**

cartilaginous fish Jawed fish that has a cartilage skeleton; e.g., sharks. **439**

Casparian strip Waxy, waterproof band; seals abutting cell walls of root endodermal cells, preventing water and dissolved substances from seeping through the cell walls into the vascular cylinder. **497**

catastrophism Now-abandoned hypothesis that catastrophic geologic forces unlike those of the present day shaped Earth's surface. **262**

cDNA DNA synthesized from RNA by the enzyme reverse transcriptase. **243**

cell Smallest unit with the properties of life—the capacity for metabolism, growth, homeostasis, and reproduction. **4, 56**

cell cortex Mesh of microfilaments that reinforces the plasma membrane. **72**

cell count The number of cells of a given type present in one microliter of blood. **641**

cell cycle A series of events from the time a cell forms until it reproduces. In eukaryotes, a cycle consists of interphase, mitosis, and cytoplasmic division. **144**

cell differentiation *See* differentiation.

cell junction Structure that connects a cell to another cell or to extracellular matrix; e.g., gap junction, adhering junction, tight junction. **71**

cell-mediated immune response Immune response involving cytotoxic T cells and NK cells that destroy infected or cancerous body cells. **667**

cell plate After nuclear division in a plant cell, a disk-shaped structure that forms a cross-wall between the two new nuclei. **149**

cell theory All organisms consist of one or more cells; the cell is the smallest unit of life; each new cell arises from another cell; and a cell passes hereditary material to its offspring. **55**

cell wall In many cells (not animal cells), a semirigid permeable structure around the plasma membrane. **60, 340**

central nervous system Of vertebrates, the brain and spinal cord. **555**

central vacuole A fluid-filled organelle in many plant cells. **69**

centriole A barrel-shaped structure that has a role in microtubule formation in cilia, flagella, and eukaryotic spindles. **73**

centromere Constricted region in a eukaryotic chromosome where sister chromatids are attached. **143**

cephalization During the evolution of most kinds of animals, increasing concentration of sensory structures and nerve cells at the anterior end of the body. **404, 554**

cephalopod Soft-bodied mollusk with a closed circulatory system. Moves by jet propulsion of water from a siphon; e.g., squids, octopuses, nautiluses. **417**

cerebellum Hindbrain region with reflex centers that maintain posture and smooth out limb movements. **568**

cerebral cortex Surface layer of cerebrum; it receives, integrates, and stores sensory information and coordinates responses. **570**

cerebrospinal fluid Clear extracellular fluid that bathes and protects the brain and spinal cord; contained in a system of canals and chambers. **566**

cerebrum A forebrain region concerned with olfactory input and motor responses. In mammals, it evolved into a complex integrating center. **568**

character Quantifiable, heritable characteristic or trait. **303**

character displacement Modifications of a trait of one species in a way that low-ers intensity of competition with another species; occurs over generations. **821**

charge An electrical property. Opposite charges attract; like charges repel. **22**

charophyte algae Green algal lineage most closely related to land plants. **362**

chelicerate Arthropod with four pairs of walking legs, and a head with eyes but no antennae. A horseshoe crab or arachnid. **422**

chemical bond An attractive force that arises between two atoms when their electrons interact. **25**

chemoautotroph Organism that makes its own food using carbon from inorganic sources such as carbon dioxide, and energy from chemical reactions. **118**

chemoreceptor Sensory receptor; detects dissolved ions or molecules in fluid. **578**

chiton Marine mollusk with a dorsal shell made of eight plates. **416**

chlorophyll *a* Main photosynthetic pigment in plants, algae, and cyanobacteria. **109**

chlorophyte Member of the most diverse lineage of green algae. **362**

chloroplast Organelle of photosynthesis in plants and some protists. Two outer membranes enclose a semifluid stroma. A third membrane forms a compartment that functions in ATP and NADPH formation; sugars form in the stroma. **69, 111**

choanoflagellates Protists that are the closest known living protistan relatives of animals; resemble sponge cells. **406**

chordate Animal with an embryo that has a notochord, a dorsal hollow nerve cord, gill slits in the pharynx wall, and a tail that extends past the anus. Some, none, or all of these traits persist in adults. **434**

chorion An extraembryonic membrane of amniotes; in mammals it becomes part of the placenta. Villi form at its surface and facilitate the exchange of substances between the embryo and mother. **769**

choroid A blood vessel-rich layer of the middle eye, which is darkened by the brownish pigment melanin and prevents light scattering. **588**

chromatin All of the DNA molecules and associated proteins in a nucleus. **65**

chromosome A complete molecule of DNA and its attached proteins; carries part or all of an organism's genes. Linear in eukaryotic cells; circular in prokaryotes. **65**

chromosome number The sum of all chromosomes in a cell of a given type; e.g., it is 46 in human body cells. **144**

chyme Semidigested food in the gut. **706**

chytrid A type of fungus, the only fungal group with a flagellated stage. **391**

ciliary muscle A ring-shaped muscle of the eye that encircles the lens and attaches to it by short fibers. **589**

ciliate A heterotrophic alveolate protist with cilia at its surface; also known as a ciliated protozoan; e.g., *Paramecium*. **357**

cilium, plural **cilia** Short movable structure that projects from the plasma membrane of certain eukaryotic cells. **73**

circadian rhythm Any biological activity repeated about every 24 hours. **469, 532**

circulatory system Organ system that rapidly transports substances to and from cells; typically consists of a heart, blood vessels, and blood. Helps stabilize body temperature and pH in some animals. **638**

clade A group of species that share a set of characters. **303**

cladistics Method of determining evolutionary relationships by grouping species into clades. **303**

cladogram Evolutionary tree diagram that shows a network of evolutionary relationships among clades. **303**

classical conditioning An animal's involuntary response to a stimulus becomes associated with another stimulus that is presented at the same time. **785**

cleavage Early stage of development in animals. Mitotic cell divisions divide a fertilized egg into many smaller cells (blastomeres); the original volume of egg cytoplasm does not increase. **760**

climate Prevailing weather conditions of a region; e.g., temperature, cloud cover, wind speed, rainfall, and humidity. **862**

cloaca In fish, amphibians, reptiles, and birds, opening through which digestive and urinary wastes leave the body; may also function in reproduction. **444**

clone A genetically identical copy of DNA, a cell, or an organism. **156, 243**

cloning vector A DNA molecule that can accept foreign DNA, be transferred to a host cell, and get replicated in it. **242**

closed circulatory system Organ system in which blood flows continually inside blood vessels and does not come into direct contact with tissue fluids. **638**

club fungus Fungus that produces sexual spores in a club-shaped cell; most familiar mushrooms. **396**

cnidarian A type of radially symmetrical invertebrate that makes nematocysts; has two types of epithelial tissues and a saclike gastrovascular cavity; e.g., sea anemone, jellyfish, coral. **410**

coal A nonrenewable energy source that formed more than 280 million years ago from submerged, undecayed, and slowly compacted plant remains. **378**

cochlea A fluid-filled, coiled structure in the inner ear; transduces pressure waves into action potentials. **584**

codominance Nonidentical alleles that are both fully expressed in heterozygotes; neither is dominant or recessive. **176**

codon In mRNA, a nucleotide base triplet that codes for an amino acid or stop signal during translation. *See* genetic code. **220**

coelom Of many animals, a tissue-lined cavity that lies between the gut and body wall. **405**

coenzyme An organic cofactor. **99**

coevolution The joint evolution of two closely interacting species; each species is a selective agent that shifts the range of variation in the other. **296, 382, 818**

cofactor A metal ion or a coenzyme that associates with an enzyme and is necessary for its function; e.g., NAD+. **99**

cohesion Tendency of molecules to stick together under tension; a property of liquid water. **29**

cohesion–tension theory Explanation of how water moves from roots to leaves in plants; evaporation of water from leaves creates a continuous negative pressure (tension) that pulls water from roots upward in a cohesive column. **498**

cohort A group of individuals of the same age. **804**

collecting duct In the kidney, a small tube into which many distal tubules drain, and that in turn drains into the renal pelvis. Site where the final urine concentration is determined. **727**

collenchyma Simple plant tissue; alive at maturity. Lends flexible support to rapidly growing plant parts. **478**

colon *See* large intestine.

commensalism An interspecific interaction in which one species benefits and the other is neither helped nor harmed. **818**

communication signal A social cue that is encoded in stimuli, such as the body's surface coloration or patterning, odors, sounds, and postures. **786**

community All populations of all species in a habitat. **5, 818**

companion cell Of phloem, parenchyma cell that loads sugars into sieve tubes. **479**

comparative morphology Scientific study of the body plans and structures among groups of organisms. **261**

compartmentalization In some plants, a defense response in which an attacked region becomes walled off. **468**

competition, interspecific Interaction in which the individuals of different species compete for a limited resource; suppresses population size of both species. **818**

competitive exclusion When two species require the same limited resource to survive or reproduce, the better competitor will drive the less competitive one to extinction in the shared habitat. **820**

complement A set of proteins that circulate in inactive form in blood as part of innate immunity. When activated, they destroy invaders or tag them for phagocytosis. **660**

complete digestive system A tubular digestive system; has a mouth at one end and an anus at the other. **702**

compound Type of molecule that has atoms of more than one element. **25**

compound eye Crustacean or insect eye with multiple units, each of which samples part of the visual field. **587**

concentration The number of molecules or ions per unit volume of fluid. **82**

concentration gradient Difference in concentration between adjoining regions of fluid. **82**

condensation Chemical reaction in which two molecules become covalently bonded as a larger molecule; water often forms as a by-product. **39**

conduction Of heat: the transfer of heat between two objects in contact with one another. **733**

cone cell A vertebrate photoreceptor that responds to intense light and contributes to sharp vision and color perception. **590**

conifer A type of gymnosperm adapted to conserve water through droughts and cold winters. Cone-producing woody trees or shrubs with thickly cuticled needlelike or scalelike leaves. **380**

coniferous forest Biome dominated by conifers, which tolerate cold and drought, and poor soils. **876**

conjugation Among prokaryotes, transfer of a plasmid from one cell to another. **341**

conjunctiva Mucous membrane that lines the inner surface of the eyelids and folds back to cover the eye's sclera. **588**

connective tissue Most abundant type of animal tissue. Soft connective tissues differ in the amounts and arrangements of fibroblasts and extracellular matrix. Adipose tissue, cartilage, bone tissue, and blood are specialized types. **542**

consumer Heterotroph that gets energy and carbon by feeding on tissues, wastes, or remains of other organisms. **6, 840**

continuous variation In a population, a range of small differences in a trait; result of polygenic inheritance. **180**

contractile ring A thin band of actin and myosin filaments that wraps around the midsection of an animal cell undergoing cytoplasmic division. It contracts and pinches the cytoplasm in two. **148**

contractile vacuole In freshwater protists, an organelle that collects and then expels any excess water that moves into the cell by osmosis. **355, 722**

control group In experiments, a group that is the same as an experimental group except for one variable; used as a standard of comparison. **13**

convection Transfer of heat by moving molecules of air or water. **733**

coral reef A formation consisting mainly of calcium carbonate secreted by reef-building corals. **882**

cork Component of bark; its suberized layers waterproof, insulate, and protect surfaces of woody stems and roots. **487**

cork cambium In plants, a lateral meristem that gives rise to periderm. **487**

cornea Of a human eye, the clear outer layer through which light passes on the way to the pupil. **588**

corpus callosum The thick band of nerve tracts uniting the cerebral hemispheres of higher mammals including humans. **569**

corpus luteum A glandular structure that forms from cells of a ruptured follicle after ovulation; its progesterone and estrogen secretions help thicken the endometrium in preparation for pregnancy. **749**

cortisol Steroid hormone secreted by adrenal cortex; helps maintain the blood level of glucose between meals; its level rises when the body is stressed. **610**

cotransporter Transport protein that can move two or more substances across a membrane; e.g. sodium-potassium pump. **85**

cotyledon Seed leaf; part of a flowering plant embryo. **476**

countercurrent exchange Exchange of substances by fluids that are flowing in opposing directions and are separated by a semipermeable membrane. **686**

covalent bond Chemical bond in which two atoms share a pair of electrons. **26**

craniate A chordate that has its brain inside a cranium (brain case); any fish, amphibian, reptile, bird, or mammal. **435**

critical thinking Mental process of judging information before accepting it. **11**

crossing over Process in which homologous chromosomes exchange corresponding segments during prophase I of meiosis. Puts nonparental combinations of alleles in gametes. **160**

crustacean One of the mostly marine arthropods with two pairs of antennae; e.g., a copepod, barnacle, crab, or lobster. **423**

culture Sum of behavior patterns of a social group, passed between generations by learning and symbolic behavior. **453**

cuticle Of plants, a cover of waxes and cutin on the outer wall of epidermal cells. Of annelids, a thin, flexible secreted layer. Of arthropods, a lightweight exoskeleton hardened with chitin. **70, 372**

cyanobacterium A type of prokaryotic photoautotroph; carries out photosynthesis by the noncyclic pathway and so releases oxygen. **342**

cycad A gymnosperm of subtropical or tropical habitats; many resemble palms. **380**

cyst Of many microbes, a hardy resting stage. **353**

cytokines Signaling molecules with major roles in vertebrate immunity. **661**

cytokinesis Cytoplasmic division. **148**

cytokinin A plant hormone; promotes cell division; releases lateral buds from apical dominance, inhibits senescence. **527**

cytoplasm The semifluid matrix between a cell's plasma membrane and its nucleus or nucleoid. **56**

cytoplasmic localization Accumulation of different materials in specific regions of a cell's cytoplasm. **762**

cytosine (C) A type of nitrogen-containing base in nucleotides; also, a nucleotide with a cytosine base. Base-pairs with guanine in DNA and RNA. **206**

cytoskeleton Dynamic framework of protein filaments that structurally support, organize, and move eukaryotic cells and their internal structures. Prokaryotic cells have similar protein filaments. **72**

decomposer One of the prokaryotic or fungal heterotrophs that obtains carbon and energy by breaking down wastes or remains of organisms. **840**

deletion Loss of a part of a chromosome; also, a mutation in which one or a few base pairs are lost. **192, 224**

demographics Statistics that describe a population; e.g., size, age structure. **798**

demographic transition model Model that correlates changes in population growth with stages of economic development. **812**

denature To unravel the shape of a protein or other lare biological molecule, as by high temperature or pH. **46**

dendrite In a neuron, one of the short, branching extensions that accept signals and conduct them to the cell body. **556**

dendritic cell Phagocytic white blood cell that patrols tissue fluids; presents antigen to T cells. **661**

denitrification Conversion of nitrate or nitrite to gaseous nitrogen (N_2) or nitrogen oxide (NO_2) by soil bacteria. **855**

density-dependent factor A factor that slows population growth, and either appears or worsens with crowding; e.g., disease, competition for food. **803**

density-independent factor A factor that slows population growth; its likelihood of occurring and magnitude of effect does not vary with population density. **803**

dentin A calcium-rich material similar to but harder and denser than bone that composes the principal mass of teeth. **705**

deoxyribonucleic acid *See* DNA.

dermal tissue system Tissues that cover and protect all exposed plant surfaces. **476**

dermis Skin layer beneath the epidermis; mostly dense connective tissue. **548**

desalinization The removal of salt from saltwater. **849**

desert Biome of areas where evaporation greatly exceeds rainfall, where soil is thin and vegetation sparse. **871**

desertification Conversion of grassland or irrigated or rain-fed cropland to desertlike conditions. **900**

detrital food chain Food chain in which energy flows from producers to detritivores and decomposers (rather than herbivores). **843**

detritivore Any animal that feeds on small particles of organic matter; e.g., a crab or earthworm. **840**

deuterostome A bilateral animal belonging to a lineage in which the second opening to appear on the embryo surface becomes the mouth; e.g., an echinoderm or chordate. **405**

development The process that transforms a zygote into an adult with specialized tissues and, usually, organs. **7, 462**

diaphragm Broad sheet of smooth muscle beneath the lungs; partitions the coelom into a thoracic cavity and an abdominal cavity. **689**

diatom A photosynthetic stramenopile (protist) that lives as a single cell inside a two-part silica shell. **360**

differentiation The process by which cells become specialized; occurs as different cell lineages begin to express different subsets of their genes. **230, 765**

diffusion Net movement of molecules or ions from a region where they are more concentrated to a region where they are less concentrated. **82, 464**

digestive system Body sac or tube where food is digested and absorbed, and any undigested residues expelled. Incomplete systems have one opening; the complete systems have two (mouth and anus). **702**

dihybrid experiment An experiment in which individuals with different alleles at two loci are crossed or self-fertilized; e.g., $AaBb \times AaBb$. The ratio of phenotypes in the resulting offspring offers information about dominance relationships between the alleles. **174**

dinoflagellate Alveolate protist typically having two flagella; deposits cellulose in alveoli. Heterotrophs and photoautotrophs; some cause red tides. **358**

dinosaur One of a group of reptiles that arose in the Triassic and were dominant land vertebrates for 125 million years. **443**

diploid Having two of each type of chromosome characteristic of the species ($2n$). **145**

directional selection Mode of natural selection; forms at one end of a range of phenotypic variation are favored. **282**

disease Condition that arises when the body's defenses cannot overcome infection and activities of a pathogen interfere with normal body functions. **346**

disruptive selection Mode of natural selection that favors extreme forms in the range of variation; intermediate forms are selected against. **285**

distal tubule In a kidney nephron, tube that conveys filtrate from the loop of Henle to a collecting duct. **727**

distance effect A biogeographic pattern. Islands distant from a mainland have fewer species than those closer to the potential source of colonists. **835**

DNA Deoxyribonucleic acid. Double-stranded nucleic acid twisted into a helix; hereditary material for all living organisms and many viruses. Information in its base sequence is the basis of an organism's form and function. **7, 48**

DNA chip Microscopic array of DNA fragments that collectively represent a genome; used to study gene expression. **249**

DNA cloning A set of procedures that uses living cells such as bacteria to make many identical copies of a DNA fragment. **242**

DNA fingerprint An individual's unique array of short tandem repeats. **247**

DNA library A collection of cells that hosts different fragments of foreign DNA, often representing an organism's entire genome. **244**

DNA ligase Enzyme that seals breaks in double-stranded DNA. **208**

DNA polymerase DNA replication enzyme; assembles a new strand of DNA from free nucleotides based on the sequence of a DNA template. **208**

DNA repair mechanism One of several processes by which enzymes repair broken or mismatched DNA strands. **209**

DNA replication Process by which a cell duplicates its DNA before it divides. **144**

DNA sequencing Method of determining the order of nucleotides in DNA. **246**

dominant With regard to an allele, having the ability to mask the effects of a recessive allele paired with it. **171**

dormancy Period of arrested growth. **534**

dosage compensation Theory that X chromosome inactivation equalizes gene expression between males and females. **232**

double fertilization Mode of fertilization in flowering plants in which one sperm nucleus fuses with the egg, and a second sperm nucleus fuses with the endosperm mother cell. **512**

doubling time The time it takes for a population to double in size. **801**

drug addiction Dependence on a drug, which takes on an "essential" role; follows habituation and tolerance. **563**

dry shrubland Biome of areas that get less than 25 to 60 centimeters of rain; short, multibranched woody shrubs dominate. **873**

dry woodland Biome of areas that get about 40 to 100 centimeters of rain; may have many tall trees but no dense canopy. **873**

duplication Base sequence in DNA that is repeated two or more times. **192**

eardrum Thin membrane that vibrates in response to pressure waves (sounds), thus transmitting vibrations to the bones of the middle ear. **584**

ecdysone Insect hormone with roles in metamorphosis, molting. **613**

echinoderm A radial invertebrate with some bilateral features and calcified spines or plates on the body wall; e.g., sea star. **428**

ecoregion Broad land or ocean province influenced by abiotic and biotic factors. **897**

ecosystem Community interacting with its environment through a one-way flow of energy and cycling of materials. **5, 840**

ectoderm Outer primary tissue layer of animal embryos. **546, 764**

ectotherm An animal that can stay warm mainly by absorbing environmental heat, as by basking in the sun. **444, 733**

effector Muscle (or gland) that responds to neural or endocrine signals. **466**

effector cell Antigen-sensitized B cell or T cell; active in adaptive immunity. **666**

egg Mature female gamete, or ovum. **162**

El Niño Eastward displacement of warm surface waters of the western equatorial Pacific. Recurs, alters global climates. **885**

electron Negatively charged subatomic particle that occupies orbitals around the atomic nucleus. **22**

electron transfer chain Array of enzymes and other molecules in a cell membrane that accept and give up electrons in sequence, thus releasing the energy of the electrons in small, usable increments. **101**

electron transfer phosphorylation Third stage of aerobic respiration; electron flow through electron transfer chains in inner mitochondrial membrane sets up an H+ gradient that drives ATP formation. **130**

electronegativity A measure of an atom's ability to pull electrons away from other atoms. **25**

electrophoresis Technique of separating DNA fragments by size. **246**

element A substance that consists only of atoms with the same number of protons. **22**

embryonic induction Embryonic cells produce signals that alter the behavior of neighboring cells. **764**

embryophyte Member of the clade of land plants; its eggs and embryos develop in a multicelled reproductive structure. **370**

emergent property A property of a system that does not appear in any of its component parts; e.g., cells (which are alive) are composed of many molecules (which are not alive). **5**

emigration Permanent move of one or more individuals out of a population. **800**

emulsification In the small intestine, the coating of fat droplets with bile salts so that fats remain suspended in chyme. **707**

enamel The hardest material in the body, which covers the tooth's exposed crown and reduces wear. **705**

endangered species A species endemic (native) to a habitat, found nowhere else, and highly vulnerable to extinction. **893**

endemic species A species that is confined to the limited area in which it evolved. It is more likely to go extinct than a species with a more widespread distribution. **894**

endergonic Type of reaction in which reactants have less free energy than products; requires a net energy input to proceed. **96**

endocrine gland A ductless gland that secretes hormone molecules, which typically travel in blood to target cells. **541**

endocrine system Control system of cells, tissues, and organs that interacts intimately with the nervous system; secretes hormones and other signaling molecules. **598**

endocytosis Process by which a cell takes in a substance by engulfing it in a vesicle formed from a bit of plasma membrane. **86**

endoderm Innermost primary tissue layer of animal embryos. **546, 764**

endomembrane system Series of interacting organelles between the nucleus and plasma membrane; produces lipids and proteins for secretion or insertion into cell membranes. Includes endoplasmic reticulum, Golgi bodies, vesicles. **66**

endophytic fungi One of the fungi that lives as a symbiont inside plant leaves and stems. **398**

endoplasmic reticulum (ER) Membranous organelle, a continuous system of sacs and tubes that is an extension of the nuclear envelope. Rough ER is studded with ribosomes; smooth ER is not. **66**

endoskeleton In chordates, an internal framework consisting of cartilage, bone, or both; works with skeletal muscle to position, support, and move body. **436, 619**

endosperm Nutritive tissue in the seeds of flowering plants. **384, 513**

endosperm mother cell A cell with two nuclei ($n + n$) that is part of the mature female gametophyte of a flowering plant. At fertilization, a sperm nucleus will fuse with it, forming endosperm. **512**

endospore Of certain bacteria, a resting structure enclosing a bit of cytoplasm and the DNA; resists heat, irradiation, drying, acids, disinfectants, and boiling water. When conditions favor growth, it germinates and a bacterium emerges from it. **343**

endosymbiosis An intimate, permanent ecological interaction in which one species lives and reproduces in the other's body to the benefit of one or both. **324**

endotherm An animal warmed mainly by its own metabolically generated heat. **446, 733**

energy A capacity to do work. **6, 94**

energy pyramid Diagram that depicts the energy stored in the tissues of organisms at each trophic level in an ecosystem. Lowest tier of the pyramid, representing primary producers, is always the largest. **845**

enhancer Binding site in DNA for proteins that enhance the rate of transcription. **230**

enzyme Protein or RNA that catalyzes (speeds) a reaction without being changed by it. **80, 98**

eosinophil A white blood cell specialized for combating parasites. **661**

epidermis Outermost tissue layer of plants and nearly all animals. **479, 548**

epiglottis Flaplike structure between the pharynx and larynx; its positional changes direct air into the trachea or food into the esophagus. **689**

epiphyte A plant that grows on the trunk or branch of another plant but does not withdraw nutrients from it. **377**

epistasis Interacting products of two or more gene pairs influence a trait. **177**

epithelium (epithelial tissue) Animal tissue that covers the external body surfaces and lines tubular organs and body cavities. **541**

equilibrium model of island biogeography A model describing the number of species expected to inhabit a habitat island of a particular size and distance from mainland as source of colonists. **835**

erythropoietin Kidney hormone; induces stem cells in bone marrow to give rise to red blood cells. **696**

esophagus A muscular tube between the pharynx (throat) and stomach. **705**

essential amino acid Any amino acid that an organism cannot synthesize for itself and so must obtain from food. **713**

essential fatty acid Any fatty acid that an organism cannot synthesize for itself and so must obtain from food. **712**

estrogen Major sex hormone in females. Helps oocytes mature and primes the endometrium for pregnancy; affects growth, development, and secondary sexual traits. In males, small amount has a role in sperm production. **612, 748**

estuary Partly enclosed coastal region where seawater mixes with fresh water and runoff from land, as in rivers. **880**

ethylene Gaseous plant hormone that inhibits cell division in stems and roots; also promotes abscission and fruit ripening. **527**

eudicot Flowering plant with embryos that have two cotyledons; typically has

branching leaf veins, and floral parts in fours, fives, or multiples of these. **476**

euglenoid A flagellated protist with many mitochondria. Majority are heterotrophs, others photoautotrophs. **355**

eukaryote Organism whose cells characteristically start out life with a nucleus and other membrane-enclosed organelles; a protist, plant, fungus, or animal. **8**

eukaryotic cell Type of cell that starts life with a nucleus. **56**

eukaryotic flagella *See* flagellum.

eutrophication Nutrient enrichment of a body of water; promotes population growth of phytoplankton. **857**

evaporation Transition of a liquid to a gas; requires energy input. **29, 733**

evergreen broadleaf forest Biome between latitudes 10° north and south of the equator with rainfall averages of 130 to 200 centimeters each year; tropical rain forest. **874**

evolution Change in a line of descent. **10, 262**

evolutionary tree diagram Type of diagram that summarizes evolutionary relationships among a group of species. Each branch represents a separate line of descent; each node, a divergence. **303**

exaptation Adaptation of an existing structure for a completely different purpose; a major evolutionary novelty. **296**

exergonic Type of reaction in which products have less free energy than reactants; ends with net release of energy. **96**

exocrine gland Glandular structure that secretes a substance through a duct onto a free epithelial surface; e.g., sweat gland, mammary gland. **541**

exocytosis Fusion of a cytoplasmic vesicle with the plasma membrane; as it becomes part of the membrane, its contents are released to extracellular fluid. **86**

exodermis Cylindrical sheet of cells under root epidermis of many plants. **497**

exon Nucleotide sequence that is not spliced out of RNA during processing. **220**

exoskeleton An external skeleton; e.g., the hardened arthropod cuticle. **421, 618**

exotic species Species that has become established in a new community after dispersing from its home range. **831**

experiment A test designed to support or falsify a prediction. Involves experimental and control groups. **13**

experimental group In experiments, a group of objects or individuals that display or are exposed to a variable under investigation. Experimental results for this group are compared with results for a control group. **13**

exponential growth Population increases in size by the same proportion of its total in each successive interval. **800**

external fertilization Release of gametes into water where they combine; occurs in most aquatic invertebrates, fishes, and amphibians. **741**

extinct Refers to a species that has been permanently lost from Earth. **297**

extinction Permanent loss of a species from Earth. **297**

extracellular fluid (ECF) Body fluids not in cells; e.g., plasma, interstitial fluid. **463**

extracellular matrix (ECM) Complex mixture of fibrous proteins and polysaccharides secreted by cells; supports and anchors cells, separates tissues, and has functions in cell signaling; e.g., basement membrane, bone. **70**

extreme halophile Organism adapted to a highly salty habitat; e.g., an archaean that lives in salt ponds. **344**

extreme thermophile Organism adapted to a hot habitat; e.g., an archaean that lives in a hot spring or at a hydrothermal vent. **344**

eye Sensory organ that incorporates a dense array of photoreceptors. **586**

fall overturn During the fall, waters of a temperate zone mix. Upper, oxygenated water cools, gets dense, and sinks; nutrient-rich water from the bottom moves up. **879**

fat Lipid with one, two, or three fatty acid tails attached to a glycerol. **42**

fatty acid Simple organic compound with a carboxyl group and a backbone of four to thirty-six carbon atoms; component of many lipids. Backbone of saturated types has single bonds only; that of unsaturated types has one or more double bonds. **42**

feces Digestive waste that has been concentrated by action of the colon. **710**

feedback inhibition Mechanism by which a change that results from some activity decreases or stops the activity. **100**

fermentation An anaerobic metabolic pathway by which cells harvest energy from organic molecules. *See* alcoholic fermentation and lactate fermentation. **124**

fertilization Fusion of a sperm nucleus and an egg nucleus, the result being a single-celled zygote. **162**

fetus In mammalian development, the stage after all major organ systems have formed (ninth week) until birth. **767**

fever An internally induced rise in core body temperature above the normal set point as a response to infection. **665**

fibrous root system Root system composed of an extensive mass of similar-sized roots; typical of monocots. **485**

fight–flight response Response to danger or excitement. Parasympathetic input falls, sympathetic signals increase, and adrenal glands secrete epinephrine. This readies the body to fight or escape. **565**

fin An appendage that helps stabilize and propel most fishes in water. **437**

first law of thermodynamics Energy cannot be created or destroyed. **94**

fitness The degree of adaptation to an environment, as measured by an individual's relative genetic contribution to future generations. **265**

fixation Of an allele. In a population, loss of all but one allele at a gene locus. **288**

fixed action pattern A series of instinctive movements, triggered by a simple stimulus, that continues no matter what else is going on in the environment. **784**

flagellated protozoan One of the single-celled heterotrophic protists having one or more flagella; e.g., a diplomonad. **354**

flagellum, plural **flagella** Long, slender cellular structure used for motility. Eukaryotic flagella whip from side to side; prokaryotic flagella rotate like a propeller. **60, 73, 340**

flame cells Cells at tip of the tubes of excretory organs in planarians; contain a tuft of cilia that looks like a flickering flame under a microscope. **722**

flatworm Member of a group of bilaterally symmetrical, unsegmented invertebrates having organ systems derived from three primary tissue layers, but no coelom; e.g., a planarian, fluke, or tapeworm. **412**

flower Specialized reproductive shoot of an angiosperm. **382**

fluid mosaic model A cell membrane has a mixed composition (mosaic) of lipids and proteins, the interactions and motions of which impart fluidity to it. **78**

follicle-stimulating hormone (FSH) Pituitary hormone that acts on gonads; stimulates ovarian follicle maturation in females, acts on Sertoli cells in males. **603, 745**

food chain Linear sequence of steps by which energy stored in autotroph tissues enters higher trophic levels. **841**

food web Cross-connecting food chains consisting of producers, consumers, and decomposers, detritivores, or both. **842**

foramen magnum Opening in the skull where the vertebrate spinal cord and brain connect. Its position in bipedal species is different than that of species that walk on all fours. **620**

foraminiferan Heterotrophic single-celled protist that extends its pseudopods through a perforated calcium carbonate or silica shell. Most live on the ocean floor. **356**

fossil Physical evidence of an organism that lived in the past. **261**

founder effect A form of bottlenecking. Change in allele frequencies that occurs after a few individuals establish a new population. **289**

fovea Area of the retina that is richest in photoreceptors. **590**

free energy The amount of energy that is available (free) to do work. **96**

fruit Mature ovary, often with accessory parts, from a flowering plant. **382, 516**

functional group An atom or a group of atoms covalently bonded to carbon; imparts certain chemical properties to an organic compound. **38**

fungus, plural **fungi** Type of eukaryotic heterotroph; can be multicelled or single-celled; cell walls contain chitin; obtains nutrients by extracellular digestion and absorption. **8, 390**

gallbladder Organ that stores bile from the liver; secretes bile through a duct into the small intestine. **707**

gamete Mature, haploid reproductive cell; e.g., an egg or sperm. **156**

gametophyte A haploid, multicelled body in which gametes form during the life cycle of plants and some algae. **162, 372, 508**

ganglion, plural **ganglia** Group of neuron cell bodies; may function as an integrating center for signals. **412, 554**

gap junction Cell junction that forms an open channel across the plasma membrane of adjoining animal cells; permits rapid flow of ions and small molecules from the cytoplasm of one cell to another. **540**

gastric fluid Extremely acidic mixture of secretions from the stomach lining. **706**

gastrointestinal tract The gut. Starts at the stomach and extends through the intestines to the tube's terminal opening. **705**

gastropod Member of the most diverse mollusk group. Has distinct head and a broad muscular foot that makes up most of the lower body mass. **416**

gastrula Early animal embryo with two or three primary tissue layers. **760**

gastrulation Stage of animal development; embryonic cells formed by cleavage become arranged as two or three primary tissue layers in a gastrula. **760**

gene Heritable unit of information in DNA; occupies a particular location (locus) on a chromosome. **156, 171**

gene expression Process by which the information contained in a gene becomes converted to a structural or functional part of a cell. **171, 217**

gene flow The movement of alleles into and out of a population, as by individuals that immigrate or emigrate. **289**

gene pool All of the genes in a population; a pool of genetic resources. **278**

gene therapy The transfer of a normal or modified gene into an individual with the goal of treating a genetic disorder. **254**

genetic code Set of sixty-four mRNA codons, each of which specifies an amino acid or stop signal in translation. **220**

genetic drift Change in allele frequencies in a population due to chance alone. **288**

genetic engineering Process by which deliberate changes are introduced into an individual's chromosome(s). **250**

genetic equilibrium Theoretical state in which a population is not evolving with respect to a specified gene. **279**

genetically modified organism (GMO) An organism whose genome has been deliberately modified; e.g., a transgenic organism. **250**

genome An organism's complete set of genetic material. **244**

genomics The study of genomes. The structural branch investigates the three-dimensional structure of proteins encoded by a genome; comparative branch compares genomes of different species. **249**

genotype The particular alleles carried by an individual. **171**

genus, plural **genera** A group of species that share a unique set of traits. **8**

geologic time scale Chronology of Earth's history. **270**

germ cell Animal cell that can undergo meiosis and give rise to gametes. **156**

germ layer One of the primary tissue layers in an embryo (endoderm, ectoderm, or mesoderm). **546, 760**

germination The resumption of growth of a spore or mature embryo sporophyte after dormancy, dispersal, or both. **524**

gibberellin Plant hormone; induces stem elongation, helps seeds break dormancy, has role in flowering in some species. **526**

gill A respiratory organ. In vertebrates, usually one of a pair of thin folds richly supplied with blood exchange gases with surrounding water. **416, 437, 684**

ginkgo A gymnosperm; only surviving species is a deciduous tree that has fan-shaped leaves. **380**

gland A secretory organ derived from epithelium. Hormone-secreting endocrine glands have ducts; exocrine glands are ductless. **541**

global warming Long-term increase in temperature of Earth's lower atmosphere; rising levels of greenhouse gases contribute to the increase. **853**

glomeromycete Member of a fungal group that forms mycorrhizae in which hyphae branch inside a plant cell. **393**

glomerular filtration First step in urine formation; blood pressure forces water and solutes out of glomerular capillaries into Bowman's capsule. **728**

glomerulus In a kidney nephron, a cluster of capillaries from which fluid is filtered into Bowman's capsule. **727**

glottis Opening between vocal cords. **689**

glucagon Pancreatic hormone; stimulates conversion of glycogen and amino acids to glucose when blood glucose is low. **608**

glycolysis First stage of aerobic respiration and fermentation; glucose or another sugar molecule is broken down to two pyruvate for a net yield of 2 ATP. **124**

gnetophyte A type of woody, vinelike, or shrubby gymnosperm. **380**

Golgi body Organelle of endomembrane system; enzymes inside its much-folded membrane modify polypeptide chains and lipids; the products are sorted and packaged into vesicles. **67**

gonad Primary reproductive organ in animals; produces gametes. **612, 742**

gonadotropin-releasing hormone (GnRH) A hypothalmic hormone that induces the pituitary to release hormones (LH and FSH) that act on the gonads. **745**

Gondwana Supercontinent that formed more than 500 million years ago. **273**

grassland Biome dominated by grasses and other nonwoody plants; common in interiors of continents with warm summers, cool winters, recurring fires, and 25–100 centimeters of rain. **872**

gravitropism Plant growth in a direction influenced by gravity. **530**

gray matter Portion of the brain and spinal cord that includes cell bodies and dendrites. **566**

grazing food chain Food chain in which energy flows from producers to herbivores. **842**

greenhouse effect Some atmospheric gases absorb infrared wavelengths (heat) from the sun-warmed surface, and then radiate some back toward Earth, warming it. **852**

ground tissue system Plant tissues that make up the bulk of the plant body and function in photosynthesis, structural support, storage, other tasks. **476**

groundwater Water contained in soil and in aquifers. **848**

growth Of multicelled species, increases in the number, size, and volume of cells. Of single-celled prokaryotes, increases in the number of cells. **462, 524**

growth factor Checkpoint gene product that stimulates cell division. **150**

growth hormone (GH) Pituitary hormone that promotes growth of bone and soft tissues in the young; influences metabolism in adults. **603**

guanine (G) A type of nitrogen-containing base in nucleotides; also, a nucleotide with a guanine base. Base-pairs with cytosine in DNA and RNA. **206**

guard cell One of a pair of cells that define a stoma across the epidermis of a leaf or stem. **500**

gut A sac or tube in which food is digested. Also the gastrointestinal tract from the stomach onward. **404**

gymnosperm Nonflowering seed plant; forms its seeds on exposed surfaces of spore-producing structures; a gnetophyte, cycad, ginkgo, or conifer. **380**

habitat Place where an organism or species lives; described by physical and chemical features and array of species. **818**

habituation An animal learns through experience not to respond to a stimulus that has neither positive or negative effects. **785**

hair cell Hairlike mechanoreceptor; it fires when sufficiently bent or tilted. **583**

half-life Characteristic time it takes for half of a quantity of a radioisotope to decay. **268**

haploid Having one of each type of chromosome characteristic of the species (*n*); e.g., a human gamete is haploid. **157**

hearing Perception of sound. **584**

heart Muscular pump; its contractions circulate blood through an animal body. **638**

heartwood Dense, dark, aromatic tissue at the core of older tree stems and roots. **487**

Heimlich maneuver Procedure that can dislodge an object from the trachea of a person who is choking. Upward-directed thrusts under the diaphragm force air out of the lungs and into the trachea. **690**

hemoglobin Iron-containing respiratory protein. In humans, it occurs in red blood cells and carries the most oxygen. **683**

hemostasis Process that stops blood loss from a damaged vessel by coagulation, spasm, and other mechanisms. **642**

hermaphrodite An individual with male and female reproductive organs. **409, 740**

heterotherm An animal that maintains its core temperature by controlling metabolic activity some of the time and allowing it to rise or fall at other times. **733**

heterotroph Organism that obtains carbon from organic compounds assembled by other organisms. **118**

heterozygous Having two different alleles at a gene locus; e.g., *Aa*. **171**

histone Type of protein that structurally organizes eukaryotic chromosomes. Part of nucleosomes. **143**

homeostasis The collection of processes by which the conditions in a multicelled organism's internal environment are kept within tolerable ranges. **7, 463, 540**

homeotic gene Type of master gene; its expression controls formation of specific body parts during development. **234, 766**

hominid All humanlike and human species. **452**

homologous chromosome One of a pair of chromosomes in body cells of diploid organisms; except for the nonidentical sex chromosomes, members of a pair have the same length, shape, and genes. **156**

homologous structures Similar body parts among lineages; reflect shared ancestry. **304**

homozygous Having identical alleles at a gene locus; e.g., *AA*. **171**

homozygous dominant Having a pair of dominant alleles at a locus on homologous chromosomes; e.g., *AA*. **171**

homozygous recessive Having a pair of recessive alleles at a locus on homologous chromosomes; e.g., *aa*. **171**

horizontal gene transfer Process by which a living cell acquires genes from another cell of the same or different species; e.g., by bacterial conjugation. **340**

hormone *See* animal hormone, plant hormone.

hot spot A habitat that contains many species found nowhere else and at a high risk of extinction. **896**

human Member of the genus *Homo*. **455**

human chorionic gonadotropin (HCG) Hormone first secreted by the blastocyst, and later by the placenta; helps maintain the uterine lining during pregnancy. **769**

humus Decomposing organic matter in soil. **494**

hybrid Heterozygote. Individual with two different alleles at a gene locus. **171**

hydrogen bond Attraction that forms between a covalently bonded hydrogen atom and an electronegative atom taking part in a separate covalent bond. **27**

hydrologic cycle *See* water cycle.

hydrolysis A type of cleavage reaction in which an enzyme breaks a bond by attaching a hydroxyl group to one atom and a hydrogen atom to the other. The hydrogen atom and the hydroxyl group are derived from a water molecule. **39**

hydrophilic Describes a substance that dissolves easily in water; e.g., a salt. **28**

hydrophobic Describes a substance that resists dissolving in water; e.g., an oil. **28**

hydrostatic pressure *See* turgor.

hydrostatic skeleton A fluid-filled cavity on which muscle contractions act. **410, 618**

hydrothermal vent Underwater fissure where superheated, mineral-rich water is forced out under pressure. **320, 884**

hypertonic Describes a fluid with a high solute concentration relative to another fluid. **88**

hypha, plural **hyphae** Of a multicelled fungus, a filament having chitin-reinforced walls; component of a mycelium. **390**

hypothalamus Forebrain region; a center of homeostatic control of internal environment (e.g., salt–water balance, core temperature); influences hunger, thirst, sex, other viscera-related behaviors, and emotions. **568, 602**

hypothesis, scientific Testable explanation of a natural phenomenon. **12**

hypotonic Describes a fluid with a low solute concentration relative to another fluid. **88**

immigration One or more individuals move and take up residence in another population of its species. **800**

immunity The body's ability to resist and combat infections. **660**

immunization A process that is designed to promote immunity from a disease; e.g., vaccination. **674**

implantation In mammalian pregnancy, a blastocyst burrows into uterine lining. **768**

imprinting A form of learning triggered by exposure to sign stimuli; time-dependent, usually occurs during a sensitive period while an animal is young. **784**

inbreeding Nonrandom mating among close relatives. **289**

incomplete digestive system Saclike gut; food enters and wastes leave through the same opening. **702**

incomplete dominance Condition in which one allele is not fully dominant over another, so the heterozygous phenotype is somewhere between the two homozygous phenotypes. **176**

independent assortment Theory that alleles of one gene become distributed into gametes independently of alleles of all other genes during meiosis. **174**

indicator species Any species which, by its abundance or scarcity, is a measure of the health of its habitat. **896**

induced-fit model Explanation of how some enzymes work; an active site bends or squeezes a substrate, which brings on the transition state. **98**

inflammation A local response to tissue damage that results in redness, warmth, swelling, pain; occurs when signaling molecules released by white blood cells increase blood flow to the tissue and induce phagocytes to enter it. **665**

inheritance Transmission of DNA from parents to offspring. **7**

inhibitor A hormone that slows release of another hormone. **603**

innate immunity In vertebrates, a set of general defenses against infection; recognition of pathogen-associated molecular patterns triggers phagocytosis, inflammation, and complement activation. **660**

inner ear Of vertebrates, primary organ of equilibrium and hearing; includes the vestibular apparatus and cochlea. **584**

insertion A mutation in which extra base pairs become inserted into DNA. **224**

instinctive behavior Behavior performed without having first been learned. **784**

insulin Pancreatic hormone. Its actions lower the blood level of glucose. **608**

integrator A control center that receives, processes, and stores sensory input, and coordinates the responses; e.g., a brain. **466**

integumentary exchange In some animals, gas exchange across thin, moistened skin or some other external body surface. **684**

intercostal muscles The skeletal muscles between the ribs; help change the volume of the thoracic cavity during breathing. **689**

intermediate disturbance hypothesis An explanation of community structure; holds that species richness is greatest in habitats where disturbances are moderate in intensity, frequency, or both. **829**

intermediate filament Cytoskeletal element that mechanically strengthens cell and tissue structures. **72**

internal fertilization The union of sperm and egg within the female's body. **741**

interneuron Neuron that receives input from sensory neurons and sends signals to other interneurons or to motor neurons. **554**

interphase In a eukaryotic cell cycle, the interval between mitotic divisions when a cell grows in mass, roughly doubles the number of its cytoplasmic components, and replicates its DNA. **144**

interspecific competition *See* competition, interspecific.

interstitial fluid Fluid in between cells and tissues of a multicelled body. **638**

intervertebral disk Cartilage disk that lies between adjacent vertebrae; acts as a flex point and shock absorber. **620**

intron Nucleotide sequence that intervenes between exons; excised during RNA processing. **220**

inversion Structural rearrangement of a chromosome in which part of it becomes oriented in the reverse direction. **192**

ion Atom that carries a charge because of an unequal number of protons and electrons. **25**

ionic bond Type of chemical bond; strong mutual attraction between ions of opposite charge. **26**

iris Doughnut-shaped muscle that adjusts how much light enters the eye through the pupil at its center. **586**

isotonic Describes a fluid with the same solute concentration relative to another fluid. **88**

isotopes Forms of an element that differ in the number of neutrons their atoms carry. **22**

jaw Paired, hinged cartilaginous or bony feeding structures of most chordates. **436**

joint Area of contact between bones. **624**

karyotype Image of an individual's complement of chromosomes arranged by size, length, shape, and centromere location. **187**

key innovation An evolutionary adaptation that gives its bearer the opportunity to exploit a particular environment more efficiently or in a novel way. **297**

keystone species A species that has a disproportionately a large effect on community structure, relative to its own abundance. **830**

kidney One of a pair of vertebrate organs that filter blood, remove wastes,

and help maintain the internal environment. **437, 724**

knockout experiment An experiment in which an organism is genetically engineered so one of its genes does not function. **234**

Krebs cycle The second stage of aerobic respiration; breaks down two pyruvate to CO_2 and H_2O for a net yield of two ATP and many reduced coenzymes. **128**

K-selection Selection for traits that make offspring better competitors; occurs in a population near carrying capacity. **805**

K–T asteroid impact hypothesis Idea that an asteroid impact was the cause of the mass extinction that marks the boundary between Cretaceous and Tertiary periods, 65 million years ago. **443**

La Niña Climatic event in which Pacific waters become cooler than average. **886**

labor The birth process. **776**

lactate fermentation Anaerobic pathway that breaks down glucose, forms ATP and lactate. Begins with glycolysis; regenerates NAD^+ so glycolysis continues. Net yield: 2 ATP per glucose. **133**

lactation Milk production and secretion by hormone-primed mammary glands. **776**

lake A body of standing fresh water. **878**

lancelet An invertebrate chordate, a small filter feeder with a fishlike shape. **434**

large intestine Colon. The bacteria-rich region of the vertebrate gut that absorbs water and mineral ions and also compacts undigested food residues for elimination. **705**

larva, plural **larvae** A free-living, immature stage between the embryo and adult in the life cycle of many animals. **409**

larynx Tubular airway leading to lungs; has vocal cords in some animals. **689**

lateral bud Axillary bud. A dormant shoot that forms in a leaf axil. **480**

lateral meristem Vascular cambium or cork cambium. Sheetlike cylinder of meristem inside older stems and roots. **477**

leaching Process by which water moving through soil removes nutrients from it. **495**

learned behavior Enduring modification of a behavior as an outcome of experience in the environment. **784**

lens In camera eyes, a transparent body that bends light rays so they all converge suitably onto photoreceptors. **586**

lethal mutation Mutation that drastically alters phenotype; usually causes death. **279**

lichen Symbiotic association between a fungus and a photoautotroph—an alga or cyanobacterium. **398**

life history pattern Of a species, pattern of when and how many offspring are produced during a typical lifetime. **804**

ligament A strap of dense connective tissue that bridges a skeletal joint. **624**

light-dependent reactions First stage of photosynthesis; one of two metabolic pathways (cyclic or noncyclic) in which light energy is converted to the chemical energy of ATP. NADPH and O_2 also form in the noncyclic pathway. **111**

light-independent reactions Second stage of photosynthesis; metabolic pathway in which the enzyme rubisco fixes carbon, and glucose forms. Runs on ATP and NADPH produced in the light-dependent reactions. *See also* Calvin–Benson cycle. **111**

lignin Organic compound that strengthens cell walls of vascular plants; reinforces stems and thus helps plant stand upright. **70, 372**

limbic system Centers in cerebrum that govern emotions; roles in memory. **571**

limiting factor Any essential resource that limits population growth when scarce. **802**

lineage Line of descent. **267**

linkage group All genes on a chromosome; tend to stay together during meiosis but may be separated by crossovers. **178**

lipid Fatty, oily, or waxy organic compound; often has one or more fatty acid components. **42**

lipid bilayer Structural foundation of cell membranes; mainly phospholipids arranged tail-to-tail in two layers. **57**

liver Large organ that stores glucose as glycogen and releases it as needed. Also produces bile and detoxifies some harmful substances such as alcohol. **711**

loam Soil with roughly equal amounts of sand, silt, and clay. **494**

lobe-finned fish Only bony fish having fleshy ventral fins supported by internal skeletal elements. **439**

local signaling molecule Chemical signal secreted into interstitial fluid. Has potent effects on nearby cells, but is inactivated fast; e.g., prostaglandins. **598**

locus, plural **loci** The location of a gene on a chromosome. **171**

logistic growth Population growth pattern. A population grows exponentially when small, then levels off in size once carrying capacity has been reached. **802**

loop of Henle Hairpin-shaped, tubular part of a nephron where water and solutes are reabsorbed from interstitial fluid. **727**

lung Internal respiratory organ of all birds, reptiles, mammals, most amphibians, and some fish. **437, 684**

luteinizing hormone (LH) An anterior pituitary hormone that acts on the gonads; stimulates ovulation in females, testosterone production in males. **603, 745**

lymph Interstitial fluid that has entered vessels of the lymphatic system. **654**

lymph node Lymphoid organ that is a key site for immune responses, as executed by its organized arrays of lymphocytes. **655**

lymph vascular system The portion of the lymphatic system that takes up and conducts excess tissue fluid, absorbed fats, and reclaimable solutes to blood. **654**

lysogenic pathway Viral replication mode in which viral genes get integrated into host chromosome and may be inactive through many host cell divisions before being replicated. **336**

lysosome Enzyme-filled vesicle; functions in intracellular digestion. **67**

lysozyme Antibacterial enzyme; occurs in body secretions such as mucus. **663**

lytic pathway A rapid viral replication pathway. Viral genes direct the host cell to make new virus particles, which are released when the host cell dies. **336**

macroevolution Patterns of evolution that occur above the species level. **296**

macrophage Phagocytic white blood cell that patrols tissue fluids; presents antigen to T cells. **661**

Malpighian tubule One of many small tubes that help insects and spiders dispose of wastes without losing water. **422, 723**

mammal Only amniote that makes hair and nourishes offspring with milk from the female's mammary glands. **448**

mantle Of mollusks, a tissue draped over the visceral mass. **416**

marsupial Pouched mammal. **448**

mass extinction Simultaneous loss of many lineages from Earth. **297**

mass number Total number of protons and neutrons in the nucleus of an element's atoms. **22**

mast cell White blood cell in connective tissue; factor in inflammation. **661**

master gene Gene encoding a product that affects the expression of many other genes; cascades of master gene expression often result in the completion of a complex task such as flower formation. **233**

mechanoreceptor Sensory cell that detects mechanical energy (a change in pressure, position, or acceleration). **578**

medulla oblongata Hindbrain region. Its reflex centers control respiration and other basic tasks; coordinates motor responses with complex reflexes, e.g., coughing. **568**

megaspore Haploid spore that forms in ovary of seed plants; gives rise to a female gametophyte with egg cell. **379, 512**

meiosis Nuclear division process that halves the chromosome number, to the haploid (*n*) number. Basis of sexual reproduction. **142, 156**

melanin A brownish-black pigment deposited in skin; amounts vary among ethnic groups. **549**

melatonin Pineal gland hormone that serves as part of a biological clock. **612**

memory cell Antigen-sensitized B or T cell that forms in a primary immune response but does not act immediately. Participates in a secondary response if the same antigen re-enters the body at a later time. **667**

meninges Three membranes that enclose and protect the brain and spinal cord. **566**

menopause Stage when a human female's fertility ends; menstruation ceases and secretion of sex hormones declines. **747**

menstrual cycle Approximately monthly cycle in human females of reproductive age. Hormonal changes lead to oocyte maturation and release, and prime the uterine lining for pregnancy. If pregnancy does not occur, this lining is shed and the cycle begins again. **747**

meristem Zone of undifferentiated plant cells that can divide rapidly; gives rise to differentiated cell lineages that form mature plant tissues. **476**

mesoderm Middle primary tissue layer (between endoderm and ectoderm) of most animal embryos. **546, 764**

mesophyll Type of plant tissue; photosynthetic parenchyma. **478**

messenger RNA (mRNA) Type of RNA that carries a protein-building message; intermediary between DNA and protein synthesis. **216**

metabolic pathway Series of enzyme-mediated reactions by which cells build, remodel, or break down organic molecules; e.g., photosynthesis. **100**

metabolism All the enzyme-mediated chemical reactions by which cells acquire and use energy as they build, remodel, and break down organic molecules. **39**

metamorphosis Hormone-induced growth and tissue reorganization transforms larva into the adult form. **421, 425**

metaphase Stage of mitosis during which the cell's chromosomes align midway between poles of the spindle. **146**

methanogen Any bacterium or archaean that produces methane gas. **344**

MHC marker Self-recognition protein on the surface of body cells. Triggers adaptive immune response when complexed with antigen fragments. **666**

microevolution Of a population or species, small-scale change in allele frequencies. Occurs by mutation, natural selection, genetic drift, gene flow. **279**

microfilament Cytoskeletal element that helps strengthen or change the shape of a cell. Fiber of actin subunits. **72**

microspore Walled haploid spore of seed plants; gives rise to pollen grains. **379, 512**

microsporidian Intracellular fungal parasite of aquatic habitats; only fungal group that forms flagellated spores. **393**

microtubule Cytoskeletal element involved in the movement of a cell or its components; hollow filament of tubulin subunits. **72**

microvillus, plural **microvilli** Slender extension from free surface of some cells, such as brush border cells in the small intestine; increases surface area. **541, 708**

middle ear Eardrum and ear bones that transmit air waves to the inner ear. **584**

migration Of many animals, a recurring pattern of movement between two or more regions in response to seasonal change or other environmental rhythms. **800**

mimicry Evolutionary convergence of body form; a close resemblance between species. A defenseless species may look like a well-defended one, or several well-defended species may all look alike. **824**

mineral In nutrition, an inorganic substance essential for survival and growth. **714**

mitochondrion Double-membraned organelle of ATP formation; site of second and third stages of aerobic respiration in eukaryotes. **68**

mitosis Nuclear division mechanism that maintains the chromosome number. Basis of body growth, tissue repair and replacement in multicelled eukaryotes, as well as asexual reproduction in some plants, animals, fungi, and protists. **142**

mixture Two or more types of molecules intermingled in proportions that vary. **25**

model Analogous system used to test an object or event that cannot itself be tested directly. **12**

molecular clock Method of estimating how long ago two lineages diverged by comparing DNA or protein sequences.

Assumes that neutral mutations accumulate in DNA at a constant rate. **308**

molecule Group of two or more atoms joined by chemical bonds. **4, 25**

mollusk Only invertebrate with a mantle draped over a soft, fleshy visceral mass; most have an external or internal shell; e.g., gastropods, bivalves, cephalopods. **416**

molt Periodic shedding of worn-out or too-small body structures. **407**

monocot Flowering plant with embryos that have one cotyledon; typically have parallel-veined leaves and floral parts in threes (or multiples of three). **476**

monohybrid experiment An experiment in which individuals with different alleles at one locus are crossed or self-fertilized; e.g., *Aa* × *Aa*. The phenotype ratio of the resulting offspring offers information about dominance relationships between the alleles. **172**

monomer A small molecule that is a repeating subunit in a polymer; e.g., glucose is a monomer of starch. **39**

monophyletic group An ancestor and all of its descendants. **303**

monotreme Egg-laying mammal. **448**

monsoon Wind and weather pattern that changes seasonally and is caused by differential heating of a continental interior and the nearby ocean. **867**

morphogen Master gene product; diffuses through embryonic tissues; the resulting gradient causes transcription of different genes in different parts of the embryo. **765**

morphogenesis Process by which tissues and organs form. **765**

morphological convergence Evolutionary pattern in which similar body parts evolve separately in different lineages. **305**

morphological divergence Evolutionary pattern in which a body part of an ancestor changes in its descendants. **304**

morula A cluster of sixteen cells formed by repeated cell divisions of a zygote. **768**

motor neuron Neuron that relays signals from the brain or spinal cord to muscle cells or gland cells. **554**

motor protein Type of protein that, when energized by ATP hydrolysis, interacts with cytoskeletal elements to move cell parts or the whole cell; e.g., myosin. **72**

motor unit A motor neuron and the muscle fibers that it controls. **632**

mucosa A mucus-secreting epithelium; e.g., the inner lining of the gut wall. **706**

multiple allele system Three or more alleles persist in a population. **176**

multiregional model Idea that modern humans evolved gradually from many different *Homo erectus* populations that lived in different parts of the world. **456**

muscle fatigue Decline in muscle tension when tetanic contraction is continuous. **632**

muscle fiber Long, multinucleated cell in a skeletal muscle. **626**

muscle tension Mechanical force exerted by a contracting muscle. **632**

muscle twitch A sequence of muscle contraction and relaxation in response to a brief stimulus. **632**

mutation Permanent, small-scale change in DNA. Primary source of new alleles and, thus, of life's diversity. **10, 171**

mutualism An interspecific interaction that benefits both participants. **398, 818**

mycelium, plural **mycelia** Mesh of tiny, branching, food-absorbing filaments (hyphae) of a multicelled fungi. **390**

mycorrhiza "Fungus-root." A mutualism between a fungus and plant roots. **399, 496**

myelin sheath Lipid-rich wrappings around axons of some neurons; speeds propagation of action potential. **564**

myofibril In a muscle fiber, one of many long, thin structures that run parallel with the long axis of a muscle fiber; composed of sarcomeres arranged end-to-end. **628**

myoglobin Oxygen-storing protein, most abundant in muscle. **683**

myosin An ATP-driven motor protein that moves cell components on cytoskeletal tracks. Interacts with actin in muscle to bring about contraction. **629**

myriapod Land-dwelling arthropod with two antennae and an elongated body with many segments; a millipede or centipede. **424**

natural killer cell (NK cell) Type of lymphocyte; kills infected or cancerous cells that can evade other lymphocytes. **661**

natural selection A process of evolution in which individuals of a population who vary in the details of heritable traits survive and reproduce with differing success. **10, 265, 281**

naturalist Person who observes life from a scientific perspective. **260**

nature Everything in the universe except what humans have manufactured. **4**

nectar Sweet fluid exuded by some flowers; attracts pollinators. **510**

negative feedback mechanism A major homeostatic mechanism by which some activity changes conditions in a cell or multicelled organism and thereby triggers a response that reverses the change. **466**

nematocyst Unique feature of cnidarians. Capsule that discharges a barbed or sticky thread when touched; has roles in feeding and defense. **410**

neoplasm Tumor; abnormal mass of cells that lost control over their cell cycle. **150**

nephridium, plural **nephridia** Of annelids and some other invertebrates, one of many water-regulating units that help control the content and volume of tissue fluid. **415, 723**

nephron Functional unit of the kidney; it filters water and solutes from blood, then reabsorbs adjusted amounts of both. **727**

nerve Bundles of axons enclosed in a sheath of connective tissue. **554**

nerve cord In bilateral animals, a line of communication that runs parallel with the anterior–posterior axis. In vertebrates, it develops as a hollow neural tube that gives rise to the spinal cord and brain. **412**

nerve net Nervous system of cnidarians and some other invertebrates; asymmetrical mesh of neurons. **410, 554**

nervous system Organ system that detects internal and external stimuli, integrates information, and coordinates responses. **554**

nervous tissue Animal tissue consisting of neurons and often neuroglia. **545**

neuroglial cell, plural **neuroglia** Any of the nervous tissue cells that structurally and metabolically support neurons. **545, 554**

neuromodulator Any signaling molecule that reduces or magnifies the influence of a neurotransmitter on target cells. **562**

neuromuscular junction A chemical synapse between a motor neuron and a skeletal muscle fiber. **560**

neuron A type of excitable cell; functional unit of the nervous system. **545, 554**

neurotransmitter An intercellular signaling molecule secreted by the axon endings of a neuron. **560, 598**

neutral mutation A mutation that has no effect on survival or reproduction. **279**

neutron Uncharged subatomic particle in the atomic nucleus. **22**

neutrophil Circulating phagocytic white blood cell. **661**

niche A species' unique ecological role; it is described in terms of the conditions, resources, and interactions necessary for survival and reproduction. **818**

nitrification One stage of the nitrogen cycle. Soil bacteria break down ammonia or ammonium to nitrite; then other bacteria break down nitrite to nitrate, which plants can absorb. **855**

nitrogen cycle An atmospheric cycle. Nitrogen moves from its largest reservoir (the atmosphere), then through the ocean, ocean sediments, soils, and food webs, then back to the atmosphere. **854**

nitrogen fixation Conversion of gaseous nitrogen to ammonia. **342, 496, 854**

nondisjunction Failure of sister chromatids or homologous chromosomes to separate during meiosis or mitosis. Resulting cells get too many or too few chromosomes. **194**

nonpolar Having an even distribution of charge. Two atoms share electrons equally in a nonpolar covalent bond. **27**

nonshivering heat production In response to cold stress, mitochondria in cells of brown adipose tissue release energy as heat, rather than storing it in ATP. **735**

normal flora Microorganisms that typically live on human surfaces, including the interior tubes and cavities of the digestive and respiratory tracts. **662**

notochord A rod of stiffened tissue in chordate embryos; may or may not persist as a supporting structure in the adult. **434**

nuclear envelope A double membrane that constitutes the outer boundary of the nucleus. **64**

nucleic acid Single- or double-stranded chain of nucleotides joined by sugar–phosphate bonds; e.g., DNA, RNA. **48**

nucleic acid hybridization Base-pairing between DNA or RNA from different sources. **244**

nucleoid Of a prokaryotic cell, region of cytoplasm where the DNA is concentrated. **56, 340**

nucleolus In a nucleus, a dense, irregularly shaped region where ribosomal subunits are assembled. **65**

nucleoplasm Of a nucleus, the viscous fluid enclosed by the nuclear envelope. **65**

nucleosome Smallest unit of structural organization in eukaryotic chromosomes; a length of DNA wound twice around a spool of histone proteins. **143**

nucleotide Organic compound with a five-carbon sugar, a nitrogen-containing base, and at least one phosphate group. Monomer of nucleic acids. **48, 206**

nucleus In eukaryotic cells only, organelle with an outer envelope of two pore-studded lipid bilayers; separates the cell's DNA from its cytoplasm. **22, 56**

nutrient An element or type of molecule with an essential role in an individual's survival or growth. **6, 494**

obesity Having a health-threatening excess of fat in adipose tissue. **716**

observational learning One animal acquires a new behavior by observing and imitating behavior of another. **785**

olfaction The sense of smell. **582**

olfactory receptor Chemoreceptor for a water-soluble or volatile substance. **582**

oocyte *See* primary oocyte, secondary oocyte.

oomycote *See* water mold.

open circulatory system A system where blood moves through hearts and large vessels but also mixes with interstitial fluid. **638**

operant conditioning A type of learning in which an animal's voluntary behavior is modified by the consequences of that behavior. **785**

operator Part of an operon; a DNA binding site for a repressor. **236**

operon Group of genes together with a promoter–operator DNA sequence that controls their transcription. **236**

organ Body structure composed of tissues that interact in one or more tasks. **462, 540**

organ of Corti An acoustical organ in the cochlea that transduces mechanical energy of pressure waves into action potentials. **585**

organ of equilibrium Organ that monitors a body's position and motion; e.g., human vestibular apparatus. **583**

organ system A set of organs that are interacting chemically, physically, or both, in a common task. **462, 540**

organelle Structure that carries out a specialized metabolic function inside a cell; e.g., a nucleus in eukaryotes. **62**

organic Molecule that consists primarily of carbon and hydrogen atoms; many types have functional groups. **36**

organism An individual that consists of one or more cells. **4**

osmoreceptor Sensory receptor that detects shifts in solute concentrations. **578**

osmosis Diffusion of water in response to a concentration gradient. **88**

osmotic pressure Amount of hydrostatic pressure that prevents osmosis into cytoplasm or other hypertonic fluid. **89**

osteoblast Bone-forming cell; it secretes matrix that gets mineralized. **622**

osteoclast Bone-digesting cell; it secretes enzymes that digest bone's matrix. **622**

osteocyte A mature bone cell; osteoblast that has become surrounded by its own secretions. **622**

outer ear The pinna, a skin-covered flap of cartilage projecting from the side of the head, that collects sound waves and directs them into the auditory canal. **584**

ovary In animals, a female gonad. In flowering plants, the enlarged base of a carpel, inside of which one or more ovules form and eggs are fertilized. **382, 509, 746**

oviduct Ciliated duct between the ovary and uterus; where fertilization most often occurs. Also called a Fallopian tube. **746**

ovulation Release of a secondary oocyte from an ovary. **749**

ovule In a seed-bearing plant, structure in which a haploid, egg-producing female gametophyte forms; after fertilization, it matures into a seed. **379, 509**

oxidation–reduction reaction Reaction in which one molecule accepts electrons (it becomes reduced) from another molecule (which becomes oxidized). **101**

oxyhemoglobin In red blood cells only, hemoglobin with bound oxygen. **692**

oxytocin (OT) Pituitary hormone with roles in labor, and lactation. In some mammals, also affects social behavior; e.g., pair bonding. **602**

ozone layer An atmospheric layer of high ozone concentration. **864**

pain Perception of injury. **580**

pain receptor A sensory receptor that detects tissue damage. **578**

pancreas Glandular organ that secretes enzymes and bicarbonate into the small intestine; secretes the hormones insulin and glucagon into the blood. **608**

Pangea Supercontinent that formed about 237 million years ago and broke up about 152 million years ago. **272**

parapatric speciation A speciation model in which different selection pressures lead to divergences within a single population. **295**

parasite An organism that obtains some or all the nutrients it needs from a living host, which it usually does not kill outright. **826**

parasitism Interaction in which a parasitic species benefits as it exploits and harms (but usually does not kill) the host. **818**

parasitoid A type of insect that, in a larval stage, grows inside a host (usually another insect), feeds on its tissues, and kills it. **826**

parasympathetic neuron A neuron of the autonomic nervous system. Its signals slow overall activities and divert energy to basic tasks; also works in opposition with sympathetic neurons to make small ongoing adjustments in the activities of internal organs that they both innervate. **565**

parathyroid gland One of four small glands in the back of the thyroid gland; regulates blood calcium level. **607**

parenchyma A simple plant tissue made up of living cells; has roles in photosynthesis, storage, and other tasks. **478**

partial pressure Contribution of one gas to the total pressure of a mixture of gases. **682**

passive transport Mechanism by which a concentration gradient drives the movement of a solute across a cell membrane through a transport protein; no energy input is required. **84, 464**

pathogen Disease-causing agent. **334**

pattern formation The process by which a complex body forms from local processes during embryonic development. **235, 765**

PCR Polymerase chain reaction. Method that rapidly generates many copies of a specific DNA fragment. **244**

pedigree Chart showing the pattern of inheritance of a gene in a family. **197**

pelagic province The open waters of the ocean. **884**

pellicle A thin, flexible, protein-rich body covering of some single-celled eukaryotes such as euglenoids. **354**

penis The male organ of intercourse; also has a role in urination. **743**

per capita growth rate The rate obtained by subtracting a population's per capita death rate from per capita birth rate. **800**

periderm A plant dermal tissue that replaces epidermis on older stems and roots. Consists of parenchyma, cork, cork cambium. **487**

periodic table of the elements Tabular arrangement of the known atomic elements by atomic number. **22**

peripheral nervous system All spinal and cranial nerves, branches of which extend through the body. **555**

peristalsis Recurring waves of contraction of smooth muscle that move material through a tubular organ. **705**

peritubular capillaries A set of blood capillaries that surround the tubular parts of a kidney nephron. **727**

permafrost A perpetually frozen layer that underlies arctic tundra. **877**

peroxisome Enzyme-filled vesicle that breaks down amino acids, fatty acids, and toxic substances. **67**

pH A measure of the number of hydrogen ions in a solution. pH 7 is neutral. **30**

phagocytosis "Cell eating," an endocytic pathway by which a cell engulfs particles such as microbes or cellular debris. **86**

pharynx Tube from oral cavity to the gut. In land vertebrates, it is the entrance to the esophagus and trachea. **412, 689, 705**

phenotype An individual's observable traits. **171**

pheromone Signaling molecule secreted by one individual that affects another of the same species; has roles in social behavior. **582, 598, 786**

phloem Plant vascular tissue; distributes photosynthetic products through the plant body. **372, 479**

phospholipid A lipid with a phosphate group in its hydrophilic head, and two nonpolar fatty acid tails; main constituent of cell membranes. **43**

phosphorus cycle A sedimentary cycle. Phosphorus (mainly phosphate) moves from land, through food webs, to ocean sediments, then back to land. **856**

phosphorylation Transfer of a phosphate group to a recipient molecule. **97**

photoautotroph Photosynthetic autotroph; e.g., nearly all plants, most algae, and a few bacteria. **118**

photolysis Reaction in which light energy breaks down a molecule. Photolysis of water molecules during noncyclic photosynthesis releases electrons and hydrogen ions used in the reactions, and molecular oxygen. **112**

photoperiodism Biological response to seasonal changes in the relative lengths of day and night. **532**

photophosphorylation Any light-driven phosphorylation reaction. **114**

photoreceptor A light-sensitive sensory receptor of invertebrates and vertebrates. **579**

photorespiration Reaction in which rubisco attaches oxygen instead of carbon dioxide to ribulose bisphosphate; occurs in C4 plants when stomata close and oxygen levels rise. Produces no ATP. **116**

photosynthesis The metabolic pathway by which photoautotrophs capture light energy and use it to make sugars from CO_2 and water. **6, 108**

photosystem In photosynthetic cells, a cluster of pigments and proteins that, as a unit, converts light energy to chemical energy in photosynthesis. **111**

phototropism Change in the direction of cell movement or growth in response to a light source. **531**

phylogeny Evolutionary history of a species or group of species. **303**

phytochemicals Plant molecules that are not an essential part of the human diet, but may reduce risk of certain disorders; e.g., lutein, isoflavanones. **715**

phytochrome A light-sensitive pigment that helps set plant circadian rhythms based on length of night. **532**

pigment An organic molecule that absorbs light of certain wavelengths. Reflected light imparts a characteristic color. **108**

pilus, plural **pili** A protein filament that projects from the surface of some bacterial cells. **60, 340**

pineal gland Light-sensitive, melatonin-secreting endocrine gland in the brain. **612**

pioneer species An opportunistic colonizer of barren or disturbed habitats. Adapted for rapid growth and dispersal. **828**

pituitary gland Vertebrate endocrine gland located inside the brain; interacts with the hypothalamus to control physiological functions, including activity of many other glands. Its posterior lobe stores and secretes hormones from the hypothalamus; anterior lobe makes and secretes its hormones. **602**

placenta In a placental mammal, organ that forms during pregnancy from maternal tissue and extraembryonic membranes. Allows a mother to exchange substances with a fetus but keeps their blood separate. **448, 769**

placental mammal Member of the largest mammal subgroup, the only group in which an organ (the placenta) forms and allows materials to diffuse between the bloodstreams of a mother and the embryo developing inside her uterus. **448**

placozoan Simplest known animal, with no symmetry and just two cell layers. **408**

plankton Mostly microscopic autotrophs and heterotrophs of aquatic habitats. **356**

plant A multicelled photoautotroph, typically with well-developed roots and shoots. Primary producer on land. **9**

plant hormone Signaling molecules that can stimulate or inhibit plant development, including growth. **526**

plaque On teeth, a thick biofilm composed of bacteria, their extracellular products, and saliva glycoproteins. **663**

plasma Liquid portion of blood; mainly water with proteins, sugars, and other solutes, and dissolved gases. **640**

plasma membrane Outer cell membrane; encloses the cytoplasm. **56**

plasmid A small, circular DNA molecule in bacteria, replicated independently of the chromosome. **242, 341**

plastid In plants and algae, an organelle that functions in photosynthesis or storage; e.g., chloroplast, amyloplast. **69**

plate tectonics Theory that Earth's outer layer of rock is cracked into plates, the slow movement of which rafts continents to new locations over time. **273**

platelet Cell fragment that circulates in blood and functions in clotting. **641**

pleiotropy The effect of a single gene on multiple traits. **177**

polar Having an uneven distribution of charge. Two atoms share electrons unequally in a polar covalent bond. **27**

polarity Any separation of charge into distinct positive and negative regions. **27**

pollen grain Immature male gametophyte of seed-bearing plants. **373, 508**

pollination The arrival of pollen on a receptive stigma of a flower. **512**

pollination vector Any agent that moves pollen grains from one plant to another; e.g., wind, pollinators. **510**

pollinator A living pollination vector; e.g., a bee. **382, 510**

pollutant A natural or synthetic substance released into soil, air, or water in greater than normal amounts; it disrupts natural processes because organisms evolved in its absence, or are adapted to lower levels. **864**

polymer Large molecule of multiple linked monomers. **39**

polypeptide Chain of amino acids linked by peptide bonds. **44**

polyploid Having three or more of each type of chromosome characteristic of the species. **194**

pons Hindbrain traffic center for signals between cerebellum and forebrain. **568**

population A group of individuals of the same species in a specified area. **5, 278**

population density Number of individuals of a population in a specified volume or area of a habitat. **798**

population distribution The pattern in which individuals of a population are dispersed through their habitat. **798**

population size The number of individuals that actually or potentially contribute to the gene pool of a population. **798**

positive feedback mechanism An activity changes some condition, which in turn triggers a response that intensifies the change. **467**

predation Ecological interaction in which a predator kills and eats prey. **818**

predator A heterotroph that eats other living organisms (its prey). **822**

prediction A statement, based on a hypothesis, about a condition that should exist if the hypothesis is not wrong; often called the "if–then process." **12**

pressure flow theory Theory that the flow of fluid through phloem (trans-

location) is driven by the difference in osmotic pressure between a plant's source and sink regions. **503**

prey An organism that a predator kills and eats. **822**

primary growth Plant growth from apical meristems in root and shoot tips. **477**

primary oocyte A immature egg that has not completed prophase I. **748**

primary producer An autotroph at the first trophic level of an ecosystem. **840**

primary productivity The rate at which an ecosystem's primary producers secure and store energy in tissues. **844**

primary succession A community arises and species arrive and replace one another over time in an environment that was without soil, such as a newly formed island. **828**

primary wall The first thin, pliable wall of young plant cells. **70**

primate A type of mammal; a prosimian or an anthropoid. **452**

primer Short, single strand of DNA designed to hybridize with a template; DNA polymerases initiate synthesis at primers during PCR or sequencing. **245**

prion A type of protein found in vertebrate nervous systems; becomes infectious when its shape changes. **338**

probability Chance that a particular outcome of an event will occur; depends on the total number of outcomes possible. **173**

probe Short fragment of DNA labeled with a tracer; designed to hybridize with a nucleotide sequence of interest. **244**

producer Autotroph; an organism that makes its own food using carbon from inorganic molecules such as CO_2. Most are photosynthetic. **6**

product A molecule remaining at the end of a reaction. **96**

progesterone Sex hormone secreted by ovaries and the corpus luteum. **612**

proglottid One of many tapeworm body units that bud behind the scolex. **413**

prokaryote Single-celled organism in which the DNA is not contained in a nucleus; a bacterium or archaean. **56, 339**

prokaryotic cell *See* Prokaryote.

prokaryotic chromosome Double-stranded, circular molecule of DNA together with a few attached proteins. **340**

prokaryotic fission Cell reproduction mechanism of prokaryotic cells. **340**

prolactin (PRL) Hormone that induces synthesis of enzymes used in milk production. **603, 776**

promoter In DNA, a nucleotide sequence to which RNA polymerase binds. **219**

prophase Stage of mitosis and meiosis in which chromosomes condense and become attached to a newly forming spindle. **146**

protein Organic compound that consists of one or more polypeptide chains. **44**

protists Informal name for eukaryotes that are not plants, fungi, or animals. **8, 352**

protocell A membrane-enclosed sac of molecules that captures energy, engages in metabolism, concentrates materials, and replicates itself. Presumed stage of chemical evolution that preceded living cells. **320**

proton Positively charged subatomic particle in the nucleus of all atoms. The number of protons (the atomic number) defines the element. **22**

protostome A bilateral animal belonging to a lineage characterized in part by having the first indentation to form on the early embryo's surface become the mouth; e.g., mollusks, annelids, arthropods. **405**

proximal tubule Tubular portion of a nephron closest to Bowman's capsule. **727**

pseudocoel False coelom; a main body cavity incompletely lined with tissue derived from mesoderm. **405**

pseudopod A dynamic lobe of membrane-enclosed cytoplasm; functions in motility and phagocytosis by amoebas, amoeboid cells, and phagocytic white blood cells. **73**

puberty For humans, the post-embryonic stage when gametes start to mature and secondary sexual traits emerge. **612, 743**

pulmonary circuit Cardiovascular route in which oxygen-poor blood flows to lungs from the heart, gets oxygenated, then flows back to the heart. **639**

Punnett square A diagram used to predict the outcome of a testcross. **173**

pupil An opening at the center of the iris through which light enters the eye. **588**

pyruvate Three-carbon end product of glycolysis. **124**

quadrat One of a number of sampling areas of the same size and shape used to estimate population size. **799**

radial symmetry Animal body plan with parts arranged on a central axis like the spokes of a wheel. **404**

radioactive decay Process by which atoms of a radioisotope spontaneously emit energy and subatomic particles when their nucleus disintegrates. **23**

radioisotope Isotope with an unstable nucleus; decays into predictable daughter elements at a predictable rate. **23**

radiolarian Single-celled predatory protist with pseudopods that project through its perforated silica shell. Most drift in open upper ocean waters. **356**

radiometric dating Method of estimating the age of a rock or fossil by measuring the content and proportions of a radioisotope and its daughter elements. **268**

radula Of many mollusks, a tongue-like organ hardened with chitin; used to feed. **416**

rain shadow Reduction in rainfall on the side of a high mountain range facing away from prevailing wind; results in arid or semiarid conditions. **867**

reactant Molecule that enters a reaction. **96**

reaction Process of chemical change. **96**

receptor A molecule or structure that can respond to a form of stimulation such as light energy, or to binding of a signaling molecule such as a hormone. **6, 466**

receptor protein Plasma membrane protein that binds to a particular substance outside of the cell. **80**

recessive With regard to an allele, having effects that are masked by a dominant allele on the homologous chromosome. **171**

recognition protein Plasma membrane protein that identifies a cell as belonging to *self* (one's own body tissue). **80**

recombinant DNA A DNA molecule that contains genetic material from more than one organism. **242**

rectum Last part of the mammalian gut that stores feces before their expulsion. **705**

red alga An aquatic, usually multicelled autotrophic protist with an abundance of phycobilins. **364**

red blood cell Erythrocyte; hemoglobin containing blood cell; transports oxygen. **640**

red marrow Site of blood cell formation in the spongy tissue of many bones. **622**

referred pain Pain that is experienced when the brain mistakenly interprets signals about a visceral problem as if the signals were coming from the skin or joints. **581**

reflex Simple, stereotyped movement in response to a stimulus; sensory neurons synapse on motor neurons in the simplest reflex arcs. **566**

releaser Hypothalamic signaling molecule that enhances secretion of a hormone by the anterior pituitary. **603**

renal cortex The outer portion of the kidney, just under the renal capsule, where nephrons begin. **726**

renal medulla The inner portion of the kidney, into which the nephron's loop of Henle and collecting duct extend. **726**

replacement model Idea that modern humans arose from a single *Homo erectus* population in sub-Saharan Africa within the past 200,000 years, then spread and replaced other hominids. **456**

repressor Transcription factor that blocks transcription by binding to a (eukaryotic) promoter or (prokaryotic) operator. **230**

reproduction An asexual or sexual process by which a parent cell or organism produces offspring. **7**

reproductive base The number of actually and potentially reproducing individuals of a population. **798**

reproductive cloning Technology that produces genetically identical individuals; e.g., artificial twinning, SCNT. **210**

reproductive isolation Any mechanism that prevents gene flow between populations; part of speciation. **290**

reptile Not a formal taxon; amniotes that do not have features of birds or mammals; e.g., turtle, lizard. **442**

resource partitioning Use of different parts of a resource; permits two or more similar species to coexist in a habitat. **821**

respiration The sum of physiological processes that move O_2 from surroundings to metabolically active tissues in the body and CO_2 from tissues to the outside. **682**

respiratory cycle One inhalation and one exhalation. **690**

respiratory membrane Fused-together alveolar and blood capillary epithelia and the basement membrane in between; the respiratory surface in a human lung. **692**

respiratory protein A protein with one or more metal ions that binds O_2 in oxygen-rich animal tissues and gives it up where O_2 levels are lowest; e.g., hemoglobin. **683**

respiratory surface Any thin, moist body surface that functions in gas exchange. **682**

resting membrane potential The voltage difference across the plasma membrane of a neuron or other excitable cell that is not receiving outside stimulation. **557**

restriction enzyme Type of enzyme that cuts specific base sequences in DNA. **242**

retina In vertebrate and many invertebrate eyes, a tissue packed with photoreceptors and interwoven with sensory cells. **587**

reverse transcriptase A viral enzyme that catalyzes the assembly of nucleotides into DNA, using RNA as a template. **243, 336**

Rh blood typing Method of determining whether Rh^+, a type of surface recognition protein, is present on an individual's red blood cells; if absent, the cell is Rh^-. **643**

rhizoid A rootlike absorptive structure of some bryophytes. **374**

rhizome A stem that grows horizontally underground. **376**

ribosomal RNA (rRNA) A type of RNA that becomes part of ribosomes; some catalyze formation of peptide bonds. **216**

ribosome Site of protein synthesis. An intact ribosome has two subunits, each composed of rRNA and proteins. **56**

ribozyme A catalytic RNA. **321**

riparian zone The narrow corridor of vegetation along a stream or river. **903**

RNA Ribonucleic acid. Type of nucleic acid, typically single-stranded; important in transcription, translation, and gene control; some are catalytic. *See also* ribosomal RNA, transfer RNA, messenger RNA, ribozyme. **48**

RNA polymerase Enzyme that catalyzes transcription of DNA into RNA. **218**

RNA world Hypothetical time prior to the evolution of DNA in which RNA molecules stored genetic information and catalyzed protein synthesis. **321**

rod cell Vertebrate photoreceptor that detects very dim light; contributes to the coarse perception of movement. **590**

root Typically belowground plant part that absorbs water and minerals. **476**

root hair Hairlike, absorptive extension of a young cell of root epidermis. **485, 496**

root nodule Mutualistic association of nitrogen-fixing bacteria and roots of some legumes and other plants; infection leads to a localized tissue swelling. **496**

rotifer Tiny bilateral, coelomate animal with a ciliated head; occurs mostly in freshwater or damp environments. **419**

roundworm Bilateral invertebrate with a false coelom and complete digestive system in an unsegmented body. Most are decomposers; some are parasites. **420**

r-selection Selection that favors traits that maximize number of offspring; operates when the population is well below its carrying capacity. **805**

rubisco Ribulose bisphosphate carboxylase, or RuBP. Carbon-fixing enzyme of light-independent photosynthesis reactions. **115**

ruminant Hoofed, herbivorous mammal that has multiple stomach chambers. **703**

runoff The water that flows into streams when the ground is saturated. **848**

sac fungi Fungus that produces sexual spores in sac-shaped cells. **394**

salinization Salt buildup in soil. **849**

salivary amylase An enzyme in saliva that hydrolyzes starch, breaking it into disaccharides. **705**

salt Compound that dissolves easily in water and releases ions other than H^+ and OH^-. **31**

sampling error Difference between results derived from testing an entire group of events or individuals, and results derived from testing a subset of the group. **16**

saprobe Heterotroph that extracts energy and carbon from nonliving organic matter and so causes its decay. **339**

sapwood Of an older stem or root, the moist secondary growth between the vascular cambium and heartwood. **487**

sarcomere One of many basic units of contraction along the length of a muscle fiber. It shortens by ATP-driven interactions between its parallel arrays of actin and myosin components. **628**

sarcoplasmic reticulum Specialized ER that forms flattened, membrane-bound chambers around muscle fibers; takes up, stores, and releases calcium ions. **630**

scale One of many platelike structures at the body surface of fish and reptiles. **438**

science Systematic study of nature. **11**

scientific theory Hypothesis that has not been disproven after many years of rigorous testing, and is useful for making predictions about other phenomena. **12**

sclera A dense, white, fibrous layer of the eyeball that covers most of the eye's outer surface. **588**

sclerenchyma Simple plant tissue; dead at maturity, its lignin-reinforced cell walls structurally support plant parts. **478**

seamount Extinct seafloor volcano. **884**

second law of thermodynamics Energy tends to disperse spontaneously. **94**

second messenger Molecule produced by a cell in response to binding of a hormone at the plasma membrane; relays a signal into a cell signal; e.g., cyclic AMP. **600**

secondary growth A thickening of older stems and roots at lateral meristems. **477**

secondary oocyte A haploid cell produced by the first meiotic division of a primary oocyte; released at ovulation. **748**

secondary succession A community arises and changes over time in a habitat where another community existed previously. **828**

secondary wall Lignin-reinforced wall inside the primary wall of a plant cell. **70**

seed The mature ovule of a seed plant; contains embryo sporophyte. **373, 515**

segregation Thoery that the two members of each pair of genes on homologous chromosomes separate during meiosis. **173**

selective permeability Membrane property that allows some substances, but not others, to cross. **82**

selfish herd Animal group that forms when individuals each attempt to hide in the midst of others. **790**

semen Fluid expelled from the penis during ejaculation; consists of a small volume of sperm mixed with accessory duct secretions. **743**

semiconservative replication Describes the process of DNA replication, by which one strand of each copy of a DNA molecule is new, and the other is a strand of the original DNA. **208**

semi-evergreen forest Biome in the humid tropics that includes a mix of broadleaf trees that retain leaves year round, and deciduous broadleaf trees that shed once a year in the cold or dry season. **874**

senescence Of multicelled organisms, the phase in a life cycle from maturity until death; also applies to death of parts, such as plant leaves. **534**

sensory adaptation After a while, sensory neurons stop responding to an ongoing stimulus. **579**

sensory neuron Type of neuron that detects a stimulus and relays information about it toward an integrating center. **554**

sequence The order of nucleotides in a strand of DNA or RNA. **207**

sex chromosome Member of a pair of chromosomes that differs between males and females. **186**

sexual dimorphism Having distinct female and male phenotypes. **286**

sexual reproduction Production of genetically variable offspring by gamete formation and fertilization. **156, 740**

sexual selection Mode of natural selection in which some individuals outreproduce others of a population because they are better at securing mates. **286, 788**

sexually transmitted disease (STD) Disease agent is transferred between individuals by sexual contact; e.g., syphilis, AIDS. **754**

shell model Model of electron distribution in an atom; orbitals are shown as nested circles, electrons as dots. **24**

shivering response Rhythmic tremors in response to cold. **735**

shoot Aboveground plant parts; e.g., stems, leaves, flowers. **476**

short tandem repeat Stretch of DNA that consists of many copies of a short sequence; basis of DNA fingerprinting. **247**

sieve tube Conducting tube in phloem; distributes sugars through a plant. **479**

sink In plants, any region where organic compounds are being unloaded from sieve tubes. **502**

sister chromatid One of two attached members of a duplicated eukaryotic chromosome. **142**

sister group The two lineages that emerge from a node on a cladogram. **303**

six-kingdom classification system System of classification that groups all organisms into kingdoms Bacteria, Archaea, Protista, Fungi, Plantae, and Animalia. **312**

skeletal muscle tissue Contractile tissue that is the functional partner of bone. **544**

sliding-filament model Model for how the sarcomeres of muscle fibers contract. ATP-activated myosin heads repeatedly bind actin filaments (attached to Z lines) and tilt in short power strokes that slide actin toward the sarcomere's center. **629**

slime mold An amoebozoan; amoeba-like cells that cluster into a mass, differentiate, and form reproductive structures. **365**

small intestine Part of the vertebrate gut in which digestion is completed and from which most nutrients are absorbed. **705**

smog Atmospheric condition in which winds cannot disperse airborne pollutants trapped by a thermal inversion. **864**

smooth muscle tissue Contractile tissue in the wall of soft internal organs. **545**

social parasite An animal that takes advantage of its host's behavior, thus harming it; e.g., cuckoo. **826**

soil Mixture of various mineral particles (sand, silt, clay) and decomposing organic matter (humus). **494**

soil erosion A loss of soil under the force of wind and water. **495**

soil profile Distinct soil layers that form over time in a biome. **870**

solar tracking Circadian response; a plant part changes position in response to the sun's changing angle through the day. **532**

solute A dissolved substance. **28**

solvent Substance, typically a liquid, that can dissolve other substances; e.g., water. **28**

somatic cell nuclear transfer (SCNT) Method of reproductive cloning in which genetic material is transferred from an adult somatic cell into an unfertilized, enucleated egg. **210**

somatic nervous system Portion of the peripheral nervous system that carries messages to skeletal muscles and relays information about skin and joints. **564**

somatic sensation An easily localized sensation such as warmth or touch that arises when receptors in skin, muscle, or joints are stimulated. **580**

somatosensory cortex Part of the cerebral cortex that receives sensory information from somatic nerves. **580**

somite One of many paired segments in a vertebrate embryo that gives rise to most bones, skeletal muscles of the head and trunk, and the dermis. **766**

sorus, plural **sori** Cluster of spore-forming chambers on underside of a fern frond. 377

source In plants, any region where organic compounds are being loaded into sieve tubes. 502

speciation Formation of daughter species from a population or subpopulation of a parent species; the routes vary in their details and duration. 290

species A type of organism. Of sexually reproducing species, one or more groups of individuals that potentially can interbreed, produce fertile offspring, and do not interbreed with other groups. 8

sperm Mature male gamete. 162

sphincter A ring of muscles that alternately contracts and relaxes, which closes and opens a passageway between two organs. 705

spinal cord The part of a central nervous system inside a vertebral canal. 566

spindle *See* bipolar spindle.

spleen The largest lymphoid organ, with phagocytic white blood cells and B cells; filters antigen and used-up platelets and worn-out or dead red blood cells. In embryos only, a site of red blood cell formation. 655

sponge Filter-feeding aquatic invertebrate with no symmetry, no tissues. 408

sporangium Multicellular, spore-forming structure that protects developing spores and facilitates their dispersal. 372

sporophyte Diploid, spore-producing body of a plant or multicelled alga. 162, 372, 508

spring overturn In temperate zone lakes, a downward movement of oxygenated surface water and an upward movement of nutrient-rich water in spring. 879

stabilizing selection Mode of natural selection; intermediate phenotypes are favored over extremes. 284

stamen The male reproductive part of a flowering plant; consists of a pollen-producing anther on a filament. 508

stasis Macroevolutionary pattern in which a lineage persists with little or no change over evolutionary time. 296

statolith Organelle that acts as a gravity-sensing mechanism. 530

steroid A type of lipid with four carbon rings and no fatty acid tails. 43

stem cell Self-perpetuating, undifferentiated animal cell. A portion of its daughter cells become specialized. 546

stimulus A specific form of energy that activates a sensory receptor able to detect it; e.g., pressure. 782

stoma, plural **stomata** Gap that opens between two guard cells; lets water vapor and gases diffuse across the epidermis of a leaf or primary stem. 116, 372

stomach Muscular, stretchable sac; mixes and stores ingested food and helps break it apart mechanically and chemically. 705

strain A subgroup within a prokaryotic species that can be characterized by some identifiable trait or traits. 339

stramenopile A single-celled or multi-celled protist of a group united by DNA studies; some have cells with filaments on one of two flagella; e.g., a diatom, brown alga, or water mold. 360

strobilus (strobili) Of some nonflowering plants such as horsetails and cycads, a cluster of spore-producing structures. 376

stroma The semifluid matrix between the thylakoid membrane and the two outer membranes of a chloroplast; site of light-independent photosynthesis reactions. 111

stromatolite Fossilized dome-shaped mats of aquatic photoautotrophic bacteria. 323

substrate A reactant molecule that is specifically acted upon by an enzyme. 98

substrate-level phosphorylation The direct transfer of a phosphate group from a substrate to ADP; forms ATP. 126

surface-to-volume ratio A relationship in which the volume of an object increases with the cube of the diameter, but the surface area increases with the square. 56

survivorship curve Plot of age-specific survival of a cohort, from the time of birth until the last individual dies. 804

swim bladder Adjustable flotation sac of some bony fish. 439

symbiosis Ecological interaction in which members of two species live together or otherwise interact closely; e.g., mutualism, parasitism, commensalism. 818

sympathetic neuron A neuron of the autonomic nervous system. Its signals cause increases in overall activities in times of stress or heightened awareness. Also works in opposition with parasympathetic neurons to make small ongoing adjustments in activities of internal organs they both innervate. 565

sympatric speciation A speciation model in which speciation occurs in the absence of a physical barrier; e.g., by polyploidy in flowering plants. 294

synapse Region where the axon endings of a neuron are separated by a tiny gap from the cell that the neuron signals. 560

synaptic integration The summation of excitatory and inhibitory signals that arrive at an excitable cell's input zone at the same time. 561

syndrome The set of symptoms that characterize a medical condition. 197

system acquired resistance Of some plants, a mechanism that induces cells to produce and release compounds that will protect tissues from attack. 468

systemic circuit Cardiovascular route in which oxygenated blood flows from the heart through the rest of the body, where it gives up oxygen and takes up carbon dioxide, then flows back to the heart. 639

T cell receptor (TCRs) Antigen-binding receptor on the surface of T cells; also recognizes MHC markers. 666

T lymphocyte T cell. White blood cell that coordinates vertebrate adaptive immune responses; cytotoxic T cells carry out cell-mediated responses. 661

taproot system In eudicots, a primary root and all of its lateral branchings. 485

tardigrade A tiny coelomate animal of damp and aquatic habitats; has four pairs of legs; its dried out dormant stage can survive extremely adverse conditions. 419

taste receptor A chemoreceptor that detects solutes in the fluid bathing it. 582

taxon, plural **taxa** An organism or set of organisms. 302

taxonomy Science of naming and classifying species. 302

telophase Stage of mitosis during which chromosomes arrive at the spindle poles and decondense, and new nuclei form. 146

temperate deciduous forest Biome with 50 to 150 centimeters of precipitation throughout the year, warm summers and cold winters. Dominant vegetation is trees that shed leaves in fall. 874

temperature Measure of molecular motion. 29

tendon A cord or strap of dense connective tissue that attaches a muscle to bone. 626

teratogen A toxin or infectious agent that interferes with embryo development and causes birth defects. 774

terminal bud A shoot's main zone of primary growth. 480

testcross Method of determining genotype; a cross between an individual of unknown genotype and a homozygous recessive individual. Offspring phenotypes are analyzed. 172

testis, plural **testes** One of a pair of male gonads where sperm form by meiosis; secretes the hormone testosterone. 742

testosterone A sex hormone necessary for the development and functioning of the male reproductive system. 612, 742

tetanus Motor unit response to repeated stimulation; a strong prolonged

contraction. Also, a disease caused by bacteria in which muscles stay contracted. **632**

tetrapod Vertebrate that is a four-legged walker or descended from one. **440**

thalamus Forebrain region; a coordinating center for sensory input and a relay station for signals to the cerebrum. **568**

theory of inclusive fitness The idea that genes associated with altruism are adaptive if they cause behavior that promotes the reproductive success of an altruist's closest relatives. **792**

theory of uniformity Idea that gradual repetitive processes occurring over long time spans shaped Earth's surface. **263**

therapeutic cloning Producing human embryos by SCNT. **211**

thermal radiation Emission of heat from any object. **733**

thermocline Thermal stratification in a large body of water; a cool midlayer stops vertical mixing between warm surface water above it and cold water below it. **879**

thermoreceptor Type of sensory cell that detects a change in temperature. **578**

thigmotropism Redirected growth of a plant in response to contact with a solid object; e.g., vine curling around a post. **531**

thirst center Part of the hypothalamus that promotes water-seeking behavior when osmoreceptors in the brain detect a rise in the blood level of sodium. **730**

threatened species Species likely to be endangered in the near future. **893**

three-domain system Classification system that groups all organisms into the domains Bacteria, Archaea, and Eukarya. **312**

threshold potential Of a neuron, the membrane potential at which gated sodium channels open and start an action potential. **558**

thylakoid membrane A chloroplast's inner membrane system, often folded as flattened sacs, that forms a continuous compartment in the stroma. In the first stage of photosynthesis, pigments and enzymes in the membrane function in the formation of ATP and NADPH. **111**

thymine (T) A type of nitrogen-containing base in nucleotides; also, a nucleotide with a thymine base. Base-pairs with adenine; does not occur in RNA. **206**

thymus Endocrine gland beneath the breastbone that secretes thymosins and is the site of T cell maturation. **612**

thymus gland Lymphoid organ located beneath the sternum; T cells formed in bone marrow move to it and mature under the influence of its hormonal secretions. **655**

thyroid gland Endocrine gland located at base of neck; its hormones influence growth, development, metabolic rate. **606**

thyroid-stimulating hormone (TSH) A pituitary hormone that regulates secretion of thyroid hormone. **603**

tidal volume The volume of air that flows into and out of the lungs during a a normal inhalation and exhalation. **691**

tight junction Array of fibrous proteins that joins epithelial cells; collectively, these cell junctions prevent fluids from leaking between cells in epithelial tissues. **540**

tissue In multicelled organisms, a group of cells of a specialized type interacting in the performance of one or more tasks. **462, 540**

tissue culture propagation Method in which somatic cells are induced to divide repeatedly in the laboratory. **519**

topsoil Uppermost soil layer; has the most nutrients for plant growth. **495**

torsion Twisting of the body that occurs as gastropod mollusks develop. **416**

total fertility rate (TFR) For humans, the average number of children born to a female during her lifetime. **810**

tracer A molecule with a detectable label attached; researchers can track it after delivering it into a cell or other system. **23**

trachea Windpipe; airway that connects the pharynx (throat) with the bronchi leading to the lungs. **689**

tracheal system In insects and some other land arthropods, branching tubes that start at the body surface and end near cells; role in gas exchange. **685**

tracheid Type of tapered cell in xylem, dead at maturity; its perforated wall forms part of a water-conducting tube. **478, 498**

trait A physical, biochemical, or behavioral characteristic of an individual. **7**

transcription Process by which an RNA is assembled from nucleotides using a gene region in DNA as a template. First step in protein synthesis. **216**

transcription factor Regulatory protein that influences transcription; e.g., activator, repressor. **230**

transduction Movement of DNA from one organism to another by a virus. **341**

transfer RNA (tRNA) Type of RNA that delivers amino acids to a ribosome during translation. Its anticodon pairs with an mRNA codon. **216**

transformation In prokaryotes, a type of horizontal gene transfer in which DNA is taken up from the environment. **341**

transgenic Refers to an organism that has been genetically engineered to carry a gene from a different species. **250**

transition state In a chemical reaction, the point at which reactant bonds are at their breaking point. **98**

translation At ribosomes, information encoded in an mRNA guides synthesis of a polypeptide chain from amino acids. Second stage of protein synthesis. **217**

translocation Attachment of a piece of a broken chromosome to another chromosome. Also, the movement of organic compounds through phloem. **192, 502**

transpiration Evaporative water loss from a plant. **498**

transport protein Membrane protein that passively or actively assists specific ions or molecules into or out of a cell. **80**

transposable element Small segment of DNA that can spontaneously move to a new location in the chromosomal DNA of a cell. **224**

triglyceride A lipid with three fatty acid tails attached to a glycerol backbone. **42**

trophic level All organisms that are the same number of transfer steps away from the energy input into an ecosystem. **840**

tropical deciduous forest Equatorial biome where less than 2.5 centimeters of rain falls in the dry season. Most trees shed leaves at the start of the dry season. **874**

tropism Directional growth response to an environmental stimulus. **530**

trypanosome Member of the largest subgroup of kinetoplastid protozoans; all are parasites. **355**

tubular reabsorption A process by which peritubular capillaries reclaim water and solutes that leak or are pumped out of a nephron's tubular regions. **728**

tubular secretion Transport of H+, urea, other solutes out of peritubular capillaries and into nephrons for excretion. **729**

tumor Abnormal mass of cells. Benign tumor cells stay in their home tissue; malignant ones invade other places in the body and start new tumors. *See also* neoplasm. **150**

tunicate Filter-feeding, invertebrate chordate enclosed in a baglike secreted covering as an adult. **435**

turgor Hydrostatic pressure. Pressure that a fluid exerts against a wall, membrane, or some other structure that contains it. **88**

ultrafiltration In a capillary bed, pressure generated by the beating heart forces some protein-free plasma out of a blood capillary, into interstitial fluid. **651**

upwelling Upward movement of cool water from the depths, as when winds blow surface water away from a coast. **885**

uracil (u) A type of nitrogen-containing base in nucleotides; also, a nucleotide with a uracil base. Base-pairs with adenine; occurs in RNA, not in DNA. **216**

urea Nitrogen-containing waste excreted in urine; forms in the liver when ammonia combines with CO_2. **724**

ureter A urine-conducting tube from each kidney to the urinary bladder. **726**

urethra Tube that drains the urinary bladder; opens at the body surface. **726**

urinary system Vertebrate organ system that adjusts the blood's volume and composition; rids the body of metabolic waste. **724**

urine Fluid consisting of excess water, wastes, and solutes; forms in kidneys by filtration, reabsorption, and secretion. **724**

uterus In a female placental mammal, a muscular, pear-shaped organ in which embryos are housed and nurtured during pregnancy. **746**

vaccine A preparation introduced into the body in order to elicit immunity to an antigen. **674**

vacuole A fluid-filled organelle that isolates or disposes of waste, debris, or toxic materials. **67**

vagina In female mammals, the organ that receives sperm, forms part of birth canal, and channels menstrual flow. **746**

variable In experiments, a characteristic or event that differs among individuals and that may change over time. **13**

vascular bundle Multistranded, sheathed cord of primary xylem and phloem in a stem or leaf. **481**

vascular cambium A lateral meristem that forms in older stems or roots. **486**

vascular cylinder Sheathed, cylindrical array of primary xylem and phloem in a root. **485**

vascular tissue In vascular plants, xylem that distributes water and mineral ions and phloem that distributes sugars made in photosynthetic cells. **372**

vascular tissue system All xylem and phloem in plants that are structurally more complex than bryophytes. **476**

vector An insect or some other animal that carries a pathogen between hosts; e.g., a mosquito that transmits malaria. *See also* cloning vector. **343, 826**

vegetative reproduction Growth of new roots and shoots from extensions or fragments of a parent plant; form of asexual reproduction in plants. **518**

vein In plants, a vascular bundle in a stem or leaf. In animals, a large-diameter vessel that carries blood toward the heart. **483, 648**

ventricle A chamber of the heart that receives blood from an atrium and pumps it out into arteries. **646**

venule A small blood vessel that connects several capillaries to a vein. **648**

vernalization Stimulation of flowering in spring by low temperature in winter. **533**

vertebra, plural **vertebrae** One of a series of hard bones that protects the spinal cord and forms the backbone. **436, 620**

vertebral column Backbone, a feature common to all vertebrates. **436, 620**

vertebrate Animal with a backbone. **434**

vesicle Small, membrane-enclosed, saclike organelle; different kinds store, transport, or degrade their contents. **67**

vessel member Type of cell in xylem, dead at maturity; its perforated wall forms part of a water-conducting tube. **478, 498**

vestibular apparatus In vertebrates, an organ of equilibrium in the inner ear. **583**

villus, plural **villi** One of the many fingerlike multicellular projections that increase the surface area of some tissues in an animal body; e.g., small intestinal villi. **708**

viroid Infectious RNA molecule; most infect plants. **338**

virus Noncellular infectious particle that consists of DNA or RNA, a protein coat and, in some types, a lipid envelope; it can be replicated only after its genetic material enters a host cell and subverts the host's metabolic machinery. **334**

visceral sensation Sensation arising from sensory receptors in internal organs; difficult to localize. **580**

vision Perception of visual stimuli based on light focused on a retina and image formation in the brain. **586**

visual accommodation Adjustments in a lens position or shape that focus light rays on the retina. **589**

vital capacity Maximum volume of air moved in and out of the lungs (with forced inhalation and exhalation). **690**

vitamin Any organic substance that an organism requires in trace amounts but that it generally cannot produce. Many function as coenzymes. **714**

vomeronasal organ In many vertebrates, a cluster of sensory neurons in the nasal cavity that responds to pheromones. **582**

warning coloration In many toxic species and their mimics, bright colors, patterns, and other signals that predators learn to recognize and avoid. **824**

water cycle The process by which water moves among ocean, the atmosphere, and freshwater reservoirs. **848**

water mold A stramenopile protist; most are decomposers or opportunistic parasites, of aquatic habitats; others are plant pathogens. **361**

watershed A region of any specified size in which all precipitation drains into one stream or river. **848**

water–vascular system In echinoderms, a system of tube feet connected to canals; functions in movement, food handling. **428**

wavelength Distance between crests of two successive waves of radiant energy. **108**

wax Water-repellent lipid with long fatty acid tails bonded to long-chain alcohols or carbon rings. **43**

white blood cell Leukocyte. Type of cell that functions in immune responses; e.g., macrophage, dendritic cell, eosinophil, neutrophil, basophil, T cell, B cell. **641**

white matter Portion of brain and spinal cord that includes myelinated axons. **566**

wood Accumulated secondary xylem. **486**

X chromosome inactivation Shutdown of one of the two X chromosomes in the cells of female mammals. *See also* Dosage compensation. **232**

xenotransplantation Transplant of an organ from one species into another. **253**

xylem Complex tissue of vascular plants; conducts water and solutes through tubes that consist of the interconnected walls of dead cells. **372, 478**

yellow marrow In most mature bones, a fatty tissue that fills interior cavities; can convert to blood cell-producing red marrow if necessary. **622**

yolk Protein- and lipid-rich substance in many eggs; serves as the first food source for a developing embryo. **741**

zero population growth No net increase or decrease in population size during a specified interval. **800**

zona pellucida A noncellular layer that forms around a primary oocyte. **748**

zygote Cell formed by fusion of gametes; first cell of a new individual. **157**

zygote fungi Fungal group; sexual spores are produced in a zygospore that forms after hyphae of opposite mating types meet and fuse; e.g., black bread mold. **392**

Art Credits and Acknowledgments

This page constitutes an extension of the book copyright page. We have made every effort to trace the ownership of all copyrighted material and secure permission from copyright holders. In the event of any question arising as to the use of any material, we will be pleased to make the necessary corrections in future printings. Thanks are due to the following authors, publishers, and agents for permission to use the material indicated.

TABLE OF CONTENTS **Page vii** top, © Raymond Gehman/ Corbis. **Page viii** from left, ArchiMeDes; R. Calentine/ Visuals Unlimited; © Kenneth Bart. **Page ix** from left, Hemoglobin models: PDB ID: 1GZX; Paoli, M., Liddington, R., Tame, J., Wilkinson, A., Dodson, G.; Crystal structure of T state hemoglobin with oxygen bound at all four haems. *J.Mol.Biol.*, v256, pp. 775–792, 1996; Larry West/ FPG / Getty Images; © Professors P. Motta and T Naguro/SPL/ Photo Researchers, Inc. **Page x** from left, Dr. Pascal Madaule, France; Image courtesy of Carl Zeiss MicroImaging, Thornwood, NY; Moravian Museum, Brno; © Russ Schleipman/ Corbis. **Page xi** from left, A C. Barrington Brown © 1968 J. D. Watson; P. J. Maughan; © Jürgen Berger, Max-Planck-Institut for Developmental Biology, Türbingen; © Visuals Unlimited. **Page xii** from left, Photo courtesy of MU Extension and Agricultural Information; Jonathan Blair; © Jack Jeffrey Photography. **Page xiii** from left, © Taro Taylor, www.flickr.com; © Chase Studios/ Photo Researchers, Inc. **Page xiv** from left, © Wim van Egmond/ Micropolitan Museum; Malaria illustration by Drew Berry, The Walter and Eliza Hall Institute of Medical Research; Courtesy of © Christine Evers; Smithsonian Institution Department of Botany, G.A. Cooper @ USDA-NRCS PLANTS Database. **Page xv** from left, Robert C. Simpson/ Nature Stock; © Brandon D. Cole/ Corbis; © Dave Fleetham/Tom Stack & Associates; Herve Chaumeton/ Agence Nature. **Page xvi** from left, Heather Angel; D. & V. Blagden/ ANT Photo Library; Jean Paul Tibbles. **Page xvii** from left, Courtesy of Dr. Thomas L. Rost; © J.C. Revy/ ISM / Phototake; Photo by Jack Dykinga, USDA, ARS. **Page xviii** from left, Photo © Cathlyn Melloan/ Stone/ Getty Images; Michael Clayton, University of Wisconsin, Department of Botany; © Tony McConnell/ Science Photo Library/ Photo Researchers, Inc. **Page xix** from left, © Nancy Kedersha/ UCLA/ Photo Researchers, Inc.; From Neuro Via Clinicall Research Program, Minneapolis VA Medical Center; © Will & Deni McIntyre/ Photo Researchers, Inc. **Page xx** from left, © Mitch Reardon/ Photo Researchers, Inc.; Ed Reschke; Professor P. Motta/ Department of Anatomy/ La Sapienza, Rome/ SPL/ Photo Researchers, Inc. **Page xxi** from left, © National Cancer Institute/ Photo Researchers, Inc.; © NSIBC/ SPL/ Photo Researchers, Inc.; © Dr. Richard Kessel and Dr. Randy Kardon/ Tissues & Organs/ Visuals Unlimited. **Page xxii** from left, © John Lund/ Getty Images; © O. Auerbach/ Visuals Unlimited; Microslide courtesy Mark Nielsen, University of Utah; Ralph Pleasant/ FPG/ Getty Images. **Page xxiii** from left, Gary Head; © Dan Guravich/ Corbis; © Rodger Klein/ Peter Arnold, Inc. **Page xxiv** © Professor Jonathon Slack; Courtesy of Elizabeth Sanders, Women's Specialty Center, Jackson, MS; © Lennart Nilsson/ Bonnierforlagen AB. **Page xxv** from left, © Monty Sloan, www.wolfphotography.com; © Charles Lewallen; © Doug Peebles/ Corbis. **Page xxvi** from left, Graphic created by FoodWeb3D program written by Rich Williams courtesy of the Webs on the Web project (www.foodwebs.org); NOAA; © James Randklev/ Corbis. **Page xxvii** from left, Image provided by GeoEye and NASA SeaWIFS Project; © Dr. John Hilty; © Billy Grimes.

INTRODUCTION NASA Space Flight Center

CHAPTER 1 **1.1** Courtesy of Conservation International. **Page 3** From second from top, Jack de Coningh; © Lewis Trusty/ Animals Animals; © Nick Brent; © Raymond Gehman/ Corbis. **1.2** (a) Rendered with Atom In A Box, copyright Dauger Research, Inc.; (d) © Science Photo Library/ Photo Researchers, Inc.; (e) © Bill Varie/Corbis; (f–h) © Jeffrey L. Rotman/Corbis; (i) © Peter Scoones; (j–k) NASA. **1.4** © Y. Arthus-Bertrand/ Peter Arnold, Inc. **1.5** © Jack de Coningh. **1.7** (a) clockwise from top left, © Dr. Richard Frankel; © David Scharf, 1999. All rights reserved; © Susan Barnes; © SciMAT/ Photo Researchers, Inc.; (b) left, © R. Robinson/ Visuals Unlimited, Inc.; right, © Dr. Harald Huber, Dr. Michael Hohn, Prof. Dr. K. O. Stetter, University of Regensburg, Germany; (c) above, left, clockwise from top, © Lewis Trusty/ Animals Animals; © Emiliania Huxleyi photograph, Vita Pariente, scanning electron micrograph taken on a Jeol T330A instrument at Texas A&M University Electron Microscopy center; © Carolina Biological Supply Company; © Oliver Meckes/ Photo Researchers, Inc.; Courtesy of James Evarts; right, © John Lotter Gurling/ Tom Stack & Associates; inset, © Edward S. Ross; below, left, from left, © Robert C. Simpson/ Nature Stock; © Edward S. Ross; right, © Stephen Dalton/ Photo Researchers, Inc. **1.8** (a) Photographs courtesy Derrell Fowler, Tecumseh, Oklahoma; (b) © Nick Brent. **1.9** (a) © Lester Lefkowitz/ Corbis; (b) Centers for Disease Control and Prevention; (c) © Raymond Gehman/ Corbis. **1.10** top, © Superstock. **1.11** (a) © Matt Rowlings, www .eurobutterflies.com; (b) © Adrian Vallin; (c) © Antje Schulte. **1.12** © Gary Head. **Page 17** Photo courtesy of Dr. Robert Zingg/ Zoo Zurich. **Page 18** Scientific Paper; Adrian Vallin, Sven Jakobsson, Johan Lind, and Christer Wiklund, Proc. R. Soc. B (2005 272, 1203, 1207). Used with permission of The Royal Society and the author.

Page 19 UNIT I © Wim van Egmond, Micropolitan Museum

CHAPTER 2 **2.1** © Owaki-Kulla/ CORBIS. **Page 21** From second from top, © Michael S. Yamashita/ Corbis; © Bill Beatty/ Visuals Unlimited; © R. B. Suter, Vasar College; © W. K. Fletcher/ Photo Researchers, Inc. **2.4** Courtesy © GE Healthcare. **Page 24** Left, © Michael S. Yamashita/ Corbis. **Page 25** © Hubert Stadler/ Corbis. **2.7** (a) Left, upper, Gary Head; lower, © Bill Beatty/ Visuals Unlimited. **2.10** (b) Right, © Steve Lissau/ Rainbow; (c) Right, © Dan Guravich/ Corbis. **2.12** (a) © Lester Lefkowitz/ Corbis; (b) © R. B. Suter, Vasar College. **2.13** Photos from © JupiterImages Corporation. **2.14** Left, Michael Grecco/ Picture Group; right, © W. K. Fletcher/ Photo Researchers, Inc.

CHAPTER 3 **3.1** © ThinkStock/ SuperStock. **Page 35** From top, © Tim Davis/ Photo Researchers, Inc.; © JupiterImages Corporation; Kenneth Lorenzen. **3.2** (b) © JupiterImages Corporation. **3.3** (a), © National Cancer Institute/ Photo Researchers, Inc.; (b–d) Hemoglobin models: PDB ID: 1GZX; Paoli, M., Liddington, R., Tame, J., Wilkinson, A., Dodson, G., Crystal structure of T state hemoglobin with oxygen bound at all four haems. *J.Mol.Biol.*, v256, pp. 775–792, 1996. **3.5** © Tim Davis/ Photo Researchers, Inc. **3.7** Left, © JupiterImages Corporation. **3.8** © JupiterImages Corporation. **3.11** © Kevin Schafer/ Corbis. **Page 43** Bottom left, Kenneth Lorenzen. **3.16** (a–d, bottom) PDB files from NYU Scientific Visualization Lab. **3.17** (b, right) After: *Introduction to Protein Structure*, 2nd ed., Branden & Tooze, Garland Publishing, Inc.; (c, left) PDB ID: 1BBB; Silva, M. M., Rogers, P. H., Arnone, A.; A third quaternary structure of human hemoglobin A at 1.7-Å resolution; *J Biol Chem* 267 pp. 17248 (1992); (c, right) After: *Introduction to Protein Structure*, 2nd ed., Branden & Tooze, Garland Publishing, Inc. **3.18** PDB ID: 1BBB; Silva, M. M., Rogers, P. H., Arnone, A.; A third quaternary structure of human hemoglobin A at 1.7-Å resolution; *J Biol Chem* 267 pp. 17248 (1992). **3.19** (c) © Dr. Gopal Murti/ SPL/ Photo Researchers, Inc.; (d) Courtesy of Melba Moore. **3.20** PDB files from Klotho Biochemical Compounds Declarative Database. **3.22** PDB ID:1BNA; H. R. Drew, R. M. Wing, T. Takano, C. Broka, S. Tanaka, K. Itakura, R. E. Dickerson; Structure of a B-DNA Dodecamer. Conformation and Dynamics; PNAS V. 78 2179, 1981. **Page 50** © JupiterImages Corporation.

CHAPTER 4 **4.1** Left, © JupiterImages Corporation; right, © Stephanie Schuller/ Photo Researchers, Inc. **Page 53** From top, Tony Brian and David Parker/ SPL/ Photo Researchers, Inc.; © JupiterImages Corporation; R. Calentine/ Visuals Unlimited; © ADVANCELL (Advanced In Vitro Cell Technologies; S.L.) www .advancell.com; © Dylan T. Burnette and Paul Forscher. **4.2** © Tony Brian and David Parker/ SPL/ Photo Researchers, Inc. **4.3** (a) Parke-Davis; Above, Linda Hall Library, Kansas City, MO; (b) © Michael W. Davidson, Molecular Expressions; above, © The Royal Society. **4.7** (a) Above, © JupiterImages Corporation; (b) © Geoff Tompkinson/ Science Photo Library / Photo Researchers, Inc. **4.8** (a–b, d–e) Jeremy Pickett-Heaps, School of Botany, University of Melbourne; (c) Prof. Franco Baldi. **4.9** (0.1m) Robert A. Tyrrell; (1m) © Pete Saloutos/ Corbis; (100m) Courtesy of © Billie Chandler. **4.11** (a) ArchiMeDes; (b,c) © K.O. Stetter & R. Rachel, Univ. Regensburg. **4.12** (a) Rocky Mountain Laboratories, NIAID, NIH; (b) R. Calentine/ Visuals Unlimited. **4.13** Courtesy of Roberto Kolter Lab, Harvard Medical School. **4.14** (a) Dr. Gopal Murti/ Photo Researchers, Inc.; (b) M.C. Ledbetter, Brookhaven National Laboratory. **4.16** Right, © Kenneth Bart. **4.17** (a) Don W. Fawcett/ Visuals Unlimited; (b) © Martin W. Goldberg, Durham University, UK. **4.18** (a) © Kenneth Bart; (b,d) Don W. Fawcett/ Visuals Unlimited; (e) Micrograph, Gary Grimes. **4.19** © Conner's Way Foundation, www.connersway.com. **4.20** Micrograph, Keith R. Porter. **4.21** © Dr. Jeremy Burgess/ SPL/ Photo Researchers, Inc. **4.22** (c) © Russell Kightley/ Photo Researchers, Inc. **4.23** George S. Ellmore. **4.24** Left, © Science Photo Library/ Photo Researchers, Inc.; Right, Bone Clones®, www.boneclones

top, A C. Barrington Brown © 1968 J. D. Watson; 4th from top, Courtesy of Cyagra, Inc., www .cyagra.com. **13.3** (a) Below, © Eye of Science/ Photo Researchers, Inc. **Page 205** © JupiterImages Corporation. **13.5** Left, A C. Barrington Brown © 1968 J. D. Watson; right, PDB ID: 1BBB; Silva, M. M., Rogers, P. H., Arnone, A.: A third quaternary structure of human hemoglobin A at 1.7-Å resolution. *J Biol Chem* 267 pp. 17248 (1992). **13.9, 13.10** Courtesy of Cyagra, Inc., www.cyagra.com. **Page 212** © Mc Leod Murdo/ Corbis Sygma. **Page 213** Critical Thinking #4, Shahbaz A. Janjua, MD, Dermatlas; http://www.dermatlas.org. **13.12** Journal of General Physiology 36(1), September 20, 1952: "Independent Functions of viral Protein and Nucleic Acid in Growth of Bacteriophage."

CHAPTER 14 **14.1** Right, © Vaughan Fleming/ SPL/ Photo Researchers, Inc. **Page 215** Second From top, O. L. Miller; bottom, P. J. Maughan. **14.5** (d) Model by © Dr. David B. Goodin, The Scripps Research Institute. **14.6** O. L. Miller. **Page 222** © Kiseleva and Donald Fawcett/ Visuals Unlimited. **14.14** Left, Nik Kleinberg; right, P. J. Maughan. **14.15** (a) © John W. Gofman and Arthur R. Tamplin. From *Poisoned Power: The Case Against Nuclear Power Plants Before and After Three Mile Island*, Rodale Press, PA, 1979. **Page 226** © Vaughan Fleming/ SPL/ Photo Researchers, Inc. **14.17** (b) © Dr. M. A. Ansary/ Science Photo Library/ Photo Researchers, Inc.

CHAPTER 15 **15.1** Left Courtesy of Robin Shoulla and Young Survival Coalition; right, From the archives of www.breastpath.com, courtesy of J.B. Askew, Jr., M.D., P.A. Reprinted with permission, copyright 2004 Breastpath .com. **Page 229** From top, From the collection of Jamos Werner and John T. Lis; © Jose Luis Riechmann; © Eye of Science/ Photo Researchers, Inc. **15.4** From the collection of Jamos Werner and John T. Lis. **15.5** (a–b) © Dr. William Strauss; (c) © DermAtlas, www.dermatlas.org. **15.6** © Thinkstock Images/ Jupiter Images. **15.7** (a) below, Juergen Berger, Max Planck Institute for Developmental Biology, Tuebingen, Germany; (b) © Jose Luis Riechmann. **Page 234** Fruit Fly, Photo by Lisa Starr. **15.8** (a) © Jürgen Berger, Max-Planck-Institut for Developmental Biology, Tübingen; (b) © Visuals Unlimited; (c) © Eye of Science/ Photo Researchers, Inc.; (d) right, Courtesy of Edward B. Lewis, California Institute of Technology; others, © Carolina Biological / Visuals Unlimited. **15.9** (a–e) © Maria Samsonova and John Reinitz; (f) © Jim Langeland, Jim Williams, Julie Gates, Kathy Vorwerk, Steve Paddock, and Sean Carroll, HHMI, University of Wisconsin-Madison. **15.10** PDB ID: 1CJG; Spronk, A.A.E.M., Bovin, A.M.J.J., Radha, P.K., Melacini, G., Boelens, R., Kaptien, R.: The Solution Structure of Lac Repressor Headpiece 62 Complexed to a Symmetrical Lac Operator Structure (London) 7 pp. 1483, (1999). Also PDB ID: 1LBI; Lewis, M., Chang, G., Horton, N.C., Kercher, M.A., Pace, H.C., Schumacher, M.A., Brenan, R.G., Lu, P.: Crystal structure of the lactose operon repressor and its complexes with DNA and inducer. *Science* 271 pp. 1247 (1966); lactose pdb files from the Hetero-Compound Information Centre-Uppsala (HIC-Up). **Page 237** © Lowe Worldwide, Inc. as Agent for National Fluid Milk Processor Promotion Board.

CHAPTER 16 **16.1** © Courtesy of Golden Rice Humanitarian Board. **Page 241** Top, © Professor Stanley Cohen/ SPL/ Photo Researchers, Inc.;

from third from top, Courtesy of © Genelex Corp.; Argonne National Laboratory, U.S. Department of Energy; Photo courtesy of MU Extension and Agricultural Information. **16.3** (a) © Professor Stanley Cohen/ SPL/ Photo Researchers, Inc.; (b) with permission of © QIAGEN, Inc. **16.9** Courtesy of © Genelex Corp. **16.10** Right, © Volker Steger/ SPL/ Photo Researchers, Inc. **16.11** (a) Argonne National Laboratory, U.S. Department of Energy; (b) Courtesy of Joseph DeRisa. From *Science*, 1997 Oct. 24; 278 (5338) 680–686. **Page 250** Photo Courtesy of Systems Biodynamics Lab, P.I. Jeff Hasty, UCSD Department of Bioengineering, and Scott Cookson. **16.12** (d) © Lowell Georgis/ Corbis; (e) Keith V. Wood. **16.13** (a) The Bt and Non-Bt corn photos were taken as part of field trial conducted on the main campus of Tennessee State University at the Institute of Agricultural and Environmental Research. The work was supported by a competitive grant from the CSREES, USDA titled "Southern Agricultural Biotechnology Consortium for Underserved Communities," (2000–2005). Dr. Fisseha Tegegne and Dr. Ahmad Aziz served as Principal and Co-principal Investigators respectively to conduct the portion of the study in the State of Tennessee; (b) Dr. Vincent Chiang, School of Forestry and Wood Products, Michigan Technology University. **16.14** (a) © Adi Nes, Dvir Gallery Ltd.; (b) Transgenic goat produced using nuclear transfer at GTC Biotherapeutics. Photo used with permission; (c) Photo courtesy of MU Extension and Agricultural Information. **16.15** R. Brinster, R. E. Hammer, School of Veterinary Medicine, University of Pennsylvania. **Page 254** © Jeans for Gene Appeal. **Page 255** © Courtesy of Golden Rice Humanitarian Board. **16.16** (a) Laboratory of Matthew Shapiro while at McGill University. Courtesy of Eric Hargreaves, www.pageoneuro plasticity.com.

Page 257, UNIT III © Wolfgang Kaehler/ Corbis.

CHAPTER 17 **17.1** Left © Brad Snowder; right © David A. Kring, NASA/ Univ. Arizona Space Imagery Center. **Page 259** From top, left, © Wolfgang Kaehler/ Corbis; Courtesy George P. Darwin, Darwin Museum, Down House; Phillip Gingerich, Director, University of Michigan. Museum of Paleontology; USGS. **17.2** (a,c) © Wolfgang Kaehler/ Corbis; (b) © Earl & Nazima Kowall / Corbis; (d,e) Edward S. Ross. **17.3** (a) © Dr. John Cunningham/ Visuals Unlimited; (b) Gary Head. **17.4** (a) 2006 Dlloyd; (b) © PhotoDisc/ Getty Images; (c) Courtesy of Daniel C. Kelley, Anthony J. Arnold, and William C. Parker, Florida State University Department of Geological Science. **17.5** (a) Courtesy George P. Darwin, Darwin Museum, Down House; (b) Christopher Ralling; (e) Dieter & Mary Plage/ Survival Anglia; above, Heather Angel. **17.6** (a) © John White; (b) 2004 Arent. **17.7** (a) © Gerra and Sommazzi/ www.justbirds.org; (b) © Kevin Schafer/ Corbis; (c) Alan Root/ Bruce Coleman Ltd. **17.8** Down House and The Royal College of Surgeons of England. **17.9** (a) H. P. Banks; (b) Jonathan Blair; (c) Courtesy of Stan Celestian/ Glendale Comunity College Earth Science Image Archive. **17.10** © JupiterImages Corporation. **Page 268** zircon, Courtesy of Stan Celestian/ Glendale Comunity College Earth Science Image Archive. **17.12** (a) © PhotoDisc/ Getty Images; (b,c) Lisa Starr. **17.13** (a) W. B. Scott (1894); (b) John Klausmeyer, University of Michigan Exhibit of Natural History; above, Doug Boyer in P. D.

Gingerich et al. (2001) © American Association for Advancement of Science; inset, © P. D. Gingerich, University of Michigan. Museum of Paleontology; (c) © Bruce J. Mohn; above, © P. D. Gingerich and M. D. Uhen (1996), © University of Michigan. Museum of Paleontology; inset, Phillip Gingerich, Director, University of Michigan. Museum of Paleontology; left, © Peter Timmermans / Stone/ Getty Images. **17.14** (c) © JupiterImages Corporation. **17.16** USGS. **17.17** Above, left, © Martin Land/ Photo Researchers, Inc.; right, © John Sibbick; (a–e) After A.M. Ziegler, C.R. Scotese, and S.F. Barrett, "Mesozoic and Cenozoic Paleogeographic Maps," and J. Krohn and J. Sundermann (Eds.), *Tidal Frictions and the Earth's Rotation II*, Springer-Verlag, 1983. **Page 274** © David A. Kring, NASA/ Univ. Arizona Space Imagery Center. **17.18** Lawrence Berkeley National Laboratory.

CHAPTER 18 **18.1** Left © Reuters NewMedia, Inc./ Corbis; right © Rollin Verlinde/ Vilda. **Page 277** From top, Alan Solem; Thomas Bates Smith; Peggy Greb/ USDA; © Jo Wilkins; © Photo by Marcel Lecoufle. **18.2** (a) page 278 right, © Roderick Hulsbergen/ http:// www.photography.euweb.nl; all others, © JupiterImages Corporation. (b) Alan Solem. **18.3** © Photodisc/ Getty Images. **18.6** J. A. Bishop, L. M. Cook. **18.7** Courtesy of Hopi Hoekstra, University of California, San Diego. **18.9** © Peter Chadwick/ Science Photo Library/ Photo Researchers, Inc. **18.11** Thomas Bates Smith. **18.12** (a) Courtesy of Gerald Wilkinson; (b) Bruce Beehler. **18.13** (a,b) After Ayala and others; (c) © Michael Freeman/ Corbis. **18.14** (a,b) Adapted from S. S. Rich, A. E. Bell, and S. P. Wilson, "Genetic drift in small populations of Tribolium," *Evolution* 33:579–584, Fig. 1, p. 580, 1979. Used by permission of the publisher; below, Peggy Greb/ USDA. **18.15** Left David Neal Parks; right W. Carter Johnson. **18.16** From Meyer, A., Repeating Patterns of Mimicry. *PLoS Biology* Vol. 4, No. 10, e341 doi:10.1371/journal. pbio.0040341. **Page 290** © Alvin E. Staffan/ Photo Researchers, Inc. **18.18** Left Courtesy of Dr. James French; right Courtesy of Joe Decruyenaere. **18.19** G. Ziesler/ ZEFA. **18.20** (a) © Graham Neden/ Corbis; (b) © Kevin Schafer/ Corbis; center, © Ron Blakey, Northern Arizona University; (c) © Rick Rosen/ Corbis SABA. **18.21** Poouli, Bill Sparklin/ Ashley Dayer; All others, © Jack Jeffrey Photography. **18.23** (a) © Ian Hutton; (b) Courtesy of Peter Richardson; (c) © Jo Wilkins. **18.24** Courtesy of Dr. Robert Mesibov. **18.25** © Photo by Marcel Lecoufle. **Page 298** © Lanz Von Horsten; Gallo Images/ Corbis.

CHAPTER 19 **19.1** (a,b) © Jack Jeffrey Photography; (c) © Bill Sparklin/ Ashley Dayer. **Page 301** From top, Luc Viatour; © Taro Taylor, www.flickr.com/photos/tjt195; From "Embryonic staging system for the short-tailed fruit bat, Carollia perspicillata, a model organism for the mammalian order Chiroptera, based upon timed pregnancies in captive-bred animals" C.J. Cretekos et al., *Developmental Dynamics* Volume 233, Issue 3, July 2005, Pages: 721–738. Reprinted with permission of Wiley-Liss, Inc., a subsidiary of John Wiley & Sons, Inc. **19.2** From left, Joaquim Gaspar; Bogdan; Opiola Jerzy; Ravedave; Luc Viatour. **19.5** (a) © Taro Taylor, www.flickr.com; (b) © JupiterImages Corporation; (c) © Linda Bingham. **19.6** (a) © Lennart Nilsson/ Bonnierforlagen AB; (b)

Courtesy of Anna Bigas, IDIBELL-Institut de Recerca Oncologica, Spain; (c) From "Embryonic staging system for the short-tailed fruit bat, Carollia perspicillata, a model organism for the mammalian order Chiroptera, based upon timed pregnancies in captive-bred animals" C.J. Cretekos et al., *Developmental Dynamics* Volume 233, Issue 3, July 2005, Pages: 721–738. Reprinted with permission of Wiley-Liss, Inc., a subsidiary of John Wiley & Sons, Inc.; (d) Courtesy of Prof. Dr. G. Elisabeth Pollerberg, Institut für Zoologie, Universität Heidelberg, Germany; (e) USGS. **19.7** (a) Dr. Chip Clark; (b) top, Tait/ Sunnucks Peripatus Research; bottom, © Jennifer Grenier, Grace Boekhoff-Falk, and Sean Carroll, HMI, University of Wisconsin-Madison; (c) top, Herve Chaumeton/ Agence Nature; bottom, © Jennifer Grenier, Grace Boekhoff-Falk, and Sean Carroll, HMI, University of Wisconsin-Madison; (d) top, © Peter Skinner/ Photo Researchers, Inc.; bottom, Courtesy of Dr. Giovanni Levi. **19.10** © Jack Jeffrey Photography. **19.11** From top, © Hans Reinhard/ Bruce Coleman, Inc.; © Phillip Colla Photography; © Randy Wells/ Corbis; © Cousteau Society/ The Image Bank/ Getty Images; © Jack Jeffrey Photography. **19.12** Third from top, Bill Sparklin/ Ashley Dayer; all others, © Jack Jeffrey Photography. **19.13** John Steiner/ Smithsonian Institution. **Page 314** Bill Sparklin/ Ashley Dayer. **19.17** Courtesy of Irving Buchbinder, DPM, DABPS, Community Health Services, Hartford CT.

CHAPTER 20 **20.1** Left, © Peter Menzel/ Photo Researchers, Inc.; right, Courtesy of Agriculture Canada. **Page 317** From top, Lisa Starr; From Hanczyc, Fujikawa, and Szostak, Experimental Models of Primitive Cellular Compartments: Encapsulation, Growth, and Division; www.sciencemag.org *Science* 24 October 2003; 302; 529, Figure 2, page 619. Reprinted with permission of the authors and AAAS.; © Chase Studios/ Photo Researchers, Inc.; Photo by Julio Betancourt/ U.S. Geological Survey. **20.2** Jeff Hester and Paul Scowen, Arizona State University, and NASA. **20.3** (a) Painting by William K. Hartmann. **20.5** (a) © Eiichi Kurasawa/ Photo Researchers, Inc.; (b) Courtesy of the University of Washington; (c) © Michael J. Russell, Scottish Universities Environmental Research Centre. **20.6** (a) Sidney W. Fox; (b) From Hanczyc, Fujikawa, and Szostak, Experimental Models of Primitive Cellular Compartments: Encapsulation, Growth, and Division; www.sciencemag.org *Science* 24 October 2003; 302; 529, Figure 2, page 619. Reprinted with permission of the authors and AAAS. **20.7** (a) Stanley M. Awramik; (b,c) © University of California Museum of Paleontology; (d) © Chase Studios/ Photo Researchers, Inc.; (e) Courtesy of © Department of Industry and Resources, Western Australia. **20.8** (a,b) © Bruce Runnegar, NASA Astrobiology Institute; (c) © N.J. Butterfield, University of Cambridge. **20.10** (a) © CNRI/ Photo Researchers, Inc.; (b) © Courtesy of Isao Inouye, Institute of Biological Sciences, University of Tsukuba; (c) © Robert Trench, Professor Emeritus, University of British Columbia. **20.12** Photo by Julio Betancourt/ U.S. Geological Survey. **Page 329** © Philippa Uwins/ The University of Queensland.

Page 331 UNIT IV © Layne Kennedy/ Corbis.

CHAPTER 21 **21.1** © Lowell Tindell; inset, © NIBSC/ Photo Researchers, Inc. **Page 333** From top, Dr. Richard Feldmann/ National Cancer Institute; Lisa Starr; © SciMAT / Photo Researchers, Inc.; © Savannah River Ecology Laboratory; © CAMR, Barry Dowsett/ Science Photo Library/ Photo Researchers, Inc. **21.2** (a) left, after Stephen L. Wolfe; (c) Dr. Richard Feldmann/ National Cancer Institute; (d) © CAMR/ A. B. Dowsett/ SPL/ Photo Researchers, Inc.; (e) bottom, © Dr. Linda Stannard, UCT/ SPL/ Photo Researchers, Inc.; top, © Russell Knightly/ Photo Researchers, Inc. **21.3** Kenneth M. Corbett. **21.4** © Science Photo Library/ Photo Researchers, Inc. **21.5** © NIBSC/ Photo Researchers, Inc.; (a–e) After Stephen Wolfe, *Molecular Biology of the Cell*, Wadsworth, 1993. **21.6** (a) Photo by Barry Fitzgerald, Courtesy of USDA; (b) Photo by Peggy Greb, Courtesy of USDA. **21.7** (a) © APHIS photo by Dr. Al Jenny; (b) © Lily Echeverria/ Miami Herald; (bottom) PDB ID: 1QLX; Zahn, R., Liu, A., Luhrs, T., Riek, R., Von Schroetter, C., Garcia, F. L., Billeter, M., Calzolai, L., Wider, G., Wuthrich, K.: NMR Solution Structure of the Human Prion Protein, Proc. Nat. Acad. Sci. USA 97 pp. 145 (2000). **21.9** © CNRI/SPL/ Photo Researchers, Inc. **21.11** (a) © Dr. Dennis Kunkel / Visuals Unlimited. **21.12** P. W. Johnson and J. MeN. Sieburth, Univ. Rhode Island/ BPS. **21.13** (a) © Dr. Manfred Schloesser, Max Planck Institute for Marine Microbiology; (b) P. Hawtin, University of Southampton/ SPL/ Photo Researchers, Inc.; (c) © Dr. Richard Frankel (d) Courtesy of © Dr. Rolf Müller and Dr. Klaus Gerth. **21.14** (a) © SciMAT / Photo Researchers, Inc.; (b) Dr. Terry J. Beveridge, Department of Microbiology, University of Guelph, Ontario, Canada. **21.15** © Stem Jems/ Photo Researchers, Inc. **21.16** (a) © Dr. John Brackenbury/ Science Photo Library/ Photo Researchers, Inc.; (b) Courtesy Jack Jones, *Archives of Microbiology*, Vol. 136, 1983, pp. 254–261 Reprinted by permission of Springer-Verlag.; (c) © Dr. W. Michaelis/ Universitat Hamburg. **21.17** (a) © Martin Miller/ Visuals Unlimited; (b) © Dr. Harald Huber, Dr. Michael Hohn, Prof. Dr. K.O. Stetter, University of Regensburg, Germany, (c) © Savannah River Ecology Laboratory; (d) © Alan L. Detrick, Science Source/ Photo Researchers, Inc. **21.18** © WHO, Pierre-Michel Virot, photographer; inset, © Sercomi/ Photo Researchers, Inc. **21.19** © Camr/ B. Dowsett/ SPL/ Photo Researchers, Inc. **21.20** © Photodisc/ Getty Images.

CHAPTER 22 **22.1** Left, Malaria illustration by Drew Berry, The Walter and Eliza Hall Institute of Medical Research; right, AP Images. **Page 351** From top, D. P. Wilson/ Eric & David Hosking; © Dr. Dennis Kunkel/ Visuals Unlimited; © Frank Borges Llosa/ www.frankley.com; © Lewis Trusty/ Animals Animals; Courtesy Microbial Culture Collection, National Institute for Environmental Studies, Japan; Courtesy of www.hiddenforest.co.nz. **22.2** (a) © Astrid Hanns-Frieder michler/SPL/ Photo Researchers, Inc.; (b) © Dr. David Phillips/ Visuals Unlimited; (c) Science Museum of Minnesota; (d) D. P. Wilson/ Eric & David Hosking; (e) Steven C. Wilson/ Entheos. **22.4** (a,b) Dr. Stan Erlandsen, University of Minnesota; (c) © Dr. Dennis Kunkel/ Visuals Unlimited. **22.5** (a) After Prescott et al., *Microbiology*, third edition; (b) © Oliver Meckes/ Photo Researchers, Inc. **22.6** Left, P. L. Walne and J. H. Arnott, *Planta*, 77:325–354, 1967. **22.7** (a) Courtesy of Allen W. H. Bé and David A. Caron; (b) © Wim van Egmond. **22.8** © Ric Ergenbright/ Corbis. **22.9** Gary W. Grimes and Steven L'Hernault. **22.10** (a) Courtesy James Evarts; (b,c) Redrawn from V. & M. Pearse and M. & R. Buchsbaum, *Living Invertebrates*, The Boxwood Press, 1987. Used by permission. **22.11** (a) © Dr. David Phillips/ Visuals Unlimited; (b) © Lexey Swall/Staff from article, "Deep Trouble: Bad Blooms" October 3, 2003 by Eric Staats. **22.12** Left, © Frank Borges Llosa/ www.frankley.com; right, © Wim van Egmond/ Micropolitan Museum. **22.13** Left, © Sinclair Stammers/ Photo Researchers, Inc.; (b) right, © London School of Hygiene & Tropical Medicine/ Photo Researchers, Inc.; (c) right, © Moredum Animal Health, Ltd./ Photo Researchers, Inc.; (e) Micrograph Steven L'Hernault. **22.14** Homer, Photography by Gary Head. **22.15** (a) Greta Fryxell, University of Texas, Austin; (b) © Wim van Egmond/ Visuals Unlimited. **22.16** Right, © Lewis Trusty/ Animals Animals; Left, from T. Garrison, *Oceanography: An Invitation to Marine Science*, Brooks/Cole, 1993. **22.17** © Garo/ Photo Researchers, Inc. **22.18** Heather Angel. **22.19** (a) © Susan Frankel, USDA-FS; (b) Dr. Pavel Svihra. **22.20** Right, Courtesy of Professeur Michel Cavalla. **22.21** (a) Courtesy of Professor Astrid Saugestad; (b) © Lawson Wood/ Corbis; (c) Montery Bay Aquarium. **22.22** (a) © Wim van Egmond; (b) Courtesy Microbial Culture Collection, National Institute for Environmental Studies, Japan. **22.23** © PhotoDisc/ Getty Images. **22.24** © Wim van Egmond. **22.25** (a) © M I Walker/ Photo Researchers, Inc.; (b) © Edward S. Ross; (c) Courtesy of www.hiddenforest .co.nz. **22.26** (a–c) Carolina Biological Supply Company; (d) Courtesy Robert R. Kay from R. R. Kay, et al., *Development*, 1989 Supplement, pp. 81–90, © The Company of Biologists Ltd., 1989. **Page 366** Photo by James Gathany, Centers for Disease Control.

CHAPTER 23 **23.1** (a) © Karen Carr Studio/ www.karencarr.com; (b) © T. Kerasote/ Photo Researchers, Inc. **Page 369** From top, Courtesy of © Professor T. Mansfield; National Park Services, Paul Stehr-Green; © David C. Clegg/ Photo Researchers, Inc.; Photo by Scott Bauer/ USDA. **23.2** (b) Courtesy of © Christine Evers. **23.3** © Christopher Scotese, PALEOMAP Project. **23.4** (a) © Reprinted with permission from Elsevier; (b) Patricia G. Gensel. **23.5** (b) After E.O. Dodson and P. Dodson, *Evolution: Process and Product*, Third Ed., p. 401, PWS. **23.6** (a) Michael Clayton/ University of Wisconsin, Department of Botany; (b) Courtesy of © Professor T. Mansfield. **23.7** (a) © Andrew Syred/ Photo Researchers, Inc.; (b) © M.I. Walker/ Wellcome Images. **23.8** (a) Courtesy of © Christine Evers; (b) Gary Head. **Page 374** © University of Wisconsin-Madison, Department of Biology, Anthoceros CD. **23.9** (a,b) © Wayne P. Armstrong, Professor of Biology and Botany, Palomar College, San Marcos, California; (c) National Park Services, Martin Hutten; (d) National Park Services, Paul Stehr-Green. **23.10** Top center, © Jane Burton/ Bruce Coleman Ltd.; art, Raychel Ciemma. **23.11** (a) © Fred Bavendam/ Peter Arnold, Inc.; (b) © John D. Cunningham/ Visuals Unlimited. **23.12** (a) © Ed Reschke/ Peter Arnold, Inc.; (b) © Gerald D. Carr; (c) © Colin Bates, www.coastalimageworks .com; (d) Courtesy of © Christine Evers. **23.13** © A. & E. Bomford/ Ardea, London; art, Raychel Ciemma. **23.14** (a) © S. Navie; (b) © David C. Clegg/ Photo Researchers, Inc.; (c) © Klein Hubert/ Peter Arnold, Inc. **23.15** Right, © PaleoDirect.com. **23.16** Field Museum of Natural History, Chicago (Neg. #7500C); inset, © Brian Parker/ Tom Stack & Associates. **Page 379**

Institute of Human Origins; (b) Kenneth Garrett/ National Geographic Image Collection; (c) Louise M. Robbins; (d) Kenneth Garrett/ National Geographic Image Collection. **26.36** Jean Paul Tibbles. **26.37** © John Reader/ Photo Researchers, Inc. **26.38** (a) © Pascal Goetgheluck/ Photo Researchers, Inc.; (b,c) © Peter Brown. **26.40** © Christopher Scotese, PALEOMAP Project. **Page 458** P. Morris/ Ardea London.

CHAPTER 27 **27.1** Left, Star Tribune/ Minneapolis-St. Paul; right, © VVG/ Science Photo Library/ Photo Researchers, Inc. **Page 461** From top, © CNRI/ SPL/ Photo Researchers, Inc.; © Erwin & Peggy Bauer/ www.bciusa.com; G. J. McKenzie (MGS). **27.2**, Right From top, Courtesy of Charles Lewallen, Dartmouth Electron Microscope Facility; Photo Courtesy of Prof. Alison Roberts, University of Rhode Island. **27.3** Left, Art by Lisa Starr with © 2000 PhotoDisc, Inc; right, upper, © CNRI/ SPL/ Photo Researchers, Inc.; lower, Dr. Robert Wagner/ University of Delaware, www.udel.edu/Biology/ Wags. **27.4** Left © Montana Pritchard/ Getty Images Sport; right, © Darrell Gulin/ The Image Bank/ Getty Images. **27.5** (a) Courtesy of the National Park Service, www.nps.gov; (b) © Cory Gray; (c) Heather Angel; (d) © Biophoto Associates/ Photo Researchers, Inc. **27.6** (a) © Geoff Tompkinson/ SPL/ Photo Researchers, Inc.; (b) © Erwin & Peggy Bauer/ www.bciusa .com. **27.8** Right, © VVG/ Science Photo Library/ Photo Researchers, Inc. **27.9** Right, © Niall Benvie/ Corbis. **27.10** Left, © Kennan Ward/ Corbis; right, G. J. McKenzie (MGS). **27.11** Frank B. Salisbury. **Page 471** Upper right, Courtesy of © Christine Evers. **27.13** Courtesy of Dr. Kathleen K. Sulik, Bowles Center for Alcohol Studies, the University of North Carolina at Chapel Hill. **27.14** © John DaSiai, MD/ Custom Medical Stock Photo.

Page 473 UNIT V © Jim Christensen, Fine Art Digital Photographic Images.

CHAPTER 28 **28.1** Left © Michael Westmoreland/ Corbis; right © Reuters/ Corbis. **Page 475** From top, © Darrell Gulin/ Corbis; Courtesy of Dr. Thomas L. Rost; © Biodisc/ Visuals Unlimited; © David W. Stahle, Department of Geosciences, University of Arkansas; © iStockphoto.com/ mjutabor. **28.3** (a) from left, © Bruce Iverson; © Ernest Manewal/ Index Stock Imagery; Courtesy of Dr. Thomas L. Rost; © Franz Holthuysen, Making the invisible visible, Electron Microscopist, Phillips Research; (b) from left, Mike Clayton/ University of Wisconsin Department of Botany; © Darrell Gulin/ Corbis; Gary Head; Courtesy of Janet Wilmhurst, Landcare Research, New Zealand. **28.5** © Donald L. Rubbelke/ Lakeland Community College. **28.7** (a) © Dr. Dale M. Benham, Nebraska Wesleyan University; (b) D. E. Akin and I. L. Risgby, Richard B. Russel Agricultural Research Center, Agricultural Research Service, U.S. Dept. Agriculture, Athens, GA; (c) Kingsley R. Stern. **28.8** © Andrew Syred/ Photo Researchers, Inc. **28.9** George S. Ellmore. **28.10** (d) Above, © M. I. Walker/ Photo Researchers, Inc.; below, Gary Head. **28.11** (a) Center, Ray F. Evert; right, James W. Perry; (b) Center, Carolina Biological Supply Company; right, James W. Perry. **28.12** (c), (d) left, Benjamin de Bivort; (d) center, Miguel Bugallo; right, Sigman. **28.13** © Kenneth Bart. **28.14** (a) © N. Cattlin/ Photo Researchers, Inc.; (c) C. E. Jeffree,

et al., *Planta*, 172(1):20–37, 1987. Reprinted by permission of C. E. Jeffree and Springer-Verlag; (d) © Jeremy Burgess/ SPL/ Photo Researchers, Inc. **28.15** (a) Courtesy of Dr. Thomas L. Rost; (b) Gary Head. **28.16** (a) left, © Biodisc/ Visuals Unlimited; right, after Salisbury and Ross, *Plant Physiology*, Fourth Edition, Wadsworth; (b) above, © Brad Mogen/ Visuals Unlimited; below, © Dr. John D. Cunningham/ Visuals Unlimited. **28.17** (a) © Brad Mogen/ Visuals Unlimited; (b) © Dr. John D. Cunningham/ Visuals Unlimited; (c) Michael Clayton/ University of Wisconsin, Department of Botany. **28.20** (b) © Peter Gasson, Royal Botanic Gardens, Kew. **28.21** (a) © Peter Ryan/ SPL/ Photo Researchers, Inc.; (b) © Jon Pilcher; (c) © George Bernard/ SPL/ Photo Researchers, Inc. **28.22** (a) NOAA; (b) © David W. Stahle, Department of Geosciences, University of Arkansas. **28.23** (a) Michael Clayton/ University of Wisconsin, Department of Botany; (b) © Dinodia Photo Library/ Botanica/ Jupiter Images; (c) © iStockphoto.com/ mjutabor; (d) © Eric Sueyoshi/ Audrey Magazine; (e) © Chase Studio/ Photo Researchers, Inc.; (f) © Chris Hellier/ Corbis. **Page 490** © Darrel Plowes. **Page 491** Critical Thinking, #1, left, Edward S. Ross; right, © Ian Young, www.srgc.org.uk; #3, left, Edward S. Ross; right, Courtesy of Jeff Hutchison, University of Florida Center for Aquatic and Invasive Plants.

CHAPTER 29 **29.1** (a) © OPSEC Control Number #4 077-A-4; (b) © Billy Wrobel, 2004. **Page 493** From top, Photo courtesy of Stephanie G. Harvey, Georgia Southwestern State University; © Wally Eberhart/ Visuals Unlimited; © Jeremy Burgess/ SPL/ Photo Researchers, Inc.; © J.C. Revy/ ISM/ Phototake. **Page 494** © JupiterImages Corporation. **29.2** William Ferguson. **29.3** Photo courtesy of Stephanie G. Harvey, Georgia Southwestern State University. **29.4** (a) Courtesy of Mark Holland, Salisbury University; (b) Photo courtesy of Iowa State University Plant and Insect Diagnostic Clinic; (c) © Wally Eberhart/ Visuals Unlimited; (d) Mark E. Dudley and Sharon R. Long; (e) NifTAL Project, Univ. of Hawaii, Maui. **29.5** Above, Micrograph Chuck Brown. **29.6** (a) Alison W. Roberts, University of Rhode Island; (b,c) H. A. Core, W. A. Cote, and A. C. Day, *Wood Structure and Identification*, 2nd Ed., Syracuse University Press, 1979. **29.7** Left, The Ohio Historical Society, Natural History Collections. **29.8** (a) Micrograph by Ken Wagner/ Visuals Unlimited, computer-enhanced by Lisa Starr; (b–e) Courtesy of E. Raveh. **29.9** Left Don Hopey/ *Pittsburgh Post-Gazette*, 2002, all rights reserved. Reprinted with permission; right, © Jeremy Burgess/ SPL/ Photo Researchers, Inc. **29.10** (a) © James D. Mauseth, MCDB; (b) © J.C. Revy/ ISM/ Phototake. **29.11** Martin Zimmerman, *Science*, 1961, 133:73–79, © AAAS. **Page 504** Photo by Keith Weller, ARS, Courtesy of USDA.

CHAPTER 30 **30.1** (a) © Alan McConnaughey, www.flickr.com/photos/engrpiman; (b) Courtesy of James H. Cane, USDA-ARS Bee Biology and Systematics Lab, Utah State University, Logan, UT. **Page 507** From top, © Robert Essel NYC/ Corbis; Dartmouth Electron Microscope Facility; Photo by Stephen Ausmus, USDA, ARS; Photo by Peggy Greb, USDA, ARS. **30.2** (a) © Robert Essel NYC/ Corbis. **30.4** (a) Ravedave; (b) Courtesy of Joe Decruyenaere; (c) Joaquim Gaspar; (d) Harlo H. Hadow; (e) © Jay F. Petersen. **30.5** (a) Photo by Jack Dykinga, USDA, ARS; (b,c) Thomas Eisner, Cornell University. **30.6** Left John

Alcock/ Arizona State University; right Merlin D. Tuttle, Bat Conservation International. **30.7** (a) © David Goodin; (b) John Alcock, Arizona State University. **30.9** (a) Dartmouth Electron Microscope Facility; (b) © Susumu Nishinaga/ Photo Researchers, Inc. **30.10** (a) left, © Michael Clayton, University of Wisconsin; top right, Raychel Ciemma; bottom right, © Michael Clayton, University of Wisconsin; (b) Dr. Charles Good, Ohio State University, Lima; (c,d) Michael Clayton, University of Wisconsin. **30.12** (a) © T. M. Jones; (b) R. Carr; (c) © JupiterImages Corporation; (d) © Robert H. Mohlenbrock © USDA-NRCS PLANTS Database; (e) © Trudi Davidoff, www.WinterSown.org; (f) © iStock-photo.com/ Greggory Frieden. **30.13** (a) © Richard H. Gross; (b) © Andrew Syred/ SPL/ Photo Researchers, Inc.; (c) Photo by Stephen Ausmus, USDA, ARS. **30.14** © Darrell Gulin/ Corbis. **30.15** (a) © Richard Uhlhorn Photography; (b,c) Photo by Peggy Greb, USDA, ARS. **Page 520** Above, Courtesy of Caroline Ford, School of Plant Sciences, University of Reading, UK; below, © James L. Amos/ Corbis. **30.16** Gary Head. **30.17** © Steven D. Johnson. **Page 521** Critical Thinking #1, Edward S. Ross.

CHAPTER 31 **31.1** Michael A. Keller / FPG/ Getty Images. **Page 523** From top, Herve Chaumeton/ Agence Nature; © Andrei Sourakov and Consuelo M. De Moraes; © Cathlyn Melloan/ Stone/ Getty Images. **31.2** © Dr. John D. Cunningham/ Visuals Unlimited. **31.3** Left Barry L. Runk / Grant Heilman, Inc.; right © James D. Mauseth, MCDB. **31.4** Herve Chaumeton/ Agence Nature. **31.5** © Sylvan H. Wittwer/ Visuals Unlimited. **31.6** Left © mepr; right © Robert Lyons/ Visuals Unlimited. **31.9** (a) Courtesy of Dr. Consuelo M. De Moraes; (b–d) © Andrei Sourakov and Consuelo M. De Moraes. **31.10** (a) Michael Clayton, University of Wisconsin, Department of Botany; (b,c) © Muday, GK and P. Haworth (1994) "Tomato root growth, gravitropism, and lateral development: Correlations with auxin transport." *Plant Physiology and Biochemistry* 32, 193–203, with permission from Elsevier Science. **31.11** Micrographs courtesy of Randy Moore from "How Roots Respond to Gravity," M. L. Evans, R. Moore, and K. Hasenstein, *Scientific American*, December 1986. **31.12** Photo © Cathlyn Melloan/ Stone/ Getty Images. **31.13** Gary Head. **31.14** Cary Mitchell. **31.17**, **31.18** (a) © Clay Perry/ Corbis; (b) © Eric Chrichton/ Corbis. **31.19** Eric Welzel/ Fox Hill Nursery, Freeport, Maine. **31.20** Left © Roger Wilmshurst; Frank Lane Picture Agency/ Corbis; right © Adrian Chalkley. **31.21** Larry D. Nooden.

Page 537 Unit VI, © Kevin Schafer.

CHAPTER 32 **32.1** © Bryce Richter/ Courtesy of University of Wisconsin-Madison. **Page 539** Second from top, Ed Reschke; bottom, © Gregory Dimijian/ Photo Researchers, Inc. **32.4** From top, © Ray Simmons/ Photo Researchers, Inc.; Ed Reschke/ Peter Arnold, Inc.; © Don W. Fawcett. **32.5** (a) © John Cunningham/ Visuals Unlimited; (b,c) Ed Reschke (d) © Science Photo Library/Photo Researchers, Inc.; (e) © Michael Abbey/ Photo Researchers, Inc.; (f) © University of Cincinnati, Raymond Walters College, Biology. **32.6** Gary Head. **32.7** © Science Photo Library/ Photo Researchers, Inc. **32.8** Above, © Tony McConnell/ Science Photo Library/ Photo Researchers, Inc.; (a,b) Ed Reschke; (c)

Nielsen, University of Utah; Ralph Pleasant/ FPG/ Getty Images; Dr. Douglas Coleman, The Jackson Laboratory. **40.4** (a) © W. Perry Conway/ Corbis; (a,b art) Adapted from A. Romer and T. Parsons, *The Vertebrate Body*, Sixth Edition, Saunders Publishing Company, 1986. **40.7** After A. Vander et al., *Human Physiology: Mechanisms of Body Function*, Fifth Edition, McGraw-Hill, 1990. Used by permission. **40.8** (a) Microslide courtesy Mark Nielsen, University of Utah. **40.9** (d) Right, © D. W. Fawcett/ Photo Researchers, Inc.; Art, After Sherwood and others. **40.11** (b) National Cancer Institute. **40.13** USDA, www.mypyramid.gov. **Page 713** From left, Ralph Pleasant/ FPG/ Getty Images; © Photodisc/ Getty Images; Ralph Pleasant/ FPG/ Getty Images; © Paul Poplis Photography, Inc./ Stockfood America; © Photodisc/ Getty Images. **40.14** Gary Head. **40.15** Dr. Douglas Coleman, The Jackson Laboratory. **Page 718** Simon Law, www.flickr.com/people/sfllaw.

CHAPTER 41   41.1 Page 720, © Archivo Iconografico, S.A./ Corbis; page 721, © Ed Kashi/ Corbis. **Page 721** Bottom, © David Parker/ SPL/ Photo Researchers, Inc. **41.4** Left © Cabisco/ Visuals Unlimited. **41.5** © Susumu Nishinaga/ Photo Researchers, Inc.; above, © Stephen Dalton/ Photo Researchers, Inc. **41.8** Left, Gary Head; right, © Tom McHugh/ Photo Researchers, Inc. **41.12** Evan Cerasoli. **41.14** (a) © Bob McKeever/ Tom Stack & Associates; (b) © S. J. Krasemann/ Photo Researchers, Inc. **41.15** © David Parker/ SPL/ Photo Researchers, Inc. **41.16** (a) © Dan Guravich/ Corbis; (b) Bettmann/ Corbis. **Page 736** © Lawrence Lawry/ Photo Researchers, Inc. **41.17** Logo, USDA.

CHAPTER 42   42.1 © AFP/ Getty Images. **Page 739** Top, © Rodger Klein/ Peter Arnold, Inc.; bottom, © Western Ophthalmic Hospital/ Photo Researchers, Inc. **42.2** (a) © Biophoto Associates/ Photo Researchers, Inc.; (b) © Rodger Klein/ Peter Arnold, Inc.; (c) © Staebler/ Jupiter Images. **42.3** (a) Frieder Sauer/ Bruce Coleman, Ltd; (b) Matjaz Kuntner; (c) © Ron Austing/ Frank Lane Picture Agency/ Corbis; (d) © Doug Perrine/ seapics.com; (e) Carolina Biological Supply Company; (f) Fred McKinney/ FPG /Getty Images; (g) Gary Head. **42.5** (b) © Ed Reschke. **Page 747** © AJPhoto/ Photo Researchers, Inc. **42.10** (e) Right, © Lennart Nilsson/ Bonnierforlagen AB. **42.12** © Marilyn Houlberg. **42.13** (c) Above, Courtesy of Elizabeth Sanders, Women's Specialty Center, Jackson, MS. **42.15** © Heidi Specht, West Virginia University. **42.16** (a) © Dr. E. Walker/ Photo Researchers, Inc.; (b) © Western Ophthalmic Hospital/ Photo Researchers, Inc.; (c) © CNRI/ Photo Researchers, Inc. **42.17** (a) © David M. Phillips/ Visuals Unlimited; (b) © CNRI/ SPL/ Photo Researchers, Inc.; (c) © John D. Cunningham/ Visuals Unlimited. **42.18** © Todd Warshaw/ Getty Images. **Page 756** © Laura Dwight/ Corbis.

CHAPTER 43   43.1 © Dana Fineman/ Corbis Sygma. **Page 759** Top, Dr. Maria Leptin, Institute of Genetics, University of Koln, Germany; second from bottom, © Lennart Nilsson/ Bonnierforlagen AB. **43.3** (b–e) Carolina Biological Supply Company; (f) Left and center, © David M. Dennis/ Tom Stack & Associates, Inc.; right, © John Shaw/ Tom Stack & Associates. **43.7** Above, Carolina Biological Supply Company; below, Dr. Maria Leptin, Institute of Genetics, University of Koln, Germany. **43.8** (a,b) After S.

Gilbert, *Developmental Biology*, Fourth Edition; (c) © Professor Jonathon Slack. **43.10** Left, Peter Parks/ Oxford Science Films/ Animals Animals. **43.11** (a) © David M. Parichy; (b,c) © Dr. Sharon Amacher. **43.12** Adapted from L.B. Arey, *Developmental Anatomy*, Philadelphia, W.B. Saunders Co., 1965. **43.16** © Lennart Nilsson/ Bonnierforlagen AB. **43.18** Left, © Zeva Oelbaum/ Corbis; right, James W. Hanson, M.D. **Page 777** © Dana Fineman/ Corbis Sygma.

Page 779 Unit VII, © Minden Pictures.

CHAPTER 44   44.1 © Scott Camazine. **Page 781** From top, Reprinted from *Trends in Neuroscience*, Vol. 21, Issue 2, 1998, L.J.Young, W. Zuoxin, T.R. Insel, "Neuroendocrine bases of monogamy," Pages 71–75, ©1998, with permission from Elsevier Science; © Kevin Schafer/ Corbis; © B. Borrell Casals/ Frank Lane Picture Agency/ Corbis; © Australian Picture Library/ Corbis. **44.2** (a) Eugene Kozloff; (b) Stevan Arnold. **44.4** Left, © Robert M. Timm & Barbara L. Clauson, University of Kansas; (a,b) Reprinted from *Trends in Neuroscience*, Vol. 21, Issue 2, 1998, L.J.Young, W. Zuoxin, T.R. Insel, "Neuroendocrine bases of monogamy," Pages 71–75, ©1998, with permission from Elsevier Science. **44.5** (a) Eric Hosking; (b) © Stephen Dalton/ Photo Researchers, Inc. **44.6** © Nina Leen/ TimePix/ Getty Images; inset, © Robert Semeniuk/ Corbis. **44.7** © Professor Jelle Atema, Boston University. **44.8** © Bernhard Voelkl. **44.9** Left, Robert Maier/ Animals Animals. **44.10** (a) © Tom and Pat Leeson, leesonphoto.com; (b) © Kevin Schafer/ Corbis; (c) © Monty Sloan, www.wolfphotography.com. **44.11** © Stephen Dalton/Photo Researchers, Inc. **44.12** (a) John Alcock, Arizona State University; (b,c) © Pam Gardner/ Frank Lane Picture Agency/ Corbis; (d) © D. Robert Franz/ Corbis. **44.13** Michael Francis/ The Wildlife Collection. **44.14** (a) © B. Borrell Casals/ Frank Lane Picture Agency/ Corbis; (b) © Steve Kaufman/ Corbis; (c) © John Conrad/ Corbis. **44.15** (a) © Tom and Pat Leeson, leesonphoto.com; (b) John Alcock, Arizona State University; (c) © Paul Nicklen/ National Geographic/ Getty Images. **44.16** © Jeff Vanuga/ Corbis. **44.17** © Steve Bloom/ stevebloom.com. **44.18** © Eric and David Hosking/ Corbis. **44.19** (a) © Australian Picture Library/ Corbis; (b) © Alexander Wild; (c) © Professor Louis De Vos. **44.20** (a) Kenneth Lorenzen; (b) © Peter Johnson/ Corbis; (c) © Nicola Kountoupes/ Cornell University. **Page 794** © Lynda Richardso/ Corbis.

CHAPTER 45   45.1 © David Nunuk/ Photo Researchers, Inc. **Page 797** From top, © Amos Nachoum/ Corbis; © David Scharf, 1999. All rights reserved; © G. K. Peck; © Joe McDonald/ Corbis; © Don Mason/ Corbis. **45.2** (a) © Amos Nachoum/ Corbis; (b) © Corbis; (c) A. E. Zuckerman/ Tom Stack & Associates. **45.3** E. R. Degginger; inset, Jeff Fott Productions/ Bruce Coleman, Ltd. **45.4** (a) © Cynthia Bateman, Bateman Photography; (b) © Tom Davis. **45.5** © Jeff Lepore/ Photo Researchers, Inc. **45.6** © David Scharf, 1999. All rights reserved. **45.7** (a) © G. K. Peck; (b) © Rick Leche, www.flickr.com/photos/rick_leche. **45.9** © Charles Lewallen. **45.10** (a) © Joe McDonald/ Corbis; (b) © Wayne Bennett/ Corbis; (c) Estuary to Abyss 2004. NOAA Office of Ocean Exploration. **45.11** (a,b) Upper David Reznick/ University of California—Riverside; computer enhanced by Lisa Starr; (a,b) lower Hippocampus Bildarchiv; (c)

Helen Rodd. **Page 807** © Bruce Bornstein, www.captbluefin.com. **45.13** Left, © Mark Harmel/ Photo Researchers, Inc.; right, AP Images. **45.14** NASA. **45.17** Left © Adrian Arbib/ Corbis; right, © Don Mason/ Corbis. **45.19** © Polka Dot Images/ SuperStock. **Page 814** U.S. Department of the Interior, U.S. Geological Survey. **45.20** © Reinhard Dirscherl/ www.bciusa.com.

CHAPTER 46   46.1 (a) Photography by B. M. Drees, Texas A&M University. Http://fireant.tamu; (b) Scott Bauer/ USDA; (c) USDA. **Page 817** From top, © Martin Harvey, Gallo Images/ Corbis; © Nigel Jones; © Pat O'Hara/ Corbis; © Pierre Vauthey/ Corbis Sygma. **46.2** Left, Donna Hutchins; (a) © B. G. Thomson/ Photo Researchers, Inc.; (b) © Len Robinson, Frank Lane Picture Agency/ Corbis; (c) Martin Harvey, Gallo Images/ Corbis. **46.3** Upper, Harlo H. Hadow; lower, Bob and Miriam Francis/ Tom Stack & Associates. **46.4** © Thomas W. Doeppner. **46.5** © Pekka Komi. **46.6a** (a) Left, © Michael Abbey/ Photo Researchers, Inc.; right, © Eric V. Grave/ Photo Researchers, Inc. **46.7** Stephen G. Tilley. **46.8** Upper, © Joe McDonald/ Corbis; lower, left, © Hal Horwitz/ Corbis; right, © Tony Wharton, Frank Lane Picture Agency/ Corbis. **46.9** © W. Perry Conway/ Corbis. **46.10** Left, © Ed Cesar/ Photo Researchers, Inc.; right, © Robert McCaw, www.robertmccaw.com. **46.11** (a) © JH Pete Carmichael; (b) Edward S. Ross; (c) W. M. Laetsch; (d) © Nigel Jones. **46.12** (a–c) Edward S. Ross; (d) © Nigel Jones. **46.13** (a,b) Thomas Eisner, Cornell University; (c) David Burdick/ NOAA; (d) © Bob Jensen Photography. **46.14** (a) MSU News Service, photo by Montata Water Center; (b) © Karl Andree. **46.15** Left, © The Samuel Roberts Noble Foundation, Inc.; right, Courtesy of Colin Purrington, Swarthmore College. **46.16** © C. James Webb/ Phototake USA. **46.17** © Peter J. Bryant/ Biological Photo Service. **46.18** (a) © Richard Price/Getty Images; (b) © E.R. Degginger/ Photo Researchers, Inc. **46.19** (a) © Doug Peebles/ Corbis; (b) © Pat O'Hara/ Corbis; (c,d) © Tom Bean/ Corbis; (e) © Duncan Murrell/ Taxi/ Getty Images. **46.20** (a) R. Barrick/ USGS; (b) USGS; (c) P. Frenzen, USDA Forest Service. **46.21** (a,c) Jane Burton/ Bruce Coleman, Ltd.; (b) Heather Angel. **46.22** (a) © Pr. Alexande Meinesz, University of Nice-Sophia Antipolis; (b) © Angelina Lax/ Photo Researchers, Inc.; right, © The University of Alabama Center for Public TV. **46.23** © John Carnemolla/ Australian Picture Library. **46.24** After W. Dansgaard et al., *Nature*, 364:218–220, July 15, 1993; D. Raymond et al., *Science*, 259:926–933, February 1993; W. Post, *American Scientist*, 78:310–326, July–August 1990. **46.25** © Pierre Vauthey/ Corbis Sygma. **Page 836** John Kabashima.

CHAPTER 47   47.1 Left, © C. C. Lockwood/ Cactus Clyde Productions; right, Diane Borden-Bilot, U.S. Fish and Wildlife Service. **Page 839** From top, © D. A. Rintoul; Graphic created by FoodWeb3D program written by Rich Williams courtesy of the Webs on the Web project (www.foodwebs.org); NASA's Earth Observatory; USDA Forest Service, Northeastern Research Station. **47.3** Right top, © Lloyd Spitalnik/ lloydspitalnikphotos.com; right bottom, © Van Vives; all others, © D. A. Rintoul. **47.4** from left, top row, © Bryan & Cherry Alexander/ Photo Researchers, Inc.; © Dave Mech; © Tom & Pat Leeson, Ardea London Ltd.; 2nd row, © Tom Wakefield/ Bruce Coleman, Inc.; © Paul J. Fusco/ Photo Researchers, Inc.; © E. R.

Degginger/ Photo Researchers, Inc.; 3rd row, © Tom J. Ulrich/ Visuals Unlimited; © Dave Mech; © Tom McHugh/ Photo Researchers, Inc.; mosquito, Photo by James Gathany, Centers for Disease Contro; flea, © Edward S. Ross; 4th row, © Jim Steinborn; © Jim Riley; © Matt Skalitzky; earthworm, © Peter Firus, flagstaffotos.com .au. **47.5** Left, Courtesy of Dr. Chris Floyd; right, Graphic created by FoodWeb3D program written by Rich Williams courtesy of the Webs on the Web project (www.foodwebs.org). **47.6** NASA. **47.9** Upper, U.S. Department of the Interior, National Park Service; lower, Gary Head. **47.10** Jack Scherting, USC&GS, NOAA. **47.12** USDA Forest Service, Northeastern Research Station. **47.14** Lisa Starr after Paul Hertz; photograph © Photodisc/ Getty Images. **47.15, 47.16** Lisa Starr and Gary Head, based on NASA photographs from JSC Digital Image Collection. **47.17** © Yann Arthus-Bertrand/ Corbis. **47.19** © Jeff Vanuga/ Corbis. **47.20** © Frederica Georgia/ Photo Researchers, Inc. **47.21** Art, Gary Head and Lisa Starr; photograph, © Photodisc/ Getty Images. **47.22** Fisheries & Oceans Canada, Experimental Lakes Area. **Page 858** U.S. Department of Transportation, Federal Highway Administration. **47.24** Courtesy of NASA's Terra satellite, supplied by Ted Scambos, National Snow and Ice Data Center, University of Colorado, Boulder.

CHAPTER 48 **48.1** © Hank Fotos Photography; inset, © Wolfgang Kaehler/ Corbis. **Page 861** From second from top, NASA; © James Randklev/ Corbis; Jack Carey; Raghu Rai/ Magnum Photos. **48.5** NASA. **48.7** (a) Adapted from *Living in the Environment* by G. Tyler Miller, Jr., p. 428. © 2002 by Brooks/Cole, a division of Thomson Learning; (b) © Ted Spiegel/ Corbis. **48.8** After M. H. Dickerson, "ARAC: Modeling an Ill Wind," in *Energy and Technology Review*, August 1987. Used by permission of University of California Lawrence Livermore National Laboratory and U.S. Dept. of Energy. **48.9** NASA. **48.10** Left, © Sally A. Morgan, Ecoscene/ Corbis; right, © Bob Rowan, Progressive Image/ Corbis. **48.12** NASA. **48.13** Above, © Yves Bilat, Ardea London Ltd.; below, © Eagy Landau/ Photo Researchers, Inc. **48.15** (a,b) Courtesy of Jim Deacon, The University of Edinburgh; (c) Jeff Servos, US Fish & Wildlife

Service; (d) Bill Radke, US Fish & Wildlife Service. **48.16** (a) Ray Wagner/ Save the Tall Grass Prairie, Inc.; (b) © Tom Bean Photography; (c) Jonathan Scott/ Planet Earth Pictures. **48.17** (a) © John C. Cunningham/ Visuals Unlimited; (b) Jack Wilburn/ Animals Animals; (c) AP Images; (d) © Richard W. Halsey, California Chaparral Institute. **48.18** left, © James Randklev/ Corbis; all others, © Randy Wells/ Corbis. **48.19** Upper, © Franz Lanting/ Minden Pictures; inset, Edward S. Ross; lower, Hans Renner; inset, © Adolf Schmidecker/ FPG/ Getty Images. **48.20** (a) © Thomas Wiewandt/ ChromoSohm Media Inc./ Photo Researchers, Inc.; (b) © Raymond Gehman/ Corbis; inset, Donna Dewhurst, US Fish & Wildlife Service. **48.21** © Darrell Gulin/ Corbis; inset, Thomas D. Mangelsen/ Images of Nature. **48.22** © Pat O'Hara/ Corbis. **48.24** Jack Carey. **48.26** Ocean Arks International. **48.27** (a) © Annie Griffiths Belt/ Corbis; (b) © Douglas Peebles/ Corbis. **48.28** (a) © Nancy Sefton; (b) Courtesy of J. L. Sumich, *Biology of Marine Life*, 7th ed., W. C. Brown, 1999; (c) © Paul A. Souders/ Corbis. **Page 882** © Sea Studios/ Peter Arnold, Inc. **48.29** (a) C. B. & D. W. Frith/ Bruce Coleman, Ltd.; (b) Douglas Faulkner/ Sally Faulkner Collection; (c) © Douglas Faulkner/ Photo Researchers, Inc.; (e) lionfish, Douglas Faulkner/ Sally Faulkner Collection; all others, © John Easley, www.johneasley.com. **48.30** © Dr. Ray Berkelmans, Australian Institute of Marine Science. **48.31** (b) NOAA. **48.32** (a) Courtesy of © Montery Bay Aquarium Research Institute; (b) © Peter Herring/ imagequestmarine.com; (c) Image courtesy of NOAA and MBARI; (d–f) © Peter Batson/ imagequestmarine.com. **48.35** NASA Goddard Space Flight Center Scientific Visualization Studio. **48.36** (a) CHAART, at NASA Ames Research Center; (b) © Eye of Science/ Photo Researchers, Inc.; (c) Courtesy of Dr. Anwar Huq and Dr. Rita Colwell, University of Maryland; (d) Raghu Rai/ Magnum Photos. **Page 888** © Jose Luis Pelaez, Inc./ Corbis.

CHAPTER 49 **49.1** U.S. Navy photo by Chief Yeoman Alphanso Braggs. **Page 891 Top,** © George M. Sutton/ Cornell Lab of Ornithology; From third from top, NOAA; Bureau of Land Management. **49.3** Mansell Collection/ Time, Inc./ Getty Images. **49.4** © George M. Sutton/ Cornell Lab of Ornithology. **49.5** Jeffrey

Sylvester/ FPG/ Getty Images. **49.6** (a) © Dr. John Hilty; (b) Joe Fries, U.S. Fish & Wildlife Service. **49.8** © Billy Grimes. **49.10** NOAA. **49.11** Image provided by GeoEye and NASA SeaWIFS Project. **49.12** Claire Fackler/ NOAA. **49.13** © PhotoDisc/ Getty Images. **49.14** Bureau of Land Management. **Page 904** © Dan Guravich/ Corbis.

Appendix V Hemoglobin models: PDB ID: 1GZX; Paoli, M., Liddington, R., Tame, J., Wilkinson, A., Dodson, G., Crystal structure of T state hemoglobin with oxygen bound at all four haems. *J.Mol.Biol.*, v256, pp. 775–792, 1996.

Appendix VI Electron transfer chains: PDB ID: 1A70; Binda, C., Coda, A., Aliverti, A., Zanetti, G., Mattevi, A., Structure of the mutant E92K of [2Fe-2S] ferredoxin I from Spinacia oleracea at 1.7 Å resolution. *Acta Crystallogr.*, Sect.D, v54, pp. 1353–1358, 1998. PDB ID: 1AG6; Xue, Y., Okvist, M., Hansson, O., Young, S., Crystal structure of spinach plastocyanin at 1.7 Å resolution. *Protein Sci.*, v7, pp. 2099–2105, 1998. PDB ID: 1ILX; Vasil`ev, S., Orth, P., Zouni, A., Owens, T.G., Bruce, D., Excited-state dynamics in photosystem II: insights from the x-ray crystal structure. *Proc.Natl.Acad.Sci.*, USA, v98, pp. 8602–8607, 2001. PDB ID: 1Q90; Stroebel, D., Choquet, Y., Popot, J.-L., Picot, D., An Atypical Haem in the Cytochrome B6F Complex, *Nature*, v426, pp. 413–418, 2003. PDB ID: 1QZV; Ben-Shem, A., Frolow, F., Nelson, N., Crystal structure of plant photosystem I, *Nature*, v426, pp. 630–635, 2003. PDB ID: 1IZL; Kamiya, N., Shen, J.-R., Crystal structure of oxygen-evolving photosystem II from Thermosynechococcus vulcanus at 3.7-Å resolution, *Proc.Natl.Acad.Sci.*, USA, v100, pp. 98–103, 2003. PDB ID: 1GJR; Hermoso, J.A., Mayoral, T., Faro, M., Gomez-Moreno, C., Sanz-Aparicio, J., Medina, M., Mechanism of coenzyme recognition and binding revealed by crystal structure analysis of ferredoxin-NADP+ reductase complexed with NADP+., *J.Mol.Biol.*, v319, pp. 1133–1142, 2002. pdb ID: 1C17; Rastogi, V.K., Girvin, M.E., Structural changes linked to proton translocation by subunit c of the ATP synthase., *Nature*, v402, pp. 263–268, 1999. PDB ID: 1E79; Gibbons, C., Montgomery, M.G., Leslie, A.G., Walker, J.E., The structure of the central stalk in bovine F(1)-ATPase at 2.4 Å resolution., *Nat. Struct.Biol.*, v7, pp. 1055–1061, 2000.